Part C

Building on the previous two parts of this Springer Handbook, which have dealt with the fundamental concepts and equations that underpin experimental fluid mechanics and the measurement of primary quantities, respectively, Part C addresses experimental fluid mechanics from an application point of view. According to the application, often unique and specific forms of equipment, experimental procedure, or analysis and interpretation of results have been developed. It is the purpose of Part C to present a selection of such application areas. In particular, measurements of non-Newtonian flows, turbulence, flow visualization, wall-bounded flows, surface topology, turbomachines, hydraulics, aerodynamics, atmospheric and oceanographic measurements, combustion diagnostics and electrohydrodynamic systems are presented in this section.

Part D

This final part of the Springer Handbook is a reference source about signal and data processing techniques commonly encountered in fluid mechanics. These topics have been complemented by a section discussing data acquisition by imaging detectors, a topic becoming increasingly important for optical measurement techniques. These are all subjects that in their development are not naturally associated with fluid mechanics; hence Part D attempts to collect information from many diverse sources and present it conveniently to the fluid mechanics researcher. Topics covered in this part include fundamental topics of signal- and data-processing transforms (Fourier, Hilbert, wavelet, etc.), proper orthogonal decomposition and stochastic estimation. This is followed by a discussion of estimator expectation and variance and the influence of noise on these quantities. The Cramér–Rao lower bound (CRLB) is introduced and developed for several common signal processing examples from fluid mechanics. Imaging detectors and measures of their performance are then discussed in detail before closing with a chapter on image processing and motion analysis, two topics especially relevant to the particle image velocity (PIV) measurement technique.

Springer Handbook
of Experimental Fluid Mechanics

Springer Handbooks provide a concise compilation of approved key information on methods of research, general principles, and functional relationships in physical sciences and engineering. The world's leading experts in the fields of physics and engineering will be assigned by one or several renowned editors to write the chapters comprising each volume. The content is selected by these experts from Springer sources (books, journals, online content) and other systematic and approved recent publications of physical and technical information.

The volumes are designed to be useful as readable desk reference books to give a fast and comprehensive overview and easy retrieval of essential reliable key information, including tables, graphs, and bibliographies. References to extensive sources are provided.

Springer Handbook
of Experimental Fluid Mechanics

Cameron Tropea, Alexander L. Yarin, John F. Foss (Eds.)

With DVD-ROM, 1240 Figures and 123 Tables

Editors:

Cameron Tropea
Fachgebiet Strömungslehre und Aerodynamik
Technische Universität Darmstadt
Petersenstraße 30,
64287 Darmstadt, Germany

Alexander Yarin
Department of Mechanical and Industrial Engineering
University of Illinois at Chicago
842 W. Taylor St.,
Chicago, IL 60607-7022, USA

John F. Foss
Mechanical Engineering
Michigan State University
A118 Engineering Research Complex,
East Lansing, MI 48824-1326, USA

Library of Congress Control Number: 2007931998

ISBN: 978-3-662-49162-1   e-ISBN: 978-3-540-30299-5

This work is subject to copyright. All rights reserved, whether the whole or part of the material is concerned, specifically the rights of translation, reprinting, reuse of illustrations, recitation, broadcasting, reproduction on microfilm or in any other way, and storage in data banks. Duplication of this publication or parts thereof is permitted only under the provisions of the German Copyright Law of September, 9, 1965, in its current version, and permission for use must always be obtained from Springer-Verlag. Violations are liable for prosecution under the German Copyright Law.

Springer is a part of Springer Science+Business Media

springer.com

© Springer-Verlag Berlin Heidelberg 2007

The use of designations, trademarks, etc. in this publication does not imply, even in the absence of a specific statement, that such names are exempt from the relevant protective laws and regulations and therefore free for general use.

Product liability: The publisher cannot guarantee the accuracy of any information about dosage and application contained in this book. In every individual case the user must check such information by consulting the relevant literature.

Typesetting and production:
LE-TEX Jelonek, Schmidt&Völckler GbR, Leipzig
Senior Manager Springer Handbook: Dr. W. Skolaut, Heidelberg
Typography and layout: schreiberVIS, Seeheim
Illustrations: Hippmann GbR, Schwarzenbruck
Cover design: eStudio Calamar Steinen, Barcelona
Cover production: WMXDesign GmbH, Heidelberg
Printing and binding: Stürtz GmbH, Würzburg

Printed on acid free paper

SPIN 10934340   89/3180/YL   5 4 3 2 1 0

Preface

The purpose of this Springer Handbook is to provide comprehensive support to the experimental fluid mechanics community, for planning, executing, and interpreting experiments. This purpose is addressed by organizing the handbook into four parts: Part A (Chaps. 1 and 2) addresses the motivation for experiments and the equations that build the foundations for experimental work; Part B (Chaps. 3–8) examines the measurement of, and measurement techniques used for, all primary quantities appearing explicitly in the governing equations; Part C (Chaps. 9–21) presents topics related to a specific application area or technique and; Part D (Chaps. 22–25) is meant to serve as a reference in questions regarding signal and data acquisition and processing.

Cameron Tropea

Experimental fluid mechanics comprises a very large number of topics and special application areas and in undertaking such a handbook project a selection must necessarily be made. In making this selection the editors have attempted to cover as completely as possible the most fundamental concepts and most frequently employed measurement techniques and fluid behaviors. Those topics that have not been included in this first edition will perhaps find a place in a future edition.

The editors of this Springer Handbook would like to heartily thank the contributing authors, who have all captured the spirit of this handbook and made significant improvements to our original concept. Furthermore, a special thanks goes to Dr. Werner Skolaut at Springer for his untiring efforts in assembling the final version of all the manuscripts and coordinating the production process. And finally, thanks go to Ms. Claudia Rau and her team at LE-TeX Jelonek, Schmidt & Vöckler GbR for their skillful preparation of manuscripts into final production form.

John Foss

Inevitably there will be misprints in this handbook and readers are invited to bring these to the attention of the Editors via e-mail. Similarly, suggestions for improvements in a second edition are also very welcome.

July 2007

Cam Tropea
John Foss
Alex Yarin

Alexander Yarin

List of Authors

David M. Admiraal
University of Nebraska-Lincoln
Department of Civil Engineering
Nebraska Hall W348
Lincoln, NE 68588-0531, USA
e-mail: dadmiraal2@unl.edu

Ronald J. Adrian
Arizona State University
Ira A. Fulton School of Mechanical
and Aerospace Engineering
P.O. Box 876106
Tempe, AZ 85210, USA
e-mail: rjadrian@asu.edu

Nuri Aksel
Universität Bayreuth
Department of Applied Mechanics
and Fluid Dynamics
Universitätsstr.
95440 Bayreuth, Germany
e-mail: tms@uni-bayreuth.de

Yiannis Andreopoulos
The City College of the City University of New York
Department of Mechanical Engineering
New York, NY 10031, USA
e-mail: andre@ccny.cuny.edu

Roger E.A. Arndt
University of Minnesota
Saint Anthony Falls Laboratory
2 Third Ave. SE,
Minneapolis, MN 55414, USA
e-mail: arndt001@umn.edu

Marc J. Assael
Aristotle University
Chemical Engineering Department
Thessaloniki, 54124, Greece
e-mail: assael@auth.gr

Howard A. Barnes
University of Wales Aberystwyth
Institute of Mathematical and Physical Sciences
Penglais Aberystwyth
Ceredigion, SY23 3BZ, UK
e-mail: howard.barnes@ntlworld.com

Terry D. Blake
8 Hazely
Tring, HP23 5JH, UK
e-mail: terrydblake@btinternet.com

Eberhard Bodenschatz
Max Planck Institute for Dynamics
and Self-Organization
Laboratory of Fluid Dynamics, Pattern Formation
and Nanobiocomplexity (LFPN)
Am Faßberg 11
37077 Göttingen, Germany
e-mail: eberhard.bodenschatz@ds.mpg.de

Jean-Paul Bonnet
Université de Poitiers, ENSMA, CNRS
Laboratoire d'Etudes Aérodynamiques
43 rue de l'aérodrome
86036 Poitiers, France
e-mail: Jean-paul.bonnet@univ-poitiers.fr

Michael P. Brenner
Harvard University
School of Engineering and Applied Science
29 Oxford Street
Cambridge, MA 02139, USA
e-mail: brenner@deas.harvard.edu

Ronald J. Calhoun
Arizona State University
Mechanical and Aerospace Engineering
P.O. Box 876106
Tempe, AZ 85287-6106, USA
e-mail: Ron.Calhoun@asu.edu

Antonio Castellanos
Universidad de Sevilla
Avda. Reina Mercedes s/n
Sevilla, 41013, Spain
e-mail: *castella@us.es*

Angelo A. Cavone
NASA Langley Research Center
ATK Space Systems, Inc.
Mail Stop 493
Hampton, VA 23681, USA
e-mail: *angelo.a.cavone@nasa.gov*

Teresa K. Chereskin
University of California-San Diego
Scripps Institution of Oceanography
9500 Gilman Drive
La Jolla, CA 92093-0230, USA
e-mail: *tchereskin@ucsd.edu*

Geneviève Comte-Bellot
Ecole Centrale de Lyon
Centre Acoustique
36 avenue Guy de Collongue
69134 Ecully, France
e-mail: *genevieve.comte-bellot@ec-lyon.fr*

Joël De Coninck
Université de Mons-Hainaut
Centre for Research in Molecular Modelling
20 Place du Parc
7000 Mons, Belgium
e-mail: *joel.de.coninck@galileo.umh.ac.be*

Laurent Cordier
CNRS, Université de Poitiers
Laboratoire d'Etudes Aérodynamiques
43, route de l'aérodrome
86036 Poitiers, France
e-mail: *Laurent.Cordier@univ-poitiers.fr*

Alvaro Cuerva
Universidad Politécnica de Madrid
Instituto Universitario de Microgravedad,
Aeronáuticos
Plaza del Cardenal Cisneros 3
Madrid, 28040, Spain
e-mail: *acuerva@idr.upm.es*

Werner J.A. Dahm
The University of Michigan
Department of Aerospace Engineering
3056 FXB 2140
Ann Arbor, MI 48109-2140, USA
e-mail: *wdahm@umich.edu*

Joël Delville
CNRS, Université de Poitiers – CEAT
Laboratoire d'Etudes Aérodynamiques
47, route de l'Aérodrome
86036 Poitiers, France
e-mail: *joel.delville@lea.univ-poitiers.fr*

Andreas Dreizler
Technische Universität Darmstadt
Fachgebiet für Energie- und Kraftwerkstechnik
Petersenstraße 30
64287 Darmstadt, Germany
e-mail: *dreizler@ekt.tu-darmstadt.de*

John K. Eaton
Stanford University
Department of Mechanical Engineering
Building 500
Stanford, CA 94305, USA
e-mail: *eaton@vk.stanford.edu*

Volker Ebert
Universität Heidelberg
Physikalisch Chemisches Institut
Im Neuenheimer Feld 253
69120 Heidelberg, Germany
e-mail: *volker.ebert@pci.uni-heidelberg.de*

Yasuhiro Egami
Institute of Aerodynamics and Flow Technology
German Aerospace Center
Bunsenstr. 10
37073 Göttingen, Germany
e-mail: *yasuhiro.egami@dlr.de*

Rolf H. Engler
Institute of Aerodynamics and Flow Technology
German Aerospace Center, Experimental Methods
Bunsenstr. 10
37073 Göttingen, Germany
e-mail: *rolf.engler@dlr.de*

Marie Farge
Ecole Normale Supérieure
Laboratoire de Météorologie Dynamique
(IPSL-CNRS)
24, rue Lhomond
75231 Paris, France
e-mail: farge@lmd.ens.fr

Harindra J.S. Fernando
Arizona State University
Mechanical and Aerospace Engineering
Tempe, AZ 85287-9809, USA
e-mail: j.fernando@asu.edu

Uwe Fey
Institute of Aerodynamics and Flow Technology
German Aerospace Center
Bunsenstr. 10
37073 Göttingen, Germany
e-mail: Uwe.Fey@dlr.de

John F. Foss
Michigan State University
Mechanical Engineering
A118 Engineering Research Complex
East Lansing, MI 48824-1326, USA
e-mail: foss@egr.msu.edu

Marcelo H. García
University of Illinois at Urbana-Champaign
Ven Te Chow Hydrosystems Laboratory, Department of Civil and Environmental Engineering
205 North Mathews Avenue
Urbana, IL 61801-2352, USA
e-mail: mhgarcia@uiuc.edu

Klaus Hannemann
German Aerospace Center (DLR)
Institute of Aerodynamics and Flow Technology, Spacecraft Section
Bunsenstraße 10
37073 Göttingen, Germany
e-mail: Klaus.Hannemann@dlr.de

Lutz Heymann
Universität Bayreuth
Lehrstuhl für Technische Mechanik
und Strömungsmechanik, Fakultät
für Angewandte Naturwissenschaften
95440 Bayreuth, Germany
e-mail: lutz.heymann@uni-bayreuth.de

Bruce Howe
University of Washington
Applied Physics Laboratory
1013 NE 40th Street
Seattle, WA 98105-6698, USA
e-mail: howe@apl.washington.edu

Wolf-Heinrich Hucho
Gärtnereiweg 3
86938 Schondorf, Germany
e-mail: HuchoWHH@t-online.de

Klaus Hufnagel
Technische Universität Darmstadt
Fluid Mechanics and Aerodynamics
Petersenstr. 30
64287 Darmstadt, Germany
e-mail: k.hufnagel@aero.tu-darmstadt.de

Bernd Jähne
University of Heidelberg
Research Group Image Processing,
Interdisciplinary Center for Scientific Computing
Im Neuenheimer Feld 368
69120 Heidelberg, Germany
e-mail: Bernd.Jaehne@iwr.uni-heidelberg.de

Markus Jehle
University of Heidelberg
Research Group Image Processing,
Interdisciplinary Center for Scientific Computing
Im Neuenheimer Feld 368
69120 Heidelberg, Germany
e-mail: Markus.Jehle@iwr.uni-heidelberg.de

Joseph Katz
The Johns Hopkins University
Department of Mechanical Engineering
3400 N. Charles St.
Baltimore, MD 21218, USA
e-mail: katz@jhu.edu

Damien Kawakami
University of Minnesota
Mechanical Engineering
10550 Nicollet Ave S.
Bloomington, MN 55420, USA
e-mail: kawa0036@umn.edu

Saeid Kheirandish
Universität Karlsruhe (TH)
Institut für Mechanische Verfahrenstechnik und Mechanik
Bauing. Kolleg. III, Fasanengarten,
Gotthard-Franz-Straße 3
76131 Karlsruhe, Germany
e-mail: Saeid.Kheirandish@mvm.uni-karlsruhe.de

Michael Klar
University of Heidelberg
Research Group Image Processing,
Interdisciplinary Center for Scientific Computing
Im Neuenheimer Feld 368
69120 Heidelberg, Germany
e-mail: Michael.Klar@iwr.uni-heidelberg.de

Joseph C. Klewicki
University of New Hampshire
Department of Mechanical Engineering
Durham, NH 03824, USA
e-mail: Joe.klewicki@unh.edu

Manoochehr M. Koochesfahani
Michigan State University
Department of Mechanical Engineering
A131, Engineering Research Complex
East Lansing, MI 48824, USA
e-mail: koochesf@egr.msu.edu

Tomasz A. Kowalewski
Polish Academy of Sciences
Institute of Fundamental Technological Research
(IPPT PAN), Department of Mechanics and Physics of Fluids
Swietokrzyska 21
Warszawa, 00-049, Poland
e-mail: tkowale@ippt.gov.pl

Eric Lauga
Massachusetts Institute of Technology
Department of Mathematics
77 Massachusetts Avenue
Cambridge, MA 02139-4307, USA
e-mail: lauga@mit.edu

Joseph W. Lee
NASA Langley Research Center
Advanced Sensing and Optical Measurements Branch
Mail Stop 493
Hampton, VA 23681, USA
e-mail: Joseph.W.Lee@nasa.gov

Jacques Lewalle
Syracuse University
Mechanical Engineering
149 Link Hall
Syracuse, NY 13244, USA
e-mail: jlewalle@syr.edu

Phillip Ligrani
University of Oxford
Department of Engineering Science
Parks Road
Oxford, OX1 3PJ, UK
e-mail: phil.ligrani@eng.ox.ac.uk

Abraham Marmur
Technion-Israel Institute of Technology
Department of Chemical Engineering
Haifa 32000, Israel
e-mail: marmur@technion.ac.il

Jan Martinez Schramm
Institute of Aerodynamics and Flow Technology
Department Spacecraft
Bunsenstraße 10
37073 Göttingen, Germany
e-mail: Jan.Martinez@dlr.de

Ivan Marusic
University of Melbourne
Victoria, Australia
e-mail: imarusic@unimelb.edu.au

Beverley J. McKeon
California Institute of Technology
Graduate Aeronautical Laboratories,
Division of Engineering and Applied Science
1200 E. California Blvd. MC 301-46
Pasadena, CA 91125-4600, USA
e-mail: *mckeon@caltech.edu*

Gareth H. McKinley
Massachusetts Institute of Technology
Department of Mechanical Engineering
77 Massachusetts Avenue
Cambridge, MA 02139, USA
e-mail: *gareth@mit.edu*

Charles Meneveau
The Johns Hopkins University
Department of Mechanical Engineering
3400 N. Charles Street
Baltimore, MD 21218, USA
e-mail: *meneveau@jhu.edu*

Wolfgang Merzkirch
Universität Essen
Lehrstuhl für Strömungslehre
Universitätsstraße 2
45141 Essen, Germany
e-mail: *wolfgang.merzkirch@uni-essen.de*

James F. Meyers
NASA Langley Research Center
Advanced Sensing and Optical Measurements
Branch
Mail Stop 493
Hampton, VA 23681, USA
e-mail: *james.f.meyers@nasa.gov*

Scott C. Morris
University of Notre Dame
Department of Aerospace and Mechanical
Engineering
109 Hessert Laboratory
Notre Dame, IN 46556, USA
e-mail: *s.morris@nd.edu*

John A. Mullin
Northrop Grumman Space Technology
One Space Park
Redondo Beach, CA 90278, USA
e-mail: *John.mullin@ngc.com*

Klaas te Nijenhuis
Delft University of Technology
Department of Polymer Materials
and Polymer Engineering
Julianalaan 136
2628 BL, Delft, The Netherlands
e-mail: *k.tenijenhuis@tnw.tudelft.nl*

Holger Nobach
Max Planck Institute for Dynamics
and Self-Organization
Am Faßberg 17
37077 Göttingen, Germany
e-mail: *holger.nobach@nambis.de*

Jeffrey A. Odell
University of Bristol
Department of Physics
Tyndall Avenue
Bristol, BS8 1TH, UK
e-mail: *Jeff.odell@bristol.ac.uk*

Nicholas T. Ouellette
Max Planck Institute for Dynamics
and Self-Organization
Laboratory of Fluid Dynamics, Pattern Formation
and Nanobiocomplexity (LFPN)
Am Faßberg 11
37077 Göttingen, Germany
e-mail: *nicholas.ouellette@ds.mpg.de*

Ronald L. Panton
The University of Texas
Mechanical Engineering Department
1 University Station C2200
Austin, TX 78712-0292, USA
e-mail: *rpanton@mail.utexas.edu*

Eric R. Pardyjak
University of Utah
Department of Mechanical Engineering
Salt Lake City, UT 8411, USA
e-mail: *pardyjak@eng.utah.edu*

Marc Perlin
University of Michigan
Naval Architecture and Marine Engineering
2600 Draper Rd.
Ann Arbor, MI 48109-2145, USA
e-mail: *perlin@umich.edu*

Marko Princevac
University of California Riverside
Department of Mechanical Engineering
Bourns Hall A315
Riverside, CA 92521, USA
e-mail: marko@engr.ucr.edu

Alberto T. Pérez
Universidad de Sevilla
Departamento de Electrónica
y Electromagnetismo, Facultad de Física
Avda. Reina Mercedes s/n
Sevilla, 41012, Spain
e-mail: alberto@us.es

Giovanni Paolo Romano
University of Roma "La Sapienza"
Department Mechanics and Aeronautics
Via Eudossiana 18
00184 Rome, Italy
e-mail: romano@dma.ing.uniroma1.it

William S. Saric
Texas A&M University
Aerospace Engineering
College Station, TX 77843-3141, USA
e-mail: saric@tamu.edu

Fulvio Scarano
Delft University of Technology
Aerospace Engineering Department
Kluyverweg 1
2627 HS, Delft, The Netherlands
e-mail: f.scarano@tudelft.nl

Günter Schewe
German Aerospace Center (DLR)
Institut für Aeroelastik
Bunsenstraße 10
37073 Göttingen, Germany
e-mail: guenter.schewe@dlr.de

Kai Schneider
Université de Provence
Centre de Mathématiques et d'Informatique
39, rue F. Joliot-Curie
13453 Cedex 13 Marseille, France
e-mail: kschneid@cmi.univ-mrs.fr

Richard Schodl
German Aerospace Center (DLR)
Institute of Propulsion Technology
Engine Measurement Systems
Linder Höhe
51147 Köln, Germany
e-mail: richard.schodl@dlr.de

Christof Schulz
Universität Duisburg-Essen
Institut für Verbrennung und Gasdynamik
Lotharstrasse 1
47057 Duisburg, Germany
e-mail: christof.schulz@uni-duisburg.de

Stephen Spiegelberg
Cambridge Polymer Group
52-R Roland Street
Boston, MA 02129-1234, USA
e-mail: steve@campoly.com

Victor Steinberg
Weizmann Institute of Science
Department of Physics of Complex Systems
Rehovot 76100, Israel
e-mail: victor.steinberg@weizmann.ac.il

Howard A. Stone
Harvard University
Division of Engineering and Applied Sciences
Cambridge, MA , USA
e-mail: has@deas.harvard.edu

Stephanus A. Theron
Technion – Israel Institute of Technology
Department of Mechanical Engineering
Haifa 3200, Israel
e-mail: therons@tx.technion.ac.il

Cameron Tropea
Technische Universität Darmstadt
Fachgebiet Strömungslehre und Aerodynamik
Petersenstraße 30
64287 Darmstadt, Germany
e-mail: ctropea@sla.tu-darmstadt.de

Oğuz Uzol
Middle East Technical University
Department of Aerospace Engineering
Ankara, 06531, Turkey
e-mail: uzol@ae.metu.edu.tr

Petar V. Vukoslavčević
University of Montenegro
Department of Mechanical Engineering
Cetinjski put bb
Podgorica, 8100, Montenegro
e-mail: *petarv@cg.ac.yu*

Manfred H. Wagner
Technische Universität Berlin
Polymertechnik/Polymerphysik
Fasanenstr. 90
10623 Berlin, Germany
e-mail: *manfred.wagner@tu-berlin.de*

William A. Wakeham
University of Southampton
Highfield
Southampton, SO17 1BJ, UK
e-mail: *vice-chancellor@soton.ac.uk*

James M. Wallace
University of Maryland
Department of Mechanical Engineering
College Park, MD 20742, USA
e-mail: *wallace@eng.umd.edu*

Jerry Westerweel
Delft University of Technology
Department of Mechanical Engineering
Leeghwaterstraat 21
2628 CA, Delft, The Netherlands
e-mail: *J.Westerweel@wbmt.tudelft.nl*

Charles H.K. Williamson
Cornell University
Mechanical and Aerospace Engineering
144 Upson Hall
Ithaca, NY 14853, USA
e-mail: *cw26@cornell.edu*

Jürgen Wolfrum
Ruprecht-Karls-Universität Heidelberg
Physikalisch-Chemisches Institut
Im Neuenheimer Feld 253
69120 Heidelberg, Germany
e-mail: *wolfrum@urz.uni-heidelberg.de*

Martin Wosnik
Alden Research Laboratory
30 Shrewsbury St.
Holden, MA 01520, USA
e-mail: *mwosnik@aldenlab.com*

Haitao Xu
Max Planck Institute for Dynamics
and Self-Organization
Department of Fluid Dynamics, Pattern Formation,
and Nanobiocomplexity
Am Faßberg 11/Turm 2
37077 Göttingen, Germany
e-mail: *haitao.xu@ds.mpg.de*

Alexander L. Yarin
University of Illinois at Chicago
Department of Mechanical
and Industrial Engineering
842 W. Taylor St.
Chicago, IL 60607-7022, USA
e-mail: *ayarin@uic.edu*

Eyal Zussman
Technion – Israel Institute of Technology
Department of Mechanical Engineering
Haifa 3200, Israel
e-mail: *meeyal@tx.technion.ac.il*

James H. Duncan
University of Maryland
Department of Mechanical Engineering
3181 Martin Hall
College Park, MD 207, USA
e-mail: *duncan@eng.umd.edu*

Daniel G. Nocera
Massachusetts Institute of Technology
Department of Chemistry
77 Massachusetts Ave.
Cambridge, MA 02139-4307, USA
e-mail: *nocera@mit.edu*

Contents

The authors of the different Sections can be identified by consulting the *Detailed Contents* or the *About the Authors* part.

List of Abbreviations	XXI
Nomenclature	XXV

Part A Experiments in Fluid Mechanics

1 Experiment as a Boundary-Value Problem
Ronald L. Panton, Saeid Kheirandish, Manfred H. Wagner 3
- 1.1 Thermodynamic Equations 3
- 1.2 Kinematic Equations 5
- 1.3 Balance Laws and Local Governing Equations 6
- 1.4 Balance Laws and Global Governing Equations 8
- 1.5 Constitutive Equations 10
- 1.6 Navier–Stokes Equations 11
- 1.7 Discontinuities in Density 11
- 1.8 Constitutive Equations and Nonlinear Rheology of Polymer Melts 13
- **References** 30

2 Nondimensional Representation of the Boundary-Value Problem
John F. Foss, Ronald L. Panton, Alexander L. Yarin 33
- 2.1 Similitude, the Nondimensional Prototype and Model Flow Fields ... 34
- 2.2 Dimensional Analysis and Data Organization 42
- 2.3 Self-Similarity 57
- **References** 82

Part B Measurement of Primary Quantities

3 Material Properties: Measurement and Data
William A. Wakeham, Marc J. Assael, Abraham Marmur, Joël De Coninck, Terry D. Blake, Stephanus A. Theron, Eyal Zussman 85
- 3.1 Density 85
- 3.2 Surface Tension and Interfacial Tension of Liquids 96
- 3.3 Contact Angle 106
- 3.4 Viscosity 119
- 3.5 Thermal Conductivity and Thermal Diffusivity 133
- 3.6 Diffusion 147
- 3.7 Electric and Magnetic Parameters of Liquids and Gases 158
- **References** 169

4 Pressure Measurement Systems
Beverley J. McKeon, Rolf H. Engler 179
  4.1 Measurement of Pressure with Wall Tappings 180
  4.2 Measurement of Pressure with Static Tubes 185
  4.3 Hardware and Other Considerations 187
  4.4 Pressure-Sensitive Paint (PSP) 188
  References 209

5 Velocity, Vorticity, and Mach Number
*Beverley J. McKeon, Geneviève Comte-Bellot, John F. Foss,
Jerry Westerweel, Fulvio Scarano, Cameron Tropea, James F. Meyers,
Joseph W. Lee, Angelo A. Cavone, Richard Schodl,
Manoochehr M. Koochesfahani, Daniel G. Nocera, Yiannis Andreopoulos,
Werner J.A. Dahm, John A. Mullin, James M. Wallace,
Petar V. Vukoslavčević, Scott C. Morris, Eric R. Pardyjak, Alvaro Cuerva* 215
  5.1 Pressure-Based Velocity Measurements 216
  5.2 Thermal Anemometry 229
  5.3 Particle-Based Techniques 287
  5.4 Molecular Tagging Velocimetry (MTV) 362
  5.5 Vorticity 382
  5.6 Thermal Transient Anemometer (TTA) 434
  5.7 Sonic Anemometry/Thermometry 436
  References 446

6 Density-Based Techniques
Wolfgang Merzkirch 473
  6.1 Density, Refractive Index, and Optical Flow Visualization 473
  6.2 Shadowgraphy 474
  6.3 Schlieren Method 476
  6.4 Moiré Deflectometry 478
  6.5 Interferometry 480
  6.6 Optical Tomography 485
  References 485

7 Temperature and Heat Flux
*Tomasz A. Kowalewski, Phillip Ligrani, Andreas Dreizler, Christof Schulz,
Uwe Fey, Yasuhiro Egami* 487
  7.1 Thermochromic Liquid Crystals 488
  7.2 Measurements of Surface Heat Transfer Characteristics
      Using Infrared Imaging 500
  7.3 Temperature Measurement via Absorption, Light Scattering
      and Laser-Induced Fluorescence 515
  7.4 Transition Detection by Temperature-Sensitive Paint 537
  References 553

8 Force and Moment Measurement
Klaus Hufnagel, Günter Schewe 563
8.1 Steady and Quasi-Steady Measurement 564
8.2 Force and Moment Measurements in Aerodynamics
and Aeroelasticity Using Piezoelectric Transducers 596
References 615

Part C Specific Experimental Environments and Techniques

9 Non-Newtonian Flows
*Klaas te Nijenhuis, Gareth H. McKinley, Stephen Spiegelberg,
Howard A. Barnes, Nuri Aksel, Lutz Heymann, Jeffrey A. Odell* 619
9.1 Viscoelastic Polymeric Fluids 619
9.2 Thixotropy, Rheopexy, Yield Stress 661
9.3 Rheology of Suspensions and Emulsions 680
9.4 Entrance Correction and Extrudate Swell 720
9.5 Birefringence in Non-Newtonian Flows 724
References 732

10 Measurements of Turbulent Flows
*Giovanni Paolo Romano, Nicholas T. Ouellette, Haitao Xu,
Eberhard Bodenschatz, Victor Steinberg, Charles Meneveau, Joseph Katz* 745
10.1 Statistical Eulerian Description of Turbulent Flows 746
10.2 Measuring Lagrangian Statistics in Intense Turbulence 789
10.3 Elastic Turbulence in Viscoelastic Flows 799
10.4 Measurements for Large-Eddy Simulations 830
References 849

11 Flow Visualization
Wolfgang Merzkirch 857
11.1 Aims and Principles of Flow Visualization 857
11.2 Visualizations of Flow Structures and Flow Direction 859
11.3 Visualization of Free Surface Flows 868
References 869

12 Wall-Bounded Flows
Joseph C. Klewicki, William S. Saric, Ivan Marusic, John K. Eaton 871
12.1 Introductory Concepts 871
12.2 Measurement of Wall Shear Stress 875
12.3 Boundary-Layer Stability and Transition 886
12.4 Measurements Considerations in Non-Canonical Flows 896
References 902

13 Topological Considerations in Fluid Mechanics Measurements
John F. Foss 909
13.1 A Companion Document 909
13.2 Utilization of Topological Considerations for Flow Field Analyses 910
References 918

14 Flow Measurement Techniques in Turbomachinery
Oğuz Uzol, Joseph Katz 919
14.1 Background On Turbomachinery Flows 919
14.2 Non-Optical Measurement Techniques 921
14.3 Optical Measurement Techniques 933
14.4 Concluding Remarks 950
References 951

15 Hydraulics
Roger E.A. Arndt, Damien Kawakami, Martin Wosnik, Marc Perlin, James H. Duncan, David M. Admiraal, Marcelo H. García 959
15.1 Measurements in Cavitating Flows 959
15.2 Wave Height and Slope 1009
15.3 Sediment Transport Measurements 1020
References 1033

16 Aerodynamics
Wolf-H. Hucho, Klaus Hannemann, Jan Martinez Schramm, Charles H.K. Williamson 1043
16.1 Ground Vehicle Aerodynamics 1043
16.2 Short-Duration Testing of High Enthalpy, High Pressure, Hypersonic Flows 1081
16.3 Bluff Body Aerodynamics 1125
References 1146

17 Atmospheric Measurements
Harindra J.S. Fernando, Marko Princevac, Ronald J. Calhoun 1157
17.1 Point Measurements 1159
17.2 Dispersion Measurements 1167
17.3 Remote Sensing 1169
17.4 Satellite Measurements 1175
References 1178

18 Oceanographic Measurements
Bruce Howe, Teresa K. Chereskin 1179
18.1 Oceanography 1179
18.2 Point Measurements 1182
18.3 Lagrangian Techniques 1188
18.4 Remote Sensing 1192

18.5	Measurement Systems	1203
18.6	Experiment Case Studies	1208
	References	1214

19 Microfluidics: The No-Slip Boundary Condition
Eric Lauga, Michael P. Brenner, Howard A. Stone 1219

19.1	History of the No-Slip Condition	1220
19.2	Experimental Methods	1222
19.3	Molecular Dynamics Simulations	1226
19.4	Discussion: Dependence on Physical Parameters	1228
19.5	Perspective	1234
	References	1235

20 Combustion Diagnostics
Christof Schulz, Andreas Dreizler, Volker Ebert, Jürgen Wolfrum 1241

20.1	Basics	1242
20.2	Laser-Based Combustion Diagnostics	1243
20.3	Experimental Data Devoted to Validation of Numerical Simulations and Modeling	1244
20.4	Application of Laser-Based Techniques	1247
20.5	Conclusions	1299
	References	1300

21 Electrohydrodynamic Systems
Antonio Castellanos, Alberto T. Pérez 1317

21.1	Equations	1318
21.2	Fluid Statics and Dynamics in EHD	1320
21.3	Experimental Methods in EHD	1322
21.4	Conductivity	1323
21.5	Mobility	1327
21.6	Electric Field Measurement: Kerr Effect	1328
21.7	Velocity	1329
21.8	Visualization	1330
	References	1331

Part D Analysis and Post-Processing of Data

22 Review of Some Fundamentals of Data Processing
Holger Nobach, Cameron Tropea, Laurent Cordier, Jean-Paul Bonnet,
Joël Delville, Jacques Lewalle, Marie Farge, Kai Schneider, Ronald J. Adrian 1337

22.1	Fourier Transform	1337
22.2	Correlation Function	1342
22.3	Hilbert Transform	1344
22.4	Proper Orthogonal Decomposition: POD	1346

	22.5	Conditional Averages and Stochastic Estimation	1370
	22.6	Wavelet Transforms	1378
	References		1395

23 Fundamentals of Data Processing
Holger Nobach, Cameron Tropea 1399

	23.1	Statistical Principles	1399
	23.2	Stationary Random Processes	1401
	23.3	Estimator Expectation and Variance	1402
	23.4	Signal Noise	1406
	23.5	Cramér–Rao Lower Bound (CRLB)	1408
	23.6	Propagation of Errors	1416
	References		1417

24 Data Acquisition by Imaging Detectors
Bernd Jähne 1419

	24.1	Definitions	1419
	24.2	Types of Detectors	1420
	24.3	Imaging Detectors	1421
	24.4	Performance of Imaging Sensors	1426
	24.5	Camera Selection	1435
	References		1436

25 Data Analysis
Bernd Jähne, Michael Klar, Markus Jehle 1437

	25.1	Image Processing	1437
	25.2	Motion Analysis	1464
	References		1488

Acknowledgements	1493
About the Authors	1495
Detailed Contents	1513
Subject Index	1531

List of Abbreviations

3-D	three-dimensional
4-D MRV	4-D magnetic resonance velocimetry

A

A/D	analog-to-digital
ABS	acoustic bubble spectrometer
ABS	acoustic backscatter
acac	acetylacetonate
ACF	autocorrelation function
ACN	acetonitrile
ADC	analog-to-digital converter
ADCP	acoustic Doppler current profiler
AFM	atomic force microscopy
AFTRF	axial flow turbine research facility
AGW	adaptive Gaussian windowing
AMODE-MST	acoustic mid-ocean dynamics experiment-moving ship tomography
APE-HKE	available potential energy-to-horizontal kinetic energy
ARD	atmospheric re-entry demonstrator
ASFM	acoustic scintillation flow meter
ASTM	American Society for Testing and Materials
ATK–GASL	Alliant Techsystems Inc. – General Applied Science Laboratories
AUV	autonomous undersea vehicles
AVHRR	advanced very high-resolution radiometer
AVP	absolute velocity profiler
AXBT	air-expendable bathythermograph

B

BBO	Basset–Boussinesq–Oseen
BC	boundary condition
BCCE	brightness change constraint equation
BL	boundary layer
BT	bathythermograph

C

CAD	crank angle degree
CARS	coherent anti-Stokes Raman spectroscopy
CARS	coherent anti-Stokes Raman scattering
CCA	constant current anemometer
CCD	charge-coupled device
CD	cyclodextrin
CFD	computational fluid dynamic
CFK	carbon-fiber-reinforced plastic
CFT	continuous Fourier transform
CLSM	confocal laser scanning microscopy
CMC	critical micelle concentration
CMD	count median diameter
CMD	count mean diameter
CMOS	complementary metal oxide semiconductor
CP	cone-and-plate
CR	constraint release
CRDLAS	cavity-ring-down laser-absorption spectroscopy
CRLAS	cavity ring-down laser absorption spectroscopy
CRLB	Cramér–Rao lower bound
CRV	crew return vehicle
CS	coherent structures
CSM	cavitation susceptibility meters
CSR	controlled-shear-rate
CSS	controlled-shear-stress
CT	Couette–Taylor
CTA	constant temperature anemometer
CTAB	cetyltrimethyl ammonium bromide
CTD	conductivity–temperature–depth/pressure
CUBRC	Calspan - University at Buffalo Research Center
CVA	constant voltage anemometer

D

DAS	direct absorption spectroscopy
DBP	di-n-butyl phthalate
DBR	distributed Bragg reflector
DE	Doi and Edwards
DEHS	di-ethyl-hexyl-sebacat
DFB	distributed feedback
DFG	difference-frequency generation
DFT	discrete Fourier transform
DFWM	degenerate four-wave mixing
DGV	Doppler global velocimetry
DIDSON	dual frequency identification sonar
DIN	Deutsches Institut für Normung
DL	diode laser
DLR	German Aerospace Center
DMF	dimethylformamide
DMSO	dimethylsulfoxide
DNS	direct numerical simulation
DOE	diffractive optical element
DPIV	digital PIV
DPV	Doppler picture velocimetry
DR	dynamic range
DSNU	dark signal nonuniformity
DSPIV	dual-plane particle image velocimetry

E

EAST	Electric Arc Shock Tunnel
EBCCE	extended brightness change constraint equation
EDM	electric discharge machining
EGR	exhaust gas recirculation
ELAC	elliptical aerodynamic configuration
ELIF	excimer-laser-induced fragmentation
EM-CCD	electron-multiplying CCD
EMI	electromagnetic interference
EMVA	European Machine Vision Association
ESA	European Space Agency
ESTEC	European Space Research and Technology Center
ETW	European transonic wind tunnel

F

FARLIF	fuel–air ratio by laser-induced fluorescence
FBRM	focused beam reflectance measurement
FFT	fast Fourier transform
FFW	finite fringe width
FIR	far-infrared
FITCD	fluorescent-conjugated dextran
FMS	frequency modulation spectroscopy
FOBS	fiber-optic backscatter
FPN	fixed pattern noise
FTR	Fourier-transform rheology

G

GBCCE	generalized brightness change constraint equation
GEO	geostationary Earth orbit
GIFTS	geosynchronous imaging Fourier-transform spectrometer
GOES	geostationary operational environmental satellites
GPS	global positioning system
GUM	guide of uncertainties in measurement

H

HDG	high-pressure windtunnel
HEG	High Enthalpy Shock Tunnel Göttingen
HEM	horizontal electrometer
HF	high-frequency
HIEST	high enthalpy shock tunnel
HITRAN	high-resolution transmission molecular absorption
HPIV	holographic particle image velocimetry
HPR	heave–pitch–roll
HTV	hydroxyl tagging velocimetry
HWA	hot-wire anemometry

I

IAPWS	International Association for the Properties of Water and Steam
IC	initial condition
IC	internal combustion
ICCD	intensified CCD
ICET	international cavitation erosion test
IEP	isoelectric point
IES	inverted echosounder
IFW	infinite fringe width
IGV	inlet guide vanes
IMET	improved meteorological packages
IRT	infrared thermography
ISC	intersystem crossing
ISL	Institute of Saint Louis
ISO	International Organization for Standards
ISS	International Space Station
ITTC	international towing tank conference
IVC	iodine vapor cell
IVK	Institut für Verbrennungsmotoren und Kraftfahrwesen

J

JAXA	Japanese Aeronautics Exploration Agency
JFTA	joint frequency–time analysis
JIS	Japanese Industrial Standards

L

L2F	laser two-focus velocimetry
LAOS	large-amplitude oscillatory shear
LAS	laser absorption spectroscopy
LCO	limit-cycle oscillation
LD	laser Doppler
LDA	laser Doppler anemometry
LDPE	low-density polyethylene
LDV	laser Doppler velocimetry
LED	light-emitting diodes
LEI	laser-enhanced ionization
LENS	large energy national shock
LES	large-eddy simulation
LFM	laser frequency monitor
LIF	laser-induced fluorescence
LII	laser-induced incandescence
LIM	local intermittency measure
LIPA	laser-induced photochemical anemometry
LISST	laser in situ scattering and transmissometry
LPT	Lagrangian particle tracking
LSS	laser speckle strophometry
LT	laser transit
LTV	laser transit velocimetry
LWIR	long-wavelength infrared

M

MARIN	Maritime Research Institute Netherlands
MC	modulus-compensating
MC	methylene chloride
MDA	minimum detectable absorption
MDPR	mean depth of erosion penetration rate
MEMS	microelectromachined sensors
MFI	melt flow index
MHT	multiple hypothesis tracker
ML	maximum-likelihood
MLE	maximum-likelihood estimator
MMH	mixed metal hydroxide
MRA	multiresolution analysis
MRI	magnetic resonance imaging
MSACA	most stable apparent contact angle
MSF	molecular stress function
MT	montmorillonite
MTV	molecular tagging velocimetry
MW	Maxwell–Wiechert
MWIR	mid-wavelength infrared
MZI	Mach–Zehnder interferometer

N

NACA	National Advisory Committee for Aeronautics
NAL	National Aeronautics Laboratory
NASA	National Aeronautics and Space Administration
Nd:YAG	neodymium-doped yttrium aluminum garnet
NEE	noise-equivalent exposure
NETD	noise equivalent temperature difference
NIR	near-infrared
NMA	monomethylacetamide
NMR	nuclear magnetic resonance
NMT	monomethyltryptamine
NO	nitric oxide

O

OBS	optical backscatter
ODE	ordinary differential equations
OH	hydroxide,weg
OPG	optical parametric generation
OPO	optical parametric oscillator
OTV	ozone tagging velocimetry

P

PAA	polyacrylic acid
PAA	polyacrylamide
PACA	practical advancing contact angle
PAH	polycyclic aromatic hydrocarbon
PBS	phosphate buffer solution
PCI	peripheral component interface
PCL	polycaprolactone
PDA	phase Doppler anemometry
PDE	partial differential equations
PDF	probability density function
PDI	polydispersity index
PDMS	polydimethylsiloxane
PDV	planar Doppler velocimetry
PET	poly(ethyleneterephthalate)
PETW	pilot facility of ETW
PHANTOMM	photoactivated non-intrusive tracking of molecular motion
PIB	polyisobuthylene
PIT	phase inversion temperature
PIV	particle image velocimetry
PLIF	planar laser-induced fluorescence
PM	polarization modulation
PMMA	polymethylmethacrylate
POCS	projection onto convex sets
POD	proper orthogonal decomposition
PP	plate–plate
PPI	plan–position indicator
PRCA	practical receding contact angle
PRNU	photoresponse nonuniformity
PS	polarization spectroscopy
PSD	particle size distribution
PSD	power spectral density
PSF	point spread function
PSP	pressure-sensitive paint
PTV	particle tracking velocimetry
PU	polyurethane
PUV	pressure and two components of horizontal current
PVA	polyvinyl alcohol
PVDF	polyvinylidene fluoride
PWM-CTA	pulse-width-modulated constant-temperature anemometer

R

RANS	Reynolds-averaged Navier–Stokes equations
RASS	radio acoustic sounding systems
RELIEF	Raman excitation plus laser-induced electronic fluorescence
REMPI	resonance-enhanced multiphoton ionization
REMUS	remote environmental monitorung units
RET	rotational energy transfer
RFI	radio frequency interference
RH	relative humidity
RHI	range–height indicator
RIC	relative information content
RMS	root-mean-square

S

S+H	sample-and-hold
SAOS	small-amplitude oscillatory shear
SAR	synthetic aperture radar
SAT	sonic anemometer/thermometer
SCR	selective catalytic reduction
SEE	saturation-equivalent exposure
SER	sentmanat extension rheometer
SFA	surface force apparatus
SFG	sum-frequency generation
SG	specific gravity
SGS	subgrid-scale stress
SHG	second-harmonic generation
SI	spark ignition
SNCR	selective non-catalytic reduction
SNR	signal-to-noise ratio
SODAR	sound detection and ranging
SOFAR	sound fixing and ranging
SPL	sound-pressure level
SR	spatial resolution
SSD	sum-of-squared differences
SVD	singular value decomposition
SWIR	short-wavelength infrared

T

TACA	theoretical advancing contact angle
TCFB	two-color flow birefringence
TDC	top dead center
TDLAS	tunable diode laser absorption
TEM	transmission electron microscopy
THF	tetrahydrofuran
TLC	thermochromic liquid crystals
TOPEX	topography experiment
TPT	thermographic phosphor thermography
TR	time resolution
TRCA	theoretical receding contact angle
TS	Tollmien–Schlichting
TSP	temperature-sensitive paint
TWG	transonic wind tunnel Göttingen

U

UV	ultraviolet
UVW	three components of velocity

V

VAO	Versuchsanstalt für Wasserbau Obernach
VCSEL	vertical-cavity surface-emitting laser
VET	vibrational energy transfer
VPI	Virginia Polytechnic Institute

W

WMS	wavelength modulation spectroscopy

X

XBT	expendable bathythermograph
XPP	extended pom-pom

Nomenclature

Experimental fluid mechanics draws on numerous disciplines in addition to fluid mechanics itself: rheology, physics, electromagnetic theory, optics, electronics, signal processing, data processing, etc. Each discipline and community has developed its own nomenclature and conventions and inevitably there exist many conflicting designations when one attempts to assemble all of these conventions into one handbook. Therefore, we have instructed authors to adhere as closely as possible to the skeleton nomenclature given below and to note explicitly in their respective articles any deviations or extensions thereof. Authors of articles dealing with constitutive equations, material properties and non-Newtonian flows were asked to follow, as far as possible, the nomenclature given in J.M. Dealy: Official nomenclature for material functions describing the response of a viscoelastic fluid to various shearing and extensional deformations, J. Rheol. **39**(1), 253–265 (1995). Furthermore, all authors were asked to prepare their manuscripts using SI units, a review of which is provided below, following the nomenclature.

List of Symbols

Vectors and tensors are written bold
Complex quantities carry an underscore

f	Frequency (Hertz)
m	Mass
$\dot{m}$	Mass flux
$\boldsymbol{n}$	Outer unit normal vector
p	Pressure
t	Time
T	Temperature
$\boldsymbol{u}$ or $\boldsymbol{v}$	Velocity vector
x_1 or x	Cartesian position coordinate
x_2 or y	Cartesian position coordinate
x_3 or z	Cartesian position coordinate
γ	Ratio of specific heats (C_P/C_V)
δ	Boundary-layer thickness
μ	Dynamic viscosity
ν	Kinematic viscosity
ρ	Density
σ	Surface tension
$\boldsymbol{\sigma}$ or σ_{ij}	Stress tensor
τ	Deviatoric stress
τ_w	Wall shear stress
ω	Circular frequency (rad/s)
$\boldsymbol{\omega}$	Vorticity vector
Λ	Circulation
Ψ	Stream function

Operators

$\overline{}$	Average of ensembles, time average
$\langle\ \rangle$	Spatial average
Im{}	Imaginary part
Re{}	Real part
$\mathcal{F}${}	Fourier transform
$\hat{}$	Estimator
E[]	Expectation
b^2[]	Bias
var[]	Variance
∇	Nabla
grad	Gradient of scalar: ∇
div	Divergence of vector; $\nabla\cdot$
curl or rot	Rotation vector: $\nabla\times$
Δ	Laplace: ∇^2

International System of Units (SI)

Base and Supplementary SI Units

Quantity	Name of Unit	Symbol
Length	meter	m
Mass	kilogram	kg
Time	second	s
Electric current	ampere	A
Thermodynamic temperature	kelvin	K
Luminous intensity	candela	cd
Amount of substance	mole	mol
Supplementary units		
Plane angle	radian	rad
Solid angle	steradian	sr

Multiplying Factors

Multiple and submultiple	Prefix	Symbol
10^{18}	exa	E
10^{15}	peta	P
10^{12}	tera	T
10^{9}	giga	G
10^{6}	mega	M
10^{3}	kilo	k
10^{2}	hecto	h
10	deka	da
10^{-1}	deci	d
10^{-2}	centi	c
10^{-3}	milli	m
10^{-6}	micro	μ
10^{-9}	nano	n
10^{-12}	pico	p
10^{-15}	femto	f
10^{-18}	atto	a

Derived SI Units

Quantity	Name(s) of unit	Unit symbol or abbreviation where differing from basic form	Unit expressed in terms of basic or supplementary units
Area	square meter		m^2
Volume	cubic meter		m^3
Frequency	hertz, cycle per second	Hz	s^{-1}
Density, concentration	kilogram per cubic meter		kg/m^3
Velocity	meter per second		m/s
Angular velocity	radian per second		rad/s
Acceleration	meter per second squared		m/s^2
Angular acceleration	radian per second squared		rad/s^2
Volumetric flow rate	cubic meter per second		m^3/s
Force	newton	N	$kg\ m/s^2$
Surface tension	newton per meter, joule per square meter	$N/m, J/m^2$	kg/s^2
Pressure	newton per square meter, pascal	N/m^2, Pa	kg/ms^2
Viscosity, dynamic	newton-second per square meter	$N\ s/m^2$, Pl	kg/ms
Viscosity, kinematic	poisseuil		m^2/s
Thermal and mass diffusivity	meter square per second		m^2/s
Work, torque, energy, quantity of heat	joule, newton-meter, watt-second	$J, N\ m, W\ s$	$kg\ m^2/s^2$
Power, heat flux	watt, joule per second	W, J/s	$kg\ m^2/s^3$
Heat flux density	watt per square meter	W/m^2	kg/s^3
Volumetric heat release rate	watt per cubic meter	W/m^2	kg/ms^{-3}

Quantity	Name(s) of unit	Unit symbol or abbreviation where differing from basic form	Unit expressed in terms of basic or supplementary units
Heat transfer coefficient	watt per square meter degree	W/m^{2}deg	kg/s^{2}deg
Latent heat, enthalpy (specific)	joule per kilogram	J/kg	m^2/s^2
Capacity rate	watt per degree	W/deg	kg m^2/s^3 deg
Thermal conductivity	watter per meter degree,	W/m deg J m/s m^2 deg	kg m/s^3 deg
Mass flux, mass flow rate	kilogram per second		kg/s
Mass flux density, mass flow rate per unit area	kilogram per square meter second		kg/m^{2}s
Mass-transfer coefficient	meter per second		m/s
Quantity of electricity	coulomb	C	A s
Electromotive force	volt	V, W/A	kg m^2/A s^3
Electric resistance	ohm	Ω, V/A	kg m^2/A^2 s^3
Electric conductivity	ampere per volt meter	A/V m	A^2 s^3/kg m^3
Electric capacitance	farad	F, A s/V	A^3 s^4/kg m^2
Magnetic flux	weber	Wb, V s	kg m^2/A s^2
Inductance	henry	H, V s/A	kg m^3/A^2 s^2
Magnetic permeability	henry per meter	H/m	kg m/A^2 s^2
Magnetic flux density	tesla, weber per square meter	T, Wb/m^2	kg/A s^2
Magnetic field strength	ampere per meter		A/m
Manetomotive force	ampere		A
Luminous flux	lumen	lm	cd sr
Luminance	candela per square meter		cd/m^2
Illuination	lux, lumen per square meter	lx, lm/m^2	cd sr/m^2

Non-Dimensional Numbers

Re	Reynolds number	St	Strouhal number
Ma	Mach number	Fr	Froude number
Pr	Prandtl number	Nu	Nusselt number

Subscripts

max	maximum	z or 3	Cartesian coordinates
min	minimum	⊥	Perpendicularly polarized
x or 1	Cartesian coordinates	∥	Parallel polarized
y or 2	Cartesian coordinates		

Superscripts

′	Fluctuating quantity in time
*	Complex conjugate
T	Transpose

Physical and Mathematical Constants

c	Speed of light	g	Gravity
e	2.718281828 ...	i	Imaginary unit
h	Planck's constant	π	3.141592653 ...

Functions

arg	Argument	max	Maximum
cos	Cosine function	min	Minimum
cosh	Hyperbolic cosine function	sin	Sine function
exp	Exponential function	sinh	Hyperbolic sine function
int	Integer	sgn	Signum function
ln	Logarithmis function, base e	tan	Tangent function
log	Logarithmic function, base 10	tanh	Hyperbolic tangent function

Part A Experiments in Fluid Mechanics

1 Experiment as a Boundary-Value Problem
Ronald L. Panton, Austin, USA
Saeid Kheirandish, Karlsruhe, Germany
Manfred H. Wagner, Berlin, Germany

2 Nondimensional Representation of the Boundary-Value Problem
John F. Foss, East Lansing, USA
Ronald L. Panton, Austin, USA
Alexander L. Yarin, Chicago, USA

The objective of Part A is to establish the fundamental concepts and equations that underlie experimental fluid mechanics. The first chapter, Sects. 1.1 through 1.8, addresses both the governing equations and the constitutive equations for Newtonian and non-Newtonian fluids. Chapter 2 provides the systematic bases for model testing and the scaling of experimental results. Sections 2.1.1 through 2.1.6 derive similarity parameters (Reynolds number, Froude number, etc.) from the governing equations and the boundary conditions.

Dimensional analysis (Sect. 2.2) provides a rational approach for the organization and interpretation of experimental data. Section 2.3, covering self-similarity, documents known flow fields that exhibit this condition (for example, an axisymmetric jet in which $\bar{u}/u_c = f(r/r_{1/2})$ and $u_c r_c = $ constant) and provides guidance on what other flows may exhibit this behavior. The encyclopedic presentation of examples will allow the reader to comprehend the universal features of both complete and incomplete self-similarity.

1. Experiment as a Boundary-Value Problem

A fluid flow experiment is an attempt to isolate a part of the world and measure flow and thermodynamic properties. A fluid is defined as a material that deforms continuously if a shear stress is applied. An internal flow situation has walls bounding the flow, but an inflow and outflow position must be controlled. An external flow problem has a uniform flow far from the body of interest. In both situations the state of flow at the boundary is controlled. In the mathematical representation of the flow, the flow conditions on the boundary are specified. This is the nature of the governing physics. If the boundary conditions depend on time the flow situation in the entire region must be specified at the initial time.

In what follows the major physical laws are outlined. In most cases tensor calculus in symbolic form is employed. Scalars are lightface type, vectors are boldface type, and tensors are boldface capitals. However, in cases where confusion is possible with tensor multiplications, index notation is employed. Scalars are then without an index, vectors have one index and tensors have two or more indices.

1.1	**Thermodynamic Equations**	3
	1.1.1 Thermodynamics	4
1.2	**Kinematic Equations**	5
1.3	**Balance Laws and Local Governing Equations**	6
	1.3.1 Continuity	6
	1.3.2 Linear Momentum and Related Equations	6
	1.3.3 Angular Momentum	7
	1.3.4 Energy	7
	1.3.5 Entropy	7
1.4	**Balance Laws and Global Governing Equations**	8
	1.4.1 Regions	8
	1.4.2 Leibnitz and Gauss Theorems	8
	1.4.3 Volume	8
	1.4.4 Mass	8
	1.4.5 Linear Momentum	8
	1.4.6 Total Energy	9
	1.4.7 Thermal Energy	10
	1.4.8 Mechanical Energy	10
	1.4.9 Entropy	10
1.5	**Constitutive Equations**	10
1.6	**Navier–Stokes Equations**	11
	1.6.1 Incompressible Flows	11
1.7	**Discontinuities in Density**	11
	1.7.1 Normal Surface Discontinuity	11
	1.7.2 Fluid–Solid Boundary	12
	1.7.3 Interfaces with Surface Tension	12
1.8	**Constitutive Equations and Nonlinear Rheology of Polymer Melts**	13
	1.8.1 Classical Theories	13
	1.8.2 Convected Derivatives and Differential Equations	16
	1.8.3 Microstructural Theories	17
	1.8.4 Conclusions	29
References		30

1.1 Thermodynamic Equations

The properties of a continuum are defined by an imaginary experiment where a region of volume V with characteristic length L is imagined to contain molecules. At a given position the volume is reduced around that position as indicated by the limit process $L \to 0$. A typical molecule, denoted by the subscript i, has a mass m_i and an instantaneous velocity $\mathbf{v}_i$. The density is the sum of mass over all molecules in the region divided by the volume as the limit is taken. Although $L \to 0$ is indicated, it cannot become so small that fluctuations occur because only a few molecules are present.

$$\rho = \lim_{L \to 0} \frac{\sum m_i}{V} . \qquad (1.1)$$

The mass-averaged velocity is a vector average of the molecular velocities and mass. This is appropriate to measure the momentum:

$$\boldsymbol{v} = \lim_{L \to 0} \frac{\sum m_i \boldsymbol{v}_i}{\sum m_i} . \quad (1.2)$$

If the substance has several chemical species, $n^{(k)}$ moles in the region, a molar averaged velocity for each species k is

$$\boldsymbol{V}^{(k)} = \lim_{L \to 0} \frac{\sum \boldsymbol{v}_i^{(k)}}{n^{(k)}} . \quad (1.3)$$

Such a velocity is useful in diffusion problems. The internal energy (per unit mass) due to random translational motions of the molecules is

$$e = \lim_{L \to 0} \frac{\sum \frac{1}{2} m_i (\boldsymbol{v}_i - \boldsymbol{v}) \cdot (\boldsymbol{v}_i - \boldsymbol{v})}{\sum m_i} . \quad (1.4)$$

The total internal energy includes other molecular motions such as vibrations, and configuration energies. The properties above are well defined whether or not the substance is in thermodynamic equilibrium.

1.1.1 Thermodynamics

It is assumed that the bulk motion of the substance does not affect the thermodynamic state. All thermodynamic variables of a simple compressible substance are described by a fundamental law that gives the entropy $s = s(\rho, e)$ or in another form $e = e(s, \rho)$. Each substance has its own entropy function, however, all functions obey the fundamental differential equation of thermodynamics.

$$e = e(s, \rho) \quad (1.5)$$

$$de = T\,ds - p\,d(\rho^{-1}), \quad (1.6)$$

the thermodynamic pressure is defined by

$$p(s, \rho) \equiv \left.\frac{\partial e}{\partial (\rho^{-1})}\right|_s , \quad (1.7)$$

and the temperature is given by

$$T(s, \rho) \equiv \left.\frac{\partial e}{\partial s}\right|_\rho . \quad (1.8)$$

Other thermodynamic properties follow from their definitions, for example the enthalpy $h = e + p/\rho$.

Two equations of state are equivalent to the fundamental law of a substance. The first equation of state is of the form

$$p = p(\rho, T) \quad (1.9)$$

or

$$\rho = \rho(p, T) . \quad (1.10)$$

It is equivalent to specify the compressibility coefficient functions:

$$\alpha(p, T) \equiv \left.\frac{1}{\rho}\frac{\partial \rho}{\partial p}\right|_T , \quad (1.11)$$

$$\beta(p, T) \equiv -\left.\frac{1}{\rho}\frac{\partial \rho}{\partial T}\right|_p . \quad (1.12)$$

Integration of these functions will reproduce $\rho = \rho(p, T)$.

The second equation of state is that for energy:

$$e = e(\rho, T) . \quad (1.13)$$

The important derivative function here is the specific heat (per unit mass) at constant volume:

$$c_\mathrm{v}(\rho, T) \equiv \left.\frac{\partial e}{\partial T}\right|_\rho . \quad (1.14)$$

The other function $\partial e / \partial \rho|_T$ is related to the state equation $\rho = \rho(p, T)$ by thermodynamic theory. In summary, the functions $p = p(\rho, T)$ and $c_\mathrm{v} = c_\mathrm{v}(\rho, T)$ describe the thermodynamics of a substance.

Often the enthalpy $h = e + p/\rho$ is used in preference to the internal energy. The important derivative function here is the specific heat (per unit mass) at constant pressure:

$$c_\mathrm{p}(p, T) \equiv \left.\frac{\partial h}{\partial T}\right|_p . \quad (1.15)$$

The other function $\partial h / \partial p|_T$ is related to the state equation $\rho = \rho(p, T)$ by thermodynamic theory. Alternatively, the functions $p = p(\rho, T)$ and $c_\mathrm{p} = c_\mathrm{p}(p, T)$ describe the thermodynamics of a substance.

There are special approximations of importance: the perfect gas, ideal gas, and incompressible fluid. For a perfect gas the state equations are:

$$p = \rho R T , \quad (1.16)$$

$$e = c_\mathrm{v}(\rho, T) T . \quad (1.17)$$

Alternatively, $h = c_\mathrm{p}(\rho, T) T$. A further restriction to an ideal gas gives simpler forms,

$$e = c_\mathrm{v}(T) T \quad (1.18)$$

$$h = c_\mathrm{v}(T) T + p/\rho \quad (1.19)$$

$$c_\mathrm{p}(T) = c_\mathrm{v}(T) + R \quad (1.20)$$

where R is the specific gas constant.

An incompressible fluid has thermodynamic variables that are independent of the density. The fundamental equation is

$$s = s(e), \quad (1.21)$$
$$de = T\,ds. \quad (1.22)$$

As before, the temperature is defined by

$$T \equiv \left.\frac{\partial e}{\partial s}\right|_{\rho}. \quad (1.23)$$

Pressure is a mechanical variable that is independent of the thermodynamic state. The first equation of state does not exist, and the second equation of state is

$$e = e(T) = c_v(T)T. \quad (1.24)$$

The enthalpy is a mixture of thermodynamic and mechanical properties, $h = e + p/\rho$.

1.2 Kinematic Equations

A fluid particle is an imaginary collection of fluid that locally follows the fluid velocity. Due to random molecular motions a fluid particle does not consist of the same molecules for all time. There are two mathematical viewpoints with different independent space and time variables. The dependent variables, the thermodynamic properties and characteristics of the continuum motion (velocity vorticity, strain rate, etc.) are instantaneous concepts and are the same from either viewpoint. The Lagrangian view can be thought of as a history of a certain fluid particle. Independent variables in the Lagrangian viewpoint are the initial position of the particle and the time, x^0 and $\hat{t}$. The particle position vector $\boldsymbol{r}$ is

$$\boldsymbol{r} = \tilde{\boldsymbol{r}}(x^0, \hat{t}), \quad (1.25)$$

It follows from this that the particle velocity is

$$\boldsymbol{v} = \frac{\partial \tilde{\boldsymbol{r}}}{\partial \hat{t}}. \quad (1.26)$$

and that the particle acceleration is

$$\boldsymbol{a} = \frac{\partial \boldsymbol{v}}{\partial \hat{t}}. \quad (1.27)$$

Alternately, the Eulerian viewpoint is based on a fixed position in space and time. The independent variables are $\boldsymbol{x}$ and t. In this viewpoint the particle position is

$$\boldsymbol{r} = \boldsymbol{r}(\boldsymbol{x}, t) = \boldsymbol{x}. \quad (1.28)$$

Equating $\boldsymbol{r}$ and time provides the connection between the two viewpoints.

$$\boldsymbol{x} = \tilde{\boldsymbol{r}}(x^0, \hat{t}), \quad (1.29)$$
$$t = \hat{t}. \quad (1.30)$$

For a given particle x^0, the particle path, with time as a parameter, is

$$\boldsymbol{x}_p = \tilde{\boldsymbol{r}}(x^0, t). \quad (1.31)$$

Choosing a different initial particle x^0 gives a different particle path.

Next, consider a line of particles given by an equation $x^0 = x^0(a)$ where a is a parameter that varies over some range, say $0 \leq a \leq 1$, where $a = 0$ is the beginning of the line and $a = 1$ is the end of the line. For given time t, a streak line of particles originally at x^0 are at

$$\boldsymbol{x}_{\text{str}} = \tilde{\boldsymbol{r}}(x^0(a), t). \quad (1.32)$$

A line of marked particles would move through the flow according to (1.31).

Another important concept is that of a streamline. For given time t, a streamline is everywhere tangent to the velocity

$$\boldsymbol{v} \times d\boldsymbol{x}_{\text{stm}} = 0,$$

or

$$\frac{dx}{u} = \frac{dy}{v} = \frac{dz}{w}. \quad (1.33)$$

The second version above refers to a coordinate system x, y, z with velocity components u, v, w.

Any dependent property f can be expressed in Eulerian variables $f = f_E(\boldsymbol{x}, t)$ or in Lagrangian variables $f = f_L(x^0, \hat{t})$. A Lagrangian time rate of change, the rate of change following a fluid particle, is given by

$$\frac{\partial f_L}{\partial \hat{t}}. \quad (1.34)$$

The particle velocity and acceleration are

$$v_i = \frac{\partial \tilde{r}}{\partial \hat{t}} = \frac{dx_i}{dt}, \quad (1.35)$$
$$a_i = \frac{\partial v_i}{\partial \hat{t}}. \quad (1.36)$$

By using the chain rules of differentiation of a composite function one finds that the Eulerian representation of a Lagrangian time derivative is

$$\frac{\partial f_L}{\partial \hat{t}} = \frac{\partial f_E}{\partial t} + \boldsymbol{v} \cdot \nabla f_E . \tag{1.37}$$

The right-hand side is called the Stokes derivative, the substantial derivative, or the material derivative.

$$\frac{df_E}{dt} \equiv \frac{\partial f_E}{\partial t} + v_i \frac{\partial f_E}{\partial x_i} .$$

This offers a physical interpretation for this combination of terms.

In addition to the translational velocity, every point in the fluid has a vorticity. Given a fluid particle P at x, the solid-like rotation of a near neighbor at P' is the angular velocity $\boldsymbol{\omega}(x)/2$. By definition

$$\boldsymbol{\omega} = \nabla \times \boldsymbol{v} . \tag{1.38}$$

The rate of strain at P is defined as

$$S_{ij} = \frac{1}{2}(\partial_i v_j + \partial_j v_i); \quad S = \frac{1}{2}\left[(\nabla \boldsymbol{v}) + (\nabla \boldsymbol{v})^T\right]. \tag{1.39}$$

Consider two particles P at x and a near neighbor at P' at a distance ds in direction of a unit vector $\boldsymbol{n}$ from P. The strain velocity per unit distance is the strain vector $\boldsymbol{d}$.

$$\frac{d\boldsymbol{v}_{\text{strain}}}{ds} = \boldsymbol{d} = \boldsymbol{n} \cdot \boldsymbol{S} . \tag{1.40}$$

A small material region has a volumetric rate of expansion

$$\lim_{L \to 0} \frac{1}{V_{MR}} \frac{dV_{MR}}{dt} = \nabla \cdot \boldsymbol{v} . \tag{1.41}$$

This is of course zero for an incompressible flow.

1.3 Balance Laws and Local Governing Equations

1.3.1 Continuity

The law for conservation of mass yields the *continuity* equation, referring to the fact that the underlying assumption of a continuum is required.

$$\frac{1}{\rho}\frac{d\rho}{dt} + \nabla \cdot \boldsymbol{v} = 0 . \tag{1.42}$$

The fractional rate of change of density following a fluid particle and the rate of expansion of the material region form a balance. Another viewpoint, from a fixed point in space, gives a balance of the local rate of change of density and the divergence of the flux of fluid into the point:

$$\frac{\partial \rho}{\partial t} + \nabla \cdot (\rho \boldsymbol{v}) = 0 . \tag{1.43}$$

The continuity equation is used to derive a relation between the substantial derivative of any fluid property f_E and the local and convective derivatives of f_E observed at a fixed Eulerian location.

$$\rho \left(\frac{\partial f_E}{\partial t} + \boldsymbol{v} \cdot \nabla f_E \right) = \frac{\partial (\rho f_E)}{\partial t} + \nabla \cdot (\rho \boldsymbol{v} f_E) . \tag{1.44}$$

1.3.2 Linear Momentum and Related Equations

The linear momentum per unit mass, $\boldsymbol{v}$, responds to surface and volumetric forces. The surface stress $\boldsymbol{R}$ on an area with outward normal $\boldsymbol{n}$ is the force per unit area of the outside substance upon the inside substance. The variation of this stress with surface direction is given by $\boldsymbol{n}$ and the variation with location is given by a stress tensor $\boldsymbol{T}(x, t)$:

$$\boldsymbol{R} = \boldsymbol{n} \cdot \boldsymbol{T} . \tag{1.45}$$

The stress is divided into a viscous tensor and pressure by subtracting the thermodynamic pressure.

$$\boldsymbol{T} = -p\boldsymbol{\delta} + \boldsymbol{\tau} . \tag{1.46}$$

The trace of the stress tensor forms a mechanical pressure force:

$$p_m = -\frac{1}{3}\text{tr}(\boldsymbol{T}) = -\frac{1}{3}\sum_i T_{ii} . \tag{1.47}$$

The Stokes assumption equates these pressures, $p = p_m$.

The gravity force per unit volume is $\rho \boldsymbol{g}$ where $\boldsymbol{g}$ is a constant scalar magnitude g times a unit vector in the gravity direction. If Z is the height above a horizontal reference plane,

$$\boldsymbol{F}_g = -\rho g \nabla Z . \tag{1.48}$$

Two equivalent forms of the momentum are from the perspective of a fixed point in space or from the perspective of a material particle

$$\frac{\partial(\rho\boldsymbol{v})}{\partial t} + \nabla\cdot(\rho\boldsymbol{v}^2) = -\nabla p + \nabla\cdot\boldsymbol{\tau} + \rho\boldsymbol{g},$$

$$\rho\frac{d\boldsymbol{v}}{dt} = -\nabla p + \nabla\cdot\boldsymbol{\tau} + \rho\boldsymbol{g}. \quad (1.49)$$

An equation governing the mechanical energy or kinetic energy per unit mass is obtained by the dot product of velocity and the momentum equation.

$$\frac{\partial(\rho\frac{1}{2}v^2)}{\partial t} + \nabla\cdot\left(\rho\boldsymbol{v}\frac{1}{2}v^2\right)$$
$$= -\boldsymbol{v}\cdot\nabla p + \rho\boldsymbol{v}\cdot\boldsymbol{g} + \boldsymbol{v}\cdot(\nabla\cdot\boldsymbol{\tau}). \quad (1.50)$$

An equation governing the moment of momentum, $\boldsymbol{r}\times\boldsymbol{v}$, is obtained by $\boldsymbol{r}$ cross the momentum equation:

$$\frac{\partial(\rho\varepsilon_{ijk}r_j v_k)}{\partial t} + \partial_p[\rho v_p(\varepsilon_{ijk}r_j v_k)]$$
$$= -\varepsilon_{ijk}r_j\partial_k p + \rho\varepsilon_{ijk}r_j g_k + \varepsilon_{ijk}r_j\partial_p\tau_{pk}, \quad (1.51)$$

where any origin is permitted for $\boldsymbol{r}$.

The vorticity $\boldsymbol{\omega} = \nabla\times\boldsymbol{v}$ is governed by a equation formed by $\nabla\times$ the momentum equation. The equation is

$$\frac{d\boldsymbol{\omega}}{dt} = -\boldsymbol{\omega}\nabla\cdot\boldsymbol{v} + \boldsymbol{\omega}\cdot\nabla\boldsymbol{v} + \frac{1}{\rho^2}\nabla\rho\times\nabla p$$
$$- \frac{1}{\rho^2}\nabla\rho\times\nabla\cdot\boldsymbol{\tau} + \frac{1}{\rho}\nabla\times\nabla\cdot\boldsymbol{\tau}; \quad (1.52)$$

a derivation of this equation may be found in [1.1, 2].

1.3.3 Angular Momentum

Conservation of angular momentum is a distinct physical law from linear momentum. The net internal angular momentum per unit mass is $\tilde{\boldsymbol{a}}$. This occurs if the molecules were spinning in a preferred direction. Angular momentum crossing an imaginary surface by molecular transport (diffusion) would produce a surface couple $n_j\Omega_{ji}$. One could also propose an external physical process ρG_i that would impart angular momentum directly to the individual particles. Conservation of total angular momentum, $\boldsymbol{r}\times\boldsymbol{v}+\tilde{\boldsymbol{a}}$, leads to the equation

$$\frac{\partial(\rho\varepsilon_{ijk}r_j v_k + \rho\tilde{a}_i)}{\partial t} + \partial_p[\rho v_p(\varepsilon_{ijk}r_j v_k + \tilde{a}_i)]$$
$$= -\varepsilon_{ijk}r_j\partial_k p + \rho\varepsilon_{ijk}r_j g_k$$
$$+ \partial_p(\varepsilon_{ijk}r_j\tau_{pk}) + \rho G_i + \partial_j\Omega_{jk}. \quad (1.53)$$

Subtracting the moment of the momentum equation derived earlier yields a relation governing internal angular momentum

$$\frac{\partial(\rho\tilde{a}_i)}{\partial t} + \partial_p(\rho v_p\tilde{a}_i) = \varepsilon_{ijk}\tau_{jk} + \rho G_i + \partial_j\Omega_{jk}. \quad (1.54)$$

It is usually assumed that the molecular angular momentum is randomly distributed so that $\tilde{\boldsymbol{a}} = 0$ and $\boldsymbol{G}$ and $\boldsymbol{\Omega}$ are zero. Then $\varepsilon_{ijk}\tau_{jk} = 0$ and $\boldsymbol{\tau}$ must be symmetric. Symmetry of $\boldsymbol{\tau}$ will be assumed in the balance of Sect. 1.1.

1.3.4 Energy

The conservation law for total energy leads to

$$\frac{\partial}{\partial t}\left[\rho\left(e + \frac{1}{2}v^2\right)\right] + \nabla\cdot\left[\rho\left(e + \frac{1}{2}v^2\right)\right]$$
$$= -\nabla\cdot\boldsymbol{q} - \nabla\cdot(\boldsymbol{v}p) + \rho\boldsymbol{v}\cdot\boldsymbol{g} + \nabla\cdot(\boldsymbol{\tau}\cdot\boldsymbol{v}). \quad (1.55)$$

Here the *heat flux* vector $\boldsymbol{q}$ accounts for the transport of energy by microscopic mechanisms.

Subtracting the kinetic energy equation yields the thermal energy

$$\frac{\partial}{\partial t}\rho e + \nabla\cdot(\rho\boldsymbol{v}e) = -\nabla\cdot\boldsymbol{q} - p\nabla\cdot\boldsymbol{v} + \Phi. \quad (1.56)$$

The symbol $\Phi \equiv \boldsymbol{\tau}:\nabla\boldsymbol{v} = \tau_{ij}\partial_j v_i$ represents the viscous dissipation that creates thermal energy from kinetic energy.

The thermal energy equation with temperature as the dependent variable is derived using general thermodynamic relations. It is

$$\rho c_p(T,p)\frac{dT}{dt} = -\nabla\cdot\boldsymbol{q} + \Phi + \beta T\frac{dp}{dt}. \quad (1.57)$$

1.3.5 Entropy

The fundamental equation expressing the second law of thermodynamics is

$$\rho\frac{ds}{dt} = -\nabla\cdot\frac{\boldsymbol{q}}{T} - \frac{1}{T^2}\boldsymbol{q}\cdot\nabla T + \frac{1}{T}\Phi. \quad (1.58)$$

The terms $-\nabla\cdot\frac{\boldsymbol{q}}{T} - \frac{1}{T^2}\boldsymbol{q}\cdot\nabla T = -\frac{1}{T}\nabla\cdot\boldsymbol{q}$ are written as two terms to separate the reversible ($-\nabla\cdot\frac{\boldsymbol{q}}{T}$) and irreversible ($-\frac{1}{T^2}\boldsymbol{q}\cdot\nabla T$) effects of heat transfer. The viscous term is irreversible.

1.4 Balance Laws and Global Governing Equations

1.4.1 Regions

Global laws are integrals of the local laws over a chosen region or, on the other hand, may be postulated as basic truths from the start. An arbitrary region, designated as AR, has a specified velocity w at each point on the surface. For a material region, MR, the surface velocity is the local fluid velocity, $w = v$. A fixed region, FR, is one with $w = 0$ everywhere. It is sometimes useful to consider a region with a surface velocity that is constant in space $w = W(t)$. Such a region has a constant volume and is designated VR. The fluid velocity relative to the VR is

$$u = v - W. \qquad (1.59)$$

A region enclosing a rocket and following it through space is a volume region, VR. The velocity of the rocket is $W(t)$ and the rocket engine discharges gases from the region.

Elementary thermodynamic texts do not have a uniform notation for regions. The arbitrary region (AR) defined above might be called an *open system*, a *deformable control volume*, a *control volume*, or some combination of these terms. The fixed region (FR) defined above might be called an *open system*, or simply a *control volume*. The material region might be called a *system*, a *constant mass system*, or a *closed system*.

1.4.2 Leibnitz and Gauss Theorems

For a region with arbitrary surface velocity w, the Leibnitz theorem is

$$\frac{dI_{ij}}{dt} = \frac{d}{dt} \int_{V_{AR}} f_{ij}(x_k, t) \, dV$$

$$= \int_{AR} \frac{\partial f_{ij}(x_k, t)}{\partial t} \, dV + \int_{AR} n_m w_m f_{ij}(x_k, t) \, dS, \qquad (1.60)$$

and the Gauss theorem is

$$\int_{V_{AR}} \partial_i f_{jk}(x_l, t) \, dV = \int_{AR} n_i f_{jk}(x_k, t) \, dS. \qquad (1.61)$$

A global law is derived by letting f_{ij} in the Leibnitz theorem be the quantity of interest in the local equation, substituting the local equation for $\partial f_{ij}/\partial t$ and converting as many volume integrals as possible into surface integrals by the Gauss theorem.

1.4.3 Volume

The volume of the region changes with time according to

$$\frac{dV_{AR}}{dt} = \int_{AR} n \cdot w \, dS. \qquad (1.62)$$

Here $n \cdot w$ is the normal velocity of the surface of the control region.

1.4.4 Mass

Conservation of mass for an arbitrary region is

$$\frac{dM_{AR}}{dt} = \frac{d}{dt} \int_{AR} \rho \, dV = - \int_{AR} n \cdot (v - w) \, \rho \, dS. \qquad (1.63)$$

Here all velocities are absolute velocities with respect to an inertial frame. For a material region, MR, $v = w$, and (1.63) becomes $\frac{dM_{MR}}{dt} = \frac{d}{dt} \int_{MR} \rho \, dV = 0$.

For a volume region, VR, with fluid velocity u with respect to the moving region;

$$\frac{dM_{VR}}{dt} = - \int_{VR} n \cdot u \rho \, dS. \qquad (1.64)$$

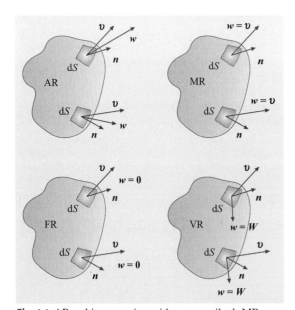

Fig. 1.1 AR, arbitrary region with w prescribed; MR, material region with w equal the local fluid velocity v; FR, fixed region with w equal zero; VR, volume region with w equal to a constant in space $W(t)$

In the rocket example the mass changes because of the relative velocity of the gases leaving the rocket motor.

1.4.5 Linear Momentum

For an arbitrary region

$$\frac{d}{dt}\int_{AR}(\rho v)\,dV = -\int_{AR}\rho\mathbf{n}\cdot(\mathbf{v}-\mathbf{w})\,\mathbf{v}\,dS$$

$$+ \int_{AR_{\text{solid}}} \mathbf{n}\cdot\mathbf{T}\,dS + \int_{AR_{\text{fluid}}} \mathbf{n}\cdot\boldsymbol{\tau}\,dS$$

$$- \int_{AR_{\text{fluid}}} \mathbf{n} p\,dS + \int_{AR_{\text{fluid}}} \rho\mathbf{F}_g\,dV. \quad (1.65)$$

Moving volume region

$$\frac{d}{dt}\int_{VR}(\rho\mathbf{u})\,dV + M_{VR}\frac{d\mathbf{W}}{dt}$$

$$= -\int_{VR}\rho(\mathbf{n}\cdot\mathbf{u})\mathbf{u}\,dS + \int_{VR_{\text{solid}}}\mathbf{n}\cdot\mathbf{T}\,dS$$

$$+ \int_{VR_{\text{fluid}}}\mathbf{n}\cdot\boldsymbol{\tau}\,dS - \int_{VR_{\text{fluid}}}\mathbf{n} p\,dS + \int_{VR_{\text{fluid}}}\rho\mathbf{F}_g\,dV. \quad (1.66)$$

Recall that he fluid velocity relative to the VR is $\mathbf{u} = \mathbf{v} - \mathbf{W}$.

1.4.6 Total Energy

For the global energy equations the surface integrals are split into regions where the surface cuts a solid and where it cuts a fluid. Shaft work arises from solid surfaces that are cut by the control surface. Work of a rotating shaft, which involves the tangential velocity, v_t, or translating shaft, which involves the normal velocity, v_n, is described by this term.

$$\dot{W}_{\text{Shaft}} = \int_{\text{Solid surfaces}} (\mathbf{n}\cdot\mathbf{T})\cdot\mathbf{v}\,dS$$

$$= \int_{\text{Solid surfaces}} (\mathbf{n}\cdot\mathbf{T})\cdot(\mathbf{v}_n + \mathbf{v}_t)\,dS$$

$$= \dot{W}_{\text{Shaft normal}} + \dot{W}_{\text{Shaft rotary}}. \quad (1.67)$$

In fluid regions the stress tensor is decomposed into pressure and viscous parts,

$$T_{ij} = -p\delta_{ij} + \tau_{ij}. \quad (1.68)$$

The total energy equation is

$$\frac{d}{dt}\int_{AR}\rho\left(e + \frac{1}{2}v^2 + gZ\right)dV$$

$$= -\int_{AR}\rho\mathbf{n}\cdot(\mathbf{v}-\mathbf{w})\left(e + \frac{1}{2}v^2 + gZ\right)dS$$

$$+ \dot{W}_{\text{Shaft normal}} + \dot{W}_{\text{Shaft rotary}}$$

$$+ \int_{\text{Fluid surfaces}}(\mathbf{n}\cdot\boldsymbol{\tau})\cdot\mathbf{v}\,dS$$

$$- \int_{\text{Fluid surfaces}} p\mathbf{n}\cdot\mathbf{v}\,dS - \int_{AR}\mathbf{n}\cdot\mathbf{q}\,dS. \quad (1.69)$$

This equation can be expressed in many forms. A popular form is to introduce the concepts of moving boundary work and flow work for the pressure work term.

$$-\int_{\text{Fluid surfaces}}\rho\mathbf{n}\cdot\mathbf{v}\left(\frac{p}{\rho}\right)dS$$

$$= -\int_{\text{Fluid surfaces}}\rho\mathbf{n}\cdot(\mathbf{v}-\mathbf{w})\left(\frac{p}{\rho}\right)dS$$

$$- \int_{\text{Fluid surfaces}}\rho\mathbf{n}\cdot\mathbf{w}\left(\frac{p}{\rho}\right)dS$$

$$= \dot{W}_{\text{Flow work}} + \dot{W}_{\text{Boundary work}}. \quad (1.70)$$

Inserting these concepts into (1.69) allows the flow work to be incorporated into the convective term and the enthalpy, $h = e + p/\rho$ to be identified.

$$\frac{d}{dt}\int_{AR}\rho\left(e + \frac{1}{2}v^2 + gZ\right)dV$$

$$= -\int_{AR}\rho\mathbf{n}\cdot(\mathbf{v}-\mathbf{w})\left(h + \frac{1}{2}v^2 + gZ\right)dS$$

$$+ \dot{W}_{\text{Shaft normal}} + \dot{W}_{\text{Shaft rotary}}$$

$$+ \dot{W}_{\text{Boundary}} + \int_{\text{Fluid surfaces}}(\mathbf{n}\cdot\boldsymbol{\tau})\cdot\mathbf{v}\,dS$$

$$- \int_{AR}\mathbf{n}\cdot\mathbf{q}\,dS. \quad (1.71)$$

1.4.7 Thermal Energy

The global thermal energy equation may be found by integrating the local thermal energy equation

$$\frac{d}{dt} \int_{AR} \rho e \, dV$$

$$= -\int_{AR} \rho n_i (v_i - w_i) e \, dS$$

$$+ \int_{AR} \tau_{ij} \partial_i v_j \, dV - \int_{AR} n_i q_i \, dS$$

$$+ \int_{AR} p \partial_i v_i \, dV . \tag{1.72}$$

1.4.8 Mechanical Energy

Subtracting the thermal energy from the total energy yields the final general form for the compressible flow in an arbitrary region. Here the total head $(1/2v^2 + gZ + p/\rho)$ appears. Note that some disciplines reserve the term *head* for items with the dimension of length, i.e., $(1/2gv^2 + Z + p/g\rho)$

$$\frac{d}{dt} \int_{AR} \rho \left(\frac{1}{2} v^2 + gZ\right) dV$$

$$= -\int_{AR} \rho \mathbf{n} \cdot (\mathbf{v} - \mathbf{w}) \left(\frac{1}{2} v^2 + gZ + \frac{p}{\rho}\right) dS$$

$$+ \dot{W}_{\text{Shaft normal}} + \dot{W}_{\text{Shaft rotary}} + \dot{W}_{\text{Boundary}}$$

$$+ \int_{\text{Fluid surfaces}} (\mathbf{n} \cdot \boldsymbol{\tau}) \cdot \mathbf{v} \, dS$$

$$+ \int_{AR} \Phi \, dV - \int_{AR} p \nabla \cdot \mathbf{v} \, dV \tag{1.73}$$

The term representing viscous dissipation is usually replaced by defining a head loss h_l.

$$\dot{m} g h_l \equiv \int_{FR} \tau_{ij} \partial_i v_j \, dV$$

$$= \int_{FR} \Phi \, dV \text{ for symmetric } \boldsymbol{\tau} . \tag{1.74}$$

1.4.9 Entropy

An exact expression of the second law of thermodynamics is

$$\frac{d}{dt} \int_{AR} \rho s \, dV$$

$$= -\int_{AR} \rho n_i (v_i - w_i) s \, dS - \int_{AR} \frac{1}{T} n_i q_i \, dS$$

$$- \int_{AR} \frac{1}{T^2} q_i \partial_i T \, dV + \int_{AR} \frac{1}{T} \tau_{ij} S_{ji} \, dV . \tag{1.75}$$

Here the last two terms are irreversible effects and always positive; neglecting them leads to an inequality well known in thermodynamics.

1.5 Constitutive Equations

Many practically important fluids obey constitutive relations for a Newtonian fluid with Fourier heat conduction. The Newtonian relationship for the stress rate of strain contains two viscosity coefficients

$$\tau_{ij} = \lambda \partial_k v_k \delta_{ij} + 2\mu S_{ij} \tag{1.76}$$

Inserting the Stokes assumption, $\lambda = -2\mu/3$, yields

$$\tau_{ij} = -\frac{2}{3} \mu \partial_k v_k \delta_{ij} + 2\mu S_{ij} . \tag{1.77}$$

The Fourier conduction law is

$$q_i = -\kappa \partial_i T \tag{1.78}$$

For completeness Fick's diffusion law for a binary mixture relates the diffusion flux and the concentration gradient with the binary diffusion coefficient as the proportionality constant.

$$\mathbf{j}_A = \rho_A (\mathbf{v}_A - \mathbf{v}) = -\rho \mathcal{D}_{AB} \nabla x_A . \tag{1.79}$$

1.6 Navier–Stokes Equations

The Navier–Stokes equations for a compressible flow may be considered as the continuity equation together with the momentum and energy equations for a Newtonian fluid:

$$\rho\left[\left(\frac{\partial \bm{v}}{\partial t} + \bm{v}\cdot\nabla\bm{v}\right)\right] = -\nabla p + \rho\bm{g} - \nabla\left(\frac{2}{3}\mu\nabla\bm{v}\right)$$
$$+ 2\nabla(\mu S),$$
$$S \equiv \frac{1}{2}\nabla\bm{v} + \frac{1}{2}(\nabla\bm{v})^{\mathrm{T}},$$
$$\rho c_{\mathrm{p}}(T,p)\frac{\mathrm{d}T}{\mathrm{d}t} = -\kappa\nabla^2 T + \Phi + \beta T\frac{\mathrm{d}p}{\mathrm{d}t}.$$
(1.80)

1.6.1 Incompressible Flows

For incompressible flows the density is approximately constant and transport coefficients are approximately constant. These are consistent assumptions at low Mach numbers (a characteristic velocity divided by the speed of sound) with adiabatic walls or isothermal walls with small temperature differences. The equations take the form

$$\nabla\bm{v} = 0, \tag{1.81}$$
$$\frac{\partial \bm{v}}{\partial t} + \bm{v}\cdot\nabla\bm{v} = -\frac{1}{\rho}\nabla p + \bm{g} + \nu\nabla\cdot\nabla\bm{v},$$
$$\nabla\cdot\nabla\bm{v} = \nabla^2\bm{v}, \tag{1.82}$$
$$\frac{\mathrm{d}\bm{\omega}}{\mathrm{d}t} = -\bm{\omega}\cdot\nabla\bm{v} + \nu\nabla^2\bm{v}. \tag{1.83}$$

Two related equations govern the enstrophy and the pressure:

$$\frac{\mathrm{d}}{\mathrm{d}t}\left(\frac{1}{2}\omega^2\right) = \omega_i\omega_j S_{ji} + \nu\partial_j\partial_j\left(\frac{1}{2}\omega^2\right)$$
$$- \nu\partial_j\omega_i\partial_j\omega_i, \tag{1.84}$$
$$\frac{1}{\rho}\nabla^2 p = \bm{v}\cdot\nabla^2\bm{v} + \bm{\omega}\cdot\bm{\omega} - \nabla^2\left(\frac{1}{2}v^2\right). \tag{1.85}$$

1.7 Discontinuities in Density

1.7.1 Normal Surface Discontinuity

Consider a surface discontinuity in the fluid. This might be thought of as a region with a finite thickness in the limit as the thickness approaches zero. The surface moves with velocity $\bm{W}$. Figure 1.2 is a typical example that depicts a shock wave caused by a blunt body moving at supersonic speed with respect to the surrounding fluid. Fluid on one side is called fluid A and that on the other fluid B. The unit normal vector $\bm{n}_A$ points from the discontinuity into fluid A, and $\bm{n}_B$ points from the discontinuity into fluid B. Assume that there is no mass, momentum or energy within the discontinuity and that the tangential velocity component is unchanged, $v_{At} = v_{Bt}$. The mass flow across the discontinuity is conserved.

$$\rho_A \bm{n}_A\cdot(\bm{v}_A - \bm{W}) + \rho_B \bm{n}_B\cdot(\bm{v}_B - \bm{W}) = 0. \tag{1.86}$$

The normal vectors may be replaced, $\bm{n}_A = -\bm{n}_B = \bm{n}$. The tangential momentum equation yields a balance of shear forces.

$$(\bm{n}_A\cdot\bm{\tau}_A)_t + (\bm{n}_B\cdot\bm{\tau}_B)_t = 0. \tag{1.87}$$

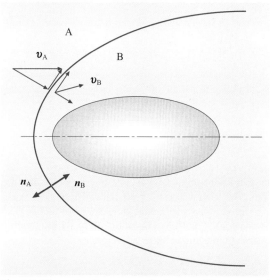

Fig. 1.2 Discontinuity across a shock wave. The tangential velocity is unchanged. The normal velocity is decreased

Let the normal component of the fluid velocity be $\bm{n}_A\cdot\bm{v}_A = v_{An}$. The normal direction momentum equation is

$$[\rho_A \cdot n_A(v_A - W)v_{An} + p_A - n_A \cdot (n_A \cdot \tau_A)]$$
$$+ n_B \cdot [\rho_B n_B \cdot (v_B - W)v_{Bn} + p_B - n_B \cdot (n_B \cdot \tau_B)]$$
$$= 0. \tag{1.88}$$

If e_t is the total (kinetic plus internal) energy, conservation of energy requires that

$$[\rho_A e_{tA} n_A \cdot (v_A - W)v_{An} + p_A v_{An}$$
$$- n_A \cdot (\tau_A \cdot v_A) + n_A \cdot q_A]$$
$$+ [\rho_B e_{tB} n_B \cdot (v_B - W)v_{Bn} + p_B v_{Bn}$$
$$- n_B \cdot (\tau_B \cdot v_B) + n_B \cdot q_B] = 0. \tag{1.89}$$

Because the discontinuity surface contains no mass, the curvature and the time dependence of W have no effect on the local validity of the equations above.

1.7.2 Fluid–Solid Boundary

The central issue here is the condition on the tangential velocity. In many situations the viscous no-slip condition is adequate. However, some situations require a more-refined approach. The analysis above covers the case of a porous or vaporizing wall without slip. Let u_s be the slip velocity of the fluid along the wall. Slip is often expressed as a slip length, defined as

$$\beta = \frac{u_s}{du/dy|_0} = \frac{u_s}{\dot{\gamma}}, \tag{1.90}$$
$$\dot{\gamma} \equiv du/dy|_0. \tag{1.91}$$

Because the interactions of gasses with solids and liquids with solids are different, these fluids must be dealt with separately.

In gases an important parameter is the Knudsen number, which compares the mean free path length λ with the flow length scale h:

$$\mathrm{Kn} = \frac{\lambda}{h} = \sqrt{\frac{\gamma \pi}{2}} \frac{M}{\mathrm{Re}}. \tag{1.92}$$

The relation with the Mach and Reynolds numbers is often useful. A first-order equation (derived from kinetic theory) for the slip is Maxwell's equation.

$$u_s = \frac{2 - \sigma_v}{\sigma_v} \mathrm{Kn} h \dot{\gamma}. \tag{1.93}$$

Here σ_v is the tangential momentum accommodation coefficient. The Knudsen number becomes large for a large object in a rarefied gas, or a very small object at atmospheric pressure.

Slip in liquids is only observed at small scales. The slip length is on the order on 100 nm (0.1 μm) Experimental results have been correlated as a power law

$$\beta = A \dot{\gamma}^B,$$
$$u_s = A \dot{\gamma}^{B+1}. \tag{1.94}$$

The constant B is about $1/2$. Unfortunately, the dimensions in the constants in the expressions above interact. Changing B changes the dimensions of A. More-detailed comments are in Chap. 19.

1.7.3 Interfaces with Surface Tension

Interfaces with two thermodynamic phases or immiscible substances may require that we postulate a surface tension property. Surface tension, force per length, is the two-dimensional world analogue to pressure. However, it is taken as a thermodynamic property of the substances and the temperature. The curvature of the surface is important. Let R_1 and R_2 be the principle radii of curvature of the surface. The curvature is

$$2\mathcal{H} = -\left(\frac{1}{R_1} + \frac{1}{R_2}\right). \tag{1.95}$$

Conservation of mass leads to the same equation as for normal discontinuities. The momentum equation contains two surface tension effects. One effect is from the curvature of the surface and another from a possible variation of the surface tension along the surface.

$$\rho n \cdot (v_A - w)v_A - \rho n \cdot (v_B - w)v_B$$
$$= [n \cdot \tau_A - n p_A] - [n \cdot \tau_B - n p_B] - \nabla_{(s)} \sigma - \sigma n 2\mathcal{H}. \tag{1.96}$$

Surfaces without mass crossing the interface have simplified expressions. In addition to the unit vector n normal to the surface, let the vectors b and t be orthogonal unit vectors within the surface. The momentum equations in these three directions are

$$n\text{-direction}: 0 = (\tau_{nn_A} - \tau_{nn_B}) - [p_A - p_B]$$
$$- 2\sigma \mathcal{H}, \tag{1.97}$$
$$n\text{-direction}: 0 = (\tau_{ntA} - \tau_{ntB}) - \frac{d\sigma}{dt}, \tag{1.98}$$
$$b\text{-direction}: 0 = (\tau_{nbA} - \tau_{nbB}) - \frac{d\sigma}{db}. \tag{1.99}$$

Contact lines, where two surfaces meet, require special treatment and if the contact line moves slip is required.

1.8 Constitutive Equations and Nonlinear Rheology of Polymer Melts

Constitutive equations describing the nonlinear rheological behavior of polymer melts have been a subject of focus due to their importance in designing and optimizing polymer processing as well as their analytical role in providing a logical picture of the molecular structure of the polymer. They are also needed to obtain closed system of the continuity and momentum equations describing flow of such materials. Constitutive equations can be classified according to their scale of work (continuum mechanics or microstructural), their method of formulation (integral or differential), or according to their approach towards time-deformation separability (separable or nonseparable) [1.3]. Most of the current theoretical work in rheology is devoted to the development of precise constitutive equations with parameters that are in some way or other obtainable through the microstructural properties of polymer melts and solutions as well as other non-Newtonian fluids. The present chapter presents an overview of the constitutive equations derived from continuum mechanics, and their evolution to the development of microstructural integro-differential equations.

Among the microstructural constitutive equations for polymer solutions and melts, the tube model has proven to be successful in predicting the linear rheological behavior of linear polymer melts and solutions, but is incapable of predicting the strain hardening behavior observed both in linear and in long-chain branched melts [1.4]. The continuum-based models, on the other hand, are merely phenomenological definitions of the solution or melt behavior and provide little or no information on the structure of the polymer molecule. It is the goal of this chapter to set the boundaries of the most recent microstructural constitutive equations and discuss the potentials of tackling the yet unsolved problems in nonlinear rheology of polymer melts.

Most recent developments which are based on modifications of the tube model have resulted in considerable progress in nonlinear viscoelastic theories and can predict strain hardening in linear polydisperse polymer melts with reasonable quantitative precision [1.5, 6]. However, a significant discrepancy arises when comparing the strain hardening of linear polydisperse melts to that of long-chain branched polymer melts since the latter show a considerably steeper onset of strain hardening [1.7, 8]. Reversing double-step strain measurements reveal another difference in melt behaviors since long-chain branched melts exhibit totally reversible behavior of the Kay-Bernstein-Kearsley-Zapas (K-BKZ) type up to considerable large deformations, whereas polydisperse linear polymers show early irreversible deformations [1.9].

In the present chapter we will discuss the main features of the evolution from the classical constitutive equations based on continuum mechanics to the microstructural theories which were developed in recent years. It is emphasized that in the present chapter, the rheological constitutive equations for concentrated polymer solutions and melts are in focus. In addition, in Sect. 1.2 and Sect. 1.3 of the handbook a number of rheological constitutive equations for slurries, gels, suspensions and emulsions is introduced and discussed in detail.

1.8.1 Classical Theories

The General Viscous Fluid

The main concern of many practical computational problems, e.g., in polymer processing is to find a suitably formulated method to calculate the flow rate. In this case and under certain circumstances the elasticity effects (and consequently the normal stress behavior in simple shear flow) may be neglected. Assuming incompressibility, and that the stress tensor $\boldsymbol{\sigma}$ only depends on the instantaneous condition of the rate of deformation tensor $\boldsymbol{D}$ (and not on its time derivatives), and assuming that the extra stress tensor $\boldsymbol{\tau} = \boldsymbol{\sigma} + p\boldsymbol{I}$ and $\boldsymbol{D}$ be coaxial (i.e., they have the same directions of principal axes), the state of stress can be described by:

$$\boldsymbol{\sigma} = -p\boldsymbol{I} + 2\eta\left(II_{\mathrm{D}}, III_{\mathrm{D}}\right)\boldsymbol{D}, \tag{1.100}$$

where II_{D} and III_{D} are the second and third invariants of the rate of deformation tensor. (The first invariant I_{D} is zero due to the assumption of incompressibility.)

Clearly, (1.100) reduces to the Newtonian fluid provided $\eta = \eta_0$ is independent of the invariants of $\boldsymbol{D}$:

$$\boldsymbol{\sigma} = -p\boldsymbol{I} + 2\eta_0\boldsymbol{D}. \tag{1.101}$$

For simple shear flows, the third invariant III_{D} vanishes identically. It is commonly assumed that III_{D} is not very important in other flows, and hence it is customary to omit the dependence on III_{D}, leading to the standard form of the general viscous fluid:

$$\boldsymbol{\sigma} = -p\boldsymbol{I} + 2\eta\left(II_{\mathrm{D}}\right)\boldsymbol{D}. \tag{1.102}$$

The later equation allows a three-dimensional representation of scalar deformation-rate-dependent flow laws.

A popular form of such flow laws has been proposed by *Ostwald* and *de Waele* [1.3]:

$$\eta = m II_D^{(n-1)/2}, \qquad (1.103)$$

hence:

$$\boldsymbol{\sigma} = -p\boldsymbol{I} + 2m II_D^{(n-1)/2} \boldsymbol{D}, \qquad (1.104)$$

where m is a temperature-dependent parameter and n determines whether the fluid is shear thinning ($n < 1$) or shear thickening ($n > 1$). However, like most phenomenological constitutive equations, (1.104) is only valid for a certain range of shear rates and fails at small shear rates. Experimental observations suggest that the viscosity finally reaches a (Newtonian) plateau at lower shear rates whereas (1.104) predicts infinite values for n<1. Other models to compensate for this shortcoming such as *Carreau–Yasuda model* [1.3],

$$\frac{\eta - \eta_\infty}{\eta_0 - \eta_\infty} = \frac{1}{(1 + \lambda^a |II_D|)^{\frac{n-1}{a}}} \qquad (1.105)$$

are in use, but their application needs a choice of extra parameters that yield no insight into the microstructural aspects of the fluid.

The Rubber–Like Liquid Theory

Based on early observations of rubber elasticity, *Lodge* [1.10] proposed a polymer melt to be a network of temporary junctions or entanglements that are created

1. at a constant rate independent of deformation or deformation rate;
2. in an isotropic state even under deformation.

This, along with the assumption of affine deformation led to the rubber-like liquid constitutive equation for the stress tensor $\boldsymbol{\sigma}$ in the form

$$\boldsymbol{\sigma} = -p\boldsymbol{I} + \int_{-\infty}^{t} \left[m(t-t') \boldsymbol{C}_t^{-1}(t') \right] dt', \qquad (1.106)$$

where p denotes the isotropic pressure contribution, $\boldsymbol{C}_t^{-1}(t')$ is the relative Finger strain tensor and $\boldsymbol{I}$ is the unit tensor. The memory function $m(t-t')$, which for simplicity is often expressed by a discrete relaxation spectrum (g_i, λ_i) [1.9],

$$m(t-t') = \sum_{i=1}^{N} \frac{g_i}{\lambda_i} e^{-(t-t')/\lambda_i}, \qquad (1.107)$$

is the time derivative of the relaxation modulus $G(t)$

$$m(t-t') = \frac{\partial G(t-t')}{\partial t'}. \qquad (1.108)$$

The discrete relaxation modes g_i and λ_i are free parameters of the theory, and have to be determined by suitable linear viscoelastic experiments. Based on the concept of reptation [1.4], considerable progress has been made in recent years in relating $G(t)$ to molar mass, molar mass distribution, and topology (linear or branched) of polymer melts, but its numerical precision depends on numerous model assumptions, especially in the case of randomly long-chain branched melts [1.11, 12]. This makes the experimental determination from dynamic mechanical analysis still the most reliable method to obtain $m(t-t')$ Chap. 9.

Later it was shown that the rubber-like liquid equation fails at large deformations [1.13], and severely overpredicts the stresses of polymer melts. This will be discussed in the following section when introducing the evolution of the K-BKZ-type constitutive equations.

K-BKZ and Related Equations

A large group of rheological equations of state for nonlinear viscoelastic behavior can be deduced as special cases of the K-BKZ equation [1.14].

The generalized relation between stress and finite strain for rubber elasticity is:

$$\boldsymbol{\sigma} = f(\boldsymbol{B}) = -p\boldsymbol{I} + g_1(I, II)\boldsymbol{B} + g_2(I, II)\boldsymbol{B}, \qquad (1.109)$$

where $\boldsymbol{B} = \boldsymbol{B}_n(t)$ is the Green deformation tensor describing the deformation from the natural, stress-free state $t' = n$ to the deformation state at time t, and the relation to the relative Finger strain tensor is given by

$$\boldsymbol{B}_n(t) = \boldsymbol{C}_t^{-1}(n). \qquad (1.110)$$

The functions g_1 and g_2 are material functions of the first and second invariant of $\boldsymbol{B}$ or $(\boldsymbol{C}^{-1})$,

$$I = I_B = \mathrm{tr}(\boldsymbol{B}) \qquad (1.111)$$

and

$$II = II_B = \frac{1}{2}\left[I_B^2 - \mathrm{tr}(\boldsymbol{B}^2) \right]. \qquad (1.112)$$

The values of invariants of the Green (and Finger) tensor for shear, uniaxial, equibiaxial and planar deformations are summarized in Table 1.1.

If the stress can be derived from an energy potential $W(I, II)$, the rubber is called hyperelastic. In the case of time-dependent viscoelastic materials, this concept can be generalized to an energy potential $w(I, II, t-t')$, which depends not only on the invariants I and II of the relative Finger tensor $\boldsymbol{C}_t^{-1}(t')$, but also on the time difference $t-t'$. The strain energy W is then given by

$$W = \int_{-\infty}^{t} [w(I, II, t-t')] \, \mathrm{d}t' \,. \tag{1.113}$$

From W, the K-BKZ equation is then derived as [1.14]:

$$\boldsymbol{\sigma} = -p\boldsymbol{I} + \int_{-\infty}^{t} \left(2\frac{\partial w}{\partial I} \boldsymbol{C}^{-1} - 2\frac{\partial w}{\partial II} \boldsymbol{C} \right) \mathrm{d}t' \,, \tag{1.114}$$

$\boldsymbol{C} = \boldsymbol{C}_t(t')$ is called the relative Cauchy tensor. By choosing a suitable potential w, (1.114) can be adapted to describe the time-dependent deformation behavior of a general viscoelastic material. Separating the effects of time and deformation will lead to a separable K-BKZ equation of the form:

$$\boldsymbol{\sigma} = -p\boldsymbol{I} + \int_{-\infty}^{t} m(t-t') \left(h_1 \boldsymbol{C}^{-1} + h_2 \boldsymbol{C} \right) \mathrm{d}t' \,. \tag{1.115}$$

If the functions $h_1(I, II)$ and $h_2(I, II)$ are not expressible as derivatives of a potential, (1.115) is called a separable Rivlin–Saywers equation [1.3], although the term K-KBZ equation is often (wrongly) also used in this case. The rubber-like liquid theory of Lodge mentioned in the previous section is recovered from (1.115) with h_1 and h_2 being equal to 1 and 0, respectively, and w is given by

$$w(I, II, t-t') = m(t-t')(I-3) \,. \tag{1.116}$$

The main problem is now reduced to obtaining a reasonable expression for w for real polymer melts in a wide range of deformations and deformation rates, a task that remained largely elusive.

However, nonlinear shear and extensional stress growth experiments on many polymer melts for sufficiently large deformations proved time-deformation separability to be valid [1.3]. This, along with neglecting

Table 1.1 First and second invariants of the Green or Finger tensor in terms of the shear strain γ and Hencky strain ε

Type of flow	I	II
Simple shear	$3+\gamma^2$	$3+\gamma^2$
Uniaxial	$e^{2\varepsilon} + 2e^{-\varepsilon}$	$2e^{\varepsilon} + e^{-2\varepsilon}$
Equibiaxial	$e^{-2\varepsilon} + 2e^{\varepsilon}$	$2e^{-\varepsilon} + e^{2\varepsilon}$
Planar	$1 + e^{-2\varepsilon} + e^{2\varepsilon}$	$1 + e^{2\varepsilon} + e^{-2\varepsilon}$

the relative Cauchy tensor by setting $h_2 = 0$ in (1.109) led to the so-called Wagner I equation [1.3, 15–17]:

$$\boldsymbol{\sigma} = -p\boldsymbol{I} + \int_{-\infty}^{t} m(t-t')h(I, II)\boldsymbol{C}^{-1} \, \mathrm{d}t' \,, \tag{1.117}$$

where $h(I, II) \leq 1$ is called the damping function, which expresses the survival probability of network strands regarding nonlinear deformations [1.16, 17]. A first approximation proposed for the damping function was:

$$h(II_B) = e^{-\sqrt{II-3}} \,, \tag{1.118}$$

which, for unidirectional shear flows, will lead to the simplified form:

$$h(\gamma) = e^{-n\gamma} \,. \tag{1.119}$$

Using the fitting exponent $n = 0.143$ for low-density polyethylene (LDPE) melts, (1.119) performs quite well for shear deformations as large as $\gamma = 13$. For larger deformations a combination of two exponential functions of the form

$$h(\gamma) = a e^{-n_1 \gamma} + (1-a) e^{-n_2 \gamma} \,, \tag{1.120}$$

was suggested, which extended the predictable deformation range up to $\gamma = 30$ [1.15].

However, comparison to experimental evidence has shown that recoverable deformations, e.g. in reversing shear or elastic recoil experiments [1.18–20], are overpredicted by the Wagner I equation. In order to overcome this defect, the damping function was replaced by a functional $H(t, t')$ of the deformation (the so-called Wagner II equation [1.3]),

$$\boldsymbol{\sigma} = -p\boldsymbol{I} + \int_{-\infty}^{t} m(t-t')H(t, t')\boldsymbol{C}^{-1} \, \mathrm{d}t' \,. \tag{1.121}$$

The functional was chosen in such a way that it always behaves as a decreasing function [1.18, 21],

$$H(t, t') = \min [h(I, II)] \,. \tag{1.122}$$

Using this functional and a linear combination of the strain invariants,

$$L = \alpha I + (1-\alpha) II \,, \tag{1.123}$$

and a damping function $h(L)$ of the form

$$h(L) = a e^{-n_1 \sqrt{L-3}} + (1-a) e^{-n_2 \sqrt{L-3}} \,, \tag{1.124}$$

reasonable quantitative agreement between theory and experiment for the nonlinear behavior both in shear and uniaxial extensional flow of many industrial (polydisperse) polymer melts could be obtained by the use of four nonlinear material parameters. Later *Papanastasiou* et al. proposed a damping function of the following sigmoidal form [1.22]

$$h(I, II) = \frac{\alpha}{(\alpha - 3) + \beta I + (1 - \beta) II} . \quad (1.125)$$

The Wagner *I* and *II* equations have the disadvantage of neglecting the second normal stress difference in shear flow, which in reality is small and negative, but nonzero. This is related to the fact that the experimentally derived strain functions summarized so far cannot be generalized to other types of deformations, even if they perform well in predicting the melt behavior under a certain flow regime. Samarkus et al. have exemplified this problem by showing that the coefficients of Wagner *II* equation obtained by shearing experiments could not predict planar extensional behavior [1.23].

For the Doi–Edwards model [1.4], to be discussed in detail in Sect. 1.8.3 of this chapter, a simple analytical approximation for the corresponding potential w (the so-called Currie approximation [1.24]) was found,

$$w \approx \frac{5}{2} \ln\left(\frac{J-1}{7}\right) , \quad (1.126)$$

with a generalized invariant J

$$J = I + 2\sqrt{II + \frac{13}{4}} . \quad (1.127)$$

This leads to

$$h_1 = \frac{5}{J-1}, \quad h_2 = \frac{-2h_1}{J-1}, \quad (1.128)$$

thereby predicting a negative second normal stress difference in shear flow. However, as the Doi–Edwards model does not account for chain stretching (Sect. 1.8.3), the predictive power of (1.128) concerning the rheology of industrially important polydisperse polymer melts is limited.

Finally, Wagner and Demarmels proposed an ansatz for the two strain functions h_1 and h_2 which represents a special form of a Rivlin–Sawyers equation describing a wide range of deformation types [1.25]. Starting with the ratio β of second (N_2) to first (N_1) normal stress differences from a shear experiment,

$$\beta = \frac{N_1}{N_2} = \frac{h_2}{h_1 - h_2} , \quad (1.129)$$

which coincides with the corresponding relation for the normal stresses in planar experiments and was assumed to be strain-independent, they proposed

$$h_1 = (1 + \beta) \, h(I, II) ,$$
$$h_2 = \beta \, h(I, II) ,$$
$$h(I, II) = \left[1 + a\sqrt{(I-3)(II-3)}\right]^{-1} , \quad (1.130)$$

which resulted in good agreement with experimental data in uniaxial, biaxial and planar deformations of a polydisperse isobutene melt.

Despite their qualitative success in describing the polymer melt behavior, constitutive equations based on classical continuum mechanics lack insight into the relation between rheology and the structure of polymers and cannot be used as predictive tools. This calls for a better understanding of the microstructure of polymer melts, subject of the following section, to develop more realistic, yet feasible models.

1.8.2 Convected Derivatives and Differential Equations

A major part of the differential constitutive equations consists of generalizations of the Maxwell model and possesses the character of continuum formulations based merely on phenomenological observations. The Maxwell model was initially utilized to describe the behavior of viscoelastic materials in the linear deformation range by considering a viscoelastic material to behave as a linear combination of a spring with spring constant g and a dashpot with constant viscosity η_0. The scalar stress σ in the material then obeys the linear first-order differential equation:

$$\sigma + \lambda \dot{\sigma} = \eta_0 \dot{\gamma} , \quad (1.131)$$

where $\dot{\gamma}$ is the scalar shear rate, and $\lambda = \eta_0/g$ is the relaxation time. (1.131) can be generalized to the three-dimensional case and to large deformations,

$$\boldsymbol{\sigma} + \lambda \stackrel{\nabla}{\boldsymbol{\sigma}} = 2\eta_0 \boldsymbol{D} , \quad (1.132)$$

by introduction of the *upper convected derivative* of the stress tensor $\stackrel{\nabla}{\boldsymbol{\sigma}}$ as

$$\stackrel{\nabla}{\boldsymbol{\sigma}} = \dot{\boldsymbol{\sigma}} - (\nabla \boldsymbol{v})^T \cdot \boldsymbol{\sigma} - \boldsymbol{\sigma} \cdot (\nabla \boldsymbol{v}) . \quad (1.133)$$

This description solves the problem of frame invariance by introducing a frame of reference that is *convected* and deformed with material lines.

The upper convective Maxwell model (1.132), is the one-mode differential equivalent of Lodge's (1.106) with an exponentially fading memory. Both integral and differential versions of Lodge's equation were successful in providing a qualitative prediction of the primary normal stress difference in shear and strain hardening in extension. Replacing the upper convective derivative in (1.126) by the *lower convected* form,

$$\overset{\Delta}{\sigma} = \dot{\sigma} - (\nabla v) \cdot \sigma - \sigma \cdot (\nabla v)^T , \qquad (1.134)$$

the lower convective Maxwell model is obtained. In the integral version, this is equivalent to replacing the Finger strain tensor by the Cauchy tensor C as the deformation measure in (1.106) with the exponentially fading memory. The lower convective Maxwell model predicts a negative second normal stress difference of the same magnitude as the first normal stress difference in shear and no strain hardening in extension, which is not in agreement with experiment. It also has no molecular basis. Equation (1.133) and (1.134) are not the only frame-invariant time derivatives. Oldroyd recognized that infinitely many could be constructed, and he proposed generalizations of the convected Maxwell models by allowing for higher-order terms to appear, e.g., in the form of the eight-constant Oldroyd equation, later followed by Giesekus and others. For details the reader is referred to [1.3]. However, experience with these purely continuum-mechanics-based equations teaches that frame invariance itself is not restrictive enough to keep the number of terms manageable, and that some molecular insight is needed.

1.8.3 Microstructural Theories

Theories from Continuum Mechanics and Their Microscopic Equivalents

In microscopic terms, stress in polymeric systems originates from orientation and extension of entropic springs, which can be thought of as e.g. representing molecular strands between entanglements. An isotropic distribution of molecular strands normalized with respect to their equilibrium length can be described by an isotropic distribution of unit vectors u. Assuming affine deformation, the inverse relative deformation gradient F^{-1} transforms a unit vector u into a deformed vector u',

$$u' = F^{-1} \cdot u . \qquad (1.135)$$

The Finger strain tensor can then be expressed as

$$C^{-1} = 3\langle u'u' \rangle \qquad (1.136)$$

where $\langle \ldots \rangle$ denotes an integral over an isotropic distribution of unit vectors before deformation,

$$\langle \ldots \rangle = \frac{1}{4\pi} \int \ldots \, d\Omega , \qquad (1.137)$$

where $d\Omega$ is an infinitesimal solid angle, and the integration is over the surface of a unit sphere. This is depicted schematically in Fig. 1.3.

A similar scheme can be used to find a microscopic representation of the Cauchy strain tensor C: If n represents an isotropic distribution of unit surfaces, and if these are assumed to be deformed affinely, the deformation gradient F transforms a unit surface vector n into the deformed surface vector n',

$$n' = n \cdot F . \qquad (1.138)$$

The Cauchy strain tensor C can then be expressed as

$$C = 3\langle n'n' \rangle . \qquad (1.139)$$

Considering further that the invariants I and II are equivalent to

$$I = 3\langle u'^2 \rangle , \qquad (1.140)$$
$$II = 3\langle n'^2 \rangle , \qquad (1.141)$$

where u' and n' represent the lengths of the vectors u' and n' respectively,

$$u' = \sqrt{u' \cdot u'} ,$$
$$n' = \sqrt{n' \cdot n'} , \qquad (1.142)$$

the separable Rivlin–Sawyers (1.115) can be expressed in microscopic terms as:

$$\sigma = -pI + \int_{-\infty}^{t} m(t-t') \Big[H_1\big(\langle u'^2 \rangle, \langle n'^2 \rangle\big) \langle u'u' \rangle$$
$$+ H_2\big(\langle u'^2 \rangle, \langle n'^2 \rangle\big) \langle n'n' \rangle \Big] dt' \qquad (1.143)$$

where the strain functions H_1 and H_2 converge to $H_1 + H_2 = 3$ in the linear viscoelastic limit [1.9].

However, while (1.143) expresses the strain measure in terms of the primitive quantities u and n, it is by no means guaranteed that this is the most appropriate representation of the strain measure when taking into

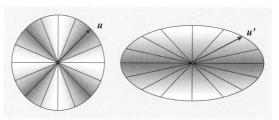

Fig. 1.3 Affine deformation of unit vectors: graphic representation of the Finger strain tensor

account topological constraints of the macromolecular chains.

The Tube Model of Doi and Edwards (DE)

The kinetic theory of *Doi* and *Edwards* [1.4] models intermolecular interaction for concentrated systems of monodisperse linear polymer chains by the tube concept: the mesh of constraints caused by surrounding chains confines the molecular chain laterally to a tube-like region. Relaxation occurs by two mechanisms: *chain retraction* by equilibration along the tube contour, which is supposed to be a fast process governed by the Rouse time τ_R of the chain with τ_R proportional to the square of the molar mass, and *chain diffusion* by reptation out of the tube with a reptation or disengagement time τ_d proportional to the third power of the molar mass. As for high values of the molar mass, chain retraction is fast compared to chain diffusion, this model explains naturally the experimentally observed time-deformation separability of the nonlinear relaxation modulus for times greater than the equilibration time.

Assuming that the diameter of the tube is not changed by deformation, or equivalently that the tension in the deformed polymer chain is equal to its equilibrium value, Doi and Edwards derived [by use of the independent alignment assumption (IAA)] a single integral constitutive equation of the form [1.4]:

$$\sigma = -p\boldsymbol{I} + \int_{-\infty}^{t} m(t-t') S_{\mathrm{DE}}^{\mathrm{IA}}(t,t') \, \mathrm{d}t' \; . \quad (1.144)$$

Here, $S_{\mathrm{DE}}^{\mathrm{IA}}$ denotes the strain measure of the Doi–Edwards (DE) model with the independent alignment assumption,

$$S_{\mathrm{DE}}^{\mathrm{IA}} = \frac{15}{3} \left\langle \frac{\boldsymbol{u}'\boldsymbol{u}'}{u'^2} \right\rangle = 5 S \; , \quad (1.145)$$

where $S = S(t,t')$ is the relative second-rank orientation tensor, and $\boldsymbol{u}'$ denotes deformed unit vectors according to (1.135). The DE model with the independent alignment assumption assumes that stress is created by an affine rotation of tube segments.

A rigorous derivation for stress relaxation after a step deformation leads to a somewhat different strain measure,

$$S_{\mathrm{DE}} = \frac{15}{4} \frac{1}{\langle u' \rangle} \left\langle \frac{\boldsymbol{u}'\boldsymbol{u}'}{u'} \right\rangle \quad (1.146)$$

and to a nonlinear integro-differential equation for the stress tensor in general flows. However, the strain measure S_{DE} is often used simplistically instead of $S_{\mathrm{DE}}^{\mathrm{IA}}$ in the single integral (1.144).

$S_{\mathrm{DE}}^{\mathrm{IA}}$ and S_{DE} produce different predictions for the second normal stress difference (with $S_{\mathrm{DE}}^{\mathrm{IA}}$ being in better agreement with available experimental data), and consequently also different predictions in biaxial deformations, whereas their difference seems to be negligible for the case of uniaxial extension [1.26].

For monodisperse polymer melts and solutions, the Doi–Edwards strain measures seem to give an acceptable description of material behavior in step-shear experiments for times greater than the equilibration time. For fast deformations (as defined below) and for polydisperse linear and branched polymer melts, on the other hand, although time-deformation separability often works over most or even the entire (experimentally attainable) time range, the measured stresses in shear and extensional flows are often much higher than predicted by the Doi–Edwards strain measure [1.3, 13, 27, 28].

Models with Pre-averaged Chain Stretch

The DE Theory with Chain Stretch. The DE theory, despite its deficiency in predicting chain stretch, established the theoretical foundation for further studies on nonlinear deformations of polymer melts.

The DE constitutive equation in shear is applicable only to deformation rates $\dot{\gamma}$ that are smaller than the reciprocal Rouse time τ_R of the chain, i.e., for flows with Deborah numbers $\mathrm{De} = \dot{\gamma}\tau_R < 1$. The first attempt to generalize the DE equation for faster flows was by applying a pre-averaged stretch ratio into the DE constitutive equation when $\mathrm{DE} > 1$ [1.4, 29]. Assuming the equilibrium chain length to be $\bar{L}$ and its value at time t

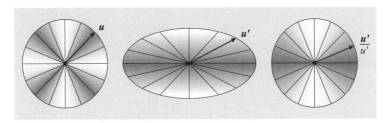

Fig. 1.4 The DE theory with independent alignment assumption (IAA): the stress tensor is the result of a mere change in chain orientation, sometimes called an *affine rotation*

to be $L(t)$, the stress tensor is given by

$$\boldsymbol{\sigma} = -p\boldsymbol{I} + \frac{G_N^0}{\bar{L}^2} \left\langle \int_0^{L(t)} \mathrm{d}s L(t) (\boldsymbol{uu}) \right\rangle, \quad (1.147)$$

where $\boldsymbol{u} = \boldsymbol{u}(s,t)$ describes the orientation at time t of a chain segment at position s along the chain, and the average is taken over all chains of the system. For fast deformations, this can be approximated by

$$\boldsymbol{\sigma} = -p\boldsymbol{I} + \frac{G_N^0}{\bar{L}^2} \left\langle \int_0^{L(t)} \mathrm{d}s L(t) (\boldsymbol{uu}) \right\rangle$$

$$\approx -p\boldsymbol{I} + G_N^0 \left(\frac{L(t)}{\bar{L}}\right)^2 \left\langle \frac{1}{L(t)} \int_0^{L(t)} \mathrm{d}s\, (\boldsymbol{uu}) \right\rangle$$

$$= -p\boldsymbol{I} + G_N^0 \left(\frac{L(t)}{\bar{L}}\right)^2 \bar{\boldsymbol{S}}(t), \quad (1.148)$$

where $\bar{\boldsymbol{S}}$ represents the orientation tensor of the ensemble of chains. Equation (1.148) turns into an exact relation for step-strain deformations.

Equation (1.148) is then approximated by a single integral equation with the stretch ratio $\lambda = (L(t)/\bar{L})$ left outside the integration,

$$\boldsymbol{\sigma}(t) = -p\boldsymbol{I} + \lambda^2(t) \int_{-\infty}^{t} m(t-t') \boldsymbol{S}_{\mathrm{DE}}(t,t') \mathrm{d}t'$$

$$= -p\boldsymbol{I} + \frac{15}{4}\lambda^2(t) G_N^0 \bar{\boldsymbol{S}}(t). \quad (1.149)$$

We call this the DE equation with pre-averaged chain stretch. It created the need to develop a proper evolution equation for $\lambda(t)$. This was proposed for the first time by *Pearson* et al. [1.29] by noticing that the rate of chain stretch can be obtained by assuming a balance between the frictional force on the chain, and the spring force created in the chain by stretching,

$$\xi \left(\boldsymbol{\kappa} : \bar{\boldsymbol{S}} L - \frac{\mathrm{d}L}{\mathrm{d}t}\right) = \frac{3k_B T}{\bar{L}^2}(L - \bar{L}), \quad (1.150)$$

where ξ is the friction coefficient, and the first term on the left hand side of (1.150) represents an affine deformation; $\boldsymbol{\kappa} = \nabla \boldsymbol{v}$ is the velocity-gradient tensor, k_B Boltzmann's constant and T temperature. By dividing by the equilibrium length $\bar{L}$ and introducing the Rouse time, $\tau_R = kM^2$, this leads to an evolution equation for the stretch $\lambda(t)$ of the form:

$$\frac{\mathrm{d}\lambda}{\mathrm{d}t} = \boldsymbol{\kappa} : \bar{\boldsymbol{S}}\lambda - \frac{1}{\tau_R}(\lambda - 1). \quad (1.151)$$

Noticing the fact that the chain cannot undergo infinite elongation, another restriction was later added to (1.151) by assuming an upper level of chain stretch, $\lambda_{\max}$. This led to modifying the right hand side of (1.151) in a fashion that would accommodate this requirement [1.30]:

$$\frac{\mathrm{d}\lambda}{\mathrm{d}t} = \boldsymbol{\kappa} : \bar{\boldsymbol{S}}\lambda - \frac{1}{\tau_R}(\lambda - 1) c(\lambda), \quad (1.152)$$

with

$$c(\lambda) = \frac{\lambda_{\max}}{\lambda_{\max} - \lambda}. \quad (1.153)$$

The newly introduced parameter $\lambda_{\max}$ determines the finite stretching limit of the chain and is reported by *Fang* et al. to be between 3 and 6 depending on the polymer system considered [1.31]. The latter work also introduces an alternative for $c(\lambda)$ in (1.153) by assuming a new $c(\lambda)$ that is derived from an entropy expression,

$$c(\lambda) = \frac{3Z\lambda_{\max}^2(\lambda + 1)}{\lambda(\lambda_{\max}^2 - \lambda^2)}, \quad (1.154)$$

where Z is the number of entanglement segments per chain. This led to somewhat better agreement with experimental data for shear experiments, but it offers only qualitative agreement with experimental results for start-up and steady-state values of extensional viscosities.

The Pom-Pom Model. The concept of a pre-averaged chain stretch was further used by *McLeish* and *Larson* [1.32] to account for the nonlinear rheology of a multi-arm model polymer initially proposed by *McLeish* [1.33]. The so-called pom-pom model is an extension of an H-shaped molecule having a backbone of molecular weight M_a and q arms of molecular weight M_b at each end (the only branching points on the backbone). The backbone is assumed to be stretched by the deformation until the tension in the backbone equals the sum of the equilibrium tensions of the dangling arms, which occurs when $\lambda = q$. Figure 1.5 shows schematically a typical pom-pom molecule and the process of backbone stretch and branch retraction into the tube of the backbone for a large enough deformation.

In the simplified version of the model [1.34], the orientation contribution to the stress from the dangling arms is neglected so that the dominant contribution arises

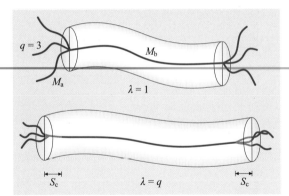
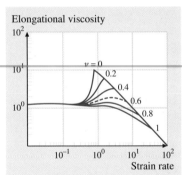

Fig. 1.5 Schematic presentation of a pom-pom molecule

Fig. 1.6 Steady viscosity in uniaxial tension for $0 < \nu < 1$ with pom-pom parameters: $q = 5$, $M_a = 5M_e$ and $M_b = 15M_e$ (after [1.35], with permission)

from the orientation and stretch of the backbone only. The stress tensor is then obtained from (1.149) as:

$$\boldsymbol{\sigma}(t) = -p\boldsymbol{I} + \frac{15}{4}\lambda^2(t)G_N^0\phi_b^2\bar{\boldsymbol{S}}(t), \quad (1.155)$$

where ϕ_b indicates the mass fraction of the backbone. While the evolution of backbone orientation is assumed to be dominated by a single orientation relaxation time τ_b,

$$\bar{\boldsymbol{S}} = \int_{-\infty}^{t} \tau_b^{-1} e^{-\frac{t-t'}{\tau_b}} \boldsymbol{S}_{DE}(t,t')\,dt', \quad (1.156)$$

backbone stretch depends on a stretch relaxation time τ_s and is derived from a similar stretch evolution equation as given by (1.141) as long as $\lambda < q$,

$$\frac{d\lambda}{dt} = \lambda\boldsymbol{\kappa}:\bar{\boldsymbol{S}} - \frac{1}{\tau_s}f(\lambda), \quad (1.157)$$

while otherwise $\lambda = q$ is assumed. $f(\lambda)$ was originally chosen to have a linear form,

$$f(\lambda) = \lambda - 1, \quad (1.158)$$

which causes the steady-state extensional viscosities to undergo a sharp maximum as shown for the case $\nu = 0$ in Fig. 1.6 This was in contrast to experimental results and unrealistic since it suggested a sudden retraction of the arm only at the moment when $\lambda = q$. In fact, relaxation of an arm can have a significant effect on the overall relaxation also before a maximum stretch is reached and very small retractions of the arms into the tube can effectively reduce the drag from the arms. To account for this effect *Blackwell* et al. [1.35] introduced *local branch point displacement* and changed $f(\lambda)$ to

$$f(\lambda) = (\lambda - 1)e^{\nu(\lambda - 1)} \quad \text{with } 0 < \nu < 1. \quad (1.159)$$

Using this new evolution equation for the backbone stretch, Blackwell et al. showed that a moderate value of ν smoothes the steady-state extensional viscosity plotted against the strain rate, as shown in Fig. 1.6, which is equivalent to a gradual arm retraction instead of an instant withdrawal.

Although the original pom-pom model was introduced by making use of the DE orientation tensor in its integral form (1.146), the orientation tensor was soon approximated by a differential evolution equation [1.32] which would allow more feasible numerical applications. The differential pom-pom equation has entirely overshadowed the original pom-pom idea since its introduction. Although the scope of the present work is integral constitutive equations, we give a short account of the differential version here because of its widespread application in numerical simulations of polymer processing [1.36]. However, we emphasize the fact that the differential approximation departs both quantitatively and qualitatively a great deal from the original pom-pom idea by causing enhanced shear thinning and neglecting the 2nd normal stress difference in shear flow [1.37, 38].

The multimode pom-pom model is another prevalent development in the theory introduced by *Inkson* et al. [1.34] in order to account for the multiple branching effects which are present in long-chain branched industrial polymer melts such as LDPE, and which are treated by the concept of the so-called seniority/priority effects. According to this model, each branching is assumed as having further branches on it so that it acts as the backbone of the next branching generation from its other end [1.39]. This leads to the introduction of multiple relaxation times representative of various branching levels, and a stress tensor that is summed over the entire range of relaxation modes. The set of equations of the multimode differential pom-pom model is therefore

Stress: $\quad \boldsymbol{\sigma} = 3\sum_i g_i\lambda_i^2\bar{\boldsymbol{S}}_i, \quad (1.160)$

Orientation: $\bar{S}_i = \dfrac{A_i}{\mathrm{tr}(A_i)}$

$$\frac{\partial}{\partial t} A_i = \kappa \cdot A_i + A_i \cdot \kappa^\mathrm{T} - \frac{1}{\tau_{\mathrm{b},i}}(A_i - I) \,, \tag{1.161}$$

Stretch: $\dfrac{\mathrm{d}\lambda_i}{\mathrm{d}t} = \lambda_i \kappa : \bar{S}_i - \dfrac{1}{\tau_{\mathrm{s},i}}(\lambda_i - 1)\mathrm{e}^{\nu_i(\lambda_i - 1)}$

$$\text{for}\quad \lambda \le q_i \,. \tag{1.162}$$

Here, for *each relaxation mode i*, four unknown parameters are required. These parameters are:

- *Backbone orientation time*: $\tau_{\mathrm{b},i}$;
- *Fractional relaxation modulus*: g_i ;
- *Number of pom-pom arms*: q_i ;
- *Backbone stretch orientation time*: $\tau_{\mathrm{s},i}$.

While $\tau_{\mathrm{b},i}$ and g_i represent the linear viscoelastic properties of the polymer melt and are obtained from linear viscoelastic experiments, q_i and $\tau_{\mathrm{s},i}$ are nonlinear parameters that are fitted to appropriate nonlinear material functions. This means that for any linear viscoelastic mode, there are two nonlinear parameters.

Figure 1.7 shows results for uniaxial extensional and transient shear viscosities of LDPE International Union of Pure and Applied Chemistry (IUPAC) A. The theoretical curves represent a nine-mode pom-pom as presented by *Inkson* et al. [1.34]. Figure 1.8 shows the nonlinear rheological behavior of a densely branched LDPE modeled with a six-mode pom-pom [1.40].

Further modifications of the differential pom-pom model were proposed by *Verbeeten* et al. [1.41]. By use of a nonlinear, Giesekus-type argument they succeeded in rewriting the pom-pom equation by excluding the finite extensibility condition and the discontinuity associated with it, and introduced a nonzero second normal stress difference. No need to say, this phenomenological description which is also known as the extended pom-pom (XPP) model, despite its computational desirability [1.42], has totally diverted from the original microstructural pom-pom idea and offers little insight into the polymer structure.

Models with Varying Tube Diameter
The original picture of the tube model proved to be a promising and flexible tool for prediction of the rheological behavior of polymer melts. However, as shown in the previous section, assuming a pre-averaged tube stretch will in practice demand numerous simplifying assumptions and a large number of nonlinear parameters as a consequence of pre-averaging. An alternative to circumvent pre-averaging is the assumption of a strain-dependent tube diameter as first suggested by *Marrucci* and *de Cindio* [1.43]. They assumed that the tube is deformed affinely during deformation and the tube volume remains constant, both in contrast to the classical DE theory, which resulted in a single integral constitutive equation of the form

$$\sigma = -pI + \frac{5}{4} \int_{-\infty}^{t} m(t-t') \left\langle \frac{u'u'}{u'} \right\rangle \mathrm{d}t' \,, \tag{1.163}$$

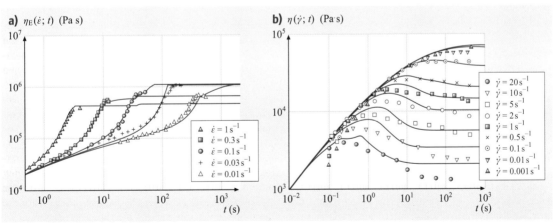

Fig. 1.7a,b Results of a nine-mode pom-pom model analysis for a standard IUPAC LDPE A with molecular weight $M_\mathrm{w} = 300\,000$ and molecular weight distribution (MWD) of 17.6 (after [1.34], with permission): (**a**) Elongational viscosity; (**b**) transient shear viscosity

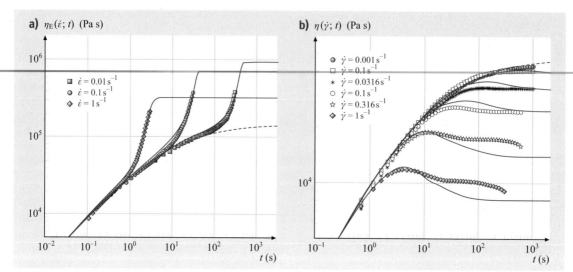

Fig. 1.8a,b Results of a six-mode pom-pom model for an LDPE with $M_w = 235\,500$ and MWD of 17.1 (after [1.40], with permission from Macromolecules 35:10091. Copyright 2002, American Chemical Society): **(a)** elongational viscosity; **(b)** transient shear viscosity

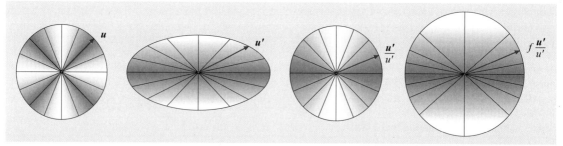

Fig. 1.9 MSF theory: stress is the result of both orientation and isotropic stretch of tube segments by a factor f

which produced better predictions for the extensional behavior of a poly(methyl methacrylate) (PMMA) melt. It was based on their work that Wagner et al. [1.44–47] adopted the concept of varying tube diameter to show that the work of the stress tensor can be correlated to the change of free energy at the molecular scale [1.7].

The Molecular Stress Function (MSF) Model for Linear Melts. In the molecular stress function (MSF) model of Wagner and coworkers [1.7, 44–52], tube stretch is caused by the *squeeze* of the surrounding polymer chains, leading to a reduction of the tube diameter a from its equilibrium value a_0. Taking into account that the tube diameter a represents the mean field of the surrounding chains and its associated strain energy, it is assumed that the tube diameter is independent of the orientation of tube segments. This is depicted schematically in Fig. 1.9.

The stress is then given by

$$\sigma(t) = -p\mathbf{I} + \int_{-\infty}^{t} m(t-t') f^2 \mathbf{S}_{DE}^{IA}(t,t')\,dt', \quad (1.164)$$

where the molecular stress function f is the inverse of the relative tube diameter,

$$f(t,t') = a_0/a(t,t'). \quad (1.165)$$

In contrast to (1.149), the tube stretch in (1.164) does not only depend on the observation time t, but also on the strain history, i.e., for time-dependent strain histories, the tube stretch varies along the tube. The dependence on t and t' is dropped in the following.

Note that while $\mathbf{S}_{DE}^{IA}$ is related directly to the deformation history via (1.145), no a priori dynamics of the internal variable f is prescribed in the MSF model.

Rather, f^2 is assumed to be directly related to the strain energy stored in the polymeric system, and is determined as solution of an evolution equation derived from an energy balance argument [1.7].

Based on prior work of *de Gennes* [1.53], and *Marrucci* et al. [1.54, 55], the molecular stress function f for linear melts is related to a strain-energy function w_{MSF} of the form

$$\frac{w_{\text{MSF}}}{3k_B T} = (f^2 - 1). \tag{1.166}$$

Neglecting dissipative constraint release, i.e., considering the hyperelastic limit, the power input of the stress tensor into the polymer system is equal to the increase of the strain energy by tube deformation [1.7]. f^2 is found as solution of the evolution equation (with velocity gradient κ and plateau modulus G_N^0)

$$\frac{1}{3k_B T}\frac{dw_{\text{MSF}}}{dt} = \kappa : \frac{\sigma}{5G_N^o} = f^2\left(\kappa : S\right) \tag{1.167}$$

to be

$$f^2 = e^{\langle \ln u' \rangle_0}, \tag{1.168}$$

i.e., f^2 is an exponential of the orientational free energy $3k_B T \langle \ln u' \rangle_0$. $\frac{d}{dt}$ indicates the material time derivative.

Note that by use of (1.168), the strain energy function of (1.166) can be expressed as

$$\frac{w_{\text{MSF}}}{3k_B T} = \langle \ln u' \rangle_0 + f^2 - \ln f^2 - 1, \tag{1.169}$$

i.e., as the sum of the orientational free energy and the stretch energy. The part of the strain energy due to chain stretch has the desired properties, namely a minimum at equilibrium ($f^2 = 1$) and a quadratic dependence on f in the vicinity of equilibrium.

Predictions of the MSF model are in excellent agreement with the onset of strain-hardening in uniaxial, equibiaxial and planar extension of polydisperse linear polymer melts (the so-called LMSF model), as exemplified in Fig. 1.10 and Fig. 1.11 [1.7].

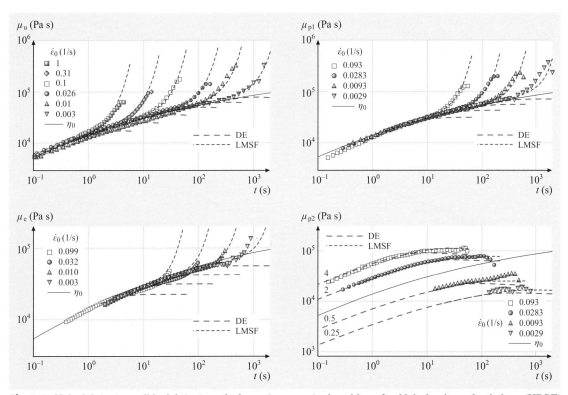

Fig. 1.10 Uniaxial (μ_u), equibiaxial (μ_e), and planar (μ_{p1}, μ_{p2}) viscosities of a high-density polyethylene (HDPE) at $T = 150\,°C$. Viscosities are normalized with respect to the zero-shear viscosity. Comparison of experimental data (*symbols*) to predictions of DE and LMSF (zero-parameter) models (after [1.7])

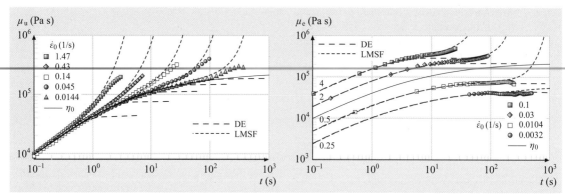

Fig. 1.11 Uniaxial (μ_u) and equibiaxial (μ_e) viscosities of a polystyrene (PS) melt. Viscosities are normalized with respect to the zero-shear viscosity. Comparison of experimental data (*symbols*) to the predictions of the DE and LMSF (zero-parameter) models.

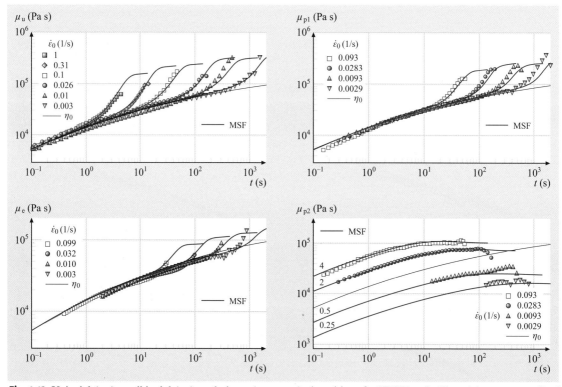

Fig. 1.12 Uniaxial (μ_u), equibiaxial (μ_e), and planar (μ_{p1}, μ_{p2}) viscosities of a HDPE melt. Viscosities are normalized with respect to the zero-shear viscosity. Comparison of experiment to predictions of the MSF model with dissipative constraint release. $f_{\max}^2 = 49$ [1.7]

Now dissipative constraint release (CR) is introduced as a dissipative process [1.7], which modifies the energy balance of tube deformation, and leads to a strain-dependent evolution equation for the molecular stress function of the form

$$\frac{\mathrm{d}f^2}{\mathrm{d}t} = f^2 \left[(\boldsymbol{\kappa} : \boldsymbol{S}) - \frac{1}{f^2 - 1} \mathrm{CR} \right] . \quad (1.170)$$

Constraint release is considered to be the consequence of different convection mechanisms for tube orientation and tube-cross section, and for constant-strain-rate flows can be expressed as

$$\mathrm{CR} = a_1(f^2-1)^2\sqrt{\boldsymbol{D}^2:\boldsymbol{S}} \\ + a_2(f^2-1)^2\sqrt{|\boldsymbol{W}\cdot\boldsymbol{D}:\boldsymbol{S}|} \quad (1.171)$$

with $\boldsymbol{D}$ and $\boldsymbol{W}$ being the rate of deformation and rate of rotation tensor, respectively. The nonlinear material parameters verify $a_1 \geq 0$ and $a_2 \geq 0$. Note that in extensional flows, constraint release depends only on the parameter a_1, while in simple shear flow, both the parameters a_1 and a_2 are of relevance. The evolution equation for the molecular stress function of linear melts in extensional flows is given by

$$\frac{\mathrm{d}f^2}{\mathrm{d}t} = \dot{\varepsilon}f^2\Big[S_{11}+mS_{22}-(1+m)S_{33}-a_1(f^2-1) \\ \times\sqrt{S_{11}+m^2S_{22}+(1+m)^2 S_{33}}\Big] \quad (1.172)$$

where the parameter m ($-1/2 \leq m \leq 1$) describes the type of extensional flow, and $\dot{\varepsilon}$ is the largest extension rate. S_{ii} are the components of the orientation tensor $\boldsymbol{S}$. At large strains, a maximum $f^2 = f^2_{\max}$ is reached and $\mathrm{d}f^2/\mathrm{d}t = 0$. Hence, the parameter a_1 can be expressed in terms of $f^2_{\max}$ as

$$a_1 = \frac{1}{f^2_{\max}-1} ; \quad (1.173)$$

$f^2_{\max}$ governs the steady-state value of the viscosity in extensional flows, and corresponds to the maximum of storable elastic energy. It is the only nonlinear material parameter of the theory for describing polymer melt rheology of linear polymers in irrotational flows. The level of agreement between experiments in different extensional deformation modes and theory for a linear polyethylene (PE) melt is demonstrated in Fig. 1.12.

The evolution equation of the molecular stress function for shear flow is given by

$$\frac{\mathrm{d}f^2}{\mathrm{d}t} = \dot{\gamma}f^2\Big[S_{12}-\frac{1}{2}\frac{f^2-1}{f^2_{\max}-1}\sqrt{S_{11}+S_{22}} \\ -\frac{a_2}{2}(f^2-1)\sqrt{|S_{11}-S_{22}|}\Big] \quad (1.174)$$

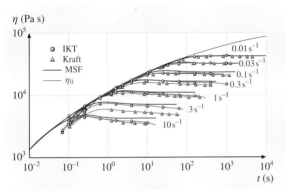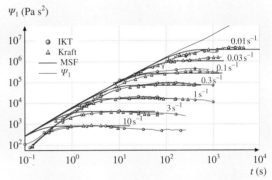

Fig. 1.13 Shear viscosity η and first normal stress function Ψ_1 of a HDPE melt. Comparison of experimental data (*symbols*) to predictions of the MSF model with dissipative constraint release. $f^2_{\max} = 49$ and $a_2 = 2.3$. For details see (after [1.7])

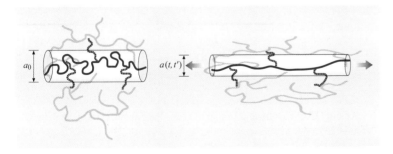

Fig. 1.14 Tube segment of a long-chain branched polymer molecule before and after deformation: one chain segment is stretched, while side-chain segments are compressed [1.49, 50]

and comparison of the predictions to experimental data of the start-up of steady shear flow for the same linear PE melt is shown in Fig. 1.13.

Note that although dissipative constraint release is a rate process, integration of (1.172) and (1.174) leads to a molecular stress function f which is deformation dependent [1.7].

The MSF Model for Long-Chain Branched Melts. The simplest model of a tube section of a long-chain branched macromolecule containing β entanglements consists of one chain segment representing one entanglement oriented in the direction of the tube (the *backbone* of the macromolecule), and one or more side chains representing $\beta - 1$ entanglements (Fig. 1.14). Note that a side chain can contain more than one chain segment, depending on the length of the side chain relative to the entanglement length. Thus, according to this model, chain segments fall into two distinct categories: either they belong to the backbone and are stretched by deformation, or they do not belong to the backbone and are compressed by deformation [1.49].

When the tube is stretched, one segment is extended, while $\beta - 1$ are compressed, leading to a total strain energy of

$$\frac{w_{\text{MSF}}}{3k_{\text{B}}T} = \frac{1}{\beta}(f^2 - 1) + \frac{(\beta - 1)}{\beta}\left(1 - \frac{1}{f^2}\right). \quad (1.175)$$

Note that in the vicinity of $f^2 = 1$, this strain energy function is well behaved, as (1.175) (with $\beta = 1$) reduces to (1.166).

The parameter β has values $\beta \geq 1$, with $\beta = 1$ for linear melts. For $\beta = 2$ (the so-called QMSF model), excellent agreement with experimental data of a long-chain branched (radiation-crosslinked) polypropylene (PP) melt is found (Fig. 1.15). Note that the increase in elongational viscosity is steeper for long-chain branched melts than for linear melts.

Introducing again constraint release as a nonlinear dissipative process, which modifies the energy balance of tube deformation, leads to a strain-dependent evolution equation for the molecular stress function of the form

$$\frac{df^2}{dt} = \frac{\beta f^2}{1 + \frac{\beta - 1}{f^4}}\left[(\kappa : S) - \frac{1}{f^2 - 1}\text{CR}\right]. \quad (1.176)$$

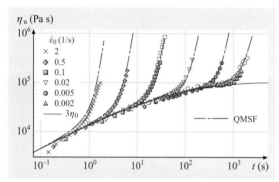

Fig. 1.15 Uniaxial viscosity η_u of a long-chain branched PP melt [1.7]. Comparison of experimental data (*symbols*) to predictions of the MSF model with $\beta = 2$ (QMSF model)

The evolution equation for the molecular stress function in constant-strain-rate extensional flows is then

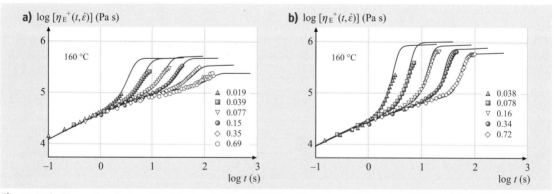

Fig. 1.16a,b Elongational viscosity data (*symbols*) of LDPE melts and predictions by MSF model [1.49]. Parameters indicate elongation rate in units of s^{-1}. (a) LDPE produced by tubular process (tubular o): $\beta = 2$ and $f_{\text{max}}^2 = 30$; (b) LDPE produced by an autoclave process (autoclave O): $\beta = 4$ and $f_{\text{max}}^2 = 80$

given by

$$\frac{df^2}{dt} = \dot{\varepsilon}\frac{\beta f^2}{1+\frac{\beta-1}{f^4}}\left[S_{11}+mS_{22}-(1+m)S_{33}\right.$$
$$\left.-\frac{f^2-1}{f_{\max}^2-1}\sqrt{S_{11}+m^2S_{22}+(1+m)^2S_{33}}\right]$$
(1.177)

and in constant-shear-rate flow by

$$\frac{df^2}{dt} = \dot{\gamma}\frac{\beta f^2}{1+\frac{\beta-1}{f^4}}\left[S_{12}-\frac{1}{2}\frac{f^2-1}{f_{\max}^2-1}\sqrt{S_{11}+S_{22}}\right.$$
$$\left.-\frac{a_2}{2}(f^2-1)\sqrt{|S_{11}-S_{22}|}\right].$$
(1.178)

The enhanced slope of elongational viscosity of long-chain branched polymer melts in comparison to linear melts is caused by the fact that a significant percentage of the chain segments of a long-chain branched molecule is compressed by elongational flow (the *side chains*), and only part of the chain segments is stretched (the *backbone*). In the multi-chain segmental MSF model described here, for one chain segment stretched, $\beta-1$ chain segments are compressed. While for LDPE melts produced by the tubular polymerization processes typically values of $\beta=2$ are found, more highly branched autoclave LDPE melts show values of $\beta=3$ and even of $\beta=4$ (Fig. 1.16) [1.49].

The level of agreement between experiments in different extensional deformation modes and in start-up of steady shear flow and theory for a tubular LDPE melt with $\beta=2$ is demonstrated in Fig. 1.17 and Fig. 1.18.

Comparison of MSF Model Predictions to Elongational and Shear Rheology of Model Branched Polystyrene Melts. It is difficult if not impossible to derive the parameter β from the topology of randomly branched LDPE; therefore β in (see (1.175)) was treated as a fit parameter (the only one in the hyperelastic limit). However, from an analysis of the nonlinear rheology of comb shaped model polystyrene melts investigated by *Hep-*

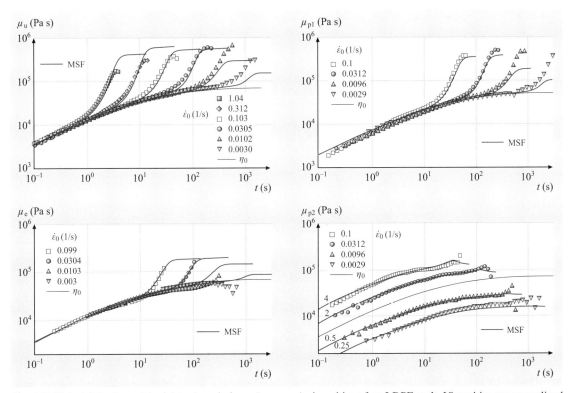

Fig. 1.17 Uniaxial (μ_u), equibiaxial (μ_e), and planar (μ_{p1}, μ_{p2}) viscosities of an LDPE melt. Viscosities are normalized with respect to the zero-shear viscosity. Comparison of experiment to predictions of the MSF model with dissipative constraint release. $f_{\max}^2 = 100$ (after [1.7])

perle and *Münstedt* [1.50, 56], it was found that indeed β, as derived from the topology of these model melts by assuming stretch of the backbone chain and compression of the side chains, is in quantitative agreement with experimental evidence seen in uniaxial extension [1.50]. The parameter β can simply be obtained as ratio of the number average molar mass of the grafted polymer, M_n, to the number average molar mass $M_{n,bb}$ of the backbone, which can be expressed in terms of the number average mass fraction $\Phi_{n,br}$ of grafted side chains,

$$\beta = \frac{M_n}{M_{n,bb}} = \frac{1}{1 - \Phi_{n,br}}. \tag{1.179}$$

For linear polymers, naturally $\beta = 1$ is obtained from (1.179).

As exemplified in Fig. 1.19, agreement between predicted and observed slopes of the elongational viscosity after inception of strain hardening is excellent for all model branched polystyrene melts investigated. Within the experimentally accessible window of elongation rates, time–strain separability of the measured elongational viscosities is observed. Also, as far as a maximum strain-hardening could be determined, the data are compatible with the implicit assumption of the MSF model that the material parameter f_{max}^2 is the same for all relaxation times of the terminal zone of the relaxation spectrum.

The shear damping function of model branched PS melts was measured by nonlinear shear relaxation experiments [1.50, 56]. Although the shear strain range investigated was limited to $\gamma < 5$, this is the impor-

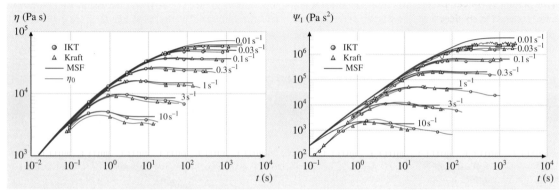

Fig. 1.18 Shear viscosity η and first normal stress function Ψ_1 of an LDPE melt. Comparison of experimental data (*symbols*) to predictions of the MSF model with dissipative constraint release $a_2 = 0.036$. For details see [1.7]

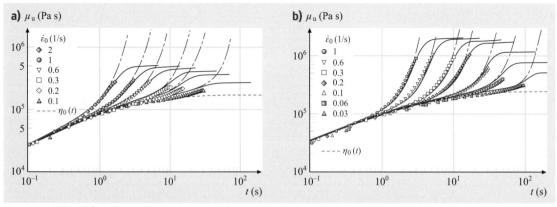

Fig. 1.19a,b Comparison of elongational viscosity data (*symbols*) of two branched PS melts to predictions (*lines*) of MSF theory. Viscosities are normalized with respect to the zero-shear viscosity; $\eta_0(t)$ indicates the start-up zero-shear viscosity [1.50, 56]. **(a)** PS 80-0.6G-22: $\Phi_{n,br} = 0.14$, $\beta = 1.2$; *dashed line*: $f_{max}^2 \to \infty$, *solid line*: $f_{max}^2 = 25$; **(b)** PS 70-3.2G-22: $\Phi_{n,br} = 0.5$, $\beta = 2.0$; *dashed line*: $f_{max}^2 \to \infty$, *solid line*: $f_{max}^2 = 80$

tant shear strain range determining the shear stress in steady shear-rate flows. As is well known, branching also has significant influence on the shear strain behavior, although the effect is usually much smaller than in extensional flows [1.50, 56]: melts with high side-chain mass fractions show substantially less shear damping than melts with low side-chain mass fractions, and for shear strains up to 5, their shear damping functions are close to the hyperelastic or K-BKZ limit, i.e., the dissipative effect of constraint release is very small Fig. 1.20a. This agrees with earlier investigations of *Wagner* and *Ehrecke* [1.9] demonstrating that an LDPE melt shows reversible (or K-BKZ) behavior in double-step shear strain experiments in contrast to a (linear) polyisobutene melt, which showed *irreversible* behavior. With decreasing side chain mass fraction, the shear damping behavior of the model branched polystyrene melts approaches the behavior of linear polystyrene, and the influence of the parameter a_2 describing the additional dissipative constraint release due to rotational flow becomes important (Fig. 1.20b).

1.8.4 Conclusions

A survey of the most recent and most often used constitutive equations in nonlinear rheology shows the amount of progress made in the recent decades. It has shown that phenomenological constitutive integral equations can function as a rational basis of microstructural theories, but their relation to polymer structure remained elusive. Constitutive equations based on the tube model with pre-averaged chain stretch have been predominantly utilized by use of a differential approximation for the orientation tensors, and have been considered as appropriate for modeling of complex flows. The main problem of equations with pre-averaged chain stretch is the need to use a large number of nonlinear parameters, thereby losing any insight into the relation to polymer topology, and sometimes rendering the models too tedious even for a computer simulation.

Integral models with varying tube diameter have shown more flexibility in predicting polymer melt behavior in typical rheological flows, and have proven to be capable of describing the rheological behavior of a wide variety of polymers. The microstructural MSF model, which is based on the variable tube diameter idea, has been the latest among such constitutive equations with excellent predictive capabilities for modeling the nonlinear extensional and shear behavior of both linear and long-chain branched industrially important polymers. The concept of a strain-dependent tube diameter, which decreases with increasing deformation, explains consistently the strain hardening of linear as well as of long-chain branched polymer melts [1.7, 49, 50]. The steeper slope of the elongational viscosity after inception of strain hardening for branched melts in comparison to linear melts is due to the fact that in branched melts only a fraction (*the backbone*) of chain segments is stretched, while side chains are compressed [1.49, 50]. Long-chain branched polymer melts show reversible or *K-BKZ* behavior in double-step shear strain experiments, because dissipative constraint release occurs only at higher shear strains, in contrast to linear melts, where dissipation starts already at smaller shear strains [1.50].

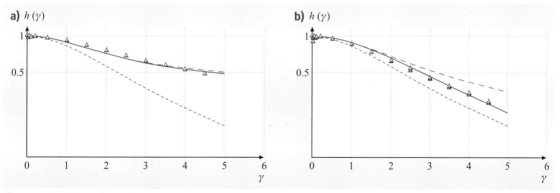

Fig. 1.20a,b Comparison of shear damping function data (*symbols*) of two PS melts to the predictions of the MSF theory [1.50, 56]; the *upper dash-dotted line* is the prediction assuming no constraint release (i.e., $f_{max}^2 \to \infty$, $a_2 = 0$); the *lower dashed line* indicates the predictions of the Doi–Edwards IAA theory, i.e., $f_{max}^2 \equiv 1$ [1.50]. (**a**) Branched PS 70-3.2G-22: the *full line* is the prediction of the MSF model with the parameters $\beta = 2$, $f_{max}^2 = 80$ and $a_2 = 0$; (**b**) linear PS-r-95: the *full line* is the prediction of the MSF model with the parameters $\beta = 1$, $f_{max}^2 = 6$ and $a_2 = 0.4$

Challenges in nonlinear rheology remaining are, to mention just a few, the relations between macromolecular architecture of homopolymers and the nonlinear parameters f_{max}^2 and a_2 of the MSF theory, as well as modeling the nonlinear rheology of blends of linear and long-chain branched polymers.

References

1.1 S.I. Green: *Fluid Vorticies* (Kluwer Academic, Dordrecht 1995)

1.2 D.D. Joseph: *Fluid Dynamics of Viscoelastic Liquids* (Springer, Berlin 1990)

1.3 R.G. Larson: *Constitutive Equations for Polymer Melts and Solutions* (Butterworths, London 1988)

1.4 M. Doi, S.F. Edwards: *The Theory of Polymer Dynamics* (Oxford Univ. Press, Oxford 1986)

1.5 M.H. Wagner: *Challenges in Nonlinear Rheology of Linear and Long-Chain Branched Polymer Melts* (Proc. XIVth Int. Congr. On Rheology, Korea 2004)

1.6 T.C.B. McLeish, S.T. Milner: Entangled dynamics and melt flow behavior of branched polymers, Adv. Polym. Sci. **143**, 195–256 (1999)

1.7 M.H. Wagner, P. Rubio, H. Bastian: The molecular stress function model for polydisperse and polymer melts with dissipative convective constraint release, J. Rheol. **45**, 1387–1412 (2001)

1.8 H. Bastian: *Non-linear viscoelasticity of linear and long-chain-branched polymer melts in shear and extensional flows*, Ph. D. Thesis (Universität Stuttgart, Stuttgart 2000), http://elib.uni-stuttgart.de/opus/volltexte/2001/894

1.9 M.H. Wagner, P. Ehrecke: Dynamics of polymer melts in reversing shear flows, J. Non-Newtonian Fluid Mech. **76**, 183–197 (1998)

1.10 A.S. Lodge: Constitutive equations from molecular theories for polymer solutions, Rheol. Acta. **7**, 379–392 (1968)

1.11 E. van Ruymbeke, R. Keunings, V. Stéphenne, A. Hagenaars, C. Bailly: Evaluation of reptation models for predicting the linear viscoelastic properties of linear entangled polymers, Macromolecules **35**, 2689–2699 (2002)

1.12 A.L. Frischknecht, S.T. Milner, A. Pryke, R.N. Young, R. Hawkins, T.C.B. McLeish: Rheology of three-arm asymmetric star polymer melts, Macromolecules **35**, 4801–4820 (2002)

1.13 M.H. Wagner, J. Meissner: Network disentanglement and time-dependent flow behavior of polymer melts, Macromol. Chem. **181**, 1533–1550 (1980)

1.14 C.W. Macosko: *Rheology, Principles, Measurements and Applications* (VCH, New York 1994)

1.15 H.M. Laun: Description of the non-linear shear behavior of a low density polyethylene melt by means of an experimentally determined strain dependent memory function, Rheol. Acta. **17**, 1–15 (1978)

1.16 M.H. Wagner: Analysis of time-dependent non-linear stress-growth data for shear and elongational flow of a low-density branched polyethylene melt, Rheol. Acta. **15**, 136–142 (1976)

1.17 M.H. Wagner: Prediction of primary normal stress difference from shear viscosity data using a single integral constitutive equation, Rheol. Acta. **16**, 43–50 (1977)

1.18 M.H. Wagner, S.E. Stephenson: The spike strain test for polymeric liquid and its relevance for irreversible destruction of network connectivity by deformation, Rheol. Acta. **18**, 463–468 (1979)

1.19 M.H. Wagner, S.E. Stephenson: The irreversibility assumption of network disentanglement in flowing polymer melts and its effects on elastic recoil predictions, J. Rheol. **23**, 489–504 (1979)

1.20 K. Osaki, S. Kimura, M. Kurata: Relaxation of shear and normal stresses in double-step shear deformations for a polystyrene solution. A test of Doi-Edwards theory for polymer rheology, J. Rheol. **25**, 549–562 (1981)

1.21 M.H. Wagner: A constitutive analysis of uniaxial elongational flow data of a low-density polyethylene melt, J. Non-Newtonian Fluid Mech. **4**, 39–55 (1978)

1.22 A.C. Papanastasiou, L.E. Scriven, C.W. Macosko: An integral constitutive equation for mixed flows: viscoelastic characterization, J. Rheol. **27**, 387–410 (1983)

1.23 T. Samurkas, R.G. Larson, J.M. Dealy: Strong extensional and shearing flows of a branched polyethylene, J. Rheol. **33**, 559–578 (1989)

1.24 P.K. Currie: Constitutive equations for polymer melts predicted by the Doi-Edwards and Curtiss-Bird kinetic theory models, J. Non-Newtonian Fluid Mech. **11**, 53–68 (1982)

1.25 M.H. Wagner, A. Demarmels: A constitutive analysis of extensional flows of polyisobutylene, J. Rheol. **34**, 943–958 (1990)

1.26 O. Urakawa, M. Takahashi, T. Masuda, N.G. Ebrahimi: Damping functions and chain relaxation in uniaxial and biaxial elongation: comparison with the Doi-Edwards theory, Macromolecules **28**, 7196–7201 (1995)

1.27 B.J.R. Scholtens, P.J.R. Leblans: Nonlinear viscoelasticity of noncrystalline EPDM rubber networks, J. Rheol. **30**, 313–335 (1986)

1.28 M.H. Wagner: The nonlinear strain measure of polyisobutylene melt in general biaxial flow and its comparison to the Doi-Edwards model, Rheol. Acta. **29**, 594–603 (1990)

1.29 D.S. Pearson, A. Kiss, L. Fetters, M. Doi: Flow-induced birefringence of concentrated polyisoprene solutions, J. Rheol. **33**, 517–535 (1989)

1.30 G. Ianniruberto, G. Marrucci: A simple constitutive equation for entangled polymers with chain stretch, J. Rheol. **45**, 1305–1318 (2001)

1.31 J. Fang, M. Kröger, H.M. Öttinger: A thermodynamically admissible reptation model for fast flows of entangled polymers: II. Model predictions for shear and extensional flows, J. Rheol. **44**, 1293–1316 (2000)

1.32 T.C.B. McLeish, R.G. Larson: Molecular constitutive equations for a class of branched polymers: the pom-pom polymer, J. Rheol. **42**, 81–110 (1998)

1.33 T.C.B. McLeish: Molecular rheology of H-polymers, Macromolecules **21**, 1062–1070 (1988)

1.34 N.J. Inkson, T.C.B. McLeish, O.G. Harlen, D.J. Groves: Predicting low density polyethylene melt rheology in elongational and shear flows with pom-pom constitutive equations, J. Rheol. **43**, 873–896 (1999)

1.35 R.J. Blackwell, T.C.B. McLeish, O.G. Harlen: Molecular drag-strain coupling in branched polymer melts, J. Rheol. **44**, 121–136 (2000)

1.36 R.G. Owens, T.N. Phillips: *Computational Rheology* (Imperial College Press, London 2002)

1.37 P. Rubio, M.H. Wagner: Letter to the Editor: A note added to "Molecular constitutive equations for a class of branched polymers: The pom-pom model", J. Rheol. **43**, 1709–1710 (1999)

1.38 P. Rubio, M.H. Wagner: LDPE melt rheology and the pom-pom polymer, J. Non-Newtonian Fluid Mech. **92**, 245–259 (2000)

1.39 R.J. Blackwell, O.G. Harlen, T.C.B. McLeish: Theoretical linear and non-linear rheology of symmetric treelike polymer melts, Macromolecules **34**, 2579–2596 (2001)

1.40 P.J. Doerpinghaus, D.G. Baird: Accessing the branching architecture of sparsely branched metallocene-catalyzed polyethylenes using the pompom constitutive model, Macromolecules **35**, 10087–10095 (2002)

1.41 W.M.H. Verbeeten, G.W.M. Peters, F.P.T. Baaijens: Differential constitutive equations for polymer melts: The extended pom-pom model, J. Rheol. **45**, 823–843 (2001)

1.42 N. Clemeur, R.P.G. Rutgers, B. Debbaut: On the evaluation of some differential formulations for the pom-pom constitutive model, Rheol. Acta. **42**, 217–231 (2003)

1.43 G. Marrucci, B. de Cindio: The stress relaxation of molten PMMA at large deformations and its theoretical interpretation, Rheol. Acta. **19**, 68–75 (1980)

1.44 M.H. Wagner, J. Schaeffer: Constitutive equations from Gaussian slip-link network theories in polymer melt rheology, Rheol. Acta. **31**, 22–31 (1992)

1.45 M.H. Wagner, J. Schaeffer: Rubbers and Polymer melts: Universal aspects of non-linear stress-strain relations, J. Rheol. **37**, 643–661 (1993)

1.46 M.H. Wagner: The non-linear strain measure of polymer melts and rubbers: A unifying approach, Makromol. Chem. Macromol. Symp. **68**, 95–108 (1993)

1.47 M.H. Wagner, J. Schaeffer: Assessment of non-linear strain measures for extensional and shearing flows of polymer melts, Rheol. Acta. **33**, 506–516 (1994)

1.48 M.H. Wagner, J. Schaeffer: Nonlinear strain measures for general biaxial extension of polymer melts, Rheol. Acta. **36**, 1–26 (1992)

1.49 M.H. Wagner, M. Yamaguchi, M. Takahashi: Quantitative assessment of strain hardening of LDPE melts by MSF model, J. Rheol. **47**, 779–793 (2003)

1.50 M.H. Wagner, J. Hepperle, H. Münstedt: Relating molecular structure of model branched polystyrene melts to strain-hardening by molecular stress function theory, J. Rheol. **48**, 489–503 (2004)

1.51 M.H. Wagner, S. Kheirandish, M. Yamaguchi: Quantitative analysis of melt elongational behavior of LDPE/LLDPE blends, Rheol. Acta **44**, 198–218 (2005)

1.52 M.H. Wagner, S. Kheirandish, K. Koyama, A. Nishioka, A. Minegishi, T. Takahashi: Modeling strain hardening of polydisperse polystyrene melts by molecular stress function theory **44**, 235–243 (2005)

1.53 P.G. de Gennes: Reptation of polymer chain in the presence of fixed obstacles, J. Chem. Phys. **55**, 572–579 (1971)

1.54 G. Marrucci, N. Grizzutti: The free energy function of the Doi-Edwards theory: Analysis of instabilities in stress relaxation, J. Rheol. **27**, 433–450 (1983)

1.55 G. Marrucci, J.J. Hermans: Non-linear viscoelasticity of concentrated polymeric liquids, Macromolecules **13**, 380–387 (1980)

1.56 J. Hepperle: *Einfluss der molekularen Struktur auf rheologische Eigenschaften von Polystyrol- und Polycarbonatschmelzen* (Shaker, Aachen 2003)

2. Nondimensional Representation of the Boundary-Value Problem

Given that an experiment can be considered to represent a physically realizable boundary value (bv) problem and given that the derived measurements are to represent aspects of the solution to the bv problem, it is rational to extend this understanding such that a maximum amount of information can be obtained from a given experiment. The first portion Sect. 2.2.1 establishes the bases for obtaining information regarding the flow associated with a prototype (the object/flow of actual interest) from measurements made in a model study. This section focuses on the large class of flows for which a Newtonian fluid and its governing equations establish the model-to-prototype information exchange.

Dimensional analysis Sect. 2.2.2 provides a complement to Section 2.1 with a less structured – and therefore a more flexible – approach to problems that extend beyond those readily addressed by the Sect. 2.2.1 material. The important issue of collecting experimental results in non-dimensional groups is addressed in Sect. 2.2.2.

The discussion of self-similarity Sect. 2.2.3 addresses the immense compaction of experimental data that is made possible for those flows that exhibit this property. The bases for, and utilization of, self-similarity are explored in detail.

2.1	**Similitude, the Nondimensional Prototype and Model Flow Fields**	34
	2.1.1 Governing Equations – Newtonian and Incompressible	34
	2.1.2 Boundary Conditions	35
	2.1.3 Initial Conditions	36
	2.1.4 Parameters that Influence the Solution to the Boundary-Value and/or the Initial-Value (BV/IV) Problem	36
	2.1.5 Governing Equations – Newtonian and Compressible	38
	2.1.6 Flows for Which U and L May Not Be Apparent	39
2.2	**Dimensional Analysis and Data Organization**	42
	2.2.1 Variables, Function List, and Extra Information	42
	2.2.2 Dimensions and Scale Ratios	43
	2.2.3 Natural Scales and Repeating Variables	43
	2.2.4 Π Theorem	45
	2.2.5 Example with Rank Less than the Number of Dimensions	45
	2.2.6 Example with Redundant Dimensions	46
	2.2.7 Anatomy of a Nondimensional Variable	47
	2.2.8 Nonuniqueness of Scales	48
	2.2.9 Reference	48
	2.2.10 Scales Chosen for Experimental Purposes	48
	2.2.11 Nondimensional Variables Interpreted as Physical Ratios	50
	2.2.12 Scales Found from Boundary Conditions and Equations	50
	2.2.13 Limiting Cases	51
	2.2.14 Singular Perturbations	52
	2.2.15 Overlap Behavior and Composite Expansions	52
	2.2.16 Common Scales and Nondimensional Parameters	55
2.3	**Self-Similarity**	57
	2.3.1 General Causes of Self-Similar Behavior in Certain Situations in Fluid Mechanics and Heat Transfer	57
	2.3.2 Implications of Self-Similarity in Experimental Studies	58
	2.3.3 Particular Examples of Self-Similar Navier–Stokes Flows	59
	2.3.4 Particular Examples of the Boundary Layer Flows	62
	2.3.5 Gas Dynamics: Strong Explosion	79
	2.3.6 Free-Surface Flows	80
References		82

2.1 Similitude, the Nondimensional Prototype and Model Flow Fields

An important class of experiments in fluid mechanics is that in which a model study of a prototype is to provide reliable information on the predicted flow properties of the associated prototype. It is this class of experiments in which the concept and principles of *similitude* are used to ensure a reliable transfer of information from the model study to the prototype. Specifically, dimensional measurements from the model (e.g., quantities such as pressure, velocity, angular speed, aerodynamic drag and lift) are used to predict the numerical values for the same quantities that would be present in the prototype flow field.

The fundamental basis for similitude is firmly grounded in the description of an experiment as a *boundary, and possibly an initial, value problem*. If the model flow can be made to represent exactly the same boundary (initial) value problem, then, of course, the *solution* to the model and prototype problems must be the same. The solution, as noted above, is represented by the experimental data acquired from the well-defined experiment. Since an *exact* solution often cannot be ensured one can substitute an *adequate* solution in order to gain useful predictions of the prototype's flow field.

This section of the Handbook is to provide guidance on how these *in principle* concepts can be transformed to *in practice* guidelines for representative application areas.

2.1.1 Governing Equations – Newtonian and Incompressible

Linearly viscous (Newtonian) fluids (air, water, oils, etc.) are widely present in fluid mechanics applications. Their ubiquity serves as part of the basis for the first restriction (Newtonian fluids in an incompressible flow environment) that is considered in this section.

The governing equations (with this restriction) are known for prototype and model flows. If the boundary and, if appropriate, initial conditions can also be made identical for the two flows, then identical solutions can be expected. This expectation is typically experienced in an experimental environment. Interesting exceptions can, and do, occur. For example, if one is investigating a *steady-state flow*, the observed values may show a strong sensitivity to the starting conditions (the initial transient period) for a given experiment. If this condition occurs, then different steady-state behaviors may exit as a result of differences in the initial transient period. Obviously, if the flow is turbulent, then only the appropriately averaged results can be predicted for the prototype from the measured model values.

The designation of an *incompressible* flow for the prototype and, therefore, for the model is one that includes many applications. The term *incompressible* must, of course, apply to the fluid. The common phrase *incompressible flow* indicates a flow in which $D\rho/Dt = 0$ is adequately met. For gases, this is often characterized by the Mach number ≤ 0.3, where $\rho/\rho_0 \leq 0.956$ for an isentropic flow at this Mach number. Liquid flows form one segment of this class of flows, although dissolved gases that can lead to *water hammer* and explosions in a submerged liquid environment are two examples in which the incompressible assumption would not be physically appropriate. Low-speed gas flows are a second (and large) segment of this class. There are only two attributes of the flow field to be characterized for an incompressible flow: the velocity $V(x, y, z, t)$ and the pressure $p(x, y, z, t)$ fields, where Cartesian coordinates are used to indicate the spatial and temporal dependencies of the dependent (V, p) variables. Cartesian coordinates are used for symbolic convenience. The derived results are applicable to any coordinate system.

It is affirmed that the analyst will be able to identify a *characteristic* length (L) and velocity (U) to *scale* the problem. In some cases the selections for U and L will be apparent. The flow past an airfoil that is not influenced by the conditions at its lateral ends is advanced as an example of the apparent choices for U and L (Fig. 2.1). In other cases (Sect. 2.1.6), L and U may have to be *created* from other characteristic properties of the prototype flow field.

Using U and L, the governing equations can be made nondimensional (for an incompressible flow) as

$$\frac{DV^*}{Dt^*} = -\nabla^* p_k^* + \frac{1}{\text{Re}} \nabla^{*2} V^*, \qquad (2.1)$$

$$\nabla^* V^* = 0, \qquad (2.2)$$

$$\text{Re} = \frac{UL}{\nu}, \qquad (2.3)$$

where $p_k^* = (p + \rho gh)/\rho U^2$, $V^* = V/U$ and $x^* = x/L$, $y^* = y/L$, $z^* = z/L$. Note, $\nu = \mu/\rho$ is the kinematic viscosity.

The term p_k clearly combines the (static) pressure and the gravitational body-force term as expressed by the elevation, h, above a datum plane. The ability to express the net force effect caused by ∇p and by $\nabla \rho gh$

Fig. 2.1 Streamlines from a time averaged LES calculation of the flow past a Valeo CD airfoil as abstracted from the work of Moreau et al. [2.1]. Courtesy of D. R. Neal

permits this useful combination. Two other aspects of this form of the equations are noteworthy.

1. Any other body force (electrical, magnetic, etc.) that *can* be expressed using the gradient operator can be incorporated with (p and ρgh), and
2. If the flow of a uniform density fluid exists in a submerged environment, then the portion of the pressure field that balances the body force effect need not be explicitly considered in the problem formulation. Specifically, the hydrostatic variation of pressure ($\partial p/\partial z = -\rho g$) does not contribute to the dynamics of the flow in such a submerged environment.

Examples wherein the electrical forces act as surface and not body forces include electrohydrodynamics and *leaky* dielectrics.

Identical governing equations for model and prototype are, therefore, obtained if the Reynolds number (Re) is the same for the two flows. This is a necessary (if viscous effects are present) but not a sufficient condition for similitude. It is also a condition that, in general, is easily satisfied when the similitude experiment is established.

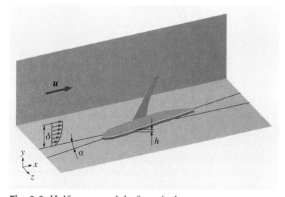

Fig. 2.2 Half-span model of an airplane

2.1.2 Boundary Conditions

Identical – or *adequately identical* – boundary conditions is often the greatest challenge for ensuring the same boundary-value problem between model and prototype. As a simple example, consider that the airfoil shape of Fig. 2.1 is the mid-span shape of a wing that

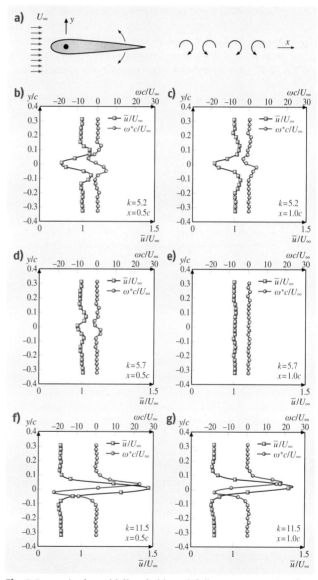

Fig. 2.3a–g A sinusoidally pitching airfoil as an example of an externally imposed reduced frequency. Note: i) the mean streamwise velocity ($\bar{u}$ and the transverse vorticity ($\bar{\omega}_z$) for $A = 2$ degrees oscillation amplitude are shown. ii) $k = \pi C f / U_0$

is attached to an airplane fuselage. If the wing were to be tested at model scale, then not only would the span-to-mid-chord-length ratio need to be represented but the flow at the wing root should also be accurately represented. This would require a full model of the airplane in order to ensure that the flow at the attachment location of the wing to the fuselage is properly represented. In order to make the model as large as possible for a given wind tunnel, a half model (attached to the tunnel wall as shown in Fig. 2.2) is often used in order to allow a larger length ratio: L_m/L_p in the model study. (Note: $[()_m = ()_{model}$, $()_p = ()_{prototype}]$.) A challenging aspect of such a model study is to ensure that important surface features: material roughness, (vents, hinges, etc.) are represented to proper scale. If the chord length is C (C now replaces the generalized L), then the rivet head diameter (d) and its protrusion (δ) should be the same relative size: (d/C) and (δ/C), for the model and the prototype. These details can be relaxed only if it is shown that such length ratios do not influence the flow parameters of interest. A challenge in wind tunnel studies is also presented by the bounding tunnel walls. If the imposed straight streamlines of the tunnel are different from the free streamlines (same nondimensional distance from the model), then the model-to-prototype common boundary conditions constraint would be violated. Adaptive tunnel walls and open test sections are two approaches to address this inherent *conflict* between using as large a model as possible for a given tunnel cross section.

2.1.3 Initial Conditions

If the flow field is one that temporally evolves from an initial condition, then the characterization of the initial conditions, in a manner that mimics that of the boundary conditions, must be established. The nondimensional representation of the evolving time will be

$$t^* = \frac{tU}{L}. \tag{2.4}$$

2.1.4 Parameters that Influence the Solution to the Boundary-Value and/or the Initial-Value (BV/IV) Problem

Nondimensional Time

There are two types of nondimensional times that must be considered in experimental fluid mechanics: the reduced frequency (ω_R) and Strouhal number (St).

Consider a flow field in which an external forcing function provides a *controlling influence* on the flow field. For example, the airfoil of Fig. 2.1 might be operated with a sinusoidal pitching motion where its angle of attack: $\alpha(t)$, is described as

$$\alpha(t) = \alpha_0 \sin \omega t ;$$

this flow field is shown schematically in Fig. 2.3. A necessary condition for similitude between model and prototype will be that the imposed time variation is similar. Namely, the model and prototype must have equal reduced frequencies

$$\omega_R|_m = \left(\frac{\omega_0 L}{U_\infty}\right)_m = \omega_R|_p = \left(\frac{\omega_0 L}{U_\infty}\right)_p. \tag{2.5}$$

The nondimensional α_0 is itself a similarity parameter between the model and prototype. Note that, in this case, the characteristic length L is designated as the chord length C and the characteristic velocity U is designated as U_∞.

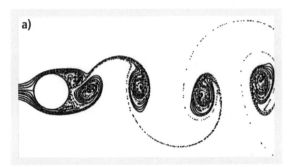

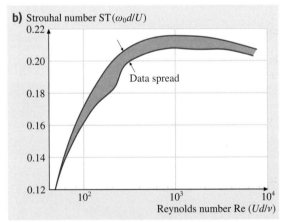

Fig. 2.4a,b Vortex shedding from a cylinder as an example of an intrinsic Strouhal number (**a**) A representation of the vortex shedding behind a cylinder (**b**) Strouhal number versus Reynolds number [2.2]

In contrast to the imposed (ω_0) time scale, which is characterized as a reduced frequency, the Strouhal number (St) represents a derived (or flow-field-dependent) time scale as in the case of *vortex shedding*. Here, the shedding frequency (f) of, for example, the flow past a cylinder (Fig. 2.4) represents a nondimensional quantity that is dependent upon the governing parameter for such a flow. Namely,

$$S_T = \frac{fd}{U_0} = \text{function(Re)},$$

since, as stated above, the Re is the controlling parameter in the equation of motion.

Froude Number

Consider a flow in which the gravitational body force influences the velocity field. Specifically, consider a flow in which the gradient of ($\rho g h$) must be considered in the BV problem. Figure 2.5, in which an airfoil serves as a hydrofoil near the free surface, provides a specific example.

Let H represent the immersion depth to the 1/4 chord location of the hydrofoil. It is considered to be apparent that the flow over the hydrofoil will be influenced by this immersion depth if H/C is sufficiently small. H/C then becomes a length scale ratio that defines one of the boundary conditions of the BV problem.

The relative body force effect is, for this BV problem, expressed by the nondimensional parameter:

$$\text{Froude no.} = \text{FR} = \frac{\rho U_0^2}{\rho g C} = \frac{U_0^2}{g C} \qquad (2.6)$$

for this ρ = constant (below the free surface) BV problem.

For completeness, it is noted that h, the elevation above a datum plane, was made nondimensional with the chord length as (h/C) in order to isolate the Froude number as a *governing parameter* in this BV problem.

The relative influence of the gravitational body force is expressed by two parameters: H/C and (U_0^2/gC). It is instructive to note that a large value of H/C renders U^2/gC irrelevant. Physically, it is observed that a planar free surface above the submerged hydrofoil indicates that the hydrofoil will not experience a dynamically significant influence on its pressure distribution from the body force term.

It is quite difficult to satisfy *both* Re and FR matching in an application where both play important roles. The hydrofoil of Fig. 2.5 provides a relevant example. Specifically, for matched Froude numbers:

$$\left.\frac{U^2}{gC}\right|_m = \left.\frac{U^2}{gC}\right|_p$$

and matched Reynolds number values,

$$\left.\frac{UC}{\nu}\right|_m = \left.\frac{UC}{\nu}\right|_p,$$

the combined constraints require that

$$\frac{U_m^2}{U_p^2} = \frac{C_m}{C_p} = \frac{\nu_m}{\nu_p}\frac{U_p}{U_m}$$

or

$$\frac{U_m^3}{U_p^3} = \frac{\nu_m}{\nu_p}. \qquad (2.7)$$

It is apparent that liquids with the indicated ratio of ν values would be difficult to find (in bulk) if U_m and U_p were very different.

The practical solution, in those cases (e.g., surface ships) where the FR is the dominant parameter, is to match the FR and to make corrections for the Re mismatch.

Densimetric Froude Number

Consider that the approach airflow, for the airfoil of Fig. 2.1, experiences a strong temperature increase as a result of absorbing thermal energy from a heated airfoil. It can clearly be expected that the wake of the airfoil will be influenced by these elevated temperatures. Specifically, the wake fluid would be lifted by the buoyancy provided by the surrounding ambient temperature fluid acting on the heated wake fluid.

The control parameter that ensures similarity between model and prototype would be the densimetric Froude number

$$\text{FR}|_D = \frac{(\Delta\rho)gC}{\rho_0 U_0^2}, \qquad (2.8)$$

where $\Delta\rho$ would express a characteristic density change in the airflow and ρ_0 would be the density of the approach flow. [Since, for a perfect gas, $p = \rho RT$ and

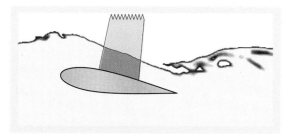

Fig. 2.5 A near-free-surface hydrofoil flow

the absolute pressure can be taken to be a constant, the $\Delta\rho/\rho_0$ value could be obtained from a characteristic temperature change as $-\Delta\rho/\rho_0 = \Delta T/T_0$ if $\Delta T/T_0$ is sufficiently small: $(\Delta T/T_0) \ll 1$.] Again, it is noted that the elevation term, h, is made nondimensional with C in this example.

As noted before, it would not be expected that a condition of complete similarity would be obtained. If density effects are dominant, then Re corrections would have to be made in the model results while the densitometric Froude number was held constant between model and prototype.

Weber Number

Surface tension effects, the presence of mechanical stresses that are present at the gas–liquid or the gas–liquid–solid boundary, can be an important factor in establishing the behavior of a liquid flow bounded by a gaseous medium. The basic property that describes the effect of the surface tension is the parameter σ or the *surface* tension force per unit length.

It is important to understand that σ is introduced in order to *preserve* the continuum mechanics understanding that the pressure is isotropic. That is, that *pressure* is unchanged in the neighborhood of the free surface whereas the normal stresses in the plane of, and those normal to, the gas–liquid interface are not equal. If the physical problem under consideration admits the introduction of a single quantity, σ, to describe the complex physical chemistry (Sect. 3.1.2) at the gas–liquid interface, then the Weber number (We) will be represented in the problem description as

$$\text{We} = \frac{\rho U^2 L}{\sigma}. \quad (2.9)$$

If the prototype involves significant surface tension effects as well as viscous effects, then the challenge would be to equate both parameters as:

$$\rho \frac{UL}{\mu}\bigg|_m = \rho \frac{UL}{\mu}\bigg|_p$$

and

$$\rho \frac{U^2 L}{\sigma}\bigg|_m = \rho \frac{U^2 L}{\sigma}\bigg|_p$$

or

$$\frac{\mu U}{\sigma}\bigg|_m = \frac{\mu U}{\sigma}\bigg|_p, \quad (2.10)$$

where the indicated ratio is referred to as the capillary number. Hence the liquid materials and the velocities would have to be selected to satisfy the capillary number constraint if a condition of similitude were to be established.

If the σ, μ and U values cannot be so manipulated, the experimentalist can attempt to *bracket* the correct condition or to make analytical adjustments to the data to gain an approximation to the complete similitude condition.

2.1.5 Governing Equations – Newtonian and Compressible

The Lagrangian (or material) derivative of the density of a fluid dynamic particle (fixed mass element) can be used to define a compressible flow succinctly. Namely,

$$\frac{D\rho}{Dt} \neq 0 \quad (2.11)$$

denotes a compressible flow.

Thompson [2.3] provides an excellent introduction to similitude considerations for compressible flows; this reference is recommended for a more thorough exposition of the compressible flow issues considered in this section.

A fundamental statement of the conservation of mass for an Eulerian cube can be written as

$$0 = \frac{\partial \rho}{\partial t} + \nabla \cdot \rho \mathbf{V} \quad (2.12)$$

from which

$$\nabla \cdot \mathbf{V} = -\frac{1}{\rho}\frac{D\rho}{Dt} \quad (2.13)$$

follows.

Thompson makes use of (2.13) to develop a nondimensional representation of $\nabla \cdot \mathbf{V}$ as {[2.3] (3.46)}

$$\nabla \cdot U = \frac{c_0^2}{c^2}\left(\frac{M_0^2}{2} U \cdot \nabla U^2 - U \cdot \tilde{G}\right)$$
$$+ \frac{c_0^2}{c^2}\frac{l_0}{c_0 l_0}\left(\frac{M_0}{2}\frac{\partial U^2}{\partial \tau} - \frac{\rho_0}{\rho}\frac{\partial \tilde{P}}{\partial \tau}\right)$$
$$- \frac{\rho_0 c_0^2}{\rho c^2}\frac{M_0^2}{\text{Re}_0}\left[U \cdot \nabla^2 U\right.$$
$$+ \left(\frac{\mu_v}{\mu} + \frac{1}{3}\right) U \cdot \nabla(\nabla \cdot U)\right]$$
$$+ \frac{1}{\text{Re}_0}\frac{T_0}{v_0}\left(\frac{\partial v}{\partial T}\right)_p \left(\frac{u_0^2}{c_p T_0}\tilde{\gamma} + \frac{1}{\text{Pr}}\nabla^2 \tilde{T}\right),$$
$$(2.14)$$

where the nondimensional values are defined as {[2.3] (3.45)}

$$U = \frac{u}{u_0}, \qquad \tilde{G} = \frac{Gl_0}{c_0^2}, \qquad X = \frac{x}{l_0},$$

$$\tilde{P} = \frac{P}{\rho_0 c_0 u_0}, \qquad \tau = \frac{t}{t_0}, \qquad \tilde{\gamma} = \frac{l_0^2}{\mu u_0^2}\gamma,$$

$$\tilde{T} = \frac{T}{T_0}, \tag{2.15}$$

and the parameters {[2.3] (3.47)}

Mach number: $\quad M_0 \equiv \dfrac{u_0}{c_0},$

Reynolds number: $\quad \mathrm{Re}_0 \equiv \dfrac{\rho_0 u_0 l_0}{\mu},$

Prandtl number: $\quad \mathrm{Pr} \equiv \dfrac{\mu c_p}{k}, \tag{2.16}$

arise naturally in the subject equation. The nomenclature for this subsection is adapted from Thompson's text. As noted, that presentation is strongly recommended for those interested in compressible flows and the present nomenclature change will facilitate the use of that reference. For completeness, it is noted that $V = u$, $L = l_0$, and $g = G$ when comparing the present nomenclature to that of Thompson. Note that G represents the body force and γ represents the dissipation or degradation of kinetic energy to the thermal form.

It is apparent that the addition of compressibility adds greatly to the complexity of the similitude considerations. It is also apparent that mutually satisfying the Mach number and Reynolds number constraints for a given model study will be essentially impossible as shown by the following.

Substitute the Mach-number-based velocity ratio into the Reynolds number as

$$\frac{u_0|_m}{u_0|_p} = \frac{\rho_0 l_0|_p}{\rho_0 l_0|_m} \frac{\mu_m}{\mu_p} = \frac{c_0|_m}{c_0|_p}.$$

Isolate the length ratio as

$$\frac{l_0|_m}{l_0|_p} = \frac{\rho_0|_p}{\rho_0|_m}\left(\frac{\mu_m}{\mu_p}\frac{c_0|_p}{c_0|_m}\right). \tag{2.17}$$

Consider the representative application of a high-speed aircraft ($M_0 = 3.5$) that is to be evaluated in a model study. A length ratio of 50:1 would be typically enforced by the available tunnel size. The bracketed term, the right-hand side of (2.17), will be sensitive to the model-to-prototype temperature ratio. To decrease (μ_m/μ_p), the model study temperature would need to be reduced since, for a gas, $\partial\mu/\partial T > 0$. Given the *low* temperature of the prototype environment (at altitude) this is a difficult condition to achieve. To increase $c_0|_m/c_0|_p$ would require the opposite condition: an elevated temperature of the model flow.

A pressurized tunnel would address the $(\rho_0|_p/\rho_0|_m)$ ratio but this is at the price of mechanical complexity and limited capability to match the required length ratio.

The practical solution is to match the Mach number values (independent of the length ratio) and attempt to address Reynolds number issues via computational corrections.

2.1.6 Flows for Which *U* and *L* May Not Be Apparent

Some flows, unlike the airfoil example discussed above, do not offer an apparent length or velocity scale. This section provides examples that suggest strategies to extract fabricated length and/or velocity scales in such circumstances. The fabricated scales then allow the experimentalist to directly utilize the parameters as described above.

A *U*-Tube Flow

Figure 2.6 shows a U-shaped tube with, at time $t \leq 0$, an elevated liquid column on one side. The fluid viscosity and density are ν and ρ and the tube diameter is D. The initial elevation difference is $h(0)$. If a model study

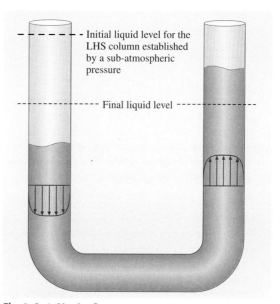

Fig. 2.6 A U-tube flow

were to be made of the prototype, it is apparent that the Reynolds number should be used as the similarity parameter with D as the length scale.

Since the flow will start with the release of the *holding pressure* (either negative on the high side or positive on the low side) there is no velocity for $t \leq 0$. The elevation difference can, however, be used to express a characteristic velocity as

$$U = \sqrt{gh} \quad (2.18)$$

from which

$$\text{Re} = \frac{UD}{\nu} = \frac{\sqrt{gh(0)}D}{\nu}$$

would provide a necessary condition for model-to-prototype similarity.

If surface tension were important in the prototype, then the Weber number would also have to be matched as

$$\text{We} = \frac{\rho U^2 L}{\sigma} = \frac{\rho g h(0) D}{\sigma}.$$

It is interesting to observe that, if We is important for the prototype, then

$$\frac{\sigma_m}{\sigma_p} = \frac{\{\mu\,[gh(0)]^{1/2}\}_m}{\{\mu\,[gh(0)]^{1/2}\}_p}. \quad (2.19)$$

That is, σ_m and σ_p must satisfy this additional constraint. If We is not important for the prototype, the experimentalist would need to ensure that it is also not important in the model study.

A Single Stream Shear Layer

One of the many possible examples of a flow without a defined length scale is that of a single stream shear layer. Figure 2.7 presents a schematic representation of such a flow which, in its idealized form, draws fluid from $y = -\infty$ by the entrainment action of the *infinitely wide* primary flow whose velocity scale is U_0.

Morris and *Foss* [2.4] provide a detailed examination of the developing region in which the separating turbulent boundary layer gradually loses its identity and a self-preserving single stream shear layer is established. As emphasized in that reference, and in the numerous prior studies cited by these authors, this is a flow without a defined length scale.

By convention and because it provides a well-defined experimental length scale, the momentum thickness at

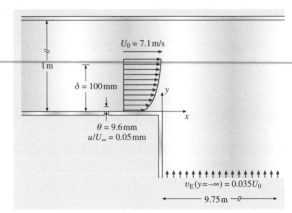

Fig. 2.7 The momentum thickness at $x = 0$ for a single stream shear layer: the flow's *length scale*

$x = 0$ is utilized as the reference condition for both the separating boundary layer and the evolving shear layer. Specifically, in the cited study, $R_{\theta(0)} = U_0 \theta(0)/\nu$ was 4860 at $x = 0$ and the self-preservation condition for the shear layer was established by $x/\theta(0) = 200$. The required length to establish self-preservation is a function of R_θ, as discussed in detail in that reference.

For completeness, it is worth noting that a planar jet (width w) is a flow *with* a defined length scale. In that flow, the role of $\theta(0)$ becomes one of a characterizing length scale ratio, $\theta(0)/w$, but the Reynolds number would be characterized as

$$\text{Re} = \frac{wU_0}{\nu},$$

if the jet, and not the separating shear layer at the sides of the jet, was the focus of the investigation. It can be readily appreciated that the jet flow, especially the region near the jet exit, will be influenced by the momentum thickness at the jet exit and, if turbulent, the relevant characteristics of the turbulent motion at the jet exit. These features constitute the influence of the boundary conditions on the solution.

However, as discussed by *Yarin* in Sect. 2.3, at sufficiently large streamwise locations the idealization that the jet nozzle width is effectively zero and the exit velocity is effectively infinite (in such a manner that the *line-source momentum flux*, $\dot{M}$, describes each downstream plane's momentum flux) leads to a condition of self-similarity. An extension to this modeling idealization is that all flows with the same $\dot{M}$ value and sufficient streamwise distance will have the same flow fields given the same distance from the apparent origin.

A Spinning Disc

Consider a flat and spinning disc whose diameter extends far beyond the central region of experimental interest. A flow from the central region outward will be established by the spinning motion. There is no apparent length scale in this problem albeit the viscous effects (i.e. Re) obviously play an important role in the resulting velocity field.

The angular speed, ω, combined with the fluid kinematic viscosity, ν, can be combined to form a velocity scale

$$U \sim \sqrt{\omega \nu}, \tag{2.20}$$

as well as a length scale

$$L \sim \sqrt{\frac{\nu}{\omega}}. \tag{2.21}$$

Not surprisingly, with these definitions, the Re value becomes unity:

$$\text{Re} = \frac{\sqrt{\omega \nu}\sqrt{\nu/\omega}}{\nu} = 1 \tag{2.22}$$

and, as a result, all nondimensional and nonturbulent velocity fields are identical. That is, $V^*(r^*, z^*)$ will be the same for any ν and ω, where

$$V^* = \frac{V(r,z)}{U} \tag{2.23}$$

and

$$r^* = \frac{r}{L}, \quad z^* = \frac{z}{L}. \tag{2.24}$$

A turbulent flow established by the rotating discs will have the same scaling parameters and its time-averaged quantities will, similarly, be identical at the same (r^*, z^*) positions.

This example problem is also a celebrated example of a clever analytical solution. It was carried out by *von Kármán* [2.5] and is presented in detail by *White* [2.6]. The latter reference does not make explicit use of the above result that Re = 1; however, this result is fully compatible with White's presentation. Specifically, from (3-183) of [2.6]

$$v_r = r\omega F(z^*), \quad v_\theta = r\omega G(z^*),$$
$$v_z = \sqrt{\omega \nu} H(z^*), \quad p = \rho \omega \nu P(z^*). \tag{2.25}$$

It is evident, using the symbols of (2.20–2.24) above, that White's equations (3-183) are compatible with the message of this section. Specifically, with the normalization provided by (2.20)

$$v_z^* = H(z^*) \quad \text{and} \quad p^* = P(z^*),$$

and with the added normalization provided by (2.21),

$$v_r^* = r^*(Fz^*) \quad \text{and} \quad v_\theta^* = r^*(Gz^*). \tag{2.26}$$

It is useful to note that the von Kármán solution (2.25, 26) is a good example of the self-similarity solutions that are considered in Sect. 2.3.

The viewpoint expressed in this presentation does differ with White in a substantive manner with regard to the cited experimental result (from the work of *Kobayashi* et al. [2.7]) that instabilities occur for

$$\text{Re} = \frac{\omega r^2}{\nu} = 8.8 \times 10^4$$

with the added result that turbulence is observed at Re = 3.2×10^5.

The disagreement can be stated as follows: *It is not correct to state these values as a Reynolds number since* Re $\equiv 1$ *in the understanding of this section. Rather, the quoted values are simply nondimensional* (radii)^2 *and their universal values attest to the present statement that all such flows are in a condition of similitude with unit Re.*

To carry the argument further for its instructive value, consider that two experimentalists were to make measurements using *large* but different disc diameters. Large, in this context, would be an r_0 value whose r_0^* is much larger than 3.2×10^5. If the experimentalists used the same ω value and their large, but finite, disc sizes (r_0) to specify a Reynolds number as they examined the issue of *when does turbulence occur*, they would conclude that the transition value for r/r_0 would be different in the two experiments.

It is instructive, in this example, that the similarity parameters for an idealized ($r_0 \to \infty$) experiment yield a more useful result than the similarity parameters ($U_0 = r_0 \omega$, $L = r_0$) that would be suggested by the physical experiment.

Finally, for this example, it is considered to be evident that the von Kármán solution will be invalid as the observation location approaches the coordinates $r \to r_0$ for $z \ll r_0$ and for smaller r as z increases (where r_0 is the radius of the experimental plate).

2.2 Dimensional Analysis and Data Organization

The organization and interpretation of experimental data is an essential activity that requires careful thought. There is not a specific recipe for organizing data. Ingenuity, insight, analysis, and creative ideas are needed. However, there are some concepts and methods that will provide guidance. One basic principle is that the simplest relationship will have the fewest number of variables. Another basic principle is that the simplest relationship will use nondimensional variables. *Dimensional analysis* is the basic procedure to produce nondimensional variables, but it needs to be refined and supplemented. Section 2.2 will present the elements of dimensional analysis followed by several examples that illustrate exceptional cases and supplemental techniques. The examples demonstrate the general features of dimensional analysis. They provide results that could in some cases be obtained from familiar analyses of fluid mechanics. The agreement between the dimensional analysis results and those from general fluid mechanics will give confidence in the ability of dimensional analysis to give valuable results in complex new problems.

The presentation of numerical results should have the most extensive validity. This is not always done. Consider the following formula to predict the flow rate of natural gas through a pipeline. Based on experimental data the following is an algebraic curve fit used in industry.

$$Q_0 = 737 \frac{T_0}{p_0} d^{2.53} \left(\frac{p_1^2 - p_2^2}{L_m G T_f} \right)^{0.510} \quad (2.27)$$

where Q_0 is the standard cubic feet per day, T_0 is the reference temperature in °R, p_0 is the reference pressure in psia, d is the pipe dimeter in inches, T_f is the fluid temperature in °R, L_m is the pipe length in miles, p_1 is the initial pressure in psia, p_2 is the final pressure in psia, G specific gravity referenced to air at 60 °F and 14.696 psia.

The constants in (2.27) implicitly have units raised to irrational powers. For example, $d^{2.53}$ must be compensated for by a length dimension with power of -2.53. It would be inappropriate to use this formula for a CO_2 pipeline. Such dimensional formulas exist in most applied fields and are quite useful. However, from a scientific viewpoint a simpler and universally valid formula could be developed, one that in principle could be applied to CO_2. The data used to produce (2.27) could have been organized into a more universal form.

2.2.1 Variables, Function List, and Extra Information

A physical experiment, event, or situation is in principle described quantitatively by a mathematical function. It is a mathematical necessity to have one dependent variable. The remaining variables are independent variables or parameters as we choose to regard them. Whether a variable is independent or a parameter is a physical choice. A correct list of variables requires knowledge of physics and is sometimes not an easy task. After a variable list is established, one should employ known physical equations and extra assumptions that reduce the number of variables.

As an example consider the question of the pressure drop between two stations, denoted as 1 and 2, in a straight pipe when the flow is incompressible. The variables are p pressure, z elevation, ρ density, ν kinematic viscosity, Q volume flow rate, g acceleration of gravity, D pipe diameter, ε relative roughness of pipe wall, L length of pipe between 1 and 2. Consider the second pressure as the dependent variable.

$$p_2 = f(p_1, \rho, Q, D, L, \varepsilon, \nu, g, z_1, z_2). \quad (2.28)$$

We have already used some knowledge of the physics in making this list. Velocity is not included because it is given by the flow rate divided by the cross-section area. The temperature of the fluid is not included because heat transfer effects are separated from flow effects in incompressible flow.

The problem can be simplified if we observe the kinetic energy equation (1.71) for incompressible flow. It is written in average values as

$$\left(p_2 + \rho \frac{1}{2} V_2^2 + \rho g z_2 \right) - \left(p_1 + \rho \frac{1}{2} V_1^2 + \rho g z_1 \right)$$
$$= -\int \Phi \, dA \, dx = h_L. \quad (2.29)$$

To be more precise one sometimes uses a coefficient in front of V_1^2 to account for the shape of the velocity profile. Here Φ is the viscous dissipation. The integral over the cross section dA and pipe length dx represents the head loss h_L between stations 1 and 2. We assume the flow in the straight pipe is fully developed so that the velocities at 1 and 2 are the same. Moreover, the dissipation depends on the velocity profile, which is unchanged along the length and does not depend on

z or p. Comparing (2.28) and (2.29) we write

$$(p_2 + \rho g z_2) - (p_1 + \rho g z_1) = f_1(\rho, Q, D, L, \varepsilon, \nu). \tag{2.30}$$

Furthermore, we propose that the dissipation should increase directly with the length of the pipe. To be precise $\int \Phi \, dA \, dx = L \int \Phi \, dA$.

$$(p_2 + \rho g z_2) - (p_1 + \rho g z_1)$$
$$= L f_2(\rho Q D \varepsilon \nu),$$

$$\frac{\mathcal{P}_2 - \mathcal{P}_1}{L} \equiv \frac{(p_2 + \rho g z_2) - (p_1 + \rho g z_1)}{L}$$
$$= f_2(\rho, Q, D, \varepsilon, \nu). \tag{2.31}$$

Thus, based on physics and guesswork the number of variables has been reduced from 11 to six. Moreover it will not be necessary to test at different values of p_1 or different changes in elevation, or even different lengths of pipes.

2.2.2 Dimensions and Scale Ratios

The basis of simplifying mathematical relationships that contain physical variables is dimensional analysis. We may define two categories of variables. Things that are counted are dimensionless. For instance, the number of molecules or the number of people in a room is dimensionless. Things that are measured require that we essentially compare the item to a scale that has a defined unit. An example here is the length of a room or its floor area. The size of a scale unit is arbitrary. There are no universal natural size units that are relevant to all physical processes. Dimensional analysis employs the fact that a physical event is independent of the scale units used to measure the variables.

Consider a length variable that has a size l when measured with a certain length unit. If a new length unit is used the value becomes $\hat{l}$. The *dimension symbol L* is the ratio of the new unit to the old unit

$$\hat{l} = lL. \tag{2.32}$$

In a similar way a time unit could be changed with scale unit ratio T or a mass unit could be changed with a scale unit ratio M

$$\hat{t} = tT, \tag{2.33}$$
$$\hat{m} = mM. \tag{2.34}$$

We could also imagine a force unit with a scale change ratio F.

$$\hat{f} = fF. \tag{2.35}$$

The symbols F, M, L, and T stand for scale change ratios for an imaginary process of changing units.

It is a fact of experience that three dimensions are sufficient to measure any physical quantity. These are called primary dimensions and may be taken as M, L, and T. Equivalently we could take F, L, and T as *primary dimensions*. However, the choice of four primary dimensions is redundant and requires, in general, that we expand the list of variables to include a *unifying dimensional constant*. The choice F, M, L, and T requires the unifying constant g_c with dimensions ML/FT^2. In principle, equations such as Newton's second law should be written with the constant included.

$$\mathcal{F} = \frac{ma}{g_c} \tag{2.36}$$

Here $\mathcal{F}$ is force and a is the acceleration. Current custom is to write equations as if a consistent system of units is employed. The dimensional constant is required in the English Engineering system, but not in the international SI system.

All other physical variables have dimensions that are combinations of the primary dimensions raised to some power. For example a velocity is

$$\hat{v} = \frac{d\hat{l}}{d\hat{t}} = \frac{d(lL)}{d(tT)} = \frac{L}{T} \frac{dl}{dt} = L^1 T^{-1} v. \tag{2.37}$$

To be precise, the value of a variable in new units $\hat{x}$ is related to the size in old units x by

$$\hat{x} = x M^a L^b T^c. \tag{2.38}$$

Bridgman [2.8] first derived this equation. Not all scale units will change in this manner. A pressure measured in decibels does not fit this relation. The Richter scale for earthquake intensity or the Rockwell hardness scale for metals are other examples of units that do not obey Bridgman's equation. However, all physical concepts can be measured with primary units so that they obey the Bridgman equation.

2.2.3 Natural Scales and Repeating Variables

As an introduction to the essential elements of dimensional analysis consider the problem of the incompressible flow at speed V over a cylinder of diameter D. The fluid density is ρ and its viscosity is μ. The question is what is the frequency ω of the *von Kármán vortex street*? The proposed mathematical function is

$$\omega = f(V, D, \rho, \mu). \tag{2.39}$$

It is useful to form a *dimensional matrix* that contains the coefficients a, b, c of Bridgman's (2.38) for each variable. For instance the frequency equation is

$$\hat{\omega} = \omega T^{-1}. \tag{2.40}$$

The powers $0, 0, -1$ are placed in the matrix.

	ω	V	D	ρ	μ
M	0	0	0	1	1
L	0	1	1	-3	-1
T	-1	-1	0	0	-1

Here we have chosen M, L, and T as the primary dimensions.

Consider the variables V, D, and ρ. What combination of these variables will produce a variable with the dimension of length? The obvious answer is D. What combination of these variables will produce a variable with the dimension of time? The answer is D/V. What combination of these variables will produce a variable with the dimension of mass? The answer is ρD^3. Instead of measuring mass quantities in units of kilograms, we will use the value of ρD^3 as the unit. This is a natural unit for this physical phenomenon. Length, time, and mass units that are associated with the physical event are

$$L_{event} = D,$$
$$T_{event} = \frac{D}{V},$$
$$M_{event} = \rho D^3. \tag{2.41}$$

The frequency variable has the dimension of $1/T$ and the viscosity variable has the dimensions of mass/(length-time).

$$\omega \left[\frac{1}{T}\right], \quad \mu \left[\frac{M}{LT}\right]. \tag{2.42}$$

Multiplying frequency and viscosity to eliminate the appropriate units produces nondimensional variables:

$$\Pi_1 = \omega\left(\frac{D}{V}\right) = \frac{\omega V}{D}, \quad \Pi_2 = \frac{\mu D(D/V)}{\rho D^3} = \frac{\mu}{\rho D V}. \tag{2.43}$$

One can observe that Π_1 is the reduced frequency and Π_2 is the reciprocal of the Reynolds number that was introduced in Sect. 2.1. This method of producing nondimensional variables is known as the method of scales. The mass, length and time units that society has defined in order to quantify physical variables are arbitrary and are not inherent to the physical event under study. We can form mass, length, and time units from variables associated with the event. Using these as units essentially produces nondimensional variables of a universal magnitude.

Let us look at the method of scales in a more general context. Consider a physical experiment that has n dimensional variables. There is one dependent variable that we write on the left-hand side

$$x_n = f(x_1, x_2, x_3, \ldots, x_{n-1}). \tag{2.44}$$

A given experiment might have several dependent variables of interest, but we should deal with them and their functions one at a time. A physical process is independent of the units used to make the measurements. Thus, in the rescaled units the same function holds

$$\hat{x}_n = f(\hat{x}_1, \hat{x}_2, \hat{x}_3, \ldots, \hat{x}_{n-1}). \tag{2.45}$$

Let us use three primary dimensions. If the scale units were changed, the first three variables would change according to

$$\hat{x}_1 = x_1 M^{a_1} L^{b_1} T^{c_1},$$
$$\hat{x}_2 = x_2 M^{a_2} L^{b_2} T^{c_2},$$
$$\hat{x}_3 = x_3 M^{a_3} L^{b_3} T^{c_3}. \tag{2.46}$$

All other x variables would change by similar relations. a, b, c are known coefficients of the Bridgman equations.

Consider the following question, is there a combination of x_1, x_2, x_3 in (2.46) that, when raised to some powers, α, β, γ, has the dimension of mass? The choice of the first three variables in the list is arbitrary. The variables can be renumbered or relisted as we see fit. The variables chosen as x_1, x_2, x_3 are called repeating variables. Let the combination that produces a mass scale y_M be

$$y_M = x_1^{\alpha_M} x_2^{\beta_M} x_3^{\gamma_M}. \tag{2.47}$$

After a change in scale units the value would be

$$\hat{y}_M = (\hat{x}_1)^{\alpha_M} (\hat{x}_2)^{\beta_M} (\hat{x}_3)^{\gamma_M}. \tag{2.48}$$

Inserting the relations (2.47) and (2.48) into (2.41) yields

$$\hat{y}_M = (\hat{x}_1)^{\alpha_M} (\hat{x}_2)^{\beta_M} (\hat{x}_3)^{\gamma_M}$$
$$= x_1^{\alpha_M} x_2^{\beta_M} x_3^{\gamma_M} M^{\alpha_M a_1 + \beta_M a_2 + \gamma_M a_3}$$
$$\times L^{\alpha_M b_1 + \beta_M b_2 + \gamma_M b_3} T^{\alpha_M c_1 + \beta_M c_2 + \gamma_M c_3}. \tag{2.49}$$

However, by design we want y_M to have the dimension of mass. Thus, we require that

$$\hat{y}_M = y_M M^1 L^0 T^0. \tag{2.50}$$

Comparing (2.50) and (2.49) shows that that the powers of M, L, and T obey the relations

$$\alpha_M a_1 + \beta_M a_2 + \gamma_M a_3 = 1,$$
$$\alpha_M b_1 + \beta_M b_2 + \gamma_M b_3 = 0,$$
$$\alpha_M c_1 + \beta_M c_2 + \gamma_M c_3 = 0. \qquad (2.51)$$

For given choices of x_1, x_2, and x_3 the coefficients a, b and c are known, and values α_M, β_M and γ_M are to be found. Cramer's rule says that a unique solution of the nonhomogeneous equation set is possible if the determinate of the coefficient matrix is not zero.

$$\det \begin{vmatrix} a_1 & a_2 & a_3 \\ b_1 & b_2 & b_3 \\ c_1 & c_2 & c_3 \end{vmatrix} \neq 0 \qquad (2.52)$$

In other words the *rank of the coefficient matrix* must be three. The rank is the size of the largest square submatrix that has a nonzero determinate. If the variables x_1, x_2, x_3 are chosen so that the rank of the dimensional matrix is three we will be able to form a mass scale from them.

We can continue to form a combination variable with the dimension of length using the same x_1, x_2, x_3. The algebra problem to find another set of coefficients α_L, β_L, γ_L is the same problem as (2.50) except that the inhomogeneous right hand side is now 0, 1, 0. A solution is still guaranteed because the coefficient matrix (2.51) is unchanged and it rank is still three. Moreover, the same process can be repeated to produce a time variable. Bridgman's equations for the length and time units are

$$\hat{y}_L = y_L M^0 L^1 T^0, \qquad (2.53)$$
$$\hat{y}_T = y_T M^0 L^0 T^1. \qquad (2.54)$$

The critical issue is that the choice of *repeating variables* x_1, x_2, x_3 is valid for all three problems if the rank of their dimensional matrix is three. If so, x_1, x_2, x_3 are said to be dimensionally linearly independent, and they may be organized into combinations that have dimensions of mass, length, and time.

2.2.4 Π Theorem

Consider any variable in the function (2.44) except x_1, x_2, and x_3. For convenience call it x_4. The Bridgman equation for x_4 is

$$\hat{x}_4 = x_4 M^{a_4} L^{b_4} T^{c_4}. \qquad (2.55)$$

Dividing (2.55) by (2.49, 53) and (2.54) yields the nondimensional Π variable

$$\hat{\Pi}_4 \equiv \frac{\hat{x}_4}{(\hat{y}_M)^{a_4}(\hat{y}_L)^{b_4}(\hat{y}_T)^{c_4}}$$
$$= \frac{x_4 M^{a_4} L^{b_4} T^{c_4}}{(y_M M)^{a_4}(y_L L)^{b_4}(y_T T)^{c_4}}, \qquad (2.56)$$
$$\frac{\hat{x}_4}{(\hat{y}_M)^{a_4}(\hat{y}_L)^{b_4}(\hat{y}_T)^{c_4}} = \frac{x_4}{(y_M)^{a_4}(y_L)^{b_4}(y_T)^{c_4}},$$
$$\hat{\Pi}_4 = \Pi_4. \qquad (2.57)$$

The variable Π_4 is nondimensional and its size is universal in the sense that it is independent of the measuring units.

The Π theorem tells us how many nondimensional variables are needed for a given problem with dimensional variables. The Π theorem can be stated as follows.

Theorem 2.1
If the physical problem is described by a function of n dimensional variables

$$x_n = f(x_1, x_2, x_3, \ldots, x_{n-1})$$

and the rank of the dimensional matrix is r, then the function may be reorganized into nondimensional variables that are $n - r$ in number

$$\Pi_{n-r} = f(\Pi_1, \Pi_2, \ldots, \Pi_{n-r-1}).$$

Obviously r cannot be greater that the number of dimensions, but it can be less. It is not a unique solution because a multiplication of two variables say $\Pi_1 \Pi_2 = \Pi_3$ is a nondimensional variable while any power $\Pi^\alpha = \Pi_4$ is also a valid nondimensional variable. Any answer that uses all the dimensional variables and has $n - r$ nondimensional variables is valid. A rigorous proof of the Π theorem can be found in [2.9].

2.2.5 Example with Rank Less than the Number of Dimensions

As an example where the rank of the dimensional matrix is smaller than the number of dimensions, consider a shock wave in a perfect gas. Let the thermodynamic state in front of the *shock wave* be specified by p_1, ρ_1 and the type of gas denoted by the specific heat ratio γ. The state of motion is fixed by the initial velocity V_1 that is measured from a coordinate system fixed to the wave. Our interest is in the pressure after the shock p_2. With some experience in fluid mechanics we know that the function describing this problem has the form

$$p_2 = f_1(p_1, \rho_1, V_1, \gamma), \qquad (2.58)$$

The dimensional matrix has three dimensions.

	p_2	p_1	ρ_x	V_1	γ
M	1	1	1	0	0
L	-1	-1	-3	1	0
T	-2	-2	0	-1	0

Try p_1, ρ_1 and V_1 as repeating variables. They will not work because the determinate of the coefficients of p_1, ρ_1 and V_1 is zero. Using p_2 or γ as a repeating variable will not correct the situation. The determinate of any 3×3 square submatrix is zero, so the rank is not three. However, the determinate of any 2×2 square submatrix is nonzero, hence the rank is two. The Π theorem predicts that the relation (2.58) can be reorganized with $5 - 2 = 3$ nondimensional variables.

Since the problem does not have three linearly independent variables, the method of scales or the other standard textbook methods does not work and we must reorganize. The purpose of the reorganization is to place zeros in one of the rows or columns of the dimensional matrix. To do this we will consider p_2/p_1 as a variable instead of p_2. This is equivalent to dividing (2.57) by p_1.

$$\frac{p_2}{p_1} = \frac{1}{p_1} f_1(p_1, \rho_1, V_1, \gamma) = f_2(p_1, \rho_1, V_1, \gamma). \qquad (2.59)$$

The dimensional matrix for f_2 is

	p_2/p_1	p_1	ρ_x	V_1	γ
M	0	1	1	0	0
L	0	-1	-3	1	0
T	0	-2	0	-1	0

One can check that the rank is still two. The Π theorem still guarantees that we can form three nondimensional variables for f_2. We note from the dimensional matrix that p_1 and ρ_1 can only occur in the combination p_1/ρ_1 and the function must have a simpler form.

$$\frac{p_2}{p_1} = f_3\left(\frac{p_1}{\rho_1}, V_1, \gamma\right). \qquad (2.60)$$

The dimensional matrix for the f_3 problem is

	p_2/p_1	p_1/ρ_1	V_1	γ
M	0	0	0	0
L	0	2	1	0
T	0	-2	-1	0

The rank is now one, so there are $4 - 1 = 3$ nondimensional variables for the f_3 problem. By inspection of the dimensional matrix, a solution will contain the combination $V_1/\sqrt{(p_1/\rho_1)}$. An answer is

$$\frac{p_2}{p_1} = f_4\left(\frac{V_1}{\sqrt{p_1/\rho_1}}, \gamma\right). \qquad (2.61)$$

The dimensional matrix for f_4 is all zeros and (2.61) contains only nondimensional variables. We have used all the dimensional variables of (2.58) and produced three nondimensional variables. Equation (2.61) is a simpler relation as predicted by the Π theorem. Those familiar with compressible flow of a perfect gas will recognize that $\sqrt{p_1/\rho_1} = \sqrt{\gamma RT/\gamma} = a/\sqrt{\gamma}$, where a is the speed of sound. Thus, the nondimensional variable $V_1/\sqrt{(p_1/\rho_1)}$ is the Mach number Ma divided by $\sqrt{\gamma}$.

2.2.6 Example with Redundant Dimensions

Previously we noted that the choice F, M, L, and T requires the *unifying dimensional constant* g_c with dimensions ML/FT^2. In principle, equations like Newton's second law should include the constant:

$$\mathcal{F} = \frac{ma}{g_c}. \qquad (2.62)$$

A change in the scale units would change the value of g_c. For example

$$g_c = 32.2 \frac{\text{lb}_\text{m}\text{ft}}{\text{lb}_\text{f}\text{s}^2} = 980 \frac{\text{Kg}_\text{m}\text{m}}{\text{Kg}_\text{f}\text{s}^2}.$$

Consider a problem

$$x_n = f(x_1, x_2, x_3, \ldots, x_{n-1}). \qquad (2.63)$$

In the $FMLT$ system the function is

$$x_n = f(x_1, x_2, x_3, \ldots, x_{n-1}, g_c). \qquad (2.64)$$

The dimensional matrix includes the dimensional constant g_c.

	x_1	x_2	...	x_n	g_c
F					-1
M					1
L					1
T					-2

The additional dimension increases the rank by one and the unifying constant increases the number of variables by one. There is no effect on the Π theorem's prediction of the number of nondimensional variables. There are other examples of dimensional unifiers. If we have angles and use the degree as a unit there would be a dimensional unifier $2\pi/360$ with units of 1/degree. Next, we will illustrate the use of extra dimensions where temperature denoted by Θ is used. The unifying constant is R with dimensions $L^2/(\Theta T^{-2})$. Again consider a *shock wave* in a perfect gas. This time the initial thermodynamic state will be specified by initial pressure p_1, and temperature T_1. The pressure after the shock is again the dependent variable. The physical function includes R (i.e., the constant $R = p/(\rho T)$ for a perfect gas) as a variable. This compensates for the use of Θ as a dimension

$$p_2 = f(p_1, V_1, T_1, \gamma, R). \quad (2.65)$$

One finds the dimensional matrix as

	p_2	p_1	T_1	V_1	R	γ
M	1	1	0	0	0	0
L	-1	-1	0	1	2	0
T	-2	-2	0	-1	-2	0
Θ	0	0	1	0	-1	0

The rank of this matrix is three (the determinant of any four variables is zero). Thus, there are not four variables that are dimensionally independent. From the Π theorem we must have $6 - 3 = 3$ nondimensional variables. If we produce three nondimensional variables and use all variables in the function list, we have a solution to the problem. Inspection of the dimensional matrix shows that p_2/p_1 can be one variable because pressures are the only variables with mass dimensions. Another combination that must occur must is RT_1 because they are the only variables with temperature dimensions. Hence, the functional form must be

$$\frac{p_2}{p_1} = f(RT_1, V_1, \gamma). \quad (2.66)$$

The dimensional matrix is now

	p_2/p_1	RT_1	V_1	γ
M	0	0	0	0
L	0	2	1	0
T	0	-2	-1	0
Θ	0	0	0	0

Checking the Π theorem shows that the rank is one and there should be $4 - 1 = 3$ nondimensional variables. Thus, we have not altered the problem by leaving out p_1 by itself. The final list of nondimensional variables is

$$\Pi_1 = \frac{p_2}{p_1}, \quad \Pi_2 = \gamma, \quad \Pi_3 = \frac{V_1}{\sqrt{RT_1}}. \quad (2.67)$$

We have satisfied the Π theorem because all dimensional variables have been used in producing three nondimensional variables. This is the same result arrived at in (2.61) since $RT_1 = p_1/\rho_1$.

2.2.7 Anatomy of a Nondimensional Variable

The formulation of nondimensional variables based on the Π theorem leads to non-unique answers. Mathematically the various forms are equally valid. However, some forms are more useful than others. Let us consider the anatomy of a nondimensional variable. Specifically, we consider a nondimensional variable (other than a parameter). The general form is

$$\Pi = \frac{x - x_{\text{Ref}}}{x_{\text{Scale}}}. \quad (2.68)$$

In physical events a variable has a natural *reference*, which may be zero, and often a preferred *scale* unit. It is desirable that the nondimensional variable be a reasonable numerical size, say 0.01–100. In addition to variables, another physical interpretation of some nondimensional numbers is that of a parameter. In the case of a parameter the nondimensional variable is often interpreted as a ratio of physical quantities or scales.

$$\Pi = \frac{x_{\text{Scale A}}}{x_{\text{Scale B}}}. \quad (2.69)$$

For example, the Reynolds number is often interpreted as a ratio of inertia effects to viscous effects. A fuller discussion of physical interpretation of parameters is given later.

The origin of a coordinate system is arbitrary, so position variables always have a natural reference. Sometimes we implicitly include the reference in stating the problem. For example, in flow over a flat plate we take the distance variable from the leading edge. This sets the distance coordinate reference implicitly. In compressible flow the absolute pressure is important and the reference pressure must be zero. In incompressible flows the pressure in the flow field increases in the same amount that the pressure at a reference location p_∞ increases. Thus, the correct nondimensional pressure is of the form

$$\Pi_p = \frac{p - p_\infty}{p_{\text{Scale}}}. \tag{2.70}$$

Whenever a quantity is governed by only gradient ∇x terms, a reference is arbitrary since $\nabla(x + x_{\text{Ref}}) = \nabla x$.

2.2.8 Nonuniqueness of Scales

Scale units in a nondimensional variable depend on the choice of *repeating variables*. Consider the following problem discussed by White [2.10] (attributed to Professor Jacques Lewalle) and by Çengel and Cimbala [2.11]. A body falls under gravity g from a height position S_0 above the ground with an initial vertical velocity V_0. The height of the body S at any time t is under study. The initial vertical velocity V_0 will be considered as a positive quantity so there will be two problems, one problem for V_0 upward and another for V_0 downward. The dimensional function is

$$S = f(t, V_0, g, S_0). \tag{2.71}$$

There are five variables and the rank of the dimensional matrix is two so there must be three nondimensional variables. The form of these depends on the choice of repeating variables. White gives the following options for nondimensional functions (the exact answer is of course known; $S = S_0 + V_0 t - (1/2)gt^2$). These variables are related as power products of the original Π variables; $S^{**} = S^* \alpha$, $t^{**} = t^* \alpha$, and $t^{***} = t^*/\sqrt{\alpha}$. As remarked earlier if Π_1 and Π_2 are nondimensional, then Π_3 formed by and product of powers, $\Pi_3 = \Pi_1^n \Pi_2^m$, is also nondimensional. It is not necessary that a problem have a single set of repeating variables. For example, another relation $S^* = f_4(t^{**}, \alpha)$ is possible if we would substitute $t^* = t^{**}/\alpha$ into $S^* = f_1(t^*, \alpha)$. The physical meaning of the scales is:

S_0 Initial distance above the ground;

V_0^2/g Distance measure of the trajectory. If V_0 is upward, S maximum is $V_0^2/2g$ above S_0;

S_0/V_0 Time to travel S_0 at speed V_0;

V_0/g Time measure of the trajectory. If V_0 is upward, V_0/g is the time to reach S maximum;

$\sqrt{S_0/g}$ Time to fall distance S_0 if V_0 is zero is $\sqrt{2S_0/g}$.

The parameter α or $\sqrt{\alpha}$ can be interpreted as ratios of these times or distances.

2.2.9 Reference

If a variable has a natural *reference* it reduces the number of variables by one. Essentially we add a little physical knowledge to the problem and simplify the result. In the falling body problem of Sect. 2.2.8 the origin of the coordinate can be placed at the initial particle position without affecting the physics. Equivalently we can consider $S - S_0$ as the dependent variable. This will give an even simpler answer than those of Sect. 2.2.8

$$S - S_0 = f(t, V_0, g). \tag{2.72}$$

The dimensional matrix is now

	$S - S_0$	t	V_0	g
M	0	0	0	0
L	1	0	1	1
T	0	-1	-1	-2

With a rank of two there will be $4 - 2 = 2$ nondimensional variables (Table 2.2). Recognizing that nondimensional variables have natural references produces a sharper result.

The exact answer is

$$\hat{S} = \pm \hat{t} - \frac{1}{2}\hat{t}^2 \begin{cases} + & \text{if } V_0 \text{ is upward}, \\ - & \text{if } V_0 \text{ is downward}. \end{cases} \tag{2.73}$$

Table 2.1

Repeating variables	Distance variable	Time variable	Parameter	Function form
S_0, V_0	$S^* = \frac{S}{S_0}$	$t^* = \frac{t}{S_0/V_0}$	$\alpha = \frac{gS_0}{V_0^2}$	$S^* = f_1(t^*, \alpha)$
g, V_0	$S^{**} = \frac{S}{V_0^2/g}$	$t^{**} = \frac{t}{V_0/g}$	$\alpha = \frac{gS_0}{V_0^2}$	$S^{**} = f_2(t^{**}, \alpha)$
S_0, g	$S^{***} = S^* = \frac{S}{S_0}$	$t^{***} = \frac{t}{\sqrt{S_0/g}}$	$\alpha = \frac{gS_0}{V_0^2}$	$S^{***} = f_3(t^{***}, \alpha)$

Table 2.2

Repeating variables	Distance variable	Time variable	Parameter	Function form
g, V_0	$\hat{S} = \frac{S - S_0}{V_0^2/g}$	$\hat{t} = \frac{t}{V_0/g}$	None	$\hat{S} = \hat{f}(\hat{t})$

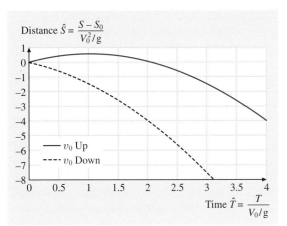

Fig. 2.8 Simplest nondimensional representation of the falling body solution

Now we have a function of only two variables. The solution is displayed in Fig. 2.8. In this figure the position of the ground is $\hat{S}_{\text{ground}} = -\alpha = -gS_0/V_0^2$.

2.2.10 Scales Chosen for Experimental Purposes

It is often desirable to choose the dependent and independent variables to match the variables being changed in the experiment. Then the repeating variables are chosen from the remaining variables. As an example consider the problem of determining the volume of an earth crater produced by an explosive charge. The explosion is at the surface of a soil that lacks cohesion; a soil like sand. A related problem is the crater formed by impact of a projectile or meteor. The dependent variable of interest is the crater volume V. Characteristics of the explosive charge are the energy per unit mass of the charge e_c, the mass of the charge m_c, and the density of the charge ρ_c. The acceleration of gravity is g, soil density ρ, and the angle of repose Φ. It requires some knowledge of the process to produce the proper list of variables. Sometimes extra variables are included only to be shown by experiments to be extraneous. Functionally, the problem is

$$V = f(e_c, m_c, \rho_c, g, \rho, \Phi). \tag{2.74}$$

The dimensional matrix is

	V	e_c	m_c	ρ_c	g	ρ	Φ
M	0	0	1	1	0	1	0
L	3	2	0	-3	1	-3	0
T	3	-2	0	0	-2	0	0

In choosing the repeating variables, note that a repeating variable may appear in several of the final nondimensional variables. Thus, we avoid V because we want it to appear in only one Π variable. It is the dependent variable. In the tests a centrifuge is used to effectively change the acceleration of gravity. Hence, we would like to use g as a major test independent variable and have it occur in only one Π group. Thus it should not be a repeating variable. Also the soil density ρ will not be varied and we would like to isolate its effect into one Π variable. Since Φ is already nondimensional, the repeating variables should be chosen from e_c, m_c, and one of the densities, say ρ_c. The determinant of these three variables (e_c, m_c, ρ_c) is nonzero so the rank is three. Hence, e_c, m_c, and ρ_c are dimensionally independent and may be formed into scales for the MLT units as follows:

$$M \sim m_c, \quad L \sim \left(\frac{m_c}{\rho_c}\right)^{1/3},$$

$$T \sim e_c^{-1/2} \left(\frac{m_c}{\rho_c}\right)^{1/3}. \tag{2.75}$$

The remaining variables that were not used in scales, V, ρ, g, and Φ are $m - r$ in number. This is just the number of nondimensional variables predicted by the Π theorem. They are taken one at a time and made non dimensional by eliminating MLT as required. We have used all the variables in the original list and produced $n - r$ nondimensional variables. Hence, the Π theorem is satisfied. In nondimensional form the final function is

$$\frac{V}{m_c/\rho_c} = f\left[\frac{g}{e_c}\left(\frac{m_c}{\rho_c}\right)^{1/3}, \frac{\rho}{\rho_c}, \Phi\right]. \tag{2.76}$$

As intended, variations in g can be used to simulate variations in the explosion strength e_c. If it is arranged so that variable occurs in only one Π group, the effect can be assessed by changing not the variable itself, but by changing another variable in that Π group.

From a series of experiments with Φ constant, *Schmidt* and *Housen* [2.12] found that for explosions (2.75) may be represented by the power relation

$$\frac{V}{m_c/\rho_c} = 0.218 \left(\frac{\rho}{\rho_c}\right)^{0.002} \left[\frac{g}{e_c}\left(\frac{m_c}{\rho_c}\right)^{1/3}\right]^{-0.464}. \tag{2.77}$$

The dependence on ρ/ρ_c is negligible according to the experiments. This is another useful aspect of dimensional analysis. The effect of soil density has been shown to be unimportant with experiments where only g and ρ_c are varied.

2.2.11 Nondimensional Variables Interpreted as Physical Ratios

Next, we try and reorganize the results of the explosion problem of Sect. 2.2.10 into a form that has a physical interpretation. We can consider a nondimensional variable as the ratio of two, dimensional, physical quantities. Let us divide the quantities in the explosion problem into two groups; those that characterize the soil and crater, ρ, V, and g, and those that characterize the charge, m_c, ρ_c and e_c. Listed below are physical concepts that are characterized by parameter groups for the charge and the crater. The mass of material from the crater is ρV, the size of the crater is characterized by $V^{1/3}$ the time to move the soil in a gravity field g is scaled by $V^{1/6}g^{-1/2}$ and so on.

Quantity	Dimensions	Crater	Charge
Mass	M	ρV	m_c
Length	L	$V^{1/3}$	$(m_c/\rho_c)^{1/3}$
Time	T	$(V/g^3)^{1/6}$	$(m_c/\rho_c)^{1/3}e_c^{-1/2}$
Volume	L^3	V	m_c/ρ_c
Density	ML^{-3}	ρ	ρ_c
Velocity	LT^{-1}	$g^{1/2}V^{1/6}$	$e_c^{1/2}$
Acceleration	LT^{-2}	g	$e_c(\rho_c/m_c)^{1/3}$
Force–weight	MLT^{-3}	$\rho g V$	$e_c(\rho_c/m_c^2)^{1/3}$
Momentum	MLT^{-1}	$\rho g^{1/2}V^{7/6}$	$e_c^{1/2}m_c$
Energy	ML^2T^{-2}	$\rho g V^{4/3}$	$e_c m_c$

We ignore the ρ/ρ_c dependence as it is negligible in (2.77). Let us replace the power 0.464 by a rational fraction. After some trial and error choose 3/7 (which is equal to 0.428). This will leave the variable m_c with a power of one. A little algebra shows that (2.74) is roughly equivalent to

$$\frac{\rho_c V^{7/6}g^{1/2}}{m_c e_c^{1/2}} = \frac{\text{Characteristic momentum of crater}}{\text{Characteristic momentum of charge}}$$
$$= 0.218^{7/6} = 0.169. \quad (2.78)$$

So the results indicate that the momentum ratio is almost, but not quite, constant for the explosion crater formation process.

2.2.12 Scales Found from Boundary Conditions and Equations

In many cases we know at least some of the equations and boundary conditions that govern a situation. They can be used as a guide in forming nondimensional variables. The object is to define variables so that the boundary conditions and equations contain pure numbers or, if that is not possible, a minimum number of nondimensional parameters. As an illustration we consider the flow in a slot with a moving upper wall, the Couette–Poiseuille problem. The lower wall will be $y = 0$ and the upper wall $y = h$. Two effects drive the flow. The upper wall is sliding with velocity V_0 and there is a constant pressure gradient $\mathrm{d}p/\mathrm{d}x$. The dependent variable is the velocity profile $v(y)$.

$$v = v(y, h, V_0, \mu, \mathrm{d}p/\mathrm{d}x) \quad (2.79)$$

The differential equation and boundary conditions that govern the flow are

$$0 = -\frac{\mathrm{d}p}{\mathrm{d}x} + \mu \frac{\mathrm{d}^2 v}{\mathrm{d}y^2}, \quad v(y=0) = 0,$$
$$v(y=h) = V_0. \quad (2.80)$$

The Π theorem applied to (2.79) predicts that the problem is governed by three nondimensional variables. We have already assumed a reference value for y by setting the lower wall to zero. From the boundary conditions it is obvious that variables of reasonable size will be

$$y^* = \frac{y}{h}, \quad v^* = \frac{v}{V_0}. \quad (2.81)$$

Substituting for the dependent variable and transforming the independent variable gives

$$0 = -\frac{\mathrm{d}p}{\mathrm{d}x} + \mu \frac{\mathrm{d}^2(v^* V_0)}{\mathrm{d}y^{*2}} \left(\frac{\mathrm{d}y^*}{\mathrm{d}y}\right)^2,$$
$$0 = -\frac{h^2}{\mu V_0}\frac{\mathrm{d}p}{\mathrm{d}x} + \mu \frac{\mathrm{d}^2 v^*}{\mathrm{d}y^{*2}}. \quad (2.82)$$

The problem now contains a nondimensional parameter

$$\mathcal{P} = -\frac{h^2}{\mu V_0}\frac{\mathrm{d}p}{\mathrm{d}x}. \quad (2.83)$$

In nondimensional variables the problem is

$$0 = \mathcal{P} + \frac{\mathrm{d}^2 v^*}{\mathrm{d}y^{*2}}, \quad v^*(y^* = 0) = 0,$$
$$v^*(y^* = 1) = 1. \quad (2.84)$$

We expect a nondimensional answer $v^*(y^*, \mathcal{P})$ with three nondimensional variables as the Π theorem predicts. Sometimes, nondimensionalizing the equations results in fewer nondimensional variables, however, in this case it does not.

2.2.13 Limiting Cases

If a parameter is very large, or very small, the other nondimensional variables must have the proper scales. Consider a nondimensional relationship that contains a parameter and we are interested in the behavior as the parameter takes on an extreme value. Say the relation is

$$\Pi_1 = f(\Pi_2, \Pi_3) \tag{2.85}$$

and $\Pi_3 \to \infty$ is of interest. It makes no difference whether the extreme value is 0 or ∞ as we could redefine the parameter as $\hat{\Pi}_3 = 1/\Pi_3$. It is important that the nondimensionalized dependent variable remain bounded and nonzero in this limit; $0 < \Pi_1 < \infty$. Often the nondimensional variable needs to be redefined in order to remain bounded. We will consider three examples.

In the Couette–Poiseuille flow problem of (2.38) we nondimensionalized the velocity by the wall velocity V_0. What if the cases we are interested in have a very small V_0. This corresponds to very large value of the parameter $\mathcal{P}$. Thus small V_0 actually means small compared to the quantity $-(h^2/\mu)\,dp/dx$. This latter quantity is a measure of the maximum velocity caused by the pressure gradient. The parameter $\mathcal{P}$ has the physical interpretation as the ratio of the velocity caused by the pressure gradient to the velocity caused by the wall motion. In the limit $\mathcal{P} \to \infty$ the variable $v^*(y^*, \mathcal{P})$ becomes unbounded. The velocity v^* has been nondimensionalized by the wall velocity which is improper when the pressure gradient dominates the flow. A hint of the difficulty is that the differential equation in (2.84) has an infinite term as $\mathcal{P} \to \infty$. The correct dependent variable for small wall velocities is

$$\hat{v} = \frac{v}{-\frac{h^2}{\mu}\,dp/dx} = \frac{v^*}{\mathcal{P}}. \tag{2.86}$$

In terms of this variable the problem is

$$0 = 1 + \frac{d^2\hat{v}}{dy^{*2}}, \quad \hat{v}(y^* = 0) = 0,$$

$$\hat{v}(y^* = 1) = \frac{1}{\mathcal{P}} \to 0. \tag{2.87}$$

A properly nondimensionalized variable is bounded in the limit.

The next example is a little more complex. Consider the falling body problem of Sect. 2.2.8. What if the initial velocity V_0 is zero? This does not fundamentally change the problem, but the previously used nondimensional variables of (2.70) are not appropriate as they all involve V_0.

$$\hat{S} = \hat{f}(\hat{t}), \quad \hat{S} = \frac{S - S_0}{V_0^2/g}, \quad \hat{t} = \frac{t}{V_0/g}. \tag{2.88}$$

Another way to look at this problem is to ask what is the behavior of $\hat{S} = \hat{f}(\hat{t})$ as $\hat{t} \to \infty$? Applying the Π theorem to (2.72) without V_0

$$S - S_0 = f(t, g) \tag{2.89}$$

shows that only one nondimensional variable is needed. We could find this in the usual way or we can ask ourselves what combination of $\hat{S}$ and $\hat{t}$ will eliminate V_0. If the initial velocity V_0 is zero the nondimensional variable is

$$\tilde{S} = \frac{\hat{S}}{\hat{t}^2} = \frac{S - S_0}{gt^2} = \text{const}. \tag{2.90}$$

In fact, we see from the exact answer (2.73) that as $\hat{t} \to \infty$, $\hat{S}/\hat{t}^2 = -1/2$.

For a final example we consider the pressure. In compressible flow the pressure has a thermodynamic role, as well as a mechanical role, and the absolute pressure occurs in the equation of state. The nondimensional pressure has a zero reference and the scale is some reference pressure. In incompressible flow the scale unit depends on the flow situation. Consider an incompressible flow with a reference velocity U_0 and a reference length L. What is the proper pressure scale p_S for a nondimensional pressure

$$p^* = \frac{p - p_{\text{Ref}}}{p_S}. \tag{2.91}$$

Let us write the x-direction momentum equation and nondimensionalize the velocities with U_0 and the lengths with L.

$$\rho\left(u\frac{\partial u}{\partial x} + v\frac{\partial u}{\partial y}\right) = -\frac{\partial p}{\partial x}$$

$$+ \mu\left(\frac{\partial^2 u}{\partial x^2} + \frac{\partial^2 u}{\partial y^2}\right),$$

$$\rho U_0^2\left(u^*\frac{\partial u^*}{\partial x^*} + v^*\frac{\partial u^*}{\partial y^*}\right) = -p_S\frac{\partial p^*}{\partial x^*} + \frac{\mu U}{L}$$

$$\times \left(\frac{\partial^2 u^*}{\partial x^{*2}} + \frac{\partial^2 u^*}{\partial y^{*2}}\right). \tag{2.92}$$

There are three possibilities. If the inertia and pressure terms are dominant, the pressure scale is $p_S = \rho U_0^2$ and the viscous term will be small if the Reynolds number $\rho U L / \mu = \mathrm{Re}$ is large. If the viscous and pressure terms are dominant, the pressure scale is $p_S = \mu U_0 / L$ and the inertia term will be small if the Reynolds number Re is small. If all terms are important then either scale is useful and the Reynolds number is moderate.

2.2.14 Singular Perturbations

In Sect. 2.2.13 the dependent variable Π_1 as required to be well behaved as the parameter Π_3 approached a limit.

$$\Pi_1 = f(\Pi_2, \Pi_3) \quad \text{as} \quad \Pi_3 \to 0 \quad \text{or} \quad \infty. \tag{2.93}$$

The situation for this section is more complex in that the dependent variable Π_1 is well-behaved for a range of the Π_2 variable, but is ill behaved for another range of Π_2. Mathematically this type of behavior is called a *singular perturbation*. One set of nondimensional variables is not appropriate for the whole space region. Wall boundary layers are the common example.

Here we consider the *fluctuations of pressure on the wall* under a turbulent boundary layer. The wavenumber k of a fluctuation of wavelength λ is

$$k = \frac{1}{2\pi} \cdot \lambda. \tag{2.94}$$

The spectral density $\phi(k)$ is a function such that its integral gives the mean square pressure.

$$\langle p'^2 \rangle = \int \phi(k) \, \mathrm{d}k. \tag{2.95}$$

The fluid is characterized by a density ρ and kinematic viscosity ν while the turbulent boundary layer has a thickness δ and a characteristic velocity u_* (this is the friction velocity). The spectrum is the dimensional function

$$\phi = \phi(k, \rho, \nu, \delta, u_*). \tag{2.96}$$

The boundary layer has two length scales: the thickness of the turbulent region is δ and the viscous scale is ν/u_*. The largest fluctuations scale with δ while ν/u_* measures the smallest possible fluctuations. The physics of the turbulent fluctuations at these two different scales is different. Low-wavenumber fluctuations are inviscid and high-wavenumber fluctuations are viscous. Straightforward dimensional analysis with inviscid parameters ρ, u_*, δ as repeating variables produces the nondimensional form

$$\frac{\phi}{\rho^2 u_*^4 \delta} = \phi\left(k_1 \delta, \frac{u_* \delta}{\nu}\right). \tag{2.97}$$

For convenience we define the nondimensional variables as

$$\Phi = \frac{\phi}{\rho^2 u_*^4 \delta}, \quad K = k_1 \delta, \quad \mathrm{Re} = \frac{u_* \delta}{\nu}. \tag{2.98}$$

The Reynolds number Re is the ratio of the boundary layer thickness to the viscous length scale. In terms of these variables the result is

$$\Phi = \Phi(K, \mathrm{Re}). \tag{2.99}$$

For high Reynolds numbers

$$\Phi_0 = \Phi_0(K) \equiv \Phi(K, \mathrm{Re} \to \infty). \tag{2.100}$$

In the limit $\mathrm{Re} \to \infty$ the spectrum is finite for low wavenumbers (frequencies), however, it approaches zero for high wavenumbers.

To obtain the spectrum function valid for the high wavenumbers we must rescale both the dependent and independent variables

$$\Phi^* = \frac{\phi}{\rho^2 u_*^3 \nu} = \Phi \, \mathrm{Re}, \quad k^* = \frac{k_1}{u_*/\nu} = \frac{K}{\mathrm{Re}}. \tag{2.101}$$

The proper wavenumber rescaling, say for instance K/Re, $K/\mathrm{Re}^{1/2}$, or $K/\mathrm{Re}^{2/3}$, depends on the physics of the process. This must be determined by analysis or experiment.

For high wavenumbers the correct nondimensional form is

$$\Phi^* = \Phi^*(k^*, \mathrm{Re}), \tag{2.102}$$

$$\Phi_0^* = \Phi_0^*(k^*) \equiv \Phi^*(k^*, \mathrm{Re} \to \infty). \tag{2.103}$$

This nondimensional form scales the wavenumber by the viscous length and is valid only for high wavenumbers. Thus, depending on the range of the independent wavenumber variable, there are two different nondimensional forms that are of order one.

2.2.15 Overlap Behavior and Composite Expansions

Singular perturbation problems have one set of nondimensional variables for small values of the independent variable, and another set for high values of the independent variable. These independent variables are related by the parameter that is taking on a limit. One might expect, and it is indeed true, that there is a range of moderate values of the independent variable where either set of variables is valid. This is called the overlap region. The behavior of the functions in the *overlap region* is called the common part. The common part depends only on the amount of rescaling needed to make the dependent variable well behaved.

If no rescaling of the dependent variable is needed then the common part is a constant. Ordinary boundary layers fall into this category. If the rescaling of the dependent variable is a power law of the parameter, then the common part is also a power law in the independent variable. In the *pressure spectrum* problem of Sect. 2.2.14 the rescaling was Re^{-1}. The corresponding overlap behavior is

$$\Phi_0 \sim \frac{1}{K} \quad \text{as} \quad K \to \infty, \tag{2.104}$$

$$\Phi_0^* \sim \frac{1}{k^*} \quad \text{as} \quad k^* \to 0. \tag{2.105}$$

These are really the same equation because of the relations (2.98)

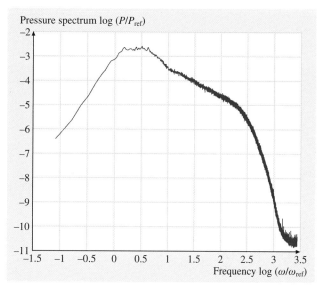

Fig. 2.9 Experimental measurement of the pressure spectrum of the turbulent atmosphere (after [2.13])

$$\Phi_0 \sim \frac{1}{K},$$
$$\Phi_0 \, Re \sim \frac{Re}{K},$$
$$\Phi_0^* \sim \frac{1}{k^*}. \tag{2.106}$$

Figure 2.9 shows experimental data taken on the atmospheric boundary layer at a very high Reynolds number. The frequency axis is equivalent to the wavenumber axis since they are connected by a convective velocity $\omega = kU_c$. The overlap region where the spectrum falls as a minus one power is obvious. This spectrum is a combination of an inviscid spectrum $\Phi_0(K)$ and a viscous spectrum $\Phi_0^*(k^*)$. Sketches of the $\Phi_0(K)$ and $\Phi_0^*(k^*)$ functions are given in Fig. 2.10. At a lower Reynolds number the drop off of the viscous part of Fig. 2.9 would occur at a lower frequency. If the Reynolds number is low enough, the overlap region disappears.

Another subtle type of overlap is where a defect law matches to a single function law. In this instance the behavior is logarithmic. A more complete description of overlap behavior is found in *Panton* [2.9, 14].

Experiments cannot be conducted at infinite values of a parameter, say the Reynolds number. Singular perturbations have a different mixture of the two different functions as the parameter changes. The importance of this mixing depends on the exact nature of the functions.

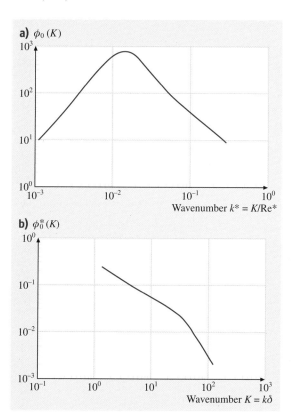

Fig. 2.10 (a) The pressure spectrum function $\Phi_0(K)$ valid for low wavenumbers. **(b)** The pressure spectrum function $\phi_0^*(k^*)$ valid for high wavenumbers

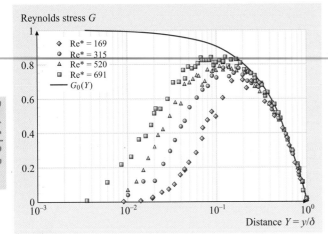

Fig. 2.11 Experimental measurement of the Reynolds stress across a pipe for several Reynolds numbers in the outer distance variable Y (after [2.15])

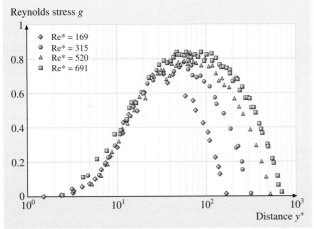

Fig. 2.12 Experimental measurement of the Reynolds stress across a pipe for several Reynolds numbers in the inner distance variable y^+ (after [2.15])

For laminar boundary layers it is not very important. For turbulent wall layers is not too important for the velocity profile. A situation where explicit variation of the parameter needs to be accounted for in the data analysis is the *Reynolds stress in pipe flow*. For pipe (or channel) flow the nondimensional Reynolds stress $\langle uv \rangle$ is a function of the distance from the wall $Y = y/R$ and the Reynolds number $\text{Re} = u_* R/\nu$.

$$-\frac{\langle uv \rangle}{u_*^2} = G(Y, \text{Re}) \,. \tag{2.107}$$

Figure 2.11 shows experimental data for the Reynolds stress in turbulent pipe flow. For high Reynolds numbers the *outer* form of this relation is

$$G_0(Y) = G(Y, \text{Re} \to \infty) \,. \tag{2.108}$$

Analysis shows that the exact equation for G_0 is

$$G_0 = 1 - Y \,. \tag{2.109}$$

Away from the wall, $Y > 0.2$, the data in Fig. 2.11 follow (2.109), but near the wall, $Y < 0.2$, the data do not follow $G_0 = 1 - Y$. At the wall $G_0(0) = 1$ is invalid and a rescaled independent variable is required. The proper independent variable scaling near the wall is

$$y^+ = \frac{y u_*}{\nu} = Y \text{Re} \,. \tag{2.110}$$

Near the wall the proper *inner* representation of the Reynolds stress is a different relation.

$$-\frac{\langle uv \rangle}{u_*^2} = g(y^+, \text{Re}) \,. \tag{2.111}$$

Figure 2.13 shows the experimental data in the *inner* representation. For high Reynolds numbers the *inner* form of this relation is

$$g_0(y^+) = g(y^+, \text{Re} \to \infty) \,. \tag{2.112}$$

Table 2.3 Length scales

Lengths	Formula	Interpretation
Flow region size	L	Characteristic length scale of flow pattern
Viscous diffusion distance	$\sqrt{\nu t}$	Distance viscous shear stress or vorticity diffuses in time t
Mean free path length	λ	Typical distance a gas molecule travels between collisions
Thermal conduction distance	$\sqrt{\alpha t}$	Distance thermal energy is conducted in time t
Capillary length	$\sqrt{\frac{\sigma}{\rho g}}$	Length scale for interface shape in gravity field
Kolmogorov	$\left(\frac{\nu^3}{\varepsilon}\right)^{1/4}$	Size of smallest turbulent eddy

Table 2.4 Velocity scales

Velocity	Formula	Interpretation
Flow	U	Velocity imposed in a boundary specification
Sound	a	Speed of an acoustic wave in a still medium
Viscous diffusion	$\sqrt{\nu/t}$	Velocity of the diffusion of viscous region at time t
Friction velocity	$u_* \equiv \sqrt{\tau_0/\rho}$	Convenient scale in wall turbulence
Thermocapillary	$\frac{1}{\mu}\partial\sigma/\partial x$	Velocity from surface tension gradient opposed by viscosity
Kolmogorov	$(\nu\varepsilon)^{1/4}$	Characteristic velocity fluctuation of smallest turbulent eddy

Near the wall, $y^+ < 20$, the data collapse to a single curve and follow (2.112), but away from the wall, $y^+ > 20$, the data do not follow $g_0 = g_0(y^+)$.

The overlap region between the inner and outer functions has a constant common part $G_{\mathrm{CP}} = G_0(Y \to 0) = 1 = g_0(y^+ \to \infty)$.

At finite Reynolds numbers the Reynolds stress does not fit either representations for all distances. For this reason G_0 and g_0 are not experimentally observed. A composite expansion is a combination of the two functions into a single uniformly valid representation. A composite expansion contains the first effects of Reynolds number. An additive *composite expansion* is

$$-\frac{\langle uv \rangle}{u_*^2} = g_0(y^+) + G_0\left(Y \to \frac{y^+}{\mathrm{Re}}\right) - G_{\mathrm{CP}}$$

$$= g_0(y^+) + (1-Y) - 1 = g_0(y^+) - \frac{y^+}{\mathrm{Re}}.$$

(2.113)

Since we know $G_0(Y) = 1 - Y$ and $G_{\mathrm{CP}} = 1$, there is only one unknown function $g_0(y^+)$ in (2.113). Solving this equation allows the data to be *corrected* for Reynolds number effects

$$g_0(y^+) = -\left.\frac{\langle uv \rangle}{u_*^2}\right|_{\mathrm{data}} + \frac{y^+}{\mathrm{Re}}.$$

(2.114)

Figure 2.12 shows data processed according to (2.114). Experimental data at a variety of Reynolds numbers is plotted in Fig. 2.13, where, within the experimental uncertainty, it falls onto a single curve. The line on Fig. 2.13 is drawn also considering channel flow experiments and direct numerical simulations.

The composite expansion (2.113) is plotted in Fig. 2.13, which shows that the Reynolds number effect is well represented by a composite expansion.

2.2.16 Common Scales and Nondimensional Parameters

Nondimensional parameters can be interpreted as a ratio of dimensional properties. The dimensional properties of the ratio can be length scales, time scales, velocity scales, forces, fluxes, energies, or other physical concepts. The

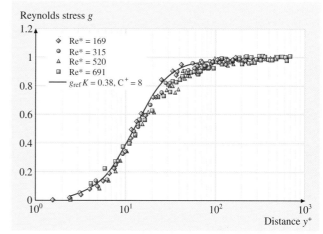

Fig. 2.13 Experimental measurement of Reynolds stress represented as the inner function $g(y^+)$ (after [2.15])

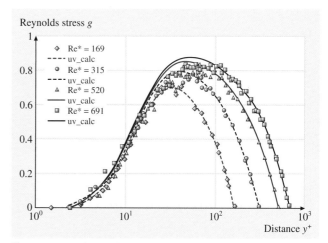

Fig. 2.14 Experimental measurement of Reynolds stress compared with the composite expansion (2.113) where g_0 is a curve fit of Fig. 2.12 (after [2.15])

Table 2.5 Nondimensional parameters

Name	Symbol	Definition	Comparison ratio
Biot number	Bi	$\frac{h}{\kappa/L}$	Convection heat transfer/conduction heat transfer
Bond number	Bo	$\frac{g(\rho-\rho_f)L^2}{\sigma}$	Buoyancy force/surface tension force (geometric length/capillary length)^2
Capillary number	Ca	$\frac{\mu V}{\sigma}$	Viscous effect/surface tension effect
Cavitation number	Cav	$\frac{p-p_v}{\rho V^2}$	Pressure difference from vapor pressure/dynamic pressure
Drag coefficient	C_D	$\frac{F_D}{1/2\rho V^2 A_x}$	Drag force/dynamic pressure times cross section area (for aircraft planform area)
Eckert number	Ec	$\frac{V^2}{c_p \Delta T}$	Kinetic energy/enthalpy change
Ekman number	E	$\frac{\nu}{f_R L^2}$	Viscous force/Coriolis force (Coriolis frequency $f_R = 2\sin\theta\Omega$ for earth rotation)
Fourier number	F	$\frac{\alpha t}{L^2}$	Heat conduction rate/energy storage rate
Friction coefficient	C_f	$\frac{\tau}{1/2\rho V^2}$	Shear stress/dynamic pressure
Friction factor	f	$\frac{h_L D/L}{1/2 V^2}$	Head loss (viscous dissipation) in pipe of length D/incoming kinetic energy
Froude number	Fr	$\frac{V^2}{gL}$	Kinetic energy/gravity potential Inertia force/gravity force
Grashof number	Gr	$\frac{g\alpha\Delta T L^3}{\nu^2}$	Bouyancy force/viscous force
Head loss coefficient	K	$\frac{h_L}{1/2 V^2}$	Head loss (viscous dissipation)/incoming kinetic energy
Knudsen number	Kn	$\frac{\lambda}{L}$	Mean free path/flow length
Lift coefficient	C_L	$\frac{F_L}{1/2\rho V^2 A_x}$	Lift force/dynamic pressure times cross section area (for aircraft planform area)
Mach number	M	$\frac{V}{a}$	Velocity/speed of sound
Marangoni number	Ma	$\frac{L}{\mu\alpha}\partial\sigma/\partial x$	Thermocapillary flow/thermal conduction
Nusselt number	Nu	$\frac{hL}{k}$	Nondimensional heat convection coefficient
Peclet number	Pe	$\frac{VL}{\alpha}$	= Re Pr bulk heat transfer/conduction heat transfer
Prandtl number	Pr	$\frac{\mu c_p}{k}$	Viscous diffusion effect/thermal diffusion effect
Pressure coefficient (Euler number)	C_p	$\frac{p-p_{Ref}}{1/2\rho V^2}$	Pressure change/dynamic pressure
Rayleigh number	Ra	$\frac{g\alpha\Delta T L^3}{\nu k}$	Modified Grashof number Gr Pr
Reynolds number	Re	$\frac{VL}{\nu}$	Inertia effects/viscous effects
Richardson number	Ri	$\frac{g\alpha\Delta T L}{V^2}$	Bouyancy force/inertia force
Rossby number	Ro	$\frac{V}{f_R L}$	Rotation time/flow time (Coriolis frequency $f_R = 2\sin\theta\Omega$ for earth rotation)
Stanton number	St	$\frac{h}{\rho c_p V}$	Heat transfer/thermal capacity of fluid
Strouhal number	St	$\frac{fL}{V}$	Frequency/(flow time)$^{-1}$
Weber number	We	$\frac{\rho V^2 L}{\sigma}$	Dynamic pressure/surface tension

exact problem under consideration and the definition of the nondimensional parameter determine the specific interpretation. For example, the *Reynolds number* is a ratio of an inertia *force* to a viscous force. The inertia effect is proportional to ρU^2 (twice the kinetic energy per unit volume) and the shear stress proportional to $\mu U/L$

$$\text{Re} = \frac{\rho UL}{\mu} = \frac{\rho U^2}{\mu U/L}. \tag{2.115}$$

However, in boundary layers and entrance length problems the shear stress is proportional to $\mu U/\delta$. Here δ is the thickness of the viscous region and is $\delta \sim \sqrt{\nu L/U}$.

Hence

$$\text{Re} = \frac{UL}{\nu} = \frac{L^2}{\nu L/U} = \left(\frac{L}{\delta}\right)^2. \tag{2.116}$$

A high Reynolds number indicates that the layer is thin compared to the length L. In the boundary layer the shear stress and inertia force are of equal importance. Nevertheless, the Reynolds number is some measure of viscous effects and inertia effects. Table 2.3 gives some length scales and Table 2.4 velocity scales that are typically defined. In Table 2.5 the reader will find common nondimensional parameters and a typical interpretation.

2.3 Self-Similarity

Certain flows lend themselves to some idealization which consists of excluding a length, time or velocity scale. This is the first sign that self-similar behavior is expected. In many important cases self-similar asymptotics *attract* real flow patterns even though the latter are *polluted* by some minor *non-ideal* details. Then, self-similar asymptotics are realizable experimentally and *attract* the initially non-self-similar flow fields. Being recognized, self-similarity can relatively readily be found analytically or numerically. Moreover, the very recognition of flow self-similarity, even without finding the corresponding solution, enables one to detect important general features in the pile of raw experimental (or numerical) data, dramatically reduces the volume of the experimental work needed for flow characterization, and makes data processing highly effective. In the present chapter the self-similarity approach is elaborated through particular examples of viscous Navier–Stokes flows, the boundary layer and gas dynamics flows, as well as a flow with the free surface. In most cases both hydrodynamic and thermal flow fields are considered and the general procedure that establishes self-similarity is demonstrated.

2.3.1 General Causes of Self-Similar Behavior in Certain Situations in Fluid Mechanics and Heat Transfer

Hydrodynamic and heat transfer problems are generally described by partial differential equations (PDEs), which implies that their solutions depend on several variables (say, spatial coordinates and/or time). The equations of motion responsible for the hydrodynamic part are nonlinear (for example, the Navier–Stokes equations), or reduce to linear equations in certain simple cases. The thermal balance equation is typically linear, but even in the simplest cases it can be solved only after the fluid mechanical part of the problem has been disposed of and the flow field established. As a rule, solution of linear and nonlinear PDEs is a much more complicated task than that of ordinary differential equations (ODEs) where functions of a single variable are sought. Fortunately, in certain cases of practical importance PDEs with the corresponding initial and boundary conditions can be reduced to ODEs subject to related conditions.

Such a reduction is possible if the problem at hand lends itself to some idealization. For example, one can consider:

- a semi-infinite depth of a fluid (instead of a finite depth) in contact with a horizontally moving infinite plate,
- a pointwise vortex core (instead of a finite one) in viscous fluids,
- fluid suction from a wedge into an infinitesimally thin (instead of a finite) slit at the wedge tip,
- viscous fluid flow, forced or due to natural convection, along a semi-infinite (instead of a finite) plate,
- jets issuing from a pointwise nozzle or buoyant plumes generated by a pointwise heat source (instead of finite ones),
- gas flows due to pointwise explosions in air (instead of finite-size ones),
- capillary waves generated by impact of a pointwise droplet or a stick (instead of finite ones) onto a thin layer of fluid.

In all these cases the idealization consists of excluding a length scale (or possibly a time or velocity scale) from the problem [2.16]. The latter is the first sign that self-similarity can exist in the given problem and its description can be reduced to ODEs. There are, however, certain restrictions. For example, there is no guarantee that the physical parameters of interest tend to finite nonzero values anywhere in the flow domain at the limit corresponding to the length scale (in fact, one of the given parameters) tending to zero. Where such finite nonzero limits exist, we are dealing with complete self-similarity (also called self-similarity of the first kind), which can be fully established by means of dimensional analysis [2.17]. Only such cases are considered in the present chapter. The situation, however, can be more complicated when the above mentioned limit tends to zero or infinity, or tends to no limit and demonstrates power-law asymptotics. In such cases, we are dealing with incomplete self-similarity (or self-similarity of the second kind). Then the exponents of the power-law asymptotics (scaling) cannot generally be found by applying dimensional analysis, but rather as the eigenvalues of the relevant eigenvalue problems. Nevertheless, the asymptotics could be fully determined only via matching the power-law *tails* to the fully non-self-similar part of the solution (which is not always possible) [2.17].

As mentioned above, self-similarity is easily recognizable and relatively readily found analytically or numerically. It also enables one to detect the instructive invariant features sometimes buried under the pile of raw experimental or numerical data. Moreover, the very recognition of intrinsic self-similarity in a certain case, even without finding the corresponding solution, dramatically reduces the volume of experimental data needed for characterizing the flow, as is shown in brief in the next subsection.

2.3.2 Implications of Self-Similarity in Experimental Studies

The idea of disregarding the detailed flow structure at certain points (for example, at the exit of a *small* nozzle releasing a jet) inevitably leads to localized singularities. At the origin of a jet flow the velocity becomes infinite, while for example the momentum flux remains finite. In spite of the singularity, such a self-similar solution is useful if the actual flow tends to the self-similar one far from the nozzle. If such is the case, we can conclude that in the far field the memory of the initial details at the nozzle exit has already faded and only such an integral characteristic as the momentum flux is important. One could actually argue that the self-similar solution should not necessarily *attract* any jet flow originating from a near-field zone, since this has not been rigorously proven so far. In other words, it is not proven that the flow field in a real non-self-similar jet will tend to the flow field corresponding to the idealized self-similar

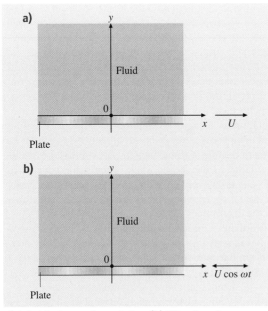

Fig. 2.15 (a) Suddenly accelerated plate **(b)** Vibrating plate

situation. However, in most fluid-mechanical cases involving boundary layers near a solid wall or free flows (as in jets and plumes), self-similar asymptotics *attract* the flow pattern even though the latter is *initially polluted* by some minor details, and as soon as the memory of the latter has faded (as in the far field of the jet) self-similar behavior is expected. The very fact that self-similar behavior is realizable experimentally shows that it *attracts* the initially non-self-similar flow fields.

Self-similar behavior can also be disrupted by the presence of unaccounted-for distant walls. A jet, for example, is never fully free, since certain confinements are present. Nevertheless, actual flow structures at sufficient distances from such walls agree fairly well in many cases with the idealized self-similar ones. This makes them valuable intermediate asymptotics of a real flow field. The self-similar solutions can be described as a tiny *island* in the *ocean* of non-self-similar flows but, their influence extends far into the latter.

The fact that the flows in the far field of, say, a steady turbulent submerged axisymmetric jet are expected to be self-similar has major implications regarding the measurement strategy in experimental study of the jet. Self-similarity implies that the longitudinal velocity in the jet is given by $u = u_{ax}(x)\Phi(\eta)$, where $u_{ax}(x)$ is the axial velocity, and the self-similar variable $\eta = \text{const} \cdot y \cdot x^\beta$, β being a known exponent. An experimenter has to measure only $u_{ax}(x)$ and the velocity profile in a single jet cross section $x = x_*$ and the values of the self-similar variable η_* in this cross section are derived from y using x_*. The measured velocity values u are also normalized by $u_{ax}(x_*)$, and the function $\Phi(\eta_*) = u/u_{ax}(x_*)$ is then established in the cross section $x = x_*$. However, self-similarity also implies that the same function describes the normalized velocity profile in all other cross sections of the jet in the far field, hence, no additional measurements are needed there. The fact that the flow is expected to be self-similar (which can be established a priori even without solving the equations of motion) permits drastic reduction of the volume of measurements.

In the following subsections the self-similarity approach will be elaborated through particular examples, beginning with the simplest ones. In Sect. 2.3.3 several examples of viscous Navier–Stokes flows are considered. Section 2.3.4 deals with flows in boundary layers, including forced and natural convection as well as laminar and turbulent flows. In Sect. 2.3.5 a self-similar flow in gas dynamics is shown, while in Sect. 2.3.6 self-similarity in a free surface flow is discussed. In all cases, the general procedure which establishes self-similarity is demonstrated. Where possible, reference is made only to the monographs in which the original works can be found.

2.3.3 Particular Examples of Self-Similar Navier–Stokes Flows

Suddenly Accelerated Flat Plates (Stokes's First Problem) and the Corresponding Thermal Problem on a Suddenly Cooled Slab

Consider an infinite plate at $y = 0$, in contact with a viscous fluid at $y \geq 0$ (Fig. 2.14a). Both plate and fluid were at rest at $t < 0$, while at the initial time $t = 0$ the plate begins to move in the x-direction with a velocity $U = \text{const}$. The kinematic viscosity of the fluid is ν. Pressure is uniform in the fluid domain, and fluid entrainment by the plate is described by the reduced Navier–Stokes equations with the appropriate initial and boundary conditions: the no-slip condition at $y = 0$ and the condition that the fluid is always at rest at infinity [2.18]. Namely,

$$\frac{\partial u}{\partial t} = \nu \frac{\partial^2 u}{\partial y^2} \tag{2.117}$$

with the initial conditions

$$t = 0; \quad u = 0 \tag{2.118}$$

and boundary conditions

$$t > 0 : \begin{cases} y = 0, & u = U, \\ y = \infty, & u = 0. \end{cases} \tag{2.119}$$

The assumption that the fluid domain is semi-infinite, $0 \leq y \leq \infty$, is an idealization of real situations where the domain is sufficiently large but finite in the y-direction. Due to the idealization the problem does not contain any given length scale. To normalize the problem (2.117–2.119) any *arbitrary* length scale L can be used, whereas the time scale becomes L^2/ν, where the dimensionality of ν is $[\nu] = \text{m}^2/\text{s}$. The velocity scale can certainly be taken as U. The dimensionless problem reads

$$\frac{\partial \bar{u}}{\partial \bar{t}} = \frac{\partial^2 \bar{u}}{\partial \bar{y}^2} \tag{2.120}$$

with the initial and boundary conditions

$$\bar{t} = 0, \quad \bar{u} = 0,$$

$$\bar{t} > 0 : \begin{cases} \bar{y} = 0 & \bar{u} = 1 \\ \bar{y} = \infty & \bar{u} = 0, \end{cases} \tag{2.121}$$

where t, y and u are rendered dimensionless by L^2/ν, L and U, respectively. Any solution of (2.120, 121) should have the form $\bar{u} = f(\bar{t}, \bar{y})$, such that the arbitrary length scale L is automatically absent in the final result. This is the case with the function

$$f(\bar{t}, \bar{y}) = F(\eta), \quad \eta = \frac{\bar{y}}{\bar{t}^{1/2}}, \tag{2.122}$$

since for any L, $\eta = y/(\nu t)^{1/2}$. Any dependence of f on $\bar{t}$ and $\bar{y}$, other than that of (2.122), does not exclude an arbitrary L which should not appear in the solution. The solution (2.122) is called self-similar and depends on the self-similar variable η. The fact of self-similarity, as established above, obviates the need to solve an equation. Once it has been established, a particular function F can be found experimentally using a time series for the velocity u at a certain location y. On the other hand, in the present case the self-similar solution can easily be found after (2.122) has been substituted in (2.120) and the resulting ordinary differential equation solved with the boundary conditions as per (2.121). The corresponding self-similar solution reads

$$\frac{u}{U} = 1 - \text{erf}\left(\frac{y}{2(\nu t)^{1/2}}\right), \tag{2.123}$$

where

$$\text{erf}(z) = \frac{2}{\sqrt{\pi}} \int_0^z e^{-\zeta^2} d\zeta \tag{2.124}$$

is the error function.

It is emphasized that, if a gap of a finite width W filled with fluid ($0 \leq y \leq W$) is considered, a length scale $L = W$ should be taken. No self-similarity would arise in that case. The self-similar solution serves, however, as a reasonable approximation of such a case at times when the presence of the wall at $y = W$ is still not strongly felt by the flow, i.e., at t significantly shorter than $W^2/(4\nu)$. During this time interval the self-similar solution represents itself as an intermediate asymptotics of the non-self-similar situation [2.17].

A kindred thermal problem arises in the case where material occupying a semi-infinite space $0 \leq y \leq \infty$ and having an initial temperature T_i is brought at $t = 0$ at $y = 0$ into contact with a *refrigerator* of temperature $T_0 < T_i$. The self-similar solution for the temperature field T reads

$$\frac{T - T_0}{T_i - T_0} = \text{erf}\left(\frac{y}{2(\alpha t)^{1/2}}\right), \tag{2.125}$$

where α is the thermal diffusivity of the material in question, $[\alpha] = \text{m}^2/\text{s}$.

Lack of Self-Similarity in the Case of a Vibrating Flat Plate (Stokes's Second Problem)

The reason for lack of self-similarity can be more subtle than in the case discussed in the preceding subsection. Consider, for example, a vibrating plate in contact with a viscous fluid (Fig. 2.14b). The problem is given by (2.117–2.119) with the no-slip condition replaced by

$$y = 0, \quad u = U\cos\omega t, \tag{2.126}$$

where U and ω are the amplitude and frequency of the vibration. In spite of the fact that the fluid domain is semi-infinite, a given length scale arises from the given parameters as $L = \sqrt{\nu/\omega}$. Then, self-similarity is not expected. In such cases, typically a numerical solution should be sought, but in the present case an analytical non-self-similar solution is available [2.16, 18]

$$\frac{u}{U} = e^{-ky}\cos(\omega t - ky), \tag{2.127}$$

where $k = [\omega/(2\nu)]^{1/2} = 2^{-1/2}L^{-1}$. This shows how the flow field depends on t and y separately, rather than on a self-similar variable.

Vorticity Diffusion

The vorticity transport equation is obtained by applying the curl operator to the Navier–Stokes equation. In the case of planar motion shown in Fig. 2.15, the vorticity transport equation takes the form [2.19]

$$\frac{\partial w}{\partial t} = \frac{\nu}{r}\frac{\partial}{\partial r}\left(r\frac{\partial w}{\partial r}\right). \tag{2.128}$$

It is assumed here that the flow depends only on the radial coordinate r, while being circularly symmetric. The vorticity $\mathbf{w} = \mathrm{curl}\,\mathbf{v}$, $\mathbf{v}$ being the velocity vector, has a single nonzero component, which is the one normal to the flow plane in Fig. 2.15. Its magnitude is denoted by w. Consider the evolution of an initially pointwise vorticity line normal to the flow plane and located at $r = 0$ (Fig. 2.15). Its strength is characterized by the initial circulation Γ_0, which implies that the velocity distribution at $t = 0$ is given by

$$t = 0, \quad v_\theta = \frac{\Gamma_0}{2\pi r} \tag{2.129}$$

with the subscript θ denoting the azimuthal direction.

The initial velocity distribution is related to the initial vorticity distribution about the vortex

$$t = 0, \quad r > 0 \quad w = 0. \tag{2.130}$$

The boundary condition for the vorticity reads

$$r \to \infty \quad w \to 0. \tag{2.131}$$

As a spatial length scale an arbitrary L should be taken, since in the present idealization of a pointwise vortex no cross-sectional radius is given. The time scale becomes L^2/ν. The corresponding vorticity scale is Γ_0/L^2, since $[\Gamma_0] = \mathrm{m}^2/\mathrm{s}$. Any solution of (2.128) should have the dimensionless form

$$\bar{w} = f(\bar{t}, \bar{r}) \tag{2.132}$$

with $\bar{w} = w/(\Gamma_0/L^2)$, $\bar{t} = t/(L^2/\nu)$, $\bar{r} = r/L$, which yields the dimensional vorticity

$$w = \frac{\Gamma_0}{L^2}f(\bar{t}, \bar{r}). \tag{2.133}$$

The only particular form of the dimensionless function f that permits automatic absence of an arbitrary length scale L in the final result is

$$f(\bar{t}, \bar{r}) = \frac{1}{\bar{t}}F(\eta), \quad \eta = \frac{\bar{r}}{\bar{t}^{1/2}}. \tag{2.134}$$

Then (2.133) and (2.134) yield

$$w = \frac{\Gamma_0}{\nu t}F(\eta), \quad \eta = \frac{r}{(\nu t)^{1/2}}. \tag{2.135}$$

The function F can be found if (2.134) is substituted into (2.128) and the resulting ordinary differential equation solved. It has the form

$$F = \frac{1}{4\pi}\exp\left(-\frac{r^2}{4\nu t}\right). \tag{2.136}$$

The corresponding velocity field becomes

$$v_\theta = \frac{\Gamma_0}{2\pi r}\left[1 - \exp\left(-\frac{r^2}{4\nu t}\right)\right]. \tag{2.137}$$

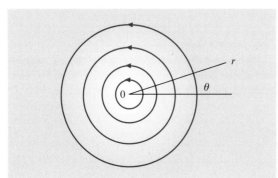

Fig. 2.16 Vorticity diffusion. The initially pointwise vortex is located at $r = 0$

The vorticity field is obviously self-similar, since

$$\frac{w}{\Gamma_0/(\nu t)} = F(\eta), \qquad (2.138)$$

which permits expression of the vorticity magnitude at a distance r_2 and time t_2 via its magnitude at any other location r_1 at time t_1.

In reality, a vortex would have a finite core of radius r_0, the flow inside of which can be visualized as solid-state rotation with an angular rate w_0. At the core boundary the angular velocity is continuous, thus $w_0 r_0 = \Gamma_0/(2\pi r_0)$. This yields $2\pi r_0^2 w_0 = \Gamma_0$. The self-similarity found holds in a real situation in the far field, where $r \gg r_0$. On the other hand, as $r_0 \to 0$ while $\Gamma_0 = \mathcal{O}(1)$, the angular rate in the core is $w_0 \to \infty$.

As usual, the fact that the vorticity field of an initially pointwise vortex in a viscous fluid should be self-similar can be (and has been) established without recourse to the solution of the governing equation (2.128) itself, merely by consideration of the problem as posed. The equation should be solved, or the experimental measurements should be undertaken, only when a particular form of the vorticity distribution is needed. The experiments, however, are simplified dramatically by the fact that due to the self-similarity expected beforehand, the measurements can be confined to a single location r. It would be instructive to do an experiment with a small-diameter rod mounted in a lathe.

Flow in a Wedge

Planar flow in a wedge is shown in Fig. 2.16a. In the approximation we accept, the slit width at the wall intersection is negligibly small, and a pointwise sink or source is located at point O. In fact, this approximation corresponds to the far field where coordinates r are much larger than the slit width. The no-slip boundary conditions at the wedge walls also do not introduce any length scale. The viscous fluid involved is characterized by its kinematic viscosity ν, the sink/source strength per unit depth is given by Q, and the wedge semi-angle is α. The stream function of the flow ψ in the general case should depend on r and on the polar angle φ. Since in the present case ν, Q and ψ have the same dimensionality m^2/s, and no length scale is given due to the assumption of a pointwise sink/source, the stream function can be taken as [2.16, 20]

$$\psi = \nu f\left(\frac{r}{L}, \varphi, \frac{Q}{\nu}\right). \qquad (2.139)$$

In the solution any arbitrary length scale L used for nondimensionalization should disappear. This can be achieved only in the case where $f(r/L, \varphi, Q/\nu) = f_1(\varphi, Q/\nu)$, i.e.,

$$\psi = \nu f_1\left(\varphi, \frac{Q}{\nu}\right). \qquad (2.140)$$

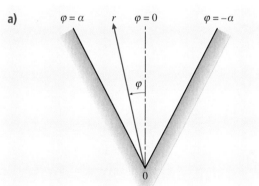

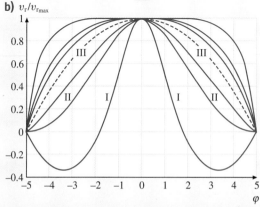

The velocity components are

$$v_r = \frac{1}{r}\frac{\partial \psi}{\partial \varphi}, \qquad v_\theta = -\frac{\partial \psi}{\partial r}. \qquad (2.141)$$

Therefore

$$v_r = \frac{\nu}{r} F\left(\varphi, \frac{Q}{\nu}\right), \qquad v_\theta = 0, \qquad (2.142)$$

where $F = f_1'$, the prime denoting differentiation with respect to φ. A solution of the Navier–Stokes equations yields F, when (2.142) are substituted therein. From the solutions, a function F satisfying the no-slip conditions at the wedge walls can be found. This yields F as a solution for the flow in the wedge. It can be expressed with the aid of the elliptic functions sn and cn. The velocity profiles for $\alpha = 5°$ are shown

Fig. 2.17a,b Viscous flow in a wedge. (**a**) Sketch, (**b**) the velocity profiles

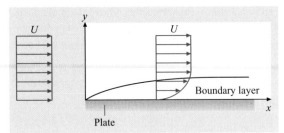

Fig. 2.18 Boundary layer near a flat plate

in Fig. 2.16b in the self-similar form $v_r/(v_r)_{\max}$, where the maximal velocity value (corresponding to $\varphi = 0$) is $(v_r)_{\max} = (\nu/r)F(0, Q/\nu)$. It is emphasized that the fact of self-similarity was established in (2.142) without any recourse to the solution of the Navier–Stokes equations, just by considering the governing physical parameters.

The velocity profiles in Fig. 2.16b are quite peculiar. Those corresponding to flows in a diverging channel (with $Q > 0$) can contain regions of reverse flow toward the source (where $v_r < 0$, as for example for profile I). The borderline diffuser profile where reverse flows disappear is denoted II. The velocity profile III is $v_r/(v_r)_{\max} = 1 - (\tan\varphi/\tan\alpha)^2$, which is the Poiseuille profile inserted in the wedge. It is located as a border line between the flows in diverging ($Q > 0$) and converging ($Q < 0$) channels. Above profile III, only confuser-like flows exist (for them v_r is, in fact, the velocity magnitude).

2.3.4 Particular Examples of the Boundary Layer Flows

Flow and Heat Transfer in Laminar Boundary Layers Near a Flat Wall

A uniform laminar flow which encounters a semi-infinite plate parallel to it adjusts itself to the no-slip condition at the plate surface and develops the boundary layer as shown in Fig. 2.17. The dynamic and thermal problems are described in the framework of the boundary layer equations [2.18]

$$\frac{\partial u}{\partial x} + \frac{\partial v}{\partial y} = 0,$$
$$u\frac{\partial u}{\partial x} + v\frac{\partial u}{\partial y} = \nu\frac{\partial^2 u}{\partial y^2},$$
$$u\frac{\partial T}{\partial x} + v\frac{\partial T}{\partial y} = \alpha\frac{\partial^2 T}{\partial y^2}, \quad (2.143)$$

with u and v being the longitudinal and transverse velocity components, and α the thermal diffusivity. If we assume a uniform temperature T_w at the plate surface and a uniform temperature of the flow at infinity T_∞, the complete set of boundary conditions reads

$$y = 0 \quad u = v = 0, \quad T = T_w$$
$$y = \infty \quad y = U, \quad T = T_\infty. \quad (2.144)$$

The dynamic part of problem (2.143) (Blasius flow) reduces to a single equation for the stream function ψ ($u = \partial\psi/\partial y$, $v = -\partial\psi/\partial x$)

$$\frac{\partial\psi}{\partial y}\frac{\partial^2\psi}{\partial x\partial y} - \frac{\partial\psi}{\partial x}\frac{\partial^2\psi}{\partial y^2} = \nu\frac{\partial^3\psi}{\partial y^3}. \quad (2.145)$$

The viscous length scale ν/U is typically so small (of the order of 10^{-5}–10^{-4} m) that it cannot be considered natural for a plate. No length scale is given, since the plate is idealized as being semi-infinite. Therefore, an arbitrary scale L is used to render the longitudinal coordinate x dimensionless. As usual in a boundary layer, the transverse coordinate y is rendered dimensionless by $L/\mathrm{Re}^{1/2}$, $\mathrm{Re} = UL/\nu$ being the Reynolds number. Then the corresponding scale of the stream function $\Psi = (UL\nu)^{1/2}$ follows from its definition, while (2.145) becomes

$$\frac{\partial\bar\psi}{\partial\bar y}\frac{\partial^2\bar\psi}{\partial\bar x\partial\bar y} - \frac{\partial\bar\psi}{\partial\bar x}\frac{\partial^2\bar\psi}{\partial\bar y^2} = \frac{\partial^3\bar\psi}{\partial\bar y^3}, \quad (2.146)$$

where the dimensionless parameters are denoted by bars. A solution has to be of the form $\bar\psi = f(\bar x, \bar y)$, i.e.

$$\psi = (UL\nu)^{1/2} f\left[\frac{x}{L}, y\left(\frac{U}{L\nu}\right)^{1/2}\right], \quad (2.147)$$

where f is a dimensionless function.

In the absence of a given length scale, only functions of a special, self-similar type, $f(\bar x, \bar y) = \bar x^{1/2} F(\bar y/\bar x^{1/2})$ can be admitted. This automatically reduces (2.147) to the form

$$\psi = (U\nu x)^{1/2} F(\eta), \quad \eta = y\left(\frac{U}{\nu x}\right)^{1/2}, \quad (2.148)$$

where L is absent. Substitution of the self-similar solution (2.148) in the dynamic part of the problem (2.143) and (2.144) yields the following boundary value problem for determining F

$$F''' + \frac{1}{2}FF'' = 0,$$
$$F(0) = F'(0) = 0,$$
$$F'(\infty) = 1, \quad (2.149)$$

where the primes denote differentiation with respect to η. The solution of (2.149) was found numerically

Table 2.6 The function $F(\eta)$ for the boundary layer along a semi-infinite flat plate

$\eta = y\sqrt{\frac{U}{\nu x}}$	F	$F' = \frac{u}{U}$	F''
0	0	0	0.33206
0.2	0.00664	0.06641	0.33199
0.4	0.02656	0.13277	0.33147
0.6	0.05974	0.19894	0.33008
0.8	0.10611	0.26471	0.32739
1.0	0.16557	0.32979	0.32301
1.2	0.23795	0.39378	0.31659
1.4	0.32298	0.45627	0.30787
1.6	0.42032	0.51676	0.29667
1.8	0.52952	0.57477	0.28293
2.0	0.65003	0.62977	0.26675
2.2	0.78120	0.68132	0.24835
2.4	0.92230	0.72899	0.22809
2.6	1.07252	0.77246	0.20646
2.8	1.23099	0.81152	0.18401
3.0	1.39682	0.84605	0.16136
3.2	1.56911	0.87609	0.13913
3.4	1.74696	0.90177	0.11788
3.6	1.92954	0.92333	0.09809
3.8	2.11605	0.94112	0.08013
4.0	2.30576	0.95552	0.06424
4.2	2.49806	0.96696	0.05052
4.4	2.69238	0.97587	0.03897
4.6	2.88826	0.98269	0.02948
4.8	3.08534	0.98779	0.02187
5.0	3.28329	0.99155	0.01591
5.2	3.48189	0.99425	0.01134
5.4	3.68094	0.99616	0.00793
5.6	3.88031	0.99748	0.00543
5.8	4.07990	0.99838	0.00365
6.0	4.27964	0.99898	0.00240
6.2	4.47948	0.99937	0.00155
6.4	4.67938	0.99961	0.00098
6.6	4.87931	0.99977	0.00061
6.8	5.07928	0.99987	0.00037
7.0	5.27926	0.99992	0.00022
7.2	5.47925	0.99996	0.00013
7.4	5.67924	0.99998	0.00007
7.6	5.87924	0.99999	0.00004
7.8	6.07923	1.00000	0.00002
8.0	6.27923	1.00000	0.00001
8.2	6.47923	1.00000	0.00001
8.4	6.67923	1.00000	0.00000
8.6	6.87923	1.00000	0.00000
8.8	7.07923	1.00000	0.00000

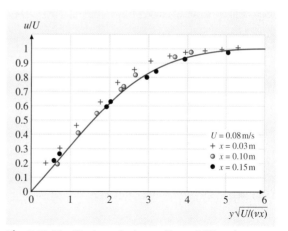

Fig. 2.19 The Blasius velocity profile: *solid line*; the experimental data – *symbols*

(Table 2.6). The corresponding self-similar velocity profile $u = UF'$ is given in Fig. 2.18 versus experimental data. It is clearly seen that in the self-similar coordinates measurements in three different cross sections $x = 0.03$, 0.1 and 0.15 m collapse onto the theoretical Blasius profile represented by the solid line. Since $F''(0) = 0.332$, the friction coefficient $c_f = \mu \partial u / \partial y|_{y=0} / (1/2\rho U^2)$, μ being the fluid viscosity, is given by

$$c_f = \frac{0.664}{\mathrm{Re}_x^{1/2}}, \quad \mathrm{Re}_x = \frac{Ux}{\nu}. \qquad (2.150)$$

The thickness of the boundary layer is $\delta \approx 5(\nu x/U)^{1/2}$.

The flow considered in the present section is incompressible. Generalization to the compressible case could be achieved using the well-known Dorodnitsyn–Howarth transformation [2.21].

In the incompressible case the solution of the corresponding thermal problem should be sought in the form

$$\theta(\eta) = \frac{T(\eta) - T_w}{T_\infty - T_w}. \qquad (2.151)$$

It yields

$$\theta(\eta) = \frac{\int_0^\eta \exp\left[-\frac{\mathrm{Pr}}{2}\int_0^\zeta F(\xi)\,\mathrm{d}\xi\right]\mathrm{d}\zeta}{\int_0^\infty \exp\left[-\frac{\mathrm{Pr}}{2}\int_0^\zeta F(\xi)\,\mathrm{d}\xi\right]\mathrm{d}\zeta}, \qquad (2.152)$$

where $\mathrm{Pr} = \nu/\alpha$ is the Prandtl number.

The dimensionless heat transfer coefficient h is nothing but the local Nusselt number Nu_x

$$\mathrm{Nu}_x = \frac{hx}{k} = \theta'(0)\mathrm{Re}_x^{1/2}, \qquad (2.153)$$

where k is the thermal conductivity of the fluid. The analytical expressions approximating the results following from (2.152) read

$$\theta'(0) = \begin{cases} 0.564\mathrm{Pr}^{1/2}, & \mathrm{Pr} < 0.05, \\ 0.332\mathrm{Pr}^{1/3}, & 0.6 \leq \mathrm{Pr} \leq 10, \\ 0.339\mathrm{Pr}^{1/3}, & \mathrm{Pr} > 10. \end{cases} \quad (2.154)$$

It is emphasized that the fact that the velocity and temperature fields in the laminar boundary layer near a semi-infinite plate are self-similar could be (and has been) established without solution of the governing equations, merely by consideration of the problem as posed.

Flow and Heat Transfer in Jets

First consider a laminar axisymmetric submerged jet [2.18, 22, 23] issuing from a pointwise nozzle (Fig. 2.19). The assumption that the nozzle is pointwise excludes a length scale (the nozzle radius) from the set of given parameters and should result in self-similarity. It is a legitimate assumption dealing only with the far-field zone of the jet, where it has already spread significantly compared to the nozzle size. In the near-field zone close to the nozzle, such simplification is impossible and a fully non-self-similar velocity field should be tackled experimentally or numerically. The dynamic and thermal problems in question involve the boundary-layer equations

$$\frac{\partial u y}{\partial x} + \frac{\partial v y}{\partial y} = 0,$$

$$u \frac{\partial u}{\partial x} + v \frac{\partial u}{\partial y} = \frac{\nu}{y} \frac{\partial}{\partial y}\left(y \frac{\partial u}{\partial y}\right),$$

$$u \frac{\partial T}{\partial x} + v \frac{\partial T}{\partial y} = \frac{\alpha}{y} \frac{\partial}{\partial y}\left(y \frac{\partial T}{\partial y}\right), \quad (2.155)$$

with x and y being the axial and radial coordinates, and u and v the longitudinal and radial velocity components, respectively.

The flow is axisymmetric, the fluid at infinity is at rest with a given temperature T_∞, namely

$$y = 0, \qquad v = \frac{\partial u}{\partial y} = \frac{\partial T}{\partial y} = 0,$$
$$y = \infty, \qquad u = 0, \ T = T_\infty. \quad (2.156)$$

The first two equations (2.155) with the boundary conditions (2.156) show that the momentum flux is constant along the jet

$$2\pi \int_0^\infty \rho u^2 y \, dy = J_x = \mathrm{const}, \quad (2.157)$$

where ρ is the fluid density, and J_x is a given momentum flux.

If we assume that the nozzle exit of the jet had a uniform velocity profile $u = u_0 = \mathrm{const}$, the momentum flux is $J_x = \pi y_0^2 \rho u_0^2$, where y_0 is the nozzle radius. Since in the present case $y_0 \to 0$, we conclude that $u_0 \to \infty$, whence $J_x = \mathcal{O}(1)$.

The dynamic equations (2.155) yield a single equation for the stream function ψ ($u y = \partial \psi / \partial y$, $v y = -\partial \psi / \partial x$), which reads

$$\frac{1}{y} \frac{\partial \psi}{\partial y} \frac{\partial^2 \psi}{\partial x \partial y} - \frac{\partial \psi}{\partial x}\left(-\frac{1}{y^2}\frac{\partial \psi}{\partial y} + \frac{1}{y}\frac{\partial^2 \psi}{\partial y^2}\right)$$
$$= \nu \left(\frac{1}{y^2}\frac{\partial \psi}{\partial y} - \frac{1}{y}\frac{\partial^2 \psi}{\partial y^2} + \frac{\partial^3 \psi}{\partial y^3}\right). \quad (2.158)$$

No length scale is given, since the jet is idealized as issuing from a pointwise nozzle. Therefore, an arbitrary length scale L is again used to render x dimensionless. As the velocity scale we take some U. Then the radial coordinate y is rendered dimensionless by $L/\mathrm{Re}^{1/2}$ with $\mathrm{Re} = UL/\nu$. The integral condition (2.157) in the dimensionless form becomes

$$\int_0^\infty \bar{u}^2 \bar{y} \, d\bar{y} = 1 \quad (2.159)$$

the chosen velocity scale being

$$U = \frac{J_x}{2\pi \rho L \nu} \quad (2.160)$$

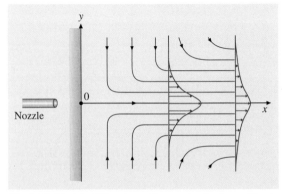

Fig. 2.20 Sketch of the streamlines and velocity profiles in the axisymmetric submerged jet

with the dimensionless parameters denoted by bars. Equation (2.160) shows that, in fact, U is not arbitrary any more, and the only arbitrary scale involved is L.

The corresponding scale of the stream function $\Psi = \nu L$ follows from its definition, while (2.158) becomes

$$\frac{1}{\overline{y}} \frac{\partial \overline{\psi}}{\partial \overline{y}} \frac{\partial^2 \overline{\psi}}{\partial \overline{x} \partial \overline{y}} + \frac{1}{\overline{y}^2} \frac{\partial \overline{\psi}}{\partial \overline{x}} \frac{\partial \overline{\psi}}{\partial \overline{y}} - \frac{1}{\overline{y}} \frac{\partial \overline{\psi}}{\partial \overline{x}} \frac{\partial^2 \overline{\psi}}{\partial \overline{y}^2}$$

$$= \frac{1}{\overline{y}^2} \frac{\partial \overline{\psi}}{\partial \overline{y}} - \frac{1}{\overline{y}} \frac{\partial^2 \overline{\psi}}{\partial \overline{y}^2} + \frac{\partial^3 \overline{\psi}}{\partial \overline{y}^3} . \qquad (2.161)$$

The solution has to be of the form $\overline{\psi} = f(\overline{x}, \overline{y})$, i.e., due to (2.160)

$$\psi = \nu L f \left[\frac{x}{L}, \frac{y}{L\nu} \left(\frac{J_x}{2\pi\rho} \right)^{1/2} \right] . \qquad (2.162)$$

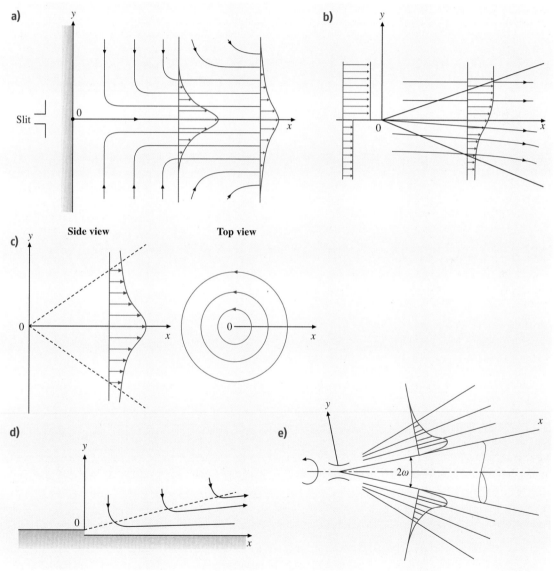

Fig. 2.21a–e Different types of jets. (**a**) Planar submerged jet; (**b**) free mixing layers; (**c**) submerged radial swirling jet; (**d**) planar wall jet; (**e**) slightly swirling jet propagating along a cone

Table 2.7 Self-similar laminar jets

Notation:
$G = \int_s \rho u \, ds$: mass flow rate; $J_x = \int_s \rho u^2 \, ds$: momentum flux; $M = \int_s \rho r u w \, ds$: moment-of-momentum flux;
$E = \frac{1}{2} \int_s \rho u^3 \, ds$: kinetic energy flux; $Q = \int_s \rho c_p u \Delta T \, ds$: excess heat flux;
$K = \int_s \rho u^2 \left(\int_s \rho u \, ds \right) ds$: first integral invariant of the wall jets; $N = \int_s \rho r u w \left(\int_s \rho u \, ds \right) ds$: second integral invariant of the wall jets;
$K_T = \int_s \Delta T^{3/2} \, ds$ – third (thermal) integral invariant of the wall jets; $K_T^1 = \int_s \rho c_p \Delta T \left(\int_s \rho u \, ds \right) ds$: same for $\Pr = 1$.
Also, $ds = (2\pi r)^k \, dy$ denotes an area element of the jet cross section, where the distance from the symmetry axis is $r \equiv y$ in the jets symmetric about the Ox axis, and $r \equiv x$ in those symmetric about the Oy axis; in the case of jets about a cone $r \equiv x \sin \omega$, 2ω being the angle at the cone apex. In the planar problems $k = 0$, in the axisymmetric ones $k = 1$.

Remarks:
In items *Differential equations*, *Boundary conditions*, etc.: I represents the dynamic problem; II the thermal one.
In items *Boundary conditions*, *General structure of the self-similar solutions*, etc.: IIa represents the thermal problem with symmetric boundary conditions for temperature; IIb that with asymmetric ones.
For wall jets: IIa represents the thermal problem for the case $T_w = T_\infty$ with T_w being a constant wall temperature; IIb that with an insulating wall; IIc that with an isothermal wall with $T_w = \mathrm{const} \neq T_\infty$.

Flow type: Planar submerged jet

Differential equations	I	$u \partial u / \partial x + v \partial u / \partial y = \nu \partial^2 u / \partial y^2$; $\partial u / \partial x + \partial v / \partial y = 0$
	II	$u \partial T / \partial x + v \partial T / \partial y = \alpha \partial^2 T / \partial y^2$
Boundary conditions	I	$v(x, 0) = \partial u / \partial y|_{y=0} = 0$; $u(x, \pm\infty) = 0$
	IIa	$\partial T / \partial y|_{y=0} = 0$, $T(x, \pm\infty) = T_\infty$
	IIb	$T(x, +\infty) = T_1$, $T(x, -\infty) = T_2$
General structure of the self-similar solutions	I	$\frac{u}{u_m} = F'(\eta)$, $u_m = Ax^\alpha$, $\eta = Byx^\beta$
	IIa	$\frac{T - T_\infty}{T_m - T_\infty} = \theta(\eta)$, $T_m - T_\infty = \Gamma x^\gamma$
	IIb	$\frac{T - T_2}{T_1 - T_2} = \theta(\eta)$
Exponents in the self-similar solutions	I	$\alpha = -\frac{1}{3}$, $\beta = -\frac{2}{3}$
	IIa	$\gamma = -\frac{1}{3}$
	IIb	$(\gamma = 0)$
Constants in the self-similar solutions	I	$A = \frac{1}{2} \sqrt[3]{\frac{3 J_x^2}{4 \rho^2 \nu}}$, $B = \frac{1}{2} \sqrt[3]{\frac{J_x}{6 \rho \nu^2}}$
	IIa	$\Gamma = \frac{Q}{c_p} \sqrt[3]{\frac{2}{9 \rho^2 \nu J_x}} \left[\int_{-\infty}^{+\infty} (F')^{\Pr+1} d\eta \right]^{-1}$
	IIb	$(\Gamma = T_1 - T_2)$
Integral invariants	I	$J_x = \int_{-\infty}^{+\infty} \rho u^2 \, dy = \mathrm{const}$
	IIa	$Q = \int_{-\infty}^{+\infty} \rho c_p u (T - T_\infty) \, dy = \mathrm{const}$
	IIb	–
Self-similar equations	I	$F''' + 2(FF'' + F'^2) = 0$
	IIa	$\theta'' + 2\Pr(F\theta' + F'\theta) = 0$
	IIb	$\theta'' + 2\Pr F\theta' = 0$
Self-similar boundary conditions	I	$F(0) = 0$, $F'(0) = 1$, $F'(\pm\infty) = 0$
	IIa	$\theta'(0) = 0$, $\theta(\pm\infty) = 0$
	IIb	$\theta(+\infty) = 1$, $\theta(-\infty) = 0$

Table 2.7 (continued)

Self-similar solutions	I	$F = \tanh\eta$, $F' = 1/\cosh^2\eta$
	IIa	$\theta(\eta) = (F')^{\text{Pr}} = 1/\cosh^{2\,\text{Pr}}\eta$
	IIb	$\theta(\eta) = \left[\int_{-\infty}^{\eta}(\cosh\eta)^{-2\,\text{Pr}}\,d\eta\right]\left[\int_{-\infty}^{+\infty}(\cosh\eta)^{-2\,\text{Pr}}\,d\eta\right]^{-1}$
Integral characteristics	I	$G = \sqrt[3]{36\rho^2\nu J_x x}$, $E = \frac{J_x}{30\mu}\sqrt[3]{6\rho\nu^2 J_x^2\frac{1}{x}}$
	IIa	–
	IIb	$Q = \frac{1}{2}c_p(T_1 - T_2)\left(\int_{-\infty}^{+\infty}F'\theta\,d\eta\right)\sqrt[3]{36\rho^2\nu J_x x}$

Flow type: free mixing layers

Differential equations	I	$u\,\partial u/\partial x + v\,\partial u/\partial y = \nu\,\partial^2 u/\partial y^2$; $\partial u/\partial x + \partial v/\partial y = 0$
	II	$u\,\partial T/\partial x + v\,\partial T/\partial y = \alpha\,\partial^2 T/\partial y^2$
Boundary conditions	I	$u(x,+\infty) = u_1$; $\partial u/\partial y|_{y=\pm\infty} = 0$; $u(x,-\infty) = u_2$
	IIa	–
	IIb	$T(x,+\infty) = T_1$, $T(x,-\infty) = T_2$
General structure of the self-similar solutions	I	$\frac{u}{u_m} = F'(\eta)$, $u_m = u_1 = \text{const}$, $\eta = Byx^\beta$
	IIa	–
	IIb	$\frac{T-T_2}{T_m-T_2} = \theta(\eta)$, $T_m = T_1 = \text{const}$, $T_2 = \text{const}$
Exponents in the self-similar solutions	I	$\alpha = 0$, $\beta = -\frac{1}{2}$
	IIa	–
	IIb	$(\gamma = 0)$
Constants in the self-similar solutions	I	$(A = u_1)$, $B = \frac{1}{2}\sqrt{\frac{u_1}{\nu}}$
	IIa	–
	IIb	$(\Gamma = T_1 - T_2)$
Integral invariants	I	–
	IIa	–
	IIb	–
Self-similar equations	I	$F''' + 2FF'' = 0$
	IIa	–
	IIb	$\theta'' + 2\,\text{Pr}\,F\theta' = 0$
Self-similar boundary conditions	I	$F'(+\infty) = 1$, $F''(+\infty) = 0$; $F'(-\infty) = m = \frac{u_2}{u_1}$, $F''(-\infty) = 0$
	IIa	–
	IIb	$\theta(+\infty) = 1$, $\theta(-\infty) = 0$
Self-similar solutions	I	$\frac{u}{u_1} = F'(\eta) = 1 + \frac{1}{2}(m-1)(1-\text{erf}\,\eta)$, $\left(\text{erf}\,\eta = \frac{2}{\sqrt{\pi}}\int_0^\eta e^{-\xi^2}\,d\xi\right)$
	IIa	–
	IIb	$\theta = \frac{T-T_2}{T_1-T_2} = \frac{1}{2}[1 + \text{erf}(\eta\sqrt{\text{Pr}})]$
Integral characteristics	I	–
	IIa	–
	IIb	–

Table 2.7 (continued)

Flow type: axisymmetric submerged jet: swirling if $w \neq 0$; no swirling if $w = 0$		
Differential equations	I	$u\partial u/\partial x + v\partial u/\partial y = \nu \frac{1}{y}\partial/\partial y\,(y\partial u/\partial y)$; $\quad \frac{\rho w^2}{y} = \partial p/\partial y$; $\quad u\partial w/\partial x + v\partial w/\partial y + \frac{vw}{y} = \nu\left[\frac{1}{y}\partial/\partial y\,(y\partial w/\partial y) - \frac{w}{y^2}\right]$; $\partial/\partial x(yu) + \partial/\partial y(yv) = 0$
	II	$u\partial T/\partial x + v\partial T/\partial y = \alpha\frac{1}{y}\partial/\partial y\,(y\partial T/\partial y)$
Boundary conditions	I	$v(x,0) = w(x,0) = \partial u/\partial y|_{y=0} = \partial w/\partial y|_{y=0} = 0$, $\quad u(x,\infty) = w(x,\infty) = 0$, $\quad p(x,\infty) = p_\infty$
	IIa	$\partial T/\partial y|_{y=0} = 0$, $\quad T(x,\infty) = T_\infty$
	IIb	–
General structure of the self-similar solutions	I	$\frac{u}{u_m} = \frac{F'(\eta)}{\eta}$, $\quad \frac{w}{w_m} = \Phi(\eta)$, $\quad \frac{p-p_\infty}{p_m-p_\infty} = P(\eta)$; $\quad u_m = Ax^\alpha$, $\quad w_m = Cx^\varepsilon$, $\quad p_m - p_\infty = \rho Dx^\delta$, $\quad \eta = Byx^\beta$
	IIa	$\frac{T-T_\infty}{T_m-T_\infty} = \theta(\eta)$, $\quad T_m - T_\infty = \Gamma x^\gamma$
	IIb	–
Exponents in the self-similar solutions	I	$\alpha = -1$, $\quad \beta = -1$, $\quad \varepsilon = -2$, $\quad \delta = -4$
	IIa	$\gamma = -1$
	IIb	–
Constants in the self-similar solutions	I	$A = \frac{3J_x}{8\pi\rho\nu}$, $\quad B = \sqrt{\frac{3J_x}{8\pi\rho\nu^2}}$, $\quad C = \frac{3M_x}{32\pi\rho\nu}\sqrt{\frac{3J_x}{8\pi\rho}}$, $\quad D = \frac{9}{2048}\frac{M_x^2 J_x}{\pi^3\rho^3\nu^4}$
	IIa	$\Gamma = \frac{(2\Pr+1)Q}{8\pi\rho\nu c_p}$
	IIb	–
Integral invariants	I	$J_x = 2\pi\int_0^\infty \rho u^2 y\,dy = \mathrm{const}$, $\quad M_x = 2\pi\int_0^\infty \rho uwy^2\,dy = \mathrm{const}$
	IIa	$Q = 2\pi\int_0^\infty \rho c_p u(T-T_\infty)y\,dy = \mathrm{const}$
	IIb	–
Self-similar equations	I	$\left(F'' - \frac{F'}{\eta}\right)' + \left(\frac{FF'}{\eta}\right)' = 0$; $\quad P' = \frac{\Phi^2}{\eta}$; $\quad \Phi'' + \frac{1+F}{\eta}\Phi' + \frac{\eta F' + F - 1}{\eta^2}\Phi = 0$
	IIa	$(\eta\theta')' + \Pr(F\theta)' = 0$
	IIb	–
Self-similar boundary conditions	I	$\left.\frac{F'}{\eta}\right|_{\eta=0} = 1$, $\quad \left.\frac{F}{\eta}\right|_{\eta=0} = 0$, $\quad \Phi(0) = 0$; $\quad \left.\frac{F'}{\eta}\right|_{\eta=\infty} = 0$, $\quad \Phi(\infty) = 0$; $\quad P(\infty) = 0$
	IIa	$\theta'(0) = 0$, $\quad \theta(\infty) = 0$
	IIb	–
Self-similar solutions	I	$F(\eta) = \frac{\frac{1}{2}\eta^2}{1+\frac{1}{8}\eta^2}$, $\quad \frac{F'}{\eta} = \frac{1}{\left(1+\frac{1}{8}\eta^2\right)^2}$, $\quad \Phi = \frac{\eta}{\left(1+\frac{1}{8}\eta^2\right)^2}$, $\quad P(\eta) = \frac{1}{\left(1+\frac{1}{8}\eta^2\right)^3}$
	IIa	$\theta(\eta) = \left(\frac{F'}{\eta}\right)^{\Pr} = \frac{1}{\left(1+\frac{1}{8}\eta^2\right)^{2\Pr}}$
	IIb	–
Integral characteristics	I	$G = \int_0^\infty 2\pi\rho uy\,dy = 8\pi\mu x$; $\quad E = \pi\int_0^\infty \rho u^3 y\,dy = \frac{\pi}{\mu}\left(\frac{3J_x}{8\pi}\right)^2 \left(\int_0^\infty \frac{F'^3}{\eta^2}\,d\eta\right)\frac{1}{x}$
	IIa	–
	IIb	–
Flow type: submerged radial swirling jet		
Differential equations	I	$u\partial u/\partial x + v\partial u/\partial y = \nu\partial^2 u/\partial y^2$; $\quad u\partial w/\partial x + v\partial w/\partial y + \frac{uw}{x} = \nu\partial^2 w/\partial y^2$; $\quad \partial/\partial x(xu) + \partial/\partial y(xv) = 0$
	II	$u\partial T/\partial x + v\partial T/\partial y = \alpha\partial^2 T/\partial y^2$

Table 2.7 (continued)

Boundary conditions	I	$v(x,0) = 0$, $\partial u/\partial y|_{y=0} = \partial w/\partial y|_{y=0} = 0$, $u(x,\pm\infty) = w(x,\pm\infty) = 0$
	IIa	$\partial T/\partial y|_{y=0} = 0$, $T(x,\pm\infty) = T_\infty$
	IIb	$T(x,+\infty) = T_1$, $T(x,-\infty) = T_2$
General structure of the self-similar solutions	I	$\frac{u}{u_m} = F'(\eta)$, $\frac{w}{w_m} = \Phi(\eta)$; $u_m = Ax^\alpha$, $w_m = Cx^\varepsilon$, $\eta = By x^\beta$
	IIa	$\frac{T-T_\infty}{T_m-T_\infty} = \theta(\eta)$, $T_m - T_\infty = \Gamma x^\gamma$
	IIb	$\frac{T-T_2}{T_1-T_2} = \theta(\eta)$
Exponents in the self-similar solutions	I	$\alpha = -1$, $\beta = -1$, $\varepsilon = -2$
	IIa	$\gamma = -1$
	IIb	$(\gamma = 0)$
Constants in the self-similar solutions	I	$A = \frac{1}{4}\sqrt[3]{\frac{9J_x^2}{2\pi^2\rho^2\nu}}$, $B = \frac{1}{2}\sqrt[3]{\frac{3J_x}{4\pi\rho\nu^2}}$, $C = \frac{3M_y}{8\pi\rho}\sqrt[3]{\frac{4\pi\rho}{3\nu J_x}}$
	IIa	$\Gamma = \frac{Q}{2\pi\rho c_p J_3}\sqrt[3]{\frac{4\pi\rho}{3\nu J_x}}$, $\left(J_3 = \int\limits_{-\infty}^{+\infty} F'\theta\,d\eta\right)$
	IIb	$(\Gamma = T_1 - T_2)$
Integral invariants	I	$J_x = 2\pi x \int\limits_{-\infty}^{+\infty} \rho u^2\,dy = \text{const}$, $M_y = 2\pi x^2 \int\limits_{-\infty}^{+\infty} \rho u w\,dy = \text{const}$
	IIa	$Q = 2\pi x \int\limits_{-\infty}^{+\infty} \rho c_p u(T - T_\infty)\,dy = \text{const}$
	IIb	–
Self-similar equations	I	$F''' + 2(FF')' = 0$, $\Phi'' + 2(F\Phi)'' = 0$
	IIa	$\theta'' + 2\Pr(F\theta)' = 0$
	IIb	$\theta'' + 2\Pr F\theta' = 0$
Self-similar boundary conditions	I	$F(0) = 0$, $F'(0) = 1$, $F'(\pm\infty) = \Phi(\pm\infty) = 0$; $\Phi'(0) = 0$
	IIa	$\theta'(0) = 0$, $\theta(\pm\infty) = 0$
	IIb	$\theta(+\infty) = 1$, $\theta(-\infty) = 0$
Self-similar solutions	I	$F = \tanh\eta$, $F' = 1/\cosh^2\eta$; $\Phi = F' = 1/\cosh^2\eta$
	IIa	$\theta(\eta) = (F')^{\Pr} = (\cosh\eta)^{-2\Pr}$
	IIb	$\theta(\eta) = \left[\int\limits_{-\infty}^{\eta}(\cosh\eta)^{-2\Pr}\,d\eta\right]\left[\int\limits_{-\infty}^{+\infty}(\cosh\eta)^{-2\Pr}\,d\eta\right]^{-1}$
Integral characteristics	I	$G = 2\pi x \int\limits_{-\infty}^{+\infty} \rho u\,dy = 2\sqrt[3]{6\pi^2\rho^2\nu J_x}\,x$, $E = \pi x \int\limits_{-\infty}^{+\infty} \rho u^3\,dy - \frac{3}{20}\frac{J_x}{\pi\mu}\sqrt[3]{\frac{4}{3}\pi\rho\nu^2 J_x^2}\frac{1}{x}$
	IIa	–
	IIb	$Q = c_p(T_1 - T_2)\sqrt[3]{6\pi^2\rho^2\nu J_x}\,x \int\limits_{-\infty}^{+\infty} F'\theta\,d\eta$

Flow type: planar wall jet

Differential equations	I	$u\partial u/\partial x + v\partial u/\partial y = \nu\partial^2 u/\partial y^2$; $\partial u/\partial x + \partial v/\partial y = 0$
	II	$u\partial T/\partial x + v\partial T/\partial y = a\partial^2 T/\partial y^2$
Boundary conditions	I	$u(x,0) = v(x,0) = 0$; $u(x,\infty) = 0$
	IIa	$T(x,0) = T_\infty$, $T(x,\infty) = T_\infty$
	IIb	$\partial T/\partial y|_{y=0} = 0$, $T(x,\infty) = T_\infty$
	IIc	$T(x,0) = T_w$, $T(x,\infty) = T_\infty$

Table 2.7 (continued)

General structure of the self-similar solutions	I	$\frac{u}{u_m} = F'(\eta)$, $u_m = Ax^\alpha$, $\eta = Byx^\beta$
	IIa	$\frac{T-T_\infty}{T_m-T_\infty} = \theta(\eta)$, $T_m - T_\infty = \Gamma x^\gamma$
	IIb	$\frac{T-T_\infty}{T_m-T_\infty} = \theta(\eta)$, $T_m - T_\infty = \Gamma x^\gamma$
	IIc	$\frac{T-T_\infty}{T_w-T_\infty} = \theta(\eta)$ ($\Gamma = T_w - T_\infty$)
Exponents in the self-similar solutions	I	$\alpha = -\frac{1}{2}$, $\beta = -\frac{3}{4}$
	IIa	$\gamma = -\frac{1}{2}$
	IIb	$\gamma = -\frac{1}{4}$
	IIc	($\gamma = 0$)
Constants in the self-similar solutions	I	$A = \sqrt{\frac{K}{4\rho^2 \nu J_1}}$, $B = \frac{1}{2}\sqrt[4]{\frac{K}{4\rho^2\nu^3 J_1}}$, $\left(J_1 = \int_0^\infty FF' d\eta\right)$
	IIa	$\Gamma(\text{Pr}) = \left[\int_0^\infty \theta^{3/2} d\eta (K_T B)^{-1}\right]^{4/3}$
	IIb	$\Gamma = \frac{Q}{c_p}\sqrt[4]{\frac{J_1}{4\rho^2 \nu K}} \left(\int_0^\infty F'\theta d\eta\right)^{-1}$
	IIc	($\Gamma = T_w - T_\infty$)
Integral invariants	I	$K = \int_0^\infty \rho u^2 \left(\int_0^y \rho u \, dy\right) dy = \int_0^\infty \rho u \left(\int_y^\infty \rho u^2 \, dy\right) dy = \text{const}$
	IIa	$K_T = \int_0^\infty (T-T_\infty)^{3/2} dy = \text{const}$
	IIb	$Q = \int_0^\infty \rho c_p u (T-T_\infty) dy = \text{const}$
	IIc	–
Self-similar equations	I	$F''' + FF'' + 2F'^2 = 0$
	IIa	$\frac{1}{\text{Pr}}\theta'' + F\theta' + 2F'\theta = 0$
	IIb	$\theta'' + \text{Pr}(F\theta)' = 0$
	IIc	$\theta'' + \text{Pr}\, F\theta' = 0$
Self-similar boundary conditions	I	$F(0) = F'(0) = 0$; $F'(\infty) = 0$
	IIa	$\theta(0) = 0$, $\theta(\infty) = 0$
	IIb	$\theta'(0) = 0$, $\theta(\infty) = 0$
	IIc	$\theta(0) = 1$, $\theta(\infty) = 0$
Self-similar solutions	I	$\eta = \frac{1}{2F_\infty} \ln \frac{F+\sqrt{FF_\infty}+F_\infty}{(\sqrt{F_\infty}-\sqrt{F})^2} + \frac{\sqrt{3}}{F_\infty}\left(\arctan\frac{2\sqrt{F}+\sqrt{F_\infty}}{\sqrt{3F_\infty}} - \arctan\frac{1}{\sqrt{3}}\right)$, $F' = \frac{2}{3}(F_\infty^{2/3} F^{1/2} - F^2)$, $F_\infty = F(\infty) = 1.7818$
	IIa	$\theta(\eta)_{\text{Pr}=1} = F'(\eta)$
	IIb	$\theta(\eta) = \exp\left[-\text{Pr}\int_0^\eta F(\xi) d\xi\right]$
	IIc	$\theta(\eta) = 1 - \left[\int_0^\eta \exp\left(-\text{Pr}\int_0^\zeta F d\xi\right) d\zeta\right]\left[\int_0^\infty \exp\left(-\text{Pr}\int_0^\zeta F d\xi\right) d\zeta\right]^{-1}$
Integral characteristics	I	$G = \int_0^\infty \rho u \, dy = F_\infty \sqrt[4]{\frac{4\rho^2 \nu K x}{J_1}}$; $J_x = \sqrt[4]{\frac{K^3}{4\rho^2 \nu J_1^3 x}} \left(\int_0^\infty F'^2 d\eta\right)$; $E = \frac{K}{4\rho J_1}\sqrt[4]{\frac{K}{4\rho^2 \nu^3 J_1 x^3}}$
	IIa	$Q = \rho c_p \frac{A\Gamma}{B} x^{-1/4} \int_0^\infty F'\theta \, d\varphi$
	IIb	–
	IIc	$Q = \sqrt[4]{\frac{4\rho^2 \nu K x}{J_1}} c_p (T_w - T_\infty) \int_0^\infty F'\theta \, d\eta$

Table 2.7 (continued)

Flow type: slightly swirling jet propagating along a cone		
Differential equations	I	$u\partial u/\partial x + v\partial u/\partial y = \nu\partial^2 u/\partial y^2$; $\rho \frac{w^2}{x}\cotan\omega = \partial p/\partial y$; $u\partial w/\partial x + v\partial w/\partial y + \frac{uw}{x} = \nu\partial^2 w/\partial y^2$; $\partial/\partial x(xu) + \partial/\partial y(xv) = 0$
	II	$u\partial T/\partial x + v\partial T/\partial y = \alpha\partial^2 T/\partial y^2$
Boundary conditions	I	$u(x,0) = v(x,0) = w(x,0) = 0$; $u(x,\infty) = w(x,\infty) = 0$; $p(x,\infty) = p_\infty$
	IIa	$T(x,0) = T_\infty$, $T(x,\infty) = T_\infty$
	IIb	$\partial T/\partial y|_{y=0} = 0$, $T(x,\infty) = T_\infty$
	IIc	$T(x,0) = T_w$, $T(x,\infty) = T_\infty$
General structure of the self-similar solutions	I	$\frac{u}{u_m} = F'(\eta)$, $\frac{w}{w_m} = \Phi(\eta)$, $\frac{p_\infty - p}{p_\infty - p_m} = P(\eta)$; $u_m = Ax^\alpha$, $w_m = Cx^\varepsilon$, $p_\infty - p_m = Dx^\delta$, $\eta = Byx^\beta$
	IIa	$\frac{T - T_\infty}{T_m - T_\infty} = \theta(\eta)$, $T_m - T_\infty = \Gamma x^\gamma$
	IIb	$\frac{T - T_\infty}{T_m - T_\infty} = \theta(\eta)$, $T_m - T_\infty = \Gamma x^\gamma$
	IIc	$\frac{T - T_\infty}{T_w - T_\infty} = \theta(\eta)$ $(\Gamma = T_w - T_\infty)$
Exponents in the self-similar solutions	I	$\alpha = -\frac{3}{2}$, $\beta = -\frac{5}{4}$, $\varepsilon = -\frac{5}{2}$, $\delta = -\frac{19}{4}$
	IIa	$\gamma = -\frac{3}{2}$
	IIb	$\gamma = -\frac{3}{4}$
	IIc	$(\gamma = 0)$
Constants in the self-similar solutions	I	$A = \sqrt{\frac{3K}{4\rho^2\nu J_1}}$, $B = \sqrt[4]{\frac{27K}{64\rho^2\nu^3 J_1}}$, $C = \frac{N}{\rho}\sqrt{\frac{3}{4\nu K J_1}}$, $D = \cotan\omega\left(\frac{N}{K}\right)^2 \sqrt[4]{\frac{3}{4\rho^2\nu}\left(\frac{K}{J_1}\right)^3}$, $\left(J_1 = \int_0^\infty FF'^2 d\eta\right)$
	IIa	$\Gamma(\text{Pr}) = \left[\int_0^\infty \theta^{3/2} d\eta (BK_T)^{-1}\right]^{-2/3}$
	IIb	$\Gamma = \frac{Q}{2\pi\rho c_p \sin\omega} \sqrt[4]{\frac{3J_1}{4\rho^2\nu K}} \left(\int_0^\infty F'\theta d\eta\right)^{-1}$
	IIc	$\Gamma = T_w - T_\infty$
Integral invariants	I	$K = \int_0^\infty \rho xu^2 \left(\int_0^y \rho xu \, dy\right) dy = \text{const}$, $N = \int_0^\infty \rho x^2 uw \left(\int_0^y \rho xu \, dy\right) dy = \text{const}$
	IIa	$K_T = \int_0^\infty (T - T_\infty)^{3/2} x \, dy = \text{const}$
	IIb	$Q = 2\pi \sin\omega \int_0^\infty \rho xu c_p(T - T_\infty) dy = \text{const}$
	IIc	–
Self-similar equations	I	$F''' + FF'' + 2F'^2 = 0$; $P' = -\Phi^2$; $\Phi'' + F\Phi' + 2F'\Phi = 0$
	IIa	$\frac{1}{\text{Pr}}\theta'' + F\theta' + 2F'\theta = 0$
	IIb	$\theta'' + \text{Pr}(F\theta)' = 0$
	IIc	$\theta'' + \text{Pr}\, F\theta' = 0$
Self-similar boundary conditions	I	$F(0) = F'(0) = \Phi(0) = 0$; $F'(\infty) = \Phi(\infty) = P(\infty) = 0$
	IIa	$\theta(0) = 0$, $\theta(\infty) = 0$
	IIb	$\theta'(0) = 0$, $\theta(\infty) = 0$
	IIc	$\theta(0) = 1$, $\theta(\infty) = 0$

Table 2.7 (continued)

Self-similar solutions	I	$\eta = \frac{1}{2F_\infty}\left[\ln\frac{F+\sqrt{FF_\infty}+F_\infty}{(\sqrt{F}-\sqrt{F_\infty})^2} + 2\sqrt{3}\left(\arctan\frac{2\sqrt{F}+\sqrt{F_\infty}}{\sqrt{3F_\infty}} - \arctan\frac{1}{\sqrt{3}}\right)\right]$; $F' = \Phi = \frac{2}{3}(F_\infty^{3/2}F^{1/2} - F^2)$, $P = -\int_\eta^\infty F'^2 \, d\eta$; $F_\infty = F(\infty) = 1.7818$
	IIa	$\theta(\eta)_{\Pr=1} = F'(\eta)$
	IIb	$\theta(\eta) = \exp\left[-\Pr\int_0^\eta F(\xi)\,d\xi\right]$
	IIc	$\theta(\eta) = 1 - \left[\int_0^\eta \exp\left(-\Pr\int_0^\zeta F\,d\xi\right)d\zeta\right]\left[\int_0^\infty \exp\left(-\Pr\int_0^\zeta F\,d\xi\right)d\zeta\right]^{-1}$
Integral characteristics	I	$G = 2\pi\rho F_\infty \sin\omega \frac{A}{B} x^{3/4}$; $J_x = 2\pi\rho\sin\omega \frac{A^2}{B} x^{-3/4} \int_0^\infty F'^2 \, d\eta$; $M = 2\pi\rho \frac{AC}{B}\sin^2\omega x^{-3/4}\int_0^\infty F'^2 \, d\eta$; $E = \pi\rho\sin\omega \frac{A^3}{B} x^{-9/4}\int_0^\infty F'^3 \, d\eta$
	IIa	$Q = 2\pi\rho c_p \frac{A\Gamma}{B} x^{-3/4}\sin\omega \int_0^\infty F'\theta\,d\eta$
	IIb	–
	IIc	$Q = 2\pi\rho c_p \sin\omega \frac{A}{B}(T_w - T_\infty)x^{3/4}\int_0^\infty F'\theta\,d\eta$

Since L is not given, only functions of a special, self-similar type, $f(\bar{x},\bar{y}) = \bar{x} f_1(\bar{y}/\bar{x})$ can be admitted. The latter automatically reduces (2.162) to the following form

$$\psi = \nu x f_1\left[\frac{y}{x}\left(\frac{J_x}{2\pi\rho\nu^2}\right)^{1/2}\right], \qquad (2.163)$$

where L is absent. Obviously, any dimensionless factor could be introduced in the self-similar variable following from (2.163). For consistency with literature, we take it as $(3/4)^{1/2}$. Then the self-similar solution corresponding to (2.163) has the form

$$\psi = \nu x F(\eta), \quad \eta = \frac{y}{x}\left(\frac{3J_x}{8\pi\rho\nu^2}\right)^{1/2}. \qquad (2.164)$$

Substitution of the self-similar solution (2.164) into (2.158), the boundary and integral conditions (2.156) and (2.157) yields the following boundary-value problem

$$\left(F'' - \frac{F'}{\eta} + \frac{FF'}{\eta}\right)' = 0,$$

$$\eta = 0, \quad F = \frac{F''\eta - F'}{\eta^2} = 0,$$

$$\eta = \infty, \quad \frac{F'}{\eta} = 0,$$

$$\int_0^\infty \left(\frac{F'}{\eta}\right)^2 \eta\,d\eta = \frac{4}{3}, \qquad (2.165)$$

which has a solution

$$F = \frac{\eta^2/2}{1+\eta^2/8}, \qquad (2.166)$$

The velocity profile rendered dimensionless by the maximal (axial) velocity $u_m(x)$ reads

$$\frac{u}{u_m} = \frac{F'}{\eta} = \frac{1}{(1+\eta^2/8)^2}, \qquad (2.167)$$

where $u_m(x) = 3J_x/(8\pi\rho\nu x)$.

It is emphasized that the fact that the flow in the far-field zone of the jet should be self-similar, can be (and has been) established without solution of the governing equations, just by consideration of the problem as posed.

The thermal field in the jet also possesses an invariant, an excess heat flux along the jet $Q = \text{const}$, namely

$$2\pi\rho c_p \int_0^\infty u(T-T_\infty)y\,dy = Q, \qquad (2.168)$$

where c_p is the specific heat at constant pressure. The corresponding self-similar solution for the temperature field has the form

$$\theta(\eta) = \frac{T-T_\infty}{T_m-T_\infty} = \frac{1}{(1+\eta^2/8)^{2\Pr}}, \qquad (2.169)$$

where the maximal (axial) excess temperature $T_m - T_\infty$ is given by

$$T_m - T_\infty = \frac{(2\Pr+1)Q}{8\pi\rho\nu c_p x}. \qquad (2.170)$$

Table 2.8 Self-similar turbulent jets

Flow type: planar submerged jet			
Differential equations	I		$u\partial u/\partial x + v\partial u/\partial y = \partial/\partial y\,(\nu_T \partial u/\partial y)$; $\quad \partial u/\partial x + \partial v/\partial y = 0$, $\quad (\nu_T = KAx^{\alpha-\beta})$
	II		$u\partial T/\partial x + v\partial T/\partial y = \partial/\partial y\,(\alpha_T \partial T/\partial y)$, $\quad (\alpha_T = K_q Ax^{\alpha-\beta})$
Boundary conditions	I		$v(x,0) = 0$, $\quad \partial u/\partial y|_{y=0} = 0$; $\quad u(x, \pm\infty) = 0$
	IIa		$\partial T/\partial y|_{y=0} = 0$, $\quad T(x, \pm\infty) = T_\infty$
	IIb		$T(x, +\infty) = T_1$, $\quad T(x, -\infty) = T_2$
General structure of the self-similar solutions	I		$\frac{u}{u_m} = F'(\eta)$, $\quad u_m = Ax^\alpha$, $\quad \eta = By x^\beta \quad (B \equiv \frac{1}{a})$
	IIa		$\frac{T - T_\infty}{T_m - T_\infty} = \theta(\eta)$, $\quad T_m - T_\infty = \Gamma x^\gamma$
	IIb		$\frac{T - T_2}{T_1 - T_2} = \theta(\eta) \quad (\Gamma = T_1 - T_2)$
Exponents in the self-similar solutions	I		$\alpha = -\frac{1}{2}$, $\quad \beta = -1$
	IIa		$\gamma = -\frac{1}{2}$
	IIb		$(\gamma = 0)$
Constants in the self-similar solutions	I		$A = \sqrt{\frac{3 J_x}{8\rho\sqrt{K}}}$, $\quad B = \frac{1}{2\sqrt{K}}$
	IIa		$\Gamma = \frac{Q}{c_p}\sqrt{\frac{2}{3\rho J_x \sqrt{K}}}\left(\int_{-\infty}^{+\infty} F'\theta\,d\eta\right)^{-1}$
	IIb		$\Gamma = T_1 - T_2$
Integral invariants	I		$J_x = \int_{-\infty}^{+\infty} \rho u^2\,dy = \text{const}$
	IIa		$Q = \int_{-\infty}^{+\infty} \rho c_p u(T - T_\infty)\,dy = \text{const}$
	IIb		–
Self-similar equations	I		$F''' + 2(FF')' = 0$
	IIa		$\theta'' + 2\text{Pr}_T(F\theta)' = 0$
	IIb		$\theta'' + 2\text{Pr}_T F\theta' = 0$
Self-similar coordinate $\eta_{1/2}$ corresponding to $u/u_m = 1/2$	I		$\eta_{1/2} = 0.88$
Self-similar boundary conditions	I		$F(0) = 0$, $\quad F'(0) = 1$, $\quad F'(\pm\infty) = 0$
	IIa		$\theta(0) = 1$, $\quad \theta(\pm\infty) = 0$
	IIb		$\theta(+\infty) = 1$, $\quad \theta(-\infty) = 0$
Self-similar solutions	I		$F = \tanh \eta$, $\quad F' = 1/\cosh^2 \eta$
	IIa		$\theta(\eta) = (F')^{\text{Pr}_T} = 1/\cosh^{2\text{Pr}_T}\eta$
	IIb		$\theta(\eta) = \left[\int_{-\infty}^{\eta}(\cosh\xi)^{-2\text{Pr}_T}d\xi\right]\left[\int_{-\infty}^{+\infty}(\cosh\xi)^{-2\text{Pr}_T}d\xi\right]^{-1}$
Mass flow rate (G), kinetic energy flux (E), etc., as well as turbulent kinematic viscosity (ν_T) and thermal diffusivity (α_T)	I		$G \sim x^{1/2}$, $\quad E \sim x^{1/2}$; $\quad \nu_T \sim x^{1/2}$
	IIa		$\alpha_T \sim x^{1/2}$
	IIb		$Q \sim x^{1/2}$, $\quad \alpha_T \sim x^{1/2}$
Flow type: free mixing layers			
Differential equations	I		$u\partial u/\partial x + v\partial u/\partial y = \partial/\partial y\,(\nu_T \partial u/\partial y)$; $\quad \partial u/\partial x + \partial v/\partial y = 0$, $\quad [\nu_T = K(u_1 - u_2)x^{\alpha-\beta}]$
	II		$u\partial T/\partial x + v\partial T/\partial y = \partial/\partial y\,(\alpha_T \partial T/\partial y)$, $\quad [\alpha_T = K_q(u_1 - u_2)x^{\alpha-\beta}]$

Table 2.8 (continued)

Boundary conditions	I	$u(x,+\infty)=u_1$, $\quad \partial u/\partial y\vert_{y=\pm\infty}=0$; $\quad u(x,-\infty)=u_2$
	IIa	–
	IIb	$T(x,+\infty)=T_1$, $\quad T(x,-\infty)=T_2$
General structure of the self-similar solutions	I	$\dfrac{u}{u_m}=F'(\eta)$, $\quad u_m=u_1=\text{const}$, $\quad \eta=Byx^\beta$ $\quad \left(B\equiv\dfrac{1}{a}\right)$
	IIa	–
	IIb	$\dfrac{T-T_2}{T_1-T_2}=\theta(\eta)$
Exponents in the self-similar solutions	I	$\alpha=0$, $\quad \beta=-1$
	IIa	–
	IIb	$\gamma=0$
Constants in the self-similar solutions	I	$A=u_1$, $\quad B=\dfrac{1}{a}=\dfrac{1}{\sqrt{2K(1-m)}}$, $\quad \left(m=\dfrac{u_2}{u_1}\right)$
	IIa	–
	IIb	$\Gamma=T_1-T_2$
Integral invariants	I	–
	IIa	–
	IIb	–
Self-similar equations	I	$F'''+2FF''=0$
	IIa	–
	IIb	$\theta''+2\Pr_T F\theta'=0$
Self-similar coordinate $\eta_{1/2}$ corresponding to $u/u_m=1/2$	I	$\eta_{1/2}\approx-0.33$ (for $m=0$)
Self-similar boundary conditions	I	$F'(+\infty)=1$, $\quad F''(\pm\infty)=0$; $\quad F'(-\infty)=m$
	IIa	–
	IIb	$\theta(+\infty)=1$, $\quad \theta(-\infty)=0$
Self-similar solutions	I	$F'(\eta)=\dfrac{u}{u_1}=1+\dfrac{1}{2}(m-1)\left[1-\text{erf}(\eta+\eta_0)\right]$. ($\eta_0\approx 0.33$)
	IIa	–
	IIb	$\theta=\dfrac{1}{2}\left[1+\text{erf}(\eta+\eta_0)\sqrt{\Pr_T}\right]$
Mass flow rate (G), kinetic energy flux (E), etc., as well as turbulent kinematic viscosity (ν_T) and thermal diffusivity (α_T)	I	$\nu_T\sim x$
	IIa	–
	IIb	$\alpha_T\sim x$

Flow type: axisymmetric submerged jet: swirling if $w\neq 0$; no swirling if $w=0$

Differential equations	I	$u\partial u/\partial x+v\partial u/\partial y=\dfrac{\nu_T}{y}\partial/\partial y\,(y\partial u/\partial y)$; $\quad \dfrac{\rho w^2}{y}=\partial p/\partial y$; $u\partial w/\partial x+v\partial w/\partial y+\dfrac{vw}{y}=\nu_T\left[\dfrac{1}{y}\partial/\partial y\,(y\partial w/\partial y)-\dfrac{w}{y^2}\right]$; $\quad \partial/\partial x(yu)+\partial/\partial y(yv)=0$
	II	$u\partial T/\partial x+v\partial T/\partial y=\alpha_T\dfrac{1}{y}\partial/\partial y\,(y\partial T/\partial y)$
Boundary conditions	I	$v(x,0)=w(x,0)=\partial u/\partial y\vert_{y=0}=0$, $\quad u(x,\infty)=w(x,\infty)=0$, $\quad p(x,\infty)=p_\infty$
	IIa	$\partial T/\partial y\vert_{y=0}=0$; $\quad T(x,\infty)=T_\infty$
	IIb	–
General structure of the self-similar solutions	I	$\dfrac{u}{u_m}=\dfrac{F'(\eta)}{\eta}$, $\quad \dfrac{w}{w_m}=\Phi(\eta)$, $\quad \dfrac{p-p_\infty}{p_m-p_\infty}=P(\eta)$; $u_m=Ax^\alpha$, $\quad w_m=Cx^\varepsilon$, $\quad p_m-p_\infty=\rho Dx^\delta$, $\quad \eta=Byx^\beta$
	IIa	$\dfrac{T-T_\infty}{T_m-T_\infty}=\theta(\eta)$, $\quad T_m-T_\infty=\Gamma x^\gamma$
	IIb	–

Table 2.8 (continued)

Exponents in the self-similar solutions	I	$\alpha = -1$, $\beta = -1$, $\varepsilon = -2$, $\delta = -4$
	IIa	$\gamma = -1$
	IIb	–
Constants in the self-similar solutions	I	$A = \sqrt{\frac{3J_x}{8\pi\rho K}}$, $B = \frac{1}{a} = \frac{1}{\sqrt{K}}$, $C = \frac{3M_x}{32\pi\rho K}\sqrt{\frac{8\pi\rho}{3J_x}}$, $D = \frac{M_x^2}{32\pi\rho K^2 J_x}$ $(K = a^2)$
	IIa	$\Gamma = (\mathrm{Pr_T} + \frac{1}{2})\frac{Q}{c_p}\frac{1}{\sqrt{6\pi\rho K J_x}}$
	IIb	–
Integral invariants	I	$J_x = 2\pi \int_0^\infty \rho u^2 y \, dy = \mathrm{const}$, $M_x = 2\pi \int_0^\infty \rho u w y^2 \, dy = \mathrm{const}$
	IIa	$Q = 2\pi \int_0^\infty \rho c_p u(T - T_\infty) y \, dy = \mathrm{const}$
	IIb	–
Self-similar equations	I	$\left(F'' - \frac{F'}{\eta}\right) + \left(\frac{FF'}{\eta}\right)' = 0$; $P' = \frac{\Phi^2}{\eta}$; $\Phi'' + \frac{1+F}{\eta}\Phi' + \frac{\eta F' + F - 1}{\eta^2}\Phi = 0$
	IIa	$(\eta \theta')' + \mathrm{Pr_T}(F\theta)' = 0$
	IIb	–
Self-similar coordinate $\eta_{1/2}$ corresponding to $u/u_m = 1/2$	I	$\eta_{1/2} = 1.82$
Self-similar boundary conditions	I	$\left.\frac{F'}{\eta}\right|_{\eta=0} = 1$, $\left.\frac{F}{\eta}\right|_{\eta=0} = 0$, $\Phi(0) = 0$; $\left.\frac{F'}{\eta}\right|_{\eta=\infty} = 0$, $\Phi(\infty) = P(\infty) = 0$
	IIa	$\theta'(0) = 0$, $\theta(\infty) = 0$
	IIb	–
Self-similar solutions	I	$F(\eta) = \frac{\frac{1}{2}\eta^2}{1 + \frac{1}{8}\eta^2}$; $F'(\eta) = \frac{\eta}{\left(1 + \frac{1}{8}\eta^2\right)^2}$; $\Phi = \frac{\eta}{\left(1 + \frac{1}{8}\eta^2\right)^2}$, $P(\eta) = \frac{1}{\left(1 + \frac{1}{8}\eta^2\right)^3}$
	IIa	$\theta(\eta) = \left(\frac{F'}{\eta}\right)^{\mathrm{Pr_T}} = \left(1 + \frac{1}{8}\eta^2\right)^{-2\mathrm{Pr_T}}$
	IIb	–
Mass flow rate (G), kinetic energy flux (E), etc., as well as turbulent kinematic viscosity (ν_T) and thermal diffusivity (α_T)	I	$G \sim x$, $E \sim x^{-1}$, $\nu_T = KA = \mathrm{const}$
	IIa	$\alpha_T = K_q A = \mathrm{const}$; $\mathrm{Pr_T} = \frac{K}{K_q}$
	IIb	–

Flow type: submerged radial swirling jet

Differential equations	I	$u\partial u/\partial x + v\partial u/\partial y = \partial/\partial y\,(\nu_T \partial u/\partial y)$; $u\partial w/\partial x + v\partial w/\partial y + \frac{uw}{x} = \partial/\partial y\,(\nu_T \partial w/\partial y)$; $\partial/\partial x(xu) + \partial/\partial y(xv) = 0$
	II	$u\partial T/\partial x + v\partial T/\partial y = \partial/\partial y\,(\alpha_T \partial T/\partial y)$
Boundary conditions	I	$v(x, 0) = 0$, $\partial u/\partial y|_{y=0} = \partial w/\partial y|_{y=0} = 0$, $u(x, \pm\infty) = w(x, \pm\infty) = 0$
	IIa	$\partial T/\partial y|_{y=0} = 0$, $T(x, \pm\infty) = T_\infty$
	IIb	$T(x, +\infty) = T_1$, $T(x, -\infty) = T_2$
General structure of the self-similar solutions	I	$\frac{u}{u_m} = F'(\eta)$, $\frac{w}{w_m} = \Phi(\eta)$, $u_m = Ax^\alpha$, $w_m = Cx^\varepsilon$, $\eta = Byx^\beta$, $\left(B \equiv \frac{1}{a}\right)$
	IIa	$\frac{T - T_\infty}{T_m - T_\infty} = \theta(\eta)$, $T_m - T_\infty = \Gamma x^\gamma$
	IIb	$\frac{T - T_2}{T_1 - T_2} = \theta(\eta)$
Exponents in the self-similar solutions	I	$\alpha = -1$, $\beta = -1$, $\varepsilon = -2$
	IIa	$\gamma = -1$
	IIb	$\gamma = 0$

Table 2.8 (continued)

Constants in the self-similar solutions	I	$A = \sqrt{\dfrac{3J_x}{8\pi\rho\sqrt{2K}}}$, $B = \dfrac{1}{\sqrt{2K}}$, $C = M_y\sqrt{\dfrac{3}{8\pi\rho J_x\sqrt{2K}}}$
	IIa	$\Gamma = \dfrac{Q\sqrt[4]{2}}{c_p}\dfrac{1}{\sqrt{3\pi\rho J_x\sqrt{K}}}\left(\int_{-\infty}^{+\infty} F'\theta\,d\eta\right)^{-1}$
	IIb	$(\Gamma = T_1 - T_2)$
Integral invariants	I	$J_x = 2\pi x \int_{-\infty}^{+\infty} \rho u^2\,dy = \text{const}$, $\quad M_y = 2\pi x^2 \int_{-\infty}^{+\infty} \rho uw\,dy = \text{const}$
	IIa	$Q = 2\pi x \int_{-\infty}^{+\infty} \rho c_p u(T - T_\infty)\,dy = \text{const}$
	IIb	—
Self-similar equations	I	$F''' + 2(FF')' = 0$, $\quad \Phi'' + 2(F\Phi)' = 0$
	IIa	$\theta'' + 2\text{Pr}_T(F\theta)' = 0$
	IIb	$\theta'' + 2\text{Pr}_T F\theta' = 0$
Self-similar coordinate $\eta_{1/2}$ corresponding to $u/u_m = 1/2$	I	$\eta_{1/2} = 0.88$
Self-similar boundary conditions	I	$F(0) = 0$, $\quad F'(0) = \Phi(0) = 1$, $\quad F''(\pm\infty) = \Phi(\pm\infty) = 0$
	IIa	$\theta'(0) = 0$, $\quad \theta(\pm\infty) = 0$
	IIb	$\theta(+\infty) = 1$, $\quad \theta(-\infty) = 0$
Self-similar solutions	I	$F = \tanh\eta$, $\quad F' = 1/\cosh^2\eta$; $\quad \Phi = F' = 1/\cosh^2\eta$
	IIa	$\theta(\eta) = (F')^{\text{Pr}_T} = (\cosh\eta)^{-2\text{Pr}_T}$
	IIb	$\theta(\eta) = \left[\int_{-\infty}^{\eta}(\cosh\xi)^{-2\text{Pr}_T}\,d\xi\right]\left[\int_{-\infty}^{+\infty}(\cosh\xi)^{-2\text{Pr}_T}\,d\xi\right]^{-1}$
Mass flow rate (G), kinetic energy flux (E), etc., as well as turbulent kinematic viscosity (ν_T) and thermal diffusivity (α_T)	I	$G \sim x$, $\quad E \sim x^{-1}$, $\quad \nu_T = KA = \text{const}$ $\quad \left(K = \dfrac{a^2}{2}\right)$
	IIa	$\alpha_T = K_q A = \text{const}$
	IIb	$\alpha_T = K_q A = \text{const}$ $\quad \left(\text{Pr}_T = \dfrac{K}{K_q}\right)$

Comments:
In items *Differential equations*, *Boundary conditions*, etc.: I represents the dynamic problem; II the thermal one.
In items *Boundary conditions*, *General structure of the self-similar solutions*, etc.: IIa represents the thermal problem with symmetric boundary conditions for temperature; IIb that with asymmetric ones.

Comparison of (2.167) and (2.169) shows that

$$\frac{T - T_\infty}{T_m - T_\infty} = \left(\frac{u}{u_m}\right)^{\text{Pr}}. \tag{2.171}$$

The effective dynamic radius of the jet is

$$\delta = \frac{8.49x}{(3J_x/8\pi\rho\nu^2)^{1/2}} \tag{2.172}$$

while the effective thermal radius δ_T depends on the Prandtl number, i.e.,

$$\delta_T = \frac{28.14x}{(3J_x/8\pi\rho\nu^2)^{1/2}} \quad \text{for} \quad \text{Pr} = 0.5,$$

$$\delta_T = \delta \quad \text{for} \quad \text{Pr} = 1,$$

$$\delta_T = \frac{4.16x}{(3J_x/8\pi\rho\nu^2)^{1/2}} \quad \text{for} \quad \text{Pr} = 2. \tag{2.173}$$

There are many other types of laminar jet flows with self-similar behavior in the far-field zone. Several such cases are shown schematically in Fig. 2.20. Their self-similarity is established in a similar manner to the case of the axisymmetric submerged jet considered above. The results are compiled in Table 2.7 with those for the axisymmetric submerged jet considered above being included as a particular case without swirling, i.e., where the angular velocity component about the jet axis $w = 0$. (When swirling is present, $w \neq 0$, a radial pressure (p) distribution forms in the jet.) It is emphasized that boundary-layer theory does not account for the effect of vertical walls shown in Figs. 2.19, 2.20a on laminar jets emerging from them. The corresponding minor corrections were discussed in the literature.

All the self-similar solutions discussed in the present section hold in the far-field zones. Numerical calculations and measurements show that the corresponding non-self-similar flow structures existing in the near-field zones close to real nozzles always tend to the self-similar ones as x increases and becomes much larger than the nozzle size y_0.

The flows considered in the present section are incompressible. Generalizations to compressible cases can be realized with aid of the Dorodnitsyn–Howarth transformation [2.21, 23].

Turbulent jets also tend to a self-similar behavior sufficiently far from the nozzle (at $x \gtrsim 30y_0$, where y_0 is the nozzle size). The kinematic eddy viscosity ν_T and the turbulent thermal diffusivity α_T in the free boundary layers (in distinction from the near-wall ones) are described rather accurately within the framework of Prandtl's second semi-empirical theory of turbulence. Both ν_T and α_T are given by power laws as per

$$\nu_T = KAx^\Omega, \quad \alpha_T = K_q Ax^\Omega, \tag{2.174}$$

where K and K_q are dimensionless empirical constants, x is the axial coordinate in the jet, A is a dimensional constant (Table 2.8). A and the exponent Ω depend on the specific type of jet. The turbulent Prandtl number $\text{Pr}_T = K/K_q$ has a value close to 0.75 for fluids whose molecular Prandtl numbers span the whole range from liquid metals to oils ($\text{Pr} \approx 10^{-2}$–$10^3$).

The corresponding self-similar solutions for free (not near-wall) turbulent jets and mixing layers resemble those for laminar jets. These solutions are presented in Table 2.8, the notation following that of Table 2.7. The only unknown empirical constant K is related to a (via B), which is thus the only empirical constant to be found by comparing the result of Table 2.8 with experimental. For example, for planar submerged jets $a \approx 0.1$–0.12, for mixing layers $a \approx 0.09$–0.12 (for $m = u_2/u_1 = 0$), and for axisymmetric submerged jets $a \approx 0.045$–0.055.

In practice, self-similar solutions can be used even at distance, $x < 30y_0$. However, when a finite nozzle is replaced by a pointwise one (which, in fact, is done in such cases) a jet polar distance should be introduced [2.24]. This means that in all the self-similar solutions x should be replaced by $x - x_0$, where an appropriate value of x_0 corresponds to the polar distance.

Boundary Layers in Natural Convection

Consider briefly laminar natural convection in the boundary layer of a hot vertical wall (Fig. 2.21). If the temperature difference between wall and fluid far from it is not too large, say $T_w - T_\infty = \mathcal{O}(10\,°\mathrm{C})$, the flow and heat transfer can be described by the following fully coupled dynamic and thermal boundary layer equations [2.16, 25]

$$\frac{\partial u}{\partial x} + \frac{\partial v}{\partial y} = 0,$$

$$u\frac{\partial u}{\partial x} + v\frac{\partial u}{\partial y} = \beta g(T_w - T_\infty)\theta + \nu\frac{\partial^2 u}{\partial y^2},$$

$$u\frac{\partial \theta}{\partial x} + v\frac{\partial \theta}{\partial y} = \alpha\frac{\partial^2 \theta}{\partial y^2}, \tag{2.175}$$

where β is the thermal expansion coefficient with dimensionality $1/K$, and g is the gravity acceleration; the normalized excess temperature $\theta = (T - T_\infty)/(T_w - T_\infty)$. The solutions of (2.175) are subject to the following boundary conditions

$$y = 0, \quad u = v = 0, \quad \theta = 1,$$
$$y = \infty, \quad u = 0, \quad \theta = 0. \tag{2.176}$$

In the case of a semi-infinite plate a length scale for x is not given, and can be taken arbitrarily as some L, since the scale $(\nu^2/g)^{1/3}$ is too small to be taken as a natural scale of the problem ($\approx 10^{-4}$ m). A velocity scale is not given either, but can be expressed via L as $U = [L\beta g(T_w - T_\infty)]^{1/2}$. The length scale in the y-direction, as usual in boundary layers, is

$$\delta = \frac{L}{(LU/\nu)^{1/2}} = \left[\frac{L\nu^2}{\beta g(T_w - T_\infty)}\right]^{1/4} \sim L^{1/4}. \tag{2.177}$$

The fact that an arbitrary L should disappear, similarly to the case considered in the section on *Flow and Heat*

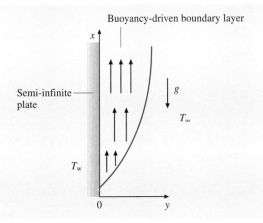

Fig. 2.22 Laminar boundary layer near vertical semi-infinite plate. The plate is kept isothermal at a temperature $T_w > T_\infty$

Transfer in Laminar Boundary Layers Near a Flat Wall in Sect. 2.3.4, leads to the self-similar solution of the following form

$$\psi = 4\nu C x^{3/4} F(\eta), \qquad \eta = C \frac{y}{x^{1/4}},$$

$$C = \left[\frac{\beta g (T_w - T_\infty)}{4\nu^2}\right]^{1/4}, \qquad \theta = \theta(\eta), \qquad (2.178)$$

where ψ is the stream function ($u = \partial\psi/\partial y$, $v = -\partial\psi/\partial x$).

As before, the self-similarity can be (and has been) established without recourse to the solutions of the governing equations, just from the given data. However, if a detailed flow and thermal structure is needed, the functions $F(\eta)$ and $\theta(\eta)$ should be found, either using the self-similar equations following from (2.175)

$$F''' + 3FF'' - 2F'^2 + \theta = 0,$$
$$\theta'' + 3\Pr F\theta' = 0,$$
$$F(0) = F'(0) = 0, \quad \theta(0) = 1,$$
$$F'(\infty) = \theta(\infty) = 0 \qquad (2.179)$$

Table 2.9 The function $H(\Pr)$

Pr	H(Pr)
Pr → 0	0.600 Pr$^{1/2}$
0.01	0.0570
0.72	0.357
1.0	0.401
2.0	0.507
5.0	0.675
7.0	0.754
10	0.826
10^2	1.55
10^3	2.80
10^4	5.01
Pr → ∞	0.503 Pr$^{1/4}$

or experimentally. (The fact that the flow is self-similar allows one to make measurements in a single cross section of the boundary layer.) Equation (2.179) were solved numerically; the results are shown in Fig. 2.22 where the vertical velocity distribution across the boundary layer [$\tilde{u}(\eta) = u/(4\nu C^2 x^{1/2}) = F'(\eta)$] and

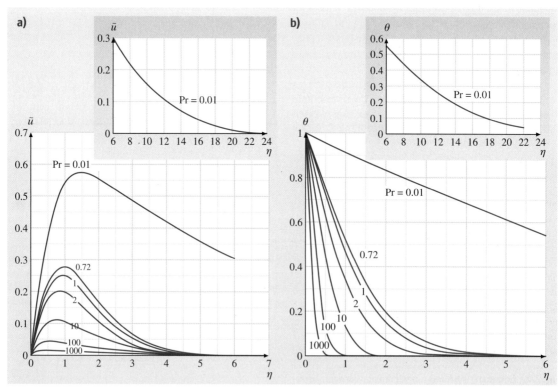

Fig. 2.23a,b Vertical velocity and temperature variation across the boundary layer for flow over an isothermal, vertical wall: **(a)** Vertical velocity, **(b)** temperature

the temperature $\theta(\eta)$ are plotted for a range of the Prandtl number. The local heat transfer coefficient $h(x)$ is calculated accordingly. Its normalized form, the Nusselt number $\text{Nu} = kx/k$ where k is the fluid thermal conductivity, is given by

$$\text{Nu} = H(\text{Pr})\text{Gr}_x^{1/4}. \tag{2.180}$$

The Grashof number is $\text{Gr}_x = \beta g x^3(T_w - T_\infty)/\nu^2$, and the function $H(\text{Pr})$ is given in Table 2.9.

Planar and axisymmetric plumes rising vertically above a heated horizontal cylinder or a sphere with a constant heat release also manifest self-similarity of the velocity and temperature distributions in the far-field zone. In a sense, they are kindred to the submerged jets. The corresponding self-similar solutions can be found in [2.25, 26].

2.3.5 Gas Dynamics: Strong Explosion

Gas dynamics also provides us with examples of important self-similar solutions. One such example is a strong explosion of a charge in a gas [2.16, 19, 27–29]. The gas is assumed to be compressible and inviscid and its motion after the explosion spherically symmetric. The corresponding set of gas dynamics equations reads

$$\frac{\partial v}{\partial t} + v\frac{\partial v}{\partial r} + \frac{1}{\rho}\frac{\partial p}{\partial r} = 0, \tag{2.181}$$

$$\frac{\partial \rho}{\partial t} + \frac{\partial \rho v}{\partial r} + 2\frac{\rho v}{r} = 0, \tag{2.182}$$

$$\frac{\partial}{\partial t}\left(\frac{p}{\rho^\gamma}\right) + v\frac{\partial}{\partial r}\left(\frac{p}{\rho^\gamma}\right) = 0. \tag{2.183}$$

Equation (2.181) represents the momentum balance, (2.182) is the continuity equation, and (2.183) the energy equation, r being the radial (spherical) coordinate, t the time, v the radial velocity of gas, ρ the density, p the pressure and γ the ratio of the specific heat at constant pressure to the specific heat at constant volume. At $t=0$ the explosive instantaneously releases an energy E_0 of dimensionality $[E_0] = J$ and a spherical shock wave propagates outwards, subdividing the gas into an infinite region ahead of the shock wave with unperturbed constant values of the pressure and density p_1 and ρ_1, and a region behind the shock wave where the flow is described by (2.181–2.183). The standard jump conditions should be satisfied at the shock front.

To make the gas flow self-similar, we have to assume that the explosion is pointwise. In other words, the charge is considered to be negligibly small compared to the relevant distances r. This, however, does not mean that no length scale is involved, since one can be constructed as $\ell = (E/p_1)^{1/3}$. However, in megaexplosions, (e.g., nuclear ones) pressurization by the shock wave is typically so strong that the pressure ahead of the shock wave p_1 can be neglected. The length scale ℓ is then lost and the problem becomes self-similar. An arbitrary length scale L can then be used, while the corresponding time scale becomes $(\rho_1 L^5/E)^{1/2}$. When the radial coordinate r and time are rendered dimensionless by these scales, $\bar{r} = r/L$, $\bar{t} = t/(\rho_1 L^5/E)^{1/2}$, the current position of the shock wave r_2, the pressure, gas velocity and density distributions behind it can be presented in the following form

$$r_2 = L f_1(\bar{t}, \gamma), \tag{2.184}$$

$$p = \frac{E}{L^3} f_2(\bar{r}, \bar{t}, \gamma), \tag{2.185}$$

$$v = \left(\frac{E}{L^3 \rho_1}\right)^{1/2} f_3(\bar{r}, \bar{t}, \gamma), \tag{2.186}$$

$$\rho = \rho_1 f_4(\bar{r}, \bar{t}, \gamma). \tag{2.187}$$

Since the final results should not contain L, the dimensionless functions $f_1 - f_4$ in (2.184–2.187) require the form

$$f_1(\bar{t}) = F_1(\gamma)\bar{t}^{2/5}, \tag{2.188}$$

$$f_2(\bar{r}, \bar{t}, \gamma) = \frac{1}{\bar{t}^{6/5}} \tilde{F}_2\left(\frac{\bar{r}}{\bar{t}^{2/5}}, \gamma\right), \tag{2.189}$$

$$f_3(\bar{r}, \bar{t}, \gamma) = \frac{1}{\bar{t}^{3/5}} \tilde{F}_3\left(\frac{\bar{r}}{\bar{t}^{2/5}}, \gamma\right), \tag{2.190}$$

$$f_4(\bar{r}, \bar{t}, \gamma) = \tilde{F}_4\left(\frac{\bar{r}}{\bar{t}^{2/5}}, \gamma\right). \tag{2.191}$$

From (2.184) and (2.188) we see that

$$\frac{r}{r_2} = \frac{\bar{r}}{F_1(\gamma)\bar{t}^{2/5}}, \tag{2.192}$$

which shows that the ratio r/r_2 defines, in fact, the self-similar variable of the distributions (2.189–2.191). Namely, we take

$$\eta = \frac{\bar{r}}{\bar{t}^{2/5} F_1(\gamma)} = \frac{r \rho_1^{1/5}}{t^{2/5} E^{1/5}} \frac{1}{F_1(\gamma)} \tag{2.193}$$

and present (2.189–2.191) as

$$p = \frac{E^{2/5} \rho_1^{3/5}}{t^{6/5}} F_2^*(\eta, \gamma), \tag{2.194}$$

$$v = \frac{E^{1/5}}{\rho_1^{1/5} t^{3/5}} F_3^*(\eta, \gamma), \tag{2.195}$$

$$\rho = \rho_1 F_4^*(\eta, \gamma). \tag{2.196}$$

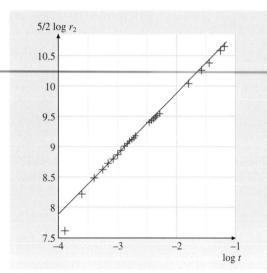

Fig. 2.24 Experimental results, shown by *crosses*, lie on a line inclined at 45° to the coordinate axes which confirms the scaling of (2.203)

It can be shown [2.19] that the pressure, gas velocity and density at the shock wave front are given by

$$p_2 = \frac{8\rho_1}{25(\gamma+1)} \left(\frac{E}{\rho_1}\right)^{2/5} \frac{1}{t^{6/5}}, \quad (2.197)$$

$$v_2 = \frac{4}{5} \frac{1}{(\gamma+1)} \left(\frac{E}{\rho_1}\right)^{1/5} \frac{1}{t^{3/5}}, \quad (2.198)$$

$$\rho_2 = \frac{\gamma+1}{\gamma-1}\rho_1. \quad (2.199)$$

Equations (2.197–2.199) enable us to present (2.194–2.196) in the following self-similar form

$$\frac{p}{p_2} = F_2(\eta, \gamma), \quad (2.200)$$

$$\frac{v}{v_2} = F_3(\eta, \gamma), \quad (2.201)$$

$$\frac{\rho}{\rho_2} = F_4(\eta, \gamma) \quad (2.202)$$

and according to (2.184) and (2.188) the position of shock wave is given by

$$r_2 = F_1(\gamma)\left(\frac{E}{\rho_1}\right)^{1/5} t^{2/5}. \quad (2.203)$$

The scaling $r_2 \sim t^{2/5}$ or $(5/2)\log r_2 \sim \log t$ is, indeed, supported by the experimental data (Fig. 2.23). The functions F_2–F_4 from (2.200-2.202) were found analytically [2.19, 27, 29]. The results are shown in Fig. 2.24 and in Table 2.10 for the case of $\gamma = 1.4$ (air). The corresponding temperature distribution, rendered dimensionless by that at the front of the shock wave T_2, can be found as $T/T_2 = (p/\rho)/(p_2/\rho_2)$ (where the gas is assumed to be ideal).

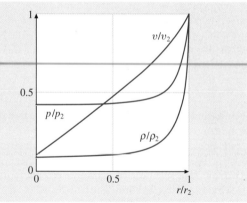

Fig. 2.25 Self-similar distribution of pressure, gas velocity and density behind the shock wave front

2.3.6 Free-Surface Flows

As an example of self-similar behavior arising in free-surface flows, consider patterns of capillary waves propagating over the free surface of a thin liquid film

Table 2.10 Self-similar distributions of pressure, gas velocity and density behind the shock wave front

$\frac{r}{r_2}$	$\frac{p}{p_2}$	$\frac{v}{v_2}$	$\frac{\rho}{\rho_2}$
1	1	1	1
0.9913	0.9109	0.9814	0.8379
0.9773	0.7993	0.9529	0.6457
0.9622	0.7078	0.9237	0.4978
0.9342	0.5923	0.8744	0.3241
0.9080	0.5241	0.8335	0.2279
0.8747	0.4674	0.7872	0.1509
0.8359	0.4272	0.7397	0.0967
0.7950	0.4021	0.6952	0.0621
0.7493	0.3856	0.6496	0.0379
0.6788	0.3732	0.5844	0.0174
0.5794	0.3672	0.4971	0.0052
0.4560	0.3656	0.3909	0.0009
0.3600	0.3655	0.3086	0.0002
0.2960	0.3655	0.2538	0.0000
0.2000	0.3655	0.1714	0.0000
0.1040	0.3655	0.0892	0.0000
0.0000	0.3655	0.0000	0.0000

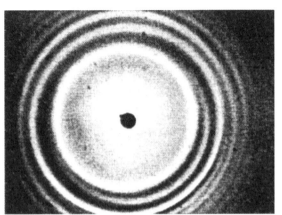

Fig. 2.26 Top view of a pattern of capillary waves taken 5 ms after impact of a copper stick onto an ethanol film of 1.4 mm thickness. The whole picture covers an area of about 25 mm × 35 mm (courtesy of Cambridge University Press)

from the point where it was impacted normally by a tiny droplet or a stick (Fig. 2.25) [2.30]. For scales of the order of several mm the gravity effect on the waves is negligibly small, and for time scales of the order of several ms viscosity effects can also be neglected. The dynamics of the wave propagation is then governed by the following set of quasi-one-dimensional equations [2.30]

$$\frac{\partial rh}{\partial t} + \frac{\partial rhV}{\partial r} = 0, \qquad (2.204)$$

$$\rho h \left(\frac{\partial V}{\partial t} + V \frac{\partial V}{\partial r} \right) = \sigma \frac{\partial}{\partial r} \left(h \frac{\partial^2 h}{\partial r^2} \right). \qquad (2.205)$$

Equation (2.204) is the continuity equation and (2.205) represents the momentum balance, which describes the competition of the inertial forces and surface tension. The waves are assumed to be axisymmetric and propagating outwards along the radial coordinate r. The fluid velocity is denoted by V, the film thickness by h, the density by ρ, the surface tension coefficient by σ, and time by t. If we consider the waves as small perturbations propagating over a liquid layer initially at rest, and of unperturbed thickness h_0, then, linearizing (2.204) and (2.205) for small perturbations of the film thickness $\chi(r, t)$, such that

$$h = h_0[1 + \chi(r, t)], \qquad (2.206)$$

we obtain the following equation for χ

$$\frac{\partial^2 \chi}{\partial t^2} + \frac{a^2}{r} \frac{\partial}{\partial r} \left(r \frac{\partial^3 \chi}{\partial r^3} \right) = 0. \qquad (2.207)$$

The parameter a combines all the given physical parameters as per

$$a = \left(\frac{\sigma h_0}{\rho} \right)^{1/2} \qquad (2.208)$$

and fully determines the wave propagation: $[a] = \mathrm{m}^2/\mathrm{s}$.

The initial impactor is assumed to be pointwise, which means that we are considering the wave pattern at distances r much larger than its diameter. Since no length scale is given, an arbitrary one, L can be used. The dimensionless wave pattern should be of the form

$$\chi = f\left(\frac{r}{L}, \frac{t}{L^2/a} \right). \qquad (2.209)$$

However, in the final result an arbitrary length scale L should automatically disappear, which means that the function f in (2.209) should depend not on its two variables separately but on their specific combination, namely,

$$\eta = \frac{r/L}{\left[t/(L^2/a) \right]^{1/2}} = \frac{r}{(at)^{1/2}}. \qquad (2.210)$$

The corresponding self-similar wave pattern $\chi = F(\eta)$ should satisfy (2.207), which yields

$$F'''' + \frac{1}{\eta} F''' + \frac{\eta^2}{4} F'' + \frac{3}{4} \eta F' = 0. \qquad (2.211)$$

The solution should correspond to the following initial perturbation

$$t = 0, \quad \chi = 4\pi S \frac{\delta(\eta)}{2\pi \eta}, \qquad (2.212)$$

where S (dimensionless) corresponds to the impact intensity, and $\delta(\eta)$ is the delta function.

Using the solution of (2.211), we find that for large η the axisymmetric waves are described by the following expression

$$\chi = \frac{2S}{\Gamma\left(\frac{1}{4}\right)} \frac{1}{\eta^{3/2}}$$

$$\times \left[\cos\left(\frac{1}{4}\eta^2 + \frac{1}{8}\pi \right) + \sin\left(\frac{1}{4}\eta^2 + \frac{1}{8}\pi \right) \right],$$

$$(2.213)$$

where the gamma function of the argument $1/4$ equals to 3.6256. The corresponding fluid velocity is given by

$$V = \left(\frac{a}{t}\right)^{1/2} \frac{S}{\Gamma\left(\frac{1}{4}\right)} \frac{1}{\eta^{1/2}}$$
$$\times \left[\cos\left(\frac{1}{4}\eta^2 + \frac{1}{8}\pi\right) + \sin\left(\frac{1}{4}\eta^2 + \frac{1}{8}\pi\right)\right].$$
(2.214)

Equation (2.213) with $\eta = r/(at)^{1/2}$ agrees pretty well with experimental data. Moreover, it shows that wave patterns similar to those of Fig. 2.25 shot at different time moments can be collapsed onto a single self-similar pattern given by (2.213), which indeed happens [2.30].

It is emphasized that self-similarity of the wave pattern in the far-field zone has been established without solution of the governing equation. The detailed wave structure has been found afterwards, theoretically and experimentally.

References

2.1 S. Moreau, D. Neal, Y. Khalighi, M. Wang, G. Iaccarino: Validation of unstructured-mesh LES of the trailing edge flow and noise of a Controlled-Diffusion airfoil, *Studying Turbulence Using Numerical Simulation Databases–XI: Proceedings of the Summer Programm 2006*, Center for Turbulence Research, Stanford University
2.2 M.C. Potter, J.F. Foss: *Fluid Mechanics* (Great Lakes, Wildwood 1982) p. 448
2.3 P.A. Thompson: *Compressible-Fluid Dynamics* (McGraw-Hill, New York 1972)
2.4 S.C. Morris, J.F. Foss: Turbulent boundary layer to single-stream shear layer: the transition region, J. Fluid Mech. **494**, 187–221 (2003)
2.5 T. von Kármán: Über laminare und turbulente Reibung, Z. Angew. Math. Mech. **1**, 233–252 (1921)
2.6 F.M. White: *Viscous Fluid Flow*, 2nd edn. (McGraw-Hill, New York 1991)
2.7 R. Kobayashi, Y. Kohama, C. Takamadate: Spiral vortices in boundary layer transition regime on a rotating disk, Acta Mech. **25**, 71–82 (1980)
2.8 P.W. Bridgman: *Dimensional Analysis* (Yale Univ. Press, New Haven 1922)
2.9 R.L. Panton: *Incompressible Flow*, 3rd edn. (Wiley, New York 2005)
2.10 F.M. White: *Fluid Mechanics*, 5th edn. (McGraw-Hill, New York 2003)
2.11 Y.A. Çengel, J.M. Cimbala: *Fluid Mechanics* (McGraw-Hill, New York 2006)
2.12 R.M. Schmidt, K.R. Housen: Problem solving with dimensional analysis, The Industrial Physicist **1**, 21 (1995)
2.13 J.C. Klewicki, H. Miner: *Wall pressure structure at high Reynolds number*, Division Fluid Dynamics Meeting, Bull. Am. Phys. Soc. (APS, College Park 2002), FH 001
2.14 R.L. Panton: Review of wall turbulence described by composite expansions, Appl. Mech. Rev. **58**, 1–36 (2005)
2.15 J.M.J. den Toonder, F.T.M. Nieuwstadt: Reynolds number effects in a turbulent pipe flow for low to moderate Re, Phys. Fluids **9**(11), 3398 (1997)
2.16 L.D. Landau, E.M. Lifshitz: *Fluid Mechanics*, 2nd edn. (Pergamon, Oxford 1987)
2.17 G.I. Barenblatt: *Similarity, Self-Similarity and Intermediate Asymptotics* (Cambridge Univ. Press, Cambridge 1996)
2.18 H. Schlichting: *Boundary Layer Theory*, 8th edn. (Springer, Berlin, Heidelberg 2000)
2.19 L.I. Sedov: *Similarity and Dimensional Methods in Mechanics*, 10th edn. (CRC, Boca Raton 1993)
2.20 L. Rosenhead: *Laminar Boundary Layers* (Clarendon, Oxford 1963)
2.21 K. Stewartson: *The Theory of Laminar Boundary Layers in Compressible Fluids* (Clarendon, Oxford 1965)
2.22 S.I. Pai: *Fluid Dynamics of Jets* (van Nostrand, Toronto, New York 1954)
2.23 L.A. Vulis, V.P. Kashkarov: *Theory of Viscous Liquid Jets* (Nauka, Moscow 1965)
2.24 G.N. Abramovich: *The Theory of Turbulent Jets* (MIT, Cambridge 1963)
2.25 Y. Jaluria: *Natural Convection Heat and Mass Transfer* (Pergamon, Oxford 1980)
2.26 Y.B. Zel'dovich: Limiting laws of freely rising convective currents. In: *Selected Works of Ya.B. Zel'dovich, Chemical Physics and Hydrodynamics*, Vol. 1, ed. by J.P. Ostriker (Princeton Univ. Press, Princeton 1992)
2.27 L.I. Sedov: Propagation of strong blast waves, Prikl. Mat. Mekh. **10**(2), 241–250 (1946)
2.28 G.I. Taylor: The formation of a blast wave by a very intense explosion, Proc. Roy. Soc. London A **201**, 159–186 (1950)
2.29 J. von Neumann: The point source solution. In: *Collected Works*, Vol. VI, ed. by A.J. Taub (Pergamon, Oxford 1963)
2.30 A.L. Yarin, D.A. Weiss: Impact of drops on solid surfaces: self-similar capillary waves, and splashing as a new type of kinematic discontinuity, J. Fluid Mech. **283**, 141–173 (1995)

Part B Measurement of Primary Quantities

3 Material Properties: Measurement and Data
William A. Wakeham, Southampton, UK
Marc J. Assael, Thessaloniki, Greece
Abraham Marmur, Haifa, Israel
Joël De Coninck, Mons, Belgium
Terry D. Blake, Tring, UK
Stephanus A. Theron, Haifa, Israel
Eyal Zussman, Haifa, Israel

4 Pressure Measurement Systems
Beverley J. McKeon, Pasadena, USA
Rolf H. Engler, Göttingen, Germany

5 Velocity, Vorticity, and Mach Number
Beverley J. McKeon, Pasadena, USA
Geneviève Comte-Bellot, Ecully, France
John F. Foss, East Lansing, USA
Jerry Westerweel, Delft, The Netherlands
Fulvio Scarano, Delft, The Netherlands
Cameron Tropea, Darmstadt, Germany
James F. Meyers, Hampton, USA
Joseph W. Lee, Hampton, USA
Angelo A. Cavone, Hampton, USA
Richard Schodl, Köln, Germany
Manoochehr M. Koochesfahani, East Lansing, USA
Daniel G. Nocera, Cambridge, USA
Yiannis Andreopoulos, New York, USA
Werner J.A. Dahm, Ann Arbor, USA
John A. Mullin, Redondo Beach, USA
James M. Wallace, College Park, USA
Petar V. Vukoslavčević, Podgorica, Montenegro
Scott C. Morris, Notre Dame, USA
Eric R. Pardyjak, Salt Lake City, USA
Alvaro Cuerva, Madrid, Spain

6 Density-Based Techniques
Wolfgang Merzkirch, Essen, Germany

7 Temperature and Heat Flux
Tomasz A. Kowalewski, Warszawa, Poland
Phillip Ligrani, Oxford, UK
Andreas Dreizler, Darmstadt, Germany
Christof Schulz, Duisburg, Germany
Uwe Fey, Göttingen, Germany
Yasuhiro Egami, Göttingen, Germany

8 Force and Moment Measurement
Klaus Hufnagel, Darmstadt, Germany
Günter Schewe, Göttingen, Germany

The objective of Part B is to provide specific information to the reader on the following primary quantities: material properties Chap. 3, flow-field properties (Chap. 4 pressure, Chap. 5 velocity and vorticity, Chap. 6 spatial density variations, and Chap. 7 temperature) and forces and moments (Chap. 8). Chapter 3 is focused on providing quantitative information for the material properties, the sources of this information and the associated confidence levels for the given data. Chapters 4 through 8 provide comprehensive guidance to the reader on: i) the objectives, ii) the available equipment, iii) utilization techniques, and iv) post-processing of the primitive information for the stated quantities.

3. Material Properties: Measurement and Data

The density of a fluid is defined as the mass of the fluid per unit volume. Some methods of measuring the density of the fluid with high precision determine the variation of one quantity in this ratio when the other is fixed and a further state variable is altered. Other methods make use of the effect of the fluid density on the position or motion of a rigid body contained within it. Examples of these instruments are described in this chapter for operation over a wide range of thermodynamic states.

- 3.1 **Density** 85
  - 3.1.1 Piezometers 86
  - 3.1.2 Bellows-Type Densimeters 87
  - 3.1.3 Vibrating-Element Densimeters 88
  - 3.1.4 Buoyancy-Type Densimeters 91
  - 3.1.5 Density Reference Values 95
  - 3.1.6 Tables of Density Values 96
- 3.2 **Surface Tension and Interfacial Tension of Liquids** 96
  - 3.2.1 Surface Tension of Pure Liquids 96
  - 3.2.2 Surface Tension of Liquid Solutions . 98
  - 3.2.3 Interfacial Tension 99
  - 3.2.4 Implications of Surface and Interfacial Tension on Liquid–Fluid Systems 100
  - 3.2.5 Measurement of Surface Tension and Interfacial Tension 101
  - 3.2.6 Surface Tension Values for Liquids .. 105
- 3.3 **Contact Angle** 106
  - 3.3.1 The Equilibrium Contact Angle 106
  - 3.3.2 Dynamic Contact Angle 112
- 3.4 **Viscosity** 119
  - 3.4.1 Oscillating-Body Viscometers 119
  - 3.4.2 Vibrating Viscometers 122
  - 3.4.3 Torsional-Crystal Viscometer 124
  - 3.4.4 Capillary Viscometers 125
  - 3.4.5 Falling-Body Viscometers 127
  - 3.4.6 Viscosity Reference Values 131
  - 3.4.7 Tables of Viscosity Values 132
- 3.5 **Thermal Conductivity and Thermal Diffusivity** 133
  - 3.5.1 Transient Methods for Thermal Conductivity 134
  - 3.5.2 Steady-State Methods for Thermal Conductivity 138
  - 3.5.3 Light-Scattering Methods for Thermal Diffusivity 141
  - 3.5.4 Thermal Conductivity Reference Values 146
  - 3.5.5 Tables of Thermal Conductivity Values 147
- 3.6 **Diffusion** 147
  - 3.6.1 Diffusion in Liquids 149
  - 3.6.2 Diffusion in Gases 154
  - 3.6.3 Diffusion Reference Values 156
  - 3.6.4 Tables of Diffusion Coefficient Values 157
- 3.7 **Electric and Magnetic Parameters of Liquids and Gases** 158
  - 3.7.1 Introduction 158
  - 3.7.2 Dielectric Constant 159
  - 3.7.3 Electric Conductivity 160
  - 3.7.4 Broadband Measurement of the Conductivity and Dielectric Constant 166
- **References** 169

3.1 Density

The density ρ is defined as the mass of unit volume of a substance under prescribed conditions. The density varies with pressure and temperature, the variation with respect to both variables being much greater in gases than liquids.

The specific gravity (SG), also termed the relative density, is determined by dividing the density of the substance by the density of a standard substance obtained under the same conditions of temperature and pressure. In particular, the usual definitions are

$$\text{Liquid SG} = \frac{\text{density of liquid}}{\text{density of water}} \quad (3.1)$$

$$\text{Gas SG} = \frac{\text{density of gas}}{\text{density of air}} \quad (3.2)$$

Commonly accepted sets of reference conditions are

- normal temperature and pressure (NTP), usually taken as the temperature at 0.00 °C and a pressure of 760 mmHg, and
- standard temperature and pressure (STP), usually taken as the temperature at 15.00 °C and a pressure of 0.101325 MPa.

Despite the apparent simplicity of its definition, the accurate measurement of the density of fluids is complex and many novel techniques have been developed. Special care is required in measurements at high pressures in either phase while the hazards associated with operation in the gas phase, when the stored energy in a system is large, add to the complexity of experimental work considerably. In this chapter, modern techniques for the measurement of density will be presented, and emphasis will be given to those techniques that cover very wide temperature and pressure ranges, with low uncertainty.

3.1.1 Piezometers

Densimeters that employ magnetic-suspension or vibrating-element techniques, which will be discussed later in this chapter, now provide more-convenient methods for achieving a low level of uncertainty in the measurement of density. However, piezometers have been used extensively in the past because of their simplicity and high accuracy when used with care. Piezometers can be divided into three categories:

1. devices that measure the amount of mass or amount of substance within a fixed volume,
2. devices that determine the change in pressure effected by a change in volume, and
3. devices that utilize one or more expansions from one volume to another.

Fixed-Volume Piezometers

In fixed-volume devices, the mass or amount of substance in a known volume is measured at a temperature and pressure that are determined independently. To determine the mass of the sample the container is often weighed directly with and without the sample present. Alternatively, if the fluid is a gas, it may be allowed to expand into a much larger volume so that the final conditions are near ambient, when a relatively simple equation of state can be used to evaluate the amount of substance present.

The instrument shown in Fig. 3.1, made out of Pyrex glass, was employed for measurements of the saturated density of liquid refrigerants up to 10 MPa pressure with an estimated uncertainty of 0.3% [3.1]. It consists of a vessel connected to a capillary. The capillary is used to define the exact volume of the liquid, thus giving high resolution when filling the piezometer. Following evacuation of the piezometer, it is weighed. The sample is then introduced, degassed, and the device disconnected from the filling line. The mass of the sample was obtained by weighing again, while its volume by observing the liquid level in the tube.

Hwang and coworkers [3.2] developed a continuously weighed piezometer, based on an earlier apparatus developed by *Machado* and *Street* [3.3]. Hwang's piezometer consists of a weight measurement system, an isothermal bath with a temperature control and measurement system, a sample pressurizing system, and a high vacuum system. The sample is introduced into the piezometer vessel of known volume via a flexible capillary feed line. The mass of the piezometer plus fluid is then determined with an electronic force balance from which it is suspended at all times. In general, fixed volume devices are capable of density measurements with an uncertainty of ±0.1% or better, depending upon the fluid properties.

Fixed-volume cells have been used to great effect over a wide range of conditions. *Kubota* et al. [3.4] employed a fixed-volume cell charged with a known

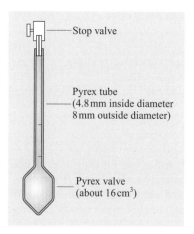

Fig. 3.1 Fixed-volume piezometer (after [3.1])

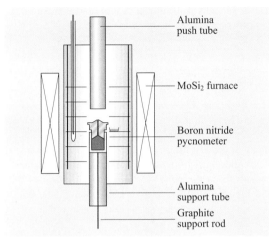

Fig. 3.2 High-temperature pycnometer, developed by *Sato* et al. [3.5]

mass of sample for the measurement of the density of refrigerants up to 100 MPa pressure. *Sato* et al. [3.5] operated a high-temperature pycnometer (Fig. 3.2) made of boron nitride, for the measurement of the density of molten silicon. In this case, the pycnometer containing a cylindrical sample of solid silicon was supported by a graphite rod and installed at the appropriate position on an alumina support tube (Fig. 3.2) inside the furnace. Then the upper part (cap) of the pycnometer had been partially inserted into the lower part during assembly. After the silicon melted, the apparatus was temporarily evacuated and the alumina push tube was lowered slowly to push the upper part of the pycnometer into position in the lower part. This caused the excess of molten silicon now in the pycnometer to overflow, leaving it exactly full and containing a defined volume. The mass of the sample is determined by subsequent weighing outside of the furnace. *Sato* et al. [3.5] reported density measurements up to 1850 K with an estimated uncertainty of better than 0.5%. However, at such high temperatures the difficulties of measurement generally mean that the discrepancies between investigators exceed their mutual quoted uncertainties so that it is not yet possible to confirm the estimate of error in this case.

Variable-Volume Piezometers

Variable-volume devices are characterized by sample cells that change in volume during the experiment. The change in volume is usually achieved using bellows (see next section) or pistons. Some older devices employed mercury as a liquid piston (e.g., *Goodwin* et al. [3.6]) but environmental and safety considerations now discourage the use of mercury in large quantities. New designs for piston devices have recently been employed [3.7]. However, the variable volume devices are generally less accurate than other methods, with uncertainties of about $\pm 1\%$, and thus are not described further here.

Expansion Piezometers

The basic principle of expansion devices is that the sample is expanded from one volume into a second volume (usually evacuated), and the ratio of the original volume to the final volume establishes the ratio of densities before ρ_0 and after expansion ρ_f with

$$\frac{\rho_0}{\rho_f} = \frac{V_a + V_b}{V_a} = r, \quad (3.3)$$

where, V_a is the volume occupied by the fluid before the expansion, $(V_a + V_b)$ is the volume after the expansion and r is the cell constant. The device may utilize either a single expansion (large r) or a series of expansions.

Single expansion devices employ volume ratios r ranging from 50 to 1000 so that a pressure near atmospheric results from the expansion. The final molar density can then be calculated from the temperature and pressure using a simple virial equation of state. Once the final density is known, the original density ρ_0 can be calculated from (3.3), or the amount of substance by multiplying the final density by the total volume of the system. Single expansion devices have been recently described by *Duarte* et al. [3.8].

The most common multiple expansion method is that developed by *Burnett* [3.9], recently successfully applied by *Stouffer* [3.10]. These techniques have the advantage that neither mass nor volume need to be measured directly. Only pressure and temperature are measured before and after the expansion from a single volume into the combination of the original volume and an additional one. Although the technique is usually not an absolute technique, the uncertainty in the measurement of the density is often better then $\pm 0.05\%$.

3.1.2 Bellows-Type Densimeters

Bellows-type densimeters are cells that contain and enclose the sample fluid. The entire cell, or at least a part of it, is a flexible bellows, which transmits the pressure on the outside of the cell to the test fluid. The pressure can be exerted and measured on the outside of the flexible cell through another pressure transmitting fluid which itself can be pressurized by a piston pump. The linear movement of the end of a bellows of constant cross section is measured to determine the compression

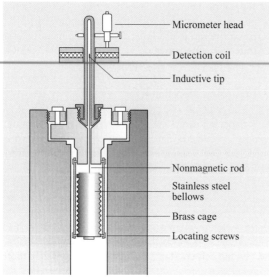

Fig. 3.3 High-pressure bellows-type pycnometer, developed by *Dymond* et al. [3.11, 12]

of the fluid in the measuring cell caused by the applied pressure.

The instrument shown in Fig. 3.3, is a bellows-type densimeter, employed by *Dymond* et al. [3.11, 12], for the measurement of the density of hydrocarbons up to 600 MPa with an estimated uncertainty of 0.1%. A known amount of liquid sample is introduced into the sealed metal bellows. In this case the movement of the bellows' end is determined using the magnetic induction between an inductive tip mounted on a rod that is itself attached to the end of the bellows and an external detector coil. Measurement of the movement of the detector coil along the tube to secure a constant induction serves to determine the position of the end of the bellows. Bellows densimeters have been employed by various investigators. *Iso* and *Uematsu* [3.13] employed such a densimeter for the measurement of the density of refrigerants up to 10 MPa pressure.

3.1.3 Vibrating-Element Densimeters

If a solid, elastic body containing (or surrounded by) a fluid is driven to oscillate, then the frequency of resonance of the solid/fluid assembly will depend upon the properties of the fluid and, in particular its density ρ. The resonant frequency will also depend upon the properties of the solids involved and, in particular, upon a modulus K of the solid oscillator of mass M, whose elastic distortion yields the oscillatory behavior. As a result, for any form of solid/fluid oscillator of this kind we can write an expression for the resonant frequency f as [3.14]

$$f = \sqrt{\frac{K}{(M+k\rho)}}, \qquad (3.4)$$

where k is characteristic of the particular arrangement of the oscillator. Evidently, if it is possible to calculate, or measure, K and k independently then measurement of the resonant frequency of the oscillation can yield the density of the test fluid.

A common feature of all types of vibrating element densimeters is the need to excite the vibration of the solid element and to observe its resonant frequency. There are two mechanisms usually employed for this purpose [3.14].

Magnetic drive of the vibrating element can be achieved by means of small coil assemblies mounted within or upon the vibrating element and forced into motion by application of electric currents in fixed driving coils. Detection of motion can be accomplished in the same way by the same or a different set of coils. If a steady drive is desired, the signals picked up by the sensors are amplified and fed back as a drive to maintain the disturbing forces on the vibrating body of the meter. This mechanism can actually be made to self-tune to the resonant frequency of the oscillator automatically if the resonance is sufficiently sharp. The main advantage of the magnetic drive and pickup systems is that they are noncontact methods; they use conventional copper windings and they are usable within the temperature range of -200 to $+200$ °C.

Alternatively, there is a wide range of piezoelectric materials that allow the direct application of an electric current to a solid so as to causes an elastic distortion. These materials demonstrate good temperature characteristics and have the advantage of being low in cost. They have a relatively high impedance, making the signal conditioning circuitry easy, unlike the circuitry for electromagnetic sensors. Piezoelectric drives are mechanically affixed to the vibrating body by adhesives. Therefore, attention must be paid to the careful placement of the mount in order to reduce the strain experienced by the piezo element owing to the thermal and pressure stresses to which the element is subjected during operation over ranges of temperature and pressure. A number of different types of densimeters have been developed that utilize vibrating elements and the most common ones are [3.14]:

1. Vibrating-tube densimeters are suitable for operation with a wide range of fluids from gases to liquids

as well as to slurries. Commercial instruments using this principle are widely available and are frequently used in research laboratories and industrial monitoring. The mode of operation is based on the transverse vibration of a single circular section tube constrained to vibrate in a single plane.
2. Vibrating-cylinder densimeters are ideal for liquified gas products or refined liquids.
3. Tuning-fork densimeters make use of the natural frequency of low-mass tuning forks. In some cases, the fluid is taken into a small chamber in which the electromechanically driven forks are situated, or in other cases the fork is inserted directly into the liquid.

In the vibrating-tube densimeter the basic measurement to be made is of the resonant frequency of the tubular oscillator filled with the fluid whose density is to be measured. The mass of fluid inside the oscillator determines the change of the resonant frequency of oscillation from that obtained when the tube is filled with another fluid or even in vacuo. Thus, by measuring this change of frequency it is possible to obtain the mass that fills the tube volume and therefore the density of the sample. In practice, the configuration of the tubular oscillator is generally chosen in the form of a 'V' or a 'U' so as to define precisely a single mode of oscillation. For that reason it is not usual to seek to perform absolute measurements with the densimeter using a fundamental theory. Instead one begins with the equation that derives from the simple phenomenological theory that arises from (3.4) so that

$$\rho = A\tau^2 - B \, , \tag{3.5}$$

where τ is the measured period of vibration for the tube when filled with a fluid of density ρ, and A and B calibration parameters of the densimeter, which may be functions of pressure and temperature. The values of A and B have to be obtained by the use of two fluids of known density, ρ_1 and ρ_2, as calibrants under the same conditions of temperature and pressure. Equation (3.5) can then be rewritten to eliminate A and B, in the form

$$\rho = \frac{(\rho_1 - \rho_2)\left(\tau^2 - \tau_2^2\right)}{\left(\tau_1^2 - \tau_2^2\right)} + \rho_2 \, , \tag{3.6}$$

where τ_1 and τ_2 are the periods of vibration of the tube when filled with calibrants 1 and 2, respectively, at the same temperature and pressure as the sample liquid. For the best vibrating tube densimeters the dependence on temperature and pressure of the two instrument constants is rather small and, if modest accuracy is sought, it is possible to treat them as constants. However, for the highest accuracy it is always best to use the full form of (3.6) at each thermodynamic state.

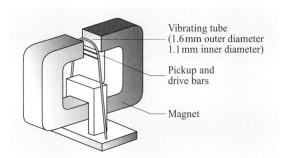

Fig. 3.4 Vibrating-tube densimeter, developed by *Hynek* et al. [3.15]

Vibrating-tube densimeters have been employed by a large number of investigators. A very small sample of the applications is listed here. *Hynek* et al. [3.15] constructed a vibrating-tube densimeter (Fig. 3.4) for measurements with corrosive solutions at temperatures up to 723 K and pressures up to 40 MPa, with an uncertainty of about 0.5%. A similar vibrating U-tube was developed by *Blencoe* et al. [3.17] for measurements of the density up to 200 MPa. Similar devices have been employed for the measurement of the density of refrigerants, for example, *Defibaugh* and *Morrison* [3.18] employed such an instrument for the measurement of the density of refrigerant R22 from 253 to 373 K and up to 6.2 MPa with an estimated uncertainty of 0.05%. *Sousa* et al. [3.19] measured the density of refrigerant R142b from 293 to 403 K and pressures up to 17 MPa.

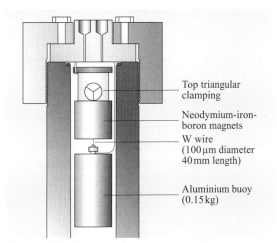

Fig. 3.5 Vibrating-wire densimeter, developed by *Padua* et al. [3.16]

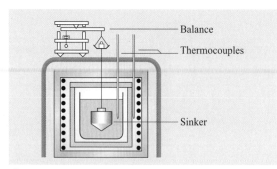

Fig. 3.6 Buoyancy-type densimeter

Their estimated uncertainty was 1% for the liquid phase and 3% for the vapor phase.

A special category of vibrating-element densimeters is the vibrating-wire densimeter. Such an instrument is shown in Fig. 3.5. In this case, the vibrating element comprises a thin metallic wire of radius R and length $2L$ with the top end fixed and a solid cylindrical weight of volume V_w suspended from the lower end. The wire is placed in a uniform permanent magnetic field generated by two rare-earth magnets. The motion of the wire is driven by the application of an oscillatory current in the wire. Variation of the frequency of the driving current with the aid of an impedance analyzer, or in some other way, can be used to find the resonant frequency. The presence of the sample fluid around the vibrating wire contributes to the change in the resonant frequency

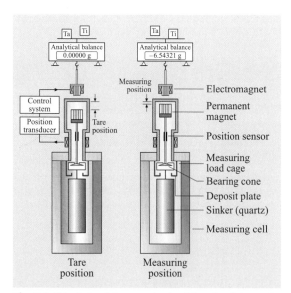

Fig. 3.7 Principle of the new single-sinker densimeter, developed by *Klimeck* et al. [3.20]

from that observed when the wire vibrates in vacuo. The largest single effect of the fluid is a result of the buoyancy force acting on the suspended weight. However, there is a minor effect caused by the fluid flow around the vibrating wire. According to the theory of such a device, treated in detail elsewhere [3.21, 22], the resonant frequency of the wire can be expressed in terms of the density of the fluid as well as its viscosity, together with the radius of the wire, the density of its material, the volume of the suspended weight and the density of the material from which the weight is constructed. Many of these quantities are amenable to direct, independent measurement but some are not. In particular, the measurements of the wire radius and of its density are usually not accomplished with sufficient accuracy by independent means. For that reason these two quantities are usually determined by calibration using two fluids of known density and viscosity. This needs to be done only at one thermodynamic state since changes with temperature and pressure are readily evaluated for other conditions.

The first attempts to build densimeters based on the vibrating-wire sensor and designed according to the theory of the vibrating wire took place in the group of *Wakeham* at Imperial College, London, around 1990 [3.23], where the most recent version of the hydrodynamic model has also been formulated. A second version of the vibrating-wire densimeter was built at the Instituto Superior Tecnico, Lisbon, initially as a pure densimeter [3.16]. This design is shown in Fig. 3.5. A 100 μm-diameter tungsten wire was employed, while a top triangular clamping ensured the predominance of the first vibrational mode of motion. The uncertainty in the density measurements achieved with this instrument is ±0.05% at the 95% confi-

Table 3.1 Reference values for the density ($kg\,m^{-3}$) of water and steam

| | Pressure (MPa) | | | |
T (°C)	0.1	1	5	10
0.1	999.8	1000	1002	1005
10	999.7	1000	1002	1004
25	997.0	997.5	999.2	1001
50	988.0	988.4	977.0	992.3
75	974.8	975.1	977.0	979.2
100	0.5897	958.8	960.6	962.9
200	0.4603	4.854	867.3	870.9
300	0.3790	3.876	22.05	715.3
500	0.2805	2.824	14.58	30.48
700	0.2227	2.233	11.30	22.94

Table 3.2 Density ρ (kg m^{-3}) of some n-alkanes in the liquid phase

		P (MPa)	273.15 K	298.15 K	323.15 K	348.15 K	373.15 K
n-Pentane	C$_5$H$_{12}$	0.101	645.7	621.2	595.5	568.0	537.7
		5	650.8	627.5	603.4	577.8	549.4
		10	655.6	633.4	610.5	586.3	559.3
		25	668.4	648.5	628.3	606.9	582.2
		50	685.8	668.2	650.5	631.5	608.6
n-Hexane	C$_6$H$_{14}$	0.101	677.2	654.9	631.7	607.3	581.4
		5	681.6	660.2	638.2	615.3	591.0
		10	685.8	665.2	644.2	622.5	599.6
		25	697.2	678.5	659.6	640.4	620.1
		50	713.1	696.4	679.8	662.9	645.0
n-Heptane	C$_7$H$_{16}$	0.101	700.5	679.5	657.9	635.6	612.1
		5	704.5	684.2	663.5	642.3	620.3
		10	708.3	688.7	668.8	648.6	627.7
		25	718.8	700.8	682.7	664.6	646.2
		50	733.7	717.4	701.4	685.4	669.3
n-Octane	C$_8$H$_{18}$	0.101	718.7	698.7	678.2	657.2	635.3
		5	722.4	703.0	683.3	663.2	642.6
		10	726.0	707.2	688.2	668.9	649.3
		25	736.0	718.5	701.1	683.8	666.3
		50	750.2	734.3	718.7	703.4	688.2
n-Nonane	C$_9$H$_{20}$	0.101	733.1	713.9	694.2	674.1	653.4
		5	736.6	717.9	699.0	679.7	660.1
		10	740.1	721.9	703.5	685.0	666.3
		25	749.6	732.6	715.7	698.9	682.1
		50	763.3	747.8	732.5	717.6	702.9
n-Decane	C$_{10}$H$_{22}$	0.101	745.1	726.2	707.2	687.8	668.0
		5	748.5	730.1	711.7	693.1	674.2
		10	751.8	733.9	716.0	698.0	680.0
		25	760.9	744.2	727.6	711.2	695.0
		50	774.3	758.9	743.8	729.2	714.9
n-Undecane	C$_{11}$H$_{24}$	0.101	754.8	736.5	718.0	699.2	680.0
		5	758.1	740.2	722.3	704.2	685.8
		10	761.3	743.9	726.4	708.9	691.2
		25	770.2	753.8	737.6	721.5	705.5
		50	783.2	768.1	753.3	738.9	724.8
n-Dodecane	C$_{12}$H$_{26}$	0.101	763.4	745.3	727.0	708.5	689.6
		5	766.6	748.9	731.1	713.2	695.1
		10	769.7	752.4	735.1	717.7	700.3
		25	778.4	762.1	745.9	729.8	714.0
		50	791.2	776.0	761.2	746.7	732.6

dence level. The design was later adapted to measure the viscosity simultaneously along with a more refined theory [3.24]. The latest vibrating-wire instrument of this type was built at the University Blaise-Pascal. *Clermont-Ferrand*, France, incorporating a number of improvements [3.25].

Table 3.3 Density ρ (kg m^{-3}) of some alkenes in the liquid phase

		P (MPa)	273.15 K	298.15 K	323.15 K	348.15 K	373.15 K
Benzene	C$_6$H$_6$	0.101	899.8	873.6	846.8	819.2	790.8
		5	903.4	877.7	851.6	824.9	797.6
		10	906.9	881.8	856.3	830.4	804.0
		25	916.7	892.9	869.1	845.1	820.8
		50	931.1	909.1	887.2	865.3	843.3
Toluene	C$_7$H$_8$	0.101	885.8	862.3	838.7	814.8	790.5
		5	889.1	866.2	843.2	820.1	796.8
		10	892.4	870.0	847.6	825.2	802.8
		25	901.6	880.4	859.5	839.0	818.8
		50	915.3	895.7	876.6	858.1	840.3
Ethylbenzene	C$_8$H$_{10}$	0.101	883.4	862.6	840.8	817.9	793.9
		5	886.7	866.3	845.1	823.1	800.0
		10	889.8	869.9	849.3	828.0	805.9
		25	898.8	880.1	860.9	841.3	821.3
		50	912.1	894.9	877.5	860.0	842.3
o-Xylene	C$_8$H$_{10}$	0.101	898.0	876.5	854.9	833.2	811.5
		5	900.9	879.8	858.7	837.7	816.6
		10	903.9	883.1	862.5	842.0	821.6
		25	912.2	892.4	872.9	853.8	835.0
		50	924.6	906.1	888.1	870.6	853.7
m-Xylene	C$_8$H$_{10}$	0.101	879.6	860.0	839.4	817.6	794.5
		5	882.8	863.7	843.7	822.7	800.6
		10	886.0	867.4	848.0	827.7	806.5
		25	894.9	877.5	859.5	841.0	821.9
		50	908.1	892.3	876.1	859.7	843.0
p-Xylene	C$_8$H$_{10}$	0.101	875.6	855.8	834.9	812.9	789.5
		5	878.9	859.5	839.2	818.0	795.6
		10	882.0	863.1	843.4	822.9	801.5
		25	890.9	873.2	855.0	836.2	816.9
		50	904.1	888.0	871.5	854.8	837.8
Mesitylene	C$_9$H$_{12}$	0.101	880.7	860.7	840.2	819.2	797.6
		5	883.8	864.1	844.2	823.9	803.2
		10	886.8	867.6	848.1	828.4	808.5
		25	895.3	877.1	859.0	840.8	822.8
		50	907.9	891.2	874.6	858.3	842.4

3.1.4 Buoyancy-Type Densimeters

The buoyancy method basically makes use of Archimedes' principle. A suspended sinker (Fig. 3.6) with a known mass and volume is attached to a fine wire, and is totally immersed in the sample liquid. A precision force balance is used to measure the force to support the sinker. Once the mass, volume, and supporting weight of the sinker are known, the density of the liquid can be calculated. The principle is rather straightforward for liquids at atmospheric pressure where containment of the liquid sample is simple and attachment of a thread to a force balance causes no breach of a sealed container. However, even in these simple circumstances, some corrections need to be made for the force exerted by surface tension on the suspension wire and for the volumetric thermal expansion coefficient of the sinker. When used with great care buoyancy-type densimeters can yield results of great accuracy.

An advanced version of the buoyancy technique is the magnetic suspension system. *Klimeck* et al. [3.20] developed an advanced single-sinker densimeter em-

Table 3.4 Density ρ (kg m^{-3}) of some n-alcohols in the liquid phase

		P (MPa)	273.15 K	298.15 K	323.15 K	348.15 K	373.15 K
Methanol	CH$_4$O	0.101	810.0	786.5	762.5	737.6	710.8
		5	814.0	791.3	768.2	743.9	717.3
		10	818.0	796.0	773.6	749.8	723.5
		25	828.9	808.7	787.9	765.4	739.5
		50	844.6	826.4	807.5	786.4	760.8
Ethanol	C$_2$H$_6$O	0.101	806.3	785.0	763.1	739.6	713.8
		5	809.9	789.2	767.8	744.8	719.1
		10	813.4	793.2	772.3	749.7	724.2
		25	823.1	804.3	784.6	762.9	737.7
		50	837.2	820.0	801.9	781.2	756.1
1-Propanol	C$_3$H$_8$O	0.101	819.1	799.5	779.0	756.8	732.4
		5	822.5	803.4	783.5	761.8	737.7
		10	825.8	807.2	787.8	766.6	742.8
		25	835.0	817.7	799.6	779.6	756.4
		50	848.5	832.9	816.3	797.6	775.1
1-Butanol	C$_4$H$_{10}$O	0.101	824.6	805.8	786.2	765.2	742.3
		5	827.7	809.4	790.3	769.9	747.5
		10	830.8	812.9	794.4	774.5	752.5
		25	839.4	822.8	805.5	786.8	765.8
		50	852.2	837.1	821.4	804.1	784.1
1-Pentanol	C$_5$H$_{12}$O	0.101	828.9	811.0	792.1	772.1	750.4
		5	831.9	814.3	796.0	776.5	755.4
		10	834.8	817.7	799.8	780.9	760.2
		25	842.9	826.9	810.4	792.6	773.1
		50	855.1	840.5	825.5	809.2	790.9
1-Hexanol	C$_6$H$_{14}$O	0.101	833.0	815.3	797.1	778.2	758.2
		5	835.8	818.5	800.8	782.4	763.0
		10	838.6	821.7	804.5	786.6	767.6
		25	846.4	830.6	814.5	797.8	780.1
		50	858.1	843.6	829.1	813.9	797.6
1-Heptanol	C$_7$H$_{16}$O	0.101	836.4	819.1	801.1	782.3	762.5
		5	839.1	822.2	804.6	786.3	767.1
		10	841.7	825.2	808.1	790.3	771.6
		25	849.3	833.8	817.8	801.2	783.7
		50	860.6	846.4	831.8	816.7	800.7
1-Octanol	C$_8$H$_{18}$O	0.101	838.8	821.7	804.1	785.5	766.0
		5	841.4	824.7	807.5	789.5	770.5
		10	844.0	827.7	810.8	793.3	774.8
		25	851.4	836.0	820.2	803.9	786.7
		50	862.4	848.3	833.9	819.0	803.3

ploying a magnetic suspension system (Fig. 3.7). The magnetic suspension coupling consists of an electromagnet, a permanent magnet, a position transducer, and a control system. The electromagnet is attached at the underfloor weighing hook of a commercial analytic balance. Inside the coupling housing there is a permanent magnet to which the sinker to be weighed is linked by means of a load coupling and decoupling de-

Table 3.5 Density ρ (kg m^{-3}) of some refrigerants in the liquid phase

		P (MPa)	248.15 K	273.15 K	298.15 K	323.15 K	348.15 K
R22	CHClF$_2$	5	1375	1299	1214	1113	975.0
		10	1386	1315	1238	1149	1043
		25	1416	1355	1290	1217	1140
R32	CH$_2$F$_2$	5	1148	1070	980.7	865.0	
		10	1162	1088	1006	910.1	857.7
		25	1194	1128	1058	982.3	1019.8
R124	C$_2$HClF$_4$	5	1518	1450	1380	1303	1202
		10	1526	1462	1402	1338	1247
		25	1547	1496	1454	1408	1330
R125	C$_2$HF$_5$	5	1447	1349	1236	1086	
		10	1466	1376	1279	1167	
		25	1510	1436	1359	1282	
R134a	C$_2$H$_2$F$_4$	5	1386	1312	1229	1135	1017
		10	1399	1328	1252	1168	1075
		25	1432	1368	1302	1234	1166
R141b	C$_2$H$_3$ClF$_2$	5	1328	1287	1244	1197	1147
		10	1333	1294	1253	1209	1163
		25	1345	1311	1276	1239	1200
R152a	C$_2$H$_4$F$_2$	5	1022	971.3	914.8	850.7	774.1
		10	1030	981.9	929.4	871.9	808.5
		25	1051	1008	962.8	915.6	867.3

vice. The upper part of the coupling housing which separates the permanent magnet from the electromagnet is manufactured of a magnetically neutral metal, namely copper beryllium. To achieve the freely suspended state of the permanent magnet, its absolute position is detected by a position sensor and controlled via a proportional–integral–differential controller. By means of a superimposed set-point controller and an additional control system, several vertical motions of the permanent magnet are generated automatically. In this way, soft up- and downward movements of the permanent magnet can be realized, and via the load coupling and decoupling device the solid quartz glass cylinder working as sinker can be coupled and decoupled.

In the tare position, the permanent magnet is suspended at a relatively large distance from the top of the coupling housing, the sinker is decoupled from the permanent magnet, and the balance can be tared to zero. In order to achieve the measuring position, the electronic control unit of the magnetic suspension coupling brings the permanent magnet closer to the top of the coupling housing. This means that the bearing cone also moves upwards and takes the measuring load cage with which the sinker is connected. In this way, the sinker is coupled with the balance and can be weighed.

In order to measure the density of the fluid in the measuring cell, the sinker is coupled and decoupled several times (changes between the tare and measuring position), so that the buoyancy force upon the sinker can be more accurately determined by averaging. Then, the density of the fluid can be determined from the simple relation,

$$\rho = \frac{(m_\text{s} - m_\text{s,fluid})}{V_\text{s}(T, P)} . \tag{3.7}$$

In this equation, m_s is the true mass of the sinker (weighed in the evacuated measuring cell), $m_\text{s,fluid}$ is the apparent mass of the sinker (weighed in the fluid-filled measuring cell), and $V_s(T, P)$ is the temperature and pressure dependent volume of the sinker.

This instrument was successfully employed for many measurements over a very wide range of conditions with an uncertainty of better than $2 \times 10^{-4} \rho$.

The group in Bochum, headed by Prof. W. Wagner, also has other types of densimeter based upon similar principles. Here, we only mention two other very successful designs. The two-sinker densimeter developed by *Kleinrahm* and *Wagner* [3.26] in 1986 automatically compensated all incidental effects (such as the zero-point shift of the balance, buoyancy forces on auxiliary

Table 3.6 Density ρ (kg m^{-3}) of some gases

		P (MPa)	273.15 K	298.15 K	323.15 K	348.15 K	373.15 K
Argon	Ar	0.101	1.778	1.628	1.502	1.394	1.300
		5	91.75	82.71	75.39	69.36	64.30
		10	189.1	168.1	151.7	138.6	127.8
Hydrogen	H$_2$	0.101	0.089	0.082	0.075	0.070	0.065
		5	4.23	3.89	3.60	3.35	3.13
		10	8.09	7.46	6.92	6.46	6.06
Nitrogen	N$_2$	0.101	1.246	1.141	1.052	0.997	0.911
		5	62.44	56.47	47.64	47.64	44.26
		10	124.4	111.6	101.5	93.3	86.5
Oxygen	O$_2$	0.101	1.424	1.304	1.203	1.116	1.041
		5	73.67	66.36	60.48	55.63	51.55
		10	152.2	135.1	121.8	111.2	102.6
Carbon monoxide	CO	0.101	1.246	1.141	1.052	0.977	0.911
		5	62.94	56.82	51.89	47.82	44.40
		10	126.1	112.8	102.3	93.9	86.9
Carbon dioxide	CO$_2$	0.101	1.970	1.802	1.660	1.540	1.436
		5		130.52	104.46	90.21	80.48
		10		379.6	230.6	186.9	
Sulfur dioxide	SO$_2$	0.101	2.908	2.649	2.435	2.254	2.099
Hydrogen sulfide	H$_2$S	0.101	1.532	1.4	1.29	1.195	
		5				85.78	
Methane	CH$_4$	0.101	0.715	0.655	0.604	0.56	0.522
		5	40.05	35.3	31.73	28.91	26.61
		10	89.6	75.7	66.4	59.5	54.1
Ethane	C$_2$H$_6$	0.101	1.351	1.235	1.137	1.054	0.982
		5		91.8	70.97	60.53	
		10		314.7	223.6	158.98	
Propane	C$_3$H$_8$	0.101	2.004	1.826	1.678	1.554	

devices, adsorption effects, surface tension, etc.) that reduce the accuracy of the density measurement when only a single sinker is employed. In 2002, the same group developed an absolute viscometer–densimeter [3.27] for measurements on gases, that operates with a 0.15–0.4% uncertainty in viscosity and 0.02–0.05% in density.

Various other investigators employed densimeters based upon the same principle. *Masui* [3.28] employed an optical sensing system fed by a fibre optic as a feedback control to stabilize the buoy support. *Masui* [3.28] measured the density of toluene from 298 to 423 K and pressures up to 30 MPa, with an estimated uncertainty of 0.025%. A similar magnetic suspension densimeter was employed by *Toscani* et al. [3.29] for density measurements of liquids and liquid mixtures. This instrument covered a temperature range from 295 to 400 K at pressures up to 100 MPa, with an uncertainty of ±0.2%. *Okada* et al. [3.30], developed a magnetic densimeter, in which the float consists of a hollow glass body containing a soft-iron core.

3.1.5 Density Reference Values

The density of water is still widely employed as a liquid density standard. The equation developed by *Wagner* and *Pruß* [3.31] was adopted by the International Association for the Properties of Water and Steam (IAPWS) in 1995. The IAPWS formulation for the thermodynamic properties of ordinary water substance for both scientific and general use is called IAPWS 1995 (IAPWS-95) [3.32]. It represents all of the thermodynamic properties of water from the melting line (251.2 K at 209.9 MPa) to a temperature of 1273 K and pressures up to 1 GPa. In this entire range IAPWS 95 represents the most accurate measurements to within the experimental uncertainty. Values of density at specific temperature and pressure points are given at Table 3.1.

3.1.6 Tables of Density Values

In Tables 3.2–3.6, the density of commonly encountered fluids is given for engineering purposes, as a function of temperature and pressure. The fluids and the temperature and pressure conditions chosen are the same for the density, viscosity and thermal conductivity discussed in the present section and in Sects. 3.4 and 3.5.

Values for the liquid density are based on a large collection of experimental data critically assessed (n-alkanes [3.33], n-alkenes [3.34], n-alcohols [3.35], refrigerants [3.36]). The uncertainty of the quoted liquid density values is much better than ±0.3%. Values for the gas-phase density have been obtained from corresponding-states software [3.37] with an estimated uncertainty better than ±0.5%.

3.2 Surface Tension and Interfacial Tension of Liquids

This chapter presents the concepts of surface and interfacial tensions, their dependencies, measurement, and prediction. It starts with an explanation of the surface tension of pure liquids and its temperature dependence. Then, the surface tension of solutions is presented, and useful correlations for the dependence of surface tension on the solute concentration are suggested. Next, the concept of interfacial tension associated with two dense phases is explained, and its relationship with the surface tensions of the two phases is discussed. The Young–Laplace equation that correlates the pressure difference across an interface with interfacial tension and curvature is presented. Various methods for measurement of surface and interfacial tension are discussed, based on the theoretical background previously explained. Finally, selected surface tension values are tabulated, and a method for predicting surface tension from other thermodynamic properties is shown.

3.2.1 Surface Tension of Pure Liquids

Observations Intuitive Concepts, and Definitions

It is well known that small liquid drops are almost spherical even under the influence of gravity, while large drops are distorted from sphericity under the effect of gravity. Also, many types of small solid particles (and insects such as striders) can float or move on water even if their density is higher than that of water, while large particles of such density readily pass through the water surface. Clearly, therefore, there exists a force that acts on particles in general, which becomes more pronounced as they become smaller. A complementary well-known fact is that, as particles become smaller, the ratio of their surface area to their volume increases. For example, the surface area of a sphere of radius R is proportional to R^2, while its volume is proportional to R^3; therefore, the ratio of surface area to volume is proportional to $1/R$, which strongly increases as R decreases. Thus, it is reasonable to suspect that the force acting on small particles is associated with their interfaces.

Thinking from a molecular point of view, the net force acting on a molecule near a surface must be different from that acting on a molecule deep in the bulk, simply because of symmetry considerations. This is schematically demonstrated in Fig. 3.8. Thus, the energy of a molecule near a surface must be different from that of the same molecule in the bulk. Based on this general picture, Gibbs developed a formalism that defines surface energy as the difference between the actual energy of a system and the sum of the energies of its components had there been no interface between them. In Gibbs' approach, the interface is considered a mathematical surface of no thickness, in line with macroscopic observations. The surface energy per unit area of a liquid surface in contact with vapor (or, in general, with a gas) is its surface tension. If the liquid is in contact with a dense phase, such as another immiscible liquid or a solid, the corresponding terms that are used are *interfacial energy* and *interfacial tension*. The latter will be

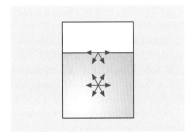

Fig. 3.8 A schematic comparison of the forces acting on a molecule near a surface, and a molecule in the bulk

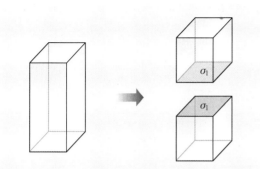
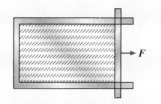

Fig. 3.9 An imaginary experiment that creates two new surfaces by breaking an infinite liquid body into two parts, and removing them to a large distance from each other

Fig. 3.10 Top view of a liquid film on a frame with a movable side. A force acting on the movable side is required to increase the surface area of the film

used below as a general term, whenever both interfacial tension and surface tension are implied. For completeness, it should be mentioned that Bakker and van der Waals developed a different approach, in which the interface is considered as a thin, but three-dimensional region. For practical purposes, the two approaches lead to the same conclusions.

It may also be useful to regard surface energy as the energy needed to create a surface. In the imaginary experiment shown in Fig. 3.9, an infinite body of a liquid is separated into two parts that are removed from each other to a sufficiently large distance, so their interaction energy is negligible. The energy per unit interfacial area needed for this separation is termed the *work of cohesion* E_c since it overcomes the cohesion between the molecules of a single species. It is equal to twice the surface tension of the liquid σ_l since two new surfaces were created in this experiment. Thus,

$$\sigma_l = \frac{E_c}{2}. \quad (3.8)$$

Another imaginary experiment may shed a different light on the surface tension concept. Figure 3.10 shows a thin frame with three fixed sides and a movable bar as a fourth side. If a liquid film (such as a soap film) is formed on this frame, then a force F has to be applied to the movable bar in order to increase the surface area of the film. The work done by this force is Fd, where d is the path length. This work is transformed into surface energy $2\sigma_l L d$, where L is the width of the frame (the factor 2 stands for the two sides, top and bottom, of the liquid film). Thus

$$\sigma_l = \frac{F}{2L}. \quad (3.9)$$

This result shows that surface tension can also be interpreted as a force per unit length. This interpretation is useful as a basis for some of the methods for surface tension measurement.

Based on the above definitions, the units in which surface tension is measured should be J/m^2 or N/m. Actually, for convenience, surface tension is usually expressed in mN/m (millinewtons per meter). The reason for this choice is that this unit is equivalent to dynes/cm that had been used for many years. Thus, by adopting the units of mN/m, the numerical values of surface tension were unchanged by the transition to SI units. Typical surface tension values for regular liquids are in the range of ≈ 14 to ≈ 73 mN/m at about room temperature. The higher end of this range corresponds to water. Liquid metals have much higher surface tension: mercury, for example, has a surface tension of ≈ 486 mN/m at room temperature.

In most practical applications, the surface tension of a pure liquid is considered independent of the size of the system. Thus, the surface tension of a small drop of a liquid is assumed the same as for a large drop. A priori, based on molecular interactions reasoning, it can be argued that this constancy may not hold for very small drops. This indeed is the case, however it was experimentally demonstrated that a drop has to be almost molecular in size in order to observe size dependence of its surface tension.

Temperature Dependence

At the critical temperature T_c there is no distinction between a liquid and its vapor. In other words, no interface exists between them. Thus, the surface tension of a liquid at the critical temperature must be, by definition, zero. This observation indicates that, in general, for most liquids, surface tension decreases as the temperature is increased. Since no exact theory seems to exist, the quantitative dependence of surface tension on the absolute temperature T (in K) can be estimated by semi-empirical, approximate expressions. A useful expression is the Eötvös equation, which reads

$$\sigma_l V^{2/3} = c(T_c - T), \qquad (3.10)$$

where V is the molar volume in m^3, c is a constant, which for many liquids has the value of approximately 2.1×10^{-4} J/K, and σ_l is measured in mN/m.

Another useful expression is the van der Waals–Guggenheim equation

$$\sigma_l = \sigma^0 \left(1 - \frac{T}{T_c}\right)^n, \qquad (3.11)$$

where σ^0 is a constant typical to the liquid, in the same units as σ_l, and n equals 11/9 for many organic liquids. The usefulness of these two expressions lies in their approximate universality in terms of the values of c or n.

3.2.2 Surface Tension of Liquid Solutions

Concepts and General Examples

In many practical processes, solutions rather than pure liquids are employed. Therefore, understanding surface tension of solutions is essential. In principle, solutes may increase the surface tension or decrease it. According to a fundamental, thermodynamic theory developed by Gibbs, the effect of a solute on surface tension depends on the tendency of the solute to concentrate at the solution–air interface. If the solute tends to concentrate at the interface more than in the bulk, an increase in its concentration will decrease the surface tension.

Figure 3.11 shows typical dependencies of surface tension on concentration in aqueous systems. Figure 3.11a shows the effect of a typical electrolyte (NaCl) dissolved in water. As can be clearly seen, the addition of the electrolyte increases the surface tension beyond that of water. However, very large concentrations are required for a relatively small increase in surface tension. Clearly, the electrolyte prefers to be in the bulk rather than at the surface. Figure 3.11b shows a typical effect of a soluble organic liquid in water (propanol). The surface tension gradually decreases from that of pure water to that of the solute. The decrease is steeper at lower concentrations of the solute and shallower at higher concentrations. The fact that the surface tension decreases indicates that the organic solute prefers to concentrate at the interface. This is understandable from a molecular point of view, because a molecule of an organic materials always has a hydrophobic (water-fearing) part in it, which tends to stay away from water as much as possible.

When an organic molecule has a polar part together with a dominant hydrophobic part, the solubility in the

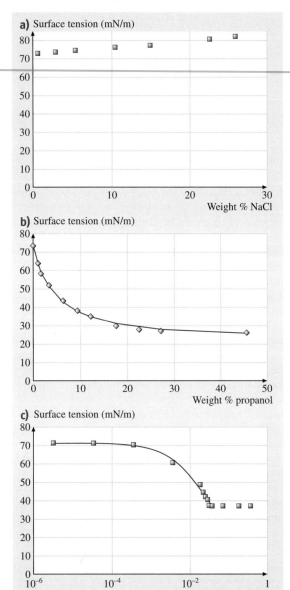

Fig. 3.11a–c Typical surface tension dependence on solute concentration in aqueous solutions: (**a**) an electrolyte (NaCl); (**b**) a soluble organic liquid (propanol, the curve represents a fit of the Connors and Wright equation); (**c**) a surfactant (CTAB in PBS, the curve represents a fit of the Szyszkowski equation)

bulk is small, and the molecule tends to adsorb mostly at the surface. Such amphiphilic molecules are referred to as *surface-active agents* or *surfactants*. Figure 3.11c shows a typical dependence of the surface tension on concentration for a surfactant (cetyl trimethyl ammonium bromide in phosphate buffer solution, CTAB in PBS). Two features are very prominent in this figure:

1. the surfactant concentration that leads to a meaningful reduction in surface tension is orders of magnitude smaller than for a soluble organic material; and
2. above a certain concentration the surface tension remains constant.

The latter is called the *critical micelle concentration* (CMC), since at this and higher concentrations of the surfactant, its molecules in the bulk aggregate in structures called *micelles*. This aggregation is an alternative to adsorbing at the liquid-air surface. The hydrophobic parts of the molecules are hidden inside the micelles in order to minimize their interaction with water, instead of being exposed to air at the interface. A spherical micelle is shown schematically in Fig. 3.12. Micelles exist in many geometrical forms, a topic which is extensively covered in textbooks and in the research literature, and which is of the utmost importance in biology (cell membrane formation by lipids). It is also interesting to note that *reverse micelles* can form in organic liquids, where the polar parts of the surfactant molecules will be *hidden* from the solvent.

Useful Surface Tension Correlations

Unfortunately, the theory regarding surface tension of solutions is not yet sufficiently developed to yield exact, general equations. Therefore, semi-empirical correlations need to be used. Two useful correlations are presented below, one that is useful for very low concentrations, and the other for relatively high concentrations.

For low concentrations, the von Szyszkowski equation may be used

$$\sigma_l = \sigma_0 - RT\Gamma_\infty \ln\left(1 + \frac{C}{b}\right), \quad (3.12)$$

where σ_0 is the surface tension of the pure solvent, R is the universal gas constant, C is the molar concentration of the solute, and Γ_∞ and b are two empirical constants to be determined by the best fit to experimental data. In the simplified theory underlying this equation, Γ_∞ is the saturation surface concentration (mols per unit area) of the adsorbed molecules at the surface. The curve in Fig. 3.11c shows an example of a fit of this equation to experimental data. The fit is excellent all the way to the CMC.

At higher concentrations, an equation developed by *Connors* and *Wright* [3.38] may be employed

$$\sigma_l = \sigma_0 - (\sigma_0 - \sigma_s)\left(1 + \frac{\beta(1-x)}{1-\alpha(1-x)}\right)x, \quad (3.13)$$

where σ_s is the surface tension of the pure solute, x is the molar fraction of the solute, and α and β are the two empirical constants to be determined. It is clear that at $x = 0$ this equation indeed predicts the surface tension to be that of the pure solvent, σ_0, and at $x = 1$ that of the pure solute, σ_s. Figure 3.11b shows that this equation can fit data very well for intermediate concentrations.

3.2.3 Interfacial Tension

Work of Adhesion

The concept of interfacial tension can be demonstrated and understood by following an imaginary experiment, similar to that used for introducing the surface tension concept. Suppose, as shown in Fig. 3.13, that two semi-infinite bodies of different materials, a and b, are separated from each other to a sufficiently large distance, so their interaction energy is negligible. At the beginning of the process, the interfacial energy per unit interfacial area is σ_{ab}, while at the end the sum of the surface energies per unit area is $(\sigma_a + \sigma_b)$. The energy per

Fig. 3.12 A schematic of a spherical micelle

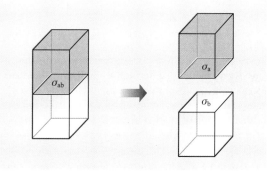

Fig. 3.13 An imaginary experiment that creates two new surfaces by separating two semi-infinite bodies of different materials at the interface between them, and removing them to a large distance from each other

unit interfacial area needed to achieve this separation, E_a, is termed the *work of adhesion*, and the interfacial tension can be expressed as

$$\sigma_{ab} = \sigma_a + \sigma_b - E_a . \tag{3.14}$$

Thus, if one can independently calculate E_a, the interfacial tension can be calculated from the individual surface tensions.

Calculation of Interfacial Tension from Surface Tensions

Unfortunately, an exact, general calculation of the work of adhesion is not yet possible. Therefore, some useful approximations have been developed over the years. *Girifalco* and *Good* [3.39] presented the first useful correlation:

$$E_a = 2\phi\sqrt{\sigma_a \sigma_b} , \tag{3.15}$$

where ϕ is a dimensionless correction factor, covering for the uncertainty in the equation. Values of ϕ for liquid–fluid systems were measured to be in the range of 0.6–1.2, so the uncertainty in this equation is meaningful.

Fowkes [3.40] developed another approach. In his approach, surface tension consists of two contributions: that of dispersion (London–van der Waals) forces and that of other sources, such as hydrogen bonding, for example:

$$\sigma = \sigma^d + \sigma^h . \tag{3.16}$$

Here, σ is surface tension in general, the superscript d stands for dispersion (London–van der Waals) forces, and the superscript 'h' stands for hydrogen bonding. In addition, the work of adhesion between two non-similar phases (e.g., an aqueous phase and a nonpolar phase) depends only on the nonpolar interactions between them. Therefore,

$$E_a = 2\sqrt{\sigma_a^d \sigma_b^d} \tag{3.17}$$

Owens and *Wendt* [3.41] generalized this approach by considering the surface tension to include, in general, also a polar (nondispersive) contribution: $\sigma = \sigma^d + \sigma^p$, where the superscript 'p' stands for the polar contributions. Then, the work of adhesion becomes

$$E_a = 2\sqrt{\sigma_a^d \sigma_b^d} + 2\sqrt{\sigma_a^p \sigma_b^p} . \tag{3.18}$$

By comparing (3.18) with (3.15), an expression for the Girifalco–Good correction factor can be derived

$$\phi = \sqrt{\frac{\sigma_a^d \sigma_b^d}{\sigma_a \sigma_b}} + \sqrt{\frac{\sigma_a^p \sigma_b^p}{\sigma_a \sigma_b}} . \tag{3.19}$$

This equation can be used to roughly estimate the value of ϕ. For water, $\sigma^d/\sigma \approx 0.3$, so for the case of an interface between water and a completely nonpolar liquid ($\sigma^p = 0$), $\phi = 0.3^{1/2} \approx 0.55$. This value agrees well with the lowest end of the experimentally measured ϕ. However, the highest value of ϕ according to (3.19) is 1, which is lower than the experimentally derived value of about 1.2. Thus, while (3.19) is probably a good approximation, it is still not sufficiently accurate. Other correlations and approaches that are based on various contributions to surface tension can also be found in the literature.

In order to elucidate the surface tension components of a polar liquid (say liquid a), its interfacial tension with a nonpolar liquid (b) should be measured. In this case, $\sigma_b^p = 0$, $\sigma_b^d = \sigma_b$, and $\phi = \sqrt{\sigma_a^d/\sigma_a}$. Thus, calculation of ϕ from interfacial tension measurement yields σ_a^d. Obviously, it is advisable to repeat the measurement with a few nonpolar liquids and get an average result. For example, the nonpolar liquids n-hexane, n-heptane, n-octane and n-decane all have a ϕ of 0.55 with water at 20 °C. Consequently, $0.55 = \sqrt{\sigma_w^d/\sigma_w} = \sqrt{\sigma_w^d/72.8}$, or $\sigma_w^d \cong 22$ mN/m and $\sigma_w^p \cong 50.8$ mN/m. Benzene has a surface tension of 28.9 mN/m and a ϕ of 0.72 with water at 20 °C. Therefore, by (3.19), $0.72 = \sqrt{\frac{22\sigma_b^d}{72.8 \cdot 28.9}} + \sqrt{\frac{50.8(28.9 - \sigma_b^d)}{72.8 \cdot 28.9}}$, and it turns out that $\sigma_b^d \cong 27.5$ mN/m. As expected, this value is quite close to the surface tension itself.

3.2.4 Implications of Surface and Interfacial Tension on Liquid–Fluid Systems

Surface Curvature

The effect of surface and interfacial tension in liquid–fluid systems is closely related to the curvature of their interfaces. Therefore, the basic principles of curvature definition are explained in the following.

For a two-dimensional sufficiently smooth curve, the local radius of curvature is the radius of the circle drawn using three infinitesimally close points on the curve (Fig. 3.14). Radii of curvature may be defined as negative or positive, depending on whether the curve is concave or convex. For a three-dimensional surface, the mean curvature H at a given point is defined by the following average

$$2H = \frac{1}{R_1} + \frac{1}{R_2} ; \tag{3.20}$$

R_1 and R_2 are the principal two-dimensional radii of curvature of the curves of intersection between the sur-

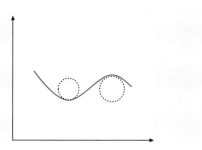

Fig. 3.14 Examples of the local radius of curvature of a two-dimensional curve (the two local radii of curvature shown in the figure have opposite signs)

face and two perpendicular planes (the principal cross sections, Fig. 3.15). It is important to realize that H is invariant to the orientation of the two planes, as long as they are perpendicular to each other.

For example, the mean curvature of a sphere is $1/R$, where R is the sphere radius, since in this case $R_1 = R_2 = R$. For a cylinder, it is convenient to choose one of the perpendicular planes to be parallel to the axis of the cylinder (so that $R_1 = \infty$) and the other perpendicular to the axis (so that $R_2 = R$, the cylinder radius). Thus, for a cylinder $H = 1/(2R)$.

Pressure Difference across Curved Interfaces

For large liquid–fluid systems, the interfaces of which have zero curvature, the pressure at equilibrium must be equal on the two sides of the interface. However, when the interfacial curvature is nonzero, a pressure difference across the interface must exist at equilibrium. The magnitude of this pressure is given by the Young–Laplace equation

$$P_i - P_0 = 2\sigma_{lf} H = \sigma_{lf}\left(\frac{1}{R_1} + \frac{1}{R_2}\right), \quad (3.21)$$

where P_i is the pressure at the side of the interface for which the radius of curvature is defined, P_o is the pressure at the other side of the interface, and σ_{lf} is the interfacial tension between the liquid and the fluid. For simple shapes, such as a bubble or a drop, P_i and P_o can be interpreted as the inside and outside pressure, respectively.

Thus, for example, for a spherical body

$$P_i - P_0 = \frac{2\sigma_{lf}}{R}. \quad (3.22)$$

It is important to notice that this equation does not differentiate between a fluid (e.g. air) bubble and a liquid drop: the pressure is always higher inside the curved body, whether it is a bubble or a drop. For a cylindrical body the pressure difference is given by

$$P_i - P_0 = \frac{\sigma_{lf}}{R}. \quad (3.23)$$

3.2.5 Measurement of Surface Tension and Interfacial Tension

Force Methods

Some surface tension measurement methods use force measurements, as described below. These methods employ the concept of surface tension (or interfacial tension) being a force per unit length.

Drop Weight. The idea underlying this method is that a drop remains attached to a capillary as long as the interfacial tension force balances its weight. This is shown

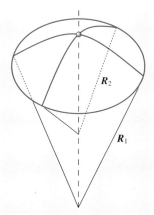

Fig. 3.15 Intersection curves of a three-dimensional surface with two perpendicular planes

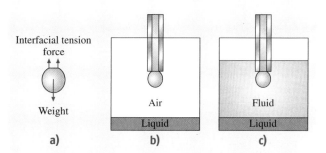

Fig. 3.16a–c The drop weight method: (**a**) the approximate force balance; (**b**) surface tension measurement; (**c**) interfacial tension measurement (the fluid is a lighter surrounding liquid)

schematically in Fig. 3.16a. According to this simplistic argument, the interfacial tension force at detachment equals the weight of the detached drop W_d. The force due to interfacial tension is assumed to be $2\pi r \sigma_{lf}$, where r is the radius of the capillary (interfacial tension is considered as force per unit length). Thus, in principle, weighing a drop and measuring the radius of the capillary enable the calculation of the interfacial tension.

However, the detachment process is much more complex than assumed above. Actually, the detachment does not occur at the line of contact between the liquid and the capillary, but through the formation of a narrowing neck in the drop itself. Therefore, the static picture of a balance between gravity and the interfacial tension force needs to be complemented. The dynamic process of detachment is rather complicated, and can be described only with the help of rather sophisticated numerical simulations. Thus, the practical solution to the problem is to add a correction factor, f_W, to the static balance equation:

$$W_d = 2\pi r \sigma_l f_W \,. \tag{3.24}$$

The correction factor turns out to depend on the dimensionless ratio between the radius of the capillary and the cubic root of the volume V of the drop: $f_W = f_W(r/V^{1/3})$. Figure 3.17 shows an approximate curve for f_W. Since the detachment process is dynamic, (3.24) may need also to be corrected for the viscosity of the liquid, if it is very high. Information about the effect of viscosity is still incomplete. Still less is known on a significant effect of the elastic stresses developing in non-Newtonian viscoelastic polymer solutions (even in the most dilute ones).

From a practical point of view, the drop weight method is a convenient, inexpensive method that can give good interfacial tension results. It may be especially useful when relative changes in interfacial tension are more important than very accurate, absolute values. The drop weight is calculated as an average of a sufficiently large number of drops. An automatic drop counter is helpful. The radius of the capillary can be accurately measured, however attention should be given to find out whether the drop hangs on the outer perimeter of the capillary or the inner one. The correction factor is best estimated by calibration of the actual experimental system with liquids of known surface tension. The drop weight method can be used to measure surface tensions (Fig. 3.16b) as well as interfacial tensions (Fig. 3.16c). For the latter purpose, the capillary tip is dipped into the fluid.

The Ring Method. The ring method is described schematically in Fig. 3.18. The principle behind it is somewhat similar to the one underlying the drop weight method: one measures the force needed to detach a wire ring from a liquid–fluid interface. The ring is dipped into the liquid, and then removed until detachment from the liquid occurs. The maximum force measured in this case F_r is the sum of the weight of the ring W_r and the interfacial tension force that acts on the inner as well as outer perimeter of the ring. Since the thickness of the wire of the ring is very small compared with its radius, the two perimeters are considered to be of the same radius R_r. As in the case of the drop weight method, the detachment process is complex, therefore a correction factor f_r is required in order to calculate the exact interfacial tension from the simplistic force balance. Thus

$$F_r = W_r + 4 f_r \pi R_r \sigma_{lf} \,. \tag{3.25}$$

f_r depends on two dimensionless ratios: $f_r = f_r(R_r/V_r^{1/3}, R_r/r_r)$, where V_r is the meniscus volume

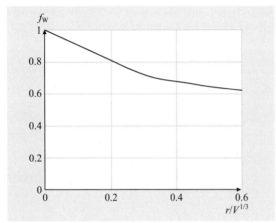

Fig. 3.17 An approximate curve for f_W, the correction factor for the drop weight method

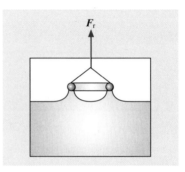

Fig. 3.18 A schematic of the ring method

(liquid carried by the ring above the bulk level), and r_r is the cross-sectional radius of the wire.

One of the main advantages of this method is that it can be easily automated and calibrated to give a surface tension reading. The f_r dependence is then an integral part of the instrument software. The main disadvantage of this method is its sensitivity to the cleanliness of the wire (usually a platinum wire). Also, the wire has to be perfectly planar in order to avoid distortions of the interface that may interfere with the measurement and calculations. The ring method may be used for measuring interfacial tensions as well as surface tensions.

Wilhelmy Plate. The Wilhelmy plate method is similar to the ring method, however a thin, vertical plate is used instead of a ring (Fig. 3.19). The main advantage is that the force balance at detachment does not need a correction factor, because of the simplicity of the plate geometry and its thinness:

$$F_p = W_p + p_p \sigma_{lf} . \quad (3.26)$$

In this equation, F_p is the force measured by the balance, W_p is the weight of the plate, and p is the perimeter of its cross section. The Wilhelmy plate method is useful for measuring interfacial tensions as well as surface tensions.

Shape Methods

The shape methods for surface tension measurement take advantage of the fact that the shape of a drop at equilibrium is determined by a balance between external forces (e.g., gravity) and surface or interfacial tension. A drop used in these methods may be hanging from a capillary (*pendant drop*) or on top of a horizontal solid surface (*sessile drop*). When external forces are negligible, the drop must be spherical, independently of interfacial tension. However, when external forces are sufficiently large to distort the shape of the drop from

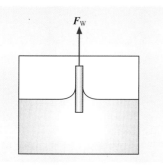

Fig. 3.19 The Wilhelmy plate method

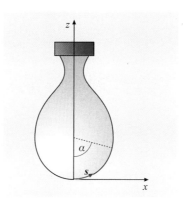

Fig. 3.20 A pendant drop with the variables used in the calculation of its shape

sphericity, the details of its shape depend also on interfacial tension. Recording the shape enables the calculation of interfacial tension from (3.21).

Drop Shape. In this method, the shape of a drop is recorded, digitally analyzed, and compared with theoretical calculations. The drop must be sufficiently large, so that the effect of gravity on its shape is meaningful. The drop must also be axisymmetric, in order to enable comparison with available theoretical calculations. The objective of the comparison with theory is to find the interfacial tension value that leads to the best fit between the recorded and calculated shapes.

For these calculations, (3.21) is transformed in the following way for the case of a pendant drop (Fig. 3.20). Very similar arguments apply to sessile drops. First, the pressure difference at each point across the interface is expressed as

$$P_i - P_o = (P_i - P_o)_{z=0} - \Delta \rho g z , \quad (3.27)$$

where x and z are the coordinates for describing the drop shape (Fig. 3.20), the point $(x = 0, z = 0)$ is the drop apex, $(P_i - P_o)_{z=0}$ is the pressure difference at the drop apex, $\Delta \rho$ is the density difference between the liquid and the fluid, and g is the gravitational acceleration.

Then, for an axisymmetric drop, the expressions for the radii of curvature become

$$\frac{1}{R_1} = \frac{(d \sin \alpha)}{dx} , \quad (3.28)$$

where R_1 is the principal radius of curvature in a drop cross section that includes the z-axis and

$$\frac{1}{R_2} = \frac{(\sin \alpha)}{x} , \quad (3.29)$$

where R_2 is the principal radius of curvature in a plane perpendicular to the above cross section, at the point (x,

z). α is the angle between the drop axis and the normal to the drop interface (Fig. 3.20). When (3.27–3.29) are introduced in (3.21), the result is a first-order differential equation in α

$$\frac{d\sin\alpha}{dx} = \frac{(P_i - P_o)_{z=0}}{\sigma_{lf}} - \frac{\Delta\rho g z}{\sigma_{lf}} - \frac{\sin\alpha}{x} . \quad (3.30)$$

Due to the symmetry at the apex, $R_1 = R_2$. Therefore

$$\frac{(P_i - P_o)_{z=0}}{\sigma_{lf}} = \frac{2}{b} , \quad (3.31)$$

where b is the radius of curvature at the apex. Thus, (3.30) finally reads

$$\frac{d\sin\alpha}{dx} = \frac{2}{b} - \frac{\Delta\rho g z}{\sigma_{lf}} - \frac{\sin\alpha}{x} . \quad (3.32)$$

However, there are still two independent variables, x and z. To solve this problem, the integration is done along the generatrix of the drop interface, introducing the arc length from the apex, s, as variable. The variables x and z are related to s by (Fig. 3.20)

$$dx = \cos\alpha\, ds \quad (3.33)$$

and

$$dz = \sin\alpha\, ds . \quad (3.34)$$

The arc length should also be introduced as a variable in (3.32), where $d\sin\alpha/dx$ is replaced by $d\alpha/ds$. Then, (3.32–3.34) constitute a set of three differential equations that are solved by routine numerical integration methods. The drop shape method has also been automated.

Rotating Cylinder. In this method, the shape of a drop is determined by a balance between interfacial tension and centrifugal force. A bubble of a fluid is introduced into a horizontal capillary that is filled with a liquid (the density of the liquid is higher than that of the fluid). The capillary rotates around its axis (Fig. 3.21). The centrifugal force increases pressure in the denser liquid, which squeezes the fluid bubble. In the absence of interfacial tension, the fluid bubble would have become a very thin and long cylinder. However, interfacial tension tends to keep the bubble as spherical as possible. The equilibrium between these two forces determines the actual shape of the bubble.

If, for simplicity, one assumes the bubble to resemble a cylinder of radius r_b (and finite length), then minimization of the system energy (the sum of the kinetic and interfacial energies) leads to the following approximate equation

$$\sigma_{lf} \cong \frac{(\omega^2 \Delta\rho r_b^3)}{4} , \quad (3.35)$$

where ω is the angular velocity of rotation. This equation can be employed to estimate the usefulness of the method. In one assumes an angular velocity of $10^2\,\mathrm{s}^{-1}$, density difference of $0.2 \times 10^3\,\mathrm{kg/m^3}$, a bubble radius of 10^{-3} m, the resulting interfacial tension is 0.5 mN/m. Thus, this method may be used for the measurement of very low interfacial tensions.

In order to be able to measure the bubble radius while the capillary is rotating, a stroboscopic light is synchronized with the frequency of rotation. Thus, the bubble appears frozen. The main technical difficulty is indeed stabilizing the rotating capillary in such a way that the frozen bubble is amenable to exact measurement of its size. This stabilization is more difficult the higher the frequency of rotation. Therefore, this method is especially useful for low interfacial tensions, for which the rotation frequency does not need to be very high.

Maximum Bubble Pressure

This method is based on a clever way of using (3.21) without the need to measure the radius of curvature of the interface directly. The idea underlying the method is related to the pressure variations inside a bubble that is growing at the tip of a capillary. At the beginning of

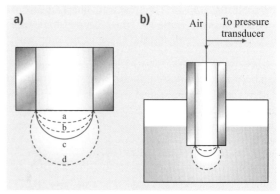

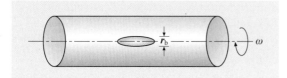

Fig. 3.21 The rotating-cylinder method

Fig. 3.22a,b The maximum bubble pressure method: Stages in the growing of a bubble

Table 3.7 Selected surface tension values of organic liquids

Group	Material	Surface tension at 20 °C [mN/m]
Acetylenes	1-hexyene	20.98
	1-tridecyne	27.56
Acids (organic)	Formic	37.67
	Isobutyric	25.04
Aldehydes	Acetaldehyde	21.18
	2-furaldehyde	43.76
Alcohols	Ethyl	22.39
	1-decanol	28.88
Alkyl halides	Chloromethane	16.2
	1-iodohexadecane	32.73
Amines	Trimethylamine	13.97
	Phenylhydrazine	45.56
Benzene and its alkyl derivatives	Benzene	28.88
	1-phenyldecane	30.97
Esters, aromatic	Ethyl α-campholanate	27.64
	Benzyl benzoate	45.95
Esters, carboxylic	Isopropyl acetate	22.30
	Methyl acetoacetate	33.09
Ethers	Ethyl methyl ether	15.93*
	Anisole	35.70*
Ketones	Acetone	26.67
	Benzophenone	44.05
Olefins	1-pnetene	16.00
	1-octadecene	28.49
Organosilicon compounds	Tetramethylsilane	12.85
	Tetrapropyl silicate	23.58
Paraffins	Pentane	16.05
	Eicosane	28.87
Perfluoro compounds	Perfluoropentane	9.89
	Perfluorocyclopentane	11.12

* value interpolated between 15 °C and 25 °C

Table 3.8 Selected surface tension values of inorganic liquids

Liquid	Surface tension at 20 °C [mN/m]
Bromine	41.8
Carbon disulfide	32.32
Hydrogen peroxide	75.87
Mercury	486.5
Water	72.88

the process of blowing a bubble, the bubble volume is small, the radius of curvature is large (Fig. 3.22a, stage a, dashed curve), and the pressure is, consequently, low. As the blowing process proceeds and the bubble grows, the radius of curvature *decreases* (Fig. 3.22a stage b, dashed curve), and the pressure *increases*. From a geometric point of view, the lowest possible radius of curvature of the bubble (when it is sufficiently small to be spherical) is the inside radius of the capillary (Fig. 3.22a stage c, solid curve). At this point, the pressure is highest. From this point on, an increase in the bubble volume must be associated with an increase in the radius of curvature (Fig. 3.22a stage d, dashed curve), therefore with a decrease in pressure. Thus, when the pressure is highest (for sufficiently small bubbles), the radius of curvature of the bubble must equal that of the capillary. By identifying the point of maximum pressure, surface tension can be calculated from the measured maximal pressure difference ΔP_{max} and the known capillary radius r_c

$$\sigma_l = \frac{\Delta P_{max} r_c}{2} . \qquad (3.36)$$

If the bubble is sufficiently large to be distorted by gravity, a corresponding correction must be made to (3.36), based on calculations of the exact shape of the bubble.

From a practical point of view, the system involves blowing gas bubbles into a sample liquid, and continuously measuring the pressure inside the bubble (Fig. 3.22b). This method is more suited to measure surface tension than interfacial tension. The bubbles are continuously blown, and the maximum pressure is recorded over many bubbles. It is important to blow the bubbles as slowly as practically possible, in order to 1. obtain equilibrium values of surface tension, and 2. avoid interactions between successive bubbles, which may interfere with the measurement. The method was neglected for a long time, since accurate pressure transducers were too expensive. Since their price has turned reasonable, this method has become a useful option.

3.2.6 Surface Tension Values for Liquids

Typical, Selected Values

Tables 3.7 and 3.8 present typical, selected surface tension values, taken from the review paper by *Jasper* [3.42]. Effort has been made to include the highest and lowest values of each group, in order to demonstrate the possible value range.

Data on the dependencies of surface tension on temperature for numerous liquids can be found in the book by *Vargaftik* et al. [3.43].

Estimation of Surface Tension

There are a few methods for predicting the surface tension of a liquid from its properties. One of the successful methods for liquids that do not contain hydrogen bonds is the corresponding-state method:

$$\sigma_l = P_c^{2/3} T_c^{2/3} Q \left(1 - \frac{T}{T_c}\right)^{11/9} . \quad (3.37)$$

In this equation, σ_l is given in mN/m, T_c and P_c are the critical temperature and pressure of the liquid in K and Pa, respectively, and T is the temperature.

$$Q = 5.553 \times 10^{-5} \left[1 + \frac{\frac{T_b}{T_c} \ln(P_c/1.013 \times 10^5)}{1 - T_b/T_c}\right] - 1.293 \times 10^{-4} . \quad (3.38)$$

In this equation, T_b is the normal boiling point (i.e., the boiling point at 1.013×10^5 Pa). Q has the dimensionality of $[\text{kg}^{1/2}\text{m}/(\text{s K})]^{2/3}$, however its numerical value is adjusted to yield the surface tension in mN/m. Equation (3.37) and (3.38) may predict surface tension to within a few percent.

3.3 Contact Angle

3.3.1 The Equilibrium Contact Angle

This section presents the various definitions of equilibrium contact angles, their measurement and interpretation. It starts with the contact angle on an ideal solid surface, its calculation, and the assessment of the solid surface tension from the value of the ideal contact angle. Then, the complexity of contact angles on real surfaces, which are rough and chemically heterogeneous, is explained. The phenomenon of hysteresis and the concepts of the advancing, receding, and most stable contact angles are presented and discussed. The conditions for meaningful measurement of contact angles are explained, and methods for their interpretation are presented.

The Ideal Contact Angle

The Young Equation. Figure 3.23 shows a typical wetting system consisting of a drop on a solid surface. This system contains three interfaces, therefore is characterized by three interfacial tensions: liquid–fluid σ_{lf}, solid–liquid σ_{sl}, and solid–fluid σ_{sf} (out of these, only σ_{lf} is directly measurable). The contact angle θ is defined as the angle between the tangent to the liquid-fluid interface and the tangent to the solid interface at the contact line between the three phases. By convention, the contact angle is measured on the liquid side (rather than on the fluid side). In many practical situations, the fluid is a gas.

The relationship between the contact angle and the interfacial tensions in the system is based on the pioneering publication by Young in 1805:

$$\cos\theta_Y = \frac{\sigma_{sf} - \sigma_{sl}}{\sigma_{lf}} . \quad (3.39)$$

In this equation, the subscript Y indicates the contact angle predicted by the Young equation. This equation was developed for the case of an ideal solid surface, which is defined as smooth, rigid, chemically homogeneous, insoluble and non-reactive. Therefore, this contact angle is referred to as the *ideal contact angle*. It is important to emphasize that this relationship depends only on the chemical nature of the three phases, and is independent of gravity. The latter may affect the shape of the drop, but not in the close proximity of the contact line, thus not its contact angle. The Young contact angle represents the state of the drop, which has the minimal Gibbs energy. It is important to note that the Gibbs energy versus contact angle curve for an ideal surface has only a single minimum at θ_Y. In other words, an ideal solid surface is characterized by a single value of the contact angle.

In principle, the three interfacial tensions may be influenced by each other at the contact line. This is

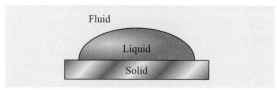

Fig. 3.23 A typical wetting system

due to the effect that one phase may have on the interaction between the other two phases. For example, the molecules of the solid may interfere with the interaction between the liquid and the fluid, thus affecting the value of the liquid-fluid interfacial tension very close to the contact line. This possibility was recognized by Gibbs, who suggested that this three-phase mutual interaction be accounted for by a *line tension*. The value of line tension has been a controversial issue for many years, however it is clear now that it may affect the contact angle of drops only when they are very small (much less than 1 micron). Therefore, line tension will not be referred to any more in the present discussion.

In order to predict the equilibrium contact angle, using (3.39), one needs to know all three interfacial tensions. However, σ_{sl} and σ_{sf} cannot be directly measured. Thus, the best that can be done to reduce the number of unknowns is to use the correlations between interfacial tensions and surface tensions that were introduced in Sect. 3.2. Since in most applications and, especially, in most measurements the fluid is a gas, σ_f is practically zero. Therefore, the solid–fluid interfacial tension σ_{sf} is actually the surface tension of a solid, σ_s. For simplicity and clarity of presentation, it is assumed throughout this chapter that the fluid is a gas.

Under this assumption, introducing, for example, (3.14) and (3.15) from Sect. 3.2 into (3.39), leads to

$$\cos\theta_Y = -1 + 2\phi\sqrt{\frac{\sigma_s}{\sigma_l}}. \tag{3.40}$$

It should be remembered, however, that σ_s, the surface tension of the clean solid, is also not directly measurable. Therefore, an equation such as (3.40) can be used in either of two ways (assuming that the value of ϕ is known):

1. to calculate the θ_Y, using σ_s as a free parameter;
2. to assess σ_s based on measurements of θ_Y.

A few numerical examples that illustrate the predictions of θ_Y by this equation are summarized in Table 3.9, assuming, for simplicity, that $\phi = 1$. The issue of the value of ϕ will be discussed later in more detail.

Example 1 represents the case of a water drop on a typical nonpolar plastic surface. The contact angle is relatively high, in agreement with typical data for water on polyethylene, for example. Example 2 represents a nonpolar liquid, such as octane, on a nonpolar plastic such as polyethylene. This case is character-

Table 3.9 Numerical predictions of (3.40) for various cases (assuming $\phi = 1$)

Example	σ_l [mN/m]	σ_s [mN/m]	$\cos\theta_Y$	θ_Y [°]
1	73	22	0.098	84.4
2	22	22	1	0
3	73	400	3.68	0

ized by a zero contact angle, and is usually termed *complete wetting*. Equation (3.40) shows that complete wetting always results when the surface tension of the liquid is approximately equal to that of the solid (depending on the exact value of ϕ). Equation (3.40) shows that the contact angle is higher than zero when the surface tension of the liquid is approximately higher than that of the solid, as demonstrated in the previous example 1. This case is termed *partial wetting*.

Example 3 demonstrates the case of a liquid spreading on a solid, whose surface tension is much higher than that of the liquid. For example, water spreading on a very clean metal surface. The interesting point is that the value of $\cos\theta_Y$ calculated from (3.40) appears to be higher than 1. This mathematical impossibility has a very simple physical explanation, as follows. Equation (3.39) or its derivatives (3.40) predict where the local minimum in energy should be, within the physically possible range of contact angles, which is $0-180°$. However, whenever the value of $\cos\theta_Y$ is supposed by this equation to be higher than 1, it means that such a local minimum does not exist. Rather, the minimal energy occurs at the border of the contact angle range, $\theta_Y = 0°$. Similarly, if $\cos\theta_Y$ appears to be lower than (-1), the minimum in energy occurs at the other border, at $\theta_Y = 180°$.

It should also be noted that a clean solid surface, when exposed to air, may adsorb components from the air, so its *effective* surface tension may be lower than that of the clean solid. In some cases, the solid may adsorb the vapor of the liquid to such an extent that its surface tension is much lowered. Systems for which complete wetting is expected to occur (based on the surface tension of the clean solid), but actually are characterized by partial wetting (because of the adsorption of vapor), were termed *autophobic* by Zisman and his collaborators.

Assessing the Surface Tension of a Solid from the Ideal Contact Angle. Equation (3.40), or a similar equation, would have enabled the calculation of the surface tension of a solid, had the value of ϕ been known. However, ϕ

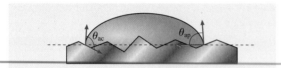

Fig. 3.24 The actual and the apparent contact angle on a rough surface

itself depends on the surface tension of the solid and on that of the liquid, as demonstrated, for example, by (3.19) in Sect. 3.2. When this equation is introduced into (3.40), one gets

$$(1+\cos\theta_Y)\sigma_l = 2\left(\sqrt{\sigma_s^d\sigma_l^d} + \sqrt{\sigma_s^d\sigma_l^d}\right). \quad (3.41)$$

Assuming that the surface tension of the liquid and its components are known, this equation involves two unknowns: σ_s^d and σ_s^p. In order to solve for both unknowns, another contact angle measurement needs to be made with a different liquid on the same solid. Then, a system of two equations enables the calculation of the two unknowns:

$$(1+\cos\theta_{Y1})\sigma_{l1} = 2\left(\sqrt{\sigma_s^d\sigma_{l1}^d} + \sqrt{\sigma_s^d\sigma_{l1}^d}\right). \quad (3.42)$$

$$(1+\cos\theta_{Y2})\sigma_{l2} = 2\left(\sqrt{\sigma_s^d\sigma_{l2}^d} + \sqrt{\sigma_s^d\sigma_{l2}^d}\right). \quad (3.43)$$

In order to get meaningful results, it is very important to correctly choose the liquids for the contact angle measurements. It turns out that it is best if one of the liquids is nonpolar, and if the liquids are as dissimilar as possible.

Contact Angles on Real Surfaces

Actual and Apparent Contact Angles. The previous section outlined the procedures for calculating either the contact angle on an ideal solid, or the surface tension of a solid from the measured ideal contact angle. However, in reality, solid surfaces are seldom ideal; they are usually rough and chemically heterogeneous to some extent. In these cases, there is a need to distinguish between actual and apparent contact angles. Figure 3.24 demonstrates these definitions for a rough surface. The actual contact angle is the angle between the tangent to the liquid-fluid interface and the actual, local surface of the solid. The apparent contact angle is the angle between the tangent to the liquid-fluid interface and the line that represents the nominal solid surface, as seen macroscopically.

Figure 3.24 clearly demonstrates that the difference between the two angles may be very large. It turns out, that the actual contact angle equals the Young contact angle, if line tension is negligible. So, the actual contact angle is the one needed for the assessment of surface tension of solid surfaces, or as a boundary condition for theoretical calculations. However, a method to routinely measure the actual contact angle has not been developed yet. The contact angle that is currently amenable to measurement is the apparent one. Therefore, the main problem that needs to be solved is the correlation between the measurable, apparent contact angle and the ideal one.

Contact Angle Hysteresis. When apparent contact angles are measured on real surfaces, which may be rough or chemically heterogeneous or both, it becomes clear that there exists a range of practically stable, apparent contact angles. This is in contrast to the prediction by the Young equation of a single contact angle on an ideal surface. Experimentally, when the drop volume is increased, the contact line appears to be pinned, while the contact angle increases (Fig. 3.25a). The apparent contact angle eventually reaches a maximum value, which is termed the *advancing contact angle*. If the drop volume is further increased, the contact line advances. Therefore, the motion of the contact line is sometimes described as a *stick-slip* motion. Similar phenomena occur when the drop volume is decreased (Fig. 3.25b): the contact line appears to be pinned, while the contact angle decreases until it reaches a minimal value called the receding contact angle; further reduction in the drop volume causes the contact line to recede. The difference between the advancing and receding contact angles, which is termed the hysteresis range, may be very large. Thus, contact angle hysteresis is a major problem in the interpretation of contact angles and the assessment of the surface tension of a solid.

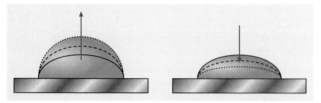

Fig. 3.25a,b Contact angle hysteresis: (**a**) increasing the drop volume increases the apparent contact angle up to the advancing contact angle; (**b**) decreasing the drop volume decreases the apparent contact angle down to the receding contact angle

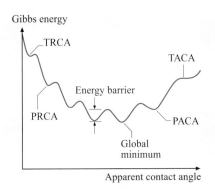

Fig. 3.26 A schematic of the Gibbs energy curve for a real surface with the following features: the global minimum, the theoretical advancing contact angle (TACA), a practical advancing contact angle (PACA), the theoretical receding contact angle (TRCA), a practical receding contact angle (PRCA), a potential barrier

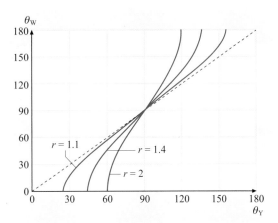

Fig. 3.27 The Wenzel versus Young contact angles for various roughness ratios. The *dashed line* corresponds to an ideal smooth surface with $r = 1$

In contrast to the Gibbs energy curve for a drop on an ideal solid surface, the energy curve for a real surface is characterized by multiple minima points, as demonstrated schematically in Fig. 3.26. Thus, many metastable apparent contact angles exist. In principle, the system tends to get to the most stable state, which is defined by the global minimum (Fig. 3.26). However, in-between the local minima in the Gibbs energy, local maxima exist. Therefore, in order to move from one local minimum to the next, the drop has to overcome an energy barrier (defined as the energy difference between a local minimum and an adjacent local maximum, Fig. 3.26). It is important to note that the energy barrier increases as the drop gets nearer to the global minimum.

Based on this theoretical picture, it is clear that the theoretical advancing contact angle is the highest angle for which there is a local energy minimum (Fig. 3.26). Similarly, the theoretical receding contact angle is the lowest angle for which there exists a local energy minimum. In practice, the system is always subject to some energy input from the environment, for example, via vibrations caused by the drop impact, or by the environment of the system. This energy input may enable overcoming energy barriers up to a certain level. Therefore, the advancing contact angle in practice is somewhat lower than the theoretical advancing contact angle (Fig. 3.26). Likewise, the practical receding contact angle is higher than the theoretical one.

The advancing and receding contact angles are measurable quantities. To some extent, they depend on the drop volume, so they should be measured for sufficiently large drops. An experimental indication for the drop being sufficiently large is minimal *stick-slip* behavior of the contact line. In other words, as the drop volume is increased or decreased, the drop retains the contact angle at its advancing or receding value, respectively, without featuring *stick-slip*. However, the correlation between the advancing, receding and the Young contact angle is not known yet. In some cases, especially when only a comparative study is required, the advancing contact angle serves as a convenient, reproducible measure. More information can be obtained from the most stable contact angle (at the global minimum in the Gibbs energy), as described in the following.

The Most Stable Apparent Contact Angle on Rough Surfaces. When a drop spreads on a rough surface, the actual solid–liquid interfacial area is higher than the nominal (projected) interfacial area. This has to be accounted for when the minimum in the Gibbs energy of the system is sought after. The surface roughness is characterized for this purpose by the roughness ratio r which is the ratio between the actual and nominal surface area of the solid. Thus, for a smooth surface $r = 1$, and for a rough surface $r > 1$. In 1936, Wenzel developed the following equation for the apparent contact angle θ_W on a rough surface

$$\cos\theta_W = r\cos\theta_Y . \tag{3.44}$$

This equation has an interesting practical prediction. If $\theta_Y < 90°$ (good wettability of a smooth surface of the same chemistry), roughness enhances wetting, i.e. $\theta_W < \theta_Y$. If, however, $\theta_Y > 90°$, roughness diminishes wetting, namely $\theta_W > \theta_Y$. This is demonstrated in Fig. 3.27 for a few values of r.

It is important to recognize that the Wenzel equation is based on the assumption that the liquid completely penetrates into the roughness grooves (Fig. 3.28a). This wetting situation on rough surfaces is termed *homogeneous wetting*. Under some roughness conditions, especially when roughness is high, this may not be the case: air bubbles may be trapped in the roughness grooves, underneath the liquid (Fig. 3.28b). The latter situation is referred to as *heterogeneous wetting* on rough surfaces, and will be described in detail below.

It turns out, that the Wenzel equation is an approximation, which becomes better as the drop becomes larger in comparison with the scale of roughness. The question of how large the drop should be in order for the Wenzel equation to apply has not yet been fully answered. However, based on some simulations and preliminary experimental data, it seems that if the drop is larger than the roughness scale by two to three orders of magnitude, the Wenzel equation applies. This is reasonable from an experimental point of view, since typical roughness is of the order of magnitude of microns, while typical drops are of the order of magnitude of millimeters.

The Wenzel contact angle represents the most stable contact angle on a rough surface, namely the contact angle that is associated with the global minimum in the Gibbs energy of the system. The methods for measuring the Wenzel contact angle will be described below. Once its value is known, the Young contact angle can be calculated using (3.44), assuming that the roughness ratio is also known. From the Young contact angle, the surface tension of the solid can be calculated as described above for an ideal solid surface. For example, the measurements and calculations can be repeated for two different liquids, then the surface tension of the solid can be calculated from (3.42) and (3.43).

The Most Stable Apparent Contact Angle on Chemically Heterogeneous Surfaces. On a chemically heterogeneous solid surface, the surface tension varies from one spot to the other. Accordingly, the Young contact angle has a different, local value at each spot. Therefore, the characterization of chemically heterogeneous surfaces is more complex than that of rough surfaces. In general, the surface can be characterized by a properly averaged apparent contact angle.

The most stable apparent contact angle on a chemically heterogeneous surface θ_C is given by the Cassie equation, which was published in 1948 for the case of a surface with only two different chemistries

$$\cos\theta_C = x_1 \cos\theta_{Y1} + (1 - x_1)\cos\theta_{Y2}. \qquad (3.45)$$

In this equation, x is the area fraction characterized by a given chemistry, and the subscripts 1 and 2 indicate the two different surface chemistries. This equation can be generalized to state that the cosine of the Cassie contact angle is the weighted average of the cosines of all the Young contact angles that characterize the surface. The weighted averaging is done according to the area fraction of each chemistry.

Like in the case of the Wenzel equation, the Cassie equation is also an approximation that becomes better when the drop size becomes larger with respect to the scale of chemical heterogeneity. According to preliminary simulations, a size ratio of two to three orders of magnitude seems to be sufficient.

The Most Stable Apparent Contact Angle in Heterogeneous Wetting on Rough Surfaces. As mentioned above, under some roughness conditions, air bubbles may be trapped in the roughness grooves, under the liquid (Fig. 3.28b). In this case, the solid surface may be considered chemically heterogeneous, and the Cassie equation (3.45) may be applied:

$$\cos\theta_{CB} = fr_f \cos\theta_Y - (1 - f). \qquad (3.46)$$

This equation was developed by Cassie and Baxter (CB), considering air to be the second chemistry in (3.45). In (3.46), θ_{CB} is the CB apparent contact angle, f is the fraction of the projected area of the solid surface that is wet by the liquid, and r_f is the roughness ratio of the wet area. The fraction f in this equation plays the role of x_1 in (3.45), and the contact angle of the liquid

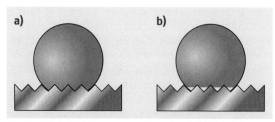

Fig. 3.28a,b Wetting on a rough surface: (**a**) homogeneous wetting – the liquid penetrates into the roughness grooves; (**b**) heterogeneous wetting – air bubbles are trapped in the roughness grooves, underneath the liquid

with air is considered to be 180°, since the shape of a sufficiently small drop in air is very close to spherical. When $f = 1$, $r_f = r$, and the CB equation turns into the Wenzel equation.

The transition from homogeneous wetting (Wenzel equation) to heterogeneous wetting (CB equation) was theoretically analyzed by *Marmur* [3.44]. It was found to be dependent not only on the roughness ratio, but also on the specific geometry, in terms of the second derivative of $(r_f f)$ with respect to f. It was also found that heterogeneous wetting on rough surfaces may lead to super-hydrophobicity, which is defined as a state of a very high contact angle ($\gtrsim 160°$) and very low roll-off (slip) angle. The latter is the inclination angle at which a drop slips from or rolls-off a solid surface. It turns out that the wetted area, i.e. the solid-liquid contact area, is much smaller in heterogeneous than in homogeneous wetting, even when the apparent contact angles are the same. Thus, it may be assumed that the force that holds the liquid to the solid is much smaller in heterogeneous wetting, thus leading to smaller roll-off angles.

Contact Angle Measurement and Interpretation. The most common method of measuring contact angles is by taking a side-view picture of a drop and evaluating the contact angle from this picture. The evaluation can be made either by direct measurement of the angle (preferably averaging the angles at the two sides of the drop), or by fitting a curve to the drop profile and calculating the angle from this curve. The fitting of the curve may be done either by using a polynomial or by using the solution to the Young–Laplace equation [as given by (3.32–3.34) of Sect. 3.2]. For many practical cases, the drop is sufficiently large for gravity to distort its shape from being spherical. Thus, fitting the shape by a circular segment is usually not satisfactory.

This method is simple, straightforward, and amenable to automation. However, the main disadvantage of this method is the lack of testing for the symmetry of the drop. If the drop is not axially symmetric, the measurement of the contact angle is questionable. Therefore, it is essential also to take a top view of the drop. If this picture of the drop assures axisymmetry, then the calculations based on the side-view are meaningful. Alternatively, it is possible to take only a top-view of the drop, and use the maximum drop diameter, drop volume, and the surface tension of the liquid to calculate the contact angle from the Young–Laplace solution for the drop shape. The drop volume is obtained via its weight and density.

In terms of interpretation, two approaches may be taken:

1. the hysteresis approach, and
2. the most stable apparent contact angle (MSACA) approach.

In the hysteresis approach, one measures the advancing and receding contact angles, θ_a and θ_r, and tries to estimate from them either the most stable contact angle θ_{ms} or the Young contact angle. The measurement of the advancing contact angle is done by increasing the drop volume until the highest possible contact angle is reached. Similarly, the receding contact angle is measured by decreasing the drop volume until the lowest possible contact angle is reached. Three methods of interpretation have been suggested within this approach. *Decker* et al. [3.45] suggested taking an average of the contact angles themselves:

$$\theta_{ms} = \frac{(\theta_a + \theta_r)}{2} . \tag{3.47}$$

Andrieu et al. [3.46] proposed an average of the cosines

$$\cos \theta_{ms} = \frac{(\cos \theta_a + \cos \theta_r)}{2} . \tag{3.48}$$

The third method is applicable only to rough surfaces, and aims at elucidating the Young contact angle. *Kamusewitz* et al. [3.47] suggested the following steps:

1. measuring the advancing and receding contact angles for a series of surfaces of the same chemistry but varying degrees of roughness;
2. drawing the advancing and receding contact angles versus the hysteresis range $(\theta_a - \theta_r)$;
3. fitting the best straight lines for the two sets of data in step 2; and
4. getting the Young contact angle from the extrapolated values of these two lines to zero hysteresis range.

In the MSACA approach, one tries to directly measure the most stable contact angle. The basic underlying idea is that by properly vibrating the surface, the drop may overcome the energy barriers and get to its most stable state. The practical problem is how to identify this most stable state. It was theoretically proven that a sufficiently large drop must be axisymmetric on a horizontal, real surface. The opposite statement has not been proven, but a working hypothesis based on it may be applied: a drop becomes more axisymmetric as it is vibrated and approaches the most stable state. Thus, the measurement procedure involves vibrating the surface while monitoring the symmetry of the drop by viewing it from above; once the drop becomes symmetric, its contact angle is measured as described above.

Unfortunately, none of the above approaches has been sufficiently studied and substantiated. Also, no standard surfaces for comparing and calibrating contact angle measurements seem to exist. However, careful performance of contact angle measurement according to the above description and explanation may yield useful information. In particular, it should be emphasized that the following two principles should be applied in all cases and methods:

1. the drop should be sufficiently large, at least two orders of magnitude larger than the scale of roughness or chemical heterogeneity; and
2. the drop should be axisymmetric at the time of measurement.

3.3.2 Dynamic Contact Angle

Here, the concept of the dynamic contact angle, its origin and significance are introduced and current theoretical interpretations are briefly surveyed. The general experimental techniques used in its measurement are then discussed in detail with particular reference to the method by which a liquid drop is applied to a flat solid surface and allowed to spread. Other methods are also outlined to illustrate some of the experimental problems and their solution.

Introduction
Significance of the Dynamic Contact Angle. Many industrial and material processing operations require a liquid to be spread onto a solid. Examples include coating, painting, printing, plant protection, gluing and lubrication. The liquid may be paint, ink, insecticide, adhesive or some other liquid, and may be Newtonian or rheologically complex. Similarly, the solid may have a surface that is smooth or rough, uniform or chemically heterogeneous. It may be shaped as a sheet or a fibre or have some more complicated shape, and it may be porous. Evidently the properties of the materials involved can vary widely.

Apart from the fundamental problem of whether a given solid is wetted by the liquid in question, which is discussed in Sect. 3.3.1, many of the practical applications require a precise knowledge of how the rate of wetting affects the process. In particular, it is often helpful to know just how fast a liquid will wet or can be made to wet a given area of the solid surface. Such information is useful in process optimisation. The underlying phenomena are also relevant to our understanding of many other processes such as oil recovery from porous rocks and ground-water flow.

The dynamic behaviour of a liquid front moving across a solid has been studied extensively over the past several decades. A variety of different configurations have been examined, but most studies have been restricted to one or more of the following:

- drops spreading on a flat substrate – relevant to inkjet printing, the spraying of liquids such as paint or insecticide, etc.;
- drops moving down an inclined plane – relevant to droplet runoff;
- liquid penetration into capillary tubes or between parallel plates – relevant to flow in porous media;
- solid substrates, such as plates, cylinders, flexible tapes or fibres being drawn into or out of liquids – useful for fundamental coating studies;
- more complex configurations such as those involved in industrial coating processes, e.g., curtain and roller coating.

In simple cases, the main parameters required to quantify the dynamics of wetting are the relative velocity at which the liquid moves across the solid, i. e., the contact-line velocity v, and the dynamic contact angle θ_D, i. e., the angle formed between the moving liquid interface and the solid surface. The dynamic contact angle is the key boundary condition for the wetting process. Significantly, the experimentally observed dynamic angle generally differs from the static contact angle θ_S and may refer to either an advancing (wetting) or a receding (dewetting) interface. Since solid surfaces are often rough or chemically heterogeneous, even equilibrium contact angles may not be single-valued, but will depend on whether the interface has been advanced or recessed a phenomenon known as contact angle hysteresis (Sect. 3.3.1). On such surfaces, contact lines tend to pin, and when they do eventually move they do so in an unsteady way. Such factors complicate both the measurement and the interpretation of the contact angle. This is especially true on surfaces that swell or reorganise in some way on contact with the liquid.

Origin of the Dynamic Contact Angle. In forced wetting or forced dewetting, the contact line is made to move by application of an external force. In such cases, a single functional relationship is expected between θ_D and

v for any given system under a given set of conditions. On changing the system or the experimental conditions (e.g., the flow rate in a coating experiment) the precise form of this relationship may change. Recent studies seem to suggest that the form of the relationship depends on the flow field in a fairly complex way [3.48, 49], so that θ_D may not be a material property of the system at any level. Nevertheless, it is generally observed that advancing angles increase while receding angles decrease with increasing rates of steady contact-line displacement. In other words the contact angle depends on both speed and direction of displacement, i.e., it is velocity dependent. This is shown schematically in Fig. 3.29 for a system that also exhibits contact angle hysteresis. While anomalies have been observed [3.50], the relationship between θ_D and v is usually monotonic.

On the other hand, if we deposit a drop of liquid onto a solid it will tend to spread spontaneously under capillary forces alone (spontaneous wetting). Under these transient conditions, the instantaneous dynamic contact angle will relax, decreasing from 180° at the moment of contact towards its static value. At the same time, the contact-line velocity will decrease from its initial value to zero at equilibrium [3.51]. The reverse situation is observed if we forcibly spread a liquid on a surface that it wets only partially and then allow it to break up and retract into individual droplets (spontaneous dewetting). In this case, the contact angle will increase from its value on rupture towards its static value. Because of contact angle hysteresis, the final, static values may differ.

Since both forced and spontaneous wetting and dewetting are examples of moving contact lines, it should be possible to describe them in some equivalent way. Since the processes occur at a finite rate, possibly with associated changes in the shape of the liquid, but certainly with changes in the wetted area, the wetting processes must be dissipative. Indeed, the fact that the observed dynamic contact angle differs from its equilibrium value is evidence of this. Several attempts have been made in the literature to explain the observed behaviour, however these boil down to essentially two approaches, which differ from each other mostly in their consideration of the effective dissipation channel.

One of these two approaches, commonly known as the hydrodynamic theory, emphasises the dissipation due to viscous flows within the slowly-moving wedge of liquid near the contact line [3.52–57]. Changes in the observed dynamic contact angle are then ascribed to viscous bending of the liquid interface in this mesoscopic region. The microscopic angle θ_m is usually assumed to retain its static value θ_S.

The other approach, which originates from the Frenkel/Eyring view of flow as a stress-modified molecular rate process, discards dissipation due to viscous flow and focuses instead on that occurring in the immediate vicinity of the moving contact line due to the process of attachment or detachment of fluid particles to or from the solid surface [3.50, 58, 59]. According to this view, the channel of dissipation is effectively the dynamic friction associated with the moving contact line [3.60], and the microscopic contact angle is velocity dependent and identical with the experimentally observed angle. This approach is usually termed the molecular-kinetic theory.

A full discussion of these theories is beyond the scope of this chapter, but it is helpful to outline the basic equations and give some examples of the magnitude of the relevant parameters. In its most simple form, the equation describing the change in the dynamic contact angle due to viscous bending may be written in terms of the capillary number $\text{Ca} = (\mu v/\sigma)$ as

$$\theta_D^3 - \theta_S^3 = 9\text{Ca} \ln\left(\frac{L}{L_m}\right), \quad \theta_D < 3\frac{\pi}{4}, \quad \theta_m = \theta_S, \tag{3.49}$$

where μ and σ are, respectively, the dynamic viscosity and surface tension of the liquid and L and L_m are, respectively, appropriately chosen macroscopic and microscopic length scales. Setting $L = 10\,\mu\text{m}$, which is the approximate distance from the contact line at which the contact angle can be measured, and $L_m = 1\,\text{nm}$, i.e., the order of molecular size, then $\ln(L/L_m)$ is estimated to

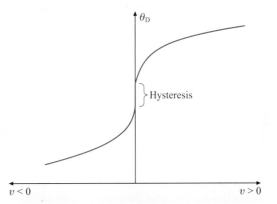

Fig. 3.29 Schematic representation of the velocity dependence of the contact angle, illustrating the behavior of the dynamic advancing and receding contact angles for a system that also exhibits contact angle hysteresis. If hysteresis is present, very low steady contact-line velocities may not be experimentally accessible

be of the order of 10. Experimental values vary widely, though values of about 10 have been found frequently for liquids that completely wet the solid. Much larger values have sometimes been reported for non-wetting liquids. These are usually thought to be non-physical, since they would appear to involve sub-molecular microscopic length scales.

For small drops and small angles (3.49) leads to simple scaling laws for the base radius R and the instantaneous contact angle as a function of time t:

$$R(t) \sim t^{1/10}, \quad (3.50)$$
$$\theta_D(t) \sim t^{3/10}. \quad (3.51)$$

These relationships have been widely confirmed for completely wetting liquids.

According to the contrasting viewpoint provided by the molecular-kinetic theory, the motion of the contact line is determined by the statistical dynamics of the molecules within its immediate vicinity. The key parameters are κ^0, the equilibrium frequency of the random molecular displacements occurring along the contact line, and λ the average distance of each displacement. In the simplest case, λ is supposed to be the distance between adsorption sites on the solid surface. The driving force for the wetting line to move in a given direction is taken to be the out-of-balance surface tension force that arises when wetting equilibrium is disturbed: $\sigma(\cos\theta_S - \cos\theta_D)$. The resulting equation for the contact-line velocity is then

$$v = 2\kappa^0\lambda \sinh\left[\sigma(\cos\theta_S - \cos\theta_D)\frac{\lambda^2}{2k_B T}\right], \quad (3.52)$$

where k_B is the Boltzmann constant and T the temperature. If the argument of the sinh function is small, (3.52) reduces to its linear form

$$v = \kappa^0\lambda^3\sigma\frac{(\cos\theta_S - \cos\theta_D)}{k_B T}$$
$$= \frac{1}{\zeta}\sigma(\cos\theta_S - \cos\theta_D), \quad (3.53)$$

where $\zeta = k_B T/\kappa^0\lambda^3$ is the coefficient of contact-line friction. This has the same units as dynamic viscosity. While the experimentally determined values of λ are usually of molecular dimensions, those of κ^0 can vary widely. Some examples are listed in Table 3.10. Consistent with theory, experimental values of ζ are always larger than the viscosity of the liquid (Table 3.10), and appear to increase both with viscosity and the strength of solid–liquid interactions.

Crucially, (3.52) and (3.53) predict scaling laws that differ from those predicted by (3.49):

$$R(t) \sim t^{1/7}, \quad (3.54)$$
$$\theta_D(t) \sim t^{3/7}. \quad (3.55)$$

Comparison with experiment therefore offers the opportunity of assessing which mode of dissipation may be occurring.

Despite their fundamentally different bases, both models have been shown to work fairly well for experimental liquid/solid systems, though problems remain. As it seems likely that both types of dissipation occur simultaneously, combined theories have been suggested [3.61–63]. For example in the case of a spreading drop, it has been predicted and apparently confirmed that the wetting-line friction regime precedes the viscous regime, which becomes dominant only as the contact angle becomes small [3.63, 64]. However, it is also important to mention the continuum hydrodynamics approach adopted by *Shikhmurzaev* [3.63]. This accommodates dissipation through standard hydrodynamic channels, but also exploits non-equilibrium thermodynamics to describe dissipation due to the solid/liquid interfacial formation process. One consequence of this approach is that the microscopic dynamic contact angle is coupled directly to the flow. Analytical expressions can be obtained for certain simplifying conditions, such as small capillary number, which are very effective in describing the experimental results found in the literature [3.65, 66].

The reason for briefly discussing these theoretical interpretations in a chapter devoted to experimental methods is two-fold. First, experimental results, unless used in a purely descriptive way, will have to be interpreted in terms of a model of some sort, though it is important not to prejudge the observations in terms of any given model. Secondly, it is wise to ensure that sufficient ancillary measurements are made to enable the results to be interpreted as fully as possible. For example, in specifying the liquid, it will usually be essential to have sufficiently accurate measurements of its viscosity (assuming this to be Newtonian, which is not always the case with industrially interesting liquids), surface tension and density (Sects. 3.1, 3.2 and 3.4). Similarly, if comparisons are to be made between different flows, then it is important to determine parameters such as geometry and flow rate that properly characterize them. It may also be desirable to determine other factors such as the roughness of the solid surface. In this latter case, techniques such as profilometry, optical scanning mi-

Table 3.10 Examples of the values of parameters obtained by applying the molecular-kinetic theory to experimental data for various systems

System	μ [Pa s]	σ [mN/m]	θ_S [°]	λ (nm)	κ^0 [s^{-1}]	ζ [Pa s]
Water on PET [3.50]	0.001	72.4	82	0.36	8.6×10^9	0.01
16% glycerol in water on PET [3.50]	0.0015	69.7	72.5	0.46	3.6×10^9	0.012
86% glycerol in water on PET [3.50]	0.104	65.8	65	0.46	3.5×10^7	1.2
Di-n-butyl phthalate on PET [3.51, 67]	0.196	34.3	< 7	1.8	1.1×10^5	6.4
Silicone oil on glass [3.50, 68]	0.958	21.3	0	0.79	2.3×10^5	35.9
Silicone oil on glass [3.50, 68]	98.8	21.7	0	0.79	2.3×10^3	3580

croscopy or, ultimately, atomic force microscopy may be applied.

Useful reviews concerning the dynamic contact angle and its interpretation can be found in the references cited above, especially [3.50, 56, 57, 65]. For further reading, *Dussan* [3.69] provides a perceptive survey of the older literature, covering both theoretical and experimental aspects, while *Blake* and *Rushak* [3.70] set the dynamic contact angle in the context of modern coating processes.

Generic Methods
General Principles and Techniques. The general principles of dynamic contact angle measurement can be illustrated by describing the experimental techniques used to study a liquid drop spreading on a flat solid substrate. Obviously, these techniques can equally be applied to the measurement of static contact angles, as discussed in Sect. 3.3.1.

The drops are usually supplied via a microsyringe, the needle of which is ground at right angles. Drop volumes of the order of 1 μl or larger are typical. If appropriate to the problem under investigation, useful measurements can be made with much smaller droplets, e.g., 100 pl [3.71], but these require special deposition techniques. For convenience, the solid substrate should be mounted on a three-axis translational stage. Both microsyringe and translation stage can be motorized and computer controlled for repeat measurements. If required, the stage can be thermostatically heated and the whole system enclosed in an environmental chamber. Sufficient time must be allowed to equilibrate the system before commencing measurements.

The profiles of the drops are easily captured using a high-resolution black and white digital video camera equipped with a suitable macro lens or long working distance microscope having the necessary magnification. The video camera is connected to a PC, which enables the images to be processed in real time and/or stored for subsequent processing. To get clear, sharp profiles, proper illumination is crucial. Diffuse uniform back-lighting seems to be best for routine video imaging of droplet profiles, whereas a collimated light source (Koehler illumination) is better for very precise measurements in the vicinity of the wetting line or for very small drops. Some means of adjusting the intensity of the light is desirable, such as a proprietary control unit or neutral density filters.

Existing set-ups are capable of capturing objects with sizes ranging from a fraction of a millimetre (such as ink-jet droplets) to about one centimetre. A typical image of a sessile drop is shown in Fig. 3.30. A computer program using suitable edge-detection algorithms and contour fitting then finds the contact angles from the profile. Typically, a full profile is discretised into about 1500 points and the best parameters are calculated in a few seconds.

Deposited drops are not always axially symmetric, so it is helpful to divide the data set for each drop into four parts, specifically the left and right side of the drop and left and right side of its reflection in the substrate. For each part we can calculate the best parameters to fit the Young–Laplace equation (Sect. 3.2) for the capillary pressure drop across a curved interface. For a drop in a gravitational field in the z direction, the equation may be written as

$$\Delta \rho g z = \sigma \left[\frac{z''}{\left(1 + z'^2\right)^{1/2}} + \frac{z'}{x \left(1 + z'^2\right)^{1/2}} \right] , \tag{3.56}$$

where $z' = dz/dx$, $z'' = d^2z/dx^2$, $\Delta \rho$ is the difference in density between the drop and the air and g the acceleration due to gravity. In doing this, we are assuming implicitly that the Young–Laplace equation can describe the shape of non-equilibrium interfaces. This is true if the capillary number based on the contact-line velocity is small, but even if this approximation is not strictly valid, the procedure will usually give an acceptable fit to the experimental data. See [3.72] for an experimental investigation of the influence of viscous deformation on the shape of a liquid meniscus during the immersion of a vertical cylinder into a pool of liquid.

To calculate the contact angle between the liquid and the substrate, it is vital to locate the exact position of the contact line. To find this with high precision, one should expand or contract the calculated profiles of the drop and its reflection, until they intersect each other symmetrically about the plane of the solid. The angle of the resulting curve with this plane is then easily calculated. Using this procedure, it is possible to calculate the left and right contact angles independently. Any small difference between the two angles is an indication of the precision of the measurement and/or the non-uniformity of the solid surface. This technique also allows one to edit the profile to eliminate anomalies. For example, it allows one to leave the needle used to deposit the liquid inside the drop while capturing the image; any deformation associated with the needle is edited out afterwards. To minimise the effect on the precision of the measurements, the diameter of the needle should be less then 1/3 of the diameter of the drop. With the needle inside the drop, it is possible to measure both advancing and receding contact angles by adding or subtracting liquid.

Another possibility is to set the focus on one contact region, rather than the whole drop. In this way, a higher magnification can be used, but information about the rest of the profile is lost. The profile of the part of the drop near the solid and its reflection can be fitted by a simple curve, such as a circular arc or a straight line. Both kinds of fits are satisfactory, and result in systematic differences in the contact angle of no more than 2°. One of the main advantages of this approach is that it can be fully automated and applied to geometries other than a spreading drop. It is particularly useful in forced wetting experiments, such as those involving plunging tapes (Fig. 3.31a) [3.66, 73] or coating processes (Fig. 3.31b) [3.49, 74, 75] where the wetting line remains more-or-less stationary in the frame of observation. With plunging tapes, a clearer image of the meniscus is obtained if the tape is curved slightly across its width at the point of entry into the liquid, creating what amounts to a plunging cylinder. By viewing along a tangent to the curve the contact angle can be measured at a single point rather than across the whole width of the tape [3.66]. Further details concerning dynamic contact angle measurements in these more specialised situations can be obtained from the references given.

A standard video system captures 50 or 60 images per second, giving one image every 20 or 17 ms. However, high-speed cameras and recording systems with rates up to at least 1000 frames per second are readily available and are especially helpful in determining the contact angle in the early stages of spreading [3.71]. Even higher effective framing rates are attainable using stroboscopic methods and suitable triggering. With such techniques, a temporal resolution of 1 µs is relatively straightforward and is especially helpful in studying rapidly evolving phenomena such as droplet deposition. However, high-speed imaging necessitates very high data transfer rates and may pose storage problems. Applied to dynamic contact angle measurements, they are more suitable if used with circular-arc or straight-line fitting techniques rather than those involving fitting entire profiles.

Computer programs for controlling drop deposition, image-capture and data processing can be custom written using standard methods. For image processing public domain packages such as NIH image are freely available [3.76]. Alternatively, all aspects of control and contact angle measurement can be implemented using high-level commercial software such as LabView or other scientific programming language, paying special attention to the quality of the edge detection algorithm. Commercial contact angle apparatus incorporating many of the features described above can also be purchased, but care should be taken to ensure they fully meet experimental requirements.

In practice, the solid/liquid systems that can be analyzed using the techniques described above have to meet the following requirements. First, the time of spreading should fall within the time frame of the equipment. This means that the contact angle should not change too much between consecutive images. Drop spreading exhibits the highest velocities just af-

Fig. 3.30 Image of the profile of a sessile drop of water on an oxidized silicon wafer chemically grafted with octadecyltrichlorosilane. Here the equilibrium contact angle is 105°

ter deposition. Secondly, the first angle that can be measured clearly has to differ significantly from the equilibrium angle. If this is not the case, the dynamic behaviour is too fast to be captured. For example in inkjet printing most of the spreading process is over within 0.1 s [3.71]. Thirdly, the contact angles have to be larger then about 5°. Below this value, the angle cannot be distinguished from zero. Special techniques are required such as the optical interference methods used to study thin liquid films. It follows also that angles greater than about 175° cannot be distinguished from 180°.

Example. Figure 3.32 shows the results obtained on depositing 0.5 to 1 µl droplets of di-*n*-butyl phthalate (DBP) onto poly(ethyleneterephthalate) (PET) [3.51]. The dynamic contact angle is plotted as a function of time after deposition. The density, dynamic viscosity and the surface tension of the DBP (Fisons, SLR grade, 99%) were respectively $1.04\,\mathrm{g\,cm^{-3}}$, $19.6\,\mathrm{mPa\,s}$ and $34.3\,\mathrm{mN/m}$ at the temperature of the experiment (21 °C). The PET was provided by Kodak Ltd. as a flexible and transparent 35 mm tape. When fixed to a rigid solid (in this case a glass slide), the surface is flat. The PET has a low roughness and shows a homogeneous, moderately low energy surface giving small contact angle hysteresis with organic liquids. The DBP droplets eventually spread to achieve a static advancing contact angle of less than 7°.

Repeatability of the experimental curves is excellent except for the first few data. If we adjust the starting time, the curves collapse perfectly. This means that the experiment is fully reproducible, except for the initial condition. This is not a problem, as all the equations can be expressed in terms of velocity relative to the initial time. The standard deviation of the data indicates the error on the individual measurements to be to be less than 1°.

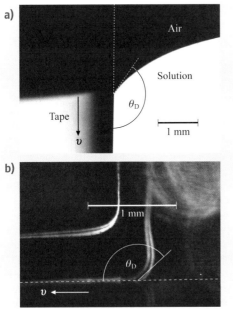

Fig. 3.31 (a) Image of a polyester tape entering a pool of aqueous glycerol at 0.05 m/s [3.70]. The tape is bent into a slight curve (to the left). By viewing along a tangent to the curve, the contact angle is measured at a single point rather than across the whole width of the tape. The air appears dark due to internal reflection. The dynamic contact angle through the liquid was determined to be $(143\pm2)°$. (b) Flow visualization of curtain coating showing the liquid impinging onto the moving substrate and being coated to the left as a uniform layer [3.49]. The various interfaces are marked by light scattered from streams of hydrogen bubbles, which are reflected in the interfaces and therefore appear doubled. The contact angle through the liquid was determined to be $(140\pm5)°$

Data Analysis. Figure 3.33 shows advancing dynamic contact angle data for DBP on PET obtained using the plunging tape method [3.51, 67]. At the lowest experimental velocity, the contact angle is about 16°. The contact angle increases with increasing velocity. Air entrainment is seen when the contact angle is close to 180° ($v \approx 52\,\mathrm{cm/s}$). This kind of curve is typical of the steady-state wetting behaviour observed for Newtonian liquids [3.50]. In forced wetting experiments, such as this, one obtains the dynamic contact angle directly as a function of wetting velocity, allowing direct theoretical comparison by standard curve-fitting techniques.

On the other hand, with spreading drops, some analysis is required to obtain the data in this direct form or to compare the results with theory. Only then can the resulting parameters be related to the physical characteristics of the system. To do this, one can make use of the scaling laws given in Sect. 3.1.2, however a more thorough analysis is preferable. For a fixed set of flow conditions, all the equations describing the different theoretical models can be expressed in general terms as

$$v = \frac{\mathrm{d}R}{\mathrm{d}t} = f(\theta_\mathrm{D}, \theta_\mathrm{S}, \sigma, \mu \ldots P_1, P_2 \ldots) \,, \quad (3.57)$$

where f is an independent or at most a weak function of the base radius R, and P_1, P_2, etc. stand for the theory-specific parameters of interest, e.g., L/L_m for

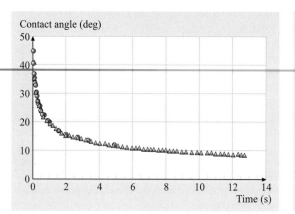

Fig. 3.32 Dynamic contact angle as a function of time for droplets of DBP on PET. The plot shows two sets of data: (*circles*) spreading followed for 5 s, and (*triangles*) spreading followed for 13 s (after [3.51])

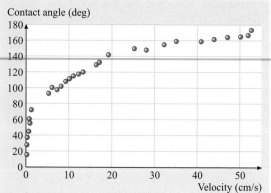

Fig. 3.33 Dynamic contact angle as a function of contact-line velocity for a PET tape plunging into a pool of DBP (after [3.51, 67])

the hydrodynamic theory, or either ζ or $\kappa^0 \lambda$ and $\lambda^2/k_B T$ for the molecular-kinetic theory. In addition, we have the following relation between R and the instantaneous contact angle θ at any time t:

$$R = \left(\frac{3V}{\pi}\right)^{1/3} \frac{\sin\theta}{(2 - 3\cos\theta + \cos^3\theta)^{1/3}}. \quad (3.58)$$

Here, it is assumed that the drop has the shape of a spherical cap (Bond number $\mathrm{Bo} = \Delta\rho g V^{2/3}/\sigma \ll 1$) and that the volume of the drop V is constant (the liquid has low volatility). Differentiating this equation yields

$$\frac{dR}{dt} = -\left(\frac{3V}{\pi}\right)^{1/3} \frac{(1-\cos\theta)^2}{(2 - 3\cos\theta + \cos^3\theta)^{4/3}} \frac{d\theta}{dt}. \quad (3.59)$$

Finally, on combining (3.53) and (3.55), we obtain

$$\frac{d\theta}{dt} = -\left(\frac{\pi}{3V}\right)^{1/3} \frac{(2 - 3\cos\theta + \cos^3\theta)^{4/3}}{(1-\cos\theta)^2} \\ \times f(\theta_D, \theta_S, \sigma, \mu, P_1, P_2 \ldots). \quad (3.60)$$

For a given system and initial conditions [e.g., $\theta(t=0) = 180°$], (3.60) may be solved by a Runge–Kutta algorithm [3.77] to find the curve $\theta(t)$. This theoretical curve is then compared with the experimental data. The total difference between both curves is minimized by adjusting the values of the parameters P_1, P_2, etc. using the downhill Simplex or Levenberg–Marquard methods [3.77]. Although this provides one with the best-fit parameters related to the experimental data, it does not give any indication of the expected errors on these values. Indeed, the calculated total difference is merely a measure of the quality of the fit, not its reproducibility. Furthermore, if the system is overdimensioned (e.g., if the parameters P_1, P_2, etc. are linearly dependent), the error on one or more parameters can diverge to infinity. To analyse the reproducibility of the fits, one may use the so-called bootstrap method [3.77]. With this powerful method, the original experimental data are used as the basis for a Monte Carlo simulation. From the original data set, 37% (i.e., $1/e$) randomly chosen points are replaced by duplicates. The duplicate points are chosen according to a normal distribution function, with the original data as average and the expected error on each datum as the standard deviation. In this way, we replace the original data set with a new one, for which the corresponding parameters are then calculated using the method described above. Successive cycles result in a simulated set of values for each parameter. If enough cycles are used (> 100), these sets turn out to be normally distributed, providing one with the mean value and standard deviation for each parameter.

Other Techniques

In this chapter, the spreading drop technique has mainly been used to illustrate the generic methods of dynamic contact angle measurement. However, several others, such as those based on plunging tapes and coating visualisation, have been touched on, as they are very powerful methods of studying forced wetting and are therefore prominent in the current literature. For similar reasons, two other approaches also deserve specific mention.

The first involves calculating the dynamic contact angle from the force exerted on a plate or fibre as it immersed or withdrawn from a pool of liquid [3.78]. Suitable equipment is available commercially or can be adapted from a Wilhelmy plate surface tension apparatus (Sect. 3.2.5). Useful data can be obtained over a good range of low speeds (typically 10^{-4}–1 cm/s) provided the solid is available in a suitable form. The surface tension of the liquid must be determined in a separate experiment, e.g., using a completely wetted plate ($\theta_S = 0$).

The second approach is based on direct or indirect measurements of the dynamic contact angle during capillary flow. In principle, the angle can be determined indirectly from measured parameters (such as the meniscus velocity, distance of penetration, and/or the pressure drop down the tube) provided the flow can be accurately modelled [3.79]. But, this is difficult and the most reliable methods are usually based on direct microscopic observation of the moving meniscus [3.68, 80, 81]. Both the Wilhelmy plate and capillary flow methods can be used in steady-state or transient modes.

Finally, mention should be made of simulation methods in which the "experiments" are done using computational techniques such as molecular dynamics [3.60, 82]. Traditionally, these are not seen as experimental methods, however as computer power increases, such techniques will find increasing value where true physical experiments are currently very difficult or essentially impossible.

3.4 Viscosity

In this section viscometers for the measurement of the viscosity of Newtonian fluids over a wide range of thermodynamic states with high precision are described. In the first three types of viscometers considered, a solid body is forced to perform oscillations while immersed in the fluid. The characteristics of the oscillation are determined by the viscosity of the fluid. In the two other types of viscometers non-oscillatory flow is employed, in one case by the fluid flow through a capillary tube, and in the other a body of revolution falling under gravity through the fluid. Each has its merits, which are discussed.

The dynamic viscosity μ of a fluid is a measure of its tendency to dissipate energy when it is disturbed from equilibrium by a velocity field $\boldsymbol{v}$, which distorts the fluid at a rate $\dot{\varepsilon}_{ij}$ given by

$$\dot{\varepsilon}_{ij} = \frac{1}{2}\left(\frac{\partial v_i}{\partial x_j} + \frac{\partial v_j}{\partial x_i}\right) . \tag{3.61}$$

For viscous Newtonian fluids the viscosity is defined by the constitutive equation

$$\sigma_{ij} = -P\delta_{ij} + \mu\left(\frac{\partial v_i}{\partial x_j} + \frac{\partial v_j}{\partial x_i} - \frac{2}{3}\delta_{ij}\frac{\partial v_k}{\partial x_k}\right) . \tag{3.62}$$

In this equation, σ_{ij} are the instantaneous stresses P the pressure and δ_{ij} is the Kronecker symbol. In (3.62) it is assumed, as usually done, that the so-called volume (dilatational) viscosity is zero. The viscosity depends on the thermodynamic state of the fluid and is usually specified by the pairs of variables (T, P) or (T, ρ) for a pure fluid, to which must be added a composition dependence in the case of mixtures. While (3.62) incorporates the viscosity it does not immediately suggest means of measurement because it is impossible to measure local shear stresses. It should further be pointed out that the type of rheometers that will be discussed undoubtedly will have a shear rate that is not constant – thus, the measured moment of the shear stress is measured at one location but the rate of strain is not known at the same location. Thus, methods of measurement of the viscosity must be based on the determination of some integral effect of the stresses amenable to precise measurement in a known flow field. Inevitably, the imposition of a shear field generates small pressure differences and dissipation causes local temperature gradients, both of which change slightly the reference thermodynamic state to which a measurement is assigned, from the initial, unperturbed, equilibrium state. The reference state for the measurement will be obtained by averaging so that it is important that the system is disturbed as little as possible from equilibrium during measurement.

3.4.1 Oscillating-Body Viscometers

Oscillating-body viscometers consist of an axially symmetric body suspended from a torsion wire so that the body performs oscillations in the fluid about its axis of symmetry. The oscillator can be a disk (oscillating freely or between fixed plates), a sphere or a cup. The fluid usually surrounds the oscillator but, in the case of the sphere or the cup, it can be inside them. The suspension wire is elastic and the system is gently rotated to start a motion. The fluid exerts a viscous drag on the oscillator causing the angular frequency of the oscillation ω and the damping decrement Δ of the resulting simple harmonic motion to be different from those in vacuum, ω_0 and Δ_0. The magnitude of the change depends on the viscosity and on the density of the fluid in addition to the physical characteristics of the oscillator. Measurements of ω, ω_0, Δ, and Δ_0 can give the viscosity of the fluid with very low uncertainty, if the density is known from an independent determination. The only measurements necessary in this technique are those of mass, length and time, in addition to those of the pressure and temperature, all of which can be made with a very high resolution. It is relatively straightforward to make the induced perturbation small, so that a near-equilibrium state is maintained.

The characteristic equation for the torsional motion of any oscillating body viscometer can be expressed [3.83, 84] as

$$(s+\Delta_0)^2 + 1 + D(s) = 0. \tag{3.63}$$

Here, $D(s)$ is the torque on the body, calculated from solutions for the fluid flow close to the oscillating body. It is therefore the torque $D(s)$ that is characteristic of the particular type of oscillating viscometer employed.

Apart from roots that are associated with the initial oscillatory behavior of the body, the roots associated with the long-time behavior [3.83, 84] can be written in the form

$$s = \pm(i-\Delta)\theta \tag{3.64}$$

where i is the imaginary unit and

$$\theta = \frac{\tau_0}{\tau}. \tag{3.65}$$

Here, τ is the period of oscillation in the fluid and the subscript zero denotes the same quantity in vacuo. The motion associated with these roots, characterized by an angular displacement a of an axially symmetric body, is

$$a(t) = a_0 \mathrm{e}^{-\Delta\theta\omega_0 t} \cos(\theta\omega_0 t), \tag{3.66}$$

and this is the motion observed experimentally after an initial transient of about two periods.

Since the characteristic equation is complex, two equations, the real and imaginary parts are obtained from (3.63).

$$2\theta(\Delta_0 - \theta\Delta) + \mathrm{Im}[D(s)] = 0, \tag{3.67}$$

$$(\Delta_0 - \theta\Delta)^2 + 1 - \theta^2 + \mathrm{Re}[D(s)] = 0. \tag{3.68}$$

Equation (3.68), for the real part, can be used to determine viscosity but it requires a higher precision in measurement to achieve the same accuracy in the viscosity as can be attained using the imaginary part of the equation. Consequently (3.67) for the imaginary part has usually been preferred.

Before proceeding it is important to introduce a natural length scale that appears in oscillatory systems, namely the boundary-layer thickness defined as

$$\delta = \left(\frac{\mu}{\rho\omega_0}\right)^{1/2}. \tag{3.69}$$

This natural length scale is important in the selection of the dimensions of oscillating-body viscometers for particular purposes as the following three sections describing specific oscillating-disk, oscillating-cup and oscillating-sphere viscometers will show.

Oscillating–Disk Viscometers

One of the first very successful oscillating-disk viscometers was developed by *Kestin* and *Leidenfrost* [3.85] in 1959, to measure the viscosity of gases near room temperature, in the pressure range 0.1–60 MPa.

The basic design of such a viscometer, consists of a disk of radius R, thickness d and moment of inertia I, oscillating between two parallel fixed plates at distances b_1 and b_2 from its surfaces. Figure 3.34 shows a development by *Vogel* [3.86] of the early disk viscometers of Kestin and his collaborators. The configuration between two fixed plates has generally been preferred to a free disk, because the presence of the two parallel fixed plates tends to increase the viscous drag on the disk, and hence produce a decrement that is easier to measure with high accuracy. The plates also reduce the likelihood of influence from spurious free convective flows within the bulk of the fluid. The characteristic equation for the torque $D(s)$ for this configuration is [3.84]

$$D(s) = \frac{\pi R^4 \mu s}{I b \omega_0}\left[C_N + \frac{s}{3}\beta_1\beta_2 - \frac{s^2\beta}{90}(\beta_1^2+\beta_2^2) \right.$$
$$\left. + \frac{s^3\beta}{945}(\beta_1^2+\beta_2^2) + \ldots\right], \tag{3.70}$$

where

$$\beta_1 = \frac{b_1}{\delta}, \qquad \beta_2 = \frac{b_2}{\delta}, \tag{3.71}$$

$$\beta = \frac{b}{\delta}, \qquad b = \frac{2b_1 b_2}{(b_1+b_2)}. \tag{3.72}$$

Equation (3.70) is valid if

$$b_1 + b_2 + d \ll R \quad \text{and} \quad b_1 + b_2 + d \ll \delta. \tag{3.73}$$

These conditions are readily met in the dilute and moderately dense gas states, so that (3.70) can be employed as the basis of the evaluation of the viscosity of the fluid. The quantity C_N is an instrumental constant that depends only on the linear dimensions of the oscillatory system. It is unity for a disk of infinite radius and zero thickness. Expressions can be found for it for real geometries in the literature [3.83, 84]. The instrument constant can thus be evaluated independently from the characteristics of the assembled instrument. This allows the formulation of the final working equation for the disk instrument to be written as

$$C_N = \left[\frac{2I}{\pi \rho b R^4}(\theta\Delta - \Delta_0) + f_1\theta\Delta\right]\beta^2 \\ + f_2\theta^2(3\Delta^2 - 1)\beta^4 + f_3\theta\Delta(\Delta^2 - 1)\beta^6 , \quad (3.74)$$

where the coefficients f_1, f_2 and f_3 only depend on the distances d_1 and d_2 between the disk and the plates

$$f_1 = \frac{1}{6}\left[\left(\frac{d_1}{d_2}+1\right) + \left(\frac{d_2}{d_1}+1\right)\right] , \quad (3.75)$$

$$f_2 = \frac{1}{720}\left[\left(\frac{d_1}{d_2}+1\right)^3 + \left(\frac{d_2}{d_1}+1\right)^3\right] , \quad (3.76)$$

$$f_3 = \frac{1}{7560}\left[\left(\frac{d_1}{d_2}+1\right)^5 + \left(\frac{d_2}{d_1}+1\right)^5\right] . \quad (3.77)$$

Equation (3.74) is a cubic equation in the square of β and the coefficients depend only on measurable quantities. The instrument can be used to obtain absolute measurements of the viscosity by solution for β for measured values of the frequency and damping of oscillations in the fluid and in vacuo, if C_N is known independently since from (3.69) and (3.72)

$$\mu = \frac{\omega_0 \rho b^2}{\beta^2} . \quad (3.78)$$

It is essential for the accuracy of the instrument that a complete parallel alignment of the fixed plates and the disk, as well as their flatness, is achieved. The theoretical expression for C_N does not take into account the small amount of additional damping introduced by the stem and mirror attached to the disk in order to observe the oscillations and permit measurement of Δ, Δ_0, ω and ω_0. Kestin et al. [3.87] have shown that the value of C_N calculated from experiments in five different gases was within 0.1% of its theoretical value. The effect of the additional damping is therefore small but not always negligible.

Although (3.74) and (3.78) permit absolute measurements to be made, the conditions on their validity, namely the inequalities (3.73), make them inapplicable for very small boundary-layer thicknesses, such as occur in liquids and dense gases. In those cases an alternative formulation of the working equation has been developed in which an edge-correction factor is defined for an instrument. This edge-correction factor is a function only of the boundary-layer thickness so that it can be determined by calibration with fluids of known properties thus making relative measurements of the viscosity of other fluids, possible.

There has been a large series of oscillating-disk viscometers. It is worthwhile identifying some of the very successful instruments. As already mentioned, Kestin and Leidenfrost [3.85] developed the first instrument to measure the viscosity of gases near room temperature, in the pressure range 0.1–60 MPa with an uncertainty of about 0.1%. This instrument was employed over a wide range of conditions, including measurements near the critical point of carbon dioxide [3.88]. Di Pippo et al. [3.89] and Kestin et al. [3.90] designed a similar viscometer for high temperatures, up to 973 K, and low-pressure gases, with an uncertainty ranging from 0.1–0.3% at the highest temperatures. Oltermann [3.91]

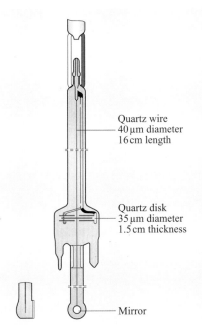

Fig. 3.34 Oscillating-disk quartz viscometer developed by *Vogel* (after [3.86])

also described an oscillating-disk viscometer for operation up to 673 K and 25 MPa in water near its critical point, with an estimated uncertainty of 1%. *Kestin* et al. [3.92] described an instrument to measure the viscosity of aqueous solutions of ionic salts at temperatures up to 573 K and pressures up to 35 MPa. It was operated in a relative mode with an estimated uncertainty of 0.3–0.5%. Finally, one of the most successful such instruments which is still in operation today is the quartz oscillating-disk viscometer (Fig. 3.34) designed by *Vogel* [3.86] for measurements on gases and vapors up to 650 K, with an uncertainty of 0.1–0.3%.

Oscillating-Cylinder Viscometers

The thick disk or cylinder configuration consists of a cylindrical body oscillating in an essentially infinite fluid. *Torklep* and *Oye* [3.93] have described an oscillating-cylinder viscometer designed for absolute measurements of low viscosity fluids up to temperatures of 1200 °C. The uncertainty quoted was 0.1% for water, where special efforts were made to achieve high accuracy, rising to 1% for routine measurements employing time-saving procedures.

Oscillating-Cup Viscometers

An oscillating-cup viscometer consists of an axially symmetric cylinder (cup) of inner radius R and height H, with the fluid contained inside. *Beckwith* and *Newell* [3.94] obtained the expression for $D(s)$ for the cases of a filled or partially filled cup. In the case of the filled cup, $D(s)$ is given by

$$D(s) = \frac{s^{3/2}}{I}\left(A\xi + B\frac{\xi^2}{s^{1/2}} + C\frac{\xi^3}{s} + D\frac{\xi^4}{s^{3/2}}\right),$$
(3.79)

where I is the moment of inertia of the cup itself, and $\xi = \delta/R$. The coefficients in the above equation are

$$A = 4 + \frac{2R}{H},$$
(3.80)

$$B = -6 - \frac{32R}{\pi H},$$
(3.81)

$$C = \frac{3}{2} + \frac{18R}{H},$$
(3.82)

$$D = 3.2 - \frac{16R}{\pi H}.$$
(3.83)

There are many successful oscillating-cup viscometers. *Grouvel* et al. [3.95] measured the viscosity of mercury in the range 20–260 °C with an uncertainty of 1.5%, while *Tippelskirch* et al. [3.96] employed an oscillating-cup viscometer to measure the viscosity of mercury from 470 to 1250 °C and to pressures up to 100 MPa. The uncertainty of this instrument was about 1% at the lower temperatures. *Abe* et al. [3.97] employed successfully such an instrument to measure the viscosity of molten salts in the temperature range 800–1200 °C. Finally, *Knapstad* et al. [3.98] measured the viscosity of many hydrocarbons up to 150 °C with an uncertainty of 0.5%.

Oscillating-Sphere Viscometers

In this type of viscometer, the fluid either fills a suspended, spherical shell or surrounds a suspended, solid sphere. Accurate expressions exist for the viscous torque in either case but the viscometer has not been as successful as have other types. This is partly because of the difficulty of manufacturing a sufficiently precise spherical shell, but also because the device offers little advantage over the simpler cylindrical viscometer. *Dumas* et al. [3.99] and *Brockner* et al. [3.100] employed oscillating-sphere viscometers to measure the viscosity of molten salts. However the uncertainty in the results achieved is higher than that obtained by oscillating-disk or oscillating-cup viscometers.

3.4.2 Vibrating Viscometers

The essential characteristic of the oscillatory viscometers discussed in the previous section was the observation of the effect of the viscosity of the fluid on the damping of the free, torsional motion of a rigid, solid body either immersed within it or surrounding it. In the case of the vibrating-wire viscometers, the oscillations involve periodic distortions of the solid body itself, which is in contact with the fluid whose viscosity is under investigation. The main advantages of vibrating viscometers are that the instruments are mechanically simpler than those of oscillatory bodies, and that the volume of fluid required for their use is much smaller, making operation at extremes of higher pressures and temperature easier.

Vibrating-Wire Viscometer

The vibrating-wire viscometer is a particularly simple form of vibrating viscometer, which has successfully been employed over a wide range of conditions. Indeed, it has played an important role in the investigation of the superfluid character of liquid helium [3.101].

A thin, circular-section wire of radius R, density ρ_s, subject to a tension, is constrained to be stationary at either end. It is surrounded by an infinite volume of the fluid of interest and performs oscillations transverse to its axis in a single plane containing the axis. If the os-

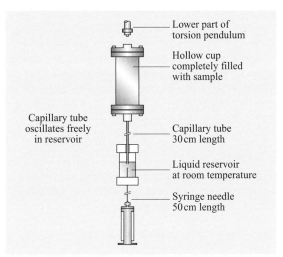

Fig. 3.35 Oscillating-cup viscometer developed by *Knapstad* et al. [3.98]

cillation is induced by an initial deflection at time $t = 0$ and subsequently allowed to decay, then the displacement of any point on the wire ξ ultimately conforms to a damped, simple harmonic motion [3.102]

$$\xi_a = A\,e^{-\Delta\omega t} \sin(\omega t + \phi)\,, \tag{3.84}$$

in which ω is the angular frequency and Δ the logarithmic decrement, and the subscript a indicates the asymptotic situation. At short times the displacement is affected by transients that cause a departure from (3.84)

$$\xi = F(t) + \xi_a(t)\,. \tag{3.85}$$

However, it has been shown [3.102] that the transient terms $F(t)$ decay rapidly compared with the decay time of the harmonic motion so that omission of the first few oscillations of an experiment is sufficient to eliminate them from consideration.

Provided that the design of the instrument is carried out to satisfy the constraints

$$\varepsilon \ll \Omega \ll \frac{1}{\varepsilon^2}\,, \tag{3.86}$$

where ε is the fractional amplitude of the motion, $\varepsilon = \xi_{\max}/R$ and

$$\Omega = \frac{\rho \omega R^2}{\mu} \tag{3.87}$$

and

$$\omega \varepsilon \frac{R}{C} \ll 1\,, \tag{3.88}$$

where C is the sonic velocity in the fluid, then its motion should correspond to that of (3.84). The decrement and frequency of the motion are related to the viscosity μ, and density ρ, of the fluid by the equation [3.102]

$$\Delta = \frac{(\rho/\rho_w)k' + 2\Delta_0}{2[1 + (\rho/\rho_w)k]}\,, \tag{3.89}$$

where k and k' are given by

$$k = -1 + 2\mathrm{Im}(A)\,, \tag{3.90}$$
$$k' = 2\,\mathrm{Re}(A) + 2\Delta\,\mathrm{Im}(A)\,, \tag{3.91}$$

where

$$A = (i - \Delta_0)\left(1 + \frac{2K_1(s)}{sK_0(s)}\right) \tag{3.92}$$

and

$$s = [(i - \Delta)\Omega]^{1/2} \tag{3.93}$$

and K_0, K_1 are modified Bessel functions. Finally, Δ_0, denotes the logarithmic decrement of the wire's oscillations in vacuo which can be determined by direct measurement. Equation (3.87) and (3.89–3.93) can be used to calculate the viscosity of the fluid from the logarithmic decrement and the frequency of oscillation in the fluid and the logarithmic decrement in vacuo, provided that the density of the fluid is known.

Only one correction to the working equations given above is significant ($> 0.01\%$) and that arises from the fact that it is necessary to enclose the fluid within a solid wall. A correction for this effect has been derived [3.102, 103] and it is rather straightforward to allow for the presence of the boundary.

In the vibrating-wire viscometer [3.104, 105] shown in Fig. 3.36, the vibrations of the metallic wire are stimulated by a pulse of current through the wire in a permanent magnetic field. The observation of the motion of the wire uses measurements of the signal induced across the wire as it oscillates perpendicular to the same field. In the particular instrument shown, samarium–cobalt alloy permanent magnets provide the 1 T magnetic field required. The wire, at its lower end carries a central weight W. This weight is connected through a balancing mechanism to an outer stainless-steel weight. The balancing mechanism acts as a fulcrum for the two weights. The volumes of the two weights are chosen so that the net effect of changing the density of the surrounding liquid results in a negligible effect upon the tension in the wire and thus the frequency of oscillation. At the same time the difference in their masses produces a constant tension in the wire. The way of supporting the two weights also produces a preferred plane of oscillation for the wire from among the degenerate set. The signal induced in the vibrating wire is observed with a bridge, and the out-of-balance signal, amplified by 30 000 times, is then interrogated with an analogue-to-digital (A/D) converter coupled to a computer. Representation of the decaying harmonic signal allows the determination of the frequency and decrement. This process is repeated in vacuo to provide all the information necessary to evaluate the viscosity using (3.87) and (3.89–3.93).

The uncertainty in the viscosity measurements made with such viscometers operated in an absolute manner is better than ±3%. Most of the uncertainty arises from those in the diameter of the vibrating wire and its density if they are determined independently. These two parameters can be determined with a greater precision through measurements on a fluid of known viscosity. When operated in the relative manner suggested by this calibration, the accuracy of viscosity measurements can be improved to one of ±0.5%, limited only by the uncertainty in the standard reference values for the viscosity. The precision in the viscosity measurements for liquids is about ±0.1%. For measurements in gases, the uncertainty is about ±1% when operated in an absolute way, and better than ±0.5% in a relative manner. Vibrating-wire viscometers have successfully been operated at Thessaloniki [3.104], London [3.105], Rostock [3.106] and Lisbon [3.107] over a wide range of conditions. A similar type of viscometer, but with the magnets outside the pressure vessel, was operated in Amsterdam [3.108]. The same types of viscometer have been operated in a continuous mode as well as the transient mode described here. In that case the resonant characteristics of the wire's vibrations have been measured with an impedance analyzer. These measurements have essentially the same accuracy as those performed in a transient mode but they have been used to make simultaneous measurements of density and viscosity on a single fluid, which has obvious advantages.

3.4.3 Torsional-Crystal Viscometer

The torsional piezoelectric-crystal viscometer was introduced by *Mason* [3.109] in 1947. It has several important advantages over other viscometers. It can cover a very large viscosity range, from about $10-10^5$ μPa s, permitting measurements in a wide range of fluid states, ranging from low-pressure gases to liquids near freezing. The motion of the crystal is exceedingly small although of high frequency.

The basic principle of the technique is the following. If a cylinder of piezoelectric material is cut with its axis along the x-axis (electric), and a sinusoidal voltage is applied to four electrodes at its quadrants, the cylinder will vibrate with a nearly pure internal torsional motion. When the crystal is immersed in a fluid, a shear wave will be produced in the fluid and will be rapidly attenuated as it moves normal to the surface of the crystal. The viscous drag, exerted by the fluid on the surface of the crystal, changes the crystal's resonant frequency, its conductance at resonance, and its bandwidth at resonance from their values in vacuo. The equation, most frequently employed in recent work [3.110]

$$\mu\rho = \pi f_e \left(\frac{M}{S}\right)^2 \left(\frac{\Delta f_e}{f_e} - \frac{\Delta f_0}{f_0}\right)^2, \qquad (3.94)$$

relates Δf_e, the crystal's bandwidth at resonance to the viscosity–density product of the fluid $\mu\rho$. Here, M, S and f_e are the mass, surface area, and resonant frequency of the crystal, respectively, while the subscript zero refers to the equivalent quantities in vacuo. In common with others this working equation neglects departures from pure torsion in the motion of the crystal and in the motion of the fluid.

Crystalline quartz is usually selected as the piezoelectric material because of its superior physical

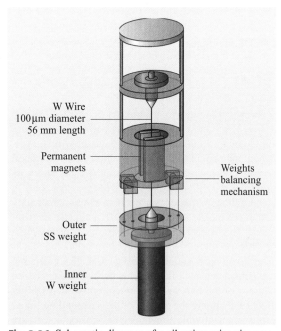

Fig. 3.36 Schematic diagram of a vibrating-wire viscometer for liquids developed by *Assael* et al. [3.104] and *Oliveira* [3.105]

properties. Crystal preparation includes obtaining the highest-quality starting material, orienting the cylinder axis along the x-axis (electric), grinding and polishing, locating the z-axis (optic) properly, sawing surface grooves at the midplane and parallel to the z-axis or plating electrodes on the crystal, measuring the crystal dimensions, and cleaning with appropriate solvents. Most torsional crystals have had electrodes plated directly on the surface of the crystal. However, several transducers [3.111, 112] have had a gap containing the test fluid between the surface of the crystal and the electrodes. An example of this type is shown in Fig. 3.37. This viscometer gives small repeatable bandwidths in vacuo. A typical crystal has a length of 5 cm, a diameter of 0.3 cm and a 39.5 kHz resonant frequency, which yields a fluid boundary layer of about 10^{-4} cm, whereas the gap containing the fluid is about 5×10^{-2} cm.

Torsional-crystal viscometers are in regular use today by *Haynes* [3.111], *Diller* and *Frederick* [3.112], *Bode* [3.113], *dos Santos* [3.114] and others. *Diller* and *Frederick* [3.112] have succeeded in expanding the temperature range of application up to 600 K. The estimated uncertainty of measurements made with these viscometers is about 2%.

3.4.4 Capillary Viscometers

Capillary viscometers are the most extensively used instruments for the measurement of viscosity, especially for the liquid phase. They have the advantage of simplicity of construction and operation. They are in regular use in many countries, for standard measurements in support of industrial investigations of the viscosity of liquids at atmospheric pressure. Hence the first section describes such instruments, while the second section describes capillary viscometers for gases and liquids at high pressures.

The principle of the capillary viscometer is based on the Hagen–Poiseuille equation of fluid dynamics (first formulated by Hagen in 1839) and its alteration for practical viscometry by *Barr* [3.115] in order to include the so-called kinetic-energy correction and the end correction. The resulting equation expresses the viscosity of a fluid flowing through a thin capillary, in terms of the capillary radius a the pressure drop along the tube ΔP the volumetric flow rate Q and the length of the tube L as

$$\mu = \frac{\pi a^4 \Delta P}{8 Q (L + na)} - \frac{m \rho Q}{8 \pi (L + na)}, \tag{3.95}$$

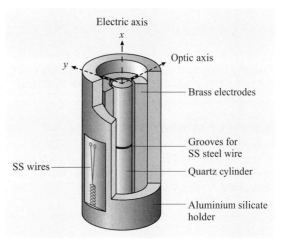

Fig. 3.37 Torsional-crystal viscometer developed by *Diller* and *Frederick* [3.112]

where n is the end-correction factor and m is the kinetic-energy correction factor with $L \gg a$. The above equation is derived on the assumptions that:

1. the capillary is straight with a uniform circular cross section,
2. the fluid is incompressible and Newtonian, and
3. the flow is laminar and there is no slip at the capillary wall.

The correction factors n and m reflect the fact that in a practical viscometer two chambers must be placed at either end of the capillary in order to measure the pressure drop. Thus, for example, the parabolic velocity distribution characteristic of most of the flow can only be realized some distance downstream from the inlet of the capillary. *Kestin* et al. [3.116] obtained theoretical values for the two correction factors in the range of Reynolds numbers $0.5 \leq \text{Re} \leq 100$, as

$$m = m_0 + \frac{8n}{\text{Re}}, \tag{3.96}$$

where

$$m_0 = 1.17 \pm 0.03 \quad \text{and} \quad n = 0.69 \pm 0.04. \tag{3.97}$$

There are also a number of experimental determinations of m and n. The values for m proposed by *Swindells* et al. [3.117] are

$$m = 1.12 \text{ to } 1.16 \quad \text{for} \quad 106 \leq \text{Re} \leq 648, \tag{3.98}$$

while *Kawata* et al. [3.118] reported

$$m = 1.08 \text{ to } 1.16 \quad \text{for} \quad 46 \leq \text{Re} \leq 1466, \tag{3.99}$$

$$n = 0.79 \text{ to } 0.88 \quad \text{for} \quad \text{Re} < 0.14. \tag{3.100}$$

There are a number of other secondary corrections to the working equation that can be found in literature. The most important ones are the effect of coiled capillaries [3.119], especially for gas phase measurements, the consequences of the nonuniformity of the capillary cross section [3.115] and the effects of surface tension in the case of the study of liquids.

Capillary Viscometers for Liquids

The capillary method has been employed for the absolute determination of the viscosity of primary standard liquids. The present viscosity standards within the International Organization for Standards (ISO), the American Society for Testing and Materials (ASTM), the Japanese Industrial Standards (JIS), the Deutsches Institut für Normung (DIN) are based on work by *Swindells* et al. [3.117] in 1952 with an absolute capillary viscometer. This group of workers determined the viscosity of distilled water at 20 °C and at atmospheric pressure. A schematic diagram of this viscometer is shown in Fig. 3.38. The viscometer itself was mounted in a thermostat bath together with a differential mercury manometer. The viscometer capillary, made of high quality glass tubing, had a radius of 0.2 mm, and was fitted into a glass chamber at either end. The flow of water through the capillary was generated by a piston–cylinder arrangement indicated as the main injector.

The mercury injector in the auxiliary bath forced mercury into the entrance chamber of the capillary and displaced water through it. The manometer consisted of a mercury-filled U-tube. The work of *Swindells* et al. [3.117], was characterized by great care in every aspect of measurement, to ensure the highest possible precision. The final value of the viscosity of water had an estimated uncertainty of better than 0.1%. *Kawata* et al. [3.118] described an absolute viscometer, employing a horizontal capillary, in order to check the value produced by *Swindells* et al. [3.117]. The accuracy claimed for all of these measurements has since been challenged and greater uncertainty bounds allowed but the absolute value reported by *Swindells* et al. remains intact.

For precise, measurements of the viscosity of liquids relative to this standard value, the capillary master viscometer shown in Fig. 3.39 is employed. The principle of each of the suspended-level viscometers shown in Fig. 3.39, is the measurement of the time taken for the meniscus of the sample liquid to fall from one timing mark to another in one arm of the viscometer. The marks are generally either side of a *timing* bulb and define the volume of liquid that flows during an experimental run under the influence of gravity. A master viscometer has a longer capillary, a larger timing bulb of oval shape and a cylindrical lower bulb compared with a routine viscometer. These differences serve to minimize the various errors, and render negligible the effect of the second term in (3.95).

For a capillary master viscometer, if $Q = V/t$, where t denotes the time required for volume V (in the timing bulb) to flow through the capillary, and $\Delta P = \rho g h$, where h is the mean effective height of the liquid column, then (3.95) becomes

$$\mu = \frac{\pi a^4 g h t}{8(L+na)V} - \frac{mV}{8\pi(L+na)t} \qquad (3.101)$$

or

$$\mu = c_1 t - \frac{c_2}{t}, \qquad (3.102)$$

where c_1 and c_2 are obtained from (3.101).

Routine capillary viscometers are used for many industrial measurements because they are very simple to use and permit rapid operation while yielding results of adequate accuracy. Some variations of such routine viscometers from among the many variants are shown in Fig. 3.39. For these viscometers (3.102) remains valid but the constants c_1 and c_2 are determined by calibration with fluids of known viscosity rather than from theory.

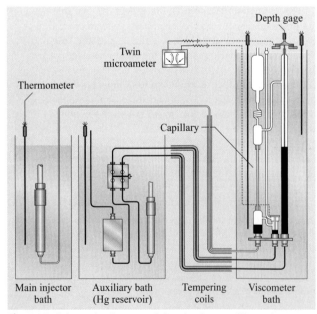

Fig. 3.38 Schematic diagram of the absolute capillary viscometer for water developed by *Swindells* et al. [3.117]

Capillary Viscometers for High Pressures

The capillary method is also suitable for application to gases and liquids at high pressures. However, in addition to the factor discussed in the previous section, a number of special precautions have to be considered. For example for measurements on gases, the measurement of the flow rate is difficult and corrections for compressibility and for slip at the capillary wall have to be considered. In addition, measurements at high pressures require a high-pressure construction for the whole system, a manometer to measure a small pressure difference in the presence of a high total pressure etc.

Assuming the fluid to be an ideal gas and the pressure drop between both ends of the capillary small, then the correction for compressibility is simple, and (3.95) becomes

$$\mu = \frac{\pi a^4(P_1 - P_2)}{8Q_1(L+na)} \frac{(P_1+P_2)}{2P_1}$$
$$- \frac{\rho_1 Q_1}{8\pi(L+na)}\left(m + \log \frac{P_1}{P_2}\right), \quad (3.103)$$

where P_1 and P_2 are the pressures at the inlet and the exit of the capillary and Q_1 is the volumetric flow rate at entry. The slip correction, proposed by *Knudsen* for such circumstances, modifies the working equation to [3.83]

$$\mu = \frac{\pi a^4(P_1 - P_2)}{8Q_1(L+na)} \frac{(P_1+P_2)}{2P_1}\left(1 + \frac{4\zeta}{a}\right), \quad (3.104)$$

where the slip correction ζ is calculated as

$$\zeta = 2.16\mu \sqrt{\frac{4}{\pi(P_1+P_2)\rho}}. \quad (3.105)$$

For high-pressure and/or high-temperature measurements, coiled capillaries are sometimes employed in order to secure a sufficiently long flow tube while maintaining uniformity of temperature. This kind of arrangement modifies the basic working equation because of the introduction of new velocity components into the flow. In most circumstances the effects can be treated as a correction to the basic equation so long as appropriate care is taken. A successful gas capillary designed by Michels and operated by *Trapeniers* et al. [3.119] for measurements on gases up to 100 MPa, is shown in Fig. 3.40.

Nagashima et al. [3.120] employed a closed-circuit capillary viscometer for measurements of the viscosity of water and heavy water in the liquid and vapor phase at temperatures in the range 50–500 °C and pressures up to 80 MPa. A similar apparatus was described by *Rivkin* et al. [3.121] for measurements in the critical region of water. *Guevara* et al. [3.122] employed a capillary viscometer for gases up to 2000 K while *Ejima* et al. [3.123] applied a capillary viscometer to measurement of the viscosity of molten salts up to 1200 K.

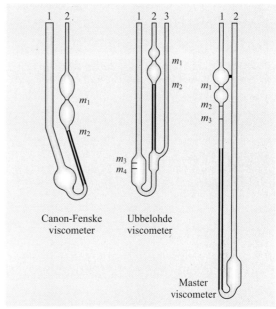

Fig. 3.39 Capillary master and routine viscometers

3.4.5 Falling-Body Viscometers

Falling-body viscometers involve the use of the time of free-fall of an object under the influence of gravity through the fluid of interest to determine the viscosity. Generally, the falling body is a body of revolution and has most commonly been a sphere or a cylinder. Generally, the instruments are not among those of the highest accuracy because it is rather difficult to ensure that the instrument operates in line with a theory of it. However, exceptionally, such instruments have been used for accurate measurements including work in Japan on a new standard reference viscometer [3.83]. More usual instruments of the same type have a number of advantages for operation at very high pressures that make them useful for routine measurements in industry.

Falling-Sphere Viscometer

The principle of a falling-sphere, or falling-ball, viscometer is based on Stokes' law. This law results from the creeping-flow Stokes equations valid for very low Reynolds numbers. For that reason this type of viscometer is usually applicable to the precision measurement of

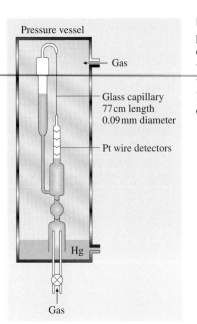

Fig. 3.40 High-pressure gas capillary viscometer developed by *Trapeniers* et al. [3.119]

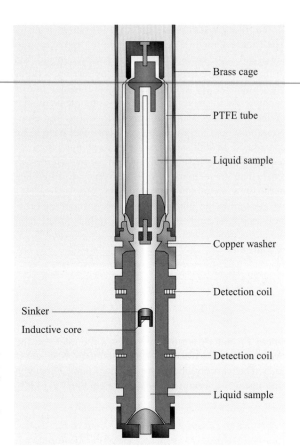

Fig. 3.41 Falling-cylinder viscometer developed by *Isdale* [3.124] and *Irvine* [3.125]

the viscosity of highly viscous liquids or measurements under high pressure. In routine operation the viscometers are used in a relative manner using a calibration of the instrument with standard liquids. For a sphere of radius a, falling through a distance L, the viscosity of the fluid is given as

$$\mu = \frac{2a^2(\rho_s - \rho)gt}{9L} f_w, \qquad (3.106)$$

where ρ_s is the density of the sphere, and t the fall time. This equation is valid provided the motion is very slow, the fluid is infinite, incompressible and Newtonian, and there is no slip between the fluid and the sphere's surface. Also the ball must fall vertically without rotation. The correction factor f_w accounts for the effect of the inevitable presence of a wall containing the fluid. It is usually obtained by calibration with fluids of known viscosity. If R is the radius of the cylindrical container and provided a/R and Re are sufficiently small (Re $\ll$ 1), *Faxen* [3.126] has derived theoretically the following correction factor

$$f_w = 1 - 2.109 \left(\frac{a}{R}\right) + 2.09 \left(\frac{a}{R}\right)^3 - 0.95 \left(\frac{a}{R}\right)^5. \qquad (3.107)$$

To measure the fall time, at atmospheric pressure a glass cylinder is often employed to contain the fluid because it allows optical observation of the fall time in a transparent fluid. *Flude* and *Daborn* [3.127] performed measurements using a laser Doppler effect and attained a precision of $\pm 0.07\%$ in the time. When the liquid is opaque, the fall velocity can be measured by various techniques for example by recording the inductance changes in a coil wound on the fall tube owing to the passage of the ball. For high-pressure measurements

Table 3.11 Reference values for the viscosity of noble gases at a pressure of 0.1 MPa

T (°C)	Viscosity (μPa s)				
	Helium	Neon	Argon	Krypton	Xenon
25	19.86	31.76	22.62	25.39	23.09
100	23.16	37.06	27.32	31.22	28.84
200	27.35	43.47	32.85	38.06	35.91
300	31.28	49.50	37.83	44.28	42.38
400	35.04	55.00	42.35	49.99	48.32
500	38.60	60.19	46.63	55.34	53.84

Table 3.12 Viscosity μ (μPa s) of some n-alkanes in the liquid phase

		P (MPa)	273.15 K	298.15 K	323.15 K	348.15 K	373.15 K
n-Pentane	C_5H_{12}	0.101	286.4	220.3	174.8	141.0	114.3
		5	301.1	232.4	185.3	150.5	122.8
		10	316.1	244.4	195.6	159.7	130.8
		25	361.6	279.9	225.0	184.6	151.6
		50	439.8	338.7	271.4	222.1	181.3
n-Hexane	C_6H_{14}	0.101	384.7	295.7	237.8	196.5	164.5
		5	403.8	310.9	251.0	208.6	176.0
		10	423.5	326.3	264.1	220.4	187.0
		25	483.8	372.8	302.9	254.5	217.5
		50	590.4	452.4	367.1	308.8	264.3
n-Heptane	C_7H_{16}	0.101	530.9	392.5	307.4	249.8	207.6
		5	557.9	412.5	323.6	264.0	220.6
		10	586.0	433.1	340.1	278.1	233.3
		25	673.6	496.0	389.6	319.6	269.7
		50	831.7	606.6	474.0	388.2	327.6
n-Octane	C_8H_{18}	0.101	754.3	531.6	401.9	318.6	260.5
		5	795.5	559.9	423.4	336.4	276.1
		10	838.5	589.2	445.5	354.4	291.7
		25	974.3	680.4	513.0	408.3	337.1
		50	1225	844.7	631.4	500.0	411.9
n-Nonane	C_9H_{20}	0.101	995.5	677.5	498.3	387.1	312.4
		5	1052	714.2	524.9	408.2	330.2
		10	1112	752.4	552.5	429.8	348.2
		25	1298	872.4	637.4	495.0	401.6
		50	1649	1092	788.7	607.9	491.1
n-Decane	$C_{10}H_{22}$	0.101	1364	886.6	630.0	477.3	378.2
		5	1445	936.9	664.7	503.5	399.6
		10	1531	989.4	700.7	530.5	421.3
		25	1805	1156	813.0	613.2	486.6
		50	2328	1465	1016	758.8	598.2
n-Undecane	$C_{11}H_{24}$	0.101	1738	1098	761.5	565.9	441.8
		5	1844	1161	803.8	596.8	466.3
		10	1956	1228	847.8	628.7	491.2
		25	2318	1439	985.7	727.1	566.7
		50	3010	1835	1238	902.3	697.3
n-Dodecane	$C_{12}H_{26}$	0.101	2319	1404	941.2	681.4	521.9
		5	2466	1488	994.7	719.1	550.7
		10	2622	1576	1051	758.1	580.2
		25	3129	1859	1227	879.4	670.2
		50	4112	2395	1554	1098	827.8

sapphire optical windows are sometimes set into the pressure vessel to permit observation of the motion.

Falling-Cylinder Viscometer

An alternative realization of a falling-body viscometer makes use of a right circular cylinder falling under the influence of gravity along the axis of a coaxial cylindrical tube. This arrangement has the advantage that it is possible to secure a low Reynolds number for the fluid flow by choosing a suitable geometry, or by reducing the cylinder's mass by making it hollow.

For a cylinder of radius r_1 and length L_s, in a tube of radius r_2, the viscosity is given by [3.124]

$$\mu = \frac{(1-\rho/\rho_s)}{A} t \, , \qquad (3.108)$$

Table 3.13 Viscosity μ (μPa s) of some alkenes in the liquid phase

		P (MPa)	273.15 K	298.15 K	323.15 K	348.15 K	373.15 K
Benzene	C_6H_6	0.101	857.6	593.8	435.4	334.0	265.0
		5	897.6	621.1	455.4	349.6	278.0
		10	939.5	649.5	476.0	365.6	291.0
		25	1072	737.9	539.5	414.0	329.8
		50	1314	897.8	651.8	497.5	394.7
Toluene	C_7H_8	0.101	761.7	538.2	408.5	326.6	271.1
		5	793.5	560.1	425.2	340.5	283.5
		10	826.5	582.8	442.4	354.7	296.0
		25	929.9	652.9	494.9	397.3	333.1
		50	1117	777.5	586.4	470.2	395.0
Ethylbenzene	C_8H_{10}	0.101	862.6	614.5	467.6	373.5	308.9
		5	900.0	640.8	487.8	390.4	324.0
		10	939.0	668.1	508.7	407.7	339.3
		25	1062	753.1	573.3	460.5	385.3
		50	1286	906.1	687.5	552.4	464.0
o-Xylene	C_8H_{10}	0.101	1091	749.8	559.8	444.4	369.1
		5	1137	780.5	582.5	462.8	385.1
		10	1186	812.5	606.1	481.7	401.5
		25	1338	912.3	678.9	539.8	451.3
		50	1617	1093	808.3	641.3	537.0
m-Xylene	C_8H_{10}	0.101	772.1	562.0	434.1	350.4	291.9
		5	804.9	585.7	452.8	366.2	306.2
		10	839.0	610.3	472.1	382.4	320.7
		25	946.1	686.8	531.5	431.9	364.4
		50	1141	824.3	636.6	518.1	439.1
p-Xylene	C_8H_{10}	0.101	798.2	583.3	452.1	365.9	305.5
		5	831.4	607.4	471.1	382.1	320.2
		10	866.0	632.4	490.7	398.7	335.1
		25	974.2	709.9	551.1	449.1	379.7
		50	1171	848.6	657.4	536.4	455.6
Mesitylene	C_9H_{12}	0.101	941.2	657.5	494.1	392.1	324.1
		5	983.8	686.8	516.3	410.4	340.3
		10	1028	717.4	539.3	429.2	356.8
		25	1169	813.2	610.9	487.4	407.3
		50	1429	987.5	739.6	590.3	495.5

where

$$A = \frac{2\pi L_s L_T}{mg\left\{\ln(r_2/r_1) - \left[(r_2^2 - r_1^2)/(r_2^2 + r_1^2)\right]\right\}}.$$

(3.109)

Here, ρ is the density of the liquid and ρ_s the density of the material of the cylinder, while L_s is the distance traveled by the falling body of mass m.

There are many difficulties in a practical instrument, because it is impossible to ensure that the falling body and its tube are perfectly cylindrical and that the former falls along the axis of the latter. A number of different designs have been developed to try to ensure correct operation that include a wide variety of shapes for the front face of the falling body itself. Pins protruding from the falling body perpendicular to its length have been employed in the past, to ensure concentricity. However, they were unsuccessful as the influence of friction between the pins and the wall was high.

An alternative approach is to design the falling body as a self-centering device. This was done successfully by

Table 3.14 Viscosity μ (μPa s) of some n-alcohols in the liquid phase

		P (MPa)	273.15 K	298.15 K	323.15 K	348.15 K	373.15 K
Methanol	CH_4O	0.101	772.2	568.3	408.1	312.3	271.5
		5	787.6	581.7	418.7	320.7	278.4
		10	803.0	595.0	429.1	328.9	285.1
		25	847.5	633.0	458.6	351.6	303.4
		50	917.4	691.8	503.2	385.2	330.0
Ethanol	C_2H_6O	0.101	1699	1104	719.5	494.4	380.3
		5	1743	1134	739.8	508.4	390.6
		10	1787	1164	760.2	522.3	400.7
		25	1920	1254	819.5	562.1	429.2
		50	2141	1401	914.0	624.1	472.6
1-Propanol	C_3H_8O	0.101	3535	1989	1141	827.2	854.8
		5	3653	2062	1185	858.6	885.7
		10	3774	2135	1229	890.4	916.6
		25	4145	2361	1362	985.1	1007
		50	4790	2749	1589	1143	1153
1-Butanol	$C_4H_{10}O$	0.101	4503	2524	1382	833	622.1
		5	4706	2644	1447	870	647.5
		10	4919	2769	1515	909	673.3
		25	5587	3160	1726	1026	750.1
		50	6812	3874	2104	1231	878.7
1-Pentanol	$C_5H_{12}O$	0.101	7203	3393	1706	1046	812.5
		5	7608	3581	1795	1094	844.9
		10	8037	3780	1887	1144	877.8
		25	9418	4415	2178	1297	976.1
		50	12057	5614	2713	1567	1141
1-Hexanol	$C_6H_{14}O$	0.101	8893	4303	2206	1263	809.3
		5	9451	4554	2319	1318	839.3
		10	10045	4820	2437	1374	869.7
		25	11979	5675	2810	1549	960.2
		50	15756	7311	3499	1855	1111
1-Heptanol	$C_7H_{16}O$	0.101	14217	5712	2658	1473	946.0
		5	15133	6041	2787	1533	978.7
		10	16110	6388	2921	1594	1012
		25	19301	7506	3345	1782	1110
		50	25568	9643	4123	2110	1273
1-Octanol	$C_8H_{18}O$	0.101	19606	7145	3240	1811	1182
		5	20866	7545	3392	1882	1222
		10	22208	7969	3551	1955	1263
		25	26598	9330	4050	2179	1385
		50	35222	11926	4962	2571	1588

Isdale [3.124], *Irvine* [3.125], *Glen* [3.128] and others, and measurements of the viscosity of hydrocarbons at very high pressures were reported. The measurements were performed on a relative basis with an estimated uncertainty of 2–3% in the final values.

3.4.6 Viscosity Reference Values

The internationally agreed standard for viscosity, ISO/TR 3666:1998, is the viscosity of water at 20 °C and atmospheric pressure (0.101325 MPa), and its approved

Table 3.15 Viscosity μ (μPa s) of some refrigerants in the liquid phase

		P (MPa)	248.15 K	273.15 K	298.15 K	323.15 K	348.15 K
R22	$CHClF_2$	5	292.1	224.0	173.4	131.8	94.2
		10	304.8	236.1	185.8	145.2	111.1
		25	341.0	269.0	217.0	175.4	142.3
R32	CH_2F_2	5	229.5	167.3	124.0	89.4	
		10	240.1	176.3	132.6	98.6	82.5
		25	269.0	199.0	152.2	116.1	112.4
R124	C_2HClF_4	5	529.2	369.8	274.3	210.8	159.2
		10	549.5	389.8	296.3	234.1	180.3
		25	610.6	448.2	356.6	292.8	228.2
R125	C_2HF_5	5	310.2	223.9	161.5	112.2	
		10	328.6	240.7	178.9	132.8	
		25	379.3	283.2	217.8	170.6	
R134a	$C_2H_2F_4$	5	405.7	295.2	219.4	161.8	114.6
		10	430.5	314.5	236.4	178.6	133.7
		25	503.6	368.6	280.8	218.3	172.0
R141b	$C_2H_3ClF_2$	5	829.5	577.5	429.8	333.3	265.5
		10	858.3	602.2	451.9	353.1	283.9
		25	946.8	677.5	518.1	410.5	335.4
R152a	$C_2H_4F_2$	5	337.8	230.9	173.7	134.8	103.3
		10	355.5	243.6	185.1	146.8	117.5
		25	407.9	279.3	215.4	176.0	147.8

value is

$$\mu = 1.0016 \, \text{mPa s} \,. \tag{3.110}$$

This value has an estimated relative uncertainty of $\pm 0.17\%$. This is based on the value of 1.0019 mPa s reported by *Swindells* et al. [3.117] in 1952, which was also the basis of ISO/TR 3666:1977. The small difference in value is due to the difference between the ITS-48 and ITS-90 temperature scales.

The temperature dependence of the viscosity of water at atmospheric pressure in the temperature range $0.01 – 100\,°\text{C}$, is given by the following recommended correlation [3.129]

$$\log \frac{\mu(\theta)}{\mu(20\,°\text{C})} = \frac{\theta}{116 - \theta} \times \left(1.2378 - 1.303 \times 10^{-3}\theta + 3.06 \times 10^{-6}\theta^2 + 2.55 \times 10^{-8}\theta^3\right), \tag{3.111}$$

where $\theta = 20 - T(°C)$. The estimated uncertainty of (3.111) is better than $\pm 0.1\%$.

In the case of the viscosity of gases, nitrogen is usually employed as a standard, since it is readily available with high purity and is also inexpensive to obtain. The following value is recommended [3.83] for the viscosity of nitrogen at 25 °C and atmospheric pressure:

$$\mu = (17.710 \pm 0.016) \, \mu\text{Pa s} \,. \tag{3.112}$$

Recommended values for the viscosity of the noble gases at a pressure of 0.1 MPa are shown in Table 3.11. They have been calculated using a combination of experimental data and available theory and their estimated uncertainty is $\pm 0.2\%$ in the range 25–200 °C, and $\pm 0.4\%$ in the range 200–500 °C.

For use at higher pressures, up to 30 MPa, and at a temperature of 25 °C, the viscosity of nitrogen is represented by the equation [3.83]

$$\mu = 0.17763 \times 10^{-4} + 0.86870 \times 10^{-8} \rho + 0.14240 \times 10^{-9} \rho^2 \,, \tag{3.113}$$

where μ is measured in Pa s and ρ in kg/m^3. Other secondary viscosity reference fluids are available in the literature [3.83].

3.4.7 Tables of Viscosity Values

In Tables 3.12–3.16, the viscosity of commonly encountered fluids is given for use in engineering calculations, as a function of both temperature and pressure. The val-

Table 3.16 Viscosity μ (μPa s) of some gases

		P (MPa)	273.15 K	298.15 K	323.15 K	348.15 K	373.15 K
Argon	Ar	0.101	21.26	22.73	24.14	25.50	26.82
		5	22.43	23.73	25.01	26.26	27.49
		10	24.32	25.29	26.32	27.39	28.48
Hydrogen	H_2	0.101	8.33	8.86	9.36	9.85	10.32
		5	8.40	8.92	9.42	9.90	10.36
		10	8.50	9.00	9.49	9.96	10.42
Nitrogen	N_2	0.101	16.54	17.62	18.66	19.66	20.62
		5	17.46	18.40	20.25	20.25	21.15
		10	18.89	19.59	20.34	21.12	21.91
Oxygen	O_2	0.101	19.12	20.48	21.79	23.05	24.28
		5	20.18	21.39	22.58	23.75	24.89
		10	21.90	22.80	23.77	24.77	25.78
Carbon monoxide	CO	0.101	16.47	17.61	18.70	19.75	20.77
		5	17.44	18.43	19.41	20.37	21.32
		10	18.99	19.70	20.49	21.30	22.13
Carbon dioxide	CO_2	0.101	13.96	15.14	16.28	17.38	18.46
		5		17.95	18.46	19.21	20.04
		10			29.35	23.64	23.16
Sulfur dioxide	SO_2	0.101	11.87	12.97	14.06	15.15	16.24
Hydrogen sulfide	H_2S	0.101	11.76	12.86	13.94	15.01	
		5				17.26	
Methane	CH_4	0.101	10.34	11.14	11.92	12.67	13.39
		5	11.49	12.11	12.75	13.39	14.04
		10	13.84	13.88	14.17	14.58	15.05
Ethane	C_2H_6	0.101	8.69	9.44	10.17	10.88	11.57
		5			13.47	13.26	13.51
		10			38.33	24.83	19.37
Propane	C_3H_8	0.101	7.61	8.28	8.93	9.57	

ues quoted are therefore not intended to be the best possible values for the viscosity of a particular fluid but are intended to be of sufficient accuracy to be useful. The fluids and the temperature and pressure conditions chosen are the same as chosen for the density and thermal conductivity in Sects. 3.1 and 3.4.

Values for the liquid viscosity are based on a large collection of experimental data correlated by a semi-empirical hard-spheres-based procedure available in the literature (n-alkanes [3.130], n-alkenes [3.131], n-alcohols [3.132], refrigerants [3.133]). The uncertainty of the quoted liquid viscosity values is much better than ±5%. Values for the gas-phase viscosity have been obtained from corresponding-states software [3.37], itself based upon experiment; it is estimated that the uncertainty is less than 2%.

3.5 Thermal Conductivity and Thermal Diffusivity

The thermal conductivity is defined by Fourier's law. In one group of instruments described in this chapter the use of a time-dependent heating perturbation is applied to the fluid, and the time dependence of the temperature in the fluid is measured and related to the thermal conductivity. In the second group a continuous and constant heat source establishes a steady temperature distribution in the fluid whose measurement again yields the thermal conductivity of the fluid. The instruments have different attributes which make them suitable for particular applications.

The thermal conductivity of a fluid measures its propensity to dissipate energy, when disturbed from equilibrium by the imposition of a temperature gradient. For isotropic fluids the thermal conductivity λ is defined by Fourier's law

$$\boldsymbol{Q} = -\lambda \nabla T, \tag{3.114}$$

where $\boldsymbol{Q}$ is the instantaneous flux of heat, which is the response of the medium to the instantaneous temperature gradient. The thermal conductivity depends on the thermodynamic state of the fluid. Because it is impossible to measure local fluxes and local gradients, (3.114) cannot be employed directly to measure the thermal conductivity of a fluid.

The rate of heat propagation through a fluid is described by its thermal diffusivity a defined as

$$a = \frac{\lambda}{\rho C_p}, \tag{3.115}$$

where ρ is the fluid density and C_p its isobaric heat capacity. In some techniques, the thermal conductivity is obtained indirectly from the measured thermal diffusivity, provided the density and the isobaric heat capacity of the fluid are known.

The starting point for the formulation of the working equations to measure the thermal conductivity of a fluid is the equation of energy conservation which, for a Newtonian, viscous, isotropic and incompressible fluid, with temperature-dependent properties, can be written [3.134]

$$\rho \frac{\mathrm{d}U}{\mathrm{d}t} = -\nabla \cdot \boldsymbol{Q} - P(\nabla \cdot \boldsymbol{v}) - \boldsymbol{\sigma} : \boldsymbol{\varepsilon}, \tag{3.116}$$

where U is the internal energy, t the time, P the hydrostatic pressure, $\boldsymbol{v}$ the hydrodynamic velocity of the fluid, $\boldsymbol{\sigma}$ the stress tensor, ρ the density and $\boldsymbol{Q}$ the heat flux vector. The notation $\mathrm{d}/\mathrm{d}t$ represents the substantive derivative [3.134]. On the assumption that the temperature perturbation is small and that a local-equilibrium thermodynamic state exists, (3.116) can be transformed to

$$\rho C_v \frac{\mathrm{d}U}{\mathrm{d}t} - T \left(\frac{\alpha_p}{k_T}\right) \left(-\alpha_p + k_T \frac{\mathrm{d}P}{\mathrm{d}T}\right) \frac{\mathrm{d}T}{\mathrm{d}t}$$
$$= -\nabla \cdot \boldsymbol{Q} + \phi, \tag{3.117}$$

where C_v is the isochoric heat capacity, α_p the isobaric expansion coefficient, k_T the isothermal compressibility and $\phi = \boldsymbol{\sigma} : \boldsymbol{\varepsilon}$ is the rate of internal energy increase owing to viscous dissipation. Assuming that $k_T(\mathrm{d}P/\mathrm{d}T) \ll \alpha_p$, (3.117) can be written

$$\rho C_p \frac{\mathrm{d}T}{\mathrm{d}t} = -\nabla \cdot \boldsymbol{Q} + \phi. \tag{3.118}$$

A general solution of (3.118) is not possible; thus it is necessary to apply a number of further restrictions before it can be employed as the basis of determinations of thermal conductivity. It is further assumed that fluid movements are avoided, so that $\boldsymbol{v} = 0$ and consequently $\phi = 0$. It is therefore necessary to make measurements of the thermal conductivity in such a way that the effect of convection is negligible even if it is unavoidable. Assuming further that the radiative heat flux is rendered negligible and employing Fourier's law (3.114) then, for an isotropic fluid with a temperature-independent thermal conductivity, density and heat capacity, (3.118) can be written

$$\rho C_p \frac{\partial T}{\partial t} = \lambda \nabla^2 T. \tag{3.119}$$

Equation (3.119) is the basis for all experimental methods for the measurement of the thermal conductivity.

In the last 30 years a variety of experimental methods have been developed, both for liquid or gaseous phases, to cover a wide range of thermodynamic states. These methods, according to the use of (3.119) can be classified in two main categories:

- Transient or unsteady-state techniques, in which the full (3.119) is used and the principal measurement is the temporal history of the fluid temperature,
- Steady-state techniques, for which $\partial T/\partial t = 0$, and (3.119) reduces to $\lambda \nabla^2 T = 0$, which can be integrated for a given geometry.

3.5.1 Transient Methods for Thermal Conductivity

The main difficulty in performing accurate measurements of the thermal conductivity of fluids lies in the realization of two of the conditions above. It should be possible to isolate the conduction process from other mechanisms of heat transfer.

The imposition of a temperature gradient in a compressible fluid in the gravitational field of the earth inevitably creates a state of motion (natural convection) so that pure conduction in a fluid is very difficult to achieve. The success of transient techniques for the measurement of the thermal conductivity of fluids is based on the fact that the characteristic time for the acceleration of the fluid by buoyancy forces is much longer than the propagation time of a temperature wave originated by a strong and localized temperature gradient. In this section two transient techniques are described: the transient hot-wire technique applicable over a wide range of conditions and an interferometric technique especially suited to the critical region.

The advantages of the transient hot-wire technique are that it permits the user to obtain the thermal conductivity by the use of an exact working equation resulting from a careful mathematical model of the instrument as well as to eliminate convective contributions to the heat transfer from the measurement. The transient hot-wire technique is an absolute technique and the instruments based on its principle are considered primary instruments and are capable of providing the lowest uncertainty possible at present. The uncertainty of the measurements has been confirmed by performing measurements in low-density noble gases, for which an exact molecular theory exists [3.135]. It has been applied successfully in most regions of the phase diagram, except very close to the critical point where the temperature gradients used are too large to maintain a state sufficiently near equilibrium, owing to large fluctuations in density and long-range correlations. It is in this particular region that the interferometric technique is singularly appropriate because it has the unique advantage that the nearer the critical point is approached the smaller can be the applied temperature gradient. Thus, the two transient techniques are complimentary.

Transient Hot-Wire Technique

A transient thermal conductivity measurement is one in which a time-dependent perturbation, in the form of a heat flux, is applied to a fluid initially at equilibrium. The thermal conductivity is obtained from an appropriate working equation relating the observed response of the temperature of the fluid to the perturbation. In principle, one can devise a wide variety of techniques of this kind. However, the only geometrical arrangement applied over a wide range of conditions is one in which the perturbing heat flux is applied by means of electrical dissipation in a thin, cylindrical wire as a step function. In this case the wire is itself used as the thermometer to monitor the temperature rise of the fluid at its interface.

The temperature rise in the fluid at a distance r from the wire at a time t, can be defined

$$\Delta T(r, t) = T(r, t) - T_0 \,. \tag{3.120}$$

It can easily be shown [3.136] that for a cylindrical wire of radius r_0, and for small values of the term $(r_0^2/4at)$,

$$\Delta T(r_0, t) = \frac{q}{4\pi\lambda} \left[\ln\left(\frac{4at}{r_0^2 \gamma}\right) + \left(\frac{r_0^2}{4at}\right) + \ldots \right] \,, \tag{3.121}$$

where $\Delta T(r_0, t)$ is the transient temperature rise of the fluid at the wire surface, λ and a the thermal conductivity and thermal diffusivity of the fluid, respectively, q the heat input power per unit length, and $\gamma = 0.577216$ is the Euler–Mascheroni constant. In the ideal model, (3.121) describes the temperature rise of the wire in contact with the fluid at its surface. In practice a real instrument departs from the ideal model in a number of respects and analytical corrections have been developed [3.136] for the departure of a practical instrument from the ideal one. The two major additive corrections to (3.121) that need to be applied in practice are:

1. The heat capacity correction ΔT_{hc} significant only at short experimental times [3.137],

$$\begin{aligned}
\Delta T_{\mathrm{hc}} &= \frac{2q^{1/2}}{\pi^2 r_0^2} \int_0^\infty \left(1 - \mathrm{e}^{-a_{\mathrm{w}} u^2 t}\right) J_1(ur_0) \\
&\quad \times \left[J_0\left(\sqrt{\frac{a_{\mathrm{w}}}{a}} ur_0\right) \phi(u) - Y_0\left(\sqrt{\frac{a_{\mathrm{w}}}{a}} ur_0\right) \phi(u) \right] \\
&\quad \times \left\{ u^3 \left[\phi^2(u) + \psi^2(u)\right]\right\}^{-1} \mathrm{d}u \,,
\end{aligned} \tag{3.122}$$

with

$$\phi(u) = \lambda_{\mathrm{w}} a^{1/2} J_1(ur_0) J_0\left(\sqrt{\frac{a_{\mathrm{w}}}{a}} ur_0\right) \\
- \lambda a_{\mathrm{w}}^{1/2} J_0(ur_0) J_1\left(\sqrt{\frac{a_{\mathrm{w}}}{a}} ur_0\right) \,, \tag{3.123}$$

$$\psi(u) = \lambda_{\mathrm{w}} a^{1/2} J_1(ur_0) Y_0\left(\sqrt{\frac{a_{\mathrm{w}}}{a}} ur_0\right) \\
- \lambda a_{\mathrm{w}}^{1/2} J_0(ur_0) Y_1\left(\sqrt{\frac{a_{\mathrm{w}}}{a}} ur_0\right) \,. \tag{3.124}$$

In these expressions, r_0 is the wire radius and the subscript 'w' refers to wire properties. Furthermore, J_0 and J_1 denote Bessel functions of first kind, of order zero and one, respectively, while Y_0 and Y_1 express Bessel functions of second kind, of order zero and one.

2. The outer boundary correction, ΔT_{ob}, significant only at long experimental times

$$\Delta T_{\text{ob}}(r_0, t) = \frac{q}{2\pi\lambda} \ln \frac{b}{r_0} + \frac{q}{2r_0\lambda} \sum_{n=1}^{\infty} e^{-a\alpha_n^2 t} \times \frac{J_0^2(b\alpha_n) \left[J_0(r_0\alpha_n) Y_1(r_0\alpha_n) - Y_0(r_0\alpha_n) J_1(r_0\alpha_n) \right]}{\alpha_n \left[J_1^2(r_0\alpha_n) - J_0^2(b\alpha_n) \right]} , \quad (3.125)$$

where b is radius of the cylindrical fluid enclosure and α_n are the positive roots of the equation

$$J_1(r_0 x) Y_0(bx) - Y_1(r_0 x) J_0(bx) = 0 . \quad (3.126)$$

Healy et al. [3.136] proposed approximate expressions of (3.122) and (3.125) valid for large values of $(4at/r_0^2)$. These expressions together with (3.121) form a consistent set in order to calculate the thermal conductivity from the measured temperature rise. The application of this methodology to liquids and gases at moderate pressures has provided many reliable thermal-conductivity data over the last two decades. Unfortunately, the analytical corrections proposed by *Healy* et al. [3.136] proved to be inadequate [3.137] for the description of experiments in the gas phase at low densities, where fluids exhibit exceedingly high thermal-diffusivity values.

To overcome these difficulties, a numerical finite-element method was proposed by *Assael* et al. [3.137], in order to solve the complete set of energy-conservation equations that describe the heat-transfer experimental processes. The choice of this particular numerical method was dictated by the high accuracy the method exhibits in computational heat transfer problems. The energy equations to be solved are two coupled partial differential equations, one for the wire, $0 < r \geq r_0$:

$$\left(\rho c_p \right)_{\text{w}} \frac{\partial T_{\text{w}}}{\partial t} = \lambda_{\text{w}} \nabla^2 T_{\text{w}} - \frac{q}{\pi r_0^2} , \quad (3.127)$$

and one for the fluid, $r_0 \geq r < \infty$:

$$\left(\rho c_p \right) \frac{\partial T}{\partial t} = \lambda \nabla^2 T . \quad (3.128)$$

On the wire/fluid interface both the temperature and the heat flux are considered to be continuous. This means that for $r = r_0$,

$$\lambda_{\text{w}} \left(\frac{\partial T_{\text{w}}}{\partial r} \right)_{r=r_0} = \lambda \left(\frac{\partial T}{\partial r} \right)_{r=r_0} \quad (3.129)$$

and

$$T_{\text{w}}(r_0, t) = T(r_0, t) . \quad (3.130)$$

This set of equations was solved subject to a suitable set of initial conditions on a straight wire [3.137]. As it can be seen from the conditions (3.129) and (3.130), the problem is one-dimensional with respect to the radial direction. *Assael* et al. [3.137] employed a forward difference (Euler) scheme coupled with a modification of the Gaussian elimination method (LU decomposition method). In practice, experimental means are employed to yield a finite segment of a wire that behaves as if it were part of an infinite wire. This allows the numerical solution of the differential equations to be used iteratively to determine the thermal conductivity and diffusivity of the fluid that yields the best match between the experimental and calculated temperature rise of this finite segment of wire.

Since the technique was first employed by *Stalhane* and *Pyk* [3.138] in 1931 to measure the thermal conductivity of powders, there have been significant improvements in the practical realization of the technique. In modern instruments the wire sensor acts both as the heat source and as a thermometer. Rapid development of analogue and digital equipment as well as of computer-driven data-acquisition systems have meant that precise measurements of transient electrical signals can be made quickly. Thus, it has become possible to measure the resistance change taking place in the hot wire as a consequence of its temperature rise with a precision better than 0.1%. Furthermore, instead of a single wire, two wires identical in all respects except for length, are employed. This allows a practical and automatic means of compensating for the finite length of the wires. For electrically insulating fluids platinum has usually been employed as the heating wire and sensing thermometer because of its chemical stability and resistance/temperature characteristics. In order to allow measurements of electrically conducting fluids, tantalum wires are often employed, because tantalum upon electrolytic oxidation forms tantalum pentoxide on its surface, which is an electrical insulator. Together with a more systematic approach to the theory it has been possible to provide instruments with an uncertainty of ±0.5%.

The theory assumes that the heat source is straight and stationary. To ensure that this is true a mechanism must be used to take up the thermal expansion of the wire as it undergoes transient heating. At the same time the expansion of the wire and its supports must not impose undue stresses during temperature change of the whole assembly. The simplest means of doing this is illustrated in Fig. 3.42 where the wires are tensioned by small weights. In Fig. 3.43 the wires are pretensioned between fixed supports but the supports of the wires and the wires themselves are made of the same material, tantalum; so that the wires are kept under constant tension even when the temperature changes.

To register with high accuracy the resistance change of the wire, and thus its temperature change, an electronic type of bridge is usually employed. Recent bridges [3.140] allow measurements from 20 μs after the initiation of heating, resulting in a very large number of measurements within 1 s. At the same time they are characterized by a temperature resolution of 5 mK and a time resolution of 1 μs.

The transient hot-wire technique is unique when compared with other techniques of measuring thermal conductivity. The mathematical model of the technique permits a detailed assessment of its precision and verifies the absence of modes of heat transfer other than conduction. The technique has successfully been employed over a very wide range of pressures and temperatures, but not near the critical point for the reasons already discussed. Transient hot-wire instruments have been successfully operated in many laboratories (in London [3.141], Thessaloniki [3.137, 139], Lisbon [3.142], Tokyo [3.143], Boulder [3.144] and other places). In the best instruments, the uncertainty of the technique is in the region of ±0.3 to ±0.5%, although measurements performed prior to the development of the complete theory in 1976 are expected to have uncertainties larger than ±0.5%.

Interferometric Technique

At the critical point the thermal conductivity of the fluid diverges, its thermal conductivity becomes infinite and its thermal diffusivity becomes zero. This divergence is a consequence of the ever-increasing length scale of correlations among the molecules as the critical point is approached, and its effect extends over a wide region around the critical point. In this region a number of special techniques for the study of thermal conductivity have been developed and we describe just one that has had the distinction of having been operated on the Earth [3.145] and in space [3.146].

The principle of the technique is quite straightforward. An infinitesimally thin, uniform source of heat q is located at the junction of a two semi-infinite materials, a solid and a fluid. Initiation of the heat flux at time $t = 0$ causes the temperature of the fluid to rise according to

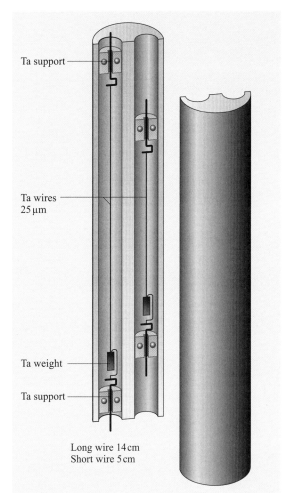

Fig. 3.42 Wires with weights arrangement developed by *Assael* et al. [3.139]

the equation

$$\Delta T(z, t) = \frac{2q}{\lambda} \sqrt{at}\, \text{ierfc}\left(\frac{z}{2}\sqrt{at}\right), \quad (3.131)$$

where z is the distance from the heat source ($z = 0$) measured along the normal to it into the fluid domain, and ierfc denotes the integrated complementary error function. The temperature distribution in the fluid at any instant can be determined by means of the effect of the temperature change on the density and refractive index n of the fluid using optical interferometry. In the simplest approach an interference pattern is produced in the fluid at a uniform temperature in a plane perpendicular to the heater surface by the superposition of two laser beams. Any subsequent nonuniformity of temperature induced

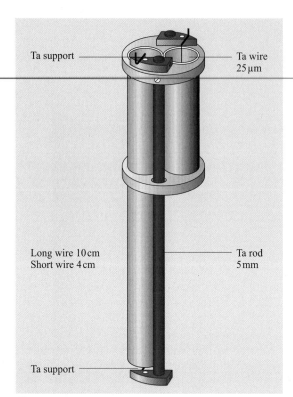

Fig. 3.43 Wires and support made of same material design developed by *Jawad* [3.147]

by the transient heating perturbs the fringe pattern in a manner described by (3.131) and $|dn/dT|$. Thus, in principle, measurements of the perturbation can be used to determine the thermal diffusivity of the fluid a. In the work of *Becker* and *Grigull* [3.145] this was achieved by means of holographic interferometry. The fringes in the cell (Fig. 3.44) at a state of uniform temperature are produced as a hologram and consequently subtracted

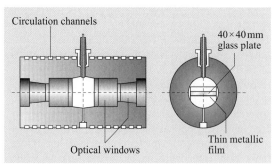

Fig. 3.44 Interferometric cell developed by *Becker* and *Grigull* [3.145]

from those produced in any other state, at any instant during the transient heating.

Using the theory of ideal interferometry, in which all light paths are assumed straight, the fringe order k is given by

$$k = 2\sqrt{at} B \, \text{ierfc}\left(\frac{z}{2}\sqrt{at}\right), \quad (3.132)$$

where

$$B = \frac{ql}{\lambda \Lambda}\left|\frac{\partial n}{\partial T}\right| \quad (3.133)$$

for light of wavelength Λ, when l is the distance through the fluid over which the temperature disturbance takes place.

Thus, measurements of the position of fringes of various orders at any instant of time, following the initiation of transient heating, can be used to determine the thermal diffusivity a by nonlinear regression, and consequently the thermal conductivity from a knowledge of the fluid density and heat capacity.

The main advantage of the technique is that it is possible to employ temperature rises of the order of 10^{-4} K because the sensitivity of the technique increases as the critical point is approached. As with any measurement near the critical point a high degree of overall temperature stability and uniformity of the order of 10^{-6} K is essential and the nature of the cell required lends itself to the fulfillment of these conditions.

3.5.2 Steady-State Methods for Thermal Conductivity

In the case of the steady-state methods employed for the measurement of the thermal conductivity, (3.119) reduces to

$$\lambda \nabla^2 T = 0. \quad (3.134)$$

In these methods, the heat flux necessary to maintain a constant temperature difference is measured. The two surfaces are usually formed by concentric cylinders or by two parallel plates. All such instruments are characterized by quite simple working equations and the difficulties in their use arise from the painstaking alignment of the two surfaces, because defects in their alignment make a first-order contribution to the instrument's uncertainty. Attention must also be given to the avoidance of significant contributions from other modes of heat transfer such as natural convection or radiation.

Coaxial-Cylinder Technique

The coaxial-cylinder technique is a steady-state method that measures the heat exchange by conduction between two concentric cylindrical surfaces separated by a small gap filled with the fluid sample, each of these surfaces being maintained at constant temperature. The technique has an intermediate position between the transient hot-wire and parallel-plate techniques as far as the secondary heat flows are concerned. The instrumentation is somewhat more complex than that for the hot-wire cells, but less complex than that for the parallel-plate apparatus. The reduction of heat losses from non-isothermal sample boundaries is easier to realize with a system involving heat transmission between a heat source and a concentric heat sink separated by the fluid than with a planar cell. However, special care must be taken in the perfect machining of the two cylindrical surfaces, which must be highly polished. The same care must be taken in the centering of the two coaxial cylinders. The heat transfer arising from convection in the coaxial-cylinder system can be made negligible, except near the critical point of the fluid, by employing a small annular gap (typically 0.2–0.3 mm). Furthermore, the radiant heat transfer can be strongly reduced by choosing a low-emissivity material from which to make the cylinders. Silver is ideal for this purpose since it has a low emissivity, a high thermal conductivity and a good resistance to chemical agents. One of the main advantages of the coaxial-cylinder technique is its versatility. Almost any fluid can be investigated, whether an electrical conductor or not.

The basic principle of the technique assumes a thin layer of a homogeneous fluid with a uniform thermal conductivity λ enclosed between two coaxial cylinders of infinite length. The external radius of the inner cylinder is r_1, and the internal radius of the external cylinder is r_2. We assume that the heat flux is uniformly generated in the inner cylinder and propagates radially through the test sample to the heat sink (the outer cylinder), in steady-state conditions. Then, the temperatures of the external surface of the inner cylinder and of the internal surface of the outer cylinder will be respectively T_1 and T_2. Employing (3.134), the amount of heat transferred by conduction per unit time and per unit length through the fluid layer is given by

$$Q = \frac{2\pi\lambda}{\log(r_2/r_1)} (T_1 - T_2) \ . \tag{3.135}$$

The thermal conductivity is obtained from the above equation by measuring the heat flux Q passing through the test sample and the temperature difference ($T_1 - T_2$). The value of the thermal conductivity determined corresponds to that at the average temperature ($T_1 + T_2$)/2.

In practice, the length of the cylinders is not infinite and the heat transfer through their ends must be considered.

In the classic coaxial-cylinder cell (Fig. 3.45), the end pieces (guard cylinders) are maintained at exactly same temperature as the inner cylinder emitter, so that all of the heat is transferred radially and (3.135) can be applied to obtain the thermal conductivity of the fluid.

Alternatively, if the end pieces are maintained at the same temperature as the inner surface of the outer cylinder, the thermal conductivity of the fluid is obtained from the equation

$$Q = \frac{\lambda}{C} (T_1 - T_2) \ , \tag{3.136}$$

where Q is the total amount of heat generated in the emitter and C represents a geometric instrument constant. The constant C depends just upon the geometry of the coaxial cylinders. It can therefore be determined from measurements of the heat flux when the annular gap is filled with a standard fluid of known thermal conductivity. Alternatively, it is possible to use the fact that there is a complete analogy between the equations of heat conduction and electrostatics so that a measurement of the electrical capacitance of the coaxial cylinder system when filled with a fluid of known dielectric constant also yields the constant C [3.148, 149].

In the measurement of the thermal conductivity, a number of corrections have to be made to obtain the thermal conductivity of the test specimen. Some of these corrections are related to the special features of the apparatus, for instance, the eccentricity between the internal and external cylinders; some are due to the fluid under investigation such as the heat transfer by convection. Others are related both to the instrument design and the fluid, the heat transfer by radiation including the absorption of radiation, or the temperature jumps between the walls of the cell and the test gas.

The coaxial-cylinder instrument shown in Fig. 3.45 [3.148, 149] incorporates two coaxial cylinders made completely out of silver. This instrument was used to measure several thermal conductivity of several samples, namely carbon dioxide, steam, ammonia, ethene, ethane, propane and many others [3.150] with special emphasis on the critical region. The large vertical extent of the cylinders (see dimensions in Fig. 3.45) means that the critical point can only be realized at one horizontal plane but measurements in the vicinity of the critical region

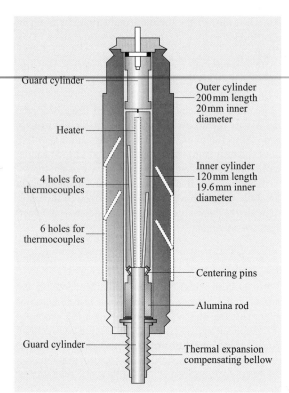

Fig. 3.45 Coaxial-cylinder cell developed by *LeNeindre* [3.148] and *Tufeu* [3.149]

were obtained successfully. The estimated uncertainty of these data is better than ±2% [3.150].

Coaxial-cylinder instruments have been successfully operated and reported in literature by many investigators, such as *Vargaftik* and *Smirnova* [3.152], *Vines* [3.153], *Ziebland* and *Burton* [3.154], *Yata* et al. [3.155], *Bailey* and *Kellner* [3.156] and others.

Parallel–Plate Technique

In the parallel-plate technique for measuring the thermal conductivity, the fluid under investigation is confined between two horizontal plates. The system is heated from above so that the upper plate has a higher temperature than the lower plate. It follows from (3.134), that, if the upper and lower plates are maintained at constant, uniform temperatures T_1 and T_2 respectively, the heat flow across an area A and a fluid layer of thickness d, is

$$Q = \lambda A \frac{(T_1 - T_2)}{d} . \tag{3.137}$$

In this solution it is assumed that the thermal conductivity of the fluid is constant, that the plate dimensions are infinite and that the heat flux is in one dimension.

In practice the upper plate is surrounded by a guard plate, sufficiently close to the upper plate to eliminate the distortion of the temperature profile at the edges of the latter. Equation (3.137) shows that it is very important to measure accurately the area A and the thickness d apart from the temperature difference since they all enter the working equation in first order. The plates must also be perfectly parallel in order that the working (3.137) can be used. Since none of these conditions can ever be satisfied exactly, there are several corrections that must be considered. In practice, the guard plate is separated from the upper plate by a distance, smaller than the thickness d. Thus, A is not the area of the upper plate but an effective area for heat transfer.

Generally, the heat transfer between the plates is not only effected by conduction through the fluid layer, but also by convection and radiation. In addition, parasitic heat losses between the plates need to be taken into account. The effects of convection can be made negligibly small (although not eliminated) by taking great care to align the plates in a horizontal position and by using a small distance d. In most cases a correction is necessary for radiative heat transfer between the plates. To reduce the radiation correction the plate surfaces must have low emissivities. For this purpose the plate surfaces are polished and protected against oxidation, sometimes by coating with nickel, chrome, silver or silver dioxide. Finally, a correction has to be made for the reference state of the fluid usually taken as the average of the lower- and upper-plate temperatures. A description of all these corrections can be found in the literature [3.151, 157].

The parallel-plate instrument shown in Fig. 3.46 [3.151] was employed in the measurement of the ther-

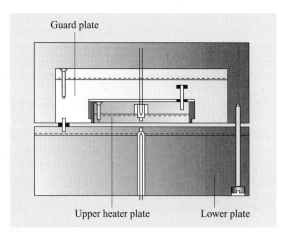

Fig. 3.46 Parallel-plate instrument (after [3.151])

mal conductivity of carbon dioxide in the liquid and dense gas states, with special attention to the critical region. In general, the parallel-plate technique makes it possible to perform accurate measurements of the thermal conductivity of fluids over a wide range of temperatures and pressures. The technique has successfully been employed from liquid-helium temperatures (*Ubbink* and *de Haas* [3.158], *Grenier* [3.159], *Roder* and *Diller* [3.157]) up to 500 °C (*Nuttall* and *Ginnings* [3.160], *Amirkhanov* and *Adamov* [3.161]) and at a pressure from a few mmHg up to 250 MPa (*Michels* et al. [3.162]). The classical work of *Sengers* and his coworkers [3.162] in the critical region demonstrated what can be achieved with this instrument given great care.

3.5.3 Light-Scattering Methods for Thermal Diffusivity

In the last 30 years there have been rapid developments in the application of optical techniques to the measurement of some of the transport properties of fluids. These techniques are applicable to a wide range of optically transparent fluids and have proved particularly valuable near the critical point of a fluid because, unlike the classic methods of measurement, it is not necessary to impose a macroscopic gradient on the fluid. In this section we consider two optical techniques. The bulk of our description of these techniques is associated with the underlying theory of the experiments because the physical aspects of the instrumentation are generally commercially available optical components assembled for particular purposes. First the application of photon-correlation spectroscopy to the measurement of the thermal diffusivity of fluids will be described. Secondly, a more recent optical technique that makes use of forced-Rayleigh scattering of light to determine the thermal diffusivity of fluids, such as molten salts, under extreme conditions, will be presented.

Photon-Correlation Spectroscopy Technique

A fluid in thermodynamic equilibrium is continuously subject to fluctuations of its thermodynamic variables. The fluctuations, although they are microscopic, are described statistically by the same equations that govern the corresponding macroscopic processes [3.163]. For example, the lifetime of entropy fluctuations is determined by the thermal diffusivity a of the bulk fluid. Light passing through a fluid is scattered by microscopic fluctuations of the local dielectric constant brought about by local fluctuations of the thermodynamic variables. The spectrum of the light scattered by such fluctuations has three peaks, namely the Rayleigh line and a symmetric pair of Brillouin lines. The Rayleigh line is the result of quasi-elastic scattering from fluid elements with entropy fluctuations and, in mixtures, also contains a component scattered from concentration fluctuations. Photon-correlation spectroscopy (dynamic light-scattering spectroscopy or light-beating spectroscopy), which operates in the time domain, is a high-resolution technique for measuring the widths of each of the lines of this scattered spectrum.

Use of photon-correlation spectroscopy to measure the thermal diffusivity and mass diffusion coefficient has the advantage that the measurements are made on a sample in thermal equilibrium so that no corrections for macroscopic temperature gradients are needed, and convective effects can be avoided. Because the transport information derives from sampling microscopic thermodynamic fluctuations with a laser beam, the total volume of material needed to make a measurement can be very small. Correspondingly, the time needed to reach equilibrium can be kept small, and experiment run times can also be short depending on the desired accuracy of the measurement. The technique is intrinsically absolute in nature, not requiring extensive calibration. It is, of course, particularly well suited to measurements of the thermal diffusivity in the critical region when fluctuations become more important. The major limitation of the technique has been inability to achieve a measurement uncertainty better than 5–10%, although recently [3.164] the uncertainty of the technique has improved to about 2.5%.

We now briefly examine the details of the theory of the method since they reveal what measurements may most easily be made. Incident light with wavenumber k (in the scattering medium) is scattered at an angle θ by fluctuations with a wavenumber q according to the Bragg condition

$$q = 2k \sin\left(\frac{\theta}{2}\right). \tag{3.138}$$

The width of the scattered Rayleigh peak is determined by the diffusivities associated with thermal fluctuations in the fluid. For the entropy fluctuations this is the thermal diffusivity a whereas for the concentration fluctuations, the determining diffusivity is the mutual diffusion coefficient D. In the time domain one works with the autocorrelation function of the electric field E of the scattered light, defined by

$$G^{(1)}(\tau) = \langle E^*(\tau)E(0)\rangle. \tag{3.139}$$

In a one-component fluid $G^{(1)}(\tau)$ decays exponentially with delay time τ, as

$$G^{(1)}(\tau) = A_s \, e^{-aq^2\tau} , \qquad (3.140)$$

where q is the wavenumber defined in (3.138), where the exponential contains the thermal diffusivity. Evidently, if it were possible to observe the autocorrelation function the thermal diffusivity could be evaluated directly. In actual photon-correlation experiments, the detectors of the scattered light are photomultiplier tubes, which respond to the intensity of the scattered light and not simply to its electric vector. Thus, when the scattered signal is processed by a digital correlator, it is the intensity correlation function

$$G^{(2)}(\tau) = \langle I(\tau)I(0) \rangle \qquad (3.141)$$

that is obtained. The intensity is simply related to the electric field vector because

$$I(t) = |E(t)|^2 = E^*(t)E(t) , \qquad (3.142)$$

in which the electric field $E(t)$ is a linear combination of all contributions to the scattered electric field at the detector. In all but the simplest of cases, the number of terms which arises from expanding (3.141) is very large. Special circumstances are employed to render the number of terms tractable.

In heterodyne measurements [3.165–167], which are most suitable for small intensities, the scattered light is mixed coherently with a static light source at the incident wavelength, usually light reflected from a window of the sample cell. In this case, the static field is added to the scattered fields at the detector, and (3.141) for a single-component fluid becomes

$$G^{(2)}(\tau) = (I_0 + I_s)^2 + I_s^2 \, e^{-2\tau/\tau_s} + 2I_0 I_s \, e^{-\tau/\tau_s} , \qquad (3.143)$$

where I_0 is the intensity of the static scattering, I_s is the intensity due to entropy fluctuations and τ_s is the decay time associated with the entropy fluctuations related to the thermal diffusivity a

$$a = \frac{1}{(q^2 \tau_s)} . \qquad (3.144)$$

Fitting the heterodyne equations to extract the decay time τ_s, it is usual to arrange the condition $I_0 \gg I_s$ so that (3.143) simplifies to

$$G^{(2)}(\tau) = (I_0 + I_s)^2 + 2I_0 I_s \, e^{-\tau/\tau_s} . \qquad (3.145)$$

Equation (3.145) forms the basis of the usual photon-correlation spectrometer measurements.

A typical arrangement is shown in Fig. 3.47. The heterodyne arrangement is ensured by using a strong reference beam generated by a beam splitter which is then combined with the Rayleigh-scattered light at the photomultiplier. Here, q is obtained from the equation [3.164]

$$q = \frac{4\pi n}{\lambda_0} \sin\left(\frac{\theta_s}{2}\right) , \qquad (3.146)$$

where n is the sample refractive index, λ_0 the laser wavelength in vacuum, and θ_s the angle between the direction of observation and the incident laser beam within the sample (Fig. 3.47). With these values taken from experiment, and with the decay time extracted from the measured correlation function, the thermal diffusivity a can be calculated from (3.144).

Recent publications [3.164] on the measurement of the thermal conductivity of toluene with a photon-correlation spectrometer showed an uncertainty of $\pm 2.5\%$, mostly attributed to errors in the angle measurement and the photon statistics, which can be improved by greater duration of the experiments.

Forced Rayleigh-Scattering Technique

The forced Rayleigh-scattering technique was initiated independently by *Eichler* et al. [3.168] and *Pohl* et al. [3.169] in 1973. *Eichler* and coworkers developed it for the determination of the thermal diffusivity of organic liquids and ruby crystals. More recently *Nagasaka* and *Nagashima* and their coworkers [3.170, 171] reworked the theory of forced Rayleigh scattering and applied it to the measurement of the thermal diffusivity of organic liquids, liquid crystals and, particularly, high-temperature molten salts (above 1000 °C).

The term *forced Rayleigh scattering* was created by analogy with spontaneous Rayleigh scattering, which we saw above is based on the scattering effect of statistical thermodynamic fluctuations in a fluid. If such weak and random fluctuations are replaced by stronger and more coherent excitations from a laser-induced grating, the forced scattering (or diffraction) of a probing beam becomes much stronger and coherent. The technique has also been called *laser-induced dynamic grating*.

The principle of the forced Rayleigh scattering method, is illustrated by reference to Fig. 3.48.

Two pulsed, high-power laser beams of equal wavelength and equal intensity intersect in an absorbing sample at an angle θ. They generate an optical interference fringe pattern whose intensity distribution is spatially sinusoidal. Following partial absorption of the laser light, this interference pattern induces a corresponding temperature distribution in the x-direction

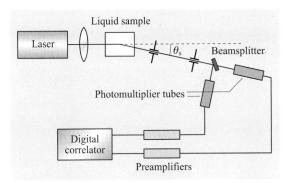

Fig. 3.47 Schematic diagram of a photon-correlation spectrometer

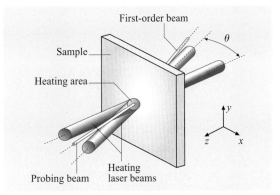

Fig. 3.48 Principle of forced Rayleigh scattering

of the sample. After the heating process, the excited temperature distribution decays exponentially owing to heat conduction and the decay of the distribution can be examined by using a non-absorbed probe-laser beam which is diffracted by the grating created by the influence of the temperature distribution on the refractive index of the fluid. It is assumed that heat conduction takes place in the x-direction only and this is permissible if the following conditions are satisfied. First the grating period Λ should be much smaller than the sample thickness d. Secondly, Λ should be sufficiently small compared with the light absorption length and finally Λ should be small compared with the diameter of the heated area. The simplest mathematical description of forced Rayleigh scattering is then provided by rewriting the thermal balance equation (3.119) in one dimension as

$$\frac{\partial T}{\partial t} = a \frac{\partial^2 T}{\partial x^2} \qquad (3.147)$$

subject to the initial spatially periodic temperature distribution produced by the interference of two laser beams.

$$T = T_0 + \Delta T_0 (1 + \cos qx) \quad \text{at } t = 0. \qquad (3.148)$$

Here, T is the temperature, t the elapsed time after heating, a the thermal diffusivity of the sample, T_0 the initial temperature, ΔT_0 the initial spatial temperature amplitude, and $q = 2\pi/\Lambda$ is the wavenumber of the interference pattern. The relation between the angle of inclination of the heating beams and the grating period is

$$\Lambda = \frac{\lambda_h}{2\sin(\theta/2)} \approx \frac{\lambda_h}{\theta} \quad (\theta \approx 0), \qquad (3.149)$$

where λ_h is the wavelength of the heating beam. The solution to (3.147) for the temperature distribution is

$$T(x,t) = T_0 + \Delta T_0 \left[1 + \cos(qx) \exp\left(\frac{-t}{\tau}\right) \right], \qquad (3.150)$$

which demonstrates that the spatial temperature amplitude decays exponentially with the relaxation time characteristic of heat conduction τ

$$\tau = \left(\frac{1}{a}\right)\left(\frac{\Lambda}{2\pi}\right)^2. \qquad (3.151)$$

As indicated earlier, the spatially periodic temperature distribution produces a corresponding refractive-index distribution. In turn this acts as an optical phase grating for a low-power probing laser beam of a wavelength λ_p not absorbed in the sample. According to the theory of diffraction, if the sample is thin enough, the first-order diffracted laser beam intensity I_1 is proportional to the square of the maximum phase variation of the grating ϕ so that for a probing laser beam of intensity I_p

$$I_1 = \frac{I_p \phi^2}{4}. \qquad (3.152)$$

Table 3.17 Reference values for the thermal conductivity of noble gases at a pressure of 0.1 MPa

T (°C)	Thermal conductivity (mWm^{-1}K^{-1})				
	Helium	Neon	Argon	Krypton	Xenon
25	155.3	49.24	17.67	9.451	5.482
100	181.1	57.84	21.36	11.63	6.852
200	213.9	67.43	25.59	14.18	8.534
300	244.7	76.79	29.60	16.50	10.07
400	274.1	85.34	33.14	18.64	11.49
500	302.0	93.39	36.50	20.64	12.81

Table 3.18 Thermal conductivity λ (mW m^{-1} K^{-1}) of some n-alkanes in the liquid phase

		P (MPa)	273.15 K	298.15 K	323.15 K	348.15 K	373.15 K
n-Pentane	C_5H_{12}	0.101	125.1	113.7	102.4	91.3	80.5
		5	127.8	116.8	105.9	95.3	84.6
		10	130.4	119.7	109.2	98.8	88.3
		25	137.6	127.6	117.8	107.9	97.3
		50	147.8	138.6	129.4	119.7	108.8
n-Hexane	C_6H_{14}	0.101	128.4	118.9	109.8	100.9	92.1
		5	130.7	121.5	112.8	104.4	96.1
		10	133.0	124.1	115.7	107.6	99.6
		25	139.3	131.0	123.3	116.0	108.7
		50	148.4	140.9	134.0	127.4	120.6
n-Heptane	C_7H_{16}	0.101	132.8	123.5	114.5	105.9	97.4
		5	134.9	125.8	117.2	108.9	100.9
		10	137.0	128.2	119.8	111.8	104.1
		25	142.9	134.5	126.8	119.4	112.4
		50	151.4	143.7	136.6	130.0	123.5
n-Octane	C_8H_{18}	0.101	137.8	128.4	119.5	110.8	102.4
		5	139.8	130.7	122.0	113.6	105.5
		10	141.8	132.9	124.4	116.3	108.5
		25	147.4	139.0	131.0	123.6	116.4
		50	155.7	147.8	140.5	133.7	127.1
n-Nonane	C_9H_{20}	0.101	140.1	130.9	122.1	113.5	105.1
		5	142.1	133.1	124.4	116.1	108.0
		10	144.0	135.1	126.7	118.6	110.8
		25	149.3	140.9	132.9	125.3	118.1
		50	157.3	149.4	142.0	135.0	128.3
n-Decane	$C_{10}H_{22}$	0.101	143.5	134.2	125.3	116.6	108.3
		5	145.4	136.3	127.5	119.1	111.0
		10	147.2	138.3	129.7	121.5	113.6
		25	152.5	143.9	135.7	128.0	120.7
		50	160.4	152.2	144.5	137.4	130.6
n-Undecane	$C_{11}H_{24}$	0.101	145.0	135.9	127.0	118.3	110.0
		5	146.8	137.8	129.1	120.7	112.5
		10	148.6	139.8	131.2	123.0	115.0
		25	153.7	145.2	137.0	129.2	121.7
		50	161.4	153.3	145.5	138.2	131.2
n-Dodecane	$C_{12}H_{26}$	0.101	147.9	138.6	129.5	120.7	112.2
		5	149.7	140.5	131.6	123.0	114.7
		10	151.5	142.4	133.6	125.2	117.1
		25	156.5	147.7	139.3	131.2	123.5
		50	164.1	155.7	147.6	140.0	132.8

The interrelation between ϕ and the maximum temperature amplitude of (3.150) can be expressed by

$$\phi = \left(\frac{2\pi d}{\lambda_p}\right)\left(\frac{\partial n}{\partial T}\right)\Delta T_0 \exp\left(\frac{-t}{\tau}\right), \qquad (3.153)$$

where n is the refractive index of the sample. Substituting (3.153) into (3.152), the thermal diffusivity of the sample can be obtained from the following equation

$$a = -\left(\frac{1}{2}\right)\left(\frac{\Lambda}{2\pi}\right)^2\left(\frac{d\log I_1}{dt}\right). \qquad (3.154)$$

Table 3.19 Thermal conductivity λ (mW m^{-1}K^{-1}) of some alkenes in the liquid phase

		P (MPa)	273.15 K	298.15 K	323.15 K	348.15 K	373.15 K
Benzene	C_6H_6	0.101	153.1	143.7	134.0	124.1	114.2
		5	154.9	145.6	136.1	126.5	116.9
		10	156.6	147.5	138.2	128.8	119.4
		25	161.5	152.9	144.1	135.2	126.4
		50	169.0	160.9	152.7	144.5	136.2
Toluene	C_7H_8	0.101	138.5	130.0	121.9	114.4	107.2
		5	139.9	131.6	123.8	116.4	109.6
		10	141.4	133.3	125.6	118.4	111.8
		25	145.7	137.9	130.6	124.0	118.0
		50	152.1	144.8	138.1	132.0	126.7
Ethylbenzene	C_8H_{10}	0.101	135.6	128.6	121.7	115.1	108.6
		5	137.0	130.2	123.5	117.1	110.9
		10	138.4	131.7	125.3	119.1	113.2
		25	142.5	136.1	130.1	124.5	119.2
		50	148.6	142.8	137.4	132.4	127.9
o-Xylene	C_8H_{10}	0.101	142.1	134.6	127.6	121.2	115.3
		5	143.5	136.0	129.2	123.0	117.4
		10	144.8	137.5	130.8	124.8	119.4
		25	148.7	141.7	135.4	129.7	124.8
		50	154.6	148.0	142.2	137.1	132.8
m-Xylene	C_8H_{10}	0.101	133.9	127.5	121.2	115.0	108.9
		5	135.3	129.1	123.0	117.0	111.2
		10	136.7	130.6	124.7	118.9	113.4
		25	140.7	135.0	129.6	124.4	119.5
		50	146.8	141.6	136.8	132.3	128.2
p-Xylene	C_8H_{10}	0.101	132.2	125.7	119.4	113.1	107.0
		5	133.6	127.3	121.1	115.1	109.3
		10	135.0	128.8	122.8	117.0	111.5
		25	138.9	133.1	127.6	122.4	117.5
		50	144.9	139.7	134.7	130.2	126.0
Mesitylene	C_9H_{12}	0.101	135.6	129.1	122.9	117.1	111.5
		5	136.9	130.6	124.6	119.0	113.7
		10	138.2	132.1	126.2	120.8	115.8
		25	142.1	136.3	130.9	126.0	121.6
		50	148.0	142.6	137.8	133.5	129.9

Accordingly, the thermal diffusivity is determined by measuring the time dependence of I_1 and the grating period Λ. The optical instrumentation required to carry through these measurements is standard and does not merit description here.

The actual experimental conditions may differ from those described. There are several secondary effects that must be considered, for example in many fluids it is necessary to add a dye to the fluid so that it absorbs sufficient light and this may affect the property measured. It is also necessary to consider the fact that the heat conduction process is not in practice one-dimensional; these and other effects have been dealt with in the literature [3.172]. By comparison with conventional techniques for the measurement of the thermal diffusivity, the forced Rayleigh technique has several following advantages: it is contact-free, it is characterized by the short duration of the measurement time (< 1 ms), small temperature rise (< 0.1 K), small sample volume (< 10 mm^3), and it is applicable to anisotropic

Table 3.20 Thermal conductivity λ (mW m^{-1}K^{-1}) of some n-alcohols in the liquid phase

		P (MPa)	273.15 K	298.15 K	323.15 K	348.15 K	373.15 K
Methanol	CH$_4$O	0.101	206.1	203.8	192.5	176.2	158.2
		5	208.7	206.9	195.9	179.5	161.2
		10	211.3	210.0	199.1	182.8	164.0
		25	218.5	218.4	208.2	191.6	171.8
		50	229.2	230.7	221.1	204.1	182.6
Ethanol	C$_2$H$_6$O	0.101	174.9	166.7	156.5	145.0	132.6
		5	177.0	169.0	159.0	147.5	135.0
		10	179.1	171.3	161.5	150.0	137.3
		25	185.0	177.8	168.3	156.8	143.6
		50	193.8	187.4	178.3	166.7	152.7
1-Propanol	C$_3$H$_8$O	0.101	159.1	156.1	150.2	141.5	130.5
		5	160.9	158.2	152.5	144.1	133.1
		10	162.8	160.3	154.9	146.6	135.5
		25	167.9	166.2	161.4	153.4	142.2
		50	175.7	175.0	171.0	163.3	151.9
1-Butanol	C$_4$H$_{10}$O	0.101	159.3	155.0	147.0	135.9	122.5
1-Pentanol	C$_5$H$_{12}$O	0.101	163.5	155.1	143.1	128.6	112.8
1-Hexanol	C$_6$H$_{14}$O	0.101	167.5	154.5	139.1	122.5	106.0

materials. The technique has successfully been applied to the measurement of the thermal diffusivity of molten salts [3.173] although its potential is greater.

3.5.4 Thermal Conductivity Reference Values

Toluene and water have been proposed as primary standard reference liquids. Under the auspices of the International Union of Pure and Applied Chemistry (IUPAC) Subcommittee on Transport Properties, *Nieto de Castro* et al. [3.174] recommended the following values as primary data:

for toluene at 298.15 K and 0.1 MPa

$$\lambda = (0.1311 \pm 0.0013) \, \text{W} \text{m}^{-1} \text{K}^{-1} \tag{3.155}$$

and for water at 298.15 K and 0.1 MPa

$$\lambda = (0.6067 \pm 0.0061) \, \text{W} \text{m}^{-1} \text{K}^{-1} \, . \tag{3.156}$$

The temperature dependence of the thermal conductivity is represented by the following equations, where $T^* = (T/298.15 \, \text{K})$, and $\lambda^* = [\lambda(T)/\lambda(298.15 \, \text{K})]$:

for toluene

$$\lambda^* = 1.68182 - 0.682022 T^* \tag{3.157}$$

at $230 \, \text{K} \leq T \leq 360 \, \text{K}$,

$$\lambda^* = 1.45210 - 0.224229 T^* - 0.225873 T^{*2} \tag{3.158}$$

at $189 \, \text{K} \leq T \leq 360 \, \text{K}$, for water

$$\lambda^* = -1.26523 + 3.70483 T^* - 1.43955 T^{*2} \tag{3.159}$$

for $274 \, \text{K} \leq T \leq 360 \, \text{K}$.

The maximum deviation of the primary experimental data from (3.157) is 1.3%, from (3.158) is 1.5% and from (3.159) is 1.1%. Other secondary thermal conductivity reference liquids are available in the literature [3.167].

Recommended values for the thermal conductivity of the noble gases are shown in Table 3.17. They have been calculated using available theory and the corresponding viscosity data and their estimated uncertainty is $\pm 0.3\%$ in the range 25–200 °C, and $\pm 0.5\%$ in the range 200–500 °C.

For higher pressures up to 30 MPa and at a temperature of 27.5 °C, the thermal conductivity of argon is represented by the equation [3.175]

$$\lambda = 17.751 + 21.402 \times 10^{-3} \rho + 27.247 \times 10^{-6} \rho^2 \, , \tag{3.160}$$

Table 3.21 Thermal conductivity λ (mW m^{-1} K^{-1}) of some refrigerants in the liquid phase

		P (MPa)	248.15 K	273.15 K	298.15 K	323.15 K	348.15 K
R22	CHClF$_2$	5	111.6	98.8	86.8	76.2	65.8
		10	114.2	102.0	90.9	81.7	74.7
		25	120.8	110.1	100.5	93.3	89.8
R32	CH$_2$F$_2$	5	171.8	147.7	127.4	112.0	
		10	176.8	153.3	134.0	121.2	129.0
		25	189.7	166.7	148.6	137.9	165.6
R124	C$_2$HClF$_4$	5	87.5	78.9	70.5	63.6	57.6
		10	88.7	80.9	73.5	67.9	62.8
		25	92.2	86.0	81.0	77.5	73.3
R125	C$_2$HF$_5$	5	81.7	75.2	66.6	55.6	
		10	84.4	78.9	71.7	63.4	
		25	91.3	87.6	82.4	76.6	
R134a	C$_2$H$_2$F$_4$	5	104.9	94.6	84.6	74.1	62.3
		10	107.6	97.8	88.6	79.3	69.9
		25	114.8	105.9	98.0	90.6	83.8
R141b	C$_2$H$_3$ClF$_2$	5	108.5	101.1	93.6	86.0	78.2
		10	109.5	102.5	95.5	88.3	81.1
		25	112.4	106.5	100.6	94.4	88.2
R152a	C$_2$H$_4$F$_2$	5	127.8	116.8	106.1	94.6	81.3
		10	130.7	120.3	110.4	100.3	89.3
		25	138.2	129.0	120.9	113.0	104.9

where λ is measured in mW/(m K) and ρ in kg/m^3. Other secondary thermal conductivity reference fluids are available in the literature [3.167].

3.5.5 Tables of Thermal Conductivity Values

In Tables 3.18–3.22, the thermal conductivity of commonly encountered fluids is given for engineering purposes as a function of temperature and pressure. The fluids and the temperature and pressure conditions chosen are the same for the density and viscosity in Sects. 3.1 and 3.4.

Values for the liquid thermal conductivity are based on a large collection of experimental data correlated by a semi-empirical hard-spheres-based procedure available in the literature (n-alkanes [3.176], n-alkenes [3.177], n-alcohols [3.178], refrigerants [3.179]). The uncertainty of the quoted liquid thermal conductivity values is much better than ±5%. Values for the gas-phase thermal conductivity have been obtained from corresponding-states software [3.37] based on experimental data and they have an estimated uncertainty better than 2%. It is important to stress that the values listed for individual materials are not intended to be the best known values; they are, rather, a set of values that are consistent with those listed for other properties with sufficient accuracy to be useful.

3.6 Diffusion

The diffusion processes in liquids and gases are characterized by vastly different time scales; the process in liquids being some 10^5 times slower than in low-density gases. As a result, the types of instruments used to measure the diffusion coefficient differ dramatically between the two phases. We describe just some of the most successful among the many techniques that have been used in each phase concentrating upon optical techniques for liquids. We consider also two classical techniques for the gas phase. Just one technique is common to both gas and liquid phases, that relying upon the phenomena of Taylor dispersion.

The diffusion coefficient of a mixture of two or more chemical species is a measure of its tendency to produce

Table 3.22 Thermal conductivity λ (mW m^{-1}K^{-1}) of some gases

		P (MPa)	273.15 K	298.15 K	323.15 K	348.15 K	373.15 K
Argon	Ar	0.101	16.64	17.79	18.89	19.95	20.98
		5	19.47	20.34	21.21	22.08	22.95
		10	22.68	23.13	23.69	24.32	25.00
Hydrogen	H_2	0.101	172.46	183.61	194.36	204.70	214.80
		5	177.70	188.50	198.80	208.90	218.80
		10	182.00	193.20	203.20	213.05	222.60
Nitrogen	N_2	0.101	24.90	26.58	28.21	29.80	31.34
		5	28.25	29.60	32.34	32.34	33.70
		10	31.84	32.76	33.80	34.92	36.07
Oxygen	O_2	0.101	25.54	27.52	29.46	31.36	33.23
		5	28.83	30.49	32.16	33.84	35.52
		10	32.57	33.74	35.05	36.44	37.90
Carbon monoxide	CO	0.101	24.89	26.71	28.48	30.20	31.89
		5	28.43	29.89	31.38	32.87	34.36
		10	32.28	33.26	34.39	35.50	36.87
Carbon dioxide	CO_2	0.101	16.25	18.09	19.94	21.80	23.65
		5		26.30	25.96	26.76	27.93
		10			55.01	37.48	34.66
Sulfur dioxide	SO_2	0.101	9.84	11.00	12.20	13.43	14.68
Hydrogen sulfide	H_2S	0.101	16.39	18.08	19.81	21.55	
		5				29.22	
Methane	CH_4	0.101	32.75	36.08	39.51	43.04	46.68
		5		38.90	41.46	44.34	47.44	50.74
		10		47.69	48.26	50.06	52.44	55.21
Ethane	C_2H_6	0.101	19.65	22.55	25.62	28.86	32.26
		5			38.31	36.89	38.49
		10			75.26	61.66	52.42
Propane	C_3H_8	0.101	15.49	18.03	20.75	23.62	

entropy when it is disturbed from equilibrium by the imposition of gradient of the chemical potential of each species. As for the other transport coefficients, the diffusion coefficient is defined as the proportionality constant between a flux and a driving force. However, unlike the other transport processes, the diffusive flux of molecules has been defined with respect to a number of different frames of reference, and the driving force has also been expressed in a variety of alternative ways. There are therefore a variety of different diffusion coefficients in use. In this work, we choose the definition that leads to a phenomenological equation most closely related to the equation which describes the diffusion process in an experiment. We also confine ourselves to binary systems for simplicity. Consequently, we employ the gradient of molar density of a species for the driving force and consider molar fluxes of the two species, J_1 and J_2, with respect to a volume-fixed frame of reference defined by

$$v_1 J_1 + v_2 J_2 = 0, \tag{3.161}$$

where v_1 and v_2 are the partial molar volumes of the components of the mixture. Fick's law gives the molar fluxes, at any instant relative to this frame of reference as

$$J_1 = -D_{12} \left(\frac{\partial C_1}{\partial x}\right)_t, \tag{3.162}$$

$$J_2 = -D_{21} \left(\frac{\partial C_2}{\partial x}\right)_t, \tag{3.163}$$

where C_1 and C_2 are the molar densities of the two components. The coefficients D_{12} and D_{21} are the diffusion coefficients for the mixture, and from the definition

$$v_1 C_1 + v_2 C_2 = 1 , \quad (3.164)$$

it follows that

$$D_{12} = D_{21} . \quad (3.165)$$

When $(\partial C_i/\partial x)_t$ changes with time, but the diffusion coefficient does not vary with concentration during the experiment, then

$$\frac{\partial C_i}{\partial t} = D_{12} \left(\frac{\partial^2 C_i}{\partial x^2} \right) , \quad i = 1, 2 . \quad (3.166)$$

The diffusion coefficient defined in this way is known also as the interdiffusion coefficient or mutual diffusion coefficient, and depends parametrically on the thermodynamic state of the fluid, which is characterized by the variables (T, P, C_i) or (T, ρ, C_i). In this section it will be referred to simply as the *diffusion coefficient*.

The diffusion coefficient, as defined above, must be distinguished from

1. the self-diffusion coefficient that refers to the diffusional motion in a single-component fluid, and is usually studied by techniques such as nuclear magnetic resonance, and
2. the intra-diffusion coefficient, or tracer diffusion coefficient, which characterizes the diffusion of each of the components i, j in an otherwise uniform mixture of two or more components where the component under study, i, is chemically identical with component j but can be distinguished by some label such as its isotopic form.

The work presented here is divided into two main sections: the first dealing with measurements in the liquid phase, and the second with measurements in the gas phase because for this property there are some substantial differences in techniques between the two phases.

3.6.1 Diffusion in Liquids

In this section, techniques that are able to yield measurements of the diffusion coefficients in liquids with a small uncertainty on a reasonable timescale (≤ 1 d), will be discussed. Thus, the methods which will be presented are the Diaphragm-cell technique, the Taylor-dispersion technique, and the Rayleigh and Gouy interferometric techniques. Other techniques can yield higher accuracy but take extreme precautions or a larger investment of time than can be justified for routine use.

Diaphragm-Cell Technique
The diaphragm cell is the simplest method of determining diffusion coefficients with an uncertainty of about 1%. It is very versatile and has been used for a wide range of temperatures and pressures. The essential features of a typical cell, of the type first employed by *Stokes* [3.180], are shown in Fig. 3.49. Each bulb contains a glass stirrer, always in contact with the glass diaphragm. The stirrers are rotated by means of external rotating magnets. For a measurement, of the diffusion coefficient of potassium chloride in aqueous solution, for example, the salt solution is placed in the bottom bulb and water in the top. After a certain period, samples are taken from both bulbs for analysis.

In the case of the diaphragm cell, the diffusion process is assumed to be one-dimensional but the characteristics of the diaphragm are not known so that application of (3.166) and its solution are heuristic. If an analysis of the compositions of the two samples in the top and bottom bulbs is conducted after a diffusion time t then the diffusion coefficient D_{12} is obtained from the set of equations [3.181, 182]

$$D_{12} = \left(\frac{1}{\beta t} \right) \log \left[\frac{(C_B^0 - C_T^0)(1 - \lambda/6)}{(C_B - C_T)} \right] , \quad (3.167)$$

with

$$\beta = \left(\frac{A}{l} \right) \left(\frac{1}{V_B} + \frac{1}{V_T} \right) \left(1 - \frac{\lambda}{6} \right) , \quad (3.168)$$

and

$$\lambda = \frac{2 V_D}{(V_B + V_T)} . \quad (3.169)$$

Here, the subscripts 'B' and 'T' denote the bottom and top compartments, while the superscript 0 denotes initial concentrations. A is the effective area of the membrane, l its effective length and V_D the liquid content in the membrane (note that A and l are not simply related, as the effective area of the membrane could be the same for a whole range of membranes while their length could be increased). Excellent measurements with this technique were carried out by *Woolf* and *Tilley* [3.183] on aqueous solutions of potassium chloride.

Taylor-Dispersion Technique
The Taylor-dispersion technique originated by Sir Geoffrey *Taylor* in 1953 [3.184], provides a means whereby

rapid measurements of diffusion coefficients can be made with moderate accuracy over a wide range of conditions. The method is a dynamic chromatographic technique that has few of the limitations of the other techniques of measuring diffusion coefficients.

In an idealized Taylor-dispersion experiment a narrow pulse of solute is injected near the axis into a long uniform tube of length L and radius R, in which solvent is flowing in a slow, laminar manner. As the pulse is carried through the tube, it spreads owing to the combined action of convection in the axial direction and molecular diffusion in the radial direction. The peak center, or the maximum, continues to move at the mean velocity of the laminar profile. Eventually, the peak elutes from the end of the long tube, where a suitable detector is employed to measure the radially averaged concentration profile as a function of time.

In cylindrical coordinates, the continuity equation for a species in terms of its molar concentration C, at fixed point (r, x) is written

$$D_{12}\left(\frac{\partial^2 C}{\partial r^2} + \frac{1}{r}\frac{\partial C}{\partial r} + \frac{\partial^2 C}{\partial x^2}\right) = u(r)\frac{\partial C}{\partial x} + \frac{\partial C}{\partial t}, \quad (3.170)$$

where $u(r)$ is the axial velocity of the flow relative to laboratory coordinates. The diffusion coefficient is assumed constant which is valid if the concentration gradient is small. It is further assumed that there is no chemical reaction occurring, the fluid density is constant, and that the fluid is in laminar flow with the familiar parabolic velocity profile for Newtonian fluids,

$$u(r) = 2\bar{u}\left[1 - \left(\frac{r}{R}\right)^2\right], \quad (3.171)$$

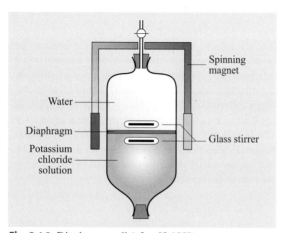

Fig. 3.49 Diaphragm cell (after [3.180])

where $\bar{u}$ is the mean flow velocity. It is, however, more convenient to measure the concentration distribution relative to an axial coordinate z, which moves with the mean speed of flow $\bar{u}$. Then the velocity in this frame of reference is

$$v(r) = u(r) - \bar{u} = \bar{u}\left[1 - 2\left(\frac{r}{R}\right)^2\right]. \quad (3.172)$$

Taylor showed that, by choosing appropriate experimental conditions, the effects of longitudinal diffusion upon the injected pulse may be neglected [3.185]. Therefore, neglecting the axial dispersion term and considering convection across a plane which moves at $\bar{u}$, (3.167) reduces to

$$\frac{\partial^2 C}{\partial r^2} + \frac{1}{r}\frac{\partial C}{\partial r} = \frac{R^2}{D_{12}}\frac{\partial C}{\partial t} + \frac{R^2 \bar{u}}{D_{12}}\left[1 - 2\left(\frac{r}{R}\right)^2\right]\frac{\partial C}{\partial z}, \quad (3.173)$$

where $z = x - \bar{u}t$. Although (3.173) cannot be solved directly for the concentration perturbation, it is readily solved for the spatial moments of the distribution C at a particular time. In particular, *Aris* [3.186] has shown that, after a sufficient time, the concentration perturbation averaged over a cross section of the tube has a Gaussian distribution along the length of the tube. The variance of the distribution is related to the dimensions of the tube, the velocity of the flow, and the diffusion coefficient of the fluid mixture. Indeed, this result formed the basis of Taylor's original measurements of diffusion coefficients.

It is experimentally more convenient to monitor the cross-section-averaged concentration distribution at an axial position $z = L$ as a function of time [3.187, 188]. In this case, the first moment of the temporal distribution $\bar{t}_{id}$ is given by [3.187]

$$\bar{t}_{id} = \frac{L}{\bar{u}}(1 + 2\zeta), \quad (3.174)$$

whereas the variance of the temporal distribution, which is no longer Gaussian, is [3.187]

$$\sigma_{id}^2 = \left(\frac{L}{\bar{u}}\right)^2 \left(8\zeta^2 + 2\zeta\right), \quad (3.175)$$

where

$$\zeta = \frac{\bar{u} R^2}{48 D_{12} L}, \quad (3.176)$$

and the subscript 'id' denotes the ideal experimental arrangement. These equations lead to a final working equation for the evaluation of the diffusion coefficient from the measured temporal moments of the distribution in the form [3.187]

$$D_{12} = \frac{R^2}{24\bar{t}_{id}} \frac{\left(1 + \frac{4\sigma_{id}^2}{\bar{t}_{id}^2}\right)^{1/2} + 3}{\left(1 + \frac{4\sigma_{id}^2}{\bar{t}_{id}^2}\right)^{1/2} + \frac{2\sigma_{id}^2}{\bar{t}_{id}^2} - 1} \, . \quad (3.177)$$

The principle of Taylor dispersion was first applied to the measurement of diffusion coefficients in gases by *Giddings* et al. [3.189]. However, the uncertainty of the measurements was very large. The development of very sensitive refractive-index detectors for liquid chromatography encouraged the development of the method, for use in liquids by *Ouano* [3.190] and *Wakeham* et al. [3.191, 192]. Figure 3.50 shows the instrument employed by *Alizadeh* and *Wakeham* [3.187, 188] for their measurements of the diffusion coefficients of *n*-alkanes in the temperature range 20–80 °C. They showed that diffusion coefficients with an uncertainty of ±1% may be obtained in an experiment lasting only one hour.

Rayleigh Interferometric Technique

The most precise methods for measuring interdiffusion coefficient in two- or three-component liquid systems have been the Rayleigh and Gouy optical interferometric methods. Both have been employed to obtain values of binary diffusion coefficients with a typical uncertainty of about 0.2% and a precision of 0.1–0.2%. There are many possible practical arrangements that can be used in experiments, but most workers have chosen the *free-diffusion* case described below. Whatever arrangement is adopted the process of diffusion produces a time-dependent distribution of refractive index or refractive index gradient in the mixing fluid that enables the diffusion process to be followed.

Free diffusion in a vertical column starts from an infinite sharp boundary between two uniform solutions of two species of different concentrations. The free-diffusion experiment is stopped before concentration changes are observed at the top or bottom of the cell [3.193]. The initial concentrations of the two solutions are normally chosen to be only slightly different so that the diffusion coefficient can be taken as a constant. In such a case, the closed-form solutions of the diffusion equation given below can be used. The diffusion (3.166), in a binary system with a constant diffusion coefficient, can be rewritten for one solute as

$$\frac{\partial C}{\partial t} = D_{12}\left(\frac{\partial^2 C}{\partial Z^2}\right) . \quad (3.178)$$

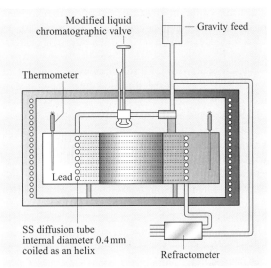

Fig. 3.50 Schematic diagram of a Taylor-dispersion instrument employed for diffusion coefficient measurements in liquids developed by *Alizadeh* et al. [3.187] and *Wakeham* [3.188]

Here C is the molar concentration, t the time elapsed from the start of the experiment, and Z the vertical distance from the starting boundary (positive in the direction of higher density and, usually, higher concentration). The solution to this equation for free diffusion is

$$C = \overline{C} + \frac{\Delta C}{2}\mathrm{erf}\left(\frac{y}{\sqrt{D}}\right) , \quad (3.179)$$

where

$$y = \frac{Z}{2\sqrt{t}} , \quad (3.180)$$

$$\overline{C} = \frac{C_T + C_B}{2} , \quad (3.181)$$

and

$$\Delta C = C_B - C_T . \quad (3.182)$$

In the above equations the subscripts 'B' and 'T' refer to the initial bottom and top solutions, respectively.

In the usual case, the refractive index n can be described by a linear function of the solute concentration

$$n = \overline{n} + R(C - \overline{C}) , \quad (3.183)$$

so that for free diffusion in a binary system

$$n = \overline{n} + \frac{\Delta n}{2}\mathrm{erf}\left(\frac{y}{\sqrt{D_{12}}}\right) , \quad (3.184)$$

where Δn is the difference of refractive index of the top and bottom solutions initially. This distribution of refractive index can be revealed by Rayleigh interferometry.

A schematic view of a Rayleigh interferometer for the measurement of the diffusion coefficient in liquids [3.194] is shown in Fig. 3.51. A monochromatic light source from a vertical slit is focused onto the camera plane by the main lens. The light passes through two parallel slits one in front of each of the two channels of a Rayleigh-type diffusion cell, and then through a horizontal cylindrical lens. If both channels contain fluids of the same refractive index, the interference pattern observed is determined solely by the superposition of the diffraction envelopes generated by the two slits. In this way, equally spaced fringes are produced within the diffraction envelope known as reference fringes. The zeroth-order fringe, which corresponds to equal optical path lengths for the two beams is located at the center of the envelope. If the refractive index of one channel is increased the interference fringes that are conjugate to the channel move sideways within the diffraction envelope and the displacement of any given fringe is proportional to the refractive index difference between the two channels. If, to this system one adds a horizontal, cylindrical lens so that the cells are imaged on the camera plane and also makes one channel a cell in which free diffusion takes place according to (3.179), accompanied by the refractive index changes of (3.184), then different vertical positions in the cell are subject to different refractive index changes so that the resulting fringe pattern will be that shown in Fig. 3.52. In this case, each fringe is shifted by an amount depending upon the extent of diffusion at each vertical position from a corresponding reference fringe. The reference fringes can themselves be obtained continuously from interference in regions of both cells where there is no diffusive perturbation such as the ends of the cell.

Measurements of the fringe shifts at a vertical position Z with respect to the initial boundary position may be used to determine the diffusion coefficient. If j denotes the j-th minimum in the fringe pattern about its center and J is the total number of fringes which corresponds to the refractive-index difference between the two initial solutions in the diffusion cell, then (3.184) leads to the result

$$(2j - J) = \text{erf}(y^*) . \qquad (3.185)$$

Now

$$y^* = \frac{Z}{2\sqrt{D_{12}t}} \qquad (3.186)$$

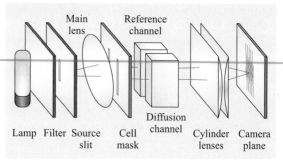

Fig. 3.51 Schematic view of Rayleigh interferometer [3.194]

so that, if (3.185) and (3.186) are written for both the j-th and $(J-j)$-th fringes, it is possible to obtain the result that for an interferogram photographed at time t:

$$D_{12} = \frac{\left[(X_{J-j} - X_j)/2y^*\right]^2}{4M^2 t} , \qquad (3.187)$$

where X_j and X_{J-j} are the fringe shifts for the j-th and $J-j$-th minima in the pattern on the photographic plate, y^* is the appropriate solution of (3.185) and M is the magnification factor of the optical system.

The Rayleigh interferometric technique has been employed for a number of precise measurements of diffusion coefficient by Sundelöf [3.195], Longsworth [3.196] and Svensson [3.197] among others. In recent years, laser light sources have considerably

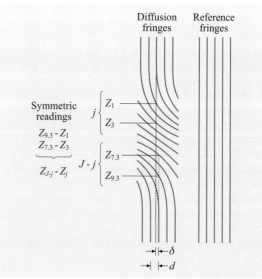

Fig. 3.52 Schematic diagram of Rayleigh interference fringes

simplified the design of Rayleigh interferometers and its modifications [3.198, 199]. This is because the coherence of the laser light removes the need for careful matching of the optical paths of the two beams which is necessary to observe interference fringes with ordinary light sources. In general, the Rayleigh interferometric technique is a very good method for measuring diffusion coefficients at moderately low (0.01 molar) to high concentrations. It is also characterized by a simple run procedure and theory, but it has seldom been applied far from ambient temperature because of the difficulties of ensuring stable optical paths in large temperature gradients.

Gouy Interferometric Technique

As already discussed, fringe positions in Rayleigh patterns yield the refractive index at corresponding levels of the diffusion cell. Fringe positions in Gouy fringe patterns are a Fourier transform of the refractive-index gradient $\partial n/\partial z$ and are related to the symmetrical positions of the gradient about the position of the maximum gradient.

Figure 3.53 illustrates a common configuration for a Gouy apparatus. The instrument is physically simpler than the Rayleigh one shown in Fig. 3.51; there is only one lens, and besides the cell, the only other optical components are the lamp, filter and slit. However, the theory of the device is more complicated.

In the Gouy system any light ray that passes through the diffusion-affected region of the free-diffusion cell will be deflected away from its point of intersection with the photographic plate in the absence of diffusion. The amount of this displacement is, to a first-order approximation, proportional to the refractive index gradient at the position in the diffusion cell at the height where the light crosses the cell. Interference fringes then arise at the photographic plate because rays that follow paths through different parts of the diffusion cell may have the same point of intersection with the photographic plate.

A set of *Gouy* [3.200] fringe patterns is illustrated in Fig. 3.54 together with a set of Rayleigh reference fringes. The Rayleigh fringes are formed by using light passing through regions of the diffusion cell undisturbed by diffusion and they are used to determine the position of the undeviated image of the slit on the photographic plate. Here $Y_0, Y_1, \ldots, Y_j$ are the distances from the undeviated slit image to the outermost fringe minimum ($j = 0$), next outer minimum ($j = 1$), etc. The distance C_t is the maximum Y-position that light would reach according to ray optical theory [3.201]. Equations for the analysis of Gouy interferometric fringe patterns

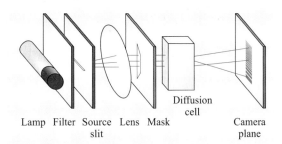

Fig. 3.53 Schematic view of Gouy interferometer (after [3.200])

to determine binary diffusion coefficients were derived independently by *Kegeles* and *Gosting* [3.201] and *Coulson* et al. [3.202]. According to Kegeles and Gosting, the positions of fringe minima Y_j are related to a parameter z_j by

$$Y_j = C_t \exp\left(-z_j^2\right). \tag{3.188}$$

The parameters z_j may be evaluated from the equations

$$f(z_j) = \text{erf}(z_j) - \left(2\frac{z_j}{\sqrt{\pi}}\right) \exp\left(-z_j^2\right), \tag{3.189}$$

$$f(z_j) = \frac{(j+Z_j)}{J}. \tag{3.190}$$

Here Z_j is a quantity for fringe j, which is calculated for fringe minima from wave optics [3.203]; it approaches 3/4 for large j. The binary diffusion coefficient is related to C_t by

$$D_{12} = \frac{(J\lambda b)^2}{4\pi C_t^2 t}, \tag{3.191}$$

where b is the optical distance from the center of the cell to the camera plane.

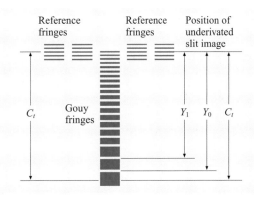

Fig. 3.54 Schematic diagram of Gouy interference fringes

Gouy interferometry has been the most widely used technique for the study of diffusion in liquids. Many careful experiments can be found in literature [3.201, 203]. As with most interferometric techniques there have been few applications far from ambient conditions.

3.6.2 Diffusion in Gases

The study of diffusion in gases began in 1863 with *Graham* [3.205], who was not only a pioneer in the field but also one its most ingenious experimentalists. Indeed, his experiments were performed even before the formulation of Fick's law. Since his work, numerous attempts have been made to devise new methods and to improve existing methods of measurement. A survey given by *Marrero* and *Mason* [3.206] summarizes the most important developments until that date and a more recent monograph by *Dunlop* et al. [3.207] provides a comprehensive review of experimental techniques.

The purpose of this section is to discuss widely used and accurate methods rather than to give an extensive overview of all the available techniques. As in the previous section on diffusion in liquids, the attention is focused only on the measurement of the mutual diffusion coefficient in binary systems.

Closed-Tube Technique

The closed-tube technique originally developed by *Graham* [3.205] in 1863 was employed by *Loschmidt* [3.208] to determine the diffusion coefficient for 10 gases in the temperature range 252–293 K. This type of instrument has therefore become known as the Loschmidt cell. In Fig. 3.55, a schematic diagram of a high-pressure Loschmidt cell employed by *Shankland* and *Dunlop* [3.204] is shown. The cell was constructed in two antisymmetric halves joined about a central pivot. The cell could be operated at pressures up to 2 MPa without leakage. At the start of an experiment the two halves of the cell were filled with different pure gases to the same total pressure. After thermal equilibrium was attained, the diffusion process was begun by rotating the upper half-cell about the central pivot to bring the two sections into coincidence.

Assuming that the diffusion coefficient D_{12} is independent of the mixture composition, for a total tube length L, the solution of (3.166) is [3.209, 210]

$$C_1(z,t) = \sum_{n=0}^{\infty} A_n \cos\left(\frac{n\pi z}{L}\right) \exp\left(-\frac{n^2\pi^2 D_{12} t}{L^2}\right),$$

(3.192)

where the coefficients A_n depend on the initial concentrations

$$A_0 = \frac{b}{L} C_{1B} + \left(\frac{L-b}{L}\right) C_{1T}$$

(3.193)

and for $n \geq 1$,

$$A_n = \left[\frac{-2(C_{1T} - C_{2B})}{n\pi}\right] \sin\left(\frac{n\pi b}{L}\right).$$

(3.194)

In these equations, b is the position of the boundary between the two cells, and

$$C_1(z,0) = C_{1T}, \qquad b \leq z \leq L,$$
$$C_1(z,0) = C_{1B}, \qquad 0 \leq z \leq b.$$

(3.195)

There are two different ways in which this basic solution may be employed for the measurement of the diffusion coefficient. In the first method the concentrations are measured at two planes normal to the longitudinal axis of the cells which are equidistant from the initial boundary. Then, all of the even terms in the summation of (3.192) disappear. Further simplifications can be made if the distance of the detection planes from the initial boundary is chosen to be $L/3$ because then the equation – truncated to include just the first term with

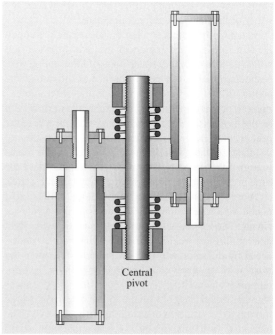

Fig. 3.55 High-pressure Loschmidt cell developed by *Shankland* and *Dunlop* [3.204]

negligible error – becomes

$$\Delta C_1(t) = 2A_1 \cos\left(\frac{\pi}{6}\right) \exp\left(-\frac{\pi^2 D_{12} t}{L^2}\right) \quad (3.196)$$

and the diffusion coefficient can be determined from measurements of ΔC_1 as a function of time by a least-squares regression technique. In this case the measurements of concentrations at the selected planes can be done by a variety of in situ means that have included interferometric means and thermal conductivity sensors.

In an alternative method of using the Loschmidt cell, the diffusion process is stopped after a time t_m before mixing is complete [3.210]. If the initial boundary is formed at $z = b = L/2$, and the two halves of the cell can be isolated from one another after the time t_m (Fig. 3.55), the average concentration in each half of the cell can be measured. If the average concentration of species 1 in the upper half of the cell at time t_m is denoted by $\langle C_{1T} \rangle$ and that in the lower half of the cell by $\langle C_{1B} \rangle$ these two quantities can be evaluated by averaging the concentration distribution of (3.192) over the length of each half-cell at time t_m, so that

$$\frac{\langle C_{1T} \rangle - \langle C_{1B} \rangle}{C_{1T} - C_{1B}} = \frac{8}{\pi^2} \sum_{n=0}^{\infty} \frac{\exp\left[-(2n+1)^2 \pi^2 D_{12} t_m / L^2\right]}{(2n+1)^2} , \quad (3.197)$$

which enables D_{12} to be determined from measurements of $\langle C_{1T} \rangle$, $\langle C_{1B} \rangle$, t_m and L. This equation can further be simplified by the proper choice of time t_m [3.210]. The method of composition analysis in this type of instrument can be any means that has adequate precision for the gaseous systems under study, including mass spectrometry.

In the first approach, the exact nature of the initial experimental conditions do not influence the evaluation of the diffusion coefficient from $\Delta C_1(t)$ as a function of t. This approach depends on the time interval between concentration measurements but not on the absolute time at which these measurements are made. On the other hand, the second method depends upon the measured time from the start of the diffusion process, as well as upon the initial concentrations in the two halves of the cell prior to a measurement.

Two-Bulb Technique

The two-bulb technique is the most widely used method for determining the diffusion coefficients of gases [3.211]. The basic arrangement for a two-bulb cell consists of two chambers of relatively large volume joined by a small-bore, small-volume diffusion tube. Initially, the two chambers are filled with fluid mixtures of different composition at the same pressure which are allowed to approach a uniform composition by means of diffusion through the tube. In an ideal model of this type of instrument, it is assumed that the diffusion coefficient of the gas mixture is independent of composition, the gas mixtures are ideal, so that there is no volume change upon mixing or heat of mixing, and the transient temperature rises due to Dufour effects are insignificant [3.181]. It is also assumed that the concentration gradient is confined to the connecting tube whereas the composition within each bulb remains uniform at all times. In addition, the pressure is assumed to be uniform throughout the cell, so that viscous effects are negligible, and high enough to minimize free-molecular (Knudsen) diffusion [3.211].

Following these assumptions, (3.166) can be solved [3.181] to produce the working equation

$$C_{1T}(t) - C_{1B}(t) = [C_{1T}(0) - C_{1B}(0)] \exp\left(\frac{-t}{\tau}\right), \quad (3.198)$$

where

$$\tau = \left[\left(1 - \frac{V_c}{3(V_B + V_T)}\right) \frac{A D_{12}}{L} \left(\frac{1}{V_B} + \frac{1}{V_T}\right)\right]^{-1}. \quad (3.199)$$

In the above equations, the subscripts 'B' and 'T' refer to the bottom and top cell volumes, V_c the volume of the connecting tube of cross-sectional area A and length L. Therefore the diffusion coefficient can be obtained from measurements of the difference between the concentrations of one species in the top and bottom bulbs as a function of time, together with the dimensions of the diffusion cell.

The two-bulb cell shown in Fig. 3.56 was employed by *van Heijningen* et al. [3.212] for the measurement of the diffusion coefficients of monatomic gas mixtures in the temperature range 65–300 K. The estimated uncertainty of the measurements was ±0.5%. To monitor the concentration changes they employed thermistors, in the same way as *Yabsley* and *Dunlop* [3.211] in their measurements of the diffusion coefficient of helium–argon and helium–oxygen mixtures near room temperature. *Taylor* and *Cain* [3.213] monitored the composition changes by withdrawing samples of gas for mass spectrometric analysis at specific time intervals during the diffusion process. Their estimated uncertainty was ±2%.

Taylor–Dispersion Technique

The Taylor-dispersion technique described in the case of diffusion in liquids has also been applied to the measurement of the diffusion of gases. The principles of the application of the technique to gases are identical to those for liquids so that no new description is necessary here. However, it is necessary to point out one important difference between the application of the theory of the method to gases and liquids which derives from the fact that in the dilute gas phase the diffusion coefficient is approximately 10^5 times larger than in the liquid phase. As a consequence the principal contribution to the dispersion of a pulse of solute gas in gas-phase measurements is the direct molecular diffusion in the axial direction and not the Taylor-dispersion contribution arising from radial diffusion.

For the same reason the dimensions of the diffusion tubes employed for measurements in the dilute gas phase [3.214] are quite different from those employed in the liquid phase. Generally, a diffusion tube with a length of a few meters and a diameter of a few millimeters is employed for measurements in the dilute gas state. When combined with the low viscosity of the gas this means that it is very difficult to satisfy all of the conditions [3.210] necessary for absolute measurements. Thus, although the technique yields a gas-phase diffusion coefficient within a period of 15 min, and is readily applied over a wide range of temperature, the uncertainty of the results has generally been modest [3.210].

More recently the same technique has been applied to measure the diffusion coefficients of gases at elevated pressure with particular emphasis on the supercritical state [3.215, 216]. Under these conditions the diffusion coefficient of the system is much closer to that characteristic of the liquid phase so that the conditions of the theory are more easily satisfied and the apparatus is essentially identical to that employed for liquids. The Taylor-dispersion technique is particularly suitable for high-pressure applications because the element to be pressurized is simply a cylindrical-section tube and because the time required for a diffusion measurement can be retained within reasonable bounds (about one hour) without a loss of uncertainty [3.215, 216].

3.6.3 Diffusion Reference Values

The interdiffusion coefficient of aqueous solutions of potassium chloride is recommended as the reference standard [3.181] for the liquid phase. The values given in

Table 3.23 Integral diffusion coefficients for aqueous potassium chloride solutions at 25 °C

C_1 (mol dm^{-3})	$\overline{D}_{12}(C_1) \times 10^9$ (m^{2}s^{-1})	C_1 (mol dm^{-3})	$\overline{D}_{12}(C_1) \times 10^9$ (m^{2}s^{-1})
0.001	1.973	0.060	1.890
0.002	1.966	0.070	1.886
0.003	1.961	0.080	1.882
0.004	1.956	0.090	1.878
0.005	1.953	0.100	1.874
0.006	1.949	0.200	1.857
0.007	1.947	0.300	1.850
0.008	1.944	0.400	1.848
0.009	1.941	0.500	1.848
0.010	1.939	0.600	1.849
0.020	1.923	0.700	1.850
0.030	1.911	0.800	1.852
0.040	1.903	0.900	1.855
0.050	1.896	1.000	1.858

Table 3.24 Calibration data for gas-phase diffusion coefficients at a pressure of 0.101325 MPa

System	$D_{12}(0) \times 10^5$ m^{2}s^{-1}	a_1	a_2
He–Ar	7.344	0.0846	1.4825
He–N$_2$	7.067	0.0676	1.4883
He–O$_2$	7.469	0.0564	1.1270
He–CO$_2$	6.029	0.0905	2.3952
N$_2$–Ar	2.034	0.0041	0.0

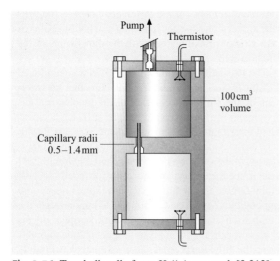

Fig. 3.56 Two-bulb cell of *van Heijningen* et al. [3.212]

Table 3.25 Diffusion coefficient D_{12} ($\times 10^{-9} \text{m}^2\text{s}^{-1}$) of some hydrocarbons in supercritical CO_2

		299.15 K 9.0 MPa	299.15 K 10.5 MPa	303.15 K 10.5 MPa	305.65 K 10.5 MPa	308.15 K 10.5 MPa
n-Pentane	C_5H_{12}	13.7	13.0	14.1	15.7	17.6
n-Hexane	C_6H_{14}	13.7	13.0	14.0	15.4	17.6
n-Heptane	C_7H_{16}	13.4	12.6	14.0	15.4	17.1
n-Octane	C_8H_{18}	13.2	12.4	13.5	15.2	17.1
n-Nonane	C_9H_{20}	12.9	12.1	13.3	14.9	16.7
n-Decane	$C_{10}H_{22}$	12.5	11.8	13.0	14.6	16.1
n-Undecane	$C_{11}H_{24}$	12.2	11.5	12.7	14.5	16.1
n-Dodecane	$C_{12}H_{26}$	11.3	11.0	12.0	13.9	15.7
n-Tetradecane	$C_{14}H_{30}$	9.53	9.41	10.9	12.4	14.1
Acetone	CH_3COCH_3	14.3	14.2	15.2	16.2	17.6
Benzene	C_6H_6	14.3	14.0	15.0	15.8	17.4

Table 3.26 Diffusion coefficient D_{12} ($\times 10^{-9} \text{m}^2\text{s}^{-1}$) of some organic solutes in n-hexane

	213.2 K	233.2 K	253.2 K	273.2 K	299.2 K	313.2 K	333.2 K
Benzene	1.21	1.80	2.57	3.43	4.70	5.53	6.96
o-Difluorobenzene	1.10	1.60	2.41	3.35	4.46	5.26	6.61
p-Difluorobenzene	1.10	1.72	2.49	3.36	4.63	5.42	6.62
1,2,4-Trifluorobenzene	1.18	1.72	2.33	3.20	4.40	5.20	6.34
1,2,3,5-Tetrafluorobenzene	1.05	1.67	2.27	3.34	4.40	4.95	6.48
1,2,4,5-Tetrafluorobenzene	1.05	1.68	2.22	3.16	4.40	5.17	6.20
Pentafluorobenzene	1.06	1.70	2.26	2.98	4.00	4.83	5.95
Hexafluorobenzene	1.01	1.62	2.16	2.84	4.05	4.62	5.81
Octafluorobenzene	0.88	1.40	1.83	2.47	3.50	4.07	4.98

Table 3.23 are the integral diffusion coefficient defined as

$$\overline{D}_{12}(\overline{C}_1) = \frac{1}{\overline{C}_1} \int_0^{\overline{C}_1} D_{12} \, dC_1 , \qquad (3.200)$$

where D_{12} is the true *differential* diffusion coefficient and C_1 is the concentration of species 1 in one of the two chambers of a diaphragm cell [3.217]. The values in Table 3.23 have been derived from the data of *Woolf* and *Tilley* [3.183].

For gas mixtures recommended values of the diffusion coefficients of five mixtures at 300 K and 0.101325 MPa can be calculated from

$$D_{12} = D_{12}(0)\left(1 + \frac{a_1 x_2}{1 + a_2 x_2}\right) . \qquad (3.201)$$

Here, x_2 represents the mole fraction of the heavier component and values of $D_{12}(0)$, a_1 and a_2 are listed in Table 3.24. Equation (3.201) represents the experimental data by *Dunlop* et al. [3.218]; the estimated uncertainty of these data is about $\pm 0.1\%$.

Temperature dependencies of diffusion coefficients in the range 50–1000 K are given by *Kestin* et al. [3.219].

3.6.4 Tables of Diffusion Coefficient Values

In Tables 3.25–3.27, the interdiffusion coefficient for some commonly encountered systems is given for en-

Table 3.27 Diffusion coefficient D_{12} ($\times 10^{-9} \text{m}^2\text{s}^{-1}$) of some organic solutes in toluene

	299.2 K	323.2 K	348.2 K
n-Hexane	2.50	3.21	
n-Decane	1.84	2.52	3.22
n-Tetradecane	1.46	1.98	2.61

gineering purposes as a function of temperature and pressure. The data in Table 3.25 have been obtained by the Taylor dispersion technique with an estimated uncertainty of 3% [3.220]. Measurements presented in Tables 3.26 and 3.27 have an estimated uncertainty of 2.5% [3.221].

3.7 Electric and Magnetic Parameters of Liquids and Gases

This section describes measurements of the dielectric constant and electrical conductivity of fluids, as well as magnetic susceptibility. In the introduction some relevant electromagnetic parameters are discussed. Throughout the chapter the term permittivity will always be equivalent to the electrical permittivity, conductivity to the electrical conductivity, and permeability to the magnetic permeability.

3.7.1 Introduction

The speed of all electromagnetic radiation (the speed of light) in vacuum is the same for all the frequencies and is denoted by c. The value of c is 2.99792458×10^8 m/s. The speed of light c can be expressed via the permittivity ε_0 and permeability μ_0 of free space, as

$$c^2 = \frac{1}{\varepsilon_0 \mu_0}, \tag{3.202}$$

where $\varepsilon_0 = 8.85418782 \times 10^{-12}$ F/m and $\mu_0 = 4\pi \times 10^{-7}$ N/A^2 (in SI units).

In other substances, however, the speed of electromagnetic radiation is a function of the radiation frequency and of the material properties. The factor by which the speed of a particular frequency of the electromagnetic radiation is changed relative to c when it travels inside a material is called the refractive index of the material at that particular frequency. That is, if v is the phase velocity of radiation of a particular frequency in a specific material, then the refractive index n is given by

$$n = \frac{c}{v}. \tag{3.203}$$

The refractive index for a number of materials and aqueous solutions is given in Tables 3.28 and 3.29, respectively. The dependence of the refractive index of a material on frequency (except in vacuum, where all frequencies travel at c) is seen in the effect known as dispersion. This is the division of white light into its constituent spectral colors, such as when it travels through a prism, and is the cause of chromatic aberration in lenses. If the refractive indices of two materials are known for a given frequency, then one can compute the angle by which radiation of that frequency will be refracted as it moves from the first material into the second using Snell's law. The refractive index is typically larger than one. The speed of light in air, for example, is slightly less than c. In denser media, such as water and glass, light can slow down to fractions of c such as $0.75c$ and $0.67c$.

The characteristic permittivity ε and permeability μ of a medium together determine the phase velocity of electromagnetic radiation through that medium according to

$$v^2 = \frac{1}{\varepsilon \mu}. \tag{3.204}$$

Thus the refractive index can be written as:

$$n^2 = \frac{\varepsilon \mu}{\varepsilon_0 \mu_0}. \tag{3.205}$$

Also

$$\varepsilon = \varepsilon_r \varepsilon_0, \tag{3.206}$$

$$\mu = \mu_r \mu_0, \tag{3.207}$$

where ε_r is the relative permittivity or dielectric constant and μ_r is the relative permeability of the material. For nonmagnetic materials like, for example, most polymer solutions, μ_r is unity and thus the square of the refractive index is equal to the dielectric constant.

On the atomic level, the deceleration of the electromagnetic radiation as it enters a material may be attributed to the continuous process of absorption and emission of photons as they interact with the atoms of the material. Between atoms, the photons travel at the speed c, as in vacuum. As they interact with the atoms, they are absorbed and re-emitted, which results in a slight delay. On a sufficiently large scale the delay manifests itself as an overall reduction in the speed of light in the material. The absorption and emission process can be thought of as the electric field of a photon creating an oscillating force on the charges of each atom (primarily the electrons). This oscillation of charges itself causes the radiation of an electromagnetic field, which is slightly out of phase compared to that of the original photon,

Table 3.28 Some representative values of the refractive index at the wavelength of $\lambda = 589$ nm (yellow sodium light). Unless stated otherwise, the temperature is 298.15 K. Melting point is abbreviated by m.p.

Substance	n
Air	1.0002926
Bromine	1.661
Carbon dioxide	1.00045
Diamond	2.419
Glass	1.5–1.9
Glycerin	1.4729
Helium	1.000036
Sodium chloride at m.p. (1123.15 K)	1.408
$Na_2O \cdot SiO_2$ (solid)	1.52
Rock salt	1.516
Water (liquid) (293.15 K)	1.333
Water (solid) (273.15 K)	1.31

Table 3.29 Refractive index of several aqueous solutions at the wavelength of 589 nm (yellow sodium light) at 293.15 K and a weight concentration of 20% unless otherwise indicated

Solute	n
Acetic acid - CH_3COOH	1.3472
Acetic acid - CH_3COOH (100%)	1.3716
Ammonia – NH_3	1.3440
Ammonium chloride – NH_3CL	1.3708
Ammonium sulfate – $(NH_3)_2SO_4$	1.3677
Barium chloride – $BaCl_2$	1.3664
Calcium chloride – $CaCl_2$	1.3839
Cesium chloride – CsCl	1.3507
Citric acid – $(HO)C(COOH)_3$	1.3598
Ethanol – CH_3CH_2OH	1.3469
Ethanol – CH_3CH_2OH (100%)	1.3614
Formic Acid – HCOOH	1.3437
Glycerol – $CH_2OHCHOHCH_2OH$	1.3572
Glycerol – $CH_2OHCHOHCH_2OH$ (100%)	1.4735
Hydrochloric acid – HCl	1.3792
Magnesium chloride – MgCl	1.3859
Methanol – CH_3OH	1.3381
Methanol – CH_3OH (100%)	1.3290
Potassium bromide – KBr	1.3591
Silver nitrate – $AgNO_3$	1.3574
Sodium chloride – NaCl	1.3684
Sulfuric acid – H_2SO_4	1.3576

leading to a slight retardation of the field and an apparent delay in the photon's travel. Sometimes the refractive index is defined as a complex number, with the imaginary part representing the absorption of the material. This representation is particularly useful when analyzing the propagation of electromagnetic waves through metals.

The magnetic permeability μ used in physics and engineering is the degree of magnetization of a material in response to a magnetic field and is defined as the ratio of magnetic flux density B (also called magnetic induction) to the magnetic field strength H

$$\mu = \frac{B}{H}. \quad (3.208)$$

Although magnetic permeability is related in physical terms most closely to electric permittivity, it is probably easier to think of permeability as of a sort of *resistance to magnetic flux*; just as those materials with high electrical conductivity let electric current through easily, materials with high permeabilities allow magnetic flux through more easily than others.

Sometimes, for non-ferromagnetic substances the permeability is so close to μ_0 that the magnetic susceptibility χ is used:

$$\mu_r = 1 + \chi. \quad (3.209)$$

The magnetic susceptibility is also defined as the ratio of magnetization M to magnetic field strength H. The magnetization is defined as the total vector sum of the magnetic moments of all the atoms in a given volume V divided by that volume. From Table 3.30, it is clear that the permeabilities of common diamagnetic and paramagnetic materials do not differ substantially from that of free space. In fact, to all intents and purposes, the magnetic properties of such materials can be safely neglected ($\mu_r = 1$). Measurements of the magnetic susceptibility are described in detail in [3.222–226].

3.7.2 Dielectric Constant

The dielectric constant ε_r is an expression of the extent to which a material concentrates electric flux, and is the electrical equivalent of the relative magnetic permeability μ_r. As the dielectric constant increases, the electric flux density (the displacement current) increases. In electromagnetism, the permittivity ε of a medium is also introduces as the ratio D/E of the electric displacement D to the electric field strength E when an external field is applied to the substance. Similarly to the magnetic susceptibility the electric susceptibility is defined as

$$\chi_e = \varepsilon_r - 1, \quad (3.210)$$

Table 3.30 Magnetic susceptibilities of some ferro-, para- and diamagnetic materials at different temperatures. For temperatures other than room temperature, the value is given next to the material name. The following notation is used: (l) liquid, (g) gas, (s) solid. The values were taken from the sources indicated in the table

Material	χ
Acetone (C_3H_6O) (l) [3.227]	-5.7803×10^{-06}
Aluminum (s) [3.228]	2.2×10^{-5}
Ammonia (g) [3.228]	-1.06×10^{-5}
Antimony (l, 903.78 K) [3.229]	-5.74409×10^{-7}
Bismuth (l, 544.55 K) [3.229]	-6.63033×10^{-6}
Bismuth (s) [3.228]	-1.67000×10^{-4}
Cadmium (l, 673.15 K) [3.230]	-1.63049×10^{-3}
Chlorine (Cl_2) (l, 238.65 K) [3.231]	-1.11623×10^{-5}
Copper (s) [3.228]	-9.8×10^{-6}
Diamond (s) [3.228]	-2.2×10^{-5}
Ethanol (C_6H_6O) (l) [3.227]	-3.55392×10^{-6}
Germanium (l, 1211.67 K) [3.232]	6.25289×10^{-6}
Glycerol ($C_3H_8O_3$) (l) [3.227]	-9.82884×10^{-6}
Hydrogen (H_2) (l, 20.3 K) [3.231]	-2.29758×10^{-6}
Hydrogen (g, 1 atm) [3.228]	-2.1×10^{-9}
Indium (l, 673.15 K) [3.230]	-8.26742×10^{-4}
Indium (l, 429.75 K) [3.229]	-5.02831×10^{-6}
Iron (s) [3.228]	3.0×10^{3}
Lead (l, 600.62 K) [3.232]	-1.1773×10^{-5}
Lead (l, 600.61 K) [3.229]	-8.97515×10^{-6}
Mercury (l) [3.231]	-2.84026×10^{-5}
MnZn (Fe_2O_4)$_2$ (s) [3.228]	2.5×10^{3}
Nitrogen (g, 1 atm) [3.228]	-5.0×10^{-9}
Oxygen (O_2) (l, 90 K) [3.231]	3.44981×10^{-3}
Oxygen (g, 1 atm) [3.228]	2.09×10^{-6}
Potassium (l, 673.15 K) [3.230]	2.07219×10^{-4}
Silicon (s) [7]	-3.7×10^{-6}
Silver (l, 1234.93 K) [3.232]	-2.64816×10^{-5}
Silver (l, 1234.93 K) [3.229]	-2.57661×10^{-5}
Sulphuric acid (H_2SO_4) (l) [3.231]	-9.1473×10^{-6}
Terbium (s) [3.228]	9.51×10^{-2}
Tin (l, 505.08 K) [3.232]	-3.1157×10^{-6}
Tin (l, 505.08 K) [3.229]	-1.31758×10^{-6}
Toluene (C_7H_8) (l) [3.227]	-7.69998×10^{-6}
Tungsten (s) [3.228]	6.8×10^{-5}
Water (l) [3.231]	-9.04015×10^{-6}
Water (l) [3.228]	-9.0×10^{-6}

where the electric susceptibility χ_e, is also defined as the ratio of polarization P to the electric field strength E. In general, the dielectric constant can be defined as a complex number, with the real part expressing reflective surface properties (Fresnel reflection coefficients), and the imaginary part expressing the radio absorption coefficient k_v.

When an electric field is applied to a medium, an electric current propagates. The total current propagating in a real medium is in general composed of two parts: a conduction current J_c and a displacement current J_d. A perfect dielectric is a material that shows displacement current only. In the case of a leaky dielectric medium (i.e., when conduction currents are not negligible) the total current density is

$$J_{\text{tot}} = J_c + J_d = \sigma E + i\omega\varepsilon E = i\omega\varepsilon^* E , \quad (3.211)$$

where σ is the specific conductivity of the medium, and the complex permittivity ε^* is defined as

$$\varepsilon^* = \varepsilon_r - i\frac{\sigma}{\varepsilon_0\omega} , \quad (3.212)$$

where $i = \sqrt{-1}$, and $\omega = 2\pi f$ is the angular frequency.

The permittivity of liquids in the radio frequency and microwave regions can also be presented by the Debye equation [3.233, 234]:

$$\varepsilon^* = \varepsilon' + i\varepsilon'' , \quad (3.213a)$$

where

$$\varepsilon' = \varepsilon_\infty + \frac{\varepsilon_s - \varepsilon_\infty}{1 + \omega^2\tau^2} , \quad (3.213b)$$

$$\varepsilon'' = \frac{(\varepsilon_s - \varepsilon_\infty)\omega\tau}{1 + \omega^2\tau^2} . \quad (3.213c)$$

ε_s is the permittivity measured in a static field or at low frequencies where no relaxation effects occur, and ε_∞ is a parameter describing the permittivity in the high-frequency limit. τ is the relaxation time for molecular orientation. The values of the dielectric constant for a number of liquids and gases are given in Table 3.31.

3.7.3 Electric Conductivity

The electric conductivity is a measure of how well a material accommodates the transport of electric charge (the detailed physical mechanisms of conductivity in liquids are discussed in Chap. 24). Conductance is an electrical phenomenon where a material contains movable particles of electricity. When a difference of electrical potential is applied across a conductor, its movable charges flow, and an electric current appears. A conductor such as a metal has high conductivity, and an insulator like glass, or vacuum, has low conductivity.

Table 3.31 The dielectric constants ε_r measured in static fields or at low frequencies where no relaxation effects occur

Solvent	ε_r	T (K)
Acetone[1]	20.7	298.15
Acetonitrile[1]	36.7	298.15
Ammonia[1] (239 K)	22	298.15
Ammonia[2]	16.61	293.2
Argon[2]	1.3247	140
Benzene[1]	2.27	298.15
Bromine[2]	3.1484	297.9
Bromine trifluoride[2]	106.8	298.2
Chlorine[2]	2.147	208.0
Dimethyl acetamide[1]	37.78	298.15
Dimethyl sulfoxide[1]	46.7	298.15
Dioxan[1]	2.21	298.15
Ethanol	24.3	298.15
Ethylene diamine[1]	12.9	298.15
Hydrogen[2]	1.2792	13.52
Hydrogen chloride[2]	14.3	158.9
Hydrogen cyanide[1] (289 K)	118.3	298.15
Hydrogen fluoride[2]	83.6	273.2
Iodine[2]	11.08	391.25
Krypton[2]	1.664	119.8
Oxygen[2]	1.5684	54.478
Ozone[2]	4.75	90.2
Phosphorus[2]	4.096	307.2
Pyridine[1]	12	298.15
Selenium[2]	5.44	510.65
Sulfur[2]	3.4991	407.2
Sulfuric acid[1]	101	298.15
Water	80.1	293.15
Xenon[2]	1.88	161.35

[1] denotes the values adopted from [3.235] and [3.236],
[2] denotes the values adopted from [3.237–239]

A semiconductor has a conductivity that may vary with conditions, such as exposure to certain frequencies of light.

Electric currents in electrolytes are flows of electrically charged ions. For example, if an electric field is applied to a solution of salt NaCl, which dissociates to Na$^+$ and Cl$^-$, the sodium ions will move towards the negative electrode (anode), and the chlorine ions will move towards the positive electrode (cathode). If the conditions are right, redox reactions will take place, which release electrons from the chlorine, and allow electrons to be absorbed into the sodium. In water ice and in certain solid electrolytes, flowing protons constitute the electric current.

The specific electrical conductivity σ is the reciprocal of the specific electrical resistivity measured in SI units in Ohm m (Ω m). The corresponding units of σ are S/m (S = Ω^{-1} stands for Siemens). Table 3.32 shows the value of σ measured for a number of substances. It is equal to the ratio of the current density J to the electric field strength E, as defined in the previous section. The latter also applies to the electrolytic conductivity of fluids.

Since in ordinary nonmetallic liquids the electric charge and current are generally related with dissolved ions, charge may be induced in poorly conducting liquids even though equilibrium net charge initially is absent. In electrohydrodynamics most work until the 1960s focused on the behavior of perfect or good conductors (mercury or water), or almost perfect dielectrics (apolar liquids such as benzene). That began to change following studies of poorly conducting liquids – leaky dielectrics [3.240].

The leaky dielectric can be modeled by the Navier–Stokes equations to describe fluid motion and an equation of charge conservation employing its Ohmic

Table 3.32 Representative values of specific conductivities

Substance	σ (1/Ωcm)	T (K)
Copper	5.8×10^5	293.15
Iron	1.1×10^5	273.15
Lead	4.9×10^5	273.15
Lithium chloride (melt)	6.221	983.15
Mercury	1.1×10^4	293.15
Molten oxide CaO · SiO$_2$	0.8	2023.15
Molten oxide K$_2$O · 2SiO$_2$	1.5	2023.15
Molten oxide Li$_2$O · SiO$_2$	5.5	2023.15
Molten oxide Li$_2$O · 2SiO$_2$	2.5	2023.15
Molten oxide Na$_2$O · SiO$_2$	4.8	2023.15
Molten oxide Na$_2$O · 2SiO$_2$	2.1	2023.15
Potassium chloride (0.01%) in water	1.3×10^{-2}	298.15
Potassium chloride (melt)	2.407	1145.15
Sodium chloride (solid)	1×10^{-3}	1073.15
Sodium chloride (melt)	3.9	1173.15
Sodium at melting point (m.p.)	1.04×10^5	370.98
Sodium chloride at m.p.	3.58	1074.15
Sulphuric acid (0.4%) in water	7.5×10^{-1}	291.15
Water	4×10^{-8}	291.15
Xylene	1×10^{-19}	298.15

conductivity. Electromechanical coupling occurs only at the interfaces where charge, carried to the interface by conduction, results in stresses of electric origin different from those present in perfect dielectrics or perfect conductors. With perfect conductors or dielectrics with no embedded charges, the stresses of the electric origin are perpendicular to the interface and alterations of the interfacial shape combined with the interfacial tension serve to balance the electric stress. Leaky dielectrics are different because free charges accumulated on the interface modify the field, and in particular, produce shear stresses. Viscous flow develops to provide stresses to balance the action of the shear stress components resulting from the tangential electric field acting on the interfacial charge.

The leaky dielectric model arises naturally through a scale analysis. Under static conditions, electric and magnetic phenomena are independent since their fields are uncoupled [3.241]. Insofar as the characteristic time for electrostatic processes is large compared to that for magnetic phenomena, the electrostatic equations furnish an accurate approximation. When external magnetic fields are absent, magnetic effects can be ignored completely. From Maxwell's equations, the characteristic time for electric phenomena τ_c can be identified as the ratio of electric permittivity and conductivity,

$$\tau_c = \frac{\varepsilon}{\sigma} \,. \tag{3.214}$$

For magnetic phenomena the characteristic time τ_M is the product of the magnetic permeability, conductivity and the square of the characteristic length

$$\tau_M = \mu \sigma l^2 \,. \tag{3.215}$$

Transport process time scales τ_P arise from viscous relaxation, diffusion, oscillation of an imposed field, or motion of a boundary. Slow processes are defined as those where $\tau_P \geq \tau_c \gg \tau_M$. The second inequality can be rearranged to give $(\varepsilon_r/\mu_r)^{1/2} \varepsilon_0/\sigma \gg l \, (\varepsilon_0\mu_0)^{1/2}$, and since $c = (\varepsilon_0\mu_0)^{-1/2}$, the right-hand side of this inequality is extremely small for leaky dielectric systems, for example most polymer solutions. On the other hand, if $\tau_P \gg \tau_c$, the liquid could be considered a perfect conductor, while for $\tau_P \sim \tau_c$ it is a leaky dielectric (a poor conductor), where interaction of the electric and hydrodynamic fields is still very important in spite of the fact that the fluid is electrically neutral in the bulk and the excessive charge accumulates only at the interfaces.

Ion Mobility

In general, ions are responsible for the transport of the electric charge in fluids. There are two aspects of ionic motion. First, there is the individual aspect. This concerns the dynamic behavior of separate ions. These ionic motions are basically random in direction and speed. Second, ionic motions have a group aspect that is of particular significance when more ions move in certain directions than in the others, and produce a drift, or flux, of ions. A flux of ions can come about in three ways. If there is a difference in the concentration of ions in different regions of the electrolyte (a leaky dielectric), the resulting concentration gradient produces a flow of ions. This phenomenon is termed diffusion (Sect. 3.6). If there are differences in the electrostatic potential at various points in the electrolyte, then the resulting electric field produces an additional flow of charge (and thus, of the ions) in the direction of the electric field. This is termed migration or conduction. Finally, if a difference in pressure, density, or temperature exists in various parts of the electrolyte, then the liquid begins to move as a whole or parts of it move relative to the other parts. This is an ordinary hydrodynamic flow, which, however, results in ion/charge convection [3.236]. There are of course fluids where the electric charge is transported by electrons and holes irrespective of the motion of the molecules/ions in the fluid. This is characteristic of metallic fluids such as mercury or molten alloys. The electric charge carriers in a fluid can consist of both ions and electrons. Examples of such fluids are solutions of conducting polymers like MEH-PPV [poly(2-methoxy, 5-(2'-ethyl-hexoxy)-1, 4-phenylene-vinylene)] [3.242, 243], polypyrrole [3.244] and polyaniline doped with d,l-camphorsulfonic acid [3.245].

Ions, like electrons, do not move at the speed of light when carrying charge from one point to another. Ions in solution participate in random (Brownian) motion in which they change momentum as a result of collisions with the other molecules and ions. Statistical bias in the motion of ions without electric field is the result of diffusion due to inequalities in the ion numbers in different regions. *Robinson* and *Stokes* [3.246] note that in the case where there are no other forces, it may be convenient to consider the electric field as representing a force. In this discussion we will concentrate on the electric field as the driving force for ion drift.

A Simplified Picture of Ionic Motion under the Influence of an Applied Electric Field. Under the influence of an external force like those corresponding to an applied electric field, ion motions are affected because of the fact that the ions are charged. Hence, the imposition of an electric field singles out one direction in space for preferential ionic movement. The walk is no longer perfectly random – the ions drift. Were an ion is completely isolated (in vacuum), it would accelerate indefinitely until it will collide with an electrode. In an electrolyte solution, an ion very soon collides with some other ion or with a solvent molecule that crosses its path. The ion changes directions; however, the electric field imparts to the ion a direction.

The initial velocity of the ion can be ignored precisely because it is random and therefore does not contribute to the preferred motion (drift) of the ion. The applied electric field imparts a sure component to the random velocities of the ion. This extra velocity component is in the direction of the electric force vector $\boldsymbol{F}$ and is called the drift velocity.

From Newton's second law we have the acceleration as

$$\frac{d\boldsymbol{v}}{dt} = \frac{\boldsymbol{F}}{m}, \quad (3.216)$$

where m is the ion mass.

The drift velocity ($\boldsymbol{v}_d$) is estimated as the product of the acceleration and the average time between collisions τ,

$$\boldsymbol{v}_d = \frac{d\boldsymbol{v}}{dt}\tau = \frac{\tau}{m}\boldsymbol{F}. \quad (3.217)$$

The flux of ions is related to the drift velocity in the following way

$$\text{Flux} = \text{Concentration of ions} \times \text{Drift velocity}. \quad (3.218)$$

Thus, if $\boldsymbol{F}$ is an electric force that induces conduction, then this equation is the molecular basis of the fundamental relation used in the macroscopic view of conduction, i.e.

$$\text{Flux} \propto \text{Electric field}. \quad (3.219)$$

The expression (3.217) reveals the condition under which the proportionality between the drift velocity (flux) and electric field breaks down. It is essential that in a collision an ion does not preserve any part of its extra velocity component arising from the electric force. If it did, then the actual drift velocity would be greater than that calculated by (3.217) because there would be a cumulative carryover of the extra velocity from collision to collision. Thus, every collision must eliminate all traces of the force-derived extra velocity, and the ion must start afresh to acquire the additional velocity. This condition can be satisfied only if the drift velocity, and therefore the field, is sufficiently small.

The proportionality constant τ/m in (3.217) is referred to as the (generalized) absolute mobility M_{abs} because it is an index of how mobile the ions are

$$M_{\text{abs}} = \frac{\tau}{m} = \frac{\boldsymbol{v}_d}{\boldsymbol{F}}. \quad (3.220)$$

The force acting on an ion in an electric field ($\boldsymbol{E}$) is equal to the charge of the ion times the field at the point where the ion is situated

$$\boldsymbol{F} = z_i e_0 \boldsymbol{E}, \quad (3.221)$$

where e_0 is the electron charge, and z_i the valence of the ion. In the literature, mobilities are usually defined as the ratio of $\boldsymbol{v}_d$ to $\boldsymbol{E}$, which introduces the conventional or electrical mobility

$$M_e = \frac{\boldsymbol{v}_d}{\boldsymbol{E}} = M_{\text{abs}} z_i e_0. \quad (3.222)$$

Though the two types of mobilities are closely related, it must be stressed that the concept of absolute mobility is more general because it can be used for any force that determines the drift velocity of ions and not only the electric force used in the definition of electrical mobilities.

The Relation between the Equivalent Conductivity and the Absolute Mobility of an Ion. The motion of an isolated body is obviously governed by Newton's second law, but in dealing with the motion of ions it is not usually necessary to consider the acceleration unless electrical fields of very high intensities or frequencies are involved. Under normal conditions, the ions are almost instantaneously accelerated to the point where their motion is limited by the viscous drag of the solvent, and all the energy supplied by the electric field is dissipated by the viscous forces. The ions thus move with a constant limiting or terminal velocity, which for all reasonably small fields is directly proportional to the applied field. This is of course the reason for the validity of Ohm's law for electrolytes subjected to ordinary electric fields, and for the fact that the conductivities of ion-containing liquids have no simple relation to the ion masses. There is, for example, little difference between the ionic conductivities of electrolytes containing chloride and iodide ions even though the latter have nearly four times the mass of the former.

From (3.218) and (3.220), the flux of ions depends on the ion's absolute mobility M_{abs}, where the absolute mobility is defined as the ion's speed due to a unit force acting on it [3.247]. In (3.221), the electric force is expressed as acting per ion but may be more conveniently expressed as that acting per mole of ions. Thus M_{abs} can relate the limiting speed v_d to the magnitude of the force per mole f

$$v_d = M_{abs} f . \quad (3.223)$$

In the case of an electric field, we can express the force per mole as f_e

$$f_e = z_i F_a E , \quad (3.224)$$

where z_i is the valence and F_a is the Faraday number and $F_a = e_0 N_A$ (Coulomb/mol), N_A is Avogadro's number. Thus $z_i F_a$ is the charge per mole of ions and

$$v_d = M_{abs} z_i F_a E , \quad (3.225)$$

or

$$v_d = M_e E , \quad (3.226)$$

as in the case where all forces but E can be neglected. Then

$$M_e = M_{abs} z_i F_a . \quad (3.227)$$

Neither M_{abs} nor M_e can be conveniently measured but they can be related to the equivalent ionic conductivity λ, which can be measured. To find this relationship, consider a column in the electrolyte solution of length l and of uniform cross-sectional area A. If there is a potential difference V between the ends of the column, it will drive a current I of one ion species through the column. If G is the conductance of the column for that ion species, the current will be given by Ohm's law as

$$I = GV . \quad (3.228)$$

The conductance G depends on length and area and is proportional to the specific ionic conductivity σ, so

$$G = \frac{\sigma A}{l} . \quad (3.229)$$

The equivalent conductivity λ is obtained by dividing the specific conductivity σ by the ionic concentration in *equivalents*

$$\lambda = \frac{\sigma}{z_i C_i} , \quad (3.230)$$

where C_i is the concentration in mol/m^3 and $z_i C_i$ the concentration in *equivalents*. When the ion concentration tends to zero in a solution, λ tends to λ^0, the limiting equivalent conductivity. The electric field driving the current is the voltage gradient

$$|E| = \frac{V}{l} . \quad (3.231)$$

From (3.228) to (3.231), we obtain the electric current

$$I = \lambda z_i C_i A |E| . \quad (3.232)$$

We can also relate the current to the drift velocity. Since the electric charge per unit volume is, $z_i C_i F_a$, the charge in the column considered is $\rho_e = z_i C_i F_a A l$. This charge moves out of the column in the time $l/|v_d|$. Thus the rate of passage of electric charge, which is actually the electric current, is

$$I = \frac{C_i z_i F_a A l}{l/|v_d|} = C_i z_i F_a A |v_d| . \quad (3.233)$$

Equating the two expressions for the current, (3.232) and (3.233), we have

$$v_d = \frac{\lambda |E|}{F_a} . \quad (3.234)$$

From (3.226) and (3.234) we express the electric mobility M_e in terms of λ as

$$M_e = \frac{\lambda}{F_a} , \quad (3.235)$$

whereas from (3.225) and (3.235) the absolute mobility M_{abs} is related to λ as

$$M_{abs} = \frac{\lambda}{z_i F_a^2} \quad \text{(for force per ion)} , \quad (3.236a)$$

$$M_{abs} = \frac{\lambda N_A}{z_i F_a^2} \quad \text{(for force per mole)} . \quad (3.236b)$$

The ratio between M_{abs} of an ion X to M_{abs} of K$^+$ (potassium ion) is defined as the relative mobility $M_{\text{rel }X}$ of ion X. Hence, the relative mobility of ion X of valence z_X is given by

$$M_{\text{rel }X} = \frac{\lambda_X}{z_X \lambda_K} , \quad (3.237)$$

since z is $+1$ for K$^+$. Both mobilities must be taken at the same temperature. For a monovalent ion Y, its relative mobility will simply be given by

$$M_{\text{rel }Y} = \frac{\lambda_Y}{\lambda_K} , \quad (3.238)$$

where λ_Y is the equivalent conductivity of Y at the same temperature as for λ_K. Table 3.33 contains the values of the limiting equivalent conductivity λ^0, the relative mobility M_{rel}, the absolute mobility M_{abs} and the electric mobility M_e for a number of ions in water at 298.15 K. Table 3.34 shows the dependence of the limiting equivalent conductivity on temperature for a number of ions in water. Tables 3.35 and 3.36 contain the values of λ^0 for a number of ions in protic and aprotic solvents, respectively. Protic solvents are strong hydrogen donors. For large concentrations the influence of the ions on each other because of proximity must be taken into account.

The Viscous Force Action on an Ion in Solution – The Stokes–Einstein Relation. If asserting the similarity between a macroscopic sphere moving in an incompressible fluid and a particle (ion) moving in a solution, then the calculations for viscous force acting on a macroscopic sphere can be used for that acting on an ion. The value of the viscous force depends on several factors – the velocity v and diameter $d = 2r$ of the sphere (an ion) and the viscosity μ and density ρ of the medium. When the hydrodynamic conditions are such that the Reynolds number (3.239) is much smaller than unity, the viscous force opposing the sphere is given by Stokes' law (3.240).

$$\mathrm{Re} = \frac{|v|d\rho}{\mu}, \tag{3.239}$$

$$F_{\mathrm{Stokes}} = 6\pi r \mu v. \tag{3.240}$$

The net driving force due to the concentration gradient and the external applied force (electric field) produces a steady-state diffusion/conduction flux of ions J for which one can imagine a drift velocity of the diffusing particles. Since the drift velocity is a steady-state velocity, the net driving force (F_{net}) must be opposed by an equal resistive force which can be taken to be the Stokes viscous force. Hence,

$$-F_{\mathrm{net}} = 6\pi r \mu v_{\mathrm{d}}. \tag{3.241}$$

with v being equal to the drift velocity v_{d}.

If we, for example, consider a bath containing a polymer solution in an electric field between two electrodes (like in the electrospinning of polymer solutions), we can estimate the drift velocity of the charge carriers, which is mainly ionic impurities (depending on the polymer and solvent). Using the following values: $|E| = 1\,\mathrm{g}^{1/2}/(\mathrm{cm}^{1/2}\,\mathrm{s})$ (300 V/cm); $e_0 = 4.8 \times 10^{-10}\,(\mathrm{g}^{1/2}\,\mathrm{cm}^{3/2})/\mathrm{s}$ $(1.6 \times 10^{-19}\,\mathrm{C})$; $z_i = +1$; $\mu = 10^{-2} - 10\,\mathrm{g}/(\mathrm{cm}\,\mathrm{s})$ (1–1000 cP) and $r = 10^{-8}\,\mathrm{cm}$ (typical for small ions such as Cl^-), $|v_{\mathrm{d}}|$ is of the order of 1 cm/s to 10^{-3} cm/s. The value of $|v_{\mathrm{d}}|$ has been estimated from (3.241) with $|F_{\mathrm{net}}| = e_0|E|$. For small ions r is the ionic radius, correct to the order of magnitude [3.236]. The solvation radii [3.246] are one to two orders larger. With the above values and $\rho \sim 1\,\mathrm{g/cm}^3$, one can easily see that $\mathrm{Re} \ll 1$.

From (3.220, 222) and (3.241) we have

$$M_{\mathrm{abs}} = \frac{1}{6\pi r \mu}, \tag{3.242}$$

$$M_{\mathrm{e}} = \frac{z_i e_0}{6\pi r \mu}. \tag{3.243}$$

The mobility given by (3.243) is also called the Stokes mobility.

The Einstein relation between the ion diffusion coefficient D and the absolute mobility M_{abs} is one of the most important relations relevant for the diffusion of ions [3.236]

$$D = M_{\mathrm{abs}} k_{\mathrm{B}} T = \frac{M_{\mathrm{e}} k_{\mathrm{B}} T}{z_i e_0}, \tag{3.244}$$

where k_{B} is the Boltzmann constant and T the temperature.

With the help of (3.242), (3.244) yields

$$D = \frac{k_{\mathrm{B}} T}{6\pi r \mu}. \tag{3.245}$$

Equation (3.245) is the Einstein–Stokes relation and it links the processes of diffusion and viscous flow.

The real question centers on the applicability of Stokes' law to microscopic ions moving in a structured medium in which the surrounding particles are roughly of the same size as the ions.

Mobility Measurements. The *natural* conductivity of dielectric liquids (leaky dielectrics) is generally very small. Therefore, in order to make the measurement of charge mobility easier, it is necessary to enhance the normal charge density in a controlled manner, usually by some form of transient external excitation. The mobility of charge carriers is defined as its drift velocity per unit of electric stress (3.226). An estimate of mobility is achieved by a time-of-flight method, which requires a measure of the time necessary for the charge to travel a known distance in the liquid under the influence of a uniform electric field. A detailed description of various experimental techniques of measurement can be found in [3.248]. A general arrangement for mobility measurements is illustrated in Fig. 3.57 [3.249]. Excess charge is created at the emitter electrode E. By applying the appropriate polarity of voltage V, ions of one sign are swept to the collector electrode C. The emitter electrode is maintained at a high potential relative to earth. The grid pair AB and DF act as electrical shutters, or gates, to allow the passage of carriers across the drift space BD. The gates are arranged to open or close with the frequency of an alternating current (AC) voltage applied to them.

Table 3.33 Experimental values of the limiting equivalent ionic conductivity λ^0, the relative ion mobility M_{rel}, the absolute ion mobility M_{abs} and the electrical ion mobility M_e for a number of different ions in water at 298.15 K. Most of the values were taken from [3.253] and [3.254]. See also [3.246]

Ion	Valence	λ^0 [(cm Ω equiv.)$^{-1}$]	M_{rel}	M_{abs} (mol s g$^{-1} \times 10^9$)	M_e (cm$^{3/2}$g$^{-1/2} \times 10^3$)
Acetate	−1	40.866	0.556	4.3907	0.4236
Br	−1	78.1	1.0626	8.3912	0.8095
Cl	−1	76.35	1.0388	8.2032	0.7914
ClO$_4$	−1	67.326	0.916	7.2336	0.6979
F	−1	55.4	0.7537	5.9523	0.5742
HCO$_3$	−1	44.4675	0.605	4.7777	0.4609
I	−1	76.8	1.0449	8.2515	0.7961
NO$_3$	−1	71.46	0.9722	7.6778	0.7407
OH	−1	198.3	2.698	21.3057	2.0555
Ag	1	61.9	0.8422	6.6506	0.6416
Cs	1	77.2	1.0503	8.2945	0.8002
H	1	350.0805	4.763	37.6133	3.6287
K	1	73.5	1.0	7.8970	0.7619
Li	1	38.6	0.5252	4.1473	0.4001
Na	1	50.1	0.6816	5.3828	0.5193
NH$_4$	1	73.5	1.0	7.8970	0.7619
Rb	1	77.8365	1.059	8.3629	0.8068
SO$_4$	−2	80.0	0.5442	4.2977	0.8292
Ba	2	63.798	0.434	3.4273	0.6613
Ca	2	59.5	0.4048	3.1964	0.6167
Cd	2	54.39	0.37	2.9219	0.5638
Co	2	54.39	0.37	2.9219	0.5638
Cu	2	53.655	0.365	2.8824	0.5562
Fe	2	54.39	0.37	2.9219	0.5638
Hg	2	63.651	0.433	3.4194	0.6598
Mg	2	53.0	0.3605	2.8472	0.5494
Mn	2	53.508	0.364	2.8745	0.5546
Ni	2	49.539	0.337	2.6613	0.5135
Pb	2	70.56	0.48	3.7905	0.7314
Sr	2	59.388	0.404	3.1904	0.6156
Zn	2	52.773	0.359	2.8350	0.5470
Fe	3	67.914	0.308	2.4323	0.7040
Gd	3	67.2525	0.305	2.4086	0.6971
La	3	69.678	0.316	2.4954	0.7222

If this frequency is changed continuously, the number of ions reaching C is a maximum when their transit time between the gates is equal to, or an integral multiple of, the period of the pulses. The variation in the collector current is shown in Fig. 3.57b, where the transit time is given by the reciprocal of the difference in frequencies corresponding to adjacent current maxima. The amplitude of the oscillations tends to fall as the frequency of the gate voltage is raised with the result that the sensitivity of the method is decreased as the drift distance is increased and the transit time shortened. *Meyer* and *Reif* [3.250] used this method for fields up to 25 kV/m in liquid helium. *Schynders* et al. [3.251, 252] used this method to determine the electron mobilities in liquid ar-

Table 3.34 Experimental values of the limiting equivalent conductivity λ^0 of ions in water at various temperatures. The units are $(cm\Omega equiv.)^{-1}$. For a more complete set of data [3.246]

Ion	Valence	Temperature (K)			
		273.15	288.15	298.15	308.15
Br$^-$	−1	42.6	63.1	78.1	94
Cl$^-$	−1	41	61.4	76.35	92.2
I$^-$	−1	41.4	62.1	76.8	92.3
Cs$^+$	1	44	63.1	77.2	92.1
K$^+$	1	40.7	59.6	73.5	88.2
Li$^+$	1	19.4	30.2	38.6	48
Na$^+$	1	26.5	39.7	50.1	61.5
Ca^{++}	2	31.2	46.9	59.5	73.2

Table 3.35 Limiting ionic conductivities λ^0 $(cm\Omega equiv.)^{-1}$ in protic solvents at 298.15 K [3.255]. MeOH – Methanol, EtOH – Ethanol, PrOH – Propane alcohol, BuOH – Butane alcohol, HCOOH – Formic acid

Ion	Valence	MeOH	EtOH	PrOH	BuOH	HCOOH
Ag	+1	50.07				
Br	−1	56.43	23.88	12.22	8.23	28.3
Cl	−1	52.09	21.87	10.45	7.76	26.52
Cs	+1	61.33	26.46			
I	−1	62.62	27.0	13.81	9.52	–
K	+1	47.78	22.2	6.88		23.99
Li	+1	39.08	17.07		8.1	19.36
Na	+1	45.08	20.37	8.35		20.97
NH$_4$	+1				6.68	27.01
NO$_3$	−1	61.13	–	–	–	–

Table 3.36 Limiting ionic conductivities λ^0 $(cm\Omega equiv.)^{-1}$ in some aprotic solvents at 298.15 K [3.255]. DMF – dimethylformamide, DMSO – dimethylsulfoxide, NMT – monomethyltryptamine, NMA – monomethylacetamide, ACN – acetonitrile

Ion	Valence	DMF	DMSO	NMT	NMA	ACN
Ag	+1	35.2				86.0
Br	−1	53.6	24.1	62.9	11.72	100.7
Cl	−1	55.1	24.4	62.7	10.6	
Cs	+1		16.1			87.3
I	−1	52.3	23.8		13.42	102.1
K	+1	30.8	14.7		7.28	83.6
Li	+1	25.0	11.4		5.65	
Na	+1	29.9	14.54		7.19	76.9
NH$_4$	+1	38.7				
NO$_3$	−1	57.3				106.4

electrodes are measured at different frequencies (f) and different distances (L) with the aid of a spectrum analyzer [3.256]. The results of the typical used to find σ and ε_r are shown in Fig. 3.59 for an aqueous solution of polyethylene oxide. In order to determine the values of σ and ε_r, a model, consisting of resistors and capacitors that simulates the impedance of the test fluid is constructed. For the above solution the model shown in Fig. 3.60 is appropriate. The impedance of the circuit shown in Fig. 3.60 is given by

$$Z = \frac{R}{1 + i\omega RC}, \qquad (3.246)$$

where R is resistance, C capacitance and ω the angular frequency. The real and imaginary parts of (3.246), given in (3.247), should correspond to the measured values of

gon. Additional information on mobility measurements can be found in Chap. 23 of the Handbook.

3.7.4 Broadband Measurement of the Conductivity and Dielectric Constant

In order to measure the conductivity σ and dielectric constant ε_r of a fluid, the apparatus shown in Fig. 3.58 can be used. It consists of a glass syringe with two brass pistons as the electrodes. Electrode 1 is connected to a linear translation stage that enables accurate setting of the distance L between the electrodes. Electrode 2 is fixed close to a small hole in the glass syringe through which the test fluid can escape when the distance between the electrodes is changed.

The real Re(Z) and imaginary Im(Z) components of the complex impedance (Z) of the volume between the

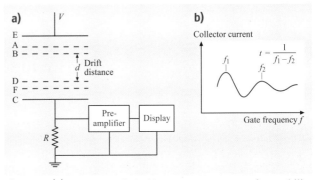

Fig. 3.57 (a) Diagram of double gate arrangement for mobility measurement, and (b) collector current as a function of the frequency of the voltage on the gates (after [3.249])

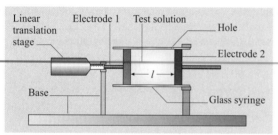

Fig. 3.58 An apparatus for measuring the electric conductivity and dielectric constant of a fluid

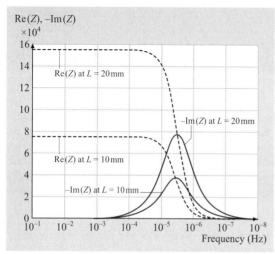

Fig. 3.59 A plot of the real Re(Z) and imaginary Im(Z) parts of the complex impedance of the fluid between the electrodes of the apparatus in Fig. 3.58. The fluid between the electrodes was an aqueous solution of polyethylene oxide: PEO (molecular weight = 4×10^6) at 1% weight concentration in ethanol/water (40/60)

Re(Z) and Im(Z):

$$\mathrm{Re}(Z) = \frac{R}{1 + \omega^2 R^2 C^2}, \quad (3.247\mathrm{a})$$

$$\mathrm{Im}(Z) = \frac{-\omega R^2 C}{1 + \omega^2 R^2 C^2}, \quad (3.247\mathrm{b})$$

The equations for Re(Z) and Im(Z) as a function of ω are fitted to the experimental data shown in Fig. 3.59 in order to find the values of R and C. With the values of R and C known, σ and ε_r can be calculated from

$$\sigma = \frac{L}{RS}, \quad (3.248\mathrm{a})$$

$$\varepsilon_r = \frac{CL}{\varepsilon_0 S}, \quad (3.248\mathrm{b})$$

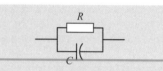

Fig. 3.60 A model that simulates the impedance of the test fluid in Fig. 3.59

where S is the surface area of the electrodes and L is the length of the sample fluid. Table 3.37 shows the values of σ and ε_r for a number of polymer solutions and pure solvents (leaky dielectrics) obtained by this method. The following notation is used: PAA for polyacrylic acid, PVA for polyvinyl alcohol, PU for polyurethane, PCL for polycaprolactone, MC for methylene chloride, DMF for dimethylformamide, and THF for tetrahydrofuran.

The electrical conductivity is a function of the electron and ion motion. In the solutions and solvents tested ion diffusion mainly determines the conductivity of the solution and the electron conductivity is much smaller. When measuring the parameters of an electrically con-

Table 3.37 Representative values of ε_r and σ obtained with the method described in this section

Polymer	ε_r	σ (mS/m)
2% PEO $M_w = 6 \times 10^5$ in ethanol/water (40/60)	67.09	0.85
2% PEO $M_w = 10^6$ in ethanol/water (40/60)	66.71	0.81
1% PEO $M_w = 4 \times 10^6$ in ethanol/water (40/60)	66.12	1.102
2% PEO $M_w = 10^6$ in water	81.96	9.43
6% PAA $M_w = 2.5 \times 10^5$ in ethanol/water (40/60)	79.5	24.47
6% PVA $M_w = 10^4$ in ethanol/water (50/50)	65.99	3.73
6% PU tecoflex in THF/ethanol (50/50)	16.75	0.093
8% PCL $M_w = 8 \times 10^4$ in acetone	25.2	0.142
10% PCL $M_w = 8 \times 10^4$ in MC/DMF (75/25)	18.55	0.191
Distilled water	88.75	0.447
Ethanol (95%)	24.55	0.0624
Acetone	20.7	0.0202
Ethanol/water (40/60)	69.47	0.150
MC/DMF (40/60)	29.82	0.505
MC/DMF (75/25)	21.3	0.273
THF/ethanol (50/50)	15.79	0.037

ducting medium (a leaky dielectric), the migration of ions in the high-frequency field causes energy to be drawn from the circuit, damping the oscillation response. In circuit terms, the electrodes and the volume between them are analogous to a capacitor with a resistor in parallel. When the resistance across the electrodes is large, the capacitance derived from the frequency response is close to the true capacitance of the medium between the electrodes. As the resistance across the electrodes is reduced, due to increased ionic conductivity, the oscillator is damped and the oscillation frequency falls. The apparent capacitance thus appears greater than the true capacitance. The effect of ionic conductivity on the permittivity measurement can be neglected below $\sigma = 0.8\,\text{S/m}$ [3.257]. Some other methods of measurement of the dielectric constant and electric conductivity are described in [3.258–262] and in Chap. 23 of the Handbook.

References

3.1 C. Yokohama, S. Takahashi: Saturated liquid densities of 2,2-dichloro-1,1,1-trifluoroethane (HCFC-123), 1,2-dichloro-1,2,2-trifluoroethane (HCFC-123a), 1,1,1,2-tetrafluoroethane (HFC-134a) and 1,1,1-trifluoroethane (HFC-143a), Fluid Phase Equilibr. **67**, 227–240 (1991)

3.2 W.R. Lau, C.-A. Hwang, H.B. Brugge, G.A. Iglesias-Silva, H.A. Duarte-Garza, W.J. Rogers, K.R. Hall, J.C. Holste, B.E. Gammon, K.N. Marsh: A continuously weighed pycnometer for measuring fluid properties, J. Chem. Eng. Data **42**, 738–744 (1997)

3.3 J.R.S. Machado, W.B. Street: Equation of state and thermodynamic properties of liquid methanol from 298 to 489 K and pressures to 1040 bars, J. Chem. Eng. Data **28**, 218–223 (1983)

3.4 H. Kubota, Y. Tanaka, T. Makita, H. Kashiwagi, M. Noguchi: Thermodynamic properties of 1-chloro-1,2,2,2-tetrafluoroethane (R124), Int. J. Thermophys. **9**, 85–101 (1988)

3.5 Y. Sato, T. Nishizuka, K. Hara, T. Yamamura, Y. Waseda: Density measurement of molten silicon by a pycnometric method, Int. J. Thermophys. **21**, 1463–1471 (2000)

3.6 A.R.H. Goodwin, C.H. Braddsell, L.S. Toczylkin: (P, ρ, T) of liquid n-octane obtained with a spherical pycnometer at temperatures of 298.03 K and 313.15 K and pressures in the range 0.7 MPa to 32 MPa, J. Chem. Thermodyn. **28**, 637–646 (1996)

3.7 V.N. Belonenko, V.M. Troitsky, Y.E. Belyaev, J.H. Dymond, N.F. Glen: Application of micro (P, V, T) apparatus for measurement of liquid densities at pressures up to 500 MPa, J. Chem. Thermodyn. **32**, 1203–1219 (2000)

3.8 C.M.M. Duarte, H.J.R. Guedes, M. da Nunes Ponte: (P, V − m, T) measurements on liquid and gaseous mixtures near the critical point − I. (xenon + ethane), J. Chem. Thermodyn. **32**, 877–889 (2000)

3.9 E.S. Burnett: Compressibility determinations without volume measurements, J. Appl. Mech. A **3**, 136–140 (1936)

3.10 C.E. Stouffer: *Densities of mixtures of carbon dioxide and hydrogen sulfide from 200 to 450 K to 23 MPa by the Burnett-isochoric method*. Ph.D. Thesis (Texas A&M University, College Station, Texas, USA 1992)

3.11 J.D. Isdale, J.H. Dymond, T.A. Brawn: Viscosity and density of n-hexane-cyclohexane mixtures between 25 and 100 °C up to 500 MPa, High Temp.-High Press. **11**, 571–580 (1979)

3.12 J.H. Dymond, J.D. Isdale, N.F. Glen: Density measurement at high pressure, Fluid Phase Equilibr. **20**, 305–314 (1985)

3.13 A. Iso, M. Uematsu: Thermodynamic properties of 1,1-difluoroethane in the super-critical and high-density regions, Physica A **156**, 454–466 (1989)

3.14 J.G. Webster: *Mechanical Variables Measurement: Solid, Fluid and Thermal* (CRC, London 2000)

3.15 V. Hynek, M. Obsil, V. Majer, J. Quint, J.-P.E. Grolier: A vibrating-tube flow densitometer for measurements with corrosive solutions at temperatures up to 723 K and pressures up to 40 MPa, Int. J. Thermophys. **18**, 719–732 (1997)

3.16 A.A.H. Padua, J.M.N.A. Fareleira, J.C.G. Calado, W.A. Wakeham: A vibrating-wire densimeter for liquids at high pressures: The density of 2,2,4-trimethylpentane from 298.15 to 348.15 K and up to 100 MPa, Int. J. Thermophys. **15**, 229–243 (1994)

3.17 J.G. Blencoe, S.E. Drummond, J.C. Seitz, B.E. Nesbitt: A vibrating-tube densimeter for fluids at high pressures and temperatures, Int. J. Thermophys. **17**, 179–190 (1996)

3.18 D.R. Defibaugh, G. Morrison: Compressed liquid densities and saturation densities of chlorodifluoromethane (R22), J. Chem. Eng. Data **37**, 107–110 (1992)

3.19 A.T. Sousa, C.A. Nieto de Castro, R. Tufeu, B. Le Neindre: Density of 1-chloro-1, 1-difluoroethane (R142b), High Temp.-High Press. **24**, 185–194 (1992)

3.20 J. Klimeck, R. Kleinrahn, W. Wagner: An accurate single-sinker densimeter and measurements of the (P, ρ, T) relation of argon and nitrogen in the temperature range from (235 to 520) K at pressures up to 30 MPa, J. Chem. Thermodyn. **30**, 1571–1588 (1998)

3.21 T. Retsina, S.M. Richardson, W.A. Wakeham: The theory of a vibrating-rod densimeter, Appl. Sci. Res. **43**, 127–143 (1986)

3.22 A.R.H. Goodwin, K.N. Marsh, W.A. Wakeham (Eds.): *IUPAC Experimental Thermodynamics Vol. VI. Measurement of the Thermodynamic Properties of Single Phases* (Elsevier, Amsterdam 2003)

3.23 M. Dix, J.M.N.A. Fareleira, Y. Takaishi, W.A. Wakeham: A vibrating wire densimeter for measurements in fluids at high pressures, Int. J. Thermophys. **12**, 357–370 (1991)

3.24 A.A.H. Padua, J.M.N.A. Fareleira, J.C.G. Calado, W.A. Wakeham: Electromechanical model for vibrating-wire instruments, Rev. Sci. Instrum. **69**, 2392–2399 (1998)

3.25 F. Audonnet, A.A.H. Padua: Simultaneous measurements of density and viscosity of n-pentane fom 298 K to 383 K and up to 100 MPa pressure using a vibrating-wire instrument, Fluid Phase Equilibr. **181**, 147–161 (2001)

3.26 R. Kleinrahm, W. Wagner: Measurement and correlation of the equilibrium liquid and vapor densities and the vapor pressure along the coexistance curve of methane, J. Chem. Thermodyn. **18**, 739–760 (1986)

3.27 C. Evers, H.W. Losch, W. Wagner: An absolute viscometer-densimeter and measurements of the viscosity of nitrogen, methane, helium, neon, argon, and krypton over a wide range of density and temperature, Int. J. Thermophys. **23**, 1411–1439 (2002)

3.28 R. Masui: Development of a magnetic suspension densimeter and measurement of the density of toluene, Int. J. Thermophys. **23**, 921–935 (2002)

3.29 S. Toscani, P. Figuiere, H. Szwarc: A magnetic suspension apparatus to measure densities of liquids as a function of temperature at pressures up to 100 MPa. Application to n-heptane, J. Chem. Thermodyn. **21**, 1263–1277 (1989)

3.30 M. Okada, N. Uemastsu, K. Watanabe: Orthobaric liquid densities of trichloro-fluoromethane, dichlorodifluoromethane, chlorodifluoromethane, 1,1,2-trichloro-trifluoroethane, 1,2-dichlorotetrafluoroethane, and of the azeotropic mixture of (chlorodifluoromethane + chloropentafluoroethane) between 203 and 463 K, J. Chem. Thermodyn. **18**, 527–543 (1986)

3.31 W. Wagner, A. Pruß: The IAPWS formulation 1995 for the thermodynamic properties of ordinary water substance for general and scientific use, J. Phys. Chem. Ref. Data **31**, 387–535 (2002)

3.32 http://www.iapws.org/

3.33 M.J. Assael, J.H. Dymond, D. Exadaktilou: An improved representation of the density of n-alkane liquid densities, Int.J. Thermophys. **15**, 155–164 (1994)

3.34 M.J. Assael, J.H. Dymond, P.M. Patterson: Correlation and prediction of dense fluid transport coefficients – V. Aromatic Hydrocarbons, Int. J. Thermophys. **13**, 895–905 (1992)

3.35 M.J. Assael, J.H. Dymond, S.K. Polimatidou: Correlation and prediction of dense fluid transport coefficients – VI. n-Alcohols, Int. J. Thermophys. **15**, 189–201 (1994)

3.36 M.J. Assael, J.H. Dymond, S.K. Polimatidou: Correlation and prediction of dense fluid transport coefficients – VII. Refrigerants, Int. J. Thermophys. **16**, 761–772 (1995)

3.37 Supertrapp Ver. 2.0, Computer Software Package for the calculation of the transport properties of nonpolar fluids and their mixtures (N.I.S.T., Gaithersburg 1998)

3.38 K.A. Connors, J.L. Wright: Dependence of surface tension on composition of binary aqueous-organic solutions, Anal. Chem. **61**, 194–198 (1989)

3.39 L.A. Girifalco, R.J. Good: A theory for the estimation of surface and interfacial energies – I. Derivation and application to interfacial tension, J. Phys. Chem. **61**, 904–909 (1957)

3.40 F.M. Fowkes: Dispersion force contributions to surface and interfacial tensions, contact angles, and heats of immersion. In: *Contact Angle, Wettability, and Adhesion*, Adv. Chem. Ser., Vol. 43, ed. by R.F. Gould (ACS, Washington 1964) pp. 99–111

3.41 D.K. Owens, R.C. Wendt: Estimation of the surface free energy of polymers, J. Appl. Polym. Sci. **13**, 1741–1747 (1969)

3.42 J.J. Jasper: The surface tension of pure liquid compounds, J. Phys. Chem. Ref. Data **1**, 841–1009 (1972)

3.43 N.B. Vargaftik, Y.K. Vinogradov, V.S. Yargin: *Handbook of Physical Properties of Liquids and Gases: Pure Substances and Mixtures* (Begell House, New York 1996)

3.44 A. Marmur: Wetting on hydrophobic rough surfaces: to be heterogeneous or not to be?, Langmuir **19**, 8343–8348 (2003)

3.45 E.L. Decker, S. Garoff: Using vibrational noise to probe energy barriers producing contact angle hysteresis, Langmuir **12**, 2100–2110 (1996)

3.46 C. Andrieu, C. Sykes, F. Brochard: Average spreading parameter on heterogeneous surfaces, Langmuir **10**, 2077–2080 (1994)

3.47 H. Kamusewitz, W. Possart, D. Paul: The relation between Young's equilibrium contact angle and the hysteresis on rough paraffin wax surfaces, Colloid. Surf. A Physiochem. Eng. Aspects **156**, 271–279 (1999)

3.48 T.D. Blake, A. Clarke, K.J. Ruschak: Hydrodynamic assist of dynamic wetting, AIChE J. **40**, 229–242 (1994)

3.49 T.D. Blake, M. Bracke, Y.D. Shikhmurzaev: Experimental evidence of non-local hydrodynamic influence on the dynamic contact angle, Phys. Fluids **11**, 1995–2007 (1999)

3.50 T.D. Blake: Dynamic contact angles and wetting kinetics. In: *Wettability*, ed. by J.C. Berg (Marcel Dekker, New York 1993) pp. 251–309

3.51 M.J. de Ruijter, J. De Coninck, T.D. Blake, A. Clarke, A. Rankin: Contact angle relaxation during the spreading of partially wetting drops, Langmuir **13**, 7293–7298 (1997)

3.52 O.V. Voinov: Hydrodynamics of wetting, Fluid Dyn. **11**, 714–721 (1976)

3.53 E.B.V. Dussan: The moving contact line: the slip boundary condition, J. Fluid Mech. **77**, 665–684 (1976)

3.54 L.H. Tanner: The spreading of silicone oil drops on horizontal surfaces, J. Phys. D **12**, 1473–1484 (1979)

3.55 R.G. Cox: The dynamics of the spreading of liquids on a solid surface, J. Fluid Mech. **168**, 169–194 (1986)

3.56 P.G. de Gennes: Wetting statics and dynamics, Rev. Modern Physics **57**, 827–863 (1985)

3.57 S.F. Kistler: Hydrodynamics of wetting. In: *Wettability*, ed. by J.C. Berg (Marcel Dekker, New York 1993) pp. 311–429

3.58 B.W. Cherry, C.M. Holmes: Kinetics of the wetting of surfaces by polymers, J. Colloid Interface Sci. **29**, 174–176 (1969)

3.59 T.D. Blake, J.M. Haynes: Kinetics of liquid/liquid displacement, J. Colloid Interface Sci. **30**, 421–423 (1969)

3.60 M.J. de Ruijter, T.D. Blake, J. De Coninck: Dynamic wetting studied by molecular modeling simulations of droplet spreading, Langmuir **15**, 7836–7847 (1999)

3.61 P.G. Petrov, P.G. Petrov: A combined molecular-hydrodynamic approach to wetting kinetics, Langmuir **8**, 1762 (1992)

3.62 F. Brochard-Wyart, P.G. de Gennes: Dynamics of partial wetting, Adv. Colloid Interface Sci. **39**, 1–11 (1992)

3.63 M. de Ruijter, G. Oshanin, J. De Coninck: Droplet spreading: partial wetting regime revisited, Langmuir **15**, 2209–2216 (1999)

3.64 M. de Ruijter, M. Charlot, M. Voué, J. De Coninck: Experimental evidence of several time scales in drop spreading, Langmuir **16**, 2363–2368 (2000)

3.65 Y.D. Shikhmurzaev: Moving contact lines in liquid/liquid/solid systems, J. Fluid Mech. **334**, 211–249 (1997)

3.66 T.D. Blake, Y.D. Shikhmurzaev: Dynamic wetting by liquids of different viscosity, J. Colloid Interface Sci. **253**, 196–202 (2002)

3.67 M. de Ruijter: *A microscopic approach to partial wetting: statics and dynamics*. Ph.D. Thesis (University of Mons-Hainaut, Mons-Hainaut 1996)

3.68 R.L. Hoffman: A study of the advancing interface – 1. Interface shape in liquid-gas systems, J. Colloid Interface Sci. **50**, 228–241 (1975)

3.69 E.B.V. Dussan: On the spreading of liquids on solid surfaces: static and dynamic contact lines, Annu. Rev. Fluid Mech. **11**, 371–400 (1979)

3.70 T.D. Blake, K.J. Ruschak: Wetting: static and dynamic contact lines. In: *Liquid Film Coating*, ed. by S.F. Kistler, P.M. Schweizer (Chapman Hall, London 1997) pp. 63–97

3.71 A. Clarke, T.D. Blake, K. Carruthers, A. Woodward: Spreading and imbibition of liquid droplets on porous surfaces, Langmuir **18**, 2980–2984 (2002)

3.72 E. Ramé, S. Garoff: Microscopic and macroscopic dynamic interface shapes and interpretation of dynamic contact angles, J. Colloid Interface Sci. **177**, 234–244 (1996)

3.73 R. Burley, B.S. Kennedy: An experimental study of air entrainment at a solid/liquid/gas interface, Chem. Eng. Sci. **31**, 901–911 (1976)

3.74 A. Clarke: The application of particle tracking velocimetry and flow visualisation to curtain coating, Chem. Eng. Sci. **50**, 2397–2407 (1995)

3.75 T.D. Blake, A. Clarke, E.H. Stattersfield: An investigation of electrostatic assist in dynamic wetting, Langmuir **16**, 2928–2935 (2000)

3.76 NIH Image: http://rsb.info.nih.gov/nih-image/ or for further possibilities see http://www.efg2.com/Lab/Library/ImageProcessing/

3.77 W.H. Press, S.A. Teukolsky, W.T. Vetterling, B.P. Flannery: *Numerical Recipes in Fortran*, 2nd edn. (Cambridge Univ. Press, Cambridge 1992)

3.78 R.A. Hayes, J. Ralston: Forced liquid movement on low energy surfaces, J. Colloid Interface Sci. **159**, 429–438 (1993)

3.79 E.V. Gribanova, L.I. Molchanova: Dependence of wetting angle on rate of meniscus movement, Kolloidn. Zh. **40**, 217–223 (1978)

3.80 J.W. Berube, C.B. Schriver, D.H. Chittenden: Photomicrography of a rapidly moving meniscus, Ind. Eng. Chem. Fundam. **12**, 236–239 (1973)

3.81 M. Fermegier, P. Jenffer: An experimental investigation of the dynamic contact angle in liquid-liquid systems, J. Colloid Interface Sci. **146**, 226–241 (1991)

3.82 G. Martic, F. Gentner, D. Sevano, D. Coulon, J. De Coninck, T.D. Blake: A molecular dynamics simulation of capillary imbibition, Langmuir **18**, 7971–7976 (2002)

3.83 W.A. Wakeham, A. Nagashima, J.V. Sengers (Eds.): *Experimental Thermodynamics, Measurement of the Transport Properties of Fluids*, Vol. III (Blackwell, Oxford 1991)

3.84 G.F. Newell: Theory of oscillating type viscometer – V. Disk oscillating between fixed plates, Z. Angew. Math. Phys. **10**, 160–175 (1959)

3.85 J. Kestin, W. Leidenfrost: *Thermodynamic and Transport Properties of Gases, Liquids and Solids*, ed. by Y.S. Touloukian (ASME McGraw-Hill, New York 1959)

3.86 E. Vogel: Konstruktion eines Quarzglas-Schwingscheibenviskosimeters und Messungen an Stickstoff und Argon, Wiss. Z. Univ. Rostock **21**, 169–179 (1972), in German

3.87 J. Kestin, S.T. Ro, W.A. Wakeham: Viscosity of isotopes of hydrogen and their intermolecular potential, J. Chem. Soc. Faraday Trans. I **68**, 2316–2323 (1972)

3.88 J. Kestin, O. Korfali, J.V. Sengers: Density expansion of the viscosity of carbon dioxide near the critical temperature, Physica **100A**, 335–348 (1980), 335–348

3.89 R. Di Pippo, J. Kestin, J.W. Whitelaw: A high temperature oscillating-disk viscometer, Physica **32**, 2064–2078 (1966)

3.90 J. Kestin, H.E. Khalifa, W.A. Wakeham: Viscosity of 5 gaseous hydrocarbons, J. Chem. Phys. **66**, 1132–1134 (1977)

3.91 G. Oltermann: *Messung der Viskosität von Wasserdampf in der Nähe des kritischen Zustandes*. Ph.D. Thesis (University of Hannover, Hannover 1977), in German

3.92 J. Kestin, R. Paul, I.R. Shankland, H.E. Khalifa: A high-temperature, high-pressure oscillating-disk viscometer for concentrated ionic solutions, Ber. Bunsenges. Phys. Chem. **84**, 1255–1265 (1980)

3.93 K. Torklep, H.A. Oye: An absolute oscillating-cylinder (or cup) viscometer for high pressures, J. Phys. E **12**, 875–885 (1979)

3.94 D.A. Beckwith, G.F. Newell: Theory of oscillating-type viscometers – II. Oscillating-cup, J. Ang. Math. Phys. **8**, 450–465 (1957)

3.95 J.M. Grouvel, J. Kestin, H.E. Khalifa, E.U. Franck, F. Hensel: Viscosity of liquid mercury from 20 °C to 260 °C along the saturation line, Ber. Bunsenges. Phys. Chem. **81**, 338–344 (1977)

3.96 H.V. Tippelskirch, E.U. Franck, F. Hensel, J. Kestin: Viscosity of fluid mercury to 1529 K and 100 bar, Ber. Bunsenges. Phys. Chem. **79**, 889–897 (1975)

3.97 Y. Abe, O. Kosugiyama, A. Nagashima: Viscosity measurements of NaCl in the temperature range 1083 to 1473 K, Ber. Bunsenges. Phys. Chem. **84**, 1178–1184 (1980)

3.98 B. Knapstad, P.A. Skjolsvik, H.A. Oye: Viscosity of pure hydrocarbons, J. Chem. Eng. Data **34**, 37–43 (1989)

3.99 D. Dumas, K. Grjotheim, B. Hogdahl, H.A. Oye: Theory of oscillating bodies and its utilization for determination of high-temperature viscosities, Acta Chem. Scand. **24**, 510–530 (1970)

3.100 W. Brockner, K. Grjotheim, K. Ohta, H.A. Oye: High temperature viscometer for liquids – 2 Viscosity of alkali chlorides, Ber. Bunsenges. Phys. Chem. **79**, 344–347 (1975)

3.101 J.T. Tough, W.D. McCormick, J.G. Dash: Viscosity of liquid He II, Phys. Rev. **132**, 2373–2378 (1963)

3.102 T. Retsina, S.M. Richardson, W.A. Wakeham: The theory of a vibrating-rod viscometer, Appl. Sci. Res. **43**, 325–346 (1987)

3.103 S.S. Chen, M.W. Wambsganaa, J.A. Jendrzejcyk: Added mass and damping of a vibrating rod in confined viscous fluids, Trans. ASME, J. Appl. Mech. **43**, 315–332 (1976)

3.104 M.J. Assael, M. Papadaki, M. Dix, S.M. Richardson, W.A. Wakeham: An absolute vibrating-wire viscometer for liquids at high pressures, Int. J. Thermophys. **12**, 231–244 (1991)

3.105 C.M.B.P. Oliveira: *Viscosity of liquid hydrocarbons at high pressure*. Ph.D. Thesis (London University, London 1991)

3.106 J. Wilhelm, E. Vogel: Viscosity measurements on gaseous propane, J. Chem. Eng. Data **46**, 1467–1471 (2001)

3.107 J.L.G.C. da Mata, J.M.N.A. Fareleira, C.M.B.P. Oliveira, F.J.P. Caetano, W.A. Wakeham: A new instrument to perform simultaneous measurements of density and viscosity of fluids by a dual vibrating-wire technique, High Temp.-High Press. **33**, 669–676 (2001)

3.108 P.S. van der Gulik, R. Mostert, H.R. van den Berg: The viscosity of methane at 25 °C up to 10 kBar, Physica **151A**, 153–165 (1988)

3.109 W.P. Mason: Measurement of the viscosity and shear elasticity of liquids by means of a torsionally vibrating crystal, Trans. ASME **69**, 359–370 (1947)

3.110 B. Welber: Damping of a torsional oscillating cylinder in liquid helium at various temperatures and densities, Phys. Rev. **119**, 1816–1822 (1960)

3.111 W.M. Haynes: Viscosity of gaseous and liquid argon, Physica **67**, 440–470 (1973)

3.112 D.E. Diller, N.V. Frederick: Torsional piezoelectric crystal viscometer for compressed gases and liquids, Int. J. Thermophys. **10**, 145–157 (1989)

3.113 B. Bode: *Entwicklung eines Quarzviskosimeters für Messungen bei hohen Drucken*. Ph.D. Thesis (Technical University Clausthal, Clausthal 1984)

3.114 F.J.V. dos Santos, C.A. Nieto de Castro: Viscosity of toluene and benzene under high presure, Int. J. Thermophys. **18**, 367–382 (1997)

3.115 G. Barr: *A Monograph of Viscometry* (Oxford Univ. Press, Oxford 1931)

3.116 J. Kestin, M. Sokolov, W.A. Wakeham: Theory of capillary viscometers, Appl. Sci. Res. **27**, 241–264 (1973)

3.117 J.F. Swindells, J.R. Coe Jr., T.B. Godfrey: Absolute viscosity of water at 20 °C, J. Res. Nat. Bur. Standards **48**, 1–31 (1952)

3.118 M. Kawata, K. Kurase, K. Yoshida: *Realization of a viscosity standard*, Proceedings of 5-th International Congress on Rheology, Vol.1 (University of Tokyo Press, Tokyo 1969)

3.119 N.J. Trappeniers, A. Botzen, H.R. van den Berg, J. van Oesten: The viscosity of neon between 25 °C and 75 °C at pressures up to 1800 atm, Corresponding states for the viscosity of the noble gases up to high densities, Physica **30**, 985–996 (1964)

3.120 A. Nagashima, I. Tanishita: Viscosity measurement of water and steam at high temperature and high pressures, Bull. JSME **12**, 1467–1478 (1969)

3.121 S.L. Rivkin, A.Y. Levin, L.B. Izrailevskii, K.G. Kharistonov: Experimental investigation of the viscosity of heavy water at temperatures of 200–375 °C and pressures up to 500 Bar, Thermal Eng. **19**, 130–134 (1972), English translation of Teploenergetika

3.122 F.A. Guevara, B.B. McInteer, W.E. Wageman: High temperature viscosity ratios for hydrogen, helium, argon and nitrogen, Phys. Fluids **12**, 2493–2505 (1969)

3.123 T. Ejima, K. Shimakage, Y. Sato, H. Okuda, N. Kumada, A. Ishigaki: Viscosity measurements of alkali chlorides with capillary viscometers, Nihon-Kagaku-Kaishi **6**, 961–968 (1982)

3.124 J.D. Isdale: *Viscosity of simple liquids including measurement and prediction at elevated pressure*. Ph.D. Thesis (University of Strathclyde, Strathclyde 1976)

3.125 J.B. Irving: *Viscosity measurements at pressures up to 14000 bar using an automatic falling cylinder viscometer*. Ph.D. Thesis (University of Glasgow, Glasgow 1980)

3.126 H. Faxen: Die Bewegung einer starren Kugel längs der Achse eines mit zäher Flüssigkeit gefüllten Rohres, Arkiv. Matematik. Astron. Fys. **17**, 1–28 (1923), in German

3.127 M.J.C. Flude, J.E. Daborn: Viscosity measurements by means of falling spheres compared with capillary viscometers, J. Phys. E **15**, 1312–1321 (1982)

3.128 N.F. Glen: *Viscosity coefficient measurement at elevated pressure*. Ph.D. Thesis (University of Glasgow, Glasgow 1983)

3.129 J. Kestin, M. Sokolov, W.A. Wakeham: Viscosity of liquid water in range 8 to 150 °C, J. Phys. Chem. Ref. Data **7**, 941–948 (1978)

3.130 M.J. Assael, J.H. Dymond, M. Papadaki, P.M. Patterson: Correlation and prediction of dense fluid transport coefficients – I. n-Alkanes, Int. J. Thermophys. **13**, 269–281 (1992)

3.131 M.J. Assael, J.H. Dymond, P.M. Patterson: Correlation and prediction of dense fluid transport coefficients – V. Aromatic hydrocarbons, Int. J. Thermophys. **13**, 895–905 (1992)

3.132 M.J. Assael, J.H. Dymond, S.K. Polimatidou: Correlation and prediction of dense fluid transport coefficients – VI. n-Alcohols, Int. J. Thermophys. **15**, 189–201 (1994)

3.133 M.J. Assael, J.H. Dymond, S.K. Polimatidou: Correlation and prediction of dense fluid transport coefficients – VII. Refrigerants, Int. J. Thermophys. **16**, 761–772 (1995)

3.134 R.B. Bird, W.E. Stewart, E.N. Lightfoot: *Transport Phenomena* (Wiley, New York 1960)

3.135 G.C. Maitland, M. Rigby, E.B. Smith, W.A. Wakeham: *Intermolecular Forces: Their Origin and Determination* (Clarendon, Oxford 1981)

3.136 J.J. Healy, J.J. de Groot, J. Kestin: The theory of the transient hot-wire method for measuring the thermal conductivity, Physica **82C**, 392–408 (1976)

3.137 M.J. Assael, L. Karagiannidis, N. Malamataris, W.A. Wakeham: The transient hot-wire technique: A numerical approach, Int. J. Thermophys. **19**, 379–389 (1998)

3.138 B. Stalhane, S. Pyk: New method of measuring thermal conductivity coefficients, Teknisk Tidskrift **61**, 389–397 (1931)

3.139 M.J. Assael, L. Karagiannidis, W.A. Wakeham: Measurements of the thermal conductivity of R11 and R12 in the temperature range 250–340 K at pressures up to 30 MPa, Int. J. Thermophys. **13**, 735–751 (1992)

3.140 M.J. Assael, K. Gialou: A transient hot-wire instrument for the measurement of the thermal conductivity of solids up to 590 K, Int. J. Thermophys. **24**, 667–675 (2003)

3.141 A.M.F. Palavra, W.A. Wakeham, M. Zalaf: Thermal conductivity of n-pentane in the temperature range 306–360 K and at pressures up to 0.50 GPa, Int. J. Thermophys. **8**, 305–315 (1987)

3.142 U.V. Mardolcar, J.M.N.A. Fareleira, C.A. Nieto de Castro, W.A. Wakeham: Measurements of the thermal conductivity of argon: a test of the accuracy of the transient hot wire instrument, High Temp.-High Press. **17**, 469–488 (1985)

3.143 Y. Nagasaka, A. Nagashima: Absolute measurements of the thermal conductivity of electrically conducting liquids by the transient hot wire method, J. Phys. E **14**, 1435–1440 (1981)

3.144 R.A. Perkins, A. Laesecke, C.A. Nieto de Castro: Polarized transient hot-wire thermal conductivity measurements, Fluid Phase Equilibr. **80**, 275–285 (1992)

3.145 H. Becker, U. Grigull: Interferometry of transparent phase objects, especially with a high interference fringe density, illustrated by an example from heat transfer, Wärme Stoffübertrag. **10**, 233–244 (1977)

3.146 W.A. Wakeham, J.P.M. Trusler, R.J. de Bruijn, R.J.J. van Diest, T.D. Karapantsios, A.C. Michels: Heat transport near the critical region of a pure fluid in microgravity. In: *Proceedings 2-nd European Symposium Fluids in Space, Naples*, ed. by A. Viviant (Giannini & Figli Publisher, Naples 1996) pp. 29–36

3.147 Sh.H. Jawad: *Thermal conductivity of polyatomic gases*. Ph.D. Thesis (London University, London 1998)

3.148 B. Le Neindre: *Contribution a l'etude experimentale de la conductivite thermique de quelques fluides a haute temperature et a haute pression*. Ph.D. Thesis (Paris University, Paris 1969)

3.149 R. Tufeu: *Etude experimentale en fonction de la temperature et de la pression de la conductivite thermique de l' ensemble des gaz rares et des melanges helium-argon*. Ph.D. Thesis (Paris University, Paris 1971)

3.150 B. Le Neindre, Y. Garrabos, R. Tufeu: Thermal conductivity in supercritical fluids, Ber. Bunsenges. Phys. Chem. **88**, 916–920 (1984)

3.151 A. Michels, J.V. Sengers, P.S. van der Gulik: The thermal conductivity of carbon dioxide in the critical region – II. Measurements and conclusions, Physica **28**, 1216–1237 (1962)

3.152 N.B. Vargaftik, Y.V. Smirnova: On the temperature dependancy of water vapour heat conductivity, Zh. Tekh. Fiz. **26**, 1221–1231 (1956), in Russian

3.153 R.G. Vines: Measurement of thermal conduction of gases at high temperature, J. Heat Transfer **82C**, 48–52 (1960)

3.154 H. Ziebland, J.T.A. Burton: The thermal conductivity of nitrogen and argon in the liquid and gaseous states, Brit. J. Appl. Phys. **9**, 52–59 (1958)

3.155 J. Yata, T. Minamiyama, K. Kajimoto: Thermal conductivity of water and steam at high temperatures and pressures. I. Experimental apparatus and results with water up to 200 °C, B. JSME **22**, 1220–1226 (1979)

3.156 B.J. Bailey, K. Kellner: The thermal conductivity of argon near the critical point, Brit. J. Appl. Phys. **18**, 1645–1647 (1967)

3.157 H.M. Roder, D.E. Diller: Thermal conductivity of gaseous and liquid hydrogen, J. Chem. Phys. **52**, 5928–5970 (1970)

3.158 J.B. Ubbink, W.J. de Haas: Thermal conductivity of gaseous He, Physica **10**, 465–470 (1943)

3.159 G. Grenier: Thermal conductivity of liquid helium, Phys. Rev. **83**, 598–603 (1951)

3.160 R.L. Nuttall, D.C. Ginnings: Thermal conductivity of nitrogen from 50 degrees to 500 degrees and 1 to 100 atmospheres, J. Res. Nat. Bur. Standards **58**, 271–2788 (1957), Research Paper No. 2760

3.161 K.I. Amirkhanov, A.P. Adamov: Heat transfer of steam in near critical and subscritical state; experimental data on heat transfer of steam in range of pressures between 200–400 kg/sq cm and at temperatures of 350–450 °C obtained by use of method of plane horizontal layer, Teplogenertika **10**, 69–72 (1963)

3.162 A. Michels, J.V. Sengers, L.J.M. van de Klundert: The thermal conductivity of argon at elevated gas densities, Physica **29**, 149–160 (1963)

3.163 L.L. Fabelinskii: *Molecular Scattering of Light* (Plenum, New York 1968)

3.164 K. Kraft, M. Matos Lopes, A. Leipertz: Thermal diffusivity and thermal conductivity of toluene by photon correlation spectroscopy: A test of the accuracy of the method, Int. J. Thermophys. **16**, 423–432 (1995)

3.165 P.N. Pusey: *Photon Correlation Spectroscopy and Velocimetry* (Plenum, New York 1976)

3.166 B. Saleh: *Photoelectron Statistics with Applications to Spectroscopy and Optical Communication* (Springer, Berlin, Heidelberg 1978)

3.167 W. Wakeham, A. Nagashima, J.V. Sengers (Eds.): *Experimental Thermodynamics, Measurement of the Transport Properties of Fluids*, Vol. III (Blackwell, Oxford 1991)

3.168 H.J. Eichler, G. Salje, H. Stahl: Thermal diffusion measurements using spatially periodic temperature distributions induced by laser light, J. Appl. Phys. **44**, 5383–5388 (1973)

3.169 D.W. Pohl, S.E. Schwartz, V. Irniger: Forced Rayleigh scattering, Phys. Rev. Lett. **31**, 32–35 (1973)

3.170 T. Hatakeyama, Y. Nagasaka, A. Nagashima: Measurement of the thermal diffusivity of liquids by the forced Rayleigh scattering method, Proc. ASME-JSME Thermal Engineering Joint Conference, Honolulu, Vol. 1 (USA 1987) pp. 311–318

3.171 Y. Nagasaka, A. Nagashima: Measurement of the thermal diffusivity of molten KCl up to 1000 °C by the forced Rayleigh scattering method, Int. J. Thermophys. **9**, 923–932 (1988)

3.172 Y. Nagasaka, T. Hatakeyama, M. Okuda, A. Nagashima: Measurement of the thermal diffusivity of liquids by the forced Rayleigh scattering method. Theory and experiment, Rev. Sci. Instrum. **59**, 1156–1168 (1988)

3.173 Y. Nagasaka, N. Nakazawa, A. Nagashima: Experimental determination of the thermal diffusivity of molten alkali halides by the forced Rayleigh scattering method – I. Molten, LiCl, NaCl, KCl, RbCl, and CsCl, Int. J. Thermophys. **13**, 555–573 (1992)

3.174 C.A. Nieto de Castro, S.F.Y. Li, A. Nagashima, R.D. Trengove, W.A. Wakeham: Standard reference data for the thermal conductivity of liquids, J. Phys. Chem. Ref. Data **15**, 1073–1086 (1986)

3.175 J. Kestin, R. Paul, A.A. Clifford, W.A. Wakeham: Absolute determination of the thermal conductivity of the noble gases at room temperature up to 35 MPa, Physica **100A**, 349–369 (1980)

3.176 M.J. Assael, J.H. Dymond, M. Papadaki, P.M. Patterson: Correlation and prediction of dense fluid transport coefficients – I. n-Alkanes, Int. J. Thermophys. **13**, 269–281 (1992)

3.177 M.J. Assael, J.H. Dymond, P.M. Patterson: Correlation and prediction of dense fluid transport coefficients – V. Aromatic hydrocarbons, Int. J. Thermophys. **13**, 895–905 (1992)

3.178 M.J. Assael, J.H. Dymond, S.K. Polimatidou: Correlation and prediction of dense fluid transport coefficients – VI. n-Alcohols, Int. J. Thermophys. **15**, 189–201 (1994)

3.179 M.J. Assael, J.H. Dymond, S.K. Polimatidou: Correlation and prediction of dense fluid transport coefficients – VII. Refrigerants, Int. J. Thermophys. **16**, 761–772 (1995)

3.180 R.H. Stokes: An improved diaphragm-cell for diffusion studies, and some tests of the method, J. Am. Chem. Soc. **72**, 763–767 (1950)

3.181 J. Kestin, W.A. Wakeham: *CINDAS Data Series on Material Properties. Transport Properties of Fluids, Thermal Conductivity, Viscosity and Diffusion Coefficient*, Vol. I (Hemisphere, New York 1988)

3.182 R. Mills, L.A. Woolf: *The Diaphragm Cell* (Australian National Univ., Canberra 1968)

3.183 L.A. Woolf, J.F. Tilley: Revised values of integral diffusion coefficients of potassium chloride solutions for the calibration of diaphragm cells, J. Phys. Chem. **71**, 1962–1963 (1961)

3.184 G.I. Taylor: Dispersion of soluble matter in solvent flowing slowly through a tube, Proc. R. Soc. London Ser. A **219**, 186–203 (1953)

3.185 G.I. Taylor: Diffusion and mass transport in tubes, Proc. R. Soc. London Ser. B **67**, 857–869 (1954)

3.186 R. Aris: On the dispersion of a solute in a fluid flowing through a tube, Proc. R. Soc. London Ser. A **235**, 67–77 (1956)

3.187 A. Alizadeh, C.A. Nieto de Castro, W.A. Wakeham: The theory of the Taylor dispersion technique for liquid diffusivity measurements, Int. J. Thermophys. **1**, 243–284 (1980)

3.188 W.A. Wakeham: Diffusion coefficient measurements by the chromatographic method, Discuss. Faraday Soc. **15**, 141–150 (1980)

3.189 J.C. Giddings, S.L. Seager: Rapid determination of gaseous diffusion coefficients by means of gas chromatography apparatus, J. Chem. Phys. **33**, 1579–1582 (1960)

3.190 A.C. Ouano: Difusion in liquid systems – I. Simple and fast method of measuring diffusion constant, Ind. Eng. Chem. Fundam. **11**, 268–277 (1972)

3.191 K.C. Pratt, D.H. Slater, W.A. Wakeham: Rapid method for determination of diffusion coefficients of gases in liquids, Chem. Eng. Sci. **28**, 1901–1903 (1973)

3.192 K.C. Pratt, W.A. Wakeham: Mutual diffusion coefficient of ethanol-water mixtures. Determination by a rapid new method, Proc. R. Soc. London Ser. A **336**, 393–406 (1974)

3.193 H.S. Carslaw, J.C. Jaeger: *Conduction of Heat in Solids* (Oxford Univ. Press, New York 1947)

3.194 H.J.V. Tyrrell, K.R. Harris: *Diffusion in Liquids* (Butterworths, London 1984)

3.195 L.O. Sundelöf: Determination of distributions of molecular size parameters by a convolution procedure. A computational and experimental study with special reference to diffusion at low concentrations, Arkiv Kemi **25**, 1–65 (1966)

3.196 L.G. Longsworth: Diffusion measurements at 1° of aqueous solutions of amino acids peptides and sugars, J. Am. Chem. Soc. **74**, 4155–4159 (1952)

3.197 H. Svensson: On the use of Rayleigh-Phulpot-Cook interference fringes for the measurement of diffusion coefficients, Acta Chem. Scand. **5**, 72–84 (1951)

3.198 S. Claesson, H. Matsuda, L.O. Sundelöf: New boundary formation technique for free-diffusion in liquids, Chem. Scripta **6**, 94–96 (1974)

3.199 F.R. McLarnon, R.H. Muller, C.W. Tobias: Reflection effects in interferometry, Appl. Opt. **14**, 2468–2472 (1975)

3.200 G. Gouy: Sur de nouvelles franges d'interfirence, Comptes Rend. **90**, 307–309 (1880)

3.201 G. Kegeles, L.J. Gosting: The theory of an interference method for the study of diffusion, J. Am. Chem. Soc. **69**, 2516–2523 (1947)

3.202 C.A. Coulson, J.T. Cox, A.G. Ogston, J.S.L. Philpot: A rapid method for determing diffusion constants in solutions, Proc. R. Soc. London Ser. A **192**, 382–402 (1948)

3.203 L.J. Gosting, L. Onsager: A general theory for the Gouy diffusion method, J. Am. Chem. Soc. **74**, 6066–6074 (1952)

3.204 I.R. Shankland, P.J. Dunlop: Pressure dependence of the mutual diffusion coefficients of the binary systems N_2+Ar, N_2+O_2, O_2+Ar, and $Ar+Kr$ at 300 and 323 K, Physica **100A**, 64–84 (1980), in German

3.205 T. Graham: On the molecular mobility of gases, Phil. Trans. R. Soc. **153**, 385–405 (1863)

3.206 T.R. Marrero, E.A. Mason: Gaseous diffusion coefficients, J. Phys. Chem. Ref. Data **1**, 3–110 (1972)

3.207 P.J. Dunlop, K.R. Harris, D.J. Young: Experimental methods for studying diffusion in gases, liquids, and solids. In: *Physical Methods of Chemistry*, Vol. VI, ed. by B.W. Rossiter, R.C. Baetzold (Wiley-Interscience, New York 1992)

3.208 J. Loschmidt: Experimental-Untersuchungen über die Diffusion von Gasen ohne poröse Scheidewunde, Sitzber. Mathem.-Natur. Clas. Kaiserl. Akad. Wiss. Wien **62**, 468–478 (1870)

3.209 J. Crank: *The Mathematics of Diffusion* (Clarendon, Oxford 1956)

3.210 W. Wakeham, A. Nagashima, J.V. Sengers (Eds.): *Experimental Thermodynamics, Measurement of the Transport Properties of Fluids*, Vol. III (Blackwell, Oxford 1991)

3.211 M.A. Yabsley, P.J. Dunlop: Study of 2-bulb method for measuring diffusion coefficients of binary gas mixtures, Physica **85A**, 160–164 (1976)

3.212 R.J.J. van Heijningen, A. Feberwee, A. van Costen, J.J.M. Beenakker: Determination of the diffusion coefficient of the system N_2-H_2 as a function of temperature and concentration, Physica **32**, 1649–1662 (1966)

3.213 W.L. Taylor, D. Cain: Temperature dependence of the mutual diffusion coefficients of He-Ar, Ne-Ar and Xe-Ar from 350 to 1300 K, J. Chem. Phys. **78**, 6220–6227 (1983)

3.214 W.A. Wakeham, D.H. Slater: Binary diffusion coefficients of homologous species in argon, J. Phys. B **7**, 297–306 (1974)

3.215 G.M. Schneider: Phase equilibria in fluid systems, Ber. Bunsenges. Phys. Chem. **88**, 841–848 (1984)

3.216 A. Wilsch, R. Feist, G.M. Schneider: Capacity ratios and diffusion coefficients of low-volatility organic compounds in supercritical carbon dioxide from supercritical chromatography, Fluid Phase Equilibr. **10**, 299–306 (1983)

3.217 G.C. Maitland, E.B. Smith: Critical reassessment of viscosity of eleven common gases, J. Chem. Eng. Data **17**, 150–167 (1972)

3.218 P.S. Arora, I.R. Shankland, T.N. Bell, M.A. Yabsley, P.J. Dunlop: Use of precise binary diffusion coefficients to calibrate 2-bulb cell instead of using standard end-correction for connecting tube, Rev. Sci. Instrum. **48**, 673–674 (1977)

3.219 J. Kestin, K. Knierim, E.A. Mason, B. Najafi, S.T. Ro, W.A. Wakeham: Equilibrium and transport properties of the noble gases and their mixtures at low density, J. Phys. Chem. Ref. Data **13**, 229–303 (1984)

3.220 S. Umezawa, A. Nagashima: Measurement of the diffusion coefficients of acetone, benzene and alkane in supercritical CO_2 by the Taylor dispersion method, J. Supercrit. Fluids **5**, 242–250 (1992)

3.221 M.A. Awan, J.H. Dymond: Transport properties of nonelectrolyte liquid mixtures – X. Limiting mutual diffusion coefficients of fluorinated benzenes in n-hexane, Int. J. Thermophys. **17**, 759–769 (1996)

3.222 P. Frick, S. Khripchenko, S. Denisov, D. Sokoloff, J.F. Pinto: Effective magnetic permeability of a turbulent fluid with macroferroparticles, Eur. Phys. J. B **25**, 299–402 (2002)

3.223 J. de Vicente, G. Bossis, S. Lacis, M. Guyot: Permeability measurements in cobalt ferrite and carbon iron powders and suspensions, J. Magn. Magn. Mater. **251**, 100–108 (2002)

3.224 I. Hrianca, I. Malaescu: The rf magnetic permeability of statically magnetized ferrofluids, J. Magn. Magn. Mater. **150**, 131–136 (1995)

3.225 J.S. Brooks, G.O. Zimmerman, R. Meservey: Apparatus for the measurement of the magnetic susceptibility of liquids in high magnetic fields, Rev. Sci. Instrum. **54**(9), 1234–1237 (1983)

3.226 M.B. Stout: *Basic Electrical Measurements*, 2nd edn. (Prentice-Hall, Englewood Cliffs 1960)

3.227 D.R. Lide (Ed.): Diamagnetic susceptibility of selected organic compounds, Handbook of Chemistry and Physics. In: *Handbook of Chemistry and Physics*, 84th edn. (CRC, Boca Raton 2004)

3.228 University of Surrey, Department of Electronics and Physical Sciences (2002) http://www.ee.surrey.ac.uk/workshop/advice/coils/mu/#mur

3.229 M. Matsuura, S. Takeuchi: Thermoelectric power and magnetic susceptibility of liquid silver alloys, Trans. TIM **17**, 707–716 (1976)

3.230 S. Takeda, S. Tamaki: Magnetic susceptibilities of liquid Cd-Na and In-Na alloys, J. Phys. Soc. Jpn. **58**(4), 1484–1485 (1989)

3.231 D.R. Lide (Ed.): Magnetic susceptibility of elements and inorganic compounds. In: *Handbook of Chemistry and Physics*, 84th edn. (CRC, Boca Raton 2004)

3.232 P. Terzieff, X. Tsuchiya: Magnetic susceptibility of liquid Ag-Ge, Ag-Sn and Ag-Pb, J. Phys. **13**, 3573–3582 (2001)

3.233 D.P. Fernandez, Y. Mulev, A.R.H. Goodwin, J.M.H.L. Sengers: A database for the static dielectric constant of water and steam, J. Phys. Chem. Ref. Data **24**, 33–69 (1995)

3.234 U. Kaatze: Complex permittivity of water as a function of frequency and temperature, J. Chem. Eng. Data **34**, 371–374 (1989)

3.235 P. Turq, J. Barthel, M. Chemla: *Transport, Relaxation and Kinetic Processes in Electrolyte Solutions* (Springer, Berlin, Heidelberg 1992)

3.236 J. O'M. Bockris, A.K.N. Reddy: *Modern Electrochemistry*, Vol. 1 (Plenum, New York 1970)

3.237 C. Wohlfahrt: *Static dielectric constants of pure liquids and binary liquid mixtures*, Landolt-Börnstein New Ser., Vol. IV/6 (Springer, Berlin, Heidelberg 1991)

3.238 K.N. Marsh (Ed.): *Recommended Reference Materials for the Realization of Physicochemical Properties* (Blackwell, Oxford 1987)

3.239 C. Wohlfart: Permittivity (dielectric constant) of liquids. In: *Handbook of Chemistry and Physics*, 84th edn., ed. by D.R. Lide (CRC, Boca Raton 2004)

3.240 D.A. Saville: Electrohydrodynamics, The Taylor Melcher leaky dielectric model, Annu. Rev. Fluid Mech. **29**, 27–64 (1997)

3.241 R.P. Feynman, R.B. Leighton, M. Sands: *The Feynman Lectures on Physics*, Vol. 2 (Addison Wesley, Palo Alto 1964)

3.242 D. Braun, A.J. Heeger: Visible light emission from semiconducting polymer diodes, Appl. Phys. Lett. **52**(18), 1982–1984 (1991)

3.243 G. Gustafsson, Y. Cao, G.M. Treacy, F. Klavetter, N. Colaneri, A.J. Heeger: Flexible light-emitting-diodes made from soluble conducting polymers, Nature **357**, 477–479 (1992)

3.244 M.N. Simon, B.Y. Lin, H.S. Lee, T.A. Skotheim, J.S. Wall: Conducting polymer films as EM substrates, Proc. XIIth Int. Congress for Electron Microscopy, Vol. 1 (San Francisco Press, San Francisco 1990) pp. 290–291

3.245 A.C. MacDiarmid, W.E. Jones, I.D. Norris, J. Gao, A.T. Johnson, N.J. Pinto, J. Hone, B. Han, F.K. Ko, H. Okuzaki, M. Llaguno: Electrostatically-generated nanofibers of electronic polymers, Synth. Met. **119**, 27–30 (2001)

3.246 R.A. Robinson, R.H. Stokes: *Electrolyte Solutions*, 2nd edn. (Butterworths, London 1959)

3.247 A. Newman: Ion transport in roots: measurement of fluxes using ion-selective microelectrodes to characterize transporter function, Plant Cell Environment **24**, 1–14 (2001)

3.248 I. Adamczewski: *Ionization, Conductivity and Breakdown in Dielectric Liquids* (Taylor Francis, London 1969)

3.249 T.J. Gallagher: *Simple Dielectric Liquids, Mobility, Conduction, and Breakdown* (Clarendon, Oxford 1975)

3.250 L. Meyer, F. Reif: Mobility of He ions in liquid helium, Phys. Rev. **110**, 279–280 (1958)

3.251 H. Schynders, L. Meyer, S.A. Rice: Electron mobilities in liquid argon, Phys. Rev. Lett. **15**, 187–180 (1965)

3.252 H. Schynders, L. Meyer, S.A. Rice: Electron drift velocities in liquefied argon and krypton at low electric field strengths, Phys. Rev. **150**, 127–145 (1966)

3.253 J.A. Dean: *Lange's Handbook of Chemistry*, 15th edn. (McGraw-Hill, New York 1999)

3.254 P. Vanysek: Ionic conductivity and diffusion at infinite dilution. In: *Handbook of Chemistry and Physics*, 83rd edn., ed. by D.R. Lide (CRC, Boca Raton 2002)

3.255 R.A. Horne: *Water and Aqueous Solutions, Structure, Thermodynamics and Transport Processes* (Wiley-Interscience, New York 1972)

3.256 R. Heinrich, S. Bonisch, D. Pommerenke, R. Jobava, W. Kalkner: Broadband measurement of the conductivity and the permittivity of semiconducting materials in high voltage XLPE cables, 8. Int. Conference on Dielectric Materials, Measurements and Applications, Edinburgh, IEE Conf. Publ. **473**, 212–217 (2000)

3.257 D.A. Robinson, C.M.K. Gardner, J. Evans, J.D. Cooper, M.G. Hodnett, J.P. Bell: The dielectric calibration of capacitance probes for soil hydrology using an oscillation frequency response model, Hydrol. Earth Syst. Sci. **2**(1), 111–120 (1998)

3.258 J.B. Hasted: *Aqueous Dielectrics* (Chapman Hall, London 1973)

3.259 R.H. Cole, P. Winsor IV: Fourier transform dielectric spectroscopy. In: *Fourier, Hadamard and Hilbert Transforms in Chemistry*, ed. by A.G. Marshall (Plenum, New York 1982) pp. 183–206

3.260 J.B. Hasted, D.M. Ritson, C.H. Collie: Dielectric properties of aqueous ionic solutions. Part I and II, J. Chem. Phys. **16**(1), 1–11 (1948)

3.261 R.J. Hodgkinson, N.J. Eastman, J. Favaron, H. Owen: Contact and contactless electrical conductivity measurements on the liquid semiconductor systems In-Te and Sb-Tc, J. Phys. C **15**, 4147–4153 (1982)

3.262 P.C. Fannin, S.W. Charles, D. Vincent, A.T. Giannitsis: Measurement of the high-frequency complex permittivity and conductivity of magnetic fluids, J. Magn. Magn. Mater. **252**, 80–82 (2002)

4. Pressure Measurement Systems

Measurements of the steady pressure in a fluid flow may be required to determine other thermodynamic properties, to determine forces on a body due to the pressure distribution over it, or in order to determine the dynamic head and flow velocity (for further details on the latter see Sect. 5.1. Pressure is a scalar representation of molecular activity, a measure of the nondirectional molecular motions. Thus it must, by definition, be measured by a device at rest relative to the flow. Whilst the common practice in the fluid mechanics community is to denote the pressure as *static* (as opposed to the coordinate-dependent *total pressure*, Sect. 3.1), this terminology introduces a fundamental redundancy.

In practice, pressure is commonly measured both at walls and in the freestream using the types of measurement device shown in Fig. 4.1 connected to a transducer of suitable sensitivity and range. The orifice of a small wall tapping represents a simple way to obtain the pressure impressed on the wall by the external flow. So-called static pressure tubes approximate the local fluid pressure in the freestream if the disturbance presented to the flow can either be accounted for or is not large to begin with. However this can only ever be strictly true for steady laminar flow due to the normal velocity component introduced when a flow becomes turbulent. Measurement of freestream pressure is one of the hardest challenges in fluid mechanics.

This chapter addresses measurement of pressure using wall tappings (Sect. 4.1) and static pressure tubes (Sect. 4.2), and especially errors due to the intrusive flow presence of real, finite-sized devices and calibrations to correct for these. *Bryer* and *Pankhurst* [4.1] and *Chue* [4.2] provided seminal monographs on the general topic of pressure probes in 1971 and 1975, respectively, which give detailed descriptions of measurement devices, coverage of the background to the various corrections and a survey of older data. The topic is covered here more concisely, with a view to practical use by the engineer, and with reference to modern literature. The reader is referred to *Bryer* and *Pankhurst* [4.1] and *Chue* [4.2] for further details on most sections.

In more recent years a further method for obtaining pressure on the surface of a wind tunnel model has been developed, based on pressure sensitive paints (PSP). The introduction of PSP provides a method to measure the pressure on the surface of a model directly without the transducers and tubing associated with conventional means. A paint, the luminescence of which is dependent on air pressure, is applied to the surface of a wind tunnel model and the pressure distribution is obtained from the images produced by proper illumination. In Sect. 4.4 the basics of PSP are discussed and further subsections address in detail different paints, paint application procedures, imaging systems and image processing. In discussing the achievable accuracy of PSP techniques, both the spatial and temporal resolution is examined. The thermal sensitivity of the paint dye is introduced and this is closely linked to temperature-sensitive paints (TSP), as discussed in Chap. 7, Sect. 7.4.

4.1	**Measurement of Pressure with Wall Tappings**...............	180
	4.1.1 Cavity Shape, Connection and Alignment...............	181
	4.1.2 Finite-Area Effects...............	181
	4.1.3 Effect of Compressibility...............	182
	4.1.4 Effect of Finite Depth...............	183
	4.1.5 Condition of the Orifice Edge.........	184
	4.1.6 Correction for Distance from Measuring Point...............	184
4.2	**Measurement of Pressure with Static Tubes**...............	185
	4.2.1 Effect of Geometry...............	185
	4.2.2 Effect of Hole Location...............	186
	4.2.3 Directional Sensitivity...............	186
	4.2.4 Effect of Turbulence...............	186

4.3 Hardware and Other Considerations ... 187
4.4 Pressure-Sensitive Paint (PSP) ... 188
4.4.1 Basics of PSP ... 188
4.4.2 Paints ... 190
4.4.3 Imaging Systems ... 200
4.4.4 Processing ... 204
4.4.5 Applications ... 206
4.4.6 Concluding Remarks ... 208
References ... 209

4.1 Measurement of Pressure with Wall Tappings

A wall tapping, or piezometer, is a simple means of obtaining pressure at the wall, p_w, in a wall-bounded flow, but one that requires some subtlety in many flows. For example, accurate determination of the pressure distribution on a scale model in a small-cross-section high-Mach-number wind tunnel may be complicated by changes in the flow field due to the diameter of the tappings (dictated by manufacturing or response time constraints, Sect. 4.1.4), which may be large compared to the boundary-layer thickness (which changes with the streamwise location).

The finite size of tappings that can be reliably and smoothly manufactured may be sufficiently large to induce an error in the measured pressure, such that $p_{wm} = p_w + \Delta p_w$. Dimensional analysis shows that, for a pressure tapping of a given geometry in a zero-pressure-gradient flow (or where the tapping diameter is small compared with the scale of pressure variation), Δp nondimensionalised with the wall shear stress τ_w is a function of the following variables:

$$\Pi = \frac{\Delta p}{\tau_w}$$
$$= f\left(\frac{d_s u_\tau}{\nu}, \frac{d_s}{D}, M, \frac{l_s}{d_s}, \frac{d_c}{d_s}, \frac{\epsilon}{d_s}\right) \quad (4.1)$$

and, of course, the (laminar or turbulent) condition of the wall-bounded flow. Here d_s is the tapping (orifice) diameter, $u_\tau = \sqrt{\tau_w/\rho}$ is the friction velocity, D is the flow lengthscale, M is the Mach number (the ratio of local velocity to the local speed of sound), l_s is the depth of the orifice, d_c is the diameter of the cavity behind the orifice, ϵ is the root-mean-square height of burrs on the edge of the tapping orifice, ρ is the fluid density and ν is the kinematic viscosity (Fig. 4.1a). The true pressure at the wall, p_w, is then given by

$$p_w = p_{mw} - \Pi \tau_w . \quad (4.2)$$

The complexity of the flow local to the tapping means that analytical or numerical solutions for the pressure error are at present available only for very low Reynolds numbers and/or two-dimensional geometries, e.g. [4.3]. The majority of the experimental data is for a turbulent flow over the orifice, and comparisons of experiments reveal significant scatter between results, probably due to the difficult nature of the experiments (a pressure error that is of the same order as the experimental uncertainty) and extrapolation of the *true* pressure for quantification of the error. In what follows, we tackle the effect of each of the nondimensional parameters in (4.1) on the measurement error.

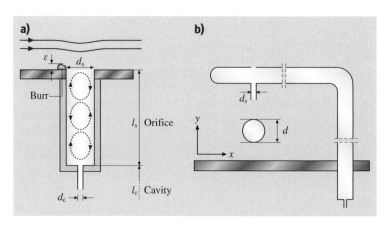

Fig. 4.1a,b Determination of steady pressure: **(a)** wall tapping; **(b)** static tube

4.1.1 Cavity Shape, Connection and Alignment

First we consider the effect of orifice shape. While the traditional geometry employed for measurement of pressure at the wall is the straight-edged, circular cylinder drilled perpendicular to the plane of the wall (as in Fig. 4.1a), researchers have experimented with several other geometries, including slot-type (noncircular) [4.2], angled [4.4] and radiused or chamfered-edge circular tappings ([4.4], Sect. 4.1.5). In addition, various connector geometries from the tapping to the transducer have been designed, dictated by experimental constraints, manufacturing (i.e. drilling) techniques and even with the goal of reducing the pressure error. The efforts of *Allen* and *Hooper* [4.5] in seeking to define a commercial standard for tapping geometry through the investigation of geometries including recessed, conical, countersunk and protruding orifice tubes were essentially inconclusive, although they did yield several useful scaling trends.

The difference between the pressure measured by cylindrical and slot-type tappings at supersonic speeds is of the order of $\pm 1\%$ (Douglas Aircraft company tests reported by *Chue* [4.2]), with implications for appropriate correction of measurements from tappings that are drilled out-of-round.

The directional sensitivity of tappings and effect of alignment of the tapping with the wall normal direction was investigated by *Rayle* [4.4], who found a zero error when the tapping centreline was angled $30°$ downstream with respect to the outward wall normal, increasing/decreasing as the centreline was rotated upstream/downstream, respectively.

The analysis of *Ducruet* and *Dyment* [4.6] identifies an effect of radius of curvature r_b of the wall containing the orifice, of obvious importance when $d/r_b \sim \mathcal{O}(1)$. While this suggests that there will be a difference in the errors induced by tappings in pipe, flat plate boundary layers/channels and bodies of revolution, etc., this effect has not been identified in the majority of the experimental literature and results from flat and curved surfaces have been used together in an attempt to obtain general scaling laws.

The below-surface geometry of the tapping (i.e. the cavity geometry and ratio d_s/d_c) will also have an effect. Chue suggests that differences between the experimental results of *Livesey* et al. [4.7] (a *pinhole* design, with $d_s/d_c = 1/14$ and $l_s \ll l_c$) and *Shaw* [4.8] (cylindrical tappings with $d_s/d_c = 1/2$ and l_s large) must lie in the wall fittings, all other parameters being equal. This is discussed further in Sect. 4.1.4.

In summary, while the current knowledge concerning straight-edged, cylindrical tappings will be reviewed here, use of other geometries will most likely require in situ calibration over the expected range of flow conditions to obtain wall pressure measurements with high confidence.

4.1.2 Finite-Area Effects

Given a deep tapping with smooth edges, the finite size (diameter) of the tapping causes local curvature of the streamlines and a complicated system of vortices within the cavity (and potentially a stagnation point on the downstream wall), as sketched in Fig. 4.1a and observed by *Miyadzu* [4.9] and *Ray* [4.10]. The effect of a finite tapping diameter on the measured static pressure is expressed in the dependence of the pressure error on both $d_s^+ = d_s u_\tau/\nu$ and d_s/D, the ratios of tapping diameter to viscous scale and flow lengthscale, respectively:

$$\Pi = f\left(d_s^+, \frac{d_s}{D}\right). \tag{4.3}$$

Measurements of Π are extremely difficult to make due to the small magnitude of the pressure differences, the sensitivity to manufacturing tolerances, the need to isolate only either d_s^+ or d_s/D (especially hard in boundary layers where both the friction velocity and the dominant lengthscale, the displacement thickness δ^*, change with streamwise location) and the difficulty of establishing the *zero-error* condition (by extrapolation [4.8] or use of a flush surface pressure transducer [4.11]), where $\Pi \to 0$ as $d^+ \to 0$. Studies include those by *Miyadzu* [4.9], *Ray* [4.10], *Franklin* and *Wallace* [4.11] and *Ducruet* and *Dyment* [4.6] on the static hole error in a plane surface (i.e. beneath a flat plate turbulent boundary layer or in channel flow), *Fuhrmann* [4.12] in an axisymmetric turbulent boundary layer and *Allen* and *Hooper* [4.5], *Rayle* [4.4], *Shaw* [4.8] and *Livesey* et al. [4.7] in pipes. However the results are quite scattered due to variations in the exact geometries of the tappings under test.

For tappings that are *small* with respect to the flow lengthscale D, the results of *Shaw* [4.8], obtained in a pipe for $0.008 < d_s/D < 0.1$, $25 < d_s^+ < 800$ and $1.5 \leq l_s/d_s \leq 6$ ($M < 0.2$) had been used as the standard for the variation of pressure error with diameter for deep holes. However more recently *McKeon* and *Smits* [4.13] also explored the pressure error in pipe flow, extending the range of Reynolds numbers by changing d_s^+ without making d_s/D large ($0.0020 \leq d_s/D \leq 0.0184$,

$d_s^+ \leq 8000$) with $M \leq 0.07$ and $l_s/d_s = 4$ with effectively infinite connection. They found that *Shaw*'s [4.8] results masked the dependence of the error on d_s/D.

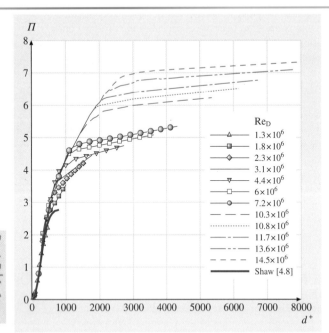

Fig. 4.2 Effect of hole size on pressure error, Π, in pipe flow (After *McKeon* and *Smits* [4.13]). Each curve corresponds to the error found from a set of wall tappings with $0.002 \leq d_s/D < 0.02$ at one pipe Reynolds number, i.e., the variation in local Reynolds number d_s^+, is achieved at each pipe Reynolds number by changing d_s/D

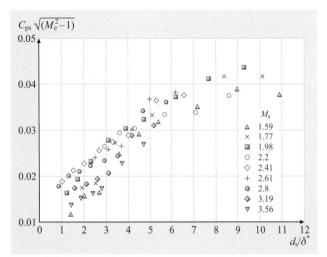

Fig. 4.3 Effect of compressibility on pressure corrections for large holes (After *Rainbird* [4.14])

Curve fits to their data are replicated in Fig. 4.2, showing the effect of increasing the pipe Reynolds number on the pressure error for the same set of different diameter tappings. While Π is a function of tapping Reynolds number alone alone for low d_s^+, the influence of d_s/D becomes increasingly important, even for tappings with small diameters. At the same d_s^+, the error for a small tapping exceeds that for a large one (although note that the pipe Reynolds number will be much larger for the small tapping).

The intuitively simple proposal of Livesey et al. to relate the pressure error to observed streamline deflection through an integral of the dynamic pressure at a distance proportional to d_s from the surface gives a curve similar to those in Fig. 4.2 (but not exactly the same).

For large tappings and high tapping Reynolds number, the data indicate [4.13, 14] that the pressure error tends to a function of d_s/D alone, hence plots of the pressure coefficient $C_{ps} = \Delta p/(1/2\rho U^2)$ versus d_s/D would be more appropriate. This has been addressed in the compressible, high-Reynolds-number regime, where experiments have often been performed with d_s/D orders of magnitude larger than in the incompressible case because of limitations on manufacturing smooth small orifice tappings and thin boundary layers.

In summary, the error introduced into the measurement of pressure at the wall by a tapping of finite diameter in turbulent flow is positive for deep holes and increases with the ratio of the hole diameter to the viscous scale, d_s^+ (for a given tapping diameter d_s/D), but decreases as the ratio to the flow lengthscale, d_s/D, increases (for a given d_s^+). Of course the error will be averaged out in the measurement of pressure gradient in steady internal flows since the flow lengthscale does not change; however for boundary-layer pressure gradients the effect may be significant. Note that Rainbird suggests that the error, Π, for large holes in laminar flow does not tend to a high-Reynolds-number asymptote and can become very large, $\mathcal{O}(50)$.

In a flow with a rapid spatial variation of pressure, there will be an additional effect of spatial averaging of the pressure measured by a tapping that is large compared with the lengthscale of pressure variation.

4.1.3 Effect of Compressibility

The effect of compressibility has not often been isolated experimentally, particularly due to the competing effects of putting small tapping diameters into small models while requiring rapid response times due to generally short run times in higher-Mach-number facilities.

Rainbird performed measurements with large tappings ($0.7 \leq d_s/\delta^* \leq 10.9$) normal to the axis of a cone under a zero-pressure-gradient boundary layer at Mach numbers external to the boundary layer $1.59 \leq M_e \leq 3.56$. When plotted in the form $C_{ps}/\sqrt{M_e^2-1}$ versus d_s/δ^* for high Reynolds numbers only ($d_w^+ > 1000$ based on wall conditions), as in Fig. 4.3, the data demonstrate a general trend of increasing error with decreasing Mach number and increasing d_s/δ^* (at least for the smaller tapppings). *Rainbird* [4.14] proposed that for $d_s/\delta^* > 10$ approximately, the calibration coefficient is given by

$$C_{ps} \approx \frac{0.04}{\sqrt{M_e^2-1}} \,. \tag{4.4}$$

Note that *Plentovich* [4.15] recorded the generation of Mach waves by finite-sized tappings on a National Advisory Committee for Aeronautics (NACA) $65_1\,613$ airfoil at $0.6 \leq M_\infty \leq 0.8$ and high chord Reynolds number (also inconclusive results concerning the efficiency of fitting a porous metal plug in the tapping flush with the orifice in reducing the measurement error). Finally, *Ducruet* [4.16] has suggested that the effect of Mach number is more important for laminar than turbulent boundary layers.

4.1.4 Effect of Finite Depth

The depth of the tapping cavity l_s is defined here as the distance between the plane of the wall and either the cavity behind the orifice or the connection to the transducer, with diameter d_c and depth l_c, as in Fig. 4.1a. This parameter impacts on the extent of the system of eddies that is set up within the tapping and has been shown to affect the magnitude of the pressure error in a complex fashion (that also involves the relative cavity width with respect to the orifice, d_c/d_s). As such, several authors have sought to eliminate the depth from the problem by defining a minimum *deep* tapping l_s/d_s ratio beyond which the error ceases to change. A consensus appears to be $l_s/d_s \approx 2$, e.g. [4.5, 8, 9].

For shallower tappings, *Ray* [4.10] investigated tappings with $0.1 \leq l_s/d_s \leq 1.75$ and varying connection diameter d_c in a low-aspect-ratio duct and proposed that the error could be represented as follows for $1.7 < d_s^+ < 31.6$:

$$\Pi = f\left(\frac{l_s}{d_s}\right)\sqrt{d_s^+} \tag{4.5}$$

with

$$f\left(\frac{l_s}{d_s}\right) = 0.25 \quad \text{for} \quad l_s/d_s = 1.75\,,$$
$$= 0.54 \quad \text{for} \quad l_s/d_s = 0.1\,. \tag{4.6}$$

Livesey et al. found a similar trend of increasing error with hole depth for tappings with a large d_c, and a deep hole limit of $l_s/d_s \approx 7.5$. For this d_c configuration and low l_s/d_s, the sign of the error changes with increasing d^+ (Fig. 4.4), as also noted by *Miyadzu* [4.9] for

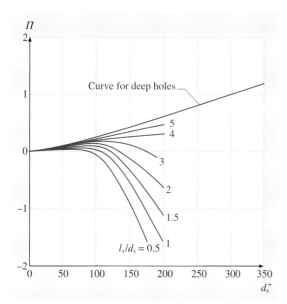

Fig. 4.4 Variation of pressure error Π with relative tapping depth l_s/d_s for pinhole-type tappings with $d_c = 14d_s$ (After *Livesey* [4.7])

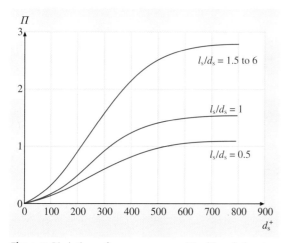

Fig. 4.5 Variation of pressure error Π with relative tapping depth l_s/d_s for narrow tappings with $d_c = 2d_s$ (After *Shaw* [4.8])

$l_s/d_s = 0.4$. Seeking to eliminate the error, Livesey et al. proposed that tappings with $l_s/d_s = 2$ and $d_c/d_s = 14$ have $\Pi \sim 0$ for $0 < d_s^+ < 300$, at least for their experimental setup.

Shaw's investigation also included shallower tappings, in the range $0.5 \le l_s/d_s \le 6$ (with $d_c/d_s = 2$). He found that the pressure error was independent of the depth-to-diameter ratio for $l_s/d_s > 1.5$, but that the error steadily decreased for shallower tappings. The errors for tappings with $d_s^+ < 750$ and depth $l_s/d_s = 0.5$, 1 and ≥ 1.5 are shown in Fig. 4.5 (although note that the effect of d_s/D detailed in Sect. 4.2.2 may also apply to these results).

4.1.5 Condition of the Orifice Edge

Two aspects of the condition of the orifice edge have been shown to affect the pressure error by altering the flow field in and around the tapping: the size of any burrs remaining after drilling [4.5, 8] and the magnitude of the radius on the edge of the hole (either imposed or caused by drilling, sanding or polishing [4.4, 5, 8]). A smooth upstream surface approaching the tapping is assumed.

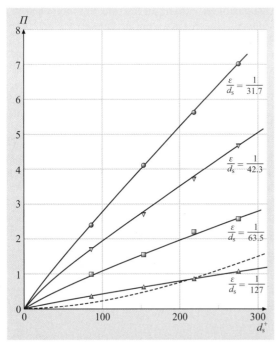

Fig. 4.6 The effect of the condition of the tapping orifice edge: the variation of the pressure error Π with relative burr height ϵ/d_s (After *Shaw* [4.8]). The dashed line is the small-tapping, zero-burr error

The effects of the form of the orifice edge on the pressure error can be summarised [4.4] as an increasing positive error with increasing edge radius and increasing negative error with increasing depth of chamfer (for a constant chamfer angle of 82°). Errors of up to -1% relative to the *dynamic* pressure measured in the straight-edged case have been observed (see also the pictorial summary by *Chue* [4.2]). *Allen* and *Hooper* [4.5] confirm this trend with edge radius and suggest that a radius of $d_s/4$ or smaller gives no detectable error. These authors also found errors of up to -0.4% of the mean velocity head for burrs hanging over/into the tapping (i.e. not protruding into the flow). Large negative errors were also observed for orifice tubes that protruded into the flow, increasing with the distance of protuberance from the pipe wall.

Shaw investigated the effect of burr heights in the range $1.57 \times 10^{-4} < \epsilon/d_s \le 3.4 \times 10^{-2}$ on deep tappings ($l_s/d_s = 4$), with $d_s/D = 0.03175$ by varying the feed rate of the drill while maintaining the same drill speed. His results, shown here in Fig. 4.6, demonstrate that the burr height can cause an error that exceeds the one due to due to finite area for small holes (a maximum value of $\Pi \approx 8$ was observed for the largest burr at the highest $d_s^+ \approx 275$, but the error appears to increase linearly with d_s^+), and hence extreme care must be taken with the drilling process. Note also that a large speck of dust resting on the edge of a tapping could cause a similar effect, hence there is reason to clean out tappings before making sensitive measurements. Several authors have discussed the best method to produce small tappings without burrs or radiused edges (or out-of-round, Sect. 4.1.1). *Franklin* and *Wallace* [4.11] experimented with drilling, lapping and reaming techniques for the orifice and polishing for the surrounding surface, settling on a complicated method that used tapered plugs to fill the hole and grinding to take off any burrs and finish the surface (these plugs subsequently reamed the tappings on removal). A Talysurf optical profiler was used to inspect the orifice edges. *Zagarola* [4.17] used a 32 *times* optical level to inspect orifice edges holes after the surrounding pipe surface has been honed and polished.

Figure 4.7a,b show magnifications of two tappings from the study of *McKeon* and *Smits* [4.13]. These were drilled from the polished measurement surface inwards, thus reducing the expected level of burring, however the tapping in (a) was discarded due to the obvious burring ($\epsilon/d_s = 0.63 \times 10^{-3}$) although the error due to the burring was estimated to be negligible, while (b) was accepted and used in the experiments.

4.1.6 Correction for Distance from Measuring Point

Obtaining the pressure in the internal part of a turbulent flow from a measurement at the wall requires correction for the difference between local and wall pressure introduced by turbulent fluctuations. For boundary layers at high Reynolds number with thickness δ,

$$\frac{p_\mathrm{w} - p}{\tau_\mathrm{w}} = \frac{\overline{v'^2}}{u_\tau^2} + \frac{1}{u_\tau^2} \int_0^y \left(U \frac{\partial V}{\partial x} + V \frac{\partial V}{\partial y} \right) \mathrm{d}y, \quad (4.7)$$

which will depend on the degree of spatial variation of the flow (and where is has been assumed that the streamwise gradient of Reynolds shear stress $\partial \overline{u'v'}/\partial x$ is negligibly small). In pipe flow with radius R,

$$\frac{p_\mathrm{w} - p}{\tau_\mathrm{w}} = \frac{\overline{v'^2}}{u_\tau^2} + \frac{1}{u_\tau^2} \int_r^R \left(\frac{\overline{w'^2} - \overline{v'^2}}{r} \right) \mathrm{d}r. \quad (4.8)$$

Fig. 4.7a,b Magnified images of wall tapping with $d_\mathrm{s} = 2.381$ mm. (**a**) Rejected due to burring ($\epsilon/d_\mathrm{s} = 0.63 \times 10^{-3}$) and (**b**) Accepted (After *McKeon* and *Smits* [4.13])

This latter pressure gradient has been measured indirectly by hot-wire anemometry (*Sandborn* [4.18]) and directly by static pressure tube (*Patterson* et al. [4.19]) for low to intermediate pipe Reynolds numbers. These measurements both found that the integral term on the right in (4.8) is negligible, and that the pressure difference $(p_\mathrm{w} - p)/\tau_\mathrm{w} \approx 1$ away from the wall and $(p_\mathrm{w} - p)/\tau_\mathrm{w} \to 0$ near the wall.

4.2 Measurement of Pressure with Static Tubes

A static tube such as that shown in Fig. 4.1b may be used to obtain the freestream pressure. This is a hard measurement to make, since strictly the probe should be at rest relative to the flow in order to measure the scalar quantity that is local pressure. In practice, rather than using a flying probe (with an unknown, complex trajectory in turbulent flow), a slender tube is aligned parallel to the incoming flow to minimise disturbance to the flow and the pressure is measured by several tappings in the same radial plane (to avoid the effect of local pressure gradients) downstream of the nose geometry. The measurement location is removed from both the probe tip and probe stem, and will be a point at which the surface pressure equals, or can be related by calibration to the undisturbed freestream value. Recall the pressure coefficient

$$C_\mathrm{ps} = \frac{p_\mathrm{m} - p}{\frac{1}{2}\rho U^2} \quad (4.9)$$

to account for the difference between the measured pressure and the true pressure at the measurement location. *Chue* [4.2] notes that the pressure measured by static tubes is essentially independent of Reynolds number in the range $3000 \leq \mathrm{Re_d} \leq 53\,000$, such that for most applications in which a static tube would be employed viscosity does not influence the reading.

Many of the results from the section on wall tappings carry over to errors introduced by the orifices in a static tube (which will be large compared with the boundary layer developing on the probe), but some additional effects should be considered. Note that, although much experimental work has been performed on Pitot-static probes (Sect. 5.1.3), negligible mutual interference between the measurements of total head and (static) pressure [4.1] means that some results can be applied equally well to static tubes.

Standard tip shapes include ellipsoidal and hemispherical designs for subsonic measurements. While cylindrical probes with rounded tips will primarily be considered here, other static pressure probes include a short-head design (where the static pressure holes are located at the downstream position corresponding exactly to the undisturbed static pressure), a spherical form, a static wedge and the static disc (which has the advantage of being largely insensitive to flow direction in the plane of the disc, but has been shown to demonstrate large variations in calibration function with Reynolds number and a nonlinear yaw calibration that should preclude its use in turbulent flow [4.20]).

4.2.1 Effect of Geometry

In general, probes that resemble a body of revolution with a rounded nose are used for subsonic flow, while an ogival or conical section may be preferable for supersonic flow. The latter design performs better by keeping the bow shock attached to the probe and minimising the disturbance to the flow. The static pressure is obtained from shock tables.

As the freestream Mach number approaches unity, the probe geometry will dictate the development of local shocks in the region of flow acceleration around the probe tip. With increasing Mach number, the shocks move back along the probe, passing over the tappings (at which point the measured static pressure will drop abruptly) followed by locally supersonic flow. Before the shock passes, an additional error may be introduced into the measured pressure due to the interaction of the shock with a laminar boundary layer. The effect may be reduced by fitting a trip ring to the nose of the probe to promote a turbulent boundary layer that is more resistant to the passage of the shock, or eliminated by using a fine-nosed probe for high-subsonic freestream flows ($M \leq 0.8$, [4.1]).

Allen and *Hooper* [4.5] investigated probes of differing geometries, concluding that the dependence of the behaviour on the mean velocity distribution and the angular sensitivity (of all designs except a sphere) implied that individual calibrations were required for all but the well-documented *standard* design of Fig. 4.1b.

Note that the Venturi effect caused by the blockage due to the probe body may also influence the pressure measured using a static tube in confined flows.

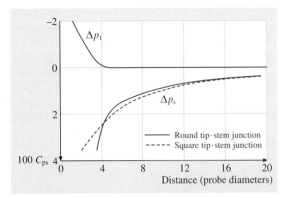

Fig. 4.8 Percentage error in the measured pressure, i.e., $100 C_{ps}$, due to the effect of tip (Δp_{tip}) and stem (Δp_{stem}) on pressure measurement by static tube (or Pitot static probe) (After *Pope* and *Harper* [4.21])

4.2.2 Effect of Hole Location

The downstream location of the holes relative to the probe tip and stem will influence the measured pressure if the separation from either is too small. Flow in the immediate vicinity of the tip undergoes a local acceleration and a tapping in this region will record a pressure that is significantly lower than the true value for several probe diameters d downstream of the tip (although note that this distance will have some dependence on tip geometry). Figure 4.8 shows a typical variation of the percentage error, $100 C_{ps}$, induced by the tip flow (labelled Δp_{tip}) in terms of the separation of the static hole and the tip in probe diameters. *Liepman* and *Roshko* [4.22] suggest that the holes should be $10-15d$ downstream of the nose to eliminate this error.

The effect of the stem on the pressure reading is in the opposite sense: approaching flow experiences a pressure rise due to local deceleration. This effect is more persistent (Δp_s as shown for two stem/probe geometries and tappings in the plane of the stem axis in Fig. 4.8).

Probes may be designed such that the tip and stem errors cancel, as in so-called Prandtl probes. *Pope* [4.21] suggests that for this type of probe the errors will cancel if the static holes are located $6d$ downstream of the tip and $8d$ upstream of the stem.

The tappings will normally have a depth of order $\mathcal{O}(d/2)$, greater than twice the orifice diameter, d_s, such that the deep hole tapping correction will be applicable (Sect. 4.1.2). Note that the generally shallow depth of the cavity behind the orifice implies that the measurements will be particularly sensitive to burrs, both internal and external.

4.2.3 Directional Sensitivity

By the nature of the quantity static tubes are designed to measure, they are quite directionally sensitive. Flow around ellipsoidal nose designs separates later than around hemispherical tips and thus the former are less sensitive to yaw (in the sense of smaller errors for moderate yaw $< 15°$, but note that at larger yaw angles the error increases rapidly and exceeds the hemispherical nose value). The use of several holes around the circumference of the tube decreases this sensitivity, but in any case accuracy in the region of 1% in C_{ps} can be obtained for a yaw of less than 5° in turbulent flow, or 2% for yaw less than 26° in the plane normal to the hole axes for a round-nosed tube [4.1]. If the incoming flow is laminar, probes exhibit a narrower angle of approximately constant C_{ps}, followed by a faster increase in error for larger yaw angles [4.20].

4.2.4 Effect of Turbulence

While the effects of turbulence on measurements made with static tubes remain somewhat unclear, it can be determined that the influence will depend on the distribution of turbulent energy between the components (only the normal components, $\overline{v'^2}$ and $\overline{w'^2}$, contribute for a correctly aligned probe), eddy size in relation to the probe diameter (or more specifically to the separation of static tappings) and turbulence intensity.

Proposals for the form of the error have been given for eddies that are small (and uncorrelated at the measurement locations) [4.23, 24] and large (such that a simple quadratic yaw response is appropriate) [4.25]:

$$C_{ps} = +\frac{1}{2}\frac{(\overline{v'^2} + \overline{w'^2})}{U^2}, \quad d/D \text{ large}$$
(Fage [4.24])

$$= -\frac{1}{2}\frac{(\overline{v'^2} + \overline{w'^2})}{U^2}, \quad d/D \text{ small}$$
(Bradshaw and Goodman [4.25], after Toomre).

(4.10)

However *Christiansen* and *Bradshaw* [4.20] indicate that $C_{ps} = 0 \pm 2\%$ for ellipsoidal-tipped probes in the range of eddy-size-to-probe-diameter ratio that would be used in common practice. They also noted that the error should be expected to increase as the square of turbulent intensity by (4.10).

4.3 Hardware and Other Considerations

The accuracy of pressure measurements depends critically on the sensitivity of the transducer employed. For high precision, fast response times and a digital output, diaphragm-type transducers are the preferred choice, although the proportionality of sensitivity to the full-scale reading means that high-range transducers give poor accuracy in flows with low velocities. In addition, the effect of temperature on zero drift must be considered, such that calibration in situ using reference pressures may be required. Liquid manometers tend to be larger and slower to respond to input pressures.

For measurements at multiple locations, multichannel devices such as the older mechanical scanning Scanivalve systems or newer multisensor units permit sequential and simultaneous measurements, respectively. Alternatively, a bank manometer may be used.

The exact configuration of the probes and accompanying hardware, including the transducer, traversing system, position encoder etc., will depend on the flow under consideration: in a large, low-speed wind tunnel, they may all be mounted inside the test section downstream of the measurement location, whereas for confined flows only the probe(s) will be inserted into the flow and connectors will lead outside to the transducer.

The connections between probes and transducers are commonly made using flexible, plastic tubing (where the exact material may be selected based on the temperature requirement) when moderate pressures are involved, or metal tubing for higher pressures. Care should be taken that the tubing is sufficiently thick to avoid kinking due to the path between probe and transducer and/or failure due to the pressure difference across the wall. The response time of the system may be optimised by minimising the volume between probe and transducer without reducing it so much that the flow required to adjust to a new pressure experiences significant resistance [4.1].

Techniques for the measurement of steady pressure using wall tappings and static tubes have been described here. For information on the determination of velocity magnitude and direction from pressure measurements using Pitot tubes, Pitot static probes, multihole probes and yawmeters the reader is referred to Sect. 5.1 and *Chue* [4.2].

4.4 Pressure-Sensitive Paint (PSP)

4.4.1 Basics of PSP

General

Wind-tunnel tests are designed to give aerodynamicists information about the performance of a model when subjected to airflow. Such testing is vital in the development of new aircraft, cars, etc. in the prediction of performance and manoeuvrability, and in the identification and resolution of aerodynamic problems. Aspects of interest include structural loading, aerodynamic efficiency, boundary layer and transition effects, and the validation of computational fluid dynamic (CFD) codes. Pressure measurements made on the surface of wind tunnel models play an important role, particularly in the development of wing design and when covering the complete model surface, forces and moments. Since first basic tests in the early 1990 other applications of pressure and temperature measurements (PSP/TSP) have been performed, such as PSP/TSP tests on the blades of turbo machines or transition measurements in cryogenic wind tunnels, which requires TSP for very low temperatures or very low concentration of oxygen for PSP measurements.

Conventionally, surface pressures are measured [4.26] using hundreds of pressure taps. These are connected to mechanical or electronic scanning systems mounted within the model or, for highest accuracy, large secondary standard pressure transducers mounted externally to the model. These orifices can influence the flow over the model and introduce measurement errors. The introduction of pressure-sensitive paint (PSP) provides a method [4.27] to measure the pressure on the surface of a model directly without the transducers and tubing. A paint, the luminescence of which is dependent on air pressure, is applied to the surface of a wind tunnel model and the pressure distribution is obtained from the images produced by proper illumination [4.28–98]. With PSP technology the potential exists for a considerable saving in the cost of model design and manufacture.

Objectives

Herein we describe the theory and practice of optical pressure measurement (PSP), summarising the work of researchers around the world in many application areas and draw on our own group's experience in the industrial wind tunnel community.

Section 4.4.2 describes the theory and practice of pressure-sensitive paint Section 4.4.3 provides an understanding of paint excitation and imaging systems, and Sect. 4.4.4 the necessary image processing techniques. Section 4.4.5 discusses some applications where optical pressure measurement is already in use.

Concepts

The method is based on the phenomenon of deactivation of photoexcited molecules of organic luminophores by oxygen molecules (quenching). The ability of oxygen to quench the luminescence of organic luminophores was discovered by *H. Kautsky* and *H. Hirsch* in 1935 [4.26]. Certain materials are luminous when excited by the correct light wavelength. This luminescence can be quenched by the addition of another material. In 1919 *Stern* and *Volmer* [4.99] published a paper describing the physics behind this phenomenon and a set of equations that model it. In 1980 *Peterson* and *Fitzgerald* [4.100] used a luminescent dye sensitive to oxygen quenching to visualise a jet of oxygen flowing over a surface.

In 1985 *Pervushin* et al. [4.101] used oxygen quenching to measure the pressure of air on the surface of wind tunnel models. This was the first time a pressure-sensitive paint was used as a tool for aerodynamic research. Subsequent advances in imaging technology, notably high-resolution charge-coupled device (CCD) cameras and digital image processing, have made this technique accessible to more users. Finally a coating that can be applied to the model surface like an ordinary paint is necessary. A comparison of accuracy is usually carried out against measurements performed with conventional pressure orifices. However, various problems have to be considered during the covering of pressure tapped models with PSPs. The orifices may be protected by covering them with small pieces of an ordinary adhesive tape before PSP application, but after removing these pieces sharp edges of PSP occur in the vicinity of pressure tap orifices (Fig. 4.9a). The orifices may also

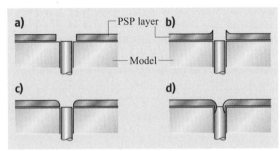

Fig. 4.9a–d Paint formation around a pressure tap orifice at different means of orifice protection

be protected by inserting wire pieces of an appropriate diameter before PSP application, but the PSP then forms hillocks around the orifices (Fig. 4.9b). The alternative way to protect the pressure tap orifices from being covered by the PSP is to blow air through the taps during PSP application, but this creates craters of PSP around the orifices (Fig. 4.9c). Moreover, blowing air can be performed only if the other sides of the pressure taps are accessible. Experience of numerous experiments shows that spraying of PSP as a thin polymer layer does not damage pressure taps without any orifice protection and does not changes their shape significantly (Fig. 4.9d) since the total thickness of PSP is usually 20–60 µm.

Deformations of the model and the model support take place under aerodynamic loads in the airflow. These deformations cause deformation and displacement of the model image acquired in the airflow. A set of markers must be established on the PSP surface to be able to correct for the displacement and deformation of model image during image processing [4.27]. Characteristic points of the model image may be used as markers, but the restricted number of these does not allow adequate correction of the model image. It is more reasonable to generate the necessary number of these markers, uniformly distributed on the model surface, as clearly identifiable points on the PSP surface; markers can be made using luminescent points on the model surface with very high luminescence intensity, but some may then be lost during image recording since the measurement system is adjusted to obtain a luminescence signal as close to the maximum signal of the image detector as possible. Thus, it is preferable to make markers as contrasted, dark points on the PSP-coated surface. They may be made by covering some points of the model surface with small pieces of an ordinary adhesive tape during active layer application and removing these after drying of the active layer. The number and position of markers required depend on the processing algorithm, while their size depends on the spatial resolution of the image detector and has to allow a precise enough measurement of their position on the model's image. Knowledge of the correct coordinates of the markers on the model surface allows transformation of the bidimensional pressure distribution on the model image to the three-dimensional pressure distribution on the real model geometry.

The deformation of the model and the model support also causes a change of the excitation light intensity distribution on the model surface. This change must be corrected in the case of the luminescence intensity acquisition to avoid additional errors during pressure reconstruction. For this purpose, a few spots of reference luminescent material, insensitive to pressure and temperature, may be placed on the model surface in regions of less interest. The number of spots depends on the processing algorithm, while their size depends on the spatial resolution of the image detector and has to allow a precise enough measurement of their luminescence intensity. If a change of the excitation light intensity does not exceed a few percent, a restricted number of luminescent references allows precise enough determination of the correction function for the excitation light intensity.

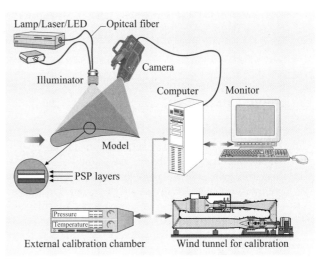

Fig. 4.10 Schematic of pressure-sensitive paint measurement system

The painted model may be stored for an extended period of time, being protected from external light, dust and contact with any undesirable materials (oil, solvents, other polymeric materials, etc.). Illumination of the paint initiates photochemical reactions inside the active layer that causes photodegradation of its components and thus the loss of its pressure-sensitive properties under an extended period of illumination. The best way to keep the model is to place it in an opaque box, and store it in complete darkness at 20–80% relative humidity.

The schematic of the test setup in Fig. 4.10 shows all the essential optical and electrical components of a PSP system. The system consists of various illumination devices, twin CCD cameras or filter wheels for pressure and reference images (because of the use of binary paint, with pressure-sensitive molecules and reference molecules without pressure and temperature dependency), a local image and data-acquisition system and an external calibration chamber, if possible, to pressurise the wind tunnel, the test section can also be used for full-size model calibration. In order to be applicable to the widest possible range of wind tunnel tests modular and highly sophisticated acquisition and processing subsystems are required. User-friendly software packages have already been developed by various users.

4.4.2 Paints

General Description

Figure 4.13 shows a basic schematic of the processes of luminescence and oxygen quenching. A pressure-sensitive paint consists of a dye held in an oxygen-permeable binder. The dye absorbs light, and the energy is used to shift an electron from one part of the molecule to another. The former part of the molecule gains a positive charge with a negative charge on the latter and these are stabilised and held apart. For ruthenium-based paints the excitation wavelength is in the blue ($\lambda = 450$ nm) and the emission wavelength in the red ($\lambda = 620$ nm), For pyrene-based paints the excitation wavelength is in the ultraviolet (UV, $\lambda = 340$ nm) and the emission wavelength is in the blue ($\lambda = 450$ nm). The molecule becomes a tiny battery that will discharge a photon by the process of luminescence after a characteristic lifetime. The paint is seen to glow. The battery can also be discharged without luminescence by a collision with oxygen. The higher the oxygen partial pressure the less the paint glows. The intensity of the luminescence gives a measure of the partial pressure of oxygen and hence the pressure of the air.

If the paint is illuminated with a pulse of light the luminescence will decay exponentially with a characteristic lifetime. This lifetime is also quenched by oxygen. The higher the oxygen pressure the shorter the lifetime as the dye molecules loses energy by collision with oxygen instead of luminescence.

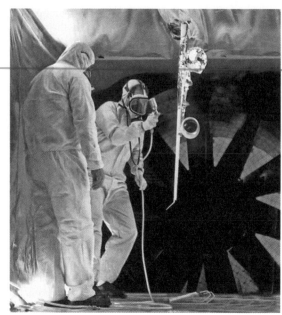

Fig. 4.12 PSP paint spraying in the covered test section itself

Pressure-Sensitive Paint Realisations

To utilise the phenomenon of dye-molecule quenching by oxygen to create pressure-sensitive sensors it is necessary to solve three closely related problems.

1. It is necessary to find such dye molecules and to place them in such a binder that quenching collisions between oxygen and dye molecules become probable during the lifetime of an excited dye molecule. If these collisions are improbable, no effect of oxygen pressure on luminescence will be observed. On the contrary, if large numbers of these collisions take place, no luminescence can be seen. This means that both very low- and very high-sensitivity paints do not allow highly accurate pressure measurements and that there is an optimal pressure sensitivity of the PSP that allows a minimum pressure measurement error.
2. Dye molecules should be attached to the model surface and the PSP layer should withstand both normal and tangential loads of the airflow. To increase the luminescent signal, and thus the accuracy of the measurements, the number of dye molecules must be as large as possible but these molecules must not affect each other. There are two different methods to attach dye molecules to the model surface: to spread them into a polymer binder or to adsorb them on a microporous surface.

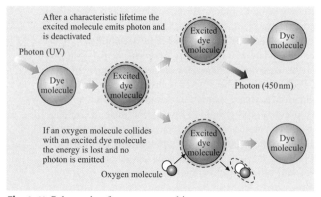

Fig. 4.11 Schematic of oxygen quenching

Table 4.1 Historical review of paint development

Luminophore	Binder	Mechanical properties	Ref.	Year
Trypaflavin	Silicagel	Powder	[4.101–103]	1964
Pyrene, coronene, p-terphenyl	Mineral oil, apiezon, decaline, glycerol, silicagel, alumina, molecular sieves	Viscous liquid, powder	[4.104]	1971
Anthracene	solution	Liquid	[4.105]	1974
Platinum group metal complexes; porphyrin and phthalocyanine complexes of VO^{+-}, Cu^{++}, Zn^{++}, Pt^{++}, Pd^{--}; dimeric Rh, Pt, Ir complexes.	Plexiglas, polystyrene, polycarbonate, resins, polyvinylchloride, latex, teflon, polypropylene, polyvinylidene, fluoride, silicon rubbers	Films	[4.106]	1986
$[Ru(Ph_2phen)_3](ClO_4)_2$	GE RTV SILASTIC 118	Film	[4.106]	1986
PtOEP, PdOEP	Polysterene	Film	[4.107]	1995
$[Ru(Ph_2phen)_3](ClO_4)_2$	Polystyrene	Film	[4.108]	1995
$Ru(bipy)_3^{++}$	Zeolite+silicone	Paint, film	[4.109]	1995
$[Ru(Ph_2phen)_3](ClO_4)_2$ $Ru(bipy)_3^{++}$	Silicagel+silicone	Paint, film	[4.109]	1995

3. To provide optimal pressure sensitivity the dye and the binder of the PSP should be chosen to provide optimum quenching conditions. The polymer binder or the microporous surface on the one hand restrict the probability of the oxygen–dye collisions that determine the pressure sensitivity of a luminescent pressure sensor and on the other hand affect dye molecules by changing their luminescence characteristics. The oxygen permeability of the binder also affects the time response of the PSP to a pressure change, so highly permeable polymers and absorbers are preferable to decrease the PSP response time. Also, the probability of oxygen luminophore collisions increases the oxygen permeability of the binder, so the luminophore lifetime should be of an appropriate range to create a PSP using a highly permeable binder. Chemical immobilisation of dye molecules in the polymer binder or on the microporous surface to provide the required dye concentration while preventing interaction of the dye molecules with each other is also a rather complex problem. A further problem is the technology for the application of a selected luminophore–binder system to the model surface. To use a PSP in aerodynamic research it should be relatively thin, smooth and should adhere sufficiently well to the model surface. To perform accurate pressure measurements with pressure-sensitive sensors it is also very important to provide the same PSP characteristics at all points on the model surface.

Oxygen Sensors. There exist a number of luminescent oxygen sensors for medicine, biology, environmental monitoring and industrial process control. Some of the most important are presented in Table 4.1.

As we will see later these systems are widely used in pressure measurements. The first oxygen sensors were liquids or powders and could not be used in aerodynamic research. Analysis of the latest publications shows that the most promising luminophores are platinum or palladium porphyrins (PtOEP, PdOEP) and ruthenium complexes such as tris-4,7-diphenyl-1,10 phenathroline ruthenium(II) ($[Ru(Ph_2phen)_3,]^{++}$) or tris-2,2'-bipyridyl ruthenium(II) ($Ru(bipy)_3$) and the most preferable binder is silicone rubber.

Dyes. Pressure-sensitive paints appear to be divided into three families, using ruthenium polypyridyls, platinum/palladium porphyrins or pyrene derivatives for dyes. Figure 4.13 shows the structure of these dyes. The ruthenium compounds are excited in the blue, glow red and are very photostable. However, they are difficult to incorporate into polymer systems and have a low sensitivity to oxygen. The porphyrin compounds may be excited in the UV or the green and produce a red luminescence. They have long luminescent lifetimes and are very sensitive to oxygen but often have low signal strength at atmospheric pressure. The pyrene derivatives are UV excited, have luminescence in the blue and have the best thermal stability, but suffer from photo-degradation.

Binders. The key to a successful pressure-sensitive paint lies in the choice of an appropriate binder. The binder should be

- stable, with a long shelf life
- compatible with and adhere to the materials in wind tunnel models
- safe to apply
- quick to cure without the application of high temperatures
- aerodynamically smooth
- highly permeable to oxygen
- safely and easily removable with no damage to the model

Researchers tend to use dimethylsiloxane polymers, as these have high oxygen permeability. These polymers are used widely by industry and come in a variety of forms. Room-temperature-curing polymers use a catalyst, often atmospheric moisture, to polymerise [4.110]. These rubbers contain powerful adhesives that bind to many substrates. They may be self-levelling but cannot be polished. Other thermally activated polymers cure to form a hard surface that can be polished flat. However, the application of heat is detrimental to wind tunnel models covered by such binders.

Solvents. A solvent is used to dissolve the dye and the binder. It also thins the paint so that it can be sprayed and evaporates to leave a homogenous layer of paint. Most pressure-sensitive paints use chlorinated organic solvents such as dichloromethane and trichloroethane. Dichloromethare is the active constituent of paint stripper and allows the dye to dissolve by expanding the polymer. The use of alcohol has also been reported [4.109].

Manufacture and Spraying

The order in which the components are mixed is very important during PSP manufacture. The relative concentrations of dye, polymer and solvent can also change the paint characteristics. Low dye concentration will reduce the signal from the paint while high dye concentration causes the paint to self-quench, reducing the paint luminescence.

Pyrene- as well as ruthenium-based paints were developed in combination with special binders to minimise their polymerisation time, permeability and lack of photodegradation.

For normal paint, which means for transonic use, we mainly use pyrene-based paint as well in a polymer binder as in anodised aluminium (AA). The function of the polymer binder has been explained above, and Fig. 4.12 of the structure of AA from *Sakaue* [4.111] should give an impression. It is possible to create very small holes like an open sponge in a thin aluminium surface. For anodisation processing with H_2SO_4 is first necessary, followed by treatment with H_3PO_4.

It is possible to put pressure-sensitive dyes into these holes. Oxygen has to penetrate via diffusion from the top into the holes. Compared with other so-called *open systems* the response time is very short (50 μs to 1 ms) and therefore this type of paint is a good candidate for unsteady PSP.

An electrochemical process is applied to obtain an anodised aluminum layer, which is formed on an aluminum anode in an electrolytic solution. A self-ordered structure is obtained using this anodisation at a constant direct current in the electrolyte at a constant voltage (i. e. electrolyte temperature).

There are a huge number of micropores on the surface. This anodised aluminum is formed using dilute sulfuric acid as electrolyte. The anodising voltage was 20 V, and the current density was 12.5 mA/cm^2. The anodising time was about 30 minutes. The diameter of the micropores is proportional to the anodising voltage. The depth of the anodised aluminum is proportional to the anodising time, with a constant voltage and a constant current density.

In order to create uniformly distributed micropores as shown in Fig. 4.12, it is necessary to conduct careful pre- and post treatments. The porous anodised aluminum layer is optically transparent if the diameter of micropore is much less than the wavelength of light. The anodised aluminum layer is also formed on aluminum alloys that contain Mg, Fe, Cr or Si, though the structure of the micropores is somewhat disordered [4.112].

Nonperiodic unsteady pressure is difficult to measure using the PSP technique, even though several attempts have been made to measure an unsteady pressure field [4.113, 114]. The most serious problem with unsteady measurement is that the signal from the PSP is inevitably weak due to the short exposure time. As for periodically fluctuating measurements, accurate pressure data can be obtained using phase-locking methods. In such a case, several snapshots at a given phase angle are superposed in a single image until sufficient light is accumulated. However, today the superposition technique is no longer applied to nonperiodic phenomena.

Temperature and *humidity* effects on the PSP signal are other issues. The temperature effect causes substantial errors in all PSP measurements. On the other hand, sensitivity to humidity is not negligible in the

AA-PSP, because the anodised aluminum layer has a hydrophilic nature. The moisture content in the layer alters the oxygen concentration in the AA-PSP. We carefully examine the acquisition and processing of PSP images to eliminate these errors.

Another *open system* has been developed by German Aerospace Center (DLR) [4.115], at first for cryogenic use, which is sprayable like the well-known steady paints. One new type of pressure-sensitive dye is palladium tetra-(pentafluorophenyl)porphine (PdTFPP) embedded in a porous binder of poly(trimethyl-silyl-propyne) [poly(TMSP)] – for very high oxygen permeability.

Schanze [4.116] published a paint formulation based on the dye tris(4,7-diphenyl-phenanthroline) ruthenium (d) dichloride (RuDPP) in poly(dimethyl siloxane). The paint is made by dissolving 3 mg of RuDPP and 1 g of DMS-D33 in 4 ml dichloromethane solvent. The paint is applied with an airbrush and allowed to dry for 24–48 h in low-humidity conditions.

A US patent [4.108] describes many different ways of producing pressure-sensitive paints, both with and without silica gel encapsulated in rubber. For example, the dyes pyrene and perylene were dissolved in Dow 732 silicon rubber. The paint is made by first dissolving 1 g Dow 732 silicon rubber in 7 g isooctane solvent. The resulting solution is shaken and placed in an ultrasonic cleaner for 15 min. 1.5 mg of pyrene and 1.8 mg perylene are added and allowed to dissolve. The paint is applied with an airbrush and allowed to cure at ambient temperature and humidity.

Kavandi et al. [4.117] published a paint formulation based on the dye platinum octaethylporphyrin (PtOEP) from Porphyrin Products, USA, in a poly(dimethyl siloxane) solution GP-197. The paint is made by mixing 0.1 g of PtOEP in one litre of GP-197 solution. The dye quickly dissolves to give a dark-red solution. The paint is applied by airbrush and cured at $100\,^\circ$C for 2 h. *McLachlan* et al. [4.118] mixed a fine powdered rutile-grade TiO_2 into this paint to improve the signal strength 2.5 times by increased scattering.

Spraying Procedure

PSP is usually applied to a wind tunnel model by an adhesive layer or spraying using professional airbrushes like those used by car painters. The selection of the nozzle diameter as a function of the actual air pressure is very important.

Pressurised air for the system must be clean and free of compressor oil. The airbrush can be cleaned by various solvents, such as acetone. The uniformity is judged by eye and spraying will be stopped when a good colour is seen. In most cases a neutral blue pigment [4.99] is added to the final layer and thickness can be controlled easily on a white background layer by the intensity of the blue colour.

Various instruments for measuring the coating thickness can be used after curing. A typical total thickness for steady measurements is 40–60 µm, although for fast paints (response time < 1 ms) the typical thickness is about 5 µm. It is common practice to fill all the screw holes and other wrinkles before a model enters a wind tunnel. Some of these surface fillers fluoresce significantly under PSP lighting, which changes the paint calibration above the filler and limits the exposure time of the carriers. The fillers themselves may also emit various wavelengths or may shrink during PSP spraying because of the various solvents in the paints. It is becoming increasingly popular to coat the model directly in the test section [4.119] (Fig. 4.14) to minimise the installation time.

A very stable filler for most existing paints was found from the Loctite Company (Fig. 4.15).

When spraying in closed rooms ventilation with absolutely dust-free fresh air is necessary. Also personal protection with a mask and an overall must be provided and safety precautions taken.

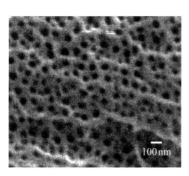

Fig. 4.13 Surface of modified aluminum (anodised aluminum)

Fig. 4.14 Various fillers for stabilisation tests using PSP subsequently

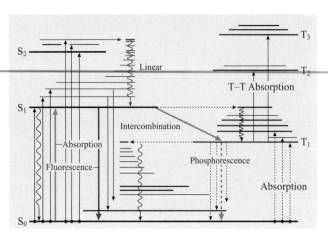

Fig. 4.15 Triplet and singlet levels and intersystem crossing

the photophysics of dye molecules. The theory uses the concept of quantisation of allowed molecular excitation states and electron spin. Figure 4.16 shows an idealised energy-state (Jablonski) diagram of a molecule. The vertical axis represents the molecular energy. The boxed arrows show the electron spin. For states with electron spins in opposite directions the term singlet is used. For states with electron spins in the same direction the term triplet is used.

Parker [4.120] describes the energy transfer mechanisms for molecules and notes that nearly all molecules have an even number of electrons [4.99, 120].

The electrons populate a ground, or lowest-energy, state (S_0) in pairs, with the spin of the electrons opposed. However, if one of the electrons is promoted to an upper energy state (S_1, T_1) the spin of the electrons can be in the same or opposite directions.

The solid arrows in Fig. 4.16 show radiative absorption and emission between states. The bold horizontal lines correspond to the excited states of the molecule. The fine lines correspond to the vibrational energy states of the molecule between which transitions occur by absorption and emission of long-wavelength radiation indicated by the wavy lines. This process is known as internal conversion. Radiationless intersystem crossing (a) occurs from the lowest-energy state of a singlet level (S_1) to one of the upper vibrational levels of the lowest triplet level (T_1). From here the molecule quickly looses its excess vibrational energy. The rate of intersystem crossing can successfully compete with the rate of fluorescence in some molecules.

At first sight it appears that radiationless intersystem crossing (b) $T_1 \rightarrow S_0$ should be just as fast as possible. In fact it occurs at least 10^8 times slower and is the reason that phosphorescence can be observed.

It should be pointed out that all pressure-sensitive molecules are believed to constitute a health hazard. The toxicity of the other ingredients like the solvents is not well known. In all cases, it is always recommended to avoid skin contact and inhalation during spraying.

Theory

Photophysics for the Intensity-Based System. The theory of pressure-sensitive paint luminescence is found in

$\longrightarrow$ Radiative decay
$- \rightarrow$ Radiationless decay
k_{IC} = internal conversion $S_1 \rightarrow S_0$
k'_{ISC} = intersystem crossing $S_1 \rightarrow S_0$
$k_Q[Q]$ = quenching
k_P = phosphorescence and
k_F = fluorescence
k_{ISC} = intersystem crossing $S_1 \rightarrow S_0$
$k_{thermal}[T]$ = thermal deactivation

Rate Factors and Quenching Processes. Figure 4.15 shows the relevant intramolecular decay step mechanisms for pressure-sensitive paint, each characterised by its own rate constant. Each excited state is characterised [4.121] by its lifetime, given by

$$\tau = \frac{1}{\sum k_i}, \qquad (4.11)$$

where k_i is the rate constant for a generic molecular process that causes the disappearance of the excited state.

The quantum yield for each luminescence process is the ratio between the number of photons produced to the number of photons absorbed. For example the quantum

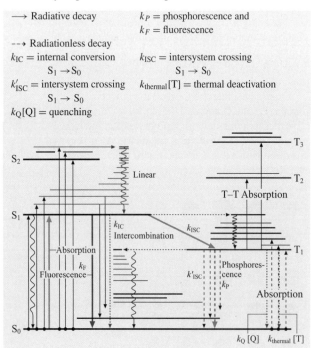

Fig. 4.16 Schematic of possible energy paths

yield Φ_P for phosphorescence from the lowest excited triplet state is:

$$\Phi_P = \eta_{ISC}\eta_P, \quad (4.12)$$

where η_{ISC} is the efficiency with which the triplet state is populated by light absorption and η_P is the efficiency of phosphorescence from the triplet state. These efficiencies are computed by determining the ratio of the desired energy path to the sum of all energy paths. By inspecting Fig. 4.17

$$\eta_{ISC} = \frac{k_{ISC}}{k_{ISC} + k_F + k_{IC}} \quad (4.13)$$

and ignoring thermal effects $k_{thermal}(T)$ and quenching $k_Q[Q]$ to simplify this step

$$\eta_P = \frac{k_P}{k'_{ISC} + k_P} = k_P \tau_{T1}, \quad (4.14)$$

where τ_{T1} is the lifetime of the excited triplet state.

When the lifetime of the excited state is sufficiently long the excited molecule may have the chance to encounter a molecule with which to interact to allow deactivation without radiation. This is called bimolecular quenching. If $k_Q[Q]$ is the quenching rate and $[Q]$ is the concentration of the quencher then (4.14) can be modified to

$$\eta'_P = \frac{k_P}{k'_{ISC} + k_P + k_Q[Q]}. \quad (4.15)$$

Stern–Volmer Relations. Stern and Volmer [4.99] noted that, if the ratio of the die luminescent intensity with a quencher I and without a quencher I_0 was plotted against the quencher concentration, then a straight line was obtained. By definition the quantum efficiencies are directly proportional to the luminescent intensity. Taking the ratio of (4.12) with and without the quencher and substituting in (4.14) and (4.15) gives the equation of a straight line

$$\frac{I_0}{I} = \frac{\eta_{ISC}\eta_P}{\eta_{ISC}\eta'_P} = \frac{\frac{k_P}{k'_{ISC}+k_P}}{\frac{k_P}{k'_{ISC}+k_P+k_Q[Q]}} = 1 + \frac{k_Q[Q]}{k'_{ISC}+k_P}. \quad (4.16)$$

The lifetime of the phosphorescent state is computed using (4.11) by inspection of Fig. 4.17, again ignoring thermal effects

$$\tau = \frac{1}{k'_{ISC} + k_P + k_Q[Q]}. \quad (4.17)$$

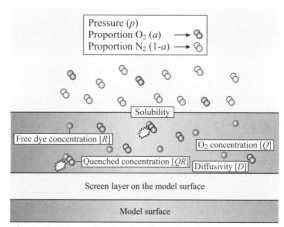

Fig. 4.17 Schematic of paint photochemistry

Taking the ratio of the lifetimes with a quencher τ and without quenching τ_0 reveals another Stern–Volmer equation:

$$\frac{\tau_0}{\tau} = \frac{\frac{1}{k'_{ISC}+k_P}}{\frac{1}{k'_{ISC}+k_P+k_Q[Q]}}$$
$$= 1 + \frac{k_Q[Q]}{k'_{ISC}+k_P} = 1 + \tau_0 k_Q[Q]. \quad (4.18)$$

Comparing (4.16) with (4.18) and noting that (4.16) assumes that η_{ISC} is unchanged between the two measurements of I_0 and I it is seen that the intensity and lifetime ratios follow the same equation:

$$\frac{I_0}{I} = \frac{\tau}{\tau_0} = 1 + \tau_0 k_Q[Q]. \quad (4.19)$$

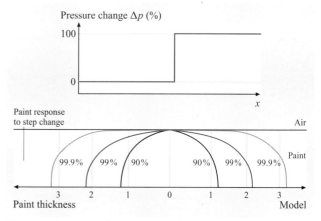

Fig. 4.18 Response of pressure-sensitive paint to spatial step change in pressure

Photochemistry. Wolfbeis [4.122] predicts the basic behaviour of immobilised indicators (dye molecules held in a matrix) that are quenched by a dissolved gas. A schematic of this process is shown in Fig. 4.18. An equilibrium is established between the concentrations of the quencher molecules [Q], unquenched indicator molecules [R] and the associated quenched complex [QR]. Assuming a 1:1 stoichiometry, the equilibrium is described by

$$K_E = [QR]/[Q][R], \quad (4.20)$$

where K_E is the equilibrium constant.

The total concentration [C] of the immobilised indicator is the sum of the free and complexed molecules

$$[C] = [R] + [QR]. \quad (4.21)$$

If the total amount of indicator remains constant it follows that rearranging (4.20) and substituting into (4.21) gives

$$[C] = [R](1 + K_E[Q]). \quad (4.22)$$

If the dissolved gas can quench the luminescence from the indicator then the luminescence I will be proportional to the concentration of the unquenched indicator. If there is no dissolved gas then [R] = [C] and the luminescence I_0 will be proportional to the concentration of the indicator. Substituting I and I_0 into (4.22) yields an equation equivalent to the Stern–Volmer equation [4.99]:

$$\frac{I_0}{I} = 1 + K_E[Q]). \quad (4.23)$$

The coefficient K_E can be expanded to be the product of the lifetime τ_0 of the indicator in the absence of quenching gas and the bimolecular quenching constant k_Q by comparison to (4.19).

Wolfbeis noted that k_Q is related to the collisional frequency of quenching by a modified Smoluchowski equation

$$k_Q = \frac{4\pi \gamma N D r}{1000}, \quad (4.24)$$

where γ is a factor that accounts for the quenching efficiency of a collision, N is Avagadro's number, D is the diffusion coefficient of the quencher (the indicator is immobilised) and r is the collision radius. The term $N/1000$ converts molarity into molecules per cubic centimetre. For pressure-sensitive paint the concentration of the quenching gas in the polymer [Q] is equal to the product of the solubility S of the gas and the partial pressure of gas outside the polymer (Henry's law). It should be noted that Henry's law only holds for sparingly soluble gases at low partial pressures. This assumption is used to simplify the analysis. Real systems tend to have a second-order relationship between solubility and partial pressure [4.123].

The total pressure p of the air outside the polymer is equal to the sum of the partial pressures of its constituents. Equation (4.23) can be rewritten:

$$\frac{I_0}{I} = 1 + Kp, \quad K = \tau_0 \frac{4\pi \gamma N D r}{1000} Sa, \quad (4.25)$$

where a is the volume fraction of oxygen in the air, and K is the pressure sensitivity of the system and has units of reciprocal pressure.

Why Oxygen Quenching? Ranby and Rabek [4.124] give an introduction to the structure of molecular oxygen. Oxygen exists in the atmosphere predominantly as a diatomic structure and is paramagnetic. In this way oxygen differs from other permanent gases. This property is interpreted as being due to its two outer electrons having parallel spins. This uncoupled electron pair classifies oxygen as a triplet in its lowest-energy state.

The next two higher-energy states are singlets and are located 22.5 and 37.5 kcal/mol above the triplet state. This makes the excitation of oxygen relatively easy. Oxygen can successfully quench red dye luminescence by the transfer of excess energy from the dye to the oxygen molecules. *Turro* [4.125] notes that ground-state molecular oxygen is a general and efficient quencher of both singlet and triplet states of organic molecules. The process of quenching can occur in many ways; it is not even dependent on the luminophore, giving 22.5 kcal/mol as the triplet O_2 will catalyse intersystem crossing from the singlet to the triplet state.

Characteristics

Thermal Sensitivity. The quantum efficiency and lifetime of radiation from a dye are both temperature dependent [4.126]. The efficiency of $T_1 \to S_0$ phosphorescence is critically dependent on temperature and is observed to decrease by several orders of magnitude between cryogenic and room temperatures.

This is due to the long lifetime of the triplet state that allows the possibility of thermally activated, radiationless transition with a rate $k_{thermal}$ (Fig. 4.17). The rate constant for thermal deactivation can be modelled by the Arrhenius formula

$$k_{thermal} = k_0 + A \exp(-\Delta E / k_B T) \quad (4.26)$$

where k_0 and A are constants, ΔE is the energy for the deactivation mechanism, k_B is Boltzmann's constant and T is the temperature (Kelvin).

The actual thermal sensitivity of a dye can be very complex [4.125] with (4.11) containing the sum of several competing processes. However, both I_0 and τ_0 can be empirically modelled using

$$\log\left(\frac{1}{I_0} - \frac{1}{I_0'}\right) \propto -\frac{1}{T} \quad \text{and}$$

$$\log\left(\frac{1}{\tau_0} - \frac{1}{\tau_0'}\right) \propto -\frac{1}{T}, \quad (4.27)$$

where I_0 and τ_0 are the intensity and lifetime at temperature T (Kelvin) and I_0' and τ_0' are the intensity and lifetime extrapolated to 0 K, all with no quenching.

The pressure sensitivity of the system K is dependent on τ_0 and also on the product of the diffusion and solubility constants, which is defined [4.127] as the permeability of the matrix P

$$P = DS. \quad (4.28)$$

The permeability temperature dependence can be represented by

$$P = P_0 \exp(-E_M/RT), \quad (4.29)$$

where E_M is the activation energy of permeation, P_0 is the exponential premultiplier, R is the gas constant and T is the temperature in Kelvin.

This analysis suggests that the production of a temperature-insensitive paint is unlikely. The lifetime falls with increasing temperature while the permeability increases. This characteristic can be used to produce a paint with K that is insensitive to temperature over a limited range, but at the expense of the thermal sensitivity of I_0 and τ_0. Choosing a dye with a low τ_0 sensitivity to temperature will increase the thermal effect on K.

Pervushin et al. [4.101] produced a paint with a luminescent intensity that remains constant over a limited temperature interval, although the thermal sensitivity of K is not quantified.

Schanze et al. [4.116] concluded that for ruthenium dyes the temperature dependence is dominated by the nonradiative decay rate of the dye molecule. However, the temperature dependence of the oxygen diffusion rate also plays a strong role. They concluded that to minimise the temperature sensitivity of a paint it is necessary to design a binder to have a low activation energy for the oxygen diffusion rate.

Spatial Resolution. The diffusion of oxygen into a layer of pressure-sensitive paint is the direct analogy of heat flowing into a solid. There are many publications [4.128] which explain the models in more detail. The flow of heat and the diffusion of oxygen are both modelled by the diffusion equation

$$\frac{\partial^2 [Q]}{\partial x^2} + \frac{\partial^2 [Q]}{\partial y^2} + \frac{\partial^2 [Q]}{\partial z^2} = \frac{1}{D}\frac{\partial [Q]}{\partial t}. \quad (4.30)$$

Figure 4.19 shows the distribution of oxygen in the paint layer due to a step change in pressure across the surface of the paint. It can be modeled using the diffusion equation expanded into three dimensions with $\partial[Q]/\partial t = 0$:

$$\frac{\partial^2 [Q]}{\partial x^2} + \frac{\partial^2 [Q]}{\partial y^2} + \frac{\partial^2 [Q]}{\partial z^2} = 0. \quad (4.31)$$

Equation (4.31) indicates that the spatial variation of oxygen is only dependent on geometry, and in the case of PSP, on the thickness of the paint. *Bukov* et al. [4.129] have determined that the paint will have reached 99% of the pressure step change across the surface of the paint in five times the paint thickness.

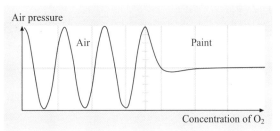

Fig. 4.19 Passage of pressure wave through an air–paint interface

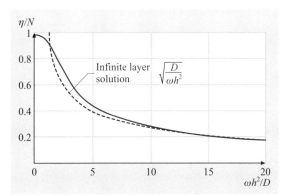

Fig. 4.20 Transfer function for pressure-sensitive paints

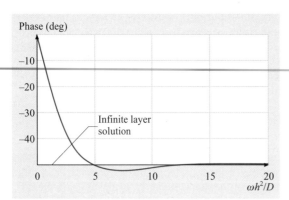

Fig. 4.21 Phase-dependent transfer function for pressure-sensitive paints

Response Time. Figure 4.20 shows a pressure wave of frequency ω impinging on the surface of a layer of pressure-sensitive paint. The passage of the pressure wave into the paint can be modelled by the diffusion equation

$$\frac{\partial^2 [Q]}{\partial x^2} = \frac{1}{D}\frac{\partial [Q]}{\partial t}, \qquad (4.32)$$

where [Q] is the concentration of quencher molecules in the paint and D is the diffusion coefficient. The solution to the diffusion equation is not trivial, but in this case is identical to the periodic flow of heat into a conducting solid [4.130]

$$[Q](x,t) = \cos\left(\omega t - \frac{x}{\delta}\right)\exp\left(-\frac{x}{\delta}\right),$$
$$\text{with}\quad \delta^2 = \frac{2D}{\omega}, \qquad (4.33)$$

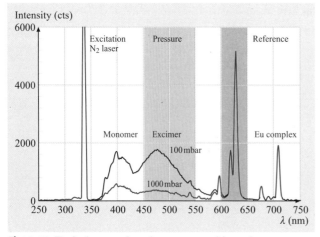

Fig. 4.22 Typical pyrene-based pressure-sensitive paint spectrum

where δ is often called the skin depth of the oscillation.

The diffusion into a finite thickness h of PSP can be modelled by applying the boundary condition $\partial [Q]/\partial x = 0$ at $x = h$. As the frequency of modulation is increased, the skin depth reduces and a smaller active layer of paint is able to respond.

Equation (4.33) can be integrated over x to find the number of modulated quencher molecules n in the layer of paint. The amplitude and phase of n is plotted against the frequency factor in Fig. 4.22. In this figure, n has been normalised against the total number N of quenching molecules in the paint thickness. As ω increases the solution tends to the infinite-layer case.

This analysis concludes that the paint does not have to be thin to detect pressure fluctuations. As the frequency increases the effective sensing layer reduces in thickness. The imaging system will detect a reducing modulation on top of an increasing offset. Reducing the paint thickness does not increase the signal strength; it reduces the often observed offset. Counterintuitively, the fluctuating pressure data from thick paint would be easier to analyse than from a thin paint layer as the transfer function will be less sensitive to frequency and paint thickness.

Carroll et al. [4.131] and *Engler* [4.132] investigated experimentally the response of two pressure-sensitive paints to sinusoidal and sawtooth pressures waves. Their results agree well with this analysis.

Borovoy et al. [4.133] note that the time for a paint to achieve 99% of its response to a step change in pressure is:

$$t_{\text{step}} = 12h^2/\pi^2 D . \qquad (4.34)$$

Carroll et al. [4.104] computed the diffusion of oxygen after a step change as an infinite series of exponential functions. They found that this could be simplified to a series of two exponential functions. Several paints with varying thickness were tested to verify the dependence on thickness and diffusion rate.

Engler [4.132] conducted tests looking at the response of two paints to modulated pressures and found that OPTROD LTD. type 18E was limited to 1 Hz and OPTROD LTD. type 21-Fl could be used to measure up to 20 Hz.

Inspection of (4.25) reveals that the ideal high-bandwidth paint would utilise a binder with a high diffusion rate but a low solubility. This would keep the permeability of the polymer and the Stern–Volmer coefficient unchanged.

Spectra. The spectrum of a pressure-sensitive paint is largely due to the type of dye used. Figure 4.23 shows the spectra from a typical UV-excited phosphorescent pyrene-based dye. The emission peak is at a longer wavelength (440 nm) than the absorption peak (450 nm). The shift between the peaks is a function of the difference between the singlet and triplet energy levels. For fluorescent dyes the absorption and emission spectra can significantly overlap, which is not the case for pyrene-based dyes. The excimer peak in the range 500 ± 50 nm is very pressure sensitive whereas the peak at 400 ± 10 nm is highly temperature sensitive. In general it is desirable for the spectral separation to be large so that broadband light filters can be used to separate the excitation light source output from the paint's luminescence.

As can be seen in Fig. 4.23, two different dyes were used: a pressure-sensitive one and a reference dye which is not pressure or temperature sensitive. The illumination for this dye is the same as for the pyrene dyes, but the emission is in the far red wavelength about 630 ± 10 nm with a high peak. Therefore it is easy to separate it from the pressure peak in the blue.

These dyes are only sensitive to the illumination intensity and are therefore an ideal reference component for illumination correction, because there is no chance of creating a stable and homogeneous illumination on the model's surface.

Also in these spectra the influence of two different pressures, red (100 mbar) and green (1000 mbar), is clearly visible and illustrates that a much higher intensity is given at 100 mbar than at 1000 mbar.

Intrusion. Pressure-sensitive paint is an intrusive technique. The finite thickness of the paint alters the profile of the aerofoil and the surface of the paint never has the same surface finish as the underlying material. These effects lead to differences in the pressure distribution on the model surface and will alter the aerodynamic drag.

These problems are not unique to pressure-sensitive paint. There has been much research on the effect of surface finish on aircraft drag. A detailed assessment of the drag processes was investigated and documented by *Hoemer* [4.134]. It should be noted that cryogenic testing of a temperature-sensitive paint (TSP) by *Fey* et al. [4.135], with a roughness of $< 0.5\,\mu\text{m}$ Ra was accepted by the European transonic wind tunnel (ETW) because this TSP could be polished after spraying and the paint created no additional surface effects like turbulent wedges at the leading edge.

Sources of Error

Bukov et al. [4.129] described many of the errors found in the PSP measurement technique. Typical errors associated with the paint include:

- Nonlinearity of the PSP calibration due to second-order solubility effects requires a second-order fit to reduce errors to 0.2%.
- Spatial variation of the PSP calibration gives a 0.3% error caused by microheterogeneity of the polymerisation. This can be corrected by pixelwise instream calibration in pressurised wind tunnels. This has also been reported by *Torgerson* et al. [4.136] as a more significant error for low-speed flows.
- Temperature sensitivity of the PSP calibration can be a significant error and corrections must be applied. Depending on the PSP dye the error is between 0.3–5%/°C.
- Temperature hysteresis is dependent on the type of polymer used and has been noted by *Jules* et al. [4.137] to be an irreversible effect. Between $-20\,°C$ and $80\,°C$ flow or model surface temperature there is no significant error for the DLR02 paint, but for temperatures between $100\,°C$–$120\,°C$ there is also an irreversible effect.
- Pressure hysteresis is caused by the irreversible solubility of oxygen in a polymer and is explained by *Grate* and *Abrahams* [4.138]. Elastic polymers have the lowest hysteresis.
- Spatial resolution is limited by sideways oxygen diffusion in the paint layer, while average time resolution is limited by the paint thickness and diffusion rate. For steady flow using the intensity method this is no problem, but it causes phase errors when measuring dynamic pressure signals.

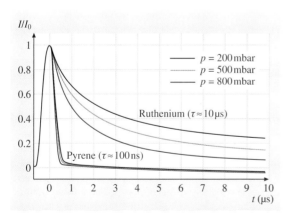

Fig. 4.23 Principle of lifetime measurement for two different dyes

- Paint ageing. Paints are now more stable than five years ago, when paints could change by as much as 5% per day due to slow polymerisation, slowing down after complete polymerisation. The paint's calibration changes because the environment around the individual dye molecules changes at different rates. This can leave some luminous molecules that are unable to be quenched. The ratio between these insensitive molecules and the sensitive molecules introduces a standing offset in the luminescent signal and changes the linearity of the Stern–Volmer plot [4.139]. In 1996 *Zharkova* et al. [4.109] found that their paint had a very short life and became insensitive to pressure 15 d after spraying. DLR's experience [4.115] using their own DLR02 paint, is that about 10% of sensitivity is lost after half a year storage in a dark box at 20–80% relative humidity at room temperature 20–40 °C.
- Photobleaching is proportional to the intensity of the light source and depends on the type of dye used, and may be significant for some paints.

4.4.3 Imaging Systems

Radiometric Imaging

Radiometric imaging is the simplest and most widely used technique for acquiring pressure-sensitive paint images. The surface is illuminated with a continuous light source and the paint's luminescence is normally detected by a cooled, charge-coupled device (CCD), slow scan camera [4.140].

The Stern–Volmer equation (4.35) predicts that the intensity of the fluorescence is inversely proportional to the partial pressure of oxygen. Equation (4.35) describes the ratio IR of the intensity I of paint at pressure p to the intensity I_0 at zero pressure and has been rewritten as

$$IR = \frac{I}{I_0} = \frac{1}{1+kp}, \qquad (4.35)$$

where k is the pressure sensitivity of the paint (measurable by calibration).

This equation requires an estimate of the intensity of the paint at zero pressure. This will vary over the surface as the illumination is not uniform and the thickness of the dye on the surface depends on the application method. A wind-off reference image is taken at atmospheric pressure to measure these variations and the Stern–Volmer equation is modified accordingly [4.117]. The pressure image is produced from a ratio between this image and a wind-on image.

Any wind-on model displacement will make direct pixel-by-pixel computation impossible and the registration between the two images has to be corrected by image transformation [4.141]. Various correction methods, also applicable to real three-dimensional wind tunnel models, are developed in most PSP groups.

The error in the illumination intensity (ξ) due to the movement of the model can be approximated by:

$$\zeta \approx \frac{2l^2}{r^2+l^2}\frac{\varepsilon_\mathrm{r}}{r} - \frac{2l^2}{r^2+l^2}\frac{\varepsilon_\mathrm{l}}{l} - \frac{2rl}{r^2+l^2}\Phi, \qquad (4.36)$$

where r is the distance between the lamp and the model, l is the length of the model, ε_r is the distance of movement from the lamp, ε_l is the distance of movement laterally and Φ is the angular rotation of the model under load.

The most sensitive factor of (4.36) for sting-mounted models is the angular rotation. Sting-bending rotation is corrected by computing the model movement from the measured loads, and pitching the model to the wind-on angle of attack before taking the reference image.

This is quite successful for flat models illuminated using a point source. However, practical light sources tend to deliver a cone of light. This leads to significant changes in illumination over the surface of the model and any model movement will cause pressure measurement errors in the illumination gradient. Highly curved areas of the model, such as leading edges, are similarly affected. The most successful system for correcting for model movement has been demonstrated by *Engler* [4.132, 142]. Here the model attitude and displacement are directly measured and the model is moved by a multidimensional sting actuator to match the reference image to the wind-on image within two pixels. The model is now in the same illumination field for both images. This technique cannot be applied to all wind tunnels. The model deformation can be corrected by the DLR software. The deformation occurs due to the cantilever arrangement of the wings and other surfaces bending under aerodynamic loads.

Lifetime Measurement

Lifetime measurement illuminates the paint with a pulse of light and measures the response with a gated detector. This was one of the first methods [4.101] used to image pressure-sensitive paint with a film camera exposed at various intervals after illumination by a flash lamp. The method has been used to make fluorescence images of microscope slides [4.143] using a scanning laser system and has been adapted for aerodynamics by *Davies* [4.144], *Engler* et al. [4.142] and *Burns* and *Sul-*

livan [4.145]. Figure 4.24 shows the behaviour of the paint when illuminated with a pulse of light. The intensity of the paint J_{pulse} can be modelled, to first order, by an exponential decay:

$$J_{\text{pulse}} = J_0 \exp(-t/\tau), \quad (4.37)$$

where J_0 is the intensity at time $t = 0$. The lifetime of the paint τ is dependent on pressure in the same way as intensity to first order. Equation (4.19) has been rewritten as

$$\frac{\tau_0}{\tau} = 1 + Kp, \quad (4.38)$$

where τ_0 is the lifetime of the paint at zero pressure and Kp is the pressure sensitivity. Figure 4.24 shows the exponential decay curves for ruthenium- and pyrene-based paints. Since the decay time for ruthenium is in the range 2–10 μs it must be pointed out that for pyrene-based paint the decay time is in the 50–100 ns range and the exponential function is nearly monoexponential. To compute the lifetime at least two measurements of intensity must be made on the decay curve. The measurement of the intensity cannot be performed instantaneous and must be integrated using a time gate (Fig. 4.25). By making the gates the same width t_{gate} (4.37) can be used to predict the relation of the ratio of the two intensities LR to pressure

$$LR = \frac{\int_{t_{\text{gate}}}^{2t_{\text{gate}}} J_{\text{pulse}} \, dt}{\int_0^{t_{\text{gate}}} J_{\text{pulse}} \, dt} = \exp\left[-(1 + Kp)\frac{t_{\text{gate}}}{\tau_0}\right]. \quad (4.39)$$

Often there is no need for a wind-off reference image using this technique. This overcomes the problems associated with model movement and the radiometric method. An experiment by DLR [4.142] gives a good impression of a direct comparison between the intensity and lifetime methods.

Fluorescent Lifetime Imaging

Holmes [4.146] has developed a fluorescent lifetime imaging method which uses a modulated light source to illuminate the paint. He developed an area measurement system based on arrays of light-emitting diodes and a solid-state phase-sensitive camera. The response of the paint can be modelled by using the analogy of a low-pass filter built around a resistor and capacitor. The paint is able to store light in the same way as a capacitor stores charge. The paint emits light in the same way as the capacitor loses charge through the resistor. Because this very specific technique has up to now only be used by him, details can be taken from this report [4.146].

Light Sources

The purpose of a PSP light source is to illuminate the paint at the correct wavelength. The light source must produce sufficient light to enable an image to be taken in a reasonable time frame for the experiment. It is also helpful, but not absolutely necessary, that the light source emits little power in the luminescence band of the paint. Stacks of interference and absorption filters achieve the desired spectrum at the expense of efficiency. The light source must be stable, as any illumination fluctuation is magnified at least four times in the final pressure image.

For paint formulations that require ultraviolet excitation, possible light sources would be based on Hg-vapour

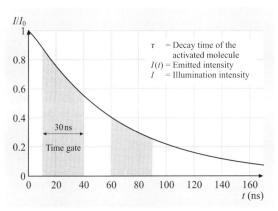

Fig. 4.24 Two-gate technique for pyrene-based paint

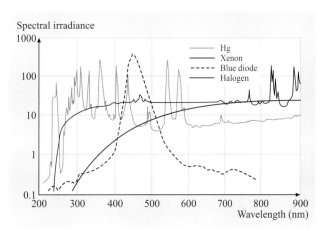

Fig. 4.25 Spectral output of tungsten, mercury and xenon lamps and a blue diode

or xenon-arc lamps, or on an excimer laser. Short-wavelength UV light is not directly visible and is not transmitted by ordinary glass. For paints that are excited in the visible band (usually 400–500 nm), xenon-arc, halogen, argon laser or light-emitting diodes are possible sources (Fig. 4.26).

Laser light sources have obvious benefits in terms of narrow-band, intense illumination. However, lasers have distinct drawbacks as they are heavy, expensive, and must be transported and handled with care. Proper laser safety procedures are essential. The use of lasers requires extensive safety precautions that add complexity to the test schedule. Excimer output energy typically fluctuates less from pulse to pulse. Mirror systems in some designs are sensitive to vibration and temperature change.

Halogen lamps darken less with time than ordinary lamps, but the halogen cycle requires a specific temperature range to operate efficiently. Cooling air circulating around the bulb can block the halogen cycle and shorten the useful life of the lamp. Also, small variations in the operating voltage can have a large effect on the blue end of the spectrum. *Carroll* et al. [4.131] report that a stabilised constant-current power supply can reduce ripple to 0.04%. Optical feedback is often required to improve long-term stability.

The optical output of a mercury-arc or xenon-arc lamp depends on the operating temperature. The lamp must be allowed to achieve thermal equilibrium before taking images, which can take several seconds. Very useful especially for wind tunnels are optical fibers which are directly connected to the arc lamps. Liquid fibers are available for wavelengths from the UV up to the IR, meaning that thermal effects can be reduced to a minimum and electrical problems with high-power devices can also be excluded.

LEDs make excellent light sources for pressure-sensitive paint. Arrays of diodes can be used to illuminate the model from many directions. They are small, easy to mount, produce little heat, and in sufficient numbers they can produce a fairly uniform, tailored, distribution field. Their disadvantage is the change in the emitted wavelengths when they are heated in the wind tunnel, especially in low-speed tunnels where air coolers are not installed.

DLR and THE FRENCH AEROSPACE LAB (ONERA) [4.147] initially used liquid optical fibers, which can be connected to lamps as well as to lasers to transmit the light to the working section. Filters in front of xenon-arc lamps can select the right wavelength for very different paints on the market. From the viewpoint of illumination users are independent of paint selection, because various filter sets can be used.

Detectors

Photographic film was the first detector [4.101] used for pressure-sensitive paint. However the inherent non-linearity of the emulsions, the sensitivity to processing procedures and the long processing time means that film is seldom used in wind tunnels today. The advantage of photographic film is that it does not require expensive and bulky instrumentation. This has made it the choice for flight trials [4.148, 149].

Wind tunnel researchers use both electronic cameras and scanning detectors to view the paint. Full-field, cooled, scientific-grade, CCD cameras are used for global imaging applications. Photomultipliers and avalanche photodiodes are used for scanning spot applications. Recently phase-sensitive CCD cameras have become available for making global fluorescent lifetime images.

La Belle and *Garvey* [4.150] give an introduction to the properties of CCD cameras suitable for pressure-sensitive paint. CCD cameras come in three varieties: interline, frame-transfer and full-field devices. Interline devices are designed for video picture applications; half of the CCD imaging area is covered by pixels, which slightly reduces their sensitivity and resolution but they have the eminent advantage that they do not need mechanical shutter. After a selected integration time the image will be shifted to the covered area. DLR [4.151] developed in cooperation with the OMT camera company a special black-and-white PSP version of this device for 12- and 14-bit operation. Frame-transfer devices are designed for high frame rates. Full-field devices are designed for scientific imaging applications and are optimised for low-noise operation by slow-

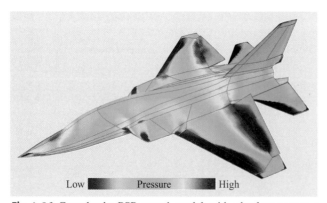

Fig. 4.26 Completely PSP-coated model with absolute pressure distribution to calculate loads

Table 4.2 Observed camera properties

Camera	Run	Median image value	Median intensity ratio	Standard uncertainty	Standard deviation of ratio images
12 bit	1	3780	0.31	0.0019	0.009
16 bit	2	50 063	0.30	0.0008	0.003
Gain = 1	3	13 662	0.299	0.0004	0.0009
Gain = 4	4	10 220	0.284	0.0004	0.0017

ing the scan rate, but they work with a mechanical shutter.

An important factor that controls the use of camera type is the necessity to capture images rapidly in order to increase the run rate and decrease the required tunnel occupancy time. This generated interest in the possibility of trading off analog-to-digital converter (ADC) bit depth for faster image download in the PSP cameras. Therefore signal strength, noise, and measurement uncertainty are compared.

Pressure-sensitive paint (PSP) systems have historically incorporated scientific-grade cameras with 16 bit analog-to-digital converters (ADC). Such high resolution was always considered necessary to resolve small pressure differences, particularly those encountered in low-speed testing. The system has matured such that we now find ourselves conducting more production-oriented tests, with a concomitant requirement to improve overall productivity. A careful accounting of the system revealed a bottleneck in image acquisition caused by the time required to transfer the image data off the camera. This generated an interest in the possibility of trading off ADC bit depth for faster image download.

Exploring this possibility, the specifications of a number of commercially available cameras were collected by *Brown* et al. [4.152], from which their theoretical performance and effective ADC resolution could be compared. The results are shown in Table 4.2. The current standard PSP camera is the Photometrics CH350^{TM}, shown in the first column. It has an image upload time of 6 s. This compares with times of 2 and 0.12 s for the Photometrics PXL^{TM} camera and the IDT sharpVISION^{TM} 1500EX, which have ADC resolutions of 14 and 12 bits, respectively.

When taking exposures under typical operational conditions, one must also take into account the performance of the charge-coupled devices (CCD). Lower quantum efficiencies will require longer exposure times to accumulate electrons, and higher full-well capacities will require longer times to accumulate maximum signal in order to take full advantage of the ADC resolution. One must also consider the effects of read and shot noise, which reduce the signal-to-noise ratio and can significantly reduce the effective resolution. Taking these factors into account, the effective resolution of the 16 bit Photometrics CH350 camera is actually less than 11 bits, as shown in Table 4.2. The 12 bit resolution of the IDT 1500 is actually a little less than 10. The difference in resolution of these two cameras is more like one or two bits instead of four. Moreover, even though the IDT camera requires almost three times the exposure time of the CH350 camera, its total time per image is less by a factor of six. Perhaps maximum ADC resolution is not optimum, especially considering the high purchase price for high-resolution cameras.

To understand these tradeoffs better, comparative tests were performed on the cameras listed in Table 4.2. Each camera was used to acquire calibration data on a PSP coupon. By using a common coupon, the true performance of each camera in terms of pressure resolution and signal-to-noise ratio could be determined.

Shot noise describes the random variations in accumulating charge in a well of the CCD. This is described by the Poisson probability distribution, which has a standard deviation equal to the square root of the total amount of accumulated charge.

The quantity being accumulated is photons (ph), which are converted to pholoelectrons by the quantum efficiency η, which is then converted to counts by the gain G:

$$\sigma_{\text{shot}} = (\text{ph})^{1/2} \,,$$
$$\sigma_{\text{shot}} = (I\eta/G)^{1/2} \eta/G \,,$$
$$\sigma_{\text{shot}} = (I\eta/G)^{1/2} \,. \tag{4.40}$$

In this way, we can estimate the total random uncertainty in a raw image due to read and shot noise as:

$$U_I^2 = \sigma_{\text{read}}^2 + \sigma_{\text{shot}}^2 \,. \tag{4.41}$$

Read noise is independent of the signal level, while shot noise increases with the square root of the signal. Using these equations and the median image values and median intensity ratios, the computed image statistics are shown in Table 4.2.

The uncertainty for the 12 bit camera should be about twice that of the 16 bit camera. The standard deviation of the ratio of the images, showing the data scatter, are represented in Table 4.2.

Noise was estimated by computing the standard deviation of the residuals, that is, the differences between the pixel data and the least-squares line. The signal level was quantified by computing the range of the least-squares line, namely the maximum minus the minimum value. The results are shown in Table 4.3.

The variation in intensity ratio across the coupon is on the order of 0.01. For this signal level, the 12 bit camera has a signal-to-noise ratio of 1, while the 16 bit camera has a signal-to-noise ratio of 6. Looking at it in another way, the 12 bit camera has an observed noise level of 0.009 in intensity ratio, which corresponds to 0.16 psi. The 16 bit camera has an observed noise level of 0.002 in intensity ratio, corresponding to 0.037 psi. The 12 bit IDT camera can discern pressure variations that are only four times larger than we can currently see with the 16 bit CH350 camera.

With the gain set to 1, the PXL camera has an observed noise level of 0.0007 in intensity ratio, which corresponds to 0.0013 psi. When the gain is set to 4, the observed noise level is increased to 0.0014, corresponding to 0.023 psi. The PXL camera can discern smaller pressure variations when the gain is set to 1 than 4.

The 12 bit camera should not replace the 16 bit camera. The resolution of the 12 bit camera is not high enough to discern actual fluctuations on the order of 0–2 psi. The random uncertainty due to read and shot noise is also up to three times as high for the 12 bit camera and the signal is lost in the noise.

The impact of gain on the PXL cameras has been quantified. These cameras should be used with the gain set to 1, because they have a higher signal-to-noise ratio and the camera can discern smaller pressure fluctuations.

Filters

Light filters are used to separate the luminescence from the excitation light source. Filters can be split into two categories:

- *Interference filters* select a band of light through a complex process of constructive and destructive interference. They consist of a substrate onto which varying thicknesses and types of chemical layers are vacuum deposited in such a fashion that the transmission of certain wavelengths is enhanced, while other wavelengths are either reflected or absorbed. *Bandpass filters* only transmit light in a defined spectral band. The peak central wavelength and spectral width can be controlled very tightly. *Edge filters* only transmit light above (long pass) or below (short pass) a certain wavelength. They are designed to exhibit a sharp transition over a narrow wavelength range and are frequently known as *hot* and *cold mirrors*. Short-pass filters tend to only have a limited stop band and also transmit long wavelengths.

- *Colour glass filters* are used for applications that do not need precise control over wavelengths and transmission intensities. They operate by selective absorption within the glass to give different spectral characteristics and are relatively inexpensive compared to interference filters.

Table 4.3 Camera signal properties

Camera	Run	Signal	$S_{residual}$	S/N ratio
12 bit	1	0.0098	0.0085	1.2
16 bit	2	0.0104	0.0017	6.3
Gain = 1	3	0.0013	0.0007	1.9
Gain = 4	4	0.0013	0.0014	0.9

The key filter characteristic is the ratio of transmission to blocking. The integral over wavelength is computed to compare different filters.

Filters may not be operated above 60 °C and must be cooled to remove the heat caused by the incident and absorbed radiation. All filters are sensitive to the angle of incidence of the incoming light. For interference filters the peak transmission wavelength will decrease for angles away from normal, while the bandwidth and transmission characteristics generally remain unchanged. For colour glass filters increasing the incidence angle increases the transmission path, reducing the transmission efficiency.

Excitation light sources require short-pass filters. Researchers often use a stack of filters to first remove the IR spectrum then block the spectral region of the paint luminescence. Bandpass filters tend to be used on the detector to block the excitation light source.

4.4.4 Processing

All pressure-sensitive paint results are generated from two or more measurements or images. The form of the processing is highly dependent on the experimental method. *Stilwell Bowen* [4.153] recommends that any data-reduction methodology should be user independent. The following processes are often performed on the pressure data.

Remove Self-Illumination Effects

Self-illumination is the reflectance of luminescence from one part of a model to another. It can occur on complex model structures. The signal from any point on the model is the sum of all the rays reflected from other points. *Bukov* et al. [4.129] found that associated pressure errors can be as large as 8% without correction. This forced them to test individual components separately. *Ruyten* describes the processing involved to make these corrections [4.154]. *Engler* et al. [4.151] found that in the most cases the typical error produced by analysing the models in parts is about 3%, but the influence of the test section with its windows and polished walls can create an additional 3–6% error. Therefore it is recommended to cover the test section with absorbing paint or self-adhesive layers, as done by these authors in various wind tunnels with great success.

Calibration

Three types of pressure calibration are routinely performed for pressure-sensitive paint:

- A priori methods either calibrate a coupon of paint or the complete model in a pressure chamber.
- In situ calibration uses the pressure-sensitive paint to interpolate between pressure taps.
- Direct calibration of the complete model in the wind tunnels test section, which is only possible when the tunnel can be pressurised.

The advantage of a priori calibration is that the paint is a standalone transducer that can be used on any model. However, the absolute level of pressure can be difficult to compute due to the many bias errors that affect the measurement. *Engler* [4.132] achieved an accuracy of ±1 mbar for a paint calibrated externally and for the same paint calibrated by changing the pressure of the tunnel without flow. *Engler* [4.132] controlled the temperature to ensure that thermal errors were reduced by using the paint at its minimum sensitivity.

Mebarki [4.155] proposed a general formulation of the Stern–Volmer law where the calibration coefficients are temperature dependent. This a priori allows different temperature conditions to be taken into account between the reference and the wind-on measurements.

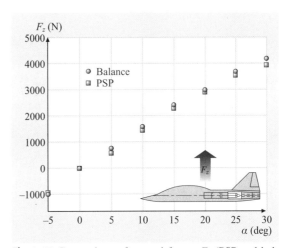

Fig. 4.27 Comparison of normal forces F_z (PSP and balance)

The major sources of bias errors in all calibrations are thermal and illumination errors. These have forced many researchers to apply different calibrations to separate parts of the model. This makes the data-reduction process very user dependent and no longer traceable.

Analysis

Most researchers present their data as calibrated pressure images and chord-wise distributions compared to pressure taps. DLR [4.151] have presented real 360° PSP results from the transonic wind tunnel HST (High Speed Windtunnel) of DNW (Deutsch/Niederländischer Windkanal) at Amsterdam (Fig. 4.27). Three-dimensional

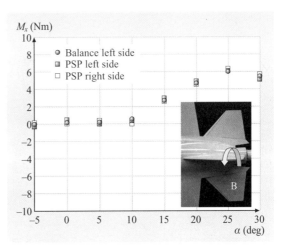

Fig. 4.28 Comparison of the bending moment M_x of the model (PSP and balance)

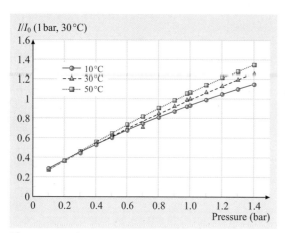

Fig. 4.29 Typical intensity calibration for a pressure-sensitive paint

wind-tunnel models showing the pressure distribution were obtained by pressure-sensitive paint using eight cameras simultaneously. The complete pressure integration around the model surface gives the loads. However, detailed comparison with internal balances shows the accuracy which is possible today.

The goal for industrial pressure-sensitive paint is to produce fully qualified, traceable pressure data in a form suitable for input into aircraft design procedures. For the measurements of Fig. 4.27 a structured grid was used to transform the results from pixel values to the knot points of the grid. For each knot point the coordinates x, y, z and C_p value exist for the complete model. All data were given to the clients in ASCII as well as Tecplot format.

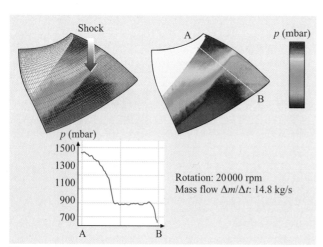

Fig. 4.30 Pressure distribution on a turbine blade

The pressure maps were used for later integration to calculate the total loads of the model (Fig. 4.28) and of single components like the horizontal tail (as shown in Fig. 4.29). The main goal of this test was to obtain detailed information about the accuracy of the PSP data.

The PSP method can be an excellent tool to detect the sources of separation on the model's surface, because balances give information averaged over the complete model, or model parts equipped with balances.

Asymmetric pressure distributions on the wings as obtained by means of PSP made these effects obvious. As far as the quantitative data are concerned the difference between hinge moment balances and PSP is about $\approx 5\%$ in moments for symmetric flow using lenses with a large zoom factor (Fig. 4.29). Online PSP data [4.151] provide a fast overview of the pressure field over the very wide flight envelope of the aircraft, both in terms of speed and angle of attack. This helps reduce the effort, costs and time required for numerical calculations of the pressure field and the loads considerably.

Thermal Compensation

Figure 4.30 shows a typical pressure and temperature calibration for a pressure-sensitive paint, where thermal effects can be seen. The main problem with any thermal correction technique is that the precision of the temperature measurement must be accurate enough that the correction process does not introduce errors into the pressure measurement.

This problem has yet to be adequately solved. *Morris* and *Donovan* [4.156] used a temperature-sensitive paint to generate a thermal map to compensate the pressure-sensitive paint. The two paints were applied and data acquired in separate runs. This technique has been particularly useful in the visualisation of turbine blades as separate blades can be sprayed and the data acquired in the same run.

Researchers have succeeded in combining two dyes in the same binder to make pressure and temperature measurements [4.157]. This method will probably yield the best results as both dyes are in close proximity to one an other and should be at the same temperature. *Henne* [4.119] realised this and used infrared thermography images to correct for thermal errors in the pressure-sensitive paint method.

4.4.5 Applications

Subsonic–Transonic Flow

There have been numerous papers written on the use of pressure-sensitive paint in flows between Mach $= 0.4$

and 1.4. Error analysis predicts that this is probably the flow regime where pressure-sensitive paint is the most effective due to the large changes in pressure and relatively small thermal effects. This flow regime also matches the envelope of many military and civil models in industrial wind tunnels. Pressure-sensitive paint is expected to reduce aircraft development costs by reducing the manufacturing costs and time. Most major developers have now conducted trials using pressure-sensitive paint, although many do not publish the results due to customer confidentiality [4.81–98, 158–175].

Supersonic Flow

Some of the first pressure-sensitive paint results [4.101] were made at Mach = 3.0 on a sphere, a blunt cone and a wedge. Thermal errors were kept to a minimum by choosing paint with a low thermal sensitivity in the temperature range of the experiment.

Early work by *Morris* et al. [4.140] showed that their paints adhered to the model in the regions of highest shear stress in a Mach = 2.0 flow, even though the paint could be damaged by touch. *McLachlan* et al. [4.157] gave a detailed account of a test at Mach = 1.6–2.0. The thermal errors meant that an a priori calibration could not be used, as the temperature was not measured. The best accuracy of 0.017 C_p was achieved by only computing pressure values for single chord lines and using in situ calibration at this time.

Paints were developed at DLR for tests up to Ma = 6.0 for temperatures between −40 °C and +80 °C. These paints resist all the demanding conditions of such a flow, although condensation remains a problem, as the quenching process is constrained and does not then give correct pressure results.

Hypersonic Flow

Borovoy et al. [4.133] published the results of their findings at Mach = 6.0 and 19–74 bar in a shock wind tunnel with a 40 ms run time. At these speeds thermal effects cause the greatest errors.

Images were taken 8–18 ms after the start of a run so that the paint had time to react to pressure changes, but before thermal changes affected the paint characteristics. The PSP results were within 10% of the prediction when compared to computational fluid dynamics (CFD).

Jules et al. [4.137] also reported that thermal errors significantly affected the response of paints tested in the hypersonic regime. They found that beyond a certain temperature threshold the paint intensity degradation became irreversible. *Morris* and *Donovan* [4.156] used a temperature-sensitive paint in a separate run to measure the surface temperature of the model and correct the pressure results.

Low-Speed Flow

Pressure-sensitive paint is an absolute pressure transducer and is not suited to low-range differential pressures. Performance can only be improved by increasing the precision of the measuring instrument. For CCD cameras this equates to reducing the shot noise, which requires averaging sequences of images. *Sajben* [4.176] showed theoretically that C_p measurement errors increase at low Mach numbers due to the small pressure differences generated. However, *Morris* et al. [4.177] have shown that low-speed measurements are possible by careful experiment design. Valid pressure measurements have been obtained at 20 m/s on a NACA 00 12 airfoil model with an error of 0.12 in C_p [4.178].

Torgerson et al. [4.136] managed a pressure resolution of 0.01 psi. They found that model movement gave significant errors and used two-dye paints and phase-sensitive laser scanning to remove the need for reference images. They were also limited by spatial variations of the paint characteristic caused by the microheterogeneity of the polymer structure.

Verhaagen [4.179] used pressure-sensitive paint to compare the vortex structure on a double delta wing to CFD predictions.

In the paper by *Engler* [4.180] describing cooperative work between DLR and ONERA various tests are described from car investigations with the advantage of a hard-shell model (no distortion between wind-off and wind-on image) to obtain high-pressure resolution. They also describe the tests as low speed, but the main problem is temperature correction, because normal low-speed wind tunnels have no coolers and the temperature changes during the runtime of a complete set of a alpha-polar requires about 30 min and already creates temperature changes of up to 20 °C. *Henne* [4.119] solved this problem using infrared cameras for pixel-by-pixel temperature correction.

Turbine Measurement

The measurement of pressures on turbine blades has traditionally been difficult. The use of pressure-sensitive paint offers a potential solution but there are problems that need to be solved. These include: poor optical access, large thermal gradients, the need to *freeze* rapidly moving blades and the problems of generating reference images. There appear to be few papers published in the open literature, but several outstanding experiments were presented at the Fifth Pressure-Sensitive Paint Workshop held at Arnold Engineering Development Center (AEDC), USA in May 1997.

It is also known that TsAGI have made significant advances measuring rotating machinery [4.181]. *Trinks* [4.182] investigated the shock location on turbomachines as a function of the mass flow. The rotation of the blades was 20 000 rpm at a temperature of about 40 °C.

Two-dimensional flow measurements have been made on the end wall of a cascade wind tunnel. *Burns* and *Sullivan* [4.145] used a scanning laser to excite PSP on a fan blade. A photomultiplier tube detected the paint decay and the pressure was computed by a lifetime measurement.

In Flight

McLachlan et al. [4.118] imaged, from the inside, the pressure distribution of a pressure plotted window coated with a UV-excited paint. The experiment was self-contained and mounted under an F-104G aircraft flying between Mach $= 1.0$ and 1.6. A standard 8 bit video camera was used. The experiment had to be done at night to minimise ambient light intrusion. An in situ calibration was used due to avoid photodegradation. An accuracy of 35 psf in the range of 400 psf was achieved at Mach $= 1.6$. At Mach $= 1.3$ thermal effects dominated the results.

Houck et al. [4.148] conducted a flight test to image the pressure distribution on a practice bomb coated with a blue-light-excited paint. The experiment used a conventional film camera and custom flash light sources as a self-contained unit attached to an adjacent aircraft hard point. The film was processed and digitised by projection onto a scientific-grade CCD camera. Luminescent reference markers were used to correct for illumination fluctuation.

Unsteady Measurements

Investigations of unsteady processes are becoming increasingly important in aerodynamic. A large amount of experience exists for steady measurements, but the industry and some aerodynamic areas such as aeroelasticity and turbomachinery need new, fast measurement techniques like the nonintrusive and planar PSP method, also to optimise their own CFD calculations. The first steps in paint development have been taken, with short response times in the region of 0.1 ms to 50 µs.

Winslow et al. [4.113] analysed the response time of a PSP luminescence which obeys a linear Stern–Volmer relation. According to their theory, the time to reach 90% of the total response is:

$$t_{90\%} = 0.85 \frac{h^2}{D}, \qquad (4.42)$$

where h denotes the thickness of the PSP layer and D the gas diffusivity in the layer. The flow in the micropore is close to free-molecular flow, because the diameter of the micropore in the anodised aluminum is less than the mean free path of gas molecules. The diffusion process in such a case is called Knudsen diffusion. The gas diffusivity of the Knudsen diffusion is

$$D = d\sqrt{\frac{8RT}{9\pi}}, \qquad (4.43)$$

where d denotes the pore diameter, R the gas constant for a unit mass, T the absolute temperature, and π the ratio of the circumference of a circle to its diameter. The gas constant for air is 287 J/(kg K). Note that the diffusion coefficient is linearly proportional to d.

The response of the hydrophobic AA-PSP to a step change in pressure was tested by using a shock tube. Such a shock tube can produce a sharp, discontinuous pressure change (normal shock) with a rise time of less than a microsecond. The PSP sample was placed at the end of the tube. A photomultiplier tube was used to measure the luminescent intensity.

In terms of the hardware, the main problem is still data acquisition using fast complementary metal oxide semiconductor (CMOS) CCD cameras and fast data storage. The first steps have mainly been made in paint development. After the successful development of *conventional* PSP, industrial partners are now encouraging the investigation of these more-complex problems.

4.4.6 Concluding Remarks

Pressure-sensitive paints have been developed since the early 1980s. A number of theoretical concepts are involved, but in the late 1990s the technique reached sufficient maturity to be used as a tool in industrial wind tunnels.

Pressure-sensitive paint is an absolute transducer that converts a unit of pressure into a unit of light. The paint can be calibrated in situ or a priori. There are several methods to acquire the data: radiometric, lifetime and fluorescent lifetime imaging etc.. The raw data requires image processing to produce pressure results.

Optical pressure measurement has been used in production testing for subsonic and supersonic up to cryogenic flows [4.183–187]. Developers are beginning to solve the problems involved in low-speed and hypersonic applications. Satisfactory thermal compensation techniques are in progress but data reduction is still very user dependent.

The goal for industrial pressure-sensitive paint is to produce fully qualified, traceable pressure data in a form suitable for input into aircraft design procedures.

References

4.1   D.W. Bryer, R.C. Pankhurst: *Pressure Probe Methods for Determining Wind Speed and Flow Direction*, Technical Report (Her Majesty's Stationery Office, London 1971), .

4.2   S.H. Chue: Pressure probes for fluid measurement, Prog. Aerosp. Sci. **16**, 147–223 (1975)

4.3   A. Thom, C.J. Appelt: The pressure in a two-dimensional static hole at low Reynolds numbers, Rep. Memor. Aero. Res. Council **3090** (1957)

4.4   R.E. Rayle Jr.: An investigation of the influence of orifice geometry on static pressure measurements, MS Thesis (M.I.T., Boston 1949)

4.5   C.M. Allen, L.J. Hooper: Piezometer investigation, ASME Trans. **54**(HYD-54-1) (1932)

4.6   C. Ducruet, A. Dyment: The pressure-hole problem, J. Fluid Mech. **142**, 251–267 (1984)

4.7   J.L. Livesey, J.D. Jackson, C.J. Southern: The static-hole error problem: an experimental investigation of errors for holes of varying diameters and depths, Aircraft Eng. **34**, 43–47 (1962)

4.8   R. Shaw: The influence of hole dimensions on static pressure measurements, J. Fluid Mech. **7**, 550–564 (1960)

4.9   A. Miyadzu: Über den Einfluss der Bohrungen auf die Druckanzeige, Ingen.-Arch. **7**, 35–41 (1936), in German

4.10   A.K. Ray: On the effect of orifice size on static pressure reading at different Reynolds numbers, Ing.-Arch. **24**(3) (1956), , Engl. transl. Aero. Res. Counc. unclassified 'A3' report 18829, FM 2479, EA 527, TP 498

4.11   R.E. Franklin, J.M. Wallace: Absolute measurements of static-hole error using flush transducers, J. Fluid Mech. **42**, 33–48 (1970)

4.12   G. Fuhrmann: Theoretische und experimentelle Untersuchungen an Ballonmodellen, Ph.D. Thesis (Göttingen University, Göttingen 1912)

4.13   B.J. McKeon, A.J. Smits: Static pressure correction in high Reynolds number fully developed turbulent pipe flow, Meas. Sci. Tech. **13**, 1608–1614 (2002)

4.14   W. J. Rainbird: Errors in measurement of mean static pressure of a moving fluid due to pressure holes. In: Quart. Bull. Div. Mech. Eng. Nat. Aero. Est. DME/NAE #3, (Nat. Res. Council, Ottawa, 1967)

4.15   E.B. Plentovich: Status of orifice induced pressure error studies, AIAA **84-0245** (1984)

4.16   C. Ducruet: A method for correcting wall pressure measurements in subsonic compressible flow, J. Fluids Eng. **113**, 256–260 (1991)

4.17   M.V. Zagarola: Mean-flow scaling of turbulent pipe flow, Ph.D. Thesis (Princeton University, Princeton 1996)

4.18   V.A. Sandborn: *Experimental evaluation of momentum terms in turbulent pipe flow*, Tech. Note, Vol. 3266 (NACA Lewis Flight Prop. Lab., Cleveland 1955)

4.19   G.K. Patterson, W.J. Ewbank, V.A. Sandborn: Radial pressure gradient in turbulent pipe flow, Phys. Fluids **10** (1967)

4.20   T. Christiansen, P. Bradshaw: Effect of turbulence on pressure probes, J. Phys. E **14**, 992–997 (1981)

4.21   A. Pope, J.J. Harper: *Low-speed Wind Tunnel Testing* (Wiley, New York 1966)

4.22   H.W. Liepmann, A. Roshko: *Elements of Gas Dynamics* (Wiley, New York 1957)

4.23   S. Goldstein: A note on the measurement of total head and static pressure in a turbulent stream, Proc. Roy. Soc. London A **155**(886), 570–575 (1936)

4.24   A. Fage: On the static pressure in fully-developed turbulent flows, Proc. Roy. Soc. A **155**(886), 576–596 (1936)

4.25   P. Bradshaw, D.G. Goodman: The effect of turbulence on static-pressure tubes, Rep. Memor Aero. Res. Council **3527** (1968)

4.26   B.L. Welsh, P.R. Ashill: *Pressure Measurement Techniques in Use at the Royal Aerospace Establishment*, von Karman Institute Lecture Series (1989)

4.27   R.H. Engler, K. Hartmann, B. Schulze: A new optical pressure measurement system (OPMS) ICIASF' 91 Record, Rockville (1991) pp. 163–170

4.28   O.S. Wolfbeis, M.J.P. Leiner: Recent progress in optical oxygen sensing, SPIE Proc. **906**, 42–48 (1988)

4.29   A. Baron, M. Gouterman: 100 µs images of unsteady surface flow using fast responding pressure sensitive paint, Proc. 7th Int. Symp. Flow Visualization, Seattle (1995)

4.30   D. Lubbers, N. Opitz: Quantitative fluorescence photometry with biological fluids and gases, Adv. Exp. Med. Bin. **75**, 441–448 (1976)

4.31   M. P. Gouterman, I. L. Kavandi, J. Gallery, J. B. Callis: Surface pressure measurement by oxygen quenching of luminescence, European Patent Application 0 472 243 A2, August (1991)

4.32 K. Asai, H. Kanda, J.P. Sullivan: Boundary layer transition detection in a cryogenic wind tunnel using luminescent paint, J. Aircraft **34**(l), 34–42 (1997)

4.33 J. H. Bell, B. G. McLachlan: Image registration for luminescent paint sensors, AIAA J. (1993) paper 93-0178

4.34 A. Orlov, V. Radchenko, N. Sadovskii, I. Truyanovsky: Luminescent pressure sensitive composition, European Patent Application 0 558 771 Al, March (1992)

4.35 Y. Mebarki, M. C. Merienne, Y. Le Sant: Application d'un revetement luminescent a deux composes pour la mesure de la pression et de la temperature, Societe Francais des Thermiciens (SFT), Journee d'Etudes (1997)

4.36 J.R. Lakowicz, et al.: Fluorescent lifetime imaging, Anal. Biochem. **202**, 316–330 (1992)

4.37 C. G. Morgan: Measurement of luminescence, US Patent 5,459,323 (1995)

4.38 D. M. Oglesby, C. K. Puram, B. T. Upchurch: Optimisation of measurements with pressure sensitive paints, NASA TM (4695), June (1995)

4.39 A. P. Bukov, et al.: Application of luminescent quenching for pressure field measurements on the model surface, Proc. Wind Tunnels Wind Tunnel Test Tech. Conf., Royal Aeronautical Society, Southampton (1992)

4.40 T. Araki, H. Misawa: Light emitting diode-based nanosecond ultraviolet light source for fluorescence lifetime measurements, Rev. Sci. Inst. **66**, 2 (1995)

4.41 Y. Le Sant, M-C. Merienne: An image resection method applied to mapping techniques, IEEE ICIASF Record Wright-Patterson AFB, 46.1–46.8 (1995)

4.42 J. W. Holmes: Fluorescent lifetime imaging for pressure sensitive paint, GARTEUR AD(AG-21) Final Report, HWA/CN/RD/98006/1 (1998)

4.43 R.M. Dowgwillo, M.J. Morris, J.F. Donovan, M.E. Benne: Pressure sensitive paint in transonic wind-tunnel testing of the F-15, J. Aircraft **33**, 1 (1996)

4.44 R. M. Dowgwillo, et al.: The application of the pressure sensitive paint technique to high speed wind tunnel testing of a fighter aircraft configuration with complex store loadings, AIAA J. (1994) paper 94–1932

4.45 J. P. Crowder: Flow visualisation in flight testing, AIAA J. 90-1273 5th Biannual Flight Testing Conference (1990)

4.46 R. H. Engler, K. Hartmann, I. Troyanovski, A. Vollan: Description and assessment of a new optical pressure measurement system (OPMS) demonstrated in the high speed wind tunnel of DLR in Göttingen, DLR-FB 92-24, 1–66 (1992)

4.47 M.C. Merienne, Y. Mebarki: Contribution to the study of wing-nacelle interaction, Proc. CEAS Wind Tunnels and Wind Tunnel Test Tech. Conf., Cambridge (1997)

4.48 M. Kammeyer, C. Kelble, J. Donovan, M. Benne, T. Kihlken: Recent improvements in pressure-sensitive paint measurement accuracy at, Boeing, AIAA 2002-2907, 22nd AIAA Aerodynamic Measurement Technology and Ground Test Conference, St. Louis (2002)

4.49 M. Kammeyer, J. Donovan, C. Kelble, M. Benne, T. Kifilken, J. Felter: Accuracy assessment of a pressure-sensitive paint measurement system, AIAA 2002-0530, 40th AIAA Aerospace Sciences Meeting and Exhibit, Reno (2002)

4.50 T. Gadella, T. Jovin, R. Cegg: Fluorescence lifetime imaging microscopy (FLIM): Spatial resolution of microstructures on the nanosecond time scale, J. Biophys. Chem. **48**, 221–239 (1993)

4.51 C. Klein: Einsatz einer druckempfindlichen Beschichtung (PSP) zur Bestimmung des momentanen Druckfeldes von Modellen im Windkanal. Dissertation (University Göttingen), DLR-Forschungsbericht 97-55 (1997)

4.52 R.H. Engler, M.C. Merienne, C. Klein, Y. Le Sant: Application of PSP in low speed flows, Aerosp. Sci. Technol. **6**, 313–322 (2002), DLR Göttingen and ONERA Meudon (2002)

4.53 Y. Iijima, Y. Egami, A. Nishizawa, K. Asai, U. Fey, R. H. Engler: Optimization of temperature-sensitive paint formulation for large-scale cryogenic wind tunnels, 20th ICIASF'03 Record, Göttingen, (2003) 70–76

4.54 H. Sakaue, J.W. Gregory, J.P. Sullivan, S. Raghu: Porous pressure sensitive paint for characterizing unsteady flowfields, AIAA J. **40**, 1094–1098 (2002)

4.55 K. Asai, H. Kanda, C. T. Cunningham, R. Erausquin, J. P. Sullivan: Surface pressure measurement in a cryogenic wind tunnel by using luminescent coating, International Congress on Instrumentation in Aerospace Simulation Facilities '97 Record (1997) pp. 105–114

4.56 Y. Amao, K. Asai, I. Okura: Photoluminescent oxygen sensing using palladium tetrakis(4-carboxyphenyl)porphyrin self-assembled membrane on alumina, Anal. Commun. **36**, 170–180 (1999)

4.57 M. Kameda, N. Tezuka, T. Hangai, K. Asai, K. Nakakita, Y. Amao: Adsorptive pressure-sensitive coatings on porous anodized aluminium, Meas. Sci. Technol. **15**, 489–500 (2004)

4.58 V. Mosharov, V. Radchenko, A. Orlov: Binary pressure paint: A lot of problems, Presented at the 7th Annual Pressure Sensitive Paint Workshop, Purdue Univeristy, West Lafayette (1999)

4.59 R. H. Engler: Development of pressure sensitive paint, Presented at the 7th Annual Pressure SensitivePaint Workshop, Purdue Univeristy, West Lafayette (1999)

4.60 J. Crafton, S. Fonov, E. Jones, L. Goss, C. Carter: Bi-luminophore pressure-sensitive paint development, 8th Annual Pressure Sensitive Paint Workshop, NASA Langley Research Center (2000)

4.61 M. Kameda, T. Tabei, K. Nakakita, H. Sakaue, K. Asai: Image measurements of unsteady pressure fluctuation by a pressure-sensitive coating on porous anodized aluminum, Meas. Sci. Technol. (2005)

4.62 J. W. Gregory: Porous pressure-sensitive paint for measurement of unsteady pressures in turbomachinery, Proc. 42nd AIAA Aerospace Sci. Meeting Exhibit, AIAA-2004 294 (2004)

4.63 Y. Sakamura, M. Matsumoto, T. Suzuki: High framerate imaging of surface pressure distribution using a porous pressure-sensitive paint, Meas. Sci. Technol. **16**, 759–765 (2005)

4.64 T. Kawakami, T. Tabei, M. Kameda, K. Nakakita, K. Asai: Unsteady pressure-field measurements by a fast-responding PSP on porous anodized aluminum, Proc. 11th Int. Symp. Flow Visualization (2004) Paper No. 217

4.65 H. Sakaue, T. Tabei, M. Kameda: Hydrophobic monolayer coating on anodized aluminum pressure-sensitive paint, Sens. Actuat. (2005) (in preparation)

4.66 H. Masuda, K. Fukuda: Ordered metal nanohole arrays made by a two-step replication of honeycomb structures of anodic alumina, Science **268**, 1466–1468 (1995)

4.67 H. Masuda, F. Hasegawa, S. Ono: Self-ordering of cell arrangement of anodic porous alumina formed in sulfuric acid solution, J. Electrochem. Soc. **144**, L127–L130 (1997)

4.68 H. Masuda, K. Yada, A. Osaka: Self-ordering of cell configuration of anodic porous alumina with large-size pores in phosphoric acid solution, Jpn. J. Appl. Phys. **37**, L1340–L1342 (1998)

4.69 M. Saito, Y. Shiga, M. Miyagi, K. Wada, S. Ono: Optical transmittance of anodically oxidized aluminum alloy, Jpn. J. Appl. Phys. **34**, 3134–3138 (1995)

4.70 T. Hangai, M. Kameda, K. Nakakita, K. Asai: Time response characteristics of pyrene-based pressure-sensitive coatings on anodic porous alumina, Proc. 10th Int. Symp. Flow Visualization (2002) Paper No. F0269

4.71 I. Gursul: Recent development in delta wing aerodynamics, Aeronautical J. **108**, 437–452 (2004)

4.72 D.G. Mabey: Some aspects of aircraft dynamic loads due to flow separation, Prog. Aerospace Sci. **26**, 115–151 (1989)

4.73 N. Chandrasekharan, M. Hammer, L. Kelly, L. A. Mattes: A paradigm shift for pressure sensitive paints, The 8th Pressure Sensitive Paint Workshop, NASA Langley Research Center (2000)

4.74 Y. Sakamura, Y. Kidoh, T. Suzuki, M. Matsumoto: Time resolved pressure measurements in shock-induced flows usinga pressure-sensitive paint, Proc. 23rd Int. Symp. Shock Waves, Fort Worth (2001) pp. 456–462

4.75 R. Crites, M. Benne: Emerging technology for pressure measurements in wind tunnels – pressure sensitive paint, AIAA 95-0106, Presented at the 33rd Aerospace Sciences Meeting, Reno (1995)

4.76 J. H. Bell: Accuracy limitation of lifetime-based pressure sensitive paint measurements, 19th Int. Congr. Instrum. Aerospace Simulation Facilities, Cleveland (2001)

4.77 L. P. Goss, D. D. Trump, B. Sarka, L. N. Lydick, W. M. Baker: Multi-dimensional time-resolved pressure-sensitive-paint techniques: A numerical and experimental comparison, AIAA-2000-0832, Presented at the 38th Aerospace Sciences Meeting, Reno (2000)

4.78 T. Liu, S. D. Torgerson, J. P. Sullivan, R. Johnston, S. Fleeter: Rotor blade pressure measurement in a high speed axial compressor using pressure and temperature sensitive paints, AIAA-97-0162, Presented at the 35th Aerospace Sciences Meeting, Reno (1997)

4.79 A. G. Davies: Temperature compensated PSP measurements on a 2D wing in transonic flow, Presented at the 6th Pressure-Sensitive Paint Workshop, Seattle (1998)

4.80 J.W. Holmes: Analysis of radiometric, lifetime and fluorescent lifetime imaging for pressure sensitive paint, Aeronaut. J. **2306**, 189–94 (1998)

4.81 W. M. Baker: Recent experiences with pressure sensitive paint testing AIAA-2001-0306, Presented at the 39th Aerospace Sciences Meeting, Reno (2001)

4.82 M. E. Sellers: Application of pressure sensitive paint for determining aerodynamic loads on a scale model of the F-16C AIAA-2000-2528, Presented at the 21st Aerodynamic Measurement Technology and Ground Testing Conference, Denver (2000)

4.83 W.M. Ruyten: Self-Illumination Calibration Technique for Luminescent Paint measurements, Rev. Sci. Instrum. **68**(7), 3452–57 (1997)

4.84 S. Ponomarov, M. Gouterman: Ideality of Pressure-Sensitive Paint, I. Platinum Tetra(penta-fluorophenyl)porphine in Fluoroacrylic Polymer, J. Appl. Polym. Sci. **77**(8), 2795 (2000)

4.85 M. Lyonnet, B. Deleglise, G. Grenat, A. Bykov, V. Mosharov: The two-component PSP investigation on a civil aircraft model in S2MA wind tunnel, AGARD Conf. Proc. CP-601: Adv. Aerodyn. Meas. Technol., 81st Fluid Dyn. Panel Sym, Seattle, pp 30–1 – 30–8. Neuilly -sur-Seine, France: AGARD (1998)

4.86 V. Mosharov, V. Radchenko, S. Fonov: Luminescent pressure sensors in aerodynamic experiment, Moscow: Cent. Aerohydrodyn. Inst. (TsAGI). CWA Int. Corp. (1997) p. 151

4.87 T. Liu, B.T. Campbell, S.P. Burns, J.P. Sullivan: Temperature- and pressure-sensitive luminescent paints in aerodynamics, Appl. Mech. Rev. **50**(4), 227–246 (1997)

4.88 K. Nakakita, M. Kurita, K. Mitsuo: Development of the pressure-sensitive paint measurement for large wind tunnels at japan aerospace exploration agency, 24th Congress of the International Council of the Aeronautical Sciences, Yokohama (2004) ICAS2004-3.2.2

4.89 J. H. Bell: Applications of pressure sensitive paint to testing at very low flow speeds, 42nd AIAA.

Aerospace Sciences Meeting & Exhibit, Reno (2004) AIAA-2004-0878

4.90 K. Mitsuo, Y. Iijima, A. Nishizawa, K. Nakakita, K. Asai: Application of PSP. Measurement to low speed wind tunnel testing using an automobile model, Proc. MOSAIC Workshop, Tokyo (2003) pp.70–71

4.91 Y. Mébarki, K. R. Cooper: Aerodynamic testing of a generic automotive model with pressure sensitive paint, 10th Int. Symp. Flow Visualization, Kyoto (2002) ISFV-2002-F0120

4.92 Y. Le Sant, F. Bouvier, M. C Merienne, J. L Peron: Low speed tests using PSP at ONERA, 39th AIAA Aerospace Sciences Meeting & Exhibit, Reno (2001) AIAA 2001-0555

4.93 Y. Shimbo, R. D. Mehta, B. J. Cantwell: Vortical flow field investigation using the pressure sensitive paint technique at low speed, 35th AIAA Aerospace Science Meeting, Reno (1997) AIAA 97-0388

4.94 Y. Shimbo, N. Komatsu, K. Asai: Pressure sensitive paint application at large production wind tunnels, 22nd International Congress of Aeronautical Science, Harrogate (2000) ICAS2000-3.3.3

4.95 C. Y. Huang, H. Sakaue, J. W. Gregory, J. P. Sullivan: Molecular sensors for MEMS, 40th Aerospace Sciences Meeting & Exhibit, Reno (2002) AIAA2002-0256

4.96 C. Y Huang, J. P Sullivan: Flow visualization and pressure measurement in micronozzles, 11th Int. Symp. Flow Visualization, Univ. of Notre Dome (2004)

4.97 A. Davies, D. Bedwell, M. Dunleavy, N. Brownjohn: Pressure sensitive paint limitations and solutions, Proc. 17th International Congress on Instrumentation in Aerospace Simulation Facilities, Pacific Grove (1997) pp.11–21

4.98 M. Hamner, B. Campbell, T. Liu, J. P Sullivan: A scanning laser system for temperature and pressure sensitive paint, AIAA J. (1994) 94-0728

4.99 V.O. Stern, M. Volmer: Über die Abklingungszeit der Fluoreszenz, Phys. Z. **20**, 183 (1919), in German

4.100 J.I. Peterson, R.V. Fitzgerald: Rev. Sci. Instrum. **51**, 670 (1980)

4.101 M.M. Ardasheva, L.B. Nevskii, G.E. Pervushin: Measurement of pressure distribution by means of indicator coatings, J. Appl. Mech. Techn. Phys. **2**, 469–474 (1985)

4.102 M. Brenci: Fibre optic optrodes for chemical sensing, Proc. Opt. Fibre Sensors (1993)

4.103 M. E. Sellers, I. A. Brill: Demonstration test of pressure sensitive paint in the AEDC 16ft transonic wind tunnel, AIAA paper 94-2481, 18th AIAA Ground Testing Conference (1994)

4.104 W. Holmes: The relevance of pressure sensitive paint to aerodynamic research, J. Fluoresc. **3**(3), 179–183 (1994)

4.105 B. Carroll, M. Morris: Step response of pressure sensitive paint, AIAA J. **34**(3), 521–526 (1996)

4.106 The Boeing Aircraft Company, St Louis (2006)

4.107 J. W. Holmes: Pressure sensitive paint measurements in the DRA 8 ft × 8 ft high speed wind tunnel (GARTEUR Version) DRA/AS/HWA/TR95051/1, October (1995)

4.108 S. D. Schwab, R. M. Levy: Pressure sensitive paint formulations and methods, US Patent 5,359,887, Nov. 1 (1994)

4.109 G. M. Zharkova, A. I. Maksinov, A. A. Pavlov, V. M. Khachaturyan: Pressure visualization on aerodynamic surface by the method of luminescent coating, Proc. 6th Int. Symp. on Flow Visualization Yokohama (Springer, Heidelberg 1992)

4.110 F. Rodriguez: *Principles of Polymer Systems*, 2nd edn. (McGraw-Hill, New York 1987)

4.111 H. Sakaue, J.P. Sullivan: Time response of anodized aluminum pressure-sensitive paint, AIAA J. **39**, 1944–1949 (2001)

4.112 N.A. Winslow, B.F. Carroll, A.J. Kurdila: Model development and analysis of the dynamics of pressure-sensitive paints, AIAA J. **39**, 660–666 (2001)

4.113 R. H. Engler: Pressure sensitive paints in quantitative wind tunnel studies, Proc. 11th Int. Symp. Flow Visualization, University of Notre Dame, South Bend (2004)

4.114 T. Liu, J.P. Sullivan: *Pressure and Temperature Sensitive Paint* (Springer, Heidelberg 2004)

4.115 R. H Engler: PSP and TSP for different Wind Tunnels and Flow Facilities, The 8th International Symposyium on Fluid Control, Measurement and Vizualization, Chengdu (2005)

4.116 K.R. Schanze, B.F. Carroll, S. Korotkevitch, M. Morris: Temperature dependence of pressure sensitive paints, AIAA J. **35**(2), 306–310 (1997)

4.117 J. Kavandi: Luminescent barometry in wind tunnels, Rev. Sci. Instrum. **61**, 11 (1990)

4.118 B. G. McLachlan, J. H. Bell: Flight testing of a luminescent surface pressure sensor, NASA Tech. Memorandum 103970 (1992)

4.119 U. Henne: Application if the PSP technique in low speed range wind tunnels, STAB Workshop, Göttingen (2005)

4.120 C. Parker: *Photoluminescence of Solutions* (Elsevier, Amsterdam 1968), Chapter 1C

4.121 N.J. Turro: *Modern Molecular Photochemistry* (Univ. Science Books, New York 1991)

4.122 O. Wolfbeis: Fibre-optical fluor sensors in analytical and clinical chemistry, Chem. Anal. **77**, 129 (1988)

4.123 A. Vollan, L. Alati: A new optical pressure measurement system, 14th ICIASF Congress (1991) p. 3

4.124 B. Ranby, J.F. Rabek: *Singlet Oxygen: Reactions with Organic Compounds and Polymers* (Wiley, New York 1978)

4.125 A. Juris, V. Balzani, F. Barigelld: Ru (II)polypyridine complexes: Photophysics, photochemistry, electrochemistry, and chemo luminescence, Coord. Chem. Rev. **84**, 85277 (1988)

4.126 F. Lythe: Die luminescence of Tris (Bipyridyl) ruthenium (11) chloride, J. Am. Chem. Soc. **91**(2), 131–137 (1969)

4.127 S. Pauly: Permeability and diffusion data. In: *Polymer Handbook*, ed. by J. Brandrup, E.H. Immergut (Wiley Interscience, New York 1989)

4.128 J.R. Welty: *Engineering Heat Transfer* (Whiley, New York 1974) pp. 102–114

4.129 A. Bukov, V. Pesetsky: Optical surface pressure measurements: Accuracy and application field evaluation, Proc. AGARD CP535, Brussels, 73rd Fluid Dynamics Panel, October (1993) Paper 24

4.130 E.U. Condon, H. Odishaw: *Handbook of Physics* (McGraw Hill, New York 1958)

4.131 B. Carroll, A. Winslow, J. Abbitt, K. Schanz, M. Morris: Pressure sensitive paint application to a sinusoidal pressure fluctuation, IEEE ICIASF Record Wright–Patterson AFB (1995) 35/1–35/6

4.132 R. H. Engler: Further developments of pressure sensitive paints (OPMS) for non flat models in steady transonic flow and unsteady conditions IEEE ICIASF Record Wright–Patterson AFB (1995) 33/1–33/8

4.133 V. Borovoy, A. Bukov, V. Mosharov, et al.: Pressure sensitive paint in shock wind tunnel, IEEE ICIASF Record Wright–Patterson AFB (1995) 34/1–34/4

4.134 S. F. Hoemer: Fluid–dynamic drag, Library on Congress Catalog Card Number 64–19666 (1965) chapters 2 and 5

4.135 U. Fey, R. H. Engler Y. Egami, Y. Iijima, K. Asai, U. Jansen, J.Quest: Transition detection by temperature sensitive paint at cryogenic temperatures in the European Transonic Wind Tunnel (ETW), ICIASF'03 Record, Göttingen (2003) pp. 77–88

4.136 S. D. Torgerson, T. Liu, J. P. Sullivan: Use of pressure sensitive paints in low speed flows, AIAA–96–2184,19 AIAA Advanced Measurement and Ground Testing Technology Conference (1996)

4.137 K. Jules, M. Carbonaro, S. Zemsch: Application of pressure sensitive paints in hypersonic flows, NASA Tech. Memorandum 106824, February (1995)

4.138 J. W. Grate, M. H. Abrahams: Solubility properties of siloxane polymers for chemical sensors, Pacific Northwest Fibre Optic Sensors Workshop 3–4, SPIE 2574:71–77 (1995)

4.139 K. Goswami, S. Klainer: Fibre optic chemical sensor for the measurement of partial pressure of oxygen, SPIE **990** (1988)

4.140 M. J. Morris, et al.: Aerodynamic applications of pressure sensitive paint, AIAA J. (1992) paper 92–0264

4.141 J. F. Donovan, et al.: Data analysis techniques for pressure–and temperature-sensitive paint, AIAA J. (1993) paper 93–0176

4.142 C. Klein, R. H. Engler: First results using the new DLR PSP system–Intensity and lifetime measurements, 43.1–43.9 Proc. CEAS Wind Tunnels Wind Tunnel Test Tech. Conf. Cambridge (1997)

4.143 A. Draeijer, R. Sanders: Fluorescence lifetime imaging: a new tool in confocal microscopy. In: *Handbook of Confocal Microscopy* (Plenum, New York 1995)

4.144 A. G. Davies: Recent developments in pressure sensitive paint measurements using the BAe system, 28.1–28.11 Proc. CEAS Wind Tunnels Wind Tunnel Test Techn. Conf. Cambridge (1997)

4.145 S. P. Burns, I. P. Sullivan: The use of pressure sensitive paints on rotating machinery, IEEE ICIASF Record Wright –Patterson AFB (1995) 32.1–32.14

4.146 J. W. Holmes: Analysis of radiometric, lifetime and fluorescent lifetime imaging, J. Roy. Aeronaut. Soc. (1998) Paper 2306

4.147 R. C. Crites: Pressure sensitive paint technique, Measurement Techniques Lecture Series 1993–05, von Karman Institute for Fluid Dynamics (1993)

4.148 S. W. Houck, R. G. Hepp, M. I. Morris, M. E. Benne: Pressure sensitive paint flight test, IEEE Aerospace Applications Conf., Aspen Co 4:241–252 (1996)

4.149 C. A. Fuentes, J. D. Abbitt: Development of a film-based pressure sensitive paint technique, AIAA 96–2933, 32nd AIAA/ASME/SAE/ASEE Joint propulsion conference, Lake Buena Vista, FL (1996)

4.150 R. D. La Belle, S. D. Garvey: Introduction to high performance CCD cameras, IEEE ICIASP Record Wright–Patterson AFB (1995)

4.151 R. H. Engler, U. Fey, U. Henne, C. Klein, W. E. Sachs: Pressure sensitive paints and temperature sensitive paints in quantitative wind tunnel studies, J. Visualization (2004) paper 04–053

4.152 N. Brown, M. E. Benne, M. E. Kammeeyer: Factors influencing camera selection for the boing pressure sensitive paint system, Proc. 42nd AIAA Aerospace Sciences Meeting and Exhibit (2004) AIAA-2004-294

4.153 S. S. Bowen: Comparison of motion estimators for an intensity variant image sequence, SPIE Image and Video Processing H 2182 (1994)

4.154 W.M. Ruyten: Self-illumination calibration technique for luminescent paint measurements, Rev. Sci. Instrum. **68**(7), 3452–3457 (1997)

4.155 Y. Mebarki: Peintures sensibles a lapression: Application en soufflerie aerodynamique, Thesis (University of Lille, Lille 1997)

4.156 M. J. Morris, J. F. Donovan: Application of pressure- and temperature sensitive paints in high speed flows, AIAA J. 94–2231, 25th AIAA Fluid Dynamics Conference (1994)

4.157 B.G. McLachlan, J.H. Bell, H. Park, et al.: Pressure-sensitive paint measurements on a supersonic high-sweep oblique wing model, J. Aircraft **32**(2), 470–483 (1995)

4.158 J. Holmes: Analysis of radiometric, lifetime and fluorescence lifetime imaging, Aeronautical J. **2306**, 189–194 (1998)

4.159 L. Goss, D. Trump, B. Sarka, L. Lydick, W. Baker: Multi-dimensional time-resolved pressure-sensitive-paint techniques: A numerical and experimental comparison (2000) AIAA 2000–0832

4.160 L. Coyle, M. Gouterman: Correcting Lifetime Measurements for Temperarure, Sens. Actuat. B **61**, 92–99 (1999)

4.161 J. Hradil, C. Davis, C. Mongey, C. McDonagh, B. D MacCraith: Temperature-corrected pressure-sensitive paint measurements using a single camera and a dual-lifetime approach, Meas. Sci. Technol. **13**, 1552–1557 (2002)

4.162 K. Mitsuo, Y. Egami, H. Suzuki, H. Mizushima, K. Asai: Development of lifetime imaging system for pressure-sensitive paint (2002) AIAA 2002–2909

4.163 K. Mitsuo, K. Asai, A. Takahashi, H. Mizushima: Advanced lifetime PSP imaging system for simultaneous pressure and temperature measurement (2004) AIAA 2004–2188

4.164 A. Watkins, J. Jordan, B. Leighty, J. Ingram, D. Ogelsby: Development of next generation lifetime PSP imaging system, Proc. 20th Int. Congr. Instrum. Aerospace Simulation Facilities, Gottingen (2003) pp. 372–377

4.165 R. H. Engler: PSP/Acoustic circulation method for surface pressure and flow field investigation around a delta wing, Proc. 21th Int. Congr. Instrum. Aerospace Simulation Facilities, Sendai (2005)

4.166 B. Schulze. C. Klein: Light emitting surfaces of wind tunnel models for excitation of pressure sensitive paint, Proc. 21th Int. Congr. Instrum. Aerospace Simulation Facilities, Sendai (2005)

4.167 W. Ruyten: Assimilation of physical chemistry models for lifetime analysis of pressure-sensitive paint (2004) AIAA 2004–0880

4.168 W. Ruyten, M. Sellers: Lifetime analysis of pressure-sensitive paint PtTFPP in FIB′ (2004) AIAA (2004)–0881

4.169 W. Ruyten: Optimization of three-gate lifetime pressure-temperature-sensitive paint measurements (2004) AIAA 2004–2190

4.170 L. Goss, G. Jones, J. Crafton, S. Fonov: Temperature compensation in time-resolved pressure measurements, Proc. 11th Int. Symp. Flow Visualization, University of Notre Dame (2004)

4.171 E. Puklin, B. Carlson, S. Gouin, C. Costin, E. Green, S. Ponomarev, H. Tanjii, M. Gouterman: Ideality of Pressure Sensitive Paint. I. Platinum Tetra(pentafluorophenyl)porphine in Fluoroacrylic Polymer, J. Appl. Polym. Sci. **77**(8), 2795–2804 (2002)

4.172 R. H Engler: DLR intensity and lifetime systems, PSP Workshop Pressure Sensitive Paint Workshop, NASA Langley Research Center (2000)

4.173 D. Gebbie, M. Reeder, C. Tyler, V. Fonov, J. Crafton: PSP-based experimental investigation of a blended wing body aircraft (2005) AIAA 2005–4719

4.174 J.H. Bell, et al.: Surface Pressure MeasurementsUsing Luminescence Coatings, Annu. Rev. Fluid Mech. **33**, 155–206 (2001)

4.175 S. Fonov, R. H. Engler, C. Klein, S. Mihailov, V. Mosharov, V. Kulesh, V. Radchenko, E. Schairer: Investigations of the pressure fields on the oscillating wings by pressure sensitive paint, Proc. 11th DGLR-Fach-Symposium der AG STAB vom 10–12 Nov. Technischen Universität, Berlin (1998)

4.176 M. Sajben: Uncertainty estimates for pressure sensitive paint measurements, AIAA J. **31**(11), 2105–2110 (1993)

4.177 M. J. Morris: Use of pressure-sensitive paint in low-speed flows, IEEE ICIASF Record Wright-Patterson AFB (1995)

4.178 O.C. Brown, R.D. Mehta, B. T. Cantwell: Low-speed flow studies using the pressure sensitive paint technique, 81th AGARD conference, Seattle (1997)

4.179 N. G. Verhaagen, L. N. Jenkins, S. B. Kern, A. E. Washburn: A study of the vortex flow, over a 76/40 deg double delta wing, NASA Contractor Report 195032, ICASE Report 95–5 (1995)

4.180 W. Ruyten, C. Fisher: On the effects of reflected light in luminescent paint measurements, 38th Aerospace Sciences Meeting and Exhibit, AIAA-2000-0833 (2000)

4.181 V. E. Mosharov, V. N. Radchenko, S. D. Fonov: Luminescent pressure sensors in aerodynamic experiments, published privately. Contact S. D. Fonov, TsAGI, Zhukovsky, 140160, Moscow reg., USSR or M. Osin, RUKAR, Russia (1994)

4.182 O. Trinks: Entwicklung und Einsatz einer Fluoreszenz-Lebensdauer-Methode zur Bestimmung instationärer Strömungsvorgänge an Verdichterschaufeln unter Verwendung druckempfindlicher Beschichtungen, Dissertation, Universität Göttingen (2000)

4.183 Fonov, L. Goss, J. Jones, V. Crafton, M. Fonov: New method for surface pressure measurements, 43rd AIAA Aerospace Science Meeting, Reno (2005) AIAA-2005-1029

4.184 R. H. Engler: Pressure sensitive paint in quantitative wind tunnel studies CEAS/KATnet Conference on Key Aerodynamic Technologies, Hilton Bremen (2005)

4.185 C. Klein, W. E. Sachs, U. Henne, R. H. Engler, A. Wiedemann, R. Konrath: International vortex flow experiment 2 (VFE-2), Experimental Pressure Distribution on the 65° Delta Wing Configuration using PSP, (Paper in preparation for the 44th AIAA Congress, Reno (2006))

4.186 M. Kurita, K. Nakakita, K. Mitsuo, S. Watanabe: Data processing of pressure-sensitive paint for industrial wind tunnel testing, 24th AIAA Aerodynamic Measurement Technology and Ground Testing Conference, Portland (2004) Paper 2004–2189

4.187 M. Benne, R. Kammeyer, J. Donovan, M. Rueger, J. Harris, D. Morgenroth, E. Green: General strategy for the development of an improved pressure-sensitive paint system, 2nd AIAA Aerodynamic Measurement Technology and Ground Test Conference, St. Louis (2002) AIAA 2002–2906

5. Velocity, Vorticity, and Mach Number

The objective of this chapter is to provide a comprehensive statement of the experimental methods that can be used to transduce the velocity and its companion quantity: vorticity ($\nabla \times \overline{u}$). Velocity measurements can be understood to represent *spatially integrated* and *pointwise* values. Thermal transient anemometry (Sect. 5.6) and sonic anemometers (Sect. 5.7) represent the former. Pressure-based velocity measurements (Sect. 5.1), thermal anemometry (Sect. 5.2), and particle-based techniques (Sect. 5.3) represent the latter. In addition, particle image velocimetry (PIV, Sect. 5.3.2), planar Doppler velocimetry (Sect. 5.3.3), and molecular tagging velocimetry (Sect. 5.4) also provide spatial distributions of the pointwise measurements for the *instant* at which the image is formed. The vorticity measurements rely on some form of the above pointwise measurements. A general overview of optical methods is presented in Sect. 5.5.1.

5.1	**Pressure-Based Velocity Measurements**	216
	5.1.1 Measurement of Total Pressure Head with Pitot Tubes	218
	5.1.2 Dynamic Head from Separate Measurements of Total and Static Pressures	226
	5.1.3 Direct Measurement of Dynamic Head (Combined Pitot-Static and Other Probes)	226
	5.1.4 Measurement of Dynamic Head and Flow Direction (Multihole Probes)	228
5.2	**Thermal Anemometry**	229
	5.2.1 Introduction	229
	5.2.2 Sensors	231
	5.2.3 Anemometer Electronics	258
	5.2.4 Calibration Procedures in Subsonic Flows	266
	5.2.5 Measurement of Velocity and Temperature Fluctuations	273
	5.2.6 Calibration Procedures in Compressible Flows	278
	5.2.7 Special Techniques	279
	5.2.8 A Comprehensive Technique for X-Array Calibration and Data Processing	283
5.3	**Particle-Based Techniques**	287
	5.3.1 Tracer Particles	287
	5.3.2 Laser Doppler Technique	296
	5.3.3 Particle Image Velocimetry	309
	5.3.4 Doppler Global Velocimetry	342
	5.3.5 Laser Transit Velocimetry	353
5.4	**Molecular Tagging Velocimetry (MTV)**	362
	5.4.1 The Photochemistry of MTV: Molecular Tracers and Chemical Mechanisms	363
	5.4.2 The Experimental Implementation of MTV: Tagging Methods, Detection and Processing	373
	5.4.3 Examples of MTV Measurements	377
	5.4.4 Summary and Conclusions	382
5.5	**Vorticity**	382
	5.5.1 Optical Techniques in Strophometry – Vorticity Measurements Methods	383
	5.5.2 High-Resolution Dual-Plane Stereo PIV (DSPIV)	400
	5.5.3 Measurements of the Vorticity Vector and Other Velocity Gradient Tensor-Based Turbulence Properties	408
	5.5.4 Transverse Vorticity Measurements with a Four-Sensor Hot-Wire Probe	429
5.6	**Thermal Transient Anemometer (TTA)**	434
	5.6.1 Operational Description	434
	5.6.2 Representative Results	435
5.7	**Sonic Anemometry/Thermometry**	436
	5.7.1 Definition	436
	5.7.2 Measurement Principles	437
	5.7.3 Device Characteristics, Accuracy, and Limitations	439
	5.7.4 Data-Acquisition Requirements	444
	5.7.5 Use and Calibration Procedures	444
	5.7.6 Manufacturers and Costs	445
	5.7.7 Device Comparison	445
References		446

5.1 Pressure-Based Velocity Measurements

Measurement of dynamic head, in conjunction with use of the Bernoulli equation, is perhaps the most commonly used means of determining fluid velocity in a steady, mid-to-high velocity flow. The fabrication of suitable pressure probes is simple and inexpensive, the transducer range may be selected to suit the flow under consideration and a differential measurement of total and static head is sufficient for incompressible flows, in which the density is known. It is, however, an intrusive technique, and one which requires some subtlety for accurate results over a range of flow types and Reynolds numbers using finite-sized probes. With correct use and under appropriate corrections high-accuracy velocity measurements may be obtained.

Governing Equations

The technique rests on the general (5.1), known as the *Bernoulli equation*, which is valid along a single streamline in steady, inviscid flows where the continuum approximation holds.

$$\frac{dp}{\rho} + U\, dU = 0 \,. \qquad (5.1)$$

Here U is the local mean magnitude of the velocity along the streamline, ρ is the local fluid density and p is the local pressure. Equation (5.1) may be integrated along the streamline to give the general form for compressible flow of an ideal gas:

$$\frac{p_0}{p} = \left(1 + \frac{\gamma-1}{2} M^2\right)^{\frac{\gamma}{\gamma-1}}, \qquad (5.2)$$

where p_0 would be the pressure in the fluid after an isentropic, adiabatic deceleration along the streamline to a zero-velocity condition. This quantity is commonly called the *stagnation* or *total* pressure despite its non-scalar nature (a dependence on coordinate system) and consequently the true (scalar) pressure p is often referred to as the *static pressure*.

The pressure–density relation

$$\frac{1}{2}\rho U^2 = \frac{1}{2}\gamma M^2 p \qquad (5.3)$$

has been utilized in writing (5.2), where the Mach number $M = U/a$, the speed of sound is $a = \sqrt{\gamma p/\rho}$, and γ is the ratio of specific heats. For small Mach numbers, binomial expansion gives

$$p_0 - p = \frac{1}{2}\rho U^2 \left(1 + \frac{M^2}{4} + \frac{(2-\gamma)M^4}{24} + \ldots\right) \qquad (5.4)$$

and the familiar incompressible ($M=0$) form

$$p_0 - p = \frac{1}{2}\rho U^2 \,, \qquad (5.5)$$

where $1/2\rho U^2$ is termed the dynamic head or dynamic *pressure*. Thus, for incompressible flow, measurements of density, and total and static pressures can be used to obtain the local velocity.

Note that truncation of the binomial expansion of (5.2) causes an error in the determination of dynamic head of 0.2% for $M \leq 0.09$ if only the first term of the expansion is retained (i.e., the incompressible Bernoulli equation), for $M \leq 0.55$ if two terms are used and through $M = 1$ for three or more terms (assuming zero measurement error).

In compressible flow, the continuum approximation may be violated if the ratio of the molecular mean free path λ to the characteristic flow dimension (here the diameter of the probe d), i.e., the Knudsen number $\text{Kn} = \lambda/d$, is sufficiently large that boundary interactions instead of inter-molecular collisions determine relaxation times. Using the Chapman–Enskog solution of the Boltzmann equations to obtain the viscosity, the conditions under which the continuum approximation holds ($\text{Kn} < 0.01$) can be written in terms of the Mach number and Reynolds number based on d as

$$\text{Kn} = \frac{\lambda}{d} = \left(\frac{\pi\gamma}{2}\right)^{1/2} \frac{M}{\text{Re}_d} \ll 1 \,. \qquad (5.6)$$

Only continuum flow in an ideal fluid will be considered here. For a detailed discussion of pressure-based velocity measurement in non-continuum gases, see *Chue* [5.1, Sect. 1.8].

In the case of a continuum supersonic freestream, a probe of finite size will cause the formation of a detached shock (Fig. 5.1d), hence a total pressure probe will measure the reduced stagnation pressure p_{02} in the decelerated subsonic flow of region 2. Because of the non-isentropic nature of the compression through the shock wave, the Bernoulli equation cannot be used. The freestream Mach number M_∞ can be found by means of the Rayleigh supersonic pitot formula and a measurement of the freestream static pressure p_∞:

$$\frac{p_\infty}{p_{02}} = \left[\frac{2\gamma M_\infty^2 - (\gamma-1)}{\gamma+1}\right]^{1/(\gamma-1)}$$
$$\times \left[\frac{\gamma+1}{2} M_\infty^2\right]^{-\gamma/(\gamma-1)} . \qquad (5.7)$$

If the upstream reservoir pressure is known and the flow is isentropic up to the shock, normal shock relations give

$$\frac{p_{0\infty}}{p_{02}} = \left[\frac{2\gamma M_\infty^2 - (\gamma-1)}{\gamma+1}\right]^{1/(\gamma-1)}$$
$$\times \left[\frac{2 + (\gamma-1)M_\infty^2}{(\gamma+1)M_\infty^2}\right]^{\gamma/(\gamma-1)}, \quad (5.8)$$

or, of course, for downstream pressure measurements (5.2) may be used. If required, U_∞ can be determined from (5.3). Thus both the freestream Mach number and velocity can be obtained more easily from separate measurements of the stagnation pressure p_{02} and freestream static pressure than from a difference measurement of dynamic head in compressible flow.

Measurement Techniques

The measurement technique selected will depend on the flow under consideration, and potentially the ability to physically place an apparatus into the flow. In confined flows or flows with large velocity gradients (such as wall-bounded flows) where probe size may be limited by spatial resolution concerns, and where the variation of pressure with distance from the wall is known or negligible, a Pitot (impact) tube may be used with a static pressure tapping at the wall, or a static pressure tube. (Measurements of static pressure have been addressed in Chap. 4 *Pressure of Measurement Systems*, and it will be assumed that all static pressure measurements have been acquired and corrected using the techniques described therein.) In free flow, a Pitot-static tube may be used to obtain reasonably closely collocated measurements of static and total pressures, with the limitation that these probes are relatively large and thus may present blockage issues. Larger, multihole probes may be used to determine both the magnitude and direction of the velocity vector.

For all methods of determining the velocity, measurement errors will be introduced unless:

- measuring devices are small enough that there is negligible disturbance to the incoming flow because of the presence of the device and any associated support or traversing mechanism (i. e., blockage and interference are negligible, all measurement devices are *small*),
- the devices are correctly aligned to the flow velocity vector within the range of yaw acceptance (note that this condition may be violated in a highly turbulent flow),
- the stagnation pressure device is small enough that the assumption of deceleration of a single streamline holds (approximately),
- the stagnation pressure device Reynolds number is sufficiently large that the assumption of isentropic, inviscid streamline deceleration is valid,
- the dimensions of the device are negligible compared to the incoming flow lengthscales, i. e., the pressure can be considered approximately uniform across the measurement face,
- measurements are averaged over a sampling time selected to be long enough to give convergence of the reading to the true velocity, having accounted for the response of the probe (where, in general, the smaller the orifice, the longer the response time),
- the flow is free of particulates, condensation, etc., which could adversely affect the device readings.

Well-designed measurement devices will minimize the effects of violating these conditions.

In addition, the accuracy of the measurements will depend on the accuracy of the transducer. The sensitivity of the velocity to errors in the differential measurement of stagnation and static pressures for compressible and incompressible flows are given in (5.9) and (5.10), respectively:

$$\frac{dU}{U} = \frac{1}{\gamma M_\infty^2} \frac{p\, dp_0 - p_0\, dp}{p_0 p}, \quad (5.9)$$
$$\frac{dU}{U} = \frac{1}{2} \frac{d(p_0 - p)}{p_0 - p}. \quad (5.10)$$

Devices to measure the (static) pressure have been considered in Chap. 4: static pressure wall tappings (Sect. 4.1) and static pressure tubes (Sect. 4.2). That chapter also addresses suitable hardware for determining velocity from pressure measurements in Sect. 4.3. Here the use of Pitot/impact tubes is described for the measurement of total pressure in Sect. 5.1.1, and in conjunction with a separate static pressure device to obtain the dynamic head in Sect. 5.1.2, while Sect. 5.1.3 covers combined probes such as the Pitot-static tube for the local measurement of dynamic pressure, and Sect. 5.1.4 introduces multihole probes from which both the velocity direction and magnitude may be determined. Some fundamental ideas will carry over from Chap. 4 and between sections, especially the correction methods for measurements of steady pressure. It has been noted in Chap. 4 that there is subtlety in the order of application of the various corrections to measured data, and this must be emphasized when the velocity is to be determined from the dynamic head, the difference between two measured quantities, p_0 and p, that may not be the true values in the flow. In situ calibration may be required to maximize probe accuracy. The reader is referred to *Bryer* and *Pankhurst* [5.2] and *Chue* [5.1] for further details on most sections.

The convention of aligning x with the mean velocity direction and y with the direction of traverse of the probe (i.e., parallel to the probe support axis, Fig. 5.1) will be adopted. Reynolds decomposition will be used to write $u_x = U + u'$, $u_y = V + v'$ and $u_z = W + w'$, where U, V, and W are the mean (time-averaged) component velocities and a prime denotes a fluctuating velocity.

5.1.1 Measurement of Total Pressure Head with Pitot Tubes

Consider the probes of Fig. 5.1a–c for the measurement of total pressure p_0 commonly deemed *Pitot probes* after *Henri de Pitot* [5.3] (see also Sect. 5.1.3 for details of other total pressure devices). In what follows, it will be assumed that the only errors are introduced into measurement of p_0, i.e., p and ρ are measured accurately. In general, the results will hold for both impact-type and Pitot-static probes (Sect. 5.1.4), with the flow around the latter more severely disturbed due to the larger thickness (a higher ratio of inner to outer diameter, $\theta = d_{in}/d$) required to accommodate both static and stagnation pressure connections (Fig. 5.7c).

For a Pitot probe of a particular geometry, dimensional analysis shows that, given a completely accurate measurement of static pressure, the dynamic pressure measured by a transducer attached to the probe $p_{0m} - p$, will be related to the true velocity as a function of the following variables

$$C_p = \frac{p_{0m} - p}{\frac{1}{2}\rho U^2}$$

$$= f\left(\tilde{\phi}, \mathrm{Re}_d, M, \frac{\overline{u'^2}}{U^2}, \frac{\overline{v'^2}}{U^2}, \frac{\overline{w'^2}}{U^2}, \alpha, \frac{y}{d}\right), \quad (5.11)$$

where $\tilde{\phi}$ is some measure of the incoming flow angle relative to the yaw sensitivity of the probe, Re_d is the local probe Reynolds number, M is the incoming Mach number, α relates the lengthscale of incoming shear to the probe diameter, y/d considers proximity to a surface and the components of relative turbulent intensity are given by $\overline{u'^2}/U^2$, etc.; the effect of turbulence scale relative to the probe can be considered in the yaw behav-

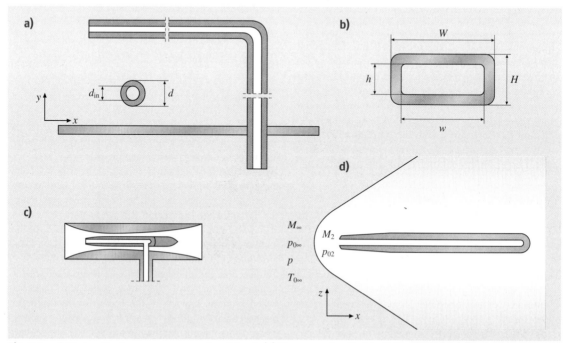

Fig. 5.1a–d Probes for the measurement of total head: (**a**) blunt-nosed cylindrical Pitot probe; (**b**) flattened Pitot probe; (**c**) Kiel probe; and from shock relationships: (**d**) Pitot probe in supersonic flow with known freestream static pressure p_∞ or reservoir pressure $p_{0\infty}$

ior. If $C_p = 1$ then the true total pressure is measured by the probe (and Bernoulli's equation holds uncorrected), but in general the magnitude of each of the parameters and their influence on the velocity determined from the measurements must be assessed.

Probe Selection and the Effect of Geometry

The finite size of practical Pitot probes violates the conditions of strict applicability of the Bernoulli equation. Rather than measuring the total pressure due to the stagnation of a single streamline, the probe will measure a uniform pressure imposed by the flow pattern at its nose. Depending on the probe Reynolds number (see the section on this topic below), a stagnation streamsurface will form inside or in front of the probe and the resulting flow pattern will determine the measured total head. If the probe is not fully aligned with the incoming flow, a further complexity to this tip flow is introduced and must be accounted for.

Spatial averaging in inhomogeneous flows may be limited by by using small-diameter probes to approximate the single-streamline condition, and the disturbance of the incoming streamlines due to the presence of the probe may be reduced through appropriate design of the probe and its mounting/traversing mechanism. However the dimensions of the probe may be limited by structural constraints and the poor temporal response of small-orifice designs. Blockage considerations may further constrain the dimensions of the probe, such that d/D must be small (Sect. 5.1.2).

The probe tip geometry will influence the flow in the region of the stagnation point and hence also impact on the measurement error. Standard probe designs include a cylindrical square-ended design (Fig. 5.1a) and a flattened cylinder (appropriate for measurement close to a surface) (Fig. 5.1b). For multiple simultaneous measurements rakes of probes may be used (although note that care must be taken to avoid mutual interference, especially in supersonic flow).

The effect of the ratio of inner to outer probe diameters θ will be detailed later, however historically $\theta = 0.6$ has been used by many investigators and thus many of the Pitot corrections have been formulated for this type of probe.

Directional Sensitivity

A Pitot probe may experience an incoming flow that is not parallel to its centerline either through physical misalignment to the predominant flow direction (which may be eliminated by careful experimental setup) or because of large variations in the angle of a turbulent or unsteady flow. The directional sensitivity of the probe will depend on the tip shape, probe diameter ratio θ, and the Mach and Reynolds numbers, and may be different in the yaw (z) and pitch (y) directions because of the presence of the probe support, Fig. 5.1 (note also that flattened probes are particularly sensitive to pitch because of their geometry). The zero-yaw condition corresponds to the maximum pressure reading (of interest when the direction of the flow is also to be determined, Sect. 5.1.3) and for commonly used designs $C_p \approx 1$ for a wide angular range, such that exact alignment with the flow direction is not critical to accurate measurements.

For simple square-ended probes, *Liepmann* and *Roshko* [5.4] suggest that an accuracy of $\approx 1\%$ of dynamic head ($C_p \geq 0.99$) may be obtained for yaw angles up to $\pm 20°$, decreasing with decreasing relative orifice size d_{in}/d and increasing rounding of the nose. *Bryer* and *Pankhurst* [5.2, 5] note that the acceptable yaw angle also increases slightly with increasing Mach number.

Chue [5.1] has summarized the yaw characteristics of several commonly used nose shapes (shown here in Fig. 5.2). The Kiel probe (Fig. 5.1c) gives the best yaw response by a factor of approximately two.

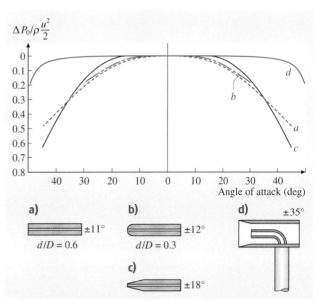

Fig. 5.2a–d Yaw sensitivity of commonly used Pitot probe tip designs: (**a**) square-ended Pitot probe with $\theta = 0.6$; (**b**) round-nosed Pitot/Pitot-static probe with $\theta = 0.3$; (**c**) Pitot probe with tapered tip, suitable for supersonic measurements; (**d**) Kiel probe (after *Chue* [5.1])

Probe Reynolds Number: The Effect of Viscosity

For real, viscous fluids, the assumption of inviscid (isentropic) flow on which the Bernoulli equation relies must be revisited. When the probe Reynolds number $Re_d = dU_{local}/\nu$ is low (where U is the local flow velocity and ν is the kinematic viscosity), in particular, viscosity may affect the flow around the probe tip such that measured total pressure p_{0m} differs significantly from the true impact pressure since the stagnation streamsurface may extend forward of the probe tip. This is commonly corrected by means of a calibration process, since there is not a simple extension of the Bernoulli equation to a viscous fluid.

Early measurements were taken with the round, square-ended design of Fig. 5.1a by *Barker* [5.8], *Hurd* et al. [5.7] and *MacMillan* [5.9]. For incompressible flow and $Re_d \to 0$ the effect of viscosity is to increase the measured impact pressure (and hence also measured velocity) above the true value, i.e., $C_p > 1$. There is some discussion in the literature of whether or not C_p ever falls below zero in the mid-Reynolds number range [5.7] or whether this is an artefact of experimental method [5.9]. The exact form of C_p appears to be sensitive to the mode of calibration, the condition of the probe tip (square ends reduce C_p relative to spherical/hemispherical tip values) and the ratio of inner to outer probe diameter (large θ reduces C_p).

Chue suggested that d_{in} is the more-appropriate dimension to use at low probe Reynolds number, $Re_{d_{in}} < 10$, and under this scaling it appears that there is reasonably good collapse for most probe tip de-

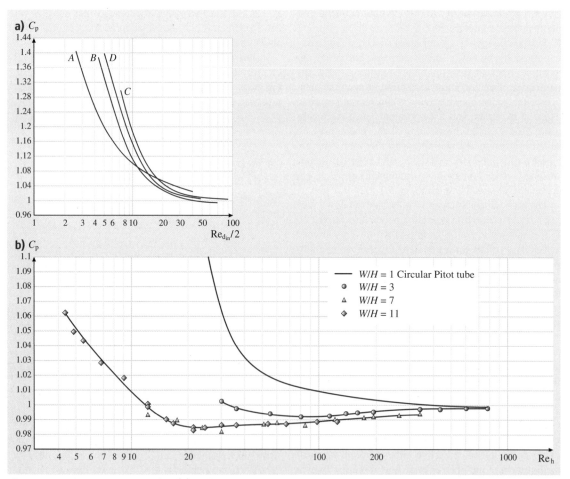

Fig. 5.3a,b Viscous correction for: **(a)** cylindrical probes: A. *Sherman* [5.6], B. *Hurd* et al. [5.7], C. *MacMillan*, D. *Barker* [5.8] (after *MacMillan* [5.9]); **(b)** flattened, boundary-layer-type probes (after *MacMillan* [5.10])

signs to the low-Reynolds-number numerical results of *Lester* [5.11]. For vanishingly small Reynolds number, inertial effects are negligible and the pressure distribution becomes independent of density, $C_p = A/\text{Re}_d$.

Figure 5.3a shows the variation of C_p calibrations with $\text{Re}_{d_{in}}$ for cylindrical probes and square [5.6] and round [5.7–9] noses.

With reasonable accuracy, the effect of viscosity at low probe Reynolds number may be represented by

$$C_p = \frac{4.14}{\text{Re}_{d_{in}}} \quad \text{Re}_{d_{in}} < 1 \quad \text{Hurd et al. [5.7]}$$
$$= 1 + \frac{2.5}{\text{Re}_{d_{in}}^{1.25}} \quad 1 < \text{Re}_{d_{in}} < 10 \quad \text{Lester [5.11]}, \tag{5.12}$$

where the results of *Hurd* et al. [5.7] have been reformulated in terms of the inner probe diameter and a curve fit to *Lester*'s [5.11] data has been suggested. For higher Reynolds numbers,

$$C_p = \frac{\text{Re}_d}{1.07\text{Re}_d - 7.4} \quad 12 < \text{Re}_d < 60$$
$$\text{\textit{Mikhailova} and}$$
$$\text{\textit{Repik} [5.12, 13]}$$
$$= 1 + \frac{10}{\text{Re}_d^{1.5}} \quad \text{Re}_d > 30$$
$$\text{\textit{Zagarola} [5.14]}$$
$$= 1 \pm 2\% \quad \text{Re}_d > 600$$
$$\text{\textit{Chue} [5.1]}. \tag{5.13}$$

For a hemispherical nose design, *Merriam* and *Spaulding* [5.15] have shown that the diameter ratio does not affect C_p for $0.2 < \theta < 0.74$. The results of *Sherman* [5.6] for a round-nosed probe at low Reynolds number are shown as curve A in Fig. 5.3a. For $\text{Re}_d > 20$ they agree well with the theoretical solution of *Homann* [5.16] for the stagnation region of a sphere:

$$C_p = 1 + \frac{6}{\text{Re}_d + 0.455\sqrt{\text{Re}_d}}. \tag{5.14}$$

The effect of viscosity is perhaps more often important for square-ended flattened probes (Fig. 5.1b) since this kind of probe is often used close to surfaces where the velocity may be low. Common probe dimensions lie in the region of $0.3 < h/H < 0.6$ and $3 < W/H < 11$. The results of *MacMillan* [5.10], shown here in Fig. 5.3b, reveal that a single C_p versus $\text{Re}_{d_{in}}$ curve is appropriate for aspect ratio $W/H \geq 7$. C_p falls below unity for increasing non-axisymmetry. However note that h/H was varied for these experiments so the effect of θ has not been taken into account. *Mikhailova* and *Repik* [5.13] attempted to relate the calibration curves for flattened probes to those for circular probes by defining a complex equivalent diameter with a theta-dependent coefficient (not given here).

For compressible flow, a calibration coefficient for the total pressure alone C_μ may be defined

$$C_\mu = \frac{p_{0m} - p_0}{\frac{1}{2}\rho U^2} = C_p - \frac{p_0 - p}{\frac{1}{2}\rho U^2}. \tag{5.15}$$

In subsonic compressible flow the effect of compressibility should be to decrease the calibration coefficient below its incompressible value, as a superposition of viscous and compressibility effects [from (5.4)]. However, the available subsonic data of *Sherman* [5.6] show negligible effect of compressibility, thus the incompressible form, $C_\mu = C_p - 1$, assessed at the appropriate Reynolds number, should be applied to subsonic, compressible flow measurements [5.1].

For supersonic flow, the calibration coefficient is more appropriately written in terms of the (subsonic) flow conditions downstream of the probe-induced bow shock (Fig. 5.1d):

$$C_\mu = \frac{p_{02m} - p_{02}}{\frac{1}{2}\rho U_2^2} \tag{5.16}$$

and evaluated as for subsonic flow. (If there is strong shear then the bow shock may no longer be normal to the stagnation streamline and the pressure measured by the probe can no longer be simply expressed.) An estimate of the velocity behind the shock U_2 for the Reynolds number at which the correction must be evaluated can be obtained using the Rayleigh supersonic Pitot formula [(5.7) with the measured stagnation pressure p_{02m}] for M_∞ and the normal shock relation of (5.17) for the Mach number behind the shock

$$M_2 = \left(\frac{1 + \frac{\gamma-1}{2}M_\infty^2}{\gamma M_\infty^2 - \frac{\gamma-1}{2}} \right)^{\frac{1}{2}}. \tag{5.17}$$

The velocity follows after a temperature measurement:

$$U_2 = \frac{M_2\sqrt{\gamma R T_{02}}}{\sqrt{\frac{T_{02}}{T_2}}} = \frac{M_2\sqrt{\gamma R T_{02}}}{\left(1 + \frac{\gamma-1}{2}M_2^2\right)^{\frac{1}{2}}}. \tag{5.18}$$

Note that, for an adiabatic flow, $T_{02} = T_{01}$ so that the freestream temperature measurements may be used.

Direct experimental calibrations of p_{02m}/p_{02} versus Re_d in general retain the influence of Mach number and/or probe shape. *Chue* [5.1] has proposed the approximate curve of Fig. 5.4 for hemispherical, flat, and internally chamfered probes, noting that the Mach number does not seem to influence the calibration within the scatter of the available data. For $\mathrm{Re}_{d\infty} > 200$ the effect of viscosity is small. Note, however, that $\mathrm{Re}_{d\infty}$ is the freestream Reynolds number, which will not capture the subtleties of the flow downstream of the shock wave. This calibration should be treated as approximate. A somewhat improved collapse may be obtained by using p_{02m}/p_{02} versus $\mathrm{Re}_{d2}\sqrt{\rho_2/\rho_\infty}$, where Reynolds number Re_{d2} is assessed downstream of the shock, as the abscissa [5.17].

Effect of Turbulence

In a turbulent flow, the impact pressure at a point in the flow includes a contribution from the turbulent fluctuating velocity that increases the reading above its true value in the following way [5.18]:

$$p_{0m} = p + \frac{1}{2}\rho\left(U^2 + \overline{u'^2} + \overline{v'^2} + \overline{w'^2}\right), \quad (5.19)$$

where the overbar denotes a time average and quasi-steady flow is assumed. Although the detailed nature of the turbulence and the size of the characteristic scale relative to the probe diameter probably influences the measured stagnation pressure to some degree, *Christiansen* and *Bradshaw* [5.19] suggest that (5.19) is appropriate and *Tavoularis* and *Szymczak* [5.20] demonstrated negligible effect of characteristic scale (at least for low turbulence intensity)

$$\frac{\sqrt{\overline{u'^2}}}{U} < 10\%.$$

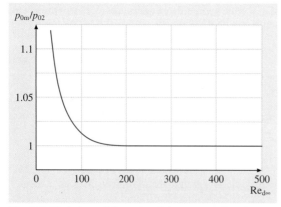

Fig. 5.4 Variation of the ratio of the measured to the true stagnation pressure downstream of the probe-induced shock versus the upstream Reynolds number (after *Chue* [5.1])

However for large-scale fluctuations and coherent structures, the instantaneous yaw angle must be considered in the response. *Zagarola* [5.14] included the response of the probe to normal fluctuations, i. e., the directional sensitivity, through the constant K_t, which was proposed to have a value $K_t \approx 0.3$ by *Ozarapoglu* [5.21].

$$p_{0m} - p = \frac{1}{2}\rho U_m^2$$
$$= \frac{1}{2}\rho\left(U^2 + \overline{u'^2}\right) + \frac{1}{2}\rho(1-K_t)\left(\overline{v'^2} + \overline{w'^2}\right). \quad (5.20)$$

For $\overline{v'^2}, \overline{w'^2} \ll \overline{u'^2}$, (5.20) can be simplified to give

$$U = \sqrt{U_m^2 - \overline{u'^2}} \quad (5.21)$$

or (for low turbulence intensity $\overline{u'^2}/U_m^2$)

$$\frac{U - U_m}{U_m} = -\frac{\overline{u'^2}}{2U_m^2}. \quad (5.22)$$

For example, *Dickinson* [5.22] proposed that this form is suitable for $y^+ = yu_\tau/\nu > 50$ in turbulent pipe flow.

Thus, correction for the effect of turbulence requires either simultaneous measurement of the fluctuating velocity, e.g., by hot-wire anemometry (which most probably negates the need for pressure-based velocity diagnostics), or the use of a known or empirical form of the distribution of $\overline{u'^2}$, $\overline{v'^2}$, and $\overline{w'^2}$.

Special attention should be paid to whether or not other corrections applied to the measurements implicitly include the effect of turbulence, and the magnitude of the turbulence correction relative to other corrections required. In a pipe with diameter D, *McKeon* et al. [5.23] found that, for Pitot probes with $d/D = 0.002-0.014$ and high Reynolds numbers ($d^+ < 2500$), the turbulence correction was an order of magnitude smaller than other corrections and could be neglected, however *Monty* [5.24] found that this was not the case for Pitot-static tubes with $d/D = 0.015$ at lower Reynolds numbers.

Effect of Shear

If a Pitot probe is to be used in a free shear flow, two additional effects on the impact pressure must be considered: the nonlinear effect of spatial integration of the inhomogeneous flow field across the probe face and displacement of the stagnation streamline induced by the finite size of the probe. Both are consequences of the violation of the single-streamline assumption and both lead to an indicated velocity that is higher than the true value corresponding to the location of the centerline of the probe (i.e., an apparent deflection of incoming streamlines towards the region of lower velocity). The magnitude of this effect can be assessed either in terms of an implied error in probe position $\Delta y = y - y_{\text{cl}}$, the offset of the probe centerline relative to the position of the streamline under measurement, or as the difference between measured and true velocity, ΔU, at y_{cl}, where

$$\frac{\Delta U}{U(y_{\text{cl}})} = \frac{U(y_{\text{cl}} + \Delta y) - U(y_{\text{cl}})}{U(y_{\text{cl}})}$$
$$= 2\alpha \frac{\Delta y}{d} - \beta \frac{(\Delta y)^2}{d^2} + \ldots \quad (5.23)$$

The shear parameters α and β are given by

$$\alpha = \frac{d}{2U(y_{\text{cl}})} \left.\frac{\mathrm{d}U}{\mathrm{d}y}\right|_{\text{cl}}, \quad (5.24)$$

$$\beta = -\frac{1}{2} \frac{d^2}{U(y_{\text{cl}})} \left.\frac{\mathrm{d}^2 U}{\mathrm{d}y^2}\right|_{\text{cl}}, \quad (5.25)$$

and the subscript 'cl' implies that the variables are assessed on the probe centerline. In boundary-layer flows both α and β takes their maximum values near the wall and are necessarily zero in the freestream, and $\alpha = 0$ on the centerline of a pipe. Figure 5.5a,b shows the variation of α and β, respectively, for a cylindrical probe of 0.3 mm diameter in pipe flow at $\mathrm{Re}_D = \overline{U}D/\nu = 75 \times 10^3$ and 4×10^6, where $\overline{U}$ is the bulk velocity [5.25].

Integration of a linear velocity profile across a circular probe yields the following expression for the apparent offset of the probe centerline due to spatial integration, Δy_{int}:

$$\frac{\Delta y_{\text{int}}}{d_{\text{in}}} = \frac{1}{2\alpha}\left(\sqrt{1 + \frac{\alpha^2}{4}} - 1\right). \quad (5.26)$$

Thus for a probe with $\theta = 0.6$, this offset can be seen to have a maximum value of $\Delta y_{\text{int}}/d = 0.035$ when $\alpha = 1$, falling with decreasing α. This effect is usually small compared to the other shear-induced corrections, and hence is often neglected.

In addition, account must be made for a displacement of the effective stagnation streamline in the direction of the region of lower velocity due to the presence of a finite-sized probe in a shear flow. In practice, account is taken of this effect by applying a displacment correction to the probe position $\Delta y/d$, which is intuitively a function of the local shear α. Theoretical approaches that consider a sphere in potential flow due to *Hall* [5.26]

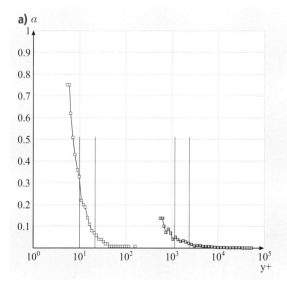
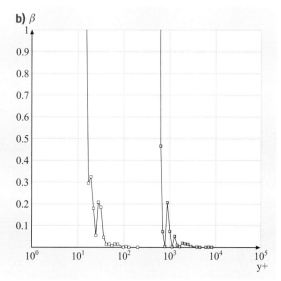

Fig. 5.5a,b The variation of shear parameters (**a**) α (5.24) and (**b**) β (5.25) in pipe flow. *Open symbols*: $\mathrm{Re}_D = 75 \times 10^3$ and *closed symbols* $\mathrm{Re}_D = 4 \times 10^6$ (after *McKeon* [5.25])

and *Lighthill* [5.27] give expressions of the form

$$\frac{\Delta y}{d} = A_1 \tanh(A_2 \alpha) = C_1 \alpha - C_2 \alpha^3 + O(\alpha^5) \,. \tag{5.27}$$

However early experimental work [5.28, 30–32] led to the conclusion that a constant displacement correction,

$$\frac{\Delta y}{d} = \varepsilon_d \tag{5.28}$$

was appropriate, with $\varepsilon_d = 0.08$–0.18 for cylindrical probes (apparently independent of θ, $Re_{d_{in}}$, and the velocity gradient). The effect of θ can be taken into account using the expression of *Young* and *Maas* [5.31]):

$$\frac{\Delta y}{d} = 0.131 + 0.082\theta \,. \tag{5.29}$$

More recently, *Tavoularis* and *Szymczak* [5.20] proposed $\varepsilon_d = 0.16$ for $\theta = 0.6$ and $\alpha \geq 0.03$, based on experiments covering $0.002 \leq \alpha \leq 0.14$. These authors noted that, as $\alpha \to 0$, so does the error in U_m, even though the positional error remains constant. For rectangular probes, *Quarmby* and *Das* [5.29] suggest $\varepsilon_d = \Delta y/H = 0.19$, in qualitative agreement with early work by *Young* and *Maas* [5.31].

However the most commonly used version of this correction is due to *MacMillan* [5.28], who investigated the effect of shear on cylindrical probes with $\theta = 0.6$ in the pipe Reynolds number range $Re_D = (20-100) \times 10^3$ and found

$$\varepsilon_d = 0.15 \,. \tag{5.30}$$

In many situations (or at least over the range of shear parameter in which Pitot probes are commonly used) this correction is sufficient and simple to apply.

Intuition, however, suggests that the full correction must be a function of α [5.1, 23, 33], such that it tends to zero in uniform flow (consider the problem that arises on a pipe centerline with a constant ε_d). *McKeon* et al. [5.23] proposed the following, empirical form (the *Princeton displacement correction*):

$$\frac{\Delta y}{d_p} = 0.15 \tanh\left(4\sqrt{\alpha}\right) \,. \tag{5.31}$$

Equation (5.31) takes a similar form to the potential flow theory of (5.27) and has been demonstrated to provide the best collapse between data taken with probes of different diameter (in pipe flow), and for larger values of α than the MacMillan correction. As such it is recommended that this correction be used if high measurement accuracy is required.

A comparison of the different displacement corrections for cylindrical and square Pitot probes is given in Fig. 5.6. It should be noted that the accuracy of the corrected data depends critically on the error in probe location and the practical difficulty of separating out shear and turbulence effects in experiment. In addition, a correction for wall proximity must be applied for $y/d < 2$.

Effect of Wall Proximity

When a Pitot probe is used near a solid boundary an additional correction is required. Consider a probe resting on a wall in uniform flow: the resultant streamline deflection is in the opposite sense to that due to shear, i.e., towards the region of higher velocity. The cumulative effect for a probe in proximity to a wall in shear flow such that $d/D \ll 1$, then, is a reduction in the velocity-gradient-induced displacement correction.

MacMillan [5.28] suggested that measurements can be corrected for the proximity of the wall by application of the following (Reynolds-number-independent) adjustment to the velocity:

$$\frac{\Delta U}{U} = 0.015 \exp\left[-3.5\left(\frac{y}{d} - 0.5\right)\right] \,, \tag{5.32}$$

which should be used in addition to the displacement correction (5.30) for $y/d < 2$. Although the MacMil-

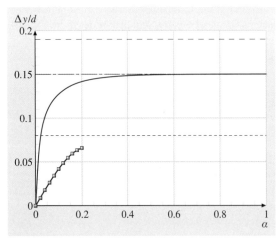

Fig. 5.6 The variation of different suggestions for the displacement correction with shear parameter α. *Solid line*: Princeton displacement correction [5.23]; *dot-dash line*: MacMillan correction [5.28]; *symbols*: Lighthill (potential theory) [5.27]; *long dashes*: correction for rectangular probes (*Quarmby* and *Das* [5.29]); and *short dashes*: lower limit of constant corrections ϵ_d in the literature ($0.08 \leq \epsilon_d \leq 0.18$ for cylindrical probes)

lan corrections take the simplest form of the available formulations, they have been shown to overcorrect data taken in high-Reynolds-number pipe flow [5.23]. For flattened probes, the results of *Quarmby* and *Das* [5.29] indicate that (5.32) may be used with reasonable confidence when the largest probe dimension, i.e., the probe width H, is used instead of the circular probe diameter (although note that these authors found some Reynolds number dependence of the correction).

The experiments of *McKeon* et al. demonstrated that data within one probe diameter of the wall could not be used with the same confidence as further away from the wall. They proposed the *Princeton wall correction* for the effect of both shear and wall proximity in the near-wall region, based on the Preston probe pressure data of [5.34]. Assuming that a probe resting on the wall reads the true pressure at a distance from the wall of

$$\frac{y}{d} = \frac{1}{2} + \frac{\delta_w}{d}, \qquad (5.33)$$

δ_w is the required wall correction, which will depend on probe Reynolds number. The correction is best given by

$$\frac{\delta_w}{d} = 0.150 \quad \text{for } d^+ < 8$$
$$= 0.120 \quad \text{for } 8 < d^+ < 110$$
$$= 0.085 \quad \text{for } 110 < d^+ < 1600, \qquad (5.34)$$

and should be applied for $1 < y/d < 2$ only. Further from the wall the free shear displacement correction of (5.31) should be used. When tested on pipe flow data taken with probes with four different diameters, excellent collapse was demonstrated for $10 < d^+ < 4000$.

Note that nearer the wall, $y/d \leq 1$, neither wall correction gives appropriate collapse of data from different diameter probes. In this region the accuracy of measurements made using Pitot probes degrades significantly and errors in probe position become critically important. This is especially unfortunate since near-wall measurements are often used to determine wall shear stress or

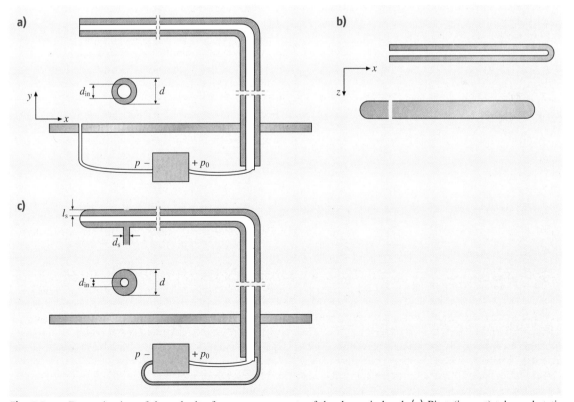

Fig. 5.7a–c Determination of the velocity from measurements of the dynamic head: (**a**) Pitot (impact) tube and static pressure wall tapping; (**b**) Pitot probe and static tube; (**c**) Pitot-static probe

slope of the mean velocity profile (Fig. 5.9). Experimentalists should be aware of the limitations of this measurement technique in the very near-wall region, particularly at high Reynolds numbers where the probe diameter becomes large relative to the viscous scale.

5.1.2 Dynamic Head from Separate Measurements of Total and Static Pressures

Separate measurements of total and static pressures from a Pitot probe (above) and either a wall tapping or static tube (Chap. 4) can be used to determine the Mach number or velocity from the Bernoulli equation [(5.2) and (5.5), respectively]. These configurations are shown in Fig. 5.7a,b using a single transducer to obtain the dynamic head. It has been noted above that, while a differential measurement to give the dynamic pressure is most suitable for incompressible flow, it is the ratio p_0/p that is required to obtain the Mach number (in which case the total and static pressure measurements must separated).

Additional corrections may need to be made when separate measuring devices are used for the total and static pressures.

The probe body and strut and the static pressure device must be selected and located such that there is negligible interference effect between them. *Hubbard* [5.35] considered the blockage effects of the strut on a wall pressure tapping in a pipe with diameter $D = 300$ mm. The relative mean velocity error $(1/B)\Delta U/U$, where $B = 2d/(\pi D)$, for the strut alone (under full extension to the pipe centerline) in terms of streamwise and azimuthal separation of strut and tapping is replicated here in Fig. 5.8. This represents the worst-case error since the blocking mass will be a maximum on the centerline for a retracting strut. Note that the sign of the induced error depends on the streamwise separation of the tapping from the probe strut and that, in practice, the measured pressure is often obtained by T-ing together two or more tappings offset in the spanwise direction from the impact tube streamwise measurement location.

The decrease in measured static pressure caused by the blockage because of the probe body must also be taken into account, varying from zero when the probe tip is far downstream of the location of static pressure measurement to a value that can be calculated from the Venturi effect when it is several diameters upstream [5.35].

In general, measurements with the Pitot tube in proximity to the wall require special attention, as indicated at the end of the previous section. The dramatic effect of the wall tapping correction together with the Pitot displacement and wall corrections on a high-Reynolds-number pipe mean velocity profile [5.25] is shown in Fig. 5.9, where $y^+ = yu_\tau/\nu$ and $U^+ = U/u_\tau$, and it has been shown [5.23] that there is a change to the inferred value of the von Kármán constant κ in the logarithmic law of the wall in pipe flow of the order of 5% compared to the analysis with no corrections.

5.1.3 Direct Measurement of Dynamic Head (Combined Pitot-Static and Other Probes)

Combined dynamic pressure probes may be used to obtain the flow velocity in situations where the complexity or size of separate measuring devices is undesirable. For unconfined flows, the Pitot-static probe of Fig. 5.7c is commonly used, utilizing coaxial tubing to combine a Pitot probe inside a static tube. Well-designed Pitot-static probes may be used to obtain accuracies of the order of 0.1% of dynamic head [5.36] or down to the limit of transducer sensitivity [5.2].

In general, the errors that arise when using these probes are a simple combination of the effects described

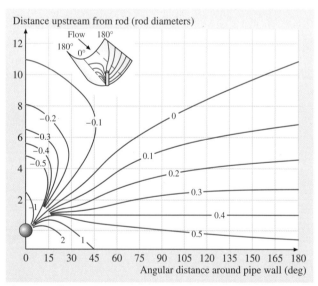

Fig. 5.8 Effect of blockage due to the probe support strut on the measurement of the static pressure by a tapping in a pipe wall. The worst case for the strut (under full extension to the pipe centreline) in relative mean velocity error $(1/B)\Delta U/U$, where $B = 2d/(\pi D)$, is given in terms of the streamwise and azimuthal separation of strut. (After *Hubbard* [5.35])

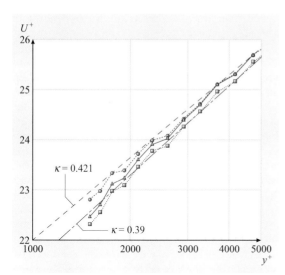

Fig. 5.9 Mean velocity profiles from *McKeon* [5.25] at $Re_D = 10 \times 10^6$ under the *MacMillan* [5.28] displacement correction. *Shaw* [5.37] static pressure correction (*squares*). *McKeon* and *Smits* [5.38] static pressure correction (*triangles*), with MacMillan wall term (*circles*). Log laws with $U^+ = 1/\kappa y^+ + B$ and $\kappa = 0.421$ (*dashed line*) and $\kappa = 0.390$ (*dot-dashed line*)

for the separate measurement of total and static pressure above and in Chap. 4. Due to the complexity of design these probes tend to have a larger diameter than equivalent Pitot probes and a lower ratio of inner to outer diameter θ. Consequently, use near walls is not recommended due to a large wall-proximity effect, and care must be taken in flows with large velocity gradients (shear). The influence of viscosity on the dynamic pressure through the measurement of total head is as detailed in Sect. 5.1.1, but the impact of turbulence will depend on the correlation between the total and static pressure fluctuations.

The effect of yaw is shown in Fig. 5.10, which gives the percentage error in measured dynamic head, $100 C_{pm}$, where $C_{pm} = (p_{0m} - p_m)/(1/2\rho U^2)$, due to errors in the measured total and static heads with yaw angle for a standard Pitot-static probe. It should be noted that the zero-yaw case no longer corresponds to the highest recorded velocity since there is a balance between the competing effects on the total and static pressures. The effects of pitch and yaw have been shown to be the same within approximately ±1% [5.2].

Chebbi and *Tavoularis* [5.39] have demonstrated that a severe reduction in Reynolds number (in this case over four orders of magnitude to a low Reynolds number

$Re_d = 0.14$) also dramatically increases the yaw sensitivity of a Pitot-static probe with a rounded tip and $\theta = 0.28$, reducing the range of negligible error by an order of magnitude. In general, it is unlikely that the Reynolds number effect is important, since $Re_d = 0.14$ lies well outside the normal Reynolds number range in which Pitot probes are used.

The tip and stem influence the measured dynamic head through the static pressure as detailed in Sect. 4.2.2 and shown in Fig. 4.8. In the Prandtl probe design, a cancellation of the errors occurs through the location of the static pressure holes; tip-to-static-hole and static-hole-to-stem separations of $6d$ and $8d$, respectively, are used in a standard modified Prandtl design.

Other types of combined probe include the compact, cantilever-type Pitot cylinder, wedges, and cones. In the former, cylindrical design, coaxial tubes are used in conjunction with static and total tappings located at the same lengthwise distance along the cylinder, but separated circumferentially (and the outer cylinder is sealed with a rounded tip). The probe is inserted transverse to the incoming flow and rotated such that the total pressure tapping is aligned with the incoming flow. The relationship between the angular location of the static pressure tappings and local static pressure is determined by calibration since the flow may deviate from the ideal potential-flow solution due to the influence

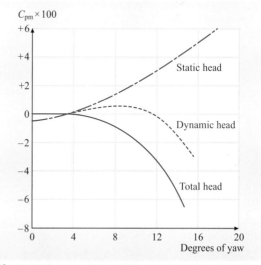

Fig. 5.10 Yaw sensitivity of Pitot-static probes (after *Pope* and *Harper* [5.36]), demonstrating the percentage error in the measured dynamic pressure, i.e., $100 C_{pm}$, due to the effect of angular misalignment on the total, static, and dynamic head readings

of the tapping size or Reynolds and Mach number effects (shocks will occur in the accelerating flow over the front half of the cylinder for freestream Mach numbers $0.55 \leq M < 1.0$). The total pressure reading is independent of Reynolds number for $Re_d > 4000$, but a function of the separation of the tapping from the probe tip when this is less than five probe diameters [5.1]. *Chue* [5.1] states that the accuracy of the cantilever probe is 5% of the impact pressure per degree of yaw, implying that this kind of probe is most useful for higher flow velocities. [Note that, with two static tappings symmetrically situated around the cylinder circumference from the total tapping, the local flow direction can also be obtained by the null-reading method (Sect. 5.1.4) provided that the Mach number is low enough to preclude shocks on the cylinder surface.]

Mach number can be determined in an isentropic supersonic flow using the relationship between reservoir pressure and surface pressure on a wedge (or cone) via the oblique (conical) shock relationships [5.4].

For more details of other probe designs, the reader is referred to *Bryer* and *Pankhurst* [5.2], and *Chue* [5.1].

5.1.4 Measurement of Dynamic Head and Flow Direction (Multihole Probes)

So-called *multihole* probes are often used to obtain the flow direction, the static pressure and the dynamic head in flows where the incoming angle is unknown and/or has a large variation in the flow domain, for example, in turbomachinery and wake flows. Probes typically consist of a central, stagnation pressure hole or tube, downstream from which a number of pairs of static pressure holes/tubes are symmetrically displaced, with equal and opposite angles between the orifice faces and the flow. Figure 5.11a,b shows conical five- and seven-hole probes; other typical designs include hemispherical and wedge arrangements.

Multihole probes may be used in either *null-reading* (for symmetric probes) or *calibrated* modes. In the former method, the angular orientation of the probe is altered about the yaw and/or pitch axes until both static holes in a pair give the same pressure reading and the stagnation hole is aligned with the flow direction. The flow angle can then be determined from the rotations required to null the probe, which (when appropriately connected to a set of transducers and with simple probe calibrations) then gives measurements of total, static and dynamic heads, and hence the velocity vector. This method is extremely time-consuming and relies on the assumption that the flow is locally symmetric about the probe (a constraint either on probe manufacturing tolerances or the size of the probe relative to the lengthscale of any shear in the flow), however in this case the Reynolds and Mach number dependence of the measurement can be simply obtained. Note that the sensitivity of this type of probe is designed to give a more pronounced variation with angle than could be obtained by using the yaw response of the types of probe described earlier in the chapter.

Alternatively, the probe can be calibrated by forming single-valued dimensionless pressure ratios (which are a function of angle and independent of Reynolds number) in flows where either a suitable rotation mechanism is not available or leads to movement of the tip of the probe during the nulling process. For use in flows where the flow angle may lead to separation over one or more tappings (approximately 30° for most probes [5.40]), the calibrations are typically subdivided based on where the highest pressure is measured. (In practice, the calibrations become double-valued when the angle is large enough to give a significant separation in the tip region.) *Sumner* [5.41] has compared the accuracy of measurements for which the calibrations are found by interpolation and curve-fit polynomials between calibration points for seven-hole probes, finding that the former gave better results in the presence of flow separation and that results were sensitive to the grid of calibration points used.

If there is angular variation in only one plane, a three-hole Cobra probe may be used, consisting of a stagnation tube flanked by two in-plane static pressure tubes. *Chue* [5.1] proposes that the flow direction within a boundary layer may be determined to within 0.2° by the calibration method if the pitching angle does not exceed 6°.

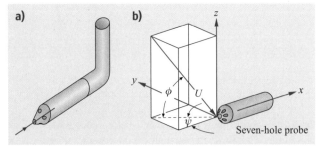

Fig. 5.11a,b Schematic of (**a**) a five-hole truncated-cone probe, and (**b**) a conical seven-hole probe, demonstrating the angular determination of the velocity vector

A five-hole probe, Fig. 5.11a, may be used to determine the angular variation in both pitch and yaw directions with a magnitude of less than 30° [5.42]. At larger angles, where the five-hole probe becomes ineffective for angle determination by calibration, a seven-hole probe, Fig. 5.11b, may be more suitable (although note that the nulling method becomes prohibitively time-consuming for this type of probe). For a conical design, the cone angle is typically 30° or 45°, with the tappings spaced at 60° intervals.

Zilliac [5.40] suggests that a calibrated seven-hole probe offers a measurement accuracy of 1% in velocity and 0.5° in angle (comparable to a five-hole probe in nulling mode) also noting that since six tubes pack closely around the central stagnation tube there is only a small concomitant increase in probe size. In subsonic compressible flows, the calibration method may be used at flow angles up to 70° to determine flow angle with accuracy ±2° and Mach number to ±0.04 with 95% certainty [5.43]. *Takahashi* [5.44] has investigated the suitability of seven-hole probes for unsteady flow measurements.

While the use of multihole probes is ubiquitous in the aerospace and turbomachinery communities, there are serious limitations on these techniques that preclude their use for highly accurate measurements. The sources of error are many, including most seriously the effects of instantaneous yaw and all-component fluctuation intensity in turbulent flow, the large disturbance presented by the probe (the connections for multiple pressure lines mean that multihole probes are necessarily relatively large) and the effects of flow separation. It is not clear how the empirical corrections described in earlier sections of this chapter apply to the complex flow fields induced around a probe tip that is not aligned with the flow. While an attempt is often made to address these issues during the calibration process, the requirement for both known flow angle and velocity probably implies a laminar or at least low turbulence calibrating flow. Strictly speaking, calibration in a nonturbulent freestream justifies use of the probe only in a similar, nonturbulent flow.

In addition, multihole probes are not suitable for use in flows where there is a significant gradient in total pressure or in which the Mach number is sufficiently high that there is the potential for asymmetric shocks to form in front of the probe, or over only some of the tappings. (The effect of the latter phenomenon may be mitigated by reducing the separation between small tappings and/or using probe shapes, such as a wedge, for which analytic solutions exist.)

Further details on multihole probes may be found in *Bryer* and *Pankhurst* [5.2].

This section has described devices for the determination of the magnitude of the velocity (or Mach number), assuming at least approximate alignment of probes with the incoming flow. In general, the flow angle may be determined from Pitot and static probes through yaw calibrations if the flow conditions are otherwise known, or multihole yawmeters may be rotated until the same readings are obtained by holes on either side of the probe centerline, the *null-reading* method. For further information on these techniques and appropriate hardware, the reader is referred to *Chue* [5.1].

5.2 Thermal Anemometry

Hot-wire anemometry is a versatile technique that can be used for the measurement of velocity and temperature fluctuations in the time domain for investigations in turbulent flows. With appropriate designs hot-wire anemometry is useful over wide speed ranges from low subsonic to high supersonic flows. The method is capable of detecting turbulent perturbations with a large dynamic response because of the very small hot-wire thermal inertia and its correction in the anemometer. The hot-wire can be operated in three popular methods: constant current, constant temperature, and the new constant voltage. The response of the sensor and that of the associated operating electronic circuit are analyzed. Also included are calibration techniques and approximations for low turbulence levels.

5.2.1 Introduction

Thermal anemometry is used when one wishes to measure rapidly varying velocities with good spatial and time resolution. An example is given in Fig. 5.12 for a supersonic boundary layer. The method relies on the changes in heat transfer from a small heated sensor exposed to the fluid in motion. The sensor is made with a material whose electric resistivity depends on the temperature. It usually has the shape of a small cylindrical wire or a thin film. Experiments can be made in gases, from low subsonic to high supersonic velocities, and transparent, opaque, and even electrically conducting liquids. Instantaneous wall shear stresses can also be measured using sensors mounted flush to the surface. The measurement of other flow properties such as temperature, density, composition, is also part of thermal anemometry. Some aspects of these property measurements are considered in the present chapter.

Heat is introduced in the sensor by Joule heating, and is lost primarily by forced convection. The heating rate is $R_w I_w^2$, where I_w is the current intensity in the sensor and R_w the sensor resistance when heated. The cooling rate is of the form $(T_w - T_a)\phi_{conv}(U \ldots)$, where T_w is the temperature of the wire when heated, and T_a is the temperature of the wire at the same location when unheated. In liquids and low subsonic gas flows, T_a is simply the temperature of the surrounding fluid. When compressibility effects are present, T_a becomes the recovery temperature T_r of the sensor. The function $\phi_{conv}(U \ldots)$ mainly depends on the fluid velocity normal to the wire. The role of the other parameters will be discussed in Sects. 5.2.3 and 5.2.4, and for the moment the forced convection effect is simply written $\phi_{conv}(U)$. A small fraction of the heat supplied to the wire is, however, conducted towards the supports holding the sensor, resulting in a nonuniform temperature distribution along its length. The quantities T_w or R_w are in reality spatial averages over the sensor. This problem is neglected for the moment, but it will be considered later in this section.

For a steady flow, the heat balance between the heating and the cooling rates has the simple form

$$R_w I_w^2 = (T_w - T_a)\Phi_{conv}(U) . \qquad (5.35)$$

The velocity U can be deduced from the electrical quantities R_w and I_w, if T_w can be related to R_w and similarly T_a to R_a, where R_a is the resistance of the unheated sensor. Such relations exist for most sensors. Often an assumed linear dependence is adequate to describe $R(T)$

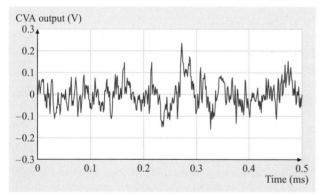

Fig. 5.12 Time trace of a turbulent signal provided by hot-wire anemometry based on experiments conducted with a constant-voltage anemometer (CVA) in a Mach 2.3 boundary layer

around the operating point, so that

$$R_w = R_0[1 + \chi(T_w - T_0)] \quad \text{and}$$
$$R_a = R_0[1 + \chi(T_a - T_0)] , \qquad (5.36)$$

where T_0 is a reference temperature, R_0 the resistance of the sensor at that temperature, and χ the temperature coefficient of resistivity of the wire material around T_0, and T_0 is often chosen close to T_a. The value of χ is positive for metallic materials. For example, for tungsten, $\chi \simeq 4.0 \times 10^{-3}\,\text{K}^{-1}$, resulting in 100% increase of the sensor resistance when the operating temperature T_w is increased by 250 K over T_a. The value of χ is negative for thermistors. For polysilicons, χ can be positive or negative, depending on the material used for the doping and its concentration.

A significant parameter that controls the operation of the sensor is the relative difference in temperature, $(T_w - T_a)/T_a$, or the relative difference in resistance, $(R_w - R_a)/R_a$. The latter is preferred because of its direct deduction from electrical quantities. It is simply denoted a_w and called the overheat ratio of the sensor,

$$a_w = \frac{R_w - R_a}{R_a} . \qquad (5.37)$$

When needed, the relative difference in temperature can be expressed in terms of a_w as

$$\frac{T_w - T_a}{T_a} = a_w \frac{1 + \chi(T_a - T_0)}{\chi T_a} . \qquad (5.38)$$

In practice, for air or other gases, a_w is in the range 0.05–1.0. Beyond this value, the sensor could be permanently deformed or oxidized. For example, platinum has an excellent resistance to oxidation, but is naturally soft and weak. Its tensile strength can be increased by the addition of iridium or rhodium. On the other hand, tungsten, which presents a high tensile strength, oxidizes above 600 K. For water or other fluids, a_w is limited to around 0.10 because of the occurrence of vapor bubbles from boiling.

The simple balance (5.35) holds for steady flows only. If a flow is unsteady, which is always the case if the flow is turbulent, thermal energy is unsteadily stored in the sensor, and the instantaneous heat balance becomes

$$m_w c_w \frac{dT_w}{dt} = R_w I_w^2 - (T_w - T_a)\Phi_{conv}(U) , \qquad (5.39)$$

where m_w is the mass of the sensor and c_w is the specific heat of the sensor material. Conduction effects are not included in (5.39).

When the sensor is an element of an appropriate electrical system, the change in resistance can be used to generate a measurable signal. Three operating modes are possible.

- The current intensity in the sensor is maintained constant, $I_w = $ constant, which leads to the constant-current anemometer (CCA). Any change of U creates a change in R_w and the measurable signal is directly this resistance change.
- The resistance of the sensor is maintained constant, $R_w = $ a constant, which leads to the constant-resistance anemometer, usually called a constant-temperature anemometer (CTA) because of (5.36). The measurable signal when a change in U occurs is then the change in current intensity I_w to be fed to the sensor to fulfill the imposed condition.
- The voltage across the sensor is maintained constant, $V_w = R_w I_w = $ constant, which leads to the constant-voltage anemometer (CVA). The measurable signal is then the change in current intensity I_w to be fed to the sensor under that new imposed condition.

The main difference between these three operating modes is linked to the handling of thermal inertia of the sensor, represented by the left-hand side of (5.39). In usual applications, the frequencies of the flow fluctuations to be measured are much higher than the natural frequency of the sensor, so that electronic compensation is needed. In a CCA and CVA, this is achieved by a first-order high-pass filter integrated in the amplifying unit. In a typical CCA application, the filter's response is tuned to compensate exactly the thermal lag of the sensor. The overall bandwidth is then only limited by the amplifier's characteristics, mainly its gain–bandwidth product. With a CVA, the thermal lag is partially compensated during measurements and is fully compensated when postprocessing the data. This permits high productivity for large-bandwidth applications because no adjustment is required when the experiment is running. In a CTA, the temperature of the sensor is maintained constant by a feedback loop, so that its thermal inertia is, in principle, automatically compensated. In this case, the maximum bandwidth is limited by the amplifier's properties and some characteristics of the practical setup. Full compensation can be made after carefully tuning the circuit. These points will be analyzed in Sect. 5.2.3.

In any of these three operating modes, the temperature T_a is a major parameter, as it modifies the difference $T_w - T_a$. In fact, changing T_a is almost the same as changing T_w. This point is most important when using a CTA, which maintains $R_w = $ const, hence $T_w = $ const, whatever T_a. For example, in a subsonic flow, if the temperature T_a of the incoming flow increases at fixed U, the current intensity needed by the sensor decreases as $T_w - T_a$ decreases. A benefit, however, is that thermal anemometry can be used to measure temperature changes occurring in turbulent flows.

The Section is organized as follows. Section 5.2.2 describes the nature and shape of typical sensors, and their geometrical disposition relative to the incident flow, in order to obtain specific flow characteristics. The steady heat-transfer laws expressing the cooling rate of the sensors are then indicated for various fluids and flow conditions. They are most useful when interpreting the electrical signals available at the anemometer output. For unsteady flows, the thermal lag of the sensor is an inherent limiting parameter and some estimates of this lag are indicated. Finally, two important topics are considered: the heat loss towards the supports holding the sensor and the spatial resolution of the sensing element.

Section 5.2.3 deals with the electronic circuits used in the CCA, CTA, and CVA. Treatment of the thermal lag of a sensor is examined with special attention. For example, it is shown how constant-voltage operation reduces the time constant compared to constant-current operation. The frequency response as well as the electronic noise of anemometers are analyzed and compared. Effects of connecting cables are also considered.

Section 5.2.4 refers to calibration procedures in low subsonic flows including the effect of fluid temperature drifts, and Sect. 5.2.5 deals with the measurement of temperature fluctuations.

Section 5.2.6 refers to calibration procedures in high supersonic flows.

Section 5.2.7 presents some special methods, such as the use of the time of flight between sensors or the use of pulse-width-modulated wires, and also the benefit of a flying anemometer when the main flow direction is not well defined, for example in separated zones or near jet edges.

Thermal anemometry has been used for many years, with many improvements over the years. Accordingly, the literature is very rich. Acknowledgement of previous works will often be made and some comprehensive books or reviews on the subject have been contributed by *Corrsin* [5.45], *Bradshaw* [5.46], *Hinze* and *Freymuth* [5.47], *Blackwelder* [5.48], *Perry* [5.49], *Fingerson* [5.50], *Lomas* [5.51], *Bruun* [5.52], *Lekakis* [5.53], and *Comte-Bellot* [5.54–56].

5.2.2 Sensors

Materials and Sensor Geometries

The sensing elements used in thermal anemometry are either wires or films. Conventional wires are typically 0.5–5 µm in diameter and nominally 0.5–2 mm long. Films are about 0.1 µm thick and deposited on fine cylinders of quartz, about 25–50 µm in diameter, on quartz wedges, or on thin acetate or kapton foils. Furthermore, a very thin quartz coating deposited on the sensor provides both protection against a hostile environment and isolation when operating in a conductive medium. The micromachined hot wires usually have a flat trapezoidal cross section, which can be as small as $0.5 \times 0.5\,\mu\text{m}^2$. Their lengths are in the range 1–200 µm. The most common materials for wires and films are listed in Table 5.1.

The upper part of Table 5.1 refers to the classical and easy-to-manipulate metallic elements. Their electric resistances at ambient temperature, $T_a \simeq 293\,\text{K}$, are in the range 3–10 Ω. The temperature resistivity coefficient χ is a very important parameter and needs to be controlled since the wire manufacturing process can influence the value. Handbooks generally indicate values only corresponding to macroscopic amounts of pure material, for example, $\chi = 0.0045\,\text{K}^{-1}$ for tungsten, whereas the measured values for hot-wires can be over a wide range 0.0036–0.0047 K^{-1}. Very useful references on the subject are available in *Van Dijk* and *Nieuwstadt* [5.65].

The lower part of Table 5.1 refers to the wires and films more recently developed using micromachining technology. Their electric resistances are much larger, around 0.5–5 kΩ, meaning that the classic electronic circuitry of the CCA, CTA, or CVA should be redesigned to meet that new requirement.

Incidentally, it is interesting to measure the true wire diameter, for example with an electron microscope. Indeed, *Lemay* and *Benaïssa* [5.66] reported wire diameters which are 15–20% higher than the values stated by the manufacturer for 0.5 and 1 µm Pt–10%Rh wires. *Dersken* and *Azad* [5.67] observed that, even for specially built wires (platinum plated, tungsten core, gold plated ends), a 2.84–3.85 µm range was observed for a nominal value of 2.5 µm, and a 4.82–5.77 µm range for a nominal value of 5 µm. These differences in diameter translate into variations of the cold resistance R_a of the wire and can also be detected in that way.

Regarding the sensor dimensions, one has to prefer large aspect ratios. For hot-wires, this ratio is defined by l/d, where l is the wire length and d the wire diameter. A large aspect ratio, above 300, permits a wire to be cooled by the velocity component normal to it, with negligible heat transfer to its supports. For a fixed aspect ratio, a smaller diameter obviously leads to a shorter wire, resulting in more-local measurements, and a higher-frequency response. However, fine wires are expensive and brittle, and therefore reserved to turbulence measurements close to walls, where fine turbulent scales have to be resolved.

Table 5.1 Physical properties of sensor materials at 293 K (approximate values for MEMS polysilicons doping, P is phosphor and B boron)

Material	Resistivity (Ωm)	Temperature coefficient of resistivity χ (K^{-1})	Density ρ (kg/m^3)	Specific heat c (J kg^{-1}K^{-1})	Thermal conductivity k (W m^{-1}K^{-1})	Ref.
Copper	1.6×10^{-8}	$+4.0 \times 10^{-3}$	8900	385	400	[5.57]
Nickel	7.0×10^{-8}	$+6.0 \times 10^{-3}$	8900	438	90	
Platinum	1.1×10^{-7}	$+3.9 \times 10^{-3}$	21500	130	70	[5.58]
Pt/10% Rh	1.9×10^{-7}	$+1.7 \times 10^{-3}$	19900	150	40	[5.59]
Silver	1.6×10^{-8}	$+3.8 \times 10^{-3}$	10500	235	428	[5.57]
Tungsten	6.0×10^{-8}	$+4.5 \times 10^{-3}$	19300	140	170	[5.58]
Silicon	2.3×10^3	-7.5×10^{-2}	2330	705	148	[5.60]
Polysilicon						
High P doping	5×10^{-5}	$+1.0 \times 10^{-3}$			50	[5.61]
Low P doping	5×10^{-5}	-2.5×10^{-3}				
B doping	2×10^{-5}	$+8.5 \times 10^{-4}$			60	[5.61, 62]
B doping $2 \times 10^{16}/\text{cm}^2$	5×10^{-5}	$+8.0 \times 10^{-4}$				[5.63, 64]

At its ends, the sensor is connected to fine but sufficiently strong supports that act only as the electrical conductor and mechanical support. For conventional hot-wires, these elements are usually called *prongs* or *needles*, and *stem* respectively. The two parallel prongs are usually 5–10 mm long and taper down to around 100 μm near the wire. The prongs are epoxy-fitted inside the stem, whose diameter can be as small as 1 mm. To decrease the aerodynamic perturbations brought by these elements, longer prongs, up to 20 mm, were suggested by *Comte-Bellot* et al. [5.68] and *Strohl* and *Comte-Bellot* [5.69]. Specially miniaturized probes were developed by *Willmarth* and *Sharma* [5.70] and *Ligrani* and *Bradshaw* [5.71].

For microelectromechanical sensors (MEMS), the successive connecting elements are called *shanks, beam,* and *handle*. According to *Jiang* et al. [5.72], the two parallel supporting shanks are 100 μm long, 20 μm wide, and 0.5 μm thick. The silicon beam is 1 mm long, 200 μm wide, and 75 μm thick. The handle is thick, around 500 μm. Similar values are reported by *Ebefors* [5.64] for a three-sensor probe.

Often the sensor does not span the entire distance between the needles. The inactive part close to the ends of the sensor should have a low electrical resistance in order not to participate in the sensor response. This part is often called the *stub*. For platinum wires extruded by the Wollaston process, the silver sheath is etched away to provide the active part, and preserved near the two ends. The same procedure holds for tungsten wires, which are copper-plated before soldering. Superconducting endings have also been suggested by

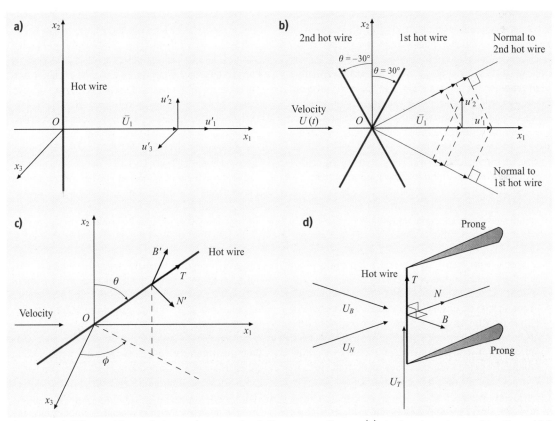

Fig. 5.13a–d Wire positions relative to the space-fixed $O_{x_1 x_2 x_3}$ coordinates: (**a**) single wire set normal to O_{x_1}, which is also $\bar{U}_1$; (**b**) X-wires in the $O_{x_1 x_2}$ plane with $\theta = 30°$ for the first wire and $\theta = -30°$ for the second wire; (**c**) long wire with yaw angle θ and pitch angle ϕ in spherical coordinates; N', B', T is the wire-fixed frame, N' corresponds to an increment in θ, B' to an increment in ϕ, and T is along the wire [a long wire is cooled by the resulting normal velocity $U_n = (U_{N'}^2 + U_{B'}^2)^{1/2}$]; (**d**) short wire with its supporting prongs, N, B, T is the prong-fixed frame. The effective cooling velocity is given by (5.172) and (5.173)

Castaing et al. [5.73] for experiments in helium II at 4 K. For polysilicon sensors, various doping concentrations are used for the active and nonactive parts, as suggested by *Jiang* et al. [5.74]. Let us recall that common dopants are phosphorus and boron, and that heavier doping results in lower resistivity by increasing the number of charge carriers. The average grain size of boron-doped polysilicon layers can be estimated from atomic force microscopy (AFM), and a clear illustration is given by *McConnell* et al. [5.61].

Sensors Setup Relative to the Flow

Identification of the principal mean flow direction is an important prerequisite for hot-wire or hot-film measurements. This direction, when known, is usually taken as the x_1-coordinate axis and corresponds therefore to $\overline{U}_1$. Three basic positions can be given to a wire, as illustrated by Fig. 5.13: Fig. 5.13a the wire can be placed in the $O_{x_1 x_2}$ plane and normal to x_1; Fig. 5.13b the wire can be placed in the $O_{x_1 x_2}$ plane and yawed on x_2 by an angle θ, hence inclined on O_{x_1} by $\pi/2 - \theta$; Fig. 5.13c the wire can be outside the $O_{x_1 x_2}$ plane and inclined relative to this plane with a pitch angle ϕ. The angles θ and ϕ are the standard spherical coordinate angles. These wire positions are defined relative to the space-fixed coordinates $O_{x_1 x_2 x_3}$. For short wires, auxiliary angles referring to the probe itself are introduced and are defined in Fig. 5.13d.

Use of these sensor positions, or combinations of these positions, leads to the following possibilities:

- Single hot-wires placed normal to $\overline{U}_1$ mainly respond to the instantaneous longitudinal velocity U_1 when the velocity fluctuations are of small amplitude. This instantaneous velocity can be written as $U_1 = \overline{U}_1 + u_1'$ using the Reynolds decomposition, where u_1' is the longitudinal velocity fluctuation. The transverse fluctuations u_2' and u_3' behave as second-order terms in this setup and can be neglected. Obviously, $\overline{U}_1$ should be different from zero, otherwise the transmitted signal suffers rectification, since the hot-wire is just as much cooled by the flow coming from the right-hand side as from the left.
- Cylindrical hot-films work similarly to single hot-wires, but they are more robust when experimenting in water, oils, or liquid metals and even in slightly dirty gases. A thin quartz coating deposited on sensors prevents hydrolysis of the neighboring liquid.
- X wires are made of two inclined wires lying in the $O_{x_1 x_2}$ plane and often placed symmetrically relative to $\overline{U}_1$, as in the letter X, i.e., θ (first wire) $= -\theta$ (second wire). Figure 5.13b illustrates this case for $\theta = 30°$. In practice, the midpoints of both wires are slightly separated along O_{x_3} to reduce aerodynamic and thermal interferences [5.75]. For small-amplitude velocity fluctuations, the wire responses are

$$\overline{U}_1 \cos\theta + u_1' \cos\theta - u_2' \sin\theta \quad \text{for the 1st wire}$$
$$\overline{U}_1 \cos\theta + u_1' \cos\theta + u_2' \sin\theta \quad \text{for the 2nd wire}$$
(5.40)

and the sum and difference of the fluctuating parts of these signals give $u_1'(t)$ and $u_2'(t)$. If the wires are not perfectly symmetric relative to $\overline{U}_1$, a weighted sum and a weighted difference provide $u_1'(t)$ and $u_2'(t)$. In extreme cases, it is possible to place one of the wires normal to the flow. This wire thus directly gives $u_1'(t)$ and an appropriate combination of its response with that of a slanted wire gives $u_2(t)$ [5.76]. Note that u_1' and u_2' are in the plane defined by the X wires. The other transverse fluctuating component u_3' together with u_1', can be obtained by rotating the X-wires by 90° around the direction of $\overline{U}_1$. These classic setups are described in most textbooks (*Hinze* [5.47], *Bruun* [5.52], and *Bailly* and *Comte-Bellot* [5.77]). After proper calibration, an X probe can also be used in a two-dimensional (2-D) flow with unknown mean velocity directions (Sect. 5.2.4). A two-dimensional flow means that $U_3 \simeq 0$. The instantaneous velocity vector can be obtained, as well as the two mean components, $\overline{U}_1$ and $\overline{U}_2$, and the velocity fluctuations, u_1' and u_2', even when they are large. However, the incident velocity vector should always give positive velocity components relative to each wire to avoid spurious signals due to rectification. This condition defines the correct approach *quadrant*. The term *quadrant* comes from the use of two mutually orthogonal wires. In practice the acceptance angle is less than 90° because of the nearby needles supporting the wires.
- V probes work similarly to X probes. They permit measurements in close proximity to a wall. A cantilevered V probe has even been suggested by *Hishida* and *Nagano* [5.78, 79].
- Triple-sensor probes provide information about the instantaneous velocity vector at a point, hence the three velocity components, U_1, U_2, U_3 simultaneously, after proper manipulation of the three wire outputs. $\overline{U}_1$, $\overline{U}_2$, and $\overline{U}_3$ can then be obtained, along with the three velocity fluctuations u_1', u_2', and u_3'. Velocity component relative to each wire should always stay positive, and this defines the correct

approach *octant* to be respected. The term *octant* comes from the use of three mutually orthogonal wires. In practice the acceptance cone is smaller because of the prongs holding the wires. A fourth hot wire can be added to the probe to enlarge the acceptance cone. A recent survey is given by *van Dijk* and *Nieuwstadt* [5.80].

- Multiple hot wires, from 4 to 12, can provide one or more components of the vorticity vector as discussed by *Wallace* and *Foss* [5.81].
- Arrays of around 10 single wires permit to scan the free edges of turbulent flows, as pioneered by *Sunyach* [5.82] for jets using thermal marking. The technique was applied to boundary layers by *Chen* and *Blackwelder* [5.83] and adapted to micromachined sensors by *Jiang* et al. [5.74]. *Kang* et al. [5.84] achieved space–time correlation of active-grid-generated turbulence with an array of X-wires. *Delville* et al. [5.85] used two rakes of 12 X-probes each to analyze the large structures of a plane mixing layer on the basis of a proper orthogonal decomposition (POD). *Citriniti* and *George* [5.86] used an array of 138 single wires to reconstruct the velocity field of a round jet.
- Wall films permit one to obtain the instantaneous wall shear stress $\tau_w = \mu \partial U_1 / \partial x_2$. Large arrays of wall films can be made as these sensors are non intrusive. Around 100 hot-films are most useful when investigating transition on flat or curved surfaces, separation bubbles over airfoils, or shock oscillations in transonic regimes. Laboratory experiments are reported by *Hall* et al. [5.87] using CTA, and flight experiments by *Moes* et al. [5.88] using CVA.

Many types of hot-wire probes are available commercially nowadays, but probes can be fabricated in most laboratories. Valuable advice is given by *Lomas* [5.51], e.g., for welding and soft-soldering wires on fine needles. Regarding films, even if the very first ones were successfully made by a paint and fire technique, the sensors are now made by specialized manufacturers by vapor deposition or atomic sputtering. They are reproducible within 1% and this greatly reduces the labor of calibrating large arrays. For polysilicon sensors, only a few laboratories are currently equipped to make them, but mass production is possible as a single silicon wafer can encompass a large number of sensors.

Because hot-wire anemometers have to compete with laser anemometers which are advantageously located outside the flows, hot-wire probes should be as non-intrusive as possible. In low subsonic flows, minia-turized hot-wire probes are recommended. *Willmarth* and *Sharma* [5.70] and *Ligrani* and *Bradshaw* [5.71] reported experiments with wires having lengths as small as 100 μm and diameters of 0.5 μm. For larger probes, prongs can be placed at 45° to the flow, as suggested by *Comte-Bellot* et al. [5.68], and *Strohl* and *Comte-Bellot* [5.69]. This location permits compensation of the flow deceleration due to the blocking effect of the stem – which is maximal for a stem aligned with $\overline{U}_1$, and the flow acceleration occurring between the prongs – which is maximal for prongs normal to $\overline{U}_1$. Estimates of the aerodynamic perturbations were also obtained by *Adrian* et al. [5.89] on the basis of an inviscid irrotational flow around the probe elements. Historically, *Hoole* and *Calvert* [5.90] were the first to reveal the aerodynamic perturbation resulting from a probe by comparing the wire response for a probe (prongs and stem) facing the flow or held normal to the flow, with the wire position kept unchanged.

In supersonic flows, the prongs are always aligned with the flow to withstand shock waves. Slightly slack wires with a drop of rubber cement at the soldering points provide satisfactory protection against vibration and fatigue failure as recommended by *Kovasznay* [5.91]. Slack is introduced when soldering the wire, or later if a tiny screw permits the change of prong separation, as suggested by *Bonnet* [5.92]. Control of the adjustment is made by inspecting the spectra which should not contain the sharp peaks associated with strain-gauging above 100 kHz. Measurements in transonic flows remain however very difficult due to large fluctuating aerodynamic loads and temperature fluctuations.

For MEMS sensors, strong aerodynamic perturbations were reported by *Jiang* et al. [5.74] when changing the yaw and pitch angles of their probes, in a similar way as *Hoole* and *Calvert* [5.90] did with conventional probes. Additional investigations would be of considerable interest, as the shanks, beams and handle are relatively large and close to the sensors in that new type of technology [5.93].

Steady Convective Heat Loss Laws in Subsonic Flows

The simplified heat balance (5.35) is now rewritten including all the flow and wire parameters except the aspect ratio, which will be considered later in this section. Using dimensionless analysis, this balance is of the form

$$\mathrm{Nu} = \mathrm{Nu}\left(M, \mathrm{Re}, \mathrm{Gr}, \mathrm{Pr}, \gamma, \frac{T_w - T_a}{T_a}\right), \quad (5.41)$$

where M is the flow Mach number, Nu, Re, and Gr are the Nusselt, Reynolds and Grashoff numbers, respectively, based on the wire diameter

$$M = \frac{U}{c}, \quad \text{Nu} = \frac{R_w I_w^2}{\pi l k (T_w - T_a)},$$

$$\text{Re} = \frac{\rho U d}{\mu}, \quad \text{and} \quad \text{Gr} = g\beta(T_w - T_a)d^3 \left(\frac{\rho}{\mu}\right)^2. \tag{5.42}$$

The physical properties involved for the fluid are the thermal conductivity (k), the density (ρ), the Newtonian coefficient of dynamic viscosity (μ), the thermal coefficient of expansion at constant pressure (β), the ratio of the heat capacities at constant pressure and constant volume ($\gamma = C_p/C_v$), and the Prandtl number (Pr = $\mu C_p/k$). The incident flow velocity is simply noted U instead of U_1 or U_n. The temperature at which the physical properties are taken will be discussed later in this section.

Reference to a Nusselt number shows that conduction terms and boundary conditions, hence temperature gradients, prevail in the energy equation. Indeed, for hot-wires, the Reynolds number is usually in the range $2 \leq \text{Re} \leq 40$, before the onset of vortex shedding.

Several cases of flows are of interest as follows.

Heat-Transfer Laws for Subsonic Air Flows. M and γ obviously have no effect, Pr is fixed, for air Pr $\simeq 0.70$ over a large temperature range, and T_a is the ambient fluid temperature. Furthermore, forced convection often prevails – the criterion for neglecting buoyancy effect is examined at the end of this section. Equation (5.41) simplifies therefore to

$$\text{Nu} = \text{Nu}\left(\text{Re}, \frac{T_w - T_a}{T_a}\right). \tag{5.43}$$

Several heat-transfer relationships for forced convection have been reported for very long wires (l/d in the range 2000–8000) (e.g., by *McAdams* [5.94], *Van der Hegge Zijnen* [5.95], *Collis* and *Williams* [5.96], *Davis* and *Fisher* [5.58], *Bradbury* and *Castro* [5.97], *Kramers* [5.98]). The McAdams analysis already concerned 13 data sets in 1950, and in a more-recent survey in 1972, *Andrews* et al. [5.99] listed about 40 experiments. All these laws have the form

$$\text{Nu} = (A_N + B_N \text{Re}^n)\left(\frac{T_w + T_a}{2}\right)^m, \tag{5.44}$$

where A_N, B_N, n, and m are constant. There are, however, some tiny differences:

1. the power for the Reynolds number n varies slightly around $n \simeq 0.50$,
2. a temperature loading depending on m is introduced or not, with a maximum value of $m \simeq 0.17$; it is attributed to the flow distortion around the wire caused by the addition of heat, and
3. the physical properties of the fluid are evaluated at either the incoming flow temperature T_a or at the *film* temperature $T_f = (T_w + T_a)/2$.

In what follows, the distinction is made by using the subscripts a and f respectively.

The most-favored correlation formulae are due to *Collis* and *Williams* [5.96] and *Bradbury* and *Castro* [5.97]

$$\text{Nu}_f = \left(0.24 + 0.56\text{Re}_f^{0.45}\right)T_f^{0.17}$$

for $0.02 \leq \text{Re}_f \leq 44$

$$\text{Nu}_f = 0.48\text{Re}_f^{0.51} T_f^{0.17}$$

for $44 \leq \text{Re}_f \leq 140$ \hfill (5.45)

and *Kramers* [5.98]

$$\text{Nu}_a = 0.39 + 0.51\text{Re}_a^{0.50} \quad \text{for } 2.5 \leq \text{Re}_a \leq 1475. \tag{5.46}$$

We recall that the temperature dependence of the physical properties of air around $T_a = 293$ K may be expressed by the power laws [5.100, 101]

$$\frac{\rho(T)}{\rho(T_0)} = \left(\frac{T}{T_0}\right)^{-1}$$

$$\frac{\mu(T)}{\mu(T_0)} = \left(\frac{T}{T_0}\right)^m$$

$$\frac{k(T)}{k(T_0)} = \left(\frac{T}{T_0}\right)^n. \tag{5.47}$$

with $m \simeq 0.765$ and $n \simeq 0.885$. Power laws are convenient when taking logarithmic derivatives of heat-transfer laws, especially in supersonic flows.

The Collis and Williams or Kramers laws concern the average Nusselt number around the wire. An extensive survey of the local values of the Nusselt number for hot wires was conducted by *Dennis* et al. [5.102], using both experimental data and numerical predictions. The highest value is at the forward stagnation point, and its relative importance increases almost linearly with the Reynolds number. Similar results for large cylinders are available in textbooks, e.g., by *Eckert* and *Drake* [5.103].

Regarding MEMS sensors, very preliminary heat-transfer laws are indicated by *Jiang* et al. [5.72]. Surprisingly, the exponent n in U^n depends on the anemometer type, $n = 1.0$ for CCA and $n = 0.6$ for CTA. Further research is definitely needed to clarify this point. However, an exponent n larger than the $n \simeq 0.50$ usually

Table 5.2 Physical properties of usual fluids and substrates for hot-films at 293 K

Fluid/substrate	Prandtl number Pr	$C_p/C_v = \gamma$	Density ρ (kg/m³)	Specific heat C_p (J kg⁻¹K⁻¹)	Thermal conductivity k (W m⁻¹K⁻¹)	Dynamic viscosity μ (kg m⁻¹s⁻¹)
Air	0.71	1.40	1.2	1005	0.0257	1.82×10^{-5}
Helium	0.70	1.66	0.17	5180	0.157	1.86×10^{-5}
Water	7.0		1000	4180	0.60	1.00×10^{-3}
Mercury	0.022		13600	140	9.3	1.55×10^{-3}
Quartz			2650	710	6–11	
Kapton			1420	1090	0.10–0.35	
Perspex			1190	1470	0.21	
Tufnol			1320	1500	0.29	
Plywood			580	2500	0.13	

admitted for circular cylinders could be possible, because the rectangular or trapezoidal cross section of this new type of sensor gives rise to shed eddies and separated zones. In textbooks, exponents in the range 0.60–0.75 are reported for elongated hexagonal cylinders, although for larger Reynolds numbers [5.104].

Equations (5.46) and (5.36) lead to the practical form

$$\frac{R_w I_w^2}{R_w - R_a} = A + BU^{0.50} \quad (5.48)$$

with

$$A = 0.39 \frac{\pi l k_a}{\chi R_0} \quad \text{and} \quad B = 0.51 \frac{\pi l k_a}{\chi R_0} \left[\frac{\rho_a d}{\mu_a}\right]^{0.50}. \quad (5.49)$$

Equation (5.48) is known as King's law. It concerns the wire itself, not the anemometer outputs. In order to retrieve King's law from anemometer outputs, all the resistances present in the circuits have to be taken into account, for example, the top bridge resistance in the arm containing the sensor in a CTA circuit.

In King's law, A and B are improperly called *constants*. Obviously they are not; they depend on the wire diameter and on the fluid temperature. If the temperature dependencies of the physical parameters (5.47) are introduced in (5.49), one obtains

$$\frac{dA}{A} = 0.886 \frac{dT_a}{T_a}$$
$$\frac{dB}{B} = (0.886 - 0.50 - 0.38)\frac{dT_a}{T_a} = 0.006\frac{dT_a}{T_a}. \quad (5.50)$$

Coefficient B therefore varies far less with temperature than coefficient A. This result is particularly interesting at large values of U when A can be neglected compared to $BU^{0.50}$. *Uberoi* and *Corrsin* [5.105] confirmed these trends when experimenting on hot jets.

Regarding buoyancy effects, a criterion that permits one to ensure that forced convection prevails has been established by *Collis* and *Williams* [5.96] for a cross-flow

$$\text{Re}_f > \text{Gr}_f^{1/3}. \quad (5.51)$$

For example, taking $T_a \simeq 293$ K, $a_w \simeq 1$, $g = 9.81$ ms⁻² and $\beta = 1/273$, and using (5.47) and Table 5.2, this criterion gives $U \geq 0.07$ m/s whatever the wire diameter.

When a mixed convection regime occurs, heat exchange laws become complicated. *Hatton* et al. [5.106] suggested an explicit correlation based on a vectorial addition of the forced and natural Reynolds numbers

$$\text{Nu}_f = \left(0.384 + 0.581 \text{Re}_{\text{eff}}^{0.439}\right) \left(\frac{T_f}{T_a}\right)^{0.154}, \quad (5.52)$$

where

$$\text{Re}_{\text{eff}}^2 = \text{Re}_f^2 \left(1 + 2.06\frac{\text{Ra}^{0.418}}{\text{Re}_f}\cos\theta_z + 1.06\frac{\text{Ra}^{0.836}}{\text{Re}_f^2}\right), \quad (5.53)$$

where Ra is the Rayleigh number, $\text{Ra} = \text{Gr} \cdot \text{Pr}$, and θ_z is the angle from the vertically upward direction of the forcing flow.

Heat-Transfer Laws for Other Gases in Subsonic Flows. *Andrews* et al. [5.99], and *Pitts* and *McCaffrey* [5.107] at the National Bureau of Standards carried out extensive studies of heat exchanges for hot wires and hot films using various gases such as argon, helium, methane, propane and carbon dioxide, in the Reynolds number range below and just near vortex shedding. For all gases, except helium, the law valid for air was approximately verified once the proper temperature dependencies of the gas molecular properties were taken into account. The Prandtl number effect is found to be relatively weak. The case of helium, for which accommodation effects are strong, is singular and a special correction was elaborated.

Heat-Transfer Laws for Liquid Flows. Equation (5.43) now involves the Prandtl number, for example $Pr = 7.1$ for water, 0.0225 for mercury, 0.01 for molten sodium, and in the range 20–5000 for oils. These fluids are useful either in industrial heat exchangers or in laboratory experiments to model the Earth's outer core. According to *Hill* and *Sleicher* [5.110, 111] who revisited many previous measurements done for long cylinders, the heat-transfer law that *Kramers* [5.98] suggested is particularly useful

$$\mathrm{Nu}_a = 0.42 \mathrm{Pr}^{0.20} + 0.57 \mathrm{Pr}^{0.33} \mathrm{Re}_a^{0.50} \ . \quad (5.54)$$

Buoyancy effects in liquids have also been analyzed. *Gebhart* and *Pera* [5.112] considered a large range of Prandtl numbers of 6.3–63. The criterion for forced convection to prevail is similar to that of *Collis* and *Williams* [5.96]

$$\mathrm{Re}_f > \mathrm{Gr}_f^{1/n} \quad \text{where } n \text{ is between 2 and 3} \ . \quad (5.55)$$

Regarding the heat transfer laws, *Gebhart* and *Pera* [5.112] indicated a set of curves for the Nusselt number in terms of both the Péclet and Rayleigh numbers ($Pe = Re \cdot Pr$, $Ra = Gr \cdot Pr$).

All the above results hold only for Newtonian fluids. Otherwise, very different phenomena are observed. Investigations made with cylindrical hot films, by *Smith* et al. [5.113] or *James* and *Acosta* [5.114] among others, showed that the Nusselt number decreases when the polymer concentration increases, at a given Reynolds number. Moreover, a cylinder normal to the flow is cooled less than a cylinder oblique to the flow. The explanation given by *Lumley* [5.115] invoked the expansion of polymer molecules in pure strain (stagnation region, hot-film set normal to the flow) so that the viscosity increases, the boundary layer on the film becomes thicker and the heat transfer is reduced.

Heat-Transfer Laws in Compressible Flows

Kovasznay [5.116], *Morkovin* [5.100], *Morkovin* and *Phinney* [5.117] and also *Laufer* and *McClellan* [5.59] pioneered the possibility of using hot-wires for supersonic flows. For a wire normal to the flow, a detached bow shock forms in front of the wire. Subsequently, the wire is in a subsonic flow, and the heat-transfer laws mentioned above regain some importance. The fluid-mechanics equations also correlate the flow parameters upstream and downstream of the shock. Especially, momentum and total temperature are conserved through the shock. Figure 5.14 illustrates the shock which exists in front of a wire for two different flows, a Mach 3 air flow and a Mach 10 helium flow. In the latter case, the bow shock becomes much thicker and forms closer to the wire. However, hot-wire anemometry remains possible in these hypersonic flows. This situation has been particularly investigated by *Spina* and *McGinley* [5.108]. The strong needles supporting the wire also create shocks, but these are attached to the needle tips, and their effect on the wire can be reduced if the inactive regions, even if short, exist at both ends of the active part of the wire.

Heat-transfer laws at high velocities have received considerable attention, for Mach number up to 6, and Reynolds numbers up to 400. Many experimental data were obtained and analyzed by *Kovasznay* [5.116], *Laufer* and *McClellan* [5.59], and *Spangenberg* [5.118], who reported very complete data tables, *Morkovin* and *Phinney* [5.117], and *Dewey* [5.119, 120], who drew many valuable curves, *Baldwin* et al. [5.121], *Gaviglio* [5.122], *Barre* et al. [5.123], and *Stainback* and *Nagabushana* [5.124], with a large bibliography, and *Spina* and *McGinley* [5.108] for gases other than air.

Regarding the temperature to be used for the physical parameters μ and k entering the Reynolds and Nusselt numbers, one can think of using the conditions behind the detached shock wave, ρ_2, U_2 and T_2, as being *the apparently free stream* for the wire, as suggested by *Kovasznay* [5.116] and used by *Laufer* and *McClellan* [5.59]. Here the subscript 2 refers to values after the shock, as usual in compressible flows, not to a lateral velocity component. On the other hand, the total temperature T_t was suggested by *Baldwin* et al. [5.121]

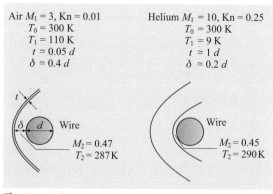

Fig. 5.14 A hot-wire probe in a supersonic flow. Two examples: Mach 3 air flow, Mach 10 helium flow (after *Spina* and *McGinley* [5.108] and *Elizarova* et al. [5.109])

or *Dewey* [5.119, 120]. Of course, the temperature T_2 is close to T_t as can be deduced from the basic relations

$$T_t = T_\infty \left(1 + \frac{\gamma - 1}{2} M^2\right) = T_2 \left(1 + \frac{\gamma - 1}{2} M_2^2\right) \quad (5.56)$$

and

$$M_2^2 = \frac{M^2 + 2/(\gamma - 1)}{2\gamma M^2/(\gamma - 1) - 1}, \quad (5.57)$$

where T_∞ is the freestream static temperature before the shock, also denoted by T_1 or simply T (this should not be confused with the temperature existing outside a boundary layer, this external zone being often called *freestream*). For example, differences between T_t and T_2 are 6.6%, 3.8%, and 3.2% for $M = 2$, 4, and 6, respectively. The temperature T_t also has the advantage of being a constant in adiabatic shock flows and very nearly constant across most boundary-layer and wake flows. Furthermore, $\rho_2 U_2 = \rho_\infty U_\infty$. These remarks lead us to choose the references in the following way. Density and velocity are based on freestream conditions, also called static conditions. Viscosity and thermal conductivity are based on the total temperature T_t, also called the stagnation or reservoir temperature.

The temperature difference entering into the Nusselt number makes use of the recovery temperature T_r, which is the temperature of the unheated wire placed in the flow. It therefore resembles the temperature T_a of the incoming flow at subsonic velocities. It is the temperature above which an additional cooling by the flow is possible. The useful dimensionless numbers are therefore

$$\mathrm{Nu}_t = \frac{R_w I_w^2}{\pi l k_t (T_w - T_r)}, \quad \mathrm{Re}_t = \frac{\rho_\infty U_\infty d}{\mu_t},$$

$$M = M_\infty = \frac{U_\infty}{\sqrt{\gamma R T_\infty / m}}, \quad (5.58)$$

where the subscript 't' refers to an evaluation at the stagnation temperature, the subscript '∞' at the freestream static temperature, and the subscript 'r' at the recovery temperature. $R = 8314 \,\mathrm{J\,kg^{-1}\,K^{-1}}$ is the universal gas constant, and m the molar mass of the gas considered: $R/m = 287$ for air, $R/m = 2078$ for helium.

A general heat-transfer law in supersonic flow is therefore of the form

$$\mathrm{Nu}_t = \mathrm{Nu}_t(M, \mathrm{Re}_t, \gamma, \Theta). \quad (5.59)$$

The ratio γ is usually fixed in an experiment ($\gamma = 1.4$ for air, $\gamma = 1.66$ for helium). Θ represents the wire temperature loading, denoted $(T_w - T_r)/T_r$ by *Kovasznay* [5.116] or *Smits* et al. [5.125], and T_w/T_r by *Morkovin* [5.100]. Some attempts to extract Θ from the Nusselt number have been made in special cases, e.g., by *Kovasznay* [5.116] for CCA, and by *Smits* et al. [5.125], *Spina* and *McGinley* [5.108], or *Weiss* et al. [5.126] for CTA. Keeping the general form (5.59) is much more rewarding (Sect. 5.2.6).

Concerning the wire Reynolds number Re_t, let us be precise in that it is connected to the unit Reynolds number of a wind tunnel, $\mathrm{Re}_{\mathrm{WT}} = \rho_\infty U_\infty / \mu_\infty$, by

$$\mathrm{Re}_t = \mathrm{Re}_{\mathrm{WT}} d \frac{\mu_\infty}{\mu_t}$$

$$= \mathrm{Re}_{\mathrm{WT}} d \left(1 + \frac{\gamma - 1}{2} M^2\right)^{-0.765}. \quad (5.60)$$

The wire resistance R_r at the recovery temperature T_r, is also of special interest. Since an electric current, even if very small, could modify the resistance R_r to be measured, it is advisable to obtain R_r from the plot of the measured hot-wire resistance R_w with the electric power injected into the wire at different overheats and extrapolating the curve towards zero dissipation [5.127]. It is also helpful to have some knowledge of the recovery factor η and the ratio of T_r to T_t, which depends on both the Mach and the Reynolds numbers for air flows, as

$$\eta = \frac{T_r}{T_t} \quad \eta = \eta(M, \mathrm{Re}_t). \quad (5.61)$$

Some physical insight of η is obtained when introducing the Knudsen number. A Knudsen number expresses the continuous or rarefied flow conditions. *Stalder* et al. [5.128], using the kinetic theory of gases to relate density, viscosity and sound speed in terms of molecular velocity and molecular mean free path, arrived at the following very useful expression

$$\mathrm{Kn}_\infty = \frac{\text{molecular mean free path}}{\text{wire diameter}} \simeq \sqrt{\frac{\pi \gamma}{2}} \frac{M}{\mathrm{Re}_\infty}, \quad (5.62)$$

where $\mathrm{Re}_\infty = \rho_\infty U_\infty d / \mu_\infty$ is the freestream Reynolds number for the wire. Here the freestream conditions prevail as a Knudsen number expresses a geometric ratio. The continuous regime is for $\mathrm{Kn}_\infty \leq 0.01$. Slip flow is characterized by $0.01 \leq \mathrm{Kn}_\infty \leq 1$ and free molecular flows exist for $\mathrm{Kn}_\infty \geq 1$. Let us observe that, as η is related to M, Re_t, and R_∞, it is superfluous to introduce a Knudsen number in the Nusselt equation (5.59).

Values for η are reported by *Laufer* and *McClellan* [5.59], *Morkovin* [5.100], and *Baldwin* et al. [5.121]. They are also available in *Spina* and *McGinley* [5.108], *Gaviglio* [5.122], *Dewey* [5.120], and

Stalder et al. [5.128]. In that work the recovery factor is denoted by r and is defined by $T_r = T\{1 + [(\gamma - 1)/2]rM^2\}$, but r can be connected to η by (5.56).

With some simplifications, η is roughly constant, $\eta \simeq 0.97 \pm 0.01$ as long as $M \leq 2$ and $Kn_\infty \leq 0.2$, and η rises sharply with M and Kn_∞ afterwards, reaching values of around 1.15. For these η values in excess of one, *Stalder* et al. [5.128] invoked the small amount of energy per molecule transported from a body for the case of free-molecule flow. For marked slip flows, *Dewey* [5.119] could normalize η in terms of two analytical functions that approach the free-molecule flow and the continuum flow respectively. That expression is also available in *Eckert* and *Drake* [5.103]. These results present even more interest nowadays because of a large incentive for hypersonic flights [5.108, 129].

Plots giving Nu_t in terms of Re_t, with the additional Knudsen number Kn_∞ as a parameter, were prepared by *Baldwin* et al. [5.121] as early as 1960. Similar plots were drawn and supplemented with new data by *Dewey* [5.120], and referenced afterwards by *Fingerson* and *Freymuth* [5.50] and *Eckert* and *Drake* [5.103] among others.

The Baldwin–Dewey diagram is reproduced in Fig. 5.15. In the upper part of the figure, where the values of the Nusselt number are large, the curves are close to straight lines, $Nu_t \sim Re_t^{1/2}$ with an additive constant slightly decreasing with the Mach number. The Knudsen number is less than 0.01, the fluid acts as a continuum and the heat exchange is large. In the lowest part of the figure, for large Knudsen numbers and small Nusselt numbers, the Mach number dependence is strongly represented as an important parameter. In this regime, Nu_t becomes proportional to Re_t rather than $Re_t^{1/2}$; detailed values are available in *Dewey* [5.119]. When doing experiments, it is thus important to locate the M, Re_t, and Kn_∞ values in these diagrams, and to estimate the most appropriate exponent n for the Re_t number dependence.

A great simplification occurs when $M \geq 1.2$: the Nusselt number becomes independent of the Mach number, and the main remaining parameter is the Reynolds number. Furthermore, if the Reynolds number stays reasonably high, $Re_t \geq 20$, the flow is not too far from continuum conditions, and the slope relating the Nusselt number to the Reynolds number is close to $1/2$.

All the heat-transfer laws we have just listed constitute useful guidelines in different occasions: to interpret calibration curves, see Sects. 5.2.4, 2.6, to obtain the wire sensitivity coefficients to small turbulent fluctuations, see the following sections, and to estimate the time constant of a wire.

Wire Sensitivity Coefficients for Subsonic Flows and Small Fluctuations

Fluctuations whose amplitude do not exceed 10% are usually assimilated as *small fluctuations*. The fluctuating wire response may then be obtained by a first-order Taylor expansion around a mean operating point. Fluctuations are assumed to follow the tangent to the response curve rather than the response curve itself. The slopes which appear are then conveniently expressed in terms of the dimensionless ratios (fluctuation)/(mean value). The reason is that these ratios are directly connected to

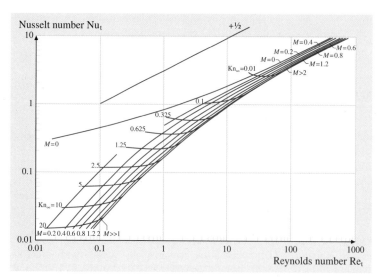

Fig. 5.15 The Baldwin–Dewey diagram for heat loss from hot wires in compressible flows. The Nusselt and Reynolds numbers are defined by (5.58) (after [5.120, 121])

the logarithmic derivatives of the heat transfer laws. This is the standard procedure in compressible flows, but for introductory purposes it is useful to treat two subsonic cases:

1. a single normal wire placed in a non-isothermal flow, and
2. a yawed wire placed in an isothermal flow.

Single Normal Wire in Non-isothermal Flows. In a subsonic incident flow with velocity and temperature fluctuations u'_1 and θ'_a around an operating point, $\overline{U}_1, \overline{T}_a$, the response of a single normal wire can be expressed as

$$\frac{r'_w}{\overline{R}_w} = S^{CC}_{U_1} \frac{u'_1}{\overline{U}_1} + S^{CC}_{T_a} \frac{\theta'_a}{\overline{T}_a}$$

for constant-current (CC) operation,

$$\frac{i'_w}{\overline{I}_w} = S^{CT}_{U_1} \frac{u'_1}{\overline{U}_1} + S^{CT}_{T_a} \frac{\theta'_a}{\overline{T}_a}$$

for constant-temperature (CT) operation,

$$\frac{i'_w}{\overline{I}_w} = S^{CV}_{U_1} \frac{u'_1}{\overline{U}_1} + S^{CV}_{T_a} \frac{\theta'_a}{\overline{T}_a}$$

for constant-voltage (CV) operation, (5.63)

where the factors S are the sensitivity coefficients. The superscript (CC, CT, or CV) denotes the way the hot-wire is operated and the subscript (U_1 or T_a) the nature of the flow parameter. For example,

$$S^{CT}_{U_1} = \frac{\overline{U}_1}{\overline{I}_w} \left(\frac{\partial \overline{I}_w}{\partial \overline{U}_1} \right) \quad \text{at constant } \overline{R}_w \text{ and } \overline{T}_a$$ (5.64)

or

$$S^{CV}_{T_a} = \frac{\overline{T}_a}{\overline{I}_w} \left(\frac{\partial \overline{I}_w}{\partial \overline{T}_a} \right) \quad \text{at constant } \overline{V}_w \text{ and } \overline{U}_1.$$ (5.65)

Note that (5.63) and the following equations in this section refer only to the wire. The sensitivity coefficients based on the anemometer outputs are considered in Sect. 5.2.3 as the components of the electronic circuits are implied in their expressions.

The S coefficients introduced in (5.63) are obtained by taking the derivative of (5.48), in which $U^{0.50}$ can be more generally written U^n to permit some deviation from the simple square-root law.

First, we assume that A and that B do not depend on T_a. This gives

$$\frac{dR_w}{\overline{R}_w} + 2\frac{dI_w}{\overline{I}_w} - \frac{\overline{R}_w}{\overline{R}_w - \overline{R}_a} \frac{dR_w}{\overline{R}_w}$$
$$= n \frac{B\overline{U}_1^n}{A + B\overline{U}_1^n} \frac{dU_1}{\overline{U}_1} - \frac{\overline{R}_a}{\overline{R}_w - \overline{R}_a} \frac{dT_a}{\overline{T}_a}. \quad (5.66)$$

Inserting into (5.66) the $R(T)$ relation specified by (5.36) and the wire overheat ratio $\overline{a}_w$ defined by (5.37) yields

$$-\frac{1}{\overline{a}_w} \frac{dR_w}{\overline{R}_w} + 2\frac{dI_w}{\overline{I}_w}$$
$$= n \frac{B\overline{U}_1^n}{A + B\overline{U}_1^n} \frac{dU_1}{\overline{U}_1} - \frac{\chi \overline{T}_a}{1 + \chi(\overline{T}_a - T_0)} \frac{dT_a}{\overline{T}_a}. \quad (5.67)$$

This equation is advantageously valid no matter how the wire is operated. The constraint relative to the chosen type of anemometer can now be introduced, i. e.,

$$\frac{dI_w}{\overline{I}_w} = 0 \text{ for CC}, \quad \frac{dR_w}{\overline{R}_w} = 0 \text{ for CT},$$
$$\frac{dR_w}{\overline{R}_w} + \frac{dI_w}{\overline{I}_w} = 0 \text{ for CV}. \quad (5.68)$$

Finally, noting the fluctuations $r'_w = dR_w$, $i'_w = dI_w$, $u'_1 = dU_1$, and $\theta'_a = dT_a$, yields

$$S^{CC}_{U_1} = -\overline{a}_w \frac{nB\overline{U}_1^n}{A + B\overline{U}_1^n},$$

$$S^{CC}_{T_a} = \frac{\chi \overline{T}_a}{1 + \chi(\overline{T}_a - T_0)},$$

$$S^{CT}_{U_1} = \frac{1}{2} \frac{nB\overline{U}_1^n}{A + B\overline{U}_1^n},$$

$$S^{CT}_{T_a} = -\frac{1}{2\overline{a}_w} \frac{\chi \overline{T}_a}{1 + \chi(\overline{T}_a - T_0)},$$

$$S^{CV}_{U_1} = \frac{\overline{a}_w}{1 + 2\overline{a}_w} \frac{nB\overline{U}_1^n}{A + B\overline{U}_1^n},$$

$$S^{CV}_{T_a} = -\frac{1}{1 + 2\overline{a}_w} \frac{\chi \overline{T}_a}{1 + \chi(\overline{T}_a - T_0)}. \quad (5.69)$$

Figure 5.16 shows the dependence of the coefficients S as a function of the mean wire overheat $\overline{a}_w$, for $n = 1/2$. One can observe the very large sensitivity of a wire to temperature for CT operation at small overheats. Regarding the S^{CV} sensitivities, which are smaller than the S^{CC} or S^{CT} values, one should remember that the S given by (5.69) refer only to the wire. The effect of the complete electronic circuits will be examined in Sect. 5.2.3.

Equation (5.67) also permits one to derive the relations that exist between the wire sensitivities of the three anemometers, whatever the turbulent fluctuation considered

$$S^{CT} = -\frac{1}{2\bar{a}_w} S^{CC}, \quad S^{CV} = -\frac{1}{1+2\bar{a}_w} S^{CC},$$

$$S^{CV} = \frac{2\bar{a}_w}{1+2\bar{a}_w} S^{CT}. \tag{5.70}$$

If a temperature loading factor m is introduced, see (5.44), a similar development can be conducted. Equation (5.67) then takes the modified form

$$-\frac{1}{\bar{a}_w}(1+\epsilon)\frac{\mathrm{d}R_w}{R_w} + 2\frac{\mathrm{d}I_w}{I_w}$$
$$= n\frac{B\overline{U}_1^n}{A + B\overline{U}_1^n}\frac{\mathrm{d}U_1}{\overline{U}_1}$$
$$-\frac{\chi \overline{T}_a}{1+\chi(\overline{T}_a - T_0)}(1+\epsilon)\frac{\mathrm{d}T_a}{\overline{T}_a}, \tag{5.71}$$

where $\epsilon = m\bar{a}_w(1+\bar{a}_w)/(2+\bar{a}_w)$. The sensitivity coefficients, denoted by a star with the temperature load included, become

$$S_{U_1}^{CC\star} = S_{U_1}^{CC}(1+\epsilon)^{-1}, \quad S_{T_a}^{CC\star} = S_{T_a}^{CC}$$
$$S_{U_1}^{CT\star} = S_{U_1}^{CT}, \quad S_{T_a}^{CT\star} = S_{T_a}^{CT}(1+\epsilon)$$
$$S_{U_1}^{CV\star} = S_{U_1}^{CV}\left(1 + \frac{\epsilon}{1+2\bar{a}_w}\right)^{-1},$$
$$S_{T_a}^{CV\star} = S_{T_a}^{CV}(1+\epsilon)\left(1 + \frac{\epsilon}{1+2\bar{a}_w}\right)^{-1}. \tag{5.72}$$

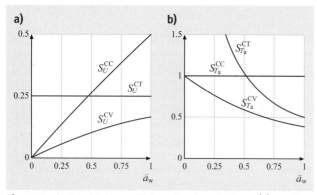

Fig. 5.16a,b Sensitivity coefficients in subsonic flows, (**a**) to velocity fluctuations, (**b**) to ambient temperature fluctuations, for a wire operated in constant-current, constant-temperature, or constant-voltage mode

Corrective terms appear more or less on all anemometers, but they are at most $2m/3$ when $\bar{a}_w = 1$, and even smaller for a CV-operated wire.

Finally, if the dependence of A and B on the fluid temperature is introduced, as suggested by (5.50), a similar development can be conducted and yields, for example,

$$S_{T_a}^{CC} = \frac{\chi \overline{T}_a}{1+\chi(\overline{T}_a - T_0)} - 0.80\bar{a}_w\frac{A}{A + B\overline{U}^n}. \tag{5.73}$$

The difference with the value indicated in (5.69) is therefore negligible at high velocities and low overheat ratios.

Yawed Wire in Isothermal Flows. For a wire yawed at an angle θ on x_2, see Fig. 5.13, the velocity cooling the wire is now $U_1 \cos\theta - U_2 \sin\theta$. The sensitivity coefficient to U_1 and U_2 can be defined for small-amplitude fluctuations and obtained by deriving (5.48) with respect to U_1 and U_2 for a fixed θ. This gives

$$\frac{r'_w}{R_w} \quad \text{or} \quad \frac{i'_w}{I_w}$$
$$= S_{U_1}(\text{yawed wire})\frac{u'_1}{U_1} + S_{U_2}(\text{yawed wire})\frac{u'_2}{U_1} \tag{5.74}$$

with

$$S_{U_1}(\text{yawed wire}) = S_{U_1}(\text{normal wire}) \tag{5.75}$$

and

$$S_{U_2}(\text{yawed wire}) = -S_{U_1}(\text{yawed wire}) \tan\theta. \tag{5.76}$$

Wire Sensitivity Coefficients
for Compressible Flows and Small Fluctuations
In a compressible flow, a single wire placed normal to the incident flow responds to changes in three fundamental variables, velocity U, density ρ, and total temperature T_t. However, other triplets of variables can be used and they will also be examined.

Use of ρ, U, T_t. Morkovin [5.100] established the expressions of the sensitivity coefficient to ρ, U, and T_t variables by taking the logarithmic derivative of the general equation (5.59). For newcomers to the field, *Bradshaw* [5.46] gave a clear view for a CT-operated wire and the benefit of using log derivatives. The algebra for the derivation is straightforward, but rather long.

One has to use the definitions of Nu_t, Re_t and M given by (5.58), and two additional basic aerodynamic relations

$$M^2 = \frac{U^2/c_t^2}{1-(\gamma-1)U^2/2c_t^2}$$

$$\frac{dM}{M} = \frac{1}{\alpha}\left(\frac{dU}{U} - \frac{1}{2}\frac{dT_t}{T_t}\right) \quad (5.77)$$

with

$$c_t^2 = \gamma \frac{RT_t}{m}, \quad \text{and} \quad \alpha = 1 + \frac{\gamma-1}{2}M^2. \quad (5.78)$$

Morkovin treated the CC and CT cases separately, but a unified approach is possible, which can moreover encompass the CV case as well. This general equation is

$$-\frac{1}{\overline{A}_w}\frac{d\overline{R}_w}{\overline{R}_w} + 2\frac{d\overline{I}_w}{\overline{I}_w}$$
$$= (X)\frac{d\overline{\rho}}{\overline{\rho}} + (XX)\frac{dU}{U} + (XXX)\frac{dT_t}{T_t} \quad (5.79)$$

with

$$(X) = \frac{\overline{Re}_t}{\overline{Nu}_t}\frac{\partial \overline{Nu}_t}{\partial \overline{Re}_t} - \frac{1}{\overline{a}_w}\frac{\overline{Re}_t}{\overline{\eta}}\frac{\partial \overline{\eta}}{\partial \overline{Re}_t}$$

$$(XX) = \frac{\overline{Re}_t}{\overline{Nu}_t}\frac{\partial \overline{Nu}_t}{\partial \overline{Re}_t} + \frac{1}{\alpha}\frac{\overline{M}}{\overline{Nu}_t}\frac{\partial \overline{Nu}_t}{\partial \overline{M}}$$
$$\quad - \frac{1}{\overline{a}_w}\frac{\overline{Re}_t}{\overline{\eta}}\frac{\partial \overline{\eta}}{\partial \overline{Re}_t} - \frac{1}{\overline{a}_w}\frac{1}{\alpha}\frac{\overline{M}}{\overline{\eta}}\frac{\partial \overline{\eta}}{\partial \overline{M}}$$

$$(XXX) = n_t + 1 - \frac{1+\overline{A}_w}{\overline{A}_w}\chi' - m_t\frac{\overline{Re}_t}{\overline{Nu}_t}\frac{\partial \overline{Nu}_t}{\partial \overline{Re}_t}$$
$$\quad - \frac{1}{2\alpha}\frac{\overline{M}}{\overline{Nu}_t}\frac{\partial \overline{Nu}_t}{\partial \overline{M}} + m_t\frac{1}{\overline{a}_w}\frac{\overline{Re}_t}{\overline{\eta}}\frac{\partial \overline{\eta}}{\partial \overline{Re}_t}$$
$$\quad + \frac{1}{2\alpha}\frac{1}{\overline{a}_w}\frac{\overline{M}}{\overline{\eta}}\frac{\partial \overline{\eta}}{\partial \overline{M}}$$

where χ' is the exponent of a power law for the temperature dependence of the wire resistance

$$\frac{R_w}{R_0} = \left(\frac{T_w}{T_0}\right)^{\chi'}, \quad \begin{array}{l} \chi'=1.15 \text{ when } \chi=0.0041\,\text{K}^{-1}, \\ \chi'=1.05 \text{ when } \chi=0.0036\,\text{K}^{-1} \end{array}$$
(5.80)

and is hence advantageously close to one. The χ' values are deduced from χ and (5.36) by a simple Matlab routine; m_t and n_t denote the power exponents for the temperature dependence of viscosity and thermal conductivity, already introduced in (5.47), hence $m_t \simeq 0.765$, and $n_t \simeq 0.885$. Finally, $\overline{A}_w$ is a generalized overheat parameter defined by

$$\overline{A}_w = \frac{1}{2}\frac{\overline{I}_w}{\overline{R}_w}\left(\frac{\partial \overline{R}_w}{\partial \overline{I}_w}\right) \quad \text{at constant flow}. \quad (5.81)$$

The factor $\overline{A}_w$ depends on the wire overheat ratio $\overline{a}_w$ and its determination is examined later in this section.

The interesting characteristic of (5.79) is that the left-hand side concerns the wire, and the right-hand side concerns the flow, once the wire overheat ratio $\overline{a}_w$ is chosen. This form is similar to that which was developed earlier for subsonic flows (5.67). It therefore permits the treatment of any anemometer type, CCA, CTA or CVA. The appropriate constraint, listed in (5.68), has to be taken into account. This leads to the following expressions for the sensitivity coefficients at the wire level

$$S_\rho^{CC} = -\overline{A}_w(X), \qquad S_\rho^{CT} = \frac{1}{2}(X),$$

$$S_\rho^{CV} = \frac{\overline{A}_w}{1+2\overline{A}_w}(X),$$

$$S_U^{CC} = -\overline{A}_w(XX), \qquad S_U^{CT} = \frac{1}{2}(XX),$$

$$S_U^{CV} = \frac{\overline{A}_w}{1+2\overline{A}_w}(XX),$$

$$S_{T_t}^{CC} = -\overline{A}_w(XXX), \qquad S_{T_t}^{CT} = \frac{1}{2}(XXX),$$

$$S_{T_t}^{CV} = \frac{\overline{A}_w}{1+2\overline{A}_w}(XXX). \quad (5.82)$$

Obviously, the quantities (X), (XX) and (XXX) are complex, but a few remarks are helpful:

1. The leading term is the derivative of the Nusselt number with respect to the Reynolds number. This term has already been obtained at low subsonic speed. Moreover, if the temperature dependence of the physical properties are neglected, i. e., $m_t = n_t = 0$, and if the wire overheat stays less than about 0.50, then $\overline{A}_w = \overline{a}_w$, and the relations (5.69) are recovered.
2. The recovery factor η can often be taken as a constant, unless one works in a rarefied gas. The derivatives of η relative to the Mach and Reynolds numbers can then be neglected. *Morkovin* [5.100] estimated that the error is often less than 6%.
3. When the Mach number is greater than 1.2, the Baldwin–Dewey plots [5.119, 121], sketched in Fig. 5.15, show that the heat transfer no longer depends on the Mach number. Furthermore, there is the simple relation $S_\rho = S_U = \hat{S}_{\rho U}$ concerning the sensitivity coefficients.
4. When the Mach number is in the range 0.5–1.2, one has to estimate the log derivative of the Nusselt number with respect to the Mach number, in addition to the usual Nusselt dependence on the

Reynolds number. Using Fig. 5.15, one can appreciate this Mach number effect, by noting the Nusselt number change along a parallel to the ordinate axis at the chosen Reynolds number. For example, for $Re_t \simeq 20$ and around $M = 0.90$, the Nusselt number decreases by about 15%. Incidentally, the values reported by *Morkovin* [5.100] in his Fig. VI-2 are much too high. *Baldwin* et al. [5.121] also referred to an earlier work by *Scadron* and *Warshawsky* [5.130] and pointed out that a possible correlation could be $Nu_t = 0.431 Re_t^{0.50}(1 + M^2)^{-0.50}$.

Use of ρU, M, T_t. This ρU, M, T_t triplet is the favorite for experimentalists because of the direct presence of the Mach number, which is the key parameter of any experiment. The sensitivity coefficients fitting that new group, noted $\hat{S}$, can be related to the S defined by (5.82). Writing the differential response for any anemometer, we have on one hand

$$\frac{r'_w}{R_w} \text{ or } \frac{i'_w}{I_w}$$
$$= \underbrace{\frac{\overline{\rho}}{\overline{R}_w} \frac{\partial \overline{R}_w}{\partial \overline{\rho}}}_{S_\rho} \frac{d\rho}{\overline{\rho}} + \underbrace{\frac{\overline{U}}{\overline{R}_w} \frac{\partial \overline{R}_w}{\partial \overline{U}}}_{S_U} \frac{dU}{\overline{U}} + \underbrace{\frac{\overline{T}_t}{\overline{R}_w} \frac{\partial \overline{R}_w}{\partial \overline{T}_t}}_{S_{T_t}} \frac{dT_t}{\overline{T}_t}$$
(5.83)

and on the other hand

$$\frac{r'_w}{R_w} \text{ or } \frac{i'_w}{I_w} = \underbrace{\frac{\overline{\rho U}}{\overline{R}_w} \frac{\partial \overline{R}_w}{\partial (\overline{\rho U})}}_{\hat{S}_{\rho U}} \frac{d(\rho U)}{\overline{\rho U}} + \underbrace{\frac{\overline{M}}{\overline{R}_w} \frac{\partial \overline{R}_w}{\partial \overline{M}}}_{\hat{S}_M} \frac{dM}{\overline{M}}$$
$$+ \underbrace{\frac{\overline{T}_t}{\overline{R}_w} \frac{\partial \overline{R}_w}{\partial \overline{T}_t}}_{\hat{S}_{T_t}} \frac{dT_t}{\overline{T}_t} .$$
(5.84)

In (5.84), $d(\rho U)$ can be developed and dM transformed using (5.77), hence

$$\frac{r'_w}{R_w} \text{ or } \frac{i'_w}{I_w} = \hat{S}_{\rho U} \left(\frac{d\rho}{\overline{\rho}} + \frac{dU}{\overline{U}} \right)$$
$$+ \hat{S}_M \frac{1}{\alpha} \left(\frac{dU}{\overline{U}} - \frac{1}{2} \frac{dT_t}{\overline{T}_t} \right) + \hat{S}_{T_t} \frac{dT_t}{\overline{T}_t}$$
(5.85)

and identifying with (5.83) gives

$$\hat{S}_{\rho U} = S_\rho$$
$$\hat{S}_M = \alpha(S_U - S_\rho)$$
$$\hat{S}_{T_t} = S_{T_t} + \frac{1}{2}(S_U - S_\rho) .$$
(5.86)

The most interesting result is that, when $M \geq 1.2$ and $\eta \simeq 1$, the derivatives involving M and η in (X) and (XX) are negligible, and (5.86) simply become

$$\hat{S}_{\rho U} = S_\rho = S_U , \quad \hat{S}_M = 0 , \quad \hat{S}_{T_t} = S_{T_t} . \quad (5.87)$$

In fact, this result seems to hold already in transonic flows. Using a CC-operated wire, *Barre* et al. [5.123] and *Dupont* and *Debiève* [5.131] did a complete survey in the freestream of a wind tunnel for $0.7 \leq M \leq 2.3$, $4 \leq Re_t \leq 20$, $0 \leq \bar{a}_w \leq 0.50$. Using a CT-operated wire, *Horstman* and *Rose* [5.132] reported experiments in a transonic boundary layer and found that $S_\rho \simeq S_U$ holds even for $Re_t \leq 20$ and $a_w \geq 0.50$. Similar results were obtained by *Jones* et al. [5.133] who investigated the Mach number range $0.05 \leq M \leq 1.0$ in a specially designed probe calibration tunnel [5.134].

In these conditions, a hot-wire response simplifies to

$$\frac{r'_w}{R_w} \text{ or } \frac{i'_w}{I_w} = \hat{S}_{\rho U} \frac{d(\rho U)}{\overline{\rho U}} + \hat{S}_{T_t} \frac{dT_t}{\overline{T}_t} . \quad (5.88)$$

Operating a normal wire at different overheats modifies the relative ratio between $\hat{S}_{\rho U}$ and $\hat{S}_{T_t}$. With one wire, one can obtain the variances and the correlation coefficient for the fluctuations $d(\rho U)$ and dT_t. This requires a minimum of three values of $\bar{a}_w$, but a larger number, around ten, allows the use of a polyfit technique and yields greater accuracy. This approach is used in many supersonic experiments [5.127, 135]. An automated stepping procedure to change $\bar{a}_w$ was developed by *Walker* et al. [5.136], *Weiss* et al. [5.126], or *Norris* and *Chokani* [5.137]. With two wires, placed in the close vicinity of each other, the instantaneous values of $\rho U(t)$ and $T_t(t)$ can be obtained and recorded [5.124, 136, 138]. These quantities are most useful when comparisons are made with turbulence predictions, because the density-weighted averages introduced by Favre are of common use in compressible flows [5.77].

The separation of ρ and U fluctuations remains difficult, however, even when several wires are used, because of the inherent equality $S_\rho = S_U$, which makes the reduction matrix singular. Nevertheless, in some cases, it would be beneficial to know the velocity fluctuations themselves, for example when comparing hot-wire anemometry to Doppler laser velocimetry or to particle image velocimetry. Physical assumptions are therefore needed.

For example, in a boundary layer, the strong Reynolds analogy, suggested by *Morkovin* [5.139], *Gaviglio* [5.122, 140], or *Smith* and *Smits* [5.138] is most useful. It states that when a fluid particle leaves the wall region where hot and slow motions prevail it carries with it a positive temperature fluctuation $T' > 0$, hence $\rho' < 0$, and a negative longitudinal velocity fluctuation $u' < 0$, so that the correlation between ρ' and u' is high and positive

$$\overline{\rho' u'} \simeq 0.80 \sqrt{\overline{\rho'^2}} \sqrt{\overline{u'^2}} \, . \tag{5.89}$$

In addition, the assumption of a constant stagnation temperature and constant static pressure permits one to relate the root-mean-square (rms) values of ρ' and u' by

$$\frac{\sqrt{\overline{\rho'^2}}}{\overline{\rho}} = (\gamma - 1) M^2 \frac{\sqrt{\overline{u'^2}}}{\overline{U}} \, . \tag{5.90}$$

The Kovasznay Modes. Another triplet of fundamental variables are the modes introduced by *Kovasznay* [5.91] and extensively used by *Morkovin* [5.100]: the vorticity mode ω, the entropy (spotiness) mode σ, and the acoustic mode π. The sensitivity coefficients of a wire to each of these modes, denoted by a double hat, can also be expressed in terms of the sensitivity coefficients S given by (5.82), independently of the anemometer type, by

$$\hat{S}_\sigma = -S_\rho + \alpha S_{T_t} \, ,$$
$$\hat{S}_\omega = S_U + \alpha (\gamma - 1) M^2 S_{T_t} \, ,$$
$$\hat{S}_\pi = S_\rho + \frac{n_x}{M} S_U + \alpha (\gamma - 1)(1 + n_x M_x) S_{T_t} \, , \tag{5.91}$$

where n_x is the direction of the sound wave.

Use of the Kovasznay modes allows reasonable assumptions to be made. For example in the freestream of wind tunnels, sound waves predominate, due to sound radiation from the supersonic, turbulent, boundary layer on the nozzle walls [5.126]. Temperature spottiness, leading to an entropy mode, comes from incomplete mixing in upstream reservoirs of high-speed wind tunnels.

The General Overheat $\overline{A}_w$. The parameter $\overline{A}_w$, which expresses the temperature load of the wire, is easily obtained from $\overline{R}_w$ and $\overline{I}_w$ measurements at different overheat values $\overline{a}_w$, if the flow conditions are unchanged, see (5.81) and *Morkovin* [5.100]. A series of 10–12 values is needed to accurately compute the derivative involved in the definition of $\overline{A}_w$.

Most importantly, the factor $\overline{A}_w$ does not depend on the anemometer type. For example, it is possible to compute $\overline{A}_w$ in the *Smits* et al. experiments [5.125], although this was not explicitly done by the authors. Indeed, the $S_{T_t}^{\text{CT}}$ coefficient given by (5.82) should be the same as that which can be deduced from any other explicit relation. In particular, *Smits* et al. [5.125] wrote

$$\text{Nu}_t = A f(\tau) + B g(\tau) \text{Re}_t^n \, , \tag{5.92}$$

with

$$f = 1 - 0.65 \tau \, , \quad g = 1 - 0.085 \tau \, ,$$
$$\tau = \frac{(T_w - T_r)}{T_t} \, . \tag{5.93}$$

Therefore, we have on one hand, from (5.82)

$$S_{T_t}^{\text{CT}} = n_t + 1 - \frac{1 + \overline{A}_w}{\overline{A}_w} \chi' - n m_t \frac{B g \text{Re}_t^n}{A f + B g \text{Re}_t^n} \tag{5.94}$$

and on the other hand, from (5.92) and (5.93)

$$S_{T_t}^{\text{CT}} = n_t - \frac{\eta}{\tau} - (\eta + \tau) \frac{A f' + B g' \text{Re}_t^n}{A f + B g \text{Re}_t^n}$$
$$- n m_t \frac{B g \text{Re}_t^n}{A f + B g \text{Re}_t^n} \, , \tag{5.95}$$

where $\tau = \eta \overline{a}_w$ and $\eta + \tau = \eta (1 + \overline{a}_w)$. After identification, the choice of $\text{Re}_t = 85$, and the use of the fitted values $A = 0.27$, $B = 0.35$ and $n = 0.55$, one obtains for $\overline{A}_w$

$$\frac{1}{\overline{A}_w} = -1$$
$$+ \frac{1}{\chi'} \left[\frac{1}{\overline{a}_w} - \eta (1 + \overline{a}_w) \frac{0.518}{4.30 - 0.48 \overline{a}_w} + 1 \right] \, . \tag{5.96}$$

Equation (5.96) advantageously makes $\overline{A}_w$ an explicit function of $\overline{a}_w$. Figure 5.17 indicates the results deduced from this equation, assuming $\chi' = 1$ and $\eta \simeq 0.95$. Other experimental data by *Morkovin* [5.100], *Comte-Bellot* and *Sarma* [5.127], or *Weiss* et al. [5.141] are also reported, concerning CC-, CT-, or CV-operated wires. Very clearly, $\overline{A}_w \simeq \overline{a}_w$ as long as $\overline{a}_w \leq 0.50$, and $\overline{A}_w > \overline{a}_w$ when $\overline{a}_w \geq 0.50$.

The increase of $\overline{A}_w$ with $\overline{a}_w$ corresponds to a decrease of the Nusselt number. This decrease is made explicit by *Smits* et al. [5.125] by means of the two functions f and g. The fitted values of A and B already used above give a relative decrease of Nu_t from 1.0 to 0.88 when $\overline{a}_w$ increases from 0 to 1. *Spina* and *McGinley* [5.108] also observed a very strong Nu_t variation in a Mach 10 helium tunnel, with $g = 1 - 0.50 \tau$. *Weiss* et al. [5.126] in the freestream of a Mach 2.54 wind tunnel found $f \simeq 1 - 0.72 \tau$ and $g \simeq 1 - 0.18 \tau$.

The temperature load that has just been considered appears to act in a way which is opposite to that we have indicated for subsonic flows when the film temperature is used (5.45). Indeed expressing the physical properties at the film temperature $T_f = (T_w + T_a)/2$ rather than at T_a leads to $\mathrm{Nu}_f < \mathrm{Nu}_a$ and the need to adjust the factor $(T_f/T_a)^{0.17}$.

Practical Values of the Sensitivity Coefficients for Supersonic Flows. For sufficiently large Mach and Reynolds numbers and a continuum flow, the sensitivity coefficients can be obtained from (5.86) and (5.82) with the appropriate simplifications listed in point 3. after (5.82). Hence

$$(X) = (XX) = \frac{1}{2} \quad \text{and}$$

$$(XXX) = n_t + 1 - \frac{1 + \overline{A}_w}{\overline{A}_w}\chi' - \frac{1}{2}m_t . \quad (5.97)$$

In practice, one has only to measure A_w according to (5.81) or to use Fig. 5.17. For air $n_t = 0.885, m_t = 0.765$, and for a tungsten wire $\chi' \simeq 1.10$.

The sensitivity coefficients that can be expected in compressible flows are indicated in Fig. 5.18. For CC- and CT-operated wires, the curves agree with experimental data collected by *Gaviglio* [5.122], *Bonnet* [5.92], *Smits* et al. [5.125], and *Smits* and *Dussauge* [5.142]. Similar results have been established by *Kosinov* [5.143]. Extension to the CV-operated wire is therefore permitted from (5.82) on the same basis. In these data, convective heat transfer rates are important. When this is not the case, discrepancies may occur, because of thermal conduction end losses, as indicated by *Ko* et al. [5.144] among others.

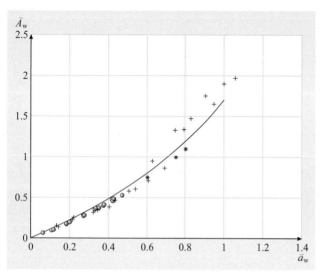

Fig. 5.17 The factor $\overline{A}_w$ as a function of the overheat ratio $\overline{a}_w$: *stars*: CCA *Morkovin* [5.100], *circles*: CCA and CVA *Comte-Bellot* and *Sarma* [5.127], *pluses*: CTA and CVA, *Weiss* et al. [5.141], *line*: (5.96)

Time Constant for Small Fluctuations

When thermal energy is unsteadily stored in a wire, the instantaneous heat balance (5.39) has to be rewritten as

$$m_w c_w \frac{dT_w}{dt} = (R_w - R_a)I_w^2 - \Phi_{\mathrm{conv}}(T_w - T_a) + R_a I_w^2 , \quad (5.98)$$

where m_w is the wire mass, and c_w is the heat capacity of the material constituting the wire. As a consequence the wire does not respond instantaneously to the forcing function Φ_{conv}. The difference between the response of an *ideal* wire, devoid of heat capacity, and the response of a CC-, CT-, or CV-operated wire, is therefore essential. Moreover, the amplitude of the turbulent fluctuation is an important parameter. We therefore consider the case of small-amplitude fluctuations in this section and the general case in the next section.

Constant-Current Operation. The thermal balance of an ideal wire having the same cold resistance R_a and heated with the same current intensity I_w as the real wire is

$$0 = (R_w^\star - R_a)I_w^2 - \Phi_{\mathrm{conv}}(T_w^\star - T_a) + R_a I_w^2 , \quad (5.99)$$

where $T_w^\star$ and $R_w^\star$ are the instantaneous temperature and resistance of this ideal wire, respectively. Substituting

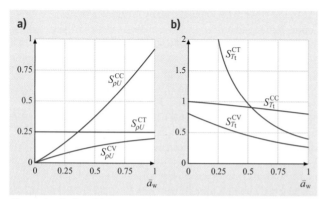

Fig. 5.18a,b Sensitivity coefficients in supersonic flows when $M > 1.2$, (**a**) to velocity or mass flux fluctuations, (**b**) to total temperature, for a wire operated in constant-current, constant-temperature, or constant-voltage mode

Φ_{conv} from (5.98) into (5.99), and using the $R(T)$ relation specified by (5.36) gives

$$\frac{m_{\text{w}} c_{\text{w}}}{\chi R_0} \frac{1}{I_{\text{w}}^2} \frac{1}{R_{\text{a}}} \frac{dR_{\text{w}}}{dt} + \frac{R_{\text{w}} - R_{\text{a}}}{R_{\text{w}}^\star - R_{\text{a}}} = 1. \quad (5.100)$$

The resistance R_{w} is therefore related to $R_{\text{w}}^\star$ by a first-order linear differential equation, where one coefficient, $R_{\text{w}}^\star - R_{\text{a}}$, is a random function of time.

For small-amplitude fluctuations, i. e.,

$$R_{\text{w}} = \overline{R}_{\text{w}} + r'_{\text{w}} \quad \text{with } r'_{\text{w}} \ll \overline{R}_{\text{w}} \quad \text{and}$$
$$R_{\text{w}}^\star = \overline{R}_{\text{w}}^\star + r'^\star_{\text{w}} \quad \text{with } r'^\star_{\text{w}} \ll \overline{R}_{\text{w}}^\star, \quad (5.101)$$

a first-order expansion of (5.100) gives $\overline{R}_{\text{w}} = \overline{R}_{\text{w}}^\star$ and

$$M_{\text{w}}^{\text{CC}} \frac{dr'_{\text{w}}}{dt} + r'_{\text{w}} = r'^\star_{\text{w}} \quad \text{with}$$
$$M_{\text{w}}^{\text{CC}} = \frac{m_{\text{w}} c_{\text{w}}}{\chi R_0} \frac{1}{I_{\text{w}}^2} \frac{\overline{R}_{\text{w}} - R_{\text{a}}}{R_{\text{a}}}. \quad (5.102)$$

The response to a step function is an exponential, and the simplest notion of a time constant M_{w}^{CC} applies. The amplitude attenuation is therefore given by

$$\left[1 + \omega^2 (M_{\text{w}}^{\text{CC}})^2\right]^{-1/2},$$

or $-20\,\text{dB/decade}$, or $-6\,\text{dB/octave}$.

Often, M_{w}^{CC} is simply denoted M_{w}. Furthermore, the wire Nusselt number can be introduced in (5.102) in place of I_{w}^2. This leads to the compact form

$$M_{\text{w}}^{\text{CC}} = (1 + \overline{a}_{\text{w}}) \frac{d^2}{4} \frac{\rho_{\text{w}} c_{\text{w}}}{k_{\text{a}}} \frac{1}{\text{Nu}_{\text{a}}}. \quad (5.103)$$

For $a_{\text{w}} \simeq 0$, the expression (5.103) was already obtained by *Scadron* and *Warshasky* [5.130] when working on thermocouples. Nu_{a} can be measured or estimated using a heat-transfer law, e.g., Kramers law valid for air or liquids (5.46). The wire characteristics are listed in Table 5.1 and the fluid and/or substrate properties in Table 5.2. For example, $M_{\text{w}}^{\text{CC}} \simeq 0.50\,\text{ms}$ for a $5\,\mu\text{m}$ tungsten wire, operated at $\overline{a}_{\text{w}} = 0.80$ in a $U = 50\,\text{m/s}$ air flow. The amplitude attenuation is then around $-30\,\text{dB}$ at $10\,\text{kHz}$. Direct measurements of M_{w}^{CC} are discussed in Sect. 5.2.3.

Regarding the dependence of M_{w}^{CC} on the wire diameter at a given velocity, (5.103) leads to the approximate form $M_{\text{w}}^{\text{CC}} \sim d^{3/2}$.

For a cylindrical hot-film sensor, the frequency response was analyzed by Lowell and Patton, as early as 1955, and by *Kidron* [5.145]. When the bulk of the cylinder responds as a whole, $\rho_{\text{w}} c_{\text{w}}$ can be replaced by $\rho_{\text{s}} c_{\text{s}}$ in (5.103), where the subscript 's' refers to the substrate. Useful properties for the substrate are indicated in Table 5.2, especially the thermal conductivity k_{s}. For example, $M_{\text{film}}^{\text{CC}} \simeq 4.0\,\text{ms}$ for a $25\,\mu\text{m}$ cylindrical hot-film, with a quartz core, operated at $\overline{a}_{\text{w}} = 0.80$ in a $U = 50\,\text{m/s}$ air flow. This film needs therefore an important frequency correction. In water, the situation is greatly improved because of a substantial increase of the heat transfer rate due to water's larger thermal conductivity. For example, $M_{\text{film}}^{\text{CC}} = 0.17\,\text{ms}$ for a $50\,\mu\text{m}$ film placed in a $U = 2\,\text{m/s}$ water flow and operated at $\overline{a}_{\text{w}} = 0.10$.

When hot-films are placed on wedges, the response becomes more complicated because part of the thermal energy through the wire leaks towards the substrate. *Bellhouse* and *Schultz* [5.146] investigated a one-dimensional case where the heat loss takes place along a normal to the substrate. They clearly pointed out that the dynamic sensitivity cannot be obtained from a steady flow calibration. *Brison* et al. [5.147] considered a two-dimensional case, where the thermal transfer may also take place in the longitudinal direction. They showed, in particular, that a film should not be placed too close to the stagnation point of a wedge, because the high transfer rates to the external flow which prevail in that region serve as a heat sink for the substrate and therefore for the film.

Constant-Temperature Operation. In principle, since R_{w} is maintained constant whatever the flow conditions, the time constant concept is not required for an idealized CTA. However, since the condition of constant resistance can only be maintained by an appropriate electronic circuit, the way this circuit handles that function is an essential aspect of the actual CTA. This topic is considered in Sect. 5.2.3.

Constant-Voltage Operation. The *real* wire now has to respect $R_{\text{w}} I_{\text{w}} = V_{\text{w}}$, and the *ideal* wire $R_{\text{w}}^\star I_{\text{w}}^\star = V_{\text{w}}$, with $V_{\text{w}} = \text{constant}$. Equation (5.98) keeps its form, and (5.99) becomes

$$0 = (R_{\text{w}}^\star - R_{\text{a}}) I_{\text{w}}^{\star 2} - \Phi_{\text{conv}}(T_{\text{w}}^\star - T_{\text{a}}) + R_{\text{a}} I_{\text{w}}^{\star 2}. \quad (5.104)$$

Combining (5.98) and (5.104) yields

$$\frac{1}{V_{\text{w}}^2} \frac{m_{\text{w}} c_{\text{w}}}{\chi R_0} \frac{dR_{\text{w}}}{dt} = \frac{1}{R_{\text{w}}} - \frac{1}{R_{\text{w}}^\star} \frac{R_{\text{w}} - R_{\text{a}}}{R_{\text{w}}^\star - R_{\text{a}}}. \quad (5.105)$$

For small-amplitude fluctuations, a first-order expansion gives $\overline{R}_w = R_w^\star$, $\overline{I}_w = I_w^\star$, and

$$M_w^{CV} \frac{dr'_w}{dt} + r'_w = r'^\star_w \quad (5.106)$$

with, for the new time constant,

$$M_w^{CV} = \frac{\rho_w c_w}{\chi R_0} \frac{1}{\overline{I}_w^2} \frac{(\overline{R}_w - R_a)}{2\overline{R}_w - R_a} = \frac{M_w^{CC}}{1 + 2\overline{a}_w} . \quad (5.107)$$

Equation (5.107) clearly shows that M_w^{CV} is smaller than M_w^{CC}. This result was anticipated by *Sarma* [5.148] and established by *Comte-Bellot* [5.55] and *Comte-Bellot* and *Sarma* [5.127]. *Sarma* [5.149] and *Reimann* et al. [5.150] explained that, if R_w decreases due to flow, I_w increases, and this increased I_w partially restores the hot-wire resistance. These small M_w^{CV} values constitute a noticeable benefit in favor of the CV operation. For example, for the 5 μm tungsten wire considered above, M_w^{CV} goes down from 0.50 ms to 0.19 ms at $\overline{a}_w = 0.80$. Figure 5.19 reports results obtained in supersonic conditions.

Dynamic Nonlinearity: Large Fluctuations
Constant-Current Operation. Equation (5.100) cannot be linearized when the turbulent fluctuations present large amplitudes. *Corrsin* [5.45] pointed out that higher harmonics are generated by *parametric excitation* because of the coefficient $(R_w^\star - R_a)$, which is a random function of time. Using Fourier series, he computed the amplitude and phase of the second harmonic that appears for a sinusoidal input. Extension to random signals featuring turbulence was made by *Comte-Bellot* and *Schon* [5.151] on an analog computer. Obviously, the electronic compensation circuit of a CCA cannot discriminate between the true turbulence components and the wrong higher harmonics due to the wire behavior. As a consequence, errors on odd turbulence moments are large, such as on the skewness factors which are third-order moments. Nowadays, (5.100) can easily be solved by a Matlab procedure. Results are presented in Sect. 5.2.3.

Constant-Temperature Operation. Similarly to Sect. 5.2.2, the role of the electronic circuit is crucial, see Sect. 5.2.3.

Constant-Voltage Operation. For large-amplitude fluctuations, the full equation (5.105) has to be considered and higher harmonics could be generated. A systematic analysis has however not yet been performed. See also Sect. 5.2.3.

Sensors of Finite Length – End Losses
A real hot-wire is held by supports that are at ambient or near-ambient temperature, see Fig. 5.20. Thermal conduction can therefore take place axially along the wire towards the supports, resulting in a marked temperature maximum near the middle of the wire. This behavior was clearly illustrated by *Champagne* et al. [5.152] who deduced the temperature profile along a wire from its infrared emission. Very simply, one can observe a wire under a magnifying glass while gradually increasing the heating current intensity. The central part of the wire is the first to turn red.

The partial differential equation governing the temperature T at a location ξ_3 along the wire or the stubs is determined by the local heat balance between the

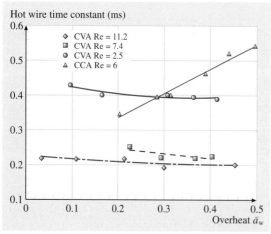

Fig. 5.19 Time constant for a CCA- and CVA-operated wire in supersonic flows (after *Comte-Bellot* and *Sarma* [5.127]). At similar Re and $\overline{a}_w$, $M_w^{CV} < M_w^{CC}$ (5.107)

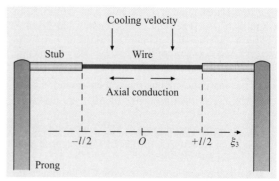

Fig. 5.20 Sketch of a short wire with stubs and prongs

ohmic heating, the convective cooling, and the axial heat conduction. This balance, which is presented in most reviews on hot-wire anemometry [5.49, 51, 52], may be written

$$\rho c \frac{\pi d^2}{4} \frac{\partial T}{\partial t}$$
$$= R_1 I_w^2 - \mathrm{Nu}_1 \pi k_a (T - T_a) + \frac{\pi d^2}{4} k \frac{\partial^2 T}{\partial \xi_3^2} . \quad (5.108)$$

This is an extension of (5.39), the axial conduction towards the supports now being introduced. The subscript 'w' has been dropped to generalize (5.108) for a point located on the wire as well as on the stubs. Hence, d is the wire diameter d_w when ξ_3 concerns the wire, and the stub diameter d_s when ξ_3 concerns one of the two stubs, and similarly for the thermal conductivity k, the specific heat c, and the Nusselt number, which involves a Reynolds number based on the wire (or stub) diameter. The subscript '1' refers to a value per unit length (not to be confused with the top resistance of the Wheatstone bridge). The subscript 'w' is kept only for I_w as the current intensity is the same through the wire and the stubs.

The spatial average of R_1 over the wire and the stubs constitutes the input to the anemometer whatever its type, CCA, CTA, or CVA. This average is noted $\langle R \rangle$ and is given by

$$\langle R \rangle = \int_{-s-l/2}^{+s+l/2} R_1(\xi_3) \, \mathrm{d}\xi_3 . \quad (5.109)$$

It is therefore very important to examine how T and R_1 are distributed along the wire and the stubs, when the main parameters are changed, such as changes in the material, the wire aspect ratio, the relative importance of the stubs, the flow velocity normal to the wire, the flow temperature, and the current intensity through the wire.

Equation (5.108) can be rewritten in a compact form as

$$\tau \frac{\partial T}{\partial t} = \frac{\partial^2 T}{\partial \xi_3^2} - C(T - T_a) + D , \quad (5.110)$$

where

$$\tau = \frac{\rho c}{k} , \quad C = 4 \left(\mathrm{Nu}_1 \frac{k_a}{k} - \frac{\chi R_{01}}{\pi k} I_w^2 \right) \frac{1}{d^2} ,$$
$$D = 4 R_{a1} \frac{1}{\pi k} I_w^2 \frac{1}{d^2} . \quad (5.111)$$

R_{01} is the wire (or stub) resistance per unit length at the reference temperature T_0. Similarly R_{a1} is the wire (or stub) resistance per unit length at the flow temperature T_a, and $\mathrm{Nu}_1 = R_1 I_w^2 / \pi k_a (T - T_a)$ is the local Nusselt number per unit length. Locally, Nu_1 is assumed to be that of a long wire, as indicated for subsonic and supersonic flows in Sect. 5.2.2. The incident velocity therefore occurs in the factor C, the incident flow temperature in the factor D, and the wire current as well as the wire diameter in the two factors C and D.

The basic case of a wire directly soldered onto relatively massive prongs, which can be assumed to stay at temperature T_a, can be treated analytically when the incident flow is steady, uniform and isothermal. This was investigated very early by *Betchov* [5.153], *Corrsin* [5.45] or *Davies* and *Fisher* [5.58]. The solution of (5.110) that verifies the boundary conditions $T = T_a$ for $\xi_3 = \pm l/2$ gives the temperature profile $T(\xi_3)$

$$T(\xi_3) - T_a = \frac{D}{C} \left(1 - \frac{\cosh \sqrt{C} \xi_3}{\cosh \sqrt{C} l/2} \right) . \quad (5.112)$$

One can observe from (5.110) that the factor D/C is the temperature difference $(T - T_a)$ which would be established for an infinitely long wire. The space average of $T(\xi_3)$ along the wire is

$$\langle T \rangle - T_a = \frac{D}{C} \left(1 - \frac{\tanh \sqrt{C} l/2}{\sqrt{C} l/2} \right) . \quad (5.113)$$

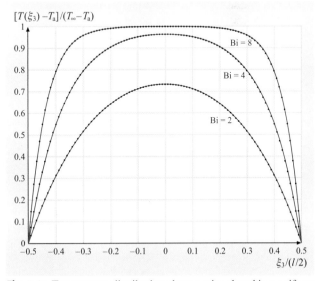

Fig. 5.21 Temperature distribution along a wire placed in a uniform perpendicular flow for different Biot numbers $\mathrm{Bi} = \sqrt{C} l/2$ with C defined by (5.111). Both ends are assumed to be at the ambient temperature T_a

Table 5.3 Steady end losses for different wires and flows, according to (5.114) when $\langle a \rangle = 0.80$

Velocity (m/s)	Material	l (mm)	d (μm)	l/d	Re_t	Nu_t	R_{01} (Ω/m)	I_w (mA)	$Bi = \sqrt{Cl}/2$	ϵ_{th} (%)
Air										
30	Pt	1	2.5	400	5	1.5	22400	25	3.7	7.5
30	Pt	2	5	400	10	2.0	5600	57	8.4	6.5
30	Pt	1	5	200	10	2.0	5600	57	4.2	13
30	W	1	2.5	400	5	1.5	11200	33	4.7	12
30	W	2	5	400	10	2.0	2800	75	5.4	11
30	W	1	5	200	10	2.0	2800	75	2.7	20
60	W	2	5	400	20	2.7	2800	87	6.2	9
10	W	2	5	400	5	1.5	2800	66	2.3	20
[a] $M = 2.5$[a]	W	1.2	5	240	20	1.5	2800	65	2.8	20
[b] $M = 6$	W	1.2	5	240	7	0.7	2800	60	1.1	38
Helium										
[c] $M = 10$	W	1.2	5	240	15	1	2800	135	5.4	10

The supersonic conditions are: [a] $T_t \simeq 293$ K and $p_t \simeq 1 \times 10^5$ Pa [5.141]; [b] $T_t \simeq 450$ K and $p_t \simeq 9 \times 10^5$ Pa [5.129]; and [c] $T_t \simeq 300$ K and $p_t \simeq 25 \times 10^5$ Pa [5.108]

Figure 5.21 shows the temperature profile along a wire, according to (5.112) for different values of $\sqrt{Cl}/2$, and the corresponding mean values given by (5.113). For the anemometer, the change in $\langle R \rangle$, which is related to the change in $\langle T \rangle$ by $\langle R \rangle = R_a + \chi R_0(\langle T \rangle - T_a)$, is the quantity of interest.

The heat loss q to both supports can be deduced from (5.112)

$$q = -2k_w \frac{\pi d^2}{4} \frac{\partial T}{\partial \xi_3}\bigg|_{\xi_3 = l/2}$$

$$= 2k_w \frac{\pi d^2}{4} D \frac{\tanh \sqrt{Cl}/2}{\sqrt{C}} \quad (5.114)$$

and the fraction ϵ_{th} relative to the input ohmic heating is

$$\epsilon_{th} = \frac{q}{\langle R \rangle I_w^2} = \frac{1}{1 + \langle a \rangle} \frac{\tanh \sqrt{Cl}/2}{\sqrt{Cl}/2}, \quad (5.115)$$

where $\langle a \rangle = (\langle R \rangle - R_a)/R_a$.

The dimensionless quantity $\sqrt{Cl}/2$ is the Biot number Bi. It expresses the ratio between the heat-transfer coefficient of a solid at its surface and the heat conductance to the center of the solid (it is most important in food processing). Here it is based on the half-wire length $l/2$ and should be as large as possible to favor the cooling of the wire by the incident flow. Referring to the definition of C (5.111), one sees that $\sqrt{Cl}/2$ depends on l/d as well as on the flow conditions and the wire diameter. In practice one therefore has to consider both l/d and $\sqrt{Cl}/2$. In other words, a *long* wire involves more than the unique condition $l/d \to \infty$.

Estimates of ϵ_{th} are given in Table 5.3 for different wires and flows. Use is made of the physical data listed in Table 5.2, of Kramers' law (5.46) for subsonic flows, and of the Baldwin–Dewey diagram for supersonic flows, Fig. 5.15. Table 5.3 clearly shows that end losses become large when the Nusselt number decreases. This is the case for low-density flows. *Dewey* [5.119], *Davies* and *Fisher* [5.58], and *Lord* [5.154] studied this interesting case, even carrying out experiments in a vacuum. *Lord* [5.154] established a correction which can be applied during calibration. It was later used with success by *Ko* et al. [5.144]. *Pitts* and *McCaffrey* [5.107] also did a careful analysis of end losses for tungsten wires, $d = 4$ μm, $l = 1.25$ mm, and 0.1 μm-thick platinum films on quartz cylinders, $d = 50$ μm, $l = 1$ mm, in air and many other gases for a Reynolds number range of 2–35. The end losses were analyzed in terms of the A and B variations in King's law. These results confirm that Nu_∞ for a wire of infinite length is always less than Nu_m for a wire of finite length. The fractional decrease of A from A_∞ to A_m is very important, on the order of 50%; the fractional decrease of B stays around 5%.

When stubs exist between the wire and the prongs, the topic is conveniently investigated using a numerical technique to solve (5.108) or (5.110). *Morris* and *Foss* [5.155] studied a 1 mm active region of a 5 μm tungsten filament with 50 μm copper-plated stubs, where the velocity was 7.5 m/s and $\langle a \rangle = 0.7$. They found that the temperature at the end of the sensor was 13.2 K above ambient and the mean wire temperature was 210 K above ambient. *Li* et al. [5.156] studied longer stubs and different wires, and obtained in some cases a temperature 60 K above ambient for the stubs, close to that measured by *Champagne* et al. [5.152].

Recently, *Löfdahl* et al. [5.157] computed the temperature distribution along a MEMS-fabricated hot-wire sensor. The wire length was 400 μm, its cross section was 2 μm × 2 μm, and the supports were 100 μm long with a cross section of 2 μm × 20 μm. A nonuniform temperature distribution along the supports was clearly observed, with a rise of around 30 K. In addition, the convective heat transfer from the wire was found to constitute about 50–60% of the total heat generated. A conventional hot-wire has a corresponding value of 80–85%.

Fingerson and *Freymuth* [5.50] reported results for cylindrical hot-films. They found, very interestingly, that steady end losses are strongly decreased, roughly by a factor of 16, compared to tungsten wires with the same aspect ratio. Indeed, the main path for heat towards the supports is the conductive metallic film deposited on the quartz core, and this film is very thin, hence precluding heat losses. Cylindrical hot-film sensors can therefore be used with relatively low aspect ratios, on the order of 20–50.

When velocity, temperature or current intensity fluctuations are present, the problem becomes more complex, and a small-perturbation analysis is generally adopted. *Betchov* [5.153] used an analytical approach for velocity and current changes for a CC operated wire. Similarly, *Bremhorst* et al. [5.158] considered velocity, ambient temperature and current fluctuations. *Lord* [5.159] proposed a methodical approach for both a CC- and a CT-operated wire. *Beljaars* [5.160] and *Li* [5.161] considered a CT-operated wire and calculated the new dynamic response using a numerical technique. For velocity fluctuations, a 5% step down is found around 1000 Hz. This frequency is roughly the inverse of the time scale of the stub. *Perry* [5.49] reported differences between velocity fluctuations and temperature fluctuations, as well as between hot and cold wires. More results are given in Sect. 5.2.5.

Spatial Resolution

The velocity fluctuations that occur in a turbulent field vary in space from one point to another. The length scales of these variations cover a continuous spectrum from the largest scales of size L, where kinetic energy is produced, down to the smallest ones, of the size of the Kolmogorov scale η, where kinetic energy is dissipated into heat. A perfect flow analysis would require a very short wire able to respond to the smallest scales of the flow. Instead, a wire has a non-negligible length and responds to an averaged value of the turbulent fluctuations over its length. The wire response will thus depend strongly on its length l as well as on the energy contents at all scales, from L to η.

For free turbulent flows, such as jets or wakes, the scale L is roughly the lateral dimension of the flow, hence the simple relation $l \ll L$ usually holds. On the other hand, η can be very small as is related to L by $\eta/L \simeq \text{Re}_L^{-3/4}$ [5.77]. For example, for a jet with $\overline{U} = 20$ m/s, $u'/\overline{U} \simeq 20\%$ and $L = 0.10$ m, one has $\eta \simeq 0.05$ mm. Obviously, $\eta \ll l$ and the wire drastically smooths out the fine turbulent velocity fluctuations.

For the inner part of wall turbulent flows, L becomes of the order of the wall distance and η approaches L in the viscous sublayer [5.77, 162]. Thus, $l \simeq L \simeq \eta$, and measurements will become erroneous for all structure sizes. Furthermore turbulence is highly anisotropic in the near-wall region.

In what follows, the first part deals with isotropic turbulence where analytical developments can be conducted. This is an important case as local isotropy is a reasonable assumption for many flows at high Reynolds numbers. In addition, some guidelines are obtained regarding the ratios L/l and η/l. In a second part, the wall region with its high anisotropy is considered. This case requires systematic experiments and/or the use of databases provided by direct numerical simulations (DNS) of the flow.

Spatial Resolution of Hot-Wires in Homogeneous Isotropic Turbulence.
Spatial Resolution for a Single Normal Wire in Isotropic Turbulence. For a single wire normal to O_{x_1} and parallel to the x_3 direction, the wire reading attributed to the wire midpoint O' is given by

$$u_1^m(O') = \frac{1}{l} \int_{-l/2}^{l/2} u_1(0, 0, \xi_3) \, \mathrm{d}\xi_3 \, . \tag{5.116}$$

The ξ_i refer to coordinates centered at O' with ξ_3 along the wire. The superscript $()^m$ denotes the mapping due to the wire. The sensitivity coefficient along the wire is assumed to be constant, and the tangential and binormal cooling of the wire is neglected. The mean square value of u_1^m is thus given by the double integral

$$\overline{u_1^{2m}}(O') = \frac{1}{l^2} \iint_{\mathcal{D}} \overline{u_1(0,0,\xi_3) u_1(0,0,\xi_3')} \, d\xi_3 \, d\xi_3' ,$$

(5.117)

where $\mathcal{D}$ is the domain $-l/2, \leq \xi_3, \xi_3' \leq l/2$. Equation (5.117) can be transformed by introducing $r_3 = \xi_3' - \xi_3$ and the transverse correlation function $g(r_3)$ between longitudinal velocity fluctuations at two different points on the wire. One of the integrals can then be computed and (5.117) simplifies into

$$\overline{u_1^{2m}}(O') = \overline{u_1^2} \frac{2}{l^2} \int_0^l (l - r_3) g(r_3) \, dr_3 .$$

(5.118)

This result was first obtained by *Dryden* et al. [5.163], who detailed all the cross terms in (5.117), and later by *Frenkiel* [5.164] who directly processed the equation as indicated above. Incidentally the method is similar to that used by *Taylor* in his study of particle diffusion through turbulence [5.77]. Equation (5.118) shows the role of the velocity correlation drop along the wire, even when measuring $\overline{u_1^2}$, which is mostly associated with large structures. This can be illustrated by using a composite correlation function that includes a transverse large scale L_g and a transverse Taylor scale λ_g connected to L_g by $\lambda_g/L_g = (15/\text{Re}_L)^{1/2}$ [5.77]. Writing $g(r_3)$ as

$$g(r_3) = \frac{1}{2} \left[\exp\left(-\frac{r_3}{L_g}\right) + \exp\left(-\frac{r_3^2}{\lambda_g^2}\right) \right] ,$$

(5.119)

(5.118) gives $\overline{u_1^{2m}}(O')/\overline{u_1^2} = 0.91, 0.86$ and 0.74 for $\lambda_g = 2l, l$ and $l/2$, respectively, when $L_g = 4l$. Clearly the correlation drop over the wire length significantly affects the measurements.

The spatial resolution of a wire is of great concern when measuring spectra. The Fourier decomposition of the turbulent field around the wire midpoint O' is given by

$$u_i(\xi_1, \xi_2, \xi_3)$$
$$= \iiint \hat{u}_i(k_1, k_2, k_3)$$
$$\times \exp[i(k_1\xi_1 + k_2\xi_2 + k_3\xi_3)] \, dk_1 \, dk_2 \, dk_3 ,$$

(5.120)

where k_1, k_2, k_3 are the wavevector components and $\hat{u}_i$ is the Fourier contributions in the sense of generalized functions. The subscript i is $i = 1, 2$ or 3. The Fourier frame has coordinate axes parallel to the space-fixed coordinates O_{x_1,x_2,x_3}. The wire response according to (5.116) becomes

$$\hat{u}_1^m(k_1, k_2, k_3) = \frac{\sin(k_3 l/2)}{k_3 l/2} \hat{u}_1(k_1, k_2, k_3) \quad (5.121)$$

and the spectral tensor $\Phi_{11}^m(k_1, k_2, k_3)$ mapped by the wire is

$$\Phi_{11}^m(k_1, k_2, k_3) = \left[\frac{\sin(k_3 l/2)}{k_3 l/2}\right]^2 \Phi_{11}(k_1, k_2, k_3) ,$$

(5.122)

where $\Phi_{11}(k_1, k_2, k_3)$ represents the exact spectral tensor relative to the longitudinal velocity fluctuations. For isotropic turbulence the general form of the spectral tensor is

$$\Phi_{ij}(k_1, k_2, k_3) = \frac{E(k)}{4\pi k^4} \left(k^2 \delta_{ij} - k_i k_j\right) , \quad (5.123)$$

where $E(k)$ is the kinetic energy spectrum of turbulence

$$\frac{\overline{u_i^2}}{2} = \int_0^\infty E(k) \, dk \quad \text{with } k = \left(k_1^2 + k_2^2 + k_3^2\right)^{1/2} .$$

(5.124)

The one-dimensional spectrum $E_{11}^m(k_1)$ obtained from the response of the wire is the integral of $\Phi_{11}^m(k_1, k_2, k_3)$ over k_2 and k_3

$$E_{11}^m(k_1)$$
$$= \int_0^\infty \int_0^\infty \left[\frac{\sin(k_3 l/2)}{k_3 l/2}\right]^2 \frac{E(k)}{4\pi k^4} \left(k^2 - k_1^2\right) dk_2 \, dk_3 .$$

(5.125)

This is related to the usual power spectral density of a signal recorded in time by the Taylor hypothesis, which links k_1 to the frequency f, $|k_1| = 2\pi f/\overline{U}_1$. Equation (5.125) can be evaluated numerically once $E(k)$ is given, as suggested by *Wyngaard* [5.165]. However, the former approach by *Uberoi* and *Kovasznay* [5.166] sheds light on the physics behind the errors. Indeed, one of the integrals in (5.125) can be analytically computed whatever the form of $E(k)$. This is done by using polar coordinates in the k_2–k_3 plane, with the radius $\sigma = (k_2^2 + k_3^2)^{1/2}$ and the angle ϕ between k_2 and σ, and by introducing y such that $y = k_3 l/2 = \sigma l \sin\phi/2$, hence

$\mathrm{d}\phi = \mathrm{d}y[(\sigma l/2)^2 - y^2]^{-1/2}$. This yields

$$E_{11}^{\mathrm{m}}(k_1) = 4 \int_0^\infty \left[\frac{E\left(\sqrt{k_1^2 + \sigma^2}\right)}{4\pi(k_1^2 + \sigma^2)^2} \sigma^3 \right.$$

$$\left. \times \int_0^{\sigma l/2} \left(\frac{\sin y}{y}\right)^2 \frac{\mathrm{d}y}{[(\sigma l/2)^2 - y^2]^{1/2}} \right] \mathrm{d}\sigma \,. \tag{5.126}$$

Using the new variable $k = (\sigma^2 + k_1^2)^{1/2}$, which gives $k\,\mathrm{d}k = \sigma\,\mathrm{d}\sigma$ at fixed k_1, (5.126) can be written as

$$E_{11}^{\mathrm{m}}(k_1) = \frac{1}{2} \int_{k_1}^\infty W \frac{E(k)}{k^3}(k^2 - k_1^2)\,\mathrm{d}k \,, \tag{5.127}$$

where the weighting function W is given by

$$W\left(\sigma\frac{l}{2}\right) = \frac{\pi}{2} \int_0^{\sigma l/2} \left(\frac{\sin y}{y}\right)^2 \frac{\mathrm{d}y}{[(\sigma l/2)^2 - y^2]^{1/2}}$$

with $\sigma\dfrac{l}{2} = \sqrt{(k^2 - k_1^2)}\dfrac{l}{2}$. \hfill (5.128)

When $W \simeq 1$, (5.127) provides the correct expression of the one-dimensional longitudinal velocity spectrum [5.77]

$$E_{11}(k_1) = \frac{1}{2} \int_{k_1}^\infty \frac{E(k)}{k^3}(k^2 - k_1^2)\,\mathrm{d}k \,. \tag{5.129}$$

The graph of the weighting function W is drawn in Fig. 5.22. As long as $[\sqrt{(k^2 - k_1^2)}l/2] \leq 1$, $W \simeq 1$ and subsequently $W \simeq [\sqrt{(k^2 - k_1^2)}l/2]^{-1}$. The value at which W starts to be less than one depends on l as well as on k_1 and k. This zone and its border in the k–k_1 plane are sketched in Fig. 5.23. The spectrum $E(k)$ is also indicated on the right of the figure along with the construction which shows the spectral zone of $E(k)$ for which $W < 1$ at given l and k_1.

Wyngaard [5.165] used a Pao spectrum for $E(k)$

$$E(k) = C_K \epsilon^{2/3} k^{-5/3} \exp\left(-\frac{3}{2}C_K(k\eta)^{4/3}\right), \tag{5.130}$$

where C_K is the Kolmogorov constant, $C_K \simeq 1.5$, and ϵ is the dissipation rate of kinetic energy. Figure 5.24 illustrates the errors for a Pao spectrum and for different wire lengths expressed in terms of the Kolmogorov scale η. These values agree with those of Wyngaard when $k_1 l$ is transformed into $k_1 \eta$ for the abscissa. However, the Pao spectrum describes only the high-wavenumber contributions. In comparison, the Karman–Saffman–Pao form describes an entire spectrum whose inertial range is given by the $k^{-5/3}$ law [5.77]. Its expression is

$$E(k) = 1.45 \frac{u'^2}{k_e} \frac{(k/k_e)^4}{[1 + (k/k_e)^2]^{17/6}}$$

$$\times \exp\left[-\frac{9}{4}(k\eta)^{4/3}\right]. \tag{5.131}$$

The maximum of $E(k)$ occurs at $\sqrt{12/5}k_e$ or below, depending on k_e. Moreover, one can compute the integral length scale L using the classic relation for isotropic turbulence

$$L = \frac{3\pi}{4} \frac{\int_0^\infty k^{-1} E(k)\,\mathrm{d}k}{\int_0^\infty E(k)\,\mathrm{d}k} \,. \tag{5.132}$$

Hence L, which depends on k_e, is adjustable. For example, L/η takes the values 3.2, 4.5, 7.9, 12.8 and 22, for $k_e\eta = 1, 1/2, 1/5, 1/10, 1/20$, respectively. Graphs comparing the Pao and the Karman–Saffman–Pao spectra are available in *Ewing* et al. [5.167].

Figure 5.25 shows the attenuation that affects $E_{11}^{\mathrm{m}}(k_1)$ for a given wire length and different values of the integral length scale L, for a Karman–Saffman–Pao

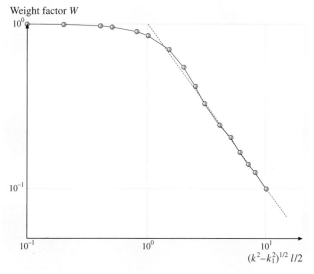

Fig. 5.22 Spatial resolution: the weighting function W for a single wire and isotropic turbulence (5.128)

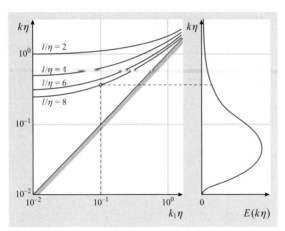

Fig. 5.23 Curves limiting the $W = 1$ and $W < 1$ zones for different wire lengths l. Under each curve $W = 1$ and above each curve $W < 1$. The dotted lines between this plot and $E(k)$ show the highest wavenumber for which $W = 1$ at a selected $k_1 \eta$ when $l/\eta = 8$

spectrum. It is clear that the attenuation is large at small values of L.

In this analysis some simplifications were introduced, such as the constant value of the wire sensitivity along its length. Attempts to introduce this additional parameter were conducted by *Elsner* et al. [5.168]. The increased sensitivity in the central section of the wire is beneficial since it reduces the apparent wire length.

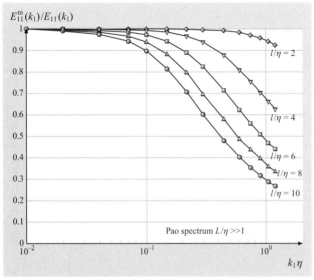

Fig. 5.24 Spatial resolution: the effect of wire length on the one-dimensional velocity spectra, for a Pao spectrum

Spatial Resolution for Two Parallel Wires in Isotropic Turbulence. Two wires of length l, normal to Ox_1, parallel to Ox_3 and separated along Ox_2 by a distance s_2, are currently used to obtain the derivative $\partial u_1/\partial x_2$. The spatial resolution of this array was considered very early by *Frenkiel* [5.164] who established the errors affecting the rms values and more recently by *Ewing* et al. [5.167] who considered the spatial filtering of the probe taking into account changes in the spectral model.

An analysis similar to that conducted in the section above is again possible. The Fourier decomposition is made relative to the midpoint O' of the array, hence giving for the wires (denoted by I and II)

$$\hat{e}_\mathrm{I}^\mathrm{m} = \frac{1}{l} \int_{-l/2}^{l/2} \mathrm{d}\xi_3 \iiint \hat{u}_1(k_1,k_2,k_3) \exp\left(-\mathrm{i}k_2 \frac{s_2}{2}\right)$$
$$\times \exp(-\mathrm{i}k_3 \xi_3)\,\mathrm{d}k_1\,\mathrm{d}k_2\,\mathrm{d}k_3$$

$$\hat{e}_\mathrm{II}^\mathrm{m} = \frac{1}{l} \int_{-l/2}^{l/2} \mathrm{d}\xi_3' \iiint \hat{u}_1(k_1,k_2,k_3) \exp\left(+\mathrm{i}k_2 \frac{s_2}{2}\right)$$
$$\times \exp(-\mathrm{i}k_3 \xi_3')\,\mathrm{d}k_1\,\mathrm{d}k_2\,\mathrm{d}k_3 \,. \qquad (5.133)$$

This yields for the one-dimensional spectrum $D_{1,2}^\mathrm{m}$ of $(\hat{e}_\mathrm{I} - \hat{e}_\mathrm{II})/s_2$, which approaches the lateral derivative $\partial u_1/\partial x_2$ of the longitudinal fluctuations

$$D_{1,2}^\mathrm{m}(k_1) = \iint \left[\frac{\sin(k_3 l/2)}{k_3 l/2}\right]^2 \left[\frac{\sin(k_2 s_2/2)}{k_2 s_2/2}\right]^2$$
$$\times \Phi_{11}(k_1,k_2,k_3)\,\mathrm{d}k_2\,\mathrm{d}k_3 \,. \qquad (5.134)$$

A similar result was obtained by *Ewing* et al. [5.167]. The same integration procedure as before yields

$$D_{1,2}^\mathrm{m} = \frac{1}{4} \int_{k_1}^{\infty} \mathrm{WD}\,\frac{E}{k^3}\left(k^2 - k_1^2\right)^2 \mathrm{d}k \,, \qquad (5.135)$$

where the weighting function, denoted by WD, reads

$$\mathrm{WD}\left(\sigma\frac{l}{2}\right) = \frac{4}{\pi}\frac{1}{(\sigma l/2)^2} \int_0^{\sigma l/2} \left[\left(\sigma\frac{l}{2}\right)^2 - y^2\right]^{1/2}$$
$$\times \left(\frac{\sin y}{y}\right)^2 \left(\frac{\sin \Delta}{\Delta}\right)^2 \mathrm{d}y \qquad (5.136)$$

with

$$\sigma\frac{l}{2} = \sqrt{\left(k^2 - k_1^2\right)}\frac{l}{2} \quad \text{and}$$

$$\Delta = \frac{s_2}{l}\left[\left(\sigma\frac{l}{2}\right)^2 - y^2\right]^{1/2}$$

and y as introduced earlier for (5.126). When WD $= 1$ the exact spectrum is recovered:

$$D_{1,2}(k_1) = \frac{1}{4} \int_{k_1}^{\infty} \frac{E}{k^3} (k^2 - k_1^2)^2 \, dk \,. \quad (5.137)$$

Equation (5.136) takes into account both the length l of the two parallel wires (assumed to be identical) and the lateral separation s_2 of the wires. Figure 5.26 shows the weighting function WD for $s_2/l = 0.5$ and 1.0, and also the case where the wire length is neglected. It is clear that this simplification is not acceptable. Both the l and s_2 effects have to be considered. For further estimates, approximate curves can be used. In the case of Fig. 5.26, with the wire length included, WD may simply be represented by

$$\text{WD} \simeq \exp\left[-0.15 \left(\sigma \frac{l}{2}\right)^2\right] \quad \text{for } \frac{s_2}{l} = 0.5$$

$$\text{WD} \simeq \exp\left[-0.24 \left(\sigma \frac{l}{2}\right)^{2.7}\right] \quad \text{for } \frac{s_2}{l} = 1.0 \,.$$

$$(5.138)$$

Figure 5.27 illustrates the errors on $D_{1,2}^m$ for $s_2/l = 0.50$ (a usual case) and $l/\eta = 2 - 4 - 6$. A von Karman–Saffman–Pao spectrum is used with $k_e \eta = 0.10$, hence $L/\eta = 12.8$. One of the main results is that measurements of the lateral derivatives require a correction even at low k_1 values. This point was already stressed by *Wyngaard* [5.169], *Antonia* et al. [5.170], *Antonia* and *Mi* [5.171], and *Ewing* et al. [5.167]. The last of these works provides clear plots of the joint effects of the wire length and separation for a Lin spectrum which rolls off faster than a Pao spectrum in the dissipation range. As practical information, an attenuation of the order of 10% appears on the transverse velocity derivatives when $l/\eta \simeq s_2/\eta \simeq 3$.

Spatial Resolution for an X Array in Isotropic Turbulence.
For X-wires, let us consider the case of two identical wires of length l, placed in the Ox_1x_2 plane, inclined at 45° to Ox_1, and separated by s_3 along Ox_3. Wire I should respond to $u_1 - u_2$, and wire II to $u_1 + u_2$ (Fig. 5.13). When taking into account the exact geometry of the probe, and making use of the Fourier decomposition, the responses are in reality

$$u_1^m - u_2^m$$
$$= \iiint \exp\left(+\frac{ik_3 s_3}{2}\right)$$

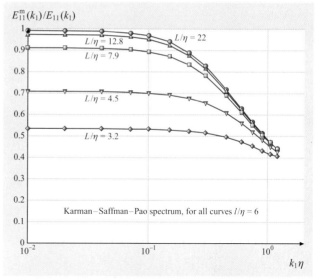

Fig. 5.25 Spatial resolution: the effect of the integral length scale L included in a Karman–Saffman–Pao spectrum on the one-dimensional velocity spectra, for a wire of fixed length

$$\times \frac{1}{l} \int_{-l/2}^{l/2} \exp[i(k_1 \xi_1 + k_2 \xi_2)] \, ds (\hat{u}_1 - \hat{u}_2) \, dk_1 \, dk_2 \, dk_3 \,,$$

$$u_1^m + u_2^m$$
$$= \iiint \exp\left(-\frac{ik_3 s_3}{2}\right)$$
$$\times \frac{1}{l} \int_{-l/2}^{l/2} \exp\left[i(k_1 \xi_1' + k_2 \xi_2')\right] ds' (\hat{u}_1 + \hat{u}_2) \, dk_1 \, dk_2 \, dk_3$$

$$(5.139)$$

or, after integration by ds and ds' along each wire,

$$u_1^m - u_2^m = \iiint (\hat{u}_1 - \hat{u}_2) \frac{\sin\left[(k_1 + k_2)l/(2\sqrt{2})\right]}{(k_1 + k_2)l/(2\sqrt{2})}$$
$$\times \exp\left(+\frac{ik_3 s_3}{2}\right) dk_1 \, dk_2 \, dk_3 \,,$$

$$u_1^m + u_2^m = \iiint (\hat{u}_1 + \hat{u}_2) \frac{\sin\left[(k_1 - k_2)l/(2\sqrt{2})\right]}{(k_1 - k_2)l/(2\sqrt{2})}$$
$$\times \exp\left(-\frac{ik_3 s_3}{2}\right) dk_1 \, dk_2 \, dk_3 \,. \quad (5.140)$$

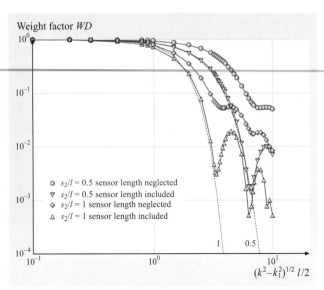

Fig. 5.26 Spatial resolution: the weighting function WD for two parallel wires and isotropic turbulence (5.136)

As above, spatial averaging along the wires reduce the readings by roughly the same amount even if different wavevectors enter into the wire responses: $k_1 + k_2$ for the first wire and $k_1 - k_2$ for the second. In addition, there is a mutual effect between u_1 and u_2. This effect is most easily shown when neglecting the wire length effect. Indeed, in Fourier space, the sum and difference of

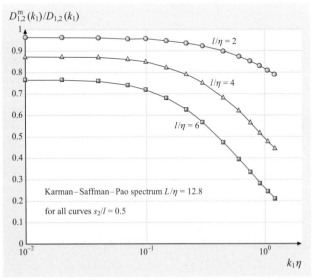

Fig. 5.27 Effect of wire length and wire separation on the one-dimensional $\partial u_1 / \partial x_2$ spectra

(5.140) yields

$$\hat{u}_1^m = \hat{u}_1 \cos\left(\frac{k_3 s_3}{2}\right) - i\hat{u}_2 \sin\left(\frac{k_3 s_3}{2}\right),$$

$$\hat{u}_2^m = \hat{u}_2 \cos\left(\frac{k_3 s_3}{2}\right) - i\hat{u}_1 \sin\left(\frac{k_3 s_3}{2}\right), \quad (5.141)$$

which clearly shows that $\hat{u}_2$ enters into $\hat{u}_1^m$ and that $\hat{u}_1$ enters into $\hat{u}_2^m$. The measured spectra are therefore

$$\Phi_{11}^m = \cos^2\left(\frac{k_3 s_3}{2}\right) \Phi_{11} + \sin^2\left(\frac{k_3 s_3}{2}\right) \Phi_{22}$$

$$\Phi_{22}^m = \cos^2\left(\frac{k_3 s_3}{2}\right) \Phi_{22} + \sin^2\left(\frac{k_3 s_3}{2}\right) \Phi_{11}.$$

(5.142)

Numerical 2-D integration is here required to treat different Φ_{ij} forms. *Wyngaard* [5.165] used a Pao spectrum and an X-probe with $l/\eta = 1, 3$ and 10 and $s_3/\eta = 0.5$ and 1. He found that the lateral component u_2' is hardly affected by the wire separation and that the longitudinal component u_1' can be increased for small wire separations. It therefore appears that accurate measurements of the longitudinal component would best be made using a single wire, reserving an X-array for the lateral component.

Some Experimental Results. Many experiments have been conducted in grid turbulence or in flows with characteristics close to isotropy, such as jets, far wakes, central zones of channel/pipe flows, and upper parts of the atmospheric boundary layer. The objective was either to improve the probe design or to correct measurements when inappropriate probes were used.

Wyngaard [5.165] conducted experiments in atmospheric turbulence and also in a curved mixing layer, with $\eta \simeq 0.06$ mm and $l/\eta = 2, 4$ and 6. The corrections he developed for the isotropic case appear to provide reasonable support for his data analysis.

Derksen and *Azad* [5.67] found that the isotropic correction of Wyngaard was also satisfactory for measurements in a pipe flow at $x_2/R = 0.10$ from the wall, $x_2^+ = 150$, with around 15 different wires with $d = 3.6-5.4 \,\mu\text{m}$, $l = 0.3-2.8$ mm, and $l/\eta = 3-20$.

Elsner et al. [5.168] reported systematic measurements in grid turbulence for spectra and kinetic-energy dissipation. In those experiments $\eta = 0.25$ mm, and $l = 0.5-3.9$ mm. The mean dissipation rate for longitudinal fluctuations ϵ_1 was equivalently obtained from the second moment of the one-dimensional spectra and from the time derivative of the longitudinal velocity fluctuations. Some typical errors are -30% if $l/\eta = 10$ and

-10% if $l/\eta = 2$. A convenient linear regression line of ϵ_l versus the wire length was obtained, giving the true turbulent energy dissipation at an extrapolated zero wire length.

Zhou et al. [5.173] conducted extensive investigations of the kinetic and thermal dissipation rates in grid turbulence, with $\eta = 0.32 – 0.73$ mm. Several configurations of X-wires and pairs of parallel wires were used with $l = 0.5$ mm, $l/d \simeq 200$, s_2/η and s_3/η in the range 2.8–7.8. After correction, the theoretical value $\overline{(\partial u_1/\partial x_2)^2} = 2/15$, with all terms normalized by the Kolmogorov scales, was recovered except for wire separations smaller than 3η, where aerodynamic and thermal interferences were claimed to be significant.

Antonia and *Mi* [5.171] considered a heated jet and also revisited the case of atmospheric turbulence studied by *Wyngaard* [5.165]. The corrections implying isotropy appeared to provide reasonable support for the analysis.

Browne et al. [5.174] considered the far wake of a cylinder and used an X wire made of two wires whose separation can be accurately set and measured. The conditions were $\eta = 0.45$ mm, $l = 0.75$ mm, and $s_3/\eta = 0.38–7.2$. They observed that $\overline{u_1^2}$ decreases and that $\overline{u_2^2}$ increases when the wire separation increases: a result which differs from *Wyngaard* [5.165]. However, in that type of flow, isotropy is only approached at high wavenumbers as pointed out by *Antonia* et al. [5.170]. The large-scale structures of the flow exhibit strong organization which could affect the results.

Spatial Resolution in Anisotropic Turbulence. Analytical developments are obviously not possible in highly anisotropic turbulent fields, such as the wall region of turbulent boundary layers. Systematic experimental investigations or computer simulations are therefore needed. An advantage of computer simulations is that the cooling of the wire by the three velocity components can be included.

Ligrani and *Bradshaw* [5.175] reported extensive measurements in the region $x_2^+ = 5–100$ in a boundary layer whose freestream velocity was 7.3 m/s with wire diameters in the range $d = 0.62–5$ μm with $l/d = 260–600$ and $l^+ = 3.3–60$. The superscript '+' denotes normalization by the wall variables, i.e., friction velocity u_τ and kinematic viscosity ν. The greatest effect concerns the maximum value of $(\overline{u_1^2})^{1/2}/u_\tau$ which was found to decrease from 2.8 to 2.0 when the wire length increased. Skewness and flatness factors were less affected. The probability density function of the longitudinal velocity was also investigated; a slight decrease was observed at large negative u_1 values for long wires. A stronger effect was noted by *Johansson* and *Alfredsson* [5.176] in a channel flow. Large negative $u_1(t)$ amplitudes correspond to ejection events that are naturally at small scales and hence are strongly affected by spatial averaging.

Klewicki and *Falco* [5.177] made an extensive compilation of previous works regarding the statistics of small-scale structures in boundary layers. In addition, they performed measurements using a probe for which the separation s_2 between the two parallel sensors could be varied systematically. In particular, the variance of $\partial u_1/\partial x_2$ in the buffer layer, at $x_2^+ = 38$ or 53, was found to be attenuated by only about 5% for $s_2/\eta \simeq 3$ but the attenuation increased rapidly to about 35% for $s_2/\eta \simeq 6$. In these experiments $\overline{U}_1 = 0.60–23$ m/s in the freestream and the Reynolds number based on the momentum thickness was $R_\theta = 1010–4850$.

Moin and *Spalart* [5.178] launched the use of direct numerical simulation (DNS) databases to evaluate X-wire responses. These early computations concerned a channel flow with Re = 3300 and $\text{Re}_h^+ = 180$. However, only one wire separation, $s_3/\eta = 1$, was used and wires were assimilated to points.

Suzuki and *Kasagi* [5.172] extended the computations of Moin and Spalart to different types of probes. In these computations $\text{Re}_h^+ = 150$, $l^+ = \sqrt{2}(1-10)$, and $s_3^+ = 1-10$. A single wire placed near the maximum streamwise turbulent intensity was found to give a ra-

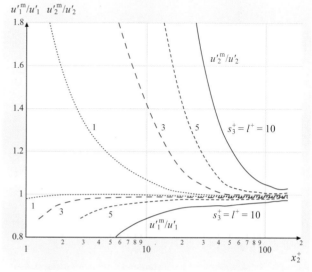

Fig. 5.28 Spatial resolution of an X-wire close to a wall: errors on the rms values of u_1' and u_2' (after *Suzuki* and *Kasagi* [5.172])

tio between the measured and exact value of 0.95 for $l^+ = 10$, and 0.90 for $l^+ = 20$. Figure 5.28 reports the corresponding ratios for u'_1 and u'_2 for conventional X-probes (wires inclined at $\pm 45°$) and several l^+ and s_3^+ values. The u'_2 component appears to be much more affected than u'_1, which is probably a strong anisotropy effect as this tendency is the opposite of that reported by Wyngaard.

Antonia et al. [5.179, 180] analyzed the lateral velocity derivatives obtained with two parallel wires in a channel flow, both experimentally and numerically. The conditions were $Re_h = 3300$, $Re_h^+ \simeq 180$, $d = 1.3$ μm, $l = 0.3$ mm, $l/d = 230$, and $\eta = 0.2$ mm at $x_2^+ = 2$. In the wall region, for x_2^+ in the range 5–40, only a wire spacing of $s_3^+ = 2$–4 was found to be adequate.

Close to walls, many other characteristics are also of great interest such as the components of the vorticity vector. Such investigations require the use of special probes made of between four and nine sensors. Their responses are extensively analyzed by *Wallace* and *Foss* [5.81] and later on in this Handbook. In addition to the spatial resolution regarding the velocity fluctuations in the probe volume, the velocity gradient effect, and the aerodynamic perturbations have also to be considered.

Finally, the additional cooling of sensors by nearby walls is also an important topic. It is considered later in this Handbook.

5.2.3 Anemometer Electronics

Constant-Current Anemometer (CCA)

Historically, CCA was successfully used as early as 1920 to investigate wind velocity and wind direction in the atmosphere by *Huguenard* et al. [5.181] and *Simmons* and *Bailey* [5.182]. Its fast response compared to wind vanes or cup anemometers was a strong motivation for its adoption. Flows through pipes or orifices were also attempted by *King* [5.183], *Thomas* [5.184], and *Richardson* [5.185]. The use of several wires simultaneously was of great benefit to treat large directional changes and even flow reversals. The extension to flows with rapid time changes was made possible by the pioneering work of *Dryden* and *Kuethe* [5.186] who showed that an RC circuit could be incorporated into the amplifier to compensate for the thermal inertia of the wire.

The simplest circuit for a CCA is shown in Fig. 5.29. VS is a stable direct-current (DC) low-noise voltage source, and R_B a large ballast resistor that keeps the current I_w in the wire fairly constant despite changes in the wire resistance. An adjustable resistor placed in series with R_B permits the adjustment of I_w, and the potential drop across a precision resistor also in series with the hot-wire yields I_w. A more-functional circuit places the wire in a Wheatstone bridge. The bridge allows the measurement of the resistance of the cold wire, adjustment of the wire overheat, and also, possibly, the correction of the cable inductance by adding an appropriate capacitance or inductance in a specific arm of the bridge, similar to a CTA. A regulated constant-current supply can also be used in place of a voltage source; very stable ones are commercially available.

The amplifier, which has to be high gain and ultralow noise, is always made of several stages, and the RC compensating circuit with a gain increasing with frequency often covers two stages. The frequency responses offered by commercial CCA systems can reach 300 kHz. Some higher-frequency responses, up to 500 kHz, have recently been reported by *Repkov* et al. [5.187]. These authors also added to their circuit a microprocessor which permits measurements in a short-time-duration wind tunnel with protection against wire burn-out.

The compensated CCA output signal $e'^\star(t)$ is retrieved from the measured signal $e'(t)$ according to (5.102)

$$e'^\star(t) = e'(t) + M_w^{CCA} \frac{de'}{dt} \quad \text{with } M_w^{CCA} = M_w^{CC}.$$

(5.143)

The adjustment of M_w^{CCA} is done in situ, using a square-wave-intensity signal superposed onto the heating current. *Dryden* and *Kuethe* [5.186] and more recently *Perry* [5.49] have shown that step changes in U, T_a, or I_w are equivalent as long as the fluctuations stay small. The correct adjustment is obtained when a square output is restored. Multiple time records and

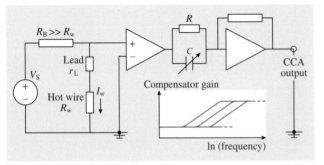

Fig. 5.29 Block diagram of a constant-current anemometer (CCA). RC is the adjustable compensating circuit integrated in a high-gain amplifier

phase averaging greatly improve the accuracy of the adjustment.

The static sensitivity coefficients S^{CCA} expressed at the CCA output are such that

$$\left(\frac{e'}{\overline{E}}\right)_{CCA} = S_{U_1}^{CCA} \frac{u'_1}{\overline{U}_1} \quad \text{and}$$

$$\left(\frac{e'}{\overline{E}}\right)_{CCA} = S_{T_a}^{CCA} \frac{\theta'_a}{\overline{T}_a} . \tag{5.144}$$

These S^{CCA} are easily connected to the wire sensitivity coefficients S^{CC} defined in Sect. 5.2.2. For a CCA working without a bridge, there is the simple relation

$$S^{CCA} = S^{CC} . \tag{5.145}$$

For a CCA involving a bridge, with R_1 in the top arm on the side of the hot-wire, one has

$$S^{CCA} = S^{CC} \frac{\overline{R}_w}{\overline{R}_w + r_L + R_1} . \tag{5.146}$$

The electronic noise of a CCA circuit is easy to estimate when the wire is directly connected to the amplifier input through a cable lead of resistance r_L. Neglecting the stray noise picked up through the inductance and capacity of the wire and its lead (twisted cables are advised, as well as separate circuits for the wire heating and the transmitted wire signal), the noise sources are the thermal noise of the wire and lead, and the input voltage noise of the amplifier. The signal-to-noise ratio

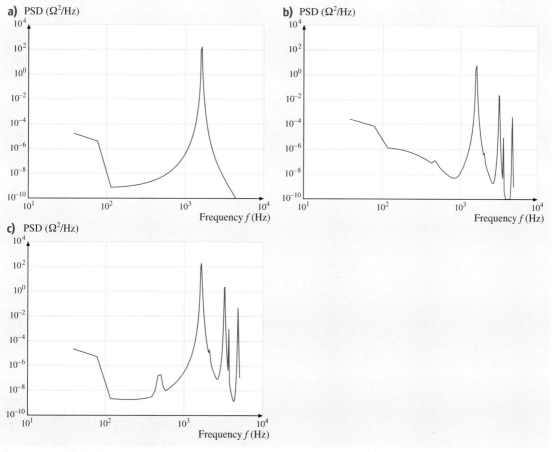

Fig. 5.30a–c Higher harmonics generated by parametric excitation in a CCA: (**a**) PSD of the input signal $R_w^\star$, (**b**) PSD of the solution R_w of (5.100), (**c**) PSD of the linearly corrected CCA output r'_w based on (5.102). $R_w^\star = 10 + \sqrt{2}\cos\omega t$, $M_w\omega = 5$, skewness factors are 0.0002, 0.0200, and -0.2724 for $R_w^\star$, R_w, and r'_w, respectively

(SNR) at frequency f is therefore of the form

$$\text{SNR}|^{\text{CCA}} = \frac{S_w^{\text{CCA}}}{2\pi f M_w^{\text{CCA}}} \frac{\overline{E}}{\hat{e}_n} \frac{\hat{u}_1}{\overline{U}_1}, \quad (5.147)$$

where $\hat{e}_n$ is the total resulting noise in V/$\sqrt{\text{Hz}}$, as usual in noise analysis, and $\hat{u}_1$ the turbulence spectrum in ms^{-1}/$\sqrt{\text{Hz}}$. In $\hat{e}_n$, the most important term is the amplifier noise.

If the wire is placed in a Wheatstone bridge, the SNR estimate is more complicated because it involves all the resistances constituting the bridge. *Kidron* [5.188] and *Freymuth* [5.189] did the complete analysis, and managed to show that a CCA circuit behaves similarly to a CTA circuit. Expression (5.157), which is indicated later on in Sect. 5.2.3, can then be used. *Bestion* et al. [5.190] made systematic comparisons for supersonic flows and observed no significant differences between SNR$|^{\text{CCA}}$ and SNR$|^{\text{CTA}}$.

Finally, when the incident fluctuations are large, the higher harmonics generated by parametric excitation at the wire level (Sect. 5.2.2), are not at all suppressed by the compensating circuit. Even worse, they are amplified as if they were true turbulence. Figure 5.30 represents the power spectral density (PSD), for the input signal $R_w^\star$, for the solution R_w of (5.100), and for the CCA output signal $Q(t)$ corrected by applying the linear thermal lag correction defined by (5.143). In that assumed corrected CCA output $Q(t)$, the second harmonic is of the order of 10^{-2} relative to the fundamental, and overly large negative values on the velocity skewness appear, of the order of -0.27 for amplitude fluctuations of 15% and $M_w^{\text{CCA}}\omega \simeq 5$. These results agree with those obtained by *Comte-Bellot* and *Schon* [5.151] with an analog computer.

Bremhorst and *Graham* [5.191] also considered the case of large velocity changes. However, their approach remained based on a first-order system, only taking into account the M_w^{CCA} changes with the velocity. Second and higher harmonics could not therefore be observed in their analysis.

Constant-Temperature Anemometer (CTA)

A typical CTA circuit is shown in Fig. 5.31. The two vital elements are the Wheatstone bridge and the feedback amplifier. The sensor is placed in one arm of the bridge. The voltage difference across the bridge is a measure of the change in wire resistance with the incident velocity or with the flow temperature, and it constitutes the input to the operational amplifier. The designed amplifier has an output current feedback to the top of the bridge, and hence to the wire to restore its resistance. For example, the top bridge voltage E increases when the flow velocity increases or when the fluid temperature decreases, hence increasing the current intensity in the wire. The feedback amplifier usually consists of several stages of voltage gain and a unity-gain current booster.

The resistance ratio R_2/R_1 is called the bridge ratio. It is often chosen larger than one, e.g., 5, 10, or 20. The amplifier current is thus mainly introduced into the active arm containing the wire rather than into the opposite passive arm. Furthermore, the top resistance R_1 is often chosen to be large, $R_1 = 10\,\Omega$, $20\,\Omega$, or even $50\,\Omega$, to increase the top bridge voltage E and its fluctuations, and to ease their acquisition. However, for any asymmetric bridge the Wheatstone bridge accuracy is decreased, as explained in basic textbooks [5.192]. The wire overheat is set up by adjusting resistance R_3.

In principle, since R_w is maintained constant, the thermal lag of the wire is automatically suppressed (5.39). Such an advantage was recognized as very useful at an early stage, by *Ziegler* in 1934 [5.193], *Weske* in 1943 [5.194], or *Ossofsky* in 1948 [5.195]. However, it was not until the mid 1960s that reliable CTA devices became available with the development of high-

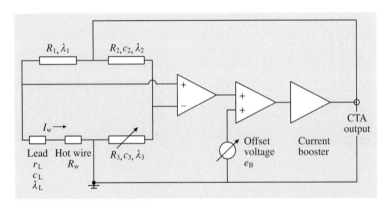

Fig. 5.31 Block diagram of a constant-temperature anemometer (CTA). The wire overheat is adjusted by modifying resistance R_3. The adjustable capacities and inductances placed in arms 2 and 3 of the bridge, c_2, λ_2, c_3, λ_3, and the offset voltage e_B, make the CTA stable

quality solid-state operational amplifiers. Many circuits were then suggested, some even without a Wheatstone bridge [5.196], and many aspects of the control theory involved in the CTA response were analyzed over the years by *Freymuth* in 1967 [5.197], *Davis* and *Davies* in 1968 [5.198], *Perry* and *Morrison* in 1971 [5.199], *Wood* in 1975 [5.200], *Freymuth* again in 1977 and 1998 [5.201, 202], and *Watmuff* in 1995 [5.203]. Possibilities offered by recent computer-controlled CTA are described by *Fingerson* [5.204] or *Jorgensen* [5.205].

Let us now analyze the CTA performance in more detail. Concerning the frequency response, it is first rewarding to examine the case of an ideal CT anemometer, without any additional circuits, as done by *Hinze* [5.47]. This is just a first-order system approach. From the change of the wire resistance, the error signal to the amplifier input can be computed taking into account the time constant of the wire and the amplifier response involving a closed loop. This yields for the time constant of the wire under CT operation

$$M_w^{CTA} = \frac{M_w}{1 + 2G\bar{a}_w \bar{R}_w g_{tr}}, \quad (5.148)$$

where g_{tr} is the transconductance of the bridge, G the amplifier gain, M_w the time constant of the wire alone, and $\bar{a}_w$ the overheat ratio. *Comte-Bellot* [5.54, 206] computed a similar expression, which shows the roles of the resistances R_1, R_2, and R_3 of the bridge

$$M_w^{CTA} = \frac{M_w}{1 + 2G\bar{a}_w \frac{\bar{R}_w}{\bar{R}_w + R_1} \frac{R_2}{R_2 + R_3}}. \quad (5.149)$$

The time constant in constant-temperature operation is therefore much smaller than that in constant-current operation, as G can be on the order of 1000. However, the wire overheat ratio $\bar{a}_w$ has to be sufficiently high, and a value of $\bar{a}_w = 0.80$ is often advised by the manufacturers. At low overheat ratios, when $\bar{a}_w \to 0$, the time constant of the anemometer system approaches that of the wire M_w, and this precludes measurements as no compensation is provided by the CTA circuit. Expression (5.149) also shows that a symmetric bridge yields the smallest possible value for M_w^{CTA}, thus permitting the best frequency response.

A realistic analysis of the CTA must, of course, go beyond that of a first-order system. The reason is that many parameters play a role in the dynamic response, and all are essential to the CTA stability and high-frequency response. It is useful to list them:

1. the time response of the hot-wire of the form $(1 + M_w s)^{-1}$ with a Laplace transform ($s = i\omega$); this supposes small-amplitude fluctuations [5.45] (Sect. 5.2.2),
2. the resistance of the hot-wire prongs, around $0.50\,\Omega$, and the resistance r_L of the connecting cable, around $0.20\,\Omega/m$,
3. the inductance of the coaxial connecting cable λ_L, around $0.40\,\mu H/m$,
4. the capacitance of the coaxial connecting cable c_L, around $60\,pF/m$,
5. the bridge resistances,
6. the trim inductors introduced in the bridge,
7. the trim capacitors introduced in the bridge,
8. the roll-off frequency and gain of the feedback amplifiers, the gain can be frequency dependent,
9. the offset value applied to the last feedback amplifier.

Let us remark that λ_L has a greater effect than c_L, as $1 + i\omega\lambda_L/R_w \simeq 1 + 0.3i$ whereas $1 - i\omega R_w c_L \simeq 1.00$, for a 5 m-long cable and a frequency of 100 kHz, using the values indicated in 3 and 4 in the above list. Furthermore, λ_L can be balanced out by a capacitor c_2 placed in parallel to resistor R_2 in the opposite arm of the bridge, and such that $c_2 = \lambda_L/R_1 R_3$ according to the alternating-current (AC) equilibrium condition for a Maxwell bridge [5.192]. Another possibility would be to place an inductance in series with the top resistance R_1, then $\lambda_1 = R_2\lambda_L/R_3$, or in series with resistance R_3, then $\lambda_3 = R_2\lambda_L/R_1$. Trim capacitors are easier to manufacture, so that the solution using c_2 is usually preferred.

Freymuth [5.197, 201, 207, 208] made significant contributions taking into account points 1, 7, 8 and 9. For small-amplitude fluctuations, he showed that a CTA is well represented by a third-order system with a differential equation of the form

$$\frac{M_w M''}{G}\frac{d^3 e'}{dt^3} + \frac{M_w}{G}\left(M' - G\frac{\bar{R}_w}{\bar{R}_w + R_1}R_2 c_2\right)\frac{d^2 e'}{dt^2}$$
$$+ \frac{M_w}{G}\frac{e_B}{\bar{E}}\frac{de'}{dt} + e'$$
$$= s_u u' + s_{sq} sq', \quad (5.150)$$

where M_w is the time constant of the wire, M' and M'' are time constants that characterize the frequency response of the feedback circuit (M' in s, and M'' in s^2), G is the amplifier gain, e_B is the offset voltage, c_2 is a capacitance in the arm opposite to the wire, s_u is the dimensional sensitivity to a velocity change, and s_{sq} is the sensitivity to a square wave applied to the bridge (the derivative of sq is also involved), e' is the voltage fluctuation at the top of the bridge, $\bar{E}$ is the mean voltage at the top of the bridge, and t is the time. Let us recall that a first-order system exhibits an exponential decay, a second-order system an oscillating decay, and a third-order system both types of decay simultaneously.

For a given amplifier and a given wire working in fixed conditions, M', M'', and M_w are fixed. Therefore, the only way to adjust the CTA dynamic response is to act on G, e_B, and c_2. In particular e_B, which is around 5–10 mV, is present in a damping term. *It is also advisable to make all adjustments at the expected maximum flow velocity. This is because M_w decreases when U increases, hence the damping term in front of de'/dt decreases, and the system may oscillate.*

Freymuth [5.201] also analyzed the CTA response for a sinusoidal forcing velocity and for an electronic step voltage and showed that the best optimized cut-off frequency, f_c is given by

$$f_c = \frac{1}{2\pi}\left(\frac{G}{M'' M_w}\right)^{1/3} \tag{5.151}$$

and that the transient response to an electronic step voltage should present a high-amplitude pulse followed by an undershoot without oscillation. In practice, this electronic test is performed by applying a square-wave signal to the bridge; Fig. 5.32 illustrates this response for a wire and a film. Recently, *Khoo* et al. [5.209] and *Teo* et al. [5.210] reported many results for wires and films, using either voltage or velocity perturbations.

The time constants M' and M'' in (5.150) were interpreted by *Freymuth* [5.211] in terms of a two-stage feedback amplifier with each stage characterized by a first-order frequency response. The gains of these stages were denoted by G_1 and G_2, their first-order time constants by M_1 and M_2, and their gain–bandwidth products were such that $GBP_1 = G_1/2\pi M_1$ and $GBP_2 = G_2/2\pi M_2$. Then, $G = G_1 G_2$, $M' = M_1 + M_2$, and $M'' = M_1.M_2$, and f_c becomes

$$f_c = (f_w GBP_1 GBP_2)^{1/3}. \tag{5.152}$$

For a wire frequency f_w that is usually of the order of 100 Hz and gain–bandwidth products of the order of 3 MHz, $f_c \simeq 100$ kHz. This is quite close to the maximum CTA frequency announced by manufacturers. The offset voltage e_B is also clearly applied to the second amplifier.

Watmuff [5.203] performed a systematic investigation of all points 1 to 9. Using Laplace transforms, he started with a model deliberately of higher order than is anticipated, with a transfer function of the form

$$P(s) = P_7 s^7 + P_6 s^6 + P_5 s^5 + P_4 s^4 \\ + P_3 s^3 + P_2 s^2 + P_1 s + P_0. \tag{5.153}$$

He used a compact matrix notation and conducted computer simulations to explore the behavior of representative configurations. The polynomial form for the poles of (5.153) is really seventh order when all points 1 to 9 are taken into account. It becomes fifth order when points 4 and 7 are not taken into account. Interestingly, Watmuff observed that his fifth-order system has only three dominant modes, a pair of complex poles and a simple one. The two higher simple modes remain well behaved and are beyond the frequency range of interest. This aspect confirms therefore *Freymuth*'s results [5.201, 202].

Watmuff [5.203] also provided a good understanding of the square wave test used to optimize a CTA. In particular, he showed that a satisfactory response cannot be obtained by adjusting the offset when the cable inductance is too high. Another interesting result concerns the subminiature wires that often lead to CTA instabilities.

Weiss et al. [5.212] suggested the use of the Fourier transform of the square-wave response to obtain the amplitude of the transfer function of the CTA, and to correct

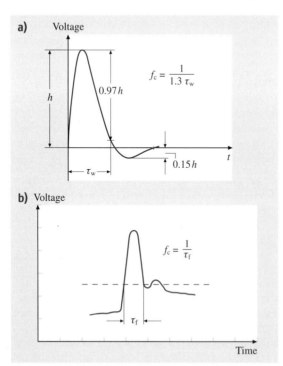

Fig. 5.32a,b The square-wave test response of a hot-wire probe (**a**) and hot-film probe (**b**) operated by a CTA (after *Bruun* [5.52])

the measured spectra during postprocessing. Figure 5.33 illustrates how the frequency response can be substantially extended, up to around 300 kHz. Even the case of small overheat ratios can be treated satisfactorily. The technique was used by *Weiss* et al. [5.141, 213] in a Mach 2.5 boundary layer and allowed the corrected CTA spectra to agree with the CCA and CVA measurements. This was not the case in previous investigations [5.127, 214].

The sensitivity coefficients S^{CTA} expressed at the CTA output (top of the bridge) are defined by

$$\left(\frac{e'}{\overline{E}}\right)_{CTA} = S^{CTA}_{U_1} \frac{u'_1}{\overline{U}_1} \quad \text{and} \quad \left(\frac{e'}{\overline{E}}\right)_{CTA} = S^{CTA}_{T_a} \frac{\theta'_a}{\overline{T}_a} \,. \tag{5.154}$$

for velocity or temperature fluctuations. They are such that

$$S^{CTA} = S^{CT} \tag{5.155}$$

whatever the cable resistance r_L and the top bridge resistance R_1. Indeed one has the simple equation

$$I_w = \frac{E}{(R_1 + R_w + r_L)} \,, \quad \text{which gives} \quad \frac{dI_w}{I_w} = \frac{dE}{E} \,. \tag{5.156}$$

This simplification should not let one believe that any cable can be used, as cable capacitance and inductance have to be taken into account in the CTA dynamic response. Optimized cables are therefore often advised by CTA manufacturers.

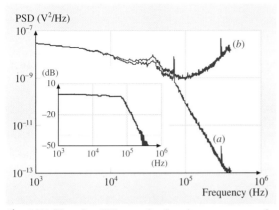

Fig. 5.33 Use of a CTA transfer function to measure the freestream power spectra in a Mach 2.5 wind tunnel: (*a*) original spectrum, (*b*) corrected spectrum, insert corresponding to bridge frequency response. The CTA noise in f^2 is only visible on curve (*b*) (after *Weiss* et al. [5.212])

Because $e_B \neq 0$, the Wheatstone bridge is not perfectly balanced. *Weiss* [5.215] analyzed how this could affect S^{CTA}_U and $S^{CTA}_{T_a}$. He showed that both S^{CTA} values are lower than those of a perfectly balanced bridge, and that the degree of discrepancy increases with increasing offset voltage and decreasing overheat ratio. However, a direct calibration procedure may take this effect into account.

The electronic noise of a CTA has been analyzed by *Kidron* [5.188] and *Freymuth* [5.189]. The signal-to-noise ratio (SNR), at frequency f, may be expressed as

$$\text{SNR}|^{CTA} = \frac{1}{2} \frac{1}{2\pi f M_w} \frac{m \overline{a}_w}{(1+m)^2} \frac{\overline{E}}{\hat{e}_n} \frac{\hat{u}}{\overline{U}} \,, \tag{5.157}$$

where m denotes the ratio R_1/R_w, $\hat{u}$ the spectrum of the fluctuating velocity u', and $\hat{e}_n$ the spectrum of the equivalent noise e'_n of the bridge and amplifier occurring at the amplifier input. The spectrum $\hat{u}$ is expressed in $\text{ms}^{-1}/\sqrt{\text{Hz}}$ and the spectrum $\hat{e}_n$ in $\text{V}/\sqrt{\text{Hz}}$. Some adaptation has been made from the original paper [5.201] to transform the M^{CTA}_w value, which involved the bridge, into M_w, which characterizes the wire itself. Equation (5.157) shows that a high SNR requires $m \simeq 1$ in addition to a high $\overline{a}_w$ value.

Regarding the CTA output noise spectrum itself, which is the square of the denominator in (5.157), a growth in f^2 is expected. This behavior was observed only recently as experiments require very low-turbulence freestream wind tunnels and CTA with very large bandwidths. *Saddoughi* and *Veeravalli* [5.216] reported experiments in subsonic flows, and *Weiss* et al. [5.212] [5.141] in the freestream of a Mach 2.5 wind tunnel. The f^2 growth was followed by a final decay which was very abrupt as the high-frequency amplitude response of a CTA circuit described by (5.150) varies as f^{-3}, which corresponds to a Butterworth filter of third order [5.201].

Although the arguments presented in this section mostly concern hot-wires, they can apply to hot-films in the case of a bulk response of the film and its quartz core. When this is not the case, the high-frequency f^{-3} law for hot-wires becomes a $f^{-2/5}$ law for hot-films, according to the analysis performed by *Freymuth* [5.217] and the experimental data collected by *Fingerson* and *Freymuth* [5.50] and *Fingerson* [5.204].

Last but not least, *Freymuth* [5.217] carried out a nonlinear analysis of the CTA response. He used the full equation (i.e., not linearized) for the amplifier and the wire responses. This led him to a nonlinear third-order equation with variable coefficients that he solved

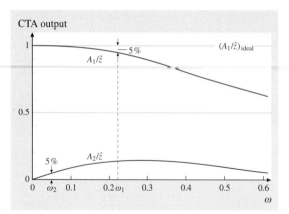

Fig. 5.34 Normalized fundamental component $A_1/\hat{z}$ and second harmonic $A_2/\hat{z}$ of a CTA output for an input function $z = \hat{z}\sin(\omega t)$ with $\hat{z} = 0.4$ (after *Freymuth* [5.217])

by iteration. Figure 5.34 shows the second harmonic A_2 generated in the case of a relatively large sinusoidal change in the flow velocity. Since the numerical estimate of Freymuth is rather complex, the problem could be revisited because the velocity signal and the electronic signal could differ from each other when nonlinearities are included.

Constant-Voltage Anemometer (CVA)

The block diagram of a CVA is shown in Fig. 5.35. This recent technique was conceived and developed by *Sarma* [5.148, 218]. Prototypes have been tested in several investigations mostly dealing with high-speed transitional and turbulent boundary layers, in the laboratory [5.127, 150, 219–221] and in flight using remote control [5.88].

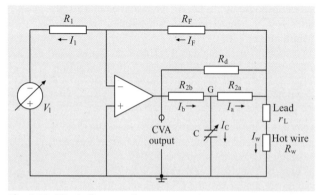

Fig. 5.35 Block diagram of a constant-voltage anemometer (CVA). The output signal is large due to the large resistance $R_{2a} + R_{2b} = R_2 \simeq 100\,\Omega$

The sensor, either a wire or a film, is placed in the feedback loop of the operational amplifier. The voltage across it, $V_w = R_w I_w$, can be adjusted and measured, and once set it is independent of the hot-wire resistance R_w. Any change in wire resistance due to fluid flow produces a change of current in the wire, whose only path is through resistance R_2. The anemometer output V_s is closely related to the voltage drop in resistor R_2. The circuit parts involving R_{2a}, R_{2b}, and C with $R_{2a} + R_{2b} = R_2$ permit a linear frequency increase of the CVA output and therefore the thermal lag compensation of the wire. The time constant T_C associated with this compensation is

$$T_C = \frac{R_{2a} R_{2b}}{R_2} C \,. \tag{5.158}$$

The Ohm–Kirchhoff laws applied to the network give

$$\begin{aligned}
-R_1 I_1 + V_1 &= 0 \,, \\
V_w - R_F I_F &= 0 \,, \\
I_F &= I_1 \,, \\
V_w &= R_w I_w \,, \\
V_s - R_{2b} I_b &= V_G \,, \\
V_G - R_{2a} I_a &= V_w \,, \\
V_G - \frac{I_C}{\mathrm{i} C \omega} &= 0 \,, \\
I_b &= I_a + I_C \,, \\
I_a &= I_w + I_F \,.
\end{aligned} \tag{5.159}$$

An infinite gain–bandwidth product is assumed for the operational amplifier, the resistor R_d, which is large compared to R_2, is neglected. The analysis including lead resistance r_L will be considered later on in this text. From (5.159) one can easily deduce

$$V_w = V_1 \frac{R_F}{R_1} \tag{5.160}$$

which shows that the wire voltage V_w is indeed constant, as V_1 is a stable reference voltage. The adjustment of V_w is done by adjusting V_1, manually or in a automated way by a PC. After some algebra, one can also obtain from (5.159) the CVA output signal V_s at frequency ω

$$\frac{V_s}{V_w} = 1 - \frac{R_2}{R_{2a}} + R_2 \left(\frac{1}{R_w} + \frac{1}{R_F} + \frac{1}{R_{2a}} \right)(1 + \mathrm{i}\omega T_C) \tag{5.161}$$

with T_C given by (5.158). Also,

$$\overline{V}_s = \left[1 + R_2 \left(\frac{1}{R_w} + \frac{1}{R_F} \right) \right] V_w \,, \tag{5.162}$$

permitting one to deduce $\overline{R}_w$ from $\overline{V}_s$ by

$$\overline{R}_w = R_2 \left[\frac{\overline{V}_s}{\overline{V}_w} - \left(1 + \frac{R_2}{R_F}\right) \right]^{-1}. \quad (5.163)$$

The output change v'_s due to a flow change at frequency ω, is obtained from (5.161) by derivation with respect to R_w, if the fluctuation amplitudes are small. Taking also into account the wire thermal lag M_w^{CVA} in CV operation, this gives

$$v'_s = -R_2 \overline{I}_w \frac{r'_w}{R_w} \frac{1 + i\omega T_C}{1 + i\omega M_w^{CVA}}$$

with $M_w^{CCV} = M_w^{CV}$. $\quad (5.164)$

The factor $R_2 \overline{I}_w$ clearly shows that the CVA output is very large, as resistance R_2 can be chosen to be large, of the order of 100 Ω. A factor of 20 can therefore be gained compared to a CCA, and a factor of 2 compared to a CTA with a 50 Ω top resistance. The factor $1 + i\omega T_C$ shows the increase in gain of the CVA output with frequency. Resistor R_d, which has been neglected, has a damping role to limit that increase. Details are available in *Sarma* [5.148] and *Kegerise* and *Spina* [5.222, 223]. In addition, a high-order low-pass Butterworth filter is placed just before the CVA output, and below its frequency limit, around 650 kHz, the CVA fluctuating output $e'(t)$ is identical to v'_s, and for the mean values, $\overline{E} = \overline{V}_s$.

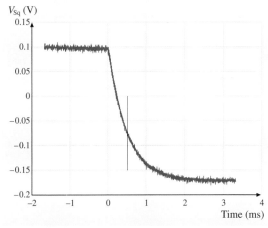

Fig. 5.36 Time trace of a 16 Hz square-wave signal applied to a 5 μm tungsten wire placed in the potential core of a $U = 15$ m/s jet with $\overline{a}_w = 0.25$. The curve shows the quality of the exponential fit. The *vertical line* corresponds to a 63% reduction and indicates the M_w^{CVA} value, 0.48 ms in the present case

As a whole, the system is of order one and the gain demand in CV operation is smaller than in CC operation, as $M_w^{CVA} < M_w^{CCA}$, so that a CVA can be stable and have a high frequency response.

The time constant M_w^{CVA} is measured in situ using a square-wave test and an auxiliary circuit developed by *Sarma* and *Lankes* [5.224]. An example is shown in Fig. 5.36.

In order to correct for M_w^{CVA}, one can select a T_C value which provides for the exact $1 + i\omega M_w^{CVA}$ compensation in (5.164). On current prototypes around 40 values are available [5.225]. However, a drawback is that the bandwidth of the CVA decreases when T_C increases, as the gain–bandwidth product of an operational amplifier is fixed. Another approach is therefore preferred. It consists of making a partial hardware correction in the CVA during the experiments, with a fixed T_C value, and to complete the correction later, during data processing, as suggested by *Kuppa* et al. [5.226] and *Sarma* et al. [5.225]. Denoting these two signals by $e'_{raw}(t)$ and $e'_{corr}(t)$, respectively, *Comte-Bellot* and *Sarma* [5.127] explained the relation to apply to $e'_{raw}(t)$ to restore $e'_{corr}(t)$ in terms of the i-th and $(i-1)$-th samples acquired at a sampling frequency of f_s

$$e'_{corr}(i) = \frac{e'_{raw}(i) + f_s \{ M_w^{CVA} [e'_{raw}(i) - e'_{raw}(i-1)] + T_C e'_{corr}(i-1) \}}{1 + T_C f_s}. \quad (5.165)$$

This two-step procedure has several advantages:

1. it increases the productivity during data acquisition,
2. it ensures a constant frequency bandwidth whatever the flow conditions or wire overheat, and
3. it permits slightly different M_w^{CVA} values to be tested within experimental accuracy (±2%).

A point, however, not yet investigated is how high the sampling frequency has to be in order to allow a first-order scheme. Further investigation on this topic is encouraged.

The CVA output sensitivity coefficients to flow changes, denoted by S^{CVA}, can be obtained from (5.162) and (5.164) and related to the wire sensitivity S^{CV} in CV operation. Also one has to note that $r'_w/\overline{R}_w = -i'_w/\overline{I}_w$. For example, this gives for velocity fluctuations

$$\left(\frac{e'_{corr}(t)}{\overline{E}}\right)_{CVA} = S_U^{CVA} \frac{u'(t)}{\overline{U}_1}$$

with $S_U^{CVA} = \left[1 + \overline{R}_w \left(\frac{1}{R_2} + \frac{1}{R_F}\right)\right]^{-1} S_U^{CV}$.

$\quad (5.166)$

The factor in the brackets is advantageously close to 1.0, as $R_2 = 100\,\Omega$, $R_F = 300\,\Omega$, and $\overline{R}_w \simeq 10\,\Omega$. The correction is therefore rarely needed.

The lead resistance r_L slightly modifies the sensitivity coefficient given by (5.166) as well as the time constant given by (5.164) because the constant voltage condition is $(R_w + r_L)I_w = $ constant and not simply $R_w I_w = $ constant. This topic has been investigated by *Comte-Bellot* et al. [5.227] who established the following useful compact form, which is valid for velocity fluctuations as well as for temperature fluctuations,

$$S^{\mathrm{CVA}} \text{ (with lead)} = \mathrm{LS} \times \mathrm{LM} \times S^{\mathrm{CV}} \qquad (5.167)$$

with

$$\mathrm{LS} = \left[1 + (\overline{R}_w + r_L)\left(\frac{1}{R_2} + \frac{1}{R_F}\right)\right]^{-1}$$
$$\times \left[1 + \frac{r_L}{\overline{R}_w(1 + 2\overline{a}_w)}\right]^{-1} \qquad (5.168)$$

(the CVA circuit resistances R_2 and R_F have been included in LS) and

$$\mathrm{LM} = \left(1 + \frac{r_L}{\overline{R}_w}\right)\left[1 + \frac{r_L}{\overline{R}_w(1 + 2\overline{a}_w)}\right]^{-1}. \qquad (5.169)$$

Very advantageously, $\mathrm{LS} \times \mathrm{LM} \simeq 1$ because LS is slightly greater than one, and LM slightly less than one. The correction can therefore be often omitted.

Because of the constant-voltage operation the capacitance of the cable does not introduce any limitation on the dynamic response and also does not pick up extraneous noise, as pointed out by *Sarma* [5.148]. Inductance may also have a similar neutral behavior. Indeed the CVA has consistently exhibited superior noise immunity over CTA and CCA in tests in high radiofrequency interference (RFI) and electromagnetic interference (EMI) environments in hypersonic wind tunnels as reported by *Lachowicz* et al. [5.220] and *Blanchard* et al. [5.228]. This advantage makes it possible for fast setup of the CVA. Only the lead resistance of the cable needs to be considered, as presented above.

The electronic noise of a CVA prototype was analyzed by *Weiss* and *Comte-Bellot* [5.221]. Equation (5.159) are in this case completed with the Nyquist–Johnson noise sources of all resistances, and with the input voltage and current noise of the operational amplifier. Experiments were also conducted with different carbon resistances and tungsten wires operating with no flow, at the setting $T_C = 0.10\,\mathrm{ms}$ generally used. The rms noise value was found to decrease as the wire resistance increased, hence permitting an upper noise limit to be determined from experiments on an unheated wire. Regarding the noise power spectra, an increase with the square of the frequency was observed, as in other anemometers. An explicit relation was obtained for the signal-to-noise ratio, $\mathrm{SNR}|^{\mathrm{CVA}}$ at frequency f

$$\mathrm{SNR}|^{\mathrm{CVA}} = \frac{1}{2\pi f M_w^{\mathrm{CVA}}} \frac{R_{2a}}{R_{2a} + \overline{R}_w} \frac{\overline{a}_w}{2(1 + 2\overline{a}_w)} \frac{V_w}{\hat{e}_n} \frac{\hat{u}_1}{\overline{U}_1} \qquad (5.170)$$

with $\hat{e}_n$ the noise spectrum in $\mathrm{V}/\sqrt{\mathrm{Hz}}$, e.g., $\hat{e}_n \simeq 2.7 \times 10^{-9}\,\mathrm{V}/\sqrt{\mathrm{Hz}}$ in [5.221], and $\hat{u}_1$ the turbulent spectrum also in $\mathrm{V}/\sqrt{\mathrm{Hz}}$. It was also shown that $V_w = R_F V_1/R_1 \pm e'_n$, denoting the fluctuations of e_n by e'_n. The value of R_{2a}, which is proprietary to the manufacturer, can be assumed to be around $R_2/2$, and hence still large relative to R_w, so that $R_{2a}/(R_{2a} + \overline{R}_w) \simeq 1$. Equation (5.170) is therefore easy to use. It permits the estimation of SNR values when investigating the low-energetic fine-scale turbulent structures or the background perturbations of wind tunnels.

Finally, for large-amplitude fluctuations, the higher-order harmonics generated by parametric excitation at the wire level (5.105), are compensated as if they were turbulence. A suggestion to reject them would be to deduce $R_w(t)$ from $V_s(t)$, and then $R^\star(t)$ by inverting (5.105) [5.229]. Nowadays time data are acquired at high sampling frequencies and this procedure could lead to satisfactory results. Attempts in that direction would be most useful. A similar procedure could also work for a CCA, but it has never been suggested, despite the long history of the CCA. Also, as M_w^{CVA} is less than M_w^{CCA}, the inversion of (5.105) could be done with greater accuracy.

5.2.4 Calibration Procedures in Subsonic Flows

The calibration of a hot-wire probe associated with its anemometer establishes the relationship between the anemometer output voltage and the magnitude and direction of the incident velocity vector. A quiescent flow is the best to quickly record accurate data. In air, the potential region of a plane or circular jet, or the freestream of a wind tunnel are often chosen. Nowadays, the flow velocity and probe angle are easily controlled by a personal computer (PC), and the anemometer outputs are acquired in an automated way.

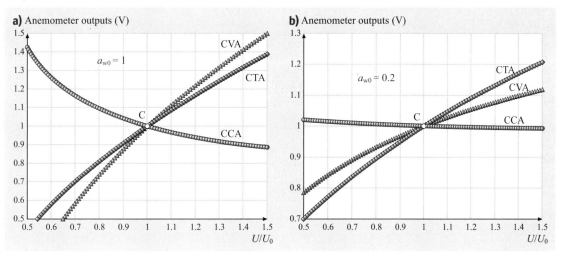

Fig. 5.37a,b Comparison of the CCA, CTA, and CVA response curves to velocity changes: **(a)** for $\bar{a}_w = 1.0$, **(b)** for $\bar{a}_w = 0.20$, at the reference velocity $U_0 = 10$ m/s. The vertical voltage shifts (in V) are 0.831, 4.157, and 8.314 for **(a)**, and 0.288, 2.208, and 4.800 for **(b)**, for the CCA, CTA and CVA respectively. For the CTA, $R_1 = 40\,\Omega$, for the CVA $R_2 = 100\,\Omega$

Preliminaries

The responses of anemometers generally exhibit nonlinearities for large velocity or fluid temperature changes. Part of the nonlinearity comes from the wire and the other part comes from the anemometer electronics. Although both effects are included in calibration curves, a hint of the shape of the curves is useful prior to the systematic acquisition of calibration data. This can be achieved by using a simple heat-loss law for the wire, such as King's law (5.48). Figure 5.37 reports Matlab simulations for a single wire placed normal to a flow, for the three types of anemometer, CCA, CTA and CVA, and two mean wire overheat values, $\bar{a}_w = 0.20$ and 1.0, at the reference velocity $U_0 = 10$ m/s. All anemometer outputs are translated along the y-axis in order to coincide at C. The curvature inherent to each curve is hence

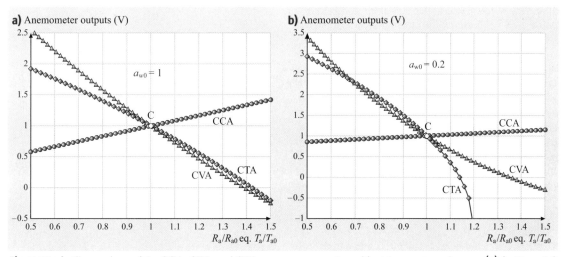

Fig. 5.38a,b Comparison of the CCA, CTA, and CVA response curves to ambient temperature changes: **(a)** for $\bar{a}_w = 1.0$, **(b)** for $\bar{a}_w = 0.20$, at the reference temperature $T_{a0} = 293$ K, $R_{a0} = 5\,\Omega$. The vertical voltage shifts and the values of R_1 and R_2 are the same as in Fig. 5.37

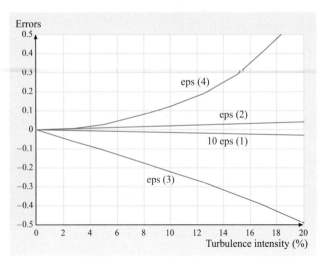

Fig. 5.39 Errors made by linearizing the anemometer calibration curve around the operating point as a function of the turbulence level. Example of a CTA; eps(1) to eps(4) are defined by (5.171)

$$\text{eps}(3) = \frac{\overline{e'^3}}{\left(\overline{e'^2}\right)^{3/2}}, \quad \text{eps}(4) = \frac{\overline{e'^4}}{\left(\overline{e'^2}\right)^2} - 3 \quad (5.171)$$

with

$$e'(t) = E(t) - \overline{E} \quad \text{and}$$

$$S_{U_1} = \frac{1}{4} \frac{B\sqrt{U_0}}{A + B\sqrt{U_0}} = 0.215 \, .$$

The errors that affect the mean or the rms values are small, less than 1%, and are hence negligible. On the other hand, the flatness factors and more so the skewness factors show large errors. Their measurement therefore requires the use of the full calibration curve. This means that the calibration curve has to be inverted with accuracy at each sampling point used in the data-acquisition routine to deduce the velocity from the anemometer output.

Results from the application of a time-varying sinusoidal input signal u' is also instructive. The higher harmonics that necessarily appear have low power content, for example 10^{-2} and 10^{-4} down from the fundamental. Errors on the PSD spectra are therefore small. Double correlation functions, which are Fourier transforms of PSDs, are also expected to be free of error. However, the skewness factors are in error, giving for example eps(3) $\simeq -0.23$ because of the cross-product, which involves the square of the fundamental and the first harmonic.

Static nonlinear effects can be similarly simulated for the two other types of anemometers, the CCA and CVA. For the operating point considered above ($U_0 = 10 \, \text{m/s}$, $a_{w0} = 1.0$), at a Gaussian turbulence level of 10% for all the three anemometers, eps(1) and eps(2) are both less than 1%, the higher harmonics have the same low relative order of magnitude as indicated above, eps(3) values are -0.22 for CTA, -0.34 for CVA and $+0.47$ for the CCA, and eps(4) values are 0.12 for CTA, 0.28 for CVA and 0.48 for CCA. For these higher moments, inversion of the calibration curve is therefore recommended whatever the type of anemometer.

Calibration of a Single Wire Normal to the Flow

For a single normal wire, only the incident velocity has to be changed. It is often measured by a Pitot tube connected to a pressure transducer. It is common practice to carry out around 20–30 readings over the selected velocity range. The description of the calibration curves can be made by using

1. power relationships, close to King's law or a similar law with added terms

preserved and permits a direct visual appraisal of the nonlinear effects. For example, at $\bar{a}_w = 1.0$, the CVA is seen to be able to keep an almost linear response over a $\pm 10\%$ velocity change. A satisfactory linear response also appears on the CTA, but strong nonlinearities occur in a CCA.

The wire response to a change in R_a, which corresponds to a change in the fluid temperature T_a, can similarly be investigated; results are presented in Fig. 5.38. An exact linear response appears with a CCA and a satisfactory one for a CVA whatever $\bar{a}_w$. On the contrary, a CTA presents a strong nonlinear response at low overheat ratios, already visible at $\bar{a}_w = 0.20$.

Figure 5.39 shows for a CTA the errors that exist if only the tangent to the velocity calibration curve at the operating point is used to interpret the measurements. The operating point is such that $U_0 = 10 \, \text{m/s}$, $a_{w0} = 1.0$, and $E_0 = 4.157 \, \text{V}$ (with $R_1 = 40 \, \Omega$ for the top bridge resistance). The synthetic input turbulence u' is a random Gaussian noise, with 10^6 values, its mean and rms values are 0 and 1.0, and the skewness and flatness factors are 0 and 3, respectively. The turbulence level is varied between 0 and 20%. The errors considered are defined by

$$\text{eps}(1) = \frac{\overline{E} - E_0}{E_0},$$

$$\text{eps}(2) = \left(\frac{\sqrt{\overline{e'^2}}}{\overline{E}} - S_{U_1} \frac{\sqrt{\overline{u'^2}}}{U_0} \right) \left(\frac{\sqrt{\overline{e'^2}}}{\overline{E}} \right)^{-1},$$

2. simple or extended polynomial fits
3. spline fits.

Bruun et al. [5.231] compared the accuracy of the different techniques. Technique 3 seems the best. It relies on a flexible strip (a drafting spline) that provides a smooth curve through and between points on the graph. The data range is subdivided into subintervals, and each interval is fitted with a cubic spline.

To ensure smoothness the empirical fit must provide a monotonic relationship between the anemometer output and the velocity including the derivatives of all orders. For example, for a CTA or CVA, the fit must verify $d^n E_{\text{output}}/dU_1^n > 0$ for $n = 1$, < 0 for $n = 2$, and > 0 for $n = 3$, and so on. The calibration curve can then easily and accurately be inverted to deduce the velocity in the flow under investigation from the anemometer output.

For small turbulence levels, the local slope to the calibration, at the mean operating point, gives the sensitivity coefficient $S_{U_1}^{\text{CCA}}$, $S_{U_1}^{\text{CTA}}$, $S_{U_1}^{\text{CVA}}$, defined in Sect. 5.2.3.

For very low velocities the reference velocity using pressure differentials is not accurate enough. The probe is therefore displaced to a quiet environment at a known speed. A rotating arm was used by *King* [5.183, 232], a horizontal translating carriage by *Baille* [5.233], a sliding carriage with adjustable inclination by *Tewari* and *Jaluria* [5.234] to study directional effect of buoyancy. *Khoo* et al. [5.235] kept the probe fixed and placed it in the air gap between a top rotating disc and a bottom stationary disc. Their setup permits direct calibration under the cooling influence of a nearby wall.

De Haan [5.236] investigated the use of acoustic waves for the *dynamic* calibration of a hot-wire. An acoustic standing wave was superposed on the mean velocity. Experiments were conducted at a frequency of 2.5 kHz for a mean speed in the range 6–18 m/s for 18 wires differing in length, diameter and stub importance. The rms velocities measured by the wire, and based on a *static* calibration, were found to agree to within 4% with those deduced from the microphone readings.

In the absence of mean velocity, the wire can be placed at an acoustic velocity antinode within a resonant tube, but due to the rectified signal transmitted by the anemometer, data processing must only be performed on the incoming velocity maxima. The acoustic pressure at the pressure antinode is measured by a microphone flush-mounted at the wall. Acoustic velocity u' and pressure p' are simply related by $u' = p'/\rho_0 c_0$, where ρ_0 and c_0 are the density and speed of sound, respectively. The velocity is changed by adjusting the level of the acoustic source, which is usually a loudspeaker.

Calibration in liquids is mostly performed in a towing tank or in a large circular channel with a rotating arm. Many setups are described in the *Lomas* book [5.51]. A moving container filled with the test liquid and displaced along the wire was used by *Dring* and *Gebhart* [5.237]. This method allows the easy experimentation of different liquids at different temperatures.

Calibration of X-Wires in Subsonic Flow
For an X-probe, an additional angular calibration of each wire with respect to its yaw angle θ is needed at different velocities. This is done by placing the X-probe on a small rotating rig, with the probe center kept unchanged [5.238].

For low turbulence intensities, the θ range can be limited to $\pm 10°$. Then S_{U_1} (yawed wire) and S_θ (yawed wire) defined in Sect. 5.2.3 can be evaluated and used. Weighted sum and difference of the two anemometer outputs then give $u'_1(t)$ and $u'_2(t)$. Furthermore, the $u'_2(t)$ signal is almost unaffected by the nonlinear wire response to U_1 [5.239].

For high turbulence levels, a larger θ range is needed, and the interval $\pm 35°$ is commonly used. An example of a calibration map, borrowed from *Abdel* et al. [5.230], is given in Fig. 5.40. Similar plots are reported by *Lueptov* et al. [5.240] or *Johnson* and *Eckelmann* [5.241] among others.

The calibration data can be used to create a look-up matrix that contains, in discrete form, the relation-

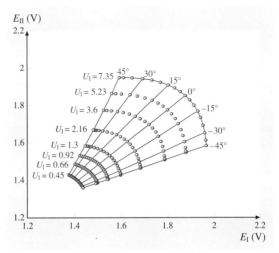

Fig. 5.40 Typical yaw angle calibration of a CTA-operated X-probe at different velocities U_1 (m/s) (after *Abdel* et al. [5.230])

ship between the two anemometer outputs, E_I and E_II, and the velocity components, $U_1(t)$ and $U_2(t)$, or their equivalents in plane polar coordinates, $|U|(t)$ and $\theta_U(t)$, for the velocity magnitude and flow angle, respectively. This method makes no assumption regarding the wire geometry or the wire response. When conducting turbulence measurements, the reconstruction of the velocity field, $U_1(t)$ and $U_2(t)$, needs bilinear interpolation schemes with high-resolution grids. This important topic is discussed by *Lueptow* et al. [5.240], *Chew* and *Ha* [5.242], *van Dijk* and *Nieuwstadt* [5.80]. Historically, *Willmarth* and *Bogar* [5.243], and *Willmarth* [5.244] pioneered the method and showed that a unique velocity vector in the plane of the X-wire can be obtained and followed in time. Very recently *Burattini* and *Antonia* [5.245] compared different X-wire calibration schemes and concluded that the look-up-table approach is more reliable than the effective-angle calibration scheme.

The calibration data can be also expressed, for each wire, in terms of a relationship involving King's law. The concept of an effective velocity, U_eff, is therefore widely used to represent the velocity cooling the wire

$$U_\text{eff}^2 = U_\text{N}^2 + U_\text{B}^2 + k^2 U_\text{T}^2 \,. \tag{5.172}$$

U_N and U_T are defined in Fig. 5.13d. The factor k takes into account the effects due to the finite wire length and the prong orientation with respect to the flow, usually $k \simeq 0.10\text{--}020$ [5.52]. *Chew* and *Ha* [5.238] studied more precisely how k depends on the velocity and on the probe angle. *Bruun* and *Tropea* [5.246] examined the yaw effect on A, B, and n in King's law. When reducing data for turbulence measurements, two equations have to be resolved at each time to recover the velocity vector. This computation is easier to perform when one considers the variables $|U|(t)$ and $\theta_U(t)$. The ratio of the two equations, one for each wire, gives $\theta_U(t)$. Subsequently, either equation gives $|U|(t)$. The procedure is greatly simplified if the X-probe is made of two mutually orthogonal wires, as U_N for the first wire becomes U_T for the second wire, and U_N for the second wire becomes U_T for the first wire. However, there are always four solutions to the system, and only the correct one must be kept, the one in the approach quadrant (Sect. 5.2.2). This approach is mostly used with constant-temperature anemometers and details are reported by *Kawall* et al. [5.247], *Hishida* and *Nagano* [5.78,79], *Bruun* [5.52], or *Lekakis* et al. [5.248] among others. A comparison between the two approaches, look-up matrix and effective velocity, was made by *Browne* et al. [5.249] and they found that the effective velocity gave surprisingly good results even in high-intensity turbulent flows.

Calibration of Three- and Four-Wire Probes in Subsonic Flows

The probe calibration now requires data collection for each wire, with respect to the velocity U_1, yaw angle θ, and pitch angle ϕ. So the amount of data to collect and save is very large. Examples are available in *Lakshminarayana* and *Davino* [5.250] for turbomachinery flows or in *Pompeo* and *Thomann* [5.251] for wall turbulence. When conducting turbulence measurements, the data are exploited using the lookup-table method or the concept of effective cooling velocity.

Van Dijk and *Nieuwstadt* [5.80] recently compared the two approaches and strongly supported the lookup-table method as a very good tool to process hot-wire measurements, since rapid computer processing is readily available. Among the pioneers of the techniques, let us cite *Pailhas* and *Cousteix* [5.252], and *Döbbeling* et al. [5.253]. Special surfaces fitting the calibration data were computed by *Samet* and *Einav* [5.254] and applied to measurements in a swirling jet issuing into a coflowing stream.

The effective velocity approach was, however, used successfully by *Lekakis* et al. [5.248] who experimented with orthogonal and non-orthogonal probes, and by *Kawall* et al. [5.247], *Lekakis* et al. [5.248], or *Gieseke* and *Guezennec* [5.255]. In three-dimensions (3D), the effective cooling velocity also contains the binormal component U_B, hence U_eff is given by

$$U_\text{eff}^2 = U_\text{N}^2 + h^2 U_\text{B}^2 + k^2 U_\text{T}^2 \tag{5.173}$$

with $h \simeq 1.02\text{--}1.05$ [5.52]. The three equations provided by the anemometer outputs are now solved in two steps. Firstly, the ratios of the second and third equations by the first equation give the yaw and pitch angles for the velocity vector. Then, either equation gives the velocity magnitude. The resolution is much more direct if the three wires constituting the probe are mutually orthogonal. In any case, eight solutions are numerically possible, and only one is correct, the one in the approach octant. Addition of a fourth wire increases the acceptance cone as shown by *Lemonis* and *Dracos* [5.256] and by *Döbbeling* et al. [5.257] who could extend the acceptability range to a hemisphere. The acceptance angle was also analyzed by *Lavoie* and *Pollard* [5.258] who reported convincing results in the near field of a turbulent jet. When using the effective cooling velocity approach, there is also the need to refer to the matrix ω_{ij}, which relates the wire-fixed frame $O'_{x'_i}$, i.e.,

$(N'B'T)$, to the space-fixed frame O_{x_j}, i.e., (O_{x_1,x_2,x_3}). This is not difficult, just cumbersome. From Fig. 5.13, one has

$$\omega_{ij} = \cos\left(O'_{x'_i}, O_{x_j}\right)$$
$$= \begin{cases} \cos\theta\sin\phi - \sin\theta + \cos\theta\cos\phi, \\ \cos\phi + 0 - \sin\phi, \\ \sin\theta\sin\phi + \cos\theta + \sin\theta\cos\phi. \end{cases} \quad (5.174)$$

Hence, the velocity components in the wire-fixed frame, $U_{N'}$, $U_{B'}$, U_T, are related to those in the space-fixed frame, U_1, U_2, U_3, by

$$U_{N'} = U_1\cos\theta\sin\phi - U_2\sin\theta + U_3\cos\theta\cos\phi$$
$$U_{B'} = U_1\cos\phi + 0 - U_3\sin\phi$$
$$U_T = U_1\sin\theta\sin\phi + U_2\cos\theta + U_3\sin\theta\cos\phi \quad (5.175)$$

and conversely, U_1, U_2, U_3 are related to $U_{N'}$, $U_{B'}$, U_T by

$$U_1 = U_{N'}\cos\theta\sin\phi + U_{B'}\cos\phi + U_T\sin\theta\sin\phi$$
$$U_2 = -U_{N'}\sin\theta + 0 + U_T\cos\theta$$
$$U_3 = U_{N'}\cos\theta\cos\phi - U_{B'}\sin\phi + U_T\sin\theta\cos\phi. \quad (5.176)$$

It is interesting to note that these relations are similar to those used when studying the 3-D turbulence spectra [5.77], with a wavenumber vector playing the same role as a vector coinciding with the hot-wire. Finally, a simple 2-D matrix permits the expression of a rotation around the wire to pass from the prong-frame (N B T) to the wire-frame (N', B', T), with T common to the two frames. This additional degree of freedom has to be kept for each wire as the prongs location relative to the wire is important [5.68, 90].

Wall Sensors in Subsonic Flows

Wall sensors are most conveniently calibrated by using the wall region of a fully developed pipe or channel flow, where the measured pressure gradient can serve as a direct measure of the wall shear stress. Examples are given by *Cousteix* and *Juillen* [5.259], *Haritonidis* [5.260] and *Cook* [5.261, 262] for metallic films, and by *Huang* et al. [5.263] and *Lin* et al. [5.62] for MEMS. Turbulent boundary layers can also be used. The wall sensor is calibrated against a reference method (skin friction balance, oil-film interferometry) or well established wall laws [5.264]. A properly calibrated Preston tube can also be used [5.265]. The flow between a fixed plate and a rotating plate was developed by *Brown* and *Davey* [5.266] as a very convenient and inexpensive calibration device. The Stewartson solution for the mean flow and the wall shear was used and adopted also by *Khoo* et al. [5.235].

For metallic elements, the calibration curves are observed to be of the form

$$\frac{R_w I_w^2}{R_w - R_a} = A + B\tau_w^{1/3}. \quad (5.177)$$

Such a reference equation was analytically established by Lévêque in 1928 and by Fage and Falkner in 1931, and revisited by *Knudsen* and *Katz* in 1958 [5.104], *Cousteix* and *Jullien* [5.259], or *Bailly* and *Comte-Bellot* [5.77]. A numerical heat-transfer model was recently developed by *Cole* and *Beck* [5.267]. The main assumption is always that the thermal boundary layer of the heated sensor lies entirely within the linear region of the velocity profile, usually called the viscous sublayer. Experiments supporting this are those of *Brown* and *Davey* [5.266], *Reichert* and *Azad* [5.268] or *Alfredsson* et al. [5.269] among others. Coefficient B depends on the probe and flow characteristics, and the additive constant A depends mainly on the heat conduction leakage to the substrate. In air it can represent as much as 50% of electric power injected [5.270]. Very interestingly, *Sarma* and *Moes* [5.271] suggested a method to measure in situ the heat loss to the substrate for a multiple hot-film array operated by CVAs. In water, leakage to the substrate is much less severe. Recently, *Mangalam* [5.272] reported many successful experiments conducted with CVAs on autonomous underwater vehicles.

For miniature MEMS devices, (5.177) is not supported by direct measurements. *Lin* et al. [5.62] reported laws where the electric power varies as $\tau_w^{0.67}$ for nitride-based sensors and as $\tau_w^{0.85}$ for parylene diaphragm sensors. The authors invoked an additional heat-transfer mechanism and developed an appropriate model for it. *Ruedi* et al. [5.273] also reported significant difficulties regarding the repeatability of calibration sets and explored various reasons for them, such as the drift of ambient temperature or quite small overheat ratios for the sensor. Even after reduction of these effects, the power dissipated by the film was closer to a linear dependance on τ_w than to the classic $\tau_w^{1/3}$ law.

Regarding the frequency response of the hot-film, *Cook* [5.262] reported very interesting measurements in an oscillating flow generated by controlling the area of a sonic throat. They used a quartz-substrate gauge, a glue-on gauge with a thin flexible substrate, and a surface film with a cavity underneath. Interestingly, a laminar case was first studied [5.261] and they could observe that the skin friction preceded the freestream velocity by 45°, as theoretically established by *Lighthill* [5.274]. For turbulent flows, the phase advance went down to around 15–20°. This experiment was one of those simulated by *Cole* and *Beck* [5.267]. Their numerical approach permits to simulate different thermal diffusivities for the substrate and different ratios D/a, where D is the substrate depth and a is the half-length of the hot-film in the flow direction, for example $D/a \simeq 6$ for comparison with the experimental data of Cole and Beck. To this day, the highest frequency achieved is only around 20 Hz, the limitation being due to the maximum number of time steps in the computation.

In hot-film arrays, an individual calibration of the sensors is rarely needed as commercial sensors are manufactured with nearly identical characteristics. In special cases, for example to study the flow around a cylinder, *Desgeorges* et al. [5.275] individually calibrated 251 hot films simply by rotating the cylinder to bring each sensor successively to the same angular position relative to the flow.

Mean Fluid Temperature Drifts in Subsonic Flows

In many flows, the mean temperature $\overline{T}_a$ changes from day to day because of the environment, or during an experiment because of the heat released by the machines powering the wind tunnels, or from point to point in non-isothermal flows such as hot jets. The velocity calibration obtained at temperature $\overline{T}_{a0}$ is not valid at another temperature $\overline{T}_a$. The reason is that the curves expressing the relationship between the anemometer output and the velocity are shifted down for a CTA or a CVA and up for a CCA when $\overline{T}_a > \overline{T}_{a0}$, and vice versa when $\overline{T}_a < \overline{T}_{a0}$. For the CTA, *Koppius* and *Trines* [5.276] collected many data for a long wire $l/d \approx 770$ over a temperature range of 283–353 K. *Nitsche* and *Haberland* [5.265] and *Thünker* et al. [5.277] considered hot-wires and hot-films, and they could combine a refrigerated flow and a heated flow to make measurements over a very large temperature range, 250–425 K. These extreme flow conditions are typical of thermo and cryo wind tunnels.

There are several ways to account for the temperature difference $\overline{T}_a - \overline{T}_{a0}$:

1. Velocity calibration curves can be collected at a number of different fluid temperatures, covering the range of interest. All the curves are stored and the adequate curve is used in the course of turbulence measurements, which require the local temperature $\overline{T}_a$ to be independently measured. A thermocouple is sufficient for a global temperature drift of a wind tunnel. For non-isothermal flows, local measurements are needed and they require a *cold* hot-wire, usually operated in a CCA mode with a small ($\simeq 1$ mA) current. The procedure has to be repeated for every wire in a multiple-wire probe. Interpolation procedures also have to be developed to cover intermediate temperatures. An example is given by *Meyer* [5.278] for two wires making an X-probe and a third cold wire.

2. When only one velocity calibration curve has been recorded, at temperature $\overline{T}_{a0}$, only an analytical correction can be attempted to recover it from measurements carried out at T_a. The correction is essentially based on King's law. For a CTA, *Lemonis-Dracos* [5.256] expressed the correction procedure as

$$\overline{E}_{\text{corr}}^{\text{CTA}} = \overline{E}_{\text{meas}}^{\text{CTA}} \left[\frac{(\overline{T}_w - T_{a0})}{(\overline{T}_w - \overline{T}_a)} \right]^{1/2}, \quad (5.178)$$

which comes directly from the dependence $\overline{T}_w^2 \sim (\overline{R}_w - \overline{R}_a)$ at fixed $A + B\overline{U}_1^n$ in King's law. *Benjamin* and *Roberts* [5.279] revisited this correction for a large temperature change, 300–500 K, and found that the physical properties of the fluid which could affect A and B could be ignored, and even that the approximate form deduced from (5.178) by a first-order expansion

$$\overline{E}_{\text{corr}}^{\text{CTA}} \simeq \overline{E}_{\text{meas}}^{\text{CTA}} \left[1 - \gamma_c \frac{(\overline{T}_{a0} - \overline{T}_a)}{(\overline{T}_w - \overline{T}_a)} \right]$$

with $\gamma_c = 0.50$ (5.179)

performs even better than (5.178). One can also obtain (5.179) using the sensitivity coefficient $S_{T_a}^{\text{CTA}} = -1/(2\overline{a}_w) = -\overline{R}_a/2(\overline{R}_w - \overline{R}_a) \simeq -\overline{T}_a/2(\overline{T}_w - \overline{T}_a)$ defined in Sects. 5.2.2 and 5.2.3. *Meyer* [5.278] examined the case of yawed wires and found that $\gamma_c \simeq 0.45$ when the yaw angle reaches around $\pm 30°$. *Cimbala* and *Park* [5.280] also made use of King's law, and added other factors to take into account a DC offset and a gain. A relatively small temperature range was inspected, 300–308 K. *Kostka* and *Vasanta Ram* [5.281] and *Vasanta Ram* [5.282] operated in the temperature range 293–333 K, using wires of four different aspect ratios: 718, 306, 248 and 106. Their reduced data show that the overall Nusselt number Nu_{a0}, with physical properties evaluated at a fixed temperature, $\overline{T}_{a0} = 313\,\text{K}$, decreases linearly with the temperature difference $\overline{T}_a - \overline{T}_{a0}$.

3. *Hollasch* and *Gebhart* [5.283], working with liquids, suggested establishing the correction by changing the wire temperature T_w rather than the fluid temperature T_a.
4. Automatic compensation of ambient-temperature drift can be introduced in the electronic circuit. This has mostly been developed for CTA. Three approaches have been analyzed by *Drubka* et al. [5.284]: the constant-overheat mode, the constant-resistance-difference mode, the constant-output-voltage mode. All of them, however, require an additional temperature-compensating probe, placed in the flow and integrated in the Wheatstone bridge, and complex electronics. The simple King's law with constant coefficients is also assumed.
5. The need for an additional sensor is eliminated by *Sarma* and *Comte-Bellot* [5.285] in a CVA setup. An automated procedure permits to obtain in situ the heated and cold resistances of a hot-wire in each measurement cycle. The *high* V_w step, with $I_w \simeq 100\,\text{mA}$, permits the turbulence to be recorded and M_w^{CVA} to be measured, while the *low* V_w step, with $I_w \simeq 1\,\text{mA}$, yields R_a and T_a. Fig. 5.41 shows a typical set of calibration curves obtained for different fluid temperature T_a and the possibility to reduce all the different calibration curves into a unique calibration curve by using the ratio

$$\text{PDR} = \frac{R_w I_w^2}{(R_w - R_a)}, \quad (5.180)$$

which involves the power injected in the wire and the difference in the hot and cold resistance of the wire. The PDR is therefore in essence a Nusselt number.

Truzzi et al. [5.286] used this method to measure the unsteady flow in the near-orifice region of synthetic jets. The method requires only one hot-wire. This is a definite advantage over previous methods established for CCA or CTA where an additional temperature sensor is needed.

5.2.5 Measurement of Velocity and Temperature Fluctuations

In subsonic non-isothermal flows, velocity fluctuations are accompanied by fluctuations in the fluid temperature. A single hot-wire placed normal to the mean incident velocity therefore has a response of the form

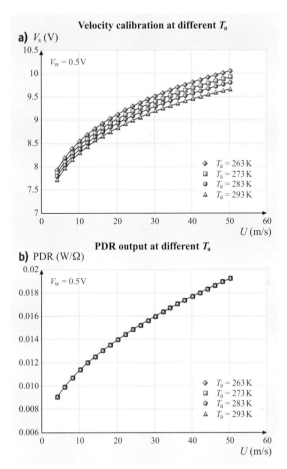

Fig. 5.41a,b Calibration curves collected at different fluid temperatures T_a for a CVA, and correction made by using the PDR ratio defined by (5.180)

$$E(t) = f\left[\overline{U}_1 + u_1'(t), \overline{T}_a + \theta_a'(t)\right], \quad (5.181)$$

where $E(t)$ is the anemometer output, $\overline{U}_1 + u_1'$ the incident longitudinal velocity and $\overline{T}_a + \theta_a'$ the incident fluid temperature. For small fluctuations, (5.181) can be rewritten

$$\frac{e'(t)}{\overline{E}} = S_{U_1}\frac{u_1'}{\overline{U}_1} + S_{T_a}\frac{\theta_a'}{\overline{T}_a}, \quad (5.182)$$

where the factors S are the sensitivity coefficients of the chosen anemometer to velocity and temperature. As early as 1947, *Corrsin* [5.287] pointed out the possibility to adjust the wire overheat $\overline{a}_w$ to modify the relative value of the ratio S_{T_a}/S_{U_1} in a constant-current anemometer and, hence, the relative weight of the velocity and temperature contributions in (5.182). A set of independent algebraic equations was obtained and its resolution provided the turbulence levels $\overline{u_1'^2}$, $\overline{\theta_a'^2}$ and the velocity–temperature correlation $\overline{u_1'\theta_a'}$. This method is still in use. Furthermore, in the limit $\overline{a}_w \to 0$, the corresponding *cold* wire acts as a resistance temperature sensor and gives the temperature fluctuations. In practice, one has to see how small $\overline{a}_w$ can be set compared with the anemometer noise. This also leads to the use of very fine wires, down to 0.25 μm. With two wires operated simultaneously, one can dispose of two equations similar to (5.181) or (5.182); adequately combining them gives the instantaneous fluctuations $u_1'(t)$, $\theta_a'(t)$, and the mean values $\overline{U}_1$, $\overline{T}_a$. Most of these techniques can also be extrapolated to higher combinations of sensors, such as a X-wire associated with a cold wire to obtain $\overline{u_1'^2}$, $\overline{u_2'^2}$, $\overline{\theta_a'^2}$, $\overline{u_1'u_2'}$, $\overline{u_1'\theta_a'}$, and $\overline{u_2'\theta_a'}$. In this section we examine the most important topics related to these techniques.

The Multiple Overheat Ratio Method with One Wire

From the sensitivities S_{U_1} and S_{T_a} given in Sects. 5.2.2 and 5.2.3 one easily obtains the ratio S_{T_a}/S_{U_1}

$$\frac{S_{T_a}}{S_{U_1}} = -\frac{2}{\overline{a}_w} \frac{B\sqrt{\overline{U}_1}}{A + B\sqrt{\overline{U}_1}} \chi \overline{T}_a [1 - \chi(\overline{T}_a - T_0)]$$

$$\simeq -\frac{2}{\overline{a}_w} \quad (5.183)$$

independent of the anemometer type. The approximate value $-2/\overline{a}_w$ is for $A \ll B\sqrt{\overline{U}_1}$, $\overline{T}_a \simeq T_0$ and $\chi \simeq 1/\overline{T}_a$. This clearly shows that adjusting the overheat ratio $\overline{a}_w$ permits modification of the ratio S_{T_a}/S_{U_1}. Squaring (5.182), using the approximate sensitivity ratio $-2/\overline{a}_w$ and averaging, yields

$$\frac{1}{S_{T_a}^2} \frac{\overline{e'^2}}{\overline{E}^2} = \frac{\overline{\theta_a'^2}}{\overline{T}_a^2} - \overline{a}_w \frac{\overline{u_1'\theta_a'}}{\overline{U}_1 \overline{T}_a} + \frac{\overline{a}_w^2}{4} \frac{\overline{u_1'^2}}{\overline{U}_1^2}. \quad (5.184)$$

The anemometer type appears only in the left-hand side of this equation through the sensitivity coefficient S_{T_a}. A minimum of three values of the overheat ratio $\overline{a}_w$ is required to extract $\overline{\theta_a'^2}/\overline{T}_a^2$, $\overline{u_1'^2}/\overline{U}_1^2$, and $\overline{u_1'\theta_a'}/\overline{U}_1\overline{T}_a$. More values permit the data to be smoothed and yield greater accuracy, as is done in supersonic flows where temperature is an inherent variable to be determined.

The Cold Wire Method

In the limit $\overline{a}_w \to 0$, the wire operates as a resistance sensor and advantageously isolates the temperature fluctuation θ_a'. However, different problems have to be examined: the residual velocity contamination, the time constant, end effects, and electronic noise.

Using King's law (5.48), and assuming $R_w \simeq R_a$ except in the difference $R_w - R_a$, the condition on $\overline{a}_w$ shows that the sensitivity ratio at a fixed velocity varies as the inverse square of the current I_w through the wire

$$\frac{S_{T_a}}{S_{U_1}} \simeq -2\left(A + B\sqrt{\overline{U}_1}\right)\frac{1}{I_w^2} \quad \text{for a CCA}, \quad (5.185)$$

and as the inverse square of the wire voltage V_w

$$\frac{S_{T_a}}{S_{U_1}} \simeq -2\left(A + B\sqrt{\overline{U}_1}\right)\frac{R_a^2}{V_w^2} \quad \text{for a CVA}. \quad (5.186)$$

A CTA cannot operate a cold wire because its response is strongly nonlinear at low overheats and its feedback loop also loses effectiveness (Sect. 5.2.3).

Equation (5.185) clearly shows that very small current intensities have to be used. Accordingly, the wire output signal is small and a fine wire is preferred. Tungsten wires are commercially available with $d = 3.2$ and 4.3 μm and platinum and platinum–rhodium wires with $d = 2.5$, 1.0, 0.63, and even 0.25 μm. The current to realize a *cold* wire is around 0.20 mA for $d = 2.5$ μm, 0.15 mA for $d = 1.0$ μm, 0.10 mA for $d = 0.63$ μm and 0.05 mA for $d = 0.25$ μm. The overheat ratio is then less than 0.01. For a wire length of around 0.5 mm, the cold wire resistance is of the order of 10 Ω for $d = 2.5$ μm, 60 Ω for $d = 1.0$ μm, 150 Ω for $d = 0.63$ μm, and 1200 Ω for $d = 0.25$ μm. Extrafine wires are reserved for low-speed laboratory experiments requiring high-frequency responses and low noise. Special circuits have also to be developed to accept such large values of the wire resistance. Data are reported among others by *Yeh* and *Van Atta* [5.288], *Fulachier* [5.289], *Paranthoen* et al. [5.290], *Browne* et al. [5.291], *Mydlarski* and *Warhaft* [5.292], and *Tagawa* et al. [5.293] and *Mestayer* and *Chambaud* [5.294].

The wire resistance changes with flow temperature are small and they require high amplification factors or the use of special bridges. A first attempt by *Yeh* and *Van Atta* [5.288] made use of a high-frequency carrier as in AC strain gauge bridges. *Lin* and *Lin* [5.295] developed a special bridge, and *Tavoularis* [5.296] used a reference voltage source. Output signals on the order -0.10 V/K were obtained. Present commercial systems permit outputs up to -0.5 V/K. The coefficient χ that links resistance and temperature also has to be accurately known. *Van Dijk* and *Nieuwstadt* [5.65] examined different techniques, including the common one that makes use of an oven.

The electronic noise of the circuit and the signal-to-noise ratio should also be estimated prior to measurements. These values will be those which finally determine the current intensity through the wire. The use of a low-pass filter that keeps only the significant part of the signal is beneficial, for example, a low-pass filter set at 5 kHz was used by *Browne* et al. [5.291] and *Lemay* and *Benaïssa* [5.66]. When a known frequency prevails, such as in a pulsed jet or in an acoustic resonating cavity, a narrow-band pass filter can be associated to that known frequency. *Huelsz* and *Ramos* [5.300] reported results with a ±5 Hz filter around 130 Hz in an acoustic refrigerator.

Values of the time constant M_w of *cold* wires are reported in Fig. 5.42. The measurement of M_w at small overheat can be obtained in different ways.

- Use of the time step response of the wire to an externally injected square wave signal, as is done for hot-wires. This is advantageously made in a low-turbulence environment. Some difficulties can be encountered due to the electronic noise.
- Measurement of the frequency response of the wire to a sine-wave injection, then $M_w = 1/(2\pi f_c)$, where f_c is the frequency at which a -3 dB reduction occurs in the wire response.
- Use of the response of the wire placed alternatively in a hot/cold flow at a known frequency. *Antonia* et al. [5.297] placed the cold wire downstream of a larger-diameter cylinder that was pulsed at different frequencies with a short-duration voltage pulse. *Paranthoen* et al. [5.290] vibrated the wire across a cold/hot interface.
- Extrapolation towards $\bar{a}_w = 0$ of M_w data obtained at slightly higher $\bar{a}_w$ values. This can be a way to get around the electronic noise that precludes accurate measurements. Preliminary CVA data were successful using the interval $\bar{a}_w = 0.08$–0.20.
- Radiative heating of the sensor, as suggested by *Kidron* [5.145]. The radiant flux should be well focussed on the wire itself, avoiding stubs for platinum wires etched from Wollaston or prongs for bare tungsten.

Estimates of M_w can be obtained from (5.103), although established for wires with infinite aspect ratio. Letting $\bar{a}_w \to 0$ gives

$$M_w = \frac{\rho_w c_w d^2}{4 k_a \mathrm{Nu}_a} \,. \tag{5.187}$$

Usually, for a regular *hot* wire, $M_w \sim d^{3/2}$ because $B_N \mathrm{Re}^{1/2}$ is dominant compared to A_N in the Nusselt law (5.44). For a *cold* wire A_N can play a role, as the wire is far less sensitive to the flow, thus the law $M_w \sim d^2$ can also be expected. Both the $d^{3/2}$ and d^2 laws are reported in Fig. 5.42. The former seems to fit the experimental data better irrespective of the presence or absence of stubs and the influence of the wire aspect ratio.

The spatial resolution of a cold wire relative to the temperature field is also an important topic to be considered. For isotropic turbulence, it can be investigated similarly to that was made for the velocity field (Sect. 5.2.2). Indeed *Uberoi* and *Kovasznay* [5.166] entitled their work *field mapping by the wire* without reference to the nature of the turbulent fluctuation. Only the spectral tensor $\Phi_{\theta_a}(k_1, k_2, k_3)$ and the three-dimensional spectrum $G(k)$ adapted to a scalar field have to be used. They are such that

$$\Phi_{\theta_a}(k_1, k_2, k_3) = \frac{G(k)}{k} \quad \text{with } \overline{\theta_a^2} = \int_0^\infty G(k) \, \mathrm{d}k \,. \tag{5.188}$$

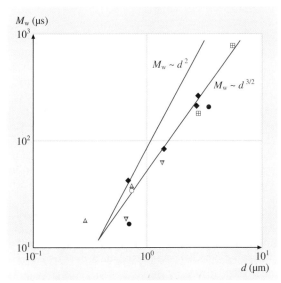

Fig. 5.42 Time constant of cold wires for $\overline{U}_1 \simeq 5$ m/s, *dagger*: Antonia et al. [5.297], *open circle*: Browne et al. [5.291], △ *LaRue* et al. [5.298], *upwards triangle*: Lemay and Benaïssa [5.66], *filled circle*: Tagawa et al. [5.293] *plus*: Fulachier and Dumas [5.299] *plus in circle*: CVA experiments (see references for the wire material, l/d ratio, and the presence or absence of stubs)

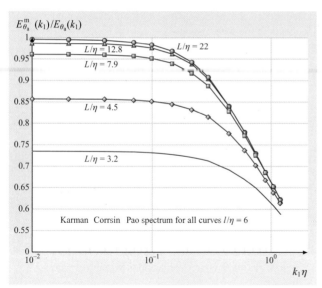

Fig. 5.43 Spatial resolution of a cold wire for the one-dimensional temperature spectra

The measured one-dimensional spectrum is therefore

$$\Phi_{\theta_a}^m(k_1) = \frac{1}{2}\int_{k_1}^{\infty} W \frac{G(k)}{k}\,dk \qquad (5.189)$$

with W still given by (5.128). When $W = 1$ the form established by *Kovasznay* et al. [5.301] for $\Phi_{\theta_a}^m(k_1)$ is

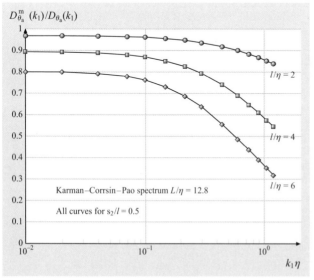

Fig. 5.44 Spatial resolution of two cold wires for the one-dimensional $\partial\theta_a/\partial x_2$ spectra

recovered. In what follows a Karman–Corrsin–Pao spectrum is used for $G(k)$, being very flexible to fit laboratory or atmospheric data. Within a numerical factor (omitted for simplicity and of no consequence for the spatial resolution error based on the ratio of measured and exact values) $G(k)$ is

$$G(k) \sim \frac{\overline{\theta_a^2}}{k_e'} \frac{(k/k_e')^2}{\left[1+(k/k_e')^2\right]^{11/6}} \exp\left[-\frac{9}{4\,\mathrm{Pr}}(k\eta)^{4/3}\right]. \qquad (5.190)$$

Moreover, one can compute the integral length scale L_{θ_a} using the relation

$$L_{\theta_a} = \frac{\pi}{2} \frac{\int_0^{\infty} k^{-1} G(k)\,dk}{\int_0^{\infty} G(k)\,dk} . \qquad (5.191)$$

Here, k_e' is adjusted in order to have $L_{\theta_a}\eta = 3.2, 4.5, 7.9, 12.8$ and 22, as for the velocity field examined in Sect. 5.2.2. This leads to $k_e'\eta = 0.99, 0.465, 0.183, 0.092$, and 0.045, respectively. These values are very close to those used for the velocity spectra, $k_e\eta = 1, 1/2, 1/5, 1/10$, and $1/20$. For atmospheric data, inner and outer scales based on the lower and upper limits of the $k^{-5/3}$ law are often preferred to k_e' and η, but the form of (5.190) still holds.

Figure 5.43 reports errors concerning the one-dimensional temperature spectra for different values of $L/\eta = L_{\theta_a}/\eta$ and a fixed wire length, $l = 6\eta$. A unity Prandtl number has been chosen. A smaller value, such as $\mathrm{Pr} = 0.70$ for air, would just lead to smaller errors as the exponential tail of $G(k)$ reduces the presence of small scales more strongly. On the whole, the errors for the temperature field appear slightly smaller than for the velocity field. A possible reason stems from the fact that Fourier components can be aligned with the wavevector for the temperature field, rather than only normal to it as for the velocity field, hence decreasing the impact of the k_3 components.

The errors on the lateral derivatives of the temperature field measured using two parallel cold wires can be similarly obtained. The measured spectrum taking into account both the wire length and the wire separation is found to be

$$D_{\theta_a,2} = \frac{1}{4}\int_{k_1}^{\infty} \mathrm{WD}\frac{G(k)}{k}(k^2 - k_1^2)\,dk \qquad (5.192)$$

with WD still given by (5.136). Results are reported in Fig. 5.44. Again, the errors appear to be smaller than the errors concerning a velocity field. For the longitudinal

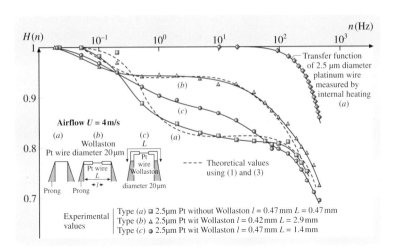

Fig. 5.45 Influence of Wollaston length and stubs on the transfer function $H(f)$ for a *cold* wire (after *Paranthoen* et al. [5.290])

derivatives, deduced from one-dimensional spectra using a single wire and the Taylor hypothesis, errors will be those that exist in the one-dimensional spectra weighted by k_1^2.

For anisotropic turbulence, no investigation has been conducted yet. Some progress using numerical simulations would be most welcome. Also, the case of *hot* wires with a mixed velocity–temperature response would be worth considering.

Finally, end losses for *cold* wires have been investigated by several researchers. As only a small part of the ohmic heating is convected away by the flow, thermal conduction towards the ends of the wire is large. Most important is the dynamic effect pointed out by *Perry* et al. [5.302], *Paranthoen* et al. [5.290, 303], and *Tsuji* et al. [5.57], with a low plateau appearing in the wire transfer function. This effect depends strongly on the aspect ratio of the wire and on the way in which the wire is bonded at its ends, in particular whether stubs are used or not (Fig. 5.45). Long stubs noticeably improve the wire response. In Paranthoen's results the wire aspect ratio is around 190, deliberately chosen small to point out the effect. Usually, fine wires present a higher aspect ratio, around 400 for a 1 μm wire that is 0.40 mm long.

The Multiple Overheat Ratio Method with Two Wires

Two wires, labeled I and II, simultaneously operated at two different overheats and closely located, give the following responses

$$E_\mathrm{I} = f_\mathrm{I}(U_1, T_\mathrm{a}),$$
$$E_\mathrm{II} = f_\mathrm{II}(U_1, T_\mathrm{a}), \qquad (5.193)$$

for large turbulent changes, or

$$\left(\frac{e'}{E}\right)_\mathrm{I} = (S_{U_1})_\mathrm{I}\frac{u_1'}{U_1} + (S_{T_\mathrm{a}})_\mathrm{I}\frac{\theta_\mathrm{a}'}{T_\mathrm{a}},$$
$$\left(\frac{e'}{E}\right)_\mathrm{II} = (S_{U_1})_\mathrm{II}\frac{u_1'}{U_1} + (S_{T_\mathrm{a}})_\mathrm{II}\frac{\theta_\mathrm{a}'}{T_\mathrm{a}}, \qquad (5.194)$$

for small fluctuations. In the former case, a complete database permits one to obtain U_1 and T_a. The technique is similar to that indicated for X-wires (Sect. 5.2.4). Direct readings, as suggested by *van Dijk* and *Nieuwstadt* [5.65] or polynomial fits as suggested by *Artt* and *Brown* [5.304] can be used. In the latter case, the properly weighted sum and difference of the responses give $u_1'(t)$ and $\theta_\mathrm{a}'(t)$. A *cold* wire can be used as wire II, hence simplifying the data processing.

Calibration in Velocity and Temperature

The calibration of a wire for joint velocity and temperature measurements relies on the recording of the anemometer output while varying first the velocity at a fixed temperature and then the temperature at a fixed velocity. This requires a wind tunnel equipped with controlled heating and cooling systems. An open jet with a temperature range of 280–320 K, and a velocity range of 2–50 m/s can cover most of the subsonic needs. The probe is placed in the potential core of the jet. Obtaining smooth calibration curves or accurate look-up tables requires sufficient data points, as explained in Sect. 5.2.4 for two velocity components.

Instead of U_1 and T_a, other doublets of variables can be used. The Pitot tube reading $h = \rho_\mathrm{a} U_1^2/2$ and the thermocouple indication T_a are directly provided by an experiment; hence they are convenient measurements for this purpose. This was suggested by *Comte-Bellot* and *Mathieu* [5.305] when working on heated wall jets. The sensitivity coefficients associated with the variables h and T_a, denoted by $\hat{S}$, can be related to S by referring to U_1 and T_a. Writing the differential response for any anemometer, we have on one hand

$$\frac{r'_w}{R_w} \text{ or } \frac{i'_w}{I_w}$$
$$= \underbrace{\left(\frac{\overline{U}_1}{\overline{R}_w}\frac{\partial \overline{R}_w}{\partial \overline{U}_1}\right)_{cstT_a}}_{S_{U_1}} \frac{dU_1}{\overline{U}_1} + \underbrace{\left(\frac{\overline{T}_a}{\overline{R}_w}\frac{\partial \overline{R}_w}{\partial \overline{T}_a}\right)_{cstU_1}}_{S_{T_a}} \frac{dT_a}{\overline{T}_a}$$
(5.195)

and on the other hand

$$\frac{r'_w}{R_w} \text{ or } \frac{i'_w}{I_w}$$
$$= \underbrace{\left(\frac{\overline{h}}{\overline{R}_w}\frac{\partial \overline{R}_w}{\partial \overline{h}}\right)_{cstT_a}}_{\hat{S}_h} \frac{dh}{\overline{h}} + \underbrace{\left(\frac{\overline{T}_a}{\overline{R}_w}\frac{\partial \overline{R}_w}{\partial \overline{T}_a}\right)_{csth}}_{\hat{S}_{T_a}} \frac{dT_a}{\overline{T}_a}.$$
(5.196)

The increment dh can be developed in (5.196). For an incompressible fluid with thermal expansion obeying the Boussinesq assumption, and hence at constant surrounding pressure, one has $\partial \rho / \overline{\rho} = -\partial T_a / \overline{T}_a$, and therefore

$$\frac{r'_w}{R_w} \text{ or } \frac{i'_w}{I_w} = \hat{S}_h \left(-\frac{dT_a}{\overline{T}_a} + 2\frac{dU_1}{\overline{U}_1}\right) + \hat{S}_{T_a}\frac{dT_a}{\overline{T}_a} .$$
(5.197)

Identifying with (5.195) gives

$$S_{U_1} = 2\hat{S}_h ,$$
$$S_{T_a} = \hat{S}_{T_a} - \hat{S}_h .$$
(5.198)

Another *doublet* is the mass flow rate $\rho_a U_1$ and the fluid temperature T_a. The former quantity is, for example, measured using a Brooks meter where a known fraction of the incident flow is derived and passed through a capillary tube where it is heated. The mass flow rate is deduced from the temperature rise and the heat capacity of air. The fluid temperature is measured with a thermocouple, placed in the main duct before the junction. The corresponding sensitivity coefficients, denoted by $\tilde{S}$, can be derived in a similar way as above, which gives

$$S_{U_1} = \tilde{S}_h ,$$
$$S_{T_a} = \tilde{S}_{T_a} - \tilde{S}_{\rho_a U_1} .$$
(5.199)

Sets of calibration curves or look-up tables will then use h and T_a, or $\rho_a U_1$ and T_a, as their parameters.

5.2.6 Calibration Procedures in Compressible Flows

Wires in Compressible Flows

A complete calibration procedure for high subsonic and supersonic flows, which would be similar to that in subsonic flows, is not easy to perform. One would need to change ρ, U and T_t or ρU, M and T_t independently, and this is only possible in specialized facilities. *Jones* [5.134] described such a facility which covers the ranges: $M = 0.05-1.0$, $P_t = 0.2 \times 10^5 - 1.0 \times 10^5$ Pa, and $T_t = 848-1098$ K for a 5.7 cm nozzle exit diameter. *Bonnet* and *Knani* [5.306] built a 1 cm supersonic calibration jet that permits the ranges $M = 1.3-3$, $P_t = 3 \times 10^5 - 14 \times 10^5$ Pa, and $T_t = 250-280$ K.

In practice, changes of the density ρ and mass flow ρU are made by acting on the total pressure P_t [5.122]. The Mach number M is changed by using different nozzles, as reported by *Bonnet* [5.92]. *Laufer* and *McClellan* [5.59] used a servo-driven stainless-steel flexible-plate nozzle. *Gaviglio* placed the hot-wire at different positions along the axis of a long supersonic nozzle [5.122].

Modification of T_t is less simple, because it implies an adjustment of P_t to keep ρU constant through the nozzle. Indeed, the mass flow rate for a nozzle of Mach number M and area A is given by [5.307]

$$A\rho U = AMP_t \left(\frac{\gamma}{rT_t}\right)^{1/2}$$
$$\times \left(1 + \frac{\gamma-1}{2}M^2\right)^{-(\gamma+1)/2(\gamma-1)} \quad (5.200)$$

where $r = R/m$ (5.58). Equation (5.200) shows that, if T_t is increased, P_t should also be increased. Often this adjustment is not possible, and the sensibility coefficient relative to T_t is computed using the term (XXX) in (5.82). This requires knowledge of n_t and m_t for the fluid, χ' for the wire, and $\overline{A}_w$ for the operating conditions. This method was used by many authors, *Rose* [5.135], *Ko* et al. [5.144], and *Gaviglio* [5.122]. Results concerning the sensitivity coefficients in supersonic flows are given in Sect. 5.2.2 and Fig. 5.18.

Calibration of inclined hot-wires in supersonic flows is usually attempted using the same procedure as in subsonic flow, by yawing the probe through an angle of about $\pm 10°$. Very satisfactory results are described by *Konrad* and *Smits* [5.308] who developed special probes. However, some difficulties were indeed encountered in earlier experiments. Wire overheat effects were reported by *Bonnet* and *Knani* [5.306]. *Fernando* et al. [5.309] investigated the problem of flow interference, where the shock waves from one wire and the wires' supports interfere with the flow on the other wire. The possible effect of the bowed shape of the wire is invoked by *Smits* and *Muck* [5.310]. Some density variation along the wire is also possible, as studied by *Reshotko* and *Beckwith* [5.311] and revisited by *Smits* and *Muck* [5.310]. The simple cosine law of the subsonic flows was therefore not so obvious for supersonic conditions, and careful analysis was advised prior to systematic measurements.

Rose and *Johnson* [5.312] preferred to use closely matched yawed wires, at approximately ±45° to the flow direction and to measure the instantaneous difference between the two signals to obtain u'_2. The procedure is similar to that suggested by (5.40) in subsonic flows, as ρ' and θ'_t disappear in the difference. The wires were considered matched when their visual appearances were similar and their cold resistances were within 5% of each other. In principle, this direct measure of u'_2 should be very pertinent for comparison with laser Doppler velocimetry (LDV) [5.313].

Attention should also be paid to the errors due to large fluctuations, as developed in Sect. 5.2.4. Since calibrations are difficult in compressible flows, sensitivity coefficients obtained with small perturbations in incompressible flows or subsonic flows are often used for the purpose (5.97). This implies that only rms, the PSD, or double correlation functions can be reasonably investigated.

Wall Sensors in High-Speed Flows

A direct calibration of wall films can also be made against a Preston tube in high subsonic and supersonic conditions. *Bradshaw* [5.314] and *Spina* et al. [5.315] analyzed the several assumptions which have to be fulfilled, and most importantly the existence of a logarithmic region. These two review articles also offer reliable skin-friction formulae that can be used whenever the global characteristics of a flow are known.

A comparison between wall hot-film and other experimental techniques for wall shear-stress measurements, such as laser interferometry, liquid crystals, or infrared imaging, was made by *Hall* et al. [5.87]. The Mach range was from 1.5–2.5 and the unit Reynolds number was $3.3 \times 10^6 - 13.1 \times 10^6$ /m. The thin films provided the most reliable and quantitative indicator of the details of the transition process.

Uncalibrated sensors are usually sufficient to detect boundary-layer transition, flow separation, or shock oscillation on airfoils, because of the important magnitude changes occurring on the anemometer signals. This is an advantage when doing in-flight experiments. The very stable CVA appears particularly well adapted to this objective. Experiments are reported by *Moes* et al. [5.88] over a Mach number range of 0.68–0.80 at an altitude of 6000 m, with a 45 Ni hot-film array, the anemometers being tuned in situ to take into account the 15% decrease of the *cold* resistance compared to its value at ground level.

5.2.7 Special Techniques

The Flying Hot-Wire

One of the main restrictions of stationary hot-wire anemometry is its inability to provide measurements when reversing flow occurs. This is overcome by displacing a probe, usually an X-wire, through the flow to be investigated. At a time t the probe is assumed to be at a known position x_p, y_p, z_p and to move at a known velocity U_p, V_p, W_p. The probe responds to the relative velocity, U_r, V_r, W_r, and the flow velocity, U, V, W is connected to the measured velocity U_r, V_r, W_r by

$$U = U_p + U_r \quad V = V_p + V_r \quad W = W_p + W_r \,. \tag{5.201}$$

The probe velocity should be sufficient to ensure that all signals correspond to a velocity vector remaining within the approach quadrant of the X wire probe.

The name of *flying* probe comes from the investigation of atmospheric turbulence using a probe attached to an aeroplane, as reported by *Payne* and *Lumley* [5.316], and *Sheih* et al. [5.317]. In the laboratory the probe motion can be of various types:

- Linear, as developed by *Panchapakesan* and *Lumley* [5.318, 319] to study the external zones of air or helium jets, or by *Kelso* et al. [5.320] to investigate a flow past a fence,
- Circular, as used by *Coles* and *Wadcock* [5.321] to study the flow past an airfoil, by *Cantwell* and *Coles* [5.322] to analyze the near wake of a cylinder, by *Walker* and *Maxhey* [5.323] to investigate grid turbulence, and by *Hussein* et al. [5.324] to study turbulent jets. This last work reported interesting comparisons between a stationary hot-wire, a flying hot-wire and LDV measurements,
- Curvilinear, as developed by *Thomson* and *Whitelaw* [5.325] and *Al-Kayiem* and *Bruun* [5.326]

with bean-shaped trajectories to approach separation bubbles over airfoils.

The flying hot-wire technique requires the development of an electronic timer circuit to obtain a cyclic operation, which includes routines to initiate the motion, to acquire data during the effective passage, to move the probe back to its initial position, possibly by another path than that used for measurements, and to respect a rest position so that all flow disturbances disappear. The data from each sweep then permit to obtain phase averages for every measuring point. A full calibration of the X-wire probe is required. Many practical details of the implementation of a flying hot-wire system are available in *Bruun* [5.52].

Nowadays, flying hot-wire anemometry is in competition with laser Doppler anemometry and particle image velocimetry in laboratory experiments. However, it might keep its place for the study of atmospheric turbulence. For example, the work by *Otten* et al. [5.327] is most interesting as the aircraft speed is in the range 100–200 m/s, hence involving compressible flow around the probe.

Three-Wire Pulsed Anemometers

When heat is rapidly produced in a wire by an electric voltage pulse of a few microseconds, a tracer of heated air is introduced into the flow. This thermal wake can be detected by a sensor acting as a resistance probe placed downstream of the heated wire. Furthermore, two wires placed on either side of the pulsed wire permit the determination of the velocity direction and to solve the forward–reverse ambiguity of the flow.

Figure 5.46 illustrates a pulsed probe. Usually the two sensor wires are parallel and their axes are perpendicular to the axis of the pulsed wire (Fig. 5.46a). This permits the thermal wake to impact either on one of the sensors even in the presence of turbulence. The sensors, kept parallel and perpendicular to the pulsed wire, can also be slightly inclined onto the pulsed wire. It was shown that this setup improves the directional response of the global probe. *Jaroch* [5.328], *Castro* and *Haque* [5.329], and *Castro* and *Dianat* [5.330] suggested setting the angle to about 30°. Sensor wires parallel to the pulsed wire Fig. 5.46b are appropriate when the thermal wake remains along a known line and downstream of one of the two sensors. This is possible for separated laminar flows or acoustic fields. Close to a wall where strong velocity gradients exist the pulsed wire has to be placed parallel to the wall. The nearby wall also permits the development of *through-wall* probes where all the prongs supporting the wires stem from the wall, as reported by *Castro* and *Dianat* [5.330].

Sketches and pictures of probes are available in *Bradbury* and *Castro* [5.331], *Westphal* et al. [5.332], *Jaroch* and *Dahm* [5.333], and *Castro* and *Dianat* [5.330] among others. A 5–10 μm tungsten wire is often used for the pulsed wire to sustain the thermal loading, while 2–5 μm wires are used for the sensors. The wire length is in the range 6–10 mm, and the spacing s is in the range 1–2 mm. The working velocity range is 1–10 m/s. The repetition frequency of the pulse is rather low, around 50 Hz, because the pulsed wire is required to cool completely between pulses.

The time of flight, T, of the wake from the pulsed wire to one of the sensing wires has to be linked to the magnitude of the velocity $|U|$ and to the yaw angle θ. The velocity vector is supposed to lie in the plane of the sensors. This requires a preliminary probe orientation, comparable to the use of X-wire probes (Sect. 5.2.2). There is also an acceptance angle for the pulsed probe which is connected to the length l of the sensors and the spacing s between the pulsed wire and either one of the sensors. Obviously, the maximum possible yaw angle is $\theta_{max} = \tan^{-1}(l/2s)$. With $l = 10$ mm and $s = 1$ mm, this leads to $\theta_{max} \simeq 75°$.

The velocity magnitude $|U|$ and the time of flight T could be thought of as simply connected through a convective process, which would give

$$T = \frac{s}{|U|\cos\theta} \quad \text{and} \quad U_1 = \frac{s}{T}. \tag{5.202}$$

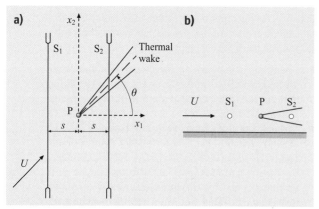

Fig. 5.46a,b Sketch of a three-wire pulsed probe. P is the pulsed wire, S_1 and S_2 are the sensors detecting the thermal wake: (**a**) general set up with sensors perpendicular to the pulsed wire, (**b**) special set up for laminar or acoustic flows

Calibrations show that the cosine law is approximately obtained at fixed values of $|U|$ [5.329, 334]. Regarding the time of flight T, more-complex expressions than (5.202) are reported at fixed θ. For example *Handford* and *Bradshaw* [5.334] or *Castro* and *Haque* [5.330] indicated

$$U = a\frac{1}{T} + b\frac{1}{T^2} + c\frac{1}{T^3} + d\frac{1}{T^4}. \quad (5.203)$$

The reason for this difference is that many physical mechanisms other than convection are involved in the generation, development, and reception of the thermal wake.

- The turbulent dispersion brought due to all three turbulent velocity fluctuations enlarge the wake. An estimate necessarily involves the statistical properties of the turbulence field. Very roughly, the assumption of a short-time dispersion gives a displacement in the i direction $\delta_i \simeq u_i' T \simeq 0.2$ mm for a turbulence intensity around 20%, for $U = 10$ m/s and $T = 0.1$ ms. The viscous diffusion of the wake adds a small widening of the wake, on the order of 0.02 mm as given by $\delta \simeq (\nu T)^{1/2}$ for Pr = 1.
- When the thermal wake hits the sensor, the impact only concerns a small fraction of the length of the wire, around 0.2 mm as estimated above, when the sensors are normal to the pulsed wire. The thermal conduction along the axis of the sensor wire becomes important, with heat conducted towards both the supporting prongs. This problem is more complicated than that considered in Sect. 5.2.2 where the whole wire was submitted to a flow. This topic has mostly been considered by *Bradbury* and *Castro* [5.331].
- Close to a wall the contours of constant temperature after the release of a heat *puff* are additionally distorted by the velocity shear as they move downstream. *Castro* and *Dianat* [5.330] established analytical results for a uniform shear flow. These contours are ellipses, highly elongated along the O_{x_1} direction by up to 0.4 mm when arriving at the sensing probe, and slightly tilted up towards the rapidly moving fluid. As a consequence, only a smeared temperature front arrives on the sensor. Incidentally, these contours are very similar to those established by *Tennekes* and *Lumley* [5.162] for the dispersion of particles in a uniform shear flow.
- Finally, the thermal inertia of the sensor introduces a time delay in its response. *Bruun* [5.52] established an interesting relation that gives the time t_{max} at which the maximum amplitude of the signal occurs, taking into account the relative importance of the thermal lag of the sensor M_S to that of the pulsed wire M_P

$$t_{max} = \frac{s}{U} + \frac{M_P M_S}{M_P - M_S} \log \frac{M_P}{M_S}. \quad (5.204)$$

To achieve the best measurement of the time of flight, *Handford* and *Bradshaw* [5.334] advise working on the signal slope, in order to differentiate the time signal provided by the sensor. The flight time is then deduced from the point at which the slope value passes through a preset level. This method was found to be nearly independent of the flow speed.

The number of flight times that are necessary to achieve a prescribed accuracy depends strongly on both the turbulence characteristics that are studied, mean velocities or Reynolds stresses, and on the level and scale of turbulence present. An order of magnitude is 2000 flight times, which gives a total acquisition time of around 40 s.

During measurements, the most appropriate orientation of the pulse-wire probe within the flow also has to be selected. Alignment with the streamwise direction $\theta = 0$ provides accurate measurements of $\overline{U}_1$ and $\overline{u_1'^2}$. Other angles θ are introduced to obtain $\overline{U}_2$, $\overline{u_2'^2}$, and $\overline{u_1'u_2'}$. Pitch angles are also sometimes needed. Many details are given by *Castro* and *Cheun* [5.335], and *Jaroch* [5.328].

The pulsed probes have proved their capabilities in the highly recirculating zones close to obstacles. *Bradbury* [5.336] obtained complete streamline patterns in the very near wake of a flat plate placed normal to a 7.5 m/s flow. *Castro* and *Haque* [5.329] added a central splitter and investigated the normal and tangential Reynolds stresses and the momentum balance. Experiments in an axisymmetric free jet were reported by *Jaroch* [5.328]. The measurements of $\overline{U}_1$ and $\overline{u_1'^2}$ with single hot-wire and different types of pulsed-wire probes agreed very well. Some scatter appeared on other quantities such as $\overline{U}_2$, $\overline{u_2'^2}$, or $\overline{u_1'u_2'}$. To reduce this, he advised the use of pulsed-wire probes with the largest and most symmetric acceptance cone. *Venas* et al. [5.337] conducted experiments in wall jets and reported satisfactory agreement with laser Doppler anemometer data.

One-Wire Pulsed Anemometers

Another technique to measure flow velocity uses the response of a wire first heated with a very short electrical pulse and then left to relax and cool. The velocity can be deduced from the time history of the evolution, which is a function of the velocity component U nor-

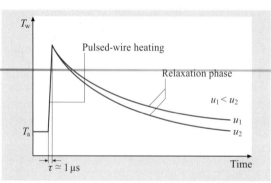

Fig. 5.47 Time evolution of the pulsed wire temperature during pulse and relaxation periods, for two speeds, U_1 and U_2 with $U_1 < U_2$: pulse duration 1 μs, current intensity when pulsing $I_{wp} = 1$ A. The pulse heats the wire by around 50 K. (After *Mathioulakis* et al. [5.339])

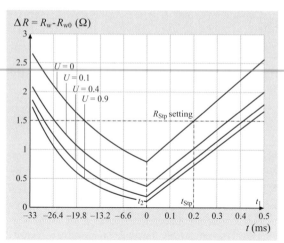

Fig. 5.48 Resistance variation in one balanced cooling–heating cycle for the heat pulse integrator method. Pulse duration 20 μs, current intensity when pulsing $I_{wp} = 0.5$ A. Note the asymmetric shape of the curve. (After *Li* and *Gow* [5.341])

mal to the wire. This idea was introduced by *Calvet* and *Liousse* [5.338] in 1971 for free convection flows, and received a renewal of attention after 1994 through the works of *Mathioulakis* et al. [5.339], *Grignon* et al. [5.340], *Li* and *Gow* [5.341], *Foss* et al. [5.342], *Foss* and *Hicks* [5.343], *Morris* and *Foss* [5.155], and *Hicks* et al. [5.344].

In the first set of papers, by Calvet, Liousse, Mathioulakis and Grignon, the wire is at ambient temperature before the pulse triggering, and after the pulse the wire is left to relax back to the ambient temperature. Figure 5.47 illustrates the temperature evolution of the wire for two incident velocities.

In that case, the characteristic time τ_1 for the temperature rise during the pulse period is obtained by neglecting the convection term in the heat balance of the wire given by (5.98). This leads to

$$\tau_1 = \frac{\pi d^2}{4} \rho_w c_w l \frac{1}{R_a \chi I_{wp}^2}, \quad (5.205)$$

where I_{wp} is the constant current applied during the pulse.

During the relaxation period, the characteristic time τ_2 is the time M_w^{CCA} given by (5.103). For clarity, this equation is rewritten, as it shows that the velocity dependence appears through a Nusselt number

$$\tau_2 = (1 + a_w) \frac{d^2}{4} \frac{\rho_w c_w}{k_a} \frac{1}{\mathrm{Nu}_a}, \quad (5.206)$$

where I_w is the constant floor current through the wire. Obviously $\tau_1 \ll \tau_2$. The temperature rise is sharp and almost linear, and the temperature relaxation is close to an exponential law. In their experiments the authors advantageously used a long wire, $l/d = 700$, with $d = 5$ μm and $l = 3.5$ mm to avoid end losses. The pulse frequency was around 1 kHz, which appeared sufficient for the free convection flows investigated, as the velocity was in the range 0.1–1 m/s. The current I_w was set low enough that the wire was *cold* between pulses and the ambient temperature could be measured.

The velocity U is obtained from (5.206) using (5.45) or (5.46). A regression was applied to the $R_w(t)$ variation during relaxation. Fitting the curve by an exponential law was found possible despite some discrepancies at short times because of thermal diffusion close to the hot wire. *Breton* [5.345] and *Grignon* et al. [5.340] developed analytical models to investigate the interactions that take place between convection, diffusion, and conduction at the wire level.

In the work of Li and Gow, the wire is not allowed to reach the ambient condition before it is reheated. The wire then works as a *heat integrator*, although a balanced cooling–heating cycle is permitted as illustrated in Fig. 5.48. Several parameters of interest were also investigated: the pulse width, the length of the interpulse period compared to the relaxation time of the wire, and the time it takes for the wire to reach some imposed high temperature. In Fig. 5.48 this set temperature point is noted Stp. As an application, the case of a 35 mm cylinder made of Tufnol and placed in a 5–10 m/s flow was considered. The cylinder was pierced to allow the

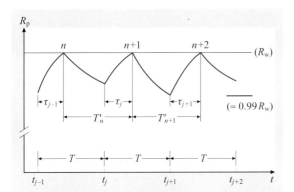

Fig. 5.49 Sketch of the heating and relaxing time traces for a one-wire pulsed probe, during three cycles. A ceiling value R_w is imposed, T is the clock period which imposes reheating. The heating and relaxing curves differ from one cycle to the next due to a velocity change. (After *Foss et al.* [5.342])

installation of 32 equally spaced wires mounted normal to the circular section. The wires, 20 μm platinum, were almost touching the cylinder surface, hence their cooling by the flow was close to that of hot-wire wall sensors. The pulse frequency was around 100 Hz. The results they found revealed good agreement with the survey conducted by *Bellhouse* and *Schultz* [5.146].

In the works by Foss, Hicks and coworkers, the wire is almost not allowed to cool between pulses, roughly by only 0.01%. The pulsing frequency is therefore very high, around 100 kHz in their most recent developments. The heating and relaxing time traces show different concavities, but their durations become roughly the same order because the convection by the flow becomes important during both the heating and relaxing periods. Figure 5.49 illustrates three cycles. Because of the imposed ceiling resistance R_w and the duration of the puls,e which has to change when the velocity varies, their system has been called a *pulse-width-modulated constant-temperature anemometer* (PWM-CTA). Block diagrams and details for the time acquisitions and connection with the velocity are available in *Foss et al.* [5.342]. Calibration of the system has been performed up to around 30 m/s and mean velocity profiles have been acquired in the shear layer of a subsonic jet.

A PWM-CTA has distinctive features. Fundamentally, it is a digital anemometer although the heat transfer and power dissipation processes that govern its behavior are analogue in character. Unlike a conventional CTA it does not require an A/D converter to transfer its output for further processing. Also, a PWM-CTA does not need a feedback loop, which is unstable in a conventional CTA. The electronic noise per se is not a relevant issue, but the comparator that is used to define the times τ_j in a cycle results in an inherent uncertainty. Further developments of these important topics are presented later in this Handbook.

5.2.8 A Comprehensive Technique for X-Array Calibration and Data Processing

Calibration

This section presents a calibration/processing strategy that is intended to provide an efficient and an accurate procedure to recover Q and γ from E_1 and E_2 (the slant-wire voltages), where Q is the velocity magnitude in the plane of the X and γ is the angle of the flow (in the plane defined by the X-array) (Fig. 5.50). This section provides an alternative approach to that presented in Sect. 5.2 of the Handbook. It is noted that the use of analytic expressions ensures the monotonicity of the calibration data: $\partial E/\partial Q$ and $|\partial E/\partial \gamma|$ are uniformly positive over the calibrated range of Q and γ. E_1 and E_2 are the output voltages from the two slant wires of the array. The velocity and angle ranges are to define a *complete calibration* such that the Q and γ values of the investigated flow field are adequately included in the calibration data.

An example of a *complete* calibration data set would be one for which 13 angles ($\gamma = +36°, +30°\ldots 0°, \ldots -30°, -36°$) were used to position the probe axis with respect to the calibration stream and a series of velocity values were obtained at each angle. These data could be fit with (using the CTA as an example):

$$E_j^2 = A_j(\gamma) + B_j(\gamma) Q^{n_j(\gamma)} \qquad (5.207)$$

for the 13 γ values (where $j = 1, 2$ for the two wires of the X-array).

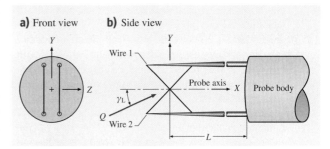

Fig. 5.50 Definition sketch for an X-array

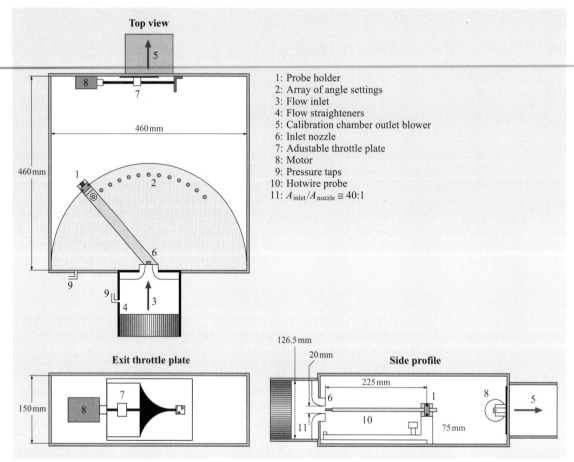

Fig. 5.51 Transient flow calibration device

The effective velocity formulation [(5.172) of Sect. 5.2.4] *collapses* the angle information of (5.207) into the single coefficient k. Experience [5.346] shows that A and n are distinctly not constant over all angles even for an X-array that meets the stringent *Strohl* and *Comte-Bellot* [5.69] geometric conditions. These earlier results are confirmed by the data of this section. The deviations in the A, B, and n values are most pronounced when the velocity approaches tangency with respect to the sensor. As a result, the use of an effective velocity (which implicitly assumes constant A and n values with the variable B reflected in the k value) cannot provide as precise a fit to the calibration data over all γ values as that provided by (5.207).

Equations of the form (5.207) also ensure the monotonicity noted above.

The requirement for an extensive body of calibration data can be effectively met using a *controlled transient* calibration tunnel to obtain the data in a minimum time. Figure 5.51 shows a representative configuration for such a calibration device. Figure 5.52 presents the transient data for each slant wire. Note that the *jagged* traces are a result of acquiring the voltages during the transient opening of the throttle plate shown in Fig. 5.51.

The analytic form of (5.207) will *smooth* the jagged data into a monotonically increasing distribution as shown in Fig. 5.53. Specifically, the processed data of Fig. 5.52 are presented in the form of $E(Q, \gamma)$ for the range $2 \leq Q \leq 18$ m/s in Fig. 5.53. It is apparent that a smooth calibration data set is achieved. The stepper-motor-controlled throttle plate was driven for a period of 40 s to develop the velocity range of 2–18 m/s. The complete calibration (13 angles) can nominally be executed in 15 min.

The A, B, and n values at each γ are selected to provide the minimum standard deviation between the

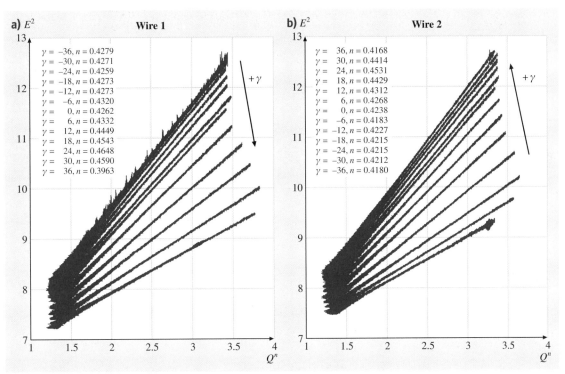

Fig. 5.52a,b Initial calibration data for the two slant-wire sensors. (a) $E_1^2 = f_1(Q^n; \gamma)$. (b) $E_2^2 = f_2(Q^n; \gamma)$

measured and the analytical values (5.208). Figure 5.54 presents $A(\gamma)$, $B(\gamma)$, and $n(\gamma)$; it is apparent that all three coefficients should be allowed to vary with respect to the calibration angle γ.

The velocity magnitudes are obtained from the pressure differences between the tap that is forward of the nozzle and the tap in the receiver chamber. (It has been confirmed that the upstream tap provides a true value for the stagnation pressure.) This use of the Bernoulli equation is predicated upon a *steady-state* condition. Specifically, that

$$\int_{S_1}^{S_2} \frac{\partial V}{\partial t} \, ds \ll \frac{1}{2}\left[V^2(S_2) - V^2(S_1)\right]$$

$$+ \frac{1}{\rho}[p(S_2) - p(S_1)]. \qquad (5.208)$$

This assumption has been evaluated by comparing the results of:

- a steady-state calibration – a 10 s acquisition at each velocity setting, with
- the results of three different transient conditions: 40, 50, 60 s durations for the acquisition period.

Figure 5.55 shows these data for the representative condition of wire 2 at $\gamma = 18°$. A fixed value of ΔE^2 has been added to the E^2 values of the transient data in order to separate these distributions on the figure. It is apparent that the transient condition does not violate the use of the Bernoulli equation. Obviously, a much longer calibration period would be required if a steady-state calibration protocol were used for all Q and γ values. These data also verify the use of (5.207) to average the unsteady oscillations properly about the correct mean values.

Processing Algorithm

Flow field measurements, of course, invert the calibration process. Voltages (E_1 and E_2) are measured and the Q and γ values are sought. The recommended algorithm can best be described graphically. The computational method described in this section was developed by *Browne* et al. [5.249] and independently by *Morris* [5.347]. It is also readily implemented via software postprocessing of the time series of voltage values.

Consider a voltage pair: $E_1 = 2.734$ V and $E_2 = 2.653$ V. From Fig. 5.53, it is apparent that these voltage values will intersect the 13 curves, resulting in

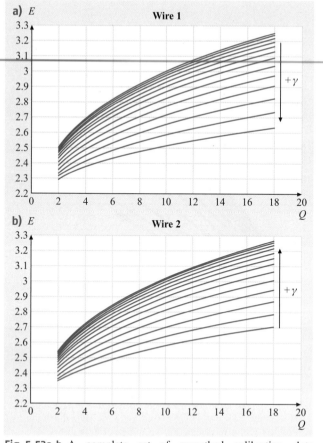

Fig. 5.53a,b A complete set of smoothed calibration data for $0 \leq Q \leq 18\,\text{m/s}$ and $-36 \leq \gamma \leq 36°$. (a) $E_1 = g_1(Q;\gamma)$. (b) $E_2 = g_2(Q;\gamma)$.

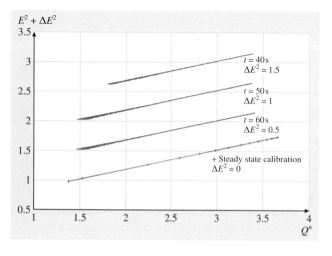

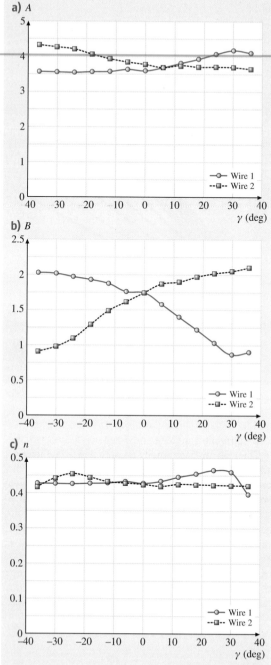

Fig. 5.54a–c Best-fit A, B, and n values for the slant-wire sensors 1 and 2

Fig. 5.55 Confirmation of the transient velocity calibration protocol. Note: these data are for slant-wire $\gamma = 18°$ ◀

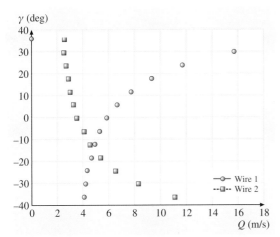

Fig. 5.56 Q and γ values that are separately compatible with the measured E_1 and E_2 values

26 discrete velocity values. The 13 (Q, γ) values for each sensor can then be plotted on common axes, as shown in Fig. 5.56. Consider that smooth curves are used to interpolate between the discrete data points from each sensor.

It is apparent that the true (Q, γ) values for the measured (E_1, E_2) values are defined by the intersection of the two curves. Computationally, one could fit a second- or third-order polynomial through the three or five points closest to the intersection point to determine the intersection values. As shown in Fig. 5.57, both interpolation curves provide a precise indication of the (Q, γ) values that correspond to the (E_1, E_2) values. The inferred velocity magnitudes differ by 0.6% and the angles differ by 0.11°. These values are within the tolerances one would expect from an X-array measurement.

Summary
The $A(\gamma)$, $B(\gamma)$, and $n(\gamma)$ values of Fig. 5.54 indicate that an accurate calibration requires a complete calibration of an X-array probe. If one adopts the practice of acquiring a complete calibration *before* and *after*

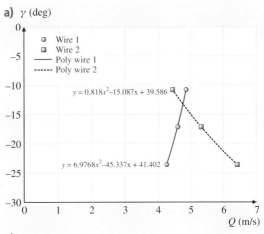

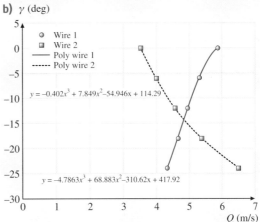

Fig. 5.57a,b Derived (Q, γ) values from the measured (E_1, E_2) values: (**a**) three-point second-order fit ($Q = 4.88$ m/s, $\gamma = -14.6°$), (**b**) five-point third-order fit. ($Q = 4.85$ m/s, $\gamma = -14.4°$)

a data set (to confirm the integrity of the data), then an efficient as well as a complete calibration process is advantageous. The transient calibration described above provides these attributes.

5.3 Particle-Based Techniques

5.3.1 Tracer Particles

In this Section four optical measurement techniques for flow velocity based on tracer or seed particles in the flow will be described: laser Doppler anemometry/velocimetry (LDA, LDV), particle image velocimetry (PIV), Doppler global velocimetry (DGV), and laser transit velocimetry (LTV). These techniques rely on the presence of particles in the flow that not only follow all flow velocity fluctuations but are also sufficient in number to provide the desired spatial or temporal resolution of the measured flow velocity. Of course it is desirable that these tracer particles are at the same time good light-scattering centers, since this improves the signal quality for a given incident laser power. The fact that flow tracking demands small particles and light scattering improves with increasing particle size already suggests the need for optimization when choosing tracer particles. The following section deals therefore with the tracer particle issues common to all of the measurement techniques, in particular the particle motion in flows is examined, the scattering properties of particles, their generation and their introduction into the flow.

Particle Motion in Flows

Within the approximation of low Reynolds number, the equation of motion of a small spherical particle immersed in a fluid flow is given as

$$\frac{4}{3}\pi a^3 \rho_p \frac{d\mathbf{v}}{dt}$$
$$= \underbrace{\frac{4}{3}\pi a^3 \rho_f \frac{D\mathbf{V}}{Dt}}_{\text{non-inertial force}} + \underbrace{\frac{4}{3}\pi a^3 (\rho_p - \rho_f)\mathbf{g}}_{\text{net body force}}$$
$$\underbrace{- 6\pi \mu a \left[(\mathbf{v}_p - \mathbf{U}) - \frac{1}{6}a^2 \nabla^2 \mathbf{U}\right]}_{\text{quasi-steady viscous force}}$$
$$\underbrace{- 6\pi \mu a^2 \int_0^t \frac{d}{d\tau}\left[(\mathbf{v}_p - \mathbf{U}) - \frac{1}{6}a^2 \nabla^2 \mathbf{U}\right] \frac{d\tau}{\sqrt{\nu(t-\tau)}}}_{\text{time history (Basset) force}}$$
$$\underbrace{- \frac{2}{3}\pi a^3 \frac{d}{dt}\left[(\mathbf{v}_p - \mathbf{U}) - \frac{1}{10}a^2 \nabla^2 \mathbf{U}\right]}_{\text{added mass force}} + \underbrace{\mathbf{L}}_{\text{lift force}} ,$$

(5.209)

where $a \, (= 1/2d_p)$ is the tracer particle radius, $\mathbf{U}$ is the velocity of the surrounding fluid, and $\mathbf{g}$ is the gravitational acceleration. The lift force originates from the finite circulation around the sphere. It can either be due to an intrinsic rotation of the sphere immersed in a uniform flow (Magnus effect) or to rotational flow such as when a particle is immersed in a shear layer. In the latter case the fluid exerts a net force in the direction of the velocity gradient. This equation is known as the Basset–Boussinesq–Oseen (BBO) equation and details of its

Table 5.4 Commonly used tracer particles for PIV experiments

Fluid	Material	Diameter (μm)	Density (kg/m³)
Air	DEHS	1–3	10^3
–	Glycol–water solution	1–3	10^3
–	Vegetable oil	1–3	10^3
–	TiO$_2$	0.2–0.5	$1-4 \times 10^3$
Water	Latex	5–50	10^3
–	Sphericell	10–100	$0.95-1.05 \times 10^3$
–	Silver-coated hollow glass spheres	30–100	$> 10^3$

derivation can be found in *Crowe* et al. [5.348]. Solutions for this equation have been given by *Hjemfelt* and *Mockros* [5.349] and by *Chao* [5.350]. A more recent discussion of this equation and solutions for turbulent flow has been given by *Mei* [5.351] and a summary of tracer particle issues in general can be found in *Melling* [5.352].

For very small particle tracers as used in LDA or PIV the first part of the quasi-steady viscous term (Stokes drag) dominates the right-hand side of the equation. In the approximation that $D\mathbf{U}/Dt = d\mathbf{v}_p/dt$ (5.209) allows the difference between the particle velocity $\mathbf{v}_p$ and that of the surrounding fluid $\mathbf{U}$ to be estimated as

$$\mathbf{v}_p - \mathbf{U} = \frac{2}{9}\frac{a^2(\rho_p - \rho_f)}{\mu}\frac{d\mathbf{v}_p}{dt} . \quad (5.210)$$

The velocity difference $\mathbf{v}_p - \mathbf{U}$ is referred to as the slip velocity. Clearly from (5.210) the choice of neutrally buoyant particles $(\rho_p - \rho_f)/\rho_f = 0$ leads to particle tracers that accurately follow the flow. While this condition can be easily satisfied for liquid flows, it cannot be achieved so easily in gas flows, where a typical value of the density ratio is $\rho_p/\rho_f = O(10^3)$, and therefore particles with a smaller diameter ($0.5 \, \mu\text{m} < d_p < 5 \, \mu\text{m}$) are employed. Larger particles used in liquid flows approximating the neutral buoyancy condition can be selected from a wide choice of materials, for instance hollow glass spheres or polyamide particles of relatively large diameter ($5 \, \mu\text{m} < d_p < 50 \, \mu\text{m}$) can be used. Table 5.4 summarizes seeding particle properties in relation to different flow experiments.

From the above relations in the case that $\rho_p/\rho_f \gg 1$ (gas flows), a single exponential decay law is commonly used to model the particle response to a stepwise variation in the flow velocity. The characteristic response time τ reads

$$\tau = \frac{2}{9}a^2\frac{\rho - p}{\mu}. \qquad (5.211)$$

The particle response time should be kept smaller than the smallest time scale of the flow. Moreover in high-speed flows, especially if shock waves are present, exact flow tracing can never be achieved. At least within limited regions downstream of shock waves (Fig. 5.59), where the flow decelerates abruptly, the particle tracers decelerate along an exponential decay curve (Fig. 5.60). The thickness of a shock wave occurring at sea level atmospheric conditions is of the order of $0.1\,\mu\text{m}$. Further information regarding the response of particles across shock waves can be found in *Dring* [5.354] or *Tedeschi* et al. [5.355].

The fidelity of the flow tracers in turbulent flows is quantified by the particle Stokes number S_k, defined as the ratio between τ and the characteristic flow time scale τ_f. The most critical conditions are met when particle tracers are immersed in turbulent flows with a high Reynolds number, where a wide range of turbulent scales is present. From a practical point of view it can be stated that the condition $S_k < 0.1$ returns an acceptable flow tracing accuracy with errors below 1%.

Among the requirements for the choice of the seeding material, health and safety aspects have to be taken into account. The material should not be hazardous or toxic if inhaled, it should not be corrosive or reactive when in contact with parts of the flow facility or other instrumentation. Finally, seeding material that naturally evaporates leaving minimum residues introduces less contamination of the facility and its optical interfaces.

Scattering Properties of Tracer Particles

Small particles are required to fulfil the fluid-mechanical requirements of tracers. However, as an opposing requirement, the particles should scatter enough light in order to be visible. Typical particle dimensions are on the order of a micrometer for gas (i.e., air) flows, and tens of micrometers for liquid (i.e., water) flows. This means

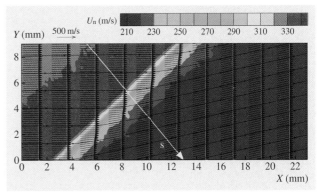

Fig. 5.59 PIV measurement of the flow across a planar shock wave. (After *Scarano* and *van Oudheusden* [5.353])

that the scattering of light by these particles occurs in the so-called Mie regime, where the particle diameter d_p is larger than the light wavelength λ. Another approach is to use tracer particles that contain a fluorescent dye; the light that is absorbed by the dye is emitted at a longer wavelength, which makes it possible to distinguish tracer particles from other objects (e.g. interfaces).

Mie Scattering. The particle scattering cross section depends on the particle diameter, the wavelength of the

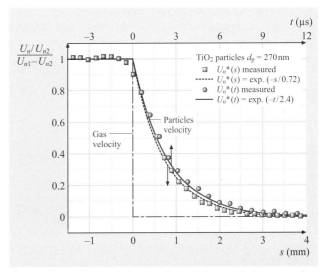

Fig. 5.60 (Particle) velocity profile and time evolution normal to the shock wave. Across shock waves the particle velocity lag is not the only source of error. Often the velocity measurement is affected by the limited spatio-temporal resolution of the measurement technique. This appears as a smoothing effect of the velocity profile at the shock location. (After *Scarano* and *van Oudheusden* [5.353])

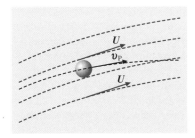

Fig. 5.58 Velocity discrepancy between the particle and the surrounding fluid velocity

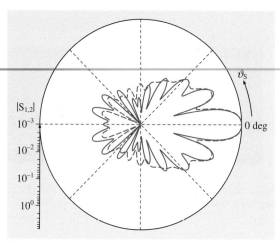

Fig. 5.61 Scattering function for a Mie parameter value of 10 computed for a water droplet in air ($m = 1.334$); *solid line*: perpendicular polarization, *dashed line*: parallel polarization (after *Albrecht* et al. [5.356])

light, and the refractive index of the particle (m) (relative to the refractive index of the surrounding medium). The Mie parameter $x_M = \pi d_p / \pi$ governs the scattering amplitude. The scattering cross section shows a very strong variation as a function of the scattering angle (υ_s). A typical Mie scattering diagram is shown in Fig. 5.61. Most of the light is scattered in the forward direction, and an appreciable amount of light is scattered in the backward direction; at a scattering angle of 90°, e.g., for single-camera PIV with the optical axis normal to the light-sheet plane, the scattering amplitude is generally very low. This is why PIV needs a much stronger light source than LDA (which operates in near-forward or near-backward scattering mode). For very small tracer particles (of less than a few μm) the light scattering cross section also depends on the direction of polarization relative to the scattering direction. As most continuous wave lasers used for LDA and frequency-doubled Nd:YAG lasers used for PIV emit polarized light, one should find the appropriate orientation of the doubling crystal of the laser.

In the Mie regime ($d_p > \lambda$) the scattering cross section is roughly proportional to the d_p^2 (i.e., the particle surface area), but when the particle diameter becomes less than the wavelength of light ($x_M < 3$) the scattering cross section is proportional to d_p^4 (Fig. 5.62). This implies a practical lower limit for tracer particles to a fraction of micrometer.

The scattering cross section also depends on the particle material refractive index relative to the refractive index of the surrounding medium. For small liquid droplets and solid particles in a gas (e.g., air) the relative change in refractive index is about 1.3–1.5, but in a liquid it is about 1.2 or less; this is why LDA and PIV in liquid flows generally requires larger tracer particles in comparison to air. The use of hollow glass spheres in liquids combines the advantages of (near-)neutral buoyancy and a large change in refractive index at the internal glass/bubble interface. The reflectivity of particles can also be improved by means of a metal coating. Silver-coated glass beads have a high reflectivity, but apart from the high specific gravity (about that of water) it is unknown whether the coating remains intact when suspended in water.

Fluorescent Tracer Particles. Tracer particles can be labeled with a fluorescent dye (e.g., Rhodamine); the

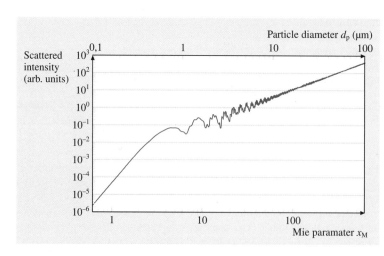

Fig. 5.62 Scattering intensity as a function of the Mie parameter x_M computed for a water droplet in air ($m = 1.333$, $\lambda_w = 514.5$ nm) at a scattering angle of $\vartheta_s = 30°$ and a circular receiver aperture of radius 10 mm (after *Albrecht* et al. [5.356])

Table 5.5 Summary of fluorescent dye properties (from *Rottenkolber* et al. [5.358])

Attribute	Rhodamin 6G	Rhodamin B	Sulforhodamin	DCM
Maximum absorption wavelength (nm)	524	544	529	481
Maximum emission wavelength (nm)	≈ 580	≈ 600	≈ 650	≈ 650
Quantum yield	High	Medium	Low	High
Lifetime (e^{-2}-decay) (ns)	6.4	2.5	5.8	1.6
Lifetime (99%-decay) (ns)	14.5	8.4	12.3	5.2
Onset (ps)	< 25	< 18	< 35	< 14

dye absorbs the incident light, which is then re-emitted at a longer wavelength. For example, when the green (532 nm) light of a Nd:YAG laser illuminates tracer particles doped with Rhodamine, the particles not only scatter the laser light, but also emit orange–red (550–680 nm) light. This makes it possible to separate the tracer particles from other scattering objects, such as walls, bubbles, droplets and other particles, by means of an appropriate optical filter that blocks the light with the wavelength of the incident light (viz., light sheet) and passes the fluorescent light. Fluorescent tracer particles are commonly used for PIV in multiphase flows, in near-wall measurements, and in microscopic measurements [5.357].

The emission of fluorescent light is isotropic, so there is no dependence of particle image intensity as a function of viewing angle with respect to the light sheet orientation (see above). The fluorescent intensity depends on the amount of fluorescent dye in the tracer particle; this is generally proportional to the particle volume, which implies that the brightness decreases in proportion to d_p^3 (as opposed to Mie scattering, which decreases in proportion to d_p^2). The duration of the fluorescence is typically ≈ 10 ns (viz., instantaneous), although prolonged exposure induces photo-bleaching that reduces the fluorescence intensity. Also, the fluorescent light is incoherent, which implies that such particles are unsuitable for holographic recording approaches.

PIV can also be combined with instantaneous thermometry by using tracer particles that contain liquid crystals. When combined with a color camera one can measure the local temperature from the color of the light scattered by these tracer particles [5.360, 361].

Fluorescent tracer particles have also been successfully used for laser Doppler measurements, the main issue being the fluorescent onset time and lifetime, since these will determine the dynamic range of detectable velocities. Using ultrashort laser pulses and time-resolved fluorescence spectroscopy, *Rottenkolber* et al. [5.362] have determined the characteristics of four common fluorescent dyes, as summarized in Table 5.5.

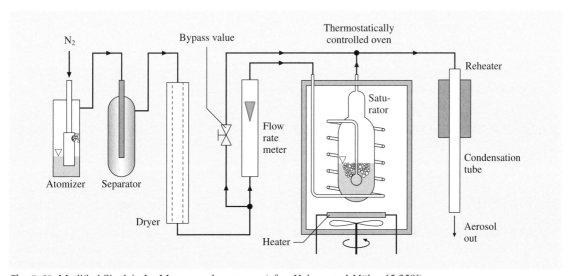

Fig. 5.63 Modified Sinclair–La Mer aerosol generator (after *Helsper* and *Mölter* [5.359])

Particle Generation

Some of the best suited particles for laser Doppler and PIV measurements, such as polymers, hollow SiO_2 microballoons or metallic-coated spheres, are quite complex to fabricate but are commercially available. Since these particles are not always inexpensive, they are better suited for closed loop systems. When larger quantities of particles are required, more economical solutions are sought and dedicated particle generators offer a good solution, some of which are also commercially available. In many cases however, the user provides his own particle generation hardware; details of several such solutions are provided below.

Particle generation can be divided according to solid and liquid particles. Solid particles or powders are used to seed liquid flows or in applications such as combustion, where liquid droplets would evaporate. In either case a monodispersed size distribution and a spherical particle shape are desirable, the latter condition being automatically fulfilled by small liquid droplets.

Droplet Generation

Condensation. Droplet generation through condensation can be spontaneous in a highly saturated vapor stream; however, the necessary control of size and concentration is really only achievable if the vapor pressure, the condensation temperature, and the seed nuclei concentration are all controlled. Such a generator, which is a modification of the classical *Sinclair* and *La Mer* [5.363] generator, is pictured in Fig. 5.63, adapted from *Helsper* and *Mölter* [5.359] and commercially available as a monodispersed aerosol generator (Palas MAGE-2000).

Fog generators, as are used in theatrical productions or discotheques, have also been used for particle seeding or flow visualization in wind tunnels. These devices also

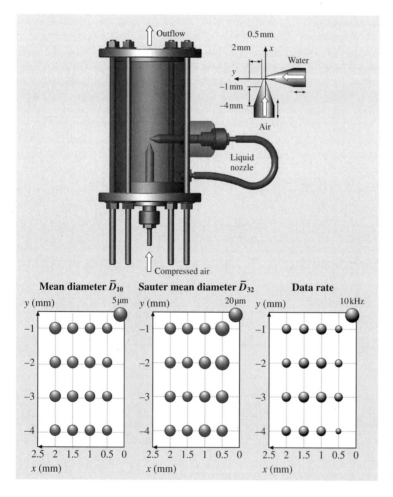

Fig. 5.64 A twin-fluid atomizer showing data rate, mean diameter and Sauter mean diameter as a function of nozzle alignment

work on the principle of condensation. Oil is evaporated on a heated plate and condenses to fine droplets in the gas flow; however, the size distribution is wide and the delivery rate is not as well controlled. These devices do not work well in very dry atmospheres, since the fog immediately evaporates. The longevity of the droplets can be controlled through the choice of fluid and its vapor pressure. Such devices are also commercially available and are a good choice where high particle rates are required, for example in large wind tunnels.

Atomization. By far the most common method of seeding gas flows is through liquid atomization. Of the many atomizer types available [5.364] the common nebulizer type used in inhalation devices is the most suitable. This is known as a twin-fluid or air-assist atomizer and draws liquid from a reservoir into a high-speed gas jet, where the liquid stream undergoes prompt atomization. Since this type of atomization results in a polydisperse spray, an impactor or separator is used to remove large droplets from the flow stream.

The droplet size depends primarily on the atomizing airflow rate and on the liquid used. Typical mean particle sizes range from $0.2\,\mu m$ using DEHS (di-ethyl-hexyl-sebacat) to $4-5\,\mu m$ with water. Fluids like olive oil or water with alcohol lie between. Particle rates as high as $10^{13}\,m^3/s$ are possible from a single nozzle (AGF 2.0, Palas GmbH), although a cascade of nozzles can also be used to vary concentration at almost constant flow rate (six-jet atomizer model 9306, TSI Inc.). Where necessary, bypass air can be used to dilute the flow to concentration levels appropriate for the particular experiment.

For many applications, the common inhalation or medication nebulizer offers an economical solution and can be easily obtained through medical suppliers. These units are available with reservoirs ranging from 20 cl to 1 l. An impactor plate is often built into such devices and a mean particle diameter of about $(4\pm2)\,\mu m$ with water can be expected.

The actual construction of such twin-fluid atomizers is quite varied, several designs being suitable for construction in ones own laboratory. One such design is pictured in Fig. 5.64. A high-speed airstream from a central nozzle creates a low static pressure and draws liquid from a reservoir into the air jet, where it is atomized. The alignment of the air nozzle with the liquid nozzle influences the number density of the droplets but not the drop size distribution, as illustrated in Fig. 5.64 for water and various operating conditions.

A popular atomizer design is the Laskin nozzle, as described in detail by *Echols* and *Young* [5.365] and as shown in Fig. 5.65. High-speed submerged air jets are created by small holes (1 mm) in the nozzle tube (10 mm diameter). These exit the nozzle tube just below a collar (16 mm diameter), which has 2 mm holes tangent to the nozzle. The exact location of these holes is quite critical for proper functioning of the unit. The dimensions of the nozzle tube and the collar are not as critical. There should be sufficient clearance between the bottom of the nozzle tube and the bottom of the container, at least 25 mm. The air holes should be no more that 25 mm below the free surface of the liquid.

The number of particles is controlled through the number of activated nozzles. The droplet size distribution is only slightly dependent on the feed air pressure but does depend on the liquid used, oils leading to a smaller size distribution than water. Dilution air can be used to adjust the final output concentration or to achieve isokinetic introduction of the particles into the

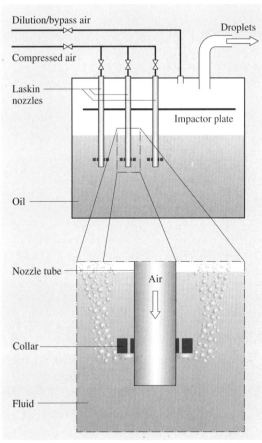

Fig. 5.65 Schematic of a Laskin nozzle unit for droplet seeding

flow stream. Laskin nozzles exhibit mean droplet diameters of about 1 µm with vegetable oils. Such atomizers can also be designed to withstand back-pressures of several bar, necessary when delivering tracer particles into flows under pressure.

A highly monodispersed stream of droplets can be generated by exploiting the inherent instabilities of a laminar liquid jet, as first analyzed by *Rayleigh* [5.366]. Although the droplet quantity is far below that typically required for flow seeding, this generator is in widespread use as a standard with which phase Doppler instruments can be aligned and adjusted. Such devices are commercially available (e.g., TSI model 3450) but also exist in a wide variety of custom designs [5.367].

The principle of this generator is pictured in Fig. 5.66a. Rayleigh showed that a round liquid jet of diameter D will disintegrate into droplets due to surface tension forces. This instability exists for wavelengths in the range 3.5–$7D$, whereby the wavelength $\lambda = 4.508 D$ exhibits unstable behavior. If the liquid jet is now externally perturbed at a wavelength in this range, the instabilities will lock on to this wavelength and a highly periodic stream of mono-dispersed droplets results. The size of each droplet is given directly by

$$d_\mathrm{p} = \left(\frac{6\dot{v}}{\pi f}\right)^{1/3}. \qquad (5.212)$$

Since both the volume flow rate $\dot{v}$ and the excitation frequency f can be exactly controlled, the drop size is adjustable to within micron accuracy.

Solid Particle Generation

Atomization. Solid particles can be produced by atomization of suspensions using techniques described in the previous section. After atomization, the liquid or solvent evaporates, leaving a solid particle. Typical suspensions are latex particles (polysterol or polyvinyltoluol) with very narrow size distributions. Such particles are usually available in 10% solutions with sizes of 10–20 µm. They are made in a polymerization process, which results in standard deviations in size of only a few percent. The suspension must be diluted before atomization such that the probability of obtaining more than one latex particle per droplet becomes very low. Also this method of aerosol generation has been accepted as a standard. Other solid particles have been atomized in this manner, in particular TiO_2 and ZrO_2 in a 5% ethyl alcohol suspension for investigations in flames [5.368].

From Powders. There are several methods of obtaining a flow of tracer particles from a powder. In all cases, however, coagulation tends to reduce the number concentration and shift the size distribution towards larger particles. Thus, it is often important to use dry delivery gas and also dry, hydrophobic powders. Two different dispersion devices will be briefly described.

The first of these is the rotary brush seeder, pictured in Fig. 5.67a. In this device a packed column of powder is fed into a rotating brush, which lifts off the top layer of powder and transports it into a high-speed stream (170 m/s) of carrier gas, usually air. This high-speed gas stream also provides sufficient energy to break up larger coagulate particles further. The rate of particle delivery is controlled by the bulk powder feed rate, the brush rotation rate, and the volume flow rate of carrier gas. Typically, powders are quartz dust, calcium, glass spheres, and titanium dioxide.

A second method of dispersing a bulk powder into a gas stream is the cyclone aerosol generator, pictured in Fig. 5.67b. *Glass* and *Kennedy* [5.369] have used such

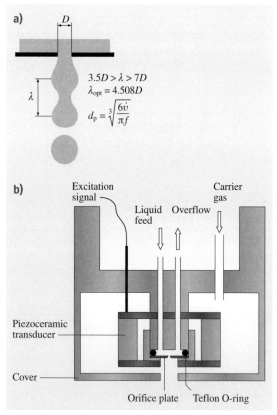

Fig. 5.66a,b Monodisperse droplet generation using the Rayleigh instability. (**a**) Principle of Rayleigh instability of a round liquid jet; (**b**) vibrating orifice generation head

a device for dispersing 0.1–1 μm Al_2O_3 particles. *Anderson* et al. [5.370] have used this design for tracer particle generation for PIV studies. The cyclone acts not only as a dispersion method but also as a particle size separator, depending on the diameter of the outlet tube.

A summary of various solid aerosol materials commonly used for seeding purposes is given in Table 5.6.

Introducing Particles into the Flow

Achieving optimum flow seeding is generally acknowledged as the most difficult part of LDA and PIV experiments. Once the flow is uniformly seeded with required concentration, the experiment has a high chance of success. The seeding procedure consists of different steps, which may vary depending on the flow facility. For water flows in a confined environment or within closed-loop flow the seeding procedure consists in dispersing the seeding material in the fluid. It is best practice to increase the seeding concentration gradually from a low level until reaching the desired one. If seeding exceeds the optimum level, multiple scattering and opacity compromise the setup of the experiment. For air flows the seeding material first needs to be entrained in the working fluid and then inserted upstream of the measurement region. Particles of a given diameter can be selected from a polydisperse particle distribution in the seeded fluid through centrifugation or making use of impact plates. The procedure of dispersing the seeded fluid within the stream is equally important. The seeding distribution system should not perturb the flow and at the same time should provide a homogeneous seeding level in the flow. This is usually accomplished with two-dimensional rakes of orifices placed upstream of screens and meshes in the wind tunnel settling chamber. In some cases a homogeneous seeding concentration cannot be achieved because of physical processes in the flow affecting the fluid density (e.g., compressible flows, flames). A homogeneous flow seeding does not show the underlying structure of the flow. Figure 5.68a is a PIV recording of the flow around a square cylinder developing as a von Kármán wake behind the model. The darker and brighter stripes are due to the light refraction and reflection through the transparent model. Conversely, in high-speed flows inhomogeneous seeding distribution is unavoidable due to the combined effect of density variations and strong inertia forces locally centrifuging seeding particles from the core of vortices. This is illustrated in Fig. 5.68b, where a Mach 2 stream develops around the model of a rocket propelled with a supersonic jet [5.372].

With atomized droplets, their lifetime before complete evaporation is of particular interest, since this

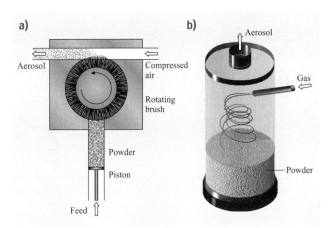

Fig. 5.67a,b Aerosol generation from powder. (**a**) Operation principle of a rotary brush seeder. (**b**) A cyclone aerosol generator

determines how far upstream the particle injection can be placed. The evaporation rate can be controlled through the choice of liquids. For instance, a 5–10% addition of glycerin to water can extend the droplet lifetime tremendously.

Deposition of particles on walls is one major obstacle to tracer-based measurement techniques in large wind tunnels, since the cost of cleaning such an installation can be enormous. For this reason water droplets are often the preferred tracer; however, other liquids exist which leave almost no deposits and are nontoxic, e.g., DEHS. Window deposition is also a limiting factor for measurements in motored and fired engines. For this reason the window design should afford quick and easy removal/cleaning. It is not unusual to experience measurement times of only tens of seconds before window cleaning is again necessary, meaning that external systems must be used to keep all engine components and fluids at operating temperatures. Recessed windows or protecting air curtains in front of windows have both

Table 5.6 Properties of various solid seeding materials (after *Agarwal* and *Johnson* [5.371])

Material	Density (g/cm^{-3})	Index of refraction	Melting point (°C)
DOP	0.98	1.49	
PSL	1.05	1.59	
NaCl	2.16	1.54	801
Al_2O_3	3.96	1.79	2015
TiO_2	4.26	2.6–2.9	1750
SiC	3.2	2.6	2700
ZrO_2	5.6	2.2	2980

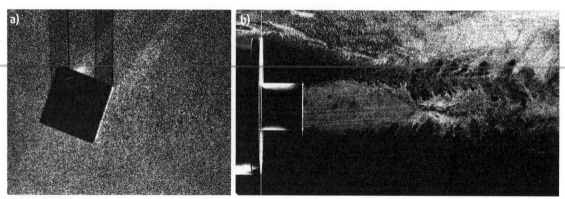

Fig. 5.68a,b Example of homogeneous seeding density in a low-speed bluff-body wake (**a**) and inhomogeneous seeding in a high-speed supersonic flow (**b**). (After *Scarano* et al. [5.372])

been used to avoid window dirtying; however, these measures are not always possible without disturbing the measured flow field.

Recommendations

A common erroneous tendency is to increase the seeding level beyond optimum in order to compensate for a drop in the signal-to-noise ratio. Such a drop can be due to several effects that should be checked before deciding to increase the seeding level:

1. contamination of optical interfaces for illumination and imaging
2. misalignment of receiving optics (LDA) or between the light sheets or incorrect (too long) time separation between pulses (PIV)
3. excessive seeding concentration causing medium opacity
4. blurred particle images due to incorrect setting of the imaging focal plane (PIV).

5.3.2 Laser Doppler Technique

Measurement Principles

The laser Doppler technique, most frequently referred to as a laser Doppler anemometer (LDA) or laser Doppler velocimeter (LDV) is a widespread technique for pointwise velocity measurements in a flow with fluid markers. Some decisive advantages of the technique over more-conventional probe-based techniques can be mentioned: non-intrusiveness, directional sensitivity, high spatial and temporal resolution, and high accuracy. A first laser Doppler instrument was presented by Cummins et al. and by *Yeh* and *Cummins* in 1964 [5.373], in which an optical configuration, subsequently known as the reference-beam mode, was introduced. The currently used dual-beam mode was introduced almost simultaneously by *Lehmann* [5.374] and *vom Stein* and *Pfeifer* [5.375], the latter of which were granted a patent. In this configuration, two in-going laser beams crossing at an intersection angle were used to form a measurement volume and the scattered light from both beams was collected on a single detector. These were the basic innovations leading to the present-day laser Doppler instrument.

The basic principle of the technique invokes the Doppler effect twice, once when the incident laser light, characterized by the wavelength λ_b and frequency f_b (subscript 'b' for beam), impinges on a moving tracer particle and once when light with a frequency f_p (subscript 'p' for particle) is scattered from the moving tracer particle and received by a stationary detector with the frequency f_r (subscript 'r' for receiver). This is illustrated in Fig. 5.69, and the corresponding relation for the detected frequency of scattered light is given by

$$f_r = f_b \frac{\left(1 - \frac{v_p e_b}{c}\right)}{\left(1 - \frac{v_p e_{pr}}{c}\right)} \approx f_b + \frac{v_p(e_{pr} - e_b)}{\lambda_b}$$
$$\text{for } |v_p| \ll c, c = f_b \lambda_b, \quad (5.213)$$

where c is the speed of light in the medium surrounding the particle and v_p is the particle velocity. The second term in (5.213) contains the Doppler shift of the incident wave frequency. The difference of the normal vectors appears when the direction of propagation of the incident and scattered wave differs. The Doppler shift is directly proportional to this difference and to the velocity of the particle. For typical flow systems and for receiver placements that maximize the magnitude of the vector different, the Doppler shift is of the order 1–100 MHz, which compared to the frequency of laser light (approximately 10^{14} MHz) is very small and virtually impossible to resolve directly. One exception is direct detection through the use of frequency-dependent absorption cells, leading to the Doppler global velocimeter described in Sect. 5.3.4.

The most common approach to circumvent the necessity of resolving the Doppler shift directly is known as the dual-beam configuration, in which a measurement volume is formed at the intersection of two incident waves and the scattered waves are detected with a single detector, as illustrated in Fig. 5.70a. Applying (5.213) to each of the input beams, the frequencies at the detectors can be written

$$f_1 = f_b + \frac{v_p(e_{pr} - e_1)}{\lambda_b} , \quad f_2 = f_b + \frac{v_p(e_{pr} - e_2)}{\lambda_b} .$$
(5.214)

The difference (beat) frequency f_D is obtained through the optical mixing of waves with frequencies f_1 and f_2 on the detector. This frequency is commonly called the Doppler frequency and, using the vector relation illustrated in Fig. 5.70b, can be expressed as

$$f_D = f_1 - f_2 = +\frac{v_p(e_1 - e_2)}{\lambda_b}$$
$$= \frac{2\sin(\Theta/2)}{\lambda_b}|v_p|\cos\alpha = \frac{2\sin(\Theta/2)}{\lambda_b}v_{p\perp} ,$$
(5.215)

where $v_{p\perp}$ is the particle velocity component perpendicular to the beam bisector. Noteworthy is the fact that the difference frequency is independent of the receiver position and that it is linearly proportional to the velocity component in the x direction. Furthermore, the intersection angle Θ and the wavelength can be determined very accurately from the optical arrangement, eliminating the need for a dedicated calibration of the device. Equation (5.215) is therefore the direct link between the frequency of modulated light intensity at the detector and the desired flow velocity component.

The fringe model is an alternative explanation of the laser Doppler principle, in which the wavefronts of two coherent beams interfere in the intersection volume, forming interference fringes with a spacing Δx, as illustrated in Fig. 5.71. Tracer particles passing through the volume will scatter modulated light according to the passing of the constructive and destructive interference fringes, yielding the Doppler frequency at the

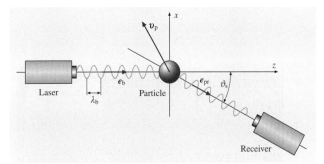

Fig. 5.69 Defining geometry for applying the Doppler effect in the laser Doppler technique

detector. This frequency is determined for each individual tracer particle passing through the volume, yielding flow velocity readings in time.

The fringe model clearly indicates the need to maintain high fringe contrast to yield a maximum modulation in the resulting signal from the receiver. This contrast is best when the intensity and polarization direction of the two incident beams are equal, and of course that the light is spatially and temporally coherent – the main reasons for using a laser as a light source.

One further requirement is to distinguish the sign of the particle velocity, since the Doppler frequency alone does not contain this information, as indicated in Fig. 5.72a. The same signal will be obtained for a par-

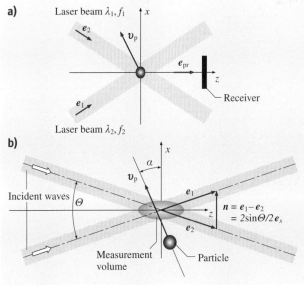

Fig. 5.70 (a) Dual-beam configuration. **(b)** Vector relations relevant to determining the Doppler frequency

ticle crossing the fringes in either direction. To achieve directional sensitivity the frequency of one of the incident beams is shifted by an amount f_s compared to the other beam, which results in a moving fringe pattern. A tracer particle with zero velocity in the measurement volume will result in a received frequency of f_s, while any movement of the particle will result in a frequency larger or smaller than f_s, depending on the direction relative to the fringe movement. This is illustrated in Fig. 5.72b.

The magnitude of the shift frequency must be chosen such that the highest expected velocity in the negative x direction, i.e., in the direction of the moving fringes, is still significantly smaller that the velocity of the fringes (Δv). How this achieved is outlined next.

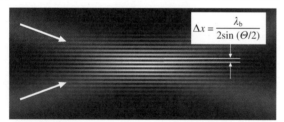

Fig. 5.71 Interference pattern in the measurement volume – fringe model

Optical Systems

There exist a very large number of different optical configurations to realize the laser Doppler technique, the choice of which depends largely on factors specific to the intended application. These include the available optical access, possible particle seeding, velocity range, required spatial and temporal resolution, number of velocity components to be measured, etc.

Fundamentally the laser Doppler technique can work with the detector placed at any scattering angle with respect to the incident beams; however the discussion in Sect. 5.3.1 indicated that the scattered light intensity from tracer particles is usually much stronger in the forward direction. This suggests that the receiving optics should be placed on the opposite side of the measurement volume to the transmitting optics. While this is desirable for better signal strength, the available optical access does not always allow this. Often a receiving optics integrated into the transmitting optics is more convenient to position and at the same time requires only one optical access to the measurement point. This fundamental distinction between forward and backscatter is illustrated schematically in Fig. 5.73, whereby also sidescatter arrangements are sometimes used for limiting the measurement volume in its length.

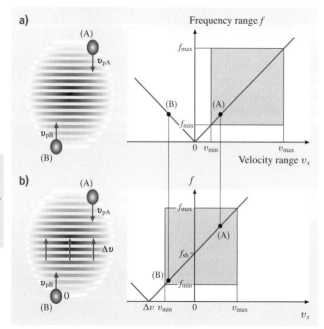

Fig. 5.72a,b Measured frequencies for particles crossing the measurement volume in positive (A) and negative (B) x directions: (a) without frequency shift; and (b) with frequency shift

Practically most laser Doppler systems are now realized with a fiber-optic link between the bulk optical components of the system and the measurement probe. A typical system is illustrated in Fig. 5.74, showing a Bragg cell used for splitting the laser beam into two beams, one shifted in frequency compared with the other (for directional sensitivity). A color-splitting prism follows, resulting in two or three colors from an all-lines laser to be separated. The use of two or three colors allows systems to be configured to measure two or three velocity components of the tracer particle movement simultaneously. The individual beams (up to six) are then transmitted to a measurement head via single-mode

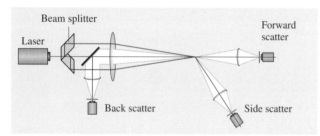

Fig. 5.73 Possible optical configurations of a laser Doppler system

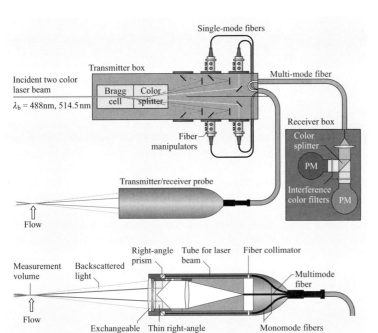

Fig. 5.74 Two-color, four-beam laser Doppler system, suitable for measuring two velocity components. The green ($\lambda = 488$ nm) and blue ($\lambda = 514.5$ nm) lines of an argon-ion laser have been used

polarization-preserving fibers, since both coherency and polarization are prerequisites for interference in the measurement volume. Fiber links of 5 or 10 m are not uncommon; hence only the probe head must be mounted and traversed to position the measurement volume in the flow field.

The Bragg cell can be understood as an acoustic diffraction grating from which the zero and first-order beams are used. Typical frequency shifts of the first-order beam are 40 MHz or 80 MHz, which then also becomes the frequency shift or the frequency of the moving fringes in the intersection volume. Particles passing through the volume will then result in signal frequencies that are smaller or larger, depending on their direction. Some laser Doppler systems employ an electronic down-mixing of this signal to shift the frequencies to lower ranges, more compatible with signal processing electronics. As an example, for a Bragg cell operated at 40 MHz and a mixing with a reference signal of 35 MHz, an effective frequency shift of 5 MHz is achieved.

The scattered light from tracer particles passing through the measurement volume is collected through a lens system and focused onto a photodetector using

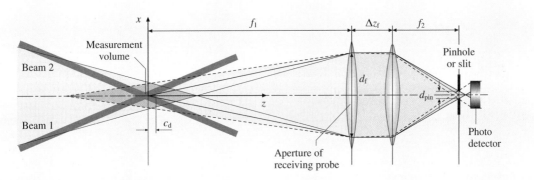

Fig. 5.75 Configuration of the receiving optics

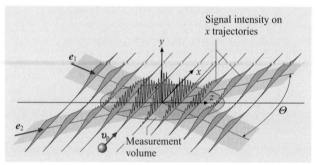

Fig. 5.76 Signal generation in the measurement volume

shown in Fig. 5.75. Using this lens arrangement, the measurement volume is imaged onto a pinhole, slit or fiber. This pinhole acts as an aperture and can limit the effective dimensions of the measurement volume, either in diameter (forward or backscatter) or in length (side scatter). The magnification of the lens system is given as $\beta = -f_2/f_1$ (where f_1 and f_2 are the focal lengths of the lens system shown in Fig. 5.75), which together with the pinhole diameter determines the observation area at the beam intersection plane.

a receiving optics. The photodetector, typically a photomultiplier or photodiode, converts the scattered light intensity into an electrical signal with proportional voltage. The receiving optics may be integrated into a single probe together with the transmitting optics, as shown in Fig. 5.74; however the optical characteristics are separate and can be represented by the two-lens system

The spatial resolution achieved with the laser Doppler technique depends both on the optical system and on the signal-processing electronics. The size of the measurement volume created at the intersection of the two incident beams depends on the degree of beam focussing. Laser beams are Gaussian beams and, when focused with a lens, they exhibit a beam waist, the diameter (d_{w1}) of which depends on the focal length of the lens (f), the wavelength (λ_b), and the diameter of the beam at the beam waist before the lens (d_{w0}):

$$d_{w1} = \frac{4 f \lambda_b}{\pi d_{w0}}. \tag{5.216}$$

A Gaussian beam exhibits a Gaussian intensity profile in its cross section; hence there is no distinct border between illuminated and non-illuminated at the edge of the beam. The diameter given in (5.216) refers to the diameter at which the intensity has fallen to e^{-1} of the maximum intensity on the beam axis.

In most laser Doppler systems the focal length of the front lens in the transmitting optics determines the standoff or working distance between the probe and the measurement position. This distance is often dictated by the application at hand and by the available optical access. However this focal length can also be applied with (5.216) to give the dimensions of the measurement volume. The nominal size of the measurement volume can be approximated as an ellipsoid with half-axes and volume given as

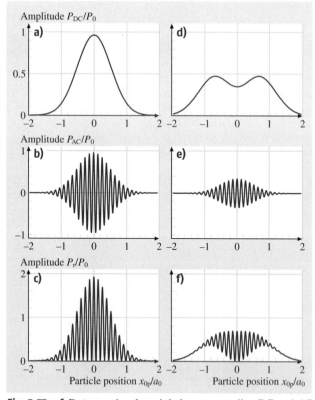

Fig. 5.77a–f Detector signals and their corresponding DC and AC parts for two different particle trajectories ($v_p = e_x v_{px}$, $S_{1r} = S_{2r}$). **(a–c)** $r_{p0} = (0, 0.1b_0, 0.1c_0)^T$, **(d–f)** $r_{p0} = (0, 0.2b_0, 0.7c_0)^T$

$$a_0 = \frac{d_{w1}/2}{\cos \Theta/2}, \quad b_0 = \frac{d_{w1}}{2}, \quad c_0 = \frac{d_{w1}/2}{\sin \Theta/2},$$
$$V_0 = \frac{4}{3}\pi a_0 b_0 c_0, \tag{5.217}$$

and the number of interference fringes in the volume will be

$$N_{fr} = \frac{a_0}{\Delta x} = \frac{2 d_{w1} \tan(\Theta/2)}{\lambda_b}. \tag{5.218}$$

Typically the diameter of the measurement volume will range between 40 μm and 200 μm and the length will be 5–10 times larger, depending on the intersection angle. If smaller measurement volumes are desired, beam expansion must be used before the front lens to increase d_{w0}, which decreases d_{w1}. The length of the measurement volume may also be truncated by using a side-scatter receiving arrangement in combination with a pinhole aperture in front of the photodetector, as indicated above. However, the volume in which tracer particles can be detected can still be larger or smaller than the nominal measurement volume. To underline the difference between the detection volume and the measurement volume, the signal generation is examined in more detail.

As illustrated in Fig. 5.76, the signal generation in the measurement volume depends on the trajectory of the particle. The trajectory influences both the low-frequency (DC/pedestal) and modulated (AC/Doppler frequency) portion of the signal. This is shown for two exemplary trajectories in Fig. 5.77. The amplitude of the signal will depend on particle size. For particles significantly larger than the fringe spacing Δx, the modulation, or visibility, will also decrease even if the fringe contrast is 100%. This occurs because larger particles will always result in several fringes being *seen* by the detector. Whether a particle will be detected of not will depend on the detection criteria invoked in the signal processing step, which may include an amplitude or a signal-to-noise threshold on the DC and/or AC portion of the signal. Hence, the effective spatial resolution of the laser Doppler system is a complex function of many optical and electronic parameters.

This discussion also underlines the trade-off in selecting the size of tracer particles. Small particles follow flow fluctuations more accurately and result in higher modulation levels in the Doppler signal; on the other hand they also result in lower signal amplitudes ($\propto d_p^2$ in the Mie scattering range), lower signal-to-noise ratios and in the worst case may not even be detected.

Using different colors of a laser source, two or three measurement volumes can be overlaid but with differing fringe orientation. By distinguishing the signals from the different beam pairs through color filtering in the receiving optics, two or three velocity components can be simultaneously acquired. An optical arrangement suitable for three velocity components is shown in Fig. 5.78. Note that the three velocity components may not all be orthogonal to one another and may require a transformation into the probe-fixed or laboratory coordinate system.

The performance of the laser Doppler system will depend largely on the optical system. Therefore any strategy for maximizing the signal-to-noise ratio and the rate of Doppler signals should begin with the optical system. The main influencing factors include

- Input laser power
- Incident intensity (for given laser power a function of measurement volume diameter)

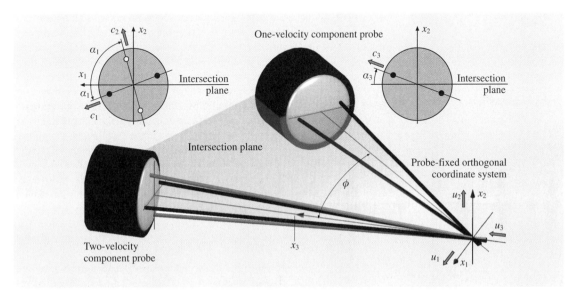

Fig. 5.78 Three-velocity component laser Doppler system, showing measured velocity components and probe-fixed, orthogonal coordinate system

- Complete intersection of the two beams, equal intensity, equal polarization, coherence
- Particle size
- Receiving aperture and alignment of the receiving optics.

In some cases these factors are choices representing conflicting requirements and additional diagnostic equipment may be necessary to achieve the best performance.

Signal Processing

Each tracer particle passing through the measurement volume results in a signal at the output of the photodetector. The signal processor has the task of detecting when a signal is present and estimating various signal parameters – Doppler frequency, arrival time, signal length, signal-to-noise ratio. Such a time series of particle signals is shown in Fig. 5.79, where a high-pass filter was used to first remove the DC part of the signal. Note that the signals arrive irregularly in time, a feature requiring special consideration in the data processing step. These signals are often referred to as burst signals, or Doppler bursts.

For the temporal resolution of the system, the number of detected or processed signals per unit time is of importance. Assuming a random and homogeneous spatial distribution of particles in the flow with concentration $\overline{n}_p$, the mean number of particles in the measurement volume is

$$\overline{N}_{p0} = \overline{n}_p V_0 = \frac{4}{3}\pi a_0 b_0 c_0 \overline{n}_p \tag{5.219}$$

and the mean signal rate (number of signals per unit time) for particle trajectories through the measurement volume parallel to the x axis is

$$\dot{N}_0 = p b_0 c_0 \overline{n}_p \overline{v}_x . \tag{5.220}$$

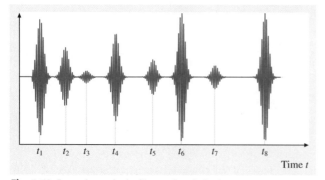

Fig. 5.79 Irregular arrival of burst signals in time

Increasing the particle concentration indefinitely does not continue to improve the temporal resolution; firstly because at high particle loadings the injection of particles start to influence the flow, but also because the signal processing may not be reliable in estimating the signal frequency when multiple particles are present simultaneously in the measurement volume. Using Poisson statistics, which is valid for random spatial distributions of particles, three concentration ranges can be distinguished:

- $\overline{N}_{p0} < 0.1$: single realization with a probability of 99.5%
- $\overline{N}_{p0} \geq 0.1$: multiple particle signals, in which the superposition of random amplitudes deteriorate signal quality and may lead to a lower signal rate
- $\overline{N}_{p0} > 10$: a quasi-continuous signal.

Signal processors operate most effectively in the single realization mode, which places upper limits on the allowable mean particle concentration for a given measurement volume size. Smaller measurement volumes tolerate higher mean concentrations. At the same time, the same laser power distributed over a smaller measurement volume will result in higher incident intensity and higher signal amplitudes and quality. However, reducing the measurement volume indefinitely will eventually result in fewer particles being seen, reducing the temporal resolution of the system.

Superimposed on the Doppler signal is noise, including stochastic noise from the photodetector and electronics (shot noise, Johnson noise, dark current) as well as from the scattering process itself. The laser may introduce signal noise, especially laser diodes. Noise can arise from unwanted reflections associated with the flow rig or other particles. Poor grounding of the experimental apparatus or poorly designed electronics may also lead to ground noise. Except for the last source, noise contributions in the system are usually considered to be spectrally white, referring to the fact that the total noise power is distributed evenly over all frequencies up to the upper bandwidth of the system. The power of the signal fluctuations σ_s^2 divided by the power of the noise fluctuations σ_n^2 is known as the signal-to-noise ratio (SNR) and is generally expressed in decibels

$$\frac{\text{SNR}}{\text{dB}} = +10 \log \left(\frac{\sigma_s^2}{\sigma_n^2} \right) . \tag{5.221}$$

In Fig. 5.80, a laser Doppler signal, a noise signal and the summation of the two in time, spectral and correlation domain is illustrated. It becomes obvious from Fig. 5.80 that the power spectral density (PSD) or the autocorrelation function (ACF) offer excellent means to monitor SNR and to determine whether a particle signal is present or not (Sect. 25.4). The presence of noise in the signal can have both direct and indirect effects on the measurement quantities. In the worst case, the scattering signal from the particle may not be detected at all or a completely wrong frequency could be measured from a noisy signal. However, noise may also affect estimates of the signal duration, generally leading to an overestimation, which then influences the derived flow quantities in the data processing. In any case, noise increases the variance of the signal frequency estimates, regardless of which processing scheme is employed. Thus, noise at the signal processing stage essentially determines the lowest resolvable level of measurable turbulence in a flow.

This level is known as the Cramèr–Rao lower bound (CRLB), and is discussed further in Sect. 25.5, as are further aspects about the estimation of noise and SNR of a signal.

The noise contribution to the signal can be decreased and the SNR increased by bandpass filtering the signal before processing; however to do this the cut-off frequencies of the filter must be carefully chosen so as not to suppress signals with widely varying Doppler frequencies due to turbulent flow fluctuations. The advantage of employing bandpass filters is that a higher particle detection rate can be achieved and the variance of the frequency estimates may decrease. Otherwise this variance will falsely appear as turbulent flow fluctuations. Nevertheless, as mentioned above, the best strategy for

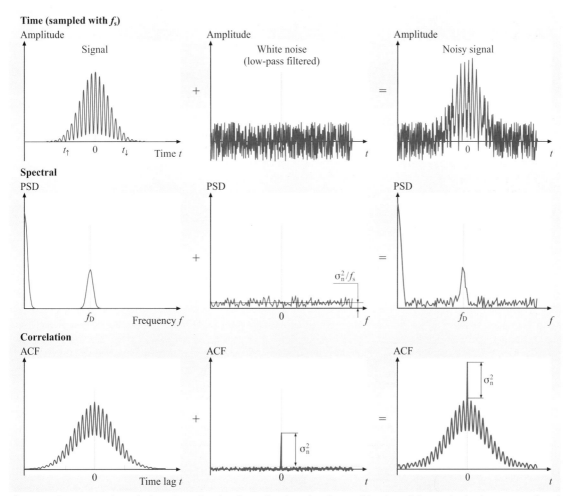

Fig. 5.80 Representation of a laser Doppler signal, a noise signa,l and a combination of the two in the time, spectral, and correlation domains

achieving a maximum signal-to-noise ratio is to design and adjust the optical system properly.

The processing of Doppler signals has seen a long history of various methods since the first instruments; however the high speed and low cost of present-day electronics has resulted in commercial processor units that exclusively use either the autocorrelation or power spectrum for both signal detection and signal parameter estimation. The main signal parameter to be estimated from each Doppler burst is the frequency f_D and, as shown in Sect. 25.5.1 of this Handbook, several estimators exist which come very close to achieving the Cramèr–Rao lower bound for the signal frequency. In fact with some a priori knowledge about the signal frequency range, and with judicious use of filters, the CRLB can even be exceeded. Modern commercial processors are able to achieve these limits.

Other signal parameters to be estimated for each burst include the arrival time, the length of the burst, and the signal-to-noise ratio. The arrival time is required for reconstruction of the velocity time series and also for checking for coincidence between signals from different channels of a two- or three-velocity component system. Coincidence is registered when a signal is detected on each channel within a given time separation, typically 10–30% of the total signal duration. This coincidence is important when the velocity values are used to estimate Reynolds stress terms, in which case only velocity component values definitely coming from the same particle should be considered.

The length of the burst, sometimes referred to as transit or residence time, is a quantity required in the data processing stage for estimating mean and fluctuating flow velocities. However the burst length may also be used as a postvalidation criterion. The burst duration can be compared with the measurement volume size divided by the measured velocity. If these do not agree within sufficient bounds, the particle is rejected from entering further statistics. Similarly, the signal-to-noise ratio can be used either as an online validation or as a postvalidation criterion. Modern signal processors allow these quantities and their acceptable limits to be adjusted through software.

With the widespread availability of fast analog-to-digital convertors and processors, customized signal-processing schemes are easily implemented, which may estimate additional signal parameters. In particular, model-based signal processing is attractive for special applications. One example is the recent use of the laser Doppler technique to estimate particle acceleration [5.376].

Data Processing

As opposed to signal processing, whose task is to estimate certain characteristic parameters from a given signal, data processing uses these estimated parameters from many signals to derive certain flow-related quantities. Two basic properties unique to laser Doppler data must be considered. First, the sample times of the flow velocity, given by the particle arrival times in the measurement volume, are irregular. This is in contrast to many other measurement instruments, which use a regular time sampling of an analog signal (analog-to-digital converter). The second important feature of laser Doppler data is that not only is the velocity sample time random, but that the short-term rate of random particle arrivals will generally be correlated with the measured flow quantity. This correlation will be referred to as the *particle-rate/velocity* correlation and without careful attention to this fact, estimators can be biased, i.e., their expectation can be incorrect.

The assumption of a homogeneous and random spatial distribution of tracer particles in the fluid leads to an exponential distribution of the distances between particles [5.377]. For a constant convective velocity, this results in an exponential distribution of interarrival times through any volume in space. Therefore, a good model of the interarrival time probability density distribution (PDF) is

$$p(\Delta t_i) = \dot{N} \exp\left[-\dot{N}(t_i - t_{i-1})\right], \quad (5.222)$$

where $\dot{N}$ is the mean particle rate and t_i and t_{i-1} are the arrival times of consecutive particles i and $i-1$, respectively. The striking feature of the distribution is that the most probable time between two particles is zero. Even at modest mean particle arrival rates, particles will quite often appear in rapid succession. Either the signal processor must be able to evaluate the signals online or suitable input buffering must be available to avoid loss of information and to prevent processor *dead time*. On the other hand, the fact that velocity information is often available over the very short time spans of consecutive particles suggests that, in principal, information about very high-frequency velocity fluctuations is contained in the data. This is in strong contrast to data sampled at equal time intervals, for which the sampling theorem applies and for which no information above the Nyquist frequency $f_{Nyq} = 1/(2\Delta t_s)$ is available, where Δt_s is the sampling interval. In fact, with randomly sampled data, there is no equivalent to the sampling theorem or the Nyquist frequency and with suitable estimators it is possible to achieve alias-free and unbiased estimates of signal power at frequencies far exceeding the mean particle arrival rate, as discussed below.

Arrival-time information is sometimes used in data processing as a basis for validation. Knowing the particle velocity and the dimensions of the measurement volume, an estimate of mean transit times can be made. Measured arrival times lying far below these estimates indicate that the signal processor may be delivering more than one velocity value per particle. Since moment estimators of velocity fluctuations generally assume the single realization condition, such multiple values per particle are unacceptable. These can often be recognized as peaks in the probability density function of the interarrival time, which also indicate suitable thresholds for validating each individual velocity before further processing.

Estimation of Moments. The particle-rate/velocity correlation is the main physical reason for requiring special moment estimators for laser Doppler data. The particle rate through the measurement volume is determined by the volume flux of fluid through the measurement volume and this is, in general, correlated with the measured velocity component. Therefore, the sample rate of the velocity also increases with velocity. For a given observation time, higher velocities will be sampled more frequently than lower velocities and a simple arithmetic mean of all samples will be positively biased over the true time mean of the velocity. The degree of bias will depend on how strong the correlation between the particle arrival rate and the measured velocity component is.

A correct estimator for the mean of the u flow velocity component must therefore weight each sample with a factor g, which is inversely proportional to the conditional probability density of a particle arrival at a time t_i, given the velocity u_i,

$$\hat{\bar{u}} = \frac{\sum_{i=1}^{N} u_i g_i}{\sum_{i=1}^{N} g_i}, \qquad (5.223)$$

where the index i refers to the arrival time t_i and the hat (^) signifies that this is only an estimation of the mean.

One possible weighting factor would be

$$g_i = \frac{1}{\sqrt{u_i^2 + v_i^2 + w_i^2}}, \qquad (5.224)$$

which uses the magnitude of the vector velocity and assumes a spherical measurement volume. For this, a three-velocity component laser Doppler system is necessary.

Alternatively, the residence time (or transit time) of the particle τ_i can be used as a weight factor, since this will be inversely proportional to the vector velocity magnitude [5.378].

$$g_i = \tau_i \, . \qquad (5.225)$$

This is only possible if the signal processor provides residence time information and if it is reliable.

An estimator for the second moment is given in a similar manner

$$\hat{\sigma}_u^2 = \frac{\sum_{i=1}^{N} \left(u_i - \hat{\bar{u}}\right)^2 g_i}{\sum_{i=1}^{N} g_i}, \qquad (5.226)$$

as are estimators for joint moments

$$\overline{u'v'} = \frac{\sum_{i=1}^{N} \left(u_i - \hat{\bar{u}}\right)\left(v_i - \hat{\bar{v}}\right) g_i}{\sum_{i=1}^{N} g_i}. \qquad (5.227)$$

The last estimator assumes that the u and v velocities are available at the same instant in time, meaning time coincidence is demanded during the acquisition. For independent time series, i.e., data collection without coincidence, this equation must be modified to

$$\overline{u'v'} = \frac{\sum_{i=1}^{N}\sum_{j=1}^{M} \left(u_i - \hat{\bar{u}}\right)\left(v_i - \hat{\bar{v}}\right) g_{ui} g_{vi}}{\sum_{i=1}^{N} g_{ui} g_{vi}}. \qquad (5.228)$$

As an example, if a residence time weighting is being used, g_{ui} corresponds to the residence time of the i-th particle, found from the u component signal.

The issue of choosing an appropriate moment estimator is certainly less critical now than in the past, simply because almost all commercial signal processors now provide reliable estimates of the transit time for each particle passage through the measurement volume. Transit-time weighting is the recommended estimator for all measurement situations involving spatially homogeneous particle seeding.

The necessity to use a weighted estimator can be checked by cross-correlating the measured velocity magnitude with the interarrival time fluctuations between particles. This is especially reliable if all three

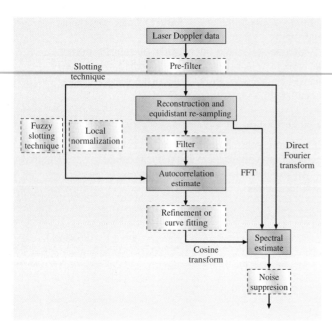

Fig. 5.81 Power spectral density estimation methods for laser Doppler data

Estimation of Turbulent Velocity Spectra. In contrast to the rather straightforward computation of power spectral density functions for equally sampled data, as described in detail in Sect. 24.1, now the input data is randomly sampled in time and no obvious equivalent to the fast Fourier transform (FFT) is available as a computational algorithm. However, as anticipated at the beginning of this discussion, some estimators are able to exploit the random sampling to achieve estimates of turbulent kinetic energy at frequencies much higher than the mean particle arrival rate. The most common of these include

- direct transform
- slot correlation followed by a cosine transform
- reconstruction with equidistant resampling and FFT.

The relation between these techniques is pictured schematically in Fig. 5.81. In each of the algorithmic routes, additional steps (shown as dashed boxes in Fig. 5.81) are possible, representing various enhancements of these basic algorithms.

A comprehensive review and evaluation of these techniques are given in *Benedict* et al. [5.379]. They come to the conclusion that two main algorithms for estimation of the PSD and autocorrelation function (ACF) can be recommended: the fuzzy slotting technique in combination with local normalization [5.380], and the refined sample-and-hold (S+H) reconstruction [5.381].

velocity components are available. A weighted estimator is only required if a significant correlation exists between these quantities.

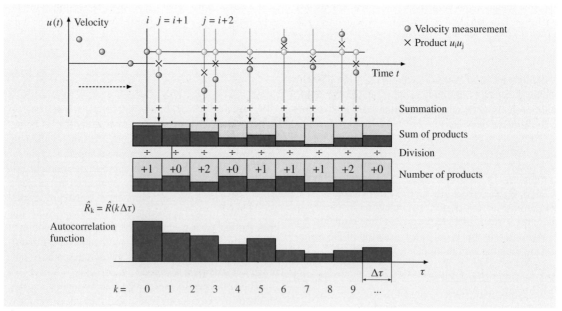

Fig. 5.82 Time series of laser Doppler velocity data and slot correlation with equidistant lag time bins $\Delta\tau$

The Slotting Technique. The slotting technique, generally credited to *Mayo* et al. [5.382], is pictured in Fig. 5.82 and computed as

$$\hat{R}_k = \hat{R}(k\Delta\tau) = \frac{\sum_{i=1}^{N}\sum_{j=J}^{N} u_i u_j b_k(t_j - t_i)}{\sum_{i=1}^{N}\sum_{j=J}^{N} b_k(t_j - t_i)},$$

$$b_k(t_j - t_i) = \begin{cases} 1 & \text{for } \left|\frac{t_j - t_i}{\Delta\tau} - k\right| < \frac{1}{2} \\ 0 & \text{otherwise} \end{cases} \quad (5.229)$$

with the velocity samples $u_i = u(t_i)$ and $u_j = u(t_j)$. The velocity product of all sample pairs with time separations falling within a given lag-time bin is added to that bin's sum as another estimate of the ACF for that lag time. After processing all possible signal pairs, each sum is divided by the number of accumulated products in that bin. The original algorithm of Mayo uses $J = 1$, which means that all sample pairs occur for k and $-k$. Calculating only positive lag times ($k \geq 0$) (5.229) can be modified to $J = i$. Note that the first ACF coefficient ($\tau = 0$) also contains self-products. With $J = i$ the self-products are over-represented compared to $J = 1$. Therefore, an additional weighting factor of $1/2$ is often used for self-products when ($J = i$). However, since noise in each of the velocity estimates is unavoidable, the first slot will be increased by the variance of the noise. Since the ACF coefficient at $\tau = 0$ normally also corresponds to the variance of the process, this variance will be overestimated using the estimator given by (5.229). This also leads to a biased PSD estimate. Using only cross-products ($J = i + 1$) leads to a spectral estimate with a noise-independent expectation. In fact, only the estimation variance increases in the case of noise. However, the first slot can also be under-represented due to processor dead times. This again can lead to a bias of the $k = 0$ ACF coefficient, and thus the PSD estimate. The ACF coefficient estimates are usually considered valid for time lags in the middle of each bin.

A one-sided PSD estimator is computed from the slot correlation by taking its discrete cosine transform

$$\hat{S}_j = \hat{S}(f_j) = \hat{S}\left(\frac{j}{2K\Delta\tau}\right)$$
$$= 2\Delta\tau\left[\hat{R}_o + 2\sum_{k=1}^{K-1} \hat{R}_k \cos(2\pi f_j k \Delta\tau) + (-1)^j \hat{R}_K\right], \quad (5.230)$$

where K is the index of the maximum time lag of the ACF and is chosen by the user.

A severe limitation of the standard slotting technique is its high variance (roughly constant up to moderate lag times), which leads to poor estimates of turbulence spectra. In order to reduce the variance of the slotting technique, *van Maanen* and *Tummers* [5.383] employ an ACF normalized by a variance estimate particular to each slot, called the local normalization. This results in an estimate of the correlation coefficient

$$\hat{\rho}_k = \hat{\rho}(k\Delta\tau)$$
$$= \frac{\sum_{i=1}^{N}\sum_{j=J}^{N} u_i u_j b_k(t_j-t_i)}{\sqrt{\left[\sum_{i=1}^{N}\sum_{j=J}^{N} u_i^2 b_k(t_j-t_i)\right]\left[\sum_{i=1}^{N}\sum_{j=J}^{N} u_j^2 b_k(t_j-t_i)\right]}} \quad (5.231)$$

and is used as the basis for the cosine transform. The corresponding one-sided, real PSD estimator is

$$\hat{S}_j$$
$$= \hat{S}(f_j) = \hat{S}\left(\frac{j}{2K\Delta\tau}\right)$$
$$= 2\hat{\sigma}_u^2 \Delta\tau\left[1 + 2\sum_{k=1}^{K-1} \hat{\rho}_k \cos(2\pi f_j k \Delta\tau) + (-1)^j \hat{\rho}_K\right], \quad (5.232)$$

where $\hat{\sigma}_u^2$ is an estimate of the velocity variance. While (5.231) has been shown to have significantly lower variance for small lag times than (5.230) normalized by $\hat{R}_0$, the variance at large lag times is unchanged.

Another method for reducing variance in the slotting technique has been dubbed the fuzzy slotting technique by *Nobach* et al. [5.381]. In this estimator, a lag-product weighting scheme is defined as

$$b_k(t_j - t_i) = \begin{cases} 1 - \left|\frac{t_j - t_i}{\Delta\tau} - k\right| & \text{for } \left|\frac{t_j - t_i}{\Delta\tau} - k\right| < 1 \\ 0 & \text{otherwise} \end{cases} \quad (5.233)$$

and is used instead of the top-hat function in the original algorithm (5.229). This estimator allows lag products to contribute to two slots simultaneously and weights lag products that lie close to the slot centers more heavily. The combination of this fuzzy slotting technique with local normalization is one of the recommended methods of computing the ACF.

Reconstruction with FFT. Reconstruction methods create equidistantly spaced time series by resampling according to various interpolation schemes, thereby allowing a FFT to be used in making PSD or ACF estimates. The most common scheme by far is the sample-and-hold (zeroth-order, S+H) reconstruction. This is the simplest of the polynomial class of reconstruction algorithms and is depicted schematically in Fig. 5.83. It can be written as

$$u^{(S+H)}(t) = u(t_i) \quad \text{for } t_i \le t < t_{i+1}$$
$$\text{and } i = 1, \ldots, N, \quad (5.234)$$

where N is the total number of samples in a given block. The reconstruction can be performed either over the entire data set or with single data blocks. The equidistant resampling with time steps of Δt_s is performed by

$$u_i^{(S+H)} = u^{(S+H)}(i \Delta t_s) \quad \text{for } i = 0, \ldots, N_R - 1 \quad (5.235)$$

and leads to a data set of N_R samples that can be processed by a Fourier transform.

The Fourier transform with the imaginary unit i is given by

$$\underline{U}_n^{(S+H)} = \sum_{i=0}^{N_R-1} u_i^{(S+H)} \exp\left(-2\pi i \frac{in}{N_R}\right)$$
$$\text{for } n = 0, \ldots, N_R - 1 \quad (5.236)$$

and leads to the full block PSD

$$\hat{S}_n^{(S+H)} = \frac{\Delta t_s}{N_R} |\underline{U}_n^{(S+H)}|^2 \quad \text{for } n = 0, \ldots, N_R - 1 \quad (5.237)$$

and, through the inverse FFT

$$\hat{\underline{X}}_k^{(S+H)} = \frac{1}{N_R} \sum_{i=0}^{N_R-1} \hat{S}_i^{(S+H)} \exp\left(2\pi i \frac{ik}{N_R}\right)$$
$$\text{for } k = 0, \ldots, N_R - 1 \quad (5.238)$$

to the full block ACF

$$\hat{R}_k^{(S+H)} = \frac{1}{\Delta t_s} \hat{X}_k^{(S+H)} \quad \text{for } k = 0, \ldots, N_R - 1. \quad (5.239)$$

To reduce the variance of the final PSD estimate, only $K + 2 < N_R$ values of the ACF are used for further calculations. K is the maximum desired lag time for the ACF. In the refinement step outlined below, $K + 1$ values of $\hat{R}_k$ are required, and one further sample in the ACF is required for the subsequent Fourier transform to the PSD, i.e., $K + 2$ samples in total.

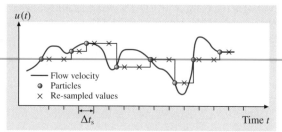

Fig. 5.83 The concept of zeroth-order reconstruction (S+H) and resampling at equal time intervals

Unquestionably the work of *Adrian* and *Yao* [5.384] was a major step forward in understanding the fundamental content of an S+H reconstructed signal. They derived an expression for the expectation of the PSD

$$k = 0, \ldots, N_R - 1 \quad E[\hat{S}^{(S+H)}(\omega)]$$
$$= \underbrace{\frac{1}{1 + \omega^2/N^2}}_{\text{filter}} \left[S(\omega) + \underbrace{\frac{2\sigma_u^2}{N^3 \lambda_u^2}}_{\text{step noise}} \right], \quad (5.240)$$

where $E[\hat{S}^{(S+H)}(\omega)]$ is the expectation of the spectral estimate and $S(\omega)$ is the true spectrum (σ_u^2 is the variance of the velocity fluctuations and λ_u is the Taylor microscale). The second term in parentheses was termed step noise and corresponds to the spectral contribution necessary to account for the step-like jumps in an S+H signal. This contribution vanishes with the inverse of the third power of the data rate $\dot{N}^{-3}$. The factor in front of the parentheses, operating on both the true spectrum and the step noise, is a first-order, low-pass filter with a cut-

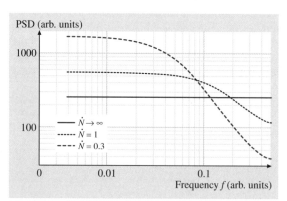

Fig. 5.84 The PSD of a white-noise process computed using a S+H reconstruction and resampling with FFT PSD estimation

off frequency of $\dot{N}/(2\pi)$. This was subsequently named the particle-rate filter and, at higher frequencies, dominates the spectrum. Actually the step noise is a form of aliasing in which the signal energy above the filter cut-off frequency is distributed evenly over the entire spectrum.

One result from *Adrian* and *Yao* [5.384] is reproduced here because of the clarity it brings to how all reconstruction methods effect spectral estimators. Simulated laser Doppler data from a white-noise process are submitted to a S+H reconstruction, resampling and a conventional FFT PSD estimation. The result is shown in Fig. 5.84. The effects of both the additive step noise and the low-pass filter are clearly evident at low particle rates; indeed the completely falsified spectrum for the lowest data rate begins to uncannily resemble that of turbulence. The conclusion of Adrian and Yao was that such spectra are reliable only up to a frequency of $\dot{N}/(2\pi)$, a substitute for the Nyquist frequency rule concerning regularly sampled data. Their assumption that the ACF of turbulent velocity fluctuations is exponential was not instrumental to this conclusion. Of course the computed spectrum below the cut-off frequency is still falsified by the step noise or any other aliased energy from higher-frequency components in the signal.

Even the nature of the interpolation curve in the reconstruction does not alter the basic limitation of the particle-rate filter. This was demonstrated by *Müller* et al. [5.385], who applied numerous other reconstruction schemes to simulated and measured laser Doppler data. These included higher-order polynomials, projection onto convex sets (POCS), fractal reconstruction, and the so-called Shannon reconstruction. Although the reconstructed signal in the time domain was visually more appealing than an S+H signal in many cases, the spectral content was altered surprisingly little.

Nobach et al. [5.386] developed a refinement that cancels the particle-rate filter effect associated with S+H reconstruction. The approach is to derive an expression for the resampled ACF in terms of the true ACF. The relation is then inverted to improve the ACF estimation. The inversion is given by

$$\hat{R}_k = \begin{cases} \hat{R}_0^{(SH)} & \text{for } k = 0 \\ (2c+1)\hat{R}'^{(SH)}_k - c\left(\hat{R}'^{(SH)}_{k-1} + \hat{R}'^{(SH)}_{k+1}\right) & \\ & \text{for } k = 1, \ldots K \end{cases}$$

$$c = \frac{\exp\left(-\dot{N}\Delta t_s\right)}{\left[1 - \exp\left(-\dot{N}\Delta t_s\right)\right]^2}, \qquad (5.241)$$

where $\hat{R}$ is the refined ACF estimate based on the ACF of the reconstructed and resampled time signal $\hat{R}^{(SH)}$. A full derivation of this relation is given by *Nobach* et al. [5.381]. The PSD follows from a cosine transform of $\hat{R}$. In principle, a refinement can be derived for any reconstruction algorithm, but for the S+H reconstruction the refinement becomes the very simple algorithm given above and is effective enough that the advantages of other reconstruction schemes become negligible.

5.3.3 Particle Image Velocimetry

What is PIV?

Most common fluids, such as air and water, are homogeneous and optically transparent. This implies that the motion of the fluid is not readily visible; this is experienced by the modern traveler who takes a view through an airliner window and is oblivious of the air rushing at speeds around 250 m/s just outside the window. Once the fluid has been marked, for example by using smoke, small particles or a dye, the fluid motion and structure of the flow become visible; methods for flow visualization are described by *Merzkirch* [5.387], and the book by *Van Dyke* [5.388] contains excellent examples of flow visualization examples providing insight in the deep physical aspects of fluid motion. Yet, visualization methods are predominantly qualitative, and in general fluid markers are introduced at specific locations. On the other hand, single-point methods (e.g., *hot-wire anemometry* and *laser-Doppler anemometry*) provide a quantitative and accurate sample of the flow at a given point. They are however unable to capture the instantaneous flow organization, also known as *coherent flow structures*, observed in turbulent flows [5.389–391]. With the development of digital camera technology and affordable digital image processing it became possible to develop new measurement methods that could simultaneously provide the instantaneous spatial flow field description and a quantitative result [5.392]. One of the most successful measurement methods that has emerged in the past two decades is *particle image velocimetry* (PIV) [5.393, 394].

The working principle of PIV is schematically described in Fig. 5.85. The different aspects of a PIV measurement are discussed in greater detail in the remainder of this chapter. The principle of PIV is based on the measurement of the displacement of small tracer particles that are carried by the fluid during a short time interval. The tracer particles are sufficiently small that they accurately follow the fluid motion and do not alter the fluid properties or flow characteristics (Sect. 5.3).

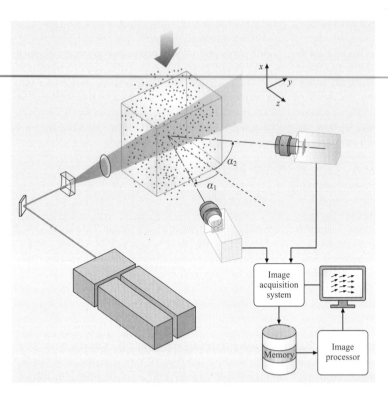

Fig. 5.85 Schematic of a typical PIV measurement system

The tracer particles are illuminated by means of a thin light sheet generated from a pulsed light source (usually a double-head pulsed laser system), and the light scattered by them is recorded onto two subsequent image frames by a digital imaging device, typically a CCD camera. By means of stereoscopic imaging it is possible to determine the three components of the flow velocity within the planar field defined by the light sheet. The simplified one-camera system suffers from the restriction that the optical axis must be aligned in the direction normal to the light-sheet plane. Moreover such systems only yield two velocity components within the measurement plane and are therefore only suitable for the investigation of flows with a negligible out-of-plane velocity component. PIV is essentially a non-intrusive measurement method; however it requires optical access for both the delivery of the light sheet and recording of the images. In order to obtain robust unbiased measurements over the flow domain, it is important that the tracer particles are homogeneously distributed within the observed flow region. This is a marked difference with respect to flow visualization methods, where fluid trac-

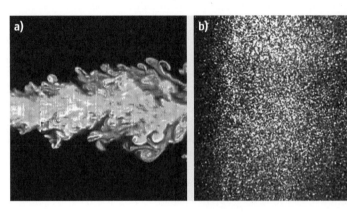

Fig. 5.86a,b Turbulent jet experiment: flow visualization of the jet by means of a dye added to the jet fluid (**a**); PIV image of the same jet where both the jet fluid and the surrounding fluid are homogeneously seeded with tracer particles (**b**)

ers are typically introduced at very specific locations, as shown in the example of Fig. 5.86.

The recorded images are typically processed offline on a digital computer. This consists essentially of a cross-correlation analysis of the particle-image patterns in small subdomains, or *interrogation regions*, between the first and second image frame. Because of the stereoscopic viewing and the finite thickness of the light sheet, the observed particle-image displacement is slightly different for each camera, which depends on the magnitude and direction of the local velocity. The three-component displacement vector is reconstructed from the displacements observed by the two cameras. The particle-image pattern displacement divided by the image magnification and the time delay between the laser light pulses yields the local fluid velocity. This process is repeated for the entire image domain, which yields the instantaneous velocity in a planar cross section of the observed flow. Sophisticated interrogation methods have been developed that make it possible to interrogate images at high spatial resolution and to determine displacements with high accuracy. The PIV analysis typically yields 10^3–10^5 velocity measurements per image (depending on the digital image format and interrogation resolution) with a typical measurement accuracy of about 1% [5.397].

PIV is a measurement method able to partially determine the instantaneous velocity gradient tensor $\partial \boldsymbol{u}/\partial \boldsymbol{x}$ within the plane defined by the light sheet. This implies that PIV measurement can be used to determine the instantaneous out-of-plane component of the vorticity, i.e., the curl of the velocity field. This has been a major breakthrough in experimental methods that has made it possible to investigate aspects of flows experimentally that were previously impossible, with particularly interesting applications for turbulent flows where quantitative information on coherent flow structures could be inferred. In this respect PIV measurements made it possible to investigate turbulent flows from a perspective other than that of the one-point statistics, retrieving the deterministic approach to the study of turbulence and restoring the concepts of dynamical organization and emphasizing the role of coherent structures, which were not straightforward when using previous experimental methods. For example, Fig. 5.87 shows the observation of a so-called *hairpin vortex* in wall-bounded turbulent flow by means of classical hydrogen-bubble visualization and the computed out-of-plane vorticity from a PIV measurement; the same figure shows an example of an LDA measurement obtained in the same region where the heads of hairpin vortices would occur, which evidently does not reveal the turbulent flow structure.

An additional contribution of the PIV technique to experimental research is the capability to determine the flow velocity at a large number of points simultaneously, which allows a significant reduction of the time needed to acquire experimental data in wind-tunnel tests [5.398]. An example is PIV measurement of the wake behind an airplane model in a towing tank [5.399], as illustrated in Fig. 5.88. The light-sheet position is at

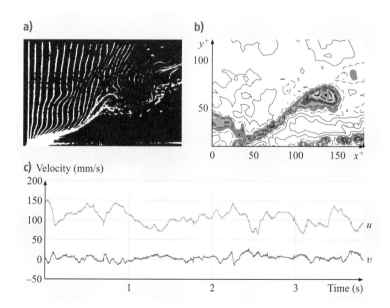

Fig. 5.87a–c Coherent structures in wall-bounded flows. (**a**) *Hairpin vortex* visualized by means of hydrogen bubbles (after *Kim* et al. [5.395]); (**b**) measured out-of-plane vorticity for a hairpin vortex measured with PIV (*Westerweel* [5.396]); (**c**) time trace of a two-component LDA measurement near the wall in a turbulent boundary layer (courtesy of A. Schwarz-van Maanen)

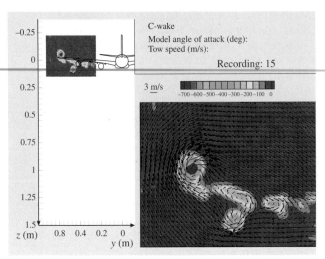

Fig. 5.88 The measured vorticity behind an airplane model in a towing tank (*Scarano* et al. [5.399])

The extension of the PIV technique into a three-dimensional measurement method has been achieved by different approaches. The most popular is the high-speed scanning method [5.400, 402] where the quasi-instantaneous three-dimensional flow structure is reconstructed from PIV recordings obtained at high repetition rate over parallel planes slightly shifted along the normal direction (Fig. 5.89a).

The instantaneous three-dimensional measurement of the flow can also be achieved by holographic recording [5.403] of an illuminated volume and the particle depth position is obtained by the reconstruction of the 3-D object from its hologram. A more-recent approach for three-dimensional measurements makes use of several viewing directions of an illuminated volume in the flow domain and a three-dimensional scattered light field reconstruction algorithm based on digital tomography [5.404]. A typical result is shown in Fig. 5.89b.

In the past decades PIV has evolved from a method that was only applied in specialized laboratories into a versatile off-the-shelf experimental tool [5.394] with a spatial resolution and accuracy that can be used for the detailed investigation of complex unsteady and three-dimensional flows. With state-of-the-art equipment it is possible to measure from large-scale supersonic flows [5.405], down to microscopic-scale low-speed flows [5.357] and even in living organisms [5.406, 407]. With the use of multiple-camera systems it is possible to combine PIV with other optical diagnostics such as *laser-induced fluorescence* (LIF), which makes it pos-

a fixed location in the laboratory system of reference, while the aircraft model is towed through the measurement plane and beyond, releasing trailing vortices. Typically such vortex wakes are quite persistent and extend over very long distances in comparison with the wingspan. Measurements with single-point methods (such as HWA or LDA) would require a passage of the model for each measurement location, while a PIV measurement captures the wake vorticity field time history at a given station within a single experimental run.

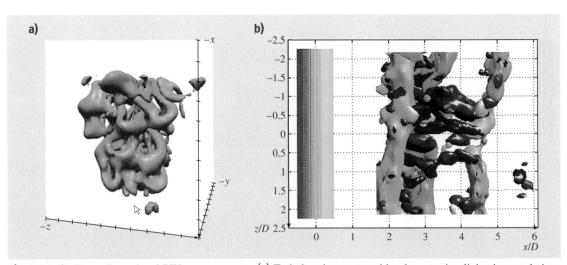

Fig. 5.89a,b Three-dimensional PIV measurements: (**a**) Turbulent jet measured by the scanning light-sheet technique and visualized with a vorticity isosurface (after *Hori* and *Sakakibara* [5.400]); (**b**) circular cylinder wake measured with tomographic PIV and visualized by color-coded vorticity components isosurfaces (after *Scarano* et al. [5.401])

sible to obtain simultaneous velocity and concentration or temperature measurements [5.408]. The use of fluorescent markers or advanced analysis methods allows an easy extension of PIV in multiphase flows [5.409–413].

Optical Systems: Seeing the Flow
Illumination.
Light Source Requirements. The measurement principle of PIV relies on a finite observation time Δt used to detect the particle motion. The particle tracers need to be illuminated and observed twice within the time separation. The first requirement for a light source suited for PIV experiments is a short duration of illumination δt. A practical criterion for the choice of the maximum duration of each illumination pulse is that particle images do not appear as streaks but rather as circular dots. This is obtained when the (imaged) particle displacement within the pulse duration (or during the imager exposure time for continuous illumination) is significantly smaller than the size of the particle image itself (see the section on imaging for the particle image size):

$$\delta t \ll \frac{d_\tau}{vM}. \tag{5.242}$$

The second requirement is that the illuminated particles are distributed within a thin sheet such that they can be imaged in focus and their position in depth is dictated by the light sheet. A third requirement is that the intensity of the light source must allow the scattered light from the seeding particles to be detected by digital imaging devices. The required pulse energy E is proportional to the linear dimension L of the area of interest. Typically $E \approx 100$ mJ is sufficient to illuminate an area with $L = 10$ cm in air flows, while about twice that area is possible in water flows. The planar illumination requires that the thickness of the light sheet is significantly smaller than its height and width (in the order of 1%).

Lasers are used as illumination sources since they can produce a pulsed, collimated and monochromatic light beam that can be easily shaped into a thin light sheet (see the next section). The most common device used for PIV experiments is the solid-state frequency-doubled Nd:YAG laser that emits light with a wavelength of 532 nm. It produces pulse energies ranging between 10 mJ and 1 J. With its very short pulse duration (5–15 ns) this instrument is practically suited to illuminate flows without any limit on the flow speed. The standard architecture of a PIV laser consists of two separate lasers firing independently at the required pulse separation. Therefore the time separation can be freely optimized for the experimental conditions, primarily the flow speed and the imaging magnification.

However its repetition rate ranges between 10 and 50 Hz, posing the major limitation of Nd:YAG-based systems in performing time-resolved experiments except for very low-speed flows ($v < 0.2$ m/s). In the last decade, important developments in the area of solid-state lasers have improved significantly the performance of diode-pumped Nd:YLF lasers for high-repetition-rate illumination. Currently Nd:YLF lasers are capable of emitting pulses of energy of 10–40 mJ at a repetition rate of 1–5 kHz. Their pulse duration is about 10 times longer than Nd:YAG lasers.

Light-Sheet Formation Optics. The circular-cross-section beam delivered by the light source is shaped into

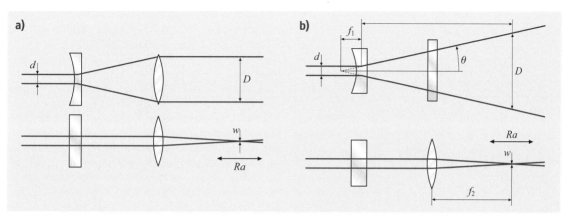

Fig. 5.90a,b Light-sheet formation optics: **(a)** constant sheet width. **(b)** linearly expanding sheet width. The Rayleigh length Ra is the length over which the light sheet has a minimum thickness w (denoted as the *waist*)

a thin sheet by means of cylindrical and spherical lenses. Common arrangements are illustrated in Fig. 5.90. In the first case (Fig. 5.90a) a cylindrical lens expands the beam along the vertical direction followed by a spherical lens that compensates the effect of the first lens. The resulting system is equivalent to a beam expander along the vertical direction. In the horizontal direction the cylindrical lens has no effect on the beam and the spherical lens will focus it at a distance equal to the focal length. This arrangement is particularly suited to achieve uniform illumination along the propagation direction. It is, however, not practical in cases for which large regions of space have to be illuminated unless a large-format spherical lens is available. The scheme illustrated in Fig. 5.90b applies for the illumination of large areas and in general when uniform illumination is not strictly required. In this case the desired sheet width can be obtained at an appropriate distance L as a function of the focal length of the cylindrical lens f_1. Systems with more than two lenses allow more-complete control of the light sheet at the cost of increased complexity.

To minimize the variation of the light sheet properties within the area of interest it is convenient to design the light-sheet optics in such a way that the beam waist does not fall within the measurement region. In fact the spatial rate of change of sheet thickness and energy density is highest around the waist. As a consequence several experimental parameters such as out-of-plane particle motion and particle brightness vary largely within the measurement area, limiting the possibility to optimize the experiment. In particular, an arrangement with the waist outside (possibly beyond) the measurement region is strictly required when performing stereoscopic measurements with a large velocity component across the light-sheet thickness.

Imaging. A schematic of the PIV optical configuration is shown in Fig. 5.91. The light sheet has a finite thickness Δz_0 and is assumed to be uniform along the x and y directions. An image of the tracer particles in the light sheet is formed by means of a lens on the surface of an image sensor (usually a CCD or CMOS sensor array). The lens is characterized by its focal length f, aperture number $f^\#$ (given by the focal length divided by the aperture diameter), and image magnification M_0, defined as the ratio of the image distance Z_0 and object distance z_0. The diameter d_τ of the image of a small tracer particle with diameter d_p in the light sheet is given by [5.393, 415]

$$d_\tau \cong \left(d_s^2 + M_0^2 d_p^2\right)^{1/2},$$
$$\text{with } d_s = 2.44(1 + M_0)f^\# \lambda, \qquad (5.243)$$

where d_s is the *diffraction-limited spot* diameter and $M_0 d_p$ the geometric image diameter. The diffraction-limited spot results from the finite resolution of the optical system due to diffraction effects; for a point source ($d_p \to 0$) or distant object ($M_0 \to 0$) the light captured by the objective is spread over a small spot also known as the *Airy* disc, with diameter d_s, surrounded by diffraction rings of decreasing brightness [5.415]. For all practical purposes in PIV, the light distribution in the Airy disc is well approximated by a Gaussian intensity distribution with an e^2 diameter of $0.74 d_s$. The first expression in (5.243) would be exact when the diffraction-limited spot and geometrical-optics particle image follow a Gaussian intensity distribution.

For typical optical parameters in PIV, e.g., $M_0 \approx 0.2$, $f^\# \approx 4$, $\lambda \approx 0.5\,\mu m$, $d_p \approx 10\,\mu m$, $d_s \approx 6\,\mu m$, it is found that $d_s \gg M_0 d_p$, so that $d_\tau \approx d_s$. In other words, the diffraction limit generally dominates the particle-image formation, and the particle-image diameter is quite uniform despite variations in d_p.

Equation (5.243) dictates the lower limit for the particle image diameter and applies for an aberration-free lens with a circular aperture. In practice this means that it is valid for sufficiently large $f^\#$, since for very small $f^\#$ aberrations decrease the performance of lens systems; it is good practice always to close the aperture of a lens by one or two f-stops, and to use a lens within its designed magnification range; different lens designs exist, e.g., for high-magnification imaging (viz., *macro*lenses) or for fixed magnification and working distances (*telecentric* lenses).

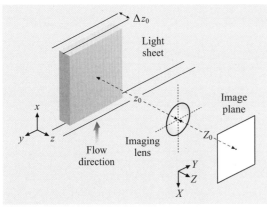

Fig. 5.91 Schematic of the optical configuration for PIV imaging. (After *Westerweel* [5.414])

Moreover, the particle-image diameter follows (5.243) for particle images in focus, i.e., when the light-sheet thickness Δz_0 is smaller than the focal depth δz of the optical system (Fig. 5.91) given by [5.393]

$$\delta z \cong 4\left(1+\frac{1}{M_0}\right)^2 f^{\#2}\lambda \,. \tag{5.244}$$

For the above example the focal depth of the imaging system is about 1 mm. In the case of μPIV the focal depth will in general be much smaller than the thickness of the illuminated volume, and the recording contains both focused particle images as well as out-of-focus particle images, which degrade the measurement precision. In this case the measurement volume is not defined by the thickness of the light sheet, but by the *correlation depth*, i.e., the depth along the optical path over which (defocused) particle images contribute significantly to the measured displacement [5.416].

Given the rather small focal depth and the strong dependence of the measurement error on particle-image diameter, it is important that the optical system is properly focused. To adjust the lens by hand while judging the focus from a monitor is often not accurate and impossible to verify. It is good practice to consider the *image contrast*, defined as the standard deviation of the image gray value divided by the mean image gray value, as a function of the focal-plane position; the position with maximum image contrast is the optimal focus position.

Stereoscopic Imaging. The use of stereoscopic imaging by means of two cameras makes it possible to reconstruct the three-component displacement vector in the plane of the light sheet. A review of stereoscopic PIV and related methods is given by *Prasad* [5.417]. The two generic configurations are the *translation method* and the *angular method* (Fig. 5.92). In the translation method the object and image planes are parallel, so that the magnification is constant, but disadvantages are that

- it is restricted to rather small viewing angles (up to about 15°), and
- the off-axis viewing is prone to optical aberrations.

The angular method allows much larger viewing angles and paraxial imaging, but the image magnification is no longer constant. This leads to a perspective deformation of the image and varying resolution over the image domain. In recent years practical and accurate methods have been developed for image reconstruction in angular stereoscopic viewing.

For an ideal imaging system with parallel object and image planes and with a thin light sheet (i.e., $\Delta z_0 \ll \delta z$), a particle at position (x, y, z) within the light sheet ($z \approx z_0$) is projected onto a position (X, Y, Z_0) with: $X/x = Y/y = M_0$. For angular stereoscopic viewing it is necessary to tilt the backplane of the camera with respect to the optical axis of the lens, according to the diagram shown in Fig. 5.93, so that all particle images are in focus; this is the so-called *Scheimpflug* configuration [5.417, 418]. Since the object plane and image plane are not parallel, the image magnification varies over the

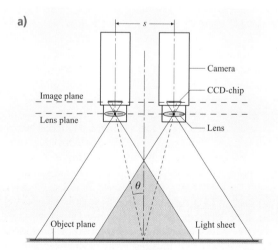
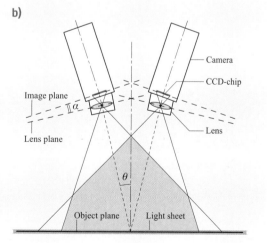

Fig. 5.92a,b Optical configurations for stereoscopic imaging: the translation method (**a**) and angular displacement method (**b**)

field of view. As a result there is a perspective deformation of the image as shown in Fig. 5.96. The geometric back-projection of (X, Y) in the image data onto (x, y) in the object plane is given by

$$x = \frac{fX \sin\alpha}{M_0 \sin\theta(X \sin\alpha + fM_0)},$$
$$y = \frac{fX}{X \sin\alpha + fM_0}, \quad (5.245)$$

where the angles α and θ are defined in Fig. 5.93. This method requires accurate values for the optical and geometric parameters; inaccuracies can lead to significant systematical errors in the PIV results. A generalized approach is to use a mapping function instead [5.417, 420]

$$x = \frac{a_0 + a_1 X + a_2 Y + a_3 X^2 + a_4 Y^2 + a_5 XY + \ldots}{1 + a'_1 X + a'_2 Y + \ldots},$$
$$y = \frac{b_0 + b_1 X + b_2 Y + b_3 X^2 + b_4 Y^2 + b_5 XY + \ldots}{1 + b'_1 X + b'_2 Y + \ldots}, \quad (5.246)$$

where the parameter vectors a, a', b, and b' are determined by means of a calibration target. When $a' = b' = 0$ the parameter vectors a and b can be found from an arbitrary number of calibration points by means of solving a set of linear equations. For correcting perspective image distortion (in which straight lines are imaged as straight lines) it is in principle sufficient to use second-order polynomials.

The three-component displacement vector $(\Delta x, \Delta y, \Delta z)$ for a symmetric angular configuration (i.e., $\theta_1 = \theta_2 = \theta$) is reconstructed from the two in-plane displacements $(\Delta x_l, \Delta y_l)$ and $(\Delta x_r, \Delta y_r)$ observed by the *left* and *right* camera, respectively,

$$\Delta x = \frac{\Delta x_1 - \Delta x_2}{2 \sin\theta}, \quad \Delta y = \frac{\Delta y_1 + \Delta y_2}{2},$$
$$\Delta z = -\frac{\Delta x_1 + \Delta x_2}{2 \cos\theta}, \quad (5.247)$$

for $|\Delta x_1|, |\Delta y_1|, |\Delta x_2|, |\Delta y_2| \ll z_0 \sin\theta$ (Fig. 5.94). It is noted that the expression for Δz in (5.247) implies that, in a single-camera configuration (with the optical axis perpendicular to the light-sheet plane), the in-plane components of the displacement are contaminated by the out-of-plane displacement; the error is proportional to $\Delta z \tan\eta$, where η represents the viewing angle with respect to the optical axis. Given typical values for $\Delta x, \Delta z \approx 0.25$ mm and $\eta \approx 6°$ (for a point near the edge of the field of view) yields an error of up to 10% of the in-plane displacement.

A generalized method to obtain the out-of-plane component is by means of so-called *three-dimensional calibration* or *Soloff method* [5.421], where an image of the calibration target is recorded at two or more positions along the z-coordinate, and the relation between the three-dimensional position x in the object domain and the two-dimensional position X in each image plane is

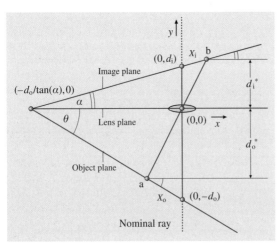

Fig. 5.93 Scheimpflug condition for tilted object and image planes

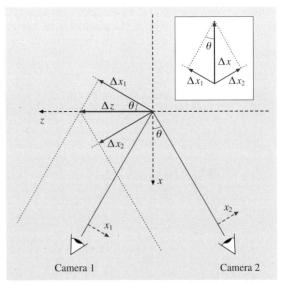

Fig. 5.94 The principle of stereoscopic PIV in which an out-of-plane displacement Δz is reconstructed from the observed motion in two cameras. In this diagram the cameras are located on opposite sides of the light sheet. (After *van Doorne* et al. [5.419])

Fig. 5.95 Three generic configurations for the light sheet and cameras in stereoscopic PIV. The *arrow* indicates the direction of propagation of the light sheet

fitted to a polynomial $F(x)$. The particle-image displacement ΔX is then given by: $\Delta X \cong (\nabla F) \cdot \Delta x$, [5.417]:

$$\begin{pmatrix} \Delta X_l \\ \Delta Y_l \\ \Delta X_r \\ \Delta Y_r \end{pmatrix} \cong \begin{pmatrix} F^l_{1,1} & F^l_{1,2} & F^l_{1,3} \\ F^l_{2,1} & F^l_{2,2} & F^l_{2,3} \\ F^r_{1,1} & F^r_{1,2} & F^r_{1,3} \\ F^r_{2,1} & F^r_{2,2} & F^r_{2,3} \end{pmatrix} \begin{pmatrix} \Delta x \\ \Delta y \\ \Delta z \end{pmatrix} \quad (5.248)$$

with $F_{i,j} = \partial F_i / \partial x_j$. The set of equations in (5.248) is solved by means of a least-squares optimization method.

Equation (5.247) shows that the random measurement error for Δz relative to the random errors in Δx and Δy is inversely proportional to $\tan \theta$. Hence, for $\theta = 45°$ the errors in the in-plane and out-of-plane components are of the same order. When stereoscopic viewing is applied for a liquid flow (viz., water) the light rays are refracted at the water–air interface. Given Snell's law, a stereoscopic viewing angle of 45° in air reduces to only

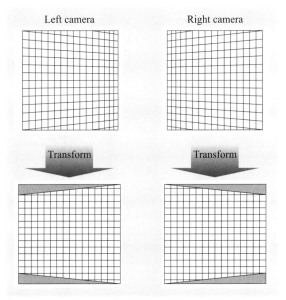

Fig. 5.96 Perspective deformation due to the angular configuration in stereoscopic viewing

32° (with a refractive index of $n = 1.33$) for water, implying that an effective 45° stereoscopic viewing angle in water is obtained for 49°, which is close to the critical angle (for $n = 1.33$), making this arrangement not viable. It is therefore common to use prisms such that the optical axis is not refracted when it crosses the wall of the test section and large viewing angles in liquids are attainable without strong image distortion [5.417].

When using a calibration-based stereoscopic reconstruction, it is important that the position of the calibration target accurately coincides with the light sheet; deviations lead to so-called *registration errors* and artefacts in the measurement results [5.419, 422]. A small deviation can be corrected by means of a procedure based on the so-called *disparity map* obtained by cross-correlating the simultaneous particle images recorded from the two views [5.418, 423].

Another aspect that can strongly influence the performance of a stereoscopic PIV system is the orientation of the cameras with respect to the light sheet. In most applications with restricted optical access it will be practical to position both cameras on one side of the light sheet; with full optical access one can position the cameras on opposite sides of the light sheet. Three generic configurations can be identified, as shown in Fig. 5.95. For the first configuration the scattering angle is around 90° for both cameras. In view of the scattering diagram shown it would be favorable to use forward or backward scattering, as is achieved in the second configuration. The level of scattering has usually increased, although it can be quite uneven between the two cameras, as forward scattering is generally an order of magnitude stronger than backward scattering. In the third configuration both cameras are positioned so that light is scattered in a forward direction and at equal scattering levels; this configuration would be favorable when working with low-power lasers or very small tracer particles, provided that sufficient optical access is available.

Non-Ideal Imaging. For non-ideal imaging systems the particle images can be distorted as a result of optical aberrations, and straight lines in the object plane become curved lines in the image plane, i.e., nonuniform imaging. Examples of optical aberrations are *coma* and *astigmatism*. Coma usually occurs for nonparaxial imaging, while astigmatism occurs when the light rays pass at significant angles through different refractive media (e.g., when no prisms are applied for stereoscopic viewing in a liquid). Nonuniform imaging is generally classified as *barrel* or *pincushion* deformation. Optical aberrations and image nonuniformity can usually be

Fig. 5.97 Stereoscopic PIV setup for measurements in a pipe. A: pipe; B: water-filled rectangular box; C: water-filled prisms; D: cameras on Scheimpflug mounts (after *van Doorne* and *Westerweel* [5.419])

2. image *enhancement* accentuates certain image features (e.g., improvement of image contrast, suppression of background illumination).

The interrogation analysis of PIV images by means of spatial correlation is based on the requirement that the image statistical properties are *homogeneous*, i.e., the image mean and standard deviation are spatially uniform and the spatial correlation is a function of the difference of two spatial locations. This implies:

1. homogeneous seeding
2. uniform illumination
3. uniform image background.

In many practical situations these requirements may not be fully met: for large flow facilities and high-speed flows it is difficult to maintain a homogeneous seeding; the Gaussian intensity profile of laser beams and the use of an uncollimated light sheet imply gradual variations of the local illumination intensity; PIV measurements near wall and other objects usually suffer from nonuniform backgrounds due to reflections. Some of these non-ideal aspects in PIV images can be compensated by means of image-processing methods, which are briefly reviewed here.

avoided when lenses are used within the design specifications. A small degree of image nonuniformity can be corrected by using higher-order polynomials as mapping functions (5.246).

Optical aberrations and image nonuniformity also occur when the flow is observed through curved windows, for example when imaging a flow in a cylindrical pipe. This can become very strong when such a pipe is filled with a liquid. In that case the optical distortions and image nonuniformity can be reduced considerably by enclosing the pipe section at the measurement location with a box that is filled with the same liquid. It is also beneficial to replace the pipe wall locally with a thin-walled pipe section.

As an example, Fig. 5.97 shows the optical configuration for stereoscopic PIV measurements in a pipe. The cameras are on Scheimpflug mounts on opposite sides of the light sheet in the forward scattering direction, and the setup includes a water-filled box with a thin-walled pipe section and prisms to minimize optical distortions.

Digital Image Processing. Digital image processing encompasses all computer operations on digital images. For a full account of all possible methods one is referred to the many text books that have been written on this subject [5.424, 425]. In general two approaches are distinguished in image processing:

1. image *restoration* attempts to *repair* undesirable effects due to the imaging (e.g., perspective distortion, image blur due to defocusing), whereas

Low-Pass and High-Pass Filtering. Straightforward image-processing operations are low- and high-pass filtering. These can be implemented by means of moving average or autoregressive methods. Alternatively, these filters can also be implemented in the Fourier domain making use of the duality of multiplication and convolution [5.426]. These operations are linear, which implies that subtracting a low-pass filtered image from its original image is equivalent to a high-pass filtered image. In general, these methods are *lossy* (see later), which means that information is lost. For example, high-pass filtering generally removes a slowly varying background image, but also reduces the effective particle-image diameter, which can lead to increased discretization errors (peak locking); also, high-pass filtering does not remove sharp edges present in background images. An effective approach for the elimination of background noise and sharp edges is based on the subtraction of the statistical minimum intensity locally from each image. This procedure can be applied under the constraint that the background noise is constant during the acquisition of the series of recordings.

Histogram. In general the gray values in a PIV image are not equally distributed. The typical gray value histogram will have a dominant peak that represents the dark image background and an exponentially decaying part that represents the particle images (Fig. 5.98). In general the number of gray values (typically 4096 gray values for a 12 bit digitizer) by far exceeds the number of gray values a human can detect, which is about 60 [5.425]. Therefore only the brightest particle images are usually observed, whereas the actual number of particle images that contribute to the spatial correlation can be much larger than what would be judged on visual inspection of PIV images. To enhance the visibility of weak particle images a nonlinear relation between intensity and pixel gray value can be applied with the aim to achieve a more-uniform distribution over the gray values. This is referred to as *histogram equalization* [5.425].

Min/Max Filtering. A method to remove a slowly varying background image and to normalize the image contrast is *min/max* filtering [5.396]. The image is filtered in two ways:

1. each pixel is replaced by the *smallest* value in an $M \times N$-pixel neighborhood (i.e., a local minimum filter), and the result is passed through an $M \times N$-pixel moving-average filter;
2. the same operation is carried out but now for the *largest* value in an $M \times N$-pixel neighborhood (this results also passes an $M \times N$-pixel moving-average filter).

The results are the so-called lower and upper *envelopes* of the pixel gray values. The lower envelope is subtracted from the original image, and the result is divided by the (scaled) difference between the upper and lower envelope. The result represents an image with zero background and homogeneous image contrast. The optimal filter size depends on the particle-image diameter and seeding density, and can be found by optimizing the fraction of valid displacement measurements [5.428].

Image Compression. A typical image format in PIV is 1024×1024 px with 12 bit dynamic range (i.e., 4096 gray levels). Hence, a single stereoscopic PIV recording requires about 8 MB of storage. Recording a large time series easily takes up several gigabytes, and a full experiment can consume several hundreds of gigabytes. Standard compression algorithms, such as Joint Photographic Experts Group (JPEG), are *lossy* compression methods, in which fine image details are lost. As PIV images are completely filled with small-scale objects (viz., particle images), the use of lossy compression methods always implies a loss of performance [5.429], i.e., an increase in the number of spurious vectors and an increase of the random noise error (due to an increase of the effective particle-image diameter). Lossless compression methods, such as Lempel–Ziv–Welch (LZW), are better suited for PIV image data, although the compression ratio is modest due to the fact that there is little to no structure in PIV images (which typically appear as *white-noise* images). As a matter of fact, the best compression is PIV interrogation itself, which reduces an 8 MB stereoscopic image set to about 50 kB of displacement data.

Particle-Image Detection. Particle tracking methods and super-resolution PIV (Sect. 5.4) depend on the detection of individual particle images. As particle images generally appear as bright dots on a dark background, a common approach is to use the condition of local maximum above a selected grey value threshold: all pixels above the given threshold represent particle images. The number of detected particle images depends on the threshold value: when this threshold value is too high a significant number of particle images remain undetected, whereas a value that is too low will result in the spurious detection of background noise as particle images. In addition, nonuniform illumination makes it difficult to use a *single* threshold value for the entire image, unless image processing has been used to normalize the image (see before). For uniform illumination and monodisperse tracer particles the grey value histogram would be bimodal, and the optimal threshold value would lie in between the histogram peaks that represent the particle images and image background [5.430].

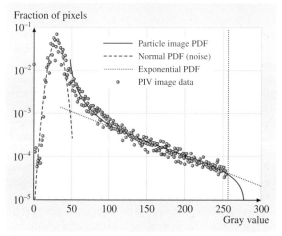

Fig. 5.98 Histogram of pixel gray values for an 8 bit digital PIV image (after *Westerweel* [5.427])

However, in practice illumination is not uniform (mainly due to the Gaussian profile of the light sheet), and the histogram will *not* be bimodal; instead it has a single background peak and an exponential tail that represents the particle images (Fig. 5.98). Inevitably, this means that the number of detected particle images depends on the threshold and that there will be a bias towards bright particle images, i.e., large tracer particles (that may have a significant slip with respect to the fluid velocity) and tracer particles that occur in the central (i.e., bright) section of the light sheet. The latter implies that a threshold reduces the effective light-sheet thickness.

Image Interrogation: Inside the Computer

Velocity from Tracer Motion. In PIV the motion of a fluid is visualized by the motion of small tracer particles added to the fluid. These tracer particles constitute a *pattern* that can be used to evaluate the fluid motion. If the tracer density is very low (i.e., the distance between distinct particles is much larger than the displacement) then it is very easy to evaluate the displacement from individual tracer particles. This mode of operation is generally referred to as *low-image-density* PIV, or *particle tracking velocimetry* (PTV). However, for most PTV algorithms the amount of information that can be retrieved from an image is very limited. By increasing the concentration of tracer particles, the displacement becomes *larger* than the particle-image spacing, and it is no longer possible to identify matching pairs unambiguously. This mode of operation is generally referred to as *high-image-density* PIV.

Low-Image-Density PIV. An intuitive approach for interrogating PIV images is by matching individual particle-image pairs. Consider the simplest interrogation algorithm, i.e., *nearest-neighbor matching*: for each particle image in frame 1 a matching particle image is searched within an interrogation domain in frame 2 (Fig. 5.99). The probability of identifying the *correct* particle-image pair is given by the probability P_0 that no *other* particle images occur within a circle with an area $A = \pi \Delta X^2$. The number of particle images that appears within a given area is a Poisson process (given that the tracer particles have a homogeneous random distribution over the flow volume), with the mean number of particle image per unit area given by: $C\Delta z_0 / M_0^2$, where C is the tracer number density, Δz_0 the light-sheet thickness, and M_0 the image magnification. Then, the probability P_0 is given by

$$P_0 = F_O e^{C\Delta z_0 / M_0^2 \pi \Delta X^2}, \quad (5.249)$$

where F_O is the probability that the tracer particle leaves the light sheet between the two exposures. In order to achieve a success rate P_α for the interrogation of the frames, one can use (5.242) to determine an upper limit for the displacement ΔX (assuming $F_O = 1$):

$$\Delta X < \delta \cdot \sqrt{\frac{4}{\pi} \ln\left(\frac{1}{P_\alpha}\right)},$$

$$\text{with } \delta \cong 0.5 \left(\frac{C\Delta z_0}{M_0^2}\right)^{-1/2}, \quad (5.250)$$

where δ is the mean distance between neighboring particle images. It is easily verified that an acceptable interrogation success rate implies that $\Delta X \ll \delta$, e.g., $\Delta X/\delta < 0.26$ for $P_\alpha = 95\%$. This implies that the interrogation algorithm is only successful when the *image density*, defined as

$$N_I = \left(\frac{C\Delta z_0}{M_0^2}\right) A \quad \text{with } A = \pi \Delta X^2, \quad (5.251)$$

is considerably smaller than one; the phrase *low-image-density PIV* has been coined to categorize PTV methods.

The obtained velocity vectors are randomly distributed over the image domain and rather sparse: if one would subdivide an image into interrogation domains with area A, only a fraction $N_I \ll 1$ would actually contain a particle-image pair. For example, for an 8 px displacement range and a 95% interrogation success rate, the number of measured velocity vectors would be about one thousand. More-advanced PTV interrogation algorithms make use of iterative approaches where the particle images are paired based on a velocity estimate obtained by a previous cross-correlation analysis. This approach is known as *super-resolution* analysis in the sense that information is extracted at a resolution higher than that of the interrogation windows [5.431]. In this case a much higher vector density can be achieved.

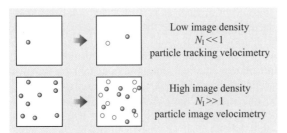

Fig. 5.99 Interrogation strategies for low image density and high image density

The limiting principle in this case is that the displacement differences within the interrogation window must always be smaller than a given parameter related to the particle-image spacing – as explained above for the nearest-neighbor matching algorithm.

High-Image-Density PIV. As explained above, one cannot rely on a simple particle-image matching algorithm to measure displacements that are larger than the mean particle-image spacing. Hence, to increase the number of samples that can be extracted it is necessary to use an approach that is capable of matching particle-image pairs when the displacement ΔX is *larger* than the mean particle-image spacing δ (Fig. 5.99). Instead, it is necessary to use a statistical approach to identify the most probable match of particle images in the two frames.

Spatial Correlation Analysis. Consider an interrogation domain in frames 1 and 2 that contains a substantial number of particle images, as shown in Fig. 5.100a. For a given particle image in the first frame each particle image in the interrogation domain in frame 2 is a possible match candidate and each pair represents a possible displacement with equal likelihood; this is represented in the displacement histogram in Fig. 5.100b where each peak represents a possible displacement and all peaks have equal amplitude to represent that each matched pair has an equal probability. This procedure is then repeated for all particle images in the interrogation domain in frame 1, and the probabilities in the displacement histogram are superimposed. Hence, the displacement peak for each of the matching pairs will soon be dominant over the displacement peaks for unrelated pairs, as shown in Fig. 5.100b.

Already two important requirements for PIV interrogation are evident:

1. the displacement of the particle images within the interrogation domain should be (nearly) uniform;
2. the interrogation domain should contain a considerable number of particle image pairs, i. e., $N_I \gg 1$; this is why this interrogation strategy is also referred to as *high-image-density PIV*.

The important difference with respect to low-image-density PIV is that the interrogation in any location normally contains sufficient particle images to obtain a valid measurement. In general, the number of velocity data extracted from high-image-density PIV is an order of magnitude larger than for low-image-density PIV, i. e., about 10^4 interrogations/image. In addition the velocity data is arranged on a regular grid, as opposed to the random data in low-image-density PIV, which is more convenient in view of post-interrogation processing of the data.

The statistical analysis to identify the most likely particle-image pairing is equivalent to find the particle-image displacement by means of a spatial cross-correlation $R(s)$ of the interrogation images $W_1(x)I_1(x)$ and $W_2(x)I_2(x)$:

$$R(s) = \int_A W_1(x)I_1(x)W_2(x+s)I_2(x+s)\,\mathrm{d}x \, ,$$

(5.252)

where W_1 and W_2 are windowing functions that define the interrogation domains in the images I_2 and I_2, respectively.

Spatial Correlation Analysis. A theoretical analysis of PIV interrogation using spatial correlation was initially developed by *Adrian* [5.432] for double-exposure continuous images obtained from the photographic recording technique. Later this theory was extended to image pairs with cross-correlation. *Westerweel* [5.396, 414, 433] adapted the theory for analysis of digital images.

To understand the operating principle of the spatial correlation analysis, the images I_1 and I_2 are split into mean and fluctuating parts, i. e., $I_1(X) = \langle I_1 \rangle + I'_1(X)$ and similarly for I_2, where $\langle \cdots \rangle$ denotes an ensemble average over images under identical circumstances but with different particle-image patterns. Also, it is implicitly assumed that the particle images are distributed homogeneously so that $\langle I_1 \rangle$ is a constant. After substitution in (5.244), $R(s)$ can be written as

$$R(s) = R_C(s) + R_F(s) + R_D(s) \, ,$$

(5.253)

where R_C is the spatial cross-correlation of the mean image intensities, R_F the cross-correlation of $\langle I_1 \rangle$ and $I'_2(X)$ and vice versa, and R_D the cross-correlation of the fluctuating images intensities $I'_1(X)$ and $I'_2(X)$. This is

Fig. 5.100 Histogram analysis for finding the most probable displacement

also illustrated in Fig. 5.101. In practice this means that the terms R_C and R_F can be eliminated by subtracting the mean image intensity from the interrogation images. The remaining term R_D can be split into mean and fluctuating parts, where the averaging is taken over an ensemble of tracer patterns for a given (fixed) velocity field $\boldsymbol{u}(X, t)$:

$$R_D(s) = \langle R_D(s) | \boldsymbol{u} \rangle + R'_D(s), \quad (5.254)$$

where $\langle R_D(s) | \boldsymbol{u} \rangle$ is commonly referred to as the *displacement-correlation peak* and $R'_D(s)$ as the *random correlation term*. Figure 5.102 shows typical examples of $\langle R_D(s) | \boldsymbol{u} \rangle$ and $R'_D(s)$. Evidently $\langle R_D(s) | \boldsymbol{u} \rangle$ shows a single peak at a location $s = s_D$, which coincides with the local particle-image displacement within the interrogation domain (provided that the velocity field within the interrogation domain is uniform).

However, it is only possible to identify the correct displacement-correlation peak when its amplitude is larger than the highest random correlation peak in $R'_D(s)$.

If this is not the case, the interrogation analysis yields a *spurious* value for the particle-image displacement.

The ratio of the highest correlation peak and that of the second highest correlation peak is a measure of the *detectability* of the displacement-correlation peak, indicated by D_0, and can be considered as a lower limit of the signal-to-noise ratio.

Theoretical analysis shows that for a (nearly) uniform velocity field the quantity $\langle R_D(s) | \boldsymbol{u} \rangle$ is given by

$$\langle R_D(s) | \boldsymbol{u} \rangle \sim N_I F_O(\Delta z) F_I(s) F_\tau(s - s_D), \quad (5.255)$$

where N_I is the *image density*, F_O the *loss of correlation due to out-of-plane motion* of the tracer particle (i.e., tracer particles that enter or leave the light sheet between the two exposures), F_I the *loss of correlation due to in-plane motion* of the tracer particles (i.e., tracer particles that enter or leave the interrogation domain between the two exposures), and F_τ the particle-image self-correlation. Given that the light sheet has a Gaussian intensity profile along the z-coordinate, and that the in-

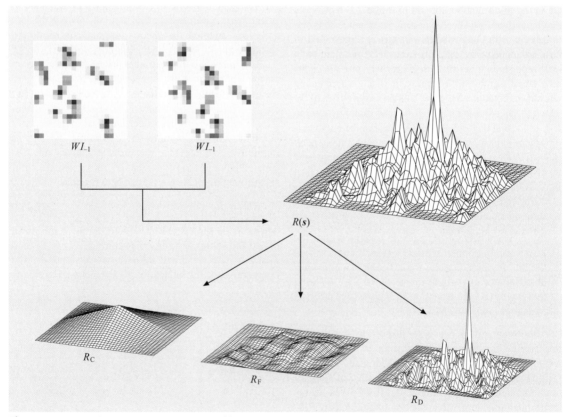

Fig. 5.101 The spatial cross-correlation $R(s)$ of interrogation images WI_1 and WI_2, (*top-hat* windowing) when split into mean and fluctuating parts, can be separated into three terms: R_C, R_F, and R_D

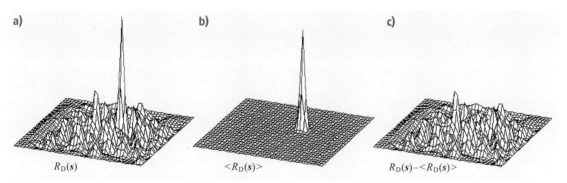

Fig. 5.102a–c Separation of the displacement-correlation term R_D (**a**) into mean (**b**) and fluctuating terms (**c**)

terrogation windows are identical (i.e., $W_1 = W_2 = W$) and uniform with an area $A = D_I \times D_I$, F_O and F_I are given by

$$F_O(\Delta z) = \exp\left(-8\frac{\Delta z}{\Delta z_0}\right), \quad \text{and}$$

$$F_I(s_X, s_Y) = \left(1 - \frac{|s_X|}{D_I}\right)\left(1 - \frac{|s_Y|}{D_I}\right), \quad (5.256)$$

for $|s_X|, |s_Y| < D_I$. For identical particle images that have an (approximate) Gaussian shape with an e^{-2} width d_τ, the particle-image self-correlation F_τ is given by

$$F_\tau(s) = \exp\left(-4\frac{s^2}{d_\tau^2}\right) \quad \text{with } s^2 = s_X^2 + s_Y^2. \quad (5.257)$$

Note that F_τ has a Gaussian shape with an e^{-2} width d_τ. The expressions (5.247) and (5.249) can be used to evaluate the effect of the image density and in-plane and out-of-plane motion on the location and shape of the displacement-correlation peak.

Image Density. Figure 5.103 shows the effect of a variation in the image density N_I on the spatial correla-

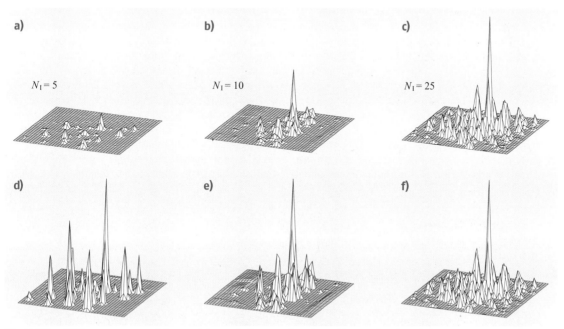

Fig. 5.103a–f The appearance of the spatial correlation as a function of image density, viz. $N_I = 5$, 10, and 25. Graphs (**d–f**) are identical to graphs (**a–c**), but normalized with respect to the maximum peak height

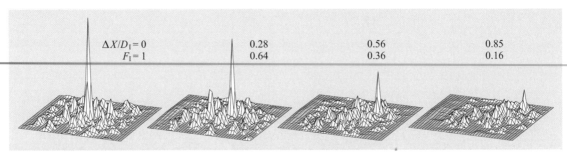

Fig. 5.104 The spatial correlation for increasing in-plane displacement (at constant image density)

tion, while the in-plane and out-of-plane displacements remain constant. As predicted by (5.247) the peak amplitude is directly proportional to N_I, i.e., the signal strength increases when the number of particle images in the interrogation domain increases. It is also clear that the *noise* (i.e., the random correlations) increases in proportion to N_I. However, when the correlation is normalized with respect to the highest peak (Fig. 5.103d–f), it is clear that the *relative* noise level (defined as the highest random correlation peak) *decreases* with increasing N_I.

In-Plane and Out-of-Plane Motion. The amplitude of the displacement-correlation peak decreases for increasing in-plane displacement. Examples of the spatial correlation for increasing in-plane displacements are shown in Fig. 5.104. The peak amplitude is proportional to the fraction of the particle images that remains within the interrogation domain; this fraction is equal to the value of $F_I(s)$ (5.249). Note that the amplitude of the displacement-correlation peak decreases for increasing displacement, while the amplitude of the noise peaks remains at the same level. Hence, for increasing in-plane displacement, the probability that the displacement peak is correctly identified reduces. Later it is shown that the location of W_2 can be shifted over (s_X, s_Y), so that the number of particle images within the interrogation domain is maximized and $F_I \approx 1$ (Sect. 5.4.2). An alternative way to achieve $F_I = 1$ is to use a larger window W_2 such that all particle image present in W_1 are also included in W_2.

Out-of-Plane Motion. Figure 5.105 shows the spatial correlation for a fixed image density and in-plane displacement, but for increasing out-of-plane displacement. The peak amplitude decreases for increasing $|\Delta z|$ and is proportional to $F_O(\Delta z)$ as defined in (5.248) (for a light sheet with a Gaussian intensity profile). Unlike the effect of in-plane motion, it is not possible to reinterrogate the image pairs and compensate for the out-of-plane motion; in practice this means that loss of correlation due to out-of-plane motion is a major limiting aspect of a planar PIV measurement.

It should also be noted that in Fig. 5.105 it was assumed that the light sheet is *identical* (perfectly overlapped) for both exposures. In practice this may not always be the case when a double-cavity laser is used; when the two laser beams do not perfectly overlap, the value of Δz in Fig. 5.105 should be replaced by $\Delta z + z^*$, where z^* is the light-sheet misalignment. In PIV measurements where the principal flow direction is perpendicular to the light-sheet plane a deliberate offset can be used to maximize F_O (this is very similar to the window offset method, although it cannot be executed in

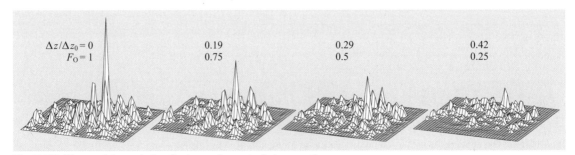

Fig. 5.105 The spatial correlation for increasing out-of-plane motion

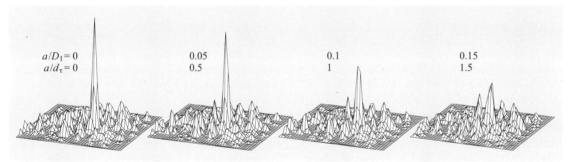

Fig. 5.106 The spatial correlation for an increasing velocity difference within the interrogation window

a postrecording phase). A practical procedure to achieve the highest correlation in relation to the light sheets overlap is to perform online cross-correlation while displacing one light sheet and monitor the height of the correlation peak.

Velocity Gradient. The effect of a local velocity gradient in the measurement field was ignored in the present theory. An extension to arbitrary flow fields is not trivial. Instead it can be assumed that there is a *small* local variation of the velocity field within the interrogation domain; this variation can be due to a large-scale gradient in the in-plane or out-of-plane fluid motion, or a local variation at a scale below the dimensions of the interrogation domain. It is then investigated to what extent such variations do not significantly affect the theory developed for purely uniform fluid motion.

Figure 5.106 shows the spatial correlation for increasing local variation $a = M_0|\Delta u|\Delta t$ within the interrogation domain of the displacement field. For increasing a the peak amplitude *decreases* while the *width* of the peak *increases*, and the total volume under the correlation peak remains (approximately) constant. The loss of correlation F_Δ due to a (small) variation of the displacement field is given by

$$F_\Delta(a) = \exp\left(-\frac{a^2}{d_\tau^2}\right). \qquad (5.258)$$

The effects of a local variation of the displacement field are negligible when a is smaller than d_τ. For larger values of a it can happen that the displacement-correlation peak excessively broadens and splits up into several correlation peaks, and an unambiguous estimation of the displacement is no longer possible. Hence, $a/d_\tau \approx 1$ is an upper limit for which the assumption of local uniform flow holds. In the special case that the local variation of the displacement field is only due to large-scale in-plane gradients, it is possible to use window deformation methods [5.435] to recover the displacement-correlation peak for $a > d_\tau$.

PIV Design Rules. In summary, the amplitude of the displacement-correlation peak is given by

$$R(s_D) \sim N_P N_I F_I F_O F_\Delta. \qquad (5.259)$$

For cross-correlation analysis of two image frames $N_P = 1$, but in the case of *ensemble correlation*, which can be used for steady flows (primarily in μPIV), N_P is the number of superimposed correlations. The peak amplitude in relation to the highest noise peak (or signal-to-noise ratio) can be *enhanced* by increasing N_I (and N_P), but is *impeded* by in-plane and out-of-plane loss of

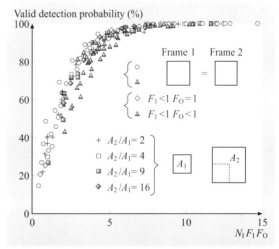

Fig. 5.107 The valid detection probability for the displacement-correlation peak as a function of the image density N_I and the in-plane and out-of-plane loss of correlation, F_O and F_I, respectively (after *Keane* and *Adrian* [5.434])

correlation and loss of correlation due to local variation of the displacement field.

The probability of obtaining a detectability D_0 defined as the signal-to-noise ratio that is larger than 1.1 as a function of the parameters N_I, F_I and F_O more or less collapses onto a single curve, as shown in Fig. 5.107. The data are based on extensive Monte Carlo simulations, and show that the detection probability for cross-correlation analysis and uniform displacements (i.e., $N_P = 1$ and $F_\Delta = 1$) reaches close to 100% when $N_I F_I F_O$ is larger than about 7.5. This result has led to the following general *design rules* for PIV measurements [5.434, 436, 437]

$$N_I > 10 |\Delta X| < \frac{1}{4} D_I,$$
$$|\Delta z| < \frac{1}{4} \Delta z_0, \quad M_0 |\Delta u| \Delta d_\tau, \quad (5.260)$$

i.e., when the interrogation domain contains at least 10 particle-image pairs, the in-plane and out-of-plane motion should be less than about one-quarter of the interrogation window size and light-sheet thickness, respectively, and local variations of the displacement within the interrogation domain should be less than the particle-image diameter. When these criteria are met at least 95% of the interrogations should return the correct particle-image displacement.

As mentioned before, some of the criteria in (5.252) can be relaxed, e.g., the use of a *window offset* eliminates the second rule and the use of *window deformation* relaxes the fourth rule (but only when in-plane gradients are dominant). Nonetheless, any advanced methods will rely on an initial interrogation result that should comply with these design rules.

Digital Correlation. The spatial correlation analysis described previously deals with *continuous* images. In practice one is dealing with *digital images*, i.e., the continuous image field is sampled at discrete locations

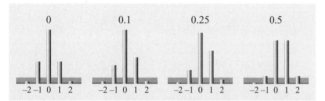

Fig. 5.108 Detailed shape of the correlation peak (in one dimension) as a function of the subpixel part of the displacement (here shown for a 1.6 pixel particle-image diameter). The *horizontal shaded region* represents the background noise level in the correlation

and the image intensity is quantified to discrete levels. The effects of spatial discretization are not important according to the Nyquist theorem [5.426] when the image sampling rate is at least twice the bandwidth of the optical system [5.415], and the intensity quantization can be ignored when the difference between the discrete levels is small compared to the intensity variations [5.425]. *Westerweel* [5.438, 439] demonstrated that for digital PIV analysis it is not necessary to sample the images at the Nyquist rate, since the information on the particle-image displacement is represented at much smaller spatial frequencies than the detailed information of the individual particle-image locations. In practice a 32×32 px interrogation domain is typically used; when the design rule in (5.252) for in-plane motion is used, this would imply an 8 px maximum displacement. If the displacement uncertainly would be determined by the pixel dimension (i.e., a ± 0.5 px error), then the relative error for digital PIV measurements would not be better than $1/8 (\approx 13\%)$. Fortunately, it is possible to obtain accurate measurements of the displacement by means of interpolating the discrete displacement-correlation peak, or so-called *subpixel interpolation* (Fig. 5.108); the typical uncertainty is $0.05-0.10$ px, which implies relative errors around 1% or better. This is sufficient for application of PIV to the investigation of a wide range of unsteady and turbulent flows, and is key to the success of *digital* PIV. However, as one is operating below the Nyquist sampling rate, a number of effects emerge that are typical for digital PIV analysis that will be discussed below.

Pixelization. The image $I(x, y)$ is discretized by the image sensor (i.e., CCD or CMOS device) that integrates the light intensity over a small area, or *pixel*:

$$I[i, j] = \iint p(x - i \cdot d_\tau, y - j \cdot d_r) I(x, y) \, dx \, dy, \quad (5.261)$$

where $p(x, y)$ is the pixel spatial sensitivity. This is usually a uniform rectangular area, but more-complex forms should be used to represent the effect of *microlenses* (where the sensitivity may be nonuniform and leakage from adjacent pixels can be incorporated). The total area of $p(x, y)$ divided by d_r^2 (for a square pixel layout) is the *pixel fill factor* β. When β is significantly smaller than one, the light sensitivity of the sensor decreases proportionally. At the same time the fill factor affects the efficiency of subpixel interpolation [5.427].

The image intensity is represented in a finite number of discrete levels. Common quantization level ranges are

256 or 4096 gray levels, i.e., 8 bit and 12 bit quantization, respectively. The quantization can be represented as additive white noise [5.425]:

$$I[i,j] = I^{\bullet}[i,j] + \varsigma[i,j] . \quad (5.262)$$

When the variance of ς is small with respect to the variance of I, the effect of quantization can be ignored, and usually the sensor noise level is larger than the quantization noise level. Therefore $I^{\bullet}$ will simply be written as I. However, an extreme situation is a *binary image* (which contains only gray levels 0 and 1), which is sometimes used to implement a special form of spatial correlation; in that case the increased noise contribution of ς needs to be compensated by a higher image density.

Discrete Correlation. For digital images $I_1[i,j]$ and $I_2[i,j]$ the discrete spatial correlation is given by

$$\phi[p,q] = \sum_i \sum_j W_1[i,j]\bigl(I_1[i,j] - \overline{I}_1\bigr) \\ \times W_2[i+p,j+q]\bigl(I_2[i+p,j+q] - \overline{I}_2\bigr) , \quad (5.263)$$

where $\overline{I}_1$ and $\overline{I}_2$ are the averaged image values over the windows W_1 and W_2, respectively. Commonly, W_1 and W_2 are identical and uniform $N \times N$-px domains that are equal to 1 inside the domain and zero outside the domain. A generalized form of the discrete correlation when W_1 and W_2 are not identical is given by Ronneberger [5.440]; if W_2 is larger than W_1 then is no longer a constant, i.e., and vice versa.

The discrete spatial correlation $\phi[p,q]$ is related to the continuous spatial correlation $R_D(s)$ by convoluting R_D with the self-correlation Φ_{pp} of $p(x,y)$ and then sampling the result at integer multiples of d_r:

$$\phi[p,q] \doteq \{\Phi_{pp} * R_D\}(p \cdot d_\tau, q \cdot d_r) , \quad (5.264)$$

(ignoring trivial normalization and/or conversion factors, e.g., conversion from light intensity to gray value).

Computation of the Spatial Correlation. The spatial correlation involves a summation over $M \times N$ multiplications for each *pixel* in the correlation domain, which has to be evaluated for a total of $2M \times 2N$ px in the correlation domain; hence, the *direct* evaluation of requires a total of $(2M \times 2N)^2$ multiplications, which implies a considerable computational effort. Fortunately, there exists a very efficient algorithm to improve the processing speed for the evaluation of the spatial correlation, i.e., the *fast Fourier transform* (FFT).

The spatial correlation in (5.255) can be computed by means of a *discrete Fourier transform* (DFT) and is *identical* to the correlation provided that the domain is at least $2M \times 2N$ to contain $\phi[p,q]$; the proof is given in Appendix. The DFT can be computed by means of the fast and efficient FFT algorithm [5.441]. For a $U \times V$ domain the FFT algorithm requires only $4UV \log_2 UV$ multiplications (when U and V are both powers of 2), which implies an enormous reduction in the computational effort. In its most common form the domain size has to be a power of 2, but generalized algorithms exist in which the domain size can be any combination of a power of prime numbers (e.g. 2, 3, 5, which captures almost any domain size). It is however not necessary to restrict the application of FFT to interrogation domains that are only a power of 2, or prime numbers; with proper *zero padding* (Appendix A) any domain size can be selected, as long as the padded domain matches the requirements for the FFT algorithm that is being used.

It is common to use an $M \times N$ correlation domain, which implies that part of the correlation that falls outside the domain is folded back into the (periodic) domain. This only causes significant errors when the correlation attains significant nonzero values outside the central $M \times N$ domain. This only occurs when the particle-image displacement $[p,q]$ is *outside* the domain: $-M/2 < p < +M/2$, $N/2 < q < +N/2$. When the experimental conditions comply with the design rules in (5.252) this does not occur, and the use of an $M \times N$ correlation domain generally implies a substantial reduction of the computational effort (of roughly a factor four), compared to a $2M \times 2N$ domain, at the cost of a small but acceptable increase of the noise level.

Subpixel Interpolation. One of the most powerful aspects of digital PIV is the ability to determine the particle-image displacement at subpixel level: under optimal conditions the particle-image displacement can be estimated with a precision better than 0.1 px. This means that the measurement error can be as low as 1% (or less) for a 32×32 px interrogation domain with an 8 px particle-image displacement (in accordance to the 1/4-rule).

The peak location at subpixel level is obtained by *interpolation* of the correlation values around the correlation maximum. Figure 5.108 shows the correlation values in a narrow peak as a function of the subpixel position of the displacement-correlation peak. Note that the value $\phi_0 \equiv \phi[m_0, n_0]$ of the correlation maximum remains practically unchanged, while the two adjacent correlation values $\phi_{-1} \equiv \phi[m_0 - 1, n_0]$ and $\phi_{+1} \equiv \phi[m_0 + 1, n_0]$ change considerably in amplitude. This means that the subpixel location of the maximum (i.e., the particle-image displacement) can be estimated from the correlation values $\{\phi_{-1}, \phi_0, \phi_{+1}\}$ for each coordinate.

The most commonly used method for the sub-pixel estimation of the displacement is the *three-point Gaussian peak fit*, where the subpixel part for each component of the displacement is given by [5.442]:

$$\varepsilon_X = \frac{1}{2} \frac{(\ln \phi_{-1} - \ln \phi_{+1})}{(\ln \phi_{-1} - \ln \phi_{+1} - 2 \ln \phi_0)} , \quad (5.265)$$

The numerator in (5.257) represents the skew of the correlation peak as also shown in Fig. 5.108. Evidently, this estimator fails when:

1. any of the considered correlation values is zero or negative, or
2. when the denominator is close to zero.

The former can occur for very narrow correlation peaks, i.e., $d_\tau/d_\mathrm{r} \ll 1$, or when then the image background is nonuniform (e.g., in the case of reflections), whereas the latter can occur for very wide correlation peaks when: $\phi_0 \approx \phi_{-1} \approx \phi_{+1}$, i.e., $d_\tau/d_\mathrm{r} \gg 1$.

In principle the correlation values ϕ_{-2} and ϕ_{+2} would also carry information with regard to the subpixel peak position, but for very narrow peaks ($d_\tau/d_\mathrm{r} \approx 2$) and finite image density ($N_\mathrm{I} \approx 10$) the correlation noise level is of the order of the correlation values ϕ_{-2} and ϕ_{+2}, so that their contribution to the estimated displacement value is practically insignificant. For larger particle images ($d_\tau/d_\mathrm{r} \gg 2$) or higher image densities ($N_\mathrm{I} \gg 10$) it would make sense to use a larger domain for the subpixel estimation. In that case the displacement is estimated by means of two-dimensional least-squares fit of a Gaussian curve over a 5×5 kernel (or larger) domain. This has the additional advantage that the fit can be generalized to accommodate elliptical peaks with arbitrary orientation (which can occur in the case of strong local velocity gradients). However, for narrow peaks that effectively cover a 3×3 px domain, there is no significant difference between the one-dimensional (1-D) fit (5.257) applied for each component and a least-squares 2-D fit.

Bias Correction. The spatial correlation $R(s)$ defined in (5.244) is a biased estimate of the true two-point ensemble correlation for a *finite* interrogation domain; this bias is proportional to the in-plane loss of correlation $F_\mathrm{I}(s)$, which skews the displacement-correlation peak and causes a bias in the estimated displacement towards the origin of the correlation domain (even when the displacement is uniform). This bias is proportional to the size of the interrogation domain and the width of the displacement-correlation peak [5.414, 434]. This bias can be corrected by using

$$\phi^*[m, n] = \frac{\phi[m, n]}{F_\mathrm{I}[m, n]} , \quad (5.266)$$

where $F_\mathrm{I}[m, n]$ is given by (5.249) for uniform and identical interrogation windows W_1 and W_2. The difference between the measured and actual image displacement using $\phi[m, n]$ and $\phi^*[m, n]$, respectively, is shown in Fig. 5.109. It is important that the peak identification is done in $\phi[m, n]$, as the unbiased correlation $\phi^*[m, n]$ has a diverging variance for large $[m, n]$ [5.443], which can cause the appearance of dominant spurious correlation peaks near the edges of the correlation domain.

Peak Locking. Reduction of the particle-image diameter relative to the pixel size can lead to an effect commonly referred to as *peak locking*. This effect becomes apparent in the displacement histogram of PIV data, as illustrated in Fig. 5.110. Evidently, peak locking significantly reduces the accuracy of PIV measurements [5.444]; although the effect on velocity flow statistics may be within acceptable limits, this effect can lead to awkward results for instantaneous spatial derivative data such as vorticity [5.445].

The effect originates from the fact that the displacement-correlation peak increasingly deviates from its assumed Gaussian shape for decreasing d_τ/d_r, i.e., the term Φ_pp in (5.256) deforms the correlation

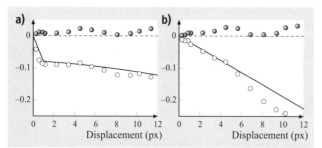

Fig. 5.109a,b The difference between the measured and actual displacement as a function of the actual displacement for uniform (**a**) and Gaussian (**b**) interrogation windows. The *open symbols* and *line* indicate the estimate using the biased correlation ϕ, while the *filled symbols* indicate the result using the unbiased estimate ϕ^* (5.259)

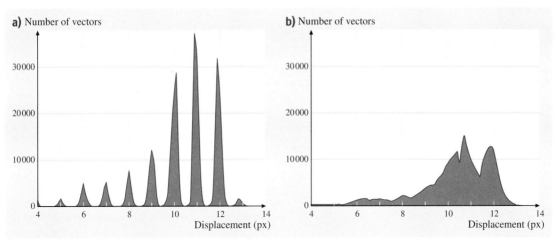

Fig. 5.110a,b PIV data from a turbulent pipe flow obtained with two different subpixel interpolation methods: (**a**) peak centroid, and (**b**) Gaussian peak fit

peak. This is illustrated in Fig. 5.111, which shows the difference between the measured and true subpixel displacements for different values of d_τ/d_r. Note that the slope of the curves near the origin in Fig. 5.111 is always negative, which implies that the measured displacement is biased towards integer pixel values.

Further analysis shows that diffraction-limited particle images are fully resolved when the particle-image diameter is at least four pixels [5.427]. This means that the continuous particle images can be fully reconstructed from the discrete samples according to the Shannon–Whittaker sampling theorem [5.426], and that peak-locking effects do not occur. Between 2 and 4 px

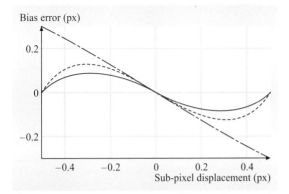

Fig. 5.111 The *bias error*, or mean difference between the measured displacement and true displacement, as a function of the subpixel fraction of the displacement for the peak centroid (*dash-dotted*), parabolic peak fit (*dashed*) and Gaussian peak fit (*solid*) (after *Westerweel* [5.396])

in diameter, the bias error is small and its magnitude depends on the pixel fill ratio [5.427]. For very small particle images, i.e., 1–2 px in diameter, the effect of peak locking becomes very strong, although the correlation peak may still contain sufficient information to recover the displacement data, i.e., ϕ_{-1} and ϕ_{+1} are sufficiently above the background noise level. It should be noted that the peak height is a function of the subpixel displacement, which enhances biasing towards integer pixel displacements. For $d_\tau/d_r < 1$ all information is retained in ϕ_0, and ϕ_{-1} and ϕ_{+1} are no longer above the background noise level, i.e., signal recovery is no longer feasible.

Numerous approaches have been suggested to eliminate the peak-locking effect, including image defocusing, (*anti-aliasing*) image filtering, alternate subpixel interpolation methods, and window subpixel reinterpolation [5.446]. Most of these methods have in common that they redistribute the image information and thus artificially increase the effective particle-image diameter. The best practice seems to be a proper choice of light source energy and optical configuration (viz., image magnification and lens aperture) to match the diffraction limit to the available image pixel resolution.

Optimization. From the preceding paragraphs it already becomes clear that the optimal particle-image diameter is somewhere between one and four pixels. Analysis shows that this is primarily determined by the information content of the estimated spatial correlation $\phi[m, n]$ and that the random error for the three-point Gaussian peak fit in (5.257) has a minimum value of about

0.05–0.10 px at $d_\tau/d_r \approx 2$ under idealized conditions. This has been confirmed by means of simulation results and controlled experiments [5.438, 439, 447].

The estimated correlation values $\phi[m, n]$ are not independent *samples* of $R_D(s)$, but are correlated over a finite domain with an area L^2 that is given by [5.414]

$$L^2 = \sum_m \sum_n \frac{\text{cov}\{\phi[m_0, n_0], \phi[m_0+m, n_0+n]\}}{\text{var}\{\phi[m_0, n_0]\}},$$
(5.267)

where $[m_0, n_0]$ is the location of the displacement-correlation peak. The value of L^2 can be interpreted as the number of correlated samples in the interrogation domain, so that $(L/M)^2$ represents the relative uncertainty level for an $M \times M$ domain [5.443]. The typical behavior of L/M as a function of d_τ/d_r is shown in Fig. 5.114, which indicates two scaling regimes:

1. for small particle images ($d_\tau/d_r < 1$) the increase in pixels decreases the uncertainty, i.e., the amount of available information in the correlation peak location increases with increasing number of pixels (or decreasing d_r);
2. for large particle images ($d_\tau/d_r > 1$) the correlation peak spreads over a larger number of pixels with decreasing d_r, but this does not increase the information content of the correlation peak and therefore does not reduce L/M.

This special property is related to the fact that the *width* of the correlation peak is directly related to the *maximum slope* of the correlation peak (which determines the estimation uncertainty), as is the case for a Gaussian-like shape [5.448, 449]. As a result the root-mean-square (rms) error is proportional to the particle-image diameter (for resolved particle images), i.e.,

$$\sigma_{\Delta X} \cong c d_\tau,$$
(5.268)

with $c \approx 0.05 - 7$ [5.393, 433, 444] This relation holds for large (Gaussian-like) particle images (i.e., $d_\tau/d_r \gg 1$) and indicates that the performance improves for decreasing particle-image diameter. However, improvement breaks down when the effect of finite pixel size becomes apparent. Figure 5.112 shows the simulation results and a theoretical prediction for the rms error as a function of the particle-image diameter relative to the pixel dimension. The theoretical result is based on the image statistical properties as the only input parameters. Also, two scaling regimes for the rms error as a function of d_τ/d_r are identified: $\sigma_{\Delta X} \sim d_\tau$ for constant *image density*, and $\sigma_{\Delta X} \sim d_\tau^2$ for constant *source density*.

Hence, the optimal particle-image diameter is determined by two opposing trends, i.e., the reduction of the random error for decreasing particle image diameter and an increase for the bias and random error for decreasing particle image diameter as a result of the finite pixel dimensions; the optimal value for d_τ/d_r is about 2 (or slightly larger) based on theoretical and simulation results, but in the literature values between 2 and 4 are reported based on practical considerations.

The rms estimation error is approximately constant as a function of the displacement, but is proportional to the displacement for $|\Delta X|/d_r < 0.5$, as shown in Fig. 5.113.

The error vanishes for zero displacement as the interrogation images WI_1 and WI_2 are identical (for zero out-of-plane motion). This property can be utilized to improve the performance of subpixel interpolation by means of window matching.

The fact that the rms error is proportional to the subpixel displacement (rather than constant) implies also a bias towards integer pixel displacement values; the error fluctuations are much smaller for near-integer pixel values, but increase for increasing subpixel displacement, causing a typical variation in the histogram

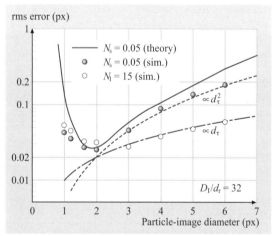

Fig. 5.112 The rms displacement error as a function of the particle image diameter (in pixel units) for 32×32 px interrogation windows and constant image density $N_I = 15$ (*open symbols*) and constant source density $N_s = 0.05$. The *solid line* represents a theoretical result, while the *dotted* and *dashed lines* represent scaling relations. (After Westerweel [5.433])

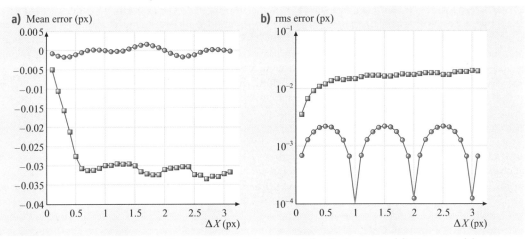

Fig. 5.113a,b Measurement error due to discretization as a function of the displacement. (**a**) Bias error; (**b**) rms error. *Squares* are relative to single-step correlation. *Circles* are relative to multistep correlation and subpixel window shift

for subpixel displacements as shown in Fig. 5.113. The vanishing of the rms error towards zero in-plane displacement (i.e., when the two interrogation images become identical) only occurs when the out-of-plane displacement is negligible. Therefore one can expect that this contribution towards peak locking will disappear for increasing out-of-plane motion, as shown by the simulation results of *Foucaut* et al. [5.450]. Hence, for flows with appreciable turbulence levels, this contribution to peak locking is usually not significant.

Dynamic Range. The lower end of the dynamic range is determined by the requirement that the corresponding displacement can be distinguished from the noise level.

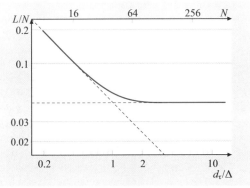

Fig. 5.114 The level of uncertainty in the estimation of the displacement from the spatial correlation as a function of the particle-image diameter (in pixels)

The minimum detectable displacement shall be denoted with ΔX_{min}. The maximum measurable displacement ΔX_{max} can be directly related to the minimum F_I criterion stating that the in-plane displacement should not exceed one-quarter of the interrogation window size D_I, i.e., $\Delta X_{max} \leq 1/4 D_\mathrm{I}$ (5.253). The velocity dynamic range therefore reads:

$$\frac{\Delta X_{max} - \Delta X_{min}}{\Delta X_{min}} = \frac{D_\mathrm{I}}{4\Delta X_{min}} - 1 \,. \quad (5.269)$$

This expression reads as a accuracy–resolution trade off. The dynamic range can be increased with a larger window size, at the expense of the spatial resolution. The dependence of ΔX_{min} on other experimental parameters has been one of the most investigated topics in PIV [5.393, 396, 442]. It is strongly dependent upon the particle image diameter (see previous section) and the type of subpixel interpolation used to determine the location of the correlation peak.

A typical value of $\Delta X_{min} = 0.1$ px can be assumed under generic measurement conditions. Following *Willert* and *Gharib* [5.442] this can be achieved using a three-point Gaussian peak-fit model, particle images of two to three pixel diameter, displacements not exceeding 10 px, and a window size of 32×32 px. With the previous considerations (5.262) yields a dynamic range in the order of 10^2 for a window size of 32×32 px.

Detectability Enhancement. The detectability defined above is directly dependent upon the number of paired particle images within the correlation windows and in general increases with the interrogation window size at the expense of the spatial resolution. In the following paragraphs a few methods are presented to enhance the detectability without the direct need to increase the size of the interrogation window. However it should be said that each of the proposed methods relies upon assumptions and is strictly valid only within specific hypotheses.

Local Correlation Averaging. Under several experimental conditions a high source density cannot be achieved for technical reasons. As a consequence analysis performed at low image number becomes critical, with the appearance of a large number of spurious correlation vectors. A first approach to increasing the number of particle pairs without increasing the window size consists of accumulating the effect of several instantaneous particle image recordings (basically summing up the images, i.e., taking $N_P > 1$). This approach is only valid when the flow is strictly steady. However the method has the limit that beyond a given point, individual particle images cannot be distinguished anymore. A better solution is therefore to average the correlation maps from the analysis of the single recordings [5.409, 451]. In this case no upper limit is found on the number of recordings that can contribute to build the correlation signal. In μPIV this procedure is very frequently adopted to average out of the particle velocity spatiotemporal fluctuations solely due to the Brownian motion of the molecules in the fluid.

Correlation Multiplication. The correlation map produced at a given location is the result of the correlation signal from each single pixel. A possible way to enhance the correlation peak height and reduce the noisy peaks is to multiply several correlation maps for a correlated value of the velocity but uncorrelated particle image pattern [5.452]. The most common approach is to multiply each correlation map by those obtained at the neighboring locations. The method can also be applied for unsteady flows, however it has not yet been ascertained whether this approach yields a net benefit in flows with a significant velocity gradient.

Phase-Only Correlation. A recent approach specific to the drop in S/N close to optical interfaces (e.g., bright walls) is to eliminate the amplitude information during the FFT-based evaluation of the cross-correlation map. The so-called DC pedestal in the correlation map can be significantly suppressed and reliable particle correlation information could be retrieved in conditions where bright objects other than particles where included in the correlation window [5.453].

Advanced Interrogation Techniques. This discussion places severe limitations on the measurement performance of PIV when solely associated to the spatial correlation operator. Advanced techniques have been introduced that allow one to overcome the limitations posed by the application of the cross-correlation operator to interrogate the images [5.454]. The first group of techniques allows the measurement dynamic range to be increased in relation to the in-plane particle motion. The second group shows the possibility of enhancing the correlation peak. Finally the discussion touches upon possible improvements to the spatial resolution.

Window Shift Method. When the in-plane particle image displacement is estimated by means of a first

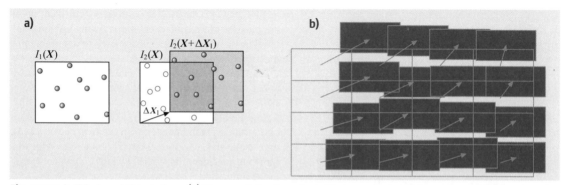

Fig. 5.115a,b Window shift technique. (**a**) The correlation is evaluated between two particles patterns shifted, accounting for the translation due to the flow motion. Shaded shifted window with the estimated displacement. (**b**) Window shift for a group of windows

correlation analysis, a relative shift between the correlation windows can be applied in order to compensate for the loss of pairs due to the average particle motion. As a result the number of particle image pairs F_I is increased. The technique can be implemented as a two-step analysis where the interrogation is repeated the second time with the window in the second exposure shifted by the value of the displacement obtained from the first interrogation [5.439]. Denoting the result from the first interrogation by ΔX_1, the subsequent interrogation will be performed with a modified form of the cross-correlation operation, which reads:

$$R_1(s) = \int_A W_1(x) I_1(x) W_2[x + \text{round}(\Delta x_1) + s] \\ \times I_2[x + \text{round}(\Delta x_1) + s] \, dx \,. \quad (5.270)$$

The argument of the windowing function W will be omitted in the remainder, assuming a uniform (top-hat) weighting. The increase of paired particle images reduces substantially the systematic underestimate of the particle displacement and marginally the rms uncertainty. The result of (5.263) is then added to the previously obtained displacement.

Multigrid Analysis. A logical consequence of the window shift technique is the multigrid approach: the interrogation process is repeated several times, with the interrogation window shifted after each interrogation. At this point the window size does need to comply with the 1/4 rule given in (5.253) related to the average particle in-plane motion. As a result the dynamic range is amplified; when the window-refinement multigrid window-shift technique is applied, (5.262) can be rewritten as:

$$\frac{\Delta X_{\max} - \Delta X_{\min}}{\Delta X_{\min}} = \frac{RD_I}{4\Delta X_{\min}} - 1 \,, \quad (5.271)$$

where D_I^0 is the interrogation window size at the first interrogation and R is the window refinement defined as the ratio between the initial and final size of the interrogation window. In conclusion, for a fixed window size, the dynamic range is increased proportionally with the refinement ratio R, which typically ranges between 2 and 6. For two-dimensional flow problems, the maximum displacement can even exceed the window linear dimension (except for the first interrogation). Other factors may limit the maximum displacement, out-of-plane particle motion and high velocity gradient being the main sources of pair loss.

Subpixel Window Shift. PIV measurements obtained with digital imaging devices may be affected by errors arising from the limited resolution of the light intensity pattern and in turn of the spatial correlation map. As discussed before the estimate of the subpixel particle image, displacement is affected by an error due to imperfect interpolation of the peak or to its distorted shape [5.396]. These effects can be corrected applying the relative shift between I_1 and I_2 with subpixel precision [5.446, 454, 455]. The expression for the spatial correlation function for the subpixel window shift is a slightly modified version of (5.263):

$$R_1(s) = \int_A W_1 \tilde{I}_1 \left(x - \frac{\Delta x}{2}\right) W_2 \tilde{I}_2 \left(x - \frac{\Delta x}{2} + s\right) dx \,, \quad (5.272)$$

where $\tilde{I}$ indicates that the image intensity values are obtained by interpolation at subpixel locations. As addi-

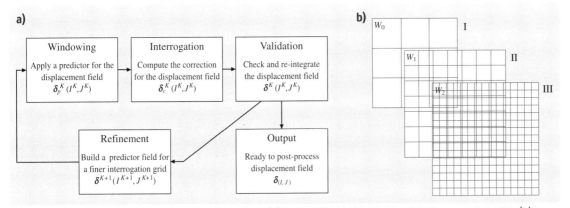

Fig. 5.116a,b Block diagram of multigrid interrogation (**a**) and graphical description of the interrogation grids (**b**)

tional advantage, the method can be implemented with second-order precision following an accurate centered interrogation scheme where the two windows are shifted backward and forward by an equal amount [5.456]. The advantage with respect to the discrete window offset technique is that the subpixel shift returns a correlation map with a fractional displacement that tends to zero, and is therefore unbiased (Fig. 5.113). The particle displacement measured after the application of the window shift is regarded as a correction term to be added to the applied shift. A drawback of the subpixel shift method is that it requires image samples at intermediate locations between pixels, obtained with image interpolation techniques at additional computational effort. Moreover the image interpolation may introduce signal degradation, especially when the particle images are barely sampled ($d_\tau/d_r < 2$).

Iterative Window Deformation. This discussion has been carried out with the assumption that the motion of particles within the interrogation window can be assumed to be uniform. This rough hypothesis is never strictly valid and in most flows of interest the velocity field may exhibit significant variations within the interrogation windows. A hypothesis with more-general validity is the flow to be represented with a linear function within the interrogation window rather than a constant value, which correspond to a local representation by truncated Taylor expansion [5.457] as shown in (5.267). The expression of the spatial correlation function is therefore further modified into (5.266).

$$R_1(s) = \int_A W_1 \tilde{I}_1 \left(x - \frac{\Delta x(x)}{2} \right) W_2 \tilde{I}_2 \left(x + \frac{\Delta x(x)}{2} + s \right) dx , \quad (5.273)$$

where the displacement applied to the intensity pattern is now a function of space and may therefore be nonuniform. Several levels of approximation may be considered depending on the truncation of the Taylor series below, however the most common choice is to stop at the first-order term

$$\Delta x_1(x) = \Delta x_1(x_0) + \nabla[\Delta x_1(x_0)](x - x_0) + \ldots \sigma(x - x_0)^2 . \quad (5.274)$$

In (5.274), x_0 denotes the position of the center of the interrogation window. It should be mpetd that the above expression can return all previous interrogation methods depending on the number of parameters used for the displacement approximation. If no parameter is considered ($\Delta x_1 = 0$) the interrogation corresponds to the basic cross-correlation. The zeroth-order truncated expansion corresponds to the window-shift method and a first-order truncation corresponds to the window deformation technique. When the spacing of the displacement vectors is smaller than the size of the window (i.e., overlap factor of 50% or higher) the displacement distribution within the window can be represented with a piecewise linear function, which corresponds to a higher-order approximation of the flow pattern within the window.

The technique can be implemented within the multigrid approach and its main advantage with respect to the window-shifting method is an increased robustness and accuracy over highly sheared flows such as boundary layers, vortices, and turbulence in general (Fig. 5.117c). A correlation map sample from a turbulent shear flow is shown in Fig. 5.118. The correlation map (Fig. 5.118a) shows several peaks of height comparable to the peak of the most probable displacement. In this situation the detection of a false peak is very likely to occur with increased measurement uncertainty along the shear layer. These spurious peaks are approximately distributed in the direction normal to the local velocity gradient, causing a broadening of the correlation signal. The correlation map obtained by the window deformation shows a more-pronounced single peak only marginally affected by broadening.

Higher-Order Correlation Methods. The interrogation process attempts a global matching of the particle images patterns between the two exposures, which may be regarded as an optimization process:

$$\max_{\Delta X} \int_A I_1[x] I_2[x + \Delta x] dA , \quad (5.275)$$

where ΔX is the vector parameter to be determined that maximizes the correlation function. Spatial cross-correlation analysis (even including the schemes with window deformation) only yields one vector from each interrogation window, corresponding to a weighted average of the particle displacement inside the window. When the optimization is extended from the displacement to its spatial derivatives the optimization space becomes multidimensional, where a maximum includes the particle displacement as well its gradient or higher-order derivatives:

$$\max_{\Delta X, \text{grad}(\Delta X), \ldots} \int_A I_1[x] I_2[x + \Delta x] dA . \quad (5.276)$$

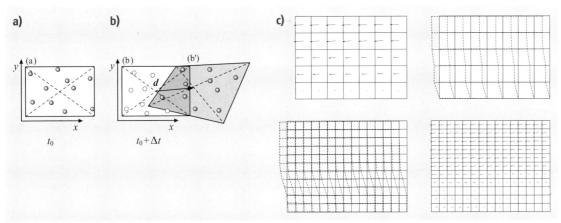

Fig. 5.117a–c Window deformation principle. **(a,b)** Tracer pattern in the first **(a)** and second **(b)** exposure (*solid circles* in **(b)** are the tracers correlated with the first exposure). **(c)** Schematic view of the multigrid window deformation method

The number of parameters involved in the optimization is given by the order of approximation of the Taylor expansion used to represent the displacement within the interrogation window. Higher-order correlation information can in principle return at each measurement cell the values of the velocity and its derivative(s) with the advantage of an increased spatial resolution. However the properties of estimators higher than first order are not yet sufficiently understood. As an example of application of higher-order correlation, the evaluation of the spatial correlation associated to the second derivatives of the displacement field can be used to reduce the filtering effect associated to the conventional spatial cross-correlation analysis [5.458] significantly.

Iterative Interrogation. Most common PIV image-analysis software nowadays performs multistep analysis of PIV recordings, which can be seen as composed out of two procedures:

1. *Multigrid analysis*, where the interrogation window size is progressively decreased. This process allows the elimination of the 1/4-rule constraint and is usually terminated when the required window size (the smallest) is applied.
2. *Iterative analysis* at a fixed sampling rate (grid spacing) and spatial resolution (window size). This process further improves the accuracy of the image deformation and enhances the spatial resolution.

The iterative analysis can be described by a predictor–corrector loop as represented schematically in Fig. 5.121. The iterative equation in its simplest form reads

$$\Delta x_{k+1} = \Delta x_k + \Delta x_{\text{corr}}\,, \tag{5.277}$$

where Δx_k indicates the result of the evaluation at the k-th iteration used as the predictor for the $k+1$-th iteration. The correction term Δx_{corr} can be intended as a residual and is the displacement vector obtained by

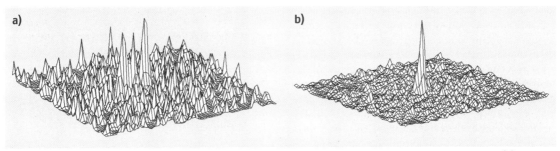

Fig. 5.118a,b Discrete spatial correlation in a turbulent shear flow. Interrogation with single-step correlation **(a)** and with the multistep window deformation technique **(b)**

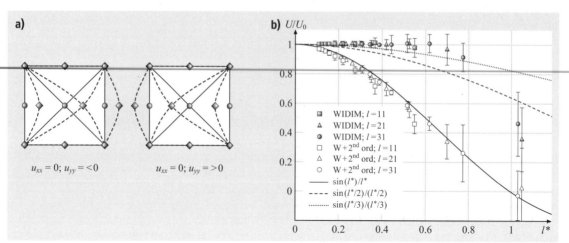

Fig. 5.119a,b Schematic representation of the second-order correlation scheme (**a**). Comparison between the second-order correlation method spatial response and the basic correlation (**b**)

cross-correlation analysis of the deformed images I_1^k, I_2^k obtained from the original images I_1, I_2 according to the relations

$$I_1^k = \tilde{I}_1\left[x - \frac{\Delta x_k(x)}{2}\right], \quad I_2^k = \tilde{I}_2\left[x + \frac{\Delta x_k(x)}{2}\right]. \tag{5.278}$$

The procedure can be reiterated several times, however two to three iterations are already sufficient to achieve a well-converged result, which allows deformation of the images and compensation for most of the in-plane particle image motion.

Stability and Spatial Resolution of Iterative Interrogation. The overall approach described above appears very logical and its simplicity makes it straightforward to implement, which probably justifies why it has been so broadly adopted in the PIV community (PIVchallenge II and III, *Stanislas* et al. [5.397]). However when such iterative interrogation is performed without spatial filtering of the velocity field, the process tends to oscillate and eventually diverge unless the image processing is interrupted at an early stage. The instability arises from the sign reversal in the sinc shape of the response function associated to the top-hat function of the interrogation window. If two almost identical images are considered (except for artificial pixel noise) the displacement field measurement after some iterations oscillates spatially with a wavelength $\Lambda_{\text{unst}} = 2/3 D_\text{I}$.

The above result is consistent with the fact that the response function r_s of a top-hat interrogation window (the common option) corresponds to the $\sin(x/D_\text{I})/(x/D_\text{I})$ function. Therefore the range with negative values of the sinc function is responsible for the amplification of the specific wavelength. The iterative process requires therefore a stabilization by means of a low-pass filter applied to the updated result (Fig. 5.121). In line with the above definitions, the update term reads

$$\Delta x_{k+1} = r_s(\Delta x_0 - \Delta x_k). \tag{5.279}$$

The iterative process stability can be achieved with the application of a spatial low-pass filter to the velocity predictor, as shown in the block diagram of Fig. 5.121.

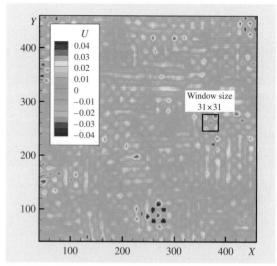

Fig. 5.120 Displacement field from iterative analysis of images with a zero-displacement field

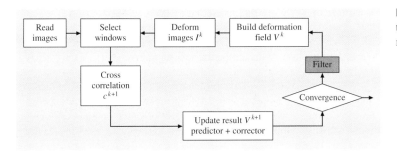

Fig. 5.121 Block diagram of the iterative image deformation interrogation method with filtered predictor

The property of the filter is to damp the growth of fluctuations at wavelengths smaller than the window size. A moving-average filter with a kernel size corresponding to that of the interrogation window is proven to be more than sufficient to stabilize the process [5.459]. The iterative equation relative to the block diagram in Fig. 5.121 can be rewritten as

$$\Delta x_{k+1} = f(\Delta x_k) + \Delta x_{\text{corr}} . \quad (5.280)$$

Introducing the spatial response of the chosen low-pass filter r_f (in the present case we assume for simplicity that $r_f = r_s$), the filtered version of the iterative equation reads

$$\Delta x_{k+1} = \Delta x_k + r_s^2 (\Delta x_0 - \Delta x_k) , \quad (5.281)$$

where the sign reversal is eliminated by taking the square of the response function. Other means of stabilization are nonlinear filtering such as least-squares spatial regressions [5.460] or directly weighting the interrogation window (e.g., pyramidal, Gaussian, or LFC technique as suggested by *Nogueira* et al. [5.461]). A numerical simulation of the sine-wave test shows that the single-step cross-correlation yields a spatial response that compares well with that of a moving-average filter. The iterative correlation conversely returns a spatial response with almost no modulation error up to wavelengths twice the size of the interrogation window. However without a spatial filter, the result tends to overshoot the actual amplitude and small-wavelength fluctuations are amplified. When a spatial regression filter is applied, such behavior is avoided, whereas the spatial resolution is still significantly better than that of the single-step cross-correlation in terms of modulation error.

Postprocessing

The interrogation analysis of PIV images yields an array of displacement vectors that represent the two or three component velocity field in the measurement volume (this is either a plane or a volume). The next step is *data validation*, i.e., the identification of erroneous displacement vectors. In general the signal dropout due to invalid data will be low, and the missing data is replaced in order to retain the regular array structure. The final step is then data evaluation, which can include the computation of derived quantities (e.g., vorticity and strain rate) and turbulence statistics (e.g., Reynolds stress), or the detection of specific flow structures.

Data Validation. If one would optimize a PIV measurement in accordance to the *recipe* presented in (5.253) one could end up with a result as shown in Fig. 5.123; the displacement field contains a number of displacement vectors that do not appear to fit with the surrounding displacement field. Such vectors occur when a random correlation peak exceeds the amplitude of the displacement-correlation peak, and therefore such vectors are commonly referred to as *spurious vectors*. These

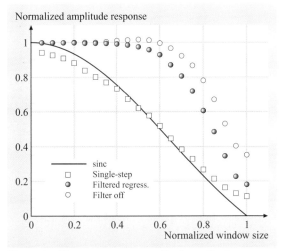

Fig. 5.122 Sine wave test: normalized amplitude response as a function of the normalized window size. *Sinc* function (*solid line*); single-step correlation (*squares*); iterative window deformation (*circles*); filtered iterative window deformation (*filled circles*)

vectors are often the result of insufficient particle images, large in-plane or out-of-plane displacements, or high spatial gradients of the velocity; other causes can be a strong background image (e.g., due to a surface reflection) or light-sheet inhomogeneity (e.g., a shadow cast by a larger object).

The aim of optimizing PIV is to maximize the probability of detecting the displacement-correlation peak. One could aim to increase the image density to achieve a near-100% valid detection rate, but this can lead to other undesirable effects, such as fluid opacity (due to multiple particle scattering) and multiphase effects (i.e., increase of the apparent fluid viscosity and particle–particle interactions). Therefore, a more-practical approach is to accept a small fraction (typically about 5%) of spurious vectors, and to rely on a post-interrogation detection, removal, and replacement procedure.

There are several approaches to identify spurious vectors. An intuitive approach is visual inspection of a plotted vector map. Such an approach is acceptable for the evaluation of a handful of PIV results (as was the case when PIV was first developed), but is impractical nowadays when PIV measurements tend to generate 10^3 vectors maps or more.

For the automated validation there are generally two approaches, which are based on

1. correlation signal quality
2. local *coherence* of the vector map, i.e., a comparison of each vector with measured displacements in adjacent interrogation regions

Both will be discussed below, but in general detection methods based on correlation signal quality are not very robust and the evaluation based on local *coherence* appears to be much more efficient.

Correlation Signal Quality. The quality of a PIV interrogation can be expressed by the amplitude of the displacement-correlation peak, which is proportional to $N_\mathrm{I} F_\mathrm{I} F_\mathrm{O} F_\Delta$. However, the amplitude also depends on the local image properties, such as mean intensity and intensity variance, which are determined by the local light level and particle image characteristics. When the correlation is normalized by the (local) variance of the image intensity, the correlation amplitude varies between 0 and 1, where 1 means perfect correlation of the particle-image pattern. However, the amplitude also depends on the local value of F_τ, which complicates the interpretation of the correlation signal amplitude; for example, when one of the interrogation images contains an unmatched bright particle image, it simply correlates with the brightest particle image in the other interrogation image, leading to a spurious displacement with near-unity correlation coefficient.

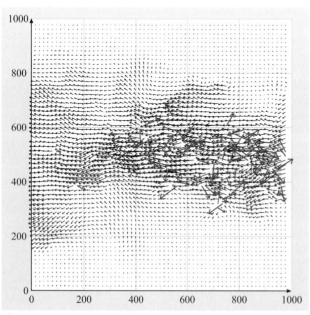

Fig. 5.123 Example of a PIV result containing spurious vectors (after *Westerweel* and *Scarano* [5.462])

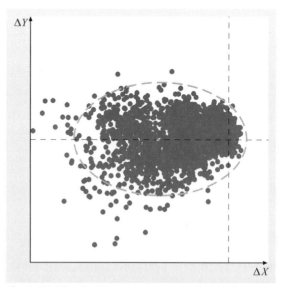

Fig. 5.124 The displacements corresponding to the vector map in Fig. 5.123. The ellipse indicates the region that contains the range of expected displacements

The signal quality is also characterized by the amplitude of the displacement-correlation peak relative to the highest random correlation peak; this ratio is generally referred to as the *peak detectability* D_0, and this measure can be regarded as a lower limit for the correlation signal-to-noise ratio. However, the detectability appears to be a reliable measure only when the image density N_I is high, while D_0 can strongly fluctuate for low N_I, leading to erroneous acceptance of spurious vectors.

Vector Field Coherence. The measured displacement field is expected to have a certain *smoothness* or *coherence*, in which spurious vectors stand out with respect to the overall or local displacement field. A first approach is to generate a scatter plot of the observed displacements; see the example in Fig. 5.124. The valid displacements appear to fall in a closed region indicated by the ellipse. Evidently, this region would be different for each flow. One can identify displacements that occur outside this region as spurious data. This type of identification represents a *global approach* in which fluctuations that are unacceptably large with respect to the mean and variance of the total flow field are discarded. However, this global approach cannot identify spurious displacements within the global fluctuation level that show up because of a strong deviation with respect to the *local* displacement field.

A local approach would be to compare each vector with the local mean displacement of the four or eight adjacent displacement vectors. A complication is that the local mean is a linear estimate that is strongly influenced by the possible presence of a spurious vector in the neighborhood of adjacent displacement vectors. This is easily avoided by using the *median* of the displacement components, which has statistical properties similar to the local mean but is quite impervious to the presence of spurious data and therefore much more robust. Hence, the *residual* r of a displacement vector U_0 is defined as

$$r = \frac{|U_0 - U_\mathrm{m}|}{\sigma}, \quad (5.282)$$

where U_m is the *median* vector of the eight vectors $\{U_1, \ldots, U_8\}$ in the 3×3 neighborhood of U_0 and σ a normalization factor; a robust estimate for σ is the median r_m of $r_i = |U_i - U_\mathrm{m}|$ (with $i = 1\ldots 8$). A small offset ε is added to r_m, i.e. $\sigma = r_\mathrm{m} + \varepsilon$, where $\varepsilon \approx 0.1$ px, to improve the performance of the evaluation for very low fluctuation levels. Now, a vector is labeled as *spurious* when r is larger than a given threshold value r_c. This approach is commonly referred to as the *median test* [5.462, 463].

The optimal value for r_c depends on the experimental details, the flow conditions and the quantities to be extracted from the data. The proper choice of r_c is important: if it is chosen too small then valid data can be labeled as spurious, and if it is chosen too large then spurious data remain undetected; these erroneous detections are generally referred to as errors of the first and second kind, respectively. It depends on the objectives of the measurement whether or not it is acceptable to have a small fraction of spurious data to remain or of valid data being rejected; this should be investigated for each measurement. In general, low-order turbulence velocity statistics are quite insensitive, while derivative data (such as the vorticity) are strongly affected, by undetected spurious data.

The normalization described by *Westerweel* and *Scarano* [5.462] makes r_c relatively independent of the flow conditions, and a typical value for r_c is 2, with larger and smaller values yielding a less- and more-stringent detection, respectively.

Data Replacement. The most common strategy is to accept that there is a small fraction of spurious vectors in a PIV result, to rely on a robust validation method for the detection of spurious data, and to replace the missing vector data. This strategy can only work when the fraction of spurious vectors is low. In general, PIV data are considered of good quality when the fraction of spurious vectors is less than about 0.95. Behind this rule of thumb there is a simple statistical argument that is explained below.

For flows with fairly homogeneous statistics for the velocity fluctuations and homogeneous seeding, the occurrence of spurious vectors is randomly distributed over the measurement domain. The median test is quite insensitive to the presence of any spurious vectors when less than one-half of the adjacent interrogations return a spurious result. Consider the four directly adjacent interrogation positions in a 3×3 neighborhood. Given a probability γ that each of these interrogations yields a spurious result, the probability of finding k spurious vectors in n interrogations is given by a binomial distribution $B(k, n; \gamma)$. The probability that the four adjacent interrogations contain no more than two spurious vectors is then given by

$$\mathrm{Prob(successful\ median\ test)} = \sum_{k=0}^{2} B(k, n = 4; \gamma). \quad (5.283)$$

Suppose it is required that the median test should not fail in the measurement domain. For a typical image of 1024 × 1024 px interrogated with 32 × 32 px interrogations regions with 50% overlap, this means that it should fail less than 1 in 4096 interrogations. Solving the above equation shows that γ should be less than 0.04, which is close to the rule of thumb value ($\gamma = 0.05$).

Estimation of Derivative Data. The vorticity and deformation are flow properties that can be directly obtained by numerical differentiation of the PIV velocity field. To obtain reliable estimates of these derivative properties it is important that the PIV measurement data is accurate, does not contain spurious data, and that the spatial resolution is adequate to perform a differentiation of the data. In the previous section it was shown that reliable and accurate velocity data can be obtained provided the flow is appropriately seeded with small tracer particles, and the images are recorded in correspondence to the optimization rules in (5.253). The spatial resolution of the PIV measurement is characterized by the dimension size of the interrogation window. Under ideal circumstances it is possible to use 16 × 16 px (or even 8 × 8 px) interrogation regions, which provide sufficient spatial resolution to determine the instantaneous velocity derivatives (mostly associated with small-scale turbulence). Usually an overlap of 50–75% is used between adjacent interrogation positions, which provides sufficient data density to estimate the deformation and vorticity.

Methods for the estimation of derivative data have been described by *Lourenco* and *Krothapalli* [5.464], *Abrahamson* and *Lonnes* [5.465], *Fouras* and *Soria* [5.466], *Raffel* et al. [5.467], and *Foucaut* and *Stanislas* [5.468]. Here we describe the basic approach for the estimation of the out-of-plane component of the vorticity that is commonly referred to as the *circulation method*.

The deformation tensor $\partial \boldsymbol{u}/\partial \boldsymbol{x}$ can be split into symmetric and antisymmetric parts. The symmetric part is referred to as the stress tensor, while the antisymmetric part (which has only three nonzero components) can be written as a vector, i.e., the vorticity $\boldsymbol{\omega}$ [5.469]. The vorticity vector is then defined as the curl of the velocity field:

$$\boldsymbol{\omega} = \nabla \times \boldsymbol{u}, \quad \text{with } \omega_z = \frac{\partial v}{\partial x} - \frac{\partial u}{\partial y}, \tag{5.284}$$

where u and v are the two in-plane components. For PIV data that are obtained at discrete locations the vorticity ω_z can be estimated from central first-order differences of the measured velocity data

$$\omega_z(i, j) \cong \frac{v(i+1, j) - v(i-1, j)}{2\Delta x} - \frac{u(i, j+1) - u(i, j-1)}{2\Delta y}, \tag{5.285}$$

where (i, j) are the indices of the interrogation grid in the x- and y-directions, respectively, and Δx and Δy are the distances between mesh points. Given that a 50% overlap is used between subsequent interrogations, the estimation of the vorticity involves data over a region with an area of $3D_\mathrm{I} \times D_\mathrm{I}$. This implies that the vorticity is computed with a resolution length scale that is larger than the resolution length scale of the velocity data. This can be compensated partially by using interrogation data with a larger overlap, but this also increases the correlation of the noise in the velocity data, which deteriorates the accuracy of the vorticity estimate.

The differentiation scheme in (5.285) tends to amplify the noise in the measured velocity data. *Landreth* and *Adrian* [5.470] propose the use of a mild low-pass filter with a Gaussian kernel to attenuate the noise while retaining the velocity signal unaffected. A more-common approach is to estimate the vorticity by means of the circulation, i.e.,

$$\omega_z = \lim_{A \to 0} \frac{1}{A} \oint_C \boldsymbol{u} \cdot \mathrm{d}\ell, \tag{5.286}$$

where C is the contour that encloses an area A (Fig. 5.125). The discrete representation of (5.286) is [5.471]:

$$\omega_z(i, j) \cong \frac{1}{4\Delta x \Delta y} \times \begin{pmatrix} \Delta y \cdot v(i+1, j) \\ + \frac{1}{2}\Delta y[v(i+1, j-1) + v(i+1, j+1)] \\ -\Delta x \cdot u(i, j+1) \\ -\frac{1}{2}\Delta x[u(i-1, j+1) + u(i+1, j+1)] \\ -\Delta y \cdot v(i-1, j) \\ -\frac{1}{2}\Delta y[v(i-1, j-1) + v(i-1, j+1)] \\ +\Delta x \cdot u(i, j-1) \\ +\frac{1}{2}\Delta x[u(i-1, j-1) + u(i+1, j-1)] \end{pmatrix}. \tag{5.287}$$

Note that this is *identical* to first applying a filter to the velocity data:

$$\tilde{u}(i,j) = \frac{1}{2}\left\{u(i,j) + \frac{1}{2}[u(i-1,j) + u(i+1,j)]\right\},$$

$$\tilde{v}(i,j) = \frac{1}{2}\left\{v(i,j) + \frac{1}{2}[v(i,j-1) + v(i,j+1)]\right\},$$

(5.288)

which are then substituted in (5.285). For a 50% overlap the vorticity estimate involves a region with an area of $4D_I \times D_I$, which is only slightly larger than the central difference scheme using unfiltered velocity data. However, the noise in the vorticity result is substantially reduced. It can be easily verified using (5.285) and (5.288) that, for statistically independent velocity data, the rms error in ω_z is reduced by a factor of $\sqrt{3/8} \approx 0.6$. For a 1% noise level in the velocity data (i.e., a 0.1 px error and a 10 px displacement) the vorticity can be estimated with an error of about 5–10% of the rms variation of the vorticity.

Appendix:
Computation of the Spatial Correlation
Using the Discrete Fourier Transform

Consider an $M \times N$ px image $I[m,n]$. This image may be considered as an infinitely large image with $I[m,n] = 0$ outside: $1 \leq m \leq M$, $1 \leq n \leq N$, i.e., *zero padding*. Let $\tilde{I}[m,n]$ be a periodic image with periods U and V in m and n, respectively, such that $\tilde{I}[m,n] = I[m,n]$, for $1 \leq m \leq U$, $1 \leq n \leq V$ with $U \geq M$, $V \geq N$. The two-dimensional discrete Fourier transform (DFT) $\tilde{F}[u,v]$ of $\tilde{I}[m,n]$ is given by

$$\tilde{F}[u,v] = \frac{1}{UV} \sum_{m=\{U\}} \sum_{n=\{V\}} \tilde{I}[m,n] \exp\left[-2\pi i \left(\frac{mu}{U} + \frac{nv}{V}\right)\right].$$

(5.289)

Here the notation of *Oppenheim* et al. [5.426] is adopted, in which $m = \{U\}$ denotes the summation of m over one period U, and similarly for $n = \{V\}$. The DFT is periodic in u and v with periods of U and V, respectively. The inverse two-dimensional DFT is defined as

$$\tilde{I}[u,v] = \sum_{u=\{U\}} \sum_{v=\{V\}} \tilde{F}[u,v] \exp\left[2\pi i \left(\frac{mu}{U} + \frac{nv}{V}\right)\right].$$

(5.290)

Let $\tilde{I}_1[m,n]$ and $\tilde{I}_2[m,n]$ be the periodic images associated with the interrogation images $I_1[m,n]$ and $I_2[m,n]$, respectively, constructed according to the definitions

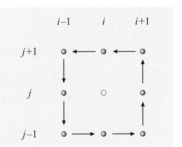

Fig. 5.125 The computation of the vorticity ω_z by means of the contour integral in (5.287) is equivalent to the central differentiation scheme in (5.285) using the filtered velocity data defined in (5.288)

above. For each of these two signals the discrete Fourier transform is given by (5.289), denoted by $\tilde{F}_1[u,v]$ and $\tilde{F}_2[u,v]$ respectively. The cross power spectral density $\tilde{S}[u,v]$ is given by the convolution of $\tilde{F}_1[u,v]$ and $\tilde{F}_2^*[u,v]$, where the asterisk denotes the complex conjugate. Substitution of (5.289) yields

$$\tilde{S}[u,v]$$
$$= \frac{1}{(UV)^2} \sum_{k=\{U\}} \sum_{l=\{V\}} \sum_{m=\{U\}} \sum_{n=\{V\}} \tilde{I}_1[k,l] \tilde{I}_2[m,n]$$
$$\times \exp\left[-2\pi i \left(\frac{m-k}{U} u + \frac{n-l}{V} v\right)\right],$$

(5.291)

which, after substitution of $r = m-k$ and $s = n-l$, can be written

$$\tilde{S}[u,v]$$
$$= \frac{MN}{(UV)^2} \sum_{r=\{U\}} \sum_{s=\{V\}} \exp\left[-2\pi i \left(\frac{ru}{U} + \frac{sv}{V}\right)\right]$$
$$\times \left\{\frac{1}{MN} \sum_{k=\{U\}} \sum_{l=\{V\}} \tilde{I}_1[k,l] \tilde{I}_2[k+r,l+s]\right\}.$$

(5.292)

The term between braces is recognized as the spatial correlation $\tilde{\phi}[r,s]$ of the two *periodic* images $\tilde{I}_1[m,n]$ and $\tilde{I}_2[m,n]$. The summations over k and l are taken over an area for which $\tilde{I}_1[k,l] = I_1[k,l]$, so that

$$\tilde{S}[u,v]$$
$$= \frac{MN}{(UV)^2} \sum_{r=\{U\}} \sum_{s=\{V\}} \tilde{\phi}[r,s] \exp\left[-2\pi i \left(\frac{ru}{U} + \frac{sv}{V}\right)\right],$$

(5.293)

with

$$\tilde{\phi}[r,s] = \sum_{\alpha=-\infty}^{+\infty}\sum_{\beta=-\infty}^{+\infty} \phi[r+\alpha U, s+\beta V], \quad (5.294)$$

where ϕ is defined in (5.256). The most important difference with the direct computation of the spatial correlation is that the DFT applies to *periodic* fields only. For images that are defined over an $M \times N$ px domain, the spatial correlation is defined over a $2M \times 2N$ px domain. This implies that the *exact* spatial correlation $\phi[r,s]$ can be retrieved when the interrogation images are padded by zeroes over a domain that is at least *twice* the size of the original signal in each direction. This is illustrated in Fig. 5.126. Hence, proper zero padding eliminates the effect that the DFT deals with periodic signals, which is usually referred to as DFT with (quasi-)nonperiodic boundaries.

In conclusion, the *exact* spatial correlation is obtained when zero padding is applied in accordance with the size of the correlation domain. So, *Fourier correlation* is not different from *direct correlation*. However, small errors are introduced when the DFT computation is done without zero padding, which can result in a significant increase of the measurement error for the subpixel displacement. Zero padding is used implicitly in the *direct* computation of the correlation, but it is unnecessary slow in comparison to the zero-padded DFT computation which can be evaluated with the fast and efficient FFT algorithm.

5.3.4 Doppler Global Velocimetry

Doppler global velocimetry (DGV), also known as planar Doppler velocimetry, is a member of a class of laser-based velocimetry techniques that use the absorption characteristics of molecular filters to determine the Doppler shift of scattered laser light. Tuning a single-frequency laser to a point midway along the side of an absorption line results in the absorption of approximately half of the light energy by the media. If the optical frequency is changed, e.g., by tuning the laser, Doppler shift in the light, etc., the amount of energy absorbed by the media changes. Thus the absorption line acts as an optical frequency-to-intensity converter, and is linear over a significant frequency range for a majority of absorption lines. Developing an optical system to monitor the instantaneous effects of this transfer function on collected scattered laser light yields a method to determine the velocity of targets passing through a specific point or points within the illuminating laser beam. This may be a single point along the beam, or an array of points within a laser light sheet as viewed by a video camera.

The direct determination of optical frequency eliminates the need to mix multiple laser light sources (e.g., reference-beam laser velocimetry, fringe-type laser velocimetry, interferometry-based laser velocimetry) or time-track individual particles (e.g., laser transit anemometry, and particle image velocimetry (PIV)). Thus any source of Doppler-shifted scattered light from a single-frequency laser may be measured to yield the velocity of the scattering object. This characteristic yields the ability to obtain average, time-dependent, or instantaneous velocity measurements from single or multiple submicron particles and even gas molecules. The limitation is the acquisition of sufficient scattered photons to activate the detector with adequate photon statistics to yield good signal-to-noise ratios. The technology also

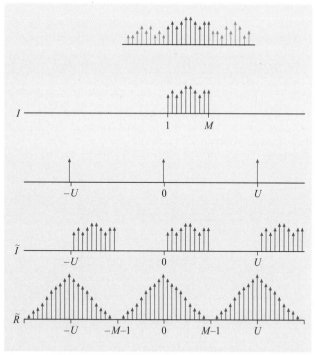

Fig. 5.126 Computation of the spatial (auto-)correlation by means of the discrete Fourier transform (DFT). *Top to bottom*: the signal, the finite domain signal (viz., interrogation image), the pulse train with a period of U, the harmonic signal $\tilde{I}$, and the spatial aut-correlation. The DFT implies a circular repetition of the original signal, so that the exact spatial correlation is only obtained when the period of the DFT domain is at least twice the signal length

allows the measurement-point dimensions to be defined by the viewing optical system, and eliminates the necessity to resolve single micron-sized particles. Finally, the ability to measure the velocity of very small particles and even molecules virtually eliminates measurement errors resulting from particle dynamics.

Note that molecular filter technology does have limitations, however. High-energy lasers and/or heavily seeded flows, scientific-grade cameras and integration techniques may be required to produce the signal levels needed to obtain accurate flow measurements. The overall measurement accuracy is dependent on the quality of the frequency content within the laser beam, the precision of the molecular filter calibration, the amplitude resolution and signal-to-noise ratio of the detector, the analog-to-digital converter, and the precision and geometry of the component placement.

Basic Principles

The shift in optical frequency of light scattered from a moving object is not only dependent on the velocity of the object, but also on the geometric configuration of the light source and the detector. If, as shown in Fig. 5.127, a laser beam is transmitted from the left to right along the vector direction $\hat{i}$, and an object passes through the beam, light will be scattered in all directions. If a detector is placed along a vector direction $\hat{o}$ to intercept a portion of the scattered light, that light will have an optical frequency shift described by

$$\Delta \nu = \frac{\nu_o \mathbf{V} \cdot (\hat{o} - \hat{i})}{c}, \quad (5.295)$$

where $\Delta \nu$ is the Doppler shift frequency, ν_o is the laser optical frequency, $\mathbf{V}$ is the velocity of the object, and c is the speed of light. The vectorial dot product in (5.295) indicates that this configuration will be sensitive to velocity along a single direction in space: the direction of $(\hat{o} - \hat{i})$ lying within the plane of $\hat{o}$ and $\hat{i}$. Thus multiple velocity components can be measured by changing the propagation direction of the laser beam ($\hat{i}$), or by placing multiple detectors about the measurement point, e.g., $\hat{o}_A$, $\hat{o}_B$, $\hat{o}_C$.

The laser must have the ability to generate a single-frequency output that is tunable to allow alignment with the chosen absorption line. The laser should also be frequency stabilized to keep the optical frequency within the dynamic range of the line. The following two examples illustrate these characteristics for the application of an iodine vapor cell. The 514.5 nm output from an argon-ion laser lies close to several candidate absorption lines. The laser must include an etalon to select a single longitudinal mode that yields bandwidths that are narrow enough (≈ 10 MHz) to provide a usable frequency resolution along the absorption line. Heating and/or tilting the etalon also provides a frequency tuning capability. Another commonly used laser, a pulsed, frequency-doubled Nd:YAG, provides the capability of making measurements nearly instantaneously. However, an injection seeder is needed to provide single-frequency operation.

The determination of the Doppler shift is obtained using a molecular absorption filter acting as an optical frequency discriminator to measure the frequency of the collected scattered light. Once found, the original laser frequency, typically determined using the same or a second molecular absorption filter [laser frequency monitor (LFM)], is subtracted from the measured frequency to yield the Doppler shift. The velocity is then obtained us-

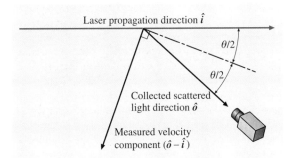

Fig. 5.127 Measurement of the velocity direction based on the orientation of the laser propagation direction and the detector location.

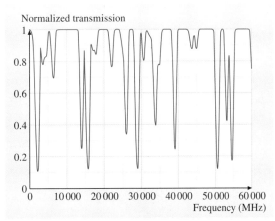

Fig. 5.128 Absorption spectra of iodine near the 514.5 nm line of an argon-ion laser (after *Forkey* et al. [5.472])

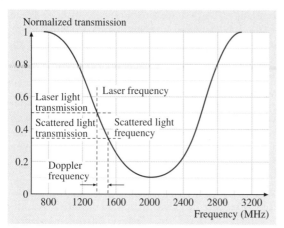

Fig. 5.129 Iodine vapor cell (IVC) transfer function

ing (5.295). The most commonly used molecular filter is an iodine vapor cell. Iodine possesses a large number of absorption lines in the green portion of the visible spectrum, which matches well with several laser systems, e.g., argon ion, doubled Nd:YAG, and doubled Yb:YAG. *Forkey* et al. [5.472], who developed a model outlining the absorption characteristics of iodine vapor with respect to optical frequency, aided the task of selecting an appropriate line. An example portion of the spectrum is shown in Fig. 5.128. A line that had a deep well with a linear slope was selected. The vapor pressure was then set to obtain the desired frequency-to-intensity characteristics.

A velocimeter can be created for flow measurement applications using these principles by projecting a single-frequency laser beam through a flow field and viewing a point(s) along the beam with a detector (Fig. 5.127). An iodine vapor cell is placed in front of the detector and the laser frequency is adjusted to a point along the edge of a linear absorption line (Fig. 5.129). When the flow is started, the collected scattered light is shifted in frequency based on the Doppler effect and the transmissivity through the iodine vapor changes (Fig. 5.129). Measurement of this change yields the information needed to determine the velocity of the scattering media. However, it also includes effects of variation in scattering intensity related to particle number density and size distribution along with inherent spatial variations in the laser beam intensity. Using a second detector to sample the collected light prior to the iodine vapor cell can eliminate these effects. This serves as the normalizing reference signal needed to determine the effects of the vapor transfer

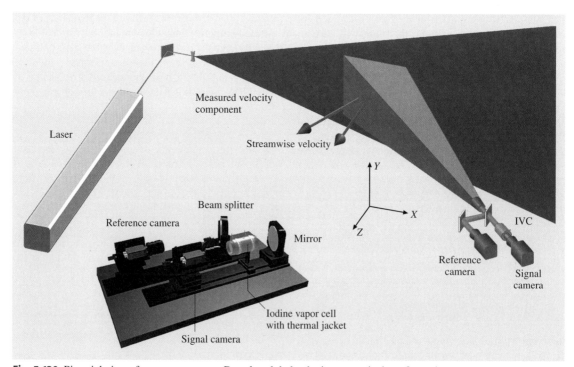

Fig. 5.130 Pictorial view of a one-component Doppler global velocimeter optical configuration

function on the collected scattered light, as illustrated for a single-component configuration in Fig. 5.130.

Molecular filter technology has been used in a variety of configurations to provide flow measurements in diverse applications. Although the primary application has been wind-tunnel testing, significant efforts have been directed toward internal flow applications in engines, and even in medical drug-delivery systems. The technology has been used to provide the *third* component measurement in laser transit anemometry and particle image velocimetry systems to eliminate the need for multiple viewing ports in internal flow applications.

Configurations and Applications

The original concept was proposed and demonstrated in the laboratory by *Komine* [5.473] and *Komine* et al. [5.474] at the Northrop Research and Technology Center. The technology was transferred to the NASA Langley Research Center in 1991, where *Meyers* and *Komine* [5.475] conducted the first wind-tunnel demonstration. The technology was advanced toward a practical wind-tunnel measurement system by capturing the video frames digitally and redirecting the laser light sheet to three orthogonal positions within the same measurement plane to obtain sequential three-component measurements [5.476, 477]. Development continued at Langley toward the goal of increasing measurement accuracy along with determining the limits of the technology through wind-tunnel testing; see *Meyers* [5.478] and *Meyers* et al. [5.479].

The ability to obtain three-component velocity measurements without the need to resolve individual seed particles opened the potential to use DGV in large wind tunnels to view the flow field over large measurement planes. The first application of multiple cameras to obtain simultaneous three-component measurements was a long-focal-length application in the NASA Ames Research Center 40×80 foot national full-scale aerodynamic complex (NFAC); see *Meyers* [5.478] and *Reinath* [5.480] (Fig. 5.131). The resolved streamwise velocity component of the 463 °C flow exiting a high speed civil transport (HSCT) engine simulator is shown in Fig. 5.132 along with the velocity profile extracted from a row of pixels along the horizontal diameter. The spatial resolution was 1.25 mm at focal distances of 15.8, 15.2, and 16.8 m, respectively, for the three-camera systems. This effort continued at NASA Ames leading to rotorcraft measurements in the 80×120 foot leg of the NFAC using a pulsed Nd:YAG-based system; see *McKen-*

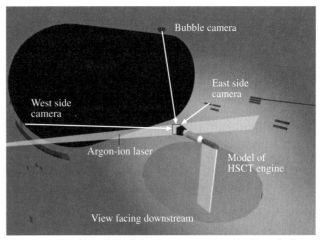

Fig. 5.131 Configuration of the three-component DGV optical system in the NASA Ames 40×80 foot NFAC to measure the flow from a high-speed jet

zie [5.481] and *McKenzie* and *Reinath* [5.482]. The long-range capabilities of DGV led to the use of several configurations in the US Air Force Wright Laboratory subsonic aerodynamic research laboratory. Primarily these efforts concentrated on the investigation of leading-edge vortex interaction with vertical stabilizers on high-performance aircraft; see *Beutner* et al. [5.483], *Elliott* and *Beutner* [5.484], and *Elliott* et al. [5.485].

A research program at ThermoTrex Corporation used various elements (cesium, potassium, and rubidium) to obtain optical frequency-to-intensity conversions that led to the development of a unique long-focal-length approach, *Kremer* et al. [5.486] and *Bloom* et al. [5.487]. One of the systems used potassium in a Faraday-cell configuration where polarization rotation was used to establish the transfer function instead of using optical energy absorption. This accomplished a factor of two increase in dynamic range, but required the use of an Alexandrite laser operating at 770 nm. This pulsed system was successful in measuring the line-of-sight velocity profile of the flow about a helicopter rotor at ranges exceeding 400 m with a measurement uncertainty of 0.6 m/s.

The ability to measure the flow velocity without resolving individual seeding particles made this technology attractive for the measurement of very high-speed flows. Research at Princeton University led to the development of filtered Rayleigh-scattering techniques that were able to measure supersonic/hypersonic flows using molecular scattering, *Miles* et al. [5.488], and *Forkey* et al. [5.472, 489]. They used a *dense* iodine vapor cell that yielded a nearly step-function response to view Doppler frequency contours as the optical frequency of the pulsed Nd:YAG laser was tuned. Conventional DGV approaches were used at NASA Langley by *Meyers* [5.478] – argon, and *Smith* et al. [5.490] and *Smith* [5.491] – Nd:YAG, Ohio State University by *Elliott* et al. [5.492] and *Clancy* et al. [5.493] – Nd:YAG, and Oxford University by *Quinlan* et al. [5.494] to investigate flows about models and in high-speed jets.

The engine measurement techniques research group of the Institute of Propulsion Technology at the DLR Research Center Köln-Porz has developed DGV systems to measure various internal flows. Applications have included jet inlet, turbine, combustor, internal combustion engine, and engine exhaust flow fields along with measurements of sprays and flames by *Röhle* and *Schodl* [5.495], *Röhle* [5.496], *Röhle* et al. [5.497], *Willert* et al. [5.498], and *Fischer* et al. [5.499]. Example results from flow measurements inside a cold combustion chamber are shown in Fig. 5.133.

A unique modification to a standard DGV configuration was developed at Cranfield University in England. *Nobes* et al. [5.500] adapted fiber-optic imaging bundles to provide three-component flow measurements along with laser frequency measurements using a single iodine vapor cell and only a pair of cameras. This approach greatly reduced system cost and complexity. It also lowered measurement uncertainty since a single iodine vapor cell could be used for all three components and the LFM function. The only disadvantage appeared to be image blurring caused by the edges of the individual fibers that limited the effective measurement spatial resolution. This approach was used by *Willert* et al. [5.501] of DLR Köln to obtain the first three-component measurements in a large cryogenic wind tunnel.

The ability to accept scattered light from multiple particles was exploited by *Kuhlman* et al. [5.502, 503] and *Kuhlman* and *Collins* [5.504] at West Virginia University, and *Crafton* et al. [5.505] at Purdue University through the development of point measurement configurations. The goal was to determine if the technology would provide continuous temporal measurements that could be interrogated for turbulence intensity and power spectral measurements. The results indicated a strong potential for measurements with better accuracy than can be obtained with fringe-type laser velocimetry.

The final category addressed is hybrid configurations that combine DGV technology with other techniques to produce systems with unique characteristics. An innovative hybrid was developed by *Förster* et al. [5.506] at DLR Köln that used DGV technology to obtain the third component in a modified laser transit anemometer system. The resulting system provided three-component point measurements within turbomachinery using a single optical access port. Taking the approach a step further, *Wernet* [5.507] from the NASA Glenn Research Center combined DGV technology with particle image velocimetry to produce a system capable of measuring three-component flow fields with a single viewing port. This work was directed toward turbomachinery and engine applications.

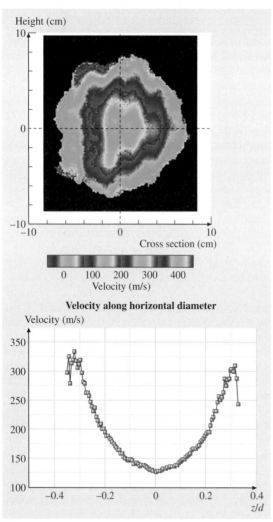

Fig. 5.132 Resolved streamwise component of velocity from the high-speed jet flow operating at 463 °C at a freestream Mach number of 0.15

System Design, Calibration, and Potential Error Sources

Although the fundamental physics for molecular-filter-based laser velocimetry has been presented along with example systems/applications to illustrate the breadth of this technology, the insight needed to develop a usable system would be difficult to obtain from this information. The purpose of this section is to highlight physics/equipment anomalies and how system design, calibration procedures, and software processing techniques can minimize the effects of the anomalies. The predicted quality of the measurements can be determined through an error analysis of a system based on an iodine vapor cell with argon-ion/frequency-doubled Nd:YAG laser sources. Although specific to DGV, many of the characteristics that will be analyzed here are common to point-based systems and systems using other molecular-filter/laser combinations.

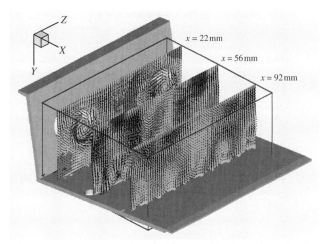

Fig. 5.133 Measurements in a double-staged isothermal combustion chamber by DLR, Köln (after *Röhle* et al. [5.497])

Laser. The search for sources of measurement uncertainty begins with the laser and the quality of its output. The laser characteristics described in the *Basic Principles* section pertain to single-frequency outputs that can be tuned to the midpoint along the edge of an iodine absorption line. The most common lasers that satisfy these requirements are continuous-wave (CW) argon-ion lasers operating at 514.5 nm, and pulsed, injection-seeded, frequency-doubled Nd:YAG lasers operating at 532 nm. It is noted that currently available continuous-wave frequency-doubled Nd:YAG lasers cannot be easily tuned and have frequency drifts of several MHz/min, and thus are not suitable. Furthermore, all lasers must be environmentally protected as any changes in ambient temperature or pressure, or vibrations in mounting structure(s) will cause them to become unstable, producing multimode operation or even cessation of lasing.

Single frequency operation is obtained in the argon-ion laser by placing a temperature-controlled etalon within the laser cavity. The etalon forms a second resonator that, if tuned to one of the resonant frequencies supported by the laser cavity, isolates a single longitudinal mode for amplification. The etalon resonant frequency is tuned by changing its temperature or by mechanically tilting it – both methods change the effective etalon thickness. The dependence of the index of refraction of the media and the component thickness on the environmental temperature results in frequency drift of the two resonant frequencies. This will reduce the laser output power as the two resonant points misalign, which may cause the output frequency to oscillate between two adjacent longitudinal modes before *hopping* to the next mode. Typically, the drift originates from the laser cavity since the etalon is normally enclosed in a temperature-controlled oven. The most efficient method to eliminate this drift is through an active feedback control system using the laser frequency measurements from the LFM as an input to control a piezoelectric transducer that actively positions a cavity mirror [5.506] to compensate for environmentally induced changes in cavity length.

The use of an injection seed laser makes it possible to operate a pulsed, frequency-doubled Nd:YAG laser in single-frequency mode. This laser-diode-pumped seed laser emits a low-power, single-frequency continuous-wave beam that is directed onto the primary laser cavity axis, thus providing a stimulating source when the primary laser is pulsed. In this manner, the primary laser acts as an amplifier of the single-frequency beam from the seed laser, thus maintaining its optical frequency characteristics. Adjusting the temperature of the seed laser will change the length of its Nd:YAG rod and thus the output frequency. The tunable range is approximately 20 GHz (an order of magnitude greater than with an argon) with a minimal loss of laser energy over multi-mode operation (the etalon in the argon laser reduces the output power by 50%). The main laser cavity resonant frequency is adjusted through a feedback-driven piezoelectric transducer mounted to the fully reflective mirror. The control function is derived from the measurement of the Q-switch build-up time – the shorter the time, the better the frequency overlap between the seed laser and the host, which improves single-frequency operation. The piezoelectric transducer is normally dithered to bracket any changes in the resonant frequency caused by thermal expansion/contraction of the main cavity. Pulse-to-pulse optical frequency variations can be as large as 70 MHz, requiring the measurement of each pulse frequency with the LFM. In addition, the optical frequency is not homogeneous throughout the laser beam cross section. Unlike a gas laser whose media index of refraction in the laser cavity is virtually the same as air, a Nd:YAG laser rod has a much larger index of refraction. Thus any microscopic variation in rod length caused by fabrication or temperature variations will effectively change the laser cavity length over the cross section, yielding a resonant frequency distribution. The theoretical distribution, based on manufacturer rod face flatness and parallelism specifications, was determined to be approximately 100 MHz, a similar range to that found experimentally. The simplest method to eliminate this error source is to obtain a wind-off measurement of optical frequency with the DGV system. The resulting mean-removed frequency image represents the variation caused by the laser. This *frequency flat field* is simply subtracted from the acquired Doppler frequency measurements to yield the unbiased images.

The laser output must be monitored to insure single-frequency operation. An optical spectrum analyzer placed in the LFM can provide a real-time reading of the optical frequency spectra of an argon-ion laser. If a clean single frequency is not found, the laser should be adjusted to reestablish a single frequency. Measurement of the Q-switch build-up time in a Nd:YAG laser provides a parameter indicating single-frequency operation. The build-up time can be determined by measuring the time between the Q-switch trigger and the occurrence of the optical pulse. When the laser is operating in single-frequency mode, this build-up time is minimized.

Iodine Vapor Cell. The absorption of light by iodine vapor is based on the temperature-dependent shape of the selected absorption line. While the location of the optical frequency at the center of a given absorption line is fixed by the atomic characteristics of iodine, the depth and width of that line are dependent on the number of iodine molecules in the optical path. For a fixed cell length the number of molecules in vapor state, and thus the line shape, is dependent on the temperature of the coldest point within the cell (Fig. 5.134). Typically the cell body is held at a high temperature and a narrow stem extending from the body is kept at a tightly controlled lower temperature – thus defining the iodine vapor pressure. While this method is acceptable for spectroscopy applications, a variation of $\pm 0.1\,°C$ of the control temperature represents a ± 3 m/s velocity uncertainty in DGV measurements. Additionally, the inability to isolate the cells completely from hostile wind-tunnel environments results in large variations in line shape because of the inherent slow response of a heating/cooling system to counteract ambient temperature changes. This is the reason for the lack of quantitative velocity measurements in early wind-tunnel applications of the technology.

The solution to the inability to control cell temperature was the construction of a vapor-limited or starved cell [5.479, 484]. After a standard iodine vapor cell has been constructed, the desired vapor pressure is set using the above approach. Once the internal pressure has stabilized, the stem is sealed at the cell body to separate excess crystalline iodine from the cell. Setting the cell to any temperature at or above the original stem temperature yields a constant cell vapor pressure since all iodine molecules will be in the vapor state once the original stem temperature is reached. Elevating the cell temperature, for example an additional $20\,°C$, provides a buffer to account for the slow response of the temperature controllers to any change in environmental temperature, and maintain a constant optical frequency-to-intensity transfer function.

Optical Components: Transmitter. The creation of a laser light sheet normally involves simply placing a cylindrical-lens system in the laser beam. However, the Gaussian and approximate Gaussian power distributions found in argon-ion and pulsed Nd:YAG lasers, respectively, can produce detector saturation due to the high-intensity scattered light from the optical axis, yet

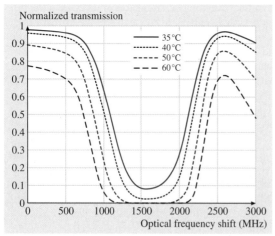

Fig. 5.134 Theoretical iodine vapor absorption line profiles as a function of cold finger temperature

be undetectable away from the axis. Additionally, the diameter of the laser beam can produce a thick light sheet that will lessen the common volume viewed by the three receivers, potentially introducing velocity errors in areas with large velocity gradients. Thus the area viewed should be set to the half-power limits of the light sheet along with focusing the sheet thickness at the center of the viewing area. If average velocity measurements are acceptable, slowly scanning an argon beam along with on-camera integration will yield good results, as the light sheet will then have a flat intensity distribution. Note, however, that rapid scanning can impose an apparent Doppler frequency shift in the measurements.

Optical Components: Receiver. The receiver optical system, shown in Fig. 5.130b, is fairly simple. However, attention must be given to the characteristics of the optical component or significant measurement errors may occur. The first element reached by the scattered light is the beam splitter. The transmission/reflection ratio is dependent on the input light polarization in all beam splitters, including the so-called nonpolarizing variety. This is significant since the polarization of light scattered by small particles is dependent on particle size, scattering angle and to a lesser extent index of refraction. Thus any differences in input polarization, e.g., calibration setup, variations in particle size distribution, etc., will change the transmission/reflection ratio imposed by the beam splitter, and effectively rescale the system transfer function in an unknown manner. Additionally, spatially dependent transfer function variations caused by any optical element, e.g., dirt, fingerprints, interference fringes, etc., will result in local deviations in the system transfer function. These effects can be greatly reduced or eliminated by operating the system normally with particle scatter, but with the laser frequency adjusted so that all optical frequencies are outside any absorption lines. The resulting reference images are normalized by the signal images to yield the *intensity flat field*. The laser frequency is then tuned to the desired location along the absorption line and data acquired. Resultant normalized signal images are then multiplied by the intensity flat field, thus negating all polarization (assuming the particle size distribution has not changed) and spatially dependent optical transmission effects.

Other suggestions regarding the optics include the use of a cube-type beam splitter, a cold-welded iodine vapor cell, and green photographic filters placed on the cameras. Typically beam splitter plates are used to divide incoming light into two optical paths. These plates have a partially reflective coating on the input side and an antireflective coating on the other. Thin plates should be avoided as they may be physically distorted by the mounting hardware, leading to a degradation of image quality. Beam splitter plates have exhibited the property of softening/blurring the transmitted image, producing artifacts in the normalized results. Thus, beam splitter cubes are recommended since they do not exhibit this characteristic. Aligning the iodine vapor cell along the signal leg optical axis can produce fringes from the interference of scattered light from cell window surfaces. These fringes are typically circular, as the low pressure within the cell will pull the centers of the windows inward. Further, hot welding the windows to the cylinder body will induce heat stress in the windows producing additional optical distortions. Cold welding the windows to the cylinder body and misaligning the iodine vapor cell a few degrees from the optical axis will greatly reduce or eliminate these optical distortions. Placing photographic green filters on the cameras attenuates contributions from other light sources, e.g., room lights, combustion products, etc., without greatly affecting the collected scattered light intensity. The filters will not eliminate contributions from these sources, but will help since any background light energy acquired will decrease the effective dynamic range of the cameras.

Optical Components: Cameras. Camera requirements are dependent on specific techniques and applications, thus the choice of camera is difficult to specify in a generic sense. The following factors should be considered:

- Spatial resolution/image size
- Imaging rate
- Pixel-well depth/amplitude resolution
- Overall signal-to-noise ratio
- Image integration/shutter control
- CW or pulsed laser applications
- Working focal distance
- Background light/environment
- External triggering/synchronization capabilities.

In general the camera/detector system must be linear with a repeatable optical intensity-to-voltage calibration since amplitude is the measured parameter. In multiple-camera configurations, all cameras must be externally synchronized using the same synchronization waveform, precluding a master/slave approach since image acquisition must be simultaneous over exactly the same time period. Overall a cooled CCD sensor with at least a 12 bit onboard digital conversion capability and a controllable electronic shutter is recommended. However, this may be more camera than is needed for a given experiment – RS-170-type cameras may be acceptable. In any case, it is advisable to measure the overall camera signal-to-noise ratio using the method outlined by *Meyers* et al. [5.479] since the signal-to-noise ratio quoted by camera manufacturers is typically based only on electronic noise, which is not necessarily sufficient for DGV applications.

Facility and Model. All surfaces that will scatter laser light should be painted either flat black or red (if a green photographic filter is used) to reduce spurious light sheets or flare. A light sheet impinging on any surface (including antireflective-coated windows once they become coated with dirt or particles) will produce some flare, which limits measurements to several millimeters above the surface. If on-surface measurements are desired, light should originate from under the surface. Treating surfaces will have the added benefit of reducing the secondary scatter caused by surface–particles–detector and particles–surface–detector paths. This source of background light can not be eliminated and will affect measurement accuracy as the Doppler shift imposed on this secondary scattered light will not be the same as the scattered light from particles passing through the light sheet [5.495].

Facility configuration must also be considered. If the opposing facility wall has a window that is not antireflective coated and the light sheet reflection cannot be directed outside all camera views, it should be orthogonal to the window, making the reflected light sheet coplanar with the original. As with any spurious light sheet, a measurement error will occur, as the additional scattered light from this sheet can not be eliminated. However, since it is coplanar, each measurement vector and its contribution can be estimated based on Mie scattering, if the particle size distribution is known.

Flow Seeding. Typically a flow measurement application using molecular filter velocimetry will involve

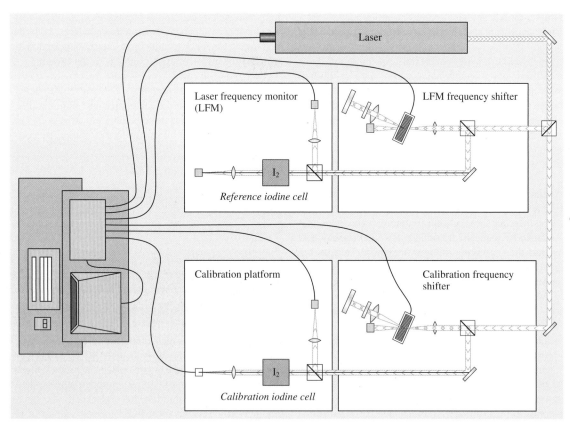

Fig. 5.135 Iodine vapor cell calibration layout using tunable Bragg cells; tuning range: 420–660 MHz, laser mode spacing: 129 MHz

seeding the flow with small particles. Particle generation techniques normally used for fringe-type laser velocimetry and PIV, such as Laskin nozzles, liquid sprays with embedded polystyrene latex, etc., typically have limited particle number densities when the plume is expanded to fill large laser light sheets. In addition, these particles tend to be fairly large and may not follow the flow with the desired fidelity. Smoke generators, such as theatrical foggers and smoke simulators used to train firefighters, will provide sufficient number densities. Care must be used if theatrical foggers are chosen as they typically generate size distributions that have extremely long tails, upwards of 20–50 μm. The recommended generator is the smoke simulator, as this device will produce 0.2–0.7 μm mineral oil particles in number densities controllable by the generator. The output can range from a barely visible wisp, to a cloud capable of filling the NASA Langley 14×22 foot subsonic tunnel in a few minutes. If testing in a supersonic flow, water condensation is a good choice as the particle size is extremely small and particles will completely fill the flow field. The presence of water, however, will affect the flow properties to a minor extent.

System Calibration: Iodine Vapor Cell. Determining an accurate optical frequency-to-transmissivity transfer function for the iodine vapor cell (IVC) is critical to the overall accuracy of DGV measurements. Example calibration procedures are outlined below.

1. Laser tuning – The output laser frequency is changed by mode-hopping (argon) or adjusting the seed laser temperature (Nd:YAG) and tracking the transfer function. This technique assumes that the change in laser frequency is a linear function of the adjustment. Through experimentation it has been determined that this approach does not provide the accuracy required for DGV applications and thus is not recommended.
2. Measuring the Doppler frequency of a rotating wheel – The IVC is placed in a DGV system that is configured to measure velocity in a single plane. The Doppler frequency profile across the wheel can be predicted from distance and rotational speed measurements. By having a Doppler shift range greater than the laser resonator frequency, the profiles may be overlapped to produce a transfer function that does not depend on knowing the absolute laser or mode hop frequencies. This approach is recommended.
3. Bragg cell frequency shift – The best accuracy has been obtained using a pair of tunable Bragg cells to change the input frequency by known amounts at each laser longitudinal mode, Fig. 5.135. The results have yielded standard deviations about a B-spline curve fit of less than 0.2% full scale. This technique is highly recommended.

Image Distortion. The determination and correction of perspective and other distortions is necessary to insure accurate overlap of the signal and reference camera images. Furthermore, accurate overlap is required to align multiple image pairs to determine standard Cartesian velocity component measurements. Standard 2-D polynomial dewarping methods using single equations to adjust an image do not account for high-spatial-frequency distortions typically present in windows and optical components. Thus a piecewise method must be used to insure accurate overlap between the camera images. Using a piecewise bilinear warping procedure [5.478], a distorted fiber optic view of a dot card target can be corrected (Fig. 5.136).

Intensity Flat Field. The intensity flat field is the amplitude calibration between the signal and reference camera images. The correction accomplishes two tasks:

1. it determines the overall beam splitter division of collected scattered light as affected by the polarization characteristics of the scattering media; and,
2. it determines any higher spatial frequency variations in the optical transfer function, e.g., dirt, interference fringes.

The intensity flat field should be obtained by acquiring a normal data image with the laser and all Doppler-shifted frequencies located outside all absorption lines.

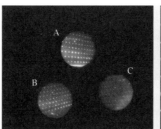 

Camera view of three imaging fibres   Dewarped image of fiber A
Internal fiber bundle diameter: 0.5 mm (30 000 fibers), dot spacing 25 mm

Fig. 5.136 Image dewarping of a dot card spatial alignment target viewed by three fiber-optic imaging systems: piecewise bilinear method

Frequency Flat Field (Crystal-Based Lasers). Crystal-based lasers, such as Nd:YAG, have a distribution of optical frequencies over the output beam cross section. This distribution is caused by microscopic variations in rod length that change the effective length of the optical cavity based on the rod index of refraction and the spatial distribution of rod length. Typically, 10 arc s of parallelism and a flatness of $\lambda/10$ can yield optical frequency variations of up to 100 MHz. The distribution of optical frequencies can be measured by obtaining Doppler frequency maps under wind-off conditions. An intensity flat field should also be taken to account for any particle size changes between wind-on and wind-off conditions. The measured frequency distribution is then subtracted from the Doppler frequency maps of the flow to negate these frequency variations.

Application Sequence

Continuing with the example of using DGV to measure a flow field, the following data-acquisition and data-processing sequence is suggested. Other molecular filter velocimetry techniques and measurement applications have a similar operating sequence.

Data Acquisition.
Image Alignment. An alignment target is placed in the measurement plane and data images acquired. A target composed of a rectangular pattern of equally spaced dots is suggested, as individual dot centroid locations are more easily and accurately determined than the intersection of lines. Further, the determination of piecewise dewarping coefficients based on the centroid locations will increase spatial accuracy of the overlaid images. Using a back-illuminated perforated panel of millimeter-sized holes with small spatial separations as the alignment target may increase this accuracy further. The resulting image would be a *star* field where centroids may be more-accurately determined using the greater contrast of bright spots on a dark background (the same principle used in particle image velocimetry).

Background Images. A background data set should be acquired with all components having the same settings that would be used when acquiring velocity data – except no particles are injected. The background contains contributions from all light sources, e.g., room lights, laser flare from walls, and models. The background image levels can then be subtracted from the data images to eliminate these sources. Secondary scattering [5.495] however, can not be eliminated, thus procedures must be implemented to eliminate or minimize any extraneous laser light.

Frequency Flat Field Images (Crystal-Based Lasers Only). Inject particles into the laser light sheet without a measurable flow to determine the laser-induced optical frequency distribution. The resulting optical frequency measurement images should be subtracted from the Doppler frequency images obtained from the flow measurements to eliminate these frequency variations. Note: a separate intensity flat field image set should be obtained as the particle scattering characteristics may not be the same as during the acquisition of flow measurements.

Intensity Flat Field Images. The facility should be operating at the desired conditions for the flow measurements and the seeding particles injected into the flow. The laser should then be frequency tuned to place all optical frequencies outside of the absorption line (insure that an adjacent line is not entered). An alternate method for crystal-based lasers is to operate the laser multimode. The resulting normalized images represent the transfer function of the optical system without the effects of the iodine vapor. The normalized images obtained from the flow measurements should be multiplied by this transfer function to eliminate the spatially dependent influences of the optical elements.

Flow Measurement Images. Measurement accuracy is strongly dependent on the amplitude and spatial distribution of the collected scattered light in the data images. Higher measured amplitudes will lower the significance of noise sources, such as photon, electronic, and quantizing noise. The more uniform the flow seeding, the lower the effect that minor spatial *misalignments* of the signal image with the reference image will have on the accuracy of the normalization process. The signal amplitude can be effectively increased in CCD detectors by allowing them to integrate the incoming collected scattered light over a time period needed to charge the capacitors to a high level [5.495–497]. Integration has the added benefit of averaging laser speckle noise along with any motion of the particle stream that would reduce spatial amplitude gradients at the stream boundary [5.479]. If integration is used, whether on the detector or in the computer, electronic noise becomes more significant. Thus a cooled camera should be used to lower these noise levels. This approach also limits DGV to average or phase-locked measurements – instantaneous measurements have the potential for too great a level of measurement uncertainty even when a high-power, pulsed laser is used. It is noted, however, that point measurement configurations using photomultipliers as detectors in conjunction with high-speed digitizers can be used for temporal measurements. These detectors are far more sensitive and have lower noise levels than cameras.

Data Monitoring. Unlike fringe-type laser velocimetry or particle image velocimetry, molecular filtered velocimetry does not have a definitive parameter that determines if a measurement is acceptable in real time. The sensitivity of the technology to single-frequency stability of the laser, uniformity of seeding, and signal strength minimums require constant monitoring of many parameters. In addition, sample data should be processed in near real time to determine if the velocity measurements are reasonable, and as an indicator of potential problems.

Summary

Molecular filter velocimetry has characteristics that make the technique advantageous for many fluid-mechanical investigations. These include high-speed flows, highly three-dimensional flows, cyclic flows, flows with a large dynamic range in velocity, and for the specific case of point Doppler velocimetry, temporal measurements such as turbulence power spectra. From an installation viewpoint, these techniques provide long focal distance, large measurement area capabilities and, through the use of fiber optics, measurements of internal flows in confined structures.

As with any measurement technique, however, there are limitations. The technique requires careful monitoring of the laser and components, heavier seeding than other laser-based methods, and no in situ indicators that the measurements will be accurate. Additionally, since the measurement accuracy is an absolute value, low-speed flow measurements may not be as accurate as other methods. Finally, although instantaneous measurements can be obtained, their use for turbulence statistics would require great care to insure minimal error.

5.3.5 Laser Transit Velocimetry

This section starts with a short overview of the history of the development of laser transit velocimetry (LTV) also known as laser two-focus velocimetry (L2F). The advantageous properties of LTV that qualify this technique for high-speed and turbomachinery flows and for near-wall measurement applications are explained. The fundamentals are then treated, describing the general principal, the statistical data evaluation and the prospects for the measurement of turbulence statistics. Next, a state-of-the-art LTV system is considered, presenting details about the optical setup, the measurement procedure and the data-processing and evaluation methods. Subsequently, three-component LTV techniques are treated with the main emphasis being placed on the recently developed three-component Doppler-L2F velocimeter. In this hybrid system the L2F principle is combined with a Doppler-frequency shift analysis by which the velocity component in the direction of the optical axis is determined. The unique measurement system is described in detail and examples of applications are presented.

In 1968, very soon after the laser Doppler (LD) method became known, an alternative optical flow velocity measuring method was introduced by *Thompson* [5.508]. This method, referred to as the *tracer particle fluid velocimetry*, was the forerunner of the kind of optical measuring devices known today as laser transit (LT) or laser two-focus (L2F) velocimeters. The idea was to measure the time of flight of tracer particles carried with the fluid passing through two separated, parallel, highly focused laser beams (Fig. 5.137). When Thompson compared his method with the known LD method, he saw the advantages of a simpler optical setup, simpler data processing, and the employment of lower-power lasers to detect the scattered light of the small tracer particles.

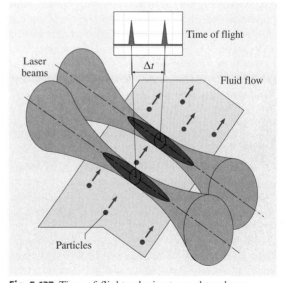

Fig. 5.137 Time-of-flight velocimeter probe volume

In Thompson's system the two beams in the probe volume had a waist diameter d of 40 μm and a separation s of a few millimeters. Photomultipliers adjusted to the probe volume generate pulse-type signals when detecting the scattered light of particles crossing the laser beams. An oscilloscope internally triggered by the photomultipliers' voltage pulses was used for the time-of-flight measurements. However, successful measurements could only be obtained when the orientation of the plane containing the two beams was adjusted parallel to the mean flow direction. In this case the probability of successful dual beam transits of the particles attains its maximum value. Under these conditions multiple traces integrated on the oscilloscope screen show two bright, pulse-type signals, one on the left side of the screen, which starts the oscilloscope trace and a second one separated from the first by $\overline{\Delta t}$ indicating the successful time-of-flight measurements (Fig. 5.138). Due to the integration the detected value $\overline{\Delta t}$ represents the mean value of the time-of-flight measurements, which together with the known separation s of the beams provides the magnitude of the mean flow velocity component perpendicular to the beams

$$\overline{u}_\perp = \frac{s}{\overline{\Delta t}}. \qquad (5.296)$$

Knowing the orientation $\overline{\alpha}$ of the plane containing the two beams the mean velocity vector components $\overline{\alpha}$ and $\overline{u}_\perp$ in the plane perpendicular to the beam axis are determined.

The brightness of the second signal on the oscilloscope screen is a measure of the probability of successful dual-beam crossings. The level of brightness reaches its maximum value when the angle α of the beam's plane is correctly adjusted to the mean flow direction.

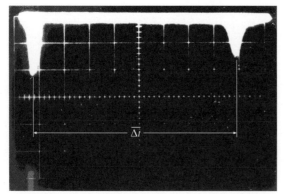

Fig. 5.138 Time-of-flight signals superimposed on an oscilloscope

The detectability of the second signal and the adjustment of the beam's plane is quite easy in flows of low turbulence intensities where the flow direction is very uniform. However, in cases where the flow turbulence intensity increases and exceeds values of about 10%, detectability of the second signal disappears almost completely, rendering this time-of-flight measuring method inapplicable.

Due to this limitation Thompson's method did not achieve much importance. Only a few authors [5.509, 510] followed his proposal and applied similar optical setups to flows where the turbulence intensities were low, the mean flow direction was well predictable and the beams plane could be set to a fixed orientation.

Then in 1973 a paper was published [5.511] describing how this laser transit velocimetry technique could also be operated in turbulent flows. It showed how the collected data could be processed to determine the two components of the mean flow velocity, turbulence intensities, Reynolds shear stresses, and also higher-order moments of the fluctuating velocity quantities.

This knowledge about its applicability to turbulent flows changed matters dramatically and, as of that point in time, this technique underwent rapid development.

While the optical arrangements proposed have not undergone significant modifications, considerable development has taken place on the signal-processing electronics. Two different development directions were initiated. One group of researchers [5.510, 512–521] applied correlators and photon correlators and a second group used time-of-flight measurements and statistical data analysis [5.522–536]. Theoretically the correlation analysis should be the superior technique, especially at high data rates (< 200 kHz), but this superiority has not been proved in practice to date where the data rate is comparatively low (< 20 kHz). As can be shown [5.520, 528, 537] under these conditions there is complete agreement between the measured cross correlogram and the time-of-flight histogram. Because of its availability in the area of nuclear physics and its simplicity the time-of-flight measurement method has attained much more importance. This technique is therefore treated in this article. The output of both techniques, either the correlogram or the histogram requires further data processing.

Many authors have theoretically analyzed the LT measurement method in detail [5.510, 512, 513, 515, 520, 521, 523, 525, 526, 531, 538–544]. Most of the main error sources are known today and they can be estimated quantitatively as a function of the probe volume geometry.

Over the years LT velocimetry has gained importance in the experimental fluid flow analysis. Due to its high signal-to-noise level resulting from the highly focused laser beams LT velocimetry extends the range of applicability of LD systems and can therefore be considered a supplementary method.

Its specific fields of application are: high-speed flows, applications with limited optical access where backscatter arrangements with low signal level are required, and applications in narrow flow channels, where background flare reduction is important. Apart from these positive features the LT velocimetry technique is in practice limited to applications in flows with turbulence intensities smaller than 30%, and hence cannot be applied in, e.g., combusting and mixing flows, where the turbulence intensities by far exceed this value.

Several LT systems are on the market and some industrial and research organizations have built their own system for in-house use. LT velocimetry has been successfully applied to long-range wind speed analysis, heat exchangers, wet steam flows, water pumps, diesel engines, plasma flows, and standard and cascade wind tunnels. Nevertheless the LT velocimetry has found its widest application in the field of turbomachinery flow investigation. Numerous papers have been published producing results of measurements in centrifugal and axial compressors as well as turbines. A rather complete overview is given in [5.532, 540]. These detailed LT velocimetry data have contributed a great deal to our improved understanding of turbomachinery internal flow. This is especially true with regard to rotor flow investigations, which are almost inaccessible for conventional measuring techniques. The comparison with theoretical data has resulted in improved mathematical models and design procedures.

Laser Transit Velocimetry
Applied to Turbulent Flows

In turbulent flows the velocity vector at any location is considered as a random variable, which under stationary conditions can be described by a three-dimensional probability density function that depends generally on all three components of the velocity vector. As the LT velocimeter as well as the usual two-component LD systems are only capable of measuring the velocity components in the plane perpendicular to the optical axis, in the ideal case the measured data can only approach the two-dimensional probability density function, which can be derived from the three-dimensional function by integration over the on-axis velocity component. To be able to perform this integration, both systems must also be capable of detecting the respective velocity components from such flow vectors that have a significant on-axis component. Therefore the axial length of the probe volumes must be long enough. Since this requirement can often not be fulfilled sufficiently in practice the systems with a limited integration range require a roughly perpendicular alignment of the optical axis with respect to the expected mean flow velocity. Assuming that the LT velocimeter can be operated in such a way that the measured data can be approached by a two-dimensional probability density function, the mean values of any velocity function defined in the plane perpendicular to the optical axis, e.g., the mean velocity components, turbulence intensities, and Reynolds shear stresses, can be deduced [5.532].

Let us consider a typical LT velocimeter probe volume, as shown in Fig. 5.139. The beams waist diameter d is about $10\,\mu\text{m}$, the beam separation s is about $0.2{-}0.4\,\text{mm}$. The plane containing the two beams – the plane of beams – is rotatable and its angular position is indicated by the value α. Two photomultipliers are adjusted to the beams and deliver the start and stop signal for the time-of-flight measurements. Due to the given ratio of beam diameter and separation a certain angular capture range $\Delta\alpha$ is defined, within which particles can move and cross the two beams. When at a location in a turbulent flow the beam's plane is adjusted to a certain angle α; within the rather wide angular range of fluctuating flow directions only those particles that move along the plane within the range $\Delta\alpha$ can pass through both the beams and deliver two successive scattered light pulses. The time elapsed between the pulses yields the component of the particle velocity perpendicular to the optical axis.

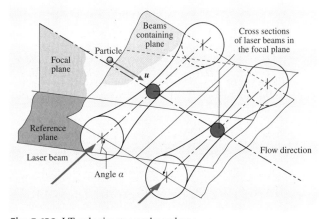

Fig. 5.139 LT velocimeter probe volume

When particles move outside this small angular range $\Delta\alpha$, they may still start a time measurement but then either there will be no stop signal within the preselected time range or a different particle will stop the measurement, hence generating incorrect measurements.

To distinguish between correct and incorrect time-of-flight measurements many measurements at the chosen position of the plane of beams are necessary. In Fig. 5.140, as an example, a probability histogram of such collected data is plotted. Due to its random distribution the incorrect measurements are represented by the baseline of nearly constant frequency level. The net distribution, which exceeds the baseline frequency, is shaped like a Gaussian curve and represents the correct time-of-flight measurements. The necessity of collecting multiple measurement events is not a real drawback since the statistical analysis in any case requires a large number of data, which determines the measurement accuracy.

By setting the plane containing the beams to other, slightly different, angles and by accumulating time-of-flight data until the preset number of started measurement events has been achieved, more velocity histograms can be constructed. This is possible as long as the chosen angles of the plane are within the range of the velocity angle fluctuations of the turbulent flow. The result of this measurement procedure is a two-dimensional probability histogram (Fig. 5.141) as a function of time-of-flight and angle settings. These data are the input data for a rather sophisticated evaluation procedure, based on nonlinear regression fits of an ideal probability density function of the flow velocity to the measured histogram [5.540, 541]. Mean values of the velocity components and turbulence intensities are the typical output, but Reynolds shear stresses and higher-order moments can also be computed.

The accuracy is mainly determined by the number of collected data, the background noise resulting from laser light reflection from surfaces in the vicinity of the probe volume, the flow turbulence, and the selected beam separation. In the case of low turbulence intensities ($< 5\%$) the accuracy of the velocity magnitude and angle determination is very high (measurement error $< 1\%$ of the flow magnitude and less than $0.2°$ of the flow angle), better than can be achieved by LDA systems, although the measurement error increases with flow turbulence. When turbulence intensities exceed 30% the LT velocimeter is practically not applicable because of difficulties in distinguishing between correct and incorrect measurement events.

Optics

The schematic assembly of the LT velocimeter that pioneered the first application in turbulent flows [5.511] is shown in Fig. 5.142. An argon laser was used as a light source. The linear polarized laser beam was transformed by a quarter-wave plate into circular polarization. In this way the following rotatable Rochon prism splits the laser beam into two beams of different linear polarization but

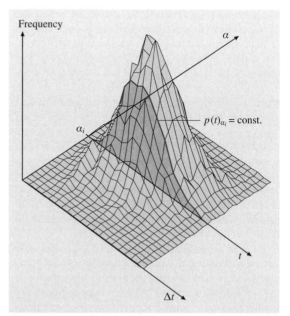

Fig. 5.141 Two-dimensional histogram of measured time-of-flight data collected at different angle α settings of the plane of beams

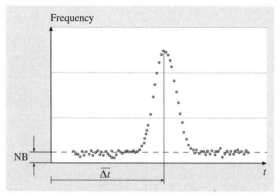

Fig. 5.140 Time-of-flight data histogram collected at a given setting of the plane of beams (incorrect measurements are indicated by the noise band NB)

of equal intensity, independent of the angular position of the Rochon prism. One output beam follows the direction of the incident laser beam while the other is deflected from that direction by an angle δ (compare detail A).

As the focal point of the following lens L1 is located at the point of division, both beams leave the lens parallel to each other and are focused in the second focal plane.

The two beams are then relayed by the inner part of lens L2 to the measuring volume, which is to be placed in the selected measuring position in a flow field (e.g., in a wind tunnel) (compare detail B).

Particles that traverse the beams in this area each emit scattering light pulses and the part which is scattered back is detected by the outer area of lens L2 and sent via a spatial filtering device to two photodetectors, each assigned to a beam in the measuring volume.

The spatial filtering device serves to keep the background radiation, which arises in narrow channels as a result of reflection of the laser beams off the channel walls, to a minimum. The use of microscope optics makes the spatial filtering very effective. Due to the enlargement the two very small backscattered beams can be adjusted very precisely to the two holes of the aperture.

This optical setup for two-component measurements has not changed significantly over the years as its optical design was already rather optimized. It was rather emerging technologies such as fibres and laser diodes that initiated the development of improved LT systems, which were easier and more reliable to handle and operate [5.540].

Electronics

When the photomultipliers detect the scattered light pulses from particles crossing the assigned laser beam in the probe volume, voltage output pulses are generated. To compensate for losses in the signal transfer chain (Fig. 5.143) the photomultiplier output signal must first be amplified. Fast amplifiers are required as, especially at high velocities, the scattered light pulses become very short – pulse widths of only a few nanoseconds.

Depending on the velocity and the size of the particles and the position where they cross the beams the amplitude of the scattered light pulse varies greatly. At low speeds many photons are integrated and strong Gaussian-shaped pulses are generated, whereas at very high speeds only a few single photons are transferred to the photomultipliers. Because of this high dynamic range of pulse amplitude and pulse width signal filters are required by which an optimized signal can be attained when the time-constant adjustment is matched to the specific experimental conditions. The range of the adjustable time constants of the filters is typically 2 ns to 1 µs. With this filter technique the signal-to-noise ratio can be essentially improved, hence allowing a more-effective reduction of background noise.

Discriminators are implemented to transform these signals with varying amplitude into uniform pulses,

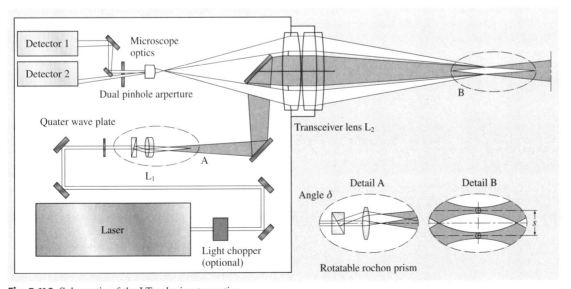

Fig. 5.142 Schematic of the LT velocimeter optics

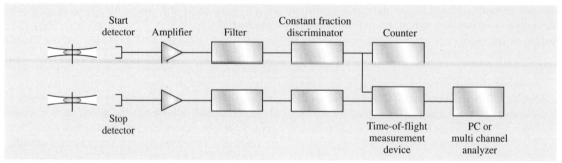

Fig. 5.143 Schematic of the LT velocimeter electronics

which are needed as input signals for precise time-of-flight measurements.

When the output pulses of the discriminator are generated, by employing simple leading-edge triggers, timing errors may occur due to the varying amplitude of the input pulses (Fig. 5.144). Therefore constant-fraction discriminators are preferably implemented, which operate with a special triggering method that allows the output pulses to be generated exactly at the instant of the maximum of the input signal, independent of the amplitude. The constant-fraction technique can only work precisely if an internal delay time is adjusted according to the time constant of the signal. Since this time constant is defined by the aforementioned signal filters the setting of filters and constant-fraction discriminator is combined.

The output signals from the discriminator are the input signals for the further time-of-flight analysis. This can either be performed by cross-correlation techniques or by direct time-of-flight measurements in combination with statistical data analysis.

In the early stages of development of the LT technique time-to-amplitude converters and multichannel analyzers were used (these are very common for time measurements in nuclear physics). Nowadays, fast counters are in use for time-of-flight measurements and statistical analysis is completely PC-based with an integrated custom-developed, dedicated PC card.

A normalization of the measured histogram taken at each angle setting of the plane of the beams is necessary to enable the construction of the desired two-dimensional histogram. The reference value for this normalization is the number of measurement events started. These are indicated by valid start signals delivered by the time-of-flight measuring system and acquired by a preset counter. After having reached the preset number, which has to be selected by the operator depending on the actual measurement conditions, the counter interrupts the data acquisition. The straightforward way of operation is to set the preset number to the same value for all angle settings of the beam's plane.

The measured two-dimensional histogram of data is then used as the input for further data evaluation.

Experimental Operation

The probe volume geometry of a LT velocimeter, the waist diameter d, and the beam separation s are determined by the diameter D of the incident laser beam, the splitting angle δ of the Rochon prism, and the focal length f_1 of the lens L1 (Fig. 5.142), assuming that the imaging scale of the lens system L2 is 1:1.

The Gaussian waist diameter ($1/e^2$ diameter) is

$$d = \frac{4}{\pi}\lambda\frac{f_1}{D}, \qquad (5.297)$$

where λ is the wavelength of the laser beam, and the beam separation is

$$s = f_1 \delta. \qquad (5.298)$$

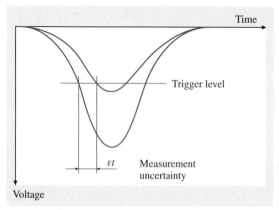

Fig. 5.144 Measurement uncertainty as a function of varying signal amplitude

The angular range $\Delta\alpha$ in the LT system that is capable of detecting successful dual-beam crossings is only a few degrees and is given by

$$\Delta\alpha \sim \frac{d}{s} = \frac{4\lambda}{\pi\delta D} \, . \tag{5.299}$$

The sensitive axial length of the probe volume l depends on the laser power P, the beam waist diameter d, the diameter d_p of the particles considered, the effectiveness of spatial filtering, and the amplitude discrimination of the photomultiplier signals.

For a laser power P of 0.5 W and particles with diameters of 0.5–1 µm it turns out from practical experience that the axial length l is about 10–40 times the beam waist diameter d. That means, for a typical diameter of $d = 10\,\mu\text{m}$,

$$l \approx 0.1\text{–}0.4\,\text{mm} \, . \tag{5.300}$$

While in the case of LD systems the fringe spacing (and this determines the probe volume diameter) is matched to the expected velocity magnitude it is very advantageous in the case of LT system to match the beam separation s to the expected turbulence intensities [5.532]. In this way an LT system can be optimized with respect to the measuring accuracy and time required to perform the measurements. Assuming a constant beam diameter of $d = 10\,\mu\text{m}$, a beam separation of $s = 0.35$ mm is recommendable for turbulence intensities of 3% and the separation should be changed to $s = 70\,\mu\text{m}$ when the flow turbulence reaches 16%. With the reduced beam separation at higher turbulence intensities the corresponding angular capture range $\Delta\alpha$ increases from $\Delta\alpha = 1.5°$ to $\Delta\alpha = 8.0°$ so that the number of successful dual beam crossings (and with it the measurement time) becomes equal in the two different flow cases. Additionally, with the wider angular range $\Delta\alpha$ the angular step width of the beam's plane settings can also be extended so that in both flow cases the total number of angle settings can also be kept constant and likewise the total measurement time. With respect to the measurement accuracy, 10–02 angle settings placed within the angular fluctuation range of the turbulent flow are sufficient. In practice more angle settings are often needed to make the angular scanning range wide enough so that the unknown flow angle at a new measurement position most probably falls within this range. The measured data give a clear indication whether the range was correct and sufficiently wide. Otherwise measurements must be repeated with a modified angular setting range.

Applications in periodically fluctuating flows, e.g., rotor flow analysis in turbomachines, require special

Fig. 5.145 3C-L2F optical head

synchronizing electronics to enable phase-resolved velocity measurements. Either a one-per-revolution or a blade-frequency signal is needed as an input. The synchronizer divides the blade period periodically in up to 64 windows. When a measurement event is indicated by a valid time-of-flight measurement, at that instant of time the window number assigned to the measurement event is identified and the measurement data are then directed to a corresponding storage. After the data collection all 64 storages are filled with data. The data, collected in each individual storage, determine the two-dimensional probability histogram, which is required for further data evaluation. The result is the velocity information (mean value and turbulence intensities) distributed in 64 circumferential positions over the blade period. In principle, measurement can be performed in a single, selected blade passage, however in practice – because of the saving in terms of measurement time – the data from all blade passages are mixed together and further calculations provide representative mean values.

When the rotor blades move through the probe volume, strong light reflections occur and may destroy the

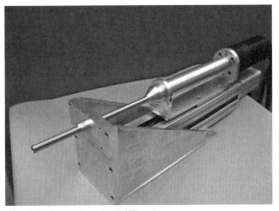

Fig. 5.146 3C-Doppler L2F probe

photomultiplier. In order to avoid this problem either the laser beam is switched off during the short blade-passage time by means of Pockels or Bragg cells, or the photomultipliers are gated by changing the polarization of the high voltage between the cathode and first dynode. During this blade-passage time – a small part of the blade period – the data collection is interrupted.

Three-Component Systems

To date the two-component LT technique has been best suited to overcoming serious access problems, since it is a back-scattering technique with a confocal optical beam path. This is the main reason why the two-component LT technique is well established in the experimental flow analysis of turbomachines, where the optical access problems are most severe.

This statement is not valid when three-component measurements are considered. Standard three-component velocimeters consist of two separate optical units – one or two velocity-component measuring devices – which are angularly displaced and operate on a common probe volume. For reasons of measurement accuracy of the velocity components the displacement angle should be at least 30°, which leads to a solid angle of greater than 40° necessary for optical access. This condition cannot be fulfilled in most turbomachinery applications.

A specially designed three-component LT system [5.540, 545] is on the market today and in use at various European institutions. This system (Fig. 5.145) is set up with two independent two-component LT systems with a tube-type optical head construction. The two tubes are mounted on a mechanical rotation unit inclined to the rotational axis at an angle of 7.5°. The location where the laser beams intersect is on the rotational axis and determines the probe volume. From the two-component data of the two independent LT systems the three velocity components can be deduced. Since the system is designed to adjust to the maximum sensitivity of the radial component measurement automatically, the solid angle required for optical access is only 20°, remarkably smaller than that needed for a standard three-component arrangement.

Even though this three-component system is well suited for turbomachinery measurements in general there are extreme applications where either this small

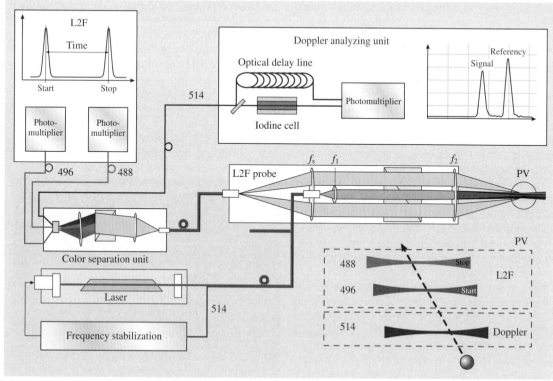

Fig. 5.147 Schematic of the 3C-Doppler L2F device

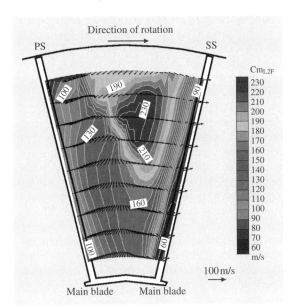

Fig. 5.148 Absolute velocity distribution measured at 0% meridional length just in front of the blades

solid angle for optical access is still too large or where not enough space is available to place and operate the optical head. In these cases only two-component systems can be used because they only need an 11° solid angle for optical access.

Recently a new three-component system was developed, called the 3C-Doppler-L2F, which operates with the same confocal optical setup as the two-component LT system, thus enabling three-component measurements even under difficult conditions, e.g., in centrifugal compressors, to be carried out. This technique combines the principle of the LT method with the principle of the Doppler global velocimetry [5.546, 547]. From the time-of-flight data, the standard two-component measurements provide the velocity vector component in the plane perpendicular to the optical axis. An additional frequency analysis of the scattered light determines the Doppler frequency shift, which represents the velocity component along the optical axis. The Doppler shift is measured by using the frequency-dependent absorption of iodine.

The system was developed for an application in a transonic centrifugal compressor and designed in the shape of an optical probe with an outer diameter of 25 mm and a probe throw of about 100 mm. The 3C-Doppler-L2F probe is shown in Fig. 5.146.

The setup of the 3C-Doppler-L2F device is shown in Fig. 5.147. A frequency-stabilized argon-ion laser with an intracavity etalon is used as a light source. While operating in multicolor mode, the Ar^+ laser was frequency stabilized using the green line ($\lambda = 514$ nm). The laser was fibre linked to the probe head. Lens f_1 collimates the laser beam emerging from the fibre end and guides the multicolor beam to a dispersion prism where the various colors experience different angular deflections, so that with the aid of lens f_2, parallel beams with differing colors and separations are focused in the probe volume (PV).

The multicolor light scattered by the particles traversing these beams is collected by the outer area of lens f_2 and sent through the same dispersion prism, where the various colors are again deflected such that the light produced at the various places in the probe volume is projected on a single focal point by lens f_3 and thus may be coupled into a multimode receiving fibre, which additionally functions as a aperture for spatial filtering. The receiving fibre guides the multicolor scattered light to a color-separation unit where the various colors of the detected light are separated and launched into the three assigned fibres.

The scattering light pulses from the 488 nm and 496 nm laser beams in the probe volume are guided to a L2F processing unit, where the start and stop signals for the time-of-flight measurements are generated. The signals are processed by standard commercially available L2F signal processors and provide the two-velocity component $v_\perp$ and angle α in the plane perpendicular to the optical axis.

The velocity component v_Z along the optical axis is deduced from the Doppler-frequency shift analysis using the scattering light pulses from the 514 nm laser beam in the probe volume. In this Doppler analyzing unit the 514 nm scattering light is collimated and split by a nonpolarizing beam splitter into two beams of equal intensity, One beam is guided through an iodine cell, providing signal light pulses while the other beam is guided through a multimode fibre that serves as an optical delay line. Along this path the reference light pulses are transmitted. Both pulses are detected by a single photomultiplier with a constant time delay determined by the fibre. The amplitude ratio of these pulses is proportional to the transmission $T(v)$ of the iodine cell, from which the scattered light frequency can be derived. The measurement uncertainty of the v_Z-component is ± 0.5 m/s.

The 3C-Doppler L2F system was applied to a centrifugal compressor in order to measure the flow field in the rotating flow channels of a transonic splitter blade impeller of advanced design [5.548, 549]. These flow data are very much desired for the validation of 3-D Navier–Stokes numerical simulations.

In Fig. 5.148 the velocity distribution between two main blades measured just in front of the rotor leading edge is plotted. The three-component absolute velocities are represented by the vector plot of the velocity components within the measurement plane and the color-coded out-of-plane velocity component CmL2F. The triangularly shaped area in the central upper part of the plot is the region where the relative intake velocity is supersonic. The steep gradient on the left side of this region indicates the location of a compression shock where the supersonic relative velocity suddenly changes to subsonic. The splitter blades, which are centered between two main blades (Fig. 5.148), are not shown in this plot since their leading edge is located further downstream at 30% of the meridional length. Their upstream influence on the flow at the location in front of the rotor, however, is presumably indicated by the in-plane flow vectors pointing upwards between the two main blades. Another successful application in a low-pressure turbine demonstrates the outstanding capability of the three-component Doppler L2F system.

Outlook

Nowadays planar velocimetry systems such as the particle image velocimetry and global Doppler velocimetry are common, as they are very efficient and therefore cost saving. These planar systems in comparison to point measurement techniques also supply additional information such as visualization of flow structures, and for this very reason are very much in use today. However for planar systems the problem of optical accessibility is more severe and cannot be solved in some applications. Therefore we are convinced that point measurement techniques, and LD and LT velocimetry, will continue to be of importance in the future, especially in turbomachinery flow research.

5.4 Molecular Tagging Velocimetry (MTV)

Molecular tagging velocimetry (MTV) is a whole-field optical technique that relies on molecules that can be turned into long-lifetime tracers upon excitation by photons of appropriate wavelength. These molecules are either premixed or naturally present in the flowing medium (i.e., unseeded applications). Typically a pulsed laser is used to *tag* the regions of interest, and those tagged regions are interrogated at two successive times within the lifetime of the tracer. The measured Lagrangian displacement vector provides the estimate of the velocity vector. This technique can be thought of as essentially a *molecular* counterpart of particle image velocimetry (PIV), and it can offer advantages compared to particle-based techniques where the use of seed particles is not desirable, difficult, or may lead to complications. Figure 5.149 illustrates one implementation of the technique where the particular tracer used is a water-soluble phosphorescent supramolecule. A planar grid of intersecting laser beams, formed from a pulsed excimer laser (20 ns pulse width, 308 nm wave-

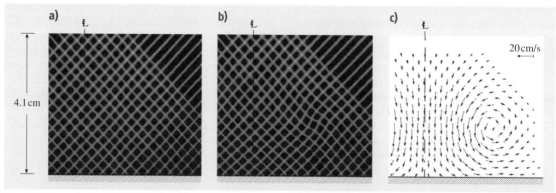

Fig. 5.149a–c Typical MTV image pairs and the resultant velocity field [5.550]. The flow shown is from a vortex ring impacting on a flat wall at normal incidence. The axis of symmetry is indicated by the dashed lines. (**a**) The grid imaged 1 μs after the laser pulse. (**b**) The same grid imaged 8 ms later. (**c**) The velocity field derived from (**a**) and (**b**)

length), turns on the luminescence of the supramolecules that are premixed in a water flow of a vortex ring approaching a solid wall at normal incidence [5.550]. The displacement of the tagged regions is determined, in this case, using a direct spatial correlation method. The conventional planar imaging shown in Fig. 5.149 provides information on two components of the velocity vector, the projection onto the viewed plane. Stereo imaging can produce the complete three components of the velocity vector [5.551].

While the earliest use of MTV can be traced back at least three decades, this technique has seen significant advances over the past 15 years [sometimes under alternate titles such as laser-induced photochemical anemometry (LIPA) and flow tagging velocimetry]. The ability to non-intrusively tag the molecules in a flowing medium and observe their subsequent evolution offers other possibilities besides velocimetry in fluid flows. Depending on the tracer properties and the method of implementation, molecular tagging can be used for flow visualization, simultaneous quantification of velocity and scalar (concentration or temperature) fields, and novel methods in the study the Lagrangian evolution of scalar fields, entrainment, and mixing. As we describe the various elements involved in using molecular tagging techniques, and in particular molecular tagging velocimetry, we draw attention to several review articles that have appeared in the past [5.552–5.555], in addition to a special issue of *Measurement Science and Technology* on this topic [5.556].

5.4.1 The Photochemistry of MTV: Molecular Tracers and Chemical Mechanisms

A molecular complex is suitable for molecular tagging applications if its lifetime as a tracer is long enough relative to the flow convection time scale to allow sufficient displacement of the tagged regions during the interrogation time. The photophysical properties of the tracer, in turn, dictate the wavelength and number of photon sources needed to create the tracer (i.e., the tagging process) and those needed for interrogation.

All MTV techniques are based on the chemistry of molecules in electronic excited states. Normally residing in dark or ground states, molecules enter excited states when they absorb energy from the environment. Electronically excited molecules are generally very reactive, eventually returning either to their original ground state or to a new ground-state molecule. The departure of a molecule from its excited state may occur in the absence of a reacting partner (called an intramolecular decay process) or with the help of a second molecule (called an intermolecular decay process), as shown schematically in Fig. 5.150.

Figure 5.150a depicts the intramolecular processes that govern the decay of an isolated excited state molecule to its ground state. Competing photon emission (radiative) and thermal (nonradiative) relaxation pathways are described by the radiative (k_r) and nonradiative (k_{nr}) rate constants, respectively. The radiative rate constant is an intrinsic property of the molecule and it reflects the probability of the excited state emitting a photon of a given frequency; k_r is directly related to the ability of the molecule to absorb light, described by the oscillator strength or the molar absorptivity constant [5.557–5.559]. k_{nr} encompasses all decay pathways that do not lead to photon emission. Nonradiative decay most typically entails the conversion of an excited state's electronic energy into high-energy vibrations of the ground-state molecule [5.560–5.562]. Vibrational relaxation to the equilibrated ground state molecule (M) is accompanied by the concomitant release of heat ($k_{nr}^{(M)}$); alternatively, the excess vibrational energy may be manifested in the creation of a different ground-state molecule ($k_{nr}^{(P)}$), called the photoproduct. The

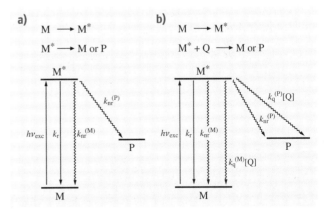

Fig. 5.150 (a) Simple excited-state diagram for intramolecular decay processes of a molecule in an excited state M*. The radiative rate constant k_r and nonradiative rate constant $k_{nr}^{(M)}$ describe the conversion of M* to its ground state M by photon and thermal emission, respectively. $k_{nr}^{(P)}$ is the rate constant describing the return of M* to a new molecule P. (b) The excited state diagram including intramolecular and intermolecular decay processes of a molecule in an excited state M*. The quenching rate constants $k_q^{(M)}$ and $k_q^{(P)}$ describe the intermolecular conversion of M* by quencher Q to its ground state M or a new ground-state molecule P

overall nonradiative decay rate constant k_{nr} is simply $k_{nr} = k_{nr}^{(M)} + k_{nr}^{(P)}$.

The interplay between k_r and k_{nr} determines the fundamental properties of molecules in excited states [5.557, 563]. The intensity of the luminescence I_0 is directly dependent on the luminescence efficiency, usually called the emission quantum yield Φ_e, which is the ratio of the number of photons emitted per photons absorbed. The emission quantum yield is directly related to k_r and k_{nr} by the following expression

$$I_0 \sim \Phi_e = \frac{k_r}{k_r + k_{nr}} = k_r \tau_0 , \qquad (5.301)$$

where τ_0 is the observed lifetime of the electronic excited state, and

$$\tau_0 = \frac{1}{k_r + k_{nr}} . \qquad (5.302)$$

Nonradiative relaxation pathways usually dominate (i.e., $k_{nr} \gg k_r$) and, as can be seen from (5.301), molecules under this condition will remain dark upon excitation. However, when thermal emission is inefficient with regard to photon emission (i.e., $k_r \gg k_{nr}$), excited states will literally light up or luminesce.

Two types of luminescence are of significance to MTV techniques: fluorescence and phosphorescence. In the ground state, virtually all molecules reside in electronic states in which the electrons are paired. For this situation, a molecule is said to be in a singlet ground state. Quantum-mechanical rules dictate that the most probable transitions to excited states will be to other singlet excited states. Fluorescence is the radiative process that accompanies the relaxation of a molecule from a singlet excited state to its singlet ground state. Because singlet-to-singlet transitions are quantum-mechanically allowed, they occur with a high probability. Hence k_r is large, ranging from 10^8 to 10^{10} s^{-1}. From the perspective of (5.301) and (5.302), there are two consequences of such large values of k_r. Nonradiative decay channels k_{nr} are generally not competitive with k_r (i.e., $k_r \gg k_{nr}$). As deduced directly from (5.301), the emission intensity of a fluorescent excited state is therefore intense. Moreover, the lifetime of a fluorescent excited state will also approach that of k_r^{-1} (5.302); thus fluorescent lifetimes are short, on the order of 1–100 ns with most fluorescent lifetimes being $\approx$ 2–10 ns.

Conversely, phosphorescence is long-lived because the transitions between the excited state and ground state are forbidden. In a phosphorescent excited state, the electrons are triplet-paired or aligned with each other. Here the absorption process occurs between the singlet ground state and an excited state of different spin. Quantum-mechanically this is a forbidden transition, and hence k_r is small. The radiative decay rates can therefore approach seconds and the phosphorescence lifetimes can be large (owing to the small k_r). However, the phosphorescence intensity from the spin forbidden excited state may be significantly weaker than fluorescence because k_{nr} will become competitive with k_r on the micro- to millisecond time scale. Moreover, because the absorption cross section of the triplet state is small, it is usually not produced by direct absorption. Alternatively, the triplet excited state is usually populated from a correspondent singlet state by the intersystem crossing (ISC) process. Because the triplet has a lower energy than the singlet, the wavelength for phosphorescence is significantly red-shifted from the absorption profile. Thus, whereas the wavelength difference between the absorption and emission profile (known as the Stokes shift) may be relatively small for fluorescence, a large shift between the wavelengths of absorption and phosphorescence is typically observed.

Equations (5.301) and (5.302) only consider the intramolecular decay processes of the excited-state molecule in the absence of external reactants. Because electronically excited molecules are highly energetic, they are susceptible to intermolecular reactions in which M* physically or chemically interacts with species in its environment (called quenchers Q) [5.564] to again return to ground state M or to another molecule P (Fig. 5.150b). When quenching processes are present, (5.301) and (5.302) must be modified by adding $k_q[Q]$ to their denominators,

$$I \sim \Phi_e = \frac{k_r}{k_r + k_{nr} + k_q[Q]} , \qquad (5.303)$$

$$\tau = \frac{1}{k_r + k_{nr} + k_q[Q]} , \qquad (5.304)$$

where k_q is the quenching rate constant and the concentration of the reacting partner [Q] accounts for the bimolecular nature of the quenching process. It follows from this formalism that quenching pathways are dissipative and their presence will diminish the luminescence intensity and shorten the excited-state lifetime. The Stern–Volmer relation quantitatively defines the attenuation in luminescence lifetime and intensity under quenching conditions as

$$\frac{I_0}{I} = \frac{\tau_0}{\tau} = 1 + \tau_0 k_q [Q] , \qquad (5.305)$$

where I_0, I, and τ_0, τ are the luminescence intensity and lifetime in the absence and presence of Q, respectively. Owing to the short lifetimes of singlet excited states, significant concentrations (typically 0.01 molar or greater) of Q are required to quench fluorescence. This is not the case for phosphorescence. The long lifetimes of phosphorescent excited states makes them especially susceptible to quenching at extremely small concentrations of Q.

Most of the imaging techniques based on molecular tagging may be understood in the context of the simple relations defined by (5.301–5.305). Below is a description of the chemistry behind these techniques.

The Different Mechanisms of MTV

The four basic mechanisms that encompass current MTV techniques are shown in Fig. 5.151. Sometimes referred to as laser-induced photochemical anemometry (LIPA), mechanism A describes measurements based on the image produced by a photochromic dye and is the only MTV technique that relies on measuring absorbance. Light excitation produces a high-energy form of the dye, which is usually more strongly absorbing than the ground-state molecule M (usually from transparent to opaque); the dye molecules in the flow are therefore tagged by the darkened image. Mechanism B describes the Raman excitation plus laser-induced electronic fluorescence (RELIEF) technique. In a RELIEF experiment, a high-energy form of the molecule, M′ (specifically vibrationally excited oxygen), is also responsible for creating the image. But unlike in LIPA, M′ emits photons upon subsequent irradiation. The RELIEF image is therefore revealed by detecting luminescence rather than absorbance. Mechanism C relies on the production of a luminescent molecule P upon excitation of a ground-state molecule M. For the technique with the moniker photoactivated non-intrusive tracing of molecular motion (PHANTOMM), a laser dye is produced. Variants of the technique arise from the production of different P species: ozone in the oxygen tagging velocimetry (OTV) technique, hydroxyl radicals in the hydroxyl tagging technique (HTV), and two techniques based on the photorelease of NO. In each case, a luminescent image is produced upon excitation with a *write* laser and the image is detected by irradiating with a subsequent *read* laser pulse. Within mechanism C we also include the reverse approach such as photobleaching where, instead of releasing a luminescent tracer, a non-luminescent species is produced from fluorescent dyes, thereby creating a *negative* image. Mechanism D is the most straightforward of the MTV techniques. An excited-state molecule M* is produced by excitation. An image is detected by directly monitoring the long-lived phosphorescence from M* upon its radiative return to its ground state M. Details of each of the MTV techniques are presented next.

Mechanism A: MTV by Absorbance. Originally introduced by *Miller* [5.565] for visualizing flows, photochromic chemicals were extensively used first by Hummel and his group for velocity measurements [5.566–569] and subsequently were more-thoroughly applied to the MTV technique by *Falco* and *Chu* [5.570] under the acronym LIPA. In a photochromic process, laser excitation of molecule M produces an electronically excited molecule M*, which then relaxes by nonradiative decay ($k_{nr}^{(M)}$ and $k_{nr}^{(P)}$). Whereas most of the vibrational energy is released as heat ($k_{nr}^{(M)}$), some vibrational energy is channeled into the making and/or breaking of bonds ($k_{nr}^{(P)}$) to produce a high-energy form

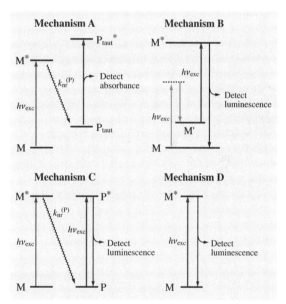

Fig. 5.151 The different mechanisms of MTV. M and M* designate a molecule in its ground and electronic excited state, respectively, and P indicates the formation of a new molecule. Consistent with Fig. 5.150, *solid arrows* describe radiative transitions and *wavy arrows* indicate nonradiative transitions. The excitation lines are color-coded to emphasize the number of different frequencies typically needed for each experiment. The different diagrams summarize the photochemistry and photophysics that give rise to the various MTV techniques under the various acronyms of LIPA based on photochromism (type A), RELIEF (type B), PHANTOMM, OTV, HTV (type C), and direct luminescence (type D)

of the molecule P_{taut}; in chemical terms, this high-energy molecular species is called a tautomer (the same molecular framework connected by a different bonding pattern). This tautomeric process is adaptable to MTV because the absorption spectra of M and P_{taut} are different. The color change accompanying the M-to-P_{taut} conversion can be detected by exciting with a white-light source. In the context of MTV description, the long-lifetime tracer in this case is the newly produced P_{taut}, which can persist for several seconds to minutes.

The photochromism of commercially available substituted benzyl pyridines and spiropyran dyes have been used for most velocimetry measurements to date. The structure of these dyes and their overall photochemistry are depicted in Fig. 5.152. The nature of the X_i substituent on either tracer may be varied to tune the precise spectroscopic and photophysical properties of the dye. In the benzyl pyridine and spiropyran systems, M is colorless to slightly yellow, whereas P_{taut} is dark blue. This color change results from the increased conjugation (the alteration of single and double bonds) of the molecular framework upon conversion to the high-energy tautomer. For the molecules on the left side of Fig. 5.152, electrons in the π system of the dye (those electrons associated with the alternating single and double bonds) are confined to a single six-membered ring (highlighted by the dark outline). Excitation to M* followed by subsequent nonradiative decay causes the making and breaking of bonds such that the π system is extended. The increase in the conjugation allows the electrons to delocalize over a longer part of the molecule, resulting in a red shift of the absorption spectrum from the ultraviolet to the visible spectral region as explained by quantum-mechanical particle-in-a-box formalisms [5.571]. Use of photochromic chemicals requires two photon sources. When solutions of the photochromic tracers are irradiated with a *write* laser, typically a UV laser (e.g., with a wavelength of 351 nm), P_{taut} is produced and an opaque image is generated along the path of excitation. A second light source (usually a white-light source) is then needed to see or *read* the image.

There are several advantages of molecular tagging by a photochromic tracer. The nonradiative conversion from M* to P_{taut} occurs rapidly, within the duration of a nanosecond laser pulse, and hence the image is formed promptly upon laser irradiation. The conversion of P_{taut} back to M occurs reversibly so the tracer is reusable. Moreover, P_{taut} persists for several seconds to minutes before thermally converting back to M. Thus, flows moving at slow speeds may be studied and the photochromic tracer can be used repeatedly. Though yet to be implemented in fluid flow studies, there are numerous other photochromic imaging systems offering a wide range of available chemical and spectroscopic properties. A purported drawback of this mechanism is the insolubility of these photochromic dyes in water. While it is true that most studies using photochromics have been performed in organic liquids (typically kerosene), relatively minor chemical modifications to many photochromic dyes would enable them to dissolve easily in water [5.572].

Despite these advantages (i.e., long lifetime, reusable), there is one drawback of photochromic systems that limits their overall utility for MTV. Because the image is produced by a change in absorbance, the difference between the incident and transmitted light must be measured, a task which greatly complicates data collection and analysis. Indeed, emitted light (against a black background) is more easily and accurately detected than transmitted light, and consequently images based on luminescence are more amenable to MTV. For this reason, the major advances in the MTV technique in recent years have been made with the luminescence-based imaging schemes of mechanisms B, C, and D. Nevertheless, MTV with photochromics have been used effectively in fluid-dynamics studies [5.570, 573–579].

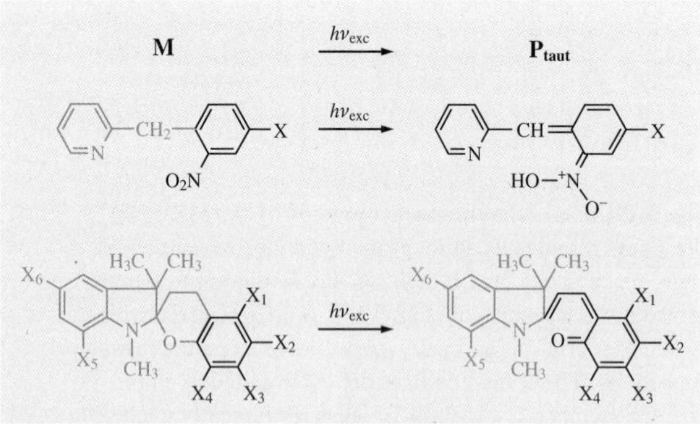

Fig. 5.152 Photochemical transformations of two popular photochromic dyes used for MTV: the substituted benzyl pyridines (*top*) and spiropyran dyes (*bottom*). The molecules are labeled with respect to the M and P_{taut} designations of Fig. 5.150 (mechanism A). The *bold lines* show the molecular framework responsible for the color change upon laser excitation

Mechanism B: MTV by Vibrational Excited-State Fluorescence. The Raman excitation plus laser-induced electronic fluorescence (RELIEF) technique was designed by *Miles* and coworkers [5.580–584] for molecular tagging in unseeded air flows. The technique is schematically described in Fig. 5.151 (mechanism B).

The oxygen in the air flow is tagged by a laser to produce oxygen in a vibrational excited state, which is long-lived and its displacement can subsequently be interrogated, or *read*, by exciting to an electronic excited state that is emissive. The mechanism is ingenious in its design because the long-lifetime tracer needed to follow the flow is derived from the vibrational excited state of oxygen; the short-lived fluorescence of the electronic excited state of oxygen is used only to interrogate the vibrationally excited molecule.

RELIEF is the most spectroscopically sophisticated MTV technique. The spectroscopic underpinning of RELIEF is based on the stimulated Raman effect [5.585], which is familiar to most as a method to Raman-shift the wavelength of light from a laser pump source. In normal Raman scattering, laser irradiation of frequency ν_0 results in the spontaneous population of a vibrational excited level. The populations are small, however, and vibrational excited states are not produced in sufficient concentration to be viable for velocimetry. At high laser irradiances (electric fields of $> 10^9$ V/m) nonlinear effects dominate, such that a substantial fraction of the exciting radiation of frequency ν_0 is converted into radiation of frequency $\nu_0 \pm n\nu$, where ν is the vibrational frequency of the molecule. Most of the power (sometimes as much as 50%) is transferred to the first Stokes line, $\nu_0 - \nu$. In the context of Fig. 5.152, the exciting nonlinear radiation $\nu_0 - \nu$ pumps the system from its ground to its first vibrational level by driving the system through a virtual excited state (indicated by the dashed lines). For oxygen, the process involves the population of the first vibrational level ($\nu = 1$) of the $^3\Sigma_g^-$ electronic ground. The formation of the long-lifetime tracer, i.e., the vibrationally excited oxygen, is rapid and occurs within the laser pulse of order 10 ns. This vibrational level can be very long-lived, approaching 27 ms at 1 atm [5.586]. However, for MTV applications, the usable lifetime is in the 100 ms range owing to collisional deactivation of vibrationally excited oxygen by water and carbon dioxide [5.587]. In most applications to date, the time delay between tagging and interrogation has been of order 10 μs or less. From the $^3\Sigma_g^-(\nu = 1)$ vibrational excited state, oxygen can be excited to its $^3\Sigma_u^-$ (or sometimes referred to as the B) excited state (responsible for the famous Schumann–Runge bands of oxygen [5.588] from which 200–400 nm fluorescence may be observed for interrogation purposes.

The requirement of three frequencies (i.e., two to tag, one to interrogate) makes the RELIEF technique challenging to implement. The power transferred to the first Stokes line at $\nu_0 - \nu$ is related exponentially to the incident laser power. The intense light at the difference frequency needed to produce sufficient concentrations of the vibrationally excited oxygen is provided by mixing the 532 nm second harmonic of a Nd:YAG laser and the 580 nm output of a dye laser, or a Raman-cell frequency shifter, also driven by the 532 nm line of the same Nd:YAG laser. For interrogation, the 193 nm output of an ArF excimer laser is needed to promote the vibrationally excited oxygen molecules into the fluorescent Schumann–Runge manifold. On the detection side of the experiment, the overall imaging of the flow is complicated by the fact that the $^3\Sigma_u^-$ fluorescent excited state of oxygen is predissociative, leading to the homolytic cleavage of the oxygen–oxygen bond [5.589]. Bond cleavage is of course a nonradiative decay process, hence the overall quantum efficiency for fluorescence is small. In addition, the electronic communication between excited and ground states is poor (owing to poor Franck–Condon factors) so that the efficiency for fluorescence is further compromised. Despite these challenges, MTV measurements with RELIEF have been successfully carried out in gas-phase, oxygenated flows [5.581–583, 590, 591].

Mechanism C: MTV by Photoproduct Fluorescence. The photoactivated non-intrusive tracking of molecular motion (PHANTOMM) method is based on tracers that would otherwise luminesce if not for the attachment of a deactivating, or caging, group. First introduced by *Lempert* et al. [5.592] for liquid-phase flows, the overall technique relies on two sources of photons. Laser excitation leads to the photolytic removal of the deactivating group to produce a fluorescent dye with high emission quantum efficiencies. In the context of MTV description, the long-lifetime tracer in this case is the uncaged laser dye, which can persist for a very long time. The laser dye is spatially produced only where the *write* laser has excited the flow. The displacement of the tagged regions is revealed by illuminating the flow with a *read* laser to reveal the fluorescence from the uncaged dye. The uncaged dye can be tracked indefinitely since the photolytic removal of the deactivating group is irreversible. For this reason, the PHANTOMM technique can accommodate very slow flow speeds and has been utilized in several novel applications [5.593–599].

An obvious question is why does the laser dye need to be photogenerated in the first place? Returning to the discussion of Sect. 5.4.2, fluorescence is short-lived; thus a fluorescent image that is directly produced by laser excitation will exist for only 1–10 ns. The glowing image is too short-lived to lead to sufficient displacement over the convection timescale of most flows. The PHANTOMM technique overcomes this limitation by uncaging the fluorescent dye. From a chemical perspective, PHANTOMM is the luminescent counterpart to LIPA. In both techniques, an initial excitation pulse is used to produce a tracer, which can be followed in time; with LIPA the tracer is absorbing and hence tracked by the detection of transmitted light, whereas in PHANTOMM the tracer is tracked by detecting emitted light.

The detection of a fluorescent image offers several advantages to the PHANTOMM technique. Because the transition between the singlet excited state and singlet ground state is allowed for strongly absorbing materials such as organic laser dye molecules, k_r approaches $10^9 \, s^{-1}$. This rate of radiative relaxation is so large that the nonradiative decay channels k_{nr} are generally not competitive (i.e., $k_r \gg k_{nr}$). From (5.301), the emission intensity of the PHANTOMM image is therefore intense. Also because fluorescence lifetimes are short [on the order of ≈ 1–10 ns (5.302)], the excited state (and hence the tagged region in the PHANTOMM technique) is not readily quenched by oxygen. This can be understood within the context of (5.305). The quenching rate for oxygen is diffusion controlled (i.e., $k_q = 10^9 \, mol^{-1} s^{-1}$), and its concentration in solution is typically 10^{-3} mol. With an excited-state lifetime of $\approx 10^{-9} \, s^{-1}$, a maximum quenching of 0.1% will be observed. Thus the fluorescence image is not affected by air, a fact long known to the practitioners of laser-induced fluorescence (LIF). The short lifetime of the fluorescence of uncaged dye offers another significant advantage for increasing signal gain. Once uncaged, a dye can be excited multiple times, allowing for multiple photons to be detected from an individual activated dye over time durations that are short with respect to fluid motion. *Lempert* et al. [5.592] have estimated that the photogenerated fluorescein dye (with its natural radiative lifetime of 4.5 ns and quantum yield of 0.90) can emit 150 photons over the duration of a conventional flashlamp-pumped dye laser (FWHM = $2 \, \mu s$), which is a time scale that is frozen with respect to fluid motion in most liquid applications.

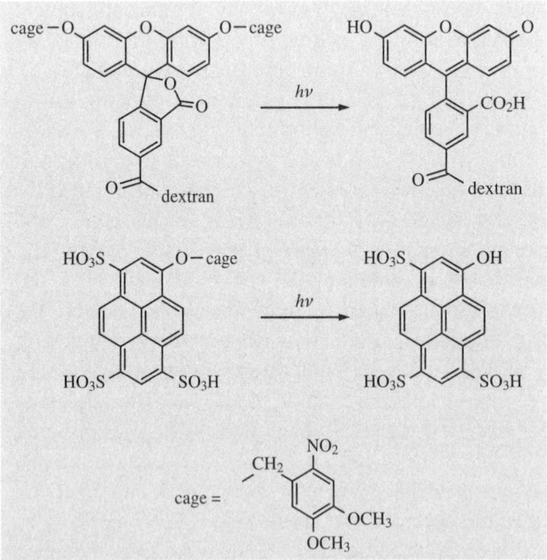

A variety of caged fluorescent dyes is currently available commercially from Molecular Probes, Inc. (Eugene, Oregon, USA). Two such tracers are shown in Fig. 5.153. In each case the dye is modified with a methoxynitroaromatic deactivating, or caging, group. The covalent attachment of the deactivating group renders the latent fluorescent dye nonemissive. The tracers can be activated with the third harmonic of a Nd:YAG laser (355 nm) to produce dyes with luminescence properties of fluorescein and pyrene. Though the precise mechanism for the photolytic removal of the deactivating group has yet to be determined, it likely occurs via well-known resonance structures involving the nitro group to yield an aldehyde through a photochemistry that appears to be initiated by π to π^* excitation of the deactivating group [5.600]. However, even though the excitation light may be absorbed within the appended group, it cannot be contained there and it is free to flow to all bonds in the molecule on vibrational time scales (sub-picosecond). The nonspecificity of the excitation process means that the photolytic removal of the group does not occur immediately upon laser excitation; specifically, the deactivating group falls off the dye on millisecond time scales [5.592], which establishes the upper limit for the flow speeds that can be investigated. Moreover, as emphasized in Fig. 5.153, the luminescent dye is produced irreversibly by cleavage of a covalent bond. Thus the overall tracer can be used only once. A significant advance to the overall approach will be creation of photoactivated dyes that can be uncaged promptly and reversibly.

Interestingly, the PHANTOMM concept, though not so named at the time, was first implemented by *Boedeker* [5.601], and further utilized by *Goss*

Fig. 5.153 Chemical structures of two PHANTOMM tracers before and after photolytic cleavage of the deactivating group. The photochemistry of the top caged tracer produces a fluorescein dye, whereas that of the bottom caged tracer yields a pyrene fluorophore

et al. [5.602], using a much simpler tracer molecule than those shown in Fig. 5.153. Water will photodissociate to OH and H radicals upon two-photon excitation at 248 nm [5.603]. Subsequent excitation with a second laser at 308 nm yields the fluorescence characteristic of the Q_1 transitions [5.604] of the OH radical. With regard to the PHANTOMM construct, the caged tracer is water, the deactivating group or cage is an H radical, and the fluorescent dye is the OH radical. The need for two-photon excitation to release the OH radical makes the technique difficult to implement. Nevertheless, Boedecker's approach permits gas-phase flows to be imaged by the PHANTOMM technique with the caveat that the OH radical tracer has a short lifetime and thus is restricted to high-speed flows. Related studies by *Goss* et al. [5.605] have expanded Boedecker's approach to permit high-speed combusting flows to be imaged by luminescence from C, C_2, and CN tracers produced by the photolysis of hydrocarbons.

Under the acronym hydroxyl tagging velocimetry (HTV), Pitz and coworkers have advanced the use of OH radical fluorescence by replacing the two-photon *write* pulse with a one-photon pulse [5.606–611]. In this adaptation, single-photon dissociation of vibrationally excited water molecules to produce the OH radical is achieved by the 193 nm excitation wavelength of an ArF excimer laser. The tagged regions are imaged with the OH radical fluorescence excited by the 248 nm wavelength of a KrF excimer laser. Though not yet implemented, the 308 nm excitation frequency of a XeCl excimer laser should also be suitable to image the OH radical fluorescence. The technique has been used to image the flow fields of combusting H_2–air flames. At lean concentrations of H_2, OH radical lifetimes approach 1 ms. For rich H_2 flames, however, the lifetime of the OH radical drops to ≈ 200 ns owing to its reaction with H_2 to produce H_2O and a H radical. The advantage of single-photon generation of OH radical from H_2O is that larger spatial regions can be tagged with single-photon excitation. HTV has recently been used in a noncombusting supersonic flow where water was sprayed upstream to provide moist air and adequate OH signal level [5.609].

In the ozone tagging velocimetry (OTV) technique of Pitz and coworkers [5.608, 612, 613] the *caged dye* is ozone, which is created from oxygen O_2 by the same reaction chemistry that produces ozone in the upper atmosphere. The basic photochemistry of the OTV technique is shown below:

$$O_2 + h\nu(\lambda = 193\,\text{nm}) \rightarrow O(^3P) + O(^3P)$$
$$O(^3P) + O_2 \rightarrow O_3,$$

$$O_3 + h\nu(\lambda = 248\,\text{nm}) \rightarrow O(^3P) + O_2^*$$
$$O_2^* \rightarrow O_2 + h\nu_{\text{fluor}}.$$

Two sources of photons are therefore needed for this approach. The 193 nm light from an ArF excimer laser impinging on air causes oxygen to dissociate into two oxygen radicals. These $O(^3P)$ radicals react with undissociated oxygen to produce ozone O_3, which is the long-lifetime tracer and the tagging reagent. After a known delay time, the position of the O_3 tag line is revealed by the application of a 248 nm pulsed laser sheet produced from a KrF excimer laser. The 248 nm excitation undoes the chemistry of the 193 nm laser line, causing ozone to photodissociate to an $O(^3P)$ radical and O_2, which fluoresces via the Schumann–Runge transitions that are accessed by the KrF interrogation laser. The long imaging lifetimes are derived from the ozone tracer, which is stable in air at room temperature. However, at increased temperatures, ozone is unstable with respect to thermal decomposition; thus, the OTV technique is not suitable for hot flows (e.g., beyond 600 K). In addition, the formation of O_3 from O_2 with the 193 nm laser pulse is not rapid and occurs over a time scale of $\approx 20\,\mu\text{s}$. In a situation similar to the uncaging of fluorescein in the PHANTOMM technique, the delay between laser tagging and the generation of enough ozone to obtain a fluorescent image with sufficient signal-to-noise ratio dictates the fastest flow speeds that can be accommodated.

Another tracer molecule that can be produced from uncaging is nitric oxide NO. The fluorescence of NO has long been used in LIF measurements [5.587]. As with all the techniques described in this section, implementation of the tracer for MTV applications requires its generation from a parent molecule. Whereas many other smallmolecular tracers exhibit only short lifetimes due to energy-transfer processes or high reactivity, NO is stable on the millisecond time scale, thus allowing relatively slow flows to be imaged. There are two caged NO molecules that have been employed to date for this MTV mechanism. NO_2 may be dissociated in the focused beam of a 308 nm XeCl laser [5.614]. The NO generated may be imaged by the fluorescence of NO, generated by its excitation with a Raman-shifted 226 nm line from a KrF excimer laser. NO may also be generated from tert-butyl nitrite seeded in an air or nitrogen flow [5.615, 616]. NO is released upon excitation with either a 193 nm ArF laser or a 248 nm KrF excimer laser. The photodissociation cross section of the tert-butyl nitrite at 248 nm is a factor of ≈ 2 smaller than compared to 193 nm, but the pulse energies of the KrF lasers are generally higher. Finally, NO may be generated by irradiating air with an intense laser beam in a method known as air photolysis and recombination tracking (APART) [5.617–619]. The tracer results from the creation of NO from reaction of N_2 and O_2 in the field of illumination of a focused ArF excimer laser beam (193 nm wavelength). This MTV scheme is attractive because the air flow may be imaged without any kind of seeding.

A different implementation of mechanism C for MTV applications relies on the laser-enhanced ionization (LEI) of an atomic species naturally present in the flow or seeded into it [5.620]. The LEI technique relies on the excitation a neutral atom or molecule to an excited state near its ionization limit. A second photon can then be used to ionize the atom or molecule, thus producing a depletion of the neutral species, which is typically fluorescent. The detected region is revealed by an attenuation of the luminescence intensity. Pertaining to the mechanism described in this section, the LEI technique yields a *negative* image, in contrast to the foregoing techniques of this section. Instead of releasing a luminescent tracer, the ionizing laser pulse produces a nonluminescent species. Detection of a quenched luminescence against a bright background complicates the implementation of MTV and limits the overall dynamic range of the technique. Most recently, the LEI technique has been improved by using the secondary laser to produce a luminescent image [5.621]. In this implementation, neutral strontium ions seeded into a flame are ionized by first resonantly exciting ($\lambda_{\text{exc}} = 460.733$ nm) neutral strontium atoms from their ground state ($5s^2$, $^1S_0$) to a Rydberg state ($5s^1 5p^1$, $^1P_1$), followed by subsequent ionization of this excited species to produce strontium ions with the 308 nm output of an excimer laser. Fluorescence from the strontium ions is detected at 407.8 nm upon excitation with a 421.552 nm *read* laser pulse. The precise excitation frequencies are produced from XeCl-pumped dye lasers. The technique is somewhat cumbersome to implement as two pump and two dye lasers are needed to produce the fluorescent image; one laser system for the production of the strontium ions and another to produce the laser-induced fluorescence image. The short lifetime of strontium ions (of the order of 60 μs due to ion–electron recombination processes) makes this approach suitable for imaging supersonic and hypersonic flows.

Yet another alternate implementation of mechanism C includes a photobleaching approach. Continuous excitation of organic dyes reduces the ability of the dye molecules to absorb photons and emit fluorescence, thereby reducing the effective concentration of active dye molecules and a reduction of fluorescence emission from the bleached regions. The detrimental effects of photobleaching have been addressed previously in quantitative LIF studies [5.622–624]. A first-order description of the photobleaching process defines the bleaching effectiveness in terms of a rate that is proportional to the excitation source photon flux and properties of the dye molecule, i.e., absorption cross section and bleaching quantum efficiency [5.622]. As a result, effective photobleaching over a short timescale requires a high-intensity laser source. In comparison with the PHANTOMM concept, this is a reverse approach where, instead of releasing a luminescent tracer, a nonluminescent species is produced from fluorescent dyes, thereby creating a *negative* image. As already described previously, however, detection of reduced luminescence against a bright background complicates the implementation of MTV and limits the overall dynamic range of the approach. Nevertheless, MTV measurements based on photobleaching have been successfully carried out [5.625–628].

Mechanism D: MTV by Direct Phosphorescence. Molecular tagging based on direct photoluminescence is the easiest technique to implement. A single laser is used to produce a luminescent excited state, which is extremely long-lived and its displacement by the flow can be imaged by simply monitoring the emanating luminescence. In the context of MTV description, the long-lifetime tracer in this case is the excited-state molecule itself. The emission of a photon returns the molecule back to its ground state where the tracer may be re-excited; thus the tracer is reusable.

The overall approach of the MTV technique based on mechanism D is similar to LIF, but the tracer has a long enough lifetime to permit sufficient displacement of the luminescence-tagged regions with the flow. The requisite long lifetimes of the tracer necessarily require the luminescence to be phosphorescence-based. This presents two obstacles to the development of MTV tracers for mechanism D. Successful imaging systems with long lifetimes must display the intrinsic property of a small k_r, but k_{nr} must be even smaller if the emission quantum yield is to approach its theoretical limiting value of unity (5.301, 302) and produce intense luminescence intensities. This requirement demands that efficient nonradiative decay pathways be eliminated in the design of imaging reagents. However, this property does not necessarily ensure a viable MTV technique; even if k_{nr} remains small with regard to k_r, long-lived excited-state molecules have enough time to react with other molecules in their environment by quenching. As discussed in the context of (5.303) and (5.304), quenching leads to decreased emission intensity and lifetime, respectively.

For MTV applications based on mechanism D, quenching is an issue pervading most measurements. Water, oxygen, and residual metals in the environment are good quenchers of luminescence. As described by (5.305), the lifetime and intensity of an excited state decrease with increasing concentration of these quenchers. Thus a primary challenge in the design of any successful imaging system is to minimize these quenching pathways. The most problematic quencher in engineering applications is oxygen, owing to its efficiency and prevalence in the environment. Oxygen quenches phosphorescent excited states by an energy-transfer mechanism. The problems with oxygen in the MTV technique can be explicitly demonstrated by considering (5.305). Typically the quenching rate for oxygen is diffusion controlled (i.e., $k_q = 10^9$ mol^{-1}s^{-1}), and its concentration in solution is typically 10^{-3} mol and in air is 10^{-2} mol. According to (5.305), the lifetime and intensity of a tracer possessing an inherent 1 ms photoluminescence lifetime (τ_0) will be attenuated by 10^3 in oxygenated solutions and will be reduced by over 10^4 in air.

Indeed, in the absence of oxygen, MTV measurements can be performed with a variety of organic phosphors. Some of the popular tracers such as biacetyl and acetone, which have often been used for flow visualization and LIF concentration measurements, can also be used as MTV tracers. The potential of biacetyl for this purpose has been known for sometime [5.629–631], and detailed multipoint measurements have been carried out with it [5.632–639]. Biacetyl (also called 2,3-butanedione) has a broad absorption spectrum with maxima at 270 and 420 nm. It is nontoxic and its relatively high vapor pressure (40 Torr at room temperature) allows a molar seeding of $\approx 5\%$. Biacetyl phosphoresces with a high quantum yield (0.15) and with a lifetime approaching 1.5 ms in nitrogen and helium flows. Nevertheless, oxygen completely quenches the green phosphorescence of biacetyl, even when present at only trace levels, rendering it impractical for general use in oxygenated flows such as air. More recently, *Lempert* and coworkers have demonstrated the acetone-based MTV approach and applied it to high-speed microjet flows [5.640, 641].

Long-lived, unquenchable tracers for MTV applications may be constructed by connecting small molecular subunits in intricate ways to lead to elaborate molecular architectures [5.642–645]. These supramolecules contain multiple sites with complementary function – a photoactive center capable of emitting visible light at one site and a docking site for the photoactive center at the other. This strategy centers on manipulating the fundamental parameters governing the luminescence intensity of a photoactive center such that molecular recognition of a constituent at the docking site causes k_r to be much greater than k_{nr} and k_q [5.646]. In this manner, it is possible to develop tracers with long-lived, bright phosphorescence that are not quenched.

Examples of such tracers are those based on lanthanide ions such as europium (Eu^{3+}) and terbium (Tb^{3+}) that have been developed for the study of two-phase flows. These ions exhibit luminescence from transitions involving electrons residing in orbitals buried deep within the ion. For this reason, the luminescent excited state is shielded from the external environment. Although oxygen will physically contact the lanthanide ion upon collision, it cannot communicate with the luminescent excited state owing to poor electronic coupling [5.647, 648] and oxygen quenching of lanthanide luminescence is not predominant. However, water is an excellent quencher of lanthanide luminescence because its O−H bonds act as good thermal receptacles for the excited-state energy of lanthanide-based active sites [5.649]. The deleterious effect of water can be circumvented by encapsulating the lanthanide ion in a molecular cage – a cryptand ligand (Fig. 5.154). The two nitrogens and five oxygens comprising the three straps of the 2.2.1 cryptand ligand occupy seven of the nine coordination sites of the lanthanide ion; the two remaining coordination sites are important

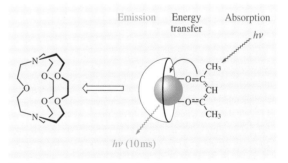

Fig. 5.154 A cryptand molecular cage encapsulates the photoluminescent Tb$_3^+$ ion (indicated by the *shaded ball*) with two remaining coordination sites occupied by light-harvesting groups, thereby allowing the photoactive Tb$_3^+$ center to be indirectly excited

for docking a light-harvesting group described below. The excited states of lanthanide ions are intrinsically bright lumophores, but they show little or no luminescence under direct irradiation because the excited states are spin-forbidden, and therefore they have low absorbance [5.650]. Docking or coordination of a light-absorbing group like acetylacetonate (acac) at the two unoccupied coordination sites opens up a conduit for energy to flow from the light-harvesting group to the lanthanide ion [5.651] and results in the green luminescence of the Tb^{3+} ion, which subsequently decays with its natural lifetime of 1.7 ms. This absorption-energy transfer-emission process [5.652] allows the lanthanide ion to be excited indirectly with intensities 10^3-10^4 times greater than in its absence. Thus this tracer emits brightly, possesses a long lifetime and its excited state is not quenched by oxygen or water. These properties have allowed successful use of this tracer in a study of two-phase flows using the MTV technique [5.552].

Interest in developing tracers with even longer lifetimes has led to a class of supramolecules called cyclodextrins (CDs). Cyclodextrins are cyclic oligosaccharides [5.653], formed from connecting six, seven, or eight D-glucose rings (a common sugar) in a head-to-tail arrangement (α-, β-, and γ-, respectively) (Fig. 5.155). The molecule is cup-shaped with its size determined by the number of sugars in the structure. The CD most often used in MTV applications [5.550] is β-CD, which is constructed from seven sugar subunits, resulting in an outer cup dimension of 15.3 Å and an inner cup dimension of 7.8 Å. Hydroxyl groups of the sugar rings encircle the outer rims of the CD cup, imparting water solubility. To increase the solubility even more, an additional glucosyl subunit is hung off of the rim (called Gβ-CD). The hydrocarbon rings of the D-glucose subunits of the cup walls define a water-repelling or hydrophobic interior that allows it to absorb water-insoluble or hydrophobic compounds. Thus the cyclodextrin is a miniature bucket that dissolves in water but will fill itself with hydrophobic compounds.

The CD cup can be filled with a lumophore such as 1-bromonaphthalene (1-BrNp). This molecule has a bright fluorescence with a 9 ns lifetime. The purpose of attaching the bromine to the molecule is to induce a crossing from the fluorescing, singlet excited state to a triplet ex-

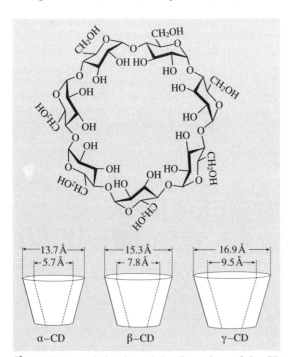

Fig. 5.155 A cyclodextrin and the dimensions of the CD cup for the α, β, and γ forms

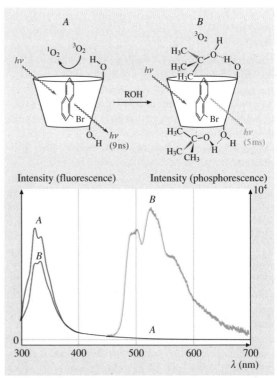

Fig. 5.156 Phosphorescence from 1-bromonaphthalene (1-BrNp) included within a CD is observed only when alcohols are present. In the absence of alcohol (*A*), only blue fluorescence is observed. Green phosphorescence is triggered upon addition of an alcohol such as tert-butanol (*B*)

cited state. The bright green phosphorescence (10 ms is the natural lifetime but the lifetime may decrease to ≈ 5 ms when intense irradiation sources are used) from 1-BrNp is efficiently quenched by oxygen, even when the lumophore is included within a CD cup, as shown in Fig. 5.156 (spectrum A) where only the short-lived blue fluorescence of the 1-BrNp is observed. The phosphorescence can be recovered when certain alcohols (indicated collectively by ROH) are added to the solution. Exhaustive equilibria, photophysical, and kinetics studies for a series of alcohols [5.654] reveal that a ternary complex $(1 - \text{BrNp} \bullet \text{G}\beta - \text{CD} \bullet \text{ROH})$ is formed among the alcohol, CD, and 1-BrNp. The mechanism that gives rise to the triggered phosphorescence by alcohols is shown in Fig. 5.156. In short, the alcohol hydrogen-bonds to the rim of the CD cup, and the aliphatic end of the alcohol flips over the hydrophobic interior of the CD. Accordingly, the alcohol acts as a lid for the CD cup, thereby shielding 1-BrNp from oxygen (Fig. 5.157). Phosphorescence enhancements can be very large, approaching $10^4 - 10^5$, depending on the fit of the alcohol lid to the top of the CD cup [5.655].

The details of the MTV implementation and other applications of the phosphorescent supramolecules described above can be found in *Gendrich* et al. [5.550]. The choice of the alcohol and its concentration strongly influence the phosphorescence lifetime and intensity. For MTV studies, cyclohexanol and neopentanol are the alcohols of choice, and the excitation sources used to date are the 308 nm of the XeCl excimer laser or the 266 nm of quadrupled Nd:YAG laser. Since phosphorescence emission is produced only when all components (Gβ-CD, 1-BrNp, and alcohol) of the ternary complex are present, it is possible to devise in addition to pure velocimetry other applications for molecular-tagging passive-scalar mixing regions, or chemical reaction interfaces, and monitor their Lagrangian evolution. In more-recent studies [5.656, 657] the original glucosyl subunits (i.e., in Gβ-CD) have been replaced by maltosyl subunits (Mβ-CD). The measured properties of both glucose- and maltose-based triplexes are quite similar and the two can be used interchangeably. The three components of these phosphorescent complexes are commercially available; the various alcohols and the lumophore (1-BrNp) are readily found in catalogs of most scientific chemical companies, and maltosyl beta CD is available from Cyclodextrin Technologies Development, Inc. (Gainesville, FL, USA) under the trade name Trappsol.

5.4.2 The Experimental Implementation of MTV: Tagging Methods, Detection and Processing

The specific complexities in the photophysics of the various molecular tracers used for MTV determine the wavelength and number of photon sources needed for tagging and those needed for interrogation. Regardless of these complexities, implementation of MTV for fluid flow studies involves certain common issues that are discussed in this section.

Tagging Methods

Molecular tagging along single or multiple parallel lines is the simplest method of tagging and has been utilized in a large fraction of MTV studies to date. The experiment is typically configured to create tagged lines perpendicular to the primary flow direction and the velocity is estimated from the displacement of these lines. Two examples of single-line tagging are shown in Figs. 5.158 and 5.159. Figure 5.158 depicts the displacement of a tagged line that is created by RELIEF (MTV mech-

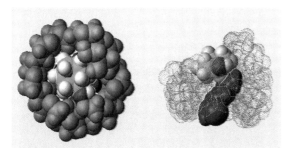

Fig. 5.157 Top and side views of computer-generated molecular model of the tracers from 1-BrNp, CD, and cyclohexanol. The CD is speckled in the side view so that the inside of the cup may be viewed. The purple molecule is 1-BrNp and the multicolored molecule is cyclohexanol. In the top view, the cyclohexanol obscures the 1-BrNp. It is this *blocking action* that prevents oxygen from quenching the phosphorescence of the 1-BrNp

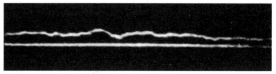

Fig. 5.158 Typical single-line RELIEF image pair taken in a turbulent free air jet with 0 μs (*lower line*), and 7 μs (*upper line*) time delay. The *line* is approximately 1.2 cm long (after [5.590], with permission of Cambridge Univ. Press)

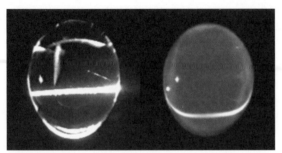

Fig. 5.159 Single-line tagging image pair inside a free-falling 5 mm-diameter water droplet using caged fluorescein [5.658]. The time delay between the initial tagging (*left*) and displaced image (*right*) is 29.5 ms

anism B) in a turbulent air jet. The resulting velocity data have been used to elucidate certain characteristics of turbulent velocity fluctuations [5.590]. Figure 5.159 illustrates the image pair from single-line tagging using caged fluorescein (MTV mechanism C). These image data have been used to characterize the internal circulation inside a falling water droplet [5.593, 594, 658]. An example of multi-line tagging using phosphorescent supramolecules (MTV mechanism D) in a water tunnel is provided in Fig. 5.160. This figure illustrates the trailing-edge region of a NACA-0012 airfoil oscillating sinusoidally at a high reduced frequency of $k = 8.8$ and amplitude of $2°$ [5.659]. The time delay between tagging and interrogation is selected to be long (about 20 ms) to allow a large displacement of the tagged lines so that the flow patterns caused by vortical structures and vortex sheets can be easily observed.

When using line tagging, it is important to recognize that this tagging method allows the measurement of only one component of velocity, that normal to the tagged line. In addition, the estimate of this velocity component has an inherent error associated with it, which is connected with the ambiguity in the unique determination of the displacements of various portions of a (continuous) tagged line. Following the analysis of *Hill* and *Klewicki* [5.660], the error can be written in the form $\Delta u/u = (v/u)(\partial u/\partial y)\Delta t$, where u is the estimated velocity component normal to the tagged line, Δu is the error in the estimated velocity, v is the flow velocity component parallel to the tagged line, and Δt is the time delay between tagging and interrogation. Clearly an a priori knowledge of the flow field is necessary in order to provide an estimate of the error. It can be observed, however, that this inherent error is identically zero only in flows where the velocity component v along the tagged line is zero (i.e., unidirectional flows) or where the velocity gradient along the line $\partial u/\partial y = 0$. In a general flow field where these constraints are not met, the error can be reduced by decreasing the delay time Δt, but it cannot be made arbitrarily small, since Δt needs to be large enough for the resulting displacement of the tagged line to be measured with adequate accuracy. While bearing these issues in mind, it is sometimes possible to take advantage of an a priori knowledge of the flow field under investigation to design the experimental parameters such that the inherent error discussed here becomes minimal compared to other measurement errors. Representative examples of fluid flow measurements obtained with line tagging methods include studies of jets [5.581–583, 590, 640, 641], boundary layers [5.660, 661], pulsatile flow through tubes [5.569], particle-laden coaxial jets [5.662], buoyancy-induced flow in directional solidification [5.663–665], and axial flow within concentrated vortex cores [5.666, 667]. MTV with line tagging has also found extensive use in microfluidic applications where the flow is primarily unidirectional [5.596, 599, 627, 668, 669].

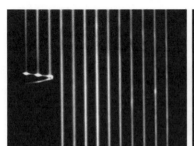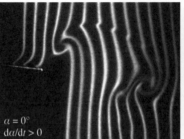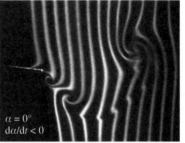

Fig. 5.160 Multiline tagging in the trailing region of a NACA-0012 airfoil pitching sinusoidally with an amplitude of $2°$ [5.659]. The tagging pattern (*left image*) right after the laser pulse is a series of parallel lines with a spacing of 3 mm (about 0.025 chord length). The displaced lines are shown at zero angle of attach during pitch-up (*middle image*) and pitch-down (*right image*)

In order to measure two components of the velocity in a plane unambiguously, the luminescence intensity field from a tagged region must have spatial gradients in two, preferably orthogonal, directions. For single-point velocimetry, this can be achieved using a pair of crossing laser beams and measuring the displacement of the beam intersection; a grid of intersecting laser beams allows multipoint velocity measurements as illustrated in Fig. 5.149. This tagging scheme, first suggested by *D'Arco* et al. [5.573] and later improved upon and utilized by *Falco* and *Chu* [5.570], is now commonly used.

The methods used to date for generating a grid of intersecting laser beams typically rely on standard optics (i.e., beam splitters, mirrors, spherical and cylindrical optics, and beam blockers). For example, the tagging pattern shown in Fig. 5.149 is generated by manipulating the main beam of an excimer laser to increase its aspect ratio using a cylindrical lens, followed by splitting the outgoing laser sheet with a 50:50 beam splitter and redirection by mirrors, and then passing them through a pair of beam blockers [5.550]. Spherical lenses are used as needed to control the spatial scaling of the entire grid pattern. The maximum density of grid intersections is limited by different factors depending on the type of molecular tracer that is used. In some cases the energy requirements to create the tag and detection sensitivity set the limit. *Miles* et al. [5.591] have reported a single intersection created with RELIEF, whereas *Pitz* et al. [5.609] recently achieved a 7×7 grid pattern with their HTV approach using OH (Fig. 5.161). When using phosphorescent tracers for MTV in both gas- and liquid-phase flows, the energy in each tagging beam is typically in the range $1-2$ mJ or less per pulse and the maximum grid density has been limited not by the available laser energy but by the pixel density of the imaging detector arrays used [5.550, 635, 670]. For 640×480 px detector arrays, tagging with a grid pattern in the range 30×30 has been achieved. Recent work with higher-resolution detectors (1280×1024 px array) tags the phosphorescent molecules in a water flow with about a 70×70 grid pattern [5.657].

It is useful to note that tagging by a nonuniform laser illumination in the form of a grid is needed when the molecular tracer is present uniformly in the fluid. This method is only a special case of a more-generalized approach to induce a spatially nonuniform luminescence intensity in a tagged region. For example, the nonuniform passive-scalar field typical of most turbulent flows can sometimes be used as a natural source of luminescence nonuniformity without the need for grid illumination, or further enhancement of the effect of grid illumination. Examples of this type are discussed in *Koochesfahani* et al. [5.553] and *Gendrich* et al. [5.550].

Detection

The most common implementation of MTV uses a single detector; the initial (or reference) tagging pattern is recorded once, usually at the beginning of the experiment, and then the *delayed* images are acquired. This approach works well as long as the initial tagging pattern remains spatially invariant throughout the experiment. Otherwise, any variations in the initial tagging pattern (e.g., due to laser-beam pointing instability, vibration of the optics, nonuniform tracer concentration, etc.) is misinterpreted as flow velocity fluctuations and reduces the measurement accuracy of the instantaneous velocity.

To minimize these potential problems, *Gendrich* et al. [5.550] implemented a two-detector imaging system for their work with phosphorescent tracers. The MTV image pairs are acquired by a pair of CCD detectors that view the same region of interest in the flow through a beam splitter. Using a reference target, the two cameras are aligned to within one pixel, and the remaining residual displacement field between the two detectors is quantified to subpixel accuracy and accounted for in subsequent data processing (for details see [5.550]). Immediately after the pulsed laser fires to tag the molecules in the flow, the first detector records an initial image of the tagged regions. After a prescribed time delay, the second detector records

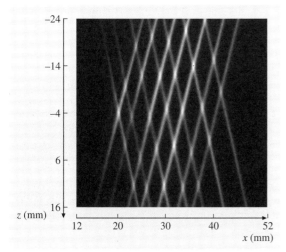

Fig. 5.161 A 7×7 grid of tagged OH lines in a supersonic flow (Mach number 2) over a cavity [5.609]

a second image of the tagged regions displaced by the flow. Such a two-image system offers advantages over the typical single-image system in that no assumption needs to be made a priori about the intensity field in a tagged region and one can properly take into account the variations in the initial tagging pattern. Two-detector imaging has been used in many MTV studies, especially those based on phosphorescent tracers [5.638, 659, 664–667, 669–675].

These one- or two-detector schemes address conventional planar imaging for obtaining information on two components of the velocity vector over the viewed plane (or one velocity component in the case of line tagging). When stereo imaging is used to obtain the out-of-plane velocity component as well (*Krüger* [5.615]; *Bohl* et al. [5.551]), the number of detectors needed increases by a factor of two. With advances in detector technology, the two-detector MTV approach can now be accomplished with a single *dual-frame* camera that allows the acquisition of two images of the tagged regions with a programmable time delay between them [5.657, 666]. This capability not only simplifies the implementation of conventional MTV imaging but also provides for a significant reduction of effort in aligning multiple cameras for stereo imaging [5.666].

Processing

A common method for finding the displacement of tagged lines or grids has been to locate the center of each line by various techniques [5.592, 660, 667]. One approach is to use the best fit to an assumed laser line shape, for example, a Gaussian intensity distribution [5.592]. Another is to use a second-order curve fit following Gaussian smoothing [5.660]. *Hill* and *Klewicki* [5.660] report the rms accuracy in determining the displacement of a tagged line to be 0.35 px. The performance of these methods for finding line centers tend to suffer when the intensity distribution of the tagged regions cannot be assumed in advance, for example, due to a nonuniform tracer distribution, difficulties associated with laser-beam transmission through a flowing medium, photobleaching effects, etc. *Sadr* and *Klewicki* [5.676] have recently explored a spline-based method and report an improvement in subpixel accuracy by about a factor of three.

Another approach for finding the displacement of tagged regions is based on a direct digital spatial correlation technique, and offers certain advantages over the traditional line-center methods. In particular, it is a more-general scheme that is independent of the specific intensity distribution within a tagged region and can accommodate arbitrary tagging patterns including those due to nonuniform scalar mixing fields. The details of this approach and its performance are described in *Gendrich* and *Koochesfahani* [5.677]. A small window, referred to as the source window, is selected from a tagged region in the earlier undelayed image, and it is spatially correlated with a larger roam window in the second delayed image. A well-defined correlation peak occurs at the location corresponding to the displacement of the tagged region by the flow; the displacement peak is located to subpixel accuracy using a multidimensional polynomial fit (see the example in Fig. 5.162). This approach is similar to that used in DPIV processing of particle image pairs and, because of the averaging process inherent in the correlation procedure, it is more robust to the presence of noise and typically more accurate than line-center methods. Based on both experiments and an extensive statistical study, it has been found that the displacement of the tagged regions can be typically determined with a 95% confidence limit of 0.1 subpixel accuracy (i.e., 95% of the displacement measurements are accurate to better than 0.1 px). This corresponds to an rms accuracy of 0.05 px, assuming a Gaussian distribution for error. For high values of image signal-to-noise ratio, the 95% confidence level can be as low as 0.015 px (0.0075 px rms). An example of the application of this procedure is provided in Fig. 5.149; the velocity vectors shown in this figure are *raw* and have not been filtered or smoothed.

The spatial correlation method described above has also been applied to processing line tagging images in an investigation of flows in microchannels [5.668, 669].

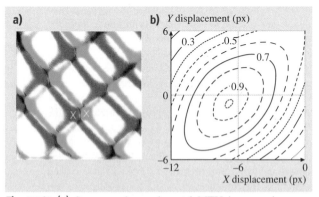

Fig. 5.162 (a) Superposed experimental MTV image pair: *green lines* are the grid at $t = 0$, *red lines* are the grid after $\Delta t = 6$ ms delay. **(b)** Correlation coefficient contours for the indicated intersection in **(a)** (after *Koochesfahani* et al. [5.553])

In this case, a one-dimensional correlation procedure is used and accuracies similar to *Gendrich* and *Koochesfahani* [5.677] are obtained. A modified version of the two-dimensional spatial correlation method of *Gendrich* and *Koochesfahani* [5.677] has been introduced by *Zheng* and *Klewicki* [5.678] where the subpixel accuracy is determined not based on the original multidimensional polynomial fit but using two decoupled polynomials. This approach offers a faster computational algorithm at the expense of reduced accuracy. Other processing methods such as the image correlation velocimetry approach of *Tokumaru* and *Dimotakis* [5.679] have also been applied to MTV image pairs [5.608].

MTV velocity data are obtained originally on an irregularly spaced measurement grid. In order to take advantage of standard data display and processing techniques, the data need to be remapped onto a regular grid with uniform spacing. One method that has been used for this purpose is based on a least-squares fit to low-order polynomials. The details of the procedure and its performance characteristics are given by *Cohn* and *Koochesfahani* [5.680]. The velocity field on a regular grid is then used to compute the vorticity field in the same manner as in DPIV data. Typically a finite-difference approach, for example, second-order central difference, is utilized [5.671, 680]. The accuracy of the measured vorticity depends on the accuracy in the velocity measurement and the data grid spacing. Vorticity measurement accuracies in the range $1-2\,\mathrm{s}^{-1}$ (95% confidence limit) have been reported [5.670, 673].

An example of MTV data obtained by the procedures described above is given in Fig. 5.163. This example shows the instantaneous whole-field measurement of two components of the velocity vector over a plane in a study of unsteady boundary-layer separation caused by the interaction of a vortex ring impinging on a solid wall. This interaction generates an unsteady adverse pressure gradient on the wall, which results in boundary-layer separation and the formation of a secondary vortex. An LIF visualization of this flow right after the formation of the secondary vortex is depicted in Fig. 5.163a (only the right half of the flow is included). Figure 5.163b shows the corresponding instantaneous MTV velocity data (1 mm grid spacing with the first grid 0.5 mm away from the wall) and the computed vorticity field. These data have sufficient resolution to study the behavior of the unsteady boundary layer on the wall and the progression of the boundary-layer separation process [5.671].

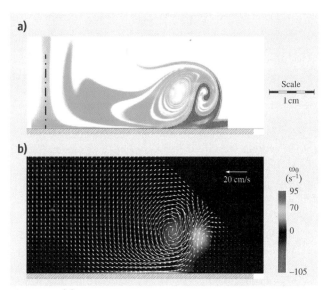

Fig. 5.163 (a) LIF image of the *right-half* of a downward-moving vortex ring impinging on a solid wall. The ring and wall-layer fluids are marked by green- and red-emitting laser dyes, respectively. (b) Velocity and vorticity fields measured using MTV (after *Koochesfahani* et al. [5.553])

5.4.3 Examples of MTV Measurements

The references cited in this work present the development and application of molecular tagging velocimetry in flows over a wide range of speeds, from liquid-phase microfluidic flows with speeds that are measured in μm/s to gas-phase flows at supersonic speeds. The scope of the measurements covers a range from the instantaneous profile of one component of velocity vector along a tagged line to whole-field three-component velocity data over a plane obtained with stereo imaging. Some of the flows that have been successfully investigated include pulsatile flow in tubes, electro-osmotically and pressure-driven microfluidics, internal circulation in droplets, unsteady boundary-layer separation, vortex flows and mixing enhancement, convective flows in directional solidification, flows during intake and compression inside motored internal combustion (IC) engines, flows in engine testing facilities, high-temperature reacting flows, free and wall-bounded turbulent flows, and highly three-dimensional vortex flows with strong out-of-plane motions where the primary flow direction is normal to the tagged plane.

In this section we provide several examples to highlight some of the measurement capabilities that have been achieved with molecular tagging velocimetry based on phosphorescent tracers (MTV mechanism D). These examples consider only velocimetry applications. The unique properties of some molecular tracers have led to novel methods for studying scalar mixing in turbulent flows using caged fluorescent dyes [5.681, 682] and high-sensitivity temperature measurements with phosphorescent supramolecules [5.656]. In addition, MTV has been extended to multivariable mapping such as simultaneous velocity–concentration [5.674] and velocity–temperature [5.657, 683, 684] experiments.

Boundary-Layer Resolved Measurements of Leading-Edge Separation on a Pitching Airfoil

When an airfoil dynamically pitches to high angles of attack the viscous boundary layer near the leading edge eventually separates and causes catastrophic events such as dynamic stall. The evolution of the flow within the boundary layer at the onset of separation has been captured with molecular tagging velocimetry [5.672], providing the first detailed map of the events that occur within the boundary layer near the surface of a pitching airfoil. The results shown in Fig. 5.164 are from measurements carried out in a 60×60 cm water tunnel using a 12 cm chord NACA-0012 airfoil pitching at a constant rate (nondimensional pitch rate $= 0.1$) from 0 to 60° angle of attack. The chord Reynolds number is about 12 000. Since the range of spatial scales in this problem is too large to be captured in a single field of view, these measurements are carried out over multiple fields of view with decreasing size. Repeatability of this flow allows the data to be compiled into a single finely resolved data set for each 0.25° change in angle of attack.

The nondimensional vorticity field computed from this data set is shown in Fig. 5.164 at four angles of attack near the onset of leading-edge separation. These results illustrate the ability of MTV to image fluid flows near walls. Among the details that are resolved in these measurements are the occurrence of a thin reversed-flow region near the airfoil surface, and the eruption of the boundary-layer vorticity away from the wall in a highly localized manner in both time and space. Complementary two-dimensional Navier–Stokes computations have also been carried out at the flow conditions of these experiments. A conclusion from these experimental data is that the process of boundary layer separation occurs over a shorter time scale, and is more eruptive, than that captured by the computations.

Measurement of Buoyancy-Driven Convective Flow During Unidirectional Solidification

Solidifying a binary alloy under off-eutectic conditions is often accompanied by convection in the melt. The convection mechanisms caused by solutal and thermal forces for such conditions can produce inhomogeneities and imperfections such as solute-rich channels in the final fully solidified ingot casting. While the convective phenomena involved in the formation of these imperfections are not completely understood, it is generally accepted that they are best described in terms of complex dynamic fields and that most experimental methods applied to date are poorly suited to measure these fields. One means of learning more about these phenomena has been the use of transparent analogs of metallic alloys. One such analog is the binary aqueous ammonium chloride (NH_4Cl–H_2O) system. Some of the complex features of convective phenomena associated with solidification in the aqueous ammonium chloride include: the vertical growth of a solid–liquid interface with concurrent early development of numerous fine structures of salt fingers followed by the appearance of a small number of plumes ejecting fluid jets from channels in the mushy zone of the growing dendritic crystal mass. These features are illustrated in the shadowgraph images in Fig. 5.165 [5.664, 665].

Measurement of this flow field by particle-based techniques can be problematic since the presence of particles could interfere with the solidification process and create unwanted nucleation sites. In addition, the mushy

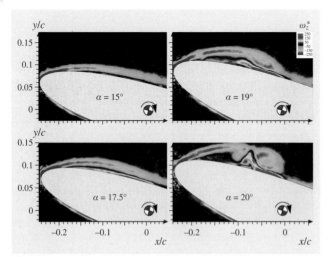

Fig. 5.164 Evolution of the vorticity field at the onset of leading-edge separation on an airfoil pitching to high angles of attack [5.672]

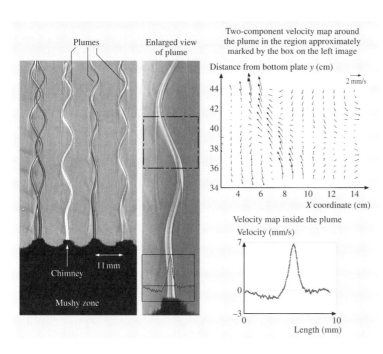

Fig. 5.165 Shadowgraph image of organized plumes above an advancing solidification front and the velocity fields measured by MTV [5.664, 665]. The velocity profile at the bottom of plume is measured using line tagging

zone acts as a very fine-scale porous medium and traps the particles that are contained within the liquid above. As a result, the fluid within the plume is often devoid of particles. The MTV approach using phosphorescent supramolecules have been used to measure the velocity field within and around the plumes [5.664, 665]. The portion of the flow near the bottom of the plume is mostly unidirectional (vertically upward flow) and single line tagging is sufficient to map the velocity profile within the plume. Farther up, however, two-component velocity maps are obtained using molecular tagging by an intersecting grid of laser lines. A unique aspect of these measurements is that the time delay between tagging and interrogation needs to be long due to the slow flow speeds involved. The 60 ms delay time used in these measurements is about a factor of 12 longer than the phosphorescence lifetime (i.e., $1/e$ point) of the tracer used, resulting in a significant reduction of phosphorescence emission intensity and necessitating the use of an intensified camera for detection.

Figure 5.165 shows the measured instantaneous velocity profile across a single plume as it exits the chimney, and the velocity map around the plume farther up. The upward jet-like velocity field within the plume is captured along with a nonsymmetric downward flow around the plume. Note the peak velocity is only 7 mm/s and the width of the plume is about 2 mm.

Measurements in a *Steady Flow Rig* Model of an IC Engine

The steady flow rig configuration is commonly used in the internal combustion (IC) engine research community to study the fundamentals of the intake flow. The particular geometry used here consists of a quartz cylinder of radius $R_0 = 41$ mm, placed axisymmetrically around a nozzle with a valve body placed axisymmetrically inside the jet nozzle. In this case the flow exiting through the valve opening, which simulates the intake flow into an IC engine geometry, is in the form of an

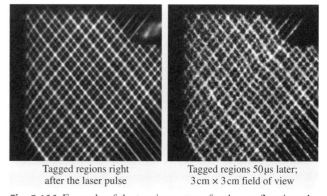

Fig. 5.166 Example of the tagging pattern for the gas flow into the steady flow rig (after *Stier* and *Koochesfahani* [5.635]). Images were acquired by a gated image-intensified camera

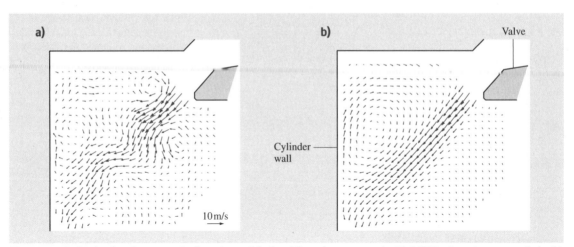

Fig. 5.167a,b Intake flow into a steady flow rig model of an IC engine measured with MTV using nitrogen seeded with biacetyl (after *Stier* and *Koochesfahani* [5.635]). (**a**) The instantaneous velocity field derived from the image pair in Fig. 5.166 using a spatial correlation procedure. (**b**) Time-averaged velocity field based on 320 instantaneous realizations

annular jet. In this study, the valve opening (valve lift) is set at $L = 9$ mm and the maximum intake speed is about 10 m/s. The instantaneous accelerations in the shear layer at the interface between the intake jet and adjacent fluid can be as high as 5000 g, making it difficult to rely on the results of particle-based techniques. The details of this work can be found in *Stier* and *Koochesfahani* [5.635].

Figure 5.166 shows a 3×3 cm field of view in the nitrogen/biacetyl flow being investigated and the regions tagged by a grid pattern generated from an excimer laser ($\lambda = 308$ nm, 20 ns pulse). Part of the valve body and the left wall of the cylinder are visible in the picture. The maximum flow speed in the annular jet entering the cylinder is about 10 m/s. Also shown is an example of the later image of the tagged regions after a 50 μs delay. For this time delay, the maximum displacement of tagged regions is about 8 px (≈ 500 μm). Image pairs such as those in Fig. 5.166 are used to determine the instantaneous radial and axial velocity components in this flow field. An example of the instantaneous velocity field and the structure of the intake flow in this geometry are shown in Fig. 5.167 along with the average velocity field based on 320 realizations. The instantaneous flow map shows a highly unsteady intake annular jet, which has an undulating appearance with opposite sign large scale vortical structures on its two sides. The mean flow map indicates a large-scale region of recirculation in the upper left corner of the engine cylinder, a feature typical of an IC engine flow field. These data have also been used to derive other properties of the flow such as the instantaneous and average vorticity fields and velocity fluctuations [5.635].

Velocity Field During Late Compression in a Motored IC Engine

One of the main obstacles in optimizing combustion in gasoline-fueled internal combustion engines is the large cycle-to-cycle variation in in-cylinder flow and mixing characteristics. Cycle-to-cycle variability puts constraints on the lean limits of combustion. In this study velocity field data are obtained using MTV during late compression of an internal combustion engine, the most critical time of the four-stroke cycle. Such data are highly sought after since the state of the

Fig. 5.168 The optically accessible research engine and a typical MTV grid in the nitrogen/biacetyl flow within the engine cylinder (after *Koochesfahani* et al. [5.639])

flow just before the firing of the spark plug directly influences the subsequent combustion and emission production.

The measurements are conducted in an optically accessible motored research engine shown in Fig. 5.168. The cylinder is made from quartz and the flat-head piston face is modified for optical access through a quartz window. A typical MTV grid generated within the engine cylinder is also shown in Fig. 5.168. Measurements are made at late compression at a position of 270 crank angle degree (CAD), as the piston approaches the top dead

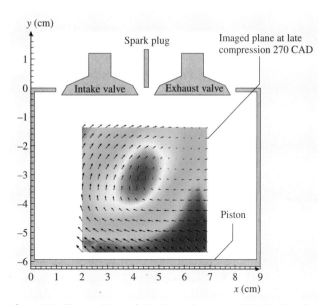

Fig. 5.169 The position of the imaged plane at 270 CAD (after *Goh* [5.638])

center (TDC) of the engine, with the engine running at 600 rpm. For the conditions described here, and the engine compression ratio of nearly 10, the gas temperature can reach a value as high as 600 K. The measurements

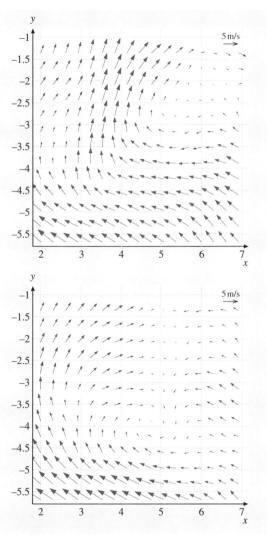

Fig. 5.170 Two instantaneous realizations at 270 CAD showing the large cycle-to-cycle variability of the flow field (after *Goh* [5.638])

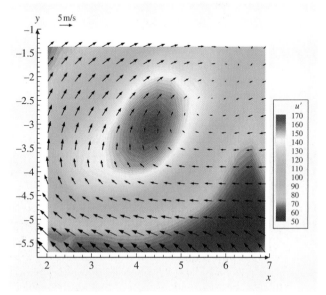

Fig. 5.171 Maps of the ensemble-averaged velocity field and the rms fluctuation of the horizontal component of the velocity u' (in cm/s). Colors denote rms fluctuation level (after *Goh* [5.638])

consist of 500 independent realizations of the velocity map at the same crack angle (270 CAD) at the mid-tumble-plane. The position of the imaged plane over which the velocity maps are obtained is given in Fig. 5.169. The details of this work can be found in *Goh* [5.638].

Two samples of the instantaneous velocity field, for two different engine cycles, are shown in Fig. 5.170. It is clear the flow pattern is significantly different between these two realizations, an indication of the large cycle-to-cycle variability of the flow. The maps of the ensemble-averaged velocity field and the rms fluctuation of the horizontal velocity component are computed from 500 such realizations (Fig. 5.171). As expected for a flow with large cycle-to-cycle variation, the mean velocity field shows little resemblance to the instantaneous field; the local fluctuation level can be higher than the mean by several hundred percent. The data shown in Figs. 5.170 and 5.171 are the first cycle-resolved velocity measurements that use molecular tracers rather than particles to determine flow velocities in a piston-engine assembly. Such measurements have been used to advance the understanding and quantification of cycle-to-cycle variability in an internal combustion engine, and to find methods for reducing it [5.638].

5.4.4 Summary and Conclusions

Molecular tagging velocimetry has seen significant advances as a result of improvements in laser and detector technologies, imaging techniques, data analysis methods, and chemical design and synthesis of novel molecular structures. These advances are expected to continue. The MTV approach has continuously evolved and its use is steadily increasing in both fundamental flow studies and applied engineering measurements.

5.5 Vorticity

Vorticity ($\nabla \times \vec{V}$) is recognized as a primitive variable of considerable interest in fluid dynamics. However, its time-resolved measurement has significantly lagged behind the development of velocity and pressure measurement techniques. The necessity to obtain a measure of the *curl of the velocity*, that is, to accurately measure velocity differences over very small distances, has been a formidable challenge. Two presentations of hot-wire probe techniques are presented to guide the reader in the selection of a suitable measurement technique for a given application.

It is also appropriate to note the capacity to obtain well-resolved, single images of the vorticity component normal to the image plane using molecular tagging velocimetry is presented in Sect. 5.4. Particle image velocimetry (Sect. 5.3.3) techniques can also give a measure of the vorticity although their spatial resolution does not typically allow as fine a resolution of the scales of the vorticity field.

Most of the experimental work carried out in the past on turbulent flows has focused on acquiring velocity information on various types of flow configurations, which has been used to enhance our understanding of the underlying physics of these flows. The velocity field, however, is not well suited for defining and identifying organized structures in time-dependent vortical flows because the streamlines and pathlines are completely different in two different inertial frames of reference. Better understanding of the nature of turbulent structures and vortex motions of turbulent flows, particularly in the high-wavenumber region, often requires spatially and temporarily resolved measurements of gradients of fluid-mechanical quantities such as velocity or pressure. The availability of such measurements will lead to information on important quantities such as the structure function of velocity or pressure, the vorticity, the rate-of-strain tensor and its matrix invariants, and the dissipation rate of turbulent kinetic energy.

Lighthill [5.685] in his wide-ranging introduction to boundary-layer theory is credited as being the first researcher to emphasize the role of vorticity for the better understanding of fluid-dynamics phenomena. His work provided an extensive description of vorticity dynamics in a variety of flows by using vorticity as a primary variable.

Vorticity characterizes the rotation rate of a fluid particle. In the case of constant-density incompressible flows, vorticity is acquired by a pressure gradient introduced at a physical surface. The pressure gradient at the surface is balanced by the stress gradient, which is related to the vorticity flux entering the flow. Thus the existence of vorticity generally indicates that viscous effects are important. In turbulent flows vorticity is the key to improving our understanding of the mechanisms that control the motion of the large-scale structures and the mechanisms by which irrotational fluid lumps gain rotational motion through the action of viscosity and thereby entrain ambient fluid into the flow field.

5.5.1 Optical Techniques in Strophometry – Vorticity Measurements Methods

Although measurements of instantaneous vorticity, particularly in unsteady or separated turbulent flows, are very much in demand, such data are scarce at present due to the extreme difficulties involved in such measurements. In most of the available techniques Taylor's hypothesis is invoked, the validity of which has not yet been justified for the case of separated flows. Recent techniques in hot-wire and hot-films anemometry can measure several components of the vorticity vector [5.686–693]. Hot-wire techniques have been extended to measure three-dimensional vorticity in vortex-dominated flows such as those over delta wings [5.689] and in compressible turbulence with or without the presence of shock waves [5.694–696]. However, these techniques are intrusive and invoke Taylor's hypothesis. In addition, hot-wires and hot-films are insensitive to the large changes in velocity direction that are present in separated flows and flows with high turbulence intensity.

In the context of the present work the term *strophometry* is defined as the science and technology involved in the measurement of fluid rotation, being a derivative of the Greek word strophe (στροφή) that literally means rotation. This definition is extended to include measurements of any spatial gradient of a given quantity or one or more velocity gradients.

Background Information

For an observer traveling with the local velocity at a given point inside the flow, the geometry of the instantaneous flow pattern at some point in the neighborhood can be described, to first order, using the terminology of the critical-point theory of *Perry* and *Chong* [5.697]. According to this theory, the local flow topology can be classified according to the nature of the eigenvalues λ of the velocity gradient tensor $A_{ij} = \partial v_i / \partial x_j$, which are the roots of the cubic equation

$$\det(A_{ij} - \lambda \delta_{ij}) = \lambda^3 + P\lambda^2 + Q\lambda + R = 0 \, ,$$

where the coefficients P, Q, and R are the invariants of the tensor since their values are unchanged by rotation of the coordinate frame. These invariants are $P = -\text{tr}(A)$, $Q = 1/2[P^2 - \text{tr}(A^2)]$ and $R = -\det(A)$.

The velocity gradient tensor (A_{ij}) can be decomposed, not uniquely, into a symmetric and antisymmetric part. Specifically, these two parts are the rate-of-strain tensor, $S_{ij} = 1/2(\partial v_i / \partial x_j + \partial v_j / \partial x_i)$ and the-rate-of-rotation tensor, $R_{ij} = 1/2(\partial v_i / \partial x_j - \partial v_j / \partial x_i) = 1/2 \varepsilon_{ijk} \Omega_k$, where Ω_k is the vorticity vector; Ω_k is defined as the curl of the velocity vector

$$\Omega_k = \varepsilon_{ijk} \frac{\partial v_i}{\partial x_j} \tag{5.306}$$

and ε_{ijk} is the alternating unit vector. At each point in the flow field we may imagine a spherical fluid particle whose motion can be decomposed into translation, expansion, shearing, and rotation. The vorticity can be interpreted as just twice the instantaneous solid-body-like rotation rate of the fluid particles along the principle axes in the fluid where there exists no shear deformation. We may alternatively interpret the vorticity as the circulation per unit area of a surface perpendicular to the vorticity field.

Why is vorticity so important in understanding the dynamics of turbulent fluid flow? Insight into this may be gained by looking at the equation describing the transport of vorticity [5.696]

$$\frac{d\Omega_i}{dt} = S_{ik}\Omega_k - \Omega_i S_{kk} + \varepsilon_{iq\eta} \frac{1}{\rho^2} \frac{\partial \rho}{\partial x_q} \frac{\partial p}{\partial x_\eta}$$
$$+ \varepsilon_{iq\eta} \frac{\partial}{\partial x_q} \left(\frac{1}{\rho} \frac{\partial \tau_{\eta j}}{\partial x_j} \right) \, . \tag{5.307}$$

This transport equation indicates that, for an observer moving with the flow, the rate of change of the vorticity component Ω_i of a particle is provided by four dynamically significant processes, namely that of stretching or compression and reorientation or tilting by the strain S_{ik}, vorticity generation through dilatation, baroclinic generation through the interaction of pressure and density gradients, and viscous effects expressed by the viscous stress term. The term $S_{ik}\Omega_k$ consists of one stretching or compression component and two tilting components. This term has been illuminated mostly through numerical work and appears to play an important role in the various physical processes involved in turbulent flows. If the viscous term can be ignored when its magnitude is small, then the change of vorticity of a fluid element in a Lagrangian frame of reference can be entirely attributed to vortex stretching and/or tilting, to dilatational effects, and to baroclinic torque.

In the case of barotropic fluids, the pressure does not enter directly into the equation, unlike the transport equation for momentum, leading to some simplification in interpretation and computation. The reason for this is that the pressure acts through the center of mass of particles and does not produce rotation; rotation is produced by the viscous shearing stresses acting tangentially at the surface of the particles. However, the pressure gradients associated with the velocity field are necessary for the diffusion of vorticity into the flow at solid surfaces. The diffusion of the vorticity is a relatively slow process governed by the same diffusivity constant, the kinematic viscosity, as that for momentum. This is shown by considering the momentum equation evaluated at the wall, where the velocity vector is identically zero because of the nonslip conditions. In this case the momentum equation is reduced to $\partial p/\partial x_i|_\text{wall} = \rho \partial \tau_{ki}/\partial x_k|_\text{wall}$ which is equivalent to

$$\left.\frac{\partial p}{\partial x_i}\right|_\text{wall} = -\mu \varepsilon_{ijk} \left.\frac{\partial \Omega_k}{\partial x_j}\right|_\text{wall}. \tag{5.308}$$

The term $-\partial \Omega_k/\partial x_j|_\text{wall}$ is the vorticity flux density, a term first introduced by *Lighthill* [5.685], who defined it for two-dimensional flows by analogy to the Fourier heat conduction as $-\nu \partial \Omega_k/\partial x_j|_\text{wall}$. The term vorticity flux describes the rate of vorticity production *at the wall*, which then enters the flow. In that sense it is more important to know the amount of vorticity entering the flow than the amount of vorticity at the wall. Figure 5.172 depicts the mechanism of vorticity change of a small fluid element moving toward the wall. A pressure gradient tangential to the boundary causes the fluid to change its rotation due to the non-slip conditions at the wall. A fluid element with initial vorticity $\Omega_\text{in} < 0$ enters the positive-pressure-gradient wall region and acquires a positive vorticity change $\Delta\Omega_y$ that changes its vorticity to the output vorticity $\Omega_\text{out} > 0$. According to *Andreopoulos* and *Agui* [5.698] the vorticity change shed from the solid boundary and sensed at a nearby location inside the flow at a distance Δy from the wall, to a first-order approximation, will be $\Delta\Omega_z = -(\partial \Omega_z/\partial y)_\text{w}\Delta y = \frac{1}{\mu}(\partial p/\partial x)_\text{w}\Delta y$ and $\Delta\Omega_x = -(\partial \Omega_x/\partial y)_\text{w}\Delta y = -\frac{1}{\mu}(\partial p/\partial z)_\text{w}\Delta y$. Since $\Delta y > 0$ always, the signs of $\partial p/\partial x$ and $\partial p/\partial z$ determine whether the wall acts as a source or sink of vorticity. Spatially and time-resolved measurements of the vorticity flux terms $\partial \Omega_x/\partial y$ and $\partial \Omega_z/\partial y$ in a two-dimensional boundary layer have been obtained in the work by *Andreopoulos* and *Agui* [5.698].

Physics of the Structure Function

There is a remarkable similarity and considerable overlap in the way velocity or pressure gradients are estimated and the way the structure functions of velocity or pressure are obtained, although they are defined quite differently. The velocity or pressure gradients, $\partial v_i/\partial r_j$ or $\partial p/\partial r_j$, respectively, are local quantities defined at one point inside the flow field, whereas the transfer function is defined as the difference between the values of velocity or pressure at two points of the flow field. Specifically the structure function is defined, before time averaging, as $\Delta v_i = v(r + \Delta r_j) - v(r)$ for velocity and $\Delta p = p(r + \Delta r_j) - p(r)$ for pressure.

For small separations Δr_j and after considering Taylor's expansion, these relations become $\Delta v_i = v(r + \Delta r_j) - v(r) = \partial v_i/\partial r_j \Delta r_j$ and $\Delta p_i = p(r + \Delta r_j) - p(r) = \partial p/\partial r_j \Delta r_j$, respectively.

In practice, the velocity or pressure gradients are obtained from the value of the corresponding structure function as the separation distance Δr_j tends to a very small value. Thus, the velocity or pressure gradients defined at one point are estimated from the difference of velocity or pressure values at two different but nearby points. In that respect the velocity or pressure gradients and the structure functions of velocity or pressure are interrelated and the underlying physics behind their behavior may be the same.

Let us assume that the differential $\Delta Q = Q(r + \Delta r) - Q(r)$ of a quantity $Q(r)$ is required to be esti-

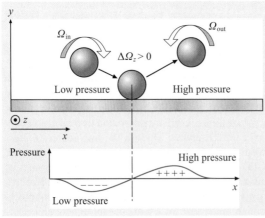

Fig. 5.172 Lighthill's mechanism of vorticity change at the wall beneath a flow

mated at a position r. In this case, two measurements of the quantity Q at close locations r and $r+\Delta r$, are usually attempted. ΔQ is a fluctuating quantity and if we consider the time average $\overline{\Delta Q^2}$ then the structure function is defined as

$$\overline{\Delta Q^2} = \overline{[Q(r+\Delta r) - Q(r)]^2}$$
$$= \overline{Q(r)^2}\left[(1-\rho)^2 + 2\rho(1-R_{\Delta r})\right], \quad (5.309)$$

where ρ is the ratio $\rho^2 = \overline{Q^2(r+\Delta r)}/\overline{Q^2(r)}$ and $R_{\Delta r}$ is the correlation coefficient $R_{\Delta r} = R_{r,r+\Delta r} = \frac{\overline{Q(r)Q(r+\Delta r)}}{[\overline{Q^2(r)}]^{1/2}[\overline{Q^2(r+\Delta r)}]^{1/2}}$, not to be confused with the rate-of-rotation tensor.

Thus for the rms of a gradient we obtain

$$\left(\frac{\overline{\Delta Q^2}}{\Delta r^2}\right)^{1/2}$$
$$= \frac{[\overline{Q(r)^2}]^{1/2}}{\Delta r}\left[(1-\rho)^2 + 2\rho(1-R_{\Delta r})\right]^{1/2}. \quad (5.310)$$

The quantity Q could be, for instance, one component of the velocity vector or pressure or even one vorticity component. The relation (5.310) indicates that the rms of the quantity $\Delta Q/\Delta r$ depends, among other quantities, on $(1-\rho)$ and $(1-R_{\Delta r})$. The first quantity $(1-\rho)$ is a measure of the inhomogeneity of the turbulence intensities, while the second one $(1-R_{\Delta r})$ indicates the decorrelation between the two fluctuations of Q at r and $r+\Delta r$. It can be argued that, for small separations Δr, in most flows either $2\rho(1-R_{\Delta r}) \gg (1-\rho)^2$ or $\rho \approx 1$ is valid. In low-speed incompressible boundary layers, for instance, the ratio $2\rho(1-R_{\Delta r})/(1-\rho)^2$ is close to 150 at a location just outside the viscous sublayer for $\Delta r = \Delta x_2^+ = 16$, where x_2 is the wall normal direction. In another example: isotropic turbulence behind rectangular grids $\rho \approx 1$ [5.694]. It is, therefore, reasonable to assume that $(\overline{\Delta Q^2}/\Delta r^2)^{1/2} = [\overline{Q(r)^2}]^{1/2}/\Delta r[2\rho(1-R_{\Delta r})]^{1/2}$, which suggests that most contributions to $\Delta Q/\Delta r$ come from the uncorrelated parts of the two signals, $Q(r)$ and $Q(r+\Delta r)$. Thus the rms of $(\overline{\Delta Q^2}/\Delta r^2)^{1/2}$ between two perfectly correlated signals $(R_{\Delta r} = 1)$ is vanishing. In turbulent flows it is known that the existence of small-scale turbulence decreases $R_{\Delta r}$. In that context there will always be a nonzero $(\overline{\Delta Q^2}/\Delta r^2)^{1/2}$ since $R_{\Delta r} < 1$ for $\Delta r > 0$. On the other hand two sinusoidal signals with a phase shift between them φ, i.e., $y_1 = A_1 e^{i\omega t}$ and $y_2 = A_2 e^{i(\omega t + \varphi)}$, will also result in a correlation coefficient $R_{\Delta r} < 1$ since $R_{\Delta r} = \cos\varphi$. In all cases other than $\varphi = 0$, $R_{\Delta r} < 1$ and $(\overline{\Delta Q^2}/\Delta r^2)^{1/2}$ reaches nonzero values. For $\varphi = 180°$ $(1-R_{\Delta r})$ reaches a maximum value of 2 and therefore $(\overline{\Delta Q^2}/\Delta r^2)^{1/2} = 2(\overline{Q(r)^2})^{1/2}/\Delta r$. From this example it can be concluded that contributions to $\Delta Q/\Delta r$ come mostly from the uncorrelated parts of the flow or from the out-of-phase events. Small-size eddies represented in the high-wavenumber part of the spectrum are known to be weakly correlated with each other and therefore in the present context they represent contributions to $\Delta Q/\Delta r$ from the uncorrelated parts of the flow.

The limit of (5.310) when Δr approaches zero is also of interest in the present considerations since it is related to the potentially available spatial resolution. In this case $R_{\Delta r}|_{\Delta r \to 0} = 1$ and $\rho|_{\Delta r \to 0} = 1$ and thus the limit appears to become indefinite. After applying the Hôpital rule the following relation can be obtained $\lim_{\Delta r \to 0}(\overline{\Delta Q^2}/\Delta r^2) \approx -[\overline{Q(r)^2}]\{\partial^2 R/\partial(\Delta r)^2 - [\partial\rho/\partial(\Delta r)]^2\}$ where $\partial^2 R/\partial(\Delta r)^2$ is the curvature of the cross-correlation function R, at the origin. Thus a Taylor microscale λ can be defined as $(\partial^2 R/\partial(\Delta r)^2)_{\Delta r \to 0} = -1/\lambda^2$ and therefore $(\overline{\Delta Q^2}/\Delta r^2)_{\Delta r \to 0} \approx [\overline{Q(r)^2}]/\lambda^2\{1 + \lambda^2[\partial\rho/\partial(\Delta r)]^2\}$.

General Requirements

The measurement of one or more components of vorticity in turbulent flows of research and technical interest has been a long-held, but elusive, goal of fluid-mechanics instrument developers and researchers. The review article by *Wallace* and *Foss* [5.81] provides a detailed account of the available techniques, as of that date, used to measure vorticity and compare existing data. The majority of the techniques are based on thermal anemometry or optical methods. The present Section will address issues related to optical techniques, which possess several advantages and some limitations with respect to those based on thermal anemometry. Optical techniques are non-intrusive, are sensitive to large changes in velocity direction and most do not require any calibration. These characteristics make them ideal for use in separated flows and flows with large turbulence intensities where the use of intrusive methods is prohibited or where recirculating effects are present. Most of these optical techniques are based on detecting the light scattered by particles imbedded in the flow. In that respect, seeding the flow with appropriate particles that follow the fluid motion very closely is critical (Sect. 5.3.1). It should be noted here that, for gas flows in which seed must be added and in which the flow does not recirculate, adding particles can be far more intrusive than a hot-wire probe.

There are several requirements that determine the performance of each technique beyond its basic idea and the principles upon which it is based.

Formal Derivation in Implicit Formulations. There is a need for a formal definition of the measured quantities, either the vorticity components Ω_k or velocity gradient tensor terms A_{ij}. Specifically, there are two common ways to derive vorticity. In the first method, measurements of velocity components at nearby locations are obtained and used in a finite-difference scheme to derive vorticity. In the second method measurements of microcirculation around a small area are obtained and vorticity is computed as circulation density.

The first definition is based on a Taylor's series expansion of velocity in a nearby location, which to a first-order approximation becomes $v_i(r_j + \Delta r_j) = v_i(r_j) + \Delta r_j \partial v_i / \partial r_j + \ldots$

For r_j in the y-direction and $\Delta r_j = (0, \Delta y, 0)$ this becomes $v_i(y + \Delta y) \approx v_i(y) + \Delta y \partial v_i / \partial y$, which yields the following relation to compute the velocity gradient $\partial v_i / \partial y \approx [v_i(y + \Delta y) - v_i(y) / \Delta y]$. This relation will also define the uncertainties in the measurement of the velocity gradient.

The second method of estimating vorticity is due to *Foss* and co-workers [5.699, 700] who developed a hot-wire technique for the measurement of the transverse vorticity component. The technique involves the measurement of the microcirculation Γ around a small, finite domain of area ΔS normal to the flow. The estimated vorticity appears to be computed through the relation $\langle \Omega_z \rangle = \Gamma / \Delta S$.

These two methods are not the only ones in use. There is no need to have a computational formalism for vorticity when it is estimated explicitly. For instance, direct estimate of vorticity through measurements of fluid rotation does not require vorticity computation through an implicit scheme [5.701].

Spatial and Temporal Resolution. The spatial and temporal resolution of any probe affects the accuracy of measurements of time-dependent velocity gradients A_{ij} or vorticity components Ω_k. This resolution depends on the optical arrangement of the technique, the hardware used to generate the measuring probe volume and to process the signals, and the particles used as light scatterers and their arrival rate in the measuring locations. Thus the technique has a typical characteristic length scale L_p that is usually associated with the separation distance between the sensing areas and a time scale T_p, which depends mostly on the speed of data acquisition or burst processor. The technique should be able to resolve the small scales that exist in a given flow, which appear to be the Kolmogorov viscous scales $\eta = \nu^{3/4} \varepsilon^{-1/4}$ for length and $T_\varepsilon = \nu^{1/2} \varepsilon^{-1/2}$ for time, where ε is the dissipation rate of turbulent kinetic energy and ν is the kinematic viscosity. L_p and T_p should be less than η and T_ε, respectively, in order to capture contributions from the smallest eddies, which are responsible for most of the content of vorticity or velocity gradient since small eddies are uncorrelated [see relation (5.310)]. *Wyngaard* [5.702] was the first to evaluate the problem of spatial resolution of hot-wire-based vorticity techniques and found that no attenuation occurred for $L_p / \eta = 1$, while substantial attenuation of vorticity was evident for $L_p / \eta > 3.3$. The review by *Wallace* and *Foss* [5.81] concludes that reasonable estimates of velocity gradients can be obtained without attenuation for $L_p / \eta < 2$–4.

Although L_p has to be as small as possible it can not be minimized independently from other parameters such as the speckle noise. Errors due to noise in the measurements amplify as the length scale decreases. There is certainly an optimum length intermediate between attenuation and amplification of noise.

Determining η requires a good estimate of the turbulent kinetic energy dissipation rate E, which is very difficult to measure directly in a laboratory. Previous studies obtained the value of E by making assumptions to reduce some of the terms in the expression of the total dissipation rate [5.694]:

$$E = \tau_{ij} S_{ij} = 2\mu S_{ij} S_{ij} - \frac{2}{3} \mu S_{kk} S_{kk} \,. \tag{5.311}$$

The second term on the right-hand side of this relation represents the additional contribution of compressibility to the dissipation rate of kinetic energy. This term disappears in the case of incompressible flow. Since this term is always positive, the negative sign in front may erroneously suggest that compressibility reduces dissipation. This is incorrect because the term $S_{ij} S_{ij}$ also contains contributions from dilatation effects, which can be revealed if one considers that

$$S_{ij} S_{ij} = \frac{1}{2} \Omega_k \Omega_k + \frac{\partial v_i}{\partial x_j} \frac{\partial v_j}{\partial x_i} \,. \tag{5.312}$$

The second term on the right-hand side represents the inhomogeneous contribution in the case of incompressible flows. In the case of compressible flows, terms related to dilatation can be extracted, and the instantaneous total dissipation rate then becomes

$$E = \mu \Omega_k \Omega_k + 2\mu \left(\frac{\partial v_i}{\partial x_j} \frac{\partial v_j}{\partial x_i} - S_{kk} S_{kk} \right) + \frac{4}{3} \mu S_{kk} S_{kk} \,. \tag{5.313}$$

The third term on the right-hand side describes the direct effects of compressibility, i.e., dilatation on the dissipation rate. The first and the last terms on the right-hand side of the last relation are quadratic with positive coefficients and positive signs and are therefore always positive. The second term on the right-hand side indicates the contributions to the dissipation rate by the purely nonhomogeneous part of the flow. Its time-averaged contribution disappears in homogeneous flows. This term, in principle, can obtain negative values and thus can reduce the dissipation rate. This does not violate the second law of thermodynamics as long as the total dissipation remains positive at any point in space and time.

The estimates that are usually provided for E, particularly those based on indirect methods, may be very misleading. An experiment with poor spatial resolution will result in attenuation of all measured turbulent quantities including the dissipation rate, which will indicate high η and therefore high resolution.

It should be mentioned here, however, that there is increasing evidence provided by *Tsinober* et al. [5.693] and *Honkan* and *Andreopoulos* [5.689] that the Taylor microscale λ should be used to determine the spatial resolution in vorticity measurements. It has been shown in Sect. 5.3, that in the limit of vanishing separation Δx_j between the measurement locations, the rms of the fluctuations of one velocity gradient $\partial v_i/\partial x_j$ is given by $\lim_{\Delta x_j \to 0} \overline{(\Delta v_i/\Delta x_j)}^2 = \overline{u_i^2}/\lambda_{ij}^2 \{1 + \lambda_{ij}^2 [\partial \rho/\partial(\Delta x_j)]^2\}$, where λ_{ij} is the Taylor microscale. If the gradient, $\partial \rho/\partial \Delta x_j$, which indicates the degree of variation of the rms of velocity fluctuations, is much smaller than $1/\lambda_{ij}^2$ then the relation

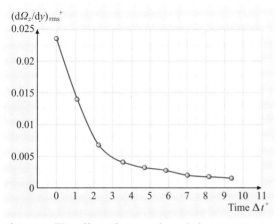

Fig. 5.173 The effect of temporal resolution on rms vorticity flux are made nondimensional by inner wall scaling (after *Andreopoulos* and *Agui* [5.698])

$\lim_{\Delta x_j \to 0} \overline{(\Delta v_i/\Delta x_j)}^2 = \overline{u_i^2}/\lambda_{ij}^2$ provides the best estimate for ideal resolution when Δx_j tends to zero. In that respect the ratio L_p/λ_{ij} should be less than 1.

Although the statistics of fluctuations of individual velocity components do not require increased temporal resolution, cross-correlation/shear stress and velocity gradients/vorticity measurements require a high degree of coincidence of particles at different locations inside the probe volume. This can be achieved through a high rate of particles, which results in practically continuous signals. The resolution in time basically depends on the rate of arrivals of particles in the probe volume and on the data-processing rate of the hardware.

An example of the effect of temporal resolution on the statistical results of vorticity flux is obtained from the data of *Andreopoulos* and *Agui* [5.698]. Vorticity flux is computed by (5.308) by subtracting two pressure signals obtained at very close locations that are about 0.8 mm or $\Delta x^+ = 6$ apart. Artificial sampling times Δt were produced by skipping by various numbers equally sampled data digitized at rates substantially faster than the Kolmogorov frequency $1/T_\varepsilon$. The results plotted in Fig. 5.173, show a substantial dependence of the rms of vorticity flux on the temporal resolution Δt.

As the Reynolds number of the flow increases its spatial and temporal scales decrease and therefore, if the size of the measuring volume and data-acquisition rate remain unchanged, it is expected that the spatial and temporal resolution of the measuring technique will deteriorate.

Validation. Eventually the developed technique has to be validated in a real flow that will provide a good test bed for evaluation of its performance. In the past, velocity gradient/vorticity measurement techniques were tested in laminar flows that possessed known rotational flow characteristics very close to those predicted by theory. Although such tests are adequate to demonstrate the basic principles of the techniques, they are insufficient in evaluating or predicting the performance of the probe in a real turbulent flow where a wide range of flow scales is present and resolution issues are an important part of the overall assessment. In that respect, it may be better to test the technique in a canonical turbulent flow such as a two-dimensional, zero-pressure-gradient boundary layer over a smooth wall, a two-dimensional shear layer, or isotropic homogeneous grid-generated turbulence. The bulk and turbulent characteristics of canonical flows are well known and existing data are well documented and adequately tabulated. Unfortunately, velocity gradient/vorticity data are not always available and therefore a comparison with previous velocity gradient/vorticity data is not possible, except in some cases of incompressible boundary-layer flows, free shear layer flows or grid turbulence-generated flows where such data exist. The preceding taxonomy of flows also includes compressible grid-generated turbulence interaction with shock waves.

Existing Techniques and Their Classification

The inherent difficulties of velocity gradient/vorticity measurements, which require high spatial and temporal resolution in turbulent flow fields, have only recently been substantially overcome.

Laser speckle strophometry (LSS) or particle image velocimetry (PIV) techniques are non-intrusive but cannot in general provide adequate time-history information. In addition slight out-of-plane motion of the scattering particles due to flow three-dimensionality results in different speckle patterns (speckle decorrelation). In particle image velocimetry, crossing of the flowing particles paths in highly fluctuating flows may cause ambiguity in the detection scheme. PIV is also limited by spatial resolution loss due to averaging from the correlation schemes. Although the resolution of a typical CCD camera may typically be $1\,\text{k} \times 1\,\text{k}$ px, subregions of 32×32 px known as interrogation areas are usually considered to validate a local velocity measurement. According to *Arik* [5.703] in a typical PIV setup approximately 10 particles are needed to validate a velocity vector measurement and 10 velocity vectors are needed to evaluate a vorticity vector. Thus approximately 100 particles are needed to evaluate a vorticity vector, a number which results in 30 to 40 particles for each of the three vorticity components.

Lourenco et al. [5.704] used PIV techniques to obtain one-component vorticity measurements in a jet flow. *Meinhart* and *Adrian* [5.705] used PIV to measure the near-wall vortical structure of low-speed boundary layers with reasonable spatial resolution.

Laser speckle strophometric methods have been developed by *Breyer* et al. [5.706], *Kriegs* et al. [5.707], and *Kriegs* and *Staude* [5.708]. In these techniques the speckle pattern, which is the result of the diffraction of a plane wave produced by a random grating, is basically affected by the flow velocity gradient while the translation of the grating caused by the constant flow velocity will not affect the pattern.

These authors applied their double-pulse strophometric technique to obtain direct measurements of the quantities $\partial v/\partial x + \partial v/\partial y$, $\partial v/\partial x - \partial v/\partial y$ and $\partial v/\partial z$ in a channel flow with Reynolds number based on the channel width in the range of 2250–6700. This information was used to obtain statistics of the $\partial v/\partial y$ gradient.

Holographic particle image velocimetry (HPIV) was used by *Zhang* et al. [5.709] to obtain velocity gradient measurements in a square duct flow with moderate resolution. A polarization-based approach with dual-plane particle image velocimetry (DSPIV) has been used by *Kähler* [5.710], *Ganapathisubramani* et al. [5.711] and *Hu* et al. [5.712] to obtain vorticity measurements in boundary-layer flows and in a lobed jet mixer. A two-frequency DSPIV was applied in jet flows to measure all nine components of the velocity gradient tensor with resolution of about six Kolmogorov length scales [5.713]. These recent works based on PIV techniques have demonstrated the potential to provide measurements of vorticity with excellent spatial resolution. At the moment, their temporal resolution is rather limited, which does not however affect the process of statistical averaging. In the following a description of optical techniques which have been used to measure one velocity gradient/vorticity at a single point with reasonable success will be provided. These techniques can be classified as direct or indirect, depending on whether vorticity is obtained from the measurement of fluid particles rotation or from the measurement of individual velocity gradients inside the flow field.

One of the first direct measurements of vorticity was obtained by *Frish* and *Webb* [5.701] by measuring the solid rotations of tiny transparent spherical beads with

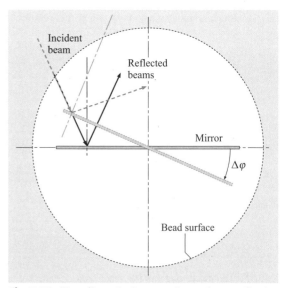

Fig. 5.174 The effect of mirror rotation on beam reflection as in the technique by *Frish* and *Webb* [5.701]

a diameter of 25 μm, each containing embedded planar crystal mirrors specially developed for this purpose, suspended in a refractive-index-matched liquid.

The flows and the beads were illuminated by a laser light and the rate of rotation of the reflected light, which is proportional to vorticity, was measured directly. Figure 5.174 depicts the effect of a rotating mirror on the direction of the reflected beam. If the mirror is rotated by an angle $\Delta\varphi$ within an infinitesimally small time Δt then its angular velocity will be $\omega = \Delta\varphi/\Delta t$ and the in-cident and reflected angles of the light beams are α_i and α_r, respectively, then each angle will change by the same amount $\Delta\varphi$, and the angle between the reflected and incident beams will change by $2\Delta\varphi$. For a fixed $\Delta\varphi$ the measurement of Δt can lead to the determination of the rate of rotation, which is half of the vorticity. The spatial resolution of the technique, which is determined by the size of the beads, is considered to be very good. However, the complexity of this technique and the requirement for matched refractive index prohibit its use in air.

Most of the techniques used to obtain indirect measurements of vorticity take advantage of the Doppler effect and have configurations similar to those used in laser Doppler anemometry (LDA). In typical dual-beam configurations for velocity measurements, two laser beams cross each other to form the measuring/probe volume with the fringe pattern, as shown in Fig. 5.175a. Receiving optics can be aligned in any direction since the Doppler frequency shift Δf_D is independent of the k_s direction. The velocity component normal to the fringe orientation $v_i(x_j)$ can be obtained by measuring $\Delta f_D = \mathbf{V} \cdot (\mathbf{k}_{01} - \mathbf{k}_{02})/2\pi$, where $\mathbf{k}_{01}$ and $\mathbf{k}_{02}$ are the wavenumber vectors along the two incident beams forming the measurement volume.

A natural extension of the traditional single-point LDA measurement of one velocity component to obtain one velocity gradient is shown in Fig. 5.175b, where two independent probe volumes are formed at a distance Δx_j apart to measure two velocity components in the same direction, $v_i(x_j)$ and $v_i(x_j + \Delta x_j)$, respectively. The velocity gradient $\partial v_i/\partial x_j$ can be estimated through the finite-difference relation $\partial v_i/\partial x_j \approx \Delta v_i/\Delta x_j \approx v_i(x_j + \Delta x_j) - v_i(x_j)/\Delta x_j$. For one-component vorticity measurements, simultaneous measurements of the two velocity gradients are required.

This configuration has been used by *Lang* and *Dimotakis* [5.714] and *Lang* [5.715] to obtain measurements of one component of vorticity in a two-dimensional shear layer. Two probe volumes, separated by $\Delta y = 1.9$ mm, were aligned in the y-direction to measure the $\partial u/\partial y$ gradient and another two probe volumes separated by $\Delta x = 1.9$ mm were aligned in the x-direction to measure the $\partial V/\partial x$ gradient. Thus eight beams were required to measure the spanwise vorticity component $\Omega_z = \partial V/\partial x - \partial U/\partial y$, which was estimated as $\Omega_z = (V_2 - V_1)/\Delta x - (U_2 - U_1)/\Delta y$. Mean and rms values of Ω_z were obtained. The lack of estimates for the dissipation rate of turbulent kinetic energy prevented direct evaluation of the spatial resolution of this technique. A similar arrangement has been used by *Romano* et al. [5.716] to measure velocity structure functions

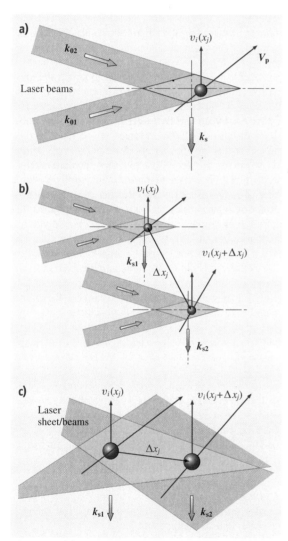

Fig. 5.175 (a) Typical fringe pattern in LDA. **(b)** Configuration for the measurement of one velocity gradient $\partial U_i/\partial x_j$. **(c)** Configuration for the measurement of the cross-correlation of velocity components

in a channel flow where two completely different LDA systems were used.

Instead of generating two distinct probe volumes to obtain two velocity estimates at nearby locations, an alternative method is to form a single probe volume with focus at two distinct points inside the volume. The advantage of the latter approach is that fewer optical components are required. It has already been used for cross-correlation measurements in space [5.719, 720] and for simultaneous velocity measurements at several points by using an elongated control volume generated by the intersection of two *sheet* beams formed by passing the beams through two cylindrical lenses (Fig. 5.175c) as described by *Nakatani* et al. [5.721]. The light intensity is reduced significantly in this arrangement of two interfering sheets and therefore the signal-to-noise ratio is deteriorated.

The LAVOR (laser vorticity probe) method proposed by *Agui* and *Andreopoulos* [5.717, 718] uses two parallel laser beams focused with a long-focal-length lens to generate a moderately narrow probe volume defined by the region of intersection of the two beams. The velocity at two different locations inside the probe volume is measured. There are several advantages to this approach over the *sheet* beam approach. These include fewer optical components, simple alignment of the optics, smaller windows for optical access, and better use of laser intensity.

A schematic drawing of the principle is shown in Fig. 5.176. The measurement locations within the probe volume are defined by imaging the entire probe volume onto the optical fibers located on the opposite side of the receiving optics. Therefore, scattered light will be collected from two distant locations within the probe volume. Scattered light from the two points on the probe volume is designed to enter the acceptance cone of the optical fibers. The distance Δx_j between the two measurement points can be adjusted to match the spatial resolution requirements. No magnification through the lenses was used in this application and therefore the distance between the two fibers $\Delta x_{j,f}$ was the same as the distance Δx_j. The distance of the fibers $\Delta x_{j,f}$ can be of the order of 1 mm and, for the case where the two fibers are very close together, the distance between the two points in the probe volume will be on the order of 0.3–1 mm. The value of Δx_j sets the spatial resolution of the probe and has to be compared to the Kolmogorov length scale of the turbulent flow under investigation $\eta = (\nu^3/\varepsilon)^{1/4}$, where ε is the dissipation rate of turbulent kinetic energy, if the smallest viscous scales are be resolved [5.702, 722]. It should be noted that, although the resultant probe volume is large, the spatial resolution is defined by the intersection of the probe volume and the image of the optical fiber ends. The aforementioned range of the spatial resolution is not unrealistic; it is based on the experimental results of *Johns* et al. [5.720] and *Fraser* et al. [5.719]. The authors in the first reference succeeded in making spatial correlation measurements between velocity components at different locations as close as 70 μm and without *crosstalk* between the scattered light by the two particles present at the measurement volume.

The uniqueness of the LAVOR technique for vorticity measurements lies in the use of

1. a small control volume imaged onto the core of a fiber
2. fiber optics that allows high spatial resolution and
3. fewer optical components.

An extremely fast digital data-acquisition device was used, which allows *instantaneous* processing of the burst signal with a high mean rate of 78 000 bursts/s by using a record length of 64 points to digitize the individual

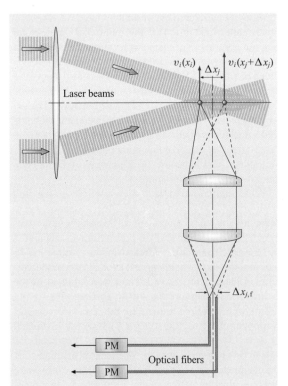

Fig. 5.176 Principle of measurement in the technique by *Agui* and *Andreopoulos* [5.717, 718]. PM: photomultiplier

bursts. This value determines the temporal resolution of the new technique. In addition to vorticity, measurements of velocity components are also obtained at the same time so that an additional mapping of the velocity flow field is not necessary.

The implementation of the previously described idea for measurements of vorticity is shown in Figs. 5.177a,b. For measurement of one vorticity component, the determination of the two velocity gradients $\partial v_2/\partial x_1$ and $\partial v_1/\partial x_2$ is required. Therefore, the probe volumes have to be oriented appropriately to coincide with the x_1–x_2 plane. It should be emphasized that the measured vorticity Ω_3 is invariant with rotation of both probe volumes about the x_3-axis, which is normal to the x_1–x_2 plane. Scattered light collected through the receiving optics enters the fibers through a module located on the focal plane. The fiber-optics module with the arrangement of the four fiber cable ends is shown in Fig. 5.177c. It should be noted that the x_1–x_2 plane coincides with the focal plane of the arrangement of the receiving optics.

An appreciation of the advantages of the LAVOR technique can be obtained if one compares it with the arrangement adopted by *Romano* et al. [5.716] to measure velocity transfer functions in channel flows that required the use of instrumentation, equipment, and optical components equivalent to two different LDA systems.

A second category of single-point velocity gradient/vorticity measurement techniques exists that are not based on the interference of two laser beams and the scattering light of a particle moving through the fringes realized on the surface of the photodetector. This category includes techniques that use laser Doppler signals from two particles illuminated by two different light beams. *Durst* et al. [5.723] in panel 3.26 of that book show that the introduction of phase variation with time caused by the movement of light-scattering particles will yield signals that originate from different particles illuminated by different light beams (Fig. 5.178a). An equation for the light intensity at the surface of the photomultiplier has been derived for the case of two beams that have the same wavelength, which shows that the intensity will vary with time and cause the photodetector to respond to the Doppler frequency, which is a function of the relative velocity of the two particles if their position is known. If the two beams are parallel, the directions of the two scattering waves are the same, and the photodetector is orientated in the same direction and positioned halfway between the two beams then the Doppler shift Δf_D is proportional to the velocity difference. No measurements have been presented to validate the proposed technique.

The idea of heterodyning of the scattered light from two particles was explored by *Hanson* [5.724] and more recently by *Ötügen* et al. [5.725] and *Yao* et al. [5.726]. Hanson's heterodyning arrangement uses an optical filter to define the relative separation of two particles, which requires the fabrication of miniature spatial gratings and rings with geometries tailored to the specific flows under investigation. His technique, which has been tested in a cylindrical laminar Couette flow, is insensitive to the sign of velocity gradient.

The technique by *Ötügen* et al. [5.725], shown schematically in Fig. 5.178b, resolves the sign of the velocity gradient by using a Bragg cell to shift the fre-

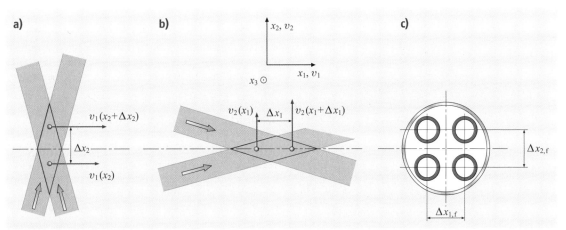

Fig. 5.177 (a) Beam orientation for the measurement of $\partial v_1/\partial x_2$. **(b)** Beam orientation for the measurement of $\partial v_2/\partial x_1$. **(c)** Fiber optics module

quency of one laser beam and has been tested in the laminar flow formed between two concentric cylinders with one of them rotating. *Yao*'s et al. [5.726] technique, which is quite similar to that of Ötügen et al., has been tested successfully in a laminar channel flow.

If only one particle is present in one of the two targeted locations, its Doppler effect will be lost within the much larger frequency of the laser beam that goes undetected by the photomultiplier. Only when both particles are present is there cancelation of the original laser frequency and is the beat frequency detected.

The technique by Ötügen et al. originally used two different arrangements to collect scattered light from the two particles in the measurement volume. The two scattered light paths were recombined by a beam combiner before they enter the photomultiplier. This technique has been further developed recently and a single arrangement of collecting optics has been used in which both particles appear to be practically on the same focal plane because of the close proximity of the two laser beams. The technique has been used to carry out measurements of vorticity in the turbulent boundary layer of a small closed-loop water tunnel [5.727] and in the turbulent shear layers behind a splitter plate placed in the same wind tunnel [5.728]. Each of the shear layers had a thickness of about 4 mm and the distances Δx and Δy between the two particles for each of the two pair of beams were $\Delta x = 0.7$ mm and $\Delta y = 1$ mm. Thus the ratio of the large flow scale to the separation distance Δx or Δy is 4–6, which is considered inadequate to resolve even the large-scale contributions to the vorticity let alone the contributions of small scales of the size of the Kolmogorov viscous scales, which was estimated to be $0.047 - 0.067$ mm. This is a shortcoming of the test flow and the size of the water-tunnel facility rather than of the technique itself, which appears to be very promising. In fact, in their boundary-layer experiments the boundary-layer thickness was about 40 mm and the separation distances were reduced to $\Delta x = 0.4$ mm and $\Delta y = 0.78$ mm. In this case, the Kolmogorov length scales appear to be of the order of $0.03 - 0.07$ mm in the region close to the wall. Thus the ratio $\Delta x / \eta$ is in the range 13–23, which can be considered reasonably small but inadequate.

In all these techniques that estimate the velocity gradient through the heterodyning of the scattered light from two different particles, the velocity field cannot be measured simultaneously with the velocity gradient, which requires separate measurements. In this respect, the inability of these methods to provide information about the velocity field at the same time as the velocity gradient measurements may be considered a drawback.

It is also evident from the description above that many of the techniques have been validated in a laminar flow environment. This is a first step in a full evaluation plan that should be followed up with testing in a turbulent flow where documentation of the flow-field quantities exists.

Particle Rates and Temporal Resolution

Although the statistics of fluctuations of individual velocity components do not require increased temporal resolution, cross-correlation/shear stress and velocity gradients/vorticity measurements require a high degree of coincidence of particles at different locations in the probe volume. This can be achieved, to a certain extent, through high rate of particles arriving at the measurement locations. If P_A ($0 \leq P_A \leq 1$) is the probability of having one particle at a given location A and P_B

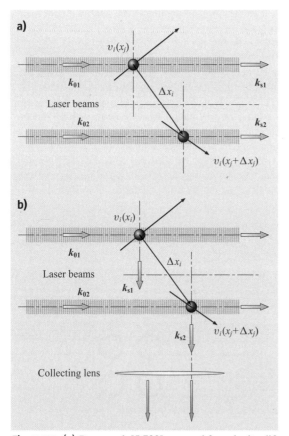

Fig. 5.178 (a) *Durst* et al. [5.723] proposal for velocity difference measurements. **(b)** *Ötügen* et al. [5.725] technique for velocity difference measurements

($0 \leq P_B \leq 1$) is the probability of having one particle at a second location B then the probability of having two particles at the same time in locations A and B is $P(A \cap B) = P_A \times P_B$, provided that these two events are independent of each other [5.729]. This results in a value of $P(A \cap B)$ that is smaller than either P_A or P_B. If the presence of four particles at the same time at four different locations A, B, C, and D is required, called coincidence in the present context, then the probability of this event is P^4 where $P = P_A = P_B = P_C = P_D$. For a typical $P = 0.5$ then $P^4 = 0.0625$, a value eight times less than P.

Increasing the particle concentration so that there is always at least one pair of particles in the measuring volume can increase the coincidence and therefore the data rate. Further increase in particles concentration may result in the presence of more than one particle inside the measuring volume, which can cause interference due to multiple light scattering. In addition, the presence of too many particles will affect the flow field itself, which may become a slurry.

As the Reynolds number of the particular flow increases, turbulence fluctuations will increase and therefore turbulent diffusion, a process that is responsible for transporting particles into the measurement volume, will increase and therefore the particle arrival rate will increase. However, the residence time of each particle will be shorter and the rate of validation will decrease because the coincidence criteria will not be met. In effect, achieving high rates of validated particles present at various locations of the flow field at the same time becomes increasingly difficult and, if this is compounded with the requirement to have continuous signal, it is close to impossible.

The resolution in time of the measurements depends basically on the rate of arrivals of particles in the probe volume and on the data processing rate of the hardware. Since most burst-type processors can process data at a rate of about 200 000 bursts/s, which is considered adequate for temporal resolution of flows up to low subsonic Mach numbers, the temporal resolution is entirely dependent on the rate of particle arrivals at the focusing points.

To demonstrate several of the typical issues involved in optical strophometry, the LAVOR technique of *Agui and Andreopoulos* [5.718] will be described in more detail than above. This technique has been used to obtain measurements of the spanwise vorticity in the near-wall region of a two-dimensional boundary layer configured in the CCNY (City College New York) low-speed wind tunnel, which has a $1.2\,\text{m} \times 1.2 \times 8.4\,\text{m}$ working section. Three different experiments were carried out at Reynolds numbers $Re_\theta = 2700$, 3900 and 5400. This test flow has been extensively investigated and data of all turbulent quantities, including three-dimensional time-resolved vorticity obtained by subminiature multi hot-wire techniques, are well documented in several publications [5.689, 730]. Thus a direct comparison of the LAVOR data with the hot-wire data of vorticity will be more meaningful since both have been obtained in the same facility under similar flow conditions.

The flow was seeded with $1-3\,\mu\text{m}$-diameter particles produced by an oil-based smoke machine that was

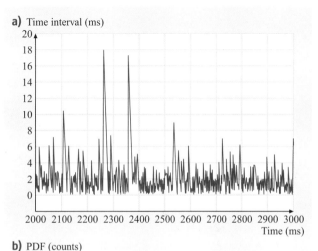

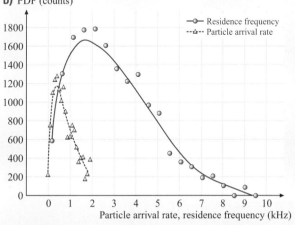

Fig. 5.179 (a) Typical time record of time intervals between successive particle arrivals (after *Agui and Andreopoulos* [5.718]). (b) Typical probability density function (PDF) of particle arrival rates (number/s), F_p and the inverse of the residence time in the probe volume $F_R = 1/T_R$ in the experiments by *Agui and Andreopoulos* [5.718]

introduced into the boundary layer through a hole in the wall about 1.8 m or 12.6δ upstream of the measuring location. The particle size, pinhole-fiber diameter, the size of the probe volume, and the separation distance between focusing points on the probe volume were the most important dimensions to be optimized in the LAVOR probe for good spatial resolution. The size of the focusing point, determined by the image of the fiber/pinhole diameter, was 100 µm, which defined the spatial resolution of velocity measurements. As mentioned earlier, the separation distance between the focusing locations on the probe volume determines the spatial resolution of the measurements of velocity gradients, i.e., vorticity. This separation in the case of the $Re_\theta = 2700$ flow was about three Kolmogorov microscale units and was based on the case that this separation distance was 1.0 mm, which appeared to be the upper bound of the estimates. This resolution is about 35% better than that in the experiments of *Balint* et al. [5.731] and *Honkan* [5.690]. As a result of improved resolution, the uncertainty in measurements of vorticity due to unresolved scales is estimated to be considerably below the 9% estimate predicted by the theory of *Ewing* and *George* [5.732].

In this boundary-layer flow, the frequency scale of the Kolmogorov-size eddies $f_\varepsilon = (\varepsilon/\nu)^{1/2}$, which can be considered the maximum frequency anticipated in the flow, is about 250 Hz [5.690, 731]. This indicates that adequate temporal resolution can be achieved with a rate of 250 bursts/s. Figure 5.179a shows a record of the time interval between successive particle arrivals T_p. The bursts produced by these particles have been validated by the burst processors. It can be seen that most of the time intervals are between 1 and 5 ms, which indicates a data rate much faster than 200 Hz. The validated particle data rate, defined as $F_p = 1/T_p$, has a probability density function (PDF), which is shown in Fig. 5.179b. It appears that the most probable value is about 450 particles per second, which is higher than the Kolmogorov frequency of the dissipative scales f_ε and about half the frequency in near-wall viscous units, u_τ^2/ν, which is approximately 1 kHz. Thus the data density in the dissipative scale is $F_p/f_\varepsilon \approx 1-2$. At the same time the particle residence time T_R is smaller than the time between particles T_p. This is also shown in Fig. 5.179b where the PDF of the frequency $F_R = 1/T_R$ is plotted for direct comparison with the PDF of particle rate F_p. The most probable value of F_R appears to be 2.2 kHz, which is substantially larger than the most probable value of the particle rate.

Time coincidence in this case was realized within a time window t_c, which in this case was taken as the inverse of the maximum frequency of the flow $1/f_\varepsilon$, usually characteristic of the dissipative range of scales. This was facilitated in the data-processing software by interpolating between successive bursts separated by a time $\Delta t < t_c$. When this time Δt was greater than t_c signal dropout occurred. Figure 5.180a shows a schematic of the interpolation algorithm used to analyze the data.

The statistical results obtained in the work of *Agui* and *Andreopoulos* [5.718] are not very sensitive to small or large changes in t_c. Figure 5.180b shows the results of a sensitivity analysis of the quantities $\overline{\Omega_3}$, and $(\overline{\omega_3^2})^{1/2}$ when the ratio of t_c/T_ε obtains values in the range 0.1–1.4. The data were obtained at $x_2^+ = 9.58$ and are presented as normalized differences from the value of the corresponding quantity Q at $t_c/T_\varepsilon = 1$, which was finally adopted, i.e., $\Delta Q/Q = [Q(t_c) - Q(T_\varepsilon)]/Q(t_c)$, where Q is either of the two previously mentioned quantities. The results show less than 4% variation in any of the quantities shown above for a 130% variation in t_c/T_ε. These typical results suggest that there is no appreciable sensitivity of $\overline{\Omega_3}$, and $(\overline{\omega_3^2})^{1/2}$ to changes in the coincidence window t_c/T_ε.

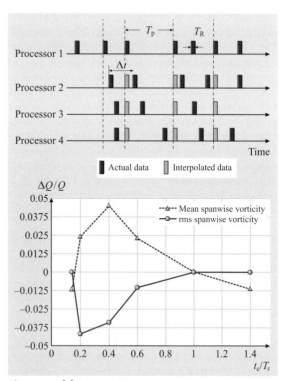

Fig. 5.180 (a) Interpolation scheme used in *Agui* and *Andreopoulos* [5.718]. *Black*: actual data; *grey*: interpolated data. (b) The effects of the coincidence window t_c on vorticity statistics (after *Agui* and *Andreopoulos* [5.718])

Uncertainties in Determining the Separation Distance for Optical Methods

In order to compute the vorticity component Ω_3 from the measured velocity differences, the distance between the two particles should be known. Although the focusing area can be determined rather well, the distance between the two particles cannot be considered as constant because the particle sizes are much smaller than the diameters of the two areas and they may not cross through the centers of the respective areas. Figure 5.181a illustrates the probe volume with two particles entering it at different locations. The actual distance between the two particles Δx_j is bounded in the following range $\Delta x_F - D_F \leq \Delta x_j \leq \Delta x_F + D_F$ where Δx_F is the fixed distance between the centers of the two fiber images and D_F is the diameter of the fiber image, which in the case of the work of Agui and Andreopoulos was about 0.1 mm, a value very close to half the maximum width of the probe volume. Although on average it is expected that $\overline{\Delta x_F} \approx \overline{\Delta x_j}$, instantaneous values of Δx_j can vary by $\pm D_F$ (at most) since particles can cross the probe volume without passing through the center of the fiber image. Thus the ratio of $\pm D_F/\Delta x_j$ can provide the lower and upper bounds of the estimate of the uncertainties involved in the determination of the separation distance between the two particles. Typically the distance Δx_j can be of the order of 1 mm or a fraction of that and can be fixed in advance. In this respect, it is expected that $D_F/\Delta x_j = 0.1$. Agui and Andreopoulos [5.718] attempted to reduce this uncertainty further in two different ways: first, by estimating the relative position of the particle entering the fiber image by taking into account its residence-time information and velocity as well as its local direction θ determined from the measurement of the other velocity component, and second, by filtering out data with residence traveling distance L_R within the image of the fibers, which is considerably less than the diameter D_F.

The first technique accounts for some variation in Δx_j. The associated procedure is better illustrated in Fig. 5.181b where only one image of the fiber is considered, which is assumed to be circular. The path of a particle entering at a point A_{in} with coordinates (x_1^{in}, x_2^{in}) and exiting the probe volume at A_{out} with coordinates (x_1^{out}, x_2^{out}) is also shown in this figure. The coordinate system has its origin at the centroid of the probe volume. Since the location and therefore the coordinates of the particle change during the residency time, a decision has to be made as to which location will be selected to compute the value of Δx_j. It appears that the most plausible location is the mid distance between the entry and exit points (x_1^m, x_2^m). Geometrical considerations indicate that $x_1^m = x_1^{in} - 1/2 L_R \tan\theta$ where L_R in the normal distance between A_{in} and A_{out} ($x_2^{in} - x_2^{out}$), which can be estimated through the residency time T_R and the measured local velocity v_2. Thus $x_2^{in} - x_2^{out} = L_R = v_2 T_R$ and the estimated offset location of the particle coordinate in the x_1 direction is

$$\delta x_j = \cos\theta \left[\frac{D_F^2}{4} + \left(\frac{L_R}{2\cos\theta} \right)^2 \right]^{1/2}$$

$$= \left(\frac{D_F^2}{4}\cos^2\theta + \frac{L_R^2}{4} \right)^{1/2}. \quad (5.314)$$

The angle θ is the local direction of the particle at A_{in} defined as $\theta = \tan^{-1}(v_1/v_2)$. The value of v_1 at this location is not known. An estimate of this velocity component is available at the origin of the coordinate system, which was obtained at the other probe volume. This estimate of v_1 can be used in evaluating the previous equation without accounting for a possible variation along the x_1.

In order to gain some insight into the contributing terms of the previous equation, Agui and Andreopoulos [5.718] computed values of the residency distance L_R and their PDF is shown in Fig. 5.181c. L_R cannot exceed the diameter of the image of the fiber within the probe volume D_F, which was about 0.1 mm. The data of Fig. 5.181c clearly show that the majority of the events take place at $L_R \approx D_F$. The data also show that there exist contributions from events with $L_R < D_F$. The PDF drops steeply for values of $L_R > D_F$. The data indicate that the useful range of L_R is between 0.001 mm $< L_R <$ 0.165 mm the upper bound of which is slightly higher that the anticipated 0.1 mm. It appears that events with values close to the two limits of this range contribute insignificantly to the statistical data.

In the second method used to reduce the uncertainty in Δx_j, events with values of L_R outside the range 0.016 mm $< \Delta x_j <$ 0.14 mm were filtered out from the statistics.

The drawback of the first technique is that the depth of field of the fiber end images has been ignored in these considerations. In addition, there is an ambiguity in determining the sign of δx_j under these assumptions. The disadvantage of the second procedure is that by filtering out events with data, which correspond to particles crossing the fiber end image at locations away from its center, the resultant time series of the processed data contain more signal dropouts than before. It appears, however, that both corrections, when applied to the acquired data, have minor effect on the statistical results. Mean and rms vorticity values are reduced by 1–2% only. The reason is that the number of events affected by the corrections is rather small in comparison to the total number of realizations. If the fixed distance between the two fibers is reduced to 0.4 mm the effect of both corrections is expected to be greater than in the case of $\Delta x_j = 1$ mm. As a result of these considerations, the data of *Agui* and *Andreopoulos* [5.718] have not been corrected to reduce the uncertainty in the fiber separation Δx_j. In addition, no corrections due to the velocity bias have been applied to their data.

Uncertainty Estimates of Velocity Gradients Measurements

Direct estimates of the uncertainties associated with the measurements of velocity gradients can be obtained by considering the propagation of the uncertainties in the measurement of each quantity involved in the process. Following the example of *Honkan* and *Andreopoulos* [5.689], a typical velocity gradient is measured by the following approximation:

$$\frac{\partial v_i^+}{\partial x_j^+} \cong \frac{v_i(x_j + \Delta x_j) - v_i(x_j)}{\Delta x_j} T_v = F, \quad (5.315)$$

where T_v is an appropriate time scale to normalize velocity gradients, which in the case of wall-bounded flows can be the viscous time scale in the near-wall region, $T_v = v/u_\tau^2$ with u_τ being the friction velocity.

For the typical test case of *Agui* and *Andreopoulos* [5.718], the uncertainties in the measurements of $v_i(x_j)$ and $v_i(x_j + \Delta x_j)$ can be assumed to be the same, $\Delta v_i(x_j) = \Delta v_i(x_j + \Delta x_j) = \Delta v$, and if the uncertainty in the estimate of Δx_j is $\Delta(\Delta x_j)$, then the relative uncertainty $\Delta F/F$ for the case of wall-bounded flows will be given by

$$\frac{\Delta F}{F} = \left\{ 2 \left[\frac{\Delta v}{v_i(x_j) - v_i(x_j + \Delta x_j)} \right]^2 + \left[\frac{\Delta(\Delta x_j)}{(\Delta x_j)} \right]^2 + 4 \left(\frac{\Delta u_\tau}{u_\tau} \right)^2 \right\}^{1/2}. \quad (5.316)$$

As can be seen from this relation, the factor 4 in front of the third term on the right-hand side makes the uncertainty in the measurement of the friction velocity u_τ one of the major contributors to the total uncertainty. A typical uncertainty of 5% is assumed in determining u_τ.

The relative uncertainty shown in (5.316) has typically been estimated in the viscous sublayer where F is maximum. *Agui* and *Andreopoulos* [5.718] suggest that a typical value of the uncertainty Δv is 1%, which corresponds to about 0.01 m/s while a typical velocity difference is $\overline{v_1(x_j) - v_j(x_j + \Delta x_j)} = u_\tau \Delta x_2^+ = 0.2$ m/s, where $\Delta x_2^+ = 1.6$ (the probe size). A typical value of the uncertainty in estimating the distance Δx_j is 2%, a value based on the results of the sensitivity analysis described earlier. Thus the uncertainty $\Delta F/F$ appears to be about 11% in the viscous sublayer. The relative un-

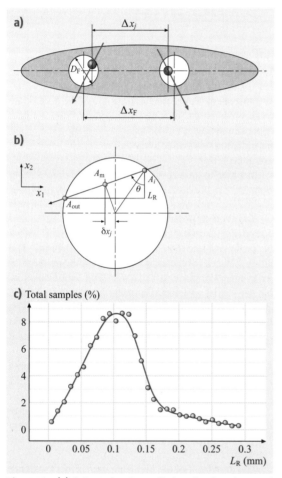

Fig. 5.181 (a) Schematic of two distinct focal regions and particle crossings on the same probe volume. (b) Geometrical details of particle crossings on half probe volume. (c) Typical probability density function of the transient/residence distance inside the probe volume in the experiments by *Agui* and *Andreopoulos* [5.718]

certainty $\Delta F/F$ increases as the distance away from the wall increases because the absolute value of F decreases.

Comparison of Mean Velocity-Gradient Data in Boundary-Layer Flows

A comparison between the mean velocity gradient and the derivative of the mean velocity profile is a good indicator of the probe's ability to measure at least one velocity gradient with sufficient accuracy. An attempt has been made here to compare several results obtained by different techniques in different facilities in boundary-layer or channel flow experiments.

Figure 5.182 shows the distribution of the mean velocity gradient $\overline{\partial U_1^+/\partial x_2^+}$ (negative $\overline{\Omega_3^+}$) across the boundary layer. Table 5.7 lists some of the bulk characteristics of the various datasets used in the present comparison. The data of *Honkan* and *Andreopoulos* [5.689], *Briassulis* et al. [5.694], and *Agui* and *Andreopoulos* [5.718] were obtained in the same wind-tunnel facility at CCNY. The data of *Agui* and *Andreopoulos* [5.718] were obtained with the optical technique LAVOR and the results shown in Fig. 5.182 are those measured at the centroid of the probe by time-averaging its instantaneous values. These data can be directly compared with the data of *Honkan* and *Andreopoulos* [5.689]. The two datasets agree very well in the buffer region while the LAVOR data appear to be slightly higher in the logarithmic region.

On the same figure the data of the mean velocity derivative are plotted; were obtained by differentiating *Spalding*'s [5.733] formula for the mean velocity distribution in the near-wall region, including the buffer zone and the logarithmic region. In addition, the data of mean velocity derivative in the logarithmic region, $1/\kappa x_2^+$, where κ is the so-called von Kármán constant, are also plotted for comparison. The data obtained by differentiating the near-wall velocity measurements of *Purtell* et al. [5.734] and *Andreopoulos* et al. [5.735] have also been plotted in the same figure. Purtell et al.'s mean velocity data show no Re_θ effect in the near-wall region and therefore the deduced mean velocity-gradient data are independent of Re_θ. The data of Andreopoulos et al. correspond to $Re_\theta = 3624$.

The LAVOR data are also compared with the hot-wire data of *Balint* et al. [5.731]. It appears that the agreement of the LAVOR data with those of Balint et al. and with Spalding's formula in the buffer region is very satisfactory. Both data sets are also in very good agreement with the data obtained from the relation of the logarithmic law in the region of its validity. The data of Purtell et al. are lower than those predicted by Spald-

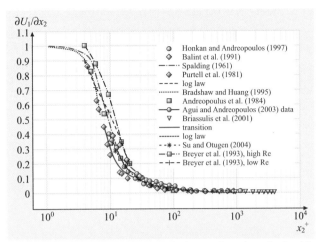

Fig. 5.182 Comparison of mean velocity gradient data obtained in boundary-layer flows by various experimental techniques

ing's correlation while the data of Andreopoulos et al. are closer to Spalding's prediction in the upper edge of the buffer layer and close to Purtell's data at the upper edge of the viscous sublayer. However, if one considers that the uncertainty introduced in deducing these data and the uncertainty band associated with Spalding's formula, which originates from a modest 5% uncertainty in the initial mean velocity measurements, the agreement of the LAVOR measurements with existing data and correlations is quite satisfactory. It should also be pointed out that Spalding's relation should be reconsidered in light of recent demonstrations of limitations of the classical law-of-the-wall for the viscous sublayer. *Bradshaw* and *Huang* [5.736], for instance, have shown that integration of the total shear stress expression yields $\partial \overline{U_1}/\partial x_2^+ = 1 - A(x_2^+)^3$, where $A = \overline{\partial u_1/\partial x_2 \partial p/\partial x_2}$. The value of this coefficient, as suggested by *Mansour* et al. [5.737], is 1.4×10^{-3} and the whole subtracting term represents the contribution of the turbulent part of the total shear stress if the latter is independent of the wall distance in the viscous sublayer. Based on this relation the velocity gradient appears to be lower than that calculated from Spalding's formula (Fig. 5.182).

The data of *Su* and *Ötügen* [5.738] obtained in a water tunnel agree reasonably well with the rest of the data. The high-Reynolds-number data of *Breyer* et al. [5.706] ($Re_h = 100\,000$) are substantially higher than the rest of the data in the near-wall region. The low-Reynolds-number data ($Re_h = 3500$) show better agreement with the majority of the data under comparison in the near-wall region. In general, most of

Table 5.7 Principal parameters of various experiments providing rms vorticity data

Datasets	Re_θ	Resolution L_p/η	L_p^+	Measurement technique	Measured quantities
[5.718]	2790	< 4	< 6	Optical	Ω_3
[5.696]	3900	5	13	Optical	Ω_3
	5400	6	19		
[5.731]	2685	6	11	Hot wire	$\Omega_1, \Omega_2, \Omega_3; S_{ij}$
[5.689]	2790	7	12	Hot wire	$\Omega_1, \Omega_2, \Omega_3; S_{ij}$, full dissipation
[5.739]	1450	6	–	Hot wire	Ω_3
[5.705]	2270	–	15	Optical, PIV	Ω_3
[5.705]	6450	–	20	Optical, PIV	Ω_3
[5.740]	1100	0.6–1.5	1.95	Hot wire	Ω_3
	2870	0.9–3.4	4.75		
	4850	1.7–7.0	7.82		
[5.741]	5×10^6	–	< 6	Hot wire	Ω_3
[5.738]	1837	17	26	Optical	Ω_3
[5.706]	(Re_h) 3500	2	–	Optical, laser speckle	$\partial v/\partial x + \partial v/\partial y, \partial v/\partial x - \partial v/\partial y,$ and $\partial v/\partial z$
	6700	4			
	100 000				

the data agree reasonably well with each other in the region $x_2^+ > 80$ while there are substantial differences in the closer region to the wall with $x_2^+ < 80$ where most of the intensive vortical activities take place. Thus the best test for the method is how well it works in the near-wall region of the boundary-layer flow.

Comparison of Fluctuating Vorticity Statistical Data in Boundary Layers

Comparison of various data sets that describe the distribution of the rms spanwise vorticity fluctuations is shown in Fig. 5.183a plotted against x_2^+ and normalized by u_τ^2/ν. The data sets under comparison are listed in Table 5.7 together with their spatial resolution estimates. In addition to these experimental data, the DNS results of *Spalart* [5.742] obtained at a rather low $Re_\theta = 1400$ and by *Kim* et al. [5.743] are also plotted for comparison. The data of *Rajagopalan* and *Antonia* [5.739] and *Meinhart* and *Adrian* [5.705] obtained by PIV techniques with a spatial resolution of about $\Delta x^+ = 20$ are also plotted on the same figure. It should be mentioned that all the data sets, with the exception of Meinhart and Adrian and Su and Ötügen, agree remarkably well in the region $x_2^+ > 60$. Considerable differences, however, seem to exist in the near-wall region. The LAVOR data of *Agui* and *Andreopoulos* [5.718] seem to be substantially higher than any of the other measurements. The data by *Klewicki* [5.740] obtained at $Re_\theta = 2870$ are reasonably close to the LAVOR data but start to deviate in the buffer region with $x_2^+ < 20$ where differences of up to 15% can be observed.

Comparison with data obtained in other facilities under different conditions is meaningful only when appropriate scaling is available. The work of *Balint* et al. [5.731] and *Honkan* and *Andreopoulos* [5.689] clearly indicated that current scaling ideas based on inner, outer, or mixed-layer parameters failed to collapse velocity gradient/vorticity data. The data of *Honkan* and *Andreopoulos* [5.689], however, have been obtained in the same wind tunnel under exactly the same flow conditions as the LAVOR measurements. This allows a direct comparison in the absence of any problem due to inadequate scaling. It appears that the LAVOR data are higher than the hot-wire data in the $x_2^+ < 60$ region. One possible reason for this behavior is the higher spatial resolution of the LAVOR technique and its non-intrusive character.

There are two more parameters to be considered for a meaningful comparison of the various data sets. First is the accuracy in the measurements of vorticity and the uncertainty in measuring u_τ, which as discussed later appears to be one of the most dominant contributors to the overall uncertainty. Unfortunately the overall issue of accuracy in the measurements of vorticity is not addressed in detail in the published literature and information on uncertainty in the measurements of u_τ is not available for the majority of datasets. Second is the effect of Re_θ which, as will be shown in the next Section, is yet to be established. There is evidence suggesting that ω_3^+-rms increases with increasing Re_θ. However, quantitatively this effect is not strong enough to explain the rather sub-

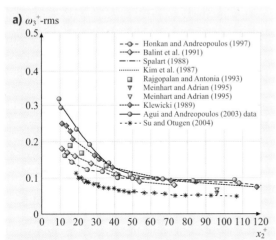

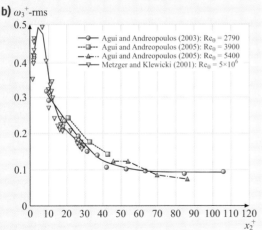

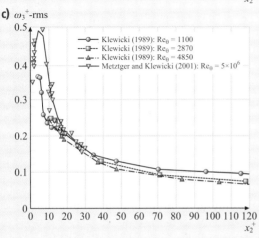

stantial differences between the various datasets shown in Fig. 5.183a. Most probably these differences are attributed to the combined effects of different accuracies involved in each of the techniques applied including those due to spatial resolution, the lack of appropriate scaling, and the effects of Re_θ.

Reynolds-Number Effects

The effect of Reynolds number on statistical properties of velocity fluctuations in turbulent flows is always of interest, since most opractical engineering flows are in the high-Reynolds-number regime and most laboratory data have been obtained in low-Reynolds-number flows. Reynolds-number effects on vorticity statistics have yet to be demonstrated experimentally. Some rms data for the quantities $\partial v/\partial x + \partial v/\partial y$, $\partial v/\partial x - \partial v/\partial y$, and $\partial v/\partial z$ obtained by *Kriegs* et al. [5.707] show no effect for Reynolds numbers of 2600–4000. In the work of *Kriegs* and *Strude* [5.708] the rms data for $\partial v/\partial x + \partial v/\partial y$ seem to increase slightly in the near-wall region, with increasing Reynolds number from 2250 to 6700, while the rms data of the quantity $\partial v/\partial z$ decrease with increasing Re.

Figure 5.183b shows the rms of the spanwise vorticity fluctuations at three different values of Re_θ obtained by *Agui* and *Andreopoulos* [5.744] by using their LA-VOR techniques. Although the spatial resolution is reduced with increasing Re_θ (Table 5.7) it remains reasonably acceptable at $Re_\theta = 3900$ and 5400. There is some evidence that indicates that ω_3-rms increases slightly with increasing Re_θ in the near-wall region. There is a modest increase of about 20% of ω_3-rms for a Re_θ increase of 39% at $x_2^+ = 30$. In the case of $Re_\theta = 5400$, no data in the near-wall region are available and therefore the Re effect shown in the case of $Re_\theta = 3900$ cannot be corroborated.

The high-Reynolds-number data of *Metzger* and *Klewicki* [5.741] obtained in the atmospheric boundary layer developed over the salt flats of the Great Salt Lake Desert are also plotted for comparison in Fig. 5.183b. These data appear to be higher than the present ones within the viscous sublayer and lower in the outer part of the buffer zone and the beginning of the logarithmic region. Thus the picture that emerges from this comparison is that, in the buffer and the viscous sublayer, the rms of the spanwise vorticity

Fig. 5.183 (a) Comparison of rms data for spanwise vorticity fluctuations obtained in boundary-layer flows by various experimental techniques. (b,c) Effects of Reynolds number on the vorticity statistics in a boundary layer. (b) After *Agui* and *Andreopoulos* [5.696]; (c) After *Klewicki* [5.740]

fluctuations increases with increasing Re_θ. This dependency is considerably reduced and reverses away from the wall.

Quite similar behavior is exhibited by the data of *Klewicki* [5.740], obtained in a wind tunnel at three different values of Re_θ. For clarity these data have been plotted in Fig. 5.183c separately to those shown in Fig. 5.183b. These data agree well with the data of *Agui* and *Andreopoulos* [5.744]. A comparison with the high-Re_θ data of *Metzger* and *Klewicki* [5.741] clearly suggests a rather strong dependency of vorticity fluctuations on Re_θ in the viscous sublayer and the buffer region.

Concluding Remarks

The importance of vorticity, velocity gradient tensor, rate-of-strain tensor, and wall vorticity flux was discussed first in the context of the structure function in the present contribution, followed by a discussion of the general requirements of strophometric techniques. Of paramount significance amongst these are the issues of the temporal and spatial resolution of the probe volume for the measuring technique and its validation in a turbulent flow field. Several optical techniques used in strophometry were described in detail. The major advantage of optical techniques over those based on hot wires is that they are non-intrusive in the flow field and are therefore suitable for separated flows or flows with relatively high variation in the pitch or yaw angles of the instantaneous velocity vector. In canonical turbulent boundary-layer flows, for instance, the rms of the pitch and yaw angles is about 9° very close to the wall [5.689], which for a Gaussian distribution with 3 rms range variation indicates a range of 0 to 54. In that respect, hot-wire techniques with sensitivity to pitch/yaw angle variation in this range can be effective in channel or boundary-layer-type flows. The non-intrusive nature of the optical techniques always has to be considered in relation to possible disturbances introduced by seeding since adding particles can alter the flow substantially.

A comparison of data measured in boundary-layer or channel flows using optical techniques with measurements obtained by other means was attempted to evaluate their performance.

The value of each of these techniques can be estimated by considering its cost, complexity, and performance. Most of the optical techniques developed to date have demonstrated capabilities in measuring only one component of the vorticity vector. Future work and development should focus on:

1. Increasing the spatial and temporal resolution of the techniques so that reliable time-dependent strophometric data can be obtained in high-Reynolds-number flows
2. Increasing the capabilities to measure more than one vorticity component
3. Decreasing the cost of hardware and level of complexity of the techniques
4. Applying present techniques to a wide range of flows.

5.5.2 High-Resolution Dual-Plane Stereo PIV (DSPIV)

One of the main motivations behind the continuing development of advanced measurement methods in fluid mechanics is the need for experimental studies of turbulent flows, where the wide range of length and time scales involved in the fluid motion creates one of the most demanding challenges in all of measurement science. Researchers in fluid mechanics are increasingly seeking to address issues regarding intermediate and small scales of inhomogeneous turbulence of the type encountered in practical turbulent shear flows. Unlike homogeneous isotropic or uniformly sheared turbulence in periodic domains, spatially developing turbulent shear flows at large Reynolds numbers cannot be readily studied by direct simulation. At the same time, new measurement capabilities are beginning to provide experimental access to flow information that had previously been accessible only by simulations. These include multiple-sensor hot-wire probes that provide single-point time-series data for various key turbulence quantities, such as the time-varying vorticity and strain rate components, as well as the enstrophy and the dissipation rate. In parallel, these data are being augmented with new multidimensional optical measurement methods that are providing two-, three- and even four-dimensional data for velocity gradient fields that are essential to understanding the intermediate and small scales of turbulent shear flows. Collectively, these techniques are beginning to provide a direct experimental measurement capability that, at least in the near term, may provide the most practical way to obtain the detailed structure, statistics, similarity and scaling properties of gradient quantities in turbulent shear flows at a level of resolution and detail that has traditionally been associated only with direct numerical simulations.

The fields of principal interest for understanding the precise structure and dynamics of inhomogeneous turbulence in spatially developing turbulent shear flows are all derived from the nine simultaneous components of the velocity gradient tensor field $\partial u_i/\partial x_j(\mathbf{x}, t)$, given by

$$\frac{\partial u_i}{\partial x_j} = \begin{pmatrix} \partial u/\partial x & \partial u/\partial y & \partial u/\partial z \\ \partial v/\partial x & \partial v/\partial y & \partial v/\partial z \\ \partial w/\partial x & \partial w/\partial y & \partial w/\partial z \end{pmatrix} \quad (5.317)$$

The components of this tensor determine the three components of the vorticity vector

$$\omega_x = \left(\frac{\partial w}{\partial y} - \frac{\partial v}{\partial z} \right),$$
$$\omega_y = \left(\frac{\partial u}{\partial z} - \frac{\partial w}{\partial x} \right),$$
$$\omega_z = \left(\frac{\partial v}{\partial x} - \frac{\partial u}{\partial y} \right),$$

the three normal components of the strain rate tensor

$$S_{xx} = \left(\frac{\partial u}{\partial x} \right),$$
$$S_{yy} = \left(\frac{\partial v}{\partial y} \right),$$
$$S_{zz} = \left(\frac{\partial w}{\partial z} \right),$$

and the three shear components of the strain rate tensor

$$S_{xy} = \frac{1}{2}\left(\frac{\partial u}{\partial y} + \frac{\partial v}{\partial x} \right),$$
$$S_{yz} = \frac{1}{2}\left(\frac{\partial v}{\partial z} + \frac{\partial w}{\partial y} \right),$$
$$S_{zy} = \frac{1}{2}\left(\frac{\partial w}{\partial x} + \frac{\partial u}{\partial z} \right).$$

The strain rate and vorticity components can be written as

$$S_{ij} \equiv \frac{1}{2}\left(\frac{\partial u_i}{\partial x_j} + \frac{\partial u_j}{\partial x_i} \right) \quad \text{and}$$
$$\omega_k \equiv \frac{1}{2}\left(\frac{\partial u_j}{\partial x_i} - \frac{\partial u_i}{\partial x_j} \right)\varepsilon_{ijk}.$$

Of these, specific interest is often directed at such gradient fields as the enstrophy $\omega_i\omega_i$, the enstrophy production rate $\omega_i S_{ij}\omega_j$, and the kinetic-energy dissipation rate $2\nu S_{ij}S_{ij}$, and in their properties at the quasi-universal intermediate and small scales of turbulent shear flows.

Numerous studies have sought to develop accurate ways of measuring these nine quantities simultaneously in turbulent flows, including single-point probe-based measurements [5.81, 687, 693] as well as various multidimensional imaging techniques. Most of the latter methods are based on particle imaging approaches. Classical particle image velocimetry (PIV) allows measurement of two in-plane velocity components, which provide access to four of the nine velocity gradient tensor components. These permit formation of three of the six components of the strain rate tensor and a single vorticity component. Stereo PIV additionally provides the out-of-plane velocity component, thereby giving access to two further velocity gradient components, however these do not provide any additional strain rate or vorticity components. Particle tracking velocimetry (PTV) provides three-component velocity fields throughout a three-dimensional volume, allowing access to all nine components of $\partial u_i/\partial x_j$, but the large particle separations needed for accurate tracking typically prevent gradient measurements at the small scales of turbulence. The most extensive velocity gradient measurements in turbulent flows to date have come from holographic particle image velocimetry (HPIV) [5.709], although the resolution in those measurements is typically larger than the smallest scales in the turbulent flow. Indirect measurements of $\partial u_i/\partial x_j$ via scalar imaging velocimetry (SIV) are based on three-dimensional laser-induced fluorescence imaging of a scalar field [5.745, 746] and inversion of the conserved scalar transport equation to obtain the underlying three-component velocity field [5.747, 748]. This allows highly resolved measurements of all nine simultaneous components of the velocity gradients at the intermediate and small scales of a turbulent flow, but requires smoothness and continuity constraints in the inversion to obtain the velocity field.

Dual-plane stereo particle image velocimetry (DSPIV) is based on two separate stereo PIV measurements that provide all three components of velocity in two parallel light-sheet planes. This allows all nine components of $\partial u_i/\partial x_j$ to be calculated from the measured velocities in the two planes, with the resolution determined largely by the thickness and spacing of the light-sheet planes and the size of the PIV correlation window. The approach requires separating the light scattered by particles in the two light-sheet planes onto two independent stereo camera pairs. A polarization-based approach, in which the two light sheets were arranged with orthogonal polarizations so that each stereo camera pair saw the scattered light from only one of the sheets, was introduced in [5.749]. This has been used to measure comparatively large-scale features of the flow in a turbulent boundary layer [5.750, 751], and the same technique has been used to investigate large-scale features of a lobed jet mixer [5.752]. A related polarization-based dual-plane approach that combines stereo PIV and conventional PIV, with continuity used to estimate the out-of-plane gradient component, has been used to study features of a turbulent boundary layer [5.753]. These studies all investigated specific aspects of these flows, and did not seek to resolve or study velocity gradients on the quasi-universal intermediate and small scales of turbulent shear flows. Moreover, maintaining the orthogonal polarization in the Mie scattered light requires the scattering particles to be spherical, and as a consequence these studies used fine liquid droplets as seed particles. This can be done in isothermal flows, but in exothermic reacting turbulent shear flows such liquid droplets do not survive and the polarization-based method cannot be used.

A two-frequency DSPIV approach [5.713, 754, 755] allows traditional solid metal oxide particles to be used as the seed, thereby permitting measurements in reacting as well as nonreacting turbulent shear flows. This approach is based on measurements in two differentially spaced light-sheet planes, using two different laser frequencies in conjunction with filters to separate the light scattered from the seed particles onto the individual stereo camera pairs. The differential spacing of the two stereo PIV planes allows resolution of all nine components of the $\partial u_i / \partial x_j$ fields down to the local inner (viscous) scale of the turbulent flow, permitting highly-resolved experimental studies of the quasi-universal intermediate and small scales of nonreacting and exothermically reacting turbulent shear flows.

DSPIV Measurement Setup

As indicated in Fig. 5.184, the two-frequency DSPIV method [5.713] is based on two simultaneous, independent stereo PIV measurements in two differentially spaced light-sheet planes. Two different light-sheet frequencies are used in conjunction with appropriate optical filters to separate the scattered light from particles seeded in the flow onto two independent stereo PIV camera pairs. The two-frequency system involves four Nd:YAG lasers, of which two are sequentially triggered to create the double pulses for the light sheets of wavelength 532 nm and the other two, also sequentially triggered at the same two instants of time, pump two pulsed dye lasers that provide the light sheets of wavelength 635 nm. Light scattered from 0.5 μm aluminum oxide seed particles is recorded on four 12 bit 1280×1024 px PIV cameras. A typical field of view of 15.5×12.5 mm is achieved using Sigma 70–300 f/4-5.6 APO macro lenses. The 532 nm and 635 nm stereo camera pairs each have narrow-band optical filters that block the scattered light from the other light sheet. The four cameras are arranged in an asymmetric angular-displacement forward–forward imaging configuration

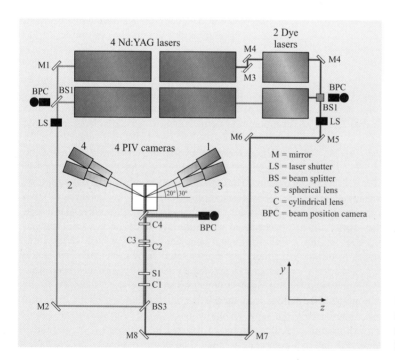

Fig. 5.184 Experimental arrangement for two-frequency dual-plane stereo PIV measurements (after [5.713])

with a 50° included angle satisfying the Scheimpflug condition. The small field of view required for high-resolution measurements and the long focal length of the camera lens necessitate large-f-number imaging to obtain sufficiently focused particle images across the entire field of view. Precise alignment of the object, lens, and image planes requires custom-designed camera mounts to maintain a coincident field of view for all four cameras. Calibration of this imaging arrangement is based on a transparent precision calibration target [5.713] that provides registration of all four cameras to a common reference frame, as well as the mapping from the object planes to each of the four image planes to correct for distortions due to the variable magnification across each image. Particle images are processed with an adaptive multipass technique using 32×32 px interrogation boxes with no overlap in the final vector fields. This provides the same in-plane resolution between vectors as the out-of-plane resolution between the planes. More-detailed descriptions of this two-frequency DSPIV setup can be found in [5.713].

Light–Sheet Characterization

The thickness, spacing, and two-axis parallelism of the two pairs of light sheets are of central importance to the resolution and accuracy of DSPIV measurements. The four laser beams are initially aligned without the sheet-forming optics to be coincident to within $\pm 25\,\mu$m over the entire optical path, as verified optically using CCD cameras, and only then formed into light sheets by inserting the spherical and cylindrical lenses. Prior to each measurement, transmitted light from a knife edge traversed across each sheet is measured with a photodiode to assess the characteristics of the four light sheets quantitatively [5.713]. An error-function fit to each resulting profile is differentiated to obtain a Gaussian sheet-normal intensity profile. The three lowest-order moments of this profile allow the sheet centerline position and the local $1/e^2$ thickness of each laser sheet to be determined. This procedure is repeated at 16 vertical positions along each sheet in the center of the field of view, as well as at the right and left edges, to characterize the four light sheets over the entire field of view, as shown in Fig. 5.185. Typical results verify actual thicknesses of $400 \pm 20\,\mu$m in the two 532 nm light sheets and $800 \pm 20\,\mu$m in the two 635 nm light sheets. Sheet noncoincidence in each light-sheet pair is typically less than $40\,\mu$m, and less than 1° of nonparallelism can be achieved between the 532 nm and 635 nm light-sheet pairs. This procedure also quantitatively confirms the $400 \pm 20\,\mu$m separation between the two light-sheet

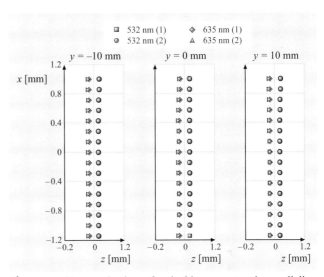

Fig. 5.185 Characterization of coincidence, two-axis parallelism and separation of the two light-sheet pairs (after [5.713])

pairs that matches the out-of-plane spatial resolution to the in-plane resolution of the measurements.

Relative Measurement Resolution

The light sheet characteristics alone do not determine the spatial resolution achieved in any dual-plane measurement. Four principal factors collectively contribute to the spatial resolution of these DSPIV measurements:

1. the light-sheet thicknesses and separation
2. the PIV correlation window size
3. the typical bad-vector replacement region size
4. the PIV smoothing filter scale.

As noted before, typical achievable laser sheet thicknesses are $\Delta_1 \approx 400\,\mu$m and $\Delta_1 \approx 800\,\mu$m in the 523 nm and 635 nm light sheets, respectively. The 32×32 px PIV correlation window, together with the pixel size and magnification ratio, effectively filters the velocity field at a scale of $\Delta_2 \approx 400\,\mu$m. From the vector-to-vector spacing, the typical bad-vector replacement scale $\Delta_3 \approx 800\,\mu$m is applied to approximately 5% of the vectors, and the median filter applied to the velocities in the final processing stage is at a scale of $\Delta_4 \approx 800\,\mu$m. Since these contributions to the net resolution are independent, the resulting velocity gradient fields are resolved to a length scale of $\Delta_{\text{net}} = (\Delta_1^2 + \Delta_2^2 + \Delta_3^2 + \Delta_4^2)^{1/2} \approx 1100\,\mu$m.

To assess the relative resolution of any such measurement, the spatial resolution Δ_{net} can be compared with the local Kolmogorov scale $\lambda_K \equiv (\nu^3/\varepsilon)^{1/4}$, defined in terms of the viscosity ν and the local mean dissipation rate ε in homogeneous isotropic turbulence and applied to turbulent shear flows via the assumption of local isotropy. For the DSPIV measurement results shown herein, at $Re_\delta = 6000$ and $30\,000$, the respective Kolmogorov scales λ_K are $210\,\mu\text{m}$ and $108\,\mu\text{m}$, giving a relative resolution of $\Delta_{\text{net}}/\lambda_K \approx 5.2$ at the lower Reynolds number. Here Re_δ is the local outer-scale Reynolds number $Re_\delta \equiv u_c \delta/\nu$, where u_c and δ are the local velocity and length scales, respectively, in the local mean velocity profile that characterize the peak mean shear. In turbulent jets, u_c and δ are, respectively, the local centerline mean velocity and local flow width. The corresponding inner (viscous) length $\lambda_\nu \approx 5.9\,\lambda_K$ is the local strain-limited viscous diffusion scale on which gradient fields in turbulent flows are concentrated by the competing effects of strain and viscous diffusion [5.756, 757]. At $Re_\delta = 6000$ and $30\,000$ the λ_ν values are $1240\,\mu\text{m}$ and $640\,\mu\text{m}$, with the resulting $\Delta_{\text{net}}/\lambda_\nu \approx 0.89$ at the lower Reynolds number.

Velocity Gradient Calculation

From the measured velocity component values (u, v, w) at each point in the two differentially spaced z-planes,

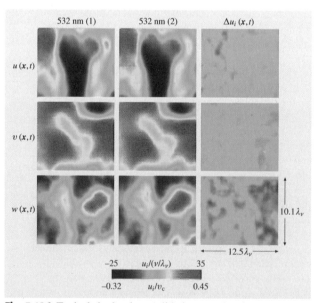

Fig. 5.186 Typical single-plane validation test showing $u_i(x, t)$ velocity component fields measured in the same light-sheet plane by both stereo camera pairs (*left* and *middle*), and differences $\Delta u_i(x, t)$ (*right*) on same color scale; note that $\lambda_\nu \approx 5.9\,\lambda_K$, as described in the text (after [5.758])

the in-plane components of the velocity gradient tensor $\partial u_i / \partial x_j$ can be obtained from central differences in the x- and y-directions, and the out-of-plane components obtained from one-sided differences along the z-direction. However, this would only make use of the measured velocities at six of the 18 points available in the $3 \times 3 \times 2$ template around each point. If instead the natural pixel-based coordinate frame is rotated by $\pm \pi/4$ along the z-axis, or by $\pm \tan^{-1}(1/2)$ along the x- and y-axes, then separate evaluations of the gradient components can be obtained in each rotated frame and projected back onto the original frame. Averaging the resulting velocity gradient components obtained from all the frames produces a differencing stencil that makes use of all 18 points around each point. This provides a reduction of random error in the gradient components without significantly degrading the net spatial resolution in the resulting derivative fields. This can be contrasted to an explicit filter applied to the velocity gradients in the original frame, which for the same error reduction would lead to a substantial degradation in resolution.

Single-Plane Validations

Although it is always possible to differentiate any measured velocity field to produce a matrix of derivative values, it is an entirely different matter to verify that these derivatives represent the true velocity gradient tensor components in the flow. Since high-resolution measurements of velocity gradients necessarily involve differences between nearly equal measured velocity values, it is possible for the resulting derivatives to be completely dominated by errors in the original measured velocities and thus not reflect the true velocity gradients to any significant accuracy. Avoiding this requires considerable effort to minimize even small errors in the velocity measurements, along with careful assessments of the relative errors in the resulting velocity differences to validate the tensor of derivative values. The most fundamental of such validation tests is based on single-plane imaging [5.758], in which both stereo camera pairs in the DSPIV system are arranged to image the same particle field in the same double-pulsed light sheet, so that $\Delta z \equiv 0$. In principle, the velocity fields from the two camera pairs should then be identical, and differences in the two independently measured velocity fields thus provide a quantitative assessment of the accuracy with which velocity gradients can be measured. Typical results from such single-plane imaging assessments are shown in Figs. 5.186 and 5.187, where the measurements are from the centerline of an axisymmetric turbulent jet at $Re_\delta = 6000$. Figure 5.186 shows the instantaneous velocity component fields $u_i(\mathbf{x}, t)$ from the two stereo camera pairs when both image the same 532 nm light sheet, as well as the differences $\Delta u_i(\mathbf{x}, t)$ between these two independent measurements, shown on the same color scale. It is apparent that the differences Δu_i are far smaller than the velocity component values themselves, reflecting the comparatively high accuracy that can be obtained in such DSPIV measurements once numerous error sources have been addressed. The distributions of the measured Δu_i values obtained over many such instantaneous realizations are shown in Fig. 5.187. Relative errors, given by the ratio of rms values of Δu_i and u_i, are found to be 6.5% for the u- and v-components and 10.5% for the w-components, with the latter being slightly larger due to the inherently higher errors in measuring the out-of-plane velocity component with stereo PIV systems.

Coincident-Plane Validations

The additional errors due to light-sheet positioning inaccuracies and the separate formation of 532 nm and 635 nm light-sheet pairs can be obtained from coincident-plane imaging tests [5.758]. In this case, the velocity component fields obtained from the two stereo camera pairs are compared when separate 532 nm and 635 nm light sheets are created and positioned with the optical alignment system to be as nearly coincident as experimentally feasible, i.e., $\Delta z \approx 0$. Typical results from such coincident-plane imaging assessments are shown in Figs. 5.188 and 5.189. In principle, the $u_i(\mathbf{x}, t)$ fields from the 532 nm and 635 nm camera pairs should again be identical, with differences being due to the errors addressed before and the additional errors introduced by the independent dual light-sheet genera-

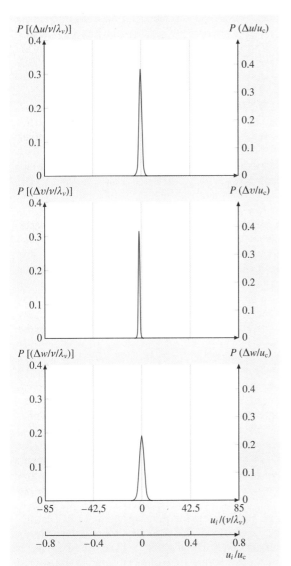

Fig. 5.187 Distributions of velocity component differences from single-plane validations tests of the type in Fig. 5.186, shown normalized on inner- and outer-scale variables (after [5.758])

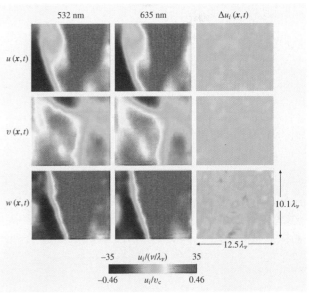

Fig. 5.188 Typical coincident-plane validations, showing $u_i(\mathbf{x}, t)$ velocity component fields measured in two coincident light-sheet planes by the two stereo camera pairs (*left* and *middle*), and differences $\Delta u_i(\mathbf{x}, t)$ (*right*) shown on the same color scale (after [5.758])

tion and positioning. It is apparent that the differences Δu_i in Fig. 5.188 are again far smaller than the velocity component values themselves, indicating that the additional errors due to separate light-sheet formation and positioning inaccuracies are substantially smaller than the errors already inherent to stereo imaging of the finite-thickness light sheets. The distributions of the measured Δu_i values from many such realizations are

shown in Fig. 5.189. In this case, the elative errors are found to be 8–9% for the u- and v-components, and 16% for the w-components, verifying that the errors due to the dual light-sheet formation and positioning procedure are measurable, but are substantially smaller than those inherent in stereo PIV imaging. The results from these coincident-plane validation tests also demonstrate that further efforts to improve the light-sheet formation and positioning procedures would provide only marginal improvements in the accuracy of the velocity gradient measurements. Recent advances in mapping calibrations for stereo PIV systems [5.759, 760] may provide a way to reduce this limiting accuracy in stereo PIV measurements.

Separated-Plane Validations

The accuracy of all nine simultaneously measured velocity gradient components $\partial u_i/\partial x_j$ can be further assessed in separated-plane validation tests [5.758, 761]. The two stereo camera pairs are now arranged to image separate 532 nm and 635 nm light-sheet pairs with a differential spacing $\Delta z \approx 400\,\mu m$ that matches the in-plane separation between adjacent vectors. An example of all three simultaneously measured instantaneous velocity component fields $u_i(\mathbf{x}, t)$ in the two differentially spaced light sheets is given at the top in Fig. 5.190. The two velocity fields should now no longer be identical due to the differential z-spacing, which gives rise to the z-derivatives in the velocity gradient components.

Figure 5.190 also shows all nine simultaneous $\partial u_i/\partial x_j$ components measured in the same flow as above. Perhaps the most stringent validation test that can be applied to any measurement of the full velocity gradient tensor is the extent to which the measured components satisfy the zero-divergence requirement from continuity in the velocity field. The resulting distribution of divergences is shown in Fig. 5.191, where the measured divergence values are normalized by the local instantaneous norm of the velocity gradient tensor, giving the relative error in the measured velocity gradients. Note that at locations where the gradients are small, this relative divergence error will be relatively large, since these errors result primarily from measurement uncertainties that are independent of the velocity gradient values. The divergence values can be compared with the distributions of the velocity gradients measured at $Re_\delta = 6000$ and 30 000, corresponding to $Re_\lambda = 44$ and 113, in the same flow, shown in Fig. 5.192. Ratios of the rms divergence errors to rms values of the measured on- and off-diagonal gradient components $\partial u_i/\partial x_j$

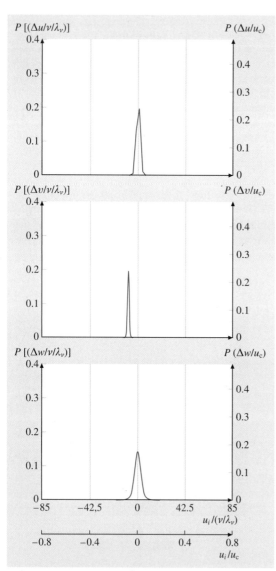

Fig. 5.189 Distributions of velocity component differences from coincident-plane validations tests of the type in Fig. 5.188, shown on inner- and outer-scale variables. Compare with Fig. 5.187 (after [5.758])

are 11.7% and 8.9%, respectively, consistent with accuracies inferred from the single- and coincident-plane validation tests.

Numerous further tests based on the similarity, scaling, and isotropy implied by classical turbulence theory allow further validation of any measurement of the velocity gradient tensor in turbulent flows [5.761]. Examples of such validations for DSPIV measurements obtained at $\mathrm{Re}_\delta = 6000$ and 30 000 are seen in distributions of the velocity gradient components in Fig. 5.192 and in the principal strain rates and vorticity components in Fig. 5.193. In each case, isotropy requires the inner-scaled distributions shown in each panel to be identical. On the semi-logarithmic axes it can be verified that such similarity extends in the measured results even to the rare-event tails of these distributions.

Velocity Gradient Field Measurements

Figure 5.194a–c show typical examples of the instantaneous enstrophy field $\omega_i\omega_i$, the enstrophy production rate field $\omega_i S_{ij}\omega_i$, and the kinetic-energy dissipation rate field $2\nu S_{ij} S_{ij}$ obtained from such DSPIV measurements at $\mathrm{Re}_\delta = 6000$ and 30 000 on the centerline of an axisymmetric turbulent jet. The relative size of each plane is given in terms of the inner (viscous) length scale λ_ν. Note that the enstrophy and dissipation rate fields are second-order gradient quantities that scale, respectively, with local outer variables as $(u_c/\delta)^2 \mathrm{Re}_\delta$ and (u_c^3/δ). In contrast, the enstrophy production rate is a third-order gradient field and thus scales as $(u_c/\delta)^3 \mathrm{Re}_\delta^{3/2}$. Distributions of each of these quantities obtained from over 1000 such instantaneous measured realizations [5.761] verify these outer-variable scalings as well as the similarity in the corresponding inner-variable scalings with ν and λ_ν, and provide further validations of the DSPIV measurement results for these fields. Fields of the type in Fig. 5.194 are of central importance to understanding the structure and dynamics of turbulence at the quasi-universal intermediate and small scales in inhomogeneous turbulent shear flows. Various aspects of these fields, including their precise scale-similarity properties [5.761], are likely to play a key role in developing improved subgrid-scale models for more-accurate large-eddy simulations of practical turbulent shear flows. To date the inability to access these fields in inhomogeneous turbulence has been a major factor hindering the development of such models. The advent of multidimensional measurement techniques such as DSPIV that are capable of giving accurate and highly resolved access to these fields is likely to be a significant step toward developing more-accurate physically based models of turbulence for such simulations.

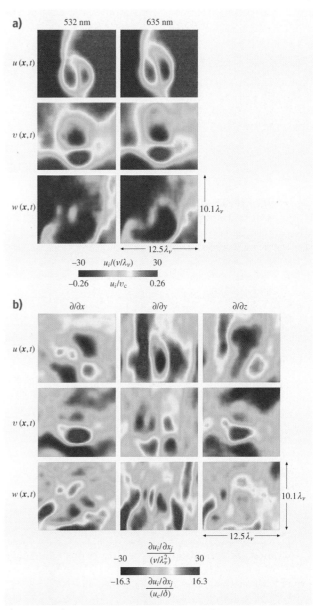

Fig. 5.190a,b Typical separated-plane DSPIV measurements, showing $u_i(x, t)$ velocity component fields measured in the two differentially spaced light-sheet planes by the two stereo camera pairs (**a**), and the resulting nine components (**b**) of the velocity gradient tensor field $\partial u_i/\partial x_j$, with normalizations on the inner- and outer-scale variables (after [5.758])

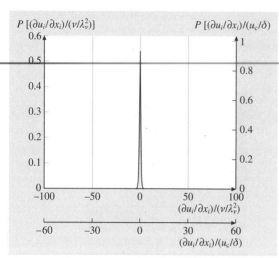

Fig. 5.191 Distribution of the measured divergence errors from separated-plane DSPIV measurements, obtained by summing the principal values of the strain rate tensor S_{ij} from the measured velocity gradient tensor field $\partial u_i/\partial x_j$, shown normalized on inner- and outer-scale variables. Compare with the range of gradient values in Fig. 5.192 (after [5.758])

Summary

DSPIV measurements of the type described here allow highly resolved direct experimental access to all nine simultaneous velocity gradient component fields $\partial u_i/\partial x_j$ on the intermediate and small scales of fully developed turbulent shear flows under conditions that are, at present, well beyond the reach of direct numerical simulations. Proper design of the DSPIV setup allows accurate measurements of the structure, statistics, similarity, and scaling of key gradient fields in such inhomogeneous shear flow turbulence, including the vorticity vector field, the strain rate tensor field, and the true kinetic-energy dissipation rate field. Proper differential spacing of the two stereo PIV light-sheet planes, as well as precise characterization of the thickness, coincidence, and two-axis parallelism of these sheets over the entire field of view, permits resolution of the velocity gradient tensor down to the local inner (viscous) length scale of the turbulent flow. Careful validations of such dual-plane measurements are possible through single- and coincident-plane tests, as well as through separated-plane measurements of the velocity divergence and further tests of the similarity, isotropy, and Re scaling in the measured gradient components and various fields derived from them. Results from such tests have established the accuracy of these fields from DSPIV measurements of the type described here. Moreover the two-frequency DSPIV method described here allows use of solid metal oxide seed particles to permit such highly resolved measurements in both nonreacting and exothermically reacting turbulent shear flows. For the foreseeable future, direct experimental measurements based on high-resolution imaging methods such as DSPIV will likely provide the only practical means for investigating the structure, statistics, similarity, and scaling properties of key gradient fields in spatially developing turbulent shear flows at a level of resolution and detail that has traditionally been associated only with direct numerical simulations.

5.5.3 Measurements of the Vorticity Vector and Other Velocity Gradient Tensor-Based Turbulence Properties

To know the vorticity vector and the strain rate tensor it is necessary to simultaneously determine the instantaneous velocity gradients of all three velocity components in the coordinate axes directions: $\partial U_i/\partial x$, $\partial U_i/\partial y$, $\partial U_i/\partial z$ ($U_1 = U$, $U_2 = V$, $U_3 = W$). Although only six velocity gradients (four cross-stream and two streamwise) have to be known to determine the vorticity vector, the four cross-stream gradients needed cannot be measured accurately with hot-wire anemometry if the other two cross-stream gradients are neglected. In order to determine the velocity gradients experimentally with a finite-difference approximation, the velocity vector has to be measured at two points closely separated in the coordinate directions. To measure three velocity vector components with hot-wire anemometry, it is necessary to use an array of at least three sensors. By separating these arrays in the y and z coordinate directions at an appropriate distance, the six cross-stream velocity gradients, $\partial U_i/\partial y$, $\partial U_i/\partial z$, can be determined with a minimum of nine sensors. Determination of the three streamwise gradients, $\partial U_i/\partial x$, using hot-wire anemometry is possible by employing *Taylor*'s [5.762] hypothesis. Depending on the number of sensors, their orientation in the probe arrays, and the position of the arrays with respect to the coordinate axes, several types of vorticity probes have been developed to date.

Geometrical Configurations of Three-Component Vorticity Probes

Nine-Sensor Probes. The first attempt to measure all three components of the vorticity vector simultaneously was made by *Wassman* and *Wallace* [5.763], who used a nine-sensor probe. It was configured in three identical arrays with three sensors in each array. To make the probe as compact as possible, the supporting prong at the center of each array was common to the three sensors. They solved the problem of crosstalk between the electrical circuits due to the common prongs, but they did not develop the probe to an operational level that permitted accurate measurements of three-dimensional velocity and vorticity fields.

T-Configuration. The nine-sensor probe idea was developed further and a version capable of simultaneously measuring all three vorticity components with reasonable accuracy was brought to operational status by *Vukoslavčević* et al. [5.687]. This design consisted of three T-configured arrays, each containing three sensors inclined at 45° to the probe axis, as shown in Fig. 5.195. Two of the sensors form a V-shaped array in the x–z plane and the third sensor is oriented at the same angle in the x–y plane. The extent of the sensing area was 1.7 mm vertically and 2.2 mm horizontally, with a distance between the array centers of approximately 1.2 mm in the y- and z-directions. The distance h between the prongs supporting each sensor was 0.5 mm. The diameters of the tungsten sensors were 2.5 μm and

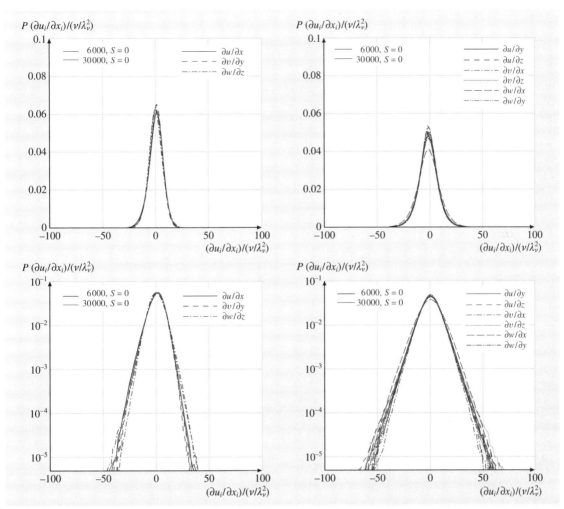

Fig. 5.192 Distributions of on-diagonal (*left*) and off-diagonal (*right*) components of inner-scaled velocity gradients $\partial u_i/\partial x_j$ measured via DSPIV in a turbulent shear flow at $\text{Re}_\delta = 6000$ and $30\,000$ ($\text{Re}_\lambda = 44$ and 113), in linear (*top*) and semilogarithmic (*bottom*) axes. Note the agreement with the similarity and scaling of isotropic turbulence (after [5.761])

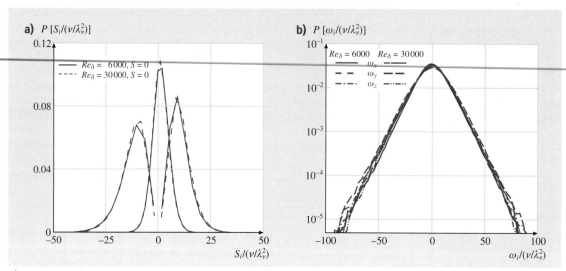

Fig. 5.193a,b Further validations of velocity gradient components $\partial u_i/\partial x_j$ measured via DSPIV in a turbulent shear flow at $Re_\delta = 6000$ and $30\,000$ ($Re_\lambda = 44$ and 113), showing the distributions of the principal strain rates (**a**) and vorticity components (**b**). Note the excellent agreement with the requirements of isotropy and inner-scale similarity from classical theory (after [5.761])

their lengths were about 0.7 mm, giving an aspect ratio of about 280.

The length of the prongs from their tip to the base where they are glued together was 25 mm, and the base diameter was about 3.2 mm. This gives a ratio of about 8, which was large enough to avoid a very serious influence on the flow at the location of the sensors by the probe shaft as indicated by the inviscid flow considerations discussed later.

Orthogonal Configuration. Honkan [5.690], Andreopoulos and Honkan [5.691] and Honkan and Andreopoulos [5.689] also constructed a nine-sensor probe capable of simultaneously measuring all three vorticity components. This probe consisted of a set of three triple-sensor arrays as sketched in Fig. 5.196. The sensors in each array are mutually orthogonal to each other. They are oriented at approximately $54.7°$ to the probe axis and placed in meridian planes at $120°$ with respect to each other. For this probe each sensor was supported by a separate pair of prongs, eliminating any possibility of electrical crosstalk between the hot-wire circuits.

The total extent of the sensing region of this probe was about 4 mm across the flow, with the array centers located at the corners of an equilateral triangle with sides approximately 2.46 mm long. The array centers were separated by 2.46 mm in the y-direction and 2.13 mm in the z-direction. The distance between the supporting prongs of each sensor was 0.8 mm. The diameters of the platinum-plated tungsten sensors was $2.5\,\mu$m, and their lengths were about 0.8 mm, giving a sensor aspect ratio of about 320.

The length of the prongs of this probe from their tips to the base where they were glued together was approximately 10 mm, while the base occupied a space approximately 5 mm in diameter. This gives a ratio of only 2, so there was a possible influence of the probe shaft on the flow at the location of the sensors.

Twelve-Sensor Probes. Tsinober et al. [5.693] constructed a 12-sensor probe with the configuration sketched in Fig. 5.197. They added an additional tungsten sensor of 2.5 mm diameter to each array of the nine-sensor T-configuration probe and used separate pairs of prongs for each sensor, making a total of 12 sensors supported by 24 prongs.

The extent of the sensing region of this probe was 2.3 mm vertically and horizontally, with distances between the array centers of 1.4 mm and 0.4 mm between the supporting prongs of each sensor. The diameter of the tungsten sensors, inclined at $45°$ to the probe axis, was $2.5\,\mu$m, and their length was about 0.56 mm, giving an aspect ratio of about 224.

The length of the prongs, from their tip to the base where they are glued together, is approximately 20 mm while the base occupied a space approximately 3 mm in

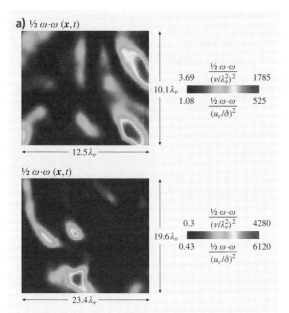
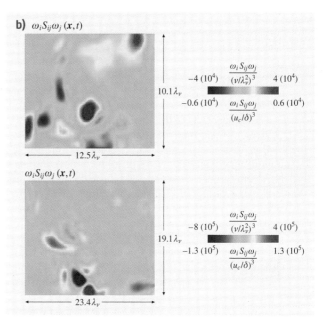

Fig. 5.194a–c Typical results from DSPIV measurements at quasi-universal intermediate and small scales of turbulent shear flows, showing (**a**) the enstrophy field, (**b**) the enstrophy-production-rate field, and (**c**) the kinetic-energy dissipation rate field obtained from measured velocity gradient components $\partial u_i/\partial x_j$ at $\mathrm{Re}_\delta = 6000$ and $30\,000$ ($\mathrm{Re}_\lambda = 44$ and 113). Values are shown normalized on inner- and outer-scale variables (after [5.761]) (**a**) Enstrophy fields at $\mathrm{Re}_\delta = 6000$ (*top*) and $30\,000$ (*bottom*) ($\mathrm{Re}_\lambda = 44$ and 113) (**b**) Enstrophy production rate fields at $\mathrm{Re}_\delta = 6000$ (*top*) and $30\,000$ (*bottom*) ($\mathrm{Re}_\lambda = 44$ and 113) (**c**) Dissipation rate fields at $\mathrm{Re}_\delta = 6000$ (*top*) and $30\,000$ (*bottom*) ($\mathrm{Re}_\lambda = 44$ and 113)

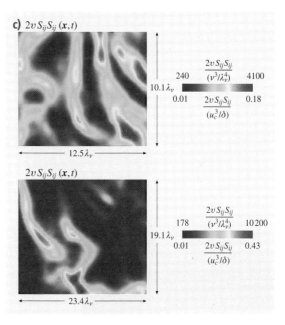

diameter, giving an aspect ratio of about 6.7. Thus there was possibly a small influence of the probe shaft on the flow at the location of the sensors.

Tsinober et al. [5.693] and *Kholmyansky* et al. [5.764] also used a 21-sensor probe consisting of five arrays of four hot-wires as well as one cold-wire positioned upstream of the central array to measure the temperature, as shown in the photograph in Fig. 5.197b. The most recent version of this probe has 25 sensors: one cold and four hot sensors for each of the five arrays. The prongs of the central array are longer than those of the surrounding arrays, so its sensors are positioned further upstream than those of the outer arrays. With this probe *Galanti* et al. [5.765] avoided having to use Taylor's hypothesis to obtain the streamwise gradients by differencing the velocity components obtained from the central array with the average of those from the four surrounding arrays. Of course, avoiding Taylor's hypothesis in this manner is at the cost of having to use this spatially averaged information for the estimation of streamwise gradients, $\partial U_i/\partial x$. It is also at the cost of the lower spatial resolution of this probe compared to probes with fewer sensors.

Nguyen [5.766] and *Marasli* et al. [5.767] used a 12-sensor probe with a similar array configuration to that of *Tsinober* et al. [5.693] but with a different arrangement of the arrays and a different data-reduction procedure. The probe design and signal data-reduction procedure are explained in detail by *Vukoslavčević* and *Wallace* [5.768].

The total extent of the sensing area of this probe was 2.20 mm in the y-direction and 2.40 mm in the z-direction, with an average distance between the array centers of approximately 1.14 mm in the y-direction and 1.32 mm in the z-direction. The distance between the supporting prongs of each sensor, inclined at 45° to the probe axis, was 0.44 mm. The diameter of the tungsten sensors was 2.5 μm, and their length was about 0.62 mm, giving an aspect ratio of about 250.

The length of the prongs from their tip to the base, where they are glued together, is approximately 33 mm while the base occupied a space approximately 3.6 mm in diameter, giving a ratio of 9. This is large enough to avoid most of the influence of the probe shaft on the flow at the probe sensors, as indicated by the inviscid flow considerations discussed later.

Eight-Sensor Probe. *Zhu* and *Antonia* [5.769], *Antonia* et al. [5.770] and *Antonia* et al. [5.771] have analyzed, constructed and used an eight-sensor probe capable of simultaneously measuring all three vorticity components. The probe consists of the four X-wire arrays shown in Fig. 5.199.

Each array can measure two velocity components providing the information necessary to determine the four cross-stream velocity gradients. In contrast to the nine- and 12-sensor configurations described above, the binormal velocity component cooling the sensors of this eight-sensor probe has to be neglected. Two arrays are positioned with horizontal orientations while the other two are in vertical orientations. The sensors in each array

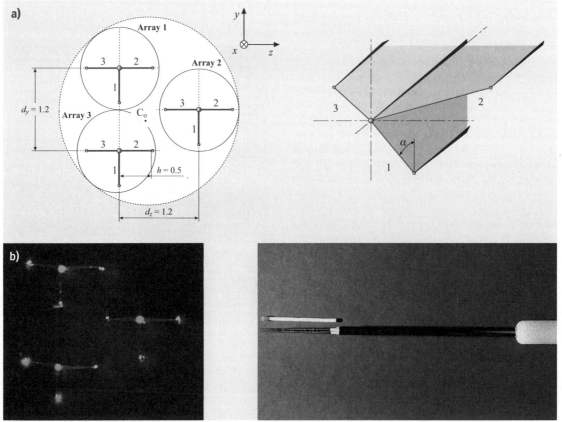

Fig. 5.195 (a) Sketch of the front view of a nine-sensor T-configuration probe and schematic view of one of its arrays. Dimensions in mm. (b) Photographs of the nine-sensor probe of *Vukoslavčević* et al. [5.687] (courtesy CUP)

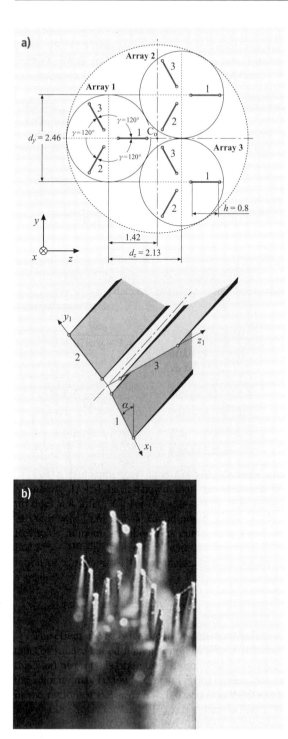

Fig. 5.196 (a) Sketch of the front view of a nine-sensor probe with orthogonally configured sensors and schematic view of one of its arrays. Dimensions in mm. (b) Photograph of an orthogonally configured nine-sensor probe of *Honkan* and *Andreopoulos* [5.689] (courtesy CUP)

The separations between the centers of the X-arrays are $d_y = 2.1$ mm in the y-direction and $d_z = 2.2$ mm in the z-direction. All the sensors were etched from a Wollaston (Pt–10%Rh) wire to an active length of 0.5 mm. The diameters of the wires were 2.5 μm, giving an aspect ratio of about 200. The extent of the sensing area of the probe was 2.8 mm vertically and 2.7 mm horizontally.

Probe Fabrication Method

Fabrication of these vorticity probes is very difficult because of their small size, complex design, and fragility. The main problem is to fit 12, 18, 24, or even more supporting prongs into the smallest feasible space, while still satisfying the accuracy as well as spatial resolution criteria. Arranging the prongs with very small distances between them may result in two difficult problems to be overcome: preventing electrical shorting between the prongs or between the prongs and the probe body and maintaining the prongs in place so that they keep the desired array geometry. Several technical solutions and various prong materials have been tried.

For the first version of the nine-sensor probe [5.687], a tungsten rod of about 0.25 mm diameter was used for the prongs. The prongs were chemically tapered from their original diameter to a tip diameter of about 50 μm. Although using a common prong for each array for this probe reduced the aerodynamic disturbances to the flow, it introduced some electrical crosstalk between the anemometer circuits. To make this crosstalk insignificant and to avoid instabilities from the anemometer bridge circuits, the common prong resistance should be less than about 0.1 Ω. To reduce the resistance without unduly increasing the common prong diameter, each common prong was first plated with copper. Then the tips of all the prongs were plated with a thin layer of nickel to facilitate welding of the tungsten sensors to the prongs. The four prongs of each array were drawn through round four-bore ceramic insulating tubes with a 1.6 mm outside diameter, and then placed in a stainless-steel tube with one flat side in order to bring the probe as close as possible to the wall. The ends of the prongs were soldered to copper stranded wire with gold-plated connectors on the other end. The ceramic tubes were sealed into the outer stainless-steal tube with epoxy. The prong tips were manipulated under a measuring micro-

are inclined at 45° to the probe axis. The separation between the two inclined wires in each X-array is 0.6 mm.

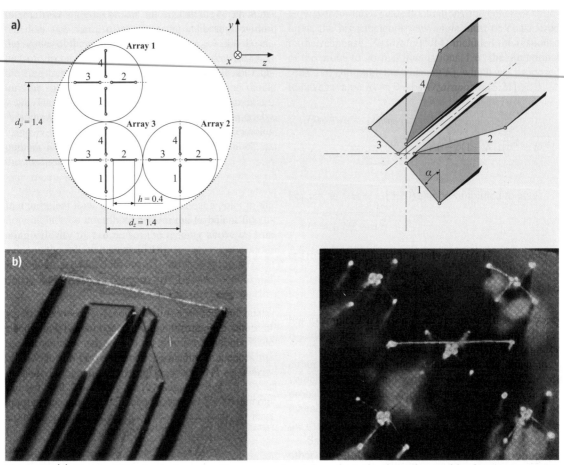

Fig. 5.197 (a) Sketch of the front view of a 12-sensor probe and a schematic view of one of its four-sensor arrays. Dimensions in mm. (b) Photograph of one array and of the 21-sensor probe of *Tsinober* et al. [5.693] (courtesy CUP)

scope into their proper geometric positions within about a ±0.01 mm tolerance. After nickel plating the prongs, the 2.5 μm-diameter unsheathed tungsten sensor wires were welded to the tips. The relative position of the prong tips and the sensor wires are shown in Fig. 5.195.

The main problem with the design of the nine-sensor probe with common central prongs was to make these central prongs small enough to avoid flow blockage but also to have resistances close to zero to avoid crosstalk between the sensors and circuit instabilities. It was difficult to reach these goals simultaneously.

To avoid this problem, *Tsinober* et al. [5.693] used four separated tiny prongs, put together to form the central prongs, in place of common prongs for each array of their 12-sensor vorticity probe. The prongs were made of tungsten wire, tapered at the tip to a 20 μm diameter, insulated by Teflon, glued together with a tip separation of 80 μm and plated with nickel or tin. They were positioned using 30 μm microspacers. A special manipulator was used to carry and position short pieces of sensor wire for welding to the tips of the prongs.

The tungsten prongs were replaced with stainless-steel prongs of the same dimension (0.25 mm) for the 12-sensor probe of *Vukoslavčević* and *Wallace* [5.768], shown in Fig. 5.198, in order to avoid both the crosstalk problem caused by common prongs and the tedious plating procedure. The resistance of a stainless-steel prong is well over the critical value of 0.1 Ω, even when it is plated to a reasonable thickness. Therefore, in place of the common prongs, four separate prongs were used in the center of each array. These prongs were tapered to a tip diameter of less than 50 μm and then drawn through four-bore ceramic tubes of 1.23 mm outside diameter.

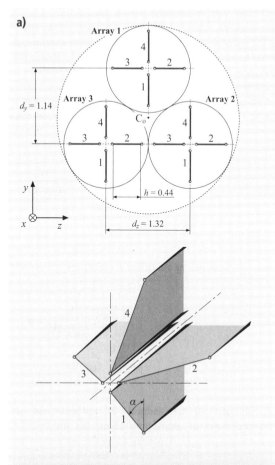

Fig. 5.198 (a) Sketch of the front view of a twelve-sensor probe of *Vukoslavčević* and *Wallace* [5.768] and a schematic view of one of its four-sensor arrays. (b) Photograph of the twelve-sensor probe of *Vukoslavčević* and *Wallace* [5.768] (shown in *Bernard* and *Wallace* [5.772], courtesy Wiley)

around these ceramic tubes using constant-diameter fishing line as a spacer. The whole assembly was drawn into a plastic tube of 3.6 mm outside diameter and then placed in a stainless-steal tube of 4.6 mm outside diameter. Although the number of prongs was doubled compared to the nine-sensor probe, the probe dimensions were only slightly increased. This was made possible using thinner ceramic tubes for the central prongs and placing the outside prongs around them. The prongs were positioned under a measuring microscope with an accuracy of ±0.01 mm, which was the most tedious part of this approach.

For the nine-sensor orthogonal probe of *Honkan* [5.690] and *Honkan* and *Andreopoulos* [5.689] a separate two-hole, 1.23 mm-diameter ceramic tube was used for each pair of prongs. Two 0.25 mm-diameter stainless-steel prongs, tapered at one end electrolytically to a diameter of 190 μm, were inserted into a ceramic tube and glued at one end. Using a special rig, two prongs were made simultaneously flat to mount the sensor at the required angle of 54.6° to the probe axis. This process was repeated for the eight other sensor elements. A second jig was used to position three of these elements to form one array of the triple orthogonal probe and the third one to put all three arrays together in the desired position. The unit was then inserted into a thin-walled brass tube. Miniature-sized gold connectors were used to connect the prongs and shielded wire leads. In this way the

Three ceramic tubes were glued together, forming the centers of each of the three arrays of the 12-sensor probe. The rest of the 12 surrounding prongs were positioned

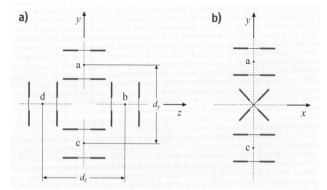

Fig. 5.199a,b Sketches of the front and side view of an eight-sensor probe of *Zhu* and *Antonia* [5.769] (courtesy of Inst. of Phys.): (a) streamwise view, and (b) cross-stream view

crosstalk problem was resolved and the plating process avoided. This method of positioning the prongs seems to be more convenient compared to the approach described above, but the dimensions of the probe body were much larger due to the use of nine separate ceramic tubes.

Spatial Resolution

Because measuring velocity gradients with hot-wire sensors to determine vorticity components requires the determination of the velocity vector at least at two points in the flow simultaneously, the spatial resolution of such probes is a serious problem that must be addressed. To measure the velocity vector accurately at a given point in a turbulent flow field, the sensors and their arrays should be as small as possible. Additionally, the distances between arrays must also be as small as possible. Unfortunately, small sensor and array sizes can result in thermal interference and inadequate sensor response. If the spacing is too small the effects of noise can overwhelm the velocity gradient signal. These issues are discussed in some detail by *Wallace* and *Foss* [5.81] and will be summarized here.

A hot-wire sensor cannot resolve motions of turbulent coherent structures that have scales smaller in size than its length [5.776]. In order to resolve the nonuniform velocity field of a vortex structure, a hot-wire probe must be much smaller than the spatial scale of that structure. This is not easy to achieve with multi-sensor vorticity probes. Whether a probe is capable of resolving the smallest structures in the flow depends on the ratio of its dimensions to the *Kolmogorov* [5.777] length scale η.

Wyngaard [5.702] analyzed the response of the Kovasznay-type probe, designed to measure the streamwise component of vorticity Ω_x. He assumed isotropic turbulence and the theoretical three-dimensional velocity spectrum of *Pao* [5.773] and investigated the attenuation due to finite sensor length L and sensor separation S. He found that the ratio of measured to the true variance $\overline{\Omega_x^2}$ decreases sharply as η/L decreased. It is essentially 1.0 for $\eta/L = 1$, and drops to 0.86 for $\eta/L = 0.32$. Serious attenuation occurs for smaller values of η/L. He concluded that the ratio of η/L should not be smaller than 0.3, as was later confirmed by *Browne* et al. [5.778]. He also found that S/L has a strong influence on the measurement results. His analysis of the effect of S/L was performed for a fixed value of $\eta/L = 0.32$. The value of $S/L = 0.7$ provided the best probe response, while wider sensor spacing resulted in attenuation in the variance of Ω_x. *Klewicki* and *Falco* [5.692] used a hot-wire probe that enabled systematic variation of the separation of two parallel sensors. Their investigation showed that, for a Reynolds number based on the momentum thickness of 1010 in a turbulent boundary layer, the variance of the velocity gradient just above the buffer layer is attenuated by 5% for $S/\eta = 3.3$.

Antonia et al. [5.775] performed parallel sensor probe analysis to investigate the spatial resolution of velocity gradient measurements. They used the DNS data of *Kim* et al. [5.743] and *Moser* et al. [5.779] at the centerline of a turbulent channel flow simulation, with $R_\tau = 180$ and 400, where $R_\tau = u_\tau b/v$, $u_\tau = \sqrt{\tau_w/\rho}$ is the friction velocity, b is the channel half width, τ_w is the wall shear stress and ρ is the fluid density. They determined the *measured* velocity gradient $\Delta u/\Delta y$ by dividing the difference between two simulated streamwise velocity component values, separated in the cross-stream direction by a distance $\Delta y = S$, by the *true* velocity gradient $\partial u/\partial y$ determined with spectral differentiation using Chebychev polynomials. The ratios of the variances of the *measured* and *true* velocity gradients are compared with previous predictions in Fig. 5.200 This ratio decreased with increasing sensor separation, as had been predicted and experimentally confirmed previously. It is clear that the sensor separation should not exceed 2–4 Kolmogorov lengths to determine the velocity gradients with reasonable accuracy.

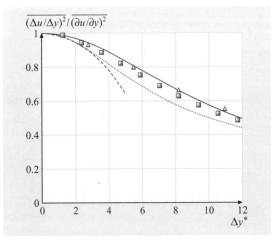

Fig. 5.200 Attenuation of the *measured* velocity gradient due to the separation distance $\Delta y^* = \Delta y/\eta$ over which the velocity difference is determined. DNS: (*triangles*) $R_\tau = 180$, (*squares*) $R_\tau = 400$; Wyngaard's analysis (*dotted line*) with *Pao*'s spectrum [5.773], (*solid line*) with DNS spectrum, (*dashed line*) truncated by *Hussein* and *George* [5.774] with Pao's spectrum. (After *Antonia* et al. [5.775])

However, decreasing the distance between the sensors causes another serious problem. The errors in the gradient measurements increase greatly for small sensor separations for three reasons. The percentage error in the velocity differences, the numerator in $\Delta u/\Delta y$ determined from the velocities measured at each of the two sensors, becomes larger with decreasing sensor separation. This is because these differences are between large numbers that approach the same value as the sensors come very close together. Furthermore, the temperature fields of each of the sensors can begin to influence the other sensor. In addition the percentage error in the determination of the separation distance, the denominator in $\Delta u/\Delta y$, also rapidly increases as the separation distance decreases. An example from *Antonia* et al. [5.775] of the errors that can creep into the finite-difference approximation of velocity gradients for very small sensor separations is strikingly clear, as shown in Fig. 5.201.

In this figure, the errors in the velocity gradient variances that were observed in the DNS analysis of the *measured* velocity gradient finite-difference approximation $\Delta u/\Delta y$ are compared to physical measurements by a real hot-wire probe with variable sensor separations $\Delta y^* = \Delta y/\eta$. For small values of $\Delta y^* < 2$, the measured variance values are increasingly overestimated, resulting in a large gradient measurement error. This error primarily depends on the accuracy of the velocity measurements as well as the accuracy of the sensor separation measurements, as explained above, and can differ from one experiment to the other.

Most of the discussion thus far has been about measurements of velocity gradients or vorticity with specific probe configurations. Therefore, for other probe configurations this discussion should only be used as a general guide. Final decisions about the probe dimensions of any other configuration depend on many probe and flow parameters that must be considered individually.

There are several characteristic lengths to be considered for probe configurations designed to measure all three vorticity components: the sensor length, the separation between the sensors within an array, and the separation between the arrays in the y- and z-directions. The choices of these lengths have to be made in a manner consistent with one of the crucial assumptions required for these probes, viz. that the velocity gradients are constant over the probe sensing volume. It is very hard to meet this condition in turbulent flows, especially in flows with high mean velocity gradients such as in the wall region of turbulent boundary layers. In this case, the dimension of the probe in the wall normal direction should be smaller than in the lateral direction.

To avoid disturbing the flow too seriously by the probe body, the distance from the tips of the prongs to their base should also be sufficiently long. A rule of thumb often used by probe designers, based on a potential flow calculation, is that this length should be about 10 times the diameter of the probe body. This requirement is not easy to meet. Prongs that are too long easily vibrate in a turbulent flow, making the probe too fragile and subject to sensor breakage. The minimum diameter of the probe body is determined by the number of prongs to be placed within it while maintaining insulation between them.

The nine-sensor probe of *Vukoslavčević* et al. [5.687] and the 12-sensor probe of *Vukoslavčević* and *Wallace* [5.768] were used in a boundary-layer flow with the same characteristics. At $y^+ = 11.2$, which was the closest measurement location that the probe could be placed from the wall, they estimated the Kolmogorov microscale to be $\eta = 0.19$ mm. Thus the ratio of the Kolmogorov microscale to the sensor length was $\eta/l = 0.27$ for the nine-sensor probe and $\eta/l = 0.31$ for the 12-sensor probe. An equivalent of the sensor separation S in the *Wyngaard* [5.776] spatial resolution analysis of the Kovasznay-type probe are the separations d_y and d_z of the array centers for the vorticity probes discussed here. For the nine-sensor probe of *Vukoslavčević* et al. [5.687], $d_y/\eta = d_z/\eta = 6.3$. For the 12-sensor

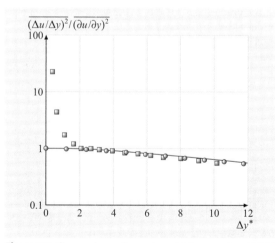

Fig. 5.201 Comparison of experimental (*squares*) and DNS (*circles*) *measured* estimates of velocity gradient variance to DNS *true* values showing the dependence on the separation distance $\Delta y^* = \Delta y/\eta$ over which the velocity difference is determined. Wyngaard's analysis with DNS spectrum (*solid line*) is also shown. (After *Antonia* et al. [5.775])

probe of *Vukoslavčević* and *Wallace* [5.768], $d_y/\eta = 6.0$ and $d_z/\eta = 6.9$.

This same 12-sensor probe was used by *Nguyen* [5.766] and *Marasli* et al. [5.767] to measure the three velocity and vorticity components simultaneously in the intermediate wake of a circular cylinder with Reynolds number $Re_d = 2000$ and $x/d = 30$. For this experiment d is the cylinder diameter and x is the distance downstream from the cylinder. At this location the Kolmogorov microscale was 0.2–0.4 mm, so that, in the worse case, this probe resolved the small scales as well as in the boundary-layer experiment of *Vukoslavčević* and *Wallace* [5.768] near the wall.

The same probe also was used by *Loucks* [5.780] in a two-stream mixing layer. Again in this experiment all the components of the velocity and vorticity vectors were measured simultaneously. Measurements were made at three Reynolds numbers based on the momentum thickness: $R_\theta = 432$ at a location before the mixing transition and at two stations downstream where the flow was self similar and $R_\theta = 1792$ and 2483, respectively. At the highest Reynolds number the Kolmogorov microscale had its smallest value of 0.4 mm. Thus in this flow $\eta/l = 0.65$, $d_y/\eta = 2.9$, and $d_z/\eta = 3.3$.

Comparing these values with the spatial resolution analysis given at the beginning of this section, it is clear that the required ratio of the Kolmogorov microscale to sensor length $\eta/l \geq 0.3$ was met in all the flows in which these probes were used. The ratios of the Kolmogorov microscale to the array separations, d_y/η and $d_z/\eta \approx 2$–4, required for good spatial resolution while maintaining reasonable accuracy, was satisfied when the 12-sensor probe was used in the mixing-layer experiment and was reasonably close to these recommended values in the wake experiment and in the boundary-layer experiment discussed above. For the positions further from the wall in the boundary layer and at other positions in the cylinder wake, η becomes larger, giving better spatial resolution. From these considerations, it seems that this 12-sensor probe was capable of resolving all but the smallest scales in the turbulent boundary-layer, wake, and mixing-layers flows in which it has been used.

The nine-sensor probe of *Honkan* [5.690] and *Honkan* and *Andreopoulos* [5.689] was used in a similar boundary-layer flow with a Kolmogorow microscale of $\eta = 0.2$ mm at $y^+ = 12.5$. Thus $\eta/l = 0.25$, $d_y/\eta = 12.30$, and $d_z/\eta = 10.65$.

The spatial resolution of the 12-sensor probe of *Tsinober* et al. [5.693] would have been slightly less than that of *Vukoslavčević* and *Wallace* [5.768] at comparable locations close to the wall in a boundary layer with a similar Reynolds number. However, the closest measurements they made in their boundary layer with their probe were further out in the logarithmic layer, where the Kolmogorov microscale was 0.5 mm.

The eight-sensor probe of *Antonia* et al. [5.770] and *Antonia* et al. [5.771] was used in decaying grid turbulence with Kolmogorov microscale values of 0.17–0.4 mm at different distances from the grid. Thus the ratio of the Kolmogorov microscale to the sensor length was in the range $0.34 < \eta/l < 0.8$, while the array center separations were in the range $5.7 < d_z/\eta < 13$, depending on the distance from the grid. The array center separations of this probe are limited by the sensor lengths and the separations of the prongs. The method of *Zhu* and *Antonia* [5.769], developed to correct velocity gradients spectra for insufficient spatial resolution, was used with this eight-sensor probe. If welded wires were used for the sensors of this design instead of Wollaston etched wires, the probe could be made at least half the size, giving better spatial resolution.

With the same sensor length and prong separation and the same type of fabrication, the nine-sensor T-configuration probe will have a better spatial resolution than the nine-sensor orthogonal-configuration probe or the 12-sensor probes. However, this is not the only design consideration. As will be discussed later, the situation is just the opposite with regard to the uniqueness domain of the probes. The uniqueness domain of a 12-sensor probe is almost double that of the nine-sensor T-configuration probe. How to trade off between these two design considerations depends on the type of flow.

Determination of Cross-Stream Velocity Gradients

Two basically different approaches have been used to determine the cross-stream velocity gradient components. One is based on the assumption that the velocity gradients can be neglected over each array, assuming that the effective velocity cooling each sensor is the same within an array, but that it changes between the arrays. In addition, for the eight-sensor probe the influence of the binormal velocity component on the sensor in each array must also be neglected to measure the other two velocity components.

The other approach is based on the assumption that a different effective velocity acts at the center of each sensor. In this case the velocity gradients over the arrays are not neglected. The only necessary assumption in this second approach is that the velocity gradient components are constant over the entire probe sensing area.

This assumption can also be applied to the eight-sensor probe, but in this case two cross-stream velocity gradients have to be neglected. The first approach is much simpler but less accurate. The second approach is more accurate but requires a rather complex mathematical procedure. Different models describing sensor cooling and relating the effective cooling velocities to the anemometer voltage outputs can be utilized with both approaches. For the approach where the velocity gradients over the arrays are neglected, a direct method can also be used in which the velocity vector is mapped to the voltage outputs for each array.

The Sensor Response Equation. A general form of the sensor response equation is

$$F(E, U_e) = 0, \quad (5.318)$$

where E is the anemometer voltage output and U_e is the velocity effectively cooling the sensor. The simplest forms of this expression are the well-known King's law

$$E^2 = A + BU_e^n, \quad (5.319)$$

and a higher-order polynomial fit,

$$\sum_{p=1}^{5} b_p E^{p-1} = U_e^2. \quad (5.320)$$

One of the most accurate expressions for the effective cooling velocity in a three-dimensional flow field is that proposed by *Jorgensen* [5.781],

$$U_e^2 = U_n^2 + k^2 U_t^2 + h^2 U_b^2, \quad (5.321)$$

where U_n, U_t, and U_b are the orthogonal components of the velocity vector with respect to the sensor coordinate system, i.e., normal, tangential, and binormal to the sensor, while k and h are the so-called yaw and pitch coefficients. Another expression,

$$U_e^2 = U_{ne}^2 + h^2 U_b^2, \quad (5.322)$$

based on the effective sensor angle, introduced by *Bradshaw* [5.782] for X-array probes, has also been successfully extended to three-velocity-component measurements. In this case, the influence of the tangent cooling component U_t is incorporated into the normal component in the plane of the sensor and its prongs U_{ne} defined with respect to an effective sensor inclination angle. This effective angle is determined in the calibration procedure. The influence of the binormal velocity component is taken into account in the same way as in Jogensen's cooling law given in (5.321).

Data–Reduction Algorithms.
Nine-Sensor Probe. *Vukoslavčević* et al. [5.687] developed a special numerical algorithm for their nine-sensor T-configuration probe that made use of King's law (5.319) and the effective cooling velocity (5.322) concept. Relating the velocity components U_{ne} and U_b in the orthogonal coordinate system to U, V, and W, the components in probe coordinate system, the following expressions for the effective cooling velocity U_{eij} of the j-th sensor in the i-th array ($i, j = 1, 2, 3$) can be derived for the sensors in the x–y and x–z planes, respectively:

$$U_{eij}^2 = (U_{ij} \cos \alpha_{eij} + V_{ij} \sin \alpha_{eij})^2 + h_{ij}^2 W_{ij}^2 \quad (5.323)$$

and

$$U_{eij}^2 = (U_{ij} \cos \alpha_{eij} \pm W_{ij} \sin \alpha_{eij})^2 + h_{ij}^2 V_{ij}^2. \quad (5.324)$$

U_{eij} can be expressed in terms of the three velocity components U_0, V_0, and W_0 at the probe center C_0 (Fig. 5.195), the six cross-stream velocity gradients (in the plane normal to the probe axis passing through the sensor centers), the distance projected on this plane between the probe center and the center of each sensor, the individual sensor effective angle α_{eij}, and the pitch coefficients h_{ij}. To do this the velocity components U_0, V_0, and W_0 were expanded to first order in a Taylor series about C_0, and the effective cooling velocity was related to the voltage output using King's law. This gave the following set of nine nonlinear algebraic equations with three velocity and six cross-stream velocity gradient component unknowns:

$$\left(\frac{E_{ij}^2 - A_{ij}}{B_{ij}} \right)^{\frac{2}{n}}$$
$$= \left[K_{ij1} \left(U_0 + C_{ij1} \frac{\partial U}{\partial z} + C_{ij2} \frac{\partial U}{\partial y} \right) \right.$$
$$\left. + K_{ij2} \left(V_0 + C_{ij3} \frac{\partial V}{\partial z} + C_{ij4} \frac{\partial V}{\partial y} \right) \right]^2$$
$$+ \left[K_{ij3} \left(W_0 + C_{ij5} \frac{\partial W}{\partial z} + C_{ij6} \frac{\partial W}{\partial y} \right) \right]^2. \quad (5.325)$$

For this equation, the cross-stream velocity component V is exchanged with W for the sensors $j = 2, 3$ (Fig. 5.195).

The 27 coefficients K_{ijk}, the nine coefficients A_{ij}, and the nine coefficients B_{ij} had to be determined experimentally from calibration. The constants K_{ij1}, by their nature, are cosines of the effective angles α_{eij}, the K_{ij2} coefficients are sines of the effective angles, while the K_{ij3} coefficients account for aerodynamic blockage (pitch coefficients h_{ij}). The C_{ijl} ($l = 1, 2, 3, 4, 5, 6$) constants can be expressed as positive or negative fractions of the projected prong spacing h for a given probe geometry. The probe was calibrated in a uniform (gradient-free) flow using a mechanism to pitch and yaw the probe in the range where it gives a unique solution. Assuming an optimal value of 0.4–0.5 for the exponent n in King's Law, varying the speed magnitude for zero pitch and yaw angle and using a least-squares fit, the coefficients A_{ij} and the products $B_{ij}(K_{ij1})^{n/2}$ were found. The coefficients K_{ij1}, K_{ij2} and K_{ij3} can be determined for the sensor in the vertical or horizontal planes by yawing or pitching the probe. The equations that have to be solved to find the unknown calibration coefficients are linear.

Although various algorithms are available, a new numerical algorithm was developed to solve the system of nine nonlinear algebraic (5.325) for the unknown velocity components at the probe center and the six velocity gradients in the cross-stream direction. The method was based on a successive iteration procedure that converged after a few iterations. The probe has to be used inside the solution uniqueness range, otherwise multiple real roots can appear. This issue will be discussed in detail below.

The advantage of this solution approach used for the nine-sensor T-configuration probe is that the velocity variation over each array does not have to be neglected and the sensor inclination angles α do not have to be measured. The calibration procedure is rather simple, but the sensors have to be positioned in the horizontal or vertical planes as accurately as possible.

In their first solution method, Honkan [5.690] and Honkan and Andreopoulos [5.689] assumed that the velocity gradients, over each array of their nine-sensor probe with an orthogonal geometrical configuration, could be neglected and that only gradients between the arrays had to be taken into account. In this way the problem is reduced to the problem of velocity measurements with a triple-sensor orthogonal probe. The velocity gradients can then be estimated by finite differences, knowing the velocity at each array center and the separation distance between the array centers. They analyzed two possibilities for defining the response of the three arrays of their nine-sensor probe: direct mapping through a look-up table and the determination of the cooling law of each sensor.

The first approach directly relates the velocity vector to the voltage outputs for each array. The data are used to form a look-up table which maps, one-to-one, each set of three voltages, E_{i1}, E_{i2}, and E_{i3}, for each array, to a corresponding velocity vector, defined by its magnitude Q and its pitch and yaw angles φ and ϑ with respect to the probe coordinate system.

$$|E_{i1}, E_{i2}, E_{i3}| \Leftrightarrow |Q, \varphi, \vartheta|. \tag{5.326}$$

The advantages of this approach are that there is no need to assume any type of cooling law for the sensors, and no knowledge of the exact array geometry is required. The disadvantage is that the method requires a large number of calibration data in order to achieve high resolution. This large number of calibration points is set by the desirability of having the voltage difference ΔE_{ik} between two successive points to be of the order of the instrument resolution in order to achieve the optimum accuracy. Various methods can be used to interpolate for cases where optimum data resolution is not obtained. The uniqueness domain can be determined using an approach proposed by Vrettos [5.783]. Of course, as mentioned above, an inherent and serious disadvantage of this approach is that it must be assumed that the velocity vector field across each array is constant.

The second data analysis approach of Honkan [5.690] and Honkan and Andreopoulos [5.689] is based on determining a cooling law describing the response of each sensor. The effective cooling velocities for the i-th array are expressed as the system of algebraic nonlinear equations:

$$U_{ei1}^2 = H_{i1}^2 U_{li}^2 + K_{i1}^2 V_{li}^2 + M_{i1}^2 W_{li}^2,$$
$$U_{ei2}^2 = H_{i2}^2 U_{li}^2 + K_{i2}^2 V_{li}^2 + M_{i2}^2 W_{li}^2,$$
$$U_{ei3}^2 = H_{i3}^2 U_{li}^2 + K_{i3}^2 V_{li}^2 + M_{i3}^2 W_{li}^2, \tag{5.327}$$

where U_l, V_l, and W_l are the velocity components in a local coordinate system x_l, y_l, z_l with its origin attached to the tip of an array and its axes along the sensors, as shown in Fig. 5.196.

These expressions are similar to Jorgensen's (5.321) cooling law. However, they are equivalent to Jorgensen's cooling law only for a special orientation of the prongs for which the velocity component along one of the sensors is normal to the second one and binormal to the third one. This was not the case for their probe.

Using the King's law fit (5.319), this system of equations can be expressed in matrix form as

$$\begin{vmatrix} [(E_{i1}^2 - A_{i1})/B_{i1}]^{2/ni1} \\ [(E_{i2}^2 - A_{i2})/B_{i2}]^{2/ni2} \\ [(E_{i3}^2 - A_{i3})/B_{i3}]^{2/ni3} \end{vmatrix} = \begin{vmatrix} H_{i1}^2 & K_{i1}^2 & M_{i1}^2 \\ H_{i2}^2 & K_{i2}^2 & M_{i2}^2 \\ H_{i3}^2 & K_{i3}^2 & M_{i3}^2 \end{vmatrix} \begin{vmatrix} U_{li}^2 \\ V_{li}^2 \\ W_{li}^2 \end{vmatrix}.$$
$$\tag{5.328}$$

The system (5.328) is linear in U_{li}^2, V_{li}^2, and W_{li}^2 and can easily be solved for each array. The coefficients A_{i1} and B_{i1} can be determined using a calibration procedure by varying the speed magnitude with zero pitch and yaw angle and assuming the coefficient n_{ij} to be in the range 0.45–0.5. The values of the matrix coefficients H_{ij}, K_{ij}, and M_{ij} can be found by appropriately pitching and yawing the probe. Only the positive values of U_{li}, V_{li}, and W_{li} can be inside the probe uniqueness range, as will be discussed below.

The velocity components U_i, V_i, and W_i in the probe coordinate system x, y, z shown in Fig. 5.196, and the velocity components U_{li}, V_{li} and W_{li} in the local coordinate systems of each array are linearly related by

$$\begin{vmatrix} U_i \\ V_i \\ W_i \end{vmatrix} = \begin{vmatrix} p_{i1} & p_{i2} & p_{i3} \\ q_{i1} & q_{i2} & q_{i3} \\ r_{i1} & r_{i2} & r_{i3} \end{vmatrix} \begin{vmatrix} U_{li} \\ V_{li} \\ W_{li} \end{vmatrix} , \quad (5.329)$$

where the coefficients p_{ij}, q_{ij}, and r_{ij} are the combinations of the cosine and sines of angle α and γ shown in Fig. 5.196. After solving the system (5.328) for U_{li}, V_{li}, and W_{li}, the velocity components U_i, V_i, and W_i can be found from the (5.329) for each array and then the six velocity gradients in the y–z plane can be calculated.

Using a similar approach to that of *Vukoslavčević* et al. [5.687], *Honkan* and *Andreopoulos* [5.689] modified this solution method so that the assumption of constant velocity over each array was no longer needed. They concluded that this was a more-accurate method, because it usually is not reasonable to assume that no velocity gradients exist over the sensing areas of the arrays while determining the gradients between array centers. This assumption would only be reasonable if the size of the arrays were much smaller than the distances between the centers of the arrays, which was not the case for any of the probes described here.

Twelve-Sensor Probe. *Tsinober* et al. [5.693] also assumed that the velocity gradients were constant over each of the three arrays of four sensors forming their 12-sensor probe. They considered each four-sensor array as a combination of four three-sensor arrays. The instantaneous velocity vectors obtained from each of the four subarrays were used to form a mean for the four-sensor array at each instant. This mean was taken to be the true value of the instantaneous velocity vector at the four-sensor array. They also used direct mapping with a look-up table to relate the velocity vector and voltage output for each array. The advantage of adding a fourth sensor to each array of the nine-sensor probe to increase the uniqueness range, as will be discussed below, was not utilized with this solution method. As was discussed above, assuming that no velocity gradients exist over the sensing areas of the arrays while determining the gradients between array centers calls into question the accuracy of this method. This is because the size of the arrays of this probe is about the same as the distances between the centers of the arrays.

The advantage of the fourth sensor, related to the uniqueness range, was discussed by *Vukoslavčević* and *Wallace* [5.768]. As in the case of the nine-sensor probe, they assumed a flow field which instantaneously varies across the sensing area of the probe so that each sensor experiences a different velocity vector. Based on their previous experience with the nine-sensor probe, they found that, for the low velocities in the near-wall region of turbulent boundary layer, the polynomial fit (5.320) and *Jorgensen*'s [5.781] expression (5.321) give better results than King's law (5.319) and the effective-angle (5.322) approach. Replacing the orthogonal velocity components U_n, U_t, and U_b in Jorgensen's expression (5.321) with the U, V, and W velocity components in the probe coordinate system x, y, z and equating it to the polynomial (5.320), the following expressions for the response of the j-th ($j = 1$–4) sensor of the i-th ($i = 1$–3) array is obtained

$$\sum_{n=1}^{5} b_{ijn} E_{ij}^{n-1}$$
$$= a_{ij1} U_{ij}^2 + a_{ij2} V_{ij}^2 + a_{ij3} W_{ij}^2$$
$$+ a_{ij4} U_{ij} V_{ij} + a_{ij5} U_{ij} W_{ij} + a_{ij6} V_{ij} W_{ij} . \quad (5.330)$$

The coefficients a_{ijk} depend on the probe geometry, the flow distortion by the prongs and thermal effects. In fact, they are products of the sines and cosines of the angles α and of the pitch and yaw coefficients, h and k, of each sensor. Dividing by a_{ij1} and taking $A_{ijk} = a_{ijk}/a_{ij1}$ and $B_{ijn} = b_{ijn}/a_{ij1}$, (5.330) becomes

$$\sum_{n=1}^{5} B_{ijn} E_{ij}^{n-1}$$
$$= U_{ij}^2 + A_{ij1} V_{ij}^2 + A_{ij2} W_{ij}^2$$
$$+ A_{ij3} U_{ij} V_{ij} + A_{ij4} U_{ij} W_{ij} + A_{ij5} V_{ij} W_{ij} . \quad (5.331)$$

None of the coefficients in this equation need to be known in advance; they can be determined in a calibration procedure. By placing the probe in a nominally irrotational and uniform calibration jet flow and pitching and yawing it, an optimal number m of different flow realizations, $U_{ij} = U_{0m}$, $V_{ij} = V_{0m}$, and $W_{ij} = W_{0m}$, can be induced. These known flow realizations and measured voltages yield a corresponding number of linear equations (5.331) with the A_{ijk} and B_{ijn} coefficients unknown. The number of equations should, of course, be equal to or greater than the number of unknown calibration coefficients for each sensor, requiring at least 10 flow realizations. The method of least squares can be used to minimize the measurement error.

As for the case of the nine-sensor probe, the velocity components U_{ij}, V_{ij}, and W_{ij} can be defined in terms of the velocity components U_0, V_0, and W_0 at the probe center C_0 (Fig. 5.198), their gradients in the plane normal to the probe axis passing through the sensor centers, and the distances c_{ij} and d_{ij} in the y- and z-directions between C_0 and the centers of each sensor.

Expanding the velocity components U_0, V_0, and W_0 to first order in a Taylor series about C_0 results in the following 12 nonlinear algebraic equations with nine velocity and cross-stream velocity gradients unknowns:

$$\sum_{n=1}^{5} B_{ijn} E_{ij}^{n-1}$$
$$= (U_0 + gU_{ij})^2 + A_{ij1}(V_0 + gV_{ij})^2$$
$$+ A_{ij2}(W_0 + gW_{ij})^2 + A_{ij3}(U_0 + gU_{ij})(V_0 + gV_{ij})$$
$$+ A_{ij4}(U_0 + gU_{ij})(W_0 + gW_{ij})$$
$$+ A_{ij5}(V_0 + gV_{ij})(W_0 + gW_{ij}), \quad (5.332)$$

where

$$gU_{ij} = c_{ij}\frac{\partial U}{\partial y} + d_{ij}\frac{\partial U}{\partial z}$$

and with U replaced by V and W for gV_{ij} and gW_{ij}.

A special numerical algorithm was developed to solve these equations. For each array, it chooses the three sensor response equations, out of the four available, that give the maximum uniqueness range. In addition to obtaining the maximum possible uniqueness range, a second advantage of this method is that knowing the exact orientation of the sensors is not required. In order to define the velocity gradient, the only thing that must be known (and with good accuracy) is the spacing between the sensor centers. These distances are determined with a measuring microscope.

Nguyen [5.766] and *Marasli* et al. [5.767] used the same response equations as those by *Vukoslavčević* and *Wallace* [5.768] but their iterative solution method employed a minimization of the least-squared error from all 12 sensors. How this method affects the uniqueness range of the probe has not been analyzed.

Eight-Sensor Probe. The data-reduction mechanism of the eight-sensor probe is the simplest of all those described here. Any of the well-known methods developed for X-arrays probes can be used in this case. The assumption of *Antonia* et al. [5.770] and *Antonia* et al. [5.771] that the velocity field is constant across each array but that it varies between arrays is not necessary. This assumption can be questionable in high-gradient flows, bearing in mind that the ratio of the sensor separations to the separations between arrays is about 0.3. Assuming a constant-gradient flow field over the whole probe sensing area will make the response equations more complex, but they can be easily solved using a successive iteration procedure similar to the one developed for the nine- and 12-sensor probes. However, it is necessary to neglect the binormal component cooling each sensor and, therefore, its gradients when only eight sensors are used.

Streamwise Velocity Gradients

Using the data obtained with the nine- or 12-sensor probes, three velocity components and six velocity gradients in the cross-stream directions can be obtained. However, only four of the cross-stream gradients can be obtained with the eight-sensor probe. The streamwise gradients have to be evaluated using an alternate method.

A common method is to apply *Taylor*'s [5.762] hypothesis that the turbulent flow pattern is *frozen* for a sufficiently short time interval as it passes the probe at a fixed point in space with a convection speed U_c. Under this hypothesis, the streamwise gradient of any instantaneous velocity component U_i can be expressed as

$$\frac{\partial U_i}{\partial x} = -\frac{1}{U_c}\frac{\partial U_i}{\partial t}. \quad (5.333)$$

The convection velocity U_c is usually taken as the local mean velocity $\overline{U}$ or as the local instantaneous velocity U, giving the following forms of Taylor's hypothesis:

$$\frac{\partial U_i}{\partial x} = -\frac{1}{\overline{U}}\frac{\partial U_i}{\partial t} \quad (5.334)$$

or

$$\frac{\partial U_i}{\partial x} = -\frac{1}{U}\frac{\partial U_i}{\partial t}. \quad (5.335)$$

Another approach is based on the assumption that an instantaneous property φ of a material fluid particle does not change along its Lagrangian path during a sufficiently small time interval so that the material derivative $D\varphi/Dt = 0$. Under this assumption the momentum equation reduces to a refined form of Taylor's hypothesis

given by

$$\frac{\partial U_i}{\partial x} = -\frac{1}{U}\left(\frac{\partial U_i}{\partial t} + V\frac{\partial U_i}{\partial y} + W\frac{\partial U_i}{\partial z}\right). \quad (5.336)$$

It is clear from the momentum equation that this expression is strictly valid only if the pressure forces negate the viscous forces, and its accuracy depends on the degree to which this is true.

The streamwise gradient $\partial U/\partial x$ can also be determined from the continuity equation for incompressible fluid flow

$$\frac{\partial U}{\partial x} = -\left(\frac{\partial V}{\partial y} + \frac{\partial W}{\partial z}\right). \quad (5.337)$$

This exact expression can be employed to test the different forms of Taylor's hypothesis used to determine the streamwise velocity gradient.

All three equations, (5.334), (5.335) and (5.336) have been analyzed by evaluating the magnitude of the terms in the momentum equation. They have been tested using experimental and numerical data for different types of turbulent flows. For a turbulent flow field that is convected locally in the x-direction by the mean streamwise velocity $\overline{U}$ ($\overline{V} \approx 0$, $\overline{W} \approx 0$, $\partial \overline{U}/\partial z = 0$), the refined Taylor hypothesis expression takes the form

$$\frac{\partial U_i}{\partial x} = -\frac{1}{\overline{U}}\left(\frac{\partial U_i}{\partial t} + v\frac{\partial u_i}{\partial y} + w\frac{\partial u_i}{\partial z} + v\frac{\partial \overline{U}_i}{\partial y}\right), \quad (5.338)$$

where u_i are the velocity fluctuations about the mean values and v and w are components in the cross-stream directions. The last three terms in this equation are small for a sufficiently small turbulence level. In that case, the refined (5.336) and classical form (5.335) of Taylor's hypothesis give similar results.

The momentum equation can be rewritten in the following form

$$\frac{\partial U_i}{\partial t} + \overline{U}\frac{\partial u_i}{\partial x} + \left(u\frac{\partial u_i}{\partial x} + v\frac{\partial u_i}{\partial y} + w\frac{\partial u_i}{\partial z}\right)$$
$$- \left[\frac{\partial}{\partial x}(\overline{uu}) + \frac{\partial}{\partial y}(\overline{uv}) + \frac{\partial}{\partial z}(\overline{uw})\right]$$
$$+ u\frac{\partial \overline{U}_i}{\partial x} + v\frac{\partial \overline{U}_i}{\partial y} + w\frac{\partial \overline{U}_i}{\partial z} + \frac{1}{\rho}\frac{\partial p}{\partial x_i}$$
$$- \nu\left(\frac{\partial^2 u_i}{\partial x^2} + \frac{\partial^2 u_i}{\partial y^2} + \frac{\partial^2 u_i}{\partial z^2}\right) = 0, \quad (5.339)$$

where p are the pressure fluctuations about the mean values. This form of the momentum equation is convenient to analyze the magnitude of the terms to be neglected in order to test the accuracy of Taylor's hypothesis for a given type of flow. An analysis of the order of magnitude of these terms is given by *Heskestad* [5.784]. He found that the terms $u_j(\partial \overline{U}_i/\partial x_j)$ are always small compared to $\overline{U}(\partial u_i/\partial x)$, and that the $(\partial/\partial x_i)\overline{u_i u_j}$ terms are also small compared to $u_j \partial u_i/\partial x_j$ if the Taylor microscale λ is quite small compared to some characteristic length L so that $\partial u_i/\partial t \gg \lambda/L$. Neglecting the viscous terms $\nu \partial^2 u_i/\partial x_k^2$, which he did on the grounds that viscous forces do not greatly affect the large-scale motion of the flow associated with its momentum, the momentum equation (5.339) reduces to the form

$$\frac{\partial u_i}{\partial x} = -\frac{1}{\overline{U}}\left(\frac{\partial U_i}{\partial t} + v\frac{\partial u_i}{\partial y} + w\frac{\partial u_i}{\partial z} + \frac{1}{\rho}\frac{\partial p}{\partial x_i}\right). \quad (5.340)$$

In the case of boundary-layer flow, where $\partial \overline{U}/\partial x \approx 0$, this expression differs from the generalized Taylor's hypothesis (5.338) only because it includes the pressure fluctuation gradient, and because *Heskestad* [5.784] considered the $v(\partial \overline{U}/\partial y)$ term in (5.338) to be negligible. *Batchelor* [5.785] estimated that, in an isotropic field, the ratio of the mean square pressure gradient to a typical mean square inertia term is inversely proportional to the Reynolds number $\text{Re}_\lambda = \lambda u/L$. If this is the case, only for $\text{Re}_\lambda > 1000$ can the pressure terms be neglected and the streamwise gradients be evaluated from the generalized Taylor hypothesis (5.336) with sufficient accuracy.

Piomelli et al. [5.786] used direct numerical simulation (DNS) and large-eddy simulation (LES) data from turbulent channel flows as well as the experimental boundary-layer data of *Balint* [5.787] to extend the numerical study of Taylor's hypothesis initiated by *Hussain* et al. [5.788]. They compared the rms of the streamwise velocity gradient obtained directly from the LES and used the classical (5.334) as well as the refined form (5.336) of Taylor's hypothesis. Very good agreement was found above the buffer layer. By comparing $\partial U_i/\partial t$ with $\overline{U}\partial U_i/\partial x$ from the LES, high phase coherence between these two signals was observed above the buffer layer as evidenced by correlation coefficient values of 0.95 or higher everywhere in this region of the flow for all three streamwise velocity gradients. The amplitudes of these two signals were also similar in this region of the flow, as evidenced by their similar rms levels. In interpreting these results, one should keep in mind that in LES only the larger-scales are computed accurately, while the smaller ones are modeled. The influence of the small scales obviously affect the velocity derivatives. To assess the importance of the neglected small scales, the rms intensities of the velocity gradients were compared to those obtained from DNS and from the experimental data of *Balint* [5.787]. Above the buffer layer, the agreement between the LES and the experimental rms values, which were determined with comparable resolution, was very good, while both sets of values are somewhat lower than those of the DNS due to insufficient resolution.

Kim and *Hussain* [5.790] investigated the propagation speed of velocity, vorticity, and pressure fluctuations from a turbulent channel flow DNS simulation. This investigation has obvious implications for Taylor's hypothesis, as they pointed out. They found, surprisingly, that the propagation velocity is very close to the local mean velocity except in the near-wall region, i.e., the viscous and buffer layers, where it has a nearly constant value of about $10\,u_\tau$. The relative lack of agreement seen by *Piomelli* et al. [5.786] between the amplitudes of $\partial U_i/\partial t$ and $\overline{U}\partial U_i/\partial x$ in and below the buffer layer, as indicated by their rms values there, can be explained by the fact that turbulence fluctuations are not convected with the mean velocity in this region near the wall. Use of a more-appropriate convection velocity in the near-wall region of bounded flows will clearly improve the ability of the Taylor's hypothesis time–space transformation to provide adequate estimates of streamwise gradients.

A comparison of the streamwise gradient $\partial U/\partial x$ evaluated using the continuity equation and different forms of Taylor's hypothesis was given by *Loucks* [5.780]. The time series, shown in Fig. 5.202, was obtained from the boundary-layer data of *Ong* [5.789] using a nine-sensor probe in the logarithmic layer at $y^+ = 89$.

The time series of the two forms of Taylor's hypothesis match extremely well in both phase and amplitude. They can hardly be distinguished from each other. They both also track fairly closely in phase and shape with the continuity equation. However, the amplitudes of the Taylor's hypothesis signals deviate from that of the continuity equation signal during some short intervals of the time series shown. *Loucks'* [5.780] further analysis of this data, as well as the data taken by a 12-sensor probe in a plane mixing layer, have shown that these differences between $\partial U/\partial x$ obtained from continuity and from Taylor's hypothesis are not an effect of the data sampling rate. We have determined that this amplitude disagreement in the boundary-layer data of *Ong* [5.789] can be much reduced by a judicious choice of the Taylor's hypothesis convection velocity, viz. by using a convection velocity of $(0.6-0.7)\,\overline{U}$ near the wall. This is consistent with the findings of *Kim* and *Hussain* [5.790] mentioned above.

To compare the phase shifts of the different signals, *Loucks* [5.780] calculated the cross-correlation coefficients shown in Fig. 5.203, where R_1, R_2 and R_3 are the cross-correlation of $\partial U/\partial x$ obtained from continuity (5.337) with the various forms of Taylor's hypothesis given by (5.334), (5.336) and with (5.336), where $\overline{U}$ was used as the convection velocity in place of U.

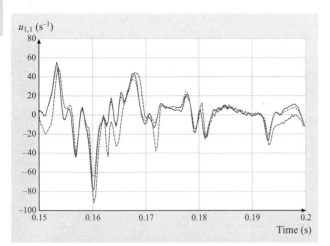

Fig. 5.202 A time series of the streamwise velocity gradient from *Loucks* [5.780] using the boundary-layer data of *Ong* [5.789] at $y^+=89$: continuity equation (5.337) (*dashed line*), Taylor hypothesis (5.334) (*solid line*), and refined Taylor hypothesis (5.336) (*dotted line*)

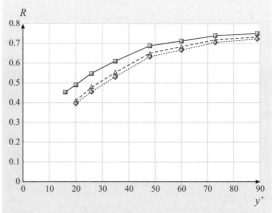

Fig. 5.203 Correlation coefficients of the streamwise velocity gradients from *Loucks* [5.780] using the boundary-layer data of *Ong* [5.789] at $y^+ = 89$: continuity (5.337) with Taylor's hypothesis (5.334) (*squares*), continuity (5.337) with refined Taylor's hypothesis (5.336) (*circles*), and continuity (5.337) with refined Taylor's hypothesis (5.336) but using $\overline{U}$ as the convection velocity (*triangles*)

Although all three coefficients are close to each other, surprisingly the highest correlation was obtained using the classical form of the Taylor hypothesis. Similar results were obtained in the mixing-layer flow.

To summarize, it can be stated that Taylor's hypothesis can be used to approximate streamwise gradients with an accuracy that improves with distance from the wall for bounded flows and with increasing mean velocity in mixing-layer flows. Adding the additional terms for the refined form of Taylor's hypothesis does not improve the accuracy. For all the flows mentioned here for which Taylor's hypothesis was examined, the Taylor microscale Reynolds number Re_λ was in the range 500–700. These are large values, but are still smaller than the criterion of *Heskestad* [5.784].

Uniqueness Range

It is well known that different fluid velocity vectors, as illustrated in Fig. 5.204, can produce the same hot-wire response. For any number of heated sensors included in a hot-wire probe, multiple solutions of the sensors' response equations can be obtained, and different velocity vectors can be determined from any specific set of anemometer output signals. This is the well-known *uniqueness* or *rectification* problem referred to by *Tutu* and *Chevray* [5.792].

Fortunately, there is a solution domain, known as the *uniqueness domain*, where the velocity vector can be uniquely determined for a given probe. The uniqueness domain of a multiple-sensor probe is shown schematically in Fig. 5.204b. The outer boundary of this domain can be described by the largest angle, φ_{cr}, defined as the angle between the probe axis, shown in Fig. 5.204b, and the velocity vector U_{cr} at the edge of the domain. Within this domain the fluid velocity vector U can be uniquely determined. The geometrical form of such a domain is usually an asymmetric conical surface for which φ_{cr} is a function of the angle θ, which specifies the orientation of the meridian plane containing U_{cr}. Its shape varies, depending primarily on the number of sensors and their configuration. However, the *uniqueness range* is more commonly represented by the symmetric cone, shown in Fig. 5.204b, the half angle ε of which is defined as the minimum value of the angle φ_{cr}. Inside this cone there is a unique solution independent of the rotation of the probe around its axis. Thus hot-wire probes should only be used in flows with velocity vectors contained within the uniqueness cone. Otherwise, at least two fluid velocity vectors will give the same output signals, and there is no way to recognize which of them really exists in a given flow. The uniqueness range of a vorticity probe is the same as the uniqueness range of its arrays.

To date, a variety of investigations about the uniqueness range of different hot-wire probe configurations and their signal interpretation procedures have been published. An illustrative graphical representation, based on the *cosine law*, of the response of an ideal, infinitely long, hot-wire sensor to possible fluid velocity vector directions was given by *Willmarth* [5.793]. In the case of a three-sensor probe, there can be up to eight intersection points between the three cylindrical surfaces that represent three-probe sensor responses.

Detailed numerical and graphical analyses of the uniqueness domain of three- and four-sensor probes have been carried out by many researchers; one such survey can be found in *Vukoslavčević* et al. [5.791]. They also analyzed the uniqueness domain and derived analytical expressions defining the uniqueness range of three-

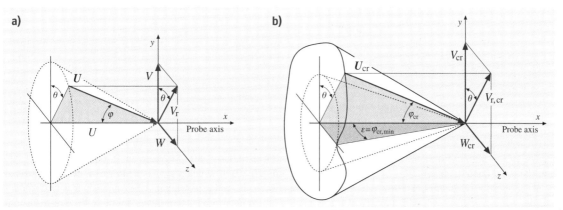

Fig. 5.204 (a) Velocity vector in Cartesian and spherical coordinates. (b) The uniqueness domain and uniqueness cone of a multisensor hot-wire probe. (After *Vukoslavčević* et al. [5.791], courtesy Inst. of Physics)

and four-sensor probes. These expressions are based on the effective angle approach (5.322) or on the effective cooling velocity (5.321) described by Jorgensen's law.

For the nine-sensor T-configuration probe, shown in Fig. 5.195, or one of its arrays, the half angle ε of the uniqueness cone corresponds to the meridian plane $\theta = 0°$ ($W = 0$) and is given by

$$\varepsilon = \varphi_{\text{cr,min}} = a\tan\left[\frac{\sin 2\alpha_e}{2h^2}\right]. \quad (5.341)$$

A maximum value of $\varepsilon = 26.56°$ can be achieved for the case of an ideal sensor response ($\alpha = \alpha_e$, $h = 1$) and a sensor inclination angle of $\alpha = 45°$. For any other value of angle α the uniqueness range will decrease. In the case of a nine-sensor orthogonal-configuration probe or one of its arrays, shown in Fig. 5.196, the half cone angle ε corresponds to meridian planes containing any of the sensors and their supporting prongs and is defined by

$$\varepsilon = \varphi_{\text{cr,min}} = a\tan\left(\frac{\sin 2\alpha_e}{h^2 + \sin^2\alpha_e}\right), \quad (5.342)$$

with a maximum value of $\varepsilon = 35.26°$ for an ideal sensor response ($\alpha = \alpha_e$, $h = 1$) and a sensor inclination angle of $\alpha = 35.26°$. As for the nine-sensor T-configuration probe, the uniqueness range will decrease for any other value of the sensor inclination angle α.

Two expressions were derived for the 12-sensor probe shown in Fig. 5.198 or for one of its four-sensor arrays. One is based on the effective-angle approach and is given by

$$\varepsilon = \varphi_{\text{cr,min}} = a\tan\left(\frac{\sqrt{2}\cos\alpha_e}{\sqrt{\sin^2\alpha_e + h^2}}\right). \quad (5.343)$$

The other is based on Jorgensen's cooling law and is given by

$$\varepsilon = \varphi_{\text{cr,min}} = a\tan\sqrt{\frac{2(\cos^2\alpha + k\sin^2\alpha)}{\sin^2\alpha + k^2\cos^2\alpha + h^2}}. \quad (5.344)$$

Both of these correspond to the same meridian planes $\theta = \pm 45°$ and $\theta = \pm 135°$ ($W = \pm V$). For the case of an ideal sensor response ($\alpha = \alpha_e$, $k = 0$, $h = 1$), they reduce to the same form

$$\varphi_{\text{cr1,2}} = a\tan\left(\frac{\sqrt{2}\cos\alpha}{\sqrt{\sin^2\alpha + 1}}\right), \quad (5.345)$$

which was first derived by *Pailhas* and *Cousteix* [5.794] and gives a value of $\varepsilon = 39.23°$ for $\alpha = 45°$.

It is therefore clear that the 12-sensor probe has the highest value of the uniqueness half cone angle with a value of $\varepsilon = 39.23°$, compared to $\varepsilon = 26.56°$ and $\varepsilon = 35.26°$ for nine-sensor T-configuration and nine-sensor orthogonal-configuration probes, respectively. In addition, in contrast to the nine-sensor configurations, this value can be further increased by decreasing the sensor inclination angle α, as can be seen in Fig. 5.205 where ε is shown for different values of the pitch coefficient h and the sensor inclination angle α_e.

If the wire angle is decreased to the value $\alpha = 35.26°$ that corresponds to the nine-sensor orthogonal probe, the uniqueness half cone angle will be $\varepsilon = 45.0°$, which is about 9° higher than for the case of the nine-sensor orthogonal probe and about 18° higher than for the nine-sensor T-configuration probe. Unfortunately, decreasing the sensor inclination angle α decreases the sensitivity of the probe to the cross-stream velocity components.

All four expressions, (5.341), (5.342), (5.343), and (5.344), clearly show that for real sensor responses the uniqueness range cone half-angle ε decreases with an increase of the pitch coefficient h. In addition, the same trend can also be seen in Fig. 5.205 for the 12-sensor probe.

In order to achieve the maximum possible value of the uniqueness range cone half-angle for the 12-sensor probe shown in Fig. 5.198, an appropriate choice of three out of the four available sensors has to be made for each array, as explained in *Vukoslavčević* and *Wallace* [5.768]. Each array of a 12-sensor probe can be viewed as a combination of two subarrays of

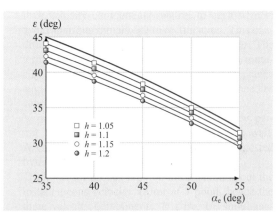

Fig. 5.205 The uniqueness cone half-angles $\varepsilon = f(h, \alpha)$ for a 12-senor probe for $h = 1$ (lines), $h = 1.05$ (open square), $h = 1.1$ (filled square), $h = 1.15$ (open circle), and $h = 1.2$ (filled circle). (After Vukoslavčević et al. [5.791], courtesy Inst. of Physics)

three-sensors in the T-configuration. For example, one subarray is formed by sensors 1, 2, and 3 in the geometrical arrangement of the arrays of the nine-sensor probe shown in Fig. 5.195, and the other is formed by sensors 2, 3, and 4 corresponding to the same three-sensor subarray configuration rotated by 180°. For a triple-sensor array, shown in Fig. 5.195, there are, in fact, two minimums of the critical angle φ_{cr}. The first is in the range $|\theta| \leq 90°$, or $V \leq 0$ and the second is in the range $90° \leq \theta \leq 180°$, or $V \geq 0$. For the combination of sensors 1, 2, and 3 the first (smaller) minimum is given by (5.341) and the second (higher) one by (5.343) and (5.344). The situation is reversed for the combination of sensors 2, 3, and 4. The higher ε corresponds to $V \geq 0$ or $90° \leq \theta \leq 180°$ and the lower one to $V \leq 0$ or $|\theta| \leq 90°$. This means that, in order to achieve the higher value for the uniqueness range cone half-angle given by (5.343) or (5.344), a choice of sensors 1, 2, and 3 has to be made for negative values of V, and sensors 2, 3, and 4 have to be chosen for positive V. However, the value of the V velocity component is not known in advance. The procedure can be started by any arbitrary initial choice and then updated depending on the sign of V. The same conclusion can be drawn from the subarray sensor combinations 1, 2, and 4 and 1, 3, and 4, together with the sign of the W velocity component.

In the case of the mapping approach using a look-up table, the only condition for which the mapping given by (5.326) is unique is given by

$$J = \begin{vmatrix} \partial E_{i1}/\partial Q & \partial E_{i2}/\partial Q & \partial E_{i3}/\partial Q \\ \partial E_{i1}/\partial \varphi & \partial E_{i2}/\partial \varphi & \partial E_{i3}/\partial \varphi \\ \partial E_{i1}/\partial \vartheta & \partial E_{i2}/\partial \vartheta & \partial E_{i3}/\partial \vartheta \end{vmatrix} \neq 0 . \quad (5.346)$$

The nine elements of this Jacobian J are the sensitivities of each sensor. The partial derivatives can be found by using the finite-difference method at the successive points obtained in a calibration procedure. If J does not change sign at any point inside a domain, a unique measurement of the velocity vector can be obtained in that domain. The first attempt to verify the uniqueness of this type for an orthogonal triple hot-wire was given by *Vrettos* [5.783]. This condition is very general and can be applied to any type of probe configuration.

Examples of Measurements with Multisensor Hot-wire Probes

The probes described above have been used to measure the vorticity components as well as other important velocity gradient-based properties of turbulence such as the strain rate $S_{ij} = \partial u_i / \partial x_j + \partial u_j / \partial x_i$, the dissipation rate $\varepsilon = 2 \langle S_{ij} S_{ij} \rangle$, and the helicity density $U_i \cdot \Omega_j$ in grid turbulence as well as in turbulent boundary layers, wakes, jets and mixing layers [5.689, 690, 693, 730, 731, 764, 766–768, 770, 771, 780, 787, 789, 795–805]. Here a few representative examples from the rich variety of the results from these measurements will be shown. In Fig. 5.206, the distribution of enstrophy, $0.5\overline{\omega_i \omega_i}$, measured in turbulent boundary layers with several of these probes is compared to the distribution obtained from a direct numerical simulation (DNS) by *Spalart* [5.742] of a turbulent boundary layer and of a turbulent channel flow by *Moser* et al. [5.806].

Here the comparison data are all from boundary-layer experiments or DNS with Reynolds numbers given in terms of the momentum thickness θ and the freestream velocity U_∞, with the exception of the channel flow DNS of Moser et al. where the Reynolds number is given in terms of half the channel width and the friction velocity u_τ. The experimental profiles show the same monotonic increase of the enstrophy as the wall is approached as seen in the DNS profiles. However, all the experimental data fall below the two DNS profiles except for the data of *Ong* and *Wallace* [5.803] after the normalizing velocity scale u_τ has been corrected. This correction, discussed thoroughly in *Ong* and *Wallace* [5.803], is appropriate, because using the Clauser method to determine u_τ, as was done for all the experimental data shown here, results in significantly overestimated values. Furthermore, because u_τ^4 is used in the normalization of $0.5\overline{\omega_i \omega_i}$, this error in the normalizing scale is greatly amplified.

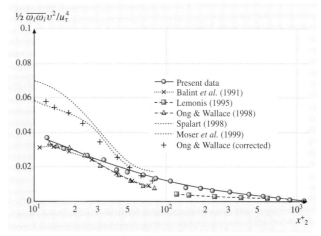

Fig. 5.206 Normalized enstrophy, $0.5\overline{\omega_i \omega_i}/u_\tau^4$, measured in turbulent boundary layers with vorticity probes and compared to simulation results. (After *Andreopoulos* and *Honkan* [5.730], courtesy CUP)

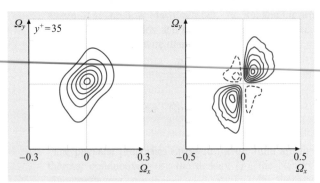

Fig. 5.207 Joint PDF and covariance integrand of the vorticity components Ω_x and Ω_y in a turbulent boundary layer at $\text{Re}_\theta = 1070$ (after *Ong* and *Wallace* [5.803], courtesy CUP)

The joint probability density function (JPDF) of the streamwise and spanwise components of vorticity $P(\Omega_x, \Omega_y)$ reveals interesting characteristics of the vortical structure of turbulent shear flows. For example, the left-hand plot of $P(\Omega_x, \Omega_y)$ in Fig. 5.207 is of data taken just above the buffer layer in the turbulent boundary-layer experiment of *Ong* and *Wallace* [5.803]. The inclination of the major axis of the JPDF into the first and third quadrants of the hodograph plane implies a preference for these vorticity components to occur simultaneously with like signs, either both plus or both minus. This is even more evident in the right-hand plot of the covariance integrand, $\Omega_x \Omega_y P(\Omega_x, \Omega_y)$, where it is seen that the occurrence of like signs of these two components (first and third quadrants) dominates the value of the covariance, $\overline{\Omega_x \Omega_y}$. In fact, as *Ong* and *Wallace* [5.803] point out, the peak values of the contours in the first and third quadrants of the covariance integrand plots can be used to find the projections on the x–y plane of the angles of orientation of the vortex filaments that most contribute to the covariance $\overline{\Omega_x \Omega_y}$ at this y^+ location. These characteristics of the JPDF and covariance integrand are consistent with the prevalent view that the bounded flows are made up of organized, but often asymmetric, *hairpin* vortices, inclined downstream in the x–y plane.

In addition to simultaneous measurement of the vorticity vector components, the fact that these probes provide good estimates of the complete velocity gradient tensor, with the exception of the eight-sensor probe, make them useful in determining several other fundamental properties of turbulence. For example, all the terms of the turbulent dissipation rate are available. Figure 5.208 shows time series obtained by *Folz* [5.801] in an experiment carried out in the near-surface layer of the atmospheric boundary layer over the Utah salt flats with a 12-sensor probe like the one described in Fig. 5.198. The full dissipation rate signal ε is compared to approximations in which

1. the inhomogeneous cross-product terms are neglected so that $\varepsilon \sim 2\nu \sum_i \sum_j (\partial u_i / \partial x_j)^2$,
2. the enstrophy is used as a surrogate so that, $\varepsilon \sim \zeta/\nu = (1/\nu)\omega_i \omega_i$, and
3. isotropy is assumed so that $\varepsilon \sim 15\nu (\partial u/\partial x)^2$.

There are significant differences in both the amplitude and phase between the time series of each of these approximations of the dissipation rate and the full dissipation rate ε.

A strong alignment of the vorticity vector with the intermediate eigenvector of the rate-of-strain tensor, α_i for $i = 2$, was observed by *Tsinober* et al. [5.693] in both turbulent grid flow and in a turbulent boundary layer at $y/\delta = 0.2$, as seen in Fig. 5.209. This striking result is in agreement with numerical simulations by *Ashurst* et al. [5.807], but its physical significance is still unexplained.

The decay of the vorticity component variances measured by *Antonia* et al. [5.770] in grid turbulence is shown in Fig. 5.210, where it is compared to an isotropic estimate of their decay. Each of the components decays at the same rate as the isotropic estimate, providing convincing evidence that the vorticity field, which is dominated by high wavenumbers, is isotropic in this grid flow.

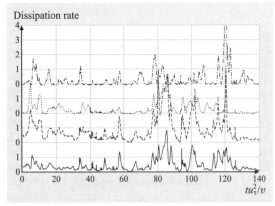

Fig. 5.208 Time series of differences between full dissipation rate, $\varepsilon = 2S_{ij}S_{ij}$ (*solid line*), and three approximations to ε at $y^+ = 45$ in the atmospheric surface layer over the Utah salt flats: homogeneous, $2\nu \sum_i \sum_j (\partial u_i / \partial x_j)^2$ (*dashed line*); entropy, ζ/ν (*dotted line*); and isotropic, $15\nu(\partial u/\partial x)^2$ (*dot-dashed line*). (After *Folz* [5.801])

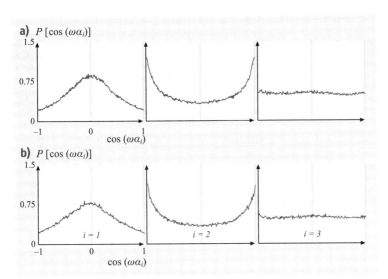

Fig. 5.209a,b Probability density functions of the cosine of the angle between the vorticity vector and the eigenvectors of the rate-of-strain tensor: $\alpha_2 < \alpha_2 < \alpha_3$; (**a**) grid flow at $x/M = 30$ and (**b**) boundary layer at $y/\delta = 0.2$. (After *Tsinober* et al. [5.693], courtesy CUP)

Summary

Only within the last 20 years has knowledge of the important properties of turbulence defined by the velocity gradient tensor, including the vorticity vector, become accessible by experimental measurements and numerical simulations. This access has enabled the examination of the structural characteristics of turbulence and the related transport processes for a rich variety of flows, as well as the testing of previously unverifiable turbulence theories. The development and use of multisensor hot-wire probes to obtain this knowledge has been an important part of this progress in turbulence research.

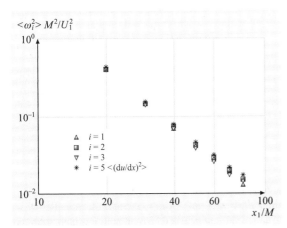

Fig. 5.210 Decay of the vorticity component variances with downstream distance in grid turbulence compared with isotropic decay. (After *Antonia* et al. [5.770], courtesy CUP)

5.5.4 Transverse Vorticity Measurements with a Four-Sensor Hot-Wire Probe

This section describes the use of a four-sensor hot wire probe that is capable of measuring time series of two components of velocity (u, v) and a single component of vorticity that is perpendicular to the measured velocities (ω_z). The original probe design and processing algorithm were developed by Foss and coworkers, and can be found in *Foss* and *Haw* [5.808], *Wallace* and *Foss* [5.81]. Measurements from this sensor are presented in, for example, *Foss* [5.809] and *Morris* and *Foss* [5.810].

A three-view drawing of the basic probe geometry is shown in Fig. 5.211. A perspective view is shown in Fig. 5.212 for clarity. Two of the sensors are parallel with

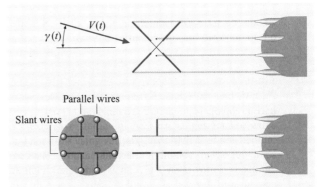

Fig. 5.211 Three-view drawing of the probe and coordinate system. Note that the wire thicknesses are not to scale

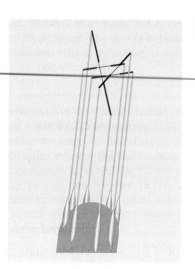

Fig. 5.212 Perspective view of the probe geometry

1–1.5 mm spacing. The remaining two sensors are configured as a standard X-wire probe, and placed through the two parallel sensors as shown. The result is a nominal measurement domain that is of the order 1 mm³ in size. The definition of the approach velocity vector is shown in the coordinate system of the probe in Fig. 5.211. The component of velocity normal to the plane shown at the top of Fig. 5.211 is assumed to be small.

The sensors are typically created using tungsten wire with a nominal 5 μm diameter. Copper plating aerodynamically isolates the 1 mm active portion of the sensor from the support prongs. The copper plating is typically hand soldered to the supports using a multi-degree-of-freedom traverse to position the sensor over the prongs. Note that a significant challenge in the manufacture of the probe is to *thread* the fourth wire through the others. That is, the first three sensors can be positioned on the support prongs without difficulty but, due to the geometry, the fourth wire cannot be positioned without threading it through the others. This can be accomplished by first fixing the fourth wire to a small support needle.

Algorithm

The time series output from the probe includes the magnitude of velocity in the *plane of the probe* as determined by the two straight wires q_1, q_2 and the angle γ determined from the X-array. Note that the symbol $q = (u^2 + v^2)^{1/2}$ is used since these sensors essentially respond to the velocity magnitudes that are perpendicular to their axes. The X-array, which could provide both u and v, is only necessary to provide the instantaneous flow angle γ. Methods for calibrating single-sensor hot-wires and X-array probes are given in Sect. 5.2. The component of vorticity to be calculated from the hot-wire signals is given by

$$\omega_z = \frac{\partial v}{\partial x} - \frac{\partial u}{\partial y} \ . \tag{5.347}$$

Rather than approximating the spatial velocity derivatives directly, an integral approach will be presented based on the circulation

$$\langle \omega_z \rangle_\mu = \frac{1}{A_\mu} \oint \boldsymbol{V} \, \mathrm{d}s \ . \tag{5.348}$$

That is, the probe is used to evaluate a spatially averaged vorticity over a defined microcirculation domain of area A_μ by evaluating this circulation integral. The objective of the processing algorithm is to use the outputs from the hot-wire sensors (q_1, q_2, γ) to create a convected domain where (5.348) can be evaluated. A schematic of the domain of integration is shown in Fig. 5.213.

The domain is defined by the streamwise ($\hat{s}$) and probe-normal ($\hat{y}$) directions. In short, the time series of the flow speed provides the definition of a convected domain. The circulation integral can be evaluated given that the two parallel wires provide the velocity on the *top* and *bottom* of the region, and the X-wire probe provides a single estimate of the vertical component at both the fore and aft sides of the domain.

The domain is created using the time series $q_1(t_n)$ and $q_2(t_n)$ with the assumption of a local (in time) Taylor's hypothesis. The height of the domain Δy is specified by the distance between the parallel wires. The length of the domain Δs is created from a convected length that is nominally the same dimension: $\Delta s \approx \Delta y$.

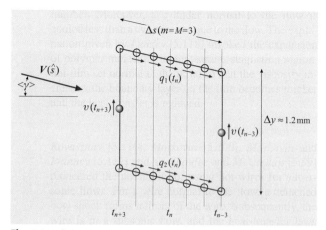

Fig. 5.213 Schematic of the microcirculation domain for the evaluation of (5.348) assuming $M = 3$

However, the definition of the convected length in the streamwise direction is limited to discrete time record values set by the acquisition frequency. Specifically, the value of Δs is evaluated at time t_n as:

$$\Delta s_n(m) = \sum_{i=n-m}^{n+m} \left[\frac{q_1(t_i)+q_2(t_i)}{2}\right] \Delta t . \quad (5.349)$$

where Δt is the inverse of the data-acquisition frequency, and the convection velocity is defined as the mean from the two parallel sensors. The size of Δs is established by increasing the dummy variable m until $\Delta s_n(m) > \Delta y$. The smallest value of m to satisfy this inequality is denoted by $m = M$.

Note that the circulation domain is centered on the time series point t_n, and is expanded M data points in both directions. The total number of data points used $(2M+1)$ for each calculation time (t_n) depends on the instantaneous flow speed. The data-acquisition frequency must be large enough that a minimum number of time series points are included for each evaluation of the integral given in (5.348). Specifically, $f_{acq} > 2MU/\Delta y$, where U is the minimum velocity expected in the time series, and M should be at least 3.

Note that the evaluation of the domain length defined in (5.349) assumes that the angle γ is constant over the time period that defines the ≈ 1 mm convected length. The variability in $\gamma(t)$ from the averaged value in this time period, $\langle \gamma_n \rangle$ (shown in Fig. 5.213), can be accounted for by including the factor $\cos[\gamma(t) - \langle \gamma_n \rangle]$ in the summation of (5.349). However, this correction is nearly unity in practice, and does not affect the definition of the convected length.

Once the microdomain has been *created* at time record t_n, the average vorticity is calculated from the circulation around the domain. The magnitude of the velocity can be projected onto the *top* of the microdomain as $q_1 \cos(\gamma - \langle \gamma \rangle)$ and then integrated along the length. The discrete representation is given by

$$\langle q_1(t_n) \rangle = \frac{1}{2M-1} \sum_{i=n-M}^{n+M} q_1(t_i) w_i \cos[\gamma(t_i) - \langle \gamma \rangle] . \quad (5.350)$$

The weighting variable is defined as $w_i = 0.5$ if $i = \pm M$ and 1 otherwise, which properly accounts for the numerical integration at the end points of the domain.

The quantity $\langle q_2(t_n) \rangle$ should be evaluated similarly for the *bottom* of the domain. Again, the cosine factors in these expressions can typically be ignored without a significant loss of accuracy, although this should be evaluated for a given measurement environment. The velocity component in the y-direction is calculated from

$$v(t_{n\pm M}) = \frac{q_1(t_{n\pm M}) + q_2(t_{n\pm M})}{2} \sin(\gamma_{n\pm M}) \quad (5.351)$$

at both the upstream and downstream sides of the domain.

For each point of the time series (n), the vorticity is calculated from:

$$\langle \omega_n \rangle = \frac{1}{(\Delta s)(\Delta y)} \oint (V \, ds)$$
$$= \frac{v(t_{n+M}) - v(t_{n-M})}{\Delta s} - \frac{\langle q_2 \rangle - \langle q_1 \rangle}{\Delta y} . \quad (5.352)$$

Following the evaluation of $\langle \omega_n \rangle$ at time t_n, the time step is incremented, and the algorithm is repeated, beginning with a new evaluation of Δs. Given the relatively high acquisition frequency required for the evaluation of the circulation domain, it is often desirable to down-sample the vorticity by skipping to the next time step as $t_{n+\langle M \rangle}$, where $\langle M \rangle$ is the average value of M that is anticipated.

Uncertainty

The ability of the probe and algorithm to recover a time series of the fluid vorticity accurately depends highly on the specific application and implementation of the technique. This section will provide a summary of possible errors and suggested methods to minimize their effects.

An important consideration for any hot-wire-based technique is proper calibration. Typically the probe is calibrated to a velocity reference that is derived from a pressure measurement as described in Sect. 5.2. With the vorticity probe, the calibration errors can be minimized by utilizing a two-step calibration procedure. First, the two parallel sensors are calibrated and fitted. The sensor with the lower standard deviation of error is then used as the master, to which the other parallel sensor and the X-wire are calibrated. This will inherently minimize calibration differences between the two parallel sensors [the difference being of obvious importance in (5.352)].

A second consideration is the effect of electronic noise on the measurement. This is of particular importance due to the relatively high-frequency response that is required. As noted earlier, the number of points needed to create the microcirculation domain should be nominally greater than six. For example, a probe with $\Delta y = 1.0$ mm and a velocity of 10 m/s requires an acquisition frequency of 60 kHz. Note that this implies the requirement that the anemometer circuitry have a frequency response at least as high. The ef-

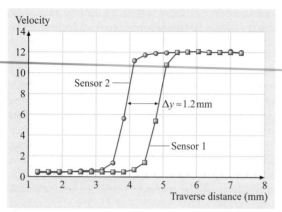

Fig. 5.214 Results from traversing the probe through a laminar shear layer. The displacement in the mean velocity curves allows the evaluation of Δy

contamination. That is, $M > 10$ should be a nominal requirement.

A source of potential error in the magnitude of the recovered vorticity is uncertainty in the distance between the parallel wires Δy. Although the support prongs may be positioned at a nominal separation, the operation of hand soldering the wires to the prongs typically results in a relatively large variability in the actual sensor placement. An effective method to determine the sensor separation is to traverse a calibrated probe through a laminar shear layer with a large mean velocity gradient. If the traversal is executed with sufficiently small spatial increments, the two parallel wires will provide identical mean velocity profiles, with a lateral shift that is equal to the probe spacing. An example is shown in Fig. 5.214. A sample probe was traversed in increments of 0.32 mm across the shear layer on one side of a calibration jet. The two sensors clearly measure an identical velocity profile that is shifted spatially, which provides an estimate of the true Δy.

As an example of the net uncertainty in the vorticity measurement, consider a probe with $\Delta y = 1.0$ mm, and a calibration range of 1–15 m/s. A typical uncertainty in the mean velocity in this range is of order 20 mm/s, with a *noise* level of about 10 mm/s. The error in the mean vorticity will then be of order $20\,\text{s}^{-1}$, with a noise level of about $10\,\text{s}^{-1}$. This clearly makes the technique unreliable for the recovery of the mean vorticity in many applications. However, many turbulent flows have fluctuations in vorticity that are in the range of several hundred inverse seconds in which case these errors amount to only a few percent of the total vorticity variance.

fects of electronic noise can be minimized by using the highest-quality anemometer circuitry available, and by using an appropriately defined analog filter to reduce the contamination of high-frequency noise that is not representative of the fluid velocity fluctuations. If an analog filter is not used, then the acquisition frequency should be set sufficiently high that the integration procedure effectively filters out any high-frequency

A final consideration for the uncertainty is the assumption of Taylor's hypothesis in the algorithm. Although viscous effects are often negligible in turbulent shear flows, strong pressure gradients will affect the sensors's response. Specifically, the algorithm described will result in a nonzero vorticity if the probe is subjected to a potential flow in which local streamline curvature is strong compared to the probe size. Although this effect is difficult to quantify, it is typically not considered to be as important as the calibration and electronic noise errors.

Sample Results

A recent investigation of the vorticity field of a single stream shear layer serves as an example of the use of the vorticity probe [5.811]. The flow field is also described in detail in *Morris* and *Foss* [5.812]. Essentially, the single stream shear layer, or *half-jet*, is a planar flow with a freestream on one side ($v_o = 7.1$ m/s), and a perpen-

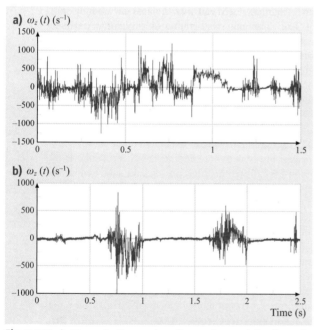

Fig. 5.215a,b Example time series of spanwise vorticity from the center of a shear layer (**a**) and the edge of a shear layer (**b**)

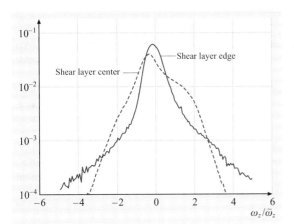

Fig. 5.216 Two examples of the vorticity PDF taken from a single stream shear layer

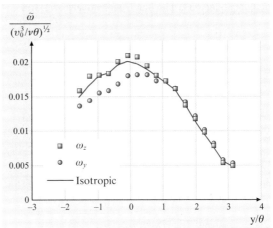

Fig. 5.217 An example profile of the rms value of two components of vorticity in a single stream shear layer. Note that the ordinate is normalized by the local momentum thickness (θ), and the vorticity is normalized by the freestream velocity (v_o), the viscosity (v), and θ. The isotropic values were evaluated from (5.353)

dicular entrainment on the low-speed side with a speed of approximately $0.035v_o$. The coordinate system of the shear layer is specified such that x represents the streamwise direction, y represents the direction perpendicular to the plane of the shear layer, and z is the direction of spanwise homogeneity.

Two example time records of the spanwise vorticity field are given in Fig. 5.215. These data were acquired at a streamwise distance of 684 times the momentum thickness of the boundary layer at separation (9.8 mm). The first time record was acquired at a position in the center of the shear layer where the flow is fully turbulent. The rms value of the vorticity time series $\tilde{\omega}_z$ was $184\,\text{s}^{-1}$, compared to the mean shear $\partial \bar{u}/\partial y = 9\,\text{s}^{-1}$.

The second of the time records was acquired at a location near the high-speed edge of the shear layer where the flow is convectively intermittent. Specifically, relatively long periods of low vorticity levels representative of the freestream are interrupted by bursts of high levels of vorticity that contain significant high-frequency content. The rms value of the vorticity time series at this location was $\tilde{\omega}_z = 50\,\text{s}^{-1}$, compared to a nominal mean shear of zero.

From the vorticity time series, a number of stochastic measures can be computed, including spectral distributions and probability distribution functions (PDF). As an example, the complete time records from the spatial locations of Fig. 5.215 were used to compute their respective distributions (Fig. 5.216). These data show a number of interesting characteristics, most notably an exponential decrease in the PDF observed as a linear region in the semilogarithmic plot for $\omega_z/\tilde{\omega}_z > 2$. Similar features have been observed in computational solutions to isotropic turbulence [5.813].

Lastly, the vorticity probe was traversed across the shear layer in order to quantify the spatial distributions of the vorticity moments. In Fig. 5.217, the rms value of vorticity, both spanwise ($\tilde{\omega}_z$) and transverse ($\tilde{\omega}_y$), are shown with the predicted values based on single-velocity-component measurements using isotropic relations. Specifically, the time series record from one of the parallel sensors was used with Taylor's frozen-flow hypothesis to estimate the spatial derivative $\partial u/\partial x$, from which the rms vorticity was calculated [5.814] as

$$\tilde{\omega}_z^2 = 5v \overline{\left(\frac{\partial u}{\partial x}\right)^2}. \tag{5.353}$$

The agreement between the expression and the directly measured rms vorticity values shown in Fig. 5.217 is generally quite good. Although (5.353) is exact only in homogeneous and isotropic turbulence, agreement between the rms value of the vorticity time series and (5.353) can serve to verify that the probe and computer algorithm are working correctly as long as the subject flow field is not too far from equilibrium.

5.6 Thermal Transient Anemometer (TTA)

The TTA has been developed to obtain the time- and space-averaged values of the temperature and the density–velocity product of the flow through a designated area which, in this presentation, will be termed a *cell*. The complete area of interest, or segment of the total area of interest, may be used to define a cell. The former would describe, for example, the separate passages of an automotive climate control system; the latter would describe, as a second example, the 4×4 cells at the aft location of an automotive radiator. See Foss et al. [5.815–817] for the initial publication of this technique.

5.6.1 Operational Description

Figure 5.218 presents the multi-X pattern of the cells in a typical TTA frame. In this image, the small-diameter (≈ 0.1–0.2 mm) tungsten wire makes 15 spans across the cell width. Figure 5.218b also shows the tether locations where the tungsten wire is secured in a teflon retaining plug. In each case, the cells are intended to provide a complete representation of the flow rate and temperature for the area of interest. The 3×4 frame of Fig. 5.218 was prepared to cover the face of an automotive radiator.

Once fabricated, the resistance of the cell is measured at ambient temperature (R_a) and this value serves as the reference condition for the temperature measurement. Specifically, given (5.355) from Sect. 5.2 and

$$R_w = R_a[1 + \chi(T_w - T_a)], \qquad (5.354)$$

the resistance of the cell (R_w) permits the temperature to be determined. Note that χ is the temperature coefficient of resistance. This is the first measurement made in the operation of the TTA.

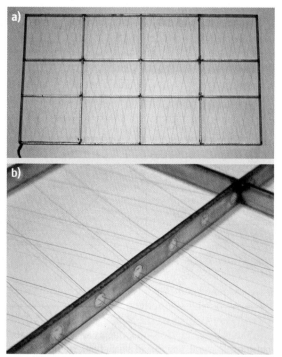

Fig. 5.218a,b A 3×4 TTA frame. (**a**) The complete frame. (**b**) A detailed view to show the tether plugs

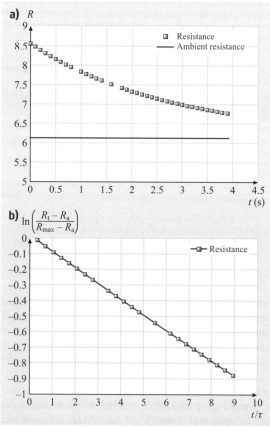

Fig. 5.219a,b $R_w(t)$ from a velocity calibration data set shown using: (**a**) algebraic coordinates, and (**b**) logarithmic–linear coordinates

The circuitry of the TTA then elevates the wire's temperature by delivering current until the wire's resistance achieves the preset value: R_{hot}. Heat transfer, as described by (5.358) in Sect. 5.2.1, results in the decrease of $R_w(t)$ from R_{hot} to R_a. Specifically, adopting a simplified lumped-parameter analysis, the temperature of the wire during the cool-down period is described as

$$mc\frac{dT_w}{dt} = -\left[hA_p(T_w - T_a) + 2k_w A_c \left|\frac{dT}{dx}\right|_{0,L}\right], \tag{5.355}$$

where m is the mass of the wire, c is its specific heat, h is the convective cooling coefficient, $A_p = \pi L d$ is the perimeter area, k_w is the sensor's thermal conductivity, $A_c = \pi d^2/4$ is the cross-sectional area of the wire, and $0, L$ show the locations where the wire is tethered to the teflon plugs.

A modeling strategy, which has to be complemented with experimental data, is to characterize the conduction terms by an arbitrary length, $L(T_a)$, and a constant multiplier, λ, such that

$$2k_w A_c \left|\frac{dT}{dx}\right|_{0,L} = \lambda k_w A_c \frac{(T - T_a)}{L}. \tag{5.356}$$

The designation $L(T_a)$ is made to recognize that the *effective conduction length* may be a function of the ambient temperature. Similarly, k_w is a known function of temperature and this will be incorporated into the conduction term of response question (5.360).

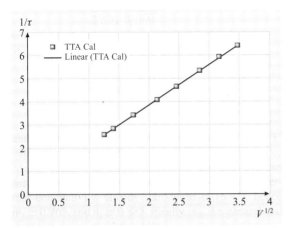

Fig. 5.220 Representative velocity calibration data for the cell of Fig. 5.219

Given the definition of the Nusselt number (Nu), the convection term can be written as

$$hA_p(T_w - T_a) = \left[\frac{Nu k_f}{d}\right](T_w - T_a). \tag{5.357}$$

The nominal temperatures and the calibration and use of the TTA in air means that the influence of the Prandtl number on Nu can be neglected. Hence, the standard relationship for a cylinder heat-transfer problem

$$Nu = C\left(\frac{Vd}{\nu}\right)^{0.5}, \tag{5.358}$$

can be used in (5.357) with the result

$$\text{convection term} = \frac{Ck_f}{d^{0.5}\nu_f^{0.5}}V^{0.5}. \tag{5.359}$$

Hence, (5.355) can be written as

$$\frac{d(T_w - T_a)}{(T_w - T_a)} = -\left\{A\left[\frac{k_w(T_a)}{k_w(T_{a-cal})}\right] + \frac{Bk_f}{\nu_f^{0.5}}V^{0.5}\right\} dt \tag{5.360}$$

or

$$\frac{T_w - T_a}{T_{hot} - T_a} = \exp\left(-\frac{t}{\tau}\right), \tag{5.361}$$

where

$$\frac{1}{\tau} = \left\{A\left[\frac{k_w(T_a)}{k_w(T_{a-cal})}\right] + \frac{Bk_f}{\nu_f^{0.5}}V^{0.5}\right\}. \tag{5.362}$$

5.6.2 Representative Results

The $R_w(t)$ values from a velocity calibration are shown in Fig. 5.219. The tungsten sensor wire diameter was 0.175 mm for these data. The standard deviation between the measured values and the exponential fit was 0.69 ms. Hence, these data provided $\tau = (293 \pm 0.69)$ ms.

Figure 5.220 shows the balance of the calibration results for the cell of Fig. 5.219. These data are displayed as suggested by (5.362). The standard deviation of the fitted equation to the separate (τ, V) data pairs is 0.0118 m/s. This *small* value attests to the veracity of (5.362) to represent the effective cooling of the sensor wire at the calibration temperature.

5.7 Sonic Anemometry/Thermometry

5.7.1 Definition

Similar to hot-wire anemometry and LDA, ultrasonic anemometry (or, as it is more commonly in the literature, sonic anemometry) is a measurement technique that is used to obtain multiple components of instantaneous velocity at a point in space.

In addition to the three velocity components, most modern sonic anemometers also provide virtual temperature (T_v) measurements. The virtual temperature is the temperature at which dry air has the same density as moist air at the same pressure [5.818]. Using this definition, a convenient form of the equation of state may be written as $P = \rho R_d T_v$, where P is total pressure of air including moisture (Pa), ρ is the total density of air (kg/m^3) and R_d is the gas constant for dry air (287 J/(kg K)) [5.819]. The virtual temperature is given by $T_v = T(1 + 0.61q)$, where T is the absolute temperature. Here, $q = M_w/(M_w + M_d)$ is the specific humidity, defined as the ratio of the mass of moist air (M_w) to the total mass of air ($M_w + M_d$).

Sonic anemometer/thermometers (SATs) are very robust: they have no moving parts, require infrequent calibration, and can operate in harsh atmospheric environmental conditions. SATs can be left out in the field for extended periods of time with little maintenance compared to hot-wire, cup, and propeller anemometers. These qualities combined with the fact that the length and time scales that are resolved with current sonic anemometer technology are quite large (compared to fast-response engineering techniques) have made this technique popular with researchers probing the atmospheric boundary layer [5.820]. While sonic anemometers are quite robust, the response of sonic anemometers could be affected by intense rain, severe contamination (dirt, dust, etc.), ice formation, and structural vibrations. All of these effects are areas of current research efforts [5.821]. Table 5.8 provides a comparison of the main characteristics of the following three traditional atmospheric boundary-layer measuring techniques: sonic, cup, and propeller anemometers.

In recent years, many researchers have used SATs in conjunction with other instrumentation to calculate covariances using so-called eddy correlation or eddy covariance techniques. The SAT allows correlations of velocity components and the virtual temperature to be readily computed. For example, useful quantities such as Reynolds stresses ($\overline{u_i' u_j'}$) and turbulent heat fluxes ($\overline{u_i' T_v'}$) are easily calculated from the time series output from the SAT. Additional physical quantities such as the vertical turbulent concentration fluxes of CO_2 and water vapor are regularly measured using this technique [5.822] For example, gas sensors can be collocated with a SAT to allow quantities such as $\overline{u_i' \chi'}$ to be calculated (where χ represents the concentration of the tracer gas). As a result of the ability to calculate such useful quantities, SAT use has spread to a variety of applications including air quality [5.823, 824], wind energy [5.825, 826], urban boundary layer [5.827, 828], forest canopy and agriculture [5.829], hydrological [5.830], and trace-gas budget studies (often related to global climate change [5.831]). Additionally, SATs have been used as an option for exploration of extraterrestrial atmospheres [5.832, 833], and low-pressure gas anemometry. There is also a flow meter technique based on sonic anemometry [5.834].

The type of sonic anemometry described in this section has been used by engineers and micrometeorologists since the late 1950s [5.835]. Historically two types of sonic anemometers have been used, pulse and continuous wave. Pulse anemometers measure time delays between ultrasonic pulses (defined as having frequencies > 20 kHz) while continuous-wave anemometers measure phase shifts [5.836–838]. Using this technol-

Table 5.8 Summary of the main operational characteristics of sonic, cup, and propeller anemometers

System	Calibration	Maintenance	Sampling frequency	Three components of wind speed?	Temperature measurement
Sonic anemometer/ thermometer	Initial	Cleaning	Up to 100 Hz	Yes (for three-path models)	Yes (sonic temperature)
	Sonics present some weaknesses that are considered in the text				
Cup anemometer	Periodic	Intense	Distance constant > 1 m	No	No
	Mobile parts, slow response, influenced by turbulence, influenced by vertical velocity components.				
Propeller anemometer	Periodic	Moderate	Distance constant > 1 m	Estimate for three-probe configurations	No
	Mobile parts, slow response, influenced by turbulence and misalignment				

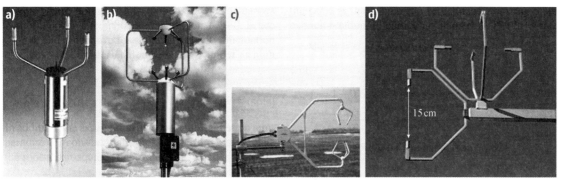

Fig. 5.221a–d Various sonic anemometer configurations: (**a**) 2-D sonic anemometer with vertical mounting base, (**b**) 3-D SAT with vertical mounting base, (**c**) 3-D SAT with horizontal mounting nonorthogonal pulse paths, (**d**) 3-D SAT with horizontal mounting and orthogonal pulse paths. Photographs are courtesy of (**a**) Vaisala, Inc. (**b**) RM Young Company, (**c**) Campbell Scientific, Inc. and (**d**) Applied Technologies, Inc.

ogy, a SAT measures a *line-averaged* velocity of the flow field along the acoustic path. The line averaging represents the main limitation to the instrument's frequency response [5.818]. Today most manufacturers use pulse-based anemometers. Figure 5.221 shows four different sonic anemometers currently in manufacture, with varying path lengths and probe configurations.

5.7.2 Measurement Principles

Most modern sonic anemometers determine velocity components along a known path length by measuring the difference in transit times of acoustic pulses sent simultaneously in opposite directions between two sensors. Multiple paths oriented at various angles allow for the measurement of multiple components of the velocity vector. A number of excellent sources exist that describe the fundamental principles of sonic anemometers [5.818, 836, 839]. For most devices, the ultrasonic pulse is generated using piezoelectric transducers that act as both the transmitter and receiver. In some particular cases, when a sonic anemometer is used in other atmospheres (i.e., gases different than air or different conditions, leading to different sound speed) the transducer technology may differ. Piezoelectric transducers present an acoustic impedance on the order of 1.2×10^7 kg/m²s, whereas the Earth's air presents an acoustic impedance of 400 kg/m²s. According to [5.832], the ratio of these values results in an acceptable attenuation of an ultrasound signal when it is transmitted from a piezoelectric sensor in the Earth's atmosphere. However, for instance in the Martian case, the impedance ratio would be unacceptable; the extremely low acoustic impedance of the Martian atmosphere (3 kg/m²s) would lead to extreme attenuation of the ultrasound signal transmitted from a piezoelectric sensor. In this case, capacitive transducers with acoustic impedances of 1000 kg/m²s can be used as an alternative to piezoelectric transducers.

The description given here is for an ultrasonic pulse-type anemometer and is adapted from *Kaimal* [5.836], *Kaimal* and *Finnigan* [5.818] and *Cuerva* and *Sanz-Andres* [5.839]. The basic principles of operation are best understood by considering a steady, uniform, 2-D flow field with an along-path velocity component of u and a velocity component normal to the path of w, as shown in Fig. 5.222. The time of flight for a pulse sent from transducer 1 to traverse to transducer 2 will be denoted by t_{12}. Similarly a pulse sent from transducer 2 to 1 will have a time of flight t_{21}. The following equation may be written using the ray vectors shown in Fig. 5.222 for a pulse being transmitted from sensor 1 to 2:

$$ct_{12}\cos\beta + ut_{12} = L,\qquad(5.363)$$

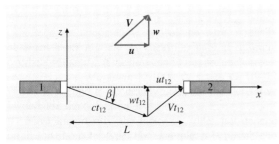

Fig. 5.222 2-D schematic of the ray vectors associated with an ultrasonic pulse being sent from the transmitter (1) to the receiver (2) in a steady, uniform velocity field

where, $\sin\beta = w/c$, c is the speed of sound, and L is the distance between transducers 1 and 2. A similar equation may be written for a pulse traveling from sensor 2 to 1. The associated times of flight are then

$$t_{12} = \frac{L}{c\cos\beta + u} \quad (5.364)$$

and

$$t_{21} = \frac{L}{c\cos\beta - u} \,. \quad (5.365)$$

The transmitted time difference is then

$$\Delta t = t_{21} - t_{12} = \frac{2uL}{c^2\cos^2\beta - u^2} \,. \quad (5.366)$$

Substituting $\sin^2\beta + \cos^2\beta = 1$ and $u^2 + w^2 = V^2$ gives

$$\Delta t = \frac{2uL}{c^2 - V^2} \,. \quad (5.367)$$

For small-Mach-number flow, $V \ll c$, the along-path wind component simplifies to

$$u = \frac{c^2}{2L}\Delta t \,. \quad (5.368)$$

Equation (5.368) requires the speed of sound to be estimated. Following [5.836], the effect of temperature and water-vapor content can be accounted for by approximating the speed of sound as $c^2 = C_T T(1 + 0.32e/p)$, where T is the absolute temperature (K), e is the vapor pressure of water in air, $C_T = 403\,\mathrm{m^2/s^2\,K}$ and p is the atmospheric pressure. The virtual temperature is approximated as $T_v = T(1 + 0.32e/p)$ so that $c^2 = C_T T_v$. This can be related to the familiar definition of the speed of sound given by $c^2 = \gamma p/\rho$, where $\gamma = C_p/C_v$ is the ratio of specific heats at constant pressure and volume, and ρ is the density of air (kg/m³). Substituting the ideal gas relationship with the virtual temperature, $p = \rho R_d T_v$, yields $c^2 = \gamma R_d T_v$.

Most SATs however, use a single path to send pulses simultaneously from sensor 1 to 2 and from sensor 2 to 1. These instruments use electronics to subtract time inverses directly, namely

$$\frac{1}{t_{12}} - \frac{1}{t_{21}} = \frac{2u}{L} \,. \quad (5.369)$$

This yields the velocity along the path without the need to calculate the speed of sound separately. The resulting equation for the velocity is simply

$$u = \frac{L}{2}\left(\frac{1}{t_{12}} - \frac{1}{t_{21}}\right) \,. \quad (5.370)$$

It should be noted that, in a real configuration, the pulses are not sent simultaneously because of physical limitations (it is technically impossible for a sensor to act as both an emitter and receiver simultaneously). In practice there is a time delay between shots (z_B) that is on the order of 0.001 s, depending on the manufacturer. The effect of this time delay between pulses must be taken into account when a sonic anemometer measures the turbulence velocity and temperature spectra [5.839].

As shown in *Kaimal* and *Finnigan* [5.818] the time-inverses method also allows the speed of sound to be calculated directly, hence providing an estimate of the virtual temperature. Squaring the sum of the inverse flight times yields

$$\left(\frac{1}{t_{12}} + \frac{1}{t_{21}}\right)^2 = \frac{4c^2}{L^2}\cos^2\beta = \frac{4c^2}{L^2}\left(1 - \frac{w^2}{c^2}\right) \,. \quad (5.371)$$

Equation (5.371) can then be solved for the speed of sound,

$$c^2 = \frac{L^2}{4}\left(\frac{1}{t_{12}} + \frac{1}{t_{21}}\right)^2 + w^2 \,. \quad (5.372)$$

Using the approximation derived above, $c^2 = C_T T_v$ yields the following expression for the virtual temperature

$$T_v = \frac{L^2}{4C_T}\left(\frac{1}{t_{12}} + \frac{1}{t_{21}}\right)^2 + \frac{w^2}{403} \,. \quad (5.373)$$

For a multicomponent anemometer, w in (5.373) can be calculated using the velocity components from the other axes of the anemometer. The exact formulation

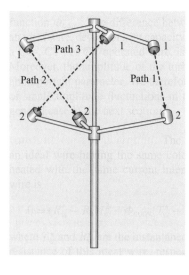

Fig. 5.223 Schematic of a commercial 3-D sonic anemometer indicating the three pathlines

Table 5.9 Pulse sequence and timing for a METEK USA-1 sonic anemometer. The activated path refers to the paths shown in Fig. 5.223, while the path sense refers to the transmitter-to-receiver communication shown in Fig. 5.222

Activated path	Path sense	Time (ms)
1	1-2	z_B 1.14
1	2-1	z_B 1.14
2	1-2	z_B 1.14
2	2-1	z_B 1.14
3	1-2	z_B 1.14
3	2-1	z_B 1.14
Electronic and computation		z_E 0.88

Table 5.10 Typical pulse firing sequence and block averaging for a gill wind research sonic anemometer. The time delay between pulses is $z_B = 0.001$ s; the data delivery speed is 168 samples/s

Path	Pulse sense	Start time	End time
Path 1	1–2	0	z_B
	2–1	z_B	$2z_B$
Path 2	1–2	$2z_B$	$3z_B$
	2–1	$3z_B$	$4z_B$
Path 3	1–2	$4z_B$	$5z_B$
	2–1	$5z_B$	$6z_B$

Mode	Averaged measurements	Final number of averaged measurements delivered (true number of samples per second)
1	8	21 (168/8)
2	3	56 (168/3)

for w depends on the geometry of the SAT axes. This temperature is actually the sonic temperature (the temperature reported by the SAT) but for most applications represents the virtual temperature quite well [5.840]. As noted by [5.841], most SATs actually determine sonic temperature as an average over multiple paths. In that case, (5.373) represents the sonic temperature obtained for one path.

For a three-path sonic anemometer (Fig. 5.223), a full sequence leading to a complete measurement of the wind-speed vector consists of at least six shots of ultrasound pulses, two pulses per acoustic path. The time required to compute the wind-speed vector from individual path measurements is limited by the time delay z_B between consecutive pulses and the time after the sixth pulse z_E required to reset the electronics. Table 5.9 shows a typical pulse sequence, which lasts $6z_B + z_E = 7.72$ ms [5.842].

In general, SATs do not provide the direct measurements from the individual pulse sequence. Instead they carry out averaging processes on them. These processes, normally called *block averaging*, are manufacturer dependent. A typical scheme for this process is included in Table 5.10.

5.7.3 Device Characteristics, Accuracy, and Limitations

Most SATs attributes vary across manufactures. Typical sensor measurement paths are approximately 0.10–0.20 m. The velocity range of most sensors is approximately ± 30 m/s with velocity accuracies in the range of ± 0.02–0.05 m/s. Temperature accuracy tends to vary much more significantly over a range of ± 0.1–2.0 °C. The sampling frequency range is typically up to about 100 Hz. Some manufactures also allow for oversampling of wind components and output of an average filtered signal at a lower frequency, as discussed above, via a block averaging process (Table 5.10).

The frequency response of a SAT is limited by the attenuation in the spatial response imposed by line averaging along a path. *Kaimal* [5.818] and *Kaimal and Finnigan* [5.818] have suggested as a *rule of thumb* that the low-pass filtering attenuation distortions are confined to spectral wavelengths $\lambda < 2\pi L$. If the paths of sensors are separated (e.g., nonintersecting paths, as in Fig. 5.221d) then, the response of the turbulent fluxes (e.g., $\overline{u_i' T_v'}$ or $\overline{u_i' u_j'}$) calculated between the sensors is compromised if $\lambda < 2\pi s$, where s is the separation distance between the midpoints of the pulse paths. For a typical SAT with intersecting sensor paths and a sensor separation of $L = 0.10$ m (e.g., Fig. 5.221c), spectral attenuation will occur for wavelengths less than about 0.63 m. Using Taylor's frozen turbulence hypothesis to express this limit in terms of frequency for $L = 0.1$ m and a typical wind speed of $u = 10$ m/s leads to a limiting frequency of $f = u/(2\pi L) \approx 15.9$ Hz.

These *rules of thumb* were developed assuming a homogeneous turbulent velocity field [5.843]. *Cuerva and Sanz-Andres* [5.839] have investigated the effect of this assumption on measurements made on more-realistic flow fields. In particular, they investigated line-averaging effects for a nonsteady, nonuniform velocity field through the sensor. They found that the flow Mach number and the time delay between pulses increases the wavelength at which spectral attenuation begins. In some cases, the combination of line averaging and geometry effects leads to overestimation of the spectra instead of spectral attenuation.

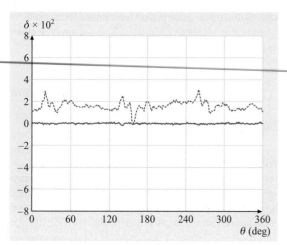

Fig. 5.224 Difference between the measured wind-speed magnitude $|u|^M$ and the actual magnitude $|u|$ as a percentage of the actual one (represented as δ) versus the incidence angle θ. The *dotted line* represents the uncorrected response (0–3%) of the anemometer and the *solid line* the corrected response after calibration in wind tunnel [5.848]

non-negligible effects of the determination of turbulence properties [5.846, 847].

The effect of aerodynamic disturbances associated with the acoustic path supporting structures on the measurement of averaged characteristics of the wind-speed vector must be considered. This effect is highly manufacturer dependent [5.849]. Figure 5.224 shows the corrections in the measurement of the average value of the total magnitude of the wind speed as a function of the angle of incidence of the wind-speed vector for a commercial sonic anemometer unit. These corrections are easily determined by a calibration process and can be implemented internally (by the manufacturer) or during postprocessing by the user.

In general, it can be stated that two phenomena:

1. aerodynamic interference in the acoustic paths due to supporting structures
2. the line-averaging process

make the sonic anemometer response non-isotropic (i.e., directional), both when measuring averaged and turbulent magnitudes. The characteristics of this directional response depend on the sensor design.

As a result of the line averaging, care must be taken when placing the sensors near surfaces. For use in the atmospheric surface layer, the minimum distance between the center of the probe and the ground has been suggested to be $z_{min} = 8\pi L$ [5.844]. *Wamser* et al. [5.845] however, showed that the maximum attenuation occurs in the vertical velocity variance near the ground and that the z_{min} restriction can be relaxed down to about $z_{min} = 4L$, or just less than one meter for most SATs over flat terrain. Additionally, depending on the application, the block averaging process can lead to

Coordinate Transformation Methods

The alignment of a SAT is of great importance since misalignment produces a deviation between the assumed probe coordinate system and the actual one. SATs are often aligned in the field such that the y-axis points northward, the x-axis eastward, and the z-axis normal to the surface. Another common horizontal alignment technique is to orient the anemometer to limit flow distortion as much as possible based on the expected winds. Alignment is typically accomplished

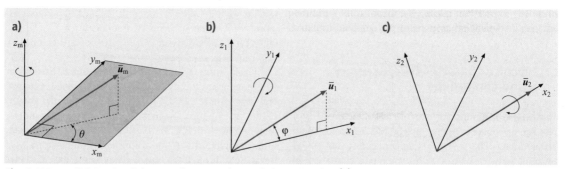

Fig. 5.225a–c Schematic of the coordinate rotation technique showing **(a)** the first rotation θ about the z_m axis (yaw), **(b)** the second rotation φ about the y_1 axis (pitch) and **(c)** the final rotation ψ about the x_2 axis (roll). $\bar{\boldsymbol{u}}_m = (\bar{u}_m, \bar{v}_m, \bar{w}_m)$, $\bar{\boldsymbol{u}}_1 = (\bar{u}_1, 0, \bar{w}_1)$, $\bar{\boldsymbol{u}}_2 = (\bar{u}_2, 0, 0)$ are the average velocity vectors referred to in the sonic anemometer coordinate system (x_m, y_m, z_m), after the first rotation with a reference system (x_1, y_1, z_1), and the second rotation with a reference system (x_2, y_2, z_2), respectively

using tools such as inclinometers, compasses, global positioning units, and levels. Unfortunately, some level of misalignment is unavoidable and causes the various components of the velocity field to be incorrectly redistributed. As noted by [5.850], this can be particularly problematic when calculating velocity covariances over sloping terrain where the SAT is usually either aligned with the vertical component normal to the slope or leveled. *Wilczak* et al. [5.850] review several correction/coordinate transformation algorithms that are useful for sonic anemometers. Following [5.818, 850] and [5.851] the most widely used coordinate transformation technique is presented below, followed by *Wilczak* et al.'s [5.850] alternative method. These methods transform velocities in the measured coordinate system into a streamline coordinate system. This operation is very useful for comparing wind-tunnel data or theory to atmospheric surface layer data. It is also particularly useful for investigating turbulent surface exchanges such as the turbulent vertical flux of momentum or a scalar (e.g., $\overline{u'_i \chi'}$ or $\overline{u'_i u'_j}$). The coordinate transformations are most appropriate over horizontally homogeneous, uniformly sloping terrain (e.g., in the inertial layer above forests, buildings, fields, etc.).

A transformation of the measured coordinate system into a streamline coordinate system is a transformation between two orthogonal coordinates systems and is shown schematically in Fig. 5.225. This is a rotation operation that can be represented in matrix form as

$$\begin{pmatrix} u_f \\ v_f \\ w_f \end{pmatrix} = A \begin{pmatrix} u_m \\ v_m \\ w_m \end{pmatrix}, \quad A = T \cdot S \cdot R, \quad (5.374)$$

where u_m, v_m, and w_m are the instantaneous velocity components in the measured sonic coordinate system, and u_f, v_f, and w_f are the final instantaneous velocity components after the transformation. As noted by *Sozzi* [5.851], a method to obtain an approximation of A is to combine one or more rotations in order to align the coordinate system axes in which the measurements where made (x_m, y_m, z_m with unit vectors i_m, j_m, and k_m) with the tangent, binormal, and principal normal of the local streamline (x_f, y_f, z_f with unit vectors i_f, j_f, and k_f) as shown in Fig. 5.226 using

$$T = \begin{pmatrix} 1 & 0 & 0 \\ 0 & \cos\psi & \sin\psi \\ 0 & -\sin\psi & \cos\psi \end{pmatrix},$$

$$S = \begin{pmatrix} \cos\varphi & 0 & \sin\varphi \\ 0 & 1 & 0 \\ -\sin\varphi & 0 & \cos\varphi \end{pmatrix},$$

$$R = \begin{pmatrix} \cos\theta & \sin\theta & 0 \\ -\sin\theta & \cos\theta & 0 \\ 0 & 0 & 1 \end{pmatrix}.$$

Here, ψ, φ, and θ are the roll, pitch, and yaw rotation angles, respectively, taken as positive counterclockwise about each axis. The principal normal is defined by the direction in which the streamline is curving most rapidly [5.818]. Often in practice the estimation procedure is further simplified with a rougher estimate that only consists of the first two rotations, or a so-called double rotation, first about the z_m axis and then about the y_1 axis.

The procedure is typically broken up into three rotation operations. If x_m, y_m, z_m is the measuring coordinate system of the sonic anemometer, the first operation is a rotation about the vertical (z_m) axis that aligns the new x-coordinate with the mean streamwise wind direction, i.e., (x_m, y_m, z_m) $\rightarrow$ (x_1, y_1, z_1) so that the mean spanwise velocity is zero:

$$\begin{pmatrix} u_1 \\ v_1 \\ w_1 \end{pmatrix} = R \begin{pmatrix} u_m \\ v_m \\ w_m \end{pmatrix}$$

or,

$$u_1 = u_m \cos\theta + v_m \sin\theta ,$$
$$v_1 = -u_m \sin\theta + v_m \cos\theta ,$$
$$w_1 = w_m , \quad (5.375)$$

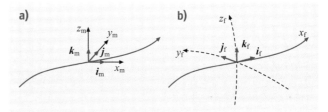

Fig. 5.226a,b Schematic of (**a**) a streamline passing through the sonic anemometer coordinate system and (**b**) the streamline coordinate system as discussed in the text (after [5.852]). i_f is tangent to the streamline, j_f is binormal to the streamline, and k_f is normal to the streamline. As noted by *Finnigan* [5.852], the z_f coordinate lines are tangent to the field of k_f vectors, the y_f coordinate lines are tangent to the field of j_f vectors, and the streamlines form the x_f coordinates

where

$$\theta = \tan^{-1}\left(\frac{\overline{v}_m}{\overline{u}_m}\right), \quad \sin\theta = \frac{\overline{v}_m}{\sqrt{\overline{u}_m^2 + \overline{v}_m^2}} \quad \text{and}$$

$$\cos\theta = \frac{\overline{u}_m}{\sqrt{\overline{u}_m^2 + \overline{v}_m^2}}. \tag{5.376}$$

In these equations, $\overline{v}_m$ and $\overline{u}_m$ are mean velocities in the measured sonic anemometer coordinate system. Therefore, the transformation is based on the choice of the mean velocity. This should be an ensemble average, however, practically it is usually estimated with time averages. After the first rotation, $\overline{v}_1 = 0$ by definition. Once θ is determined, either the time series can be rotated using (5.375) or the following equations for mean flow and turbulent stresses can be used

$$\overline{u}_1 = \overline{u}_m \cos\theta + \overline{v}_m \sin\theta,$$
$$\overline{v}_1 = -\overline{u}_m \sin\theta + \overline{v}_m \cos\theta,$$
$$\overline{w}_1 = \overline{w}_m, \tag{5.377}$$
$$\overline{u'^2_1} = \overline{u'^2_m}\cos^2\theta + \overline{v'^2_m}\sin^2\theta + 2\overline{u'_m v'_m}\sin\theta\cos\theta,$$
$$\overline{v'^2_1} = \overline{u'^2_m}\sin^2\theta + \overline{v'^2_m}\cos^2\theta - 2\overline{u'_m v'_m}\sin\theta\cos\theta,$$
$$\overline{w'^2_1} = \overline{w'^2_m}, \tag{5.378}$$
$$\overline{u'_1 v'_1} = \overline{u'_m v'_m}(\cos^2\theta - \sin^2\theta)$$
$$+ \sin\theta\cos\theta\left(\overline{v'^2_m} - \overline{u'^2_m}\right),$$
$$\overline{u'_1 w'_1} = \overline{u'_m w'_m}\cos\theta + \overline{v'_m w'_m}\sin\theta,$$
$$\overline{v'_1 w'_1} = -\overline{u'_m w'_m}\sin\theta + \overline{v'_m w'_m}\cos\theta. \tag{5.379}$$

As shown in Fig. 5.225b, the second rotation is about the y_1 axis and forces $\overline{w}_2 = 0$. The velocities are given by

$$u_2 = u_1 \cos\varphi + w_1 \sin\varphi,$$
$$v_2 = v_1$$
$$w_2 = -u_1 \sin\varphi + w_1 \cos\varphi, \tag{5.380}$$

where

$$\varphi = \tan^{-1}\left(\frac{\overline{w}_1}{\overline{u}_1}\right),$$

$$\sin\varphi = \frac{\overline{w}_1}{\sqrt{\overline{u}_1^2 + \overline{w}_1^2}} \quad \text{and}$$

$$\cos\varphi = \frac{\overline{u}_1}{\sqrt{\overline{u}_1^2 + \overline{w}_1^2}}. \tag{5.381}$$

Equations similar to (5.377–5.379) can then be easily obtained. This completes the double rotation and aligns x_2 with the mean wind direction. The last rotation or triple rotation rotates the coordinate system such that the final z-axis is normal to the mean wind trajectory [5.851] (Fig. 5.224b). According to [5.850] there are an infinite number of rotations that satisfy $\overline{v}_f = \overline{w}_f = 0$. It has been suggested that the approximation $\overline{v'_f w'_f} = 0$ can be made over flat land [5.818, 853]. If this is done, the following equations result:

$$u_f = u_2,$$
$$v_f = v_2 \cos\psi + w_2 \sin\psi,$$
$$w_f = -v_2 \sin\psi + w_2 \cos\psi, \tag{5.382}$$

where

$$\psi = \frac{1}{2}\tan^{-1}\left(\frac{2\overline{v'_2 w'_2}}{\overline{v'^2_2} - \overline{w'^2_2}}\right). \tag{5.383}$$

Since the third rotation is based on a strong (and likely incorrect for complex flow) constraint on the turbulent stress, it should be used with extreme caution. *Finnigan* [5.852] recommends that the third rotation should not be applied but that the general procedure should be used.

Planar Fit Method
As an alternative, *Wilczak* et al. [5.850] suggest a *planar-fit* technique to determine the pitch and roll angles of the anemometer, which reduces the error associated with the second and third rotations described above. The method uses average velocity vectors from an ensemble of averaging periods to define a new best-fit x–y plane using multiple linear regression. This plane is defined by a pitch and roll angle with respect to the sonic anemometer coordinate system similar to that described above. This regression forces the average vertical velocity to be zero (i. e., $\overline{w}_{pf} = 0$). First, the planar-fit velocity components are determined through a partial rotation, namely

$$\begin{pmatrix} u_{pf} \\ v_{pf} \\ w_{pf} \end{pmatrix} = \mathbf{T}(\psi_{pf}) \cdot \mathbf{S}(\varphi_{pf}) \cdot \begin{pmatrix} u_m \\ v_m \\ w_m \end{pmatrix} = \mathbf{P} \cdot \begin{pmatrix} u_m \\ v_m \\ w_m \end{pmatrix}, \tag{5.384}$$

where $\mathbf{P} = \mathbf{T}(\psi_{pf})\mathbf{S}(\varphi_{pf})$ and the planar-fit pitch (φ_{pf}) and roll (ψ_{pf}) angles are determined from a least-squares fitting of the mean velocity vectors to a single plain. The final rotated velocities ($\overline{u}_f \neq 0$, $\overline{v}_f = 0$ and $\overline{w}_f = 0$) are obtained by rotating the planar-fit velocities through the appropriate yaw angle θ_{pf} using (5.385) as

$$\begin{pmatrix} u_f \\ v_f \\ w_f \end{pmatrix} = \mathbf{R}(\theta_{pf}) \cdot \begin{pmatrix} u_{pf} \\ v_{pf} \\ w_{pf} \end{pmatrix}, \quad (5.385)$$

where

$$\theta_{pf} = \tan^{-1}\left(\frac{\overline{v}_{pf}}{\overline{u}_{pf}}\right). \quad (5.386)$$

Here, $\overline{u}_{pf}$ and $\overline{v}_{pf}$ represent average values taken for each averaging period. To determine the planar-fit pitch (φ_{pf}) and roll (ψ_{pf}) angles [5.850] consider a modified version of (5.384) for the mean velocities that includes a mean offset error vector $\mathbf{a}$ associated with the error in the measured velocities associated with the instrument

$$\begin{pmatrix} \overline{u}_{pf} \\ \overline{v}_{pf} \\ \overline{w}_{pf} \end{pmatrix} = \begin{pmatrix} p_{11} & p_{12} & p_{13} \\ p_{21} & p_{22} & p_{23} \\ p_{31} & p_{32} & p_{33} \end{pmatrix} \begin{pmatrix} \overline{u}_m - a_1 \\ \overline{v}_m - a_2 \\ \overline{w}_m - a_3 \end{pmatrix}. \quad (5.387)$$

Wilczak et al. [5.850] show that the $\overline{u}$ and $\overline{v}$ biases are usually much smaller than the $\overline{w}$ bias. Consequently, their method only contains a mean offset in the vertical velocity. This is done by setting the last equation in (5.387) to zero (i.e., $\overline{w}_{pf} = 0$) and rearranging to obtain

$$\overline{w}_m = a_3 - \frac{p_{31}}{p_{33}}\overline{u}_m - \frac{p_{32}}{p_{33}}\overline{v}_m \quad (5.388)$$

that is,

$$\overline{w}_m = b_0 + b_1 \overline{u}_m + b_2 \overline{v}_m. \quad (5.389)$$

The b coefficients are then solved for by minimizing the function S,

$$S = \sum_{i=1}^{N} \left(\overline{w}_{m,i} - b_0 - b_1 \overline{u}_{m,i} - b_2 \overline{v}_{m,i}\right)^2, \quad (5.390)$$

where $\overline{u}_{m,i}$, $\overline{v}_{m,i}$, and $\overline{w}_{m,i}$ are the components of the mean velocities measured by the sonic anemometer for each averaging period. Taking the partial derivative of (5.390) with respect to b_0, b_1, and b_2 and setting the resulting equations each equal to zero yields the following system of equations:

$$\begin{pmatrix} N & \sum_i \overline{u}_{m,i} & \sum_i \overline{v}_{m,i} \\ \sum_i \overline{u}_{m,i} & \sum_i \overline{u}_{m,i}^2 & \sum_i \overline{u}_{m,i}\overline{v}_{m,i} \\ \sum_i \overline{v}_{m,i} & \sum_i \overline{u}_{m,i}\overline{v}_{m,i} & \sum_i \overline{v}_{m,i}^2 \end{pmatrix} \begin{pmatrix} b_0 \\ b_1 \\ b_2 \end{pmatrix}$$
$$= \begin{pmatrix} \sum_i \overline{w}_{m,i} \\ \sum_i \overline{u}_{m,i}\overline{w}_{m,i} \\ \sum_i \overline{v}_{m,i}\overline{w}_{m,i} \end{pmatrix}.$$

The b_i coefficients can easily be obtained using a standard matrix-inversion technique. The planar-fit pitch and roll angles are then calculated using the following equations that are obtained by combining (5.389) and the orthogonality condition (i.e., $p_{31} + p_{32} + p_{33} = 1$) as

$$p_{31} = \frac{-b_1}{\sqrt{b_1^2 + b_2^2 + 1}} = -\cos\psi_{pf}\sin\varphi_{pf},$$

$$p_{32} = \frac{-b_2}{\sqrt{b_1^2 + b_2^2 + 1}} = -\sin\psi_{pf},$$

$$p_{33} = \frac{1}{\sqrt{b_1^2 + b_2^2 + 1}} = \cos\psi_{pf}\cos\varphi_{pf},$$

$$\sin\psi_{pf} = -p_{32},$$

$$\sin\varphi_{pf} = \frac{b_1}{\sqrt{b_1^2 + 1}},$$

$$\cos\psi_{pf} = \sqrt{\frac{b_1^2 + 1}{b_1^2 + b_2^2 + 1}},$$

$$\cos\varphi_{pf} = \frac{1}{\sqrt{b_1^2 + 1}}.$$

The planar-fit pitch and roll angles can now be substituted into (5.384) to solve for u_{pf}, v_{pf}, and w_{pf}. The last step is to solve for the final rotated velocities using (5.385).

Other Sensor Issues

While, it has been observed that thermal expansion of SATs has a negligible effect on accuracy [5.836], the sensors and supports of SATs create wakes that interfere with the flow field being measured and result in a velocity deficit error. Aerodynamic distortion on the acoustic path is not only due to wake effects but also due to blockage effects, leading to acceleration in certain parts of the acoustic path regions [5.849]. Horizontally mounted SATs (Fig. 5.221c,d) are usually pointed into the mean wind and have a restricted range of permissible wind angles to avoid the wakes of the sensor and mounting equipment. Vertically mounted SATs (Fig. 5.221a,b) are often axisymmetric but have supports that induce wake effects. Also, any mean vertical winds will cause significant flow distortion due to the base. Many manufacturers include corrections that are meant for specific anemometer orientations to account for wake-shadowing effects of the sensor supports. These usually assume that the mean flow is horizontal and are not valid for flow with significant mean verti-

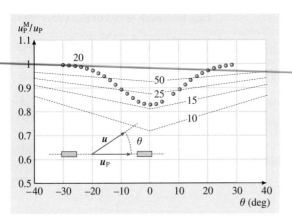

Fig. 5.227 Normalized velocity attenuation u_P^M/u_P (where u_P is the projection of the wind vector $\mathbf{u}$ onto the pulse path) from transducer shadowing in a sonic anemometer (with orthogonal probe axes) shown as a function of wind direction (θ) for various values of L/a (10–50) (after [5.818, 854]). Here, a is the diameter of the transducer, L is the separation distance between transducers

cal velocity components. Shadowing effects have also been investigated in the literature [5.849, 854–859]. Figure 5.227 is an example of the corrections that can be applied to particular SATs. With more sensors being used in environments with a nonhorizontal mean flow (e.g., urban flow and complex terrain), this topic has become an important issue. In particular it has been shown that introducing an angle of attack to the sonic anemometer can cause large errors in mean velocities and turbulent fluxes [5.860]. Additional calibration data over a full range of wind angles is necessary to alleviate this problem. It should be remarked that, in extremely poor designs, certain sonic anemometer geometries may lead to undefined measured wind speed values (i.e., one measurement corresponding to multiple wind values) as described in [5.861].

Temperature measurements from the SAT are sensitive to moisture content [5.841] and cross-stream velocity and should be corrected [5.862]. *Schotanus* et al. [5.863] and *Liu* et al. [5.841] have proposed corrections. Some manufacturers include correction algorithms that can be run online or implemented in postprocessing. *Liu* et al. [5.841] have provided correction information for several specific types of sonic anemometers including the Campbell Scientific CSAT3, the Metek USA-1 and the Gill Instrument Solent R2, R3, R31, and HS.

5.7.4 Data-Acquisition Requirements

Most SATs have data-acquisition electronics built into the instrument on a mounting arm or in an included electronics box. Communication with the electronics typically takes place through a modem-type communication protocol using a PC terminal (e.g., via an RS-232 serial port) or a data logger. Some manufacturers use proprietary digital communication protocols (e.g., SDI (synchronous digital interface) for Vaisala, Inc. and SDM (synchronous devices for measurement) for Campbell Scientific, Inc.) that give the best instrument performance. These digital communication techniques allow for external instrument triggering. In addition, most manufacturers' electronics can be programmed to output analog signals.

5.7.5 Use and Calibration Procedures

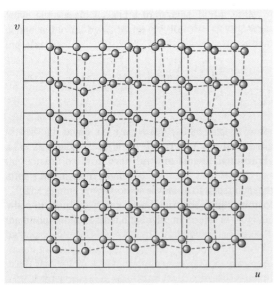

Fig. 5.228 Schematic of the results from a two-dimensional calibration. The graphical relation $f_{\mathrm{cal}}: \mathbb{R}^2 \to \mathbb{R}^2$ is presented. The grey dots represent the wind-tunnel measurements (real values of u and v), whereas the brown dots represent the values measured by the sonic anemometer (measured values of u and v)

One of the biggest advantages of SATs is that they typically only require an initial calibration. Usually, SATs only need to be recalibrated if the sensors have been disturbed (for example, causing the pulse path distance L to be changed). Some models require regular zeroing in a still/anechoic chamber [5.821, 864]. In general, prior to instrument deployment, checking the zero of

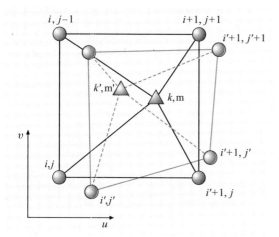

Fig. 5.229 Schematic of the correction process of a sonic measurement. The grey dots represent the real values of the wind speed determined during calibration in wind tunnel, the brown dots represent the values of u and v measured by the sonic anemometer during calibration, the grey triangle represents the values of u and v measured in the field, and the white triangle represents the values of the interpolated (corrected after calibration) u and v

the instrument in a still chamber is a good practice. If the path distance has changed, SATs are usually calibrated using a reference standard (e.g., Pitot-static or hot-wire probe) in a wind tunnel. Calibration method specifications are proposed in the literature [5.821, 864]. However, a need for a uniform systematic calibration method still exists. The calibration of a sonic anemometer involves the determination of the function $f_{\text{cal}} : \mathbb{R}^3 \to \mathbb{R}^3$ for a number of different values of wind speed magnitude $|\boldsymbol{u}|$, wind direction (or horizontal incidence angle) θ, and vertical angle of incidence α, which may give rise to a large number of calibration points. For example, a typical full calibration may consist of $N_u \times N_\theta \times N_\alpha = 10 \times 40 \times 10 = 4000$ calibration points. From these experimental relations between the real and measured wind speed vectors (u, v, w) [or equivalent vector magnitudes $(|\boldsymbol{u}|, \theta, \alpha)$] a proper fit to the function $f_{\text{cal}} : \mathbb{R}^3 \to \mathbb{R}^3$ that can be applied to correct future measurements must be determined.

The authors have explored linear interpolation methods to reduce the information contained in the calibration test. The philosophy behind this method can be more easily illustrated if a two-dimensional calibration is considered. If only the variation of two components of the wind speed are evaluated (i. e., the two horizontal components u and v), the relation between the measured and real values can be represented graphically as in Figs. 5.228 and 5.229, where the white dots represent the calibration values for the vector (u, v) and the grey dots the corresponding measured values $(u^{\text{M}}, v^{\text{M}})$. (A similar scheme with a different data process is presented in [5.857].)

An enlargement of the schematic shown in Fig. 5.228 is represented in Fig. 5.229. The white triangle represents the interpolated (corrected after calibration) value of the measured vector $(u^{\text{M}}, v^{\text{M}})$. This calibrated value is obtained in this case by a linear calibration function $f_{\text{cal}}^{\text{l}} : \mathbb{R}^2 \to \mathbb{R}^2$ so that the corrected vector is obtained as

$$(\boldsymbol{u}_{k,\text{m}})_{\text{C}}^{\text{M}} = \left(u_{k,\text{m}}^{\text{M}}, v_{k,\text{m}}^{\text{M}} \right)_{\text{C}}$$
$$= f_{\text{cal}}^{\text{l}}(\boldsymbol{u}_{k',\text{m}'}; \boldsymbol{u}_{i,j}, \boldsymbol{u}_{i',j'}, \boldsymbol{u}_{i+1,j}, \boldsymbol{u}_{i'+1,j}$$
$$\ldots \boldsymbol{u}_{i'+1,j'+1}) \,. \qquad (5.391)$$

To avoid cross contamination of the velocity components, extreme care needs to be taken during mounting and leveling SATs. Over level terrain, bubble indicators are useful for leveling. However, over sloping terrain users must determine if the sensor is to be placed parallel to the slope or oriented with respect to the gravitational vector. In addition, sensors should be placed to maximize the useful range of measurements by minimizing tower and sensor distortion by orienting the sensor into the predominant wind direction. Quality control routines are used in postprocessing to remove or mark data with winds that have been disturbed by the tower.

Finally, wet sensors, rain, snow, icing, and debris on sensors degrade the ultrasonic signal. Most manufactures have heated sensors that are intended for cold weather use to prevent icing.

5.7.6 Manufacturers and Costs

A number of manufacturers exist around the world including: Vaisala, Inc., RM Young Company, Applied Technologies, Inc., Campbell Scientific, Inc., Metek GmbH, and Gill Instruments Ltd. Anemometer cost varies significantly. 2-D SATs are less expensive than 3-D devices, typically costing $ 1000–3000. The price range of 3-D SATs is approximately $ 2000–8000. The prices generally vary with accuracy and additional sensor options and do not include data loggers or laptop computers for data storage.

5.7.7 Device Comparison

SATs provide a number of advantages for making point measurements in flows with length scales of interest that are greater than $O(1\,\mathrm{m})$ for frequencies less than $O(10\,\mathrm{Hz})$. Over this range, they provide much better response than traditional meteorological instruments such as cup-and-vane or propeller anemometers [5.820, 844, 865] (Table 5.8). SATs are much easier to operate and much more robust than traditional engineering laboratory equipment such as LDA, PIV, or hot-wire anemometry. However, for investigating high-frequency turbulence at the smallest length scales, traditional engineering laboratory equipment must be employed.

References

5.1 S.H. Chue: Pressure probes for fluid measurement, Prog. Aerosp. Sci. **16**, 147–223 (1975)

5.2 D.W. Bryer, R.C. Pankhurst: *Pressure Probe Methods for Determining Wind Speed and Flow Direction* (Her Majesty's Stationery Office, London 1971)

5.3 H. Pitot: Déscription d'une machine pour mesurer la vitesse des eaux courantes et le sillage desvaisseaux. In: *Mémoires de L'Académie des Sciences* (L'Académie des Sciences, Paris 1732)

5.4 H.W. Liepmann, A. Roshko: *Elements of Gas Dynamics* (Wiley, New York 1957)

5.5 R.C. Pankhurst, D.W. Holder: *Wind-Tunnel Technique* (Sir Isaac Pitman, London 1952)

5.6 F. S. Sherman: New experiments on impact pressure interpretation in supersonic and subsonic rarified airstreams, Tech. Note *2995*, (NACA, Washington 1953)

5.7 C.W. Hurd, K.P. Chesky, A.H. Shapiro: Influence of viscous effects on impact tubes, J. Appl. Mech. **20**(2), 254–256 (1953)

5.8 M. Barker: On the use of very small Pitot-tubes for measuring wind velocities, Proc. R. Soc. London, Vol. 101 (1922) 435–445

5.9 F.A. MacMillan: Viscous effects on Pitot tubes in shear flow, J. R. Aeronaut. Soc. **58**, 837–839 (1954)

5.10 F.A. MacMillan: Viscous effects on flattened Pitot tubes at low speeds, J. R. Aeronaut. Soc. **58**, 837–839 (1954)

5.11 W. G. S. Lester: The flow past a Pitot tube at low Reynolds numbers, Rep. Memor. *3240*, Aero. Res. Council (1961)

5.12 E.U. Repik: Measuring the velocity in a boundary layer with a total-head Pitot tube, J. Eng. Phys. Thermophys. **22**(3), 357–360 (1972)

5.13 N.P. Mikhailova, E.U. Repik: Effect of viscosity on Pitot tube readings at low flow speeds, Izv. Akad. Nauk. SSSR **1**, 136–139 (1976)

5.14 M. V. Zagarola: *Mean-flow scaling of turbulent pipe flow*. Ph.D. Thesis (Princeton University, Princeton 1996)

5.15 K. G. Merriam, E. R. Spaulding: Comparative tests of Pitot-static tubes, Tech. Note *546* (NACA, Washington 1935)

5.16 F. Homann: The effect of high viscosity on the flow around a cylinder and a sphere, Tech. Mem. *1334* (NACA, Washington 1952)

5.17 J.L. Potter, A.B. Bailey: Pressures in the stagnation regions of blunt bodies in rarified flow, AIAA J. **2**(4), 743–745 (1964)

5.18 S. Goldstein: A note on the measurement of total head and static pressure in a turbulent stream, Proc. R. Soc. London A **155**(886), 570–575 (1936)

5.19 T. Christiansen, P. Bradshaw: Effect of turbulence on pressure probes, J. Phys. E Sci. Instrum. **14**, 992–997 (1981)

5.20 S. Tavoularis, M. Szymczak: Displacement effects of square-ended Pitot tubes in shear flows, Exp. Fluids **7**, 33–37 (1989)

5.21 V. Ozarapoglu: *Measurements in incompressible turbulent flows*. Ph.D. Thesis (Université Laval, Quebec 1973)

5.22 J. Dickinson: *Turbulent Skin Friction Techniques* (Laval University, Quebec 1975)

5.23 B.J. McKeon, J. Li, W. Jiang, J.F. Morrison, A.J. Smits: Pitot probe corrections in high Reynolds number fully developed turbulent pipe flow, Meas. Sci. Technol. **14**, 1449–1458 (2003)

5.24 J. P. Monty: *Developments in smooth wall turbulent duct flows*. Ph.D. Thesis (University of Melbourne, Australia 2005)

5.25 B. J. McKeon: *High Reynolds number turbulent pipe flow*. Ph.D. Thesis (Princeton University, Princeton 2003)

5.26 I.M. Hall: The displacement effect of a sphere in two-dimensional shear flow, J. Fluid Mech. **1**, 142–162 (1956)

5.27 M.J. Lighthill: Contributions to the theory of the Pitot-tube displacement effect, J. Fluid Mech. **1**, 493–512 (1957)

5.28 F. A. MacMillan: Experiments on Pitot tubes in shear flow, Rep. Memor. *3028*, Aero. Res. Council. (1956)

5.29 A. Quarmby, H.K. Das: Displacement effect on Pitot tubes with rectangular mouths, Aeronaut. Q. **20**, 129–139 (1969)

5.30 P.O.A. Davies: The behaviour of a Pitot tube in transverse shear, J. Fluid Mech. **3**, 441–456 (1958)

5.31 A. D. Young, J. N. Maas: The behaviour of a Pitot-tube in a transverse total-pressure gradient, Rep. Memor. *1770*, Aero. Res. Council (1936)

5.32 J.L. Livesey, J.D. Jackson, C.J. Southern: The static-hole error problem: an experimental investigation of errors for holes of varying diameters and depths, Aircr. Eng. **34**, 43–47 (1962)

5.33 E.U. Repik, V.K. Kuzenkov, N.P. Mikhailova: Influence of velocity gradient and wall proximity on the readings of a Pitot tube in measurement of surface friction and velocity distribution in a turbulent boundary layer, J. Eng. Phys. Thermophys. **48**(6), 642–649 (1985)

5.34 V.C. Patel: Calibration of the Preston tube and limitations on its use in pressure gradients, J. Fluid Mech. **23**, 185–206 (1965)

5.35 C.W. Hubbard: Investigation of errors of Pitot tubes, Trans. ASME **61**, 477–492 (1939)

5.36 A. Pope, J.J. Harper: *Low-Speed Wind Tunnel Testing* (Wiley, New York 1966)

5.37 R. Shaw: The influence of hole dimensions on static pressure measurements, J. Fluid Mech. **7**, 550–564 (1960)

5.38 B.J. McKeon, A.J. Smits: Static pressure correction in high Reynolds number fully developed turbulent pipe flow, Meas. Sci. Technol. **13**, 1608–1614 (2002)

5.39 B. Chebbi, S. Tavoularis: Pitot-static tube response at very low Reynolds numbers, Phys. Fluids **3**(3), 481–483 (1991)

5.40 G.G. Zilliac: Calibration of seven-hole pressure probes for use in fluid flows with large angularity, Tech. Mem. *102200* (NASA, Washington 1989)

5.41 D. Sumner: A comparison of data-reduction methods for a seven-hole probe, J. Fluid Eng. **124**, 523–527 (2002)

5.42 C. Venkateswara Babu, M. Govardhan, N. Sitaram: A method of calibration of a seven-hole pressure probe for measuring highly three-dimensional flows, Meas. Sci. Technol. **9**, 468–476 (1998)

5.43 A.A. Gerner, C.L. Maurer, R.W. Gallington: Nonnulling seven-hole probes for high angle flow measurement, Exp. Fluids **2**, 95–103 (1984)

5.44 T.T. Takahashi: Measurement of air flow characteristics using seven-hole cone probes, Tech. Mem. *112-194* (NASA, Washington 1997)

5.45 S. Corrsin: Turbulence: Experimental methods. In: *Handbuch der Physik/Encyclopedia of Physics VIII/2*, ed. by S. Flügge (Springer, Berlin, Heidelberg 1963) pp. 524–590

5.46 P. Bradshaw: *An Introduction to Turbulence and its Measurements* (Pergamon, Oxford 1975)

5.47 J.O. Hinze: *Turbulence* (McGraw-Hill, New York 1975)

5.48 R.F. Blackwelder: Hot-wire and hot-film anemometers. In: *Methods of Experimental Physics: Fluid Dynamics*, Vol. 18, ed. by R.J. Emrich (Academic, New York 1981) pp. 259–314

5.49 A.E. Perry: *Hot-Wire Anemometry* (Clarendon, Oxford 1982)

5.50 L.M. Fingerson, P. Freymuth: Thermal anemometers. In: *Fluid Mechanics Measurements*, Vol. 4, ed. by R.J. Goldstein (Hemisphere, Washington 1983) pp. 99–154

5.51 C.G. Lomas: *Fundamentals of Hot-Wire Anemometry* (Cambridge Univ. Press, Cambridge 1986)

5.52 H.H. Bruun: *Hot-Wire Anemometry* (Oxford Scientific, Oxford 1995)

5.53 I. Lekakis: Calibration and signal interpretation for single and multiple hot-wire/hot-film probes, Meas. Sci. Technol. **7**, 1313–1333 (1996)

5.54 G. Comte-Bellot: Hot-wire anemometry, Ann. Rev. Fluid Mech. **8**, 209–231 (1976)

5.55 G. Comte-Bellot: Hot-wire anemometry. In: *Handbook of Fluid Dynamics*, ed. by R.W. Johnson (CRC, Boca Raton 1998) pp. 34.1–34.29

5.56 G. Comte-Bellot: Les méthodes de mesure physique de la turbulence, J. Phys. **37**(Suppl. C1), 67–78 (1976)

5.57 T. Tsuji, Y. Nagano, M. Tagawa: Frequency response and instantaneous temperature profile of cold-wire sensors for fluid temperature fluctuation measurements, Exp. Fluids **13**, 171–178 (1992)

5.58 P.O.A.L. Davies, M.J. Fisher: Heat transfer from electrically heated cylinders, Proc. R. Soc. London A, Vol. 280 (1964) 486–527

5.59 J. Laufer, R. McClellan: Measurements of heat transfer from fine wires in supersonic flows, J. Fluid Mech. **1**, 276–289 (1956)

5.60 J. Fraden: *Handbook of Modern Sensors* (Springer-AIP, Berlin, New York 2004)

5.61 A.D. McConnell, S. Uma, K.E. Goodson: Thermal conductivity of doped polysilicon layers, J. MEMS **0654**, 1–11 (2001)

5.62 Q. Lin, F. Jiang, X.Q. Wang, Y. Xu, Z. Han, Y.C. Tai, J. Lew, C.M. Ho: Experiments and simulations of MEMS thermal sensors for wall-shear stress measurements in aerodynamic control applications, J. Micromech. Microeng. **14**, 1640–1649 (2004)

5.63 T. Ebefors, E. Kälvesten, G. Stemme: Three dimensional silicon triple-hot-wire anemometer based on polyimide joints, IEEE Int. Workshop on Micro Electro Mechanical System, Vol. MEMS'98 (1998) 25–29

5.64 T. Ebefors: *Polyimide V-groove joints for three-dimensional silicon transducers*. Ph.D. Thesis (Royal Institute of Technology, Stockholm 2000)

5.65 A. van Dijk, F.T.M. Nieuwstadt: The calibration of (multi) hot-wire probes. 1. Temperature calibration, Exp. Fluids **36**, 540–549 (2004)

5.66 J. Lemay, A. Benaïssa: Improvement of cold-wire response for measurement of temperature dissipation, Exp. Fluids **31**, 347–356 (2001)

5.67 R.W. Derksen, R.S. Azad: An examination of hot-wire length corrections, Phys. Fluids **26**, 1751–1754 (1983)

5.68 G. Comte-Bellot, A. Strohl, E. Alcaraz: On aerodynamic disturbances caused by hot-wire probes, J. Appl. Mech. **38**, 767–774 (1971)

5.69 A. Strohl, G. Comte-Bellot: Aerodynamic effects due to configuration of X-wire anemometers, J. Appl. Mech. **40**, 661–666 (1973)

5.70 W.W. Willmarth, L.K. Sharma: Study of turbulent structure with hot wires smaller than the viscous length, J. Fluid Mech. **142**, 121–149 (1984)

5.71 P.M. Ligrani, P. Bradshaw: Subminiature hot-wire sensors: development and use, J. Phys. E Sci. Instrum. **20**, 323–332 (1987)

5.72 F. Jiang, Y.C. Tai, C.M. Ho, W.J. Li: A micromachined polysilicon hot-wire anemometer, Solid State Sensor and Actuator Workshop (1994) 264–267

5.73 B. Castaing, B. Chabaud, B. Hébral: Hot-wire anemometer operating at cryogenic temperature, Rev. Sci. Instrum. **63**, 4167–4173 (1992)

5.74 F. Jiang, Y.C. Tai, R. Karen, M. Gaustenauer, C.H. Ho: Theoretical and experimental studies of micromachined hot-wire anemometers, Intern. Electron. Devices Meeting (1994) 139–142

5.75 J.M. Österlund, A.V. Johansson: Dynamic behavior of hot-wire probes in turbulent boundary layers. In: *Advances in Turbulence*, ed. by R. Benzi (Kluwer, Dordrecht 1995) pp. 398–402

5.76 G. Comte-Bellot: Écoulement turbulent entre deux parois planes parallèles, Publ. Sci. Tech. Min. Air Paris **419**, 1–159 (1965), [previously: Contribution à l'étude de la turbulence de conduite, Ph.D. Thesis (Univ. Grenoble, Grenoble 1963), No 315, translated into English by P. Bradshaw: Turbulent flows between parallel walls, A.R.C 31609 FM 4102 (1969)]

5.77 C. Bailly, G. Comte-Bellot: *Turbulence* (CNRS, Paris 2003)

5.78 M. Hishida, Y. Nagano: Turbulence measurements with symmetrically bent V-shaped hot-wires. Part 1: Principles of Operation, J. Fluid Eng. Trans. ASME **110**, 264–269 (1988)

5.79 M. Hishida, Y. Nagano: Turbulence measurements with symmetrically bent V-shaped hot-wires. Part 2: Measuring velocity components and turbulent shear stresses, J. Fluid Eng. Trans. ASME **110**, 270–274 (1988)

5.80 A. van Dijk, F.T.M. Nieuwstadt: The calibration of (multi) hot-wire probes. 2. Velocity calibration, Exp. Fluids **36**, 550–564 (2004)

5.81 J.M. Wallace, J.F. Foss: The measurement of vorticity in turbulent flows, Ann. Rev. Fluid. Mech. **27**, 469–514 (1995)

5.82 M. Sunyach: *Contribution à l'étude des frontières d'écoulements turbulents libres*. Thèse d'Etat (Univ. Lyon, Lyon 1971), No. 1971-37 (in French)

5.83 C.H.P. Chen, R.F. Blackwelder: Large-scale motion in a turbulent boundary layer: a study using temperature contamination, J. Fluid Mech. **89**, 1–31 (1978)

5.84 H.S. Kang, S. Chester, C. Meneveau: Decaying turbulence in an active-grid-generated flow and comparisons with large-eddy simulation, J. Fluid Mech. **480**, 129–160 (2003)

5.85 J. Delville, L. Ukeiley, L. Cordier, J.P. Bonnet, M. Glauser: Examination of large-scale structures in a turbulent plane mixing layer. Part 1. Proper orthogonal decomposition, J. Fluid Mech. **391**, 91–122 (1999)

5.86 J.H. Citriniti, W.K. George: Reconstruction of the global velocity field in the axisymmetric mixing layer utilizing the proper orthogonal decomposition, J. Fluid Mech. **418**, 137–166 (2000)

5.87 R.M. Hall, C.J. Obara, D.L. Carraway, C.B. Johnson, R.E. Wright Jr, P.F. Covell, M. Azzazy: Comparisons of boundary-layer transition measurement techniques at supersonic Mach numbers, AIAA J. **29**, 865–871 (1991)

5.88 T.R. Moes, G.R. Sarma, S.M. Mangalam: Flight demonstration of a shock location sensor using constant voltage hot-film anemometry, NASA Tech. Memo **4806** (1997)

5.89 R.J. Adrian, R.E. Johnson, B.G. Jones, P. Merati, A.T.C. Tung: Aerodynamic disturbances of hot-wire probes and directional sensitivity, J. Phys. E Sci. Instrum. **17**, 62–71 (1984)

5.90 B.J. Hoole, J.R. Calvert: The use of a hot-wire anemometer in turbulent flow, J. R. Aeronaut. Soc. **71**, 511–512 (1967)

5.91 L.S.G. Kovasznay: Turbulence in supersonic flow, J. Aeronaut. Sci. **20**, 657–675 (1953)

5.92 J.-P. Bonnet J.P.: *Etude théorique et expérimentale de la turbulence dans un sillage supersonique*. Thèse d'Etat (Univ. Poitiers, Poitiers 1982), No. 355

5.93 C.M. Ho, Y.C. Tai: Micro-electro-mechanical-systems (MEMS) and fluid flows, Ann. Rev. Fluid Mech. **30**, 579–612 (1998)

5.94 W.H. MacAdams: *Heat Transmission* (McGraw-Hill, New York 1954)

5.95 B.G. Van der Hegge Zijnen: Modified correlation formulae for the heat transfers by natural and by forced convection from horizontal cylinders, Appl. Sci. Res. **6**, 129–140 (1956)

5.96 D.C. Collis, M.J. Williams: Two-dimensional convection from heated wires at low Reynolds numbers, J. Fluid Mech. **6**, 357–384 (1959)

5.97 L.J.S. Bradbury, I.P. Castro: Some comments on heat transfer laws for fine wire, J. Fluid Mech. **51**, 487–495 (1972)

5.98 H. Kramers: Heat transfer from spheres to flowing media, Physica **XII**(2-3), 61–80 (1946)

5.99 G.E. Andrews, D. Bradley, G.F. Hundy: Hot wire anemometer calibration for measurements of small gas velocities, Int. J. Heat Mass Transf. **15**, 1765–1786 (1972)

5.100 M.V. Morkovin: Fluctuations and hot-wire anemometry in compressible flows, Wind Tunnel AGARDographs Ser. **24**, 1–102 (1956)

5.101 F.M. White: *Viscous Fluid Flow* (McGraw-Hill, New York 1991)

5.102 S.C.R. Dennis, J.D. Hudson, N. Smith: Steady laminar forced convection from a circular cylinder at low Reynolds numbers, Phys. Fluids **11**, 933–940 (1968)

5.103 E.R.G. Eckert, R.M. Drake Jr: *Analysis of Heat and Mass Transfer* (Hemisphere, Washington 1987)

5.104 J.G. Knudsen, D.L. Katz: *Fluid Dynamics and Heat Transfer* (McGraw-Hill, New York 1958)

5.105 M.S. Uberoi, S. Corrsin: Spectra and diffusion in a round turbulent jet, NACA Rep. **1040** (1951)

5.106 A.P. Hatton, D.D. James, H.W. Swire: Combined forced and natural convection with low-speed air flow over horizontal cylinders, J. Fluid Mech. **42**, 17–31 (1970)

5.107 W.M. Pitts, B.J. McCaffrey: Response behaviour of hot wires and films to flows of different gases, J. Fluid Mech. **169**, 465–512 (1986)

5.108 E.F. Spina, C.B. McGinley: Constant-temperature anemometry in hypersonic flow: critical issues and sample results, Exp. Fluids **17**, 365–374 (1994)

5.109 T.G. Elizarova, I.A. Shirokov, S. Montero: Numerical simulation of shock-wave structure for argon and helium, Phys. Fluids **17**, 068101 (2005)

5.110 J.C. Hill, C.A. Sleicher: Convective heat transfer from small cylinders to mercury, Int. J. Heat Mass Transf. **12**, 1595–1604 (1969)

5.111 J.C. Hill, C.A. Sleicher: Directional sensitivity of hot-films sensors in liquid metals, Rev. Sci. Instrum. **42**, 1461–1468 (1971)

5.112 B. Gebhart, L. Pera: Mixed convection from horizontal cylinders, J. Fluid Mech. **45**, 49–64 (1971)

5.113 K.A. Smith, E.W. Merrill, H.S. Mickley, P.S. Virk: Anomalous Pitot tube and hot film measurements in dilute polymer solutions, Chem. Eng. Sci. **22**, 619–626 (1967)

5.114 D.F. James, A.J. Acosta: Laminar flow of dilute polymer solutions around circular cylinders, J. Fluid Mech. **42**, 269–288 (1970)

5.115 J.L. Lumley: Drag reduction in turbulent flow by polymer additives, J. Polym. Sci. Macromol. Rev. **7**, 263–290 (1973)

5.116 L.S.G. Kovasznay: The hot-wire anemometry in supersonic flow, J. Aeronaut. Sci. **17**, 565–584 (1950)

5.117 M.V. Morkovin, R.E. Phinney: *Extended Applications of Hot-wire Anemometry to High-speed Turbulent Boundary Layers* (John Hopkins Univ., Baltimore 1958)

5.118 W.G. Spangenberg: Heat-loss characteristics of hot-wire anemometers at various densities in transonic and supersonic flow, NACA Tech. Note **3381** (1955)

5.119 C.F. Dewey Jr: Hot wire measurements in low Reynolds number hypersonic flows, J. Am. Rocket Soc. **28**, 1709–1718 (1961)

5.120 C.F. Dewey Jr: A correlation of convective heat transfer and recovery temperature data for cylinders in compressible flow, Int. J. Heat Mass Transf. **8**, 245–252 (1965)

5.121 L.V. Baldwin, V.A. Sandborn, J.C. Laurence: Heat transfer from transverse and yawed cylinders in continuum, slip, and free molecule air flows, J. Heat Transf. Trans. ASME **82**, 77–86 (1960)

5.122 J. Gaviglio: Sur les méthodes de l'anémométrie par fil chaud des écoulements turbulents compressibles de gaz, J. Mécanique **2**, 449–498 (1978)

5.123 S. Barre, P. Dupont, J.P. Dussauge: Hot-wire measurements in turbulent transonic flows, Eur. J. Mech. B Fluids **11**(4), 439–454 (1992)

5.124 P.C. Stainback, K.A. Nagabushana: Review of hot-wire anemometry techniques and the range of their applicability. In: *Thermal Anemometry*, Vol. FED167, ed. by ASME (ASME, New York 1993) pp. 93–133

5.125 A.J. Smits, K. Hayakawa, K.C. Muck: Constant temperature hot-wire anemometer practice in supersonic flows. Part 1: The normal wire, Exp. Fluids **1**, 83–92 (1983)

5.126 J. Weiss, H. Knauss, S. Wagner: Experimental determination of the free-stream disturbance field in a short-duration supersonic wind tunnel, Exp. Fluids **35**, 291–302 (2003)

5.127 G. Comte-Bellot, G.R. Sarma: Constant voltage anemometer practice in supersonic flows, AIAA J. **39**, 261–270 (2001)

5.128 J.R. Stalder, G. Goodwin, M.O. Creager: Heat transfer to bodies in a high-speed rarefied-gas stream, NACA Tech. Note **2438** (1951)

5.129 N. Chokani, A.N. Shiplyuk, A.A. Sidorenko, C.B. McGinley: Comparison between a hybrid constant-current anemometer and constant-voltage anemometer in hypersonic flow, 34th AIAA Fluid Dynamics Conf (AIAA, 2004), AIAA Paper 2004-2248

5.130 M.V. Scadron, I. Warshasky: Experimental determination of time constants and Nusselt numbers for bare-wire thermocouples in high-velocity air streams and analytical approximation of conduction and radiation errors, NACA Tech. Note **2599** (1952)

5.131 P. Dupont, J.F. Debieve: A hot wire method for measuring turbulence in transonic or supersonic heated flows, Exp. Fluids **13**, 84–90 (1992)

5.132 C.C. Horstman, W.C. Rose: Hot-wire anemometry in transonic flow, AIAA J. **15**, 395–401 (1977)

5.133 G.S. Jones, P.C. Stainback, C.D. Harris, C.W. Brooks, S.J. Clukey: Flow quality measurements for the Langley 8-foot transonic pressure tunnel LFC experiment, AIAA Paper, Vol. 89-0150 (1989)

5.134 G.S. Jones: Wind tunnel requirements for hot-wire calibrations, AIAA Paper **94-2534** (1994)

5.135 W.C. Rose: Turbulence measurements in a compressible boundary layer, AIAA J. **12**, 1060–1064 (1974)

5.136 D.A. Walker, W.F. Ng, M.D. Walker: Experimental comparison of two hot-wire techniques in supersonic flow, AIAA J. **27**, 1074–1080 (1989)

5.137 J.D. Norris, N. Chokani: Rapid scanning of overheat ratios using a constant voltage anemometer, AIAA J. **41**, 1619–1621 (2003)

5.138 D.R. Smith, A.J. Smits: Simultaneous measurements of velocity and temperature fluctuations in the boundary layer of a supersonic flow, Exp. Therm. Fluid Sci. **7**, 221–229 (1993)

5.139 M.V. Morkovin: Effects of compressibility on turbulent flows, Colloque CNRS, Vol. 108 (1962) 367–380

5.140 J. Gaviglio: Reynolds analogies and experimental study of heat tranfer in the supersonic boundary layer, Int. J. Heat Mass Transf. **30**, 911–926 (1987)

5.141 J. Weiss, H. Knauss, S. Wagner, N.D. Chokani, G. Comte-Bellot, A. Kosinov: Comparative measurements in M=2.54 flow using constant-temperature and constant-voltage anemometers, 41st AIAA Aerospace Sci. Meeting and Exhibit (2003) , , AIAA Paper 2003-1277

5.142 A.J. Smits, J.P. Dussauge: Hot-wire anemometry in supersonic flow, AGARDograph **315**, 5.1–5.14 (1989)

5.143 A.D. Kosinov, N.V. Semionov, Y.G. Yermolaev: Disturbances in test section of T-325 supersonic wind tunnel, Russ. Acad. Sci. Novosibirsk ITAM Rep. **6-99** (1999)

5.144 C.L. Ko, D.K. McLaughlin, T.R. Troutt: Supersonic hot-wire fluctuation data analysis with a conduction end-loss correction, J. Phys. E Sci. Instrum. **11**, 488–494 (1978)

5.145 I. Kidron: Measurement of the transfer function of hot-wire and hot-film turbulence transducers, IEEE Trans. Instrum. Meas. **15**(3), 76–81 (1966)

5.146 B.J. Bellhouse, D.L. Schultz: The determination of fluctuating velocity in air with heated thin film gauges, J. Fluid Mech. **29**, 289–295 (1967)

5.147 J.F. Brison, G. Charnay, G. Comte-Bellot: Calcul des transferts thermiques entre film chaud et substrat par un modèle à deux dimensions : prévision de la réponse dynamique des sondes usuelles, Int. J. Heat Mass Transf. **22**, 111–119 (1978)

5.148 G.R. Sarma: Transfer function analysis of the constant voltage anemometer, Rev. Sci. Instrum. **69**, 2385–2391 (1998)

5.149 G.R. Sarma: Analysis of a constant voltage anemometer circuit, IEEE Instrum. Meas. Technol. Conf. **1993**, 731–736 (1993)

5.150 C.A. Reimann, N. Chokani, G.R. Sarma: Improved calibration methods for a constant voltage anemometer-operated hot-wire in hypersonic flow, 41th AIAA Aerospace Sci. Meeting and Exhibit (2003) , , AIAA Paper 2003-0976

5.151 G. Comte-Bellot, J.P. Schon: Harmoniques créés par excitation paramétrique dans les anémomètres à fil chaud à intensité constante, Int. J. Heat Mass Transf. **12**, 1661–1677 (1969)

5.152 F.H. Champagne, C.A. Sleicher, O.H. Wehrmann: Turbulence measurements with inclined hot-wires, J. Fluid Mech. **28**, 153–175 (1967)

5.153 R. Betchov: L'influence de la conduction thermique sur les anémomètres à fils chauds, Proc. K Ned. Akad. Wet., Vol. 51 (1948) 721–730

5.154 R.G. Lord: Hot-wire probe end-loss corrections in low density flows, J. Phys. E Sci. Instrum. **7**, 56–60 (1974)

5.155 S.C. Morris, J.F. Foss: Transient thermal response of a hot-wire anemometer, Meas. Sci. Technol. **14**, 251–259 (2003)

5.156 J.D. Li, B.J. McKeon, W. Jiang, J.F. Morrison, A.J. Smits: The response of hot wires in high Reynolds-number turbulent pipe flow, Meas. Sci. Technol. **15**, 789–798 (2004)

5.157 L. Löfdahl, V. Chernoray, S. Haasl, G. Stemme, M. Sen: Characteristics of a hot-wire microsensor for time-dependent wall shear stress measurements, Exp. Fluids **35**, 240–251 (2003)

5.158 K. Bremhorst, L. Krebs, D.B. Gilmore: The frequency response of hot-wire anemometer sensors to heating current fluctuations, Int. J. Heat Mass Transf. **20**, 315–322 (1977)

5.159 R.G. Lord: The dynamic behaviour of hot-wire anemometers with conduction end losses, J. Phys. E Sci. Instrum. **14**, 573–578 (1981)

5.160 A.C.M. Beljaars: Dynamic behaviour of the constant temperature anemometer due to thermal inertia of the wire, Appl. Sci. Res. **32**, 509–518 (1976)

5.161 J.D. Li: Dynamic response of constant temperature hot-wire system in turbulence velocity measurements, Meas. Sci. Technol. **15**, 1835–1847 (2004)

5.162 H. Tennekes, J.L. Lumley: *A First Course in Turbulence* (MIT Press, Cambridge 1972)

5.163 H.L. Dryden, G.B. Schubauer, W.C. Mock Jr, H.K. Skramstad: Measurements of intensity and scale of wind-tunnel turbulence and their relation to the critical Reynolds number of spheres, NACA Tech. Rep. **581** (1937)

5.164 F.N. Frenkiel: Effects of wire length in turbulence investigations with a hot-wire anemometer, Aeronaut. Q. **V**, 1–24 (1954)

5.165 J.C. Wyngaard: Measurement of small-scale turbulence structure with hot-wires, J. Phys. E Sci. Instrum. **1**(Serie 2), 1105–1108 (1968)

5.166 M.S. Uberoi, L.S.G. Kovasznay: On mapping and measurement of random fields, Q. Appl. Math. **10**, 375–393 (1953)

5.167 D. Ewing, H.J. Hussein, W.K. George: Spatial resolution of parallel hot-wire probes for derivative measurements, Exp. Therm. Fluid Sci. **11**, 155–173 (1995)

5.168 J.W. Elsner, P. Domagala, W. Elsner: Effect of finite spatial resolution of hot-wire anemometry on measurements of turbulence energy dissipation, Meas. Sci. Technol. **4**, 517–523 (1993)

5.169 J.C. Wyngaard: Spatial resolution of the vorticity meter and other hot-wire arrays, J. Phys. E Sci. Instrum. **2**(Serie 2), 983–987 (1969)

5.170 R.A. Antonia, D.A. Shah, L.W.B. Browne: Spectra of velocity derivatives in a turbulent wake, Phys. Fluids **30**, 3455–3462 (1987)

5.171 R.A. Antonia, J. Mi: Corrections for velocity and temperature derivatives in turbulent flows, Exp. Fluids **14**, 203–208 (1993)

5.172 Y. Suzuki, N. Kasagi: Evaluation of hot-wire measurements in wall shear turbulence using a direct numerical simulation database, Exp. Therm. Fluid Sci. **5**, 66–77 (1992)

5.173 T. Zhou, R.A. Antonia, L.P. Chua: Performance of a probe for measuring turbulent energy and temperature dissipation rates, Exp. Fluids **33**, 334–345 (2002)

5.174 L.W.B. Browne, R.A. Antonia, D.A. Shah: Selection of wires and wire spacing for X-wires, Exp. Fluids **6**, 286–288 (1988)

5.175 P.M. Ligrani, P. Bradshaw: Spatial resolution and measurement of turbulence in the viscous sublayer using subminiature hot-wire probes, Exp. Fluids **5**, 407–417 (1987)

5.176 A.V. Johansson, P.H. Alfredsson: Effects of imperfect spatial resolution on measurements of wall-bounded turbulent shear flows, J. Fluid Mech. **137**, 409–421 (1983)

5.177 J.C. Klewicki, R.E. Falco: On accurately measuring statistics associated with small-scale structure in turbulent boundary layers using hot-wire probes, J. Fluid Mech. **219**, 119–142 (1990)

5.178 P. Moin, P.R. Spalart: Contributions of numerical simulation data base to the physics, modeling, and measurement of turbulence, NASA Tech. Memo **100022** (1987)

5.179 R.A. Antonia, Y. Zhu, J. Kim: On the measurement of lateral velocity derivatives in turbulent flows, Exp. Fluids **15**, 65–69 (1993)

5.180 R.A. Antonia, Y. Zhu, J. Kim: Corrections for spatial velocity derivatives in a turbulent shear flow, Exp. Fluids **16**, 411–416 (1994)

5.181 E. Huguenard, A. Magnan, A. Planiol: Les appareils à fils chauds, leur application à l'étude des mouvements atmosphériques, Serv. Technol. Aeronaut. Paris **32**, 1–66 (1926)

5.182 L.F.G. Simmons, A. Bailey: A hot-wire instrument for measuring speed and direction of airflow, Phil Mag S7 **13**(3), 81–96 (1927)

5.183 L.V. King: The linear hot-wire anemometer and its application in technical physics, J. Franklin I. **181**, 1–25 (1916)

5.184 J.S.G. Thomas: The hot-wire anemometer: its application to the investigation of the velocity of gases in pipes, Phil Mag S6 **233**(39), 505–534 (1920)

5.185 E.G. Richardson: *Les appareils à fil chaud, leurs applications dans la mécanique expérimentale des fluides* (Gauthier-Villars, Paris 1934)

5.186 H.L. Dryden, A.M. Kuethe: The measurement of fluctuations of air speed by the hot-wire anemometer, NACA Rep. **320** (1929)

5.187 V.V. Repkov, A.N. Shiplyuk, A.A. Sidorenko, V.A. Lebiga, A.Y. Pak, V.N. Zinoviev: Contant current anemometer with built-in microcontroller, Int. Conf. Meth. Aerophys. Res (Nonparel, Novosibirsk 2004) 4.247–4.249

5.188 I. Kidron: The signal-to-noise ratios of constant-current and constant-temperature hot-wire anemometers, IEEE Trans. Instrum. Meas. **16**(1), 68–73 (1967)

5.189 P. Freymuth: Noise in hot-wire anemometers, Rev. Sci. Instrum. **39**, 550–557 (1968)

5.190 D. Bestion, J. Gaviglio, J.P. Bonnet: Comparison between constant-current and constant-temperature hot-wire anemometers in high speed flows, Rev. Sci. Instrum. **54**, 1513–1524 (1983)

5.191 K. Bremhorst, L.J.W. Graham: A fully compensated hot/cold wire anemometer system for unsteady flow velocity and temperature measurements, Meas. Sci. Technol. **1**, 425–430 (1990)

5.192 G. Asch: *Les capteurs en instrumentation industrielle* (Dunod, Paris 2006)

5.193 M. Ziegler: The construction of a hot wire anemometer with linear scale and negligible lag, Verh. K. Akad. Wet. **15**, 3–22 (1934)

5.194 J.R. Weske: A hot-wire circuit with very small time lag, NACA Tech. Rep. **881** (1943)

5.195 E. Ossofsky: Constant temperature operation of the hot-wire anemometer at high frequency, Rev. Sci. Instrum. **19**, 881–889 (1948)

5.196 J.C. Wyngaard, J.L. Lumley: A constant temperature hot-wire anemometer, J. Sci. Instrum. **44**, 363–365 (1967)

5.197 P. Freymuth: Feedback control theory for constant-temperature hot-wire anemometers, Rev. Sci. Instrum. **38**, 677–681 (1967)

5.198 Davis M.R., Davies P.O.A.L.: *The physical characteristics of hot-wire anemometers*, Tech. Rep. 2 (Univ. Southampton, Southampton 1968)

5.199 A.E. Perry, G.L. Morrison: A study of the constant-temperature hot-wire anemometer, J. Fluid Mech. **47**, 577–599 (1971)

5.200 N.B. Wood: A method for determination and control of the frequency response of the constant-temperature hot-wire anemometer, J. Fluid Mech. **67**, 769–786 (1975)

5.201 P. Freymuth: Frequency response and electronic testing for constant-temperature hot-wire anemometers, J. Phys. E Sci. Instrum. **10**, 705–710 (1977)

5.202 P. Freymuth: On higher order dynamics of constant-temperature hot-wire anemometers, Meas. Sci. Technol. **9**, 534–535 (1998)

5.203 J.H. Watmuff: An investigation of the constant-temperature hot-wire anemometer, Exp. Therm. Fluid Sci. **11**, 117–134 (1995)

5.204 L.M. Fingerson: Thermal anemometry, current state, and future directions, Rev. Sci. Instrum. **65**, 285–300 (1994)

5.205 F.E. Jorgensen: The computer-controlled constant-temperature anemometer. Aspects of set-up, probe calibration, data acquisition and data conversion, Meas. Sci. Technol. **7**, 1378–1387 (1996)

5.206 G. Comte-Bellot: Anémométrie à fil chaud (cas des fluides incompressibles). In: *Techniques de Mesure dans les Écoulements*, ed. by CEA/EDF (Eyrolles, Paris 1974) pp. 117–198

5.207 P. Freymuth: The effect of varying resistance ratio on the behaviour of constant-temperature hot-wire anemometers, J. Phys. E Sci. Instrum. **14**, 1373 (1981)

5.208 P. Freymuth: Second or third order control theory for constant-temperature hot-wire anemometers?, Exp. Fluids **23**, 175–176 (1997)

5.209 B.C. Khoo, Y.T. Chew, C.J. Teo, C.P. Lim: The dynamic response of a hot-wire anemometer: III. Voltage-perturbation versus velocity-perturbation testing for near-wall hot-wire/film probes, Meas. Sci. Technol. **10**, 152–169 (1999)

5.210 C.J. Teo, B.C. Khoo, Y.T. Chew: The dynamic response of a hot-wire anemometer: IV. Sine-wave voltage perturbation testing for near-wall hot-wire/film probes and the presence of low-high frequency response characteristics, Meas. Sci. Technol. **12**, 37–51 (2001)

5.211 P. Freymuth: Interpretations in the control theory of thermal anemometers, Meas. Sci. Technol. **8**, 174–177 (1997)

5.212 J. Weiss, H. Knauss, S. Wagner: Method for the determination of frequency response and signal to noise ratio for constant-temperature hot-wire anemometers, Rev. Sci. Instrum. **72**, 1904–1909 (2001)

5.213 J. Weiss, N. Chokani, G. Comte-Bellot: Constant-temperature and constant-voltage anemometer use in a Mach 2.5 flow, AIAA J. **43**, 1140–1143 (2005)

5.214 G. Comte-Bellot, G.R. Sarma, T. Faure, J.P. Dussauge, P. Dupont, J.F. Debiève: Performance studies of the constant voltage anemometer in a Mach 2.3 boundary layer, ICIASF 99 (1999) 40.1–40.9

5.215 J. Weiss: Effect of bridge imbalance on the estimation of sensitivity coefficients for constant-temperature hot-wire anemometers, Meas. Sci. Technol. **14**, 1373–1380 (2003)

5.216 S.G. Saddoughi, S.V. Veeravalli: Hot-wire anemometry behaviour at very high frequencies, Meas. Sci. Technol. **7**, 1297–1300 (1996)

5.217 P. Freymuth: Nonlinear control theory for constant-temperature hot-wire anemometers, Rev. Sci. Instrum. **40**, 258–262 (1969)

5.218 G.R. Sarma: Flow Rate Measuring Apparatus, US Patent 5,074,147 (1991)

5.219 S.M. Mangalam, G.R. Sarma, S. Kuppa, L.R. Kubendran: A new approach to high-speed flow measurements using constant voltage anemometry, 17th Aerospace Ground Testing Conf (1992), , AIAA Paper 92-3957

5.220 J.T. Lachowicz, N. Chokani, S.P. Wilkinson: Boundary-layer stability measurements in a hypersonic quiet tunnel, AIAA J. **34**, 2496–2500 (1996)

5.221 J. Weiss, G. Comte-Bellot: Electronic noise in a constant voltage anemometer, Rev. Sci. Instrum. **75**, 1290–1296 (2004)

5.222 M.A. Kegerise, E.F. Spina: A comparative study of constant-voltage and constant-temperature hot-wire anemometers, PartI: The static response, Exp. Fluids **19**, 154–164 (2000)

5.223 M.A. Kegerise, E.F. Spina: A comparative study of constant-voltage and constant-temperature hot-wire anemometers, Part II: The dynamic response, Exp. Fluids **19**, 165–177 (2000)

5.224 G.R. Sarma, R.W. Lankes: Automated constant voltage anemometer with in situ measurements of overheat and time constant of the hot-wire, Rev. Sci. Instrum. **70**, 2384–2386 (1999)

5.225 G.R. Sarma, G. Comte-Bellot, T. Faure: Software corrected hot wire thermal lag for the constant voltage anemometer featuring a constant bandwidth at the selected compensation setting, Rev. Sci. Instrum. **69**, 3223–3231 (1998)

5.226 S. Kuppa, G.R. Sarma, S.M. Mangalam: Effect of thermal inertia on the frequency response of constant voltage hot-wire anemometer and its compensation, ASME Fluid Meas. Instrum. Forum, Vol. FED 161 (1993) 67–73

5.227 G. Comte-Bellot, J. Weiss, J.C. Béra: Lead-resistance effects in a constant voltage anemometer, Rev. Sci. Instrum. **75**, 2075–2081 (2004)

5.228 A.E. Blanchard, J.T. Lachowicz, S.P. Wilkinson: NASA Langley Mach 6 quiet wind-tunnel performance, AIAA J. **35**, 23–28 (1997)

5.229 G. Comte-Bellot, J. Weiss, J.C. Béra: Control of the dynamic non-linearity in a constant voltage anemometer, Bull. Am. Phys. Soc. **50**(9), 189 (2005)

5.230 A.A. Abdel-Rahman, G.J. Hitchman, P.R. Slawson, A.B. Strong: An X-array hot-wire technique for heated turbulent flows of low velocity, J. Phys. E Sci. Instrum. **22**, 638–644 (1989)

5.231 H.H. Bruun, M.A. Khan, H.H. Al-Kayiem, A.A. Fardad: Velocity calibration relationships for hot-wire anemometry, J. Phys. E Sci. Instrum. **21**, 225–232 (1988)

5.232 L.V. King: On the convection of heat from small cylinders in a stream of fluid: determination of the convection constants of small platinum wires with applications to hot-wire anemometry, Phil. Trans. R. Soc. Lond. Ser. A **214**, 373–432 (1914)

5.233 A. Baille: *Lois de refroidissement des fils chauds aux faibles vitesses*. Ph.D. Thesis, (Univ. Aix-Marseille, Aix-Marseille 1971), (in French)

5.234 S.S. Tewari, Y. Jaluria: Calibration of constant-temperature hot-wire anemometers for very low velocities in air, Rev. Sci. Instrum. **61**, 3834–3845 (1990)

5.235 B.C. Khoo, Y.T. Chew, C.P. Lim: The flow between a rotating and a stationary disc: application to near-wall hot-wire calibration, Meas. Sci. Technol. **9**, 650–658 (1998)

5.236 R.E. De Haan: The dynamic calibration of a hot wire by means of a sound wave, Appl. Sci. Res. **24**, 335–353 (1971)

5.237 R.P. Dring, B. Gebhart: Hot-wire anemometer calibration for measurements at very low velocity, ASME J. Heat Transf. **91**, 241–244 (1969)

5.238 Y.T. Chew, S.M. Ha: The directional sensitivities of crossed and triple hot-wire probes, J. Phys. E Sci. Instrum. **21**, 613–620 (1988)

5.239 G. Comte-Bellot: Hot-wire and hot-film anemometers. In: *Measurements of Unsteady Fluid Dyn. Phen.*, ed. by B.E. Richards (Hemisphere, Washington 1977) pp. 123–162

5.240 R.M. Lueptow, K.S. Breuer, J.H. Haritonidis: Computer-aided calibration of X-probes using a look-up table, Exp. Fluids **6**, 115–118 (1988)

5.241 F.D. Johnson, H. Eckelmann: A variable angle method of calibration for X-probes applied to wall-bounded turbulent shear flow, Exp. Fluids **2**, 121–130 (1984)

5.242 Y.T. Chew, S.M. Ha: A. critical evaluation of the explicit data analysis algorithm for a crossed wire anemometer in highly turbulent isotropic flow, Meas. Sci. Technol. **1**, 775–781 (1990)

5.243 W.W. Willmarth, T.J. Bogar: Survey and new measurements of turbulent structure near the wall, Phys. Fluids **20S**, 9–21 (1977)

5.244 W.W. Willmarth: Geometric interpretation of the possible velocity vectors obtained with multiple sensors, Phys. Fluids **28**, 462–465 (1985)

5.245 P. Burattini, R.A. Antonia: The effect of different X-wire calibration schemes on some turbulence statistics, Exp. Fluids **38**, 80–89 (2005)

5.246 H.H. Bruun, C. Tropea: The calibration of inclined hot-wire probes, J. Phys. E Sci. Instrum. **18**, 405–413 (1985)

5.247 J.G. Kawall, M. Shokr, J.F. Keffer: A digital technique for the simultaneous measurement of streamwise and lateral velocities in turbulent flows, J. Fluid Mech. **133**, 83–112 (1983)

5.248 I.C. Lekakis, R.J. Adrian, B.G. Jones: Measurement of velocity vectors with orthogonal and non-orthogonal triple-sensor probes, Exp. Fluids **7**, 228–240 (1989)

5.249 L.W.B. Browne, R.A. Antonia, L.P. Chua: Calibration of X-probes for turbulent flow measurements, Exp. Fluids **7**, 201–208 (1989)

5.250 B. Lakshminarayana, R. Davino: Sensitivity of three sensor hot wire probe to yaw and pitch angle variation, J. Fluid Eng. Trans. ASME **110**, 120–122 (1988)

5.251 L. Pompeo, H. Thomann: Quadrupole hot-wire probes in a simulated wall flow, Exp. Fluids **14**, 145–152 (1993)

5.252 G. Pailhas, J. Cousteix: Méthodes d'exploitation des données d'une sonde anémométrique à quatre fils chauds, Rech. Aérospatiale **2**, 161–168 (1986)

5.253 K. Döbbeling, B. Lenze, W. Leuckel: Computer-aided calibration and measurements with a quadruple hotwire probe, Exp. Fluids **8**, 257–262 (1990)

5.254 M. Samet, S. Einav: A hot-wire technique for simultaneous measurements of instantaneous velocities in 3D flows, J. Phys. E Sci. Instrum. **20**, 683–690 (1987)

5.255 T.J. Gieseke, Y.G. Guezennec: An experimental approach to the calibration and use of triple hot-wire probes, Exp. Fluids **14**, 305–315 (1993)

5.256 G. Lemonis, T. Dracos: A new calibration and data reduction method for turbulence measurement by multihotwire probes, Exp. Fluids **18**, 319–328 (1995)

5.257 K. Döbbeling, B. Lenze, W. Leuckel: Basic considerations concerning the construction and usage of multiple hot-wire probes for highly turbulent three-dimensional flows, Meas. Sci. Technol. **1**, 924–933 (1990)

5.258 P. Lavoie, A. Pollard: Uncertainty analysis of four-sensor hot-wires and their data-reduction schemes used in the near field of a turbulent jet, Exp. Fluids **34**, 358–370 (2003)

5.259 J. Cousteix, J.C. Juillen: Jauges à fil chaud pour la mesure du frottement pariétal: réalisation, étalonnage, applications, Rech. Aérospatiale **3**, 207–218 (1982)

5.260 J.H. Haritonidis: The measurement of wall shear stress. In: *Advances in Fluid Mechanic Measurements*, Vol. 45, ed. by M. Gad-el-Hak (Springer, Berlin, Heidelberg 1989) pp. 229–261

5.261 W.J. Cook, T.H. Giddings, J.D. Murphy: Response of hot-element wall shear-stress gages in laminar oscillating flows, AIAA J. **26**, 706–713 (1988)

5.262 W.J. Cook: Response of hot-element wall shear stress gages in unsteady turbulent flows, AIAA J. **32**, 1464–1471 (1994)

5.263 J.B. Huang, F.K. Jiang, Y.C. Tai, C.M. Ho: A micro-electro-mechanical-system-based thermal shear-stress sensor with self-frequency compensation, Meas. Sci. Technol. **10**, 687–696 (1999)

5.264 H. Schlichting, K. Gersten: *Boundary Layer Theory* (McGraw-Hill, New York 2000)

5.265 W. Nitsche, C. Haberland: Ein vereinfachtes Eichverfahren für Hitzdrahtanemometer (Einzelpunkteichung), Z. Flugwiss. Weltraumforsch. **8**, 264–270 (1984), (in German)

5.266 G.L. Brown, R.F. Davey: The calibration of hot-films for skin friction measurement, Rev. Sci. Instrum. **42**, 1729–1731 (1971)

5.267 K.D. Cole, J.V. Beck: Conjugated heat transfer from a hot-film probe for transient air flow, J. Heat Transf. Trans. ASME **110**, 290–296 (1988)

5.268 J.K. Reichert, R.S. Azad: Wall shear stress measurement with a hot film in a variable temperature flow, Rev. Sci. Instrum. **48**, 341–345 (1977)

5.269 P.H. Alfredsson, A.V. Johansson, J.H. Haritonidis, H. Eckelmann: The fluctuating wall-shear stress and the velocity field in the viscous sublayer, Phys. Fluids **31**, 1026–1033 (1988)

5.270 R.I. Tanner: Theory of a thermal fluxmeter in a shear flow, J. Appl. Mech. **34**, 801–805 (1967)

5.271 G.R. Sarma, T. Moes: Demonstration of skin friction measurements featuring in situ estimation of conduction loss using constant voltage anemometers and surface hot-films, Rev. Sci. Instrum. **76**, 055102 (2005)

5.272 S.M. Mangalam: Real-time extraction of hydrodynamic flow characteristics using surface signatures, IEEE J. Oceanic Eng. **29**, 622–630 (2004)

5.273 J.D. Ruedi, H. Nagib, J. Österlund, P.A. Monkewitz: Evaluation of three techniques for wall-shear measurements in three-dimensional flows, Exp. Fluids **35**, 389–396 (2003)

5.274 M.J. Lighthill: The response of laminar skin friction and heat transfer to fluctuations in the stream velocity, Proc. R. Soc. London A, Vol. 224 (1954) 1–23

5.275 O. Desgeorges, T. Lee, F. Kafyeke: Multiple hot-film array calibration and skin friction measurement, Exp. Fluids **32**, 37–43 (2002)

5.276 A.M. Koppius, G.R.M. Trines: The dependence of hot-wire calibration on gas temperature at low Reynolds numbers, Int. J. Heat Mass Transf. **19**, 967–974 (1972)

5.277 R. Thünker, W. Nitsche, M. Swoboda: Hot-wire, hot-film and surface hot-film applications in strongly temperature loaded flows, ICIASF, Vol. 89 (1989)

5.278 L. Meyer: Calibration of a three-wire probe for measurements in nonisothermal flow, Exp. Therm. Fluid Sci. **5**, 260–267 (1992)

5.279 S.Г. Benjamin, C.A. Roberts: Measuring flow velocity at elevated temperature with a hot wire anemometer calibrated in cold flow, Int. J. Heat Mass Transf. **45**, 703–706 (2002)

5.280 J.M. Cimbala, W.J. Park: A direct hot-wire calibration technique to account for ambient temperature drift in incompressible flow, Exp. Fluids **8**, 299–300 (1990)

5.281 M. Kostka, V. Vasanta Ram: On the effects of fluid temperature on hot wire characteristics. Part 1: Results of experiments, Exp. Fluids **13**, 155–162 (1992)

5.282 V. Vasanta Ram: On the effects of fluid temperature on hot wire characteristics. Part 2: Foundations of a rational theory, Exp. Fluids **13**, 267–278 (1992)

5.283 K. Hollasch, B. Gebhart: Calibration of constant-temperature hot-wire anemometers at low velocities in water with variable fluid temperature, ASME J. Heat Transf. **94**, 17–22 (1972)

5.284 R.E. Drubka, J. Tan-atichat, H.M. Nagib: Analysis of temperature compensating circuits for hot-wires and hot-films, DISA Inf. **22**, 5–14 (1977)

5.285 G.R. Sarma, G. Comte-Bellot: Automated constant voltage anemometer for measurements with fluid temperature drifts, Rev. Sci. Instrum. **73**, 1313–1317 (2002)

5.286 G.E. Truzzi, G.R. Sarma, N. Chokani: Constant voltage anemometer operated hot wire at subsonic speeds over wide overheats in unstady flows, Rev. Sci. Instrum. **73**, 4363–4368 (2002)

5.287 S. Corrsin: Extended applications of the hot-wire anemometer, Rev. Sci. Instrum. **18**, 469–471 (1947)

5.288 T.T. Yeh, C.W. Van Atta: Spectral transfer of scalar and velocity fields in heated-grid turbulence, J. Fluid Mech. **58**, 233–261 (1973)

5.289 L. Fulachier: Hot-wire measurements in low speed heated flows, Proc. Dyn. Flow Conf. DISA, ed. by B.W. Hansen (1979) 465–487

5.290 P. Paranthoen, C. Petit, J.C. Lecordier: The effect of the thermal prong-wire interaction on the response of a cold wire in gaseous flows (air, argon, helium), J. Fluid Mech. **124**, 457–473 (1982)

5.291 L.W.B. Browne, R.A. Antonia, A.J. Chambers: Effect of the separation between cold wires on the spatial derivatives of temperature in a turbulent flow, Bound.-Lay. Meteorol. **27**, 129–139 (1983)

5.292 L. Mydlarski, Z. Warhaft: Passive scalar statistics in high-Peclet-number grid turbulence, J. Fluid Mech. **358**, 135–175 (1998)

5.293 M. Tagawa, K. Kato, Y. Ohta: Response compensation of fine-wire temperature sensors, Rev. Sci. Instrum. **76**, 094904 (2005)

5.294 P. Mestayer, P. Chambaud: Some limitations to measurements of turbulence micro-structure with hot and cold wires, Bound.-Lay. Meteorol. **16**, 311–329 (1979)

5.295 S.-C. Lin, S.-C. Lin: Study of strong temperature mixing in subsonic grid turbulence, Phys. Fluids **16**, 1587–1598 (1971)

5.296 S. Tavoularis: A circuit for the measurement of instantaneous temperature in heated turbulent flows, J. Phys. E Sci. Instrum. **11**, 21–23 (1978)

5.297 R.A. Antonia, L.W.B. Browne, A.J. Chambers: Determination of time constants of cold wires, Rev. Sci. Instrum. **52**, 1382–1385 (1981)

5.298 J.C. LaRue, T. Deaton, C.H. Gibson: Measurement of high-frequency turbulent temperature, Rev. Sci. Instrum. **46**, 757–764 (1975)

5.299 L. Fulachier, R. Dumas: Spectral analogy between temperature and velocity fluctuations in a turbulent boundary layer, J. Fluid Mech. **77**, 257–277 (1976)

5.300 G. Huelsz, E. Ramos: Temperature measurements inside the oscillatory boundary layer produced by acoustic waves, J. Acoust. Soc. Am. **103**, 1532–1537 (1998)

5.301 L.S.G. Kovasznay, M.S. Uberoi, S. Corrsin: The transformation between one- and three- dimensional power spectra for an isotropic scalar fluctuation field, Phys. Rev. **76**, 1263–1264 (1949)

5.302 A.E. Perry, A.J. Smits, M.S. Chong: The effects of certain low frequency phenomena on the calibration of hot wires, J. Fluid Mech. **90**, 415–431 (1979)

5.303 P. Paranthoen, J.C. Lecordier: Mesures de température dans les écoulements turbulents, Rev. Gen. Therm. **35**, 283–308 (1996)

5.304 D.W. Artt, A. Brown: The simultaneous measurement of velocity and temperature, J. Phys. E Sci. Instrum. **4**, 72–74 (1971)

5.305 G. Comte-Bellot, J. Mathieu: Sur la détermination expérimentale des coefficients de sensibilité aux fluctuations de vitesse et de température des anémomètres à fil chaud, CR Acad. Sci. Paris **246**, 3219–3222 (1958)

5.306 J.P. Bonnet, M.A. Knani: Calibration and use of inclined hot-wires in a supersonic turbulent wake, Exp. Fluids **6**, 179–188 (1988)

5.307 S. Candel: *Mécanique des fluides* (Dunod, Paris 1995)

5.308 W. Konrad, A.J. Smits: Reynolds stress measurements in a three-dimensional supersonic turbulent boundary layer, Proc. ASME Symp. Therm. Anemometry, Vol. FED 167 (1993) 219–230

5.309 E.M. Fernando, J.F. Donovan, A.J. Smits: The calibration and operation of a constant-temperature anemometer crossed-wire probe in supersonic flow, Proc. ASME Symp. Thermal Anemometry, Vol. FED 53 (1987) 43–49

5.310 A.J. Smits, K.C. Muck: Constant temperature hot-wire anemometer practice in supersonic flows. Part 2: The inclined wire, Exp. Fluids **2**, 33–41 (1984)

5.311 E. Reshotko, I.E. Beckwith: Compressible laminar boundary layer over a yawed infinite cylinder with heat transfer and arbitrary Prandtl number, NACA Tech. Rep. **1379** (1957)

5.312 W.C. Rose, D.A. Johnson: Turbulence in a shock-wave boundary-layer interaction, AIAA J. **13**, 884–889 (1975)

5.313 D.A. Johnson, W.C. Rose: Laser velocimeter and hot-wire anemometer comparison in a supersonic boundary layer, AIAA J. **13**, 512–515 (1975)

5.314 P. Bradshaw: Compressible turbulent shear layers, Ann. Rev. Fluid Mech. **9**, 33–54 (1977)

5.315 E.F. Spina, A.J. Smits, S.K. Robinson: The physics of supersonic turbulent boundary layers, Ann. Rev. Fluid Mech. **26**, 287–319 (1994)

5.316 F.R. Payne, J.L. Lumley: One-dimensional spectra derived from an airborne hot-wire anemometer, Q. J. R. Meteorol. Soc. **92**, 397–401 (1966)

5.317 C.M. Sheih, H. Tennekes, J.L. Lumley: Airborne hot-wire measurements of the small-scale structure of atmospheric turbulence, Phys. Fluids **14**, 201–215 (1971)

5.318 N.R. Panchapakesan, J.L. Lumley: Turbulence measurements in axisymmetric jets of air and helium. Part 1. Air jet, J. Fluid Mech. **246**, 197–223 (1993)

5.319 N.R. Panchapakesan, J.L. Lumley: Turbulence measurements in axisymmetric jets of air and helium. Part 2. Helium jet, J. Fluid Mech. **246**, 225–247 (1993)

5.320 R.M. Kelso, T.T. Lim, A.E. Perry: A novel flying hot-wire system, Exp. Fluids **16**, 181–186 (1994)

5.321 D. Coles, A.J. Wadcock: Flying-hot-wire study of flow past an NACA 4412 airfoil at maximum lift, AIAA J. **17**, 321–329 (1979)

5.322 B. Cantwell, D. Coles: An experimental study of entrainment and transport in the turbulent near wake of a circular cylinder, J. Fluid Mech. **136**, 321–374 (1983)

5.323 M.D. Walker, M.R. Maxey: A whirling hot-wire anemometer with optical data transmission, J. Phys. E Sci. Instrum. **18**, 516–521 (1985)

5.324 H.J. Hussein, S.P. Capp, W.K. George: Velocity measurements in a high-Reynolds-number, momentum-conserving, axisymmetric, turbulent jet, J. Fluid Mech. **258**, 31–75 (1994)

5.325 B.E. Thomson, J.H. Whitelaw: Flying hot-wire anemometry, Exp. Fluids **2**, 47–56 (1984)

5.326 H.H. Al-Kayiem, H.H. Bruun: Evaluation of a flying X hot-wire probe system, Meas. Sci. Technol. **2**, 374–380 (1991)

5.327 L.J. Otten, A.L. Pavel, W.E. Finley, W.C. Rose: A survey of recent atmospheric turbulence measurements from a subsonic aircraft, AIAA Paper **81-0298** (1981)

5.328 M. Jaroch: Development and testing of pulsed-wire probes for measuring fluctuating quantities in highly turbulent flows, Exp. Fluids **3**, 315–322 (1985)

5.329 I.P. Castro, A. Haque: The stucture of a turbulent shear layer bounding a reversed flow region, J. Fluid Mech. **179**, 439–468 (1987)

5.330 I.P. Castro, M. Dianat: Pulsed wire velocity anemometry near walls, Exp. Fluids **8**, 343–352 (1990)

5.331 L.J.S. Bradbury, I.P. Castro: A pulsed-wire technique for velocity measurements in highly turbulent flows, J. Fluid Mech. **49**, 657–691 (1971)

5.332 R.V. Westphal, J.K. Eaton, J.P. Johnston: A new probe for measurement of velocity and wall shear stress in unsteady, reversing flow, J. Fluid Eng. Trans. ASME **103**, 478–482 (1981)

5.333 M. Jaroch, A. Dahm: A new pulsed-wire probe for measuring the Reynolds stresses in the plane containing the main shear direction of a turbulent shear flow, J. Phys. E Sci. Instrum. **21**, 1085–1094 (1988)

5.334 P.M. Handford, P. Bradshaw: The pulsed-wire anemometer, Exp. Fluids **7**, 125–132 (1989)

5.335 I.P. Castro, B.S. Cheun: The measurement of Reynolds stresses with a pulsed-wire anemometer, J. Fluid Mech. **118**, 41–58 (1982)

5.336 L.J.S. Bradbury: Measurements with a pulsed-wire and a hot-wire anemometer in the highly turbulent wake of a normal flat plate, J. Fluid Mech. **77**, 473–497 (1976)

5.337 B. Venås, H. Abrahamsson, P.Å. Krogstad, L. Löfdahl: Pulsed hot-wire measurements in two- and three-dimensional wall jets, Exp. Fluids **27**, 210–218 (1999)

5.338 P. Calvet, F. Liousse: Mesures locales de températures, pressions et vitesses au moyen de capteurs thermorésistants chauffés par impulsions, Rev. Gen. Therm. **114**, 527–542 (1971)

5.339 E. Mathioulakis, M. Grignon, J.G. Poloniecki: A pulsed-wire technique for velocity and temperature measurements in natural convection flows, Exp. Fluids **18**, 82–86 (1994)

5.340 M. Grignon, E. Mathioulakis, P. Ngae, J.G. Poloniecki: Pulsed-wire technique for velocity measurements in natural convection flows – a numerical optimisation tool, Int. J. Heat Mass Transf. **41**, 3121–3129 (1998)

5.341 D. Li, K.S. Gow: The development of a pulsed multi-element hot-wire anemometer, Meas. Sci. Technol. **6**, 1175–1185 (1995)

5.342 J.F. Foss, D.G. Bohl, T.J. Hicks: The pulse width modulated-constant temperature anemometer, Meas. Sci. Technol. **7**, 1388–1395 (1996)

5.343 J.F. Foss, T.J. Hicks: Processing time series data (velocity, shear stress) from the pulse width modulated-constant temperature anemometer (PWM-CTA). In: *19th ICIASF IEEE*, ed. by M.P. Wernet (IEEE, Piscataway 2001) pp. 448–453

5.344 T.J. Hicks, J.K. Schwannecke, M.J. Norconk, A.M. Hellum, J.F. Foss: The evolved 100 kHz PWM-CTA. In: *20th ICIASF*, ed. by R.H. Engler (IEEE, Piscataway 2003)

5.345 J.-L. Breton: *Dissipation thermique de fils chauds soumis à des impulsions calorifique*. Thèse Spécialité (Univ. Paul Sabatier, Toulouse 1972)

5.346 J.F. Foss, R.C. Haw: Transverse vorticity measurements using a compact array of four sensors. In: *The Heuristics of Thermal Anemometry*, Vol. FED97, ed. by D.E. Stock, S.A. Sherif, A.J. Smits (ASME, New York 1990) pp. 71–76

5.347 S.-C. Morris: *The Velocity and Vorticity Fields of a Single Stream Shear Layer*, Ph.D. Thesis (Michigan State University, East Lansing 2002)

5.348 C. Crowe, M. Sommerfeld, Y. Tsuji: *Multiphase Flows with Droplets and Particles* (CRC, Boca Raton 1998)

5.349 A.T. Hjemfelt, L.F. Mockros: Motion of discrete particles in a turbulent fluid, Appl. Sci. Res. **16**, 149–161 (1966)

5.350 B.T. Chao: Turbulent transport behavior of small particles in a turbulent fluid, Österr. Ingenieurarch. **19**, 7 (1964)

5.351 R. Mei: Velocity fidelity of flow tracer particles, Exp. Fluids **22**, 1–13 (1996)

5.352 A. Melling: Tracer particles and seeding for particle image velocimetry, Meas. Sci. Technol. **8**, 1406–1416 (1997)

5.353 F. Scarano, B.W. van Oudheusden: Planar velocity measurements of a two-dimensional compressible wake, Exp. Fluids **34**, 430–441 (2003)

5.354 R.P. Dring: Sizing criteria for laser anemometry particles, J. Fluid Eng. Trans. ASME **104**, 15–17 (1982)

5.355 G. Tedeschi, H. Gouin, M. Elena: Motion of tracer particles in supersonic flows, Exp. Fluids **26**, 288–296 (1999)

5.356 H.-E. Albrecht, M. Borys, N. Damaschke, C. Tropea: *Laser Doppler and Phase Doppler Measurement Techniques* (Springer, Berlin, Heidelberg 2003)

5.357 C.D. Meinhart, S.T. Wereley, J.G. Santiago: PIV measurements of a microchannel flow, Exp. Fluids **27**, 414–419 (1999)

5.358 G. Rottenkolber, R. Meier, O. Schäfer, S. Wachter, K. Dullenkopf, S. Wittig: Phase discrimination inside a spray: LDV measurements using fluorescent seeding particles (FLDV). In: *Laser Techniques for Fluid Mechanics*, ed. by R.J. Adrian, D.F.G. Durao, F. Durst, M.V. Heitor, M. Maedo, J.H. Whiteclaw (Springer, Berlin, Heidelberg 2002) pp. 511–526

5.359 C. Helsper, L. Mölter: Erzeugung von Prüfaerosolen für die Kalibrierung von optischen partikelmessverfahren nach VDI 3491, Tech. Mess. **56**, 229–234 (1989), (in German)

5.360 D. Dabiri, M. Gharib: Digital particle image thermometry: The method and implementation, Exp. Fluids **11**, 77–86 (1991)

5.361 A. Günther, P.R. von Rohr: Influence of the optical configuration on temperature measurements with fluid-dispersed TLCs, Exp. Fluids **32**, 533–541 (2002)

5.362 G. Rottenkolber, R. Meier, O. Schäfer, S. Wachter, K. Dullenkopf, S. Wittig: Combined PDA and LDV measurements: Phase discrimination inside a spray using fluorescent seeding particles, Proc. 10th Int. Symp. Appl. Laser Technol. Fluid Mech. Lisbon (2000) , paper 31.1

5.363 D. Sinclair, V.K. LaMer: Light scattering as a measure of particle size in aerosols: The production of monodisperse aerosols, Chem. Rev. **44**, 245–267 (1949)

5.364 A. Lefebvre: *Atomisation and Sprays* (Hemisphere, New York 1989)

5.365 W.-H. Echols, J.-A. Young: Studies of air-operated aerosol generators, NRL Rep. *5929* (US Naval Res. Lab., Washington 1963)

5.366 : Proc. London Math. Soc., Vol. 10 (1878) 4

5.367 G. Brenn, F. Durst, C. Tropea: Monodisperse sprays for various purposes – their production and characteristics, Part. Part. Syst. Charact. **13**, 179–185 (1996)

5.368 N. Paone, G.M. Revel, E. Nino: Velocity measurement in high turbulent premixed flames by a PIV measurement system, Proc. 8th Int. Symp. Appl. Laser Technol. Fluid Mech. Lisbon (1996) , , paper 3.4

5.369 M. Glass, I.M. Kennedy: An improved seeding method for high temperature laser Doppler velocimetry, Combust. Flame **49**, 155–162 (1977)

5.370 D.J. Anderson, C.A. Greated, J.D.C. Jones, G. Nimmo, S. Wiseall: Fibre optic PIV studies in an industrial combustor, Proc. 8th Int. Symp. Appl. Laser Technol. Fluid Mech. Lisbon (1996) , , paper 18.4

5.371 J.K. Agarwal, E.M. Johnson: Generating aerosol for laser velocimeter seeding, TSI Quart. **VII** (1981)

5.372 F. Scarano, B.W. van Oudheusden, W.J. Bannink, M. Bsibsi: Experimental investigation of supersonic base flow plume interaction by means of particle image velocimetry, 5th Eur. Symp. Aerothermodyn. Space Vehicles Cologne (2004)

5.373 Y. Yeh, H.Z. Cummins: Localized fluid flow measurements with a He-Ne laser spectrometer, Appl. Phys. Lett. **4**, 176–179 (1964)

5.374 B. Lehmann: Geschwindigkeitsmessung mit Laser-Doppler-Anemometer Verfahren, Wissenschaftliche Berichte AERG-Telefunken **41**, 141–145 (1968), (in German)

5.375 H.D. vom Stein, H.J. Pfeifer: A Doppler difference method for velocity measurements, Metrologia **5**, 59–61 (1969)

5.376 B. Lehmann, H. Nobach, C. Tropea: Measurement of acceleration using the laser Doppler technique, Meas. Sci. Technol. **13**(9), 1367–1381 (2002)

5.377 W. Feller: *An Introduction to Probability Theory and Its Applications*, Vol. I (Wiley, New York 1971)

5.378 P. Buchhave, W.K. George Jr, J.L. Lumley: The measurement of turbulence with the laser-Doppler anemometer, Ann. Rev. Fluid. Mech. **11**, 443–504 (1979)

5.379 L.H. Benedict, H. Nobach, C. Tropea: Estimation of turbulent velocity spectra from laser Doppler data, Meas. Sci. Technol. **11**, 1089–1104 (2000)

5.380 H.R.E. van Maanen, H. Nobach, L.H. Benedict: Improved estimator for the slotted autocorrelation function of randomly sampled LDA data, Meas. Sci. Technol. **10**, L4–L7 (1999)

5.381 H.E. Nobach Müller, C. Tropea: Efficient estimation of power spectral density from laser Doppler anemometer data, Exp. Fluids **24**, 499–509 (1998)

5.382 W.T. Mayo Jr, M.T. Shay, S. Ritter: Digital estimation of turbulence power spectra from burst counter LDV data, Proc. 2nd Int. Workshop on Laser Velocimetry (Purdue Univ., 1974) 16–26

5.383 H.R.E. van Maanen, M.J. Tummers: Estimation of the auto correlation function of turbulent velocity fluctuations using the slotting technique with local normalization, Proc. 8th Int. Symp. Appl. Laser Technol. Fluid Mech. Lisbon, ed. by R.J. Adrian, D.F.G. Durao, F. Durst, M.V. Heitor, M. Maeda, J.H. Whitelaw (1996) , , paper 36.4

5.384 R.J. Adrian, C.S. Yao: Power spectra of fluid velocities measured by laser Doppler velocimetry, Exp. Fluids **5**, 17–28 (1987)

5.385 E. Müller, H. Nobach, C. Tropea: LDA signal reconstruction: application to moment and spectral estimation, Proc. 7th Int. Symp. Appl. Laser Technol. Fluid Mech. Lisbon, ed. by R.J. Adrian, D.F.G. Durao, F. Durst, M.V. Heitor, M. Maedo, J.H. Whiteclaw (1994) , , paper 23.2

5.386 H. Nobach, E. Müller, C. Tropea: Refined reconstruction techniques for LDA data analysis, Proc. 8th Int. Symp. Appl. Laser Technol. Fluid Mech. Lisbon, ed. by R.J. Adrian, D.F.G. Durao, F. Durst, M.V. Heitor, M. Maedo, J.H. Whiteclaw (1996) , , paper 36.2

5.387 W. Merzkirch: *Flow Visualization* (Academic, Orlando 1987)

5.388 M. Van Dyke: *An Album of Fluid Motion* (Parabolic, Stanford 1982)

5.389 G.L. Brown, A. Roshko: On density effects and large structure in turbulent mixing layers, J. Fluid Mech. **64**, 775–816 (1974)

5.390 B.J. Cantwell: Organized motion in turbulent flow, Ann. Rev. Fluid Mech. **13**, 457–515 (1981)

5.391 A.K.M.F. Hussain: Coherent structures and turbulence, J. Fluid Mech. **173**, 303–356 (1986)

5.392 L. Hesselink: Digital image processing in flow visualization, Ann. Rev. Fluid Mech. **20**, 421–486 (1988)

5.393 R.J. Adrian: Particle-imaging techniques for experimental fluid mechanics, Ann. Rev. Fluid Mech. **23**, 261–304 (1991)

5.394 R.J. Adrian: Twenty years of particle image velocimetry, Exp. Fluids **39**, 159–169 (2005)

5.395 H.T. Kim, S.J. Kline, W.C. Reynolds: The production of turbulence near a smooth wall in a turbulent boundary layer, J. Fluid Mech. **50**, 133 (1971)

5.396 J. Westerweel: *Particle Image Velocimetry – Theory and Application* (Delft Univ. Press, Delft 1993)

5.397 M. Stanislas, K. Okamoto, C.J. Kähler, J. Westerweel: Main results of the second international PIV challenge, Exp. Fluids **39**, 170–191 (2005)

5.398 W. Kühn, J. Kompenhans, J.C. Monnier: Full scale PIV test in an industrial facility. In: *Particle Image Velocimetry – Progress Towards Industrial Application*, ed. by M. Stanislas, J. Kompenhans, J. Westerweel (Kluwer, Dordrecht 2000) pp. 91–150

5.399 F. Scarano, C. van Wijk, L.L.M. Veldhuis: Traversing field of view and AR-PIV for mid-field wake vortex investigation in a towing tank, Exp. Fluids **33**, 950–961 (2002)

5.400 T. Hori, J. Sakakibara: High-speed scanning stereoscopic PIV for 3D vorticity measurement in liquids, Meas. Sci. Technol. **15**, 1067–1078 (2004)

5.401 F. Scarano, G.E. Elsinga, E. Bocci, B.W. van Oudheusden: Investigation of 3-D coherent structures in the turbulent cylinder wake using Tomo-PIV, 13th Int. Symp. Appl. Laser Tech. Fluid Mech., Lisbon (2006)

5.402 C. Brucker: 3-D scanning-particle-image-velocimetry: Technique and application to a spherical cap wake flow, Appl. Sci. Res. **56**, 157–79 (1996)

5.403 S.F. Herrmann, K.D. Hinsch: Light-in-flight holographic particle image velocimetry for wind-tunnel applications, Meas. Sci. Technol. **15**, 613–621 (2004)

5.404 G.E. Elsinga, B. Wieneke, F. Scarano, B.W. van Oudheusden: Tomographic particle image velocimetry, Exp. Fluids **41**, 933–947 (2006)

5.405 J. Haertig, M. Havermann, C. Rey, A. George: Particle image velocimetry in Mach 3.5 and 4.5 shock-tunnel flows, AIAA J. **40**, 1056 (2002)

5.406 J.R. Hove, R.W. Köster, A.S. Forouhar, G. Acevedo-Bolton, S.E. Fraser, M. Gharib: Intracardiac fluid forces are an essential epigenetic factor for embryonic cardiogenesis, Nature **421**, 172–177 (2003)

5.407 P. Vennemann: In vivo micro particle image velocimetry measurements of blood-plasma in the embryonic avian heart, J. Biomech. **39**, 1191–1200 (2006)

5.408 C. Fukushima, L. Aanen, J. Westerweel: Investigation of the mixing process in an axisymmetric turbulent jet using PIV and LIF. In: *Laser Techniques for Fluid Mechanics*, ed. by R.J. Adrian, D.F.G. Durao, M.V. Heitor, M. Maeda, C. Tropea, J.H. Whitelaw (Springer, Heidelberg, Berlin 2002) pp. 339–356

5.409 E. Delnoij, J. Westerweel, N.G. Deen, J.A.M. Kuipers, W.P.M. van Swaaij: Ensemble correlation PIV applied to bubble plumes rising in a bubble column, Chem. Eng. Sci. **54**, 5159–5171 (1999)

5.410 K.T. Kiger, C. Pan: PIV technique for the simultaneous measurement of dilute two-phase flows, J. Fluid Eng. **122**, 811–818 (2000)

5.411 D.A. Khalitov, E.K. Longmire: Simultaneous two-phase PIV by two-parameter phase discrimination, Exp. Fluids **32**, 252–268 (2002)

5.412 R. Lindken, W. Merzkirch: A novel PIV technique for measurements in multiphase flows and its application to two-phase bubbly flows, Exp. Fluids **33**, 814–825 (2002)

5.413 C. Poelma, J. Westerweel, G. Ooms: Turbulence statistics from optical whole field measurements in particle-laden turbulence, Exp. Fluids **40**, 347–363 (2006)

5.414 J. Westerweel: Fundamentals of digital particle image velocimetry, Meas. Sci. Technol. **8**, 1379 (1997)

5.415 J.W. Goodman: *Introduction to Fourier Optics* (McGraw-Hill, New York 1968)

5.416 M.G. Olsen, R.J. Adrian: Out-of-focus effects on particle image visibility and correlation in microscopic particle image velocimetry, Exp. Fluids **29**, S166–S147 (2000)

5.417 A.K. Prasad: Stereoscopic particle image velocimetry, Exp. Fluids **29**, 103 (2000)

5.418 C. Willert: Stereoscopic digital particle image velocimetry for application in wind tunnel flows, Meas. Sci. Technol. **8**, 353 (1997)

5.419 C.W.H. van Doorne, J. Westerweel: Measurement of laminar, transitional and turbulent pipe flow using stereoscopic-PIV, Exp. Fluids **42**, 259–279 (2007)

5.420 J. Westerweel, J. van Oord: Stereoscopic PIV measurement in a turbulent boundary layer. In: *Particle Image Velocimetry: Progress Towards Industrial Application*, ed. by M. Stanislas, J. Kompenhans, J. Westerweel (Kluwer, Dordrecht 2000) pp. 459–478

5.421 S.M. Soloff: Distortion compensation for generalized stereoscopic particle image velocimetry, Meas. Sci. Technol. **8**, 1441 (1997)

5.422 S.J.M. Coudert, J.P. Schon: Back-projection algorithm with misalignment corrections for 2D3C stereoscopic PIV, Meas. Sci. Technol. **12**, 1371 (2001)

5.423 B. Wieneke: Stereo-PIV using self-calibration on particle images, Exp. Fluids **39**, 267–280 (2005)

5.424 A. Rosenfeld, A.C. Kak: *Digital Picture Processing*, 2nd edn. (Academic, Orlando 1982)

5.425 A.K. Jain: *Fundamentals of Digital Image Processing* (Prentice-Hall, Englewood Cliffs 1989)

5.426 A.V. Oppenheim, A.S. Willsky, S. Hamid, S.H. Nawab: *Signals and Systems* (Prentice-Hall, Englewood Cliffs 1983)

5.427 J. Westerweel: Effect of sensor geometry on the performance of PIV interrogation. In: *Laser Techniques Applied to Fluid Mechanics*, ed. by R.J. Adrian, D.F.G. Durao, F. Durst, M.V. Heitor, M. Maedo, J.H. Whitelaw (Springer, Berlin, Heidelberg 2000) pp. 37–55

5.428 K.E. Meyer, J. Westerweel: Advection velocities of flow structures estimated from particle image velocimetry measurements in a pipe, Exp. Fluids **29**, S237 (2000)

5.429 C. Freek, J.M.M. Sousa, W. Hentschel, W. Merzkirch: On the accuracy of a MJPEG-based digital image compression PIV system, Exp. Fluids **27**, 310 (1999)

5.430 Y.G. Guezennec, N. Kiritsis: Statistical investigation of errors in particle image velocimetry, Exp. Fluids **10**, 138 (1990)

5.431 R.D. Keane, R.J. Adrian, Y. Zhang: Super-resolution particle imaging velocimetry, Meas. Sci. Technol. **6**, 754–768 (1995)

5.432 R.J. Adrian: Statistical properties of particle image velocimetry measurements in turbulent flow. In: *Laser Anemometry in Fluid Mechanics*, Vol. III, ed. by D.F.G. Durao, R.J. Adrian, T. Asanuma, F. Durst, J.H. Whitelaw (1988) pp. 115–129

5.433 J. Westerweel: Theoretical analysis of the measurement precision in particle image velocimetry, Exp. Fluids **29**, S3 (2000)

5.434 R.D. Keane, R.J. Adrian: Optimization of particle image velocimeters. Part I: Double-pulsed systems, Meas. Sci. Technol. **1**, 1202 (1990)

5.435 H.T. Huang, H.F. Fielder, J.J. Wang: Limitation and improvement of PIV, Part II. particle image distortion, a novel technique, Exp. Fluids **15**, 263–273 (1993)

5.436 R.D. Keane, R.J. Adrian: Optimization of particle image velocimeters. Part II: Multiple-pulsed systems, Meas. Sci. Technol. **2**, 963 (1991)

5.437 R.D. Keane, R.J. Adrian: Theory of cross-correlation analysis of PIV images, Appl. Sci. Res. **49**, 191 (1992)

5.438 J. Westerweel: Measurement of fully-developed turbulent pipe flow with digital particle image velocimetry, Exp. Fluids **20**, 165 (1996)

5.439 J. Westerweel: The effect of a discrete window offset on the accuracy of cross-correlation analysis of digital PIV recordings, Exp. Fluids **23**, 20 (1997)

5.440 O. Ronneberger: *Messung aller drei Geschwindigkeitskomponenten mit Hilfe der Particle Image Velocimetry mittels einer Kamera und zweier paralleler Lichtschnitte*, Diploma Thesis (Georg-August-Universität, Göttingen 1998)

5.441 E.O. Brigham: *The Fast Fourier Transform and Its Applications* (Prentice-Hall, Englewood Cliffs 1988)

5.442 C.E. Willert, M. Gharib: Digital particle image velocimetry, Exp. Fluids **10**, 181 (1991)

5.443 M.B. Priestley: *Spectral Analysis and Time Series* (Academic, San Diego 1992)

5.444 A.K. Prasad, R.J. Adrian, C.C. Landreth, P.W. Offutt: Effect of resolution on the speed and accuracy of particle image velocimetry interrogation, Exp. Fluids **13**, 105–116 (1992)

5.445 K.T. Christensen: The influence of peak-locking errors on turbulence statistics computed from PIV ensembles, Exp. Fluids **36**, 484 (2004)

5.446 A.M. Fincham, G. Delerce: Advanced optimization of correlation imaging velocimetry algorithms, Exp. Fluids **29**, S013–S022 (2000)

5.447 C.E. Willert: The fully digital evaluation of photographic PIV recordings, Appl. Sci. Res. **56**, 79 (1996)

5.448 N. Bobroff: Position measurement with a resolution and noise-limited instrument, Rev. Sci. Instrum. **57**, 1152 (1986)

5.449 D.P. Hart: Sparse array image correlation, Proc. 8th Int. Symp. Appl. Laser Tech. Fluid Mech. Lisbon (1996)

5.450 J.M. Foucaut, B. Miliat, N. Perenne, M. Stanislas: Characterization of different PIV algorithms using the EUROPIV synthetic image generator and real images from a turbulent boundary layer. In: *Particle Image Velocimetry: Recent Improvements*, ed. by M. Stanislas, J. Westerweel, J. Kompenhans (Springer, Heidelberg, Berlin 2004) pp. 163–185

5.451 C.D. Meinhart, S.T. Wereley, J.G. Santiago: A PIV algorithm for estimating time-averaged velocity fields, J. Fluid Eng. **122**, 285–289 (2000)

5.452 D.P. Hart: PIV error correction, Exp. Fluids **29**, 13–22. (2000)

5.453 M.P. Wernet: Symmetric phase only filtering: a new paradigm for DPIV data processing, Meas. Sci. Technol. **16**, 601–618 (2005)

5.454 F. Scarano: Iterative image deformation methods in PIV, Meas. Sci. Technol. **13**, R1–R19 (2002)

5.455 B. Lecordier: Etude de l'interaction de la propagation d'une flamme premelangee avec le champ aerodynamique, par association de la tomographie Laser et de la Velocimetrie par Images de particules. These de doctorat (Universite de Rouen, Rouen 1997)

5.456 S.T. Wereley, C.D. Meinhart: Second-order accurate particle image velocimetry, Exp. Fluids **31**, 258 (2001)

5.457 P.T. Tokumaru, P.E. Dimotakis: Image correlation velocimetry, Exp. Fluids **19**, 1–15 (1995)

5.458 F. Scarano: A super-resolution PIV interrogation approach by means of velocity 2nd derivatives correlation, Meas. Sci. Technol. **15**, 475–486 (2004)

5.459 F.F.J. Schrijer, F. Scarano: Effect of predictor-corrector filtering on the stability and spatial resolution of iterative PIV interrogation, Exp. Fluids, (in press) (2007)

5.460 F. Scarano, F.F.J. Schrijer: Effect of predictor filtering on the stability and spatial resolution of iterative PIV interrogation, 5th International Symposium on PIV, Caltech Pasadena (2005)

5.461 J. Nogueira, A. Lecuona, P.A. Rodriguez: Local field correction PIV, implemented by means of simple algorithms, and multigrid versions, Meas. Sci. Technol. **12**, 1911 (2001)

5.462 J. Westerweel, F. Scarano: A universal detection criterion for the median test, Exp. Fluids **39**(6), 1096–1100 (2005)

5.463 J. Westerweel: Efficient detection of spurious vectors in particle image velocimetry data sets, Exp. Fluids **16**, 236–247 (1994)

5.464 L. Lourenco, A. Krothapalli: On the accuracy of velocity and vorticity measurements with PIV, Exp. Fluids **18**, 421–428 (1995)

5.465 S. Abrahamson, S. Lonnes: Uncertainty in calculating vorticity from 2D velocity fields using circulation and least-squares approaches, Exp. Fluids **20**, 10–20 (1995)

5.466 A. Fouras, J. Soria: Accuracy of out of plane vorticity measurements derived from in-plane velocity field data, Exp. Fluids **25**, 409–430 (1998)

5.467 M. Raffel, C. Willert, S. Wereley, J. Kompenhans: *Particle Image Velocimetry – a practical guide* (Springer, Berlin, Heidelberg 2007)

5.468 J.M. Foucaut, M. Stansislas: Some considerations on the accuracy and frequency response of some derivative filters applied to particle image velocimetry fields, Meas. Sci. Technol. **13**, 1058–1071 (2002)

5.469 G.K. Batchelor: *An Introduction to Fluid Mechanics* (Cambridge Univ. Press, Cambridge 1967)

5.470 C.C. Landreth, R.J. Adrian: Measurement and refinement of velocity data using high image density analysis in particle image velocimetry. In: *Applications of Laser Anemometry to Fluid Mechanics*, ed. by D.F.G. Durao, R.J. Adrian, M. Maeda, F. Durst, J.H. Whitelaw (Springer, Berlin 1990) pp. 484–497

5.471 C.C. Landreth, R.J. Adrian: Impingement of a low Reynolds number turbulent circular jet onto a flat plate at normal incidence, Exp. Fluids **9**, 74–84 (1990)

5.472 J.N. Forkey, W.R. Lempert, R.B. Miles: Corrected and calibrated I2 absorption model at frequency-doubled Nd:YAG laser wavelengths, Appl. Opt. **36**(27), 6729–6738 (1997)

5.473 H. Komine: System for measuring velocity field of fluid flow utilizing a laser-Doppler spectral image converter, US Patent 4,919,536 (1990)

5.474 H. Komine, S.J. Brosnan, A.B. Litton, E.A. Stappaerts: Real-time Doppler global velocimetry, AIAA Paper, Vol. 91-0337 (1991)

5.475 J.F. Meyers, H. Komine: Doppler global velocimetry – A new way to look at velocity. In: *Laser Anemometry: Advances and Applications*, ed. by A. Dybbs, B. Ghorashi (ASME, New York 1991)

5.476 J.F. Meyers: Doppler global velocimetry – The next generation?, AIAA Paper, Vol. 92-3897 (1992)

5.477 J.W. Lee, J.F. Meyers, A.A. Cavone, K.E. Suzuki: Doppler global velocimetry measurements of the vortical flow above an F/A-18, AIAA Paper, Vol. 93-0414 (1993)

5.478 J.F. Meyers: Development of Doppler global velocimetry as a flow diagnostics tool, Measurement in Fluids and Combustion Systems, Special Issue, Meas. Sci. Technol. **6**(6), 769–783 (1995)

5.479 J.F. Meyers, J.W. Lee, R.J. Schwartz: Characterization of measurement error sources in Doppler global velocimetry, Meas. Sci. Technol. **12**, 1–12 (2001)

5.480 M.S. Reinath: Doppler global velocimeter development for large wind tunnels, Meas. Sci. Technol. **12**, 432–441 (2001)

5.481 R.L. McKenzie: Planar Doppler velocimetry for large-scale wind tunnel applications, AGARD Fluid Dynamics Panel 81st Meeting and Symposium on Advanced Aerodynamic Measurement Technology Seattle (1997) , , paper 9

5.482 R.L. McKenzie, M.S. Reinath: Planar Doppler velocimetry dapabilities at low speeds and its application to a full-scale rotor flow, AIAA Paper, Vol. 2000-2292 (2000)

5.483 T.J. Beutner, G. Elliott, A. Mosedale, C. Carter: Doppler Global velocimetry applications in large scale facilities, AIAA Paper, Vol. 98-2608 (1998)

5.484 G.S. Elliott, T.J. Beutner: Molecular Filter Based Planar Doppler Velocimetry, Prog. Aerosp. Sci. **35**, 799–845 (1999)

5.485 G.S. Elliott, T.J. Beutner, C.D. Carter: Application of planar Doppler velocimetry wind tunnel testing, AIAA Paper, Vol. 2000-0412 (2000)

5.486 R.-M. Kremer, E.-J. Korevaar, S.-H. Bloom: Laser radar device, US Patent 5,267,010 (1993)

5.487 S.H. Bloom, P.A. Searcy, K. Choi, R. Kremer, E. Korevaar: Helicopter plume detection by using an ultranarrow-band noncoherent laser Doppler velocimeter, Opt. Lett. **18**, 244–246 (1993)

5.488 R.B. Miles, J.N. Forkey, W.R. Lempert: Filtered Rayleigh scattering measurements in supersonic/hypersonic facilities, AIAA Paper, Vol. 92-3894 (1992)

5.489 J.N. Forkey, N.D. Finkelstein, W.R. Lempert, R.B. Miles: Control of experimental uncertainties in filtered Rayleigh scattering measurements, AIAA Paper, Vol. 95-0298 (1995)

5.490 M.W. Smith, G.B. Northam, J.P. Dummond: Application of absorption filter planar Doppler velocimetry to sonic and supersonic Jets, AIAA J. **34**, 434–441 (1996)

5.491 M.W. Smith: Application of a planar Doppler velocimetry system to a high Reynolds number compressible jet, AIAA Paper, Vol. 98-0428 (1998)

5.492 G.S. Elliott, M. Samimy, S.A. Arnette: A molecular filter based velocimetry technique for high speed flows, Exp. Fluids **18**, 107–118 (1994)

5.493 P.S. Clancy, M. Samimy, W.R. Erskine: Planar Doppler velocimetry: three-component velocimetry in supersonic jets, AIAA Paper, Vol. 98-0506 (1998)

5.494 N.J. Quinlan, R.W. Ainsworth, B.J. Bellhouse, R.J. Manners, S.J. Thorpe: Application of Doppler global velocimetry to supersonic gas-particle flows in drug delivery. In: *GALA 7th International Conference on Laser Anemometry – Advances and Applications*, ed. by B. Ruck (GALA, Karlsruhe 1997), , paper 21-1

5.495 I. Röhle, R. Schodl: Evaluation of the Accuracy of the Doppler Global Technique, Proc. Optical Methods and Data Processing in Heat and Fluid Flow London (1994) 155–161

5.496 I. Röhle: Three-dimensional Doppler global velocimetry in the flow of a fuel spray nozzle and in the wake region of a car, Flow Meas. Instrum. **7**, 287–294 (1996)

5.497 I. Röhle, R. Schodl, P. Voigt, C. Willert: Recent developments and applications of quantitative laser light sheet measuring techniques in turbomachinery components, Meas. Sci. Technol. **11**, 1023–1035 (2000)

5.498 C. Willert, E. Blümcke, M. Beversdorff, W. Unger: Application of phase-averaging Doppler global velocimetry to engine exhaust flows, 10th Int. Symp. Appl. Laser Tech. Fluid Mech. Lisbon (2000) , , paper 35-2

5.499 M. Fischer, J. Heinze, K. Matthias, I. Roehle: Doppler global velocimetry in flames using a new developed, frequency stabilized, tunable, long pulse YAG laser, 10th Int. Symp. Appl. Laser Tech. Fluid Mech. Lisbon (2000) , , paper 35-3

5.500 D.S. Nobes, H.D. Ford, R.P. Tatam: Three dimensional planar Doppler velocimetry using imaging fibre bundles, AIAA Paper, Vol. 2002-0692 (2002)

5.501 C. Willert, G. Stockhausen, J. Klinner, M. Beversdorff, J. Quest, U. Jansen, M. Raffel: On the development of Doppler global velocimetry for cryogenic wind tunnels, IEEE 20th ICIASF (IEEE, New York 2003)

5.502 J. Kuhlman, S. Naylor, K. James, S. Ramanath: Accuracy study of a 2-component point Doppler velocimeter (pDv), AIAA Paper, Vol. 97-1916 (1997)

5.503 J. Kuhlman, P. Collins, T. Scarberry: Two-component point Doppler velocimetry data in circular jets, Meas. Sci. Technol. **12**, 395–408 (2001)

5.504 J. Kuhlman, P. Collins: Circular jet 2-component point Doppler velocimetry (pDv) Velocity Data, AIAA Paper, Vol. 2000-2296 (2000)

5.505 J. Crafton, N.M. Messersmith, J.P. Sullivan: Filtered Doppler velocimetry: development of a point system, AIAA Paper, Vol. 98-0509 (1998)

5.506 W. Förster, G. Karpinsky, H. Krain, I. Röhle, R. Schodl: 3-component-Doppler-laser-two-focus velocimetry applied to a transonic centrifugal compressor, 10th Int. Symp. Appl. Laser Tech. Fluid Mech. Lisbon (2000) , , paper 7-2

5.507 M.P. Wernet: Planar particle imaging Doppler velocimetry, a 3-component velocity measurement technique, AIAA Paper, Vol. 2004-0022 (2004)

5.508 D.H. Thompson: A tracer particle fluid velocity meter incorporating a laser, J. Sci. Instrum. **1**, 929–932 (1968)

5.509 L.H. Tanner: Particle timing laser velocity meter, Opt. Laser Technol. **5**, 108–110 (1973)

5.510 L. Lading: Analysis of a laser correlation anemometer, Proc. 3rd Symp. Turbulence in Liquids (1973) 205–219

5.511 R. Schodl: Ein neues optisches Geschwindigkeitsmessverfahren: Laser-Zweistrahlverfahren, DFVLR-Nachrichten **12**, 506–508 (1973), , (in German)

5.512 L. Lading: Comparing a laser Doppler anemometer with a laser correlation anemometer, Proc. Conf. Engineering Uses of Coherent Optics Glasgow (1975) 19–36

5.513 L. Lading: The time-of-flight laser anemometer, AGARD CP **193**, 23 (1976)

5.514 A.E. Smart: Special problems of laser anemometry in difficult applications, AGARD Lecture Ser. **90** (1977)

5.515 W. T. Mayo Jr., A. E. Smart: Comparison of data from the transit time velocimeter with other systems now in use for velocity measurements, AED-TR-79-32, Final Report on Contract F40500-78-C-002 (1979)

5.516 W.T. Mayo Jr., A.E. Smart, T.E. Hunt: Laser transit anemometer with microcomputer and special digital electronics: measurements in supersonic flows, 8th ICIASF (1979)

5.517 K.G. Barlett, C.Y. She: Single particle correlated time-of-flight velocimeter for remote wind speed measurement, Opt. Lett. **1**, 175 (1977)

5.518 R.G.W. Brown, E.R. Pike: A combined laser Doppler and time of flight anemometer, Opt. Laser Technol. **10**(6), 317–319 (1978)

5.519 C.Y. She, R.F. Lelley: Scalling law and photon-count distribution of a laser time-of-flight velocimeter, J. Opt. Soc. Am. **72**, 365–371 (1982)

5.520 R.G.W. Brown: Velocimetry signals and data reduction in simulation and practice. In: *Photon Correlation Techniques*, ed. by E.O. Schulz-du Bois (Springer, Berlin, New York 1983)

5.521 M.N. Bross: Transit laser anemometry data reduction for flow in industrial turbomaschinery, Proc. 3rd Int. Conf. Photon Correlation Techniques in Fluid Mechanics Cambridge (1979)

5.522 R. Schodl: A laser dual beam method for flow measurement in turbomaschines, ASME Paper, Vol. 74-GT-159 (1974)

5.523 R. Schodl: Laser-two-focus velocimetry for use in aero engines. Laser optical measurement methods for aero engine research and development, AGARD-LS **90**, 4.1–4.34 (1977)

5.524 R. Schodl: Development of the laser-two-focus method for nonintrusive measurement of flow vectors particularly in turbomachines, ESA-TT **528** (1979)

5.525 R. Schodl: A laser-two-focus velocimetry for automatic flow vector measurements in the rotating components of turbomaches, Measurement Methods in Rotating Components of Turbomachinery (ASME, New York 1980)

5.526 L. Lading: Spatial Filtering in Laser Anemometry, Proc. of the 4th Symp. Turbulence in Liquids (1975)

5.527 U. Schricker: *Optimierung der Messsignalaufbereitung der Datensammlung für das Laser-Zwei-Fokus-Geschwindigkeitsmessverfahren* (Inst. F. Strahlantriebe und Turboarbeitsmaschinen, Aachen 1979)

5.528 R. Schodl: Comparison of signal processing by correlation and by pulse-fair timing in laser dual focus velocimetry, Proc. Symp. Long Range and Optical Velocity Measurements (1980)

5.529 P.W. McDonald: A comparison between measured and computed flow fields in a transonic compressor rotor, Trans. ASME J. Eng. Power **102**(4), 883–891 (1980)

5.530 L. Maretto, M. Troilo: A dual focus fiber optic anemometer for measurements in wet stream, Measuring Techniques in Transonic and Supersonic Flows in Cascades and Turbomachines (Ecole Centrale de Lyon, Ecully 1981)

5.531 L. Lading: Estimating time and time-lag in time-of-flight velocimetry, Appl. Opt. **22**(22), 3637–3643 (1983)

5.532 R. Schodl: Laser-two-focus velocimetry, Advanced Instrumentation for Aero Engine Components, AGARD-CP-399 (AGARD, Philadelphia 1986) 7

5.533 S. Yasu, T. Tamaki, S. Nagno: Measurements of flow field within an axial flow fan using a laser two focus velocimeter, Int. Gas Turbine Congress Tokyo (1983), , paper 83-Tokyo-IGTC-49

5.534 L. Fottner, J. Lichtfuss: Design of transonic compressor cascade for minimal shock losses and comparison with test results, Viscous Effects in Turbomachines, AGARD CP 351 (1983)

5.535 G. Janssens., J. Labbe: Two focus laser velocimeter applied to measurements in an experimental centrifugal compressor, ICIASF, Vol. 83 (1983) 251–254

5.536 H. Hayami, Y. Senoo, H. Ueki: Flow in the inducer of a centrifugal compressor measured with a laser velocimeter, ASME Paper, Vol. 84-GT-74 (1984)

5.537 M.M. Ross: *DFVLR/Leicester University/GEC Laser Anemometer Experiments at the High Speed Aerodynamics Laboratory, Leicester University* (Whetstone, Leicester 1978)

5.538 J.C. Erdmann: Particle statistics in transit velocimetry. In: *Photon Correlation Techniques*, ed. by E.O. Schulz-du Bois (Springer, Berlin, New York 1983) pp. 168–191

5.539 J.B. Attis: Statistical aspects of signal processing in laser transit velocimetry, Conf. Photon Correlation Techniques in Fluid Mechanics Standford (1980)

5.540 R. Schodl: Laser-two-focus velocimetry: Two and three dimensional techniques, VKI Lecture Series on Advanced Measurement Techniques, D/1998/0238/455, VKI LS 1998-06 (1998)

5.541 W. Förster: Laser-2-focus data analysis using a nonlinear regression model, 16 ICIASF Symp. Dayton (1995)

5.542 H. Selbach: Laser-2-Fokus Geschwindigkeitsmessverfahren. In: *Lasermethoden in der Strömungsmesstechnik*, ed. by B. Ruck (AT Fachverlag, Stuttgart 1990), , (in German)

5.543 J.D. Trolinger, J.J. Ginouy: Laser applications in flow diagnostics, AGARD-AG **296** (1988)

5.544 W. M. Humphreys Jr., W. E. Hunter Jr.: Estimating laser transit anemometry noise performance capabilities. Presented at the 13th ICIASF 89 in Göttingen, Germany (1989)

5.545 R. Schodl, W. Förster: A multi-color fiberoptic laser-two-focus velocimeter for 3-dimensional flow analysis, Fourth Int. Symp. Appl. Laser Anemometry to Fluid Mechanics Lisbon (1988)

5.546 I. Roehle, R. Schodl: Method for measuring flow vectors in gas flow, UK Patent GB 2 295 670 A (1995)

5.547 I. Röhle: Three dimensional Doppler global velocimetry in the flow of a fuel spray nozzle and in the wake region of a car, Flow Meas. Instrum. **7**(3/4), 287–294 (1996)

5.548 R. Schodl, W. Förster, G. Karpinsky, I. Röhle: 3-component-doppler global laser-two-focus velocimetry applied to a transonic centrifugal compressor, 10th Int. Symp. Appl. Laser Tech. Fluid Mech. Lisbon (2000)

5.549 H. Krain: High pressure ratio centrifugal compressor with transonic flow, Proc. 3rd ASME/JSME Conference San Francisco, FEDSM99-7801 (1999) 9

5.550 C.P. Gendrich, D.G. Bohl, M.M. Koochesfahani: Whole-field measurements of unsteady separation in a vortex ring/wall interaction, AIAA Paper, Vol. 97-1780 (1997)

5.551 D. Bohl, M. Koochesfahani, B. Olson: Development of stereoscopic molecular tagging velocimetry, Exp. Fluids **30**, 302–308 (2001)

5.552 R.E. Falco, D.G. Nocera: Quantitative multipoint measurements and visualization of dense solid-liquid flows using laser induced photochemical anemometry (LIPA). In: *Particulate Two-Phase Flow*, ed. by M.C. Rocco (Butterworth-Heinemann, London 1993) pp. 59–126

5.553 M.M. Koochesfahani, R.K. Cohn, C.P. Gendrich, D.G. Nocera: Molecular tagging diagnostics for the study of kinematics and mixing in liquid phase flows, Proc. Eighth Int. Symp. Appl. Laser Tech. Fluids Mech. Lisbon, Vol. I (1996) 1–12

5.554 M.M. Koochesfahani: Molecular tagging velocimetry (MTV): Progress and applications, AIAA Paper, Vol. 1999-3786 (1999)

5.555 W.R. Lempert, S.R. Harris: Molecular tagging velocimetry. In: *Flow Visualization – Techniques and Examples*, ed. by A.J. Smits, T.T. Lim (Imperial College Press, London 2000) pp. 73–92

5.556 M.M. Koochesfahani (Ed.): Special Feature: Molecular Tagging Velocimetry, Meas. Sci. Technol. **11**(9), 1235–1300 (2000)

5.557 V. Balzani, V. Carassiti: *Photochemistry of Coordination Compounds* (Academic, London 1970)

5.558 L.S. Forster: Photophysical processes – energy levels and spectra. In: *Concepts of Inorganic Photochemistry*, ed. by A.W. Adamson, P.D. Fleischauer (Wiley-Interscience, New York 1975) pp. 1–35

5.559 R.P. Wayne: *Principles and Applications of Photochemistry* (Oxford Univ. Press, Oxford 1980)

5.560 J. Jortner, S.A. Rice, R.M. Hochstrasser: Radiationless transitions in photochemistry, Adv. Photochem. **7**, 149–309 (1969)

5.561 K.F. Freed: Radiationless transitions in molecules, Acc. Chem. Res. **11**, 74–80 (1978)

5.562 S.H. Lin (Ed.): *Radiationless Transitions* (Academic, New York 1980)

5.563 N.J. Turro: *Modern Molecular Photochemistry* (Benjamin/Cummings, Menlo Park 1978)

5.564 V. Balzani, L. Moggi, M.F. Manfrin, F. Bolletta: Quenching and sensitization processes of coordination compounds, Coord. Chem. Rev. **15**, 321–433 (1975)

5.565 Miller S.: *Photochemical reaction for the study of velocity patterns and profiles*. B.A.Sc. Thesis (University of Toronto, Ontario 1962)

5.566 A.T. Popovich, R.L. Hummel: A new method for nondisturbing turbulent flow measurement very close to a wall, Chem. Eng. Soc. **22**, 21–25 (1967)

5.567 F. Frantisak, A. Palade de Iribarne, J.W. Smith, R.L. Hummel: Nondisturbing tracer technique for quantitative measurements in turbulent flows, Ind. Eng. Chem. Fund. **8**, 160–167 (1969)

5.568 R.L. Hummel: Nondisturbing flow measurement using a photochromic tracer technique. High speed photography, Proc. SPIE **97**, 302–307 (1976)

5.569 M. Ojha, R.S.C. Cobbold, K.W. Johnston, R. Hummel: Pulsatile flow through constricted tubes: an experimental investigation using photochromic tracer methods, J. Fluid Mech. **203**, 173–197 (1989)

5.570 R.E. Falco, C.C. Chu: Measurement of two-dimensional fluid dynamic quantities using a photochromic grid tracing technique, Proc. SPIE **814**, 706–710 (1987)

5.571 E. Merzbacher: *Quantum Mechanics* (Wiley, New York 1970)

5.572 V.N. Yurechko, Y.S. Ryazantsev: Fluid motion investigation by photochromic flow visualization technique, Exp. Therm. Fluid Sci. **4**, 273–288 (1991)

5.573 A. D'Arco, J.C. Charmet, M. Cloitre: Nouvelle technique de marquage d'ecoulement par utilisation de molecules photochromes, Rev. Phys. Appl. **17**, 89–93 (1982)

5.574 C.C. Chu, Y.Y. Liao: A quantitative study of the flow around an impulsively started circular cylinder, Exp. Fluids **13**, 137–146 (1992)

5.575 C.C. Chu, C.T. Wang, C.H. Hsieh: An experimental investigation of vortex motions near surfaces, Phys. Fluids **5**(3), 662–676 (1993)

5.576 P. Zalzal, M. Ojha, C.R. Ethier, R.S.C. Cobbold, K.W. Johnston: Visualization of transitional pipe flow using photochromic tracer method, Phys. Fluids **6**, 2003–2010 (1994)

5.577 C.C. Chu, C.T. Wang, C.C. Chang, R.Y. Chang, W.T. Chang: Head-on collision of two coaxial vortex rings: experimental and computation, J. Fluid Mech. **296**, 39–71 (1995)

5.578 G.G. Couch, K.W. Johnston, M. Ohja: Full-field flow visualization and velocity measurement with a photochromic grid method, Meas. Sci. Technol. **7**(9), 1238–1246 (1996)

5.579 H. Park, J.A. Moore, O. Trass, M. Ojha: Laser photochromic velocimetry estimation of the vorticity and pressure field – two-dimensional flow in a curved vessel, Exp. Fluids **26**, 55–62 (1999)

5.580 R. Miles, C. Cohen, J. Conners, P. Howard, S. Huang, E. Markovitz, G. Russell: Velocity measurements by vibrational tagging and fluorescent probing of oxygen, Opt. Lett. **12**(11), 861–863 (1987)

5.581 R.B. Miles, J.J. Connors, E.C. Markovitz, P.J. Howard, G.J. Roth: Instantaneous profiles and turbulence statistics of supersonic free shear layers by Raman Excitation plus Laser-Induced Electronic Fluorescence (RELIEF) velocity tagging of oxygen, Exp. Fluids **8**, 17–24 (1989)

5.582 R.B. Miles, J.E. Connors Markovitz, P. Howard, G. Roth: Instantaneous supersonic velocity profiles in an underexpanded sonic air jet by oxygen flow tagging, Phys. Fluids **1**(2), 389–393 (1989)

5.583 R.B. Miles, D. Zhou, B. Zhang, W.R. Lempert, Z.S. She: Fundamental turbulence measurements by RELIEF flow tagging, AIAA J. **31**(3), 447–452 (1993)

5.584 R.B. Miles, W.R. Lempert: Quantitative flow visualization in unseeded flows, Ann. Rev. Fluid Mech. **29**, 285–326 (1997)

5.585 D.A. Long: *Raman Spectroscopy* (McGraw-Hill, New York 1977)

5.586 R. Frey, J. Lukasik, J. Ducuing: Tunable Raman excitation and vibrational relaxation in diatomic molecules, Chem. Phys. Lett. **14**(4), 514–517 (1972)

5.587 R.L. McKenzie: Progress in laser spectroscopic techniques for aerodynamic measurements: an overview, AIAA J. **31**, 465–477 (1993)

5.588 G. Herzberg: *Molecular Spectra and Molecular Structure: Spectra of Diatomic Molecules*, 2nd edn. (van Nostrand, New York 1950)

5.589 M.W. Feast: Emission Schumann–Runge O2 bands, Nature **162**, 214–215 (1948)

5.590 A. Noullez, G. Wallace, W. Lempert, R.B. Miles, U. Frisch: Transverse velocity increments in turbulent flow using the RELIEF technique, J. Fluid Mech. **339**, 287–307 (1997)

5.591 R.B. Miles, J. Grinstead, R.H. Kohl, G. Diskin: The RELIEF flow tagging technique and its application in engine testing facilities and for helium–air mixing studies, Meas. Sci. Technol. **11**(2), 1272–1281 (2000)

5.592 W.R. Lempert, K. Magee, P. Ronney, K.R. Gee, R.P. Haughland: Flow tagging velocimetry in incompressible flow using photo-activated nonintrusive tracking of molecular motion (PHANTOMM), Exp. Fluids **18**, 249–257 (1995)

5.593 S.R. Harris, W.R. Lempert, L. Hersh, C.L. Burcham, A. Saville, R.B. Miles, K. Gee, R.P. Haughland: Quantitative measurements on internal circulation in droplets using flow tagging velocimetry, AIAA J. **34**(3), 449–454 (1996)

5.594 S.R. Harris, R.B. Miles, W.R. Lempert: Observations of fluid flow produced in a closed cylinder by a rotating lid using the PHANTOMM (Photo-Activated Non Intrusive Tracking of Molecular Motion) flow tagging technique, Proc. Eighth Int. Symp. Appl. Laser Tech. Fluid Mech. Lisbon (1996) 1–9

5.595 M. Biage, S.R. Harris, W.R. Lempert, A.J. Smits: Quantitative velocity measurements in turbulent Taylor-Couette flow by PHANTOMM flow tagging, Proc. Eighth Int. Symp. Appl. Laser Tech. Fluid Mech. Lisbon (1996) 1–8

5.596 P.H. Paul, M.G. Garguilo, D.J. Rakestraw: Imaging of pressure and electrokinetically driven flows through open capillaries, Anal. Chem. **70**, 2459–2467 (1998)

5.597 W.R. Lempert, S.R. Harris: Flow tagging velocimetry using caged dye photo-activated fluorophores, Meas. Sci. Technol. **11**(9), 1251–1258 (2000)

5.598 J.S. Park, H.J. Kim, K.D. Kihm: Molecular tagging fluorescence velocimetry (MTFV) for Lagrangian flow field mapping inside evaporating meniscus: potential use for micro-scale applications, J. Flow Visual. Image Process. **8**, 177–187 (2001)

5.599 D. Sinton, X. Xuan, D. Li: Thermally induced velocity gradients in electroosmotic microchannel flows: the cooling influence of optical infrastructure, Exp. Fluids **37**, 872–882 (2004)

5.600 J.A. McCray, D.R. Trentham: Properties and uses of photoreactive caged compounds, Ann. Rev. Biophys. Chem. **18**, 239–270 (1989)

5.601 L.R. Boedeker: Velocity measurements by H2O photolysis and laser-induced fluorescence of OH, Opt. Lett. **14**(10), 473–475 (1989)

5.602 L. Goss, T. Chen, D. Trump, B. Sarka, A. Nejad: Flow tagging velocimetry using UV-photodissociation of water vapor, AIAA Paper, Vol. 91-0355 (1991)

5.603 G. Meijer, J.J. ter Meulen, P. Andresen, A. Bath: Sensitive quantum state selective detection of H2O and D2O by (2+1)-resonance enhanced multiphoton ionization, J. Chem. Phys. **85**, 6914–6922 (1986)

5.604 R.J. Exton, M.E. Hillard: Raman Doppler velocimetry – A unified approach for measuring molecular-flow velocity, temperature, and pressure, Appl. Opt. **25**, 14–21 (1986)

5.605 L. Goss, T. Chen, D. Schommer, A. Nejad: Laser diagnostics for velocity measurements in supersonic combusting environments, AIAA Paper, Vol. 92-0008 (1992)

5.606 J.A. Wehrmeyer, L.A. Ribarov, D.A. Oguss, F. Batliwala, R.W. Pitz, P.A. DeBarber: Flow tagging velocimetry for low and high temperature flowfields, AIAA Paper **99-0646** (1999)

5.607 J.A. Wehrmeyer, L.A. Ribarov, D.A. Oguss, R.W. Pitz: Flame flow tagging velocimetry with 193 nm H2O photodissociation, Appl. Opt. **38**, 6912–6917 (1999)

5.608 R.W. Pitz, J.A. Wehrmeyer, L.A. Ribarov, D.A. Oguss, F. Batliwala, P.A. DeBarber, S. Deusch, P.E. Dimo-

5.608 takis: Unseeded molecular flow tagging in cold and hot flows using ozone and hydroxyl tagging velocimetry, Meas. Sci. Technol. **11**(9), 1259–1271 (2000)

5.609 R.W. Pitz, M.D. Lahr, Z.W. Douglas, J.A. Wehrmeyer, S. Hu, C.D. Carter, K.-Y. Hsu, C. Lum, M.M. Koochesfahani: Hydroxyl tagging velocimetry in a supersonic flow over a cavity, Appl. Opt. **44**(31), 6692–6700 (2005)

5.610 L.A. Ribarov, J.A. Wehrmeyer, R.W. Pitz, R.A. Yetter: Hydroxyl tagging velocimetry (HTV) in experimental airflows, Appl. Phys. B **74**, 175–183 (2002)

5.611 L.A. Ribarov, J.A. Wehrmeyer, S. Hu, R.W. Pitz: Multi-line hydroxyl tagging velocimetry measurements in reacting and nonreacting experimental flows, Exp. Fluids **37**, 65–74 (2004)

5.612 R.W. Pitz, T.M. Brown, S.P. Nandula, P.A. Skaggs, P.A. DeBarber, M.S. Brown, J. Segall: Unseeded velocity measurement by ozone tagging velocimetry, Opt. Lett. **21**(10), 755–757 (1996)

5.613 L.A. Ribarov, J.A. Wehrmeyer, F. Batliwala, R.W. Pitz, P.A. DeBarber: Ozone tagging velocimetry using narrowband excimer lasers, AIAA J. **37**(6), 708–714 (1999)

5.614 C. Orlemann, C. Schulz, J. Wolfrum: NO-flow tagging by photodissociation of NO2. A new approach for measuring small-scale flow structures, Chem. Phys. Lett. **307**(1–2), 15–20 (1999)

5.615 S. Krüger, G. Grünefeld: Stereoscopic flow-tagging velocimetry, Appl. Phys. B **69**, 509–512 (1999)

5.616 S. Krüger, G. Grünefeld: Gas-phase velocity field measurements in dense sprays by laser-based flow tagging, Appl. Phys. B **70**, 463–466 (2000)

5.617 N.J. Dam, R.J.H. Klein-Douwel, N.M. Sijtsema, J.J. ter Meulen: Nitric oxide flow tagging in unseeded air, Opt. Lett. **26**, 36–38 (2001)

5.618 N.M. Sijtsema, N.J. Dam, R.J. Klein-Douwel H, J.J. ter Meulen: Air photolysis and recombination tracking: A new molecular tagging velocimetry scheme, AIAA J. **40**(6), 1061–1064 (2002)

5.619 W.P.N. van der Laan, R.A.L. Tolboom, N.J. Dam, J.J. ter Meulen: Molecular tagging velocimetry in the wake of an object in supersonic flow, Exp. Fluids **34**, 531–533 (2003)

5.620 P. Barker, A. Thomas, H. Rubinsztein-Dunlop, P. Ljungberg: Velocity measurements by flow tagging employing laser-enhanced ionisation and laser induced fluorescence, Spectrochim. Acta B **50**, 1301–1310 (1995)

5.621 H. Rubinsztein-Dunlop, B. Littleton, P. Barker, P. Ljungberg, Y. Malmsten: Ionic strontium fluorescence as a method for flow tagging velocimetry, Exp. Fluids **30**, 36–42 (2001)

5.622 M.-M. Koochesfahani: *Experiments on turbulent mixing and chemical reactions in a liquid mixing layer*. Ph.D. Thesis (California Institute of Technology, Pasadena 1984)

5.623 J.P. Crimaldi: The effect of photobleaching and velocity fluctuations on single-point LIF measurements, Exp. Fluids **23**, 325–330 (1997)

5.624 G.R. Wang, H.E. Fiedler: On high spatial resolution scalar measurement with LIF. Part 1: Photobleaching and thermal blooming, Exp. Fluids **29**, 257–264 (2000)

5.625 J. Rička: Photobleaching velocimetry, Exp. Fluids **5**, 381–384 (1987)

5.626 K.F. Schrum, J.M. Lancaster, S. Johnston E, S.D. Gilman: Monitoring electroosmotic flow by periodic photobleaching of a dilute, neutral fluorophore, Anal. Chem. **72**, 4317–4321 (2000)

5.627 B.P. Mosier, J.I. Molho, J.G. Santiago: Photobleached-fluorescence imaging of microflows, Exp. Fluids **33**, 545–554 (2002)

5.628 G.R. Wang: Laser induced fluorescence photobleaching anemometer for microfluidic devices, Lab. Chip. **5**, 450–456 (2005)

5.629 B. Hiller, R.A. Booman, C. Hassa, R.K. Hanson: Velocity visualization in gas flows using laser-induced phosphorescence of biacetyl, Rev. Sci. Instrum. **55**, 1964–1967 (1984)

5.630 H.S. Lowry: Velocity measurements using the laser-induced phosphorescence of biacetyl, AIAA Paper, Vol. 87-1529 (1987)

5.631 J.B. Liu, Q. Pan, C.S. Liu, J.R. Shi: Principles of flow field diagnostics by laser induced biacetyl phosphorescence, Exp. Fluids **6**, 505–513 (1988)

5.632 H.S. Hilbert, R.E. Falco: Measurements of flows during scavenging in a two-stroke engine, SAE Tech. Paper **910671** (1991)

5.633 B. Stier, M.M. Koochesfahani: Molecular tagging velocimetry in gas phase and its application to jet flows, ASME Paper, Vol. FEDSM97-3687 (1997)

5.634 B. Stier, M.M. Koochesfahani: Whole field MTV measurements in a steady flow rig model on an IC engine, SAE Tech. Paper **980481** (1998)

5.635 B. Stier, M.M. Koochesfahani: Molecular tagging velocimetry (MTV) measurements in gas phase flows, Exp. Fluids **26**(4), 297–304 (1999)

5.636 H.G. Hascher, M. Novak, K. Lee, H. Schock, H. Rezaei, M.M. Koochesfahani: An evaluation of IC-engine flow with the use of modern in-cylinder measuring techniques, AIAA Paper, Vol. 98-3455 (1998)

5.637 E.M. Thurlow, J.C. Klewicki: Experimental study of turbulent Poiseuille–Couette flow, Phys. Fluids **12**(4), 865–875 (2000)

5.638 A.-C.-H. Goh: *Active flow control for maximizing performance of spark-ignited stratified charge engines*. M.S. Thesis (Michigan State University, East Lansing 2001)

5.639 M.M. Koochesfahani, A.C. Goh, H.J. Schock: Molecular Tagging Velocimetry (MTV) and its automotive applications. In: *The Aerodynamics of Heavy Vehicles: Trucks, Busses, and Trains*, Lecture Notes in Applied and Computational Mechanics, Vol. 19, ed. by R. McCallen, F. Browand, J. Ross (Springer, Heidelberg, Berlin 2004) pp. 143–155

5.640 W.R. Lempert, N. Jiang, S. Sethuram, M. Samimy: Molecular tagging velocimetry measurements in supersonic microjets, AIAA J. **40**, 1065–1070 (2002)

5.641 W.R. Lempert, M. Boehm, N. Jiang, S. Gimelshein, D. Levin: Comparison of molecular tagging velocimetry data and direct simulation Monte Carlo simulations in supersonic micro jet flows, Exp. Fluids **34**, 403–411 (2003)

5.642 Z. Pikramenou, J.-A. Yu, R.B. Lessard, A. Ponce, P.A. Wong, D.G. Nocera: Luminescence from supramolecules triggered by the molecular recognition of substrates, Coord. Chem. Rev. **132**, 181–194 (1994)

5.643 M.A. Mortellaro, D.G. Nocera: A turn-on for optical sensing, Chem. Tech. **26**, 17–23 (1996)

5.644 M.A. Mortellaro, D.G. Nocera: A supramolecular chemosensor for aromatic hydrocarbons, J. Am. Chem. Soc. **118**, 7414–7415 (1996)

5.645 D.G. Nocera: Not just another pretty shape, New Sci. **149**, 24–27 (1996)

5.646 C.M. Rudzinski, D.G. Nocera: Buckets of light, Opt. Sensors Switches **7**, 1–90 (2001)

5.647 V. Balzani, F. Bolletta, F. Scandola: Vertical and "nonvertical" energy transfer processes – a general classical treatment, J. Am. Chem. Soc. **102**, 2152–2163 (1980)

5.648 J.F. Endicott: Manipulation of superexchange couplings, Acc. Chem. Res. **21**, 59–66 (1988)

5.649 W. Horrocks, M. Albin: Lanthanide ion luminescence in coordination chemistry and biochemistry, Prog. Inorg. Chem. **30**, 1–104 (1983)

5.650 J.-C.G. Bünzli: Luminescent probes. In: *Lanthanide Probes in Life, Chemical and Earth Sciences*, ed. by J.-G.C. Bünzli, G.R. Choppin (Elsevier, Amsterdam 1988) pp. 219–293

5.651 J.-A. Yu, R.B. Lessard, D.G. Nocera: Direct observation of intramolecular energy transfer from a Î²-diketonate to terbium(III) ion encapsulated in a cryptand, Chem. Phys. Lett. **187**, 263–268 (1991)

5.652 V. Balzani, F. Scandola: *Supramolecular Photochemistry* (Ellis-Horwood, Chichester 1991)

5.653 G. Wenz: Cyclodextrins as building blocks for supramolecular structures and functional units, Angew. Chem. Int. Edit. **33**, 803–822 (1994)

5.654 A. Ponce, P.A. Wong, J.J. Way, D.G. Nocera: Intense phosphorescence triggered by alcohols upon formation of a cyclodextrin ternary complex, J. Phys. Chem. **97**, 11137–11142 (1993)

5.655 W.K. Hartmann, M.H.B. Gray, A. Ponce, D.G. Nocera: Substrate induced phosphorescence from cyclodextrin lumophore host-guest complexes, Inorg. Chim. Acta **243**, 239–248 (1996)

5.656 H. Hu, C. Lum, M.M. Koochesfahani: Molecular tagging thermometry with adjustable temperature sensitivity, Exp. Fluids **40**, 753–763 (2006)

5.657 H. Hu, M.M. Koochesfahani: Molecular tagging velocimetry and thermometry technique and its application to the wake of a heated cylinder, Meas. Sci. Technol. **17**, 1269–1281 (2006)

5.658 S.-R. Harris: *Quantitative measurements in a lid driven, cylindrical cavity using the PHANTOMM flow tagging technique*. Ph.D. Thesis (Princeton University, Princeton 1999)

5.659 M.M. Koochesfahani, D. Bohl: Molecular tagging visualization and velocimetry of the flow at the trailing edge of an oscillating airfoil, Proc. 10th Int. Symp. Flow Visualization Kyoto (2002) , , paper F0453

5.660 R.B. Hill, J.C. Klewicki: Data reduction methods for flow tagging velocity measurements, Exp. Fluids **20**(3), 142–152 (1996)

5.661 J.C. Klewicki, R.B. Hill: Laminar boundary layer response to rotation of a finite diameter surface patch, Phys. Fluids **15**(1), 101–111 (2003)

5.662 R. Sadr, J.C. Klewicki: Flow field characteristics in the near field region of particle-laden coaxial jets, Exp. Fluids **39**, 885–894 (2005)

5.663 K. Wirtz, M.M. Koochesfahani, J.J. McGrath, A. Benard: Molecular tagging velocimetry applied to buoyancy-driven convective phenomena during solidification, ASME Paper, Vol. HTD-361-4 (1998) 103

5.664 C. Lum, M.M. Koochesfahani, J.J. McGrath: Measurements of the velocity field with MTV during the solidification of an alloy analog with mushy region, ASME/IMECE2001 Paper, Vol. HTD-24222 (2001)

5.665 C. Lum: *Velocity field measurement for a unidirectional solidification of an ammonium chloride (NH4 Cl) solution using molecular tagging velocimetry (MTV)*. M.S. Thesis (Michigan State University, East Lansing 2001)

5.666 D.-G. Bohl: *Experimental Study of the 2-D and 3-D Structure of a Concentrated Line Vortex Array*. Ph.D. Thesis (Michigan State University, East Lansing 2002)

5.667 D.G. Bohl, M.M. Koochesfahani: MTV measurements of axial flow in a concentrated vortex core, Phys. Fluids **16**(11), 4185–4191 (2004)

5.668 D. Maynes, A.R. Webb: Velocity profile characterization in sub-millimeter diameter tubes using molecular tagging velocimetry, Exp. Fluids **32**, 3–15 (2002)

5.669 C. Lum: *An experimental study of pressure- and electroosmotically-driven flows in microchannels with surface modifications*. Ph.D. Thesis (Michigan State University, East Lansing 2005)

5.670 R.-K. Cohn: *Effect of forcing on the vorticity field in a confined wake*. Ph.D. Thesis (Michigan State University, East Lansing 1999)

5.671 C.P. Gendrich, M.M. Koochesfahani, D.G. Nocera: Molecular tagging velocimetry and other novel applications of a new phosphorescent supramolecule, Exp. Fluids **23**(5), 361–372 (1997)

5.672 C.-P. Gendrich: *Dynamic stall of rapidly pitching airfoils: MTV experiments and Navier–Stokes simulations*. Ph.D. Thesis (Michigan State University, East Lansing 1998)

5.673 R.K. Cohn, M.M. Koochesfahani: Vorticity field evolution in a forced wake, Proc. 1st Int. Symp. Turbulence and Shear Flow Phenomena Santa Barbara, ed. by S. Banerjee, J.K. Eaton (1999) 901–906

5.674 M. Koochesfahani, R. Cohn, C. MacKinnon: Simultaneous whole-field measurements of velocity and concentration fields using combined MTV and LIF, Meas. Sci. Technol. **11**(9), 1289–1300 (2000)

5.675 R. Sadr, J.C. Klewicki: An experimental investigation of the near-field flow development in coaxial jets, Phys. Fluids **15**(5), 1233–1246 (2003)

5.676 R. Sadr, J.C. Klewicki: A spline-based technique for estimating flow velocities using two-camera multiline MTV, Exp. Fluids **35**, 257–261 (2003)

5.677 C.P. Gendrich, M.M. Koochesfahani: A spatial correlation technique for estimating velocity fields using Molecular Tagging Velocimetry (MTV), Exp. Fluids **22**(1), 67–77 (1996)

5.678 Q. Zheng, J.C. Klewicki A fast data reduction algorithm for molecular tagging velocimetry: the decoupled spatial correlation technique, Meas. Sci. Technol. **11**(9), 1282–1288 (2000)

5.679 P.T. Tokumaru, P.E. Dimotakis: Image correlation velocimetry, Exp. Fluids **19**, 1–15 (1995)

5.680 R.K. Cohn, M.M. Koochesfahani: The accuracy of remapping irregularly spaced velocity data onto a regular grid and the computation of vorticity, Exp. Fluids **29**, S61–S69 (2000)

5.681 J.E. Guilkey, K.R. Gee, P.A. McMurtry, J.C. Klewicki: Use of caged fluorescent dyes for the study of turbulent passive scalar mixing, Exp. Fluids **21**, 237–242 (1996)

5.682 L. Hansen, J.E. Guilkey, P.A. McMurtry, J.C. Klewicki: The use of photoactivatable fluorophores in the study of turbulent pipe mixing: effects of inlet geometry, Meas. Sci. Technol. **11**(9), 1235–1250 (2000)

5.683 S.L. Thomson, D. Maynes: Spatially resolved temperature measurements in a liquid using laser induced phosphorescence, J. Fluid Eng. **123**, 293–302 (2001)

5.684 H. Hu, M.M. Koochesfahani: A novel technique for quantitative temperature mapping in liquid by measuring the lifetime of laser induced phosphorescence, J. Visual. Jpn. **6**(2), 143–153 (2003)

5.685 M.J. Lighthill: Chap. II. In: *Laminar Boundary Layer*, ed. by L. Rosenhead (Oxford University, Oxford 1963)

5.686 J.M. Wallace: The measurement of vorticity in turbulent flows, 9th Biannual Symposium on Turbulence (1984)

5.687 P. Vukoslavcevic, J.M. Wallace, J. Balint: The velocity and vorticity vector fields of a turbulent boundary layer, Part I. Simultaneous measurement by hot-wire anemometry, J. Fluid Mech. **228**, 25–51 (1991)

5.688 A. Honkan, J. Andreopoulos: Experimental investigation of time-resolved vorticity in two dimensional turbulent boundary layer flows, AIAA Paper, Vol. 93-2910 (1993)

5.689 A. Honkan, Y. Andreopoulos: Vorticity, strain rate and dissipation characteristics in the near wall region of turbulent boundary layers, J. Fluid Mech. **350**, 29–26 (1997)

5.690 A. Honkan: *An experimental study of the vortical structure of turbulent flows*. Ph.D. Thesis (City University of New York, New York 1994)

5.691 Y. Andreopoulos, A. Honkan: Experimental techniques for highly resolved measurements of rotation, strain and dissipation-rate tensors in turbulent flows, Meas. Sci. Technol. **7**, 1462–1476 (1996)

5.692 J.C. Klewicki, R.E. Falco: On accurately measuring statistics associated with small-scale structure in turbulent boundary layers using hot wire probes, J. Fluid Mech. **219**, 119–142 (1990)

5.693 A. Tsinober, E. Kit, T. Dracos: Experimental investigation of the field of velocity gradients in turbulent flows, J. Fluid Mech. **242**, 169–192 (1992)

5.694 G. Briassulis, J. Agui, Y. Andreopoulos: The structure of weakly compressible grid turbulence, J. Fluid Mech. **432**, 219–283 (2001)

5.695 Y. Andreopoulos, J.H. Agui, G. Briassulis: Shock wave-turbulence interactions, Ann. Rev. Fluid Mech. **32**, 309–345 (2000)

5.696 J.H. Agui, G. Briassulis, Y. Andreopoulos: Studies of interactions of a propagating shock wave with decaying grid turbulence: velocity and vorticity field, J. Fluid Mech. **524**, 143–195 (2005)

5.697 A.E. Perry, M.S. Chong: A description of eddying motions and flow patterns using critical point concepts, Ann. Rev. Fluid Mech. **19**, 125–155 (1987)

5.698 J. Andreopoulos, J.H. Agui: Wall-vorticity flux dynamics in a two-dimensional turbulent boundary layer, J. Fluid Mech. **309**, 45–84 (1996)

5.699 J.F. Foss: Accuracy and uncertainty in transverse vorticity measurements, Bull. Am. Phys. Soc. **21**, 1237 (1976), (abstract)

5.700 J.F. Foss: Advanced techniques for transverse vorticity measurements, Proc. Biennial Symp. Turbulence (1981) 208–218

5.701 M.B. Frish, W.W. Webb: Direct measurements by Optical Probe, J. Fluid Mech. **107**, 173–185 (1981)

5.702 J.C. Wyngaard: Spatial resolution of the vorticity meter and other hot-wire arrays, J. Phys. E Sci. Instrum. **2**, 983–987 (1969)

5.703 E.B. Arik: Current status of particle image velocimetry and laser Doppler anemometry instrumentation. In: *Flow at Ultra High Reynolds and Rayleigh Numbers: A Status Report*, ed. by R.J. Donnelly, K.R. Sreenivasan (Springer, Berlin, New York 1998) pp. 138–158

5.704 L. Lourenco, A. Krothapalli: The role of photographic parameters in Laser-Speckle of particle image displacement velocimetry, Exp. Fluids **5**, 23–31 (1987)

5.705 C.D. Meinhart, R.J. Adrian: Measurements of the zero-pressure-gradient turbulent boundary layer using particle image velocimetry, AIAA Paper, Vol. 95-0789 (1995)

5.706 H. Breyer, H. Kriegs, U. Schmidt, W. Staude: The measurement of velocity gradients in fluid flow by

laser light scattering, Part 1: Mean gradients, Exp. Fluids **15**, 200–208 (1993)
5.707 H. Kriegs, R. Schulz, W. Staude: The measurement of velocity gradients in fluid flow by laser light scattering, Part 2: Statistical properties of gradients, Exp. Fluids **15**, 240–246 (1993)
5.708 H. Kriegs, W. Staude: A laser-pulse technique for the measurement of time-resolved velocity gradients in fluid flow, Meas. Sci. Technol. **6**, 653 (1995)
5.709 J. Zhang, B. Tao, J. Katz: Turbulent flow measurement in a square duct with hybrid holographic PIV, Exp. Fluids **23**, 373–381 (1997)
5.710 C.J. Kähler: Investigation of the spatio-temporal flow structure in the buffer region of a turbulent boundary layer by means of multiplane stereo PIV, Exp. Fluids **36**, 114–130 (2004)
5.711 B. Ganapathisubramani, E.K. Longmire, I. Marusic, S. Pothos: Dual-plane PIV technique to resolve complete velocity gradient tensor in a turbulent boundary layer, Exp. Fluids **32**(2), 222–231 (2004)
5.712 H. Hu, T. Saga, T. Kobayashi, N. Taniguchi, M. Yasuki: Dual-plane stereoscopic particle image velocimetry: System setup and its application on a lobed jet mixing flow, Exp. Fluids **31**, 277–293 (2001)
5.713 J.A. Mullin, W.J.A. Dahm: Dual-plane stereo particle image velocimetry (DSPIV) for measuring velocity gradient fields at intermediate and small scales of turbulent flows, Exp. Fluids **38**, 185–196 (2005)
5.714 D.B. Lang, P.E. Dimotakis: Vorticity measurements in a two-dimensional shear-layer, Bull. Am. Phys. Soc. **29**(9), 1556 (1984)
5.715 D.-B. Lang: *Laser Doppler velocity and vorticity measurements in turbulent shear layers*. Ph.D. Thesis (Caltech, Pasadena 1985)
5.716 G.P. Romano, R.A. Antonia, T. Zhou: Evaluation of LDA temporal and spatial velocity structure functions in a low Reynolds number turbulent channel flow, Exp. Fluids **27**, 368–377 (1999)
5.717 J.H. Agui, J. Andreopoulos: Development of a New LASER Vorticity Probe-LAVOR. Fluids Engineering Division of ASME, Int. Symp. Laser Anemometry, Vol. FED191, ed. by T.T. Huang, M.V. Ötügen (1994) 11–19
5.718 J.H. Agui, Y. Andreopoulos: A new laser vorticity probe – LAVOR: its development and validation in a turbulent boundary layer, Exp. Fluids **34**, 192–205 (2003)
5.719 R.A. Fraser, P.G. Felton, F.V. Bracco, D.A. Santavica: Preliminary turbulence length scale measurements in a motored IC engine, SAE Tech. Paper **860021** (1986)
5.720 R.J.R. Johns, G.F. Pitcher, E. Winkelhofer: Measurement of spatial correlations in a turbulent flow with a two-dimensional backscatter LDA system, 3rd Int. Symp. Applications of Laser Anemometry to Fluid Mechanics Lisbon (1986)
5.721 N. Nakatani, M.E. Tokita, M.E. Mocgowa, T. Yamada: Simultaneous measurement of flow velocity variations at several points with multi-point LDV, Int. Conf. Laser Anemometry Manchester (1986)
5.722 A.A. Kolmogorov: The Local Structure of Turbulence in Incompressible Viscous Fluid for Very Large Reynolds Numbers, CR Akad. Sci. SSSR **30**, 301–306 (1941)
5.723 F. Durst, A. Melling, J.H. Whitelaw: *Principles and Practice of Laser-Doppler Anemometry* (Academic, New York 1981)
5.724 S.G. Hanson: Application of the laser gradient anemometer (LGA) for fluid flow measurements, 2nd Symp. Applications of Laser Anemometry to Fluid Mechanics Lisbon (1984)
5.725 M.V. Ötügen, G. Su W-J; Papadopoulos: A New Laser-Based Method for Strain Rate and Vorticity Measurements, Meas. Sci. Technol. **9**, 267–274 (1998)
5.726 S. Yao, P. Tong, B. Ackerson: Proposal and testing of a fiber optic based measurements of flow vorticity, Appl. Opt. **40**(24), 4022–4027 (2001)
5.727 V.P. Stepaniuk, C. Tarau, M.V. Ötügen, W.-J. Su: Evaluation of a laser velocity gradient probe and measurements in a boundary layer, Exp. Fluids **36**, 510–513 (2004)
5.728 V.P. Stepaniuk, W.-J. Su, C. Tarau, M.V. Ötügen: Measurements in turbulent shear layers using a laser vorticity probe, AIAA Paper, Vol. 2004-1223 (2004)
5.729 L. Blank: *Statistical Procedures for Engineering, Management and Science* (McGraw-Hill, New York 1980)
5.730 Y. Andreopoulos, A. Honkan: An experimental study of the dissipative and vortical motion in turbulent boundary layers, J. Fluid Mech. **439**, 131–163 (2001)
5.731 J. Balint, J.M. Wallace, P. Vukoslavcevic: The velocity and vorticity vector fields of a turbulent boundary layer, Part 2 Statistical Properties, J. Fluid Mech. **228**, 53–86 (1991)
5.732 D. Ewing, W.-K. George: Private communication (1993)
5.733 D.B. Spalding: A single formula for the law of the wall, J. Appl. Mech. **28**, 455–458 (1961)
5.734 L.P. Purtell, P.S. Klebanoff, F.T. Buckley: Turbulent boundary layer at low Reynolds number, Phys. Fluids **24**(5), 802–811 (1981)
5.735 J. Andreopoulos, F. Durst, Z. Zaric', J. Jovanovic: Influence of Reynolds number on Characteristics of Turbulent Wall Boundary Layers, Exp. Fluids **2**, 7–16 (1984)
5.736 P. Bradshaw, G.P. Huang: The law of the wall in turbulent flow, Proc. R. Soc. London A, Vol. 451 (1995) 165–188
5.737 N.N. Mansour, J. Kim, P. Moin: Reynolds-stress and dissipation-rate budgets in a turbulent channel flow, J. Fluid Mech. **194**, 15–44 (1988)
5.738 W.-J. Su, M.-V. Ötügen: Private communication (2004)
5.739 S. Rajagopalan, R.H. Antonia: RMS spanwise vorticity measurements in a turbulent boundary layer, Exp. Fluids **14**, 142–144 (1993)

5.740 J.-C. Klewicki: *On the interactions between the inner and outer region motions in turbulent boundary layers*. Ph.D. Thesis (Michigan State University, East Lansing 1989)

5.741 M.M. Metzger, J.C. Klewicki: A comparative study of near-wall turbulence in high and low Reynolds number boundary layers, Phys. Fluids **13**(3), 692–701 (2001)

5.742 P.R. Spalart: Direct simulation of a turbulent boundary layer up to R_θ = 1410, J. Fluid Mech. **187**, 61–98 (1988)

5.743 J. Kim, P. Moin, R. Moser: Turbulence statistics in fully developed channel flow at low Reynolds number, J. Fluid Mech. **177**, 133–166 (1987)

5.744 J.-H. Agui, Y. Andreopoulos: Reynolds number effects on vorticity statistics (2005) (in preparation)

5.745 W.J.A. Dahm, K.B. Southerland, K.A. Buch: Direct, high resolution, four-dimensional measurements of the fine scale structure of Sc >> 1 molecular mixing in turbulent flows, Phys. Fluids **3**, 1115–1127 (1991)

5.746 W.J.A. Dahm, L.K. Su, K.B. Southerland: A scalar imaging velocimetry technique for four-dimensional velocity field measurements in turbulent flows, Phys. Fluids **4**, 2191–2206 (1992)

5.747 L.K. Su, W.J.A. Dahm: Scalar imaging velocimetry measurements of the velocity gradient tensor field at the dissipative scales of turbulent flows, Part I: Assessment of errors, Phys. Fluids **8**, 1869–1882 (1996)

5.748 L.K. Su, W.J.A. Dahm: Scalar imaging velocimetry measurements of the velocity gradient tensor field at the dissipative scales of turbulent flows, Part II: Experimental results, Phys. Fluids **8**, 1883–1906 (1996)

5.749 C.J. Kähler, J. Kompenhans: Multiple plane stereo PIV: Technical realization and fluid-mechanical significance, Proc. 3rd Int. Workshop on PIV Santa Barbara (1999)

5.750 C.J. Kähler, M. Stanislas, J. Kompenhans: Spatio-temporal flow structure investigation of near-wall turbulence by means of multiplane stereo particle image velocimetry, Proc. 11th Int. Symp. Appl. Laser Tech. Fluid Mech. Lisbon (2002)

5.751 C.J. Kähler: Investigation of the spatio-temporal flow structure in the buffer region of a turbulent boundary layer by means of multiplane stereo PIV, Exp. Fluids **36**, 114–130 (2004)

5.752 H. Hu, T. Saga, T. Kobayashi, N. Taniguchi, M. Yasuki: Dual-plane stereoscopic particle image velocimetry: System setup and its application on a lobed jet mixing flow, Exp. Fluids **31**, 277–293 (2001)

5.753 B. Ganapathisubramani, E.K. Longmire, I. Marusic, S. Pothos: Dual-plane PIV technique to resolve complete velocity gradient tensor in a turbulent boundary layer, Proc. 12th Int. Symp. Appl. Laser Tech. Fluid Mech. Lisbon (2004)

5.754 J.A. Mullin, W.J.A. Dahm: Highly-resolved three-dimensional velocity measurements via dual-plane stereo particle image velocimetry (DSPIV) in turbulent flows, AIAA Paper, Vol. 2002-0290 (2002)

5.755 J.A. Mullin, W.J.A. Dahm: Direct experimental measurements of velocity gradient fields in turbulent flows via high-resolution frequency-based dual-plane stereo PIV (DSPIV), Proc. 12th Int. Symp. Appl. Laser Tech. Fluid Mech. Lisbon (2004)

5.756 K.A. Buch, W.J.A. Dahm: Experimental study of the fine-scale structure of conserved scalar mixing in turbulent flows, Part 1. Sc >> 1, J. Fluid Mech. **317**, 21–71 (1996)

5.757 K.A. Buch, W.J.A. Dahm: Experimental study of the fine-scale structure of conserved scalar mixing in turbulent flows, Part 2. Sc $\approx$ 1, J. Fluid Mech. **364**, 1–29 (1998)

5.758 J.A.W.J.A. Mullin & Dahm: Dual-plane stereo particle image velocimetry measurements of velocity gradient tensor fields in turbulent shear flow. Part 1: Accuracy assessments, Phys. Fluids **18**(3), 035101 (2006)

5.759 D.C. Bjorkquist: Stereoscopic PIV calibration verification, Proc. 11th Int. Symp. Appl. Laser Tech. Fluid Mech. Lisbon (2002)

5.760 B. Wiener: Stereo-PIV using self-calibration on particle images, Proc. 5th Int. Symp. Particle Image Velocimetry Busman (2003)

5.761 J.A. Mullin, W.J.A. Dahm: Dual-plane stereo particle image velocimetry measurements of velocity gradient tensor fields in turbulent shear flow. Part 2: Experimental results, Phys. Fluids **18**(3), 035102 (2006)

5.762 G.I. Taylor: Production and disipation of vorticity in a turbulent fluid, Proc. R. Soc. London **164**, 15–23 (1938)

5.763 W.W. Wassman, J.M. Wallace: Measurement of vorticity in turbulent shear flow, Bull. Am. Phys. Soc. **24**, 1142 (1979)

5.764 M. Kholmyansky, A. Tsinober, S. Yorish: Velocity derivatives in the atmospheric turbulent flow at Re_λ =104, Phys. Fluids **13**, 311–314 (2001)

5.765 B. Galanti, G. Gulitsky, M. Kholmyansky, A. Tsinober, S. Yorish: Adv. in Turbulence X. In: *Proc. 10th European Turb. Conf*, ed. by H.I. Andersson, P.Å. Krogstad (CIMNE, Barcelona 2004)

5.766 P.-N. Nguyen: *Simultaneous measurements of the velocity and vorticity vector fields in the turbulent near wake of a circular cylinder*. Ph.D. Thesis (Univ. of Maryland, College Park 1993)

5.767 B. Marasli, P. Nguyen, J.M. Wallace: A calibration technique for multiple-sensor hot- wire probes and its application to vorticity measurements in the wake of a circular cylinder, Exp. Fluids **15**, 209–218 (1993)

5.768 P. Vukoslavcevic, J.M. Wallace: A 12-sensor hot-wire probe to measure the velocity and vorticity vectors in turbulent flow, Meas. Sci. Technol. **7**, 1451–1461 (1996)

5.769 Y. Zhu, R.A. Antonia: Spatial resolution of a 4-X-wire vorticity probe, Meas. Sci. Technol. **7**, 1492–1497 (1996)

5.770 R.A. Antonia, T. Zhou, Y. Zhu: Three-component vorticity measurements in a turbulent grid flow, J. Fluid Mech. **374**, 29–57 (1998)

5.771 R.A. Antonia, P. Orlandi, T. Zhou: Assessment of a three-component vorticity probe in decaying turbulence, Exp. Fluids **33**, 384–390 (2002)

5.772 P.S. Bernard, J.M. Wallace: *Turbulent Flow: Analysis, Measurement and Prediction* (Wiley, Hoboken 2002)

5.773 Y.N. Pao: Structure of turbulent velocity and scalar fields of large wave numbers, Phys. Fluids **8**, 1063–1075 (1965)

5.774 H.J. Hussein, W.K. George: Influence of wire spacing on derivative measurements with parallel hot-wire probes, Forum on Turbulent Flow, Vol. 94 (1990) 121–124

5.775 R.A. Antonia, Y. Zhu, J. Kim: On the measurement of lateral velocity derivatives in turbulent flows, Exp. Fluids **15**, 65–69 (1993)

5.776 J.C. Wyngaard: Measurement of small-scale turbulence structure with hot-wires, J. Phys. E Sci. Instrum. **1**, 1105–1108 (1968)

5.777 A.N. Kolmogorov: Dissipation of energy in locally isotropic turbulence, CR Acad. Sci. URSS **32**, 19–21 (1941)

5.778 L.W.B. Browne, R.A. Antonia, D.A. Shah: Selection of wires and wire spacing for X- wires, Exp. Fluids **6**, 286–288 (1988)

5.779 R.D. Moser, J. Kim, N.N. Mansour: Direct numerical simulation of turbulent channel flow up to Re=590, Phys. Fluids **11**, 943–945 (1989)

5.780 R.-B. Loucks: *An Experimental Investigation of the Velocity and Vorticity Fields in a Plane Mixing Layer*. Ph.D. Thesis (Univ. of Maryland, College Park 1998)

5.781 F.E. Jorgensen: Directional sensitivity of wire and fiber film probes, DISA Inf. **11**, 31–37 (1971)

5.782 P. Bradshaw: *An Introduction to Turbulence and Its Measurements* (Pergamon, New York 1971)

5.783 N. Vrettos: *Experimentelle Untersuchung der Ausbreitung von Kulturmschwaden im Windkanal*. Studienarbeit (Univ. of Karlsruhe, Karlsruhe 1984) (in German)

5.784 G. Heskestad: A generalized Taylor hypothesis with application for high Reynolds number turbulent shear flows, Trans. ASME **87**, 735–739 (1965)

5.785 G.K. Batchelor: Pressure fluctuations in isotropic turbulence, Proc. Cambridge Phil. Soc., Vol. 47 (1951) 359

5.786 U. Piomelli, J.L. Balint, J.M. Wallace: On the validity of Taylor Hypothesis for wall bounded flows, Phys. Fluids **3**, 609–611 (1988)

5.787 J.-L. Balint: *Contribution de l'Etude de la Structure Tourbillonaire d'une Couche Limite Turbulente au Moyen d'une Sonde a Neuf Fils Chaude Measurant le Rotationnel*. Docteur d'Etat en Sciences (Ecole Centrale de Lyon, Lyon 1986), in French

5.788 F. Hussain, K. Jeong, J. Kim: Structure of turbulent shear flows, Proc. Summer Prog. Ctr. Turb. Res. Stanford Univ./NASA Ames (1987) 273–290

5.789 L. Ong: *Visualization of turbulent flows with simultaneous velocity and vorticity measurements*. Ph.D. Thesis (Univ. of Maryland, College Park 1992)

5.790 J. Kim, F. Hussain: Propagation velocity of perturbations in turbulent channel flow, Phys. Fluids **3**, 695–706 (1993)

5.791 P. Vukoslavcevic, D. Petrovic, J.M. Wallace: An analytical approach to the uniqueness problem of hot-wire probes to measure simultaneously three velocity components, Meas. Sci. Technol. **15**, 1848–1854 (2004)

5.792 N.K. Tutu, R. Chevray: Cross wire anemometer in high intensity turbulence, J. Fluid Mech. **71**, 785–800 (1975)

5.793 W.W. Willmarth: Geometrical interpretation of the possible velocity vectors obtained with multiple sensor probes, Phys. Fluids **28**, 462–465 (1985)

5.794 G. Pailhas, J. Coustex: Methode d'explotation des donnes d'une sonde anemometrique a quatre fils chauds, Rech. Aerospatiale **2**, 161–168 (1986)

5.795 E. Kit, A. Tsinober, J.L. Balint, J.M. Wallace, E. Levich: An experimental study of helicity related properties of turbulent flow past a grid, Phys. Fluids **30**, 3323–3325 (1988)

5.796 J.M. Wallace, J.L. Balint, L. Ong: An experimental study of helicity density in turbulent flow, Phys. Fluids **4**, 2013–2026 (1992)

5.797 S. Menon: *A study of velocity and vorticity statistics in an axisymmetric jet*. M.S. Thesis (Univ. of Houston, Houston 1993)

5.798 J.J. Gorski, J.M. Wallace, P.S. Bernard: The enstrophy equation budget of bounded turbulent shear flows, Phys. Fluids **6**, 3197–3199 (1994)

5.799 L. Ong, J.M. Wallace: Local isotrophy of the vorticity field in a boundary layer at high Reynolds number. In: *Advances in Turbulence*, Vol. V, ed. by R. Benzi (Springer, Berlin, Heidelberg 1995) pp. 392–397

5.800 G.-C. Lemonis: *An experimental study of the vector fields of velocity and vorticity in turbulent flows*. Ph.D. Thesis (Swiss Federal Institute of Technology, Zurich 1995)

5.801 A. Folz: *An experimental study of the near-surface turbulence in the atmospheric boundary layer*. Ph D. Thesis (Univ. of Maryland, College Park 1997)

5.802 A. Tsinober, L. Shtilman, H. Vaisburd: A study of vortex stretching and enstrophy generation in numerical and laboratory turbulence, Fluid Dyn. Res. **21**, 477–494 (1997)

5.803 L. Ong, J.M. Wallace: Joint probability density analysis of the structure and dynamics of the vorticity field of a turbulent boundary layer, J. Fluid Mech. **367**, 291–328 (1998)

5.804 B. Galanti, A. Tsinober: Self-amplification of the field of velocity derivatives in quasi-isotropic turbulence, Phys. Fluids **12**, 3097–3099 (2001)

5.805 N. Li: *Passive scalar dispersion in a turbulent mixing layer*. Ph.D. Thesis (Univ. of Maryland, College Park 2004)

5.806 J.-N. Kim: *Wirbelstärkemessungen in einer turbulenten Scherschicht*. Dr.-Ing. Thesis (Tech. Univ. Berlin, Berlin 1989)

5.807 W.T. Ashurst, A.R. Kerstein, R.A. Kerr, C.H. Gibson: Alignment of vorticity and scalar gradient with strain rate in simulated Navier-Stokes turbulence, Phys. Fluids **30**, 2343–2353 (1987)

5.808 J.F. Foss, R.C. Haw: Transverse vorticity measurements using a compact array of four sensors, ASME **FED 97**, 71–76 (1990)

5.809 J.F. Foss: Vorticity considerations and planar shear layers, Exp. Therm. Fluid Sci. **8**, 260–270 (1994)

5.810 S.C. Morris, J.F. Foss: Vorticity spectra in high Reynolds number anisotropic turbulence, Phys. Fluids **17**, 088102 (2005)

5.811 S.-C. Morris: *The velocity and vorticity fields of a single stream shear layer*. Ph.D. Thesis (Michigan State University, East Lansing 2002)

5.812 S.C. Morris, J.F. Foss: Boundary layer to shear layer - the transition region, J. Fluid Mech. **494**, 187 (2003)

5.813 K. Sreenivasan, A. Juneja, A. Suri: Scaling properties of circulation in moderate Reynolds number turbulent wakes, Phys. Rev. Lett. **75**, 433 (1995)

5.814 G. Batchelor: *Homogeneous Turbulence* (Cambridge Univ. Press, Cambridge 1954)

5.815 J.F. Foss, J.K. Schwannecke, A.R. Lawrenz, M.W. Mets, S.C. Treat, M.D. Dusel: The thermal transient anemometer, Meas. Sci. Technol. **15**, 2248–2255 (2004)

5.816 J.F. Foss, J.A. Peabody, M.J. Norconk, A.R. Lawrenz: Ambient temperature an free stream turbulence effects on the thermal transient anemometer, Meas. Sci. Technol. **17**, 2519–2526 (2006)

5.817 J.F. Foss: Thermal transient anemometer having sensing cell assembly, US Patent 7051599 (2006)

5.818 J.C. Kaimal, J.J. Finnigan: *Atmospheric Boundary Layer Flows: Their Structure and Measurement* (Oxford Univ. Press, New York 1994)

5.819 J.D. Anderson: *Modern Compressible Flow: With Historical Perspective*, 2nd edn. (McGraw-Hill, New York 1990)

5.820 J. Wyngaard: Cup, propeller, vane, and sonic anemometers in turbulence research, Ann. Rev. Fluid Mech. **13**, 399 (1981)

5.821 VDI-Guideline (Ed.): Turbulence measurement with sonic anemometers, VDI *3786*, Part 12 (VDI, Düsseldorf 1994)

5.822 M. Aubinet, A. Grelle, A. Ibrom, U. Rannik, J. Moncrieff, T. Foken, A.S. Kowalski, P.H. Martin, P. Berbigier, C. Bernhofer, R. Clement, J. Elbers, A. Granier, T. Grunwald, K. Morgenstern, K. Pilegaard, C. Rebmann, W. Snijders, R. Valentini, T. Vesala, A.H. Fitter, D.G. Raffaelli: Estimates of the annual net carbon and water exchange of forests: the EUROFLUX methodology, Adv. Ecol. Res. **30**, 113 (2000)

5.823 B. Duan, C.W. Fairall, D.W. Thomson: Eddy correlation measurements of the dry deposition of particles in wintertime, J. Appl. Meteorol. **27**, 642 (1988)

5.824 B.B. Hicks, D.R. Matt, R.T. McMillen: A micrometeorological investigation of surface exchange of O_3, SO_2 and NO_2: A case study, Bound.-Lay. Meteorol. **47**, 127 (1988)

5.825 A. Cuerva, S. López-Díez, D. Bercebal: Higher level descriptions of sites and wind turbines by means of principal component analysis, Proc. European Wind Energy Conference (James James Sciences, London 1999) 663

5.826 A. Fragoulis: The complex terrain wind environment and its effects on the power output and loading of wind turbines, Aerospace Sciences Meeting and Exhibit, AIAA (1997) 33–40

5.827 M.W. Rotach, S.-E. Gryning, E. Batchvarova, A. Christen, R. Vogt: Pollutant dispersion close to an urban surface: The BUBBLE tracer experiment, Meteorol. Atmos. Phys. **87**, 39 (2004)

5.828 N. Kato, T. Ohkuma, J.R. Kim, H. Marukawa, Y.J. Niihori: Full scale measurements of wind velocity in two urban areas using an ultrasonic anemometer, Wind Eng. Aerodyn. **88**, 67–78 (1992)

5.829 J. Finnigan: Turbulence in plant canopies, Ann. Rev. Fluid Mech. **32**, 519–571 (2000)

5.830 R.M. Petrone, J.S. Price, J.M. Waddington, H. Von Waldow: Effects of a changing surface cover on the moisture and energy exchange of a restored peatland, J. Hydrol. **295**, 198–210 (2004)

5.831 X. Lee, W. Massman, B. Law (Eds.): *Handbook of Micrometeorology: A Guide for Surface Flux Measurement and Analysis* (Kluwer Academic, Dordrecht 2004)

5.832 D. Banfield, R. Dissly, A.D. Toigo, P.J. Gierasch, W.R. Dagle, D. Schindel, D.A. Hutchins, B.T. Khuri-Yakub: Mars acoustic anemometer - eddy fluxes, Proc. 6th Int. Conf. Mars (2003) 3144

5.833 A. Cuerva, A. Sanz-Andres: Sonic anemometry of planetary atmospheres, J. Geophys. Res. Planets **108**, 5029 (2003), (DOI:10.1029/2002JE001944)

5.834 J.G. Webster: *Instrumentation and Sensors Handbook* (CRC, Boca Raton 1999)

5.835 V.-E. Suomi: Energy budget studies and development of the sonic anemometer for spectrum, AFCRC Tech. Rep. *56-274* (University of Wisconsin, 1957)

5.836 F. Dobson, L. Hasse, R. Davis: *Air-Sea Interaction - Instruments and Methods* (Plenum, New York 1982)

5.837 H.L. Fox, Bolt, Beranek, Newman (Eds.): AFCRL 68-0180 (1968)

5.838 S.E. Larson, F.W. Weller, J.A. Businger: Continuous wave sonic anemometer for investigation of atmospheric turbulence, J. Appl. Meteorol. **18**, 562–568 (1979)

5.839 A. Cuerva, A. Sanz-Andres: On the sonic anemometer measurement theory, J. Wind Eng. Aerodyn. **88**, 25–55 (2000)

5.840 J.C. Kaimal, J.E. Gaynor: Another look at sonic thermometry, Bound.-Lay. Meteorol. **56**, 401–410 (1991)

5.841 H. Liu, G. Peters, T. Foken: New equations for sonic temperature variance and buoyancy heat flux with an omnidirectional sonic anemometer, Bound.-Lay. Meteorol. **100**, 459 (2001)

5.842 A. Cuerva, A. Sanz-Andres, J. Navarro: On multiple-path sonic anemometer measurement theory, Exp. Fluids **34**, 345–357 (2003)

5.843 J.C. Kaimal, J.C. Wyngaard, D.A. Haugen: Deriving power spectra from a three-component sonic anemometer, J. Appl. Meteorol. **7**, 827–837 (1968)

5.844 H. Lenschow: *Probing the Atmospheric Boundary Layer* (Am. Meteorol. Soc., Boston 1986)

5.845 C. Wamser: The frequency response of sonic anemometers, Bound.-Lay. Meteorol. **84**, 231 (1997)

5.846 K. Henjes, P.K. Taylor, M.J. Yelland: Effect of pulse averaging on sonic anemometer spectra, J. Atmos. Ocean. Technol. **16**, 181–184 (1999)

5.847 A. Cuerva, A. Sanz-Andres: Steps to reach a safe scenario with sonic anemometers as standard sensor for wind measurements in wind energy, Proc. European Wind Energy Conference Nice (James James, London 1999) 649

5.848 A. Cuerva, A. Sanz-Andrés, S. Franchini: The role of ultrasonic anemometry in wind engineering. In: *Fourth European African Conf. Wind Engineering*, ed. by J. Náprstek, C. Ficher (ITAM AS CR, Prague 2005), , paper 152

5.849 A. Wieser, F. Fielder, U. Corsmeier: The influence of the sensor design on wind measurements with sonic anemometer systems, J. Atmosph. Ocean. Technol. **18**, 1585–1608 (2001)

5.850 J.M. Wilczak, S.P. Oncley, S.A. Stage: Sonic anemometer tilt correction algorithms, Bound.-Lay. Meteorol. **99**, 127–150 (2001)

5.851 R. Sozzi, M. Favaron: Sonic anemometry and thermometry: theoretical basis and dataprocessing software, Environ. Softw. **11**, 259–270 (1996)

5.852 J.J. Finnigan: A re-evaluation of long-term flux measurement techniques, Part II: Coordinate systems, Bound.-Lay. Meteorol. **113**, 1–41 (2004)

5.853 R.T. McMillan: An eddy correlation technique with extended applicability to non-simple terrain, Bound.-Lay. Meteorol. **43**, 231 (1988)

5.854 J.C. Wyngaard, S.-F. Zhang: Transducer-shadow effects on turbulence spectra measured by sonic anemometer, J. Atmosph. Ocean. Technol. **2**, 548 (1985)

5.855 J.A. Murray, J.C. Klewicki, C.A. Biltoft: An experimental investigation of sonic anemometer wake blockage effects, ASME, Vol. FED 161 (1993) 43

5.856 A. Grelle, A. Lindroth: Flow distortion by a solent sonic anemometer: Wind tunnel calibration and its assessment for flux measurement over forest field, J. Atmosph. Ocean. Technol. **11**, 1529 (1994)

5.857 C. Kraan, W.A. Oost: A new way of anemometer calibration and its application to a sonic anemometer, J. Atmosph. Ocean. Technol. **6**, 516 (1989)

5.858 D. Heinemann, D. Langner, U. Stabe, H. Waldl (Ed.), R. Watson: Measurement and correction of ultrasonic anemometer errors and impact on turbulence measurements, European Wind Energy Conference Dublin (1998) 409

5.859 N.G. Mortensen, J. Højstrup: The solent sonic - response and associated errors, Symp. Meteorology Observation and Instrumentation (Am. Meteorol. Soc., Boston 1992) 501

5.860 M.K. van der Molen, J.H.C. Gash, J.A. Elbers: Sonic Anemometer cosine response and flux measurement: II The effect of introducing an angle of attack dependent calibration, Agric. For. Meteorol. **122**, 95 (2004)

5.861 A. Cuerva, A. Sanz-Andres, O. Lopez: Singularities and un-definitions in the calibration functions of sonic anemometers, J. Atmosph. Ocean. Technol. **21**, 1868 (2004)

5.862 J. Wyngaard: Transducer-shadow effects on turbulence spectra measured by sonic anemometers, J. Atmosph. Ocean. Technol. **2**, 548 (1985)

5.863 P. Schotanus, F.T.M. Nieuwstadt, H.A.R. De Bruin: Temperature measurement with a sonic anemometer and its application to heat and moisture fluxes, Bound.-Lay. Meteorol. **26**, 81 (1983)

5.864 ISO: *Meteorology-Sonic anemometer/thermometers – Acceptance test methods for mean wind measurements*, Vol. 16622, 1st edn. (ISO, 2001)

5.865 S. Cervenka (Ed.): Selection of an Anemometer For Measurement of Wind Turbulence, ECN-I-92-029 DE92 557290 (Energieonderzoek Centrum Nederland, 1992)

6. Density-Based Techniques

The methods dealt with in this section are based on changes of fluid density; hence its index of refraction. As a result of these changes, optical phase and, coupled with it, direction of propagation of a light wave transmitted through the flow are altered in comparison to the properties of the incident light. The available signal can be presented in planar form, i.e., as a flow picture, and the methods are often referred to as *optical flow visualisation*, because the changes in index of refraction are detected and measured by optical techniques. The obtainable information is integrated along the whole path of the light in the fluid field (*line-of-sight methods*) and, in a three-dimensional (3D) object field, special techniques for interpreting the signal pattern are necessary (*tomography*) in order to provide local data values of the quantity to be determined, e.g., density. Four major groups of experimental methods can

6.1	Density, Refractive Index, and Optical Flow Visualization	473
6.2	Shadowgraphy	474
6.3	Schlieren Method	476
6.4	Moiré Deflectometry	478
6.5	Interferometry	480
6.6	Optical Tomography	485
References		485

be distinguished: shadowgraphy, schlieren technique, moiré techniques, and interferometry. The fluid mechanical problem areas to which these optical measuring techniques can be applied are compressible flow, convective heat transfer, mixing and mass transfer, combustion, and flows with density stratification.

6.1 Density, Refractive Index, and Optical Flow Visualization

The methods dealt with in this section are essentially those that provide information on the state of flow using light, which is transmitted through the flow field as depicted in Fig. 11.1 of Chap. 11. Optical phase and, coupled with it, the direction of propagation of the transmitted light wave are altered compared to the properties of the incident wave if the fluid's refractive index varies in the flow. The refractive index n is related to the fluid density through the Clausius–Mosotti equation [6.1], which for a gas reduces to the simpler form of the Gladstone–Dale equation

$$n - 1 = K\rho \qquad (6.1)$$

with ρ being the gas density and K the *Gladstone–Dale constant*, which has the dimension of $1/\rho$, is specific for a gas, and depends weakly on the wavelength of light used [6.1]. In compressible flow of an ideal gas the density ρ is a function of the Mach number [6.2], and, for these flows, the information obtainable with the methods to be discussed in this section is therefore a measure of the Mach number or flow velocity.

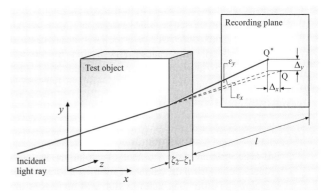

Fig. 6.1 Refractive deflection of a light ray in an object field (flow) with varying refractive index (caused by varying fluid density)

Table 6.1 Methods of optical flow visualisation

Method	Signal form	sensitive to changes of
Shadowgraph	Displacement Δ	$\partial^2\rho/\partial x^2, \partial^2\rho/\partial y^2$
Schlieren, moiré	Deflection angle ε	$\partial\rho/\partial x, \partial\rho/\partial y$
Speckle photography	Deflection angle ε	$\partial\rho/\partial x, \partial\rho/\partial y$
Shearing interferometry	Phase difference	$\partial\rho/\partial x, \partial\rho/\partial y$
Reference beam interferometry	Phase difference	ρ

The methods dealt with below collect and present the signal in planar form, i.e., as a flow picture, and they are often referred to as *optical flow visualisation*, because the aforementioned changes in optical phase are detected and measured by optical techniques. As for any *line-of-sight method* the obtainable information is integrated along the whole path of the light in the fluid field and, in a three-dimensional (3-D) object field, special techniques for desintegrating the signal pattern are necessary in order to provide local data values of the quantity to be determined, e.g., density. This desintegration, with the aim of reconstructing the 3-D object from a number of two-dimensional (2-D) data sets, requires the transmission of light waves through the object in different directions and is called optical tomography (Sect. 6.6).

The propagation of a light wave or ray through an inhomogeneous transparent object field, here a flow with variable fluid density, can be analysed by applying Fermat's principle (see [6.1] and references therein). As a result of the variation of the refractive index, the light ray is deflected from its undisturbed, original direction. The signal evidencing this deflection in a recording plane (Fig. 6.1) is either the displacement with the two components Δ_x, Δ_y or the deflection angle with its two components ε_x, ε_y; z is the direction of the incident light ray. In addition to the light deflection, the optical phase of the wave is different from the phase that the wave would have in the recording plane in the absence of the object. This difference in optical phase (or its equivalent expressed as an optical path length) can also be transformed into a visible signal. The signal patterns are collected in the recording plane, e.g., with a photographic or electronic camera. By means of the analysis of wave propagation in the refractive index field, one can relate the observable signal to the refractive index or density distribution in the flow under study.

According to the different signal forms (displacement Δ, deflection angle ε, optical phase difference) one can distinguish between three different groups of methods of optical flow visualisation. They also differ systematically in the relationship between the observable or measurable signal and the fluid density (Table 6.1).

Besides compressibility there may be other reasons for alterations of the fluid density in a flow, the most common being temperature differences as the result of heat transfer processes. Another cause is differences of concentration values in an inhomogeneous fluid with several components; the related fluid-mechanical process is mixing, which applies to both gaseous and liquid flows. Both mixing and heat transfer cause combustion processes to exhibit variations in density, so that they are often the object of optical flow visualisation. Finally, flow processes of a fluid that, in equilibrium, has a natural vertical density gradient (*stratified flow*) can be investigated with the methods discussed here. In some of these fields the techniques of optical flow visualisation were developed independently and with exclusive applicability to the respective field [6.3–5].

For a quantitative measurement of the density or one of its derivatives the exact relationship between refractive index and density must be known. Values of the Gladstone–Dale constant appearing in (6.1) or relationships applying either to pure or salted water (used for experiments with stratified flows) are included in [6.1].

6.2 Shadowgraphy

In its simplest form, the shadowgraph does not need any optical component, and the effect can therefore be observed in many situations outside of a laboratory. An essential feature of the method is the use of a point-shaped light source. The light diverging from this source is transmitted through the test object, and a shadow pattern caused by an inhomogeneous density field can be observed in a vertical plane at a distance l behind the object [6.1, 6]. More common for laboratory experiments is an arrangement with parallel light through the flow, which might be bounded by viewing windows (Fig. 6.2). In order to be able to record a shadowgraph of reduced size with a camera, the recording plane (photographic film, chip of an electronic camera) is focused by means

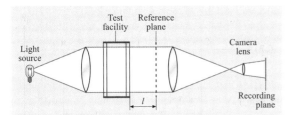

Fig. 6.2 Shadowgraph setup with parallel beams through the test object

of the camera lens onto the plane at a distance l from the object. In this way, the object is not in focus in the shadow picture.

When passing through the flow field under study, the individual light rays are refracted and bent out of their original path as indicated in Fig. 6.1, so that the light intensity in the plane of observation is altered with respect to the undisturbed case. By means of an analysis of the light propagation in the refractive index field of the object [6.1], one derives for the relative changes of light intensity in the plane of observation

$$\frac{\Delta I}{I} = l \int_{\zeta_1}^{\zeta_2} \left(\frac{\partial^2}{\partial x^2} + \frac{\partial^2}{\partial y^2} \right)(\ln n)\, dz\,, \qquad (6.2)$$

where I is the light intensity, l is explained in Fig. 6.2, z is the optical axis and direction of light propagation, and x and y are the coordinates in planes normal to the optical axis, $n(x, y, z)$ is the refractive index in the flow; $\Delta I/I(x, y)$ is the observable signal, i.e., a distribution of shades or values of grey in the recording plane; ζ_1 and ζ_2 are the entrance and exit coordinates, respectively, of the light rays transmitted through the test object. The integral evidences the property of all *light-of-sight* methods: The information on the quantity of interest, here density, is integrated along the light path in the object.

Equation (6.2) is derived with the assumption that the displacements of a light ray as shown in Fig. 6.1, Δ_x and Δ_y, are small, e.g., in comparison to l (*weak refraction*). By applying the Gladstone–Dale formula, (6.1), it becomes evident that the shadowgraph is sensitive to changes of the second derivative of the (gas) density. From this it follows that the shadowgraph is not a method suitable for a direct quantitative measurement of density, since such an evaluation would require one to perform a double integration of the signal data. However, owing to its simplicity, the method is convenient for obtaining a quick survey of flow fields with varying fluid density.

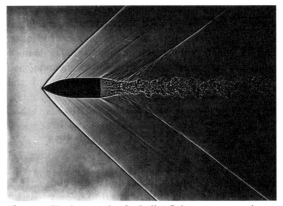

Fig. 6.3 Shadowgraph of a bullet flying at supersonic velocity (courtesy Deutsch-Französisches Forschungsinstitut, ISL, St. Louis, France)

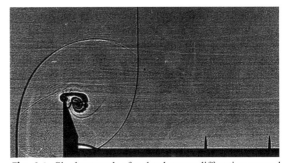

Fig. 6.4 Shadowgraph of a shock wave diffracting around a vertical wedge; vortex formation at the apex of the wedge (courtesy Deutsch-Französisches Forschungsinstitut, ISL, St. Louis, France)

An indirect determination of the density distribution in a flow was proposed for cases when a numerical solution of the flow field is available: by inserting the computed density distribution $\rho(x, y, z)$ into (6.2), one may construct a shadow pattern, and it is common to conclude that the numerical solution is correct if the computational and the experimental shadowgraphs look alike [6.7, 8]. Such procedures, that are taken as *validation* of the computational results, are also known for the methods described in the following sections. It appears that any conclusion about the validity of the numerical result from a similarity of the computed and experimentally visualised flow patterns must be done very carefully, because small deviations of the two patterns might still be associated with large differences in the respective density distributions.

The most drastic change in the second derivative of n or ρ occurs in a shock wave. The shadowgraph is most

6.3 Schlieren Method

The German word *Schliere* designates an inhomogeneity or disturbance in a transparent medium, most often with reference to a glass plate or window. The credit for having introduced the schlieren method as a tool for visualising density inhomogeneities in a flow is usually given to Toepler, as outlined in the reviews of *Schardin* [6.9] and *Settles* [6.6], and the method is often referenced as the *Toepler schlieren method*, in order to distinguish this technique from other optical methods that are also associated with the word *schlieren* (see below). A system with parallel light through the test object, e.g. compressible flow, is considered (Fig. 6.5). An image of the light source is formed in the plane of the knife edge, which is placed in the focal plane of the second lens, called the *schlieren head*. Here, the knife edge is perpendicular to the plane of the figure, and the light source is either point-shaped or a narrow slit parallel to the knife edge. The camera lens serves to form an image of the test object on the recording plane (photographic film, chip of an electronic camera). Thus, the test object is in focus, and the formation of shadow effects (Sect. 6.1) is avoided.

If the light source is cutting off part of the image of the light source, the light intensity with which the recording plane is illuminated will be reduced. Let a be the reduced height of the light source image as seen in the direction of the optical axis (Fig. 6.6). An optical analysis [6.1, 6, 9] then delivers for the distribution of the relative light intensity in the recording plane

$$\frac{\Delta I}{I} = \frac{f_2}{a} \int_{\zeta_1}^{\zeta_2} \frac{1}{n} \frac{\partial n}{\partial y} dz . \qquad (6.3)$$

As in (6.1), $n(x, y, z)$ is the refractive index in the fluid flow, $I(x, y)$ is the light intensity in the recording plane, f_2 is the focal length of the schlieren head, and z is the direction of the optical axis. Note the integration of the information along the trajectories of the light in the test object, i.e. from ζ_1 to ζ_2. Equation (6.3) is derived under the assumption of very small angles of light deflection, ε (Fig. 6.1). If the system is applied to a gas flow, then $n \approx 1$, and with the Gladstone–Dale constant K, the equation can be written as

$$\frac{\Delta I}{I} = \frac{K f_2}{a} \int_{\zeta_1}^{\zeta_2} \frac{\partial \rho}{\partial y} dz \qquad (6.4)$$

from which it follows that the schlieren method is sensitive to changes of the density derivative normal to the knife edge, here $\partial \rho / \partial y$, because the knife edge in Fig. 6.5 is assumed to extend in the x-direction. By turning the knife edge (and the slit source) by 90° around the optical axis, one measures the component $\partial \rho / \partial x$.

From (6.3, 4) it follows that, for a given density distribution in the flow, the relative change in light intensity

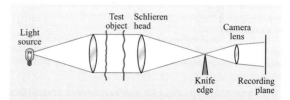

Fig. 6.5 Schlieren setup with parallel light through the test field

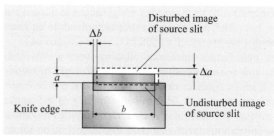

Fig. 6.6 Image of a light source of size $a \times b$ in the focal plane of the schlieren head, as seen in the direction of the optical axis; shift of the light source by Δa and Δb, respectively, caused by light deflection in the refractive index object

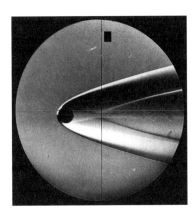

Fig. 6.7 Schlieren photograph of a sphere flying at supersonic velocity [6.1]

and therefore the contrast in the recording plane is larger the smaller the ratio a/f_2. Usually, in a schlieren setup it is not possible to replace the schlieren head during an experiment, and f_2 might be taken as fixed. There are several reasons why a cannot be reduced arbitrarily, the most important being diffraction, whose influence on the image quality must be avoided. Another reason results from the demand that the system should be capable of measuring light deflections in positive and negative directions, in Fig. 6.1 expressed by $+\varepsilon_y$ and $-\varepsilon_y$ when the knife edge is in the x-direction. In the schlieren system shown in Fig. 6.5 $+\varepsilon_y$ generates an increase of the light intensity, $-\varepsilon_y$ an intensity loss. Therefore, a should be chosen such that, in the undisturbed case, the (background) intensity of illumination is reduced, and light deflections in the positive and negative direction can be distinguished by a brighter or darker illumination at the respective position in the schlieren image (Fig. 6.7). High image quality is obtained if the background intensity in the recording plane, e.g. illumination without flow, is constant throughout this plane. This requires the use of a point source or slit source with minimal slit height a.

The term *schlieren method* is limited to optical systems in which the light distribution in the image of the source (focal point of schlieren head) is somehow manipulated, here by the knife edge. This should be kept in mind if the word *schlieren* appears in connection with other methods that are based on different optical principles (*background oriented schlieren* and *synthetic schlieren* in Sect. 6.4, and *schlieren interferometer* in Sect. 6.5). The *manipulation* of light can be defined more precisely, if one follows the descriptions of the schlieren method by means of system analysis or Fourier optics [6.10–12]. In terms of these analyses the knife edge acts, in the Fourier domain, as a frequency filter cutting off part of the transmitted frequency spectrum.

A great variety of modifications are reported in the literature; these aim at increasing the diameter of the light beam transmitted through the object, allowing the study of extended test objects, increasing the optical sensitivity towards the resolution of weak density changes, introducing some kind of quantitative assessment of the signal, etc. [6.1, 6]. A closer look at the proposed modifications shows that almost none of these were not already included in Schardin's fundamental work on schlieren systems [6.9]. The most interesting extension of the classical schlieren system, whose signal pattern is a planar distribution of grey levels, is the addition of colour to the signal pattern, e.g. by means of a transparent colour filter that replaces the schlieren knife edge. In addition to the production of very decorative flow pictures, the use of

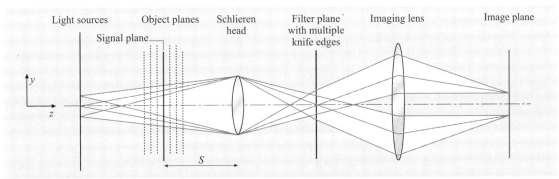

Fig. 6.8 Principle of a *sharp-focusing* schlieren system. Several light beams (*channels*, only three shown here), each originating from a source in the source plane, pass through the test field in different directions. Every channel has its knife edge in the filter plane where the schlieren head forms an image of each source. The signal plane at a distance S from the schlieren head is imaged onto the image plane. The 3-D refractive index field can be scanned by imaging different object planes onto the imaging plane

colours allows one to discriminate between the refractive deflection of light in the positive and negative directions, as well in the *x*- and *y*-directions; *Settles* [6.6, 13] presents the most complete and informative review of colour schlieren systems.

Due to the *line-of-sight* effect, in (6.2, 3, 4) expressed by the integration along *z*, information on $\partial\rho/\partial y$ or $\partial\rho/\partial x$ can be obtained directly only for a nominally two-dimensional (2-D) object with $\rho = \rho(x, y)$. Information on a three-dimensional (3-D) flow field is available by applying the techniques of optical tomography (Sect. 6.6). However, for the schlieren method, a special approach for resolving a 3-D density field is known. The *sharp-focusing schlieren system* [6.14] uses multiple off-axis point sources, with light beams or *channels* originating from each source and traversing the 3-D test field in different angular directions (Fig. 6.8). Each source has a corresponding knife edge or filter in the focal plane of the schlieren head. The image plane where all channels overlap is focused onto a particular plane in the object, normal to the *z*-direction, and the information from the overlapping signals is a measure of the desired quantity, here the fluid density gradient, in the focused object plane, while the contributions to the signal from regions outside of the focused object plane average out and reduce to noise. In practice, the focused object plane is a slice with small but finite extension in the *z*-direction, the direction of the optical axis. By means of the imaging lens, the position of the object plane, which is in focus, can be varied, and the 3-D object can thus be scanned in the *z*-direction.

Extensive information on the setting up of such a *sharp-focusing* or *multiple-source* schlieren system has been given by *Weinstein* [6.14]. Both from experiments and numerical simulations [6.12, 15] it is apparent that using five sources or channels provides a spatial resolution of the object that is barely improved by increasing this number. It cannot be expected that a satisfactory determination of the 3-D density distribution $\rho(x, y, z)$ is possible with this technique. However, it has been proven that valuable information on the spatial distribution of flow structures in compressible turbulent flows can be obtained with the *sharp-focusing* schlieren technique [6.16].

6.4 Moiré Deflectometry

Due to light deflection by a variable density field a geometrical pattern seen through this field appears deformed. The local deflections are described quantitatively by the displacements Δ_x and Δ_y shown in Fig. 6.1. Since these displacements are a measure of the respective deflection angles ε_x and ε_y, the measurement of the local displacement of the geometrical pattern, as seen in the recording plane, allows one to determine the local deflection angles and thus the derivatives of the density, $\partial\rho/\partial x$ and $\partial\rho/\partial y$, which are linked to the deflection angles by integral relationships similar to (6.4).

The deformation of a given pattern when viewed through a density field is a phenomenon of daily experience; e.g. it can be realised if an object is observed through the warm air (plume) rising from a candle flame. The use of this effect for quantitative measurement has been known for a long time [6.1]. The geometrical pattern used for this purpose can be regular or random. When a regular pattern in the form of a grid is used, the technique is known as the *moiré method*; other names are *deflection mapping* or the *Ronchi method* (with a Ronchi grid). The grid, which can be placed on either side of the test object, can be different in shape, showing, e.g., equidistant lines, or regular patterns of concentric circles or squares. In principle, one observes through the test field a pattern of known geometry, which is distorted due to the refractive deflection of the light rays in the density field (Fig. 6.9). Many variations of moiré grids or patterns have been reported, including the use of two grids, one on either side of the test object, such that a situation that resembles the fringe patterns of an interferometer can be generated (Sect. 6.5).

These moiré or Ronchi methods are based on the use of a regular and macroscale geometrical reference pattern; *macroscale* in the sense that the grid dimensions are not very small in comparison to typical scales of

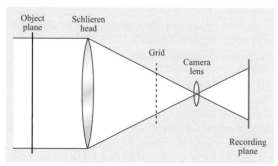

Fig. 6.9 Possible setup for deflection mapping by a Ronchi grid

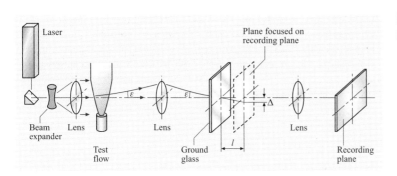

Fig. 6.10 Setup for deflection mapping by speckle photography [6.17]

the flow field to be visualised. In contrast to this situation, for the methods named *laser speckle photography*, microscale reference patterns of random geometry are used. These patterns, whose deformation caused by the refractive light deflection is studied, are generated by the optical speckle effect; for reviews see [6.17–19].

In the optical setup a ground glass placed in the light beam serves as the speckle-generating element (Fig. 6.10). Here, the test object is placed in front (left side) of the ground glass, and is imaged by means of a lens onto the plane of the ground glass, while the recording plane is focused onto a plane at a distance l from the ground glass. With a laser light source, optical speckles are generated to the right of the ground glass as a result of multiple interference of the light scattered from the elements of the ground glass. Two exposures are recorded and superimposed with the arrangement shown in Fig. 6.10: one exposure in the absence of the test object, and a second exposure with the object. Due to the refractive deflection in the second exposure, corresponding light rays from the two exposures appear separated by a displacement Δ in the plane at a distance l from the ground glass. The two components of this displacement, Δ_x and Δ_y, can be measured simultaneously in the recording plane, and with the relationship $\varepsilon \approx \Delta/l$, the two components of the density derivative, $\partial\rho/\partial x$ and $\partial\rho/\partial y$, can be determined. Since the speckle pattern as the reference pattern has a fine and evenly distributed structure, the distribution of $\Delta(x, y)$ can be measured practically at any point in the recording plane, i.e. with very high spatial distribution, in contrast to the methods using regular (macroscale) grids, where the information is only available at the position of the grid lines.

The key for a successful application of the speckle method is the evaluation of the two recorded, superimposed speckle patterns by means of digital image processing. The same procedures as used for particle image velocimetry are used here, e.g., evaluation by means of correlation techniques, in order to determine the distribution of the displacement vectors $\Delta(x, y)$ (Sect. 5.3.2). In fact, such evaluation algorithms on the basis of correlation schemes were developed for the speckle method prior to their use in particle image velocimetry. Speckle photography renders no direct visualisation of the flow with density variations; a visualisation is possible by subjecting the recorded speckle pattern to a spatial filtering process, and the result is similar to a schlieren pattern, but with much lower image quality [6.17–19].

While the speckle method uses a speckle pattern generated purposely by inserting a ground glass into the light path, the *background oriented schlieren method* makes use of a reference pattern that is naturally available, e.g., the structure of the background in the rear of the transparent object under study [6.20]. The background pattern as seen through the test region is recorded twice: once without flow, a second time with flow. The two overlapping patterns differ slightly due to the refractive deflection of the light in the test flow and, as in the case of the speckle method, the local displacement vectors $\Delta(x, y)$, which are a measure of the light deflection angles $\varepsilon(x, y)$, can be determined by digital image processing (correlation methods). This technique has practically no limit regarding the diameter of the field of view, and allows the study of large test objects outside the laboratory. Here, the term *schlieren* is used because the method gives information on the deflection angle or the density derivative, like a schlieren system; but this term is misleading here because the setup is not a schlieren system according to the definition given in Sect. 6.3. The same argument applies to a method called *synthetic schlieren* which uses, in principle, two moiré grids [6.21]. However, in contrast to the scales applicable to conventional methods, the second grid is *virtual* and employs the digitisation of the field of view and the edges of pixels to form a reference pattern.

In contrast to *classical* schlieren technique (Sect. 6.3), all moiré methods aim to provide quantitative information on the amount of light deflection in the flow with density variations. In comparison to schlieren, the optical setup is less sensitive to disturbances and therefore less costly, it is easier to investigate objects of large dimension, and the application of digital image processing for quantitative evaluation is straightforward; but the visual impression of schlieren pictures is probably much bigger; schlieren pictures, if well made, can appear more attractive than moiré records, and the optical sensitivity of the schlieren system is perhaps higher than that of a moiré setup.

6.5 Interferometry

The change in optical phase of a light wave transmitted through the flow field can be measured by instruments known as interferometers. The instruments applied to optical flow visualisation or the measurement of fluid density are exclusively *two-beam interferometers*, i.e., the visible signal pattern, called an *interferogram*, is the result of optical interference of two light waves, of which at least one is transmitted through the object. If this object wave interferes with a second wave that remains undisturbed and thus serves as a reference, one has *reference beam interferometry*. If both waves pass through the object field, but displaced by a small distance with respect to each other or *sheared*, they belong to the class of *shearing interferometry*, sometimes also called *differential* or even (and without technical reason) *schlieren interferometry*.

The principle of two-beam interferometry is explained in Fig. 6.11 by considering light rays instead of waves [6.1]. A parallel beam of light is passed through the test object along the z-direction. Each ray of the beam is considered to interfere with a conjugate ray separated from the first ray by a distance d. Fig. 6.11 is a projection of the y–z plane; the test field is bounded by the surfaces $\zeta_1(x, y)$ and $\zeta_2(x, y)$, and for the following analysis it is assumed that these surfaces are planes $z = $ const., which is no general restriction. Behind the test object, a lens or mirror, equivalent to the *schlieren head* in Sect. 6.3, focuses the light beam, and a black box, here called the *interferometer unit*, ensures that all *conjugate rays* either coincide after having passed through this unit, or are made to intersect in the recording plane, where they interfere and produce the visible interference pattern.

The difference in optical path length between two conjugate rays is

$$\Delta l = \int_{\zeta_1}^{\zeta_2} n\left(x, y + \frac{d}{2}, z\right) dz$$
$$- \int_{\zeta_1}^{\zeta_2} n\left(x, y - \frac{d}{2}, z\right) dz . \quad (6.5)$$

The optical phase difference is obtained by dividing Δl by the wavelength λ. Bright interference fringes appear in the recording plane where

$$\frac{\Delta l(x, y)}{\lambda} = 0, \pm 1, \pm 2, \ldots, \quad (6.6)$$

which therefore represents the equation of fringes in the recording plane (the integers designate the *fringe order*). Thus, $\Delta l(x, y)/\lambda$ is the signal in planar form, which serves to determine the unknown quantity n; the problem that the signal is two dimensional, whereas the quantity of interest $n(x, y, z)$ is three dimensional, was mentioned above (*line-of-sight* methods). However, (6.5) incorporates a further problem: three is only one equation for two unknowns, namely n at two different positions, $y + d/2$ and $y - d/2$. A solution to this problem is provided by an appropriated design of the *interferometer unit* in Fig. 6.11, thereby reducing the number of unknowns to only one. Two different approaches that define two different classes of interferometers are known. They are characterised by the magnitude of the ratio d/D, with D being the diameter of the field of view or the diameter of the parallel light beam in Fig. 6.11:

1. With $d/D \geq 1$, one ray of each pair of conjugate rays propagates outside the test field, where it remains undisturbed; this is the class of *reference beam interferometers*. One of the two integrals in (6.5) that

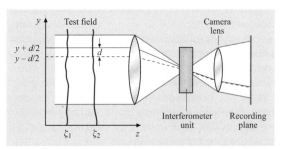

Fig. 6.11 Principle of a two-beam interferometer with parallel light through the test field

describes the optical path covered by the *reference beam* reduces to a constant.

2. For $d/D \ll 1$, both rays traverse the test field, where they are separated or *sheared* by the small distance d. It is common to call such systems *shearing interferometers*. Developing (6.6) into a Taylor series around the value of $n(x, y, z)$ and taking into account only the linear terms (the quadratic terms vanish), because d is a small quantity, results in

$$\frac{d}{\lambda} \int_{\zeta_1}^{\zeta_2} \frac{\partial}{\partial y} n(x, y, z) \, dz = 0, \pm 1, \pm 2, \ldots \quad (6.7)$$

as the equation of fringes for a shearing interferometer. It is seen that the signal is proportional to the first derivative of the refractive index or the gas density, as is the case for schlieren systems (6.3, 4), which explains the name *schlieren interferometer*, occasionally used for this class of instruments. Another name, *differential interferometer*, results from the appearance of the derivative in (6.6). The derivative $\partial n / \partial x$ can be measured by rotating the system around the optical axis (z-axis) by 90°.

Equations 6.6 and 6.7 apply to alignment of the interferometer for which a uniform test field (with $n = \text{const.}$ or $\partial n / \partial y = \text{const.}$) appears uniformly illuminated, i.e., no fringes are seen in the field of view. This case is called the *infinite fringe width* (IFW) alignment. Both interferometers can be aligned in a different way, so that a system of equidistant, parallel interference fringes appears for the uniform test field. A density variation in the flow will distort this regular fringe pattern, and the deviation or shift of a fringe from its undisturbed position is then the signal and a measure for the density variation. An alignment with this signal form is called the *finite fringe width* (FFW) alignment. A fringe shift by one fringe width (a distance S of the fringes in the undisturbed case) is equivalent to a difference in optical path length of one wavelength. It follows that a fringe shift ΔS at a point (x, y) measured in terms of the width S is

$$\frac{\Delta S(x, y)}{S} = \frac{1}{\lambda} \int_{\zeta_1}^{\zeta_2} n(x, y, z) \, dz - \text{const.} \quad (6.8)$$

for the reference interferometer, and

$$\frac{\Delta S(x, y)}{S} = \frac{d}{\lambda} \int_{\zeta_1}^{\zeta_2} \frac{\partial}{\partial y} n(x, y, z) \, dz \quad (6.9)$$

for the shearing interferometer. The experimenter can select either the IFW or FFW alignment; the choice is made according to the special experimental conditions.

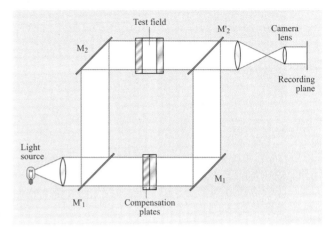

Fig. 6.12 Principal setup of a Mach–Zehnder interferometer; M_1, M_2 are mirrors; M'_1, M'_2 are beam splitters

The classic reference beam instrument is the Mach–Zehnder interferometer (MZI), which has been in use since the end of the 19th century. It is designed to allow a wide separation of test and reference beam, i.e., $d/D > 1$ (Fig. 6.12). The *interferometer unit*, mentioned in Fig. 6.11, is here the second beam splitter M'_2, while the first beam splitter M'_1 separates the incident beam into the test and reference beam. If all mirrors and beam splitters, M_1, M_2, M'_1, M'_2, are aligned exactly parallel, one has the IFW mode. Tilting one of the plates, usually M'_2, provides the FFW mode. For flow studies, particularly in a wind tunnel, a large-diameter field of view D is desired. This results in tight requirements on the optical and mechanical precision of the mirrors, beam splitters, test section windows, and compensation plates (which compensate for the optical path length experienced by the test beam in the test section windows). These requirements made the MZI a very expensive instrument, difficult to handle, and sensitive to mechanical disturbances or vibrations. The invention of holography opened a new route to develop an attractive alternative to the MZI: holographic interferometry.

The key to holographic interferometry is that it is possible to store on one holographic plate the information of two (or more) light waves and to simultaneously release, or reconstruct, the originally separate information in each [6.5, 22]. The reconstructed waves coincide spatially and can interfere if the conditions for optical coherence are fulfilled. In a holographic interferometer two consecutive holographic exposures are taken through the field of interest, most often the first exposure without flow, and the second in the presence of the flow with varying fluid density. Here, the first

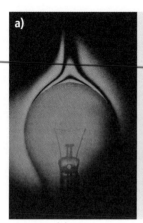 

Fig. 6.13 (a) Mach–Zehnder interferogram visualising the plume rising from a hot light bulb (IFW mode). **(b)** Holographic interferogram visualising the temperature fields outside and inside the hot bulb (IFW mode)

exposure is the reference wave, while the second exposure carries information about the flow and is therefore the test wave. Upon simultaneous reconstruction of the two holographic recordings, one produces an interfering wave pattern equivalent to the interfering test and reference beams in the MZI. The difference between the two approaches is that in the MZI the reference and test beams exist simultaneously but are separated in space, whereas in a holographic interferometer the two beams coincide spatially but are separated in time, because they are recorded at different instants of time. From this spatial coincidence follows an important advantage of holographic interferometry compared to Mach–Zehnder interferometry: disturbances in the optical components of the setup, which cause distortions of the MZI fringe system, are cancelled out in a holographic interferometer because they affect the test and reference beam in the same way and at the same position. This is exemplified in Fig. 6.13. The Mach–Zehnder interferogram of a hot light bulb visualises the plume rising in the air above the bulb, but nothing is seen inside the bulb because the glass is low quality, and thus does not allow interference of light rays that pass through the bulb. In the holographic interferogram the disturbances caused by the bulb glass are compensated, and interference fringes visualise the temperature field inside the bulb.

In a typical setup of a holographic interferometer (Fig. 6.14) a laser beam is separated by a beam splitter into two beams. One of them, after being expanded, traverses the test field as a collimated beam and then strikes the holographic plate in the recording plane. The second beam is fed outside the test field where it remains undisturbed and, after expansion by a second beam expander, it also strikes the holographic plate. This second beam is needed as the holographic reference beam, and its role must not be confused with that of the reference beam in the MZI. The holographic reference beam is also needed to reconstruct the object from the hologram [6.22].

A number of modifications of the holographic setup shown in Fig. 6.14 are described in the literature, among which some allow the recording of real-time holographic recordings by means of cinematography. A particular modification of the setup is the generation of a certain degree of three-dimensional information. For this purpose it is necessary to introduce a ground glass as a light diffuser into the optical path, designated by D in Fig. 6.14. The diffuser scatters the light into various directions, such that a continuum of beam directions traverses the test field and, as a consequence, the reconstructed scene may be examined under different viewing angles. In this way, different interferograms can be produced from the same holographic exposure for each viewing angle. However, the total angle of possible viewing directions is limited, and one holographic interferogram of this

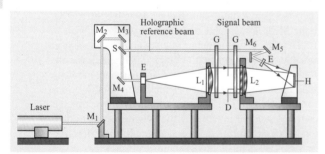

Fig. 6.14 Setup of a holographic interferometer [6.1]. M_1–M_6 mirrors; L_1, L_2 lenses; S beam splitter; E beam expander; G glass windows of test section; D diffuser (ground glass); H holographic plate

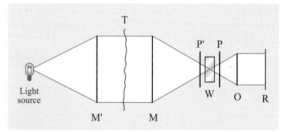

Fig. 6.15 Setup of a shearing interferometer using a Wollaston prism W and parallel light through the test field T; M, M' are lenses or spherical mirrors; P, P' polarisers; O imaging lens; R recording plane

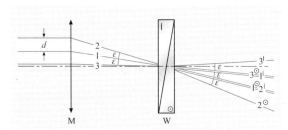

Fig. 6.16 Shearing of the light beam by the Wollaston prism W; M is the schlieren head (Fig. 6.15)

Fig. 6.17 Mach–Zehnder interferogram of Bénard convection in a 2-D liquid fluid cell; fringes are curves of constant temperature (W. Merzkirch)

kind is usually not sufficient for a complete tomographic study of the test field; see Sect. 6.6.

The dependence of a shearing interferometer's signal on the (first) derivative of the density (6.7, 9) makes these devices interesting for studies in convective heat transfer. A great variety of shearing interferometers are known; the most common setup applied in flow studies uses a Wollaston prism as the *interferometer unit* [6.1]. The setup sketched in Fig. 6.15 is similar to that of a schlieren system; the knife edge is simply replaced by the combination of the Wollaston prism with two polarisers, one in front and one behind the prism, and the two polarisers rotated by 90° with respect to each other. The similarity with the schlieren system is another reason for the name *schlieren interferometer*. A Wollaston prism splits each incident ray into two diverging rays separated by an angle ε, which is given as a constant of the system (Fig. 6.16). The two separated rays are linearly polarised, with polarisation directions normal to each other, in Fig. 6.16 characterised by the symbols $\otimes$ and $|$. It can be seen that each ray $i\otimes$ coincides with a ray $(i+1)|$, and these two rays (or waves) can interfere due to the presence of the second polariser behind the prism. The first polariser in front of the prism (Fig. 6.15) simply provides equal intensity of the conjugate rays. The two interfering rays traverse the test field at different positions, separated by the distance d (Fig. 6.16). The quantity d is the product of the focal length of the lens (or *schlieren head*) M and the separation angle ε (with $\varepsilon \ll 1$).

If the centre of the prism coincides with the focal point of M (Fig. 6.15), and if there is a homogeneous density distribution in the test field, all interfering rays will not exhibit a phase difference, and the field of view is free of interference fringes (*infinite fringe width* alignment, IFW). A shift of the prism from this central position along the optical axis (z-direction) results in the formation of parallel, equidistant fringes (*finite fringe width* alignment, FFW). It can be shown [6.1] that the fringe width in the FFW mode is

$$S = \frac{\lambda f_2}{\varepsilon w} \qquad (6.10)$$

with λ being the wavelength of the used light source, f_2 the focal length of M, ε the separation angle of the Wollaston prism, and w the shift of the prism from the central position along the optical axis. The direction of the measured density gradient [y in (6.9)] is normal to the undisturbed fringes in the FFW mode. A vertical shift of the prism causes no change in the fringe width; it only changes the order of the fringes appearing in the field of view [6.1].

In a shearing interferogram solid contour lines, e.g. the edge of a solid body or the frame of the field of view, appear in the form of a double image or gray band of width $d\cos\alpha$, where α is the angle between the direction of shear [y in (6.9)] and the normal to the wall. The exact position of the wall edge is in the middle of the gray band. A shock wave causing a density jump is also depicted as a band of width $d\cos\alpha$.

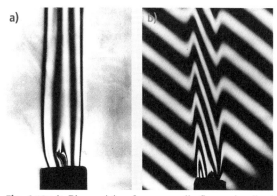

Fig. 6.18a,b Plume rising from a candle flame visualised by a Wollaston prism shearing interferometer; (**a**) IFW mode; note the double image of the candle contour line; (**b**) FFW mode (W. Merzkirch)

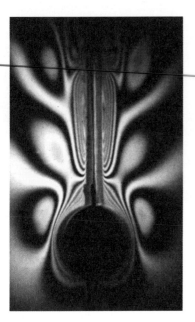

Fig. 6.19 Formation of an internal gravity wave in density-stratified saltwater visualised by a shearing interferometer; near field around a solid cylinder towed in the downward direction (W. Merzkirch)

Typical examples of interferograms are presented in Figs. 6.17, 6.18, 6.19. They include a reference beam interferogram of a two-dimensional flow field taken in the IFW mode such that the fringes are curves of constant density (here constant temperature), and shearing interferograms in the IFW and FFW modes, one taken with a *white* light source and therefore being coloured since the position of the fringes then depends on the wavelength λ (6.7).

In an interferogram the information on optical path length or optical phase, and therefore on fluid density, is given along the interference fringes. As expressed by (6.5) and (6.7) this information is available in discontinuous form, because the fringes are separated by a path length difference of a wavelength or by a phase difference of 2π. A number of methods have been communicated for automated evaluation of an interferogram, and have been continuously improved along with the progress made in digital image processing (e.g. [6.24]). With these methods it is possible to interpolate the information in the space between two interference fringes. The result of such an interpolation is less precise, the lower the number of fringes in the field of view, which means that the measurement accuracy will be low for weakly refractive test objects.

A nearly continuously distributed measurement of optical phase or optical path length in the recording plane is possible with *digital phase-stepping interferometry*, a combination of holographic interferometry and computer-based signal evaluation [6.23]. Here, the interferogram's phase is manipulated in a known manner such that it can be displayed as a gray scale and determined very accurately at a large number of points in the image plane. During the reconstruction process the holographic reference beam (Fig. 6.14) is translated step by step, with each step being a fraction of a wavelength. In this way, the constant appearing in (6.5), when this equation applies to a reference beam interferometer, is altered within fractions of a wavelength, or the constant optical phase of the interferometric reference beam is altered, step by step, by fractions of 2π. These stepwise alterations result in changes of the image intensity, which can be recorded by a digital camera and calibrated for further evaluation in the form of a lookup table [6.25]. Quantitative optical information in the space between two *regular* interference fringes is now available as the result of a physical process and not only by numerical interpolation and, hence, the sensitivity of interferometric methods for weakly refractive objects is increased by at least one order of magnitude (Fig. 6.20). While the resolution of a conventional interferometer is usually given as 1/10 of a wavelength (this value must be converted into values of density with the equations given above and with application to a particular object), the resolution of digital phase stepping is estimated to be between 1/100 and 1/1000 of a wavelength [6.23].

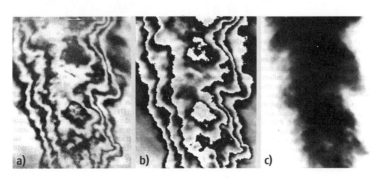

Fig. 6.20 (a) Reference beam interferogram of a transitional helium jet; **(b)** phase display in the form of a digital interferogram; **(c)** gray scale display of absolute phase of **(a)** (courtesy Prof. D. W. Watt, University of New Hampshire, USA [6.23])

6.6 Optical Tomography

As noted in Sect. 6.1 line-of-sight methods integrate the optical information on the density distribution in the object field along the light path in the fluid. Techniques for desintegrating the signal pattern and providing local data values of the 3-D density (or density gradient, respectively) are called optical tomography. These techniques are based on the possible transformation of the z–y Cartesian coordinate system (the x coordinate is not considered here for simplicity) into a z–α coordinate system, with α being the angle of the direction under which a point z, y is seen from the origin $(0, 0)$. If one restricts the field of interest to values $z \geq 0$, then all values α cover a range of 180°. The idea is that the light beam of the optical system is directed through the object field in multiple directions α (*projections*) in order to *reconstruct* the density function $\rho(z, y)$ from the multiple of signal patterns obtained for each transmission direction α. That this is possible becomes evident from Fig. 6.21, where the object field is discretised into a finite number of elements, each including a constant value of the density (or gradient, respectively) $\rho_i (i = 1 \ldots N)$; the values ρ_i are the unknowns that must be determined from a set of N equations. With the discretisation the integral, e.g. in (6.5), is replaced by a summation; each projection provides a number of equations $n < N$ that link the signal pattern recorded for the projection α_j with n unknowns. The projections in different directions α_j serve to produce a sufficient number (N) of independent equations.

A number of algorithms for optical tomography have been described in the literature; see, particularly [6.26]. The aim of most approaches is to provide high 3-D resolution of the reconstruction with a minimum number of projections. In analogy to the formerly mentioned z–α coordinate system, a total angular range of projections of 180° is needed to reconstruct a random 3-D object field. Any symmetry in the object reduces the number of necessary projections and, possibly, the total angular

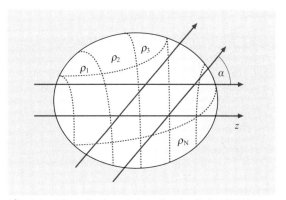

Fig. 6.21 Discretisation of the refractive index field into N elements with constant value of the density ρ_i; variation of projections for optical tomography

range to be covered. The simplest case is an axisymmetric object, which can be reconstructed with only one projection using the Abel inversion as the reconstruction algorithm [6.1].

The majority of tomographic methods apply to optical interferometry, e.g. [6.27, 28]. A complex and elaborate optical setup is needed for the investigation of unsteady flows because the interferograms of all projections must be recorded simultaneously [6.29]. An accurate 3-D reconstruction requires a dense distribution of the signal in the recording planes. Since the signal density in an interferogram is limited, attempts have been made to use the high data rate available with optical speckle densitometry (Sect. 6.4) for more-precise tomographic reconstructions of flows with varying fluid density [6.30]. Finally, it should be noted that tomography can provide the spatial resolution necessary for evidencing turbulent large-scale structures, but not yet, at this state of the development, for resolving turbulent fine scales.

References

6.1 W. Merzkirch: *Flow Visualization*, 2nd edn. (Academic, Orlando 1987)
6.2 F. Peters: A compact presentation of gasdynamic fundamentals, Forsch. Ingenieurw. **68**, 111–119 (2003)
6.3 F.J. Weinberg: *Optics of Flames* (Butterworth, London 1963)
6.4 W. Lauterborn, A. Vogel: Modern optical techniques in fluid mechanics, Annu. Rev. Fluid Mech. **16**, 223–244 (1984)
6.5 F. Mayinger, O. Feldmann (Eds.): *Optical Measurements*, 2nd edn. (Springer, Berlin, Heidelberg 2001)
6.6 G.S. Settles: *Schlieren and Shadowgraph Techniques* (Springer, Berlin, Heidelberg 2001)
6.7 W. Schöpf: A new way of analyzing the shadowgraph method, J. Flow Vis. Image Proc. **4**, 179–187 (1997)
6.8 G.B. Brassington, J.C. Patterson, M. Lee: A new algorithm for analyzing shadowgraph images, J. Flow Vis. Image Proc. **9**, 25–51 (2002)

6.9 H. Schardin: Die Schlierenverfahren und ihre Anwendungen, Ergeb. Exakten Naturwiss. **20**, 303–439 (1942), in German

6.10 H. Wolter: Schlieren, Phasenkontrast und Lichtschnittverfahren. In: *Handbuch der Physik*, Vol. 24, ed. by S. Flügge (Springer, Berlin, Heidelberg 1956) pp. 555–645

6.11 C.A. Lopez: Numerical simulation of a schlieren system from the Fourier optics perspective, AIAA Paper **94-2618** (1994)

6.12 A. Hanenkamp, W. Merzkirch: Investigation of the properties of a sharp-focusing schlieren system by means of Fourier analysis, Opt. Lasers Eng. **44**, 159–169 (2006)

6.13 G.S. Settles: Colour-coding schlieren techniques for the optical study of heat and fluid flow, Int. J. Heat Fluid Flow **6**, 3–15 (1985)

6.14 L.M. Weinstein: Large-field high-brightness focusing schlieren system, AIAA J. **31**, 1250–1255 (1993)

6.15 S.H. Collicott, T.R. Salyer: Noise-reduction properties of a multiple-source schlieren system, AIAA J. **32**, 1683–1688 (1994)

6.16 S. Garg, G.S. Settles: Measurements of a supersonic turbulent boundary layer by focusing schlieren deflectometry, Exp. Fluids **25**, 254–264 (1998)

6.17 W. Merzkirch: Density-sensitive whole-field flow measurement by optical speckle photography, Exp. Thermal Fluid Sci. **10**, 435–443 (1995)

6.18 K.D. Kihm: Laser speckle photography technique applied for heat and mass transfer problems, Adv. Heat Transf. **30**, 255–311 (1997)

6.19 N.A. Fomin: *Speckle Photography for Fluid Mechanics Measurements* (Springer, Berlin, Heidelberg 1998)

6.20 M. Raffel, H. Richard, G.E.A. Meier: On the applicability of background oriented optical tomography for large scale aerodynamic investigations, Exp. Fluids **28**, 477–481 (2000)

6.21 S.B. Dalziel, G.O. Hughes, B.R. Sutherland: Whole-field density measurements by "synthetic" schlieren, Exp. Fluids **28**, 322–235 (2000)

6.22 C.M. Vest: *Holographic Interferometry* (Wiley, New York 1979)

6.23 D.W. Watt, C.M. Vest: Digital interferometry for flow visualization, Exp. Fluids **5**, 401–406 (1987)

6.24 H.H. Bartels-Lehnhoff, P.H. Baumann, B. Bretthauer, G.E.A. Meier: Computer-aided evaluation of interferograms, Exp. Fluids **16**, 46–53 (1993)

6.25 T.A.W.M. Lanen, C. Nebbeling, J. L.van Ingen: Digital phase-stepping holographic interferometry in measuring 2-D density fields, Exp. Fluids **9**, 231–235 (1990)

6.26 G.T. Herman: *Image Reconstruction from Projections* (Academic, New York 1980)

6.27 H. Philipp, T. Neger, H. Jäger, J. Woisetschläger: Optical tomography of phase objects by holographic interferometry, Measurement **10**, 170–181 (1992)

6.28 K. Muralidhar: Temperature field measurement in buoyancy-driven flows using interferometric tomography, Annu. Rev. Heat Transf. **12**, 265–375 (2002)

6.29 B. Timmerman, D.W. Watt: Tomographic high-speed digital holographic interferometry, Meas. Sci. Technol. **6**, 1270–1277 (1995)

6.30 T.C. Liu, W. Merzkirch, K. Oberste-Lehn: Optical tomography applied to speckle photographic measurement of asymmetric flows with variable density, Exp. Fluids **7**, 157–163 (1989)

7. Temperature and Heat Flux

Thermochromic liquid crystals (TLCs) can be applied for thermographic measurements of heat transfer and temperature in fluid mechanics, delivering important quantitative full-field data for comparison with and validation of numerical simulations. Thin coatings of TLCs at surfaces are utilized to obtain detailed heat transfer data for steady or transient processes. Application of TLC tracers allows instantaneous measurement of the temperature and velocity fields for two-dimensional cross sections of flows. These methods are based on computerized true-color analysis of digital images for temperature measurements and modified particle image velocimetry, which is used to obtain the flow field velocity. In this Chapter, the advantages and limitations of liquid-crystal thermography are discussed, followed by several examples of thermal flow field measurements.

The use of infrared thermography for nonintrusive measurement of spatially resolved surface heat transfer characteristics is described for five different measurement environments, including situations where large gradients of surface temperature are present. In the first of these, measurements are made on the surface of a therapeutic biomedical patch, where the quantity of interest is the time-varying spatially resolved surface temperature. For the other situations, the measured temperature distributions are used to deduce quantities such as the surface Nusselt numbers on the surface of a swirl chamber, the effectiveness of surface adiabatic film cooling downstream of individual shaped film cooling holes, the surface heat flux reduction ratio downstream of two rows of film cooling holes placed on a model of the leading edge of an airfoil, and thermal boundary condition information for numerical predictions of the heat transfer characteristics on the surface of a passage with an array of rib turbulators. In all of these situations, in situ calibration procedures are employed in which the camera, imaging, and data-acquisition systems are all calibrated together in place within the experimental facility as the infrared measurements are obtained. This requires separate, simultaneous, and independent measurements of surface temperatures, and produces spatially resolved results from infrared images with high levels of accuracy and resolution.

7.1	**Thermochromic Liquid Crystals**...............	488
	7.1.1 Materials......................................	489
	7.1.2 Illumination	491
	7.1.3 Acquisition and Calibration	493
	7.1.4 Examples	496
	7.1.5 Conclusions	499
7.2	**Measurements of Surface Heat Transfer Characteristics Using Infrared Imaging** ...	500
	7.2.1 Introduction and Background........	500
	7.2.2 Chapter Organization....................	501
	7.2.3 Infrared Cameras...........................	501
	7.2.4 Overall Physical Arrangement	502
	7.2.5 In Situ Calibration	503
	7.2.6 Measurement of Surface Nusselt Numbers	503
	7.2.7 Measurement of Surface Adiabatic Film Cooling Effectiveness	506
	7.2.8 Measurement of Surface Heat Flux Reduction Ratio	508
	7.2.9 Transient Surface Temperature Measurements	510
	7.2.10 Boundary-Condition Information for Numerical Predictions..............	512
	7.2.11 Summary and Conclusions............	515
7.3	**Temperature Measurement via Absorption, Light Scattering and Laser-Induced Fluorescence**	515
	7.3.1 Overview......................................	516
	7.3.2 Non-Resonant Techniques	516
	7.3.3 Resonant Techniques....................	523
	7.3.4 Nonlinear Techniques	535
	7.3.5 Conclusions	537

7.4 Transition Detection by Temperature-Sensitive Paint 537	7.4.4 Transition Detection by Means of TSP 548
7.4.1 Introduction 537	7.4.5 Comparison with Other Methods 551
7.4.2 Surface Heat Transfer Processes 543	
7.4.3 Temperatures in Laminar and Turbulent Boundary Layer 545	References 553

7.1 Thermochromic Liquid Crystals

The description of transient temperature fields is one of the most sought after goals of experimental fluid mechanics. Unfortunately, performing accurate temperature measurements is not an easy task. Point measurements, which are the most common, usually give insufficient information if complex configurations are investigated. Full-field measurements, although not as accurate, may offer greater confidence for comparison with the numerical results. One such full-field technique is based on the use of *thermochromic liquid crystals* (TLCs). Their use to study heat transfer, flow visualization, and thermal mapping is playing an increasing role in the compilation of experimental temperature-related data in several areas of applications, competing with or replacing traditional infrared thermography [7.1].

Liquid crystals are highly anisotropic fluids that exist between the phase boundaries of the solid phase and the conventional, isotropic liquid phase [7.2]. Their temperature sensitivity is based on the property of some cholesteric, chiral-nematic liquid-crystal materials. This phase is typically composed of molecules arranged in a stack of very thin layers with the director in each layer twisted with respect to those above and below. An important characteristic of the cholesteric mesophase is the pitch, which is defined as the distance over which the director rotates a full turn in the helix structure of the nematic layers. The helical structure of the liquid-crystal material is responsible for its ability to selectively reflect light of wavelengths related to the pitch length. As the incident light passes through a TLC material, the periodic variation of the refractive index (Sect. 6.1) modulates the light polarization and causes interference within the multilayered material. According to Bragg's law, a characteristic light wavelength is singled out depending on the light scattering angle, refractive index and structure periodicity. Hence, a single color will be reflected when the molecular pitch is equal to the corresponding wavelength of light in the visible spectrum. Temperature variation or stress executed on the sample will change the director orientation in successive molecular layers, which modifies the pitch length. This makes it possible to use liquid crystals as temperature indicators, where detection of the color change in reflected light leads to quantitative information. The side effect, stress sensitivity of the liquid crystals, is successfully used in fluid mechanics to detect shear or pressure changes [7.3]. However, for temperature measurements this should be minimized to avoid ambiguity of information given by color changes.

Bulk TLC material is normally clear, or slightly milky in appearance. It shows a change of color to red in response to an increase of temperature, followed by yellow, green, blue, violet, finally turning colorless again at higher temperatures. The molecular transition responsible for the selective light reflection is limited to a narrow range of temperature variation, defined for a specific TLC material as the *color play range*. The color changes of TLCs are repeatable and reversible as long as the material is not physically or chemically damaged. This being the case, TLCs can be calibrated accurately with proper care and used as temperature indicators. They modify incident white light and display color with a wavelength that is related to temperature. Beyond its *clearing point temperature*, a TLC material is a transparent liquid, simplifying its use as an additive to paints, filters or warning labels.

Pure liquid-crystal materials are thick, viscous liquids, and are greasy and difficult to deal with under most heat transfer laboratory conditions. TLC materials are also sensitive to mechanical stress. A microencapsulation process that encloses small portions of liquid-crystal material in polymeric material has been introduced to solve problems with stress sensitivity and chemical deterioration.

In the past liquid crystals have been extensively applied in the visualization of entire, either steady-state or transient temperature fields on solid surfaces. Since quantifying color is a difficult and somewhat ambiguous task, application of TLCs was initially largely qualitative; application of color photography or interference filters for color analysis was tedious and inaccurate. Quantitative and fast temperature measurements were only brought about with the adoption of the charge-coupled device (CCD) color camera and digital image processing. The rapid development of hardware and software image-processing techniques has made the use of inexpensive systems, real-time, transient, full-field temperature measurements using TLCs possible.

Thermochromic liquid crystals can be painted onto a surface [7.4–11] or suspended in the fluid to render the temperature distribution visible [7.12–16]. By disseminating the liquid-crystal material throughout the flow, TLCs not only become classical tracers for flow visualization, but simultaneously, minute thermometers monitoring the local fluid temperature (Fig. 7.1).

The typical diameter of TLC tracers used in flow measurements is 20–50 μm. As the density of TLCs is close to that of water, they are well conveyed in liquid flows. The response time of TLC materials is 3–10 ms, sufficiently fast for typical thermal problems in fluids. Application of TLCs as tracers facilitates instantaneous measurement of temperature and velocity fields [particle image velocimetry (PIV); Sect. 5.3.2] in a two-dimensional cross section of the flow. It is a unique method of combining full-field temperature and velocity measurements [7.17–20]. In the following we give some details on application of TLCs for fluid mechanics measurements, mostly gained through our own experience *playing with colors* during the last 20 years.

7.1.1 Materials

Generally there are three different forms of TLCs available: unsealed pure cholesteric material, slurries of encapsulated TLCs, and mechanically protected thin liquid-crystal films. TLCs can be purchased in one of these forms with the chemical composition matched to the temperature variation of the problem. Typically they have a color play range of about 0.5–40 °C, associated with clearing point temperatures of −30 °C to above 120 °C. The proper choice of material depends on the application. Its working temperature (clearing point) should match the expected temperature range. Selection of narrow-band TLCs offers accurate resolution of temperature but only over a very narrow interval. This is convenient in detecting transient changes of temperature, where the passage of a single, well-defined isotherm through the investigated media can easily be detected. Broadband TLCs give rather qualitative information, discriminating hot and cold regions with very little accuracy of measured temperature values. It is worth noting that most TLC producers define the temperature response of their products using a null angle between illumination and observation. These must be modified according to Bragg's law for other configurations, which results in a shift of the clearing point and shrinkage of the bandwidth of the TLC's color response. Table 7.1 shows the basic properties of a few TLC sam-

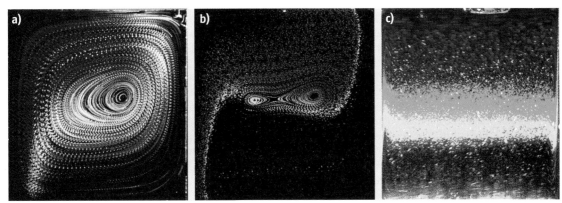

Fig. 7.1a–c Natural convection in a differentially heated box. Three different flow regimes are visualized using unencapsulated liquid-crystal tracers: (**a**) Low Rayleigh number (10^4): temperature difference between walls 4 °C, TLC material TM107, box size 38 mm; (**b**) Intermediate Rayleigh number (8×10^4): temperature difference 16 °C, TLC material TM445, box size 38 mm; (**c**) High Rayleigh number (1.3×10^8): temperature difference 34 °C, TLC material TM317, box size 80 mm. Images (**a**) and (**b**) were obtained from color slides exposed 20 times every 15 s; image (**c**) was created from 20 digital images taken every 200 ms and added in a computer memory

Table 7.1 Thermochromic liquid crystals applied to flow measurements. Nominal red clearing point T_{start} and temperature range ΔT according to catalogue data

Company	Symbol	T_{start} (°C)	ΔT (°C)	Form	Comments
Hallcrest	BM 250/R0C 10W /S33	0	10	Encapsulated	$\Delta T \approx 5\,°C$, heavy tracers
Hallcrest	BM 100/R90F 2W /S33	32.5	2	Encapsulated	
Hallcrest	BM 100/R90F10W /S33	32.5	10	Encapsulated	
Hallcrest	BM R29C 4W /S33	29	4	Encapsulated	
Hallcrest	BM 100/R20C10W /S33	20	10	Encapsulated	
Hallcrest	BM 100/R6C12W /S33	6	12	Encapsulated	
Hallcrest	BM 100/R29C4W /S33	29	4	Encapsulated	
Hallcrest	BM /R96C6W	96	6	Liquid	
Hallcrest	BM /R60C6W	59.8	6	Encapsulated	
Hallcrest	BN /R70C6W	69.5	6	Liquid	
BDH	TM 445 (R17 C6W)	17	6	Liquid	Strong colors, $\Delta T \approx 4\,°C$
BDH	TM 446 (R37 C6W)	37	6	Liquid	Strong colors, $\Delta T \approx 4\,°C$
BDH	TM 317	21	20	Liquid	Strong colors, $\Delta T \approx 4\,°C$
BDH	TM 107 (R27 C6W)	27	8	Liquid	Strong colors, $\Delta T \approx 3\,°C$
Merck	TM 912	−2	10	Liquid	Strong colors
Merck	TCC 1001 (27C-31C)	27	4	Encapsulated	Very stable suspension

ples used as tracers with some practical remarks about their usage.

Liquid-crystal organic compounds degrade very easily when exposed to chemical contamination and to ultraviolet light. Slight chemical contamination may shift the color play range over several degrees Celsius or even completely remove the TLC temperature sensitivity. In our experience, apparently negligible sources of contamination such as the use of polyvinyl tubes for providing the suspension of liquid crystals appear to produce surprisingly strong effects on the TLC response characteristics. As the effect of chemical deterioration is usually difficult to predict, it is recommended to use unsealed TLCs only in water and its solutions with glycerol. For all other flow configurations protection through encapsulation is necessary. Despite this inconvenience in most of our experiments pure, unsealed TLCs material was used to produce TLC tracers. There are two main reasons for this. Pure material offers a very good signal-to-noise ratio for color evaluation. For large enough tracers and for glycerol or water as a carrier liquid, the intensity of light scattered by unsealed TLCs (Mie scattering band) beyond their clearing point is negligibly small. Hence, tracers are practically invisible for temperatures outside their color play range. This practically cancels any secondary light scattering, improves color evaluation, and allows for deeper optical penetration in the flow. The second reason for using pure TLCs material, which is very important in our experience, is their flexibility in matching the color play range. In experiments where phase changes takes place (freezing, boiling) the color play of TLCs must exactly match the desired temperature range. Usually it is not possible or very expensive to obtain TLCs manufactured exactly to that range, and even then experimental conditions may shift the resulting color response, making new, expensive matching of the ordered TLCs necessary. The same result can be achieved simply in the laboratory using unsealed material by mixing two samples of TLCs with clearing points above and below the desired working temperature. The mixture obtained will exhibit temperature sensitivity in the range proportional to its composition.

Pure TLC material is commercially available as a thick, viscous liquid. It can be dissolved in several organic solvents (e.g., diethylether) and sprayed onto the solid surface or directly into the fluid to be investigated. It is also possible to emulsify them into the liquid by strong mixing with the carrier fluid. When applying unsealed TLCs as tracers, it is important to achieve a homogeneous, very dilute suspension of small droplets. The size of the tracers should be minimized to avoid buoyancy effects and guarantee that tracers follow the flow pattern. However, they should be large enough to be detectable and, what is even more important in the case of TLCs thermography, to guarantee a strong enough signal-to-noise ratio of reflected light for color evaluation.

Several techniques have been investigated to achieve this target. Mixing diethylether solutions of TLCs with a hot carrier liquid is a relatively simple and efficient method. First, a thin layer of the solution is left at the liquid surface until the ether has completely evaporated. Then, by mechanical stirring, a tiny microlayer of pure TLC is broken into a suspension of microdroplets. However, the resulting droplets vary quite substantially in size and an additional selection procedure (sedimentation) is necessary to remove the large ones. The method was improved using piezoceramic droplet generator to disperse the ether TLC solution in a 2 m-high fall tower [7.21]. The ether solvent in droplets with a diameter of 0.4 mm completely evaporates during their flight, and pure TLC material can be collected in the carrier liquid. The apparatus allows a suspension of almost perfectly monodisperse TLC droplets with a diameter of 50 μm to be obtained, evidently improving the quality of tracer images (Fig. 7.1a,b). A less tedious but in most cases satisfactory method of dispersing pure TLC materials in liquids is turbulent mixing of a preheated carrier liquid doped with a small amount of hot TLC material. Using this method, the TLC droplet size is 10–50 μm, forming a uniform colorful mist when observed by a CCD camera.

Perhaps the most common method employed for flow visualization experiments is the application of encapsulated TLCs, which are commercially available as slurries of polymer capsules in water. Each microcapsule, of size 50–150 μm, contains approximately 40% (by weight) of enclosed TLC material. These are custom formulated for the required color change properties. It is possible to control the buoyancy (apparent specific gravity) characteristics of the microcapsules within limits by varying the composition of the TLC mixture, and the microcapsule diameter and wall properties. The TLC slurry can be added directly to the carrier fluid. Theoretically, it should be possible to use such microcapsules in any liquid that is not aggressive to the encapsulation material. Since the slurry carrier is water, in practice it seems that they can only be mixed with aqueous solutions of glycerol, ethylene glycol, and other similar low-molecular-weight polyhydric alcohols. Seeding with dry particles is not possible. Polymeric shells enclosing TLC materials are very fragile. Any attempt to remove the TLC particles from the slurry by drying or filtering leads to their damage and loss of their content. Perhaps the only nonaqueous carrier liquid successfully used with encapsulated TLCs is silicon oil. Usually encapsulated TLCs produce less-saturated colors. Light diffused by the capsule creates an additional white background in addition to the selective reflection by the liquid crystals, decreasing the saturation of the analyzed colors and the overall transparency of the carrier liquid. Therefore the potential advantages of using encapsulated TLCs, the chemical and mechanical protection of the TLC material, is diminished by the reduced flexibility in matching size and composition, their worse optical properties, and the much higher unit price.

For surface temperature measurements, the unencapsulated material (unsealed liquid) is applied to a clear plastic sheet and sealed with a black backing coat to form a prepackaged assembly. Commercially temperature-indicating devices using a thin film of liquid crystal sandwiched between a transparent polyester sheet and a black absorbing background are available. If chemical or mechanical protection is not a serious problem, it is possible to deposit pure material directly onto the black surface. Evaporation of an ether solution of TLCs has been successfully used to cover interrogated surfaces with a very uniform, thin microlayer of thermochromic material.

Application of TLCs sheets and films is limited to measurements of the heat transfer coefficient and temperature distribution over areas that are difficult to access with other complementary techniques such as infrared (IR) thermography. TLC films have an advantage over IR techniques in all configurations where direct monitoring of long-wavelength infrared radiation is obstructed, for example by a water film falling over the surface in question. However, obtaining quantitative measurements with TLC films is not simple and in many cases impossible due to artefacts produced by nonuniform illumination of the interrogated object. Additional sources of uncertainty are heat flux resistance occurring between the surface and the thermochromic foil or unknown variations of the effective thickness of the thermosensitive paints sprayed over the surface.

7.1.2 Illumination

TLCs show colors by selectively reflecting incident white light. The observed color depends on the TLC temperature and also on the reflection angle, measured relative to the incident light direction. The effects are different in thin films and paints compared to tracers suspended in liquids. In both cases two basic rules apply:

1. the light source should have smooth and stable spectral characteristics, and

2. the angle between the observer (camera) and the direction of incident light should be kept fixed across the whole monitored area.

To fulfil the first condition, collimated light from strong halogen lamps or xenon flash tubes is practically the only choice. For flow illumination, strong and relatively easily fabricated light sources are assembled using linear 1000 W halogen lamps with a tungsten filament spanning a 150 mm tube. Lamps are switched on for only a short time (0.5–5 s) to avoid excessive heating and to extend the lifetime of the filaments. A filament preheating circuit is used to speed up switching of the lamp to full power. High-energy xenon discharge lamps are employed for short illumination times. Such lamps consist of a 150 mm-long tube connected to a battery of condensers and can deliver as much as 1 kJ of energy during a 1 ms pulse. Repetition of the light pulses is relatively slow (several seconds). The total thermal load and condenser charging time are both factors limiting the repetition rate. In practice, two sets of condensers with an electronic switch can be used to allow two light pulses to be emitted from the same tube within approximately 200 ms. Such a configuration is useful when applying TLC tracers for simultaneous temperature and velocity measurements (particle image thermometry (PIT) and velocimetry (PIV); for the latter see Sect. 5.3.2).

For surface temperature measurements special care is necessary to avoid specular light reflections form the surface covered by the TLCs. Sometimes polarization filters may help to minimize this effect. Application of diffuse light, often preferred for the illumination of complex surface geometries, introduces uncertainty in the color–temperature relation and limits the accuracy of measurements due to variation of the illumination/observation angle over the investigated area. To avoid additional heating by irradiating light, infrared filters and short light pulses should be applied. Use of fluorescent tubes is not advised as the illumination they produce is pulsating in time and spectrally discontinuous.

Incident light should be collimated to achieve well-defined conditions for Bragg reflection of a single wavelength. This is especially important for tracers, where reflected light is observed directly. For TLCs sealed between multilayered films, internal light reflections partially diminish their sensitivity to the observation angle variation. In some surface temperature studies, variation of the lighting/viewing angle by up to 30° is often accepted as a compromise between illumination uniformity and good color resolution [7.4]. However, for quantitative measurements such a compromise is not acceptable and, as pointed out by *Sabatino* et al. [7.10], only a calibration over the full viewing area can help to remove viewing-angle bias from surface temperature measurements. In practice this requires a calibration curve for each point (image pixel) of the analyzed surface. Such a technique, which is easy to perform when using digital recording, takes into account most sources of uncertainty in the perceived hue, including reflections, nonuniformity of the color response characteristics of the CCD pixels, lenses, and optical windows.

Good collimation of panchromatic (white) light, which is necessary for quantitative temperature measurements, is not an easy task. Optical passages including multiple slits, parabolic mirror and cylindrical lenses are used to extract a 2 mm-thick 100 mm-wide sheet of light from a halogen or flash tube, with the sheet collimation preserved over a distance of about 10 cm. The cost of this is strong attenuation of the light intensity, leaving little left from a 1000 W light source. To resolve the difficulties of light collimation it is possible to apply a bundle of three monochromatic light sources, such as lasers or laser diodes operating at blue, green and red colors. However, an additional problem arises with equalizing the color balance at every illuminated point. Hence, in practice lasers (HeNe) are applied for the illumination of TLCs only to extract single isotherms for calibration purposes.

A procedure has been developed to resolve the question of the single-wavelength accuracy of the method and its sensitivity to variations in lighting and viewing angle [7.22]. A suspension of unsealed TLCs (TM317) in glycerol was produced using the aforementioned ether-solvent-based dispersing method. Calibration of the suspension was performed in a small cubic container (38 mm internal size) made of Plexiglas, with top and bottom walls made of black, oxidized metal. Both metal walls were kept at a constant temperatures, 28.7 °C for the top and 26.6 °C for the bottom, ensuring stably stratified thermal field in the liquid. The temperature in the cavity varied linearly with height, with small deviations in the vicinity of the Plexiglas side walls. The liquid was illuminated by a chopped 0.5 mm-diameter laser beam from a 5 mW HeNe laser, passing through the center cross section of the cavity (Fig. 7.2a). The scattered light was observed in the perpendicular direction ($\phi = 90°$) with an optical system equipped with a focusing lens and a photodiode. Scattered light was collected from a region approximately 1 mm in size, selected in the middle

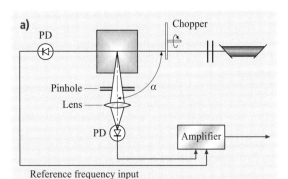

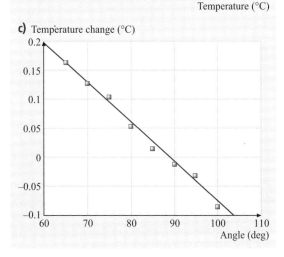

Fig. 7.2a–c Intensity of HeNe laser light scattered by suspension of unencapsulated TLC tracers (TM317) kept in a stable vertical temperature gradient in a 38 mm Plexiglas box: (**a**) schematic of the apparatus, (**b**) effect of temperature variation, (**c**) effect of the angle of observation on the measured temperature shift

red-band sensitivity of TLCs. A strong peak in the signal was observed (Fig. 7.2b) when the condition for matching the laser wavelength with the TLCs characteristic pitch was reached. The near-Gaussian shape of the signal, with a half-amplitude width corresponding to a 0.2 °C temperature change, indicates the nominal sensitivity of the TLC sample. This value corresponds to the ideal conditions of 633 nm-wavelength evaluation of isotherms, and for the given experimental configuration defines the minimum error of quantitative temperature measurements in the red part of the color play range of the TLC sample. The dependence of the reflecting angle on the temperature shift was evaluated by varying the position of the detector on the arc centered at the interrogated point of liquid. Figure 7.2c illustrates the linear decrease of the detected light intensity, with a slope corresponding to a change of 0.07 °C in the recorded temperature for each 10° change in the viewing angle. This test underlines the importance of preserving a fixed angle between the illuminated plane (light sheet plane) and the camera, and keeping the viewing angle of the lens small. For example, observing a 5 cm-wide area with a CCD camera in the perpendicular direction, the observation distance should be about 400 mm to keep the viewing angle below 4°.

7.1.3 Acquisition and Calibration

The calibration procedure is the most tedious and delicate element of TLCs thermography. Due to the unstable properties of TLCs and the large sensitivity of their color response to experimental conditions, the same illumination and recording system should be used for calibration and measurements. This guarantees that color bias due to external optical effects is taken into account. A typical experimental setup used for quantitative thermography consists of an illumination set, a three-chip CCD color camera and a 24 bit frame grabber. A commercially available standard three-CCD camera can be used, but its color response should be verified and properly balanced. The electronically generated color bars are useful to balance the gain and contrast of each of three color channels. Features that compensate for the nonlinear color response of the human eye, automatic gain, and white

of the beam. The scattered light intensity was monitored using a differential, phase-locked voltmeter. Moving the cell up and down with a micrometer screw, liquid layers with linearly varying temperature were interrogated. The intensity of the scattered light (reflected by the TLCs) remained almost constant until the temperature-matched

balance, usually set by default in standard cameras, should be deactivated.

For flow measurements, the investigated cavity is illuminated with a thin sheet of white light from the aforementioned halogen lamp or xenon flash tube and observed in the perpendicular direction. The color images, typically of 768 × 564 pixels, can be easily acquired using a color frame grabber installed in a personal computer equipped with a peripheral component interconnect (PCI) or Firewire (IEEE 1394) interface. The setup (Fig. 7.3) additionally consists of a computer-controlled system of three stepper motors, which allow the automated acquisition of several horizontal and vertical planes across the flow within one or two minutes. Hence, for slowly varying flow structures, a transient recording of three-dimensional flow features is possible. Recording of the transient flow patterns and temperature fields is performed periodically. Typically every 10–300 s, a short series of images is acquired and stored on the hard disk of the computer for later evaluation. In the case of TLC tracers their displacement can be analyzed using the PIV technique (Sect. 5.3.2, and full flow velocity fields are obtained [7.18].

Temperature measurements using TLCs are based on color analysis of the images and needs appropriate calibration. This is eased by using preprocessed color information from a three-CCD camera, which splits light into three basic components: red, green and blue (RGB). This process is known as trichromic decomposition. Each of the three color components is usually recorded as a separate 8 bit intensity image. Numerous methods of subtracting color information from a trichromic RGB signal exist [7.6]. The most straightforward is to convert the RGB trichromic decomposition to another trichromic decomposition based on *hue* (color value), *saturation*, and *intensity* (HSI). Such a decomposition is common in image analysis and also serves as a natural means of converting color images to their black-and-white representation. Classical conversion of an RGB color space to an HSI decomposition is based on three simple relations. Light intensity I (or brightness) is defined as the sum of its three primary components:

$$I = \frac{\sqrt{(R^2 + G^2 + B^2)}}{\sqrt{3}} \qquad (7.1)$$

with R, G and B being the intensities of the red, green and blue components, respectively.

Using an 8 bit representation, the maximum intensity is equal to 255. The saturation S represents color purity, i.e., the relative value of the reminder after subtracting the amount of colorless (white) light:

$$S = 255 \left(1 - \frac{\min(R, G, B)}{I}\right) . \qquad (7.2)$$

In our representation pure colors have saturation equal to 255. The hue H relates to the dominant color and is usually obtained from the algebraic or trigonometric relation between the two dominant primary colors. To make use of the 8 bit signal dynamics and limiting ourselves to the spectral colors [7.18], the following formula is used to calculate the hue value:

$$H = \begin{cases} 63 + [(G' - R')63]/(G' + R') & \text{for } B' = 0 , \\ 189 + [(B' - G')63]/(G' + B') & \text{for } R' = 0 , \end{cases} \qquad (7.3)$$

where $R' = R - \min(R, G, B)$, $G' = G - \min(R, G, B)$, and $B' = B - \min(R, G, B)$.

Temperature is determined by relating hue to a temperature calibration function. This is the most critical stage of TLC-based thermography. Light refracted by TLCs is not monochromatic, even if the observed sample has a uniform temperature. Color depends on observation angle, the scattering properties of the TLCs material, the color and refractive index of the carrier liquid, and may also vary with the size of the particles. Additional factors, such as the color of the light source, the color transmission properties of the acquisition system, as well as reflected and ambient light modify the resulting color information. The observed color may also depend on light intensity. Very careful calibration is therefore necessary to obtain quantitative information. It is advised to generate a separate calibration curve for each small area of the investigated surface and use these to normalize the measured color information.

For the best effects the calibration procedure should be performed with an identical experimental arrangement to that used for measurements. To perform the

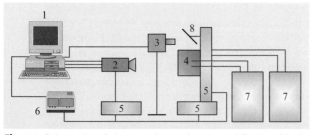

Fig. 7.3 Schematic of the experimental system. PC (1) with the acquisition card controlling the color camera (2), halogen lamp (3) and three stepping motors (5) using a driver (6). The temperature in the cavity (4) is controlled by two thermostats (7); the mirror (8) used to direct the light sheet

calibration, an interrogated surface or liquid in a cavity with a suspension of TLC tracers must be kept at a constant and uniform temperature. Temperature stability and uniformity better than 0.1 °C is required for TLCs with narrow color bands. For measurements of temperature fields in liquids this is achieved by keeping the walls of the experimental cavity at a constant temperature and continuously mixing the liquid with a magnetic stirrer. Data for calibration are obtained by adjusting the liquid temperature in small increments (usually 0.3 °C steps) and acquiring several sequences of images for future processing. Compared to surface thermography, the use of TLCs suspension for flow investigations yields additional problems. The color images of the flow are discrete, i.e., they represent a noncontinuous cloud of points and their color usually varies from point to point. This is not only caused by slight variations in the TLC temperature responses but also indicates local fluctuations in the fluid temperature present due to the limited thermal diffusivity of liquids. To minimize this effect, calibration is usually performed for small areas of about 30×30 pixels, extracted in the vicinity of the temperature sensors from each image and averaged for several subsequent images.

Hue evaluation is performed for each pixel of the resulting averaged image, under constraints of minimum and maximum pixel intensity and minimum saturation. Only good pixels, i.e., those having an intensity and saturation above a given threshold level, are used to build a hue matrix, which is smoothed using a 5×5 median filter followed by a low-pass filter. An average hue value is then calculated for each image and used as a reference point for the calibration procedure. The procedure can be repeated for several parts of the image to create a position-dependent temperature–hue relation. For the full color range of TLCs about 20–30 calibration points are collected at gradually increasing liquid temperatures. A smooth and unique hue–temperature relation is established by fitting a polynomial of degree 6–10. After removing outliers, a final fit is performed to obtain a smooth temperature–hue function (Fig. 7.4). The accuracy of the measured temperature depends on the color (hue) value as a direct consequence of the nonlinearity of the curve. The relative error is 3–10%, and is based on the temperature range defined by the TLC color play limits. It is estimated by calculating the uncertainty

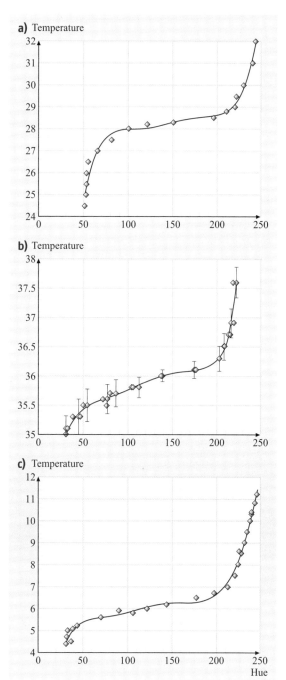

Fig. 7.4a–c Temperature versus hue calibration points and high-order polynomial fits for three different TLCs dispersed as tracers in a flow. (**a**) Encapsulated tracers TCC1001 in water; (**b**) Suspension of unencapsulated TLCs TM446 in water, error bars indicate estimated accuracy; (**c**) Mixture of two TLC types (TM445 and TM912) used to produce a suspension of liquid crystals matching the desired temperature range

of single-point measurements from several images of the same flow. For the TLCs used to obtain the curve in Fig. 7.4b (TM446), an absolute accuracy of 0.15 °C results for low temperatures (red–green color range) and 0.5 °C for high temperatures (blue color range). The most sensitive color region is the transition from red to green, which occurs over a temperature change of less than 1 °C. Additional, quantitative information beyond the calibration curve can be obtained by careful analysis of the remaining two components of the color decomposition. These can be used to analyze the temperature in areas of the TLC clearing point, which exhibits a gradual fall in pixel intensity and the saturation of colors. Quantitative evaluation of these areas is difficult but possible, for example using neural networks to include complete image information in the color analysis [7.23].

Obtaining full calibration curves prior to each experimental run is not only tedious but also limits rapid repetition of different experimental configurations. Such repetitions are often necessary, for example when different TLCs are used to shift the analyzed temperature range. Observing the shape of different calibration curves taken under similar conditions we may find large similarities. Therefore in some cases it is possible to assume the existence of a general calibration curve, independent of the active range. A calibration procedure based on such a *master* curve would require only a few measurements to produce a useful calibration over an entire active range. Using a nondimensional representation of temperature *Hay* and *Hollingsworth* [7.6] obtained repeatability of their regression curves with an acceptable accuracy for TLCs painted onto a heated surface.

Similarities are also observed for the regression curves obtained for TLC tracers (Fig. 7.4). For reproducible experimental conditions simple stretch and shift operations may be used to transform one of them to another. Such a procedure can be applied to account for small changes of TLC response, either due to aging or when a new mixture is prepared.

7.1.4 Examples

In the following we illustrate several applications of TLC tracers to study the natural convection of liquids in closed cavities, and also with a phase change (freezing of water). In conjunction with the experimental program, numerical simulations of the problem were performed using finite-difference models of the Navier–Stokes and energy equations [7.13, 18, 24, 25]. The significance of the full-field temperature and flow measurements for the verification of the numerical results becomes evident by a direct comparison of the evaluated and predicted fields.

The first experiment deals with natural convection in a cubic cavity with two vertical, isothermal walls kept at different temperatures. This flow configuration resembles a popular benchmark, an idealized case of the flow in a cubical cavity used to test numerical solutions of the Navier–Stokes equations [7.26]. The characteristic recirculating flow is generated in the cavity by the temperature gradients between the two opposite metal walls of the cube. In addition to its theoretical interest, this type of convective flow has numerous potential applications, among which probably the most popular

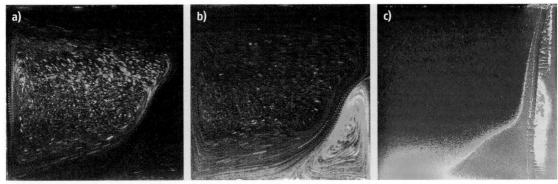

Fig. 7.5a–c Visualization of natural convection in water close to the freezing point, using a hot (left) wall at a temperature of 10 °C and a right wall at a temperature of 0 °C for (**a**) and (**b**), −10 °C for (**c**). Three types of liquid-crystal tracers were used to indicate the temperature variation: (**a**) encapsulated TLCs BM100/R6C12W showing details of flow and temperature for *normal* circulation (on the left); sum of 10 images taken every 1 s; (**b**) TLC mixture (TM445 and TM912) active at lower temperatures, visualizing the complex structure of *abnormal* circulation; sum of 20 images; (**c**) TLC mixture with a composition matched to the freezing point

is that of double glazing. Initially our interest was directed towards understanding the flow in the vertical central plane of the cavity. For this purpose, the observations of flow patterns and temperature fields were performed for several systems with increasing Rayleigh number [7.13, 14]. The flow structure in the differentially heated cavity strongly depends on the Rayleigh number. At small Rayleigh number (Ra $< 10^3$) the flow is dominated by conduction, which is seen in the form of vertical isotherms across the cavity. In the parameter range analyzed (Ra $= 2 \times 10^4 – 10^5$), both convection and conduction are important. At the lower end of the range, heat transfer due to convection begins to play a significant role, generating a vertical temperature gradient in the center of the cavity. The horizontal temperature gradient is positive everywhere. The streamlines correspond to a single roll located at the center of the cavity. This can be well observed in the multiply exposed photographs of liquid-crystal tracers conveyed by the flow of glycerol (Fig. 7.1a). At higher Rayleigh numbers (Ra $> 6 \times 10^4$), the horizontal temperature gradient becomes locally negative in some regions. This causes horizontal elongation of the streamlines and the development of a second roll in the core (Fig. 7.1b). By further increase of the Rayleigh number, a new flow regime is observed (Fig. 7.1c), with a strongly stratified temperature distribution in the center and thin thermal boundary layers at both isothermal side walls. This regime after increasing Rayleigh number above 10^8 leads to a transition to turbulence. Such a transition, easily identified using full-field temperature visualization and quantification, can be directly compared with numerical simulations of the corresponding cases.

Flow visualization via liquid crystals allows the observation of the flow structure and identification of variation of the temperature field. Even without quantitative data, the general shape of the isotherms can be deduced from the color photographs. However, despite the undoubted beauty of multiply exposed color photographs of liquid-crystal tracers, they have little value for quantitative image analysis. The lack of visible particles in large regions of the flow and their individual modulation of color are the main obstacles to computerized analysis. Hence, digital images of flow taken to implement particle image velocimetry and thermometry analysis exhibit a rather dense, uniform crowd of small, singly exposed colorful dots, representing a fine dispersion of the liquid-crystalline material in the flow. The human eye may still easily analyze the color distribution of such images (Fig. 7.5c), but particle displacements can be detected only by computational means.

The behavior of natural convection of water in the vicinity of the freezing point shows an interesting feature in the typical configuration with differentially heated walls. It is mainly due to the strongly nonlinear temperature dependence of the density function, which has an extremum at 4 °C. The competing effects of positive and negative buoyancy force result in a flow with two distinct circulations. There is a *normal* clockwise circulation, where the water density decreases with temperature (upper-left cavity region) and an *abnormal* convection with the opposite density variation and counterclockwise rotation (lower-right region). As mentioned, TLCs allow

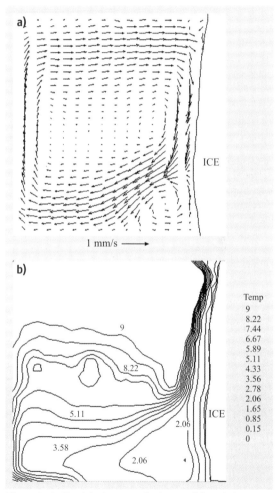

Fig. 7.6a,b Particle image velocimetry (**a**) and thermometry (**b**) applied to measure the velocity and temperature fields for natural convection in water freezing in a differentially heated cavity. Temperatures applied: 10 °C (on the left wall) and −10 °C (on the right wall)

the detection of temperature over a limited range of their color play regime. In some cases it is necessary to obtain precise information over a wider range of temperature variation, a task that is impossible to achieve with high precision for a single brand of TLCs. For steady or reproducible flow configurations it is possible to repeat the same experiment using different types of TLCs, as illustrated in Fig. 7.5. Regions of low temperature are visualized in Fig. 7.5a using TLCs operating in the temperature range 0–4 °C, and the same experiment is repeated in Fig. 7.5b to elucidate the temperature pattern in the left part of the cavity, using other types of TLCs.

To investigate natural convection with a phase change, one of the isothermal walls was held at a temperature of $T_c = -10\,°C$. As this is below the freezing temperature of water, ice is formed at the cold wall (Fig. 7.5c). Initially uniform growth of the ice layer is quickly modified by the two colliding *normal* and *abnormal* circulations. In the upper part of the cavity, convective heat transfer from the hot wall is limited by the abnormal circulation, separating it from the freezing front. Figure 7.6 demonstrate the ability of TLCs to resolve simultaneously the transient variation of the velocity and temperature fields. These data, which were compared with numerical simulations performed for the freezing problem [7.24], led us to discover significant discrepancies and indicated directions to improve the model [7.27–29].

In the second solidification experiment flow in a cube with only one isothermal wall was investigated. The top isothermal wall was kept at a low temperature and the other five walls were nonadiabatic, allowing a heat flux from the fluid surrounding the cube. The temperature at the internal surfaces of the cube adjusted itself depending on both the flow and the heat flux through and along the walls. The lid-cooled cubic cavity was selected to investigate the convective flow with and without a phase change (freezing of water at the top wall). When the phase change occurs, the configuration resembles to some extent a directional solidification in a Bridgman furnace used for crystal growth (Fig. 7.7). Physically this configuration bears some similarity to the Rayleigh–Bénard problem. The symmetry of the enclosure imposes a strong downward flow along the vertical axis of symmetry. However, before a stable final flow structure is achieved, several oscillatory changes in its pattern are observed [7.30, 31]. The initial flow instabilities are seen well in the temperature and velocity fields visualized by TLCs in the box and could be confirmed in the numerical simulations. The formation of ice has been studied by decreasing the lid temperature to $-10\,°C$. A complicated flow pattern with azimuthally varying structure was also reproduced in the star-like ice surface [7.32]. It was found that the creation of the ice layer at the lid has a stabilizing effect on the flow. This follows from the symmetry of the solid ice surface, which imposes the direction and character of the flow, eliminating the instabilities observed in the pure convective case. There is also a density inversion under the lid that decelerates the main jet and limits convective heat flux in that region [7.33].

Another example of the investigated configurations illustrates the simulation of flow characteristics accompanying the casting processes (Fig. 7.8). For this purpose

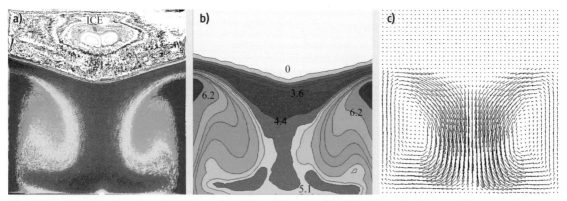

Fig. 7.7a–c Ice crystal growing from the top in a lid-cooled cavity in a Plexiglas cube immersed in an external water bath at a temperature of 20 °C. The isothermal metal lid has a temperature of −10 °C. Unencapsulated mixture of thermochromic liquid crystals (TM445 and TM912) used as tracers to evaluate both the temperature and velocity flow fields. (**a**) Superposition of five recorded images taken every 400 ms, (**b**) the evaluated temperature, and (**c**) the velocity fields

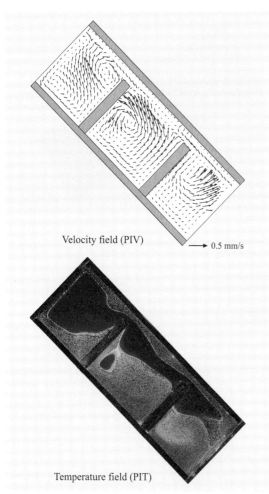

Fig. 7.8 Velocity field and temperature distribution visualized for the cavity inclined at 45°. Two cold isothermal walls (upper and lower) are responsible for sudden cooling of the fluid. This changes the color of the seeded liquid crystals (TM446) from blue (hot) to red (cold regions)

a rectangular cavity with two isothermal walls was filled with pressurized hot, viscous liquid (glycerol). The main features of the experiment, such as flow acceleration and deceleration at obstacles, a free surface flow and a sudden increase of the fluid viscosity on cooling, are typical for melt solidification in a mould. In contrast to a real casting, use of TLCs as tracers enables full-field measurement of the temperature and velocity fields. Quantitative transient data for the velocity and temperature fields were collected to verify and validate the numerical models used for typical casting problems. The main aim of the investigation was to create an experimental benchmark for the mould-filling problem [7.25, 34].

A similar rectangular, inclined box filled with water was used to simulate the up-slope, down-slope flow transition, typical for natural convection of air in valleys. Despite the differences in scales, some atmospheric phenomena, such as updraughts typical of the morning transition, down-slope front propagation during the evening transition, and nocturnal stable stratification can be visualized well in the laboratory experiment. To investigate up-slope flow phenomenon a negative temperature gradient is set up between the lower and upper wall, simulating solar heating of the ground. For transition and down-slope flow studies the temperature of the walls is reversed by switching the coolant settings. TLC tracers changing the color of refracted light with temperature indicate the onset of flow instabilities with periodically rising plumes and hot fluid ejections, analogous to those observed for diurnal circulation on long, sloping surfaces (Fig. 7.9). These periodical ejections could be identified well in the experimentally measured velocity fields and reproduced in the numerical simulation. After reversing the temperature gradient a transition front was observed both in the temperature field and the velocity field. The front builds up about 1 min after the reversal of temperature of the walls at a distance of 2 cm from the lower edge of the slope, and disappears about 5 min later, when regular nocturnal stratification of the temperature and the flow pattern develops [7.35].

7.1.5 Conclusions

Application of TLC tracers to laboratory studies of thermally driven flows appears to be, in most cases, the only option for acquiring full-field information about the temperature and velocity fields in liquids. The noninvasive character of the method and its relative simplicity enables a valuable tool for the full-field verification and validation of numerical results for small-scale laboratory configurations. We found that a large improvement in the quality and reliability of numerical simulations could be obtained by validating and tuning methodologies using information obtained from the full-field flow measurements described [7.36].

However, despite its apparent simplicity, TLC thermography requires tedious and frequently repeated calibrations to offer high-quality quantitative measurements. This is especially true for applications involving flow field measurements. Further investigations both in improving TLC stability and the robustness of color evaluation procedures are necessary.

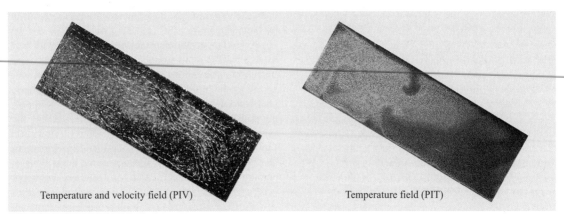

Fig. 7.9 Laboratory simulation of a diurnal up-slope flow, temperature and velocity measured with TLCs suspended in water. *Left*: unencapsulated tracers TM107 used to evaluate temperature and velocity fields in the cavity inclined at 30°; *right*: encapsulated tracers (TCC1001) visualizing the temperature field in the cavity inclined at 20°

7.2 Measurements of Surface Heat Transfer Characteristics Using Infrared Imaging

7.2.1 Introduction and Background

The use of infrared thermography has increased dramatically throughout the world over the past 15 years for the measurement of surface characteristics for a variety of research investigations involving convective heat transfer phenomena. The technique is especially viable and useful because it gives spatially resolved surface temperature distributions non-intrusively, even when large gradients of surface temperature are present. This can be done with high levels of accuracy and resolution, when the infrared imaging system and acquisition procedures are properly calibrated, and the nuances of camera behavior and infrared imaging are properly taken into account. In many research situations, the surface temperature distribution itself is of interest. However, its measurement is often one step in the determination of a variety of other spatially varying quantities that may be of interest, such as surface heat transfer coefficient distributions, nondimensional quantities containing surface heat transfer coefficients, surface adiabatic effectiveness, overall thermal effectiveness distributions, and thermal boundary condition information for numerical predictions of arrangements that are measured experimentally.

One critical step in the use of infrared thermography is the calibration of the images in some way so that local surface temperatures can be determined from the different shades, gray scales, or colors within infrared images. This is possible analytically, but requires knowledge of a variety of physical characteristics such as the surface emissivity, the effective atmospheric transmittance, the temperatures of surrounding walls, the atmospheric temperature, and other quantities. Because of the complexity and limited accuracy that is possible with such an approach [7.37], most investigators use some form of in situ calibration instead. Such an in situ calibration procedure was described and used as early as 1968 by *Czysz* and *Dixon* [7.38] for surface measurements using thermographic phosphors. Details of in situ calibration procedures employed by other investigators for specific measurement apparatus and environments using infrared thermography are described by *Meyers* et al. [7.39], *Gartenberg* and *Roberts* [7.40], *Wendt* [7.41], *Westby* [7.42], *Martiny* et al. [7.43], *Sargent* et al. [7.44], *Schulz* [7.37], and others. In the *Meyers* et al. [7.39] study, film cooled combustor walls are investigated under transient conditions. During heating of the combustor, these investigators recorded data simultaneously from both the infrared camera scanner and thermocouples that were embedded in the walls of the chamber. These data were then used for a best-fit approximation of the calibration parameters of the camera. *Westby* [7.42] calibrates prior to measurements by stepping a test article through a series of independently measured, steady-state temperatures, as infrared images of the article are recorded simultaneously. *Sargent* et al. [7.44], and *Schulz* [7.37] employ steady-state calibration procedures, by matching the measured local surface temperatures to gray scale values from black-and-white infrared images. Additional details of these calibration procedures are described later in the chapter.

When in situ calibrations are employed to obtain accurate spatially resolved surface temperature distribution data, infrared thermography is used for a variety of research problems, applications, and measurement environments with a variety of physical arrangements. For example, *Scherer* et al. [7.45], and *Scherer* and *Wittig* [7.46] describe measurements of convective heat transfer distributions on surfaces downstream of a backward-facing step, and downstream of a normal high-momentum jet, respectively. *Gartenberg* and *Roberts* [7.40] and *Wendt* [7.41] describe the use of infrared thermography for measurements in a variety of aeronautical and aerodynamic flows. *Westby* [7.42] uses related techniques for heat transfer measurements in rarefied flows.

Since the 1980s, and even more so in the 1990s, infrared thermography has been widely used for a variety of turbomachinery investigations and applications [7.37, 39, 44, 47–57]. Of these investigations, *Martiny* et al. [7.47], *Sen* et al. [7.50], *Gritsch* et al. [7.52, 53], *Yuki* et al. [7.51], *Sweeny* and *Rhodes* [7.54], *Schulz* [7.37], *Bell*, et al. [7.55], and *Furukawa* and *Ligrani* [7.56] investigate film cooling environments. *Meyers* et al. [7.39], and *Martiny* et al. [7.48] consider film and effusion cooled combustor walls. *Hedlund* et al. [7.49] and *Sargent* et al. [7.44] present measured results from the surfaces of swirl chambers used for internal cooling of turbine airfoil leading edges, and *Won* et al. [7.57] present measured results from a rib turbulator surface, used for the internal cooling of the mid-span regions of gas turbine blades and vanes.

Advances in measuring techniques and image analysis is continuing, with recent advances in these areas described by *Driggers* et al. [7.58]. These investigators also analyze some of the problems associated with laboratory measurement of sampled infrared imaging systems, and with processing the resulting images. Infrared thermography measurements made in a variety of other thermal systems, including on the surfaces of a passage with a rotating disc and within a rotating channel, are described by *Astarita* et al. [7.59], and *Cardone* et al. [7.60]. More recently, *Hindle* et al. [7.61] employed a fiber-based near-infrared absorption tomography system to measure distributions of hydrocarbons in the cylinder combustion chamber of an internal combustion engine. This technique takes advantage of the hydrocarbon absorption of $1.7\,\mu\mathrm{m}$ radiation using newly developed solid-state all-optoelectronic components.

These different investigations illustrate the variety of measurement environments where infrared thermography is employed to measure spatially resolved quantities on a surface or over a volume of interest. In every case, measured temperature distributions are used to deduce other quantities such as surface heat transfer coefficients, surface Nusselt numbers, surface film cooling effectiveness, or hydrocarbon distributions.

7.2.2 Chapter Organization

Infrared cameras are described in Sect. 7.2.3, the overall physical measurement arrangement is described in Sect. 7.2.4, and in situ calibration procedures are presented and discussed in Sect. 7.2.5. Subsections 7.2.6–7.2.10 then describe five different quantities measured and determined using infrared thermography, for five different physical situations. Specific procedures and approach details are given for measurement of these different quantities, which can be used in a variety of other physical situations and measurement environments. Included are specific examples and applications of in situ calibration of infrared thermography systems, and different approaches for capturing and processing infrared camera images to obtain useful information. A summary and conclusions are finally presented in Sect. 7.2.11.

7.2.3 Infrared Cameras

There are two general types of infrared thermography systems, scanning systems and staring systems. Scanning systems measure surface radiation point by point. Oscillating mirrors and rotating prisms are then used to transfer object points to a single infrared detector. Staring systems use infrared sensor arrays or focal-plane arrays. With this arrangement, every element of the object is transferred to a corresponding sensor element of the focal plane. Because of the considerable progress that has been achieved with infrared sensor arrays in recent years (up to 3×10^5 elements, each $24\,\mu\mathrm{m}$ by $24\,\mu\mathrm{m}$ in size), the majority of infrared cameras that are used today are staring systems.

Infrared camera systems generally operate over one of three different wavelength regions or bands of the infrared wavelength band, as it exists as part of the electromagnetic spectrum. These three bands are:

1. the short-wavelength infrared (SWIR) or near-infrared (NIR) band, which extends from 0.4 μm to 2–3 μm,
2. the mid-wavelength infrared (MWIR) or mean-infrared band, from 2–3 μm to 5 μm, and
3. the long-wavelength infrared (LWIR) or far-infrared (FIR) band, which extends from 8 μm to 12 μm.

The wavelength band employed depends on the application and measurement requirements. For example, the NIR or SWIR wavelength band is generally employed for laser applications. The MWIR and LWIR bands are generally used for thermal imaging and temperature detection, with important differences in the technologies employed in cameras that operate over these two different wavelength bands. Mid-wavelength infrared (MWIR) band cameras generally produce higher-quality images with better resolution and better magnitudes of noise equivalent temperature difference (NETD) compared to LWIR cameras. In contrast, LWIR cameras often operate with an uncooled focal-plane array of detectors, which makes them less expensive.

The different wavelength bands and the spectral emissivity are important when determining surface temperatures from measured spectral radiation. The relations, based on Planck's laws, for infrared energy levels also apply when the number of photons is quantified. This is important because numerous infrared cameras use photon detectors as sensing elements. The detectors most often used in the SWIR band are short mercury cadmium telluride, or indium gallium arsenide (InGaAs). The detectors most often used in the MWIR band are indium antimonide (InSb), or mercury cadmium telluride. The detectors most often used in the LWIR band are microbolometers (MB), gallium arsenide (GaAs), or mercury cadmium telluride. The typical detector size for all three wavelength bands is 30×30 μm for a 320×256 focal plane array, and 25×25 μm for a 640×512 focal plane array. Each element in each focal-plane array contributes to one pixel value in the infrared image. Many commercial infrared camera systems also include elaborate and advanced software, and data-processing capabilities to process digital images. This is accomplished either with real-time or digital data-acquisition systems. Many advanced cameras also have high frame rate and windowing capabilities with 14 bit processors. The resulting dynamic range images can then be acquired using real-time video reorders, with frames rates of 25 Hz, 30 Hz, or 60 Hz.

7.2.4 Overall Physical Arrangement

A typical test setup arrangement for measuring surface temperatures is shown in Fig. 7.10 [7.37]. Here, the incident radiation onto the infrared camera I_{meas} is comprised of:

1. the radiation from the test surface $I_{T,\text{obj}}$, which is attenuated by the atmospheric transmittance τ,
2. the radiation from the surroundings $I_{T,\text{amb}}$ (i.e., the flow channel and the side walls), and
3. the radiation that has been reflected from the test surface and also attenuated by the atmospheric surroundings $I_{T,\text{atm}}$.

Note that the atmosphere itself may also be emitting radiation into the camera. The test surface emission depends upon its emissivity ε because such surfaces are not perfect black bodies. If radiation from the atmosphere itself is neglected, according to *Schulz* [7.37], the measured radiation is then given by the relation

$$I_{\text{meas}} = \tau \varepsilon I_{T,\text{obj}} + \tau(1-\varepsilon)I_{T,\text{amb}} + (1-\tau)I_{T,\text{atm}} \ .$$

The effective atmospheric transmittance consists of the transmittance of the hot gas, the infrared translucent window, and the surrounding air. In the test arrangements described in this chapter, the windows are either sapphire, sodium chloride, or zinc selenide, which are used because of their high transmittance in the appropriate wavelength range and their high temperature resistance.

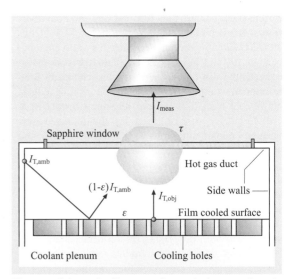

Fig. 7.10 Schematic diagram of a typical measuring situation using infrared thermography to analyze surface temperatures (after [7.37], used by permission)

Overall, the physical arrangement shown in Fig. 7.10, which includes this variety of physical phenomena described, is quite complicated. To overcome these complications and to obtain accurate measurement of surface temperature distributions, many investigators employ in situ calibration procedures, such as the ones that are briefly described in the next section.

7.2.5 In Situ Calibration

There are a variety of approaches to calibrating infrared camera systems to obtain thermal data. One of the most accurate means to calibrate infrared camera systems for measurement involves in situ calibration procedures. With this approach, the camera, imaging, and data-acquisition systems are all calibrated together in place within the experimental facility, and the calibrations are generally conducted simultaneously as the measurements are obtained.

In situ calibration procedures, like the ones described by *Sargent* et al. [7.44], are now widely used throughout the worldwide engineering community. Here, the infrared images are calibrated at the same times as the measurements are made by using thermocouples for simultaneous surface temperature measurements. Surface temperatures are thus measured at discrete locations (using thermocouples) at the same moment as each infrared image is recorded. With this approach, the relationship between infrared image gray scales and surface temperatures is determined for each infrared image. This relationship is a second- of third-order polynomial where temperature is a function of gray scale value. With this calibration polynomial, recorded digitized data, and location maps of thermocouple positions, surface temperature distributions over the entire test surfaces are determined from spatially resolved distributions of gray scale values recorded with the infrared camera. Magnitudes of spatially resolved convective heat transfer coefficients, Nusselt numbers, or surface effectiveness values are then determined from these and other measurements.

Similar calibration procedures are used for all five of the situations described in this chapter. In every case, some camera output parameter, such as the infrared image gray scale value, is correlated to temperature. These temperature values are determined from some other, independent means to measure local temperatures on the same surface from which the infrared images are obtained. In most cases, thermocouples are used for these local temperature measurements, which are recorded simultaneously as the infrared images of the surfaces are recorded.

Questions regarding a thermocouple's accuracy arise when applying thermocouple-based correction procedures to infrared thermography. A thermocouple always only senses the temperature of its junction. One error that may then result is the ability of a thermocouple junction to accurately measure the surrounding temperature of the material within which it is embedded, as well as the temperatures of nearby surfaces (from which an infrared image is obtained). This ability depends upon the thermocouple size, location, mounting arrangement, position, materials, wire lead arrangement and orientation, hole depth, relative temperatures, and materials and the material within which it is mounted. When transient measurements are undertaken, this task is further complicated by the thermal inertial influences of the thermocouple as well as the surrounding mounting substrate.

According to *Schulz* [7.37], the accuracy of a thermocouple reading also depends upon its calibration curve, and upon its exact position in the surface and its orientation with respect to temperature gradients. The latter is especially important when high surface temperature gradients are present. Even very small discrepancies between the thermocouple position and the location of the according pixel of the infrared image can lead to pronounced temperature differences. For the calibrations used in investigations by *Schulz* [7.37], *Gritsch* et al. [7.52, 53], *Furukawa* and *Ligrani* [7.56], *Won* et al. [7.57], and others, only those thermocouples located in approximately uniform temperature areas are employed, while thermocouples located in regions of large temperature gradient are excluded from the calibration procedure.

In some cases, additional refinements and alterations to in situ calibration procedures are employed [7.37]. These include extensive numerical analysis of infrared data to reduce the root-mean-square error values of temperature measurement differences and errors. For the investigations described by *Schulz* [7.37], these are based on the relationship between the measured radiation and surface temperature, as produced by the AGEMA Thermovision 900 camera and software.

7.2.6 Measurement of Surface Nusselt Numbers

Here, the use of infrared thermography to measure spatially resolved distributions of surface heat transfer coefficients, and surface Nusselt numbers on the interior surface of a swirl chamber is described. Swirl chambers are used in a wide variety of applications because of the heat and mass transfer increases they provide. For example, swirl chambers are employed in heat exchangers, automobile engines, furnaces, biomedical devices, and devices used for heating and cooling of metal ingots. They are also used to fractionate solid particulates suspended and transported in liquids and gases, to enhance mixing processes in combustion chambers, and for spray-drying applications used with atomizers.

In the investigation of *Hedlund* et al. [7.49], the heat transfer coefficient h is given by

$$h = \frac{\dot{q}''}{(T_w - T_i)},$$

where $\dot{q}''$ is the surface heat flux, T_w is the local surface temperature, and T_i is the mixed mean air temperature at the inlet of the swirl chamber. The surface Nusselt number is then determined using $\text{Nu} = hD/k$, where k is the air thermal conductivity, and D is the inner diameter of the main cylinder of the swirl chamber. For the results that are presented from *Hedlund* et al. [7.49], a constant surface heat flux boundary condition is applied to the test surfaces, and the inlet mixed mean air temperature T_i is also invariant as each experimental data set is obtained. Thus, variations of h and Nu are linked to spatial variations of surface temperature T_w, which are measured with infrared thermography.

Swirl Chamber for Heat Transfer Measurements

A schematic of the swirl chamber used for heat transfer measurements is shown in Fig. 7.11, including the coordinate system. The acrylic cylinder of the swirl chamber contains 30 copper–constantan thermocouples, and its inner surface is adjacent to the airstream. Each of these thermocouples is located 0.051 cm below this surface to provide measurements of local surface temperatures, after correction for thermal contact resistance and the temperature drop through the 0.051 cm thickness of acrylic. Acrylic is chosen because of its low thermal conductivity ($k = 0.16\,\text{W/mK}$ at $20\,°C$) to minimize axial and circumferential conduction along the test section, and thus minimize smearing of spatially varying temperature gradients along the test surface. Acrylic also works well for infrared imaging because of its surface emissivity value.

Surface Nusselt Number Measurement Procedures

Infrared imaging is used to measure spatially resolved surface temperature variations along the inner surface of the swirl chamber. To do this, the infrared radiation emitted by the heated interior surface of the swirl chamber is captured using a VideoTherm 340 infrared imaging camera [7.62], which operates at infrared wavelengths of $8–14\,\mu\text{m}$. Temperatures, measured using the 30 calibrated copper–constantan thermocouples distributed along the swirl chamber surface adjacent to the flow, are used to perform the in situ calibrations simultaneously as the radiation contours from surface temperature variations are recorded.

This is accomplished as the camera views the test surface through a custom-made, cylindrical zinc selenide window, which transmits infrared wavelengths

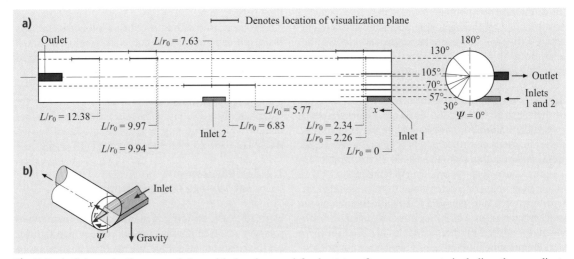

Fig. 7.11a,b Schematic diagrams of the swirl chamber used for heat transfer measurements including the coordinate system (**b**) and arrangement of inlet and outlet ducts (**a**) [7.49]

between 6 and 17 μm. Reflection and radiation from surrounding laboratory sources are minimized using an opaque shield that covers the camera lens and the zinc selenide window. Frost build-up on the outside of the window (due to the fact that the air inside the swirl chamber is cooled to less than 1 °C) is eliminated using a small heated airstream from a hair dryer. The window is located on a segment of the swirl chamber that is either rotated or relocated axially so that the camera can view different portions of the interior surface of the swirl chamber. At least two, and as many as five, thermocouple junction locations are present in any field viewed by the camera. The exact spatial locations and pixel locations of these thermocouple junctions and the coordinates of a 12.7×12.7 cm field of view are known from calibration maps obtained prior to measurements. During this procedure, the camera is focused, and rigidly mounted and oriented relative to the test surface in the same way as when radiation contours are recorded.

With these data, grey scale values at pixel locations within videotaped images from the infrared imaging camera are readily converted to temperatures. Examples of such calibration data are shown in Fig. 7.12 for three different window camera views of the swirl chamber test surface. Because such calibration data depend strongly on camera adjustment, the same brightness, contrast, and aperture camera settings are used for all three data sets. Each calibration data set shown in Fig. 7.12 is curve-fit to a second- or third-order polynomial equation. The slight difference between the three calibration data sets is because the transmissivity of the zinc selenide window and the emissivity of the swirl chamber test surface both exhibit small variations as the camera angle, camera and test surface orientation, surface temperatures, or experimental conditions are changed. The in situ calibration approach rigorously and accurately accounts for these variations.

For these measurements, the images from the infrared camera are recorded as 8 bit gray scale images on commercial videotape using a video recorder. Images are then digitized using software, operated on a desktop personal computer (PC). Subsequently, software is used to perform coordinate transformations to correct for nonrectangular, stretched, or distorted recorded images because of camera perspective or because the camera lens orientation is not normal to the curved target surface. This software also converts each of the 256 possible gray scale values to temperature at each pixel location using calibration data, and determines values of local Nusselt numbers. Contour plots of local surface temperature and Nusselt number (in *unrolled* planar, Cartesian coordinates) are prepared using graphics software. Each individual image covers a 250×250 pixel area.

Local Surface Nusselt Number Results

The surface Nusselt number distribution measured just downstream of the second inlet of the swirl chamber is shown in Fig. 7.13b. Included is the enhanced infrared image used to determine this distribution, presented in Fig. 7.13a. These data are given for $Re = 19\,400$ ($Re_D = 7205$) and $T_i/T_w = 0.85$. The location of the data, based on the coordinate system shown in Fig. 7.11, is $x/r_0 = 7.0\text{--}8.0$ and $\psi = 0\text{--}50°$. Here, Re is the Reynolds number based on the inlet duct characteristics, Re_D is the Reynolds number based on the swirl chamber main cylinder characteristics, T_i is the air mixed-mean temperature at the swirl chamber inlet, T_w is the local surface temperature, x is the axial distance measured from the swirl chamber end face, r_0 is the radius of the large cylinder comprising the swirl chamber, and ψ is the circumferential angle measured from the swirl chamber vertical plane. The surface Nusselt number variations in Fig. 7.13b show clear evidence of an array of Görtler vortex pairs, which develop in the chamber as the flow exits the second inlet passage of the chamber. In Fig. 7.13a, these are indicated by the collection of light and dark stripes, which correspond to vortex pair trajectories along the concave inner surface of the chamber. This image is as recorded directly from the camera with some enhancement, but with no corrections for camera angle and perspective. The surface temperature increases in Fig. 7.13a as image regions become lighter and whiter. Lower surface temperatures then coincide

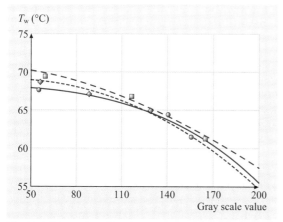

Fig. 7.12 Example calibrations of infrared images used for different window orientations and different camera views into the swirl chamber [7.49]

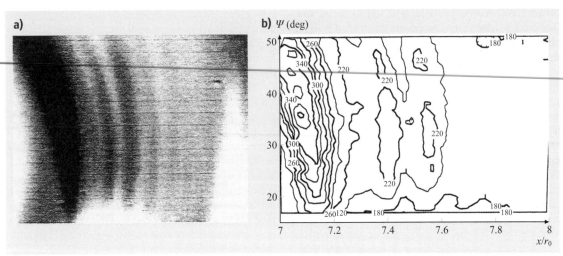

Fig. 7.13a,b Surface Nusselt number variations due to Görtler vortex pairs in the swirl chamber for Re = 19 400 (Re$_D$ = 7205) and $T_i/T_w = 0.85$ at $x/r_0 = 7.0$–8.0 and $\psi = 0$–50° just downstream of the second inlet [7.49]: (**a**) enhanced infrared image, (**b**) Nusselt number contours

with higher local Nusselt numbers, and higher surface temperatures coincide with lower local surface Nusselt numbers [7.49].

The spatially resolved surface Nusselt numbers determined from this infrared image are shown in Fig. 7.13b. In spite of the relatively high Reynolds number of this flow, important Nusselt number variations due to the Görtler vortex pairs are evident, where higher and lower local Nusselt numbers correspond to vortex pair downwash and upwash regions, respectively [7.49]. The high Nusselt number region located at $x/r_0 = 7.0$–7.2 in Fig. 7.13 is due to the shear layer near the edge of the inlet jet. This shear layer contains vortex structures that also contribute to surface Nusselt number enhancements. Such results illustrate the excellent accuracy and spatial resolution obtained with the infrared imaging and measurement techniques employed in this particular investigation [7.49].

7.2.7 Measurement of Surface Adiabatic Film Cooling Effectiveness

In this section, the use of infrared thermography to measure another quantity of interest to heat transfer researchers, the surface adiabatic film cooling effectiveness, is described. Such data are useful in film cooling investigations, and are needed to determine the performance and protection provided by different film cooling hole configurations. Effective performance of the films produced by these configurations is important for cooling blades and vanes in the turbines of gas turbine engines. Such cooling maintains acceptable metal temperatures, and in doing so, maintains blade and vane durability, reliability, and lifetimes for safe and efficient gas turbine engine operation. In efforts to design film cooling holes with improved performance, recently attention has been devoted to contouring the holes employed [7.52, 53, 55, 56].

Surface adiabatic film cooling effectiveness distributions from holes with diffuser-shaped expansions are described here because these are believed to improve film cooling performance relative to holes that are cylindrical [7.55, 56]. This is accomplished as the increased cross-sectional area reduces the exit velocity and thus the momentum flux of the film as it exits the holes. The penetration of the film jet into the main flow is then reduced, which gives higher overall film cooling effectiveness. The lateral expansion of such holes also causes improved lateral spreading of the film concentrations which leads to more-uniform coverage of the test surfaces with protective film cooling air [7.55, 56].

One parameter that quantifies film cooling performance is the adiabatic film cooling effectiveness, η [7.56]. For high-speed flows, where compressibility and viscous dissipation are non-negligible, this quantity is given by

$$\eta = \frac{(T_{aw} - T_{r\infty})}{(T_{oc} - T_{r\infty})},$$

where T_{aw} is the surface adiabatic temperature, $T_{r\infty}$ is the free-stream recovery temperature, and T_{oc} is the coolant stagnation temperature [7.56]. As such, the adiabatic effectiveness η is the ratio of actual surface temperature reduction due to the cooling film, and the maximum possible temperature reduction given by the temperatures of the coolant and mainstream gas. In the investigations of *Gritsch* et al. [7.52, 53], *Bell* et al. [7.55], and *Furukawa* and *Ligrani* [7.56], $T_{r\infty}$ and T_{oc} are maintained constant during a given test, and the spatial variation of surface adiabatic temperature T_{aw} is measured using infrared thermography. In most test arrangements, a material with low thermal conductivity is employed to produce an adiabatic test surface where T_{aw} is measured. In the results from *Gritsch* et al. [7.52, 53], which are presented to illustrate adiabatic effectiveness determination using infrared thermography here, data for three different hole configurations are presented, each for three different film cooling blowing ratios. The blowing ratio M is the ratio of mass flux provided by the coolant at the inlet of the cooling holes, and the mass flux in the free stream of the mainstream gas.

Experimental Arrangements and Apparatus for Surface Effectiveness Measurements

The single film cooling hole arrangements employed by *Gritsch* et al. [7.52, 53] are shown in Fig. 7.14. The three configurations are: cylindrical, fan-shaped, and laid-back fan-shaped. The exit-to-entry ratio for the fan-shaped, and laid-back fan-shaped holes are 3.0 and 3.1, respectively. For the infrared thermography measurements downstream, the bottom test wall is equipped with thermocouples to measure local surface temperatures, which are also employed for in situ calibration of the infrared images. The surface is also painted with a black color with a high, temperature-independent emissivity to ensure good radiative properties. The camera observes this test surface through a sapphire window located in the top wall of the test section channel. All results are obtained under steady-state thermal-equilibrium conditions.

The infrared camera employed to obtain these data is an AGEMA Thermovision 900 device with two indium antimonide (InSb) detectors [7.63]. The detector within this scanner operates in the 2–5.6 μm spectral range. To insure low-noise operation, the detector is cooled to approximately 80 K using a Stirling motor. In contrast to liquid-nitrogen-cooled scanners, the Stirling-cooled infrared scanner allows for arbitrary observations in multiple directions and unlimited operation, which is useful when infrared thermography is employed in laboratory

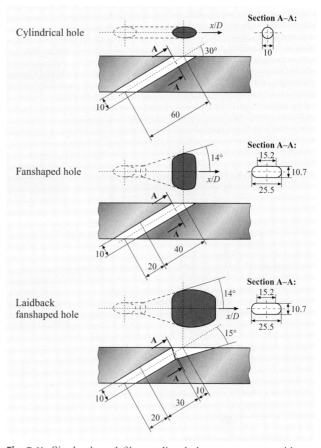

Fig. 7.14 Single shaped film cooling hole arrangements with expanded exits employed in the investigations by *Gritsch* et al. (after [7.37, 52, 53], used by permission)

conditions. Each infrared image produced consists of 136 lines, with each line consisting of 272 samples. The corresponding image acquisition frequency is 15 Hz, which can be doubled by using half the number of scan lines.

Local Surface Adiabatic Film Cooling Effectiveness Results

Figure 7.15 shows adiabatic surface film cooling effectiveness distributions for the cylindrical, fan-shaped, and laid-back fan-shaped holes for blowing ratios M of 0.5, 1.0, and 1.5, where x represents the streamwise distance along the test surface, z represents the spanwise distance along the test surface, and D is the entrance diameter of the film cooling holes. In this multi-part figure, the film produced by the cylindrical hole detaches from the surface at a blowing ratio M of 1.5, result-

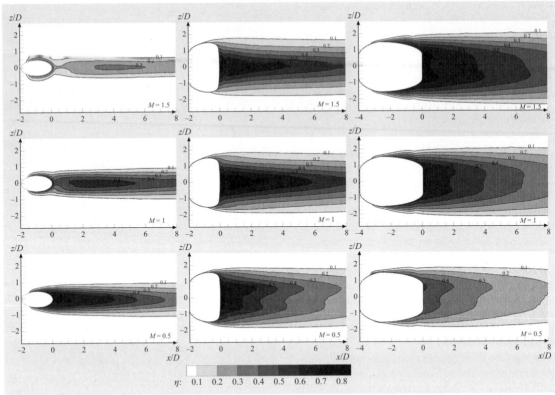

Fig. 7.15 Surface adiabatic film cooling effectiveness distributions obtained using the hole arrangements shown in Fig. 7.14 from investigations by *Gritsch* et al. [7.37, 52, 53]. Surface distributions are given for different hole arrangements and different film cooling blowing ratios (used by permission)

ing in relatively low values of adiabatic effectiveness. Decreasing the blowing ratio increases the local effectiveness magnitudes downstream of the cylindrical holes because the penetration of the jet into the main flow is reduced [7.37, 52, 53]. For all three blowing ratios, lateral spreading of the film downstream of the cylindrical hole is very poor. Significantly higher magnitudes of adiabatic surface effectiveness are present downstream of the fan-shaped, and laid-back fan-shaped holes in Fig. 7.15. Lateral spreading of the film on the test surface is also improved for these hole configurations compared to the cylindrical hole, however, most of the coolant is still ejected along the centerline, resulting in pronounced lateral effectiveness gradients, especially at the higher blowing ratios. Note that a small separation region is present just downstream of the fan-shaped holes for higher blowing ratios, which is indicated by a local decrease in adiabatic film effectiveness. The centerline effectiveness magnitude for the laid-back fan-shaped hole is somewhat reduced compared to the distribution produced by the fan-shaped hole, although the lateral spreading is improved [7.52, 53].

7.2.8 Measurement of Surface Heat Flux Reduction Ratio

Increases in heat transfer coefficients due to film cooling injection can be combined with adiabatic effectiveness values (like those described in the previous section) to determine the heat transfer rates, and consequently, the net heat flux reduction Δq_r produced by film cooling. From *Sen* et al. [7.50], Δq_r is defined as

$$\Delta q_r = 1 - \frac{h_f(T_{aw} - T_w)}{h_0(T_{r\infty} - T_w)}.$$

Here, the subscript '0' denotes no film cooling, and the subscript 'f' denotes with film cooling. A nondimensional surface temperature can be defined using:

$$\theta = \frac{(T_{r\infty} - T_w)}{(T_{r\infty} - T_{oc})}.$$

Combining these equations with the defining equation for adiabatic film cooling effectiveness η then gives

$$\Delta q_r = 1 - \left(\frac{h_f}{h_0}\right)\left(1 - \frac{\eta}{\theta}\right) .$$

From this equation, it is evident that the net heat flux reduction can be determined from measurements of adiabatic film cooling effectiveness and the heat transfer coefficient ratio, h_f/h_0. The net heat flux reduction Δq_r is positive if film cooling gives a reduction in the heat transfer rate along the test surface. However, in some cases, injecting the coolant film into the flow along a surface increases the heat transfer coefficient, $h_f/h_0 > 1$, and negative values of Δq_r. In order to produce a net decrease in heat flux to the surface, the adiabatic effectiveness must be large enough to compensate for increases in the heat transfer coefficient ratio h_f/h_0 [7.50, 51].

Experimental Facility for Surface Heat Flux Reduction Measurements

Figure 7.16 shows a schematic diagram of the leading-edge test facility employed for heat flux reduction measurements using infrared imaging, from *Yuki* et al. [7.51]. This leading edge matches configurations for vane airfoil leading edges like those used in industrial gas turbine engines. The arrangement is a half-section leading edge. The test section is placed in the wind tunnel with a suction slot immediately in front of the leading edge to control the location of the stagnation line. The constant-radius leading edge contains two rows of holes, which are angled laterally at 90° with respect to the bulk flow direction. The two rows of holes are positioned 0° and 25° relative to the horizontal, at x/d locations of 0 and 3.5. Within the plane of each row of holes, the holes are angled 20° with respect to the test surface. The diameter of all holes is 6.35 mm. Each row contains nine holes with a spanwise hole spacing of 7.64 hole diameters. The rows are arranged so that holes in adjacent rows are staggered with respect to each other. The leading-edge thickness is 25 mm giving hole length-to-diameter ratios of 12. To minimize conduction error along the surface, the leading edge is made of extruded polystyrene foam with a thermal conductivity of 0.027 W/mK. The width of the suction slot and the amount of suction are adjusted so that the stagnation line is located at $x/d = 0$, and the stagnation streamline is parallel to the approaching mainstream direction. During heat transfer measurements, the mainstream velocity is set at 10 m/s, giving a Reynolds number based on the effective leading-edge diameter of 65 000 [7.51].

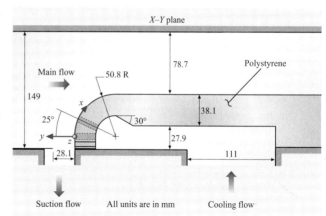

Fig. 7.16 Schematic diagram of a leading-edge test facility for heat flux reduction measurements using infrared imaging from *Yuki* et al. (after [7.51], used by permission)

Infrared Imaging Procedures

Surface temperature data are collected using an Inframetrics 600L infrared camera [7.64] mounted 45° relative to the horizontal direction, and pointed towards the center of curvature of the leading edge. A window made of sodium chloride is used to allow maximum transmission of radiation emitted from the surface to the camera. The camera is calibrated by comparing camera-detected temperatures to values measured with type E ribbon thermocouples mounted at several positions along the test surface. According to *Yuki* et al. [7.51], the uncertainty magnitude of surface temperature measurements, determined from the curve-fit error for the linear calibration equation, is ±0.2 °C. The spatial resolution of the infrared measured surface temperatures is 0.5 times one film cooling hole diameter, where discrete pixel levels are based on integrated averages over a 3×3 mm surface area. This resolution is adequate to resolve measured temperature gradients except within one-half of one diameter of the film cooling holes because of the steep gradients of surface temperature at these locations. Mainstream and plenum temperatures are measured using type E thermocouples with an experimental uncertainty of ±0.1 °C.

To measure surface heat transfer coefficients, a sheet of 25 μm-thick stainless steel is attached to the leading edge to provide a constant heat flux surface. The 597×102 mm sheet covers the entire curved portion of the leading edge plus approximately 10 mm extra upstream and downstream to avoid end effects. Electrical current is directed through the plate using copper busbars attached on either end. Elliptically shaped holes,

1.2 mm wider than the film cooling holes, are cut into the steel sheet at the injection hole locations. To determine power input, the voltage across the stainless-steel plate and the current is determined from voltage measurements across a 3.33 mΩ shunt resistor connected in series with the heat flux plate. The total heat rate from the test plate is determined by subtracting the analytically determined radiative heat flux from the total heat flux. Conduction losses are estimated to be negligible. Local heat transfer coefficients are then determined from the convective heat flux, local surface temperature, and mainstream temperature. During testing the nominal plate temperature is 45 °C and the nominal mainstream temperature is 27 °C [7.51].

Net Heat Flux Reduction Results

The measured heat flux reduction distribution on the leading-edge test section is shown in Fig. 7.17 [7.51]. These data are given for a nondimensional temperature θ of 0.6, a density ratio of 1.0, and a momentum flux ratio of 3.5. For these data, the stagnation line is located at $x/d = 0$, where x represents the streamwise distance along the test surface, z represents the spanwise distance along the test surface, and d is the entrance diameter of the film cooling holes. Figure 7.17 shows that $\Delta q_r > 0$ over most of the surface, indicating improved performance relative to operation with no film cooling. Net heat flux reduction magnitudes are maximum at locations directly beneath the largest coolant concentrations, where adiabatic effectiveness magnitudes are maximum. However, downstream of the second row of holes, there is an elongated region where the net heat flux reduction is negative with values as low as $\Delta q_r = -0.4$. This region corresponds to locations along the test surface where heat transfer coefficients are augmented from local mixing near the edge of the coolant jets. Such data further illustrate the ability of infrared imaging to provide detailed, spatially resolved data on critical components in cooling applications subject to high-temperature gas environments.

7.2.9 Transient Surface Temperature Measurements

Here, the use of infrared imaging to acquire transient, spatially resolved surface temperature distributions at different times is described. The technique is applied to biomedical therapeutic patch testing.

Infrared Imaging and Data Acquisition Procedures

Spatially resolved temperature distributions along a patch test surface are determined using infrared imaging in conjunction with thermocouples, digital image processing, and in situ calibration procedures. To accomplish this, the infrared radiation emitted by the heated patch surfaces is captured using a VideoTherm 340 infrared imaging camera [7.62], which operates at infrared wavelengths from 8 μm to 14 μm. Each patch, as it is tested, is mounted on a flat sheet of acrylic. Temperatures, measured using calibrated, miniature copper–constantan thermocouples distributed along the patch surface being measured, which is adjacent to still air, are used to perform the in situ calibrations simultaneously as the radiation contours from surface temperature variations are recorded. Because the patches are exposed to natural convection induced by laboratory air, no window to transmit infrared energy is required. Six thermocouple junction locations are present in the infrared field viewed by the camera. The exact spatial locations and pixel locations of these thermocouple junctions and the coordinates of a 1.3 × 1.3 cm field of view are known from calibration maps obtained prior to the measurements. During this procedure, the camera is focused, and rigidly mounted and oriented relative to the test surface in the same way as when radiation contours are recorded. Voltages from the thermocouples are acquired using a Hewlett Packard 44422T data-acquisition card installed in a Hewlett Packard 3497A data-acquisition

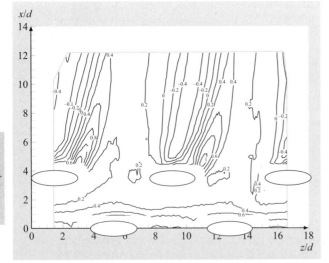

Fig. 7.17 Heat flux reduction distribution on the leading-edge test section from *Yuki* et al. [7.51] for $\theta = 0.6$, density ratio of 1.0, and momentum flux ratio of 3.5. The stagnation line is located at $x/d = 0$ (used by permission)

control unit, which is controlled by a Hewlett Packard A4190A series computer.

With these data, gray scale values at pixel locations within videotaped images from the infrared imaging camera are readily converted to local surface temperature values. Because such calibration data depend strongly on camera adjustment, the same brightness, contrast, and aperture camera settings are used to obtain the experimental data. The in situ calibration approach rigorously and accurately accounts for these variations.

Images from the infrared camera are recorded as 8 bit gray scale directly into the memory of a PC using a frame-grabber video card, and associated software. Three sets of 60 frames are recorded at a rate of one frame per second. All 180 resulting images are then ensemble-averaged to obtain the final gray scale data image. This final data set is then imported into software to convert each of the 256 possible gray scale values into the local Nusselt number at each pixel location using calibration data. Each individual image covers a 300 × 300 pixel area.

Transient Surface Temperature Results

Figure 7.18 shows infrared images acquired 1, 6, and 60 min from the beginning of testing of one biomedical therapeutic patch. Here, X and Y represent the coordinates in two directions along the patch surface. Infrared camera images are shown in the left-hand column, with associated surface temperature contours presented in the right-hand column. Notice that the highest temperatures are located near the center of the patch for the 1 and 6 min data sets. As time proceeds, temperature levels drop near the center of the patch, and the outer portions of the patch heat up as energy is conducted away from the patch center. The time variations of the maximum temperature and spatially averaged patch temperature are shown in Fig. 7.19. The interesting feature of these data is the time of the maximum temperature,

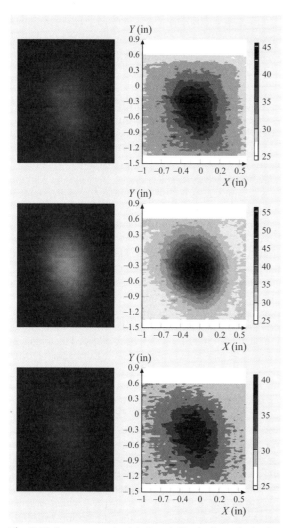

Fig. 7.18 Images acquired 1, 6, and 60 min from the beginning of testing of a biomedical therapeutic patch acquired to determine spatial and temporal surface temperature variations. Infrared camera images are shown in the left-hand column, with associated surface temperature contours presented in the right-hand column

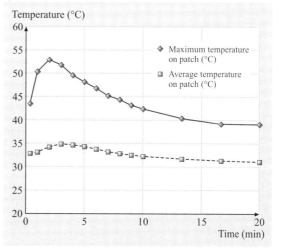

Fig. 7.19 Time variations of maximum temperature and average temperature from tests of the biomedical therapeutic patch

which is about 2 min after the test begins. Overall, such data illustrate the ability of infrared imaging, in situ calibration procedures, and digital image processing to provide spatially resolved, time-varying data for specific applications.

7.2.10 Boundary-Condition Information for Numerical Predictions

In this section, the use of infrared thermography to provide thermal boundary-condition information for numerical predictions is described. The test surface that is analyzed contains a collection of rib turbulators that are angled with respect to the flow stream. Overall, the vortices and the accompanying secondary flows produced by these rib turbulators aid convective processes for heat transfer augmentation by:

1. increasing secondary advection of fluid between the central parts of the channel and regions near the wall, and
2. producing regions with high three-dimensional shear and high magnitudes of turbulence production over much of the channel cross section, thereby substantially increasing turbulence transport levels in all three coordinate directions [7.65].

In many cases, remarkable enhancements of local and spatially averaged surface heat transfer rates are possible with rib turbulators, in spite of the lower local Nusselt numbers at certain locations along the ribbed surfaces. This is aided by the wide collection of secondary flow phenomena produced by the rib turbulators, which include flow recirculation zones both upstream and downstream of individual ribs, shear layer reattachment, pairs of counter-rotating flow cells, and different collections of small- and large-scale vortex pairs [7.65].

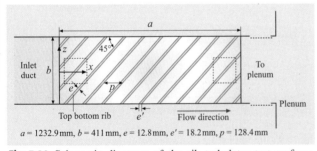

Fig. 7.20 Schematic diagrams of the rib turbulator test surfaces, including coordinate system [7.57]

In the discussion that follows, the Nusselt number Nu is given by

$$\mathrm{Nu} = h \frac{D_\mathrm{h}}{k} \,,$$

where h is the heat transfer coefficient, D_h is the channel hydraulic diameter, and k is the molecular thermal conductivity of air. The heat transfer coefficient is then based on the flat projected area and is determined using

$$h = \frac{\dot{q}''}{(T_\mathrm{w} - T_\mathrm{m})}$$

where $\dot{q}''$ is the local surface heat flux, T_w is the local wall temperature, and T_m is the time-averaged, local mixed mean temperature. Nu_0 is then the baseline Nusselt number in a smooth channel with no rib turbulators.

Experimental Facility and Test Surface for Heat Transfer Measurements

Figure 7.20 shows the geometric details of the rib turbulator test surface, including the rib turbulator geometry. Prior to this ribbed turbulator test section is a 411×103 mm inlet duct that is 1219 mm in length. This is equivalent to 7.4 hydraulic diameters (where the hydraulic diameter is 164.7 mm). Two trips are employed on the top and bottom surfaces of the inlet duct, just upstream of the test section, which follows with the same cross section dimensions. The duct exits to another duct, and then to a 0.60 m square plenum. Figure 7.20 shows that a total of 13 ribs or rib segments are used on the top wall and on the bottom wall of the rib turbulator test section. As mentioned, these are arranged with $45°$ angles with respect to the streamwise flow direction, such that the ribs on opposite walls of the channel are parallel and aligned with respect to each other. Each rib is 12.8 mm high and has a square cross section. The ratio of the rib height to the hydraulic diameter is 0.78, the rib pitch-to-height ratio is 10, and the blockage provided by the ribs is 25% of the channel cross-sectional area [7.57, 65]. The top wall of the test section also has two cut-out regions (one at the upstream end and one at the downstream end) where a zinc selenide window can be installed to allow the infrared camera to view a portion of the test surface on the bottom wall. When this window is not in use, inserts with ribs (which exactly match the adjacent rib turbulators on the top wall) are used in its place. Also identified in Fig. 7.20 is the test section coordinate system employed for the study. Note that the Y coordinate is directed normal to the bottom wall [7.57].

All exterior surfaces of the facility are insulated with Styrofoam ($k = 0.024$ W/mK), or two to three layers of 2.54 cm-thick, Elastomer Products black neoprene foam insulation ($k = 0.038$ W/mK) to minimize heat losses. Calibrated copper–constantan thermocouples are located between the three layers of insulation located all around the test section to determine conduction losses. Between the first layer and the 3.2 mm-thick acrylic test surfaces are custom-made Electrofilm etched-foil heaters (each encapsulated between two thin layers of Kapton) to provide a constant heat flux boundary condition on the flat horizontal, bottom surface of the ribbed test plate. The acrylic surfaces, which are adjacent to the airstream, contain 35 copper–constantan thermocouples, which are placed within the ribs as well as within flat portions of the test surfaces between the ribs. Each of these thermocouples is located 0.051 cm below the surface to provide measurements of local surface temperatures, after correction for thermal contact resistance and the temperature drop through the 0.051 cm thickness of acrylic. Acrylic is chosen because of its low thermal conductivity ($k = 0.16$ W/mK at 20 °C) to minimize streamwise and spanwise conduction along the test surfaces, and thus minimize smearing of spatially varying temperature gradients along the test surfaces. Energy balances performed on the heated test surfaces then allow the determination of the local magnitudes of the convective heat flux.

The mixed-mean stagnation temperature of the air entering the test section is measured using five calibrated copper–constantan thermocouples spread across its cross section. To determine this temperature, thermocouple-measured temperatures are corrected for thermocouple wire conduction losses, channel velocity variations, as well as for the differences between the stagnation and recovery temperatures. All measurements are obtained when the test facility is in a steady state.

Local Nusselt Number Measurement

To determine the surface heat flux (used to calculate heat transfer coefficients and local Nusselt numbers), the convective power levels provided by the etched foil heaters are divided by flat test surface areas. Spatially resolved temperature distributions along the bottom rib turbulator test surface are determined using infrared imaging in conjunction with thermocouples, energy balances, and in situ calibration procedures [7.44, 57]. To accomplish this, the infrared radiation emitted by the heated interior surface of the channel is captured using a VideoTherm 340 infrared imaging camera [7.62], which operates at infrared wavelengths from 8 μm to 14 μm. Temperatures, measured using the calibrated, copper–constantan thermocouples distributed along the test surface adjacent to the flow, are used to perform the in situ calibrations simultaneously as the radiation contours from surface temperature variations are recorded.

This is accomplished as the camera views the test surface through a custom-made zinc selenide window, which transmits infrared wavelengths between 6 and 17 μm, located on the top wall of the test section. Eleven to 13 thermocouple junction locations are usually present in the field viewed by the infrared camera. The exact spatial locations and pixel locations of these thermocouple junctions and the coordinates of a 12.7 × 12.7 cm field of view are known from calibration maps obtained prior to the measurements [7.57].

Images from the infrared camera are recorded as 8 bit gray scale directly into the memory of a PC computer using a frame-grabber video card, and associated software. Three sets of 60 frames are recorded at a rate of one frame per second. This is done after the rib turbulator flows are established to obtain data when the test surface is in a steady-state condition. All of the 180 resulting images are then ensemble-averaged to obtain the final gray scale data image. This final data set is then imported into software to convert each of the 256 possible gray scale values to local Nusselt number values at each pixel location using calibration data. Each resulting image then covers a 300 × 300 pixel area. Thermal conductivity in the Nusselt number is based on the average of the local wall temperature and the temperature of the air at the upstream inlet [7.57].

Conduction Analyses Within the Test Surface

Three-dimensional conduction along and within the test surface is determined using version 6.0 of the ANSYS numerical code. To accomplish this, a portion of the test surface, with one rib segment, is modeled using approximately 17 480 numeric node elements. In addition, a constant heat flux boundary condition (determined from values used in the experiments) is imposed on the back side of the 3.2 mm-thick acrylic test surface. Local surface temperatures, measured experimentally using infrared thermography, are used for the thermal boundary condition on portions of the test surface next to the airstream, including the top of the rib segment and the flat parts of the test surface around the rib. Temperatures are also imposed as the thermal boundary condition on the vertical parts of the rib segment, which are determined by interpolation of values measured at the edge of the rib top and at the corner (between each rib side and flat parts of the test surface) [7.57].

Ordinarily, without the conduction analysis applied, the same heat flux produced by the etched-foil heater is assumed to leave the test surface next to the airstream at the rib top and along the flat parts of the test surface around the rib. This means that no heat is assumed to leave the vertical side walls of the rib, and that all of the thermal power going into the test surface leaves entirely from the horizontal surfaces next to the airstream. The ANSYS version 6.0 conduction analysis is used to determine the nonuniform variations of the surface heat flux into the airstream, which are actually present, including along the vertical sides of the rib segment. As mentioned, this is accomplished using experimentally measured values as thermal boundary conditions around the domain solved by the numerical code.

Results with and Without Test Surface Conduction Analysis

The results presented in Fig. 7.21a for $Re_H = 17\,000$ (where Re_H is the Reynolds number based on the channel height) are again given in Fig. 7.21b, but with variable surface heat flux and surface conduction analysis applied. Comparison of these two figures thus provides information on the effects of three-dimensional conduction along and within the test surface. The results under these experimental conditions are chosen to illustrate these effects, because variations due to conduction are qualitatively similar to those observed at other Reynolds numbers. Note that the results in the latter figure cover smaller ranges of X/D_h and Z/D_h values, where X represents the streamwise distance along the test surface, Z represents the spanwise distance along the test surface,

and D_h is the channel hydraulic diameter. A comparison of the data in Fig. 7.21a,b shows that the Nu/Nu_0 ratio changes due to three-dimensional conduction are most apparent on the top of the rib segment, and on the flat base regions immediately adjacent to the rib segment. Such three-dimensional conduction effects are limited, in part, because of the high-quality construction and performance of the heated test surface, within which conduction is one-dimensional over most of the volume [7.57].

The Nu/Nu_0 data in Fig. 7.21a are obtained by assuming that all of the heat flux into the rib segment leaves entirely from the top of the rib (with none leaving

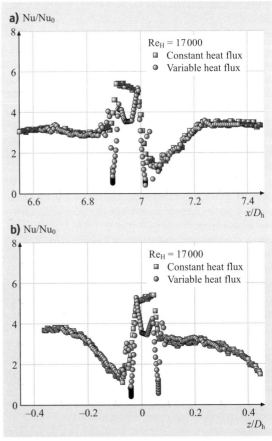

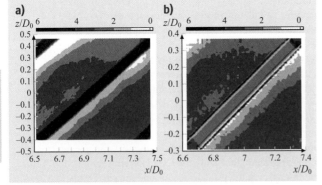

Fig. 7.21a,b Local Nusselt number ratio Nu/Nu_0 distribution along the rib turbulator test surface for $Re_H = 17\,000$ and $T_{oi}/T_w = 0.94$ [7.57]. (**a**) With constant surface heat flux and no surface conduction analysis applied. (**b**) With variable surface heat flux and surface conduction analysis applied

Fig. 7.22a,b Variable surface heat flux, local Nusselt number ratios Nu/Nu_0 along the rib turbulator test surface for a Reynolds number $Re_H = 17\,000$ and $T_{oi}/T_w = 0.94$. Data are given for constant surface heat flux (no surface conduction analysis) and for variable surface heat flux (with surface conduction analysis) [7.57]. (**a**) At constant $z/D_h = 0.0$ as x/D_h varies. (**b**) At constant $x/D_h = 7.0$ as z/D_h varies

from the sides). When three-dimensional conduction is present and a portion of the heat transfer leaves through the vertical sides of the rib, the surface Nu/Nu_0 distribution in Fig. 7.21b is produced. As before, the darker diagonal regions with relatively high Nu/Nu_0 ratios are again present along the tops of the ribs in this figure. Relative to constant heat flux values in Fig. 7.21a, Nusselt number ratios in Fig. 7.21b are then lower in the middle part of the rib top, with higher Nu/Nu_0 values near the upstream and downstream edges of the rib top. Nusselt number ratios on the flat surfaces just adjacent the ribs are also lowered in corners and then increased (compared to constant heat flux values) along lines that are approximately parallel to the rib segment [7.57].

The effects of three-dimensional conduction within and along the test surface are further illustrated by the results presented in Fig. 7.22a,b [7.57]. In both cases, local Nusselt number ratios Nu/Nu_0 for constant surface heat flux (no surface conduction analysis) and for variable surface heat flux (with surface conduction analysis) are presented for a Reynolds number Re_H of 17 000 and T_{oi}/T_w of 0.94. In the first of these figures, local Nu/Nu_0 data are given as they vary with X/D_h for constant $Z/D_h = 0$. In the second of these figures, Nu/Nu_0 data are given as they vary with Z/D_h for constant $X/D_h = 7.0$. Both figures show variable heat flux Nu/Nu_0 ratios that are significantly lower than the constant heat flux values in the central part of the rib. As mentioned, this is partially due to heat transfer from the sides of the ribs, which also causes local Nu/Nu_0 increases, determined with a variable heat flux thermal boundary condition, to be higher than values measured with constant heat flux, near the edges of the rib top. Other variations of importance in Figs. 7.22a and 7.22b are present on the flat surfaces near the edges of the ribs, where variable flux Nu/Nu_0 values are locally lower than constant flux Nu/Nu_0 values over very small areas [7.57]. As such, these results illustrate the usefulness of infrared thermography to obtain high-quality spatially resolved surface temperature data for use as thermal boundary-condition information in numerical prediction codes.

7.2.11 Summary and Conclusions

Results from several recent investigations [7.49, 51, 52, 57] have been presented to illustrate the variety of measurement environments where infrared thermography is employed to measure spatially resolved quantities over surfaces of interest. In one of these studies, measurements were made on the surface of a therapeutic, biomedical patch, where the quantity of interest is time-varying spatially resolved surface temperature. For the other situations, the measured temperature distributions are used to deduce other quantities such as surface Nusselt numbers on the surface of a swirl chamber [7.49], surface adiabatic film cooling effectiveness downstream of shaped film cooling holes [7.52], surface heat flux reduction ratio downstream of two rows of film cooling holes placed on an airfoil leading-edge model [7.51], or thermal boundary-condition information for numerical predictions of heat transfer characteristics on the surface of a passage with an array of rib turbulators [7.57]. The corresponding results illustrate the excellent accuracy and spatial resolution obtained with infrared imaging, and the other measurement techniques employed in each particular investigation.

To achieve high accuracy, in situ calibration procedures are required. In situ calibration procedures, like those described by *Sargent* et al. [7.44], are now widely used throughout the worldwide engineering community. With this approach, the camera, imaging, and data acquisition systems are all calibrated together in place within the experimental facility as the infrared measurements are obtained, using simultaneous, separate and independent means to measure surface temperatures. Infrared thermography then gives spatially resolved surface temperature distributions non-intrusively, even when large gradients of surface temperature are present. This can be done with high levels of accuracy and resolution, provided the infrared imaging system and acquisition procedures are properly calibrated, and the nuances of camera behavior and infrared imaging are properly included within the measurement procedures.

7.3 Temperature Measurement via Absorption, Light Scattering and Laser-Induced Fluorescence

Temperature distributions are of major interest in the description of reactive and nonreactive gaseous flow systems, where they play a crucial role in chemistry (reaction rates, equilibrium state) and physics (heat transfer, fluid dynamics, droplet evaporation). Since the first development of lasers in the 1960s, laser spectroscopy has become an important tool in fundamental and applied research in natural sciences such as physics, chemistry, biology and medicine. The unique properties of laser light enable selective and quantitative probing of many physical and chemical parameters with high temporal and spatial resolution, and the optical nature of laser techniques allows access to enclosed systems with minimum perturbation of the system under study. Laser techniques have therefore found broad application in fluid mechanics research. Here, they allow the remote measurement of physical (pressure, temperature), chemical (species composition) and gas dynamical (flow velocities, turbulence structures) properties. A number of fundamentally different laser-based techniques have been developed and applied for temperature measurements in gas-phase flows. Detailed discussions can be found in the review articles of *Stricker* [7.66] and *Laurendeau* [7.67]. Thermometry principally relies either on the temperature dependence of the total number density (e.g., Rayleigh scattering, spontaneous Raman scattering) or the temperature dependence of the population of different rotational, vibrational, or electronic states of atomic or molecular probes [e.g., spontaneous Raman scattering, laser-induced fluorescence (LIF), or coherent anti-Stokes Raman scattering (CARS)]. However, there is to date no versatile, accurate, and robust technique that can be applied to a wide range of experimental conditions. The selection of a specific method depends on the actual problem. The following paragraphs will therefore present different approaches to laser-based temperature measurements.

7.3.1 Overview

CARS thermometry typically provides the highest accuracy, but it allows pointwise measurements only. It will be briefly explained in Sect. 7.3.4. The other techniques can in principle be used for imaging. Rayleigh scattering (Sect. 7.3.2) provides relatively strong signals that can be used to determine temperature as long as the local gas composition, and hence the local effective Rayleigh scattering cross-section, is known [7.68]. In many practical situations, however, Rayleigh thermometry suffers from strong elastic scattering off surfaces and particles; under these conditions filtered Rayleigh scattering [7.69–71] has been successfully used to separate surface scattering from the Doppler-broadened gas-phase signal. This requires stable narrow-band lasers that coincide with narrow molecular or atomic absorption lines used as notch filters. Spontaneous Raman scattering (Sect. 7.3.2) is frequently used for temperature evaluation in one-dimensional, resolved multispecies measurements [7.72]. It is typically too weak for two-dimensional imaging. It has, however, been demonstrated for instantaneous concentration imaging with long-pulse flash-lamp-pulsed dye lasers as high-energy light sources [7.73]. Laser-induced fluorescence (Sect. 7.3.3) provides strong signals and has frequently been used for pointwise and imaging thermometry either with species that naturally occur within the gas mixture or tracers added to the flow. While point measurements are based on probing several ground states (*multi-line* thermometry), instantaneous imaging measurements typically rely on quasi-simultaneous excitation at two wavelengths for temperature evaluation (*two-line* or *two-color* thermometry). In special cases where the concentration of the fluorescing species is invariant in space and time the fluorescence intensity can be used directly for temperature measurements (*single-line* thermometry). Absorption spectroscopy (Sect. 7.3.3) has been widely applied for temperature measurements using tunable diode lasers in the infrared wavelength region. While all these techniques access local temperatures, absorption techniques necessarily accumulate information along the line of sight of the laser path through the measurement volume. In homogeneous media signal interpretation from two or multiple lines or the evaluation of line-shapes is straight forward. In systems with strong concentration and temperature inhomogeneities within the interaction path results must be critically interpreted. Spatially resolved measurements are possible via multipath absorption tomography.

7.3.2 Non-Resonant Techniques

Physical Background of Rayleigh and Spontaneous Raman Scattering

If a photon of energy $\hbar\omega_0$ that is non-resonant with any allowed single- or multiphoton transition interacts with a molecule with a certain probability, elastic or inelastic light scattering occurs. This process is sketched in Fig. 7.23. The more probable elastic scattering process

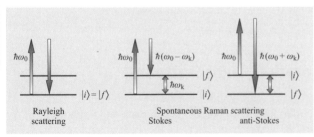

Fig. 7.23 Energy-level diagram of Rayleigh and spontaneous Raman scattering

is termed *Rayleigh* scattering whereas the much weaker inelastic process is called spontaneous *Raman* scattering. After a Rayleigh scattering process the molecule returns to its original quantum state. In contrast, Raman scattering is associated with a net energy exchange between photon and molecule. The energy transfer might have increased or decreased the rotational (denoted by J) and/or vibrational quantum state (denoted by v) of the molecule. Due to energy conservation scattered Raman photons contain correspondingly decreased or increased energy $\hbar(\omega_0 \mp \omega_k)$ with $\hbar\omega_k$ being the energy difference between the initial $|i\rangle$ and final $|f\rangle$ quantum states involved in the process ($|i\rangle$ and $|f\rangle$ denote wavefunctions and are solutions of the corresponding Schrödinger equation). In analogy to the postulate of *Stokes* [7.74] that fluorescence is shifted towards longer wavelengths relative to the exciting wavelength, red-shifted Raman scattered photons $\hbar(\omega_0 - \omega_k)$ are termed *Stokes* lines and blue-shifted photons $\hbar(\omega_0 + \omega_k)$ *anti-Stokes* lines.

In a semiclassical view where only molecules but not the incident light are treated as quantum objects, Rayleigh and Raman scattering result from a dipole moment induced by the incident electromagnetic field. When a molecule is exposed to an alternating monochromatic electric field the resulting forces push electrons and nuclei back and forth. As a consequence the molecular system behaves analogously to an oscillating dipole. The frequency ω_0 of the alternating electric field and the oscillating dipole frequency are equal and in phase. According to the laws of electrodynamics, the molecule emits light of frequency ω_0, giving rise to Rayleigh scattering.

The flexibility of this forced intramolecular motion depends on the chemical species, temperature and quantum state. It is expressed by the polarizability tensor $\boldsymbol{\alpha}$, which is second rank. The internal degrees of freedom such as rotation or vibration modulate the polarizability. As a consequence, the induced dipole moment oscillates additionally on side bands $\omega = \omega_0 \mp \omega_k$ but with an arbitrary phase difference to the incident radiation. Herein, ω_k denotes a discrete frequency characterizing the internal motion.

For a monochromatic electromagnetic field, as provided by a laser beam, $\boldsymbol{E} = \boldsymbol{E}_0 \cos \omega_0 t$, the linear induced dipole moment can be expressed by

$$p^{(1)} = \boldsymbol{\alpha} \cdot \boldsymbol{E}. \qquad (7.4)$$

The aforementioned variation of the polarizability tensor $\boldsymbol{\alpha}$ with internal motions such as discrete molecular vibrations can be expanded in a Taylor series. For the example of a normal vibration k with its coordinate Q_k and in the limit of the electrical harmonic approximation that neglects powers higher than the first, the polarizability tensor may be expressed by

$$\boldsymbol{\alpha}_k = \boldsymbol{\alpha}_0 + \boldsymbol{\alpha}'_k Q_k. \qquad (7.5)$$

The components of the new tensor $\boldsymbol{\alpha}'_k$ consist of the derivatives of the tensor components $(\alpha_{ij})_0$ with respect to the normal coordinate $Q_k : (\alpha'_{ij})_k = (\partial \alpha_{ij}/\partial Q_k)_0$. For harmonic motion the normal coordinate can be expressed by $Q_k(t) = Q_{k_0} \cos \omega_k t$, where Q_{k_0} denotes the normal coordinates amplitude. By inserting $Q_k(t)$ into (7.5) the time-dependent polarizability tensor reads

$$\boldsymbol{\alpha}_k = \boldsymbol{\alpha}_0 + \boldsymbol{\alpha}'_k Q_{k_0} \cos(\omega_k t). \qquad (7.6)$$

Equation (7.4) can be reformulated by using (7.6) and one finally obtains

$$p^{(1)} = \boldsymbol{\alpha}_0 \boldsymbol{E}_0 \cos \omega_0 t + \frac{1}{2} \boldsymbol{\alpha}'_k \boldsymbol{E}_0 Q_{k_0} \cos(\omega_0 - \omega_k)t$$
$$+ \frac{1}{2} \boldsymbol{\alpha}'_k \boldsymbol{E}_0 Q_{k_0} \cos(\omega_0 + \omega_k)t. \qquad (7.7)$$

The first term in (7.7) defines Rayleigh scattering. The condition to observe Rayleigh scattering is that the polarizability tensor $\boldsymbol{\alpha}_0$ is nonzero. In practice this is fulfilled for any molecule. The second and third term in (7.7) define Stokes and anti-Stokes Raman scattering, respectively. For Raman activity the derivative of the polarizability with respect to the normal coordinate $\boldsymbol{\alpha}'_k$ must be nonzero at the equilibrium position.

To derive an expression for the individual intensities of Rayleigh, Stokes, and anti-Stokes Raman scattering, the quantum-mechanical transition moments $\Phi(\alpha)$ must be calculated for a specific experimental arrangement. These transition moments replace $\boldsymbol{\alpha}$ in (7.4). In the following cartesian coordinates are applied and it is assumed that the incident monochromatic electromagnetic radiation is linearly polarized along the z-axis ($E_{0,x} = 0$, $E_{0,y} = 0$, $E_{0,z} \neq 0$) and is propagating along

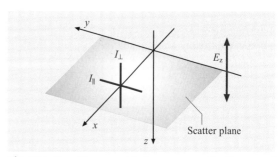

Fig. 7.24 Specific geometrical arrangement for Raman and Rayleigh scattering

Table 7.2 Space-averaged components of the transition moments for Rayleigh and ro-vibronic Raman scattering. The factor K_v is given by either $K_v = (v+1)\frac{h}{4\pi(\omega_0 - \omega_k)}$ for Stokes transitions or by $K_v = v\frac{h}{4\pi(\omega_0 + \omega_k)}$ for anti-Stokes transitions with v being the vibrational quantum number. The Placzek–Teller coefficients b_{J_f, J_i} are listed in [7.75]

Process	Branch	ΔV	ΔJ	$\overline{\langle f|\alpha_{zz}|i\rangle^2} + \overline{\langle f|\alpha_{yz}|i\rangle^2}$
Rayleigh		0	0	$(a)_0^2 + \frac{7}{45} b_{J,J}(\gamma)_0^2$
Rot.–vib. Raman	O	±1	−2	$K_v \frac{7}{45} b_{J,J-2}(\gamma')_0^2$
Rot.–vib. Raman	Q	±1	0	$K_v\left[(a')_0^2 + \frac{7}{45} b_{J,J}(\gamma')_0^2\right]$
Rot.–vib. Raman	S	±1	+2	$K_v \frac{7}{45} b_{J,J+2}(\gamma')_0^2$

the y-axis (Fig. 7.24). Detection takes place along the x-axis perpendicularly to the excitation light propagation ($\theta = \frac{\pi}{2}$). In this specific arrangement the scatter plane is spanned by the x and y axes. In space-fixed coordinates the polarizability tensor exhibits nonzero off-trace components. Rayleigh- and Raman-scattered light following the oscillation of an induced dipole moment therefore consists of two polarization components that are aligned perpendicular and parallel to the scattering plane (*depolarization*). Consequently, only two transition moments must be considered. For Rayleigh scattering, where strictly no net energy exchange between molecule and radiation takes place, the initial and final state are equal, whereas the initial and final states are different for Raman scattering. The transition moment relevant for Rayleigh and Raman scattering in the arrangement depicted in Fig. 7.24 is expressed by

$$\Phi(\alpha) = \langle f|\alpha_{zz}|i\rangle^2 + \langle f|\alpha_{yz}|i\rangle^2 , \quad (7.8)$$

where α_{zz} and α_{yz} are either the tensor elements of the polarizability α_0 (Rayleigh) or rather the derivative of the polarizability with respect to the normal coordinate of the intermolecular motion α'_k (Raman). The two components of the induced dipole moment for the transition $f \leftarrow i$ can be accordingly expressed by

$$p^{(1)}_{f \leftarrow i, y} = \langle f|\alpha_{yz}|i\rangle E_{0,z} ,$$
$$p^{(1)}_{f \leftarrow i, z} = \langle f|\alpha_{zz}|i\rangle E_{0,z} . \quad (7.9)$$

Following the laws of electrodynamics, the power of radiation emitted by an oscillating dipole with oscillation frequency ω can be expressed, with $T_{\perp,\parallel}$ being the polarization-dependent transmissivity of the detection optics, by:

$$I(\theta, T_{\perp,\parallel})_{f \leftarrow i} = T_{\perp,\parallel}\frac{c}{32\varepsilon_0\pi^2}\omega^4 \left(p^{(1)}_{f \leftarrow i}\right)^2 \sin^2\theta . \quad (7.10)$$

The angle θ between exciting radiation and detection direction in the specific arrangement considered here is $\pi/2$. Consequently, $\sin^2\theta = 1$. For Rayleigh scattering the oscillation frequency is given by $\omega = \omega_0$. For Stokes and anti-Stokes Raman scattering the frequencies are correspondingly $\omega = \omega_0 \mp \omega_k$. Exploiting (7.9), the polarization-dependent scattering intensities read

$$I_\perp(\theta, T_\perp)_{f \leftarrow i} = T_\perp \frac{c}{32\varepsilon_0\pi^2}\omega^4 \langle f|\alpha_{zz}|i\rangle^2 |E_{0,z}|^2 ,$$
$$I_\parallel(\theta, T_\parallel)_{f \leftarrow i} = T_\parallel \frac{c}{32\varepsilon_0\pi^2}\omega^4 \langle f|\alpha_{yz}|i\rangle^2 |E_{0,z}|^2 . \quad (7.11)$$

Assuming that $T_\perp = T_\parallel = 1$, the total intensity from a single spatially fixed molecule is given by the sum of both polarization components

$$I(\theta)_{f \leftarrow i} = \frac{c}{32\varepsilon_0\pi^2}\omega^4$$
$$\times \left(\langle f|\alpha_{zz}|i\rangle^2 + \langle f|\alpha_{yz}|i\rangle^2\right) |E_{0,z}|^2 . \quad (7.12)$$

For an ensemble consisting of $N_{C,tot}$ molecules of species C with $N_{v,J}$ being in a specific initial quantum state $|i\rangle = |v, J\rangle$, the Boltzmann fraction relates these two numbers by $f_{v,J}(T) = N_{v,J}/N_{C,tot}$ with T being the temperature. Because molecule movements in a low-density gas are random, in an actual measurement space-averaged transition moments (subsequently denoted by overbars) need to be considered. The space averages of the squares of the transition moments may be expressed in terms of the invariants of the derived polarizability tensor components associated with the k-th normal mode. Commonly, space invariants are expressed by $(a)_0$ and $(\gamma)_0$ for Rayleigh scattering, and by $(a')_0$ and $(\gamma')_0$ for Raman scattering. Table 7.2 summarizes the space-averaged components of the transition moments for Rayleigh and Raman scattering.

The Rayleigh/Raman scattering intensity then reads

$$I(\theta)_{f \leftarrow i} = \frac{c}{32\varepsilon_0\pi^2}\omega^4 \left(\overline{\langle f|\alpha_{zz}|i\rangle^2} + \overline{\langle f|\alpha_{yz}|i\rangle^2}\right)$$
$$\times |E_{0,z}|^2 f_{v,J}(T) N_{C,tot} . \quad (7.13)$$

This expression can be reformulated by introducing a temperature-dependent differential cross-section $\partial\sigma/\partial\Omega(T)$ by

$$\frac{\partial\sigma}{\partial\Omega}(T) = \frac{c}{32\varepsilon_0\pi^2}\omega^4 \times \left(\overline{\langle f|\alpha_{zz}|i\rangle^2} + \overline{\langle f|\alpha_{yz}|i\rangle^2}\right) f_{v,J}(T). \tag{7.14}$$

An actually measured Rayleigh or Raman signal is due to the number of photons S collected from a specific probe volume. By using (7.13) and (7.14) S is given by

$$S_{C,f\leftarrow i}(\theta) = \frac{I(\theta)}{\hbar\omega} = \frac{1}{\hbar\omega}\frac{\partial\sigma}{\partial\Omega}|E_{0,z}|^2 N_{C,\text{tot}}\Omega\eta. \tag{7.15}$$

where Ω is the solid angle of the collection optics and η is the overall detection efficiency (transmissivity of optics, quantum efficiency of detector). Note that both factors should be derived from calibration measurements.

For a reliable simulation of Raman signals it is imperative to calculate the transition frequency $\omega = \omega_0 \mp \omega_k$ correctly. In general, anharmonic corrections and deviations from the rigid-rotator assumption need to be applied. Depending on the molecule under consideration this can already be problematic especially for other than diatomic molecules [7.76].

The Boltzmann fraction $f_{v,J}(T)$ can be calculated as a function of the term values E_i (in wavenumbers) and degeneracy factors g_i by

$$f_{v,J}(T) = \frac{g_i \exp\left(\frac{hcE_i(v,J)}{kT}\right)}{\sum_i g_i \exp\left(\frac{hcE_i(v,J)}{kT}\right)}$$
$$= \frac{g_i}{Z(T)} \exp\left(\frac{hcE_i(v,J)}{kT}\right) \tag{7.16}$$

with $Z(T)$ being the temperature-dependent partition function. This implies the use of correct term values and once again the need for anharmonic and non-rigidity corrections. The space-averaged transition moments so far have been discussed in the framework of the electrical harmonic approximation (compare (7.5)). For high temperatures, such as in chemically reacting flows, electrical anharmonicities [higher orders in (7.5)] need to be considered as well.

The temperature dependence of the measured Rayleigh and Raman intensities in (7.15) is given through the Boltzmann fraction $f_{v,J}(T)$ and the number density $N_{C,\text{tot}}(T,p)$. The latter can be expressed in terms of the total molar gas number density n, the mole fraction x_c of species C and the avogadro number N_A: $N_{C,\text{tot}} = x_c n N_A$. By assuming the validity of the ideal gas law, the total number density $N = nN_A$ is a function of pressure p and temperature T. In the following sections we discuss how these temperature dependencies can be exploited for thermometry. Note, that in contrast to thermometry based on fluorescence, the Rayleigh and Raman scattering processes are not disturbed by intermolecular collisions.

Thermometry by Rayleigh Scattering

Rayleigh-scattered light is in phase with the incident radiation. This can cause interference by scattering from different molecules. However, this is prevented in the gases considered here due to the random motion of the molecules. Furthermore, in Rayleigh scattering all chemical species contained in a finite-sized probe volume give rise to a measured signal. Therefore, contributions of each chemical species C as expressed in (7.15) sum up to a total Rayleigh-scattering signal, resulting in

$$S_{\text{ray}}(\theta) = \frac{1}{\hbar\omega}\left(\frac{\partial\sigma}{\partial\Omega}\right)_{\text{ray}}|E_{0,z}|^2 N\Omega\eta. \tag{7.17}$$

Here, an effective Rayleigh cross-section $(\partial\sigma/\partial\Omega)_{\text{ray}} = \sum_C x_C (\partial\sigma/\partial\Omega)_{C,\text{ray}}$ is used. In general, $(\partial\sigma/\partial\Omega)_{C,\text{ray}}$ is approximately independent of temperature. Then, as is obvious from (7.17), Rayleigh scattering can be used to assess the total number densities N if the gas composition is known. Therefore, for spatially-homogeneous (measured) pressure, the temperature can be derived via the ideal gas law.

Rayleigh thermometry is a technique relying in many practical applications on single laser shots [7.77]. Thereby a high temporal resolution much shorter than typical Kolmogorov time scales in turbulent (reactive) flows can be accessed. With relatively little experimental effort Rayleigh scattering can be applied in an imaging setup. A sketch of a typical setup is shown in Fig. 7.25.

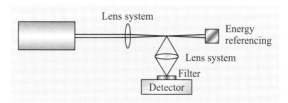

Fig. 7.25 Schematic of the experimental setup for Rayleigh scattering

Rayleigh thermometry has therefore been widely applied to turbulent systems, e.g., in turbulent flames [7.68, 78] or internal combustion engines [7.79]. Inferring number densities from scattering data, however, requires information about local effective Rayleigh cross-sections and therefore the local gas composition, which is usually unavailable in inhomogeneously mixed or reactive systems [7.80]. In combustion, special fuel blends have been proposed that provide similar total cross-sections for unburned and burned gases [7.68, 81], enabling Rayleigh thermometry even in turbulent combustion systems. Alternatively, Raman scattering can be applied simultaneously with Rayleigh scattering to measure the gas composition [7.72]. This is, however, limited spatially to one dimension [7.82].

Rayleigh thermometry suffers from problems with background scattering (e.g., off walls or windows); in the presence of particles such as soot in flames; this technique is therefore not feasible at all. Here, a filtered Rayleigh-scattering technique has been applied [7.69–71]. This approach makes use of the fact that gas-phase molecules has a wider Doppler width due to their thermal motion (≈ 0.1–$0.3\,\mathrm{cm}^{-1}$) than scattering off surfaces (walls, particles, droplets). If a single-longitudinal-mode laser is used, the signal of surface-scattered light can be rejected by narrow-band ($< 0.05\,\mathrm{cm}^{-1}$) filters (usually molecular filters like io-

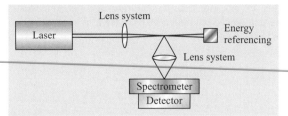

Fig. 7.27 Schematic of the experimental setup for Raman scattering

dine or atomic vapor filters such as mercury contained in quartz cells). Increasing temperature causes a larger Doppler width of the Rayleigh-scattered light originating from the gas phase. This line broadening causes an increasing signal, bypassing the narrow-band filter. The principle of filtered Rayleigh scattering thermometry is shown in Fig. 7.26. Exact control of the laser wavelength, spectral profile, absorption profile of the molecular or atomic line filter, and knowledge of line broadening is required.

Thermometry by Raman Scattering

A sketch of a Raman spectrometer is shown in Fig. 7.27.

Each chemical species exhibits a specific red and blue shift (Stokes and anti-Stokes lines) according to the energy difference $\hbar\omega_k$ between the initial and final states involved to the process. Table 7.3 summarizes averaged relative normalized differential Raman cross-sections for Q-branches $[(\partial\sigma/\partial\Omega)_Q\,(T=300\,\mathrm{K})]$ and Raman frequency shifts for combustion-related molecules. Averaged differential Raman cross-sections

Table 7.3 Raman shifts and averaged normalized differential Raman cross-sections for the Q-branch of some combustion-related molecules at room temperature. The absolute Raman cross-section of the Q-branch of nitrogen is $(5.05 \pm 0.1) \times 10^{-48} \,(\omega_0/2\pi c - 2331\,\mathrm{cm}^{-1})\,\mathrm{cm}^6/sr$. Data are taken from [7.83]

Molecule	Raman shift $\omega_k/2\pi c$ (cm^{-1})	$(\partial\sigma/\partial\Omega)_Q$ ($T = 300$ K)
$CO_2\,(\nu_2)$	1285.0	0.75
$CO_2\,(\nu_1)$	1388.0	1.13
O_2	1555.0	1.04
CO	2143.0	0.93
N_2	2331.0	1.0
$CH_4\,(\nu_1)$	2914.0	8.55
$CH_4\,(\nu_2)$	3017.0	5.7
H_2O	3652.0	3.51
H_2	4155.2	3.86

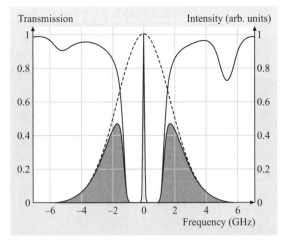

Fig. 7.26 Molecular line-shape of Rayleigh-scattered light at elevated temperature (*dashed line*), transmission profile of an atomic vapor such as mercury (*solid line*) and spectral laser emission (peak at 0 GHz). The spectral overlap of the atomic notch and Doppler-broadened Rayleigh profiles result in the spectral wings transmitted to the detector (*hatched area*) (after [7.84])

scale nonlinearly with temperature. For molecular hydrogen Fig. 7.28 shows the theoretical temperature dependence of the differential Q-branch cross-sections normalized to its value at 300 K. Depending on the theoretical approach [7.76], the temperature dependence, especially at temperatures exceeding 1000 K, can differ by a few percent, depending on the molecule.

In contrast to pure Rayleigh scattering, signals from different chemical species C show species-specific Raman frequency shifts and therefore must be spectrally dispersed before detection. The advantage of Raman thermometry is that dispersed Raman spectra contain all the necessary information to deduce the gas temperature. This is in contrast to Rayleigh thermometry or laser-induced fluorescence (LIF) schemes, where additional information on gas composition, and collisional energy-transfer pathways in the case of LIF, is required for a thorough analysis. Therefore, despite the rather small scattering cross-sections, which need to be overcome by high excitation laser photon fluxes and low-noise detection components, Raman thermometry can be an important alternative. Note that, in addition to temperature, the gas composition can be evaluated as well (simultaneous Raman/Rayleigh scattering [7.72, 82, 85]). Raman scattering is therefore the only multiscalar diagnostic based on a single excitation frequency.

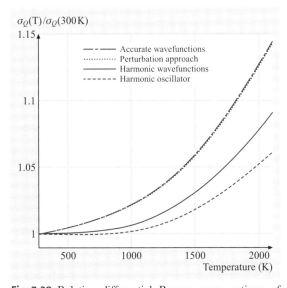

Fig. 7.28 Relative differential Raman cross-sections of molecular hydrogen as a function of temperature. Dependent on the theoretical description for temperatures exceeding 1000 K the differences between the calculated cross-sections differ by a few percent (after [7.76])

Local temperatures can be extracted from Raman spectra in different ways. For the case of high spectral dispersion, single rotational or ro-vibronic lines might be resolved. Raman spectra can be simulated and fitted to the measured spectra [7.76, 82]. However, in addition to (7.15), line widths must be considered as well. Raman lines might be pressure and Doppler broadened. In addition, the limited spectral resolution of the spectrometer broadens the measured lines. Therefore, each allowed transition needs to be convoluted with an effective line-shape and a spectrum is received from the superposition of all relevant allowed transitions. Figure 7.29 shows schematically the ro-vibronic N_2 Raman bands at various temperatures. In this example the effective line width is broad compared to the spectral separation of adjacent transitions. This leads to an overlap of ro-vibronic Raman lines, resulting in a single Raman band up to approximately 400 K and additional hot bands at temperatures above 500 K. It is obvious that Raman bands and spectral intensities are strongly temperature dependent.

A second way to extract temperatures from Raman scattering is to determine the number densities of each chemical component present in the probe volume. To do this it is assumed that the gas composition is comprised of only Raman-active species at sufficiently high concentrations to allow accurate detection and Stokes shifts into spectral regions that are actually monitored. Then the number density of all chemical components can be added to find the overall number density in the probe volume under consideration. With the validity of the ideal gas law assumed, as in Rayleigh thermometry, the total number density in connection with the (measured) pressure can be used to determine a local temperature.

The third way to determine temperatures from Raman scattering is to perform measurements of both Stokes- and anti-Stokes-shifted Raman lines (or bands). In the limit of the harmonic approximation and identical degeneracy factors for both quantum states, the measured ratio of Stokes and anti-Stokes lines is given by

$$\frac{S_{\text{Stokes}}}{S_{\text{anti-Stokes}}} = \left(\frac{\omega_0 - \omega_k}{\omega_0 + \omega_k}\right)^3 \exp\left(\frac{\hbar c \omega_k}{kT}\right). \quad (7.18)$$

The intensity ratio must be a value which can be measured with sufficient signal-to-noise ratio. This requires a sufficiently high population in the excited state from which anti-Stokes Raman scattering can take place. Near room temperature Raman lines shifted in the range 0–500 cm^{-1} are useful, whereas at around 1000 K spectral shifts up to 2000 cm^{-1} should be used.

Practical Considerations

In general Rayleigh scattering can be applied at single points [zero-dimensional (0-D)], along lines [one-dimensional (1-D)], or in two-dimensional (2-D) planes. By combining at least two parallel-orientated 2-D planes, quasi-three-dimensional applications are possible as well. As depicted in Fig. 7.25, for Rayleigh thermometry an exciting light source (most commonly a laser), a detector and some optical components such as a bandpass filter in front of the detector and lens systems are needed. In the case of 0-D and 1-D applications the laser might be focussed simply by a long-focal-length spherical lens, in the case of 2-D applications the laser is formed into a light sheet by appropriate spherical and cylindrical lenses.

Raman scattering experiments can also be performed in 0-D, 1-D or 2-D. Two-dimensional applications, however, are only feasible for flows containing just a few main components, as each component needs an individual detector equipped with a suitable bandpass filter. For the general case, where many components occur in the flow such as hydrocarbon/air flames, Raman scattering can be applied in practice only up to 1-D. In this case along the probe volume Raman bands stemming from the various chemical components are spectrally dispersed before detection (Fig. 7.27).

In the following some practical aspects on hardware components and issues related to calibration are discussed. However, this paragraph is far from complete and for more-detailed information the reader is referred to the literature and textbooks such as [7.72, 76, 86].

Laser System. Many fluid-mechanical problems are characterized by turbulence; hence, exposure times of single temperature measurements should be sufficiently short to monitor frozen states of the turbulent flow under investigation. The shortest time scales in turbulent flows are characterized by Kolmogorov scales [7.87]. Taking an industrial-type nozzle for combustion applications as an example [7.88], Kolmogorov time scales can be estimated assuming homogeneous isotropic turbulence. Typical Kolmogorov time scales are on the order of few microseconds. Consequently, single temperature measurements should resolve at least one microsecond. In practice, this requirement presupposes the usage of pulsed lasers in the sub-microsecond regime. Due to their high photon fluxes, reliability, and ease of operation solid-state, quality-switched lasers are increasingly used for Rayleigh or Raman thermometry, replacing flash-lamp-pumped dye [7.89] or excimer lasers [7.90, 91]. Among the class of solid-state lasers, the Nd:YAG laser is currently the workhorse in many laboratories. Typical specifications are pulse widths of 10 ns, repetition rates ranging from 10 to 100 Hz, and pulse energies in the fundamental at 1064 nm up to 2.5 J. The frequency conversion up to the fourth harmonic at 266 nm is easily possible due to the high peak intensities. Generally spoken, highest pulse energies as possible are favored but in practice optical breakdown phenomena set a limit. Optical breakdown occurs by absorption processes most often at remaining dust particles that exist as trace components in many flows. During a breakdown high light intensities are emitted that can damage photodetectors such as image intensifiers or CCD devices. To lower the probability of optical breakdown, pulse stretchers are used to lower the intensity but maintain a high pulse energy [7.72].

The choice of the excitation wavelength (laser emission wavelength) depends on the specific application. In general, the Rayleigh and Raman cross-sections increase nonlinearly with decreasing excitation wavelength [compare (7.14)]. However, laser-induced fluorescence interference, especially with Raman scattering, should be avoided [7.92]. As the electronic resonances of many molecules are located in the ultraviolet (UV) region [7.93, 94], the higher scattering cross-sections in the UV can be exploited only for flows consisting of specific components that are non-resonant with the excitation wavelength. In combustion research, UV excitation has been exploited for the investigation of hydrogen/air flames [7.95]. For hydrocarbon/air flames, partially oxygenized or polycyclic hydrocarbons may be formed during the oxidation process, giving rise to signals that interfere with Raman bands [7.92]. Therefore, and at the expense of decreased cross-sections, an excitation laser wavelength in the visible spectral region needs to be used. This loss can be partly compensated by higher laser pulse energies and higher detector quantum efficiencies in the visible spectral region.

Detection. Prior to detection, the signal radiation is dispersed in the case of Raman scattering, most commonly using a Czerny–Turner spectrometer, or is filtered by a narrow-bandwidth interference filter to reject spurious light in the case of Rayleigh scattering.

Due to the small Raman cross-sections, signal intensities especially in Raman thermometry are rather low. For this reason a low-noise detector, large solid angle of detection optics and high quantum efficiency are mandatory. For 0-D applications photomultiplier tubes (PMT) have been commonly used. At the exit plane of a suitable spectrometer for each chemical component at least one PMT monitors individual species concentrations. For 1-D Raman applications a 2-D array detector is needed (space and wavelength directions). For this reason CCD detectors are commonly used. The lowest noise level combined with the highest quantum efficiency is currently delivered by backside-illuminated CCD array detectors. If high temporal resolution is required and some background luminosity is present, as is the case in flames, an additional fast shutter assembly is needed. As exemplified in [7.96], a fast mechanical shutter with exposure times on the order of $10\,\mu s$ can be built from rotating slit wheels.

In the case of Rayleigh scattering, the requirements for low-noise high-quantum-efficiency detectors are less stringent compared to Raman scattering because of the much larger cross-sections. For 0-D applications PMTs are the best solution, whereas for 1-D or 2-D applications intensified CCD array detectors (ICCD) are best suited. The intensifier allows a short-gated detector, with very efficient spurious-light suppression, to be used. The filter in front of the detector might be either an interference bandpass filter, or a notch filter (Fig. 7.26) in the case of filtered Rayleigh scattering.

Calibration. Spectroscopic methods are in general not self-calibrated. Several factors such as probe volume, the solid angle of detection, transmission through optical elements, or the detector quantum efficiency depend very specifically on the individual experimental setup and may in part vary with wavelength. For this reason calibration measurements must be performed. Taking Rayleigh thermometry as an example, a measured temperature can be expressed relative to a calibration condition on the basis of (7.17) and the ideal gas law by

$$T = \frac{S_{\mathrm{ray}}}{S_{\mathrm{ray,ref}}} \frac{(\partial\sigma/\partial\Omega)_{\mathrm{ray,ref}}}{(\partial\sigma/\partial\Omega)_{\mathrm{ray}}} \frac{|E_{0,z,\mathrm{ref}}|^2}{|E_{0,z}|^2} \frac{p_{\mathrm{ref}}}{p} T_{\mathrm{ref}} \,. \tag{7.19}$$

Resolution. The spatial resolution achieved in any optical diagnostic method depends on the laser beam diameter and imaging quality of the optical components that image the probe volume onto the detector. Beam diameters can be reduced using rather short-focal-length focussing lenses and operating the laser in the TEM$_{00}$ mode. Depending on the specific configuration, spot sizes of $200\,\mu m$ or less can easily be realized. Using well-designed lenses, imaging of $50\,\mu m$ spots on the detection side is possible as well. In typical experimental arrangements of single-shot Rayleigh thermometry, a probe volume size of $\approx (100\,\mu m)^3$ can therefore routinely be achieved. This might be not true for Raman thermometry in general, where optical breakdown often limits the laser spot diameter to values above $200\,\mu m$.

In contrast to the very high temporal resolution (sub-microsecond regime using quality-switched laser pulses), which is often below the Kolmogorov time scales for many practical turbulent flows, the length scales of the smallest structures of these flows, given by *Batchelor* [7.97], might not be resolved spatially. The situation is even worse for fluids with Schmidt numbers much larger than 1. This limited spatial resolution, although much better than typical intrusive probe techniques such as thermocouples, can act similar to a low-pass filter and may influence the measurement of local temperatures and, even worse, local temperature gradients in the case of 1-D or 2-D applications.

7.3.3 Resonant Techniques

Temperature from Absorption Measurements
Absorption spectroscopy is one of the oldest techniques for the non-intrusive investigation of gaseous media. The subject of interest in laser absorption spectroscopy (LAS) is mostly the absolute concentration or absorber density of one or more atomic or molecular species that is present in a mixture [7.98, 99]. Apart from that, LAS is able to determine simultaneously physical boundary conditions such as temperature, velocity, pressure or mass flux etc. to which the absorbers are exposed [7.100].

The basic setup of an absorption spectrometer can be described as a radiation source, which emits collimated light with an initial intensity $I_0(\nu)$ through the sample under investigation towards a radiation detector. Part of the radiation, $dI(\nu)$, is absorbed by atomic or molecular constituents along each path increment dl in a simple one-step process. The light intensity $I(\nu)$ behind the sample is monitored to determine the total loss and is used as a quantitative measure for the number of absorbers along the beam path. The transition frequency ν at which the absorption occurs is characteristic of the absorbing substance, so that a simultaneous species identification (a spectral fingerprint) is possible. Integration of the light losses over the total absorption path l leads, under the assumption of a spatially homogeneous absorber distribution and a monochromatic light source, to the Lambert–Beer equation

$$I(\nu, l) = I_0(\nu) \exp\left[-\kappa(\nu)Cl\right], \tag{7.20}$$

where C is the absorber concentration, and $\kappa'(\nu) = \kappa(\nu)C$ is the linear molar absorption coefficient [7.98].

At thermal equilibrium the fractional population $f_{v,J}(T)$ of level $i = |v, J\rangle$ with energy E_i is described by the Boltzmann equation (7.16). In an equilibrium state of particle ensemble and radiation field the upward and downward radiative transition rates must be equal and the Einstein coefficients are related by the following equation:

$$\frac{A_{21}}{B_{21}} = \frac{8\pi h v_{12}^3}{c^3}. \tag{7.21}$$

At low laser intensities the total amount ΔI absorbed at the transition frequency v_{12} by any upward transitions from E_1 to E_2 is then described as:

$$-\Delta I(v_{12}) = I_0(v_{12}) N_1 \Delta l \frac{A_{21} c^2}{8\pi v_{12}^2}. \tag{7.22}$$

Absorption-Line Parameters: Line Strength and Shape. In order to remove its line-shape dependence, $\kappa(v)$ is often expressed as a product of a frequency-independent line-strength factor, S, called the integrated absorption coefficient, and a normalized line-shape function $\phi(v - v_{12})$ with the half-width γ centered at v_{12}:

$$\kappa(v) = S\phi(v - v_{12}); \quad \int_{-\infty}^{\infty} \phi(v - v_{12}) \, dv = 1;$$

$$S = \int_{-\infty}^{\infty} \kappa(v) \, dv. \tag{7.23}$$

With these relations we can rewrite the Lambert–Beer relation as

$$I(v, l) = I_0(v) \exp\left[-S\phi(v) N_1 l\right] \tag{7.24}$$

and remove the line-shape dependence by integrating over the entire absorption line, so that the total spectral area S occupied by the absorption line reflects only the absorption strength and the number density of absorbers in the initial level E_1. In this case the relation between the Einstein coefficient for spontaneous emission, A_{21}, and S becomes, $S = (A_{21} N_1 \frac{g_2}{g_1} v_{12}^2) \frac{c^2}{8\pi}$ [7.101], which shows that the integrated absorption coefficient is only proportional to the absorber specific constants and the number of atoms in the initial state.

Temperature Dependencies. In any combustion-related sensor for molecular species it is important to recognize that S is a strong function of temperature. If the pressure remains constant during a temperature change the absorber density will change as T^{-1}. In addition, the fractional population $f_{v,J}$ (7.16) at the initial energy level changes as a function of temperature and energy of the probed state. A change in temperature of several hundred degrees can cause variations in the integrated absorption coefficient over two orders of magnitude. On the other hand, this strong temperature dependence can be used to measure the temperature itself by taking the ratio of the integrated absorption of at least two suitable transitions [7.100, 102].

Line shapes and Broadening. The finite lifetimes of all excited states lead to natural or radiative line broadening, which is described by a Lorentzian profile with a width γ_N. In addition, the thermal motion of the absorbers with mass M in a gas at thermal equilibrium at a temperature T causes a velocity-dependent Doppler shift in the transition frequency, resulting in a distribution of transition frequencies, which is modeled by a Gaussian line-shape function of width γ_D. Furthermore, collisional broadening is generated by frequent perturbations of the absorber through collisions with other gas molecules, which results in a shortening of the lifetime of the excited state and in a shift of the energy levels involved. This line-shape is often sufficiently described by a Lorentzian function. To take into account the effect of different collision partners separate foreign-gas and self-broadening coefficients γ_{L_i} are introduced and empirically derived expressions for the pressure and temperature dependence of the half-width are used and can be combined to give

$$\gamma_{\text{Lorentz}} = \left(\gamma_{L_s} \frac{p_s}{p_{\text{tot}}} + \gamma_{L_f} \frac{p_f}{p_{\text{tot}}}\right) \left(\frac{T_0}{T}\right)^n \tag{7.25}$$

with T_0 a reference and T the actual temperature. p_i and p_{tot} are the partial and total pressures, respectively; n reflects the temperature dependence with kinetic gas theory yielding a power law of $n = 0.5$. Finally, at intermediate pressures, when none of the aforementioned mechanisms dominates, the experimental line-shape is described by a convolution of the Gauss and Lorentz profiles, the so-called Voigt function, which generally cannot be expressed in an analytical form. Approximations and further literature on line-shapes can be found in [7.98, 103, 104].

LAS can be applied to any absorber that has either a permanent dipole moment or an allowed electronic transition. In the first case nearly any heteronuclear molecule relevant to combustion such as H_2O, CO, CO_2, CH_4, NO, NO_2, NH_3 [7.99, 102, 105, 106] can be detected via strong ground-state rotational–vibrational transitions in the mid-infrared or via weaker overtone and combination bands in the near-infrared (0.75–3 μm) spectral region. Even stronger electronic transitions of all atoms, ions, molecules, and free radicals important in combustion, e.g., H, O, N, O_2, NO, CH_4, CH, OH, NH_2, CH_3, HO_2 [7.98, 107, 108] are accessible in the visible (350–750 nm) and most often in the ultraviolet part (< 350 nm) of the spectrum.

Long-Path Absorption Techniques. To increase detection sensitivities in absorption measurements the path length of the sample-traversing beam must be increased. White- or Herriot-type mirror arrangements are often employed at the cost of spatial resolution. Another line-of-sight absorption technique that has emerged recently as a species-specific combustion diagnostic tool is cavity ring-down laser absorption spectroscopy (CRLAS) [7.109] to monitor reactants, intermediates and products with very high sensitivity ($< 10^{-6}$ fractional absorption). In CRLAS a laser beam is coupled into a linear cavity formed by two concave mirrors with high reflectivity (> 99.98%). The intensity decay in the cavity is given by $dI/dt = -ITc/L$, where L is the length of the cavity with mirrors of transmissivity T. This leads to the exponential solution $I(t) = I_0 \exp(-Ttc/L)$ with the decay constant $\tau = L/Tc$. One can define a total round-trip loss of $\Gamma = 1 - \exp(-2L/c\tau)$, since the round-trip time in the cavity is $\tau_{rt} = 2L/c$ [7.109]. The intensity of the trapped light pulse is monitored in time via transmission through the output mirror on successive passes inside the cavity. Every absorption species along its multiple trips between the mirrors due to the probed species increases the overall photon decay rate of the light pulse

$$\frac{1}{\tau} = \frac{[T + \Lambda + \kappa(\nu)h]c}{L} \qquad (7.26)$$

from that of the empty resonator with τ_{rt}. In (7.26) $T + \Lambda$ are cavity losses due to mirror transmission (T), and due to nonspecific absorption and scattering (Λ). Therefore, the decay time is a quantitative measure of the absorption strength, and hence the concentration of the species inside the cavity. Because only the decay time is relevant the method also has the advantage of being insensitive to pulse-to-pulse energy fluctuations of the laser source. However, any other intensity losses should be avoided (scattering, surface reflection, bulk material absorption etc.), as they reduce the beam intensity inside the cavity and thereby the decay time.

Intermediates in hydrocarbon combustion, such as hydroxyl, methylene and methyl radicals, have been detected in laminar flames [7.110, 111] and low-pressure plasmas [7.112], respectively, using CRLAS or intracavity absorption (ICLAS) with ultraviolet light to stimulate electronic transitions. The exceptionally high signal-to-noise level in such measurements leads to very accurate, path-integrated temperature determinations with accuracies of less than 2% in atmospheric pressure flames. Using CRLAS, HCO [7.113], hydrocarbons (CH_4, CH_3), water, and CO_2 have also been detected quantitatively in their characteristic infrared spectrum around 3 μm with very high spectral resolution (0.007 cm^{-1}) in low-pressure (50 mbar) laminar methane/air flames using an optical parametric oscillator [7.114], and ICLAS has made possible the detection of CH_2 in flames [7.111]. Besides combustion diagnostics CRLAS is also increasingly being applied in other fields of physical chemistry including high-resolution spectroscopy [7.115] and chemical kinetics [7.116].

Absorption Thermometry. For temperature measurements based on the ratio of absorption strengths, the peak absorption or the integrated area of at least two absorption features of a target molecule must be known. From this data the absolute temperature can be calculated using the Boltzmann law (7.16). The technique is frequently described in the literature [7.117]. In an optical path with spatially homogeneous properties in thermal equilibrium the integrated absorbance areas a_1 and a_2 (cm^{-1}) of the two transitions

$$a_i = \int_{-\infty}^{\infty} -\ln\left(\frac{I}{I_0}\right) d\nu \qquad (7.27)$$

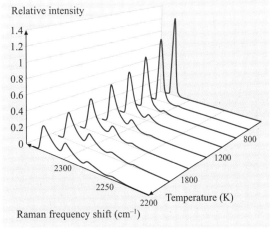

Fig. 7.29 Relative Raman scattering intensities of molecular nitrogen for different temperatures [7.86]

are measured. I and I_0 represent the transmitted and the incident light intensities, respectively. The integration is over frequency ν (cm^{-1}). Each area can be expressed as

$$a_i = S_i(T) pxl, \quad (7.28)$$

where x and T (in K) are the absorber mole fraction and temperature, respectively; p (in bar) is the pressure and the path length is given by l (in cm). The line strengths S_i (cm^{-2} bar^{-1}) can be expressed as

$$S_i(T) = S_i(T_0) \frac{T_0}{T} \frac{Z(T_0)}{Z(T)} \frac{1 - \exp\left(-\frac{hc\nu_{0,i}}{k_B T}\right)}{1 - \exp\left(-\frac{hc\nu_{0,i}}{k_B T_0}\right)}$$
$$\times \exp\left[-\frac{hc}{k_B} E_i'' \left(\frac{1}{T} - \frac{1}{T_0}\right)\right], \quad (7.29)$$

where T_0 is a reference temperature, k_B is the Boltzmann constant, Z is the absorber's partition function, and $\nu_{0,i}$ (in cm^{-1}) and E_i'' (in cm^{-1}) are the frequency and lower-state energy of the i-th transition, respectively. The ratio R of the areas a_1/a_2 can therefore be reduced to a function of the absorber temperature T only:

$$R \equiv \frac{a_1}{a_2} = f(T)$$
$$\approx \frac{S_1(T_0)}{S_2(T_0)} \exp\left[-\frac{hc}{k_B}(E_1'' - E_2'')\left(\frac{1}{T} - \frac{1}{T_0}\right)\right]. \quad (7.30)$$

Typically a_1 and a_2 are measured and one can derive the temperature by solving (7.30). If the two transitions are chosen in the right way as explained below the ratio R is a unique function of temperature over several 100 K, as demonstrated in Fig. 7.30.

In principle, there are different ways to detect integrated absorbance areas. A common approach is to use tunable, spectrally narrow-band light sources and tune the wavelength over the absorption features to detect entire line-shapes. The tuning approach has some advantages in contrast to simple peak absorption measurements since one can use the integrated areas of the peaks to derive temperature more accurately. Additionally, species concentration can be measured and one is able to determine other physical parameters such as pressure (due to pressure broadening) from the lineshape.

Alternatively, spectrally broadband light can be transmitted through the region of interest and the signal can be detected by wavelength dispersion. In this way, several absorption peaks appear on the wavelength axis, with center frequencies corresponding to molecular transitions.

Tunable Diode Laser Absorption Spectroscopy (TDLAS). Laser diodes are small, easy to handle and provide powerful laser light at almost any desired wavelength in the near and mid-IR [7.100]. Diodes originally manufactured for telecommunication applications are available at low prices. Distributed feedback (DFB) laser diodes emit narrow-band [full-width at half-maximum (FWHM) 10^{-4} cm^{-1}] laser light that can be wavelength tuned either by varying temperature or current. The temperature tuning range is typically several nm wide, but only slowly adjustable. Current tuning can be done very fast and the tuning range is typically a few cm^{-1}. This is sufficient to tune the laser diode over an absorption line (typically 0.15 cm^{-1} FWHM at atmospheric pressure) within one millisecond or faster.

A typical setup for direct absorption TDLAS is shown in Fig. 7.31. The simplest sensor for TDLAS thermometry contains one fiber-coupled laser diode that is tuned over two absorption features that are spectrally close but not overlapping [7.119]. The laser light is led through a flexible fiber and pitched across the region of

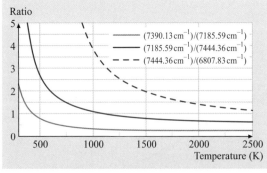

Fig. 7.30 The peak ratio and the integrated absorbance area ratio is a unique function of temperature. Shown are three different ratios from the four water transitions used in [7.118]

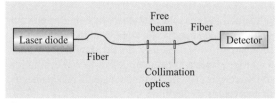

Fig. 7.31 TDLAS setup with a tunable laser diode, the system to measure within the free beam path, and the detector

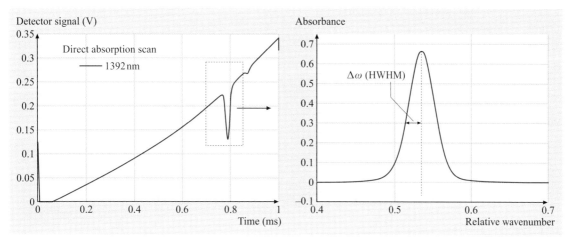

Fig. 7.32 The raw signal (*left*) is transformed into a Voigt peak [7.120]

interest. A detector collects the intensity of the transmitted light over time. The temporal variation in wavelength is observed by simultaneously monitoring the laser tuning with an etalon. The area of the Voigt-shaped absorption peak is integrated to get the integrated absorbance area a_i as shown in Fig. 7.32. A part of the laser beam can be coupled out and used as reference signal. A baseline is then fit to the scan and the peak absorbances or the integrated areas a_i of the Voigt-shaped peaks are computed to derive temperature from their ratios.

Line Selection. Two-line thermometry has a high temperature sensitivity over a limited temperature range (several 100 K) that depends on the corresponding ground-state energies of the two transitions used. In principle, the two peaks should have a large difference in ground-state energies to make sure the ratio of the detected absorptions changes strongly with temperature. But the absorbance of the two lines should be of the same order of magnitude (0.25 < ratio < 4) to prevent one line from dominating the ratio [7.121]. For low temperatures, lines with low ground-state energies should be used and for flame temperatures, transitions originating from higher ground-state energies are necessary. If higher accuracy over a wider temperature range is desired, several laser diodes can be multiplexed into one fiber to access more transitions with different ground-state energies. A sensor with four laser diodes has demonstrated excellent temperature sensitivity in the range 300–2500 K by using multiple ratios [7.118]. The multiplexing approach also enables temperature and several species concentrations to be monitored simultaneously, e.g., fuel and water in an engine [7.122]. The drawback of the multiplexed sensor is that the system becomes more complex since the different wavelengths of the transmitted light must be separated and detected by multiple detectors. In a 600 MW power plant, CO, H_2O and gas temperature were measured in [7.123].

Spectral positions, ground-state energies and further information about more then 1.7 million absorption lines of 37 molecules can be found in the high-resolution transmission molecular absorption/high-temperature spectroscopic absorption parameters HITRAN/HITEMP database [7.124]. Water vapor is present in moist air and is one of the main products in hydrocarbon combustion and is therefore present in many practical systems. It has a rich absorption spectrum in the mid-IR that can be probed with diode lasers. This is the reason why a combination of several water transitions is often used for temperature measurements in flow and combustion processes.

Wavelength Modulation Spectroscopy (WMS) with 2f Detection. In the simple TDLAS direct absorption approach the sensitivity is limited to absorbances of typically > 1%. This is due to the limited signal-to-noise (S/N) ratio since the small absorption signal has to be separated from the huge 100% transmission signal and a baseline has to be taken from a reference channel or from a fit to the wings of the wavelength scan. If one wishes to use peaks with lower absorbance, the laser tuning (at, e.g., 1 kHz) can be overlayed with a fast modulation frequency (of, e.g., 500 kHz). Then, the signal is detected with a lock-in amplifier to separate the phase-coupled signal from noise. When we detect the second harmonic of the modulated signal as shown in

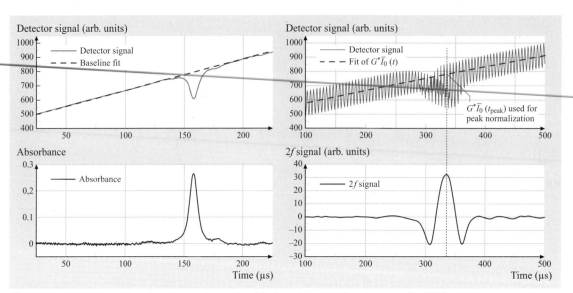

Fig. 7.33 Comparison of the direct absorption method (*left*) and the WMS technique with $2f$ detection (*right*) [7.125]

Fig. 7.33, the signal sits on a zero background, making a zero-absorption baseline unnecessary.

The gain in S/N ratio is especially good for weak lines, as shown in Fig. 7.34. This gain in sensitivity comes at the expense of a more complex sensor and a more difficult interpretation of the signal. For thermometry it is sufficient to take the ratio of the peak intensities and not integrated areas. This makes the evaluation of the WMS technique with 2f detection straightforward [7.126].

Laser-Induced Fluorescence

Laser-induced fluorescence (LIF) is the process of spontaneous emission from an excited electronic state populated upon absorption of a laser photon. Large cross-sections mainly in the visible and ultraviolet spectral range enable species detection down to the sub-ppm concentration level. Many combustion-relevant species such as OH, O_2, NO, CH, CN, NH, C_2 can be accessed selectively. Detailed treatments are given in the textbooks of *Eckbreth* [7.86], *Kohse-Höinghaus* and *Jeffries* [7.77], and in the review articles of *Kohse-Höinghaus* [7.127] and *Daily* [7.128]. LIF can be understood as two subsequent steps. Absorption of (typically) one photon leads to the population of a (typically electronically) excited state in the respective atom or molecule. After a certain lifetime τ this excited species can relax into a lower-lying state by emitting a fluorescence photon. Fluorescence typically competes with alternative processes that lead to a depopulation of the excited state (quenching, photodissociation, ionization) without the emission of photons. The ratio of fluorescence rate to excitation rate is given by the fluorescence quantum yield ϕ.

Equation (7.31) describes the LIF intensity I_{LIF} for one species in the regime of weak excitation as

$$I_{\text{LIF}} = c I_{\text{laser}} N(p,T) f_{v,J}(T) B_{ik} \Gamma(p,T) \phi \qquad (7.31)$$

with c being a parameter dependent on the specific experimental setup. The LIF intensity I_{LIF} is proportional

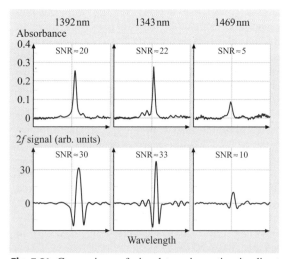

Fig. 7.34 Comparison of signal-to-noise ratios in direct absorption (*upper row*) and wavelength modulation with second-harmonic detection (*lower row*) [7.125]

to the number density of the excitable molecules in the probed volume V, which is determined by the number density of the respective species $N(p, T)$ times the temperature-dependent Boltzmann fraction $f_{v,J}$ giving the population of the initial level i. The Einstein B_{ik} coefficient describes the absorption probability for transition $i \to k$. Within the linear regime, the fluorescence intensity depends linearly on the laser intensity I_{laser} and the spectral overlap $\Gamma(p, T)$ of the laser profile and the absorption line. All these factors determine the number density of excited molecules and, therefore, the absorption part of the LIF process. The fluorescence quantum yield ϕ in (7.32) gives the ratio of the spontaneous emission rate (from level k) versus the total rate of (radiative and nonradiative) relaxation processes:

$$\phi = \frac{A_{ki}}{\sum_j A_{kj} + Q_k(p, T) + P_k} \,. \tag{7.32}$$

ϕ therefore depends on the rate of spontaneous emission (given by the Einstein A_{ki} coefficient) divided by the sum of the rates of all depopulation processes of the excited state (spontaneous emission from state k to all possible ground-state levels: ΣA_{kj}, quenching $Q_k(p, T)$ and pre-dissociation P_k). The detection efficiency of fluorescence photons depends on the observed solid angle $\Omega/4\pi$, the transmission of the optical system ε and the response of the detector η. The effects of polarization are discussed in [7.129].

When quantifying signal intensities obtained from LIF one usually has to account for numerous temperature-dependent effects. The resulting overall temperature dependence in turn is the basis for LIF thermometry. The spectral overlap factor $\Gamma(p, T)$ can be calculated from spectra simulations which are available for the most important combustion-relevant species [7.130, 131]. This is mainly important in high-pressure applications where collisional broadening causes the individual rotational lines to blend, yielding absorption features that are spectrally broader than the laser line. At atmospheric pressure and above the denominator in the fluorescence quantum yield ϕ is dominated by fluorescence quenching ($Q \gg A$). For a large number of colliders quenching rates, e.g., for OH and NO, have been determined as a function of temperature and pressure, enabling the development of simulation models [7.132, 133]. Therefore, quenching can be quantified as long as the local gas composition and temperature are known. Additional losses from the excited state can occur due to pre-dissociation P_k and photoionization. Besides these processes other collisionally induced energy-transfer processes must be considered. Vibrational (VET) and rotational energy transfer (RET) populate excited levels within the excited state, which can subsequently fluoresce. This especially gains importance if levels with significantly different effective fluorescence lifetimes are involved. Furthermore, fluorescence signal may be shifted to different spectral regions compared to the direct transitions from the laser-populated level. The implications of these effects are discussed in further detail in [7.86, 129].

With increasing laser intensity a transition from the linear LIF response [as given in (7.31)] towards a saturation regime (with I_{LIF} independent of I_{laser}) is observed. Saturation occurs through downward pumping of population from the excited state. At the same time, the laser-coupled ground-state levels can be depleted because of too slow population equilibration via RET from neighboring states. The overall effect can be modeled based on the rate equations of the individual processes. For simple systems with constant laser intensity, the system (that is, its state populations) reaches a steady state after some characteristic time τ_{ss}. If the temporal laser pulse width τ_{laser} is much larger than τ_{ss}, the steady-state solution may be used to describe the overall LIF process. This yields relatively simple *nontransient LIF models* [7.130, 134]. *Transient LIF models*, on the other hand, involve solving the fully time-dependent rate equations [7.129, 135]. For many practical applications, simple nontransient models yield an adequate description of the LIF process. This is true in particular for nanosecond laser excitation and high-pressure environments.

LIF Thermometry: Theory. For LIF thermometry, different methods have been established that involve different experimental and data evaluation procedures: single-line techniques (where only a single ground state is probed), two-line techniques (where two ground states are probed and the temperature is inferred from the ratio of two LIF intensities), and multi-line techniques (where three or more ground states are probed and temperature is inferred from a Boltzmann plot or by fitting spectral simulations). Theoretical analysis of line-shape, quenching, energy transfer, and noise effects for the different techniques have been performed by several authors [7.129, 136–139].

Particularly relevant in this context are of course the temperature-dependent terms. The dominant temperature influence in (7.31) is given by the Boltzmann fraction (7.16) of the species population in the laser-excited ground state. The temperature dependence is due to both the exponential factor for the particular transition and the partition function $Z(T)$. The spectral overlap fraction Γ depends on temperature via collisional broadening and shifting effects. The quenching rate Q depends on temperature due to both the varying collisional frequency and quenching cross-sections. Finally, the number density N of the target LIF species depends on concentration, pressure, and temperature via the ideal gas law.

In *single-line LIF thermometry*, an overall temperature dependence of the LIF signal can be calculated from (7.31) if the spectroscopic data (term energies for Boltzmann fraction, collisional broadening, quenching) are known. The laser intensity I_{laser} is readily measured. It is, however, evident from (7.31) that the number density of the target species N must be known. This is the case in nonreactive flows where concentrations are constant (e.g., O_2 in the inlet stroke of internal combustion engines [7.140], or NO seeded in constant amounts to a flow [7.138]) or when concentrations can be derived from chemical kinetic simulations (as shown, e.g., for NO in a low-pressure flame [7.141]). Finally, the calculated relative temperatures need to be calibrated at a known temperature.

Two-line LIF thermometry is much more frequently used. In this approach, two LIF measurements are performed from different ground states. The fluorescence ratio R from the two measurements is then given by

$$R_{12} = \frac{I_{\text{LIF},1}}{I_{\text{LIF},2}} = \frac{c_1 I_{\text{laser},1} g_1 \exp\left(-\frac{\varepsilon_1}{k_B T}\right) B_1 \Gamma_1(p,T) \frac{A_1}{[A_1+Q_1(p,T)]}}{c_2 I_{\text{laser},2} g_2 \exp\left(-\frac{\varepsilon_2}{k_B T}\right) B_2 \Gamma_2(p,T) \frac{A_2}{[A_2+Q_2(p,T)]}} \quad (7.33)$$

with ε_i being the term values in Joules. By taking the ratio the species concentration, which is the major unknown factor in most applications, cancels out. This makes the two-line techniques applicable to reacting systems. It should be noted that the partition function $Z(T)$ also cancels out in the ratio.

Commonly, more assumptions about the remaining temperature-dependent factors are made: Q is assumed to dominate over A (which is usually the case in atmospheric or high-pressure systems); if Q is the same for the excited states in both measurements (by choosing either a molecule for which Q does not depend on the rotational quantum number such as NO, or by choosing transitions that excite into the same upper state), then Q also cancels out: the ratio R_{12} becomes independent of quenching processes. If, finally, the temperature dependence of the overlap coefficients and the influence of laser energy fluctuations are neglected, then the fluorescence ratio is given by a Boltzmann relation only,

$$R_{12} \propto \exp\left(-\frac{\Delta\varepsilon_{12}}{k_B T}\right) \quad (7.34)$$

depending on the temperature T, the ground-state energy difference $\Delta\varepsilon_{12}$, and the Boltzmann constant k. Equation (7.34) reduces the temperature dependence of the LIF ratio to the temperature dependence of the relative ground-state populations, which represents the most important temperature influence in the two-line technique. Nevertheless, additional influence of the overlap fraction $\Gamma(p,T)$ and quenching $Q(p,T)$ may become important depending on the specific target species.

The constants c_1 and c_2 in (7.33) summarize the efficiency of the detection system. If they are the same for both excitation wavelengths, quantitative temperature measurements are in principle possible without calibration if the spectroscopic constants of the two transitions are known. In many cases, however, known temperature data is necessary to calibrate the c_1/c_2 ratio.

The precision of any thermometry technique depends on the sensitivity of the LIF signal strength(s) on temperature. In a two-line technique, high sensitivity of the fluorescence ratio R_{12} on temperature is desired, i.e., a large change in R_{12} with temperature. Mathematically, this can be expressed as the second derivative being zero:

$$\frac{d^2 R_{12}}{dT^2} = 0 \Leftrightarrow T = \frac{\Delta\varepsilon_{12}}{2k} . \quad (7.35)$$

Depending on the temperature range to be investigated in the two-line technique, this equation gives the condition for the ground-state energy differences that yield the highest temperature sensitivity. For much higher or lower temperatures, the sensitivity may decrease severely. A two-line technique is therefore most sensitive within a certain temperature range only. Table 7.4 compares the sensitivities of different NO two-line thermometry methods used in flames.

Multi-line LIF thermometry offers high temperature sensitivity over a wide temperature range by probing a number of ground states with different term energies. In a basic approach we can assume, similar to the two-line technique, that the main temperature influence of the LIF signal arises from the ground-state population only. Furthermore it is assumed that the influence of temperature on the overlap fraction and quenching is equal for all transitions, c is equal for all transitions, and

Table 7.4 Comparison of the temperature sensitivities of different NO two-line LIF thermometry methods in flames

Reference	Excited NO lines	Maximum temperature sensitivity $\Delta \varepsilon_{12}/2k$ (K)
Tamura et al. [7.142]	A – X(0, 0)O_{12}(1.5) and A – X(0, 0)O_{12}(19.5)	498
Tsujishita et al. [7.143]	A – X(0, 0)Q_2(17.5) and A – X(0, 0)Q_2(27.5)	566
Bessler et al. [7.144]	A – X(0, 0)Q_1 & P_{21}(33.5) and A – X(0, 2)O_{12}(5.5)	1420
Bell et al. [7.145]	A – X(0, 0)R_1 & Q_{21}(21.5) and A – X(0, 2)O_{12}(8.5–10.5) band head	2300

$A \ll Q$ is valid. Then, for any transition i, (7.31) can be rewritten as

$$\ln\left(\frac{I_{\text{LIF},i}}{I_{\text{laser},i}} \frac{g_i}{A_i B_i}\right) = -\frac{1}{T}\frac{\varepsilon_i}{k} - \ln\left(\frac{Z(T)}{cN(T)\Gamma(p,T)Q(p,T)}\right), \quad (7.36)$$

where the last term is constant for all transitions. The signals $I_{\text{LIF},i}$ and the laser intensity $I_{\text{laser},i}$ are measured, and g_i, A_i and B_i are spectroscopic constants. By plotting the left term of (7.36) versus ε_i for different transitions, the temperature can be derived from the slope of the plot, which is $-1/k_B T$. This is called a *Boltzmann plot* (Fig. 7.35).

In many practical applications there are a number of effects that complicate this simple analysis:

- The assumption that Q is constant for all transitions does not always hold. In OH, for example, Q is known to depend on the rotational quantum number J [7.136].
- Transitions may overlap, so that the LIF signal does not arise from a single ground state. This is the case, e.g., for NO, especially at elevated pressures. The overlap fractions Γ of many transitions must then be considered simultaneously.
- In many practical applications, background signals are present, such that $I_{\text{total}} = I_{\text{LIF}} + I_{\text{background}}$. The simple Eqs. (7.31–7.36) then cannot be applied; instead, corrections for the background contribution are necessary.
- The assumption of weak laser excitation that was used for deriving (7.31) may not hold. LIF saturation and rotational energy transfer may have significant effects on thermometry techniques [7.129, 146].
- Especially for larger molecules, the spectroscopy is complex and often only poorly understood.

Accurate LIF thermometry techniques need to include a full description of these effects. This can either be done by multi-temperature calibration measurements (yielding an empirical description of the temperature dependence of the signals) or by using detailed spectral simulations. For NO-LIF thermometry a multi-line scanning approach [7.147] that simultaneously considers the effects of line-shape and background contribution based on spectra simulations has been developed [7.131].

Target Molecules for LIF-Based Thermometry. The choice of the target molecule for LIF thermometry influences the accessible temperature range, versatility of the technique, experimental approach, and spectroscopic evaluation procedure. Temperature measurements have been performed using atomic fluorescence tracers such as indium [7.148–150], diatomic tracers such as OH [7.136, 139, 143, 151–155], NO [7.137, 138, 142, 143, 145, 147, 153, 155–168], O_2 [7.140, 151, 169–171], iodine [7.172, 173], CH [7.163] and C_2 [7.163], and organic molecules such as ketones [7.174–177] or toluene [7.178–180]. The availability of these molecules differs in reactive systems. The presence of OH and hot O_2 is limited to regions of reactive or hot gases like behind the flame front in combustion processes; the indium technique requires the activation of a precursor within a flame front; CH and C_2 are short-lived combustion intermediates; organic molecules that can be added to the fuel are only available in unburned gases and are therefore limited to temperature measurements in unburned mixtures.

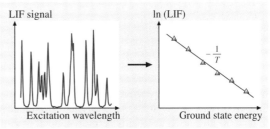

Fig. 7.35 Section of a typical LIF excitation scan (here NO at ≈ 225.3 nm) and the resulting Boltzmann plot. The slope of the Boltzmann plot gives the negative inverse temperature

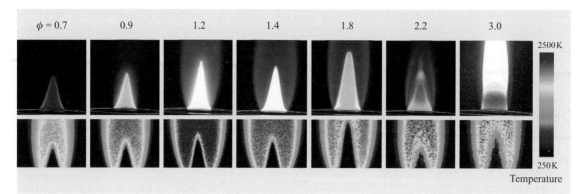

Fig. 7.36 *Top*: time-averaged flame luminosity. *Bottom*: temperature fields measured with NO multi-line LIF thermometry in laboratory flames at different equivalence ratios ϕ [7.147]

Nitric oxide (NO), seeded into reactive flow systems, has the advantage for combustion-related thermometry to be present in both the unburned and post-flame gases (Fig. 7.36). It is stable at room and elevated temperatures and therefore has the potential of making a wide temperature range accessible for LIF thermometry. NO seeding is easy to perform and does not require difficult procedures like those required for indium. Furthermore, NO provides strong fluorescence signals. Therefore, only low seeding concentrations, which do not influence the system under study, are necessary for thermometry ($\approx 100-1000$ ppm); since NO is produced in many combustion systems, seeding may not be necessary at all. In contrast to OH, fluorescence quenching in the NO $A^2 \Sigma^+$, $v' = 0$ state is insensitive to rotational quantum number [7.181–183]. Therefore, it is not required that both laser-excited transitions populate the same excited rotational level.

Single-line techniques have dominantly been applied to turbulent flows seeded with NO [7.157–159]. *Tamura* et al. used detailed chemistry simulations to calculate local NO concentrations for the interpretation of a single-line thermometry technique [7.142]. Two-line techniques have been applied to turbulent flows [7.137, 139, 160, 164, 165] and low-pressure [7.142] and atmospheric-pressure [7.143, 144, 184] flames (Table 7.4). Almost all of this work employed different rotational transitions within the $A - X(0, 0)$ vibrational bands. Vibrational temperatures have been measured with transitions in the NO $A - X(0, 0)$ and $(0,2)$ bands [7.144]. Multi-line techniques have been applied in stable low-pressure [7.155], atmospheric-pressure [7.143, 162] and high-pressure. [7.161, 185] flames, sooting flames [7.147], and spray systems [7.167], and in an arcjet reactor [7.163]. Comparisons were performed among NO single-, two- and multi-line techniques [7.138, 142, 162]. NO thermometry has also been compared to LIF methods using other target molecules [7.143, 153, 155, 163].

Whereas mono- and diatomic species typically provide line spectra, both in absorption and fluorescence emission, in the UV that resolve individual rotational transitions, in larger molecules the high density of states causes blending of many transitions into continuous absorption and emission bands. Furthermore, intramolecular energy transfer processes (intersystem crossing, internal conversion) open up in addition to those mentioned above. The variation of the efficacy of these energy-transfer paths as a function of vibrational excitation, temperature and pressure, can result in a temperature-dependent shift of the absorption and emission spectra.

Organic molecules such as aromatics and ketones are frequently used for observing mixing processes between fuel and air in combustion systems [7.186]. The original application was fuel concentration measurements. For many of these substances that are often used as fluorescent tracers pronounced temperature dependence has been found. The temperature dependence of LIF signals from organic molecules is usually a detrimental effect for quantitative fuel concentration measurements. However, the temperature dependence can also be exploited to measure temperature and even to measure temperature and fuel concentration simultaneously.

For a few larger organic molecules the temperature behavior of absorption cross-sections and fluorescence quantum yields is well enough understood to use them for gas-phase measurements. Acetone has been most thoroughly studied and has been used for temperature measurements in flows [7.175, 187, 188]. The understanding of 3-pentanone fluorescence has also evolved to a state [7.187, 189] that makes this tracer useful not only for fuel concentration but also for simultaneous temperature measurements, i.e., in internal combustion engines [7.176, 188]. The temperature effects in toluene LIF have been investigated [7.178, 180, 190]. It appears that toluene is also suitable as a tracer for measuring temperature at least in oxygen-free environments [7.179]. The combination of simultaneous temperature and fuel concentration measurements is particularly interesting for combustion systems because it provides fuel number densities, fuel concentrations and equivalence ratios (based on the assumption that oxygen is evenly distributed within the air) in fuel/air mixtures prior to ignition [7.188].

Thurber et al. [7.175] determined temperature fields (350–500 K range) in a flow passing a heated cylinder using single-line excitation of acetone at 248 nm. In this nonreacting flow the mole fraction of acetone in a nitrogen stream was constant. Independence from the requirement of constant mole fraction is achieved when two-line excitation schemes are employed. This was first demonstrated for 3-pentanone by *Großmann* et al. [7.187] and then expanded for applications with acetone by *Thurber* et al. [7.175, 191]. They also compared single- and two-line acetone thermometry and assesses the conditions (for homogeneous distribution of tracers) where one of the two techniques has advantages in terms of temperature precision and signal-to-noise ratio [7.175].

The absorption spectrum of 3-pentanone exhibits a temperature-induced shift towards longer wavelengths of about 10 nm per increase of 100 K (Fig. 7.37). Upon excitation in this region, fluorescence is emitted between 330 and 550 nm, with a spectral distribution that is almost independent of the absorbed wavelength. This spectral shift of the absorption, albeit undesired for concentration measurements with single-wavelength excitation, can be used for measuring temperature. After excitation at two different wavelengths (248 and 308 nm)

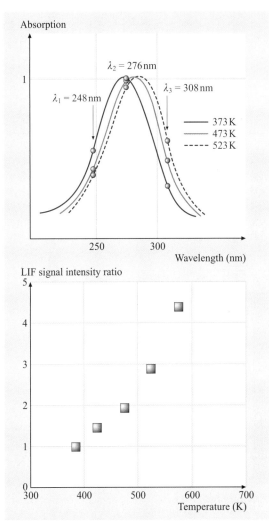

Fig. 7.37 Temperature dependence of the absorption spectrum of 3-pentanone (*top*) and the resulting temperature effect (*bottom*) found by taking the ratio of the signals from two excitation wavelengths (248 and 348 nm), which is the basis for thermometry based on two excitation colors and a single detection color [7.188]

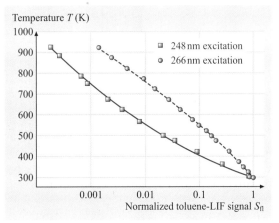

Fig. 7.38 Calibration curves for the determination of temperature from the toluene LIF signal in a homogeneously seeded nitrogen gas flow at atmospheric pressure (*Symbols*: experimental data, *lines*: fit)

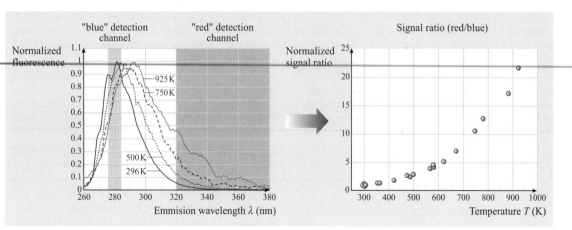

Fig. 7.39 Scheme for two-color toluene LIF

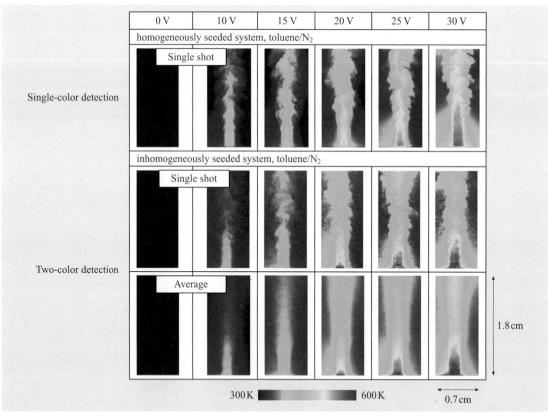

Fig. 7.40 Temperature measurements based on toluene LIF. From *left* to *right* increasing heating power of the injected gas jet that is injected into room-temperature nitrogen [7.177]

the ratio of the LIF intensities reflects the local temperature; Fig. 7.37 shows data for 308 nm [7.192]. In imaging measurements the local temperature can be evaluated for each pixel from the relative signal intensities. This technique has been applied to measurements in an optically-accessible two-stroke engine [7.176, 188].

The temperature dependence of the toluene LIF signal can be used to visualize temperature distributions in homogenously seeded flows using a one-laser one-camera setup based on the strong variation of the fluorescence quantum yield with temperature (Fig. 7.38). In flows with inhomogeneous toluene concentration distribution (e.g., mixing studies) the LIF signal depends on both the local tracer number density and the local temperature. In this case, one can take advantage of the red shift of the emission spectrum with increasing temperature (Fig. 7.39). A two-color detection technique, using single-color excitation and two-color detection then allows temperature measurements.

A demonstration of toluene LIF thermometry based on single- and two-color detection is shown in Fig. 7.40 [7.177]. A heated gas jet is injected through a 1.7 mm orifice into room-temperature nitrogen. The jet and the surrounding atmosphere is seeded with 0.5–2% toluene. The first row of images relies on spatially invariant toluene composition while the second row displays images where the toluene concentration in the jet and the surrounding area is different. The first two lines shows single-shot images, the third row displays temperature measurements that are averaged over subsequent measurements. Images are taken with 248 nm excitation and detection in the spectral bands shown in the left frame in Fig. 7.39.

7.3.4 Nonlinear Techniques

Nonlinear wave-mixing phenomena are usually not observed at the low light intensities of incoherent thermal radiators. However, the development of high-power tunable laser systems has triggered the study of numerous new nonlinear optical phenomena in liquid, solid and gaseous samples. The attained signal intensities in molecule-specific resonantly enhanced higher-order wave mixing processes can be large due to their strong dependence on the intensities of all incoming beams, the concentration of the sample molecule and the interaction length (e.g., the effective spatial extent in the sample where the signal is produced inside the interaction region of all beams). The energy level schemes of multibeam techniques most often established in combustion diagnostics today are summarized in Fig. 7.41 together with sketches of the experimental setups. The nonlinear susceptibilities are enhanced whenever one or more of the frequencies of the in-

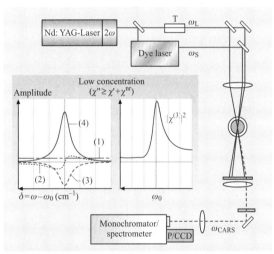

Fig. 7.42 Experimental arrangement of a CARS spectrometer. The dye laser can be either a narrow-band laser for scanning CARS or a broadband laser for single-shot CARS thermometry. In the case of scanning CARS a photomultiplier tube is used at the exit plane of the monochromator; in the case of a broadband dye laser a CCD camera is used in connection with a spectrometer. The *insets* show a typical CARS line-shape function for an isolated transition at low concentration of the species under consideration; it is composed of the constant nonresonant background (1), the interference contributions with χ_{nr} (2, 3) and the Lorentzian-shaped resonant part (4) of the third-order susceptibility $\chi^{(3)}$

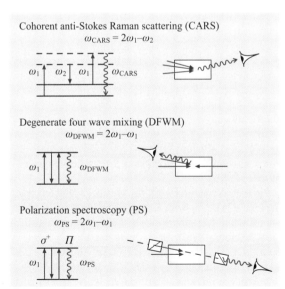

Fig. 7.41 *Left*: energy-level diagrams of various optical four-wave-mixing techniques. *Right*: schematic experimental arrangements of laser beams and direction of signal detection

teracting laser beams coincide with allowed single- or multiphoton transitions. In the macroscopic picture the induced higher-order polarization terms arise from the nonlinear mixing of the input waves and give rise to the radiation of the same or various new sum- and/or difference frequency combinations of the frequencies of the input waves. This produces a coherent signal beam in a direction determined by the phase-matching condition

$$k_{sig} = \sum k_j , \qquad (7.37)$$

where $|k_j| = n_j \omega_j / c$ is the magnitude of the corresponding beam wave vector. Through focusing of the beams good spatial resolution transverse to the beam propagation direction can be obtained with sample volumes of several tens of μm^3 without sacrificing signal intensity due to the larger crossing angles. Each of these techniques can be used for thermometry. However, CARS has emerged as the most robust technique even under harsh conditions such as gas turbine combustors [7.193].

Coherent Anti-Stokes Raman Scattering (CARS)

coherent anti-Stokes Raman scattering (CARS) [7.86] belongs to the class of nonlinear, coherent processes. A typical experimental sketch is presented in Fig. 7.42. The signals of coherent processes are generally strong and emitted in a laser-like beam, but the signal generation depends nonlinearly on the concentration of the probe molecules and is also intrinsically a nonlinear function of the exciting laser intensities. CARS therefore imposes stronger demands on the laser source such as to stability, beam quality, and mode structure, and since two laser beams are involved, on alignment. For thermometry in air-breathing flames, N_2 is used preferentially as the probe molecule because of its large concentration and the complete spectroscopic database that is available. The CARS technique allows only point measurements; simultaneous thermometry of an extended line or plane (as in LIF or Raman) is not possible. However, due to its high accuracy, CARS is often used to define a temperature standard for the investigation and validation of other thermometry methods.

In CARS experiments (Fig. 7.41) one fixed frequency (pump) and one broadband (Stokes) laser beam in the visible with a frequency difference equal to a Raman-allowed transition are aligned and focused onto the sample to create, in a four-wave mixing process, a coherent beam at the anti-Stokes frequency $(2\omega_1 - \omega_2)$,

well separated spatially and spectrally from all incoming beams. The CARS radiation is then dispersed in a spectrometer and detected with a CCD camera (Fig. 7.42). In the unsaturated regime the CARS intensity at frequency $\omega_{CARS} = \omega_3 = 2\omega_1 - \omega_2$ for an isolated transition [7.86] is (7.38)

$$I_{CARS} = I_3 = \frac{\omega_3^2}{n_1^2 n_2 n_3 c^4 \varepsilon_0^2} (lN\Delta_j)^2 I_1^2 I_2 \left| \chi_{CARS}^{(3)} \right|^2 , \qquad (7.38)$$

with

$$\chi_{CARS}^{(3)} = K_j \frac{\Gamma_j}{2\Delta\omega_j - i\Gamma_j} + \chi_{nr} ;$$

$$K_j = \frac{\omega_3^2 (4\pi)^2 n_1 \varepsilon_0 c^4}{n_2 \hbar \omega_2^4 \Gamma_j} \left(\frac{\partial \sigma}{\partial \Omega} \right)_j ; \qquad (7.39)$$

I_{CARS} is dependent on higher powers of the pump (1) and Stokes (2) laser intensities as well as on the square of the number density difference $N\Delta_j$ between the lower and upper Raman levels of the j-th transition. Equation (7.38) is written for perfect phase matching of the beams, i.e., when (7.37) holds and the corresponding signal intensity is maximized. Γ_j is the total broadening rate of the transition and $\Delta\omega_j = \omega_1 - \omega_2 - \omega_j$, the detuning from the Raman transition frequency. The spectral structure of the CARS signature over thermally populated levels is contained in the expression of the third-order susceptibility $\chi^{(3)}$ (7.39), which is therefore an accurate measure of the gas temperature. This is exploited for thermometry applications through least squares fitting of computer-generated to experimental CARS spectral shapes, if proper account is taken of the various physical effects contributing to the CARS spectral signature, such as coherences in the CARS pump beams [7.194, 195], saturation and collisional effects [7.196, 197] or the noise characteristics when broadband Stokes lasers are employed [7.198].

In their pioneering work *Taran* et al. [7.199] recorded the first CARS spectra of molecular hydrogen in an atmospheric pressure Bunsen flame using a narrow-band scanning dye laser as a Stokes beam. *Eckbreth* et al. [7.193] introduced single-pulse temperature and species measurements by using a broadband Stokes dye laser that covers Raman shifts of the whole investigated spectral branch. Species concentrations can be determined from either the integrated CARS line intensity or its spectral shape [7.200, 201] through calibration measurements and spectral modeling, respectively. For low concentrations the (real) nonresonant contribution χ_{nr} in the susceptibility expression (7.39) is significant and limits the detection sensitivity due to interference with the resonant contribution. To illustrate this behavior, the real and imaginary susceptibility components determining the CARS line-shape are depicted in the inset of Fig. 7.42 for an assumed low concentration of the resonant species in a nonresonant buffer gas.

Due to the $\Delta v = 1$, $\Delta J = 0$ selection rules in vibrational Q-branch Raman spectra, individual lines are closely spaced and one needs different laser dyes to cover all the Stokes wavelengths for detecting different species. Alternatively, *Aldén* et al. and *Seeger* et al. excited the pure rotational transitions to enable temperature and species concentration measurements to be made in what is called the single-shot broadband rotational CARS (RCARS) technique [7.202–204]. The 4-dicyanmethylene-2-methyl-6-(p-dimethylaminostyryl)-4H-pyran (DCM) laser dye covers most of the rotational Stokes shift even for spectra at 1000 K or higher and multiple species can be detected within the same spectral range as well. Since both the pump and Stokes beam in the corresponding level scheme (Fig. 7.41) originate from the same Stokes laser, additional spectral noise averaging takes place, which makes this technique less susceptible to mode noise in the laser sources. Problems in this technique are stray light from the pump beam as well as a lower temperature measurement accuracy at high temperatures.

7.3.5 Conclusions

In this section various spectroscopic methods that enable non-intrusive temperature measurements have been outlined. The main focus here was on gas-phase applications. In general, no single technique for thermometry is appropriate for all problems. Depending on the demands such as the required resolution, accuracy, and precision, and parameters such as the density, gas composition, and temperature range, different techniques exhibit specific benefits. Common to all the methods discussed is the application of lasers as a light source and sensitive low-noise detectors. Low-signal techniques such as Raman scattering or nonlinear techniques such as CARS have thereby recently become feasible. For other techniques such as absorption or fluorescence spectroscopy these improvements have led to dramatic improvements in sensitivity and selectivity. The various techniques differ in their temporal and spatial characteristics from time averaging to instantaneous detection and from point (0-D), line (1-D) or planar (2-D) measurements.

Most of the instrumentation necessary for laser-based thermometry is commercially available. The different methods were not only developed for laboratory environments but have already proved their applicability in harsh environments in many cases. Problems in fluid mechanics, engineering science and others have been tackled using laser diagnostics with great success. Spectroscopy is therefore a prominent example of the transfer of fundamental science to the solution of technical tasks.

7.4 Transition Detection by Temperature-Sensitive Paint

7.4.1 Introduction

The method of temperature-sensitive paint (TSP) encompasses the visualization and quantitative evaluation of the temperature on a body's surface with the help of a temperature-sensitive coating. This coating has to be excited by narrow-band, incident light and temperature-sensitive emitted light is detected with the help of photosensitive devices (i.e., CCD cameras). In contrast to thermometric measurements using point sensors such as thermocouples or resistance temperature detectors, this method enables the mapping of a complete three-dimensional (3-D) surface at once. An overview of temperature measurement in general is given, for example, by *Childs* et al. [7.205].

Short Historical Overview
Nowadays, there may be the impression that the TSP technique is a rather new method that had emerged at the same time as the pressure-sensitive paint (PSP) method in the 1980s. But this holds true (if at all) for the *name TSP* only. The technique itself has been known (and in use) since the 1940s and is well documented in the literature in a variety of papers dealing with *thermographic phosphors* or *phosphor thermometry*. Moreover, the principle of phosphorescence has been known much longer and was largely the domain of academia until the introduction of the fluorescent lamp in the 1930s. In parallel to this, there was increased interest in luminescing materials and the temperature dependency of such *phosphors* soon played an important rule, since the

degradation of light output with increasing temperature represents a drawback for lamp manufacturing. Therefore, thermal characterization was an important task for those studying phosphors and it was early suggested by *Neubert* [7.206] to use phosphors for temperature measurements. The thermographic *method* originated with *Urbach* in 1949 [7.207], who additionally published the characteristics of a number of thermally sensitive phosphors known at that time.

Phosphor thermometry as a tool for 2-D surface temperature measurement on a wind-tunnel model was reported for the first time by *Bradley* in 1953 [7.208]. He used a half-wedge which was partly covered by a phosphor-coated plastic sheet to study the boundary layer in a supersonic flow (Mach number M = 2.3), using a temperature-sensitive coating provided by Urbach. In the following decades, the development of rockets and space reentry vehicles had shown the need for heat transfer measurements on wind-tunnel models, especially in supersonic and hypersonic testing. Therefore, the use of temperature-sensitive paints was concentrated on inorganic, thermographic phosphors capable of being applied in the high temperature range [7.209].

Additionally to wind tunnel testing, the TSP method with the use of thermographic phosphors was applied to a variety of other applications. For example, Monaweck and McGonnagle [7.210] used the method of phosphor thermometry in 1956 for nondestructive testing of reactor fuel elements, where the heat transfer from the fuel to the coolant was investigated. Ongoing development of the *thermographic phosphor thermography* (TPT) has extended its application range to temperature field measurement on electronic circuits, isolation checks, to medicine, or turbine and combustion testing, to name just a few examples. Furthermore, research has been undertaken to use thermographic phosphors to compensate the temperature dependency of pressure-sensitive paints [7.211].

With the advent of modern infrared (IR) camera techniques, an alternative method has been made available for heat transfer measurement on wind-tunnel models, as long as combustion is not involved. Since in turbine testing the radiation of combustion is superimposed on the surface temperature radiation of the turbine blade under investigation, it is problematical to use the IR technique in that case. The use of a thermographic phosphor, on the other hand, makes it possible to filter out the surface temperature distribution on the blade by detecting the phosphor's fluorescence intensity in a certain wavelength range only, which is different from the radiation caused by combustion.

Therefore, thermographic phosphors are a commonly used tool in turbine and combustion testing (see, for example, [7.212]), whereas their use in wind tunnels is not as widespread, since the IR technique is applicable here. Things have changed since that time, whereby now cryogenic wind tunnel testing has become more and more important.

With the advent of modern cryogenic wind tunnels like the National Aeronautics and Space Administration (NASA) Langley national transonic facility (NTF) or Europe's new counterpart, the European transonic wind tunnel (ETW), it was for the first time possible to achieve full-scale flight Reynolds numbers at transonic speeds with subscale models. By using nitrogen as the test gas instead of air, and cooling the flow by continuous injecting of liquid nitrogen, it is possible to get 5 to 6 times higher Reynolds numbers in cryogenic wind tunnels compared to so-called *warm* transonic wind tunnels. Additionally, by pressurizing the wind tunnel it is possible to increase the Reynolds number further. In high-speed testing by use of cryogenic wind tunnels, measurements are usually performed with unfixed transition. Hence, the detection of the laminar-to-turbulent boundary-layer transition nowadays has become of major interest in parallel to the increased use of cryogenic wind tunnels in the design phase of transport aircraft.

If a temperature difference is given between the laminar and turbulent part of a boundary layer which is undergoing transition, a thermographic method can be applied to detect the transition line. The use of IR cameras for surface temperature detection in cryogenic testing is rather difficult because of the rapid decrease in black-body radiation intensity and the shift to longer wavelengths, at least for temperatures less than 200 K. The method of thermographic phosphors was a promising alternative for transition detection measurement in cryogenic wind tunnels, using *phosphors* with the highest temperature sensitivity in the cryogenic wind-tunnel temperature regime.

With the appearance of the PSP method in the 1980s, the term *TSP* had been established in parallel, now meaning the application of a thermographic paint in wind tunnel testing not exclusively based on phosphors but on organic substances as well. The term *thermographic phosphor thermography* (TPT) is most often found in the literature in relation to turbine testing, combustion or hypersonic flow/heat transfer measurement. TSP can be interpreted as a general category which includes thermographic phosphors as a subset. *CryoTSP* describes the use of temperature-sensitive paints mainly for transition detection in cryogenic wind tunnels.

At the end of this chapter, the application of the TSP technique is described with respect to transition detection in cryogenic wind tunnels. For the same purpose it can be used in *warm* wind tunnels as well, simply by applying another paint. Using TSP for measuring absolute temperatures on a model surface in a cryogenic wind tunnel is of less interest since the temperature of the model is determined by the adjusted flow temperature, thus having no real aerodynamic relevance as opposed to heat transfer measurement in hypersonic wind tunnels. However, it is possible to perform absolute temperature measurements by cryoTSP the same way.

A comprehensive review on thermographic phosphors, including a more-detailed historical overview, is given in *Allison* and *Gillies* [7.214]. Pioneering work in the development of high-sensitivity paints for application to warm and cryogenic wind tunnels was carried out by *Campbell* et al. [7.215–217], *McLachlan* et al. [7.218], *Asai* and co-workers [7.219] or *Popernack* et al. [7.220]. In the following, the TSP method will be sometimes compared with other 2-D thermographic methods applicable to wind tunnel testing, namely infrared thermography (IRT) and thermography by use of thermochromic liquid crystals (TLC). A more-detailed comparison of the three techniques with respect to their suitability for transition detection is given in [7.221]. The summarized results of this comparison are shown in a table at the end of this section.

Thermography by the Use of Thermographic Phosphors

Thermographic phosphors mostly consist of a binder and a luminescent crystalline phosphor. Usually, the phosphor is a transition metal compound or a rare earth compound, such as yttrium oxysulfide or vanadate (V). Oxysulfides and phosphates singly doped with europium (Eu), terbium (Tb) or dysprosium (Dy) are also used. Here, the second dye provides the temperature sensitivity of the phosphor, which can be adjusted by tuning the doping rate. Thermographic phosphors can be applied to a surface as paint, consisting of thermostable binder mixed with the crystalline phosphor and sprayed with an airbrush, for example. In general, a wide variety of different sensors are known, and they cover a wide temperature range, starting from cryogenic temperatures up to 2000 K.

As an example, the phosphor formulation $Y_2O_2S:0.15\%$ Eu^{3+} as described in [7.213] will be addressed here. The luminescent rare-earth atom (here Eu) is excited by UV light and exhibits a temperature-dependent spectral distribution of fluorescent light, as shown in Fig. 7.43. The spectrum shows discrete lines in the visible range. The energy distribution within the different spectral lines is dependent on temperature, i.e. different temperatures leading to different colors of the emitted light, starting from orange–white at 290 K and going to red when the temperature increases to 420 K [7.213]. If the temperature remains below 570 K, the coating will survive and the sensor can be used repeatedly. A benefit of thermographic phosphors compared to liquid crystals, for example, is that viewing angle, pressure (or strain), and UV light intensity do not affect the emission color.

A thermographic phosphor (or a temperature-sensitive paint in general) does not necessarily exhibit a line spectrum as shown in Fig. 7.43. Wide-band emission spectra can be observed for temperature-sensitive molecules as well. Possible methods for evaluation of TSP thermograms may differ somehow depending on whether the emission spectrum is a line or continuous spectrum, but in principle they are comparable. A well-known method of temperature measurement by means of TSP is the so-called *intensity method*, where the total energy of the temperature-dependent emission light of the paint is used to evaluate temperatures. This method will be described in more detail later.

In general, the decay time (or lifetime) of fluorescence/phosphorescence is also sensitive to temperature (moreover, it is the *reason* for the temperature-dependent intensity), and in some applications evaluation of the lifetime of emission instead of the integral intensity will be more accurate. Nevertheless, it is more difficult to get complete surface thermograms based on lifetime meas-

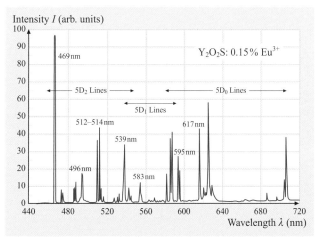

Fig. 7.43 Spectral distribution of emission light for a thermographic phosphor with line spectrum [7.213]

urement since one has to use intensified CCD cameras for direct 2-D evaluation, or a 1-D scanning system based on a photomultiplier.

In lifetime measurement, one has to excite the sensor (i. e., the paint) by a short light pulse [generated by a laser or a high-power light-emitting diode (LED) system]. Pulse duration is of the order of nanoseconds and after being excited, the TSP emits light whose intensity has a characteristic, exponential decay time (the lifetime τ), which can range from microseconds up to milliseconds. Figure 7.44, for example, shows the lifetime for the phosphor $Mg(F)GeO_6$ excited by a 266 nm pulsed laser. Starting from ambient conditions up to 550 K it shows temperature dependency of lifetime (note the logarithmic scale) and especially for $T > 400\,°C$, it shows high sensitivity, possibly making it a good sensor for that temperature range.

The method of thermographic phosphors is widely used in turbine, combustion and heat transfer measurement [7.222]. TSP in general up to now is not often used in low-speed or (warm) transonic wind-tunnel testing since infrared camera systems can be applied. However, the advantage of the TSP method compared to IRT is the large choice of optics, which may support a larger number of applications. Furthermore, high-speed cameras can be used, which enable detection of short-duration signals, this is still being a challenge with IRT.

Based on the two physical properties (temperature-sensitive *spectral distribution* and *lifetime*), at least three methods of thermographic imaging using phosphors can be defined: First, one can measure the color of emission light by the use of red–green–blue (RGB) cameras for phosphors showing line spectra. However, since the color of the light of these molecules changes less dramatically than for liquid crystals (TP: orange–white to red within 100 K. LC: red to blue within 1 K for narrow-band, or within 20 K for wide-band, liquid crystals) this technique shows rather high uncertainty. Secondly, one can use two different black-and-white (b/w) cameras equipped with optical filters, cutting out only two lines of the (multiline) emission spectrum of the thermographic phosphor (Fig. 7.43). Then, temperature calibration can be carried out by forming ratios of the two images (which have to be aligned accurately to allow for the different viewing angles). Thirdly, one optical light-gathering system and a beam splitter can be used with the two cameras to ensure the same viewing angle (but it is a rather old-fashioned method).

Another possibility is to relate images under temperature variations with a reference image at constant temperature, comparable to PSP evaluation (Sect. 4.2.2) where a *wind-OFF* image (constant pressure) is related to a *wind-ON* image (with a given aerodynamic pressure distribution). This method works well when the temperature-sensitive paint has a wide band emission rather than a line spectrum (intensity method).

Since the application of thermographic phosphors in aerodynamics was in the beginning mainly developed for hypersonic wind tunnels, the temperature increase was quite large and fast. Therefore, TPT development was focused on image processing and heat flux data reduction rather than on temperature measurement accuracy. One important matter is the insulating property of the paint, which is an issue for heat flux application. This is even the reason why NASA has developed an impressive technique that includes casting the model with the same binder as used for the paint [7.223].

Apart from phosphors, organic sensors can also be used for paint-based temperature measurement, but they can be destroyed by heat at higher temperatures and therefore cannot be used in hypersonic wind tunnels or combustion testing. However, this difference is more due to the material properties of the organic sensors than their functional characteristics; excitation wavelengths (UV or visible range), emission spectrum (line spectrum or wide band), wavelength shift ($\lambda_{em} - \lambda_{exc}$), and suitability for intensity or lifetime method are comparable for both kinds of temperature-sensitive paints. There is furthermore a difference in the solubility of the temperature-sensitive compound in a suited binder: phosphors tend to cluster more than organic sensors and it is more difficult to obtain a homogenously emitting paint. In the following, we will use the term temperature-

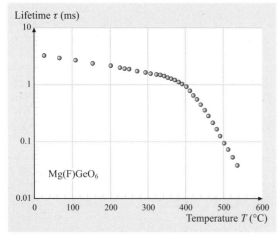

Fig. 7.44 Lifetime change with temperature for a thermographic phosphor with high sensitivity for $T > 400\,°C$

sensitive paint (TSP) for paints that are based either on phosphors or on organic molecules.

Temperature Range, Sensitivity, Spatial Resolution and Time Response of TSP

Existing TSP formulations cover the temperature range $80\,\text{K} < T < 2000\,\text{K}$, with the bandwidth of a single TSP being around 100 K. A comprehensive survey of different temperature-sensitive paint formulations is given in [7.224]. Some temperature-sensitive molecules show their highest sensitivity at cryogenic conditions, say, $100\,\text{K} < T < 200\,\text{K}$ (Fig. 7.45). The accuracy of TSP is typically around 0.1–0.8 K. As a rule of thumb, for the transition detection measurement without referencing and calibration one can clearly resolve about 0.5 K when using scientific CCD cameras with a 12 bit dynamic range. Sensitivity may be enhanced by using low noise, 16 bit CCD cameras or pixel binning. By performing a calibration of the TSP on the model under investigation (which requires temperature sensors to be in contact with the paint), and by averaging several images, thermograms with relative accuracy of less than 0.1 K can be acquired with the TSP technique. This is still more than with modern IR cameras, where up to 0.02 K changes can be resolved.

Spatial resolution is mainly determined by the CCD camera and optical detection system and therefore can by very high; this is a benefit of TSP compared to IRT. For example, thermograms of $4000 \times 2700\,\text{pixels}^2$ with 14 bit dynamic range are state of the art. Nevertheless, observation of a TSP layer with too high a magnification may show some *grainy* structures caused by clustering of the sensor molecules. In conventional wind-tunnel testing (imaging a complete wing) this will not be a problem.

TSP has two characteristic timescales: the luminescent lifetime and the thermal diffusion timescale. The luminescent lifetime can be about $1\,\mu\text{s}$ to 1 ms, which is mostly beyond the necessary exposure time for a typical TSP image in large wind-tunnel testing (approx. 1 s). However, in small wind tunnels where distances are centimeters rather than meters, the use of a high-energy light source together with a paint showing high quantum efficiency enables exposure times of the order of milliseconds when using high-sensitivity CCD cameras. The response time of a TSP *layer* is dependent upon the boundary conditions of the heat transfer in a specific application; the thermal diffusion time for a thin TSP coating is given by b^2/α_T, where b is the coating thickness and α_T the thermal diffusivity of TSP [7.224]. In general, the thermal diffusion time is much larger than the luminescent lifetime and therefore thermal diffusion limits the time response of temperature-sensitive paints.

Sensor and Measurement System

The principle of temperature-sensitive paint measurement is based on the *thermal quenching* process of the luminescence of the sensor molecules incorporated into a binder matrix. Luminescence is a general term for both fluorescence and phosphorescence and the emitting molecules are called *luminophores*. A luminophore is activated by absorption of photons of incident light in a certain wavelength range, bringing the molecule from the ground electronic to some excited electronic states. The excited electrons return to the ground state by a combination of radiative and radiationless processes and emission occurs due to the radiative process. The wavelength of the emitted light is shifted with respect to the excitation light, appearing at longer wavelengths (Stokes shift). Ruthenium complexes, for example, can be used as luminophores which are excitable by blue light (420–550 nm) and emit at (590–740 nm), i.e., red light (Fig. 7.46). Or europium complexes exist which can be excited around (340 ± 10) nm (i.e., by UV light), showing emission around 615 nm.

The excited states of the luminophore can be deactivated by interaction with other components of the system. For example, they can be deactivated by collisions with oxygen molecules (*oxygen quenching*), which is the principle behind *pressure-sensitive paint* (PSP), described elsewhere in this book (Sect. 4.2.2). However, to enable oxygen quenching, the binder matrix has to be

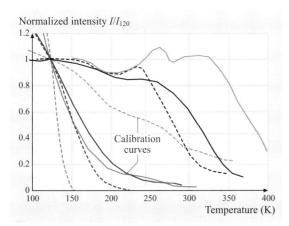

Fig. 7.45 Change of emission light intensity I with temperature for different TSPs. Some paints (brown curves) are very sensitive in the cryogenic temperature range $100 < T < 200$ K (data taken from *Liu* and *Sullivan* [7.224])

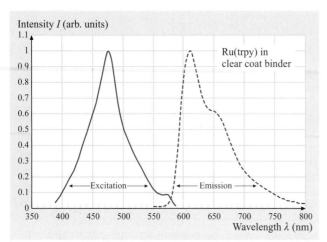

Fig. 7.46 Excitation and emission spectrum of cryoTSP based on Ru(trpy) as the sensor molecule, incorporated into a clear coat binder

penetrated by oxygen molecules, making it necessary to use a noncommercial, special binder that typically leads to a less smooth surface than for temperature-sensitive paint.

Another process for deactivation of excited states is so called *thermal quenching*: the quantum efficiency of luminescence decreases with increasing temperature because of the increased frequency of collisions. This means that the luminescent intensity of a TSP layer decreases with an increase in temperature, dark areas in an image meaning higher temperatures and bright areas meaning lower temperatures. This is in contrast to an IR result image, where high intensity denotes high temperature.

In comparison to PSP, the binder for TSP does not need to be permeable to oxygen, i.e., conventional, transparent binders like polyurethane work well. This gives the added benefit that the surface can be finished to high smoothness by polishing, which is important especially in cryogenic testing. On the other hand, TSP uses the same measurement system as PSP, i.e., same cameras and lamps (but maybe different optical filters) and the evaluation process of TSP images is nearly the same as for PSP (intensity or lifetime method).

To prevent fast temperature equilibration on metallic model surfaces by heat conduction, they have to be covered by a thermally insulating paint layer before applying the TSP coating. In addition, the color of this insulating layer should be white to serve as a screening layer, giving a more-homogenous intensity distribution and higher signal-to-noise ratio (the same holds true for PSP).

A modern TSP measurement system consists (besides the paint) of at least one scientific CCD camera with optics and an excitation light source. Using high-energy xenon flash lamps as illumination sources gives the benefit to choose excitation wavelengths by applying different filters, making *one lamp* the light source for *different luminophores*. Also (high-energy) LED arrays can be used for excitation, but with the restriction that only one wavelength range is available. For luminophores which have to be excited by UV light < 360 nm (i.e., europium complexes), the use of LEDs is hardly possible because of the present lack of efficient UV LEDs. However, this may change within the near future, as higher-power LEDs become available with their wavelength range extended to the shorter UV wavelengths (at the time of writing, 365 nm represents the short-wave limit for high-power LEDs).

To obtain a TSP thermogram of a surface using the intensity method, one has to form the ratio of two different images: a reference image taken at constant or known temperature distribution on the surface divided by a second image taken after the temperature distribution has changed somehow. With a prior measured temperature calibration for the TSP paint, one can determine absolute temperatures on the surface of interest. For example Fig. 7.47 shows the temperature variation along a line cut on the upper surface of a wing tested in the ETW cryogenic wind tunnel [7.225]. The wing image Fig. 7.47a shows the raw data of TSP intensity where inhomogeneous excitation light distribution is still seen (brighter in the mid-span area, less bright near root and tip). The resulting temperature field was evaluated by dividing the raw image Fig. 7.47a by a reference image at constant temperature. This results in the ratio image Fig. 7.47b. Line cuts within the temperature field of Fig. 7.47b are shown in Fig. 7.47c. In this example, the temperature difference between the laminar (bright areas) and turbulent part (dark grey areas) of the boundary layer was made large enough to be seen clearly in the raw image Fig. 7.47a. If the temperature difference is much smaller, it is hard to detect in the raw image but may nevertheless be clearly pronounced in the (somehow amplified) ratio image.

Some notes on the use of *intensity monitors* should be made. In pressure-sensitive paint (PSP) testing by means of the intensity method, it is common to use a *binary* paint that consists of the pressure-sensitive luminophore and a pressure-*insensitive* luminophore as an intensity monitor (often called a *reference dye*). Both are incorporated into an oxygen-permeable binder. By normalizing the pressure-sensitive emission light intensity (selected by filter A) with the pressure-insensitive reading of the intensity monitor dye (selected by filter B), the resulting intensity distribution is independent of inhomogeneous light distribution caused by angular dependent excitation or by variations in paint thickness. Normally, these deviations are also canceled out by division of *wind-ON* and *wind-OFF* image of pressure-sensitive emission light only. But since the model may eventually move and deform between *wind-OFF* (no loading) and *wind-ON* (high loading) conditions, inhomogeneous intensity distribution is not canceled out completely in PSP testing. Hence, a reference dye (intensity monitor) is needed to attain high accuracy. Additionally, an intensity monitor accounts for absolute differences in excitation light intensity between *wind-ON* and *wind-OFF* image, caused by time fluctuations of the flash energy when using xenon flash lamps, or by varying excitation light sources in general.

In TSP testing, the reference and run images are usually taken under *wind-ON* conditions (tunnel is running) and a second luminophore for normalizing the TSP intensity is not strictly necessary. However, if the absolute excitation light intensity changes between the reference and run images, this can still lead to an error in the evaluated TSP temperature. Therefore, very stable light sources should be chosen (LED systems, xenon lamps with a feedback unit) when performing accurate temperature measurements by means of TSP without an intensity monitor. For transition detection it is sufficient to divide a reference image (stationary wind-tunnel conditions, temperature constant) by a run image (acquired during a temperature change in the flow) without calibration, since one is interested in a good enhancement of the transition lines rather than in absolute surface temperatures of laminar and turbulent regions on the wing.

7.4.2 Surface Heat Transfer Processes

For transition detection measurement in wind tunnels by means of TSP, a large enough temperature difference between areas of laminar and turbulent boundary layer states has to be established. If the naturally given difference in adiabatic wall temperature is too small to be seen with a temperature-sensitive paint (as is in cryogenic testing), the laminar-to-turbulent temperature difference has to be enhanced artificially [7.219]. There exist different methods for augmentation of the adiabatic wall temperature difference, and for all of them heat transfer from the wind-tunnel flow to the model substrate plays an important rule. Hence, before explaining the

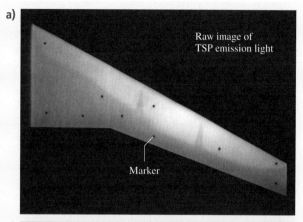
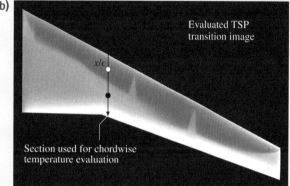
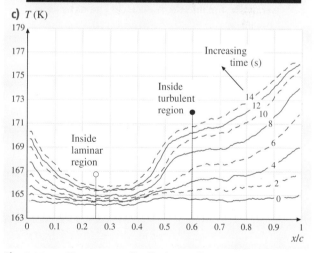

Fig. 7.47a–c Temperature distribution on line cuts crossing laminar and turbulent state of the boundary layer. Flow is from top to bottom. (a) Location of the line cut on upper wing surface. (c) Absolute temperature along line cuts at different time steps. T changes with time because of the changing flow temperature in the wind tunnel

various methods, some basic concepts of heat transfer shall first be mentioned. Three basic heat transfer processes take place near a solid wall and influence the thermal signature that is used to visualize the boundary-layer transition: heat radiation, heat conduction and heat convection.

Heat Transfer by Means of Radiation

All materials *radiate* thermal energy in amounts determined by their temperature, where the energy is removed by photons of light in the infrared and visible portions of the electromagnetic spectrum. Consider a wing section under specific test conditions in a wind tunnel, as shown in Fig. 7.48; $\dot{q}_\mathrm{rad}$ is the radiative heat flux per unit area (heat flux density) from the wing's surface, which can be written as

$$\dot{q}_\mathrm{rad} = \varepsilon \sigma T_\mathrm{W}^4 \ . \tag{7.40}$$

This represents the well-known Stefan–Boltzmann law, with T_W being the wall temperature, σ denotes the Stefan–Boltzmann constant ($\sigma = 5.6703 \times 10^{-8}$ W/m² K⁴) and ε is the emissivity, a *surface* rather than a *material* property. The constant ε denotes the ratio of actual heat radiated by a body to the maximum possible heat radiation, i.e., the radiation that would occur for a certain temperature if the body were an idealized blackbody radiator. It is dependent on the material the body is made of, surface finishing (rough or shiny), and surface color. Fortunately, many technical surfaces have an ε in the range 0.92–0.98 (painted surfaces, wood, glass, carbon fiber- or glass-fiber-reinforced plastics) whereas blank metallic surfaces have ε of the order 0.05. For known ε, surface temperature distributions can be deduced from radiation measurements with the help of infrared (IR) cameras.

Equation (7.40) shows that the reading of an IR sensor scales with T^4, meaning that IR thermography is very useful at temperatures above 300 K but becomes difficult at cryogenic conditions. Moreover, an IR sensor must be cooler than the object under investigation so as not to be affected by its own heat radiation, which would be a strong source of noise. Therefore IR sensors or sensor matrices have to be cooled down with liquid nitrogen or Peltier elements to temperatures as low as 77 K to obtain high sensitivity and a high signal-to-noise ratio. In cryogenic testing, the sensor must be cooler than the object under investigation, i.e., a liquid-helium-cooled device has to be operated, since the flow is already near liquid-nitrogen temperatures.

Heat Transfer by Means of Conduction

Conduction is heat transfer by means of molecular or atomic movement (vibration) within a material (solid body, liquid, or gas) without any net motion of the material as a whole. Conduction is the only process which enables heat flux inside a wind-tunnel model if it is completely solid. The heat flux per unit area inside a solid material between two locations separated by the distance d is given by

$$\dot{q}_\mathrm{cond} = \frac{\lambda(T_\mathrm{hot} - T_\mathrm{cold})}{d} \ , \tag{7.41}$$

where the constant λ denotes the thermal conductivity of a model material (this is a simplified 1-D approach). Compared to radiation, where the absolute temperature itself determines the radiated heat flux, now the temperature difference between two locations is the scaling factor.

Heat Transfer Caused by Convection

Convection involves the transfer of heat by the motion and mixing of *macroscopic* portions of a fluid. The term natural, or *free*, convection is used if this motion and mixing is caused by density variations resulting from temperature differences within the fluid, whereas the term *forced* convection pertains motion and mixing caused by an outside force, as, for example, the flow driven by a wind-tunnel fan. Heat transfer by convection is more difficult to analyze than heat transfer by conduction because no single, constant property of the heat transferring medium (such as thermal conductivity λ) can be defined to describe this mechanism. Heat transfer by convection depends on parameters like fluid velocity, density, viscosity, turbulence level, surface roughness, geometrical parameters, type of flow (single or two phase) and state of flow (laminar or turbulent). In practice, analysis of heat transfer by convection is treated empirically (by direct observation and meas-

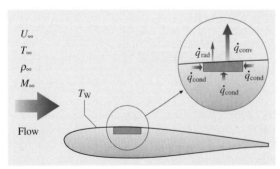

Fig. 7.48 Heat transfer processes on a wind-tunnel model's surface

urement). For example, the heat transferred per unit area from a hot isothermal plate to a forced airstream is given by Newton's law

$$\dot{q}_{\text{conv}} = \alpha(T_W - T_\infty) ,\qquad(7.42)$$

where α is the (forced) convective heat transfer coefficient, which depends upon the physical properties of the fluid as well as on the physical situation, and which is different for free convection.

For convective heat transfer in wind-tunnel testing, the turbulent convective heat transfer coefficient α is of the order of several 100. It is related to the plate's Nusselt number Nu and also depends on the Prandtl number Pr. For zero pressure gradient ($\partial p/\partial x = 0$), $Pr = 1$, and T_W = constant there is a linear relationship between the wall friction coefficient c'_f and the heat transfer coefficient α (Reynolds analogy):

$$\alpha = \frac{1}{2}c'_f u \frac{\lambda}{\nu} ,\qquad(7.43)$$

where u represents the flow velocity outside the boundary layer, and ν the kinematic viscosity of the medium. Since the wall friction in the turbulent region of the boundary layer is around 10 times higher than in the upstream laminar region, there will be different temperatures in laminar and turbulent portions of a boundary layer, when convective cooling or heating of the model by the incoming flow occurs.

7.4.3 Temperatures in Laminar and Turbulent Boundary Layer

The main principle of transition detection using thermographic methods is visualization of a temperature difference between the laminar and turbulent state of a model's boundary layer (BL) that is undergoing transition. This temperature difference occurs either naturally, caused by the different recovery factors for laminar and turbulent boundary-layer flow, or it has to be added artificially using the method of *temperature steps*, or using electrical external or internal heating of the surface. IR thermography, because of the available high-sensitivity sensors (temperature resolution up to 0.02 K), may be able to detect the small temperature differences created by the recovery process in warm, transonic wind tunnels or in flight measurements. However, if the model or wing has a blank metallic and/or shiny surface (i. e., the emissivity ε is very small and the surface can mirror radiation from heat sources surrounding the model or wing), a special paint or foil must be applied to improve emissivity. In pressure driven blow-dow type wind tunnels a temperature difference naturally occurs because flow temperature rapidly decreases due to the expension of the gas at the beginning of the test run. For transition detection measurement in continuously driven, cryogenic wind tunnels, one needs to perform an *artificial enhancement* of the laminar-to-turbulent temperature difference because there is only a small temperature difference caused by the recovery factor.

Temperature Differences Arising from the Recovery Factor

When the fluid is slowed down from its free-stream velocity U_∞ to $U = 0$ at the wall it *recovers* some heat energy from kinetic energy, which leads to an increase in temperature on a model's surface. If the surface (for example the upper side of a wing) is thermally well insulated, approximately no heat flux occurs between model and boundary layer ($\dot{q}_{\text{cond}} = 0$), and the increase in temperature within the boundary layer leads to an increased, so-called adiabatic wall temperature (or recovery temperature) T_{aw}. The change in wall temperature is described by the recovery factor R, which depends on the physical properties of the fluid (i. e., air in common wind tunnels and nitrogen (N_2) in cryogenic wind tunnels), and on flow conditions. The adiabatic wall temperature can be written [7.226, 227]

$$T_{\text{aw}} = T_\infty + R\frac{U_\infty^2}{2c_{\text{p,air}}} = T_\infty\left(1 + R\frac{\gamma-1}{2}M_\infty^2\right) ,\qquad(7.44)$$

with $c_{\text{p,air}}$ being the specific heat of air at constant pressure (which is very similar to c_{p,N_2}), γ the isentropic exponent and M_∞ denotes the Mach number of the undisturbed wind tunnel flow. γ is nearly constant at lower temperatures or pressures and can be set to 1.4 without much lack of accuracy. It is obvious that there is a different recovery factor for a laminar boundary layer in comparison to a turbulent boundary layer and the two can be estimated to $R_{\text{lam}} \approx 0.84$ and $R_{\text{tur}} \approx 0.9$, respectively (for a Prandtl number of $Pr = 0.71$). Thus, the difference in adiabatic wall temperature on a wing's surface which is subject to a partly laminar and partly turbulent boundary layer can be written as:

$$\begin{aligned}\delta T_{\text{aw}} &\equiv T_{\text{aw,tur}} - T_{\text{aw,lam}} \\ &= T_\infty(R_{\text{tur}} - R_{\text{lam}})\left(\frac{\gamma-1}{2}M_\infty^2\right) \\ &\approx 0.012 T_\infty M_\infty^2 .\end{aligned}\qquad(7.45)$$

An estimate for δT_{aw} is shown in Table 7.5 for different M_∞ and T_∞. Note that here δT_{aw} represents an up-

Table 7.5 Maximum possible adiabatic wall temperature differences caused by the recovery factor

M_∞	T_∞ (K)	δT_{aw} (K)
0.2	1000	0.48
0.2	300	0.15
0.2	100	0.05
0.8	1000	7.7
0.8	300	2.3
0.8	100	0.77

per limit for the possible temperature difference, which holds true only for a perfectly insulated model surface. Even though the model surface may be well insulated, true adiabatic conditions can never be reached, i.e., a certain amount of heat exchange between flow and model will always take place. Therefore, real δT values are smaller than shown in Table 7.5.

Temperature Difference Produced by the Method of Temperature Steps

Since in boundary layer flow the convective heat transfer coefficient α_{tur} is larger than α_{lam}, a temperature *change* in the outer flow will be seen faster on the surface underneath the turbulent boundary layer, leading to a temperature difference δT between turbulent and laminar regions for a certain amount of time. When the change stops, this difference will vanish in time because of heat exchange by internal conduction and outer convection. There is competition between the generation of a temperature difference by means of convective heat transfer (if the flow temperature differs from the model temperature) and internal equilibration by conduction. Therefore, a metallic model must be thermally well insulated against the flow, for example by the use of thermally insulating paint or foils.

More effective is the use of non-metallic models, manufactured from carbon-fiber- or glass-fiber-reinforced plastic, wood, glass or suchlike (note however that glass or wood models are rarely found in modern wind-tunnel testing). In practice, one has to apply the temperature change in the oncoming flow quickly (if possible), this being denoted a *temperature step*, which can either be positive or negative. In cryogenic testing it is very easy to generate a temperature step by increasing the amount of (normally continuously) injected nitrogen, leading to a negative gradient $-dT/dt$ [7.225]. Furthermore one can stop the continuous injection of nitrogen, leading to an increase in flow temperature $+dT/dt$, caused by the heat input of the fan (at least for the higher Mach numbers > 0.4).

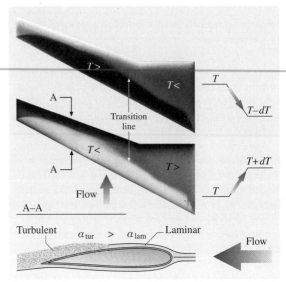

Fig. 7.49 Method of temperature steps for transition detection measurement by means of TSP, suitable for cryogenic wind tunnels

By having a sufficiently large temperature difference δT between turbulent and laminar regions on a model surface, the transition line can be seen as the borderline between light and dark areas in the evaluated TSP image (Fig. 7.49).

Temperature Differences Produced by Electrical Heating

Another method for artificially generating temperature differences between model surface and flow is to heat up the model or single parts of a model (for example, the wing, or slats and flaps). This can be achieved by classical heating foils, where the heating elements of thin electrically resistant material (thin copper strips) are embedded in Kapton foils that can be mounted on, or beneath but close to the surface. Figure 7.50 shows the application of such heating foils mounted in the leeward side of a half-model slat. Transition detection on this slat was performed successfully in a low speed, cryogenic wind tunnel [7.228].

Nowadays the highly developed technology of carbon-fiber-reinforced plastic (CFK) has enabled the use of more-homogeneous and more-efficient heating layers applicable to models made of glass-fiber-reinforced plastic or of carbon-fiber-reinforced plastic [7.229]. In the case of carbon fiber, a single sheet of carbon with electrical connections which function as an electrical heater is laminated with the other sheets

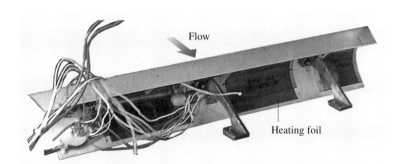

Fig. 7.50 Conventional heating foils implemented into the slat of a half model. Opposite side of the slat is painted with TSP

of carbon fiber in such a way that the carbon sheet has a very small distance to the surface to ensure that most of the heat is conducted to the surface and not to the inner parts of the model. Figure 7.51 shows a schematic drawing of the application of a heating layer of carbon inside a wing section. Thermographic investigation of the surface temperature distribution of such models equipped with carbon heating sheets has also shown very good homogeneity of the surface temperature for large heated areas and high temperatures [7.229].

Additional Methods for Establishing a Boundary-Layer Temperature Difference

In some cryogenic facilities a temperature difference can be established by injecting warm, gaseous nitrogen in addition to the continuously injected, liquid nitrogen. Furthermore, an artificial temperature difference can also be established by adjusting the complete model to a temperature different from the flow temperature. This can be realized by heating or cooling the model outside the test section (in a separate chamber) and moving it to the test section just before acquisition of transition images begins (model transport method [7.228]). Although this enables the implementation of positive as well as negative temperatures difference betweens model and flow, it is hardly practicable in most industrial-scale wind tunnels since it takes some time to move the model. Thermographic images may thereby become rather expensive by wasting tunnel time for model transport. Moreover, special care has to be taken to prevent humidity from entering the test section with the model during transport. However, sometimes this is the only method available if temperature steps cannot be generated some other way and the existing model cannot be equipped with heating foils.

Some workers have used special infrared radiators that can be used for heating the model from outside, but, in general, and compared with the aforementioned methods, these radiators produce only a small increase in temperature with poor homogeneity and less stability in time [7.229].

Finally, some blow-down wind tunnels exist (for example the Ludwieg-type cryogenic wind tunnel DNW-KRG in Göttingen), which can provide a suitably large temperature step within milliseconds. Since the flow is driven by a large pressure difference between a storage tube and a low-pressure vessel, the gas (nitrogen) expands quickly when the so-called *fast shutter valve* is opened. Thus, a temperature drop is created automatically during operation making such wind tunnels well suited for transition detection by means of TSP, also at ambient temperatures.

Influence of Temperature Difference on Boundary Layer Instability

It is well known that a temperature gradient inside the boundary layer changes the stability properties, at least for the Tollmien–Schlichting (TS) instability. This property was used as a method for laminar flow control by

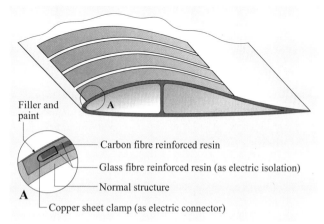

Fig. 7.51 Implementation of carbon-fiber-reinforced resin heating layer into a carbon-fiber-reinforced plastic model (courtesy K. de Groot)

cooling a model in air, or heating a model in water. Following *Reshotko* [7.230], as cited by *Saric* [7.231], the boundary-layer equation near the wall can be written as

$$\mu \frac{\partial^2 u}{\partial y^2} = \rho V_0 \frac{\partial u}{\partial y} + \frac{\mathrm{d}p}{\mathrm{d}x} - \frac{\mathrm{d}\mu}{\mathrm{d}T} \frac{\partial T}{\partial y} \frac{\partial u}{\partial y}, \quad y \approx 0, \tag{7.46}$$

with u being the flow velocity inside the boundary layer, y the wall normal coordinate, p the pressure, and μ the dynamic viscosity. A fuller boundary layer (i.e., a more-negative $\partial^2 u/\partial y^2$) usually results in lower disturbance growth so that anything that makes the left-hand side of (7.46) more negative will stabilize the boundary layer [7.231]. Thus, model cooling in air ($\partial T/\partial y > 0$, $\partial \mu/\partial T > 0$), which is equivalent to heating the flow, will result in a more-stable boundary layer. On the other hand, cooling the flow (or heating the model) makes the boundary layer more unstable. This fact has to be taken into account when performing transition detection by the method of temperature steps, especially when using negative steps $-\mathrm{d}T/\mathrm{d}t$ in the flow or using electrical heating of the model.

In cryogenic wind tunnel testing in ETW, it is possible to realize $+\mathrm{d}T/\mathrm{d}t$ as well as $-\mathrm{d}T/\mathrm{d}t$ with the same rate of heating or cooling for Mach numbers above 0.4. In a cryoTSP test campaign in 2003, the mentioned influence of temperature steps was investigated, leading to rule-of-thumb criteria for a suitable selection of $\mathrm{d}T/\mathrm{d}t$ [7.225]. Additionally, during a temperature step the Reynolds number changes slightly (caused by the changing viscosity) and it can be observed that the location of transition moves in time during image acquisition, at least for Reynolds numbers in the range where x/c_{trans} strongly depends on Re (Re < 3 Mio).

7.4.4 Transition Detection by Means of TSP

Infrared thermography is well established in flight tests, and may be easier to use in warm wind tunnels, but TSP is to be recommended for cryogenic testing. Hence, the term *cryoTSP* has become established. Nevertheless, since some industrial wind tunnels are already equipped with a PSP system, it may also be a good opportunity to perform transition detection with the TSP technique in a *warm* wind tunnel by using the same cameras and lamps as for PSP but different filter for lamps together with a TSP suited for the wind-tunnel temperature range. The following discussion on cryoTSP also holds in principle for TSP at higher temperatures.

At the end of this section, results of transition detection measurements obtained using the DLR mobile PSP/TSP system [7.225, 232] will be presented. A cryoTSP paint originally developed by the Japanese Aeronautics Exploration Agency (JAXA) (former National Aeronautics Laboratory (NAL) of Japan) was used [7.233]. Transition detection was performed [7.234, 235] in different, industrial cryogenic wind tunnels such as the European transonic wind tunnel (ETW), the pilot facility of ETW (PETW), the low-speed cryogenic wind tunnel of the German–Dutch wind tunnels in Cologne (DNW-KKK) or the cryogenic Ludwieg-tube wind tunnel in Göttingen (DNW-KRG).

Required Properties for TSP in Cryogenic Testing

It has to be stated that, in comparison to laboratory (small-scale) wind-tunnel testing, a cryoTSP suited for large-scale, industrial wind tunnels needs to have the following additional properties: *high luminescent intensity* (The intensity of excitation light varies inversely proportionally to the square of the distance between the model and the light source. To apply TSP in large-scale wind tunnels, the TSP formulation must provide very high brightness.); *coating robustness at cryogenic temperatures* (In general, polymer coating on a large metallic surface is subject to cracking when cooled down to cryogenic temperatures, or warmed up after the test. This is due to a large difference in thermal contraction between the TSP and the substrate, especially for aluminium models.); *smooth surface* (At cryogenic temperatures, boundary layers on model surfaces are very thin, especially in the leading edge region. To prevent premature transitions, surface roughness of the coating must be as small as possible.) JAXA has successfully optimized a cryoTSP with respect to these properties, this being tested in laboratory-scale wind tunnels beforehand [7.233]. The paint is based on a metal-organic compound (abbreviation Ru(trpy)) implemented into a polyurethane binder, and has a wide-band excitation and emission spectrum (Fig. 7.46).

Figure 7.52 shows a calibration curve for the cryoTSP used here in the range $100\,\mathrm{K} < T < 300\,\mathrm{K}$. At first glance, the calibration curve suggests that this TSP shows a nearly constant sensitivity in the range $110\,\mathrm{K} < T < 170\,\mathrm{K}$ and becomes less sensitive for $T > 170\,\mathrm{K}$. However, for transition detection things are different; the reason lies in the special technique of *temperature steps* used in cryogenic wind tunnels. When performing transition detection by the use of temperature steps, one starts at a given, constant flow temperature,

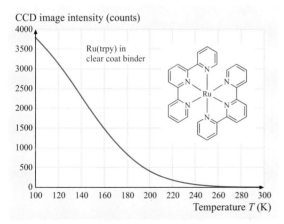

Fig. 7.52 Decrement of intensity with increasing temperature for Ru-based cryoTSP

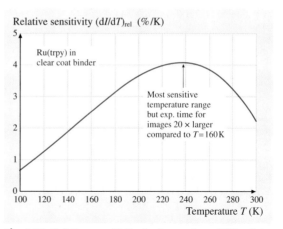

Fig. 7.53 Relative sensitivity for the used cryoTSP paint

say $T_{start} = 170$ K. Then a positive step $+dT/dt$ of (say) 0.5 K/s is applied until the flow temperature increases to 180 K. No temperature smaller than 170 K for this working point exists, and there is no need to use the full dynamic range shown in Fig. 7.52, where the curve starts at 100 K. Exposure time of the camera, aperture of the objective lenses, flash frequency, or absolute intensity of the excitation light source can be adjusted in such a way that 3000 intensity counts are measured at 170 K, instead of 1000 counts at 170 K as given in Fig. 7.52 (12 bit dynamic range camera = 4095 intensity counts at maximum). This increases the steepness of the calibration curve for this working point, making the system more sensitive.

Thus, for transition detection by the method of temperature steps it is more convenient to refer to the *relative sensitivity* of the measurement system given by

$$\left(\frac{dI}{dT}\right)_{rel} = \left(\frac{dI(T)}{dT} \frac{1}{I(T)} \times 100\right) \quad [\%/K] \quad (7.47)$$

to get an estimate for the resolvable temperature differences at a given working point (I = intensity counts on CCD image, T = flow temperature). Figure 7.53 shows $(dI/dT)_{rel}$ for the cryoTSP used here, which exhibits its highest relative sensitivity around 240 K. This fact has been amply demonstrated in different cryogenic wind-tunnel tests using the JAXA paint, showing good transition images even at 240 K. However, the image exposure time for $T = 240$ K is around 20 times larger compared to that for 160 K (Fig. 7.52), which mostly limits the practical application range of this paint to $T < 240$ K in large, continuously driven, industrial wind tunnels such as ETW.

DLR has since improved the described cryoTSP by implementing a second luminophore into the paint, showing high sensitivity in the range 240 K $< T <$ 320 K in parallel to high intensity [7.236]. This second, europium-based luminophore has to be excited by UV light that is clearly separated from the blue light necessary for excitation of the Ru(trpy) luminophore (working range: 100 K $< T <$ 240 K). On the other hand, the emitted wavelength of this europium complex is in the same range as emission light for Ru(trpy) (600 nm $< \lambda_{em} <$ 650 nm). Using this two-component cryoTSP (2C-cryoTSP), it is possible to perform transition detection in the complete range 100 K $< T <$ 320 K by the use of a single paint, with the same camera–filter assembly, but by changing the lamp filter, or by using two different lamps (for example xenon lamp with UV filter in addition to a blue high-power LED).

Especially on metallic wind-tunnel models, before applying the TSP it is necessary to spray a carrier paint which serves as a thermal insulation (thickness about 100 μm), reducing the heat flux from flow to model (or vice versa). Additionally, this paint should be white to serve as a screening layer for the TSP, giving a more homogenous background intensity distribution. The same screen layer can be used for IR measurement on a metallic model but must be different for liquid crystals, for instance, since they require a black background to absorb the light which penetrates the TLC layer.

Results of Transition Detection by cryoTSP

An example for transition detection on a full model 3D wing is shown in Fig. 7.54. This cryoTSP measurement [7.225, 234] was conducted in the ETW wind tunnel using a 12 bit dynamic range CCD camera with

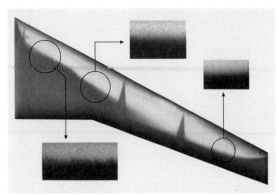

Fig. 7.54 Result image of transition detection on a full model wing tested in ETW. $Re = 6$ Mio, $M = 0.79$, $C_L = 0.5$, $T = 170$ K, tunnel pressure $p = 120$ kPa. Flow is from top to bottom

1280×1024 pixel^2 spatial resolution, and the method of temperature steps. Compared to IR images (typical resolution 320×240 pixel^2), the high spatial resolution of the CCD image allows us to highlight details of the transition line even when the complete wing is recorded by the camera as one image. In the transition image Fig. 7.54 turbulent areas are represented as dark regions and laminar areas appear bright.

Figure 7.55 shows a cryoTSP transition image of measurement in the DNW-KRG. Since this Ludwieg-type wind tunnel is operated as a blow-down wind tunnel, no artificial temperature step or model heating was applied, and high-quality result images could be gathered with a short exposure time of 0.4 s. Besides the transition line, the signatures of (strong) vortices located in the junction area of the main airfoil and the two cylindrical *noses* could be clearly seen.

Transition detection on a thick 2D profile is shown in Fig. 7.56. A carbon-fiber-reinforced plastic model was tested at low Mach numbers in the DNW-KKK low-speed cryogenic wind tunnel. Temperature *steps* were created by adding warm nitrogen to the tunnel flow.

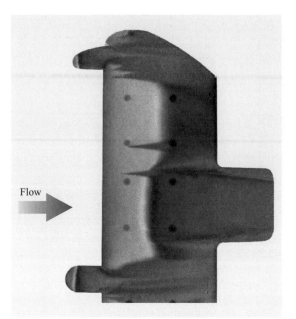

Fig. 7.55 Transition detection in the Ludwieg type DNW-KRG cryogenic wind tunnel. $Re = 4$ Mio, $M = 0.8$, $T = 200$ K, $p = 260$ kPa

Since carbon fiber has a low thermal conductivity and moderate heat capacity, results could be obtained even at the rather high temperature of $T = 258$ K and low attainable gradients by blowing in warm gas. The two examples shown in the figure refer to a 30×30 cm^2 area located in the mid-span region of a 2.40 m-long model (chord length $= 30$ cm). The left image in Fig. 7.56 shows transition for a negative angle of attack ($\alpha = -5°$), caused by a laminar separation bubble (bright vertical line in the middle). In the right image, transition takes place near the nose (sharp, vertical line), while flow separation can also be seen (rolling line in the middle). Note here (for $\alpha = 15°$): dark area = turbulent; small, somewhat brighter area near nose = laminar; bright area to the right = separated flow. The darker area near the

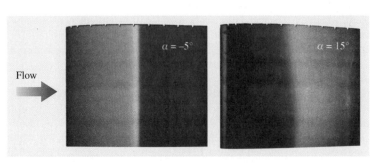

Fig. 7.56 TSP result images of a thick 2-D wing tested in the DNW-KKK wind tunnel. $Re = 2.5$ Mio, $M = 0.25$, $T = 258$ K, $p = 100$ kPa

nose in the left image in Fig. 7.56 ($\alpha = -5°$) indicates a more-effective heating here (flow was warmer than model). This is caused by a higher wall shear stress near the leading edge because of the thin boundary layer. The same holds true for the dark leading edge area in Fig. 7.54.

Finally, Fig. 7.57 shows a result of transition detection on a half model slat, tested at low Mach numbers in the DNW-KKK cryogenic wind tunnel [7.228]. We applied the heating foil method to establish a temperature difference between flow and slat (Fig. 7.50). Nevertheless the signatures of the three heating foils can still be seen in the TSP image (Fig. 7.57b), the transition line being clearly visible as the borderline between the darker, laminar region near the leading edge and the brighter turbulent region downstream. Here, the turbulent part of the boundary layer is cooled by the flow, thus appearing brighter in the image. The laminar *breakthrough* on the bottom part of the slat in Fig. 7.57b is caused by an upstream strake mounted on the model engine.

7.4.5 Comparison with Other Methods

As already mentioned, other well-established thermographic methods such as infrared thermography (IRT) or liquid-crystal thermography (TLC) can be also used for transition detection. The benefits and drawbacks of the various methods are presented in Table 7.6 (the method of thermographic phosphors (TPT) was included in the general classification of TSP). Pros and cons are listed with respect to their use in large, industrial wind tunnels. However, certain drawbacks in large-scale testing may not necessarily also be the case in laboratory-scale testing. This should be kept in mind.

The features of the different methods were compared primarily with respect to their use as a transition detection tool in Table 7.6. Their usefulness with respect to absolute temperature measurement was not analyzed in detail, since for transition detection all methods do not require a sensor calibration. However, the suitability for absolute temperature measurement is strongly related to temperature difference measurements used in transition detection. Moreover, thermographic methods can be used for the measurement of (conductive and convective) heat fluxes and wall shear stresses (because of the Reynolds analogy). The suitability of the different methods for such measurements has not been treated here.

The frame rates mentioned in the table have to be considered as *typical* image frequencies of the thermography *system*. For example, modern focal-plane array IR cameras typically have frame rates of 50 Hz, i.e., they

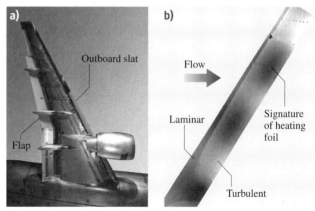

Fig. 7.57a,b Transition detection on high-lift devices in half model testing in DNW-KKK. Re = 2 Mio, $M = 0.18$, $T = 205$ K, $p = 100$ kPa. (a) Half model prepared for test. (b) Transition patterns on inner part of the outboard slat (seen in the middle of (a))

operate at video frequency. However, by binning of image pixels their frame rate can be increased up to about 500 Hz. 12 bit dynamic range CCD cameras typically used for the TSP method offer frame rates of around 10 Hz, which may be extended to around 20 Hz by binning of the CCD pixels. In large wind tunnels, exposure times usually are greater than 0.1 s (i.e., frame rates < 10 Hz result) caused by the large distances between the lamp/camera and model. The lifetime of the TSP luminophores is of the order of microseconds but incorporated into a polyurethane binder, the response time of a TSP *layer* is of the order of 5–20 ms (depending on the thickness). The response time for thermographic liquid crystals is < 6 ms but incorporated into paint or polymer it is around 100–200 ms, thus limiting the frame rate for TLC in the table to 5 Hz.

Further development of these methods is rapid and improvements in temperature sensitivity, ease of use, absolute temperature range, bandwidth, and spatial resolution will be made in the future. For example, to the author's knowledge, chemists in the UK are performing research for liquid crystals applicable in the cryogenic temperature range. However, since LCs available today are not suitable for cryogenic temperatures, their useful temperature range starts at $-30\,°\mathrm{C}$ in Table 7.6. Future IR cameras will have spatial resolution comparable to today's standard CCD cameras, in addition to increased frame rates. However, they will still be rather expensive. In the field of temperature-sensitive paint, DLR, for example, will further improve the existing cryoTSP formulation and will develop new TSPs for warm wind tunnels and turbine testing at ambient temperatures.

Table 7.6 Comparison of different methods for thermographic imaging in wind tunnel testing. (IRT: infrared thermography, TSP: temperature-sensitive paint, TPT: thermographic phospor thermometry, TLC: thermochromic liquid crystals, NB: narrow band, WB: wide band)

Method	Principle	Operation range (K)	Bandwidth (K)	Typical frame rate	Max. resolution (K)	Suitability for transition detection	Benefits	Drawbacks
IRT	Detection of radiated heat	200–1400	≈ 200	50 Hz	0.02	• Good in ambient and high temperature testing • Standard cameras not suitable for cryogenic testing (< 200 K)	• Ready to use in most applications • No surface preparation for nonmetalic models necessary • Various cameras available • Good temperature sensitivity without calibration • High temperature and time resolution	• IR system comparatatively expensive • Special optics, windows and cameras necessary • Small spatial resolution
TSP (incl. TPT)	Change of luminescent light intensity and lifetime of fluorescence/ phosporescence	80–2000	≈ 100	10 Hz	0.1	• Good for cryogenic testing and high temperature testing (using phospors) • Not much used at ambient temperatures	• Covers all temperatures given in wind tunnel testing • High spatial resolution • No *special* optics, windows or cameras necessary • Same equipment as for PSP can be used • Very smooth surface finishing possible	• Application of paint layer needed • Less temp. resolution compared to IRT • Need for excitation light source and filters for lamp and camera
TLC	Color change of scattered light	240–390	0.5–2 (NB) 5–30 (WB)	5 Hz	0.1	• At ambient conditions with small changes in absolute flow temperatures	• Various TLC formulations available • Excitation with white light • No special optics, windows or cameras necessary	• Relatively small bandwidth • Sensitive to illumination and viewing angle

Finally, progress in the development of modern, high-sensitivity, high-dynamic-range CCD cameras as well as in high-intensity excitation light sources (for example high-energy visible-range LEDs and UV LEDs) will expand the application range for thermographic imaging by means of TSPs.

References

7.1 G.M. Carlomagno: Heat transfer measurements by means of infrared thermography. In: *Measurement Techniques VKI Lect Series 1993-05 Rhode-Saint-Genese* (VKI, Rhode-Saint-Genese 1993) pp. 1–114

7.2 S.M.J. Straley: Physics of liquid crystals, Rev. Mod. Phys. **46**(4), 617–704 (1974)

7.3 P. Bonnett, T.V. Jones, D.G. Donnell: Shear-stress measurement in aerodynamic testing using cholesteric liquid crystals, Liquid Cryst. **6**, 271–280 (1989)

7.4 S.A. Hippensteele, L.M. Russell, F.S. Stepka: Evaluation of a method for heat transfer measurements and thermal visualisation of a heater elements and liquid crystals, Trans. ASME **105**, 184–189 (1983)

7.5 D.K. Hollingsworth, A.L. Boehman, E.G. Smith, R.J. Moffat: Measurement of temperature and heat transfer coefficient distributions in a complex flow using liquid crystal thermography and true-color image processing, ASME J. Heat Transfer **123**, 35–42 (1989)

7.6 J.K. Hay, D.K. Hollingsworth: A comparison of trichromic systems for use in the calibration of polymer-dispersed thermochromic liquid crystals, Exp. Thermal Fluid Sci. **12**, 1–12 (1996)

7.7 H. Babinsky, J.A. Edwards: Automatic liquid crystal thermography for transient heat transfer measurements in hypersonic flow, Exp. Fluids **21**, 227–236 (1996)

7.8 J. Stasiek: Thermochromic liquid crystals and true color image processing in heat transfer and fluid-flow research, J. Heat Mass Transfer **33**, 27–29 (1997)

7.9 R.T. Kukreja, S.C. Lau: Distribution of local heat transfer coefficient on surfaces with solid and perforated ribs, Enhanced Heat Transfer **5**, 9–21 (1998)

7.10 D.R. Sabatino, T.J. Praisner, C.R. Smith: A high-accuracy calibration technique for thermochromic liquid crystal temperature measurements, Exp. Fluids **28**, 497–505 (2000)

7.11 T.J. Praisner, D.R. Sabatino, C.R. Smith: Simultaneously combined liquid crystal surface heat transfer and PIV flow-field measurements, Exp. Fluids **300**, 1–10 (2001)

7.12 W. Hiller, T.A. Kowalewski: Simultaneous measurement of the temperature and velocity fields in thermal convective flows. In: *Flow Visualization IV*, ed. by C. Veret (Hemisphere, Paris 1987) pp. 617–622

7.13 W.J. Hiller, S. Koch, T.A. Kowalewski: Three-dimensional structures in laminar natural convection in a cube enclosure, Exp. Therm. Fluid Sci. **2**, 34–44 (1989)

7.14 W.J. Hiller, S. Koch, T.A. Kowalewski, D.G. de Vahl, M. Behnia: Experimental and numerical investigation of natural convection in a cube with two heated side walls. In: *Proc. of IUTAM Symp.*, ed. by H.K. Moffat, A. Tsinober (Cambridge Univ. Press, Cambridge 1990) pp. 717–726

7.15 D. Dabiri, M. Gharib: Digital particle image thermometry: The method and implementation, Exp. Fluids **97**, 77–86 (1991)

7.16 H.G. Park, D. Dabiri, M. Gharib: Digital particle image velocimetry/thermometry and application to the wake of a heated circular cylinder, Exp. Fluids **30**, 327–338 (2001)

7.17 W.J. Hiller, S. Koch, T.A. Kowalewski, P. Mitgau, K. Range: Visualization of 3-D natural convection. In: *Proc. 6th Int. Symp. Flow Visualization*, ed. by Y. Tanida, H. Miyashiro (Springer, Berlin, Heidelberg 1992) pp. 674–678

7.18 W.J. Hiller, S. Koch, T.A. Kowalewski, F. Stella: Onset of natural convection in a cube, Int. J. Heat Mass Transfer **36**, 3251–3263 (1993)

7.19 S. Koch: *Berührungslose Messung von Temperatur und Geschwindigkeit in freier Konvektion, Dissertation*, Mitteilungen aus dem MPI, Vol. 108 (Univ. Göttingen, Göttingen 1993)

7.20 T.A. Kowalewski, A. Cybulski, M. Rebow: Particle image velocimetry and thermometry in freezing water. In: *8th Int. Symp. Flow Visualization*, ed. by G.M. Carlomagno, I. Grant (ODE, Edinburgh 1998) pp. 24.1–24.8, CD ROM

7.21 K. Range: *Temperaturmessungen auf den Innenwänden eines Konvektionsbehälters mit Flüssigkristallen, Diplom*, MPIFS Bericht, Vol. 101 (Univ. Göttingen, Göttingen 1992)

7.22 W.J. Hiller, S. Koch, T.A. Kowalewski: Simultane Erfassung von Temperatur- und Geschwindigkeitsfeldern in einer thermischen Konvektionsströmung mit ungekapselten Flüssigkristalltracern. In: *2D-Meßtechnik DGLR-Workshop DGLR-Bericht 88-04* (DGLR, Bonn 1988) pp. 31–39, in German

7.23 E. Fornalik, Y. Yamamoto, W. Chen, K. Nakabe, K. Suzuki: Visualization of heat transfer enhancement regions modified by the interaction of inclined impinging jets into crossflow. In: *Image Processing Methods in Applied Mechanics Euromech 406 IPPT Reports 4/1999*, ed. by T.A. Kowalewski, J. Kompenhans, W. Kosinski (IPPT PAN, Warsaw 1999) pp. 95–98

7.24 T.A. Kowalewski, M. Rebow: Freezing of water in the differentially heated cubic cavity, Int. J. Comp. Fluid Dyn. **11**, 193–210 (1999)

7.25 T. Michalek, T.A. Kowalewski: Experimental model of mould filling flow. In: *Eurotherm 69 Heat and Mass Transfer in Solid-Liquid Phase Change Processes*, ed. by B. Šarler, D. Gobin (Polit. Nova Gorica, Lubljana 2003) pp. 61–68

7.26 T. Michalek, T.A. Kowalewski, B. Saler: Natural convection for anomalous density variation of water – numerical benchmark, Prog. Comput. Fluid Dyn. **5**, 158–170 (2005)

7.27 M. Giangi, F. Stella, T.A. Kowalewski: Phase-change problems with free convection: fixed grid simulation, Comp. Vis. Sci. **2**, 123–130 (1999)

7.28 M. Giangi, T.A. Kowalewski, F. Stella, E. Leonardi: Natural convection during ice formation: numerical simulation versus experimental results, Comp. Assisted Mech. Eng. Sci. **7**, 321–342 (2000)

7.29 J. Banaszek, Y. Jaluria, T.A. Kowalewski, M. Rebow: Semi-implicit FEM analysis of natural convection in freezing water, Num. Heat Transfer A **36**, 449–472 (1999)

7.30 T.A. Kowalewski, A. Cybulski: Experimental and numerical investigations of natural convection in freezing water, Proc. Int. Conf. Heat Transfer Change of Phase, Mechanics **61**(2), 7–16 (1996)

7.31 T.A. Kowalewski, A. Cybulski: Natural convection with phase change, IPPT Rep. **8**, 1–58 (1997), (in Polish)

7.32 A.Y. Gelfgat, P.Z. Bar-Yoseph, A. Solan, T.A. Kowalewski: An axisymmetry breaking instability of axially symmetric natural convection, Int. J. Trans. Phenomena **1**, 173–190 (1999)

7.33 C. Abegg, D.G. de Vahl, W.J. Hiller, S. Koch, T.A. Kowalewski, E. Leonardi, G.H. Yeoh: Experimental and numerical study of three-dimensional natural convection and freezing in water. In: *Proc. 10th Int. Heat Transfer Conf.*, Vol. 4, ed. by G.F. Hewitt (Taylor Francis, New York 1994) pp. 1–6

7.34 T.A. Kowalewski, A. Cybulski, T. Sobiecki: Experimental model for casting problems. In: *Computational Methods and Experimental Measurements*, ed. by Y.V. Esteve, G.M. Carlomagno, C.A. Brebia (WIT, Southampton 2001) pp. 179–188

7.35 K. Dekajlo, T.A. Kowalewski, H.J.S. Fernando: Experiments on up-slope to down-slope transition in an inclined box filled with water. In: *ICTAM04 CD-ROM Proceedings IPPT PAN*, ed. by W. Gutkowsky, T.A. Kowalewski (Springer, Dordrecht 2004)

7.36 T.A. Kowalewski: Experimental validation of numerical codes in thermally driven flows. In: *Advances in Computational Heat Transfer*, ed. by G.D. de Vahl, E. Leonardi (Begel House, New York 1998) pp. 1–15

7.37 A. Schulz: Infrared thermography as applied to film cooling of gas turbine components, Meas. Sci. Technol. **11**, 948–956 (2000)

7.38 W.P. Dixon, P. Czysz: The use of thermographic phosphors for heat transfer measurement in hypersonic wind tunnel. In: *Society of Photo-Optical Engineers 11th Annual Symposium* (1966)

7.39 G. Meyers, J. Van der Geest, J. Sanborn, F. Davis: Comparison of advanced cooling concepts using color thermography. In: *AIAA 3rd Applied Aerodynamics Conf. AIAA Paper AIAA-85-1289* (1985)

7.40 E. Gartenberg, A.S. Roberts: Twenty-five years of aerodynamic research with infrared imaging, J. Aircr. **29**(2), 161–171 (1992)

7.41 J. Wendt: Infrared thermography. In: *The Second Joint Europe/US Short Course in Hypersonics* (1989)

7.42 M.F. Westby: *Heat transfer measurements using infrared thermography in rarefied flows*, Technical Memorandum W1 (Defense Research Agency, Farnborough 1992)

7.43 M. Martiny, R. Schiele, M. Gritsch, A. Schulz, S. Wittig: In situ calibration for quantitative infrared thermography, QIRT '96, Stuttgart, Eurotherm Seminar **50**, 3–8 (1996)

7.44 S.R. Sargent, C.R. Hedlund, P.M. Ligrani: An infrared thermography imaging system for convective heat transfer measurements in complex flows, Meas. Sci. Technol. **9**(12), 1974–1981 (1998)

7.45 V. Scherer, A. Pfeiffer, S. Wittig: *Bestimmung der Wärmeübergangszahlen in abgelösten Strömungen mit Hilfe einer Infrarotkamera*, DGLR Workshop, 2D-Meßtechnik (DGLR, Bonn 1988) pp. 245–253

7.46 V. Scherer, S. Wittig: The influence of the recirculating region: a comparison of the convective heat transfer downstream of a backward-facing step and behind a jet in a cross-flow. In: *ASME Winter Annual Meeting* (ASME, Fairfield 1989), Paper 89-GT-59

7.47 M. Martiny, A. Schulz, S. Wittig: Full coverage film cooling investigations: adiabatic wall temperatures and flow visualization. In: *Winter Annual Meeting* (ASME, Fairfield 1995), Paper 95-WA/HT-4

7.48 M. Martiny, A. Schulz, S. Wittig: Mathematical model describing the coupled heat transfer in effusion cooled combustor walls. In: *Int. Gas Turbine Aeroengine Congr. Exhibition* (ASME, Fairfield 1997), Paper 97-GT-329

7.49 C.R. Hedlund, P.M. Ligrani, H.-K. Moon, B. Glezer: Heat transfer and flow phenomena in a swirl chamber simulating turbine blade internal cooling, ASME Trans. J. Turbomachin. **121**(4), 804–813 (1999)

7.50 B. Sen, D.L. Schmidt, D.G. Bogard: Film cooling with compound angle holes: heat transfer, ASME Trans. J. Turbomachin. **118**(4), 800–806 (1996)

7.51 U.M. Yuki, D.G. Bogard, J.M. Cutbirth: Effect of coolant injection on heat transfer for a simulated turbine airfoil leading edge. In: *Int. Gas Turbine Aeroengine Congr. Exhibition* (ASME, Fairfield 1998), Paper 98-GT-431

7.52 M. Gritsch, A. Schulz, S. Wittig: Adiabatic wall effectiveness measurements of film cooling holes with expanded exits, ASME Trans. J. Turbomachin. **120**, 549–556 (1998)

7.53 M. Gritsch, A. Schulz, S. Wittig: Heat transfer coefficient measurements of film cooling holes with expanded exits. In: *Int. Gas Turbine Aeroengine Congr. Exhibition* (ASME, Fairfield 1998), Paper 98-GT-28

7.54 P.C. Sweeny, J.F. Rhodes: An infrared technique for evaluating turbine airfoil cooling designs. In: *Int. Gas Turbine Aeroengine Congr. Exhibition* (ASME, Fairfield 1999), Paper 99-GT-142

7.55 C.M. Bell, H. Hamakawa, P.M. Ligrani: Film cooling from shaped holes, ASME Trans. J. Heat Transfer **122**(2), 224–232 (2000)

7.56 T. Fukukawa, P.M. Ligrani: Transonic film cooling effectiveness from shaped holes on a simulated turbine airfoil, AIAA J. Thermophys. Heat Transfer **16**(2), 228–237 (2002)

7.57 S.Y. Won, N.K. Burgess, S. Peddicord, P.M. Ligrani: Spatially-resolved surface heat transfer for parallel rib turbulators with 45 degree orientations including test surface conduction analysis, ASME Trans. J. Heat Transfer **126**(2), 193–201 (2004)

7.58 R.G. Driggers, C. Webb, S.J. Pruchnic, C.E. Halford, E.E. Burroughs: Laboratory measurement of sampled infrared imaging system performance, Opt. Eng. **38**(5), 852–861 (1999)

7.59 T. Astarita, G. Cardone, G.M. Carlomagno: IR heat transfer measurements in a rotating channel, QIRT '96, Stuttgart, Eurotherm Seminar **50**, 147–152 (1996)

7.60 G. Cardone, T. Astarita, G.M. Carlomagno: Heat transfer measurements on a rotating disc. In: *5th Int. Symp. Transport Phenomena and Dynamics of Rotating Machinery A* (1994) pp. 663–672

7.61 F.P. Hindle, S.J. Carey, K. Ozanyan, D.E. Winterbone, E.E. Clough, H. McCann: Measurement of gaseous hydrocarbon distribution by a near-infrared absorption tomography system, J. Electron. Imaging **10**(3), 593–600 (2001)

7.62 I.S.I. Group Inc., SWI: Southwest Infrared, VideoTherm Camera Systems, 1521 Eastridge Dr. NE, Albuquerque, NM 87112-4508 USA (http://www.swinfrared.com)

7.63 AGEMA Infrared Systems, ENEA Embedded Technology, 12760 High Bluff Drive, San Diego, CA 92130 USA (http://www.enea.com/907.epibrw)

7.64 Flir Systems Inc., 25 Esquire Road, North Billerica, MA 01862 USA (http://www.flirthermography.com/contact/worldmap.asp)

7.65 P.M. Ligrani, M.M. Oliveira, T. Blaskovich: Comparison of heat transfer augmentation techniques, AIAA J. **41**(3), 337–362 (2003)

7.66 W.P. Stricker: Measurements of temperature in laboratory flames and practical devices. In: *Applied Combustion Diagnostics*, ed. by K. Kohse-Höinghaus, J.B. Jeffries (Taylor Francis, New York 2002)

7.67 N.M. Laurendeau: Temperature measurements by light-scattering methods, Prog. Energy Combust. Sci. **14**, 147–170 (1988)

7.68 R.W. Dibble, R.E. Hollenbach: Laser Rayleigh thermometry in turbulent flames, Proc. Combust. Inst. **18**, 1489–1499 (1981)

7.69 J.N. Forkey, N.D. Finkelstein, W.R. Lempert, R.B. Miles: Demonstration and characterization of filtered Rayleigh scattering for planar velocity measurements, AIAA J. **34**, 442–448 (1996)

7.70 D. Hoffman, K.-U. Münch, A. Leipertz: Two-dimensional temperature determination in sooting flames by filtered Rayleigh scattering, Opt. Lett. **21**, 525–527 (1996)

7.71 R.B. Miles, J.N. Forkey, W.R. Lempert: Rayleigh scattering measurements in supersonic/hypersonic facilities, AIAA Paper **92-3894** (1992)

7.72 R.S. Barlow, C.D. Carter, R.W. Pitz: Multiscalar diagnostics in turbulent flames. In: *Applied Combustion Diagnostics*, ed. by K. Kohse-Höinghaus, J.B. Jeffries (Taylor Francis, New York 2002) pp. 384–407

7.73 D.F. Marran, J.H. Frank, M.B. Long, S.H. Stårner: Intracavity technique for improved Raman/Rayleigh imaging in flames, Opt. Lett. **20**, 791–793 (1995)

7.74 G.G. Stokes: *Mathematical and Physical Papers* (Cambridge Univ. Press, Cambridge 1880)

7.75 D.A. Long: *The Raman Effect: A Unified Treatment of the Theory of Raman Scattering by Molecules* (Wiley, New York 2002)

7.76 D. Geyer: *1D Raman/Rayleigh experiments in a turbulent opposed-jet*, Ph.D. Thesis (TU Darmstadt, Darmstadt 2004)

7.77 K. Kohse-Höinghaus, J.B. Jeffries (Eds.): *Applied Combustion Diagnostics* (Taylor Francis, New York 2002)

7.78 T. Landenfeld, A. Kremer, E.P. Hassel, J. Janicka, T. Schäfer, J. Kazenwadel, C. Schulz, J. Wolfrum: Laserdiagnostic and numerical studies of strongly swirling natural-gas flames, Proc. Combust. Inst. **27**, 1023–1030 (1998)

7.79 C. Schulz, V. Sick, J. Wolfrum, V. Drewes, M. Zahn, R. Maly: Quantitative 2D single-shot imaging of NO concentrations and temperatures in a transparent SI engine, Proc. Combust. Inst. **26**, 2597–2604 (1996)

7.80 J. Warnatz, U. Maas, R.W. Dibble: *Combustion*, 3rd edn. (Springer, Berlin, Heidelberg 2001)

7.81 V. Bergmann, W. Meier, D. Wolff, W. Stricker: Application of spontaneous Raman and Rayleigh scattering and 2D LIF for the characterization of a turbulent $CH_4/H_2/N_2$ jet diffusion flame, Appl. Phys. B **66**, 489–502 (1998)

7.82 D. Geyer, A. Kempf, A. Dreizler, J. Janicka: Scalar dissipation rates in isothermal and reactive turbulent opposed-jets: 1D-Raman/Rayleigh experiments supported by LES, Proc. Combust. Inst. **30**, 681–689 (2005)

7.83 H.W. Schrötter: Linear Raman spectroscopy: A state of the art report. In: *Nonlinear Raman Spectroscopy and Its Chemical Applications*, NATO Adv. Study Inst. Ser. C, Vol. 93, ed. by W.J. Kiefer, D.A. Long (Kluwer, Dordrecht 1982)

7.84 J. Zetterberg, Z.S. Li, M. Afzelius, M. Aldén: Two-dimensional temperature measurements in flames using filtered Rayleigh scattering at 254 nm. In: *European Combustion Meeting* (The Combustion Institute, Orléans 2003)

7.85 W. Meier, R.S. Barlow, Y.-L. Chen, J.-Y. Chen: Raman/Rayleigh/LIF measurements in a turbulent $CH_4/H_2/N_2$ jet diffusion flame: experimental techniques and turbulence-chemistry interaction, Combust. Flame **123**, 326–343 (2000)

7.86 A.C. Eckbreth: *Laser Diagnostics for Combustion Temperature and Species*, 2nd edn. (Gordon Breach, Amsterdam 1996)

7.87 S.B. Pope: *Turbulent Flows* (Cambridge Univ. Press, Cambridge 2000) p. 771

7.88 C. Schneider, A. Dreizler, J. Janicka: Fluid dynamical analysis of atmospheric reacting and isothermal swirling flows, Flow Turbulence Combust. **74**, 103–127 (2005)

7.89 W. Meier, S. Prucker, M.H. Cao, W. Stricker: Characterization of turbulent H_2/N_2/air jet diffusion flames by single-pulse spontaneous Raman scattering, Combust. Sci. Technol. **118**, 293–312 (1996)

7.90 R.W. Pitz, J.A. Wehrmeyer, J.M. Bowling, T.S. Cheng: Single pulse vibrational Raman scattering by a broadband KrF excimer laser in a hydrogen-air flame, Appl. Opt. **29**, 2325–2332 (1990)

7.91 A. Brockhinke, P. Andresen, K. Kohse-Höinghaus: Quantitative one-dimensional single-pulse multi-species concentration and temperature measurements in the lift-off region of a turbulent H_2-air diffusion flame, Appl. Phys. B **61**, 533–545 (1995)

7.92 W. Meier, O. Keck: Laser Raman scattering in fuel-rich flames: background levels at different excitation wavelengths, Meas. Sci. Technol. **13**, 741–749 (2002)

7.93 G. Herzberg: *Molecular Spectra and Molecular Structure*, Spectra of Diatomic Molecules, Vol. 1 (Krieger, Malabar 1989)

7.94 G. Herzberg: *Molecular Spectra and Molecular Structure*, Electronic Spectra and Electronic Structure of Polyatomic Molecules, Vol. 3, 2nd edn. (Krieger, Malabar 1991)

7.95 M.M. Tacke, D. Geyer, E.P. Hassel, J. Janicka: A detailed investigation of the stabilization point of lifted turbulent diffusion flames, Proc. Combust. Inst. **27**, 1157–1165 (1998)

7.96 R.S. Barlow, P.C. Miles: A shutter-based line-imaging sytem for single-shot Raman scattering measurements of gradients in mixture fraction, Proc. Combust. Inst. **28**, 269–277 (2000)

7.97 H. Tennekes, J. Lumley: *A First Course in Turbulence* (MIT Press, Cambridge 1972)

7.98 V. Ebert, J. Wolfrum: Absorption spectroscopy. In: *Optical Measurements-Techniques and Applications*, ed. by F. Mayinger (Springer, Berlin, Heidelberg 1994) pp. 273–312

7.99 D.S. Baer, R.K. Hanson, M.E. Newfield, N.K.J.M. Gopaul: Multiplexed diode-laser sensor system for simultaneous H_2O, O_2, and temperature measurements, Opt. Lett. **19**, 1900–1902 (1994)

7.100 M.G. Allen: Diode laser absorption sensors for gas dynamic and combustion flows, Meas. Sci. Technol. **9**, 545–562 (1998)

7.101 A.P. Thorne: *Spectrophysics* (Chapman Hall, London 1988)

7.102 E.R. Furlong, D.S. Baer, R.K. Hanson: Combustion Control and Monitoring using a Multiplexed Diode-Laser Sensor System, Proc. Combust. Inst. **26**, 2851–2858 (1996)

7.103 A. Thorne, U. Litzén, S. Johannson: *Spectrophysics* (Springer, Berlin, Heidelberg 1999)

7.104 I.I. Sobel'man, L.A. Vainshtein, E.A. Yukov: *Excitation of Atoms and Broadening of Spectral Lines* (Springer, Berlin, Heidelberg 1995)

7.105 V. Ebert, T. Fernholz, C. Giesemann, H. Pitz, H. Teichert, J. Wolfrum, H. Jaritz: Simultaneous diode-laser-based in-situ detection of multiple species and temperature in a gas-fired power plant, Proc. Combust. Inst. **28**, 423–430 (2000)

7.106 S. Schäfer, M. Mashni, J. Sneider, A. Miklos, P. Hess, V. Ebert, K.-U. Pleban, H. Pitz: Sensitive detection of methane with a 1.65 µm diode laser by photoacoustic and absorption spectroscopy, Appl. Phys. B **66**, 511–516 (1998)

7.107 C.A. Taatjes, D.B. Oh: Time-resolved wavelength modulation spectroscopy measurements of HO_2 kinetics, Appl. Opt. **36**, 5817–5821 (1997)

7.108 K. Kohse-Höinghaus, D.F. Davidson, A.Y. Chang, R.K. Hanson: Quantitative NH_2 concentration determination in shock-tube laser-absorption experiments, J. Quant. Spectrosc. Radiat. Transfer **42**, 1–17 (1989)

7.109 J.J. Scherer, J.B. Paul, A. O'Keefe, R.J. Saykally: Cavity ringdown laser absorption spectroscopy: History, development, and application to pulsed molecular beams, Chem. Rev. **97**, 25 (1997)

7.110 S. Cheskis, I. Derzy, V.A. Lozovsky, A. Kachanov, D. Romanini: Cavity ring-down spectroscopy of OH radicals in low pressure flames, Appl. Phys. B **66**, 377–381 (1998)

7.111 S. Cheskis, I. Derzy, V.A. Lozovsky, A. Kachanov, F. Stoeckel: Intracavity laser absorption spectroscopy detection of CH_2 radicals in hydrocarbon flames, Chem. Phys. Lett. **277**, 423 (1997)

7.112 P. Zalicki, Y. Ma, R.N. Zare, J.R. Dadamio, E.H. Wahl, T.G. Owano, C.H. Kruger: Methyl radical measurement by cavity ring-down spectroscopy, Chem. Phys. Lett. **234**, 269–274 (1995)

7.113 J.J. Scherer, D.J. Rakestraw: Cavity ring-down laser absorption spectroscopy detection of formyl (HCO) radical in a low pressure flame, J. Chem. Phys. **265**, 169–176 (1997)

7.114 J.J. Scherer, K.W. Aniolek, N.P. Cernansky, D.J. Rakestraw: Determination of methyl radical concentrations in a methane/air flame by infrared cavity ringdown laser absorption spectroscopy, J. Chem. Phys. **107**, 6196–6203 (1997)

7.115 Y. He, M. Hippler, M. Quack: High-resolution cavity rind-down absorption spectroscopy of nitrous oxide and chloroform using near-infrared cw diode laser, Chem. Phys. Lett. **289**, 527–534 (1998)

7.116 T. Yu, M.C. Lin: Kinetics of the C_6H_5 + NO association reaction, J. Phys. Chem. **98**, 2105–2109 (1994)

7.117 S.T. Sanders, J. Wang, J.B. Jeffries, R.K. Hanson: Diode-laser absorption sensor for line-of-sight gas

7.117 temperature distributions, Appl. Opt. **40**, 4404–4415 (2001)

7.118 D.W. Mattison, S.T. Sanders, L. Ma, K.M. Hinckley, J.B. Jeffries, R.K. Hanson, C.M. Brophy: Pulse detonation engine characterization and control using tunable diode-laser sensors, J. Propulsion Power **19**, 568–572 (2003)

7.119 X. Zhou, J.B. Jeffries, R.K. Hanson: Development of a fast temperature sensor for combustion gases using a single tunable diode laser, Appl. Phys. B **81**, 711–722 (2005)

7.120 J.T.C. Liu, G.B. Rieker, J.B. Jeffries, R.K. Hanson, M.R. Gruber, C.D. Carter, T. Mathur: Near infrared diode laser absorption diagnostic for temperature and water vapor in a scramjet combustor, Appl. Opt. **44**, 6701–6711 (2005)

7.121 V. Nagali, R.K. Hanson: Design of a diode-laser sensor to monitor water vapor in high-pressure combustion gases, Appl. Opt. **36**, 9518–9527 (1997)

7.122 S.T. Sanders, J.A. Baldwin, T.P. Jenkins, D.S. Baer, R.K. Hanson: Diode-laser sensor for monitoring multiple combustion parameters in pulse detonation engines, Proc. Combust. Inst. **28**, 587–594 (2000)

7.123 H. Teichert, T. Fernholz, V. Ebert: Simultaneous in-situ measurements of CO, H_2O and gas temperatures in a full-sized coal-fired power plant by near-infrared diode lasers, Appl. Opt. **42**, 2043–2051 (2003)

7.124 L.S. Rothman, D. Jacquemart, A. Barbe, D.C. Benner, M. Birk, L.R. Brown, K. Carleer, C. Chackerian Jr., Chance, L.H. Coudert, V. Dana, V.M. Devi, J.-M. Flaut, R.R. Gamache, A. Goldmann, J.-M. Hartmann, K.W. Jucks, A.G. Maki, S.T. Mandin, S.T. Massie, J. Orphal, A. Perrin, C.P. Rinsland, M.A.H. Smith, J. Tennyson, R.N. Tolchenov, R.A. Toth, J. Vander Auwera, P. Varanasi, G. Wagner: The HITRAN 2004 molecular spectroscopic database, J. Quant. Spectrosc. Radiat. Transfer **96**, 139–204 (2004), (http://cfa-www.harvard.edu/HITRAN/)

7.125 G.B. Rieker, J.T.C. Liu, J.B. Jeffries, R.K. Hanson, T. Mathur, M.R. Gruber, C.D. Carter: Diode laser sensor for gas temperature and H_2O concentration in a scramjet combustor using wavelength modulation spectroscopy, AIAA Paper **2005-3710** (2005)

7.126 J.T.C. Liu, J.B. Jeffries, R.K. Hanson: Wavelength modulation absorption spectroscopy with 2f detection using multiplexed diode lasers for rapid temperature measurements in gaseous flows, Appl. Phys. B **78**, 503–511 (2004)

7.127 K. Kohse-Höinghaus: Laser techniques for the quantitative detection of reactive intermediates in combustion systems, Prog. Energy Combust. Sci. **20**, 203–279 (1994)

7.128 J.W. Daily: Laser induced fluorescence spectroscopy in flames, Prog. Energy Combust. Sci. **23**, 133–199 (1997)

7.129 E.W. Rothe, Y. Gu, A. Chryssostomou, P. Andresen, F. Bormann: Effect of laser intensity and of lower-state rotational energy transfer upon temperature measurements made with laser-induced predissociative fluorescence, Appl. Phys. B **66**, 251 (1998)

7.130 J. Luque, D.R. Crosley: *LIFBASE: Database and spectral simulation for diatomic molecules* (SRI International, Menlo Park 1999), MP-99-0099, (www.sri.com/cem/lifbase)

7.131 W.G. Bessler, C. Schulz, V. Sick, J.W. Daily: www.lifsim.com). In: *3rd Joint meeting of the US sections of The Combustion Institute* (Combustion Institute, Chicago 2003) pp.1–6, Paper PI05

7.132 R. Kienle, M.P. Lee, K. Kohse-Höinghaus: A scaling formalism for the representation of rotational energy transfer in OH A in combustion experiments, Appl. Phys. B **63**, 403–418 (1996)

7.133 P.H. Paul, J.A. Gray, J.L.D. Jr., J.W.T. Jr.: A model for temperature-dependent collisional quenching of NO $A^2\Sigma^+$, Appl. Phys. B **57**, 249–259 (1993)

7.134 W.G. Bessler, M. Hofmann, F. Zimmermann, G. Suck, J. Jakobs, S. Nicklitzsch, T. Lee, J. Wolfrum, C. Schulz: Quantitative in-cylinder NO-LIF imaging in a realistic gasoline engine with spray-guided direct injection, Proc. Combust. Inst. **30**, 2667–2674 (2005)

7.135 J.W. Daily, T.B. Settersten, W.G. Bessler, C. Schulz, V. Sick: A computer code to simulate laser excitation and collision dynamics in nitric oxide. In: *4th Joint Meeting of the U.S. Sections of the Combustion Institute* (The Combustion Institute, Philadelphia 2005)

7.136 R. Cattolica: OH rotational temperature from two-line laser-excited fluorescence, Appl. Opt. **20**, 1156–1166 (1981)

7.137 K.P. Gross, R.L. McKenzie: Singe-pulse gas thermometry at low temperatures using two-photon laser-induced fluorescence in NO-N_2 mixtures, Opt. Lett. **8**, 368–370 (1983)

7.138 M.P. Lee, B.K. McMillin, R.K. Hanson: Temperature measurements in gases by use of planar laser-induced fluorescence imaging of NO, Appl. Opt. **32**, 5379–5396 (1993)

7.139 J.M. Seitzman, R.K. Hanson: Two-line planar fluorescence for temporally resolved temperature imaging in a reacting supersonic flow over a body, Appl. Phys. B **57**, 385–391 (1993)

7.140 A. Roller, A. Arnold, M. Decker, V. Sick, J. Wolfrum, W. Hentschel, K.-P. Schindler: Non-intrusive temperature measurements during the compression phase of a DI Diesel engine, SAE Tech. Paper Ser. **952461** (1995)

7.141 M. Tamura, P.A. Berg, J.E. Harrington, J. Luque, J.B. Jeffries, G.P. Smith, D.R. Crosley: Collisional quenching of CH (A), OH A, and NO (A) in low-pressure hydrocarbon flames, Combust. Flame **114**, 502–514 (1998)

7.142 M. Tamura, J. Luque, J.E. Harrington, P.A. Berg, G.P. Smith, J.B. Jeffries, D.R. Crosley: Laser-induced fluorescence of seeded nitric oxide as a flame thermometer, Appl. Phys. B **66**, 503–510 (1998)

7.143 M. Tsujishita, A. Hirano, M. Yokoo, T. Sakuraya, Y. Takeshita: Accurate thermometry using NO and OH laser-induced fluorescence in an atmospheric pressure flame, JSME Int. J. Ser. B **42**, 119 (1999)

7.144 W.G. Bessler, F. Hildenbrand, C. Schulz: Two-line laser-induced fluorescence imaging of vibrational temperatures of seeded NO, Appl. Opt. **40**, 748–756 (2001)

7.145 J.B. Bell, M.S. Day, J.F. Grcar, W.G. Bessler, C. Schulz, P. Glarborg, A.D. Jensen: Detailed modeling and laser-induced fluorescence imaging of nitric oxide in a NH_3-seeded non-premixed methane/air flame, Proc. Combust. Inst. **29**, 2195–2202 (2002)

7.146 J.W. Daily, E.W. Rothe: Effect of laser intensity and lower-state rotational energy transfer upon temperature measurements made with laser-induced fluorescence, Appl. Phys. B **68**, 131–140 (1999)

7.147 W.G. Bessler, C. Schulz: Quantitative multi-line NO-LIF temperature imaging, Appl. Phys. B **78**, 519–533 (2004)

7.148 J.E. Dec, J.O. Keller: High speed thermometry using two-line atomic fluorescence, Proc. Combust. Inst. **21**, 1737–1745 (1986)

7.149 C.F. Kaminski, J. Engström, M. Aldén: Quasi-instantaneous two-dimensional temperature measurements in a spark ignition engine using 2-line atomic fluorescence, Proc. Combust. Inst. **27**, 85–93 (1998)

7.150 J. Engström, J. Nygren, M. Aldén, C.F. Kaminski: Two-line atomic fluorescence as a temperature probe for highly sooting flames, Opt. Lett. **25**, 1469–1471 (2000)

7.151 P. Andresen, A. Bath, W. Groger, H.W. Lulf, G. Meijer, J.J. ter Meulen: Laser-induced fluorescence with tunable excimer lasers as a possible method for instantaneous temperature field measurements at high pressures: checks with an atmospheric flame, Appl. Opt. **27**, 365–378 (1988)

7.152 A. Arnold, B. Lange, T. Bouché, Z. Heitzmann, G. Schiff, W. Ketterle, P. Monkhouse, J. Wolfrum: Absolute temperature fields in flames by 2D-LIF of OH using excimer lasers and CARS spectroscopy, Ber. Bunsenges. Phys. Chem. **96**, 1388–1392 (1992)

7.153 B.K. McMillin, J.M. Seitzman, R.K. Hanson: Comparison of NO and OH planar fluorescence temperature measurements in scramjet model flowfields, AIAA J. **32**, 1945–1952 (1994)

7.154 B. Atakan, J. Heinze, U.E. Meier: OH laser-induced fluorescence at high pressures: spectroscopic and two-dimensional measurements exciting the A-X (1,0) transition, Appl. Phys. B **64**, 585–591 (1997)

7.155 A.T. Hartlieb, B. Atakan, K. Kohse-Höinghaus: Temperature measurement in fuel-rich non-sooting low-pressure hydrocarbon flames, Appl. Phys. B **70**, 435–445 (2000)

7.156 R.L. McKenzie, K.P. Gross: Two-photon excitation of nitric oxide fluorescence as a temperature indicator in unsteady gasdynamic processes, Appl. Opt. **20**, 2153–2165 (1981)

7.157 J.M. Seitzman, G. Kychakoff, R.K. Hanson: Instantaneous temperature field measurements using planar laser-induced fluorescence, Opt. Lett. **10**, 439–441 (1985)

7.158 K.P. Gross, R.L. McKenzie: Measurements of fluctuating temperatures in a supersonic turbulent flow using laser-induced fluorescence, AIAA J. **23**, 1932–1936 (1985)

7.159 K.P. Gross, R.L. McKenzie, P. Logan: Measurement of temperature, density, pressure, and their fluctuations in supersonic turbulence using laser-induced fluorescence, Exp. Fluids **5**, 372–380 (1987)

7.160 B.K. McMillin, J.L. Palmer, R.K. Hanson: Temporally resolved, two-line fluorescence imaging of NO temperature in a transverse jet in a supersonic cross flow, Appl. Opt. **32**, 7532–7545 (1993)

7.161 A.O. Vyrodov, J. Heinze, M. Dillmann, U.E. Meier, A.W. Stricker: Laser-induced fluorescence thermometry and concentration measurements on NO A-X (0,0) transitions in the exhaust gas of high pressure CH_4/air flames, Appl. Phys. B **61**, 409–414 (1995)

7.162 M. Yorozu, Y. Okada, A. Endo: Two dimensional rotational temperature measurement by multiline laser induced fluorescence of nitric oxide in combustion flame, Opt. Rev. **3**, 293–298 (1996)

7.163 E.A. Brinkman, G.A. Raiche, M.S. Brown, J.B. Jeffries: Optical diagnostics for temperature measurement in a dc arcjet reactor used for diamond deposition, Appl. Phys. B. **64**, 689 (1997)

7.164 P.C. Palma, T.J. McIntyre, A.F.P. Houwing: PLIF thermometry in shock tunnel flows using a Raman-shifted tunable excimer laser, Shock Waves **8**, 275–284 (1998)

7.165 W.M. Ruyten, M.S. Smith, L.L. Price, W.D. Williams: Three-line fluorescence thermometry of optically thick shock-tunnel flow, Appl. Opt. **37**, 2334–2339 (1998)

7.166 G. Dilecce, M. Simek, M. Vigliotti, S. De Benedictis: Fast LIF approach to NO rotational temperature and density measurement: Application to a gas-dynamic expansion, Appl. Spectrosc. **54**, 824–831 (2000)

7.167 H. Kronemayer, W. Bessler, C. Schulz: Gas-phase temperature measurements in evaporating sprays and spray flames based on NO multiline LIF, Appl. Phys. B **81**, 1071–1074 (2005)

7.168 W.G. Bessler, C. Schulz, T. Lee, D.I. Shin, M. Hofmann, J.B. Jeffries, J. Wolfrum, R.K. Hanson: Quantitative NO-LIF imaging in high-pressure flames. In: *Optical and Laser Diagnostics*, ed. by C. Arcoumanis, K.T.V. Grattan (Institute of Physics, Bristol, Philadelphia 2003) pp. 107–114

7.169 M.P. Lee, P.H. Paul, R.K. Hanson: Quantitative imaging of temperature fields in air using planar laser-induced fluorescence of O_2, Opt. Lett. **12**, 75–77 (1987)

7.170 G. Laufer, R.L. McKenzie, D.G. Fletcher: Method for measuring temperatures and densities in hypersonic wind tunnel air flows using laser-induced O_2 fluorescence, Appl. Opt. **29**, 4873–4883 (1990)

7.171 M.S. Smith, L.L. Price, W.D. Williams: Laser-induced fluorescence diagnostics using a two-line excitation method, AIAA J. **31**, 478–482 (1993)

7.172 T. Ni-Imi, T. Fujimoto, N. Shimizu: Method for planar measurement of temperature in compressible flow using two-line laser-induced iodine fluorescence, Opt. Lett. **15**, 918–920 (1990)

7.173 A. Kido, S. Kubota, H. Ogawa, N. Miyamoto: Simultaneous measurements of concentration and temperature distributions in unsteady gas jets by an iodine LIF method, SAE Tech. Paper Ser. **980146** (1998)

7.174 F. Großmann, P.B. Monkhouse, M. Ridder, V. Sick, J. Wolfrum: Temperature and pressure dependences of the laser-induced fluorescence of gas-phase acetone and 3-pentanone, Appl. Phys. B **62**, 249–253 (1996)

7.175 M.C. Thurber, F. Grisch, R.K. Hanson: Temperature imaging with single- and dual-wavelength acetone planar laser-induced fluorescence, Opt. Lett. **22**, 251–253 (1997)

7.176 S. Einecke, C. Schulz, V. Sick, A. Dreizler, R. Schießl, U. Maas: Two-dimensional temperature measurements in an SI engine using two-line tracer LIF, SAE Tech. Paper Ser. **982468** (1998)

7.177 M. Luong, W. Koban, C. Schulz: Novel strategies for imaging temperature distribution using Toluene LIF, J. Phys. **45**, 133–139 (2005)

7.178 W. Koban, J.D. Koch, V. Sick, N. Wermuth, R.K. Hanson, C. Schulz: Predicting LIF signal strength for toluene and 3-pentanone under engine-related temperature and pressure conditions, Proc. Combust. Inst. **30**, 1545–1553 (2005)

7.179 W. Koban, C. Schulz: Toluene as a tracer for fuel, temperature and oxygen concentrations, SAE Tech. Paper Ser. **2005-01-2091** (2005)

7.180 W. Koban, J.D. Koch, R.K. Hanson, C. Schulz: Toluene LIF at elevated temperatures: Implications for fuel/air ratio measurements, Appl. Phys. B **80**, 147–150 (2005)

7.181 H. Zacharias, J.B. Halpern, K.H. Welge: Two-photon excitation of NO ($A^2\Sigma^+$; $v' = 0,1,2$) and radiation lifetime and quenching measurements, Chem. Phys. Lett. **43**, 41–44 (1976)

7.182 G.F. Nutt, S.C. Haydon, A.J. McIntosh: Measurement of electronic quenching rates in nitric oxide using two-photon spectroscopy, Chem. Phys. Lett. **62**, 402–404 (1979)

7.183 I.S. McDermid, J.B. Laudenslager: Radiative lifetimes and electronic quenching rate constants for single-photon-excited rotational levels of NO ($A^2\Sigma^+$, $v' = 0$), J. Quant. Spectrosc. Radiat. Transfer **27**, 483–492 (1982)

7.184 N. Sullivan, A. Jensen, P. Glarborg, M.S. Day, J.F. Grcar, J.B. Bell, C. Pope, R.J. Kee: Ammonia conversion and NO_x formation in laminar coflowing nonpremixed methane/air flames, Combust. Flame **131**, 285–298 (2002)

7.185 W.G. Bessler, C. Schulz, T. Lee, D.I. Shin, M. Hofmann, J.B. Jeffries, J. Wolfrum, R.K. Hanson: Quantitative NO-LIF imaging in high-pressure flames, Appl. Phys. B **75**, 97–102 (2002)

7.186 C. Schulz, V. Sick: Tracer-LIF diagnostics: Quantitative measurement of fuel concentration, temperature and air/fuel ratio in practical combustion situations, Prog. Energy Combust Sci. **31**, 75–121 (2005)

7.187 F. Großmann, P.B. Monkhouse, M. Ridder, V. Sick, J. Wolfrum: Temperature and pressure dependencies of the laser-induced fluorescence of gas-phase acetone and 3-pentanone, Appl. Phys. B **62**, 249–253 (1996)

7.188 S. Einecke, C. Schulz, V. Sick: Measurement of temperature, fuel concentration and equivalence ratio fields using tracer LIF in IC engine combustion, Appl. Phys. B **71**, 717–723 (2000)

7.189 J.D. Koch, R.K. Hanson: Temperature and excitation wavelength dependencies of 3-pentanone absorption and fluorescence for PLIF applications, Appl. Phys. B **76**, 319–324 (2003)

7.190 W. Koban, J.D. Koch, R.K. Hanson, C. Schulz: Absorption and fluorescence of toluene vapor at elevated temperatures, Phys. Chem. Chem. Phys. **6**, 2940–2945 (2004)

7.191 M.C. Thurber, B.J. Kirby, R.K. Hanson: Instantaneous imaging of temperature and mixture fraction with dual-wavelength acetone PLIF, AIAA Paper **98-0397** (1998)

7.192 N.P. Tait, D.A. Greenhalgh: PLIF imaging of fuel fraction in practical devices and LII imaging of soot, Ber. Bunsenges. Phys. Chem. **97**, 1619 (1993)

7.193 A.C. Eckbreth: CARS thermometry in practical combustors, Combust. Flame **39**, 133–147 (1980)

7.194 R.J. Hall: Intensity convolutions of CARS spectra, Opt. Commun. **52**, 360–366 (1985)

7.195 R.E. Teets: Accurate convolutions of coherent anti-Stokes Raman spectra, Opt. Lett. **9**, 226–228 (1984)

7.196 M. Péalat, M. Lefebvre, J.-P.E. Taran, P.L. Kelley: Sensitivity of quantitative vibrational coherent anti-Stokes Raman spectroscopy to saturation and stark shifts, Phys. Rev. **38**, 1948–1965 (1988)

7.197 M.L. Koszykowski, R.L. Farrow, R.E. Palmer: Calculation of collisionally narrowed coherent anti-Stokes Raman spectroscopy spectra, Opt. Lett. **10**, 478–480 (1985)

7.198 S. Kröll, M. Aldén, T. Berglind, R.J. Hall: Noise characteristics of single shot broadband Raman-resonant CARS with single- and multimode lasers, Appl. Opt. **26**, 1068–1073 (1987)

7.199 P.R. Regnier, J.P. Taran: On the possibility of measuring gas concentrations by stimulated anti-Stokes scattering, Appl. Phys. Lett. **23**, 240–242 (1973)

7.200 W.A. England, J.M. Milne, S.N. Jenny, D.A. Greenhalgh: Application of CARS to an operating chemical reactor, Appl. Spectrosc. **38**, 867–876 (1984)

7.201 A.C. Eckbreth, R.J. Hall: CARS concentration sensitivity with and without nonresonant background suppression, Combust. Sci. Technol. **25**, 175–192 (1981)

7.202 M. Schenk, T. Seeger, A. Leipertz: Time-resolved CO_2 thermometry for pressures as great as 5 MPa by use of pure rotational coherent anti-Stokes Raman scattering, Appl. Opt. **44**, 6526–6536 (2005)

7.203 P.-E. Bengtsson, L. Martinsson, M. Aldén: Dual-broadband rotational CARS measurements in an IC engine, Proc. Combust. Inst. **25**, 1735–1742 (1994)

7.204 P.-E. Bengtsson, M. Aldén: Soot-visualization strategies using laser techniques, Appl. Phys. B **60**, 51–59 (1995)

7.205 P.R.N. Childs, J.R. Greenwood, C.A. Long: Review of temperature measurement, Rev. Sci. Instrum. **71**(8), 2959–2978 (2000)

7.206 P. Neubert: Device for indicating the temperature distribution of hot bodies, US Patent No. 2071471 (1937)

7.207 F. Urbach, N.R. Nail, D. Pearlman: The observation of temperature distributions and of thermal radiation by means of non-linear phosphors, J. Opt. Soc. Am. **39**(12), 1011–1019 (1949)

7.208 L.C. Bradley: A temperature-sensitive phosphor used to measure surface temperatures in aerodynamics, Rev. Sci. Instrum. **24**(3), 219–220 (1953)

7.209 G.M. Buck: Surface temperature/heat transfer measurement using a quantitative phosphor thermography system. In: *29th Aerospace Sciences Meeting AIAA Paper 91-0064* (1991)

7.210 J.H. Monaweck, W.J. McGonnagle: Thermal testing of reactor fuel elements. In: *Proc. Symp. Nondestructive Test Field Nuclear Energy*, ed. by D.R. Green (ASTM, West Conshohocken 1957), ASTM Special Tech. Publ. 352-357

7.211 S.W. Allison, M.R. Cates, D.L. Beshears: A survey of thermally sensitive phosphors for pressure sensitive paint applications, Proc. ISA 46th Annual, ISA **397**, 29–37 (2000)

7.212 J.P. Feist, A.L. Heyes: The characterization of Y2O2S:Sm powder as a thermographic phosphor for high temperature applications, Meas. Sci. Technol. **11**, 942–947 (2000)

7.213 Y. Le Sant, J.L. Edy: Phosphor thermography technique in hypersonic wind tunnels: First results. In: *15th ICIASF* (1993)

7.214 S.W. Allison, G.T. Gillies: Remote thermometry with thermographic phosphors: Instrumentation and application, Rev. Sci. Instrum. **68**(7), 2615–2650 (1997)

7.215 T. Liu, B.T. Campbell, J.P. Sullivan: Thermal paints for shock/boundary layer interaction in inlet flows, AIAA Paper **92-3626** (1992)

7.216 B.T. Campbell, T. Liu, J.P. Sullivan: Temperature measurement using fluorescent molecules. In: *6th International Symposium on Application of Laser Technique* (1992)

7.217 B.T. Campbell, T. Liu, J.P. Sullivan: Temperature sensitive fluorescent paint systems, AIAA Paper **94-2483** (1994)

7.218 B.G. Mc. Lachlan, J.H. Bell, J. Gallery, M. Gouterman, J. Callis: Boundary layer transition detection by luminescent imaging, AIAA Paper **93-0177** (1993)

7.219 K. Asai, H. Kanda, T. Kunimasu, T. Liu, J.P. Sullivan: Detection of boundary-layer transition in a cryogenic wind tunnel by using luminescent paint, AIAA Paper **96-2185** (1996)

7.220 T.G. Popernack, L.R. Owens, M.P. Hamner, M.J. Morris: Application of temperature sensitive paint for detection of boundary layer transition, ICIASF Proc. (1997)

7.221 U. Fey, K. De Groot, Y. Le Sant: Thermography as a tool in wind tunnel testing, DLR Rep. **IB 224-2007A09** (2007)

7.222 J.P. Hubner, B.F. Carroll, K.S. Schanze: Heat transfer measurements in hypersonic flow using luminescent coating techniques, AIAA Paper **2002-0741** (2002)

7.223 I. Le Sant: Private communication, ONERA (2006)

7.224 T. Liu, J.P. Sullivan: *Pressure and Temperature Sensitive Paint* (Springer, Berlin, Heidelberg 2005)

7.225 U. Fey, R.H. Engler, Y. Egami, y. Iijima, K. Asai, U. Jansen, J. Quest: ETW. In: *20th Int. Congress on Instrumentation in Aerospace Simulation Facilities, ICIASF 2003 Record* (2003) pp. 77–88

7.226 W. Banks, C.P. van Dam, H.J. Shiu, G.M. Miller: Visualization of in-flight flow phenomena using infrared thermography. In: *9th Int. Symposium on Flow Visualization* (2000) pp. 1–11, 24 (NASA TM-2000-209027)

7.227 J. Sullivan, J.W. Gregory, C.Y. Huang, H. Sakaue: Flow visualization applications of luminescent paints. In: *10th Int. Symposium on Flow Visualization* (2002)

7.228 Y. Egami, U. Fey, C. Klein, M. Sitzmann, J. Wild: Transition detection on high-lift devices in DNW-KKK by means of temperature-sensitive paint. In: *12th Int. Symp. on Flow Visual.* (2006)

7.229 K. De Groot: Private communication, DLR (2006)

7.230 E. Reshotko: *Laminar Flow Control, Special course on stability and transition of laminar flows*, AGARD Rep., Vol. 709 (VKI, Brussels 1984)

7.231 W.S. Saric, H.L. Reed: Toward practical laminar flow control – Remaining challenges. In: *34th AIAA Fluid Dynamics Conference* (2004)

7.232 U. Fey, Y. Egami, C. Klein: Using cryoTSP as a tool for transition detection and instability examination at high Reynolds numbers. In: *Notes on Numerical Fluid Mechanics and Multidisciplinary Design*, Vol. 96, ed. by C. Tropea, S. Jarkirlic, H.-J. Heinemann, R. Henke, H. Hönlinger (Springer, Berlin, Heidelberg 2007)

7.233 Y. Iijima, Y. Egami, A. Nishizawa, K. Asai, U. Fey, R.H. Engler: Optimization of temperature-sensitive paint formulation for large-scale cryogenic wind tunnels. In: *20th Int. Congress on Instrumentation in*

Aerospace Simulation Facilities (ICIASF), ICIASF 2003 Record (2003) pp. 70–77

7.234 U. Fey, Y. Egami, R.H. Engler: High Reynolds number transition detection by means of temperature sensitive paint. In: *44th AIAA Aerospace Sciences Meeting and Exhibit AIAA Paper 2006-514* (2006)

7.235 U. Fey, Y. Egami, R.H. Engler: Using cryoTSP as a tool for transition detection. In: *Int. Council of the Aeronautical Sciences* (2006)

7.236 Y. Egami, U. Fey, J. Quest: Development of new two-component TSP for cryogenic testing. In: *45th Aerospace Sciences Meeting and Exhibit* (2007)

8. Force and Moment Measurement

Measurement of steady and fluctuating forces acting on a body in a flow is one of the main tasks in wind tunnel experiments. In aerodynamic testing, strain gauge balances will usually be applied for this task as, particularly in the past, the main focus was directed on the measurement of steady forces. In many applications, however, balances based on piezoelectric multicomponent force transducers are a recommended alternative solution. Contrary to conventional strain gauge balances, a piezo balance features high rigidity and low interference between the individual force components. High rigidity leads to very high natural frequencies of the balance itself, which is a prerequisite for applications in unsteady aerodynamics, particularly in aeroelasticity. Moreover for measurement of extremely small fluctuations, the possibility exists to exploit the full resolution independently from the preload.

Concerning the measurement of small, steady forces, the application of piezo balances is restricted due to a drift of the signal at constant load. However, this problem is not as critical as generally believed since simple corrections are possible.

The aim of this chapter is to give an impression of the possibilities, advantages and limitations offered by the use of piezoelectric balances. Several types of external balances are discussed for wall-mounted models, which can be suspended one-sided or twin-sided. Additionally an internal sting balance is described, which is usually applied inside the model. Reports are given on selected measurements performed in very different wind tunnels, ranging from low-speed to transonic, from short- to continuous running time and encompassing cryogenic and high pressure principles. The latter indicates that special versions of our piezo balances were applied down to temperatures of $-150\,°C$ and at pressures of up to 100 bar.

The projects span from a wing/engine combination in a low-speed wind tunnel to flutter tests with a swept-wing performed in a transonic wind tunnel, and include bluff bodies in a high pressure and cryogenic wind tunnel, as well. These tests serve as examples for discussing the fundamental aspects that are essential in developing and applying piezo balances. The principle differences between strain gauge balances and piezo balances will also be discussed.

8.1	**Steady and Quasi-Steady Measurement**	564
	8.1.1 Basics	564
	8.1.2 Basic Terms of Balance Metrology	568
	8.1.3 Mounting Variations	570
	8.1.4 Strain Gauges	572
	8.1.5 Wiring of Wheatstone Bridges	573
	8.1.6 Compensation of Thermal Effects	575
	8.1.7 Compensation of Sensitivity Shift	576
	8.1.8 Strain Gauge Selection	578
	8.1.9 Strain Gauge Application	578
	8.1.10 Materials	579
	8.1.11 Single-Force Load Cells	580
	8.1.12 Multicomponent Load Measurement	582
	8.1.13 Internal Balances	582
	8.1.14 External Balances	586
	8.1.15 Calibration	590
8.2	**Force and Moment Measurements in Aerodynamics and Aeroelasticity Using Piezoelectric Transducers**	596
	8.2.1 Basic Aspects of the Piezoelectric Force-Measuring Technique	598
	8.2.2 Typical Properties	602
	8.2.3 Examples of Application	607
	8.2.4 Conclusions	614
References		615

8.1 Steady and Quasi-Steady Measurement

The key measurement system in a wind tunnel is the multicomponent force and moment measurement instrumentation. More than 70% of the tests in a wind tunnel require some kind of force measurements. Historically the instruments were purely mechanical and their mechanism resembled balances for weighing; hence the use of the term *balance* today. Today these balances are often based on transducers or are constructed out of a single piece of metal, on which strain gauges are applied. All balances must have a minimum of one sensing element for every component to be measured. The strain sensor usually is a resistance foil strain gauge, but semiconductor gauges are also used. Illustrated in Fig. 8.1 is one of the first wind-tunnel balances built by Gustav Eiffel in 1907. The man on the upper gallery was responsible for balancing the lift generated by the airfoil in the tunnel below and simultaneously recording its associated lift.

8.1.1 Basics

Balance Types
Balance types are distinguished by the number of force/moment components that are measured simultaneously – one to six are possible – and the location at which they are placed. If they are placed inside the model they are referred to as *internal balances* and if they are located outside the model or the wind tunnel, they are referred to as *external balances*.

Fig. 8.1 Gustav Eiffel's wind tunnel with lift balance

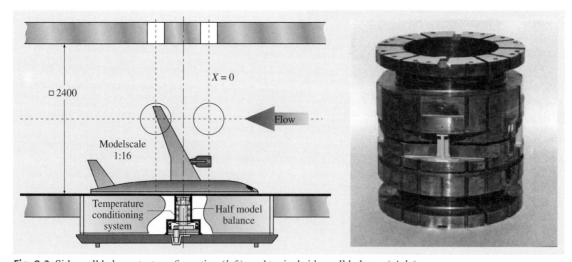

Fig. 8.2 Side-wall balance test configuration (*left*), and typical side-wall balance (*right*)

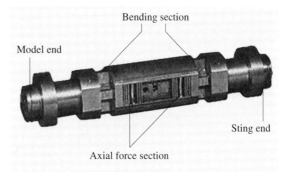

Fig. 8.4 Typical internal balance

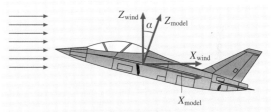

Fig. 8.5 Various possible coordinate systems for a wind tunnel and model

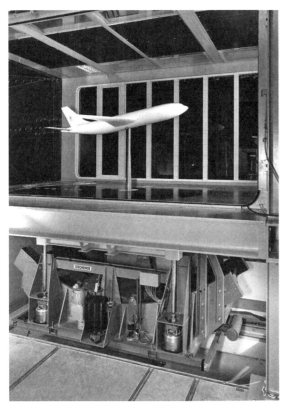

Fig. 8.3 External balance and support

External Balances

Two types of external balances exist. The first is the one-piece external balance (Fig. 8.3), which is constructed from one single piece of material and is equipped with strain gauges. Such balances are also referred to as *sidewall balances* (Fig. 8.2) as used in half-model tests.

The second type of external balance comprises single force transducers which are connected by a framework. Such a design can be built very stiff but needs more space compared to the one-piece design. However, there is usually plenty of space available around the wind tunnel for such a balance, and so the construction can be optimized with respect to measurement requirements, such as optimized sensitivity, stiffness and decoupling of load interactions.

Internal Balances

There is limited space inside the model itself, so internal balances have to be relatively small in comparison to external balances. There are two main types of internal balances. The monolithic type (Fig. 8.4), in which the balance body consists of a single piece of material, is designed in such a way that certain areas are primarily stressed by the applied loads. The other internal balance type uses small transducers that are orientated with their sensing axes in the direction of the applied loads. Such a balance is combined into a solid structure. A balance measures the total model loads and therefore is placed at the center of gravity of the model and is generally constructed from one solid piece of material.

Loads

In this chapter the word *load* will be used to describe both the applied forces and moments. The task of a balance is to measure the aerodynamic loads, which act on the model or on components of the model itself. In total there are six different components of aerodynamic loads, three forces in the direction of the coordinate axes and the moments around these axes (Fig. 8.5). These components are measured in a certain coordinate system, which can be either fixed to the model or to the wind tunnel. For the measurement of loads on model parts such as rudders, flaps and missiles, normally fewer than six components are required.

Definition of Coordinate Systems

One possible coordinate system is fixed to the wind tunnel – the wind axis system – and is aligned to the

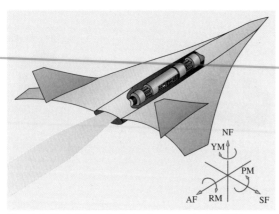

Fig. 8.6 Definition of wind axis system in the USA

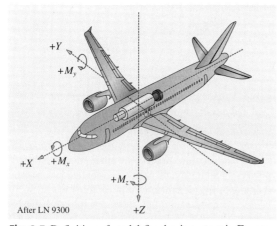

Fig. 8.7 Definition of model-fixed axis system in Europe

Table 8.1 Definition of positive axis direction

Balance axis system	Name of component	European	USA
		Positive direction	Positive direction
X	Axial force	In flight	In wind
Y	Side force	To starboard	To starboard
Z	Normal force	Down	Up
M_x	Rolling moment	Roll to starboard	Roll to starboard
M_y	Pitching moment	Turn up	Turn up
M_z	Yawing moment	Turn to starboard	Turn to starboard

The European axis system is consistent with the right-hand system and the definition is based on a standard given by DIN-EN 9300 or ISO 1151. A balance which always stays fixed in the tunnel, and relative to the wind axis system, always gives the pure aerodynamic loads on the model.

In the case of the model-fixed axis system, the balance does not measure the aerodynamic loads directly. The loads acting on the model are given by the balance and the pure aerodynamic loads must then be calculated from these components by using the correct yaw and pitch angles. The difference between the American and European definitions of the positive direction remains the same in this case.

Specification of Balance Load Ranges

Before a balance can be designed, the specifications of the load ranges and the available space for the balance are required. This is a challenging step prior to the design of a balance since cost and accuracy considerations must be made long before the first tests are performed.

The maximum combined loads specify the load ranges for the balance design. The maximum design loads of a balance are defined in various manners. For example, if several loads act simultaneously, then the load range must be specified as the *maximum combined load*. If the maximum load acts alone, then the load range is defined as the *maximum single load*. Usually such single loads do not exist in wind-tunnel tests and combined loads must be expected. Such combined loads stress the balance in a much more complicated manner and therefore deserve very careful attention. The stress analysis of the balance has to take into account this situation. Furthermore, the combination of two loads usually re-

main flow direction. The lift force is generally defined as the force on the model acting vertically to the main flow direction whereas the drag is defined as the force acting in the main flow direction. This definition is common all over the world. However, the definition of the positive direction of the forces is not universal. Whereas lift (normal force) and drag (axial force) are defined positive in the USA (Fig. 8.6), in Europe (Fig. 8.7) weight and thrust are defined as positive in the wind axis system.

To form a right-hand axis system, the side force in the USA has to be positive in the starboard direction. The definitions of the positive moments do not follow the sign rules of the right-hand system. The pitching moment is defined as positive turning right around the y-axis, but yawing and rolling moments are defined positive turning left around their corresponding axes. This makes this system inconsistent in a mathematical sense.

Table 8.2 Determination of maximum combined loads for an external balance

Test type	q	A	Wind axis system C_x	C_y	$C\ldots$	Drag	Side force	$F\ldots$, $M\ldots$	max α, β, γ	Model axis system F_x	F_y	F_z	M_x	M_y	M_z
	(Pa)	(m^2)				(N)	(N)	(N), (N m)	(deg)	(N)	(N)	(N)	(N m)	(N m)	(N m)
Full model															
Half model															
etc.															

quires that the balance carries higher loads. To determine which balance can be used for a given setup, the balance manufacturer provides the test engineer with loading diagrams that define the maximum load combinations for various available models.

In the case of external balances the available space is usually not a limiting factor. External balances have been used for several decades in wind tunnels and therefore specifications of the design load ranges are orientated more to the capabilities of the wind tunnel itself and the associated model setups such as half-model and full-model tests, or the possibility of aircraft, car and building testing. For the design of an external balance, load ranges of the principle balance configuration must first be defined. Two different options are possible.

The first option is to mount the turntable inside the weighbridge. In this case the balance stays in the wind axis system and therefore the balance always measures the wind loads. The disadvantage of this option is that the whole turntable mechanism has to be mounted on the metric side of the balance such that the balance is preloaded by the weight of this mechanism.

The second option is to mount the whole balance on a turntable. In this case some components stay in the wind axis system and some stay in the model axis system, so a calculation of the aerodynamic loads from the balance loads is necessary. For example, in the case of a full model setup, as shown in Fig. 8.9, the balance will always measure the aerodynamic loads when the angle of attack changes. In a half-model setup, as shown in Fig. 8.11, the balance will move with the model when the angle of attack changes and will measure the loads in the model axis system.

This makes the determination of the balance load ranges rather difficult. Therefore it is useful to fill out a table where the maximum loads for the different test setups are first calculated in the wind axis system. In order to do this some assumptions of the aerodynamic coefficients and the model size must be made. By assuming values for the maximum angle of attack or yaw angle, the maximum loads in the model axis system can be calculated.

From such a table the maximum of each component can be taken as the maximum load for the balance. Naturally this leads to a balance with rather high load ranges and for some cases the load range could be too high to measure with high resolution. However, if a certain test requires a high resolution and this test is mostly performed in the tunnel, it is generally better to accept the lower load range to obtain higher resolution and higher accuracy. These considerations must be made for all components to ensure a balance with the best

Table 8.3 Determination of maximum combined load for an internal balance

Test type	q	A	Wind axis system C_x	C_y	$C\ldots$	Drag	Side force	$F\ldots$, $M\ldots$	max α, β, γ	Model axis system F_x	F_y	F_z	M_x	M_y	M_z
	(Pa)	(m^2)				(N)	(N)	(N), (N m)	(deg)	(N)	(N)	(N)	(N m)	(N m)	(N m)
Transporter															
Landing															
Cruise															
Combat															
Other															

fit of the load range for the normal operation of the tunnel.

For an internal balance the available space is a major concern. The available space for balances is restricted by the diameter of the fuselage. As transport aircraft become larger, the scale for models also become larger, complicating matters since the cross sections of wind tunnels have not grown at the same rate as the aircraft. As a result the available diameter for internal balances has become smaller.

For combat aircraft the loads in relation to the available space inside the model are very high. This is the case since wind-tunnel tests for this type of aircraft are mostly performed in pressurized wind tunnels to obtain the correct Mach number. The high static pressure leads to relatively high loads on the model in comparison to tests at atmospheric pressure.

These two effects lead to higher specific loads on the balance, making it much more complicated to develop a balance with high accuracy. Therefore the definition of the load ranges and the definition of the available space must be performed very carefully to ensure a high-quality balance design.

The specification for an internal balance should therefore be made related to the model and the loads on this model during the tests, and not on the tunnel capabilities themselves. If these specifications are not carefully performed, the internal balances will provide insufficient sensitivity and accuracy for the tests.

Unlike an external balance, where some of the components are always fixed to the wind axis system, the internal balance always measures the loads in the model axis system and therefore the lift and drag are always a combination of the axial and normal forces. Because of this, the angle of attack at which maximum loading occurs must be taken into account. For combat aircraft maximum loads can act at an angle of attack as high as 40°, whereas the maximum forces for a clean transport aircraft occur in and around an angle of attack of 15°.

Because an internal balance is mounted inside the model and does not change in orientation relative to the model itself, no maximum single load occurs. For different test setups (transporter, combat, high lift, cruise condition, etc.) different maximum combined loads occur, so it is also useful to prepare a table to determine the maximum combined loads on the balance.

Maximum Combined Load:
Maximum Single Load

The definition of the maximum load can differ between a combined load acting simultaneously, termed the maximum combined load, and a maximum load acting alone, defined as the maximum single load. The maximum single load forms a load trapeze, which does not automatically cover the test requirements (Fig. 8.8). The maximum combined load specifies the load range for the balance design. Usually single loads do not occur in wind-tunnel testing. Combined loads stress the balance in a much more complicated way, so the stress analysis of the balance has to take this situation into account. With the combination of only two loads the balance usually can carry higher loads than defined by the maximum combined loading. To estimate which load can be carried for a given situation the balance manufacturer provides a loading diagram (load rhombus) which allows the test engineer to judge whether the load combination of the planned test is within the limits of the balance.

Which one of the aforementioned loads is used as the *full-scale load* depends on the balance manufacturer's philosophy. The user should specify the definition of the full-scale load, because by definition the relative uncertainty of the balance can vary by a factor of two without any difference in the absolute uncertainty. This is discussed in detail in the following section.

8.1.2 Basic Terms of Balance Metrology

In this section brief definitions are given of the terms used. Most of these terms are defined in some international standard such as the *Guide of uncertainties in measurement (GUM)* [8.1] or in the case of the United States, the American Institute of Aeronautics and Astronautics (AIAA) standards documented in *Assessment of Experimental Uncertainty with Application to Wind*

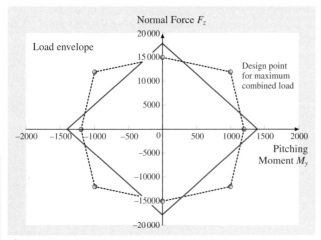

Fig. 8.8 Load rhombus and load trapeze

Tunnel Testing [8.2] and the *Calibration and Use of Internal Strain Gauge Balances with Application to Wind Tunnel Testing* [8.3].

Full-Scale Load or Output. Full-scale load or output can be interpreted in several ways. The most obvious definition is the maximum combined load or the signal corresponding to this load. It is also possible to define the full-scale load or output as the maximum of the single loads. The maximum single loads are usually much higher than the maximum combined loads, so uncertainties related to these values are much smaller than those related to the maximum combined loads. This has to be taken into account if balances from different manufactures are compared. However, for a particular wind-tunnel test the maximum of the single loads seldom occur and the use of maximum combined loads or signals as the full-scale reference yields much more realistic information about the accuracy of the balance.

Systematic Errors or Bias of an Instrument. Systematic errors are repeatable errors that occur at every measurement. If the cause of a systematic error can be detected, it can be eliminated by calibration or compensation. If the source of a systematic error cannot be detected under repeatable conditions it is then simply accepted as the difference between the measured value and the true value.

Random Errors. Random errors are defined as the difference between the mean value of an infinite number of measurements under repeatable conditions and a single measurement. The value of the random error is equal to the difference between the total error minus the systematic error.

Resolution. Resolution is the smallest value that can be detected by a balance. Similarly it can be defined as the smallest detectable difference between two loads. The maximum resolution should be in the range of 0.005% of the full-scale load. For a normal wire strain gauge, resolution in this range is not a problem and the limits of the resolution are given by the measuring equipment. Good equipment today offers a resolution of less than 0.0003% of the full-scale load. Effectively the more money one spends on the equipment, the better the obtainable resolution.

Repeatability. If a certain balance loading is repeated after a certain amount of time, the repeatability is measured as the difference in the two signals. This is a very important characteristic for a balance, since many tests in a wind tunnel compare different model configurations. Most often wind-tunnel tests cannot exactly reproduce conditions found in reality. This means that, if the test itself does not represent reality yet the engineers attempt to compare the difference between two designs under wind-tunnel test conditions, the extrapolation of results to reality can be very challenging. Repeatability can be as good as the resolution but is usually worse. The repeatability of a good balance is in the range of 0.005% of the full-scale load. Repeatability also depends on time, so it is distinguished between short-term repeatability, for example between two wind-tunnel runs, and long-term repeatability, which is for example the difference between two complete wind-tunnel installations. Both short- and long-term repeatability are of equal importance since the loads on a new aircraft as well as the loads on an older aircraft design must be measured with the same accuracy as the differences between the new model configurations. The main challenge is often the discrepancy in the flow parameters between the model and reality, especially the Reynolds number. Most of the uncertainty in the force prediction on real aircraft is then caused by this Reynolds-number gap; therefore relative measurements to the model of an older aircraft design are used to reduce this level of uncertainty.

Interactions and Interference. One of the major systematic errors of a wind-tunnel balance is caused by the interactions or interference of the model with surrounding components. For example, the sensor measuring the axial force might also pick up the loads from other components. This additional signal can be several percent of the measured axial force signal. As mentioned above, such a systematic error can be corrected through proper calibration.

Accuracy. The term accuracy is very broad and generally describes the agreement between the measured data, in this case the wind-tunnel test data, and the true value that will be measured in the flight test after wind-tunnel testing is completed. There are many influencing factors, such as the measurement of static and dynamic pressure, wall and sting interactions, the precision of the model geometry, flow quality, angle of attack and balance uncertainty. Since there are so many potential errors, it becomes difficult to specify the accuracy for a wind-tunnel test. Balance uncertainty alone must be taken into account through balance calibration.

Absolute Error and Uncertainty. The absolute error is defined as the difference between the real load acting on the model (the true value) and the load detected by the balance. Normally the true value is unknown; therefore in practice a *conventional true value* is used. Besides all the error sources in the instrumentation and the balance, the error of the load measurement is strongly influenced by the balance calibration, since in this process the relation between the signals of the balance and the true loads are determined. According to the *Guide to the Expression of Uncertainties in Measurement* [8.1], often abbreviated as GUM, the absolute error is defined as the uncertainty of the measurement and can be given as the experimental standard deviation of the whole process. For a given wind tunnel test the uncertainty of the measured data is not only dependent on the measurement of the loads, but also influenced by the measurement of the angle of attack, the dynamic pressure, the model geometry, the flow quality and all the instrumentation which are used to determine the aerodynamic performance and associated derivatives. Regarding the desired precision of the test result, minimum requirements for the load measurement can be estimated from flight test data of an existing aircraft.

Accuracy Requirements. The requirement placed on accuracy is a function of the price of the balance. If a balance is specifically designed and built for one particular set of wind-tunnel tests, it should be only as accurate as those tests require. This is however not usually the case. The high cost of balance construction usually restricts a given tunnel to a small number of balances, which should ideally cover the range of test capabilities for that particular tunnel. An estimate of the maximum required accuracy for the development of transport aircraft [8.4, 5] can be made by defining which differences between a developed aircraft and the new aircraft are to be investigated. The usual outcome of such an accuracy requirement study is that the balance must be able to measure a difference of 1–2 drag counts, which is equal to an accuracy requirement of the balance of better than 0.07% to 0.1% of the full-scale load. This value is a global one yet should be specified for every new test setup.

8.1.3 Mounting Variations

When using external balances, there are many different possibilities for mounting the models. For example, a full model aircraft can be supported by a three-sting arrangement (Fig. 8.9) or a central sting (Fig. 8.10). For cars and buildings more than three supports are sometimes necessary. The external balance support system should provide for as many arrangements as possible.

The struts on the wing support the model at the quarter-chord positions and carry most of the load, except for the pitching moment, which is balanced by the strut at the tail of the model. All links to the model are jointed around the y-axis, so that a vertical movement of the tail strut allows for easy variation of the angle of attack. Yaw angles are set by moving the whole setup on the turntable.

Using the central-sting mounting the model is fixed inside the fuselage in all directions. The adjustment of the angle of attack is done through y-axis mechanism inside the fuselage. All loads must be carried by this

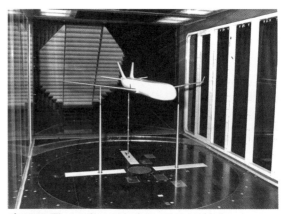

Fig. 8.9 Three-sting mounting on external balance

Fig. 8.10 Centre-sting mounting on external balance

Fig. 8.11 Semi-span models on external balance

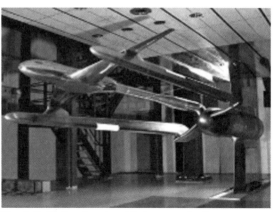

Fig. 8.14 Twin-sting rig with dummy tail sting

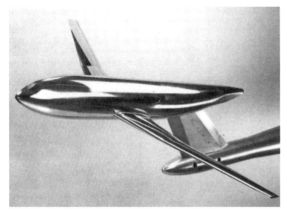

Fig. 8.12 Tail sting with fin attachment on lower side

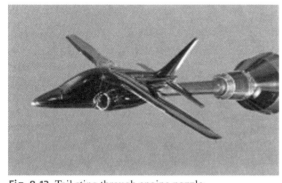

Fig. 8.13 Tail sting through engine nozzle

support and therefore it must be very rigid to minimize dynamic movements of the model during testing.

Semi-span models (Fig. 8.11) are used to increase the effective Reynolds number of the tests by increasing the geometry of the model. Besides the higher Reynolds number, the lager model size makes models with variable flaps and slats much easier to construct, thus semi-span models are often used for testing of take-off and landing configurations.

The most common setup using an internal balance is the back-sting arrangement (Fig. 8.13), in which the sting is attached through the after body of the fuselage. In some cases the vertical fin (Fig. 8.12) is also used to support the model.

For tests that require free flow around the after body two-sting setups (Fig. 8.14) can be used. In such cases one balance is needed inside each sting. To determine the influence of the tail sting two measurements are performed, one with a dummy tail sting and one without the dummy tail sting in place.

Today the support of the model by wires is seldom used anymore (Fig. 8.15). In such cases the balance must be an overhead external balance and the model hangs from the balance through wires. To keep the model stabile in the tunnel, the system must be preloaded by weights (Sp), which are usually dampened in a water basin under the test section. The advantage of such a model support is the very low interference on the flow around the model.

In Fig. 8.15 the model hangs upside down in the tunnel, minimizing the preload. This is beneficial since the balance cannot measure loads smaller than the preload. Sometimes modern wind tunnels test the model in an upside-down position in order to preload the balance in the lift direction. This way the lift generated by the model contributes to the preload such that the balance signal does not pass through zero as in a normal setup. In such a way the additional nonlinearities associated with the zero-load regime can be avoided.

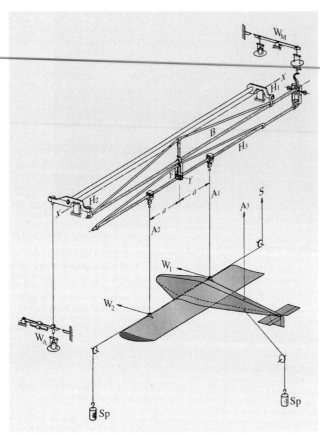

Fig. 8.15 Wire-supported model on overhead external balance

Mounting Interference

The aerodynamic loads on the model itself are always affected by the presence of the model mounts. The mounting loads themselves are subtracted from the model loads by performing tests without the model in place. The second effect to be considered is the influence of the mounts on the flow field around the model and the influence of the model on the flow field around the mounts. A complete separation of the effects is not possible. Therefore it is not possible to eliminate the influence of the model–mount interference completely. Several methods for the compensation of model–mount inference are described in [8.6].

8.1.4 Strain Gauges

Strain Gauge Fundamentals

The basic technique to measure forces with any kind of wind-tunnel balance is the measurement of the strain on an elastic spring that is deformed by the aerodynamic loads acting on the wind-tunnel model. In this chapter the fundamentals of strain measurement and strain sensors are described. For wind-tunnel balances two major types of strain sensors are used. The most commonly used is the wire strain gauge sensor. Also of importance is the semiconductor strain gauge.

The wire strain gauge is based on an electromechanical effect developed by *W. Thomson (Lord Kelvin)* in 1856. Thomson measured the electrical resistance of a metal wire and found that it could be correlated to the strain in the wire while stressed.

This effect was subsequently used by *E. Simmons* (Caltech) and *A.C. Ruge* (MIT) in 1938 in the development of the wire strain gauge. Simmons was the first to build a force transducer based on the wire strain gauge technique while Ruge used his wire strain gauges to perform experimental stress analyses. Ruge's strain gauge was very successful since it was cheap and easy to handle. Industry needed many of them such that in 1952 a technique was patented to produce the foil strain gauge in great numbers. No longer was a wire glued onto a carrier foil. Rather a thin metal foil was glued onto the carrier and the contour of the wire was etched out of the metal foil by a photochemical process. This technique is still used today to produce the foil strain gauge sensors, as it produces very precise sensors with high resolution at a low price [8.7, 8].

The physical principle of a wire strain gauge is that a change in electrical resistance is produced when a strain is applied to the gauge. The electrical resistance of a wire can be written as:

$$R = \frac{\rho l}{A}, \tag{8.1}$$

where R is the resistance of wire, l the length of the gauge grid, A the cross section of the wire and ρ the specific electric resistance.

The specific electric resistance is given as:

$$\rho = \frac{2m v_0 A l}{N_0 e^2 \lambda}, \tag{8.2}$$

where m is the mass of an electron, v_0 the velocity of the electrons, N_0 the number of free electrons, e the charge of an electron and λ the free wavelength of the electrons. With the above equation for the specific electric resistance, the resistance of a wire can be formulated as:

$$R = \frac{2m v_0 l^2}{N_0 e^2 \lambda}. \tag{8.3}$$

The relative change of the resistance of the wire is given by the following partial differential equation:

$$\frac{dR}{R} = \underbrace{\frac{2\,dl}{l}}_{\text{Influence of size}} + \underbrace{\frac{dv_0}{v_0} + \frac{dm}{m} - \frac{dN_0}{N_0} - \frac{d\lambda}{\lambda} - \frac{2\,de}{e}}_{\text{Piezoresistive effect}}. \quad (8.4)$$

The relative change of the length is related to the strain ($\varepsilon = dl/l$). With subsequent substitution into (8.4):

$$\frac{dR}{R} = 2\varepsilon + \frac{dv_0}{v_0} + \frac{dm}{m} - \frac{dN_0}{N_0} - \frac{d\lambda}{\lambda} - \frac{2\,de}{e}. \quad (8.5)$$

The relative changes in the mass (dm/m) as well as the changes in the charge of the electrons (de/e) are equal to zero such that the sensitivity (k) of a strain gauge is defined as the relative change of the resistance divided by the strain. Then the sensitivity of a wire is given by the following equation:

$$k = \frac{dR}{R\varepsilon} = 2 + \frac{1}{\varepsilon}\left(\frac{dv_0}{v_0} - \frac{dN_0}{N_0} - \frac{d\lambda}{\lambda}\right). \quad (8.6)$$

From the above equation it can be seen that the sensitivity of a wire is equal to two plus a term which itself is dependent on the strain. From real strain gauges it is known that the gauge factor is around two and constant. So for an ideal grid material the second term of (8.6) must be equal to zero such that the gauge factor remains at a value of two. For the application of strain gauge balances such gauges with a gauge factor of two are used:

$$\frac{dR}{R} = k\varepsilon. \quad (8.7)$$

8.1.5 Wiring of Wheatstone Bridges

The relative changes of the electric resistance of a strain gauge are very small. For example a signal from a 1 mV/V strain gauge with a resistance of 350 Ω would be on the order of 0.0875 Ω. Such a signal must then be measured with an accuracy of 8.75 µΩ or better. For direct measurements this would require a resistance measurement instrument with a resolution of more than 40×10^6 parts. Such an instrument even today does not exist. In 1843, *Charles Wheatstone* developed a method to directly measure the small relative changes of an electrical resistance, the so-called *Wheatstone bridge*. Four resistors of the same nominal resistance value are arranged as shown in Fig. 8.16. The layout consists of two parallel pairs of resistors in series.

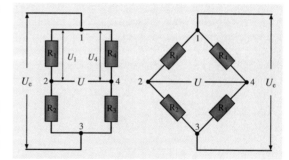

Fig. 8.16 Wheatstone bridge circuit

There is no common rule for the numbers of the resistors in Fig. 8.16. However, all equations that follow in this section are based on the numbering system found in the above figure and are valid if the following criteria are fulfilled.

1. The order of the numbers are arranged either in the clockwise or counterclockwise direction.
2. The excitation voltage is applied between points 1 and 3 and the output signal (U) is measured between points 2 and 4.

When excitation voltage is applied at points 1 and 3 the resistors R_1 and R_2 as well as resistors R_3 and R_4 each form a voltage divider. The voltages U_1 and U_4 can be described by the following equations:

$$U_1 = \frac{R_1}{R_1 + R_2}U_e \quad \text{and} \quad U_4 = \frac{R_4}{R_3 + R_4}U_e. \quad (8.8)$$

The difference between these two partial voltages is the voltage U, which is given by

$$U = U_1 - U_4 = \left(\frac{R_1}{R_1 + R_2} - \frac{R_4}{R_3 + R_4}\right)U_e. \quad (8.9)$$

The relative change of the voltage U to the excitation voltage U_e is

$$\frac{U}{U_e} = \left(\frac{R_1}{R_1 + R_2} - \frac{R_4}{R_3 + R_4}\right), \quad (8.10)$$

$$\frac{U}{U_e} = \frac{R_1 R_3 - R_2 R_4}{(R_1 + R_2)(R_3 + R_4)}. \quad (8.11)$$

It can be seen that the ratio of these two voltages, U/U_e, is proportional to the ratios of the resistances, but $U = 0$ if:

1. every resistor has the same value, i.e., $R_1 = R_2 = R_3 = R_4$,
2. the ratio $R_1/R_2 = R_4/R_3$.

If these criteria are satisfied then the bridge is termed *balanced*.

In the case of strain gauge measurements the resistors R_1 to R_4 are the strain gauges themselves, including the resistance changes of the gauges by the strain. For this case the resistors R_1 to R_4 in (8.11) are replaced by the initial resistance plus a resistance change ΔR_1 to ΔR_4 such that (8.11) can be written

$$\frac{U}{U_e} = \left(\frac{R_1 + \Delta R_1}{R_1 + \Delta R_1 + R_2 + \Delta R_2} - \frac{R_4 + \Delta R_4}{R_3 + \Delta R_3 + R_4 + \Delta R_4} \right) . \quad (8.12)$$

The relationship between the ratio of the voltages U/U_e and the ratio of the resistances $\Delta R_i/R_i$ is nonlinear. The nonlinearity can be around 0.1% if only one resistance changes. This is too large an error for a precision force transducer and therefore these nonlinearities cannot be neglected if only one strain gauge is used in the force transducer. This is the reason why most strain-gauge-based transducers use four active gauges instead.

For the case of a *full bridge*, as it is called when four gauges are active, the nominal values of the resistors R_1 to R_4 are the same and the changes in the resistances are very small, such that higher-order terms can be neglected and the (8.12) can be simplified to a linear equation:

$$\frac{U}{U_e} = \frac{1}{4} \left(\frac{\Delta R_1}{R_1} - \frac{\Delta R_2}{R_2} + \frac{\Delta R_3}{R_3} - \frac{\Delta R_4}{R_4} \right) . \quad (8.13)$$

The relation between the relative change of the resistance and the strain was found to be $\Delta R/R = k\varepsilon$ in (8.7). Now the linear relation between the relative change of the bridge output and the strain becomes:

$$\frac{U}{U_e} = \frac{1}{4} (\varepsilon_1 - \varepsilon_2 + \varepsilon_3 - \varepsilon_4) . \quad (8.14)$$

This equation is the basis for all strain-gauge-based measurement.

In this equation it can be seen that the relative resistance change or the strain in the bridge arms ε_2 and ε_4 is subtracted from the relative changes in the bridge arms ε_1 and ε_3. We now define tension to be a positive change of strain, e.g., a positive relative change of the resistance. If we form a full bridge with gauges that do have all the same tension ($\Delta R_1 = \Delta R_2 = \Delta R_3 = \Delta R_4$) no output will be measured at this bridge. As a consequence, for a force transducer we always need the strain situation to consist of two tension and two compression gauges. Such a configuration favors bending- or torsion-type transducers in force balances.

Influence of Bridge Wires

The influence of the bridge wiring can be explained by adding an imaginary wire resistance R_{wi} to every arm of the bridge (Fig. 8.17).

If it is assumed that the gauge resistance does not change ($\Delta R_i = 0$), then (8.13) can be rewritten as:

$$\frac{U}{U_e} = \frac{1}{4} \left(\frac{R_{w1}}{R_1} - \frac{R_{w2}}{R_2} + \frac{R_{w3}}{R_3} - \frac{R_{w4}}{R_4} \right) . \quad (8.15)$$

From this equation it can immediately be observed that the wires have no influence if their lengths are absolutely identical ($R_{w1} = R_{w2} = R_{w3} = R_{w4}$). This is difficult to achieve. Another possibility is to have two pairs of wires with equal lengths, i.e., $R_{w1} = R_{w2}$ and $R_{w3} = R_{w4}$. In this case the bridge is not unbalanced by the influence of the wiring such that the bridge zero output (no resistance changes in the gauges) is close to zero.

Compensation of Bridge Zero Output

An unloaded force transducer in its normal mounting position and without any load applied should have no output. Mostly this cannot be achieved even if the gauge and the wire resistances are the same, because some output is generated by the weight of the metric end of the transducer itself, causing deformation in the transducer. However, the primary reason for the bridge zero output is the difference in the nominal resistance of the strain gauges.

Looking at a data sheet of a normal foil strain gauge, the bandwidth of the initial resistance is as high as 0.4% of the nominal value. For example in the worst case the zero output of a bridge with a nominal resistance of $350\,\Omega$ is $4\,\text{mV/V}$, which is much higher than the signal achievable for a full-scale transducer (usually $2\,\text{mV/V}$).

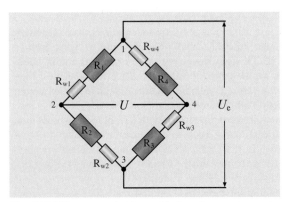

Fig. 8.17 Wheatstone bridge with imaginary wire resistances R_{wi}

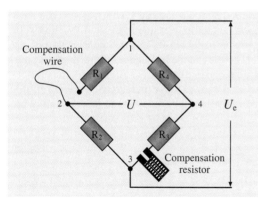

Fig. 8.18 Zero output compensation with constantan resistors

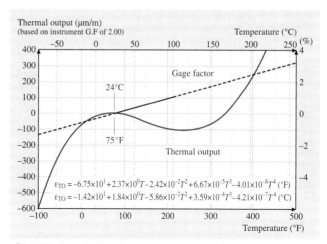

Fig. 8.19 A typical strain gauge apparent strain curve

If a strain gauge amplifier can be applied in the range up to 6 mV/V, or if the offset can be set to zero, then all problems would be resolved. However, with high-precision amplifiers this is not achievable and the zero output must be compensated to values lower than 5% of the nominal output. In order to do this the transducer or the balance must be set to its normal working position. In this position the output of the bridge must be measured and a compensation resistance must be calculated according to the following equation:

$$\Delta R_{\text{comp}} = \frac{U_0}{U_e} R_{\text{nominal}}, \qquad (8.16)$$

where R_{nominal} is the nominal resistance of the bridge and U_0/U_e is the zero output of the bridge. If a constantan wire for zero compensation is used, the length of the constantan wire must be calculated as follows:

$$L_{\text{const}} = \frac{U_0}{U_e} \frac{R_{\text{nominal}}}{R_{\text{wire}}}, \qquad (8.17)$$

where R_{wire} is the resistance per unit length of the constantan wire. Alternatively, commercially available adjustable resistors can be inserted into the bridge. In which bridge arm the compensation resistor must be installed depends on the sign of the zero output. If the zero output is negative the compensation resistor must be added into the arm which generates positive signals in the bridge and vice versa if the zero output is positive.

8.1.6 Compensation of Thermal Effects

Compensation of Zero Drift

Zero drift is defined as the change in the bridge zero output at different stable temperature conditions. It is important to have stable uniform temperature conditions, since temperature gradients inside the transducer can cause a much higher bridge output than the zero drift itself. Under stable conditions the zero drift is caused by the change in the resistance of the grid, the different heat expansion coefficients of the foil and the base material on which they are bonded. The cause of this temperature sensitivity is a combination of the direct resistance change as a function temperature as well as a real strain, which is applied to the gauge though no load is acting on the transducer. For a single gauge this output is called the *apparent strain*. Apparent strain curves are provided by the strain gauge manufacturer; a typical data sheet can be seen in Fig. 8.19.

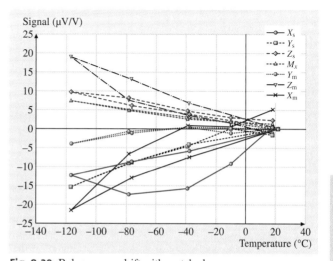

Fig. 8.20 Balance zero drift with matched gauges

For a single gauge this apparent strain is usually much higher than the zero drift of a full bridge, because at a stable uniform temperature level all four strain gauges have nearly the same apparent strain, thus canceling the effect out. This is another very important reason to use only full bridges for a general transducer or wind-tunnel balance. In Fig. 8.20 the zero drift of the balance bridges are shown where *matched gauges* are used for each bridge. This means that for every bridge four gauges were selected out of a large number to ensure that they have nearly the same apparent strain curve. This in turn minimizes the zero drift to $0.15\,\mu\text{V}/(\text{V}\,\text{K})$. In this case selection was necessary, since the balance is operated over a large temperature range. Even for operating temperatures around ambient conditions such a selection process minimizes the zero drift to very low values.

Using matched gauges in a Wheatstone bridge is a complicated method to minimize zero drift. If the operating temperature is in the range of $-10\,°\text{C}$ to $60\,°\text{C}$, compensation with the addition of temperature-sensitive wires or prefabricated temperature-sensitive resistors is much easier. Copper wires or copper resistors are mostly used in this case.

The first step in such a compensation setup is to perform a temperature run with the unloaded balance in the normal setup position. Each temperature level has to be very stable and the temperature distribution must be uniform throughout the balance. This cannot always be easily accomplished, however every temperature gradient inside the balance causes internal mechanical stresses and their associated output signals do not belong to the zero drift of the bridge. Gradient effects are not repeatable if the model changes and a correction that includes such internal gradient effects will fail later when the setup is again changed.

After obtaining the zero drift for each bridge there are two methods to follow on with.

The first possibility is the determination of a compensation resistor using the following equations:

$$\left(\frac{\Delta R}{\Delta T}\right)_{\text{comp}} = \frac{U}{U_e} \frac{1}{\Delta T_{21}} R_{\text{nominal}},\qquad(8.18)$$

$$R_{\text{comp}} = \left(\frac{\Delta R}{\Delta T}\right)_{\text{comp}} \alpha_{\text{comp}},\qquad(8.19)$$

where ΔT_{21} is the difference in temperature between the lowest and the highest measured temperatures, $\Delta U/U$ is the change of the bridge signal due to the temperature change, R_{nominal} is the nominal bridge resistance, α_{comp} [1/K] is the coefficient of the resistance change of the material which is used for compensation and R_{comp} is the required resistance for the compensation. The compensation resistor must be included into the Wheatstone bridge wiring as shown in Fig. 8.18 and the resistor must be cut according to the instructions of the manufacturer.

The second possibility is to install a temperature-sensitive wire instead of a prefabricated resistor. If this method is selected then in a third step the length of the compensation wire must be calculated:

$$L_{\text{comp}} = \left(\frac{\Delta R}{\Delta T}\right)_{\text{comp}} \alpha_{\text{comp}} R_{\text{wire}},\qquad(8.20)$$

where L_{comp} is the length and R_{wire} [Ω/m] is the resistance per unit length of the compensation wire. Materials such as copper, nickel or Balco® are most often used. For optimal results the materials data should be used from the supplier and not from handbook data. The compensation wire must be installed into the bridge circuit as shown in Fig. 8.18. The compensation resistor must be added into the bridge arm which has a lower resistance when the temperature changes.

8.1.7 Compensation of Sensitivity Shift

Sensitivity shift is defined as the change in sensitivity of a transducer as a function of temperature. There are three causes for the change in sensitivity of a strain-gauge-based force transducer. The first is the change of the geometry of the transducer body caused by thermal expansion or contraction. The second cause is the change of the transducer stiffness due to Young's modulus as a function of the temperature. The third cause of the change in sensitivity is the change in the gauge factor as a function of temperature.

To demonstrate this point, the sensitivity shift is calculated for a cantilever beam with length l and rectangular cross section of $b \times h$, loaded by a force F as shown in Fig. 8.21.

The sensitivity of this transducer is defined as the ratio of the signal ($\Delta U/U$) to the applied force (F)

$$\frac{\Delta U/U}{F} = \frac{k\varepsilon}{F}.\qquad(8.21)$$

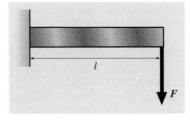

Fig. 8.21 Cantilever beam as a representative force transducer

The strain ε under the strain gauge is given by the equation: $\varepsilon = \sigma/E$, where σ is the mechanical stress under the strain gauge and E is the Young's modulus for the cantilever beam material. The stress σ can be calculated by using the following equations: $\sigma = M_b/W_b$, $M_b = Fl_{DMS}$ and $W_b = bh^2/6$, where l_{DMS} is the distance from the tip of the beam to the location of the strain gauge.

Applying the above equations, (8.21) becomes

$$\frac{\Delta U/U}{F} = \frac{k}{E}\frac{6l_{DMS}}{bh^2}. \qquad (8.22)$$

On the right-hand side of (8.22) all parameters are a function of the temperature. It can be seen now that the sensitivity shift depends on the gauge factor, the Young's modulus and the geometry of the transducer. The change in the dimensions of the cantilever beam can be written $\Delta l_i = l_i \alpha_T \Delta T$, such that the dimensions after the change in temperature are $l_i(T) = l_i + \Delta l_i$. Applying this to all geometric parameters in (8.22), the equation for the sensitivity temperature dependence becomes:

$$\frac{\Delta U/U}{F} = \frac{k(T)}{E(T)}\frac{1}{(1+\alpha_T \Delta T)^2}\frac{6l_{DMS}}{bh^2}. \qquad (8.23)$$

For the correction of the sensitivity shift there are two major results given by (8.23).

First, compensation is only possible if the gauge factor changes in the same direction as the Young's modulus of the cantilever material. This requirement can only be fulfilled by strain gauges using *Karma*® or an equivalent alloy, because the Young's modulus decreases as the temperature rises. Constantan behaves in the opposite way, so that it is impossible to compensate the sensitivity shift of a strain gauge transducer with constantan gauges.

Second, a perfect compensation can only be performed if the geometric effect is also compensated by the gauge factor variation. The consequence of this is that, for transducers with different geometries, different gauge factor drifts are necessary. The main difference appears if a bending-stress-type versus a shear-stress-type transducer is used. The influence of the geometry on a shear-stress-type transducer is bigger than that on a bending-stress-type transducer.

Typically the strain gauge manufacturers offer modulus-compensating (MC) strain gauges for the most common applications. This is done in order to compensate the main sensitivity shift caused by the Young's modulus change. The disadvantage of an MC gauge is that no self-temperature-compensation option can be chosen at the same time.

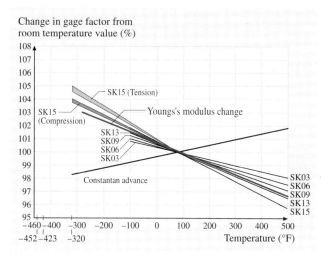

Fig. 8.22 Gauge factor shift of Karma and constantan

Another possibility is to change the excitation voltage in the same direction as the sensitivity shift change. In (8.21) it can be seen that the relative change of the signal stays constant. To achieve this, a temperature-sensitive resistor is integrated into the excitation voltage line of the bridge near the strain gauges, as shown in Fig. 8.23.

To determine the length of the compensation wire (l_{wire}) the following equation is used:

$$l_{wire} = \left(\frac{S_0}{S_T} - 1\right)\frac{RA}{\alpha_R \Delta T \rho}, \qquad (8.24)$$

where

S_0 [mV/V] = signal measured with force F acting at the reference temperature T_0 [K] and excitation voltage U_e [V],

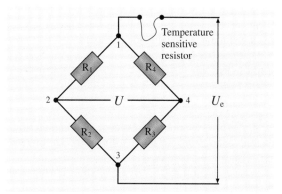

Fig. 8.23 Wheatstone bridge with sensitivity-shift compensation

S_T [mV/V] = signal measured with the same force F acting at the temperature T [K] and excitation voltage U_e [V],
R [Ω] = nominal resistance of the bridge,
A [m^2] = cross section of the wire,
α_R [1/K] = thermal expansion coefficient of the wire material,
ΔT [K] = temperature difference between the two tests,
ρ [Ω m] = specific electrical resistance of the wire material.

8.1.8 Strain Gauge Selection

The strain gauge selection process is very important in the design phase of a force transducer or balance. Many problems down the road can be avoided through the careful selection of strain gauges. The selection of the optimum strain gauge for a given transducer is an iterative process, moving between the requirements of the transducer and the characteristics of the available strain gauges.

As mentioned in Sect. 8.1.7, when using strain gauge balances gauges with a chrome–nickel alloy (*Karma*®, *Modco*®) should always be selected to achieve good sensitivity shift compensation. The gauges should be ordered from the same batch, which means they were manufactured with the same process and on the same sheet. Additionally they should be preselected to have a minimum difference of less than 0.05% in their nominal resistance.

The size and type of stress field to be measured by the gauge are the first determining factors for the gauge selection. To minimize the influence of stress concentrations, low-gradient strain fields are required. Usually the areas that fulfill this requirement are relative small, and so small gauges must be used. Sometimes on small balances even the smallest gauge does not fit on such an area without the need to cut the carrier foil.

On the other hand the excitation voltage should be as high as possible to get a high output signal. However, high excitation voltages will heat up the gauge and this will produce associated errors. Strain gauge manufacturers provide the diagrams that can be used to determine the maximum excitation voltage for a combination of the selected strain gauge area and spring material. In general, transducer materials with high heat conductivity and a gauge pattern with a large grid area allow higher excitation voltages.

It is always good practice to choose gauges with a high nominal resistance. The self-heating effect is relative small since at a constant excitation voltage the current through the bridge is lower and therefore the heating power of the gauge must be lower as well. Another advantage is the lower influence of the bridge wire thermal effects on the signal caused by the high ratio of bridge wire resistance to bridge resistance. Lastly, high-resistance gauges are small and fit into strain fields with small strain gradients. A disadvantage of high-resistance gauges is that the wires that are need to compensate zero, zero drift and sensitivity shifts must have relative high resistances too. This requires particularly thin wires that are not easy to handle.

With regard to gauge patterns, double gauges are preferred since their application is relatively simple and the two gauges sit perfectly parallel to each other. As the backing foil, fibre-reinforced epoxy laminate is the best choice for a wide temperature range and a high fatigue life. The solder dots on the gauge should be either pre-soldered or copper plated, since the chrome–nickel grid alloy needs an aggressive soldering flux in order to bind. This makes the attachment of the wires challenging and increases the risk of damage during wiring of the bridge. Welded lead wires out of copper–beryllium enables a higher fatigue life than those with copper.

8.1.9 Strain Gauge Application

The bonding of strain gauges onto the balance surface at first glance seems like a relatively simple process, however to achieve perfect adhesion requires a lot of experience. The strain gauge manufacturers themselves offer courses where all these bonding techniques can be learned. As well they provide precise instructions on how to use their adhesives and accessories. For the application of a balance or force transducer the use of adhesives supplied by the manufacturers is recommended, even if these are much more expensive than comparable ones on the market. Also important is not to apply such adhesives after their expiry date. Solvents and cleaning fluids must not be contaminated.

A study using experimental and numerical methods showed that most of the nonlinear interactions of an internal wind-tunnel balance are caused by the alignment errors of the strain gauges themselves. Consequently, near-perfect strain gauge positioning should be strived for. To minimize the errors associated with wire quality as well as to ensure durability, the wires should be as thick as possible. Stranded copper wire should be used in place of solid copper wire. As insulation *Teflon*® of *Kapton*® is preferable.

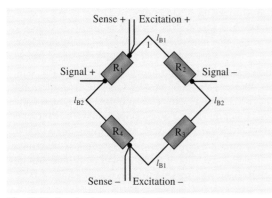

Fig. 8.24 Lead-wire connection locations

To minimize the resistance change associated with the wiring itself, either wires with equal lengths or two pairs of equal-length wire inside the bridge circuit must be used. The location of the lead wire connections can be seen in the following circuit diagram (Fig. 8.24).

8.1.10 Materials

Force transducers as well as balances are always constructed using metal. A list of often-used materials with their main physical and mechanical properties is given in Table 8.4. The selection of the material depends on the stress level and on the available space for the transducer. For high stress levels, high-strength tool steels, maraging steels, copper–beryllium or titanium are excellent materials. For low stress levels aluminum is a good choice. The material properties have a strong influence on the hysteresis, creep and repeatability characteristics of the force transducer. For the best results the yield strength of the material should be as far as possible away from the stress acting in the measurement section. Usually yield stress is three to five times higher than the stress felt by the strain gauge itself.

The mechanical properties that are of importance for the balance measurements, such as creep and hysteresis, are often not provided by the manufacturer. These properties depend on the heat treatment and the aging process. Sometimes to minimize hysteresis and creep it is advantageous to use a material with as low a tensile stress as possible. Again compromises must be made and often material tests must be made in-house. The machining process must be integrated into the heat treatment process to minimize internal stresses by forging and machining of the material. For maraging steel the integration of the manufacture into the aging process is shown in Fig. 8.25.

The time interval for each thermal step in the process depends on the material and the thickness of each part. For the best heat treatment results the instructions of the material supplier should be carefully followed. The steels that are used must be machined in an annealed condition. It is very important that electric discharge machining (EDM) is performed at this stage since EDM removes tensile stresses on and under the surface of the material. Such tensile stresses reduce the fatigue life of the balance under dynamic loads significantly. To further minimize this effect, the erosion period at the end of the EDM should be short.

The aging process will cause shrinking of the material. The dimensions of the joints, which have tight tolerances, must have an offset to allow final machining after the aging process. Grinding has nearly the same effect as EDM, so in areas with stress concen-

Table 8.4 Typical balance materials

Material	Short name	Material number	Yield stress	Tensile stress	Young's modulus	Shear modulus	Density	Coefficient of heat expansion
			(N/mm^2)	(N/mm^2)	(N/mm^2)	(N/mm^2)	(kg/m^3)	(μm/mK)
Maraging	X2 Ni Co Mo 18 8 5	1.6359	1760	1830	186000	71400	7920	11.6
Maraging	X2 Ni Co Mo 18 9 5	1.6354	1910	2010	191000	74600	8080	10.3
Maraging	X2 Ni Co MoTi 18 12 4	1.6356	2300	2400	190000	74600	8020	11.7
Maraging	X2 Ni Co Mo 18 8 3	1.6357	1430	1495	181300	68000	7920	9.0
Stainless	17-4-PH	1.4548	1170	1310	190000	75000	7780	10.0
Stainless	PH 13–8 Mo	1.4534	1150	1300	190000	75000	7760	10.0
Titanium	TI AL 6 V 4	3.7164	1000	1070	110000	43000	4430	8.6
Cu–Be	Cu Be 2	2.1247	1100	1500	123000	44000	8260	17.9
Aluminum	Al Cu Mg Mn	3.1354	300	430	72400	27600	2800	23.0
Aluminum	Al Zn Mg Cu Cr	3.4364	450	530	71000	27000	2800	23.0

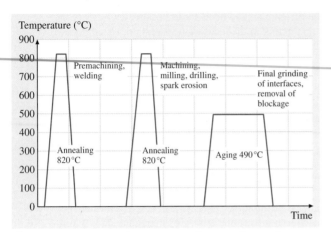

Fig. 8.25 Balance manufacture process for maraging steel

trations no grinding should be performed after the aging process.

Most steel used for balances can be machined in the annealed condition. Other materials, such as copper–beryllium, must be machined in their hardened state. The aging process of copper–beryllium causes high internal stresses, which is one reason why copper–beryllium has a high ultimate strength. The result of the aging process is the formation of large deformations, which must be corrected by machining. The advantages of copper–beryllium are the high heat conductivity, which minimizes temperature gradients inside the balance, and the low hysteresis and creep of the material.

The titanium alloy TiAl 6V4 is used in the aircraft industry for some highly loaded parts that cannot be built from aluminum. The heat conductivity is very low which forces careful cooling during machining. No heat treatment is necessary after the delivery of the material but it is very important to machine the material carefully to avoid internal stresses. Stress relief treatment can be done between 540 °C and 650 °C for about two hours, however resulting deformations may occur.

8.1.11 Single-Force Load Cells

In order to determine the forces on the model, the measurement of strain on a deforming body is needed. The strain on this body can be caused either by bending, shear or tensile stresses. In this section the different types of bending-, shear- and tensile-type single load cells are first discussed. Subsequently combinations of these types to a multicomponent force transducer will be discussed.

Bending-Type Load Cells

Load cells based on the measurement of the bending stress are the most frequently used. The main advantage of such load cells is that a bending moment always produces a positive and a negative stress of the same magnitude. This facilitates the application of a full Wheatstone bridge with a linear output and a reasonable signal magnitude. The simplest load cell is the cantilever beam, as shown in Fig. 8.26. Using this simple type of measuring beam as an example, the method to determine the dimensions of the transducer will be demonstrated.

The bending stress at the strain gauge location is defined as $\sigma_B = Fl/W_B$, whereas the measured strain is defined as $\varepsilon = \sigma_B/E$. The signal for the full bridge, with two gauges on the upper side and two gauges on the lower side, can be calculated as $\Delta U/U = k\varepsilon$. For steel ($E = 200\,000\,\text{N/mm}^2$) with a gauge factor of about $k = 2$, a signal of about $2\,\text{mV/V}$ should be strived for. In such a case the strain (ε) will be on the order of $1000\,\mu\text{D}$ (microstrain = $1000 \times 10^{-6}\,\text{m/m}$) under the gauge or a stress of about $200\,\text{N/mm}^2$. If these values are assumed, the dimensions of the gauged section can be derived from the equation $W_B = bh^2/6$. Commercially built transducers usually have a nominal signal of $2\,\text{mV/V}$. Some have a higher signal, but never more than $3\,\text{mV/V}$. All transducers designed for bending will be dimensioned using these aforementioned equations. The most challenging measurements are the bending moment (M_B) along with the section modulus (W_B). In most cases the equations for the bending moment and the section modulus are given by handbook formulas. The accuracy of these formulas is sufficient enough to determine the dimensions of a single force transducer.

These equations show that the force (F) is measured from the bending moment, such that the measured force also depends on the distance between the gauges and

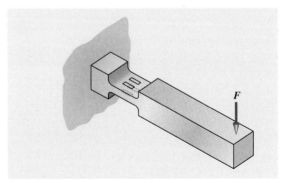

Fig. 8.26 Cantilever bending beam

the attachment point of the beam. If a force is not applied at the point where the calibration load is applied one will produce errors proportional to this misalignment. This problem can be overcome by using a different transducer design. Some possible designs are shown in Fig. 8.27. Their common design principle is that two bending beams should be coupled, forming a parallelogram. If the coupling element is stiff enough the load attachment point remains vertical, such that the transducer signal does not depend on the distance between this point and the gauge area.

Load cells that use the shear stress to determine either an applied force or a torque moment do have the same principle advantage that within the gauge area a positive and a negative shear stress is generated with the same magnitude. Therefore this type also has a linear output proportional to the applied load.

Shear-Type Load Cells

The maximum shear stress on a cantilever beam (Fig. 8.28), produced by force F, appears at the center line of the beam at $\pm 45°$, where the tension of compression stresses are nearly zero. In such a case the maximum strain also appears at $\pm 45°$, therefore to get the maximum output the gauges must be bonded as shown in Fig. 8.28. The dimensions of the beam can be calculated by determining the stresses in $\tau_{max} = Fc/A$ [N/mm^2], where F is the applied force, $A = bh$ [mm^2] is the beam cross section, and c is a form factor that depends on the shape of the beam cross section. For a rectangular cross section $b \leq h/2$ and $c = 3/2$, whereas for a circular cross section $c = 4/3$. These formulas are approximations for the centerline. The strain gauge covers a certain area around the centerline such that the integrated value measured by the strain gauge will be smaller. The signal of the transducer can be determined using the equation $\Delta U/U = k\varepsilon_{45°}$ for the full bridge output as well as the equation $\varepsilon_{45°} = \tau_{max}/2G$ for the strain under the gauge.

For a torque transducer the shear stress must be calculated using $\tau_{max} = M_t/W_p$ [N/mm^2], where M_t [Nm] is the torque moment and W_p [mm^3] is the polar section modulus.

Tension and Compression Load Cells

The main difference between shear- and bending-stress transducers and load cells using tension and compression stress measurements is that in the latter there is no positive and negative stress of equal value. If the tension stress $\sigma_{tension} = F/A$ is defined as positive, the negative compression stress becomes $\sigma_{comp} = \nu F/A$, termed the Poisson stress. For metals $\nu = 0.3$ while the compres-

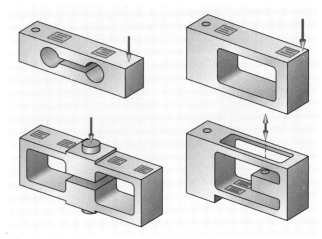

Fig. 8.27 Parallelogram-type load cells

sion stress is only about $1/3$ of the tension stress. Using a Wheatstone bridge with four gauges applied, as shown in Fig. 8.30, the output of such a transducer related to the force (F) will be nonlinear. This is not a big disadvantage as long as these nonlinearities are taken into account through calibration.

For standard load cells a nonlinear characteristic is not very common, however for applications in wind-tunnel testing they are sometimes advantageous since they do not require much space. Another advan-

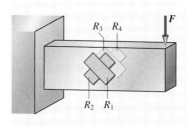

Fig. 8.28 Shear-type load cells

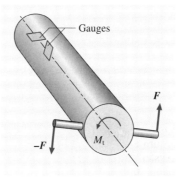

Fig. 8.29 Torque transducer

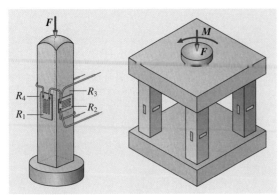

Fig. 8.30 Tension-type load cell (*left*) and cage of tension and compression columns (*right*)

tage of such a transducer is that the strain gauges are placed very close to each other, thus remaining at the same temperature. This minimizes the zero-drift and temperature-gradient sensitivity of the transducer. In some wind-tunnel balances such tension and compression columns can be arranged in a cage. In such a way it is possible to measure tension and compression as well as the moments acting on the metric side of the transducer [8.9].

8.1.12 Multicomponent Load Measurement

The fundamental criterion of all multicomponent load measurements is that all the loads acting on a model must be separated into single components as best as possible. In order to measure each single component at least one associated transducer per component must exist. This criterion can be best fulfilled in external wind-tunnel balances, where plenty of space is available to separate the load by decoupling rods with flexures. Sting balances are much smaller and so this criterion can be only partially fulfilled. Consequently, interactions or interferences in the measurement of the different components must exist.

The interactions inside a balance are systematic errors that can be determined through calibration. Therefore it is often necessary to measure more components in order to separate the errors during calibration. Another reason for having more measurement sections than components is to measure the errors caused by temperature gradients inside the balance. Temperature gradients cause deflections of the metric part against the nonmetric part, which in turn are measured by the strain gauges. This is also a systematic error which can be separated through calibration. In order to extract the loads from the balance signals, the following equation must be resolved:

$$F = \bar{E}S, \qquad (8.25)$$

where F is the force vector, $\bar{E}$ is the evaluation matrix and S the signal vector.

Through calibration one obtains an evaluation matrix whose elements take all the interactions and systematic errors into consideration. For a balance with no interactions, the six sensitivities for each component are the diagonal of the evaluation matrix while all other elements become zero. For a balance where nonlinearities up to the third order are considered, the evaluation matrix consists of $6 \times 33 = 198$ elements [8.10].

8.1.13 Internal Balances

Two general groups of internal balances exist. The first group consists of the so-called box balances. These can be constructed from one solid piece of metal or can be assembled out of several parts. Their main characteristic is that their outer shape most often appears cubic, such that the loads are transferred from the top to the bottom of the balance. The second type of internal balance is termed a sting balance. These balances have a cylindrical shape such that the loads are transferred from one end of the cylinder to the other in the longitudinal direction. Both one-piece and multi-piece sting balances exist [8.11].

Sting Balances

Internal sting balances are divided into two different groups: the force-balance type and the moment-balance type. If the bridge output is directly proportional to one load component then these balances are termed *direct read balances*. Typically for all groups, the axial force and rolling moment are directly measured with one bridge. The measurements of lift force and pitching moment or for side force and yawing moment are done in different ways characterizing each group.

Moment-type balances and force-type balances have one main feature in common: the lack of a direct output

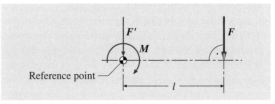

Fig. 8.31 Separation of force and moment

proportional to lift/pitch and side force/yaw. The signals that are proportional to each of these loads must then be calculated by summing or subtracting the signal from one another, before being fed into the data-reduction process. The advantage is that the associated concentrated wiring on each section is much less sensitive to temperature effects [8.12].

Force Balances. This type of balance uses two measurement sections placed in both the forward and the aft section of the balance. In these measurement sections a forward and aft force is measured most often through tension and compression transducers. These forward and aft force components are used to calculate the resulting force in the plane as well as a moment around the axis (perpendicular to the measurement plane). An example of a typical force balance is shown in Fig. 8.32.

Moment-Type Balances. Moment-type balances have a bending moment measuring section in the front as well as in the aft regions of the balance (S_1 and S_2 in Fig. 8.33).

The measurement of the two bending moments (S_1 and S_2) is used to obtain a signal that is proportional to the force in the measurement plane and a second one that is proportional to the moment around the axis (perpendicular to the measurement plane). The stress distribution shows that the moment M_y (M_z) is proportional to the sum of S_1 and S_2. However, the force F_y (F_z) is proportional to the difference in the signals S_1 and S_2.

To measure the rolling moment (M_x) one bending section must be applied with shear stress gauges to detect the shear stress τ. The most complicated part of the balance is the axial force section, which consists of flexures and a bending beam to detect axial force. These flexures enable axial movement whilst carrying the other loads.

Direct-Read Balances. A direct-read balance can be categorized as either a force-balance type or as a moment-balance type. Instead of measuring a force or a bending moment at each section separately, half bridges on every section are directly wired to a moment bridge while the other set of half bridges are directly wired to a force bridge. Thus the difference between direct-read balances and the other types is only in the wiring of the bridges. The disadvantage of such a wiring is the length of the wires from the front to the aft ends. Temperature changes inside these wires cause errors in the output signals.

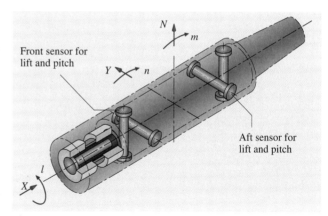

Fig. 8.32 Force balance with tension transducers in forward and aft sections (courtesy of Able Corp.)

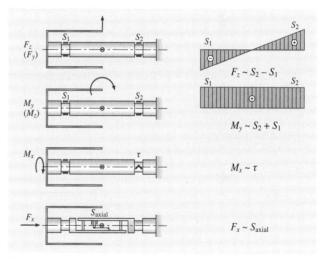

Fig. 8.33 Workings of a moment-type balance

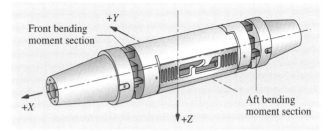

Fig. 8.34 Moment-type balance

Box Balances

The main difference between box balances and sting balances are the model and sting attachment area (Fig. 8.35). The load transfer in such balances is from

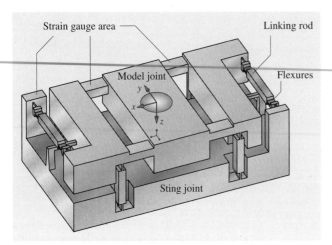

Fig. 8.35 Mono-piece box balance

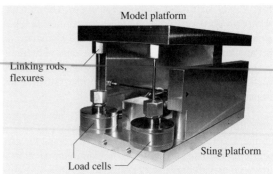

Fig. 8.36 Small box balance with load cells

top to bottom along the vertical z-axis. Therefore these balances use a central sting arrangement, as shown in Fig. 8.10 for the case of an airplane configuration. The mono-piece balance is constructed from one single piece of material. The advantage of this relatively expensive manufacturing process is the low hysteresis and good creep behavior, which are basic requirements for good balance repeatability. Multi-piece box balances are built from several parts, which can in turn be manufactured separately. The load transducers can either be integrated into the structure or separate load cells can be used instead. This enables a parallel manufacturing process with a final assembly at the end, in turn making the whole process quicker than that for a mono-piece balance. Box-type balances are considered to be internal balances, but have actually more in common with semi-span balances. In particular their temperature-sensitivity behavior is similar to that of semi-span balances.

Principle Design Equations

All internal balances measure the moment and the force in two different sections along the x-axis. The distance between the two sections defines the separation of the signal between force and moment. In the ideal case, half of the signal should be proportional to the force and the other half should be proportional to the moment.

To obtain the same output for the force and the moment, the distance $l = l_1 + l_2$ between the two sections has to be calculated. For a moment-type balance these relations can be described by the following equations

$$\sigma_1 = \frac{M_{B_1}}{W_1} = \frac{M + Nl_1}{W_1},$$

$$\sigma_2 = \frac{M_{B_2}}{W_2} = \frac{M - Nl_2}{W_2}, \quad W_1 = W_2, \quad l_1 = l_2,$$
(8.26)

where σ_1 and σ_2 are the stresses in sections 1 and 2, respectively. These stresses are caused by the moment M and the force N. W_1 and W_2 are the section moduli. The sum as well as the difference of these two stresses are

$$\sigma_1 + \sigma_2 = M\left(\frac{1}{W_1} + \frac{1}{W_2}\right) + N\left(\frac{l_1}{W_1} - \frac{l_2}{W_2}\right)$$
$$= \frac{2M}{W},$$
(8.27)

$$\sigma_1 - \sigma_2 = M\left(\frac{1}{W_1} - \frac{1}{W_2}\right) + N\left(\frac{l_1}{W_1} + \frac{l_2}{W_2}\right)$$
$$= \frac{Nl}{W}.$$
(8.28)

In (8.27) it can be seen that the sum of the stresses is only proportional to the moment M, and in (8.28) it can be seen that the difference of the stresses is only proportional to the force N. Consequently the sum of the bridge signals S_1 and S_2 is proportional to the moment M while the difference of the bridge signal S_1 and S_2 is

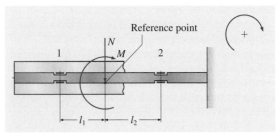

Fig. 8.37 Force and moment acting on a sleeve over the reference point

proportional to the force N

$$\sigma_1 + \sigma_2 \approx \Delta U_M ,$$
$$\sigma_1 - \sigma_2 \approx \Delta U_N . \quad (8.29)$$

At this point the ratio of the sum and the difference of the signals can be calculated

$$\frac{\Delta U_M}{\Delta U_N} = \frac{2M}{Nl} . \quad (8.30)$$

For a given moment M and a given force N, the aim is to have the ratio of the signal equal to one. The distance l between the measuring sections can be calculated in the following manner:

$$l = \frac{2M}{N} . \quad (8.31)$$

This equation is also valid for force-type balances. Therefore for the same force and moment output, there exists an optimum distance between the measuring sections; thus the length of the balance can be determined in this way. If the required load combination results in a length that does not fit into the model, then the distance between the measuring sections must be compromised in such a way that the signals for the force or the moment are smaller than the other.

Another problem appears when the distance between lift and pitch in the x–z plane is different from the distance between the side force and yaw in the y–x plane. One solution is to use different measuring sections, but this will enlarge the total length of the balance and is usually avoided.

For a single-test setup normally the balance length can be easily optimized for the model. However, the balance load range definition is more often a compromise between requirements for different test setups, where the maximum loads for all tests form the envelope of the combined-load range specification. In such cases a good compromise for the balance dimensions can seldom be found.

Specific-Load Parameter. Before designing the bending section it is good to determine first whether the balance will have high loading or not. Loading in this case refers to the ratio between the loads and the available volume for the balance. This ratio expresses the magnitude of the stress level inside the balance before starting with the calculation. This ratio is also referred to as the *specific-load parameter*

$$S_{\text{round}} = \frac{N + L/2 + M}{D^3} ,$$

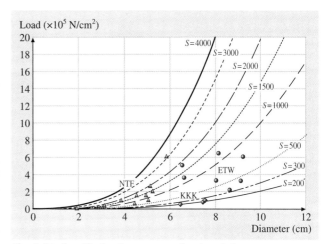

Fig. 8.38 Specific-load parameter of various balances

$$S_{\text{rectangular}} = \frac{N + L/2 + M}{1,7 BH} , \quad (8.32)$$

where L is the active length of the balance without interfaces, D is the diameter, and B and H are the maximum available width and height of the balance. The first equation is used when the main cross section of the balance is circular whereas the second is used when the main cross section is rectangular. Experience shows that highly loaded balances have a specific-load parameter greater than $1000 \, \text{N/cm}^2$. In this case the highly stressed areas occur not only in the area where strain gauges are applied, but also in the flexures of the axial force elements.

Bending Section Design. After the distance of the measuring section is fixed, the bending sections must be designed such that the output and full scale are on the order of $0.5-3 \, \text{mV/V}$. The design output depends on the data-acquisition equipment to be used. Some systems have a maximum input and therefore it must be guaranteed that no overflow occurs. This is the case even for a single maximum load, which can be up to 100% higher than the maximum combined load. In most cases, for maximum output of the design loading, the sum of the full-scale output of the force and the full-scale output of the moment are the limiting factors. Usually the maximum full-scale output is around $1.5 \, \text{mV/V}$.

In order to determine the geometry of the bending section it is sufficient to use handbook formulas. If a careful calculation is made the accuracy of the predicted full-scale output will be $\pm 10\%$. Since the real gauge factor at this stage is not yet known, it is good to

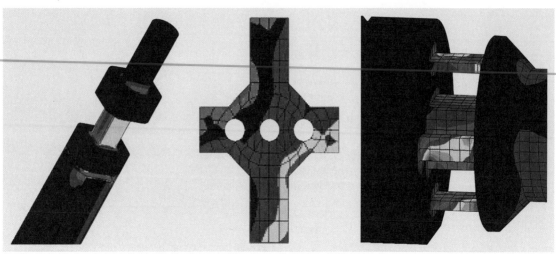

Fig. 8.39 Massive rectangular geometry (*left*), massive cross geometry (*center*) and cage with five beams (*right*)

use $k = 2$ by default. Normally the real gauge factor is somewhat higher and so the final output of the section is usually higher than calculated.

To design the optimum cross section many different designs are used. Some commonly used geometries are shown in Fig. 8.39.

Possible Geometries for Bending Section Design. On the surface of the massive cross sections (cross, rectangular or octagonal in shape) the surface stresses caused by the bending moment are measured. In a cage design the tension and compression stresses in the single beams are measured. This design is preferable when the specific-load parameter is lower than $300 \,\mathrm{N/cm^2}$. During the bending section calculation the measurement sections for five components must be designed.

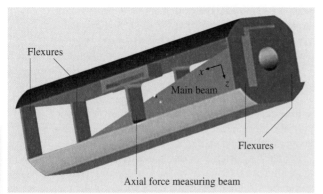

Fig. 8.40 Axial force section (the *central part* on the *left* main beam is cut out)

These components are: lift, pitch, side force, yaw and roll.

Axial Force Section Design. The axial force measurement requires much attention in the design and construction stages. Most of the cost for an internal balance is attributed to the axial force system. Usually the axial force is very small in comparison to the other forces and moments such that the sensitivity of this measurement must be extremely high. This in turn makes the axial force sensitive to interactions of the other components.

In Fig. 8.40 the components of a typical axial force section are shown. The left-hand and right-hand sides of the balance are connected by the flexures. These flexures carry the loads of all five components but are relatively flexible in the axial force direction. Most of the axial force (more than 60%) is supported by the cantilever beam in the middle of the balance.

The axial force measurements are best made in the middle of the axial force section, since the mechanical interactions are minimal in this position. For the compensation of thermal effects, in particular temperature gradients, it is better to have four rather than two axial force measuring beams placed near the flexures.

8.1.14 External Balances

Balances have been used since the beginning of aerodynamic testing. One of the first balances was built by *Otto Lilienthal*. In Fig. 8.42 Lilienthal's apparatus for the measurement of lift can be seen. The apparatus is a simple combination of a wind tunnel and a balance. Two airfoils of the same shape were fixed to a rotating arm at a certain angle of attack. Two cables were wrapped around a pulley in the center and at the other end of the cables two weights were connected. When the weights were released they fell to the ground and the arm began to rotate. During rotation the airfoils generated lift and this lift was then counterbalanced by the weights in the middle, thus measuring the aerodynamic lift.

The principle of counterbalancing weights for the measurement of aerodynamic forces on wind-tunnel models was used in many external wind-tunnel balances. Even today such external balances use weigh beams to measure the aerodynamic forces. With the advent of the strain gauge, load cells replaced these weigh beams. Subsequently the balances became stiffer, faster and more accurate. Modern external balances are the most accurate multicomponent force-measuring devices in use.

External balances can be separated into two groups, large external balances and semi-span (or half-model) balances. The latter are much more compact and are built either from one piece of material or are assembled from various. Large external balances are conversely built using a steel framework and separate load cells.

Sidewall Balances

The term sidewall balance is often used to refer to a compact external balance. Such a balance is often used for half-model testing. For semi-span model testing the bal-

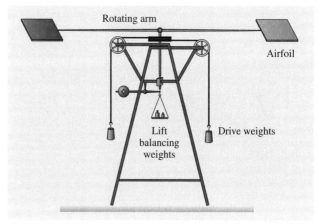

Fig. 8.41 Lilienthal's airfoil test rig with lift balance

ance has only five components since measuring the side force on a semi-span model is illogical. This very significant difference between a half-model balance and the external compact balance affords a completely different design.

The principle design of a semi-span balance is relatively simple. There are two massive parts, the earth end and the model end, both of which are connected by a flexure system and the measuring beams. The design of semi-span balances varies in the flexure system

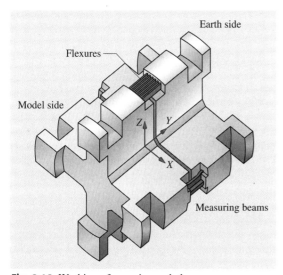

Fig. 8.42 Working of a semi-span balance

Fig. 8.43 Simple half-model balance

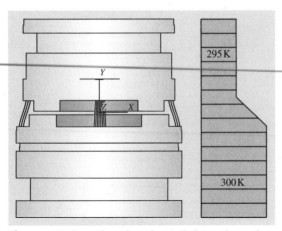

Fig. 8.44 Deformation of semi-span balance due to heat flux

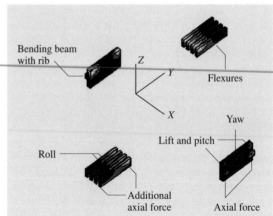

Fig. 8.45 Flexure design for half-model balance

and the form of the main body. Most have simpler flexure systems than that shown in Fig. 8.42, with only one flexure in every corner and circular cross section for the main body as shown in Fig. 8.43.

A common problem with all designs is the enormous temperature sensitivity;, which is even a problem with very slight temperature changes. Such changes occur in most tunnels, which often become progressively warmer with time. The model, with its large surface, is exposed to the tunnel air temperature and reacts very quickly whereas the balance on the outside stays at room temperature. Thus a heat flow is generated through the balance, in which the model and the flexure system are the paths of highest resistance for the heat flow. The result of this heat flow is a temperature gradient inside the balance causing a deformation of the flexures.

In Fig. 8.44 the top of the balance is cooler than the bottom, such that the upper region shrinks relative to the lower part. The stiffness of the main parts is much higher than the stiffness of the measuring flexures such that the shrinking is compensated by deformation of the flexures. This deformation in turn is measured by the strain gauges. As can be seen in Fig. 8.44, the flexures on opposite sides bend in opposite directions. If the sensitivity of both flexures is equal, then the sum of the signals is zero and the temperature gradient effect is canceled out. Therefore the aim in such a case is to build a symmetric balance with flexures of equal sensitivity but on opposite sides. The best compensation for all temperature-related effects is obtained by applying strain gauges bridges on every flexure, measuring each one separately as is typical for moment-type balances [8.13].

Design of Side-Wall Balances

As described previously, a side-wall balance consists of two massive parts plus flexures. Therefore most attention must be concentrated on the flexures and the distance between them. The simplest arrangement consists of four flexures with exactly the same shape, all of which are placed symmetrically around the center of the balance. To adjust the distribution for a force and a moment acting in one plane the same basic results (8.25, 26, 27, 28, 29, 30, 31) can be used to determine the appropriate distance between the flexures.

To adjust the sensitivity of the single component to the specified-load range a more-sophisticated flexure arrangement, as shown in Fig. 8.45, can be used. To calculate the flexure cross sections the formulas for the cage design of an internal balance are nearly identical.

Large External Balances

In contrast to semi-span balances, all large external balances are constructed from discrete parts. Outside of the wind-tunnel test section there is usually much more space available. This makes it possible to build up a large structure, enabling a better decoupling of the various components. Moreover, for such a design precision load cells can be ordered from several suppliers. The load cells define the main characteristics of the external balance such as repeatability, resolution and stability. If the load cell range is well matched to the balance load requirement the maximum resolution of the balance will exactly match the resolution of the load cell. This is usually on the order of 300 000 parts if precision electronic equipment is used. The repeatability of the balance is determined by the load cell. Furthermore, the rigidity of the balance will be at least as good as the resolution of

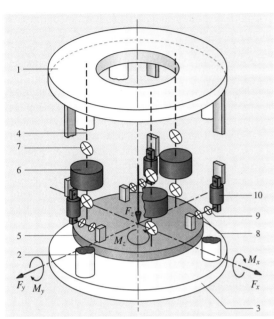

Fig. 8.46 Typical configuration of an external balance (for explanation of numbers, see text)

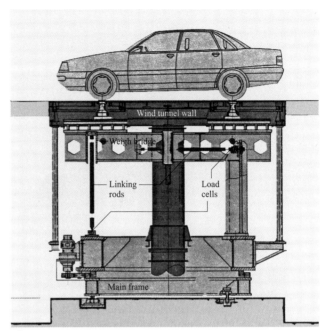

Fig. 8.47 External balance for an automotive wind tunnel (courtesy of Schenck)

the load cell. The large dimensions of the structure make the balance sensitive to temperature changes. Short-term changes in temperature must be prevented such that the whole balance is kept at a stable temperature level. For the case of a conventional wind tunnel this can be easily achieved.

There exist many different types of external balances since they are most often designed for individual wind tunnels with specific requirements. Examples of such designs are platform-, yoke- and pyramidal-type balances [8.14]. The differences between these types of balances are characterized by the design of the structure and the arrangement of the load cells. This arrangement in turn determines the signal combination for the different load components.

The design principle for external balances is nearly always the same (Fig. 8.46). The metric part of the balance is the so-called weighbridge (5). This weighbridge is connected via rods (4, 8) to the load cells (6, 10). To protect the load cells from unwanted moments, flexible knuckles (7, 9) are used on both ends of the connecting rod. These knuckles should be always elastic elements with no ball bearings. Such bearings must be avoided since any friction inside causes hysteresis. The load cells themselves are connected to the main frame (1, 2, and 3), which is itself the nonmetric side of the balance.

For a six-component balance, six load cells are needed. The vertical ones (6) measure the lift force, pitching moment and rolling moment. The horizontal load cells measure side force, drag and yawing moment.

An example in which the weighbridge is located directly under the tunnel wall is that of an automotive wind tunnel (Fig. 8.47). The cars are fixed on pads that are linked through the tunnel wall to the weighbridge of the balance without any contact with the tunnel floor.

This configuration allows for the installation of a moving belt inside the wind-tunnel floor, eliminating the floor boundary layer thus producing more-realistic boundary conditions.

Normally the entire balance is mounted on a turntable such that the balance moves together with the model, as shown in Fig. 8.48. In the case of aircraft half models and automotive testing the model axis system and the balance axis system are always identical and the wind forces and moments must be calculated by transforming the balance loads into the wind axis system.

For full model aircraft testing, the angle-of-attack setting mechanism is integrated into the balance weighbridge such that the balance stays in the wind axis system while the model alone changes its angle of attack. The measured loads in this case are the aero-

Fig. 8.48 An example of a small external balance (courtesy of Schenck)

dynamic loads in the wind axis system. For a change of yaw angle, however, the balance is moved together with the model such that the side loads must again be transformed into the wind axis system. This has to be taken into account in the balance data-reduction software. Sometimes clients need to have the side load data in the model axis system, although all the loads for the longitudinal direction should be given in the wind axis system. In some external-balance constructions the mechanism for change in angle of attack and yaw angle are integrated into the weighbridge, allowing the balance to be fixed to the earth and to the wind axis system.

8.1.15 Calibration

The primary task of the calibration is to measure the relationship between the balance signal and the measured load. In order to obtain this relationship or sensitivity, the balance must be loaded with precise single calibration loads. Such loads are often referred to as true loads. These loads can either be applied exactly in the direction of the model axis system, or alternatively if this cannot be achieved, then the exact direction of the load must be precisely measured. One of main requirements of true loads is that they are traceable to an international standard.

Another requirement is that the direction of the load during calibration is always the direction of the reference axis system. This requirement can be achieved by either realigning the loading axis system to the model axis system or by measuring the relative movements (translation and rotation) between the axis systems, thus transforming the loads into the model axis system. In the former case, a certain orientation procedure must be repeated after every change in load, whereas in the latter case, as well as the load measurement, the rotation and translation must be measured.

Since several measurements must be performed the uncertainty of the true load increases. These errors cannot be avoided thus the uncertainty of the true load application must be one order of magnitude lower than the required absolute uncertainty of the balance.

The signals of a wind-tunnel balance are more or less influenced by many systematic errors (see Sect. 8.1.2). One of the major systematic errors is the interaction of all loads on a strain gauge bridge. This means that, if a certain load combination is applied to the balance, each bridge will react to every load. This is a very undesirable situation since the original intention of the design is that each bridge should react only to one load. The elimination of the interaction is the other major task of calibration. To do so the balance must be loaded with load combinations, mainly with combinations typical of the desired wind-tunnel tests.

From all this data are the relation between signals and loads must be extracted. As mentioned in Sect. 8.1.12, the fundamental relationship between the loads and the signals of the balance is

$$F = \bar{E} \times S, \qquad (8.33)$$

where F is the unknown load vector and S the signal vector. The task of the calibration data reduction is to determine the elements of matrix $\bar{E}$, known as the evaluation matrix. This matrix is used in the tunnel software to determine the aerodynamic loads from the measured balance data.

During calibration the load vector is known and the signals are measured, so in this situation the dependent variable is the signal and the corresponding equation is:

$$S = \bar{K} \times F. \qquad (8.34)$$

Herein the matrix $\bar{K}$ is called the calibration matrix. The usual way to obtain the evaluation matrix from the calibration matrix is to invert the calibration matrix:

$$\bar{E} = \bar{K}^{-1}. \qquad (8.35)$$

As long as the calibration matrix is a 6×6 matrix it is called a linear matrix and the inversion of this linear matrix is trivial. The matrix is referred to as linear since there is only a linear relation between the interactions and the load. If the requirements on the accuracy of a multicomponent force transducer are not very high and the mechanical decoupling of the load measurement sections is adequate, such a linear matrix is suitable to describe the interactions between the components. This means that for lower accuracy requirements in some external balances a linear matrix is adequate to eliminate the interaction effects.

In reality the characteristics of the interactions are more complicated and nonlinear. This nonlinearity may cause a few-percent difference in the linear characteristic. However, the requirement of an internal wind-tunnel balance is an overall accuracy better than 0.1%, therefore nonlinear effects on direct sensitivity and interactions must be taken into account. Thus matrices on the order of 6×21 to 6×33 are used. Unfortunately, a complex 6×33 calibration matrix cannot be inverted easily to obtain the evaluation matrix. For such a calculation a numerical algorithm must be used.

Description of Interactions

Mathematically the interactions can be understood as the elements of the matrices that are not placed on the diagonal. The matrix (8.33) can be written as a sum of the terms:

$$F_i = \sum_{j=1}^{6} A_{ij} S_j + \sum_{j=1}^{6} \sum_{k=j}^{6} B_{ijk} S_j S_k + \sum_{i=1}^{6} C_{ij} S_j^3 \;. \tag{8.36}$$

This equation takes all elements of the 6×33 matrix into account. A_{ij} represents 36, B_{ijk} 126 and C_{ij} 36 coefficients. The six elements A_{ij} with $i = j$ are the direct sensitivities for the six loads.

Linear Interaction. In (8.36) the elements A_{ij} for $i \neq j$ are the coefficients of the linear interaction. As already described a linear interaction is the influence of a load component that is not to be measured with the transducer. For example, the side force generates a signal in a transducer that is used to measure tension and compression.

The bending stress caused by the side force will also affect the strain gauges since they are strained by the bending too. Another cause for the sensitivity of the side force is misalignment of the strain gauge itself. If they are not perfectly aligned in the vertical and horizontal directions they become increasingly sensitive to the strain in the other directions.

Second-Order or Product Interactions. In (8.36) the elements B_{ijk} are the coefficients of the nonlinear interaction. In the case of $j \neq k$ the interactions are called product interactions. In the case of $j = k$ they are called second-order interactions and describe the second-order nonlinearity of the sensitivity.

Product interactions are described as the sensitivity of a transducer related to the product of the load not to be measured. Take the example of an axial force acting on the parallelogram section of a balance producing a deformation dx (Fig. 8.50). If an additional normal force is loaded an additional deformation dx will be measured by the stain gauges on the flexure. The additional signal is proportional to the product of F_z and F_x.

Third-Order Interactions. In (8.36) the elements C_{ij} are the coefficients of the third-order nonlinearity. Third-order nonlinearity is taken into account only for the

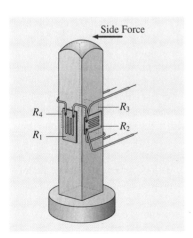

Fig. 8.49 Side-force interaction

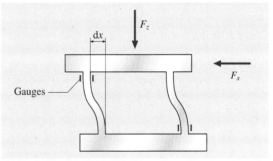

Fig. 8.50 Product interaction of F_x and F_z

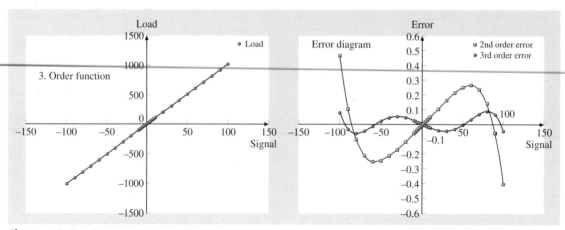

Fig. 8.51 Calibration curve and error diagram

third-order term of the loads and not for any other load combination. The main justification for taking a third-order interaction into account is the nonlinearity about zero when the load changes from positive to negative and vice versa.

The above function is a typical calibration curve of a single load cell. Though the plotted data seems to show a linear behavior, the error diagram indicates that, for a second-order fit, the error increases to up to five times higher than for a third-order fit. The error in this case is defined as the difference between the measured data and the value of the fit function used.

Verification and Accuracy

Since there are so many possible sources, a challenging task is the determination of the uncertainties or the accuracy of a balance. Primary sources include the balance itself, the measurement equipment, the calibration procedure and the loads applied. Error sources can be divided into two groups. The first group consists of the uncertainties that are generated by the calibration procedure itself. Take the example of a dead weight which at first seems to be the best true calibration load. Even a precisely measured weight has some small uncertainties associated with it. When this load is applied to the calibration sleeve of the balance it has to be aligned to the geodetic axis system. This alignment must be controlled by another measurement device, which also has uncertainties. For a moment the distance to the reference center must be measured and so we produce yet another uncertainty of the true load. The second group of errors contains all the uncertainties that are caused by the balance and the measurement equipment.

To calculate the uncertainty from every source would be a rather challenging task, so usually the back-calculated data are used to determine the uncertainty of a balance. The back-calculated data are described as the data obtained from the calculated loads using the balance evaluation matrix and the signal measured during calibration. The difference between the calibration load and the back-calculated load is assumed as the uncertainty of the balance. This data includes all uncertainties of the balance and the calibration process. Normally for a precise calibration the uncertainties caused by the true loads, as well as the uncertainties in the calibration process, are one order of magnitude lower than the uncertainties of the balance itself.

The difference between loading points is also called the residual load. These residuals contain both random errors and systematic errors (bias). Applying a statistical analysis to this data may neither be a physical nor a mathematical solution to describe the quality of the balance. Nonetheless the balance user may find such information useful in order to predict the balance accuracy.

For a proper calibration, between 100–1000 calibration points should be performed, producing one residual load for every calibration loading. This residual load data includes a lot of information but does not provide a good overview of the balance quality. Using a statistical analysis of all the residual loads will however provide some significant quality factors. Widely accepted quality factors include:

1. Minimum and maximum residual loads,
2. Mean value of the residual loads,

3. Experimental standard deviation of the residual loads.

These values can either be provided in engineering units or in a percentage related to the full-scale value of the component. The mean value of the residual loads (ΔF_{rm}) is given as:

$$\Delta F_{\text{rm}} = \sum_{i=1}^{n} \frac{\Delta F_{\text{ri}}}{n} \,. \tag{8.37}$$

Subsequently the experimental standard deviation can be calculated from the following equation:

$$S = \sqrt{\frac{\sum_{i=1}^{n} (\Delta F_{\text{ri}})^2}{n-1}} \,. \tag{8.38}$$

The analysis of the residual loads gives the balance user information about the accuracy class of the instrumentation under certain environmental conditions. All the conditions that were kept constant during calibration, such as temperature and humidity, may have another influence on the accuracy of the measurement in the tunnel itself. Thus the effects of temperature must be corrected or calibrated separately. Another source of error not often taken into account is creep. Creep effects in the balance must be tested separately as well.

All these tests provide information about the accuracy of the balance. However, a few problems remain. Firstly, how to formulate the requirements for the accuracy in a specification? Secondly, how to compare these requirements with the calibration data? One suggestion to solve these problems was made by *Ewald* and *Graewe* in the early 1980s and is briefly described in [8.14]. An equation was formulated that would take into account the error influenced by the number and the magnitude of the interactions:

$$\delta_i = A F_{i\,\text{max}} \left(a_i + b_i \sum_{n=1; n \neq i}^{6} \left| \frac{F_i}{F_{i\,\text{max}}} \right| \right), \tag{8.39}$$

where δ_i is the allowed residual load, $F_{i\,\text{max}}$ is the maximum combined load of component i, and F_i are the loads applied during calibration. The factor A is the general accuracy factor and the accuracy factor a_i is for the individual component i; b_i is a weighting of the interaction. The customer can specify the factors A, a_i and b_i and after calibration the balance manufacturer has to verify if all the residual loads are below these specifications. A more global verification is to calculate the factors A, a_i and b_i from the calibration data and to compare them with the specifications.

All other influences on the accuracy of a balance can be integrated by using the error propagation law. For the influence of the temperature the (8.39) is rewritten as

$$\delta_i = A F_{i\,\text{max}} \left[\left(a_i + b_i \sum_{n=1; n \neq i}^{6} \left| \frac{F_i}{F_{i\,\text{max}}} \right| \right)^2 c_i^2 \right.$$
$$\left. + (d_i + \Delta T_1)^2 + (e_i + \Delta T_2)^2 \right]^{\frac{1}{2}} . \tag{8.40}$$

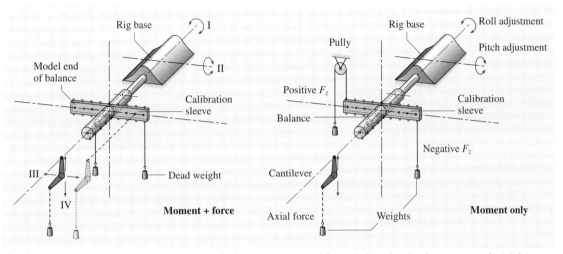

Fig. 8.52 Manual calibration with weights producing: a moment and force (*left*) and a simple moment only (*right*)

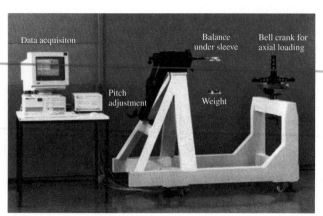

Fig. 8.53 Manual calibration rig

Calibration Principles

To obtain the aforementioned data several principles are used during calibration. The first one, most often used in manual calibration, is shown in Fig. 8.52. Here the balance is fixed with the nonmetric end to the rig base. On the balance metric or model end a calibration sleeve is mounted. This sleeve represents the model, on which the reference axis system is fixed. The calibration loads are hung on sleeves in different directions. To generate pure forces the loads must be applied directly in the axis of the reference system. To generate pure moments a positive and a negative load, as shown in the right-hand side of Fig. 8.52, must be applied to the sleeve, otherwise the loading will generate a force plus a moment as in the left-hand side of Fig. 8.52.

To generate the respective positive and negative forces, hanging weights via pulleys or bell cranks is necessary. Bell cranks are preferred, especially if they are supported with an elastic bearing. The use of pulleys is less desirable as they rely on ball bearings that have relatively high levels of friction. Such high friction causes hysteresis effects. All such mechanical devices are a source of error.

For some calibration data-reduction processes it is not necessary to apply pure moments during calibration. For these cases it is sufficient to load the balance only with a simple combination of moments and forces. On the other hand, during wind-tunnel tests pure loads normally do not occur and therefore it is much better to apply load combinations. Such complex combinations always occur during wind-tunnel testing and provide the best match between the calibration data and the actual wind-tunnel usage. Some data-reduction software is not capable of handling combined loading data and in this case some pure force and moment loadings must be performed to obtain the main sensitivity of each component separately.

In order to generate side forces, produced using positive normal forces without a pulley, the balance inside the calibration sleeve must be rotated by 90°, 180° and 270° relative to the normal position. With this measure normally most of the loads and load combinations can be generated, except for the case of axial force, which must be applied using a bell crank.

For the case of a geodetic axis system, the weight acts in exactly the vertical direction so it is very important to

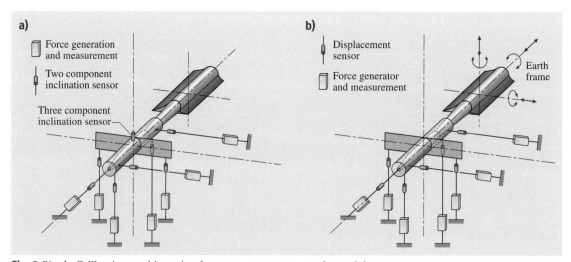

Fig. 8.54a,b Calibration machine using force generators as opposed to weights

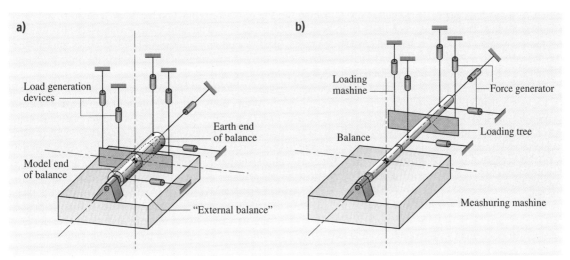

Fig. 8.55a,b Calibration machine with model end fixed to the reference system

realign the calibration sleeve to the geodetic axis system after every change of loading. This is performed to keep the reference axis system exactly in the direction of the load. For this realignment a pitch and roll adjustment in the rig base is necessary.

In the above Fig. 8.53, a manual calibration rig for small balances up to 1500 N of normal force is shown.

Calibration Methods. For automatic calibration there are several possible methods. The first method is similar to the manual one in that the metric end of the balance is fixed to a calibration sleeve and loads are applied to this sleeve. The weights and however and are replaced by force generators. Each generator has to measure the force that it is generating with a very precise load cell. Under a certain load or combination of loads the calibration sleeve will move in some direction and from the force vector related to the reference axis system only the value of the force will be known. To get the correct relative direction of the force vector and the orientation of the calibration sleeve and the orientation of the force-generator axis must be measured with the same accuracy as the force itself (Fig. 8.54a).

The accuracy of the true loads therefore is determined by two different measurement systems and both the force and the displacement measurement systems. This makes it very difficult to get a high precision for the true or calibration loads. Nearly the same principle as above mentioned can be achieved without a realignment system and which replaces the need for an orientation measuring system for the sleeve (Fig. 8.54b). However

and in this case the precision of the realignment system must be also very high and the disadvantage of a second measuring system will still exist.

A second method for automatic calibration fixes the metric end of the balance to the reference system of the measuring machine and the sting end is then con-

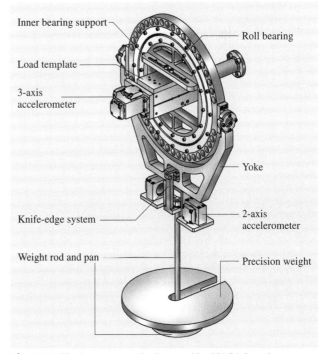

Fig. 8.56 Single vector method as used by NASA Langley

nected to a loading sleeve (Fig. 8.55a) on a loading tree (Fig. 8.55b). The mounting of the metric end of the internal balance onto the metric end of the reference measuring system and labeled as the external balance in Fig. 8.55 and has the great advantage that the reference system of both systems are always the same even when the balance is loaded.

For this method no displacement measuring system is needed and therefore the precision of the true loads is determined by the precision of the load measurement of the reference balance. In this manner the additional uncertainty of the displacement measuring system is avoided.

A final calibration method and namely the single load method and eliminates the uncertainty of the force measurement system by using a single weight as the true load and which is applied over a cardanic suspension to a sleeve which is fixed on the metric end of the balance. This single load remains always vertical. The earth end is fixed to a moving system and which moves the balance to different positions. By changing the orientation of the sleeve and balance and different load vectors are generated with only one load applied.

Additionally the orientation of the sleeve is measured very precisely by a 3-axis accelerometer system and so the orientation of the sleeve-axis system is well known relative to the single load vector. This method eliminates the force-measurement system and uses only a displacement-measurement system to determine the different load vectors. It is also possible to fix the metric end to the moving system. In that way the orientation of the moving system must be measured precisely to obtain the load vector relative to the single vertical load.

8.2 Force and Moment Measurements in Aerodynamics and Aeroelasticity Using Piezoelectric Transducers

Measurement of steady and fluctuating forces acting on bodies in a flowing fluid is the main objective in wind-tunnel experiments. In aerodynamic testing strain gauge balances will usually be applied for this task as, particularly in the past, the main focus was directed on the measurement of steady forces in attached flow, i.e., time-averaged forces acting on wind-tunnel models at rest. Moreover, in most cases the elastic compliance of the balance was neglected and the wind-tunnel models themselves were seen as rigid. With increased interest in understanding and investigating unsteady phenomena, caused for example by flow separation or oscillating models and increased importance of aeroelastic effects, the need for alternative solutions arose. In addition increasing size and lightweight construction principles applied to structures such as aircraft or bridges are coupled with a corresponding larger influence of elasticity. Thus wind-tunnel experiments must also take this into account. The entire force measuring system should be furthermore as rigid as possible, combined with the ability to resolve the unsteadiness of the occurring phenomena. Owing to their inherent stiffness, piezoelectric measuring systems fulfill these requirements very well, although there are restrictions concerning long-time measurement of small, steady forces.

In general, for multicomponent force measurements, one attempts to design the balance as stiff as possible. High rigidity leads to low interferences between the individual force/moment components and to a high natural frequency of the entire force-measuring system. Thus, all balances described in the following are based on three-component piezoelectric force transducers. Usually wind-tunnel balances are based on an arrangement of more or less elastic elements by which strain gauges are applied such that a passive measuring system is formed. In this case, the force measurement is referred to as a strain measurement, where the strain is a consequence of a force measured by a sensor – the strain gauge. A piezo balance, however, works as an active measuring system, where the force is measured directly as a consequence of the deformation of the quartz itself. That means the elastic element and the sensor are identical – the quartz. Consequently, the necessary measuring deflections of a strain gauge balance are one to two orders of magnitude larger than in the case of a piezo balance. This is due to the fact that the relatively low rigidity is a principal property of strain gauge balances.

Two examples will explain the importance of the rigidity of a balance. First, not only for static aeroelastic investigations but also in general, the air loads and the associated deformations of a wind-tunnel model under test are of major interest. Obviously, a relatively elastic balance at the root of a wind-tunnel model would change the model deformations significantly, namely depending on flow parameters. In Sect. 8.2.3 an example will be presented, where an elastic half-span model of a swept wing is compared with its rigid equivalent. The comparison concerns the behavior of the slope of the force coefficients depending on the Mach number, which is quite different. Again such measurements postulate a stiff balance at the root.

The other examples concern measurements of unsteady forces in general. In this case, the lowest natural frequency of the force-measuring system should be much higher than the frequency range of interest, i.e., vortex shedding – or buffet frequencies. Keeping in mind that, in principle, the balance/model system is an accelerometer comprising the model as a seismic mass and the more or less stiff sensor as the spring, one can imagine that the requirement of a very high natural frequency is much easier to meet using transducers, which are inherently stiff.

In particular, time-series analysis, i.e., the determination of RMS values, spectra, Strouhal numbers, correlation functions etc., needs force/moment data that are nearly free of interference (cross talk) or resonance effects caused by the measuring system itself. To demonstrate that piezo balances are able to cope with these challenges, several examples will be presented in Sect. 8.2.3, involving flow around bluff bodies or wing models, which were at rest or oscillating.

For investigation of small force fluctuations with a large steady preload, a fundamental advantage of piezoelectric measurement systems is that the charge generated by the steady preload can be short-circuited. Thus, for the detection of extremely small fluctuations, the possibility exists to exploit the full resolution of the measuring system independently of the preload.

On the other hand they have the disadvantage that only quasistatic measurements are possible, which is probably the main reason behind the scepticism towards the piezo measuring technique. The decay of the charge and fault currents in the charge amplifier cause a zero-point drift. However this is only essential when measuring small forces over long time periods. Nevertheless, the problem is not as critical as generally believed, since the drift is nearly linearly in time. This problem has been investigated and is discussed in detail.

An early application of piezoelectric force transducers in a large wind tunnel was performed by *Bridel* [8.15]. The mechanical design of his half model balance for the low-speed wind tunnel at Eidgenössische Technische Hochschule (ETH) Zürich was similar to a corresponding strain gauge balance, but he replaced the force-measuring elements by one-component piezoelec-

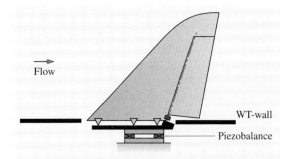

Fig. 8.57 Typical arrangement for force measurement on wall-mounted models using an external platform balance. The model is a fin including a rudder, attached to a massive plate, which is fixed to the balance

tric transducers. The first balance applied by the author was for measurements of steady and unsteady forces on two-dimensional bluff bodies in the high-pressure wind tunnel in Göttingen, which was based on multicomponent force transducers and is described in [8.16–18]. *Cook* [8.19] reported unsteady measurements in building aerodynamics using a piezoelectric platform balance.

In the last 25 years, piezo balances have been applied in many projects performed in close cooperation with colleagues from research and industry, i.e., the Office National d'Etudes et de Recherches Aérospatiales (ONERA), European Aeronautic Defence and Space (EADS), Airbus etc. [8.20, 21]. The largest wind tunnel where piezo balances were applied several times for half model testing was the French transonic wind tunnel, S2, in Modane.

The objective of this section is to give an impression of the possibilities, advantages and limitations offered by the use of piezoelectric balances. Several types of external balances are discussed for wall-mounted models, which can be suspended one-sided or twin-sided.

Figure 8.57 should give an impression of a typical test setup for an external balance measurement at the wall of a wind tunnel. A model, a half wing (here a fin), a car or a building is then attached to a measuring platform such that the cutting forces and moments between the root of the model and a reference base of the wind tunnel can be measured.

Additionally an internal balance is described, which is usually located inside the model under test and was developed by *Psolla-Bress* et al. [8.22].

Reports are given on selected measurements performed in very different wind tunnels ranging from low-speed to transonic flow, from cryogenic to high-pressure principles and from short time to continuous running. The projects span a wing/engine combination in a low-speed wind tunnel via flutter tests with a swept wing performed in a transonic wind tunnel to bluff bodies in a high-pressure and cryogenic wind tunnel, respectively. These tests serve as examples for the discussion of the fundamental aspects that are essential in developing and applying piezo balances. The principle differences between strain gauge and piezo balances will be discussed.

8.2.1 Basic Aspects of the Piezoelectric Force-Measuring Technique

The piezoelectric effect can be shortly explained as follows. If a crystal without a center of symmetry is composed of ions, then a mechanical deformation of the crystal results in an electric charge on its surface and vice versa (the inverse piezoelectric effect). The reason is that the individual negative grid points (ions) and the positive ones are shifted against each other, resulting in a polarization of the entire crystal (Fig. 8.58). This behavior is typically of quartz (SiO_2) or special ceramics, materials that are usually used for the design of transducers. For high-accuracy measurements, quartz is preferred because it is superior compared to ceramics in terms of long-term stability, linearity, temperature behavior and lack of hysteresis and pyroelectric effects. Depending on how the crystal is cut in reference to its crystallographic axis, distinguishing for example, between a longitudinal and shear effect, it will exhibit different piezoelectric coefficients and consequently different sensitivities. Depending on the measurement problem, one effect may be more appropriate then the other, which will become obvious in the description of multicomponent transducers.

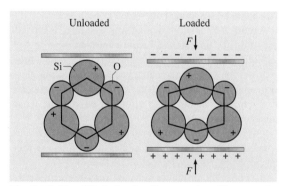

Fig. 8.58 Simplified sketch of a crystal to illustrate the shift of the negative and positive charges against each other due to mechanical deformation (the piezoelectric effect)

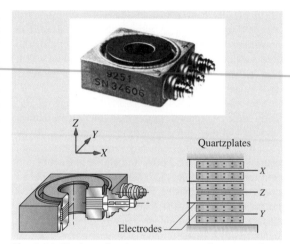

Fig. 8.59 Design of a three-component force transducer (after Kistler). In detail, for every component there is a pair of quartz plates, which are arranged in such a way that the sensitivity is doubled [8.23]

Force Transducer

Typical piezoelectric force transducers have the form of a washer, in which a quartz ring is mounted between two steel plates in the transducer case. If the quartz exhibits the longitudinal piezoelectric effect, then we have a one-component force transducer for axial direction. In the following balance applications we will use three-component transducers; in this case a washer consists of three stacked quartz rings, mechanically operating in series under prestress (Fig. 8.59). Two rings are shear sensitive and are aligned to the measurements of both orthogonal components of a shear force in an arbitrary direction. The third ring is pressure sensitive and measures the compression force on the transducer. As mentioned the quartz packets are prestressed and protected by the rust-proof, tightly welded transducer case. Typical technical data of such transducers, which were applied with the herein described balances, are listed below (Kistler type 9067(9252)). The size is about 56 mm (24 mm) squared. The values in the brackets belong to the smaller transducer (9252).

Shear forces: F_x, F_y: ±20 kN (2.5)
Axial force*: F_z: ±20 kN (5)
Rigidity: C_x, C_y: 3500 N/μm (300)
                C_z: 8000 N/μm (1000)
Sensitivity: E_x, E_y: 8 pC/N (shear effect)
                E_z: −4 pC/N (longitudinal effect)
*standard prestress = 160 kN (25)

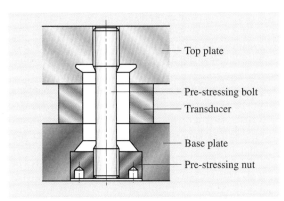

Fig. 8.60 Assembly of a three-component force transducer (after [8.24]). The mechanical rigidity of the prestressing bolt (force shunt) is small (10%) compared with the rigidity of the transducer

The possible range for the axial force F_z depends on the prestress with which the transducer is assembled, a point which will be discussed in the next chapter. The sensitivity is a property of the quartz itself and not of the transducer design, and thus is nearly independent of the size of the transducer and the orientation of the quartz disks when considering shear effects. The criterion for the selection of the transducer is the available space, the required range, the desired rigidity and weight.

Special versions of the force transducers were manufactured by Kistler for applications down to cryogenic temperatures ($T = -150°$) or pressures up to 100 bar, described in Sects. 8.2.2, 2.3. It should be noted that such force transducers were applied both in water and down to helium temperatures (4 K).

Integration and Arrangement for Multicomponent Force and Moment Measurements

Figure 8.60 shows the typical assembly of such an individual load washer. The force transducer is prestressed by an elastic bolt between a base plate and a top plate so that the shear forces can be transmitted by friction. As mentioned, the prestress can have values up to 160 kN, which is required in order that shear forces up to ±10% of the prestress can be transmitted by friction. In order to measure all six components of a force and a moment, a spatial arrangement of at least three multicomponent force transducers is necessary. Experience showed that a platform with four transducers are superior because of its higher stiffness. Figure 8.61 illustrates the principal scheme, which is the same in most platform balances.

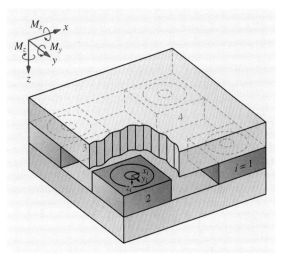

Fig. 8.61 Scheme of a multicomponent platform balance based on four three-component force transducers

The lowest eigenfrequency of such a platform is usually given by the bending of the top plate or the base plate.

It is obvious that, for applications in wind tunnels, such a platform can be installed out of the flow behind the wind tunnel wall, as shown in Fig. 8.57.

In Fig. 8.62 we see a platform balance in more detail, attached to a wing/engine combination. This example shall serve as a paradigm to explain the essential points concerning the design and construction of external piezoelectric balances, typically mounted outside the wind-tunnel walls on a turntable or solid base. The bal-

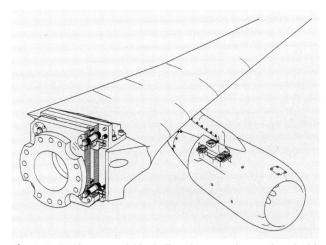

Fig. 8.62 Half-span model including ejector engine, equipped with two piezo balances: one at the root of the wing and the second integrated in the engine

ance integrated in the engine, also shown in Fig. 8.61, is described in *Schewe* [8.25] and *Triebstein* et al. [8.26]. In this case as well four multicomponent transducers are sandwiched between the pylon and the core element of the engine model.

Observing the platform balance at the root of the wing, it is obvious that it corresponds with the scheme of four multicomponent transducers (Fig. 8.62) that are prestressed by elastic bolts between two solid plates. When designing and applying piezoelectric force measuring platforms it is important to consider the force transmission in the balance. This is the junction between the balance and the model, and the attachment of the balance at a device of the wind tunnel (a turntable, for example). When designing the test setup and the balance, one must consider deformations of the solid plates within the balance. Keeping in mind that the measuring deformations are only of the order of microns, one can imagine that even the smallest deformations can produce interference effects between the individual components. Concerning the unavoidable deformations, the balance should be designed in such a way that the deformations that occur are antisymmetric. Such deformations of opposite sign can compensate for each other since the resultant force or moment components are composed (i.e., are the sum) of several (individual) signals. In other words, deformations that fulfill certain symmetry conditions can be seen as internal forces of the balance, which are not present in the resultant values, i.e., composed signals. Thus, in order to achieve a high degree of symmetry, the arrangement of the four load washers should be quadratic, if possible. Further, the base plate of the balance is not directly attached to the more or less plane surface of the test setup. A special flange between the two provides decoupling between the deformations caused by the attachment and the base plate of the balance. On the other side, the wing is screwed to the balance in such a way that the attachment screws are arranged symmetrically and as close as possible to the prestressing bolts of the balance.

In general, a multicomponent piezo balance must be as accurately designed and constructed as high-fidelity optic equipment.

This type of balance has been applied in several variations, mainly differentiated by size. A special application is described later in Sect. 8.2.3, where two such balances, attached at each end of a two-dimensional (2D) airfoil model in a bilateral suspension, were used. In this case, both balances could be connected in parallel, such that they acted electrically as one balance.

A further special version of this type of platform balance was designed for application in cryogenic conditions. This balance, called a *cryo-balance* in the following is also based on four modified transducers (type 9252) as described in Sect. 8.2.2. The parts of the balance are made of special steel for cryogenic applications called *austenitic chrome nickel steel*.

If three-dimensional bodies are to be investigated, for example complete models of aircrafts or delta wings, then a rear sting support for the model is necessary, usually attached at a turning device to vary the angle of attack (Fig. 8.63). In this case, the force measurement is performed by a balance located at the end of the sting and inside the model. This type of balance, called an internal balance, usually has the form of a long circular

Fig. 8.63 Opened test section of the transonic wind tunnel. The rear sting support of a delta wing model is visible. The internal piezo balance is mounted inside the model [8.22]

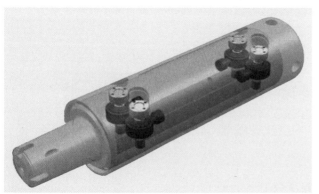

Fig. 8.64 Internal six-component piezo balance developed in cooperation by DLR and Kistler (after [8.22])

pipe segment and is difficult to design due to the small available space.

Psolla-Bress et al. [8.22] developed in cooperation with Kistler such a six-component sting balance also based on four piezoelectric force transducers manufactured by Kistler (type 9017B). These load washers have a circular case with a diameter of 16 mm. The schematic illustration of the balance in Fig. 8.64 shows that the arrangement of the transducers and the principle of operation corresponds to the platform balances described below. The outer annulus of the circular balance at which the model is fixed corresponds to the force conducting top plate and the counterpart acts as the base plate, which is attached at the support of the test section.

Details can be found in [8.22].

Comparison Between Piezoelectric and Strain Gauge Balance for a High-Pressure Wind Tunnel

Given the force measurements in the previously mentioned high-pressure wind tunnel, we have a good opportunity to compare both principles, since there are two external balances for performing three-component measurements on two-dimensional airfoils or bluff bodies.

The principal difference between the active piezoelectric measuring system and the passive strain gauge system was already discussed in the introduction.

Figures 8.65 and 8.66 further illustrate the fundamental differences between these two measurement systems. Figure 8.65 depicts the principle of a strain gauge balance for three-component measurements on a two-dimensional airfoil model. The model is fixed at both ends by a force conducting plate supported by three elastic ring elements. The ring elements, on which

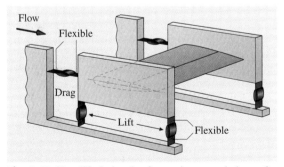

Fig. 8.65 Simplified sketch of a strain gauge balance for airfoil measurements. The strain gauges are fixed onto the individual rings (sketch by K.E. Möller)

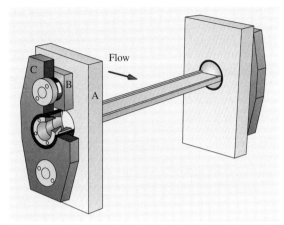

Fig. 8.66 Schematic drawing of the piezo balance for a high-pressure wind tunnel; (A) wall of the test section, (B) three-component load washer (Kistler 9067), (C) force-conduction top plate, the four load washers are prestressed by elastic bolts between A and C such that the shear forces can be transmitted by friction

the strain gauge sensors are applied, represent the force transducers. When considering the drag direction (x), it is obvious that a sufficient measuring deflection in the ring element for drag also requires a corresponding flexibility of the lift measuring elements in the x-direction and vice versa. In other words, the necessary measuring deflections for every component require a mechanical decoupling of the components that are perpendicular to the desired one. This requirement consequently reduces the desired rigidity of the entire system. In reality, the setup is highly sophisticated using six high-end force transducers manufactured by Hottinger, but the mechanical decoupling remains necessary in any case. This balance is described by *Schaake* [8.27].

For comparison we consider the piezo balance in Fig. 8.66 which was the first balance developed, and has been applied by *Schewe* [8.16] since 1980. At first glance, there are no elastic elements, but two rather stiff multicomponent load washers (B) at each side. With the help of elastic bolts and the two common top plates (C), two pairs of load washers are pressed to each vertical wall (A) of the test section. A modification of the force transducer for use up to 100 bar will be described in Sect. 8.2.3. The circular ends of the freely suspended model are passed through the wind-tunnel walls (A) and the ends are clamped in the holes of the force conducting top plate by a ring locking assembly. This detail is noteworthy as it provides rotational symmetry of the strain distribution in the force conducting top plate. The test section is the basis for the force measurements as the force is introduced over both force conducting top plates (C). This means that the force components, drag (F_x) and lift (F_z), which act as shear forces on the load washers, are transmitted by friction.

It is obvious that mechanical decoupling is not necessary since each of the four load washers is able to measure three components, where here the axial component is not relevant. The possibility to forego the mechanical decoupling further enhances the rigidity of the balance, leading to the favorable property that, applying for example a circular cylinder, the natural frequency is determined by the rather high bending frequency of the model. In the case of the strain gauge balance, the elastic elements act as springs; consequently the balance in combination with the model and attachment mass determine the natural frequency, which is typically rather low.

An important difference between both principles concerns the ability to measure force components of very different magnitude simultaneously. For example, this may be the case when investigating airfoils in attached flow, where the lift is roughly two orders of magnitude larger than the drag force. For this reason the drag force elements of strain gauge balances are typically more sensitive than the lift force elements. In other words, there is the possibility to adapt the individual force ranges by varying the corresponding sensitivities, whereas in the case of multicomponent piezoelectric transducers this possibility does not exist, as the sensitivities in one element are fixed. This small disadvantage can be partly balanced by selecting a range of higher sensitivity in the charge amplifier.

Processing of Signals

Referring to the platform balance (Fig. 8.61) and the balance for the high-pressure wind tunnel it is obvious that the four load washers deliver 12 signals. The proper combination of these signals allows, in principle, all six components of the resultant force and moment to be determined with respect to the geometric center of the four transducers.

Considering first the simpler, two-dimensional case of measurements in the high-pressure wind tunnel, where only a three-component measurement of drag, lift and pitch moment is required. In this case, only shear forces on the load washers are relevant. Since the sensitivity of all elements to the shear forces is equal, related components can be wired electrically parallel and routed to a charge amplifier. Parallel wiring of the four components in the lift direction yields the resultant lift, while the pitch moment is obtained from the difference between the two lower drag components and the two upper drag components. Total drag corresponds to the sum of the lower and the upper drag force. Finally three electric signals are produced for a three-component measurement.

Considering a three-dimensional problem, for example a half model attached to a platform balance or an internal sting balance, in which all six components of the resultant force and moment are desired. In this case the number of charge output signals is reduced to eight, when the corresponding individual components with a common line of action have been paralleled directly.

The charge amplifier converts electrical charge into a proportional electric voltage. It is, in principle, an inverting direct-current (DC) amplifier and integrates the input current $i = dQ/dt$, which is created by the changes of charge at the quartz sensor. This is made possible by an amplifier with capacitive feedback, which has the highest possible internal gain and insulation resistance at the input [8.23]. In particular for quasistatic measurements, a high resistance R at the input is a prerequisite for low drift effects. Thus, all cables and plugs between the transducer and the charge amplifier must have an insulation resistance of higher than $R \approx 10^{13}\,\Omega$. This value is more than three orders of magnitude higher than that of ordinary bayonet Neill–Concelman (BNC) cables.

For quasi-steady measurements, the outputs of the three or eight charge amplifiers (for example, Kistler 5007) are evaluated by a personal computer equipped with an analogue/digital (A/D) conversion system. Apart from the calculation of the resultant forces and moment, a simple drift correction procedure, which will be discussed in Sect. 8.2.2, is used.

8.2.2 Typical Properties

Calibration Tests (Interference Effects)

As an example of calibration behavior and interference effects, we use the external balance of the high-pressure wind tunnel, which is referred to as a HDG balance in the following. In the quasistatic mode of the charge amplifiers the static calibration of the balance was performed with weights for the drag force F_x, lift force F_z and pitch moment M_y. Figure 8.67 shows an example of a calibration curve for the lift F_z, including interference on the drag F_x.

The charge Q_z [nC] is plotted against the calibration weight, which was applied in the direction of the F_z component. The weight was varied over two orders of magnitude and it is evident that the linearity of $Q_z(F_z)$ is

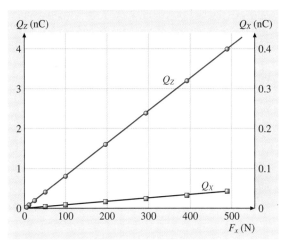

Fig. 8.67 Example of a calibration curve for the lift force F_z and the interference from the z- to the x-component. $Q_x(F_z)$ is enlarged by a factor of 10. The interference $Q_x(F_z)$ is approximately linear ($dQ_x/dQ_z = 1\%$)

quite good. The scale of the right-hand ordinate, indicating interference on the x-component $Q_x(F_z)$, is enlarged by a factor of 10. Nevertheless, we see that the interference effects on the x-component are small and nearly linear. A quantitative measurement of the interference is the ratio of the two slopes (i.e., the sensitivities for both components) $Q_z(F_z)$ and $Q_x(F_z)$, which amounts to $dQ_x/dQ_z = 1\%$. All calibration results for a multicomponent force-measuring device can be collected in the calibration matrix E_{ij}, which gives a summary of the static properties of the system. The following relation holds:

$$Q_i = E_{ij} F_j \quad E_{ij} = \frac{\partial Q_i}{\partial F_j}$$

with $i, j = 1, 2, 3$ for a three-component calibration. Q_i are the measured charges and F_j are the desired components of the force or pitch moment. The matrix elements $E_{ij} = dQ_i/dF_j$ are the sensitivities, determined by calibration using weights for the individual components.

In Fig. 8.67, for example, we determined $E_{22} = dQ_z/dF_z = 8.117\,\mathrm{pC/N}$ and $E_{12} = dQ_x/dF_z = 0.085\,\mathrm{pC/N}$. With a circular cylinder as a test model, we obtained the following calibration matrix:

$$\begin{pmatrix} Q_x \\ Q_z \\ Q_m \end{pmatrix} = \begin{pmatrix} 8.127 & 0.085 & -0.065 \\ 0.025 & 8.117 & 0.071 \\ -0.036 & 0.066 & 8.187 \end{pmatrix} \begin{pmatrix} F_x \\ F_z \\ M'_y \end{pmatrix} \quad (8.41)$$

The matrix elements of the last column (E_{i3}), belonging to the calibration of pitch moment, are normalized with the length $l = 0.2\,\mathrm{m}$ ($M'_y = M_y/l$). Thus, all matrix elements have the same dimension pC/N. The length l corresponds to the lever arm used in the calibration of the pitch moment. Under ideal conditions, all elements E_{ij}, $i \neq j$ that represent the interferences must be zero, while the diagonal elements ($i = j$) represent the sensitivities for the individual components $E_{11} = dQ_x/dF_x$, $E_{22} = dQ_z/dF_z$ and $E_{33} = dQ_m/dM'_y$.

Nevertheless, it is evident that interference elements with $i \neq j$ are at least two orders of magnitude smaller than the diagonal elements. Furthermore, the sensitivities (i.e., the slopes) for both force components F_x and F_z are identical within the accuracy possible in the drawings ($E_{11}/E_{22} = 1.0012$).

Dynamic Behavior and Resolution

The dynamic behavior can also be demonstrated by means of the high-pressure wind tunnel balance for the following reasons: first, it can be shown that, despite the extensive arrangement of the transducers, there are no eigenfrequencies below the rather high natural bending frequency of a circular cylinder model; second, we have the possibility to compare with a strain gauge balance.

The eigenfrequencies were obtained by the Fourier transform (Fig. 8.68) of the impulse response from the balance signals $F_x(t)$ and $F_z(t)$. Because of the high rigidity of the quartz elements themselves, the lowest eigenfrequency is determined by the bending of the

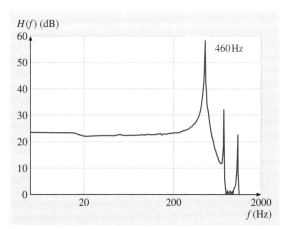

Fig. 8.68 Transfer function for the x-component obtained by Fourier transform of the impulse response, taken using the circular cylinder in the external balance of the HDG

model that connects the two parts of the balance. In this special case of the circular cylinder, the lowest eigenfrequency is $f = 460$ Hz for the x-component. Due to the different clamp conditions, the eigenfrequency is higher for the z-component with a value of $f = 535$ Hz. Finally, Fig. 8.68 shows that there are no significant resonance frequencies lower than $f = 460$ Hz. This value is high enough, considering that in our case typical vortex shedding frequencies of bluff bodies are lower than 200 Hz.

Schaake [8.27] applied in the high-pressure wind tunnel the external strain gauge balance for investigating two-dimensional bluff bodies and found resonance effects around 100 Hz caused by the balance itself. The resonance amplitudes were so high that the balance had to be blocked at high Reynolds numbers, which are coupled with high loads.

Here the undesired situation that the resonances of the force measuring system lie in the frequency range of interest occurs, in this case in the frequency regime of the vortex shedding caused by the bluff bodies.

The next example concerns the cryo-balance, which was mounted at the calibration setup. In addition a large body with rectangular cross-section, a simplified model of a building, was attached to the balance with a minimum bending frequency of 16.4 Hz. The low bending frequency was achieved by a compliant junction between the balance and the model.

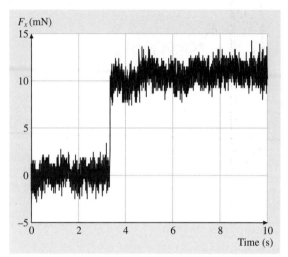

Fig. 8.70 Demonstration of the high resolution of the piezo balance in its range of highest sensitivity. A 1 g weight was hung from the balance, and the significant jump was caused by unloading the weight

A very simple experiment shall demonstrate the static and dynamic behavior using a small weight.

In Fig. 8.69, we see a zoom of the time history $F_x(t)$ for a situation where a 1 g metal weight was dropped on the model; the height of fall was approximately 0.02 m. After a high-frequency response due to the impact of the weight, the excitation of the natural frequency of the cylinder is obvious. Additionally the rather small change in the mean value could be resolved. For the calculation of the mean values of F_x before and after the impact, the entire measuring time of 10 s was used.

In order to demonstrate the limit of resolution at optimal conditions, the 1 g (≈ 10 mN) weight was hung from the balance without a model, and a jump was then caused by unloading the weight. Figure 8.70 shows the corresponding jump in the time function of the balance signal F_x, demonstrating that the resolution is even better than 0.01 N. The peak-to-peak value of the noise of the entire system is roughly 3–4 mN. For this measurement the range of the charge amplifier was switched to highest sensitivity. Finally it should be noted again that these experiments can be performed independent of the static preload and that the results are also typical of balances based on the larger force transducer (type 9067), as the sensitivities are equal.

In order to provide an example illustrating the attainable measuring range, we consider the HDG balance or a platform balance, which are based on four of the larger transducers (9067). The typical maximum load

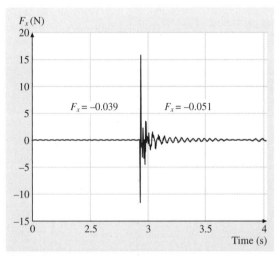

Fig. 8.69 Zoom of the time history for a situation where a weight of 1 g was dropped on the model, which was attached at the balance (height 0.02 m). The excitation of the natural frequency is obvious. Also the jump in the mean values of F_x before and after the impact is measured

for each component is at least $F_i \approx 10$ kN. Thus, the measuring range extends over six orders of magnitude, since the threshold for dynamic measurements is as low as 0.01 N. Experience shows, as does the later result in Fig. 8.71, that quasistatic measurements are possible down to below $F \approx 1$ N. Consequently, in quasistatic mode a measuring range of four orders of magnitude is attainable.

Reproducibility

In order to get a measure for the reproducibility, a calibration cycle was repeated 10 times over a time period of a few hours under normal laboratory conditions. For this test, a platform balance for half-span models, based on the larger transducer (type 9067), was used. One calibration cycle was performed with calibration weights of 5, 10, 20, 35 and 50 kg, lasting roughly 15 min. Table 8.5 shows the results of the measurements. The second column displays the sensitivity of the x-component E_x in [pC/N], determined from the slope of the calibration curves. The third column singles out one measuring point of the calibration cycle, the voltage U_x corresponding to a load of 50 kg.

Calculating the mean and the standard deviation we find for the sensitivity of the x-component:

$$E_x = (7.8634 \pm 0.0005) \text{ pC/N} . \tag{8.42}$$

The ratio between the difference of the peak values and the full-scale range can be seen as a measure of the reproducibility. These values can be derived from the

Table 8.5 Results of 10 calibration cycles repeated under normal laboratory conditions

Cycle No	E_x (pC/N)	U_x (V/50 kg)
1.	7.8637	1.9287
2.	7.8638	1.9287
3.	7.8625	1.9283 min.
4.	7.8634	1.9285
5.	7.8630	1.9285
6.	7.8633	1.9284
7.	7.8632	1.9285
8.	7.8631	1.9285
9.	7.8644	1.9288 max.
10.	7.8636	1.9285
Mean	7.8634	1.9285
Stand. dev.	0.0005	0.0002

second column: largest deviation $U' = U_{x\text{max}} - U_{x\text{min}} = 0.5$ mV, full-scale range $(x_{12} + x_{34})$ each $10 \text{ V} \Rightarrow U_{\text{FS}} = 20$ V.

Thus the reproducibility has a value of:

$$\frac{U'}{U_{\text{FS}}} = 0.025\text{\textperthousand} . \tag{8.43}$$

These results are also, at least, representative of other external balances and demonstrate that it is possible to attain good accuracy if one is able to hold the ambient conditions constant.

Behavior Under Cryogenic Conditions

As mentioned in the introduction, a piezoelectric balance was built for the cryogenic Ludwieg tube in Göttingen (KRG) in such a way that the balance itself is totally in the cryogenic environment. This fact is not self-evident as such a balance has to withstand a large amount of thermal stresses when the temperature is changed over the entire range of the cryogenic facility.

One motivation for this activity was to clarify the question of to what extent force measurements could be made with piezoelectric transducers under cryogenic conditions.

At our request, a special version of transducer type 9252 was manufactured by Kistler for applications at cryogenic temperatures. The modification concerns the insulators in the plugs, by default made from silicone, which were replaced by ceramic ones such that application down to cryogenic temperatures ($T = -150°$) and pressures up to 10 bar was more feasible. The modification mainly ensures that no humidity diffuses into the transducer case during the unavoidable large temperature changes.

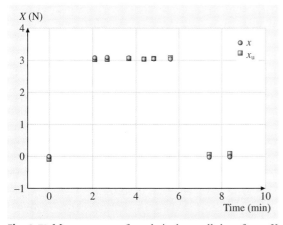

Fig. 8.71 Measurement of a relatively small drag force X on a building model, taken in the boundary layer wind tunnel of the University of Florence using the cryo-balance ($u = \text{const} = 4.3$ m/s. Wind off at $t = 0$ and $t > 7$ m). X_u: uncorrected values, i.e., without drift correction

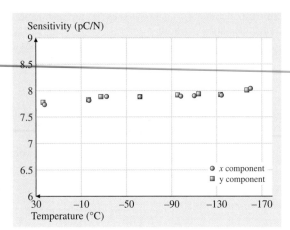

Fig. 8.72 Determination of the sensitivity of the shear components F_x and F_y of the cryo-balance depending on temperature

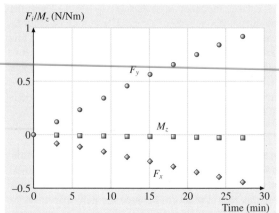

Fig. 8.73 Demonstration of the drift behavior in the range of highest sensitivity of the entire force measuring system under ideal environmental conditions

Before application in the wind tunnel, the balance with the modified multicomponent transducers (Sect. 8.2.1) was tested and calibrated in a cryostat down to a temperature of $T = -160°$. The low temperature was produced by injection of vaporized nitrogen. For the connection between the force transducer in the cryostat and the charge amplifier outside, the usual cables for piezoelectric sensors could be used.

In the pre-tests we found no significant deviation in the behavior of the balance concerning drift or electrical noise of the signals compared with normal conditions.

The calibration was conducted using weights, which were appended outside of the cryostat via cables. Concerning the axial component F_z, it was found that the cool-down from $24°$ to $-169°$ led to a contraction of the prestressing bolts, resulting in a 1.5 kN increase of the prestressing force, which is one order of magnitude smaller than the standard prestress. This effect was caused by the different thermal expansion coefficients of the transducer and the prestressing bolt.

Both shear components F_x and F_y showed good linearity compared to that at normal conditions, and the slopes, i.e., the sensitivities, as a function of temperature are displayed in Fig. 8.72. It is obvious that the sensitivity changes slightly with temperature. This sensitivity shift is caused by the fact that the corresponding piezoelectric coefficient (d_{11}), which is responsible for the shear effect, increases slightly with decreasing temperature [8.23].

In this context, it should be mentioned that, at normal conditions, arrangements of force transducer are also sensitive to temperature changes during a measurement cycle. This problem concerns, in particular, the direction of the prestressing bolts. Thus, in any case the balances must be protected against changes in thermal distribution, at least while running a measurement.

Measuring Time for Quasistatic Measurements

As mentioned, the piezoelectric measuring system is an active one and generates an electrical quantity. Herein lies a fundamental disadvantage, namely that only quasistatic measurements are possible. A time constant $T = R_g C_g$, characteristic of the exponential decline of the charge signal, results from the entire input capacity C_g and the finite insulation resistance R_g, posed by the transducer, cord, plugs and amplifier input etc. Further fault currents in the charge amplifier, which are fortunately independent of the load, cause the zero point to drift. Thus the combined error is smaller for larger charges on the quartz, meaning that the effective measuring time increases as the loading and the sensitivity increase. Fortunately, the drift is dominated by the fault currents in the input devices of the charge amplifier, which are nearly constant, leading to a linear drift depending on time at the output. The sign of the drift can be positive or negative. A simple linear correction for every test point in time can be computed as follows. For each measurement, the raw data, integrated over time, and the corresponding time are stored. Knowing the zero point before the measurement (flow speed $U = 0$) and after the measurement (where the flow speed is zero again), a correction can be computed by linear interpolation.

In order to give an impression of the drift behavior under extreme conditions, the time function of the three components of the cryo-balance were recorded, while the balance was attached to the calibration setup, i.e., without load and under good environmental laboratory conditions. In addition, the cables and plugs were verified as having an overall resistance of at least $R \approx 10^{13} \, \Omega$. In another (unfavorable) condition, the system was switched in the range of highest sensitivity, where the drift effect was maximal. Figure 8.73 shows, firstly, that the drift was rather linear for all three components, and secondly, that at least under good environmental conditions the drift effects are not as extensive as generally believed. In other words, the results show the probable lower reachable limit when components such as plugs and cables are selected and the test conditions are nearly optimal.

In order to provide an impression of the drift error at small loads, an example is presented in Fig. 8.71, taken in the boundary-layer wind tunnel of the University of Florence. A simplified building model was mounted on the cryo-balance as shown in the sketch in Fig. 8.57. This figure shows a measurement of a relatively small drag force of $X = 3.1$ N at a low flow speed $u = \text{const} = 4.3$ m/s over time t. The airflow was started at $t = 0$ min and switched off for $t > 7$ min. The index u (X_u) denotes values without drift correction. Looking at the square symbols (uncorrected) at $t = 0$ min, we see a small deviation in the negative direction, caused by switching to the operate mode. At the end of the measurement run at $t = 8.5$ min, there is also a small deviation but in the positive direction. The small difference in the outputs between the start and end is caused by a drift in the positive direction. After correction, the corresponding points (o) lie on the zero line.

Although the wind-tunnel speed is not exactly constant, we can state that the effect of the drift of the zero point is not as extensive as generally believed.

8.2.3 Examples of Application

Bluff Bodies in a High-Pressure Wind Tunnel
The flow around a circular cylinder is a classic problem of fluid dynamics, which exhibits strong Reynolds-number effects, mirrored in drastic variation of the drag coefficient and Strouhal number. A second classic bluff-body case is the sharp-edged square-section cylinder, which has a high drag coefficient that is nearly independent of the Reynolds number.

Thus, both sections were selected as examples concerning force measurements performed in the high-pressure wind tunnel (HDG). The flow speed ranged from 2 m/s to 38 m/s; the pressure could be increased

up to 100 bar and the test section measures 0.6×0.6 m^2. Since even steady loads in this low-speed wind tunnel can vary over four orders of magnitude merely by changing the flow parameters, the results give an impression of the large dynamic range of a piezo balance. Indeed, this property of the wind tunnel and the deduced requirement of a large dynamic range motivated the author to design and built his first piezo balance [8.16].

The principle setup of the balance was described in Sect. 8.2.1 (Fig. 8.66). We selected the larger three-component force-measuring elements (Kistler, type 9067), because of their large load range. A special difficulty when measuring in the high-pressure wind tunnel is the fact that the model, the sensors and the balance are located in the high-pressure section, while it is desirable to have the electronic equipment outside at atmospheric pressure. Since one of the primary prerequisites for quasistatic measurements is that an insulation resistance of $R \approx 10^{13} \, \Omega$ is reached for all connections between the force element and charge amplifier, a special solution was necessary for the crossover from the high- to the atmospheric-pressure sections. Because this requirement is difficult to satisfy with hermetic connectors, the necessary cables were inserted into flexible pressure-resistant hoses, which were pressure-seal connected at the transducer and then laid through the wind tunnel

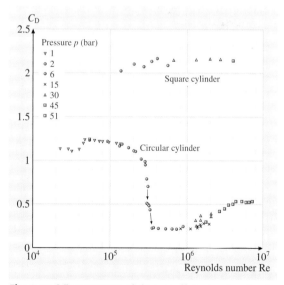

Fig. 8.74 Measurement of drag coefficients of a circular and a square cylinder in a high-pressure wind tunnel. Due to the large dynamic range of the balance the individual Re number ranges could be overlapped by merely changing the flow parameters

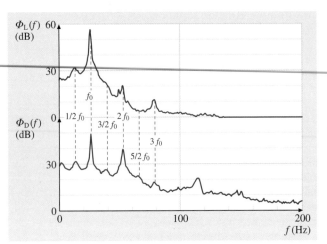

Fig. 8.75 Power spectra of the lift and drag fluctuations on a square cylinder at an angle of incidence of $\alpha = 10°$. Even the very low intense sub- and superharmonics can be resolved. The distance of the peak height is up to 50 dB

wall and connected to the charge amplifier outside of the high-pressure section. In this way, atmospheric pressure was maintained within the element, as well as in the corrugated hoses, so that an increase of static pressure in the wind tunnel acts as a preload on the element. Since the zero point in the sensor/charge-amplifier system can be chosen arbitrarily by short-circuiting the charge, the preload in the axial direction has no significant influence on the accuracy of the measurement.

This modification of the force transducer for application at high pressure up to 100 bar was also manufactured by Kistler.

Such modified transducers were also applied in several water tanks in Hamburg to measure the forces on ship models and marine structures [HSVA Jochmann]. A detailed description of the balance and many results concerning flow around more or less bluff bodies can be found in [8.17, 18, 28–30].

As an example of a quasisteady measurement, Fig. 8.74 shows drag coefficients of a circular cylinder and a square cylinder as a function of the Reynolds number. The flow speed was increased to 40 m/s and the pressure up to 51 bar. In the case of the circular cylinder, the smallest drag force was 0.8 N at the lowest Reynolds number. The largest drag force was about 2000 N for the square cylinder at $Re \approx 5 \times 10^6$. Since only about 20% of the range of the balance was used, it was demonstrated that, in principal, steady measurement over four orders of magnitude was possible. As for fluctuating forces, the dynamic range is extended to

six orders of magnitude because the resolution is about 0.01 N.

Figure 8.75 depicts a typical dynamic measurement, which is representative of the ability to resolve and detect even very weak nonlinear effects in a global force measurement. The figure shows the power spectra of the lift – and drag fluctuations on a square cylinder at an angle of incidence of $\alpha = 10°$. In this state, the free separated shear layer is probably more or less regularly attaching to the side wall of the square cylinder, producing sub- and superharmonic peaks. It is surprising that this localized effect can be found and resolved in a global, i. e., integrated measurement. In addition, apart from the fundamental Strouhal peak and the sub- and superharmonics, there are significant peaks at 3/2 and 5/2 of the vortex shedding frequency [8.31]. The dynamic is reflected in the distance between the spectral densities, which is up to 50 dB.

Further applications concerning force measurement on a bridge section and an airfoil at a high angle of attack can be found in *Schewe* [8.29].

Typical Half-Span Model in a Conventional Low-Speed Wind Tunnel

The wing/engine combination described in Sect. 8.2.1 was tested in the low-speed wind tunnel $(3 \times 3 \, \text{m}^2)$ in Göttingen. The project comprised force and pressure distribution measurements. The influence of the thrust of the ejector engine on the aeroelastic and aerodynamic behavior of the wing/engine combination was of particular interest. For this reason, in the engine model four multicomponent force transducer were installed to measure separately the global forces on the engine. More about the motivation for the project, a detailed description of the test setup, and the results of the pressure measurements can be found in *Triebstein* et al. [8.26].

Figure 8.76 shows as an example of the steady normal force and its RMS value for angles from $-4°$ up to $14°$, taken at a flow speed of 60 m/s. In this case the balance was not fixed to the wind tunnel, but rotated with the model, thus the normal force on the wing and not the lift was measured. The results demonstrate that the measurement of steady forces is possible with sufficient accuracy. The angle $\alpha = 0$ was measured twice – at the beginning and at the end of the test run. Within drawing accuracy, both measuring points overlap.

Referring to the span of the measurement range, it should be mentioned that, at the maximum normal force of more than 1000 N, only about 6% of the range of the balance is used, meaning that this balance can also

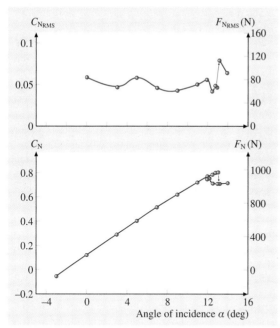

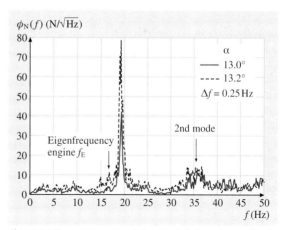

Fig. 8.76 Normal-force coefficient and corresponding root-mean-square (RMS) value depending on the angle of incidence of a wing/engine combination in a low-speed wind tunnel. $C_N(\alpha)$ is not a straight line since the balance is rotated with the wing. After the discontinuous drop of the steady normal force at $13°$, the RMS jumps to nearly double its previous value

Fig. 8.77 Power spectra of the normal force immediately before ($\alpha = 13°$) and at the onset ($\alpha = 13.2°$) of flow separation. Concerning the balance, there are no resonance peaks caused by the force-measuring system itself

be applied at much higher loads (i.e., in the transonic range) without any changes being necessary.

As an example of an unsteady measurement, in the upper part of Fig. 8.76 the RMS value of the normal force is presented, being nearly constant below the onset of flow separation. After the discontinuous drop of the steady normal force at $13°$, the RMS jumps to nearly double its previous value. One has to bear in mind that inertia forces are included.

The RMS of the bending moment was also measured, and showed similar behavior. All RMS values were obtained by integrating the corresponding power spectra. Figure 8.77 shows two examples of normal force spectra immediately before and at the onset of the flow separation. The dotted line represents the spectrum taken at $\alpha = 13.2°$, while the solid line represents the spectrum taken at $13°$. The spectra are dominated by a rather narrow peak at the bending frequency of the wing. The bending frequency is excited by the noise-like random contributions that are inherent to the flow. Particularly after the onset of flow separation, the intensity of the fluctuating aerodynamic loads, and consequently the bending vibration of the wing, increase drastically. Hence, the RMS jumps to nearly double its previous value in the normal force and bending moment. When considering these unsteady force measurements, it is important to remember that the signals contain several components. In addition to the aerodynamic loadings, contributions from elastic and inertia forces due to the vibration are also present.

Finally, it can be stated that the changes of the flow field at the onset of separation lead to a small frequency shift of the bending frequency. This frequency change, Δf, represents a small contribution to the stiffness, caused by aerodynamic effects.

As for the balance, the spectrum shows that there are no significant peaks that could be produced by the balance itself.

Aeroelastic Experiments:
Oscillating Models in a Transonic Wind Tunnel

Aeroelasticity studies the interplay between elastic structures in an air stream and aerodynamic forces and moments. The interaction between aerodynamic and structural forces may lead to complicated nonlinear, static and dynamic effects, for example the feared flutter oscillations. Thus knowledge of the steady and unsteady forces and moments is essential. Therefore three different setups for aeroelastic experiments at airspeeds up to the transonic regime have been developed. These setups are usually applied in the $1\,\text{m} \times 1\,\text{m}$ adaptive test section of the transonic wind tunnel in Göttingen.

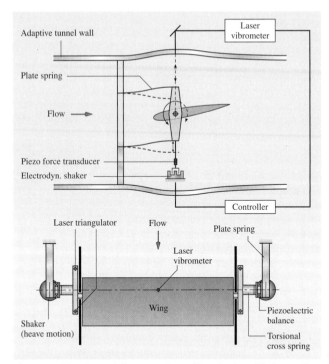

Fig. 8.78 Sketches of the setup for self-excited oscillations of an airfoil with two degrees of freedom. A piezoelectric balance is installed on each side between the bending and the torsion spring. A laser vibrometer, the electrodynamic exciter and the controller form the flutter control system. The exciter force is also measured using piezoelectric force transducers

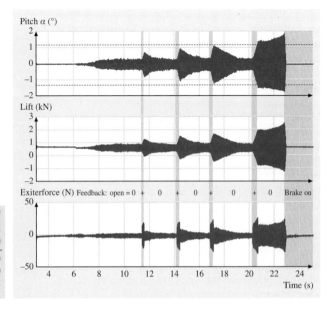

The first setup sketched in Fig. 8.78 is designed for investigating flutter phenomena of airfoil models. It is symmetrically built from blade- and cross-springs for bilateral elastic suspension. To measure the steady and unsteady forces, a piezoelectric balance is installed on each side between the bending and the torsion spring. Each balance corresponds to the platform balances, consisting of four multicomponent elements (type 9252) as described for half models in Sect. 8.2.1. In the case of flutter or forced motion the balance oscillates in the heaving direction. Therefore the smaller transducers were selected due to their lower weight and the advantage gained by reducing the moving masses.

Figure 8.78 also shows a schematic representation of the flutter control system used to dampen or excite oscillations of the model. A laser vibrometer was used to measure heaving motion of the airfoil oscillations. The heave-speed signals were fed into an operational amplifier and then into one electrodynamic exciter at each side. The excitation acts directly on the bending springs in the heave direction, whose connecting box forms the basis for the piezoelectric balance. Thus the excitation forces do not act directly on the piezoelectric balance, but rather appear only indirectly in the balance signals since the heave motion of the airfoil induces inertial forces. Forced heave oscillations can be driven by replacing the electrodynamic exciters by stronger hydraulic linear actuators. The exciter force F_{ex} itself can be measured directly with the one-component piezoelectric force sensor (Kistler 9301A), built into the heave rod.

Figure 8.79 shows an instructive example of a flutter measurement using a symmetrical National Advisory Committee for Aeronautics (NACA) 0012 airfoil taken at the flutter boundary. The state of the system is critical, i.e., in the transition range between stability and instability. We will see that it is a hysteretic-encumbered transition (subcritical bifurcation), which was recorded at a mean angle of attack of $\alpha = 1.1°$. The curves plotted in the Fig. 8.79 are: (a) the angle of attack $\alpha(t) - \alpha_{mean}$, (b) the lift force $L(t)$, and (c) the heave-rod force F_{ex} of the excitation, measured with the piezoelectric transducer. First we see in the offset that the mean lift is 0.7 kN. Second the time functions demonstrate how the

Fig. 8.79 Artificial initiation of self-excited flutter oscillations in a hysteresis regime (subcritical bifurcation). The initial amplitudes are set for different temporal lengths via positive feedback. The limit amplitude of $\alpha = 1.2°$ acts as a repeller. Depending on whether or not the initial amplitudes lie below or above this threshold value, the oscillations will be damped or excited

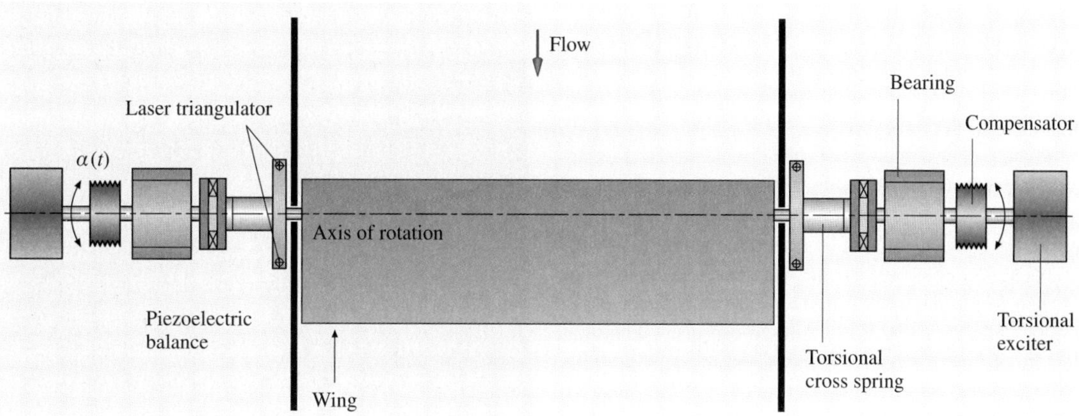

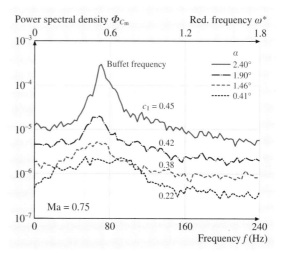

Fig. 8.80 Pitch test setup for forced torsion oscillations. The installation of both optional torsion springs (cross springs) gives the wing a torsional degree of freedom for investigating torsional flutter. The piezoelectric balances are between the hydraulic actuators and cross springs

system reacts to disturbances of different strengths or durations. Such a disturbance occurred when the control device was switched to positive feedback for a short time. The period of excitation of the system can be clearly seen in the curve for $F_{ex}(t)$, where the amplitudes shoot up at the indicated time points at $t \cong 11.3$, 14.2, 16.5, and 20.0 s. It is also obvious that the time span of the disturbances increases with increasing time. Whereas the flutter oscillations decay after the first three short bursts of excitation, they diverge after the fourth, the longest of the excitations. This behavior indicates that we must be in the hysteresis range of a subcritical bifurcation, i. e., the last disturbance was sufficient to lift the system above an unstable limit cycle, which then acted as a repeller to cause the increase in the amplitudes. At this point, we should note that the forces needed to influence the system are about two orders of magnitude smaller than the lift forces that occur, even though $F_z(c)$ still contains inertial forces.

Finally the example shows that it is possible to perform measurement of steady and unsteady forces under very difficult conditions, i. e., while the balance is oscillating.

By use of the second test setup (Fig. 8.80), the aerodynamic effect of forced pitching motions of a model may be investigated. In this case, two torsional hydraulic actuators work with a 180° phase shift and force the pitching motion of a two-side suspended model in the air stream. It is successfully used for example in testing airfoil models equipped with piezoelectrically driven flaps for dynamic stall control. The piezoelectric actuators use the inverse piezoelectric effect and are described by *Schimke* et al. [8.32]. This test setup can be upgraded with a torsional degree of freedom to investigate free pitch oscillations of the model. To measure the steady and unsteady forces, the same piezoelectric balances described above can be installed on each side between the hydraulic actuators, the model, and the optional cross-springs, respectively.

The ability to investigate unsteady phenomena such as buffet oscillations in transonic flow is demonstrated

Fig. 8.81 Power spectra of moment fluctuations for different angles of attack for a fixed supercritical airfoil. With increasing α a peak at the buffet frequency of $\omega^* = 0.56$ appears. The corresponding steady lift coefficients c_l are also included in Fig. 8.81, where spectra of the moment fluctuations are presented. In this case, the cross-springs were not installed, the model with supercritical section (NLR 7301) was at rest, and the angle of incidence α was varied using the hydraulic actuators.

These spectra show what happens when α is increased while the Mach number is held constant at Ma $= 0.75$. Above an angle of attack of $\alpha = 1.5°$, peaks appear in the spectra that can be traced back to self-excited shock oscillations (buffet) and, therefore, because of the absence of the necessary degrees of freedom for oscillations, have purely fluid-dynamical origins. The nondimensional buffet frequency has a value of $\omega^* = 0.55$, which corresponds to a frequency of $f = 72.7\,\text{Hz}$. Simultaneously, the mean lift was measured and the values are included in the figure. For this type of measurement, the high stiffness of the twin-sided piezo balance is a significant prerequisite. The test setups and the experiments are described in more detail by *Schewe* et al. [8.33].

The third test setup, sketched in Fig. 8.82, shows a wing model mounted on a turntable device such that the wing meets the sidewall with a negligible gap. The turntable device allows the model's angle of attack around an axis perpendicular to the sidewall of the test section to be adjusted. Furthermore, the model can be forced by means of a hydraulic rotation actuator to

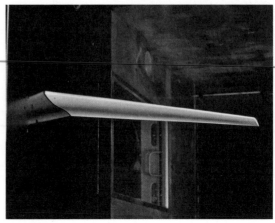

Fig. 8.83 Elastic wing models in the adaptive test section of a transonic wind tunnel

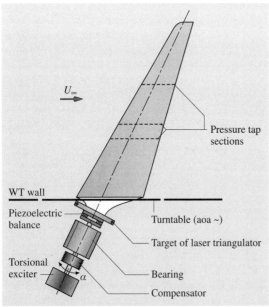

Fig. 8.82 Test setup for oscillating half models in a transonic wind tunnel. The piezo balance is located at the root of the wing

perform pitch oscillations around the spar axis plotted in Fig. 8.82. Laser triangulators are used to measure the pitch according to the spar axis. Also here, a piezoelectric platform balance is used to measure the root loads outside of the test section. The time-averaged signal of the balance allows the global aerodynamic loads to be determined, while the unsteady part of the balance signal represents the sum of unsteady air loads and inertia forces.

Figure 8.83 shows the half-span model of the *Aerostabil* project [8.34] mounted in the adaptive test section of the transonic wind tunnel in Göttingen (TWG). There are two models with the same outer geometry, representing the outer part of a transport aircraft wing with a supercritical airfoil section. The first model was designed conventionally, i.e., as rigid as possible, while the second was elastically scaled corresponding to the prototype. The aim was to study static aeroelastic effects, which have often been ignored in the past. Figure 8.84 shows the influence of the elasticity of the swept wing on the dependence of the lift curve slope $\partial C_L / \partial \alpha$ and the pitching-moment slope curve on the Mach number at a constant angle of incidence at the model root and constant Reynolds number. Significant differences due to elastic deformation can be seen as the lift curve slope is reduced up to 16% at Ma $= 0.83$. For the highest Mach numbers, the regions on the wing with flow separation are probably larger in the case of the rigid model, resulting in smaller values of the lift curve slope.

If the elasticity of a half-span model is designed in an appropriate way, flutter tests can be performed in this test setup, as can be seen in Fig. 8.85. We see a snapshot of the half wing twisted and bent by the steady air loads; in

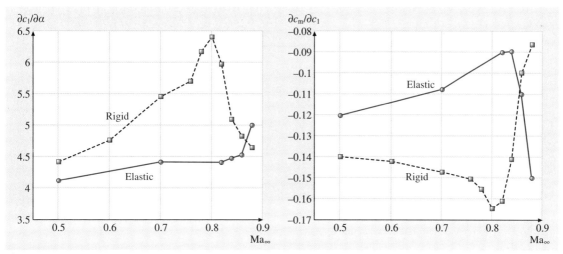

Fig. 8.84 Comparison of the lift-curve slope $\partial 2c_l/\partial\alpha$ and $\partial 2c_m/\partial c_l$ for the rigid and the elastic model as a function of Mach number

addition the wing is undergoing flutter, i. e., a limit-cycle oscillation (LCO) at a Mach number of Ma = 0.865.

Circular Cylinder in a Cryogenic Ludwieg Tube
The flow around a circular cylinder and its dependence on the Reynolds number is one of the most important problems and favorite test cases in fluid dynamics. Thus, this simple geometry was selected for the first application of the piezoelectric balance under cryogenic conditions [8.35].

The measurements were performed in the cryogenic Ludwieg tube in Göttingen (KRG), which is a blow-

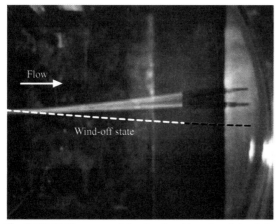

Fig. 8.85 Half-span model twisted and bent by air loads in the state of a limit cycle oscillation (LCO). Such measurements postulate a rigid balance

down wind tunnel for high-Reynolds-number research in transonic flow and is described by *Rosemann* [8.36]. The Mach number range is $0.28 < \text{Ma} < 0.95$ and high Reynolds numbers up to more than 10^7 are achieved by cooling down the fluid medium to nitrogen temperature and by pressurizing up to 10 bar. The test section measures $0.4 \times 0.35\,\text{m}^2$ and the effective measuring time is roughly 0.5 s.

The test setup is similar to others using two-dimensional models such as airfoils or cylinders. Behind each wall of the rectangular test section there is a cryo platform balance, attached to a solid base at the test section. Similar to the high-pressure wind tunnel, the circular ends of the freely suspended circular cylinder are passed through the wind tunnel walls and the ends are clamped in the holes of the force conducting top plate of each balance by a ring locking assembly. In this case, both balances can also be connected in parallel, thus electrically they act as one balance, yielding three signals for drag, lift and moment.

Figure 8.86 shows a typical result for the drag force as a function of time, measured on a smooth circular cylinder (aspect ratio 11.4) at extreme values of the operating range of the facility. The intention was to get the highest possible Reynolds number in nearly incompressible flow. Thus the measurement was taken at the lowest temperature of $T = -150\,°\text{C}$, the highest pressure possible $p_0 = 10\,\text{bar}$ and the lowest attainable Mach number of Ma = 0.28, resulting in a Reynolds number of $\text{Re} = 5.8 \times 10^6$. The curve clearly shows that, with the onset of flow, the drag in-

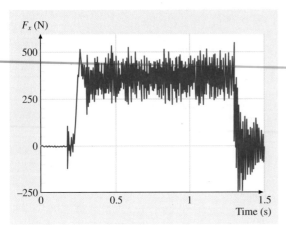

Fig. 8.86 Drag as a function of time for a circular cylinder, measured with a piezo balance in the cryogenic Ludwieg tube at extreme conditions (Re $= 5.8 \times 10^6$, Ma $= 0.28$, $T = -150\,°C$, $p_0 = 10$ bar). The drag coefficient is $c_d = 0.53$, taken in the time window Δt between 0.8 and 0.9 s

creases rapidly up to a certain level, remains there as long as flow continues, then decreases quickly to zero when flow ceases. The drag coefficient is $c_d = 0.53$, and a narrow peak in the spectrum of the lift fluctuations (Fig. 8.87) generated by the Karman vortex street, leads to a Strouhal number of St $= 0.25$. In order to substantiate the force measurements, pressure measurements at the shoulder of the cylinder ($\varphi = 90°$) were performed with a pressure sensor (Kulite). The lower spectrum in Fig. 8.87 displays the behavior of the fluctuating pressure and thus confirms that the corresponding peak in the force spectrum is indeed due to the Karman vortex street at the same Strouhal number.

The RMS value of the lift fluctuations for this case amounts to $c'_1 = 0.20$, which is more than double the value in other measurements [8.17]. The reason is that, at the lowest Mach number and with this low flow velocity, the low vortex shedding frequency ($f_V = 465$ Hz) was quite near the natural frequency of the cylinder ($f_n = 435$ Hz), leading to superelevation due to resonance effects.

The measured drag coefficient and the Strouhal number are in good agreement with the result in the high-pressure wind tunnel [8.17].

The strong oscillations in the drag signal, in particular after the flow has ceased ($t > 1.3$ s) may have many causes. One could be the mentioned proximity to the vortex resonance range. Although the vortex-induced drag fluctuations have double the value of f_{Vortex}, a higher-

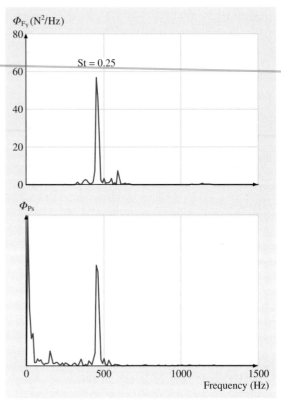

Fig. 8.87 *Upper*: power spectrum of the lift fluctuations corresponding to above measurement. *Lower*: power spectrum of the pressure fluctuations at the shoulder of the circular cylinder ($\varphi = 90°$)

order resonance due to nonlinear interaction may be responsible for the significant oscillations in the drag signal. In any case the rapid decrease of the flow velocity acts as a transient or jump excitation on the cylinder. This effect was not observed when both frequencies were well separated.

Nevertheless, it is interesting to note that, in spite of the pulse operating mode of the tunnel, the steady and unsteady processes can be measured very well.

8.2.4 Conclusions

Based on the applications described before, the typical positive properties of a piezo balance can be summarized as follows:

- High rigidity, which leads to small static deformations and to high natural frequencies
- Low interferences, typically lower than 1%

- High resolution (< 0.01 N), which is independent of the preload
- A large dynamic range; dynamic: six orders of magnitude, quasistatic: up to four orders of magnitude possible
- Applications in difficult ambient conditions such as cryogenic conditions, very high pressures and water, are possible.

The properties of piezoelectric measuring systems that can be disadvantageous when the generated charges are small are:

- Restriction to quasistatic measurements imposed by drift of the zero point, although this is not as extensive as generally believed and drift corrections are possible
- Sensitivity to temperature changes during the measurement, in particular in the direction of the prestressing bolts

Finally, in the field of aircraft aerodynamics, where the flow is attached and interest is focused on drag measurements with a resolution of drag counts ($\Delta C_d = 0.0001$), the application of strain gauge balances is suggested, since so far piezoelectric measuring systems cannot guarantee accuracy for very small steady values.

In all problems where the flow is more or less separated, for example in bluff body flows or situations around stall or when the model is oscillating, a high-end drag measurement has less significance and the disadvantages of piezo balances are more than outweighed by its inherent rigidity, which leads to well-known positive consequences.

References

8.1 ISO: *Guide to the Expression of Uncertainties in Measurements* (International Organization for Standardization, Geneva 1995)

8.2 AIAA: *Assessment of Experimental Uncertainty with Application to Wind Tunnel Testing*, AIAA S-071A-1999 (AIAA, Reston 1999)

8.3 AIAA: *Calibration and Use of Internal Strain Gauge Balances with Application to Wind Tunnel Testing*, AIAA R-091-2003 (AIAA, Reston 2003)

8.4 B. Ewald, G. Krenz: The accuracy problem of airplane development force testing in cryogenic wind tunnels, AIAA Paper 86-0776, *Aerodynamic Testing Conference Palm Beach* (AIAA, 1986)

8.5 B. Ewald: Balance Accuracy and Repeatability as a Limiting Parameter in Aircraft Development Force Measurements in Conventional and Cryogenic Wind Tunnels, *AGARD FDP Symposium, Neapel, September 1987* (AGARD, 1987)

8.6 B.C. Carter: *Interference Effects of Model Support Systems*, AGARD Rep. R 601 (AGARD, 1973)

8.7 F.G. Tatnall: *Tatnall on Testing* (American Society of Metals, Metals Park 1966)

8.8 K. Hoffmann: *An Introduction to Measurement using Strain Gages* (Hottinger Baldwin, Darmstadt 1989), Company Print

8.9 Measurement Group: *Strain Gage Based Transducers* (Measurement Group, Raleigh 1988), Company Print

8.10 B. Ewald: Multi-component force balances for conventional and cryogenic wind tunnels, Meas. Sci. Technol. **11**, 81–94 (2000)

8.11 K. Hufnagel, B. Ewald: Force testing with internal strain gage balances, AGARD FDP Special Course, AGARD R-812. In: *Advances in Cryogenic Wind Tunnel Technology* (AGARD, 1996)

8.12 M. Dubois: *Feasability Study on Strain Gage Balances for Cryogenic Wind Tunnels at ONERA, Cryogenic Technology Review Meeting*, NLR Amsterdam (1982)

8.13 K. Hufnagel: A new Half-Model-Balance for the Cologne-Cryogenic-Wind-Tunnel (KKK), *The Second International Symposium on Strain Gauge Balances, Mai 1999, Bedford* (TU Darmstadt, Darmstadt 1999)

8.14 A. Pope, K.L. Goin: *High-Speed Wind Tunnel Testing* (Krieger, New York 1978)

8.15 G. Bridel: *Untersuchung der Kraftschwankungen bei einem querangeströmten Kreiszylinder*, Dissertation, Nr. 6108 (ETH Zürich, Zürich 1978), in German

8.16 G. Schewe: A multicomponent balance consisting of piezoelectric force transducers for a high-pressure windtunnel, Techn. Messen **12**, 447–452 (1982), in German

8.17 G. Schewe: On the force fluctuations acting on a circular cylinder in crossflow from subcritical to transcritical Reynolds numbers, J. Fluid Mech. **133**, 265 (1983)

8.18 G. Schewe: Force measurements in aerodynamics using piezo-electric multicomponent force transducers. *Proc. 11th ICIASF '85 Record, Stanford Univ* (IEEE, New York 1985) p. 263

8.19 N.J. Cook: A sensitive 6-component high-frequency-range balance for building Aerodynamics, J. Phys. E **16**, 390–393 (1983)

8.20 H. Hönlinger, J. Schweiger, G. Schewe: The use of aeroelastic windtunnel models to prove structural design methods, Proc. No. 403 of the 63rd SMP Meeting of AGARD, Athens, Greece (AGARD, Neuilly-sur-Seine 1986) pp. 9-1–9-15

8.21 H. Zingel: *Measurement of steady and unsteady airloads on a stiffness scaled model of a modern transport aircraft wing*. Proc. Int. Forum on Aeroelasticity and Structural Dynamics, Aachen, DGLR-Bericht 91-06 (DGLR, Bonn 1991) p. 120

8.22 H. Psolla-Bress, H. Haselmeyer, A. Hedergott, G. Höhler, H. Holst: *High roll experiments on a delta wing in transonic flow*, Proc. 19th ICIASF 2001 Record, Cleveland, Ohio, Aug. 27-30 2001 (IEEE, New York 2001) p. 369

8.23 G. Gautschi: *Piezoelectric Sensorics* (Springer, Berlin, Heidelberg 2002)

8.24 J. Tichy, G. Gautschi: *Piezoelektrische Meßtechnik* (Springer, Berlin, Heidelberg 1980), in German

8.25 G. Schewe: Beispiele für Kraftmessungen im Windkanal mit piezoelektrischen Mehrkomponentenmeßelementen, Z. Flugwiss. Weltraumforsch. **14**, 32–37 (1990), in German

8.26 H. Triebstein, G. Schewe, H. Zingel, S. Vogel: Measurements of unsteady airloads on an oscillating engine and a wing/engine combination, J. Aircr. **31**(1), 97 (1994)

8.27 N. Schaake: *Querangeströmte und schiebende Zylinder bei hohen Reynoldszahlen, Dissertation* (Univ. Göttingen, Göttingen 1995), DLR-Report No FB 95-37 (DLR, Götzingen 1995)

8.28 G. Schewe: Sensitivity of transition phenomena to small perturbations in flow round a circular cylinder, J. Fluid Mech. **172**, 33 (1986)

8.29 G. Schewe: Reynolds-number effects in flow around more-or-less bluff bodies, J. Wind Eng. Ind. Aerodyn. **89**, 1267 (2001)

8.30 G. Schewe: *Reynolds-number-effects and their influence on flow induced vibrations*, Proc. Structural Dynamics Eurodyn 2005 (Millpress, Paris 2005) p. 337

8.31 G. Schewe: Nonlinear flow-induced resonances of an H-shaped section, J. Fluid Struct. **3**, 327–348 (1989)

8.32 D. Schimke, P. Jänker, V. Wendt, B. Junker: *Wind tunnel evaluation of a full scale piezoelectric flap control unit*, Proc. 24th European Rotorcraft Forum, Marseille, 15-17. Sept. 1998 (Organizer, 1998)

8.33 G. Schewe, H. Mai, G. Dietz: Nonlinear effects in transonic flutter with emphasis on manifestations of limit cycle oscillations, J. Fluid Struct. **18**, 3 (2003)

8.34 G. Dietz, G. Schewe, F. Kießling, M. Sinapius: *Limit-Cycle-Oscillation Experiments at a Transport Aircraft Wing Model*, Proc. Int. Forum Aeroelasticity and Structural Dynamics 2003, Amsterdam (Netherlands Association of Aeronautical Engineers, 2003)

8.35 G. Schewe, C. Steinhoff: Force measurements on a circular cylinder in a cryogenic-Ludwieg-tube using piezoelectric transducers, Exp. Fluids **42**(3), 489–494 (2007)

8.36 H. Rosemann: The Cryogenic Ludwieg-Tube-Tunnel at Göttingen, AGARD Special Course. In: *Cryogenic wind tunnel technology* (AGARD, Neuilly-sur-Seine 1997)

Part C Specific Experimental Environments and Techniques

9 Non-Newtonian Flows
Klaas te Nijenhuis, Delft, The Netherlands
Gareth H. McKinley, Cambridge, USA
Stephen Spiegelberg, Boston, USA
Howard A. Barnes, Ceredigion, UK
Nuri Aksel, Bayreuth, Germany
Lutz Heymann, Bayreuth, Germany
Jeffrey A. Odell, Bristol, UK

10 Measurements of Turbulent Flows
Giovanni Paolo Romano, Rome, Italy
Nicholas T. Ouellette, Göttingen, Germany
Haitao Xu, Göttingen, Germany
Eberhard Bodenschatz, Göttingen, Germany
Victor Steinberg, Rehovot, Israel
Charles Meneveau, Baltimore, USA
Joseph Katz, Baltimore, USA

11 Flow Visualization
Wolfgang Merzkirch, Essen, Germany

12 Wall-Bounded Flows
Joseph C. Klewicki, Durham, USA
William S. Saric, College Station, USA
Ivan Marusic, Victoria, Australia
John K. Eaton, Stanford, USA

13 Topological Considerations in Fluid Mechanics Measurements
John F. Foss, East Lansing, USA

14 Flow Measurement Techniques in Turbomachinery
Oğuz Uzol, Ankara, Turkey
Joseph Katz, Baltimore, USA

15 Hydraulics
Roger E.A. Arndt, Minneapolis, USA
Damien Kawakami, Bloomington, USA
Martin Wosnik, Holden, USA
Marc Perlin, Ann Arbor, USA
James H. Duncan, College Park, USA
David M. Admiraal, Lincoln, USA
Marcelo H. García, Urbana, USA

16 Aerodynamics
Wolf-H. Hucho, Schondorf, Germany
Klaus Hannemann, Göttingen, Germany
Jan Martinez Schramm, Göttingen, Germany
Charles H.K. Williamson, Ithaca, USA

17 Atmospheric Measurements
Harindra J.S. Fernando, Tempe, USA
Marko Princevac, Riverside, USA
Ronald J. Calhoun, Tempe, USA

18 Oceanographic Measurements
Bruce Howe, Seattle, USA
Teresa K. Chereskin, La Jolla, USA

19 Microfluidics: The No-Slip Boundary Condition
Eric Lauga, Cambridge, USA
Michael P. Brenner, Cambridge, USA
Howard A. Stone, Cambridge, USA

20 Combustion Diagnostics
Christof Schulz, Duisburg, Germany
Andreas Dreizler, Darmstadt, Germany
Volker Ebert, Heidelberg, Germany
Jürgen Wolfrum, Heidelberg, Germany

21 Electrohydrodynamic Systems
Antonio Castellanos, Sevilla, Spain
Alberto T. Pérez, Sevilla, Spain

Building on the previous two parts of this Springer Handbook, which have dealt with the fundamental concepts and equations that underpin experimental fluid mechanics and the measurement of primary quantities, respectively, Part C addresses experimental fluid mechanics from an application point of view. According to the application, often unique and specific forms of equipment, experimental procedure, or analysis and interpretation of results have been developed. It is the purpose of Part C to present a selection of such application areas. In particular, measurements of non-Newtonian flows, turbulence, flow visualization, wall-bounded flows, surface topology, turbomachines, hydraulics, aerodynamics, atmospheric and oceanographic measurements, combustion diagnostics and electrohydrodynamic systems are presented in this section.

9. Non-Newtonian Flows

Rheological constitutive equations of non-Newtonian liquids are discussed in detail in Chap. 1.8. In the present chapter they are used in the discussion of measurement techniques intended to establish an appropriate constitutive equation of a given liquid and attribute values for its material parameters. As is widely done in rheology, shear viscosity (not necessarily constant) is always denoted by η; only for Newtonian liquids, where $\eta = $ constant, is it denoted by μ as elsewhere in this Handbook.

9.1 **Viscoelastic Polymeric Fluids** 619
    9.1.1 Measurements in Shear Flow 627
    9.1.2 Rheogoniometers and Rheometers . 635
    9.1.3 Elongational Flows 646

9.2 **Thixotropy, Rheopexy, Yield Stress** 661
    9.2.1 A History of Thixotropy.................. 661
    9.2.2 Description of Thixotropic Phenomenon 665
    9.2.3 Typical Thixotropic Experiments 669
    9.2.4 Semi-Empirical Phenomenological Theories Used to Fit Experimental Data........ 670
    9.2.5 The Breakdown and Build-Up of Isolated Flocs........................... 674
    9.2.6 Examples of Systems and Studies from the Literature...................... 674
    9.2.7 Overall Conclusions...................... 679

9.3 **Rheology of Suspensions and Emulsions** 680
    9.3.1 Preliminaries 680
    9.3.2 Suspensions 683
    9.3.3 Emulsions 711

9.4 **Entrance Correction and Extrudate Swell**............................. 720
    9.4.1 Correction for Entrance Effect: Bagley Correction 720
    9.4.2 Extrudate Swell or Die Swell 721
    9.4.3 Conclusions 723

9.5 **Birefringence in Non-Newtonian Flows** .. 724
    9.5.1 The Molecular Origin of Birefringence 725
    9.5.2 Techniques for Birefringence Measurement...... 725
    9.5.3 Relation Between Birefringence and Molecular Strain 729
    9.5.4 Optical Properties of Macromolecules in Solution: Effects of Macroform and Microform Birefringence 731
    9.5.5 Example Calculation of the Theoretical Birefringence for Stretched Molecules of Atactic Polystyrene 731
    9.5.6 Conclusions 732

References 732

9.1 Viscoelastic Polymeric Fluids

An introduction into the viscoelastic phenomenon of polymeric materials in their fluid state is presented. First, the three principal measurement methods are described: stress relaxation, creep, and dynamic mechanical measurements. This is followed by the description of the Maxwell model, the Voigt–Kelvin model and the Burgers model. Finally, the Boltzmann superposition principle is introduced and applied to various flows.

For the measurement of viscoelastic properties of polymeric fluids in principle three methods are available: stress relaxation, creep, and dynamic mechanical measurements.

Stress Relaxation. In a stress relaxation experiment a constant shear deformation γ_0 is applied; the resulting stress is time dependent and decreases as a function of time. The shear modulus or relaxation modulus is de-

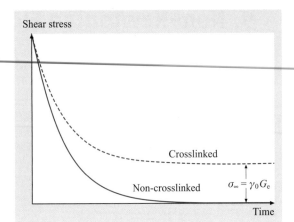

Fig. 9.1 Schematic representation of stress relaxation for viscoelastic and for a rubber-like materials

fined as the proportionality between the time dependent stress, $\sigma(t)$ and the imposed deformation:

$$G(t) = \frac{\sigma(t)}{\gamma_0} . \tag{9.1}$$

In Fig. 9.1 the stress relaxation is schematically shown for a viscoelastic material (non-crosslinked) and for a rubber-like (crosslinked) material. For the viscoelastic fluid the stress decreases to zero (fading memory), whereas for the rubber-like material the stress relaxes to a value $\gamma_0 G_e$, where G_e is the rubber shear modulus.

Creep. In a creep experiment a constant shear stress σ_0 is applied; the resulting deformation, i. e., the creep, is time dependent and increases as a function of time. The shear compliance is defined as the proportionality between the time dependent shear deformation $\gamma(t)$ and the imposed stress σ_0:

$$J(t) = \frac{\gamma(t)}{\sigma_0} . \tag{9.2}$$

In Fig. 9.2 creep is schematically shown for a viscoelastic and a rubber-like material. For a viscoelastic fluid the long-time creep increases linearly with time, with a slope equal to σ_0/η, whereas for the rubber-like material the creep increases to a constant value $\sigma_0 J_e$, where J_e is the rubber shear compliance, equal to $1/G_e$.

Dynamic Mechanical Measurements. In dynamic mechanical measurements a sinusoidal shear deformation with radian frequency ω is applied (Fig. 9.3). For linear viscoelastic materials the stress response is in the steady state also sinusoidal, but is out of phase with the strain:

$$\gamma(t) = \gamma_0 \sin(\omega t) . \tag{9.3}$$

$$\sigma(\omega, t) = \sigma_0 \sin(\omega t + \delta)$$
$$= \gamma_0 [G' \sin(\omega t) + G'' \cos(\omega t)] , \tag{9.4}$$

where: δ is the loss angle and

$$G' \equiv \frac{\sigma_0}{\gamma_0} \cos \delta \quad \text{and} \quad G'' \equiv \frac{\sigma_0}{\gamma_0} \sin \delta . \tag{9.5}$$

G' is called the storage modulus, representing the elastic properties, whereas G'' is called the loss modulus, representing the viscous properties. For ideal rubbers $G' = G_e$ and $\delta = 0$, whereas for Newtonian liquids $G'' = \eta \omega$ and $\delta = \pi/2$.

The dynamic moduli can also be defined in complex notation:

$$\gamma^* = \gamma_0 \exp(i\omega t) \quad \sigma^* = \sigma_0 \exp[i(\omega t + \delta)]$$
$$\text{and} \quad G^* \equiv \frac{\sigma^*}{\gamma^*} = \frac{\sigma_0}{\gamma_0} \exp(i\delta) = |G^*| \exp(i\delta)$$
$$\tag{9.6}$$

where:

$$G^* = G' + iG'' \quad \text{and} \quad i = \sqrt{-1} . \tag{9.7}$$

The complex viscosity is defined as:

$$\eta^* \equiv \frac{\sigma^*}{d\gamma^*/dt} = \frac{\sigma^*}{-i\omega\gamma^*}$$
$$= -\frac{iG^*}{\omega} = \frac{G''}{\omega} - i\frac{G'}{\omega} = \eta' - i\eta'' \tag{9.8}$$

with:

$$\eta' = \frac{G''}{\omega} \quad \text{and} \quad \eta'' = \frac{G'}{\omega} . \tag{9.9}$$

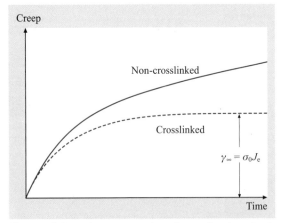

Fig. 9.2 Schematic representation of creep for viscoelastic and for a rubber-like materials

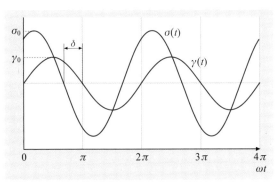

Fig. 9.3 Dynamic experiment: sinusoidal stress and deformation δ (rad) out of phase

The absolute values of the complex modulus and viscosity are:

$$|G^*| = \sqrt{G'^2 + G''^2} \quad \text{and} \quad |\eta^*| = \sqrt{\eta'^2 + \eta''^2}. \tag{9.10}$$

In Fig. 9.4 the complex material properties are demonstrated in the complex plane for $\omega = 1/2$ rad.

Mechanical Models. In order to describe the viscoelastic behavior of polymeric fluids use is made of mechanical models consisting of an elastic Hookean spring with spring constant G and a viscous Newtonian dashpot with viscosity η. The relationships between stress and deformation are, for a spring and dashpot,

$$\sigma = G\gamma \quad \text{and} \quad \sigma = \eta \frac{d\gamma}{dt}, \tag{9.11}$$

respectively.

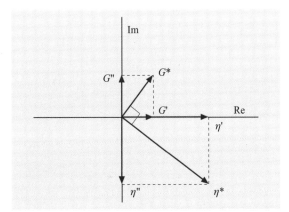

Fig. 9.4 The dynamic moduli and viscosities in the complex plane

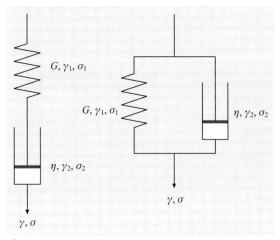

Fig. 9.5 Maxwell (*left*) and Voigt–Kelvin models (*right*)

In the Maxwell model, the spring and dashpot are linked in series while in the Voigt–Kelvin model they are linked parallel (Fig. 9.5). The Maxwell model is appropriate to describe stress relaxation, and the Voigt–Kelvin model to describe creep.

For the Maxwell model the relationships between deformations and between stresses are:

$$\gamma = \gamma_1 + \gamma_2 \quad \text{and} \quad \sigma = \sigma_1 = \sigma_2 \tag{9.12}$$

and for the Voigt–Kelvin model they are

$$\gamma = \gamma_1 = \gamma_2 \quad \text{and} \quad \sigma = \sigma_1 + \sigma_2. \tag{9.13}$$

Stress Relaxation of a Maxwell Model. If at time $t = 0$ a Maxwell model has a constant strain of γ_0 suddenly

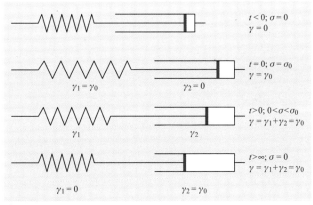

Fig. 9.6 Schematic representation of the stress relaxation of a Maxwell model

imposed, the spring is momentarily stretched to γ_0. The stress needed for this deformation is

$$\sigma_0 = G\gamma_0 \, . \tag{9.14}$$

Due to this stress the plunger of the dashpot starts to move (Fig. 9.6). As a result the stress starts to decrease gradually.

The relationships between the stress and deformation are

$$\sigma_1 = G\gamma_1 \quad \text{and} \quad \sigma_2 = \eta \frac{d\gamma_2}{dt} \, , \tag{9.15}$$

$$\sigma_1 = \sigma_2 = \sigma \, , \tag{9.16}$$

and

$$\gamma_1 + \gamma_2 = \gamma_0 \, . \tag{9.17}$$

By differentiation of (9.17) with respect to time:

$$\frac{d\gamma_1}{dt} + \frac{d\gamma_2}{dt} = \frac{d\gamma_0}{dt} = 0 \tag{9.18}$$

and by subsequent combination with (9.15) we obtain

$$\frac{1}{G}\frac{d\sigma}{dt} + \frac{\sigma}{\eta} = 0 \, . \tag{9.19}$$

Integration of (9.19) with the boundary value $\sigma(t=0) = \sigma_0$

$$\int_{\sigma_0}^{\sigma(t)} \frac{d\sigma}{\sigma} = -\frac{G}{\eta} \int_0^t dt \tag{9.20}$$

yields

$$\sigma(t) = \sigma_0 \exp(-t/\tau) \, , \tag{9.21}$$

where

$$\sigma_0 = G\gamma_0 \quad \text{and} \quad \tau = \frac{\eta}{G} \, , \tag{9.22}$$

where τ is the relaxation time of the Maxwell model, i.e., the time needed for relaxation of the stress from σ_0 to σ_0/e (i.e., approximately $0.37\sigma_0$, Fig. 9.7a). The shear modulus is defined as

$$G(t) \equiv \frac{\sigma(t)}{\gamma_0} = G \exp(-t/\tau) \, . \tag{9.23}$$

Upon plotting $\sigma_{\text{red}} = \sigma(t)/\sigma_0 = \exp(-t/\tau)$ versus $\log t$ a curve is obtained with a strong decrease from 1 to 0 around $t = \tau$, according to the results in Table 9.1.

Hence, the stress $\sigma(t)$ decreases *stepwise* within a few decades from approximately σ_0 to zero, which manifests viscoelastic liquid-like behavior. This is demonstrated in Fig. 9.7a, where the reduced relaxation modulus $G_{\text{red}}(t) = G(t)/G = \sigma(t)/\sigma_0$ is plotted versus log time, for two Maxwell models with relaxation times of $\tau = 1$ s and $\tau = 10\,000$ s, respectively.

In general, the decrease of the stress is not as sharp as shown in Table 9.1 and in Fig. 9.7a. One could imagine that this is the result of the presence of more relaxation times. We can describe this phenomenon by a so-called Maxwell–Wiechert (MW) model, i.e., N Maxwell elements linked in parallel, each with its own spring

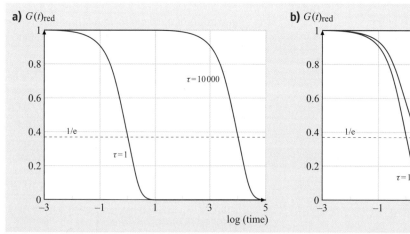

Fig. 9.7 (a) Semi-logarithmic plot of the reduced relaxation modulus $G_{\text{red}}(t) = G(t)/G = \sigma(t)/\sigma_0$ for two Maxwell elements, with relaxation times of 1 s and 10 000 s, respectively (b) Semi-logarithmic plot of the relaxation of the reduced stress σ/σ_0 for a Maxwell–Wiechert model, with relaxation times of 1 s and 10 000 s and spring constants G_1 and $G_2 = 0.5G_1$; the results of (a) are also shown

Table 9.1 Fast decrease of $\exp(-t/\tau)$ in two decades of time around $t/\tau = 1$

t/τ	0.01	0.1	1	10
$\log(t/\tau)$	−2	−1	0	1
$\exp(-t/\tau)$	0.990	0.904	0.365	0.000

constant G_i and relaxation time τ_i, for which the shear modulus can be found to be

$$G(t) = \sum_{i=1}^{N} G_i \exp(-t/\tau_i) \,. \tag{9.24}$$

The number of Maxwell elements needed in the MW model depends on the accuracy desired to describe the viscoelastic behavior.

If there are two distinct relaxation times far away from each other (e.g., 1 s and 10 000 s), then in a plot of $\sigma(t)$ versus $\log t$ the stress decreases in two steps (Fig. 9.7b): first a step from $\sigma_0 = \gamma_0(G_1 + G_2)$ to $\sigma = \gamma_0 G_2$ at around $t = 1$ s and subsequently a step from $\sigma = \gamma_0 G_2$ to $\sigma = 0$ at around $t = 10\,000$ s.

For high-molecular-weight polymers the modulus also decreases in two steps from the glassy state, at approximately 3×10^9 N/m^2, via the pseudo-rubber plateau, at approximately 10^5 N/m^2, to the fluid state. However the decrease is much less sharp than the decreases shown in Fig. 9.7b. The viscoelastic behavior of these polymers can be described by a (large) number of relatively short relaxation times around the glass rubber transition and another (large) number of relatively long relaxation times around the decrease from the pseudo-rubber plateau to the fluid state. The pseudo-rubber plateau is the result of the presence of entanglements in high-molecular-weight polymers.

Creep of a Voigt–Kelvin Model. When a stress σ_0 is suddenly applied to a Voigt–Kelvin model, we have:

$$\frac{d\sigma_0}{dt} = \frac{d\sigma_1}{dt} + \frac{d\sigma_2}{dt} = 0$$

$$= G\frac{d\gamma_1}{dt} + \eta\frac{d^2\gamma_2}{dt^2} = G\frac{d\gamma}{dt} + \eta\frac{d^2\gamma}{dt^2} \,. \tag{9.25}$$

The boundary values are:

$$\gamma(t=0) = 0 \quad \text{and} \quad \gamma(t=\infty) = \sigma_0/G \,. \tag{9.26}$$

Hence, the solution of the differential equation reads

$$\gamma(t) = \frac{\sigma_0}{G}[1 - \exp(-t/\tau)] = \sigma_0 J[1 - \exp(-t/\tau)] \,, \tag{9.27}$$

where the characteristic time, now called the retardation time, is again $\tau = \eta/G$, and $J = 1/G$ is the compliance of the spring.

In Fig. 9.8 the reduced creep, i.e., $\gamma_{\text{red}} = \gamma(t)/\gamma_\infty = \gamma(t)G/\sigma_0$, is shown for a viscoelastic system with a retardation time of 1 s. The retardation time is equal to the time needed to reach a value of $1 - 1/e \approx 0.63$, as illustrated in Fig. 9.8. Upon comparing this figure with Fig. 9.2 it becomes clear, that it illustrates the creep of a crosslinked polymer, a rubber-like material. Hence, the Voigt–Kelvin model describes a viscoelastic solid and is not able to describe the creep behavior of viscoelastic fluids.

Creep of a Burgers Model. The creep behavior of viscoelastic fluids can be described in a better way with the Burgers model, i.e., a Maxwell model linked in series with a Voigt–Kelvin model (Fig. 9.9). The following equations are available:

$$\gamma = \gamma_1 + \gamma_2 + \gamma_3 \quad \text{and} \quad \sigma = \sigma_1 + \sigma_2 \,, \tag{9.28}$$

$$\gamma_1 = \sigma/G \,, \quad \gamma_2 = \frac{\sigma}{G_1}[1 - \exp(-t/\tau_1)] \quad \text{and}$$

$$\frac{d\gamma_3}{dt} = \frac{\sigma}{\eta} \quad \text{where} \quad \tau_1 = \eta_1/G_1 \,. \tag{9.29}$$

The solution of these equations yields, for a Burgers model subjected to a constant stress σ_0, the following

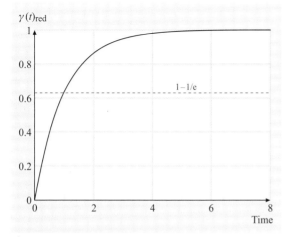

Fig. 9.8 Creep for a Voigt–Kelvin model with a retardation time of 1 s

expression for creep

$$\gamma(t) = \sigma_0 \left\{ \frac{1}{G} + \frac{1}{G_1}[1-\exp(-t/\tau_1)] + \frac{t}{\eta} \right\}$$
$$= \sigma_0 \left\{ J + J_1[1-\exp(-t/\tau_1)] + \frac{t}{\eta} \right\} = \sigma_0 J(t),$$
(9.30)

where the time-dependent compliance is given by

$$J(t) = J + J_1[1-\exp(-t/\tau_1)] + \frac{t}{\eta} \quad (9.31)$$

and where $J = 1/G$ and $J_1 = 1/G_1$, i.e., the compliances of the two springs, and the retardation time is $\tau_1 = \eta_1/G_1$.

It is clear that creep of a Burgers model is the sum of the creep of a Maxwell model and of a Voigt–Kelvin model. As demonstrated in Fig. 9.10, the values of J and J_1 can be obtained by extrapolation of the linear part that is obtained after long times, well after the retardation time, where the slope is equal to $1/\eta$.

For polymer melts, J and J_1 are of the order of 3×10^{-10} m^2/N and 10^{-5} m^2/N, both nearly independent of temperature and molecular weight, whereas the

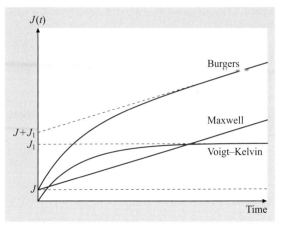

Fig. 9.10 Creep of a Burgers model

viscosity η is strongly dependent on both the temperature and molecular weight.

Boltzmann Superposition Principle. If at time $t = 0$ a viscoelastic system is suddenly subjected to a deformation γ_0 and at time $t = t_1$ suddenly to an extra deformation γ_1 then for $t \geq t_1$ the stress will be given by

$$\sigma(t \geq t_1) = \gamma_0 G(t) + \gamma_1 G\left(\overline{t-t_1}\right), \quad (9.32)$$

where the bar above $t - t_1$ means that G is a function of $t - t_1$. This principle is called the Boltzmann superposition principle, which is only appropriate to describe linear behavior. In general the stress after stepwise deformations $\Delta \gamma_i$ at times t_i is given by

$$\sigma(t) = \sum_i \Delta \gamma_i G\left(\overline{t-t_i}\right). \quad (9.33)$$

If the deformation changes continuously then the sum changes into an integral

$$\sigma(t) = \int_{-\infty}^{t} G\left(\overline{t-t'}\right) \frac{\partial \gamma(t')}{\partial t'} \, \mathrm{d}t' \quad (9.34)$$

where t' is the running time or time in the prehistory, and t is the actual time.

Examples of the Application of the Boltzmann Superposition Principle.
Steady Shear Flow. Steady shear flow means that the shear flow was started at $t' = -\infty$ and is still going on at time $t' = t$. The constant shear rate is then given by

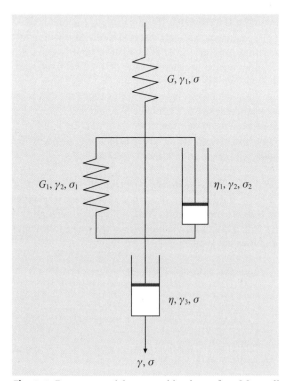

Fig. 9.9 Burgers model, a combination of a Maxwell model and a Voigt–Kelvin model in series

$\dot{\gamma}_0$, hence, $\partial \gamma / \partial t' = \dot{\gamma}_0$. Equation (9.34) yields:

$$\sigma(t) = \dot{\gamma}_0 \int_{-\infty}^{t} G\left(\overline{t-t'}\right) dt' \tag{9.35}$$

Upon substitution of $s = t - t'$ we obtain

$$\sigma(t) = \dot{\gamma}_0 \int_{0}^{\infty} G(s) \, ds = \sigma_\infty , \tag{9.36}$$

which appears to be independent of t.

If the viscoelastic behavior is given by a sum of Maxwell models, i.e., by an MW model, then we obtain

$$\sigma(t) = \dot{\gamma}_0 \int_{0}^{\infty} \sum_i G_i \exp(-s/\tau_i) \, ds$$

$$= \dot{\gamma}_0 \sum_i G_i \int_{0}^{\infty} \exp(-s/\tau_i) \, ds = \dot{\gamma}_0 \sum_i G_i \tau_i . \tag{9.37}$$

Hence, the viscosity is equal to

$$\eta = \frac{\sigma_\infty}{\dot{\gamma}_0} = \int_{0}^{\infty} G(s) \, ds = \sum_i G_i \tau_i = \sum_i \eta_i . \tag{9.38}$$

Starting Steady Shear Flow at Time $t = 0$. Upon starting steady shear flow at time $t = 0$ the stress at time t can be calculated as follows

$$\sigma(t) = \dot{\gamma}_0 \int_{0}^{t} G\left(\overline{t-t'}\right) dt' = \dot{\gamma}_0 \int_{0}^{t} G(s) \, ds \tag{9.39}$$

so that the time-dependent viscosity $\eta^+(t)$ is equal to

$$\eta^+(t) = \frac{\sigma(t)}{\dot{\gamma}_0} = \int_{0}^{t} G(s) \, ds . \tag{9.40}$$

Upon substitution (9.24) we obtain

$$\eta^+(t) = \sum_{i=1}^{N} G_i \tau_i [1 - \exp(-t/\tau_i)] . \tag{9.41}$$

For $t \to \infty$ we have

$$\eta = \lim_{t \to \infty} \eta^+(t) = \sum_{i=1}^{N} G_i \tau_i . \tag{9.42}$$

From (9.40) and (9.41) it becomes clear that the stress increases gradually to $\dot{\gamma}_0 \eta$, as demonstrated schematically in Fig. 9.11a. It is also clear that the stress is proportional to the shear rate $\dot{\gamma}_0$. Hence, if the shear stress is divided by the shear rate then the various curves are identical, while at $t \to \infty$ the viscosity has a constant value. This is only in agreement with practice if the shear rate is small. For higher shear rates the stresses are not proportional anymore with the shear rate and for polymeric fluids $\sigma(t)/\dot{\gamma}_0$ becomes smaller or can even show a maximum value, the so-called stress overshoot (Fig. 9.11b). As a result the corresponding steady-state value after long times has a smaller value so that the viscosity decreases with increasing shear rate.

Sinusoidal Deformation. Upon substitution (9.3) into (9.34) we obtain

$$\sigma(t) = \gamma_0 \omega \int_{-\infty}^{t} G\left(\overline{t-t'}\right) \cos(\omega t') \, dt' \tag{9.43}$$

which, upon substitution of $s = t - t'$, is transformed into

$$\sigma(t) = \gamma_0 \omega \int_{-\infty}^{t} G(s) \cos[\omega(t-s)] \, ds \tag{9.44}$$

or

$$\sigma(t) = \gamma_0 \omega \sin(\omega t) \int_{0}^{\infty} G(s) \sin(\omega s) \, ds$$

$$+ \gamma_0 \omega \cos(\omega t) \int_{0}^{\infty} G(s) \cos(\omega s) \, ds . \tag{9.45}$$

Comparing this result with that of (9.4), we can conclude:

$$G'(\omega) = \omega \int_{0}^{\infty} G(s) \sin(\omega s) \, ds , \tag{9.46}$$

$$G''(\omega) = \omega \int_{0}^{\infty} G(s) \cos(\omega s) \, ds . \tag{9.47}$$

Upon substituting a sum of Maxwell models, i.e., (9.24), we obtain

$$G'(\omega) = \sum_i G_i \frac{\omega^2 \tau_i^2}{1 + \omega^2 \tau_i^2} , \tag{9.48}$$

$$G''(\omega) = \sum_i G_i \frac{\omega \tau_i}{1 + \omega^2 \tau_i^2} . \tag{9.49}$$

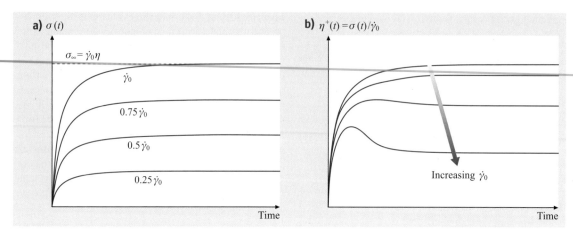

Fig. 9.11 (a) Increase of stress after starting steady shear flow at zero time $\dot{\gamma} = \dot{\gamma}_0$ =constant. **(b)** Non-Newtonian shear flow schematically shown

From (9.48) and (9.49) it follows, for very low angular frequencies, that

$$\lim_{\omega \to 0} \frac{G'}{\omega^2} = \sum_i G_i \tau_i^2 = \Psi_1 , \qquad (9.50)$$

$$\lim_{\omega \to 0} \frac{G''}{\omega} = \sum_i G_i \tau_i = \eta . \qquad (9.51)$$

Ψ_1 (i.e., the first normal stress coefficient) and η are material functions that are defined in steady shear rheology (Sect. 9.1.1). From (9.48) and (9.49) it follows that, upon plotting $\log G'$ and $\log G'$ versus $\log \omega$ at low frequencies, straight lines are obtained with slopes of 2 and 1, respectively. From these regions Ψ_1 and η can be determined. Another important conclusion is that there is a correspondence between steady shear flow behavior and oscillatory flow behavior.

For a Maxwell model with a relaxation time of 1 s the reduced dynamic moduli, i. e., $G'_{\text{red}} = G'/G$ and $G''_{\text{red}} = G''/G$, are plotted versus angular frequency in Fig. 9.12 on double logarithmic scales. A maximum in G'' arises where the G' and G'' curves cross each other. At low frequencies straight lines are obtained with slopes of 2 and 1, respectively. At high frequencies G' becomes constant, equal to G, whereas G'' decreases with a slope of -1. Both curves cross at $\omega\tau = 1$.

In Fig. 9.13 analogous results are shown for a Maxwell–Wiechert model with one relaxation time $\tau_1 = 10\,000$ s and a corresponding spring constant G and one relaxation time $\tau_2 = 1$ s with a corresponding spring constant of $100G$, hence $G'_{\text{red}} = G'/(101G)$ and $G''_{\text{red}} = G'/(101G)$. In this case two plateaus in the storage modulus and two maxima in the loss modulus are present.

This result is qualitatively in agreement with the dynamic moduli of high-molecular-weight polymers, where in general two transitions are present: the high-frequency transition is the glass–rubber transition (from approximately 3×10^9 to 10^5 N/m²) while the low-frequency transition is the rubber-flow transition. Both transitions are attended with maxima in the loss modulus. Of course the transitions are much less sharp.

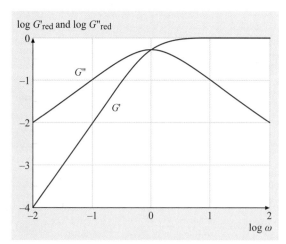

Fig. 9.12 Double logarithmic plot of the reduced dynamic moduli of a Maxwell model with one relaxation time of 1 s, versus the angular frequency

Creep of a Burgers Model. The Boltzmann superposition principle holds not only for stress relaxation but also for creep experiments. It reads, for discrete changes in

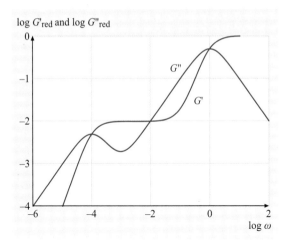

Fig. 9.13 Double logarithmic plot of the reduced dynamic moduli of a Maxwell–Wiechert model with two relaxation times of 10 000 s and 1 s, versus the angular frequency. The corresponding spring constants are G and $100\,G$

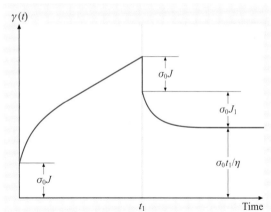

Fig. 9.14 Creep and recovery for a Burgers model

the stress,

$$\gamma(t) = \sum_i \Delta\sigma_i J\left(\overline{t-t_i}\right) \tag{9.52}$$

and for a continuously changing stress

$$\gamma(t) = \int_{-\infty}^{t} J\left(\overline{t-t'}\right) \frac{\partial \sigma(t')}{\partial t'} \, dt' \; . \tag{9.53}$$

If a Burgers model is subjected to a constant stress at time $t=0$, which is removed at time t_1, then the Boltzmann superposition can be used in the following way: removing a stress at time t_1 is equivalent to continuing the stress σ_0 and adding a stress $-\sigma_0$ at time t_1. Hence, for the deformation, the creep recovery, results:

$$\gamma(t>t_1) = \sigma_0 J(t) - \sigma_0 J(\overline{t-t_1}) \tag{9.54}$$

or

$$\gamma(t>t_1) = \sigma_0\left(J_1\{\exp[-(t-t_1)/\tau_1] - \exp(-t/\tau_1)\} + \frac{t_1}{\eta}\right) . \tag{9.55}$$

For $t \gg \tau_1$ this reduces to:

$$\gamma(t>t_1) = \sigma_0\left(J_1\{\exp[-(t-t_1)/\tau_1]\} + \frac{t_1}{\eta}\right) \tag{9.56}$$

and for $t-t_1 \gg \tau_1$ the final or remaining deformation is

$$\gamma(\infty) = \frac{\sigma_0 t_1}{\eta} \; , \tag{9.57}$$

thus providing an alternative method to determine the viscosity η (see Fig. 9.14).

Conclusions. This introduction covers the subjects stress relaxation, creep, and dynamic mechanical measurements of polymer melts and solutions. In order to describe the viscoelastic behavior of those materials the well-known mechanical models were introduced: (a) the Maxwell model, which is able to describe the stress relaxation of polymer melts qualitatively, (b) the Voigt–Kelvin model, which is able to describe the creep behavior of rubber-like polymers qualitatively, and (c) the Burgers model, which is able to describe qualitatively the creep of polymer melts and rubber-like polymers. The Boltzmann superposition principle was introduced to describe the application of experiments with time-dependent deformation or with time-dependent stress. Examples of this principle were given for various experiments.

9.1.1 Measurements in Shear Flow

In this Section measurements in shear flow will be discussed. It will start with a short introduction to the mechanics of the measurement of the viscoelastic properties of polymeric fluids. This will be followed by a description of the methods to determine the time-dependent compliance with the aid of creep measurements. Most of the discussion concerns the determination of the frequency-dependent dynamic moduli with the aid of dynamic mechanical measurements. Both compliance and dynamic moduli are illustrated with presentations from literature.

We will first discuss creep, whereas most of the discussion will deal with dynamic mechanical measurements. In shear flow measurements the sample is in general confined to the instrument, e.g., between two cylinders or between a cone and a plate. Accordingly, the translation into viscoelastic properties of the force/linear displacement ratio or of moment-of-force/angular displacement ratio will depend on the shape of the sample via a so-called shape factor. If a force F or a moment of force T_M (or torque T) is applied to a system the resulting strain/stress ratio is equal to

$$\frac{\gamma}{\sigma} = b_F \cdot \frac{x}{F} \quad \text{or} \quad \frac{\gamma}{\sigma} = b_T \cdot \frac{\phi}{T_M}, \qquad (9.58)$$

where b is a geometric shape factor (b_F in m, b_T in m^3), x is the linear displacement (in m), and ϕ is the angular displacement (in radians).

Examples of shape factors for various geometries, as shown in Fig. 9.15, are given in Table 9.2 [9.1, 2], where lines e and f are of course not appropriate for polymeric liquids, but for solid polymers.

Creep Tests

In tests to determine the compliance a specimen is subjected to a sudden, constant stress σ_0 and the resulting deformation, $\gamma(t)$, is measured as a function of time. The time-dependent compliance $J(t)$ then follows from

$$J(t) = \frac{\gamma(t)}{\sigma_0}. \qquad (9.59)$$

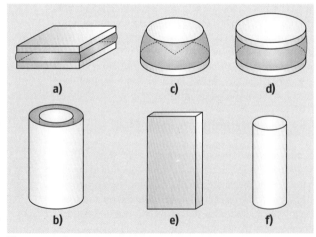

Fig. 9.15a–f Various geometric shapes for the measurement of viscoelasticity (see text) (After [9.1])

If after some time t_0, the stress is removed, the creep will be partly recovered, as was shown in Sect. 9.1:

$$\gamma(t \geq t_0) = \sigma_0 \left[J(t) - J(t - t_0) \right]. \qquad (9.60)$$

In Fig. 9.16 two arrangements for creep measurements are schematically shown: a sandwich construction and a cone-and-plate construction. In the sandwich construction the force is applied to the sample by a weight that is connected via a pulley to the surface A of the upper plate. This results in a shear stress F/A. The upper plate moves over a distance $x(t)$ and this causes a shear of the sample equal to $\gamma(t) = x(t)/d$. Accordingly, the compliance is equal to

$$J(t) = \frac{\gamma(t)}{\sigma} = \frac{Ax(t)}{Fd}. \qquad (9.61)$$

In the cone-and-plate construction the force F is transferred into a moment of force $T_M = Fr$, where r is the radius of the cylinder the string is wrapped round. Due to the moment the cone will rotate over an angle $\phi(t)$, resulting in a time-dependent shear of the sample. The constant shear stress in the sample is equal to

$$\sigma = \frac{3T_M}{2\pi R^3} \qquad (9.62)$$

and the time-dependent shear in the sample is

$$\gamma(t) = \frac{\phi(t)}{\Delta\Theta}. \qquad (9.63)$$

Accordingly, the compliance can be calculated as

$$J(t) = \frac{\gamma(t)}{\sigma} = \frac{2\pi R^3 \phi(t)}{3 T_M \Delta\Theta}. \qquad (9.64)$$

Dynamic Mechanical Tests

If a viscoelastic material is deformed sinusoidally (Sect. 9.1) according to

$$\gamma = \gamma_0 \sin(\omega t) \qquad (9.65)$$

the shear stress is given by

$$\begin{aligned}
\sigma(t) &= \sigma_0 \sin(\omega t + \delta) \\
&= \sigma_0 \left[\sin(\omega t) \cos\delta + \cos(\omega t) \sin\delta \right] \\
&= \gamma_0 [G' \sin(\omega t) + G'' \cos(\omega t)] \\
&= G'\gamma + \frac{G''}{\omega} \frac{d\gamma}{dt},
\end{aligned} \qquad (9.66)$$

where

$$G' \equiv \frac{\sigma_0}{\gamma_0} \cos\delta \quad \text{and} \quad G'' \equiv \frac{\sigma_0}{\gamma_0} \sin\delta. \qquad (9.67)$$

Accordingly, the storage and loss moduli, G' and G'', can be calculated easily by determining the amplitudes of strain γ_0 and stress σ_0 and the phase angle δ.

Table 9.2 Various geometric shape factors

Geometry	Geometric shape factor	Units
a) sandwich construction	A/d	m
b) rotation between concentric cylinders	$4\pi L r_i^2 r_0^2 / (r_0^2 - r_i^2)$	m^3
c) cone and plate	$2\pi R^3 / (3\Delta\Theta)$	m^3
d) parallel plates	$\pi R^4 / d$	m^3
e) torsion of a bar with rectangular cross section	$cd^3 f(c/d)/(16L) \quad c/d > 1$	m^3
f) torsion of a bar with circular cross section	$\pi R^4 / (2L)$	m^3

where
A = area of the sample in contact with the plates (m²),
d = thickness of the sample or torsion bar (m); in Sect. 9.3 this notation is changed by h to distinguish from particle diameter,
c = width of the torsion bar (m),
$f(c/d)$ = function of c/d, with numerical values varying between 2.25 and 5.33,
L = height of cylinder or bar (m),
R = radius of bar or plate (m),
r_i = radius of inner cylinder (m),
r_0 = radius of outer cylinder (m),
$\Delta\Theta$ = angle between cone and plate (rad), assumed to be small.

Classification of Techniques. There are various methods to determine the dynamic moduli, as shown in Table 9.3.

Dynamic Techniques. In general, when a shearing force F is applied, for example, to the upper side of the sliding plate of mass m in the sandwich construction in Fig. 9.15a, the equation motion of the plate is given by

$$F - A\sigma_{sh} = m \frac{d^2 x}{dt^2}, \qquad (9.68)$$

where A is the area that is in contact with the fluid in the gap.

If the force is an oscillating shearing force with amplitude F_0, the equation of motion becomes:

$$F_0 \sin(\omega t) - A\sigma_{sh} = m \frac{d^2 x}{dt^2}. \qquad (9.69)$$

With $b = A/d$ and $\gamma = x/d$, and using (9.66) we obtain for oscillatory flow

$$F_0 \sin(\omega t) = m \frac{d^2 x}{dt^2} + b \frac{G''}{\omega} \frac{dx}{dt} + (bG' + c) x. \qquad (9.70)$$

In (9.70) we additionally introduced an elastic constant for the instrument c. A similar equation can be derived for a rotational device, where x and m have to be replaced

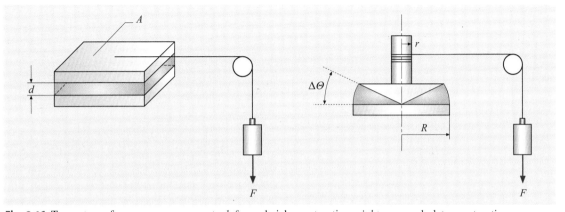

Fig. 9.16 Two set-ups for creep measurements; *left*: sandwich construction; *right* cone-and-plate construction

Table 9.3 Classification of experimental dynamic techniques

Experimental technique	Frequency (Hz)
Free damped vibrations	10^{-2}–10
Forced vibrations: resonance	10^{-2}–10^4
Forced vibrations: nonresonance	10^{-5}–10^2
Wave propagation	1–3×10^9

by θ, i.e., the angle of torsion, and I, i.e., the moment of inertia of the rotor, respectively. In some cases an extra viscous term has to be added to (9.70).

For the solution of (9.70) we distinguish three cases:

1. $F_0 = 0$: free vibrations,
2. $F_0 \neq 0$ and $m\omega^2 \ll bG' + c$: forced vibrations (nonresonance),
3. $F_0 \neq 0$ and $m\omega^2 \approx bG' + c$: forced vibrations (resonance),

where (2) and (3) are written for a sliding device. For a rotational device $m\omega^2$ should be replaced by $I\omega^2$.

Free Vibrations. The solution of (9.70) for $F_0 = 0$ for a rotational instrument is:

$$\theta = \theta_0 \exp\left[-bG''t/(2I\omega_e)\right] \cos(\omega_e t) \qquad (9.71)$$

provided

$$bG' + c > \frac{b^2 G''^2}{4I\omega_e^2},$$

with $\quad \omega_e^2 = \dfrac{4I(bG'+c) - b\eta'^2}{4I^2}, \qquad (9.72)$

and where θ_0 is the initial value of θ.

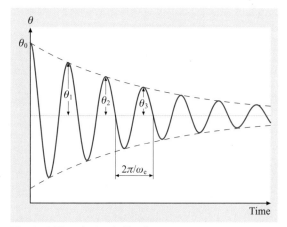

Fig. 9.17 Free damped vibrations

Equation 9.71 is the equation for free vibrations with decreasing amplitude (Fig. 9.17). The ratio of two successive maxima is

$$\frac{\theta_n}{\theta_{n+1}} = \exp\left[\pi bG''/\left(I\omega_e^2\right)\right], \qquad (9.73)$$

whereas the so-called logarithmic decrement is given by

$$\Lambda = \ln(\theta_n/\theta_{n+1}) = \pi bG''/\left(I\omega_e^2\right) \qquad (9.74)$$

which is independent of n. Hence, a plot of log θ_n versus n (several successive θ_n are shown in Fig. 9.17) should yield a straight line with slope -0.4343Λ(log e = 0.4343). The dynamic moduli now follow from

$$G' = \frac{I\omega_e^2}{b}\left(1 + \frac{\Lambda^2}{4\pi^2}\right) - \frac{c}{b}, \qquad (9.75)$$

$$G'' = \frac{I\omega_e^2}{b} \cdot \frac{\Lambda}{\pi}. \qquad (9.76)$$

If the damping of the test piece is small, then the damping of the instrument itself (air damping, friction in the bearings, the suspension etc.) has to be taken into account. In that case (9.75) has to be replaced by

$$G'' = \frac{I\omega_e^2}{b} \frac{\Lambda}{\pi} - E\omega_e, \qquad (9.77)$$

where E is the friction coefficient.

For the determination of the dynamic moduli two kinds of torsion pendulums are available (Fig. 9.18): the normal torsion pendulum, used for hard viscoelastic

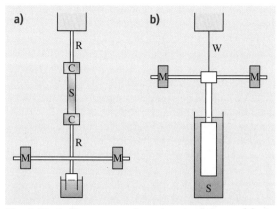

Fig. 9.18a,b Two types of torsion pendulums: (a) the normal torsion pendulum and (b) the inverted torsion pendulum. S: sample; C: clamps; W: torsion wire; M: extra adjustable masses; R: supporting rods. (After [9.3])

solids and the inverted one, appropriate for elastic liquids. The frequency range, about 0.1–100 rad/s, can be controlled by varying the moment of inertia I. This is possible by varying the masses M and their distance to the axis.

Forced Vibrations. In free vibrations experiment a defined system has a preference for only one frequency ω_e. On the other hand, if the system is driven by a sinusoidal force, then after a starting period the amplitude of the deformation remains constant. The vibration is a forced harmonic vibration. In this case the stationary solution of (9.70) is [9.2]

$$x(t) = F_0 f(\omega) \sin(\omega t - \phi) , \quad (9.78)$$

where ϕ is a phase angle, not to be confused with the loss angle δ, and

$$f(\omega) = \left[\left(bG' + c - m\omega^2\right)^2 + \left(bG''\right)^2 \right]^{-1/2} . \quad (9.79)$$

The result is a harmonic oscillation with a constant amplitude $F_0 f(\omega)$. At low frequencies the term $m\omega^2$ can be neglected. Moreover, if $c \ll bG'$, the solution reduces at low frequencies to

$$x(t) = x_0 \sin(\omega t - \delta) = \frac{F_0}{b|G^*|} \sin(\omega t - \delta) , \quad (9.80)$$

where $|G^*| = \left(G'^2 + G''^2\right)^{1/2}$.

Resonance Vibrations. With increasing frequency, $f(\omega)$ increases to a maximum value ω_0 when

$$m\omega_0^2 = bG' + c , \quad (9.81)$$

where the amplitude of the oscillation is

$$x_0^{\text{res}} = \frac{F_0}{bG''} . \quad (9.82)$$

The system is then said to be in resonance. If c is known and the values of m, ω_0 and x_0^{res} are measured, both dynamic moduli can be determined from (9.81) and (9.82) without measurement of the phase angle. Nevertheless, the value of the loss tangent can also be determined from the resonance curve, shown in Fig. 9.19. The width $\Delta\omega = \omega_2 - \omega_1$ is determined from $\tan \delta$ in the following way:

for $f(\omega)/f(\omega_0) = 0.5$, $\Delta\omega/\omega_0 = \sqrt{3} \tan \delta$, (9.83)

for $f(\omega)/f(\omega_0) = \frac{1}{2}\sqrt{2}$, $\Delta\omega'/\omega_0 = \tan \delta$. (9.84)

If the frequency remains far below the resonance frequency, the term $m\omega^2$ (or, in a rotational instrument, $I\omega^2$) can be neglected compared with bG', and the result is:

$$f(\omega) = \frac{1}{b|G^*|} \quad \text{and} \quad \tan\phi = \frac{G''}{G'} = \tan\delta . \quad (9.85)$$

It has to be mentioned, however, that for large values of the loss modulus the maximum of $f(\omega)$ is hardly perceptible. Moreover, if G' and G'' are strongly frequency dependent, a maximum will not appear at all.

Non-Resonance Vibrations. If the frequency remains far below the resonance frequency, the result for the dynamic moduli is

$$G' = \frac{F_0 \cos\phi}{bx_0} + \frac{m\omega^2}{b} - \frac{c}{b} , \quad (9.86)$$

$$G'' = \frac{F_0 \sin\phi}{bx_0} . \quad (9.87)$$

In order to determine the values of the dynamic moduli as a function of the frequency, apart from the instrument constants b, c and m (or I), determination of F_0/x_0 and the phase angle ϕ as a function of frequency is required. Note that ϕ is the measured phase angle, which is now not equal to the loss angle δ.

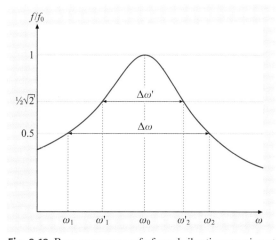

Fig. 9.19 Resonance curve of a forced vibration experiment

For hard materials dynamic measurements are in general taken on rods, bars or tubes in sinusoidal torsion or flexure. Liquids and soft viscoelastic solids are usually confined between two concentric cylinders, two cones, two plates or between a cone and a plate. In Fig. 9.20 two principles of measuring arrangements are shown on the basis of coaxial cylinders. In most types of equipment one of the two boundary surfaces is oscillated over an angle ε_1^o. The oscillation of the other boundary, over an angle ε_2^o, as well as the phase angle ϕ between the two oscillations, is measured.

For the left-hand side instrument the torque T is equal to

$$T = b \left| G^* \right| (\varepsilon_1 - \varepsilon_2) \ . \tag{9.88}$$

If the torsion bar is very stiff, ε_2^o is very small. In that case the dynamic moduli follow from

$$G' = \frac{T_0}{b\varepsilon_1^o} \cos\phi \ , \tag{9.89}$$

$$G'' = \frac{T_0}{b\varepsilon_1^o} \sin\phi \ , \tag{9.90}$$

where ε_1^o, ε_2^o and T_0 are the amplitudes of the respective oscillations. Accordingly the phase angle ϕ is equal to the loss angle δ.

In the right-hand side instrument a driving shaft is brought into oscillation ε_1. Via a torsion wire, with torsion constant c, this oscillation is transferred to the inner cylinder, which then performs an oscillation ε_2. The dynamic moduli follow from

$$G' = \frac{c}{b} \left(\frac{\varepsilon_1^o}{\varepsilon_2^o} \cos\phi - 1 \right) + \frac{I\omega^2 - c}{b} \ , \tag{9.91}$$

$$G'' = \frac{c}{b} \frac{\varepsilon_1^o}{\varepsilon_2^o} \sin\phi \ . \tag{9.92}$$

Wave Propagation. In the wave propagation technique the frequency range is enormous: from 1 Hz to 3×10^9 Hz and the technique is appropriate in the range from low-viscosity liquids to solid metals.

Two types of waves can be distinguished in the wave propagation technique, viz. shear waves and longitudinal waves.

Shear Waves. A flat plate with a large area is in contact with an isotropic viscoelastic medium of density ρ. The plate is vibrating, with an angular frequency ω, in the x-direction in its own plane, with an amplitude x_0 (Fig. 9.21). It follows for the shear wave propagating into the medium occupying the semi-space $z > 0$ that:

$$x(z, t) = x_0 \exp\left[\mathrm{i}\left(\omega t - 2\pi z/\lambda\right) - z/z_0\right] \tag{9.93}$$

where $\mathrm{i} = \sqrt{-1}$, and λ is the wavelength of the attenuated propagating wave, whose amplitude decreases with a factor e over a distance z_0 (Fig. 9.21). The attenuation α is equal to $1/z_0$. The dynamic moduli follow

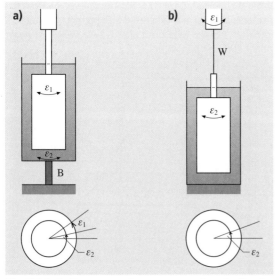

Fig. 9.20a,b Two types of nonresonant forced vibration experiments. B: torsion bar; W: torsion wire; ε_1 and ε_2: angles of rotation. (After [9.4])

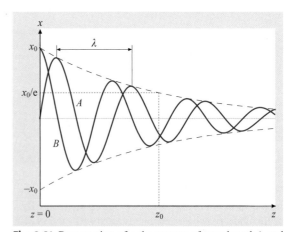

Fig. 9.21 Propagation of a shear wave of wavelength λ and attenuation $1/z_0$ in the positive, semi-infinite direction z, drawn for time $t = 2\pi n/\omega$ (*curve A*) and $t = (2n + 1/2)\pi/\omega$ (*curve B*)

from

$$G'(\omega) = \frac{\rho\omega^2(4\pi^2/\lambda^2 - 1/z_0^2)}{(4\pi^2/\lambda^2 + 1/z_0^2)^2},\quad (9.94)$$

$$G''(\omega) = \frac{\rho\omega^2 4\pi/(\lambda z_0)}{(4\pi^2/\lambda^2 + 1/z_0^2)^2},\quad (9.95)$$

$$\tan\delta = \frac{4\pi\lambda z_0}{4\pi^2 z_0^2 - \lambda^2}.\quad (9.96)$$

From these equations it follows that the dynamic moduli are determined by measuring the attenuation α and the wavelength λ (Fig. 9.21). If the damping is small, i.e., if λ/z_0 is small, the shear wave is propagated over a long distance. If, on the other hand, λ/z_0 is large, the wave will be attenuated so fast, that even the determination of the wavelength becomes difficult. According to *Ferry* [9.1] the upper limit would be $\lambda/z_0 = 3$. For this kind of experiments the frequency can be varied from 4 Hz to 5000 Hz.

In particular for very high frequencies (3000–3×10^9 Hz) use is made of reflection of propagating waves in a quartz crystal against an interface between the quartz and a thin film of a viscoelastic liquid [9.1].

Longitudinal Waves. If the flat plate similar to that of Fig. 9.21 is vibrating along the direction of its normal, i.e., in the z-direction, the medium will be compressed and extended in an oscillatory way. In this case G' and G'' have to be replaced by M' and M'', where

$$M' = K' + \frac{4}{3}G',\quad (9.97)$$

$$M'' = K'' + \frac{4}{3}G'',\quad (9.98)$$

$$M^* = K^* + \frac{4}{3}G^*,\quad (9.99)$$

where K' and K'' are the components of the complex dynamic bulk modulus K^*, whereas M' and M'' are the components of the complex dynamic longitudinal bulk modulus M^*.

Examples of Measurements on Polymer Melts

We will complete this discussion on methods for shear measurements with some results presented in the literature.

Compliance. Compliance measurements by *Plazek* and *O'Rourke* (tabulated results on page 605 in [9.1]) on polystyrene with a narrow molecular-weight distribution and a molecular weight of 600 000 are shown in Fig. 9.22. With the aid of the so-called time–temperature superposition principle [the William, Landal and Ferry (WLF) equation, see Chap. 11 in [9.1], by which measurements at a number of different temperatures are reduced to one reference temperature, in this case 100 °C, thereby increasing the frequency or time window]. Accordingly, the time window was increased to approximately 18 decades. The compliance $J(t)$, as well as $J(t) - t/\eta_0$ (where η_0 is the zero-shear viscosity) are plotted versus time. Note that at long times the slope of the compliance curve, $J(t)$, is equal to 1. In these regions the viscosity can be calculated to be approximately 2.75×10^{14} Ns/m² (i.e., just above the glass-transition temperature). The value of $J(t)$ at very short times is equal to the glass compliance $J_g = 10^{-9}$ m²/N. The value of the equilibrium shear compliance J_e follows from the curve for $J(t) - t/\eta_0$ at long times, $J_e = 1.6 \times 10^{-5}$ m²/N. Both values are in agreement with those mentioned in Sect. 9.1.

Dynamic Moduli. The dynamic moduli, G' and G'', of polystyrene Hostyrene N4000V were measured at temperatures of 140–206 °C in a Couette-type dynamic rheometer [9.5]. With the aid of the time–temperature superposition principle they were reduced to a temperature of 170 °C. In this way the frequency window was increased to approximately seven decades. Results are shown in Fig. 9.23, as $\log G'$ and $\log G''$ versus $\log \omega$ [9.6]. From a comparison of these results with

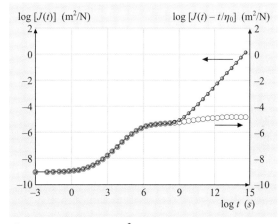

Fig. 9.22 Compliance (m²/N) of polystyrene with a narrow molecular-weight distribution with $M = 600\,000$ versus time on double logarithmic scales, according to *Plazek* and *O'Rourke* [9.1]

Fig. 9.12, it becomes clear that only a small part of the dynamic moduli was measured: the low-frequency flow behavior of the polymer and the transition to the rubber-like entanglement plateau. A much larger frequency window is needed to describe the viscoelastic behavior up to the glassy region with an instrument that is able to measure much higher moduli, up to 3×10^9 N/m^2. The dashed straight lines at low frequencies have slopes of 1 (for G'') and 2 (for G'), as was mentioned in Sect. 9.1.

In Sect. 9.1, (9.51), it was shown that the zero-shear viscosity can be determined from the loss modulus by dividing G'' by ω in the limit $\omega \to 0$. From Fig. 9.23 it follows for the zero-shear viscosity $\eta_0 = 10^5$ Ns/m^2 (the notation η_0 is introduced here instead of the η of (9.51) in Sect. 9.1). In general the loss modulus divided by the angular frequency is equal to the so-called dynamic viscosity η' (Sect. 9.1):

$$\eta' \equiv \frac{G''}{\omega} \quad \text{so that} \quad \eta_0 = \lim_{\omega \to 0} \frac{G''}{\omega} = \lim_{\omega \to 0} \eta' \,. \tag{9.100}$$

Because G' decreases much faster than G'' with decreasing frequency, we can also conclude that

$$\eta_0 = \lim_{\omega \to 0} \frac{|G^*|}{\omega} = \lim_{\omega \to 0} |\eta^*| \,. \tag{9.101}$$

It has to be mentioned that the viscosity in (9.100) and (9.101) is only equal to the zero-shear viscosity η_0, if the amplitude of the sinusoidal deformation is small enough (i.e., only for linear behavior).

In Fig. 9.24 the dynamic viscosity, η', and the absolute value of the complex viscosity, $|\eta^*|$, as calculated from Fig. 9.23, are plotted versus angular frequency on double logarithmic scales, as a dashed and full line, respectively. At low frequencies both viscosities are equal and independent of frequency: they are equal to the zero-shear viscosity of 10^5 Ns/m^2. At higher frequencies they decrease, with the dynamic viscosity decreasing faster than the complex viscosity. In Fig. 9.24 results of measurements of the viscosity in a cone-and-plate rheometer and in a capillary rheometer (Sect. 9.1) are also shown, both plotted as functions of the shear rate $\dot{\gamma}_0$. The curve $\eta = \eta(\dot{\gamma}_0)$ is decreasing as $\dot{\gamma}_0$ is increasing, which is referred to as shear thinning. It appears that in a cone-and-plate rheometer measurements were possible for shear rates of $0.001 – 0.1$ s^{-1}, whereas in the capillary rheometer measurements were possible for $3 – 3000$ s^{-1}. Most striking is the phenomenon that the transient viscosities (the symbols) follow the line of the complex viscosity versus angular frequency. Accordingly, it is experimentally shown that, over a large range of shear

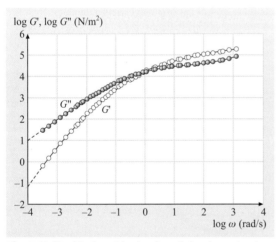

Fig. 9.23 Double logarithmic plot of the dynamic shear moduli G' and G'' versus angular frequency ω for polystyrene Hostyren N4000V, measured at temperatures varying from 140 °C to 206 °C and reduced to the reference temperature of 170 °C (see text), according to [9.6]. The measurements were taken in a Couette-type dynamic rheometer [9.5]

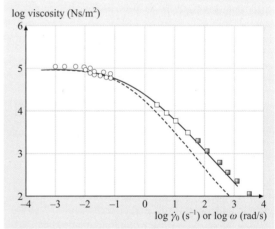

Fig. 9.24 Non-Newtonian shear viscosity $\eta(\dot{\gamma}_0)$ at 170 °C as a function of shear rate, $\dot{\gamma}_0$, for the polystyrene mentioned in Fig. 9.23, measured in a (*circles*) cone-and-plate rheometer and in a (*squares*) capillary rheometer, where the *filled squares* show the results at such high shear rates that melt fracture occurs [9.6] and the dynamic and complex viscosities, $\eta'(\omega)$ (*dashed line*) and $|\eta^*(\omega)|$ (*full line*), respectively, as functions of angular frequency, as calculated from Fig. 9.23 [9.6]

rates and angular frequencies,

$$\eta(\dot{\gamma}_0) = |\eta^*(\omega)| \quad \text{with} \quad \dot{\gamma}_0 = \omega. \tag{9.102}$$

This is called the empirical *Cox–Merz relation*.

From the transient measurements it follows first that in general it is impossible to measure the zero-shear viscosity with the aid of a capillary rheometer, because the shear rates are not low enough in this instrument. For that purpose a cone-and-plate rheometer is much more convenient, or in general a rotating rheometer. However, the best method to determine the zero-shear viscosity is measurement of the loss modulus G'', or better the dynamic viscosity η' at low frequencies. The reason for the preference for dynamic mechanical measurements is the fact that the moduli can be measured down to very low frequencies. A second point that follows from Fig. 9.24, is the high-shear-rate behavior, where it follows that $\log \eta$ is a linear function of $\log \dot{\gamma}_0$; this is called power-law behavior, which will be discussed in more detail in the next section.

Conclusion. In this Section it has been shown that there are several techniques to measure the compliance $J(t)$ of polymer melts and solutions. Many more techniques are available to determine the dynamic moduli G' and G'', however, varying in angular frequency from 0.01 to 3×10^9 Hz. Literature data for the compliance of a polymer melt are presented. Results of the dynamic moduli of polymer melts are also shown and the dynamic viscosities calculated from these measurements. It has been shown that there is a close relationship between the absolute value of the complex viscosity, $|\eta^*|$, as a function of the angular frequency, ω, and the steady shear viscosity, η, as a function of shear rate, $\dot{\gamma}_0$, known as the Cox–Merz relation.

9.1.2 Rheogoniometers and Rheometers

This Section starts with an introduction, where equations of motion are introduced in various coordinate systems, local coordinates and shearing planes are defined, and the shear rate is calculated. Subsequently three drag flow rheogoniometers are discussed, followed by pressure-driven rheometers. Shear-rate-dependent viscosity, and first and second normal stress differences and coefficients can be determined as functions of shear rate. It will be shown that measurement of the normal stress differences is possible due to the curvature of the streamlines.

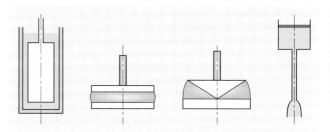

Fig. 9.25 Four commonly used rheo(gonio)meters; from *left to right*: coaxial cylinders in Couette flow, coaxial parallel plates, coaxial cone and plate and cylindrical capillary

Definition and Classification of Rheometers. A *rheometer* is an instrument that measures both *stress* and *deformation history* on a material for which the rheological constitutive equation is not known. A special case of a rheometer is a *viscometer*, which can measure only the steady shear viscosity function, $\eta(\dot{\gamma})$. According to "The British Standard Glossary of Rheological Terms" *rheogoniometers* are rheometers designed for the measurement of normal components as well as shear components of the stress tensor. Accordingly, a rheogoniometer can be used to determine material functions.

The rheological behavior of incompressible, isotropic elastic liquids can be described by σ_{21}, $N_1 = \sigma_{11} - \sigma_{22}$ and $N_2 = \sigma_{22} - \sigma_{33}$, where σ_{ij} denotes the components of the stress tensor, and N_1 and N_2 are the first and second normal stress difference, respectively. The other deviatoric components of the stress tensor are often equal to zero, due to the symmetry of the shear flow on the one hand and material isotropy on the other hand. Stationary simple shear flow is extremely important for the description of engineering processes, because this kind of flow is easily realized in laboratory and frequently occurs in (polymer) engineering. However, the measurement of elongation flow is also of great importance, because in polymer engineering these flows also play an important role Sect. 9.1.3. In practice flows are often complicated mixtures of simple shear and elongation flows.

For the measurement of shear flow many methods have been developed.

Shearing experiments have to take place in principle by moving two infinitely extended parallel plates, with the liquid of consideration in between, with a constant velocity difference with respect to each other. In practice this is of course problematic. Moreover, in this way it is only possible to determine σ_{21} and σ_{22}. For that reason we have to resort to instruments with another geometry: the streamlines do not need to be rectilinear, but they may be curved and in many cases are circular. It will become clear that, as a result of the curvature of the streamlines, normal stress differences can be determined from measurements of the normal stress σ_{22}, as a function of the position in the flowing material. An example of an approximation of infinitely extended, parallel plates is the Couette geometry, consisting of two coaxial cylinders, where the liquid under consideration is confined in the narrow gap between the two cylinders. The cylinders rotate with respect to each other with different angular velocities, and in general one is fixed. If the diameters of the cylinders are large with respect to the gap width and small with respect to their lengths, then in the gap a shear flow is obtained, which to a good approximation is comparable with the shear flow between two infinitely extended parallel plates.

In principle we distinguish two kinds of rheometers: drag flow rheometers (e.g., Couette, plate–plate, and cone–plate) and pressure-driven flow rheometers (e.g., capillary and slit rheometers). They are shown schematically in Fig. 9.25, with the exception of the slit rheometer, which will be shown later. In the three drag flow rheometers one of the two parts rotates or oscillates: the fluid in the direct vicinity of the rotating part rotates with the same angular velocity Ω, whereas the velocity of the fluid in the direct neighborhood of the fixed part will be zero (the no-slip conditions for non-Newtonian fluids are assumed to hold in these instruments; possible slip effects are discussed in Chap. 19). The fluid in between will rotate with an angular velocity decreasing from Ω to zero, which is the origin of shear and of rate of shear. In pressure-driven rheometers the flow results from a pressure above the capillary or slit and in this case it is assumed that the velocity decreases from its maximum value at the center of the capillary or slit to zero at the walls.

To be able to determine the viscoelastic properties of the liquids, we have to be familiar with local stresses and with local shear rates in the flowing liquids. For the determination of the local stresses we make use of equations of motion and for the calculation of the shear rate we first have to define shearing planes.

Equations of Motion. In order to describe the stresses in rheogoniometers we need equations that are able to express the local stresses induced by flow. For that purpose we make use of the momentum equation:

$$\rho \dot{v} = \nabla \cdot \sigma + \rho b \,, \qquad (9.103)$$

where ρ is the fluid density, v is the velocity of the fluid, ∇ is the nabla operator, σ is the stress tensor, amd b is the body force (i.e., the force per unit of mass), e.g., the gravitational acceleration. The dot over v denotes material time differentiation.

For constant viscosity (i.e., Newtonian liquids) this equation reduces to the well-known Navier–Stokes equation.

We will neglect inertial forces, i.e., $\rho \dot{v} = 0$, so that the equation is simplified to

$$\nabla \cdot \sigma + \rho b = 0 \,. \qquad (9.104)$$

In Cartesian coordinates the three projections read

$$\frac{\partial \sigma_{11}}{\partial x_1} + \frac{\partial \sigma_{21}}{\partial x_2} + \frac{\partial \sigma_{31}}{\partial x_3} = -\rho b_1 \,, \qquad (9.105)$$

$$\frac{\partial \sigma_{12}}{\partial x_1} + \frac{\partial \sigma_{22}}{\partial x_2} + \frac{\partial \sigma_{32}}{\partial x_3} = -\rho b_2 \,, \qquad (9.106)$$

$$\frac{\partial \sigma_{13}}{\partial x_1} + \frac{\partial \sigma_{23}}{\partial x_2} + \frac{\partial \sigma_{33}}{\partial x_3} = -\rho b_3 \,, \qquad (9.107)$$

in cylindrical coordinates (r, ϕ, z)

$$\frac{\partial \sigma_{rr}}{\partial r} + \frac{1}{r}\frac{\partial \sigma_{r\phi}}{\partial \phi} + \frac{\partial \sigma_{rz}}{\partial z} + \frac{\sigma_{rr} - \sigma_{\phi\phi}}{r} = -\rho b_r \,, \qquad (9.108)$$

$$\frac{\partial \sigma_{r\phi}}{\partial r} + \frac{1}{r}\frac{\partial \sigma_{\phi\phi}}{\partial \varphi} + \frac{\partial \sigma_{\phi z}}{\partial z} + \frac{2\sigma_{r\phi}}{r} = -\rho b_\phi \,, \qquad (9.109)$$

$$\frac{\partial \sigma_{rz}}{\partial r} + \frac{1}{r}\frac{\partial \sigma_{z\phi}}{\partial \phi} + \frac{\partial \sigma_{zz}}{\partial z} + \frac{\sigma_{rz}}{r} = -\rho b_z \,, \qquad (9.110)$$

and in spherical coordinates (r, ϕ, θ):

$$\frac{\partial \sigma_{rr}}{\partial r} + \frac{1}{r}\frac{\partial \sigma_{r\theta}}{\partial \theta} + \frac{1}{r\sin\theta}\frac{\partial \sigma_{r\phi}}{\partial \phi} \\ + \frac{2\sigma_{rr} - \sigma_{\theta\theta} - \sigma_{\phi\phi} + \sigma_{\theta r}\cot\theta}{r} = -\rho b_r \,, \quad (9.111)$$

$$\frac{\partial \sigma_{r\phi}}{\partial r} + \frac{1}{r}\frac{\partial \sigma_{\theta\phi}}{\partial \theta} + \frac{1}{r\sin\theta}\frac{\partial \sigma_{\phi\phi}}{\partial \phi} \\ + \frac{2\sigma_{r\phi} + \sigma_{\phi r} + (\sigma_{\theta\phi} + \sigma_{\phi\theta})\cot\theta}{r} = -\rho b_\phi \,, \quad (9.112)$$

$$\frac{\partial \sigma_{r\theta}}{\partial r} + \frac{1}{r}\frac{\partial \sigma_{\theta\theta}}{\partial \theta} + \frac{1}{r\sin\theta}\frac{\partial \sigma_{\theta\phi}}{\partial \phi} \\ + \frac{(\sigma_{\theta\theta} - \sigma_{\phi\phi})\cot\theta + 2\sigma_{r\theta} + \sigma_{\theta r}}{r} = -\rho b_\theta \,. \quad (9.113)$$

In slit rheometers use is made of the Cartesian coordinate system; for coaxial cylinders, parallel-plate, and capillary rheometers the cylindrical coordinate system is necessary, whereas for cone-and-plate rheometers the use of the spherical coordinate system is convenient.

In the description of the rheometric methods we will make use of the assumption that N_1, N_2, and σ_{21} only depend on the local value of the shear rate, independent of the curvature of the streamlines, provided the shear rate is constant in the gap. However, the local values of σ_{11}, σ_{22}, and σ_{33}, do depend on the curvature. We will come back to this point later in the description of the Couette system.

Local Coordinate System. In the common rheogoniometers the equations of motion can be greatly simplified, because in general: (a) $\partial/\partial\phi = 0$, and (b) $\sigma_{13} = \sigma_{31} = \sigma_{23} = \sigma_{32} = 0$. Here we will consider: (1) Couette flow between two coaxial cylinders, (2) flow between coaxial, parallel plates, (3) flow between coaxial cone and plate, and (4) flow in a cylindrical or rectangular capillary. It is not clear a priori what the three directions are, e.g., in some cases the 1-direction is the ϕ-direction, but it may also be the r-direction or the z-direction, depending on the geometry of the instrument of consideration. Hence, we have to define the *local coordinate system* in the following way (Fig. 9.26)

- The 1-direction is the direction of flow, i.e., the direction of the tangent of the local streamline (steady state is discussed);
- The 2-direction is the direction perpendicular to the local shearing plane (Sect. 9.1.2) and in general (but not always) positive in the direction of increasing flow rate;
- The 3-direction completes a right-handed coordinate system.

Rate of Shear. As we move on from rectilinear to consider curvilinear shear flow, we have to define *shearing planes* as planes that move stiffly during shear flow, i.e., planes where all particles maintain their mutual distance and accordingly have the same velocity. Different planes have different velocities and this is the origin of shear and rate of shear. In rectilinear shear flow between two infinitely extended parallel plates the shear planes are fluid planes parallel to the plates: different planes move with different velocities. In Couette flow the planes of shear are coaxial cylinders: different cylinders rotate with different angular velocities. In parallel-plate (or disks) rheometers they are circular disks perpendicular to the axis: different disks rotate with different angular velocities. In cone-and-plate rheometers they are coaxial cones: different cones rotate with different angular velocities. In capillary rheometers they are coaxial cylinders: different cylinders move axially with different velocities (also called telescopic flow).

The difference in (angular) velocity between the various planes of shear is a measure of the rate of shear. In rectilinear shear flow the rate of shear is easy to calculate as the velocity gradient, as will be shown. In rotating rheometers, however, the rate of shear is not as easy to calculate as in rectilinear flow, because there is a snake in the grass. If in a Couette instrument both the cylinders and the fluid rotate with the same angular velocity there will be no shear. Notwithstanding the lack of shear there is a difference in velocity between the rotating cylinders ($v = \Omega r$), thus there is a velocity gradient. Hence, we have to be careful when calculating the rate of shear.

In rectilinear shear flow the velocity and the velocity gradient are equal to (Fig. 9.27):

$$v_1 = \frac{v_\mathrm{w}}{d} x_2 \quad \text{and} \quad \frac{\mathrm{d}v_1}{\mathrm{d}x_2} = \frac{v_\mathrm{w}}{d}. \qquad (9.114)$$

The shear rate is equal to

$$\dot{\gamma} \equiv \frac{\mathrm{d}\gamma}{\mathrm{d}t} = \frac{\mathrm{d}\tan\alpha}{\mathrm{d}t} = \frac{\mathrm{d}}{\mathrm{d}t}\left(\frac{\mathrm{d}x_1}{\mathrm{d}x_2}\right) = \frac{\mathrm{d}}{\mathrm{d}x_2}\left(\frac{\mathrm{d}x_1}{\mathrm{d}t}\right)$$
$$= \frac{\mathrm{d}v_1}{\mathrm{d}x_2}, \qquad (9.115)$$

where α is the angle shown in Fig. 9.27 and $\mathrm{d}/\mathrm{d}t$ denotes material time differentiation.

Accordingly, in rectilinear shear flow the velocity gradient is equal to the shear rate. If the shear rate is

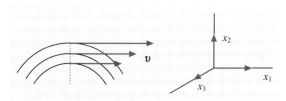

Fig. 9.26 Definition of a local coordinate system in Couette flow

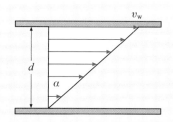

Fig. 9.27 Rectilinear shear flow

independent of time, the flow is called *stationary*, and $\dot{\gamma} = \dot{\gamma}_0 = $ constant.

In Couette geometry the velocity gradient is equal to

$$\frac{dv}{dr} = \frac{d(\Omega r)}{dr} = \Omega + r\frac{d\Omega}{dr}. \quad (9.116)$$

If the inner and outer cylinder rotate together with the fluid as a rigid system, then $\Omega = $ constant and thus $d\Omega/dr = 0$. In this case the velocity gradient is equal to Ω, whereas the shear rate is equal to zero. Hence, in the determination of the shear rate, we have to take into consideration the fact that the velocities of neighboring shearing planes contain a rigid-body rotation component, $\Omega \Delta r$, which we have to subtract from the difference in velocities (see also Fig. 9.28), in order to calculate the shear rate

$$v(r) = \Omega r \text{ and}$$
$$v(r + \Delta r) = (\Omega + \Delta\Omega)(r + \Delta r) = \Omega r + \Omega \Delta r$$
$$+ r\Delta\Omega + \text{higher-order terms} \quad (9.117)$$

so that

$$\dot{\gamma} = \lim_{\Delta r \to 0} \frac{v(r + \Delta r) - \Omega \Delta r - v(r)}{\Delta r}$$
$$= \lim_{\Delta r \to 0} r\frac{\Delta\Omega}{\Delta r} = r\frac{d\Omega}{dr}. \quad (9.118)$$

Similar considerations lead to expressions of shear rates in other geometries.

Drag Flow Rheometers

In this section we will discuss instruments with coaxial cylinders, coaxial plates, and coaxial cone and plate.

Coaxial Cylinders in Couette Flow: Rate of Shear. It was found before that the shear rate in a Couette instrument (Fig. 9.29) is equal to $\dot{\gamma} = r\,d\Omega/dr$. Because the gap is narrow, i.e., $(r_0 - r_i)/r \ll 1$, use can be made of the following approximation:

$$d\Omega/dr \approx \Delta\Omega/\Delta r = \Omega_0/(r_0 - r_i),$$

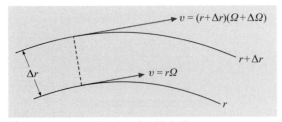

Fig. 9.28 Couette flow in a liquid shell

where $\Delta r = r_0 - r_i$ and $\Delta\Omega = \Omega(r_0) - \Omega(r_i) = \Omega_0$ so that we find for the shear rate

$$\dot{\gamma}_0 \approx r\frac{\Omega_0}{\Delta r} \approx \frac{r_0 + r_i}{2} \cdot \frac{\Omega_0}{\Delta r} = \frac{r_{\text{av}}}{\Delta r} \cdot \Omega_0 \quad (9.119)$$

with $r_{\text{av}} = (r_0 + r_i)/2$.

Hereafter it is assumed that Ω_0 is constant and thus $\dot{\gamma}$ does not depend on time.

It has to be mentioned that a still better approximation is

$$\dot{\gamma}(r) = \frac{\Omega_0}{r^2} \cdot \frac{r_0^2 r_i^2}{r_{\text{av}} \Delta r}. \quad (9.120)$$

It appears from these equations that the shear rate is practically constant in the gap.

Equations of Motion. For this instrument it follows from Fig. 9.29 that the 1-direction corresponds with the ϕ-direction, the 2-direction with the r-direction, and the 3-direction with the z-direction. We will make use of cylindrical coordinates. Many terms will vanish, because, as mentioned above: (a) in the stationary axisymmetric state there will be no ϕ dependence, so that $\partial/\partial\phi = 0$, and (b) $\sigma_{13} = \sigma_{31} = \sigma_{23} = \sigma_{32} = 0$, which in the present system means $\sigma_{\phi z} = \sigma_{z\phi} = \sigma_{rz} = \sigma_{zr} = 0$. Accordingly, from the equations of motion (9.108–9.110) with b_z being the acceleration due to gravity g, only the following terms remain:

$$\frac{\partial\sigma_{22}}{\partial r} - \frac{\sigma_{11} - \sigma_{22}}{r} = 0 \text{ or } \frac{\partial\sigma_{22}}{\partial r} - \frac{N_1}{r} = 0, \quad (9.121)$$

$$\frac{\partial\sigma_{21}}{\partial r} + \frac{2\sigma_{21}}{r} = 0, \quad (9.122)$$

$$\frac{\partial\sigma_{33}}{\partial z} = -\rho g. \quad (9.123)$$

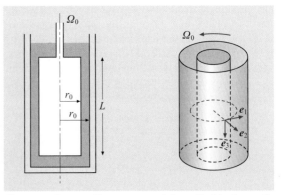

Fig. 9.29 Couette flow expressed in cylindrical coordinates, where $e_1 = e_\phi$, $e_2 = e_r$ and $e_3 = e_z$

Measurement of the First Normal Stress Difference.
From (9.122) it follows that

$$r\frac{\partial \sigma_{22}}{\partial r} = \sigma_{11} - \sigma_{22} = N_1 \ . \tag{9.124}$$

In the narrow gap of the Couette instrument

$$\frac{\partial \sigma_{22}}{\partial r} \approx \frac{\Delta \sigma_{22}}{\Delta r} = \frac{\sigma_{22}(r_0) - \sigma_{22}(r_i)}{r_0 - r_i} \tag{9.125}$$

As a result we find

$$N_1 \approx \frac{\Delta \sigma_{22}}{\Delta r} r_{av} \ , \tag{9.126}$$

so that N_1 is practically constant and can be determined *easily* by measuring σ_{22} both on the outer and inner cylinder. Experimentally it has been found that: (a) $\sigma_{22}(r_0)$ and $\sigma_{22}(r_i)$ are negative (Fig. 9.30), and (b) $\Delta \sigma_{22}$ is positive. The important conclusion is that the first normal stress difference N_1 is positive.

For the first normal stress coefficient (Sect. 9.1) we obtain:

$$\Psi_1 \equiv \frac{\sigma_{11} - \sigma_{22}}{\dot{\gamma}_0^2} = \frac{N_1}{\dot{\gamma}_0^2} = \frac{\Delta \sigma_{22}}{\Omega_0^2} \cdot \frac{\Delta r}{r_{av}} \ , \tag{9.127}$$

where (9.119) has been used.

Local Stresses in Couette Flow. In the description of the measuring methods we have made use of the assumption that N_1, N_2, and σ_{21} are only dependent on the local value of the shear rate. This means that during the flow their values are independent of the curvature of the shearing planes and the streamlines, provided that the shear rate is constant in the gap. However, the values of σ_{11}, σ_{22}, and σ_{33} do depend on the curvature and even at constant shear rate they depend on their location in the flowing liquid. Such a situation is shown in Fig. 9.30 for Couette flow.

The normal stresses σ_{11} and σ_{22} are not constant in the gap between the inner and outer cylinders, but it appears that at every point their difference $\sigma_{11} - \sigma_{22} = N_1$ is constant, independent of the location in the gap. In fact, in some cases we make use of the change of the stress components with the streamlines in order to determine N_1 and/or N_2 (Fig. 9.30). For example in a cone-and-plate instrument σ_{22} on the plate is a function of the distance to the cone axis and thus a function of the curvature of the streamlines, whereas in the present case the shear stress and the first normal stress differences are independent of this curvature.

Measurement of the Viscosity. Integration of (9.122) yields

$$r^2 \sigma_{21} = \beta = \text{constant} \ . \tag{9.128}$$

The moment of the shear force acting on a fluid cylinder in the gap ($r_i \leq r \leq r_0$), with respect to the cylinder axis reads

$$T_M(r) = Fr = \sigma_{21} 2\pi r L r = 2\pi \beta L \ . \tag{9.129}$$

The important conclusion is that T_M is independent of position in the liquid and thus T_M is also equal to the moment on the inner cylinder, which can be measured. The general result is

$$\frac{T_M}{2\pi L} = \beta = r^2 \sigma_{21} = r^2 \eta \dot{\gamma} \tag{9.130}$$

so that

$$\eta(\dot{\gamma}) \equiv \frac{\sigma_{21}}{\dot{\gamma}} = \frac{T_M}{4\pi \Omega_0 L} \frac{r_0^2 - r_i^2}{r_0^2 r_i^2} \ , \tag{9.131}$$

where use has been made of (9.120).

Hence, by control of the angular velocity of the outer cylinder and by measuring the momentum on the inner cylinder the viscosity can be determined as a function of the shear rate.

The Parallel-Plate Instrument in Torsional Flow
In this instrument the gap, of thickness z_0, between two circular, parallel plates is filled with the liquid under investigation (Fig. 9.31). The lower plate is fixed and the upper plate rotates with an angular velocity Ω_0 around the vertical axis CC' through both mid points.

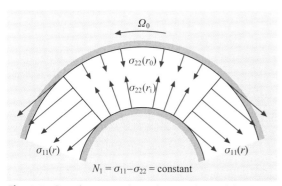

Fig. 9.30 Local stresses dependent on the position in the gap between two coaxial cylinders. Note that $\sigma_{22}(r_0)$ and $\sigma_{22}(r_i)$ are both pointing into the fluid, which means that both are negative normal stresses, and that $\sigma_{22}(r_i)$ is more negative than $\sigma_{22}(r_0)$, hence it follows that $\sigma_{22}(r_0) - \sigma_{22}(r_i) > 0$

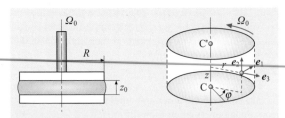

Fig. 9.31 Parallel plates with cylindrical coordinate system ϕ, z, r, with $\boldsymbol{e}_1 = \boldsymbol{e}_\phi$, $\boldsymbol{e}_2 = \boldsymbol{e}_z$ and $\boldsymbol{e}_3 = \boldsymbol{e}_r$

We consider the rotation as anticlockwise if observed from above.

The velocity is directed along the tangents on the circular streamlines, hence the ϕ-direction is the 1-direction; the 2-direction is the direction in which the velocity of the shearing planes increases, i.e., the z-direction ($z = 0$ on the bottom plate and $z = z_0$ on the upper plate). The 3-direction, i.e., the r-direction, completes a right-handed coordinate system ($r = 0$ at the axis and $r = R$ at the outer edge of the boundary). The effect of the outer boundary is neglected, which is possible if $z_0 \ll R$.

Rate of Shear. The shear rate is found to be

$$\dot{\gamma} = \lim_{\Delta z \to 0} \frac{r\Omega(z+\Delta z) - r\Omega(z)}{\Delta z} = r\frac{d\Omega}{dz}. \quad (9.132)$$

The shear rate is not constant in the gap: it increases from $\dot{\gamma} = 0$ on the axis to $\dot{\gamma}(R) = \Omega_0/z_0$ at the outer boundary. This is a big disadvantage of the parallel-plate instrument: it can be used properly only in the range where the viscoelastic properties are not shear rate dependent. In general the instrument is used only for oscillatory measurements with small amplitudes. It will be shown however, how to overcome these disadvantages.

Equations of Motion. The equations of motion in cylindrical coordinates are

$$\frac{\partial \sigma_{33}}{\partial r} = \frac{N_1 + N_2}{r}, \quad (9.133)$$

$$\frac{\partial \sigma_{12}}{\partial z} = 0, \quad (9.134)$$

$$\frac{\partial \sigma_{22}}{\partial z} = -\rho b_z = \rho g. \quad (9.135)$$

It can be proven, with the aid of (9.134), that the shear rate only depends on r and not on z. The result is

$$\dot{\gamma}(r) = r\frac{d\Omega}{dz} = r\frac{\Omega_0}{z_0} \quad (9.136)$$

so that the shear rate appears to increase linearly with distance from the axis.

Determination of the Viscosity η and $\Psi_1 - \Psi_2$ as Functions of the Rate of Shear. If the viscosity is shear rate dependent, it seems impossible to determine the *real* viscosity. The same holds for the difference of the first and second normal stress coefficients $\Psi_1 - \Psi_2$, where

$$\Psi_1 = (\sigma_{11} - \sigma_{22})/\dot{\gamma}^2, \quad \Psi_2 = (\sigma_{22} - \sigma_{33})/\dot{\gamma}^2.$$

However it can be proven that the viscosity and the difference of the first and second normal stress coefficients can be determined as functions of shear rate in a proper way, by measuring the moment T_M exerted by the flowing liquid on the upper plate, and the normal force F_n exerted by the flowing liquid on the upper plate, both as functions of the angular velocity Ω_0. This yields a method to determine σ_{21} and $N_1 - N_2$, both as functions of the shear rate $\dot{\gamma}_R = \dot{\gamma}$ at $r = R$:

$$\sigma_{21}(\dot{\gamma}_R) = \frac{T_M}{2\pi R^3}\left(3 + \frac{d\log T_M}{d\log \Omega_0}\right) \quad (9.137)$$

and

$$N_1(\dot{\gamma}_R) - N_2(\dot{\gamma}_R) = \frac{F_n}{\pi R^2}\left(2 + \frac{d\log F_n}{d\log \Omega_0}\right). \quad (9.138)$$

Hence, upon plotting T_M and F_n as functions of Ω_0 on double logarithmic scales, the $\dot{\gamma}_R$ dependent slopes $d\log T_M/d\log \Omega_0$ and $d\log F_M/d\log \Omega_0$ can be determined and thus σ_{21} and $N_1 - N_2$ can be determined as functions of $\dot{\gamma}_R$, and accordingly also the viscosity $\eta = \sigma_{21}/\dot{\gamma}$ and the difference between the first and second normal stress coefficients $\Psi_1 - \Psi_2 = (N_1 - N_2)/\dot{\gamma}^2$.

The Cone-and-Plate Instrument

In a cone-and-plate instrument the liquid under investigation is present in the gap between a cone with a large top angle (at least 170°) and a circular plate. The top of the cone, whose axis is perpendicular to the plate, is in principle positioned in the central point of the plate. In order to prevent friction between cone and plate during rotation of the cone, the top of the cone is truncated with the virtual, fictitious top corresponding to the plate center. The angle between the cone and plate, $\Delta\Theta$, is small (less than 5°). The cone rotates at an angular velocity of Ω_0, and counterclockwise when seen from above (Fig. 9.32). The shearing planes, the planes that rotate stiffly with a rate increasing from the plate to the cone itself, are coaxial cones with top angles between π and $\pi - 2\Delta\Theta$ rad. The 1-direction is the indifferent

ϕ-direction, in the 2-direction the rotational velocity of the shearing planes increases, i.e., the θ-direction, and the 3-direction is the r-direction.

Rate of Shear. The shear rate can be calculated, in a quite complicated way. The result is

$$\dot{\gamma}_0 = \sin\theta \frac{d\Omega}{d\theta} \approx \frac{\Omega_0}{\Delta\Theta} \,. \tag{9.139}$$

This means that the shear rate is almost constant in the sample under investigation ($\sin\theta$ varies approximately from 1 to 0.996). This makes the cone-and-plate instrument highly suitable for the determination of the viscosity and the normal stress coefficients as functions of shear rate.

Equations of Motion. The equations of motion in spherical coordinates can also be simplified in this case. The result, following (9.111–9.113), is:

$$\frac{\partial \sigma_{33}}{\partial r} - \frac{N_1 + 2N_2}{r} = -\rho g \cos\theta \,, \tag{9.140}$$

$$\frac{1}{r}\frac{\partial \sigma_{22}}{\partial \theta} - \frac{N_1}{r}\cot\theta = \rho g \sin\theta \,, \tag{9.141}$$

$$\frac{1}{r}\frac{\partial \sigma_{21}}{\partial \theta} + \frac{2\sigma_{21}}{r}\cot\theta = 0 \,. \tag{9.142}$$

Viscosity Measurement. Integration of (9.142) yields $\sigma_{21}(\theta) = \text{constant} = \sigma_{21}(1/2\pi) = C$, i.e., the shear stress on the plate. As a result, the moment T_M on the plate is

$$T_M = \int_0^R r\sigma_{21}\left(\frac{\pi}{2}\right) 2\pi r dr = \frac{2\pi}{3} CR^3 \quad \text{so that}$$

$$\sigma_{21} = \frac{3T_M}{2\pi R^3} \,. \tag{9.143}$$

For the viscosity we now find with the aid of (9.139)

$$\eta \equiv \frac{\sigma_{21}}{\dot{\gamma}_0} = \frac{3T_M}{2\pi \dot{\gamma}_0 R^3} \approx \frac{3T_M \Delta\Theta}{2\pi R^3 \Omega_0} \,. \tag{9.144}$$

Hence by measuring T_M as a function of Ω_0, the viscosity can be determined as a function of shear rate.

One remark has to be made, however. Equation (9.142) follows from (9.112) with the assumption that $\sigma_{\phi\theta}$ (i.e., σ_{12}) $= \sigma_{\theta\phi}$ (i.e., σ_{21}), which in general holds for polymeric systems, but not for systems such as low-molecular-weight liquid crystals. Hence, one has to be careful in the determination of the viscosity of low-molecular-weight liquid crystals in the cone-and-plate instrument. For liquid-crystal polymers, however,

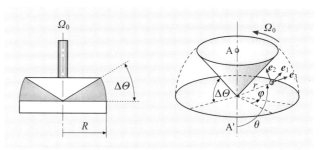

Fig. 9.32 Cone-and-plate instrument with spherical coordinates ϕ, θ, r, with $e_1 = e_\phi$, $e_2 = e_\theta$ and $e_3 = e_r$

the polymeric nature overrules the liquid-crystalline behavior, so that the cone-and-plate instrument can also be used to determine their viscosity.

Measurement of Normal Stress Differences: $N_1 + 2N_2$. Because $\cos\theta$ is approximately equal to 0, (9.140) may be approximated by

$$r\frac{\partial \sigma_{33}}{\partial r} = N_1 + 2N_2 \,. \tag{9.145}$$

The equation is only correct on the plate ($\theta = \pi/2$). Because $\sigma_{22} - \sigma_{33}$ is a function of $\dot{\gamma}$ and $\dot{\gamma}$ is not a function of r, in (9.145) $\partial\sigma_{33}/\partial r$ may be replaced by $\partial\sigma_{22}/\partial r$. Integration then eventually yields

$$\sigma_{22}(r) = \sigma_{22}(R) + (N_1 + 2N_2)\ln(r/R) \,. \tag{9.146}$$

(N.B. the fact that $\sigma_{11} - \sigma_{22}$ is a function of $\dot{\gamma}$ and thus not a function of on r, has also been used).

From this equation it follows that, upon plotting $\sigma_{22}(r)$ versus $\ln(r/R)$, a straight line will be obtained

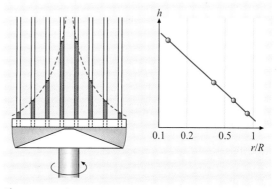

Fig. 9.33 *Left*: cone-and-plate instrument supplied with tubes for the measurement of the normal stress $\sigma_{22}(r)$ as a function of the distance to the axis. Note that the normal stress σ_{22} is negative and equal to $-\rho g h$. *Right*: height of the liquid in the tubes plotted versus $\ln(r/R)$

with a slope equal to $N_1 + 2N_2$. This normal stress $\sigma_{22}(r)$ can be measured as a function of r with pressure gauges positioned at various places on the plate (Fig. 9.33). The method with tubes mounted on the plate is only appropriate for liquids with a low viscosity, which can ascend the tubes within a reasonable time. For more-viscous liquids real pressure gauges are needed. In the figure shown the normal stresses σ_{22} are negative, because a stress is positive if the surroundings pull on the material. In this case the liquid columns push on the liquid in the gap, so that the normal stresses are negative. The slope is negative, in agreement with the observation that N_1 is positive for polymer fluids in general, whereas N_2 is in general negative and absolutely only a small fraction of N_1. In Fig. 9.34 results are shown for a 2% solution of polyisobutylene. This demonstrates that $N_1 + 2N_2$ increases with increasing shear rate. Division by $\dot{\gamma}_0^2$ yields the sum $\Psi_1 + 2\Psi_2$, which appears to be constant, independent of the shear rate.

It has to be mentioned that the device shown in Fig. 9.32, which is inverted with respect to Fig. 9.31, would result in a term $+g\cos\theta$ instead of $-\rho g\cos\theta$ in (9.140). However, because $\cos\theta$ is approximately equal to 0, we can make use of (9.140), which leads to (9.145).

The First Normal Stress Difference. The total force exerted by the flowing fluid on the plate is equal to

$$F_n = -\int_0^R \sigma_{22}(r) 2\pi r \, dr \,. \tag{9.147}$$

Integration eventually yields

$$N_1 = \frac{2F_n}{\pi R^2} \,. \tag{9.148}$$

In the derivation of (9.148) it is assumed that the free surface of the sample is spherical. In that case there are no surface stress components present, so that the term $\sigma_{33}(R)$ that arises in the integration of (9.147), is equal to 0.

Hence, by measuring the total normal force exerted by the flowing liquid on the plate as a function of shear rate the first normal stress difference N_1 and thus the first normal stress coefficient $\Psi_1 = N_1/\dot{\gamma}^2$ can be determined as a function of the shear rate.

The method presented before yields the value of $N_1 + 2N_2$, so that it is now also possible to determine the second normal stress difference N_2 and thus the second normal stress coefficient $\Psi_2 = N_2/\dot{\gamma}^2$.

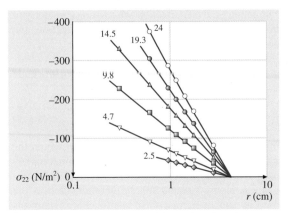

Fig. 9.34 Normal stresses $\sigma_{22}(r)$ on the plate in a cone and plate instrument ($R = 4.4$ cm), versus distance to the axis, during shear flow of a 2% solution of polyisobutylene (B200) in oppanol B1 (a Newtonian liquid with $\eta = 23.6$ mPas at $25\,^\circ\text{C}$), measured at various shear rates, as indicated in s^{-1} accoreding to [9.7]

Hole Effect in the Measurement of Normal Stresses

It is worthwhile mentioning here the so-called *hole effect*. For the measurement of normal stresses use is made of pressure gauges. In general they are mounted in the walls of the instrument. The disadvantage of this is that holes are needed for these, and in the neighborhood of a hole the streamlines will be curved in the direction of the hole. This causes an extra force due to the tension in the streamlines caused by the elasticity of the fluid. The hole pressure is the difference between the pressure that would exist at the wall if the flow were undisturbed and the measured pressure at the bottom of the hole (Fig. 9.35). Theoretically it can be shown that for a circular opening this error is:

$$\Delta P = P - P_{\text{hole}} = C_1 N_1 + C_2 N_2 \quad \text{where}$$
$$C_1 \approx 0.25 \quad \text{and} \quad C_2 \ll C_1 \,. \tag{9.149}$$

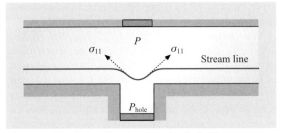

Fig. 9.35 The hole effect in viscoelastic fluids

Pressure-Driven Rheometers

In a pressure-driven rheometer the fluid is pressed from a reservoir into a capillary tube of circular cross section or into a gap of rectangular cross section (Fig. 9.36) and Fig. 9.42, respectively. The pressure drop and flow rate through this tube or gap are used to determine the viscosity. In contrast to drag flows there is an entrance region where the fluid is accelerated and an exit region where a viscoelastic fluid is liable to die swell. For such rheometers the important assumptions in the derivation of the relation between shear stress and shear rate are fully developed steady and laminar flow with no slip at the walls (i.e., $v_w = 0$).

The Capillary Rheometer. In a capillary rheometer the liquid under investigation flows through a straight tube with a circular cross section with radius R. The shear planes are coaxial cylinders with radius r. The velocity is constant in a cylindrical shear plane and decreases from the axis to the wall (for this reason this kind of flow is sometimes called telescopic flow). The streamlines are parallel with the axis and the flow is rectilinear (Fig. 9.37). The z-direction is the 1-direction, the r-direction is the 2-direction (in this case the direction where the rate of the shear planes decreases), and the ϕ-direction is the 3-direction.

Rate of Shear. The flow is rectilinear, so that the shear rate is given by

$$\dot{\gamma} = \dot{\gamma}(r) = \lim_{\Delta r \to 0} \frac{v(r + \Delta r) - v(r)}{\Delta r} = \frac{dv}{dr} \leq 0. \quad (9.150)$$

Equations of Motion. The equations of motion in cylindrical coordinates obtained from (9.108–9.110) are

$$\frac{\partial \sigma_{22}}{\partial r} + \frac{\partial \sigma_{21}}{\partial z} + \frac{N_2}{r} = 0, \quad (9.151)$$

$$\frac{\partial \sigma_{21}}{\partial r} + \frac{\partial \sigma_{11}}{\partial z} + \frac{\sigma_{21}}{r} = -\rho b_z, \quad (9.152)$$

(the third equation of motion yields the identity $0 = 0$).

Because σ_{21} is a function of $\dot{\gamma}$ and $\dot{\gamma}$ is not a function of z, it follows that $\partial \sigma_{21}/\partial z = 0$. However, this is only true far enough from the entrance of the capillary. If we neglect the entrance effect, both equations can be simplified to

$$\frac{\partial \sigma_{22}}{\partial r} + \frac{N_2}{r} = 0, \quad (9.153)$$

$$\frac{\partial \sigma_{11}}{\partial z} + \frac{1}{r}\frac{d(r\sigma_{21})}{dr} = -\rho b_z. \quad (9.154)$$

Integration of (9.154), and bearing in mind that $\partial \sigma_{11}/\partial z = $ constant, yields

$$\sigma_{21}(r) = -\frac{1}{2}r\left(\rho b_z + \frac{\partial \sigma_{11}}{\partial z}\right). \quad (9.155)$$

Two cases can be distinguished:

a) $\partial \sigma_{11}/\partial z$ is determined by gravitational forces and is comparable to ρg: the vertical viscometer for dilute solutions, e.g., the Ubbelohde viscometer, which is a subject of discussion in the Chapter on viscosity measurements in Newtonian liquids Sect. 3.4.4.
b) $|\partial \sigma_{11}/\partial z| \gg \rho g$: the rheometer for polymer melts, where $\partial \sigma_{11}/\partial z$ is significant and determined by an imposed pressure ΔP, which is the subject of the present discussion.

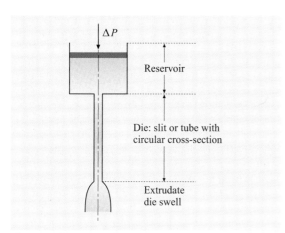

Fig. 9.36 Pressure-driven rheometer

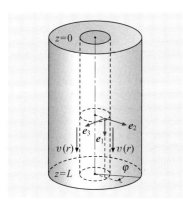

Fig. 9.37 Cylindrical coordinate system z, r, ϕ in a capillary rheometer with $e_1 = e_z, e_2 = e_r$ and $e_3 = e_\phi$

The Capillary Rheometer for Polymer Melts. In this case a high pressure ΔP is applied, so that the gravitational forces can be neglected. This means that $\sigma_{11}(0) = -\Delta P$ and $\sigma_{11}(L) = 0$, so that:

$$\frac{\partial \sigma_{11}}{\partial z} = \frac{\sigma_{11}(z=L) - \sigma_{11}(z=0)}{L} = \frac{\Delta P}{L}. \quad (9.156)$$

Because $\Delta P/L \gg \rho b_z$, it follows from (9.155)

$$\sigma \equiv \sigma(r) = -\frac{r\Delta P}{2L} < 0. \quad (9.157)$$

Again the shear stress is a linear function of r (independent of the viscosity), and maximal at the wall

$$\sigma_{\mathrm{w}} \equiv \sigma_{21}(R) = -\frac{R\Delta P}{2L} \quad (9.158)$$

and zero on the axis, so that

$$\frac{\sigma}{\sigma_{\mathrm{w}}} = \frac{\sigma_{21}(r)}{\sigma_{21}(R)} = \frac{r}{R}. \quad (9.159)$$

Velocity Profile. For *Newtonian liquids* we have

$$\eta = \frac{\sigma_{21}}{\dot{\gamma}} = -\frac{r\Delta P}{2L \, (\mathrm{d}v/\mathrm{d}r)}. \quad (9.160)$$

Integrating yields the velocity profile

$$v(r) = \frac{\Delta P}{4\eta L}\left(R^2 - r^2\right). \quad (9.161)$$

Hence, for Newtonian liquids this velocity profile is parabolic and the flow is called Poiseuille flow. In Fig. 9.38 the velocity and the shear stress profiles are shown for Poiseuille flow. Note that the shear stress is negative: the wall pulls on the flowing liquid and liquid cylinders pull on the neighboring cylinders closer to the axis.

For a *power-law fluid*, the Ostwald–de Waele power-law constitutive equation holds:

$$\sigma = K\dot{\gamma}^n \quad \text{and} \quad \eta = K\dot{\gamma}^{n-1}, \quad (9.162)$$

where n is the so-called power-law index and K is the consistency index. Then

$$\eta(\dot{\gamma}) = \frac{\sigma_{21}}{\dot{\gamma}} = -\frac{r\Delta P}{2L \, (\mathrm{d}v/\mathrm{d}r)} = K\left(\frac{\mathrm{d}v}{\mathrm{d}r}\right)^{n-1}. \quad (9.163)$$

Integrating the latter, we obtain the velocity profile

$$v(r) = \left(\frac{\Delta P}{2KL}\right)^{1/n} \frac{n}{n+1} \left(R^{1+1/n} - r^{1+1/n}\right). \quad (9.164)$$

The average velocity is

$$\langle v \rangle \equiv \frac{\int_0^R 2\pi r v(r) \, \mathrm{d}r}{\int_0^R 2\pi r \, \mathrm{d}r} = \left(\frac{\Delta P}{2KL}\right)^{1/n} \frac{n}{3n+1} R^{1+1/n} \quad (9.165)$$

and the normalized velocity profile is

$$v_{\mathrm{rel}}(r) \equiv \frac{v(r)}{\langle v \rangle} = \frac{3n+1}{n+1}\left[1 - \left(\frac{r}{R}\right)^{1+1/n}\right]. \quad (9.166)$$

Note that for $n = 1$ the Poiseuille profile for Newtonian liquids is recovered.

In Fig. 9.39 the relative velocity profiles are shown for a Newtonian liquid ($n = 1$), for a power-law liquid with $n = 1/3$, and for plug flow ($n = 0.001$). The maximum velocities on the axis are 2, 3/2, and 1, respectively.

Shear Rate at the Wall, the Rabinowitch Equation. Because the shear rate is a function of r, it seems impossible to determine the viscosity as a function of shear rate for non-Newtonian fluids. This problem can be overcome, however, in the following way. The volume flow rate Q (in m^3/s) of a liquid flowing through the tube in unit

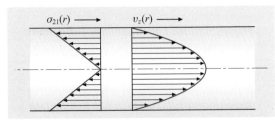

Fig. 9.38 Velocity and stress profiles for Poiseuille flow (i.e., for a liquid of constant viscosity) in straight tubes with circular cross-section: the velocity profile is parabolic and the shear stress profile is linear

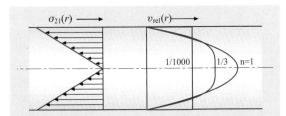

Fig. 9.39 Velocity profile in a straight tube with circular cross section, for power-law fluids with $n = 1$ (Newtonian liquid, Poiseuille flow), $n = 1/3$ and for $n = 0.001$ (plug flow)

time is

$$Q = \int_0^R v(r) \cdot 2\pi r \, dr. \quad (9.167)$$

If there is no slip at the wall, integration by parts yields

$$Q = -\int_0^R \pi r^2 \dot{\gamma}(r) \, dr. \quad (9.168)$$

It is quite easy to derive that for Newtonian liquids the relationship between the shear rate at the wall and Q is

$$\Gamma \equiv \dot{\gamma}_w = -\frac{4Q}{\pi R^3}. \quad (9.169)$$

If for non-Newtonian fluids we define the apparent shear rate at the wall as

$$\Gamma_a \equiv -\frac{4Q}{\pi R^3} \quad (9.170)$$

then the relationship between the shear rate at the wall and the volume flow rate can be derived as

$$\begin{aligned}\dot{\gamma}_w &= \frac{1}{4}\Gamma_a \left(3 + \frac{d \log |\Gamma_a|}{d \log |\sigma_w|}\right) \\ &= -\frac{Q}{\pi R^3}\left(3 + \frac{d \log Q}{d \log \Delta P}\right).\end{aligned} \quad (9.171)$$

Equation (9.171) is called the *Rabinowitch equation* already derived in 1929 (see Tadmor and Gogos in Furthter Reading) for flow of a liquid through a straight tube with a circular cross section. This shows that the shear rate

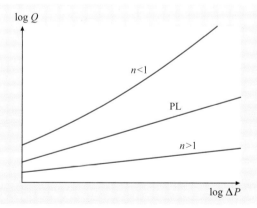

Fig. 9.40 Plot for the determination of $\dot{\gamma}_w$ for a Newtonian fluid (slope 1; denoted PL, Poiseuille); pseudoplastic shear thinning fluid (slope increasing and > 1), and dilatant shear thickening fluid (slope decreasing and < 1)

at the wall follows from the volume flow rate and from the slope of the volume flow rate Q versus the imposed pressure ΔP. Upon plotting $\log Q$ versus $\log \Delta P$, the slope can be found for every value of ΔP and/or Q. For Newtonian liquids the line obtained is a straight line with a slope of 1. On the other hand, for shear thinning pseudoplastic fluids with $n < 1$ a line is obtained with an increasing slope larger than 1. For dilatant shear thickening fluids with $n > 1$ a straight line with slope less than 1 is obtained (Fig. 9.40). In general, as it is easy to see from (9.171), since a simple derivation for the power-law liquids yields

$$\dot{\gamma}_w = -\frac{Q}{\pi R^3}\left(3 + \frac{1}{n}\right). \quad (9.172)$$

For a slit rheometer of rectangular cross section, with length L and sides W and H, where $L \gg W \gg H$ (Fig. 9.41) the shear rate at the wall is found to be:

$$\begin{aligned}\dot{\gamma}_w &= \frac{1}{3}\Gamma_a\left(2 + \frac{d\log|\Gamma_a|}{d\log|\sigma_w|}\right) \\ &= -\frac{2Q}{WH^2}\left(2 + \frac{d\log Q}{d\log \Delta P}\right) \quad \text{where}\end{aligned}$$

$$\Gamma_a \equiv -\frac{6Q}{WH^2}. \quad (9.173)$$

Measurement of the sum $2N_1 + N_2$ and of N_1. The measurements of the volume flow (or of the volumetric flow rate) may be completed with measurements of the total force F at the exit of the capillary, exerted by the fluid on the tube in the direction parallel to the axis. This force is equal to the difference in momentum of the fluid thread just before and after the exit. For molten polymers this difference is almost equal to zero. This yields an expression for $2N_1 + N_2$ at the wall of the capillary.

The mentioned force F is given by the integral of the normal stress $\sigma_{11}(r)$ over the surface:

$$F = -\int_0^R 2\pi r \sigma_{11}(r) \, dr. \quad (9.174)$$

This integral can be evaluated, yielding an expression for $2N_{1,w} + N_{2,w}$:

$$\begin{aligned}2N_{1,w} + N_{2,w} = &-\frac{F + \pi R^2 \sigma_{22}(R,L)}{\pi R^2} \\ &\times \left(2 + \frac{d\log\left|F + \pi R^2 \sigma_{22}(R,L)\right|}{d\log|\sigma_w|}\right),\end{aligned} \quad (9.175)$$

Table 9.4 Survey of possible outcomes of measurement techniques. [a] means that these measurements are not easy and in general not possible in commercially available instruments

Instrument	Shear rate	Viscosity	Normal stress differences/coefficients		
Couette	$\dot\gamma \approx$ const.	$\eta(\dot\gamma)$	$N_1(\dot\gamma)$	$\Psi_1(\dot\gamma)$	[a]
Parallel plates	$\dot\gamma = \dot\gamma(r)$	$\eta(\dot\gamma_R)$	$N_1(\dot\gamma_R) - N_2(\dot\gamma_R)$	$\Psi_1(\dot\gamma_R) - \Psi_2(\dot\gamma_R)$	
Cone and plate	$\dot\gamma =$ const.	$\eta(\dot\gamma)$	$N_1(\dot\gamma)$	$\Psi_1(\dot\gamma)$	
			$N_1(\dot\gamma) + N_2(\dot\gamma)$	$\Psi_1(\dot\gamma) + \Psi_2(\dot\gamma)$	[a]
Capillary rheometer	$\dot\gamma = \dot\gamma(r)$	$\eta(\dot\gamma_w)$	$2N_1(\dot\gamma_w) + N_2(\dot\gamma_w)$	$2\Psi_1(\dot\gamma_w) + \Psi_2(\dot\gamma_w)$	[a]
Slit rheometer	$\dot\gamma = \dot\gamma(h)$	$\eta(\dot\gamma_w)$	$N_1(\dot\gamma_w)$	$\Psi_1(\dot\gamma_w)$	

where $\sigma_{22}(R, L)$ is the normal pressure at the wall just before the exit, which can be measured. The force F can be calculated as the difference in momentum flux of the flowing fluid just before and after the exit:

$$F = \rho \int_0^R 2\pi r v_r^2(r)\, dr - \pi R_s^2 \rho v_s^2, \quad (9.176)$$

where v_s and R_s are the velocity and radius of the fluid thread after the exit, respectively ($R_s > R$ due to die swell, which causes the change in the momentum flux). From (9.175) it follows that, upon plotting $\log \left| F + \pi R^2 \sigma_{22}(R) \right|$ versus $\log |\sigma_w|$, the sum $2N_{1,w} + N_{2,w}$ can be determined and thus $2\Psi_{1,w} + \Psi_{2,w}$ as a function of $\dot\gamma_w$.

If it is assumed that $F \ll \pi R^2 \sigma_{22}(R)$ (9.175) can be reduced to:

$$2N_{1,w} + N_{2,w} = P_e \left(2 + \frac{d \log P_e}{d \log |\sigma_w|} \right)$$
$$= P_e \left(2 + \frac{d \log P_e}{d \log \Delta P} \right), \quad (9.177)$$

where $P_e = P(R, L) = -\sigma_{22}(R, L)$ is the exit pressure exerted by the flowing fluid on the wall of the capillary.

For *slit rheometers* the corresponding equation reads:

$$N_{1,w} = P_e \left(1 + \frac{d \log P_e}{d \log |\sigma_w|} \right)$$
$$= P_e \left(1 + \frac{d \log P_e}{d \log \Delta P} \right). \quad (9.178)$$

Of course it is difficult to really measure the wall pressure at the exit. For this reason wall pressures at different distances z are measured and the results are extrapolated to the exit $z = L$ (Fig. 9.42). It will be clear that measuring the wall pressure $P(z, R)$ is easier in slit rheometers than in capillary rheometers.

Conclusions Concerning the Discussed Rheogoniometers

In Table 9.1 a survey is shown of the possible results of the discussed measuring techniques. The italics in the Table mean that these properties can be measured in principle, but more-sophisticated instruments are needed than are available from manufacturers.

9.1.3 Elongational Flows

Most industrial processing of polymeric materials involves a combination of shear and extensional flow. The larger strains experienced in extensional flow fields can result in highly non-Newtonian dependencies on both strain and strain rate. As these dependencies are usually inadequately described by shear characterization alone,

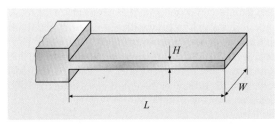

Fig. 9.41 Slit flow geometry, with $L \gg W \gg H$

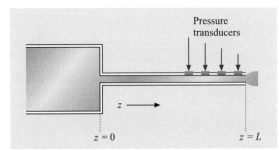

Fig. 9.42 Schematic representation of the extrapolation procedure to determine the exit wall pressure in a slit rheometer

extensive research has been conducted in the past several decades in an attempt to understand the extensional flow behavior of polymers. This research has involved a parallel effort in developing the theory of extensional flow behavior in order to determine the critical parameters and desired flow kinematics, and the experimental design of extensional rheometers that will successfully impose these kinematics.

In this Section, we discuss the current state of understanding of extensional rheology for highly viscous polymeric materials, such as polymer melts, and for low to moderately viscous materials. The state of the art in extensional rheometer design is discussed, and examples of commercially available instrumentation are presented.

List of Symbols

η: shear viscosity
η_E^+: extensional viscosity
η_E: steady-state extensional viscosity
η_{app}: apparent extensional viscosity
ε: Hencky strain
$\dot{\varepsilon}$: uniaxial strain rate
$\dot{\varepsilon}_{eff}$: effective strain rate, computed from diameter measurement
$\dot{\varepsilon}_0$: idealized strain rate
R_m: midpoint filament radius
D_m: midpoint filament diameter
D_0: initial midpoint filament diameter
L: filament length
F: tensile force
λ: longest relaxation time (Zimm)
σ: surface tension
g: gravitational constant
t: time
Tr: Trouton ratio $= \eta_E/\eta$
De: Deborah number $= \lambda\dot{\varepsilon}$
Ca: Capillary number $= \eta\dot{\varepsilon}_0 D_0/2\sigma$
Bo: Bond number $= \rho g D_0^2/4\sigma$
Ec: Elastocapillary number $= $ De/Ca

Industrial processing of polymer solutions and melts is usually a combination of shearing and extensional kinematics. Consequently, rheological characterization of materials in both a shear and an extensional flow field is necessary to accurately mathematically model these materials, detect subtle dissimilarities in composition, and to predict the processing conditions that will optimize the properties of the final product.

In most industrial processes, manufacturers desire to maximize line or web speed without compromising material properties. The shear rheological response of a complex fluid such as a polymer solution or melt that is going to be fiber-spun, blow-molded, calendared, foamed, deposited from a gun, or injected into a die or mold, will not correctly predict the processing behavior of these materials in these flow fields, which are dominated by a significant component of extensional strain. The nonlinear behavior of polymeric materials makes inference of their extensional properties from shear properties challenging, if not impossible. It is necessary to perform extensional rheology tests in addition to shear rheology to achieve a reliable characterization of the material.

Despite the recognized need for extensional measurements of polymeric fluids and melts, the development of instrumentation has proceeded slowly due to several challenges. The principal challenge is to generate a homogeneous extensional flow. To impose a deformation on a fluid, it is typically necessary to place the material in contact with a solid surface. Flow over a surface, however, will result in a shear stress, which will corrupt the extensional flow field and the resultant interpretation of the data. Thus, the flow cannot be confined, and must involve deformation in air or a sufficiently low-viscosity outer fluid.

The second challenge is that the strain history must also be known for all the fluid elements in the flow field. Strain rates must be high enough to stretch the chains (see Deborah number) and the strain high enough to stretch the chains beyond their normal radius of gyration. The high levels of strain required to achieve significant polymer chain extension and the increase in stress that ensues necessitates motion control systems with a large dynamic velocity range, a large achievable travel distance, and sensitive position control.

The large range of viscosities exhibited by polymeric materials has segregated individual extensional rheometer designs into distinct designs that are suited to a particular viscosity range, as shown in Table 9.5. High-viscosity materials (with zero-shear viscosity $\eta_0 > 1000$ Pa s) are best characterized by constant-length devices, or by constant-volume devices employing an outer fluid neutral buoyancy. Materials in the medium viscosity range of $1-1000$ Pa s are best characterized by filament stretching extensional rheometers. Materials in the low-viscosity range, which include dilute polymer solutions $0.01-1$ Pa s, are best characterized by capillary break-up rheometers, contraction flows, or opposed jet devices. More details on these devices will be provided in the next sections.

Table 9.5 Summary of extensional rheometer designs and application ranges

Instrument type	Geometry	Flow	Shear viscosity range [Pa s]	Limitations	Example ref.
Filament stretching, constant volume, medium viscosity		Uniaxial extension, constant strain rate	1–1000	Sample gripping, limited to medium with high viscosity, elastic instability	[9.16–18]
Filament stretching, constant volume, high viscosity		Uniaxial extension, constant strain rate	> 1000	Limited to low strain rates, temperature control	[9.10, 13, 14]
Filament stretching, constant length, high viscosity		Uniaxial extension, constant strain rate	> 1000	Limited to low strain rates, temperature control	[9.15, 19, 20]
Fiber spinning		Uniaxial extension	> 1	Low strain, non-uniform strain rates, pre-shear history	[9.21–24]
Capillary breakup rheometry		Uniaxial extension	0.01 – 1	Inertial and surface tension dominate at low end of viscosity, variable strain rates	[9.25–27]
Four-roll mill		Uniaxial extension	1–1000	Variable strain rates and strain histories	[9.28, 29]

As early as 1906, *Fred Trouton* verified the relationship between the uniaxial extensional viscosity and shear viscosity in Newtonian fluids by performing low-extension-rate experiments on pitch [9.8]. He

Table 9.5 (cont.)

Instrument type	Geometry	Flow	Shear viscosity range [Pa s]	Limitations	Example ref.
Entrance flows		Uniaxial extension	> 1	Variable strain rate, mixed with shear	[9.30, 31]
Sheet stretching		Biaxial planar extension	> 1000	Sample gripping, flow uniformity	[9.15]
Opposed jet		Uniaxial extension	0.01–1	Variable strain rates and strain histories, some shear	[9.32]
Planar elongation		Planar extension, constant strain rate	> 1000	Sample preparation, Planar Elongation, limited to low extension	[9.33]
Tubeless siphon		Uniaxial extension	1–1000	Pre-shear history, variable strain rate, low achievable strain rates	[9.34]

also derived theoretically the relationship between the extensional viscosity and shear viscosity of a Newtonian fluid; $\eta_E = 3\eta_0$, where η_E is the steady-state extensional viscosity. This result is a kinematic consequence of the differences in deformation between a shear flow and an extensional flow and is analogous to the relationship between the Young's modulus and the shear modulus in an incompressible isotropic elastic solid. The ratio between the extensional viscosity and shear viscosity is commonly referred to as the Trouton ratio, $\mathrm{Tr} = \eta_E/\eta_0$. Most early work was concerned with characterizing more-qualitative flow behaviors such as *spinnability* (or *Spinnbarkeit* in German) and *stickiness*. This early work is comprehensively reviewed in the only monograph on the topic by *Petrie* [9.9]. As a result of technological difficulties and theoretical misunderstandings, little definitive work in extensional rheology was performed until the 1970s, when force transducer and motion control technology was adequately refined. The initial work in the 1970s and 1980s for extensional viscosity characterization focused on instrument designs for polymeric melts [9.10–15]. The high viscosities of these materials facilitated the imposition of extensional deformations and resulted in relatively high extensional forces that were readily measured with off-the-shelf force transducers. Gripping methods employed in tensile testing of solids can often be employed on polymeric melts without compromising the flow field [9.10, 11]. The high viscosity of these materials allows the use of a surrounding stabilizing medium, such as an oil that is immiscible with the polymer, to both heat the material as well as provide neutral buoyancy, without imparting a measurable shearing stress at the oil-polymer interface [9.14]. Most testing of polymeric melts occurs at relatively low strain rates $\dot{\varepsilon}_0 \leq 1\,s^{-1}$ and moderate strain levels.

Development of instrumentation for moderate- to low viscosity materials has proceeded more slowly. The low viscosity range presents challenges in designing instrumentation that reliably impose the required kinematics and to measure the force levels that can range from milligrams and span several orders of magnitude. Very low viscosity materials, such as those used in spraying applications, present even greater challenges, where surface tension and gravitational effects are of the same order of magnitude as viscoelastic effects [9.35]. These fluids can also be influenced by inertial effects, which can result in filament break-up before deformation is complete.

The pronounced impact of polymeric additives to Newtonian fluids on jet break-up, drag reduction, and droplet formation has increased the need to characterize these materials in an extensional flow field. An example of an as-yet unresolved need in the industry is a reliable method of measuring the extensional behavior of inkjet ink. The diagram shown in Fig. 9.43 shows a schematic makeup of a printing head of an inkjet printer. Ink flowing through the head and ejected onto the surface of the printed substrate will experience three different types of extensional processes. The quest to maximize print speed while minimizing droplet trailing is facilitated by characterization of the rheological properties of the ink in the flow field dominated in this process, namely uniaxial extension. The low-shear viscosity of these materials, close to that of water, makes this class of materials extremely challenging to characterize in an extensional flow [9.36].

Through the years, a variety of instrument and test methods have been developed that achieve a certain approximation to extensional flow. In addition to uniaxial extensional flows, processes such as blow molding, squeezing flows, and drop impact involve planar and biaxial extensional flows. While these flow fields are important, this chapter is limited to uniaxial flows where the technology to characterize uniaxial extensional flow is more mature and commercially available solutions exist.

Extensional rheometer types are summarized in Table 9.5. Appropriate citations are also included for each instrument type. The accuracy of the data obtained from the types of instruments shown in Table 9.5 depends on the ability of the operator to know both the extensional strain rate and the amount of extensional strain imposed on the sample precisely. In some of the instrumentation, for example filament stretching rheometers, the strain rate and strain is uniform in the sample and constant throughout the test, allowing the calculation of a true transient extensional viscosity. In other instrumentation, for example, the fiber spinning and opposed jet geometries, a representative fluid element experiences a range of strain rates as it is elongated in the flow. To interpret the data from the latter types of instruments, operators will report an average strain rate and strain, thus yielding an *apparent* extensional viscosity. These latter instruments find utility in indexing materials and ranking formulations according to their response to an extensional flow field [9.37].

In this section, we discuss the kinematics of extensional flows from both a theoretical and an experimental viewpoint. The current state of understanding of extensional rheology is presented for highly viscous polymeric materials (i.e., polymer melts) and for low to moderately viscous materials. Currently available commercial instrumentation for extensional rheometry experiments are summarized at the end of the chapter.

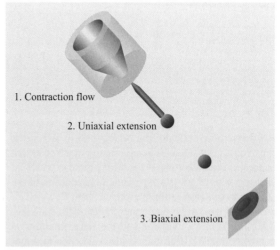

Fig. 9.43 Schematic diagram of an inkjet head, showing: (1) a contraction flow as the diameter of the channel narrows; (2) uniaxial extension followed by droplet breakup as the ink stream accelerates out of the head; and (3) biaxial extension as the ink drop impacts on the solid substrate

Kinematics of Elongational Flows
Homogeneous Versus Nonhomogeneous. A pure extensional flow is an irrotational motion in which there is no vorticity or shearing of material elements [9.38]. Such a deformation provides an extremely efficient way of orienting and elongating the underlying molecular structure of a complex fluid such as a polymer solution, fiber suspension or micellar fluid. If an extensional flow can be realized, it is thus a sensitive probe of material microstructure.

The simplest class of extensional flows is a spatially homogeneous flow, in which the velocity field is given by

$$v_x = -\frac{1}{2}(1+b)\dot{\varepsilon}x, \quad v_y = -\frac{1}{2}(1-b)\dot{\varepsilon}y,$$
$$v_z = b\dot{\varepsilon}z, \qquad (9.179)$$

where $\dot{\varepsilon}$ is the deformation rate, and b is a flow-type parameter; the Cartesian coordinate z is rendered along the stretching direction, and x and y are normal to it. Simple differentiation of these expressions shows that there are no off-diagonal or shearing contributions to the velocity gradient tensor $\partial v_j/\partial x_i$ for $i \neq j$. Uniaxial elongation corresponds to $b=0$, $\dot{\varepsilon} > 0$, biaxial extension (or uniaxial compression) corresponds to $b=0$, $\dot{\varepsilon} > 0$, and planar elongation corresponds to $b=1$.

If the flow is temporally homogeneous or steady then $\dot{\varepsilon} \neq \dot{\varepsilon}(t)$ but unsteady formulations (such as the start up of steady elongational flow $\dot{\varepsilon}(t) = \dot{\varepsilon}_0 H(t)$ where $H(t)$ is the Heaviside step function) can also be represented in this formalism.

In particular, it must be recognized, that because there is a natural time scale (the *relaxation time*) in non-Newtonian fluids, the dynamical response to an imposed extensional deformation is not instantaneous but a function of time and the deformation rate. The extensional viscosity functions can then be defined as

$$\bar{\eta}_1^+(\dot{\varepsilon}_0, t) = (\tau_{zz} - \tau_{xx})/\dot{\varepsilon}_0,$$
$$\bar{\eta}_2^+(\dot{\varepsilon}_0, t) = (\tau_{xx} - \tau_{yy})/\dot{\varepsilon}_0, \qquad (9.180)$$

where t is time.

In an uniaxial deformation ($b=0$, $\tau_{xx} = \tau_{yy}$), there is only a single material function, the uniaxial extensional viscosity, which is often denoted by $\eta_E^+ \equiv \bar{\eta}_1^+$. Most of the discussion in this chapter focuses on uniaxial flows.

The total material strain accumulated by a material element can be obtained from direct integration of (9.179). For a uniaxial elongation ($b=0$) at constant extension rate ($\dot{\varepsilon}(t) = \dot{\varepsilon}_0$), the position of a material element (identified by a label $[i]$) at time t is $Z^{[i]}(t) = Z_0^{[i]} \exp(\dot{\varepsilon}_0 t)$ and the strain between two neighboring elements is given by the natural or Hencky strain [9.38]

$$\varepsilon_H = \dot{\varepsilon}_0 t = \ln(\Delta Z(t)/\Delta Z_0), \qquad (9.181)$$

where $\dot{\varepsilon}_0$ is the idealized strain rate.

For a linear viscoelastic material described by a spectrum of discrete relaxation times λ_k and modal viscosity contributions η_k, the extensional response of a material can be solved analytically to give

$$\bar{\eta}_{\mathrm{LVE}}^+(t) = \sum_{k=1}^{N} 3\eta_k \left[1 - \exp(-t/\lambda_k)\right]. \qquad (9.182)$$

At long times $t \gg \lambda_k$ the material thus approaches a steady-state elongational viscosity $\eta_E \to \sum 3\eta_k = 3\eta_0$.

This *linear viscoelastic envelope* is an important material limit that can be used to validate the performance of a specific instrument as we show below. This material function also quantifies two important adjectives that are commonly used in describing the elongational response of materials.

Strain hardening refers to the progressive deviation of the material response function above the linear viscoelastic envelope as the Hencky strain increases such that $\eta_E^+(\dot{\varepsilon}_0, t)/\bar{\eta}_{E,\mathrm{LVE}}^+(t) > 1$.

Extensional thickening refers to an increase in the steady-state extensional viscosity (measured at long times) above the value of the steady shear viscosity such that $\eta_E(\dot{\varepsilon}_0) > \eta(\dot{\gamma})$. Because most complex fluids also exhibit shear thinning, care must be taken to compare data at equivalent deformation rates. Consideration of the second invariant of the deformation rate tensor suggests that a suitable comparison is under conditions for which $\dot{\gamma} = \sqrt{3}\dot{\varepsilon}_0$.

It is important to recognize that the majority of industrial flows do not lead to spatially homogeneous deformation fields. The material response is then a function of both time and spatial position. This dependency makes it difficult to use such a flow to measure a true material function [9.39]. A representative situation is shown in Fig. 9.44, which compares a filament stretching deformation with a fiber-spinning operation. In a filament stretching operation the displaccment of a material element of fixed identity (labeled 'A' here) is controlled to be exponential in time. The resulting Eulerian velocity field at any location (denoted '1' or '2') also changes (approximately exponentially) in time. The material element thus experiences *motion with constant stretch history* $\dot{\varepsilon}(t) = \dot{\varepsilon}_0$ and we can define the transient extensional viscosity in terms of the measured evolution in the (time-varying) tensile force and the midplane radius of the filament. By comparison, in a fiber-spinning operation, such as the Rheotens test discussed below, the flow is steady in an Eulerian sense (so that the velocity is constant at any axial location) and the tensile force is constant at any specified take-up rate. However material elements now experience a time-varying deformation history including an upstream shear, followed by a die swell at the nozzle exit and finally a spatially varying uniaxial elongation flow. As a result, it is only possible to define an apparent material function such as a *spinning viscosity* or a critical stress to break. For further details see [9.40].

Dimensionless Parameters. It is common to report fluid-dynamical measurements and material functions in terms of dimensionless parameters. For extensional rheology experiments the natural formulation is to express the *transient Trouton ratio* $\mathrm{Tr}(\varepsilon_H, \mathrm{De}) = \bar{\eta}^+(\dot{\varepsilon}_0, t)/\eta_0$ (i.e., a dimensionless extensional viscosity scaled with the zero-shear-rate viscosity) as a function of the dimensionless time or *Hencky strain* $\varepsilon_H = \dot{\varepsilon}_0 t$, and a dimensionless ratio of the material and flow time scales *Deborah number* $\mathrm{De} = \lambda \dot{\varepsilon}_0$, where λ is an appropriate measure of the relaxation spectrum λ_k of the non-Newtonian material being tested (typically either the longest time constant or an appropriately weighted *mean* value of the relaxation time). Note that the dimensionless deformation rate may also be referred to as a *Weissenberg number* for elongational flow, since the characteristic deformation rate and flow timescale are inherently related in an extensional flow; $t_{\mathrm{flow}} \sim 1/\dot{\varepsilon}_0$.

Many extensional flows and elongational rheometers involve free-surface deformations and additional material functions characterizing the relative importance of interfacial tension forces arise. Because of the relative ease of measuring shear flow properties, these are invariably referenced to viscometric properties of a fluid, such as the steady-state shear viscosity, rather than the elongational viscosity. Relevant parameters include the capillary number; a ratio of the viscous stress to capillary

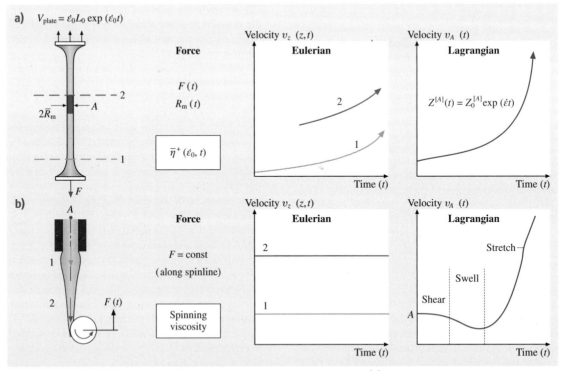

Fig. 9.44a,b Schematic of two different classes of uniaxial elongational flow; (**a**) filament stretching of a sample can be controlled to provide a constant deformation rate; (**b**) fiber spinning or the Rheotens experiment is steady in an Eulerian sense but unsteady from the point of view of a material deformation history. Z is the longitudinal axial coordinate along the filament or spin line. $\bar{\eta}^+(\dot{\varepsilon}_0, t)$ is the extensional viscosity, v_z the longitudinal velocity, and F the tensile force

pressure, $\mathrm{Ca} = \eta_0 \dot{\varepsilon}_0 R/\sigma$ (where σ is the surface tension and R is a characteristic length scale such as a thread or jet radius) and the elastocapillary number (or surface elastic parameter), $\mathrm{Ec} = \mathrm{De}/\mathrm{Ca} = \lambda\sigma/\eta_0 R$. The latter parameter plays an important role in jet break-up [9.41] and adhesive instabilities [9.42]. In deformations in which gravity plays a role it is common to report the Bond number $\mathrm{Bo} = \rho g R^2/\sigma$ as a ratio of the two dominant forces (gravity and surface tension) under static (no-flow) conditions with g being the gravitational constant. Of course any combinations and ratios of these common dimensionless parameters are also valid and have been used in the past.

Highly Viscous Materials ($\eta_0 > 1000$ Pa s). For very viscous materials such as tar or pitch (first studied by Trouton in 1906), the roles of gravity and capillarity can be safely neglected (i. e., $\mathrm{Ca} \gg 1 > \mathrm{Bo}$) and the material does not sag or neck during testing. A number of different Instron-like mechanical testing machines can then be constructed for testing such materials. These devices can be classified as *constant-sample-length* or *constant-sample-volume* instruments and they have been comprehensively reviewed by *Meissner* [9.43] and *Gupta* and *Sridhar* [9.44]; we provide additional details later.

A representative example of the measurement of the transient extensional viscosity function $\bar{\eta}^+(\dot{\varepsilon}_0, t)$ for a highly branched low-density polyethylene (LDPE) melt using the sentmanat extension rheometer (SER) universal test fixture (a constant-length device) is shown in Fig. 9.46 (taken from [9.45]). Equivalent measurements can also be performed with a constant-sample-volume device such as the filament stretching rheometer [9.46]. The linear viscoelastic response is also shown by the dark line (LVE) and provides a bounding envelope for the data at low Deborah numbers, $\mathrm{De} \ll 1$. As the imposed extension rate (and the corresponding Deborah number, $\mathrm{De} = \lambda\dot{\varepsilon}_0$) is increased, the material shows a progressively increased *strain-hardening* response and the viscosity climbs above the value expected from linear viscoelastic characterization. This strain hardening stabilizes the material against necking failure and accounts for the importance of using branched polymeric materials in film-blowing and spinline operations. The transient extensional stress growth of materials such as linear low-density polyethylene (LLDPE) that do not exhibit such pronounced strain hardening can also be measured using the SER fixture and closely follow the LVE envelope for all times and strains [9.45].

A large fraction of early research in the extensional viscosity area was focused on determining the steady extensional viscosity; however it is now recognized that in many commercial processes and flows this is not a relevant parameter as it requires prolonged extension and very large strains to be achieved; such conditions are not commonly attained in industrial operations. Determining the transient material response function and fitting the results to an appropriate constitutive model (such as those covered by *Petrie* [9.9]; *Bird* et al. [9.38] and in more-recent work such as that of *McLeish* and *Larson* [9.47]) is a preferred approach. However, if the long-time limit of the data at each deformation rate shown in Fig. 9.45 is taken as the steady-state value, then a curve of $\eta_E(\dot{\varepsilon}_0)$ can be constructed, as shown in Fig. 9.46. At very low deformation rates the extensional viscosity approaches a constant Trouton ratio corresponding to $\mathrm{Tr} = \eta_E(\dot{\varepsilon}_0)/\eta_0 \to 3$. An extensional thickening is then observed at intermediate rates as a result of molecular stretching before rate thinning leads to a power-law-like decrease in the extensional viscosity rate $\eta_E \sim \dot{\varepsilon}_0^{-0.5}$ at very high deformation rates. The molecular origin of this extensional thinning is still under debate but it may involve either: (i) the anisotropic drag experienced by the polymer molecules as they disentangle and approach full extension, or (ii) the external *pressure* resulting from the surrounding chains [9.48].

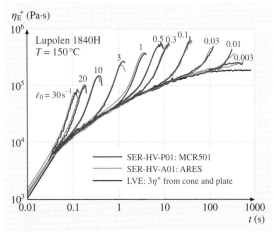

Fig. 9.45 Tensile stress growth curves at a melt temperature of 150 °C for a branched low-density polyethylene (Lupolen 1840H) over a range of Hencky strain rates $\dot{\varepsilon}_0$ of $0.003-30\,\mathrm{s}^{-1}$ generated with the SER on two different host platforms, an MCR501, and an ARES, and a plot of the linear viscoelastic envelope (LVE) taken from cone and plate measurements in start-up of steady shear flow at a shear rate of $0.005\,\mathrm{s}^{-1}$

A final area of recent interest in which the form of the extensional viscosity function for polymer melts and other viscous materials (such as hot melt adhesives) is important is the stability of polymeric threads against rupture and failure. In low-viscosity materials (discussed further below) the degree of strain hardening and its interplay with surface tension controls the formation of beads-on-a-string. In a polymer melt, surface tension is negligible (Ec $\ll$ 1) and filament failure is governed purely by the viscoelastic response of the material in extension. In the limit of high De $\gg$ 1, the filament stability can be characterized by the Considère criterion [9.49]. The principal result is typically written in terms of the tensile force in the filament; however it can also be re-expressed in terms of the transient Trouton ratio and the requirement for homogeneous extension becomes $d \ln \text{Tr}^+/d\varepsilon_H \geq 1$ ([9.50]); i.e., the Trouton ratio must increase at least exponentially fast with strain for homogeneous filament elongation. Examination of the data at the highest rates in Fig. 9.46 shows that, soon after the samples cease to strain harden exponentially with strain, the sample fails. The role of viscous stresses at moderate strain rates makes the analysis more complicated than that for a purely elastic material, but rupture events are still to be expected [9.51].

Low to Moderately Viscous Materials (0.01–1000 Pa s). Homogeneous versus Nonhomogeneous Instruments. In the past two decades, the development of extensional rheometers for low to moderately viscous materials has moved along two design paths. On the first path, the deformation kinematics are imposed on the material via the instrument, and the response of the material is monitored. On the second path, the material is deformed into an unstable configuration, but any further response is driven by the material alone. In both types of instrumentation, however, the operator must know the strain rate history of the material as well as the accumulated extensional strain on the fluid element being observed.

The round robin study in the framework of the project M1 compiled by *James* et al. [9.39] demonstrated a key fact in extensional rheometry: for non-Newtonian fluids, the extensional viscosity cannot be characterized by a single value. In this round robin, researchers reported the extensional viscosity of a dilute solution of polyisobutylene in polybutene [9.52] as a function of strain rate, but without consideration of the accumulated strain in the samples. Without the inclusion of this strain, the compiled results showed no correlation between η_E and $\dot{\varepsilon}$, and disappointing comparison

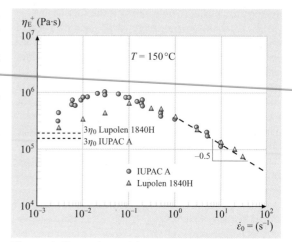

Fig. 9.46 Comparison of the steady-state extensional viscosity behavior as a function of Hencky strain rate at a melt temperature of 150 °C for two LDPE samples taken from the literature; Lupolen 1840H generated with the SER (*triangles*) and IUPAC A (*circles* from *Munstedt* and *Laun* [9.14])

between extensional rheometers. This study pushed researchers to recognize the need to accurately know and report the transient extensional viscosity as a function of both strain rate and strain. This need resulted in the design of the first generation of filament stretching rheometers.

Matta and *Tytus* [9.53] first proposed the use of a falling-plate rheometer, whereby a small volume of test fluid pinned between two circular plates is stretched as the bottom plate falls under the influence of gravity. The lack of active machine control on the movement of the bottom plate resulted in a time-varying extensional strain rate, which complicated the analysis of the data. *Sridhar* and coworkers resolved this issue by driving the two plates apart with a motion control system at user-specified velocity profiles, which allowed them to impose a deformation at a constant axial strain rate [9.17, 54]. By coupling a force transducer on one plate and stationing a laser micrometer at the midpoint of the deforming filament to monitor the filament diameter, these researchers were able to extract the transient extensional viscosity as a function of strain and strain rate for moderate to highly viscous fluids. The success of this design has led to further instrument development from a variety of groups [9.16, 17, 55–58]. The basic design for this class of instruments, termed *filament stretching extensional rheometers*, is shown in Fig. 9.47. These instruments fall under the classification of homogeneous instrumentation, given that the strain rate is spatially uniform in the fluid sample, for the most part.

Newtonian fluids were used to validate these first-generation extensional rheometers, given that the extensional viscosity is well known ($\bar{\eta} = 3\eta_0$) and is independent of both $\dot{\varepsilon}_0$ and ε_H. The early results by *Sridhar* [9.17] and *Spiegelberg* et al. [9.16] showed that for the first several units of Hencky strain, the Trouton ratio for the Newtonian test fluids exceeded the known value of Tr = 3; after being stretched to approximate $\varepsilon = 2$, the Trouton ratio approached the theoretical value of Tr = 3. Theoretical analysis in Fig. 9.59 showed that the initial flow field is dominated by shear flows resulting from the pinning condition of the fluid at the endplates [9.16]. Numerical simulation studies by *Yao* et al. [9.59] showed that the initial shear flow could be minimized by increasing the initial length of the fluid sample L relative to the initial filament diameter D. It was determined that the effects of the initial endplate shear could be minimized by utilizing an initial ratio $L/D \approx 1$. A further consequence of the fluid pinning conditions at the endplates is that a reduced amount of fluid is participating in the extensional deformation. This phenomenon, coupled with the strain-hardening behavior of most non-Newtonian fluids, usually produced a temporally-varying strain rate as determined from the midpoint diameter. *Kolte* et al. [9.60] suggested three types of experimental strategies to address the nonuniform flow exhibited with these devices:

Type 1. The endplates are separated at an exponentially increasing rate $L(t) = L_0 e^{\dot{\varepsilon}_0 t}$. The extensional viscosity is based on this imposed strain rate $\dot{\varepsilon}$ and is calculated as

$$\eta_E(t, \dot{\varepsilon}_0) = F(t)/\dot{\varepsilon}_0 A_0 e^{-\dot{\varepsilon}_0 t} \qquad (9.183)$$

where A_0 is the initial fluid area and F is the tensile force. The Hencky strain is calculated as per (9.181).

Type 2. The deformation profile is the same as in the type 1 experiment, but the data are processed differently. In the type 2 experiment, the midpoint radius R_m is measured, in addition to the force on the endplates. The varying midpoint strain rate, or *effective strain rate*, $\dot{\varepsilon}_{\text{eff}}$ is calculated as a function of time as

$$\dot{\varepsilon}_{\text{eff}}(t) = -\frac{2}{R_m}\frac{dR_m}{dt} . \qquad (9.184)$$

The extensional viscosity is calculated as

$$\eta_E(\dot{\varepsilon}_{\text{eff}}, \varepsilon) = \frac{F(t)}{\dot{\varepsilon}_{\text{eff}}\pi R_m^2(t)} . \qquad (9.185)$$

Because the strain rate varies with time due to the nonuniform flow generated by the endplates, the Hencky strain is calculated from the time integral over the strain rate based on the midpoint radius:

$$\varepsilon_{\text{eff}}(t) = \int_0^t \dot{\varepsilon}_{\text{eff}}(t)\,dt = -2\ln\left(\frac{R_m}{R_0}\right) . \qquad (9.186)$$

Type 3. In the type 3 experiment, the endplate velocity is controlled, so that the midpoint filament radius decreases as it would in an ideal flow, namely with a constant effective strain rate strain rate ($\dot{\varepsilon}_{\text{eff}} = \dot{\varepsilon}_0$), or

$$R_m(t) = R_0 e^{-\frac{1}{2}\dot{\varepsilon}_{\text{eff}} t} . \qquad (9.187)$$

The extensional viscosity and Hencky strain are calculated as in the type 1 experiment.

Researchers initially employed trial-and-error approaches to manipulate the velocity profiles to ensure a constant strain rate [9.16, 17]. *Anna* and coworkers developed a feedback system based on real-time measurements of the midpoint radius that adjusted the endplate velocity on the fly to ensure a constant strain rate [9.61], while others have developed endplate velocity profiles for type 3 experiments based on master curves developed from type 2 experiments [9.62].

These refinements, along with higher-resolution force transducers and laser micrometers, and motion control systems capable of resolving several orders of magnitude in velocity while maintaining position control on the micrometer scale, have greatly advanced

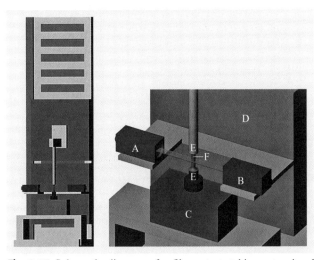

Fig. 9.47 Schematic diagram of a filament stretching extensional rheometer for moderately viscous materials. A: laser emitter; B: laser detector; C: force transducer; D: motor; E: endplates; F: fluid filament

the accuracy and repeatability of extensional viscosity measurements in fluids, as shown in the next section. The improved instrumentation has allowed researchers to conduct more-refined extensional flow experiments, such as probing individual polymer chain conformation with concurrent birefringence measurements [9.63, 64], examining step strain responses of materials [9.62, 65], and characterizing elastic instabilities in extensionally deforming fluids [9.66].

Filament stretching rheometers are typically limited to fluids exhibiting shear viscosities of approximately 1 Pa s or greater. At lower viscosities, surface tension and gravity begin to dominate the flow behavior of the fluid. It is this recognition that led *Rozhkov* and coworkers to the development of capillary break-up rheometers, which are in the class of nonhomogeneous extensional rheometers, given that the strain rate is not spatially homogeneous in the sample [9.26, 67]. Rather than trying to minimize the influence of these forces, surface tension is allowed to drive the deformation of the fluid filament, which resists the deformation through viscous and elastic forces, and gravitational forces are rendered inconsequential through small sample dimensions. In this experiment, a cylindrical fluid filament is placed between two circular endplates, similar to a filament stretching experiment. As opposed to separating the endplates at a constant strain rate while measuring force and radius, in this test a rapid axial step strain is imposed, resulting in a stretched fluid filament. The plate separation occurs at a rate faster than the longest relaxation time in the test fluid. The capillary break-up rheometer monitors the filament radius as a function of time as it thins under capillary pressure. The extensional stress is derived from the surface tension of the fluid σ, reducing the calculation of extensional viscosity to

$$\eta_{app} = \frac{\sigma/R_m}{\dot{\varepsilon}_{eff}} = \frac{\sigma/R_m}{-\frac{2}{R_m}\frac{dR_m}{dt}} = -\frac{\sigma}{2\,dR_m/dt} \,. \tag{9.188}$$

The behavior of Newtonian fluids in a capillary break-up rheometer is well predicted, and these experiments are useful for easily determining their shear viscosity if the surface tension is independently known. As discovered by *Papageorgiou* [9.68], the midpoint radius of a Newtonian filament is expected to evolve according to

$$R_m(t) = 0.0709 \frac{\sigma}{\eta_s}(t_c - t) \,, \tag{9.189}$$

where t_c is the critical break-up time of the filament. *McKinley* and *Tripathi* [9.25] showed very good correlation between the shear viscosity measured with a torsional shear rheometer and results obtained on a capillary break-up rheometer using this expression. Capillary break-up rheometry has been used to evaluate the time-dependent change in extensional viscosity of adhesives containing a volatile solvent [9.69], the extensional rheology of ink and paint [9.70], the relaxation times of non-Newtonian fluids [9.71, 72], and to probe time scales in low-viscosity fluids [9.73].

Results from Filament Stretching Rheometer Experiments. The considerable improvements made in filament stretching rheometer designs have allowed researchers to attempt another round robin comparison of results to compare and validate the basic approach. As opposed to the original M1 round robin study, this round robin proved to be much more successful. Three universities [9.74] constructed filament stretching rheometers based on the original concept by *Sridhar* [9.17] that were capable of type 3 experiments. Although the basic design was the same in all three instruments, different motion control systems, diameter sensors, and force transducers were used in the three systems. The limitations of the mechanical systems were discussed in detail in this work, and standard error propagation analysis was performed to determine the errors in the results corresponding to various stretching regimes.

In this round robin study, the researchers tested three solutions of high-molecular-weight polystyrenes dissolved in oligomeric styrene oil. The polymer concentration of all three solutions was 0.05 wt %, and the polystyrenes all had very narrow molecular-weight distributions. The material and viscometric properties, including weight-averaged molecular weight of the polystyrene (M_w), zero-shear viscosity (η_0), and Zimm relaxation time (λ), are shown in Table 9.6.

The three institutions generated master curves to determine the stretching history that would yield the optimal diameter profile required for a type 3 test [9.61, 62]. The resultant diameter profiles of the idealized stretching history showed both excellent agreement within the institutions, as well as good agreement with the ideal curve, given that the experimental diameter at each point was within 10% of the ideal curve.

One set of computed extensional viscosities is shown in Fig. 9.48. The Deborah numbers in these tests were $De = 12.0 - 17.0$, due to temperature variations and slight variations in motion control between the three institutions. However, these differences did not impact on the results, given the extremely good comparison of the three sets of results shown. The error bars at the four locations were determined from the error propagation

Table 9.6 Viscometric properties of polystyrene-based test fluids dissolved in styrene oil. The solution concentration was 0.05 wt % for all samples

Fluid	M_w [g/mol]	η_0 [Pa s]	λ [s]
SM-1	2×10^6	39.2	3.7
SM-2	6.5×10^6	46.1	31.1
SM-3	2.0×10^7	55.5	155

analysis. The results show that the refinements in instrumentation and testing technique have resulted in a very reproducible and reliable extensional rheometry test.

In addition to examining the interlaboratory reproducibility of this testing technique, Anna and co-workers thoroughly examined the ideal operating range of these rheometers. Achievable strain and strain rate regimes are limited by motor length and motor velocity. The final measurable extensional viscosity is limited by both the resolution of the force transducer and the diameter-sensing device. Highly elastic fluids can exhibit an instability at the endplates, causing partial decohesion of the material at higher levels of strain, which influences the measured extensional viscosity [9.75]. Lower-viscosity fluids are susceptible to gravitational sagging at lower strain rates, which is characterized by the Bond number (Bo) and the ratio of the Bond number to the capillary number Ca. *Anna* et al. provide critical test conditions where these conditions can occur based on fluid properties and test geometry [9.74].

Results from Capillary Break-up Extensional Rheometry. Two representative studies are presented that employ capillary break-up rheometry to indicate the utility of this technique. In the first, *Stelter* and coworkers used a capillary break-up rheometer to quantify the relaxation times and extensional viscosities of a series of ionic and nonionic polymers as a function of concentration [9.76]. This work exploited the influence of ionically charged polymer molecules on their chain flexibility to explore the effects of polymer chain flexibility on the extensional flow behavior.

When a non-Newtonian fluid is stretched in a capillary break-up rheometer, the evolution of the fluid thread diameter $D(t)$ is expected to thin in two distinct regimes, with the first regime decreasing exponentially with time t according to the expression

$$D(t) = D_0 e^{-t/3\lambda}, \quad (9.190)$$

where D_0 is the filament diameter after the initial stretch at time $t = 0$. In the second regime, the filament thins linearly according to

$$D(t) = D_0 - \frac{\sigma}{\mu_{el,t}} t, \quad (9.191)$$

where σ is the surface tension of the solution, and $\mu_{el,t}$ is the terminal extensional viscosity.

These authors prepared nonionic solutions from Praestol 2500, a linear polyacrylamide (degree of hydrolysis 3–4%, $8-10 \times 10^6$ g/mol), polyethylene oxide (PEO, $8-10 \times 10^6$ g/mol), and a branched graft copolymer composed of a carboxymethyl cellulose backbone and polyacrylamide branches (CMC-g-PAM). Two ionic solutions were prepared from a linear polyacrylamide Praestol 2540 (degree of hydrolysis 40%, 14×10^6 g/mol), and a xanthan gum solution (2×10^6 g/mol). All solutions were prepared with concentrations in water at 62–4000 ppm, resulting in values of $c[\eta]$ between 1-2, which indicates that the solutions are in the semi-dilute range; c is the polymer concentration and $[\eta]$ the intrinsic viscosity. The terminal extensional viscosity $\mu_{el,t}$ is plotted as a function of relaxation time in Fig. 9.49.

During extensional testing, each polymer system showed a dependence of the measured relaxation time and terminal elongational viscosity on solution concentration, resulting in the multiple data points shown in Fig. 9.49. The data arrange on two distinct lines. The Praestol 2500, lacking substantial hydrolysis and therefore ionic behavior, will act as a flexible polymer chain,

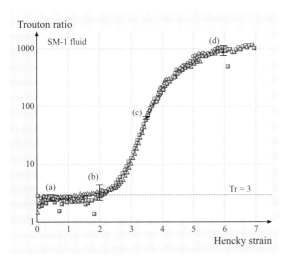

Fig. 9.48 Comparison of transient extensional viscosity results for the round robin study on fluid SM-1 from three institutions MIT (*circles*), Monash (*squares*) and Toronto (*triangles*). The error bars at strains a to d indicate the estimated error at low, moderate, and high Hencky strains (reproduced with permission after [9.74])

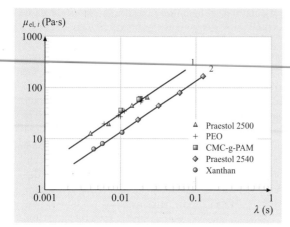

Fig. 9.49 Terminal elongational viscosity plotted as a function of relaxation time for three nonionic and two ionic semidilute aqueous solutions. Line 1 corresponds to flexible polymer chain behavior, while line 2 corresponds to rigid-like behavior. (After [9.76] with permission)

as will the polyethylene oxide. The CMC backbone of the graft copolymer is only weakly ionic, so that this copolymer will also act as a flexible chain. The repulsive behavior of the charges on the ionic polyacrylamides will prevent flexible chain motion, so that these chains will act in a rigid-like manner. The testing allowed these researchers to clearly differentiate materials exhibit-

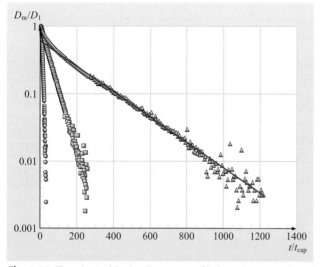

Fig. 9.50 Transient midpoint diameter profile for polystyrene-based [SM-1 (○), SM-2 (□) and SM-3 (△)]. D_1 is the initial diameter, and t_{cap} is trhe capillary time calculated as $t_{cap} = \eta_0 D_t / 2\sigma$. (After [9.65] with permission)

ing rigid-rod-like behavior and flexible chain behavior. Chain behavior will influence the spraying properties of polymer solutions.

In another study, *Anna* et al. investigated the dependence of the extensional flow properties on molecular weight with a series of dilute polystyrene solutions [9.65]. In this work, the results from capillary break-up rheometry tests conducted on the three test fluids were compared with results measured on a filament stretching rheometer. The three test fluids had the viscometric properties already shown in Table 9.6.

The evolving midpoint diameters of the three test fluids tested in the capillary break-up rheometer are shown in Fig. 9.50. Regression on the linear portion of the plot yielded relaxation times that compared well with the relaxation times determined from shear rheology and reported in Table 9.6, according to (9.190).

Conversion of the diameter versus time data to the apparent extensional viscosity versus Hencky strain with (9.187) and (9.185), respectively, allows a direct comparison between the capillary break-up rheometric data and the filament stretching data, as shown in Fig. 9.51. The differences between the two sets of curves are consequences of the different kinematics in the two devices. The best comparison between the tests occurred with SM-2, where the two types of extensional tests approached the same final steady-state Trouton ratio of $Tr \approx 5000$. It is not surprising that the transient extensional viscosities leading up to $\bar{\eta}$ in the SM-2 fluid tested in the two instruments were not similar, given that the filament stretching test was conducted at a constant strain rate, while the capillary break-up rheometer test imposes a variable strain rate set by the fluid itself. It is expected, however, that the final steady-state viscosity will be independent of the prior strain-rate history, however, which the results from SM-2 confirm. The final steady-state plateau can usually not be reached for highly elastic fluids due to endplate instabilities [9.66], which was observed in SM-3. Additionally, gravitational sagging can also limit the operating range of filament stretching rheometry. As Anna and McKinley suggest, capillary break-up rheometry provides a means of increasing the available operating range of extensional rheometry tests in highly elastic fluids [9.65].

Commercially Available Instruments. Several extensional rheometer designs have been commercialized in the past. A list of previously and currently available instruments is summarized in Table 9.7. Rheometrics commercialized an opposed jet device in the 1980s under the model name RFX. Now discontinued, the

Fig. 9.51 Comparison of transient extensional viscosity measured in a filament stretching rheometer (*solid symbols*) and apparent extensional viscosity measured with a capillary breakup rheometer (*open symbols*). The *lines* show the predictions from multimode FENE (finite extensible nonlinear elastic) calculations. [SM-1, $De = 17.0$ (●), $De \approx 2/3$ (○); SM-2, $De = 15.2$ (■), $De \approx 2/3$ (□); and SM-3 $De = 21.2$(▲), $De \approx 2/3$ (△)]. Values below a $Tr = 3$ are usually the results of gravitation sagging. (After [9.65] with permission)

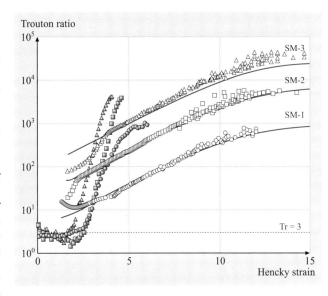

instrument suffered from the underlying problems of opposed jet devices, namely a nonhomogeneous deformation and a significant pre-shear history, both of which complicate the data interpretation [9.77]. The instrument is also limited to strains of approximately $\varepsilon = 1$, and can suffer from inertial effects [9.78]. Despite these limitations, the RFX was the only commercially available rheometer for examining low-viscosity fluids, and provided useful information for indexing materials. *Ng* and coworkers [9.79] performed opposed jet measurements on dilute solutions containing carboxymethyl cellulose, polyacrylamide, and polyisobutylene to examine their impact on drag reduction, and were able to observe differences in the solutions in extensional flows, while shear flows led to no measurable differences.

Two extensional rheometers for melts were previously available that both used the constant-length geometry previously discussed. The MXR2, developed by the National Physical Laboratory's Center for Materials Measurement and Technology and commercialized by Magna Projects, stretched polymer melt filaments by winding them onto a rotating wheel. The filaments were both heated and buoyed by an oil bath. The RME, developed by *Meissner* and coworkers and commercialized by Rheometrics [9.19], employed stationary tread-like grips that stretched a polymer melt filament at controlled rates. The filament was both heated and buoyed by gas passed through a hot porous frit. The RME could reach Hencky strains of $\varepsilon_H = 7$ and strain rates of $\dot{\varepsilon}_0 = 1\,\mathrm{s}^{-1}$. Neither the MXR2 nor the RME are currently being manufactured.

A currently available melt extensional rheometer is the Sentmanat Extension Rheometer (SER) (Xpansion Instruments, Tallmadge, OH), shown in Fig. 9.52. The SER is designed to be field-installed on commercially available shear rheometers, thus allowing them to

Table 9.7 Previously and currently available extensional rheometers

Instrument name	Company	Viscosity range [Pa s]	Flow type	Data type
Currently available				
Rheotens	Göttfert (Rock Hill, SC)	> 100	Fiber spinning	Indexer
CaBER®	Thermo (Karlsruhe, Germany)	0.01 – 10	Capillary breakup	$\eta(\dot{\varepsilon}_0, \varepsilon)$
Sentmanat extension rheometer (SER)	Xpansion Instruments (Tallmadge, OH)	> 10000	Constant-length filament stretching	$\eta(\dot{\varepsilon}_0, \varepsilon)$
Previously available				
RFX	Rheometrics Scientific	0.01-1	Opposed jet	Indexer
RME	Rheometrics Scientific	> 10000	Constant-length filament stretching	$\eta(\dot{\varepsilon}_0, \varepsilon)$
MXR2	Magna Projects & Instruments	> 10000	Constant-length filament stretching	$\eta(\dot{\varepsilon}_0, \varepsilon)$

Fig. 9.52 Rheotens filament windup rheometer for characterizing melts

Fig. 9.53 Rotating cylinders of SER melt rheometer. The *rectangle* shows the uniform texting area

Fig. 9.54 CaBER® capillary breakup rheometer. *Inset* shows a closeup of a thinning fluid filament stretched in the instrument

perform extensional rheometric experiments. The SER employs two rotating drums that wind up the polymer melt filament, imposing constant, step, or variable strain rates, depending on user choice. The torque exerted on one of the drums provides a measure of the stress in the sample.

A variant of the controlled-rate rheometer discussed above is a controlled-velocity device, embodied in the the Rheotens (Göttfert, Rock Hill, SC). The Rheotens is a fiber spinning apparatus, whereby polymer melt is pumped from an upstream tube. A set of wind-up wheels elongates the fluid, while a force transducer mounted on one of the take-up wheels monitors the resulting force exerted by stretching the sample. The Rheotens has significant pre-shear induced while the polymer melt is pumped through the upstream tube, and its extensional flow field has variations in strain rate [9.80]. As a result of these complications, the data obtained from the Rheotens is usually used to index polymer melts, rather than interpreting the data as absolute extensional rheometric data. An example of a Rheotens is shown in Fig. 9.52.

The only currently commercially available extensional rheometer for fluids is the CaBER (ThermoFisher HAAKE, Karlsruhe, Germany). An example is shown in Fig. 9.54. The CaBER is a capillary break-up rheometer, and is based on the work of *Rozhkov*, *Entov*, and coworkers [9.26, 67, 81]. In this instrument, a fluid filament is stretched rapidly between two circular endplates to a fixed level of extensional strain. After the cessation of flow, a laser micrometer monitors the axial midpoint diameter of the fluid filament as a function of time as the filament thins and breaks up under the influence of surface tension, elasticity, and viscosity. In contrast to the previous devices, the CaBER does not impose a specific deformation rate on the sample; rather, the deformation rate is set by the material. The extensional viscosity as a function of strain rate and strain can be determined from the profile of diameter versus time and knowledge of the surface tension of the test fluid. The CaBER is useful for moderate- to low-viscosity fluids, including Newtonian fluids. Once a researcher has determined that the CaBER data departs from the break-up kinematics of a Newtonian fluid using the models provided in the software, they can compare the data to other non-Newtonian models. The quality of the fit of the non-Newtonian rheological model provides insight to the type of viscoelastic behavior exhibited by the test fluid. Similar to the model fitting used in shear rheology, the optimal choice of the fluid model is up to the researcher.

9.2 Thixotropy, Rheopexy, Yield Stress

Thixotropy is a decrease of the apparent viscosity under constant shear stress or shear rate, followed by a gradual recovery when the stress or shear rate is removed. It comes about first because of the finite time taken for any shear-induced change in the microstructure – that confers the viscosity – to take place. The microstructure is brought to a new equilibrium microstructure by competition between the processes of tearing apart by stress and flow-induced collision, in a time that can be minutes. Then, when the flow ceases, Brownian motion is able to move the elements of the microstructure around slowly to more-favorable positions and rebuild the structure. This can take many hours to complete. The whole process is completely reversible.

In this Section, the history of thixotropy is reviewed. The effects of the microstructural changes on the flow properties, which result in the various manifestations of thixotropy, are described. The various mathematical descriptions of the phenomenon are summarized.

In their *Glossary of Rheological Terms*, *Barnes*, *Hutton* and *Walters* [9.82] (following the 1975 British Standards Institution definition) defined thixotropy as 'A decrease of the apparent viscosity under constant shear stress or shear rate, followed by a gradual recovery when the stress or shear rate is removed. The effect is time-dependent.'

They also defined anti-thixotropy as the precise opposite of thixotropy and said that rheopexy is a synonym of anti-thixotropy. These terms are now generally accepted in the rheological world as definitive descriptions.

Thixotropy comes about because of the finite time required for any shear- or stress-induced changes in the microstructure of structured liquids – suspensions, emulsions, polymer solutions, etc. – to take place. The microstructure of the liquid involved is brought to a new equilibrium by competition between the processes of reordering or tearing apart by stress and flow-induced collision, in a time that can typically be minutes. Then, when the applied flow or stress ceases or is removed, Brownian motion is able to move the elements of the microstructure around slowly to more-favorable positions and hence rebuild the structure. This manifests itself as an increased viscosity and can take many hours to complete. The whole process of breakdown and rebuilding is completely reversible if the basic elements are not destroyed during flow.

Non-experts have deliberately built such thixotropic behavior into some commercial products to make them usable, with the best-known examples being thixotropic paints. However, as will be shown, what is usually wanted in these cases is extreme shear thinning. However the way it is brought about usually introduces thixotropy as well, which is then almost always an irritation.

9.2.1 A History of Thixotropy

In the Beginning
Previous reviews of thixotropy have been produced by *Bauer* and *Collins* [9.83], *Mewis*, 1979 [9.84], *Cheng* [9.85] and *Godfrey* [9.86] and by the present author in 1997. Barnes [9.87] noted that the origins of thixotropy as a recognized physical phenomenon go back to 1923 when Schalek and Szegvari found that aqueous iron oxide gels 'have the property of becoming completely liquid through gentle shaking alone, to such an extent that the liquified gel is hardly distinguishable from the original sol. These sols were liquified but become solidified again after a period of time ... the change of state process could be repeated many times without any visible change in the system' [9.88,89]. The word *thixotropy* was used by *Peterfi* in 1927 [9.90,91] in the first paper that properly described the phenomenon,

combining the Greek words *thixis* (stirring or shaking) and *trepo* (turning or changing).

By 1935 *Freundlich* had published a book called *Thixotropie* [9.92] which dealt with the subject: this was the first time that the term was introduced into the title of a publication. Freundlich and coworkers soon found thixotropic effects manifested by a whole variety of systems including vanadium pentoxide sols, starch pastes, and pectin gels.

It is clear that thixotropy originally referred to the reversible changes from a flowable liquid to a solid-like gel. Previously these kinds of physical changes had only been known by changing the temperature, when such gels would melt on heating and then recooling. It was believed at the time that a new kind of phase change had been found albeit a similar behavior of various yoghurts and kefirs was long known but had not attracted the attention of rheologists.

Progress

Early work in this area in the USA is exemplified by three papers published by *McKillen* [9.93] in 1932, who reported the results of his doctoral investigations into the thixotropy of a large number of flocculated paints. He showed that the fluidity (the inverse of viscosity) as a function of time decreased in some cases by four orders of magnitude, showing an almost quadratic dependence on the rest time.

Writing in the UK in 1942, *Scott-Blair* [9.94] could still state that 'the whole subject [of thixotropy] is very new', but in the update of his textbook published in 1949, nearly 120 papers on thixotropy were cited, so things had moved on quickly. Among the examples of thixotropic materials that he cites are clays and soil suspensions, creams, drilling muds, flour doughs and suspensions, fibre greases, jellies, paints, carbon-black suspensions, and starch pastes. He also lists a number of papers on so-called *thixotrometers*, instruments specially devised to characterize the phenomenon. In this respect he raised some interesting points, among them whether thixotropy ought to be studied at constant shear rate or at constant stress, which is still a controversial question.

Scott-Blair quotes Hamaker's explanation of thixotropy as being due to the fact that 'particles can form a loose association which is easily destroyed by shaking but restores itself on standing': this explanation still stands. With our present knowledge of microstructural changes, it is probably safe to say that all materials that are shear thinning are also thixotropic, in that they will always take a finite time to bring about the rearrangements needed in the structural elements that are responsible for shear thinning. However, the number of times that these effects come within the measurable range (say > 1 s) is limited. As Scott-Blair concludes, 'If this recovery is very rapid, the phenomenon is observed as structural viscosity [shear thinning]; if slow, it is observed as thixotropy'.

In the 1930s and 1940s *Pryce-Jones* [9.95, 96] studied around 250 paints, using a so-called *thixotrometer* that he made himself [9.97]. He noted that 'It is a well-established fact that thixotropy is more pronounced in systems containing non-spherical particles', this is because they have to find themselves in the best three-dimensional (3-D) structure by rotation as well as movement, and progress from a solid gel to a freely flowing liquid due to complete microstructural breakdown (Fig. 9.55).

The full extent of thixotropy was maintained by *Bauer* and *Collins* in their 1967 review [9.83]: 'When a reduction in magnitude of rheological properties of a system, such as elastic modulus, yield stress, and viscosity, for example, occurs reversibly and isothermally with a distinct time dependence on application of shear strain, the system is described as thixotropic'. They went on to say that thixotropy was 'usually conceived as an unusual property of very special systems such as aqueous iron oxide dispersions, thixotropy in the sense described above, has been found to be exhibited by a great many and a large variety of systems. Along with the breakdown in structure, other non-rheological features change, such as conductivity and dielectric constant'.

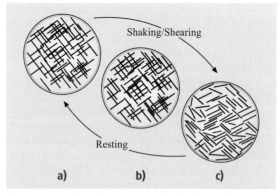

Fig. 9.55a–c Breakdown of an idealized two-dimensional (2-D) thixotropic system: (**a**) completely structured giving elastic, solid-like behavior; (**b**) partly structured giving a viscoelastic response and (**c**) completely unstructured giving a viscous, shear-thinning response

How is Thixotropy Generally Understood Today?
To guide a researcher or an engineer in deciding whether the material with which he or she is dealing is thixotropic or not, the following are various definitions offered in the current general scientific literature, scientific dictionaries, and encyclopaedias that reflect these two points of view. Some are misleading, while even the best are often incomplete. The following are a selection that illustrate the situation.

Oxford Encyclopaedic Dictionary of Physics [9.98]. 'Thixotropy: certain materials behave as solids under very small applied stresses but under greater stresses become liquids. When the stresses are removed the material settles back into its original consistency. This property is associated with certain colloids which form gels when left to stand but which become sols when stirred or shaken, due to a redistribution of the solid phase.'

Chambers Dictionary of Science and Technology [9.99]. 'Rheological property of fluids and plastic solids characterized by a high viscosity at low stress, but a decreased viscosity when an increase in stress is applied. A useful property of paints, because it makes for a thick film which is nevertheless easily worked.'

McGraw–Hill Dictionary of Scientific and Technical Terms [9.100]. 'Property of certain gels which liquefy when subjected to vibratory forces, such as ultrasonic waves or even shaking and then solidify again when left standing. Thixotropic clay: a clay which weakens when disturbed and increases in strength upon standing.'

Van Nostrand's Scientific Encyclopaedia [9.101]. 'A thixotropic fluid is a fluid whose viscosity is a function not only of the shearing stress, but also of the previous history of motion within the fluid. The viscosity usually decreases with the length of time the fluid has been in motion. Such systems commonly are concentrated solutions of substances of high molecular weight colloidal suspensions.'

Oxford Concise Science Dictionary [9.102]. 'More common, however, is the opposite effect in which the viscosity depends not only on the viscosity gradient but also on the time for which it has been applied. These liquids are said to exhibit thixotropy. The faster a thixotropic liquid moves the less viscous it becomes. This property is used in non-drip paints (which are more viscous on the brush than on the wall) and lubricating oils (which become thinner when the parts they are lubricating start to move).'

Chambers 20th Century Dictionary [9.103]. 'Thixotropy: the property of gels of showing a temporary reduction in viscosity when shaken or stirred.'

Definitions given in more-specialized dictionaries emphasize the time aspect of thixotropy.

Polymer Technology Dictionary [9.104]. 'A term used in rheology which means that the viscosity of a material decreases significantly with the time of shearing and then increases significantly when the force inducing the flow is removed.'

Polymer Science Dictionary [9.105]. 'Time-dependent fluid behavior in which the apparent viscosity decreases with the time of shearing and in which the viscosity recovers to, or close to, its original value when shearing ceases. The recovery may take place over a considerable time. This may sometimes occur with polymer systems, when molecular disentanglement increases with time of shearing.'

The definition of thixotropy in the rheological literature has changed over the years. The (American) Society of Rheology was quoted by *Reiner* and Scott-Blair in 1949 [9.106] as defined thixotropy as 'that property of a body by virtue of which the ratio of shear stress to rate of deformation (viscosity) is temporarily reduced by previous deformation'.

Some time later thixotropy was defined as 'a comparatively slow recovery, on standing, of the consistency lost through shearing'.

Gellants like Carbopol, polyacrylates and polysaccharides are used to make fuels and oxidizers (and their simulants) for rocket engines gel. The gelled products become thixotropic (see, e.g., [9.107]).

Internet Use. In general on the Internet, the word thixotropy is often very loosely defined and certainly not according to rheological orthodoxy. At best such definitions are confusing, but often they are incorrect in that no reference is made to time as a variable, the indispensable part of the proper rheological definition of thixotropy. To illustrate this point, the following (anonymous, but easily found) selection of typical definitions currently given on the Internet is cited. These appear in discussions covering a wide range of non-Newtonian liquid products. In most cases, the term thixotropy is either partly or completely confused with what we would properly define as shear thinning (Fig. 9.24 in Sect. 9.1.1).

Typical definitions:

- 'Thixotropy – the property of some gels of becoming fluid when stirred or shaken, and setting to gel again when allowed to stand.'
- 'Reversible behavior of certain gels that liquefy when they are shaken, stirred, or otherwise disturbed and reset after being allowed to stand. Thixotropy occurs in paint ... which flows freely when stirred and reverts to a gel-like state on standing.'
- 'High thixotropy materials become thinner when they are sheared.'
- 'The property of certain gels to become liquid upon being shaken or agitated and to coagulate again when left in an undisturbed condition.'
- 'Thixotropy is the property of [clay] slips becoming thicker when they are at rest, i. e., *thixotropy too high* means that the slip thickens up very quickly.'

Definitions referring to specific systems:

- 'Characteristic of a *lubricant* to become momentarily ... thinner due to mechanical action such as stirring.'
- 'The property of some *printing inks* of becoming fluid when worked and setting to a semi-solid state when at rest; the cause of some inks tending to back away from the ink fountain roller.'
- 'The thixotropic index (*of adhesives*) – a ratio of the low-shear viscosity to high-shear viscosity. For our purposes, the thixotropic index is the Brookfield viscosity at 5 rpm divided by the viscosity at 50 rpm.'
- 'The ability of wet *clay* to maintain a given shape.'
- 'The property possessed by certain *gels* of liquefying when shaken. This is the ability to resist draining off vertical surfaces while retaining spreadability under the applied stress of a brush, roller, or squeegee. Thixotropy should not be confused with viscosity ... a thixotropic liquid's viscosity is high when at rest and diminishes when stressed.'
- 'In *slip-casting*, bad drainage, brittleness, casting spots, cracking, blabbiness, pin holing are all due to too-low thixotropy, i. e., too fluid.'
- 'Property of certain materials to ... liquefy upon agitation ([*paint*] viscosity decreases upon application of shear during brushing or roller), and to stiffen to its original state when allowed to rest (viscosity increases). A product that possesses thixotropy can resist the pull of gravity.'
- 'The *ink* with good thixotropy is not good in leveling, while it is excellent in reproducibility and thick-coating.'

- 'Thixotropy... the viscosity of a substance decreases as the substance is set in motion by some mechanical action such as stirring or shaking. Thixotropy can be observed in *non-drip paints* – as the paint is being applied the viscosity drops and when the paint is on the wall the viscosity increases to its stationary value.'

Even the celebrated Encyclopaedia Britannica has a very strange definition of synovial fluid 'it is a markedly thixotropic fluid; that is, one which is both viscous and elastic'. Of course the proper definition of its rheological properties is 'viscoelastic'.

Helpful Internet Definitions

On the other hand, there are very helpful definitions. A good example is that found on the Thermo-Haake website [www.thermohaake.com/],

Definition of thixotropic flow behavior:

- Decrease of viscosity as a function of time upon shearing, 100% recovery (= regaining the original structures) as a function of time without shearing.

Determination

- Time curves – at *constant shear rate* to observe the destruction of the structures within the material, *dynamic (oscillation)* experiment or time curves after different waiting times to observe/determine the regeneration of the sample.
- Flow curves – *upwards and downwards (loop test)* at constant temperature. The hysteresis is a measure for the thixotropy.

Then the definition given in the IUPAC Compendium of Chemical Terminology 2nd Edition (1979, 51, 1217) (available online) is also very helpful:

- Thixotropy: see *work softening*.
- Work softening: the application of a finite shear to a system after a long rest may result in a decrease of the *viscosity* or the consistency. If the decrease persists when the shear is discontinued, this behavior is called work softening (or shear breakdown), whereas if the original viscosity or consistency is recovered this behavior is called thixotropy.
- Work hardening: opposite of *work softening*, in which shear results in a permanent increase of *viscosity* or consistency with time.

It is obvious then that, while most rheologists take a general view of thixotropy which covers all time effects resulting from microstructure changes, in the popular image, the older idea of a transition on shearing–resting is still often held. Strictly speaking, what is usually meant by thixotropy in that case would now be termed *extremely shear thinning*, to give near solid-like properties at rest but flow under stress, as for instance in thixo-forming using metals or the addition of so-called thixotropes to paints.

A better and extended definition of thixotropy is clearly needed, and it should contain the ideas of both considerable shear thinning (i.e., gel–fluid transition) and also time changes over and above those encountered when in its structured state the thixotropic material might be viscoelastic with its attendant time effects.

9.2.2 Description of Thixotropic Phenomenon

General Considerations

As stated above, all liquids with microstructure can show thixotropy, because thixotropy only reflects the finite time taken to move from any one state of microstructure to another and back again, whether from different states of flow or to or from rest. The driving force for microstructural change is the result of the competition between breakdown due to flow stresses, build-up due to in-flow collisions, and Brownian motion. Brownian motion is the random thermal agitation of atoms and molecules that results in elements of the microstructure being constantly bombarded, which causes them to move to a favorable position where they can – given the necessary force – attach themselves to other parts of the microstructure. Very occasionally situations arise where existing weakly attached microstructural elements – brought together by collision during shear – are slowly torn apart by the constant action of the random Brownian motion. In that case, the opposite of thixotropy is seen, i.e., anti-thixotropy (rheopexy), where flow and rest destructures the material.

The general term microstructure, as used here, while usually associated in thixotropic systems with particles, can also mean alignment of fibres, favorable spatial distribution of particles in suspensions or drops in emulsions (Sect. 9.3), or entanglement density or molecular associations in polymer solutions. All these determine the level of viscosity and elasticity (Sect. 9.1), and they all take a finite time to change from one state to another under the action of shear and/or Brownian forces. In these cases, the *maximum microstructure* is seen when the alignment and spatial distribution are random (in three dimensions) and entanglement density is at a maximum. Both these conditions result in the greatest viscous (and usually elastic) response. On the other hand, *minimum microstructure* is when there is maximum alignment with the flow of fibres; the drop or particle spatial distribution is asymmetrical in the flow direction, or there are a minimum number of entanglements or associations – all these leading to minimum viscous and elastic response.

When the time scales involved in these changes become long compared with either the response of a viscometer (or rheometer, Sect. 9.1.1), or the flow time in a particular flow geometry, we can sensibly talk about thixotropy. These time scales often range from seconds to hours, with rebuilding usually taking much longer than breakdown.

Typical Behavior

If we place a thixotropic material into a viscometer/rheometer (Fig. 9.56) and apply a constant shear rate, the measured viscosity will decrease with time, but it will eventually tend to a steady, constant value. If we then switch off the shear and allow the material to rest for a some time (without drying or any other artefacts such as sedimentation or separation occurring) and switch the shear on again, the measured viscosity will initially be higher, but will decrease and end up at the same value as that seen after the original shearing. However, the level for the original value will not necessarily be the same, because that will depend on how carefully or vigorously the material was initially loaded and how long it was left to rest before shearing.

If on the other hand, a third experiment is performed where the material is allowed to come to equilibrium and then allowed to rest for the same length of time as before, the results will be identical. If now, after equi-

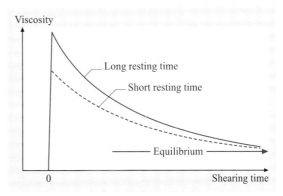

Fig. 9.56 Shearing a thixotropic liquid after short and long rest times

librium is achieved, the shear rate is instantaneously decreased to a lower value, the measured shear stress drops instantly, but thereafter it will slowly increase towards a new equilibrium. If instead of applying a given shear rate we applied a particular shear stress, then the inverse applies – the shear rate would increase as the structure breaks down and the change to another (lower) shear stress will result in a sudden decrease in shear rate followed by a further drop, Fig. 9.57.

If we now enquire what is happening on a microscale, we can imagine the picture in Fig. 9.58, where the viscosity/shear-rate behavior of a typical thixotropic material (which for the sake of argument could be a flocculated suspension) is presented. We start from a point where the microstructure at rest is a series of large flocs. Then, if the applied shear rate is increased progressively and sufficient time allowed, the floc size decreases until at a high enough shear rate, the floc has disintegrated completely into its constituent primary particles. Because large flocs *trap* more continuous-phase liquid than smaller flocs of the same particles, the viscosity is higher. The minimum viscosity is seen with individual particles.

We now imagine another experiment where the shear rate is increased stepwise from point a to end up at point b in Fig. 9.58. Instantaneously, the floc size will be that appropriate to the shear stress condition at point a, but as it experiences the higher shear rate at position a, it begins to erode, until it reaches an equilibrium size appropriate to the higher rate. This process can take some time. If now the opposite happens – the shear rate is instantaneously decreased – the individual particles (which gave the low viscosity) begin to collide and flocculate until the size of the flocs formed is appropriate to the new, lower shear rate. This process also takes time, with the build-up proceeding at a different rate to the breakdown.

Any concentrated suspension of particles is shear thinning, thus when we consider flocs, they show the phenomena as illustrated in Fig. 9.58. If we imagine that the particles in a floc are permanently glued together and thus the floc size is fixed, the flow curve of such a suspension of fixed-size flocs would follow the lines shown in the figure according to the floc size. If the floc size is now decreased (and the overall concentration maintained) but again the floc size remains unchanged when sheared, the flow curve will be lower and (for a given concentration) the degree of shear thinning decreases.

However because we are interested in systems flocculated in a secondary minimum, the size of the floc is not constant but decreases with increasing shear rate (or shear stress). Because viscosity decreases with floc size (see above), we now have a double shear-thinning effect. This means that flocculated systems are very shear thinning, see the equilibrium curve in Fig. 9.58. The extreme shear thinning also results in such flocculated, thixotropic systems appearing to have a yield stress, since the stress only decreases very slowly as the shear rate is decreased.

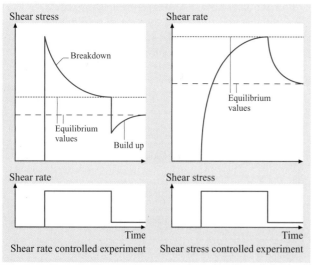

Fig. 9.57 Two kinds of step experiment and their effect on thixotropic liquids

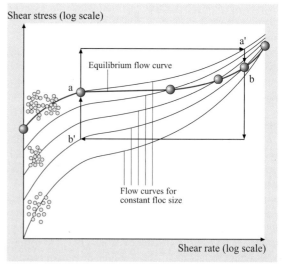

Fig. 9.58 The relationship between microstructure and (thixotropic) flow properties, illustrated for a flocculated suspension

The true steady-state behavior of a thixotropic liquid is seen both after an infinite shear time at any shear rate and shear stress of interest or infinite rest time. Of course, as all true equilibrium states are approached asymptotically, one comes close to this state after a long time rather than an infinite time, but even then, breakdown times of hours and days might be necessary to describe a very thixotropic system fully.

In a flocculated system, breakdown is towards an equilibrium situation that is governed by the balance of hydrodynamic shear stresses pulling structures apart by erosion, and of Brownian and shear forces building the structure up by collision and accretion of particles that agglomerate into flocs. The forces holding the structure together are colloidal in nature, acting over short distances (~ 10 nm) within the composite particle (floc). At rest, only the Brownian rebuilding forces are present, and these are quite small compared to the shearing, with energies of of the order of $k_B T$, where k_B is the Boltzmann constant and T is temperature. This means that the rebuilding time can be very long, since this small, random force takes a long time to rearrange large particles that, as flocs, are getting larger as they move into a more favorable structure, which is then manifested as a higher viscosity.

The typical response to a stepwise change from one steady-state condition to another in terms of the viscosity is often characterized by the so-called 'stretched-exponential' model:

$$\eta = \eta_{e,\infty} + (\eta_{e,0} - \eta_{e,\infty})\left(1 - e^{-\left(\frac{t}{\tau}\right)}\right)^r , \quad (9.192)$$

where $\eta_{e,0}$ is the viscosity at the commencement of shearing, $\eta_{e,\infty}$ is the viscosity after an infinite time, τ is a time constant and r is a dimensionless constant (which in the simplest case is unity). This equation can cope with build-up and breakdown in steps up or steps down with the values of τ and r depending on both the level and the direction, i.e., going from 0.1 to 1 s^{-1} will not have the same value of r found as going from 1 to 10 s^{-1}. The values of r vary with the conditions of the test, as well as the particular system being tested and they decrease linearly with the log of the shear rate, typically from 0.7–0.9 at 10^{-2} s^{-1} to 0.3 at 1 s^{-1} in Mewis's experiments [9.108]. He also showed that, for a typical flocculated system (fumed silica in paraffin oil), τ is a decreasing function of $\eta_{e,0}\dot{\gamma}/\eta_{\text{cont}}$, where η_{cont} is the viscosity of the continuous phase, which when taken into account properly describes the effect of temperature. (Mewis also reports optical, dielectric, and conductivity results on these systems.)

Mewis's approach was also used to describe the thixotropy of hydrophobically modified hydroxyethyl cellulose and nonassociative cellulose water solutions, see *Maestro* et al. [9.109]. The Mewis model fitted structure rebuilding experiments with an exponent r of around 2.

Heymann et al. [9.110] investigated the build-up after shearing of the yield stress of newsprint inks, with a formulation containing carbon black (Sect. 9.3). A period of pre-shearing was carried out at 1500 Pa, and then the flow curves were measured as a function of time. For inks, judged to have a yield stress, an equation similar to that above was used to describe the rebuilding of the shear stress

$$\sigma_y(t) = \sigma_{y,0} + [\sigma_{y,\infty} - \sigma_{y,0}](1 - e^{-\frac{t}{\tau}}) . \quad (9.193)$$

For these materials so described, the values of τ were found to be well over 100 s. Because of this it was judged that the recovery behavior had no relevance to the printing process, because process times are of the order of a few seconds at most.

Not only can shear break down or build up flocs, but it can also change their internal morphology. For instance, the work of *Mills* et al. [9.111] showed that shearing freshly prepared flocculated suspensions can densify individual flocs, causing reorganization within the flocs. This shows how important it is to *condition* such systems before shearing them in experiments to elucidate their thixotropic behavior. They found that, following prolonged shearing, loosely packed flocs became tightly packed and more monodisperse. This can be interpreted as a loss of structure – rheomalaxis or rheodestruction.

The experiments of *Woithers* et al. [9.112] showed how important thixotropy can be in typical flocculated systems. They examined the behavior of a depletion-flocculated polymer latex suspension and showed that the shear stress dropped significantly with time, and the effect was still present after one hour. The values of the initial and final viscosities they found were at least an order of magnitude different. They also found that the unsheared samples contained more open flocs, while shearing flocs made fresh flocs that were denser.

Viscoelasticity and Thixotropy

Viscoelastic systems sheared in their linear region show time dependency because the microstructure takes time to respond to the flow/stress (Sect. 9.1). At short times (high frequencies) structures cannot respond quickly, and we see an elastic response, while at low frequencies, the system can adjust itself continuously, i.e., it can flow, showing viscous effects (Sects. 9.1 and 9.1.1). Thus, when observed over all time (and frequency) scales, the system is viscoelastic.

For nonlinear viscoelastic – or simply inelastic but shear-thinning systems – not only does the microstructure take time to respond to the flow, but it is also changed by the flow and this change will itself take time. This is the essential difference between linear viscoelasticity and thixotropy – that while both are time effects, the former is in the linear region, where the structure responds but remains unchanged and the latter takes place in the nonlinear region where the structure is broken down by deformation as well as responding to it.

Shear thinning can occur for many reasons, e.g.,

- alignment of rod-like particles in the flow direction
- loss of entanglements in polymer solutions
- rearrangement of microstructure in suspension and emulsion flow
- breakdown of flocs

Since changes in any of these states take some time to come about (either from rest or from some other configuration), thixotropy is always (in principle) to be expected from a thinning mechanism. However thixotropy becomes significant when its time scale becomes significantly longer than the response times of instruments used to test rheology or longer than some flow time in a process, e.g., the average time for a liquid to flow through a pipe. Thixotropic time scales can be longer than viscoelastic time scales and practically important when these time scales become minutes and longer for breakdown. This will inevitably mean many more minutes or even hours for the recovery of the structural nature of most microstructural features that produce thixotropic effects (see above). It is obvious that, at conditions near the fully structured rest state, viscoelastic effects will also be seen. The typical response to a start-up experiment from a rest state shows this. Only a few theories have sought to account for this effect, which shows concurrently the breakdown/rebuilding of both viscous and elastic responses.

The picture presented becomes even more complicated when we consider the linear viscoelastic response of a rebuilding structure, where the typical storage and loss moduli – G' and G'' (Sect. 9.1) – evolve with time. This can be used as a measure of the rebuilding mode since the test is conducted at low-enough stresses/deformations that the evolving structure is unaffected by deformation. The growth of G' particularly is very sensitive to structure rebuilding. *Barut* et al. [9.113] studied an acrylic polymer solution in a mixture of solvents, with titanium dioxide (TiO_2) or a mixture TiO_2 and aluminosilicate particles. These were pre-sheared in a controlled-stress rheometer at 200 Pa for 2 min and after this ceased, the linear oscillatory properties were monitored for 10 min at strains of less than 0.1 over a frequency range of 0.03–62.8 rad s^{-1}. The form of the rebuilding curve of the storage modulus was also of a stretched-exponential form:

$$G' = G'_\infty - (G'_\infty - G'_0) \exp(-kt^p), \qquad (9.194)$$

where G'_0 is the storage modulus at the commencement of oscillations, G'_∞ is the storage modulus after an infinite time, and k and p are material parameters.

Williams and *Ren* [9.114] used an oscillatory rheometer operating over the range 250–800 Hz to examine the rebuilding of 0.045 g/ml aqueous Laponite RD dispersions (synthetic-type clay, circular discs 300 × 10 Å), again measured by G' but now derived from the phase velocity. At these high frequencies, G' is a particularly sensitive measure of the structural solid-like response of the clay. They showed that restructuring was significant over the first 10 minutes or so, but was still going on after 20 minutes. Viewed on a log scale there is a rise from an initial value at small times and a fast build up through a power-law region, eventually (as it must) flattening out at times greater than 20 minutes.

Bouda and *Mikešová* [9.115] simultaneously monitored the AC conductivity and the storage modulus G' to establish the build-up of a carbon-black network in a polyethylene melt. After an initial period where the conductivity was constant, it then increased rapidly. The time a sample with 7.4% by weight took to show the rapid increase was nearly 200 min. Similar behavior was also observed for G'. They explained the observations as resulting from the setting up of a continuous network, and used percolation theory to show that the observed behavior was explainable on the basis of diffusion-limited aggregation of small clusters of primary carbon-black particles under the action of Brownian motion. As the flocs collide and stick together they eventually form an interconnecting network as the percolation threshold is reached, at which point the electrical conductivity and the storage modulus G' rose rapidly; see also [9.116].

On the other hand, *Greener* and *Connelly* [9.117] point out how easy it is to misinterpret thixotropy loops, particularly if there is viscoelasticity present in the sample being tested. They compared the supposed thixotropic behavior of an aqueous poly-acrylamide solution with the behavior of the Wagner model (cf. Sect. 1.3) and show that all the effects are accounted for by the viscoelastic behavior alone, and not by true thixotropy.

9.2.3 Typical Thixotropic Experiments

Thixotropic/Hysteresis Loops

One of the favorite ways of measuring thixotropy is to perform a loop test; that is to say, to linearly increase the shear rate (or sometimes shear stress) from zero to a maximum value, and then to return at the same rate to zero. This test can then be repeated, until eventually an invariable loop behavior is seen; see Fig. 9.59. The area between the up and down curve is automatically measured in some computer-based rheometers as a measure of thixotropy.

However, this kind of test is to be depreciated, for although useful as a quick, qualitative test, the procedure has a number of disadvantages. First, the loop test is often carried out too quickly, and inertial effects due to the measuring head are introduced but not always recognized. (However, inertia effects can now be accounted for by some rheometer software packages.) Secondly, both shear rate and time are changed simultaneously, in a material where the flow properties are a function of both shear rate and time – as thixotropy obviously is. This is bad experimentation because the response cannot then be resolved into the separate effects from both variables from the one experiment. However the problem in interpreting the loop is even more difficult when we realize that the first part of the behavior on start-up is essentially elastic (cf. the overshoot in Fig. 9.13). As the strain becomes large this moves to nonlinear elastic response. If the behavior were just linear elasticity, then the strain in a simple loop test would increase parabolically, and the curve would be concave to start with, but soon turns over. At the same time the viscous behavior can become apparent, giving a tendency to flatten out. The viscous behavior then itself becomes nonlinear as the microstructure begins to break down at large strains. As the strain rate increases further, the liquid would like to shear-thin, but this takes time since the structure cannot adjust itself fast enough to the increasing shear rate. When the time is long enough and the structure has broken down, the down curve which will be under the up curve. Rebuilding will then begin to take place slowly. Even with an apparatus that responds perfectly to the stress and strain, interpreting the data to obtain the parameters corresponding to a model is very difficult, if not impossible.

A simpler and more-sensible test for a thixotropic liquid is performing and deriving results from stepwise experiments where the shear rate or stress is changed from one condition to another with a carefully controlled prehistory. Even so, it is impossible to eliminate the elastic response and instrument inertia.

Start-Up Experiments

Any experiment that starts from rest is another kind of thixotropic test. The typical behavior of strain- or stress-controlled experiments is shown in Fig. 9.60.

Most, if not all, thixotropic materials that have been at rest for some time show viscoelastic behavior, so the immediate response of such tests is elastic, then depending on the conditions, a thixotropic response will be seen as an overshoot in the stress in strain-controlled exper-

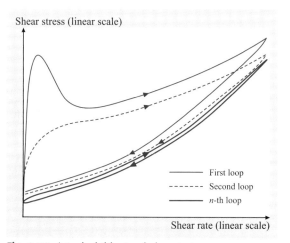

Fig. 9.59 A typical thixotropic loop test

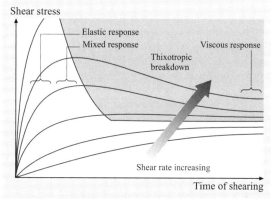

Fig. 9.60 Various regions of start-up flow

iments or an increase in the strain–time curve in creep tests (Sect. 9.1). This sometimes happens after a critical strain has been achieved. This initial elastic response giving way to a thixotropic viscous response makes the behavior quite complicated.

Artefacts Involved in Measuring Thixotropy

The greatest difficulty in understanding and modeling thixotropic materials is in knowing the effect of the – often unknown – deformation prehistory on the liquid of interest. This is particularly true in situations where a thixotropic liquid is subject to mixing and pumping, or even the seemingly simple task of filling a viscometer or rheometer. In both cases it would be interesting to be able to predict the subsequent flow even if only the initial response was known, say the initial torque on a viscometer operating at a certain shear rate, or the torque on a mixing vessel prior to the liquid being pumped out.

A number of methods have been devised to establish a consistent initial condition: fixed rest time after sample loading; pre-shear at a prescribed shear rate for a prescribed time followed by period; pre-shearing to equilibrium at a low shear rate followed by testing at a higher shear rate, etc. These eliminate the problem in characterization, but they can never completely eliminate the problem in practice, since the effects of prehistory on a previously untested sample are always unknown.

The mechanical inertia of the rotating members in viscometers and rheometers (Sect. 9.1.1) means that, in experiments where the shear rate or stress is changed quickly or instantaneously, the instrument response is delayed, and this is often mistaken for thixotropy, and even if thixotropy is present, this can complicate its measurement. Sometimes the presence of a low-compliance spring complicates the measurement because the output of the experiment relates to this spring response.

For isotropic materials, as we have seen, a typical source of thixotropy phenomenon is the breakdown of large flocs. However this also leads to the appearance of apparent slip at the wall, e.g., depletion giving large effects, see *Barnes* [9.118]. This is a case where observing the pressure gradient alone in a pressure-driven rheometer (Sect. 9.1.1) will not differentiate between or account for both effects. A detailed investigation of the flow profile will alone show what is happening. As well as pipe flows, there are many apparently simple flows where the shear rate or shear stress is not constant spatially, for instance, concentric cylinders and parallel plates (Sect. 9.1.1). These geometries are often used in the rheological characterization of thixotropic liquids, and unwittingly many workers are unaware of the complications.

9.2.4 Semi-Empirical Phenomenological Theories Used to Fit Experimental Data

Requirements of Useful Models

The ideal model to describe thixotropic behavior would start from the fact that some rheology-determining physical entity takes time to change when the flow field around it is changed or is changing. As we have already said, this might be, for example, the size of a floc, the orientation angle of an alienable particle, or the density of transient entanglements. In the simplest models, all such fundamental parameters change instantly with shear, shear rate, or shear stress. First we have to know at what rate these changes take place, and then if we know how the microstructure relates to the stress, we can predict the overall behavior. Most workers in this field have used theories to describe viscous thixotropic phenomena while only a few have attempted to describe viscoelastic effects. Representative classes of theories are described below and others in that field are noted.

Viscous Theories

Current theories for thixotropy fall into three groups: first those that use a very good description of microstructure described by a numerical value of a scalar parameter, typically λ and then use $d\lambda/dt$ as the working parameter; second those who attempt some description of the temporal change of the microstructure as for instance the number of bonds, or attempt to describe real floc architecture using fractal analysis, and third those that use time data itself on which to base a theory.

Indirect Microstructural Theories

Most workers in this area have developed mathematical theories of thixotropy. Based on some numerical scalar measure of structure, often designated by λ and using this simplistic concept, the completely built structure is represented by $\lambda = 1$ and completely broken down by $\lambda = 0$. In the simplest case of a typical, inelastic, non-Newtonian liquid with upper and lower Newtonian viscosity plateaus, $\lambda = 1$ corresponds to the zero-shear viscosity η_0 and $\lambda = 0$ corresponds to the lower asymptotic viscosity at an *infinite* shear rate η_∞, with λ taking intermediate values between.

Thixotropy is then usually introduced via the time derivative of the structure parameter $d\lambda/dt$, which is given by the sum of the build-up and breakdown terms, which in the simplest theories are only controlled by the shear rate and the current level of the structure λ. The general description of the rate of breakdown due to shearing is given by the product of the current level of structure and the shear rate raised to some power. The driving force to build up is controlled by the distance the structure is from its maximum value, raised to another power. Thus

$$\frac{d\lambda}{dt} = g(\dot{\gamma}, \lambda) = a(1-\lambda)^b - c\lambda\dot{\gamma}^d, \quad (9.195)$$

where a, b, c, and d are constants for any one system. The value of the function $g(\dot{\gamma}, \lambda)$ is negative if the system is breaking down towards equilibrium; if it is positive, it is building up to equilibrium. At equilibrium, for every value of shear rate there is a particular value of λ, which in the equation is found by setting $d\lambda/dt$ to zero.

Godfrey [9.86] summarized the development of these two-process models beginning with

- *Goodeve* and *Whitfield* in 1938 [9.119], which led to an earlier version of the above equation
- to *Moore* [9.120] with both b and d as unity,
- progressing through *Cheng* and *Evans* [9.121] with b unity, but d being non-unitary,
- through to the very general *structural kinetic model* of Mewis, with both powers taking no-unitary values, the formulation used by Lapasin's group [9.122].

The next step in this kind of approach is to relate the λ – as calculated using the equations above – to the stress σ or viscosity η in some suitable equation. As we shall see this has been done in a variety of ways which range from the Bingham equation

$$\sigma = \sigma_y + k\dot{\gamma} \quad (9.196)$$

through the Cross model to a Cross-like model containing a yield stress. Most of the differences between the theories in this area are accounted for in combinations of these structure change and structure–viscosity formulations.

Baravian et al. [9.123] proposed a modification of the Cheng and Evans approach. Their study is worth looking at in detail for two reasons. First it follows the traditional route to describe thixotropy, and second, it shows how long and seemingly involved the procedure to describe thixotropy becomes.

They postulate that the relationship between viscosity and structure is given by

$$\eta(\sigma, t) = \eta(\lambda) = \frac{\eta_\infty}{(1-K\lambda)^2}, \quad K = 1 - \left(\frac{\eta_\infty}{\eta_0}\right)^{-\frac{1}{2}}. \quad (9.197)$$

This assumes that the effect of stress on the viscosity is also accounted for by the current value of the structural parameter λ, which can be written

$$\lambda = \left[1 - \left(\frac{\eta_\infty}{\eta}\right)^{\frac{1}{2}}\right]/K, \quad (9.198)$$

where η is the current value of viscosity. Then for any system one has to find the values of η_0 and η_∞ (and thus, K), and from these, all values of η can be converted into values of λ. They assume that the relationship between the equilibrium value of the structural parameter λ and the shear stress σ is given by

$$\lambda_{\text{equil}}(\sigma) = \left[1 + \left(\frac{\sigma}{\sigma_c}\right)^p\right]^{-1}, \quad (9.199)$$

where σ is the shear stress and σ_c is its critical value.

From any one particular equilibrium position of stress, 1 Pa jumps were made up and down, and plots of $d\lambda/dt$ versus λ were made. These were of the form

$$\dot{\lambda} = \frac{1}{t_a}\left(1 - \frac{\lambda}{\lambda_{\text{equil}}}\right). \quad (9.200)$$

This kind of curve was obtained for many values of equilibrium stress. The values of the constants were then described by the equation:

$$\lambda(t) = \lambda_{\text{equil}} + (\lambda_{\text{ini}} - \lambda_{\text{equil}})\exp\left(-\frac{t}{t_a}\right), \quad (9.201)$$

where λ_{ini} is the initial value of λ.

Thus, for all values of stress, they could now define λ_{ini}, λ_{equil}, and t_a. Using the stress-up as well as stress-down data, the values of t_a were the same. This theory was used to describe a loop test very well, once instrument inertia had been accounted for.

Other variations on this theme include that of *De Kee* et al. [9.124] (following *Tiu* and *Boger* [9.125]) who described the breakdown behavior of various food systems by

$$\frac{d\lambda}{dt} = -c\dot{\gamma}^d(\lambda - \lambda_{\text{equil}})^n, \quad (9.202)$$

where c is a constant parameter.

The stress was then described by a multiple exponential-type flow law given by

$$\sigma(\dot{\gamma}) = \lambda \left[\sigma_0 + \dot{\gamma} \sum_i \eta_i \exp(-t_i \dot{\gamma}) \right]. \quad (9.203)$$

where η_i and t_i are the model parameters. The theory was used to characterize viscosity decay curves for yoghurt and mayonnaise.

Direct Structure Theories

Denny and *Brodkey* applied reaction kinetics to thixotropy via a simple scheme that described the distribution of broken and unbroken bonds [9.126]. The number of these bonds was later related to the viscosity. The forward and reverse rate constants, k_1 and k_2, represented the breakdown characteristics in flow, and the build-up kinetics, see for instance *van den Tempel* [9.127] who related the flocculated system of fat globules. Denny and Brodkey wrote the rate of structure breakdown as (cf. *Ruckenstein* and *Mewis* [9.128]):

$$-\frac{d(\text{unbroken})}{dt} = k_1(\text{unbroken})^n - k_2(\text{broken})^m, \quad (9.204)$$

and solved to give the viscosity by assuming that it is linearly proportional to the unbroken structure, with a maximum value (η_0) when completely structured and a minimum value (η_∞) when completely destructured. The rate constant k_2 is assumed to be independent of shear rate, being merely a description of Brownian collisions leading to restructuring and the rate of breakdown constant is related to shear rate by a power-law expression.

The way the well-known Cross model was derived is instructive [9.129]. Assuming that the structured liquid was made up of flocs of randomly linked chains of particles, Cross described a rate equation of the form:

$$\frac{dN}{dt} = k_2 P - (k_0 + k_1 \dot{\gamma}^m) N, \quad (9.205)$$

where N is the average number of links per chain, k_2 is a rate constant describing Brownian collision, k_0 and k_1 are rate constants for the Brownian and shear contribution to break down, P is the number of single particles per unit volume, and m is a constant less than unity. At equilibrium dN/dt is zero, so

$$N_e = \frac{k_2 P}{k_0 (1 + \frac{k_1}{k_0} \dot{\gamma}^m)}. \quad (9.206)$$

Then, assuming that the viscosity was given by a constant η_∞ plus a viscous contribution proportional to the number of bonds N_e, he derived his well-known relationship

$$\frac{\eta - \eta_\infty}{\eta_0 - \eta_\infty} = \frac{1}{1 + \frac{k_1}{k_0} \dot{\gamma}^m}, \quad (9.207)$$

where η_e is the equilibrium value of viscosity at long time of shearing (cf. (9.247) in Sect. 9.3).

Cross could have used the nonequilibrium data to derive a thixotropic model, which others did later using his model.

Lapasin et al. [9.130] used a fractal approach to describe flocculated suspensions. They argued that

$$\frac{dN_{\max}}{dt} = a(N - N_{\max}) - b\sigma^q(N_{\max} - N_{\max,\infty}), \quad (9.208)$$

where $N_{\max}$ is the number of primary particles in a floc, $N_{\max,\infty}$ is the lower limit to which $N_{\max}$ tends as the shear stress σ becomes infinite, and a and q are material constants. This can be solved for $N \gg N_{\max}$ to give

$$N_{\max} = N_{\max,\infty} + \frac{aN}{b} \sigma^{-q}, \quad (9.209)$$

which yields

$$\left(\frac{\eta}{\eta_s}\right)^{-\frac{1}{2}} = 1 - \frac{\phi}{\phi_{\max}} \frac{B + A \left(N_{\max,\infty} + \left(\frac{\sigma}{\sigma_c}\right)^{-q} \right)^{\beta+1} - 1}{B + N_{\max,\infty} + \left(\frac{\sigma}{\sigma_c}\right)^{-q} - 1}, \quad (9.210)$$

where ϕ is the volume fraction of flocs, and A and B are constants, whereas $\beta = 3/D - 1$, where D is the fractal dimension of the floc.

This equation describes the breakdown of flocs under flow, and predicts a yield stress and a high-shear-rate viscosity. It described very well the behavior of both titanium dioxide (TiO_2) and mica dispersions in steady state. (A much more-complicated theory of this kind had been proposed much earlier by *Eyring* [9.131].)

Simple Viscosity Theories

Frederickson [9.132] formulated an expression for the rate of change of fluidity Θ (the inverse of viscosity) of a non-Newtonian system as

$$\frac{d\Theta}{dt} = k_1 \dot{\gamma}^2 \left(\frac{\Theta_\infty - \Theta}{\Theta} \right) - k_2 (\Theta - \Theta_0). \quad (9.211)$$

He used this expression for steady-state flow ($d\Theta/dt = 0$), recovery at rest (i.e., zero shear rate), the sudden application of a step stress, and a loop test. Like the Cross model, this model also has higher and lower Newtonian plateaux at equilibrium, and if the quadratic dependence on shear rate is replaced a power-law relationship, they become similar.

Mewis and *Schryvers* [9.133] devised a theory that also circumvents the use of any parameter such as λ, and instead used the viscosity as a direct measure of structure. They proposed that the rate of change of viscosity rather than rate of change of structure be the viscosity difference between the steady state and current values of viscosity (not the structure difference), viz.

$$\frac{d\eta}{dt} = K[\eta_s(\dot\gamma) - \eta]^n , \qquad (9.212)$$

where η_s is the steady-state viscosity.

This integrates to give

$$\eta = \eta_{e,\infty} - (\eta_{e,\infty} - \eta_{e,0})$$
$$\times \{[(n-1)Kt(\eta_{e,\infty} - \eta_{e,0})^{n-1} + 1]^{n-1}$$
$$+ 1\}^{1/(1-n)} . \qquad (9.213)$$

In the notation $\eta_{e,0}$ and $\eta_{e,\infty}$ the initial subscript refers to the fact that the viscosities referred to are at equilibrium, while the second refers to the shear rate.

Mewis and Schryvers then made the (Newtonian) assumption that $\eta_{e,0} = \eta_{e,\infty} = \mu$, which makes the equation simpler since it is much easier to measure the eventual viscosity of the initial steady-state condition. This assumes that the viscosity at the end of the initial steady-state period as that at the beginning of the new shear-rate test, i.e., that the system is essentially between those conditions. This is reasonable under some conditions but, as Mewis's work has shown, it is only strictly true for most systems at higher shear rates. Using a value of n of $5/3$, they fitted experimental data very well for carbon black in mineral oil.

Kristen et al. [9.134] modeled the thixotropic breakdown behavior of maize starch pastes using the equation:

$$(\eta - \eta_\infty)^{1-m} = [(m-1)kt + 1](\eta_0 - \eta_\infty)^{1-m} , \qquad (9.214)$$

where η_0 and η_∞ are the asymptotic values of viscosity η (representing the fully structured and fully destructured states, respectively) measured at time t for any particular shear rate $\dot\gamma$, and k and m are material constants. This approach had been used with previous minerals and industrial suspensions [9.135]. For the starch paste they studied the data was satisfactorily described by $m = 3$.

Over the shear rate range tested, the steady-state flow curves could be described by a power-law model.

Viscoelastic Theories

Almost any viscoelastic theory can have thixotropy introduced if the particles that give the viscous and elastic responses are made to change in the way we have described for purely viscous behavior. Probably a model due to *Acierno* et al. [9.136] gives the best example of this approach. They considered a model based on a series of Maxwell elements (Sect. 9.1). The number of elements in their model was such that they could be represented by a continuous spectrum. Simplifying the model to a discrete series, it is possible to write the behavior as

$$\sigma = \sum_i \sigma_i , \quad \frac{\sigma_i}{G_i} + \theta_i \frac{d}{dt}\left(\frac{\sigma_i}{G_i}\right) = \theta_i \dot\gamma , \qquad (9.215)$$

where σ_i is the stress, G_i the modulus, and θ the relaxation time ($= \eta_i/G_i$) of the i-th element in the discrete spectrum.

Thixotropy is introduced via the well-used structure parameter λ (labeled x_i in their paper):

$$G_i = G_{0,i}\lambda_i , \quad \theta_i = \theta_{0,i}\lambda_i^{1.4} . \qquad (9.216)$$

The rate equation is then given by

$$\frac{d\lambda_i}{dt} = \frac{1-\lambda_i}{\theta_i} - \frac{a\lambda_i}{\theta_i}\left(\frac{E_i}{G_i}\right)^{\frac{1}{2}} , \qquad (9.217)$$

where E_i is the instantaneous elastic energy in the i-th element. This is the same as the Moore linear model [9.84], except that the shear rate is now replaced by the generalized expression $(E_i/G_i)^{1/2}/\theta_i$ that accounts for the elasticity as well as the viscosity. The theory gives an excellent description of most of the rheological behavior of a low-density polyethylene melt in shear and extensional transient and steady-state flow. The model is equivalent to the Moore model if reduced to the viscous case. *Shoong* and *Shen* [9.137] introduced a power-law dependence in the breakdown term, which then compares with the Cheng and Evans' inelastic model.

Quemada [9.138] developed a thixotropic model based on an explicit viscosity–structure relationship, $\eta(S)$, between the viscosity and a structural variable S. Under unsteady conditions, characterized by a reduced shear rate, $\gamma(t)$, shear-induced structural change obeys a kinetic equation (through shear-dependent relaxation times). The general solution of this equation is a time-dependent function, $S(t) = S[t, \gamma(t)]$. Thixotropy was modeled by introducing $S[t, \gamma(t)]$ into $\eta(S)$, which led directly to $\eta(t) = \eta[t, \dot{\gamma}(t)]$ without the need for any additional assumptions in the model. Moreover, whilst observation of linear elasticity requires small enough deformation, i.e., no change in the structure, larger deformations cause structural buildup/breakdown, i.e., the presence of thixotropy, and hence leads to a special case of nonlinear viscoelasticity that can be called *thixoelasticity*. Predictions of a modified Maxwell equation, obtained by using the above-defined $\eta(S)$ and assuming $G = G(0)S$ (where $G(0)$ is the shear modulus in the resting state defined by $S = 1$) were discussed in the case of start-up and relaxation tests. Similarly modified Maxwell–Jeffreys and Burger equations (Sect. 9.1) are used to predict creep tests and hysteresis loops. Discussion of model predictions mainly concerned both the effects of varying the model variables or/and the applied shear-rate conditions and comparisons with some experimental data.

9.2.5 The Breakdown and Build-Up of Isolated Flocs

Two key mechanisms of thixotropy for typical systems are floc erosion and Brownian collisions. Work has been published on these topics for diluted flocs and it is instructive to relate these studies to thixotropic build-up and breakdown of suspensions of flocs.

The diffusion rate of isolated flocs decreases significantly as their size grows. *Reynolds* and *Goodwin* [9.139] measured the diffusion coefficients of isolated flocs and found a rapid decrease of diffusion rate with floc size, with the particular value depending on the floc geometry, which they studied as linear or clustered flocs. As these are quite moderate floc sizes, it is obvious that flocs of hundreds or thousands of primary particles will move very sluggishly. The effect of primary particle size on translational diffusion coefficient was worked out by Einstein as a simple inverse dependence on size (cf. (3.203) in Sect. 3.7). However the rotational diffusion of particles scales as the inverse cube of particle size [9.140]. This behavior explains why exponential expressions for rebuilding contain a driving force such as $(1 - \lambda)$ because rebuilding starts as the floc size grows and the diffusion coefficient decreases. This means that collisions become less frequent, and as rebuilding progresses it gets slower and slower, but theoretically never stops.

The breakdown of isolated flocs in imposed shear fields has been studied by a number of workers. *Sonntag* [9.141] summarized the results as $d_f = C\dot{\gamma}^s$ where s has been measured as 0.2 or 0.5. The surface shear force experienced by an isolated floc is given by the product $d_f d \eta \dot{\gamma}$, where d_f is the floc size, d is the size of the primary particles, η is the viscosity of the continuous phase, and $\dot{\gamma}$ is the shear rate. It is this force which produces surface erosion of primary particles if it is greater than some bond shear strength between the primary particles, see *Mühle* [9.142]. This expression shows that breakdown in a given shear field is fastest for the largest sized flocs, i.e., at the shortest times as being proportional to the shear rate raised to a power. These facts are reflected in the structure breakdown criteria normally used, $d\lambda/dt \propto -\lambda \dot{\gamma}^n$.

9.2.6 Examples of Systems and Studies from the Literature

Previous Work

There are a very large number of systems that have been found to be thixotropic; previous reviews have listed many examples (see *Bauer* and *Collins* [9.83], *Mewis* [9.84], *Cheng* [9.85], and *Godfrey* [9.86]). Here a set of largely new examples is given, with special emphasis on hitherto unreported eastern European and oriental studies.

The present author (*Barnes* [9.87]) has produced extensive tabulations of work on specific thixotropic systems in the following areas:

- Thixotropic paints, inks and coatings: when coatings are applied to vertical or inclined surfaces, the time taken for rebuilding to occur can cause the material to drain. This is obviously undesirable. The thixotropic breakdown of paints is important when such paints are being put onto the brush or brushed out. The desirable properties of nondrip paint might appear quite quickly, but the paint has to be worked to make it thin enough to apply evenly.
- Thixotropic detergent systems: thixotropy in commercial detergent liquids can give rise to problems when they have to be poured from containers or poured into machines. Then dispersion can be a problem. If rebuilding is very slow, physical instability can result due to sedimentation or creaming.
- Thixotropic clay systems: clays are probably the best known examples of thixotropy, because of the extreme changes brought about by shear. A clay suspension can be shaken in a bottle, and the sound generated is almost water-like, but on standing the clay becomes completely gelled and manifests a ringing sound if tapped in a glass container. Clays such as the natural bentonites and the manufactured

and modified Laponites, because of the extensive nature of the very thin sheets from which they are made up, give a very good thickening effect without giving the unwanted viscoelastic effects found in some kinds of polymeric thickeners. However, because of the size of these clay sheets, the rate of structuring is very slow and also the difference between the fully formed structure at rest and the flowing dispersion is large.

- Thixotropic oils and lubricants: greases are thixotropic because of the flocculation of the dispersed material suspended in an oil phase. This is very important in lubricating situations where the grease has to break on shearing so that no unwanted extra drag is experienced in bearings.
- Thixotropic coal suspensions: coal–oil and coal–water suspensions show considerable thixotropy, and problems with start up of pumps and after stopping the flow can be quite severe. The pump duties required for flow of sheared suspensions are very different from start up of a rested suspension. This can result in pump failure, since start-up torques can be very high.
- Thixotropic metal slushes: if metals are sheared just below their melting point, they take on the appearance of a shear-thinning liquid. However, they have been described as thixotropic. While they might possess some thixotropy, the title is probably a misnomer because shear thinning is perhaps more important than thixotropy. In fact, as stated above, thixotropy is probably a nuisance, because what is needed is a fast-responding, very shear-thinning, liquid-like material for casting.
- Thixotropic rubber solutions: the manufacture of black rubber tyres, etc., uses carbon black. When dispersed, carbon-black particles are attracted to one another, and form a network throughout the rubber solution. The latter itself is shear thinning, and the carbon-black network also renders a degree of shear thinning, but is also thixotropic.
- Thixotropic food and biological systems: many food and biological systems are well known examples of thixotropy. For instance, flow makes them thinner, but leaving them to rest thereafter thickens them again. Thixotropy in food thickeners such as xanthan gums can cause problems in that the suspending properties given to liquids may take time to appear after shearing and this could cause some initial sedimentation or creaming of suspended material.
- Thixotropy in creams and pharmaceuticals products: creams and other personal-product and pharmaceutical materials are given *body* by using so-called thixotropes that happen to be thixotropic. Here the original meaning of thixotropy of conferring gel-like properties is still very often the controlling idea. The time effects seen using these materials are therefore again only of nuisance value.

Published Work on the Engineering Consequences of Thixotropy

Flow in Mixers. Edwards et al. [9.143] found that the behavior of a range of thixotropic materials in a series of mixers was quite easy to characterize if one assumed that the mixer behaved as a viscometer running at the same shear rate as the average shear rate in the mixer. An average shear rate for the flow in a cylindrical vessel with anchor, helical ribbon or helical screw impellers is given by the impeller rotational speed N (rev s^{-1}) times a constant depending on the impeller geometry, k, where values of k ranges from about 12 for the helical screw, to around 20 for the anchor, or 30 for the helical ribbon. They compared the torque produced by the mixer with the signal from a viscometer running at the same shear rate, both of which could halve over the course of the experiment. For salad cream, tomato ketchup, yoghurt, paint and 3 and 4% aqueous Laponite dispersions, they found that the average viscosity as a function of time in the mixer at a given impeller speed compared well with that in a viscometer running at the same shear rate. For the salad cream, tomato ketchup, and paint, the viscosities agreed to within 10%. The predicted values for 3 and 4% Laponite agreed reasonably well for the anchor and helical ribbon, but were 20–40% too low for the helical screws. This latter fact was probably due to the strongly non-Newtonian behavior and nature of the Laponite dispersions.

Flow in Pipes. When a thixotropic liquid enters a long pipe from a large vessel where it has been at rest, the development of the velocity and pressure fields in the pipe is very complicated. The large pressure involved in the start-up of flow of a thixotropic liquid can cause problems in terms of the necessary pump performance. Often cavitation can be the cause, since even though the pump could cope with sheared material, it might be unable to initiate flow of the material that has been at rest for some time. Cavitation in the liquid within the pipe can also occur.

Once flow has started, the liquid near the pipe wall is subjected to the highest shear rate and the lowest velocity; hence it is subjected to the shear for longer than the fluid flowing in the middle of the pipe. This results in a very fast and prolonged breakdown near the wall, giving a low-viscosity layer that effectively lubricates the inner, more-viscous layers (the phenomenon quite similar to the one expected for a purely Bingham fluid possessing a yield stress). If the pipe is long enough, the flow profile will evolve such that eventually the steady state profile is established. However, for short pipes the flow can be quite complicated, with a nonlinear pressure profile being a distinct possibility. Distinguishing thixotropy from a developing slip layer caused by particle depletion can be very complicated (Barnes [9.114]).

Schmitt et al. [9.144] derived an equation for the prediction of the mean value of the friction factor during the flow of a thixotropic fluid in a horizontal rectilinear cylindrical pipe. According to their equation, the pressure drop is a function of three dimensionless numbers: the generalized Reynolds number Re standing for the ratio of inertial to viscous forces for an Ostwaldian fluid, a modified memory effects number, the Deborah number $De = \theta/t_{ch}$, where θ is the characteristic memory time, and t_{ch} the characteristic flow time, and a *structural* number Se, which is correlated to the maximum breakdown of structure of the fluid. This equation is only valid for laminar flows and is based on several hypotheses described in [9.144]. The rheological model used in [9.144] is based on a structural approach, featuring a rheological state equation describing shear stress, and a structural decay equation. The fluid was stirred yoghurt and its structural parameter λ follows a second-order kinetic equation. Experimental validation of the friction factor formula showed good agreement.

Recent Publications on the Subject of Thixotropy

Much work has been presented since the present author originally reviewed the thixotropy literature in 1997 [9.87].

Chinese workers seem to be leading the field recently in studying thixotropic systems and their work deserves separate mention. First *Hou* et al. [9.145] showed a novel type of thixotropic or, as they called it, complex behavior. Similarly *Li* et al. [9.146] looked at the effect of pH on the thixotropy of ferric aluminium magnesium hydroxide/montmorillonite suspensions. They found that for relatively low ratios of Fe-Al-Mg-mixed metal hydroxide, MMH, to sodium montmorillonite (MT, 0, 0.012 and 0.051 by weight), when the pH was slowly increased, the suspension behavior changed from negative, anti-thixotropy (rheopexy), to typical thixotropy; when the ratio of MMH to MT was relatively high (0.091), rising gradually, the suspension behavior changed from 'complex' thixotropy to negative thixotropy (rheopexy) with increasing pH. In another series of similar experiments reported by *Li* et al.[9.147], thixotropy and negative thixotropy (rheopexy) were reported. This showed the complex effects of time and shear rate in influencing the microstructure. The experiments were carried out by first shearing the suspension at high shear rate to destroy the microstructure of the suspension, then monitoring the change of the viscosity or stress with time at a low shear, i.e., to study the process of structure recovery. The shear history and the shear rate strongly influenced the recovery process for MMH/MT suspensions. The system with a weight ratio (R) of MMH/MT of 0.0 and with $R = 0.013$ both show typical thixotropy in the shear-rate range $10-1022\,\mathrm{s}^{-1}$, i.e., shear rate does not change the thixotropic type behavior of the suspensions. The suspension with $R = 0.051$ showed complex thixotropy under lower shear-rate values (10 and $170\,\mathrm{s}^{-1}$), but showed negative thixotropy (rheopexy) under higher shear-rate values (511 and $1022\,\mathrm{s}^{-1}$). The system with $R = 0.091$ showed negative thixotropy (rheopexy) at lower shear-rate values (10 and $170\,\mathrm{s}^{-1}$), but shows complex thixotropy at higher shear-rate values (511 and $1022\,\mathrm{s}^{-1}$). With increasing shear rate, the viscosity of these suspensions decreases gradually, and the degree of change in viscosity similarly slows down, i.e., the thixotropic behavior becomes weaker.

Li et al. [9.148] have also reported on the influence of measuring conditions on the thixotropy of hydrotalcite-like/montmorillonite (HTlc/MT) suspensions. Three kinds of these suspensions were studied, their mass ratios of HTlc to MT, R, were 0.013, 0.051, 0.091, respectively. The HTlc/MT suspension with $R = 0.013$ showed normal thixotropy at zero time of shearing, $t(s) = 0$, but it changed to rheopexy with increasing $t(s)$. The suspension with $R = 0.051$ transformed from complex thixotropy at low shear rate or positive thixotropy at high shear rate into negative thixotropy (rheopexy) with increasing $t(s)$. The suspension with $R = 0.091$ showed complex thixotropy when the shear rate was $1022\,\mathrm{s}^{-1}$, and showed negative thixotropy (rheopexy) at $10\,\mathrm{s}^{-1}$. However, the suspension changed from negative thixotropy to the weak complex one with increasing $t(s)$ when the shear rate was 170 or $341\,\mathrm{s}^{-1}$. For all the systems, the equilibrium viscosity decreased gradually with $t(s)$ at the low shear rate, but the equilibrium viscosity increased with $t(s)$ at high shear rate because of the memory effect.

Last of all for these systems, we quote the work of *Li* et al. [9.149] on the influence of electrolytes on the thixotropic behavior of ferric aluminum magnesium hydroxide-montmorillonite suspensions [Fe-Al-Mg-MMH and Na-montmorillonite (MT)]. When sodium or magnesium chloride were added to the suspension, the thixotropy of the suspension changed from positive and complex into negative thixotropy (rheopexy), but the electrolytes did not change the thixotropic type of a rheopexic suspension. When aluminium chloride was added to positive thixotropic systems, they were transformed to a complex one, whereas a complex and negative thixotropic suspension remains unchanged, for additions less than 0.01 mol/L; when the level of added aluminium chloride was increased, all types of thixotropic systems are changed to non-thixotropic! In addition, both sodium and magnesium chloride controlled the initial viscosity, measured after cessation of intensive shearing increase, but the value of the viscosity decreased rapidly with time. The equilibrium viscosity of the suspension decreased gradually with increasing concentrations of chlorides in the suspension. With increasing concentration of aluminium chloride, the equilibrium viscosities of the positive thixotropic suspension and the complex thixotropic suspension increase at first, but later decreased, and the equilibrium viscosity of negative thixotropic (rheopexic) suspensions decreased gradually.

The thixotropic properties of mixed suspensions containing oppositely charged colloidal particles were studied by *Guo* et al. [9.150]. In particular, the rheological properties of single-component clay dispersions and clay/MMH mixed dispersions containing oppositely charged colloidal particles were investigated. As the clay concentration is increased, the clay suspensions develop from Newtonian to *yield-stress* behavior. Adding sodium chloride to clay/MMH mixed suspensions caused a decrease in the yield values and apparent viscosities. In the structural recovery measurements, the time dependence of viscoelastic properties of clay/MMH mixed dispersions showed only very small variations of the storage moduli over three hours, in contrast to unusual thixotropic properties of aqueous single-component dispersions of Laponite or MMH. The results were explained using attractive electrostatic interactions between clay and MMH particles due to their opposite charges, similar to the edge(+)/face(-) interactions in single-component clay dispersions at low pH values.

Sun et al. [9.151] investigated the rheological properties of aqueous dispersions of the aluminium–magnesium mixed-metal hydroxide (MMH), which forms solid-like structures. Special emphasis was placed on thixotropy, with structural recovery at rest after steady shear, was characterized by steady shear and small-amplitude oscillatory shear measurements. With increasing MMH concentration, the behavior changed from predominately viscous to a solid-like response. The magnitude of the storage modulus, G', increased strongly and became less dependent on frequency with increasing MMH concentration. After cessation of steady shear, the complex viscosity increased monotonically with time and even after three hours no equilibrium viscosity value was seen, while shear stress under steady rate of $10\,\mathrm{s}^{-1}$ approached the equilibrium value only after about 10 min. The recovery of the MMH suspension after cessation of steady shear was strongly affected by pre-shear history and rebuilding time, so that the greater the intensity of pre-shearing, the lower the values of the elastic moduli after pre-shearing had ceased. Conversely, the longer the rebuilding time, the larger the values of the elastic moduli. These results were similar to those obtained for clay dispersions of different ionic strength and clay concentration. Similarities in particle size and shape, though oppositely charged, and rheology between the two systems provided indirect evidence of similar rebuilding mechanism, so it was assumed that, for positively charged MMH suspension, long-range electrostatic double layers forces led to the formation of a solid-like structure.

The rheological properties of calcium-carbonate-filled polypropylene were examined by *Wang* and *Yu* [9.152] using a Rheometrics dynamic analyzer. The study included steady-shear tests, transient stress growth tests with sequential deformation history, and two-step dynamic oscillatory shear flow. Thixotropic behavior was observed in transient tests for highly filled compounds at volume loading exceeding a critical value at about 20%. The material responses of these viscoelastic thixotropic materials depended on the duration of shear as well as on the rate of shear. The effects of filler on the rheological behavior of highly filled compounds were dominant at low strain rates; however, the effects of activity of the filler were almost negligible at high strain rates because of complete breakdown of the filler network. The time scales for structural changes in filled systems often lengthened compared with the viscoelastic time constants of the unfilled melt. The magnitudes of rheological properties and the degree of hysteresis appeared to increase with increasing volume loading of filler particles. Conversely, surface treatment of fillers, which presumably reduced the interaction between the filler particles and the extent of agglomeration, resulted in major reductions of both the rheological properties and the degree of hysteresis. The diverse experimental observations were interpreted in terms of a system forming a filler network due to weak interparticle forces. The thixotropy resulting from breakdown and recovery of the filler network was dependent on the characteristic time of the individual test.

An experimental investigation was carried out by *Assaad* et al. [9.153] to determine the influence of thixotropy on the use of concrete. Five *self-consolidating* concrete (SCC) mixtures prepared with different combinations of cementitious materials and two flowable mixtures of different stability levels were assessed. The thixotropy of the fresh concretes was evaluated using the variations in apparent yield stress with time and the evolution of the structural breakdown curves. Changes to the impeller of a modified Tattersall concrete rheometer were proposed for the protocol used to assess thixotropy. Instead of the H-shaped impeller that rotates in a planetary motion, a four-bladed vane impeller rotating coaxially around the main shaft was used. This resulted in less slip in the flow of the fresh concrete and an increase in the sheared surface during rotation. Test results showed that thixotropy was not an inherent property of a typical SCC. However, the concrete exhibited a high degree of thixotropy when mixed with ternary cement containing 6% silica fume and 22% fly ash compared with similar concrete made with 4% silica fume and no fly ash. The incorporation of set-accelerating and set-retarding agents resulted in greater and lower degrees of thixotropy, respectively. In the case of one particular very flowable concrete, the addition of a viscosity-modifying additive was shown to increase thixotropy significantly compared with similar concrete made without any viscosity-modifying admixture.

Kinlock et al. [9.154] carried out a rheological study of concentrated aqueous nanotube dispersions, at concentrations at which the nanotubes interacted with each other. The dispersed nanotubes represented a high-aspect-ratio system. The dispersions were thixotropic and hence recovered their structure, and hence their viscosity, on standing.

An experimental study of the viscosity of a macroscopic suspension, i.e., a suspension for which Brownian motion can be neglected, under steady shear was published by *Voltz* et al. [9.155]. Their suspension was made up to a high packing fraction and density matched with the Newtonian continuous phase. The thixotropic behavior was characterized by a long relaxation time that was a unique function of shear. The relaxation times showed a systematic decrease with increasing shear rate. These relaxation times were larger when decreasing the shear rates, compared to those observed after increasing the shear.

Mujumdar et al. [9.156] developed a nonlinear rheological model to account for the time-dependent elastic, viscous and yielding phenomena of thixotropic materials that exhibit an apparent yield stress. A key feature of their formulation was a smooth transition from an elastically dominated response to a viscous response without a discontinuity in the stress–strain curve. The model was phenomenological and based on the kinetic processes responsible for structural changes within the thixotropic material. As such, it could predict thixotropic effects, such as stress overshoot during start-up of a steady shear flow and stress relaxation after cessation of flow. An analysis and comparison to experimental data involving oscillatory shear flow were provided to evaluate the accuracy of the model and to estimate the model parameters in a series of concentrated suspensions of silicon particles and silicon carbide whiskers in polyethylene. The data obtained with this experimental system indicated much better agreement between the theory and experiments than that obtained in earlier work by the authors.

A simple model consisting of the upper-convected Maxwell constitutive equation (cf. Sect. 1.3) and a kinetic equation for destruction and construction of structure, first proposed by Fredrickson in 1970, was used by *Bautista* et al. [9.157] to reproduce the complex rheological behavior of viscoelastic systems that also exhibited both thixotropy and rheopexy under shear flow. The model required five parameters that have a physical significance and could be estimated from rheological measurements. Several steady and unsteady flow situations were analyzed with the model. The model predicted creep behavior, stress relaxation, and the presence of thixotropic loops when the sample is subjected to transient stress cycles. The same kind of behavior has been observed with surfactant-based solutions and dispersions.

To account for thixotropic effects of typical gelled suspensions in the the paint, foodstuffs and pulp and paper areas, *Yziquel* et al. [9.158] proposed a structural network model based on a modified upper-convected Jeffreys model with a single relaxation time and a kinetic equation to describe the flow-induced microstructure evolution. Three distinct kinetic equations were tested for this purpose. The proposed model described yield and thixotropic phenomena, nonlinear viscoelastic behavior and output signal distortions observed for relatively small strain amplitude during oscillatory measurements, and overshoots observed in stress growth

Table 9.8 Details of various studies of thixotropic systems

System	Worker(s)	Description
Vesicular lava	*Bagdassarov*, *Pinkerton* [9.160]	Effect of delayed bubble deformation and recovery
Aqueous fluid gels, based on mixtures of xanthan and gellan	*Martinez-Padilla* et al. [9.161]	Effect of electrolyte addition on thixotropy
Semi-solid dairy desserts	*Tarrega* et al. [9.162]	Effect of test temperature
Aqueous pseudo-bohemite suspensions	*Li* et al. [9.163]	Effect of source of bohemite and electrolyte
Aqueous smectite suspensions	*Malfoy* et al. [9.164]	Effects of the nature of the exchangeable cation and clay concentration
Aqueous solutions of an amphoteric guar gum	*Zhou* et al. [9.165]	Effects of concentration, salts, and temperature
Ayran, a Turkish yoghurt drink	*Koksoy* and *Kilic* [9.166]	Effects of water and salt level
Semi-solid metal alloys	*Koke* and *Modigell* [9.167]	Effects of shear rate and cooling rate on microstructure
Roll coating of paints	*Lopez* and *Rosen* [9.168]	Thixotropy of paints correlated with levelling performance
Some cosmetic products	*Moise* and *Aruxandei* [9.169]	Development of equations to describe thixotropy
Concentrated yoghurt (labneh)	*Abu-Jdayil* and *Mohameed* [9.170]	The effect of storage time, 1–14 days

experiments. A comparison of model predictions and experimental data for fumed silica and coating colors was also presented. However, different model parameters had to be used to correctly predict the different flow properties indicating that a more-versatile or generalized kinetic equation needed to be developed. *Goodwin* and *Reynolds* [9.159] noted that the rheology of such flocculated dispersions is dependent on the suspension microstructure and its time evolution as well as the type and magnitude of the interparticle forces. Much current work focuses on this microstructure as recent developments in experimental techniques have made this type of information more accessible. They noted that computer simulation continued to add breadth to the understanding of the behavior of flocculated suspensions.

Last, we consider in detail two studies that used direct measurement of the flow fields in thixotropic fluids. *Corvisier* et al. [9.171] studied the velocity profile of a thixotropic fluid flow in a pipe. The axial velocity distribution was determined using a particle image velocimetry (PIV) technique (cf. Sect. 5.3.2) and ultrasonic velocity profile measurement. At the entrance section, the fluid was assumed to be in a homogeneous structural state corresponding to a high shear rate. The experimental results showed a progressive flattening of the velocity profiles as the fluid restructured as it moved down the pipe. *Raynaud* et al. [9.172] used a magnetic resonance imaging system to investigate the flow field in a concentric cylinder flow, using thixotropic and very shear-thinning colloidal suspensions. The velocity profiles between the coaxial cylinders were found to be made up of two parts: first, close to the inner cylinder the fluid is sheared at a rate larger than a critical, finite value (in contrast with the behavior of an ideal yield stress fluid), while second, where the fluid is not sheared at all close to the outer cylinder. They monitored the displacement of the critical radius in time after sudden changes of the imposed rotation velocity. They established that the apparent thixotropy of these fluids could be followed by the displacement of the interface between the sheared and unsheared regions.

We conclude our review of recent work with a general list of some other interesting thixotropic systems published by workers from a wide range of countries, together with a brief description of the work, see Table 9.8.

9.2.7 Overall Conclusions

Thixotropy comes about first because of the finite time taken for any shear-induced change in microstructure to take place. The microstructure is brought to a new equilibrium by competition between the processes of tearing apart by stress and flow-induced collision, in a time that can be minutes. Then, when the flow ceases, Brownian motion is able to move the elements of the microstructure around slowly to more-favorable positions and thus rebuild the structure. This can take many hours to complete. The whole process is completely reversible. The manifestation of the effect of the microstructural changes on the flow properties result in the various manifestations of thixotropy described in the present chapter.

9.3 Rheology of Suspensions and Emulsions

Suspensions and emulsions or more generally dispersions are disperse materials containing at least two phases of immiscible constituents. Due to their complicated physical and chemical nature, the rheological behavior of suspensions and emulsions is very complex. Nonlinear flow behavior, normal stress differences, viscoelasticity, and yield stresses are some examples of the effects that can occur in the flow processes of these materials. In this Section some essential basics of the rheological behavior of both material classes, suspensions and emulsions, are described. A classification of disperse materials is given. Common properties of suspensions and emulsions obtained in rheological measurements are discussed and some important sources of errors that may occur in rheological measurements are explained. The physical fundamentals of suspension rheology concerning the properties of the disperse and continuous phases are summarized. The mechanisms of particle–particle interactions and their consequences regarding the stability are shown. A classification of models describing the viscosity as a function of solid volume concentration and shear rate is given. Typical experimental results of the rheological behavior of suspensions both in shear and elongational flows as well as in complex flow situations are summarized. Emulsions as liquid–liquid systems show a similar flow behavior under certain circumstances. A survey of the deviating rheological behavior compared to suspensions is given. Methods of fabrication of emulsions are described. Common ground between the rheology of suspensions and emulsions complete the chapter.

9.3.1 Preliminaries

Basics and Definitions

This chapter deals with the large class of two- or multiphase materials that occur in many fields of natural and engineering sciences, industry, and daily life. The chapter is structured as follows. In the first part, we introduce the classification of disperse systems and explain some general assumptions for proper rheological measurements of such multiphase materials. In the second part, the theoretical fundamentals, which are necessary to understand the rheology of suspensions, are summarized and, based on these fundamentals, representative experimental results of the flow behavior of solid–liquid systems are illustrated and discussed. The third part deals with theoretical basics and experimental results of liquid–liquid systems, whereby we only refer to phenomena occurring additionally compared to solid–liquid systems.

Suspensions and emulsions or, more generally, dispersions are ubiquitous materials. Their appearance ranges from biological materials such as blood or cell suspensions over pharmaceutical or food products of the daily life to inorganic materials such as concrete, drilling mud, printing inks or multicomponent metal melts to name a few.

These examples show that dispersions are highly complex systems. Hence, the rheological behavior of dispersions depends on the composition of the system. The most important parameters are summarized in Table 9.9.

Knowledge of the rheological properties of disperse systems is a prerequisite for the increase of effectiveness of production and processing operations, but also for the handling of disperse products by the customer. Disperse systems consist of at least one solid or liquid phase (the disperse phase), which is dispersed in a liquid continuous phase or matrix liquid. Both phases are immiscible. In the context of this chapter only

Table 9.9 Most important properties for characterizing dispersions (modified after *Chander* [9.173])

Properties of the disperse phase
Particle size and size distribution
Particle shape and shape distribution
Surface properties
Density viscosity, viscoelasticity (emulsions)
Surface energy
Particle volume concentration

Properties of the continuous phase
Viscosity, viscoelasticity
Aqueous, nonaqueous
Dissolved substances

Properties of the interface
Electric double layer
ζ potential
Adsorption density
Thickness of the adsorbed layer
Structure of the adsorbed layer

solid–liquid and liquid–liquid dispersions will be considered. Dispersions with a dispersed fluid-like phase in a solid continuous phase (porous bodies, solid foams) and a solid or liquid phase dispersed in a gaseous continuous phase (dusts, smoke, aerosols, foams) are outside of the scope of this chapter.

Based on the state of aggregation of the disperse phase these materials are distinguished in the following way. Suspensions are materials that contain a granular solid material dispersed in a Newtonian or non-Newtonian matrix liquid. Emulsions consist of a liquid phase dispersed in a second liquid. In some cases, both a liquid and a solid phase may be dispersed in more than one liquid continuous phase.

Dispersions may be regarded as the generic term for materials consisting of discontinuities of any kind dispersed in a continuous phase of different composition or state [9.174]. Regardless of its size or shape, such a discontinuity or discrete element is called a particle, which stands both for solid, slightly deformable, or liquid discontinuities.

According to the International Union of Pure and Applied Chemistry (IUPAC) recommendations [9.175], a colloidal) dispersion is a system, in which particles of (colloidal) size of any nature (gas, liquid, solid) are dispersed in a continuous phase. This definition includes both suspensions and emulsions. The term *colloidal* is used for particles having at least one dimension in the size range 1 nm to 1 μm. Particles with mean size above 1 μm are called *noncolloidal*.

In the literature, another nomenclature can often be found. The term *dispersion* is used for a material consisting of fine insoluble or only slightly soluble solid particles with sizes larger than 1 μm distributed throughout a continuous medium [9.176], whereas emulsions are cited separately without any reference to the particle size. In this context, a suspension with colloidal solid particles is a colloidal dispersion [9.177]. The problem with this notation is that information on the particle size is necessary.

Due to the uncertainties and ambiguities of this second nomenclature, the first classification with the term *dispersion* as generic term describing multiphase materials should be preferred.

At the level of the particles, dispersions can generally be classified concerning their size, size distribution, and the shape or shape distribution of the constituents of the disperse phase. Based on the particle size, dispersions can be classified as shown in Fig. 9.61.

According to the particle size distribution (PSD), a dispersion is termed *monodisperse* if all particles are (nearly) of the same size. If more than one discrete size distribution mode occurs, the dispersion is referred to as *heterodisperse*. If only a few discrete size modes occur the dispersion is *paucidisperse*, e.g. bimodal, trimodal [9.175]. A dispersion consisting of particles of many sizes is called *polydisperse*, if less than 90% of the size distribution lies within ±5% of the average diameter $\bar{d}$, i.e.,

$$\frac{1.645 \sigma_{\mathrm{SD}}}{\bar{d}} > 0.05, \qquad (9.218)$$

where σ_{SD} is the standard deviation [9.174]. This classification is valid for single particles and not for aggregates. Due to agglomeration or coalescence, the number and size of particles can change.

A second possibility to quantify the polydispersity of a dispersion is to calculate the (weighted) polydispersity index PDI as the ratio of the volume average diameter $\bar{d}_{\mathrm{V}}$ and the number-average diameter $\bar{d}_{\mathrm{N}}$

$$\bar{d}_{\mathrm{V}} = \frac{\sum_i n_i d_i^4}{\sum_i n_i d_i^3}, \quad \bar{d}_{\mathrm{N}} = \frac{\sum_i n_i d_i}{\sum_i n_i}, \quad \mathrm{PDI} = \frac{\bar{d}_{\mathrm{V}}}{\bar{d}_{\mathrm{N}}}$$

$$(9.219)$$

with n_i as the number of particles with diameter d_i. A dispersion is called polydisperse if the relation PDI > 1.05 holds.

A third possibility to characterize the polydispersity of a dispersion is given by the so-called grade of dispersity ξ [9.178] according to

$$\xi = \frac{d_{\mathrm{V},84} - d_{\mathrm{V},16}}{2 d_{\mathrm{V},50}} \qquad (9.220)$$

with $d_{\mathrm{V},i}$ as the volume-average diameter (9.219) below which 1% of the particles lie. A dispersion can then be classified as

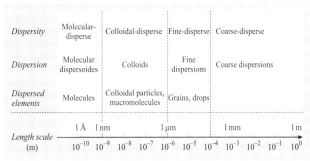

Fig. 9.61 Classification of disperse systems with respect to particle size

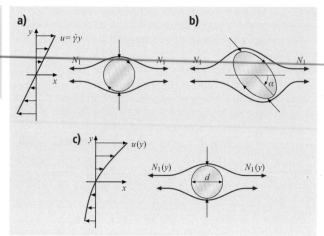

Fig. 9.62a–c Hydrodynamic forces acting on a spherical (**a**) and nonspherical particle (**b**) in a steady homogeneous shear flow and (**c**) on a spherical particle in a nonhomogeneous shear flow (After [9.179], with permission by B. G. Teubner)

- monodisperse if $\xi < 0.14$,
- quasi-monodisperse if $0.14 \leq \xi \leq 0.41$,
- polydisperse if $\xi > 0.41$.

In this chapter we use the grade of dispersity to characterize the polydispersity of a dispersion.

A formal specification of the size of particles with shapes deviating from a sphere is difficult. For this reason, some methods are developed to specify the size of nonspherical particles. In most cases, it is sufficient to reduce the size of a nonspherical particle to the diameter of an equivalent sphere concerning the mass, volume, surface or settling velocity. The sphericity has been proposed as a measure of the deviation of the particle shape from that of a sphere. Different definitions of sphericity are in use. One of the most important definition suggested by *Wadell* [9.180] relates the squared volume-equivalent diameter d_V (diameter of a sphere with the same volume as the particle) to the squared surface-equivalent diameter d_S (diameter of a sphere with the same surface as the particle) of a particle [9.176]. Various other equivalent circle diameters or shape factors can be determined from the projected area of a particle [9.176, 181–185].

It is unusual and also impossible to classify dispersions with regard to all possible particle shapes. If the particle shapes strongly deviate from a spherical shape, the flow behavior of the dispersion will be modified compared with that of hard-sphere suspensions under the same conditions. Hence, a shape factor or shape factor distribution function is necessary. Additionally, the particle size and shape distribution functions affect the applicability of the measuring geometry of the rheological measurements.

Basic Assumptions for Correct Measurement of the Macroscopic Rheological Properties of Dispersions

There are some essential and necessary requirements to be fulfilled in order to characterize dispersions rheologically in the right manner concerning the representative sampling, stability of the sample (settling, aging), migration, the stability of the flow, thermal effects, and the validity of the continuum hypothesis.

These requirements apply independently of the type of the rheological experiment (steady, transient, oscillatory) and the measuring geometry used (coaxial cylinder, cone-and-plate, torsional plate–plate, capillary, and others; Sect. 9.1.1).

In static experiments gravity forces are in competition with the Brownian forces. To estimate the *sedimentation* of samples of suspensions (of spherical particles), the ratio between gravity (settling) and Brownian forces

$$\frac{\text{gravity (settling) force}}{\text{Brownian (thermal) force}} = \frac{(\rho_\text{d} - \rho_\text{c})ga^3}{k_\text{B}T/a} \quad (9.221)$$

should be provided, where ρ_d and ρ_c are the densities of the disperse and continuous phase, g is the acceleration of gravity, a is the particle radius, k_B is the Boltzmann's constant, and T is the absolute temperature. If this ratio is greater than unity, sedimentation may occur. As shown by *Larson* [9.186], sedimentation takes place for $\Delta\rho = \rho_\text{d} - \rho_\text{c} \approx 10^3 \text{ kg/m}^3$ and particles or aggregates larger than 1 μm in radius. Prior to rheometric experiments, the tendency to sedimentation should be checked and if necessary, the experimental flow conditions should be adjusted so that during the experimental time the sedimentation does not play an important role. On the other hand, under shearing flow conditions sedimentation is always retarded.

Beyond sedimentation, particle migration under flow conditions can also falsify measurements due to the resulting gradient in solid volume concentration inside the dispersion. Normal force defects acting on particles can cause particle migration. *Böhme* [9.179] sketched the phenomenon of particle migration by illustrating the normal forces acting on a single solid particle in a non-Newtonian shear flow. In Fig. 9.62a the normal forces are equal in magnitude and of opposite sign due to the symmetry. Hence, no particle migration occurs. This is also true for a nonspherical particle (Fig. 9.62b). The spherical and nonspherical particles only rotate due to the shear force and torque. In case of a nonspherical particle, the particle rotation increases the effective hydrodynamic diameter. Magnus forces due to the rotation of particles in a potential flow do not play any role here and hence do not contribute to migration.

If the shear flow is nonhomogeneous (Fig. 9.62c), the normal stresses at the upper and lower side of the sphere can be quite different. The resulting force acts in the direction of decreasing shear rates and migration occurs. The migration can also occur in Newtonian fluids if the shear flow is nonhomogeneous [9.187]. Hence in any case nonhomogeneous shear flows should be avoided for correct measurement of viscosity, e.g., in coaxial cylinder rheometers (Sects. 9.1.1 and 9.1.2).

Particle inertia can also influence the results of transient or oscillatory experiments. If the shear viscosity of the continuous phase is too small or if the particles are too heavy or too large, they cannot follow the unsteady external flow field. With the particle Reynolds number

$$\text{Re}_\text{P} = \frac{\rho_\text{c} \dot{\gamma} a^2}{\eta_\text{c}} \tag{9.222}$$

an estimation of the inertial effects is possible. In this equation, $\dot{\gamma}$ and η_c are the shear rate and the viscosity of the continuous phase, respectively. Values of Re_P approximately greater than 10^{-1} indicate that particle inertia may be of relevance [9.188].

One of the most essential assumptions of rheometry is the existence of a hydrodynamically stable flow. During experiments, flow instabilities (e.g., secondary flows in cone-and-plate geometries or Taylor vortices in coaxial cylinder geometries) can falsify the results. Details on stability of rheometric flows can be found in the monographs of *Walters* [9.189] and *Macosko* [9.188].

An important problem may arise if temperature effects come into play. Temperature gradients in the gap of the measurement geometry, or viscous dissipation lead to density differences. This may firstly result in convective flows superimposed on the main flow in a rheometer. Secondly, temperature gradients affect the rheological properties of the continuous phase. Hence, it is important to avoid local temperature differences by tempering the measurement system. If this is not possible or the tempering is not sufficient, it is necessary to correct the data numerically (further information on this problem can be found in the monograph of *Macosko* [9.188]).

Finally, for all rheological measurements, it is necessary to ensure that the material can be regarded as a continuum. In other words, the continuum hypothesis must be fulfilled [9.190]. The macroscopic physical behavior of a fluid under rheometric considerations must be the same if it is perfectly continuous and uniform in structure. The physical quantities associated with the fluid (e.g., density, viscosity) contained within a given volume will be regarded as being spread uniformly over the whole volume. *Jeffrey* and *Acrivos* [9.191] pointed out, that it is possible to regard the dispersion as a continuum *when the length scales describing the motion of the suspension as a whole are much larger than the average size or average separation of the particles*. An essential consequence of the continuum hypothesis is that the size of the largest particle in the dispersion should be substantially smaller than the gap size of the measurement geometry. In rheology, it is a common rule that the maximum particle size must be at least one order of magnitude smaller than the smallest dimension of the measuring geometry. *Barnes* [9.192] showed that this ratio increases with increasing volume concentration of the disperse phase. Independent of the continuum hypothesis, the behavior of single particles or particle collectives in both shear and elongational flows may be of interest as shown by several publications concerning these issues in shear flows.

Aggregation, agglomeration (occurring in suspensions), or coalescence, Ostwald ripening, and creaming (occurring in emulsions) during rheological measurements influence the particle distribution function and should be prevented by implementation of appropriate actions.

If all the criteria mentioned above are fulfilled, it is possible to measure the material function of interest using the methods and devices described in Sect. 9.1.1 (shear flows), Sect. 9.1.3 (elongational flows) and Sect. 9.2 (thixotropy, rheopexy and yield stress).

The rheological nomenclature used in this chapter is based on the recommendation of the Executive Committee of the Society of Rheology [9.193].

9.3.2 Suspensions

The rheological behavior of suspensions is influenced by a very large number of parameters. Due to the fact that suspensions are multiphase materials one has to take into consideration both the properties of continuous and disperse phases, as well as the interactions within and between both phases.

This section gives a survey of the rheology of suspensions. It starts with a short general description of the properties of the components of a suspension. The stabilizing mechanisms are explained. Then a dimensional analysis is applied to show the influencing factors on the viscosity. Subsequently, the basics in measurements and possible sources of errors will be given and representative results of rheological measurements in shear and extensional flow are illustrated and discussed. For more details, we refer to the textbooks of *Larson* [9.186], *Russel* et al. [9.194], and *Morrison* and *Ross* [9.195].

Properties of the Continuous Phase

The continuous phases and their properties can be classified from different points of view – physicochemically or rheologically.

The continuous phase, consisting of water (aqueous suspension) or organic solvent (nonaqueous suspension), can be characterized by its chemical composition, dielectric constant, refractive index, or surface tension (cf. Sects. 3.7.2 and 6.1, and 3.2, respectively). An important property of the continuous phase is the polarizability. A liquid is polarizable if dissociation of the species is possible under special circumstances or if the molecules are dipole-like. By addition of surfactants, dispersants, or electrolytes, it can show strongly different properties compared to the pure liquid. The polarizability affects the type of specific interactions between the continuous and disperse phases when charged species (particles, polymers, polyelectrolytes) are added. If the continuous phase is nonaqueous, the charge properties of the particles are not so important since other mechanisms of interaction between the phases such as hydrodynamic interactions are more dominant [9.195].

The rheological properties of the continuous phase may have an essential influence on the overall rheological properties of the suspensions. If a suspension consists of particles dispersed in a Newtonian liquid, all deviations from the Newtonian behavior (nonlinearity, viscoelasticity) stem from the presence of the disperse phase.

Non-Newtonian continuous phases such as polymer melts or aqueous solvents containing surfactants or other additives modify the interparticle interactions in suspensions and emulsions. In the non-Newtonian case the problem of a locally inflated shear rate due to the particle movement becomes relevant. Due to the shear dependence a viscosity distribution may occur inside of the continuous phase. Thus, the movement of the particles during flow may be modified compared with the case of a Newtonian continuous phase. If the continuous phase is highly elastic, normal stress differences due to local shear rates (Sect. 9.1.2) can fragmentize weak aggregates into smaller species.

Properties of the Disperse Phase

The disperse phase may consist of particles, which have a variety of sizes, size distributions, shapes, densities, and surface morphologies. The surface of the particles can be chemically treated or untreated. The treatment of the particle surface influences the interparticle interactions and, hence, the rheological properties of the suspension.

In the simplest case of a dispersion, the particles are spherical, rigid, and monodisperse. In the case of deviations from the spherical shape, the aspect ratio, i.e., the ratio of the main axes of the particle, must be taken into account. If the aspect ratio is high enough, e.g., for rods, the particles align during the flow which results in a corresponding viscosity. Polydispersity of the particles may decrease the viscosity under certain circumstances (cf. Subsect. "Influence od solid volume concentration" in this section).

An additional difficulty arises if the particles are porous or if the particles aggregate. In these cases some amount of the continuous phase may be immobilized in the pores of the particles or in the interspaces between the aggregated. Thus, the amount of continuous phase around the particles decreases. Hence, the apparent solid volume concentration as well as the dispersion viscosity is then higher than expected for nonporous particles or nonaggregating dispersions.

Polymer blends consisting of immiscible components with a distinct lower viscosity of the dispersed phase show that the particles deform under flow. The steric or electrostatic layer surrounding particles can be deformable under flow (soft spheres). These effects modify the overall flow behavior of the suspension.

Brownian Motion

If the particles suspended in a low viscous continuous phase are sufficiently small (less than 1 μm), they perform a stochastic thermal movement, called Brownian motion. Nonspherical particles (rods, ellipsoids) additionally show a rotational Brownian motion.

The mechanism of translational Brownian motion can be described by a translational diffusion coefficient. For an isolated particle (or droplet) the diffusion coefficient is the ratio of the thermal energy of the particle and the (Stokes–Einstein) friction force acting at the particle which is suspended in a Newtonian continuous phase of the viscosity η_c

$$D_t = \frac{k_B T}{f_t}, \qquad (9.223)$$

where $f_t = 6\pi\eta_c a_H$ is the translational friction coefficient with a_H being the hydrodynamic radius of the particle. For a spherical particle a_H is equal to its radius a (cf. [9.196] for a survey on the hydrodynamic radii for nonspherical particles). It is seen that the diffusion coefficient is inversely proportional to the viscosity of the continuous phase and the particle size. Under the assumption that the observation time t is long enough with respect to the viscous relaxation time τ_v, i.e., $t \gg \tau_v$ with $\tau_v = \rho_c a_H^2/\eta_c$, the mean squared displacement of the particle is [9.194]

$$\langle s^2(t) \rangle = 2 D_t t \,. \tag{9.224}$$

In the case of rotational Brownian motion, the mean displacement and the translational friction coefficient must be replaced by the mean squared change of the orientation angle $\langle \phi^2(t) \rangle$ and the rotational friction coefficient f_r, respectively. Hence, a rotational diffusion coefficient $D_r = k_B T / f_r$ comes into play [9.198, 199]. Details concerning the Brownian motion can be found in the textbooks of *Russel* et al. [9.194], *van de Ven* [9.198], *Takeo* [9.196], and *Doi* and *Edwards* [9.199].

Interparticle Interactions and Stabilization

If particles in a suspension approach each other, various types of interactions occur (Fig. 9.63). These interparticle interactions are determined by the solid volume concentration. Beyond the hydrodynamic interaction, there are several other interactions. An extensive review on surface forces can be found in the monograph by *Israelachvili* [9.200]. For hydrodynamic interactions we refer to the monograph by *van de Ven* [9.198]. In the following sections we consider the influence of the hydrodynamic interaction on the suspension and emulsion viscosity not in detail but implicitly through the solid volume concentration and additionally by the Péclet number indicating the nonlinear flow behavior at higher solid volume concentrations (cf. Subsect. "Dimensional Analysis" in this section).

Hard- and Soft-Sphere Interaction. The simplest case of a suspension is that the particles behave like hard spheres. The relevant influence parameters are then viscous forces, Brownian motion, and the excluded volume of the particles. The interaction potential V_{hs}, which generally characterizes the kind and the range of interparticle forces, has the form

$$V_{hs}(r) = \begin{cases} \infty & r \leq a \\ 0 & r > a \end{cases}, \tag{9.225}$$

in the case of rigid repulsion. Here, r is the radial coordinate starting from the midpoint of the particle. An

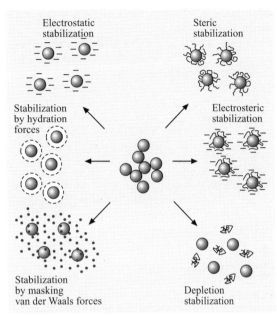

Fig. 9.63 Stabilization methods for dispersions (after [9.197], with permission by Marcel Dekker)

interacting particle causes short- and long-range effects in its neighborhood. Hence, it is (geometrically or virtually) surrounded by a layer (electric double layer, steric molecules). The effective diameter of the particle is then larger than its geometric diameter. If the particles cannot approach each other more than a certain distance, say 2δ or below, due to repulsive forces, the particle radius a must be replaced by an effective particle radius $a_{\mathrm{eff}} = a(1 + \delta/a)$ with δ as the thickness of the surrounding layer. If the surrounding layer is deformable and the particles can approach each other less than the distance 2δ, these interactions are called soft-sphere interactions. In the following, an overview of the interparticle forces is given.

Van der Waals Forces. Van der Waals forces are a combination of dispersion interaction (London), dipole–dipole interaction (Keesom) and dipole-induced dipole interaction (Debye) [9.201]. These forces are caused by a temporal asymmetry of the charge distribution around a neutral atom or molecule due to the motion of its electrons. The interaction energies between two single atoms or molecules decay as the inverse sixth power of the atom distance. The nature of the van der Waals interaction is determined by the material-specific Hamaker constant, which characterizes the relative strength of the van der Waals force between two surfaces. We consider now only two spheres 1 and 2 with given radii a_1 and a_2 (or a ratio of radii $y = a_2/a_1 \leq 1$) and with different properties, surrounded by a continuous phase 0. Hamaker [9.202] obtained for the van der Waals interaction energy V_A the relation [9.203]

$$V_A(x) = -\frac{A_{\text{eff}}}{12}\left(\frac{y}{x^2+xy+x} + \frac{y}{x^2+xy+x+y} + 2\ln\frac{x^2+xy+y}{x^2+xy+x+y}\right) \quad (9.226)$$

with $x = h_0/2a_1$ as the dimensionless surface distance (h_0 is the minimum separation between the surfaces of the spheres). The parameter A_{eff} is the effective Hamaker constant, which is calculated according as

$$A_{\text{eff}} = A_{00} - A_{01} - A_{02} + A_{12}, \quad (9.227)$$

where A_{ij} are the Hamaker constants of the disperse phase ($i = j = 0$), of the sphere 1–sphere 2 interaction ($i = 1, j = 2$), of the continuous phase–sphere 1 interaction ($i = 0, j = 1$), and of the continuous phase–sphere 2 interaction ($i = 0, j = 2$), respectively.

From (9.226) follows that the van der Waals interaction potential depends both on the diameter ratio of the spheres and on the interparticle distance. For $x \ll 1$ the interaction potential is inversely proportional to the interparticle distance. If the components 1 and 2 are identical, the Hamaker constant is positive and, hence, the van der Waals interaction is always attractive. In contrast, when the continuous phase 0 has a Hamaker constant between those of the components 1 and 2, the van der Waals interaction may be repulsive [9.203]. Hamaker constants for some inorganic materials have been calculated by *Bergström* [9.204]. A characteristic value for dimensional-analytical estimates can be given by $V_A = A_{\text{eff}}$, which is of the order 10^{-20}–10^{-19} J for the most materials.

Due to the fact that the van der Waals interaction (in the majority of cases) is attractive, it decreases the stability of suspensions by promoting the formation of aggregates.

Electrostatic Forces. There are three main mechanisms that cause the formation of charged phase interfaces in aqueous dispersions: the transition of ions from the disperse to the continuous phase, the specific adsorption of ions from the continuous phase at the surface of the particles, and the adsorption of polar surfactants at the phase interface.

These mechanisms lead to a (partial) charge separation between the particle surface and the surrounding liquid volume. Along with the thermal motion a diffused electrical double layer is generated which consists of the charged surface, neutralizing counterion, and, farther from the surface, co-ions distributed in a diffusive manner [9.203]. Stern proposed a model that assumes that, near the particle surface, some of the counterions are adsorbed (Stern plane) whereas the remaining counterions are distributed diffusively in a double layer (Gouy plane) due to thermal motion. The dependence of the electrical potential or repulsive interaction energy, respectively, on the distance from the particle surface and the electric double layer are shown in Fig. 9.64. Furthermore, the potential at a certain distance, the shear plane, is denoted ζ. *Russel* et al. [9.194] defined the shear plane as the *envelope where shear appears in the fluid adjacent to a rigid body when fluid and solid are in relative motion*.

In the simplest case the repulsive interaction energy between two identical charged spheres of the same radius a with a surface potential ψ_0 is given by [9.205]

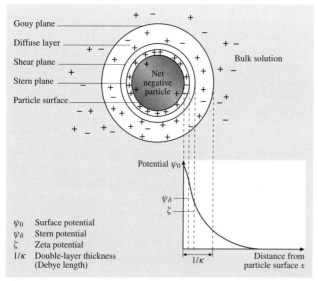

Fig. 9.64 Electrical potential and double layer surrounding a charged particle in a polar liquid (after Birdi [9.203], with permission by CRC)

$$V_e(x) = 2\pi\varepsilon_r\varepsilon_0\psi_0^2 a \begin{cases} \ln\left(1+e^{-\kappa x}\right) & \kappa a > 10, \\ e^{-\kappa x} & \kappa a < 5, \end{cases} \quad (9.228)$$

where x denotes the distance of the particle surfaces, ε_r and ε_0 are the relative permittivity of the material and the permittivity of the vacuum, and κ^{-1} is the Debye length which can be regarded as the thickness of the double layer (Fig. 9.64). It can be shown that for small distances, i.e., within the Stern plane the repulsive interaction energy decreases linearly with increasing particle distance. A characteristic value for dimensional-analytical estimates can be given by $V_e = \varepsilon_r \varepsilon_0 \psi_S^2 a$ with ψ_S^2 as the potential at a characteristic distance from the particle surface, e.g., the zeta potential.

The electrostatic forces have a stabilizing effect on the suspensions by preventing the particles from agglomeration (electrostatic stabilization).

DLVO Theory. *Derjaguin* and *Landau* [9.206] as well as *Verwey* and *Overbeek* [9.207] developed independently the first quantitative theory of interactions in dispersions – the Derjaguin–Landau-Verwey–Overbeek (DLVO) theory. The total interaction force between two particles is supposed to be a superposition of the attractive van der Waals forces and the repulsive electrostatic forces. A typical curve of the total interaction energy versus interparticle distance shown in Fig. 9.65 exhibits a maximum representing an energy barrier against agglomeration and two minima, i.e., a primary and a secondary minimum. The primary minimum stems from strong short-range repulsive forces. If the particles are small, the secondary minimum is not very distinctive. Between the primary and the secondary minimum a local maximum, i.e., a so-called electrostatic barrier may appear. If this barrier is high enough, agglomeration does not take place. For larger particles, the secondary minimum may be deep enough to cause weakly flocculated agglomerates due to the slight dominance of attractive forces. A schematic illustration of the potential energy and the corresponding structure of the suspension are depicted in Fig. 9.66. Dominant attractive forces cause aggregation or flocculation of particles and, hence, a destabilization of the suspension, whereas strong repulsive forces stabilize the suspension.

Steric Forces. An alternative to stabilize suspensions is steric stabilization. In contrast to the electrostatic stabilization, this method can be used in aqueous and nonaqueous systems. The mechanism of steric repulsion can be realized by the adsorption of organic molecules at the particle surface [9.205]. The adsorbed molecules act like a *brush*, where the strength of this polymer bridge depends strongly on the molecular weight (chain length). If the adsorbed layer has a sufficient thickness and density, the attractive forces can be reduced in a manner that aggregation or bridging flocculation is prevented.

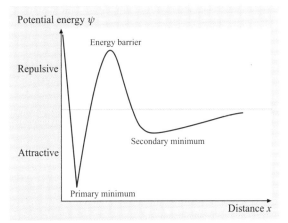

Fig. 9.65 Diagram of the potential energy between two particles in dependence on the surface distance: DLVO theory

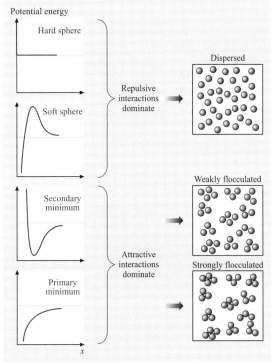

Fig. 9.66 Schematic illustration of the relationship between the total interparticle potential energy and suspension structure (after Lewis [9.205], with permission by the American Ceramic Society)

Table 9.10 Some relevant physical quantities and corresponding dimensionless groups in suspension rheology

Property	Dimensional expression	Dimensionless group
Viscosity	η	$\eta_r = \eta/\eta_c$ (relative viscosity)
Viscosity of the continuous phase	η_c	$\rho_r = \rho_d/\rho_c$ (density ratio)
Density of the disperse phase	ρ_d	$\mathrm{Re}_P = \rho_c \dot{\gamma} a^2 / \eta_c$ (particle Reynolds number)
Density of the continuous phase	ρ_c	ξ (polydispersity measure)
Polydispersity	ξ (dimensionless)	$\varphi = n(4\pi/3)a^3$ (volume concentration)
Number density	n	$\mathrm{Pe} = \eta_c \dot{\gamma} a^3 / k_B T$ (Péclet number)
Particle radius	a	δ/a (relative surface layer thickness)
Gap width of the rheometer	h	a/h (continuum parameter)
Adsorption layer thickness	δ	$t_r = t/(\eta_c a^3 k_B T)$ (relative startup time)
Time	t	$\mathrm{Ga} = \frac{(\rho_d - \rho_c)ga}{\eta_c \dot{\gamma}}$ (settling parameter)
Acceleration of gravity	g	
Thermal energy	$k_B T$	
Dispersion (van der Waals) energy	A_{eff}	$\frac{A_{\mathrm{eff}}}{\varepsilon_r \varepsilon_0 \psi_S^2 a}$
Electrostatic energy	$\varepsilon_r \varepsilon_0 \psi_S^2 a$	

If two sterically stabilized particles with the adsorption layer thickness δ approach each other, two main domains can be distinguished: a domain, where $\delta < x < 2\delta$, i.e., where the two steric layer interpenetrate, and a domain, where $x < \delta$ with an additional compression of the layers. Hence, the interparticle forces depend on both the distance x and the strength of the steric layers.

The long segments of the polymer or surfactant protruding into the continuous phase can adsorb onto the surface of neighboring particles and form a polymer bridge between the particles [9.176].

Electrosteric Forces. Polyelectrolytes as stabilizers for aqueous suspensions combine steric and electrostatic interactions (electrosteric stabilization). They have at least one type of functional groups (e.g., carboxyl or sulfonic acid groups) that can be ionized. This functional group is responsible for the electrostatic interactions. The chain of the polyelectrolyte causes steric interactions. The adsorption of polyelectrolytes is influenced by the electrochemical and physical properties of the particle surface and the continuous phase. It is interesting to note that small amounts of polyelectrolytes can neutralize the surface charge of the particle and, consequently, weak flocculation may occur. At higher amounts, the polyelectrolytes stabilize the suspension due to the long-range repulsive forces caused by the electrosteric forces.

The thickness of the polyelectrolyte layer is mainly influenced by the concentration of the electrolyte in the aqueous continuous phase, i. e., the pH value, and the ion strength (a measure of electrolyte concentration) [9.208].

Depletion Forces. Depletion forces are a result of the interactions between large colloidal particles suspended in a continuous phase that contains non-adsorbing, smaller constituents – so-called depletants (polymers, polyelectrolytes, smaller particles). Due to depletion, stabilization occurs because the particle aggregation is inhibited by unadsorbed polymer in the continuous phase. That is, depletion characterizes a situation where the concentration of depletants increases with increasing distance from the particle surface up to an equilibrium value obtained in the continuous phase. This distance, called the depletion layer thickness, has an order of two depletant diameters. The shape of the depletant molecules influences the depletion force. It is known that rod-like macromolecules, compared to spheres, increase the depletion force. In the case where the depletants are removed from the interparticle gap due to an osmotic pressure difference (exclusion of the free polymer from the interparticle regions) flocculation may occur. With increasing depletant concentration, restabilization is possible.

Dimensional Analysis

By application of dimensional analysis (Chap. 2.2) we can identify the parameters that influence the rheological behavior of the suspensions. It results in a number of dimensionless groups (dimensionless ratios of various influence parameters) which characterizes the ratio of the various influence parameters which allow the estimation of the physical relevance or irrelevance of influence parameters. In Table 9.10 the relevant physical parameters for suspensions are summarized.

From dimensional analysis it can be deduced that the following general relation holds for the dependence of the relative viscosity η_r on the dimensionless influence parameters (Table 9.10):

$$\eta_r = f\left(\varphi, \frac{\delta}{a}, \frac{a}{h}, \xi, \rho_r, t_r, \text{Ga}, \text{Pe}, \text{Re}_P, \frac{A_{\text{eff}}}{\varepsilon_r \varepsilon_0 \psi_S^2 a}\right). \tag{9.229}$$

In the following, we consider buoyant ($\rho_r \to 1$) suspensions with negligibly small particle Reynolds numbers ($\text{Re}_P \ll 1$). For suspensions that are not highly concentrated one can also neglect the influence of the interparticle forces. Hence, the functional relation for the relative viscosity of suspensions within the continuum hypothesis ($a \ll h$) reduces to

$$\eta_r = f\left(\varphi, \frac{\delta}{a}, \xi, t_r, \text{Pe}\right). \tag{9.230}$$

The continuum parameter a/h, used in this context, is similar to the Knudsen number Kn as the ratio of the molecular mean free path and the characteristic body length. Kn is applied in the field of rarefied gas dynamics or flows in microchannels. For convenience we do not change the functional symbol f. For suspensions with monodisperse unvarying particle size, (9.230) takes the form

$$\eta_r = f(\varphi, \text{Pe}). \tag{9.231}$$

in the steady state and for thin surface layers $\delta \ll a$. The Péclet number gives the ratio between the shear force and the Brownian force, or in other words, the ratio between order due to shearing and disorder due to Brownian movement. The presence of particles changes the flow behavior from linear flow behavior (Newtonian) to nonlinear flow behavior (non-Newtonian) due to the hydrodynamic interactions between the particles. Hence, with increasing concentration the nonlinear flow behavior increases, which shows the coupling between φ as a measure for hydrodynamic interaction and Pe representing the nonlinear flow behavior.

Only for highly diluted suspensions does the relative viscosity depend solely on the solid volume concentration yielding a Newtonian behavior if the continuous phase is Newtonian

$$\eta_r = f(\varphi). \tag{9.232}$$

In the literature one can often find another definition of the Péclet number:

$$\text{Pe}_{\text{eff}} = \frac{\sigma a^3}{k_B T} = \frac{\eta \dot\gamma a^3}{k_B T}. \tag{9.233}$$

In contrast to the Péclet number given in Table 9.10, this *effective* Péclet number is based on the macroscopic shear stress $\sigma = \eta \dot\gamma$ or the effective viscosity of the dispersion η, respectively.

Influence of Solid Volume Concentration

The solid volume concentration is defined as the ratio of the volume of the disperse phase V_d to the total volume

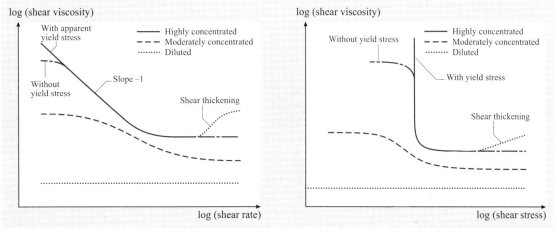

Fig. 9.67 Scheme of the shear viscosity as a function of shear rate and shear stress, respectively. The curve parameter is the solid volume concentration

$V_c + V_d$, where V_c is the volume of the continuous phase:

$$\varphi = \frac{V_d}{V_c + V_d} \ . \quad (9.234)$$

This equation holds, strictly speaking, in the case of nonporous hard spheres. In the case that the particle is surrounded by a layer, the relative layer thickness δ/a can be combined with the solid volume concentration φ to form an effective solid volume concentration

$$\varphi_{\text{eff}} = \varphi \left(1 + \frac{\delta}{a}\right)^3 , \quad (9.235)$$

resulting in a modified form of (9.231):

$$\eta_r = f(\varphi_{\text{eff}}, \text{Pe}) \ . \quad (9.236)$$

Dependent on the solid volume concentration, suspensions can be classified into three main groups: diluted, concentrated, and highly concentrated or solid suspensions [9.173, 209]. Generally, the flow behavior depends on either the shear rate or shear stress, respectively, the solid volume concentration, and the properties of the continuous phase. A sketch of the possible observations is depicted in Fig. 9.67.

Diluted suspensions of identical spherical particles, i.e., suspensions with very low solid volume concentrations, show a Newtonian behavior. In this case the distances between the particles are large enough, so that the Brownian (thermal) motion of the particles predominates over the effect of the interparticle interactions for colloidal suspensions [9.209]. *Einstein* [9.210, 211] was the first who investigated the hydrodynamic forces resulting from the motion of the continuous phase with respect to noninteracting rigid spherical particles. He derived the equation

$$\eta_r = \frac{\eta}{\eta_c} = 1 + k_1\varphi + O(\varphi^2) \quad (9.237)$$

for the viscosity of highly diluted suspensions obtained in pure shear flow, where η_c is the viscosity of the continuous phase and $k_1 = 5/2 = 2.5$ is the Einstein coefficient. The order of approximation is denoted by the symbol $O(.)$. A generalization of the Einstein coefficient is the intrinsic viscosity, which is defined as

$$[\eta] = \lim_{\varphi \to 0} \frac{\eta_r - 1}{\varphi} = \lim_{\varphi \to 0} \frac{\eta_{\text{sp}}}{\varphi} , \quad (9.238)$$

with η_{sp} as the specific viscosity. For other particle shapes the intrinsic viscosity varies (cubes: 3.1; uniaxially oriented fibres parallel to the tensile stress component: $2l/d$ with l and d as fibre length and diameter, respectively) [9.212]. Equation (9.237) shows a linear increase of the suspension viscosity with increasing solid volume concentration, however the behavior remains Newtonian. It is applicable to suspensions of spherical particles with solid volume concentrations up to 10^{-2}.

A further increase of the solid volume concentration, i.e., the transition to concentrated suspensions, causes increasing interactions of the *hydrodynamic fields* between spheres or aggregates. The interaction between two particles was described by *Batchelor* and *Green* [9.213] as well as by *Batchelor* [9.214] to extend Einstein's equation to higher solid volume concentrations:

$$\eta_r = 1 + k_1\varphi + k_2\varphi^2 + O(\varphi^3) \ . \quad (9.239)$$

In this equation, the coefficient k_2 describes the deviation from the very dilute limit of the suspension. The equation holds for solid volume concentrations up to 2×10^{-1}. This type of power series, a pseudovirial approximant, can be summarized in a general equation for the dependence of the relative viscosity on the solid volume concentration with k_i as concentration-independent expansion coefficients:

$$\eta_r = \sum_{i=0}^{N} k_i \varphi^i \ . \quad (9.240)$$

The relative viscosity in (9.240) represents both shear and elongational viscosities. Table 9.11 gives examples for the coefficients of some useful models for suspensions of monomodal spherical particles (hard spheres) based on power series expansion up to the order of 3. For details refer to the book of *Russel* et al.[9.194].

It should be mentioned that the coefficient $k_2 = 7.6$ obtained by *Batchelor* and *Green* [9.213] for the extensional flow of equal-sized spheres, is uniquely determined. However, in the case of a simple shear flow, the coefficient k_2 cannot be uniquely determined due to the occurrence of closed orbits around a reference particle. This difficulty can be overcome by introduction of some additional physical process, such as three-sphere encounters or Brownian motion, or by the assumption of some particular initial state. By allowing a superimposed Brownian motion *Batchelor* [9.214] calculated the value of the coefficient $k_2 = 6.2$. For non-Brownian suspensions *Batchelor* and *Green* [9.213] found, with some assumptions about the particular initial state, $k_2 = 5.2$.

To extend the application range to higher solid volume concentrations up to $\varphi < 0.60$, *Thomas* [9.215] proposed $k_2 = 10.05$ and replaced the third-order term

Table 9.11 Some models describing the dependence of the viscosity on the solid volume concentration for moderately concentrated suspensions

Source	N	k_0	k_1	k_2	k_3	Comment
[9.210, 211, 216]	1	1	2.5	-	-	$\varphi < 0.01$
[9.213]	2	1	2.5	7.6 (elongation)	-	$\varphi < 0.2$
				5.2 (shear)		
[9.214]				6.2 (shear)		
[9.217]	3	1	2.5	4.94	8.78	
[9.218]	3	1	2.5	6.25	15.7	$\varphi < 0.3$
[9.219]						
Low-shear limit	3	1	2.5	4 ± 2	42 ± 10	$\varphi < 0.35$
High-shear limit	3	1	2.5	4 ± 2	25 ± 7	

empirically by an exponential term

$$k_3 \varphi^3 \to 2.73 \times 10^{-3} \, e^{16.6\varphi}, \quad (9.241)$$

contrary to asymptotic power-series expansion. Many other models of this type with other coefficients are available for special suspensions.

For higher solid volume concentrations, where hydrodynamic and surface force interactions as well as many-body interactions become relevant, the power-series expansion fails since the viscosity tends to infinity at solid volume concentrations in the vicinity of the maximum possible packing fraction. Furthermore, the models given above (with the exception of the de Kruif model [9.219]) are applicable only for a Newtonian behavior or in the low shear or elongation range. Hence, another type of models has been developed taking into account the fact that the viscosity reaches a maximum value at concentrations near the maximum packing fraction. The maximum packing fraction is a characteristic scalar value that mainly depends on the particle size distribution, particle shape (distribution), type of the interactions between the disperse and continuous phases and the shear intensity. For idealized regular packing fractions of monodisperse hard spheres the values given in Table 9.12 hold.

Barnes et al. [9.82] summarized the results described in the literature for nonspherical particles. Table 9.13 reveals that the product $[\eta]\varphi_{\max}$ varies only in the range 1.31–3.77, depending on the kind of particles.

Mewis et al. [9.220] found from experimental data a maximum packing fraction of $\varphi_{\max} = 0.96$ for soft spheres, which approaches in case of infinite polydispersity $\varphi_{\max} \to 1$.

Based on experimental data *Kitano* et al. [9.221] presented the following simple equation (9.242) to calculate the maximum packing fraction for short fibres with an average length-to-diameter ($\overline{l/d}$) ratio of 6–27:

$$\varphi_{\max} = 0.54 - 0.0125 \, \overline{l/d} . \quad (9.242)$$

If the particle size distribution becomes broader, higher values of the maximum packing fraction occur because smaller particles can fill the gaps between larger particles. Under specific conditions this effect is often coupled with a decrease of the relative viscosity. Examples for bi- and trimodal suspensions are given by *Barnes* et al. [9.82].

In Fig. 9.68 the dependence of the relative viscosity on the amount of smaller particles in a bimodal suspension of various solid volume concentrations is shown. Similar results are described elsewhere [9.222–225].

Table 9.12 Maximum packing fractions for regular arrangements of monodisperse spherical particles

Type of packing	Maximum packing fraction $\varphi_{\max}$ (theoretically from geometry)
Face-centered cubic (fcc)	$\sqrt{2}\pi/6 = 0.7405$
Body-centered cubic (bcc)	$\sqrt{3}\pi/8 = 0.6802$
Random closed	0.6370
Hexagonal	$\sqrt{3}\pi/9 = 0.6046$
Simple cubic	$\pi/6 = 0.5236$
Diamond	$\sqrt{3}\pi/16 = 0.3401$

Table 9.13 Intrinsic viscosities and maximum packing fractions for suspensions with asymmetric particles (in extracts from [9.82] with permission from Elsevier)

System	$[\eta]$	φ_{max}	$[\eta]\varphi_{max}$
Glass fibres ($l/d = 21$)	6.00	0.233	1.398
Glass fibres ($l/d = 14$)	5.03	0.260	1.308
Glass rods (30 × 700 μm)	9.25	0.268	2.479
Laterite	9.00	0.350	3.150
Quartz grains (53–76 μm)	5.80	0.371	2.152
Glass fibres ($l/d = 7$)	3.80	0.374	1.421
Glass plates (100 × 400 μm)	9.87	0.382	3.770
Titanium dioxide	5.00	0.550	2.750
Ground gypsum	3.25	0.690	2.243

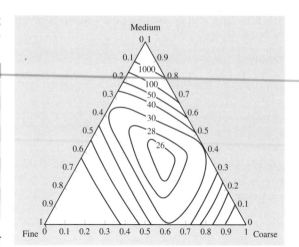

Fig. 9.69 Influence of the particle-size distribution on the relative viscosity of a trimodal suspension of spheres. Contours show values of the relative viscosity at 65% total solids (after Barnes et al. [9.82] with permission by Elsevier)

From Fig. 9.68 one can recognize that an addition of smaller particles under perpetuation of the solid volume concentration reduces the relative viscosity, especially in the range of $\varphi = 0.6$ (amount of large particles) where the relative viscosity has a minimum. This phenomenon is known as the Farris effect. The path P → Q indicates the 50-fold reduction in the relative viscosity of a suspension with $\varphi = 0.6$ if the monodispersity is changed to a bidispersity (50%/50%). The path P → S illustrates

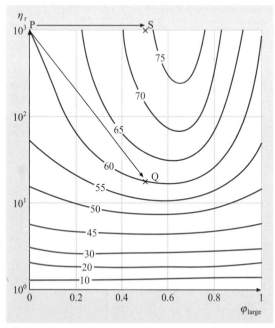

Fig. 9.68 Effect of binary particle-size fraction on the relative viscosity with total percentage solid volume concentration as parameter (particle-size ratio of 1:5) (after Barnes et al. [9.82] with permission by Elsevier)

that a 15% increase of the solid volume concentration by the addition of smaller particles without a change of the relative viscosity is possible.

In Fig. 9.69 a triangular plot of the relative viscosity of a trimodal suspension with a total solid volume concentration of 60% is shown. A relative viscosity minimum of 25 is reached at a given optimum ratio of the solid volume concentration of the components whereas the binary mixture shows a relative viscosity of 30 [9.82].

These effects give rise to the possibility to influence the viscosity of a suspension by an optimum mixing of various fractions of the disperse phase.

In the case of nonspherical particles (ellipsoids, cubes, rods, fibres) the maximum packing fraction decreases. At the same solid volume concentration, the viscosity increases compared to that of a hard-sphere suspension.

The maximum packing fraction is often obtained by fitting of equations of the type shown in Table 9.14 [e.g., (9.244)]. These models describe the dependence of the reduced viscosity on the solid volume fraction. The applicability is not restricted to suspensions with spherical particles because the parameter φ_{max} or k can be regarded as a fitting parameter.

Equations (9.243–9.245) can be formally expanded in a power series according to (9.240), where the coefficients now depend on the solid volume concentration. For small solid volume concentrations the Einstein limit (see eq. (9.237)) is reached.

Table 9.14 Viscosity–concentration models for higher solid volume concentrations

[9.228]	$\ln \eta_r = 2.5\varphi$	9.243
[9.229]	$\eta_r = f(\varphi)\left(1 - \frac{\varphi}{\varphi_{\max}}\right)^{-p(\varphi)}$	9.244
[$p = 2$, $f(\varphi) = 1$]		
[9.230]		
($p = [\eta]\varphi_{\max}$, $f(\varphi) = 1$)		
[9.231]	$\eta_r = \exp\left(\frac{2.5\varphi}{1-k\varphi}\right)$	9.245
$k = 1.35$ (face centered cubic)		
$k = 1.91$ (simple cubic)		
[9.232, 233]	$d\eta_r = \eta_r[\eta](1-k\varphi)^{-\alpha}\,d\varphi$	9.246
generalized differential form with $k = 1/\varphi_{\max}$, $[\eta] = 2.5$		
$\alpha = 0$: Arrhenius		
$\alpha = 1$: Krieger/Dougherty		
$\alpha = 2$: Mooney		

Extensive review articles on viscosity models for suspensions (including models for suspensions of non-spherical particles) are given by *Rutgers* [9.226] and *Jinescu* [9.227].

Under shear the arrangement of the particles in the fluid is modified, so that the maximum packing fraction depends on the shear rate or shear stress applied (Fig. 9.70). Thus, the (shear-dependent) maximum packing fraction has been introduced in the models (e.g., [9.219, 234, 235]).

Wildemuth and *Williams* [9.234] developed a simple model to describe the shear-dependent maximum packing fraction. They found the relation

$$\frac{1}{\varphi_{\max}(\sigma)} = \frac{1}{\varphi_{\max,0}} - \left(\frac{1}{\varphi_{\max,0}} - \frac{1}{\varphi_{\max,\infty}}\right)S(\sigma),$$

(9.247)

which describes the dependence of the maximum packing fraction on the shear stress. The values $\varphi_{\max,0}$ and $\varphi_{\max,\infty}$ are the maximum packing fractions in the low- and high-shear-rate limit. In analogy to some kinetic models to describe the time-dependent viscous (thixotropic) behavior of suspensions using parameter $\lambda(t)$ (Sect. 9.2), the function $S(\sigma)$ can be interpreted as a scalar structural parameter with values between 0 and 1 depending on the given shear stress σ. For the structural parameter the relation

$$S(\sigma) = \frac{1}{1 + K\sigma^{-m}}$$

(9.248)

has been proposed. The parameters K and m must be evaluated by fitting the experimental data. With (9.244), (9.247), and (9.248) the non-Newtonian behavior of concentrated suspensions can be modeled. A similar relation has been proposed by *Zhou* et al. [9.235].

Moderately concentrated suspensions often possess a monotonously decreasing viscosity with plateaus in the low (η_0) and high (η_∞) shear rate limit and a power-law region in between. In Table 9.15 some useful models to describe the time-independent viscous behavior of suspensions without yield stress in steady shear flows with a minimum of parameters are summarized.

If the solid volume concentration approaches the maximum packing fraction, a transition from liquid- to solid-like behavior occurs. The interactions between the particles become very intensive and a true or apparent yield stress may occur. This results in a viscosity that tends to infinity at very low shear rates or shear stresses (Fig. 9.67). The main reason for this behavior is that the free movement of particles is hindered by particles

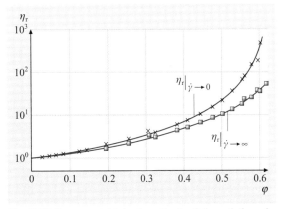

Fig. 9.70 Low- and high-shear limit of the relative viscosity in dependence on the solid volume concentration for monodisperse lattices (after *Barnes* et al. [9.82] with permission by Elsevier)

Table 9.15 Viscosity models (time independent) for suspensions without yield stress

[9.236]	$\dfrac{\eta(\varphi)-\eta_\infty(\varphi)}{\eta_0(\varphi)-\eta_\infty(\varphi)} = \dfrac{1}{1+(\sigma/\sigma_{cr}(\varphi))^{m(\varphi)}}$	9.249
	$1 \leq m(\varphi) \leq 2$	
	σ_{cr} critical shear stress	
[9.129]	$\dfrac{\eta(\varphi)-\eta_\infty(\varphi)}{\eta_0(\varphi)-\eta_\infty(\varphi)} = \dfrac{1}{1+k(\varphi)\dot\gamma^{n(\varphi)}}$	9.250
[9.237]	$\dfrac{\eta(\varphi)-\eta_\infty(\varphi)}{\eta_0(\varphi)-\eta_\infty(\varphi)} = \left(1+(\tau(\varphi)\dot\gamma)^{m(\varphi)}\right)^{\frac{n(\varphi)-1}{m(\varphi)}}$	9.251
special case: Carreau ($m=2$)	(cf. with (9.207) in Sect. 9.2)	
	$\tau(\varphi)$ characteristic material time	
	$n(\varphi)-1$ slope in the power-law region	
	$m(\varphi)$ affects the shape of the transition	
	between zero-shear viscosity and power-law region	

in their direct neighborhood. From a physical point of view, the yield stress is not a single scalar value, but it should be considered a region that characterizes the solid–liquid transition. It is influenced by the solid volume concentration and the interparticle interactions, but also by the intensity of the mechanical load and the type of the flow. The most appropriate method to measure the yield stress is to apply a shear stress ramp. Starting from a very low value, the shear stress is slowly increased and the shear strain or shear rate, respectively, is measured. To resolve the low-shear-stress range accurately, the shear stress should be given as a logarithmic ramp. Assuming a solid-like behavior below the yield stress, a sharp increase of the slope of the $\gamma(\sigma)$ curve occurs if the given shear stress approaches the yield stress. The intersection point of the asymptotes of the solid-like range and the initial liquid-like ranges characterizes the yield stress. It is important to note that the steepness of the shear stress ramp influences the solid–liquid transition. Hence, preliminary investigations are necessary to determine the critical shear stress ramp where the yield stress is independent of the ramp steepness.

Some useful models have been presented to describe the flow behavior of the so-called viscoplastic suspensions (Bingham, Herschel-Bulkley, Casson (see [9.188]). However, it should be noted that the yield stress obtained by fitting the model parameter to the experimental data is neither a unique nor a single material property. It depends on the shear rate range and even on the model used. Detailed information on the measurement and estimation of the yield stress can be found in the review articles by *Nguyen* and *Boger* [9.238] and *Barnes* [9.239].

At higher shear rates, highly concentrated suspensions with solid volume concentration of 50% and higher may show shear thickening, i.e., the viscosity increases with increasing shear rate or shear stress (Fig. 9.71). This effect is caused by a more or less sudden change in the microstructure of the suspension which could be investigated by means of scattering methods (e.g., [9.240–242]). As *Barnes* [9.243] pointed out, this shear-thickening region ranges over more than one decade of shear rate. Most of the parameters shown in Table 9.9 may control the shear-thickening behavior. Usually, the shear-thickening effect is reversible since the viscosity decreases again if the shear rate or stress decreases. Schematically, the phenomenon of shear-thickening is shown in Fig. 9.71. At higher solid volume concentrations a shear-thinning region can be observed which passes into a region with increasing viscosity at a critical shear rate $\dot\gamma_{cr}$. The critical shear rate increases and the onset of shear thickening is less severe with a broadening of the particle size distribution. Furthermore, a local viscosity maximum is reached at

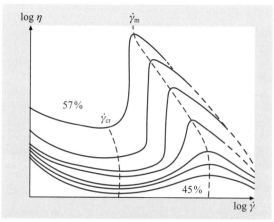

Fig. 9.71 Schematic representation of the viscosity function for shear-thickening suspensions as a function of the solid volume concentration (see text for explanation, after *Barnes* [9.82] with permission by the Society of Rheology)

a shear rate $\dot{\gamma}_m$. Both characteristic shear rates shift to lower values with increasing solid volume concentrations. It is interesting to note that the critical shear rate tends to zero if the maximum packing fraction is reached.

It should be mentioned that a correlation between the particle size and the critical shear rate exists. Based on the data in many publications, *Barnes* [9.243] gave an empirical evidence of an inverse quadratic dependence

$$\dot{\gamma}_{cr}(a) = Ka^{-2} \qquad (9.252)$$

with a value of $K \approx 4\,\mu\text{m}^2/\text{s}$ obtained from a rough fit to the available data.

Furthermore, the critical shear rate depends on the solvability of the continuous phase with regard to a stabilizer (nonaqueous suspensions) or pH value (aqueous suspensions). *Frith* et al. [9.244] indicated for sterically stabilized nonaqueous suspensions that the onset of shear thickening scales with the viscosity of the continuous phase due to the hydrodynamic effects. The critical shear rate was found to decrease with decreasing stabilizer layer thickness, i.e., if the continuous phase is a better solvent for the stabilizer. *Franks* et al. [9.245] demonstrated that the critical shear rate increases if the pH value is adjusted farther from the isoelectric point (IEP). At the (material-specific) IEP, the ζ potential is zero. This effect has been explained by an increase of the repulsive forces between the particles, which delay the onset of shear thickening. If the pH value deviates from the IEP, the addition of salt decreases the repulsive forces and, hence, decreases the critical shear rate. From these observations it can be concluded that shear thickening is not only a hydrodynamic effect but depends on the surface forces as well.

The effect of shear thickening does not only occur in steady shear experiments. *Raghavan* and *Khan* [9.246] observed so-called strain thickening in oscillatory shear flows. Strain thickening in this case means that a critical combination of a shear strain and an angular frequency exists where the complex viscosity sharply increases. Two different cases are possible: high critical strains at low frequencies and high critical frequencies at low strains. In the first case, a correlation with the results obtained in steady shear flow is possible using a modified Cox–Merz rule (the Rutgers–Delaware rule, *Doraiswamy* et al. [9.247])

$$\left|\eta^*(\hat{\gamma}\omega)\right| = \eta(\dot{\gamma})\big|_{\omega=\dot{\gamma}} \,. \qquad (9.253)$$

Compared to the original empiric Cox–Merz rule ((9.102) in Sect. 9.1.1), the angular frequency ω has been scaled with the shear strain amplitude $\hat{\gamma}$, which now describes the maximum dynamic shear rate in oscillatory shear experiments. *Gleissle* and *Hochstein* [9.248] extended the range of applicability of the Cox–Merz rule using the concept of the shear stress equivalent inner shear rate. The modified Cox–Merz rule has been applied to the high-shear region where hydrodynamic interactions dominate.

A simple rheological model in a closed form, which describes the complete viscous behavior in simple shear in a broad range of shear rates including the shear thinning and the shear thickening region, is not yet available. Here, the regions should be described separately by the models given in Table 9.15.

Rheological Measurements
Before performing the rheological measurement the following points should be clarified: a suitable measurement geometry, and the dosage and rheological preconditioning of the sample.

In Sect. 9.1.1 various measurement geometries have been described. In the following only the properties of the measurement geometries relevant to the investigation of dispersions will be explained. Details can be found in the monograph by *Walters* [9.189] or the contributions by *Powell* [9.249] on rotational rheometry, and *Mackley* and *Rutgers* [9.250] on capillary rheometry.

Choice of the Measurement Geometry – Settling, Particle Migration, and Wall Slip. The choice of the measurement geometry depends on the expected rheological properties of the suspension. Because the rheological properties are a priori unknown, preliminary measurements are absolutely necessary. However, it is possible to choose a suitable geometry based on the following information:

1. the shear rate or shear stress range of interest – stability of flow in the range of interest
2. the existence of a possible yield stress, as well as the order of magnitude of the viscosity and/or normal stress differences
3. the possibility, to prevent wall slip or, if this is not possible, to detect and correct for wall slip

The stability of flow is an important criterion for the choice of the geometry. The cone-and-plate (CP) system (cf. Fig. 9.32) is suitable for low and medium shear rates. Especially at higher concentration, the flow in CP systems becomes unstable, e.g., due to shear fracture. Starting from the free surface at the edge, the sample contracts with increasing time and shear rate, which results in a decreasing shear plane. In controlled-shear-rate (CSR) experiments, the slope of the flow curve is reduced dramatically and, hence, the viscosity decreases apparently. This results in strong shear-thinning behavior starting at a critical shear rate. In controlled-shear-stress (CSS) experiments, shear fracture is manifested as a sharp increase of the measured shear rate. Independent of the problems of sample stability, the CP system with small cone angles ($\leq 4°$) is the most preferable geometry for the investigation of suspensions because the shear rate and the shear stress are approximately constant over the whole sample [cf. (9.139) and (9.143) in Sect. 9.1.2]. Especially when investigating the yield stress in suspensions, the CP geometry is the most suitable system. Furthermore, the CP geometry allows the determination of the first normal stress difference from the axial force [cf. (9.148) in Sect. 9.1.2] which can easily be measured by modern rotational rheometers.

Similar sample instabilities can be observed in torsional plate–plate (PP) geometries (cf. Fig. 9.31). Although the main advantage lies in the simple adjustment of the shear rate by varying the gap height, the PP system is unsuitable for the rheological measurements of non-Newtonian suspensions due to the strong inhomogeneous shear field in the gap [cf. (9.136) in Sect. 9.1.2]. However, the torsional PP geometry is a necessary tool for measurements of the difference between the first and second normal stress difference [cf. (9.138) in Sect. 9.1.2], i.e., for the estimation of the second normal stress difference if the first normal stress difference is already known.

Because of the small gap height and the direction of gravity perpendicular to the gap, sedimentation or demixing effects influence the experimental results obtained in CP and PP geometries to a major degree.

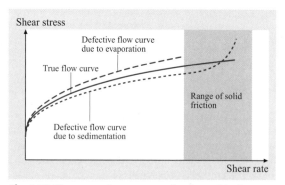

Fig. 9.72 Erroneous flow curves of a shear-thinning suspension in a coaxial cylinder system due to evaporation, sedimentation, or friction between bob and sediment

Here, demixing effects manifest themselves as an untypical time-dependent decrease of the viscosity, which in the case of a Newtonian continuous phase, tends towards the viscosity of the continuous phase. This effect could be misinterpreted as shear thinning or thixotropy (Sect. 9.2).

It is possible to estimate the influence of settling on the rheological measurements. Based on the balance between the Stokes force (cf. Sect. 3.4.5), the gravitational force, and the buoyant force an equation can be derived to estimate the time required for a single sphere to migrate over a length l

$$t_{\text{exp}} = \frac{9}{2} \frac{\eta_c l}{\Delta \rho g a^2} \qquad (9.254)$$

with $\Delta \rho = |\rho_d - \rho_c|$ and t_{exp} as experimental time (cf. Sect. 3.4.5: falling-sphere viscometer). For spheres of 10 μm diameter in water and a migration length of 1/10 of a typical gap h of 1 mm (e.g., in a PP geometry), the particle density must be within 0.2% of that of water to prevent settling during typical measurement times of 1000 s [9.188]. As seen from (9.254), the density difference is inversely proportional to the square of the particle size under unchanged conditions, i.e., if smaller particles are used, the possible density difference between particles and liquid increases quadratically. The settling time increases if the solid volume concentration is high enough that hindered settling occurs or if a superposed shear flow during measurements is realized. Generally, the relation

$$\frac{2}{9} \frac{t_{\text{exp}} \Delta \rho g a^2}{\eta_c h} \ll 1 \qquad (9.255)$$

must be fulfilled to neglect the influence of settling.

The problems discussed with CP and PP geometries can be at least partially avoided if coaxial cylinders (cf. Fig. 9.29) are used since the gravity acts parallel to the cylinder gap. Due to the length of the gap between the two cylinders, the influence of sedimentation is not as dramatic as has been described for the CP and PP geometries. However, sedimentation leads to depletion of particles in the shear gap and hence to an initial decrease of the viscosity. If the solid volume concentration is high enough, a compact layer of solid is generated at the bottom of the outer cylinder. If this layer reaches the lower edge of the inner cylinder, an additional torque occurs due to solid friction between the inner cylinder and the sediment. This observation, schematically shown in Fig. 9.72, can be misinterpreted as shear thickening or rheopexy (Sect. 9.2). Further problems may appear due to Taylor vortices obtained in coaxial cylinder systems where the inner cylinder rotates (Searle systems) and/or edge effects. An estimation of the critical shear rates or shear stresses is possible with common methods.

Another general problem is partial evaporation of the continuous phase during the measurements. Evaporation, i.e., the loss of continuous phase, leads to an increase of the solid volume concentration, which causes an increase of the relative viscosity due to the strong influence of the solid volume concentration. This effect can be minimized by carrying out the measurements in a saturated atmosphere of the continuous phase. If this is not possible, a small amount of an immiscible nonevaporating liquid can be applied to the air–sample surface to prevent evaporation.

One of the most well-known problems arising in experimental investigations of suspensions is wall slip, i.e., the violation of the no-slip condition at the walls of the measurement geometry. There are two different types of wall slip.

The first type, true wall slip, has been found with unfilled polymers [9.252]. At higher shear stresses, a relative velocity between the fluid at the wall and the wall velocity itself sets in. This effect is generally associated with the flow instabilities, e.g., during the extrusion process (stick–slip phenomenon). However, the true wall slip is only of relevance when dealing with polymers.

With moderately and highly concentrated suspensions, the second type, an apparent wall slip effect, can occur which is caused by particle migration. An initially homogeneously distributed suspension, which is subjected to non-homogeneous shear field, can demix during the experiment. This demixing effect is caused by additional forces acting at the particles, even in the case where neither inertial nor interparticle forces are relevant. The physical depletion of particles near the solid walls occurring even without flow results from a distortion of the local microstructure. Furthermore, the local isotropy caused by the Brownian motion (of colloidal particles) may be destroyed. The flow-induced forces cause a movement of the particles from regions of higher (near the walls) to regions of lower shear rates or vice versa. This effect, called particle migration, results in a solid volume concentration distribution over the sample. *Ho* and *Leal* [9.251] analyzed the problem of particle migration both in drag (Couette, see Fig. 9.27) and pressure-driven (plane Poiseuille, see Fig. 9.41) slit flow theoretically. The solid volume concentration distributions are shown in Fig. 9.73. Ho and Leal found a criterion for neglecting migration effects that compares the inertial effects and Brownian motion

$$K = \frac{\rho_c \bar{u}^2 a^4}{h k_B T} \begin{cases} < 0.1 & \text{Couette flow ,} \\ < 0.01 & \text{plane Poiseuille flow ,} \end{cases}$$

(9.256)

where $\bar{u}$ is the mean velocity in the gap (Couette flow: $\bar{u} = \frac{1}{2} u_{\max}$, plane Poiseuille flow: $\bar{u} = \frac{2}{3} u_{\max}$) and h is the gap width. Figure 9.73 makes clear that, in the Couette device where the outer cylinder rotates, the particles (spheres) move towards the centerline. In contrast, in plane Poiseuille flow a maximum solid volume concentration occurs at the distances of about $0.2h$ from the walls. The intensity of migration depends, among other things, on the mean velocity (or shear rate), particle size and the thermal energy of the particle. The danger of migration strongly increases with the particle size, since the influence of the thermal forces decreases with increasing particle size.

An extensive numerical study of particle migration in various geometries has been carried out by *Graham* et al. [9.253]. They used a modified shear-induced migration model based on the model of *Leighton* and *Acrivos* [9.254]. By using the momentum equation, mass conservation equation, and a kinetic equation for the change of the solid volume concentration, they calculated the steady-state spatial concentration profiles and the resulting integral values of pressure drop or torque.

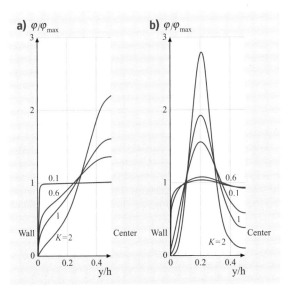

Fig. 9.73a,b Concentration distribution over the shear gap: (a) Couette flow, (b) plane Poiseuille flow. K is given by (9.256) (after *Ho* and *Leal* [9.251] with permission by Cambridge University Press)

For geometries with inhomogeneous shear fields, they found particle concentrations varying from low values at higher shear rates to higher values at lower shear rates. The surprising result of the numerical study is that, even in the case of a CP system with a practically homogeneous shear field, radial and azimuthal concentration profiles establish. As a result, the driving torque increases significantly due to the outward migration of the particles, which can be misinterpreted as a dilatancy effect. It has been shown that the particle migration results from the curvature of the CP geometry.

Allende and *Kalyon* [9.256] studied pressure-driven flows of suspensions of neutrally buoyant, noncolloidal and unimodal spheres dispersed in a Newtonian continuous phase. On the basis of the model of *Phillips* et al. [9.257] they found that the shear-induced particle migration is typically negligible for length-to-diameter (or gap) ratios of 0–50 provided that the ratios of particle-to-tube (or channel gap) radii are 5×10^{-3} and smaller. Under these conditions, the wall concentration is within 2% of the initial concentration of the suspension in the solid volume concentration range $0.1 \leq \varphi \leq 0.5$.

Experimentally, the effect of particle migration has been shown by the use of noninvasive methods. *Hartman Kok* et al. [9.258] used the total-reflection Fourier-transform infrared spectroscopy to determine the thickness of the wall slip layer in dependence of the shear rate in a PP geometry. From rheological measurements the thickness of the slip layer δ_S could be estimated [9.259] by

$$\delta_S = \frac{u_S \eta_s}{\sigma} \tag{9.257}$$

with u_S as the slip velocity (obtained from a plot of the apparent shear rate at the rim of the PP geometry as a function of the inverse gap height). They found a reasonable agreement between both methods where the thickness of the slip layer was of the order of 1 μm and nearly independent of the Péclet number (or dimensionless shear rate).

Other noninvasive methods such as nuclear magnetic resonance (NMR, *Abbott* et al. [9.260] for PP geometry, *Han* et al. [9.261] for tube flow) or laser doppler anemometry (LDA, *Jana* et al. [9.262]; Chap. 5.3.1) have been used to investigate the distribution of the solid volume concentration in various measurement geometries. *Abbott* et al. [9.260] found a particle-depleted zone near the inner cylinder of a wide gap coaxial cylinder (CC) geometry where the particle concentration at the outer wall reached the maximum packing fraction. This may lead to unsheared regions in the cylinder gap and hence to erroneous measurements. The reason for this observation is again the inhomogeneity of the shear field. The effect has been found to be irreversible and independent of the shear rate and the viscosity of the continuous phase. An essential conclusion is that wider cylinder gaps amplify the occurrence of particle migration.

To avoid wall slip caused by particle migration, it is necessary to disrupt the slip layer. This can be achieved by serrated or at least roughened measurement systems. The roughness must be larger than the slip layer thickness, so that the peaks of the surface disrupt the slip layer. On the other hand, to avoid secondary flows in the valleys of the rough surface, the roughness should not be too large.

The influence of the surface roughness on the rheological measurements of concentrated suspensions has been investigated extensively and systematically by *Aral* and *Kalyon* [9.255] for various PP geometries. The suspension used was a poly(butadiene-acrylonitrile-acrylic acid) terpolymer (PBAN) filled with glass beads of 85.4 ± 35.3 μm diameter. Various materials with surfaces of different asperities have been investigated. The authors found that wall slip generally occurs with stainless-steel plates, but could be avoided with plates made of aluminium oxide. A significant influence of the surface roughness could be found for plates of the same material. In Fig. 9.75, the influence of the surface topology on the transient shear stress obtained

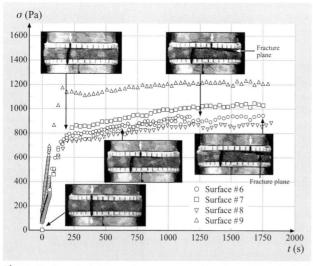

Fig. 9.74 Sample fracture: shear stress as a function of time, including photos of the sample under shear, at an apparent shear rate of 2×10^{-3} s^{-1} (after *Aral* and *Kalyon* [9.255] with permission by the Society of Rheology)

in a start-up flow with a given shear rate is shown. The number of the surface #6 to #9 corresponds to the ratio of mean surface roughness to the diameter of the plates (0.005, 0.006, 0.008, and 0.02). Obviously, the corresponding shear stress values increase with increasing surface roughness. Furthermore, some photographs are shown that depict the mechanism of wall slip and shear fracture. At short times or low shear strains ($\gamma = \dot{\gamma}t$), respectively, a homogeneous deformation of the sample occurs (see the marker lines in Fig. 9.75), whereas at higher strains an increasing shear fracture occurs. Shear fracture is caused by internal slip within the sample. This leads to a discontinuity in the velocity field occurring between two neighboring sample layers. It is interesting to note that, in most cases described by *Aral* and *Kalyon* [9.255], the absence of wall slip was coupled with the appearance of shear fracture, i.e., the formation of a fracture plane.

Both effects, i.e., wall and internal slip have also been found by *Persello* et al. [9.263] for colloidal silica dispersions using a marker technology. A result of their experiments is that the plane of shear fracture, which occurred at larger shear rates, typically moved from the upper plate at the start of the experiment to the middle of the gap if wall slip had been suppressed. The internal slip layers occurred due to a partially demixing of the dispersion, i.e., the formation of a layer of the matrix liquid acting as lubricant between the solid layers. Homogeneous deformation without wall slip and shear fracture has been found at low shear stresses. In Fig. 9.74 the effects of homogeneous sample deformation, wall slip, and shear fracture are shown schematically.

It is not surprising that flow instabilities such as wall slip can lead to unusable results. *Walls* et al. [9.264] investigated the influence of various surfaces of the test geometry on the rheological properties of silica gels. They found distinct differences of the viscosity and storage modulus curves obtained with smooth and serrated plates as shown in Fig. 9.76. From their results it can be concluded that wall slip occurs at a critical shear stress or critical shear stress amplitude. At higher stresses, it seems that wall slip disappears and both curves coincide.

Generally, it is possible to correct the effect of wall slip if a Couette device with two different inner (radii $R_{i,1}$ and $R_{i,2}$) cylinders and outer cylinders ($R_{o,1}$ and $R_{o,2}$) is used. The radii ratio $\beta = R_{o,1}/R_{i,1} = R_{o,2}/R_{i,2}$ is the same for the two geometries, whereas the gap width varies. The wall slip velocity can be calculated by using the formula of *Yoshimura* and *Prud'homme* [9.265] for the Couette device with $\beta - 1 \ll 1$ (small gaps)

$$u_S(\sigma_i) = \frac{\beta}{\beta+1} \left(\frac{\Omega_1 - \Omega_2}{R_{i,1}^{-1} - R_{i,2}^{-1}} \right), \quad (9.258)$$

where σ_i is the same shear stress at the inner cylinder for the two independent measurements with angular speeds Ω_1 and Ω_2 and the two cylinder systems 1 and 2. The true shear rate at the wall can then be obtained from

$$\dot{\gamma}_{i,a} = \dot{\gamma}_i(\sigma_i) + \frac{2u_S(\sigma_i)}{R_i}. \quad (9.259)$$

A plot of the apparent shear rate versus the inverse inner radius for a given shear stress results in a straight line. The true shear rate can be obtained from the intersection of the straight line with the ordinate. The slope of the straight line corresponds to the twofold increase of the slip velocity.

For torsional plate–plate geometries, *Yoshimura* and *Prud'homme* [9.265] proposed

$$u_S(\sigma_R) = \frac{\dot{\gamma}_{aR,1}(\sigma_R) - \dot{\gamma}_{aR,2}(\sigma_R)}{2\left(h_1^{-1} - h_2^{-1}\right)} \quad (9.260)$$

to calculate the slip velocity from two independent experiments carried out with two different gap heights h_1 and h_2 and apparent shear rates $\dot{\gamma}_{aR,1}$ and $\dot{\gamma}_{aR,2}$ at the outer edge of the plates for the same shear stress σ_R at the outer edge R. The true shear rate at the walls can then be calculated according to

$$\dot{\gamma}_{aR} = \dot{\gamma}_R(\sigma_R) + \frac{2u_S(\sigma_R)}{h}. \quad (9.261)$$

The same procedure can also be used for tube flows.

For further details on wall slip and its prevention in rheometers we refer to the review article by

Homogeneous deformation   Slip at the wall   Localization

Fig. 9.75 Homogeneous deformation, wall slip, and shear fracture (in modified form after *Persello* et al. [9.263] with permission by the society of Rheology)

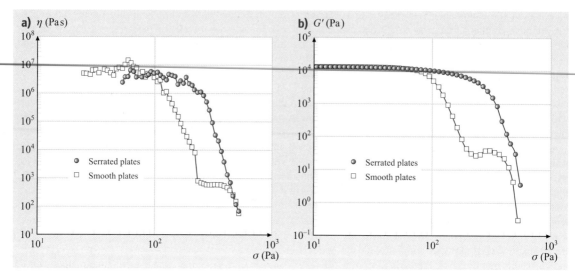

Fig. 9.76a,b Influence of the surface topology (**a**) on the viscosity function and (**b**) the storage modulus (cf. (9.5) in Sect. 9.1.1, adapted from *Walls* et al. [9.264] with permission from the Society of Rheology)

Barnes [9.266] who formulated some conditions which usually lead to significant wall slip effects, such as:

- large particles in the disperse phase (including flocs or aggregates)
- a strong dependence of viscosity on the solid volume concentration of the disperse phase
- smooth walls (sandblasted walls or profiled cones, plates or cylinders can prevent wall slip, but can lead to shear fracture)
- small flow dimensions
- usually low speeds/flow rates
- walls and particles with electrostatic charges while the continuous phase is electrically conductive.

Dosage and Rheological Preconditioning of the Sample. Preconditioning of the rheological sample is an important factor to get correct (and especially reproducible) results. If rotational rheometers are used, the suspension structure at rest can be destroyed by the filling process. Moreover, the lowering of the upper part of the measurement geometry leads to a squeeze flow of the sample. This flow process causes an intensive shear and elongation of the sample and hence a modification and an induced direction of the inner structure. Furthermore, normal or shear stresses that cannot relax even at long times occur, if the samples are highly concentrated and show a yield stress. To avoid such problems, a simple method has been developed by *Heymann* et al. [9.267] to dose and preform a pasty suspension reproducibly. They used a template with a circular opening in the middle. The diameter of the opening corresponded approximately to the diameter of the cone or plate, the thickness was nearly 90% of the maximum gap height in CP systems. The sample is put into the circular opening by a process similar to screen printing with a doctor blade. After forming the circular sample disk, the template was removed and the geometry was closed. The main advantage of this handling is a preformed sample of the thickness of the order of the gap height.

After filling, it is indispensable to use a pre-shear phase to ensure reproducible results. The influence of pre-shear on the results of rheological measurements has been studied intensively. *Heymann* et al. [9.267] analyzed the influence of steady and oscillatory pre-shear on the subsequent oscillatory experiments with suspensions of polymethylmethacrylate (PMMA) spheres dispersed in a Newtonian low-molecular-weight polydimethylsiloxane (PDMS). They found distinct differences between the storage and loss moduli obtained in shear stress controlled (CSS) amplitude sweeps after steady and oscillatory pre-shear. At low shear-stress amplitudes, the moduli were higher after oscillatory pre-shear. At higher shear-stress amplitudes, where the hydrodynamic interactions dominate, no significant differences between steady and oscillatory pre-shear could be obtained. Furthermore, the normal force, induced by the closing of the test geometry, could be reduced and the reproducibility could be enhanced by oscillatory pre-shear. A problem arises if the suspension shows yield behavior. In both cases (stationary and oscillatory pre-shear), the shear stress induced by pre-shear cannot relax completely and the subsequent measurements do not start with a stress-free initial state of the suspensions. In particular, the oscillatory option is problematic because the oscillation cannot be stopped in a well-defined way. This problem can be avoided by a suitable choice of an event control algorithm of the rheometer to set the shear stress to zero.

A dramatic influence of the pre-experimental sample preparation has also been demonstrated by *Carreau* et al. [9.269]. They analyzed the influence of the homogenization procedure prior to the rheological experiments on low-concentrated aqueous colloidal suspensions of fumed silica particles with a mean diameter of 12 nm. After manual mixing, the viscosity showed slightly shear-thinning behavior. In contrast to this observation, the influence of ultrasonic dispersion was dramatic because the viscosity was strongly shear thinning at a clearly higher level. This has been explained by the interaction between water and the silica particles based on hydrogen bridges. At rest, large aggregates are formed, which can be destroyed by shear only slightly. After the ultrasonic dispersion, the suspension consist of single particles that cause a higher viscosity level and can be structured by shear. Furthermore, they observed in controlled-shear-stress (CSS) experiments with suspensions of fumed silica particles in polypropylene glycol that the low-shear viscosity, the onset, and the importance of shear thickening decrease with increasing pre-shear stresses. This effect has been explained by a break-up of aggregates due to pre-shear.

Schmidt and *Münstedt* [9.270] investigated the rheological behavior of concentrated monodisperse suspensions. They detected a significant influence of the pre-shear time at a given shear stress on both the low-shear viscosity and the dynamic moduli. The low-shear viscosity was found to increase with increasing pre-shear time, caused by structuring effects. In contrast to the loss modulus, which has not been influenced significantly by pre-shear, the storage modulus increased markedly with increasing pre-shear times at low angular frequencies. This was interpreted as a transition from a liquid- to a solid-like viscoelastic structure of the particle network.

Summarizing all the results concerning the influence of pre-shear it is indispensable to insert a phase of rheological sample preparation before any rheological measurements are carried out with rotational rheometers (Sect. 9.1.2). Furthermore, the pre-experimental history influences the rheological behavior of suspensions markedly.

If capillary viscometers (Sect. 9.1.2) are used, the influence of various mechanical histories is not as dramatic as in the case of rotational rheometers since the sample flows from a reservoir with a larger diameter into the capillary with a distinctly smaller diameter, driven by a pressure. Due to the convergent flow field, elongational effects modify the structure of the suspension and orientation phenomena disappear.

Viscometric Flows. In this chapter, the most relevant experimental results and phenomena obtained in viscometric flows are presented and explained.

A typical example of the influence of the solid volume concentration on the shear-stress-dependent viscosity shown in Fig. 9.77 has been given by *Laun* [9.268] for an aqueous suspension of charged polystyrene-ethylacrylate copolymer spheres of 250 nm diameter. The viscosity at low solid volume concentrations $\varphi < 0.18$ remains Newtonian. Increasing the solid volume concentration, non-Newtonian shear-thinning behavior can be observed with a subsequent shear thickening region at higher solid volume concentrations. At very high solid volume concentrations a striking increase of the viscosity at lower shear stresses can be obtained, which could be an indication of an apparent, or possibly true, yield stress.

These observations could be attributed to the particle structure changes by applying rheooptical methods during the shear experiments. *Lyon* et al. [9.271] studied

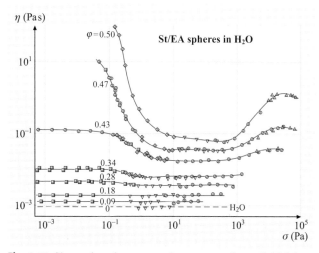

Fig. 9.77 Shear viscosity versus shear stress of a colloidal latex suspension with 250 nm particles at different solid volume concentrations at nearly the same pH values (after *Laun* [9.268] with permission by Hüthig & Wepf Verlag)

the structure formation of suspensions in a simple shear flow and found a transition from a disordered state at the beginning of the experiment to an ordered state induced by a given shear strain or time of shear. The noncolloidal monodisperse particles order in shear bands and, locally, in a hexagonal pattern due to particle migration across the streamlines of the flow.

By a combination of optical and rheological measurements, similar effects in the dependence of viscosity on the microstructure of the suspensions could be obtained by *Gondret* and *Petit* [9.272] for oscillatory shear (Fig. 9.78) as well as by *Völtz* et al. [9.274] for steady shear. They found a time-dependent decrease of the viscosity during shear, as shown for a moderately concentrated suspension of noncolloidal spheres. The decrease of the viscosity was coupled with a transition from a disordered to a periodic band-like [9.272] or hexagonal structure [9.274] of the suspension. The range of the characteristic structuring times can be some milliseconds to hours. Furthermore, it can be stated that structural modifications from a disordered state via a band-like and hexagonal to disordered structures are the reason for the complex behavior of the suspensions. These effects, dependent on the interparticle interactions and solid volume concentration, occur in both colloidal and noncolloidal suspensions.

Silbert et al. [9.275] studied concentrated, aggregated colloidal suspensions of spherical particles by numerical simulations. Among other things, they claimed, that the microstructure strongly depends on the solid volume concentration under the same shearing conditions, i.e., the same Péclet number. Starting from a semiordered phase with a coexistence of particle strings and disordered regions at $\varphi = 0.47$ they found a transition to a hexagonally packed string phase at $\varphi = 0.50$, an intermediate string/layer phase at $\varphi = 0.55$, and finally to a truly ordered layered phase at $\varphi = 0.57$.

As shown in Table 9.10, the viscosity of a suspension depends strongly on the interparticle interactions and, hence, on the kind of stabilization (hard spheres, electrostatic stabilization, steric or electrosteric stabilization). In the following, some illustrative experimental results will be given for hard spheres, charged spheres, and sterically stabilized spheres.

Shear Viscosity. In the absence of interparticle surface forces the shear viscosity of hard-sphere suspensions depends only on the solid volume concentration and the dimensionless shear rate or Péclet number (9.231). *Krieger* [9.273] investigated some model colloidal hard-sphere suspensions of polystyrene spheres in water or benzyl alcohol and metacresol. He proved that the relative viscosity data form a master curve independent of the continuous phase, as portrayed in Fig. 9.79.

A viscosity function of this type can be best described by (9.249) with $m = 1$. The critical shear stress σ_{cr}, which characterizes the shear thinning region, has been found to be dependent on the solid volume concentration. As demonstrated by *de Kruif* et al. [9.219] the critical shear stress increases initially with increasing solid volume concentration up to a value of $\varphi \approx 0.5$.

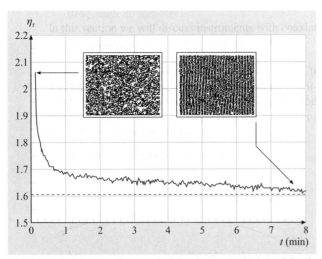

Fig. 9.78 Structuring of a suspension of noncolloidal particles in a Newtonian continuous phase with a solid volume concentration of 0.2 in an oscillatory shear flow (after *Gondret* and *Petit* [9.272] with permission by the AIP)

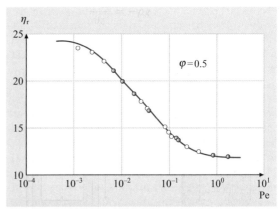

Fig. 9.79 Dimensionless viscosity function of hard sphere suspensions (after *Krieger* [9.273] with permission by Elsevier). Continuous phase: water (*line*), benzyl alcohol (*open circles*), metacresol (*filled circles*)

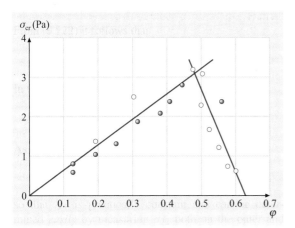

Fig. 9.80 Critical shear stress as a function of solid volume concentration for colloidal hard-sphere suspensions (after *de Kruif* et al. [9.219] with permission by the AIP)

After reaching this maximum, it decreases and reaches zero for a solid volume concentration in the vicinity of the maximum packing fraction, where the viscosity diverges. The maximum of the critical shear stress at $\varphi \approx 0.5$ is an indicator that a hard-sphere disorder–order transition is rheologically relevant [9.194]. The dependence of the critical shear stress on the solid volume concentration is shown in Fig. 9.80.

So et al. [9.277] indicated that it is possible to suppress shear-thickening effects for highly concentrated suspensions of smaller stabilized hard spheres in contrast to suspensions with nonstabilized larger particles at the same solid volume concentration. In the latter case, weak interparticle forces, i.e., attractive forces, lead to a formation of clusters and hence to a modification of the microstructure, which deviates from that of a face-centered cubic or hexagonally close packing at solid volume concentration above $\varphi = 0.5$.

Heymann et al. [9.276] studied noncolloidal suspensions of PMMA spheres dispersed in a Newtonian silicone oil. They measured the flow curves and shear stress–shear strain curves of the suspensions of various solid volume concentration in the controlled-shear-stress (CSS) mode with different logarithmic ramp times t_R. An open question in rheology of highly concentrated hard-sphere suspensions (or suspensions generally) is how the transition from the solid-like behavior at low shear strains or shear rates to the liquid-like behavior at higher mechanical loading takes place.

An example is given in Fig. 9.81. At higher shear stresses and shear rates all flow curves (Fig. 9.81a) coincide and no dependence on the ramp time is found. The shear rate depends only on the actual shear stress. Hence, the suspension shows typical liquid-like behavior at high shear stresses with terminal Newtonian behavior. At small shear stresses scattering of the data occurs. However, the same data presented in the shear strain–shear stress curves (Fig. 9.81b) unveils the problem. At low shear stresses, the curves coincide, independent of the ramp time. This indicates a unique relation between the shear stress and the shear strain. Hence, the suspension behaves like a solid at low shear stresses. Experiments with increasing and subsequently decreasing shear stresses reveal strain hysteresis in the solid-like regime, which depends on the solid volume concentra-

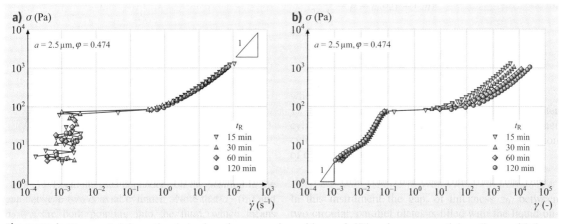

Fig. 9.81a,b Flow and deformation behavior of a suspension with $\varphi = 0.474$ and a mean particle radius a of 2.5 μm: (a) flow curve, (b) shear stress versus shear strain (after *Heymann* et al. [9.276] with permission by the Society of Rheology)

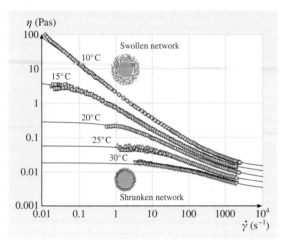

Fig. 9.82 Viscosity change due to reversible swelling of the surface layer as a function of shear rate at various temperatures (after *Senff* et al. [9.278] with permission by the American Chemical Society)

tion and the ramp time of the shear stress variation. At very low shear stresses, Hookean behavior can be speculated. By further increasing the solid volume concentration one leaves the region of suspension rheology and arrives in the region of bulk materials.

Senff et al. [9.278] analyzed the rheological behavior of sterically stabilized, temperature-sensitive core–shell lattices. The polymer layer affixed at the particle surface undergoes a volume transition as a function of the tem-

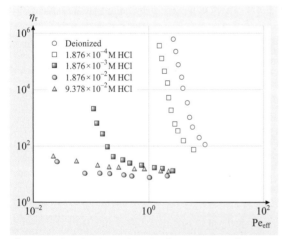

Fig. 9.83 Steady shear viscosity for aqueous polystyrene latex suspensions for $\varphi = 0.4$ at different ion strengths (after *Krieger* and *Eguiluz* [9.279] with permission by the Society of Rheology)

perature. Due to the volume change the effective size of the particles and, hence, the solid volume concentration changes. By small-angle X-ray and small-angle neutron scattering it could be shown that the spheres behave as hard spheres in both cases of shrunk and swollen surface layers. The volume change was inversely proportional to the temperature, i. e., the effective particle size increased with decreasing temperature. The modification of the solid volume concentration with the temperature can be clearly recognized from the viscosity functions. The viscosity increase over some orders of magnitudes with decreasing temperature, shown in Fig. 9.83, is solely caused by the increase of the effective solid volume concentration due to swelling of the particle surface layer.

Additionally to the influence parameters for neutral hard spheres, the charge properties of the particles and the continuous phase ($\varepsilon_r \varepsilon_0 \psi_S^2$) and the ratio of the particle radius to the Debye length (a/κ^{-1}) must be taken into account in the case of electrostatically and electrosterically stabilized suspensions.

Krieger and *Eguiluz* [9.279] examined the rheological behavior of colloidal electrostatically stabilized polystyrene lattices consisting of spherical particles with a diameter of 110 nm in water as a function of the ion strength, characterized by the concentration of hydrochloric acid. As seen in Fig. 9.84, the ion strength (or the acid concentration) strongly influences the viscosity. With decreasing ion strength, the viscosity increases dramatically and a transition from liquid-like behavior at higher ion strengths to a more solid-like behavior at lower ion strength, comparable to that of deionized water, occurs. The viscosity lies some orders of magnitude above that for uncharged hard spheres. This effect is caused by the increasing attractive forces and the formation of aggregates. Similar results were obtained for the viscosity as a function of electrosterically stabilized silica-particle suspensions by *So* et al. [9.280].

By steric stabilization or surface modification of the particles it is possible to modify the flow behavior of the suspensions. In Fig. 9.85 the flow curves of suspensions of noncolloidal untreated and surface-modified (treated) silica spheres with a mean diameter of 4.6 μm dispersed in a Newtonian silicone oil (NM1-200) are shown. The flow curves were measured with increasing (open symbols) and a subsequent decreasing shear stresses (filled symbols). The surface of the treated particles has been modified with octadecyl groups. It can be recognized that surface modification leads to a decrease of the shear rates at comparable shear stresses and solid volume concentrations. This indicates the occurrence of a yield stress due to the surface modification. Furthermore, it can be stated that at low shear stresses hysteresis between the up and down curves of the treated suspensions occurs, which is caused by the brush-like structure of the surface modification. At higher shear rates no differences between the behavior of untreated and treated suspensions can be observed due to the dominance of the hydrodynamic over the interparticle effects. The pure continuous phase shows a Newtonian behavior.

Normal Stress Differences. Experimental results concerning the first and second normal stress difference of true hard-sphere suspensions with a Newtonian continuous phase are very rare in literature due to their small magnitude and the prevalence of many troublesome artefacts.

Experimental results concerning the first and second normal stress difference in charged or sterically stabilized non-hard-sphere suspensions are also rare in the literature. Most of the results have been obtained by numerical simulation. For higher shear rates, a general relation between the first normal stress coefficient Ψ_1 (Sect. 9.1.2) and the shear rate

$$\Psi_1(\dot{\gamma}) \sim |\dot{\gamma}|^{-n} \tag{9.262}$$

is valid. *Brady* and *Bossis* [9.282] investigated the shear flow behavior of interacting charged spherical particles dispersed in a Newtonian continuous phase. From numerical results, based on the Stokesian dynamics, they found an inverse proportionality ($\Psi_1(\dot{\gamma}) \sim |\dot{\gamma}^*|^{-1}$) between the first normal stress coefficient and the dimensionless shear rate $\dot{\gamma}^* = 3\pi\eta a^2\dot{\gamma}/\varepsilon_r\varepsilon_0\psi_S^2$ for $|\dot{\gamma}^*| \to \infty$, which corresponds to the experimental results of *Gadala-Maria* [9.283]. Here, ψ_S is the surface potential for infinite separation between the particle surfaces. An initially constant first normal stress coefficient for low shear rates could not be obtained. *Schoukens* and *Mewis* [9.284] observed, for suspensions of carbon-black particles in mineral oil, which form weak particulate structures, the proportionality $\Psi_1 \sim |\dot{\gamma}|^{-3/2}$. Again, no indication for a constant first normal stress coefficient at $|\dot{\gamma}| \to 0$ could be found. Indeed, the exponent $-3/2$ shows that the anisotropic structure of the particle network is weaker and can be destroyed by shear.

In contrast to these observations, *Moan* et al. [9.285] observed a qualitatively different behavior for aqueous, highly concentrated suspensions of oblate particles, which interact through excluded volume and electrostatic potential. For the first normal stress difference N_1 and the difference between the normal stress differences

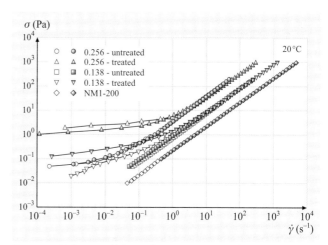

Fig. 9.84 Influence of the surface treatment on the rheological behavior of suspensions. *Open symbols*: up curve. *Filled symbols*: down curve (*Heymann* and *Aksel* [9.281]). The numbers in the legend indicate the solid volume concentration

$N_1 - N_2$, they found negative values. At low shear rates both functions decreased to a minimum and increased subsequently again. It is interesting that the second normal stress difference was negative at low shear rates and increased to positive values at higher shear rates contrary to the familiar behavior of polymers with a positive first and a smaller but negative second normal difference.

Lin-Gibson et al. [9.286] detected in nanotube suspensions a first normal stress difference which becomes negative during shear over a time period of 1000 min. By rheooptical investigations it could be shown that the development of a negative first normal stress difference is coupled with the formation of cylindrical aggregates orientated perpendicular to the flow and shear gradient direction. Similar effects were observed in emulsions too (Sect. 9.3.3).

Zarraga et al. [9.287] measured in noncolloidal suspensions of spheres in a Newtonian fluid negative values for both the first and second normal stress differences with $|N_1| < |N_2|$. The normal stress differences were proportional to the shear stress and the difference between the first and second normal stress difference, as well as the first normal stress difference were proportional to the solid volume concentration with $N_1 - N_2 \sim \varphi^3 e^{2.34\varphi}$ and $N_1 \sim \varphi^3 e^{2.34\varphi}$, respectively, in the range $\varphi = 0.35-0.5$. For noncolloidal suspensions of spheres in an elastic fluid with constant viscosity (Boger fluid), *Zarraga* et al. [9.288] measured positive first and negative second normal stress differences, as known for polymers. The normal stress differences increased with the solid volume concentration, but the ratio $|N_1/N_2|$ decreased as the solid volume concentration increased. The magnitude of N_2 at high solid volume concentrations approached the magnitude measured for the suspensions with a Newtonian continuous phase [9.287], while the magnitude of N_1 could be attributed to the viscoelasticity of the continuous phase.

A familiar normal stress behavior, known from polymers, could be observed by *Mall-Gleissle* et al. [9.290]. They investigated suspensions of glass spheres dispersed in a constant-viscosity but viscoelastic continuous phase (Boger fluid). The first normal stress difference was positive and decreased in magnitude with increasing solid volume concentration, whereas the second normal stress difference was negative, with a magnitude that increased with increasing solid volume concentration. The variation of both normal stress differences with the shear stress followed a power-law behavior.

A new method to measure normal stresses of non-Brownian suspensions at low to moderate solid volume concentration has been developed by *Singh* and *Nott* [9.291]. Due to the low normal forces at moderate solid volume concentrations the force signal was found to be very noisy. Therefore, they used pressure holes (cf. Fig. 9.35) to measure the radial component of the normal stresses in a Couette device or the axial component in a torsional PP device. They found that both the first and second normal stress coefficients are negative and $|N_1| < |N_2|$ with a strong dependence on the solid volume concentration at higher concentrations.

Summarizing the results concerning the normal stress differences, it can be stated that suspensions behave in a more-complex manner than polymer melts or solutions. It is not a priori clear how the magnitudes and the sign of the normal stress differences depend on the physicochemical properties of the suspensions. A crucial assumption for an exact measurement of the normal stress differences is an absolutely stress-free state of the sample at the beginning of the experiment. Residual stresses, induced by the closing procedure of the measurement geometry, have an essential influence of the results. One should be aware of the fact that normal stress measurements are very sensitive to errors since normal stress differences are of second order. Consequently, the sample preparation must be carried out in a much more exact way as it is necessary for the measurements of the viscosity.

Storage and Loss Modulus, Fourier-Transform Rheology. *Shikata* and *Pearson* [9.289] investigated the linear viscoelastic behavior of concentrated suspensions of submicron silica spheres in a continuous phase consisting of ethylene glycol and glycerol. Typical master curves of the storage and loss modulus for this hard-sphere suspension with a solid volume concentration of $\varphi = 0.37$ are illustrated in Fig. 9.86. The data were obtained using the time–temperature superposition [9.292]) where a temperature-dependent shift factor a_T is used to shift the curves horizontally to produce a master curve. (cf. *Shikata* and *Pearson* [9.289] for details; definitions of all the parameters shown in Fig. 9.87 are given in Sect. 9.1.1). The viscous properties dominate over the frequency range investigated. Furthermore, the loss modulus shows Newtonian regions at low and high frequencies, whereas the storage modulus reaches a plateau at high frequencies.

So et al. [9.280] analyzed the storage and loss moduli of colloidal, electrosterically stabilized silica particle suspensions in dependence on the solid volume concentration and ion strength. At a given low ion strength they found a transition from a viscously dominated behavior via a gel-like to an elastically dominated behavior with increasing solid volume concentration as shown in Fig. 9.87a. It is noteworthy that the moduli vary over more than four orders of magnitude. Furthermore, a solid-like behavior arises with increasing solid volume concentration indicated by the plateau of the loss modulus. Figure 9.87b shows the influence of the salt concentration, i.e., the ion strength. At higher salt concentrations, the suspensions show liquid-like behavior which passes to a gel-like behavior if the salt con-

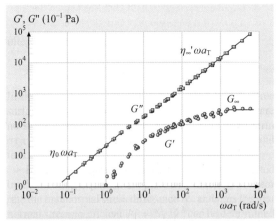

Fig. 9.85 Typical master curves of the storage and loss moduli as a function of the angular frequency for a hard-sphere suspension with $\varphi = 0.37$ (after *Shikata* and *Pearson* [9.289] with permission by the Society of Rheology)

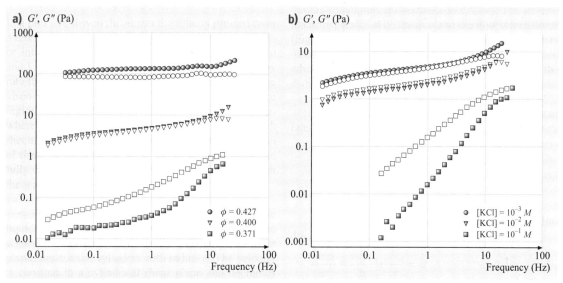

Fig. 9.86a,b Storage (G', *filled symbols*) and loss (G'', *open symbols*) moduli of an electrosterically stabilized silica particle suspension as a function of frequency (after *So* et al. [9.277] with permission by Elsevier): **(a)** dependence on the solid volume concentration at an ion strength $[\text{KCl}] = 10^{-3}$ M, **(b)** dependence on the ion strength at a solid volume concentration of $\varphi = 0.4$

centration decreases and the influence of the attractive interparticle forces increases.

A crucial problem in determining the storage and loss modulus is to meet the necessary condition of linearity between input (strain or stress) and output (stress or strain) signal. Due to the strong interactions in highly concentrated suspensions the critical shear strain or shear stress amplitudes are very small in contrast to polymers. Exceeding the critical values (intentionally in large-amplitude oscillatory shear (LAOS) or unintentionally) can lead to nonharmonic output signals and the equations for the calculation of the storage and loss moduli are no longer applicable (Sect. 9.1.1). The only way out is the analysis of the output signal concerning the occurrence of higher harmonic components (overtones) by use of the method of Fourier-transform rheology (FTR) as developed by the group of *Wilhelm* [9.293–295]. *Heymann* et al. [9.267] used this method to analyze the rheological behavior of highly concentrated suspensions. As depicted in Fig. 9.88 for a hard-sphere suspension with a solid volume concentration of $\varphi = 0.474$, the fifth harmonic in the shear stress signal occurs in the vicinity of the yield stress ($\sigma_y = 7.9$ Pa), i.e., in the range of transition from solid- to liquid-like behavior. However, the behavior of the suspension becomes already nonlinear at a shear stress amplitudes below the yield stress which is indicated by the first occurrence of the third harmonic.

Kallus et al. [9.296] studied polymer dispersions with the FTR method. An essential result of their experi-

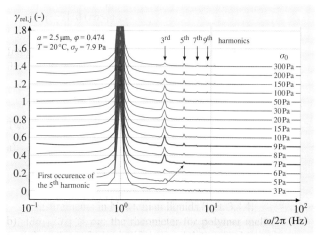

Fig. 9.87 Example of relative strain amplitude spectra obtained in oscillatory shear flow for various shear stress amplitudes of a hard-sphere suspension (for details see legend) (after *Heymann* et al. [9.267] with permission by the Society of Rheology). The curves are shifted vertically to avoid overlap. The amplitude of the shear stress σ_0 is defined in Fig. 9.3 in Sect. 9.1.1

ments is that the particle surface characteristics influence the nonlinear response significantly. The intensity of the overtones was found to be higher for dispersions of particles with a *hairy* swollen surface layer compared to a system of smooth particles at the same steady shear viscosity.

A general classification of the rheological behavior of complex fluids such as dispersions in LAOS has been given by *Hyun* et al. [9.297]. They found four main types of the fluid behavior in LAOS amplitude sweeps and pointed out that it is possible to relate the results of LAOS experiments to the microstructure of the materials. However, it must be emphasized again that the calculation of the storage and loss moduli on the basis of the linear theory is no longer valid, if overtones in the output signal occur.

In the case of flocculated colloidal suspensions *Otsubo* [9.298] proposed a rule similar to the Cox–Merz rule ((9.102) in Sect. 9.1.1), which relates the storage modulus to the first normal stress difference at low frequencies and shear rates, respectively

$$G'(\omega) = \frac{1}{2} N_1(\dot{\gamma}) \bigg|_{\omega = \dot{\gamma}} . \tag{9.263}$$

Transient Experiments – Shear Creep and Start-Up Flow. Studies on the transient behavior of suspensions were carried out, e.g., by *Heymann* et al. [9.276] and *Schmidt* and *Münstedt* [9.299]. *Heymann* et al. [9.276] investigated the creep and start-up flow behavior of non-

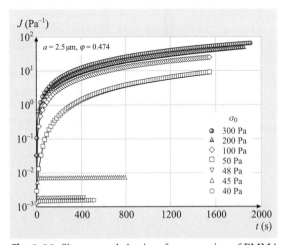

Fig. 9.88 Shear creep behavior of a suspension of PMMA spheres in a Newtonian continuous phase for various shear stresses (after *Heymann* et al. [9.276] with permission by Springer)

colloidal suspensions of monodisperse PMMA spheres (*Heymann* et al. [9.267] for details) dispersed in silicone oil. They pointed out a qualitatively different behavior of the suspensions below and above the yield stress in creep flow. Below the yield stress a typical elastic behavior has been found characterized by a time-independent creep compliance at a given shear stress. Above the apparent yield stress, a typical liquid-like behavior could be observed. The results of the creep experiment could be correlated to the start-up flow experiment. At low shear rates a linear viscoelastic behavior was found, followed by a nonlinear viscoelastic behavior with an overshoot in the shear stress output. A further increase of the shear rate led to a liquid-like behavior characterized by an instantaneous increase of the shear stress without any delay.

In Fig. 9.89, the shear compliance $J(t)$ of suspensions of (PMMA) hard spheres in a Newtonian continuous phase is shown. One can distinguish between two ranges. At shear stresses below the yield stress (in this case $\sigma_y = 48.5$ Pa) no creep could be observed. Exceeding the yield stress leads to a significant creep. It should be noted that this transition from solid- (elastic) to liquid-like behavior occurs in a small range of 48 Pa $< \sigma_0 <$ 50 Pa.

Schmidt and *Münstedt* [9.299] presented similar results for suspensions of monodisperse hydrophilic glass spheres in a Newtonian continuous phase. At low shear stresses a viscoelastic behavior has been observed in creep, whereas at high shear stresses the suspensions showed liquid-like behavior. Due to the problems described in the part dealing with the normal stress differences, transient experiments seem to be a suitable tool to investigate the viscoelastic properties of suspensions.

Most of the authors cited above investigated suspensions of spheres in various continuous phases. If the particles are nonspherical (e.g., ellipsoids, rods), additionally to the aforementioned influence parameters, the axis ratio must be taken into account to understand the flow behavior of such suspensions. For details on this subject it is referred to the monographs of *Macosko* [9.188] or *Larson* [9.186], to the contributions of *Utracki* [9.300], and to the review article of *Petrie* [9.301] on the rheology of fibre suspensions.

Elongational flows. The kinematics of elongational flows (cf. Sect. 9.1.3) is quite different from the kinematics of viscometric flows (cf. Sect. 9.1.1). Due to the exponential path–time relation, the distance between neighboring particles increases more rapidly than in viscometric flows. Hence, the collision probability between the particles is lower than in shear flows. Furthermore, particle migration, as described for shear flows, does not occur due to the kinematics.

However, experimental results obtained for suspensions in elongational flows are very rare in literature. A review of elongational rheometers applicable to special disperse and other materials has been given by *Greener* and *Evans* [9.303].

The main problem of elongational measurements with filament stretching rheometers is that the preparation of a stable sample for measurements is very difficult. Often, rods with cylindrical or rectangular cross-sectional area are used, which must be stable under gravity or in a buoyancy-neutral environment. Other devices (four-roller apparatus, opposite nozzle device) have been used for suspensions with viscosities lower than that of filled polymer melts.

Greener and *Evans* [9.302] investigated suspensions of aluminium powder dispersed in a polyisobutylene (PIB) matrix. Using a fibre filament stretching device they measured the elongational viscosity at constant strain rates. Qualitatively, they observed a similar behavior of the elongational viscosity in dependence of the extensional rate and the solid volume concentration as found for the shear properties. The Trouton ratio (the ratio of the elongational to shear viscosity at the same elongational and shear rates), was found to be approximately 3 in the low shear and strain rate range as is exactly the case for incompressible Newtonian liquids. It is interesting to note that this observation is true up to solid volume concentrations of $\varphi = 0.42$. The uniaxial elongational viscosity η_e as a function of the strain rate $\dot{\varepsilon}$ and the relative uniaxial elongational viscosity $\eta_{r,e} = \eta_e/\eta_{e,c}$ as a function of the solid volume concentration are shown in Fig. 9.90. The behavior differs from that obtained for unfilled or filled polymers. Contrary to polymers where the elongational viscosity rises from a plateau at low strain rates through a pronounced maximum and then decreases at higher strain rates, the elongational viscosity of the suspension shown here decreases monotonously with increasing strain rates.

In a more-quantitative manner, *Le Meins* et al. [9.304] observed for suspensions of colloidal polystyrene spheres in polyisobutylene (PIB) a deviation of the relative elongational viscosity from the behavior, which can be described by the Krieger–Dougherty equation (9.244). For larger particles and at low solid volume concentrations good agreement has been found, whereas for smaller particles and at higher solid volume concentrations deviations occurred. The deviations were smaller for the relative elongational viscosity than for the relative shear viscosity. These observations have been explained by dominating hydrodynamic effects occurring with larger particles, whereas with colloidal particles the interparticle interactions become relevant.

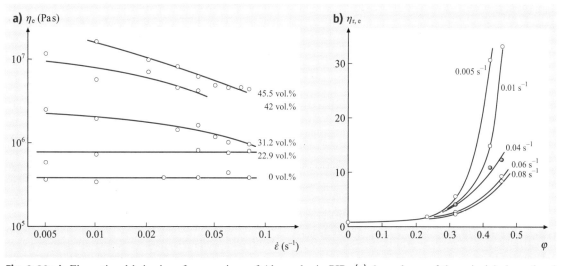

Fig. 9.89a,b Elongational behavior of suspensions of Al powder in PIB, (**a**) dependence of the uniaxial elongational viscosity η_e on the extensional strain rate $\dot{\varepsilon}$ at various solid volume concentrations, (**b**) dependence of the relative extensional viscosity on the solid volume concentration at various extensional strain rates (after *Greener* and *Evans* [9.302] with permission from the Society of Rheology)

Furthermore, a reduction in strain hardening could be observed for various elongational rates if rigid spherical polystyrene particles are added to the viscoelastic polymer matrix. The particles cannot follow the externally applied elongational field and, hence, a partial conversion form elongational to shear flow within the polymer matrix occurs.

O'Brien and *Mackay* [9.306] obtained for suspensions of kaolin suspensions elongational viscosities with elongation thinning and elongation thickening ranges depending on the particles size. Additionally, they observed particle-size-dependent Trouton ratios of the order of 50–100.

Husband et al. [9.305] measured the tensile creep behavior of suspensions containing noncolloidal limestone (CaCO$_3$) particles dispersed in a polyisobuthylene (PIB) matrix over long time periods. At low stresses they obtained a solid-like behavior indicated by an equilibrium tensile strain. At higher stresses a flow process, i.e., a liquid-like behavior, has been observed characterized by continuously increasing strain rates. The results of the tensile creep experiments are portrayed in Fig. 9.91. The tensile creep coefficient is defined as

$$\eta_{E,c}^+[t, \sigma_E(t)] = \frac{\sigma_E(t)}{\dot{\varepsilon}^-[t, \sigma_E(t)]} \qquad (9.264)$$

where $\dot{\varepsilon}^-[t, \sigma_E(t)]$ is the tensile creep rate decay function, $\sigma_E(t) = F/A(t)$ is the tensile stress, F is the force, and $A(t)$ is the time-dependent cross-sectional area of the sample. For tensile stresses $\sigma_E < 540$ Pa constant tensile strains, i.e., vanishing tensile strain rates, are reached. The tensile strains depend on the tensile stress. The vanishing tensile strain rates correspond to an increase of the tensile creep rate coefficient. This is attributed to a viscoelastic solid-like behavior. For higher tensile stresses a tensile creep rate coefficient independent of both the tensile strain and the tensile stress has been found as a characteristic of a viscoelastic liquid-like behavior. The critical tensile stress, indicating the transition between both ranges, can be interpreted as a static yield stress. The observations concerning the yield transition obtained by *Husband* et al. [9.305] in elongational flow correspond qualitatively to the results obtained by *Heymann* et al. [9.276] in viscometric flows.

Complex Flows. Due to the tensorial character of the constitutive equations (Sect. 1.5 and Sect. 1.8) the shear and elongational experiments are not sufficient to describe the material completely in every flow situation.

One of the most popular complex flow situations is squeeze flow, which combines shear and elongational flow kinematics. A defined volume of the sample is placed in a gap between two (usually circular) parallel plates. One of the plates is kept fixed and the other plate moves towards the fixed one either at a given velocity or driven by a given force. Depending on the experimental mode, the normal force, velocity, or the separation between the plates is measured. Two modes are possible: the constant-area and constant-volume techniques. Details of the modeling of the squeeze flow process have been described by *Gibson* et al. [9.307] and *Macosko* [9.188].

The kinematics of the flow field is highly complex because of two velocity components, one in radial direction ($u(r, z)$) and the other in axial direction ($w(r, z)$), where r and z are the radial and axial coordinates. In contrast to viscometric flows, where only one velocity gradient occurs, in the squeeze flow up to four velocity gradients ($\partial u/\partial z$, $\partial u/\partial r$, $\partial w/\partial z$, $\partial w/\partial r$) complicate the situation, especially for non-Newtonian materials where the rheological behavior of the material is a priori unknown. Therefore, it is impossible to obtain rheologically exactly defined material functions for non-Newtonian liquids from squeeze flow experiments.

Beyond the aforementioned difficulties, the boundary conditions have an essential influence on the kinematics of squeeze flow. If the no-slip condition at the walls holds, the flow is dominated by shear effects. The existence of wall slip leads to the dominance of

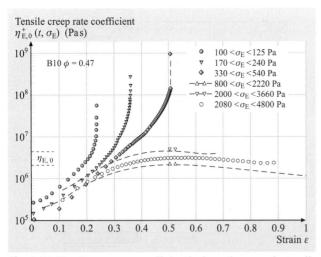

Fig. 9.90 Tensile creep rate coefficient in dependence on the tensile strain ε for a suspension with $\varphi = 0.47$; the *dashed lines* near the ordinate indicate the range of the limiting viscosity $\eta_{E,0}$ at zero extension rate (after *Husband* et al. [9.305] with permission by the Society of Rheology)

elongational effects. If it is not clear whether or not the material slips at the walls, it is not possible to obtain reliable results from the measurements.

Despite these disadvantages, squeeze flow of disperse materials is subject of many investigations. The determination of the squeeze force as a function of the gap height and estimation of the yield stress is mainly the aim of these studies.

An important problem in the investigation of suspensions in squeeze flows is demixing of the sample, i.e., filtration of the continuous phase through the *particle skeleton*. *Delhaye* et al. [9.308] found for a highly concentrated suspension of spheres in a Newtonian fluid a significant change of the radial concentration distribution in an initially homogeneously distributed sample during the experiment. Similar effects of binder migration were shown by *Poitou* and *Racineux* [9.309]. Due to the inhomogeneity of the sample, the measured squeeze force is not representative for the experiment. Other authors (e.g., *Chan* and *Baird* [9.310], *Sherwood* and *Durban* [9.311]) calculated the squeeze force as a function of the gap width for various viscoplastic models (Herschel-Bulkley, Bingham) whose parameters were obtained from definite viscometric flows. They found a reasonable agreement between the numerical simulation and the experimental data. Some contributions, dealing only with the analytical or numerical simulation of the squeeze process, are available, which consider wall slip and various models for the flow behavior [9.312–314]. However, a relation to experiments has not been established and no relevant material parameters (viscosity, yield stress) have been given. Summarizing the results available in the literature, it must be stated that the squeeze flow is not the first choice as an experiment for the measurement of rheologically relevant material functions.

9.3.3 Emulsions

Emulsions are disperse systems containing one liquid dispersed in another liquid. Both liquids are immiscible or, at least, partially immiscible. Usually, drop sizes are in a range between 10 nm and 10 μm. Emulsions with drop sizes between 10 nm and 100 nm are called microemulsions. Emulsions with drop sizes between 100 nm and 1 μm are known as miniemulsions [9.196]. Some fundamental principles of emulsion rheology have been summarized by *Sherman* [9.315], *Barnes* [9.316], *Tadros* [9.317], and *Pal* [9.318].

Generally, two fundamental types of emulsions can be distinguished: water in oil (W/O) and oil in water (O/W) emulsions. The classification depends on the kind of the dispersed (liquid 1) and continuous (liquid 2) phase. In the case of a nonaqueous continuous phase this classification is not very clear. But, in a more-general manner, an O/W emulsion is characterized by a viscosity of the disperse phase higher than that of the continuous phase, or vice versa for W/O emulsions. A further addition of a liquid 1, dispersed in a liquid 2, up to a critical volume concentration $\varphi_{1,PI}$ leads to a change of the type of emulsion, called phase inversion. In Fig. 9.91 the influence of the disperse phase 1 on the relative viscosity of emulsions is depicted. Similar to suspensions, the relative viscosity increases with increasing volume concentration of the disperse phase 1 until the maximum packing fraction is reached. At this point, a phase inversion takes place, leading to a decrease of the viscosity with increasing volume concentration of the, from now on, continuous phase 1. The continuous phase 2 below $\varphi_{1,PI}$ is now the disperse phase. Phase inversion is a typical phenomenon in emulsions that does not appear in suspensions.

In the previous sections, the rheological behavior of suspensions has been extensively described. Most of the basics described for suspensions hold also for emulsions. However, compared to suspensions, differences occur because the disperse phase consist of a second liquid with a finite viscosity and surface tension. The drops of the disperse phase can be deformed or broken-up due to shear or elongational flow. The interfacial tension influences the interactions between the two phases. Furthermore, the drop size distribution may change due to coalescence, Ostwald ripening, creaming, flocculation,

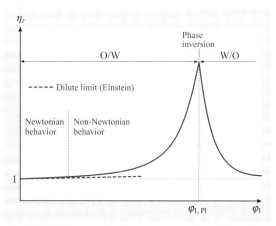

Fig. 9.91 Influence of volume concentration of phase 1 on the relative viscosity of emulsions prior and after phase inversion

or drop break-up. Additionally, the continuous phase contains surface-active agents (surfactants) that may cause increased viscosity and even non-Newtonian behavior of the continuous phase and influence the overall rheology of the emulsion.

In the following, only properties of emulsions that deviate from those of suspensions will be illustrated.

Dimensional Analysis. In addition to the dimensionless groups in suspension rheology (Sect. 9.3.1), emulsions involve the capillary number

$$\text{Ca} = \frac{\eta_c \dot{\gamma} a_0}{\sigma_\text{I}} \quad (9.265)$$

as the ratio of viscous to capillary forces (σ_I is the interfacial tension) and the viscosity ratio

$$V_\text{r} = \frac{\eta_\text{d}}{\eta_\text{c}} \, . \quad (9.266)$$

As the characteristic droplet length in (9.265) the initial droplet radius a_0 is given. In contrast to suspensions, the initial droplet radius can change at higher shear or elongational rates. Larger droplets with a weak surfactant layer can deform during shear, which is often coupled with an orientation of the droplets in the flow direction as shown in Fig. 9.94. Droplet deformation and/or break-up lead to the formation of new interfacial areas. For non-Newtonian continuous phases the definition of the capillary number according to (9.265) is not unique because the viscosity η_c is itself a function of the shear rate.

If one neglects van der Waals, electrostatic, electrosteric, and steric interactions and assumes a gap width large enough compared to the initial droplet size, the relative viscosity of emulsions can then be formulated as [9.320, 321] (cf. Table 9.10)

$$\eta_\text{r} = f\left(\varphi, V_\text{r}, \xi, \rho_\text{r}, t_\text{r}, \frac{\delta}{a_0}, \text{Ga}, \text{Ca}, \text{Pe}, \text{Re}_\text{p}\right) \, . \quad (9.267)$$

The relation can be further simplified for emulsions containing a nonvarying monodisperse phase ($\xi = \text{const.}$) of buoyant droplets ($\rho_\text{r} \rightarrow 1$) with thin surface layer thickness ($\delta/a_0 \ll 1$) and with vanishingly small droplet Reynolds number ($\text{Re}_\text{p} \rightarrow 0$) under steady-state conditions ($t_\text{r} \rightarrow \infty$) to give the relation

$$\eta_\text{r} = f(\varphi, V_\text{r}, \text{Ca}, \text{Pe}) \, . \quad (9.268)$$

The deviations in (9.268) from the analogous relation for suspensions (9.231) stems from the appearance of the viscosity ratio V_r and the capillary number Ca, which occur only in emulsions.

For high viscosity ratios ($V_\text{r} \gg 1$) and high interfacial tensions ($\text{Ca} \rightarrow 0$), emulsions with small droplets behave like suspensions (Sect. 9.3.2) as demonstrated by *Teipel* and *Aksel* [9.321]. On the other hand, for highly dilute emulsions with high interfacial tensions G. I. Taylor derived a relation for $\eta_\text{r} = f(\varphi, V_\text{r})$ which for $\varphi \rightarrow 0$ tends to the Einstein result (9.237).

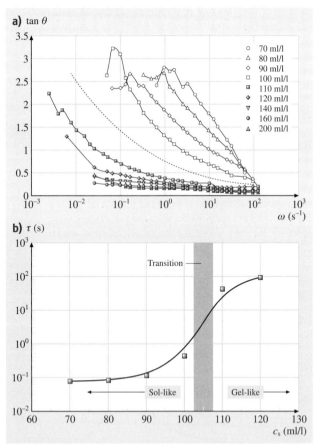

Fig. 9.92a,b Aqueous solution of Tween 85® at a temperature of 20 °C: (**a**) Loss tangent as a function of angular frequency and volume concentration (after *Teipel* et al. [9.319] with permission by Elsevier), (**b**) characteristic relaxation times τ as a function of the volume concentration of surfactant. The lines are drawn to guide the eye. The surfactant concentration is denoted by c_s

Properties of the Continuous Phase, Micellar Rheology. The continuous phase of emulsions often consists of low-molecular-weight liquids showing a Newtonian behavior. To stabilize the emulsions emulsifying agents (emulsifiers or surfactants) must be added, which lead to a modification of the rheological properties of the continuous phase. Surfactants are low-molecular-weight materials with amphiphilic properties (Chap. 3.2). Due to their amphiphilic character the surfactants can adsorb at interfaces, which results in a decrease of the interfacial or surface tension. Above a certain surfactant concentration the surfactant molecules undergo reversible aggregation to form micelles (Chap. 3.2).

A typical example has been given by *Teipel* et al. [9.319] who systematically investigated aqueous solutions of a nonionic surfactant polyoxyethylene sorbitan trioleate (Tween 85®) at various volume concentrations of surfactant. They demonstrated that with increasing surfactant concentration an increasing shear thinning effect in the viscosity function could be observed. This concentration-dependent non-Newtonian viscosity behavior is coupled with an increase of the elasticity of the micellar solutions. The oscillatory data show that, in the range of surfactant concentrations of 100–110 ml/l, a transition from a weakly elastic (sol-like) to a highly elastic (gel-like) response occurs. This transition, shown in Fig. 9.92a, is characterized by crossing the value of the loss tangent $\tan\theta = 1$ (Sect. 9.1.1). Correspondingly, the transition from sol- to gel-like behavior can also be seen in Fig. 9.92b by the characteristic relaxation time obtained from the inverse angular frequency at $\tan\theta = 1$, which increases dramatically over more than two orders of magnitude. The grade of elasticity indirectly shows the change of the internal structure of the micelles in the solution.

The rheological properties of the micellar solutions strongly depend on the conditions of preparation. In Fig. 9.93 the viscosity functions obtained under various fabrication (ϑ_F) and measurement (ϑ_M) temperatures are depicted.

Obviously, both the fabrication temperature and the measurement temperature influence the viscosity function. The low-shear viscosity increases over approximately one day if measured at a temperature than the fabrication temperature. In the high-shear-rate region the influence of the temperature preloading vanishes due to the dominance of hydrodynamic forces. These effects are caused by a structure formation process both during and after the fabrication process.

Other important properties of the continuous phase are the polarity and pH value. Both affect the charge of the droplets and the repulsive forces. Further information on the physicochemical properties of micellar solutions can be found, e.g., in the textbooks of *Larson* [9.186] or *Morrison* and *Ross* [9.195].

Properties of the Disperse Phase. In contrast to suspensions, the droplets of the disperse phase of emulsions have a finite viscosity. Hence, the viscosity ratio (9.266) must be taken into consideration. During shear (or elongation) the droplets can be deformed and/or broken up, as shown in Fig. 9.94. *Rumscheidt* and *Mason* [9.323] analyzed the droplet deformation and break-up mechanism, varying the viscosity ratio and shear rate. At low viscosity ratios Fig. 9.94a and for shear rates above a critical value, formation of a sigmoidal droplet shape and creation of small satellite droplets has been observed. At viscosity ratios of the order of 1 two identical droplets with smaller satellite droplets could be found, resulting from an elongation and an increasing contraction of the initial droplet (Fig. 9.94b). Alternatively, a droplet is stretched due to shear and a subsequent droplet break-up occurred (Fig. 9.94c). At higher viscosity ratios only droplet deformation and orientation without break-up have been observed (Fig. 9.94d). Generally, a droplet orientation was found at higher shear rates. However, it should be noted that the results described above have been found for single droplets without the influence of neighboring droplets.

The deformability of the droplets depends both on the viscosity ratio and the capillary number. *Taylor* [9.324] derived a theoretical function

$$f_1(V_r) = \frac{1 + 19V_r/16}{1 + V_r} \tag{9.269}$$

in simple shear, taking into account the viscosity ratio according to (9.266). The function $f_1(V_r)$ can vary in the

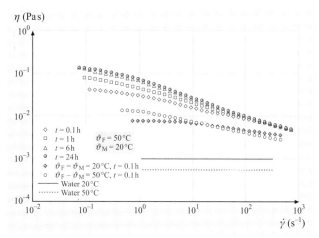

Fig. 9.93 Time dependence of the micellar solutions (Tween 85® in water) at different fabrication (ϑ_F) and measurement (ϑ_M) temperatures (after *Teipel* et al. [9.322] with permission by Elsevier)

range 1–1.88. The droplet deformation D, dependent on the viscosity ratio, is then given by

$$D = \frac{A_r - 1}{A_r + 1} = f_1(V_r)\mathrm{Ca} \qquad (9.270)$$

where A_r is the main axis ratio of ellipsoidal drops (major to minor axis length).

The effect of droplet break-up is then characterized by a critical capillary number $\mathrm{Ca_{cr}}$. Droplet break-up occurs if the capillary number exceeds a critical value, i.e., $\mathrm{Ca} > \mathrm{Ca_{cr}}$. Grace [9.325] reported some experimental findings of critical capillary numbers as a function of the viscosity ratio of the two phases for Newtonian liquids. The interpolating curve in Fig. 9.95 separates the region below where droplet break-up occurs from the region above where no droplet break-up takes place. The horizontal part of the curve represents tip-streaming break-up where small droplets are shed off from the tips of a larger sigmoidally shaped *mother drop*, as shown in Fig. 9.94a. It should be noted that *Grace* [9.325] used the dynamic surface tension to calculate the capillary number.

For a planar elongational flow, *Grace* [9.325] found that the critical capillary numbers were even smaller compared to those for shear flows at the same viscosity ratio. Furthermore, he found that it is possible to induce droplet break-up in elongational flows for viscosity ratios up to 1000. Hence, the elongational flow mechanism is more efficient for the fabrication of emulsions, even at higher viscosity ratios where shear flows fail.

Some reports have been published in the literature dealing with the problem of single droplet deformation and break-up in shear and elongational flows. *Tsakalos* et al. [9.326] investigated experimentally the deformation and break-up mechanism of a Newtonian droplet in a viscoelastic continuous phase in steady shear flow. The initially spherical droplets deform to threads during shear. They found for large capillary numbers that the deformation of threads, above a certain strain larger than 20, follows a pseudoaffine deformation for $\mathrm{Ca/Ca_{crit}} > 2.5$. Another mechanism of droplet break-up is end pinching (the ends of a stretched drop pinch off from the central filament due to drop relaxation after an abrupt change of the flow conditions). Capillary instabilities only develop when the thread diameter reaches a critical value. *Cristini* et al. [9.327] investigated the break-up phenomenon in impulsively started shear flow. They stated that the mechanism of break-up of Newtonian droplets in a Newtonian continuous phase (formation of threads with capillary break-up or end pinching) depends only on the average initial particle size. The mechanism of end pinching of Newtonian droplets in a Newtonian continuous phase during large oscillatory shear could be confirmed by *Wannaborworn* and *Mackley* [9.328] both experimentally and by numerical simulation. Furthermore, the critical capillary number was found to be higher for oscillatory shear than for steady shear, i.e., the droplets are more stable in oscillatory shear flows. At low strains, i.e., in the case of small-amplitude oscillatory

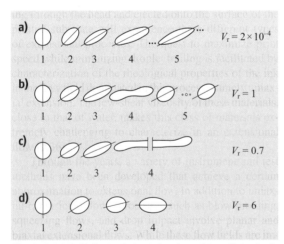

Fig. 9.94a–d Deformation and orientation of a single droplet in shear flow for increasing shear rates, indicated by the increasing numbers, at various viscosity ratios (after *Rumscheidt* and *Mason* [9.323] with permission by Elsevier)

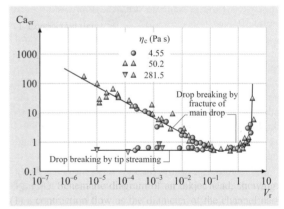

Fig. 9.95 Influence of the viscosity ratio on the critical capillary number obtained in shear flows (after *Grace* [9.325] with permission by Taylor & Francis)

shear (SAOS), *Cavallo* et al. [9.329] found a harmonic time behavior of the axis of the deformed droplets at lower strains, i.e., the response of the drop was sinusoidal, but shifted in phase with respect to the applied stress. At higher strains overtones with multiple frequencies of the basic frequency occurred. End-pinching effects have not been found in LAOS experiments. *Mighri* et al. [9.330] investigated emulsions with both phases consisting of constant-viscosity elastic (Boger) liquids in uniaxial elongational flow. They found that, for a given continuous phase, the drop deformation decreases as its elasticity increases, whereas for a given drop liquid, the drop deformation increases with increasing elasticity of the continuous phase. An overview on the dynamics of drop deformation and break-up has been given by *Stone* [9.331].

Loewenberg and *Hinch* [9.332] studied numerically the collision behavior of two deformable drops in shear flow. Their calculations indicate that drop interactions do not induce significantly subcritical capillary number break-up. In dilute emulsions, the critical Capillary number is a weak function of volume fraction because deformable drops can easily squeeze past each other.

As mentioned above, the continuous phase contains surfactants as stabilizers for the emulsion. The surfactants form a surfactant layer around the droplets. The formation of this interfacial layer modifying the droplet–droplet interactions is a kinetic process depending on the temperature, concentration of surfactants, the diffusion and adsorption properties of the surfactants. The adsorption kinetics of nonionic surfactants has been investigated by *Teipel* and *Aksel* [9.333].

If the surfactant layer is only weakly elastic it can transmit flow-induced shear stresses into the droplets and the liquid inside of the droplet can rotate, similar to the movement of a crawler. Additionally, droplet deformation occurs as shown in Fig. 9.94. With decreasing droplet radius and perfect surfactant coverage the emulsions show suspension-like behavior.

As in the case of suspensions, the viscosity of an emulsion increases with decreasing droplet size at a fixed volume concentration.

Interfacial Rheology. The mechanical properties of the interface between the continuous and disperse phases influence both the rheological properties and the stability of the emulsions. The surfactants added to facilitate the fabrication of emulsions and to stabilize them adsorb at the interface between the two phases, forming a thin film of surfactant molecules. At higher surfactant concentrations, the surfactant surface forms a compact interface with rheological properties that can differ greatly from those of the bulk phases. This can lead to viscoelastic or viscoplastic properties of the interface even though the bulk phases are Newtonian. To investigate the rheological properties of surfactant layers or films adsorbed at the interface, special rheological devices are needed. Reviews on some basic principles of interfacial rheology have been given by *Tadros* [9.317] and *Warburton* [9.334].

Generally, interfaces between two liquid phases contain monolayers of adsorbed surfactants. These layers show resistance of the surface against change of shape in shear, change of area in dilatation, and change of curvature in bending [9.334].

One of the most important mechanical properties of interfaces is the interfacial shear viscosity η_I. In contrast to the bulk viscosity η, the interfacial shear viscosity η_I has units of Pa s · m (surface Pa s). Hence, it represents not the viscosity in the volume, but the viscosity in a plane, namely in the interface. The measurement of the interfacial shear viscosity requires special devices such as torsional pendulum or rotational viscometers with special measurement geometries. Three examples are shown in Fig. 9.96a–c. The theory of the measurements of the interfacial shear viscosity with a torsional pendulum viscometer has been given by *Criddle* [9.335].

The shear device is suspended by a torsional wire and positioned at the interface. If the pendulum oscillates, the damping of the oscillation and the period due to the viscous drag of the interface is measured. A main disadvantage of the torsional pendulum viscometer is that

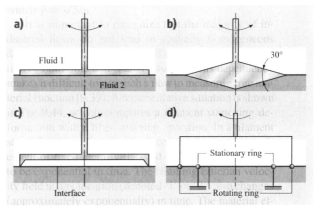

Fig. 9.96a–d Geometries used for the measurement of interfacial shear viscosity: **(a)** disk viscometer, **(b)** knife-edge disk viscometer, **(c)** ring viscometer, **(d)** concentric ring viscometer

the shear rate varies in the course of one period, which makes the measurement of non-Newtonian properties of the interface impossible. In this case, the concentric-ring viscometer, schematically shown in Fig. 9.96d, should be preferred. One ring rotates at a given angular velocity and the torsional momentum is measured at the other stationary ring. If the distance between the rings is small, the shear rate is approximately constant as in the case of the concentric cylinder viscometer (Sect. 9.1.2). In this case the non-Newtonian properties of the interface layer can be measured with more confidence.

There are other methods such as the deep-channel surface viscometer [9.317] where small particles are used as tracers to measure the surface velocities. *Petkov* et al. [9.336] developed a method to calculate the interfacial shear viscosity from the drag coefficient of a small sphere moving through an interfacial film.

To measure the dilatational surface properties three methods are used [9.317]: application of surface waves to the interface; rotation, translation or deformation of bubbles and droplets; and the maximum bubble pressure method. From these methods some information on the mechanical properties of the interface, but also on the kinetics of the surfactant adsorption at the interface, can be obtained.

Influence of Drop Volume Concentration. The deformation of the droplet surface and hence the transmission of surface tractions, causes internal circulations in the droplet, which results in a modification of the Einstein's viscosity relation (9.237)

$$\eta_r = 1 + (5/2) f_2(V_r) \varphi \,. \tag{9.271}$$

The *correction factor* $f_2(V_r)$ has been derived by *Taylor* [9.324] as

$$f_2(V_r) = \frac{\eta_d + (2/5)\eta_c}{\eta_d + \eta_c} = \frac{V_r + (2/5)}{V_r + 1} \tag{9.272}$$

for the case of infinitely diluted emulsions of two immiscible Newtonian liquids. The droplets of the dispersed phase are assumed to be perfectly spherical due to their smallness and the surface-tension effect. In the limit $V_r \to \infty$ (solid particles in a Newtonian liquid) with $f_2(V_r) = 1$, Einstein's equation for suspensions is obtained. The other limit $V_r \to 0$ describes the case of bubbles in a Newtonian liquid, leading to $f_2(V_r) = 2/5$. For finite viscosity ratios the viscosity of a comparable emulsion according to (9.271) and (9.272) is smaller than that of a suspension at the same volume concentration.

As in the case of suspensions, equations of the power-series type fail for higher volume concentrations. For this, the models summarized in Table 9.14 for suspensions of arbitrary particle shapes and shape distributions should be used because their applicability is not restricted to solid spherical particles of uniform size. *Pal* [9.337] combined Taylor's approach with the equations of *Mooney* [9.231] and *Krieger* and *Dougherty* [9.230]. Thus, he obtained two viscosity equations for volume concentrations up to 0.635:

Model 1 (Mooney type):

$$\eta_r \left(\frac{2\eta_r + 5V_r}{2 + 5V_r} \right)^{\frac{3}{2}} = \exp\left(\frac{2.5\varphi}{1 - \varphi/\varphi_{\max}} \right), \tag{9.273}$$

Model 2 (Krieger/Dougherty type):

$$\eta_r \left(\frac{2\eta_r + 5V_r}{2 + 5V_r} \right)^{\frac{3}{2}} = \left(1 - \frac{\varphi}{\varphi_{\max}} \right)^{-2.5\varphi_{\max}}. \tag{9.274}$$

Pal [9.337] fitted both models on the basis of several experimental data and found good agreement. As shown for suspensions, the maximum packing fraction $\varphi_{\max}$ is a fit parameter that depends (in this case) additionally on the type of model.

An empirical equation for the viscosity–volume concentration relation of non-Newtonian emulsions with volume concentration below a maximum packing fraction has been derived by *Pal* and *Rhodes* [9.338]

$$\eta_r(\dot{\gamma}) = (1 - K_0 K_F(\dot{\gamma})\varphi)^{-A} \tag{9.275}$$

with

$$A = 6(\eta_d - \eta_c)/[10(\eta_d + \eta_c)] \\ + (19\eta_d + 16\eta_c)/[10(\eta_d + \eta_c)] \tag{9.276}$$

as shown by *Lee* et al. [9.339]. For $V_r \gg 1$ the value $A = 2.5$ is valid. The factor K_0 accounts for the effect of hydration, i.e., the association of significant amounts of the continuous-phase liquid with the droplets, which is constant for a given emulsion. Due to the immobilization of the continuous phase, the effective dispersed-phase volume concentration increases compared to the true concentration. Furthermore, a shear-rate-dependent flocculation factor $K_F(\dot{\gamma})$, describing the morphology of the emulsion, has been introduced. The product of both factors can be interpreted as the inverse of a shear-dependent maximum packing fraction, resulting in an equation similar to the Maron/Pierce equation (9.244), however with an exponent of 2.5.

Due to the deformability of the particles the maximum packing fraction can reach values greater than those for corresponding monodisperse spherical rigid particles and the viscosity increases significantly compared to suspensions. At this stage, the occurrence of yield stresses is possible. The dependence of the viscosity on the shear rate can then be described with the models discussed in Sect. 9.3.2. A further addition of disperse-phase liquid leads to a sudden decrease of the viscosity due to the phase inversion process (Fig. 9.91).

Beyond nonlinearity, emulsions show also viscoelastic flow behavior. For this reason, models for the complex modulus G^* and complex viscosity (Sect. 9.1), respectively, have been developed. For moderately concentrated emulsions of monodisperse deformable droplets in a viscoelastic continuous phase with constant interfacial tension, *Palierne* [9.340] has shown that the complex modulus in the linear viscoelastic region is given by

$$G^* = G_c^* \left(\frac{1 + 3\varphi H/2}{1 - \varphi H} \right) , \quad (9.277)$$

where

$$H = \frac{2\left[(G_d^* - G_c^*)(19G_d^* - 16G_c^*) + (4\sigma_I/a)(5G_d^* + 2G_c^*)\right]}{(2G_d^* - 3G_c^*)(19G_d^* - 16G_c^*) + (40\sigma_I/a)(G_d^* + G_c^*)} . \quad (9.278)$$

The Palierne model underpredicts the shear modulus at higher volume concentrations of the disperse phase. It fails at volume concentrations in the vicinity of the maximum packing fraction. For vanishing interfacial tension, the model of *Kerner* [9.341] can be obtained from (9.278), which is restricted to high frequencies where the influence of the interfacial tension can be neglected. *Bousmina* [9.342] extended the Kerner model to the low-frequency range by considering the interfacial tension. For the complex modulus of the emulsions he obtained

$$G^* = G_c^* \frac{2(G_d^* + \sigma_I/a) + 3G_c^* + 3\varphi(G_d^* + \sigma_I/a - G_c^*)}{2(G_d^* + \sigma_I/a) + 3G_c^* - 2\varphi(G_d^* + \sigma_I/a - G_c^*)} . \quad (9.279)$$

The prediction of the Bousmina model deviates only slightly from that of the Palierne model.

Based on the Palierne model for the shear modulus of emulsions and the models of *Mooney* [9.231] (9.245) and *Krieger* and *Dougherty* [9.230] (9.244), *Pal* [9.343] developed models for the shear modulus of emulsions of two immiscible viscoelastic liquids taking into account that the shear modulus diverges at $\varphi \to \varphi_{max}$ in a similar way as done for the viscosity. *Pal* [9.343] found a reasonable agreement between the model predictions and experimental data.

Preparation and Stability of Samples. An overview of various methods commonly used to prepare emulsions has been given by *Morrison* and *Ross* [9.195]. The most usual mechanical equipment used to fabricate emulsions are high-shear mixers, colloid mills, homogenizers, and ultrasonic and sonic dispersers, which realize intensive steady or oscillating high-shear fields to breakup the droplets or aggregates and mix the phases. Some examples are shown in Fig. 9.97.

Besides the mechanical processes, some physico-chemical methods [9.195] exist to produce emulsions by phase inversion, phase inversion temperature (PIT), condensation or electric emulsification.

An important requirement to obtain stable emulsions is that the concentration of the emulsifier is high enough to be adsorbed at the required concentration at the newly developed interfaces in order to avoid, or at least retard, coalescence. If the surfactant at the droplet interface is inhomogeneously distributed due to shear, temperature

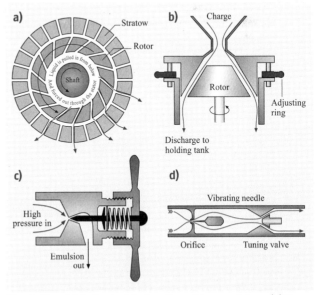

Fig. 9.97a–d Mechanical devices for making emulsions: (**a**) Rotor-stator disperser, (**b**) colloid mill, (**c**) single-stage homogenizer, (**d**) sonolator (after *Morrison* and *Ross* [9.195] with permission by Wiley)

effects or depletion, effects such as coalescence, Ostwald ripening, creaming or flocculation can destabilize the emulsion.

Coalescence is caused by the collision of two or more droplets, leading to depletion of surfactant at the interface. The collisions result in the formation of larger droplets and take place when the volume concentration of the disperse phase is high. Coalescence involves breaking of the interfacial film and is therefore irreversible. Various factors (solubility and concentration of the surfactant, pH value, salt concentration, phase-volume ratio, temperature, and properties of the film) affect the coalescence of emulsions.

Ostwald ripening only occurs in polydisperse emulsions. Collisions between two droplets may lead to one bigger droplet and one smaller droplet. Due to the material transition, small droplets become smaller and, in the extreme case, become solubilized in the continuous medium. Ostwald ripening requires a high solubility of the disperse in the continuous phase.

Due to the density differences between the two phases, a demixing process may occur. This process is called creaming [9.344] and may often be coupled with flocculation and a subsequent coalescence. The rate of creaming can be lowered by reducing the droplet size (with retention of the same interface loading of surfactant), lowering the density difference between the phases, or increasing the viscosity of the medium. In addition, the creaming rate is dependent on the volume fraction of the dispersed phase, and is usually slow in concentrated emulsions. Creaming is dominant at medium volume concentrations of 0.1–0.5 and droplet sizes of 2–5 μm. Complete demixing of the liquid phases caused by a combination of all other processes is called breaking. Further information on emulsion stability can be found in the contribution of *Walstra* [9.345].

Viscometric Flows.
Shear Viscosity. Emulsions containing small droplets can behave like suspensions. The suspension-like behavior of emulsions in shear flow has been experimentally demonstrated by *Teipel* and *Aksel* [9.321] for O/W emulsions and suspensions with the same volume concentrations of the dispersed phase, the same droplet and particle sizes, and the same viscosity for both continuous phases. The lower points in Fig. 9.98 exhibit Newtonian behavior $\eta_r = f(\varphi)$, while the upper points show slight shear-thinning behavior $\eta_r = f(\varphi, \text{Pe})$. The viscosity functions of suspensions and emulsions are nearly identical in both cases, which shows that the droplets can be regarded as rigid particles under the given shear conditions.

If the influence of droplet deformation and hence the modification of the droplet size distribution due to droplet break-up become relevant, the viscous behavior of emulsions is expected to be quite different. Consequently, the droplet size plays an important role in emulsions. Furthermore, additional viscoelastic effects may occur due to drop-shape relaxation mechanisms.

The influence of the droplet size, droplet deformation, and volume concentration of the dispersed phase on the rheological properties of the emulsions has been investigated by *Teipel* [9.346] and *Teipel* [9.319]. He confirmed the well-known effect that the viscosity of monodisperse emulsions with smaller droplets ($a = 0.5\,\mu\text{m}$) is significantly higher than that of emulsions with larger droplets ($a = 10\,\mu\text{m}$). At lower volume concentrations (40%) this effect has been observed over the whole shear-rate range investigated. However, at higher volume concentrations (up to 65%) the viscosities of the 0.5–10 μm emulsions coincide surprisingly at higher shear rates. This observation can be explained only by modification of the morphology of the 10 μm emulsion. Due to shear, the larger droplets deform and orientate in the flow direction. The effective hydrodynamic radius of the deformed and orientated droplets decreases and the viscosity tends to that of the 0.5 μm emulsion.

Pal [9.320] found that the relative viscosities of two polydisperse oil-in-water emulsions with $V_r \gg 1$

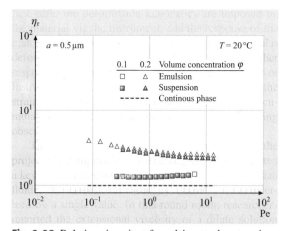

Fig. 9.98 Relative viscosity of emulsions and suspensions of monodisperse spherical particles as a function of the dimensionless shear rate (Péclet number) at identical volume concentrations of the disperse phase (after *Teipel* and *Aksel* [9.321] with permission by Wiley-VCH)

and mean droplet diameters of 6.5 μm(fine) and 32 μm (coarse), as well as 9.1 μm (fine) and 65 μm (coarse), scale with the particle Reynolds number based on the initial droplet radius. From an estimate of the Capillary numbers, he concluded that the deformation of the droplets due to shear is negligible. Only under this restriction are the results of *Pal* [9.320] applicable.

The influence of the volume concentration of the disperse phase and the viscosity ratio on the relative viscosity has been investigated by *Pal* [9.337]. In Fig. 9.99 some experimental data obtained by several authors for various emulsions ([9.337] for details) are summarized, showing the influence of the viscosity ratio for a given volume concentration. The limiting values of a constant viscosity at very low (bubbly liquid) and very high viscosity ratios (suspensions) are reached. Furthermore, the prediction of Pal's model 2 for the relative viscosity of emulsions (9.276) at a given volume concentration and a maximum packing fraction of 0.63 is depicted, which sufficiently describes the experimental trends.

Li and *Pozrikidis* [9.347] studied numerically the influence of insoluble surfactants on the rheology of dilute emulsions in Stokes flow. For the numerical calculations they assumed a viscosity ratio of $V_r = 1$ and a linear dependence of the interfacial tension on the local surfactant concentration. As an important result they discovered a decrease of the reduced interfacial stress $\sigma_{12}/\eta_c \dot{\gamma}$ with increasing capillary number, i.e., with decreasing interfacial tension. Hence, it can be concluded that, under otherwise identical conditions, the viscosity of emulsions with higher interfacial tensions is higher than for lower interfacial tensions due to the drop deformation, i.e., the emulsions behave like suspensions at higher interfacial tensions. This numerical result confirms the experimental results of *Teipel* and *Aksel* [9.321] for emulsions with $V_r \approx 60$. Details concerning the surface and interfacial tension can be found in Chap. 3.2.

Hollingsworth and *Johns* [9.348] studied the steady-state shear properties of various emulsion systems using the technique of rheo-nuclear magnetic resonance (rheo-NMR) coupled with a torque transducer. From direct measurements of the velocity maps they were able to determine the parameter of the Herschel–Bulkley model. Moreover, with the rheo-NMR technique it is possible to detect wall slip at both walls of the Couette cell used for the experiments.

Normal Stress Differences. Recently, *Montesi* et al. [9.349] investigated the shear behavior of a moderately concentrated, *attractive emulsion* of water droplets

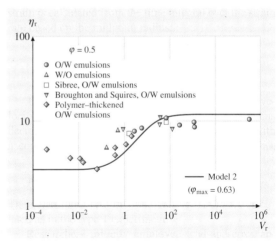

Fig. 9.99 Relative viscosity of emulsions as a function of the viscosity ratio at a given volume concentration (after *Pal* [9.337] with permission by the Society of Rheology)

dispersed in a lubricating oil ($V_r = 0.01$). They found a sharp transition from a positive to a negative first normal stress difference in a range where the shear viscosity is still constant. This observation has been explained by the formation of dough-rolling aggregates, leading to larger normal stresses in shear gradient direction compared to the normal stress in flow direction. Because this effect has also been found in other complex materials such as nanotube suspensions [9.286] it suggests some underlying rheological principle.

Li and *Pozrikidis* [9.347] calculated an increase of the first normal stress difference with increasing capillary number in shear flows, i.e., an increase of the elastic properties with decreasing interfacial tension or increasing deformation of the droplets. The negative second normal stress difference was found to increase in magnitude, reach a maximum and then start to decline with increasing capillary number. The minimum of the negative second normal stress difference occurred at capillary number $Ca = 0.4$ for all the investigated conditions.

Storage and Loss Moduli. Due to the large number of factors that influence emulsions and hence the range of types of the rheological behavior of emulsions, it is not possible to give generally valid curves for the storage and loss moduli. The behavior at various frequencies depends not only on the rheological properties of both phases, the composition of the emulsion, the droplet sizes, and the size distribution, but also on the properties of the interface between the two phases. Qualitatively, *Barnes* [9.316] pointed out some curves for the typ-

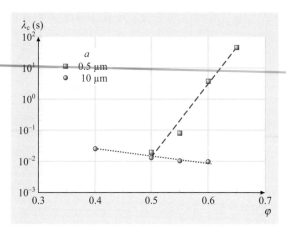

Fig. 9.100 Structural relaxation time as a function of the volume concentration for two monodisperse O/W emulsions with different droplet radii (based on results of *Teipel* [9.346]). The *dotted* and *broken lines* are drawn to guide the eye

ical behavior of emulsions in oscillatory shear flows. For low and medium concentrations the loss tangent as (ratio of the loss and storage modulus) is greater than one, indicating the dominance of the viscous component, whereas at higher concentrations the elastic component dominates. The influence of relevant parameters such as the viscosity ratio, interfacial tension or droplet sizes, and a possible solid–liquid transition have not been considered in the qualitative explanation. The statements given by Barnes have been experimentally confirmed by *Bower* et al. [9.350] for concentrated oil-in-water emulsions or by *Kulicke* et al. [9.351] for hydroxylpropylmethylcellulose-stabilized emulsions.

For flocculated W/O emulsions with a given surfactant concentration, *Pal* [9.352] showed that at low frequencies the elasticity of the emulsion is the dominating effect, whereas at higher frequencies the emulsion behaves as a viscous liquid. The crossover frequency ω_C or the characteristic structural relaxation time $\lambda_C = \omega_C^{-1}$, which characterizes the transition from the elastic to the viscous regime, depends on the volume concentration of the disperse phase at a given surfactant concentration. Contrary to the results of Pal, *Moates* et al. [9.353] observed for weakly flocculated concentrated emulsions dominating viscous properties at low and high frequencies and dominating elastic effects at medium frequencies, which indicates a double crossover in the curves of the storage and loss moduli.

Pal [9.352] found a decrease of the structural relaxation time with increasing volume concentration. In contrast to Pal, *Teipel* [9.346] observed an opposite effect in a monodisperse O/W emulsion with a droplet radius of 0.5 μm, i.e., a significant increase of the relaxation time over three decades with increasing volume concentration ($\varphi = 0.5$–0.65). For an emulsion with droplets of $a = 10$ μm a slight decrease of the relaxation time has been found, confirming the trend observed by Pal. The results are shown in Fig. 9.100.

Elongational Flows. Investigations of the elongational properties of emulsions are very few in the literature. *Anklam* et al. [9.354] used the opposed nozzle configuration to investigate the elongational behavior of W/O emulsions. They showed that for highly diluted emulsions the extensional viscosity decreases with increasing elongational rate similar to the shear behavior whereby the limiting Trouton ratio (for Newtonian liquids) of 3 seems to be reached. For higher volume concentrations an influence of the nozzle diameter and nozzle separation has been observed, which may be caused by the increasing impact of the yield stress. The effect of elongational thinning has also been found by *Plucinski* et al. [9.355] for a food emulsion.

As a final statement, one can conclude that emulsion rheology and partially suspension rheology are not completely understood and that research work in these fields is continuing.

9.4 Entrance Correction and Extrudate Swell

This section is dedicated to two effects that appear during flow in pressure-driven rheometers and also in plastic processing. The first is the entrance effect during flow from the reservoir into the capillary; it results in so-called Bagley plots. The second is so-called extrudate swell or die swell, which occurs as an increase of the diameter of the polymer extrudate with respect to the diameter of the capillary. This phenomenon is due to the elasticity of the polymer melt.

9.4.1 Correction for Entrance Effect: Bagley Correction

In the foregoing we neglected the effect of the entrance of the fluid into the capillary. The flow rate in the capillary

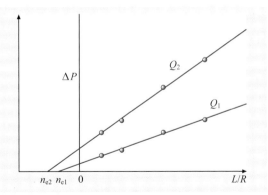

Fig. 9.101 Bagley correction plots for two flow rates: $Q_2 > Q_1$

is much higher than in the reservoir. This causes an increase of the kinetic energy, so that part of the imposed pressure ΔP in the reservoir above the capillary is used to accelerate the flowing mass in the capillary entrance region, so that

$$\Delta P = \Delta P_{\text{capillary}} + \Delta P_{\text{entrance}}. \tag{9.280}$$

For the determination of the shear stress at the wall we made use of the assumption of fully developed flow and neglected the entrance effect in $\partial \sigma_{11}/\partial z$ and $\partial \Delta P/\partial z$. The consequence of neglecting the entrance effect is an apparent lengthening of the capillary. For constant-volume flow rate the shear stress at the wall has to be corrected by increasing the length of the capillary by a term $n_e R$, where n_e is dimensionless. The end correc-

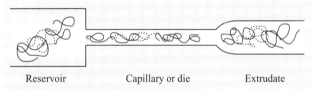

Fig. 9.103 Schematic representation of three macromolecules during flow in a capillary rheometer

tion n_e seems to be constant for a given radius of the capillary. The shear stress at the wall is now expressed as

$$\sigma_w = -\frac{R \Delta P}{2(L + n_e R)} \tag{9.281}$$

or

$$\Delta P = -2\sigma_w \left(\frac{L}{R} + n_e \right). \tag{9.282}$$

This entrance correction is called the Bagley correction. The value of the end correction n_e may be determined by measuring ΔP at constant-volume flow rate Q as a function of L/R, where preferably R is to be constant. This is shown in Fig. 9.101 for two values of the volumetric flow rate Q.

An example of a Bagley plot is shown in Fig. 9.102 for polyethylene with a melt flow index (MFI) of 2.9 at 190 °C for shear rates of 40–250 s^{-1}, as measured by Bagley himself [9.356]. It clearly shows straight lines and increasing values of n_e with increasing shear rate or increasing volume flow rate.

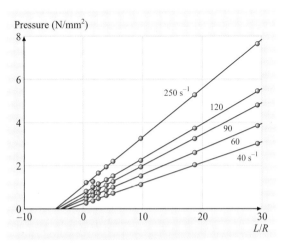

Fig. 9.102 Bagley plot for a polyethylene melt with melt flow index of 2.9 at 190 °C for shear rates of 40–250 s^{-1}, as presented by *Bagley* himself [9.356]

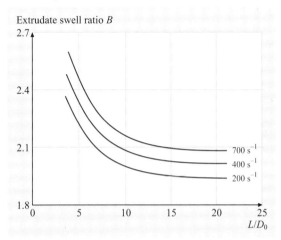

Fig. 9.104 Extrudate swell ratio as a function of L/D_0 for high-density polyethylene at 180 °C for shear rates, as indicated. After [9.357]

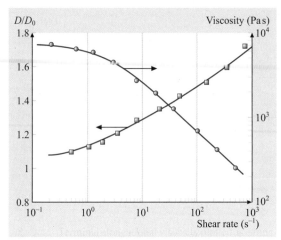

Fig. 9.105 Extrudate swell ratio and viscosity, both as functions of the shear rate for a polystyrene melt of $\overline{M}_w = 2.2 \times 10^5$ and $\overline{M}_w/\overline{M}_n = 3.1$, where $\overline{M}_w$ and $\overline{M}_n$ are the weight and number-averaged molecular weights, respectively. After [9.3]

9.4.2 Extrudate Swell or Die Swell

When a polymer melt leaves the die of a capillary rheometer (or, e.g., an extruder or injection moulding machine) its diameter increases, as was shown

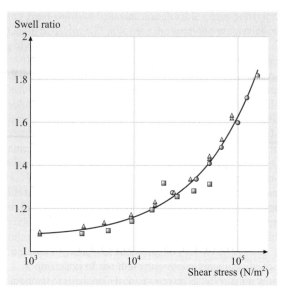

Fig. 9.107 Extrudate swell ratio of the same polystyrene melt as in Fig. 9.105, now plotted versus shear stress for temperatures of (*circles*) 160 °C; (*triangles*) 180 °C and (*squares*) 200 °C; L/D_0 ratios varied from 27 to 56. After [9.3]

schematically in Fig. 9.36. This swelling behavior is the consequence of the elastic properties of polymer melts. Let us consider a thought experiment in which

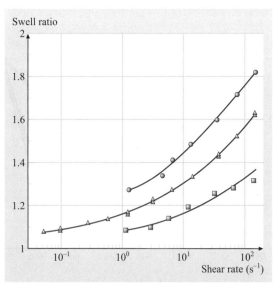

Fig. 9.106 Extrudate swell ratio of the same polystyrene melt as in Fig. 9.105, plotted versus shear rate for temperatures of (*circles*) 160 °C; (*triangles*) 180 °C and (*squares*) 200 °C; L/D_0 ratios varied from 27 to 56. After [9.3]

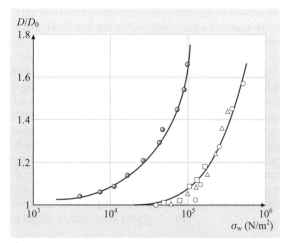

Fig. 9.108 Extrudate swell ratio as a function of shear stress at the wall for various polystyrene melts. *Filled symbols*: broad ($\overline{M}_w/\overline{M}_n = 3.1$) molecular-weight distribution polystyrene mentioned in Figs. 9.105–9.107. *Open symbols*: narrow ($\overline{M}_w/\overline{M}_n < 1.1$) molecular-weight distribution polystyrene, with $\overline{M}_w = 1.6 \times 10^5$ (After [9.3])

a crosslinked rubber is driven through a capillary. When the rubber leaves the capillary it will swell again, trying to recover its original dimensions due to elasticity. A polymer melt will swell almost like the rubber. However, upon leaving the die the original dimensions of the reservoir will not be reached, because of stress relaxation, primarily due to the coil–stretch–coil transition of macromolecules (see Fig. 9.103) and their reptation (disentanglement and subsequent re-entanglement). Die swell, or the swell ratio B, being a function of many parameters, is defined as ratio of the diameters of the extrudate and the capillary:

$$B \equiv D/D_0 = B(\dot{\gamma}_w, T, t, L/D_0, M, \text{MWD}), \tag{9.283}$$

where D_0 is the diameter of the capillary with circular cross section, D is the diameter of the extrudate at a relatively large distance from the outlet, T the is temperature, L/D_0 the length-to-diameter ratio of the capillary, M is the molecular weight of the polymer, and MWD its molecular-weight distribution (polydispersity). B can reach values of 2–4.

Many of the effects of the various parameters can be attributed to stress relaxation. In general die swell is larger when the residence time in the capillary t is shorter and when the polymer relaxation times are longer. Several of these effects are demonstrated in Figs. 9.104–9.108. In Fig. 9.104 the extrudate swell ratio is plotted versus L/D_0 for various values of the shear rate [9.357]. This shows that, at larger L/D_0 ratios, there is more time for relaxation and accordingly the swell ratio decreases. The same holds for the shear rate as the parameter: increasing the shear rate means shorter residence times and thus less time for relaxation; accordingly the swell ratio increases with increasing shear rate. In Fig. 9.105 the extrudate swell ratio and viscosity of a polystyrene melt are shown as a function of shear rate. This clearly shows again that die swell increases with shear rate, in this case even in the range where the viscosity decreases, and hence where the long relaxation times that are responsible for high die swell ratios are already felt. One could imagine that die swell ratios would even be higher if the viscosity would not be shear thinning.

In Fig. 9.106 die swell is shown as a function of shear rate for temperatures of 160 °C, 180 °C, and 200 °C for the same polystyrene melt as presented in Fig. 9.105.

The L/D_0 ratios are high, so that die swell is almost independent of this ratio. It is not surprising that the die swell ratio decreases with increasing temperature, because the relaxation times are strongly decreasing functions of temperature. If one plots the extrudate swell ratio versus some function of relaxation times one could expect a common curve. This is demonstrated in Fig. 9.107, where the extrudate swell ratio is now shown as a function of the shear stress, which is equal to shear rate × viscosity, and the viscosity is strongly dependent on the relaxation times. The points measured at the three different temperatures follow the same line over more than two decades of shear stress. Apparently, in this case the die swell ratio scales well with viscosity.

Finally in Fig. 9.108 die swell data are shown as a function of shear stress at the wall for various polystyrene melts. The filled and open symbols refer to polystyrenes with broad and narrow molecular-weight distributions, respectively. The *narrow* polymer was measured under different conditions. The *broad* polymer exhibits much higher extrudate swell ratios than the *narrow* polymer. The weight-averaged molecular weight of the *broad* polymer is approximately two times higher than that of the *narrow* polymer, which means that the viscosity is 10 times higher (the viscosity increases as $\overline{M}_w^{3.4}$). Apparently, the viscosity is not the only parameter that is responsible for the die swell phenomenon. Elasticity of the melt, as expressed by G_e or J_e or ψ_1, also seems to be of great importance [9.3].

9.4.3 Conclusions

In this Section measurements relevant to two phenomena that are observed in pressure-driven rheometers (and also in plastic processing such as extrusion or injection moulding) are discussed:

1. Entrance correction, due to accelerating forces during the flow from the reservoir into the capillary. This results in an apparent increase of the length of the capillary, which can be measured with the aid of a Bagley correction plot.
2. Extrudate swell, which occurs when the polymer leaves the capillary, and also due to elasticity of the polymer melt. This phenomenon depends on many parameters and in principle can be qualitatively related to the relaxation times. The exact description is not available yet.

9.5 Birefringence in Non-Newtonian Flows

The optical properties of transparent media, representing the interaction of the electric vector of the electromagnetic wave with electrons in the material are briefly reviewed. A distinction is drawn between optically isotropic and anisotropic media, and the molecular origin of optical anisotropy is discussed. Methods for measurement and imaging birefringence are discussed and compared, including Sénarmont, polarization modulation, and two-color birefringence techniques. Experimental issues discussed include choice of light source, windows, photodetectors, optical components, and the geometry of flow cells. Finally the origin of birefringence for the important class of flexible coil polymers is discussed, with interpretation of birefringence in terms of molecular strain, and in terms of bond polarizability, and micro- and macroform birefringence.

Light propagates through transparent media via a series of interactions with electrons, mediated by the bonds between atoms. Birefringence refers to a physical system that transmits light but is optically anisotropic, i.e., the light propagates differently in different directions; generally (but not always) this manifests itself as a directional variation of the refractive index (Chap. 3.7). In general this is due to the interaction between the electric vector of the light and the bond polarizabilities of the medium.

Many substances are isotropic, that is they have the same physical properties in all directions. This includes most fluids, most amorphous materials, and cubic crystal structures. Of specific importance here is that the bonds that mediate the passage of light either possess such symmetry or such randomness that the light propagates in the same way in all directions. Indeed normal, stress-free, single-phase fluids are intrinsically optically isotropic, even when they contain, as solutions or suspensions, molecules or particles that are themselves anisotropic. This is due to the randomization of the orientational order of the molecules or particles.

The exception, at rest, is the liquid-crystalline fluid, where above a specific concentration of anisotropic molecules, spontaneous order will emerge in the form of the liquid-crystalline phase. This ordering is long-range and cooperative, typically extending over many microns. In nematic phases this corresponds to orientational order, but in smectic liquid crystals it is both orientational and positional, and in cholesteric materials it possesses a longer-length-scale twisting structure, arising generally from the packing constraints of helical molecules. Such systems also include the important classes of surfactant and lipid mesophases, where anisotropic structures have specific biological functions, for instance in cell walls.

Non-Newtonian fluids that are isotropic at rest are of special interest here. In general the non-Newtonian flow characteristics arise from an internal structure which changes with applied flow. This may arise from orientation of suspended particles or colloidal structures (Sect. 9.2), where the particles possess an intrinsic optical anisotropy. Of particular importance is the non-Newtonian flow of flexible polymer melts and solutions (Sect. 9.1). At rest the polymer chains are locally anisotropic (in general most of the covalent bonds lie along the chain direction), but the chains exist as *drunkard's walk* random coils, either interpenetrating in melts and concentrated solutions or isolated in dilute solutions. On the scale of the wavelength of light such coils are physically and optically isotropic.

When such polymeric fluids exhibit non-Newtonian behavior it is due to a flow-induced order being imposed upon the molecules. Typically, beyond some flow-rate characteristic of the system relaxation time, the random coils become orientated preferentially, or, exceptionally, stretched by the flow field. This, coupled with the associated relaxation processes as entropy drives the coil back to its equilibrium conformation, creates the anisotropic stresses and normal forces characteristic of such fluids.

As discussed below, birefringence is a relatively straightforward phenomenon to measure. For 200 years birefringence has been used as a window into molecular processes. It has been used in conjunction with flow since 1873 when Maxwell discovered birefringence in

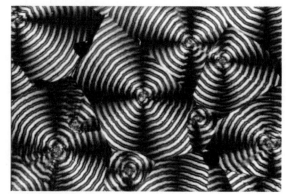

Fig. 9.109 Birefringent spherulites growing from a melt of biological thermoplastic poly(hydroxybutyrate)

flowing Canada balsam resin [9.358]. Many birefringent techniques can give fast, noninvasive information related to the microstructure and origin of non-Newtonian flow stresses. Further, normally a birefringent microscope image is available, giving information on the spatial variation of stresses and orientational order created in the flow field.

There is a problem, however. This is analogous to the phase problem is X-ray scattering: there is not a unique relationship between birefringence and induced microstructure. The same birefringence can arise from more than one molecular or phase structure, so that birefringence analysis needs to contain a model and a conjectured mode of the deformation of that model, so that calculated birefringence can be compared with observed birefringence. Sometimes, then, the correspondence is so compelling that the structure is clearly *right*. This was famously the case with spherulitic growth structures in crystallization of semicrystalline polymers [9.359] (Fig. 9.109). However, sometimes the interpretation can be contentious so that unambiguous, structures can only be assigned through a combination of birefringence studies with other techniques such as light, X-ray, or Neutron scattering.

9.5.1 The Molecular Origin of Birefringence

As discussed above, the essence of an optically anisotropic material is that its electron bonds possess directional variation. Now light propagates through the interaction of the electric E vector with electrons in the bonds, resulting in photon re-emission. The speed of this propagating wave is determined by the difference in the frequency of the wave and the characteristic frequency of the electrons. The nature of this difference determines the speed of propagation and hence the refractive index, n, for the wave (Chap. 3.7). It clearly follows that light with differing orientations of the E vector will exhibit differing refractive indices and propagation delays in the material. Specifically there will generally be an E vector direction exhibiting a maximum refractive index and another a minimum. A material or fluid of this type that displays spontaneous or flow-induced variation of refractive index is said to be *birefringent*. Dependent upon the polarization and direction of the incoming light, components of the light will propagate at different rates, recombining on emergence from the medium to give the phase and chromic effects familiar in microscope birefringent images.

In general there will be one or more directions through such a medium where the bonds (or components of bonds) perpendicular to that direction exhibit the same interaction with the E vector, which is itself perpendicular to the direction of propagation. In this special direction, the refractive index will not be a function of the direction of polarization and the medium will not exhibit birefringence. This is termed the *optic axis*. For a given direction of propagation of light there will be a direction in the fluid in which the refractive index (n_1) is a maximum (the principal axis of the birefringence) and an orthogonal direction in which the refractive index is a minimum (n_2). For this direction the birefringence is $\Delta n = (n_1 - n_2)$. Light polarized in each of these directions will propagate at different speeds (c/n) and emerge out of phase. The *retardation phase shift* (δ) results in interference when the waves recombine.

The birefringence, Δn, and δ are related by

$$\Delta n = \frac{\lambda \delta}{2\pi d}, \qquad (9.284)$$

where d is the path length through the birefringent medium, and λ is the incident wavelength.

The birefringence axes are of course related to the internal molecular structure through the intrinsic birefringence of the structural entities (from the sum of bond polarizabilities) and the average orientation of the entities (molecules, etc). The average orientation of the entities is characterized as the *director*. This will often, but not always, coincide with the principal axis of the birefringence. The molecular direction can correspond to the maximum birefringence (positive birefringence) or the minimum (negative birefringence). In a few cases (e.g., polyacrylamide) the bond polarizabilities almost exactly cancel, resulting in a non-birefringent entity.

Here we should formally mention that, if the electromagnetic frequency is comparable to one of the characteristic frequencies of the anisotropic electron vibrations, this radiation will be heavily absorbed, so that one polarization or direction may be suppressed. This phenomenon is called *dichroism*. For many materials and fluids such frequencies are in the infrared and the technique of infrared dichroism spectroscopy can indicate stress and orientational order associated with specific bond types. Beyond this, dichroic materials are the basis for the polarizer filters ubiquitously used in sun glasses and the optical systems described below.

9.5.2 Techniques for Birefringence Measurement

The polarizing microscope is shown schematically in Fig. 9.110a (the lenses have been omitted for clarity). It

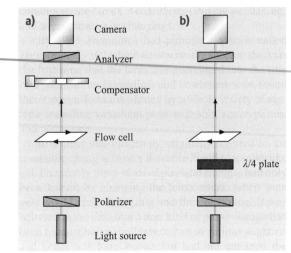

Fig. 9.110a,b The optical trains of (**a**) a polarizing microscope, (**b**) a Senarmont compensator

consists of a source, which can be monochromatic or white light, light from which passes through a polarizer to pick out a unique polarization direction, then through the medium of interest and an optional compensator and finally through an analyzer, generally orientated orthogonal to the polarizer. Such an instrument is generally semiquantitative; optically isotropic media will be seen as dark, since there is no rotation of the polarization direction and the light is stopped by the analyzer. The degree of retardation of the waves determines the perceived brightness when the emerging waves recombine. A full analysis of the propagation of polarized light through anisotropic media can be approached through Jones and Mueller matrix calculations [9.360], the intensity of the transmitted light is

$$I_s = I_0 \sin^2 2\theta \sin^2(\delta/2), \qquad (9.285)$$

where I_s is the intensity of transmitted light, I_0 is the intensity of the input beam, and θ is the angle between the incident polarization direction and the direction of the principal axis of the birefringence (maximum refractive index n_1), δ is the average retardation angle.

If θ is set to 45° the transmitted intensity becomes simply [9.361]:

$$I_s = I_0 \sin^2(\delta/2) \quad \text{or for small } \delta: \qquad (9.286)$$

$$I_s = \frac{I_0 \delta^2}{4}. \qquad (9.287)$$

For polychromatic or white light, generally colors are seen due to different interference for different wavelengths; this can be semiquantitatively interpreted through the *Michel-Lévy* chart. The compensator consists of a well-defined adjustable crystal, typically of calcite or quartz. More-precise measurements can be obtained by adjusting the compensator to provide an exactly equal and opposite retardation in the optical train, at which point the birefringence can be read from the compensator calibration.

Equation (9.287) demonstrates that the quadratic dependence of intensity on retardation and problems will arise from subtraction of scattered and stray light at low intensities due to the nonlinear dependence of δ upon I_s. A particular problem with flow birefringence is stray birefringence due to the variable stresses in the windows. These can dominate the signal, especially for solution studies, where the birefringence may be very small. For more-quantitative assessment, higher speed, greater linearity, and greater sensitivity many other techniques have been invented.

Figure 9.110b illustrates the Sénarmont compensator technique [9.362]. Now a quarter-wave plate (introducing up to an additional $\lambda/4$ retardation dependent upon rotation, but normally aligned with its high-refractive-index direction parallel to the low refractive index of the analyzer with the polarizers crossed) is included in the optical train. The analyzer is then uncrossed a little (by an angle ϕ). This can be used to compensate for stray window birefringence or to compensate out the signal birefringence as a way of measuring it.

Without flow birefringence, the transmitted intensity can be written

$$I_s = I_0 \cos^2 \varphi, \qquad (9.288)$$

where ϕ is the angle between the polarizer and analyzer.

We can see that the same intensity received by the photodiode can be induced either by a flowing solution [that is caused by retardation in the sample, (9.287)] or by uncrossing the polarizers in the absence of the sample (9.288).

Comparing (9.288) and (9.286) we can estimate the value of the retardation in a flow knowing the rotation angle of the analyzer

$$\sin^2(\delta/2) = \cos^2 \varphi. \qquad (9.289)$$

From (9.289)

$$\delta = \pi - 2\varphi. \qquad (9.290)$$

This value is used in the calculation of experimental birefringence via equation (9.285)

Substituting (9.290) into (9.284) we finally obtain for Δn:

$$\Delta n = \frac{\lambda \arcsin(\cos\varphi)}{\pi d}. \qquad (9.291)$$

A more-useful *linear* detection method [9.363], yielding much improved signal-to-noise ratio, can easily be adapted from the Sénarmont compensator. This simply involves uncrossing the polarizers to permit a background intensity to be transmitted. It is then possible to express δ as $\delta = \delta_1 + S$, where δ_1 is the constant background signal introduced by uncrossing the polarizers and S is the retardation introduced by all other optical effects. Hence from equation (9.287) we obtain:

$$I_s = \frac{I_0}{4}(\delta_1^2 + 2\delta_1 S + S^2) = \frac{I_0}{4}\delta_1^2 + \frac{I_0}{2}\delta_1 S + \frac{I_0}{4}S^2. \qquad (9.292)$$

If δ_1 is large compared to S, the S^2 term may be neglected, leaving a constant intensity plus a term linear in S. In practice the background intensity signal can be set to around 10 or more times the intrinsic signal, only limited by the stability of the light source.

Such systems work well in cases where the orientational angle θ is known or defined by the flow symmetry (e.g., in extensional flows). They can be impractical where the orientational angle is unknown or variable; this is generally the case in Couette, cone and plate, Poiseuille and other shear flows (Sect. 9.1.1). In simple shear flows, generally the principal axis of the birefringence (the director) will lie at a setting angle to the velocity and the velocity gradient vector. Techniques such as polarization modulation (PM) and two-color flow birefringence (TCFB) are capable of determining simultaneously this orientational angle and the optical retardation and birefringence.

PM [9.364] techniques cause a fast oscillation of the polarization direction, which then samples all refractive indices normal to the direction of propagation, enabling analysis to determine δ and θ simultaneously. This is commonly achieved by mechanically rotating a $\lambda/2$ plate (circular polarizer) in the beam line. There are photoelastic alternatives requiring no moving parts, such as Pockels or Faraday cells. Phase-sensitive detection techniques are employed to monitor the intensity and compare it with the incident beam, enabling correction for variations of the input intensity and an absolute determination of δ, without calibration.

The TCFB technique [9.365] is powerful, sensitive and faster than the PM techniques. The technique employs, typically, the blue (488.8 nm) and green (514.5 nm) lines of an argon-ion laser with the polarization of the two beams rotated by 45° to each other, to provide the two independent measurement required to determine δ and θ separately. However, TCFB cannot determine the sign of the orientation angle or the retardation. *Fuller* [9.366] has presented a comprehensive description of the field of optorheometry.

Light Sources

Any precise determination of birefringence requires the light source to be highly monochromatic. Further, the achievement of maximum sensitivity in possibly tiny volumes, requires high focusability of the source. These conditions point to lasers as the primary source. Most systems have utilized relatively low-powered HeNe lasers, of order 10–20 mW having λ of 632.8 nm. Optical components can be easily and economically obtained with antireflection coatings optimized for HeNe wavelengths, practically eliminating unwanted reflections.

Such lasers are cheap, reliable, and robust. They are not, however, intrinsically stable and the output intensity and power will drift considerably as the laser warms up and with ambient conditions and age of the laser. As discussed above, stability is an important requirement to achieve optimum signal-to-noise ratio and sensitivity. Stabilized HeNe lasers are available, generally operating on the principle of sensing the laser output and closed-loop feedback into a heater circuit to maintain either frequency or intensity stability. The lasers are primarily intended for metrology, so output powers tend to be at the low end of the spectrum, around 1 mW.

Increasingly solid-state lasers are available with power and performance comparable with or exceeding that of HeNe lasers. These are also intrinsically unstable with temperature, but can be feedback stabilized rather more easily than the HeNe with small Peltier heaters and coolers. They can also be an order of magnitude cheaper than similar HeNe devices. A possible drawback is that the effective source size is larger than for HeNe lasers, so that the minimum size to which the beam may be focused is limited, as is the specific achievable intensity. Such lasers can have wavelengths of operation very similar to HeNe lasers, enabling quite effective use of HeNe antireflection optics.

Windows

Clearly high-clarity low-birefringence windows are important in optorheology. But for solution studies, where the birefringent signals are commonly very small (as low as $\delta = 0.01$ nm or $\lambda \times 10^{-5}$), windows are the bane of optorheology birefringence measurements. Even though the windows may be free of stress and birefringence at rest, during flow they will inevitably contain stresses, arising from the pressure, the shear forces or the normal forces, dependent upon the geometry. These stresses will induce variable birefringence that may swamp the fluid birefringence. Part of the solution is usually to utilize thick windows. The retardation will scale with thickness for a particular stress, but the stress will generally reduce much more rapidly with window thickness. Similarly the area of the window should be kept as small as is feasible.

Photodetectors

Peltier-cooled charge-coupled device (CCD) detectors now provide excellent performance combined with speed and convenience. They can have greater than 1 megapixel resolution. Air cooling down to of the order $-40\,°C$ can reduce thermal (dark-current) noise to the level of 0.01 electrons per second per pixel. Coupled with photon efficiencies of the order of 50%, long integration times can be employed to recover even the weakest signal. Selectable *areas of interest* enable fast frame rates. At the same time 12- or 16-bit cameras can provide a wide dynamic range, which is essential for experiments that require background subtraction.

Polarizers and Waveplates

The performance of any birefringence experiment will be restricted by the performance of the polarizers. Dichroic film polarizers are designed to produce a ratio of intensity of the unwanted polarization to the wanted of order 10^{-5} (the extinction ratio). Higher ratios, up to 10^{-6}, can be obtained from birefringent components such as calcite (Glan–Taylor and Glan–Thompson prisms). These devices have the further advantages of high power throughput without damage and high transparency for the wanted polarization (up to 90%).

Temperature stability of wave plates is of prime importance, and is best achieved by the use of *zeroth-order* plates, which utilize compensating components to give exactly $\lambda/4$, rather than higher-order but much cheaper plates which give $(4n+1)\lambda/4$, with correspondingly worse stability.

Flow Cells

All optorheological experiments, with the exception of thread-line experiments, require the construction of flow cells to define and constrain the flow and provide windows through which the optical effects can be observed. These flows can be defined that which approach idealized flows as closely as possible.

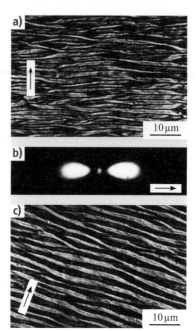

Fig. 9.111a–c Birefringence in a parallel-plate flow cell in relaxation of oriented thermotropic liquid-crystalline material ((**a**) with polars at 45° to flow direction, (**c**) with polars at 90° to flow direction), together with the accompanying light scattering pattern (**b**). The bands correspond to regions of co-operative disorientation, where the variation in brightness arises from periodic variation of the molecular director. Such banding is ubiquitous in liquid-crystalline systems, but is poorly understood

Simple shear flow will be approached through a parallel-plate or cone-and-plate apparatus (Sect. 9.1.1), usually with the light beam along the velocity gradient direction and the plates forming the windows. An alternate arrangement for simple shear is the Couette or concentric cylinders apparatus. In principle, this permits optical probes along the velocity or velocity gradient direction, but optical trains are difficult to define compared to with parallel-plate cells.

Figure 9.111 shows simple shear birefringence in a thermotropic liquid-crystalline copolymer. The flow cell is a parallel-plate cell, with polarizing optics and simultaneous light scattering. The pronounced banding texture correspond to regions of cooperative disorientation, where the variation in brightness arises from periodic variation of the molecular director. The local director can be assessed by rotation with respect to the crossed polarizer and analyzer; darkness indicates a director parallel to a principal axis. Such banding is ubiquitous in liquid-crystalline systems, but poorly understood, the banding suggests a zigzag variation in the director with respect to the shear direction, whilst the single peak in the light scattering suggests a sinusoidal variation in director.

Extensional flows can be attained through capillary entrance flows (Sect. 9.1.2), G. I. Taylor four-roll mill experiments or cross-slots (pure shear) and opposed jets. The opposed jets realize a good approximation to ideal extensional flows. If the jets impinge, then the flow will correspond to a uniaxial compression, which is equivalent to a biaxial extension. If the jets are sucking, then the flow is close to ideal uniaxial extension. The four-roll mill and jets devices permit very high strain rates coupled with the incorporation of a stagnation point. The stagnation point has a special significance, since the strain rate is finite, but the residence time is infinite, so that the fluid strain is unlimited and birefringent structures build up to close to equilibrium (maximum extension or orientation). This leads to discrete *strands* of birefringence, localized in space. Such a structure is shown in Fig. 9.112a, where the solution is just 10 ppm concentration of atactic polystyrene; retardations of less than $\lambda/1000$ can be resolved. Figure 9.112b shows the birefringent profile across the strand through the stagnation point.

Fluid mechanics presents a range of *benchmark* problems, such as flow around corners and around cylinders and spheres. Appropriate cells can be constructed to enable an optical probe to analyze birefringent patterns. In conjunction with velocimetry such approaches can visualize the stress distributions and their effects on the flow-field. Figure 9.113 shows the birefringence behind a falling ball, again the birefringence has a strand-like appearance, arising from the stagnation point behind the sphere.

Beyond well-defined flows there are flows of major industrial and commercial interest, such as porous media flow. This is generally modeled by flow through an assembly of glass spheres (ballotini). Multiple scattering makes the use of optical probes problematic, but refractive-index matching of the fluid to the glass has resulted in direct birefringence observations in random porous media [9.367].

9.5.3 Relation Between Birefringence and Molecular Strain

The orientation of the segments of a macromolecule depend on the volume and deformation of the macromolecule as a unit. Therefore birefringence demonstra-

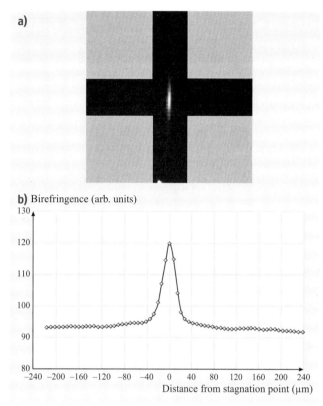

Fig. 9.112 (a) Birefringence response from a solution of 10 ppm 10.2×10^6 molecular-weight atactic polystyrene (aPS) in dioctyl phthalate (DOP). The flow corresponds to a pure shear extension along the vertical channel, coupled with a stagnation point at the center. The strain rate is $995\,\mathrm{s}^{-1}$. (b) A birefringence scan across the birefringent strand through the stagnation point

ted by polymer solutions is a function of the geometrical, hydrodynamic and optical properties of the dissolved macromolecules.

If the optical polarizability of the segments is anisotropic, their preferential orientation within the coiled molecule will result in an overall optical anisotropy of the chain (intrinsic anisotropy). According to *Peterlin* [9.368, 369] the optical anisotropy of a randomly coiled macromolecule is:

$$\gamma_1 - \gamma_2 = 3/5(\alpha_1 - \alpha_2)_i (h^2/h_0^2) v(\beta) \ ; \quad (9.293)$$

here: γ_1 and γ_2 are the principal polarizabilities of a macromolecule, α_1 and α_2 are the polarizabilities of a segment parallel and perpendicular to its length with subscript i denoting segmental anisotropy, h^2 is the mean square end-to-end distance of a deformed molecule, h_0^2 is the mean square end-to-end distance of a molecule at

Fig. 9.113 Birefringence behind a 10 mm diameter steel ball bearing falling through a 0.7% solution of 10.2×10^6 molecular weight atactic polystyrene (aPs) in dioctyl phthalate (DOP)

rest (the gyration radius squared), the factor:

$$\upsilon(\beta) = \frac{5}{3\beta}\left(1 - \frac{3\beta}{L^{-1}(\beta)}\right), \quad (9.294)$$

where $\beta = h/L$ is the degree of extension of a molecule, L is the contour length of a macromolecule, and $L^{-1}(\beta)$ is the inverse Langevin function.

The intrinsic optical anisotropy of a chain (9.293) has been derived from the average optical polarizability of the ensemble of all conformations compatible with a fixed end-to-end vector $\mathbf{h}$ [9.370]. The factor $\upsilon(\beta)$ takes into account the small nonlinearity effect on the optical anisotropy of the macromolecule as a function of the mean square end-to-end distance and varies from 1 at $\beta = 0$ to 5/3 at $\beta = 1$.

Now one has to relate the polarizability of a molecule and refractive index of an optically anisotropic medium, and also to obtain the dependence of birefringence on the degree of extension of a molecule in a hydrodynamic field. The Lorenz–Lorentz formula relates the average refractive index, n, to the specific polarizability of a medium P

$$\frac{n^2 - 1}{n^2 + 2} = \frac{4}{3}\pi P. \quad (9.295)$$

Upon differentiating, this gives

$$\frac{6n}{(n^2 + 2)^2}\,\mathrm{d}n = \frac{4}{3}\pi\,\mathrm{d}P \quad (9.296)$$

and the increment in polarizability can be related to the polarizabilities of the subunits through

$$\mathrm{d}P \approx \Delta P = \frac{MN_A}{\rho}(\alpha_1 - \alpha_2), \quad (9.297)$$

where M is the molecular weight and ρ is the density and N_A is Avogadro's number.

Kuhn and Grün assumed the validity of the above formula in respect of the separate polarizabilities of the subunits (α_1, α_2) and applied the differential equation (9.296) to the finite difference of polarizabilities (9.293), since the absolute values of differences ($\alpha_1 - \alpha_2$) are small. Rearranging equation (9.296) we get

$$\frac{\Delta n}{n} = \frac{2\pi N_A \rho}{9Mn^2}(\alpha_1 - \alpha_2)(n^2 + 2)^2, \quad (9.298)$$

The dependence of birefringence on the degree of extension of macromolecules in flow was obtained by *Peterlin* [9.371]. He considered the problem of the dynamics of macromolecules in flow and used for this purpose two models: the simple dumbbell [9.372] and the more-realistic necklace model [9.373, 374]. The difference between solutions of the diffusion equation obtained by the two models resides more in the values of numerical coefficients than in the functional dependence.

For a dumbbell model in a predominantly extensional flow, where stretching occurs in the x-direction, the specific birefringence of a solution ($\Delta n/nc$) can be written as:

$$\frac{\Delta n}{nc} = \frac{6\pi}{5h_0^2}\left\{\frac{n^2 + 2}{3n}\right\}^2(\alpha_1 - \alpha_2)\frac{N_A}{M}\upsilon(\beta)\langle\beta_x^2 - \beta_y^2\rangle, \quad (9.299)$$

and the corresponding expression for simple shear flow

$$\frac{\Delta n}{nc} = \frac{6\pi}{5h_0^2}\left\{\frac{n^2 + 2}{3n}\right\}^2(\alpha_1 - \alpha_2)\frac{N_A}{M}$$
$$\times \upsilon(\beta)\sqrt{\langle\beta_x^2 - \beta_y^2\rangle^2 + 4\langle\beta_x^2\beta_y^2\rangle^2}, \quad (9.300)$$

where β_x and β_y are square projections of vector $\boldsymbol{\beta} = \mathbf{h}/L$ on the x- and y-axes,

$$\beta^2 = \beta_x^2 + \beta_y^2 + \beta_z^2 \quad (9.301)$$

with β_z being the projection of β on the third axis z, n is the refractive index of a solution, and c is the concentration, conventionally in g/cm^3. The total polarizability $(\alpha_1 - \alpha_2)$ is given by (9.309) below.

The ratio of experimental birefringence (Δn) to the maximum achievable for fully stretched molecules (Δn_0) is

$$\frac{\Delta n}{\Delta n_0} = \frac{3}{5}\upsilon(\beta)\langle\beta_x^2 - \beta_y^2\rangle. \quad (9.302)$$

In order to plot the dependence of the dimensionless birefringence (9.302) on the degree of extension we make the following approximations. Taking into account that the deformation occurs mainly in the stretching direction x we obtain

$$\langle\beta_y^2\rangle = \langle\beta_z^2\rangle = \frac{1}{3N}, \quad (9.303)$$

where N is the number of segments in a molecule, and symmetry of the y- and z-directions is assumed (axisymmetric stretching).

From (9.301) and (9.303) it follows that

$$\langle \beta_x^2 - \beta_y^2 \rangle = \beta^2 - 1/N. \tag{9.304}$$

The function $\upsilon(\beta) = 5/3\beta[1 - 3\beta/L^{-1}(\beta)]$ can be approximated by the dependence [9.375]:

$$\upsilon(\beta) \approx \left(1 + \frac{\beta^2}{3} + \frac{\beta^4}{3}\right). \tag{9.305}$$

Using (9.302) and the approximation (9.305), we finally obtain for the dependence of dimensionless birefringence on the degree of extension

$$\frac{\Delta n}{\Delta n_0} = \frac{3}{5}\left(1 + \frac{\beta^2}{3} + \frac{\beta^4}{3}\right)\left(\beta^2 - \frac{1}{N}\right). \tag{9.306}$$

The dependence (9.306) is shown in Fig. 9.114 for $N = 10^4$, which corresponds to a molecule of atactic polystyrene with $M \approx 8 \times 10^6$ (the $1/N$ term is negligible)

An analogous result for the dependence of birefringence on degree of extension has been obtained by Treloar. His model of the uncoiling process, as well as Peterlin's dumbbell, implied that the mean molecular segmental orientation (denoted by the end-to-end separation, h) during each stage of uncoiling in the flow field, is consistent with that of a molecule in a stationary solution whose ends are fixed at a distance h apart.

Other models of uncoiling, for example, the yo-yo model of *Ryskin* [9.376] and the models of *Rallison* and *Hinch* [9.377], *Wiest* et al. [9.378], *Larson* [9.379] and *Hinch* [9.380], who predicted the formation of kinks during uncoiling, would lead to birefringence versus extension curves different from that of Peterlin and Treloar models. The models of [9.376–380] produce lower estimated molecular strain values for the same birefringence during extension in comparison to the Peterlin and Treloar result (9.306), since folded chains give the same birefringence as extended ones.

9.5.4 Optical Properties of Macromolecules in Solution: Effects of Macroform and Microform Birefringence

In the derivation of the intrinsic anisotropy (9.293) the optical influence of the solvent, which modifies the internal field and hence the effective optical properties of the segment as well as those of the whole coil, has been neglected. If the refractive index of the solvent n_s is different from that of the dissolved polymer, n_p, an additional anisotropy of the molecule arises as a result of optical interaction between separate parts of the chain [9.381]. Interaction between chain elements far removed from each other (long-range optical interaction) leads to an anisotropy of the polarizing field within the molecular coil. This anisotropy is positive in sign and is directly dependent on the shape of the molecular coil (macroform anisotropy). The difference between the two principal polarizabilities of the macromolecule (in the direction of vector h and perpendicular to it) corresponds to the macroform effect and is [9.382]:

$$(\alpha_1 - \alpha_2)_f = \left(\frac{n_p^2 - n_s^2}{4\pi n_s \rho N_A}\right)^2 \frac{M^2}{V_s}(L_2 - L_1), \tag{9.307}$$

where ρ is the density of the polymer, $V_s = 0.36 h^3$ is the volume of the molecular coil in solution (including the solvent in the molecule), and $L_2 - L_1$ is the optical shape factor, which is a function of the axial ratio of the statistical coil. The anisotropy of the polarizing field that occurs as a result of this is therefore a value averaged over the whole volume of a molecule.

The neighboring elements (monomers) in the chain are arranged in a linear order and their interactions cannot be spherically symmetrical. Such asymmetrical short-range interaction causes local anisotropy of the polarizing field, which is analogous to the average field anisotropy due to the geometrical asymmetry of the entire molecule. Thus an additional anisotropy of polarizability arises (microform anisotropy), which is also positive in sign. The principal polarizability difference corresponding to the microform effect can be written

$$(\alpha_1 - \alpha_2)_{fs} = \left(\frac{n_p^2 - n_s^2}{4\pi n_s}\right)^2 \frac{M_0 s}{\rho N_A}(L_2 - L_1)_s. \tag{9.308}$$

Here $(L_2 - L_1)_s$ is the segmental spatial asymmetry function, M_0 is the molecular weight of the monomer, and s is the number of monomers per segment.

Therefore the total polarizability difference for the macromolecule in solution $(\gamma_1 - \gamma_2)$ is the sum of three components: the segmental anisotropy, the anisotropy due to the macroscopic shape of the polymer chain (or *macroform*), and the microstructure (or *microform*) anisotropy

$$\alpha_1 - \alpha_2 = (\alpha_1 - \alpha_2)_i + (\alpha_1 - \alpha_2)_f + (\alpha_1 - \alpha_2)_{fs}. \tag{9.309}$$

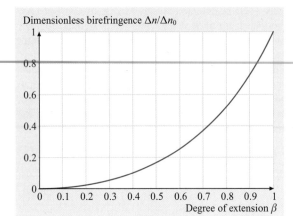

Fig. 9.114 Dependence of birefringence on the degree of extension of a polymer coil

So, the total principal polarizability difference $(\gamma_1 - \gamma_2)$ can be calculated with the help of (9.293), with $(\alpha_1 - \alpha_2)$ replacing $(\alpha_1 - \alpha_2)_i$.

9.5.5 Example Calculation of the Theoretical Birefringence for Stretched Molecules of Atactic Polystyrene

The maximum achievable birefringence calculated under the assumption that all molecules in solution are fully stretched follows from (9.298)

$$\Delta n = \frac{2\pi}{9}\left(\frac{n^2+2}{n}\right)^2 (\alpha_1 - \alpha_2)\frac{N_A c}{M_s}, \quad (9.310)$$

where $(\alpha_1 - \alpha_2)$ is determined by (9.309), and $M_s = M_0 s$ is the molecular weight of the segment. It is easily seen that, as a macromolecular coil extends in the flow and the distances between the peripheral segments increase, so the optical long-range interactions in the chain will become weaker. This means that the relative role of the macroform effect is reduced. According to Tsvetkov [9.383], the macroform effect $(\alpha_1 - \alpha_2)_f \to 0$ for fully stretched molecules. So finally we obtain

$$\Delta n_0 = \frac{2\pi}{9}\left(\frac{n^2+2}{n}\right)^2 \frac{N_A c}{M_s}$$
$$\times [(\alpha_1 - \alpha_2)_i + (\alpha_1 - \alpha_2)_{fs}], \quad (9.311)$$

where $(\alpha_1 - \alpha_2)_{fs}$ is determined by (9.308).

For atactic polystyrene in tricresylphosphate (TCP) we have [9.384]

$$n_p = 1.6, \ n_s = 1.55, \ (L_2 - L_1) = 5,$$
$$M_s = M_0 s = 822, \ (\alpha_1 - \alpha_2)_i = 145 \times 10^{-25} \text{ cm}^3. \quad (9.312)$$

The value of intrinsic segmental anisotropy of atactic polystyrene (a-PS) was determined in bromoform, which has a refractive index very close to that of a-PS molecule. Substituting these values into (9.310) and (9.311), we get for maximum achievable for a-PS in TCP: $\Delta n = 0.09 c$, where c (g/cm^3) is the concentration of the solution. The contribution of the microform effect to the total anisotropy is about 3% for this system.

9.5.6 Conclusions

Birefringence continues to provide a major insight into molecular structure; it is noninvasive, can be used in vivo and is extremely fast. The latter property makes it especially significant in non-Newtonian fluid mechanics. Careful experimentation can yield extraordinary sensitivity. Birefringence relies upon small differences in polarizability, but the *subtraction* is done at the molecular level. On the experimental scale, zero birefringence is the background upon which flow effects are superimposed. However, the relationship between birefringence and structure is ambiguous and models need to be used with caution to interpret results in terms of molecular orientation and conformation.

References

9.1 J.D. Ferry: *Viscoelastic Properties of Polymers*, 3rd edn. (Wiley, New York 1980)
9.2 R. Whorlow: *Rheological Techniques*, 2nd edn. (Ellis Horwood, Chichester 1992)
9.3 W.W. Graessly, S.D. Grasscock, R.L. Crawley: Die Swell in Molten Polymers, Trans.Soc.Rheol. **14**, 519–544 (1970)
9.4 K. te Nijenhuis: Survey of measuring techniques for the determination of the dynamic moduli. In: *Rheology*, Vol. 1, ed. by G. Astarita, G. Marucci, L. Nicolais (Plenum, New York 1980) pp. 263–280
9.5 K. te Nijenhuis, R. van Donselaar: A newly designed coaxial cylinder type dynamic rheometer, Rheol. Acta **24**, 47–57 (1985)

9.6 F.H. Gortemaker, M.G. Hansen, B. de Cindio, H.M. Laun, H. Janeschitz-Kriegl: Flow birefringence of polymer melts: application to the investigation of time dependent rheological properties, Rheol. Acta **15**, 256–267 (1976)

9.7 J. L. den Otter: Dynamic Properties of Some Polymeric Systems, Doctoral Thesis (Leiden, 1967)

9.8 F. Trouton: On the coefficient of viscous traction and its relation to that of viscosity, Proceedings of the Royal Society of London **A77**, 426–440 (1906)

9.9 C.J.S. Petrie: Elongational flows: Aspects of the behavior of model elasticoviscous fluids. In: *Research Notes in Mathematics*, ed. by Pitman (London 1979)

9.10 G.V. Vinogradov, V.D. Fikhman, B.V. Radushkevich, A.Y. Malkin: Viscoelastic and relaxation properties of a polystyrene melt in axial extension, J. Polym. Sci.: Part A-2 **8**, 657–678 (1970)

9.11 G.V. Vinogradov, B.V. Radushkevich, V.D. Fikhman: Extension of elastic fluids: Polyisobutylene, J. Polym. Sci., Part A-2 **8**, 1–17 (1970)

9.12 H. Munstedt, S. Kurzbeck, L. Egersdorfer: Influence of molecular structure on rheological properties of polyethylenes Part II. Elongational behavior, Rheol. Acta **37**, 21–29 (1998)

9.13 H. Munstedt: New universal extensional rheometer for polymer melts. Measurements on a polystyrene sample, J. Rheol. **23**, 421–436 (1979)

9.14 H. Munstedt, H.M. Laun: Elongational behaviour of a low density polyethylene melt II. Transient behavior in constant stretching rate and tensile creep experiments. Comparision with shear data. Temperature dependence of the elongational properties, Rheol. Acta **18**, 492–504 (1979)

9.15 J. Meissner, T. Raible, S.E. Stephenson: Rotary clamp in uniaxial and biaxial extensional rheometry of polymer melts, J. Rheol. **25**, 1–28 (1981)

9.16 S. Spiegelberg, D. Ables, G. McKinley: Role of end-effects on measurements of extensional viscosity in viscoelastic polymer solutions with a filament stretching rheometer, J. Non-Newt. Fluid Mech. **64**(2-3), 229–267 (1996)

9.17 V. Tirtaatmadja, T. Sridhar: Filament Stretching Device for Measurement of Extensional Viscosity, J. Rheol. **36**(3), 277–284 (1993)

9.18 R. Muller, D. Froelich: New extensional rheometer for elongational viscosity and flow birefringence measurements: some results on polystyrene melts, Polymer **26**, 1477–1482 (1985)

9.19 J. Meissner, J. Hostettler: New elongational rheometer for polymer melts and other highly viscoelastic liquids, Rheol. Acta **33**, 1–21 (1994)

9.20 M.L. Sentmanat: Dual windup drum extensional rheometer. Patent 6691569 (2004)

9.21 M.H. Wagner, B. Collignon, J. Verbeke: Rheotens-mastercurves and elongational viscosity of polymer melts, Rheol. Acta **35**, 117–126 (1996)

9.22 M.H. Wagner: Rheotens-mastercurves and drawability of polymer melts, Polym. Eng. Sci. **36**, 925–935 (1996)

9.23 W.H. Talbott, J.D. Goddard: Streaming birefringence in extensional flow of polymer solutions, Rheol. Acta **18**, 505–517 (1979)

9.24 R.C. Chan, R.K. Gupta, T. Sridhar: Fiber spinning of very dilute solutions of polyacrylamide in water, J. Non-Newt. Fluid Mech. **30**(2-3), 267–283 (1988)

9.25 G.H. McKinley, A. Tripathi: How to extract the Newtonian viscosity from capillary breakup measurements in a filament rheometer, J. Rheol. **44**(3), 653–670 (2000)

9.26 A.V. Bazilevskii, V.M. Entov, A.N. Rozhkov: Liquid Filament Microrheometer and Some of its Applications. In: *Proceedings of the 3rd European Rheology Conference* (Elsevier, New York 1990)

9.27 G.J.C. Braithwaite, S.H. Spiegelberg: *Capillary breakup to determine the extensional properties of polymeric fluids. In American Chemical Society North Eastern Regional Meeting* (American Chemical Society, Durham 2001)

9.28 P.N. Dunlap, L.G. Leal: Dilute polystyrene solutions in extensional flows: birefringence and flow modification, J. Non-Newt. Fluid Mech. **23**, 5–48 (1987)

9.29 G.G. Fuller, L.G. Leal: Flow birefringence of concentration polymer solutions in two-dimensional flows, J. Polym. Sci., Polym. Phys. **19**, 557–587 (1981)

9.30 D.M. Binding: Further consideration of axisymmetric contraction flows, J. Non-Newt. Fluid Mech. **41**, 27–42 (1991)

9.31 D.V. Boger: Viscoelastic flows through contractions, Ann. Rev. Fluid Mech. **19**, 157–182 (1987)

9.32 G.G. Fuller, C.A. Cathey, B. Hubbard, B.E. Zebrowski: Extensional viscosity measurements for low-viscosity fluids, J. Rheol. **31**, 235–249 (1987)

9.33 H.M. Laun, H. Schuch: Transient elongational viscosities and drawability of polymer melts, J. Rheol. **33**, 119–175 (1989)

9.34 M. Moan, A. Magueur: Transient extensional viscosity of dilute flexible polymer solutions, J. Non-Newt. Fluid Mech. **30**(2-3), 343–354 (1988)

9.35 A.L. Yarin: *Free liquid jets and films: Hydrodynamics & rheology. Interaction of Mechanics and Mathematics Series* (Wiley, New York 1993)

9.36 H.J. Shore, G. Harrison: The Effect of Added Polymer on the Formation of Drops Ejected from a Nozzle, Phys. Fluids **17**, 033104 (2005)

9.37 R.H. Fernando, L.-L. Xing, J.E. Glass: Rheology Parameters Controlling Spray Atomization and Roll-Misting Behavior of Waterborne Coatings, Prog. in Organic Coatings **42**, 244–248 (2001)

9.38 R.B. Bird, R.C. Armstrong, O. Hassager: *Dynamics of Polymeric Liquids. Volume 1: Fluid Mechanics*, 2nd edn. (Wiley Interscience, New York 1987)

9.39 D.F. James, K. Walters: A critical appraisal of available methods for the measurement of extensional properties of mobile systems. In: *Techniques of Rhe-*

9.40 C.J.S. Petrie: Extensional flow - A mathematical perspective, Rheol. Acta **34**, 12–26 (1995)

9.41 D.W. Bousfield, R. Keunings, G. Marrucci, M.M. Denn: Nonlinear analysis of the surface-tension driven breakup of viscoelastic fluid filaments, J. Non-Newt. Fluid Mech. **21**, 79–97 (1986)

9.42 H.K. Rasmussen, O. Hassager: The role of surface tension on the elastic decohesion of polymeric filaments, J. Rheol. **45**(2), 527–537 (2001)

9.43 J. Meissner: Experimental aspects in polymer melt elongational rheology, Chem. Eng. Commun. **33**, 159–180 (1985)

9.44 R.K. Gupta, T.E. Sridhar: Elongational Rheometers. In: *Rheological Measurement*, ed. by A.A. Collyer, D.W. Clegg (Elsevier, London 1988) pp. 211–245

9.45 M.L. Sentmanat, B. Wang, G.H. McKinley: Measuring the extensional rheology of polyethylene melts using the SER Universal Testing Platform, J. Rheol. **49**(3), 571–803 (2004)

9.46 A. Bach, H.K. Rasmussen, O. Hassager: Extensional Viscosity for Polymer Melts Measured in the Filament Stretching Rheometer, J. Rheol. **47**(2), 429–441 (2003)

9.47 T.C.B. McLeish, R.G. Larson: Molecular constitutive equations for a class of branched polymers: the Pom-Pom polymer, J. Rheol. **42**, 81–110 (1998)

9.48 H.K. Rasmussen, J.K. Nielsen, A. Bach, O. Hassager: Viscosity Overshoot in the Start-Up of Uniaxial Elongation of Low Density Polyethylene Melts, J. Rheol. **49**(2), 369–381 (2005)

9.49 A.Y. Malkin, C.J.S. Petrie: Some conditions for rupture of polymer liquids in extension, J. Rheol. **41**, 1–25 (1997)

9.50 G.H. McKinley, O. Hassager: The Considere condition and rapid stretching of linear and branched polymer melts, J. Rheol. **43**(5), 1195–1212 (1999)

9.51 Y.M. Joshi, M.M. Denn: Failure and recovery of entangled polymer melts in elongational flow. In: *Rheology Review, British Soc. Rheology*, ed. by D.M. Binding, K. Walters (2004) pp. 1–17

9.52 T. Sridhar: An overview of the project M1, J. Non-Newt. Fluid Mech. **35**, 85–92 (1990)

9.53 J. Matta, R. Tytus: Liquid stretching using a falling cylinder, J. Non-Newt. Fluid Mech. **35**, 215–229 (1990)

9.54 T. Sridhar, V. Tirtaatmadja, D.A. Nguyen, R.K. Gupta: Measurement of extentional viscosity of polymer solutions, J. Non-Newt. Fluid Mech. **40**, 271–280 (1991)

9.55 J. van Nieuwkoop, M.M.O. von Muller Czernicki: Elongation and subsequent relaxation measurements on dilute polyisobutylene solutions, J. Non-Newt. Fluid Mech. **67**, 105–124 (1996)

9.56 M.J. Solomon, S.J. Muller: Transient extensional behavior of polystyrene-based Boger fluids of varying solvent quality and molecular weight, J. Rheol. **40**, 1–19 (1996)

9.57 R.W.G. Shipman, M.M. Denn, R. Keunings: Mechanics of the falling plate extensional rheometer, J. Non-Newt. Fluid Mech. **40**, 281–288 (1991)

9.58 S. Berg, R. Kröger, H. Rath: Measurement of extensional viscosity by stretching large liquid bridges in microgravity, J. Non-Newt. Fluid Mech. **55**, 307–319 (1994)

9.59 M. Yao, G.H. McKinley: Numerical simulation of extensional deformations of viscoelastic liquid bridges in filament stretching devices, J. Non-Newt. Fluid Mech. **74**, 47–88 (1998)

9.60 M.I. Kolte, H.K. Rasmussen, O. Hassager: Transient filament stretching rheometer II: numerical simulation, Rheol. Acta **36**, 285–302 (1997)

9.61 S.L. Anna, C. Rogers, G.H. McKinley: On controlling the kinematics of a filament stretching rheometer using a real-time active control mechanism, J. Non-Newt. Fluid Mech. **87**, 307–335 (1999)

9.62 N.V. Orr, T. Sridhar: Probing the dynamics of polymer solutions in extensional flow using step strain rate experiments, J. Non-Newt. Fluid Mech. **82**, 203–232 (1999)

9.63 J.P. Rothstein, G.H. McKinley: Inhomogeneous transient uniaxial extensional rheometry, J. Rheol. **46**(4), 1419–1443 (2002)

9.64 P.S. Doyle, E.S.G. Shaqfeh, G.H. McKinley, S.H. Spiegelberg: Relaxation of dilute polymer solutions following extensional flow, J. Non-Newt. Fluid Mech. **76**(1–3), 79–110 (1998)

9.65 S.L. Anna, G.H. McKinley: Elasto-capillary thinning and breakup of model elastic liquids, J. Rheol. **45**, 115–138 (2001)

9.66 S.H. Spiegelberg, G.H. McKinley: Stress relaxation and elastic decohesion of viscoelastic polymer solutions in extensional flow, J. Non-Newt. Fluid Mech. **67**, 49–76 (1996)

9.67 A.V. Bazilevskii, V.M. Entov, M.M. Lerner, A.N. Rozhkov: Failure of polymer solution filaments, Polymer Science, Series A **39**(3), 316–324 (1997)

9.68 D. Papageorgiou: On the breakup of viscous liquid threads, Physics of Fluids **7**(7), 1529–1544 (1995)

9.69 A. Tripathi, P. Whittingstall, G.H. McKinley: Using filament stretching rheometry to predict strand formation and processability in adhesives and other non-Newtonian fluids, Rheol. Acta **39**, 321–337 (2000)

9.70 N. Willenbacher: Elongation viscosity of aqueous thickener solutions from capillary breakup elongation rheometry (CaBER). In: Proc. XIVth Int. Cong. Rheol. (2004). Seoul (S. Korea)

9.71 G. Neal, G. Braithwaite: The use of capillary breakup extensional rheology to examine concentration dependence of relaxation time. In: Society of Rheology Annual Meeting. Pittsburgh, PA (2003)

9.72 V.M. Entov, E.J. Hinch: Effect of a spectrum of relaxation times on the capillary thinning of a filament of elastic liquid, J. Non-Newt. Fluid Mech. **72**, 31–53 (1997)

9.73 L.E. Rodd, T.P. Scott, J.J. Cooper-White, and G.H. McKinley: Capillary breakup rheometry of low-viscosity elastic fluids, submitted to Applied Rheology, (2004)

9.74 S. Anna, G.H. McKinley, D.A. Nguyen, T. Sridhar, S.J. Muller, J. Huang, D.F. James: Interlaboratory comparison of measurements from filament-stretching rheometers using common test fluids, J. Rheol. **45**, 83–114 (2001)

9.75 S.L. Anna, S.H. Spiegelberg, G.H. McKinley: Elastic instability in elongating filaments, Phys. Fluids **9**(5), 10 (1997)

9.76 M. Stelter, G. Brenn, A. Yarin, R. Singh, F. Durst: Investigation of the elongational behavior of polymer solutions by means of an elongational rheometer, J. Rheol. **46**(2), 507–528 (2002)

9.77 P. Dontula, M. Pasquali, L.E. Scriven, C.W. Macosko: Can extensional viscosity be measured with opposed nozzle devices, Rheol. Acta **36**, 429–448 (1997)

9.78 D.F. James: Developments in extensional rheology, AMD **191**, 1–3 (1994)

9.79 S.L. Ng, R.P. Mun, D.V. Boger, D.F. James: Extensional viscosity measurements of dilute solutions of various polymers, J. Non-Newt. Fluid Mech. **65**, 291–298 (1996)

9.80 D.F. James: Extensional viscosity, an elusive property of mobile liquids. In: 3rd European Rheology Conference (1990)

9.81 V.M. Entov, A.L. Yarin: Influence of elastic stresses on the capillary breakup of jets of dilute polymer solutions, Fluid Dynamics **19**, 21–29 (1984)

9.82 H.A. Barnes, J.F. Hutton, K. Walters: *An Introduction to Rheology* (Elsevier, Amsterdam 1989)

9.83 W.H. Bauer, E.A. Collins: Chap. 8. In: *Rheology: Theory and Applications*, ed. by F.R. Eirich (Academic, New York 1967)

9.84 J. Mewis: Thixotropy – a general review, J. Non-Newt. Fluid Mech. **6**, 1–20 (1979)

9.85 D.C.-H. Cheng: Report Number CR 2367 (MH), Warren Spring Laboratory, UK, October 1 (1982)

9.86 J.C. Godfrey: Ph.D. Thesis, Characterisation of thixotropic fluids (University of Bradford 1983)

9.87 H.A. Barnes: Thixotropy – A review, J. Non-Newt. Fluid Mech. **70**(1/2), 1–33 (1997)

9.88 F. Schalek, A. Szegvari: Ferric oxide jellies, Kolloid Z **32**, 318–319 (1923)

9.89 F. Schalek, A. Szegvari: The slow coagulation of concentrated ferric oxide sols to reversible jellies, Kolloid Z **33**, 326–334 (1923)

9.90 T. Peterfi: Arch, Entwicklungsmech. Organ. **112**, 680 (1927)

9.91 T. Peterfi: Uber die thixotropie, Verhanitlungen 3rd. Intern. Zellto Arch. Exp. Zelif. **15**, 373–381 (1934)

9.92 H. Freunlich: *Thixotropie* (Hermann, Paris 1935)

9.93 E.L. McMillen: Thixotropy and Plasticity. 1 – The measurement of thixotropy, J. Rheol. **3**(1), 75–94 (1932)

9.94 G.W. Scott-Blair: *A Survey of General and Applied Rheology* (Pitman, London 1943)

9.95 J. Pryce-Jones: Thixotropy, JOCCA **17**, 305–375 (1934)

9.96 J. Pryce-Jones: JOCCA, Some fundamental aspects of thixotropy **19**, 295–337 (1936)

9.97 J. Pryce-Jones: Experiments on thixotropic and other anomalous fluids with a new rotation viscometer, J. Sci. Instr. **18**, 39–48 (1941)

9.98 J. Thewlis (Ed.): *Oxford Encyclopedia Dictionary of Physics* (Pergamon Press, Oxford 1962)

9.99 T.C. Collocott (Ed.): Chambers Dictionary of Science and Technology (W&R Chambers 1971)

9.100 S.P. Parker (Ed.): *McGraw-Hill Dictionary of Scientific and Technical Termsn*, 4th edn. (McGraw-Hill, New York 1989)

9.101 D.M. Considine (Ed.): *Van Nostrand's Scientific Encyclopedia*, 5th end (Van Nostrand Reinhold Company, New York 1976)

9.102 *Oxford Concise Science Dictionary*, 2nd edn. (Oxford Univ. Press, Oxford 1991)

9.103 *Chambers 20th Century Dictionary* (Chambers, London 1993)

9.104 T. Whelan (Ed.): *Polymer Technology Dictionary* (Chapman Hall, London 1994)

9.105 M.S.M. Alger (Ed.): *Polymer Science Dictionary* (Elsevier, London 1990)

9.106 M. Reiner, G.W. Scott-Blair: *Rheological Terminology in Rheology Theory and Applied* (Academic Press, New York 1967), Chap. 9

9.107 S. Rahimi, B. Natan: Thixotropic effect of inorganic gel fuels, J. Propuls. Power **16**, 1182–1184 (2000)

9.108 Unpublished report, International Fine Particle Research Institute, (1996)

9.109 A. Maestro, C. Gonzalez, J.M. Gutierrez: Shear thinning and thixotropy of HMHEC and HEC water solutions, J. Rheol. **46**(6), 1445–1457 (2002)

9.110 L. Heymann, E. Noack, L. Kampfe, B. Beckmann: Rheology of printing inks – Some new experimental results. In: *Proceedings of the XIIth International Congress on Rheology*, ed. by A. al Ait-Kadi et (Laval University, Quebec 1996) pp. 451–452

9.111 P.D.A. Mills, J. Goodwin, B.W. Grover: Shear field modification of strongly flocculated suspensions – aggregate morphology, Colloid. Polym. Sci. **269**(9), 949–963 (1991)

9.112 W. Woithers: Proc. XIIth International Congress on Rheology, ed. by A. Ait-Kadi et al. (Laval University, Quebec 1996) pp. 526–527

9.113 M. Barut, R. Lapasin, A. Zupanic, M. Zumer: Time-dependent rheological behavior of TiO2 suspensions. In: *Proceedings of the XIIth International Congress on Rheology*, ed. by A. al Ait-Kadi et (Laval University, Quebec 1996) pp. 578–579

9.114 P.R. Williams, S.R. Ren: Direct determination of the mechanical relaxation spectra of concentrated suspensions in the presence of controlled shear fields. In: *Proceedings of the XIIth International Congress*

9.115 V. Bouda, J. Mikešová: Vibrational carbon black agglomaration in polythylene melt. In: *Proceedings of the XIIth International Congress on Rheology*, ed. by A. Ait-Kadi et al. (Laval University, Quebec 1996) pp. 27–28

9.116 H.H. Winter: Evolution of rheology during chemical gelation, Progr. Colloid Polym. Sci. **75**, 104–110 (1987)

9.117 J. Greener, R. W.Connelly: The response of viscoelastic liquids to complex strain histories – the thixotropic loop, J. Rheol. **30**(2), 285–300 (1986)

9.118 H.A. Barnes: A review of the slip (wall depletion) of polymer solutions, emulsions and particle suspensions in viscometers: its cause, character and cure, J. Non-Newt. Fluid Mech. **56**, 221–251 (1995)

9.119 C.F. Goodeve, G.W. Whitfield: The measurement of thixotropy in absolute units, Trans. Farad. Soc. **34**, 511–520 (1938)

9.120 F. Moore: The rheology of ceramic slips and bodies, Trans. Brit Ceramics Soc. **58**, 470–494 (1959)

9.121 D.C.-H. Cheng, F. Evans: Phenomenological characterization of rheological behaviour of inelastic reversible thixotropic and antithixotropic fluids, Brit. J. Appl. Phys. **16**(11), 1599–1617 (1965)

9.122 A. Allesandrini, R. Lapasin, F. Sturzi: The kinetics of thixotropic behavior in clay kaolin aqueous suspensions, Chem. Eng. Commun. **17**(1-6), 13–22 (1982)

9.123 C. Baravian, D. Quemada, A. Parker: A new methodology for rheological modelling of thixotropy application to hydrocolloids. In: *Proceedings of the XIIth International Congress on Rheology*, ed. by A. al Ait-Kadi et (Laval University, Quebec 1996) pp. 779–780

9.124 D. de Kee, R.K. Code, G. Turcotte: Flow properties of time-dependent foodstuffs, J. Rheol. **27**, 581–604 (1983)

9.125 C. Tiu, D.V. Boger: Complete rheological characterization of time dependent food products, J. Texture Stud. **5**(3), 329–338 (1974)

9.126 D.A. Denny, R.S. Brodkey: Kinetic interpretation of non-Newtonian flow, J. Appl. Phys. **33**(7), 2269–2274 (1962)

9.127 M. van den Tempel: Mechanical properties of plastic-disperse systems at very small deformations, J. Coll. Sci. **16**, 284–296 (1961)

9.128 E. Ruckenstein, J. Mewis: Rheology of non-Newtonian fluids – a new flow equation for pseudoplastic systems, J. Coll. Interface Sci. **44**, 532 (1973)

9.129 M.M. Cross: Rheology of non-Newtonian fluids: a new flow equation for pseudoplastic systems, J. Coll. Sci. **20**, 417–437 (1965)

9.130 R. Lapasin, M. Grassi, S. Pricl: Viscoelastic properties of welan systems. In: *Proceedings of the XIIth International Congress on Rheology*, ed. by A. al Ait-Kadi et (Laval University, Quebec 1996) pp. 524–525

9.131 I. Park, T. Ree: Kinetics of Thixotropy of Aqueous Bentonite Suspension J, Korean Chem. Soc. **15**(6), 293–303 (1971)

9.132 A.G. Fredrickson: Model for the thixotropy of suspensions, AIChEJ **16**(3), 436–441 (1970)

9.133 J. Mewis, J. Schryvers: unpublished International Fine Particle Research Institute report (1996)

9.134 P.J. Kristensen, C.T.B. Jensen, Q.D. Nguyen: Rheological characterisation of maize starch pastes. In: *Proceedings of the XIIth International Congress on Rheology*, ed. by A. Ait-Kadi et al. (Laval University, Quebec 1996) pp. 471–472

9.135 Q.D. Nguyen, D.V. Boger: Thixotropic behavior of concentrated bauxite residue suspensions, Rheol. Acta **24**(4), 427–437 (1985)

9.136 D. Acierno, F.P. La Mantia, G. Marrucci, G. Titomanlio: Nonlinear viscoelastic model with structure-dependent relaxation-times .1. basic formulation, J. Non-Newt. Fluid Mech. **1**(2), 125–146 (1976)

9.137 D. Shoong, M. Shen: Kinetic network model for non-linear viscoelastic flow properties of entangled monodisperse polymers, J. Polym. Sci., Lett. Ed. **17**(9), 595–599 (1979)

9.138 D. Quemada: Rheological modelling of complex fluids: IV: Thixotropic and thixoelastic behaviour. Start-up and stress relaxation, creep tests and hysteresis cycles, European Physical J. Appl. Phys. **5**, 191–207 (1999)

9.139 P.A. Reynolds, J.W. Goodwin: Direct measurement of the translational diffusion-coefficients of aggregated polystyrene latex-particles, Colloid Surf. **11**(1-2), 145–154 (1984)

9.140 H.A. Barnes: *Dispersion Rheology, a Survey of Industrial Problems and Academic Progress* (Royal Soc. Chem. Industrial Div., London 1981)

9.141 H. Sonntag: Coagulation Kinetics. In: *Coagulation and Flocculation: Theory and Applications*, ed. by B. Dobias (Dekker, New York 1993), Chap. 3

9.142 K. Mühle: Floc stability in laminar and turbulent flow. In: *Coagulation and Flocculation: Theory and Applications*, ed. by B. Dobiasc (Dekker, New York 1993)

9.143 M.F. Edwards, J.C. Godfrey, M.M. Kashani: Power requirement for mixing of thixotropic liquids, J. Non-Newt. Fluid Mech. **1**(4), 309–322 (1976)

9.144 L. Schmitt, G. Ghnassia, J.J. Bimbenet, G. Cuvelier: Flow properties of stirred yogurt: Calculation of the pressure drop for a thixotropic fluid, J. Food Engin. **37**, 367–388 (1998)

9.145 W.G. Hou, D.J. Sun, S.H. Han: A novel thixotropic phenomenon – Complex thixotropic behavior, Chem. Res. Chinese Universities **13**, 86–88 (1997)

9.146 S.P. Li, W.G. Hou, X.N. Dai, J.F. Hu, D.Q. Li: The influence of pH on the thixotropy of ferric aluminum magnesium hydroxide/montmorillonite suspension, Chem. J. Chinese Universities **23**, 1763–1766 (2002)

9.147 S.P. Li, W.G. Hou, J.F. Hu, D.Q. Li: Influence of shear rate on thixotropic suspensions, J. Disp. Sci. Technol. **24**, 709–714 (2003)

9.148 S.P. Li, W.G. Hou, J.C. Xiao, J.F. Hu, D.Q. Li: Influence of measuring conditions on the thixotropy of hydrotalcite-like/montmorillonite suspension, Colloids and Surfaces A-Physicochemical and Engineering Aspects **224**, 149–156 (2003)

9.149 S.P. Li, W.G. Hou, X.N. Dai, P. Jiang: Influence of electrolytes on the thixotropy of ferric aluminum magnesium hydroxide-montmorillonite suspensions, J. Disp. Sci. Techn. **24**, 145–152 (2003)

9.150 P.Z. Guo, D.J. Sun, Z.L. Jin, C.G. Zhang: Rheological properties of mixed suspensions containing oppositely charged colloidal particles, Chem. J. Chinese Universities **24**, 1052–1055 (2003)

9.151 D.J. Sun, W.G. Hou, S.Y. Liu, B.Q. Zhang, C.G. Zhang: Thixotropic properties of aqueous dispersions of positively charged Al/Mg mixed metal hydroxide particles, Acta Chimica Sinica **59**, 163–167 (2001)

9.152 Y. Wang, M.J. Yu: Effect of volume loading and surface treatment on the thixotropic behavior of polypropylene filled with calcium carbonate, Polymer Composites **21**, 1–12 (2000), sec2-Taiwan

9.153 J. Assaad, K.H. Khayat, H. Mesbah: ACI Materials, J. Assessment of thixotropy of flowable and self-consolidating concrete **100**, 99–107 (2003)

9.154 I.A. Kinloch, S.A. Roberts, A.H. Windle: A rheological study of concentrated aqueous nanotube dispersions, Polymer **43**, 7483–7491 (2002)

9.155 C. Voltz, M. Nitschke, L. Heymann, I. Rehberg: Thixotropy in macroscopic suspensions of spheres, Phys. Rev. E **65**(5), 051402 (2002)

9.156 A. Mujumdar, A.N. Beris, A.B. Metzner: Transient phenomena in thixotropic systems, J. Non-Newton. Fluid Mech. **102**, 157–178 (2002), Special Issue SI

9.157 F. Bautista, J.M. de Santos, J.E. Puig, O. Manero: Understanding thixotropic and antithixotropic behavior of viscoelastic micellar solutions and liquid crystalline dispersions. I. The model, J. Non-Newt. Fluid Mech. **80**, 93–113 (1999)

9.158 F. Yziquel, P.J. Carreau, M. Moan, P.A. Tanguy: Rheological modeling of concentrated colloidal suspensions, J. Non-Newt. Fluid Mech. **86**, 133–155 (1999)

9.159 J.W. Goodwin, P.A. Reynolds: The rheology of flocculated suspensions, Current Opinion in Colloid & Interface Science **3**, 401–407 (1998)

9.160 N. Bagdassarov, H. Pinkerton: Transient phenomena in vesicular lava flows based on laboratory experiments with analogue materials, J. Volcanology and Geothermal Research **132**, 115–136 (2004)

9.161 L.P. Martinez-Padilla, F. Lopez-Araiza, A. Tecante: Steady and oscillatory shear behavior of fluid gels formed by binary mixtures of xanthan and gellan, Food Hydrocolloids **18**, 471–481 (2004)

9.162 A. Tarrega, L. Duran, E. Costell: Flow behaviour of semi-solid dairy desserts. Effect of temperature, Int. Dairy J. **14**, 345–353 (2004)

9.163 M.M. Li, J.P. Zhi, G.Y. Zhang, Q.L. Ma: Thixotropy comparison of different peptized pseudo-bohemites, Acta Petrologica Sinica **20**, 35–39 (2004)

9.164 C. Malfoy, A. Pantet, P. Monnet, D. Righi: Effects of the nature of the exchangeable cation and clay concentration on the rheological properties of smectite suspensions, Clays And Clay Minerals **51**, 656–663 (2003)

9.165 J.F. Zhou, L.M. Zhang, P.S. Hui: Rheological characteristics of aqueous solutions of an amphoteric guar gum derivative, Acta Physico-Chimica Sinica **19**, 1081–1084 (2003)

9.166 A. Koksoy, M. Kilic: Effects of water and salt level on rheological properties of ayran, a Turkish yoghurt drink, Int. Dairy J. **13**, 835–839 (2003)

9.167 J. Koke, M. Modigell: Flow behaviour of semi-solid metal alloys, J. Non-Newt. Fluid Mech. **112**, 141–160 (2003)

9.168 F.V. Lopez, M. Rosen: Rheological effects in roll coating of paints, Latin American Applied Research **32**, 247–252 (2002)

9.169 A. Moise, A. Aruxandei: Tire influence of temperature on the viscoplastic compound of some cosmetic products, Revista De Chimie **53**, 222–225 (2002)

9.170 B. Abu-Jdayil, H. Mohameed: Experimental and modelling studies of the flow properties of concentrated yogurt as affected by the storage time, J. Food Engin. **52**, 359–365 (2002)

9.171 P. Corvisier, C. Nouar, R. Devienne, M. Lebouche: Development of a thixotropic fluid flow in a pipe, Exp. Fluids **31**, 579–587 (2001)

9.172 J.S. Raynaud, P. Moucheront, J.C. Baudez, F. Bertrand, J.P. Guilbaud, P. Coussot: Direct determination by nuclear magnetic resonance of the thixotropic and yielding behavior of suspensions, J. Rheol. **46**, 709–732 (2002)

9.173 S. Chander: Challenges in characterization of concentrated suspensions, Colloids and Surfaces A **133**, 143–150 (1998)

9.174 V.A. Hackley, C.F. Ferraris: The Use of Nomenclature in Dispersion Science and Technology, National Institute of Standards and Technology, Technology Administration, U.S. Department of Commerce, Special Publication 960-3, Washington (2001)

9.175 D.H. Everett: Manual of symbols and terminology for physicochemical quantities and units (Appendix II/Part I), Pure. Appl. Chem. **31**, 577–638 (1972)

9.176 E. Kissa: *Dispersions – Characterization, Testing, and Measurement* (Dekker, New York 1999)

9.177 G.D. Parfitt (Ed.): *Dispersions of Powders in Liquids*, 2nd edn. (Wiley, New York 1973)

9.178 VDI-Richtlinie 3491: *Messen von Partikeln, Kennzeichnung von Partikeldispersionen in Gasen, Begriffe*

9.178 und Definitionen, VDI-Handbuch: Reinhaltung der Luft, Band 4 (VDI-Verlag, Düsseldorf 1980)
9.179 G. Böhme: Strömungsmechanik nicht-Newtonscher Fluide, 2nd edn. (Teubner, Stuttgart 2000), in German
9.180 H. Wadell: Volume shape and roundness of rock particles, J. Geol. **40**, 443–451 (1932)
9.181 A. Jillavenkatesa, S.J. Dapkunas, L.S.H Lum: Particle size characterization, National Institute of Standards and Technology, Technology Administration, U.S. Department of Commerce, Special Publication 960-1, Washington (2001)
9.182 M.H. Pahl, G. Schädel, H. Rumpf: Methods for describing particle shapes (part 1), Aufbereitungstechnik **5**, 257–263 (1973)
9.183 M.H. Pahl, G. Schädel, H. Rumpf: Methods for describing particle shapes (part 2), Aufbereitungstechnik **10**, 672–683 (1973)
9.184 M.H. Pahl, G. Schädel, H. Rumpf: Methods for describing particle shapes (part 3), Aufbereitungstechnik **11**, 759–764 (1973)
9.185 R.M. Turian, T.W. Ma, F.LG. Hsu, D.J. Sung: Characterization, settling, and rheology of concentrated fine particulate mineral slurries, Powder Technol. **93**, 219–233 (1997)
9.186 R.G. Larson: The Structure and Rheology of Complex Fluids (Oxford Univ. Press, New York 1999)
9.187 P.G. Saffmann: The lift on a small sphere in a slow shear flow, J. Fluid. Mech. **22**, 385–400 (1965)
9.188 C.W. Macosko: Rheology – Principles, Measurements, and Applications (VCH, New York 1994)
9.189 K. Walters: Rheometry (Chapman Hall, London 1975)
9.190 G.K. Batchelor: An Introduction to Fluid Dynamics (Cambridge Univ. Press, Cambridge 2000)
9.191 D.J. Jeffrey, A. Acrivos: The rheological properties of suspensions of rigid particles, AIChE J. **22**, 417–432 (1976)
9.192 H.A. Barnes: Measuring the viscosity of large-particle (and flocculated) suspensions – a note on the necessary gap size of rotational viscometers, J. Non-Newt. Fluid Mech. **94**, 213–217 (2000)
9.193 Official nomenclature for material functions describing the response of a viscoelastic fluid to various shearing and extensional deformations, J. Rheol. **39**, 253–265 (1995)
9.194 W.B. Russel, D.A. Saville, W.R. Schowalter: Colloidal Dispersions (Cambridge Univ. Press, Cambridge 1989)
9.195 I.D. Morrison, S. Ross: Colloidal Dispersions – Suspensions, Emulsions, and Foams (Wiley, New York 2002)
9.196 M. Takeo: Disperse Systems (Wiley-VCH, Weinheim 1999)
9.197 R.J. Pugh, L. Bergström: Surface and Colloid Chemistry in Advanced Ceramic Processing (Dekker, New York 1994)
9.198 T.G.M. van de Ven: Colloidal Hydrodynamics (Academic, London 1989)
9.199 M. Doi, S.F. Edwards: The Theory of Polymer Dynamics (Clarendon Press, Oxford 2003)
9.200 J. Israelachvili: Intermolecular and Surface Forces, 2nd edn. (Academic, London 1992)
9.201 H.R Kruyt: Colloidal Science Vol. I (Elsevier, Amsterdam 1952)
9.202 H.C. Hamaker: The London-van der Waals attraction between spherical particles, Physica **4**, 1058–1072 (1937)
9.203 K.S. Birdi (Ed.): Handbook of Surface and Colloid Science (CRC Press, Boca Raton 1997)
9.204 L. Bergström: Hamaker constants of inorganic materials, Adv. Colloid. Interface Sci. **70**, 125–169 (1997)
9.205 J.A. Lewis: Colloidal Processing of Ceramic, J. Am. Ceram. Soc. **83**, 2341–2358 (2000)
9.206 B.V. Derjaguin, L. Landau: Theory of the stability of strongly charged lyophobic sols and the adhesion of strongly charged particles in solutions of electrolytes, Acta Physicochim URSS **14**, 633–652 (1941)
9.207 E.JW. Verwey, J.TG. Overbeek: Theory of the Stability o Lyophobic Colloids (Elsevier, Amsterdam 1948)
9.208 O.J. Rojas, P.M. Claesson, D. Muller, R.D. Neuman: The Effect of Salt Concentration on Adsorption of Low-Charge-Density Polyelectrolytes and Interactions between Polyelectrolyte-Coated Surfaces, J. Colloid. Interface Sci. **205**, 77–78 (1998)
9.209 T.F. Tadros: Correlation of viscoelastic properties of stable and flocculated suspensions with their interparticle interactions, Adv. Colloid. Interface Sci. **68**, 97–200 (1996)
9.210 A. Einstein: Eine neue Bestimmung der Moleküldimensionen, Ann. Phys. **34**, 591–592 (1906)
9.211 A. Einstein: Eine neue Bestimmung der Moleküldimensionen, Ann Phys **19**, 289–306 (1911)
9.212 R.K. Gupta: Polymer and Composite Rheology (Dekker, New York 2000)
9.213 G.K. Batchelor, J.T. Green: The determination of the bulk stress in a suspension of spherical particles to order c^2, J. Fluid. Mech. **56**, 401–427 (1972)
9.214 G.K. Batchelor: The effect of Brownian motion on the bulk stress in a suspension of spherical particles, J. Fluid. Mech. **83**, 97–117 (1977)
9.215 D.G. Thomas: Transport characteristics of suspensions: VIII. A note on the viscosity of Newtonian suspensions of uniform spherical particles, J. Colloid. Sci. **20**, 267–277 (1965)
9.216 A. Einstein: Berichtigung zu meiner Arbeit: Eine neue Bestimmung der Moleküldimensionen, Ann Phys **34**, 591–592 (1911)
9.217 H. Eilers: Die Viskositäts-Konzentrationsabhängigkeit kolloider Systeme in organischen Lösungsmitteln, Kolloid.-Z **102**, 154–169 (1943)
9.218 E.H. Harbard: An investigation into the variation of the viscosity of a suspension with its concentration, Chem. Ind. **22**, 491–492 (1956)
9.219 C.G. de Kruif, E.MF. van Iersel, A. Vrij, W.B. Russel: Hard sphere colloidal dispersions: Viscosity as a

9.219 ...function of shear rate and volume fraction, J. Chem. Phys. **83**, 4717–4725 (1985)
9.220 J. Mewis, J.W. Frith, T.A. Strivens, W.B. Russel: Rheology of suspensions containing polymerically stabilized particles, AIChE J. **35**, 415–422 (1989)
9.221 T. Kitano, T. Kataoko, T. Shiraota: An empirical equation of the relative viscosity of polymer melts filled with various inorganic fillers, Rheol. Acta. **20**, 207–209 (1981)
9.222 D. He, N.N. Ekere: Viscosity of concentrated non-colloidal bidisperse suspensions, Rheol. Acta. **40**, 591–598 (2001)
9.223 K. Qin, A.A. Zaman: Viscosity of concentrated colloidal suspensions: comparison of bidisperse models, J. Colloid. Interface Sci. **266**, 461–467 (2003)
9.224 T. Shikata, H. Niwa: Viscoelastic behaviour of bimodal ideal suspensions, Nihon Reorojï Gakkaishi **28**, 127–135 (2000)
9.225 T. Shikata: Viscoelastic behaviour of ideal bimodal suspensions, Chem. Eng. Sci. **56**, 2957–2966 (2001)
9.226 I.R. Rutgers: Relative viscosity and concentration, Rheol. Acta. **2**, 305–348 (1962)
9.227 V.V. Jinescu: The rheology of suspensions, Int. Chem. Eng. **14**, 397–420 (1974)
9.228 S. Arrhenius: The viscosity of solutions, Biochem. J. **11**, 112–133 (1917)
9.229 S.H. Maron, P.E. Pierce: Application of Ree-Eyring generalized flow theory to suspensions of spherical particles, J. Colloid. Sci. **11**, 80–95 (1956)
9.230 I.M. Krieger, I.J. Dougherty: A mechanism for non-Newtonian flow in suspensions of rigid spheres, Trans. Soc. Rheol. **3**, 137–152 (1959)
9.231 M. Mooney: The viscosity of a concentrated suspension of spherical particles, J. Colloid. Sci. **6**, 162–170 (1951)
9.232 R.C. Ball, P. Richmond: Dynamics of colloidal dispersions, Phys. Chem. Liq. **9**, 99–116 (1980)
9.233 R.D. Sudduth: A generalised model to predict the viscosity of solutions with suspended particles I, J Appl Polym Sci **48**, 25–36 (1993)
9.234 C.R. Wildemuth, M.C. Williams: Viscosity of suspensions modeled with a shear-dependent maximum packing fraction, Rheol. Acta. **23**, 627–635 (1984)
9.235 J.Z.Q. Zhou, T. Fang, G. Luo, P.HT. Uhlherr: Yield stress and maximum packing fraction of concentrated suspensions, Rheol. Acta. **34**, 544–561 (1995)
9.236 W.J. Frith, J. Mewis, T.A. Strivens: Rheology of concentrated suspensions: Experimental investigations, Powder Tech. **51**, 27–34 (1987)
9.237 K. Yasuda, R.C. Armstrong, R.E. Cohen: Shear flow properties of concentrated solutions of linear and star branched polystyrenes, Rheol. Acta. **20**, 163–178 (1981)
9.238 Q.D. Nguyen, D.V. Boger: Measuring the flow properties of yield stress fluids, Annu. Rev. Fluid. Mech. **24**, 47–88 (1992)
9.239 H.A. Barnes: The yield stress–a review or $\pi\alpha\nu\tau\alpha\rho\epsilon\iota$ everything flows?, J. Non-Newt. Fluid Mech. **81**, 133–178 (1999)
9.240 H.M. Laun, R. Bung, S. Hess, W. Loose, O. Hess, K. Hahn, E. Hadicke, R. Hingman, F. Schmidt, P. Lindner: Rheological and small angle neutron scattering investigation of shear-induced particle structures of concentrated polymer dispersions submitted to plane Poiseuille and Couette flow, J. Rheol. **36**, 743–787 (1992)
9.241 P. d'Haene, J. Mewis, G.G. Fuller: Scattering dichroism measurements of flow induced structure of a shear thickening suspension, J. Colloid. Interface Sci. **156**, 350–358 (1993)
9.242 M.K. Chow, V.F. Zukoski: Nonequilibrium behaviour of dense suspensions of uniform particles: Volume fraction and size dependence of rheology and microstructure, J. Rheol. **39**, 33–59 (1995)
9.243 H.A. Barnes: Shear-thickening (Dilatancy) in suspensions of nonaggregating solid particles dispersed in Newtonian liquids, J. Rheol. **33**, 329–366 (1989)
9.244 W.J. Frith, P. d'Haene, R.R. Buscall, J.J. Mewis: Shear thickening in model suspensions of sterically stabilized particles, J. Rheol. **40**, 531–547 (1996)
9.245 G.V. Franks, Z. Zhou, N.J. Duin, D.V. Boger: Effect of interparticle forces on shear thickening of oxide suspensions, J. Rheol. **44**, 759–779 (2000)
9.246 S.R. Raghavan, S.A. Khan: Shear thickening response of fumed silica suspensions under steady and oscillatory shear, J. Colloid. Interface Sci. **185**, 57–67 (1997)
9.247 D. Doraiswamy, A.N. Mujumdar, I. Tsao, A.N. Beris, S.C. Danforth, A.B. Metzner: The Cox-Merz rule extended: a rheological model for concentrated suspensions and other materials with a yield stress, J. Rheol. **35**, 647–685 (1991)
9.248 W. Gleissle, B. Hochstein: Validity of the Cox-Merz rule for concentrated suspensions, J. Rheol. **47**, 897–910 (2003)
9.249 R.L. Powell: Rotational rheometry. In: *Rheological Measurement*, 2nd edn., ed. by A.A. Collyer, D.W. Clegg (Chapman Hall, London 1998), Chap. 9
9.250 M.R. Mackley, R.PG. Rutgers: Capillary rheometry. In: *Rheological Measurement*, 2nd edn., ed. by A.A. Collyer, D.W. Clegg (Chapman Hall, London 1998)
9.251 B.P. Ho, L.G. Leal: Inertial migration of rigid spheres in two-dimensional unidirectional flows, J. Fluid. Mech. **65**, 365–400 (1974)
9.252 H. Gevgilili, D.M. Kalyon: Step strain flow: Wall slip effects and other error sources, J. Rheol. **45**, 467–475 (2001)
9.253 A.L. Graham, A.A. Mammoli, M.B. Busch: Effects of demixing on suspension rheometry, Rheol. Acta. **37**, 139–150 (1998)
9.254 D.T. Leighton, A. Acrivos: The shear-induced migration of particles in concentrated suspensions, J. Fluid. Mech. **181**, 415–439 (1987)

9.255 B.K. Aral, D.M. Kalyon: Effects of temperature and surface roughness on time-dependent development of wall slip in steady torsional flow of concentrated suspensions, J. Rheol. **38**, 957–972 (1994)

9.256 M. Allende, D.M. Kalyon: Assessment of particle-migration effects in pressure-driven viscometric flows, J. Rheol. **44**, 79–90 (1999)

9.257 R.J. Phillips, R.C. Armstrong, R.A. Brown, A.L. Graham, J.R. Abbott: A constitutive equation for concentrated suspensions that accounts for shear-induced particle migration, Phys. Fluids A **4**, 30–40 (1992)

9.258 P.J.A. Hartman Kok, S.G. Kazarian, C.J. Lawrence, B.J. Briscoe: Near-wall particle depletion in a flowing colloidal suspension, J. Rheol. **46**, 481–493 (2002)

9.259 U. Yilmazer, D.M. Kalyon: Slip effects in capillary and parallel disk torsional flows of highly filled suspensions, J. Rheol. **33**, 1197–1212 (1989)

9.260 J.R. Abbott, N. Tetlow, A.L. Graham, S.A. Altobelli, E. Fukushima, L.A. Mondy, T.S. Stephens: Experimental observations of particle migration in concentrated suspensions: Couette flow, J. Rheol. **35**, 773–795 (1991)

9.261 M. Han, C.C. Kim, M.M. Kim, S.S. Lee: Particle migration in tube flow of suspensions, J. Rheol. **43**, 1157–1174 (1999)

9.262 S.C. Jana, B. Kapoor, A. Acrivos: Apparent wall slip velocity coefficients in concentrated suspensions of non-colloidal particles, J. Rheol. **39**, 1123–1132 (1995)

9.263 J. Persello, A. Magnin, J. Chang, J.M. Piau, B. Cabane: Flow of colloidal aqueous silica dispersions, J. Rheol. **38**, 1845–1870 (1994)

9.264 H.J. Walls, S. Brett Caines, A.M. Sanchez, S.A. Khan: Yield stress and wall slip phenomena in colloidal silica gels, J. Rheol. **47**, 847–868 (2003)

9.265 A. Yoshimura, R.K. Prud'homme: Wall slip correction for Couette and parallel disk viscometer, J. Rheol. **32**, 53–67 (1988)

9.266 H.A. Barnes: A review of the slip (wall depletion) of polymer solutions, emulsions and particle suspensions in viscometers: its cause, character, and cure, J. Non-Newt. Fluid Mech. **56**, 221–251 (1995)

9.267 L. Heymann, S. Peukert, N. Aksel: Investigation of the solid-liquid transition of highly concentrated suspensions in oscillatory amplitude sweeps, J. Rheol. **46**, 93–112 (2002)

9.268 H.M. Laun: Rheological properties of aqueous polymer dispersions, Angew. Makromol. Chem. **123**, 335–359 (1984)

9.269 P.J. Carreau, P.A. Lavoie, F. Yziquel: Rheological properties of concentrated suspensions. In: *Advances in the Flow and Rheology of Non-Newtonian Fluids. Rheology Series 8*, ed. by D.A. Siginer, D. De Kee, R.P. Chhabra (Elsevier, Amsterdam 1999)

9.270 M. Schmidt, H. Münstedt: Rheological behaviour of concentrated monodisperse suspensions as a function of preshear conditions and temperature: an experimental study, Rheol. Acta. **41**, 193–204 (2002)

9.271 M.K. Lyon, D.W. Mead, R.E. Elliott, L.G. Leal: Structure formation in moderately concentrated viscoelastic suspensions in simple shear flow, J. Rheol. **45**, 881–890 (2001)

9.272 P. Gondret, L. Petit: Viscosity of periodic suspensions, Phys. Fluids **8**, 2284–2290 (1996)

9.273 I.M. Krieger: Rheology of monodisperse lattices, Adv. Colloid. Interface Sci. **3**, 111–136 (1972)

9.274 C. Völtz, M. Nitschke, L. Heymann, I. Rehberg: Thixotropy in macroscopic suspensions of spheres, Phys. Rev. E **65**, 051402 (2002)

9.275 L.E. Silbert, J.R. Melrose, R.C. Ball: The rheology and microstructure of concentrated, aggregated colloids, J. Rheol. **43**, 673–700 (1999)

9.276 L. Heymann, S. Peukert, N. Aksel: On the solid-liquid transition of concentrated suspensions in transient shear flow, Rheol. Acta. **41**, 307–315 (2002)

9.277 J.H. So, S.M. Yang, J.C. Hyun: Microstructure evolution and rheological responses of hard sphere suspensions, Chem. Eng. Sci. **56**, 2967–2977 (2001)

9.278 H. Senff, W. Richtering, C. Norhausen, A. Weiss, M. Ballauf: Rheology of temperature sensitive core-shell latex, Langmuir **15**, 102–106 (1999)

9.279 I.M. Krieger, M. Eguiluz: The second electroviscous effect in polymer lattices, Trans Soc Rheol **20**, 29–45 (1976)

9.280 J.H. So, S.M. Yang, C. Kim, J.C. Hyun: Microstructure and rheological behaviour of electrosterically stabilized silica particle suspensions, Coll. Surf. A **190**, 89–98 (2001)

9.281 L. Heymann, N. Aksel: Progress and trends in rheology V. In: *Proc. 5th Europ. Conf. Rheol* (Steinkopff Verlag, Darmstadt 1998)

9.282 J.F. Brady, G. Bossis: The rheology of concentrated suspensions of spheres in simple shear flow by numerical simulation, J. Fluid. Mech. **155**, 105–129 (1985)

9.283 F. Gadala-Maria: The rheology of concentrated suspensions, Ph. D. Dissertation, Stanford University (1979)

9.284 G. Schoukens, J. Mewis: Nonlinear rheological behaviour and shear-dependent structure in colloidal dispersions, J. Rheol. **22**, 381–394 (1978)

9.285 M. Moan, T. Aubry, F. Bossard: Nonlinear behaviour of very concentrated suspensions of plate-like kaolin particles in shear flow, J. Rheol. **47**, 1493–1504 (2003)

9.286 S. Lin-Gibson, J.A. Pathak, E.A. Grulke, H. Wang, E.K. Hobbie: Elastic flow instability in nanotube suspensions, Phys Rev Lett **92**, 048302; Erratum: Phys. Rev. Lett. **92**, 239901 (2004)

9.287 I.E. Zarraga, D.A. Hill, D.T. Leighton Jr: The characterization of the total stress on concentrated suspensions of noncolloidal spheres in Newtonian fluids, J. Rheol. **44**, 185–220 (2000)

9.288 I.E. Zarraga, D.A. Hill, D.T. Leighton Jr: Normal stresses and free surface deformation in concentrated suspensions of noncolloidal spheres in a viscoelastic fluid, J. Rheol. **45**, 1065–1084 (2001)

9.289 T. Shikata, D.S. Pearson: Viscoelastic behaviour of concentrated spherical suspensions, J. Rheol. **38**, 601–616 (1994)

9.290 S.E. Mall-Gleissle, W. Gleissle, G.H. McKinley, H. Buggisch: The normal stress behaviour of suspensions with viscoelastic matrix fluids, Rheol. Acta. **41**, 61–76 (2002)

9.291 A. Singh, P.R. Nott: Experimental measurements of the normal stress in sheared Stokesian suspensions, J. Fluid. Mech. **490**, 293–320 (2003)

9.292 J.D. Ferry: *Viscoelastic properties of polymers* (Wiley, New York 1980)

9.293 M. Wilhelm: Fourier-transform rheology, Macromol. Mater. Eng. **287**, 83–105 (2002)

9.294 M. Wilhelm, D. Maring, H.W. Spiess: Fourier-transform rheology, Rheol. Acta. **37**, 399–405 (1998)

9.295 M. Wilhelm, P. Reinheimer, M. Ortseifer: High sensitivity Fourier-transform rheology, Rheol. Acta. **38**, 349–356 (1999)

9.296 S. Kallus, N. Willenbacher, S. Kirsch, D. Distler, T. Neidhöfer, M. Wilhelm, H.W. Spiess: Characterization of polymer dispersions by Fourier transform rheology, Rheol. Acta. **40**, 552–559 (2001)

9.297 K. Hyun, S.H. Kim, K.H. Ahn, S.J. Lee: Large amplitude oscillatory shear as a way to classify the complex fluids, J. Non-Newt. Fluid Mech. **107**, 51–65 (2002)

9.298 Y. Otsubo: Rheology of colloidal suspensions flocculated by reversible bridging, Chem. Eng. Sci. **56**, 2939–2946 (2001)

9.299 M. Schmidt, H. Münstedt: On the elastic properties of model suspensions as investigated by creep revovery measurements in shear, Rheol. Acta. **41**, 205–210 (2002)

9.300 L.A. Utracki: The rheology of two-phase flows. In: *Rheological Measurement*, ed. by A.A. Collyer, D.W. Clegg (Elsevier, Barking 1988), Chap. 15

9.301 C.J.S. Petrie: The rheology of fibre suspensions, J. Non-Newt. Fluid Mech. **87**, 369–402 (1999)

9.302 J. Greener, J.RG. Evans: Uniaxial elongational flow of particle-filled polymer melts, J. Rheol. **42**, 697–709 (1998)

9.303 J. Greener, J.RG. Evans: Review – Measurements of elongational flows in ceramic processing, J. Europ. Ceram. Soc. **17**, 1173–1183 (1997)

9.304 J.F. Le Meins, P. Moldenaers, J. Mewis: Suspensions of monodisperse spheres in polymer melts: particle size effects in extensional flows, Rheol. Acta. **42**, 184–190 (2003)

9.305 D.M. Husband, N. Aksel, W. Gleissle: The existence of static yield stresses in suspensions containing noncolloidal particles, J. Rheol. **37**, 215–235 (1993)

9.306 V.T. O'Brien, M.E. Mackay: Shear and elongation flow properties of kaolin suspensions, J. Rheol. **46**, 557–572 (2002)

9.307 A.G. Gibson, G. Kotsikos, J.H. Bland, S. Toll: Squeeze flow. In: Rheological measurement, ed. by A.A Collyer, Clegg DW (Elsevier, Barking 1988) Chap. 18

9.308 N. Delhaye, A. Poitou, M. Chaouche: Squeeze flow of highly concentrated suspensions of spheres, J. Non-Newt. Fluid Mech. **94**, 67–74 (2000)

9.309 A. Poitou, G. Racineux: A squeezing experiment showing binder migration in concentrated suspensions, J. Rheol. **45**, 609–625 (2001)

9.310 T.W. Chan, D.G. Baird: An evaluation of a squeeze flow rheometer for the rheological characterization of a filled polymer melt with a yield stress, Rheol. Acta. **41**, 245–256 (2002)

9.311 J.D. Sherwood, D. Durban: Squeeze flow of a power-law viscoplastic solid, J. Non-Newt. Fluid Mech. **62**, 35–54 (1996)

9.312 H.M. Laun, M. Rady, O. Hassager: Analytical solutions for squeeze flow with partial wall slip, J. Non-Newt. Fluid Mech. **81**, 1–15 (1999)

9.313 D.N. Smyrnaios, J.A. Tsamopoulos: Squeeze flow of Bingham plastics, J. Non-Newt. Fluid Mech. **100**, 165–190 (2001)

9.314 A. Matsoukas, E. Mitsoulis: Geometry effects in squeeze flow of Bingham plastics, J. Non-Newt. Fluid Mech. **109**, 231–240 (2003)

9.315 P. Sherman: *Rheological properties of emulsions*. In: Encyclopedia of Emulsion Technology, Vol. 1 (Dekker, New York 1983), Chap. 7

9.316 H.A. Barnes: Rheology of emulsions – a review, Coll. Surf. A **91**, 89–95 (1994)

9.317 T.F. Tadros: Fundamental principles of emulsion rheology and their applications, Coll Surf A **91**, 39–55 (1994)

9.318 R. Pal: Rheology of emulsions containing polymeric liquids. In: *Encyclopedia of Emulsion Technology*, Vol. 4, ed. by P. Becher (Dekker, New York 1996), Chap. 3

9.319 U. Teipel: Einfluss der Tropfengröße auf das rheologische Verhalten von Emulsionen (Influence of droplet size on the rheological behaviour of emulsions), Chemie-Ingenieur-Technik **73**, 1006–1012 (2001), in German

9.320 R. Pal: Scaling of relative viscosity of emulsions, J. Rheol. **41**, 141–150 (1997)

9.321 U. Teipel, N. Aksel: Rheologically identical behaviour of emulsions and suspensions in steady shear flow: Dimensional analysis and experimental evidence, Chem. Eng. Technol. **26**, 947–951 (2003)

9.322 U. Teipel, L. Heymann, N. Aksel: Indirect detection of structural changes in micellar solutions by rheological measurements, Coll. Surf. A **193**, 35–49 (2001)

9.323 F.D. Rumscheidt, S.G. Mason: Particle motions in sheared suspensions, XII. Deformation and burst of drops in shear and hyperbolic flow, J. Colloid. Sci. **16**, 238–261 (1961)

9.324 G.I. Taylor: The formation of emulsions in definable fields of flow, Proc. Roy. Soc. **A146**, 501–523 (1934)

9.325 H.P. Grace: Dispersion phenomena in high viscosity immiscible fluid systems and application of static mixers as dispersion devices in such systems, Chem. Eng. Comm. **14**, 225–277 (1982)

9.326 V.T. Tsakalos, P. Narvard, E. Peuvrel-Disdier: Deformation and breakup mechanism of single drops during shear, J. Rheol. **42**, 1403–1417 (1998)

9.327 V. Cristini, S. Guido, A. Alfani, J. Blawzdziewicz, M. Loewenberg: Drop breakup and fragment size distribution in shear flow, J. Rheol. **47**, 1283–1298 (2003)

9.328 S. Wannaborworn, M.R. Mackley: Experimental observation and matching numerical simulation for the deformation and breakup of immiscible drops in oscillatory shear, J. Rheol. **46**, 1279–1293 (2002)

9.329 R. Cavallo, S. Guido, M. Simeone: Drop deformation under small-amplitude oscillatory shear flow, Rheol. Acta. **42**, 1–9 (2003)

9.330 F. Mighri, A. Ajji, P.J. Carreau: Influence of elastic properties on drop deformation in elongational flow, J. Rheol. **41**, 1183–1201 (1997)

9.331 H.A. Stone: Dynamics of drop deformation and breakup in viscous fluids, Annu. Rev. Fluid Mech. **26**, 65–102 (1994)

9.332 M. Loewenberg, E.J. Hinch: Collision of two deformable drops in shear flow, J. Fluid. Mech. **338**, 299–315 (1997)

9.333 U. Teipel, N. Aksel: Studies on the adsorption of nonionic surfactants and the dripping behaviour of their solutions by the drop volume and maximum bubble pressure technique, Tenside Surf. Det. **37**, 297–309 (2000)

9.334 B. Warburton: Interfacial rheology. In: *Rheological Measurement*, 2nd edn., ed. by A.A. Collyer, D.W. Clegg (Chapman Hall, London 1998), Chap. 22

9.335 D.W. Criddle: The viscosity and elasticity of interfaces. In: *Rheology – Theory and Applications*, Vol. 3, ed. by F.R. Eirich (Academic, New York 1960)

9.336 J.T. Petkov, K.D. Danov, N.D. Denkov, R. Aust, F. Durst: Precise method for Measuring the surface shear viscosity of surfactant monolayers, Langmuir **12**, 2650–2653 (1996)

9.337 R. Pal: Novel viscosity equations for emulsions of two immiscible liquids, J. Rheol. **45**, 509–520 (2001)

9.338 R. Pal, E. Rhodes: Viscosity/concentration relationship for emulsions, J. Rheol. **33**, 1021–1045 (1989)

9.339 H.M. Lee, J.W. Lee, O.O. Park: Rheology and dynamics of water-in-oil emulsions under steady and dynamic shear flow, J. Colloid. Interface Sci. **185**, 297–305 (1997)

9.340 J.F. Palierne: Linear rheology of viscoelastic emulsions with interfacial tension, Rheol. Acta. **29**, 204–214 (1990), Erratum Rheol. Acta. **30**, 497 (1991)

9.341 E.H. Kerner: The elastic and thermoelastic properties of composite media, Proc. Phys. Soc. A **69**, 808–813 (1956)

9.342 M. Bousmina: Rheology of polymer blends: linear model for viscoelastic emulsions, Rheol. Acta. **38**, 73–83 (1999)

9.343 R. Pal: Novel shear modulus equations for concentrated emulsions of two immiscible elastic liquids with interfacial tension, J. Non-Newt. Fluid Mech. **105**, 21–33 (2002)

9.344 M.M. Robins: Emulsions – creaming phenomena, Curr. Opinion Coll. Interface Sci. **5**, 265–272 (2000)

9.345 P. Walstra: Emulsion stability. In: *Encyclopedia of Emulsion Technology*, Vol. 4, ed. by P. Becher (Dekker, New York 1996), Chap. 1

9.346 U. Teipel: Rheologisches Verhalten von Emulsionen und Tensidlösungen (Rheological behaviour of emulsions and surfactant solutions), Ph.D. Thesis, University of Bayreuth, published in: Wissenschaftliche Schriftenreihe des Fraunhofer-Institutes für Chemische Technologie, Pfinztal (Germany), in German (1999)

9.347 X. Li, C. Pozrikidis: The effect of surfactants on drop deformation and on the rheology of dilute emulsions in Stokes flow, J. Fluid. Mech. **341**, 165–194 (1997)

9.348 K.G. Hollingsworth, M.L. Johns: Rheo-nuclear magnetic resonance of emulsion systems, J. Rheol. **48**, 787–803 (2004)

9.349 A. Montesi, A.A. Peña, M. Pasquali: Vorticity alignment and negative normal stresses in sheared attractive emulsions, Phys. Rev. Lett. **92**, 058303 (2004)

9.350 C. Bower, C. Gallegos, M.R. Mackley, J.M. Madiedo: The rheological and microstructural characterisation of the non-linear flow behaviour of concentrated oil-in-water emulsions, Rheol. Acta. **38**, 145–159 (1999)

9.351 W.M. Kulicke, O. Arendt, M. Berger: Characterization of hydroxylpropylmethylcellulose-stabilized emulsions. Part II: The flow behaviour, Rheol. Acta. **276**, 1024–1031 (1998)

9.352 R. Pal: Dynamics of flocculated emulsions, Chem. Eng. Sci. **52**, 1177–1187 (1997)

9.353 G.K. Moates, A.D. Watson, M.M. Robins: Creaming and oscillation rheology of weakly flocculated concentrated emulsions, Coll. Surf. A **190**, 167–178 (2001)

9.354 M.R. Anklam, G.G. Warr, R.K. Prud'homme: The use of opposed nozzles configuration in the measurement of the extensional rheological properties of emulsions, J. Rheol. **38**, 797–810 (1994)

9.355 J. Plucinski, R.K. Gupta, S. Chakrabarti: Wall slip of mayonnaises in viscometers, Rheol. Acta. **37**, 256–269 (1998)

9.356 E.B. Bagley: End Corrections in the Capillary Flow of Polyethylene, J. Appl. Phys. **28**, 624–627 (1957)

9.357 C.D. Han, M. Charles, W. Philippoff: Rheological Implications of the Exit Pressure and Die Swell in Steady Capillary Flow of Polymer Melts. I. The Primary Normal Stress Difference and the Effect of L/D Ratio on Elastic Properties, Trans. Soc. Rheol. **14**, 393–408 (1970)

9.358 J.C. Maxwell: On double refraction of a viscous fluid in motion Proc. R. Soc. London, Ser. A **22**, 46–47 (1873)

9.359 J.K. Hobbs, D.R. Binger, A. Keller, P.J. Barham: Spiralling optical morphologies in spherulites of poly(hydroxybutyrate) J.Polym. Sci. Part B **38**, 1575–1583 (2000)

9.360 R.M.A. Azzam, N.M. Bashara: *Ellipsometry and Polarized Light* (North Holland, Amsterdam 1977)

9.361 N.H. Hartshorne, A. Stuart: *Crystals and the Polarizing Microscope* (Edward Arnold, London 1960)

9.362 E. Fredericq, C. Houssier: *Electric Dichroism and Electric Birefringence* (Clarendon Press, Oxford 1973)

9.363 C.L. Riddiford, H.G. Jerrard: Relaxation times by the Kerr effect, J. Phys. D.: Appl. Phys. **3**, 1314–1321 (1970)

9.364 G.G. Fuller, K.J. Mikkelsen: Optical rheometry using a rotary polarization modulator, J.Rheol **33**, 761–769 (1989)

9.365 A.W. Chow, G.G. Fuller: Xanthan gum two color flow birefringence, J. Rheol. **28**, 23–43 (1984)

9.366 G.G. Fuller: Optical Rheology, Annu. Rev. Fluid Mech. **22**, 387–417 (1990)

9.367 M. Muller, J. Vorwerk, P.O. Brunn: Optical studies of local flow behaviour of a non-Newtonian fluid inside a porous medium, Rheol. Acta **37**, 189–194 (1998)

9.368 A. Peterlin: Streaming birefringence of soft, linear macromolecules with finite chain length, Polymer **2**, 257–264 (1961)

9.369 A. Peterlin: Hydrodynamics of macromolecules in a velocity field with longitudinal gradient, J.Polym. Sci. Poly. Lett. **4**, 287–291 (1966)

9.370 W. Kuhn, F. Grün: Beziehungen zwischen elastischen Konstanten und Dehnungsdoppelbrechung hochelastischer Stoffe, Kolloidzscr. **101**, 248–271 (1942)

9.371 A. Peterlin: Streaming birefringence and hydrodynamic interaction, Pure and Appl. Chem. **12**, 563–586 (1966)

9.372 W. Kuhn: Über die Gestalt für den förminger Moleküle in Lösungen, Kolloidzscr. **68**, 2–15 (1934)

9.373 P.E. Rouse: A theory of the linear viscoelastic properties of dilute solutions of coiling polymers, J. Chem. Phys. **21**, 1272–1280 (1953)

9.374 B.H. Zimm, : Dynamics of polymer molecules in dilute solution: viscoelasticity flow birefringence, J. Chem. Phys. **24**, 269–278 (1956)

9.375 L.K.G. Treloar: *The Physics of Rubber Elasticity*, 3rd edn. (Clarendon Press, Oxford 1975)

9.376 G. Ryskin: Calculation of the effect of polymer additive in a converging flow, J. Fluid Mech. **178**, 423–440 (1987)

9.377 J.M. Rallison, E.J. Hinch: Do we understand the physics in the constitutive equation?, J. Non-Newt. Fluid Mech. **29**, 37–55 (1988)

9.378 J.M. Wiest, L.E. Wedgewood, R.B. Bird: On coil-stretch transitions in dilute polymer solutions, J. Chem. Phys. **90**, 587–594 (1989)

9.379 R.G. Larson: The unraveling of a polymer chain in a strong extensional flow, Rheol. Acta **29**, 371–390 (1990)

9.380 E.J. Hinch: Uncoiling a polymer molecule in a strong extensional flow Non-Newton, Fluid Mech. **54**, 209–230 (1994)

9.381 (Nauka, Moscow 1964). Translated by C. Crane-Robinson, National lending Library for Science and Technology Boston Spa. (1971)

9.382 V.N. Tsvetkov: Dynamic flow birefringence, optical anisotropy, and shape of macromolecules in solutions, J. Polym. Sci. **23**, 151–166 (1957)

9.383 V.N. Tsvetkov: The effects of shape in streaming birefringence of polymer solutions, J. Polym. Sci. **30**, 297–314 (1958)

9.384 J. Brandrup, E.H. Immergut: *Polymer Handbook*, 3rd edn. (Wiley, New York 1989)

10. Measurements of Turbulent Flows

Fluid flows in nature and technology normally depart from laminarity and are turbulent in the majority of cases, including flows around bodies such as airplanes, vehicles, ships, and in internal flows such as in ducts, turbomachines, propulsors, and even in blood circulation in the human body. Laminarity is the anomaly and not the standard. As will be shown in this chapter, the parameter which is fundamental to the transition from laminarity to turbulence is the Reynolds number, i.e., the ratio of inertial to viscous forces. In Sect. 10.1 the statistical Eulerian description of turbulent flows will be developed followed by a section on Reynolds decomposition and Reynolds equations. Section 10.1.3 finally surveys scales in turbulent flows.

In Sect. 10.2 the optical Lagrangian particle-tracking technique, capable of producing robust, single- and multiparticle Lagrangian measurements, is presented. First the image-processing algorithms used to determine the particle trajectories are discussed and then the implementation of the technique in the laboratory is described. A brief presentation of results focusing on the separation of particle pairs in intense turbulence is also given.

In Sect. 10.3 a novel type of random flow in a dilute polymer solution of a flexible high-molecular-weight polymer in two different flow setups that share the same feature of high curvature of the flow lines is discussed. In the first part of this section the hydrodynamic description of dilute polymer solution flows and the nondimensional parameters that follow from these equations to characterize these flows are presented. Variation of one of these control parameter responsible for the elastic properties of a fluid can lead to a new elastic instability in various flows that is distinguished by the presence of curvilinear trajectories. The theoretical criteria for this elastic instability in three different flows together with experimental verification are discussed. To complete the basics, the rheometric properties of the polymer solutions used and their relation to Boger fluids are given.

The first observation of elastic turbulence, in the flow between two plates, is described. Then the experimental measuring techniques used to characterize the flow are given, and a complete description of the results of measurements together with a discussion of the results is presented. Finally, the role of elastic stress, a recent theory of elastic turbulence, and comparative studies of elastic versus hydrodynamic turbulence are discussed. The last part of the section deals with the description of the elastic turbulence in a curvilinear channel or Dean flow, where a particularly detailed experiment on mixing due to elastic turbulence was conducted. A summary of the results is given finally.

Section 10.4 briefly reviews large-eddy simulations (LES) and the specific data requirements for LES (Sect. 10.4.1) and then describes the experimental methods that have been employed to obtain such data starting with arrays of point-measurement techniques (Sect. 10.4.2) and optical planar velocimetry measurement methods (Sect. 10.4.3). Sample results from the latter applied to studies of LES models are presented in (Sect. 10.4.4). The application of optical volumetric techniques for three-dimensional (3-D) velocity measurements are described in Sect. 10.4.5. Scalar fluctuation measurements using optical techniques and their applications to the study of LES variables of interest to scalar mixing and combustion are reviewed in Sect. 10.4.6.

10.1	**Statistical Eulerian Description of Turbulent Flows**..............................	746
	10.1.1 Basics of Measurements of Turbulent Flows	746
	10.1.2 Reynolds Decomposition and Equations	765
	10.1.3 Scales in Turbulent Flows	776
10.2	**Measuring Lagrangian Statistics in Intense Turbulence**	789
	10.2.1 Image Processing	789
	10.2.2 Experimental Implementation	792
	10.2.3 Turbulent Relative Dispersion	797
	10.2.4 Concluding Remarks	799

10.3 **Elastic Turbulence in Viscoelastic Flows**... 799	10.4.2 Arrays of Single-Point Instruments for Studies of SGS Dynamics........... 833
10.3.1 Basics.. 799	10.4.3 Planar Particle Image Velocimetry (PIV) for SGS Dynamics and LES....... 835
10.3.2 Elastic Turbulence in Swirling Flow Between Two Plates 804	10.4.4 Case Studies and Sample Results Using Planar PIV Measurements 836
10.3.3 Elastic Turbulence in a Curved Channel: Dean Flow 821	10.4.5 Holographic PIV Measurements of SGS Dynamics 844
10.3.4 Conclusions.............................. 829	10.4.6 Scalar Concentration Measurements for SGS Mixing and Combustion Studies 848
10.4 **Measurements for Large-Eddy Simulations** 830	**References** 849
10.4.1 Large-Eddy Simulation and Data Requirements............................ 831	

10.1 Statistical Eulerian Description of Turbulent Flows

When dealing with fluid flows in nature and technology, departure from laminarity and the presence of turbulence are normally observed; in fact, flows are turbulent in the majority of cases, such as in external flows around bodies as aeroplanes, vehicles, ships, and in internal flows as in ducts, turbomachines, propulsors, and even in blood circulation in the human body. Laminarity is the anomaly and not the standard.

10.1.1 Basics of Measurements of Turbulent Flows

Differences Between Measurements in Laminar and Turbulent Flows
As will be shown in the following, the parameter which is fundamental to the transition from laminarity to turbulence is the Reynolds number, i.e., the ratio of inertial to viscous forces. Dimensionally,

$$Re = \left|\frac{ma}{F_v}\right| = \frac{mU^2}{L}\frac{1}{\tau L^2}$$
$$= \frac{\rho L^3 U^2}{L^3 \mu U/L} = \frac{\rho U L}{\mu} = \frac{UL}{\nu},$$

where m is the mass, a the acceleration, F_v the viscous force, U a characteristic velocity, L a characteristic length, τ the shear stress, μ the dynamic viscosity coefficient, ν the kinematic viscosity coefficient, and ρ the density.

The Reynolds number is a parameter which carries information on the overall behavior of the flow field (regardless of the fact that in many cases, for example in near-wall flows, there could be differences between the characteristic lengths in different regions of the field). When the Reynolds number is below a critical value Re_C (where it would really be more correct to speak about a range of critical values, which also depends on boundary and initial conditions, BC and IC), the viscous forces are high enough to smooth instabilities in the flow (caused by unwanted small changes in BC and IC). On the other hand, above Re_C, the inertia of the flow largely overcomes the dissipative viscous effects and exponential growth of instabilities occurs; the flow rapidly becomes turbulent. The importance of the previous sentence on anomalous laminar flows becomes clearer when we consider that, owing to the small value of the kinematic viscosity of common fluids, the Reynolds number is usually large. Therefore, it is very important to evaluate the value Re_C in each set of experimental conditions. It is not straightforward to derive this value under general assumptions. Empirically, it has been found that, in external flows where the reference length is the size of the body, $(Re_C)_L \approx 3 \times 10^5$, whereas in internal flows where the reference length is the radius of the duct (orthogonal to the mean flow), $(Re_C)_\delta \approx 3 \times 10^3$. It is interesting to notice that, when considering the boundary layer over a slender body, the ratio between the length of the body (L) and the boundary-layer thickness (δ) from Prandtl theory [10.1] scales as $(Re)_L^{1/2}$; therefore, the ratio of the Reynolds number based on the size of the body over that based on the transverse characteristic length δ, scales as $L/\delta \approx (Re)_L^{1/2}$. This leads to $(Re)_L/(Re)_\delta \approx 10^2$, which is in remarkable agreement with empirical observations. Whenever $Re > Re_C$ (external or internal), the flow will start to exhibit all the features of developed turbulence.

These features establish the major differences between laminar and turbulent flows:

- extreme sensitivity to initial and boundary conditions
- unpredictability and randomness
- wide range of structures (scales) in space and time
- fully three-dimensional nature
- higher diffusion compared to the laminar case
- presence of cross-fluctuating terms among fluid-mechanics variables.

The sensitivity to initial conditions is the key point for the understanding of the behavior of turbulent flows; it is not possible, even in principle, to control the BC and IC to an arbitrarily small degree, especially if they pertain to a turbulent flow field (such as the inlet conditions for a channel). In comparison to laminar flow, a turbulent flow will exhibit a substantial (not only apparent due to our ignorance) unpredictable and random nature at any point and any time which originates from this high sensitivity to BC and IC; this also produces correlated fluctuations among the fluid-mechanics variables. These complex behaviors seem to prevent the possibility of an analytical approach when dealing with turbulent flows; in fact, from the mathematical point of view, there is still no theorem proving the existence and uniqueness of solutions to the Navier–Stokes equations in fully three-dimensional conditions (as in turbulent flows), for arbitrary time intervals, whatever the Reynolds number [10.2]. Referring to Sir Horace Lamb [10.3],

"I am an old man now, and when I die and go to Heaven there are two matters on which I hope enlightenment. One is quantum electrodynamics and the other is turbulence. About the former, I'm really quite optimistic."

there is still a lot to do in turbulence investigations (for example consider the $1 000 000 prize being offered by the Clay Mathematics Institute for a proof of the existence and uniqueness of solutions to the Navier–Stokes equations, www.claymath.org/millennium). Nonetheless, some key point is quite clear; it is sure that the fluid-mechanics equations are valid even in the turbulent regime, remembering that they are derived from fundamental conservation principles (the relation between turbulence and equations will be considered in detail in Sect. 10.1.1). Moreover, simplified mathematical descriptions of nonlinear systems reveals that unpredictability and randomness are effectively derived from deterministic equations [10.4].

The observation of a wide range of scales within turbulent flows probably has the most important practical consequences when dealing with experiments in turbulence. This has been recognized since Leonardo da Vinci's (1452–1519) famous pictures of the flowing water of the Arno River and is summarized in poetic style in the well-known verse by *Richardson* [10.5]:

"Big whorls have little whorls, which feed on their velocity; And little whorls have lesser whorls, And so on to viscosity (in the molecular sense)."

(even if for some turbulence quantity the range effectively extends well below the size where viscosity effects act). The previous description clearly points out the presence of several eddies, the interactions between them, and the effect of viscous dissipation.

Flow scales will be considered in detail in Sect. 10.1.2; as an introductory argument, for the velocity field, larger scales in space are related to the previously mentioned reference length L and time $\tau_0 = L/U$, whereas smaller scales are given by the Kolmogorov scales ($\eta = (\nu^3/\varepsilon)^{1/4}$, $\tau_\eta = (\nu/\varepsilon)^{1/2}$, where ε is the mean turbulent kinetic energy dissipation). From a simple dimensional argument, considering the balance between the inertial and dissipative terms in the turbulent kinetic energy equation [10.6], it is possible to derive that $\varepsilon \sim U^3/L$. Therefore, it is quite easy to obtain the ratios

$$\frac{L}{\eta} \sim L \left(\frac{U^3 L}{\nu^3}\right)^{1/4} = Re^{3/4},$$

$$\frac{\tau_0}{\tau_\eta} \sim \tau_0 \left(\frac{U^3}{L\nu}\right)^{1/2} = Re^{1/2},$$

i.e., the range between large and small scales $(L-\eta)/\eta$ (and similarly in time) increases as the Reynolds number to the power 0.75 (or 0.5 for time scales). For a moderate Reynolds number (10^4), this means that there are almost three orders of magnitude between the large and small space scales (and about two orders of magnitude in time); refer to Sect. 10.1.2 for further details on this.

Considering the transformed domain in wavenumber ($k = 2\pi/\ell_i$, where ℓ_i is the generic length scale), which allows one to derive the spatial distribution of energy or the energy spectrum, the large and small scales bound a range of interesting wavenumber scales in fluid mechanics from $k_L = 2\pi/L$ to $k_\eta = 2\pi/\eta$. Similarly, the distribution of energy among the different time scales is investigated by means of the energy spectrum in frequency ($f = 1/\Delta t_i$, where Δt_i is the generic time scale);

the frequency range is between $f_0 = 1/\tau_0$ and $f_\eta = 1/\tau_\eta$ (note that while f_0 is obtained as U/L, f_η is different from U/η; refer to the sections on Integral Scales and Kolmogorov Scales).

In fact, from both the numerical and experimental points of view, the existence of a wide range of scales determines the maximum and minimum dimensions (and times) to be *resolved* by a particular method (in terms of the previous example, this means that the experimental system should be able to measure over three orders of magnitude in space and two orders in time). In theory, the whole range should be detectable to allow a full description of the flow field; in practice, researchers frequently concentrate separately on intermediate–large or intermediate–small scale ranges. This limitation depends on the fact that, due to the large Reynolds numbers involved in practical investigations, the range of scales is very wide and exceeds the resolution capabilities of existing numerical codes and experimental methods (it should also be considered that the previous range has been evaluated for a one-dimensional space, while in three-dimension investigations or numerical simulations, the range must be increased as the third power).

The full three-dimensional nature of turbulent flows derives from the presence of a large range of scales which correspond to the presence of small and large eddies with shapes and strength depending on the particular flow field. These eddies also ensure more-effective diffusion (of each fluid variable) within the flow field in comparison to the laminar case in which only molecular diffusion acts.

To summarize, it is quite difficult to define turbulence univocally [10.7],

"What is turbulence? Turbulence is like pornography. It is hard to define, but if you see it, you recognize it immediately."

Prerequisites for Measurements in Turbulent Flows

Due to the previous considerations, from the point of view of measurements, even though many techniques can be used both in laminar and turbulent conditions, there are some peculiar characteristics of a measuring system to be considered when dealing with turbulent flows experimentally:

- spatial resolution
- time resolution
- the characteristic size of the system and duration of a measurement
- intrusiveness
- statistical treatment

To resolve the wide range of scales observed in turbulent flows, a measurement system should be able to identify rapid changes in fluid-mechanics variables both in space and time. To do this, the sensor (specifically the probe) of the system, must be small enough compared to the smallest spatial scale in the flow and must respond at least as quickly as the smallest time scale of the flow. The spatial resolution (SR), of a measurement system gives information about the inverse of the minimum detectable length in the flow field,

$$\text{SR} = \frac{1}{d_{\min}},$$

where $d_{\min}$ is usually closely related to the characteristic size of the sensor, although for some experimental method, there is not a physical sensor so that this size should be referred to the region which contributes to the measurement, the so-called *measurement volume*. A system with high spatial resolution is able to perform a measurement over a small length (i.e., to capture fluctuations of the considered fluid-mechanics variable over short distances). For example, in hot-wire anemometry (HWA, Sect. 5.2), the sensor is a thin wire (diameter of a few micrometers) and $\text{SR} \approx 10^5 \, \text{m}^{-1}$ (the measurement unit is length^{-1}), while in laser Doppler anemometry (LDA), (Sect. 5.3) the sensor consists of the region where the two laser beams cross (with a characteristic size of 10^{-4} m) so that $\text{SR} \approx 10^4 \, \text{m}^{-1}$ and a Pitot tube (Sect. 5.1, with a hole diameter of about 3×10^{-3} m) has $\text{SR} \approx 3 \times 10^2 \, \text{m}^{-1}$. In this sense, we can say that HWA is able to detect small moving spatial structures of the velocity field better than LDA and much better than a Pitot tube.

Similarly, the time resolution (TR), gives information on the inverse of the minimum detectable time interval in the flow field,

$$\text{TR} = \frac{1}{\Delta t_{\min}},$$

where $\Delta t_{\min}$ is the maximum time among the time interval requested by the whole measurement system to perform a measure (*measurement interval*) and the time interval for the sensor to respond to a variation of the considered fluid-mechanics variable (the *response time*, which is also connected to the size of the sensor itself). A system with high time resolution is able to take measurements separated by small time intervals (i.e., to capture fast fluctuations of the considered fluid-mechanics variable). For example, in HWA, due to the small sensor size and the fast electronics, the system takes a measurement quite rapidly (it is really an analogue system which is resampled digitally) so that TR $\approx 10^5$ s^{-1} (the measurement unit is s^{-1} or Hz), for LDA the value of Δt_{min} depends on the seeding particle inter-arrival time, which can be reduced to 10^{-4} s, so that TR $\approx 10^4$ s^{-1}, while for Pitot tubes the inertia of the manometer fluid limits the time interval to a fraction of a second, so that TR ≈ 10 s^{-1} (modern pressure transducers are able to respond quite rapidly to pressure changes but are still quite large, so that TR $\approx 10^3-10^4$ s^{-1}).

Spatial and temporal resolutions give an indication of the achievable lower limit on the scale range. On the other hand, the upper limit is related to the extension (overall size) of the measurement system in space and to the duration of the measurement in time. A single-point measurement system (such as HWA, LDA or a Pitot tube) cannot directly detect the scale spatial structure, even in homogeneous and isotropic turbulence, unless a hypothesis on the relation between behavior in space and time is assumed (see the section on Taylor's hypothesis). The detection of large-scale spatial structures can be accomplished by using multipoint systems such as using light sheets in the particle image velocimetry technique (PIV) (Sect. 5.3.2), which is then able to determine the flow behavior over a length scale d_{max} (for example, in PIV, d_{max} is the dimension of the light sheet in the test section). From the temporal point of view, in stationary flow conditions (when the large time scales are closely connected to the large space scales by Taylor's hypothesis), by performing one measurement over a time Δt_{max}, it is possible to investigate the evolution of the field up to such a time. On the other hand, in unsteady conditions (when large time scales are not simply attainable from the spatial ones), it is necessary to acquire several time sequences over a time Δt_{max} to derive the large-time-scale behavior; in this case various strategies can be employed (for example, ensemble or phase averaging as illustrated in Sect. 10.1.1).

The smallest $(d_{min}, \Delta t_{min})$ and largest $(d_{max}, \Delta t_{max})$ measurable lengths and times must be evaluated preliminarily for the measurement system under consideration and compared to the expected flow scales (for example starting from a simple evaluation in homogeneous and isotropic turbulence) to know the effective range that can be investigated. In small-scale investigations, and in unsteady conditions, it is not possible to derive the information in time from that in space by a simple transformation using the mean velocity (as stated by Taylor's hypothesis) so that both the information in space and time are requested (consider the definitions of the Kolmogorov time scale, or the Kolmogorov frequency, which do not include the mean velocity, as an example of this statement). On the other hand, for large-scale investigations, and in stationary conditions, the time and space behaviors are much more closely related by means of the mean velocity (Taylor's hypothesis).

When considering the distribution of energy among the different wavenumbers, the sampling theorem states that the energy content can be effectively detected only up to $k_{max} = \pi/d_{min}$ (and not up to $k_{max} = 2\pi/d_{min}$ as expected). Similarly, for the distribution of energy among the different frequencies, the sampling theorem states that the energy content can be effectively detected only up to $f_{max} = 1/(2\Delta t_{min})$ (and not up to $f_{max} = 1/\Delta t_{min}$ as expected). In this sense, the spatial and temporal resolutions correspond to two times the maximum wavenumber and frequency which can be detected by the measurement system. The situation is summarized in Fig. 10.1, in physical space and wavenumber domains.

The intrusiveness of the sensor within the flow field also seems to be a very important aspect of lami-

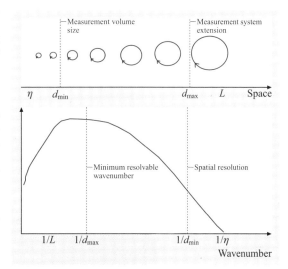

Fig. 10.1 Minimum and maximum size of the measurement system in comparison to flow structures in space and wavenumber domains

nar flow measurement. This is only partially true, due to the fact that in the presence of turbulence the insertion of a probe into the flow itself generates flow structures that modify the previous (unknown) field, whereas in laminar flow this effect is more localized (at least far from the transition regimes); thus, intrusiveness gains importance in turbulence flow measurements. In principle, it is not possible to completely avoid intrusiveness due to the requirement of interaction between the measurement system (the sensor in particular) and the fluid flow when performing experiments. In practice, this interaction should be reduced as much as possible and the effect must at least be evaluated in each case; examples are the evaluation of the interactions between intrusive sensors and boundaries in near-wall flows and the determination of spatial and temporal (wavenumber and frequency) filtering of the tracers used in optical methods (Sect. 5.3.3).

The problem of statistics is closely linked to the random nature of turbulent flows; the reader is referred to Sect. 10.1 for details of how statistics is performed. The effect of statistical accuracy and the required number of samples to attain such an accuracy are considered in the sections on Statistical Accuracy and Chap. 25. However, it is important to remark that instantaneous fields contain information on both the spatial and temporal evolution of the flow and that special care must be taken in averaging procedures (such as in the case of stationary turbulence or in homogeneous and isotropic turbulence).

Errors in experimental techniques will be not considered here because a general treatment of errors is presented elsewhere (Chapter 25) and specific measurement system errors derive from peculiar aspects of each measurement method regardless of whether laminar or turbulent conditions dominate.

Flow Variables Relevant to Turbulence and their Measurement

Once the general characteristics of a system measuring in turbulent flows have been established, it is necessary to establish which fluid-mechanics variables have to be determined for a complete description of the turbulent field. Starting from equations (Chapter 1, Sect. 10.1.1), the following fields are of interest:

- three velocity components
- pressure
- temperature
- density
- species concentration

Many other quantities are important in fluid mechanics, such as the vorticity, strain rate, stresses, dissipation, and enstrophy (which can be derived from the velocity field, although direct measurements are sometimes possible). As mentioned in the introduction, interaction terms among these variables are also of great interest in turbulence (both from the fundamental and applied points of view); these have to be determined by combined simultaneous measurements of two or more of them (examples will be given in Sect. 10.1.2).

It should also be pointed out that the equations themselves establish relations between flow variables so that one can be derived from the others (even if we are dealing with second-order partial differential equations which set severe practical limits when computing derivatives from noisy experimental data).

In these equations, other important quantities such as viscosity and other diffusion coefficients are present. Usually they are derived from constitutive relations and inserted into the equations, and empirical laws are used for their evaluation (Sect. 1.3).

For further information on the techniques and measurements mentioned above, the reader can use the following cross-references to part B of this Handbook:

- Velocity components
  - Pitot and pressure based (Sect. 5.1)
  - Thermal anemometry (Sect. 5.2)
  - Laser Doppler anemometry (Sect. 5.3.1)
  - Particle image velocimetry (Sect. 5.3.2)
  - Sonic anemometers (Sect. 5.7)
- Pressure
  - Manometers and transducers (Chap. 4)
- Temperature
  - Thermocouples (Sect. 6.1)
  - Resistive sensors (Sect. 6.1)
  - Liquid crystals (Sect. 6.3)
  - Other methods (Chap. 6)
- Density
  - Shadowgraph (Sect. 5.6)
  - Schlieren (Sect. 5.6)
  - Interferometry (Sect. 5.6)
- Concentration
  - Laser-induced fluorescence (LIF, Chap. 11)

It should be clear to experimentalists (and similarly to people involved in numerical simulations), that the large number of variables to be determined and the many potentially interesting applications mean that there are still hundreds of years' worth of interesting measurements in turbulent flows to be enjoyed. From the experimental point of view, it is also clear that efforts must be given in developing almost-nonintrusive, multipoint systems for the simultaneous measurement of various quantities in the flow field. Among others methods, this has almost been achieved for simultaneous multipoint three-component velocity (stereo PIV, 3-D particle tracking velocimetry (PTV) (Sect. 5.3.2), for combined velocity–concentration (PIV LIF , Chap. 11), velocity–temperature (liquid crystals (Sect. 6.3)) and concentration–temperature (LIF–liquid crystals) measurements.

Difficulties in measuring all of these quantities are related to:

- the need to define, determine, and verify a common *measurement volume*
- crosstalk between probes or systems
- simultaneous control and optimization of different setups
- the need for simultaneous triggering of data acquisition
- the need for a common space grid and time step definition between the various measurement systems
- the need for measurement errors of the same order of magnitude
- handling and storing large amounts of data.

To solve such problems great care and expertise must be applied; this could be the reason why advanced multipoint combined measurement systems are still on a prototype level. However, as done in the past, detailed standard procedures will be derived to achieve these requirements to a reasonable degree of accuracy and design relatively simple industrial systems.

Some additional comments should be given on the possibility of measuring pressure and density fields (of course we are still speaking about almost-nonintrusive, multipoint measurement systems). For pressure fields, high-temporal-resolution single-point measurements are already possible on the surface of bodies and on walls in general, but problems are encountered when performing such measurements with high spatial resolution (due to the probe size) within the field, i.e., far from the boundaries (due to intrusiveness of the pressure probes) and in a multipoint approach (due to the intrusiveness and interaction between the probes themselves). At these locations, simultaneous velocity–pressure correlations are also difficult to obtain. The situation is just the opposite for density field measurements, in which quite high spatial resolution in multipoint conditions can be attained far from the boundaries (albeit with low precision, i.e., with large measurement errors due to the analysis of light intensity fields), while it is quite difficult to measure close to solid surfaces (due to light reflection), especially in complex geometrical configurations (due to the almost two-dimensional nature of the measurement methods).

Some other difficulties in the measurement of turbulent flows are related to the evaluation of the derivatives that appear in the equations. Provided that the measuring system has a high spatial resolution, it is necessary to obtain derivatives from finite differences in space and time (such as those obtained as the output from numerical simulations and/or experiments). This is always a difficult task, which can be solved using interpolation algorithms coupled with evaluations of the finite differences over different grid spacings or time steps.

The substantial equivalence of many problems encountered when dealing with turbulent flows (wide range of scales, spatial and time resolution problems, errors and statistics, grid and time step problems) in both numerical simulations and experiments has already been pointed out. Many experiments using well-established measurement systems have been devoted to testing of numerical models, especially in nonstandard conditions and at high Reynolds numbers. On the other hand, well-tested numerical codes can be used to verify the quality of data obtained from advanced or nonstandard application measurement techniques. This allows the establishment of a mutual relationship between numerical simulations and experiments; an example of this is the evaluation of the turbulent kinetic-energy dissipation rate ε, which can be derived from isotropic theories (Sect. 10.1.1 and on *How to measure length and time scales*) or from numerical simulations and be used to derive dimensionless variables from measurements. This quantity, which consists of a squared sum of velocity gradients, is very difficult to evaluate from the experimental point of view with the required high spatial and temporal resolutions; on the other hand, it is quite easy to derive values for ε from numerical simulations in homogeneous and isotropic turbulence (at least at small Reynolds numbers). Thus, the numerical homogeneous and isotropic surrogate for ε can be used as a preliminary indication to evaluate flow scales and the required spatial and time resolutions in real experiments (even in inhomogeneous and anisotropic conditions).

There are large number of fields of common investigations and partnerships between numerical simulations

Lagrangian and Eulerian Descriptions

The flow field variables mentioned so far are independent of the frame reference due to Galilean invariance (at least for flow velocities much less than that of light, i.e., for all fluid-mechanics applications except for astrophysics). This statement is true for flow variables (except for the velocity itself) and their gradients, but is not true for partial derivatives in time [10.4]. Therefore if we consider all the flow at rest or in motion with a uniform velocity; in fact, the flow is not a rigid body in motion with uniform velocity and each flow element could have a velocity different from the neighbors. The variation of the fluid-mechanics quantities between different flow elements could depend on their relative position and velocity, and thus on the way in which such quantities are evaluated. Therefore, it is not trivial to consider the differences between the flow description obtained at a fixed point and the one in motion with the considered flow element; the former is referred to as Eulerian description, while the latter is known as the Lagrangian description (Sect. 10.2).

Usually, the difference between those two descriptions is clarified by considering a standing rock in a river (Eulerian frame), a leaf carried away by the flow (Lagrangian frame), and a fish moving on its own (i.e., with its own velocity). Clearly, the former undergoes variations of the fluid-mechanics quantities both in time (unsteady conditions) and from point to point (inhomogeneous conditions), while the other two exhibit variations when moving along each trajectory (a sequence of positions in time) or along different trajectories at different times, thus only in time. Such a statement can be clarified by considering the relation between the variations along a trajectory, $x(x_0, t)$, (D/Dt, substantial or Lagrangian derivative) (where x_0 is the fixed initial position) and those at a fixed point in space, x_0, ($\partial/\partial t$, the time partial or Eulerian derivative), of a fluid-mechanics variable $A[x(x_0,t), t]$ given by *Batchelor* [10.1]

$$\frac{DA}{Dt} = \left.\frac{\partial A}{\partial t}\right|_{x(x_0,t)=\text{const}}$$
$$= \left.\frac{\partial A}{\partial t}\right|_{x_0=\text{const}} + \left.\left(\frac{\partial x_i}{\partial t}\frac{\partial A}{\partial x_i}\right)\right|_{x_0=\text{const}}$$
$$= \left.\left(\frac{\partial A}{\partial t} + u_i \frac{\partial A}{\partial x_i}\right)\right|_{x_0=\text{const}},$$

where repeated indexes mean sum for $i = 1, 3$ (the sum will be omitted hereafter). The second term on the right-hand side of the last equivalence represents the transport term, which accounts for variations in space due to the motion of the fluid with velocity components u_i.

The former relation also allows one to understand how it is possible to derive one description from the other; when it is possible to know the variation in time of a fluid-mechanics variable at a given point and simultaneously the variation of the same variable between different points then we can derive the substantial derivative. This is just the case of the Lagrangian description in which the knowledge of the variables along flow trajectories allows one to derive their variations in both time and space, and thus to compute the overall variation (i.e., to derive the Eulerian description). On the other hand, given the overall variation of a quantity, it is not possible, in general, to derive the variation in time along a trajectory (i.e., to derive the Lagrangian description); the Eulerian field cannot be integrated to derive trajectories due to sensitivity to the IC (in the laminar case this would be possible). In particular, the possibility of interchange between the two descriptions depends on the specific flow field considered; the hypothesis of *Tennekes* [10.8], generalizing Taylor's hypothesis (Sect. 10.1.1, Taylor's hypothesis), establishes a direct connection between the Lagrangian and Eulerian description in isotropic turbulence, in which at large Reynolds numbers Lagrangian flow scales are larger than Eulerian flow scales (Sect. 10.2 and [10.4]).

Thus, when possible, it would be much better to attain a Lagrangian description rather than a Eulerian. However, from the experimental point of view, it is not simple to obtain fluid-mechanics variables along flow trajectories; only a few techniques allow the Lagrangian description to be obtained. Among the others, the PTV technique (Sect. 5.3.2) is now well established to derive tracer particle velocities (which under certain assumptions are representative of the flow element motions) along trajectories.

The importance of the Lagrangian approach arises in situations in which the determination of the spreading and dispersion of the flow is required (for example in pollutant dispersion or in mixing and combustion investigations) or when the history of the same flow element must be considered (for example when evaluating the deformation and stresses on blood cells in haemodynamics).

The flow variables, the derived quantities (time and space derivatives and so on), and the flow scales defined in previous paragraphs must be considered separately in the Eulerian or Lagrangian frameworks. More details on this aspect will be given in the following (Sect. 10.1.3, Sect. 10.2), but in general the evaluation of a variable in the Eulerian frame requires an average (in the sense reported in Sect. 10.1.1) at a fixed point in space or among different fixed points; on the other hand, the evaluation in a Lagrangian frame requires an average along trajectories (separating the results on the basis of the starting or arriving positions).

Alternative Approaches to Statistics

The practical consequence of the sensitivity of a turbulent flow to BC and IC is that it exhibits fluctuations in time and space. Within these fluctuations all the essential and interesting flow behaviors are embedded. Thus, the value of a fluid-mechanics variable at a given point and time (q) can be considered as the sum of a *mean* term, $\langle q \rangle$ (the way in which this term is obtained will be specified in the following of this section), plus a term containing all the information on *coherent* structures (at this stage coherent means with a well-defined pattern in the flow field, Chap. 24) in the field (large and intermediate scales) $(q')_c$, plus a random term containing all *incoherent* fluctuations (small scales), $(q')_n$,

$$q = \langle q \rangle + (q')_c + (q')_n ,$$

where $q = q(\boldsymbol{x}, t)$ is a function of the considered point ($\boldsymbol{x}$) and time (t), as are all terms on the right-hand side of the previous relation. The goal of this description is to try to separate the three contributions from each other, i.e., to determine all possible effects to which the flow variable is subjected under the influence of the flow motion. Therefore, the problem is first to perform an average of the flow variables preserving the *coherent* and *incoherent* parts and secondly to separate these two.

Unfortunately, there is no unique solution to this problem; the result depends on the criterion that is used to quantify the *coherence* or *incoherence* character of the flow variable, i.e., whether we treat them as indistinguishable from each other (no criterion), to determine the energy of the related structures, to evaluate the resemblance with sample structures defined a priori, to minimize some difference in the least-squares sense, and so on. To give an approximate idea of the problem, the most important approaches used are described in the following cross references:

- Direct averaging (Sect. 10.1.1, Sect. 10.1.2)
- Proper orthogonal decomposition (POD) (Sect. 24.4)
- Wavelets (Sect. 25.5)
- Linear stochastic estimation (LSE) (Sect. 24.6)

In this section, only direct averaging will be discussed; in this approach, all flow scales are considered as equivalent (in a Fourier sense) so that it is possible only to separate the average term from the other two in the previous decomposition, but it is not possible to distinguish between the *coherent* and *incoherent* contributions.

Statistical Domain: Ensemble Averaging, Time or Space Averaging

Once the approach to the analysis of a turbulent flow has been defined, it is necessary to specify on what domain the statistics are performed. For direct averaging, the most general way to derive averaged quantities is the ensemble average (indicated with triangular brackets); therefore, the n-th statistical moment of a fluid-mechanics variable q is given by

$$\langle q^n \rangle = \frac{\int_{-\infty}^{+\infty} q^n p(q)\, dq}{\int_{-\infty}^{+\infty} p(q)\, dq} ,$$

where $p(q)$ is the probability density function (PDF) of the variable q (Sect. 25.1 for details) and n is the order of the considered moment (Chap. 25 for details on the evaluation of statistics). Each of the previous quantities (included the PDF) is a function of the considered point and time; for example $p(q) = p(q)(\boldsymbol{x}, t)$.

In principle, the integral of the PDF over all possible values should be equal to 1; the integral is retained in the previous definition to point out that a normalization is required (especially when the integral is evaluated over a discrete set of samples). The PDF could be obtained by acquiring all possible values of the variable q at a given point and time; in practice, this means repeating the same experiment several times from the beginning (with more or less the same IC and BC), evaluating the investigated variable at the same time and position (in the general case of unsteady inhomogeneous phenomena). Such a procedure is summarized in Fig. 10.2; different values of the flow variable at the same point and time are observed for each new experiment; the mean value obtained by averaging over the different experiments is different from the one derived by a time average over a single experiment.

Two problems arise in this practical approach. Firstly, the experiments are discrete and cannot be repeated indefinitely so that the statistical treatment to determine the finite number of independent samples must be considered (Chap. 25). In this case, the PDF must be derived from a finite number of data (N_T) and the previous integral is replaced by the sum over experiments with the same PDF

$$\frac{\sum_i^N q_i^n p(q_i) \Delta q_i}{\sum_i^N p(q_i) \Delta q_i} = \frac{\sum_i^N q_i^n p(q_i) \Delta q_i}{\sum_i^N n_i}$$

(where Δq_i and n_i are the width and number of data for each bin of the N-bin histogram); if all events have the same bin width and $p(q)$, then the previous sum reduces to the usual average

$$\frac{1}{N_T} \sum_i^{N_T} q_i^n \ .$$

Secondly, the ensemble over which averaging is performed is a set of several results of the same experiment; thus, it will be necessary to repeat the same experiment systematically several times, which is a difficult problem for experimentalists. Therefore, although it is clear that ensemble averaging would be the most general way of performing averaging (without any a priori assumption about the flow field), in practice, whenever possible, it is necessary to perform the average in a much simpler way. To this end, consider the following special fluid-mechanics conditions (statistically)

- periodic flow field (in time or space)
- stationary flow field
- homogeneous flow field

for each one of which particular average procedures are possible.

In a statistically periodic flow field (period equal to T_0), the PDF is the same at time t and time $t + T_0$

$$p(q)(x, t) = p(q)(x, t + T_0)$$

(or equivalently in space at point x and $x + X_0$), so that the ensemble over which the average is performed reduces to one period of the investigated periodic phenomenon. The mathematical expressions are the same (for continuous or discrete data), but the time (or space) dependence of the results is limited to just one period. In practice, it is completely equivalent to run the experiments over N different periods or to repeat ex novo the experiments N times, providing that the BC and IC are constant (it should be required that the characteristic time, or length, scales of the phenomenon are less than T_0, Sect. 10.1.3). In this case, so-called conditional or phase averaging (i. e., averaging the acquired signal over different phases) allows one to limit the computation of $\langle q \rangle$ to times $t_0 + mT_0$, where m is an integer number and t_0 is the phase in time (and equivalently at positions $x_0 + mX_0$). Examples of such averaging are encountered in the analysis of periodic signals in time (such as for daily varying atmospheric data or for data acquired close to propellers and other rotating devices) and of periodic fields in space (such as in the wake of bluff bodies or at the outlet of jets at relatively low Reynolds numbers).

In the case of statistically stationary flow fields, the PDF is completely independent of the considered time

$$p(q)(x, t) = p(q)(x)$$

so that, statistically, each instant over a set of the ensemble should be equivalent to the others (mathematically the equivalence would also require that the statistics of the process is independent on the particular selected set of the ensemble; this is a sufficient condition. See [10.9] for further details on this argument). As depicted in Fig. 10.2, in the computation of the n-moment this allows one to replace the ensemble average by the average over a single set (i. e., in time) indicated by $\overline{q^n}$,

$$\langle q^n \rangle = \frac{\int_{-\infty}^{\infty} q^n p(q) \, dq}{\int_{-\infty}^{\infty} p(q) \, dq} = \frac{\int_0^{\infty} q^n p(q) \, dt}{\int_0^{\infty} p(q) \, dt}$$

$$= \lim_{T \to \infty} \frac{\int_0^T q^n p(q) \, dt}{\int_0^T p(q) \, dt} = \lim_{T \to \infty} \frac{1}{T} \int_0^T q^n \, dt = \overline{q^n}$$

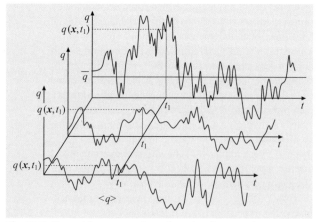

Fig. 10.2 Ensemble average along the axis of *repeated experiments* and time average along the time axis

using the hypothesis that the PDF is not dependent on time. All quantities can be functions of the spatial positions (but not of time after averaging). This is a simplified version of the ergodicity theorem, which states the equivalence between the average over the different sets of the ensemble and over time on a single set; the theorem is valid for a random process that is stationary, providing that all the statistics obtained from a set are equal to that from the others; such a process is referred as ergodic [10.9]. It must be emphasized that the simple verification of stationarity does not imply the ergodicity of a process, i.e., the replacement of the ensemble average with the time average; it is also necessary to verify that independent sets of the ensemble give the same statistics (whereas an ergodic process is automatically stationary). In practice, it is possible to divide the set into subsets and to verify the equivalence of statistics over the subsets. The widespread class of random processes that satisfy ergodicity is known as the class of Gaussian random processes with continuous power spectral density functions, i.e., without infinite-density peaks [10.9].

It is of interest to show examples of ergodic and non-ergodic processes. Consider the following unsteady non-ergodic (to be proved) periodic set

$$q(t) = Q \sin(\omega t + \varphi)$$

in which the amplitude Q and phase φ are constant over the ensemble. Ensemble averaging, at a fixed time, gives the following values for the first two statistical moments and for the correlation function (Chap. 24, Chap. 25)

$$\langle q(t) \rangle = \int_{-\infty}^{+\infty} Q \sin(\omega t + \varphi) \, p(q) dq$$
$$= Q \sin(\omega t + \varphi),$$

$$\langle q^2(t) \rangle = \int_{-\infty}^{+\infty} Q^2 \sin^2(\omega t + \varphi) \, p(q) dq$$
$$= Q^2 \sin^2(\omega t + \varphi),$$

$$\langle q(t)q(t+\tau) \rangle = \int_{-\infty}^{+\infty} Q^2 \sin(\omega t + \varphi)$$
$$\times \sin[\omega(t+\tau) + \varphi] \, p(q) dq$$
$$= Q^2 \sin(\omega t + \varphi) \sin[\omega(t+\tau) + \varphi],$$

which depends on time t (i.e., the process is unsteady). On the other hand, the average in time gives (for the same quantities)

$$\overline{q} = \lim_{T \to \infty} \frac{1}{T} \int_0^T Q \sin(\omega t + \varphi) \, dt = 0,$$

$$\overline{q^2} = \lim_{T \to \infty} \frac{1}{T} \int_0^T Q^2 \sin^2(\omega t + \varphi) \, dt$$
$$= \frac{Q^2}{2},$$

$$\overline{q(t)q(t+\tau)} = \lim_{T \to \infty} \frac{1}{T} \int_0^T Q^2 \sin(\omega t + \varphi)$$
$$\times \sin[\omega(t+\tau) + \varphi] \, dt$$
$$= \frac{Q^2}{2} \cos(\omega \tau),$$

which are of course independent on the set, but are different from the values obtained from the ensemble averaging; therefore, the process is non-ergodic. Note that, if in the time average the data are collected at the same time and phase ($\omega t + \varphi$ = constant), i.e., conditional or phase averaging is performed, the following results are obtained

$$\overline{q} = \lim_{T \to \infty} \frac{1}{T} \int_0^T Q \sin(\omega t + \varphi)_{\text{const}} dt$$
$$= Q \sin(\omega t + \varphi)_{\text{const}},$$

$$\overline{q^2} = \lim_{T \to \infty} \frac{1}{T} \int_0^T Q^2 \sin^2(\omega t + \varphi)_{\text{const}} dt$$
$$= Q^2 \sin^2(\omega t + \varphi)_{\text{const}},$$

$$\overline{q(t)q(t+\tau)} = \lim_{T \to \infty} \frac{1}{T} \int_0^T Q^2 \sin(\omega t + \varphi)_{\text{const}}$$
$$\times \sin[\omega(t+\tau) + \varphi]_{\text{const}} \, dt = Q^2$$
$$\times \sin(\omega t + \varphi)_{\text{const}}$$
$$\times \sin[\omega(t+\tau) + \varphi]_{\text{const}},$$

which are identical to the results of the ensemble averaging, so that phase averaging can be used as a surrogate of ensemble averaging.

A similar result is obtained for periodic sets with different amplitudes $Q(k)$ for each set k (all with the same phase)

$$q_k(t) = Q(k) \sin(\omega t + \varphi).$$

For the same moments as before, ensemble averaging gives

$$\langle q(t) \rangle = \int_{-\infty}^{+\infty} Q(k) \sin(\omega t + \varphi)\, p(q_k) dq_k$$
$$= \langle Q \rangle \sin(\omega t + \varphi),$$
$$\langle q^2(t) \rangle = \int_{-\infty}^{+\infty} Q^2(k) \sin^2(\omega t + \varphi)\, p(q_k) dq_k$$
$$= \langle Q^2 \rangle \sin^2(\omega t + \varphi),$$
$$\langle q(t) q(t+\tau) \rangle = \int_{-\infty}^{+\infty} Q^2(k) \sin(\omega t + \varphi)$$
$$\times \sin[\omega(t+\tau) + \varphi]\, p(q_k) dq_k$$
$$= \langle Q^2 \rangle \sin(\omega t + \varphi)$$
$$\times \sin[\omega(t+\tau) + \varphi],$$

which depend on time t (i.e., the process is unsteady). The average in time gives

$$\overline{q} = \lim_{T \to \infty} \frac{1}{T} \int_0^T Q(k) \sin(\omega t + \varphi) dt = 0,$$

$$\overline{q_k^2} = \lim_{T \to \infty} \frac{1}{T} \int_0^T Q^2(k) \sin^2(\omega t + \varphi) dt$$
$$= \frac{Q^2(k)}{2},$$

$$\overline{q(t) q(t+\tau)} = \lim_{T \to \infty} \frac{1}{T} \int_0^T Q^2(k) \sin(\omega t + \varphi)$$
$$\times \sin[\omega(t+\tau) + \varphi] dt$$
$$= \frac{Q^2(k)}{2} \cos(\omega \tau),$$

which depend on the set (except the mean value), so they are different from the values obtained from the ensemble averaging; therefore, this process is also non-ergodic.

On the other hand, if we consider the following stationary ergodic (to be proved) periodic sets with different phases $\varphi(k)$ for each set (all with the same amplitude)

$$q_k(t) = Q \sin[\omega t + \varphi(k)].$$

The ensemble averaging gives

$$\langle q \rangle = \int_{-\infty}^{+\infty} Q \sin[\omega t + \varphi(k)]\, p(q_k) dq_k$$
$$= 0,$$
$$\langle q^2 \rangle = \int_{-\infty}^{+\infty} Q^2 \sin^2[\omega t + \varphi(k)]\, p(q_k) dq_k$$
$$= \frac{Q^2}{2},$$
$$\langle q(t) q(t+\tau) \rangle = \int_{-\infty}^{+\infty} Q^2 \sin[\omega t + \varphi(k)]$$
$$\times \sin[\omega(t+\tau) + \varphi(k)]\, p(q_k) dq_k$$
$$= \frac{Q^2}{2} \cos(\omega \tau),$$

which do not depend on time t (i.e., the process is stationary). The average in time gives

$$\overline{q} = \lim_{T \to \infty} \frac{1}{T} \int_0^T Q \sin[\omega t + \varphi(k)] dt = 0,$$

$$\overline{q^2} = \lim_{T \to \infty} \frac{1}{T} \int_0^T Q^2 \sin^2[\omega t + \varphi(k)] dt$$
$$= \frac{Q^2}{2},$$

$$\overline{q(t) q(t+\tau)} = \lim_{T \to \infty} \frac{1}{T} \int_0^T Q^2 \sin[\omega t + \varphi(k)]$$
$$\sin[\omega(t+\tau) + \varphi(k)] dt$$
$$= \frac{Q^2}{2} \cos(\omega \tau),$$

which are independent of the set and equal to the values obtained from the ensemble averaging; therefore, this process is ergodic.

Lastly, consider the stationary non-ergodic (to be proved) periodic sets with different uncorrelated amplitudes and phases [$Q(k)$ and $\varphi(k)$] for each set [10.9]

$$q_k(t) = Q(k) \sin[\omega t + \varphi(k)].$$

The ensemble averaging gives

$$\langle q \rangle = \int_{-\infty}^{+\infty} Q(k) \sin[\omega t + \varphi(k)] \, p(q_k) \mathrm{d}q_k$$
$$= 0,$$
$$\langle q^2 \rangle = \int_{-\infty}^{+\infty} Q^2(k) \sin^2[\omega t + \varphi(k)] \, p(q_k) \mathrm{d}q_k$$
$$= \frac{\langle Q^2 \rangle}{2},$$
$$\langle q(t)q(t+\tau) \rangle = \int_{-\infty}^{+\infty} Q^2(k) \sin[\omega t + \varphi(k)]$$
$$\sin[\omega(t+\tau) + \varphi(k)] \, p(q_k) \mathrm{d}q_k$$
$$= \frac{\langle Q^2 \rangle}{2} \cos(\omega \tau),$$

which does not depend on time t (i.e., the process is stationary). The average in time gives

$$\overline{q} = \lim_{T \to \infty} \frac{1}{T} \int_0^T Q(k) \sin[\omega t + \varphi(k)] \mathrm{d}t$$
$$= 0,$$
$$\overline{q^2} = \lim_{T \to \infty} \frac{1}{T} \int_0^T Q^2(k) \sin^2[\omega t + \varphi(k)] \mathrm{d}t$$
$$= \frac{Q^2(k)}{2},$$
$$\overline{q(t)q(t+\tau)} = \lim_{T \to \infty} \frac{1}{T} \int_0^T Q^2(k) \sin[\omega t + \varphi(k)]$$
$$\times \sin[\omega(t+\tau) + \varphi(k)] \mathrm{d}t$$
$$= \frac{Q^2(k)}{2} \cos(\omega \tau),$$

which are dependent on the set and therefore different from the values obtained from the ensemble averaging (note that the values obtained for the correlation coefficients using ensemble ($\langle q(t)q(t+\tau)\rangle/\langle q^2\rangle$) or time ($\overline{q_k(\tau)}/\overline{q^2}$) averaging are both equal to $\cos(\omega t)$): the process is therefore non-ergodic. Thus, mathematical examples of an unsteady non-ergodic process, a steady ergodic process, and a steady non-ergodic process have been given. In practice, the previous examples could correspond to several realizations of an oscillatory behavior as in measurements around propellers (or other rotating devices) and in the atmosphere; amplitudes and phases can change from one sample of the ensemble to the other due to changes in trim at each revolution (propeller) or modifications of weather conditions from day to day (atmosphere).

Returning to the possible ways in which an average can be performed, for homogeneous flow fields, the same arguments used for stationary flows can be converted from time to space. The PDF is independent of the spatial position

$$p(q)(x,t) = p(q)(t)$$

and the ensemble average can be replaced by the average in space

$$\overline{q^n} = \lim_{V \to \infty} \frac{1}{V} \int_V q^n \mathrm{d}V$$

(all quantities can be function of time after averaging). The same considerations given before are valid in this case. Nevertheless, the assumption of homogeneity in practical fluid flow conditions is much farther from reality than that of stationarity; to this end, it is important to stress that, from the Navier–Stokes equations, it is impossible to have simultaneous homogeneity and stationarity of a flow except for a fluid at rest.

Relevant Statistical Quantities in Turbulence

By performing the average over the ensemble (when possible), in time or space, it is possible to derive all the statistics for the considered fluid-mechanics variable; see Chap. 24, 25 for further information on the evaluation of mean values, higher statistical moments and cross-moments (in turbulence, cross-moments are fundamental quantities appearing in the averaged equations, Sect. 10.1.2).

Particularly interesting are the statistics related to recurrences in space and time domains (correlation functions, Sect. 24.2) or in transformed wavenumber and frequency domains (spectral functions that are related to the former by Fourier-transform operations, Sect. 24.1). In particular, it is important to point out that the general definitions of correlation and spectral functions between two points at x (time t) and $x' = x + r$ (time $t + \tau$) are the following

$$R_{q_i q_j}(x, r, t, \tau) = \langle q_i(x,t) \, q_j(x+r, t+\tau) \rangle,$$
$$F^k_{q_i q_j}(x, k, t, \tau) = \mathfrak{Re}\{\mathfrak{I}_k[R_{q_i q_j}(x, r, t, \tau)]\},$$
$$F^f_{q_i q_j}(x, r, t, f) = \mathfrak{Re}\{\mathfrak{I}_f[R_{q_i q_j}(x, r, t, \tau)]\},$$
(10.1)

where $\mathfrak{Re}\{\cdot\}$ means the real part of the argument, and $\mathfrak{I}_f[\cdot]$ and $\mathfrak{I}_k[\cdot]$ mean the Fourier transform in the frequency and three-dimensional wavenumber domain of the quantity in brackets, respectively. The inverse Fourier-transform relations hold when passing from the spectral to the correlation functions (Wiener–Khintchine theorem, Chap. 24 [10.9]). These quantities are vectors if the two variables q_i and q_j are scalars, second-order tensors if one of the two is a vector (i.e., the flow velocity) and third-order tensors if both are vectors (i.e., if all velocity components are considered); this depends on the fact that three projections along the reference axis of the previous quantities can be derived. For example, for the correlation function, the quantity

$$R_{u_i u_j}(x, r_x, t, \tau) = \langle u_i(x,t) \, u_j(x+r_x, t+\tau) \rangle$$

is the correlation function tensor between the velocity components u_i and u_j at the points x and $x+r_x$ along the x axis of the reference system; varying i and j from 1 to 3, nine of these functions are to be evaluated in a general flow field. Of course, each one of these functions is dependent on the initial spatial position (x), the separation distance (r), the initial time (t), and the separation in time (τ).

Each one of these correlation functions reduces to the single-point cross-statistical moments (covariance) when the same point ($r=0$) and time ($\tau=0$) are considered

$$R_{q_i q_j}(\boldsymbol{x}, 0, t, 0) = \langle q_i(\boldsymbol{x}, t) \, q_j(\boldsymbol{x}, t) \rangle \,. \tag{10.2}$$

These quantities (when obtained for the fluctuating fluid-mechanics variables of a turbulent flow field) are of fundamental importance in turbulence investigations (Sect. 10.1.2); in particular, for the velocity field, these represent the Reynolds stress symmetric tensor

$$R_{u_i u_j} = \begin{pmatrix} u_1^2 & u_1 u_2 & u_1 u_3 \\ u_2 u_1 & u_2^2 & u_2 u_3 \\ u_3 u_1 & u_3 u_2 & u_3^2 \end{pmatrix} \,.$$

As a consequence of the previous single-point limit for correlation functions, the inverse Fourier transform gives for the spectral functions in the wavenumber and frequency domains

$$\int_0^\infty F_{q_i q_j}^k(\boldsymbol{x}, \boldsymbol{k}, t, 0) \, \mathrm{d}\boldsymbol{k} = R_{q_i q_j}(\boldsymbol{x}, 0, t, 0) = \langle q_i q_j \rangle \,,$$

$$\int_0^\infty F_{q_i q_j}^f(\boldsymbol{x}, 0, t, f) \, \mathrm{d}f = R_{q_i q_j}(\boldsymbol{x}, 0, t, 0) = \langle q_i q_j \rangle \,,$$

in which, as usual in transformed domains, the single-valued limit in the correlation functions becomes an integral value for the spectral functions. For $i = j$, the integral of the spectral function just gives the mean square value of the variable under consideration.

Returning to Sect. 10.1.1, the situation is simpler when the hypothesis of a stationary flow field can be assumed; in this case, the dependence on the initial time is avoided so that

$$R_{q_i q_j}(\boldsymbol{x}, \boldsymbol{r}, \tau) = \lim_{T\to\infty} \frac{1}{T} \int_0^T q_i(\boldsymbol{x}, t) \, q_j$$
$$\cdot (\boldsymbol{x}+\boldsymbol{r}, t+\tau) \, \mathrm{d}t \,,$$
$$F_{q_i q_j}^k(\boldsymbol{x}, \boldsymbol{k}, \tau) = \mathfrak{Re}\{\mathfrak{I}_k[R_{q_i q_j}(\boldsymbol{x}, \boldsymbol{r}, \tau)]\} \,,$$
$$F_{q_i q_j}^f(\boldsymbol{x}, \boldsymbol{r}, f) = \mathfrak{Re}\{\mathfrak{I}_f[R_{q_i q_j}(\boldsymbol{x}, \boldsymbol{r}, \tau)]\} \,.$$

Similarly, for the case of homogeneous flow field the dependence on initial position is avoided

$$R_{q_i q_j}(\boldsymbol{r}, t, \tau) = \lim_{V\to\infty} \frac{1}{V} \int_V q_i(\boldsymbol{x}, t) \, q_j$$
$$\cdot (\boldsymbol{x}+\boldsymbol{r}, t+\tau) \, \mathrm{d}\boldsymbol{x} \,,$$
$$F_{q_i q_j}^k(\boldsymbol{k}, t, \tau) = \mathfrak{Re}\{\mathfrak{I}_k[R_{q_i q_j}(\boldsymbol{r}, t, \tau)]\} \,,$$
$$F_{q_i q_j}^f(\boldsymbol{r}, t, f) = \mathfrak{Re}\{\mathfrak{I}_f[R_{q_i q_j}(\boldsymbol{r}, t, \tau)]\} \,.$$

The situation is also simplified when purely spatial variation are considered

$$R_{q_i q_j}(\boldsymbol{x}, \boldsymbol{r}, t) = \langle q_i(\boldsymbol{x}, t) \, q_j(\boldsymbol{x}+\boldsymbol{r}, t) \rangle \,,$$
$$F_{q_i q_j}^k(\boldsymbol{x}, \boldsymbol{k}, t) = \mathfrak{Re}\{\mathfrak{I}_k[R_{q_i q_j}(\boldsymbol{x}, \boldsymbol{r}, t)]\} \,,$$

(which can be further simplified if stationary or homogeneous flow fields are assumed, so that the dependence on t or $\boldsymbol{x}$ is avoided and averages in time or space are employed), and for the purely temporally varying case

$$R_{q_i q_j}(\boldsymbol{x}, t, \tau) = \langle q_i(\boldsymbol{x}, t) \, q_j(\boldsymbol{x}, t+\tau) \rangle \,,$$
$$F_{q_i q_j}^f(\boldsymbol{x}, t, f) = \mathfrak{Re}\{\mathfrak{I}_f[R_{q_i q_j}(\boldsymbol{x}, t, \tau)]\} \,,$$

(which also can be further simplified if stationary or homogeneous flow fields are assumed so that the dependence on t or $\boldsymbol{x}$ is avoided and averages in time or space are employed).

Is it also possible to define autocorrelation and autospectral functions by performing the previous evaluations for $q_i = q_j$ ($i = j$); for the x axis, this is

$$R_{q_i^2}(x, r_x, t, \tau) = \langle q_i(x, t) q_i(x + r_x, t + \tau) \rangle ,$$

$$F_{q_i^2}^{k_x}(x, k_x, t, \tau) = \mathfrak{Re}\left\{\mathfrak{I}_{k_x}\left[R_{q_i^2}(x, r_x, t, \tau)\right]\right\} ,$$

$$F_{q_i^2}^{f}(x, r_x, t, f) = \mathfrak{Re}\left\{\mathfrak{I}_{f}\left[R_{q_i^2}(x, r_x, t, \tau)\right]\right\} ,$$

(10.3)

and similarly for the other two axes. In this case, there are always three independent correlation functions to be determined (i.e., the projections along the three axes) for each scalar variable q_i, so that there are nine independent functions for the three components of fluid velocity. The functions in (10.3) are also known as one-dimensional correlation and spectral functions. Normalizing these quantities by the zero-separation ($r = 0$) or zero-time-delay ($\tau = 0$) single-point values ((10.2) with $i = j$), the so-called correlation coefficients are obtained. Relations between one- and three-dimensional correlation and spectral functions can be derived in general flow fields and in the case of isotropy [10.1, 4, 10].

Combined space–time and higher-order correlation and spectral functions can also be defined. In particular, it is interesting for the following (see the Sect. on multipoint equations), to consider the triple-point correlation functions between different velocity components at two points; one of those for the case of purely spatial correlation is written as

$$R_{u_i u_j, u_k}(x, r, t) = \langle u_i(x, t) u_j(x, t) u_k(x + r, t) \rangle ,$$

$$F_{u_i u_j, u_l}^{k}(x, k, t) = \mathfrak{Re}\{\mathfrak{I}_k[R_{u_i u_j, u_l}(x, r, t)]\} ,$$

(10.4)

where the comma separates the quantity evaluated at the first point from those at evaluated at the second point.

Other important multipoint high-order functions are the spatial structure functions (see also Sect. 24.2 and [10.4, 11])

$$S_{q_i q_j \ldots q_k}^n(x, r, t) = \langle [q_i(x + r, t) - q_i(x, t)]$$
$$\times [q_j(x + r, t) - q_j(x, t)]$$
$$\times [q_k(x + r, t) - q_k(x, t)] \rangle .$$

Equivalently, temporal structure functions can be defined. The purely longitudinal and transversal n-th-order velocity structure functions

$$S_{u_L}^n(x, r, t) = \langle [u_L(x + r, t) - u_L(x, t)]^n \rangle ,$$

$$S_{u_N}^n(x, r, t) = \langle [u_N(x + r, t) - u_N(x, t)]^n \rangle ,$$

(where u_L and u_N are the velocity components parallel and orthogonal to r, respectively) can be defined. These functions are often indicated by D_L and D_N, see *Monin* and *Yaglom* [10.12]. For $n = 2$, there is a simple relation between the spatial second-order structure function of the same variable and the purely spatial autocorrelation function

$$S_{q_i}^2(x, r, t) = \langle [q_i(x + r, t) - q_i(x, t)]^2 \rangle$$
$$= \langle q_i^2(x + r, t) \rangle + \langle q_i^2(x, t) \rangle$$
$$\quad - 2 R_{q_i^2}(x, r, t)$$
$$= 2\left[\langle q_i^2(x, t)\rangle - R_{q_i^2}(x, r, t)\right]$$
$$= 2\left[R_{q_i^2}(x, 0, t) - R_{q_i^2}(x, r, t)\right]$$

using (10.2) (and similarly in time). Note that the last two equivalences hold in homogeneous turbulence (steady turbulence for the time functions). Thus, the following relation exists between the second-order structure function and spectral function (using (10.3))

$$S_{q_i}^2(x, r_x, t) = 2\left(\int_0^\infty F_{q_i^2}^{k_x}(x, k_x, t) \mathrm{d}k_x\right.$$
$$\left. - \mathfrak{Re}\{\mathfrak{I}_{k_x}[F_{q_i^2}^{k_x}(x, k_x, t)]\}\right)$$
$$= 2\mathfrak{Re}\left[\int_0^\infty (1 - \mathrm{e}^{\mathrm{i}k_x r_x})\right.$$
$$\left. F_{q_i^2}^{k_x}(x, k_x, t) \, \mathrm{d}k_x\right]$$

(10.5)

using (10.2) (and similarly in time). It is important to notice that from the previous relation if a region in which a power-law wavenumber spectrum exists ($F \sim A k_x^p$), then the structure function also exhibits a power-law region ($S \sim B r_x^q$), with the following relation among exponents

$$p = -(q + 1) ,$$

(10.6)

and vice versa (see *Frisch* [10.11] and *Pope* [10.4] for the details and limitations of this relation).

It is interesting to notice two limiting behaviors for second-order structure functions; firstly, from the fact that the derivatives of a fluid-mechanics variable must be finite when the separation goes to zero (in practice, this means for separations on the order of the Kolmogorov microscale or less), it results that

$$S_{q_i}^n(x, r, t) = A r^n \quad \text{for } r \to 0 .$$

Note that this result is derived for the n-th-order moment of the same variable and that for $n = 2$ this can also be derived from properties of the autocorrelation functions for $r \to 0$ (Sect. 10.1.3). With proper normalization, i.e., using Kolmogorov length and velocity scales, this result can be written in an exact form (for example, for the longitudinal component of velocity with $n = 2$ it is $A = 1/15$ [10.12, 13]). For the case of large separation, i.e., $r \to \infty$, which in practice means separations larger than the integral scale beyond which the autocorrelation function is almost null (see the Sections on the behavior of correlation functions and integral scales) one obtains

$$S^2_{q_i}(\mathbf{x}, \mathbf{r}, t) = 2\langle q'^2_i \rangle \quad \text{for } r \to \infty \,;$$

similar relations can be obtained for higher-order structure functions [10.12, 13]).

From the point of view of how statistical quantities are obtained, it is important to point out that in the previous definitions the acquired data are assumed to be continuous while in practice, frequently, discrete data are obtained and the integrals must be replaced by sums. Moreover, it is usually assumed that the data are regularly sampled; in some circumstances this is not the case and irregularly sampled data (in time or space) are encountered (such as in LDA and PTV, see Sect. 5.3.1, 5.3.2). In this case, the problem is how to derive the statistics for such a data set; the simplest solution is the use of interpolation procedures (Sect. 27.1). Other interesting approaches, such as maximum-entropy methods, Kalman filters, and advanced correlation computation are possible; (see Chap. 27, *van Maanen* [10.14], for details).

Relevant Statistical Quantities in Turbulence: Isotropic Flows

In isotropic homogeneous flow conditions the single-point statistical moments of each fluid-mechanics variable are independent of the position and direction of the considered point; therefore fluctuation statistics (and PDFs) are the same in all directions

$$\langle q'^2_1 \rangle = \langle q'^2_2 \rangle = \langle q'^2_3 \rangle \,.$$

Moreover, all the centered statistical moments of each velocity component are equal and the cross-moments of a fluid variable with the velocity are zero independent of the velocity component involved (except the case when mean square velocity fluctuations are considered)

$$\langle u'_1 q'_i \rangle = \langle u'_2 q'_i \rangle = \langle u'_3 q'_i \rangle = 0 \,.$$

It is important to point out that equivalence of statistical moments implies isotropy, especially at large scales, but not necessarily at small scales (usually referred to as local isotropy). On the other hand, variable derivatives mainly depend on small scales; thus, the following relations between derivatives are a more-stringent test for local isotropy of the velocity field [10.12]

$$\left\langle \left(\frac{\partial u'_1}{\partial x_1}\right)^2 \right\rangle = \left\langle \left(\frac{\partial u'_2}{\partial x_2}\right)^2 \right\rangle = \left\langle \left(\frac{\partial u'_3}{\partial x_3}\right)^2 \right\rangle,$$

$$2\left\langle \left(\frac{\partial u'_1}{\partial x_1}\right)^2 \right\rangle = \left\langle \left(\frac{\partial u'_1}{\partial x_2}\right)^2 \right\rangle = \left\langle \left(\frac{\partial u'_2}{\partial x_1}\right)^2 \right\rangle$$

$$= \left\langle \left(\frac{\partial u'_1}{\partial x_3}\right)^2 \right\rangle = \left\langle \left(\frac{\partial u'_3}{\partial x_1}\right)^2 \right\rangle$$

$$= \left\langle \left(\frac{\partial u'_3}{\partial x_2}\right)^2 \right\rangle = \left\langle \left(\frac{\partial u'_2}{\partial x_3}\right)^2 \right\rangle,$$

$$\left\langle \frac{\partial u'_1}{\partial x_2} \frac{\partial u'_2}{\partial x_1} \right\rangle = \left\langle \frac{\partial u'_1}{\partial x_3} \frac{\partial u'_3}{\partial x_1} \right\rangle = \left\langle \frac{\partial u'_2}{\partial x_3} \frac{\partial u'_3}{\partial x_2} \right\rangle$$

$$= -\frac{1}{2}\left\langle \left(\frac{\partial u'_1}{\partial x_1}\right)^2 \right\rangle,$$

and for the scalar field

$$\left\langle \left(\frac{\partial q'}{\partial x_1}\right)^2 \right\rangle = \left\langle \left(\frac{\partial q'}{\partial x_2}\right)^2 \right\rangle = \left\langle \left(\frac{\partial q'}{\partial x_3}\right)^2 \right\rangle .$$

For multipoint quantities, the definitions presented in the previous section become quite simple in isotropic homogeneous flow conditions where only one correlation function is necessary for a complete description of a scalar field and only two independent correlation functions for a vector field. For the latter, in the case of the velocity field, these two independent functions are referred to as the longitudinal (f) and transverse (g) correlation functions (usually the correlation coefficients are employed); referring to (10.3), for the purely spatial correlation, considering the velocity components parallel (u_1) and orthogonal (u_2) to the direction of separation (it is now possible to take r along the direction x)

$$f(r, t) = R_{u^2_1}(r, t)/R_{u^2_1}(0, t)$$
$$= \langle u_1(\mathbf{x}, t) u_1(\mathbf{x} + \mathbf{r}, t) \rangle / \langle u^2_1(\mathbf{x}, t) \rangle ,$$
$$g(r, t) = R_{u^2_2}(r, t)/R_{u^2_2}(0, t)$$
$$= \langle u_2(\mathbf{x}, t) u_2(\mathbf{x} + \mathbf{r}, t) \rangle / \langle u^2_2(\mathbf{x}, t) \rangle . \quad (10.7)$$

Thus, each correlation among the velocity components in any direction can be expressed as a function of the previous two functions [10.10]

$$R_{u_i u_j}(\mathbf{r}, t) \equiv \langle u_i(\mathbf{x}, t) u_j(\mathbf{x}+\mathbf{r}, t)\rangle$$
$$= \langle u_i^2 \rangle \left[\frac{f(r,t) - g(r,t)}{r^2} \right.$$
$$\left. \times x_i x_j + g(r) \delta_{ij} \right], \quad (10.8)$$

where δ_{ij} is the Kronecker delta and x_i and x_j are the coordinates along the reference axis of the first and second point (separated by a distance r).

For an incompressible fluid, it is also possible to derive a relation between the longitudinal and transverse correlations [10.10]

$$g(\mathbf{r},t) = f(\mathbf{r},t) + \frac{r}{2}\frac{\partial f(r,t)}{\partial r}. \quad (10.9)$$

i.e., the longitudinal correlation can be derived from the transverse, so that only one function is required for the description of velocity correlations. The verification of the previous relation is a stringent test for isotropy in incompressible fluids.

Similarly, it is possible to greatly reduce the number of independent triple-point correlation functions (fourth-order tensors); only three of those 27 (each one along three axis) are different from zero in isotropic homogeneous flow [10.10]. Again, by considering the velocity components aligned with the separation distance (r taken along x), these three functions are

$$k(r,t) = \frac{R_{u_1^2, u_1}(r,t)}{R_{u_1^3}(0,t)} = \frac{\langle u_1^2(x,t) u_1(x+r,t)\rangle}{\langle u_1^3(x,t)\rangle},$$

$$h(r,t) = \frac{R_{u_2^2, u_1}(r,t)}{R_{u_1^3}(0,t)} = \frac{\langle u_2^2(x,t) u_1(x+r,t)\rangle}{\langle u_1^3(x,t)\rangle},$$

$$q(r,t) = \frac{R_{u_2 u_1, u_2}(r,t)}{R_{u_1^3}(0,t)}$$
$$= \frac{\langle u_2(x,t) u_1(x,t) u_2(x+r,t)\rangle}{\langle u_1^3(x,t)\rangle}.$$

For an incompressible fluid, a further reduction to only one independent triple correlation is possible [10.10]

$$h(r,t) = -\frac{k(r,t)}{2},$$
$$q(r,t) = \frac{1}{4r}\frac{\partial}{\partial r}[r^2 k(r,t)]. \quad (10.10)$$

Regarding the structure functions in homogeneous and isotropic flow conditions it is possible to derive a relation similar to (10.8), relating the second-order structure function among two velocity components separated in a generic direction to the longitudinal and transverse second-order structure functions

$$S_{u_i u_j}^2(r,t) = S_{u_2}^2(r,t)\delta_{ij}$$
$$+ \left[S_{u_1}^2(r,t) - S_{u_2}^2(r,t)\right]\frac{x_i x_j}{r^2}.$$

As for correlation functions, for incompressible fluid, it is possible to derive a relation similar to (10.9) between longitudinal and transverse second-order structure functions

$$S_{u_2}^2(r,t) = S_{u_1}^2(r,t) + \frac{r}{2}\frac{\partial}{\partial r}S_{u_1}^2(r,t),$$

so that only one scalar structure function is required for a complete description of second-order velocity differences along arbitrary directions.

According to the Kolmogorov similarity hypothesis [10.11], it is possible to derive a functional form for the second-order structure functions in isotropic homogeneous turbulence for separations within the so-called inertial range (i.e., for separations much larger than the Kolmogorov scale and much smaller than the integral scale, $\eta \ll r \ll \Lambda$) (Sect. 10.1.3),

$$S_{u_1}^2(r,t) = C_2(\varepsilon r)^{2/3}, \quad (10.11)$$

where C_2 is a constant (in theory not dependent on Reynolds number). Generalizing the previous power-law dependence to n-th order, it is possible to write

$$S_{u_1}^n(r,t) = C_n(\varepsilon r)^{n/3}. \quad (10.12)$$

The previous equation has been tested numerically and experimentally in several flow fields. Although the question of effective isotropy to be obtained is raised (i.e., large enough Reynolds numbers), it appeared that departures from Kolmogorov theory due to intermittency of small scales are present; several models have been proposed to account for the nonlinear structure function exponent behavior but none is able to explain the behavior of n-th-order structure functions over the entire range of separations (see *Frisch* [10.11] for details of this argument).

For the spectral functions, using (10.6) and (10.11), it is possible to obtain the following prediction for the behavior in the inertial range

$$F_{u_1^2}^{k_x}(k_x, t) = C_K \varepsilon^{2/3} k_x^{-5/3}, \quad (10.13)$$

which is the well known $-5/3$ power law for the inertial range in velocity spectrum of turbulence.

Relation Between Quantities Evaluated in Space and Time: Taylor's Hypothesis

It is important to point out again the differences between statistics obtained at a single point and at different points. Let us take the example of purely spatial or purely time correlation functions of a fluid-mechanics variable (see the previous two sections for details). Of course, from the experimental point of view, the former can only be evaluated by performing multipoint measurements of the variable, while the latter are obtained by measuring the evolution of the same variable in time at a single point. As already pointed out, in general, these are not uniquely linked to one another, especially when considering small vortical structures of the flow field (Sect. 10.1.3). In many circumstances, however, it is much simpler to perform measurements at a single point rather than simultaneously at several points. Therefore, it is important to establish if a relation between the behaviors in space and time holds and the degree of validity of such an approximation.

J.T. Taylor [10.15], devised an hypothesis (since then known as Taylor's hypothesis) in which time and space behaviors (along the mean direction of motion) of a fluid-mechanics variable q are simply related by the convection velocity U_c (along the same mean direction)

$$\frac{\partial q}{\partial t} + U_c \frac{\partial q}{\partial x_1} = 0 \,. \tag{10.14}$$

This hypothesis is equivalent to vanish the mean substantial derivative, i.e., to ignore in the equations of motion the transport of q along directions orthogonal to the mean flow and its diffusion in space (when q is the velocity, the pressure term is also not considered). The hypothesis seems to be a rather coarse approximation with a poor degree of validity in practical conditions; surprisingly, this is not the case and the hypothesis holds quite well in many flow conditions.

Before considering the range of validity of Taylor's hypothesis, it is important to specify in more detail the experimental meaning of the convection velocity and how the hypothesis is applied in practice. The simplest way (as in the original formulation) to apply the hypothesis is to consider the convection velocity as a constant for all flow scales, for example assuming

$$U_c = U_0 \,;$$

this means that the flow pattern is simply *frozen* at a given instant and convected over the observation point in time (this is why Taylor's hypothesis is also known as the frozen-field hypothesis). Thus, the developments in time and space are simply related by a translation from the space to the time axis and vice versa. The purely spatial autocorrelation function is evaluated through the purely time autocorrelation function

$$R_{q_i^2}(x_1, r_x = \tau U_c, t, 0) = R_{q_i^2}(x_1, 0, t, \tau) \,.$$

This relation is obtained by considering that integration of (10.14) from t to $t - \tau$, assuming U_c to be a constant, gives

$$q_i(x_1, t + \tau) = q_i(x_1 - U_c \tau, t) \,,$$

(when dealing with wavenumber spectral functions, Taylor's hypothesis allows the replacement of the wavenumber along the mean flow direction with the frequency and vice versa, $k_1 = 2\pi f/U_c$, i.e., to replace the wavenumber with the frequency spectra).

Experiments have demonstrated that this simple application of the hypothesis is not effective and that in many circumstances the convection velocity is different from the mean velocity. For example, in near-wall flows, to allow Taylor's hypothesis properly, the convection velocity should be selected between $0.7U_0$ and $0.9U_0$ depending on the distance from the wall [10.16]. Thus, the idea of a frozen field at all flow scales (i.e., regardless of the value of the local convection velocity for each flow scale) is only a first-order approximation. By performing simultaneous time correlation (at a single point) and space correlation (at several points), it is possible to show how valid the hypothesis of constant U_c is.

In Fig. 10.3, a comparison between time and space autocorrelation coefficients is given (see *Romano* [10.16] for details); these results are obtained by performing velocity measurements in a turbulent channel flow at different near-wall locations and different Reynolds numbers by means of two independent LDA systems separated along the streamwise direction (at a separation of r). The spatial correlation is transformed into a time correlation by using the local mean velocity (U_l, i.e., the velocity at the considered location) of each data set (with a constant value of convection velocity). It can be observed that near the centreline ($y^+ = 200$, where the plus indicates nondimensional wall units, refer to Chap. 12) and at the largest Reynolds number (top of Fig. 10.3), the spatial and time correlations agree reasonably well especially in the range between the Taylor and integral scales (indicated by the vertical dotted lines in the figure; see Sect. 10.1.3 for the definitions of turbulent scales). On the other hand, close to the wall ($y^+ = 20$, $y^+ = 10$) and at low Reynolds numbers (at the bottom of the figure), the space correlation overestimate the time correlation; this means that the convection velocity used to transform space into the time correlation (i.e., the local mean velocity) is too small. A better evaluation of the convection velocity, presented in the following, allows one to obtain space correlations (*dashed curves*) that are quite close to the time correlations. Note that, even with this correction, for scales smaller than the Taylor microscale there is a definite departure between the space and time correlations (as for the cases at the centreline and higher Reynolds numbers). Note also that, the larger the turbulence intensity (indicated at the bottom left corner of each plot), the greater the difference between the space and time correlations.

A possible way to evaluate the correct convection velocity from data is to compute the time delay between the time autocorrelation and the envelope of shifted space cross-correlations between two points separated by r. Such a time delay is given in Fig. 10.4, on the left for a space cross-correlation shifted by r/U_0 (i.e., using the centreline mean velocity) and on the right by r/U_l (i.e., using the local mean velocity).

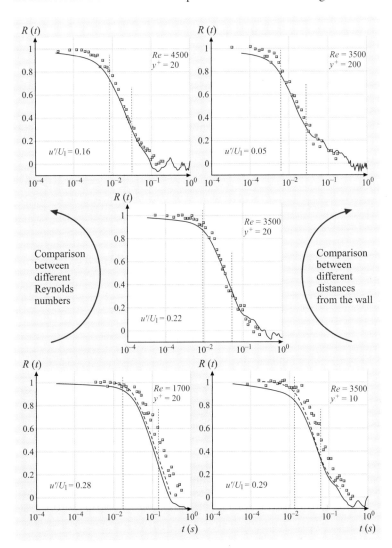

Fig. 10.3 Comparison between time autocorrelation (*continuous lines*) and space autocorrelation (*symbols*) coefficients at different Reynolds numbers and distances from the wall. *Vertical dotted lines* show Taylor and integral time scales. Longitudinal velocity fluctuations normalized by local mean velocity are also indicated. *Dashed curves* show space autocorrelation coefficients with time lag obtained using measured convection velocity rather than local mean velocity (after *Romano* [10.16])

When the centreline mean velocity shift is used, the time delay increases almost linearly; an increase in time delay with separation between points means that the effective convection velocity of turbulent structures is smaller than the centreline mean velocity. A linear increase means that this effective convection velocity is constant (for the separation between points considered); the convection velocity is just the inverse of the slope of the line. Thus, the effective convection velocity is approximately equal to the mean flow velocity at the centreline, while it decreases when moving towards the wall (plot at the bottom left); on the other hand, in comparison to the mean centreline velocity, it is smaller for larger Reynolds numbers (plot at the top left; note that data at the same Reynolds number are acquired at $y^+ = 20$, i.e., close to the wall). When the local mean velocity shift is used (plots on the right), the time delay is almost zero for large Reynolds numbers and far from the wall (indicating that the convection velocity is almost equal to the local mean velocity), whereas

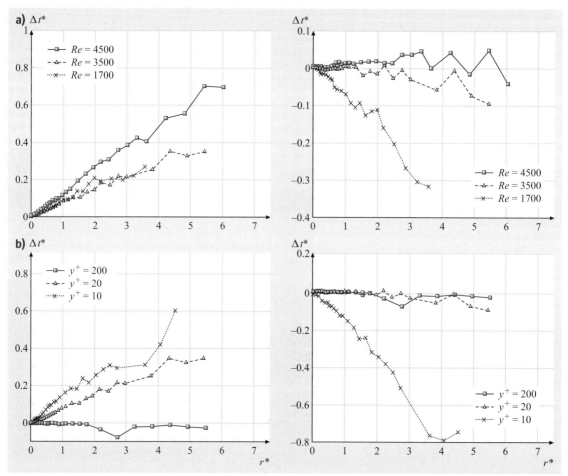

Fig. 10.4 *On the left*: dimensionless time delay (using the integral time scale) between the measured time autocorrelation and the envelope of the space cross-correlation shifted by r/U_0 (where U_0 is the mean flow velocity) versus the dimensionless separation (using the integral length scale). Different Reynolds numbers (**a**) and different distances from the wall (**b**). *On the right*: dimensionless time delay (using integral time scale) between measured the time autocorrelation and the envelope of the space cross-correlation shifted by r/U_1 (where U_1 is the local mean flow velocity) versus the dimensionless separation (using the integral length scale). Different Reynolds numbers (**a**) and different distances from the wall (**b**) (after Romano [10.16])

it decreases almost linearly for the smallest Reynolds numbers and very close to the wall (indicating that the effective convection velocity is larger that the local mean velocity).

The previous observations lead to a practical criterion for the application of Taylor's hypothesis which connects the variability of the entire flow field (which is usually valuable) to the variation of the convection velocity over the different flow scales (which is not simply valuable):

$$U_c = \begin{cases} \Lambda/\tau_0 \sim U_0 & \text{large scales} \\ \lambda/\tau_\lambda \sim U_0 f(I) & \text{intermediate scales} \\ \eta/\tau_\eta = v_\eta \sim U_0 \mathrm{Re}^{1/4} & \text{small scales} \end{cases}$$

$$\left. \begin{array}{l} (\text{order } \Lambda, \tau_0) \\ (\text{order } \lambda, \tau_\lambda) \\ (\text{order } \eta, \tau_\eta) \end{array} \right\},$$

(see Sect. 10.1.3) where $f(I)$ is a function of the turbulence intensity I (the ratio between the root-mean-square fluctuations and mean velocity) and the Reynolds number is based on the large scales. Thus, while large scales are convected by the mean flow velocity and small scales with a much lower velocity (which depends on the Reynolds number), the convection velocity of intermediate scales depends on the turbulent intensity. A practical criterion for Taylor's hypothesis to hold for intermediate scales is to have a turbulent intensity less than 20%. Many investigations have confirmed this limit (even if the hypothesis is used up to turbulent intensities as high as 30%) in a variety of flow conditions [10.16]. Figure 10.4 shows an example of the degree of validity of the hypothesis. The reader is also referred to Sect. 10.1.3 for further details on the application of Taylor's hypothesis at the different flow scales.

10.1.2 Reynolds Decomposition and Equations

Generalities

The statistical Eulerian description of turbulent flows considers the previously defined domains to perform averaging (in particular, in this section we will refer to ensemble averaging) and the decomposition into average and fluctuating terms as introduced in Sect. 10.1.1.

The decomposition in terms of two contributions (average and fluctuations) was introduced by O. Reynolds in 1894; it is substituted into the equation of motion for each fluid-mechanics variable to derive the averaged equations of motion in the presence of turbulence. This may appear a rather strange way to proceed when considering that the equations themselves do not contain any reference to laminar or turbulent conditions. Therefore, the equation of fluid mechanics, as an instantaneous representation of the flow field, must be considered fully valid for a turbulent flow (whatever the Reynolds number). The direct numerical simulation (DNS) approach effectively considers and solves the equations of fluid mechanics without any additional treatment (at least for moderate Reynolds numbers due to resolution problems on the small scales). In DNS, instantaneous fields are computed and their evolution in time and space is evaluated. A similar approach can be adopted for experimental data by measuring the instantaneous flow fields, especially when interest is focused on the generation, interaction, and dissipation of *coherent* vortical structures in turbulent flow fields. Even in this case, some sort of averaging is required to distinguish between casual and repetitive events. On the other hand, the instantaneous representation is not useful when we are interested in the average behavior of the fluid-mechanics variables (not only for mean values but for higher-order moments also) and on the average contribution of each term in the equations of motion, for example in the evaluation of energy balances in flow fields; in these cases, averaging procedure are required.

Hypothesis and Useful Relations

The essential point of Reynolds decomposition is the aforementioned separation between the mean and fluctuating parts for a fluid-mechanics variable q,

$$q = \langle q \rangle + q',$$

where triangular brackets (i.e., ensemble averaging) will be used unless otherwise specified. The fluctuating part q' averaging the previous relation has a vanishing mean value

$$\langle q' \rangle \equiv 0.$$

The other relations that have to be used before performing averaging of the equations is concerned with the properties of space and time derivatives when averaging (which is a linear operator),

$$\frac{\partial^n q}{\partial x_i^n} = \left\langle \frac{\partial^n q}{\partial x_i^n} \right\rangle + \frac{\partial^n q'}{\partial x_i^n},$$

$$\frac{\partial^n q}{\partial t^n} = \left\langle \frac{\partial^n q}{\partial t^n} \right\rangle + \frac{\partial^n q'}{\partial t^n},$$

where we should note that the exchange between average and space (or time) derivatives is strictly valid only if ensemble averaging is performed, while it is not valid for the case of space (or time) averaging; in these cases, great care must be used when considering the average turbulent equations, i.e., each term should be reconsidered separately. The following also hold:

$$\left\langle \frac{\partial^n q'}{\partial x_i^n} \right\rangle = \frac{\partial^n}{\partial x_i^n} \langle q' \rangle = 0,$$

$$\left\langle \frac{\partial^n q'}{\partial t^n} \right\rangle = \frac{\partial^n}{\partial t^n} \langle q' \rangle = 0.$$

For cross-moments

$$q_1 q_2 = \langle q_1 \rangle \langle q_2 \rangle + \langle q_1 \rangle q_2' + \langle q_2 \rangle q_1' + q_1' q_2',$$

which when averaged gives (due to the vanishing mean value of the fluctuating part)

$$\langle q_1 q_2 \rangle = \langle q_1 \rangle \langle q_2 \rangle + \langle q_1' q_2' \rangle.$$

This result is not true for three or more variables cross-moments, where additional terms are present.

For cross-moment derivatives

$$\left\langle \frac{\partial^n (q_1 q_2)}{\partial x_i^n} \right\rangle = \frac{\partial^n}{\partial x_i^n} (\langle q_1 \rangle \langle q_2 \rangle) + \frac{\partial^n}{\partial x_i^n} (\langle q_1' q_2' \rangle),$$

$$\left\langle \frac{\partial^n (q_1 q_2)}{\partial t^n} \right\rangle = \frac{\partial^n}{\partial t^n} (\langle q_1 \rangle \langle q_2 \rangle) + \frac{\partial^n}{\partial t^n} (\langle q_1' q_2' \rangle).$$

and similarly for cross-moments second-order derivatives.

Derivation of the Equations for Turbulent Flows: Single-Point Results

The equations we are dealing with are those of fluid mechanics (Chap. 1), i.e., the equations of mass conservation, balance of momentum and energy, and the equation of state. In the case of a species in the flow, the equation for species concentration conservation must also be added.

Let us start with the *conservation of fluid mass* for compressible fluid flows

$$\frac{\partial \rho}{\partial t} + \frac{\partial (\rho u_i)}{\partial x_i} = 0,$$

where ρ is the density and u_i the i-th velocity component. It is assumed that the flow variables are continuous functions within the control volume (i.e., discontinuities have to be considered in an integral rather than a differential approach [10.1]). Applying Reynolds decomposition to both the density and velocity components yields

$$\frac{\partial (\langle \rho \rangle + \rho')}{\partial t} + \frac{\partial}{\partial x_i} [(\langle \rho \rangle + \rho')(\langle u_i \rangle + u_i')] = 0$$

and evaluating the average equation (i.e., the average of all terms in the equation due to the linearity of the averaging operations)

$$\frac{\partial \langle \rho \rangle}{\partial t} + \frac{\partial}{\partial x_i} (\langle \rho \rangle \langle u_i \rangle) + \frac{\partial}{\partial x_i} \langle \rho' u_i' \rangle = 0,$$

which is different from the starting equation (even if average values are considered rather than instantaneous one) due to the presence of the cross-terms $\langle \rho' u_i' \rangle$; these terms must be considered as new variables in turbulent flows; in principle they cannot be derived from the other unknowns. This equation is valid for all flows with density changes (also temperature changes as in convective flows or in the presence of flows with different densities). For flows in which density fluctuations can be neglected, this reduces to

$$\frac{\partial \langle \rho \rangle}{\partial t} + \frac{\partial}{\partial x_i} (\langle \rho \rangle \langle u_i \rangle) = 0,$$

which is formally equivalent to the original equation for the conservation of mass, if we replace the velocity components and density with their mean values. If the density is also constant over space and time, as in incompressible fluid flows,

$$\frac{\partial \langle u_i \rangle}{\partial x_i} = 0.$$

As the divergence of the instantaneous and average velocity fields is equal to zero, the Reynolds decomposition means that the divergence of the fluctuating velocity field must also vanish

$$\frac{\partial u_i'}{\partial x_i} = 0.$$

Thus, for the case of small density fluctuations, the equation of mass conservation for the turbulent field does not add another unknown to the list of fluid-mechanics variables.

The equation for the *balance of momentum* for a Newtonian fluid (i.e., the Navier–Stokes equations) is given by (under the hypothesis of constant dynamic viscosity coefficient for a perfect gas, while for liquids the last term on the right-hand side vanishes)

$$\rho\left(\frac{\partial u_i}{\partial t}+u_j\frac{\partial u_i}{\partial x_j}\right)=-\frac{\partial p}{\partial x_i}+\rho f_i+\mu\frac{\partial^2 u_i}{\partial x_j^2}+\frac{\mu}{3}\frac{\partial}{\partial x_i}\left(\frac{\partial u_j}{\partial x_j}\right).$$

where p is the pressure, μ is the viscosity coefficient, and f_i are the components of the external mass forces (as the gravity force). The terms on the left-hand side of the equation can be grouped to form the substantial derivative. Let us assume that the viscosity coefficient and the mass forces are constant with respect to turbulent fluctuations, and apply the Reynolds decomposition to the velocity components, pressure, and density. After averaging, the previous equation becomes

$$\langle\rho\rangle\frac{\partial\langle u_i\rangle}{\partial t}+\left\langle\rho'\frac{\partial u_i'}{\partial t}\right\rangle+\langle\rho\rangle\langle u_j\rangle\frac{\partial\langle u_i\rangle}{\partial x_j}$$
$$+\langle\rho\rangle\left\langle u_j'\frac{\partial u_i'}{\partial x_j}\right\rangle+\langle u_j\rangle\left\langle\rho'\frac{\partial u_i'}{\partial x_j}\right\rangle$$
$$+\langle\rho' u_j'\rangle\frac{\partial\langle u_i\rangle}{\partial x_j}+\left\langle\rho' u_j'\frac{\partial u_i'}{\partial x_j}\right\rangle$$
$$=-\frac{\partial\langle p\rangle}{\partial x_i}+\langle\rho\rangle f_i+\mu\frac{\partial^2\langle u_i\rangle}{\partial x_j^2}+\frac{\mu}{3}\frac{\partial}{\partial x_i}\left(\frac{\partial\langle u_j\rangle}{\partial x_j}\right),$$

where the right-hand side of the equation, which contains linear terms, is the same as in the original case (if we replace the variables with their mean values), while the left-hand side contains many cross-terms. Again, if we assume that the density is constant over the ensemble (i.e., no turbulent fluctuations in the density), we obtain

$$\langle\rho\rangle\frac{\partial\langle u_i\rangle}{\partial t}+\langle\rho\rangle\langle u_j\rangle\frac{\partial\langle u_i\rangle}{\partial x_j}+\langle\rho\rangle\left\langle u_j'\frac{\partial u_i'}{\partial x_j}\right\rangle$$
$$=-\frac{\partial\langle p\rangle}{\partial x_i}+\langle\rho\rangle f_i+\mu\frac{\partial^2\langle u_i\rangle}{\partial x_j^2}+\frac{\mu}{3}\frac{\partial}{\partial x_i}\left(\frac{\partial\langle u_j\rangle}{\partial x_j}\right),$$

where only the cross-term between the velocity fluctuations is different from the original case. For the constant-density case (over space and time), using the conservation of mass for the fluctuating part, we can write the above cross-term as

$$\langle\rho\rangle\left\langle u_j'\frac{\partial u_i'}{\partial x_j}\right\rangle=\langle\rho\rangle\frac{\partial}{\partial x_j}\langle u_i' u_j'\rangle$$

which better highlights that, when compared to the starting equations, the new variable $\rho\langle u_i' u_j'\rangle$ is added (the Reynolds stress). As before, in principle, this variable cannot be obtained from other variables unless a new equation is written for it.

Applying the *conservation of energy* (total energy, including thermal and kinetic contributions, from the first law of thermodynamics), the following equation is obtained

$$\rho\frac{DU}{Dt}+\rho u_i\frac{Du_i}{Dt}=-\frac{\partial(pu_i)}{\partial x_i}+\rho f_i u_i$$
$$+\frac{\partial}{\partial x_j}\left(\sigma_{ji}u_i\right)+\rho Q-\frac{\partial q_i}{\partial x_i},$$

where U is the internal energy ($=c_V T$ for a perfect gas, where c_V is the specific heat at constant volume and T the temperature), $\sigma_{ji}=\mu\left[\left(\frac{\partial u_j}{\partial x_i}+\frac{\partial u_i}{\partial x_j}\right)-\frac{2}{3}\frac{\partial u_k}{\partial x_k}\delta_{ji}\right]$ is the viscous part of the stress tensor, Q and q are the heat fluxes exchanged by radiation and conduction, respectively, (q is usually given in terms of the Fourier constitutive relation $q_i=-k\frac{\partial T}{\partial x_i}$, where k is the fluid thermal conductivity). This equation contains an independent (from the other equations) part that is purely a balance of thermal energy

$$\rho\frac{DU}{Dt}=-p\frac{\partial u_i}{\partial x_i}+\sigma_{ji}\frac{\partial u_i}{\partial x_j}+\rho Q-\frac{\partial q_i}{\partial x_i},$$

which can also be given in terms of entropy, and a part that contains the balance of kinetic energy

$$\rho u_i\frac{Du_i}{Dt}=-u_i\frac{\partial p}{\partial x_i}+\rho f_i u_i+u_i\frac{\partial\sigma_{ji}}{\partial x_j},$$

which is not independent of the other equations of fluid mechanics, being derivable from the balance of momentum by scalar multiplication with the velocity.

When applying the Reynolds decomposition to the thermal energy balance, the following equation is obtained

$$\rho c_V\left(\frac{\partial\langle T\rangle}{\partial t}+\langle u_i\rangle\frac{\partial\langle T\rangle}{\partial x_i}+\left\langle u_i'\frac{\partial T'}{\partial x_i}\right\rangle\right)$$
$$=-\langle p\rangle\frac{\partial\langle u_i\rangle}{\partial x_i}-\left\langle p'\frac{\partial u_i'}{\partial x_i}\right\rangle+\mu\left[\frac{\partial\langle u_j\rangle}{\partial x_i}\frac{\partial\langle u_i\rangle}{\partial x_j}\right.$$
$$+\left\langle\frac{\partial u_j'}{\partial x_i}\frac{\partial u_i'}{\partial x_j}\right\rangle+\left(\frac{\partial\langle u_i\rangle}{\partial x_j}\right)^2+\left\langle\left(\frac{\partial u_i'}{\partial x_j}\right)^2\right\rangle$$
$$\left.-\frac{2}{3}\left(\frac{\partial\langle u_k\rangle}{\partial x_k}\right)^2-\frac{2}{3}\left\langle\left(\frac{\partial u_k'}{\partial x_k}\right)^2\right\rangle\right]$$
$$+\rho\langle Q\rangle+k\frac{\partial^2\langle T\rangle}{\partial x_i^2},$$

assuming that c_V, μ, k, and Q are constant, and that the density has no fluctuations over the ensemble. Under the assumption of a zero-divergence flow field, this reduces to

$$\rho c_V \left(\frac{\partial \langle T \rangle}{\partial t} + \langle u_i \rangle \frac{\partial \langle T \rangle}{\partial x_i} + \frac{\partial \langle u_i' T' \rangle}{\partial x_i} \right)$$
$$= \mu \left[\frac{\partial \langle u_j \rangle}{\partial x_i} \frac{\partial \langle u_i \rangle}{\partial x_j} + \left\langle \frac{\partial u_j'}{\partial x_i} \frac{\partial u_i'}{\partial x_j} \right\rangle + \left(\frac{\partial \langle u_i \rangle}{\partial x_j} \right)^2 \right.$$
$$\left. + \left\langle \left(\frac{\partial u_i'}{\partial x_j} \right)^2 \right\rangle \right] + \rho \langle Q \rangle + k \frac{\partial^2 \langle T \rangle}{\partial x_i^2} \,,$$

which (apart from the average terms, which are similar to the original equation) contains the cross-term $\langle u_i' T' \rangle$ similarly to the previous equations, plus the cross-derivative terms

$$\left\langle \frac{\partial u_j'}{\partial x_i} \frac{\partial u_i'}{\partial x_j} \right\rangle \,,$$
$$\left\langle \left(\frac{\partial u_i'}{\partial x_j} \right)^2 \right\rangle \,.$$

The same procedure applied to the pure kinetic-energy equation leads to

$$\rho \langle u_i \rangle \left(\frac{\partial \langle u_i \rangle}{\partial t} + \langle u_j \rangle \frac{\partial \langle u_i \rangle}{\partial x_j} + \left\langle u_j' \frac{\partial u_i'}{\partial x_j} \right\rangle \right)$$
$$+ \rho \left(\left\langle u_i' \frac{\partial u_i'}{\partial t} \right\rangle + \langle u_j \rangle \left\langle u_i' \frac{\partial u_i'}{\partial x_j} \right\rangle + \langle u_i' u_j' \rangle \frac{\partial \langle u_i \rangle}{\partial x_j} \right.$$
$$\left. + \left\langle u_i' u_j' \frac{\partial u_i'}{\partial x_j} \right\rangle \right)$$
$$= \rho f_i \langle u_i \rangle - \langle u_i \rangle \frac{\partial \langle p \rangle}{\partial x_i} - \left\langle u_i' \frac{\partial p'}{\partial x_i} \right\rangle +$$
$$\mu \left[\frac{1}{3} \langle u_i \rangle \frac{\partial^2 \langle u_j \rangle}{\partial x_i \partial x_j} + \frac{1}{3} \left\langle u_i' \frac{\partial^2 u_j'}{\partial x_i \partial x_j} \right\rangle \right.$$
$$\left. + \langle u_i \rangle \frac{\partial^2 \langle u_i \rangle}{\partial x_j^2} + \left\langle u_i' \frac{\partial^2 u_i'}{\partial x_j^2} \right\rangle \right]$$

under the assumptions of constant μ, f_i, and ρ (with respect to the ensemble). For constant density, this equation reduces to

$$\rho \langle u_i \rangle \left(\frac{\partial \langle u_i \rangle}{\partial t} + \langle u_j \rangle \frac{\partial \langle u_i \rangle}{\partial x_j} + \frac{\partial \langle u_i' u_j' \rangle}{\partial x_j} \right)$$
$$+ \rho \left(\left\langle u_i' \frac{\partial u_i'}{\partial t} \right\rangle + \langle u_j \rangle \left\langle u_i' \frac{\partial u_i'}{\partial x_j} \right\rangle + \langle u_i' u_j' \rangle \frac{\partial \langle u_i \rangle}{\partial x_j} \right.$$
$$\left. + \frac{1}{2} \frac{\partial \langle u_i'^2 u_j' \rangle}{\partial x_j} \right)$$
$$= \rho f_i \langle u_i \rangle - \frac{\partial \langle u_i \rangle \langle p \rangle}{\partial x_i} - \frac{\partial \langle p' u_i' \rangle}{\partial x_i}$$
$$+ \mu \left[\langle u_i \rangle \frac{\partial^2 \langle u_i \rangle}{\partial x_j^2} + \left\langle u_i' \frac{\partial^2 u_i'}{\partial x_j^2} \right\rangle \right]$$

Under the previous hypothesis, sometimes the third and sixth terms on the left-hand side are grouped into the term [10.17]

$$\frac{\partial \langle u_i \rangle \langle u_i' u_j' \rangle}{\partial x_j} \,,$$

which can be further separated into a part for the average kinetic energy, i.e., the kinetic energy of the mean flow ($\langle K \rangle = \langle u_i \rangle \langle u_i \rangle / 2$), which can also be obtained directly from the Navier–Stokes equations by scalar multiplication by $\langle u_i \rangle$,

$$\rho \left(\frac{\partial \langle K \rangle}{\partial t} + \langle u_j \rangle \frac{\partial \langle K \rangle}{\partial x_j} + \langle u_i \rangle \frac{\partial \langle u_i' u_j' \rangle}{\partial x_j} \right)$$
$$= \rho f_i \langle u_i \rangle - \frac{\partial \langle u_i \rangle \langle p \rangle}{\partial x_i} + \mu \langle u_i \rangle \frac{\partial^2 \langle u_i \rangle}{\partial x_j^2} \,,$$

plus a part for the fluctuating kinetic energy, i.e., the turbulent kinetic energy (TKE), $K' = \langle u_i' u_i' \rangle / 2$, which is obtained by subtracting the equation for the mean from the total kinetic-energy equation and averaging [10.10, 17]

$$\rho \left(\frac{\partial K'}{\partial t} + \langle u_j \rangle \frac{\partial K'}{\partial x_j} + \langle u_i' u_j' \rangle \frac{\partial \langle u_i \rangle}{\partial x_j} + \frac{\partial \langle K' u_j' \rangle}{\partial x_j} \right)$$
$$= - \frac{\partial \langle p' u_i' \rangle}{\partial x_i} + \mu \left\langle u_i' \frac{\partial^2 u_i'}{\partial x_j^2} \right\rangle \,.$$

The equation for kinetic energy of the mean motion contains cross-terms already recognized from the other equations for turbulent motions ($\langle u_i' u_j' \rangle$). On the other hand, the second term on the right-hand side of the TKE equation can be rewritten as

$$\mu \left\langle u'_i \frac{\partial^2 u'_i}{\partial x_j^2} \right\rangle = \mu \left(\frac{\partial}{\partial x_j} \left\langle u'_i \frac{\partial u'_i}{\partial x_j} \right\rangle - \left\langle \frac{\partial u'_i}{\partial x_j} \frac{\partial u'_i}{\partial x_j} \right\rangle \right)$$

$$= \mu \left(\frac{1}{2} \frac{\partial^2 \langle u'^2_i \rangle}{\partial x_j^2} - \left\langle \left(\frac{\partial u'_i}{\partial x_j} \right)^2 \right\rangle \right),$$

where the quantity $\varepsilon = \nu \left\langle \left(\frac{\partial u'_i}{\partial x_j} \right)^2 \right\rangle$ is usually referred to as the homogeneous TKE dissipation rate. Therefore, using the previous relations, the TKE equation can be written as (dividing by the density)

$$\frac{\partial K'}{\partial t} + \langle u_j \rangle \frac{\partial K'}{\partial x_j}$$

$$+ \langle u'_i u'_j \rangle \frac{\partial \langle u_i \rangle}{\partial x_j} + \frac{\partial \langle K' u'_j \rangle}{\partial x_j}$$

$$= -\frac{1}{\rho} \frac{\partial \langle p' u'_i \rangle}{\partial x_i} + \nu \frac{\partial^2 K'}{\partial x_j^2} - \varepsilon$$

which contains cross-terms ($\langle u'_i u'_j \rangle$, $\langle K' u_j' \rangle$, $\langle u'_i p' \rangle$) and cross-derivatives terms in ε. Note that in homogeneous turbulence the previous equation reduces to $\mathrm{d}K'/\mathrm{d}t = -\varepsilon$.

The quantity ε in the previous equation is only a part of the whole dissipation rate, so that the total dissipation rate is given by

$$\varepsilon_\mathrm{T} = \frac{1}{2} \nu \left\langle \left(\frac{\partial u'_i}{\partial x_j} + \frac{\partial u'_j}{\partial x_i} \right)^2 \right\rangle$$

$$= \nu \left[\left\langle \frac{\partial u'_j}{\partial x_i} \frac{\partial u'_i}{\partial x_j} \right\rangle + \left\langle \left(\frac{\partial u'_i}{\partial x_j} \right)^2 \right\rangle \right]$$

$$= \nu \left[\frac{\partial}{\partial x_j} \left\langle u'_i \frac{\partial u'_j}{\partial x_i} \right\rangle + \left\langle \left(\frac{\partial u'_i}{\partial x_j} \right)^2 \right\rangle \right],$$

although strictly speaking the first term on the right-hand side is not a dissipation, but rather a diffusion term. In this relation, the first part in the second square brackets (which contains three diagonal square terms plus three cross-terms) is different from zero only in imhomogeneous turbulence, where [10.12, 18]

$$\left\langle \frac{\partial u'_1}{\partial x_2} \frac{\partial u'_2}{\partial x_1} \right\rangle + \left\langle \frac{\partial u'_1}{\partial x_3} \frac{\partial u'_3}{\partial x_1} \right\rangle = -\left\langle \left(\frac{\partial u'_1}{\partial x_1} \right)^2 \right\rangle,$$

$$\left\langle \frac{\partial u'_2}{\partial x_1} \frac{\partial u'_1}{\partial x_2} \right\rangle + \left\langle \frac{\partial u'_2}{\partial x_3} \frac{\partial u'_3}{\partial x_2} \right\rangle = -\left\langle \left(\frac{\partial u'_2}{\partial x_2} \right)^2 \right\rangle,$$

$$\left\langle \frac{\partial u'_3}{\partial x_1} \frac{\partial u'_1}{\partial x_3} \right\rangle + \left\langle \frac{\partial u'_3}{\partial x_2} \frac{\partial u'_2}{\partial x_3} \right\rangle = -\left\langle \left(\frac{\partial u'_3}{\partial x_3} \right)^2 \right\rangle,$$

while the second part (which contains nine square terms) is different from zero in homogeneous turbulence and is just the quantity referred to as ε in the previous equations. In isotropic turbulence, the second part reduces to (refer to Sect. 10.1.1 on isotropy)

$$\varepsilon = 15\nu \left\langle \left(\frac{\partial u'_1}{\partial x_1} \right)^2 \right\rangle,$$

where the derivative of the axial velocity component is performed along the same direction of the velocity. Refer to Sect. 10.1.3 for how to measure the scales for further expressions of the TKE dissipation rate (Sect. 10.1.3).

The TKE dissipation rate is one of the most difficult quantities to derive in turbulence velocity measurements due to the requirement for three simultaneous velocity derivative measurements.

The final equation required to close the set of fluid-mechanics equations is the *equation of state*, i.e., an equation relating the thermodynamic variables p, ρ, and T. For an ideal gas, this law is

$$p = \rho R T$$

where $R = 287 \,\mathrm{JK/kg}$ for air is the ideal gas constant. Applying Reynolds decomposition and averaging this equation, one obtains

$$\frac{\langle p \rangle}{R} = \langle \rho \rangle \langle T \rangle + \langle \rho' T' \rangle \qquad (10.15)$$

for constant density (or temperature) over the ensemble, the last term can be neglected. Subtracting the previous equation from the instantaneous equation one obtains

$$\frac{p'}{\langle p \rangle} = \frac{\rho' T'}{\langle \rho \rangle \langle T \rangle + \langle \rho' T' \rangle}$$

$$+ \frac{T'}{\langle T \rangle + \frac{\langle \rho' T' \rangle}{\langle \rho \rangle}} + \frac{\rho'}{\langle \rho \rangle + \frac{\langle \rho' T' \rangle}{\langle T \rangle}} - \frac{1}{1 + \frac{\langle \rho \rangle \langle T \rangle}{\langle \rho' T' \rangle}}$$

$$\approx \frac{\rho' T'}{\langle \rho \rangle \langle T \rangle} + \frac{T'}{\langle T \rangle} + \frac{\rho'}{\langle \rho \rangle},$$

where the last relation holds if $\langle \rho' T' \rangle \ll \langle \rho \rangle \langle T \rangle$. In such a condition, the first term in the previous relation can also be neglected and the linear perturbation ideal gas law is obtained [10.14]. In this case no additional cross-terms are present in the equations of turbulence, while density–temperature cross-terms were present in (10.15).

In the case of a species present in the flow (say a pollutant, smoke or dye), a *conservation for the mass of the species* (concentration equation) must be added to the previous set of equations in the form of

$$\frac{\partial c}{\partial t} + \frac{\partial}{\partial x_j}(c u_j) = D \frac{\partial^2 c}{\partial x_j^2},$$

where D is the species diffusivity into the fluid (which is assumed constant), and c is the species concentration.

Applying Reynolds decomposition and averaging

$$\frac{\partial \langle c \rangle}{\partial t} + \frac{\partial \langle c \rangle \langle u_j \rangle}{\partial x_j} + \frac{\partial \langle c' u_j' \rangle}{\partial x_j} = D \frac{\partial^2 \langle c \rangle}{\partial x_j^2}$$

which, in the case of constant density (zero velocity divergence), reduces to

$$\frac{\partial \langle c \rangle}{\partial t} + \langle u_j \rangle \frac{\partial \langle c \rangle}{\partial x_j} + \frac{\partial \langle c' u_j' \rangle}{\partial x_j} = D \frac{\partial^2 \langle c \rangle}{\partial x_j^2},$$

where the cross-term $\langle c' u_j' \rangle$ is including in addition to the original equation, together with the substitution of instantaneous with average variables.

The derived equations are all those required to describe a turbulent flow; they contain terms similar to those encountered in laminar flows (providing that average rather than instantaneous variables are used) plus cross-terms between fluctuating variables that retain the effect of turbulence. Such cross-terms represent turbulent kinematic fluxes (of mass, momentum, energy or concentration) that are added to the fluxes due to the mean motions. While mean kinematic fluxes are driven by the advection velocity, for turbulent fluxes positive and negative instantaneous values are usually in balance unless a correlation between velocity and the variable fluctuation under consideration holds. Thus, a turbulent flux different from zero at a point indicates a net relation between velocity and the variable fluctuation fields. Mathematically, the cross-terms are additional unknowns to the averaged values of the fluid-mechanics variables. They cannot be derived from other fundamental principles; however, there are several possibilities to derive additional equations for these cross-terms, as will be shown in the following sections.

Derivation of the Equations for Turbulent Flows: Multi-Point Results

The preceding equations were obtained at a single-point (i. e., for statistics computed at a point). From the equations of motions it is possible to derive equations for *multipoint statistics* (refer to Sect. 10.1.1 for details on these statistical quantities). Following *Hinze* [10.10], assuming a zero-divergence velocity field, it is possible to obtain for double-point correlation (10.1)

$$\frac{\partial}{\partial t} R_{u_i' u_j'} - \frac{\partial}{\partial x_k} R_{u_i' u_k', u_j'} + \frac{\partial}{\partial x_k} R_{u_i', u_k' u_j'}$$
$$= -\frac{1}{\rho} \left(\frac{\partial}{\partial x_j} R_{u_i' p'} - \frac{\partial}{\partial x_i} R_{p' u_j'} \right) + 2\nu \frac{\partial^2}{\partial x_k^2} R_{u_i' u_j'},$$

(10.16)

where the correlations are computed between velocity component fluctuations or between velocity and pressure fluctuations. In the previous equation, as in the single-point equations, higher-order terms (triple correlations) are introduced and the solution should involve some additional constraints. In isotropic incompressible turbulence, the pressure–velocity correlations vanish [10.10], the double correlation functions can be expressed as in (10.8), using (10.7) (and similarly for triple correlations), and the previous equation reduces to (considering two points separated by a distance r)

$$\frac{\partial}{\partial t}\left(\langle u'^2 \rangle f\right) - \langle u'^2 \rangle^{3/2} \frac{1}{r^4}\frac{\partial}{\partial r}(r^4 k)$$
$$= 2\nu \langle u'^2 \rangle \frac{1}{r^4}\frac{\partial}{\partial r}\left(r^4 \frac{\partial f}{\partial r}\right),$$

(10.17)

which is known as the Karman–Howarth equation (where fluctuating variables are used). Double and triple correlation (f and k, respectively) functions depend on time t and separation r. This is the fundamental equation for two-point statistics in isotropic incompressible turbulence; the presence of derivatives makes it very difficult to compute terms of the equation from experimental data.

An equation equivalent to the Karman–Howarth equation can be derived between second- and third-order structure functions,

$$\frac{\partial}{\partial t} S_{u_1}^2 + \frac{1}{3 r^4} \frac{\partial}{\partial r}(r^4 S_{u_1}^3) = \frac{2\nu}{r^4} \frac{\partial}{\partial r}\left(r^4 \frac{\partial S_{u_1}^2}{\partial r}\right) - \frac{4}{3}\varepsilon,$$

(10.18)

where the second- and third-order structure functions depend on time t and separation r. The integral of the previous equation (over r) is known as the Kolmogorov equation [10.11]; for the stationary case,

$$S_{u_1}^3 = 6\nu \frac{\partial S_{u_1}^2}{\partial r} - \frac{4}{5}\varepsilon r.$$

Kolmogorov himself realized that in isotropic homogeneous stationary turbulence, the term containing second-order structure function is almost zero and thus the 4/5 law holds

$$S_{u_1}^3 = -\frac{4}{5}\varepsilon r ,$$

a result which justifies (10.12)

Similarly, an equation for cross scalar-velocity structure functions (for the stationary case) has been derived [10.12]

$$S_{u_1 T^2}^3 = 2\alpha \frac{\partial S_T^2}{\partial r} - \frac{4}{3}\varepsilon_T r ,$$

where ε_T and α are the energy dissipation rate and the diffusivity of the scalar field, respectively. For isotropic homogeneous stationary turbulence, the previous equation reduces to

$$S_{u_1 T^2}^3 = -\frac{4}{3}\varepsilon_T r .$$

Recently, some researchers have tried to derive equations for longitudinal and transverse correlation and structure functions in more-general flow conditions to generalize the Karman–Howarth results (10.17) and (10.18); see *Zhou* et al. [10.19], *Anselmet* et al. [10.20] and *Danaila* et al. [10.21] for details.

Similar equations can be derived in the wavenumber domain using Fourier transforms of the double and triple correlation functions (10.1) and (10.4); for isotropic incompressible turbulence one can derive [10.10]

$$\frac{\partial}{\partial t} F_{u'_i u'_j}^k = T^k - 2\nu k^2 F_{u'_i u'_j}^k ,$$

where the nonlinear spectral transfer function T^k is related to the third-order spectral function $F_{u'_i u'_j, u'_l}^k$; the previous equation is known as the Lin equation. Integration over wavenumber in the first term in the previous equation gives

$$\frac{\partial}{\partial t} \int_0^\infty F_{u'_i u'_i}^k \, \mathrm{d}k = \frac{\partial}{\partial t} \langle u_i'^2 \rangle .$$

On the other hand, integration of the last term in the previous equation gives [10.4, 10]

$$\varepsilon = 2\nu \int_0^\infty F_{u'_i u'_i}^k k^2 \mathrm{d}k , \qquad (10.19)$$

which can be also proved starting from the spectral transform of the velocity related to that of the velocity derivative. Considering the $-5/3$ law (10.13), from these relations it is also possible to obtain the wavenumber behaviors of the TKE and TKE dissipation in the inertial range

$$K' \approx C_K \varepsilon^{2/3} k^{-2/3} , \quad \varepsilon \approx \nu C_K \varepsilon^{2/3} k^{4/3} ,$$

i.e., in the inertial range, the TKE content decreases as the wavenumber increases (and the slope is negative), whereas the TKE dissipation continues to increase with wavenumber (and the slope is positive). As a consequence, the peak for TKE is at much lower wavenumbers in comparison to that of TKE dissipation.

The Problem of Closure of Equations; Hierarchy of Turbulence Equations

As shown in previous sections, the equations of fluid mechanics for a turbulent flow field retain terms similar to the original equations (using averaged variables) plus cross-terms containing the turbulent flux effects. These cross-terms are unknowns and the so-called problem of closure of turbulence equations arises; to solve the turbulent field, it is necessary to balance the number of equations with the number of unknowns. In so-called Reynolds-averaged Navier–Stokes (RANS) modeling, additional constraints on the turbulent fluxes are included into the system of equations. The zeroth-order closures simply equate the cross-terms to a constant; this is practically unsuitable for all flow field conditions. Thus, referring to first-order closures, usually known as eddy viscosity or Boussinesque closures, the turbulent kinematic fluxes of the generic fluid-mechanics variable q is given in terms of the gradient of the variable itself

$$-\langle u'_i q' \rangle = K_T^q \frac{\partial \langle q \rangle}{\partial x_i} ,$$

where K_T^q is the turbulent diffusion coefficient of the considered variable, which in general is unknown and could depend on the point considered. Algebraic or differential models aim to give useful expressions for the quantity K_T^q.

Before considering in some detail these models, let us consider the previous relation in detail for each cross-term in the equations of motion. In the equation for mass conservation the previous relation gives

$$-\frac{\partial}{\partial x_i} \langle u'_i \rho' \rangle = \frac{\partial}{\partial x_i} \left(K_T^\rho \frac{\partial \langle \rho \rangle}{\partial x_i} \right) ,$$

where K_T^ρ is the turbulent diffusion coefficient for the density. Thus, the turbulent diffusion (which is usually unknown) is related to the molecular (laminar) diffusion (which is measurable). For the equation of momentum

$$-\frac{\partial}{\partial x_i}\langle u'_i u'_j\rangle = \frac{\partial}{\partial x_i}\left[\nu_T\left(\frac{\partial \langle u_j\rangle}{\partial x_i} + \frac{\partial \langle u_i\rangle}{\partial x_j}\right)\right],$$

where ν_T is the turbulent kinematic viscosity (eddy viscosity). In the equation of thermal energy

$$-\frac{\partial}{\partial x_i}\langle u'_i T'\rangle = \frac{\partial}{\partial x_i}\left(\alpha_T \frac{\partial \langle T\rangle}{\partial x_i}\right),$$

where α_T is the turbulent thermal diffusivity. In the equation of species concentration

$$-\frac{\partial}{\partial x_i}\langle u'_i c'\rangle = \frac{\partial}{\partial x_i}\left(D_T \frac{\partial \langle c\rangle}{\partial x_i}\right),$$

where D_T is the turbulent species diffusivity. A similar result holds for the cross-terms appearing in the TKE equation.

Returning to the estimation of the eddy viscosity coefficient, it is necessary to point out that characteristic scales for turbulence are usually considered; for example for the momentum [10.4, 22],

$$\nu_T \sim u_c L_c,$$

where u_c and L_c are characteristic velocity and length scales for turbulence to be specified by modeling.

In pure *algebraic models* (zero-equation or eddy-viscosity models), the velocity and length scales are given algebraically; for example in Prandtl mixing length (l_m) theory [10.1]

$$L_c = l_m, \quad u_c = l_m \left|\frac{\partial \langle u\rangle}{\partial x_2}\right|,$$

where x_2 is taken orthogonal to the direction of the average velocity, so that

$$\nu_T = l_m^2 \left|\frac{\partial \langle u\rangle}{\partial x_2}\right|.$$

Another possibility, due to Smagorinsky [10.4], relates eddy viscosity to the mean rate-of-strain tensor $\langle s_{ij}\rangle = \frac{1}{2}\left(\frac{\partial \langle u_i\rangle}{\partial x_j} + \frac{\partial \langle u_j\rangle}{\partial x_i}\right)$,

$$\nu_T = l_m^2 (2\langle s_{ij}\rangle\langle s_{ij}\rangle)^{1/2},$$

which reduces to the Prandtl mixing length theory for unidirectional flows. The problem of these models is related to the evaluation of a proper mixing length; for example, in near-wall turbulence $l_m = \kappa y$, where y is the distance from the wall and κ is the von Karman constant, can be assumed, so that a logarithmic law for the wall is obtained, while in other flow conditions there are few arguments for guiding this choice.

In mixed *algebraic–differential models* (one-equation models), both the velocity and length scales are given in terms of the TKE, K'

$$L_c = \frac{K'^{3/2}}{\varepsilon}, \quad u_c = \sqrt{K'},$$

so that

$$\nu_T = \frac{K'^2}{\varepsilon},$$

where coefficients have been omitted. The TKE is obtained by solving the corresponding equation, while ε is obtained from L_c under mixing-length arguments (with similar limitations).

In pure *differential models* (two-equation models), the expression for eddy viscosity is the same as before, with both TKE and TKE dissipation equations to be solved by proper equations; the latter is obtained from the momentum equation by differentiation and multiplication by fluctuating derivatives so that the inertial term on the left-hand side of the equation represents the variation of homogeneous TKE dissipation, ε [10.17]

$$\frac{\partial \varepsilon}{\partial t} + \langle u_j\rangle \frac{\partial \varepsilon}{\partial x_j} = -2\nu \left\langle \frac{\partial u'_j}{\partial x_i}\frac{\partial u'_j}{\partial x_k}\right\rangle \frac{\partial \langle u_i\rangle}{\partial x_k}$$
$$- 2\nu \left\langle \frac{\partial u'_i}{\partial x_k}\frac{\partial u'_j}{\partial x_k}\right\rangle \frac{\partial \langle u_i\rangle}{\partial x_j}$$
$$- 2\nu \left\langle u'_k \frac{\partial u'_i}{\partial x_j}\right\rangle \frac{\partial^2 \langle u_i\rangle}{\partial x_k \partial x_j}$$
$$- 2\nu \left\langle \frac{\partial u'_i}{\partial x_k}\frac{\partial u'_i}{\partial x_j}\frac{\partial u'_k}{\partial x_j}\right\rangle$$
$$- 2\nu \frac{\partial}{\partial x_i}\left\langle \frac{\partial p'}{\partial x_j}\frac{\partial u'_i}{\partial x_j}\right\rangle$$
$$- \nu \frac{\partial}{\partial x_k}\left\langle u'_k \frac{\partial u'_i}{\partial x_j}\frac{\partial u'_i}{\partial x_j}\right\rangle$$
$$+ \nu \frac{\partial^2 \varepsilon}{\partial x_j^2} - 2\nu^2 \left\langle \left(\frac{\partial^2 u'_i}{\partial x_j \partial x_k}\right)^2\right\rangle.$$

These models are also known as k–ε models (or closures). Note, that, as for the equation for fluctuating TKE, the previous equation adds other unknowns in the form of triple cross-derivatives terms so that the problem is still not closed (even if applied to higher order).

This is an example of the first step of a hierarchy of equations; further equations can be derived for the triple cross-terms which would contain quadruple cross-terms and so on (similar hierarchies exist in many nonlinear systems described statistically). Numerically, this hierarchy can be truncated at some order by an approximation similar to that of Boussinesque (i. e., by relating higher-order cross-terms to lower-order terms) or with algebraic models. Another possibility is to assign initial values to the unknowns and try to find a solution iteratively, although it is not proven that this solution exists and is unique (see *Pope* [10.4], for further details).

The last equation is simplified in the case of homogeneous isotropic turbulence [10.17] to

$$\frac{\partial \varepsilon}{\partial t} + \langle u_j \rangle \frac{\partial \varepsilon}{\partial x_j} = -\frac{\varepsilon^{3/2}}{\nu^{1/2}} \frac{7}{3\sqrt{15}} S_{\partial u_1/\partial x_1} - \frac{\varepsilon^2}{K} \frac{7}{15} G ,$$

where the skewness of the derivative field,

$$S_{\frac{\partial u_1}{\partial u_1}} = \frac{\left\langle \left(\frac{\partial u_1}{\partial x_1}\right)^3 \right\rangle}{\left[\left\langle \left(\frac{\partial u_1}{\partial x_1}\right)^2 \right\rangle\right]^{3/2}} ,$$

and the quantity (known also as the palenstrophy coefficient),

$$G = \frac{\left\langle \left(\frac{\partial^2 u_1}{\partial x_1^2}\right)^2 \right\rangle}{\left[\left\langle \left(\frac{\partial u_1}{\partial x_1}\right)^2 \right\rangle\right]^2} ,$$

have been introduced. Even in this case, there are higher-order terms (contained in S and G) so that the importance of experimental verification is raised again.

The other possible approach to the closure problem is to derive an exact equation for the Reynolds stress cross-term $R_{u'_i u'_j} = \langle u'_i u'_j \rangle$; by manipulating the momentum equation, according to which we obtain [10.17]

$$\frac{\partial R_{u'_i u'_j}}{\partial t} + \langle u_k \rangle \frac{\partial R_{u'_i u'_j}}{\partial x_k}$$
$$= -R_{u'_i u'_k} \frac{\partial \langle u_j \rangle}{\partial x_k} - R_{u'_j u'_k} \frac{\partial \langle u_i \rangle}{\partial x_k} - 2\nu \left\langle \frac{\partial u'_i}{\partial x_k} \frac{\partial u'_j}{\partial x_k} \right\rangle$$
$$- \frac{\partial}{\partial x_k} \langle u'_i u'_j u'_k \rangle$$
$$- \frac{1}{\rho} \left(\frac{\partial}{\partial x_j} \langle p' u'_i \rangle + \frac{\partial}{\partial x_i} \langle p' u'_j \rangle \right)$$
$$+ \frac{1}{\rho} \left\langle p' \left(\frac{\partial u'_i}{\partial x_j} + \frac{\partial u'_j}{\partial x_i} \right) \right\rangle + \nu \frac{\partial^2 R_{u'_i u'_j}}{\partial x_k^2} , \quad (10.20)$$

which, as expected, contains higher-order terms causing a hierarchy to exist even in this case; it can be closed with iterative solutions or with further constraints similar to those already considered (but at a higher order).

Similar equations can be derived for the other turbulent fluxes, for example in the presence of species [10.17].

Let us now consider the validity of the Boussinesque hypothesis; it is important to point out that the relation itself cannot be retained as valid for at least three reasons. Firstly, if we consider the term $\langle u'_i u'_i \rangle$ and we sum from $i = 1$ to $i = 3$ (which is basically the TKE), for incompressible fluid we obtain:

$$-\langle u'_i u'_i \rangle = 2\nu_T \frac{\partial \langle u_i \rangle}{\partial x_i} = 0 ,$$

i. e., the turbulence should have null TKE, which makes no sense. This point can be overcome by the following modified expression of the Boussinesque hypothesis using the TKE,

$$-\langle u'_i u'_j \rangle = \nu_T \left(\frac{\partial \langle u_j \rangle}{\partial x_i} + \frac{\partial \langle u_i \rangle}{\partial x_j} \right) - \frac{2}{3} K' \delta_{ij} .$$

Secondly, the turbulent diffusion coefficients are not of the fluid itself but pertain to each particular flow field and also depend on the point and instant considered. Thirdly, the Reynolds and velocity deformation tensors are not aligned, i. e., their principal directions are not the same. In practice, the hypothesis is valid only if the velocity–velocity correlations are small, i. e., if we look at the phenomenon on scales that are much larger than the largest turbulent scale (Sect. 10.1.3). Nonetheless, there are several numerical codes that use the Boussinesque hypothesis; this explains why it is important to verify the hypothesis experimentally in different flow conditions.

Closure schemes have also been developed for the multipoint equations, such as the Karman–Howarth equation (10.17). It is interesting to note that the double correlation equation for the multipoint description (10.16) reduces to (10.20) when the two points coincide. The reader is referred to *Mc Comb* [10.23] for details of this argument.

It is also important to consider that, in numerical codes based on large-eddy simulation (LES), the stress tensor must also be modeled on subgrid scales while direct solution of the equations is performed at grid levels; models such as that of Smagorinsky are frequently used. Even in this case, experiments can help in evaluating the effectiveness of the models considered (*Pope* [10.4]).

Further Equations for Derived Quantities: Vorticity and Enstrophy

In many circumstances, it is useful from the conceptual and numerical points of view to consider equations for derived quantities rather than the conventional equations of fluid mechanics. In particular, it is of interest for special numerical approaches (such as vorticity–stream function methods) and for vortex dynamics to consider an equation for the *vorticity*. This can be derived by taking the vector product of the Navier–Stokes equations; for the case of incompressible flows

$$\frac{\partial \omega_i}{\partial t} + u_j \frac{\partial \omega_i}{\partial x_j} = \omega_j \frac{\partial u_i}{\partial x_j} + \nu \frac{\partial^2 \omega_i}{\partial x_j^2},$$

which allows one to eliminate the pressure term (as not in Navier–Stokes equations; this is a big advantage of vorticity-based methods). Note that a Poisson equation for the pressure can be derived by considering the scalar product of the Navier–Stokes equations [10.1]. Performing Reynolds averaging on the previous equation yields

$$\frac{\partial \langle \omega_i \rangle}{\partial t} + \langle u_j \rangle \frac{\partial \langle \omega_i \rangle}{\partial x_j} = \langle \omega_j \rangle \frac{\partial \langle u_i \rangle}{\partial x_j} + \left\langle \omega'_j \frac{\partial u'_i}{\partial x_j} \right\rangle + \nu \frac{\partial^2 \langle \omega_i \rangle}{\partial x_j^2} - \frac{\partial}{\partial x_j} \langle u'_j \omega'_i \rangle,$$

which also contains higher-order terms such as the turbulent vorticity flux (the last term on the right-hand side), and vortex stretching by the fluctuating field (the second term on the right-hand side); these must be modeled by zero- or higher-order closures as for other turbulent fluxes. Consider also that, due to the solenoidal vorticity field, the second of these terms reduces to a form similar to the first.

Another derived quantity is the *enstrophy* (i.e., the square of the vorticity $\zeta = \langle \omega'_i \omega'_i \rangle$), which is related to the vorticity as TKE is to the the velocity. Enstrophy is also closely related to the TKE dissipation rate

$$\nu \zeta = \nu \left\langle \left(\frac{\partial u'_i}{\partial x_j}\right)^2 \right\rangle - \nu \left\langle \frac{\partial u'_i}{\partial x_j} \frac{\partial u'_j}{\partial x_i} \right\rangle$$

$$= \varepsilon - \nu \left\langle \frac{\partial u'_i}{\partial x_j} \frac{\partial u'_j}{\partial x_i} \right\rangle,$$

which in homogeneous turbulence reduces to $\nu \zeta = \varepsilon$. An equation for the enstrophy can be derived as for

TKE dissipation rate [10.17]

$$\frac{\partial \zeta}{\partial t} + \langle u_j \rangle \frac{\partial \zeta}{\partial x_j} = 2 \langle \omega'_i \omega'_k \rangle \frac{\partial \langle u_i \rangle}{\partial x_k} + 2 \left\langle \omega'_i \frac{\partial u'_i}{\partial x_k} \right\rangle \langle \omega_k \rangle$$
$$- 2 \langle u'_k \omega'_i \rangle \frac{\partial \langle \omega_i \rangle}{\partial x_k} + 2 \left\langle \omega'_i \omega'_k \frac{\partial u'_i}{\partial x_k} \right\rangle$$
$$- \frac{\partial}{\partial x_k} \langle u'_k \omega'_i \omega'_i \rangle + \nu \frac{\partial^2 \zeta}{\partial x_j^2}$$
$$- 2\nu \left\langle \frac{\partial \omega'_i}{\partial x_k} \frac{\partial \omega'_i}{\partial x_k} \right\rangle,$$

which does not contain pressure terms, while still containing higher-order terms. The modeling of the enstrophy equation is similar (although not exactly the same) as that of the TKE dissipation rate equation [10.17].

The Experimental Evaluation of Turbulent Fluxes and High-Order Closures

It should be clear from the previous sections that accurate numerical modeling of turbulent flows requires the determination of the cross-terms representing turbulent fluxes and cross-derivative terms. The relations and equations derived for these cross-terms shift the problem to higher order; at some point of the hierarchy a closure approximation is required. A given closure can be valid for a specific region of the flow field, but not everywhere, and also requires changes when moving to a different flow field. Numerically, a possible solution to this problem is to give initial values to the unknowns and try to converge to a solution of the system of equations for a given flow field [10.4]. Otherwise, advanced numerical methods as direct numerical simulation (DNS) and large-eddy simulation (LES) are able to find solutions on a wide range of scales; however, the former (DNS) is still limited to low-Reynolds-number flows with quite poor resolution in time, whereas the latter (LES) also requires a sort of closure for the small scales [10.4].

Therefore, experiments must answers the question of to what extent and approximation the closure hypothesis is valid in each flow field. To do this, direct measurements of the cross and cross-derivatives terms must be performed. Let us recall those terms for the single-point equations

$$\langle \rho' u'_i \rangle, \quad \langle u'_i u'_j \rangle, \quad \langle p' u'_i \rangle,$$
$$\langle T' u'_i \rangle, \quad \langle K' u'_i \rangle, \quad \langle T' \rho' \rangle, \quad \langle c' u'_i \rangle$$
$$\left\langle \frac{\partial u'_i}{\partial x_j} \frac{\partial u'_j}{\partial x_i} \right\rangle, \quad \left\langle \left(\frac{\partial u'_i}{\partial x_j}\right)^2 \right\rangle$$

and for the multipoint case

$$\langle u'_i u'_j, u'_k \rangle ,$$

and similarly for the correlation and spectral functions. It is clear that the evaluation of the former single-point terms requires the simultaneous determination of the two fluctuating variables or velocity derivatives. This is one of the challenges for modern experimental fluid mechanics. While it is reasonable to measure multiple velocity components simultaneously (refer Sect. 5.3 on HWA, Sect. 5.3.1 on LDA, and Sect. 5.3.2 on PIV), it is not straightforward to measure multiple velocity derivatives with sufficient resolution in space (or time if Taylor's hypothesis is used). Refer to the papers by *Browne* et al. [10.24], *Antonia and Mi* [10.25], *Benedict* [10.26], *Romano and Bagnoli* [10.27] for details. It is also difficult to measure simultaneous fluctuations of velocity and temperature (or species concentration) in a global sense, unless special tracers are used (as in LIF, or liquid crystals techniques, Sect. 6.3). On the other hand, to date it is practically impossible to perform simultaneous measurements of velocity and density (or velocity and pressure) fluctuations at any point of the flow field (Sect. 5.6). In any case, advanced measurement techniques with high spatial and temporal resolutions are required for such measurements. To this point, it is important to emphasize that the determination of the velocity derivatives at one point requires measurements of the velocity at two points that are close but finitely spearated. In practice, this means two measurement systems separated by one or a few Kolmogorov lengths ($\Delta r = 2 - 3\eta$), which can easily interfere with one another unless special precautions

are taken (see *Antonia and Mi* [10.25], or *Romano and Bagnoli* [10.27] for details of the effect of spatial resolution and for experimental measurements of velocity derivatives).

The use of Taylor's hypothesis (Sect. 10.1.1) allows the replacement of velocity derivatives along the direction of motion with velocity derivatives in time

$$\left\langle \left(\frac{\partial u'_i}{\partial x_1} \right)^2 \right\rangle = \frac{1}{U_c} \left\langle \left(\frac{\partial u'_i}{\partial t} \right)^2 \right\rangle ,$$

which does not require measurements at two close points but only in time. However, even in this case the required resolution in time is high (of the order of a few Kolmogorov time scales) and noise problems arise.

Regarding derived quantities such as vorticity and enstrophy, the cross-terms to be measured are velocity–vorticity fluctuations and velocity–vorticity derivative fluctuations. The direct measurement of vorticity itself is not straightforward due to the requirements of high resolution, which are similar to highlighted previously for velocity derivatives in general. For simultaneous measurements of velocity and vorticity (and derivatives) very complex systems with multiple probes are required (see *Browne* et al. [10.24], *Antonia and Mi* [10.25] for details).

In summary, Table 10.1 holds [10.22] for the different quantities.

The experimental evaluation of turbulent fluxes and higher-order closures is the subject of a lot of papers in the recent literature on turbulence. There is a huge amount of work on this argument, and the reader is referred to *Pope* [10.4], and *Bernard and Wallace* [10.17], for details (as well as Sect. 10.4 of this book).

Table 10.1

Mean values
$\langle u_i \rangle , \langle K \rangle , \langle p \rangle , \langle T \rangle , \langle \rho \rangle , \langle c \rangle$

$\longrightarrow$ Primary numerical codes validation

Turbulent fluxes (cross-terms)
$\langle \rho' u'_i \rangle , \langle u'_i u'_j \rangle , \langle p' u'_i \rangle , \langle T' u'_i \rangle , \langle K' u'_i \rangle , \langle c' u'_i \rangle , \langle u'_i \omega'_j \rangle$

$\longrightarrow$ Turbulence models testing

Dissipation, cross-derivative terms, higher-order correlation, and spectra
$\varepsilon , \left\langle \frac{\partial p'}{\partial x_k} \frac{\partial u'_i}{\partial x_k} \right\rangle , \left\langle \frac{\partial u'_i}{\partial x_k} \frac{\partial u'_i}{\partial x_j} \frac{\partial u'_k}{\partial x_j} \right\rangle , \left\langle \frac{\partial u'_i}{\partial x_k} \frac{\partial u'_i}{\partial x_k} \right\rangle , \left\langle u'_j \frac{\partial u'_i}{\partial x_k} \frac{\partial u'_i}{\partial x_k} \right\rangle ,$
$\langle u'_i u'_j, u'_k \rangle , \langle \omega'_i \omega'_k \rangle , \left\langle \frac{\partial \omega'_i}{\partial x_k} \frac{\partial \omega'_i}{\partial x_k} \right\rangle$

$\longrightarrow$ Model refinement and novel developments

Statistical Accuracy and Number of Samples

A few words must be dedicated to the accuracy required for the determination of higher-order statistics; roughly speaking, to evaluate the variance of the n-th-order statistical moment of the variable q, $\sigma^2(\langle q^n \rangle)$, it is necessary to know the value of the $2n$-th-order moment, $\langle q^{2n} \rangle$, (see Chap. 25, and *Benedict* and *Gould* [10.28], for details). Experimentally, these higher-order moments are particularly difficult to obtain, especially when cross-moments and cross-derivative moments are involved. Therefore, approximations of the variance estimation (for example, assuming a Gaussian distribution of the data) that use lower statistical moments are employed (see Chap. 25). A useful relation for the relative error on the n-th-order moment is

$$\frac{\sigma(\langle q^n \rangle)}{\langle q^n \rangle} = \frac{\left[A\sigma^n(\langle q^2 \rangle)\right]^{1/2}}{N^{1/2} \langle q^n \rangle},$$

where $\sigma(\langle q^2 \rangle)$ is the standard deviation of the variable q. The factor A depends nonlinearly on the considered moments (approximately, it is $A=1$ for $n=1$, $A=2$ for $n=2$, $A=6$ for $n=3$, $A=100$ for $n=4$, and so on). Therefore, for a given moment (fixed n) the relative error estimation goes as the inverse square root of the number of samples. On the other hand, the error for a given number of samples (N) increases as the moment order increases (there is a factor equal to about 8 passing from $n=1$ to $n=2$ and a factor of 2.5 passing from $n=2$ to $n=4$).

Another important point, related to the way in which the statistics is obtained, is concerned with the statistical independence of the acquired data. For the previous relations concerning the accuracy of statistical moments to hold, it is necessary to ensure effective independence of the acquired data. In ensemble averaging, this is ensured by the fact that a new experiment is repeated starting from more or less the same boundary and initial conditions (Sect. 10.1.1). For time (or space) averaging, it is necessary to verify that the data are separated by at least one characteristic time (or space) scale of the phenomenon under investigation; therefore, a preliminary evaluation of such a scale is required (see Sect. 10.1.3 for details on flow scale evaluation). Nevertheless, when recurrence in time, space or transformed domains are considered (i.e., when correlation or spectral functions are evaluated), what is really under investigation is the relation (i.e., the dependence) between one piece of data and the others; in this case, the acquired data must be dependent [i.e., acquired on time (or space) intervals much lower than the characteristic time (or space) scale of the phenomenon].

10.1.3 Scales in Turbulent Flows

Generalities

The term *scale* indicates the characteristic size along one direction (if the behavior in space is considered) or duration in time (for the behavior in time) of a fluid-mechanics event that partially preserves itself in space and time. In this sense, these events could be related to the vortical structures embedded in a turbulent flow field (refer to the beginning of Chap. 10). The determination of the scales (in length and time) is an objective both for fundamental research in fluid mechanics and for many practical applications; indeed, the values of such scales allow one to derive the size, shape, and traveling times of turbulent vortical structures. From the experimental point of view, in addition to the previous goals, there is also the requirement for a practical way to determine the optimal resolutions in space or time and extension or the duration of data acquisition (see Sect. 10.1.2). Regarding the resolution of the measurement systems, the link is to the smallest space and time scales of the phenomenon. For the extension and duration of data acquisition, in theory, time or space averaging should be performed over an infinite duration or extension; in practice, the acquisition time or the volume extension will be limited by two factors:

- they should be sufficiently larger that the largest time (or space) scale of the phenomenon (by at least a factor 100);
- they should be sufficiently smaller than the characteristic scale for possible unsteady (or inhomogeneous) phenomena (by at least a factor 10).

In the second case all the considerations mentioned for phase averaging in Sect. 10.1.1 should also be taken into account. Thus, a preliminary knowledge of the time and space scales of the phenomenon is required.

Overall Behavior of Correlation Functions

As will be shown later, turbulent flow scales are derived by analysis of correlation functions (in particular of autocorrelation coefficients). Referring to purely spatial or temporal autocorrelation coefficients, the following definitions hold where, in respect to Sect. 10.1.1, we now specify that only the fluctuating part of the variable (q'_i) is used,

$$\rho_{q'^2_i}(\mathbf{x}, \mathbf{r}, t) = \frac{\langle q'_i(\mathbf{x}, t) q'_i(\mathbf{x}+\mathbf{r}, t) \rangle}{\langle q'^2_i(\mathbf{x}, t) \rangle^{1/2} \langle q'^2_i(\mathbf{x}+\mathbf{r}, t) \rangle^{1/2}},$$

$$\rho_{q'^2_i}(\mathbf{x}, t, \tau) = \frac{\langle q'_i(\mathbf{x}, t) q'_i(\mathbf{x}, t+\tau) \rangle}{\langle q'^2_i(\mathbf{x}, t) \rangle^{1/2} \langle q'^2_i(\mathbf{x}, t+\tau) \rangle^{1/2}}.$$

Therefore, it follows that

$$\rho_{q_i'^2}(\boldsymbol{x}, 0, t) = 1,$$

$$\rho_{q_i'^2}(\boldsymbol{x}, t, 0) = 1,$$

i.e., the value of the correlation function at the origin is just the mean-square value of the fluctuating part, so that the correlation coefficient is just equal to one. Note that this is not the case for the cross-correlation coefficient between two variables, which at the origin attains the value

$$\rho_{q_i' q_j'}(\boldsymbol{x}, 0, t) = \frac{\langle q_i'(\boldsymbol{x}, t) q_j'(\boldsymbol{x}, t) \rangle}{\langle q_i'^2(\boldsymbol{x}, t) \rangle^{1/2} \langle q_j'^2(\boldsymbol{x}, t) \rangle^{1/2}}.$$

It is also important to point out that the maximum degree of correlation of the variable q_i' with itself is obtained when

$$q_i'(\boldsymbol{x} + \boldsymbol{r}, t) = q_i'(\boldsymbol{x}, t)$$

for all data used in the average. On the other hand, the maximum anticorrelation is obtained when

$$q_i'(\boldsymbol{x} + \boldsymbol{r}, t) = -q_i'(\boldsymbol{x}, t)$$

and similarly for time separation. Thus, considering that

$$\langle q_i'^2(\boldsymbol{x}, t) \rangle \geq 0$$

we obtain the Schwartz inequality

$$-1 \leq \rho_{q_i'^2}(\boldsymbol{x}, \boldsymbol{r}, t) \leq 1,$$

$$-1 \leq \rho_{q_i'^2}(\boldsymbol{x}, t, \tau) \leq 1.$$

For the case of homogeneous flow field, the spatial autocorrelation coefficients (as the autocorrelation functions) are also even functions of the space separation r,

$$\rho_{q_i'^2}(x, -r, t) = \frac{\langle q_i'(\boldsymbol{x}, t) q_i'(\boldsymbol{x} - \boldsymbol{r}, t) \rangle}{\langle q_i'^2(\boldsymbol{x}, t) \rangle}$$

$$= \frac{\langle q_i'(x' + r, t) q_i'(x', t) \rangle}{\langle q_i'^2(x', t) \rangle}$$

$$= \rho_{q_i'^2}(\boldsymbol{x}, \boldsymbol{r}, t),$$

where the hypothesis of homogeneity is used in the last equivalence, where the replacement $x' = x - r \to x$ was made.

Similarly, for the case of stationary flow field, the time autocorrelation coefficients (and the autocorrelation functions) are even functions of the time interval τ,

$$\rho_{q_i'^2}(x, t, -\tau) = \rho_{q_i'^2}(x, t, \tau).$$

The limiting behavior of the autocorrelation coefficients for space separation (or time interval) going to infinity can be retrieved from the following considerations. Unless the phenomenon is laminar (which in the stationary case is equivalent to a constant value, for which the autocorrelation coefficient is always one) or periodic in space or time (which gives an oscillatory autocorrelation coefficient), it is expected that for $r \to \infty$ the samples are statistically independent and fluctuations at a point are independent of the other point (and similarly in time separation for $\tau \to \infty$). Thus, when one of the two is positive the other could be positive or negative; statistically, the average value of these negative and positive contributions vanishes (as for the case of pure noise). Therefore the limiting values of the autocorrelation coefficients are

$$\rho_{q_i'^2}(x, r \to \infty, t) = 0,$$

$$\rho_{q_i'^2}(x, t, \tau \to \infty) = 0.$$

In practice, this limit to infinity is understood to apply for separations much larger than the largest scale in the flow field. The overall shape of an autocorrelation coefficient is given in Fig. 10.5 for the case of a purely space function in homogeneous flow conditions (note that the fact that the tangent is zero at the origin has not been strictly demonstrated).

Integral Scales

At this point, it is possible to specify in greater detail the scales introduced at the beginning of Sect. 10.1, starting from the largest towards the smallest. Consider

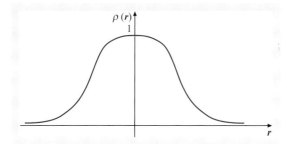

Fig. 10.5 Characteristic shape of a generic autocorrelation coefficient

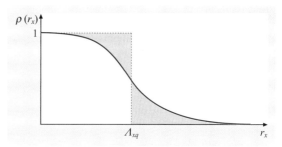

Fig. 10.6 A simple meaning for the integral scale

the purely spatial autocorrelation coefficient evaluated along the longitudinal (r_x) direction; flow scales referred to two different flow variables can be similarly introduced starting from cross-correlation coefficients. The longitudinal integral length scale of the variable q_i is defined as

$$\Lambda_{xq_i}(x,t) = \int_0^\infty \rho_{q_i'^2}(x, r_x, t)\, dr_x ,$$

which is independent of position along the axis x only in the case of a homogeneous flow field and is independent of time t only in the stationary case (note that this integral scale has dimensions of a length and can be defined even in inhomogeneous or unsteady conditions when the correlation coefficient is not symmetric).

Similarly, it is possible to define transverse integral length scales of the same variable along the directions r_y and r_z. Therefore, as for the number of independent autocorrelation functions (Sect. 10.1.1), if q is a scalar there are three independent integral length scales, while if q is a vector (with scalar components q_i, as for the flow velocity) there are nine integral length scales.

These scales are a measure of the largest structures in a flow field, i.e., of those structures that maintain a high degree of relation with themselves. For points separated by distances larger than the integral length scale there is no longer any dependence of the velocity at one point on the velocity at another. In Fig. 10.6, this is explained by considering that the previous definition of integral length scale required equivalence between the shaded areas; therefore, points separated by a distance much larger than Λ_x are practically uncorrelated ($\rho \approx 0$).

The evaluation of the integral length scales along different directions (and of different velocity components) gives information on the form of the largest structures in the flow field; for example, if the vertical and transverse integral length scales are approximately equal and smaller than the longitudinal one, the turbulent structures will appear on average in axisymmetrical elongated forms. On the other hand, if the longitudinal and transverse scales are almost equal and much higher than the vertical one, the turbulent structures have the form of disks, and so on.

In the case of a homogeneous and isotropic field, the independent autocorrelation functions reduce to just one for a scalar field and two for the velocity field [$f(r)$ and $g(r)$, described before] along arbitrary directions. Therefore, the integral length scales also reduces to one for the scalar and two for the velocity field

$$\Lambda_q(t) = \int_0^\infty \rho_q(r,t)\, dr ,$$

$$\Lambda_f(t) = \int_0^\infty f(r,t)\, dr , \quad \Lambda_g(t) = \int_0^\infty g(r,t)\, dr ,$$

where the dependence on the position along the axis is avoided and the dependence on time is avoided in the stationary case (note that in this case the autocorrelation function is symmetric). Using (10.9), it is easy to show that for incompressible homogeneous and isotropic fluid flows the following holds

$$\int_0^\infty g(r,t)\, dr = \int_0^\infty f(r,t)\, dr + \int_0^\infty \frac{r}{2} \frac{\partial f(r,t)}{\partial r}\, dr$$

$$= \int_0^\infty f(r,t)\, dr + \left. \frac{fr}{2} \right|_0^\infty$$

$$- \frac{1}{2} \int_0^\infty f(r,t)\, dr$$

$$= \frac{1}{2} \int_0^\infty f(r,t)\, dr ,$$

that is

$$\Lambda_f = 2\Lambda_g ,$$

i.e., the transverse integral length scale is one half of the longitudinal one.

For a general flow field, from the purely temporal autocorrelation coefficient, it is also possible to define an integral time scale as

$$\tau_{0q_i}(x,t) = \int_0^\infty \rho_{q_i'^2}(x, t, \tau)\, d\tau ,$$

which is independent of position along the axis (x_1) only for the homogeneous case and is independent of time t in the stationary case. There is only one integral time scale for each scalar variable: three integral time scales for the velocity, one for each component.

Using Taylor's hypothesis for the large integral scales it is possible to write

$$\tau_{0q_i}(x,t) = \frac{\Lambda_{xq_i}(x,t)}{U_c} ,$$

where U_c is the convection velocity along the axis of mean motion. This relation is a reasonable connection between the spatial and time scales, which allows the derivation of one from the other.

Taylor Microscales

An evaluation of the intermediate scales is given by considering a Taylor series expansion of the autocorrelation coefficient near the origin; for the pure spatial case with separation along the longitudinal axis this yields

$$\rho_{q'^2_i}(x, r_x, t) = \rho_{q'^2_i}(x, 0, t) + \left(\frac{\partial \rho}{\partial r_x}\right)_{r_x=0} r_x$$
$$+ \left(\frac{\partial^2 \rho}{\partial r_x^2}\right)_{r_x=0} \frac{r_x^2}{2} + O(r_x^3), \quad (10.21)$$

where $\rho_{q'^2_i}(x, 0, t) \equiv 1$, and similarly for the other axes r_y and r_z.

Neglecting the terms of order higher than r_x^2, a parabolic approximation to the autocorrelation coefficient is obtained. The separation at which the autocorrelation coefficient vanishes (i.e., the intersection of the parabola with the horizontal r_x axis) defines the so-called Taylor's microscale $\lambda_{xq_i}(x, t)$,

$$1 + \left(\frac{\partial \rho}{\partial r_x}\right)_{r_x=0} \lambda_{xq_i}(x, t) + \left(\frac{\partial^2 \rho}{\partial r_x^2}\right)_{r_x=0} \frac{\lambda_{xq_i}^2(x, t)}{2}$$
$$= 0, \quad (10.22)$$

which is a quadratic relation with unknown λ_{xq}, given the first and second derivative of the autocorrelation coefficient in the origin. This is explained graphically in Fig. 10.7; the Taylor's microscale gives indications of the size of intermediate (i.e., lower than the integral length scales) flow structures.

Even for the Taylor's microscale in space, there are three scales (one along each axis) for a scalar variable and three scales for each component of a vector.

The evaluation of Taylor's microscale is simpler when the hypothesis of a homogeneous flow field is assumed; in this case, the first-order term in the previous expansion vanishes. This can be shown by considering relation (10.21) for negative r_x values as well and

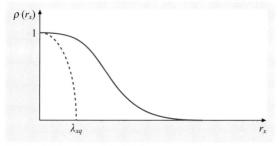

Fig. 10.7 Meaning of the Taylor microscale

subtracting and adding the two terms

$$\rho_{q'^2_i}(r_x, t) - \rho_{q'^2_i}(-r_x, t)$$
$$\equiv 0 = 2\left(\frac{\partial \rho}{\partial r_x}\right)_{r_x=0} r_x + O(r_x^3),$$

$$\rho_{q'^2_i}(r_x, t) + \rho_{q'^2_i}(-r_x, t)$$
$$= 2 + \left(\frac{\partial^2 \rho}{\partial r_x^2}\right)_{r_x=0} r_x^2 + O(r_x^4).$$

In the last relation, as the two left-hand terms are equal, it is then possible to obtain

$$\left(\frac{\partial \rho}{\partial r_x}\right)_{r_x=0} = O(r_x^2)$$

$$\left(\frac{\partial^2 \rho}{\partial r_x^2}\right)_{r_x=0} = \frac{2}{r_x^2}\left[\rho_{q'^2_i}(r_x, t) - 1\right] + o(r_x^2),$$

i.e., the first-order derivative vanishes and the second-order derivative is always nonpositive. In this case, from (10.22), the Taylor's microscale results as

$$\lambda_{xq_i}^2(t) = \frac{-2}{\left(\partial^2 \rho / \partial r_x^2\right)_{r_x=0}}$$

or equivalently

$$\rho_{q'^2_i}(r_x \to 0, t) \approx 1 - \left(\frac{r_x}{\lambda_{xq_i}}\right)^2$$

and similarly for the other axes r_y and r_z; for the stationary case the scales are not dependent on time. Regarding third-order correlations, it is possible to show that they are odd functions of the separation distance and that in isotropic turbulence the values of the function and its derivative for zero separation vanish [10.10]. Thus, for the function k, it is possible to write (10.10)

$$k(r \to 0, t) = \frac{r^3}{6}\left(\frac{\partial^3 k}{\partial r^3}\right)_{r=0}$$
$$+ \frac{r^5}{120}\left(\frac{\partial^5 k}{\partial r^5}\right)_{r=0} + O(r^7).$$

Another useful expression is obtained by considering that for a homogeneous flow

$$\left(\frac{\partial^2 \rho}{\partial r_x^2}\right)_{r_x=0} = \left[\frac{\partial^2}{\partial r_x^2}\left(\frac{\langle q'_i(x, t) q'_i(x + r_x, t)\rangle}{\langle q'^2_i(t)\rangle}\right)\right]_{r_x=0}$$
$$= -\frac{\left\langle\left(\frac{\partial q'_i(t)}{\partial r_x}\right)^2\right\rangle}{\langle q'^2_i(t)\rangle} \quad (10.23)$$

so that

$$\lambda_{xq_i}^2(t) = \frac{2\langle q_i'^2(t)\rangle}{\left\langle \left(\frac{\partial q_i'(t)}{\partial r_x}\right)^2\right\rangle} \quad (10.24)$$

and similarly for the other axes r_y and r_z (for the stationary case the scales are not dependent on time).

For homogeneous isotropic flows, the previous Taylor's microscales reduce to only one in the case of a scalar and two for the velocity field (longitudinal and transverse Taylor's microscales)

$$\lambda_q^2(t) = \frac{-2}{(\partial^2\rho/\partial r^2)_{r=0}},$$

$$\lambda_f^2(t) = \frac{-2}{(\partial^2 f/\partial r^2)_{r=0}},$$

$$\lambda_g^2(t) = \frac{-2}{(\partial^2 g/\partial r^2)_{r=0}}$$

or equivalently

$$\lambda_q^2(t) = \frac{2\langle q'^2(t)\rangle}{\left\langle\left(\frac{\partial q'(t)}{\partial x_1}\right)^2\right\rangle},$$

$$\lambda_f^2(t) = \frac{2\langle u_1'^2\rangle}{\left\langle\left(\frac{\partial u_1'}{\partial x_1}\right)^2\right\rangle},$$

$$\lambda_g^2(t) = \frac{2\langle u_2'^2\rangle}{\left\langle\left(\frac{\partial u_2'}{\partial x_1}\right)^2\right\rangle}, \quad (10.25)$$

where x_1 is taken as reference axis. Note that in these relations the two numerators are the same due to isotropy, $\langle u_1'^2\rangle = \langle u_2'^2\rangle$.

Regarding the velocity field, note that, even without the assumption of isotropy, for the case of incompressible fluid flows, it is possible to show (10.9) that

$$\frac{\partial g}{\partial r} = \frac{3}{2}\frac{\partial f}{\partial r} + \frac{r}{2}\frac{\partial^2 f}{\partial r^2}, \quad \frac{\partial^2 g}{\partial r^2} = 2\frac{\partial^2 f}{\partial r^2} + \frac{r}{2}\frac{\partial^3 f}{\partial r^3}.$$

The second equation, for $r=0$, gives

$$\lambda_f^2(t) = 2\lambda_g^2(t). \quad (10.26)$$

while in the same limit the first equation gives equivalence of first-order derivatives at the origin.

Recalling the expressions for the TKE dissipation ε for the case of homogeneous isotropic incompressible turbulence (refer to Sect. 10.1.2), the longitudinal and transverse Taylor microscales (10.25) in terms of ε are

$$\lambda_f^2(t) = \frac{30\nu\langle u_1'^2\rangle}{\varepsilon},$$

$$\lambda_g^2(t) = \frac{15\nu\langle u_2'^2\rangle}{\varepsilon}. \quad (10.27)$$

where again relation (10.26) is retrieved. The reverse relations allow one to derive the TKE dissipation rate from Taylor's microscales

$$\varepsilon = \frac{30\nu\langle u_1'^2\rangle}{\lambda_f^2(t)} = \frac{15\nu\langle u_2'^2\rangle}{\lambda_g^2(t)}. \quad (10.28)$$

This result also implies that ε is related to the correlation coefficient by

$$\varepsilon = -15\nu\langle u_1'^2\rangle\left(\frac{\partial^2 f}{\partial r^2}\right)_{r=0}$$

$$= -\frac{15}{2}\nu\langle u_2'^2\rangle\left(\frac{\partial^2 g}{\partial r^2}\right)_{r=0};$$

thus, using relations (10.3) between the autocorrelation and spectral functions and (10.7), it is also possible to write

$$\varepsilon = 2\nu\int_0^\infty F_{u_i'u_i'}^k k^2 dk,$$

which has already been obtained (10.19). Lastly, using the Karman–Howarth equation (10.17) for zero separation ($r=0$), the properties of third-order correlations, and (10.28), the simplified equation for TKE is obtained [10.4]

$$\frac{d}{dt}u_1'^2 = -10\nu\frac{u_1'^2}{\lambda_g^2} = -\frac{2}{3}\varepsilon, \quad (10.29)$$

which is equivalently obtained from the TKE equation.

Returning to (10.24) and (10.25) for the case of a *variable different from the velocity* that possesses a dissipation term (for example the temperature and its mean square value, or the species concentration), it is possible to consider the limit for the isotropic case. In particular, the temperature or species dissipation rates can be written as (see Sect. 10.1.2)

$$\varepsilon_T = \alpha\left\langle\left(\frac{\partial T'}{\partial x_j}\right)^2\right\rangle = 3\alpha\left\langle\left(\frac{\partial T'}{\partial x_1}\right)^2\right\rangle,$$

$$\varepsilon_c = D\left\langle\left(\frac{\partial c'}{\partial x_j}\right)^2\right\rangle = 3D\left\langle\left(\frac{\partial c'}{\partial x_1}\right)^2\right\rangle,$$

where, as usual, sums over repeated indexes are used and α and D are the thermal and species diffusivities). Thus, similarly to (10.27), it is possible to write the Taylor's microscales for the temperature or concentration fields for the isotropic case as

$$\lambda_T^2(t) = \frac{6\alpha\langle T'^2\rangle}{\varepsilon_T},$$

$$\lambda_c^2(t) = \frac{6D\langle c'^2\rangle}{\varepsilon_c},$$

which can also be inverted to derive the dissipation rates from Taylor's microscales.

For the time autocorrelation coefficient, the previous arguments can be repeated exactly, so that

$$1 + \left(\frac{\partial \rho}{\partial \tau}\right)_{\tau=0} \tau_{\lambda q_i}(x, t) + \left(\frac{\partial^2 \rho}{\partial \tau^2}\right)_{\tau=0} \frac{\tau_{\lambda q_i}^2(x, t)}{2} = 0$$

defines the Taylor's microscale in time $\tau_{\lambda qi}$ (only one for each scalar variable), where the dependence on position and time vanishes for homogeneous and stationary turbulence. For stationary flow field, the previous definition reduces to

$$\tau_{\lambda q_i}^2(x) = \frac{-2}{(\partial^2 \rho / \partial \tau^2)_{\tau=0}} = 2 \frac{\langle q_i'^2 \rangle}{\left\langle \left(\frac{\partial q_i'}{\partial t}\right)^2 \right\rangle},$$

i.e., the autocorrelation time coefficient close to the origin is given by

$$\rho_{q_i'^2}(x, \tau \to 0) \approx 1 - \left(\frac{\tau}{\tau_{\lambda q_i}}\right)^2.$$

For the velocity field, recalling the relations for the TKE dissipation rate in the isotropic case, we obtain (in agreement to what was previously done for the spatial separation)

$$\tau_\lambda^2(x) = \frac{30 \nu \langle u_1'^2 \rangle}{\varepsilon U_c^2},$$

where U_c is the convection velocity, and the reverse, which allows ε to be derived from Taylor's microscale in time. Thus, from the relation between the space and temporal derivatives derived from Taylor's hypothesis (Sect. 10.1.1), we obtain for the velocity component along the mean flow direction

$$\tau_\lambda \approx \frac{\lambda_f}{U_c},$$

which is a reasonable, although not exact, relation between time and space microscales.

Kolmogorov Scales

As already described in Sect. 10.1, the smallest scales are defined in terms of the TKE dissipation rate ε. For such small scales, it is necessary to distinguish clearly between the flow variables, as the definition for large and intermediate scales was based on correlation functions, while smallest scales are based on dissipation rates, which depend on the flow variable under consideration. Consider first the flow velocity, for which the smallest scales are known as the Kolmogorov length and time scales, defined as

$$\eta = \left(\frac{\nu^3}{\varepsilon}\right)^{1/4}, \quad \tau_\eta = \left(\frac{\nu}{\varepsilon}\right)^{1/2}.$$

Note that the previous definition of the space scale does not contain any information on the velocity component or the reference axis. Therefore, the assumption of isotropy is implicit in its definition; at these scales, flow structures have the same size, η, along all the axes, although the anisotropies are contained in the way in which the TKE dissipation rate is computed. It is simple to show using the above definition that a Kolmogorov velocity scale can be defined as

$$v_\eta = \frac{\eta}{\tau_\eta} = (\varepsilon \nu)^{1/4}. \tag{10.30}$$

The Reynolds number obtained using the Kolmogorov length and velocity is identically equal to 1

$$Re_\eta = \frac{v_\eta \eta}{\nu} \equiv 1,$$

thus confirming that for flow structures with characteristic size η and velocity v_η, inertial forces are balanced by viscous forces; smaller structures will be rapidly dissipated by viscosity (with a characteristic Reynolds number less than 1).

These scales can also be retrieved in another way [10.29]. Let us consider a vortical structure (eddy) with characteristic rotational velocity v_r and size r; the *eddy turnover time* is defined as

$$t_e = \frac{r}{v_r}; \tag{10.31}$$

(for large scales the relation leads to $\Lambda/\langle u'^2 \rangle^{1/2}$ [10.6]). If it is assumed that this eddy loses almost all of its energy during a turnover time, then the TKE dissipation rate will be given by

$$\varepsilon \sim \frac{\text{energy density}}{\text{turnover time}} \approx \frac{v_r^2}{(r/v_r)} = \frac{v_r^3}{r}.$$

(The eddy turnover time can also be defined in terms of the TKE and TKE dissipation $t_e = K'/\varepsilon$, so that the Reynolds number based on the velocity scale $\sqrt{K'}$ and on the length scale $t_e \sqrt{K'} = K'^{3/2}/\varepsilon$ will be $Re_T = K'^2/\nu\varepsilon$.) From the previous relation, we obtain

$$v_r \approx (\varepsilon r)^{1/3}.$$

By assuming as the characteristic velocity $v_r = v_\eta$ and as the characteristic size $r = \eta$, it is possible to obtain a first relation between these two quantities related to the TKE dissipation rate

$$v_\eta \approx (\varepsilon \eta)^{1/3} \; . \tag{10.32}$$

This relation, when applied to large scales, leads to $\varepsilon \sim (E[u])^3/\Lambda$, as described in the next section.

A second relation between those two quantities is obtained from

$$\varepsilon = 15\nu \left\langle \left(\frac{\partial u'_1}{\partial x_1}\right)^2 \right\rangle \approx \nu \left(\frac{v_\eta}{\eta}\right)^2 ,$$

that is,

$$v_\eta \approx \left(\frac{\varepsilon}{\nu}\right)^{1/2} \eta \; . \tag{10.33}$$

From (10.32) and (10.33), we obtain

$$v_\eta \eta \approx \nu \; , \quad \eta \approx \left(\frac{\nu^3}{\varepsilon}\right)^{1/4} , \quad v_\eta \approx (\varepsilon \nu)^{1/4} ,$$

so that the eddy turnover time for these structures is [from (10.31)]

$$\tau_\eta \approx \frac{\eta}{v_\eta} \approx \left(\frac{\nu}{\varepsilon}\right)^{1/2}$$

as defined before.

In the case of isotropic flow field, the useful expression for TKE dissipation rate can be used in the definitions of the scales to obtain

$$\eta = \left[\frac{\nu^2}{15\left\langle (\partial u'_1/\partial x_1)^2 \right\rangle}\right]^{1/4} ,$$

$$\tau_\eta = \left[\frac{1}{15\left\langle (\partial u'_1/\partial x_1)^2 \right\rangle}\right]^{1/2} ,$$

$$v_\eta = \left[15\nu^2 \left\langle (\partial u'_1/\partial x_1)^2 \right\rangle\right]^{1/4} ,$$

which can be used as approximate evaluations of the microscales (see Sect. 10.1.3 for different possible evaluations of ε to be used in the measurement of microscales).

As a consequence of the definition of the space and time microscales, the relation between the two is not derivable using mean flow (or convection) velocities

$$\tau_\eta \neq \frac{\eta}{U_c}$$

so that they have to be evaluated separately.

The different flow scales are summarized in Fig. 10.8 [10.30].

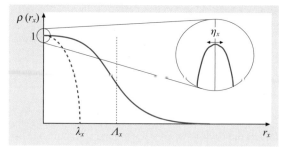

Fig. 10.8 The different flow scales

Consider now flow variables different from the velocity; in particular, for temperature or concentration fields, equations for the temperature or concentration gradients similar to those for TKE dissipation and enstrophy can be derived (see Sect. 10.1.2, Vorticity and Enstrophy). In comparison to the velocity, due to the fact that both temperature (or species concentration) and velocity fields are involved, there are four (rather than two, kinematic viscosity ν, and TKE dissipation rate ε) quantities that can be used to derive the smallest temperature or species scales; the kinematic viscosity, the TKE dissipation rate, the thermal diffusivity α (or the species diffusivity, D), and the temperature dissipation rate ε_T (or the species dissipation rate ε_c) [10.6].

In analogy with the velocity (following the arguments from (10.31)), it is possible to derive an expression for the latter variable from the definition of a temperature (or species) diffusion eddy with characteristic size (η_T), velocity ($v_{\eta T}$), time ($\eta_T/v_{\eta T}$), and temperature square fluctuation $\langle T'^2 \rangle$

$$\varepsilon_T \approx \frac{v_{\eta T} \langle T'^2 \rangle}{\eta_T} \tag{10.34}$$

and similarly for the species concentration.

Moreover, from the definition of ε_T (refer to Taylor's microscales for the temperature and concentration fields), it is possible to write in the isotropic case

$$\varepsilon_T = 6\alpha \left\langle \left(\frac{\partial T'}{\partial x_1}\right)^2 \right\rangle \approx \frac{\alpha \langle T'^2 \rangle}{\eta_T^2} , \tag{10.35}$$

and similarly for the concentration field.

Combining (10.34) and (10.35), we obtain

$$v_{\eta T} \eta_T \approx \alpha , \quad (v_{\eta c} \eta_c \approx D , \text{ for the species,})$$

which is analogous to the result for the velocity field. So far, independently of the definition of the smallest scales for the temperature or species concentration, we always

Table 10.2 Summary of scales for velocity and scalar fields in homogeneous and isotropic turbulence in which the dissipation terms can be expressed as $\varepsilon = 15\nu \left\langle \left(\frac{\partial u'_1}{\partial x_1}\right)^2 \right\rangle$ and $\varepsilon_T = 3\alpha \left\langle \left(\frac{\partial T'}{\partial x_1}\right)^2 \right\rangle$

		Integral	Taylor	Kolmogorov
Length	Velocity field	$\Lambda_f = \int_0^\infty f(r)\,dr$	$\lambda_f^2 = \frac{30\nu \langle u'^2_1\rangle}{\varepsilon}$	
		$\Lambda_g = \int_0^\infty g(r)\,dr$	$\lambda_g^2 = \frac{15\nu \langle u'^2_2\rangle}{\varepsilon}$	$\eta \approx \left(\frac{\nu^3}{\varepsilon}\right)^{1/4}$
		$\Lambda_f = 2\Lambda_g$	$\lambda_f^2 = 2\lambda_g^2$	
	Scalar field	$\Lambda_q = \int_0^\infty \rho_q(r)\,dr$	$\lambda_T^2 = \frac{6\alpha\langle T'^2\rangle}{\varepsilon_T}$	$Pr > 1\ \eta_T = \left(\frac{\alpha^2 \nu}{\varepsilon}\right)^{1/4}$
				$Pr < 1\ \eta_T = \left(\frac{\alpha^3}{\varepsilon}\right)^{1/4}$
Time	Velocity field	$\tau_{0q_i} = \int_0^\infty \rho_{q_i}(\tau)\,d\tau$	$\tau^2_{\lambda u} = 2\frac{\langle u'^2_1\rangle}{\left\langle \left(\frac{\partial u'_1}{\partial t}\right)^2\right\rangle}$	$\tau_\eta = \left(\frac{\nu}{\varepsilon}\right)^{1/2}$
	Scalar field	$\tau_{0q} = \int_0^\infty \rho_q(\tau)\,d\tau$	$\tau^2_{\lambda u} = 2\frac{\langle T'^2_1\rangle}{\left\langle \left(\frac{\partial T'_1}{\partial t}\right)^2\right\rangle}$	$Pr > 1\ \tau_{\eta T} = \left(\frac{\nu}{\varepsilon}\right)^{1/2}$
				$Pr < 1\ \tau_{\eta T} = \left(\frac{\alpha}{\varepsilon}\right)^{1/2}$

have

$$Re_{\eta_T}\,Pr = 1\,, \quad (Re_{\eta_c}\,Sc = 1, \text{ for the species})\,,$$

where the Prandtl (Pr) and Schmidt (Sc) numbers have been used [i. e., the ratio of momentum to thermal (or species) diffusivities]. These relations confirm the definition of the smallest scales in the context of temperature or species diffusion. The product $Re\,Pr$ (or $Re\,Sc$) is often reported as the Peclet number.

In contrast to the velocity field, due to the presence of temperature (or species concentration), it is not possible to obtain unique relationships for the length, velocity, and time scales. It is necessary to consider another equation for the temperature (or species) fluctuation gradients; in analogy with the vorticity gradient equation, it is possible to obtain [10.6]

$$\left\langle \left(\frac{\partial T'}{\partial x_j}\frac{\partial T'}{\partial x_i}\right)\left(\frac{\partial u'_i}{\partial x_j}+\frac{\partial u'_j}{\partial x_i}\right)\right\rangle \approx \alpha \left\langle \left(\frac{\partial^2 T'}{\partial x_j \partial x_i}\right)^2\right\rangle$$

where, for the species concentration, α is replaced by D and T' by c'. This equation can be roughly evaluated as

$$\frac{\langle T'^2\rangle}{\eta_T^2} s'_{ij} \approx \alpha \frac{\langle T'^2\rangle}{\eta_T^4}\,,$$

where

$$s'_{ij} = \frac{1}{2}\left\langle \left(\frac{\partial u'_i}{\partial x_j}+\frac{\partial u'_j}{\partial x_i}\right)^2\right\rangle^{1/2}\,;$$

thus

$$\eta_T \approx \left(\frac{\alpha}{s'_{ij}}\right)^{1/2}\,,$$

and similarly for the concentration. Once s'_{ij} has been specified for the particular range involved, the smallest length scale for the temperature (concentration) field are derived from the previous relation, and in the same way for velocity and time scales.

Two main situations hold [10.6]. When $\alpha < \nu$ (or $D < \nu$), i. e., when $Pr > 1$ (or $Sc > 1$), the fluctuating temperature dissipation is expected to occur at scales η_T smaller than η (temperature fluctuations are sensible to all possible fluctuations of velocity down to the smallest velocity scale). In this case, the quantity s'_{ij} entirely depends on the kinematic viscosity and TKE dissipation, i. e.,

$$s'_{ij} \approx \left\langle \left(\frac{\partial u'_i}{\partial x_j}\right)^2\right\rangle^{1/2} \approx \left(\frac{\varepsilon}{\nu}\right)^{1/2}\,,$$

so that [10.6, 31]

$$\eta_T = \left(\frac{\alpha^2 \nu}{\varepsilon}\right)^{1/4}\,, \quad \tau_{\eta T} = \left(\frac{\nu}{\varepsilon}\right)^{1/2}\,,$$

$$v_{\eta T} = \left(\frac{\alpha^2 \varepsilon}{\nu}\right)^{1/4}\,, \quad (10.36)$$

with α replaced by D for species concentration. Note that the smallest time scale for temperature is the same as that for velocity.

On the other hand, when $\alpha > \nu$ (or $D > \nu$), i.e., when $Pr < 1$ (or $Sc < 1$), the smallest eddies for temperature fluctuations (size of order η_T) are larger than the smallest eddies for velocity (size of order η) as not all possible fluctuations of velocity are involved in the temperature fluctuations. In this case, an evaluation of the strain rate at scales of order η_T differs from above and is smaller than $(\varepsilon/\nu)^{1/2}$. In particular, for scales $\eta_T > \eta$, s'_{ij} must be independent of the kinematic viscosity, but will depend on the specific value of η_T; this gives

$$s'_{ij} \approx \frac{\langle u'^2 \rangle^{1/2}}{\eta_T} \approx \frac{\varepsilon^{1/3}}{\eta_T^{2/3}},$$

using (10.32), so that [10.6]

$$\eta_T = \left(\frac{\alpha^3}{\varepsilon}\right)^{1/4}, \quad \tau_{\eta T} = \left(\frac{\alpha}{\varepsilon}\right)^{1/2},$$
$$v_{\eta T} = (\alpha \varepsilon)^{1/4}, \qquad (10.37)$$

with α replaced by D for concentration. In this case, in comparison to the velocity, at the smallest scales the thermal (or concentration) diffusivity replaces the kinematic viscosity. The reader is referred to Chap. 17, 18 for details of cases in which temperature does not act as a passive flow variable (convective flows). In Table 10.2, some of the previous evaluations of turbulent scales for the velocity and scalar fields are summarized.

Relations Between Flow Scales and Reynolds or Prandtl Numbers

As shown in the previous section, there are at least three classes of characteristic lengths (and times); for the velocity field, these are the integral, Taylor, and Kolmogorov scales. The first argument of this section concerns the relations among these scales; this topic was preliminary considered at the beginning of Sect. 10.1.1 with the aim of quantifying the range of flow scales. We first consider the velocity field and the possible relation between the integral and Taylor scales. From the Navier–Stokes equations, balancing inertial and viscous terms for incompressible stationary flow without external forces (i.e., neglecting the effect of pressure gradient and external forces), we obtain (see Sect. 10.1.2)

$$\langle u_i \rangle \frac{\partial \langle u_j \rangle}{\partial x_i} \approx \nu \frac{\partial^2 \langle u_j \rangle}{\partial x_i^2}; \qquad (10.38)$$

cross-terms between velocity fluctuations can be considered in terms of fluctuating velocities, but are usually much smaller than the term containing the average velocity gradient even for high fluctuations levels. Considering orders of magnitude, the previous equation can be rewritten as

$$\frac{\langle u \rangle^2}{\Lambda} \sim \frac{\nu \langle u \rangle}{\lambda^2},$$

where $\langle u \rangle$ is the characteristic velocity of the problem, Λ is the characteristic large-scale length (i.e., the integral scale that drives the inertial term), and λ is the characteristic length scale for small-scale derivatives (i.e., Taylor's microscale). Thus,

$$\frac{\Lambda}{\lambda} \sim \left(\frac{\Lambda \langle u \rangle}{\nu}\right)^{1/2} = Re^{1/2}, \qquad (10.39)$$

where Re is the Reynolds number based on the integral scale, which shows that the Taylor microscale is much smaller than the integral scale, the larger the Reynolds number.

Defining the Reynolds number based on the Taylor microscale

$$Re_\lambda = \frac{\lambda \langle u'^2 \rangle^{1/2}}{\nu},$$

it is possible to rewrite this relation between the integral and Taylor scales (10.39) as a connection between Reynolds numbers

$$Re_\lambda \sim I Re^{1/2}, \qquad (10.40)$$

where $I = \langle u'^2 \rangle^{1/2}/\langle u \rangle$ is the turbulence intensity. Thus, the Reynolds number based on the Taylor microscale is proportional to the square root of the large-scale Reynolds number. Using the definition of Re_λ, we find

$$\frac{\Lambda}{\lambda} \sim \frac{Re_\lambda}{I}.$$

Considering that, to a reasonable approximation, both the integral and Taylor microscales in time can be simply derived by the corresponding lengths using Taylor's hypothesis (see the previous section), it is possible to state that this relation between length scales also holds for time scales, i.e.,

$$\frac{\tau_0}{\tau_\lambda} \sim Re^{1/2}. \qquad (10.41)$$

Another relation between the integral and Kolmogorov scales can be derived by equating the total kinetic-energy production and dissipation rates (i.e., for the stationary, without external forces, case neglecting contributions from pressure and from fluctuating cross-terms, see Sect. 10.1.2)

$$\langle u_i \rangle \frac{\partial \left(\langle u_j \rangle \langle u_j \rangle\right)}{\partial x_i} \approx \varepsilon,$$

which can be evaluated, as an order of magnitudes, in the form

$$\frac{\langle u \rangle^3}{\Lambda} \sim \varepsilon. \qquad (10.42)$$

As already stated, this relation is useful in evaluating the TKE dissipation rate from the large-scale variable $\langle u \rangle$ and Λ.

Introducing the latter approximation into the definition of the Kolmogorov scales for length, time, and velocity (see previous section), it is possible to write

$$\frac{\Lambda}{\eta} \sim Re^{3/4}, \quad \frac{\tau_0}{\tau_\eta} \sim Re^{1/2}, \quad \frac{\langle u \rangle}{v_\eta} \sim Re^{1/4}, \qquad (10.43)$$

which give approximate evaluations of almost the entire observable range of scales in a turbulent flow from the large-scale Reynolds number.

By comparing the previous relations with those obtained between integral and Taylor scales, the ratios of the Taylor to Kolmogorov scales are also obtained

$$\frac{\lambda}{\eta} \sim Re^{1/4}, \quad \frac{\tau_\lambda}{\tau_\eta} \sim 1. \qquad (10.44)$$

The behavior of the flow length scales as a function of Reynolds number is summarized in Fig. 10.9. While the large scale is almost constant (depending on the boundaries of the flow field), the microscales decreases with the Reynolds number so that turbulence, corresponding to the fact that turbulence has been reported having the same appearance when observed for long with different enlarging factors.

More-precise relations among the flow scales can be derived under the assumption of isotropy; considering (10.28) for ε inserted into (10.42), we obtain for the ratios of the integral to the Taylor length and time scales

$$\frac{\Lambda}{\lambda_f} \sim \frac{Re^{1/2}}{30^{1/2} I}, \quad \frac{\tau_0}{\tau_\lambda} \sim \frac{Re^{1/2}}{30^{1/2} I}, \qquad (10.45)$$

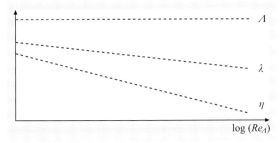

Fig. 10.9 Turbulence scales as a function of Reynolds number

where in the second relation Taylor's hypothesis has been used. These relations replace (10.39) and (10.41) in the case of isotropic flow. On the other hand, inserting the isotropic expression for ε (10.8) into the definitions of the Kolmogorov length, time, and velocity scales (see previous section), we obtain

$$\eta = \frac{\lambda_f}{30^{1/4} Re_\lambda^{1/2}}, \quad \tau_\eta = \frac{\lambda_f}{30^{1/2} \langle u_1'^2 \rangle^{1/2}},$$

$$v_\eta = \frac{30^{1/2} I \langle u \rangle}{Re_\lambda^{1/2}}. \qquad (10.46)$$

Using these relations, the connection between Reynolds numbers based on large and microscale (10.40) and Taylor's hypothesis (to convert the Taylor length microscale to time) we obtain for the ratios of the Taylor and Kolmogorov scales

$$\frac{\lambda_f}{\eta} = 30^{1/4} Re_\lambda^{1/2} \sim 30^{1/4} I^{1/2} Re^{1/4},$$

$$\frac{\tau_\lambda}{\tau_\eta} \sim 30^{1/2} I, \qquad (10.47)$$

which replace (10.44) in the case of isotropy. Considering together (10.45) and (10.47) (and using (10.46) for the velocity), it is possible to write the ratio of the integral to Kolmogorov scales for the isotropic case as

$$\frac{\Lambda}{\eta} \sim \frac{Re^{3/4}}{30^{1/4} I^{1/2}}, \quad \frac{\tau_0}{\tau_\eta} \sim Re^{1/2},$$

$$\frac{\langle u \rangle}{v_\eta} \sim \frac{Re^{1/4}}{30^{1/4} I^{1/2}}, \qquad (10.48)$$

which replace (10.43).

Consider now a flow variable different from the velocity (see also the previous section). For the temperature and species concentration, it is possible to repeat arguments similar to those involved in deriving (10.37) (considering energy and concentration equations in this case) to obtain the ratio between the integral and Taylor scales for the temperature

$$\frac{\Lambda_T}{\lambda_T} \sim (Re_T Pr)^{1/2}, \qquad (10.49)$$

where the Reynolds number based on the integral length scale for the temperature, Re_T, has been used (for the concentration field, the Schmidt number instead of the Prandtl number will appear). The result is similar to that for the velocity scales with a multiplication factor given by the Prandtl (or Schmidt) number.

For time scales, as for the velocity, assuming that Taylor's hypothesis holds, we find

$$\frac{\tau_{0T}}{\tau_{\lambda T}} \sim (Re_T Pr)^{1/2}. \qquad (10.50)$$

Table 10.3 Summary of scale ratios as functions of the Reynolds and Prandtl (or Schmidt) numbers

		Integral Kolmogorov	Integral Taylor	Taylor Kolmogorov
Length	Velocity field	$Re^{3/4}$	$Re^{1/2}$	$Re^{1/4}$
	Scalar field	$Pr > 1$, $Re^{3/4} Pr^{1/2}$	$(RePr)^{1/2}$	$Pr > 1$, $Re^{1/4}$
		$Pr < 1$, $(RePr)^{3/4}$		$Pr < 1$, $(RePr)^{1/4}$
Time	Velocity field	$Re^{1/2}$	$Re^{1/2}$	1
	Scalar field	$Pr > 1$, $Re^{1/2}$	$(RePr)^{1/2}$	$Pr > 1$, $Pr^{-1/2}$
		$Pr < 1$, $(RePr)^{1/2}$		$Pr < 1$, 1

By considering (10.42) in the definition of smallest length and time scales for temperature (or concentration) (10.36) and (10.37)), the ratios of large to small scales for the $Pr > 1$ ($Sc > 1$) case (assuming integral length and time scales for temperature, or concentration, equal to that for velocity) are evaluated as

$$\frac{\Lambda_T}{\eta_T} \sim Re_T^{3/4} Pr^{1/2}, \quad \frac{\tau_{0T}}{\tau_{\eta T}} \sim Re_T^{1/2} \quad (10.51)$$

and for the $Pr < 1$ ($Sc < 1$) case

$$\frac{\Lambda_T}{\eta_T} \sim (Re_T Pr)^{3/4}, \quad \frac{\tau_{0T}}{\tau_{\eta T}} \sim (Re_T Pr)^{1/2} \quad (10.52)$$

with the Schmidt number replacing the Prandtl number for concentration. These relations show how the range of scale for temperature (or concentration) is enhanced or restricted depending on the value of the Prandtl (or Schmidt) number. The case $Pr > 1$ ($Sc \gg 1$) is typical of water (in which $Pr \approx 7$, and $Sc > 10^3$ when a dye is used), thus exhibiting a larger (for temperature) or much larger (for concentration) range of scales in comparison to those of velocity. For a gas, $Pr \approx 1$ and $Sc \leq 1$, so that the range of scales for temperature (or concentration) is more or less the same as for the velocity. These considerations have a correspondence in the spectral behaviors of temperature (or concentration) fluctuations in comparison to those of the velocity [10.6].

By combining (10.48) and (10.49) with (10.51) or (10.52), the ratios of the Taylor to the smallest temperature (or concentration) scales can be derived; for $Pr > 1$ (or $Sc > 1$), we find

$$\frac{\lambda_T}{\eta_T} \sim Re_T^{1/4}, \quad \frac{\tau_{\lambda T}}{\tau_{\eta T}} \sim \frac{1}{Pr^{1/2}},$$

while for $Pr < 1$ (or $Sc < 1$), this reads

$$\frac{\lambda_T}{\eta_T} \sim (Re_T Pr)^{1/4}, \quad \frac{\tau_{\lambda T}}{\tau_{\eta T}} \sim 1,$$

with the Schmidt number replacing the Prandtl number for concentration.

Let us also consider the ratios of temperature (or concentration) to velocity scales. For Taylor's microscales, considering (10.39) and (10.48) for lengths, and (10.41) and (10.49) for times, we obtain

$$\frac{\lambda_f}{\lambda_T} \sim Pr^{1/2}, \quad \frac{\tau_\lambda}{\tau_{\lambda T}} \sim Pr^{1/2},$$

where the Schmidt number would replace the Prandtl number for the concentration result, assuming that the integral scales for temperature (or concentration) and velocity are almost equal, which is a reasonable approximation as both are convected by the mean velocity field. Thus, for gases, Taylor's microscales of temperature (or concentration) and velocity are almost equal, while for water Taylor's microscale for temperature is about three times smaller than that for velocity, and for dye concentration, Taylor's microscale is about a factor 30 smaller than that of the velocity.

For the smallest time scale, from the definitions (10.36), i.e., for the $Pr > 1$ ($Sc > 1$) case, we obtain

$$\frac{\eta}{\eta_T} \sim Pr^{1/2}, \quad \frac{\tau_\eta}{\tau_{\eta T}} \sim 1, \quad \frac{v_\eta}{v_{\eta T}} \sim Pr^{1/2},$$

and from (10.37), for the $Pr < 1$ ($Sc < 1$) case,

$$\frac{\eta}{\eta_T} \sim Pr^{3/4}, \quad \frac{\tau_\eta}{\tau_{\eta T}} \sim Pr^{1/2}, \quad \frac{v_\eta}{v_{\eta T}} \sim Pr^{1/4},$$

where the Schmidt number would replace the Prandtl number for the species concentration result. These relations confirm the findings already noted in Sect. 10.1.3 about the Kolmogorov scales for $Pr > 1$: the smallest scales for temperature are smaller than those for velocity (and those for concentration are much smaller when $Sc \gg 1$), whereas for $Pr \leq 1$ the smallest scales are similar. In Table 10.3, the different scale ratios for the velocity and scalars as functions of the Reynolds and Prandtl numbers are summarized.

How to Measure Length and Time Scales in a Turbulent Flow

The practical measurement of turbulent scales in flow fields leads to some potential difficulties, which are considered in this section. At the outset, it is necessary to evaluate the correlation coefficient of the flow variable under consideration in space, which requires the measurement of data at different spatial positions, and/or in time, which requires long sequences of data in time. As mentioned in Sect. 10.1.1, to derive such coefficients it is necessary to perform data sampling in space and/or time with small spatial or time separations (i.e., at high resolution); thus, the data must be correlated and not independent as required for single-point statistics (see Chap. 25).

For the evaluation of the integral scale, the integral to be performed is theoretically defined up to infinity; in practice, it is limited. One possibility is to stop at the first zero crossing, i.e., at the first space or time separation, r_{max} or τ_{max}, for which the correlation coefficient vanishes. However, in many cases the correlation coefficients start to exhibit oscillations before this zero is attained, as shown in Fig. 10.10; this is due to poor statistics (due to the fact that the number of cross-products is reduced for increasing separations) and to the random nature of the phenomenon. In this case, a second, more-convenient possibility is to define r_{max} or τ_{max} as the values for which the correlation coefficient reaches its first minimum. Another possibility is to estimate the integral scale as the value for which the correlation coefficient attains $1/e$ of its maximum (equal to 1 for zero separation), i.e., to the value expected if an exponential decay of the correlation coefficient is assumed. These three possibilities are summarized in Fig. 10.10.

For the case of a periodic phenomenon, the integral scale is related to the period (in space or time), which can be evaluated by the position of maxima in the correlation coefficients; in this case, a reasonable estimation can only be performed when the first of these maxima exceeds a value of 0.3–0.4 in the correlation coefficient.

For the Taylor microscale, it is necessary to distinguish between several possibilities. The first is connected to the determination of the microscale from the second-order derivative of the correlation coefficient at the origin. It is quite difficult to evaluate such a derivative and especially to determine how many correlation points near the origin must be used for such evaluation. An equivalent procedure is to fit a parabola to the data close to the origin and to determine the intersection with the horizontal axis (see the section on Taylor microscales and to Fig. 10.4). However, the presence of noise in the measurements adds a Dirac delta at the origin, thus preventing the determination of such a parabola; a plot of the correlation coefficient versus the square of the separation can support the evaluation of the effective number of points near the origin required for the fit to the parabola (*Romano* et al. [10.32]). For homogeneous (or stationary) flows the second-order derivative of the correlation coefficient can be replaced by the mean square derivative of the variable fluctuations (10.23). The evaluation of the mean square derivatives is itself not straightforward due to the increasing contribution of noise (see Chap. 25–Chap. 27 and sections devoted specifically to measurement systems); this procedure avoids the evaluation of the correlation coefficient and its second-order derivative.

For the velocity field, it is also possible to use the TKE dissipation rate, ε, in particular for the simplest case of isotropic flows if only a preliminary estimate of the Taylor microscale is required (see Sect. 10.1.3). The quantity ε can be evaluated from the mean square fluctuating velocity derivatives, from relatively simple estimates such as (10.42), or be evaluated from the spectra. In the following, several possible estimations of ε are given.

Another evaluation of the Taylor's microscale in time is derived from the analysis of the rate of zero-crossing of the fluctuating signal (the number of zero-crossing for unit time, N_0) [10.10]; the evaluation is given by

$$\tau_\lambda \approx \frac{\sqrt{2}}{\pi N_0}. \tag{10.53}$$

The Taylor length microscale can be derived by using Taylor's hypothesis (Sect. 10.1.1). The previous relation can be derived exactly for a sinusoidal oscillating

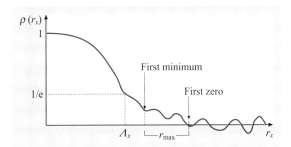

Fig. 10.10 Different evaluations for the integral length scale

signal [10.10]

$$q' = A \sin\left(\frac{2\pi t}{T_0}\right)$$

for which

$$\langle u'^2 \rangle = \frac{A^2}{2}, \quad \left\langle \left(\frac{du'}{dt}\right)^2 \right\rangle = \frac{2A^2\pi^2}{T_0^2}, \quad \dot{N}_0 = \frac{2}{T_0},$$

so that (10.53) holds exactly.

For practical evaluation of the smallest scales, it is necessary to consider that the evaluation of the TKE dissipation rate ε is required (Sect. 10.1.2). As already reported for the evaluation of the Taylor's microscale, ε can be derived in various ways. However, it is important to stress that, for general flow fields, the correct evaluation of ε requires the measurement of several mean square derivatives along different directions, which is also a challenging objective of experimental verification of advanced turbulence closure models (see Sect. 10.1.1, higher-order closures).

Here, some possible evaluations of dissipation rates are summarized (*Monin* and *Yaglom* [10.12], *George* and *Hussein* [10.18], *Pope* [10.4], *Antonia* et al. [10.33]).

For homogeneous turbulent flows, we find for velocity

$$\varepsilon = \nu \left\langle \frac{\partial u'_i}{\partial x_j} \frac{\partial u'_i}{\partial x_j} \right\rangle,$$

and for temperature

$$\varepsilon_T = \alpha \left\langle \frac{\partial T'}{\partial x_j} \frac{\partial T'}{\partial x_j} \right\rangle,$$

with D replacing α for concentration. For inhomogeneous flows, additional cross-terms are involved (see Sect. 10.1.1, single-point equations). Another expression is for exactly homogeneous flows (i.e., flows in which fluctuations of a velocity component are independent of the considered point)

$$\varepsilon_{\text{hom}} = 3\nu \left[\left\langle \left(\frac{\partial u'_1}{\partial x_1}\right)^2 \right\rangle + \left\langle \left(\frac{\partial u'_2}{\partial x_1}\right)^2 \right\rangle + \left\langle \left(\frac{\partial u'_3}{\partial x_1}\right)^2 \right\rangle \right].$$

For local homogeneous axisymmetric turbulence, useful expressions are

$$\varepsilon_{\text{axi}} = \nu\left[-\left\langle \left(\frac{\partial u'_1}{\partial x_1}\right)^2 \right\rangle + 2\left\langle \left(\frac{\partial u'_1}{\partial x_2}\right)^2 \right\rangle \right.$$
$$\left. + 2\left\langle \left(\frac{\partial u'_2}{\partial x_1}\right)^2 \right\rangle + 8\left\langle \left(\frac{\partial u'_2}{\partial x_2}\right)^2 \right\rangle \right],$$

$$\varepsilon_{T\text{axi}} = \alpha \left[\left\langle \left(\frac{\partial T'}{\partial x_1}\right)^2 \right\rangle + 2\left\langle \left(\frac{\partial T'}{\partial x_2}\right)^2 \right\rangle \right].$$

For the isotropic case (see Sect. 10.1.1 on isotropy)

$$\varepsilon_{\text{iso}} = 15\nu \left\langle \left(\frac{\partial u'_1}{\partial x_1}\right)^2 \right\rangle,$$

$$\varepsilon_{T\text{iso}} = 3\alpha \left\langle \left(\frac{\partial T'}{\partial x_1}\right)^2 \right\rangle.$$

For decaying isotropic turbulence, from (10.29), converting time derivatives into time, it is possible to derive the TKE dissipation from the fluctuating TKE decay (the same is obtained from the TKE equation)

$$\varepsilon_{\text{dec}} = -\frac{\langle u \rangle}{2} \frac{d\langle K' \rangle}{dx}.$$

An overall estimation of TKE dissipation is derived from the approximate balance between the production and dissipation of kinetic energy (see Sect. 10.1.3, on Kolmogorov scales) and similarly for temperature

$$\varepsilon \approx \frac{\langle u \rangle^3}{\Lambda}, \quad \varepsilon_T \approx \frac{\langle u \rangle \langle T \rangle^2}{\Lambda}.$$

The former result can also be reconsidered in terms of the fluctuating TKE, defining a scale Λ_ε slightly different from the integral scale:

$$\varepsilon \approx \frac{(K'/3)^{3/2}}{\Lambda_\varepsilon}.$$

From an estimate of the order of magnitude of the terms in the definition in terms of microscales (see Sect. 10.1.3 on Integral scales),

$$\varepsilon = \frac{30\nu\langle u'^2 \rangle}{\lambda_f^2}, \quad \varepsilon_T = \frac{6\alpha\langle T'^2 \rangle}{\lambda_T^2}.$$

From the integral of the energy spectrum (Sect. 10.1.1)

$$\varepsilon = 2\nu \int_0^\infty F_{u'_i u'_i}^k k^2 \, dk.$$

10.2 Measuring Lagrangian Statistics in Intense Turbulence

Even though it has been intensively studied for well over a century, fluid turbulence remains a largely unsolved problem. A full characterization of turbulence is most likely contained within the Navier–Stokes equations, but these strongly nonlinear partial differential equations continue to resist analytical solution for turbulent flows. Researchers in turbulence therefore typically turn to phenomenological models based on the 1941 hypotheses of Kolmogorov (K41) [10.34], as well as more-recent extensions [10.11]. Kolmogorov's hypotheses were originally formulated in the Eulerian framework, where flow statistics are determined at fixed spatial locations. Like all quantities in fluid mechanics, however, both turbulence and the Kolmogorov hypotheses can be cast in the Lagrangian framework, where statistics are measured along the trajectories of individual fluid elements.

Much more is known about the Eulerian characteristics of turbulence than about its Lagrangian properties, primarily because robust and precise Lagrangian experiments have historically been extremely difficult, if not impossible at high Reynolds numbers. Since a general mapping between Eulerian and Lagrangian statistics remains elusive, we cannot fully understand turbulence without a full characterization of its Lagrangian nature as well as its Eulerian side. In addition, many problems, including mixing and transport, are inherently Lagrangian [10.35].

Lagrangian experiments have been carried out for many decades in field measurements in the atmosphere and ocean using balloons and floaters, but these devices are typically too large to measure the smallest scales of the turbulence, and the measurements they produce often have significant noise and uncertainty. Over the past 15 or so years, however, Lagrangian turbulence has become the subject of laboratory experiments with the development of powerful experimental techniques based on digital imaging and signal processing, and data comparable in quality to the best Eulerian results have been measured by many groups [10.36–53]. Here, we describe an optical Lagrangian particle tracking (LPT) technique capable of producing robust single- and multiparticle Lagrangian measurements [10.54–59]. We first discuss the image-processing algorithms used to determine particle trajectories, and then describe our implementation of the technique in the laboratory. Finally, we briefly show some results from our experiments, focusing on the separation of particle pairs in intense turbulence.

10.2.1 Image Processing

Optical particle tracking is at its heart an image-processing technique and an application of machine vision. The process of converting raw images of tracer particles to Lagrangian trajectories can broadly be split into three steps. First, the particles must be located in each image. The accuracy with which the particles are found is the primary factor that determines the accuracy of the LPT system. Next, if three-dimensional resolution is desired, the two-dimensional particle positions found from the images taken by each of the multiple cameras in the system must be matched to generate the three-dimensional particle coordinates. Finally, the particle motion must be tracked in time through many images, producing particle trajectories.

Particle Identification

A typical image taken by our LPT system is shown in Fig. 10.11. Several features of this image are notable in the context of particle tracking. The image consists of bright spots, corresponding to particle images, on a dark background. It is readily apparent, however, that the particles are not uniformly illuminated, and that the image contains background noise. Additionally, some particle images overlap with one another. Overlap is unfortunately unavoidable in three-dimensional LPT, since an

Fig. 10.11 A typical image of tracer particles taken in our LPT system

entire volume is illuminated rather than simply a plane, and this makes particle identification significantly more difficult than for isolated particle images.

Figure 10.11 makes it clear, then, that a good particle-finding algorithm must be able to handle nonuniform illumination, noise, and overlap without significant loss of accuracy. Additionally, the algorithm should be computationally efficient, since hundreds of thousands of images must be analyzed to produce well-converged turbulence statistics.

After extensive tests of many types of particle-finding algorithms [10.54], including algorithms based on weighted averaging, function fitting, and neural networks, we have chosen to use an algorithm based on one-dimensional Gaussians. This algorithm is fast, provides subpixel accuracy, is resistant to moderate noise levels, and can handle overlap. We now discuss the particle-finding procedure in detail.

To identify particles, we search the image for local maxima in image intensity above a small threshold, set empirically to a level above the average background noise. To find the intensity maxima, we simply step through the image row by row and compare each pixel's intensity with those of its eight neighbors. If the pixel's intensity is greater than or equal to those of its neighbors, we assume that the center of the particle image lies within that pixel. We neglect the outer ring of pixels in the image, since particles cannot be reliably located there.

To determine the coordinates of the particle with subpixel accuracy, we first assume that the intensity profile of the light received by the detector is Gaussian. In reality, the intensity profile will be some more complex function. Near the peak, however, where the center lies, the profile can be well approximated by a Gaussian. The horizontal and vertical coordinates of the particle center are then found by fitting two independent one-dimensional Gaussians. The fitting process is simple, and noniterative for speed. By taking the local maximum pixel as well as the pixels on either side of the maximum, an analytical form for the particle center can be obtained. Let us label the coordinates of the three horizontal pixels as x_i for $i = 1, 2, 3$, where x_2 is the local maximum pixel. We can then solve the system of equations

$$I_i = I_0 \exp\left[-\frac{1}{2}\left(\frac{x_i - x_c}{\sigma_x}\right)^2\right], \quad (10.54)$$

where I_i is the intensity of pixel x_i, I_0 is the undetermined overall Gaussian intensity, x_c is the horizontal coordinate of the particle center, and σ_x is the horizontal Gaussian width, for x_c, obtaining

$$x_c = \frac{1}{2}\frac{(x_1^2 - x_2^2)\ln(I_2/I_3) - (x_2^2 - x_3^2)\ln(I_1/I_2)}{(x_1 - x_2)\ln(I_2/I_3) - (x_2 - x_3)\ln(I_1/I_2)}. \quad (10.55)$$

The vertical coordinate of the center is found analogously. Since we use digital cameras, there is a finite number of possible pixel intensities. Every possible logarithm appearing in (10.55) can therefore be precomputed and stored in a lookup table, reducing the computational cost of finding a particle center to simply a few multiplications. In contrast, if we were to fit a full two-dimensional Gaussian intensity profile to the particle image, a nonlinear, iterative fitting algorithm would be required, significantly increasing the computational cost (by at least a factor of four in our tests).

By testing this algorithm on simulated images [10.54], we estimate that the average error in the determination of the particle center is less than 0.1 pixels. In addition to its accuracy, it is also fast and robust, and is an excellent choice for the processing of images without significant background noise. If the signal-to-noise ratio becomes very poor, however, other algorithms based on neural-network pattern recognition may perform better [10.54].

Stereomatching

Once the particle positions have been determined in the two-dimensional image spaces of each camera in the LPT system, the information from all of the cameras must be combined to reconstruct the three-dimensional coordinates of the particles. For applications where the particles have distinguishing features such as shape or color, this information can be used to assist the stereomatching algorithm. In a general LPT experiment, however, the tracer particles have no such features and the only information available for stereomatching is the photogrammetric condition stating that, for each camera, the camera projective center, the particle coordinates on the image plane, and the true particle coordinates in the laboratory frame must be collinear [10.36]. Furthermore, such lines of sight drawn from all the cameras will intersect at the true location of the particle. With appropriate system calibration, which we describe in Sect. 10.2.3, simple photogrammetry can then determine the locations of the particles in three-dimensional space.

If two particles lie on the same line of sight from a single camera, ambiguities can arise. This particle shadowing will occur more frequently as the seeding density of the particles is increased. By using more cameras, however, shadowing becomes less of a problem since the amount of redundant information in the system is high. For example, *Dracos* [10.39] has shown that, for a reasonable particle seeding density, at least three cameras are needed in an LPT system.

The stereomatching algorithm we use is based on those of *Dracos* [10.39] and *Mann* et al. [10.42]. Consider a particle image p_i on one detector. As described above, we can project a line of sight from the perspective center of the camera through the particle image. We then in turn project this line of sight onto the image planes of the other cameras. Particle images on these images planes that fall within some tolerance ϵ of the projected line are considered possible matches for p_i. In this fashion, a list of possible matches for p_i is constructed for every other camera in the system. This process is then repeated for every particle image on every detector. The lists are checked for consistency, and the three-dimensional coordinates are found.

Particle Tracking

Temporal particle tracking is an example of a multidimensional assignment problem. The most general case consists of a sequence of N frames, each containing a potentially different number of particles. Solving the tracking problem requires the determination of the set of assignments between the particle positions in all temporally contiguous frame pairs so that the matched positions correspond to the same physical particle. More precisely, if we denote the i-th particle in the n-th frame by $\mathbf{x}_i^n$, we wish to find the set of links between $\mathbf{x}_i^n$ and $\mathbf{x}_j^{n+1}$, where the physical particle that was at position $\mathbf{x}_i$ in frame n is at position $\mathbf{x}_j$ in frame $n+1$, for all n, i, and j. To determine these links, we define ϕ_{ij}^n to be the "cost" of making a link between $\mathbf{x}_i^n$ and $\mathbf{x}_j^{n+1}$. If $\phi_{ij}^n = 0$, $\mathbf{x}_i^n$ and $\mathbf{x}_j^{n+1}$ refer to the same particle with perfect certainty; generally, however, $\phi_{ij}^n > 0$. The general solution to the tracking problem therefore requires minimizing

$$\Phi \equiv \sum_{n=0}^{N-1} \sum_{i=0}^{M_n} \sum_{j=0}^{M_{n+1}} \phi_{ij}^n, \quad (10.56)$$

where M_n is the number of particles in frame n. We must also allow for the possibility that some $\mathbf{x}_i^n$ or $\mathbf{x}_j^{n+1}$ will have no matches. This must be the case if $M_n \neq M_{n+1}$, and in an actual experiment corresponds to particles entering or leaving the measurement volume, or otherwise vanishing from view.

Minimizing Φ globally over all frames, as described above, is a problem of multidimensional assignment, and can be shown to be $\mathcal{NP}$-hard and therefore not tractable [10.60]. Instead, we restrict the number of frames over which we minimize Φ, in what is known as a *greedy-matching* approximation [10.61].

Greedy-matching algorithms differ in their specification of ϕ_{ij}^n. In general, however, the tracking cost is defined in terms of slowly changing variables. In the extreme case of variables that do not change, such as any distinguishing features of the particles such as color or shape, ϕ_{ij}^n could be specified as simply a binary function. In LPT, however, the particles are generally identical, and so characteristics of the motion must be used, including the particle position, velocity, and acceleration.

After testing several different algorithms using data from a direct numerical simulation (DNS) of turbulent flow [10.54], we have developed a four-frame predictive algorithm that performs well even in intense turbulence. Suppose that a particle track is partially generated up to frame n. We can then use the particle positions in frames n and $n-1$ to estimate a velocity, given by

$$\tilde{\mathbf{v}}_i^n = \frac{\mathbf{x}_i^n - \mathbf{x}_i^{n-1}}{\Delta t}, \quad (10.57)$$

where Δt is the time between frames. A possible position for the particle in frame $n+1$ is then estimated, given by

$$\tilde{\mathbf{x}}_i^{n+1} = \mathbf{x}_i^n + \tilde{\mathbf{v}}_i^n \Delta t. \quad (10.58)$$

A small search volume with an empirically determined size is constructed around the estimated position, and the particles in frame $n+1$ that fall within the volume are considered to be possible matches. For each of these $\mathbf{x}_j^{n+1}$, both a new velocity and an acceleration are estimated, given respectively by

$$\tilde{\mathbf{v}}_i^{n+1} = \frac{\mathbf{x}_j^{n+1} - \mathbf{x}_i^n}{\Delta t} \quad (10.59)$$

and

$$\tilde{\mathbf{a}}_i^{n+1} = \frac{\mathbf{x}_j^{n+1} - 2\mathbf{x}_i^n + \mathbf{x}_i^{n-1}}{\Delta t^2}. \quad (10.60)$$

Using these estimates, a possible position for the particle is estimated in frame $n+2$ for each $\mathbf{x}_j^{n+1}$ in the search volume, given by

$$\tilde{\mathbf{x}}_i^{n+2} = \mathbf{x}_i^n + \tilde{\mathbf{v}}_i^{n+1}(2\Delta t) + \frac{1}{2}\tilde{\mathbf{a}}_i^{n+1}(2\Delta t)^2. \quad (10.61)$$

A second search volume is then constructed around $\tilde{\mathbf{a}}_i^{n+1}$. Let us denote the *closest* real particle in the search volume to $\tilde{\mathbf{a}}_i^{n+1}$ by $\check{\mathbf{x}}_j^{n+2}$. The tracking cost ϕ_{ij}^n is then given by

$$\phi_{ij}^n = \left| \check{\mathbf{x}}_j^{n+2} - \tilde{\mathbf{x}}_i^{n+2} \right|. \tag{10.62}$$

In this way, the real particle in frame $n+1$ that leads to an estimated position closest to a real particle in frame $n+2$ is chosen to continue the track.

Occasionally, the same particle in frame $n+1$ will be the best match for two different particles in frame n. When such a conflict occurs, two choices are possible. The first is simply to give up and end both tracks, starting a new track with the particle in frame $n+1$. This choice guarantees that the tracking algorithm will not construct an incorrect track, but may artificially shorten the measured trajectories. The second possibility is to implement some kind of conflict resolution. One appealing possibility is to construct all of the ϕ_{ij}^n and then minimize the total cost for this frame pair, namely $\sum_{ij} \phi_{ij}^n$ [10.61]. As this is now a two-dimensional assignment problem, efficient algorithms exist to find the set of links with minimal cost [10.62]. In our tests, however, we found that this global cost minimization significantly reduced the accuracy of the tracking algorithm by frequently choosing incorrect links [10.54]. In our algorithm, we therefore use no conflict resolution and accept the tradeoff of shorter tracks for better accuracy.

The predictive algorithm we have described requires prior history of the particle trajectory to estimate future positions of the particle. For the first frame in the image sequence, however, this information is not available. In this case, we simply set $\tilde{\mathbf{v}}_i^n = 0$ and proceed as above. Prior information is also not available for particles that have newly entered the measurement volume in the middle of the image sequence. For this case, we set $\tilde{\mathbf{v}}_i^n$ to the average of the velocities of nearby particles, or to zero if there are no nearby neighbors.

Occasionally, particles will disappear from view for a short time while still remaining in the measurement volume. This occlusion can occur for many reasons. If the illumination of the particles fluctuates, the intensity of the particle image may fall below the threshold; images may also drop below threshold if some pixels on the detector have poor sensitivity or are nonfunctional. To account for the possibility of transient particle occlusion, and additionally to mitigate the effects of tracking conflicts, we have added a simple occlusion-handling system to our algorithm. If a partially constructed track finds no continuation in the next frame, the track is extrapolated using its estimated velocity and acceleration. This extrapolated position is then used in the next frame to try to find a continuation of the track. If no real continuation is found within a few frames, the track is considered to have ended, and the estimated positions are dropped.

10.2.2 Experimental Implementation

The LPT algorithm described above is general, and is applicable to any type of particle tracking experiment. Here, we describe the application of the technique in an actual experiment designed for the acquisition of Lagrangian data in high-Reynolds-number turbulence.

Flow

Central to any fluid-dynamics experiment is the flow itself. For a Lagrangian experiment where long particle tracks are desired, the flow should ideally meet several criteria. First, for an optical particle tracking experiment, the working fluid must be optically transparent. Second, the particles must remain in view for long periods of time. This is easier to achieve in practice with flows without a strong mean velocity field, though Lagrangian experiments have been conducted in wind and water tunnels [10.40, 63].

In our experiments, we have used a von Kármán swirling flow between counter-rotating disks [10.64]. 120 l of water is confined in a closed, cylindrical, plexiglass chamber 60.5 cm high and 48.3 cm in diameter. Eight round, glass windows 12.7 cm in diameter

Fig. 10.12a,b Sketch of the large-scale flow (**a**): decomposed into (**a**) pumping mode and (**b**) shearing mode

are attached around the centerline of the tank to provide optical access without lensing by the cylindrical tank walls. The top and bottom of the tank consist of hard-anodized aluminum plates. Each plate has an internal channel through which cooling water runs to remove the heat dissipated by the turbulence. Any dirt in the water is removed by pumping it through a recirculating filter loop, and bubbles are removed by pumping through a second recirculating loop open to the atmosphere.

The flow was driven by two counter-rotating disks 20.3 cm in diameter. Twelve straight vanes 4.3 cm high are attached to each disk so that the flow is forced inertially. Each disk was driven by a 1 kW direct-current (DC) motor. The large-scale flow was axisymmetric, and consisted of a pumping mode and a shearing mode, as sketched in Fig. 10.12. More details of the flow are given elsewhere [10.41,65,66]. Though the flow is anisotropic and inhomogeneous at large scales, it can generate very intense turbulence in a small amount of laboratory space. The strength of the turbulence is quantified by the Reynolds number; here, we use Re_λ, the Reynolds number based on the Taylor microscale. Re_λ, defined in isotropic turbulence as $Re_\lambda = \sqrt{15 u' L/\nu}$, where u' is the root-mean-square velocity, L is the correlation length of the velocity field, and ν is the kinematic viscosity, is an intrinsic Reynolds number based on the turbulent velocity field itself rather than on an imposed external geometry. In our experiment, we can reach $Re_\lambda \approx 1000$.

Tracer Particles

The tracking algorithm described above can accurately follow the motion of many tracer particles simultaneously in three dimensions. The typical goal of a Lagrangian particle tracking experiment is, however, to gain information on the motion of the *fluid*. The accuracy to which this is accomplished therefore depends on the extent to which the tracer particles behave like true fluid elements. Any real particle will deviate to some degree from the motion of a true fluid element due both to its inertia and to its finite size. Since turbulent flows are highly chaotic, such deviations can lead to significantly different trajectories. The particle will be a good approximation to a fluid element if its density is the same as that of the working fluid, so that it is neutrally buoyant, and if it is very small, ideally smaller than the smallest length scale in the flow, the Kolmogorov length scale $\eta \equiv (\nu^3/\epsilon)^{1/4}$, where ϵ is the mean rate of energy dissipation per unit mass. The degree to which the tracer particle behaves as a fluid element is quantified by the Stokes number, defined as

$$St = \frac{1}{18} \frac{\rho_p - \rho_f}{\rho_f} \left(\frac{d}{\eta}\right)^2, \quad (10.63)$$

where ρ_p and ρ_f are the densities of the particle and the fluid, respectively, and d is the particle diameter. In our experiments, the Stokes number ranges from 10^{-5} to 10^{-3}, and so our particles are good passive tracers [10.47,59].

Detectors

In intense turbulence, the smallest time scale in the flow, the Kolmogorov time scale $\tau_\eta \equiv \sqrt{\nu/\epsilon}$, is typically very short; at high Reynolds number in our flow, for example, τ_η is sub-millisecond. Resolving tracer particle motions on these time scales therefore requires extremely fast detectors, with speeds at least in the tens of kilohertz range if not faster. Such high imaging rates were not available in commercial cameras until very recently. In the past, therefore, we adapted the silicon strip detectors used in the vertex detectors of high-energy particle accelerators for use in Lagrangian particle tracking [10.44,47,67]. These one-dimensional detectors take images at a rate of 70 kHz, allowing full resolution of the Lagrangian acceleration. They have proved unsuitable, however, for simultaneous measurements of multiple tracers. Currently, we use the Phantom v7.1 complementary metal oxide semiconductor (CMOS) camera from Vision Research, Inc. for multiparticle experiments. At a resolution of 256×256 pixels, the Phantom v7.1 can record images at 27 kHz; as the number of sensor pixels is increased, the maximum frame rate drops. While we sacrifice some spatial resolution due to the small number of pixels in this camera, its temporal resolution allows us to resolve the Kolmogorov time scale well, and therefore to measure Lagrangian time derivatives accurately.

Illumination

With such fast cameras, the exposure time for each image is extremely short. A very intense light source is therefore required to illuminate the particles sufficiently. A pulsed source is also advantageous in a multicamera experiment since it can be used to synchronize the cameras: the detectors only record particle images while the light shines.

In our experiments, we have used two Q-switched, frequency-doubled solid-state Nd:YAG lasers for illumination, providing intense green light at 532 nm. One laser, pumped by flashlamps, achieved pulse rates of 30–70 kHz, with typical pulse widths of roughly

≈ 300 ns and a peak power of ≈ 60 W. The second laser, pumped by diode arrays, was capable of pulse rates of 10–120 kHz with pulse widths of ≈ 120 ns and a peak power of ≈ 90 W. Each laser had two Nd:YAG rods in series in the resonant cavity. The acousto-optical Q-switch both increased the power of the laser and controlled the pulse frequency. Three cameras were arranged in the forward-scattering direction from the lasers to maximize the intensity of the scattered light.

System Calibration

As discussed above, the tracking system must be calibrated before the three-dimensional positions of the particles can be determined, and the particles can be tracked. Calibration consists of determining the camera parameters and their locations in laboratory space. To determine these quantities, we use the calibration method described by *Tsai* [10.68].

Assuming a pinhole camera model, there are nine parameters that must be determined for each camera. Six of these specify the position of the camera in laboratory coordinates: three for the rotational positional and three for the translational. The additional three parameters describe the characteristics of the camera–lens system: one for the effective focal length, one for the radial distortion of the image, and one for a possible mismatch between the horizontal and vertical spacing of the pixels on the detector. We assume that the tangential distortion is negligible, which allows a significant simplification of the calibration model [10.68].

To set the model parameters we use a calibration mask consisting of a regular grid of dots mounted on a micrometer stage; images of the mask from three different cameras are shown in Fig. 10.13. The spacing of the dots is known, giving the true horizontal and vertical coordinates of the calibration dots, and the depth coordinate was measured from the micrometer on the calibration mount. By also measuring the dot positions from the camera images, the camera parameters can be determined, as described below.

Let us denote the laboratory coordinates of a particle by x_w and the coordinates in the reference frame of the camera by x_c. The mapping between these frames is given by a rotation matrix $\hat{R}$ and a translation vector T, so that

$$x_c = \hat{R} x_w + T. \quad (10.64)$$

The camera lens then projects x_c onto the two-dimensional camera image plane. Assuming a pinhole camera with an effective focal length of f_{eff}, the ideal detector coordinates x_u of the particle are given by

$$\begin{pmatrix} x_u \\ y_u \end{pmatrix} = \frac{f_{\text{eff}}}{z_c} \begin{pmatrix} x_c \\ y_c \end{pmatrix}. \quad (10.65)$$

The actual measured coordinates x_d will, however, contain some radial distortion, and so

$$x_d = (1 + k_1 r^2) x_u, \quad (10.66)$$

where $r^2 = x_d^2 + y_d^2$. Since we neglect tangential distortion, the distorted and ideal positions of the particle must lie on a line passing through the center of the image [10.68]. Therefore,

$$x_u \times x_d = 0. \quad (10.67)$$

Finally, we allow a mismatch between the horizontal and vertical spacing, so that the final measured coordinates x_p are given by

$$\begin{pmatrix} x_p \\ y_p \end{pmatrix} = \begin{pmatrix} s_x x_d \\ y_d \end{pmatrix}, \quad (10.68)$$

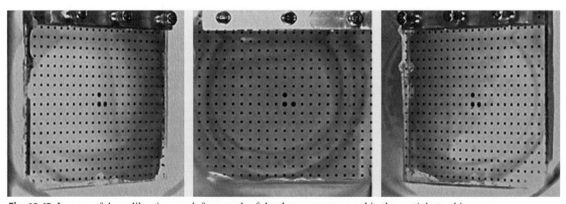

Fig. 10.13 Images of the calibration mask from each of the three cameras used in the particle tracking system

where s_x is a scale factor that is typically close to unity. Calibration therefore requires the determination of $\hat{R}$, T, f_{eff}, k_1, and s_x. By construction, we know the x_w of the dots on the calibration mask, and from the recorded camera images we can determine the x_p for each particle.

Following *Tsai* [10.68], we then write (10.67) as a set of seven linear equations, expressed as

$$\left(-x_p x_w \; -x_p y_w \; -x_p z_w \; y_p x_w \; y_p y_w \; y_p z_w \; y_p\right) \times \begin{pmatrix} T_y^{-1} R_{21} \\ T_y^{-1} R_{22} \\ T_y^{-1} R_{23} \\ T_y^{-1} s_x R_{11} \\ T_y^{-1} s_x R_{12} \\ T_y^{-1} s_x R_{13} \\ T_y^{-1} s_x T_x \end{pmatrix} = x_p \, . \quad (10.69)$$

By using least-squares fits and the orthogonality of $\hat{R}$, these equations allow the determination of the full rotation matrix, T_x, T_y, and s_x. Equation (10.66) may then be solved iteratively for the final three parameters after first estimating initial approximations of T_z and f_{eff} by setting k_1 to zero and using a least-squares fit.

After the parameters have been determined, the model can be used to map the measured particle positions into laboratory coordinates.

Temporal Differentiation

For most statistical measures in Lagrangian turbulence, velocities and accelerations must be calculated along particle trajectories. Simple finite differences, however, are not sufficient for calculating these time derivatives: the errors inherent in locating the particle centers will strongly contaminate such simple methods. Instead, we use a more-robust numerical differentiation scheme that also smooths the trajectory data. By convolving the particle tracks with a Gaussian smoothing and differentiating filter [10.49], we obtain time derivatives measured from a weighted average of many points.

The n-th-order derivative of the position in our scheme is given by

$$\frac{\mathrm{d}^n}{\mathrm{d}t^n} x(t) = \int_{-\infty}^{\infty} \tilde{x}(t-\tau) k_n(\tau) \mathrm{d}\tau \, , \quad (10.70)$$

where $\tilde{x}$ denotes the noisy measured data, and $k_n(\tau)$ is the Gaussian filtering and differentiating kernel defined by

$$k_n(\tau) = \frac{\mathrm{d}^n}{\mathrm{d}\tau^n} \left[\frac{1}{\sqrt{\pi} w} \exp\left(-\frac{\tau^2}{w^2}\right) \right] . \quad (10.71)$$

Integrating (10.70) by parts n times, the derivative in the kernel can be passed to the noisy position data; the surface term from each integration by parts vanishes since the Gaussian vanishes at infinity.

Measured tracks, however, are not infinitely long. Blindly applying (10.70), then, will produce a nonzero surface term due to the finite integration interval. We therefore truncate and renormalize the kernel before convolution with the particle trajectories.

Let us illustrate this process by considering the velocity kernel, given by (10.71) with $n = 1$. After truncation and renormalization, the kernel is given by

$$k_v(\tau) = A_v \tau \exp\left(-\frac{\tau^2}{w^2}\right) + B_v \, , \quad (10.72)$$

where the $1/(\sqrt{\pi} w)$ has been absorbed into A_v. Let us now restrict the integration domain to $[T, -T]$, where typically $(2T + 1) \approx \tau_\eta$. In principle, we must also set w; for simplicity, however, we fix $w = T/1.5$ [10.49]. To fix the constant A_v and B_v, we use the conditions that the derivative of a constant must vanish and the derivative of t must be unity. Expressed using (10.70), these conditions are given by

$$\int_{-T}^{T} k_v(\tau) \mathrm{d}\tau = 0 \quad (10.73)$$

and

$$\int_{-T}^{T} (t-\tau) k_v(\tau) \mathrm{d}\tau = 1 \, . \quad (10.74)$$

By solving these two equations simultaneously, we find that

$$A_v = \left[\frac{1}{2} w^2 \left(w\sqrt{\pi} \operatorname{erf}\left(\frac{T}{w}\right) - 2Te^{\frac{-T^2}{w^2}} \right) \right]^{-1} \quad (10.75)$$

and

$$B_v = 0 \, . \quad (10.76)$$

The acceleration kernel is defined similarly. We note that, in the discrete case, $T = 1$ corresponds to a central difference scheme.

The differentiation scheme introduced here is more accurate than a simple finite-difference algorithm, but still produces a signal that will contain errors. Here, we explicitly analyze these errors for the case of the velocity; errors in determining higher-order derivatives can be described similarly. We note that here we work in nondimensional time units.

As described above, the velocity in our scheme is given by

$$\tilde{u}(t) = \int_{-T}^{T} k_v(\tau) \tilde{x}(t+\tau) d\tau, \quad (10.77)$$

where

$$\tilde{x}(t) = x(t) + \epsilon_x(t). \quad (10.78)$$

Here, $x(t)$ is the true position of the particle and ϵ_x is the error in this measurement. We then rewrite the measured velocity as

$$\tilde{u}(t) = \int_{-T}^{T} k_v(\tau) x(t+\tau) d\tau + \int_{-T}^{T} k_v(\tau) \epsilon_x(t+\tau) d\tau. \quad (10.79)$$

The first term in this expression can be written as the sum of the true velocity $u(t)$ and an intrinsic error associated with the filtering process $\xi_f(t; T, w)$. The second term is the propagation of the uncertainty in the determination of the particle center through the filter $\xi_x(t; T, w)$. Since we fix $w = T/1.5$, both ξ_f and ξ_x are functions of time and the filter length T only.

By considering the case of $\xi_x(t; T) = 0$, i.e., perfect determination of the particle centers, we can study the uncertainty introduced by the finite frequency response of our filter. In this case, the variance of the measured velocity will be given by

$$\langle \tilde{u}(t)^2 \rangle = \left\langle \int_{-T}^{T} d\tau k_v(\tau) x(t+\tau) \right.$$
$$\left. \times \int_{-T}^{T} d\tau' k_v(\tau') x(t+\tau') \right\rangle. \quad (10.80)$$

We can write

$$\tilde{u}(t) = \int_{-T}^{T} d\tau k_v(\tau) x(t+\tau)$$
$$= \int_{-T}^{T} d\tau k_v(\tau) \left[x(t) + \int_{t}^{t+\tau} dt' u(t') \right]$$

$$= \int_{-T}^{T} d\tau k_v(\tau) \int_{t}^{t+\tau} dt' u(t'), \quad (10.81)$$

since $x(t)$ is a constant and, by definition, the convolution of the velocity kernel with a constant vanishes. Therefore, we have

$$\langle \tilde{u}(t)^2 \rangle = \int_{-T}^{T} d\tau \int_{-T}^{T} d\tau' k_v(\tau) k_v(\tau')$$
$$\times \int_{t}^{t+\tau} dt' \int_{t}^{t+\tau'} dt'' \langle u(t') u(t'') \rangle. \quad (10.82)$$

Defining $\langle \xi_f(t; T)^2 \rangle \equiv \langle u^2(t) \rangle - \langle \tilde{u}(t)^2 \rangle$, we therefore have

$$\langle \xi_f(t; T)^2 \rangle = \int_{-T}^{T} d\tau \int_{-T}^{T} d\tau' k_v(\tau) k_v(\tau')$$
$$\times \int_{t}^{t+\tau} dt' \int_{t}^{t+\tau'} dt'' \left[\langle u^2 \rangle - R(t' - t'') \right], \quad (10.83)$$

where $R(t' - t'') \equiv \langle u(t') u(t'') \rangle$ is the two-time velocity covariance and where we have assumed that the flow is stationary. The exact form of this function in turbulence, however, is not known. By turning to a DNS of low-Reynolds-number turbulence, however, we can estimate the magnitude of the error. We find that the error is less than 1% for $T < 70$, and so therefore that the error purely associated with the convolution process is negligible.

We now consider the propagated error in the determination of the particle centers. Assuming that $x(t)$ and ϵ_x are uncorrelated, the variance of $\xi_x(t; T)$ is given by

$$\langle \xi_x(t; T)^2 \rangle = \int_{-T}^{T} d\tau \int_{-T}^{T} d\tau' k_v(\tau) k_v(\tau')$$
$$\times \langle \epsilon_x(t+\tau) \epsilon_x(t+\tau') \rangle. \quad (10.84)$$

Let us now assume that the position error is δ-correlated in time and stationary, so that

$$\langle \epsilon_x(t+\tau) \epsilon_x(t+\tau') \rangle = \langle \epsilon_x^2 \rangle \delta(\tau - \tau'), \quad (10.85)$$

where now ξ_x is a function of T only. We note that this assumption is not fully realistic, since the position error will be correlated while the particle center moves within a single camera pixel. In order to compensate for this underestimation of the error, we take $\langle \epsilon_x^2 \rangle^{1/2} = 0.1$ pixels, which is probably an overestimate of the position uncertainty. With the assumption of δ-correlation, equation (10.84) is tractable, and we have

$$\langle \xi_x(T)^2 \rangle = \langle \epsilon_x^2 \rangle$$
$$\times \frac{-4(T/w) + \sqrt{2\pi} \exp(2T^2/w^2) \mathrm{erf}(\sqrt{2}T/w)}{2w^3[-2(T/w) + \sqrt{\pi} \exp(T^2/w^2) \mathrm{erf}(T/w)]^2} \quad (10.86)$$

With $w = T/1.5$, this expression reduces to

$$\langle \xi_x(T)^2 \rangle = \frac{2.11 \langle \epsilon_x^2 \rangle}{T^3}. \quad (10.87)$$

A similar analysis shows that the error in determining the acceleration scales as T^{-5}. These errors are, in general, not negligible, and must be taken into account in the calculation of turbulence statistics.

Finally, we note that it is instructive to consider how the relative error scales with turbulence parameters. Let us define α as the subpixel resolution (0.1 in our case), N_s to be the number of images taken per τ_η, N_T to be the filter length T in units of τ_η, and p to be the physical pixel size. Then the relative error in the velocity scales as

$$\frac{\langle \xi_x(T)^2 \rangle}{\langle u^2 \rangle} \sim \frac{\alpha^2}{N_s^3 N_T} \left(\frac{p}{L} \right)^2 Re_\lambda^2, \quad (10.88)$$

where L is the integral length scale. It is clear then that at high Reynolds number a simple finite-difference scheme (recall that $T = 1$ corresponds to taking central differences) is insufficient for determining the velocity accurately. For the acceleration, the problem is even more acute, and the relative error scales as

$$\frac{\langle \xi_x(T)^2 \rangle_{\mathrm{accel}}}{\langle a^2 \rangle} \sim \frac{\alpha^2}{N_s^5 N_T} Re_\lambda^3. \quad (10.89)$$

10.2.3 Turbulent Relative Dispersion

Turbulent flows abound with long-range correlations. As such, the statistics of a single Lagrangian tracer will never be sufficient to characterize turbulence fully; instead, we must turn to the joint statistics of many tracers. The simplest such multiparticle problem is that of the separation rates of pairs of fluid elements, known as turbulent relative dispersion. We have measured turbulent dispersion in our LPT experiment [10.56, 59]; here, we briefly present some of our

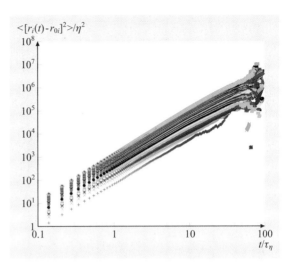

Fig. 10.14 Evolution of the mean-square particle separation at $Re_\lambda = 815$. Each set of symbols represents a bin of initial separations 1 mm ($\approx 43\eta$) wide, ranging from 0–1 mm to 49–50 mm. With both axes normalized by the Kolmogorov scales, each data curve follows a power law for over two decades in time

results as an example of the type of analysis possible with LPT.

Under the action of Brownian motion, the mean-square separation between two fluid elements in a quiescent fluid will grow linearly in time. In a tur-

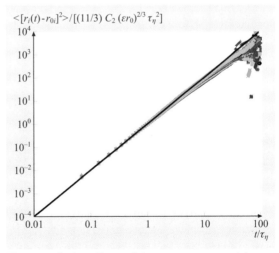

Fig. 10.15 Scale collapse of the mean-square particle separation. The same data as in Fig. 10.14 is scaled by Batchelor's constant $(11/3)C_2(\epsilon r_0)^{2/3}$ and an almost perfect collapse of the data is seen for all 50 initial separations

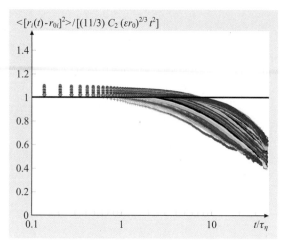

Fig. 10.16 Compensated mean-square particle separation. The time axis is scaled by τ_η. The data for different initial separations deviate from the Batchelor prediction at different times

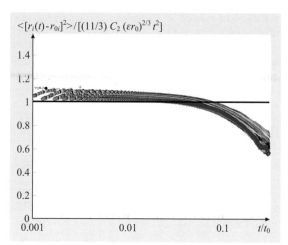

Fig. 10.17 Compensated mean-square particle separation with time scaled by t_0. Plotted in this way, there is a scale collapse in both space and time

bulent flow at sufficiently large scales, i.e., above the integral scales of the flow, pairs of fluid elements will also experience this linear Taylor diffusion [10.69]. In the inertial range, however, fluid element pairs in turbulence will separate superdiffusively, explaining why turbulent flows mix so efficiently.

Based on simple physical arguments and measurements of diffusion over a wide range of scales, *Richardson* [10.70] suggested that the mean-square pair separation should grow as t^3; subsequently, *Obukhov* [10.71] refined his work and wrote that

$$\langle r^2 \rangle = g\epsilon t^3, \qquad (10.90)$$

in the inertial range where energy neither enters nor leaves the system, and where r is the distance between the particles and g is known as the Richardson constant. Such Richardson–Obukhov scaling has proved very challenging to observe, however, and estimates of the Richardson constant span more than an order of magnitude [10.72].

In 1950, *Batchelor* [10.73] realized that while the pair retains a memory of its initial separation r_0, the mean-square separation should grow only as t^2 rather than t^3. He reasoned that this memory should persist for times on the order of the lifetime of an eddy of size r_0, given by $t_0 = (r_0^2/\epsilon)^{1/3}$. After this time, if the Reynolds number is high enough, the pair should move into the Richardson–Obukhov regime. Taking Batchelor's argu-

ments into account, we can write [10.56, 59, 74]

$$\langle [\mathbf{r}(t) - \mathbf{r}_0]^2 \rangle = \begin{cases} \frac{11}{3} C_2(\epsilon r_0)^{2/3} t^2, & t \ll t_0 \\ g\epsilon t^3, & t_0 \ll t \ll T_L \end{cases}$$
(10.91)

where C_2 is the scaling constant of the second-order Eulerian velocity structure function with a well-known

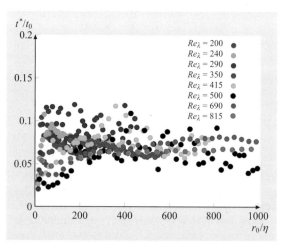

Fig. 10.18 Deviation from Batchelor's prediction; t^* measures the time when the relative dispersion data deviate by more than 5% from Batchelor's prediction. t^* is plotted against the initial separation for eight different Reynolds numbers. Irrespective of Reynolds number, $t^* = (0.071 \pm 0.009)t_0$

value of 2.13 ± 0.22 [10.75] and T_L is the integral time scale.

We have measured relative dispersion from the tracer particle trajectories in a $5 \times 5 \times 5$ cm^3 volume in the center of our experiment over a range of Reynolds numbers [10.56, 59]. In Fig. 10.14, we show our measurements of $\langle [\boldsymbol{r}(t) - \boldsymbol{r}_0]^2 \rangle$ at $Re_\lambda = 815$ with both axes normalized by the Kolmogorov scales. Figure 10.14 shows the dispersion for 50 different bins of initial separations, 1 mm wide ($\approx 43\eta$) and ranging from 0–1 mm to 49–50 mm. It is clear that, for each of these initial separations, even though the largest is approximately 70% of the integral scale L, the data follows a power law for more than two decades in time. The data, however, do not collapse for the different initial separations, suggesting that the initial separation remains a relevant parameter and that therefore the data cannot be following the Richardson–Obukhov law. Recalling Batchelor's prediction for the short-time behavior of the relative dispersion, in Fig. 10.15 we scale the same data by $(11/3)C_2(\epsilon r_0)^{2/3}$ and observe a nearly perfect collapse of the data for the various initial separations. We emphasize that the dark line drawn in Fig. 10.15 is not a fit but is rather Batchelor's predicted power law.

If we now look closer at the agreement of our data with Batchelor's prediction by compensating our experimental results by $(11/3)C_2(\epsilon r_0)^{2/3}t^2$, as shown in Fig. 10.16, we observe that the data deviate from the Batchelor prediction at times that vary with the initial separation. In order to correct for this effect, we scale time by Batchelor's $t_0 = (r_0^2/\epsilon)^{1/3}$, as shown in Fig. 10.17. This new scaling clearly collapses the data for the various initial separations much better, with each curve deviating from Batchelor's prediction at essentially the same value of (t/t_0). Our data therefore confirm both Batchelor's scaling argument and the time for which he predicted it would hold. Moreover, the value t^* for which the data deviate from Batchelor's prediction appears to be independent of Reynolds number. In Fig. 10.18, we define t^* as the time when the data deviate by 5% from Batchelor's prediction. For the entire range of Reynolds numbers tested, $t^* = (0.071 \pm 0.009)t_0$.

10.2.4 Concluding Remarks

Experimentally, Lagrangian turbulence is still very much a nascent field. It will be many years before the available Lagrangian data can rival the wealth of information we have about the Eulerian description of turbulence. Nevertheless, great progress has already been made in the development of Lagrangian experimental techniques, and these measurements are now nearly as accurate and robust as their Eulerian counterparts.

Here, we have presented a general Lagrangian particle tracking algorithm as well as our experimental implementation of the technique in an actual high-Reynolds-number flow. In addition, to demonstrate the utility of such a technique, we have shown measurements of turbulent relative dispersion, the simplest multiparticle Lagrangian problem. As Lagrangian experiments develop, we expect progress both in theory and experiment in describing more-complex multiparticle statistics and thereby probing the structure of turbulence and the nature of intermittency in great detail.

10.3 Elastic Turbulence in Viscoelastic Flows

10.3.1 Basics

This section deals with a new class of a chaotic (random) flow coined elastic turbulence that was observed in viscoelastic flows of polymer solutions [10.76]. The section is organized as follow. In the first part hydrodynamic description of dilute polymer solution flows and nondimensional parameters that follows from these equations to characterize the flows are presented. Variation of one of this control parameters that is responsible for the elastic properties of a fluid can lead to a new elastic instability in various flows, distinguished by the presence of curvilinear trajectories. Theoretical criteria for the elastic instability in three different flows together with their experimental verification are discussed. To complete the basics, the rheometric properties of polymer solutions used and their relation to Boger fluids are given. The first observation of elastic turbulence, in the flow between two plates, is described. Then experimental measuring techniques used to characterize the flow are given, and a complete description of the results of measurements together with discussion of the results are presented. Finally, the role of elastic stress, a recent theory of elastic turbulence, and comparative studies of elastic versus hydrodynamic turbulence are discussed. The third subsection deals with the description of the elastic turbulence in a curvilinear channel, or Dean flow, where a particularly detailed experiment on mixing due to elastic turbulence was conducted [10.77]. A summary of the results is given at the end of the Section.

Hydrodynamic Description of Dilute Polymer Solution Flows

Solutions of flexible high-molecular-weight polymers are viscoelastic liquids, and they differ from Newtonian fluids in many aspects [10.78]. The most striking elastic property of these polymer solutions is probably the dependence of mechanical stresses in the flow on the history of the flow. So, the stresses do not immediately become zero when fluid motion stops, but rather decay with some characteristic relaxation time λ, which can be well above a second. When a polymer solution is sufficiently dilute, its stress tensor τ can be divided into two parts, $\tau = \tau_s + \tau_p$. Here the elastic stress tensor, τ_p, is due to the polymer molecules, which are stretched in the flow, and depends on history of the flow. The first term, τ_s, is defined by the viscosity of the Newtonian solvent, η_s, and the rate of strain in the flow, $\tau_s = \eta_s [\nabla V + (\nabla V)^T]$. So, the equation of motion for a dilute polymer solution looks like

$$\frac{\partial V}{\partial t} + (V \cdot \nabla)V = -\nabla p/\rho + (\eta_s/\rho)\nabla^2 V + \nabla \tau_p/\rho, \quad (10.92)$$

where p is pressure, and ρ is the density of the fluid. One can see that τ_p enters the equation of motion linearly. The equation has a nonlinear term, $(V\nabla)V$, which is inertial in nature. The Reynolds number defines the ratio of this nonlinear term to the viscous dissipative term, $\nu \nabla^2 V$. So the degree of nonlinearity of the equation of motion can still be defined by the Reynolds number $Re = VL\rho/\eta_s$ for a polymer solution. Therefore, turbulence in fluids at high Re is a paradigm for a strongly nonlinear phenomenon in spatially extended systems [10.79, 80].

The simplest model incorporating the elastic nature of the polymer stress tensor, τ_p, is a Maxwell-type constitutive equation [10.78] with a single relaxation time λ

$$\tau_p + \lambda \frac{D\tau_p}{Dt} = \eta_p [\nabla V + (\nabla V)^T]. \quad (10.93)$$

Here $\frac{D\tau_p}{Dt}$ is a material time derivative of the polymer stress, and $\eta_p = \eta - \eta_s$ is the polymer contribution to the total viscosity η. An appropriate expression for the time derivative $\frac{D\tau_p}{Dt}$ has to take into account that the stress is carried by fluid elements, which move, rotate, and deform in the flow. The translational motion implies an advection term $(V\nabla)\tau_p$ in an appropriate expression for $\frac{D\tau_p}{Dt}$, while the rotation and deformation of the fluid particles should lead to contributions like $(\nabla V)\tau_p$ or $\tau_p(\nabla V)$ [10.78]. Therefore, along with terms linear in τ_p and V, some nonlinear terms, in which τ_p is coupled to V, should appear in the constitutive relation. A simple model equation for $\frac{D\tau_p}{Dt}$, which is commonly used for description of dilute polymer solutions, is the upper convected time derivative,

$$\frac{D\tau_p}{Dt} = \frac{\partial \tau_p}{\partial t} + (V\nabla)\tau_p - (\nabla V)^T \tau_p - \tau_p(\nabla V). \quad (10.94)$$

The equations (10.93, 94) together with the expression for τ_s constitute the Oldroyd-B model of polymer solution rheology [10.78]. One can see that nonlinear terms in the constitutive equation (10.93, 94) are all of the order $\lambda(V/L)\tau_p$. The ratio of those nonlinear terms to the linear relaxation term τ_p is given by a dimensionless expression $\lambda(V/L)$, which is usually called the Weissenberg number Wi that represents the ratio of the relaxation time to the characteristic flow time. The relaxation term τ_p is somewhat analogous to the dissipation term in the Navier–Stokes equation.

One can expect the mechanical properties of the polymer solutions to become notably nonlinear at sufficiently large Weissenberg numbers. Indeed, quite a few effects originating from nonlinear polymer stresses have been known for a long time [10.78]. So, in a simple shear flow of a polymer solution there is a difference between normal stresses along the direction of the flow and along the direction of velocity gradient. At low shear rates this normal stress difference N_1 is proportional to the shear rate squared (see Sect. 9.1.1). When flow lines are curvilinear, this gives rise to a volume force acting on the liquid in the direction of the curvature, the *hoop stress*. Therefore, if a rotating rod is inserted into an open vessel with a highly elastic polymer solution, the liquid starts to climb up the rod instead of being pushed outwards [10.81]. This phenomenon is known as *rod climbing*, or the *Weissenberg effect*. Further, in a purely extensional flow, the resistance of a polymer solution depends on the rate of extension in a strongly nonlinear fashion. There is a sharp growth in the elastic stresses when the rate of extension exceeds $1/(2\lambda)$, that is at $Wi > 1/2$. As a result, the apparent viscosity of a dilute polymer solution can increase by up to three orders of magnitude [10.82]. Both the Weissenberg effect and the growth of extensional flow resistance have been most clearly observed in viscous polymer solutions and in flows with quite low Re, where nonlinear inertial effects are insignificant.

A natural question arising here is, whether there may exist some kind of turbulent flow produced by the nonlinear polymer stresses alone, in the absence of any significant inertial effects, at low Re. An important step in this direction was made about a decade and a half ago when purely elastic instabilities were experimentally identified in a rotational flow between two plates [10.83] and in a Couette–Taylor (CT) flow between two cylinders [10.84].

Elastic Instability and its Experimental Observations in Various Curvilinear Flows

Purely elastic instabilities occur at Wi numbers of order unity and vanishingly small Re. As a result of these instabilities, secondary vortex flows are developed [10.84], and the flow resistance increases [10.83]. The analysis showed that the nonlinear mechanical properties of the polymer solution can indeed lead to a flow instability, and a simple mechanism for this purely elastic instability was proposed.

During the past decade the purely elastic instabilities in viscoelastic fluids have been a subject of many theoretical and experimental studies, which are partially reviewed in [10.85, 86]. After the pioneering work by *Larson, Muller* and *Shaqfeh* [10.84, 87], purely elastic instabilities were also found in other shear flows with curvilinear streamlines. Those included the flow between a rotating cone and a plate (Sect. 9.1.1) and the Taylor–Dean flow [10.85, 86]. The original theoretical analysis of [10.87] was refined, and more-elaborate experiments were carried out. A few new mechanisms of flow instability driven by nonlinear elastic stresses were suggested for cone-and-plate and Taylor–Dean flows. The original mechanism proposed in [10.87] was verified experimentally in [10.88].

The most thorough and detail studies were conducted on the elastic instability in Couette–Taylor (CT) flow between two coaxial cylinders. In spite of the fact that instabilities in viscoelastic fluids were studied for decades, the purely elastic instability in CT flow was first investigated both experimentally and theoretically rather recently [10.84, 87]. The mechanism for the elastic instability in the CT system suggested in [10.87], is based on the Oldroyd-B model. The primary flow in the CT system with the inner cylinder rotating (Couette flow) is a pure shear flow in the r–θ plane that generates a normal stress difference $N_1 \equiv \tau_{\theta\theta} - \tau_{rr} = 2\eta_p\lambda(\dot{\gamma}_{r\theta})^2$, and a radial force N_1/r per unit volume. Here r, θ are cylindrical coordinates in the plane perpendicular to the cylinder axis, $\tau_{\theta\theta}$ and τ_{rr} are the components of the polymer stress tensor τ_p, and $\dot{\gamma}_{r\theta}$ is the only nonzero component of the rate-of-strain tensor $\dot{\gamma}$. A secondary flow in the CT flow includes regions of elongational flow in the r-direction with $\dot{\epsilon} \equiv \partial v_r/\partial r \neq 0$. The radial extensional flow stretches the polymer macromolecular coils in the r-direction, though it is a small perturbation on the top of the azimuthal stretching produced by the primary shear flow. However, being stretched in the radial direction, the macromolecular coils become more susceptible to the basic shear flow. The coupling between the radial and shear flow leads to a further increase in shear stresses. Thus, the result is further stretching of the polymer in the θ-direction that generates additional normal stress difference. So, the elastic instability mechanism is based on the coupling between the perturbative radial elongational flow and the strong azimuthal shear flow that results in a radial force. The latter reinforces the radial flow. As pointed out in [10.88], this transition can be only a finite-amplitude (first-order) transition. The corresponding criterion for the instability is

$$K \equiv \frac{\eta_p}{\eta_s} \frac{d}{R_1} Wi^2. \tag{10.95}$$

Here d and R_1 are the gap and the inner cylinder radius, respectively. The elastic instability occurs when the parameter K exceeds a certain threshold value [10.87]. This criterion is valid at sufficiently small values of η_p/η_s and in the small-gap-ratio limit, d/R_1. A more-general expression is given in [10.87]. When the gap ratio, d/R_1, and the polymer viscosity ratio, η_p/η_s, are fixed the elastic instability is just defined by the critical Weissenberg number Wi_c.

As was suggested in [10.89], where the CT flow was discussed, there is some analogy between flow transitions driven by elasticity and inertia. So, the inertially driven Taylor instability occurs at constant Taylor number [10.79, 80], $Ta = \frac{d}{R_1} Re^2$, while the elastic instability is controlled by the parameter K from (10.95) [10.87, 88]. The Weissenberg number is defined here as $Wi = \lambda\Omega R/d$, where Ω is the angular velocity of the rotating inner cylinder. (It was termed the Deborah number in some of the original texts [10.87–89].) The Weissenberg number appears to be analogous to the Reynolds number. The geometric parameter determining curvature, d/R_1, enters the expressions for both Ta and K. Scales of time and velocity for the purely elastic flow transition are given by λ and d/λ, respectively. As was shown in [10.89] these are analogous to t_{vd} and d/t_{vd}, which define scales of time and velocity for the inertially driven flow transitions. Here t_{vd} is the viscous diffusion time, defined as $t_{vd} = d^2/\nu$.

Nevertheless, along with all those analogies, there are still some important differences between flow transitions driven by inertia and elasticity. For example, it is an experimental fact that *any* laminar flow of a Newtonian fluid becomes unstable at sufficiently high Re, and all high-Reynolds-number flows are turbulent. This includes rectilinear shear flows such as Poiseuille flow in a circular pipe and plane Couette flow, which are supposed to be linearly stable at any Re. In contrast, purely elastic flow instabilities in shear flows have only been observed so far in systems with curvilinear streamlines. All these instabilities are supposed to be driven by the hoop stress, which originates from the normal stress differences.

This difference between the inertial and elastic instabilities originates, of course, from the distinct governing equations. There are, however, some purely practical factors that can explain rather well the lack of observations of purely elastic flow transitions in rectilinear shear flows. Inertial instabilities in rectilinear shear flows of Newtonian fluids occur at quite high Reynolds numbers. Those are typically about two orders of magnitude higher than Re, at which curvilinear shear flows with large gap ratios become unstable. A priori, one may suggest that rectilinear and curvilinear shear flows would have a similar relation between Wi at thresholds of the purely elastic flow instabilities as well. The problem is, however, that while it is rather easy to generate high-Re flows with low-viscosity Newtonian fluids, it is usually impossible to reach the corresponding high values of Wi in shear flows of elastic polymer solutions. That is, there always exist rather severe practical limitations restricting nonlinearity in elastic polymer stresses in shear flows. Their molecular mechanisms have recently been elucidated in a seminal paper by Chu's group [10.90].

Polymer molecules have finite extensibility, and their relaxation time decreases when they are stretched in a shear flow. This *thinning* of the relaxation time at high Wi is usually quite a strong and well-recognized effect. In fact, different variations of the basic Oldroyd-B model have been specially developed to take into account this shear thinning [10.78]. A significant decrease of λ with shear rate $\dot{\gamma}$ renders growth of the Weissenberg number $Wi = \lambda(\dot{\gamma})\dot{\gamma}$ much slower than linear in $\dot{\gamma}$. Substantial stretching of the polymer molecules in the primary shear flow also reduces their ability for further extension and the susceptibility of the elastic stresses to flow perturbations, which is necessary for the generation of flow instabilities and secondary vortex flows [10.87]. Finally, high shear rates cause mechanical degradation of the polymer molecules, which leads to a permanent reduction of elasticity during experimental runs and a decay of λ that can be very fast at high Wi. Because of all these reasons it was found to be very difficult or even impossible to observe elastic instabilities, when expected values of Wi at the instability threshold were high. This was the case in curvilinear flows with small gap ratios [10.91], and small viscosity ratios [10.88].

The first experimental observations of the elastic instability in CT flow found Wi_c indeed rather close to the theoretical prediction [10.86, 87]. However, the observed critical modes were different from the predicted axisymmetric one. This discrepancy was resolved later both theoretically [10.92] and then experimentally [10.88]. In fact it was shown that the non-axisymmetric mode becomes unstable first, and the bifurcation is backward, as was verified experimentally. In spite of this, the instability mechanism remains as first suggested and verified in [10.87, 88].

A purely elastic instability apparently also occurs by a similar mechanism in other rotational shear flows, namely concentric cone-and-plate and plate-and-plate geometries (Sect. 9.1.1). There the secondary flow is also driven by a *hoop stress*, and by the coupling between the primary shear flow and secondary elongation flow in the axial direction. Historically a transition in fluid rheology was first observed experimentally in these geometries [10.83], and Magda and Larson suggested that its origin lies in an elastic flow instability. A complete set of numerical and experimental studies of the elastic instability in these flows was carried on in [10.91, 93, 94]. Due to the more-complicated flow structure than in CT flow, no analytic expression for the stability criterion is available in either of these cases. However, some common features for all rotational shear flows can be pointed out. For all systems the threshold of the elastic instability strongly depends on the aspect ratio, which is defined differently in each case. The instability criterion also depends on the viscosity ratio, though the functional form of this dependence was found only in CT flow. In all systems the most unstable mode is non-axisymmetric and oscillatory, and the transition is discontinuous, or first order (inverse bifurcation).

Another curvilinear flow that is relevant to further results is a flow through a curved channel driven by a streamwise pressure gradient, which was first studied for a Newtonian fluid by *Dean* [10.95]. He found that the instability is defined similarly to CT flow by $Re^2 d/R_1$ when it exceeds some critical value. The linear stability analysis for the Oldroyd-B fluid and experimental observation of the instability in the elastic Dean flow were performed by *Joo* and *Shaqfeh* [10.96–98]. It was predicted by numerical calculations and verified experimentally that the stability of the pressure-driven flow through a curved channel is defined by a criterion (in a small-aspect-ratio limit) that is rather similar to that found in CT flow (10.95) with a rather close critical value [10.97, 98]. However, in contrast to the rotational shear flows the most unstable mode is stationary and axisymmetric that shows up in a secondary toroidal vortex flow centered near the outer curved surface. The dependence of the stability threshold on the aspect ratio indicates that, like in rotational shear flows, the curvature is of crucial importance for the instability to occur. The elastic instability in the Dean flow is also driven by the *hoop stress* but the mechanism is rather different [10.97] and is related to the existence a normal stress gradient across the streamlines. The necessary condition for the instability is $\partial \tau_{\theta\theta}/\partial r > 0$, i.e., a larger hoop stress is located outward from the center of curvature. Then a small radial inward flow is reinforced by the increased hoop stress gradient. Thus, all three flow configurations with curvilinear trajectories considered in this chapter exhibit the elastic instability. Their instability criterion as well as the most unstable mode are rather well investigated and known.

Experimental Considerations to Observe Elastic Turbulence

In order to maximize the nonlinear elastic effects and to have a better opportunity to observe the elastic turbulence one has to choose experimental conditions quite carefully [10.76]. First, it is important to obtain an elastic instability at a possibly low critical Weissenberg number Wi_c. For that purpose the gap ratio and the viscosity ratio had to be possibly large. Therefore, one should use a polymer solution with a rather large η_p/η of about 1/4. [Further increase of the polymer concentration (or of η_p/η) was not very efficient, and would also complicate the solution rheology, including large shear thinning of the solution viscosity.] One should use a large aspect ratio, either in the CT flow or the swirling flow between two plates. In the swirling flow the experiments were carried out with a large gap ratio, between about 0.2 and 0.53 with the possibility to vary it.

An appropriate polymer sample for the solution that would not suffer major mechanical degradation under experimental flow conditions had to be chosen. So, the fluids most often used in experiments on elastic instabilities were so-called Boger fluids [10.99]. These fluids are highly elastic. They are obtained by solving high-molecular-weight polymers at low concentration (dilute solution) in a viscous Newtonian solvent. The Boger fluids are almost universally used as model viscoelastic fluids. Their relaxation times can be quite large, reaching seconds or even minutes, while their rheological properties are semiquantitatively described by the simple Oldroyd-B model.

The limit for extensibility of the polymer molecules had to be high compared with their typical conformations at the instability threshold. We used polyacrylamide (PAA) with a large average molecular weight of $M_w = 18 \times 10^6$ g/mol and a broad molecular-weight distribution (and low concentration of polyacrylic acid monomers). This commercial polymer sample proved to be remarkably stable with respect to mechanical degradation, which allowed us to reach high values of Wi and explore strongly nonlinear flow regimes. The high molecular weight of the polymers resulted in a large characteristic relaxation time λ even with solvents of moderate viscosity, and at a small characteristic stress, $\tau_0 = \eta_s/\lambda$. One can learn from (10.94, 95) that τ_0 sets a scale for the polymer stress at which its nonlinearity becomes significant. Therefore, the value of τ_0 together with η_p/η and d/R determine the polymer stress τ_p in the primary shear flow at the instability threshold. It is rather natural to suggest that, when polymer molecules transduce less stress, they are also less subjected to mechanical degradation. Further, molecular interpretation of the Weissenberg number in a shear flow relates it to the degree of deformation of polymer molecules from their relaxed random coil conformations. So $Wi = 1$ can be regarded as the characteristic value at which extension of polymer molecules becomes considerably larger than the size of relaxed coils. High molecular weight and flexibility of a polymer suggests a large number of Kuhn segments [10.78, 100] in the polymer chain, and a high ratio between its contour length (size, when fully extended) and the size (radius of gyration R_g) of a relaxed coil. (For a polymer molecule in a good solvent, with $R_g \sim M_w^{3/5}$, this ratio should increase as $M_w^{2/5}$ [10.100]. We would like to point out here that addition of NaCl to the solution reduces R_g.) Therefore, by using a higher-M_w polymer, we increased the domain of extensibility starting from the typical conformation at Wi_c, which opened the way for richer flow dynamics above the elastic instability threshold.

Rheometric Properties of Polymer Solutions

Several series of experiments on elastic turbulence in different experimental setups conducted during the last several years were based on the same stock (master) polymer solution of PAA ($M_w = 18 \times 10^6$ g/mol supplied by Polysciences). First we dissolved 0.9 g of PAA powder and 3 g of NaCl in 275 ml of deionized water by gentle shaking. (Addition of NaCl was necessary to improve the solubility of PAA.) Next the solution was mixed for 3 h in a commercial mixer with a propeller at a moderate speed. The idea behind this procedure was to cause mechanical degradation of PAA molecules with the highest weights, and to *cut* the high-M_w tail of the broad molecular-weight distribution of the PAA sample. In a solution with a broad distribution of polymer molecular weights the heaviest molecules, which are most vulnerable to mechanical degradation, may also make a major contribution to the solution elasticity. A possible negative effect of this is significant degradation of elasticity during experimental runs, and inconsistency in the experimental results. We found empirically that the procedure of predegradation in the mixer leads to a substantial reduction of degradation during the experiments and to substantial improvement of the consistency. Finally, 9 g of isopropanol was added to the solution (to preserve it from aging) and water was added up to 300 g. The final concentrations of PAA, NaCl and isopropanol in the stock solution were 3000 ppm, 1%, and 3%, respectively.

As a viscous Newtonian solvent for PAA, we used a solution of $\approx 65\%$ sugar (saccharose) in all experiments, if it is not mentioned specifically otherwise. Viscosity and relaxation time were measured with the same AR-1000 rheometer in a temperature-controlled narrow-gap Mooney–Ewart (Couette) geometry. The solvent viscosity was $\eta_s = 0.324$ Pa s at $12\,^\circ$C and $\eta_s = 0.114$ Pa s at $22\,^\circ$C. The PAA concentration was 80 ppm.

Solution 1, used in the early experiments at low temperature ($12\,^\circ$C), contained 1% NaCl to fix the ionic contents. Viscosity η was slowly decreasing with $\dot\gamma$, so that its shear thinning was about 7% per decade of $\dot\gamma$. At a shear rate of $\dot\gamma = 1\,\text{s}^{-1}$, corresponding to the onset of a purely elastic instability in the standard configuration (see below), η was 0.424 Pa s, and the viscosity ratio was $\eta_p/\eta = 0.235$. The polymer relaxation time, λ, was measured in oscillatory tests (Sect. 9.1) with a shear rate amplitude of $1\,\text{s}^{-1}$ in a range of angular frequencies, ω. Then $\lambda(\omega)$ at $\omega \to 0$, estimated as 3.4 s, was chosen as a representative relaxation time, λ. The diffusion coefficient for the saccharose molecules was about $D = 8.5e - 7\,\text{cm}^2/\text{s}$ [10.101].

Solution 2, used in the later experiments at higher temperature ($22\,^\circ$C), had a viscosity of $\eta = 0.138$ Pa s at a shear rate of $2\,\text{s}^{-1}$, and a viscosity ratio of $\eta_p/\eta = 0.174$. The polymer relaxation time obtained by the rheometric measurements via small oscillations using both a AR-1000 rheometer from TA Instruments and a Vilastic viscoelasticity analyzer (Vilastic-3, Vilastic Scientific, Austin, TX) as a function of shear rate showed the relaxation time scaling with *shear thinning* as $\lambda \sim \dot\gamma^{-\alpha}$, with $\alpha \simeq 0.3$ similar to that found earlier for solutions with lower-molecular-weight PAA samples [10.88], and the relaxation time in the limit of $\dot\gamma \to 0$ as $\lambda(0) = 4.7$ s.

10.3.2 Elastic Turbulence in Swirling Flow Between Two Plates

Some flow patterns observed above the purely elastic instability threshold in the curvilinear flows had a rather disordered appearance and exhibited chaotic velocity spectra. This was first mentioned in [10.88, 102] and then specifically in [10.89] for CT flow. So, it was reasonable to suggest that under some conditions a truly *turbulent* flow might be excited by elastic stresses at vanishingly small Re. This idea was explicitly stated in [10.89], where an analogy between elastic and inertial flow transitions was discussed. In fact, irregular flow patterns and growth of flow resistance in elastic polymer solutions at low Re were observed even before the purely elastic instabilities were identified [10.103]. These flow phenomena were even sometimes coined *elastic turbulence*. In all these cases, however, the term *turbulence* was used in a rather loose fashion, without a proper definition. More importantly, no quantitative data on either the flow velocity field or the spatial and temporal velocity spectra in these irregular flows were ever presented.

Although the notion of turbulence is widely used in scientific and technical literature, there is no unique, commonly accepted definition. Therefore, turbulent flow is usually identified by its main features. Turbulence implies fluid motion on a broad range of temporal and spatial scales, so that many degrees of freedom are excited in the system. There are no characteristic scales of time and space in the flow, except for those restricting the excited temporal and spatial domains from above and below. Turbulent flow is also usually accompanied by a significant increase in momentum and mass transfer. That is, the flow resistance and rate of mixing in a turbulent flow become much higher than they would be in an imaginary laminar flow with the same Re.

In recent publications [10.76, 77, 104] it was shown how these features of turbulence appeared in a highly elastic polymer solution at low Reynolds numbers in two curvilinear flows. The first quantitative experiments were done in a swirling flow between two plates with a wide gap, and the phenomenon was coined elastic turbulence. In this section we discuss the results of the experiments on the elastically driven turbulent flow in two experimental setups where the most complete studies were carried out, namely a swirling flow between two plates and an open flow in a curved channel [10.76, 77, 104–109]

Experimental Setup and Procedure

The experimental apparatus is schematically shown in Fig. 10.19.

Polymer solution was held in a stationary cylindrical cup with a flat bottom (lower plate). A coaxial rotating upper plate was just touching the surface of the fluid. The cup was mounted on top of a commercial rheometer, AR-1000 from TA instruments. The upper plate was attached to the shaft of the rheometer, which allowed precise control (within 0.5%) of its angular velocity Ω and measurements of the torque T (Ω-mode), or opposite control (within 0.5%) of T and measurements of Ω (T-mode). The average shear stress at the upper plate, τ_w, was calculated using the equation $T = \tau_w \int r \, dS$, which gave $\tau_w \equiv 3T/(2\pi R^3)$, where the integration is over the upper plate surface.

The moment of inertia of the shaft of the rheometer was $I_s \approx 14\,\mu\text{Nms}^2$ and that of the upper plate I_d was about $61\,\mu\text{Nms}^2$ for setup 1, and about $84\,\mu\text{Nms}^2$ for setup 2. The accuracy of the angular speed measurements in constant-torque mode is about 2% and the accuracy of the torque measurements in the constant-speed mode is about 1%. One has to point out here that smallness of the fluctuation rate of the angular velocity is not a sufficient criterion to have constant-speed forcing. Corresponding to the Ω-mode, $(I_s + I_d) \partial \Omega / \partial t$ should also be much smaller than typical values of the torque T.

The sidewalls of the cup were machined from a single piece of perspex, which was optically clear. The cup was circular on the inside and square on the outside in a horizontal cross section. This allowed measurements of the flow velocity in the horizontal plane by a laser Doppler velocimeter (LDV) (Sect. 5.3.1) with two crossing frequency-shifted beams. By appropriate positioning and orientation of the beam-crossing region, azimuthal (longitudinal) and radial (spanwise) velocity components, V_θ and V_r, respectively, could be measured at different r and z. Here (r, θ, z) are cylindrical coordinates. The bottom of the cup was machined of stainless steel and the temperature was stabilized at 12 °C by circulation of water below the bottom plate.

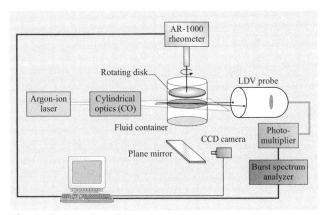

Fig. 10.19 Overview of the swirling flow apparatus

A slightly modified version of the setup was designed to photograph the flow from below, to use particle image velocimetry (PIV)(Sect. 5.3.1) and particle tracking velocimetry (PTV) techniques, and to observe mixing in the flow. A special cup of the same shape but with a transparent bottom (lower plate, made of perspex) was attached to the rheometer concentrically with the shaft but above the rheometer base, and a mirror tilted by 45° was placed under the cup, as schematically shown in Fig. 10.19. The mirror was used both to illuminate the fluid by diffuse light and to relay images of the flow to a charge-coupled device (CCD) camera. The camera was equipped with a regular video lens and mounted horizontally near the rheometer (Fig. 10.19). The source of the diffuse light was an illuminated white screen around the camera. The images were digitized by an 8-bit 512×512 frame grabber in the case of flow visualization and mixing, and at 12 bits with 640×512 pixels at 25 frame/s and 1280×1024 at 12.5 frame/s using a PixelFly digital camera (PCO, Germany) in the case of PIV and PTV. In order to provide thermal stabilization, the whole rheometer was placed in the thermally isolated box with through flow of temperature-controlled air.

In the basic setup used in the early experiments, the radii of the upper plate and the cup were $R_1 = 38$ mm and $R_2 = 43.6$ mm, respectively, and the distance between the plates was $d = 10$ mm. The configuration was similar to devices with rotating flow between two plates used in experiments on purely elastic instability [10.83, 86, 91, 94]. Its gap ratio, d/R, was significantly higher, though. In order to study the dependence of the flow conditions on the size of the system, two smaller setups, a half-size and a quarter-size system, with all the dimensions reduced by factors of two and four, respectively, were used. Each time, when dimensions of an experimental setup are nonstandard, this is stated explicitly in the text.

In the later experiments two setups were used. In setup 1 a cylindrical container of radius $R_c = 2.2$ cm and a disk with radius $R_d = 2$ cm were used, while setup 2 had $R_c = 4.9$ cm and a disk with $R_d = 4.8$ cm. The distance between the disk and the bottom plate was $d = 1$ cm in both setups. The system was illuminated laterally by a thin (30 μm in the center of the setup and about 130 μm at the edges of the setup) laser sheet through the transparent walls of the fluid container at the mid distance between the plates. The laser sheet was generated by passing a laser beam delivered by a 300 mW argon-ion laser through a block of two crossed cylindrical lenses (CO) mounted in a telescopic arrangement. The main flow investigation tool was the digital PIV technique. As flow tracers 10 μm fluorescent beads were used. We acquired 2000 pairs of flow images every 120 ms using a digital camera. The time delay between consecutive images was 40 ms. Time series of velocity fields were obtained by using a multipass PIV algorithm [10.108]. The accuracy of the method was carefully checked by running test experiments with the solvent, in the same range of mean particle displacements and under similar illumination conditions. Although the instrumental error increases more or less linearly with the mean particle displacement, it never exceeded 5% of the mean displacement. The spatial resolution was about 120 velocity vectors over 10 cm. By postprocessing the velocity fields, we obtained the profiles of the velocity components, fields of fluctuations of each velocity component, spatial spectra of the velocity fluctuations, velocity gradients and their fluctuations, structure functions of gradients, and Eulerian velocity correlation functions. The space–time measurements together with simultaneous global measurements of the flow resistance provided a rather complete description of the different flow regimes as a function of Wi [10.109]. A Lagrangian frame flow investigation was also conducted [10.107].

In a swirling flow between two plates the shear rate is quite inhomogeneous over the fluid bulk, even when the flow is laminar. So, the choice of a representative shear rate becomes somewhat arbitrary. We decided to consider the simple expression $\Omega R/d$ as a characteristic shear rate, and to define the Weissenberg number as $Wi = \lambda \Omega R/d$. The Reynolds number was defined as $Re = \Omega R d \rho / \eta$.

In order to evaluate the growth of flow resistance due to elastic instability and irregular secondary flow in the system, the average shear stress near the upper plate τ_w^{lam} in an imaginary laminar shear flow at the same Ω had to be estimated. The stress τ_w^{lam} depends on an average shear rate $\dot{\gamma}_{av}$ at the upper plate, and on the viscosity of the polymer solution η at this shear rate; an appropriate expression for it is $\tau_w^{\text{lam}} = \eta(\dot{\gamma}_{av})\dot{\gamma}_{av}$. The average shear rate $\dot{\gamma}_{av}$ was estimated from measurements of the ratio τ_w/η_0 in a laminar flow of a Newtonian fluid with a large viscosity η_0 at low Re. The shear rate in the laminar flow, calculated as $\dot{\gamma}_{av} = \tau_w/\eta_0$, was proportional to Ω, being $\dot{\gamma}_{av} = 1.12 \Omega R/d$ in the standard configuration, and gave a properly weighted average over the surface of the upper plate.

Observation of Elastic Turbulence

Flow Resistance. The first indication of a strongly nonlinear state in the swirling flow was significant growth of the flow resistance above the elastic instability threshold, which can be characterized by the ratio $\tau_w/\tau_w^{\text{lam}}$, a measure of strength of turbulence and of the turbulent resistance. The dependence of $\tau_w/\tau_w^{\text{lam}}$ on the shear rate, $\dot{\gamma} = \Omega R/d$, is shown in Fig. 10.20.

A sharp transition in the flow of the polymer solution (curve 1, black line) occurs at $\dot{\gamma} \simeq 1\,\text{s}^{-1}$ (corresponding to $Wi = 3.5$) and is manifested by a significant increase in τ_w compared to the laminar flow. The Reynolds number at the transition point is about 0.3, so that inertial effects are quite negligible there. The transition has pro-

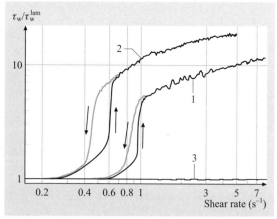

Fig. 10.20 The ratio of the average stress at the upper plate τ_w measured in the flow, to the stress τ_w^{lam} in an imaginary laminar shear flow with the same boundary conditions, as a function of the shear rate $\dot{\gamma}$. Curves 1 and 2 are for the polymer solution flow with gaps of $d = 10$ mm and 20 mm, respectively. The shear rate was gradually varied in time. *Thin black lines* represent increasing $\dot{\gamma}$; *thick gray lines* represent decreasing $\dot{\gamma}$. Curve 3 is for the pure solvent

nounced hysteresis (thick gray line, corresponding to a slow reduction of $\dot{\gamma}$), which is rather typical of a purely elastic flow instability [10.88]. The higher the shear rate, the higher $\tau_w/\tau_w^{\text{lam}}$. The latter reaches a value about 12 times larger than it would be in a laminar flow at the highest $\dot{\gamma}$. In the same range of shear rates flow of the pure solvent (curve 3) is completely laminar and the ratio $\tau_w/\tau_w^{\text{lam}}$ is unity within the resolution of the rheometer (about 1%). For a gap ratio of 0.526 the the ratio $\tau_w/\tau_w^{\text{lam}}$ reaches a value of 19 (curve 2 in Fig. 10.20). For Newtonian fluids in the same flow geometry such a growth of the flow resistance is found at Re values of about 2×10^4. For flow in a circular pipe this value of $\tau_w/\tau_w^{\text{lam}}$ is reached at $Re \simeq 10^5$, which is usually considered as a region of rather developed turbulence [10.79, 80].

In the setup with a radius of $R_1 = 4.8$ cm and an aspect ratio of 0.21 accurate power (P) measurements in the Ω-mode reveal, besides a drastic increase in the flow resistance, scalings of the average injected power $P \sim Wi^{\beta_1}$ and the rms of power fluctuations $P_{\text{rms}} \sim Wi^{\beta_2}$ with $\beta_1 \simeq 3.34 \pm 0.05$ and $\beta_2 \simeq 3.2 \pm 0.3$ (Fig. 10.21).

Here the injected power was calculated as $P = T\Omega$, and the shear thinning of the polymer relaxation time, $\lambda(\dot{\gamma})$, was taken into account. For low values of Wi in the laminar regime the injected power grows as $P_{\text{lam}} \sim Wi^{2.85}$, or quadratically with Ω, which is expected for the laminar flow. Due to the smallness of $Re < 16$, a nonlinear inertial contribution is negligible. So additional growth of the average injected power in the elastic turbulence compared to its laminar values occurs solely due to elastic stresses τ_p^{turb}. Therefore, according to the fit in Fig. 10.21 $P/P_{\text{lam}} \propto \tau_p^{\text{turb}} \propto Wi^{0.49}$.

Temporal and Spatial Velocity Fluctuations Spectra.
Temporal spectra of the azimuthal and radial components of the velocity in the horizontal plane were measured with LDV at various locations at the cell and

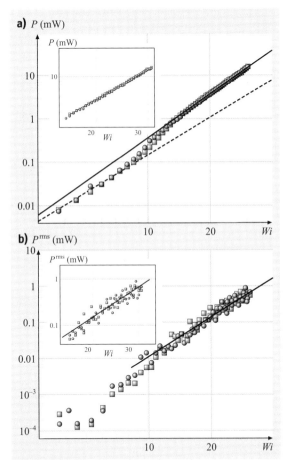

Fig. 10.21 (a) Dependence of the mean power P on the control parameter, Wi. *Squares*: increasing Wi, *circles*: decreasing Wi. The *full line* is a guide to the eye $Wi^{3.34}$. The *dotted line* is $Wi^{2.85}$. The *inset* shows a power-law fit, $Wi^{3.34\pm0.05}$. **(b)** Dependence of the rms power fluctuations P^{rms} on Wi. *Squares*: increasing Wi, *circles*: decreasing Wi. The *full lines* are power laws $Wi^{3.21}$. The *inset* shows the power-law fits $Wi^{3.21\pm0.3}$. Data were collected in setup 2 in the Ω-forcing mode

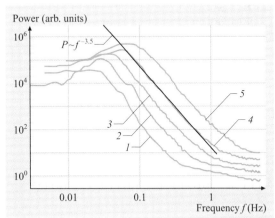

Fig. 10.22 Power spectra of the azimuthal velocity fluctuations in the standard setup at different shear rates $\dot{\gamma}$. The fluid velocity was measured by LDV in the center of the flow. The *curves* 1–5 correspond to $\dot{\gamma} = 1.25, 1.85, 2.7, 4$, and 5.9 s^{-1}, respectively (all above the transition point $\dot{\gamma} \simeq 1$). The power P of fluctuations is fitted by a power law $P \sim f^{-3.5}$ for $\dot{\gamma} = 4 \text{ s}^{-1}$ over about a decade in frequencies, f

had similar appearance [10.76, 104]. The azimuthal component of the velocity spectra measured in the center of the setup, where its average value was zero, at different constant shear rates $\dot{\gamma}$ (all above the elastic instability threshold) are shown in Fig. 10.22.

The power of the fluctuations and their characteristic frequencies increase with $\dot{\gamma}$, but the spectra remain very similar in appearance. In particular, the spectra do not have distinct peaks, and do span a broad range of about an order of magnitude in frequencies, where the power of the fluctuations decays according to a power law $P \sim f^{-\alpha}$. (Flattening of the curves at high f is due to instrumental noise.) The exponent α is about 3.5, which is much larger than the Kolmogorov exponent of 5/3 found for velocity spectra of high-Re inertial turbulence [10.79, 80].

To obtain quantitative information about the spatial structure of the fluctuating velocity field from the LDV measurements, we explored velocity spectra in various off-center points with nonzero average azimuthal velocity $\bar{V}_\theta$. The spectra of fluctuations of the radial component of velocity V_r at $\dot{\gamma} = 4\,\mathrm{s}^{-1}$ ($Wi = 13.5$) at four different radii are shown in Fig. 10.23; they look similar to the spectra in Fig. 10.22 with a similar power-law decay. They were all measured at $z = d/2$, where z is the distance from the upper plate. The root-mean-square values of the fluctuations at all four points were rather close, varying between 0.88 and 0.99 mm/s.

One can learn from Fig. 10.23 that, as the measurement point moves away from the center, the characteristic frequencies of the fluctuations become higher. The most reasonable explanation for this is growth of the average azimuthal velocity $\bar{V}_\theta$, which was 3.81 mm/s and 6.99 mm/s, at $r = 2d$ and $r = 3d$, respectively. So, the fluctuations of the velocity in time at these two points are mainly due to fluctuations in space, which are advected by the large mean flow velocity $\bar{V}_\theta$. Applying the Taylor hypothesis, we can view the spectra in time as spectra in space, with the relation between the frequency and the wavenumber given by $k = 2\pi f/\bar{V}_\theta$. Then the power-law decay regions in curves 3 and 4 imply that the fluid motion is excited in the whole corresponding ranges of k. The ranges of the spatial scales where the motion is excited span about an order of magnitude for both curves. The exponents α in the power laws $P \sim f^{-\alpha}$ (and, so, $P \sim k^{-\alpha}$) are again quite large: about 3.6 for $r = 2d$ and about 3.3 for $r = 3d$ [10.104].

In a separate experiment the spatial velocity power spectra were measured by using PIV measurements. These were obtained by averaging 2000 instantaneous spatial spectra. Although the finite spatial resolution of PIV, which limits the accessible range of wavenumbers, leads to an artificial cutoff at $k \approx 3000\,\mathrm{m}^{-1}$, a power-law decay with $k^{-3.5}$ is clearly observed

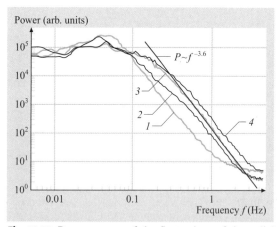

Fig. 10.23 Power spectra of the fluctuations of the radial velocity V_r at $\dot{\gamma} = 4\,\mathrm{s}^{-1}$ measured at $z/d = 0.5$ at different radii. Curves 1–4 correspond to $r = 0, r = d/2, r = 2d$ and $r = 3d$, respectively. The average flow velocities (V_θ, V_r) in mm/s are (0,0), (0.13, 0.19), (3.81, 1.17), (6.99, 0.89) for *curves* 1–4, respectively

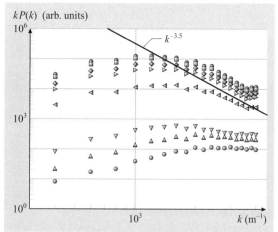

Fig. 10.24 Power spectra of fluctuations of azimuthal velocity component at different Wi. *Circles*: $Wi = 4.41$, *up triangles*: $Wi = 5.72$, *down triangles*: $Wi = 8.32$, *left triangles*: $Wi = 11.1$, *right triangles*: $Wi = 12.7$, *diamonds*: $Wi = 16$, *half filled squares*: $Wi = 18$, *empty circles*: $Wi = 19$. *Solid line* presents the power law decay $k^{-3.5}$. Data were collected in setup 1

(Fig. 10.24 [10.109]). This is rather consistent with the scaling obtained above for the velocity power spectra in the frequency domain. These large values of α imply that the power of the fluctuations decays very quickly as the size of the eddies decreases. The main contribution to fluctuations of both velocity and velocity gradients [the power of the latter should scale as $k^{-(\alpha-2)}$] should therefore be due to the largest eddies.

This also indicates that the Taylor hypothesis is applicable to this chaotic flow [10.108]. The agreement between the velocity spectra measured at a single point in the frequency domain [10.76, 104] and the directly measured spectra in the k-domain [10.109] deserves brief discussion. Following *Lumley* [10.110], the relation between the spatial spectrum, $P(k)$ and the frequency spectrum, $P(f)$ can be written as $P(k) = VP(f) - \frac{I_t^2}{2}\frac{d^2(k^2 P(k))}{dk^2} + O(I_t^4)$, where $I_t = V_\theta^{\text{rms}}/V_\theta$, $V_\theta^{\text{rms}} \equiv \overline{V_{\theta\,t}^2}^{1/2}$ and V_θ are the rms values of the fluctuations of the azimuthal velocity and the average azimuthal velocity, respectively. If $P(k) \propto k^{\alpha_2}$ and $P(f) \propto f^{\alpha_1}$, the equation above leads to: $\alpha_2 - \alpha_1 \propto \log[1 + \frac{I_t^2}{2}\alpha_2(\alpha_2+2)]/\log(k)$. If one inserts into the last equation $\alpha_2 \approx -3.6$, the difference between the exponents is (for $k \approx 1000\,\text{m}^{-1}$) as small as $\alpha_2 - \alpha_1 \approx 0.2$. Thus, the experimental resolution does not allow us to observe the difference in the scaling exponents for the spatial and temporal spectra [10.109].

Mixing in the Flow. Mixing in the flow was observed using a droplet of black ink added to the working fluid before rotation of the upper plate was started (Fig. 10.25). Using a micropipette the droplet was carefully placed near the center (at $r = 0$) at about half of the fluid depth. The ink was dissolved in a concentrated sugar syrup, to match the density of the droplet with the density of the working fluid.

Consecutive stages of mixing in the polymer solution are shown in Fig. 10.25.

Rotation of the upper plate was started abruptly at $t = 0$ at a shear rate of $\dot{\gamma} = 5.6\,\text{s}^{-1}$. It took about one minute for the irregular flow to develop after the rotation of the upper plate was started. (Development of the irregular flow was judged by growth of the flow resistance, which was saturated after about one minute [10.111].) So, no significant changes in the ink distribution occurred during the first 15 s (Fig. 10.25). After 30 s the ink spread over the surface of the lower plate by the large toroidal vortex discussed above. In the snapshots taken at later times (60, 90, and 120 s in Fig. 10.25) one can see many fine-scale structures. Those may be due either to excitation of the fluid motion on small spatial scales or significant stretching of fluid elements along their Lagrangian trajectories by randomly fluctuating large-scale eddies. The contrast of the patterns gradually decreases with time, which indicates progressing mixing. The pattern in the last snapshot, taken 8 min after the flow had been started, appears completely homogeneous. From the appearance of the mixing patterns in Fig. 10.25, the characteristic time of mixing can be estimated as 120 s, corresponding to about 30 full turns of the upper plate. The time required for mixing by molecular diffusion

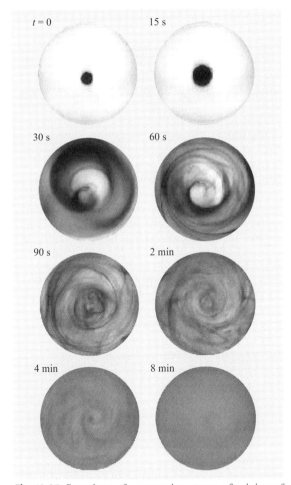

Fig. 10.25 Snapshots of consecutive stages of mixing of a droplet of ink in the polymer solution in the half-size setup, viewed from below. The area of the photographs corresponds to the area of the white upper plate. Rotation of the upper plate at $\Omega = 1.47\,\text{s}^{-1}$ ($\dot{\gamma} = 5.6\,\text{s}^{-1}$) was started suddenly at $t = 0$

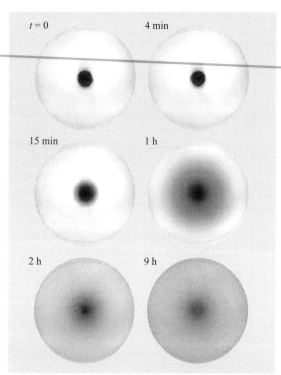

Fig. 10.26 Snapshots of consecutive stages of mixing of a droplet of ink in the pure solvent in the half-size setup, viewed from below. Rotation of the upper plate at $\Omega = 1.47\,\mathrm{s}^{-1}$ was started suddenly at $t=0$

without macroscopic flow can be estimated as dR/D, which gives a value of about 10^6 s, i.e., at least four orders of magnitude larger than the mixing time observed in the flow [10.104].

Mixing in the flow of the pure solvent at the same shear rate is shown for comparison in Fig. 10.26. One can see that the distribution of the ink remained inhomogeneous even after 9 h, although the ink became somewhat spread out the time. The Reynolds number was about 0.5, and there were some nonvanishing inertial effects in the flow [10.104].

Summarizing the experimental results, we conclude that the flow of the elastic polymer solution at sufficiently high Wi indeed has all the main features of developed turbulence stated above. The fluid motion is excited in broad ranges of frequencies and wavenumbers, both spanning about an order of magnitude. The flow is accompanied by a dramatic increase in the rate of transfer of momentum and mass. In terms of the strength of the turbulent resistance, the mixing rate, and the span of scales in space and time at which the fluid motion is excited, the observed flow can be compared to turbulence of a Newtonian fluid in a pipe at Re of about 10^5. This apparently turbulent fluid motion in the swirling flow between two plates arises at very low Re, where inertial effects are negligible, solely because of the nonlinear mechanical properties of the elastic polymer solution. We therefore call the phenomenon *elastic turbulence*. Distinctions between the elastic turbulence and the usual inertial turbulence observed in Newtonian fluids at high Re are discussed below.

Further Properties of Elastic Turbulence

Power Fluctuations and the Statistics of Injected Power. The injected power fluctuations were measured for different Wi in the elastic turbulence regime in two modes: Ω-forcing (Fig. 10.27) and T-forcing (Fig. 10.28). For each value of Wi the statistics of the power fluctuations were collected with 180 000 data points evenly sampled in time ($t_s \approx 0.038$ s). The time series of the injected power at different Wi for both modes are presented in both figures. In the elastic turbulence regime the probability distribution functions (PDFs) of the power fluctuations deviate strongly from the Gaussian distribution for both forcing modes. The PDFs in the Ω-forcing mode have a left-side skewness, while in the T-forcing mode they show a right-side skewness. Since the injected power fluctuations reflect fluctuations of the turbulent elastic stress averaged over the upper plate, one can conclude that the statistics of the elastic stress deviate strongly from Gaussian statistics and are therefore intermittent. The Wi dependencies of both P/P_{rms} (Fig. 10.21) the skewness of the PDF in elastic turbulence suggest an analogy with similar behavior in hydrodynamic turbulence, though the reasons for the effects are entirely different [10.112, 113].

Flow Structure and Velocity Profiles. To characterize the structure of the velocity field and velocity profiles flow visualization (PIV and LDV) techniques were used in various experiments and setups [10.104, 108, 109]. An impression of the flow structure in elastic turbulence can be obtained from just a few snapshots of the flow, visualized by seeding with light-reflecting flakes (1% of Kalliroscope liquid). The upper plate was black, so the bright regions indicate where the flakes are oriented parallel to the upper plate. The patterns of the polymer solution flow above the transition at $Wi = 6.5$ (Fig. 10.29a,b), and at $Wi = 13$ (Fig. 10.29c,e), appear quite irregular and exhibit structures of different sizes. One can see, however, that the structures tend to have a spiral-like form, which is probably imposed by the average azimuthal flow and circular symmetry of the setup. Further, there is a dark spot in the middle, which appears in most of the snapshots. This corresponds to the center of a large, persistent toroidal vortex, which has dimensions of the whole setup (see also below). The direction of vortical motion was inwards near the upper plate, downwards near the center and outwards near the lower plate [10.104]. The average flow velocity along the radial direction was measured by LDV at a few points, and the results agreed quite well with the presence of the large, persistent toroidal vortex. The flow of the pure solvent at the same shear rate was completely laminar, as can be seen from the snapshot in Fig. 10.29f, which appears quite uniform [10.104].

Quantitative characterization of the global flow structure and its dynamics was obtained via PIV. Instantaneous images of fields of the horizontal components of the velocity as well as averaged images over 2000 fields taken equidistant between the plates in setup 1 for several values of Wi are presented in Fig. 10.30, 31. The left column in Fig. 10.30 shows the instantaneous vector fields at three increasing values of Wi, while the right column shows the averaged vector field at the same Wi values. One can easily identify the core of the toroidal vortex, which appears in all the images above the thresh-

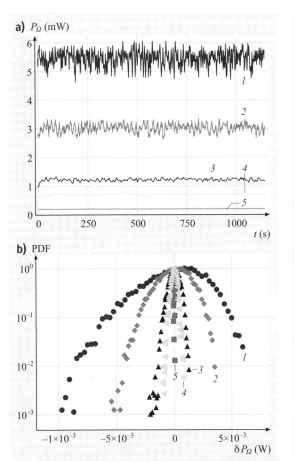

Fig. 10.27 (a) Time series (partially shown) of the injected power in the constant Ω-forcing regime for different Wi. **(b)** PDF of the power fluctuations in constant Ω forcing regime for different Wi. The curves are: (5) $Wi = 5$, (4) $Wi = 19$, (3) $Wi = 24$, (2) $Wi = 31.5$, (1) $Wi = 40$. Data were collected in setup 2

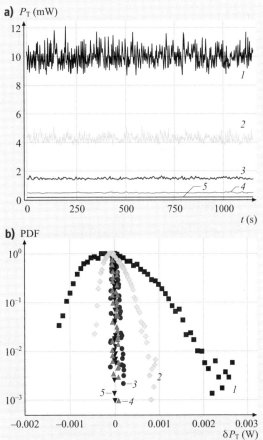

Fig. 10.28 (a) Time series (partially shown) of the injected power in constant T forcing regime for different Wi. **(b)** The PDF of the power fluctuations in the constant T-forcing regime for different Wi. The curves are: (5) $Wi = 5$, (4) $Wi = 17$, (3) $Wi = 25$, (2) $Wi = 34$, (1) $Wi = 46$. Data were collected in setup 2

old for the instability, and a spiral vortex that in addition occurs in the elastic turbulence regime. These flow structures and their reorganization can also be observed in separate presentations of the averaged azimuthal, radial components, and rms of fluctuations of the azimuthal velocity component of the velocity field at the same Wi (Fig. 10.31, 32). The profile of the average azimuthal velocity V_θ below the elastic instability exhibits a linear increase along a radius with a slope, Ω^{-1}, that corresponds to rigid-body rotation (inset in Fig. 10.33a), while there is no motion in the radial direction, $V_r \approx 0$ (Fig. 10.33b). Above the instability one can clearly see the creation of the core of the toroidal vortex at the cell center, and restructuring in the radial motion [10.109].

The large, toroidal vortex driven by the hoop stress is actually quite well known in swirling flows of elastic fluids [10.78, 114], and inhomogeneity of the shear-rate profile in the primary laminar flow has long been recognized as the origin of this phenomenon. In our system this vortex first arises as a stationary structure at low shear rates. As can be learned from Fig. 10.20 (curve 1 at $\dot\gamma < 0.75\,\text{s}^{-1}$ and curve 2 at $\dot\gamma < 0.4\,\text{s}^{-1}$), the toroidal vortex leads to some increase in the flow resistance even before the elastic instability, so we can conclude that the transition to elastic turbulence in the swirling flow between two plates is mediated by this vortex. The toroidal vortex provides a smooth, large-scale velocity field (see the right-hand panels in Fig. 10.30), which is randomly fluctuating in time (left-hand panels in Fig. 10.30), and in which the fluid and the embedded stress tensor are chaotically advected. This type of advection can create variations of the stress over a range of smaller scales, which may cause small-scale fluid motion [10.115, 116]. This would be analogous to generation of small-scale concentration variations in chaotic mixing by large fluctuating vortices.

The large-scale toroidal vortex is also responsible for the ring-shaped topology of fluctuations of the azimuthal velocity $V_\theta^{\text{rms}} \equiv \overline{V_\theta^2}^{1/2}$ (Fig. 10.32). The velocity fluctuations in the laminar regime are only due to

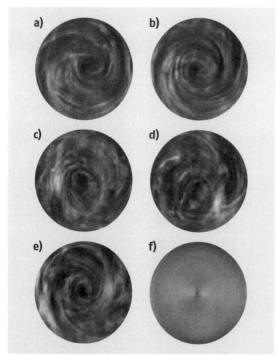

Fig. 10.29a–f Representative snapshots of the flow taken from below. The field of view corresponds to the upper plate area. The flow was visualized by seeding the fluid with light-reflecting flakes. (**a**), (**b**) the polymer solution at $Wi = 6.5$, $Re = 0.35$; (**c**), (**d**), (**e**) the polymer solution at $Wi = 13$, $Re = 0.7$; (**f**) the pure solvent at $Re = 1$

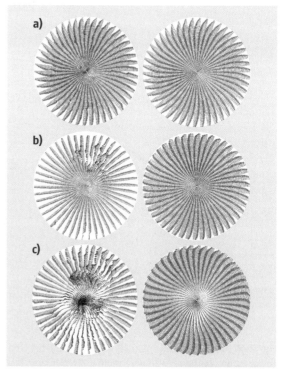

Fig. 10.30a–c Instantaneous (*left column*) and time-averaged (*right column*) velocity fields for different values Wi: (**a**) $Wi = 2.48$ (**b**) $Wi = 9.88$ (**c**) $Wi = 18.96$. Data were collected in setup 1 at the mid distance between the plates

instrumental noise, which does not exceed 5% of the average value (panel a in Fig. 10.32). The average radial velocity changes sign at about half of the radius (see Fig. 10.33b). At higher value of Wi the toroidal vortex forces a transition to elastic turbulence, which is characterized by the second flow structure reorganization to the spiral vortex (Fig. 10.30, 31) and the scaling region in P, P_{rms} (Fig. 10.21). At the same time the circular symmetry of V_θ^{rms} is broken and becomes dipolar (Fig. 10.32). The radial profiles of V_θ and V_r also change drastically. These structural transitions in the flow are also reflected in the average radial gradients of the azimuthal velocity field and its average vorticity field, which are presented in Fig. 10.34 for several values of Wi.

To learn more about the velocity field generated by elastic turbulence, we measured the average velocity and rms of the velocity fluctuations at different points. Profiles in the z-direction of the average azimuthal velocity $\overline{V}_\theta$ and of the rms fluctuations V_θ^{rms} at different flow conditions are shown in Fig. 10.35 [10.104].

The measurements were done at $r = 2d$, which is rather far from the edge of the upper plate ($R - r = 1.8d$). The profile of $\overline{V}_\theta$ in a low-Re flow of the pure solvent is an almost straight line (Fig. 10.35, curve 3). The elasticity-driven turbulent flow significantly changes the distribution of $\overline{V}_\theta$. It produces a high-shear-rate layer near the upper plate (Fig. 10.35 curve 1), and a low-shear-rate region near the middle of the gap (at $z/d = 0.5$). Such a distribution of $\overline{V}_\theta$ is reminiscent of the average velocity profiles in usual high-Re turbulence. The perturbation of the $\overline{V}_\theta$ profile becomes stronger when $\dot{\gamma}$ is increased (Fig. 10.35, curve 2). In particular, the slope of the $\overline{V}_\theta$ curve at small z/d becomes larger, which obviously corresponds to growth of τ_w/τ_w^{lam} with $\dot{\gamma}$ (Fig. 10.20).

Fluctuations of the azimuthal velocity (curve 4 in Fig. 10.35) are small near the upper plate, reach a maximum at $z/d \simeq 0.25$, and start to decrease at larger z. Again, such a distribution of V_θ^{rms} along the z-direction is reminiscent of the velocity fluctuations in turbulent flows of Newtonian fluids [10.79, 80]. The rms fluctuations reach a value of about $0.5d/\lambda$, so that the rate of deformation produced by the fluctuating velocity field is on the order of $1/\lambda$. That implies an essentially nonlinear relation between the rate of deformation and the fluctuating elastic stress. The maximal rms value of the velocity fluctuations at $\dot{\gamma} = 4\,\mathrm{s}^{-1}$ was about $1.55\,\mathrm{mm/s}$, which was about 7.5% of the upper-plate velocity (V_{max}) at $r = 2d$ and about 25% of the local value of $\overline{V}_\theta$ [10.104].

Another way to characterize a turbulent flow is to display the radial profiles (azimuthally averaged) of the turbulent intensity, defined as $I_t = V_\theta^{rms}/V_\theta$ at several values of Wi (Fig. 10.36) [10.109]. The velocity fluctuations in the laminar regime occur only due to instrumental errors and do not exceed 5% of the mean values. Above the elastic transition, I_t increases sharply but remains rather uniform at 20–30% in a peripheral region for radius ratios r/R_C of 0.2–1 (see Fig. 10.36). In the case of elastic turbulence, I_t further increases, and its dependence on Wi, presented in the inset in Fig. 10.36, exhibits a power-law scaling, $I_t \sim Wi^{0.49}$ [10.109].

The PIV measurements of the time-dependent velocity fields allowed us to calculate the average velocity gradients and vertical vorticity, and their rms fluctuations without involving the Taylor hypothesis, which is questionable in a smooth, random flow. Typical radial distribution of the velocity gradients and vorticity averaged in an azimuthal direction spatially and over 2000 images temporally is rather uniform in the bulk for all Wi but increases sharply near the wall (Fig. 10.37), while the rms of the velocity gradients and vorticity gradually increase along a radius (Fig. 10.38). The dependence of the average vertical vorticity on Wi is displayed in

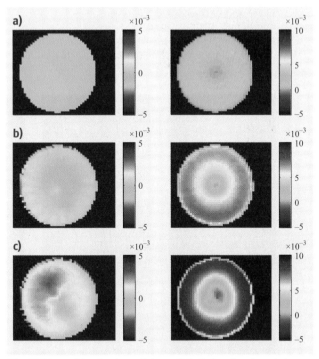

Fig. 10.31a–c Distribution of the radial (*left column*) and the azimuthal velocity components (*right column*) for different values of Wi: (**a**) $Wi = 2.48$, (**b**) $Wi = 9.88$, (**c**) $Wi = 18.96$. Data were collected in setup 1 at the mid distance between the plates

Fig. 10.37a, exhibiting a tendency to saturation at higher Wi for vorticity in the bulk and gradual growth near the wall. On the other hand, the rms of the vorticity fluctuations definitely level off at $Wi > 15$. A similar behavior is displayed by all components of the velocity gradient, which is consistent with theoretical predictions, though the saturated value $\omega_z^{rms}\lambda \approx 2.5$ is rather high [10.115, 116] (see further discussion of this issue below). The relation between the rms fluctuations of the velocity gradients and the Eulerian correlation time will be discussed in the next section.

Velocity and Velocity Gradient Fields Statistics and Wi Dependence. Characteristic probability distribution functions (PDFs) of the azimuthal component of the flow velocity in the regime of elastic turbulence are shown in Fig. 10.39. A similar PDF was found for the radial velocity component [10.104]. These distributions were obtained from LDV measurements taken at $r = 2d$, $z = 0.25d$ at a shear rate of $4\,\mathrm{s}^{-1}$. One can see that the distribution has only minor skewness and is very nearly Gaussian.

The PDF for the gradients of velocity (rates of deformation) in the azimuthal direction (and similarly for transverse), obtained from the same LDV time series, is shown in Fig. 10.40. The rates of deformation are multiplied by the relaxation time λ to make them dimensionless. The velocity gradients were estimated using the Taylor hypothesis, with smoothing over a distance of about 1.45 mm. The difference ΔV between consecutive velocity readings with even time intervals of $\Delta t = 0.22\,\mathrm{s}$ was divided by Δt and by the average velocity $V = (V_\theta^2 + V_r^2)^{1/2} = 6.6\,\mathrm{mm/s}$. One can see that the distribution of $\partial V_\theta/r\partial\theta$ (as well as $\partial V_r/r\partial\theta$) cannot be fitted by a Gaussian curve [10.104]. In contrast to the velocity distribution in Fig. 10.39, the PDF in Fig. 10.40 has a well-pronounced exponential tail, which implies significant intermittency of the velocity gradients. The situation of nearly Gaussian statistics of velocities and essentially non-Gaussian, strongly intermittent distributions of velocity gradients is actually quite typical for high-Re inertial turbulence [10.117]. Hence, the elastic turbulence also resembles high-Re inertial turbulent flows in this respect.

The statistical properties of elastic turbulence can be characterized in many ways. Besides the PDF of the local velocity and velocity gradients and the power spectrum of velocity fluctuations in the frequency and wavenumber domains, one can calculate various correlation and structure functions of the velocity and velocity gradients. Their scalings can give information on the degree of deviation from a Gaussian random field, a standard approach in hydrodynamics to quantify intermittency and to compare it with the corresponding scaling in known cases. First, let us find typical spatial and temporal correlation scales in the elastic turbulence.

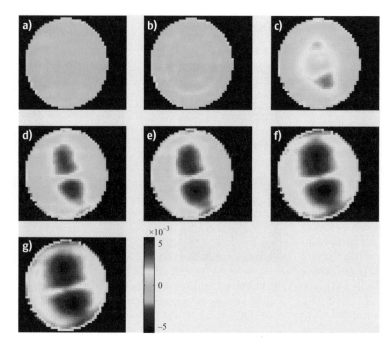

Fig. 10.32a–g Fields of fluctuations of the azimuthal velocity at different Wi: **(a)** $Wi = 8.32$, **(b)** $Wi = 9.88$, **(c)** $Wi = 11.1$, **(d)** $Wi = 12.72$, **(e)** $Wi = 13.83$, **(f)** $Wi = 17$, **(g)** $Wi = 19$. Data were collected in setup 1 at the mid distance between the plates

A typical scale on which the elastic stress is correlated can be estimated as $L = 2\pi \int P(k)\mathrm{d}k / \int k P(k)\mathrm{d}k$ and in the elastic turbulence regime one gets $L \simeq 5.9$ cm, i.e., on the order of the cell size. The Eulerian correlation time is defined as $\tau_c = \int t C(t)\mathrm{d}t / \int C(t)\mathrm{d}t$, where $C(t) = \overline{V(T)V(T+t)}/v^{\mathrm{rms}}v^{\mathrm{rms}}$ is the Eulerian correlation function. Figure 10.41 presents τ_c as a function of Wi at some radial positions in the cell. The correlation time drops significantly in the transition region and then saturates in the elastic turbulence region similar to the rms of the vorticity (velocity gradients) fluctuations. It is worth pointing out that the inverse of the saturated value of the correlation time is of the order of the saturated value of ω_z^{rms} and λ^{-1}, as expected by theory [10.115, 116].

Using PIV data one can also calculate the structure functions of the velocity gradients defined as $S_p(r) = \langle |\frac{\partial V_\theta(r_0+r)}{\partial r} - \frac{\partial V_\theta(r_0)}{\partial r}|^p \rangle_{r_0}$. Similar to inertial turbulence, let us look for a scaling in the range of scales corresponding to the power-law decay of the velocity spectra (Fig. 10.24) in the form $S_p(r) = r^{\zeta_p}$. The dependence of the scaling exponent ζ_p on the order of the structure functions p is presented in Fig. 10.42. The deviation from a linear dependence that is characteristic to the Gaussian field is clearly observed and is rather close to the analogous dependence for a passive scalar [10.118]. The relation between the passive-scalar problem and the statistics of the velocity gradient are discussed in the next subsection.

As noted in Sect. 10.3.1 on flow resistance, the comparison of the velocity power spectra in the frequency and wavenumber domains leads to the conclusion that the Taylor hypothesis is applicable in spatially smooth random-in-time flows, a representative of which is elastic turbulence. On the other hand, from Fig. 10.35 one learns that, in the cell interior at $r/R_c < 0.2$, the rms velocity fluctuation becomes much larger that the mean velocity, which casts doubt on the applicability of the Taylor hypothesis. The validity of the Taylor hypothesis in this flow was investigated experimentally in detail [10.108]. By using cross-correlation technique as well as the structure function approach, it was shown that the breakdown of the Taylor hypothesis occurs near the cell center due to strong velocity fluctuations, while the flow smoothness and the lack of scale separations in the elastic turbulence limit the quantitative applicability of the Taylor hypothesis close to the boundaries. However, the latter deficiency can be corrected by a proper choice of the advection velocity [10.108].

Role of the Elastic Stress and Theory of the Elastic Turbulence

Along with the apparent similarity in phenomenology between the elastic and inertial turbulence, there are also a few important distinctions, which are reflected in the dependence of the onset of elastic instability and the scaling of the main characteristics of elastic turbulence with system size and fluid viscosity. These results were found to be in sharp contradiction with the corresponding dependence in Newtonian fluid mechanics but quite in line with the concept of viscoelastic similarity suggested for purely elastic instability [10.88]. As was shown in [10.89], an increase in the viscosity

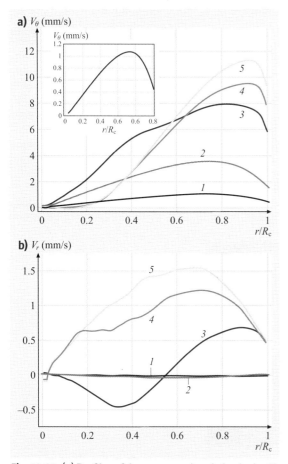

Fig. 10.33 (a) Profiles of the average azimuthal velocity V_θ for several values of Wi. The *inset* shows a typical laminar profile of the azimuthal velocity component. (b) Profiles of the average radial velocity V_r for several values of Wi. The curves are: (1) $Wi = 2.48$, (2) $Wi = 4.41$, (3) $Wi = 11.1$, (4) $Wi = 15$, (5) $Wi = 18$. Data were collected in setup 1 at the mid distance between the plates

of the solvent should lead to a transition to elastic turbulence at lower velocity, a prediction that was verified experimentally [10.104]. When the size of the setup is proportionally reduced, the velocity required for excitation of elastic turbulence should increase. This property was examined in setups twice and four times smaller than the main one [10.104]. Transition to the elastic turbulence is expected to occur at the same characteristic stress, and its dependence on Wi is supposed to be the same for all system sizes, as was also verified experimentally [10.104]. The equivalence of the statistical properties of the velocity field and the mixing patterns in elastic turbulence has also been demonstrated [10.104].

An obvious reason for the differences in the scalings in inertial and elastic turbulence is the different physical mechanisms that underlie these two kinds of turbulent motion. As is well known, the high flow resistance in high-Re inertial turbulence is due to large Reynolds stresses. The Reynolds stress tensor is defined as the average value $\rho \langle V_i V_j \rangle$, where V_i and V_j are different components of the flow velocity. In the case of elastic turbulence the Reynolds stresses are quite small, since Re is low. So, the high flow resistance in elastic turbulence is due to a large elastic stress, τ_p [10.111]. Therefore, one can suggest that in the case of elastic turbulence it would be more relevant to study the field of stresses and of rates of deformations rather than the velocity field. It would certainly be quite instructive to explore the spatial structure and temporal distribution of the elastic stress, but there is currently no technique for local measurements of τ_p in a turbulent flow. On the other hand, large-scale properties of the τ_p field can be inferred from measurements of the torque or injected power.

It is now widely accepted that the statistical properties of a random flow in the elastic turbulent regime and significant increase in the contribution of the polymer stresses to the flow resistance are associated with significant polymer stretching in a random flow. To support this theoretical vision, experimental studies of polymer stretching in a 3-D random flow between two plates were conducted in macro- and microscale setups [10.111, 119]. The first experiment was performed in a swirling flow setup with a high aspect ratio of $R/d = 1$, where $R = 30$ mm, and with a very viscous solvent, $\eta_s = 1.36$ Pa s to suppress inertia [10.111]. By experimental analysis and estimates of the contributions of the Reynolds, viscous, and elastic stresses to the shear stress on the upper plate, we found that the Reynolds stress constitutes less than 0.1% of the total stresses, and that the viscous stress is rather constant. So, as a result of a secondary random 3-D flow superimposed on a primary applied shear flow between two plates, the polymer contribution to the shear stress increases by

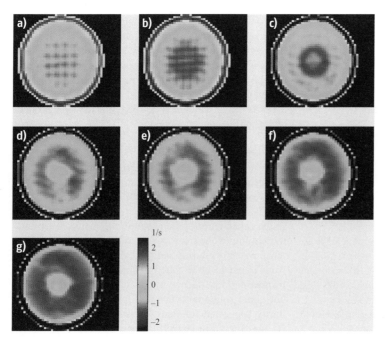

Fig. 10.34a–g The average vertical component of the vorticity ω_z at different Wi: (**a**) $Wi = 8.32$, (**b**) $Wi = 9.88$, (**c**) $Wi = 11.1$, (**d**) $Wi = 12.72$, (**e**) $Wi = 13.83$, (**f**) $Wi = 17$, (**g**) $Wi = 19$. Data were collected in setup 1 at the mid distance between the plates. The *squared pattern* visible in panels (**a,b**) is a result of the combination of the peak-locking effect and numerical differentiation and should be disregarded

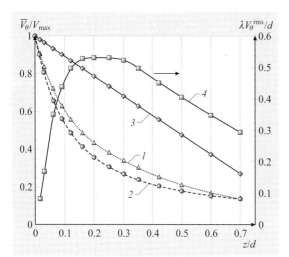
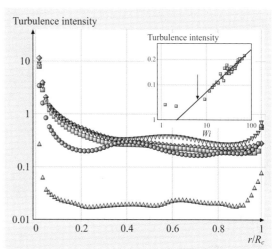

Fig. 10.35 Average azimuthal velocity $\overline{V}_\theta$ (y-axis on the *left, curves* 1–3), and rms fluctuations of the azimuthal velocity V_θ^{rms} (y-axis on the *right, curve* 4), as functions of the distance z from the upper plate. The measurements were done at $r = 2d$. The average velocities are divided by the upper plate velocity at $r = 2d$. Curve 1: polymer solution at $\dot\gamma = 2.7\,\text{s}^{-1}$; *curves* 2,4 polymer solution at $\dot\gamma = 4\,\text{s}^{-1}$ (Fig. 10.21, 22, 23, 26); *curve* 3 pure solvent at $\dot\gamma = 4\,\text{s}^{-1}$, $Re \simeq 1.2$. The rms velocity fluctuations in the polymer solution (*curve* 4) is multiplied by λ/d to make it dimensionless

Fig. 10.36 Turbulence intensity versus the reduced radial coordinate at different Wi. *Rhombus*: $Wi = 31.56$, *squares*: $Wi = 27.73$, *down triangles*: $Wi = 23.67$, *circles* $Wi = 7.54$, *up triangles*: $Wi = 2.82$. The *inset* shows the dependence of the turbulence intensity on Wi at $r/R_c = 0.85$ obtained from LDV measurements. The *full line* is a power-law fit $I_t \propto Wi^{0.49 \pm 0.06}$. The *arrow* indicates the onset of the elastic instability. The data were collected in setup 2

as much as 170 times. If one assumes linear elasticity of the flow-stretched polymer molecules (PAA), then the elastic stress causes a 13-fold polymer extension [10.111].

Thus, the first logical step in theoretical exploration of the whole problem of turbulent hydrodynamics of a dilute polymer solution was to describe a single polymer stretching in a random flow. More than 30 years ago *Lumley* [10.110] first suggested a semiquantitative theory for polymer stretching in a random flow. Since then, numerous publications on theoretical as well as numerical calculations of mean polymer extension in a stochastic flow as a function of Wi with the aim of qualitatively verifying the predicted coil-stretch transition have been published [10.120–122]. However, the statistics of a random flow was not related to the statistics of the polymer extension in these works. Recently Lumley's theory was revised, and a quantitative theory of the coil-stretch transition of a polymer molecule in a 3-D random flow was developed [10.123, 124]. The dynamics of a polymer molecule are sensitive to fluid motion at the dissipa-

tion scale, where the velocity field is spatially smooth and random in time [10.110]. On this scale polymer stretching is determined only by the velocity gradient tensor, $\nabla_i V_j$, which varies randomly in time and space: $\partial_t R_i = R_j \nabla_j V_i - R_i/\lambda(R) + \zeta_i$. Here R_i is the end-to-end vector and R is the end-to-end distance for the stretched polymer molecule, respectively, $\lambda(R)$ is the polymer relaxation time, which is R-dependent, and ζ_i is the thermal noise. At $R \ll R_{\text{max}}$ the linear regime of polymer relaxation is characterized by the polymer relaxation time $\lambda_{\text{rel}} = \lambda(0)$, where R_{max} is the maximum end-to-end stretched polymer length. At $R \gg R_g$ one can use, e.g., the FENE (finitely extensible, nonlinear elastic) model with $\lambda(R) = \lambda(0)(1 - R^2/R_{\text{max}}^2)$. In a 3-D random flow $\nabla_i V_j$ always has an eigenvalue with a positive real part, so that there exists a direction with a pure elongation flow [10.125]. The direction and rate of the elongation flow change randomly as a fluid element rotates and moves along the Lagrangian trajectory. If $\nabla_i V_j$ remains correlated within finite time intervals, the overall statistically averaged stretching of the fluid element will increase exponentially fast in time. The rate of stretching is defined by the maximal Lyapunov exponent α of the turbulent flow, which is the average

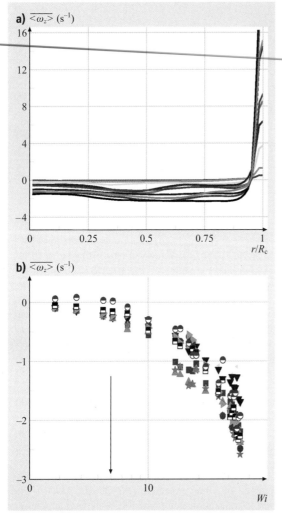

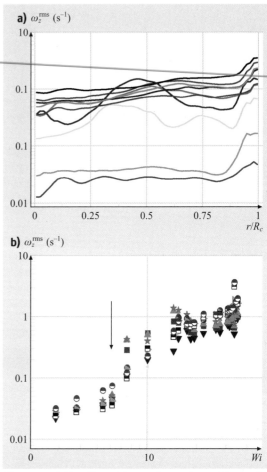

Fig. 10.37 (a) Profiles of the vorticity $\langle\omega_z\rangle$ averaged in an azimuthal direction spatially and over 2000 images temporally at different Wi: black-$Wi = 34.5$, red-$Wi = 32.3$, green-$Wi = 27.73$, blue-$Wi = 26.13$, magenta-$Wi = 19.29$, dark yellow-$Wi = 17.43$, dark blue-$Wi = 14.48$, yellow-$Wi = 10$, orange-$Wi = 5.4$, pink-$Wi = 2.82$ (b) The dependence of the average vorticity on Wi at different radial positions. Squares-$r/R_c = 0.33$, circles-$r/R_c = 0.66$, up triangles-$r/R_c = 0.4$, down triangles-$r/R_c = 0.2$, left triangles-$r/R_c = 0.1$, right triangles-$r/R_c = 0.7$, stars-$r/R_c = 0.5$, half-filled squares-$r/R_c = 0.8$, half-filled circles-$r/R_c = 0.9$. The arrow indicates the onset of the elastic instability. Data were collected in setup 2

Fig. 10.38 (a) Profiles of the rms fluctuations of the vorticity ω_z^{rms} at different Wi: black-$Wi = 34.5$, red-$Wi = 32.3$, green-$Wi = 27.73$, blue-$Wi = 26.13$, magenta-$Wi = 19.29$, dark yellow-$Wi = 17.43$, dark blue-$Wi = 14.48$, yellow-$Wi = 10$, orange-$Wi = 5.4$, pink-$Wi = 2.82$. (b) Dependence of the rms of fluctuations of the vorticity on Wi at different radial positions. Squares-$r/R_c=0.33$, circles-$r/R_c = 0.66$, up triangles-$r/R_c = 0.4$, down triangles-$r/R_c = 0.2$, left triangles-$r/R_c = 0.1$, right triangles-$r/R_c = 0.7$, stars-$r/R_c = 0.5$, half-filled squares-$r/R_c = 0.8$, half-filled circles-$r/R_c = 0.9$. The arrow indicates the onset of the elastic instability. Data were collected in setup 2

logarithmic rate of separation of two initially close trajectories, where the value of α is usually on the order

of the rms of the fluctuations of the velocity gradient,
$$\left(\frac{\partial V_i}{\partial r_j}\right)^{\mathrm{rms}} \equiv \overline{\left(\frac{\partial V_i}{\partial r_j}\right)^2}^{1/2}.$$
Stretching of a polymer molecule follows deformation of the surrounding fluid element. So the statistics

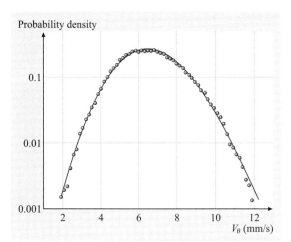

Fig. 10.39 Probability distribution functions (PDF) of the azimuthal flow velocity V_θ measured at $r = 2d, z = d/4$, and $\dot\gamma = 4\,\text{s}^{-1}$ (*circles*). The *solid line* represents the Gaussian fit with some skew

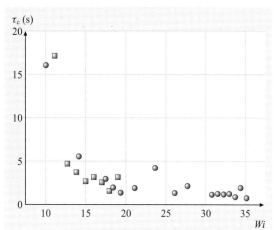

Fig. 10.41 Eulerian correlation times of the azimuthal velocity component as function of Wi. *Squares*: setup 1, *circles*: setup 2

of polymer stretching in a random, smooth flow depends critically on the value of $\lambda\alpha$ [or equivalently $\lambda\left(\partial V_i/\partial r_j\right)^{\text{rms}}$], which plays the role of a local Weissenberg number for a random flow, Wi'. According to the theory [10.110, 123], the polymer molecules should become vastly stretched if the condition $\lambda\alpha > 1$ is fulfilled, and the coil–stretch transition is defined by the relation $\lambda_{\text{cr}}\alpha = 1$, similar to that in an elongation flow with the strain rate α [10.126]. A somewhat surprising conclusion of the theory is that a generic random flow is on average an extensional flow at every point, with the rate of extension $\dot\varepsilon = \alpha$ and unlimited Henky strain. Dramatic extension of the flexible polymer molecules in the turbulent flow environment, inferred here from the bulk measurements of the flow resistance, has recently been confirmed by direct visualization of individual polymer molecules in a random flow [10.119]. According to the recent theory [10.123], the tail of the

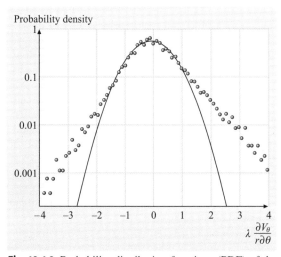

Fig. 10.40 Probability distribution functions (PDF) of the longitudinal velocity gradient, $\frac{\partial V_\theta}{r\partial\theta}$ measured at $r = 2d, z = d/4$ and $\dot\gamma = 4\,\text{s}^{-1}$ (*circles*). The velocity gradients are made dimensionless by multiplication by the relaxation time λ. The *solid line* represents a Gaussian fit

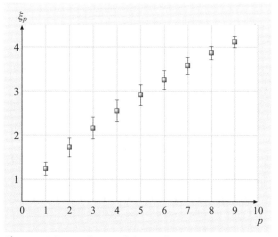

Fig. 10.42 Dependence of the scaling exponent ξ_p of the order of the structure function of the velocity gradients p

PDF of the molecular extensions is described by the power law $P(R_i) \sim R_i^{-\beta-1}$, where $\beta \sim (\lambda^{-1} - \alpha)$ in the vicinity of the transition. Positive β corresponds to the majority of the polymer molecules being unstretched. On the contrary, at $\beta < 0$ a significant fraction of the molecules is strongly stretched, and their finite size is defined by the feedback reaction of the polymers on the flow [10.123] and by the nonlinearity of molecular elasticity [10.124]. Thus, the condition $\beta = 0$ can be interpreted as the criterion for the coil-stretch transition in turbulent flows [10.123]. This criterion was quantitatively verified in the experiment by statistical analysis of both a velocity field in the Lagrangian frame (by statistical analysis of the Lyapunov exponents to define the average one) in the elastic turbulence and of a single polymer stretching in the same flow [10.119].

In the framework of the molecular theory of polymer dynamics the tensor τ_p is found to be proportional to the polymer concentration n and to the average polymer conformation tensor, $\tau_{p,ij} \sim n\lambda^{-1}\langle R_i R_j \rangle$ (if one neglects the thermal noise and uses the Hookean approximation). The growth of the elastic stresses is also evidence of significant extension of the polymer molecules in the flow. So, we can suggest that the elastic stress tensor τ_p should be the object of primary importance and interest in the elasticity-driven turbulent flow, and that it may be appropriate to view elastic turbulence as turbulence of the τ_p field.

The next crucial step towards a theoretical description of elastic turbulence was to relate the elastic stress field to the linearly decaying passive field problem [10.115, 116, 124, 127]. As was shown in [10.115], the elastic stress tensor can be rewritten as uniaxial, i.e., $\tau_{ik} = B_i B_k$, if the contribution to the elastic stress due to thermal fluctuations and polymer internal nonlinearity can be neglected. Then (10.93, 94) can be rewritten for the vector B_i in a form that is similar to the equation for the magnetic field in magnetohydrodynamics [10.116]. Then in the case of the elastic turbulence one obtains:

$$\partial_t \boldsymbol{B} + (\boldsymbol{V}\nabla)\boldsymbol{B} = (\boldsymbol{B}\nabla)\boldsymbol{V} - \boldsymbol{B}/\lambda . \tag{10.96}$$

This equation, complemented by the equation of motion rewritten as

$$\nabla P = \rho(\boldsymbol{B}\nabla)\boldsymbol{B} + \eta\nabla^2 \boldsymbol{V} , \tag{10.97}$$

and by the boundary conditions, leads to instability at $Wi > 1$, as already explained in detail. The instability eventually results in a chaotic, statistically steady dynamics. As was explained above for a single polymer stretching, a steady state occurs due to a back reaction of the stretched polymers (or the elastic stress in (10.97)) on the velocity field. Thus, stationarity of the statistics implies that the velocity gradients on a small scale should be limited by some mechanism. The saturation mechanism suggested in [10.115] assumes that both dissipative terms in (10.96, 97), namely viscous $\nu\nabla^2 V$ and relaxation B/λ, are of the same order. Then the velocity gradients become smaller when the scale decreases, and become of the same order as λ^{-1} on a small scale fixed by the stationarity condition. This means that in the chaotic flow the velocity fluctuations dominate on the scale of the system size. So, the elastic stress can be estimated as

$$\tau_p = B^2 \sim \nu \nabla_i V_j \sim \frac{\eta}{\lambda} . \tag{10.98}$$

The large-scale velocity fluctuations produce smaller-scale fluctuations in $\boldsymbol{B}$ via (10.96) that in turn induce small-scale fluctuations of the velocity via (10.97). Velocity gradients become smaller as the scale decreases, since the large-scale velocity gradient is of the order of λ^{-1}. The small-scale fields $\boldsymbol{v}$ and $\boldsymbol{B}'$ evolve passively in the large-scale fields $\boldsymbol{V}$ and $\boldsymbol{B}$, where $\boldsymbol{v}, \boldsymbol{B}' \ll \boldsymbol{V}, \boldsymbol{B}$. Thus, this problem is reduced to a linearly decaying passive field problem considered in [10.128]. In the case of elastic turbulence the equation of motion for v is dissipative, and one can neglect the gradient terms, since $\nabla \boldsymbol{B}' \sim \nabla \boldsymbol{B}$ and $\nabla \boldsymbol{v} \ll \nabla \boldsymbol{V}$.

The saturation of the rms fluctuations of the vorticity at high values of Wi observed in the data in Fig. 10.38 gives us the possibility to test which of the two theoretical mechanisms discussed above, either the feedback reaction of a molecule on the flow or the nonlinearity of molecular elasticity, primary leads to the saturation. As one finds from Fig. 10.38, $(\partial V_i/\partial r_j)^{\text{rms}} \simeq 1\,\text{s}^{-1}$. This gives for the local Weissenberg number, $Wi' = \lambda(\partial V_i/\partial r_j)^{\text{rms}} \simeq 2.5$, which exceeds the unity value predicted for the linear molecule elasticity and the feedback mechanism discussed [10.123]. So, this value indicates that the nonlinearity of molecular elasticity is mostly responsible for the saturation [10.124].

The analysis of the equations for the small-scale fluctuations of both fields leads to the power-like decaying spectrum for the elastic stress that in a spherical presentation looks like $\langle B'_i B'_j \rangle \sim F(k) \sim (k)^{-\alpha}$, where $\alpha > 3$. It is clear that the field B'_i (and therefore the field $B'_i B'_j = \nabla_i v_j$) is the passive field in the problem. The mechanism leading to the power-law spectrum for small-scale fluctuations of $\boldsymbol{B}'$ in a chaotic flow is rather general and directly related to the Batchelor scaling of a passive scalar revealed long ago [10.125]. A linear relation between small-scale fluctuations of the fields $\boldsymbol{B}'$ and

v allows the establishment of the power-law spectrum of the velocity, which in a spherical representation looks like $E(k) \sim k^{-\alpha}$, where $\alpha > 3$, in good agreement with the experimental values of $\alpha = 3.3-3.6$ (Fig. 10.22, 24). Since the velocity spectrum decays faster than k^{-3}, the elastic turbulent flow is smooth and strongly correlated on the global scale. This is the main feature of the Batchelor regime, where spatially smooth and random-in-time flow is observed [10.125]. The smoothness of the velocity field in elastic turbulence was experimentally tested by investigating the shape of the cross-correlation functions of the velocity field [10.108]. It was found that the second-order spatial derivative of the velocity field was about an order of magnitude smaller than the first-order derivative, which is a direct confirmation of the flow smoothness [10.108]. It should be emphasized that, in contrast to hydrodynamic inertial turbulence, in elastic turbulence the algebraic power spectrum decay is not related to the energy cascade and any conservation law, since the main energy dissipation occurs at the largest scales. The rapid decay of the fluctuation power with k implies a velocity field in which the main contribution to deformation and stirring (stretching and folding) on all scales comes from the randomly fluctuating velocity field at the largest scale of the system. So, it is suggested that the leading mechanism for the generation of small-scale (high-k) fluctuations in the elastic stress is advection of the fluid (which carries the stress) in this fluctuating large-scale velocity field. Hence, the fluctuating velocity field and stress tensor can both be decomposed into large- and small-scale components, and the leading mechanism for the generation of the small-scale (high-k) portions is advection by the fluctuating large-scale flow. The theory considers the elastic stress tensor as passively advected in a random velocity field, which is analogous to the concept of a passively advected vector in the magnetic dynamo theory [10.116]. Thus, the mechanism leading to the algebraic power spectra for the elastic stress and the velocity in this case is related to advection and linear decay accompanied by stretching of the fluid element carrying the elastic stress.

It follows from the stationarity of statistics of both fields that the velocity gradients $\nabla_i V_j$ are of the same order as λ^{-1} in the bulk. This leads to the formation of a boundary layer, where $\nabla_i V_j$ exceeds λ^{-1} and drops to the value in the bulk. The boundary layers at the upper plate and the walls are clearly observed in the experiment (Figs. 10.33, 35), where the flow is mainly shear dominated, and therefore, the polymers are weakly stretched [10.90, 119].

The saturation of the rms fluctuations of the vorticity (and velocity gradients) and thus of the elastic stress in the bulk that results from (10.98) and from the experimental observations naturally explains the power-law behavior of the injected power, $\frac{P}{P_{\text{lam}}} \sim Wi^{0.49}$, obtained from the plot in Fig. 10.21a. Indeed, the injected power is proportional to the torque, $P = T\Omega$. The latter is just the shear stress τ_w averaged over the upper plate. As was shown above, the flow resistance (or torque) growth above the instability threshold occurs solely due to growth of the elastic stress τ_p. In the elastic turbulence regime, however, the theory predicts $\tau_p \sim \eta/\lambda$. On the other hand, it was found that, due to shear thinning of the solution used, $\lambda \sim \Omega^{-0.3}$. Thus, the theoretically expected growth of the mean injected power with respect to its laminar value, P_{lam}, should be solely due to the elastic stress and has the following power-law scaling: $P/P_{\text{lam}} \sim \tau_p \sim Wi^{0.43}$. The latter is rather close to the experimentally observed scaling $P/P_{\text{lam}} \sim \tau_p \sim Wi^{0.49}$. Thus, this scaling is consistent with the saturation of the rms of the fluctuations of the vorticity (and the velocity gradients) in the elastic turbulence observed experimentally [10.109].

Based on this discussion, one can suggest the following scenario for the development of elastic turbulence. The polymer molecules are stretched in the primary shear flow, which leads to a large elastic stress. The elastic stress renders the primary shear flow unstable and causes an irregular secondary flow. The flow acts back on the polymer molecules, stretching them further and raising the elastic stress even more [10.111]. This makes the flow increasingly turbulent, until a kind of a saturated dynamic state is finally reached. This state implies some mutually consistent fields of average stresses and velocities, and their fluctuations, related to each other by the equation of motion and the constitutive equation.

10.3.3 Elastic Turbulence in a Curved Channel: Dean Flow

Experimental Setup and Procedure
Experimental Setup and its Fabrication. Another experimental system where elastic turbulence was studied in detail is flow in a curvilinear channel, or Dean flow. The main motivation for the experiments on Dean flow was to carry out a detailed quantitative study of mixing in elastic turbulence [10.77]. It is an open flow that allows extended continuous experimental runs with reproducible and well-controlled initial conditions, and easy collection of extensive data at different stages of mixing. Meanwhile the elastic turbulence and mixing ex-

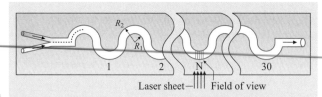

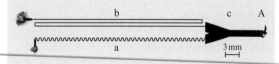

Fig. 10.43 Schematic of the curvilinear channel showing the inlet, a region of observation, and the outlet

periments were conducted in macro- and microchannels of similar geometry. The macrochannel, schematically shown in Fig. 10.43, had a uniform depth of $d = 3\,\text{mm}$, machined in a transparent bar of perspex, and was sealed from above by a transparent window.

The channel consisted of a sequence of smoothly connected half-rings with inner and outer radii of $R_1 = 3\,\text{mm}$ and $R_2 = 6\,\text{mm}$, respectively; it was square in cross section, and had quite a high gap ratio of $d/R_1 = 1$, which was intended to facilitate the development of an elastic instability at low Wi and of intensive irregular flow above the instability threshold. The macrochannels had, depending on the specific experiment, 40 to 64 repeating units, each one being $18\,\text{mm}$ in length. Because of the periodic structure of the channel, it is convenient to use the number of a segment N starting from the inlet as a discrete linear coordinate along the channel [10.77, 104].

The experiments were conducted in a microchannel consisting of 64 smoothly connected half-rings with inner and outer radii of $R_1 = 100\,\mu\text{m}$ and $R_2 = 200\,\mu\text{m}$, respectively, and a uniform thickness of $d = 100\,\mu\text{m}$. This channel has the same proportions as the macroscopic channel but its dimensions are reduced by a factor of 30 (Fig. 10.44 [10.105]). Another microchannel with 46 identical segments of the same geometry and sizes was used to measure the flow resistance (Fig. 10.45 [10.106]). The auxiliary rectilinear channel (b) has a width of $90\,\mu\text{m}$ and a total length of about $72.5\,\text{mm}$. Channel (b) and the comparator region (c) serve to make differential in situ measurements of flux versus pressure by the method described in [10.129].

The microchannel devices consist of a silicon elastomer (Sylgard 184, Dow Corning) chip sealed to a #1 microscope cover glass. The channel structure of the chip was fabricated using soft lithography [10.130]. First, a negative master mold was fabricated in UV-curable epoxy (SOTEC microsystems SU8-1070) using conventional photolithography. The epoxy was spun onto a silicon wafer at $1800\,\text{rpm}$ for $60\,\text{s}$ to create a $100\,\mu\text{m}$ layer and patterned by using a high-resolution negative transparency mask. Liquid elastomer was poured onto the mold to a thickness of $\approx 5\,\text{mm}$ and cured in an $80\,°\text{C}$ oven for $1.5\,\text{h}$. The elastomer was then peeled off the mold, trimmed to its final size, and liquid-feeding ports were punched using a 20 gauge luer stub. The patterned side of the chip was bonded to the cover glass by overnight baking in an oven at $80\,°\text{C}$.

Fig. 10.45 Photograph of the microfluidic device. The microchannels were filled with ink for better contrast

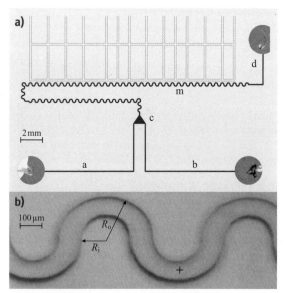

Fig. 10.44 (a) Photograph of the microfluidic device. The microchannel was filled with ink for better contrast. (b) Photograph of a section of the functional curvilinear element. The point where instantaneous flow velocity measurements (averaged over a $20 \times 20\,\mu\text{m}^2$ region) were made is marked by a *cross*

Flow Control. The flow in the microchannels was generated and controlled by pressure differences between the inlets and the outlet (Fig. 10.45). The pressures were generated hydrostatically using long vertical rails with precise rulers and sliding stages. Working liquids were kept in $30\,\text{ml}$ plastic syringes, which were held upright, open to the atmosphere and connected to the two inlets and the outlet by plastic tubing with an internal diameter of 0.76 mm. The pressure drop in the tubing was estimated to be less than 1% of the total. The two syringes feeding the inlets were attached to the sliding stages. The difference in liquid elevation between these two syringes and the outlet syringe was measured and adjusted with a precision of about 0.1 mm corresponding to 1 Pa in pressure. The dependence of the volumetric low rate Q in the curvilinear channel on the pressure difference between inlet 1 and the outlet (Fig. 10.45) was determined with a relative precision of about 0.5% using an in situ measurement technique described in [10.129]. A syringe pump (PHD 2000, Harvard Apparatus, Boston) with a 50 ml gastight Hamilton syringe was used for an absolute flow-rate calibration [10.129].

Measuring Flow Velocity. The flow velocity in a macrochannel was measured directly by LDV. Because of the small width of the channel, special effort was made to obtain high spatial resolution by reducing the region of space where the two laser beams crossed, and reduction of the distance between the interference fringes. Focusing lenses with a small focal length (about 25 mm) were used, and the angle between the beams was raised to about 90° in air (and about 60° in the liquid). As a result, the region of beam crossing was decreased to 15 µm × 15 µm × 40 µm, and the distance between the fringes was 0.44 µm [10.77, 104].

Measurements of the flow velocity in a microchannel were carried out using custom developed microscopic particle image velocimetry (micro-PIV). The polymer solution was seeded with 0.2 µm yellow–green fluorescent beads (Polysciences, Warrington), and epifluorescent imaging of the flow in the microchannel (Fig. 10.46) was made with an inverted microscope (Olympus IMT2, Warrington) and narrow-band excitation and emission filters in the dichroic filter cube. The objective was a long-working-distance lens with 20×, N.A. = 0.40, and the images were projected onto a CCD array with a 640 × 480 pixel resolution (PixelFly camera from PCO, Germany) digitized to 12 bits. Snapshots of the flow were taken with even time intervals of 40 ms, and digitally postprocessed. Images of out-of-focus particles were disregarded. The velocity field was found by cross-correlating positions of the particles in two consecutive snapshots, and the particle velocity vectors were neighbor-validated. (The calculated velocity field corresponded to the time interval between the two snapshots.) The collected time series represent velocity averaged over a 20 µm × 20 µm square region at the middle plane of the channel. Its position in the channel is equidistant from the channel walls (at $R = 150$ µm) and equidistant from the interconnections of the two half-rings at $N = 35$ [10.105, 106, 109].

Measuring Tracer Concentration Profile. The liquids to be mixed were fed into the macrochannel by two identical syringe pumps through two separate tubing lines, always at equal discharge rates. The chemical composition of the two liquids was always identical as well, with the only difference being a low concentration ($c_0 = 2$ ppm) of a fluorescent dye (fluorescein) added to one of them with the diffusion constant of the dye taken as that for the saccharose molecules, $D = 8.5 \times 10^{-7}$ cm^2/s. They were prepared from the

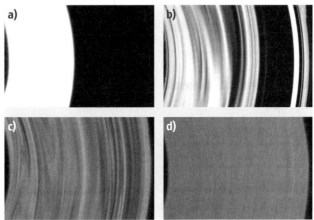

Fig. 10.47a–d Photographs of the flow taken with the laser sheet visualization at different N. The field of view is 3.07 × 2.06 mm, and corresponds to the region shown in Fig. 10.43 (rotated 90° counterclockwise). *Bright regions* correspond to high concentration of the fluorescent dye. (**a**) Flow of the pure solvent at $N = 29$; (**b**)–(**d**), flow of the polymer solution at $Wi = 6.7$ and at $N = 8, 29, 54$, respectively

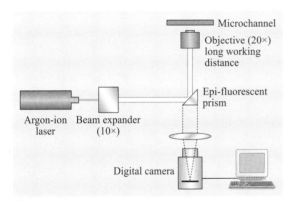

Fig. 10.46 Overview of the micro-PIV setup

same stock of a carefully filtered liquid, which was divided into two equal parts. A small amount of a concentrated solution of the dye was added to one part, while the other part was diluted by an equal amount of pure water. This method of preparation provided very good matching of the densities and refraction indices of the liquids. The channel was illuminated from one side by an argon-ion laser beam converted by two cylindrical lenses into a broad sheet of light with a thickness of about 40 μm in the region of observation. This produced a thin cut of the three-dimensional mixing pattern, parallel to the top and bottom of the channel at half of the channel depth [10.77, 104].

Fluorescent light emitted by the liquid in the direction perpendicular to the beam and to the macrochannel plane was projected onto a CCD array by a camera lens and digitized by an 8-bit, 512×512 frame grabber. Using homogeneous solutions with different amounts of dye, we found the intensity of the fluorescent light captured by the camera to be linearly proportional to the dye concentration. Therefore, the concentration of the dye was evaluated from the fluorescent light intensity (Fig. 10.47) [10.77, 104].

The experiments on mixing in microchannels were carried out by adding fluorescent dyes with low diffusivity to the solution and using them as tracers. We used a few different samples of fluorescein-conjugated dextran (FITCD, Sigma) with average molecular weights M varying from 10^4 g/mol to 2×10^6 g/mol. In spite of the relatively high molecular weight of FITCD, it did not have a measurable influence on the solution relaxation time due to the high rigidity of the polysaccharide molecules. The diffusion coefficients of the FITCD samples in water were estimated using the data in [10.131], giving values of 9.1×10^{-7}–7.4×10^{-8} cm^2/s that corresponded to a broad range of biological macromolecules. The diffusion coefficients in the solvent used were estimated under the assumption that $D \sim 1/\eta_s$, resulting in $D_1 = 6.6 \times 10^{-9}$ and $D_2 = 5.4 \times 10^{-10}$ cm^2/s for 10^4 g/mol and 2×10^6 g/mol FITCD, respectively [10.105, 106].

The concentrations of the fluorescent dyes, which were used as passive tracers in the experiments on mixing in the microchannel, were evaluated using a commercial confocal microscope (Fluoview FV500, Olympus), which was equipped with 40×, N.A.= 0.85 infinity corrected objective and a 12 bit photomultiplier. The scanning was done at a rate of 56 lines per second and 512 pixels per line, corresponding to a step of 0.18 μm per pixel [10.105].

Observation of Elastic Turbulence

Flow Resistance. The flow resistance in the microchannel was measured by measuring the volumetric flux rate, Q, in a broad range of applied pressure drops and by calculating the resistance factor $Z = \Delta p_s / Q$, where Δp_s is the pressure drop per segment. The resistance factor is a constant proportional to the viscosity for Newtonian fluids in linear, low-Re regime, and can be used as a measure of turbulent flow resistance in large channels at high Re. Figure 10.48 shows the dependence of Z on Q, after Z is divided by the resistance factor Z_0 found for a Newtonian liquid with the same viscosity. The ratio Z/Z_0 is

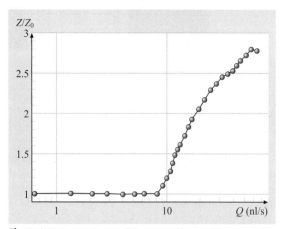

Fig. 10.48 Dependence of the normalized resistance Z/Z_0 in the flow of solution 1 through the curvilinear channel on the volumetric flow rate Q on a semilogarithmic scale

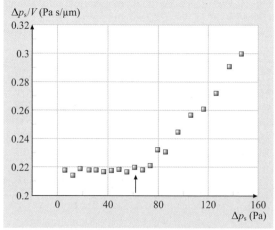

Fig. 10.49 Flow resistance, $\Delta p_s / V$ versus Δp_s. The *arrow* indicates the onset of the elastic instability

constant and equal to unity in the linear regime at low Q. At Q of about 8.5 nl/s, however, a nonlinear transition occurs; Z/Z_0 starts to grow and reaches a factor of about 2.8 at high Q. The Reynolds number was always much lower than unity, so inertial effects were always negligible [10.106].

Another way to measure the flow resistance is to measure an average flow velocity through a microchannel at various pressure drops by the micro-PIV technique. Using this method, an analog of the flow resistance, $\Delta p_s/V$, versus the pressure drop per segment, Δp_s, is shown in Fig. 10.49 [10.105]. The elastic instability occurs at $\Delta p_s \approx 50$ Pa, and the flow resistance grows up to 1.4-fold in the elastic turbulence regime.

Flow Structure, Velocity and Velocity Gradient Fields, Velocity Fluctuations Spectra, and Velocity Statistics. The flow structure and velocity flow field in the microchannel were investigated by combined micro-PIV and micro-PTV techniques for a broad range of flow rates. Above the instability threshold an initially stationary longitudinal vortex becomes time dependent. As one can see in Fig. 10.50, the vortical structure fills the whole channel cross-section and is quite spatially smooth with irregular velocity fluctuations in time [10.109]. One gets a similar impression from measurements of the transversal velocity component at several off-center locations in a macrochannel. Indeed, we found nonzero averages, which typically persisted for a few minutes and changed their sign rather randomly in time. This situation can be explained by the presence of persistent longitudinal vortical structures in the flow. These vortices probably fill the channel cross section, with their vorticity direction jumping between parallel and antiparallel to the mean flow [10.104].

The spatial smoothness of the channel flow in elastic turbulence was investigated by measuring profiles of the longitudinal and transversal velocity components by micro-PTV for different values of Wi (Fig. 10.51). One notices that, even below the transition (squares, circles, and triangles in Fig. 10.51), the profiles of the longitudinal velocity component are clearly nonparabolic and nonsymmetric. This apparent difference with respect to the case of a laminar Poiseuille flow in a straight channel has been also observed in experiments with a pure solvent (data not shown) and is confirmed by recent numerical simulations by A. Kumar (private communication) in a low-Re flow in a similar geometry [10.109].

The longitudinal velocity profile allows us to estimate the critical Weissenberg number of the elastic transition $Wi_c = \lambda(\dot{\gamma})\dot{\gamma}_c$, where the critical shear rate

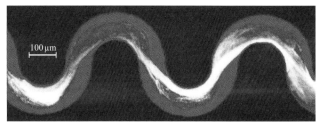

Fig. 10.50 Mid-plane horizontal confocal snapshot. The flow is seeded with 0.2 μm fluorescent spheres. The driving pressure is 120 Pa per channel segment

$\dot{\gamma}_c$ is calculated as $\dot{\gamma}_c = 2V_\theta^{\max}/d$, where $V_\theta^{\max}$ is the maximum longitudinal velocity of the profile. Indeed, at the critical pressure difference corresponding to the onset of instability, $\Delta p_s \approx 50$ Pa, one gets $\dot{\gamma}_c \simeq 4.2\,\text{s}^{-1}$, which leads to $Wi_c \simeq 13$ with $\lambda(\dot{\gamma}_c) \simeq 3.1$ s. On the other hand, one can use the time-averaged longitudinal velocity gradient across the channel, $\overline{\partial V_\theta/\partial r}$, at the transition to estimate Wi_c. Using the data in Fig. 10.54, one then obtains $Wi_c \simeq 9.3$. This critical value Wi_c is in fair agreement with the value obtained theoretically from the linear stability criterion for the aspect ratio, $d/R_1 = 1$ and $S = 0.82$, in the experiment. Then, using the instability criterion for the Dean flow one obtains $K_c \simeq 9$, which leads to $Wi_c \simeq 9$ [10.96–98], while $Re \simeq 8 \times 10^{-5}$.

PTV measurements also provide us with the possibility of measuring time series of the flow velocity at specific points in the flow. The rms fluctuations of both horizontal velocity components in the center of the microchannel as a function of Wi are shown in Fig. 10.52, 53. The fluctuations are practically absent in the laminar regime, while at $Wi_c \simeq 13$ the rms of both velocity components start to grow sharply. These measurements can be used as another indication of the transition and appearance of the elastic turbulence regime [10.109].

Figure 10.52 shows the time-averaged longitudinal velocity $\bar{V}_\theta$ as a function of the driving pressure difference. One observes that in a laminar regime the velocity growth is linear but slows above the transition. Other important information is provided in Fig. 10.45, where the average longitudinal velocity gradient across the channel at three locations as a function of Wi is presented.

The statistical properties of the velocity field are characterized by the correlation function, which is presented in Fig. 10.55 together with a typical time series of the instantaneous longitudinal velocity far above the transition. The temporal dependence has well-expressed chaotic appearance that is confirmed by the shape of the

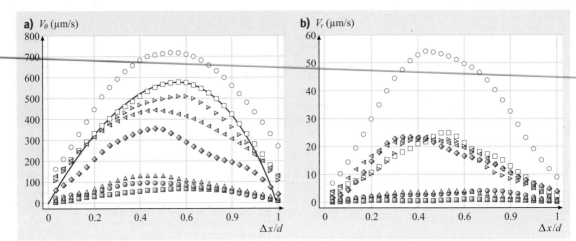

Fig. 10.51a,b Profiles of the (**a**) longitudinal velocity component, V_θ and (**b**) radial velocity component V_r across the microchannel. The symbols correspond to: *Full squares*: $Wi = 3.84$, *circles*: $Wi = 6.75$, *up triangles*: $Wi = 8.49$, *rhombus*: $Wi = 13.56$, *left triangles*: $Wi = 14.83$, *right triangles*: $Wi = 19.4$, *open squares*: $Wi = 20.15$, *open circles*: $Wi = 22$; the *full line* is the result of numerical simulations

autocorrelation function, which decays monotonically without distinct peaks [10.106].

Figure 10.56 shows the power spectra of the fluctuations of the longitudinal and transversal components of the velocity in the polymer solution flow in the macrochannel at $Wi = 6.7$. The measurements were done at $N = 12$, near the middle of the half-ring in the middle of the channel. A spectrum of the velocity fluctuations in the flow of the pure solvent at the same Q, giving just instrumental noise, is shown for comparison. The mean velocity was $\bar{V} = 6.6$ mm/s; the rms fluctuations V_{rms} were $0.09\bar{V}$ and $0.04\bar{V}$ in the longitudinal and transversal directions, respectively [10.77, 104]. The power spectra exhibit the power-law decay $P \sim f^{-\alpha}$ with an exponent $\alpha \simeq 3.3$ that is consistent with similar

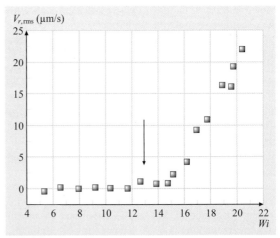

Fig. 10.52 Dependence of the rms fluctuations of the longitudinal component of the flow velocity $V_{\theta,\text{rms}}$ on Wi at the center of the microchannel. The *squares* are the instrumental error. The *arrow* indicates the onset of the elastic instability. *Inset*: dependence of the time average of the longitudinal velocity component $\overline{V_\theta}$ at the center of the microchannel on the pressure drop per channel segment Δp_s

Fig. 10.53 Dependence of the rms fluctuations of the radial component of the flow velocity $V_{r,\text{rms}}$ on Wi at the center of the microchannel. The *arrow* indicates the onset of the elastic instability

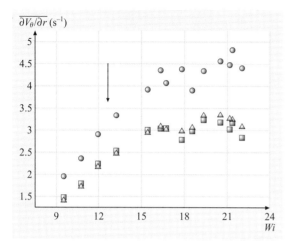

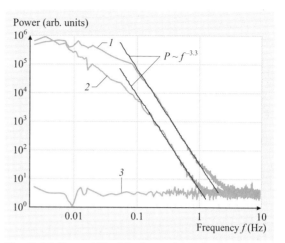

Fig. 10.54 Dependence of the average gradients of the longitudinal component of the flow velocity, $\overline{\partial V_\theta/\partial r}$ on Wi, at three different points having the same radial coordinate, $r = d/2$ but different polar angles, $\theta = -\pi/2, 0, \pi/2$. The *arrow* indicates the onset of the elastic instability. Circles: 0, squares: $\pi/2$, triangles: $-\pi/2$

Fig. 10.56 Power (P) of the fluctuations of velocity in the middle of the channel at $N = 12$ as a function of frequency f. The spectra in the polymer solution flow at $Wi = 6.7$ for the velocity components along and across the mean flow are shown by *curves* 1 and 2, respectively. *Curve* 3 shows the velocity spectrum across the mean flow for the pure solvent at the same Q

measurements in other setups and with the theoretical estimates [10.76, 77, 104, 115, 116].

Mixing in the Flow. Mixing in macrochannels was studied at quite small Reynolds numbers, reaching only 0.6 for the highest Q that was explored. Therefore, the flow of the pure solvent always remained laminar and no mixing occurred (Fig. 10.47a). The boundary separating the liquid with and without the dye was smooth and parallel to the direction of flow and only became somewhat smeared due to molecular diffusion as the liquid advanced downstream. The flow of the polymer solution was laminar and stationary up to a value of Q corresponding to Wi_c (and $Re = 0.06$), at which an elastic instability occurred. This instability led to irregular flow and fast mixing of the liquids [10.77, 104].

A few typical mixing patterns at different N in the polymer solution above the instability threshold are shown in the photographs in Fig. 10.47b–d. More insight into the structure and evolution of the mixing patterns can be obtained from space–time diagrams. Representative diagrams taken at $Wi = 6.7$ at four different N are shown in Fig. 10.57. The brightness profiles along a single line perpendicular to the channel near the middle of a half-ring (a horizontal line going through the middle of a snapshot in Fig. 10.47) were captured at even time intervals of 80 ms and plotted from top to bottom.

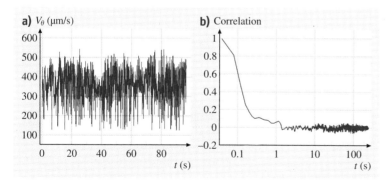

Fig. 10.55 (a) Time series of the longitudinal flow velocity V_θ at the center of the microchannel at $Wi = 20.93$. **(b)** Autocorrelation function of V_θ based on about 6000 individual velocity measurements

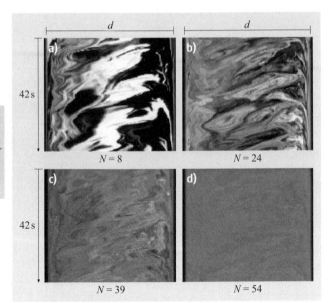

Fig. 10.57 Representative space–time diagrams of the polymer solution flow at $Wi = 6.7$ taken at different positions N along the channel

The diagrams in Fig. 10.57 share the same chaotic appearance and show features at comparable scales, but they lose contrast as liquid advances downstream and becomes progressively mixed [10.77, 104].

As illustrated by the space–time diagrams in Fig. 10.57, mixing in the polymer solution flow above

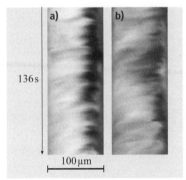

Fig. 10.59a,b Space–time plots of the FITCD distribution across the channel taken at (a) $N = 12$ and (b) $N = 18$. Confocal scanning was done along the same line across the channel in the mid-plane at equal distances from the half-ring interconnections, with even time intervals of 0.0177 s. Profiles of the FITCD concentration at consecutive moments of time are plotted from the top to bottom

the instability threshold was a random process calling for statistical analysis [10.77, 104]. A simple parameter characterizing the homogeneity of the mixture is the rms deviation of the dye concentration from its average value $\bar{c} = c_0/2$ divided by the average value itself $c_{rms} = \langle (c - \bar{c})^2 \rangle^{1/2} / \bar{c}$. A small value of c_{rms} indicates strong homogeneity and good mixing of the liquids. At the channel entrance, where the two injected liquids are perfectly separated, c_{rms} is unity, and it should become zero for a perfectly mixed liquid.

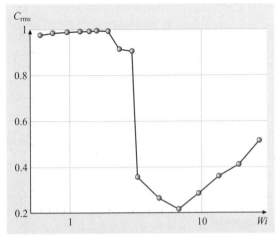

Fig. 10.58 Dependence of c_{rms} (the normalized root mean square of concentration deviations from the average) on the Weissenberg number Wi measured near the channel exit at $N = 29$ (semilogarithmic coordinates)

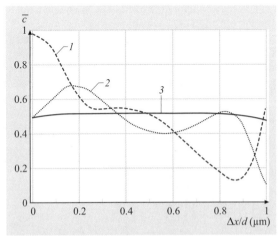

Fig. 10.60 Time average of FITCD concentration $\bar{c}$ as a function of the normalized coordinate across the microchannel at different locations downstream: (1) $N = 7$, (2) $N = 11$, and (3) $N = 41$

The dependence of c_{rms} on Wi near the exit of the channel, at $N = 29$, is presented in Fig. 10.58. The statistics of the dye concentration was evaluated from space–time diagrams similar to those in Fig. 10.57. The regions near the walls of the channel with a width of $0.1d$ were excluded from the statistics, because of possible image aberrations. In a stationary flow regime ($Wi < Wi_c$), when the concentration profile did not change in time, the brightness profiles were measured over short time intervals (about 100 s). In the regime of an irregular flow, however, the profile of concentration was strongly fluctuating. So, in order to obtain representative statistics of c, the measurements of the brightness profile were taken over quite long intervals of time (20–30 min), which typically corresponded to the total liquid discharge of about $10^3 d^3$ [10.77, 104].

The plot in Fig. 10.58 is somewhat analogous to that in Fig. 10.20, 48, 49, which show the dependence of the flow resistance on Wi. Indeed, the decrease in c_{rms} is an integral result of mass transfer produced by the irregular flow in the channel, just as growth of the flow resistance is an integral characteristic of increase of momentum transfer in elastic turbulence. The most striking feature of the plot in Fig. 10.58 is certainly an abrupt drop in c_{rms} at Wi_c, where the irregular motion of the liquid sets in.

We studied the dependence of c_{rms} on N at $Wi = 6.7$, corresponding to the greatest homogeneity of the mixture near the channel exit (Fig. 10.58), and found c_{rms} to decay exponentially with distance from the inlet with a characteristic decay length ΔN of about 15 segments [10.77, 104]. One can learn from Fig. 10.58 that, if Wi is raised above 6.7, c_{rms} starts to increase again. The most plausible explanation for this is saturation of the growth of the velocity fluctuations together with reduction of the residence time in the flow at growing Wi (and average flow velocity $\bar{V} = Q/d^2$). If the ratio between the fluctuating and average flow velocities remains constant, while they both increase, the stirring in the flow remains the same, but there is less time available for molecular diffusion, and homogeneity is reduced as a result. This situation can be quantitatively described by the growth of the Peclet number $Pe = \bar{V}d/D$. It was recently found, for a flow of a polymer solution in a channel of the same shape and at similar Wi, that the characteristic length ΔN increases as $Pe^{0.25}$ [10.105]. This suggests that c_{rms} should start increasing with Wi once growth of the velocity fluctuations has slowed down.

The typical mixing time in the channel at $Wi = 6.7$ was found to be 3–4 orders of magnitude shorter than the diffusion time, d^2/D, for the small molecules of fluorescein [10.77]. The dependence of the efficiency of mixing at the optimal flow conditions (for the 80 ppm solution it was $Wi = 6.7$) on the concentration of the polymers was surprisingly weak (although Wi_c grew quickly as the polymer concentration decreased). So, for a solution with a polymer concentration of 10 ppm ($\eta/\eta_s = 1.03$), c_{rms} values as low as 0.29 could be reached at $N = 29$. (measured at $Re = 0.065$, where inertial effects were still negligible). Excitation of irregular flow and active mixing was observed down to polymer concentrations of 7 ppm [10.77, 104].

Quantitatively similar results for mixing were observed in a microchannel [10.105, 106]. We studied in detail mixing of fluorescent-conjugated dextran (FITCD) with 2×10^6 g/mol molecular weight in the channel at $\Delta P = 124$ Pa, corresponding to a flow rate of about twice above the nonlinear transition threshold and a cross-channel space–time-averaged longitudinal velocity of about 173 µm/s. Variation of the tracer concentration profiles with time at different distances from the inlet is illustrated by the space–time plots in Fig. 10.59a,b. One can observe that the tracer concentration appears to fluctuate quite randomly without any apparent scale in time or space. Next, one can see in Fig. 10.59a, taken at $N = 12$, that the left side of the channel, where the tracer was initially injected, looks much brighter and has a much higher average concentration of the tracer. Although also noticeable in Fig. 10.59b taken further downstream, at $N = 18$, this feature is clearly weaker there.

Thus, stirring by the fluctuating velocity field seems to create a more-symmetric distribution of the tracer between the two sides of the channel. In order to validate this observation, we measured the time average of the tracer concentration $\bar{c}$ at different positions across the channel and at different N (Fig. 10.60).

One can see that the cross-channel distribution of $\bar{c}/c_0$ close to the inlet, at $N = 7$, is strongly influenced by the asymmetric conditions at the channel entrance. As one can observe from the curve at $N = 11$, however, the imprint of the initial conditions is clearly fading as the liquid advances downstream and being stirred. Further downstream, at $N = 41$, asymmetry in the tracer distribution introduced by the initial conditions has disappeared completely. Fading of the influence of the initial condition with time and restoration of symmetry in flow in a statistical sense are both distinct features of chaotic and turbulent flows. Therefore, the curves in Fig. 10.60 provide further evidence for the truly chaotic nature of the flow in the microchannel [10.105, 106].

10.3.4 Conclusions

Summarizing these experimental results, we conclude that the flow at sufficiently high Wi in curved macro- and microchannels exhibits main features that are rather similar to those observed in the swirling flow between two disks. Above the elastic instability threshold the flow of the polymer solution in the channel, firstly, exhibits two major features of turbulent flows: a major increase in the flow resistance and in the rate of mixing, and secondly, the fluid motion generates the velocity power spectra across a wide range of frequencies as well as wavenumber domains, which provides solid evidence for the turbulent character of the flow, particularly due to the absence of peaks in the spectra in Fig. 10.56 similar to those observed in the swirling flow (Fig. 10.22, 23). The spectra of both longitudinal and transversal velocity components do not exhibit any distinct peaks and have broad regions of power-law decay with an exponent of about -3.3. Since the power spectra in Fig. 10.56 were measured at a point with a high mean flow velocity (10 times higher than a characteristic fluctuating velocity), we can use the Taylor hypothesis and argue that the spectra in Fig. 10.56 actually reflect the spatial structure of the flow. Then the power-law decay region can be transferred to the spatial domain, with the power of the velocity fluctuations scaling as $P \sim k^{-3.3}$ with the wavenumber k (see the remark on this subject in the previous section). We also notice here that the exponent of -3.3 in Fig. 10.56 is very close to those measured in the flow between two plates, which varied from -3.3 to -3.6 depending on the position. So, one can suggest that the decay of the power of the velocity fluctuations with an exponent of around -3.5 is a rather general feature of elasticity-induced turbulent flows.

The functional form of the velocity power spectra, $P \sim k^{-3.3}$, suggests that the power of fluctuations of velocity gradients scales like $k^{-1.3}$. An integral of $k^{-1.3}$ diverges for $k \to 0$ and converges at $k \to \infty$. This means that the main contribution to the fluctuations of the velocity gradients and the velocity differences at all scales comes from the largest eddies, having dimensions of the whole system (the diameter of the channel or the gap between the plates). This conclusion has an immediate implication for mixing in the flow: it should result in the same type of patterns and in functionally the same statistics as in the case of a completely homogeneous flow, $V(r,t) = V_0(t) + \frac{\partial V}{\partial r}(t)(r - r(0))$, randomly varying in time (r is the position vector).

Such a flow is a realization of the so-called Batchelor regime of mixing [10.125], and the problem of the statistics of a tracer (dye) distribution in it has been solved analytically recently [10.118, 132, 133]. The Batchelor regime occurs at small scales (below the Kolmogorov dissipation scale [10.79, 117]) in usual (high-Re) turbulence, and is rather difficult to realize in the laboratory otherwise. Therefore, the elastic turbulent flow in the channel provided a very convenient experimental system for quantitative study of mixing in this regime [10.77, 104–106]. The experimental results on the correlation functions and PDF of dye concentration, and on their dependence on mixing time, agreed very well with the theoretical predictions [10.77, 105, 106]. A practical message of the experiments is that very viscous liquids can be efficiently mixed in curvilinear channels at very low flow rates by adding high-molecular-weight polymers at very low concentrations. This method of mixing, we believe, can find some industrial and laboratory applications [10.134].

10.4 Measurements for Large-Eddy Simulations

The choice of measurement technique for experimental turbulence research depends upon how turbulence is defined, in particular what type of decomposition is envisioned. In the classical Reynolds-averaged Navier–Stokes (RANS) decomposition, turbulence is defined as the deviation from the ensemble-averaged mean velocity field (Sect. 10.1.2) [10.4, 6]. For statistically stationary flows, ergodicity is often assumed and ensemble averaging is replaced with the more-practical method of time averaging. Hence for RANS, point measurement techniques such as hot wires or LDV that record long time records of turbulence signals are appropriate and have a long history of applications to turbulence research (Sect. 10.1.1). Combined with time averaging, these signals allow one to measure the mean velocity and moments of the velocity, for instance classical Reynolds stresses $\sigma_{ij}^{\text{Re}} = -\overline{u'_i u'_j}$, where an overbar denotes statistical (ensemble or time) averaging. Such data have supported model developments by providing detailed databases to which RANS simulation results can be compared, and have also provided fundamental insights into turbulent flow that have inspired and tested concepts for model developments.

In large-eddy simulation (LES) of turbulence [10.4, 135, 136], the most-often used decomposition involves spatial, three-dimensional filtering, rather than time averaging. In LES, the effects of the unresolved motions are included in the equations through new stresses (momentum fluxes) that must be parameterized by means of closure relations, and experimental data can be employed to support model development and the validation of numerical simulations. Since the type of data needed differs in significant aspects from that which is classically employed in the context of RANS, these measurement techniques are covered separately in the present section. The section briefly reviews LES and the specific data requirements for LES (Sect. 10.4.1), and then describes the experimental methods that have been employed to obtain such data starting with arrays of point measurement techniques (Sect. 10.4.2) and optical planar velocimetry measurement methods (Sect. 10.4.3). Sample results from the latter applied to studies of LES models are presented in (Sect. 10.4.4). The application of optical volumetric techniques for 3-D velocity measurements are described in (Sect. 10.4.5). Scalar fluctuation measurements using optical techniques and their applications to the study of LES variables of interest to scalar mixing and combustion are reviewed in Sect. 10.4.6.

10.4.1 Large-Eddy Simulation and Data Requirements

In LES, the objective is to solve equations for the velocity field convolved with a spatial filter at a scale Δ. The filtered velocity field is defined according to

$$\tilde{u}_i(x) = \iiint u_i(x') G_\Delta(x-x') \mathrm{d}^3 x' \,, \quad (10.99)$$

where $G_\Delta(x)$ is a filter function (here assumed to be spatially homogeneous, i.e., only a function of $x-x'$, for simplicity), of characteristic scale Δ. An example is the Gaussian filter $G_\Delta(x) = [6/(\pi\Delta^2)] \exp(-6x^2/\Delta^2)$. The decomposition of the velocity field into large (resolved or filtered) and small [unresolved or subgrid-scale stress (SGS)] scales may be performed using other types of filters [10.4, 136, 137] such as the spectral cutoff filter and the box, or top-hat, filter. The spectral cutoff filter has the desirable property that it is also a projection in the sense that twice-filtered fields are equal to single-filtered fields and it thus cleanly separates between scales. A drawback is that, when filtering spatially localized phenomena, it causes nonlocal oscillatory behavior and, since the filter has negative lobes, the resulting stress tensor does not follow some important realizability conditions [10.138]. The box filter, on the other hand, has good spatial localization but does not allow unambiguous separation between scales because of spectral overlap. The Gaussian filter has intermediate localization properties in both physical and spectral space, although it is closer to the box filter and also has spectral overlap. Convolving the Navier–Stokes equations with such filters yields the LES equations for incompressible flow

$$\frac{\partial \tilde{u}_i}{\partial t} + \tilde{u}_j \frac{\partial \tilde{u}_i}{\partial x_j} = -\frac{1}{\rho} \frac{\partial \tilde{p}}{\partial x_i} + \nu \nabla^2 \tilde{u}_i - \frac{\partial}{\partial x_j} \tau_{ij} + \tilde{f}_i \,,$$

$$\frac{\partial \tilde{u}_i}{\partial x_i} = 0 \,; \quad (10.100)$$

f_i is the external forcing and τ_{ij} is the subgrid-scale stress (SGS) tensor defined as

$$\tau_{ij} = \widetilde{u_i u_j} - \tilde{u}_i \tilde{u}_j \,. \quad (10.101)$$

This tensor arises due to the spatial filtering of the nonlinear term in the Navier–Stokes equations, and is the LES analog of the classical Reynolds stress in RANS. To highlight the analogy, we note that the classical (kinematic) Reynolds stress $\sigma_{ij}^{\mathrm{Re}}$ can also be written as $-\sigma_{ij}^{\mathrm{Re}} \equiv \overline{u'_i u'_j} = \overline{u_i u_j} - \bar{u}_i \bar{u}_j$, where an overbar still denotes statistical (ensemble or time) averaging. With an SGS model (closure), i.e., replacing τ_{ij} with an expression in terms of the filtered resolved velocity $\tilde{u}$, the LES equations may be discretized at a spatial resolution on the order of the filter scale Δ.

The most popular models for the subgrid-scale stress belong to two classes: eddy viscosity and similarity (or nonlinear) models. The first was introduced by *Smagorinsky* [10.139] and bears his name. It reads, for the deviatoric part of the stress ($= \tau_{ij} - \frac{1}{3}\tau_{kk}\delta_{ij}$)

$$\tau_{ij}^S = -2(C_s^\Delta \Delta)^2 |\tilde{S}| \tilde{S}_{ij} \,, \quad (10.102)$$

where $\tilde{S}_{ij}$ is the filtered stain rate and $|\tilde{S}| = \sqrt{2\tilde{S}_{ij}\tilde{S}_{ij}}$ is its magnitude. The Smagorinsky coefficient C_s^Δ must be specified [10.140], or it may be determined dynamically from the simulated large-scale fields as proposed by *Germano* et al. [10.141]. This procedure was based on the Germano identity [10.142],

$$L_{ij} = \tau_{ij}(\alpha\Delta) - \overline{\tau_{ij}(\Delta)} \,, \quad (10.103)$$

relating the SGS stresses at scales $\alpha\Delta$ and Δ, where $\alpha > 1$ and typically $\alpha = 2$ (an overbar denotes filtering at the scale $\alpha\Delta$). The tensor L_{ij} is the so-called resolved stress tensor defined as

$$L_{ij} = \overline{\tilde{u}_i \tilde{u}_j} - \bar{\tilde{u}}_i \bar{\tilde{u}}_j \,. \quad (10.104)$$

Substituting the Samgorinsky model for the two stress terms on the right-hand side results in

$$L_{ij} - \frac{1}{3}L_{kk}\delta_{ij} = (C_s^\Delta)^2 M_{ij},$$

where $M_{ij} = -2\Delta^2 \left[\alpha^2 \left(\frac{C_s^{\alpha\Delta}}{C_s^\Delta} \right)^2 \widetilde{|\tilde{S}|\tilde{S}_{ij}} - \widetilde{|\tilde{S}|}\widetilde{\tilde{S}_{ij}} \right].$ (10.105)

Assuming scale invariance, i.e., $C_s^{\alpha\Delta} = C_s^\Delta$, one can determine C_s^Δ based on variables available during LES. Minimizing the ensemble-averaged error between the right- and left-hand sides of this equation leads to [10.141, 143, 144]

$$\left(C_s^{\Delta,\mathrm{dyn}}\right)^2 = \frac{\langle L_{ij} M_{ij} \rangle}{\langle M_{ij} M_{ij} \rangle}.$$ (10.106)

The dynamic model has been used extensively in numerous applications of LES (see reviews in *Piomelli* [10.145] and *Meneveau and Katz* [10.146]). For an overview of several other eddy-viscosity SGS models, see *Lesieur and Metais* [10.135] and *Meneveau and Katz* [10.146].

The second type of model is related to the similarity model of *Bardina* et al. [10.147] and was originally introduced based on analysis of direct numerical simulation data. Further evidence was obtained from detailed analysis of experimental data, as reviewed in Sect. 10.4.4 below. A computationally convenient version of this model is the nonlinear model [10.148–150] that is expressed in terms of the resolved velocity gradient tensor as

$$\tau_{ij}^{\mathrm{NL}} = c_A \Delta^2 \tilde{A}_{ik} \tilde{A}_{jk}, \quad \tilde{A}_{ij} = \frac{\partial \tilde{u}_i}{\partial x_j}.$$ (10.107)

Both models may be linearly combined to form so-called mixed models [10.147]. These models also bear a resemblance to so-called deconvolution models [10.151].

For the transport of a scalar c (e.g., concentration, temperature), the filtered scalar transport equation yields SGS fluxes of the form $q_i = \widetilde{u_i c} - \tilde{u}_i \tilde{c}$ [10.152]. To quantify mixing of a conserved scalar, its variance $\widetilde{c^2} - \tilde{c}^2$ and the scalar-variance dissipation rate $\chi = \gamma[\widetilde{(\nabla c) \cdot (\nabla c)} - \nabla \tilde{c} \cdot \nabla \tilde{c}]$ are often used but must be modeled since they are not resolved in LES [10.153, 154]. In chemically reacting flows, unresolved scalar consumption rates are often written in terms of the difference $\widetilde{|\nabla c|} - |\nabla \tilde{c}|$ [10.155], which is also not resolved due to the nonlinearity inherent in the absolute value.

Studies of LES and SGS modeling based on empirical data can be divided into two broad categories *Piomelli* et al. [10.156]: a posteriori tests and a priori analysis. In a posteriori tests, the results from a simulation that utilizes a particular SGS model to be studied are compared with available data. These data can originate from direct numerical simulation (DNS) or from experiments, these being complementary approaches. DNS provides full databases with all fields resolved in three dimensions, but at limited Reynolds numbers. Experimental data may provide access to high-Reynolds-number flows in possibly complex flow conditions, but acquisition of multicomponent and multipoint velocity experimental data is challenging.

For a posteriori tests the data are processed statistically before comparing with LES output in order to provide mean velocity profiles, rms distributions, spectra, etc.. Since the LES only provides the filtered variables, for meaningful comparisons two options are available:

1. The subgrid model provides missing statistics and one may compare the statistics of the unfiltered fields.
2. The data must be filtered at a scale comparable to the LES filter scale and the statistics of the filtered fields are compared.

A posteriori tests are considered to be unambiguous tests of the combined performance of a SGS model coupled with its numerical implementation. Because of their integrated nature, they do not normally provide much insight into the detailed physics of a model and the reasons why they do, or do not, work.

In a priori analysis, one focuses upon the quantities to be modeled, such as the SGS stress or force as computed directly from data according to their definition (10.101), and compares them to model expressions that may depend upon local filtered large-scale variables. Conversely, a priori analysis allows one to focus on particular features of a model under more controlled conditions. However, a priori tests do not account for the feedback that may occur between the SGS model and the resolved scales in an actual simulation. Consequently, it has been found that good a priori behavior does not necessarily imply good performance in simulations (e.g., the similarity model). Nevertheless, by isolating specific features of the SGS physics and models, a priori studies have motivated significant fundamental advances in LES and SGS modeling.

In both a posteriori and a priori studies, one requires well-resolved data that must be filtered in space.

Since LES simulates a range of scales, and one wishes to resolve significant parts of the small-scale motions, the measurement technique must be able to resolve a significant range of length and time scales. These requirements typically imply that multipoint measurement techniques are required in experiments to support LES and SGS model developments. Experimental techniques that can record multipoint data are needed also because LES is often required to predict more than just single-point statistics, for instance spectra or spatial correlation functions.

One of the most interesting aspects of the interactions among small and large scales in LES is the rate at which energy is transferred between these scales. From the transport equations for resolved and SGS kinetic energy (e.g., *Piomelli* et al. [10.157]), this exchange can be identified as the term

$$\Pi_\Delta = -\langle \tau_{ij} \tilde{S}_{ij} \rangle \,, \tag{10.108}$$

often called the SGS kinetic-energy dissipation rate. A cascade of kinetic energy from large to small scales is characterized by positive values of Π_Δ. Many SGS stress models are calibrated by matching the modeled and real SGS dissipation, making it a central parameter that needs to be measured during a priori studies. Measurement of Π_Δ consists of evaluating as many components of the SGS stress tensor and filtered strain-rate tensor as are available from the measurement instrument, evaluating the tensor contraction, and time or ensemble averaging to obtain a statistically meaningful result. For isotropic turbulence filtered in the inertial range, the SGS dissipation is almost equal to the viscous dissipation rate [10.4, 137, 140]. As described below, often only a subset of the tensor components needed to evaluate the contractions in (10.108) are available. For instance, if two components of velocity are available in the (x_1, x_2) planes, a two-dimensional (2-D) surrogate of SGS dissipation may be evaluated according to

$$\Pi_{\Delta-2D} = -(\langle \tau_{11} \tilde{S}_{11} \rangle + \langle \tau_{22} \tilde{S}_{22} \rangle \\ + 2\langle \tau_{12} \tilde{S}_{12} \rangle) \,. \tag{10.109}$$

For isotropic turbulence these in-plane contributors constitute 7/15 of the total value. To measure the model coefficients in a priori tests, the ensemble-averaged SGS dissipation can be matched to the measured values. For instance, for the Smagorinsky model this leads to the following expression

$$(C_S^\Delta)^2 = \frac{\Pi_\Delta}{\Delta^2 \langle (2\tilde{S}_{ij}\tilde{S}_{ij})^{\frac{3}{2}} \rangle} \,, \tag{10.110}$$

in which the numerator and denominators can be measured.

10.4.2 Arrays of Single-Point Instruments for Studies of SGS Dynamics

Classical single-point measurement techniques include hot-wire anemometers (Sect. 5.2), laser Doppler anemometers (Sect. 5.3.1), sonic anemometers (Sect. 5.7), Pitot probes, etc. Using a single-point sensor only temporal filtering can be performed. Using Taylor's hypothesis this can be interpreted as one-dimensional spatial filtering in the streamwise direction. Resolving sufficient spatial resolution usually means that good high-frequency response is required (eliminating, for instance, Pitot probes from consideration).

Following this approach and using only single-point sensors (hot-wire probes), *Meneveau* [10.158] and *Meneveau and O'Neil* [10.159] analyzed data in grid turbulence to study stress–velocity correlations and the scaling of dissipation as a function of length scale, respectively, and *O'Neil and Meneveau* [10.160] considered turbulence in a cylinder wake. *Porté-Agel* et al. [10.161] studied turbulence and scalar transport in the atmospheric boundary layer using data from a single sonic anemometer. The accuracy of such one-dimensional (1-D) filtering and Taylor's hypothesis have been addressed for wall-bounded flows using DNS [10.162], LES [10.163], and field-measurement data using 2-D filtering (*Porté-Agel* et al. [10.164], see below). The results show that one-dimensional filtering does not sufficiently filter out the cross-stream variability of the turbulence. For quantitatively accurate results, these analyses show that at least two-dimensional filtering should be used.

The two-dimensional data thus required can be obtained using optical techniques based on light sheets (Sect. 10.4.3 and Sect. 10.4.4). If one is limited to point sensors, an array of point sensors arranged along a line perpendicular to the mean velocity must be used. This approach has been proposed by *Tong* et al. [10.165] and applied in *Porté-Agel* et al. [10.164, 166] and *Horst* et al. [10.167] for sonic anemometer measurements in the atmospheric boundary layer. It has also been applied to hot-wire measurements in laboratory turbulence. Compared to fully 3-D filtering, the accuracy of 2-D filtering and Taylor's hypothesis was found to be quite good based on the DNS [10.162], LES [10.163], and atmospheric field [10.168] studies.

For laboratory turbulence studies using hot-wire arrays, the spatial and temporal resolutions of each sensor is typically on the order of about 0.5 mm and 10–20 kHz, respectively. Turbulence in wind-tunnel flows typically has Kolmogorov scales on the order of 0.1–0.2 mm and advection velocities on the order of 10 m/s, and thus frequencies on the order of 50 kHz. Therefore, typical hot-wire probes under-resolve the Kolmogorov scale and Kolmogorov advection frequency in the flows considered by factors of about three to five. Also, the spacing between sensors in an array significantly exceeds the Kolmogorov scale η_K, typically by factors of five or more. Thus, these measurements do not fully resolve the viscous dissipation range of turbulence. However, just as Reynolds stresses are dominated by the large-scale turbulence fluctuations and their measurement does not require resolution of the viscous range [10.4], in LES the SGS stresses or scalar fluxes are dominated by scales near the filter scale Δ. Thus to measure the SGS momentum or scalar fluxes with reasonable accuracy, measurement resolutions of $\Delta/10$ to $\Delta/5$ are often considered to be sufficient. Thus, for instance, an experiment with spatial resolution of $5\eta_K$ allows accurate measurement of SGS fluxes at scales of 25 to $50\eta_K$, i.e., scales of interest in the inertial range of turbulence.

A complete analysis of the errors incurred when not fully resolving the entire range of scales is difficult to perform and depends upon the variables of interest. For instance, the error analysis presented in *Cerutti* and *Meneveau* [10.170] shows that, when evaluating the average value of the trace of the SGS tensor ($\langle \tau_{ii} \rangle$) from four probes separated by a distance $\Delta/2$ using a discrete box filter in the cross-stream direction, errors are on the order of 6%. This analysis is based on analytical integration of a theoretical spectrum where one assumes an ideal inertial range of turbulence. Evaluating the actual uncertainty in more complex flows with anisotropic and inhomogeneous statistics is a challenging task.

Results from a priori studies using arrays of point sensors include the works of *Cerutti* and *Meneveau* [10.170] and *Cerutti* et al. [10.171], who used four-probe hot-wire arrays in a wind tunnel to measure filtered turbulence signals and SGS stresses. The results were used to quantify so-called spectral hyperviscosity models. Later, *Kang* and *Meneveau* [10.172] used hot-wire arrays for a study of SGS anisotropy in a turbulent wake, showing that the covariance tensors of SGS quantities and filtered velocity gradients became more isotropic at small scales. However, for third-order moments associated with energy flux and SGS dissipation, the approach to isotropy was exceedingly slow. *Kang* and *Meneveau* [10.173] quantified the direct ef-

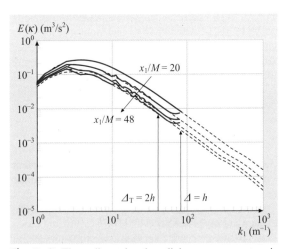

Fig. 10.61 Three-dimensional radial energy spectra in decaying isotropic turbulence. The *solid* and *dashed lines* represent the results from LES (using the dynamic Smagorinsky model) and experiments, respectively. The experimental spectra are shown in their entirety without filtering. The filter cutoff is at wavenumber $k_1 = \pi/\Delta = 78.5 \, \text{m}^{-1}$. The test-filter scale is at $k_1 = \pi/2\Delta$. (after *Kang* et al. [10.169], with permission)

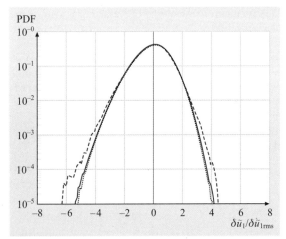

Fig. 10.62 Probability density function of the filtered velocity increment $\delta \tilde{u}_1 = \tilde{u}_1(x_1 + r) - \tilde{u}_1(x_1)$ at a displacement $r = \Delta$ and at 48 mesh sizes downstream of the active grid. The filter size was $\Delta = 0.08$ (m). The velocity increments are normalized with their root-mean-square values. *Dashed line*: experimental data; *dotted line*: Smagorinsky model; *dashed-dot line* (indistiguisheable from dotted line): dynamic Smagorinsky model; *solid line*: dynamic mixed nonlinear model (after *Kang* et al. [10.169], with permission)

fect of coherent structures in a cylinder wake on SGS dynamics and found that they had a surprisingly strong impact across scales, even at filter scales significantly smaller than the size of the coherent structures. Thus the approach to isotropy appears much slower than is assumed in most models for many practical applications of LES. *Chen* et al. [10.174] used hot- and cold-wire arrays consisting of three hot- and cold-wire probes to study a heated turbulent jet flow. Their study focused on joint statistics in the context of filtered density function methods in LES.

Array-filtered data have also been used for a posteriori tests of LES turbulence models. *Kang* et al. [10.169] generated a database from measurements in a wind tunnel (a remake of the *Comte-Bellot* and *Corrsin* [10.175] turbulence decay experiment) behind an active grid. Active grids can generate turbulence at significantly higher Reynolds number [10.176] than passive grids and hence are a useful tool for turbulence generation in LES studies, which typically focus on high-Re flows ($Re_\lambda \approx 720$ was achieved at the first measurement station of *Kang* et al. [10.169]). The energy spectra at four downstream distances from the grid, ranging from 20 to 48 grid-mesh sizes, were measured and documented for subsequent initialization of, and comparison with, LES. The data were recorded using an array of four X-wire probes with which different filter sizes could be achieved by varying the probe separation. Besides the spectra, higher-order statistics of the filtered velocity were quantified by measuring probability density functions, hyperflatness and skewness coefficients of two-point velocity increments. The data were used to study the ability of LES to reproduce both the spectral and higher-order statistics of the resolved velocity field. Specifically, the Smagorinsky, dynamic Smagorinsky, and dynamic mixed nonlinear models were considered. Overall, it is found that these various LES models predicted accurate low-order statistics and spectra of resolved scales in isotropic turbulence during the decay. For instance, Fig. 10.61 shows the time evolution of radial energy spectra deduced from the experimental data; the solid lines are the results from LES using the dynamic model. As can be seen, there is very good agreement in the range of wavenumbers resolved by the LES, except for a small pile-up of energy next to the cutoff wavenumber.

To quantify the ability of LES to reproduce higher-order statistics, probability density functions of filtered velocity increments were measured and compared to LES; the results are shown in Fig. 10.62. It can be seen that the three models underpredicted the intermittency of longitudinal velocity increments at small distances.

Turbulence in the atmospheric surface layer also poses important modeling challenges for LES. Various experimental field campaigns to deploy arrays of sonic anemometers have been undertaken. *Tong* et al. [10.163] and *Porté-Agel* et al. [10.164] performed experiments in which a horizontal array of sonic anemometers, at a single height, was used to sample the wind turbulence fluctuations. A next generation of experiments [10.166, 167, 177, 178] incorporated another horizontal array of sonic anemometers at a second height, allowing vertical gradients to be evaluated in conjunction with 2-D horizontal filtering. Such a configuration allowed for the computation of the full filtered strain-rate tensor, and comparisons of all tensor components and associated geometric alignments. A priori studies of the Smagorinsky model coefficient measured again using (10.110) were performed [10.178]. Results were used to quantify the effects of atmospheric thermal stratification and distance to the ground on the characteristic trends of the coefficient. *Kleissl* et al. [10.179] used test filtering at various scales to examine the performance of the dynamic (see [10.141]) and scale-dependent dynamic (see [10.180]) models based on the data. They found that the scale-dependent version of the model gave much better predictions of the coefficients and the trends that had been observed in *Kleissl* et al. [10.178].

10.4.3 Planar Particle Image Velocimetry (PIV) for SGS Dynamics and LES

The ideal experimental data for a priori evaluation of SGS stress models for LES must simultaneously resolve a wide range of length scales at a large number of measurement locations. Consequently, the instantaneous spatial distribution of two or three velocity components in sample planes provided by particle image velocimetry (PIV) (Sect. 5.3.2) is an ideal tool for obtaining such data. Spatial filtering of the instantaneous velocity distribution, typically using a box filter, provides the resolved flow, which can then be derived to obtain the in-plane components of the filtered velocity gradient tensor, frequently used in SGS stress models. The SGS stresses can also be calculated directly, based on the definition given in (10.101). Planar PIV provides entire planes of filtered and SGS variables, while arrays of single-point measurements described in Sect. 10.4.2 provided such data only along a single coordinate (time). 2-D maps of the local, instantaneous SGS dissipation, $(\Pi(x, t))$, defined as in (10.108) but without the averaging operator, can also be obtained from planar PIV data. Its instantaneous value can be positive, i.e., energy flowing from resolved to SGS scales or negative, i.e., backscatter of energy from subgrid to resolved scales. However, its ensemble-averaged value (10.108) is typically, but not always, positive, as described in detail in this section.

Planar PIV data only provides the in-plane components of (contributors to) the SGS dissipation. The first application of PIV data to evaluate SGS stress models was reported in *Liu* et al. [10.149] using 2-D data obtained in the far field of a round jet. Bearing in mind that these measurements were performed during the early 1990s, double-exposure film photography was the only means of obtaining PIV data that could be spatially filtered at different scales to measured trends. Their 56×56 mm negatives were scanned and converted to 5000×5000 pixels arrays, and autocorrelation analysis [10.181] was used to calculate the velocity, taking advantage of the unidirectional flow in the jet. The dissipation rate and related scales of the turbulence were estimated by calculating the ensemble-averaged, radial two-dimensional spectra of both velocity components. Fitting a Kolmogorov spectrum with a -5/3 slope line to the *inertial* part of these spectra provided the dissipation rate.

To account for the missing out-of-plane components, *Liu* et al. [10.149] used various estimates based on assumed local turbulence isotropy. Several spatial filters were used to calculate the SGS stresses and filtered velocity gradients, including:

1. A *box* or *top-hat* filter, i.e., a spatially averaged velocity over a domain with size Δ surrounding a point. The filtered area size ranged from 8×8 to 32×32 velocity vectors. Because of its simplicity, it has been the most popular method.
2. A truncated Gaussian filter.
3. A spectral cutoff filter, which inherently involved significant truncation errors in a velocity field with finite size. Consequently, spectral cutoff filters have had very limited use.

To be consistent with the data that would be available during LES, spatial velocity gradient were calculated using coarse grids, i.e., at scales of the filtered field. The results enabled direct observations of the spatial distributions of stresses in comparison to modeled values. Lack of agreement between the measured stresses and the predictions of the Smagorinsky model with a fixed coefficient was clearly evident, as also confirmed by calculation of the correlation coefficients between measured and predicted stresses. These trends confirmed prior findings based on direct numerical simulations at lower Reynolds numbers [10.147, 148].

10.4.4 Case Studies and Sample Results Using Planar PIV Measurements

To examine the relationships between flow structures at different scales, *Liu* et al. [10.149] decomposed the velocity field into logarithmic bands. Observed similarity and coherence between structures in consecutive bands led to a reformulation of the stress scale-similarity model,

$$\tau_{ij}^L = c_L L_{ij} \,, \qquad (10.111)$$

where an overbar indicates filtering at scale $\alpha \Delta (\alpha \geq 1)$, and L_{ij} is the *resolved stress* (10.104), i.e., the stress that would be obtained by filtering products of the resolved velocity components available from the LES data. For $\alpha = 1$, the scale-similarity model coincides with the *Bardina* et al. [10.147] model. The measured correlation coefficients of the similarity model with the measured stress were substantially higher than those of the eddy-viscosity closures, in agreement with visual comparisons between distributions of measured and modeled values. The model coefficient was also calculated by matching the modeled and measured SGS dissipation, i.e.,

$$C_L = \frac{\langle \tau_{ij} \tilde{S}_{ij} \rangle}{\langle L_{ij} \tilde{S}_{ij} \rangle} \,. \qquad (10.112)$$

The experimental values were $C_L \approx 1$. The first-order approximation of the similarity model, the nonlinear (or Clark model, *Clark* et al. [10.148]) model (10.107), maintained a high correlation with measured stresses. To avoid stability problems during simulations, already observed by *Bardina* et al. [10.147], these similarity models need to be combined with eddy-viscosity model, leading to a mixed model (for a review of the relevant literature, see *Meneveau and Katz* [10.146] and *Vreman* et al. [10.181]). These models also maintained the high correlation with the measured stress, but the eddy-viscosity term contributed to a positive SGS dissipation, reducing problems associated with numerical instability.

Using the PIV data of *Liu* et al. [10.149], *Meneveau and Katz* [10.182] explicitly evaluated the errors and established the accuracy with which models reproduced flow features at scales falling between the grid and test filters. The analysis demonstrated that the mixed nonlinear model yielded less error than the dynamic Smagorinsky and dynamic mixed model involving the similarity model.

Bastiaans et al. [10.183] used 2-D particle tracking velocimetry (PTV) to examine the SGS stresses, dissipation, and model predictions in a confined free convection flow generated by a transitional thermal plume. Their measured distributions of in-plane stress components and SGS dissipation rate showed regions of mean-energy backscatter within the plume, and on both sides of the impingement region on the upper wall of their facility. A priori testing of the Smagorinsky model showed reasonable qualitative agreement for the shear and wall-normal stress, except for regions located close to the wall, but their values differed within the plume. Predictions of normal SGS stress components by the dynamic model with a wide range of coefficients showed some improvements, but their values were still underestimated. For the shear stress, the dynamic model predictions had the wrong sign. However, inherently, the dynamic model provided reasonable values for the mean SGS dissipation rate. Near the no-slip wall, the predictions of Π_Δ depended on how the dynamic procedure was implemented.

Liu et al. [10.184] studied experimentally the response, evolution, and modeling of SGS stresses during rapid straining of turbulence. A unique experimental setup generated nearly isotropic turbulence with very low mean velocity in a water tank by means of four symmetrically located spinning grids. Spatially uniform rapid axisymmetric expansion, i.e., an axisymmetric stagnation-point flow, was generated by pushing two disks towards each other in the central portion of the tank. The rod connected to each disk was pushed by a cam, which was driven by precision stepping motors. In order to produce time-independent straining, the distance between the plate and the stagnation point $x_p(t)$ had to be of the form $x_p(t) = x_p(t_0) \exp[-S^*(t-t_0)]$, where S^* was the mean strain rate. Consequently the speed of the plates was $U(t) = S^* x_p(t)$. The shape of the cams was designed to produce the desired displacement.

The velocity was measured using 2-D PIV on a horizontal plane containing the axis of the plate and the stagnation point. The light sheet was generated by a copper vapor laser, and images were recorded on 35 mm film using a movie camera operating at 33 frame/s. An electro-optical image shifting technique involving a ferroelectric liquid crystal [10.185] was used to eliminate the problem of directional ambiguity in the double-exposure images. Each dataset consisted of 10 images. Results of 10 such experiments were used during the data analysis. The initial dissipation rate ε_0 and Kolmogorov scale, $\eta_0 = (\nu^3/\varepsilon_0)^{1/4}$, were again estimated from radial two-dimensional spectra of both velocity components, which had clear inertial ranges with a $-5/3$ slope. One-dimensional spectra obtained for perpendicular planes were used for confirming that local isotropy indeed existed. Subsequently, the rms values of velocity fluctuations provided the integral scale $L_0 = u'^3/\varepsilon_0$ the Taylor microscale, $\lambda = u'(15\nu/\varepsilon_0)^{1/2}$, and the Taylor microscale Reynolds number, $Re_\lambda = u'\lambda/\nu$ (=290). The dimensionless straining parameter was $S^* k_0/\varepsilon_0 = 17$, where k_0 was the turbulent kinetic energy prior to straining.

Ensemble and spatial averaging of the PIV data for each phase of the straining process provided S^*, and spatial box filtering provided in-plane SGS stresses and the velocity gradient tensor $\tilde{A}_{ij}$. When straining started, the initially isotropic turbulence became anisotropic, with the compressed τ_{11} increasing and becoming scale dependent immediately, while the extended τ_{22} responded slowly, eventually also increasing, but remaining scale independent. To elucidate the phenomena, the velocity was decomposed into a turbulent part, $u_i^T(x_i, t)$, and an applied straining part, $U_i(x_i, t)$. Accordingly, the SGS stress was decomposed into three elements: a turbulence part τ_{ij}^T, a cross-term τ_{ij}^C, and an applied straining part τ_{ij}^M, where

$$\tau_{ij}^T = \widetilde{u_i^T u_j^T} - \tilde{u}_i^T \tilde{u}_j^T \,,$$
$$\tau_{ij}^C = \widetilde{u_i^T U_j} - \tilde{u}_i^T \tilde{U}_j + \widetilde{U_i u_j^T} - \tilde{U}_i \tilde{u}_j^T \,,$$
$$\tau_{ij}^M = \widetilde{U_i U_j} - \tilde{U}_i \tilde{U}_j = \frac{\Delta^2}{12} \frac{\partial U_i}{\partial x_m} \frac{\partial U_j}{\partial x_m} \,. \qquad (10.113)$$

For spatially uniform mean straining: $\tau_{11}^M = (\Delta^2/12)\langle \tilde{S}_{11}\rangle^2$, $\tau_{22}^M = (\Delta^2/12)\langle \tilde{S}_{22}\rangle^2$, $\tau_{12}^M = 0$. Analysis of the data showed that $\langle \tau_{ij}^T \rangle \sim \Delta^{\frac{2}{3}}$, in agreement with Kolmogorov scaling, and that $\tau_{ij}^C \sim \Delta$. The data also confirmed $\tau_{ij}^M \sim \Delta^2$. At large filter scales, all the stress elements became comparable. The measured data were compared to predictions of rapid distortion theory (RDT [10.186]). Comparing the evolution of the predicted and measured normal stress components, the experimental results were more isotropic than RDT predictions. The evolution of the anisotropy tensor

$$b_{ij}^\tau = \frac{r_{ij}}{r_{kk}} - \frac{1}{3}\delta_{ij} \,,$$
$$r_{ij}(t) = \frac{[\tau_{ij}(t_i)]_{\text{spatially and ensemble-averaged}}}{[\tau_{ij}(t_0)]_{\text{spatially and ensemble-averaged}}} \,, \qquad (10.114)$$

where t_i represents a certain phase in the straining process, is shown in Fig. 10.63 below (Fig. 12 in *Liu* et al. [10.184]). As is evident, b_{ij} became increasingly isotropic with decreasing scale. However, even at the largest scale considered (280η), b_{ij} is still less anisotropic than the RDT prediction. This trend is not surprising since the measured characteristic time scale for eddies of size Δ, $T_\Delta = \varepsilon^{1/3}\Delta^{2/3}$, is comparable to the time scale of the applied straining. Thus, the basic assumption of RDT, namely that $S^{*-1} \ll T_\Delta$, does not apply to that study.

Analysis of correlation coefficients between the modeled and measured SGS stresses showed that rapid straining did not affect the high correlation values for the similarity model, and only slightly increased the low correlation coefficients for the Smagorinsky model. Another method for evaluating models was based on the squared error between the real and modeled SGS force, i.e., the divergence of the SGS stresses. The error of the Smagorinsky model was significantly larger than that of the similarity model but both were substantial. Trends of the ensemble-averaged SGS dissipation for each phase of the straining process were also compared to those calculated based on eddy-viscosity and similarity models. During straining the SGS dissipation increased (as expected) and became scale dependent, increasing with filter size. The measured SGS stress model coefficients, calculated using pairs of equations combining (10.110) and (10.112), showed that straining decreased C_L by more than 50%, and more than doubled C_S^2. The latter indicated that a model with fixed coefficients would underpredict dissipation, in contradiction to the expectation that the Smagorinsky model should overdissipate energy in highly strained flows. To explain these trends, the contributions of the turbulent, cross- and applied straining parts of the SGS stress were examined separately, and showed opposing trends. The turbulent part decreased during straining, in agreement with *McMillan and Ferziger* [10.187]. Conversely, the dissipation associated with the cross-term increased substantially, and remained high during straining. The applied straining part increased during the acceleration phase, and slightly decreased during the constant-straining stage. The opposing trends of the similarity and eddy-viscosity models during rapid straining motivated the introduction of a mixed model of the form

$$\tau_{ij}^{\text{mix}} = \sigma c_L L_{ij} - 2(1-\sigma)\left(C_S^\Delta \Delta\right)^2 |\tilde{S}|\tilde{S}_{ij} . \quad (10.115)$$

Using previously determined empirical coefficients for each model, i.e., $c_L = 1$ and $C_S^\Delta = 0.09$, the fitted value of $\sigma = 0.30$ was remarkably accurate in predicting the correct total SGS dissipation throughout the entire experiment, including the acceleration stage. The similarity term reproduced the effects of the mixed and mean terms whereas the Smagorinsky term was similar to the dissipation by the turbulent stresses. However, model coefficients that reproduced the correct amount of SGS dissipation severely underpredicted the magnitude of $\langle \tau_{ij} \rangle$ both before and during the distortion. For $\langle \tau_{11} \rangle$ there was a discrepancy of nearly a factor of 3.

Conditional averaging of the data of *Liu* et al. [10.184] was used by *Meneveau and Katz* [10.188] to study how regions of large-scale straining, rotation, and energy cascade rate were affected the SGS force and dissipation rate. The locally isotropic jet data showed that the SGS force surrounding points of large strain-rate magnitude was nearly radial. In a divergence-free velocity field, the SGS force could only affect the resolved pressure field. Being directed outwards, the SGS force decreased the resolved pressure in regions of high strain-rate magnitude. Similar trends were obtained in regions of large dissipation, but there was no effect of the resolved vorticity. In the rapidly distorted flow, the SGS force in regions of large positive dissipation decreased the resolved pressure, and opposed the mean deformation. In regions of large energy backscatter, the SGS force acted to favor the mean deformation. In terms of model performance, the mixed model fared better than the Smagorinsky or the similarity models alone. However, there were substantial discrepancies between the modeled and predicted forces.

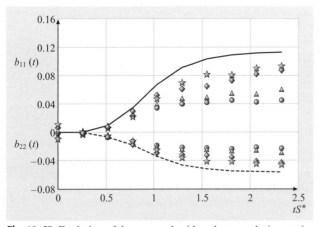

Fig. 10.63 Evolution of the mean subgrid-scale stress during straining of turbulence, as measured using planar PIV. Shown is the anisotropy tensor b_{ij} of normalized stress r_{ij}. The *solid line* and *upper set of symbols*: 11 component; *dashed line* and *lower set of symbols*: 22 component. Stars: $\Delta/\eta = 280$, rhombs: $\Delta/\eta = 80$, triangles: $\Delta/\eta = 40$, circles: $\Delta/\eta = 20$; lines: 2-D filtered RDT prediction (*Liu* et al. [10.184], with permission from CUP)

The study of the effect of rapid distortion on the dynamics of SGS stresses was expanded in *Chen* et al. [10.189, 190]. A larger experimental setup (Fig. 10.64) was used to generate a controlled planar straining–relaxation–destraining cycle of initially isotropic turbulence.

Locally isotropic turbulence at moderately high Reynolds numbers, with very low mean velocity, was again generated using four symmetrically located active grids. The uniform planar straining was applied by translating a piston vertically at prescribed velocities. The piston occupied almost the entire width of the 20 cm-wide tank. The bottom surface of the piston had rounded corners to prevent possible flow separation, which was critical when the flow was destrained. The piston was driven by a precision, programmable, linear electric actuator.

The piston maximum velocity was 1 m/s, and its maximum displacement was 76.2 cm. The sample area for PIV measurements, 9×9.5 cm, was located near the center-bottom of the tank. The images were recorded using a $2K \times 2K$ digital camera at 5 Hz, this time in a cross-correlation mode, synchronized with the piston phase. To increase the number of phases being examined to 48, data were acquired with varying initial delays. The size of the interrogation windows was 32×32 pixels, which with 50% overlap provided 121×121 vectors with a spacing of 0.7 mm. At every phase, 1000 instantaneous realizations gave a sufficient ensemble set for statistical convergence. The evolution of ensemble- and spatially averaged strain rate during the experiments is shown in Fig. 10.65. Spatial spectra at different locations confirmed the existence of local isotropy and homogeneity, and provided estimates for the dissipation rate, for the Kolmogorov scale $\eta_0 \sim 130\,\mu$m, $L_0 = 0.13$ m, $Re_\lambda \sim 400$, and $S_{max}k_0/\varepsilon_0 \sim 9.5$, where the subscript zero refers to conditions prior to straining.

The Reynolds stress components responded as expected to the applied straining (see Fig. 11 of *Chen* et al. [10.190]). During the first half of the straining phase, with increasing strain magnitude $\langle u'_1 u'_1 \rangle$, the stress in the extended direction decreased while $\langle u'_2 u'_2 \rangle$, the stress in the contracted direction, increased. Subsequently, until the end of the relaxation period, $\langle u'_2 u'_2 \rangle$ decreased gradually and $\langle u'_1 u'_1 \rangle$ remained approximately constant. At the end of the relaxation regime, the turbulence had not yet returned to an isotropic state. During destraining, the compressed $\langle u'_1 u'_1 \rangle$ increased rapidly, while $\langle u'_2 u'_2 \rangle$ continued with its decreasing trend, reaching a minimum value later than the destraining peak. A comparison of the measured stresses and

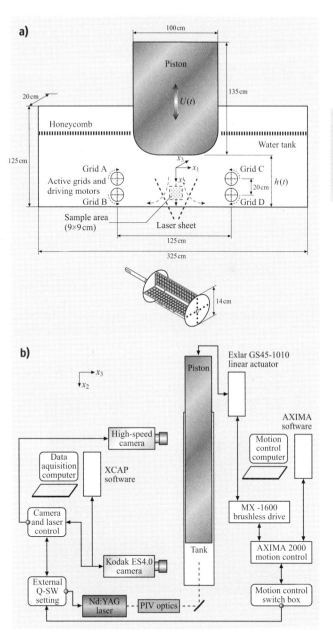

Fig. 10.64a,b Schematic diagram of facility for rapid straining and destraining turbulence experiment. (**a**) Tank and piston and a sketch of the spinning grids. (**b**) Instrumentation used to record PIV data, phase-locked with the piston motion cylce (after *Chen* et al. [10.190] with permission from CUP)

anisotropy tensor to the predictions of RDT showed that RDT gave the correct trends but overestimated $\langle u'_2 u'_2 \rangle$ and underestimated $\langle u'_1 u'_1 \rangle$, thus overpredicting the de-

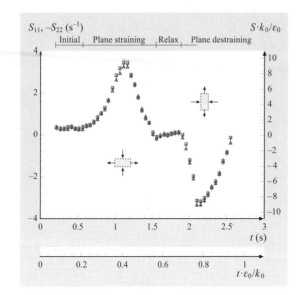

Fig. 10.65 Evolution of mean strain, spatially averaged rate S_{11} (*squares*) and S_{22} (*triangles*) during straining and destraining experiment (after *Chen* et al. [10.189], with permission from ASME). The *error bars* represent the standard deviation of spatial distribution. Measurements were performed using PIV

$$\Psi = -\langle u'_i u'_j \rangle \frac{\partial \langle u_i \rangle}{\partial x_j} \approx -\langle u'_1 u'_1 \rangle$$
$$\times \frac{\partial \langle u_1 \rangle}{\partial x_1} - \langle u'_2 u'_2 \rangle \frac{\partial \langle u_2 \rangle}{\partial x_2}, \quad (10.116)$$

well approximated from the in-plane components, followed the amplitude of the applied strain during the straining period. Once the straining stopped, Ψ vanished. During the destraining period, the production rate became negative initially, and turned to positive values, which were much smaller than those during the straining phase. The occurrence of negative production was a result of $S(t) = \partial \langle u_1 \rangle / \partial x_1 = -\partial \langle u_2 \rangle / \partial x_2$ becoming negative while $\langle u'_1 u'_1 \rangle < \langle u'_2 u'_2 \rangle$.

For homogeneous turbulence, common wisdom is that SGS variables in the LES context will tend to their RANS counterparts as the filter scale approaches the integral scale of turbulence. The PIV data obtained in the strain–destraining experiment were particularly well suited to verify such expectations for a spatially homogeneous, but temporally highly complex, flow. To perform this analysis, the SGS stress was calculated over scales ranging from 25 to $430\eta_0$. Figure 10.66 (*Chen* et al. [10.190]) shows the evolution of $b_{11}^{\tau,2D}$, the mean 2-D surrogate of the anisotropy tensor of the subgrid scales, where

$$b_{ij}^{\tau,2D} = \langle \tau_{ij} \rangle - \frac{1}{2} \langle \tau_{kk} \rangle \delta_{ij}, \quad (10.117)$$

along with the 2D surrogate of the corresponding Reynolds stress anisotropy tensor component, $b_{11}^{Re,2D}$, where

$$b_{ij}^{Re,2D} = \langle u'_i u'_j(t) \rangle / \langle u'_i u'_j(t_0) \rangle$$
$$- \frac{1}{2} \langle u'_k u'_k(t) \rangle / \langle u'_k u'_k(t_0) \rangle \delta_{ij} \quad (10.118)$$

(definitions used here differ from (10.114) since the present flow is planar while in the experiments of *Liu* et al. [10.184] it was axisymmetric). As is evident, the lag in the response of $b_{11}^{\tau,2D}$ to the applied straining increases with filter scale, i.e., the small scales respond earlier than the large scales. This trend is consistent with RDT, in contrast to the *energy cascade* process, in which the energy is first fed into the large scales, before cascading to small scales. Figure 10.66 also confirms that,

gree of anisotropy. The predicted magnitudes were only comparable to the measured data at initial times. The turbulence production rate

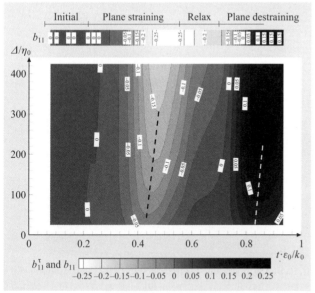

Fig. 10.66 Contour plot of the mean SGS stress anisotropy tensor component (11) as a function of dimensionless time during the straining–destraining cycle, and as function of the filter scale Δ/η. The *top bar* shows the corresponding Reynolds stress anisotropy as a function of time. Measurements were performed using planar PIV (after *Chen* et al. [10.190], with permission from CUP)

when the spatial filter scale approaches the integral scale, the response lag and magnitude of the SGS stress tend to those of the Reynolds stress. Note that for these data $L_0/\eta_0 \approx 930$, i.e., the largest filter scale, is still less than 50% of the integral scale.

A comparison between the trends of the turbulence production rate and those of $\langle \Pi_{\Delta-2D} \rangle$, presented in Fig. 10.67 (*Chen* et al. [10.190]). Prior to straining and during the relaxation period, the SGS dissipation was nearly scale independent. However, it became scale dependent during the straining and destraining periods. The SGS dissipation peak during destraining was only about 50% of the peak during straining, in spite of the fact that the amplitudes of straining and destraining were about the same. For large filter size, at the beginning of the destraining phase $\langle \Pi_{D-2D} \rangle$, became negative, i.e., there was energy backscatter, also consistent with the occurrence of negative production during the same period. This was caused by differences in the initial conditions, which for the straining phase was isotropic turbulence but for the destraining phase involved anisotropic turbulence that arose from a relaxation period that was too short to allow the turbulence to return to isotropy. When the contributions to the SGS dissipation were decomposed into turbulent, mixed, and mean straining parts, the turbulent contribution was by far the most dominant. These results further confirm that, as Δ approaches L_0, the mean SGS dissipation tends toward Ψ. The difference in their magnitudes occurs in part due to the scale gap, but mostly due to the missing out-of-plane components in $\langle \Pi_{D-2D} \rangle$. *Chen* et al. [10.190] also examined the evolution of energy spectra during the straining–destraining cycle. This analysis confirmed that small scales responded earlier than large scales to the straining. Furthermore, they also showed that RDT gave relatively good predictions during early phases of the cycle, especially at large scales, for which the underlying assumptions of RDT were better satisfied. As relaxation started, the data and predictions deviated from each other, as expected.

The evolution of the Smagorinsky coefficient C_s^2 was measured [10.189] from the PIV data (using (10.110)); C_s^2 decreased as straining or destraining started since $2\Delta^2 \langle |\widetilde{S}| \widetilde{S}_{ij} \widetilde{S}_{ij} \rangle$ responded more quickly to the applied strain than the SGS dissipation, but then increased above the equilibrium value. This deviation increased with Δ. The standard dynamic Smagorinsky coefficient, $(C_s^{\Delta,\text{dyn}})^2$, also calculated from the data but using (10.106) with $\alpha = 2$, was compared to $(C_s^\Delta)^2$. Results showed that the dynamic model overpredicted the response to straining and destraining. Furthermore, the

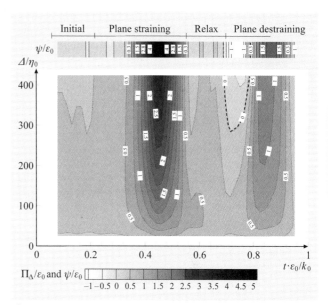

Fig. 10.67 Contour plot of mean SGS dissipation rate as a function of dimensionless time during the straining–destraining cycle, and as a function of filter scale Δ/η. The *top bar* shows the corresponding Reynolds production as function of time. The *dashed line* denotes the zero contour, meaning that in the enclosed top U-shaped region, the mean SGS dissipation is negative. (After *Chen* et al. [10.190], with permission from CUP)

scale-dependent dynamic model [10.161, 179] was also tested, now involving a second test-filter scale of 4Δ. Results did not improve and, in parts of the cycle, were even worse. Examination of the coefficients of the nonlinear dynamic mixed models showed only limited improvements over the standard dynamic Smagorinsky model results. The data were used to trace the origin of these behaviors to scale-dependent variations of the response time of turbulence to the applied straining. They showed that, even in a spatially simple (but temporally complex) flow, serious modeling challenges for LES remain.

Use of PIV data for the analysis of SGS dynamics in spatially complex flow environments with nonuniform mean flow poses further special challenges. For example, the flow structure within turbomachines involves interactions between multiple wakes generated by upstream blades with wakes and blades located downstream. Consequently, the spatial scales of the mean flow and turbulence are comparable in magnitude to the resolution of PIV measurements. Spatial filtering of such a flow field separates both the mean (phase-averaged) flow as well as the turbulence to the mean and subgrid parts, each with its own kinetic energy and fluxes.

As described in *Chow* et al. [10.191], the existence of multiple energy and flux terms may lead to confusion about the relationships between the RANS production rate (Ψ) and the SGS dissipation of kinetic energy. Their experiments were performed in a two-stage axial turbomachine flow-visualization facility containing fluid (a solution of NaI in water) with an optical index of refraction that matched that of the acrylic blades. The unobstructed view of the entire stage enabled 2-D and stereo PIV measurements in numerous sample planes. Data provided insight into the flow within turbomachines at unprecedented levels of detail, enabling the investigation of various wake–blade, wake–wake, and wake–boundary layer interactions. Such data provided a unique database for both a posteriori and a priori testing of turbulence models. The data have already been used for a posteriori testing of RANS models [10.192].

For a priori LES studies, the data were used to examine the SGS dissipation in a near-rotor wake. There the PIV data showed that the phase-averaged SGS dissipation could be negative, even when the Reynolds production rate was positive. Negative mean SGS dissipation was also measured within the stator of a centrifugal pump [10.193]. An attempt to elucidate this paradoxical result led to the identification of all the relevant kinetic-energy and energy-flux terms that resulted from spatial filtering, followed by ensemble averaging of data as one would do while analyzing the SGS dynamics. The total spatially filtered and ensemble-averaged kinetic energy, $\langle \widetilde{K} \rangle$, was decomposed into four parts,

$$\langle \widetilde{K} \rangle \equiv 0.5 \langle \widetilde{u_i u_i} \rangle = K^{\mathrm{mr}} + K^{\mathrm{ms}} + K^{\mathrm{fr}} + K^{\mathrm{fs}},$$
(10.119)

where $K^{\mathrm{mr}} = 0.5 \langle \tilde{u}_i \rangle \langle \tilde{u}_i \rangle$ and $K^{\mathrm{ms}} = 0.5(\langle \widetilde{\langle u_i \rangle \langle u_i \rangle} \rangle - \langle \tilde{u}_i \rangle \langle \tilde{u}_i \rangle)$ are the mean resolved (mr) and the mean subgrid (ms) kinetic energies, respectively; $K^{\mathrm{fr}} = 0.5 \langle \tilde{u}'_i \tilde{u}'_i \rangle$ is the fluctuating resolved (fr) kinetic energy, and $K^{fs} = 0.5(\langle \widetilde{u'_i u'_i} \rangle - \langle \tilde{u}'_i \tilde{u}'_i \rangle)$ is the fluctuating subgrid (fs) kinetic energy. Distributions of these components for the rotor near wake, using a box spatial filter of 5×5 vectors ($\approx 25\%$ of the near wake width), are presented in Fig. 10.68.

The values of $K^{\mathrm{mr}}_{\mathrm{rel}}$, the only non-Galilean invariant term, are presented in the rotor frame of reference; K^{ms} has a two-layer structure as the boundary layers on both sides of the blade extend to the wake region. This represents the kinetic energy lost when a nonuniform mean flow field is spatially filtered, and has nothing to do with the turbulence. It decays at a faster rate than the other terms, in part due to wake spreading and the use of a fixed filter size, which increases the mean resolved part at the expense of the mean subgrid part. The distribution of K^{fr}, the resolved part of the turbulent kinetic energy, is biased toward the suction side, indicating higher injections of resolved turbulence

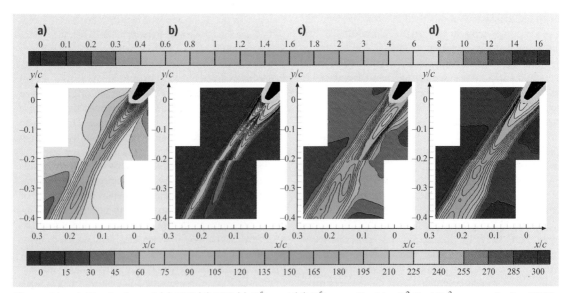

Fig. 10.68a–d Distributions of (a) $K^{\mathrm{mr}}_{\mathrm{rel}}$ (b) K^{ms} (c) K^{fr}, and (d) K^{fs} normalized by $U^2_{\mathrm{tip}} 3 \times 10^{-3}$ measured using PIV in an index-matched turbomachine facility. The *gray mask* around the blade covers points where spatial filtering overlaps with the blade (after *Chow* et al. [10.191], with permission from AIP)

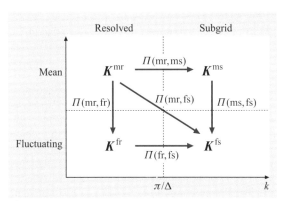

Fig. 10.69 A diagram showing the components of the kinetic energy and energy fluxes between them (after *Chow* et al. [10.191], with permission from AIP)

from the suction-side boundary layer. Regions of elevated turbulence on both sides of the rotor wake are caused by chopped wake segments of the upstream blades.

The evolution equations for each of the four kinetic-energy parts can be used to define various energy fluxes across the various regions. Specifically, it was shown in *Chow* et al. [10.191] that defining the following stresses and strain rates as $\tau_{ij}^m = \langle \widetilde{u_i} \rangle \langle u_j \rangle - \langle \tilde{u}_i \rangle \langle \tilde{u}_j \rangle$, $\tau_{ij}^f \equiv \langle \widetilde{u'_i u'_j} \rangle - \langle \tilde{u}'_i \tilde{u}'_j \rangle$, $\tau'_{ij} = \tau_{ij} - \langle \tau_{ij} \rangle$, $S'_{ij} = S_{ij} - \langle S_{ij} \rangle$ and $R^r_{ij} = -\langle \tilde{u}'_i \tilde{u}'_j \rangle$, the energy fluxes between the different kinetic energy parts are

$$\Pi(\text{mr,ms}) = -\tau_{ij}^m \langle \tilde{S}_{ij} \rangle,$$
$$\Pi(\text{mr,fs}) = -\tau_{ij}^f \langle \tilde{S}_{ij} \rangle,$$
$$\Pi(\text{mr,fr}) = R^r_{ij} \langle \tilde{S}_{ij} \rangle,$$
$$\Pi(\text{ms,fs}) = \widetilde{R_{ij} \langle \tilde{S}_{ij} \rangle} - \tilde{R}_{ij} \langle \tilde{S}_{ij} \rangle,$$
$$\Pi(\text{fr,fs}) = -\langle \tau' \tilde{S}'_{ij} \rangle. \quad (10.120)$$

In these expressions the energy fluxes are denoted as Π (donor, receiver). For clarity, the relationships between the various terms and fluxes are illustrated in Fig. 10.69, and the distributions of in-plane contributions to these flux terms are presented in Fig. 10.70. The sum of energy fluxes from the resolved to subgrid scales, i.e., in the horizontal direction in Fig. 10.69, is the ensemble-averaged SGS energy flux of (10.108)

$$\Pi_\Delta = -\langle \tau_{ij} \tilde{S}_{ij} \rangle = \Pi(\text{mr,ms}) + \Pi(\text{mr,fs}) + \Pi(\text{fr,fs}). \quad (10.121)$$

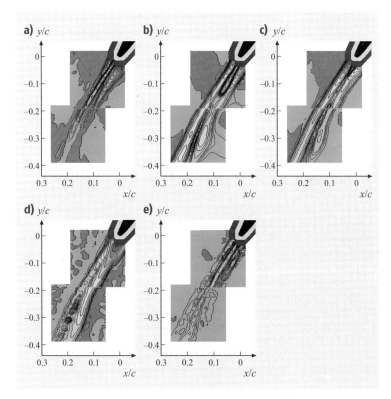

Fig. 10.70a–e Distributions of (**a**) $\Pi(\text{mr,ms})$, (**b**) $\Pi(\text{mr,fr})$, (**c**) $\Pi(\text{mr,fs})$, (**d**) $\Pi(\text{ms,fs})$, and (**e**) $\Pi(\text{fr,fs})$. All quantities are normalized by $U_{\text{tip}}^3/c^3 10^{-3}$ (after *Chow* et al. [10.191], with permission from AIP)

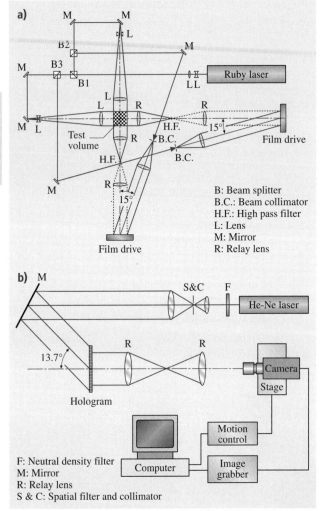

Fig. 10.71a,b Optical setup of: (**a**) the dual-view, hybrid off-axis HPIV system; and (**b**) reconstruction and scanning systems. (After *Tao et al.* [10.194], with permission from CUP)

The sum of fluxes from the mean components to the fluctuating parts, i. e., downward in Fig. 10.69, is the filtered kinetic-energy production rate,

$$\tilde{\Psi} = \widetilde{R_{ij}\langle S_{ij}\rangle} = \Pi(\text{mr,fr}) + \Pi(\text{mr,fs}) + \Pi(\text{ms,fs}) \,. \tag{10.122}$$

As is evident, $\tilde{\Psi}$ and $\tilde{P}$ have only one common term $\Pi(\text{mr,fs})$. Note that $\Pi(\text{mr,ms})$ involves only the mean flow, i. e., it has nothing to do with the turbulence, but it has a substantial impact on the SGS dissipation. The negative areas seen in Fig. 10.70 emanating from the boundary layers on both sides of the blade indicate flux from the subgrid to the resolved scales due to the growth of the wake. The highest positive flux occurs from the mean subgrid to the fluctuating subgrid kinetic energy [$\Pi(\text{ms,fs})$]. This term dominates the total production rate, i. e., most of the turbulence is produced by the subgrid mean flow. Negative production of resolved turbulence [$\Pi(\text{ms,fr})$] occurs downstream of the blade trailing edge as the boundary-layer turbulence is fed into a region with adverse velocity gradients. This analysis highlights that the production and SGS dissipation rate may have substantially different trends and magnitudes in spatially nonuniform flows. The results also highlight the usefulness of planar PIV data with good spatial resolution in elucidating important concepts at the interface between LES and RANS.

10.4.5 Holographic PIV Measurements of SGS Dynamics

Holographic particle image velocimetry (HPIV) is an experimental technique that provides three-dimensional velocity distributions and their gradients within a sample volume with extended depth. This technique consists of recording double-exposure holograms of a flow seeded with particles, and determining the 3-D velocity distributions by measuring the displacements of these particles. Until recently, holograms have been recorded only on high-resolution film, and then reconstructed optically to create a 3-D image of the original sample volume. The 3-D images are scanned to obtain 2-D slices through the 3-D volume, and then analyzed using various PIV- or PTV-based techniques. Various methods for recording holograms have been developed with optical setups ranging from simple but resolution-limited, inline holography to complex, high-resolution off-axis holography setups [10.194–201]. Achieving the 3-D vector distributions that can be filtered spatially at different scales, an ideal requirement for studying SGS dynamics, has been a challenge. Consequently, there have been very few applications of HPIV to address issues relevant to LES.

Tao et al. [10.194, 200] adapted an off-axis optical technique that had been introduced by *Zhang* et al. [10.197]. As illustrated in Fig. 10.71, this hybrid system combined the advantages of both inline and off-axis holography without having their drawbacks. In this setup, the subject beam illuminated the sample volume along the optical axis and the film recorded forward-scattered light from particles, similar to inline holography. However, a separate beam was used as the reference. A spatial high-pass filter was introduced between the sample volume and the film drive. It consisted of two identical lenses separated by twice their focal lengths and a pin installed at the focus of the first relay lens. The unscattered part of the subject beam was focused by the first relay lens and blocked by the pin. However, the light scattered from particles could still reach the film with minimum obstruction. This system reduced the speckle noise associated with inline holography, and allowed an increase in the intensity of the subject beam without overexposing the film.

This hybrid method provided the two velocity components that were perpendicular to the optical axis at high accuracy ($\approx 1\%$). The third component was significantly less accurate due to the *depth-of-focus problem*, namely that the traces of a particle persisted in out-of-plane sections over a substantial depth of about 1 mm. Several potential methods to overcome or at least alleviate this problem have been introduced, but none has succeeded in providing data in all three directions at the same level of precision in applications involving large volumes. The only available option to maintains a comparable level of accuracy was to record two perpendicular (or inclined) holograms simultaneously, as shown in Fig. 10.71a. Each hologram provided two velocity components and the 3-D data was obtained by combining the results. Using correlations, the redundant velocity component (out of the plane in the sketch) helped in precision matching of the two sets. Subsequently, *Sheng* et al. [10.201] introduced a technique that maintained the advantages of recording two orthogonal views, but required only one window and one recording system. This method was based on placing a mirror in the test section that reflected the object beam at an angle of 45°. Particles located in the volume in which the incident and reflected beams from the mirror overlapped were illuminated twice in perpendicular directions. Both views were recorded on the same hologram. Recently, (*Sheng* et al. [10.202]) showed that inline digital holographic microscopy substantially alleviated the depth-of-focus problem (but did not eliminate it) to a level that enabled measurements of 3-D particle motions at high resolution using a single view. Additional background on digital holography can be found in *Meng* et al. [10.203].

The HPIV measurements reported in *Tao* et al. [10.194, 200] were performed within a fully developed turbulent water flow in a square duct with a width H of 57 mm, and center-line mean velocity U of 2.1 m/s. The sample volume was $57 \times 57 \times 45$ mm³. The water was seeded with neutrally buoyant 20 μm

particles at a concentration of $8-12 /\text{mm}^3$. Double-exposure holograms, delayed by 60 μs, were recorded, and reconstructed using the setup of Fig. 10.71b. The illumination angle accounts for reference-beam angle and wavelength differences. A video camera equipped with a microscope objective automatically scanned the sample volume at a resolution of 4.9 μm/pixel, patched the images, enhanced them, and determined the velocity using in-house-developed PIV procedures [10.204]. The interrogation window size was 192×192 pixel $(0.93 \times 0.93 \text{ mm}^2)$. Unlike typical PIV, but similar to *Hart* [10.205], the images were compressed during acquisition and the correlation was computed directly from the compressed data. This approach maintained the high magnification without paying the penalty of a large database.

For the evaluation of SGS stress models for LES, the measurements focused on the 3-D velocity fields away from the immediate vicinity (5.25 mm) of the walls ($y^+ > 10^3$). With 65% overlap between the windows, each data set contained $136 \times 130 \times 128$ vectors with a spacing of 0.33 mm. A sample 3-D velocity map is presented in Fig. 10.72. Based on the measured spectra, the turbulent length scales were: $\eta \approx 100$ μm, $\lambda \approx 3.4$ mm and $Re_\lambda \approx 310$. The data were filtered by a 3-D, spa-

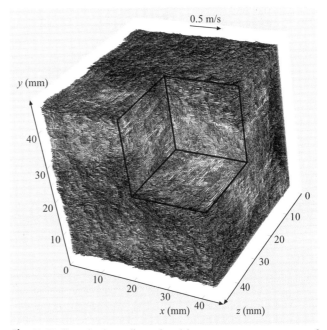

Fig. 10.72 Sample three-dimensional instantaneous vector map of turbulent flow in a square pipe measured using the setup shown in Fig. 10.71 (After *Tao et al.* [10.194], with permission from CUP)

tial box filter at $\Delta = 3.3$ mm ($= 33\eta$), i.e., in the inertial range of turbulence, which provided the SGS stresses, SGS dissipation, along with the filtered vorticity and strain-rate tensor. The uncertainty was about 2% for instantaneous velocity, and 15% for filtered velocity gradients. Data quality was evaluated by determining how well it satisfied the divergence-free condition.

Tao et al. [10.194, 200] examined the alignment trends of filtered vorticity and SGS stress components with respect to the local orientations of the eigenvectors of the filtered strain-rate tensor. Probability density functions (PDF) of scalar parameters characterizing the tensorial eigenvalue structure (following *Lund* and *Rogers* [10.206]) showed that the most probable strain state was axisymmetric extension, and the most probable SGS stress state was axisymmetric contraction. Regions of high SGS dissipation were strongly correlated with these preferred strain-rate and stress topologies. The nonlinear model showed the same trends but overpredicted the occurrence of the preferred stress state.

The relative alignment between SGS stress and strain-rate tensors is a fundamental issue in turbulence modeling since eddy-viscosity model assumes that they are aligned. During analysis, the eigenvalues of $\tilde{S}_{ij}$ were denoted as α_s, β_s and γ_s, where $\alpha_s \geq \beta_s \geq \gamma_s$, and the corresponding eigenvectors were $\boldsymbol{\alpha}_s$, $\boldsymbol{\beta}_s$ and $\boldsymbol{\gamma}_s$. Similarly, $\boldsymbol{\alpha}_{-\tau}$, $\boldsymbol{\beta}_{-\tau}$ and $\boldsymbol{\gamma}_{-\tau}$ denoted unit vectors aligned with the most compressive, intermediate, and most extensive eigenvectors of the deviatoric part of $-\tau_{ij}$, respectively. The data showed that the filtered vorticity vector was preferentially aligned with $\boldsymbol{\beta}_s$, the intermediate strain-rate eigenvector, in agreement with previous numerical and experimental data [10.207–209]. The vorticity was also preferentially aligned perpendicularly to $\boldsymbol{\gamma}_t$ but had no preferred direction in the $\boldsymbol{\alpha}_{-\tau}$–$\boldsymbol{\beta}_{-\tau}$ plane [10.200].

Three angles were needed to define the alignment of a symmetric tensor in a coordinate system defined by the eigenvectors of another symmetric tensor.

Two angles, e.g., $\phi(\boldsymbol{\alpha}_{-\tau} - \boldsymbol{\beta}_s)$ and $\cos[\theta(\boldsymbol{\alpha}_{-\tau} - \boldsymbol{\alpha}_s)]$, define the orientation of one of the eigenvectors of the SGS stress tensor ($\boldsymbol{\alpha}_{-\tau}$ in this example) in a coordinate system defined by the filtered strain-rate eigendirections. The third angle described the orientation of another SGS stress eigenvector with respect to the projection of an axis of the SGS strain rate onto a plane perpendicular to the first SGS eigenvector. This choice preserved statistical consistency, i.e., when tested against a random velocity field these variables generated a uniform 3-D joint PDF.

Figure 10.73a and b show the 3-D joint PDF of alignment between the filtered strain-rate and the SGS stress tensors obtained by analyzing nine instantaneous velocity distributions. The 3-D alignment is presented twice using different basic axial directions in order to demonstrate the dominant features; Fig. 10.73a shows the alignment of the most contracting SGS stress direction $\boldsymbol{\alpha}_{-\tau}$ better, whereas Fig. 10.73b illustrates the alignment of $\boldsymbol{\gamma}_{-\tau}$, the more extensive stress, more clearly. It is observed that $\boldsymbol{\gamma}_{-\tau}$ is preferentially aligned at 32° to $\boldsymbol{\gamma}_s$, i.e., $\cos[\theta(\boldsymbol{\gamma}_{-\tau} - \boldsymbol{\gamma}_s)] \approx 0.85$, irrespective of the other angles. From Fig. 10.73a it is evident that there are actually two distinct regions with high probability peaks. The two preferred 3-D alignments of this bimodal behavior are illustrated by inserts for clarity. The same trends were subsequently also identified in DNS and atmospheric data [10.210], indicating that these trends were not limited to the present flow geometry. Clearly, the SGS stress eigenvectors were not aligned with the filtered stain rate, in contradiction to the basic assumptions of eddy-viscosity models. Conditional averaging demonstrated that the bimodal behavior was related to the vorticity magnitude. The configuration denoted by $\alpha\beta\gamma$–$\beta\alpha\gamma$ was much more pronounced in regions of high vorticity and the configuration $\alpha\beta\gamma$–$\alpha\beta\gamma$ became equally or more dominant in regions with intermediate levels of vorticity. Some of these trends could be explained based on analysis of the nonlinear model assuming the resolved vorticity was aligned with $\boldsymbol{\beta}_s$ and that the strain-rate structure was axisymmetric expansion. This alignment trend persisted, with varying angles and peak probabilities, throughout the conditional samplings based on the magnitudes of SGS dissipation, vorticity, and strain rate [10.194].

Joint PDFs involving the measured and the nonlinear model for SGS dissipation showed significant differences in the trends, even though the nonlinear model stress eigenvectors were preferentially aligned to the same direction as those of the measured stresses. Using a mixed model, the SGS dissipation PDFs agreed better with the experimental data. However, in regions with high strain rate, high vorticity, and especially where the strain rate had a plane shearing or axisymmetric contraction topology, the mixed model still produced excessive negative SGS dissipation.

Using the same data, *van der Bos* et al. [10.211] examined the effects of SGS motions on the dynamics of $\tilde{A}_{ij}$ (10.107) based on the restricted Euler equations [10.212, 213], which was amenable to analytical solution. They showed that the tensor $\partial^2 \tau_{kj}/\partial x_i \partial x_k$, i.e., the gradient of the SGS force, appeared in the transport equation of $\tilde{A}_{ij}$. Following *Cantwell* [10.213] they examined the two invariants of $\tilde{A}_{ij}$, namely

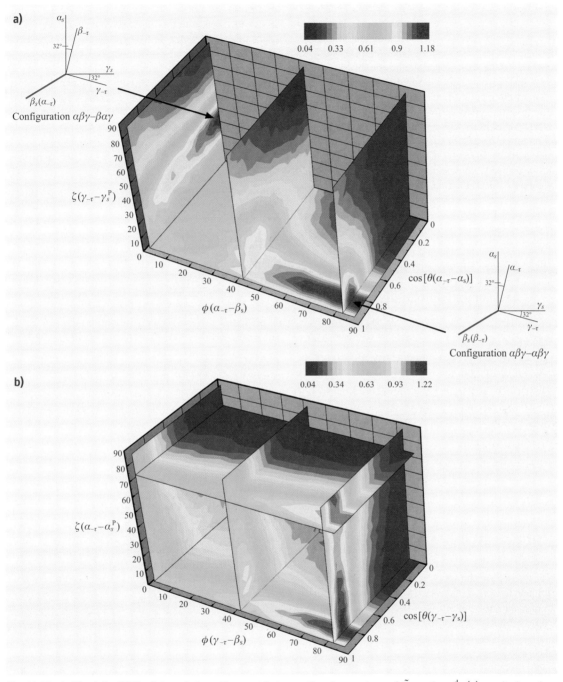

Fig. 10.73a,b The joint PDF of the relative alignment between the eigenvectors of $\tilde{S}_{ij}$ and $-\tau_{ij}^{\mathrm{d}}$: (**a**) results based on eigenvector $\boldsymbol{\alpha}_{-\tau}$ and including sketches illustrating the alignment of the SGS stress eigenvectors at the points of peak probability, and (**b**) based on $\boldsymbol{\gamma}_{-\tau}$. In the coordinate system shown, a random alignment would correspond to a uniform probability density of $(\pi/2)^{-2} \approx 0.396$. (After *Tao et al.* [10.194], with permission from CUP)

$Q_\Delta = -\tilde{A}_{im}\tilde{A}_{mi}/3$ and $R_\Delta = -\tilde{A}_{im}\tilde{A}_{mk}\tilde{A}_{ki}/3$. The effects of the SGS stress tensor on Q_Δ and R_Δ was quantified unambiguously by evaluating conditional averages that appear in the evolution equation for the joint probability distribution function of these invariants. The results showed that the SGS stresses opposed the formation of the inherent finite-time singularity along the Vieillefosse tail, which led towards $R_\Delta > 0$ and $Q_\Delta < 0$, and that the effect was significant. Trends in the SGS dissipation and other variables were quantified in different parts of the (Q_Δ, R_Δ) plane. A priori tests of the Smagorinsky, nonlinear, and mixed models showed that all reproduced the real SGS stress effect along the Vieillefosse tail, preventing the finite-time singularity, but they all failed in other regions of the (Q_Δ, R_Δ) plane. The data were also used to suggest improvements to the mixed model.

10.4.6 Scalar Concentration Measurements for SGS Mixing and Combustion Studies

Predicting the transport of scalars caused by turbulence is important for various applications such as combustion, mixing, aero-optics. As mentioned before, the prediction of subgrid variance and the scalar-variance dissipation rate are crucial ingredients in a number of LES closures for scalar transport and combustion. To measure these variables experimentally one requires well-resolved measurements of the scalar field (e.g., scalar concentration, or temperature). Several techniques for measurements of scalar fields exist. Probably the most commonly employed technique is based on laser-induced fluorescence (LIF). Planar laser-induced fluorescence (PLIF) is described in Sect. 7.4. By assuming a relationship between the recorded intensity of the emitted light and the concentration of the fluorescent dye in the flow, maps of concentration can be recorded.

Scanning a light sheet at very high speed across a volume, *Dahm* et al. [10.214] reconstructed quasi-3-D distributions of scalar concentration and followed its time evolution. Based on these data, *Cook and Riley* [10.153] analyzed various models for the subgrid scalar variance

$$\widetilde{c^2} - \tilde{c}^2 \tag{10.123}$$

and explored predictions from so-called presumed PDF models of mixing. In canonical turbulent flows the spectral characteristics of scalar distributions quite similar to those of the velocity field (including spectral slopes of -1.4 to -1.7 in the inertial–convective subrange), the scalar subgrid variance is also expected to be dominated by the largest of the unresolved scales. Thus spatial resolution is not typically required to reach the smallest scales of the scalar field (typically the Batchelor scale), but rather some fraction of the filtering scale, such as $\Delta/5$ to $\Delta/10$.

Studies of the SGS scalar fluxes

$$q_i = \widetilde{u_i c} - \tilde{u}_i \tilde{c} \tag{10.124}$$

require simultaneous measurement of the scalar concentration and velocity. Such measurements can be performed with coupled PIV and LIF (*Su and Mungal* [10.215]). Analysis of such data has recently been used to study SGS scalar fluxes in a cross-flow turbulent jet (*Sun* and *Su* [10.216]). For the study of SGS turbulent heat fluxes in the atmosphere, arrays of sonic anemometers allow simultaneous measurement of velocity and temperature, and the evaluation of SGS heat fluxes and their properties [10.166, 217]. Since scalar fluxes are expected to be dominated by the largest unresolved scales, spatial resolution down to about $\Delta/5$ to $\Delta/10$ is expected to be sufficient to resolve the most important aspects of the SGS scalar flux.

The same is not true when one wishes to evaluate the scalar dissipation rate

$$\chi = \gamma(\widetilde{(\nabla c)(\nabla c)} - \nabla \tilde{c} \nabla \tilde{c}) \tag{10.125}$$

for which scalar gradients must be resolved at the small scales (i.e., the Batchelor scale). Only very well-resolved data can be used to evaluate the scalar dissipation rate.

PLIF has also been used to measure the scalar surface density for modeling of turbulent combustion [10.218]. The surface density at the subgrid scales can be quantified by the difference

$$|\widetilde{\nabla c}| - |\nabla \tilde{c}| \tag{10.126}$$

and requires the evaluation of scalar gradients at the smallest scales before taking the absolute value and filtering. Due to the limitations of planar LIF, only two of the three gradient terms can be captured and assumptions about the third direction must be made when interpreting the results.

References

10.1 G.K. Batchelor: *An Introduction to Fluid Dynamics* (Cambridge Univ. Press, Cambridge 1967)

10.2 O.A. Ladyzenskaya: *Mathematical Theory of Viscous Incompressible Flow* (Gordon Breach, New York 1970)

10.3 S. Goldstein: Fluid Mechanics in the first half of this century, Annu. Rev. Fluid Mech. **1**, 1–29 (1969)

10.4 S.B. Pope: *Turbulent Flows* (Cambridge Univ. Press, Cambridge 2000)

10.5 L.F. Richardson: *Weather Prediction by Numerical Process* (Cambridge Univ. Press, Cambridge 1922)

10.6 H. Tennekes, J.L. Lumley: *A First Course in Turbulence* (MIT, Boston 1972)

10.7 A. Celani: *Hydrodynamic Turbulence, PhD Course* (Scuola Nazionale Fisica della Materia ISI, Torino 2004)

10.8 H. Tennekes: Eulerian and Lagrangian time microscales in isotropic turbulence, J. Fluid Mech. **67**, 561–567 (1975)

10.9 J.S. Bendat, A.G. Piersol: *Random Data: Analysis and Measurement Procedures* (Wiley Interscience, New York 1971)

10.10 J.O. Hinze: *Turbulence* (McGraw-Hill, New York 1975)

10.11 U. Frisch: *Turbulence: The Legacy of A.N. Kolmogorov* (Cambridge Univ. Press, Cambridge 1995)

10.12 A.S. Monin, A.M. Yaglom: *Statistical Fluid Mechanics: Mechanics of Turbulence* (MIT Press, Boston 1975)

10.13 G.P. Romano, R.A. Antonia: Longitudinal and transverse structure functions in a turbulent round jet: effect of initial conditions and Reynolds number, J. Fluid Mech. **436**, 231–248 (2001)

10.14 H.R.E. Van Maanen: *Retrieval of Turbulence and Turbulence Properties from Randomly Sampled Laser-Doppler Anemometry Data with Noise, PhD Thesis* (TU Delft, Delft 1999)

10.15 J.T. Taylor: The spectrum of turbulence, Proc R. Soc. Lond. A **164**, 476–490 (1938)

10.16 G.P. Romano: Analysis of two-point velocity measurements in near-wall flows, Exp. Fluids **20**, 68–83 (1995)

10.17 P.S. Bernard, J.M. Wallace: *Turbulent Flow: Analysis, Measurement and Prediction* (Wiley, New York 2002)

10.18 W.K. George, H.J. Hussein: Locally axisymmetric turbulence, J. Fluid Mech. **233**, 1–23 (1991)

10.19 T. Zhou, R.A. Antonia, L. Danaila, F. Anselmet: Approach to the 'four-fifths law' for grid turbulence, J. Turbul. **1**, 5.1–5.12 (2000)

10.20 F. Anselmet, R.A. Antonia, M. Ould Rouis: Relations between third-order and second-order structure functions for axisymmetric turbulence, J. Turbul. **1**, 3.1–3.10 (2000)

10.21 L. Danaila, R.A. Antonia, P. Burattini: Progress in studying small-scale turbulence using "exact" two-point equations, New J. Phys. **6**, 128 (2004)

10.22 C. Tropea: Post-processing of Experimental and Numerical Data. In: *von Karman Lec. Ser. 2003-03*, ed. by P. Millan, M.L. Riethmuller (von Karman Institute, Bruxelles 2003)

10.23 W.D. Mc Comb: *The Physics of Fluid Turbulence* (Clarendon, New York 1990)

10.24 L.W.B. Browne, R.A. Antonia, D.A. Sha: Turbulent energy dissipation in a wake, J. Fluid Mech. **179**, 307–326 (1987)

10.25 R.A. Antonia, J. Mi: Temperature dissipation in a turbulent round jet, J. Fluid Mech. **250**, 531–551 (1993)

10.26 L.H. Benedict: *Direct Measurements of Turbulent Dissipation Rate in Flow over a Backwards-Facing Step using LDA. Ph.D. Thesis* (North Carolina State Univ., Raleigh 1995)

10.27 G.P. Romano, F. Bagnoli: The measurement of velocity and velocity gradients in a turbulent channel by using different particle tracers. In: *Advances in Turbulence VI*, ed. by S. Gavrilakis, L. Machiels, P.A. Monkewitz (Kluwer Academic, Dordrecht 1996) pp. 525–526

10.28 L.H. Benedict, R.D. Gould: Towards better uncertainty estimates for turbulence statistics, Exp. Fluids **22**, 129–136 (1996)

10.29 M. Lesieur: *Turbulence in Fluids* (Kluwer Academic, Dordrecht 1990)

10.30 J. Mathieu, J. Scott: *An Introduction to Turbulent Flow* (Cambridge Univ. Press, Cambridge 2000)

10.31 G.K. Batchelor: *The Theory of Homogeneous Turbulence* (Cambridge University Press, Cambridge 1953)

10.32 G.P. Romano, R.A. Antonia, T. Zhou: Evaluation of LDA temporal and spatial velocity structure functions in a low Reynolds number turbulent channel flow, Exp. Fluids **27**, 368–377 (1999)

10.33 R.A. Antonia, R.J. Smalley, T. Zhou, F. Anselmet, L. Danaila: Similarity of energy structure functions in decaying homogeneous isotropic turbulence, J. Fluid Mech. **487**, 245–269 (2003)

10.34 A.N. Kolmogorov: The local structure of turbulence in incompressible viscous fluid for very large Reynolds numbers, Dokl. Akad. Nauk SSSR **30**, 301–305 (1941)

10.35 P.K. Yeung: Lagrangian investigations of turbulence, Annu. Rev. Fluid Mech. **34**, 115–142 (2002)

10.36 H.-G. Maas, A. Gruen, D. Papantoniou: Particle tracking velocimetry in three-dimensional flows – Part 1. Photogrammetric determination of particle coordinates, Exp. Fluids **15**, 133–146 (1993)

10.37 N.A. Malik, T. Dracos, D.A. Papantoniou: Particle tracking velocimetry in three-dimensional flows –

10.38 H.-G. Maas: Contributions of digital photogrammetry to 3-D PTV. In: *in: Three-Dimensional Velocity and Vorticity Measuring and Image Analysis Techniques*, ed. by Th. Dracos (Kluwer Academic, Dordrecht 1996) pp.191–207

10.39 T. Dracos: Particle tracking in three-dimensional space. In: *Three-Dimensional Velocity and Vorticity Measuring and Image Analysis Techniques*, ed. by T. by: Dracos. (Kluwer Academic, Dordrecht 1996) pp.129–152

10.40 M. Virant, T. Dracos: 3D PTV and its application on Lagrangian motion, Meas. Sci. Technol. **8**, 1539–1552 (1997)

10.41 G.A. Voth, K. Satyanarayan, E. Bodenschatz: Lagrangian acceleration measurements at large Reynolds numbers, Phys. Fluids **10**, 2268–2280 (1998)

10.42 J. Mann, S. Ott, J.S. Andersen: *Experimental Study of Relative, Turbulent Diffusion, Risø-R-1036(EN)* (Risø National Laboratory, Roskilde 1999)

10.43 S. Ott, J. Mann: An experimental investigation of the relative diffusion of particle pairs in three-dimensional turbulent flow, J. Fluid Mech. **422**, 207–223 (2000)

10.44 A. La Porta, G.A. Voth, A.M. Crawford, J. Alexander, E. Bodenschatz: Fluid particle accelerations in fully developed turbulence, Nature **409**, 1017–1019 (2001)

10.45 N. Mordant, P. Metz, O. Michel, J.-F. Pinton: Measurement of Lagrangian velocity in fully developed turbulence, Phys. Rev. Lett. **87**, 214501 (2001)

10.46 N. Mordant, J.-F. Pinton, O. Michel: Time-resolved tracking of a sound scatterer in a complex flow: Nonstationary signal analysis and applications, J. Acoust. Soc. Am. **112**, 108–118 (2002)

10.47 G.A. Voth, A. La Porta, A.M. Crawford, J. Alexander, E. Bodenschatz: Measurement of particle accelerations in fully developed turbulence, J. Fluid Mech. **469**, 121–160 (2002)

10.48 N. Mordant, E. Lévêque, J.-F. Pinton: Experimental and numerical study of the Lagrangian dynamics of high Reynolds turbulence, New J. Phys. **6**, 116 (2004)

10.49 N. Mordant, A.M. Crawford, E. Bodenschatz: Experimental Lagrangian acceleration probability density function measurement, Physica D **193**, 245–251 (2004)

10.50 N. Mordant, A. Crawford, E. Bodenschatz: Three-dimensional structure of the Lagrangian acceleration in turbulent flows, Phys. Rev. Lett. **93**, 214501 (2004)

10.51 N. Mordant, P. Metz, J.-F. Pinton, O. Michel: Acoustical technique for Lagrangian velocity measurement, Rev. Sci. Instr. **76**, 025105 (2005)

10.52 A.M. Crawford, N. Mordant, E. Bodenschatz: Joint statistics of the Lagrangian acceleration and velocity in fully developed turbulence, Phys. Rev. Lett. **94**, 024501 (2005)

10.53 B. Lüthi, A. Tsinober, W. Kinzelbach: Lagrangian measurements of vorticity dynamics in turbulent flow, J. Fluid Mech. **528**, 87–118 (2005)

10.54 N.T. Ouellette, H. Xu, E. Bodenschatz: A quantitative study of three-dimensional Lagrangian particle tracking algorithms, Exp. Fluids **40**, 301–313 (2006)

10.55 H. Xu, M. Bourgoin, N.T. Ouellette, E. Bodenschatz: High order Lagrangian velocity statistics in turbulence, Phys. Rev. Lett. **96**, 024503 (2006)

10.56 M. Bourgoin, N.T. Ouellette, H. Xu, J. Berg, E. Bodenschatz: The role of pair dispersion in turbulent flow, Science **311**, 835–838 (2006)

10.57 H. Xu, N.T. Ouellette, E. Bodenschatz: The multifractal dimension of Lagrangian turbulence, Phys. Rev. Lett. **96**, 114503 (2006)

10.58 N.T. Ouellette, H. Xu, M. Bourgoin, E. Bodenschatz: Small-scale anisotropy in Lagrangian turbulence, New J. Phys. **8**, 102 (2006)

10.59 N.T. Ouellette, H. Xu, M. Bourgoin, E. Bodenschatz: An experimental study of turbulent relative dispersion models, New J. Phys. **8**, 109 (2006)

10.60 C.J. Veenman, M.J.T. Reinders, E. Backer: Establishing motion correspondence using extended temporal scope, Artif. Intell. **145**, 227–243 (2003)

10.61 C.J. Veenman, M.J.T. Reinders, E. Backer: Resolving motion correspondence for densely moving points, IEEE T. Pattern Anal. **23**, 54–72 (2001)

10.62 F. Bourgeois, J.-C. Lasalle: An extension of the Munkres algorithm for the assignment problem to rectangular matrices, Commun. ACM **14**, 802–804 (1971)

10.63 A. Gylfason, S. Ayyalasomayajula, Z. Warhaft: Lagrangian measurements of inertial particles in wind tunnel turbulence, Bull. Am. Phys. Soc. **50**, 65 (2005)

10.64 P.J. Zandbergen, D. Dijkstra: Von Kármán swirling flows, Annu. Rev. Fluid. Mech. **19**, 465–491 (1987)

10.65 G.A. Voth: *Lagrangian acceleration measurements in turbulence at large reynolds numbers*, PhD thesis (Cornell Univ., Ithaca 2000)

10.66 A.M. Crawford: *Particle tracking measurements in fully developed turbulence: water and dilute polymer solutions*, PhD thesis (Cornell Univ., Ithaca 2004)

10.67 G.A. Voth, A. La Porta, A.M. Crawford, C. Ward, E. Bodenschatz, J. Alexander: A silicon strip detector system for high resolution particle tracking in turbulence, Rev. Sci. Instr. **12**, 4348–4353 (2001)

10.68 R.Y. Tsai: A versatile camera calibration technique for high-accuracy 3D machine vision metrology using off-the-shelf TV cameras and lenses, IEEE T. Robotic. Autom. **RA-3**, 323–344 (1987)

10.69 G.I. Taylor: Diffusion by continuous movements, Proc. Lond. Math. Soc. **20**, 196–212 (1922)

10.70 L.F. Richardson: Atmospheric diffusion shown on a distance-neighbour graph, Proc. R. Soc. Lond. A **110**, 709–737 (1926)

10.71 A.M. Obukhov: Spectral energy distribution in turbulent flow, Izv. Akad. Nauk SSSR **5**, 453–566 (1941)

10.72 B.L. Sawford: Turbulent relative dispersion, Annu. Rev. Fluid Mech. **33**, 289–317 (2001)

10.73 G.K. Batchelor: The application of the similarity theory of turbulence to atmospheric diffusion, Q. J. R. Meteorol. Soc. **76**, 133–146 (1950)

10.74 N.T. Ouellette: *Probing the Statistical Structure of Turbulence with Measurements of Tracer Particle Tracks* (Cornell Univ., Ithaca 2006)

10.75 K.R. Sreenivasan: On the universality of the Kolmogorov constant, Phys. Fluids **7**, 2778–2784 (1995)

10.76 A. Groisman, V. Steinberg: Elastic turbulence in a polymer solution flow, Nature **405**, 53–55 (2000)

10.77 A. Groisman, V. Steinberg: Efficient mixing at low Reynolds numbers using polymer additives, Nature **410**, 905–908 (2001)

10.78 R.B. Bird, C.F. Curtiss, R.C. Armstrong, O. Hassager: *Dynamics of Polymeric Liquids* (Wiley, New York 1987)

10.79 L.D. Landau, E.M. Lifschitz: *Fluid Mechanics* (Pergamon, Oxford 1987)

10.80 D.J. Tritton: *Physical Fluid Dynamics* (Clarendon, Oxford 1988)

10.81 K. Weissenberg: A continuum theory of rheological phenomena, Nature **159**, 310–311 (1947)

10.82 V. Tirtaatmadja, T. Sridhar: A filament stretching device for measurement of extensional viscosity, J. Rheology **37**, 1081–1102 (1993)

10.83 J.J. Magda, R.G. Larson: A transition occurring in ideal elastic liquids during shear flow, J. Non-Newtonian Fluid Mech. **30**, 1–19 (1988)

10.84 S.J. Muller, R.G. Larson, E.S.G. Shaqfeh: A purely elastic transition in Taylor-Coueete flow, Rheol. Acta **28**, 499–503 (1989)

10.85 R.G. Larson: Instabilities in viscoelastic flows, Rheol. Acta **31**, 213–263 (1992)

10.86 E.S.G. Shaqfeh: Purely elastic instabilities in viscometric flows, Annu. Rev. Fluid Mech. **28**, 129–185 (1996)

10.87 R.G. Larson, E.S.G. Shaqfeh, S.J. Muller: A purely elastic instability in Taylor-Couette flow, J. Fluid Mech. **218**, 573–600 (1990)

10.88 A. Groisman, V. Steinberg: Mechanism of elastic instability in Couette flow of polymer solutions: Experiment, Phys. Fluids **10**, 2451–2463 (1998)

10.89 A. Groisman, V. Steinberg: Elastic versus inertial instability in a polymer solution flow, Europhys. Lett. **43**, 165–170 (1998)

10.90 D.E. Smith, H.P. Babcock, S. Chu: Single-polymer dynamics in steady shear flow, Science **283**, 1724–1727 (1999)

10.91 J.A. Byars, A. Öztekin, R.A. Brown, G.H. McKinley: Spiral instabilities in the flow of highly elastic fluids between rotating parallel disks J, Fluid Mech. **271**, 173–197 (1994)

10.92 M. Avgousti, A.N. Beris: Non-axisymmetric modes in viscoelastic Taylor-Couette flow, J. Non-Newtonian Fluid Mech. **50**, 225–251 (1993)

10.93 A. Öztekin, R.A. Brown: Instability of a viscoelastic fluid between rotating parallel disks: analysis for the Oldroyd-B fluid, J. Fluid Mech. **255**, 473–502 (1993)

10.94 G.H. McKinley, J.A. Byars, R.A. Brown, R.C. Armstrong: Observations on the elastic instability in cone-and-plate and parallel-plate flows of a polyisobutylene Boger fluid, J. Non-Newtonian Fluid Mech. **40**, 201–229 (1991)

10.95 W.R. Dean: Fluid motion in a curved channel, Proc. R. Soc. London Ser. A **121**, 402–420 (1928)

10.96 Y.L. Joo, E.S.G. Shaqfeh: Viscoelastic Poiseuille flow through a curved channel: Anew elastic instability, Phys. Fluids A **3**, 1691–1694 (1991)

10.97 Y.L. Joo, E.S.G. Shaqfeh: A purely elastic instability in Dean and Taylor-Dean flow, Phys. Fluids A **4**, 524–543 (1992)

10.98 Y.L. Joo, E.S.G. Shaqfeh: Observations of purely elastic instabilities in the Taylor-Dean flow of a Boger fluid, J. Fluid Mech. **262**, 27–73 (1994)

10.99 D.V. Boger: A highly elastic constant-viscosity fluid, J. Non-Netonian Fluid Mech. **3**, 87–91 (1978)

10.100 M. Doi, S.F. Edwards: *The Theory of Polymer Dynamics* (Clarendon, Oxford 1988)

10.101 E. Washburn (Ed.): International Critical Tables, Vol. 5 (McGraw-Hill, New York 1929)

10.102 A. Groisman, V. Steinberg: Couette-Taylor flow in a dilute polymer solution, Phys. Rev. Lett. **77**, 1480–1483 (1996)

10.103 H.W. Giesekus: About flow stability of viscoelastic fluids, Rheol. Acta **5**, 239–252 (1968)

10.104 A. Groisman, V. Steinberg: Elastic turbulence in curvilinear flows of polymer solutions, New J. Phys. **6**, 29.1–29.48 (2004)

10.105 T. Burghelea, E. Segre, V. Steinberg: Mixing by polymers: Experimental test of decay regime of mixing, Phys. Rev. Lett. **92**, 164501-1–164501-4 (2004)

10.106 T. Burghelea, E. Segre, I. Bar-Joseph, A. Groisman, V. Steinberg: Chaotic flow and efficient mixing in a microchannel with a polymer solution, Phys. Rev. E **69**, 066305-1–066305-8 (2004)

10.107 T. Burghelea, E. Segre, V. Steinberg: Statistics of particle pair separations in the elastic turbulent flow of a dilute polymer solution, Europhys. Lett. **68**, 529–535 (2004)

10.108 T. Burghelea, E. Segre, V. Steinberg: Validity of the taylor hypothesis in a random spatially smooth flow, Phys. Fluids **17**, 103101(1–8) (2005)

10.109 T. Burghelea, E. Segre, V. Steinberg: Elastic turbulence in a swirling flow, Phys. Fluids **19**, 053104 (2007)

10.110 J. Lumley: On the solution of equations describing small scale deformations, Symp. Math. **9**, 315 (1972)

10.111 A. Groisman, V. Steinberg: Stretching of polymers in a random three-dimensional flow, Phys. Rev. Lett. **86**, 934–937 (2001)

10.112 N. Mordant, J.-P. Pinton, F. Chilla: Characterization of turbulence in a closed flow, J. Phys. II Paris **7**, 1729–1742 (1997)

10.113 J.H. Titon, O. Cadot: The statistics of power injected in a closed turbulent flow: Constant torque forcing versus constant velocity forcing, Phys. Fluids **15**, 625–640 (2003)

10.114 J.R. Stokes, L.J.W. Graham, N.J. Lawson, D.V. Boger: Swirling flow of viscoelatic fluids, Part 1, Part 2, J. Fluid Mech. **429**, 67–153 (2001)

10.115 E. Balkovsky, A. Fouxon, V. Lebedev: Turbulence of polymer solutions, Phys. Rev. E **64**, 056301-1–0563011-4 (2001)

10.116 A. Fouxon, V. Lebedev: Spectra of turbulence in dilute polymer solutions, Phys. Fluids **15**, 2060–2072 (2003)

10.117 U. Frish: *Turbulence: The Legacy of A. N. Kolmogorov* (Cambridge Univ. Press, New York 1995)

10.118 B.I. Shraiman, E.D. Siggia: Scalar turbulence, Nature **405**, 639–646 (2000)

10.119 S. Gerashchenko, C. Chevallard, V. Steinberg: Single polymer dynamics: Coil-stretch transition in a random flow, Europhys. Lett. **71**, 221–227 (2005)

10.120 N. Phan-Thien: Cone-and-plate flow of an Oldroyd-B fluid is unstable, J. Non-Newtonian Fluid Mech. **17**, 37–44 (1985)

10.121 E.S.G. Shaqfeh, D.L. Koch: Polymer stretch in dilute fixed beds of fibres or spheres, J. Fluid Mech. **244**, 17–25 (1992)

10.122 A.B. Mosler, E.S.G. Shaqfeh: The conformation change of model polymers in stochastic flow fields: Flow through fixed beds, Phys. Fluids **9**, 1222–1234 (1997)

10.123 E. Balkovsky, A. Fouxon, V. Lebedev: Turbulent dynamics of polymer solutions, Phys. Rev. Lett. **84**, 4765–4768 (2000)

10.124 M. Chertkov: Polymer stretching by turbulence, Phys. Rev. Lett. **84**, 4761–4764 (2000)

10.125 G.K. Batchelor: Small scale variation of convected quantities like temperature in turbulent fluid, J. Fluid Mech. **5**, 113–167 (1959)

10.126 P.G. de Gennes: Coil-stretch transition of dilute flexible polymers under ultrahigh velocity gradients, J. Chem. Phys. **60**, 5030–5042 (1974)

10.127 M. Chertkov: Passive advection in nonlinear medium, Phys. Fluids **11**, 2257–2262 (1999)

10.128 M. Chertkov: On how a joint interaction of two innocent partners (smooth advection and linear damping) produces a strong intermittency, Phys. Fluids **10**, 3017–3019 (1998)

10.129 A. Groisman, M. Enzelberger, S.R. Quake: Microfluidic Memory and Control Devices, Science **300**, 955–958 (2003)

10.130 Y.N. Xia, G.M. Whitesides: Soft lithography, Annu. Rev. Mater. Sci. **28**, 153–184 (1998)

10.131 Chi Wu: Laser light-scattering characterization of the molecular weight distribution of dextran, Macromolecules **26**, 3821–3825 (1993)

10.132 M. Chertkov, G. Falkovich, I. Kolokolov, V. Lebedev: Statistics of a passive scalar advected by a large-scale two-dimensional velocity field: Analytic solution, Phys. Rev. E **51**, 5609–5627 (1995)

10.133 E. Balkovsky, A. Fouxon: Universal long-time properties of lagrangian statistics in the Batchelor regime and their application to the passive scalar problem, Phys. Rev. E **6**, 4164–4174 (1999)

10.134 V. Steinberg, A. Groisman: Device and Method for Mixing Substances, US Patent No. **6,632,014 B2**, Oct. 14, 2003

10.135 M. Lesieur, O. Metais: New trends in large-eddy simulations of turbulence, Annu. Rev. Fluid Mech. **28**, 45–82 (1996)

10.136 P. Sagaut: *Large Eddy Simulation for Incompressible Flow*, 3rd edn. (Springer, Heidelberg 2005)

10.137 A. Leonard: Energy cascade in large-eddy simulations of turbulent fluid flows, Adv. Geophys. **18**, 237 (1974)

10.138 B. Vreman, B. Geurts, H. Kuerten: Realizability conditions for the turbulent stress tensor in large-eddy simulation, J. Fluid Mech. **278**, 351–362 (1994)

10.139 J. Smagorinsky: General circulation experiments with the primitive equations, Part 1: The basic experiment, Mon. Weath. Rev. **91**, 99–164 (1963)

10.140 D.K. Lilly: The representation of small-scale turbulence in numerical simulation experiments, Proc. IBM Sci. Comput. Symp. Environ. Sci. **195** (1967)

10.141 M. Germano, U. Piomelli, P. Moin, W. Cabot: A dynamic subgrid-scale eddy viscosity model, Phys. Fluids A **3**, 1760–1765 (1991)

10.142 M. Germano: Turbulence: The filtering approach, J. Fluid Mech. **238**, 325–336 (1992)

10.143 D.K. Lilly: A proposed modification of the Germano subgrid-scale closure method, Phys. Fluids A **4**, 633–635 (1992)

10.144 S. Ghosal, T.S. Lund, P. Moin, W.H. Cabot: A dynamic localization model for large eddy simulations of turbulent flow, J. Fluid Mech. **286**, 299 (1995)

10.145 U. Piomelli: Large eddy simulation: achievements and challenges, Prog. Aerosp. Sci. **35**, 335–362 (1999)

10.146 C. Meneveau, J. Katz: Scale-invariance and turbulence models for large-eddy simulation, Annu. Rev. Fluid Mech. **32**, 1–32 (2000)

10.147 J. Bardina, J. Ferziger, Reynolds: Improved subgrid scale models for large eddy simulation, AIAA Paper **80-1357** (1980)

10.148 R.A. Clark, J.H. Ferziger, W.C. Reynolds: Evaluation of subgrid models using an accurately simulated turbulent flow, J. Fluid Mech. **91**, 1 (1979)

10.149 S. Liu, C. Meneveau, J. Katz: On the properties of similarity subgrid-scale models as deduced from measurements in a turbulent jet, J. Fluid Mech. **275**, 83 (1994)

10.150 V. Borue, S.A. Orszag: Local energy flux and subgrid-scale statistics in three-dimensional turbulence, J. Fluid Mech. **366**, 1–31 (1998)

10.151 S. Stolz, N.A. Adams, L. Kleiser: An approximate deconvolution model for large-eddy simulation with application to incompressible wall-bounded flows, Phys. Fluids **13**, 997–1015 (2001)

10.152 P. Moin, K.D. Squires, W.H. Cabot, S. Lee: A dynamic subgrid-scale model for compressible turbulence and scalar transport, Phys. Fluids A **3**, 2746–2757 (1991)

10.153 A. Cook, J.J. Riley: A subgrid model for equilibrium chemistry in turbulent flows, Phys. Fluids **6**, 2868–2870 (1994)

10.154 C. Jimenez, L. Valiño, C. Dopazo: Subgrid scale variance and dissipation of a scalar field in large eddy simulations, Phys. Fluids **13**, 2433 (2001)

10.155 D. Veynante, L. Vervich: Turbulent combustion modeling, Progr. Energy Combust. Sci. **28**, 193 (2002)

10.156 U. Piomelli, P. Moin, J.H. Ferziger: Model consistency in large eddy simulation of turbulent channel flows, Phys. Fluids **31**, 1884–1891 (1988)

10.157 U. Piomelli, W.H. Cabot, P. Moin, S. Lee: Subgrid scale backscatter in turbulent and transitional flows, Phys. Fluids A **3**, 1766–1771 (1991)

10.158 C. Meneveau: Statistics of subgrid-scale stresses: Necessary conditions and experimental tests, Phys. Fluids **6**, 815 (1994)

10.159 C. Meneveau, J. O'Neil: Scaling laws of the dissipation rate of turbulent subgrid-scale kinetic energy, Phys. Rev. E **49**, 2866 (1994)

10.160 J. O'Neil, C. Meneveau: Subgrid-scale stresses and their modelling in a turbulent plane wake, J. Fluid Mech. **347**, 253–293 (1997)

10.161 F. Porté-Agel, C. Meneveau, M.B. Parlange: Some basic properties of the surrogate subgrid-scale heat flux in the atmospheric boundary layer, Bound. Layer Met. **88**, 425 (1998)

10.162 J.A. Murray, U. Piomelli, J.M. Wallace: Spatial and temporal filtering of experimental data for a-priori studies of subgrid-scale stresses, Phys. Fluids **8**, 1978–1980 (1996)

10.163 C.N. Tong, J.C. Wyngaard, S. Khanna, J.G. Brasseur: Resolvable- and subgrid-scale measurement in the atmospheric surface layer: Technique and issues, J. Atmos. Sci. **55**, 3114–3126 (1998)

10.164 F. Porté-Agel, M.B. Parlange, C. Meneveau, W.E. Eichinger, M. Pahlow: Subgrid-scale dissipation in the atmospheric surface layer: Effects of stability and filter dimension, J. Hydrometero. **1**, 75–87 (2000)

10.165 C.N. Tong, J.C. Wyngaard, J.G. Brasseur: Experimental study of the subgrid-scale stress in the atmospheric surface layer, J. Atmos. Sci. **56**(14), 2277–2292 (1998)

10.166 F. Porté-Agel, M.B. Parlange, C. Meneveau, W.E. Eichinger: A priori field study of the subgrid-scale heat fluxes and dissipation in the atmospheric surface layer, J. Atmos. Sci **58**, 2673–2698 (2001)

10.167 T. Horst, J. Kleissl, D. Lenschow, C. Meneveau, C.-H. Moeng, M.B. Parlange, P.P. Sullivan, J.C. Weil: HATS: Field observations to obtain spatially-filtered turbulence fields from crosswind arrays of sonic anemometers in the atmospheric surface layer, J. Atmos. Sci. **61**, 1655–1681 (2004)

10.168 C. Higgins, C. Meneveau, M.B. Parlange: The effect of filter dimension on the components of the subgrid-scale stress and tensor alignments in the atmospheric surface layer, J. Atmos. Ocean. Tech., **24**, 360 (2007)

10.169 H.S. Kang, S. Chester, C. Meneveau: Decaying turbulence in an active-grid-generated flow and comparisons with large-eddy simulation, J. Fluid Mech. **480**, 129–160 (2003)

10.170 S. Cerutti, C. Meneveau: Statistics of filtered velocity in grid and wake turbulence, Phys. Fluids **12**, 1143–1165 (2000)

10.171 S. Cerutti, C. Meneveau, O.M. Knio: Spectral and hyper eddy-viscosity in high Reynolds number turbulence, J. Fluid Mech. **421**, 307–338 (2000)

10.172 H.S. Kang, C. Meneveau: Universality of LES model parameters across a turbulent wake behind a heated cylinder, J. Turbul. **3**, N32 (2002)

10.173 H.S. Kang, C. Meneveau: Effect of large-scale coherent structures on subgrid-scale stress and strain-rate eigenvector alignments in turbulent shear flow, Phys. Fluids **17**, 055103 (2005)

10.174 Q. Chen, D. Wang, H. Zhang, C. Tong.: Effects of subgrid-scale turbulence on resolvable-scale velocity–scalar statistics, J. Turbul. **6**, N36 (2005)

10.175 G. Comte-Bellot, S. Corrsin: Simple Eulerian time correlation of full- and narrow-band velocity signals in grid-generated, 'isotropic' turbulence, J. Fluid Mech. **48**, 273–337 (1971)

10.176 L. Mydlarski, Z. Warhaft: On the onset of high Reynolds number grid generated wind tunnel turbulence, J. Fluid Mech. **320**, 331–368 (1996)

10.177 F. Porté-Agel, M. Pahlow, C. Meneveau, M.B. Parlange: Atmospheric stability effect on subgrid scale physics for large-eddy simulation, Adv. Water Resour. **24**, 1085–1102 (2001)

10.178 J. Kleissl, C. Meneveau, M.B. Parlange: On the magnitude and variability of subgrid-scale eddy-diffusion coefficients in the atmospheric surface layer, J. Atmos. Sci. **60**, 2372–2388 (2003)

10.179 J. Kleissl, M.B. Parlange, C. Meneveau: Field experimental study of dynamic Smagorinsky models in the atmospheric surface layer, J. Atmos. Sci. **61**(18), 2296–2307 (2004)

10.180 F. Porté-Agel, C. Meneveau, M.B. Parlange: A scale-dependent dynamic model for large-eddy simulation: Applications to a neutral atmospheric boundary layer, J. Fluid Mech. **415**, 261 (2000)

10.181 B. Vreman, B. Geurts, H. Kuerten: Large-eddy simulation of the turbulent mixing layer, J. Fluid Mech. **339**, 357–390 (1997)

10.182 C. Meneveau, J. Katz: Dynamic testing of subgrid models in LES based on the Germano identity, Phys. Fluids **11**, 1 (1999)

10.183 R.J.M. Bastiaans, C.C.M. Rindt, A.A. Van Steenhoven: Experimental analysis of a confined transitional plume with respect to subgrid-scale modelling, Int. J. Heat Mass Transfer **41**, 3989–4007 (1998)

10.184 S. Liu, J. Katz, C. Meneveau: Evolution and modeling of subgrid scale during rapid straining of turbulence, J. Fluid Mech. **387**, 281–320 (1999)

10.185 L. Bertuccioli, S. Gopalan, J. Katz: Image shifting for PIV using birefringent and ferroelectric liquid crystals, Exp. Fluids **21**, 341–346 (1996)

10.186 M.J. Lee: Distortion of homogeneous turbulence by axisymmetric strain and dilation, Phys. Fluids A **1**, 1541–1557 (1989)

10.187 O.J. McMillan, J.H. Ferziger: Direct testing of subgrid-scale models, AIAA. J. **17**, 1340 (1979)

10.188 C. Meneveau, J. Katz: Conditional subgrid force and dissipation in locally isotropic and rapidly strained turbulence, Phys. Fluids **11**, 2317 (1999)

10.189 J. Chen, J. Katz, C. Meneveau: The implication of mismatch between stress and strain-rate in turbulence subjected to rapid straining and destraining on dynamic LES models, J. Fluids Eng. **127**, 840–850 (2005)

10.190 J. Chen, C. Meneveau, J. Katz: Scale Interactions of Turbulence Subjected to a Straining-Relaxation-Destraining Cycle, J. Fluid Mech. **562**, 123 (2006)

10.191 Y.-C. Chow, O. Uzol, J. Katz, C. Meneveau: Decomposition of the spatially filtered and ensemble averaged kinetic energy, the associated fluxes and scaling trends in a rotor wake, Phys. Fluids **17**, 085102 (2005)

10.192 O. Uzol, D. Brzozowski, Y.-C. Chow, J. Katz, C. Meneveau: A Database of PIV Measurements within a Turbomachinery Stage and Sample Comparisons with Unsteady RANS, J. Turbul. **8**, N10 (2007)

10.193 M. Sinha, J. Katz, C. Meneveau: Quantitative visualization of the flow in a centrifugal pump with diffuser vanes, Part B: Addressing passage-averaged and LES modeling issues in in turbomachinery flows, J. Fluids Eng. **122**, 108–116 (2000)

10.194 B. Tao, J. Katz, C. Meneveau: Statistical geometry of subgrid-scale stresses determined from holographic particle image velocimetry measurements, J. Fluid Mech. **467**, 35–78 (2002)

10.195 H. Meng, F. Hussain: In-line recording and off-axis viewing technique for holographic particle velocimetry, Appl. Opt. **34**, 1827–1840 (1995)

10.196 D.H. Barnhart, R.J. Adrian, G.C. Papen: Phase-conjugate holographic system for high-resolution particle-image velocimetry, Appl. Opt. **33**, 7159–7170 (1994)

10.197 J. Zhang, B. Tao, J. Katz: Turbulent flow measurement in a square duct with hybrid holographic PIV, Exp. Fluids **23**, 373–381 (1997)

10.198 Y. Pu, H. Meng: An advanced off-axis holographic particle image velocimetry (HPIV) system, Exp. Fluids **29**, 184–197 (2000)

10.199 S.F. Herrmann, K.D. Hinsch: Light-in-flight holographic particle image velocimetry for wind-tunnel applications, Meas. Sci. Technol. **15**, 613–621 (2004)

10.200 B. Tao, J. Katz, C. Meneveau: Geometry and scale relationships in high Reynolds number turbulence determined from 3-D holographic velocimetry, Phys. Fluids **12**, 941–944 (2000)

10.201 J. Sheng, E. Malkiel, J. Katz: Single beam two-views holographic particle image velocimetry, Appl. Opt. **42**, 235–250 (2003)

10.202 J. Sheng, E. Malkiel, J. Katz: Digital holographic microscope for measuring three-dimensional particle distributions and motions, Appl. Opt. **45**, 3893–3901 (2006)

10.203 H. Meng, G. Pan, Y. Pu, S.H. Woodward: Holographic particle image velocimetry: From film to digital recording, Meas. Sci. Technol. **15**, 673 (2004)

10.204 G.I. Roth, J. Katz: Five techniques for increasing the speed and accuracy of PIV interrogation, Meas. Sci. Technol. **12**, 238–245 (2001)

10.205 D.P. Hart: High-speed PIV analysis using compressed image correlation, J. Fluids Eng. **120**, 463 (1998)

10.206 T. Lund, M. Rogers: An improved measure of strain state probability in turbulent flows, Phys. Fluids **6**, 1838–1847 (1994)

10.207 W.T. Ashurst, A.R. Kerstein, R.M. Kerr, C.H. Gibson: Alignment of vorticity and scalar gradient with strain rate in simulated Navier-Stokes turbulence, Phys. Fluids **30**, 2343–2353 (1987)

10.208 A. Tsinober, E. Kit, T. Dracos: Experimental investigation of the field of velocity gradients in turbulent flows, J. Fluid Mech. **242**, 169–192 (1992)

10.209 A. Vincent, M. Meneguzzi: The dynamics of vorticity tubes in homogeneous turbulence, J. Fluid Mech. **258**, 245–254 (1994)

10.210 C.W. Higgins, M.B. Parlange, C. Meneveau: Alignment trends of velocity gradients and subgrid-scale fluxes in the turbulent atmospheric boundary layer, Bound. Lay. Meteorol. **109**(1), 58–59 (2003)

10.211 F. Van der Bos, B. Tao, C. Meneveau, J. Katz: Effects of small-scale turbulent motions on the filtered velocity gradient tensor as deduced from holographic PIV measurements, Phys. Fluids **14**, 2456–2474 (2002)

10.212 P. Vieillefosse: Local interaction between vorticity and shear in a perfect incompressible fluid, J. Phys. **43**, 837–842 (1982)

10.213 B.J. Cantwell: Exact solution of a restricted Euler equation for the velocity gradient tensor, Phys. Fluids A **4**, 782–793 (1992)

10.214 W.J.A. Dahm, K.B. Southerland, K.A. Buch: Direct, high-resolution, four-dimensional measuremenets of the fine scale structure of $Sc \gg 1$ molecular mixing in turbulent flows, Phys. Fluids A **3**, 1115–1127 (1991)

10.215 L.K. Su, M.G. Mungal: Simultaneous measurements of scalar and velocity field evolution in turbulent crossflowing jets, J. Fluid Mech. **513**, 1–45 (2004)

10.216 Sun O.S., Su L.K.: Experimental assessment of scalar mixing models for large-eddy simulation, AIAA paper 2004-2550 (2004)

10.217 C. Higgins, M.B. Parlange, C. Meneveau: Turbulent heat flux and temperature gradent alignments in the lower atmosphere, Geophys. Res. Lett. **31**, L22105 (2004)

10.218 R. Knikker, D. Veynante, C. Meneveau: A dynamic flame surface density model for large eddy simulation of turbulent premixed combustion, Phys. Fluids **16**, L91–L94 (2004)

11. Flow Visualization

Most fluids, gaseous or liquid, are transparent media, and their motion remains invisible to the human eye during direct observation. Techniques allowing visualization of the flow, usually referred to as *flow visualization*, are discussed in this chapter. A great variety of such methods is known that enable one to make fluid flows visible, in the fluid mechanical laboratory, in industrial environments, and for field experiments. These methods rely mostly on the addition of a tracer material to the flowing fluid, e.g. dye or smoke, and what is then observable is merely the motion of the tracer. Differences between the motion of the tracer and that of the fluid are aimed at being minimal. Presented in this chapter are surveys of available tracer materials, techniques of introducing the tracer to the flow, techniques of proper illumination of the flow scene, methods of providing optical access to the flow, and recording of the observable information. Some of the flow visualization methods provide qualitative information on particular flow patterns, others allow to measure the flow velocity quantitatively.

11.1 Aims and Principles of Flow Visualization 857

11.2 Visualizations of Flow Structures and Flow Direction 859
 11.2.1 The Addition of Tracer Materials as Initial Condition 859
 11.2.2 Dye Lines and Contours in Liquid Flow 860
 11.2.3 Smoke Visualization in Air Flows 862
 11.2.4 Hardware for Flow Visualization Experiments: Illumination, Recording, Confinements of the Flow Regime .. 863
 11.2.5 Enhancement of Image Quality: Fluorescence, Image Processing 865
 11.2.6 Direct Visualizations of Velocity Profiles 866

11.3 Visualization of Free Surface Flows 868

References 869

11.1 Aims and Principles of Flow Visualization

Flow visualization is a tool in experimental fluid mechanics that renders certain properties of a flow field directly accessible to visual perception. It has always been believed that observation of a process pattern facilitates the development of an understanding and the subsequent analysis of such a *phenomenon*; it is unnecessary to mention that this term indicates the process' visibility. In general and under normal circumstances, most fluids, gaseous or liquid, are transparent media, and their motion must remain invisible to the human eye during direct observation, unless a technique allowing visualization of the flow is applied. A great variety of methods has been reported that enable one to make fluid flows visible, in the fluid mechanical laboratory, in industrial environments, and for field experiments.

A reader might argue that he or she has experienced seeing a flowing fluid in nature, e.g., wind, or water waves, or a flowing river. Closer inspection of what is seen in these cases shows that these flows become visible due to the natural occurrence of means equivalent to the experimental methods of flow visualization described herein.

Although the physical principles of many flow visualization methods are rather simple, a number of important discoveries were made with these techniques. A prominent example is the discovery of the existence of coherent structures in turbulent flows, which lead H. W. Liepmann to the conclusion: "It is ironical that coherent structures... were all found by the most primitive of experimental methods: flow visualization" [11.1].

Visualization of a flow requires illumination of the flow scene, and the visibility of the flow can thus be regarded as the result of the interaction of light and the flowing fluid. The light is modified due to this interaction and carries information on the state of the flow. Information on this interactive process can be obtained in two different ways (Fig. 11.1).

- One can record the light transmitted through the fluid and compare its properties with those of the incident light. The obtainable information is integrated along the whole path of the light in the fluid (*line-of-sight methods*). The properties of a light wave that are most commonly used for this way of visualization are optical phase and the direction of propagation, which change due to variations of the fluid's refractive index caused by the flow. Typical visualization methods that fall into this category are optical interferometry, schlieren, and shadowgraphy, which are discussed in Chap. 6.
- One may record the light that is scattered from a certain position in the fluid flow. A general assumption is that this light carries information on the state of the flow at the position where this light is scattered from, and that this light is not changed during its further passage through the fluid. Therefore, the recorded information is local, which is of importance if the fluid flow is three-dimensional. The intensity of the scattered light is much lower than that of the transmitted light. Visualization discussed in this chapter is based on light scattering.

The portion of light that is not transmitted further through the fluid (i.e., that is *extincted*) is reflected or scattered from the fluid particles. The intensity of light scattered from molecules is rather weak; in order to generate an intensity sufficient to be visible and recorded by *normal* means, the fluid can be seeded with tracer particles that are much bigger than the fluid molecules. What is then seen is the motion of the tracers, and particular attention must be given to whether this motion is identical to that of the fluid.

Flow visualization by means of observing the light scattered from tracer materials is mainly a qualitative method. If one resolves the light scattered from single tracer particles, the visualization becomes quantitative and allows measurement of the velocity of the scatterers. Such quantitative tracer-based techniques, sometimes

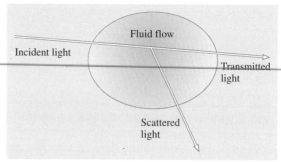

Fig. 11.1 Interaction of a light wave with a flowing fluid

referred to as *quantitative flow visualization*, are treated in Sect. 5.3. Many technical processes are governed by the interaction of a fluid flow and a solid body, and in these cases it is of great interest to know the state of the flow close to a solid wall. The interaction of the flow and the solid surface can be mechanical, thermal, or chemical; visualization methods for near-wall regimes have been developed in which the wall is coated with a material that undergoes a visible change when it interacts with the flowing fluid. Such methods for the visualization of surface flows are discussed in Chap. 7.

With the advancement of computational fluid mechanics the term *flow visualization* is also used for the visual presentation of numerical data. This kind of visualization is the result of progress made with methods that facilitate the perception of large amounts of data values, either from computation or probe measurements. In contrast to its use in these fields, the term *flow visualization* in this chapter refers to physical experimentation.

In the following information will be given on tracer materials, equipment for illumination and image recording; methods will be discussed according to the information obtainable from their application; and a number of examples will be presented for various flow situations. Review literature on flow visualization is available in the form of books that discuss methods [11.2, 3] or just show flow scenes [11.4, 5], journals specializing on this subject, e.g., [11.6], and the state of the art is regularly screened at a number of symposia, among which the International Symposia on Flow Visualization have a long tradition.

11.2 Visualizations of Flow Structures and Flow Direction

11.2.1 The Addition of Tracer Materials as Initial Condition

As stated above, the addition of tracer material ensures that sufficient light is scattered for direct observation of the flow. The materials added to the flow differ principally for the flows of liquids and gases. In order to minimize the difference in the motion of fluid and tracer it is desirable to use a neutrally buoyant tracer material; this requirement can be met in a liquid, but almost never in a gas. The most common tracers are dye in a liquid flow, most often water, and smoke in the flow of a gas, which for the present considerations, is solely air. The addition and dispersion of tracer materials generate, in principle, a two-phase fluid, and what is visualized is the motion of the dispersed phase. Here, however, it is aimed at minimizing the difference between the two phases such that fluid and tracer can be regarded to form a continuum.

What can be seen in the visualization experiment depends strongly on how and where the tracer is added, i.e., on the initial condition of the subsequent motion of the tracer material. Tracer material, released into the flow only for a short instant of time, might be collected in a flow regime where the residence time of the fluid is high, while the rest of the material is carried away with the main flow; in this way, flow structures such as vortices and regimes of separated flow become visible

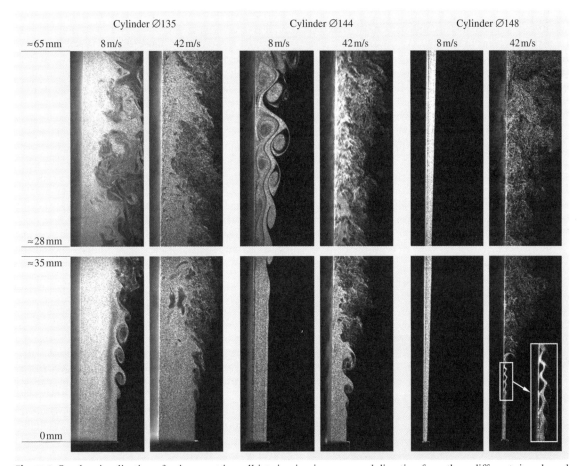

Fig. 11.2 Smoke visualization of axisymmetric wall jets issuing in an upward direction from three different ring-shaped openings (width 7.5 mm, 3 mm, 1 mm) and at two exit velocities (8 m/s, 42 m/s). Illumination is by laser light sheet. The air jets are adjacent to the wall of a vertical cylinder (courtesy of Prof. Franz Peters, Ruhr-Universität Bochum, Germany)

Fig. 11.3 Smoke visualization of the flow separated on the back of a car. Smoke is injected with a smoke pipe into the separated flow regime (courtesy of Volkswagen AG, Wolfsburg, Germany)

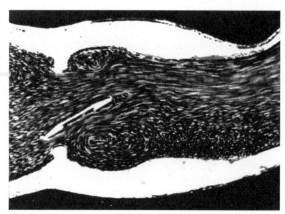

Fig. 11.5 A two-dimensional water flow model of an artificial heart valve; the length of the streaks is a measure of the local velocity of the tracer particles (courtesy Aerodynamisches Institut, RWTH Aachen, Germany)

(Figs. 11.2, 3). If the material is continuously released at a particular point in the flow, the tracer pattern visible downstream from this position will be a streamline, if the flow is stationary, and thus the flow direction can be detected (Fig. 11.4). In unsteady flow, the visible line formed by the tracers is called a streakline or filament line; this is the instantaneous locus of all fluid particles that have passed through a particular fixed point of the flow field, here the position of the tracer source. If one observes, e.g., with a long-time exposure, the trajectory that a single particle traverses in the flow field as a function of time, the result is called a particle path. Thus, the particle path contains the integrated time history of the motion of the single particle. The three curves, steamline, streakline, and particle path, coincide in steady

Fig. 11.4 Smoke line visualization of the flow over a car body in a full-scale wind tunnel (courtesy Volkswagen AG, Wolfsburg, Germany)

flow. Particular attention must be paid to the interpretation of the visible curve pattern if the observer is not at rest with respect to the coordinate system of the flow field, i.e., if relative motion exists between the object and the observer [11.2].

In such a visualization of flow structures and flow lines or direction, the tracer material can be regarded as a continuum, and the motion of individual tracer particles is not observed. At low values of tracer particle concentration in the flow, it may be possible to identify the individual moving particles, e.g., by a time exposure, and to measure their local velocity (Fig. 11.5). Apart from the possibility of producing visible flow patterns and quick surveys of the velocity distribution in an extended flow regime, this way of quantitative visualization has currently lost ground to the methods presented in Sect. 5.3.

11.2.2 Dye Lines and Contours in Liquid Flow

The marking of flow lines and contours in the flow of a liquid by means of dye can be achieved by introducing the dye into the liquid from outside (*direct injection*), or by generating the dye with an appropriate chemical reaction in the liquid. In the second case it is required that the liquid carries respective chemicals in solution, and that the dye-producing reaction is initiated at the correct position in the flow.

Directly injected dye is released either from a small ejector tube placed at the desired position in the flow or from small orifices provided in the wall of a model under investigation. In order to minimize disturbances in the flow, the ejector tube should be small and must be placed far enough upstream of the flow regime of interest. Tubes that normally serve as Pitot probes in air flows, hypodermic tubes, and syringes, with outer diameters of 1 mm or less, are often used as dye injectors. The rate at which the dye is released has to be matched to the average velocity of the liquid; formation of a dye jet at too high a release velocity must be avoided. Dyes used for flow visualization should fulfill several requirements: Besides some general properties such as being nontoxic, noncorrosive, etc., desired properties are neutral buoyancy, high stability against mixing, and good visibility.

Popular dyes for flow experiments in water are food coloring and ink. Optimal conditions can be met by arranging complete dilution of the dye in the working fluid, whose specific weight can be matched by the addition of alcohol. The undesired mixing of the dye with water, particularly if the aim is to visualize flow lines, increases with Reynolds number. Therefore, application of the method is restricted to laminar or low-Reynolds-number flows (Fig. 11.6). A number of recipes are known for stabilizing the dye filaments against mixing in such flows. Most effective is the addition of milk; the fattiness of milk apparently retards diffusion of the dye solution into the main bulk of water, and at the same time, the high reflectivity of the milk particles provides high visibility [11.8]. Another way of ensuring high optical contrast and good visibility is the use of fluorescent tracers. While ordinary dyes such as food coloring and ink require no specific provisions for illumination, fluorescent dyes

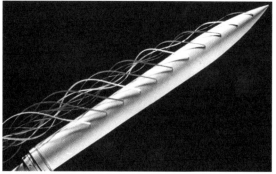

Fig. 11.6 Dye lines in the vortex flow behind a yawed cylindrical model. The fluid is water, the dye is a mixture of ink, milk, and alcohol (courtesy of Deutsch-Französisches Forschungsinstitut, ISL, St. Louis, France)

should be illuminated with the appropriate wavelengths that initiate the fluorescence. It appears that no systematic investigations on the optimization of selection of dyes and illumination have been carried out. Information on these experimental parameters is available in numerous laboratory reports and various publications. For a choice of dyes the reader is referred to Table 11.1 in [11.2].

Generating a dye without mechanically disturbing the flow using an injector is possible for the purpose of marking the contour or shear layer between two different flows that begin mixing downstream of a given position. Use is made of a dye-producing chemical reaction at the interface of the two flows. Most suitable for such a procedure is the application of pH indica-

Table 11.1 Refractive-index-matching fluids (After [11.7]); n = refractive index; ρ = density; μ = viscosity; ρ_0 and μ_0 are the corresponding values of water at 20 °C; see [11.2] for detailed references

Aqueous solutions of	n	ρ/ρ_0	μ/μ_0
Glycerin	1.33–1.47	1.0–1.26	1.0–1490
Zinc iodide	1.33–1.62		
Sodium iodide (60%)	1.5		
Sodium thiocyanate	1.33–1.48	1.0–1.34	1.0–7.5
Potassium thiocyanate	1.33–1.49	1.0–1.39	1.0–2.4
Ammonium thiocyanate	1.33–1.50	1.0–1.15	1.0–2.1
Organic liquids			
Kerosene	1.45	0.82	
Silicone oil mixture	1.47	1.03	190
Paraffin oil	1.48	0.85	
Turpentine	1.47	0.87	1.49
Soybean oil	1.47	0.93	69
Olive oil	1.47	0.92	84
Castor oil	1.48	0.96	986

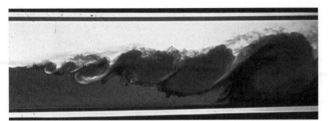

Fig. 11.7 Shear layer separating two mixing flows, visualized by means of a pH indicator reaction (courtesy Prof. R. E. Breidenthal, University of Washington, Seattle, USA; see also [11.9])

tors that can be dissolved in water and that change color upon a specific change of the pH value. The pH indicator is dissolved into one of the two streams, and the fluids in the two streams are given different pH values, below and above the critical value where the color change takes place. The mixing zone of the two fluids is visualized by the dye-producing reaction, and the structures formed in the mixing layer become visible [11.9] (Fig. 11.7).

Other dye-producing chemical reactions have been reported that are initiated along a given straight line in the flow, thus allowing the visualization of velocity profiles. These methods are discussed in section Sect. 11.2.

11.2.3 Smoke Visualization in Air Flows

The technique of marking flow lines or structures in an airstream by means of smoke is, in principle, the same as visualizing the flow pattern of a liquid by the injection of dye. Although its application is not restricted to wind tunnels, smoke visualization is a standard experimental tool for these facilities, and its technical advancement is closely related to the history of wind tunnels [11.10]. The term *smoke* is not well defined, and it can be used here in a wide sense, not only restricted to combustion products; it can also include steam, vapor, aerosols, mist. These substances cannot fulfill the requirement of being neutrally buoyant tracers, since the densities of these tracer materials are order of magnitude larger than the density of air. However, the solid or liquid particles of these materials have diameters smaller than $1\,\mu$m, thus making buoyancy effects, to a first-order approximation, negligible. The mentioned properties desired for a tracer, neutral buoyancy, stability against mixing, etc., are more difficult to meet than in the case of dyes. The basic types of producing smoke are burning or smoldering tobacco, wood, or straw; vaporizing mineral oils; producing mist as the result of the reaction of various chemical substances; and condensing steam to form a visible fog.

The generation and the use of smoke for visualization purposes is more an art than a technology, and many laboratories have developed their own recipes and equipment. Smoke and aerosol generators are also commercially available, most of them not designed primarily for laboratory experiments but for entertainment and show (*nebulisers*). The majority of smoke generators are based on the vaporization of hydrocarbon oils, particularly kerosene. A kerosene mist (smoke) generator mainly consists of a heating facility and a device in which the kerosene vapor is mixed with an airstream to form the appropriate mist [11.2]. In order to form and visualize smoke lines (Fig. 11.8), the smoke is introduced into the flow by a pipe that should be oriented parallel to the main flow. For wind-tunnel experiments it is preferable to release the smoke at the inlet of the contraction section so that the smoke lines are stabilized during acceleration of the flow in this section [11.10]. In a closed-circuit wind tunnel the smoke would soon accumulate in the tunnel system and, therefore, typical *smoke tunnels* exhaust the air into the outside, and also have high contraction ratios of the nozzle in order to keep the turbulence level low and prevent the smoke from mixing early in the airstream.

Since most wind tunnels are of the closed-circuit type, there has always been interest in using steam as the visualizing tracer material. The idea is that steam in combination with a cooling agent is introduced into the air flow, where the mixture produces a visible fog. After heat exchange with the surrounding air, the fog will disappear, thus leaving the main airstream clean in the recirculating system [11.2]. In na-

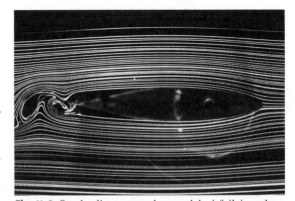

Fig. 11.8 Smoke lines around a model airfoil in a low-Reynolds-number wind-tunnel flow (courtesy of Prof. T. J. Mueller, University of Notre Dame, USA)

Fig. 11.9 Karman vortex street formed in the wind downstream of the island of Jan Mayen in the North Atlantic Sea, visible in the formation of clouds (courtesy of Sternwarte Bochum, Germany)

Fig. 11.11 Moist air rises along the slopes of Lanai Island (Hawaii), cools down with increasing height where the moisture condenses, forming a fog that visualizes the wake flow in the steady wind, here from *right* to *left* (W. Merzkirch)

ture, clouds as an accumulation of mist are the tracer material by which the motion of air can becomes visible (Fig. 11.9).

A frequently used chemical method for producing a dense white mist is based on the reaction of titanium tetrachloride, $TiCl_4$, with water. The result of the reaction is titanium dioxide, TiO_2, a white mist or smoke of high optical reflectivity. This mist can be produced in a smoke generator, and also inside the wind tunnel, i.e., without the need to introduce a smoke pipe, by depositing drops of $TiCl_4$ on the surface of a test model, which then reacts with the humidity in the air. Aerosol generators are also of great importance for velocity measurements with laser anemometers and imaging methods, on which information should be found, e.g., in Sect. 5.3.

Fig. 11.10 Open field experiment: smoke visualization of vortex formation during the start of an airplane (courtesy Prof. Cam Tropea, Technische Universität Darmstadt, Germany)

11.2.4 Hardware for Flow Visualization Experiments: Illumination, Recording, Confinements of the Flow Regime

The source, duration and form of illumination must be chosen for an experiment. Front light illumination with ordinary lamps or spot lights may be sufficient for many *simple* flow visualization studies. This situation, with the sun as the source, is also given in open field experiments (Fig. 11.10), or when observing a naturally occurring flow (Fig. 11.11). Time control of the source, either in form of slow-motion, high-speed, or stroboscopic illumination, is required if it is desired to provide information on the velocity in the flow, and it must be adjusted to the expected magnitude of the flow velocity.

In order to provide certain local information on flow structures within the flow field, illumination in the form of a light sheet has been developed and has become the most common way of illuminating the flow scene of interest. A laser is required as the light source. The laser beam is expanded to form a light sheet that defines the plane within which the flow structures are visualized. Expansion can be done with a cylinder lens or a simple glass rod, as indicated in Fig. 11.12. The thickness of the sheet is then approximately equal to the diameter of the beam emitted by the laser, i.e., approximately 1 mm. The laser beam has a Gaussian intensity profile, and in many references, where number values for the light sheet thickness are given, this thickness is not precisely defined. A very thin light sheet might be desirable for improving the spatial resolution if visual-

Fig. 11.12 Expansion of a laser beam by a glass rod (cylindrical lens) for forming a light sheet. The thickness of the sheet is controlled by a second lens

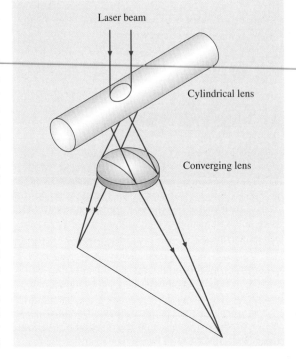

ization is quantitative with the aim of measuring flow velocities. This requires more-refined light sheet optics than a simple glass rod. A so-called telescopic arrangement consists of two spherical lenses combined with a cylinder lens [11.11]. The area of the sheet to be used effectively for the measurement and the sheet thickness can be controlled easily with one lens of the telescope. A thickness of 0.1 mm can be realized in this way. Making the sheet parallel, i.e., not expanding, by means of a large-diameter lens or a spherical mirror provides a uniform illumination intensity in the field of view. A new approach, which is particularly suitable for visualization studies of flows in microsystems, is the use of diffractive optical elements (DOE) that replace conventional optical lenses in the light sheet optics and allow the light sheet to be tailored to the specific microflow conditions [11.12]. By moving the light sheet normal to its plane one can scan the volume of the flow field and generate a kind of tomographic resolution of the three-dimensional flow structures [11.13]. This kind of *tomography* should not be confused with *optical tomography*, which is essential for resolving three-dimensional flows with varying fluid density (Chaps. 3 and 27). Light sheet optics is an essential element of setups for particle image velocimetry and related imaging systems for velocity measurements, information on which can be found in the corresponding sections, e.g., Sect. 5.3.2.

Scanning the three-dimensional (3-D) flow field with a light sheet in order to obtain 3-D information on spatial structures requires the flow to be steady during the scanning period. If this is not the case, the solution can be to use holography with single-pulse laser illumination; in this way, complete 3-D information is frozen in the hologram and can be scanned upon reconstruction of the object. Such a hologram also provides a direct 3-D view of the flow structures; see, e.g., the hologram of a jet flow in [11.15]. An alternative for producing visible 3-D information on a flow is stereo photography. This method, which employs two cameras to obtain percep-

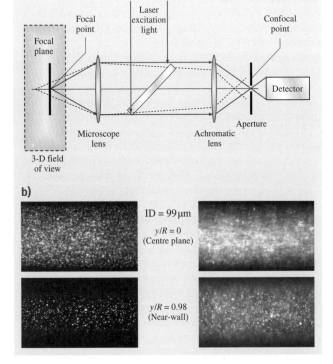

Fig. 11.13 (a) Optical setup for confocal laser scanning microscopy (CLSM). **(b)** Particle images taken in two different y-planes by the CLSM (*left column*), and by the conventional epi-fluorescence microscopy (*right column*), courtesy Prof. K. D. Kihm, University of Tennessee, USA, see also [11.14]

tion of optical depth, has been in use for several decades in various engineering applications. Such stereoscopic systems record two simultaneous, but different, views of the object. The two views are then combined, and a mathematical algorithm provides the reconstruction of the three-dimensional object field. Stereoscopic systems are of great importance for determining the three velocity components in 3-D flows by means of velocimetric imaging methods; more-detailed information is available in Sect. 5.3.2.

Confocal microscopy combined with spatial filtering and excitation of fluorescence is becoming an important means for reconstructing 3-D structures under microflow conditions [11.14]. The basic *confocal* concept is a point scanning of the laser excitation and a spatially filtered fluorescence signal emitting from the focal point onto the confocal point (Fig. 11.13a). The pinhole aperture, located at the confocal point, allows the emitted fluorescent light exclusively from the focal point (solid rays in Fig. 11.13a) to pass through the detector, and filters out the fluorescent light emitted from outside the focal point (dashed rays). This spatial filtering is the key principle behind optical-resolution enhancement by depth-wise optical slicing. The illuminating laser can scan rapidly from point to point on a single focal plane, in a synchronized way with the aperture, to complete a full-field image on the detector. The scan is repeated for multiple focal planes to reconstruct three-dimensional images. Tracer particle images obtained with confocal laser scanning microscopy (CLSM) are shown in Fig. 11.13b, compared with results of conventional epifluorescence microscopy. The raw PIV images (40 X, 0.75 NA) are shown for two selected planes of $y/R = 0$ at the center plane, and $y/R = 0.98$ near the top end of the inner surface of a 99 µm-ID microtube. The confocal microscopic images demonstrate optically sliced images with clear image definition of individual particles located within the slice thickness [11.14].

Any visualization experiment requires optical access to the flow, i.e., transparent walls for flows bounded by walls. To relate the visible information to the local position in the flow correctly it may be necessary to account for refraction of light by the viewing window. This correction is easy for a plane window, but the problem is more difficult to solve for curved optical boundaries. The problem does not exist if the refractive indices of the wall and fluid are equal. A number of liquids have been found whose refractive indices can be matched with those of glass or Plexiglas, approximately 1.5; (Table 11.1 [11.7]). Not all of these liquids are pleasant to work with, some are unstable or even unsafe, and the val-

ues of their viscosity might not be appropriate for a specific experiment. However, the number of such liquids that have tolerable properties is increasing, as a result of the growing interest in flows in porous media [11.16].

11.2.5 Enhancement of Image Quality: Fluorescence, Image Processing

Rapid diffusion of a dye into the ambient water deteriorates the image quality and makes it difficult to detect flow structures. Improvement of the imaging quality might also be desirable for flows in complex geometries, e.g., porous media. Enhancement of the visibility of flow contours can be obtained by using a fluorescent tracer. Fluorescence is induced by the corresponding light from a laser source [*laser-induced fluorescence* (LIF)] or an ultraviolet (UV) lamp, and the fluorescent radiation emitted by the tracer, at a wavelength different from that of the inducing radiation, can be isolated from the background and other disturbing light by using an optical filter in the wavelength range of the fluorescence. In this way, use is made of the inelastic scattering from the fluorescent tracer, instead of elastic scattering from *normal* tracer materials (Sect. 7.4).

A number of fluorescent dyes that can be solved in water have been described and investigated. It appears that Fluorescein is the most appropriate material to work with [11.17]. Fluorescein is highly soluble in water and has a high quantum yield of approximately 85%. The fluorescent radiation can also be used for quantitative measurements, e.g., concentration, which is not the main focus of this section. For these quantitative purposes, stability of the dye properties is desirable. Fluorescein has a weak temperature dependence; but the emitted radiation depends strongly on the pH value, which ought to be controlled and kept constant during an experiment. The absorption peak of this dye is near 490 nm, which is conveniently close to the 488 nm wavelength of an argon-ion laser, which may therefore serve as the light source for illumination. The fluorescence intensity from a dye solution is observed to diminish with time, perhaps as a result of oxidation processes. Experiments using a fluorescent dye are particularly suitable for the study of mixing processes, e.g., one fluid stream that is supposed to mix with a second stream is seeded with the dye such that the structure of the mixing zone becomes clearly visible.

Fluorescent species used for visualization experiments in gas flows are either seeded as tracers into the flow, e.g., biacetyl, O_2, I_2, NO, or are radicals produced as a result of a chemical reaction, e.g., CH, OH, C_2, CO in combustion processes. The illuminating laser must be

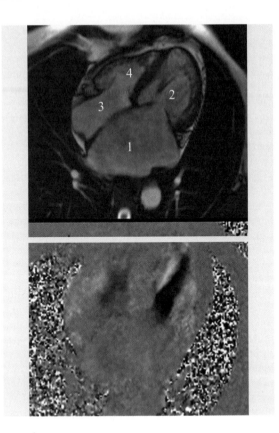

Fig. 11.14 NMR image of the heart of a patient suffering from mitral stenosis. *Top*: anatomical configuration, 1 – left auricle of increased volume, 2 – left ventricle, 3 – right auricle, 4 – right ventricle. Stenosis causes a decreased throat area when opening of the mitral valve allows the blood to flow from the left auricle to the left ventricle. *Bottom*: NMR is applied such that velocities higher than a selected threshold value appear dark, while those lower than the threshold are bright. The dark area in the left ventricle visualizes the velocity increase due to the reduced throat area (courtesy of Dr. M. Montag, Krupp-Krankenhaus, Essen, Germany)

tuned to the absorption wavelengths of the respective species, while the optical filter, which allows only the wavelength emitted from the fluorescent tracer to pass through, eliminates all the radiation generated by the combustion process. Details of reacting flows can thus be observed [11.18].

A hardware tool for the enhancement of image quality at low illumination intensity, though expensive, is the use of an image intensifier camera [11.19]. A great variety of software tools are available for improving the image quality, e.g., filtering algorithms, but it is recommended to control the physical significance of those phenomena that become visible with their application in order to avoid misinterpretation of possible numerical artifacts. For detailed information on the application of filtering algorithms, see the literature on planar velocimetry where image-processing methods are needed for the generation and presentation of experimental results (Sect. 5.3.2).

The desired visualization of flows in porous media may require more-effective means than just the method of refractive index matching mentioned in Sect. 11.2.4. Magnetic resonance imaging (MRI) has been successfully used for this purpose; it is applicable even to flows in containments with nontransparent walls, and for this reason has found intensive application for in vivo measurements in medicine, where the technique was developed (Fig. 11.14). MRI is based on nuclear magnetic resonance (NMR), which measures relaxation-weighted proton density and can also yield information on fluid velocity and temperature [11.20]. The method makes use of the dynamics of the magnetic moment of hydrogen nuclei in an externally applied magnetic field. The information obtained with an MRI system is a planar distribution of greyscale values, from which one derives, in a first step, the interstitial space in the porous medium filled with water, and thereby the geometry of the medium. Special timing of the magnetic field allows one to generate a velocity map of the fluid.

11.2.6 Direct Visualizations of Velocity Profiles

If one produces instantaneously tracers along a straight line perpendicular to the main flow direction, one may visualize local velocity profiles by observing the distortion of the line in the flow. Methods for such direct visualization of velocity profiles are known for both water and air flows. They were developed prior to the general availability of planar velocimetric methods based on the recording of tracer particle patterns and subsequent evaluation by image processing (Sect. 5.3.2), they are still in use where the equipment is available, and they are particularly useful for tutorial purposes.

The hydrogen-bubble technique is the most popular representative in this group [11.2]. It is based on the electrolysis of water; a fine wire as the cathode defines the straight line where the tracers are produced upon application of a dc voltage. Hydrogen bubbles are formed at the cathode and oxygen bubbles at the anode that is placed somewhere else in the water flow, away from the region of interest. The hydrogen bubbles are used as tracers because they are much smaller than the

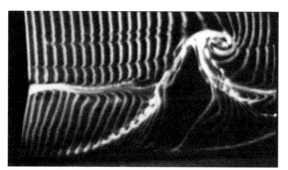

Fig. 11.15 Visualization of vortical structures in boundary layer flow by means of hydrogen-bubble time lines (courtesy Prof. C. R. Smith, Lehigh University, Bethlehem, PA, USA)

oxygen bubbles. A column-shaped row of bubbles, the instantaneous line of tracers, is formed along the wire when an electric pulse is generated; this row is carried away with the flow and deformed according to the local velocity profile. By pulsing the voltage at a constant frequency, one can produce several successive rows of bubbles, which mark the flow curves separated by a constant flow time. These curves are called time lines, and it is in this mode that the hydrogen-bubble technique is most frequently employed (Fig. 11.15).

Normal tap water may serve as the electrolytic fluid; if the water is too soft, sodium sulfate can be added in order to increase the electrolytic conductivity. A rule of thumb is that the bubble size is of the order of the diameter of the generating cathode wire. Measurement systems for this method, particularly the electric supply and control system, are commercially available. Illumination with conventional lamps at an angle of 65° between the illumination and viewing directions has been found to be most suitable. The period of time during which the bubbles can be observed in the flow is limited by the dissolution of the hydrogen bubbles in the fluid. Diffusion of the bubbles increases with Reynolds number and is very rapid in turbulent flows. The method is therefore limited to low-speed flows, maximum velocities being of the order of 20–30 cm/s. However, great achievements have been made using the hydrogen-bubble technique to study unstable and transitional boundary-layer flows.

The hardware of the thymol blue method is similar to that of the hydrogen-bubble technique [11.2]. A thin cathode wire defines the line along which the visible tracer is produced instantaneously upon application of a voltage and as the result of a sudden change of the local pH value. The production of dyes by means of pH indicators was mentioned in Sect. 11.2.2. Here, the working fluid is an aqueous solution of thymol blue which is orange–yellow in an acidic environment and turns its color to blue if the the solution becomes alkaline. This change occurs close to the cathode wire; a thin line of liquid is marked blue in the otherwise yellow fluid and moves with the flow, being deformed by the local velocity as described for the row of hydrogen bubbles. In contrast to the latter method, the applied voltage here is low in order to avoid or minimize the formation of H_2 bubbles. While the bubbles are buoyant tracers, the tracer generated with the pH indicator is non-buoyant, and this makes the thymol blue method particularly attractive for studies of flow in liquids exhibiting a natural density stratification. Furthermore, the marked fluid has the same properties, particularly size of the tracer, as the unmarked fluid, and the method is therefore also suitable for experiments with rotating flows. The comment above about diffusion of the tracer

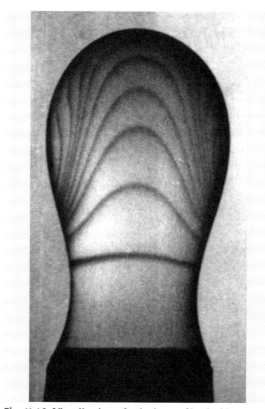

Fig. 11.16 Visualization of velocity profiles inside a growing drop by means of the photochromic dye technique (courtesy Prof. E. Marschall, University of California at Santa Barbara, USA)

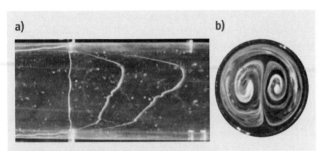

Fig. 11.17a,b Smoke wire visualization of the velocity profile in the flow through a curved pipe (**a**); this flow is known to exhibit a pair of counter-rotating vortices extending in the streamwise direction (*secondary flow*), as seen in the smoke visualization with the light sheet normal to the pipe axis (**b**). (courtesy Prof. Y. Tanida, Tokai University, Japan)

into the bulk fluid at higher Reynolds numbers applies here as well.

These two methods suffer from the presence of a disturbing electrode wire in the flow field of interest. Such mechanical disturbances are avoided if one produces the desired dye or tracer material along a straight line by a photochemical reaction with a laser beam. Photochromic materials that can be used for this purpose are known; they have the disadvantage of not being well soluble in water, and a number of organic liquids are therefore used as solvents and working fluids [11.2]. The photochromic substance absorbs UV radiation provided by a laser and becomes opaque and thus visible, while it is transparent in the unexcited state. With a short laser pulse an instantaneous line of tracers is produced along the laser beam. The reaction is reversible. The reverse transition is spontaneous but slow, and the visible tracers disappear after a given period of time that, in addition to the reverse reaction being spontaneous, can be controlled by thermal energy. The photochromic dye is a true solution and thus non-buoyant. The excitation of the reaction by UV radiation requires the use of a quartz window to introduce the laser beam into the flow. The method is applied to the study of velocity profiles in near-wall flows, and also to visualize the internal flow in drops during their formation (Fig. 11.16).

Direct visualization of velocity profiles in air flows is more difficult due to the higher degree of diffusion of possible tracers (smoke) in the flow. A frequently used method is a smoke wire that spans the flow of interest and can be heated electrically [11.2]. The wire is coated with a thin layer of oil that evaporates when the wire is heated. Application of a short electric heating pulse produces a thin line of smoke (oil mist) along the length of the wire, similar to the tracer lines in the methods described for use in liquid flows. Again, the method has been found useful for the study of near-wall flows (Fig. 11.17).

A technique for producing tracer lines in an air flow without the need to use a mechanically disturbing wire is available: the *spark tracer method* [11.2]. A high-voltage electric discharge across the mean flow direction generates a thin channel of ionized luminescing molecules that is visible and deforms, like the tracer lines in the previous cases, with the flow profile. Application is restricted to high-speed air flows.

11.3 Visualization of Free Surface Flows

The visualization of flow patterns on free water surfaces is of interest for many reasons. These flows, which include the formation and propagation of water waves, are associated with a deformation of the free surface, which is the key for a direct visualization and also quantitative measurements. It is appropriate to distinguish between large-scale deformation on the scale of water waves, and small-scale deformation or fine structure as caused, e.g., by turbulence. Water waves can be visualized by observing a regular grid immersed in the water, parallel to the free surface. To the observer, the grid pattern appears distorted as a result of the refraction of light at the wavy surface. The deformed grid pattern can be evaluated quantitatively, analogous to the application of moiré methods to the visualization of flows with varying refractive index (Chap. 6). An alternative is to observe and record the light reflected from the wavy surface, a method that makes water waves directly visible in nature (Fig. 11.18). The fine structure of such a surface, which can be caused by the interaction of the free surface with wind and turbulence, can be made visible in reflection with speckle methods that are also used for statistically analyzing turbulent tracer particle patterns [11.21]; see particularly Sect. 5.3.2.

The formation of gravitational waves of long wavelengths on the free surface of a liquid is analogous to the pattern of pressure waves in an isentropic supersonic gas flow, because these processes are governed by equations of equivalent mathematical form. This so-called

Fig. 11.18 System of waves produced by a ship moving into Seattle harbor (W. Merzkirch)

Fig. 11.19 Supercritical water flow around a wedge at zero angle of attack, visualizing a hydraulic analog of the oblique shock in supersonic flow (courtesy Prof. M. S. Wooldridge, University of Michigan, USA)

hydraulic analogy between *supercritical* flow in a water channel and supersonic gas flow enables one to visualize supersonic flow patterns in a water channel [11.2]. Although the analogy would be exact only for a gas having a ratio of specific heat of $\gamma = 2$ (which is not realistic), the method continues to attract interest, particularly for tutorial purposes [11.22] (Fig. 11.19).

References

11.1 H.W. Liepmann: The rise and fall of ideas in turbulence, Am. Scientist **67**, 221–228 (1979)
11.2 W. Merzkirch: *Flow Visualization*, 2nd edn. (Academic, Orlando 1987)
11.3 W.-J. Yang (Ed.): *Handbook of Flow Visualization* (Hemisphere, New York 1989)
11.4 M. Van Dyke: *An Album of Fluid Motion* (Parabolic, Stanford 1982)
11.5 Y. Nakayama, W.A. Woods, D.G. Clark (Eds.): *Visualized Flow* (Pergamon, Oxford 1988)
11.6 *J. Flow Visualization and Image Processing* (Begell House, New York since 1993)
11.7 R. Budwig: Refractive index matching methods for liquid flow investigations, Exp. Fluids **17**, 350–355 (1994)
11.8 H. Werle: Hydrodynamic flow visualization, Annu. Rev. Fluid Mech. **5**, 361–382 (1973)
11.9 R. Breidenthal: Structure in turbulent mixing layers and wakes using a chemical reaction, J. Fluid Mech. **109**, 1–24 (1981)
11.10 T.J. Mueller: On the historical development of apparatus and techniques for smoke visualization of subsonic and supersonic flow, AIAA Paper **80-0420-CP** (1980)
11.11 J.P. Prenel, Y. Bailly: Theoretical determination of light distributions in various laser light sheets for flow visualization, J. Flow Visual. Image Process. **5**, 211–224 (1998)
11.12 F. Peters, A. Grassmann, H.S. chimmel, B. Kley: Improving small laser light sheets by means of

11.13 J.P. Prenel, Y. Bailly: Quantitative imagery of 3-D flows: From tomographic to volumic investigations, J. Flow Visual. Image Process. **8**, 189–202 (2001)

a diffractive optical element, Exp. Fluids **35**, 4–7 (2003)

11.14 J.S. Park, C.K. Choi, K.D. Kihm: Optically sliced micro-PIV using confocal laser scanning microscopy (CLSM), Exp. Fluids **37**, 105–119 (2004)

11.15 L. Hesselink: Digital image processing in flow visualization, Annu. Rev. Fluid Mech. **20**, 421–485 (1988)

11.16 M. Stöhr, K. Roth, B. Jähne: Measurement of 3D pore-scale flow in index-matched porous media, Exp. Fluids **35**, 159–166 (2003)

11.17 D.A. Walker: A fluorescence technique of measurements of concentration in mixing liquids, J. Phys, E Sci. Instrum. **20**, 217–224 (1987)

11.18 G. Kychakoff, R.D. Howe, R.K. Hanson: Quantitative flow visualization technique for measurements in combustion gases, Appl. Opt. **23**, 704–712 (1984)

11.19 I. Grant, X. Wang: Directionally-unambiguous, digital particle image velocimetry studies using a image intensifier camera, Exp. Fluids **18**, 358–362 (1995)

11.20 M.D. Shattuck, R.P. Behringer, G.A. Johnson, J.G. Georgiadis: Convection and flow in porous media, J. Fluid Mech. **332**, 215–245 (1997)

11.21 D. Dabiri, M. Gharib: Simultaneous free-surface deformation and near-surface velocity measurements, Exp. Fluids **30**, 381–390 (2001)

11.22 S.L. Rani, M.S. Wooldridge: Quantitative flow visualization using the hydraulic analogy, Exp. Fluids **28**, 165–169 (2000)

12. Wall-Bounded Flows

Turbulent wall-bounded flows (i.e., boundary layer, pipe and channel flows) present additional measurement challenges relative to those in, say, free shear turbulent flows or grid turbulence. The physical presence of the wall and the limitations and influences it presents on the implementation of sensing technologies creates some of these challenges. Other, often more-subtle issues, however, relate to the effect that the wall has on the inherent flow dynamics. Such effects are reflected in the steep mean velocity gradient(s) in the vicinity of the surface, as well as the length and time scales of the turbulence local to the near-wall region. Regarding the latter, primary challenges are associated with the high frequencies and small scales of near-wall turbulence relative to free shear flows.

In previous Chaps. (5.2, 5.3 and 5.5.3), relatively broad discussions were provided regarding the requirements and considerations for accurate measurements of both mean and fluctuating quantities in turbulent flows. The present chapter constitutes an extension of these more-generic considerations relative to the specific case of wall-bounded turbulent flows. For the purposes of providing a background context, the initial subsection below presents a brief overview of concepts and considerations specific to wall flows. Owing to its central role in the study of the turbulent wall flows, the next subsection addresses the measurement of the wall shear stress for ca-

12.1	Introductory Concepts	871
	12.1.1 Governing Equations	871
	12.1.2 Brief Overview of Wall Flow Structure	872
	12.1.3 Scaling Ideas and Parameters	874
	12.1.4 Overview of Measurement Considerations	874
12.2	Measurement of Wall Shear Stress	875
	12.2.1 Methods for Determining the Time-Averaged Wall Shear Stress	876
	12.2.2 Time-Resolved Methods	882
12.3	Boundary-Layer Stability and Transition	886
	12.3.1 The Process of Transition for Boundary Layers in External Flows	886
	12.3.2 Nomenclature of Linear Theory	886
	12.3.3 Basic Rules for Boundary-Layer Stability Experiments	887
	12.3.4 Experimental Techniques	887
	12.3.5 Wind Tunnel Environment	890
	12.3.6 T–S Measurements	892
	12.3.7 Visualization Methods	896
12.4	Measurements Considerations in Non-Canonical Flows	896
	12.4.1 Pressure Probe Measurements	897
	12.4.2 Turbulence Measurements	898
	12.4.3 Wall Shear Stress	901
	12.4.4 Planar and Whole-Field Measurements	902
References		902

nonical boundary layer, pipe and channel flows. Considerations relative to transitional and non-canonical wall flows are presented in subsequent subsections.

12.1 Introductory Concepts

Prior to detailed discussions relative to specific measurement objectives, it is useful to provide a brief overview of background information relevant to the experimental study of wall-bounded turbulence.

12.1.1 Governing Equations

This subsection provides a brief presentation and discussion of the governing equations for the canonical turbulent boundary layer, pipe, channel flows. In each case, the development will be for incompressible, constant property flows, governed by the Reynolds averaged Navier–Stokes (RANS) equations presented previously in Sect. 10.1, and for convenience repeated here in Cartesian form for a statistically stationary flow without body forces

$$\rho \left(U \frac{\partial U}{\partial x} + V \frac{\partial U}{\partial y} + W \frac{\partial U}{\partial z} + \frac{\partial \overline{u^2}}{\partial x} + \frac{\partial \overline{uv}}{\partial y} + \frac{\partial \overline{uw}}{\partial z} \right)$$
$$= -\frac{\partial P}{\partial x} + \mu \left(\frac{\partial^2 U}{\partial x^2} + \frac{\partial^2 U}{\partial y^2} + \frac{\partial^2 U}{\partial z^2} \right), \quad (12.1)$$

$$\rho \left(U \frac{\partial V}{\partial x} + V \frac{\partial V}{\partial y} + W \frac{\partial V}{\partial z} + \frac{\partial \overline{vu}}{\partial x} + \frac{\partial \overline{v^2}}{\partial y} + \frac{\partial \overline{vw}}{\partial z} \right)$$
$$= -\frac{\partial P}{\partial y} + \mu \left(\frac{\partial^2 V}{\partial x^2} + \frac{\partial^2 V}{\partial y^2} + \frac{\partial^2 V}{\partial z^2} \right), \quad (12.2)$$

$$\rho \left(U \frac{\partial W}{\partial x} + V \frac{\partial W}{\partial y} + W \frac{\partial W}{\partial z} + \frac{\partial \overline{wu}}{\partial x} + \frac{\partial \overline{wv}}{\partial y} + \frac{\partial \overline{w^2}}{\partial z} \right)$$
$$= -\frac{\partial P}{\partial z} + \mu \left(\frac{\partial^2 W}{\partial x^2} + \frac{\partial^2 W}{\partial y^2} + \frac{\partial^2 W}{\partial z^2} \right), \quad (12.3)$$

$$\frac{\partial U}{\partial x} + \frac{\partial V}{\partial y} + \frac{\partial W}{\partial z} = 0. \quad (12.4)$$

In the above equations U, V and W are the mean velocity components in the x, y and z directions, u, v and w are the corresponding fluctuating velocity components, an overbar represents time averaging, ρ is the mass density and μ is the dynamic viscosity. Further discussion of the developments leading to (12.1), (12.2), (12.3) and (12.4) is given in Sect. 10.1.1.

Flat-Plate Boundary Layer

Two-dimensional boundary layer flow in the (x, y) plane is considered. The flow is generated by a constant free stream velocity, U_∞, flowing over a smooth flat plate located in the $y = 0$ plane of semi-infinite extent ($x \geq 0$). As is conventional the boundary layer thickness, $\delta = \delta(x)$, is taken to be the y position where $U(y) = 0.99 U_\infty$. Under these constraints and to the leading order of boundary layer approximations [12.1–3] the RANS equations reduce to,

$$\rho \left(U \frac{\partial U}{\partial x} + V \frac{\partial U}{\partial y} + \frac{\partial \overline{uv}}{\partial y} \right) = \mu \frac{\partial^2 U}{\partial y^2}, \quad (12.5)$$

$$\rho \frac{\partial \overline{v^2}}{\partial y} = -\frac{\partial P}{\partial y}, \quad (12.6)$$

$$\frac{\partial U}{\partial x} + \frac{\partial V}{\partial y} = 0. \quad (12.7)$$

Equation (12.5) indicates that the time rate of change of mean axial momentum is determined by the sum of the viscous and Reynolds stress gradients. Unlike the laminar case, (12.6) indicates that the wall normal pressure gradient is nonzero, and is determined by the gradient of the Reynolds normal stress, $\overline{v^2}$. With regard to (12.5), it is also relevant to note that additional streamwise gradient terms of the axial and wall normal velocity variances can be significant, especially under adverse pressure gradient conditions [12.2].

Fully Developed Channel and Pipe Flows

Fluid-dynamically fully developed flow in the (x, y) plane between horizontal, effectively infinite, parallel plates is considered. Flow in the x direction is generated by a constant pressure gradient, $dP/dx < 0$. The lower plate is located at $y = 0$, and the spacing between the plates is 2δ. Under these constraints, all of the terms in (12.4) are identically zero and the mean differential force balance equations for the axial and wall-normal components reduce to (e.g., [12.1–3]),

$$\rho \frac{d\overline{uv}}{dy} = -\frac{dP}{dx} + \mu \frac{d^2 U}{dy^2}, \quad (12.8)$$

$$\rho \frac{d\overline{v^2}}{dy} = -\frac{dP}{dy}. \quad (12.9)$$

Under transformation from a cylindrical to a Cartesian coordinate system, (12.8) and (12.9) can also be shown to hold for fully developed flow in a pipe [12.2]. (A particularly thorough derivation of this is also given in the dissertation by *Sahay* [12.4].) In this case, the origin is transferred from the pipe centerline to the wall, and the pipe radius R corresponds to δ. Similar to the boundary layer equations, the axial mean flow in pipes and channels is governed by a three term equation, with the mean advection term being replaced by the mean pressure gradient in the case of pipes and channels. In all cases, the wall-normal pressure gradient is determined by the wall-normal gradient of the vertical velocity variance.

12.1.2 Brief Overview of Wall Flow Structure

Physical models of the mean structure of turbulent wall flows have emerged from the extensive body of research conducted over the past century. In what follows, two views of wall flow structure are briefly presented. Broadly speaking, the first comes about from consideration of the observed properties of the mean axial velocity profile, in concert with interpretations derived from observations of the relative magnitudes of the Reynolds

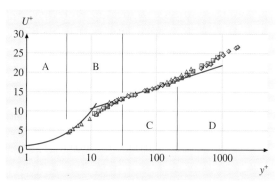

Fig. 12.1 Inner-normalized mean profiles in turbulent boundary layers and the associated layer structure: (A) viscous sublayer; (B) buffer layer; (C) logarithmic layer; (D) wake layer (Data from [12.5])

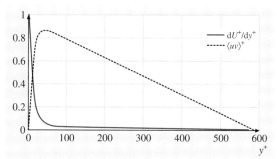

Fig. 12.2 Inner-normalized mean viscous and Reynolds shear stress profiles in turbulent channel flow $\delta^+ = 590$. (Data from [12.6])

and viscous stresses appearing in (12.5) and (12.8). The second, more recently developed interpretation, comes about by directly considering the relative magnitudes of the terms in these equations.

The prevalent view of the mean structure of turbulent wall flows [12.2, 3, 8] is largely derived from the observed properties of the mean velocity profile, along with the relative behaviors of the viscous and Reynolds stresses. Regarding the former, Fig. 12.1 shows characteristic turbulent boundary layer mean velocity profile data. In this figure the profiles have been made non-dimensional using the so-called inner variables, u_τ and ν, where $u_\tau = \sqrt{\tau_w/\rho}$ is termed the friction velocity, and ν is the kinematic viscosity. The predominant shorthand convention denotes inner variables with a superscript plus. Thus, the inner normalized wall-normal distance is denoted $y^+ = yu_\tau/\nu$. Regarding the latter, Fig. 12.2 reveals the relative magnitudes of the viscous and Reynolds stresses in turbulent channel flow. These data reveal that the magnitude of $\overline{uv}^+$ is zero at the wall, but rapidly rises to a value that is $O(1)$ by $y^+ \simeq 30$. Conversely, dU^+/dy^+ equals 1.0 at $y^+ = 0$, but diminishes to a quantity much less than $O(1)$ by $y^+ \simeq 30$. Behaviors such as these have led to the four-layer structure depicted in Fig. 12.1 and are connected to the features of the mean profile. As indicated, the viscous sublayer (layer A) extends to about $y^+ = 5$, the buffer layer (B) nominally resides between $5 < y^+ < 30$, the logarithmic layer (C) extends from near $y^+ = 30$ to $y/\delta \simeq 0.2$, and the wake layer (D) extends from the outer edge of the logarithmic layer to the outer edge of the boundary layer, $0.2 \leq y/\delta \leq 1$.

An alternative layer structure [12.7] may be derived through consideration of the relative magnitudes of the terms in the mean momentum equation. In this case, the layer structure directly reflects the dominant dynamical effects according to the mean differential statement of Newton's second law. One way to reveal this layer structure is by examining the ratio of the viscous stress gradient to Reynolds stress gradient terms in (12.5) and (12.8). Figure 12.3 schematically depicts the resulting layer structure at a fixed Reynolds number. The layer closest to the wall (layer I) is characterized by a nominal balance between mean advection and the viscous stress gradient (pressure gradient and viscous stress gradient in a pipe or channel). To a very good approximation, the adjacent layer (II) is represented by a balance between the Reynolds stress gradient (turbulent inertia) and viscous stress gradient. Across layer III (except-

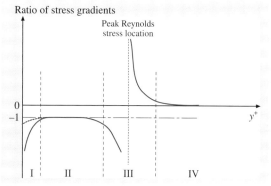

Fig. 12.3 Schematic of the structure of the boundary layer as derived from consideration of the mean momentum balance: (I) viscous stress gradient/mean advection balance layer; (II) viscous/Reynolds stress gradient balance layer; (III) Reynolds stress gradient/viscous stress gradient/mean advection balance layer; (IV) Reynolds stress gradient/mean advection balance layer (after [12.7])

ing in the immediate vicinity of the peak position of the Reynolds stress) all three terms in the mean momentum balance are of the same order, while in layer IV the mean dynamics are well represented by a balance between the Reynolds stress gradient and mean advection (or pressure gradient in a channel). The inner normalized layer widths depicted in Fig. 12.3 scale differently with Reynolds number, say $\delta^+ = u_\tau \delta/\nu$, than those in Fig. 12.1. For example, from the layer structure of Fig. 12.3 one surmises that the mean effect of viscous forces remains of dominant order much farther from the wall than indicated by the layer structure of Fig. 12.1.

12.1.3 Scaling Ideas and Parameters

Under either layer structure presented above, it is apparent that the variation of turbulent wall flow behavior with Reynolds number is characterized by a variety of scaling parameters. Relative to the theory associated with the layer structure depicted in Fig. 12.1 two independent sets of scales are required: the so-called inner and outer scales. Relative to the theory associated with the layer structure depicted in Fig. 12.3, there are three interdependent scales. In this case given any two the third may be derived. As introduced above, inner scales (e.g., length, time, velocity) are dimensionally constructed using the friction velocity and kinematic viscosity u_τ and ν, respectively. Similarly, outer scaled variables are constructed using a measure of the overall layer thickness, i.e., δ, δ^* or θ (disturbance, displacement and momentum-deficit thicknesses, respectively [12.9]), and the free stream velocity U_∞ or the deviation from the free stream velocity, $(U_\infty - U)$. (Note that in a channel flow, the maximum velocity replaces U_∞.) Nominally, the inner scales are appropriate for the characterization of the flow physics near the surface. Alternatively, outer parameters are appropriate for scaling the bulk or large scale properties of the motion centered away from the wall. In connection with this, a primary attribute of any theory of wall turbulence involves describing, as a function of Reynolds number, how the inner scaling behaviors near the wall merge with the outer scalings appropriate away from the wall. Under the theory relating to the layer structure of Fig. 12.1, this occurs in the logarithmic (or overlap) layer, where there is hypothesized to be joint validity of inner and outer scaling [12.2, 3, 8]. In the alternate view, the merging process occurs via a hierarchy of self-similar layers whose wall-normal extent asymptotically scales with the distance from the wall [12.10]. As elucidated further below, however, the physical significance of the scaling parameters are not only relevant to theoretical formulations, but are of central importance when considering the rational design of experiments.

12.1.4 Overview of Measurement Considerations

A number of important factors are worthy of careful consideration when planning for experiments of boundary layer turbulence, and particularly those in the immediate vicinity of the wall. These relate to the dynamically relevant length and time scales of the flow, the potential influences of the high mean shear near the wall on the efficacy of the measurement technique, as well as potentially detrimental effects of the wall itself on the measurement technology being employed. In addition, the convergence criteria for near-wall turbulence statistics are generally much more stringent than those in free shear flows.

Temporal and Spatial Resolution

Under general discussions of spatial and temporal resolution requirements for turbulent flow (Sect. 5.2) the length and time scales associated with Kolmogoroff scale ($\eta = (\nu^3/\epsilon)^{1/4}$, where ϵ is the mean turbulence dissipation rate) provide useful criteria relative to the size and frequency of smallest dynamical motions. That is, η itself is a measure of the smallest eddy, and for such an eddy advecting at mean velocity U an estimate of its associated circular frequency is given by $f_K = U/(2\pi\eta)$. In wall turbulence, a similar set of notions holds, but in this case the inner length scale, ν/u_τ, is typically employed as a measure of the smallest turbulent mo-

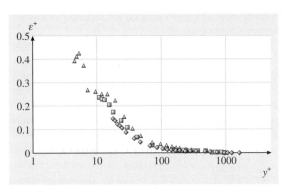

Fig. 12.4 Inner normalized profiles of the turbulence dissipation rate in zero-pressure-gradient turbulent boundary layers. In these profiles ϵ was estimated using an approximate formula based on the measurement of five velocity gradient variances [12.5]

tions. As revealed by the estimates given in Fig. 12.4, the Kolmogoroff scale is larger than the viscous scale, and depending on the distance from the wall, it can be considerably larger. Also note that u_τ^2/ν does not provide a good estimate for the highest frequency in the flow, say, for determining sampling rate. That is, while it is an inner frequency (equalling $\partial U/\partial y|_{\text{wall}}$), and does not account for the higher effective frequency associated with the advection of $O(\nu/u_\tau)$-scale eddies.

Sensor spatial resolution has proven to play a particularly important role in the study of wall turbulence, since, for example, relatively subtle Reynolds number dependencies can be masked by the competing counter-influence of decreasing spatial resolution with increasing U_∞. (Note that it is often most convenient to increase U_∞ to increase R_θ.) Important studies that have specifically examined single wire sensor resolution issues in wall-bounded flows include those by *Johansson* and *Alfredsson* [12.11], *Ligrani* and *Bradshaw* [12.12] and *Klewicki* and *Falco* [12.5]. In their study of turbulent channel flow, Johansson and Alfredsson showed considerable attenuation ($\simeq 15\%$) in the peak value of the axial intensity for nondimensional sensor lengths ranging from $4 \le \ell^+ \le 32$. They also concluded that the attenuation effect is large relative to Reynolds number dependence. In their sub-miniature hot-wire-based study of viscous sublayer turbulence in boundary layers Ligrani and Bradshaw showed very similar attenuation as in the study of Johansson and Alfredsson, and concluded that sensors having an ℓ^+ less than about 20 should be sufficient for capturing wall layer axial velocity statistics. From the analysis of a compilation of existing high resolution data and their own high-resolution boundary layer measurements, *Klewicki* and *Falco* [12.5] conclude that the peak axial velocity intensity begins to attenuate when the sensor length exceeds about eight viscous units, and that the Reynolds number dependence is smaller, but not necessarily small, in comparison with the attenuation effect. The more-recent high-resolution laser-Doppler anemometer (LDA) measurements of *DeGraaff* and *Eaton* [12.13] support these findings. Overall, relative to axial velocity statistics a common *rule of thumb* suggestion is that the characteristic sensor dimension be less than about a fifth of the distance from the wall. This is viewed as constituting a reasonable, albeit rigorously unproven, criterion.

The situation becomes more complex when considering the spatial resolution requirements for the wall normal velocity component or velocity gradients and vorticity components. For x-array hot-wire probes, effects relating to both the wire length and wire spacing must be considered, as well as whether there is a significant shear across the array [12.14]. In the case of velocity gradients and vorticity components, in principle, the objective is to attain a sensing dimension comparable to the Kolmogoroff microscale (or equivalently a small number of viscous units). In practice, however, this objective must be weighed against the effect of noise contained in signals derived from sensors that are positioned very close to each other [12.5, 15]. Regarding this issue, the results by *Folz* [12.16] based upon experiments in the atmospheric surface layer (a very high Reynolds number boundary layer) indicate that hot-wire spacings in the range of $2\pi\eta$ are optimal with regard to resolving the instantaneous gradients in the logarithmic layer.

Lastly, there are a number of measurement issues that become significant for sensors positioned in the immediate vicinity of the surface. For physical probes (e.g., Pitot tubes and hot-wire sensors) aerodynamic blockage and localized flow accelerations can become significant as the probe nears the surface [12.17]. Conduction heat transfer from a hot-wire probe to the wall is also a well-documented phenomenon, although this can be mitigated to some extent by the choosing a nonconductive surface and/or reducing the heating ratio of the wire. For optical techniques (e.g., LDV and PIV) maintaining sufficient particle seeding can pose significant challenges in the vicinity of the surface. These techniques can be influenced by unwanted light reflections from the surface as well [12.18, 19]. For LDV, the strong mean shear in and near the viscous sublayer can also cause biased measurements (Sect. 5.3.1).

12.2 Measurement of Wall Shear Stress

The wall shear stress is of central importance for both theoretical and practical reasons. When viewed as a *dynamical machine*, the primary mechanism of the turbulent boundary layer relates to its capacity to convert free-stream momentum into a shear force acting over the surface. Indeed, when viewed in this way the skin friction coefficient,

$$C_{\text{f}} = \frac{\overline{\tau}_{\text{w}}}{\frac{1}{2}\rho U_\infty^2}\,, \tag{12.10}$$

is a measure of this capacity (per unit surface area) as it constitutes the ratio of the mean wall shear stress to the axial free-stream momentum flux per unit area. Of course, from the perspective of engineering applications the accurate prediction of C_f, say as a function of R_θ, is required for the design of streamlined vehicles (e.g., aircraft and submarines) over their operating velocity ranges. Similarly, the considerations briefly outlined in Sect. 12.1.2 reveal that $\bar{\tau}_w$ (often through u_τ) constitutes a primary scaling parameter relative to theoretical treatments of wall turbulence. Lastly, the fluctuating wall shear stress has fundamental importance to the basic physics of wall-bounded flows, having, for example, relevance to improved methods for flow management and control, as well as numerical predictive methods such as large eddy simulation.

In what follows, a number of techniques for measuring the wall shear stress are discussed, and some of the key references are given for each technique. These discussions will focus on the underlying measurement principles along with the primary considerations relating to the appropriateness and application of each technique in the context of specific measurement objectives. As mentioned at the beginning of this section, many of the challenges in wall-bounded flows center on the rather stringent spatial and temporal resolution requirements for accurate near-wall measurements.

12.2.1 Methods for Determining the Time-Averaged Wall Shear Stress

Wall-flow experimental designs should carefully consider what type of wall shear stress information is required to meet the experiment objectives. That is, general techniques designed to give the spatially and temporally resolved wall shear stress can certainly also be used to obtain the time mean value. Such techniques, however, must obtain estimates of the instantaneous differential force acting over a differential area of the surface (rather than an estimate of the average force acting over a finite area), and thus their accuracy relative to estimating the mean is often diminished. It is for this reason that in cases where only the mean is required, techniques designed for this specific objective are generally preferred.

Methods Involving Measurements at the Surface

Oil film interferometry (OFI) and two pressure-based methods are considered. OFI relies on the measurement of the optically produced fringe pattern associated with light reflections from a thin film of oil on the surface over which there is a wall flow. Thus, this technique requires optical access to the facility test section. The first of the pressure-based techniques relies on the relationship between the axial mean static surface pressure variation in a duct, and its validity is restricted to the condition of fully developed flow. The second pressure-based technique, the Preston tube method, constructs a correlation between the stagnation pressure produced by the velocity profile near the surface, and thus implicitly relies on the inner-normalized universality of the mean profile over a region of the flow near the wall.

Oil–Film Interferometry. *Tanner* and *Blows* [12.20] were the first to utilize Fizeau interferometry to measure the thinning rate of an oil film deposited on a surface and subjected to a bounding flow. The basis of the technique stems from the analytical solution for flows at very low Reynolds number within the oil film. As fluid flows over the test surface, the oil film begins to thin owing to the action of the surface shear stress imparted by the bounding flow. During this thinning processes, and through the use of interferometry, the spatially and temporally varying thickness of the oil film is determined. Once the film thinning rate is known, the mean wall shear stress can be evaluated using a form of the thin-oil-film equation (12.13). Since oil film thicknesses can be measured using relatively inexpensive and simple equipment, this is often an attractive method for measuring the mean wall shear stress [12.21]. As apparent from its underlying principle, the oil-film interferometry method is restricted to smooth wall flows. In this context it is relevant to note that typical oil thicknesses are on the order of a micron, and thus for a very large number of flow

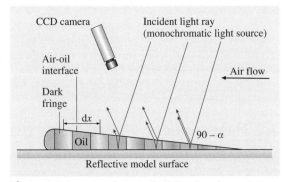

Fig. 12.5 Schematic of a typical oil-film interferometry setup (not to scale). Note that the observer viewing angle α is the deviation from normal to the surface

situations the distortion of the surface caused by the oil itself is negligible.

As indicated by the schematic of Fig. 12.5 the necessary equipment for the oil-film technique typically includes a monochromatic light source, appropriate test surface, oil film and detector (e.g., a CCD camera). While the equipment utilized by this method is fairly uncomplicated, some equipment selections will produce better images than others. Specifically, a non-laser light source such as a sodium lamp is suitable for these measurements and is actually preferable to laser light [12.21]. Additionally, *Zilliac* [12.23] has tested several surface materials and reports that polished aluminum produces poor fringes whereas polished stainless steel, mylar film, and SF 11 glass (among others) all produce satisfactory fringes. Finally, oil with a viscosity that changes little with temperature is favored since sensitivity to changes in temperature can be a source of significant measurement error [12.23]. Silicone oils are especially attractive in this regard.

In order to calculate the mean wall shear stress, the thickness of the oil film is determined using Fizeau interferometry. When a monochromatic light source illuminates the surface of a thin film, a portion of the incident light is reflected off of the oil surface while the residual portion travels through the thin film and reflects off of the surface of the test piece. As shown in Fig. 12.6, when the reflected light is collected, a series of light and dark bands appear as a result of constructive or destructive interference, depending upon the oil height at a particular position. The film thickness at the k-th (dark) fringe is given as,

$$h_k = h_0 + k\Delta h, \quad k = 0, 1, 2, \ldots, \quad (12.11)$$

where h_0 is the oil height at the zeroth fringe and the difference in height between successive fringes is given by

$$\Delta h = \frac{\lambda}{2(n^2 - \sin^2 \alpha)^{1/2}}. \quad (12.12)$$

Here, λ is the wavelength of the light source, n is the refractive index of the oil and α is the observer viewing angle. Minimizing the observer viewing angle, α, to the smallest possible values generally produces the highest-contrast fringe pattern. An example of a series of oil-film fringes, taken at various times, is shown in Fig. 12.6.

According to *Tanner* and *Blows* [12.20], the motion of the oil film is influenced by shear stress, pressure gradients, gravity, oil surface curvature and surface tension. With the aid of dimensional analysis, however, it has been shown [12.24] that for a sufficiently thin film,

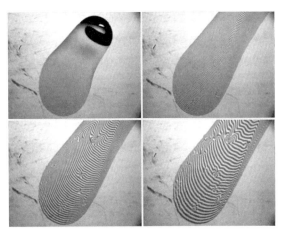

Fig. 12.6 Example fringe patterns as derived from oil film light reflections (after [12.22]). Flow is from the *upper right*

the motion of the oil is primarily due to the action of shear stress on its surface. Therefore, in most cases, the thin-oil-film equation for a two-dimensional flow reduces to

$$\frac{\partial h}{\partial t} = -\frac{1}{2\mu} \frac{\partial (\overline{\tau}_w h^2)}{\partial x}. \quad (12.13)$$

Under conditions where the shear stress is not the only dominant force influencing the motion of the oil, or if the flow is three dimensional, a more general form of the thin-oil-film equation should be used [12.21, 24].

Several methods have been developed to calculate the average wall shear stress using (12.13), and these are reviewed by *Naughton* and *Sheplak* [12.21] and *Fernholz* et al. [12.25]. Fernholz et al. derive two particularly simple methods from (12.13) under the assumption that the wall shear stress is constant. The first method is given by,

$$\overline{\tau}_w = \frac{\mu x}{ht}, \quad (12.14)$$

while the second is expressed as,

$$\overline{\tau}_w = \mu u_k \frac{2(n^2 - \sin^2 \alpha)^{1/2}}{\lambda(k + h_0/\Delta h)}, \quad (12.15)$$

where

$$u_k = \frac{\partial x}{\partial t}\bigg|_{h_k = \text{const}}. \quad (12.16)$$

While the second method appears to be more difficult to implement than the first because of the need to determine u_k (the fringe velocity), the form of the first may be deceptively simple since the height of the zeroth fringe

(h_0) must also be measured in order to calculate the height of the oil at a given location [12.26]. In the second method, however, measurements can be made that allow h_0 to be calculated from a single data set using multiple fringes [12.25]. That is, in (12.15) n, μ, α and λ are known, and u_k can be calculated from (12.16) using images of the fringe pattern taken at known times after the u_k = constant condition has been attained. Automated image-processing methodologies for determining the fringe spacing are relatively easy to implement. These generally involve finding either the leading or trailing edge of the fringe (edge detection-based method) or locating the fringe center (peak grayscale intensity-based method). Multiple fringe spacing estimates from each image can be used to construct statistically robust measurements.

While both of the above methods are based on the assumption of constant mean shear stress, errors should not be large if these methods are applied to flows with moderately varying $\overline{\tau}_w$ provided that,

$$\frac{\Delta \overline{\tau}_w}{\overline{\tau}_w} \ll \frac{\partial h}{\partial x} \qquad (12.17)$$

in the vicinity of the measurement location. It is important to check that this relation holds when using these methods since large errors may otherwise result [12.24]. Although the cases considered here are for time-invariant wall shear stress, several methods have been developed for cases where the shear stress is time dependent [12.25]. Additionally, solution methodologies exist for three-dimensional flows [12.21].

Some of the error sources and difficulties associated with this method include dust, humidity, high-shear situations, oil property variations, oil evaporation, shear stress variation, surface tension, pressure gradients and gravity effects. Fringe pattern degradation due to dust and humidity may be prevented by eliminating both from the test environment. Undesirable surface waves, that sometimes form under high-shear conditions, may also be avoided if the oil film is initially very thin [12.27]. Oil property variations may become a significant source of error and should be monitored throughout a test. Also, oil evaporation may cause the film to appear to thin faster than it actually does – this problem may be avoided by using oils having a low vapor pressure, e.g., silicone oils [12.28]. Finally, effects due to shear stress variation, surface tension, pressure gradients and gravity may be accommodated via selection of the appropriate form of the thin-oil-film equation [12.21, 24].

Mean Pressure–Gradient Method. A simple, yet highly accurate, method for measuring the average wall shear stress is available for fully developed internal flows. When the flow satisfies the conditions of being fully developed and two-dimensional (planar or axisymmetric) the mean wall shear stress $\overline{\tau}_w$ may be calculated from the measured axial pressure gradient. As elucidated in many undergraduate fluid-mechanics textbooks [12.9], a control volume analysis of a fully developed pipe or channel flow of constant cross-sectional area A_c reveals that the mean pressure drop ΔP present over a length L of the duct is related to the average wall shear stress acting over the surface area A by

$$\Delta P A_c = \int_A \overline{\tau}_w \, dA \, . \qquad (12.18)$$

For fully developed flow the surface shear stress per unit length of the duct is constant. Thus for a circular pipe

$$\overline{\tau}_w = \frac{\Delta P}{2L} \delta \qquad (12.19)$$

and in a two-dimensional channel,

$$\overline{\tau}_w = \frac{\Delta P}{L} \delta \, , \qquad (12.20)$$

where δ is either the pipe radius or channel half-height, respectively.

Harotinidis [12.29] asserts that this momentum-balance-based method is perhaps the most reliable for flows that satisfy the requisite flow conditions. Unfortunately, the applicability of this method is operationally quite limited since, in practice, it is not a trivial matter to exactly attain both the fully developed and two dimensional (axisymmetric) conditions. For example, recent results by *Lien* et al. [12.30] reveal that even in channel flows of aspect ratio greater than 12, the apparent effect of the side walls is felt in the core flow. Indeed, one may construct a physically rational argument that flow in a circular pipe is the only for which the fully developed and axisymmetric (two-dimensional) conditions can be attained to the same degree as prescribed mathematically. That is, any real channel will be finite in its span, and the side wall boundary layers will, in all likelihood, continue to grow with downstream distance. Contrary to this physical argument, *White* [12.31] provides an analysis supporting the notion that after sufficient development length the growth of the side wall boundary layers is, in fact, arrested. Interestingly, studies exploring laminar to turbulent transition cite the need for exceptionally long development lengths ($\simeq 400$ channel heights, 2δ) and especially wide channels (≥ 27 channel heights) [12.32], while high-quality turbulent flow studies are generally in the range of about 200δ and greater than 20δ for the development length and channel width, respectively. Overall, it is recommended that channel flow studies carefully verify a fully developed condition prior to utilizing the pressure drop for determining the wall shear stress.

Of course, even for fully developed pipe flow, experimental details relating to the accurate measurement of static pressure must be carefully considered. Such effects include the finite size of the pressure taps (which tends to cause an overestimation of the true pressure), as well as the quality of the tap hole relative to the internal surface of the pipe. Particularly important references in this regard include the studies of *Shaw* [12.33], *Franklin* and *Wallace* [12.34], *Ducruet* and *Dyment* [12.35], *McKeon* and *Smits* [12.36], as well as Sect. 5.1 in this handbook. Careful consideration of these effects can minimize the associated uncertainty in $\bar{\tau}_w$ to a small fraction of a percent [12.37]. Lastly, it is not advisable to use this method for flows that do not satisfy the assumptions underlying (12.19), since significant and difficult to characterize errors are likely to result [12.38].

Preston Tube Method. Though not a measurement acquired precisely at the wall, the method initially developed by *Preston* [12.39] utilizes stagnation pressure information derived from a tube placed on the wall. In some sense, this method can be considered a *single point* variant of the Clauser plot technique described below. This is because it relies on the correlation between $\bar{\tau}_w$ and the difference between the surface static pressure and the stagnation pressure produced by the portion of the mean velocity profile in the region of the flow where the so-called *law of the wall* is valid. The dimensional analysis underlying this arrives at the correlating expression

$$\frac{\overline{\Delta P}}{\bar{\tau}_w} = f\left(\frac{d^2 \bar{\tau}_w}{\rho v^2}\right), \quad (12.21)$$

where $\overline{\Delta P}$ is the mean difference between the pressure sensed by the stagnation (Preston) tube and the static pressure obtained from a nearby wall tap, and d is the outside diameter of the tube.

In any given flow, the wall thickness of the tube determines which portion of the mean profile contributes to the stagnation pressure sensed at the tube opening. For relatively thick turbulent wall flows and very small tubes, the measurement could be fully immersed in the viscous sublayer. In this case, the mean profile existing across the tube opening will be linear. Under most measurement situations, however, the finite wall thickness of a typical tube will result in the location of the tube opening being primarily in the logarithmic portion of the mean profile. It is for this situation that the majority of Preston tube calibrations have been developed. Popular among these are the calibration equations given by *Patel* [12.40], as well as the tabulated calibration given by *Head* and *Ram* [12.41] (also see *Hanratty* and *Campbell* [12.42]). Owing to the accuracy by which the mean wall shear stress can be independently determined, fully developed pipe flows have traditionally been the flow field of choice for calibrating Preston tubes.

While the Preston tube method is a simple (albeit correlation-based) technique for estimating $\bar{\tau}_w$, its limitations should be understood and some care must be taken in its implementation. Except under the condition where the tube opening is fully contained in the viscous sublayer, an important assumption implicit to the technique is that a logarithmic *law-of-the-wall* region exists in the flow considered, and that the tube opening resides in and/or below this portion of the profile. Thus, the technique is inappropriate for highly nonequilibrium wall flows, or other situations where the logarithmic portion of the mean profile has been significantly altered. This said, *Patel* [12.40] does provide some recommendations pertaining to the use of Preston tubes in boundary layer flows with axial pressure gradients. Furthermore, existing smooth wall calibrations should not be applied under rough wall situations. On the other hand, while Preston indicated that the ratio of the inner to outer diameter of the tube is important, the later study by Patel showed little sensitivity to this parameter as long as the ratio is greater than 0.2. Alignment of the Preston tube in the flow direction can be a significant source of error, especially in three dimensional boundary layers where the mean flow direction is not always easy to determine and instantaneous nonlinear yaw effects affect the measured mean value. Consistent with the behavior of Pitot tubes, errors in $\bar{\tau}_w$ of about 1% are realized for a misalignment of about $3°$. Lastly, care must be taken to make certain that the flow at the surface static pressure port is not influenced by the presence of the Preston tube on the wall.

Mean-Profile-Based Methods

Three different methods for determining the time-averaged shear stress from mean velocity profile data are considered. The von Karman integral method relies on the computation of the displacement and momentum deficit thicknesses, and thus requires measurements of the mean velocity profile. This technique is restricted to two-dimensional flows, and under some circumstances the accuracy is diminished without the measurement of streamwise gradient terms of the velocity variances. The so-called Clauser plot method is an indirect technique that relies on the observed correlation between the wall shear stress and the properties of the logarithmic mean velocity profile characteristic of wall bounded flows (Fig. 12.1). Implicit in this technique is the assumption of a universal logarithmic mean velocity profile. The mean wall gradient method involves accurately measuring the linear portion of the mean velocity profile in the immediate vicinity of the surface. This technique, however, can only be utilized in flows over smooth walls.

von Karman Momentum Integral Method. For two-dimensional developing flows, the surface stress is related to the momentum and displacement thickness by von Karman's momentum integral equation. For flows where the streamwise variation in the fluctuating velocity variances are not large, the relationship between shear stress and the properties of the mean profile is given by [12.31]

$$\frac{\bar{\tau}_w}{\rho U_\infty^2} = \frac{d\theta}{dx} + (H+2)\frac{\theta}{U_\infty}\frac{dU_\infty}{dx}. \quad (12.22)$$

Here, $H = \delta^*/\theta$ is the shape factor. Equation (12.22) includes the effects of axial pressure gradient, and thus for flat-plate boundary layer flow the last term is zero. When streamwise variations in the turbulence are non-negligible a more-complicated equation is required, and is given by [12.2]

$$\frac{\bar{\tau}_w}{\rho U_\infty^2} = \frac{d\theta}{dx} + (H+2)\frac{\theta}{U_\infty}\frac{dU_\infty}{dx}$$
$$- \frac{1}{U_\infty^2}\int_0^\delta \frac{\partial}{\partial x}(\overline{u^2} - \overline{v^2})dy$$
$$+ \text{higher-order terms}. \quad (12.23)$$

Unlike the mean pressure gradient method in ducts, the momentum integral approach does not require that the flow be fully developed. Similar to the pressure-gradient method, however, its usefulness is often diminished by the practical difficulties associated with employing it. For example, under all cases considerable care must be taken to ensure that the flow under investigation is indeed adequately planar. Additional difficulties associated with this method can also arise, in part, from the need for accurate measurements of axial velocity gradients and in some cases axial gradients of the velocity variances. In specific reference to this last point,

recent experimental evidence [12.43, 44] indicates that even under the condition of zero pressure gradient the streamwise velocity variance gradients contribute about 5% to the overall integral momentum balance.

Clauser Plot Method. Clauser [12.45] observed that the wall shear stress in turbulent boundary layers could, under equilibrium conditions, be estimated with reasonable accuracy through its correlation with the law of the wall. Equilibrium flows are those whose statistical profiles can be represented (or nearly represented) in a self-preserving form using locally determined integral parameters. Accordingly, Clauser's observation forms the basis for the method that bears his name. Essentially, the Clauser plot is a graphical method for determining wall shear stress using the properties of the time-averaged velocity profile in the logarithmic portion of the boundary layer. The attractive feature of this method is that velocity measurements in the viscous sublayer portion of the profile, that are often difficult to obtain, are not required. Conversely, however, deviations from canonical behavior may preclude the use of this method. Indeed, recent experiments strongly support the assertion that for the purposes of discerning detailed mean profile scaling behaviors an independent means of obtaining the wall shear stress is required [12.46]. Hence, this method can only be recommended for those cases where a nominally accurate $\bar{\tau}_w$ value is acceptable.

In inner normalized form, the logarithmic equation for the mean profile is given by,

$$\frac{U}{u_\tau} = \frac{1}{\kappa}\ln\left(\frac{yu_\tau}{\nu}\right) + B, \quad (12.24)$$

where, as indicated previously, the friction velocity u_τ is given by

$$u_\tau = \sqrt{\frac{\tau_w}{\rho}}, \quad (12.25)$$

and, according to the logarithmic law assumptions, κ (the von Karman constant) and B are constants. Perhaps the most often cited values for κ and B are 0.41 and 5.0, respectively [12.47], although other values are often used as well. Using equation (12.24) and the fact that the skin friction coefficient can be expressed as $C_f = 2(u_\tau/U_\infty)^2$, allows the construction of the Clauser plot. In particular, when U/U_∞ is plotted versus $U_\infty y/\nu$ a series logarithmically varying lines are generated, with each line corresponding to a specific value of C_f. Thus, as depicted in Fig. 12.7, when measured mean velocity profile data are overlaid on this graph, the wall shear stress estimate is obtained according to the best

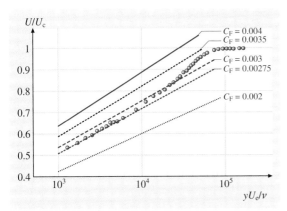

Fig. 12.7 Clauser plot with data [12.48]. Note that in the axis label U_e is used to denote the free-stream velocity

correspondence between the logarithmic portion of the velocity data and one of the parametric C_f lines.

The Clauser method is capable of producing reasonably accurate results, say ±5%. It is worth emphasizing, however, that the sole basis of this method relies on the existence and validity of the logarithmic law. Given this, the Clauser plot method should be viewed as a useful means of approximating the value of the friction velocity, but it is neither a direct nor independent method. For example, this method is clearly not appropriate for use under conditions where a well-defined logarithmic region does not exist. Flow conditions such as those with strong pressure gradients, low Reynolds numbers and separation fall into this category. Additionally, the accuracy of the method is dependent on the selection of κ and B. For flow over a flat plate, existing data support the general acceptability of the values cited. Under some circumstances, however, the values of κ and B are variable. For example, *Wygnanski* et al. [12.49] found that B exhibits some dependence on the Reynolds number while κ changes little. Additionally, *Nagib* et al. [12.44] provide compelling evidence that κ varies with pressure gradient. For these reasons, it is imperative that κ and B be chosen judiciously in applications where a Clauser plot methodology is implemented. As with all other mean-profile-based methods, another common source of uncertainty can be associated with the method of measuring the velocity profile itself. For example, significant error can be present in Pitot-tube measurements due to their sensitivity to large-amplitude velocity fluctuations, misalignment and low-Reynolds-number effects at the tube opening.

Because the original Clauser plot method was developed for smooth walls, special treatment is necessary for its use in rough wall flows. Specifically, roughness poses a challenge since it generates an "error in the origin" of the mean profile owing to the well-known downward shift of the inner normalized mean profile, e.g., [12.1,2]. In order to compensate for these challenges, *Perry* and *Li* [12.50] developed a relatively simple iterative method based on Coles formula [12.51]

$$\frac{U}{U_\infty} = 1 + \frac{1}{\kappa}\frac{u_\tau}{U_\infty}\ln\frac{z}{\delta^*} + \frac{1}{\kappa}\frac{u_\tau}{U_\infty}\ln\frac{u_\tau}{U_\infty}$$
$$+ 0.493\frac{u_\tau}{U_\infty}, \qquad (12.26)$$

where δ^* is the displacement thickness and z is the distance from the crest of the roughness elements. The procedure is then implemented as follows.

1. Using equation (12.26), generate the family of curves relating U/U_∞ to z/δ^* using different values of u_τ/U_∞, and overlay the experimental measurements on this graph.
2. From the graph, select the value of u_τ/U_∞ that corresponds to the parameterized line best approximating the data near $z/\delta^* = 1$.
3. Determine an initial estimate for e, the error in origin, from the portion of the data where $z/\delta^* \ll 1$.
4. Add this estimate for e to the distance from the crest of the element and regenerate the graph.
5. Repeat the entire procedure until a converged value of u_τ/U_∞ is returned.

Relative to the use of this procedure, it is significant to note that (12.26) is based on a wake factor of 0.55. An alternate equation could, however, be easily derived for other values. Additionally, the method of Perry and Li was developed to have the desirable trait of being fairly insensitive to the wake factor.

Viscous Sublayer Profile Method. This method can be considered direct in that it is based upon quantifying the mean gradient of the axial velocity at the wall – or more accurately, very near the wall. According to its definition, the mean shear stress produced at a point on a bounding solid surface by a flowing fluid is given by

$$\overline{\tau}_w = \mu\frac{\partial U}{\partial y}, \qquad (12.27)$$

where y is the direction that is locally normal to the surface. Thus, for a fluid with known dynamic viscosity μ the requisite task is to experimentally determine the slope of the axial velocity profile at the wall.

Attaining an accurate measure of $\overline{\tau}_w$ from this method first requires the reasonable existence of a linear region in the velocity profile, i.e., a region where

$U^+ = y^+$ in the immediate vicinity of the wall. Certainly, for canonical smooth wall-flows, this approximation is known to hold to a very good approximation [12.2, 52]. In proximity of a point of separation, for example, such a region can rationally be expected to become diminishingly small or even non-existent. Similarly, linear sublayers do not exist (in any quantifiable way) for flows that are not hydraulically smooth. For a reasonably wide range of flows over smooth walls, however, one can expect to find a region within which the mean profile exhibits the linear dependence on distance from the wall as depicted in Fig. 12.1.

Under the vast majority of flow conditions, the physical dimension of the region of linear dependence is small; typically a fraction of a millimeter. Because of this, the sensing dimension of the measurement probe must be very small as well. Although optical sensors such as laser Doppler velocimetry (LDV) and molecular tagging velocimetry (MTV) can be used to measure the sublayer profile, single element hot-wire probes are the predominant sensor of choice owing to their inherently small dimension in the plane parallel to the wall. (Recall that the diameter of a hot wire is typically 5 μm or less.) Given this, the present discussion will focus on the measurement of the sublayer mean profile using a hot-wire probe. Some of the considerations unique to the LDV and MTV techniques are discussed at the end of the next subsection.

Major challenges associated with determining $\bar{\tau}_w$ using (12.27) are (i) determining the region where the mean profile is linear, (ii) accurately and appropriately positioning the sensor in this region, and (iii) accurately calibrating the sensor for the measurements. The region of linear dependence extends from the wall to a small distance above the wall. This region is typically taken to be approximately $y^+ < 5$, although some results suggest it is even smaller [12.53]. In the vicinity of a solid surface, however, thermal anemometry is subject to conduction heat transfer and aerodynamic blockage effects (Sect. 5.2). Thus, for some portion of this region near the wall the hot-wire will produce measurements that overestimate the true velocity value. *Hutchins* and *Choi* [12.54] provide an excellent discussion of this effect relative to accurately determining the mean wall shear stress. They cite the region influenced by wall effects to be approximately $y^+ < 3.5$. For some experimental situations, however, it may be possible to confine this region even closer to the surface by reducing the heating ratio of the sensor, or by using a highly non-conductive surface material. Regarding the latter, *Chew* et al. [12.55] provide a useful study of these wall effects.

In any case, the measurements of interest must then be taken in a region above where wall effects are significant and below the upper bound of the viscous sublayer. For the numbers cited above (by Hutchins and Choi) this region is given by $3.5 < y^+ < 5.0$.

A number of methods are available for positioning the hot-wire probe at a prescribed distance away from the surface. These include the use of mechanical stops and optical displacement sensors. The present authors have also found that locating the probe through the use of a cathetometer (a traversing short-range telescope) can also be effective. In this method, the position of the surface in the immediate vicinity of the probe is located by sighting (with the cross hair of the cathetometer) the point of surface contact made with the tip of a razor blade. The Vernier scale on the traversing scope then allows the sensor position to then be determined and specified by subsequently sighting the probe tip. A repeated set of such measurements, say nine or so, provides a statistically based estimate for the probe position from the wall, as well as an estimate for the uncertainty in this measurement. As noted and discussed by Hutchins and Choi, misalignment of the wire from an orientation parallel to the plane of the surface can also lead to wall-normal positioning uncertainties. The close-up view of the probe through the cathetometer is also useful in minimizing this type of misalignment. Lastly, the probe body should be tilted slightly toward the surface such that the active wire element (tips of the prongs) are positioned to make first contact as the sensor is traversed toward the surface. This serves to minimize aerodynamic blockage effects, and avoids the situation in which the probe body runs into the surface and thus interferes with positioning the sensing wire very close to the surface.

Depending on the details of the experiment, the flow velocity within the viscous sublayer may be too small to accurately calibrate the hot wire using a Pitot tube and pressure transducer. Under such cases, special calibration devices are needed. In this regard, a particularly attractive methodology is to create a laminar flow for which the analytical solution is known. Examples of such devices include the Couette-flow-based calibrator of *Chew* et al. [12.56] and the Poiseuille device of *Yue* and *Malmstrom* [12.57].

12.2.2 Time-Resolved Methods

Relative to the number of techniques designed to measure the time-averaged wall shear stress, those capable of producing time-resolved measurements of the instantaneous wall shear stress are considerably fewer in number. Two primary reasons for this are that the correlations constituting the basis of indirect techniques are generally not instantaneously valid, and accurately quantifying instantaneous shear stress values requires sensing systems that have both a high-frequency response and are sufficiently small in size. The time-resolved methods discussed in this section are associated with either the direct measurement of the shear force acting over a small element of the surface, or are effectively instantaneous versions of the viscous sublayer profile method discussed above.

Floating-Element Sensors

Perhaps the simplest device (in theory) for measuring wall shear stress is the floating-element sensor. Floating-element sensors directly measure the shear force imparted on a movable, *floating* element portion of the wall. When subjected to flow conditions, the sensor functions either by measuring the amount of force required to keep the element in place or by correlating the displacement of the element with the applied shear force. A major positive attribute of floating-element sensors relative to sensors that indirectly measure shear stress is that no assumptions about the flow field, fluid properties or surface need to be made. Thus, unlike sensors which indirectly measure shear stress, floating-element sensors are not dependent upon the veracity of the correlating function, and therefore are not affected by the errors associated with using the simplifying assumptions needed to develop the correlation.

While floating-element sensors are simple in theory, they are, in practice, often difficult to use. *Winter* [12.59] lists several common challenges associated with the use of floating-element sensors, including:

1. The compromise between the need for an element that is sufficiently large such that the force acting on it has a magnitude that can be measured accurately, and the need for an element small enough to measure local conditions
2. The effect of gaps around the edges of the element
3. Effects due to misalignment of the element
4. Errors associated with pressure-gradient forces.

Other conditions that may influence the performance of the sensor are heat transfer, temperature changes, gravity and/or acceleration, boundary layer suction or injection, leaks, and fluctuations of normal forces (which may damage the sensor) during start-up and shutdown [12.59]. Additionally, *Harotinidis* [12.29] notes that larger sensors generally have a reduced temporal resolution, as well as, of course, reduced spatial resolution.

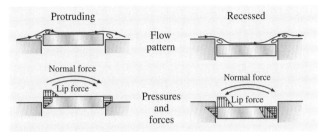

Fig. 12.8 Schematic representation of detrimental floating element misalignment errors (after [12.58])

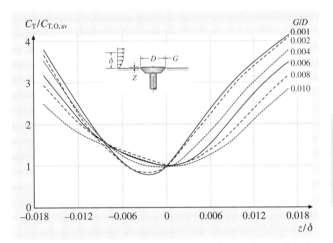

Fig. 12.9 Effects of nondimensional gap and step size on floating element drag measurements (after [12.58])

The studies by *Allen* [12.58, 60] investigated the magnitude of errors associated with gaps, misalignments and element thickness. Errors result from gaps and misalignments since these will cause additional forces under flow conditions that were not present during calibration. A summary of the types of misalignments considered by Allen is shown schematically in Fig. 12.8, while Allen's experimental findings relative to these misalignments are shown in Fig. 12.9. These findings indicate that the ratio of the gap width to the element diameter has a critical effect on the accuracy of the method. Specifically, relatively larger errors result from small gap sizes for the same vertical misalignment, and the thickness of the exposed edge of the floating element has a major impact upon the accuracy of the method. Additionally, *Acharya* [12.61] has shown that uneven pressure distributions will amplify alignment errors by disproportionately acting upon exposed edges of misaligned elements.

In an effort to circumvent the difficulties associated with the presence of gaps, *Frei* and *Thomann* [12.62] proposed filling the gap around the element with a liquid. Another solution to the gap problem, initially proposed by *Schmidt* et al. [12.63], reduces the gap size using microfabrication techniques; the review of *Löfdahl* and *Gad-el-Hak* [12.64] discusses several micro-electromechanical systems (MEMS)-based floating-element sensors that have since been produced. While the use of MEMS-based floating-element sensors is not yet widespread, *Naughton* and *Sheplak* [12.21] indicate that the errors due to pressure gradients scale favorably for MEMS sensors and that sensor misalignment problems are reduced for microsensors since the sensors are fabricated monolithically.

At its essence, the floating element sensor is a force transducer. The principle of operation for such transducers can invoke (i) single or multiple load cells [12.59], (ii) strain gauges [12.42], or (iii) displacement sensors [12.65]. In all cases, however, the output from the floating element sensor requires calibration relative to known forces. Static calibration involves the application of a series of known constant forces. Often this is most easily accomplished through the use of hanging weights. In this case, care should be taken to utilize very low friction pulleys (e.g., made of teflon) and cables made of materials that do not stretch. Depending on how easily the sensor can be removed from the test section, gravity-based calibration techniques that involve tilting the sensor have also been utilized [12.65]. For time-resolved measurements, the frequency response of the sensor must also be quantified. Under many, if not most, circumstances the floating element sensor can be treated as a one- or two-dimensional second-order spring–mass–damper system. Thus, the frequency response is estimated by applying a known impulsive force and quantifying the time it takes the sensor to return to equilibrium. In turbulent wall-bounded flows the spatial dimension of the floating element is, of course, also an important consideration. Although a comprehensive study of the effects of spatial resolution on miniaturized floating-element sensor measurements has yet to be conducted, it is rational to expect that the criteria discussed above relative to near-wall turbulence measurements are relevant. Generally, fully resolved measurements will require the characteristic dimension of the sensor to be less than about 10 viscous units. Clearly, in most flow situations MEMs sensors are required to satisfy this criterion.

Although the floating-element technique is often difficult to employ, and its accuracy is often challenging to quantify, researchers continue to make use of this method since a priori knowledge of the nature of the flow field is not necessary. Additional improvements in the accuracy and frequency response of the method are expected to follow with further development of MEMS sensors.

Instantaneous Wall Gradient Methods

Under the condition of a smooth wall, single- or multiple-point thermal anemometry based methods can be used to quantify the velocity gradient at the surface. Perhaps the most straightforward version of this involves positioning a single hot wire or a closely spaced vertical rake of wires in the viscous sublayer and acquiring time-resolved measurements of the axial velocity in the region where the velocity profile is linear. In the case where the probe is traversed toward the surface from above, the requirements for obtaining accurate measurements are essentially the same as those discussed previously for obtaining accurate mean shear stress measurements. In one variant, the prongs supporting the hot wire are embedded in a plug that can then be flush-mounted in the surface. This is sometimes termed the hot wire on the wall method [12.66]. In this case, while the operation of the hot wire is standard, its calibration must be modified. Specifically, owing to the very close proximity of the wire to the surface of the plug, it cannot be calibrated for velocity, but rather to the shear stress directly, as are flush-mounted hot-film shear stress sensors.

According to a heat transfer analysis, the static calibration equation for a shear stress sensor is given by [12.67]

$$\overline{\tau}_w^{1/3} = AE^2 + B, \qquad (12.28)$$

where E is the bridge voltage, and A and B are calibration constants. The preferred method for calibration of the sensor is in situ, since this avoids errors associated with having to reinstall and align the probe. On the other hand, this also means that another method for measuring the mean wall shear stress must be available for calibration. Popular strategies in this regard are to use either Preston tubes or the Clauser plot technique [12.68]. Owing to the fact that these techniques are correlation-based, the resulting calibrated sensor can not be viewed as an independent means for measuring $\overline{\tau}_w$. Given that the primary goal of such a sensor is to measure the τ_w fluctuations, this does not necessarily pose a major problem. Of course, if the measurement facility is a fully developed pipe or channel flow, then one may calibrate against the mean pressure drop, and thus generate an analytically well-founded calibration.

As expected, an important concern for flush-mounted shear stress sensors is their spatial resolution, and the aforementioned desirability of the sensor dimension to be less than about 10 viscous units. Perhaps an even greater concern is the loss of temporal resolution at high frequencies resultant from the thermal effects associated with the sensor substrate. Specifically, the critical issue is the heat transfer to the fluid relative to the heat transfer to the substrate [12.69]. While corrections for these effects have been developed, the empirical evidence indicates that they generally do not fully account for the attenuating effect. Thus, it is best to use fluid/substrate contributions that result in most of the heat transfer to the fluid [12.66]. For this reason, measurements in water exhibit superior performance to those in air. As with standard hot-wire anemometry, multiple wire configurations can be used to measure more than one component of velocity. In this regard, a v-array plug provides measurements of the fluctuating axial and spanwise wall velocity gradients, $\partial u/\partial y$ and $\partial w/\partial y$, respectively.

As a final note regarding flush-mounted sensors, it is also worth mentioning that in liquid flows one may also use electrochemical mass transfer probes to infer the wall shear stress fluctuations. This method relies on the correlation between the concentration gradient and velocity gradient at the surface. The development and implementation of these sensors is largely due to the extensive body of work conducted by *Hanratty* and his coworkers [12.42].

Of course, measurements of the instantaneous wall gradient are possible as long as spatially and temporally well-resolved axial velocity measurements can be accurately acquired in the viscous sublayer. In this regard, at least two optical techniques have been successfully employed.

Continuing refinement of LDV techniques have produced several systems capable of making sublayer velocity measurements in flows beyond low Reynolds number. Of course, a common characteristic of suitable systems is an especially small measurement volume. For example, in the study by *DeGraff* and *Eaton* [12.13], the dimensions (major and minor axes) of the ellipsoidal measurement volume were $65\,\mu\text{m} \times 35\,\mu\text{m}$. This was achieved by special focusing optics and using side scatter collection optics [12.70]. LDV measurements acquired close to a surface are also susceptible to negative influences associated with optical access and surface interference, as well as velocity bias associated with high shear across the measurement volume [12.18]. Regarding the latter, *Durst* et al. [12.71] have devised correction methods for the various moments of the velocity fluctuation probability density function. Owing to the low-speed flow in the sublayer, adequate seeding also requires careful attention.

Because it provides very closely spaced data instantaneously along a line, single- or multiple-line molecular tagging velocimetry (MTV), is especially well suited for wall gradient measurements. As discussed in Sect. 5.4, the basic idea underlying MTV is that fluid velocity data can be derived by tracking a pattern of excited, long lifetime, phosphors that have been mixed within the fluid and subsequently excited by laser light. Typical hardware, data acquisition and data reduction methods employed in the line version of the technique in wall-bounded flows are described in the studies by *Hill* and *Klewicki* [12.72–74] and the references therein. More-general applications of MTV and its extensions to two-component measurements are given by *Koochesfahani* and coworkers [12.75–77]. In the MTV line technique, a series of laser lines are used to excite the dissolved phosphor, and a gated camera is used to acquire the initial and time-delayed images of these lines. If the fluid is moving, these lines of excited phosphor are displaced according to the local flow velocities. The essence of the MTV technique is that velocity field data are found by tracking this displacement over a sufficiently short time duration such that a Lagrangian approximation (e.g., $u = \Delta x/\Delta t$) becomes valid.

With the line displacement and time delay between images known, the velocity may be estimated. Uncertainties are associated with the quantification of Δx, and to a lesser degree Δt. Furthermore, root-mean-square (RMS) and mean bias errors (relative to the true line location) can be associated with the line position locating algorithms. Automated image-processing-based methods do, however, allow the line displacement to be determined to within a small fraction of a pixel. The line technique is also susceptible to an error associated with the effect of having a velocity component parallel to the line. These errors result from the inability to identify and subsequently track unique fluid elements on the line. As a percentage of the instantaneous streamwise velocity this error can be expressed as

$$\frac{\Delta u}{u} = \Delta t \frac{v}{u} \frac{\partial u}{\partial y}, \tag{12.29}$$

where u, v and $\partial u/\partial y$ are instantaneous quantities. For the near-wall region of the boundary layer, *Klewicki* and *Hill* [12.73] showed that this effect causes about a 1% bias error.

For typical lens magnifications, one can obtain well over 20 measurements per millimeter along the

MTV line. This density of data is, of course, attractive when seeking to curve-fit to obtain the local slope at the surface. Owing to light reflections, however, it is often difficult to acquire data exactly at the surface. Measurements to within about $y^+ = 1$ have been reported [12.73]. The temporal resolution of the technique is limited by the camera frame rate and the laser repetition rate.

12.3 Boundary-Layer Stability and Transition

Stability and transition experimentation is no trivial task and should be undertaken by only the most serious researcher. The basic idea of an instability is that small disturbances in the flow can have large effects on the basic state, leading in some instances to additional instabilities. Small changes in an experimental set-up and measurement can introduce unanticipated disturbances that can complicate the flow or skew the interpretation of the results. When transition to turbulence proceeds through loss of stability, the process critically depends on these small effects. Thus, unlike many situations in turbulent boundary layers, measurements of stability characteristics require a special sensitivity to environmental conditions. The subsequent sections aim to highlight some of the particular details required to successfully complete a stability experiment and to advise against common mistakes throughout the process.

12.3.1 The Process of Transition for Boundary Layers in External Flows

In fluids, turbulent motion is usually observed rather than laminar motion because the Reynolds-number range of laminar motion is generally limited. The *transition* from laminar to turbulent flow occurs because of an incipient instability of the basic flow field. This instability intimately depends on subtle, and sometimes obscure, details of the flow. The process of transition for *boundary layers in external flows* can be qualitatively described using the following (albeit oversimplified) scenario.

Disturbances in the free stream, such as sound or vorticity, enter the boundary layer as steady and/or unsteady fluctuations of the basic state. This part of the process is called *receptivity* [12.78] and although it is still not completely understood, it provides the vital initial conditions of amplitude, frequency, and phase for the breakdown of laminar flow. Initially these disturbances may be too small to measure and they are observed only after the onset of an instability. A variety of different instabilities can occur independently or together and the appearance of any particular type of instability depends on Reynolds number, wall curvature, sweep, roughness, and initial conditions. The initial growth of these disturbances is described by *linear* stability theory (i.e., linearized, unsteady Navier–Stokes). This growth is weak, occurs over a viscous length scale, and can be modulated by pressure gradients, surface mass transfer, temperature gradients, etc. As the amplitude grows, three-dimensional and nonlinear interactions occur in the form of *secondary* instabilities. Disturbance growth is very rapid in this case (now over a convective length scale) and breakdown to turbulence occurs.

Since the linear stability behavior can be calculated, transition prediction schemes are usually based on linear theory. In the case of streamwise instabilities and low-disturbance environments, linear theory does very well in predicting the stability behavior. However, since the initial conditions (receptivity) are not generally known, only correlations of transition location are possible and, most importantly, these correlations must be between two systems with similar environmental conditions [12.79].

Thus, linear theory is the foundation of streamwise instabilities in low-disturbance flows. A brief review of the nomenclature of linear theory precedes the description of experimental methods. It is assumed that the reader understands the fundamental ideas of hydrodynamic stability found in [12.79–81]. Background material on transition can be found in [12.82] and [12.83].

12.3.2 Nomenclature of Linear Theory

As a reference point, start with an incompressible, isothermal flow over a flat plate with zero pressure gradient. The basic state is assumed to be locally approximated by the parallel flow $V = [U(y), 0, 0]$, where $(\hat{x}, \hat{y}, \hat{z})$ are the dimensional streamwise, wall-normal, and spanwise directions, respectively. Dependent and independent variables appearing as $\hat{q}$ are dimensional, otherwise they are dimensionless. Lengths and velocities are made dimensionless with the scales L and U_∞, respectively. Two-dimensional disturbances are superposed on the Navier–Stokes equations which are then linearized. Assuming a normal-mode disturbance of the form

$$q'(x, y, t) = q(y)\exp[i(\alpha x - \omega t)] + \text{c.c.}, \quad (12.30)$$

where q' represents a real disturbance quantity such as pressure or a velocity component. For spatially varying disturbances, the use of (12.30) results in the Orr–Sommerfeld equation (OSE) given by:

$$(D^2 - \alpha^2)^2\phi - iR[(\alpha U - \omega) \\ \times (D^2 - \alpha^2)\phi - (\alpha D^2 U)\phi] = 0, \quad (12.31)$$

where c.c. means complex conjugate, $D = d/dy$, ϕ is complex and represents the disturbance stream function, $\phi(0) = D\phi(0) = 0$, $\phi(y \to \infty) \to 0$, ω is real, $\alpha = \alpha_r + i\alpha_i$, $-\alpha_i$ is the spatial growth rate, $c = \omega/\alpha_r$ is the phase speed. Here the dimensional frequency is conserved and the length scale is $L = \sqrt{\nu\hat{x}/U_\infty}$. These give rise to the following definitions

$$R = \frac{U_\infty L}{\nu} = \sqrt{\frac{U_\infty \hat{x}}{\nu}}, \quad (12.32a)$$

$$F = \frac{\omega}{R} = \frac{2\pi f \nu}{U_\infty^2} = \text{constant}. \quad (12.32b)$$

The eigenvalue problem reduces then to finding $\alpha = \alpha(R, F)$. The locus of points for which $\alpha_i(R, F) = 0$ is called the neutral stability curve. For a given F in Blasius flow, R is double-valued. The smallest value, R_I, occurs at branch I and the largest value, R_II, occurs at branch II. Between these two Reynolds numbers, the flow is unstable. Transition depends on the measure of growth between R_I and R_II. The Reynolds number below which the flow is stable for all F is called the minimum critical Reynolds number Re_crit.

Some also have used the displacement thickness δ^* or the momentum thickness θ as the normalizing length with an attendant redefinition of the Reynolds number. All of these choices are appropriate for boundary-layer scaling. However, since no universal Reynolds number criterion appears with either R_{δ^*} or R_θ for other boundary-layer flows, the use of these scales just adds a superfluous constant. On the other hand, the use of $L = \sqrt{\nu\hat{x}/U_\infty}$ makes the dimensionless wall-normal coordinate $y = \eta$ (the Blasius similarity variable) and makes the boundary-layer Reynolds number the root of the x-Reynolds number.

In a Blasius boundary layer, R measures distance along the plate and a disturbance at the reduced frequency F is called a *Tollmien–Schlichting* (T–S) wave. Under certain conditions this wave is amplified, can interact with three-dimensional (3-D) disturbances, secondary instabilities can occur, and breakdown to turbulence can result. The generation and growth of these waves as they relate to disturbances in the basic state will be of particular interest during the experiment.

12.3.3 Basic Rules for Boundary-Layer Stability Experiments

Regardless of whether the experimental objectives are transition control, three dimensionality, secondary instabilities, nonlinear breakdown, or receptivity, the superseding rules of conducting a stability experiment are: (1) the linear problem must be correct, and (2) initial conditions must be provided for theory and computations. These rules can be considered prime directives.

Rule One
The first rule is to get the linear problem correct. Correlation of the experimental data with linear theory (in the appropriate range) ensures that the basic state is probably correct. Usually unintended weak pressure fields change the stability behavior but are not detected in the basic state measurements (Sect. 12.3.4).

Rule Two
Full documentation of physical properties, background disturbances, initial amplitudes, and spatial variations must be provided to the analyst. It is very important to measure, whenever possible, the free-stream environment (a subsection of Sect. 12.3.4 covers the details of measurements of the free-stream turbulence and sound). Any worthwhile stability experiment is going to be accompanied by a computational effort. The experimentalist needs to be able to give as many initial conditions to the analyst running the computational simulations so that an accurate comparison can be made between both methods. This includes, of course, the specification of coordinates since experiments are done in test-section coordinates while computations are done in body-oriented coordinates. The experimentalist should also heed flow symmetry requirements that the computationalist readily assumes but requires some work to achieve in the wind tunnel (see Sect. 12.3.4).

Although these seem like simple requirements, the literature has many examples of experiments that ignore these precepts. In the sections that follow, examples are discussed that illustrate the difficulty of establishing these two rules. However, all of the examples are real, correctable effects. The more advanced practitioner is referred to the *transition study group guidelines* for transition experiments [12.84].

12.3.4 Experimental Techniques

Use a Flat Plate that is Flat

For a Blasius boundary-layer experiment, a flat plate is needed; however, not all methods of manufacturing a flat plate are equally desirable. Plates originating from rolled metal are generally not recommended since the wavy-surface contour can produce a streamwise periodic pressure distribution, as was found on the original Schubauer and Skramstad plate [12.85] (now at Texas A&M). The Klebanoff flat plate [12.86] used at NBS was treated with a 1.8 m-diameter grinding disk. This option can produce satisfactory results, but is often expensive. The Saric flat plate [12.87] used at Virginia Tech and Arizona State, had a 20 mm paper honeycomb sandwiched between two 1 mm aluminum sheets in the manner that inexpensive billiard tables are fabricated. A rule-of-thumb waviness criterion for any plate intended for stability experiments is $\varepsilon/\lambda_{T-S} < 10^{-3}$, where ε is the waviness height and λ_{T-S} is the T–S wavelength. Both the Klebanoff plate and the Saric plate had a ratio of $\varepsilon/\lambda_{T-S} < 10^{-4}$.

Provide a Means for a Leading Edge

The shape of the leading edge has a large effect on the resulting flow field. *Schubauer* and *Skramstad* [12.85] used a sharp leading edge, which was drooped at a negative angle of attack to avoid separation that can occur with a sharp tip. *Klebanoff* [12.86] also chose this type of leading edge, but instead addressed the problem of separation at the tip by including a trailing-edge flap to introduce circulation and thus place the stagnation line on the test side of the plate. The difficulty with either technique is that it is difficult to simulate computationally. Another possible option is an elliptical leading edge with a trailing-edge flap, as shown in Fig. 12.10.

An ellipse with a major/minor axes ratio greater than 6:1 avoids a separation bubble on the leading edge. An ellipse has zero slope at the flat-plate intersection but has a discontinuity in curvature at that point which could be a receptivity location. The curvature discontinuity can bias an acoustic receptivity experiment so *Lin* et al. [12.88] proposed using a *modified super ellipse* whose contour follows

$$\left(\frac{\hat{y}}{b}\right)^2 + \left(\frac{a-\hat{x}}{a}\right)^m = 1, \quad m = 2 + \left(\frac{\hat{x}}{a}\right)^2. \quad (12.33)$$

Here, the origin is at the stagnation line and a and b are the major and minor axes of the ellipse. With this profile, the curvature goes continuously to zero as $\hat{x} \to a$. The aforementioned Klebanoff plate (now at

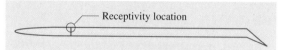

Fig. 12.10 Flat plate with a trailing-edge flap

Texas A&M), was modified by the author for receptivity experiments [12.89, 90] by machining directly on the plate, a 20:1 super ellipse on one end of the plate and a 40:1 super ellipse on the other end. Machining the leading edge directly on the plate avoids junction discontinuity issues that could also be a receptivity site.

Global Pressure Gradient

Whether one uses a blunted flat plate or a sharp flat plate at negative angle of attack, a leading-edge pressure gradient will be present and a finite distance is required for pressure recovery. Once a zero pressure gradient is obtained, the boundary-layer flow is Blasius, but referenced to a different chordwise location, $\hat{x} = \hat{x}_v$. Thus, there is a *virtual* leading edge from which the measurements and the Reynolds number must be referenced. If this is unaccounted for, it is very easy to have 20–30% errors in $\hat{x}$ (and 10–15% errors in R). For example, in order for theory to agree with the linear part of the well-known nonlinear work of *Klebanoff* et al. [12.86], one must apply a correction to $\hat{x}$ [12.91].

To be ensured of the correct streamwise location, measure the displacement thickness δ^* and then calculate the virtual location and Reynolds number with respect to the Blasius boundary-layer profile

$$\delta^*(\text{measured}) = \int_0^\infty (1 - \hat{u}/U_\infty)\,d\hat{y}, \quad (12.34)$$

where $\delta^*(\text{Blasius}) = 1.72\sqrt{\nu\hat{x}/U_\infty} \neq \delta^*(\text{measured})$. The virtual location is given by

$$\hat{x}_v = \left(\frac{\delta^*(\text{measured})}{1.72}\right)^2 \frac{U_\infty}{\nu}, \quad (12.35)$$

$$R = \left(\frac{U_\infty \hat{x}_v}{\nu}\right)^{1/2}. \quad (12.36)$$

The effects of not differentiating between the virtual and geometric locations are repeatedly demonstrated in early stability literature. Present-day researchers must be wary of which location, virtual or geometric was used to obtain transition Reynolds numbers in past literature.

Because of traverse effects and tunnel side-wall blockage, $\hat{x}_v$ may actually change with different chordwise measurements. Therefore, δ^* should be measured (and R calculated) at each location. With modern, computer-controlled experiments this is not a problem. On the other hand, it has been shown by *Klingmann* et al. [12.92] that it is possible to design the leading-edge pressure gradient on the flat plate to eliminate the virtual leading edge. Figure 12.11 shows a series of velocity profiles by the author that demonstrates a constant $\hat{x}_v$ when traverse effects are eliminated.

Local Pressure Gradient

It is difficult to measure small changes in the pressure coefficient C_p accurately. Thus, the flow may not be locally Blasius and the stability characteristics may be quite different. For example, a decrease in C_p of approximately 1% over 100 mm corresponds to a Falkner–Skan pressure gradient parameter, β, of approximately 0.1. For $\beta = +0.1$, the minimum critical Reynolds number, Re_{crit}, [based on Sect. 12.3.7] is increased by a factor of 3. In other words, the streamwise location is increased by a factor of 9.

The neutral stability curve, shown in Fig. 12.12 compares OSE, the nonparallel theory (PSE), and DNS with experiments. What is important is that OSE (dashed line), PSE (solid line), and DNS (points on the solid line) agree very well. The experiments only agree at low frequencies and high Reynolds numbers. The measured

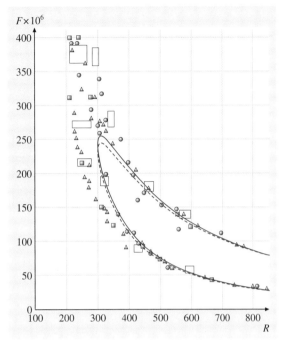

Fig. 12.12 Blasius neutral stability curve. Comparison between experiment, DNS, PSE, and OSE [12.93]

Re_{crit} is around 230 and the calculations all give a value of 300. *Saric* [12.94] conjectured that this difference is due to the extreme sensitivity to the smallest of pressure gradients (in this case adverse) that exist near branch I. This has been confirmed by *Klingmann* et al. [12.92] who designed an experiment to avoid a pressure gradient at branch I and whose data fall on the theoretical neutral stability curve. Thus, the historical discrepancy between theory and experiment has been resolved. There are other problems with the experiments and these are discussed below.

A weak adverse pressure gradient can also explain why instabilities are measured at dimensionless frequencies, $F > 250 \times 10^{-6}$, contrary to theory. The range of unstable frequencies could increase dramatically if the measurements were made in the weak adverse pressure gradient region of the recovery zone of the leading edge. Because of the low Reynolds number needed, the measurements of Re_{crit} [12.95] were conducted too close to the leading edge and too close to the disturbance source.

With extreme care one may be able to measure $\Delta C_p \approx 0.3\%$. This may not be enough. However, one could measure $U(y)$ and calculate the shape factor, $H = \delta^*/\theta$. Here, $\beta = +0.1$ corresponds to $\Delta H = 7\%$

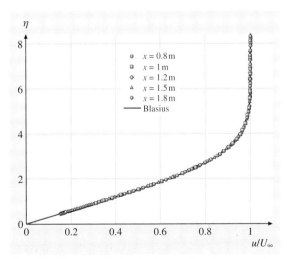

Fig. 12.11 Five measured velocity profiles at $x = 0.8$, 1.0, 1.2, 1.5, and 1.8 m superposed on the Blasius flow calculation. Streamwise location corrected for the virtual leading edge. The free-stream speed is 12 m/s

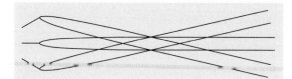

Fig. 12.13 Plate vibrations change the stagnation line

compared to $\Delta C_p = 1\%$. Thus, measurement of changes in the shape factor is more reliable than ΔC_p. One should already have $U(y)$ at each streamwise location and hence the pressure gradient can be verified at no additional work.

For Blasius flow, it is recommended that the plate be adjusted so that $H = 2.59 \pm 0.005$. Moreover, placing a boundary layer trip on the backside of the model helps avoid differential blockage problems by fixing the transition location on the non-test side. Velcro is the recommended trip since a 6 mm-high strip can excite all of the important scales.

This discussion concludes that whereas the zero-pressure-gradient case is an accepted reference test case, it is a rather sensitive and perverse test condition. The author's experience with boundary layers on wings shows that modest pressure gradients ameliorate the sensitivity to small ΔC_p. At the same time, it needs to be recognized that one should avoid measurements of Re_{crit}. Not only is this a very difficult measurement to interpret, but changes in Re_{crit} have very little to do with transition.

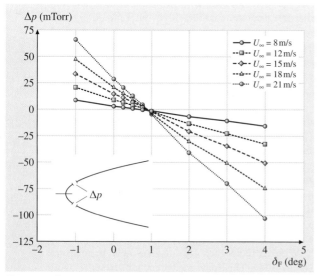

Fig. 12.14 Schematic and data of Δp measurement to achieve symmetric flow

Probe and/or Plate Vibration

If the probe support is vibrating in a direction transverse to the shear layer, the hot wire will measure different levels of the direct-current (DC) component, which in turn appear as temporal fluctuations in the alternating-current (AC) component. Carbon composites work well to stiffen a particular direction of the probe support if vibration is suspected.

Plate vibration is a very serious source of error that should be avoided at all costs. These vibrations cause oscillations in the stagnation line, as shown in Fig. 12.13, that create the initial conditions for T–S waves.

The author has used a laser vibrometer to map the vibrations of the leading edge. These studies showed vibration amplitudes on the order of one micron. If a laser vibrometer is not available, it is important to use a low-mass accelerometer. In any case, some diagnostic tool is needed to ensure that the oscillations are not in the T–S pass band.

Symmetric Flow

When providing a data base for computations of the leading-edge region, it is realistic to establish symmetric flow as an appropriate reference point. The trailing-edge flap is used to control the position of the stagnation line (the shape factor and pressure measurements will determine the plate angle). The pressure difference between the two sides of the leading edge is monitored while the trailing-edge flap angle is changed. When $\Delta p = 0$, the flow is symmetric. It is important to measure Δp in a region of large dp/dx to maximize the sensitivity of the measurement. If differential blockage is minimized and the non-test side boundary layer is tripped properly, it is possible to have the same flap setting at different speeds. Figure 12.14 shows the relationship between flap angle Δp and free-stream speed (unit Reynolds number).

12.3.5 Wind Tunnel Environment

Model Location in the Test Section

Prior to mounting the plate in the test section, all of the vortical modes must be determined. The contraction cone has the tendency to amplify the corner vortices and produce some large scale vortical motions in the test section that may take the form of those shown in Fig. 12.15. This is especially true of tunnels with contraction ratios greater than 6. This weak secondary motion is difficult to measure directly but can be observed by spanning the tunnel with a heated wire. By doing wake scans with a *cold* hot wire (no overheat) at different streamwise locations, the temperature nonuniformity can be tracked

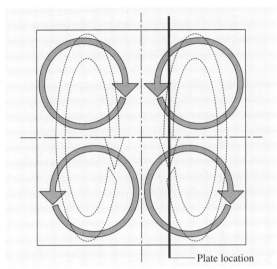

Fig. 12.15 Determination of tunnel nodes and plate placement

and any in-plane rotation can be observed. The rotational nodes can be determined and the plate placed away from these nodes. Acoustic modes will also exist in the test section and these nodes must be avoided as well.

A good rule of thumb is never to mount the plate on an axis of symmetry or at locations $1/N$ of the tunnel span where $N = 2, 3, 4$, etc. Generally, a good location is somewhere between 0.38 and 0.45 unit span.

Free Stream Disturbances: Turbulence and Sound

Ordinary wind tunnels have turbulence levels high enough to mask the appearance and growth of T–S waves. It was not until *Schubauer* and *Skramstad* [12.85], in a tunnel designed for low turbulence, that a successful boundary-layer stability experiment was conducted. It was also recognized at that time that the flight environment also had low turbulence with regard to influencing stability and transition. After the initial success of these experiments, it was recognized that something more than reducing u' fluctuations was needed to advance the knowledge base.

Unknown receptivity issues such as the roles of free-stream turbulence and sound in creating T–S waves and 3-D structures inhibit the understanding and control of transition. It is certainly clear that a naked statement of RMS streamwise fluctuations, $|u'|/U_\infty$, is not enough to describe a particular wind-tunnel environment. Free-stream disturbances are composed of rotational disturbances (turbulence) and irrotational disturbances (sound). Each plays a different role in the transition process. Cross-flow waves are very sensitive to free-stream turbulence level [12.96] while T–S waves are very sensitive to free-stream sound [12.90]. *Naguib* et al. [12.97] demonstrate a good second-order method for separating sound from turbulence that is easy to implement in real-time data acquisition.

Until we really understand the receptivity mechanisms, it is important to document the free-stream disturbance environment as completely as possible. In addition to $|u'|/U_\infty$, one should quote, in order of importance: (1) pass band and spectrum for all measurements, (2) spatial correlation measurements of all components to separate turbulence from sound, (3) flat-plate transition Reynolds number at different unit Reynolds numbers, and (4) $|u'|/U_\infty$, $|v'|/U_\infty$, and $|w'|/U_\infty$ at different positions. A general summary of flow-quality issues is found in [12.98] and a typical tunnel certification is given by [12.99].

It has been argued [12.92] that it is not necessary to have free-stream turbulence levels down to 0.04% U_∞ in order to measure T–S waves. This is a naïve statement that is only true when one knows where one is looking and one knows what one is measuring. For example, *Kendall* [12.100] has been measuring T–S waves in high-disturbance environments for years. There are two relevant points that need to be mentioned justifying a low-disturbance free stream: (1) one can always systematically increase free-stream turbulence [12.100] and study its effects and (2) different (unknown) breakdown mechanisms that are characteristic of the low-disturbance flight environment may be missed in a high-disturbance free stream. The observations of the subharmonic mechanisms [12.101, 102] fall into this category.

Another argument for low-turbulence levels can be made when streamwise vortical structures in the basic state produce a weak spanwise periodicity that is strongly susceptible to secondary instabilities. These spanwise variations were carefully documented by *Klebanoff* et al. [12.86], *Nishioka* et al. [12.103], and *Anders* and *Blackwelder* [12.104]. They strongly influence the type of breakdown to transition that is observed [12.101, 105]. These spanwise variations were not observed in the low-turbulence tunnels in Arizona, Novosibirsk, Sendai, or Stockholm. It turns out that these tunnels had slightly lower turbulence levels and that the combination of higher turbulence levels and micro surface roughness caused transient modes to grow and create the streamwise vorticity within the boundary layer. This is a good example of why it is necessary to be able to do spanwise measurements in the tunnel and within the boundary layer. Finding a turbulence level at one spanwise location does not guarantee the same turbulence level at other spanwise locations. As a general reminder, tape or junctions act similar to the micro surface roughness in that they are receptivity locations and are to be avoided on the test surface of the plate. Even though one may have a two-dimensional (2-D) roughness with an $Re_k = O(0.1)$, this is still a strong receptivity source [12.90, 106]. Therefore, in order to establish the initial conditions, one should provide the measured three-dimensional amplitude modulation within the boundary layer for comparisons with theory and computation.

12.3.6 T–S Measurements

Controlled T–S Waves – Internal Disturbance Sources

Knapp and *Roache* [12.107] tried to use the background disturbances as the source of the T–S waves and observed intermittent behavior that compromised their hot-wire measurements. It becomes necessary to fix the wave in the streamwise direction and do phase-correlated measurements. The use of an artificial disturbance source will fix the amplitude and phase at one location in order to systematically track stability and transition events. The use of a vibrating ribbon to create 2-D waves has been around since *Schubauer* and *Skramstad* [12.85]

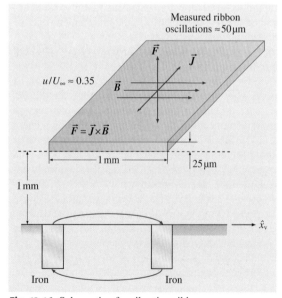

Fig. 12.16 Schematic of a vibrating ribbon

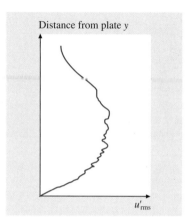

Fig. 12.17 Disturbance profile close to the ribbon

who used the idea of the Lorentz force generated by an alternating current in the ribbon in the presence of a stationary magnetic field. This is shown in Fig. 12.16. *Sreenivasan* et al. [12.108] has used the same principle on a wire in a slot. *Corke* and *Mangano* [12.102] have successfully used segmented heated wires for producing both 2-D and 3-D waves.

When the vibrating ribbon or wire is uniformly loaded, the displacement is of the form of a catenary. Therefore, a sufficiently long ribbon should be used to avoid end effects. The ribbon placement in the wall-normal direction is typically ideal (minimum displacement) if located at or near the critical layer $[U(y) = c = \omega/\alpha]$.

Even though the 2-D wave is phase correlated when using a vibrating ribbon, the interaction of this wave with the background disturbances has a random character. The Λ-vortices observed by *Saric* and *Thomas* [12.101] for different types of breakdown meandered in the span direction. Although not reported, the subharmonic measurements of *Kachanov* and *Levchenko* [12.109] were random and eyeball conditional sampling was used. The only solution is to introduce the 3-D disturbance directly with segmented heating elements [12.102]. The technique consists of using one continuous wire for the 2-D wave and a set of segmented wires, whose individual phase is controlled, for the 3-D wave.

A disturbance source such as an air jet, or heated wire, or vibrating ribbon, locally creates a disturbance that is not just a T–S wave (Fig. 12.17) but has all of the eigenmodes. A T–S wave is just one of the modes in the distribution. A relaxation distance is required to attenuate the more stable modes so that the least stable (the T–S wave) remains. If measurements are made within this relaxation distance some strong stabilizing effects may be measured. One should determine the relaxation distance downstream of the disturbance source. This would depend on the type of source used, but it should be in the range of about 10 boundary-layer thicknesses. This can be verified by first comparing $|u'(y)|$ with linear theory. Local growth rates should also be compared as a function of input amplitude. These comparisons should be documented if it is required to measure close to the disturbance source.

One would like to carry out stability measurements over a wide range of Reynolds numbers while keeping the disturbance source fixed. Unfortunately, a typical 2-D disturbance source has a finite span and *Mack* [12.110] showed that the domain of dependence of a finite span disturbance source propagates from each end toward center span at an angle of approximately 12° as shown in Fig. 12.18.

Outside of this triangular domain, the disturbance amplitude is different from linear theory. This is analogous to the boundary-condition domain of dependence in hyperbolic systems. If w is the span of the disturbance source and L is the distance in x from the source, then the centerline measurements should be made such that $L/w < 2.3$. *Ross* et al. [12.95] had a vibrating ribbon span of only 250 mm and took measurements 1 m away. Just as one is limited in the useful chord of the model due to sidewall contamination, the distance downstream of the disturbance is similarly limited. For off-centerline measurements, this value is obviously smaller.

When one attempts to study nonlinear wave interactions, the nonlinearities of the disturbance source impose different initial conditions on the nonlinear components [12.87]. For example, if one wishes to study the nonlinear interaction of waves with two frequencies f_1 and f_2, when the disturbance source, such as a vibrating ribbon, is oscillated at too high an amplitude, the disturbance source inputs $2f_1$, $2f_2$, $f_1 - f_2$, $2f_1 - f_2$, etc. into the boundary layer. As part of another difficulty, when one attempts to invoke active wave cancelation into the boundary layer through a disturbance source, a feedback signal is processed by the computer and relayed to a disturbance source. However,

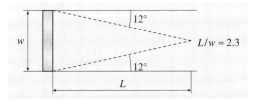

Fig. 12.18 Triangle of acceptable measurement area

the digital-to-analog (D/A) converter is a low-pass filter, the vibrating ribbon is a low-pass filter having a typical linear oscillator response, and the boundary layer is a band-pass filter/amplifier having its unique response curve. Thus the boundary-layer response is much different from the original input signal. First, one must always directly measure the disturbance-source response and the boundary-layer response in order to establish the initial conditions [12.87]. In the case of a vibrating ribbon, the disturbance source response can be measured with an end-effect inductance probe. A tailored boundary-layer response can be obtained using inverse Fourier techniques [12.111].

Controlled T–S Waves –
External Disturbance Sources

If an external sound source is used as a source of disturbance energy, say in a receptivity experiment, then the boundary-layer measurement at a particular frequency will contain probe vibrations and a component of the sound wave in addition to the T–S wave. It is easy for external sound to force, at the oscillation frequency, the mechanical system holding the hot wire. The external sound field generates a Stokes layer imbedded inside the boundary layer. All of these signals are at the same frequency and if these signals are of comparable amplitude to the T–S amplitude, one cannot obtain the usual T–S profile unless some special techniques are used to extract the T–S wave. It is for this reason that older publications with sound/stability interactions are not reliable.

This author has tried: (1) taking advantage of the exponential growth of the T–S wave so that it is larger than the background [12.106]; (2) the idea of using polar plots to separate the long-wavelength Stokes wave from the short-wavelength T–S wave [12.112]; (3) the idea of using a wavenumber spectrum using closely spaced points in the x direction [12.98]; (4) the idea of using differential surface-pressure ports [12.100]. For one reason or another, none of these techniques are satisfactory and are not recommended. The details are given in [12.113]. The major problems lie in complicated duct acoustics and reflected waves from the diffuser.

The only technique found to work is the pulsed-sound technique [12.89, 90]. The technique uses pulsed sound and is simple, effective, and lends itself to understanding the behavior of the T–S wave. From linear theory, the maximum of the T–S wave propagates at approximately one third of the free-stream speed (about 1% of the speed of the downstream-traveling sound wave). Using this fact, the traveling T–S wave can

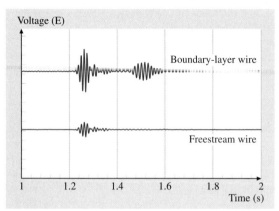

Fig. 12.19 Time traces of the free-stream wave and the boundary-layer wave

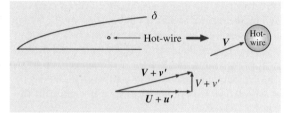

Fig. 12.20 Hot-wire measurements

be isolated from the acoustic disturbance and associated Stokes wave by sending bursts of sound into the test section. The initial sound burst is first measured and fractions of a second later, after the sound wave has passed, the slower-traveling T–S wave is measured. Figure 12.19 shows a time trace depicting the sound-burst wave sensed by hot wires in the free stream and boundary layer and the trailing T–S wave measured by the boundary-layer wire for $R = 1140$, $F = 56 \times 10^{-6}$, $f = 80$ Hz and $x = 1.8$ m. The T–S wave profile obtained with this method compares very well with OSE solutions.

There are three ways to implement this technique: (1) use the RMS amplitude of the wave packet [12.89]; (2) use the magnitude of the complex Fourier coefficient for each frequency present in the wave packet [12.90]; (3) analyze the signal in the frequency domain [12.114]. The frequency-domain approach [12.114] appears to be the best means to correctly describe the receptivity and linear amplification process of multiple-frequency signals. This is because as wave packets travel downstream, high-frequency components of the spectra, which are present initially due to the finite extent of the pulse, decay. Meanwhile, the low-frequency components in the amplified T–S pass band grow.

A feature of short sound bursts is that since they are limited in the time domain, they are extended in the frequency domain. Thus, a single sound pulse (a single frequency sine wave within an amplitude envelope) covers a wide frequency range. In many cases the pulse spectrum covers the entire T–S wavelength band. Therefore, using a pulsed-sound approach eliminates the distinction between single-frequency and broadband input.

Hot-Wire Measurements

The hot-wire anemometer is the accepted technique for the measurement of fluctuating velocities $O(10^{-3} U_\infty)$ within the boundary layer. Neither LDV nor PIV have the low-level resolution required for these measurements. Hot wires can accurately measure the streamwise and spanwise velocities (u', w') with the use of straight-wire and slant-wire pairs. Because laminar boundary layers are so thin ($\delta \approx 3\text{--}5$ mm), it is not possible to measure v' due to the span of the wire.

If the temperature of the wind tunnel undergoes changes of more than a few degrees between calibrations, temperature compensation must be used. This can be done with a simple computer solution that does both velocity and temperature compensation [12.115].

To do this, one must understand how and what data are retrieved from a hot wire. In a real boundary layer, a hot wire measures the component of velocity perpendicular to the wire as shown in Fig. 12.20.

Although the velocity is 2-D, the output signal is only a combination of U and u' because the vector sum of $U +$

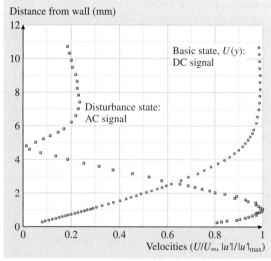

Fig. 12.21 Disturbance and basic-state profiles

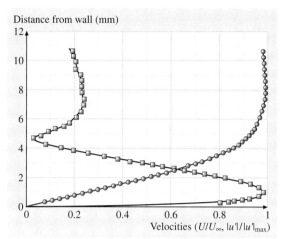

Fig. 12.22 Theory and experiment of Blasius velocity profile and T–S amplitude (after [12.94])

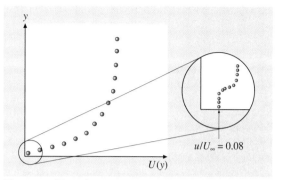

Fig. 12.23 Hot-wire measurements near the wall

V is approximately U since $V = O(1/R)$. Similarly, the AC component measures u' and not v' because u' is superposed on U. As a result, the measurement from a hot wire is u'_{rms}, which is proportional to $|D\phi|$ from the OSE. It is straightforward to separate the DC and AC signals.

When attempting to compare with theory, one commonly sees solutions of the OSE displayed in terms of the real and imaginary parts of $\phi(y)$. Since ϕ can be multiplied by any complex number, this is neither revealing nor unique. It is more meaningful to show amplitude, $|\phi|$, and phase, $\psi = \arctan(\phi_i/\phi_r)$. Using the acquired u'_{rms} data, a more-rigorous representation of results is to then plot y as a function of $|D\phi|$, where $|D\phi|$ is a positive real quantity. The resulting plot will look similar to the disturbance state in Fig. 12.21.

Once measurements are processed, they should be correlated with theory. An example with a reduced set of experimental points is shown in Fig. 12.22 [12.94].

To achieve such accuracy, a precision lead screw with anti-backlash bushings should be used in the wall normal direction. One should have the capability to make 100 measurements within the boundary layer which means step sizes on the order of 25 μm.

Wall Effects

In measuring $U(y)$ and $u'(y)$, one will need to find the wall. As the hot wire gets closer to the wall, radiation from the model removes heat from the hot wire, resulting in readings of higher velocity than is actually present. This is shown in Fig. 12.23. To compensate for this effect, stop measurements at $u/U_\infty \approx 0.1$ and for mean flow, use linear extrapolation to the wall.

Traverse Blockage

The traverse mechanism may be too large or too close to the hot wire. Moreover, a multi-wire rake may have too much local blockage. What could happen in these cases is that the weak pressure field around the probe support, although unseen in a basic-state measurement, can strongly influence T–S wave amplitude. This can be diagnosed by fixing a very small hot wire to the plate (Fig. 12.24) at a wall-normal location where say, $U/U_\infty \approx 0.3$. Establish the amplitude of a controlled T–S wave as measured by the fixed wire. Move a traverse mounted hot wire to the same location very close to the fixed wire and see if the T–S amplitude on the fixed wire has changed. This is the most sensitive and the only means for determining whether one has eliminated traverse and probe-support interference problems. One should be aware of the fact that traverses and probe supports that are quite suitable for turbulent boundary layers may not be suitable for laminar stability work.

Hot-Film Measurements

The development and application of microthin hot films [12.116, 117] have advanced their use for stability and transition measurements. These films are in the form

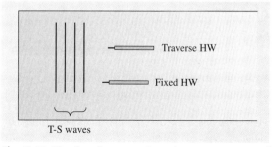

Fig. 12.24 Fixed and traverse hot-wire measurements of T–S waves

of vacuum deposited circuitry on a Kapton sheet. As many as 50 sensors can be concentrated in a small area and can be oriented in any direction. Although difficult to obtain an absolute calibration (one could use a Preston tube over the sensors), this technique is very valuable for measuring wall shear stress fluctuations. Disturbance spectra and transition location can be determined. With the use of multiple hot films, phase and group velocity directions can be determined [12.118]. The use of hot film sheets are superior to individual hot film sensors in that they provide a minimum of disturbance to the flow, are robust, and are easy to apply. This is also a superior technique for flight experiments where it may not be possible to use hot wires. In their simplest, uncalibrated use, an array of sensors would indicate an order of magnitude increase in RMS fluctuations wherever transition to turbulence occurred.

12.3.7 Visualization Methods

Visualization techniques are useful for qualitative information regarding scales and approximate transition location. For stability and transition work, they must always be supported by hot-wire or hot-film measurements.

Smoke-Wire Technique

Most smoke wires used for flow visualization have diameters in the range of 50–80 μm. This technique was modernized by *Corke* et al. [12.119]. The oil coating (toy train *smoke*) distributes itself periodically along the span of the wire and when the wire is heated it generates a short burst of smoke streaks. The computer initiates the wire voltage and the time-delayed shutter release. When used in stability experiments the smoke wire is placed near the critical layer. Much of the interesting detail is lost if the wire strays from the critical layer. The wake of the smoke wire causes a kink in the basic-state profile which alters its stability characteristics. As a result, if one examined the amplitude growth of a T–S wave in the stream direction, one would observe an almost step-like increase in amplitude downstream of the smoke wire. The amplitude could easily change by a factor of three due to the smoke wire. Thus, in contrast to its universal use in turbulent boundary layers, special care must be exercised with laminar stability.

One should always be reminded that streaklines do not correspond to streamlines in an unsteady flow. The appearance of a 3-D structure in a streakline is a historical event that is a result of the integration of the history of the smoke. A direct measurement at the location of an apparent 3-D structure may reveal something different. In the same way, visualization should always be accompanied by direct measurements. This type of visualization is good for giving scales over which you need to do the other measurements. An example of the usefulness of the smoke-wire technique is found in *Saric* [12.120].

Surface Coatings

Surface coatings have the ability to determine the approximate location of transition and only rarely something else. The author has tried them all. Shear-sensitive liquid crystals are robust but seem more useful for detecting separation than transition [12.121]. This technique is difficult to apply and introduces a nontrivial surface roughness which affects the stability characteristics. Temperature-sensitive paint (TSP) and pressure-sensitive paint (PSP) have been used for transition detection. PSP and TSP require some application skills and perhaps are best used for complicated shapes. Infrared thermography (IRT) is a non-obtrusive, successful technique requiring a sizable investment in IR cameras (see [12.122] and [12.123] for details). The author has had a great deal of success using naphthalene coatings [12.116,124,125] but health physics issues have arisen regarding the use of naphthalene, trichlorotrifluoroethane, trichloroethane, and other chemicals. For this reason, the author is reluctant to suggest their use.

12.4 Measurements Considerations in Non-Canonical Flows

Wall-bounded turbulent flows are found in a vast number of practical applications, and accordingly have been the focus of research for many years. Most studies have been on the canonical case, that is, a flat-plate turbulent boundary layer developing in a zero pressure gradient. However, most practical boundary layer flows are non-canonical, involving an increased degree of complexity. Some examples include, flow over a skewed or non-axisymmetric geometry, a boundary layer flow encountering an obstacle, or flows involving corners and junctions. In these cases, strong secondary flow and pressure gradients are often present, as is flow separation. Measurements in such flows pose particular challenges compared to the canonical boundary layer flow case, and this is the topic of this section. The discussion is limited to incompressible boundary layers that may have three-dimensionality in the mean, pressure gradients, or flows that are undergoing mild or incipient separation. In such flows the important feature of the measurement technique is that it must be able to determine the primary flow direction, as well as potentially deal with very high levels of turbulence intensity and/or instantaneous flow reversals. In these cases, perhaps more than others, the measurements must not disturb the flow.

Figure 12.25 shows sketches of typical mean velocity profiles that are being considered. Figure 12.25a corresponds to a boundary layer developing in the presence of three-dimensional pressure gradients. Pressure-driven flows are characterized by free-stream streamlines that curve. The slower fluid near the surface has less inertia than the free-stream fluid and so is turned through a larger angle, as shown in the figure. Such flows may be caused by external geometry changes in those directions or by streamline curvature of the solid body. Three-dimensional turbulent boundary layers may also be formed by shearing. An example would be an initially two-dimensional boundary layer that encounters a section of the wall that is moving at an angle different with respect to that of the free-stream direction. A review of a number of experiments describing how three-dimensionality affects the turbulent boundary layer structure is given by *Eaton* [12.126]. The second type of non-canonical turbulent boundary layers being considered, represented by Fig. 12.25b, are flows that may remain two-dimensional in the mean but involve strong streamwise pressure gradients. In the case of adverse pressure gradients this may lead to flow separation. A good review of turbulent boundary layer flows with separation is given by *Simpson* [12.127].

12.4.1 Pressure Probe Measurements

The standard method for measuring mean or average velocity is to use a Pitot static tube, as described in Sect. 5.1. Accurate use of this probe relies on aligning the probe to a known primary flow direction. *Chue* [12.128] and *Bryer* and *Pankhurst* [12.129] provide a good summary of the accuracy penalties incurred for misalignment for different Pitot-static tube designs. For complex flows, both the dynamic and static pressures are required as well as flow direction. This is most commonly achieved by using combination or multi-hole pressure probes [12.128], which typically require calibration in known flow conditions representative of those in which they will be used.

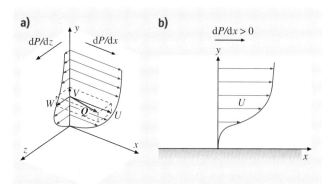

Fig. 12.25a,b Examples of non-canonical boundary layer mean velocity profiles: (**a**) pressure-driven three-dimensional boundary layer, (**b**) adverse-pressure-gradient flow with incipient separation

Many different types of probes have been proposed in the past, either for measuring flow direction, flow speed or both. Figure 12.26 shows sketches of two such devices. Figure 12.26a shows a five-hole probe, which is an example of a multi-hole probe where several small diameter tubes are axisymmetrically arranged around a central total-head tube. The outer tubes have faces that are slanted at a fixed angle as shown in the figure. Figure 12.26b shows a seven-hole probe consisting of a hemispherical tip with one central hole with the remaining holes axisymmetrically distributed at one outer radial distance from the centerline. The holes extend over the length of the probe and connect internally or by tubing to a manometer. Common alternate designs use tips that are machined to a cone or with faceted faces (*cobra* probe).

Multi-hole pressure probes require calibration, and various schemes have been proposed. See for example

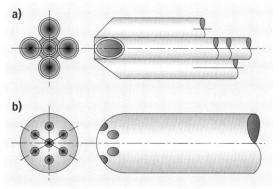

Fig. 12.26 (**a**) Schematic of five-hole probe. (**b**) Seven-hole probe with hemispherical tip

Chue [12.128], *Dominy* and *Hodson* [12.130], *Zilliac* [12.131], *Rediniotis* and *Vijayagopal* [12.132], and *Wenger* and *Devenport* [12.133]. Inaccuracies due to Reynolds-number effects can be significant and need to be considered for accurate measurements. Once calibrated, typical multi-probes have been used in flows where the primary flow direction falls within a cone with an apex angle of up to nominally 75 degrees.

Extended ranges of application, where the approximate primary flow is not known beforehand or where the flow is reversed, have been proposed. However, these usually rely on exhaustive calibrations. *Pisasale* and *Ahmed* [12.134] describe a preprocessing scheme to deal with reverse flows. *Rediniotis* and *Kinser* [12.135] describe an *omnidirectional* pressure probe that consists of a spherical tip with 18 pressure ports. The calibration database that they use is based on readings from 10 000 individual probe orientations in a steady state environment. The time average of the fluctuating pressures in a turbulent flow, when processed with the time-steady calibration information, may not accurately reflect the time-averaged flow field.

For measurements in non-canonical turbulent boundary layers, early studies (before the mid to late 1970s) used pressure probes extensively, as reviewed by *Vagt* [12.136, 137]. After which, hot wires and LDA became the preferred methods. One of the main concerns about multi-hole probes is their relative large size, and hence their intrusive influence on the flow, particularly for near-wall measurements. This is especially a concern in pressure-driven boundary layers where the imposed pressure field drives the flow. If the pressure probe and associated stems and holders are too large, then the resulting deflected streamlines may manifest in an altered pressure field and hence different flow. In an attempt to overcome these problems, some efforts have gone towards developing miniature probes, such as that of *Ligrani* et al. [12.138]. *Pompeo* et al. [12.139] used a series of pressure probes, including five-hole probes to measure mean flow profiles in a laterally strained turbulent boundary layer. *Westphal* et al. [12.140] developed a single-hole probe in which a single tube cut at 45° is rotated about its axis by a miniature stepper motor. This allows velocity magnitude and angle measurements with a significantly smaller probe, although the measurement procedure is more time-consuming.

When using multi-hole probes, one issue which should always be kept in mind is that the ultimate accuracy of the pressure probe relies on the accuracy of the calibration. Moreover, one uncertainty that remains with most existing calibration techniques is that the calibrations are performed at steady pitch and yaw angles in uniform flows, and no account of the nonlinear effects of the turbulent flow are accounted for. The level of this uncertainty remains unclear.

12.4.2 Turbulence Measurements

The hot-wire and laser Doppler anemometer (LDA) remain the methods of choice for single-point measurements of turbulence quantities, especially if time-series data are required. Comprehensive descriptions of hot-wires and LDA methods are considered in Sect. 5.2 and Sect. 5.3.1 of this handbook, respectively. Here, discussion is limited to considerations specific to the use of these techniques in non-canonical boundary layer flows.

Laser Doppler Anemometry (LDA)

LDA is well suited for non-intrusive measurements of complex flows. A three-component system with frequency shifting can, with reasonable accuracy, measure all three orthogonal components of velocity in complex flows, including regions of separated reversed flow. Successfully implementing a three component system can be difficult, not to mention expensive. Optical access issues quite often limit the type of measurement that can be made. Very-near-wall measurements are particularly challenging because of reflections and/or refraction of the laser beams due to the close proximity of the wall. Even so, successful three-component LDA measurements in three-dimensional boundary layers have been made, see for example the studies of *Driver* and *Hebbar* [12.141], and *Flack* and *Johnston* [12.142, 143].

Perhaps the most accurate LDA measurements to date for the near-wall region of a turbulent boundary layer were made by *DeGraaff* and *Eaton* [12.144]. For those experiments a high-resolution two-component system, with side-scatter collection optics, was custom built giving a measurement volume of $35\,\mu m$ in diameter, and $60\,\mu m$ in length. Such small measurement volumes are essential if one wishes to adequately resolve all the near-wall energy-containing motions in most turbulent boundary layers. For example, accurate measurement of Reynolds stresses requires the measurement volume to have a characteristic length typically less than 20 viscous wall units. This is extremely small for boundary layers in most laboratory-scale facilities at high Reynolds number. Figure 12.27, taken from [12.144], illustrates this point well and shows in physical terms the limitation of large measuring volumes (such as for a regular X-wire, or standard LDA) as Reynolds number increases in a laboratory wind tunnel.

Song and *Eaton* [12.145] used the same high-resolution LDA system to make measurements in a boundary layer that experienced strong adverse pressure gradient, separation, reattachment, and downstream redevelopment. A common problem of high-spatial-resolution LDA systems is the low data-sampling rate. Extremely high seeding levels are needed to obtain a sufficient statistical sample in a reasonable time, and it is virtually impossible to obtain high-resolution time-resolved LDA data in air flows.

Lowe and *Simpson* [12.146] have recently reported a three-color three-head optical arrangement and a higher-order processing scheme that allows simultaneous measurement of the three component velocity and the flow acceleration. Other notable studies of non-canonical boundary layer flows include that of *Simpson* et al. [12.127, 147] who made accurate measurements using a two-component system in a separating boundary layer, and made comparisons to hot-wire measurements. LDA measurements have also been extensively used for three-dimensional boundary layers. See for example *Olcmen* and *Simpson* [12.148], *Webster*, *DeGraaff* and *Eaton* [12.149], *Compton* and *Eaton* [12.150], and others as reviewed by *Eaton* [12.126].

Hot-Wire Anemometry

The hot-wire, while needing special consideration in non-canonical flows, still remains the most cost effective and simplest method for measuring turbulence flow statistics. A single wire is especially preferred for measurements in very thin boundary layers, or for measurements very close to the wall. For two-component measurements of velocity, X-wires are most often used. Three-wire probes are also commonly available commercially, and allow the measurement of all three components of velocity instantaneously. For these probes the calibration schemes are more complex, and the accuracy of the turbulence statistics are in general not as good as can be obtained using X-wires (normal and rotated at 90 degrees). Furthermore, single wires and X-wires offer the best opportunity for adequate spatial resolution. Details of three-component hot-wire systems are discussed in *Bruun* [12.151] and Sect. 5.2.

The key to successful use of the hot wire is understanding its limitations, and knowing when a reliable measurement can be made. The biggest problem facing conventional hot wires is flow ambiguity, and/or very high levels of turbulence intensity, and these will be discussed in the following.

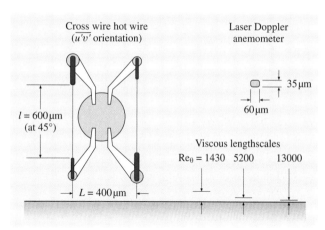

Fig. 12.27 Comparison of typical X-wire and required LDA measurement volume for near-wall boundary-layer measurements (after *DeGraaff* and *Eaton* [12.144])

High Turbulence Intensity Flows. For a single wire, the limitations due to large incident velocity vector angles can be understood by considering the response of the wire to an instantaneous flow, as shown in Fig. 12.28a. The long time-averaged velocity is $\overline{U}$, and u, v and w are the turbulent fluctuating components in the Cartesian coordinate system (x, y, z) shown in the figure. The output voltage E from an anemometer may be modeled

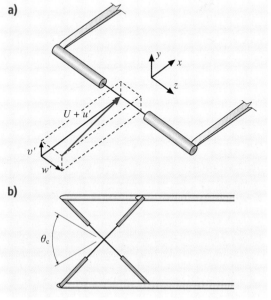

Fig. 12.28 (a) Single-normal hot-wire geometry. **(b)** Cone angle θ_c for X-wire

using a King's law formulation

$$E^2 = A + BU_e^n,\qquad(12.37)$$

where U_e is referred to as the effective velocity. A commonly used relationship for U_e is (Champagne et al. [12.153, 154])

$$U_e^2 = U_N^2 + k^2 U_L^2,\qquad(12.38)$$

where U_N is the velocity normal to the wire, and U_L is the longitudinal component of velocity along the wire. The constant k has a typical value of 0.2. Considering the velocity vector in Fig. 12.28a, and following Perry [12.155], the Pythagoras and binomial theorems are used to obtain the expression

$$U_e = (\overline{U}+u) + \frac{\overline{U}}{2}\left[\left(\frac{u}{\overline{U}}\right)^2 + \left(\frac{v}{\overline{U}}\right)^2 + k^2\left(\frac{w}{\overline{U}}\right)^2\right]$$
$$+ \text{higher-order terms}.\qquad(12.39)$$

Therefore, clearly $u/\overline{U}$, $v/\overline{U}$ and $w/\overline{U}$ need to be small in order to assume

$$U_e \approx \overline{U} + u.\qquad(12.40)$$

If the turbulence intensities are high, (12.39) may be used to estimate the possible error. *Bruun* [12.151] provides a good list of references of studies documenting uncertainties for flows with very high turbulence intensities. In reverse flows or flows with excessively high turbulence intensity, a way to successfully use a single wire is to *fly* the wire. A *flying hot wire* is a hot-wire device that moves the probe into the primary flow direction with a known bias velocity. The forward bias speed must be sufficiently fast so that the relative $\overline{U}$ is high enough to make the relative $u/\overline{U}$, etc. sufficiently small. *Cantwell* and *Coles* [12.156], *Perry* and *Watmuff* [12.157] and others (*Bruun* [12.151]) have successfully used such an approach in wake flows, and other flows with significant flow reversal. *Perry* and *Li* [12.158] used a flying hot wire in the study of rough-wall boundary layers where high turbulence intensities can also be encountered.

In the case of X-wires, an effective way to qualify the largest acceptable turbulence intensity is to define a *cone angle* [12.159] as shown schematically in Fig. 12.28b. The term cone angle, θ_c, refers to the included angle in the plane of the X-wire of an imaginary cone (assumed to be symmetric) in which the instantaneous velocity vectors fall. *Marusic* and *Perry* [12.152] quantitatively defined the cone angle as

$$\theta_c = 2(|\mu_c| + 3\sigma_c),\qquad(12.41)$$

where μ_c and σ_c are the mean and standard deviation, respectively, of the probability distribution function of the measured velocity vector angles. If θ_c exceeds a *critical cone angle* [12.152] then incorrect values of inferred Reynolds stresses will result. Determining the critical cone angle requires an independent method to accurately measure the Reynolds stresses ($\overline{u^2}$, $\overline{v^2}$, $-\overline{uv}$ etc.). *Marusic* and *Perry* [12.152] made measurements in a turbulent boundary layer developing in a strong adverse pressure gradient. They used a flying hot wire to reduce the effective θ_c and compared the measured Reynolds stresses between flying and stationary X-wires. From this, it was determined that for their X-wires (90° included angle) and calibration procedure, the critical cone angle was 40–45°. Therefore, if a stationary X-wire recorded a θ_c in excess of this value, it was determined to be unreliable. Figure 12.29 is taken from [12.152] and shows the variation of θ_c for stationary and flying hot wires in an adverse-pressure-gradient boundary layer.

The critical cone angle of a X-wire can be increased by either using a special calibration scheme to account for the large incident velocity vector angles, or by using large included angles between the wires. *Skare* and *Krogstad* [12.160] used X-wires with an apex angle of 110°, and made measurements in a very strong pressure gradient boundary layer flow on the verge of separation. *Elsberry* et al. [12.161] conducted experiments in a boundary layer with incipient separation. They adopted a calibration scheme that extended their effective critical cone angle, and made comparative measurements with LDA to confirm the accuracy of their hot-wire measurements.

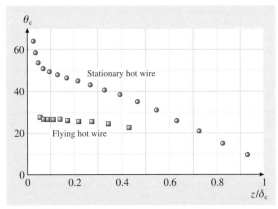

Fig. 12.29 Comparison of cone angle for stationary and flying X-wire across a turbulent boundary with strong adverse pressure gradient (after *Marusic* and *Perry* [12.152])

Unsteady, Reverse Flows. As discussed above, a flying hot wire may be used in reverse flows to remove the directional ambiguity involved. However, in many non-canonical flow applications this is not possible or convenient, due to the mechanical complexity involved with a flying apparatus, and often a continuous time series at a given point in the flow is required. Various thermal anemometry techniques have been developed over the years to address this issue, and many are reviewed by *Bruun* [12.151]. The most prominent of these is the pulsed-wire anemometer originally developed by *Bradbury* and *Castro* [12.163]. A design specific for near-wall flows was developed by *Westphal* et al. [12.162], and is shown in Fig. 12.30. The principal of operation relies on a central wire that is heated in short pulses, and upstream and downstream hot wires that operate effectively as temperature sensors. Typically the central wire is heated periodically, and the velocity is measured from the time of flight of the hot tracer fluid from the central wire to the downstream or upstream wires, giving the forward or reverse velocities respectively. Other pulsed-wire anemometer designs for near-wall flows have been proposed by *Castro* and *Dianat* [12.164] and *Devenport* et al. [12.165] and others. Limitations on the use of these probes is dependent on thermal diffusion considerations as discussed by *Schober* et al. [12.166], and in general the frequency response is limited to tens of Hz.

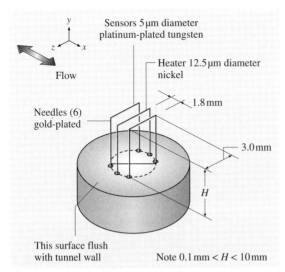

Fig. 12.30 Pulsed wire probe, suitable for measurement of near-wall velocity and wall-shear stress in reverse flows (after *Westphal* et al. [12.162])

12.4.3 Wall Shear Stress

A comprehensive review of different methods for measuring wall shear stress is given in Sect. 12.1. For non-canonical boundary layer flows the number of available techniques is greatly reduced, especially for three-dimensional boundary layers and flows with separated flow regions. For three-dimensional boundary layers, the skin friction measurement technique needs to be able to determine the principal wall shear stress magnitude and direction.

Of the direct methods, the floating element method can in principle be used for complex flows. However, its use is complicated in pressure gradients, where the device needs to be able to account for the pressure gradient contribution to the force balance across the sensor. Also, in incipient separation flows where the wall shear stress is very small, the signal-to-noise ratio is usually prohibitively small. Oil-film interferometry is perhaps the best of the direct methods for steady-state conditions. Most systems now use video cameras to record the interference fringe patterns as a function of time from an oil drop being sheared by the flow. From the two-dimensional images it is usually fairly straightforward to determine the direction of principal wall shear stress.

Of the indirect methods, several methods are available but each has its disadvantages. For example, one method is to use an equivalent of the Clauser chart, where total mean velocity Q at a wall-normal position is measured and assumed to follow a logarithmic law of the wall, thus yielding Q_τ, the total wall shear velocity. This method was used by *Compton* and *Eaton* [12.150] in their pressure-driven three-dimensional boundary layer study. This technique is not regarded as ideal as there is considerable debate as the validity of the law of the wall in complex flows.

For time-resolved wall shear stress measurements in attached flows the most common method is to use a hot wire positioned very close to the wall in the viscous sublayer. The wall shear stress, τ_0, is then known from the measured velocity U from the linear relationship

$$\tau_0 = \frac{\mu}{y} U, \qquad (12.42)$$

where y is the distance from the wall, and μ is the viscosity of the fluid. For complex flows with flow reversal, pulsed probes such as the one shown in Fig. 12.30 have been used by adopting higher-order similarity formulations ([12.162, 167]). That is, for y small

$$\tau_0 = f(U). \qquad (12.43)$$

Other techniques also based on thermal anemometry but with higher-frequency response than pulsed probes have recently been reported. These include the designs by *Spazzini* et al. [12.168] who used two electrically heated hot wires mounted above a cavity, and *Li* and *Naguib* [12.169] who used an oscillating hot-wire technique with a reported frequency response of 1.4 kHz.

12.4.4 Planar and Whole-Field Measurements

The majority of experimental studies on non-canonical turbulent boundary layers to date have been done with single-point measurements. This approach is perhaps still the best if the aim is to measure accurate turbulence statistics. However, often it is desirable to measure the velocity field simultaneously in a plane or volume, and using arrays of single-point probes is far from ideal. Fortunately, over the past decade rapid advances in digital camera and laser technology have seen techniques such as particle image velocimetry (PIV) become much more prolific. Carefully applied, PIV can be an excellent and very useful method for measuring complex boundary layer flows. Some recent studies include: Kiesow and Plesniak [12.170] who made PIV measurements in a shear-driven three-dimensional turbulent boundary layer, and *Angele* [12.171] who made measurements in a separating adverse-pressure-gradient turbulent boundary layer.

A thorough and detailed description of issues relating to PIV and its accurate use is presented in Sect. 5.3.2. For non-canonical turbulent boundary layers, and highly complex flows in general, a number of particular challenges to the accuracy of PIV are worth noting here. PIV is a technique synonymous with compromises – at least when it comes to the choice of parameters. For example, the time delay between consecutive laser sheet pulses needs to be large to minimize the error in the differencing algorithm, but not so large that the light scattering particles have had time to leave the illuminated volume. In highly complex flows, the measurement field may cover a wide range of velocities and cross-flows, and therefore an optimal choice of measurement parameters, such as laser sheet thickness and time delay between laser pulses, may only apply for a small part of the measurement field. Typically, such errors will show up prominently in measurements of the Reynolds stresses. Another common practical difficulty of using PIV in complex geometries is adequate optical access, both from the standpoint of introducing a laser sheet and taking camera images. One way to overcome this problem is to construct the flow facility out of a transparent material, and use a working fluid with the same optical refractive index as the solid surfaces. Recently, this was successfully achieved by *Uzol* et al. [12.172] in a complex axial turbo-pump flow facility. The authors constructed the rotor, stator and body completely out of transparent acrylic, and used sodium iodide, mixed with water to the right proportion, to match the refractive index of the acrylic.

Several methods exist for the measurement of instantaneous velocity field measurements in a volume, such as holographic techniques. A relatively new technique that is showing promise for whole field measurements in complex boundary layers is four-dimensional (4-D) magnetic resonance velocimetry (4-D MRV) developed by *Elkins* et al. [12.173]. The method is an adaptation of a medical magnetic resonance imaging (MRI) system and provides a non-invasive measurement of three-component mean velocity fields. *Elkins* et al. [12.173] demonstrated the technique in a complex geometry with multiple 180 degree bends. For turbulence measurements in complex geometries, several workers have also proven the effectiveness of holographic PIV. See for example, *Tao*, *Katz* and *Meneveau* [12.174] who made measurements in a turbulent duct flow, and *Konrad*, *Schroder* and *Limberg* [12.175] who made measurements in an internal combustion engine cylinder. Further advances in hologram imaging techniques will likely see the increased use of this method in the future, and possibly for non-canonical turbulent boundary-layer studies.

References

12.1 H. Schlichting, K. Gersten: *Boundary Layer Theory* (Springer, Berlin–Heidelberg 2000)
12.2 H. Tennekes, J. Lumley: *A First Course in Turbulence* (MIT Press, Boston 1972)
12.3 S.B. Pope: *Turbulent Flow* (Cambridge Univ. Press, Cambridge 2000)
12.4 A. Sahay: *The Mean Velocity and the Reynolds Shear Stress in Turbulent Channel and Pipe Flow* (Yale Univ. Press, Cambridge 1997)
12.5 J.C. Klewicki, R.E. Falco: On accurately measuring statistics associated with small scale structure in turbulent boundary layers using

hot-wire probes, J. Fluid Mech. **219**, 119–142 (1990)

12.6 R.D. Moser, J. Kim, N.N. Mansour: Direct numerical simulation of turbulent channel flow up to $R_\tau = 590$, Phys. Fluids **11**, 943–954 (1999)

12.7 T. Wei, P. Fife, J. Klewicki, P. McMurtry: Properties of the mean momentum balance in turbulent boundary layer, pipe and channel flows, J. Fluid Mech. **522**, 303–327 (2005)

12.8 A.A. Townsend: *The Structure of Turbulent Shear Flow* (Cambridge Univ. Press, Cambridge 1976)

12.9 R.W. Fox, A.T. McDonald: *Introduction to Fluid Mechanics* (Wiley, New York 2000)

12.10 P. Fife, T. Wei, J. Klewicki, P. McMurtry: Stress gradient balance layers and scale hierarchies in wall-bounded turbulent flows, J. Fluid Mech. **532**, 165–189 (2005)

12.11 A.V. Johansson, P.H. Alfredsson: Effects of imperfect spatial resolution on measurements of wall-bounded turbulent shear flows, J. Fluid Mech. **137**, 409–421 (1983)

12.12 P. Ligrani, P. Bradshaw: Spatial resolution and measurement of turbulence in the viscous sublayer using subminiature hot-wire probes, Exp. Fluids **5**, 407–417 (1987)

12.13 D.B. DeGraaff, J.K. Eaton: Reynolds number scaling of the flat plate turbulent boundary layer, J. Fluid Mech. **422**, 319–346 (2000)

12.14 S.-R. Park, J.M. Wallace: The influence of instantaneous velocity gradients on turbulence properties measured with multi-sensor hot-wire probes, Exp. Fluids **16**, 17–26 (1993)

12.15 R.A. Antonia, Y. Zhu, J. Kim: On the measurement of lateral velocity derivatives in turbulent flows, Exp. Fluids **15**, 65–69 (1993)

12.16 A.B. Folz: *An Experimental Study of the Near-Surface Turbulence in the Atmospheric Boundary Layer* (Univ. of Maryland, College Park 1997)

12.17 C.G. Lomas: *Fundamentals of Hot Wire Anemometry* (Cambridge Univ. Press, Cambridge 1986)

12.18 H.-E. Albrecht, N. Damaschke, M. Borys, C. Tropea: *Laser Doppler and Phase Doppler Measurement Techniques* (Springer, Berlin-Heidelberg 2003)

12.19 M. Raffel, C. Willert, J. Kompenhans: *Particle Image Velocimetry: A Practical Guide* (Springer, Berlin-Heidelberg 2001)

12.20 L.H. Tanner, L.G. Blows: A study of the motion of oil films on surfaces in air flow, with application to the measurement of skin friction, J. Phys. E. **9**, 194–202 (1976)

12.21 J.W. Naughton, M. Sheplak: Modern developments in shear-stress measurement, Prog. Aerospace Sci. **38**, 515–570 (2002)

12.22 J.D. Ruedi, H. Nagib, J. Osterlund, P. Monkewitz: Evaluation of three techniques for wall-shear measurements in three-dimensional flows, Expts. Fluids **35**, 389–396 (2003)

12.23 G.G. Zilliac: Further developments of the fringe-imaging skin friction technique, NASA CR-1100425 (NASA, Washington 1996)

12.24 J.L. Brown and J.W. Naughton: The thin-oil-film equation, NASA/TM 1999-208767 (NASA, Washington 1999)

12.25 H.H. Fernholz, G. Janke, M. Schober, P.M. Wagner, D. Warnack: New developments and applications of skin-friction measuring techniques, Meas. Sci. Technol. **7**, 1396–1409 (1996)

12.26 D.J. Monson, H. Higuchi: Skin friction measurement by a dual-laser-beam interferometer technique, AIAA J. **19**, 739–744 (1981)

12.27 J.D. Murphy, R.V. Westphal: The laser interferometer skin-friction meter: a numerical and experimental study, J. Phys. E. **19**, 744–751 (1986)

12.28 K.-S. Kim, G.S. Settles: Skin friction measurements by laser interferometry in shock/boundary-layer interactions, AIAA J. **28**, 133–139 (1990)

12.29 J.H. Harotinidis: The measurement of wall shear stress. In: *In: Advances in Fluid Mechanics*, ed. by M. Gad-el-Hak (Springer, Berlin-Heidelber 1989) pp. 229–261

12.30 K. Lien, J. Monty, M.S. Chong, A. Ooi: The entrance length for fully developed turbulent channel flow. In: *Proceedings of the 15th Australasian Fluid Mechanics Conference*, ed. by M. Behnia, W. Lin, G. McBain (University of Sydney, Sydney 2004)

12.31 F.M. White: *Viscous Flow* (McGraw Hill, New York 2002)

12.32 A.V. Boiko, G.R. Grek, A.V. Dovgal, V.V. Kozlov: *The Origin of Turbulence in Near-Wall Flows* (Springer, Berlin-Heidelberg 2002)

12.33 R. Shaw: The influence of hole dimensions on static pressure measurements, J. Fluid Mech. **7**, 550–564 (1960)

12.34 R.E. Franklin, J.M. Wallace: Absolute measurements of static-hole error using flush transducers, J. Fluid Mech. **42**, 33–48 (1970)

12.35 C. Ducruet, A. Dyment: The pressure hole problem, J. Fluid Mech. **142**, 251–267 (1984)

12.36 B.J. McKeon, A.J. Smits: Static pressure correction in high Reynolds number fully developed turbulent pipe flow, Meas. Sci. Tech. **13**, 1608–1614 (2002)

12.37 B.J. McKeon: *High Reynolds Number Turbulent Pipe Flow* (Princeton Univ. Press, Princeton 2003)

12.38 M.J. Walsh: Riblets. In: *In: Viscous Drag Reduction in Boundary Layers*, ed. by D.M. Bushnell, J.N. Hefner (AIAA Progress in Astronautics and Aeronautics, Reston 1989) pp. 203–254

12.39 J.H. Preston: The determination of turbulent skin friction by means of Pitot tubes, J. R. Aeronaut. Soc. **58**, 109–121 (1953)

12.40 V.C. Patel: Calibration of the Preston tube limitation on its use in pressure gradients, J. Fluid Mech. **23**, 185–208 (1965)

12.41 M.R. Head, V.V. Ram: Simplified presentation of Preston tube calibration, Aeronaut. Q. **22**, 295–300 (1971)

12.42 T.J. Hanratty, J.A. Campbell: Measurement of wall shear stress. In: *Fluid Mechanics Measurements*, ed. by R.J. Goldstein (Taylor Francis, 1996)

12.43 T.G. Johansson and L. Castillo: Near-wall measurements in turbulent boundary layers using laser Doppler anemometry, ASME FEDSM2002-31070 (ASME, Metals Park 2002)

12.44 H.M. Nagib and K.A. Chauhan and P.A. Monkewitz: Scaling of high Reynolds number boundary layers revisited, AIAA 2005-4810 (2005)

12.45 F.H. Clauser: Turbulent boundary layers in adverse pressure gradients, J. Aeronaut. Sci. **21**, 91–108 (1954)

12.46 H. Nagib and C. Christophorou and J.-D. Reidi and P. Monkewitz and J. Osterlund and S. Gravante: Can we ever rely on results from wall-bounded turbulent flows without direct measurements of wall shear stress?, AIAA 2004-392 (2004)

12.47 D. Coles: Turbulent Boundary Layers in Pressure Gradients: A Survey Lecture. In: *Proceedings of the 1968 AFOSR-IFP-Stanford Conference on Computation of Turbulent Boundary Layers*, ed. by D. Coles, H. Hirst (Stanford University, Stanford 1969)

12.48 F.S. Sherman: *Viscous Flow* (McGraw-Hill, New York 1990)

12.49 I. Wygnanski, Y. Katz, E. Horev: On the applicability of various scaling laws to the turbulent wall-jet, J. Fluid Mech. **234**, 669–690 (1992)

12.50 A.E. Perry, J.D. Li: Experimental support for the attached-eddy hypothesis in zero-pressure gradient turbulent boundary layers, J. Fluid Mech. **218**, 405–438 (1990)

12.51 D. Coles: The law of the wake in the turbulent boundary layer, J. Fluid Mech. **1**, 191–226 (1956)

12.52 A. Cenedese, G.P. Romano, R.A. Antonia: A comment on the "linear" law of the wall for fully developed turbulent channel flow, Exp. Fluids **25**, 165–170 (1998)

12.53 F. Durst, J. Jovanovic, J. Sender: LDA measurements in the near-wall region of a turbulent pipe flow, J. Fluid Mech. **295**, 305–335 (1995)

12.54 N. Hutchins, K.-S. Choi: Accurate measurements of local skin friction coefficient using hot-wire anemometry, Prog. Aerospace Sci. **38**, 421–446 (2002)

12.55 Y.T. Chew, B.C. Koo, G.L. Li: An investigation of wall effects on hot-wire measurements using a bent sublayer probe, Meas. Sci. Technol. **9**, 67–85 (1994)

12.56 Y.T. Chew, B.C. Koo, G.L. Li: A time-resolved hot-wire shear stress probe for turbulent flow: Use of laminar flow calibration, Exp. Fluids **17**, 75–83 (1994)

12.57 Z. Yue, T.G. Malmstrom: A simple method for low-speed hot-wire anemometer calibration, Meas. Sci. Technol. **9**, 1506–1510 (1998)

12.58 J.M. Allen: Systematic study of error sources in supersonic skin-friction balance measurements, NASA TN D-8291 (1976)

12.59 K.G. Winter: An outline of the techniques available for the measurement of skin friction in turbulent boundary layers, Prog. Aerospace Sci. **18**, 1–57 (1977)

12.60 J.M. Allen: Experimental study of error sources in skin friction balance measurements, J. Fluids Eng. **99**, 197–204 (1977)

12.61 M. Acharya: Development of a floating element for measurement of surface shear stress, AIAA J. **23**, 410–415 (1985)

12.62 D. Frei, H. Thomann: Direct measurements of skin friction in a turbulent boundary layer with strong adverse pressure gradient, J. Fluid Mech. **101**, 79–95 (1980)

12.63 M.A. Schmidt, R.T. Howe, S.D. Senturia, J.H. Haritonidis: Design and calibration of a microfabricated floating-element shear-stress sensor, IEEE Trans. Electron. Dev. **35**, 750–757 (1988)

12.64 L. Lofdahl, M. Gad-el-Hak: MEMS-based pressure and shear stress sensors for turbulent flows, Meas. Sci. Technol. **10**, 665–686 (1999)

12.65 W. Heuer, I. Marusic: Turbulence wall-shear stress sensor for the atmospheric surface layer, Meas. Sci. Technol. **16**, 1644–1649 (2005)

12.66 P.H. Alfredsson, A.V. Johansson, J.H. Haritonidis, H. Eckelmann: The fluctuating wall-shear stress and the velocity field in the viscous sublayer, Phys. Fluids **31**, 1026–1033 (1988)

12.67 V.A. Sanborn: *Resistance Temperature Transducers* (Metrology, Fort Collins 1972)

12.68 K. Collela, W. Keith: Measurements and scaling of wall shear stress fluctuations, Experiments in Fluids **34**, 253–260 (2003)

12.69 A.N. Menendez, B.R. Ramaprian: The use of flush-mounted hot-film gages to measure skin friction in unsteady boundary layers, J. Fluid Mech. **161**, 139–159 (1985)

12.70 D.B. DeGraaff, J.K. Eaton: A high-resolution laser Doppler anemometer: Design, qualification and uncertainty, Exp. Fluids **30**, 522–530 (2000)

12.71 F. Durst, M. Fischer, J. Jovanovic, H. Kikura: Methods to set up and investigate low Reynolds number, fully developed turbulent plane channel flows, J. Fluid Eng. **120**, 496–503 (1998)

12.72 R.B. Hill, J.C. Klewicki: Data reduction methods for flow tagging velocity measurements, Expts. Fluids **20**, 142–151 (1996)

12.73 J.C. Klewicki, R.B. Hill: Spatial structure of negative $\partial u/\partial y$ in a low R_θ turbulent boundary layer, J. Fluid Eng. **120**, 773–785 (1998)

12.74 E.M. Thurlow, J.C. Klewicki: An experimental study of turbulent Poiseuille-Couette flow, Phys. Fluids **12**, 865–875 (2000)

12.75 C.P. Gendrich, M.M. Koochesfahani, D.G. Nocera: Molecular tagging velocimetry and other

12.75 novel applications of a new phosphorescent supramolecule, Exp. Fluids **23**, 361–372 (1997)

12.76 M.M. Koochesfahani: Molecular tagging velocimetry: Progress and applications, AIAA paper no. 1999-3786 (1999)

12.77 D.G. Bohl, M.M. Koochesfahani: Molecular tagging velocimetry measurements of axial flow in a concentrated vortex core, Phys. Fluids **16**, 4185–4191 (2004)

12.78 M.V. Morkovin: On the many faces of transition. In: *Viscous Drag Reduction*, ed. by C.S. Wells (Plenum, New York 1969)

12.79 H.L. Reed, W.S. Saric, D. Arnal: Linear Stability Theory Applied to Boundary Layers, Ann. Rev. Fluid Mech. **28**, 389–428 (1996)

12.80 P.G. Drazin: *Introduction to Hydrodynamic Stability* (Cambridge Univ. Press, Cambridge 2002)

12.81 W.O. Criminale, T.L. Jackson, R.D. Joslin: *Theory and Computation in Hydrodynamic Stability* (Cambridge Univ. Press, Cambridge 2003)

12.82 E. Reshotko: Boundary-layer stability and transition, Ann. Rev. Fluid Mech. **8**, 311 (1976)

12.83 E. Reshotko: Boundary layer instability, transition, and control, AIAA 94-0001 (1994)

12.84 E. Reshotko: A Program for Transition Research, AIAA J. **13**, 261–265 (1975)

12.85 G.B. Schubauer, H.K. Skramstad: Laminar boundary-layer oscillations and transition on a flat plate. J. Res. Nat'l. Bur. Stand. 38 (1947) 251

12.86 P.S. Klebanoff, K.D. Tidstrom, L.M. Sargent: The three-dimensional nature of boundary-layer instability, J. Fluid Mech. **12**, 1 (1962)

12.87 W.S. Saric, G.A. Reynolds: Experiments on the nonlinear stability of waves in a boundary layer. In: *Laminar-Turbulent Transition*, Vol. I, ed. by R. Eppler, H. Fasel (Springer, Berlin-Heidelberg 1980)

12.88 N. Lin, H.L. Reed, W.S. Saric: Effect of leading edge geometry on boundary-layer receptivity to freestream sound. In: *Stability, Transition and Turbulence*, ed. by M.Y. Hussaini, A. Kumar, C.L. Street (Springer, New York 1992)

12.89 W.S. Saric, W. Wei, B.K. Rasmussen: Effect of leading edge on sound receptivity. In: *Laminar-Turbulent Transition Vol. IV*, ed. by R. Kobayashi (Proc. IUTAM Symp., Sendai, Japan 1995)

12.90 W.S. Saric, E. White: Influence of High-Amplitude Noise on Boundary-Layer Transition to Turbulence, AIAA 98-2645 (1998)

12.91 T. Herbert: Three-dimensional phenomena in the transitional flat-plate boundary layer, AIAA 85-0489 (1985)

12.92 B.G.B. Klingmann, A.V. Boiko, K.J.A. Westin, V.V. Kozlov, P.H. Alfredsson: Experiments on the stability of Tollmien-Schlichting waves, Eur. J. Mech. B Fluids **12**, 493 (1993)

12.93 F.P. Bertolotti, T. Herbert, P.R. Spalart: Linear and nonlinear stability of the Blasius boundary layer, J. Fluid Mech. **242**, 44 (1992)

12.94 W.S. Saric: Low-speed experiments: Requirements for stability measurements. In: *Instability and Transition*, Vol. I, ed. by Y. Hussaini (Springer, Berlin-Heidelberg 1990) p. 174

12.95 J.A. Ross, F.H. Barnes, J.G. Burns, M.A.S. Ross: The flat plate boundary layer, Part 3 Comparison of theory with experiment, J. Fluid Mech. **43**, 819 (1970)

12.96 B. Müller, H. Bippes: Experimental Study of Instability Modes in a Three-Dimensional Boundary Layer, In Fluid Dynamics of Three-Dimensional Turbulent Shear Flows and Transition, AGARD CP 438 (1988) 13-1 – 13-15

12.97 A.M. Naguib, S.P. Gravante, C.E. Wark: Extraction of turbulent wall-pressure time-series using an optimal filtering scheme, Exp. Fluids **22**, 14–22 (1996)

12.98 W.S. Saric, E. Reshotko: Review of Flow Quality Issues in Wind Tunnel Testing, AIAA 98-2613 (1998)

12.99 W.S. Saric, S. Takagi, M.C. Mousseux: The ASU Unsteady Wind Tunnel and fundamental requirements for freestream turbulence measurements, AIAA 88-0053 (1988)

12.100 J.M. Kendall: Boundary layer receptivity to freestream turbulence, AIAA 90-1504 (1990)

12.101 W.S. Saric, A.S.W. Thomas: Experiments on the subharmonic route to turbulence in boundary layers. In: *Turbulence and Chaotic Phenomena in Fluids*, ed. by T. Tatsumi (North-Holland, Amsterdam 1984)

12.102 T.C. Corke, R.A. Mangano: Resonant growth of three-dimensional modes in transitioning Blasius boundary layers, J. Fluid Mech. **209**, 93 (1989)

12.103 M. Nishioka, S. Iida, Y. Ichikawa: An experimental investigation of the stability of plane Poiseuille flow, J. Fluid Mech. **72**, 731 (1975)

12.104 J.B. Anders, R.F. Blackwelder: Longitudinal Vortices in a Transitioning Boundary Layer. In: *Laminar-Turbulent Transition Vol. I.*, ed. by R. Eppler, H. Fase (Springer, Berlin-Heidelberg 1980)

12.105 B.A. Singer, H.L. Reed, J.H. Ferziger: Effect of streamwise vortices on transition in plane channel flow, Phys. Fluids A **1**, 1960 (1989)

12.106 W.S. Saric, J.A. Hoos, R.H. Radeztsky Jr: Boundary-layer receptivity of sound with roughness. In: *Boundary Layer Stability and Transition FED-Vol. 114*, ed. by D.C. Reda, H.L. Reed, R. Kobayashi (ASME, New York 1991)

12.107 C.F. Knapp, P.J. Roache: A combined visual and hot-wire anemometer investigation of boundary-layer transition, AIAA J. **6**, 29 (1968)

12.108 K.R. Sreenivasan, S. Raghu, B.T. Chu: The control of pressure oscillations in combustion and fluid dynamical systems, AIAA 85-0540 (1985)

12.109 Yu.S. Kachanov, V.Ya. Levchenko: Resonant interactions of disturbances in transition to turbulence in a boundary layer, J. Fluid Mech. **138**, 209 (1984)

12.110 L.M. Mack: Line sources of instability waves in a Blasius boundary layer, AIAA 84-0168 (1984)

12.111 P.T. Pupator, W.S. Saric: Control of Random Disturbances in a Boundary Layer, AIAA 89-1007 (1989)

12.112 R.W. Wlezien and D.E. Parekh and T.C. Island: Measurement of acoustic receptivity at leading edges and porous strips, Appl. Mech. Rev., 43;S167 (1990)

12.113 W.S. Saric, H.L. Reed, E.J. Kerschen: Boundary-Layer Receptivity to Freestream Disturbances, Ann. Rev. Fluid Mech. Vol. **34**, 291–319 (2002)

12.114 E.B. White, W.S. Saric, R.H. Radeztsky Jr: Leading-Edge Acoustic Receptivity Measurements Using a Pulsed-Sound Technique. In: *Laminar-Turbulent Transition*, Vol. V, ed. by W. Saric, H. Fasel (Springer, Berlin-Heidelberg 2000) pp. 103–108

12.115 R.H. Radeztsky Jr., M.S. Reibert, S. Takagi: A software solution to temperature-induced hot-wire voltage drift, Proc. 3rd International Symposium on Thermal Anemometry, ASME-FED (June 1993)

12.116 J.R. Dagenhart, W.S. Saric, M.C. Mousseux, J.P. Stack: Crossflow-vortex instability and transition on a 45-degree swept wing, AIAA 89-1892 (1989)

12.117 S.M. Mangalam, D.V. Maddalon, W.S. Saric, N.K. Agarwal: Measurements of crossflow vortices, attachment-line flow, and transition using microthin hot films, AIAA 90-1636 (1990)

12.118 H. Deyhle, G. Höhler, H. Bippes: Experimental investigation of instability wave-propagation in a 3-D boundary-layer flow, AIAA J. **31**, 637 (1993)

12.119 T.C. Corke, D Koga, R. Drubka, H. Nagib: A new technique for introducing controlled sheets of smoke streaklines in wind tunnels, Proc. Int'l Congress on Instrumentation in Aerospace Simulation Facilities, IEEE Pub. 77 (1974) CH 1250-8, AES

12.120 W.S. Saric: Visualization of different transition mechanisms, Phys. Fluids **29**, 2770 (1986)

12.121 D.C. Reda: Observations of dynamic stall phenomena using liquid crystal coating, AIAA J. **29**, 308 (1991)

12.122 W.S. Saric, H.L. Reed, D.W. Banks: Flight Testing of Laminar Flow Control in High-Speed Boundary Layers, RTO-MP-AVT-111/RSM (2005)

12.123 S. Zuccher, W.S. Saric: Infrared Thermography Investigations in Transitional Supersonic Boundary Layers. Exps. Fluids. In Press (2006)

12.124 R.H. Radeztsky, Jr, M.S. Reibert, W.S. Saric: Development of stationary crossflow vortices on a swept wing, AIAA 94-2374 (1994)

12.125 R.H. Radeztsky, Jr, M.S. Reibert, W.S. Saric, S. Takagi: Effect of micron-sized roughness on transition in swept-wing flows, AIAA 93-0076 (1993)

12.126 J.K. Eaton: Effects of mean flow three dimensionality on turbulent boundary layer structure, AIAA J. **33**, 2020–2025 (1995)

12.127 R.L. Simpson: Turbulent boundary layer separation, Annu. Rev. Fluid Mech. **21**, 205–234 (1989)

12.128 S.H. Chue: Pressure probes for fluid measurements, Prog. Aerospace Sci. **16**, 147–223 (1975)

12.129 D.W. Bryer, R.C. Pankhurst: *Pressure-probe methods for determining wind speed and flow direction* (National Physical Laboratory, Teddington 1971)

12.130 R.G. Dominy, H.P. Hodson: An investigation of factors influencing the calibration of five-hole probes for three-dimensional flow measurements, J. Turbomach. **115**, 513–519 (1993)

12.131 G.G. Zilliac: Modelling, calibration and error analysis of seven-hole pressure probes, Exp. Fluids **14**, 104–120 (1993)

12.132 O.K. Rediniotis, R. Vijayagopal: Miniature multi-hole pressure probes and their neural-network-based calibration, AIAA J. **37**, 666–674 (1999)

12.133 C.W. Wenger, W.J. Devenport: Seven-hole pressure probe calibration method utilizing look-up error tables, AIAA J. **37**, 675–679 (1999)

12.134 A.J. Pisasale, N.A. Ahmed: Examining the effect of flow reversal on seven-hole probe measurements, AIAA J. **41**, 2460–2467 (2003)

12.135 O.K. Rediniotis, R.E. Kinser: Development of a nearly omnidirectional velocity measurement pressure probe, AIAA J. **36**, 1854–1860 (1998)

12.136 J.D. Vagt: Experimental techniques in three-dimensional turbulent boundary layers. In: *IUTAM Symposium: Three-dimensional turbulent boundary layers*, ed. by H.H. Fernholz, E. Krause (Springer, New York 1982)

12.137 H.H. Fernholz and E. Krause: Three-dimensional turbulent boundary layers, IUTAM Symposium, Berlin, (1982)

12.138 P.M. Ligrani, B.A. Singer, L.R. Baun: Miniature five-hole pressure probe for measurements of three mean velocity components in low-speed flows, J. Phy. E: Sci. Instrum. **22**, 868–876 (1989)

12.139 L. Pompeo, M.S.G. Bettelini, H. Thomann: Laterally strained turbulent boundary layers near a plane of symmetry, J. Fluid Mech. **257**, 507–532 (1993)

12.140 R.V. Westphal, M. Prather, and M. Toyooka: Rotatable single hole pressure probe for flow velocity and direction, AIAA-2002-3138 (2002)

12.141 D. Driver, S. Hebbar: Experimental study of a three-dimensional, shear-driven, turbulent boundary layer, AIAA J. **25**, 35–42 (1987)

12.142 K. Flack, J. Johnston: Near-wall flow in a three-dimensional turbulent boundary layer on the endwall of a rectangular bend, ASME Fluids Eng. Div. **184**, 1–19 (1994)

12.143 J. Johnston, K. Flack: Advances in three-dimensional turbulent boundary layers with an emphasis on the wall-layer regions, ASME J. Fluids Eng. **118**, 219–232 (1996)

12.144 D.B. DeGraaff, J.K. Eaton: Reynolds number scaling of the flat plate turbulent boundary layer, J. Fluid Mech. **422**, 319–346 (2000)

12.145 S. Song, J.K. Eaton: Reynolds number effects on a turbulent boundary layer with separation, reattachment, and recovery, Exp. Fluids **36**, 246–258 (2004)

12.146 K.T. Lowe, R.L. Simpson: Measurements of velocity-acceleration statistics in turbulent boundary layers, Int. J. Heat Fluid Flow **27**, 558–565 (2006)

12.147 R.L. Simpson, Y.T. Chew, B.G. Shivaprasad: The structure of a separating turbulent boundary layer. Part 1. Mean flow and Reynolds stresses, J. Fluid Mech. **113**, 23–51 (1981)

12.148 S.M. Olcmen, R.L. Simpson: An experimental study of a three-dimensional pressure-driven turbulent boundary layer, J. Fluid Mech. **290**, 225–262 (1995)

12.149 D.R. Webster, D.B. DeGraaff, J.K. Eaton: Turbulence characteristics of a boundary layer over a swept bump, J. Fluid Mech. **323**, 1–22 (1996)

12.150 D.A. Compton, J.K. Eaton: Near-wall measurements in a three-dimensional turbulent boundary layer, J. Fluid Mech. **350**, 189–208 (1997)

12.151 H.H. Bruun: *Hot-wire Anemometry. Principles and Signal Analysis* (Oxford Univ. Press, Oxford 1995)

12.152 I. Marusic, A.E. Perry: A wall-wake model for the turbulence structure of boundary layers. Part 2. Further experimental support, J. Fluid Mech. **298**, 389–407 (1995)

12.153 F.H. Champagne, C.A. Sleicher, O.H. Wehrmann: Turbulence measurements with inclined hot-wires. Part 1. heat transfer experiments with inclined hot-wire, J. Fluid Mech. **28**, 153–175 (1967)

12.154 F.H. Champagne, C.A. Sleicher: Turbulence measurements with inclined hot-wires. Part 2. hot-wire response equations, J. Fluid Mech. **28**, 177–182 (1967)

12.155 A.E. Perry: *Hot-wire Anemometry* (Clarendon, Oxford 1982)

12.156 B.J. Cantwell, D. Coles: An experimental study of entrainment and transport in the turbulent near wake of a circular cylinder, J. Fluid Mech. **136**, 321–374 (1983)

12.157 A.E. Perry, J.H. Watmuff: The phase-averaged large-scale structures in three-dimensional turbulent wakes, J. Fluid Mech. **103**, 33–51 (1981)

12.158 A.E. Perry, J.D. Li: Experimental support for the attached eddy hypothesis in zero-pressure-gradient turbulent boundary layers, J. Fluid Mech. **218**, 405–438 (1990)

12.159 A.E. Perry, K.L. Lim, S.M. Henbest: An experimental study of the turbulence structure in smooth- and rough-wall boundary layers, J. Fluid Mech. **177**, 437–466 (1987)

12.160 P.E. Skåre, P.-Å. Krogstad: A turbulent boundary layer near separation, J. Fluid Mech. **272**, 319–348 (1994)

12.161 K. Elsberry, J. Loeffler, M.D. Zhou, I. Wygnanski: An experimental study of a boundary layer that is maintained on the verge of separation, J. Fluid Mech. **423**, 227–261 (2000)

12.162 R.V. Westphal, J.K. Eaton, J.P. Johnston: A new probe for measurement of velocity and wall shear stress in unsteady, reversing flow, J. Fluids Eng. **103**, 478–482 (1981)

12.163 L.J.S. Bradbury, I.P. Castro: A pulsed-wire technique for velocity measurements in highly turbulent flows, J. Fluid Mech. **49**, 657–691 (1971)

12.164 I.P. Castro, M. Dianat: Pulsed-wire anemometry near walls, Exp. Fluids **8**, 343–352 (1990)

12.165 W.J. Devenport, G.P. Evans, E.P. Sutton: A traversing pulsed wire probe for measurements near a wall, Exp. Fluids **8**, 336–342 (1990)

12.166 M. Schober, P.E. Hancock, H. Siller: Pulsed-wire anemometry near walls, Exp. Fluids **25**, 151–159 (1998)

12.167 H.H. Fernholz, G. Janke, M. Schober, P.M. Wagner, D. Warnack: New developments and applications of skin-friction measuring techniques, Meas. Sci. Tech. **7**(10), 1396–1409 (1996)

12.168 P.G. Spazzini, G. Iuso, M. Onorato, N. Zurlo: Design, test and validation of a probe for time-resolved measurement of skin friction, Meas. Sci. Tech. **10**, 631–639 (1999)

12.169 Y. Li, A.M. Naguib: High-frequency oscillating hot-wire sensor for near-wall diagnostics in separated flows, AIAA J. **43**(3), 520–529 (2005)

12.170 R.O. Kiesow, M.W. Plesniak: Modification of near-wall turbulence structure in a shear-driven three-dimensional turbulent boundary layer, Exp. Fluids **25**, 233–242 (1998)

12.171 K.P. Angele: Pressure-based scaling in a separating turbulent APG boundary layer, In Proc. ETC9, Southampton, UK, 2002. (ed. I.P. Castro, P.E. Handcock and T.G. Thomas)

12.172 O. Uzol, Y.C. Chow, J. Katz, C. Meneveau: Unobstructed PIV measurements within an axial turbo-pump using liquid and blades with matched refractive indices, Exp. Fluids **33**, 909–919 (2002)

12.173 C.J. Elkins, M. Markl, N. Pelc, J.K. Eaton: 4D magnetic resonance velocimetry for mean velocity measurements in complex turbulent flows, Exp. Fluids **34**, 494–503 (2003)

12.174 B. Tao, J. Katz, C. Meneveau: Geometry and scale relationships in high Reynolds number turbulence determined from three-dimensional holographic velocimetry, Phys. Fluids **12**, 941 (2000)

12.175 R. Konrath, W. Schroder, W. and Limberg: Holographic particle image velocimetry applied to the flow within the cylinder of a four-valve internal combustion engine, Exp. Fluids **33**, 781–793 (2002)

13. Topological Considerations in Fluid Mechanics Measurements

The vector field variables of interest in experimental fluid mechanics include velocity, vorticity and wall shear stress. Additional vector fields of interest include the gradients of pressure and temperature. Each of these is subject to the constraints that can be placed on the isolated singular points (vector magnitude = 0) of the vector field that is projected onto a surface of interest. Identifying the relevant surfaces for a given flow and establishing, for a given identified surface, an a priori constraint on the isolated singular points of the relevant vector field, can provide experimentalists with a powerful diagnostic tool.

13.1	A Companion Document	909
13.2	Utilization of Topological Considerations for Flow Field Analyses	910
	13.2.1 A *Tabbed* Jet	910
	13.2.2 A Conical Flame Holder Geometry	911
	13.2.3 Flow Past a Small–Aspect-Ratio Protruding Cylinder	912
	13.2.4 A Vortex Ring–Wall Interaction	913
	13.2.5 An Annular Jet	915
	13.2.6 Summary	917
References		918

It is the purpose of this chapter of the handbook to enable the reader to utilize this diagnostic tool for any flow field of interest.

13.1 A Companion Document

The present author has published an article [13.1] that provides a detailed description of the steps that are required to develop and to utilize this diagnostic tool. It is expected that the interested reader will consult that reference for the supporting material that undergirds this presentation. The following items, whose focus is to develop the reader's capability to utilize this diagnostic tool, must first provide a synopsis of that more complete presentation. The following section then provides examples of the use of this diagnostic tool in the service of flow field analyses.

If the surface of interest is overlaid on a physical (no-slip) body, then the isolated singular points will satisfy the relationship

$$\chi_{\text{surface}} = \sum N - \sum S \qquad (13.1)$$

where χ is the Euler characteristic of the selected surface. The term *no-slip surface* was introduced by *Perry* and *Chong* [13.2] and [13.3] for such surfaces. Alternatively, [13.1] utilizes the term *body fitted* to emphasize the relationship of the diagnostic surface to the object of interest.

Given that any surface can be formed from a sphere into which N holes have been *punched* and/or M handles have been attached, as described in [13.1], the χ value for the surface can be established as

$$\chi_{\text{surface}} = 2 - \sum \text{holes} - 2 \sum \text{handles} . \qquad (13.2)$$

The connection between (13.1) and (13.2) is established by the condition that the Euler characteristic for a surface is equal to the sum of the indices for that surface and recognizing that a node (N) is represented by an index of $+1$ and a saddle (S) by an index of -1. [13.1] provides the identification technique to identify a node and a saddle as well as specific examples that clarify the definition of the *surfaces* to be analyzed via (13.1) and (13.2). Stated succinctly, the identification technique involves a unit vector that is, at every location of its base point, aligned with the vector under consideration. The base point of this unit vector is placed on a circle that surrounds the location of interest. The base is then caused to rotate in a clockwise direction around the circle.

- If the unit vector rotates clockwise during the circuit, then a *net* node is enclosed by the circle.
- If the unit vector rotates counterclockwise, then a *net* saddle is enclosed by the circle.

- If there is no net rotation, then there is no *net* singular point within the circle.

If a surface is selected that exists within a flow field (either fully or with boundaries that touch a physical surface), then (13.1) is generalized to

$$\chi_{\text{surface}} = 2\sum N + \sum N' - 2\sum S - \sum S'. \quad (13.3)$$

Such a surface is most easily visualized by considering an inflated balloon, gripping the balloon along *seams* that are contours in space that are tangent to the vectors, evacuating the balloon such that it collapses to a *two-sided (perhaps warped) plane*. A flattened *pancake* or *crepe* is a good simile for this surface. One or more holes may be added where the fluid can flow onto or leave the surface. (An important constraint for the vector field at a hole is: the vector field direction must be uniformly inward or outward at all points along the contour defined by the hole.) In this description, a singular point on the *seam* of the surface will appear once in the summation and, following *Hunt* et al. [13.4], the singular points on a seam are termed half-nodes or half-saddles. Singular points that are interior to the seams are *full nodes* or *full saddles* and these are counted twice in the summation as shown in (13.3). Specifically, they are counted once for each side of the collapsed sphere.

Numerous examples are provided in [13.1] as well as in the following section to clarify these concepts.

13.2 Utilization of Topological Considerations for Flow Field Analyses

In addition to the illustrative examples of the basic concepts, only two flows: *Koster* and *Müller* [13.6] plus *Zimmermann* et al. [13.7] and *Ruderich* and *Fernholz* [13.8] are considered in detail in [13.1]. The importance of the first group and the second reference is that: (i) the completeness of the inferred flow pattern was confirmed in the former, and (ii) the published flow pattern was corrected in the latter. Specifically, by applying the topological rule to different surfaces in each case, it was possible to extend the understanding of each flow beyond that provided in the original reference. The essential contribution of this Handbook section is to provide further examples that will aid the reader in the effective use of this diagnostic tool.

13.2.1 A *Tabbed* Jet

Velocity Measurements in the Tab's Wake
The addition of triangular tabs at the exit plane of a jet is an effective means of promoting rapid mixing between the jet fluid and the ambient fluid. Initial studies of the tabbed jet: *Bradbury* and *Khadem* [13.9], *Zaman*, et al. [13.10], and *Reeder* and *Samimy* [13.11], were extended by *Bohl* and *Foss* [13.12] with the addition of secondary tabs at the sides of the primary tab. Figure 13.1 shows the geometry for the primary plus secondary tabs. The Figures in this Section are taken from the MS thesis of *Bohl* [13.5]. (See *Bohl* and *Foss* [13.12] for the *open literature* source of the same information.)

The source of the strong streamwise vortex motions (that result in the near field mixing) was identified as the surface pressure distribution forward of the primary tab. The secondary tabs serve to promote this mixing by enhancing the outward trajectory of the streamwise vortex motions beyond the exit plane of the jet.

An X-array of hot-wire sensors was positioned at a regular array of y and z locations at two different downstream locations. Two time series measurements (u, v) and (u, w) were obtained over this grid of points. Using these data, the time-averaged $(\bar{v}, \bar{w})$ components of the velocity vector allowed the projected streamlines, in the planes of constant x, to be defined. A representative streamline pattern is shown in Fig. 13.2a.

The index identification technique, [13.1] and Sect. 13.1, was used to establish the indicated nodes (13.2) and saddles (13.2) that can be identified in the

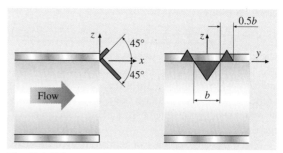

Fig. 13.1 Primary and secondary tabs (after *Bohl* [13.5])

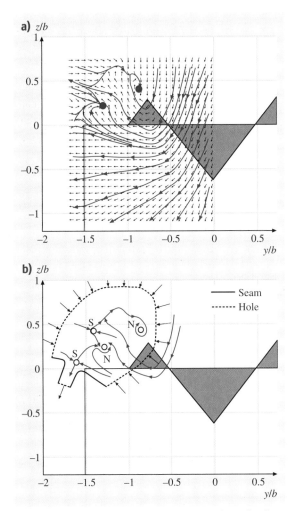

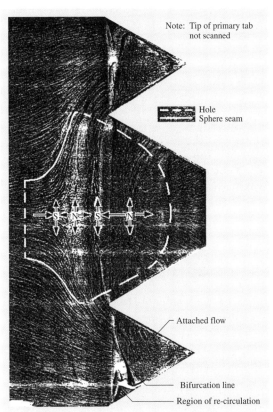

Fig. 13.2a,b Streamlines in the "collapsed surface" at $x/b = 1.2$ for the geometry of Fig. 13.1. (**a**) Measured velocity field with inferred streamlines. (**b**) Seams (*solid*) and holes (*dashed*) to analyze the flow of (**a**)

measured field. These singular points exist on the surface of the *collapsed sphere with two holes* that is also identified in this figure. Given that the collapsed sphere has two holes and given the two nodes and the two saddles, (13.3) is satisfied by the information in Fig. 13.2b. The self-consistency of these identifications gives confidence to these inferences from the experimental data.

Observations on the Approach and the Tab Surfaces

Surface streaking observations on the approach surface, plus the inclined primary tab were also made. This image

Fig. 13.3 Surface streak image (lampblack and kerosene) of the approach surface and the primary tab. The overlaid *dashed white lines* are the entry and exit holes. The seams (*solid lines*) are aligned with the limiting streamlines.

is shown in Fig. 13.3. A collapsed sphere can be identified (as shown in Fig. 13.3) with an entry and an exit hole. The two nodes and the two saddles contained within these borders are compatible with the sphere plus the two holes of Fig. 13.3. Note that a single surface formed as a disc to cover the same area could not be used since the exposed edge would not provide a unidirectional flow. (A disc is formed by adding one hole to a sphere and *flattening* the remainder of the original surface.)

A side view of the flow field's center plane can be inferred from the surface singular points (Fig. 13.4). This image, again a collapsed sphere with two holes, involves two nodes and four half-saddles that, in summation, are compatible with the $\chi = 0$ condition for (13.3). The contribution of the topological considerations is to infer the presence of the surface nodes in Fig. 13.3 as well as the nodes in Fig. 13.4 given the easily identified saddles of Fig. 13.3. The use of two surfaces to confidently

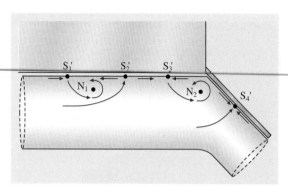

Fig. 13.4 A collapsed sphere in the center plane of the tabbed jet flow

characterize one flow field is the rule rather than an exception.

13.2.2 A Conical Flame Holder Geometry

The masters thesis work of *Holland* [13.13] examined the external flow past an axisymmetric flame holder geometry; see Fig. 13.5a for an elevation view of this axisymmetric configuration. The purpose of this investigation was to establish the baseline flow field for computation fluid dynamics (CFD) verification. The actual flame holder would involve a central tube that delivers fuel into the airstream. The flow field associated with Fig. 13.5a can be modeled with a body fitted surface as a sphere with one hole. A *sock*, which extends from the base of the conical section to the upper dashed line in Fig. 13.5a, defines this body fitted surface. Figure 13.5b identifies the one node (N_1) that satisfies the rule for this body fitted surface. There are, near the perimeter of the cone, two rings of singular points which do not contribute to the topology of this body fitted surface since they are not isolated. Their diameters d_2 and d_3 correspond to the S_2' and S_3' identifications of the collapsed surface singular points shown in Fig. 13.5c.

A collapsed-sphere surface, that can also be used to analyze this flow, is shown in Fig. 13.5a as the dashed line that includes the inlet and exiting flows. This surface has three holes. The dual ring of singular points is now involved in the counting as shown in Fig. 13.5c. The rule is satisfied as

$$\chi = 2 - 3 \text{ holes} = -1$$
$$= 2\sum N + \sum N' - 2\sum S - \sum S'$$
$$= 2(4) + 0 - 2(1) - 7 \,. \qquad (13.4)$$

It is noteworthy that the interior S_2', near the perimeter, which represents a separation location, was clearly evident in the surface streaking image. The middle half-saddle S_3' (which represents an attachment location) was quite difficult to discern. As shown in Fig. 13.5c, this middle S' location is clearly required to provide a smoothly continuous vector field.

It is left as an *interesting exercise* to consider the vector field in the presence of an exiting fuel flow which would replace the singular point S_1 in Fig. 13.5b. (Four holes lead to $\chi = -2 = 2(4) + 0 - 2(2) - 6$.)

13.2.3 Flow Past a Small-Aspect-Ratio Protruding Cylinder

Summer et al. [13.14] examined the wake region of a flow past cylinders of length (L) and diameter D with $L/D = 9, 7, 5$ and 3. The cylinders' axes were perpendicular to the surface to which they were attached. A schematic representation of this experiment is shown in Fig. 13.6.

Two representative images of the streamwise vorticity magnitudes, inferred from the (v, w) components as obtained from seven hole probe measurements, are shown in Fig. 13.7a,b for the $L/D = 5$ and $L/D = 3$ cases, respectively. It is apparent that the smaller aspect ratio case does not show the lower pair of streamwise vortex motions that are evident in Fig. 13.7a.

Figures 13.8a,b show the corresponding surfaces that can be used to represent these two flows. Namely, a sphere with one hole would form a disc which, when folded (like a taco shell) forms a two-sided semicircular disc. The straight portion of the folded disc is placed on the surface (that supports the cylinder) at the x-location of the measurements. The circular perimeter of the folded disc, which represents the hole, will experience an inflow as shown by the swirling flow pattern of the four vortex motions Fig. 13.7a and the two vortex motions Fig. 13.7b. (Alternatively, one can create the equivalent semicircular pattern with a collapsed sphere followed by opening a hole along the circumference of the semicircle.) The vectors at the perimeter (above the plate) are inward, in agreement with the spatially averaged condition, $\partial u/\partial x > 0$, for the plane of constant x in the wake.

The $L/D = 5$ vector field of Fig. 13.8a must be supplemented with inferred saddle points at lateral positions beyond the measurement domain. These are required to ensure a uniform inflow at the lateral boundary. With these additions, the topological constraint is satisfied as

$$\chi = \chi_{\text{sphere}} + 1 \text{ hole} = 1$$
$$= 2\sum N + \sum N' - 2\sum S - \sum S'$$
$$= 2(4) + 0 - 2(3) - 1 = 1 \,. \qquad (13.5)$$

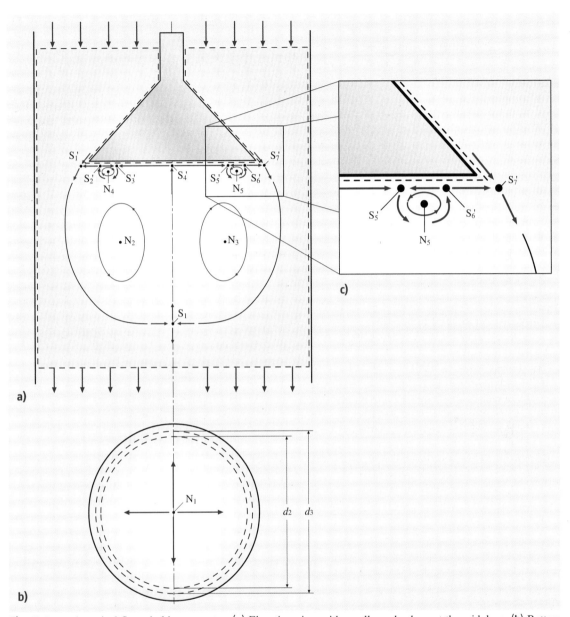

Fig. 13.5a–c A conical flame holder geometry. (**a**) Elevation view with a collapsed sphere at the midplane. (**b**) Bottom view with separation/reattachment rings. (**c**) Elevation view detail with separation/reattachment half-saddles

In contrast, the $L/D = 3$ flow field involves two streamwise vortex motions: $2N$ and three half-saddles at the lower boundary. The constraint is again satisfied as

$$\chi = 1 = 2\sum N + \sum N' - 2\sum S - \sum S'$$
$$= 2(2) + 0 - 0 - 3 = 1 \qquad (13.6)$$

13.2.4 A Vortex Ring–Wall Interaction

Gendrich et al. [13.15] utilized molecular tagging velocimetry (see Sect. 5.4 of this Handbook) to examine the kinematics of a vortex ring that impinges on a planar surface; see Fig. 13.9 for a schematic representation of this experiment. The associated velocity field, at an

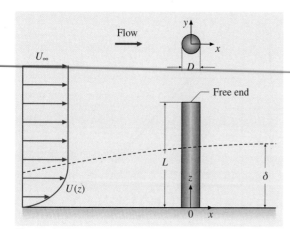

Fig. 13.6 The experiment of *Summer* et al. [13.14]

instant, is presented in Fig. 13.10. This flow shows the intriguing characteristic that, following the outward spread of the ring on the surface, a rebound effect is present in which perimeter fluid is swept back toward the ring's axis and lifted off the impact surface. A companion flow: two rings that collide with a common axis was studied by *Chu* et al. [13.16]. The present analysis would also be related to that flow.

The body fitted analysis for this flow is quite simple: a sphere with one hole forms a disc on the surface and the centered stagnation point describes the single node. The disc, a sphere with one hole (hence, $\chi = 1$), then satisfies the topological rule for both the outward, as well as the subsequent inward, flow at the disc's perimeter.

A mid-plane surface that contains the stagnation streamline provides a more interesting indication of this flow's topology. Using the domain defined by the contour in Fig. 13.11a, the surface can be described as a collapsed sphere plus three holes. That is, the χ value for Fig. 13.11a is -1. The axisymmetric nature of the vortex ring permits the streamlines, around the vortex core, to be drawn as closed contours. Seam 1, which follows this contour below the vortex, separates the inflow hole from the outflow holes. The centered stagnation point represents a half-saddle and, since the nodes of the ring vortex are excluded, the half-saddle satisfies the topological constraint as

$$\chi = -1 = 2\sum N + \sum N' - 2\sum S - \sum S'$$
$$= 0 + 0 - 0 - 1. \qquad (13.7)$$

Note that the $S' - N - S'$ pattern of the separation and reattachment, if it were included in the contour, would be self-canceling.

The instructive aspect of this example derives from the identification of seam 2 and the following considerations Fig. 13.11b.

The collapsed surface for seam 2 now encloses the node that is represented by the vortex motion and this implies the presence of two half-saddles to cancel the

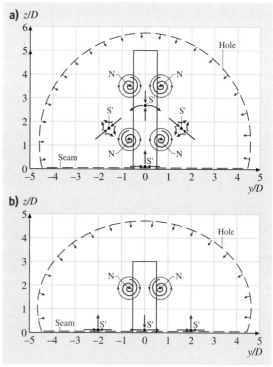

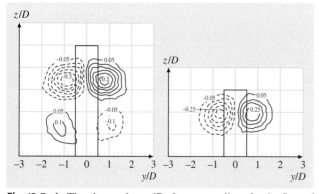

Fig. 13.7a,b The observed $x = 6D$ plane streamlines for the flow of Fig. 13.6. **(a)** $L/D = 5$; **(b)** $L/D = 3$. (After *Sumner* et al.[13.14])

Fig. 13.8a,b Topological analyses for the flows of Fig. 13.7 **(a)** $L/D = 5$; **(b)** $L/D = 3$

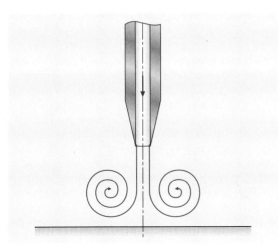

Fig. 13.9 Vortex ring–wall interaction, a schematic representation of the experiment

contribution of the node. However, in a planar representation, as viewed in the plane normal to the impact plate, it appears that the vector field is *as continuous, where seam 2 intersects the holes as it is for* seam 1.

To understand why there must be a half-saddle at the intersection of the inlet hole and the streamline that *covers* the vortex motion (see point P on Fig. 13.11b), whereas there is no half-saddle at point P for Fig. 13.11a, one can visualize the *collapsed surface* as being flattened at point P as shown in Fig. 13.12a,b. In these views, it is apparent that a half-saddle must be present at point P for the seam 2 case (Fig. 13.12b). A similar *flattening* at Q will also reveal a half-saddle for the seam 2 condition.

In contrast, a similar flattening for these same locations with seam 1 Fig. 13.12a would reveal no singular points at P and Q.

Hence, for the surface of seam 2, the rule is satisfied as:

$$\begin{aligned}\chi &= 2\sum N + \sum N' - 2\sum S - \sum S' \\ &= 2(2) + 0 - 0 - 5 \\ &= -1 \,.\end{aligned} \qquad (13.8)$$

13.2.5 An Annular Jet

Foss et al. [13.18] examined the velocity field in an annular jet formed by supporting a *streamlined plug* in the delivery tube of a round jet. The support was a cross-member attached to the perimeter of the delivery tube. The relevant features of the geometry for this flow are shown in Fig. 13.13.

The time-average separated, recirculating flow downwind from the plug is strongly three-dimensional as a result of the cross member. The present considerations will only address the vector field in a centered plane that is perpendicular to the cross-member. Particle image velocimetry (PIV) has been used to obtain 1000 images of the flow field downstream of the plug. Figure 13.14 presents the streamlines of the time-averaged field.

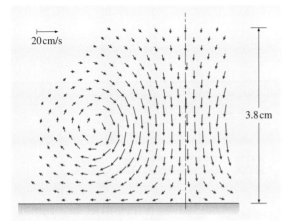

Fig. 13.10 A representative instantaneous velocity field from the experiment of Fig. 13.9. Courtesy of M. M. Koochesfahani [13.17]

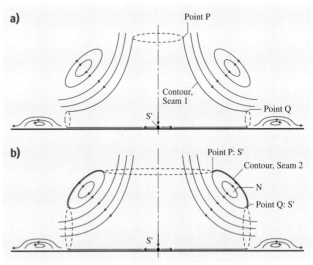

Fig. 13.11a,b The collapsed sphere plus two holes to describe the velocity field of Fig. 13.10. (**a**) A surface that excludes the right half-plane vortex (seam 1). (**b**) A surface that includes the right half-plane vortex (seam 2)

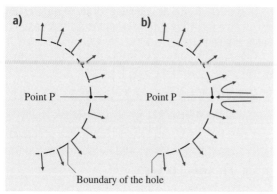

Fig. 13.12a,b A clarification of the absence/presence of a 1/2-saddle at point P in Fig. 13.11 for the surfaces of Fig. 13.11. (**a**) Absence of a half-saddle at point P for the seam 1 condition. (**b**) Presence of a half-saddle at point P for the seam 2 condition

Figure 13.15 provides a *skeleton* image of the relevant features (singular points) shown in Fig. 13.14 plus those associated with the plug itself. It also shows the domain for the topological analysis. Specifically, this collapsed sphere with two holes and one handle has an Euler characteristic of -2, which is satisfied as

$$\chi = -2 = 2\sum N + \sum N' - 2\sum S - \sum S'$$
$$= 2(3) + 0 - 2(2) - 4 \ . \qquad (13.9)$$

It is noteworthy that S'_1, S'_2 and S'_3 must be logically inferred since the PIV image cannot extend to the *near surface*. Also, S'_4 is suggested by the observed streamlines but it is not definitively shown which again represents the difficulty of PIV images being taken in the near-wall region.

The instantaneous realizations of the experiment *also* provide an opportunity to check the PIV data. In these cases, it cannot be expected that the regular (predictable) singular points of Fig. 13.15 will be observed. It is also possible that the software-generated streamlines may give a false image of the velocity field. Hence, it is of value to examine these realizations for their agreement with the appropriate topological constraint.

Figure 13.16a presents a *one-realization image* of this flow field. This figure is exceptionally busy given the intent to allow the multiple singular points to be identified. Also, its lower section has been blocked off given the ambiguities of the PIV data near the plugs' surface. The lower three holes and two seams that are shown bound the overall domain to be investigated. Note that the holes exhibit one-way flow and the seams are tangent to the streamlines. Hence, Fig. 13.16a shows a collapsed sphere with four holes. Again, the Euler characteristic is $\chi = -2$.

The terminations of the seams above the plug provide a vector field pattern that mimics that of Fig. 13.12b for the vortex ring analysis. Specifically, there are four S' values at the terminations of the two seams. Consequently, a correct vector

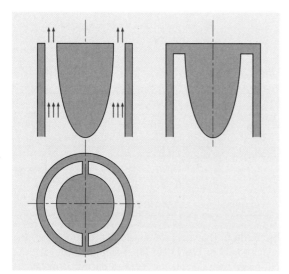

Fig. 13.13 The plug-jet geometry (after *Foss* et al.[13.18])

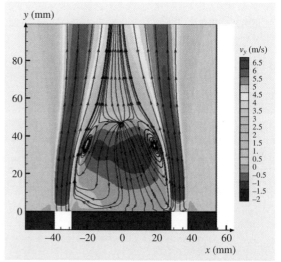

Fig. 13.14 The average velocity field (1000 realizations) in the center plane perpendicular to the support member of the plug-jet shown in Fig. 13.13. Note, x and y dimensions are in mm; V_y is the vertical velocity magnitude in m/s

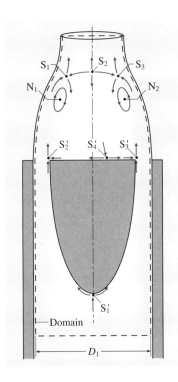

Fig. 13.15 The collapsed sphere that is used to evaluate the singular points for the flow of Fig. 13.14

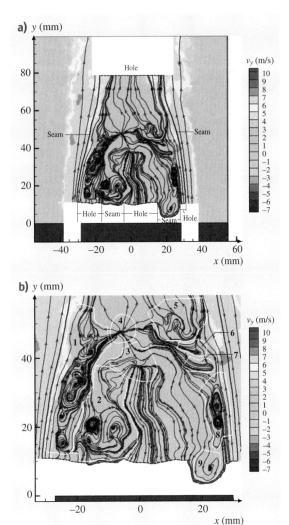

field above this lower surface will exhibit one net node.

An efficient protocol, to determine if the overall features of the vector field are correct, is to interrogate the nine subregions, identified in Fig. 13.16b, with the revolving unit vector ([13.1]). This protocol will determine if a net node or saddle exists within a given domain. It is obvious, for example, that both nodes and saddles exist within domain 1. It is, however, not readily apparent if these *cancel in pairs*.

The reader can verify that domains 1–3 and 5–9 exhibit no rotation of the unit vector as it revolves around the perimeter of those domains with the *base* of the vector on the domain boundary and its tip pointed in the direction of the streamline at the base point. In contrast, domain 4 exhibits a clockwise rotation of the unit vector as it executes that traverse. Hence, domain 4 indicates a net node.

The topological constraint is therefore satisfied as

$$\chi - 2 = 2\sum N + \sum N' - 2\sum S - \sum S'$$
$$= 2(1) + 0 - 0 - 4. \qquad (13.10)$$

Further analysis in the domains will show the self-canceling patterns of equal nodes and saddles. These considerations show that above the *lower cut-off zone* the

Fig. 13.16a,b An instantaneous realization of the velocity field for the plug-jet depicted in Fig. 13.13. (**a**) The selected domain for which reliable streamlines can be defined. Note, the lower boundary is defined by three holes and two seams at a nominal height of $x = 0.154D$. (**b**) Contour regions

indicated vector field satisfies the topological constraint and the PIV observations are justified.

13.2.6 Summary

The surface topology *rule* discussed in this section can ensure that the inferred pattern of singular points does not violate the topological constraints. As shown by the above examples and those of [13.1], the rule can also be used to extend the experimental observations by the plausible addition of singular points that are not observed. The agreement between multiple surfaces for a given flow is often required to gain confidence that the complete flow field, as typically inferred from incomplete or uncertain observations, is rational.

In addition to these positive attributes, it is appropriate to reiterate that a selected set of singular points could satisfy the rule but not represent the physical flow field. That is, the rule represents a necessary but not a sufficient constraint on the singular points of the flow field.

This section relies heavily on [13.1] for its basic elements. The examples provided herein are to give the researcher exposure to additional applications of the Rule such that its application in a new problem is readily accomplished. The half-saddles that are required for Fig. 13.11b and Fig. 13.12b (seam 2) were not recognized in [13.1]. Hence, this exposition also adds to the basic considerations covered previously.

References

13.1 J.F. Foss: Surface selections and topological constraint evaluations for flow field analyses, Exp. Fluids **37**, 883–898 (2004)

13.2 A.E. Perry, M.S. Chong: A description of eddying motions and flow patterns using critical-point concepts, Ann. Rev. Fluid Mech. **19**, 125–155 (1987)

13.3 A.E. Perry, M.S. Chong: Topology of flow patterns in vortex motions and turbulence, Appl. Sci. Res. **55**, 357–374 (1994)

13.4 J.C.R. Hunt, C.J. Abell, J.A. Peterka, H. Woo: Kinematical studies of the flows around free or surface-mounted obstacles: applying topology to flow visualization, J. Fluid Mech. **86**, 179–200 (1978)

13.5 D.G. Bohl: *An experimental study of the near field region of a free jet with passive mixing tabs*, MS Thesis (Michigan State University, East Lansing 1996)

13.6 J.N. Koster, U. Müller: Free convection in vertical gaps, J. Fluid Mech. **125**, 429 (1982)

13.7 G. Zimmermann, P. Ehrhard, U. Muller: Stationäre und Instationäre Konvektion in einer quadratischen Hele-Shaw Zelle, Primärbericht IRB-Nr **507**, 86 (1986)

13.8 R. Ruderich, H.H. Fernholz: An experimental investigation of a turbulent shear flow with separation, reverse flow, and reattachment, J. Fluid Mech. **163**, 283–322 (1986)

13.9 L.J.S. Bradbury, A.H. Khadem: The distortion of a jet by tabs, J. Fluid Mech. **70**, 801–813 (1975)

13.10 K.B.M.Q. Zaman, M.F. Reeder, M. Samimy: Control of an axisymmetric jet using vortex generators, Phys. Fluids **6**, 778–793 (1994)

13.11 M.F. Reeder, M. Samimy: The evolution of a jet with vortex-generating tabs: real-time visualization and quantitative measurements, J. Fluid Mech. **311**, 73–118 (1996)

13.12 D.G. Bohl, J.F. Foss: Near exit plane effects caused by primary and primary-plus-secondary tabs, AIAA J. **37**, 192–201 (1999)

13.13 C.M. Holland: *Exploratory investigation of the recirculation region and search for coherent motions aft of an axially mounted cone in confined flow*, MS Thesis (Michigan State University, East Lansing 1989)

13.14 D. Summer, J.L. Heseltine, O.J.P. Dansereau: Wake structure of a finite circular cylinder of small aspect ratio, Exp. Fluids **37,5**, 720–730 (2004)

13.15 C.P. Gendrich, D.G. Bohl, M.M. Koochesfahani: Whole-field measurements of unsteady separation in a vortex ring/wall interaction, 28th AIAA Fluid Dynamics Conf. 1997 (AIAA, Reston 1997)

13.16 C.-T. Chu, C.-C. Wang, R.-Y. Chang, W.T. Chang: Head-on collision of two coaxial vortex rings: experiment and computation, J. Fluid Mech. **296**, 39–71 (1995)

13.17 M.M. Koochesfahani: Private communication (2006)

13.18 J.F. Foss, R.J. Prevost, K.M. Bade, A. Levasseur: *The velocity field of an annular jet with cross-member*, ASME/JSME Joint Fluids Engineering Conf., FED SM2003-45242 (ASME, New York 2003)

14. Flow Measurement Techniques in Turbomachinery

This chapter focuses on measurement techniques that have been used during experimental investigations of turbomachinery flow fields. These techniques are not fundamentally different from those used in other external flow studies. However, implementing them within turbomachines has introduced a series of unique and specialized issues in the preparation of the experimental setup, data acquisition, and analysis procedures. This chapter provides detailed information on the methods used to address these issues, along with a comprehensive summary on how they have been implemented to investigate complex flow phenomena within turbomachinery components.

14.1 Background On Turbomachinery Flows ... 919
14.2 Non-Optical Measurement Techniques 921
    14.2.1 Data-Acquisition Techniques 921
    14.2.2 Non-Optical Measurement Techniques 928
14.3 Optical Measurement Techniques 933
    14.3.1 Applications of Laser Doppler Velocimetry (LDV) .. 934
    14.3.2 Applications of Particle Image Velocimetry (PIV) . 936
    14.3.3 Applications of Laser Two-Focus Velocimetry (L2F) 945
    14.3.4 Applications of Doppler Global Velocimetry (DGV) 948
    14.3.5 Applications of Pressure-Sensitive Paint (PSP) 950
14.4 Concluding Remarks 950
References 951

14.1 Background On Turbomachinery Flows

Turbomachines constitute a large and diverse class of devices, which are used to transfer energy either to or from a flowing stream of fluid by the dynamic action of one or more moving blade rows. They have found a wide range of applications in many engineering systems such as:

1. compressors and turbines in gas turbine engines used for power generation, propulsion of aircraft power, plants for naval surface ships, hydrofoil boats and hovercraft
2. propellers and rotors used in marine and aircraft propulsion
3. gas turbines and turbochargers used in land vehicles
4. steam turbines used in power plants
5. hydraulic turbines used in hydroelectric power plants
6. wind turbines
7. pumps used for transport of liquids, especially water, in a wide range of industrial, agricultural and residential applications
8. compressors used in industrial processes
9. pumps used in medical applications, e.g., in heart-assist devices
10. turbines and pumps used in liquid-fuel rocket engines
11. automotive torque converters, and many others.

Many books have already been published on the theory and applications of turbomachinery, and the reader should refer to them for guidance on design and analysis of components, and for information on associated fluid-mechanics and heat-transfer problems [14.1–8].

The complex flow fields within turbomachines are three-dimensional, turbulent, and inherently unsteady. Dominant unsteady phenomena occur due to the interactions of moving and stationary blades with nonuniform flow fields generated by upstream blades. For example, upstream wakes modulate the performance and modify the boundary layers on the blades, and in turn are chopped, strained, and the turbulence within them is modified as they pass through a rotor passage. Depending on applications, the boundary layers on blades range from the laminar to transitional and turbulent regimes. The associated flow fields are highly compressible within gas turbines, but often involve two phases

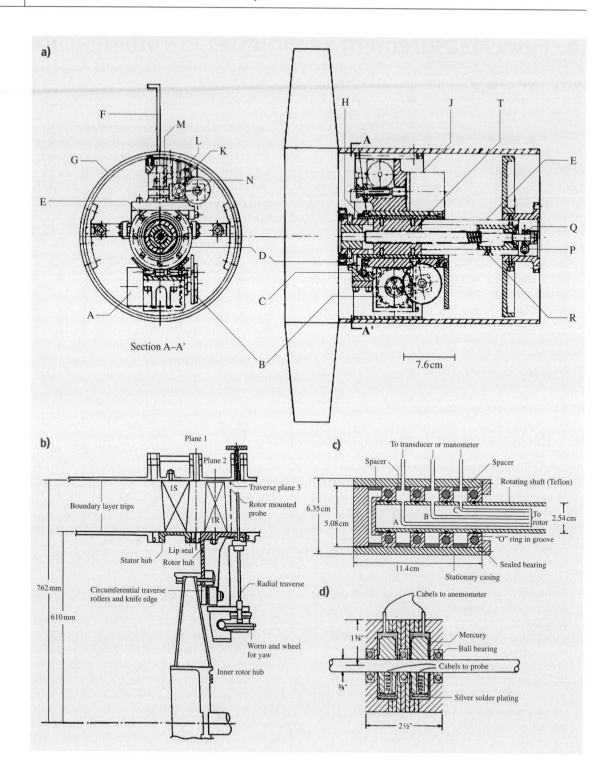

within pumps, inducers, hydroelectric turbines, and ship propellers. These complex flow phenomena affect the performance, efficiency, and range of operating conditions of these machines, as well as cause undesired phenomena, such as noise, vibrations, stall, and sometimes failure. Understanding of the flow dynamics and its effect on performance are essential for the development of more-efficient, reliable and quieter machines, as well as for developing physically meaningful and accurate prediction tools that can be incorporated into the design process. Such insight requires detailed experimental data on the mean and fluctuating components of the flow, pressure, and temperature, as well as on the resulting power, forces, torques, and vibrations. In this chapter we confine ourselves to flow and pressure measurements.

Numerous experimental studies have already investigated various aspects of the flow within turbomachinery over the years. These investigations have examined various three-dimensional, unsteady, and turbulent flow structures involving, e.g., blade–wake and wake–wake interactions, wake–boundary layer interactions, the characteristics of curved rotor wake, the structure of turbulence and deterministic stresses, secondary flows, tip vortices and tip clearance flows, rotating stall and surge, shock waves and related unsteady phenomena, wake–shock interaction, rotor–stator clocking, mixing, cavitation, and many other topics. For measuring velocity and turbulence, these studies have used hot-wire/film anemometry, laser Doppler velocimetry (LDV), laser two-focus velocimetry (L2F) and particle image velocimetry (PIV). Pressure measurements have been performed using a variety of Pitot tubes including three-, four-, five-, and seven-hole probes, surface and probe-mounted pressure transducers, and pressure-sensitive paint (PSP), the latter for mapping the instantaneous pressure distribution over surfaces. This chapter summarizes the applications of these techniques and discusses issues that are specific to turbomachines. In addition to summaries, a series of tables provide examples of applications of the various measurement techniques. Each table is dedicated to a different class of sensors and contains a brief description of objectives and type of machine involved.

14.2 Non-Optical Measurement Techniques

This category includes a variety of probes, such as single or multisensor hot-wire/film anemometers; five-hole or other types of Pitot probes and high-frequency-response pressure transducers; measurements of static pressure, wall shear stress, and surface flow visualization on blade surfaces. Although not different from those used in other aero/hydrodynamic applications, their implementation in turbomachinery investigations introduces unique issues:

1. Acquisition of data in a rotating reference frame requires complex probe traverse mechanisms and data transmission systems.
2. The flow structure is inherently unsteady with contributions from periodic variations due to the relative alignment between rotor and stator blades, large-scale instabilities occurring below the de-

Fig. 14.1a–d Two sample rotating-frame traverse mechanisms: (**a**) *Lakshminarayana* [14.9] and (**b**) *Chaluvadi* et al. [14.10]; and examples of rotating-to-stationary frame data-transfer devices: (**c**) pneumatic pressure-transfer device, *Lakshminarayana* [14.9], and (**d**) mercury slip-ring unit, *Lakshminarayana* [14.9] ◄

sign conditions such as stall and surge, as well as turbulence. Special data-acquisition and averaging procedures are needed to distinguish between these contributors in order to distinguish between them.

14.2.1 Data-Acquisition Techniques

Rotating Frame of Reference

Measurements in a rotating frame of reference require mechanisms for traversing probes in tangential, radial, axial, and *null* (probe aligned in the direction of the flow) directions. They also require means of data transmission from rotating to stationary frames, such as mercury or brush slip-ring units, pneumatic pressure-transfer devices, and wireless telemetry systems. In one of the earliest studies, *Weske* [14.11] measured pressure distributions on a rotor blade surface and traversed a Pitot probe across a rotor blade wake. He used a pulley-and-lever traverse mechanism as well as a selector switch and seal pressure-transmission system. *Gorton* and *Lakshminarayana* [14.12] used a three-sensor hot-wire to measure all three components of the mean velocity and turbulent stresses in a rotating frame of reference within a three-bladed rocket pump inducer

Table 14.1 Sample studies that have used hot-wire/film anemometry technique in investigating different aspects of turbomachinery flow fields. (S: stationary frame measurement, R: rotating frame measurement)

Hot-wire/film Author(s)	Year	Sensor type	Type of machine	Subject of study	Ref. frame
Lakshminarayana and *Poncet* [14.16]	1974	Hot-wire (X and single sensor)	Axial flow inducer	Rotor wakes	S
Gorton and *Lakshminarayana* [14.12]	1976	Hot-wire (three sensor)	Axial inducer	Mean flow and turbulence	R
Hah and *Lakshminarayana* [14.17]	1980	Hot-wire (three sensor)	Axial compressor	Effect of feestream turbulence on a rotor wake	S
Lakshminarayana et al. [14.18]	1982	Hot-wire (three sensor)	Axial fan	Effects of rotation and blade incidence on rotor wake	S
Dong and *Cumpsty* [14.19]	1990	Hot-wire (single and dual sensor)	Compressor cascade with upstream rods	Boundary layer	S
Hodson et al. [14.20]	1994	Hot-film (surface mounted)	Low pressure turbine	Unsteady boundary layer	S
Camp and *Shin* [14.21]	1995	Hot-wire, hot-film (single sensor)	Axial compressor	Turbulence intensity and length scale	S
Witkowski et al. [14.22]	1996	Hot-film (triple split)	Axial compressor	3-D wake decay and secondary flows	S
Halstead et al. [14.23]	1997	Hot-wire, hot-film (surface mounted), X hot-film probes	Axial compressor and turbine	Unsteady boundary layer	R and S
Hsu and *Wo* [14.24]	1997	Hot-wire (slanted)	Axial compressor	Unsteady wake	S
Sentker and *Riess* [14.25]	1998	Split hot-film	Axial compressor	Turbulence and unsteadiness	S
Ristic and *Lakshminarayana* [14.26]	1998	X Hot-wire	Axial turbine	3-D boundary layer	S
Prato et al. [14.27]	1998	Slanted hot-film	Axial compressor	Unsteady 3-D flow field	S
Furukawa et al. [14.28]	1998	Hot-wire	Diagonal flow rotor	Tip flow field	S
Sentker and *Riess* [14.29]	2000	Split hot-film	Axial compressor	Wake-blade interaction	S
Velarde-Suarez et al. [14.30]	2001	Hot-wire (dual sensor)	Centrifugal fan	Unsteady flow	S
Shin et al. [14.31]	2003	Hot-wire	Axial compressor	Blade boundary layers	S

passage operating in air. They did not have a rotating traverse system, and the probe was positioned manually while the inducer was not running. A rotating-probe traverse system, illustrated in Fig. 14.1a, was used by *Reynolds* et al. [14.13] to investigate the near wake of a rotor blade of an axial flow research fan. As described in detail in *Lakshminarayana* [14.9], this mechanism was mounted immediately downstream of the rotor, and allowed null and tangential traverse of a three-sensor hot-wire probe and a spherical head static pressure probe, while the rotor was in operation. Movements in the axial and radial directions could only be achieved manually, while the rotor was stationary. *Dring* and *Joslyn* [14.14] and *Joslyn* and *Dring* [14.15] reported measurements in the rotating frame of reference within an axial turbine facility. Their probes could be traversed along the tangential and radial directions while the machine was rotating. Recently, *Chaluvadi* et al. [14.10] used the rotating traverse mechanism shown in Fig. 14.1b, in a high-pressure turbine facility. This three-axis relative frame traverse mechanism had a computer-controlled stepper motor system, which allowed measurements within and at the exit of the rotor blade row.

To transmit and then measure pressure generated in the rotating frame to a stationary manometer or a transducer, early studies used pneumatic pressure-transfer devices, as shown in Fig. 14.1c [14.9]. The most common method to transmit data from embedded high-frequency pressure sensors or hot-wire signals has been

Table 14.2 Sample studies that have used various types of Pitot pressure probes in investigating different aspects of turbomachinery flow fields. (S: stationary frame measurement, R: rotating frame measurement)

Pitot pressure probe Author(s)	Year	Sensor type	Type of machine	Subject of study	Ref. frame
Weske [14.11]	1947	Pitot probe, surface pressure	Axial rotor	rotor wake, surface pressure distribution	R
Neustein [14.38]	1964	Total pressure probes	Axial pump	Flow structure	S
Dring and Joslyn [14.14]	1981	Total pressure probes (Kiel)	Axial turbine	3-D flow structure	R
Joslyn and Dring [14.15]	1992	Five-hole probe, surface pressures	Axial turbine	3-D flow structure	R
Kang and Hirsch [14.39]	1993	Five-hole probe, flow viz.	Compressor cascade	Tip clearance effects	S
Carrotte et al. [14.40]	1995	Five-hole probe	Axial compressor	OGV flow field	S
Prato et al. [14.41]	1997	Five-hole probe	Axial compressor	Mean stator flow structure	S
Doukelis [14.42]	1998	Five-hole probe	Compressor cascade (annular)	Tip clearance effects	S
Ivey and Swoboda [14.43]	1998	Three-hole cobra probe, LDV	axial compressor	Tip leakage effects	R
Dey and Camci [14.44]	2000	Five-hole probe	Axial turbine	Tip clearance flow with coolant ejection	R
Xiao et al. [14.45]	2001	Five-hole probe, surface pressures	Axial turbine	Tip clearance flow and losses	R
McLean et al. [14.46,47]	2001	Five-hole probe, Kiel probe	Axial turbine	Effects on coolant injection at upstream hub	R and S
McCarter et al. [14.48]	2001	Five-hole probe, LDV	Axial turbine	Tip clearance flow and losses	R
Coldrick et al. [14.49]	2003	Four-hole probe	Axial compressor	3-D flow structure	S
Pullan et al. [14.50]	2003	Five-hole probe	Axial turbine	Secondary flows	S
Gilarranz et al. [14.51]	2005	Five-hole probe	Centrifugal compressor	Flow structure	S

to use brush-type or mercury slip-rings. Figure 14.1d shows an earlier version of a two-channel mercury slip-ring [14.9]. Currently, a 150-channel brush-type system is being used at the axial flow turbine research facility (AFTRF) at the Pennsylvania State University [14.32]. Both brush and mercury slip rings are available commercially. Wireless telemetry systems also exist, but their implementation for unsteady pressure and velocity measurements has been rare, mostly due to their limited frequency response, which will most likely change in the future. Nevertheless, analog wireless telemetry was used in early measurements of rotor blade fluctuating and mean surface pressures [14.33, 34]. They also have been used to obtain temperature and strain data in real gas turbine engines [14.35–37]. A recently introduced digital wireless telemetry built for a high-pressure compressor rig [14.37] allowed simultaneous measurements in up to 48 channels with a bandwidth of 50 kHz. The analog version of this system supported 12 channels with a bandwidth of 30 kHz. The electrical power for both systems was supplied using a contactless induction coil.

Recent examples of measurements in a rotating frame include investigations of the tip flow field of a turbine blade with coolant ejection [14.44]; of tip clearance effects in an axial flow turbine [14.45, 48] and in an axial flow compressor [14.43]; of blade-to-blade wake variability by *Boyd* and *Fleeter* [14.52]; and of blade row–wake and vortex interactions by *Chaluvadi* et al. [14.10], [14.53]. A more-comprehensive list of references is presented in Tables 14.1–14.4, where *R* in the last column indicates that the data have been acquired in the rotating frame of reference.

Stationary Frame of Reference

Most of the experimental data have been obtained in a stationary frame of reference. Consequently, they require specialized phase-locked ensemble-averaging procedures during data analysis to distinguish between periodic and turbulent fluctuations [14.9, 16]. This ap-

Table 14.3 Sample studies that have used high-frequency-response pressure sensors in investigating different aspects of turbomachinery flow fields. (S: stationary frame measurement, R: rotating frame measurement)

High-frequency response (HFR) pressure sensors					
Author(s)	Year	Sensor type	Type of machine	Subject of study	Ref. frame
Kerrebrock et al. [14.54]	1980	Spherical HFR-pressure probe	Axial compressor	Flow structure	S
Cousins et al. [14.55]	1981	HFR-pressure transducers (on rotors)	Axial compressor	Rotating stall	R
Giannissis et al. [14.56]	1989	HFR-pressure transducers	Axial compressor	Rotating stall	S
Cherrett and Bryce [14.57]	1992	HFR-pressure probe	Axial compressor	Rotor–rotor interactions, wakes and secondary flows	S
Ainsworth et al. [14.58]	1995	HFR-pressure probe	Axial turbine	Unsteady flow	S
Fabian and Jumper [14.59]	1995	HFR-pressure transducers (surface)	Axial compressor transonic cascade with upstream rods	Unsteady pressure distributions	S
Laborde et al. [14.60]	1997	HFR-pressure transducer on wall, flow viz.	Axial pump	Tip vortex cavitation	S
Dong et al. [14.61]	1998	HFR five-hole probe	Automotive torque converter	Unsteady flow	S
Roduner et al. [14.62]	1999	HFR-pressure probe	Centrifugal compressor	Impeller exit flow	S
Kost et al. [14.63]	2000	HFR-pressure sensor (surface)	Axial turbine	Unsteady flow	S
Mailach et al. [14.64]	2001	HFR-pressure transducer (surface)	Axial compressor	Rotating instabilities	R
Tiedemann and Kost [14.65]	2001	HFR-pressure probes, L2F	Axial turbine	Effect of clocking on wake–wake interactions	S
Lohrberg et al. [14.66]	2002	HFR-pressure transducers	Hydrofoil cascade	Cavitation	S
Nohmi et al. [14.67]	2003	HFR-pressure transducer (on wall and blade)	Centrifugal pump	Cavitation	R
Leger et al. [14.68]	2004	HFR-pressure sensor (flexible and on surface)	Axial compressor	Unsteady loading	S
Spakovszky [14.69]	2004	HFR-pressure transducer (on wall)	Centrifugal compressor	Rotating stall	S
Schleer et al. [14.70]	2004	HFR-pressure probe	Centrifugal compressor	Flow structure	S
Camci et al. [14.71]	2005	HFR-pressure probe	Axial turbine	Effect of tip geometry on tip leakage flows	S

proach consists of conditionally sampling a time series of data obtained over many revolutions based on the phase or orientation of a rotor blade. A signal from a shaft encoder is typically used for synchronizing the sensor time series with the phase of the machine. For data recorded over N revolutions, the phase- or ensemble-averaged value of a certain flow variable φ at time or phase t_m in a revolution is

$$\bar{\varphi}(t_m) = \frac{1}{N} \sum_{n=1}^{N} \varphi_n(t_m) \,,$$

where φ is the instantaneous value, e.g., of the velocity, pressure or temperature, and n an index indicating a certain period/cycle. The turbulent fluctuations are defined as the difference between the ensemble-averaged value and the instantaneous value

$$\varphi'_n(t_m) = \varphi_n(t_m) - \bar{\varphi}(t_m) \,.$$

If the number of samples per period is M, the time-averaged value is

$$\tilde{\varphi} = \frac{1}{M} \sum_{m=1}^{m} \bar{\varphi}(t_m) \,.$$

Table 14.4 Sample studies that have used a combination of hot-wire/film anemometry, Pitot pressure probes and high-frequency-response pressure sensors in investigating different aspects of turbomachinery flow fields. (S: stationary frame measurement, R: rotating frame measurement)

Multiple sensors (hot-wire/film), Pitot and high-frequency pressure sensor					
Author(s)	Year	Sensor type	Type of machine	Subject of study	Ref. frame
Lockhart and Walker [14.72]	1974	Hot-wire (single sensor) pressure probe (three-hole Conrad-type yawmeter)	Axial compressor	Blade/wake interactions	S
Greitzer [14.73]	1976	Hot-wire, HFR-pressure probe	Axial compressor	Rotating stall and surge	S
Reynolds et al. [14.13]	1979	Hot-wire (three sensor), pressure probe (spherical head)	Axial fan	Rotor wakes	R and S
Gregory-Smith and Cleak [14.74]	1992	X hot-wire, five-hole probe	Axial turbine cascade	Secondary flows with inlet turbulence	S
Day [14.75]	1994	Hot-wire, HFR-pressure probe	Axial compressor	Rotating stall and surge	S
Kim and Fleeter [14.76]	1994	X hot-wire, HFR-pressure transducer	Axial compressor	Rotating stall and surge	R
Gallus et al. [14.77]	1994	Hot-wire, pressure probe (wake and blade surface)	Axial turbine	Secondary flows	R and S
Chaluvadi et al. [14.10]	2001	Hot-wire (three sensor), five-hole, three-hole and Kiel probes	Axial turbine	Blade–wake interaction	R and S
Reinmoller et al. [14.78]	2002	Hot-wire (three-sensor), five-hole probe	Axial turbine	Rotor–stator clocking	S
Sanders and Fleeter [14.79]	2002	X hot-film, HFR-pressure probe and surface transducers	Axial compressor	Blade–wake interaction and clocking	S
Johnston and Fleeter [14.80]	2002	Hot-film, HFR-pressure probe	Axial compressor	Wake–blade interaction	S
Boyd and Fleeter [14.52]	2003	Hot-wire, surface pressure (HFR)	Axial compressor	Blade-to-blade wake variability	R
Schobeiri et al. [14.81]	2003	Hot-wire (single), surface pressure	Low-pressure turbine cascade with upstream rods	Unsteady boundary layer	S
Chaluvadi et al. [14.53]	2004	Hot-wire (three sensor), five-hole probe	Axial steam turbine	Blade row–vortex interaction	R and S
Johnston and Fleeter [14.82]	2004	Hot-film, HFR-pressure probe	Axial fan	IGV-rotor potential field interaction	S

Figure 14.2a shows a sample signal from triaxial hot-wire measurements, taken from *Lakshminarayana* [14.9], and illustrates the methodology described above to calculate the ensemble-averaged and the turbulence parameters using a continuous data time series. This approach has been widely used during analysis of hot-wire/film and pressure probe signals to study various aspects of the flow field behind rotating blade rows. Some sample references are listed in Tables 14.1–14.4, and a sample application is presented in Fig. 14.2b, showing the velocity traces obtained using a hot-wire anemometer, within a compressor cascade operating downstream of a row of rotating cylindrical rods [14.19].

The so-called *phase-locked averaging technique* is another variation of the phase-dependent data acquisition and averaging methods. In this approach, the sensor is triggered based on a signal provided by the encoder to acquire data during a short but finite time δt at a fixed phase of the revolution. Different phases can be sampled by varying the delay between the shaft encoder pulse and timing of data acquisition. After collecting data over N cycles, the time delay is changed to obtain an average

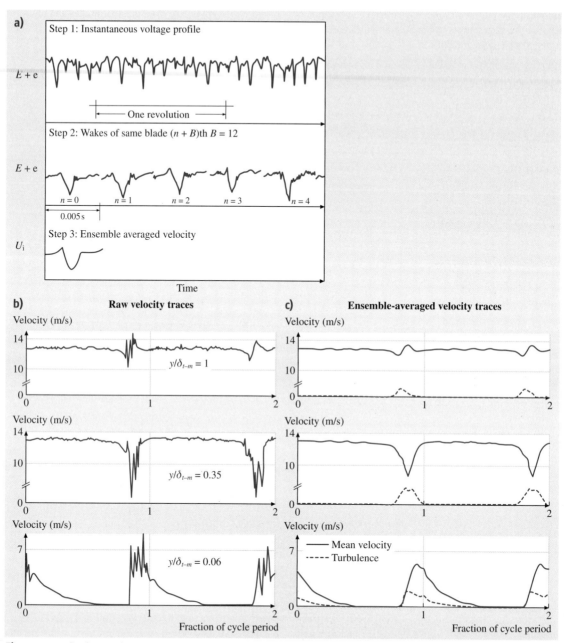

Fig. 14.2a–c Stationary-frame data acquisition and reduction procedures for hot-wire signals. (**a**) The methodology to calculate the ensemble-averaged data from continuous instantaneous time series, as explained in *Lakshminarayana* [14.9]. (**b**), (**c**) A sample application of the same measurement methodology in a compressor cascade experiment with upstream rotating rods (after *Dong* and *Cumpsty* [14.19])

Fig. 14.3a–f Examples of various hot-wire/film probes used in turbomachinery flow investigations. (**a**) *Lakshminarayana* and *Poncet* [14.16], (**b**) *Gorton* and *Lakshminarayana* [14.12], (**c**) *Witkowski* et al. [14.22], (**d**) *Sentker* and *Riess* [14.25], (**e**) *Chaluvadi* et al. [14.53]. (**f**) Surface mounted hot-film *Hodson* et al. [14.20] ▶

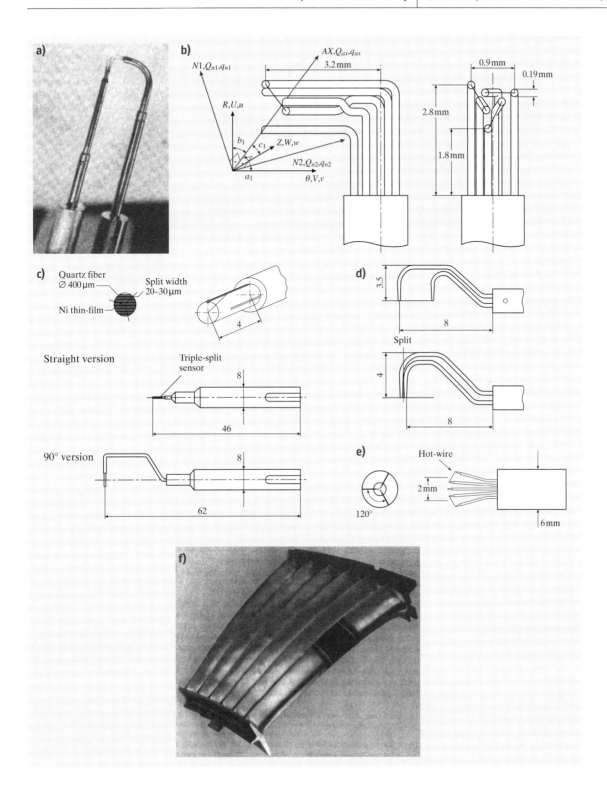

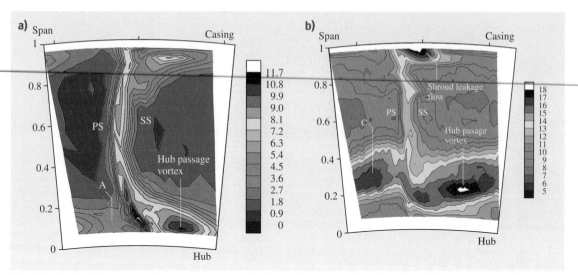

Fig. 14.4a,b Turbulence intensity contours measured using a three-sensor hot-wire behind (**a**) the stator and (**b**) the rotor of a high-pressure axial-flow steam turbine stage, with upstream delta wings used to simulate the rotor passage vortex (after *Chaluvadi* et al. [14.53])

at another phase of the rotor period. This procedure has not been used as extensively, and the reader is referred to *Lakshminarayana* [14.9] for background, and to *Hsu* and *Wo* [14.24] for an example of implementation.

Camp and *Shin* [14.21] used another approach to filter out the periodic components of a signal, i.e., those occurring at the blade-passing frequency. They used a Fourier transform to obtain the spectrum of a signal and zeroed the amplitude at the blade-passing frequency and its harmonics. The filtered Fourier transform was then transformed back into the time domain to obtain the turbulent signal. The modulated spectrum was also used for calculating the autocorrelation function directly from the Fourier coefficients, which provided the turbulent length scale. The authors claimed that these procedures were computationally efficient, and required a comparatively small amount of data. These advantages enabled them to measure the turbulence in a large number of points over a stator passage.

14.2.2 Non-Optical Measurement Techniques

Hot-Wire/Film Anemometry

Due to their small size and high-frequency response, hot-wire and hot-film anemometers have been used extensively in investigations of turbomachinery flow fields. Extensive background and technical information on these techniques can be found in *Bruun* [14.86], *Goldstein* [14.87] and other chapters in this book. A hot-wire probe consists of a short length (≈ 1 mm), fine-diameter (5 μm or less) wire made of tungsten, platinum or platinum alloys, which is attached to two prongs. Hot-film sensors are thin (about 0.1 μm) platinum or nickel films deposited on thermally insulating substrates (usually quartz), which are typically shaped as cylinders, wedges or cones. For surface measurements, hot-film sensors can be glued to the wall. In cylindrical hot-film probes, the active element is usually 25–70 μm in diameter and 1–2 mm long. Compared to hot-wire probes, the hot-film probes have a more-rigid construction, are less susceptible to contamination, and have more long-term calibration stability. Consequently, they have a lower frequency of breakage. However, it is more expensive to replace them, they have a more-complex frequency response, and their calibration may be affected by their own vortex shedding [14.86].

Early investigations in turbomachinery flows using hot-wire/film probes were mostly qualitative [14.72, 88, 89]. However, with the implementation of rotating-frame measurement techniques as well as the

Fig. 14.5a–e Sample Pitot pressure probes used in turbomachinery flow investigations: (**a**) five-hole probe *Treaster* and *Yocum* [14.83], (**b**) disk-type boundary-layer probe *Sitaram* et al. [14.84], (**c**) spherical head probe *Sitaram* et al. [14.84], (**d**) four-hole probe *Coldrick* et al. [14.49] and (**e**) seven-hole probe *Gottlich* et al. [14.85]▶

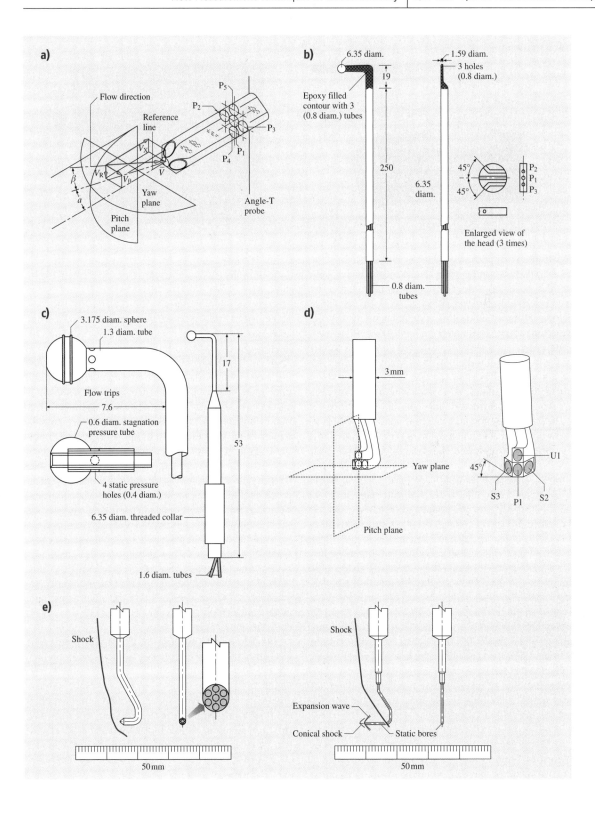

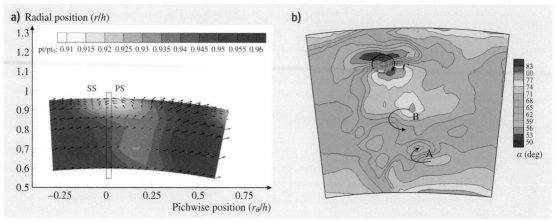

Fig. 14.6a,b Sample data obtained using five-hole probes. (**a**) Pressure-loss contours and velocity vectors downstream of an axial-flow turbine rotor. These measurements were performed in the rotating frame by *Dey* and *Camci* [14.44]. (**b**) Meridional yaw angle (α) contours at the exit of a nozzle guide vane, *Pullan* et al. [14.50]

stationary-frame acquisition and averaging methods, as described above, substantial amounts of quantitative experimental data have been collected on various aspects of the flow within turbomachines. Many different types of hot-wire and hot-film probes have been used, including single-, dual-, or triple-sensor hot-wire probes, single-element or double/triple split hot-film probes, as well as surface-mounted hot-film sensors for surface shear stress measurements. Characteristic examples of the types of probes used in investigating turbomachinery flows are presented in Fig. 14.3. Figure 14.4 shows sample turbulence intensity contours obtained within a high-pressure axial flow steam turbine stage, using the three-sensor hot-wire probe shown in Fig. 14.3e [14.53]. This specific test rig contained a stator row followed by a rotor row. This study investigated the impact of the upstream rotor passage vortex, which was simulated using delta-wing vortex generators located upstream of the stator row, on the performance of the downstream blade rows. Hot-wire measurements were performed in the stationary frame of reference and the ensemble-averaging technique was used to obtain the spatial distribution of turbulence intensity. Table 14.1 provides samples of other applications of hot-wire/hot-film anemometry, showing the types of probe and machine as well as the main topic of investigation. Studies involving hot-wire/hot-film measurements along with other sensors are summarized in Table 14.4.

Pitot Pressure Probes

Figure 14.5 shows the characteristic configurations of Pitot pressure probes that have been used in turbomachinery flow field investigations, and Tables 14.2 and 14.4 provide references for sample studies that have utilized them. A classical reference on this topic is *Sitaram* et al. [14.84]. Some sensors can only measure the total pressure, such as the wedge, spherical and flat-nosed pitot probes as well as the Kiel probe. The latter is particularly suitable in situations where the flow direction is unknown. Other sensors can measure the total and/or static pressure, such as the conventional Pitot-static (Prandtl) probe and the spherical-head total-static probe, which is insensitive to the flow direction. The three-hole cobra probe, the three-hole Conrad yawmeter probe and the five-hole probe are mostly used for measuring the flow velocity vector in addition to the total and static pressures. The thin three-hole disk-type probe can be used for two-dimensional boundary-layer measurements [14.84]. Most Pitot pressure probes inherently have a low-frequency response. Therefore, they have been used either for measurements in a rotating frame of reference or in measurements within nonrotating inlet guide vanes (IGV) and stator blade passages to obtain the mean flow characteristics.

As suggested by Table 14.2, the five-hole probe has been the most commonly used pressure probe.

Fig. 14.7a–g Sample high-frequency-response pressure probes and surface-mounted sensors used in turbomachinery flow investigations: (**a**) *Cherrett* and *Bryce* [14.57]. (**b**) *Ainsworth* et al. [14.58]. (**c**) *Kerrebrock* et al. [14.54] (**d**) *Tiedemann* and *Kost* [14.65]. (**e**) *Gossweiler* et al. [14.90]. (**f**) *Leger* et al. [14.68]. (**g**) *Ainsworth* et al. [14.91]▶

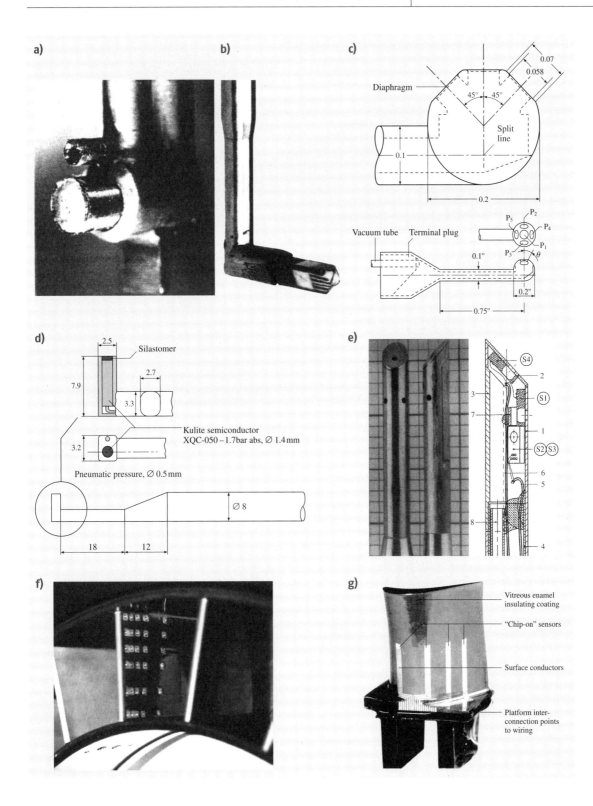

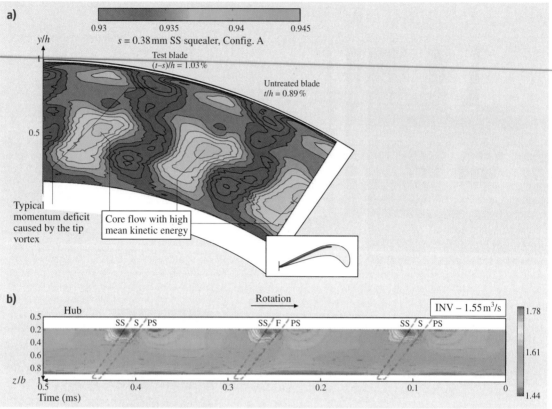

Fig. 14.8a,b Sample high-frequency-response pressure probe measurement results (**a**) P_0/P_{in} contours downstream of an axial flow turbine rotor (*Camci* et al. [14.71]) and (**b**) total pressure ration contours at the exit of the impeller of a centrifugal compressor (*Schleer* et al. [14.70])

The five measured pressures are converted to the local total and static pressures as well as the local pitch and yaw angles of the velocity vector, based on calibration. The local mean velocity magnitude is determined from the measured total and static pressures. The five-hole probe is sensitive to Reynolds number, wall vicinity, blockage, and turbulence effects. Detailed information about calibration and data-acquisition procedures for the five-hole probe can be found in *Treaster* and *Yocum* [14.83], *Lakshminarayana* [14.9] and *Sitaram* et al. [14.84]. Sample results of measurements performed using five-hole probe are presented in Fig. 14.6.

High-Frequency-Response Pressure Probes and Sensors

High-frequency pressure probes and surface pressure transducers have been used for measuring unsteady pressure fluctuations in both rotating and stationary frames of references. They typically have one or more piezoresistive (i.e., a strain-gauge attached to a diaphragm) or piezoelectric (piezoelectric crystal, typically silicon) pressure sensors, that are installed flush-mounted either within a probe or on a blade surface. The natural frequency of these sensors can extend to 150–400 kHz. Comprehensive reviews on applications of high-frequency pressure sensors to turbomachinery flow fields can be found in *Sieverding* et al. [14.92] and *Ainsworth* et al. [14.91]. Characteristic samples are listed in Tables 14.3 and 14.4. Most studies have utilized commercially available sensors produced by, e.g., Kulite, Endevco, Entran, PCB, and several others. Mounting of several sensors to a probe head to measure the total and static pressure simultaneously significantly increases the probe head size, although miniature MEMS-based transducers mounted on the tip of the probe have been introduced to remedy this problem [14.93]. Due to interference of the probe with the flow, *Kupferschmied* et al. [14.94] note that the measured pressures do not, in general, correspond to the total or static pressure of the undisturbed flow, but may be converted into them (and into flow angles and velocity) by calibrations under well-controlled flow conditions.

Figure 14.7 shows several types of probes and surface sensors that have been used in turbomachines, and Fig. 14.8 presents sample results obtained in a stationary reference frame analyzed using the ensemble-averaging technique [14.70, 71].

14.3 Optical Measurement Techniques

Optical measurement techniques are typically nonintrusive, and can be implemented within rotating blade passages without having to use rotating probes (although rotating optical probes exist). The primary challenge in applications of optical sensors is to provide appropriate optical access to the flow field within the machine. Laser Doppler velocimetry (LDV) and particle image velocimetry (PIV) have been the most commonly used techniques for measuring velocity and turbulence within turbomachines. LDV has been widely in use since the early 1970s, whereas PIV was introduced in the late 1980s, but has become popular in turbomachinery applications in the late 1990s. Another optical technique, laser two-focus velocimetry (L2F), has also been used in several studies, but less frequently. Other techniques, such as Doppler global velocimetry (DGV), are still under development. All the aforementioned techniques rely on measurements of the displacement of seed particles. In applications involving airflow, where the particles cannot be neutrally buoyant, they must

Table 14.5 Sample studies that have used laser Doppler velocimetry (LDV) technique in investigating different aspects of turbomachinery flow fields. (S: stationary frame measurement, R: rotating frame measurement)

Laser Doppler velocimetry (LDV) Author(s)	Year	Type	Type of machine	Subject of study	Ref. frame
Wisler and *Mossey* [14.95]	1973	1-D	Axial compressor	Relative flow field	S
Strazisar [14.96]	1985	1-D	Axial fan rotor	Transonic flow structure	S
Pierzga and *Wood* [14.97]	1985	1-D	Axial fan rotor	3-D flow field	S
Murthy and *Lakshminarayana* [14.98]	1986	1-D	Axial compressor	Tip flow field	S
Beaudoin et al. [14.99]	1992	2-D	Centrifugal pump	Effects of orbiting impeller	S
Stauter [14.100]	1993	3-D	Axial compressor	Tip flow field	S
Hathaway et al. [14.101]	1993	3-D	Centrifugal compressor	3-D flow structure	S
Farrell and *Billet* [14.102]	1994	–	Axial pump	Tip vortex cavitation	S
Abramian and *Howard* [14.103]	1994	1-D	Centrifugal impeller	Unsteady flow	R
Fagan and *Fleeter* [14.104]	1994	1-D	Centrifugal compressor	Flow Structure	S
Hobson et al. [14.105]	1996	2-D	Axial compressor cascade	Effect of inlet turbulence	S
Zaccaria and *Lakshminarayana* [14.106]	1997	2-D	Axial turbine	Wake/blade interaction	S
Adler and *Benyamin* [14.107]	1999	2-D	Axial turbine	Stator wake transport	S
Ristic et al. [14.108]	1999	3-D	Axial turbine	3-D flow field	S
Kang and *Thole* [14.109]	2000	3-D	Axial turbine cascade	Endwall flow structure	S
Faure et al. [14.110]	2001	2-D	Axial compressor	Flow Structure	S
McCarter et al. [14.48]	2001	3-D	Axial turbine	Tip clearance effects	S
VanZante et al. [14.111]	2002	2-D	Axial compressor	Blade/row interaction	S
Woisetschlager et al. [14.112]	2002	2-D	Axial turbine cascade	Turbine blade wakes	S
Xiao and *Lakshminarayana* [14.113]	2002	3-D	Axial turbine	Endwall flow structure	S
Matsunuma and *Tsutsui* [14.114]	2003	2-D	Axial turbine	Unsteady flow	S
Ibaraki et al. [14.115]	2003	2-D	Centrifugal compressor	Transonic flow structure	S
Gottlich et al. [14.85]	2004	2-D	Axial turbine	Stator–rotor interaction	S
Faure et al. [14.116]	2004	3-D	Axial compressor	3-D flow and turbulence structure	S

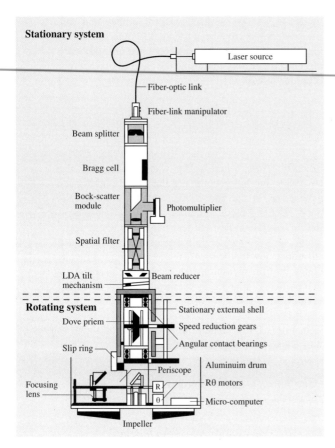

Fig. 14.9 Schematic of the stationary to rotating frame optical transfer system for relative frame LDV measurements within a centrifugal impeller (*Abramian* and *Howard* [14.103])

The operating principles of this technique are well documented in the literature [14.87, 117]. This technique attracted the attention of experimentalists in turbomachinery research soon after it was introduced in 1960s. In addition to being non-intrusive, it allowed measuring the relative velocity and turbulence fields within rotating blade rows, without having to use complex rotating probe traverse and data-transmission mechanisms. Sample studies that have used LDV for measurements within turbomachines are provided in Table 14.5.

Implementation of LDV to turbomachinery flows comes with a variety of problems. For example, due to the inherent accessibility limitations, LDV system must operate in backscatter mode, which reduces the signal-to-noise ratio by one to three orders of magnitude compared to the forward-scatter mode, depending on the particles properties [14.118]. The signal-to-noise ratio is further reduced due to reflections of the incident laser beams near end-wall regions. In addition, the three-dimensional shapes of rotor blades create shadow zones, especially near the hub regions, which necessitate use of complicated off-axis measurement systems. Another critical issue is the optical distortions to the multiple laser beams caused by the curvature of the access windows. These distortions increase the uncertainty of the measurements by deforming the measurement volume and changing the measurement location [14.101]. *Doukelis* et al. [14.119] estimated the changes in the orientation and relative position of measurement volume due to the window curvature for a three-dimensional (3-D) LDV system. They also proposed a mathematical method to correct for refractions from windows, which could be incorporated into data-acquisition procedures.

Most investigations involving LDV have been performed in a stationary frame of reference by discretizing the rotor passage period into bins, each with a finite time interval, and ensemble-averaging the results over these bins. To obtain sufficient convergence in mean velocity and turbulence parameters, one would like to increase the number of points per bin. However, increasing the bin size incorporates effects of spatial variations

be kept small enough to follow the fluid motion, especially in areas with very high pressure gradients, such as in shock waves. Consequently, the typical particle size should be around 0.5 μm to obtain a tolerable velocity lag. In applications involving liquid, the specific gravity of the particle can be better matched with that of the fluid, and significantly larger particles, in the order of 5–20 μm can be tolerated, depending on the characteristic flow scales. Non-intrusive pressure distributions measurements on blade surfaces have been performed using pressure-sensitive paint (PSP). This section introduces and provides samples of applications of these techniques in turbomachinery flow fields.

14.3.1 Applications of Laser Doppler Velocimetry (LDV)

Laser Doppler velocimetry measures the fluid velocity by detecting the Doppler frequency shift of the laser light that is scattered by small particles moving with the fluid.

Fig. 14.10a,b The LDV setup (**a**) and measured relative velocity contours within the rotor passage of a low-speed research compressor (**b**) (*Wisler* and *Mossey* [14.95]) ▶

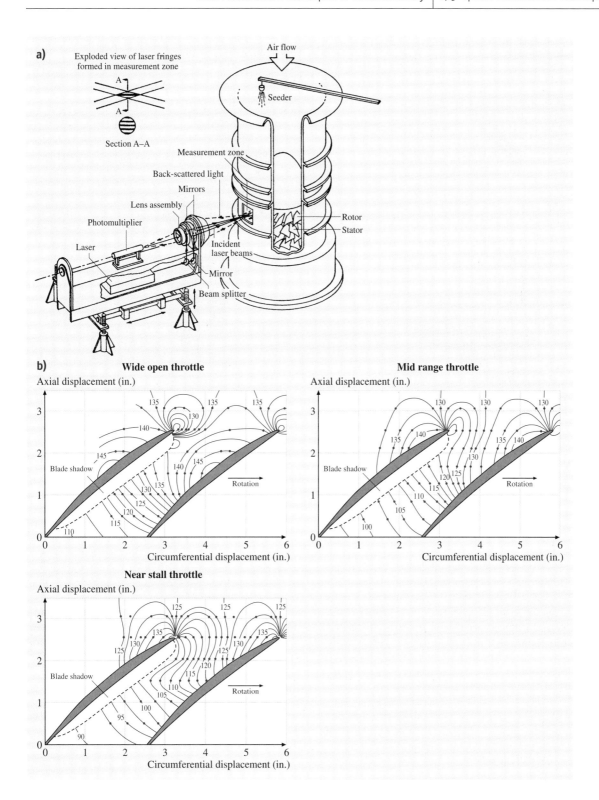

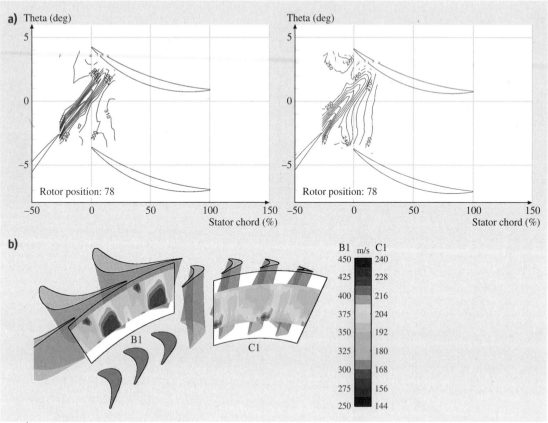

Fig. 14.11a,b Sample results from LDV measurements within (**a**) an axial compressor, rotor wake relative velocity magnitude at peak efficiency (*left*) and near stall (*right*) (*Van Zante* et al. [14.111]) and (**b**) an axial turbine, time-averaged flow field at the nozzle and rotor exits (*Gottlich* et al. [14.85])

in the flow structure within the rotor passage into the ensemble-averaged statistics. A discussion on this issue recommending a bin size of 50 points for measurements in an axial flow turbine can be found in *Ristic* et al. [14.108]. *Abramian* and *Howard* [14.103] provided detailed descriptions of mechanical and optical designs for LDV measurements in a rotating frame of reference within a centrifugal impeller. As shown in Fig. 14.9, they used a Dove prism to transfer the laser beams to the rotating frame of reference, along with a rotating periscope, which located the probe volume at any desired point within the blade passage.

As summarized in Table 14.5, one of the earliest applications of LDV to turbomachinery was described by *Wisler* and *Mossey* [14.95]. They performed velocity measurements upstream, within and downstream of the first-stage rotor row of an axial compressor, using a single-component system operating in backscatter mode. The flow was seeded by spray-atomizing a dilute water suspension of 1 μm-diameter polystyrene latex particles. Figure 14.11a,b shows their experimental setup and a sample contour plot of relative velocity within the rotor passage at 50% span. Sample results from more-recent applications within an axial compres-

Fig. 14.12a–e Sample PIV setups showing the optical access windows and laser and camera positions: (**a**) axial compressor (*Copenhaver* et al. [14.120]) WG: "Wake Generator", (**b**) axial turbine (*Lang* et al. [14.121]), (**c**) optical periscope insert for light-sheet delivery (*Balzani* et al. [14.122]), (**d**) schematic of the light-sheet probe and illumination pattern (*Wernet* et al. [14.123]), (**e**) index-matched axial pump (*Uzol* et al. [14.124]) ▶

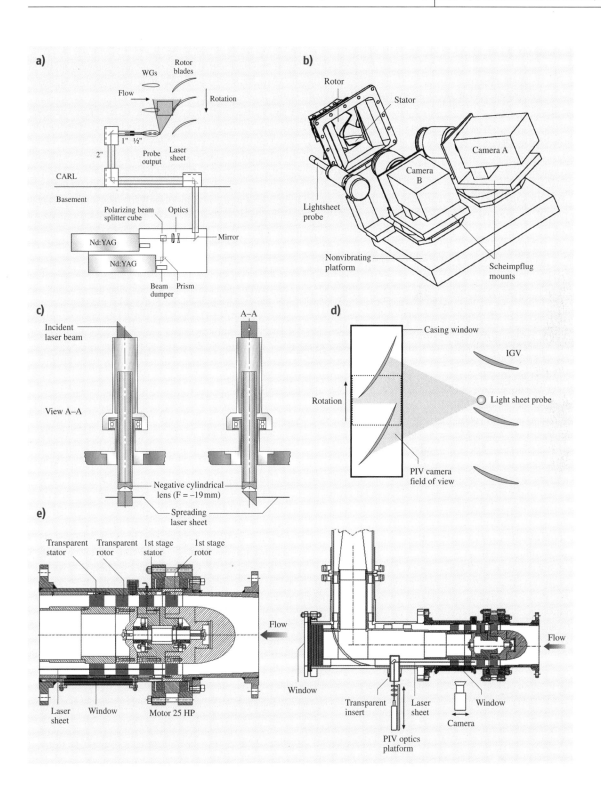

sor [14.111] and within an axial turbine [14.85] are presented in Fig. 14.12.

14.3.2 Applications of Particle Image Velocimetry (PIV)

Particle image velocimetry consists of illuminating a flow field seeded with microscopic particles with a thin light sheet, and recording a pair of images, separated by a short time interval. The velocity is determined by dividing the image into small interrogation areas, and using cross-correlation analysis to measure the displacement of particles within each area. This process provides the instantaneous distribution of two in-plane components of the velocity. Stereo imaging using two inclined cameras provides all three components of the velocity in the illuminated plane. Detailed information on technical issues associated with applications of PIV is provided in other chapters of this book. Extensive use of this technology started in the late 1980s and early 1990s [14.125–127]. Further details can be found in a series of papers published in a dedicated volume of *Measurement Science and Technology Journal* [14.128–131]) as well as in a book by *Raffel* et al. [14.132]. In this section we focus on issues relevant to applications in turbomachinery flows. Table 14.6 provides samples of applications of PIV within tu4bomachines, which includes also the types of lasers, cameras and particles used in each study. Specific components associated are discussed below. Similar to LDV, the primary challenge in applications of PIV in turbomachines are optical access and reflections from curved and rough boundaries that overwhelm the particle traces.

Components and Issues Elated to Implementation

Lasers. Similar to other applications, most of the lasers used in turbomachinery studies are frequency-doubled Nd:YAG lasers, with a wavelength of 532 nm, pulse duration of 5–10 ns, and energy of 25–350 mJ/pulse. The sheet thickness varies from about 0.5 to 3 mm. Dual-cavity Nd:YAG lasers allow short pulse-to-pulse separation times that is necessary for high-speed flow measurements. Some early studies have used pulsed copper-vapor lasers [14.133] and high-energy pulsed ruby lasers [14.134].

Image Acquisition. In early years, data were recorded on 35 mm films [14.133] or on media like Kodak Technicalpan [14.134] and then subsequently digitized. With the introduction of digital camera technology with light-sensitive charged-coupled device (CCD) chips, all PIV studies nowadays collect images in digital form. A variety of cameras with different pixel resolutions from 640 × 480 to 2048 × 2048 pixels^2 have been used. In the so-called *interline transfer* cameras, each pixel is divided to an exposed part that senses the light, and a masked part that serves as a buffer memory. Fast transfer of data from the exposed to masked sections enables the recording of two exposures on separate frames, with an interframe delay of less than 1 µs. This approach solves directional ambiguity issues, and provides a high signal-to-noise ratio in correlation analysis. Another approach to achieve short delay between frames is to use *frame-straddling*, i.e., to trigger the first laser pulse at the end of the exposure time of one frame, and the second pulse at the beginning of exposure of the subsequent frame. In a different approach, *Estevadeordal* et al. [14.135] used a two-color PIV system to resolve the directional ambiguity while investigating the flow field within an axial fan. Their first wavelength, 532 nm, was provided by a frequency-doubled Nd:YAG laser, and the second pulse at 640 nm was generated by using the Nd:YAG laser to pump a dye laser. The images were recorded using a Kodak 3k × 2k pixels digital camera with a color CCD chip.

Seeding. For measurements in axial and centrifugal air compressors and turbines, the particle size is usually less than 1 µm. The particles are generated using a variety of methods, such as commercial fog/smoke generators, atomizing glycerin and water/oil mixtures, alumina in ethanol dispersions, etc. In one of the early applications of PIV in turbomachines, which were performed in water, *Dong* et al. [14.133] used particles containing embedded fluorescent dyes that responded efficiently to green lasers. The specific gravity of these particles varied between 0.95 and 1.1, their size was in the 20–40 µm range. By filtering out the green light using a filter placed in front of the camera, they could perform measurements very close to boundaries. *Choi* et al. [14.136] used nylon 12 particles to investigate the flow field within a centrifugal impeller operating in water. Silver-coated hollow glass particles were used by *Uzol* et al. [14.124] in an axial pump facility containing a concentrated sodium iodide (NaI) solution for refractive-index-matching purposes, and by [14.137] in a centrifugal rotary blood pump.

Optical Access. The flow field in a multistage turbomachine is visually obstructed by the blades, making optical access for the laser sheet and the cameras

Table 14.6 Sample studies that have used particle image velocimetry (PIV) technique in investigating different aspects of turbomachinery flow fields.

Particle image velocimetry (PIV) Author(s)	Year	Type	Type of machine	Subject of study	Laser	Camera	Particles
Paone et al. [14.138]	1989	2-D	Centrifugal pump	Flow structure	–	–	–
Dong et al. [14.133]	1992	2-D	Centrifugal pump	Flow structure	Pulsed copper vapor and/or frequency-doubled Nd:YAG, 1 mm thickness	35 mm film and image digitizer	Fluorescent particles
Chu et al. [14.139, 140]	1995	2-D	Centrifugal pump	Unsteady flow and pressure fluctuations	Pulsed copper vapor and/or frequency-doubled Nd:YAG, 1 mm thickness	35 mm film and image digitizer	Fluorescent particles
Day et al. [14.141]	1996	2-D	Axial turbine	Effect of film cooling on flow structure	–	–	–
Tisserant and Breugelmans [14.134]	1997	2-D	Axial compressor	Blade-to-blade flow	High-energy pulsed ruby laser, 500 mJ/pulse, 1.5 mm thickness	Camera recording on Kodak Technicalpan 2415	Glycerine oil vapor mixed with air
Dong et al. [14.142]	1997	2-D	Centrifugal pump	Unsteady flow and noise	Frequency-doubled Nd:YAG, 1 mm thickness	35 mm film and image digitizer	Fluorescent particles
Balzani et al. [14.122]	2000	2-D	Axial compressor	Blade-to-blade flow	Frequency-doubled (532 nm) Nd:YAG, 2 mm thickness, 200 mJ/pulse	640 × 480 RS-170 CCD cross-correlation camera	Glycerine oil vapor mixed with air
Sinha and Katz [14.143]	2000	2-D	Centrifugal pump	Flow structure and turbulence	Nd:YAG, 25 mJ/pulse	2048 × 2048 pixels^2	Fluorescent particles
Estevadeordal et al. [14.135]	2000	2-D	Axial fan	Flow characteristics	Frequency-doubled (532 nm) and pumped-dye (640 nm) Nd:YAGs, 1 mm thickness, 60 mJ/pulse	3k × 2k Kodak color CCD	Glycerin and water mixture sub-micron particles
Wernet [14.144]	2000	2-D	Centrifugal compressor	Diffuser flow structure	Frequency-doubled (532 nm) Nd:YAG, 1 mm thickness, 200 mJ/pulse	1k × 1k pixels^2	pH-stabilized dispersion of alumina in ethanol
Wernet [14.145]	2000	2-D	Axial and centrifugal compressor	Technique and flow structure	Frequency-doubled (532 nm) Nd:YAG, 1 mm thickness, 200 mJ/pulse	1k × 1k pixels^2	Smoke and pH stabilized dispersion of alumina in ethanol
Wernet et al. [14.146]	2001	2-D	Centrifugal compressor	Surge	Frequency-doubled (532 nm) Nd:YAG, 1 mm thickness, 200 mJ/pulse	1k × 1k pixels^2	pH-stabilized dispersion of alumina in ethanol
Uzol and Camci [14.147]	2001	2-D	Axial turbine cascade	Trailing edge coolant ejection	Frequency-doubled (532 nm) Nd:YAG, 1 mm thickness, 50 mJ/pulse	1k × 1k Kodak ES 1.0	Rosco fog generator

Table 14.6 (continued)

Particle image velocimetry (PIV)								
Author(s)	Year	Type	Type of machine	Subject of study	Laser	Camera	Particles
Sinha et al. [14.148]	2001	2-D	Centrifugal pump	Rotating stall	Nd:YAG, 350 mJ/pulse	2048×2048 pixels^2	Fluorescent particles
Sanders et al. [14.149]	2002	2-D	Axial compressor	Blade-row interactions	Frequency-doubled (532 nm) Nd:YAG, 1 mm thickness, 30 mJ/pulse	$1k \times 1k$ Kodak ES 1.0	Rosco fog generator
Wernet et al. [14.123]	2002	3-D	Axial compressor	Tip clearance flow	Frequency-doubled (532 nm) Nd:YAG, 200 mJ/pulse	$1k \times 1k$ pixels^2	Rosco fog generator
Lang et al. [14.121]	2002	3-D	Axial turbine	Rotor–stator interaction	Frequency-doubled (532 nm) Nd:YAG, 2 mm thickness, 120 mJ/pulse	1280×1024 pixels^2 CCD (DANTEC 80C60)	Diethylexylsebacat particles, 0.3 µm
Uzol et al. [14.124]	2002	2-D	Axial pump	Flow structure and techniques	Frequency-doubled (532 nm) Nd:YAG, 1 mm thickness, 50 mJ/pulse	$2k \times 2k$ Kodak ES 4.0	Silver-coated hollow glass spheres
Uzol et al. [14.150]	2002	2-D	Axial pump	Unsteady flow and deterministic stresses	Frequency-doubled (532 nm) Nd:YAG, 1 mm thickness, 50 mJ/pulse	$2k \times 2k$ Kodak ES 4.0	Silver-coated hollow glass spheres
Chow et al. [14.151]	2002	2-D	Axial pump	Wake–wake interactions	Frequency-doubled (532 nm) Nd:YAG, 1 mm thickness, 50 mJ/pulse	$2k \times 2k$ Kodak ES 4.0	Silver-coated hollow glass spheres
Copenhaver et al. [14.120]	2002	2-D	Axial compressor	Wake–rotor interaction near stall	Frequency-doubled (532 nm) Nd:YAG, 1 mm thickness	$1k \times 1k$ Kodak ES 1.0	Glycerin and water mixture sub-micron particles
Estevadeordal et al. [14.152]	2002	2-D	Axial compressor	Wake–blade interactions	Frequency-doubled (532 nm) Nd:YAG, 1 mm thickness	$1k \times 1k$ Kodak ES 1.0	Glycerin and water mixture sub-micron particles
Uzol et al. [14.153]	2003	2-D	Axial pump	Deterministic stresses	Frequency-doubled (532 nm) Nd:YAG, 1 mm thickness, 50 mJ/pulse	$2k \times 2k$ Kodak ES 4.0	Silver-coated hollow glass spheres
Chow et al. [14.154]	2003	2-D	Axial pump	Rotor wake structure	Frequency-doubled (532 nm) Nd:YAG, 1 mm thickness, 50 mJ/pulse	$2k \times 2k$ Kodak ES 4.0	Silver-coated hollow glass spheres
Woisetschlager et al. [14.155]	2003	2-D	Axial turbine cascade	Turbine wake	Frequency-doubled (532 nm) Nd:YAG, 2 mm thickness, 120 mJ/pulse	1280×1024 pixels^2 CCD (DANTEC 80C60)	Diethylexylsebacat particles, 0.3 µm
Uzol et al. [14.156]	2004	3-D	Axial pump	3-D wake structure and tip vortex	Frequency-doubled (532 nm) Nd:YAG, 1 mm thickness, 120 mJ/pulse	$2k \times 2k$ Kodak ES 4.0	Silver-coated hollow glass spheres

Table 14.6 (continued)

Particle image velocimetry (PIV)							
Author(s)	Year	Type	Type of machine	Subject of study	Laser	Camera	Particles
Uzol et al. [14.157]	2004	3-D	Axial pump	Deterministic stresses	Frequency-doubled (532 nm) Nd:YAG, 1 mm thickness, 120 mJ/pulse	2k × 2k Kodak ES 4.0	Silver-coated hollow glass spheres
Liu et al. [14.158]	2004	3-D	Axial compressor	Tip flow field	Frequency-doubled (532 nm) Nd:YAG, 150 mJ/pulse	1280 × 1024 pixels^2 CCD	Smoke generator
Soranna et al. [14.159]	2004	2-D	Axial pump	Wake–boundary layer interaction	Frequency-doubled (532 nm) Nd:YAG, 1 mm thickness, 120 mJ/pulse	2k × 2k Kodak ES 4.0	Silver-coated hollow glass spheres
Lee et al. [14.160]	2004	3-D	Marine propeller	Propeller wake	Frequency-doubled (532 nm) Nd:YAG, 3 mm thickness, 125 mJ/pulse	1024 × 1024 pixels^2 CCD	Silver-coated hollow glass spheres
Sankovic et al. [14.137]	2004	2-D	Rotary blood pump	Flow structure	Frequency-doubled (532 nm) Nd:YAG, 0.37 mm thickness, 50 mJ/pulse	768 × 484 pixels^2 crosscorrelation camera	Silver-coated hollow glass spheres
Choi et al. [14.136]	2004	2-D	Centrifugal impeller	Flow structure	Frequency-doubled (532 nm) Nd:YAG, 0.5 mm thickness, 20 mJ/pulse	1300 × 1030 pixels^2 CCD	Nylon 12 particles, 30 μm
Chow et al. [14.161]	2005	2-D	Axial pump	Rotor wake, Large Eddy Simulation (LES) modelling	Frequency-doubled (532 nm) Nd:YAG, 1 mm thickness, 120 mJ/pulse	2k × 2k Kodak ES 4.0	Silver-coated hollow glass spheres
Soranna et al. [14.162]	2005	2-D	Axial pump	Wake–blade interactions	Frequency-doubled (532 nm) Nd:YAG, 1 mm thickness, 120 mJ/pulse	2k × 2k Kodak ES 4.0	Silver-coated hollow glass spheres

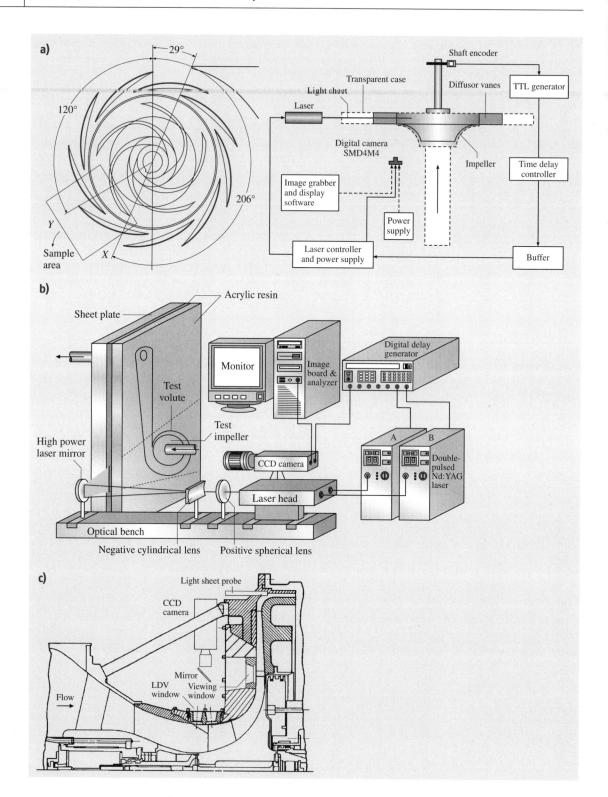

the key challenge in application of PIV in turbomachines. In addition, light reflections from the blade surfaces and end walls adversely affect the quality of images, especially near the boundaries. In applications involving air flows and realistic geometries, these problems seem to be unavoidable. As a result, most PIV data obtained in multistage turbomachines have covered limited areas, located away from boundaries and mostly between blade rows (see Table 14.6 for references). Characteristic setups for axial air turbomachines used by *Copenhaver* et al. [14.120], *Lang* et al. [14.121], *Balzani* et al. [14.122] and *Wernet* et al. [14.123] are presented in Figs. 14.12a-d. In almost all axial turbomachine applications, the laser sheet has been introduced using periscopic optical inserts (Figs. 14.12c,d), which are located upstream or downstream of the measurement domain, and attached to traverses that enable measurements at different spanwise planes. The cameras view the illuminated plane through windows installed on the outer casing, which are curved internally to match the contour of the machine.

Examples of various PIV setups for measurements within centrifugal turbomachinery, and referenced in Table 14.6, are presented in Fig. 14.13. In most applications, the sample area has been illuminated using side windows [14.136, 143] or inserted probes [14.146]. In a recent study, *Estevadeordal* et al. [14.163] introduced a fiber-optic PIV system that could be used in flows without a direct optical access, and implemented it to measure the flow within an axial turbine.

To resolve the limited access problem, *Uzol* et al. [14.124] and *Chow* et al. [14.151] have introduced a refractive-index-matched liquid facility that enabled unobstructed PIV measurements within an entire stage by matching the optical index of refraction of the blades with that of the working fluid, a concentrated solution of NaI in water ($\approx 64\%$ by weight). This method not only made it possible to obtain complete optical access to an entire stage, it also minimized the light reflection from boundaries as well, allowing high-resolution measurements of blade boundary layers. The optical setup of this system is illustrated in Fig. 14.12e. A miniature refractive-index-matching facility made entirely from transparent material was also used recently by *Sankovic*

Fig. 14.13a–c Sample PIV setups for centrifugal turbomachinery: **(a)** centrifugal pump (*Sinha* and *Katz* [14.143]); **(b)** centrifugal impeller (*Choi* et al. [14.136]); **(c)** centrifugal compressor (*Wernet* et al. [14.146]) ◄

et al. [14.137] to measure the flow structure within a centrifugal rotary blood pump.

Data Acquisition. Although high-speed PIV systems have started to appear in recent years, a common typical turbomachinery PIV system with low-frame-rate cameras cannot be used to obtain time-resolved flow field measurements. To obtain phase-locked data at specific phase of the rotor blade-passing period, a signal obtained from a shaft encoder is fed to a digital delay generator, which then triggers the lasers and the cameras. Multiple instantaneous measurements obtained at a specific phase are ensemble-averaged to obtain the phase-averaged flow and turbulence parameters. The database should be sufficiently large to obtain converged statistics. In many previous applications, 100 and 1000 samples have been shown to provide converged phase-averaged and turbulence quantities, respectively [14.164, 165]. However, it is essential to check convergence in individual applications. By averaging phase-locked data obtained over many phases, which are distributed evenly over an entire cycle, one can determine the *average-passage* or time-mean flow structure. Any averaging of a flow field introduces stresses, and time averaging of phase-locked data introduces the deterministic stresses [14.166, 167]. A detailed discussion on how to calculate the spatial distributions of deterministic stresses within a turbomachine based on PIV data obtained in discrete phases can be found in *Sinha* et al. [14.168] and *Uzol* et al. [14.150], [14.153].

Samples of Results

Figures 14.14–14.17 show selected sample results of both two-dimensional and stereoscopic PIV measurements performed within various axial and centrifugal turbomachinery facilities. A more-extended set of references is provided in Table 14.6. A two-dimensional, phase-averaged relative velocity vector map measured at 70% span within the transonic axial compressor facility at the NASA Glenn Research Center [14.145] is presented in Fig. 14.14a. Shock waves attached to the rotor leading edge of blades are clearly visible. Figure 14.14b shows the unsteady 2-D velocity field within the stator passages of the transonic research compressor facility at Purdue University [14.149]. These results were used to investigate the decay of upstream rotor wakes as they are convected through the stator passages. Figure 14.14c presents multiplane two-dimensional data, showing the spanwise variation of the rotor relative velocity within the axial compressor facility at the Von Karman Institute [14.122]. Figure 14.14d presents stereoscopic

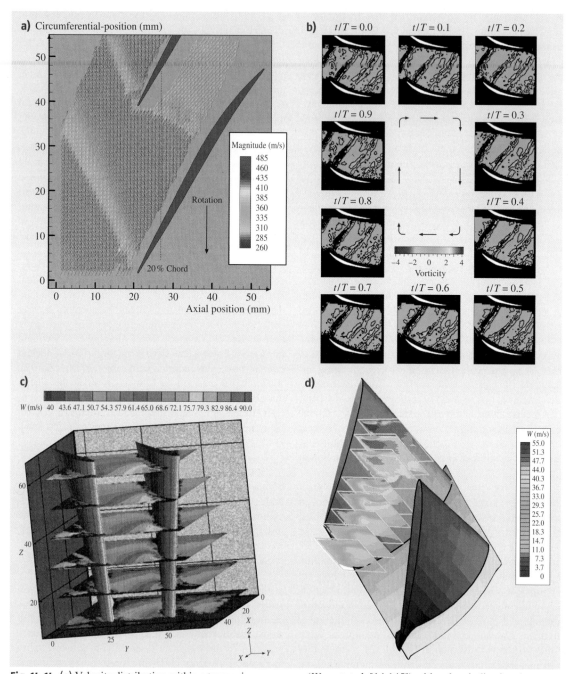

Fig. 14.14 (a) Velocity distribution within a transonic compressor (*Wernet* et al. [14.145]) with colors indicating the vector magnitude; (b) unsteady vorticity field within a stator passage (*Sanders* et al. [14.149]); (c) rotor relative velocity field from hub to tip (*Balzani* et al. [14.122]); (d) average radial velocity distribution near stall, stereoscopic measurement (*Liu* et al. [14.158])

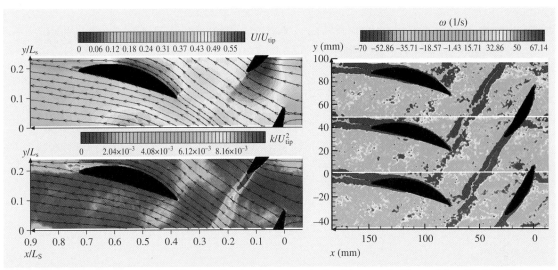

Fig. 14.15 Sample PIV data obtained in the refractive-index-matched axial pump facility at Johns Hopkins University: phase-averaged velocity field (*top left*), turbulent kinetic energy (*bottom left*) and vorticity (*right*) at mid-span within an entire stage (*Chow* et al. [14.151])

PIV data of the flow near stall conditions in the low-speed axial compressor facility at Beijing University [14.158]. These measurements provide clear views on the complex, spatially nonuniform flow within axial turbomachines. Yet, in all cases, reflection from the surfaces and limitations in access prevent coverage of the entire stage and prevent measurements close to boundaries. As noted before, in air facilities, the only ones that can examine compressibility effects, these problems are extremely difficult to resolve.

Figures 14.15 and 14.16 present sample results from the two-dimensional and stereoscopic measurements performed in the refractive-index-matched axial turbomachinery facility at Johns Hopkins University. The unobstructed view allowed detailed investigations of wake–blade, wake–wake, wake–boundary layer, and tip vortex–blade interactions in an actual multistage environment. Figure 14.15 shows characteristic distributions of phase-averaged velocity, vorticity and turbulent kinetic energy covering the entire mid-span of a second stage. Data obtained at 10 different phases was used for calculating the distributions of average-passage flow and deterministic stresses (*Uzol* et al. [14.150, 153]). High-resolution measurements (vector spacing of 117 μm) that required patching of several sample areas were used for characterization of turbulence parameters, including Reynolds and subgrid (SGS) stresses and associated energy fluxes, due to shearing and straining of wakes during passage in downstream blade rows (Figs. 14.16a,b, *Chow* et al. [14.151, 154, 161]; *Soranna* et al. [14.162, 169]). These data were also used to show that unsteady flow caused by upstream wake impingement stabilized the structure of rotor blade boundary layers (Fig. 14.16c, *Soranna* et al. [14.159]). Samples from stereoscopic measurements in multiple, closely spaced planes, spanwise variations and tip vortex–blade interactions, are presented in Fig. 14.16d.

Figure 14.17a presents sample 2-D PIV data obtained in a centrifugal pump (Fig. 14.13a) seeded with fluorescent particles and filtering of images to eliminate adverse effects of reflections [14.143]. Figure 14.17b shows sample data obtained in a high-speed centrifugal compressor (Fig. 14.13c) operating both at the design point and during surge [14.144, 146].

14.3.3 Applications of Laser Two-Focus Velocimetry (L2F)

The L2F technique measures the time of flight of small particles between two highly focused laser spots. Unlike LDV, an instantaneous realization with L2F is generally not a valid measurement, since it is not possible to ensure that the same particle passes through both laser spots. Therefore, L2F has been utilized as a statistics-based measurement device. In typical applications, one of the laser beams is rotated around the other, and many time-of-flight measurements are performed over a range

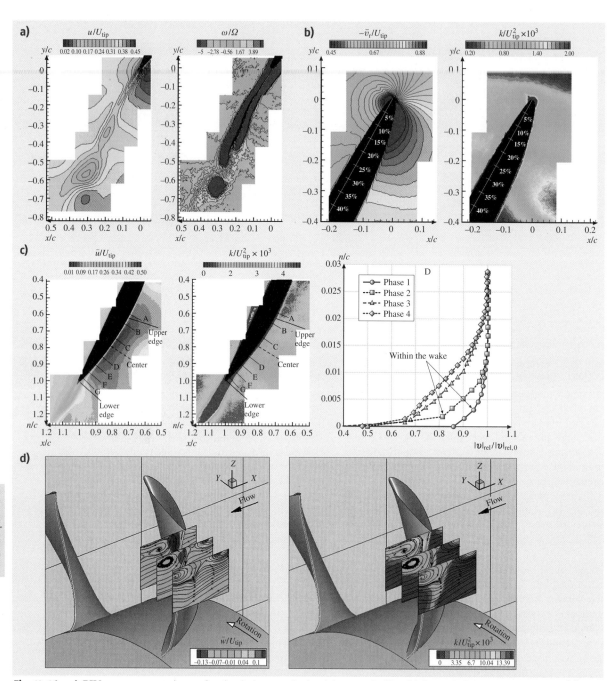

Fig. 14.16a–d PIV measurements in a refractive-index-matched axial pump facility: high-resolution measurements of (**a**) rotor wake (*Chow* et al. [14.154]), (**b**) wake–blade interaction near the rotor leading edge (*Soranna* et al. [14.162]), (**c**) wake–boundary layer interactions near the rotor trailing edge (*Soranna* et al. [14.159]) at midspan, (**d**) phase-averaged radial velocity (*left*), and turbulent kinetic energy (*right*) between 50–93% span, obtained using stereoscopic measurements (*Uzol* et al. [14.156])

Table 14.7 Sample studies that have used laser-two-focus velocimetry (L2F) technique in investigating different aspects of turbomachinery flow fields

Laser two-focus velocimetry (L2F) Author(s)	Year	Type of machine	Subject of study
McDonald et al. [14.170]	1980	Transonic compressor rotor	Flow structure
Fagan and Fleeter [14.104]	1994	Low-speed centrifugal compressor	Flow structure
Calvert and Stapleton [14.171]	1994	Transonic fan	Flow structure
Ivey and Swoboda [14.43]	1998	Low-speed axial compressor	Tip clearance effects
Kost et al. [14.63]	2000	Axial turbine	Unsteady flow
Ottavy et al. [14.172]	2001	Transonic axial compressor	Wake–shock interaction
Tiedemann and Kost [14.65]	2001	High-pressure axial turbine	Rotor–stator clocking
Schodl et al. [14.173]	2002	Transonic centrifugal compressor	3-D flow structure
Ziegler et al. [14.174]	2003	High-speed centrifugal compressor	Blade–wake interaction

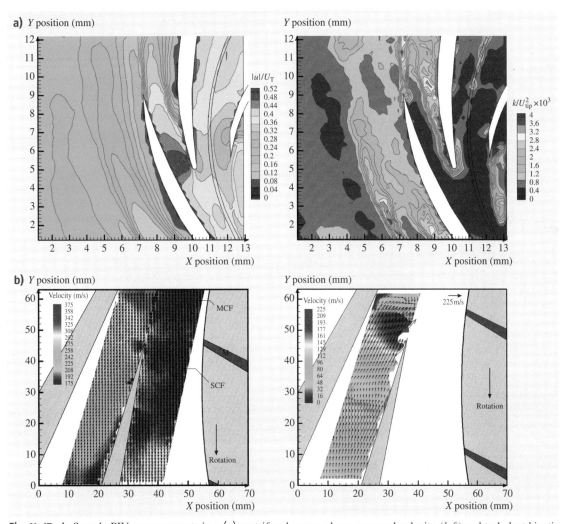

Fig. 14.17a,b Sample PIV measurements in a (**a**) centrifugal pump, phase-averaged velocity (*left*) and turbulent kinetic energy (*right*) (*Sinha* and *Katz* [14.143]), (**b**) high-speed centrifugal compressor diffuser velocity vector maps at the design point (*left*, *Wernet* [14.144]) and during a surge (*right*, *Wernet* et al. [14.146]) MCF: Main Clearance Flow, SCF: Splitter Clearance Flow

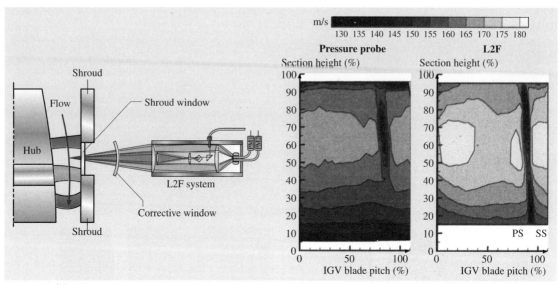

Fig. 14.18 (a) A sample L2F measurement setup within an axial compressor, and (b) corresponding data compared with three-hole pressure probe measurements (*Ottavy* et al. [14.172])

of measurement directions. The data produces a two-dimensional probability histogram, which is used to calculate the average velocity and flow angle, along with an estimate of turbulence intensity. Table 14.7 provides a list of sample studies that have used L2F to study flows within turbomachines.

Fagan and *Fleeter* [14.104] compared the results of LDV and L2F measurements performed within a large-scale, low-speed centrifugal compressor. They found that, although L2F allowed measurements closer to boundaries compared to LDV, the LDV data demonstrated more consistency, in terms of conservation of mass and a reasonable distribution of work. They concluded that LDV would be preferable to L2F due to the latter's low effective sampling rate and upper limits of turbulence intensity. Both they and *Schodl* [14.177, 178] suggest an upper limit of approximately 15% for accurate turbulence measurements.

Similar to LDV, applications of L2F in turbomachines are adversely affected by optical distortions during the passage of the laser beams through curved optical access windows. The problem is more severe for L2F since the distortions prevent the creation of acceptably focused spots, resulting in a total loss of data. *Ottavy* et al. [14.179] studied these distortions and proposed a corrective method, which restored the optical focus, and subsequently applied their setup within an axial transonic compressor [14.172]. Figure 14.18 shows their setup, along with a comparison of their L2F data to results of three-hole pressure probe measurements performed downstream of an IGV and upstream of a rotor row.

Schodl et al. [14.173] introduced a new L2F technique called Doppler laser two-focus velocimeter, and applied it to obtain data within a transonic centrifugal compressor. In this technique, the time-of-flight data was used to measure the velocity vector components in the plane perpendicular to the optical axis. The Doppler shift of the scattered light, as measured using iodine absorption (Sect. 14.3.4), was used to obtain the velocity component along the optical axis, i.e., similar to the Doppler global velocimetry (DGV) technique.

14.3.4 Applications of Doppler Global Velocimetry (DGV)

Doppler global velocimetry, also known as planar Doppler velocimetry (PDV), is a planar flow measurement technique, which is based on measuring the Doppler shift of light scattered from tracer particles. The fre-

Fig. 14.19 (a) Comparison of surface pressure distribution on a rotor suction side in a transonic axial compressor obtained using PSP (*left*) to CFD predictions (*right*) (*Navarra* et al. [14.175]); (b) Experimental setup for PSP measurements in an automotive turbocharger, and (c) pressure distributions on the compressor inlet wall and on the impeller blade (*Gregory* [14.176]) ▶

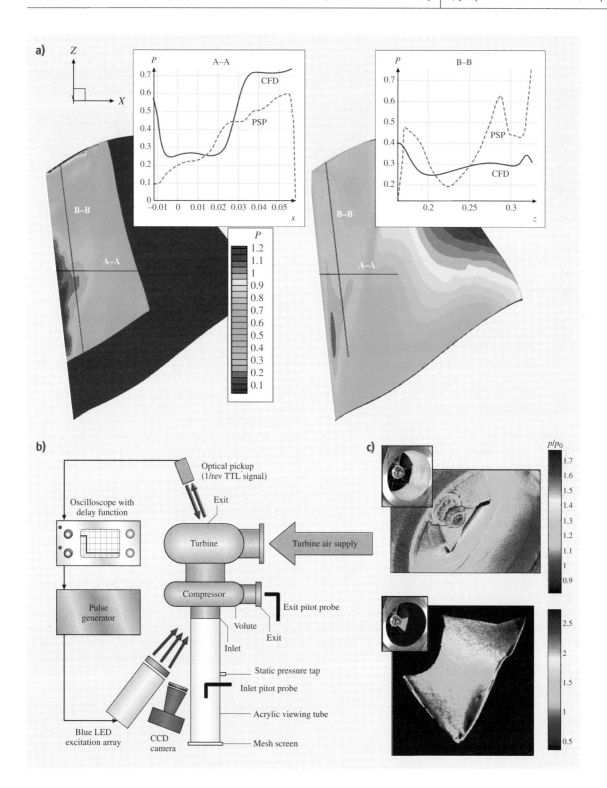

Table 14.8 Sample studies that have used pressure-sensitive paint (PSP) technique in investigating various aspects of turbomachinery flow fields

Pressure-sensitive paint (PSP)			
Author(s)	Year	Type of machine	Subject of study
Sabroske et al. [14.184]	1995	Axial compressor	Blade pressure distributions
Liu et al. [14.185]	1997	High-speed axial compressor	Blade surface pressure measurements with shocks
Engler et al. [14.186]	2000	Axial turbine	Shock movement and corner stall
Navarra et al. [14.175]	2001	Transonic axial compressor	Blade surface pressure in transonic conditions
Lepicovsky and *Bencic* [14.187]	2002	Supersonic throughflow fan	Effect of changing operating conditions
Gregory [14.176]	2004	Centrifugal compressor	Effect of inlet distortion on surface pressures

quency shift is measured by means of a frequency–intensity converter, which typically consists of measuring the attenuation of the scattered light as it passes through an iodine-vapor absorption cell [14.180]. This technique is yet to be implemented for turbomachinery flows, but setups designed for turbomachine applications have already been introduced [14.181, 182]. *Wernet* [14.183] introduced a hybrid PIV/DGV technique and proposed to use it as a viable technique that could be applied to flows with limited optical access, such as turbomachinery flow fields.

14.3.5 Applications of Pressure-Sensitive Paint (PSP)

The current PSP method is based on covering a surface with luminescent coatings that contain sensor molecules embedded in a transparent oxygen-permeable binder. When illuminated by light of appropriate wavelength, the excited sensors molecules emit light at a different wavelength, whose intensity is inversely proportional to the partial pressure of oxygen near the surface. Thus, the intensity of luminescence provides the distribution of surface static pressures. Further details about this technique can be found in a comprehensive review by *Bell* et al. [14.188]. This method is an attractive alternative to conventional surface pressure measurement techniques since pressure taps or flush-mounted transducers (Sect. 14.2.2) only provide data at discrete points, and are restricted to sites where they can be installed on the blade surfaces. It is typically impossible to instrument regions of most interest, such as thin leading edges and sharp corners. Furthermore, many sensors are required to obtain a reasonable spatial distribution, making the measurements time consuming and expensive. PSP provide the spatial distribution of surface pressures at high resolution, enabling measurements of aerodynamic loads on rotor and stator blades. It also serves as a quantitative surface visualization tool that shows the location of shocks, and boundary-layer separation and reattachment points.

Unlike external aerodynamics studies, where PSP is well established, its application to turbomachinery components is rather limited. Sample studies that have used pressure-sensitive paint in turbomachines are summarized in Table 14.8. Selecting two examples, *Navarra* et al. [14.175] used PSP to measure the pressure distribution on the suction surface of the first stage rotor of a state-of-the-art, full-scale transonic compressor. Figure 14.19a shows a comparison of their data with computational fluid dynamics (CFD) results. Recently, *Gregory* [14.176] presented results of PSP measurements in the centrifugal compressor of a Garrett T25 turbocharger, which is typically used in automotive applications. Using porous polymer/ceramic PSP they investigated the effect of inlet distortion on the pressure distribution on the surface of blades and on the wall at the inlet to the compressor. Figure 14.19b presents their experimental setup, showing the transparent inlet wall that allowed visual access, the blue light-emitting diode (LED) used to excite the PSP, and the camera that recorded the intensity of luminescence from painted surface. Figure 14.19b shows the unsteady wall pressure distribution on the inlet wall (top) and on the impeller blade (bottom).

14.4 Concluding Remarks

This chapter attempts to provide a comprehensive but brief summary of measurement techniques that have been used for studying flows within turbomachines. Being focused on techniques, we discuss issues that are unique to applications of various sensors in such a complex flow environment. For each class of sensors, a table provides examples of applications, along with sample results. We also attempt to include examples of implementation in various types of machines, e.g., axial and centrifugal compressors and pumps, large and small facilities, full-scale devices and small-scale models. However, one should bear in mind that one cannot cover more than 50 years of experimental studies of flow within turbomachines in a single chapter, or in a single book for that matter. The reader should refer to the several books mentioned in the introduction for a more-comprehensive treatment of the fluid-mechanics and heat-transfer problems within turbomachines. Besides serving as samples, the cited references can also be used as starting points for more-detailed searches on specific topics. As is evident, the level of sophistication of instruments as well as the size of database and level of detail that they can resolve has been increasing rapidly over the years, in part in response to the challenge posed by the rapid advancements in CFD tools. The ever-increasing levels of detail revealed by recent multipoint flow measurement tools enable us to elucidate and quantify complex interactions between multiple flow features, which is characteristic to turbomachines. The resulting insights are essential both for validation of predicted flows, as well as for challenging the modeling community to continue addressing discrepancies.

References

14.1 J.H. Horlock: *Axial Flow Compressors* (Kreiger Publishing Co., Melbourne FL. 1973)
14.2 J.H. Horlock: *Axial Flow Turbines* (Kreiger, Melbourne FL. 1973)
14.3 A.T. Sayers: *Hydraulic and Compressible Flow Turbomachines* (McGraw-Hill, Berkshire 1990)
14.4 E. Logan Jr.: *Turbomachinery: basic theory and applications* (Marcel Dekker, New York 1993)
14.5 B. Lakshminarayana: *Fluid Dynamics and Heat Transfer Turbomachin* (Wiley, New York 1996)
14.6 J.A. Schetz, A.E. Fuhs: *Handbook of Fluid Dynamics and Fluid Machinery* (Wiley, New York 1996)
14.7 S.L. Dixon: *Fluid Mechanics and Thermodynamics of Turbomachinery 4th edition*, 4th edn. (Butterworth-Heinemann, Woburn 1998)
14.8 D.G. Wilson, T. Korakianitis: *The Design of High-Efficiency Turbomachinery and Gas Turbines 2nd edition* (Prentice-Hall, Upper Saddle River 1998looseness1)
14.9 B. Lakshminarayana: Techniques for aerodynamic and turbulence measurements in turbomachinery rotors, J. Eng. Power **103**(2), 374–392 (1981)
14.10 V.S.P. Chaluvadi, A.I. Kalfas, M.R. Banieghbal, H.P. Hodson, J.D. Denton: Blade-Row Interaction in a High Pressure Turbine, J. Propulsion Power **17**(4), 892–901 (2001)
14.11 J.R. Weske: An investigation of the aerodynamic characteristics of a rotating axial flow blade grid, NACA Technical Note **1128** (1947)
14.12 C.A. Gorton, B. Lakshminarayana: A method of measuring the three-dimensional mean flow and turbulence quantities inside a rotating turbomachinery passage, J. Eng. Power Ser. A **98**(2), 137–146 (1976)
14.13 B. Reynolds, B. Lakshminarayana, A. Ravindranath: Characteristics of the near wake of a compressor of a fan rotor blade, AIAA J. **17**(9), 959–967 (1979)
14.14 R.P. Dring, H.D. Joslyn: Measurement of turbine rotor blade flows, J. Eng. Power **103**, 400–405 (1981)
14.15 H.D. Joslyn, R.P. Dring: Three-dimensional flow in an axial turbine: Part 1 – Aerodynamic Mechanisms, J. Turbomachin. **114**(1), 61–70 (1992)
14.16 B. Lakshminarayana, A. Poncet: A Method of Measuring Three-Dimensional Wakes in Turbomachinery, J. of Fluids Eng., J. Fluids **96**(2), 87–91 (1974)
14.17 C. Hah, B. Lakshminarayana: Freestream Turbulence Effects on the Development of a Rotor Wake, AIAA J. **19**(6), 724–730 (1980)
14.18 B. Lakshminarayana, T.R. Govindan, B. Reynolds: Effects of rotation and blade incidence on properties turbomachinery rotor wake, AIAA J. **20**(2), 245–253 (1982)
14.19 Y. Dong, N.A. Cumpsty: Compressor blade boundary layers: Part 2 – Measurements with incident wakes, J. Turbomachin. **112**(2), 231–240 (1990)
14.20 H.P. Hodson, I. Huntsman, A.B. Steele: An Investigation of Boundary Layer Development in a Multi-stage LP Turbine, J. Turbomachin. **116**(3), 375–383 (1994)
14.21 T.R. Camp, H.W. Shin: Turbulence intensity and length scale measurements in multistage compressors, J. Turbomachin. **117**(1), 38–46 (1995)
14.22 A.S. Witkowski, T.J. Chmielniak, M.D. Strozik: Experimental study of a 3D wake decay and secondary flows behind a rotor blade row of a low speed compressor stage, ASME **96-GT-415** (1996)
14.23 D.E. Halstead, D.C. Wisler, T.H. Okiishi, G.J. Walker, H.P. Hodson, H.W. Shin: Boundary layer development in axial flow compressors and turbines: Part 1 – composite picture, J. Turbomachin. **119**(1), 114–127 (1997)
14.24 S.T. Hsu, A.M. Wo: Near-wake measurement in a rotor/stator axial compressor using slanted hot-wire technique, Exp. Fluids **23**(5), 441–444 (1997)

14.25 A. Sentker, W. Riess: Measurement of unstead flow and turbulence in a low speed axial compressor, Exp. Therm. Fluid Sci. **17**, 124–131 (1998)

14.26 D. Ristic, B. Lakshminarayana: Three-dimensional blade boundary layer and endwall flow development in the nozzle passage of a single stage turbine, J. Fluids Eng. **120**, 570–579 (1998)

14.27 J. Prato, B. Lakshminarayana, N. Suryavamshi: Steady and unsteady three-dimensional flow field downstream of an embedded stator in a multi-stage axial flow compressor: Part 1 – unsteady velocity field, ASME **98-GT-521** (1998)

14.28 M. Furukawa, K. Saiki, K. Nagayoshi, M. Kuroumaru, M. Inoue: Effect of stream surface inclination on tip leakage flow fields in compressor rotors, J. Turbomach. **120**(4), 683–394 (1998)

14.29 A. Sentker, W. Riess: Experimental investigation of turbulent wake-blade interaction in axial compressors, Int. J. Heat Fluid Flow **21**, 285–290 (2000)

14.30 S. Velarde-Suarez, R. Ballasteros-Tajadura, C. Santolaria-Morros, J. Gonzalez-Perez: Unsteady flow pattern characteristics downstream of a forward-curved blades centrifugal fan, J. Fluids Eng. **123**, 265–270 (2001)

14.31 Y.H. Shin, R.L. Elder, I. Bennett: Boundary layer measurement on the blade surface of a multi-stage axial flow compressor, Proc. ASME Turbo Expo 2003 (ASME, Atlanta 2003)

14.32 C. McLean: *The Aerodynamic Effects of Wheelspace Coolant Injection into the Mainstream Flow of a High Pressure Gas Turbine*, Ph.D. Dissertation (Pennsylvania State University, University Park 2000)

14.33 W.F. O'Brien Jr., H.L. Moses, H.R. Carter: Multi-channel telemetry system for flow research on turbomachine rotors, ASME **74-GT-112** (1974)

14.34 R.W. Light: *Development of rotating to stationary data transfer system based on FM telemetry*, M.S. Thesis, Mechanical Engineering (Virginia Poly. Inst. And State Univ., Blacksburg 1975)

14.35 A. Adler: Telemetry for Turbomachinery, Mech. Eng. **101**(3), 30–35 (1979)

14.36 J.G.B. Worthy: The design, development and operation of gas turbine radio telemetry system, J. Eng. Power **103**(2), 473–479 (1981)

14.37 A. Zeisberger, L. Matziol, F. Deubert: Modern telemetry systems and rotating instrumentation, Proc. ASME Turbo Expo 2002 (Amsterdam 2002)

14.38 J. Neustein: Low Reynolds number experiments in an axial-flow turbomachine, J. Eng. Power, 257–295 (1964)

14.39 S. Kang, C. Hirsch: Experimental study on the three-dimensional flow within a compressor cascade with tip clearance: Part 1 – The Tip Leakage Vortex, J. Turbomachin. **115**(3), 444–451 (1993)

14.40 J.F. Carrotte, K.F. Young, S.J. Stevens: Measurement of the flow field within a compressor outlet guide vane passage, J. Turbomachin. **117**, 29–37 (1995)

14.41 J. Prato, B. Lakshminarayana, N. Suryavamshi: Exit flow field of an embedded stator in a multi-stage axial flow compressor, J. Propulsion Power **113**(2), 169–177 (1997)

14.42 A. Doukelis, K. Mathioudakis, K. Papailiou: The effect of tip clearance gap size and wall rotation on the performance of a high-speed annular compressor cascade, ASME **98-GT-38** (1998)

14.43 P.C. Ivey, M. Swoboda: Leakage effects in the rotor tip clearance region of a multi-stage axial compressor: Part 1 – Innovative Experiments, ASME **98-GT-591** (1998)

14.44 D. Dey, C. Camci: Development of tip clearance flow downstream of a rotor blade with coolant injection from a tip trench, Proc. International Symposium on Rotating Machinery-ISROMAC (Honolulu 2000)

14.45 X. Xiao, A.A. McCarter, B. Lakshminarayana: Tip clearance effects in a turbine rotor: Part 1 – pressure field and loss, J. Turbomachin. **123**(2), 296–304 (2001)

14.46 C. McLean, C. Camci, B. Glezer: Mainstream aerodynamic effects due to wheelspace coolant injection in a high pressure turbine stage: Part 1 - aerodynamic measurements in the stationary frame, J. Turbomachin. **123**(4), 687–696 (2001)

14.47 C. McLean, C. Camci, B. Glezer: Mainstream aerodynamic effects due to wheelspace coolant injection in a high pressure turbine stage: Part 2 - aerodynamic measurements in the rotational frame, J. Turbomachin. **123**(4), 697–703 (2001)

14.48 A.A. McCarter, X. Xiao, B. Lakshminarayana: Tip clearance effects in a turbine rotor: Part 2 - velocity field and flow physics, J. Turbomachin. **123**(2), 305–313 (2001)

14.49 S. Coldrick, P. Ivey, R. Wells: Considerations for using 3-D pneumatic probes in high speed axial compressors, J. Turbomachin. **125**(1), 149–154 (2003)

14.50 G. Pullan, J.D. Denton, M. Dunkley: An experimental and computational study of the formation of a streamwise shed vortex in a turbine stage, J. Turbomachin. **125**(2), 291–297 (2003)

14.51 J.L. Gilarranz, A.J. Ranz, J.A. Kopko, J.M. Sorokes: On the Use of Five-Hole Probes in the Testing of Industrial Centrifugal Compressors, J. Turbomachin. **127**(1), 91–106 (2005)

14.52 D.M. Boyd, S. Fleeter: Axial compressor blade-to-blade unsteady aerodynamic variability, J. Propulsion Power **19**(2), 242–249 (2003)

14.53 V.S.P. Chaluvadi, A.I. Kalfas, H.P. Hodson: Vortex Transport and Blade Interactions in High Pressure Turbines, J. Turbomachin. **126**(3), 395–405 (2004)

14.54 J.L. Kerrebrock, A.H. Epstein, W.T. Thompkins Jr.: Miniature high frequency sphere probe, Proc. Joint Fluids Eng. Gas Turbine Conf. (New Orleans 1980)

14.55 W.T. Cousins, W.F.J. OBrien, M.R. Sexton: *Dynamic pressure response with stall on axial flow compressor rotor blades*, AIAA 1981-69, 19th AIAA Aerospace Sciences Meeting 1981, St. Louis (AIAA, Reston 1981)

14.56 G.L. Giannissis, A.B. McKenzie, R.L. Elder: Experimental investigation of rotating stall in a mis-matched three-stage axial flow compressor, J. Turbomachin. **111**(4), 418–425 (1989)

14.57 M.A. Cherrett, J.D. Bryce: , Unsteady Viscous Flow in a High Speed Core Compressor, J, Turbomachinery **114**(2), 287–294 (1992)

14.58 R.W. Ainsworth, J.L. Allen, J.J.M. Batt: The development of fast response aerodynamic probes for flow measurements in turbomachines, J. Turbomachin. **117**, 625–634 (1995)

14.59 M.K. Fabian, E.J. Jumper: *Unsteady pressure distributions around compressor vanes in an unsteady, transonic cascade*, AIAA 1995-302, 33rd AIAA Aerospace Sciences Meeting 1995 (AIAA, Reno 1995)

14.60 R. Laborde, P. Chantrel: Tip clearance and tip vortex cavitation in an axial flow pump, J. Fluids Eng.. **119**, 680–685 (1997)

14.61 Y. Dong, B. Lakshminarayana, D. Maddock: Steady and unsteady flow field at pump and turbine exits of a torque converter, J. Fluids Eng. **120**, 538–548 (1998)

14.62 G. Roduner C.; Koppel P.; Kupferschmied P.; Gyarmathy: Comparison of measurement data at the impeller exit of a centrifugal compressor measured with both pneumatic and fast-response probes, J. Turbomachin. **121**(3), 609–618 (1999)

14.63 F. Kost, F. Hummel, M. Tiedmann: Investigation of the Unsteady Rotor Flow Field in a Single HP Turbine Stage, ASME Paper 2000-GT-432 (2000)

14.64 R. Mailach, I. Lehmann, K. Vogeler: Rotating instabilities in an axial compressor originating from the fluctuationg tip vortex, J. Turbomachin. **123**(3), 453–463 (2001)

14.65 M. Tiedemann, F. Kost: Some Aspects of Wake-Wake Interactions Regarding Turbine Stator Clocking, J. Turbomachin. **123**, 526–533 (2001)

14.66 H. Lohrberg, B. Stoffel, R. Fortes-Patella, O. Coutier-Delgosha, J.L. Reboud: Numerical and experimental investigation on the cavitating flow in a cascade of hydrofoils, Exp. Fluids **33**(4), 578–586 (2002)

14.67 M. Nohmi, Y. Iga, A. Goto, T. Ikahagi: Experimental and numerical study of cavitation breakdown in a centrifugal pump, ASME/JSME Joint Fluids Engineering Conference (Honolulu, 2003)

14.68 T.J. Leger, D.A. Johnston, J.M. Wolff: Flex circuit sensor array for surface unsteady pressure measurements, J. Propulsion Power **20**(4), 754–758 (2004)

14.69 Z.S. Spakovszky: Backward travelling rotating stall waves in centrifugal compressors, J. Turbomachin. **126**(1), 1–12 (2004)

14.70 M. Schleer, S.S. Hong, M. Zangeneh, C. Roduner, B. Ribi, F. Ploger, R.S. Abhari: Investigation of an inversely designed centrifugal compressor stage: Part 2 – experimental investigations, J. Turbomachin. **126**(1), 82–90 (2004)

14.71 C. Camci, D. Dey, L. Kavurmacioglu: Aerodynamics of tip leakage flows near partial squealer rims in an axial flow turbine stage, J. Turbomachin. **127**(1), 14–24 (2005)

14.72 R.C. Lockhart, G.J. Walker: The influence of viscous interactions on the flow downstream of an axial compressor stage, Proc. 2nd Int. Symp. Air Breathing Engines (Sheffield 1974)

14.73 E.M. Greitzer: Surge and rotating stall in axial flow compressors Part 2 – experimental results and comparison with theory, J. Eng. Power **98**(2), 199–217 (1976)

14.74 D.G. Gregory-Smith, J.G.E. Cleak: Secondary flow measurements in a turbine cascade with high inlet turbulence, J. Turbomachin. **114**(1), 173–183 (1992looseness1)

14.75 I.J. Day: Axial compressor performance during surge, J. Propulsion Power **10**(3), 329–336 (1994)

14.76 K.H. Kim, S. Fleeter: Compressor unsteady aerodynamic response to rotating stall and surge excitations, J. Propulsion Power **10**(5), 698–708 (1994)

14.77 H.E. Gallus, J. Zeschky, C. Hah: Endwall and unsteady flow phenomenon in an axial turbine stage, ASME **94-GT-143** (1994)

14.78 U. Reinmoller, B. Stephan, S. Schmidt, R. Niehuis: Clocking effects in a 15 stage axial turbine-steady and unsteady experimental investigations supported by numerical simulations, J. Turbomachin. **124**, 52–60 (2002)

14.79 A.J. Sanders, S. Fleeter: Rotor blade-to-blade wake variability and effect on downstream vane response, J. Propulsion Power **18**(2), 456–464 (2002)

14.80 R.T. Johnston, S. Fleeter: Airfoil row/wake interactions in a high speed axial compressor, J. Propulsion Power **18**(6), 1280–1288 (2002)

14.81 M.T. Schobeiri, B. Ozturk, D.E. Ashpis: On the physics of flow separation along a low pressure turbine blade under unsteady flow conditions, ASME Turbo Expo 2003 (ASME, Atlanta 2003)

14.82 R.T. Johnston, S. Fleeter: Three-dimensional time-resolved inlet guide vane-rotor potential field interaction, J. Propulsion Power **20**(1), 171–179 (2004)

14.83 A.L. Treaster, A.M. Yocum: The calibration and application of five-hole probes, ISA Trans. **18**(3), 23–34 (1979)

14.84 N. Sitaram, B. Lakshminarayana, A. Ravindranath: Conventional Probes for the relative flow measurement in a turbomachinery rotor blade passage, J. Eng. Power **103**(2), 406–414 (1981)

14.85 E. Gottlich, F. Neumayer, J. Woisetschlager, W. Sanz, F. Heitmeir: Investigation of stator-rotor interaction in a transonic turbine stage using laser Doppler velocimetry and pneumatic probes, J. Turbomachin. **126**(2), 297–305 (2004)

14.86 H.H. Bruun: *Hot-wire Anemometry – Principles and Signal Analysis* (Oxford Univ. Press, New York 1995)

14.87 R.J. Goldstein (Ed.): *Fluid Mechanics Measurements* (Taylor Francis, Washington 1996)

14.88 L.H. Smith Jr.: Wake dispersion in turbomachines, J. Basic Eng. Ser. D **88**(3), 688–690 (1966)

14.89 G.J. Walker, A.R. Oliver: The effect of interaction between blade rows in an axial flow compressor on the noise generated by blade interaction, J. Eng. Power **94**, 241–248 (1972)

14.90 C.R. Gossweiler, P. Kupferschmied, G. Gyarmathy: On fast response probes: Part 1-Technology, calibration and application to turbomachinery, J. Turbomachin. **117**(4), 611–617 (1995)

14.91 R.W. Ainsworth, R.J. Miller, R.W. Moss, S.J. Thorpe: Unsteady pressure measurement, Meas. Sci. Technol. **11**, 1055–1076 (2000)

14.92 C.H. Sieverding, T. Arts, R. Denos, J.F. Brouckaert: Measurement techniques for unsteady flows in turbomachines, Exp. Fluids **28**, 285–321 (2000)

14.93 L. Lofdahl, M. Gad-el-Hak: MEMS-based pressure and shear stress sensors for turbulent flows, Meas. Sci. Technol. **10**(8), 665–686 (1999)

14.94 P. Kupferschmied, P. Koppel, W. Gizzi, C. Roduner, G. Gyarmathy: Time-resolved flow measurements with fast-response aerodynamic probes in turbomachines, Meas. Sci. Technol. **11**, 1036–1054 (2000looseness1)

14.95 D.C. Wisler, P.W. Mossey: Gas velocity measurements within a compressor rotor passage using the laser doppler velocimeter, J. Eng. Power Ser. A **95**(2), 91–96 (1973)

14.96 A.J. Strazisar: Investigation of flow phenomenon in a transonic fan rotor using laser anemometry, J. Eng. Gas Turbines Power **107**, 427–435 (1985)

14.97 M.J. Pierzga, J.R. Wood: Investigation of the three-dimensional flow field within a transonic fan rotor: experiment and analysis, J. Eng. Gas Turbines Power **107**, 436–448 (1985)

14.98 K.N.S. Murthy, B. Lakshminarayana: Laser-Doppler velocimetry measurement in the tip region of a compressor rotor, AIAA J. **24**(5), 474–481 (1986)

14.99 R.J. Beaudoin, S.M. Miner, R.D. Flack: Laser velocimeter measurements in a centrifugal pump with a synchronously orbiting impeller, J. Turbomachin. **114**(2), 340–349 (1992)

14.100 R.C. Stauter: Measurement of the three-dimensional tip region flow field in an axial compressor, J. Turbomachin. **115**(3), 468–475 (1993)

14.101 M.D. Hathaway, R.M. Chriss, J.R. Wood, A.J. Strazisar: Experimental and computational investigation of the NASA low-speed centrifugal compressor flow field, J. Turbomachin. **115**(3), 527–542 (1993)

14.102 K.J. Farrell, M.L. Billet: A correlation of leakage vortex cavitation in axial flow pumps, J. Fluids Eng. **116**, 551–557 (1994)

14.103 M. Abramian, J.H.G. Howard: A rotating laser-doppler anemometry system for unsteady relative flow measurements in model centrifugal impellers, J. Turbomachin. **116**(2), 260–268 (1994)

14.104 J.R.J. Fagan, S. Fleeter: Comparison of optical measurement techniques for turbomachinery flow fields, J. Propulsion Power **10**(2), 176–182 (1994)

14.105 G.V. Hobson, B.E. Wakefield, W.B. Roberts: Turbulence amplification with incidence at the leading edge of a compressor cascade, ASME **96-GT-409** (1996)

14.106 M.A. Zaccaria, B. Lakshminarayana: Unsteady flow field due to nozzle wake interaction with the rotor in an axial flow turbine: Part 1 - rotor passage flow field, J. Turbomachin. **119**(2), 201–213 (1997)

14.107 D. Adler, R. Benyamin: Experimental investigation of the stator wake propagation inside the flow passages of an axial gas turbine rotor, Int. J. Turbo Jet Engin. **16**, 193–206 (1999)

14.108 D. Ristic, B. Lakshminarayana, S. Chu: Three-dimensional flow field downstream of an axial-flow turbine rotor, J. Propulsion Power **15**(2), 334–344 (1999)

14.109 M.B. Kang, K.A. Thole: Flowfield measurements in the endwall region of a stator vane, J. Turbomachin. **122**(3), 458–466 (2000)

14.110 T.M. Faure, G.J. Michon, H. Miton, N. Vassilieff: Laser Doppler anemometry measurements in an axial compressor stage, J. Propulsion Power **17**(3), 481–491 (2001)

14.111 D.E. Van Zante, W.-M. To, J.-P. Chen: Blade row interaction effects on the performance of a moderately loaded NASA transonic compressor stage, Proc. of ASME Turbo Expo 2002 (ASME, Amsterdam 2002)

14.112 J. Woisetschlager, N. Mayrhofer, H. Lang, B. Hampel: Experimental investigation of turbine wake flow by interferometrically triggered PIV and LDV measurements, Proc. ASME Turbo Expo 2002 (ASME, Amsterdam 2002)

14.113 X. Xiao, B. Lakshminarayana: Experimental investigation of end-wall flow in a turbine rotor, J. Propulsion Power **18**(5), 1122–1123 (2002)

14.114 T. Matsunuma, Y. Tsutsui: LDV measurements of unsteady mid-span flow in a turbine rotor at low reynolds number, Proc. ASME Turbo Expo (Atlanta, 2003)

14.115 S. Ibaraki, T. Matsuo, H. Kuma, K. Sumida, T. Suita: Aerodynamics of a transonic centrifugal compressor impeller, J. Turbomachin. **125**(2), 346–351 (2003)

14.116 T.M. Faure, H. Miton, N. Vassilieff: A laser Doppler anemometry technique for Reynolds stress measurement, Exp. Fluids **37**, 465–467 (2004)

14.117 L.E. Drain: *The laser Doppler technique* (Wiley, New York 1980)

14.118 H.B. Weyer: Optical methods of flow measurement and visualization in rotors, Proc. Joint Fluids Engineering and Gas Turbine Conference (New York 1980)

14.119 A. Doukelis, M. Founti, K. Mathioudakis, K. Papailiou: Evaluation of beam refraction effects in a 3D laser Doppler anemometry system for turbo-

14.119 machinery applications, Measure. Sci. Technol. **7**, 922–931 (1996)

14.120 W. Copenhaver, J. Estevadeordal, S. Gogineni, S. Gorrell, L. Goss: DPIV study of near-stall wake-rotor interactions in a transonic compressor, Exp. Fluids **33**(6), 899–908 (2002)

14.121 H. Lang, T. Morck, J. Woisetschlager: Stereoscopic particle image velocimetry in a transonic turbine stage, Exp. Fluids **32**(6), 700–709 (2002)

14.122 N. Balzani, F. Scarano, M.L. Riethmuller, F.A.E. Breugelmans: Experimental investigation of the blade-to-blade flow in a compressor rotor by digital particle image velocimetry, J. Turbomachin. **122**(4), 743–750 (2000)

14.123 M.P. Wernet, D.E. Van Zante, A.J. Strazisar, W.T. John, P.S. Prahst: 3-D digital PIV measurements of the tip clearance flow in an axial compressor, Proc. ASME Turbo Expo (Amsterdam, 2002)

14.124 O. Uzol, Y.-C. Chow, J. Katz, C. Meneveau: Unobstructed particle image velocimetry measurements within an axial turbo-pump using liquid and blades with matched refractive indices, Exp. Fluids **33**, 909–919 (2002)

14.125 R.J. Adrian: Multi-point optical measurements of simultaneous vectors in unsteady flow - a review, Int. J. Heat Fluid Flow **7**, 127–145 (1986)

14.126 C.E. Willert, M. Gharib: Digital particle velocimetry, Exp. Fluids **10**(4), 181–193 (1991)

14.127 R.J. Adrian: Particle imaging techniques for experimental fluid mechanics, Ann. Rev. Fluid Mech. **23**, 261–304 (1991)

14.128 Westerweel: Fundamentals of digital particle image velocimetry, Meas. Sci. Technol. **8**, 1379–1392 (1997)

14.129 H. Huang, D. Dabiri, M. Gharib: On errors of digital particle image velocimetry, Meas. Sci. Technol. **8**, 1427–1440 (1997)

14.130 A. Melling: Tracer particles and seeding for particle image velocimetry, Meas. Sci. Technol. **8**, 1406–1416 (1997)

14.131 R.J. Adrian: Dynamic ranges of velocity and spatial resolution of particle image velocimetry, Meas. Sci. Technol. **8**, 1393–1398 (1997)

14.132 M. Raffel, C. Willert, J. Kompenhans: *Particle Image Velocimetry - A Practical Guide* (Springer, Berlin 1998)

14.133 R. Dong, S. Chu, J. Katz: Quantitative visualization of the flow within the volute of a centrifugal pump: Part A – Technique, J. Fluids Eng. **114**, 390–395 (1992)

14.134 D. Tisserant, F.A.E. Breugelmans: Rotor blade-to-blade measurements using particle image velocimetry, J. Turbomachin. **119**, 176–181 (1997)

14.135 J. Estevadeordal, S. Gogineni, W. Copenhaver, G. Bloch, M. Brendel: Flow field in a low-speed axial fan: A DPIV investigation, Exp. Therm. Fluid Sci. **23**, 11–21 (2000)

14.136 Y.-D. Choi, K. Nishino, J. Kurokawa, J. Matsui: PIV measurement of internal flow characteristics of very low specific speed semi-open impeller, Exps. Fluids **37**, 617–630 (2004)

14.137 J.M. Sankovic, J.R. Kadambi, M. Mehta, W.A. Smith, M.P. Wernet: PIV investigations of the flow field in the volute of a rotary blood pump, J. Fluids Eng. **126**, 730–734 (2004)

14.138 N. Paone, M.L. Riethmuller, R.A. Van den Braembussche: Experimental investigation of the flow in the vaneless diffuser of a centrifugal pump by particle image displacement velocimetry, Exp. Fluids **1989**, 371–378 (1989)

14.139 S. Chu, R. Dong, J. Katz: Relationship between unsteady flow, pressure fluctuations and noise in a centrifugal Pump - Part A: Use of PDV data to compute the pressure field, ASME J. Fluids Eng. **117**(1), 24–29 (1995)

14.140 S. Chu, R. Dong, J. Katz: Relationship between unsteady flow, pressure fluctuations and noise in a centrifugal pump - Part B: Effects of blade-tongue interactions, ASME J. Fluids Eng. **117**(1), 30–35 (1995looseness1)

14.141 K.M. Day, P. Lawless, S. Fleeter: Particle image velocimetry measurements in a low speed two stage research turbine, AIAA 1996-2569, 32nd ASME, SAE and ASEE Joint Propulsion Conference and Exhibit 1996, Lake Buena Vista (AIAA, Reston 1996)

14.142 R. Dong, S. Chu, J. Katz: Effect of modification to tongue and impeller geometry on unsteady flow, pressure fluctuations and noise in a centrifugal pump, J. Turbomachin. **119**(3), 506–515 (1997)

14.143 M. Sinha, J. Katz: Quantitative visualization of the flow in a centrifugal pump with diffuser vanes: Part 1 – on flow structures and turbulence, J. Fluids Eng. **122**, 97–107 (2000)

14.144 M.P. Wernet: A flow field investigation in the diffuser of a high-speed centrifugal compressor using digital particle image velocimetry, Meas. Sci. Technol. **11**, 1007–1022 (2000)

14.145 M.P. Wernet: Development of digital particle imaging velocimetry for use in turbomachinery, Exp. Fluids **28**(2), 97–115 (2000)

14.146 M.P. Wernet, M. Bright, G. Skoch: An investigation of surge in a high-speed centrifugal compressor using digital PIV, J. Turbomachin. **123**, 413–428 (2001)

14.147 O. Uzol, C. Camci: Aerodynamic loss characteristics of a turbine blade with trailing edge coolant ejection Part 2: external aerodynamics, total pressure losses and predictions, J. Turbomachin. **123**(2), 249–257 (2001)

14.148 M. Sinha, Pinarbasi., J. Katz: The flow structure during onset and developed states of rotating stall within a vaned diffuser of a centrifugal pump, J. Fluids Eng. **123**, 490–499 (2001)

14.149 A.J. Sanders, J. Papalia, S. Fleeter: Multi-blade row interactions in a transonic axial compressor: Part 1

14.150 O. Uzol, Y.-C. Chow, J. Katz, C. Meneveau: Experimental investigation of unsteady flow field within a two stage axial turbomachine using particle image velocimetry, J. Turbomachin. **124**(4), 542–552 (2002)

14.151 Y.-C. Chow, O. Uzol, J. Katz: Flow non-uniformities and turbulent "hot spots" due to wake-blade and wake-wake interactions in a multistage turbomachine, ASME J. Turbomachin. **124**(4), 553–563 (2002)

14.152 J. Estevadeordal, S. Gogineni, L. Goss, W. Copenhaver, S. Gorrell: Study of Wake-Blade Interactions in a Transonic Compressor Using Flow Visualization and DPIV, J. Fluids Eng. **124**, 166–175 (2002)

14.153 O. Uzol, Y.-C. Chow, J. Katz, C. Meneveau: Average passage flow field and deterministic stresses in the tip and hub regions of a multi-stage turbomachine, J. Turbomachin. **125**(4), 714–725 (2003)

14.154 Y.C. Chow, O. Uzol, J. Katz: *On the flow and turbulence within the wake and boundary layer of a rotor blade located downstream of an IGV*, in ASME Turbo Expo 2003 International Gas Turbine Institute Conf. Atlanta (ASME, Fairfield 2003)

14.155 J. Woisetschlager, N. Mayrhofer, B. Hampel, H. Lang, W. Sanz: Laser-optical investigation of turbine wake flow, Exp. Fluids **34**, 371–378 (2003)

14.156 O. Uzol, Y.C. Chow, F. Soranna, J. Katz: 3D structure of a rotor wake at mid-span and tip regions, Proc. of 34th AIAA Fluid Dynamics Conference and Exhibit (Portland, 2004)

14.157 O. Uzol, Y.-C. Chow, F. Soranna, J. Katz, C. Meneveau: 3D measurements of deterministic stresses within a rotor-stator gap at mid-span and tip regions, Proc. of ASME Heat Transfer/Fluids Engineering Summer Conference (Charlotte, 2004)

14.158 B.J. Liu, H.W. Wang, H.X. Liu, H.J. Yu, H.K. Jiang, M.Z. Chen: Experimental investigation of unsteady flow field in the tip region of an axial compressor rotor passage at near stall condition with stereoscopic particle image velocimetry, J. Turbomachin. **126**(3), 360–374 (2004)

14.159 F. Soranna, Y.-C. Chow, O. Uzol, J. Katz, C. Meneveau: Rotor boundary layer response to an impinging wake, ASME Heat Transfer/Fluids Engineering Summer Conference (Charlotte, 2004)

14.160 S.J. Lee, B.G. Paik, J.H. Yoon, C.M. Lee: Three-component velocity field measurements of propeller wake using a stereoscopic PIV technique, Exp. Fluids **34**, 575–585 (2004)

14.161 Y.C. Chow, O. Uzol, J. Katz, C. Meneveau: Decomposition of the spatially filtered and ensemble averaged kinetic energy, the associated fluxes and scaling trends in a rotor wake, Phys. Fluids **17**(8) (2005)

14.162 F. Soranna, Y.-C. Chow, O. Uzol, J. Katz: 3D Measurements of the Mean Velocity and Turbulence Structure within the Near Wake of a Rotor Blade, Proc. ASME Fluids Engineering Division Summer Meeting and Exhibition (Houston, 2005)

14.163 J. Estevadeordal, T.R. Meyer, S.P. Gogineni: Development of a fiber optic PIV system for turbomachinery applications, AIAA, 0038 (2005)

14.164 U. Ullum, J. Schmidt, P.S. Larsen, D.R. McCluskey: Statistical analysis and accuracy of PIV data, J. Visualization **1**(2), 205–216 (1998)

14.165 O. Uzol, C. Camci: The effect of sample size turbulence intensity and velocity field on the experimental accuracy of ensemble averaged PIV measurements, Proc. 4th International Symposium on Particle Image Velocimetry (Göttingen, 2001)

14.166 J.J. Adamczyk: Model Equation for simulating flows in multistage turbomachinery, ASME Paper No. 85-GT-226 (1985)

14.167 J.J. Adamczyk, R.A. Mulac, M.L. Celestina: A model for closing the inviscid form of the average-passage equation system, ASME Paper No. 86-GT-227 (1986)

14.168 M. Sinha, J. Katz, C. Meneveau: Quantitative visualization of the flow in a centrifugal pump with diffuser vanes: Part 2 – addressing passage-averaged and LES modeling issues in turbomachinery flows, J. Fluids Eng. **122**, 108–116 (2000)

14.169 F. Soranna, Y.-C. Chow, O. Uzol, J. Katz: The effect of IGV wake impingement on the flow structure and turbulence around a rotor blade, J. Turbomachin. **128**(1), 82–95 (2006)

14.170 P.W. McDonald, C.R. Bolt, R.J. Dunker, H.B. Weyer: A comparison between measured and computed flow fields in a transonic compressor rotor, J. Eng. Power **102**, 883–891 (1980)

14.171 W.J. Calvert, A.W. Stapleton: Detailed flow measurements and predictions for a three-stage transonic fan, J. Turbomachin. **116**(2), 298–305 (1994)

14.172 X. Ottavy, I. Trebinjac, A. Vouillarmet: Analysis of interrow flow field within a transonic axial compressor: Part 1 – unsteady flow analysis, J. Turbomachin. **123**, 57–63 (2001)

14.173 C. Schodl, C. Willert, I. Roehle, J. Heinze, W. Foerster, M. Fischer, M. Beversdorff: Optical diagnostic techniques in turbomachinery, AIAA 2002-3038, 22nd AIAA Aerodynamic Measurement Technology and Ground Testing Conference (St. Louis, 2002)

14.174 K. Ziegler, E.H. Gallus, R. Niehuis: A study on impeller-diffuser interaction: Part 2 – detailed flow analysis, J. Turbomachin. **125**(1), 183–192 (2003)

14.175 K.R. Navarra, D.C. Rabe, S.D. Fonov, L.P. Goss, C. Hah: The application of pressure and temperature sensitive paints to an advanced compressor, J. Turbomachin. **123**(4), 823–829 (2001)

14.176 J.W. Gregory: Porous pressure sensitive paint for measurement of unsteady pressures in turbomachinery, Proc. 42nd Aerospace Sciences Meeting and Exhibit (ASME, Reno 2004)

14.177 Schodl R.: Development of the laser-two-focus method for non-intrusive measurement of flow vectors, particularly in turbomachines, European Space Agency Technical Translation ESA-TT-528 (1979)

14.178 Schodl R.: Laser-two-focus velocimetry. In: Advanced Instrumentation for Aero Engine Components. Papers Presented at the Propulsion and Energetics Panel 67th Symposium, AGARD Conference Proc., No. 399 (Philadelphia, 1986) p 7.1–7

14.179 X. Ottavy, I. Trebinjac, A. Vouillarmet: Treatment of L2F anemometer measurement volume distortions created by curved windows for turbomachinery applications, Meas. Sci. Technol. **9**, 1511–1521 (1998looseness1)

14.180 H. Komine: System for Measuring Velocity Field of Fluid Flow using a Laser Doppler Spectral Image Converter. US Patent No. 4919536 (1990)

14.181 H.D. Ford, R.P. Tatam: Development of extended field Doppler velocimetry for turbomachinery applications, Opt. Lasers Eng. **27**, 675–696 (1997looseness1)

14.182 I. Roehle, P. Voigt, C. Willert: Recent developments and applications of quantitative laser light sheet measuring techniques in turbomachinery components, Meas. Sci. Technol. **11**, 1023–1035 (2000)

14.183 M.P. Wernet: Planar particle imaging doppler velocimetry: a hybrid PIV/DGV technique for three-component velocity measurements, Meas. Sci. Technol. **15**, 2011–2028 (2004)

14.184 K. Sabroske, D. Rabe, C. Williams: Pressure-sensitive paint investigation for turbomachinery application, ASME **95-GT-92** (1995)

14.185 T. Liu, S. Torgerson, J. Sullivan, R. Johnston, S. Fleeter: Rotor blade pressure measurement in a high speed axial compressor using pressure and temperature sensitive paints, Proc. 35th Aerospace Sciences Meeting Exhibit (ASME, Reno 1997)

14.186 R.H. Engler, C. Klein, O. Trinks: Pressure sensitive paint systems for pressure distribution measurements in wind tunnels and turbomachines, Meas. Sci. Technol. **11**, 1077–1085 (2000)

14.187 J. Lepicovsky, T.J. Bencic: Use of pressure sensitive paint for diagnostics in turbomachinery flows with shocks, Exp. Fluids **33**, 531–538 (2002)

14.188 Bell, Schairer, Hand, Mehta: Surface pressure measurements using luminescent coatings, Ann. Rev. Fluid Mech. **33**, 155–206 (2001)

15. Hydraulics

The three segments of this chapter introduce phenomena that are of specific interest in the area of hydraulics. Where applicable (Cavitation, Sect. 15.1 and Sediment Transport, Sect. 15.3) introductory and descriptive material regarding the topic is provided. Terminology, physical examples and motivating descriptions introduce the comprehensive Cavitation subsection. Examples of the types of flows in which cavitation occurs are provided. This information sets the stage for a description of the types of facilities and instrumentation that are necessary to study the problem. Numerous photographs and descriptive sketches clarify and complement the text.

The wave height measurement segment first deals with fixed position single "point" techniques. More advanced techniques for: i) wave surface shape along a horizontal line, and ii) two-dimensional surface geometry measurements for laboratory and field observations are then described.

Following the introduction to sediment transport phenomena and terminology, the methods of measurement: manual, optical and acoustic are given detailed descriptions including calibration techniques for the latter two methods. Bed load sediment measurements: pressure difference, sediment trapping and acoustic are next described. Total load and the less common measurement techniques, plus references complete the subsection.

15.1 **Measurements in Cavitating Flows** 959
   15.1.1 Fundamentals 961
   15.1.2 Types of Cavitating Flows 967
   15.1.3 Cavitation Damage 970
   15.1.4 Facilities 972
   15.1.5 Water Quality Measurements 979
   15.1.6 Cavitation Inception 984
   15.1.7 Measurements in Cavitating Flows 988

15.2 **Wave Height and Slope** 1009
   15.2.1 Temporal Point Measurements at Fixed Geometric Positions 1009
   15.2.2 One-Dimensional Spatial Measurements 1014
   15.2.3 Two-Dimensional Spatial Measurements 1015
   15.2.4 Special Surface Elevation Measurements for Large Laboratory or Field Use 1018

15.3 **Sediment Transport Measurements** 1020
   15.3.1 A Brief Introduction to Sediment Transport 1020
   15.3.2 Methods of Measuring Suspended Sediment Transport 1022
   15.3.3 Bed Load Sediment Measurements 1029
   15.3.4 Total Load Measurements 1032
   15.3.5 Other Measurement Techniques 1032

References 1033

15.1 Measurements in Cavitating Flows

Cavitation is normally defined as the formation of the vapor phase in a liquid. The term cavitation (originally coined by R. E. Froude) can imply anything from the initial formation of bubbles (inception) to large-scale, attached cavities (supercavitation). The formation of individual bubbles and subsequent development of attached cavities, bubble clouds, etc., is directly related to reductions in pressure to some critical value, which in turn is associated with dynamical effects, either in a flowing liquid or in an acoustical field. Cavitation can be distinguished from boiling in the sense that the former is induced by the lowering of the hydrodynamic pressure, whereas the latter is induced by the raising of the vapor pressure to some value in excess of the hydrodynamic pressure. The two phenomena are related. Cavitation inception and boiling can be compared in terms of the vapor-bubble dynamics of sub-cooled and superheated liquids [15.1]. Quite often a clear distinction between the two types of phenomena cannot be made. This is especially true for cavitation in liquids other than cold water.

Any device handling liquids is subject to cavitation. Cavitation can affect the performance of turbomachinery, resulting in a drop in head and efficiency of pumps and decreased power output and the efficiency of hydroturbines. The thrust of propulsion systems can be cavitation-limited and the process can degrade the accuracy of fluid meters. Noise and vibration occur in many applications. In addition to the deleterious effects of reduced performance, noise and vibration, there is the possibility of cavitation erosion. The extent of cavitation erosion can range from a relatively minor amount of pitting after years of service to catastrophic failure in a relatively short period of time as shown in Figs. 15.1, 2. Figure 15.1 illustrates the damage that can occur on a spillway. This damage occurred after only four hours of operation. The second example is cavitation damage in a hydraulic turbine due to the formation of secondary vortices in the outlet flow. *Arndt* et al. [15.2] have identified cavitation erosion in turbines as a very common problem.

Although cavitation is normally considered to be an undesirable effect or phenomenon, it sometimes serves useful purposes. Cavitation applications include ultrasonic cleaning and the homogenization of milk. Various chemical processes are enhanced by cavitation, such as coagulation, formation of suspensions and degassing of liquids. Cavitation can be used to increase heat and mass transfer in liquids, to promote crystallization and to enhance various sonochemical reactions such as polymerization and polymer degradation. Biomedical applications include the removal of kidney stones and automated drug delivery to patients. Important new applications in the pollution control area are of interest [15.3, 4].

Fig. 15.1 Example of cavitation damage on a spillway. The extent of the damage is evident when compared with the workmen in the figure. This damage occurred after only four hours of operation

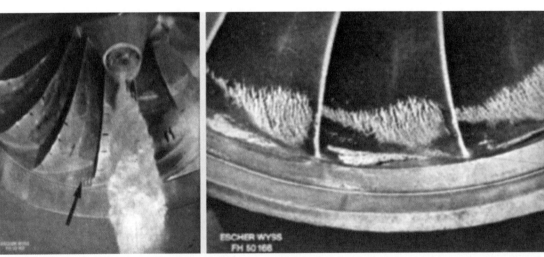

Fig. 15.2 Example of vortex cavitation induced damage in a hydroturbine. Two types of cavitation are shown on the left, a hub vortex, and a secondary vortex pattern that develops in the runner passages as illustrated by the arrow. These relative small cavitating structures are responsible for the damage shown in the *inset* (Courtesy of A. Keller)

Cavitation has been studied for many years. Leonhard Euler is credited with being the first person to postulate the possibility of cavitation in his 1754 memoir on the theory of hydraulic machines. *Osborne Reynolds* (1894) carried out what was probably the first fundamental study of the problem by observing cavitation in tubular constrictions. The introduction of propellers for marine propulsion led to an almost immediate need for cavitation research. The most significant early example of applied research is probably that of Parsons in 1897 [15.5]. His approach to a cavitation problem sets the stage for the techniques used in modern research. The HMS Turbinia was designed by him to demonstrate the application of the compound steam turbine in marine propulsion. The first trials were miserable failures. Parsons examined this problem in a water tunnel with a special stroboscope, both of his own design. He determined that the heavily loaded single screw on the vessel was cavitating and was unable to produce the required thrust. His solution was to replace the original propulsion system with three smaller turbines with the same total horsepower driving nine propellers. A speed of 32.6 knots was finally achieved, amply demonstrating the application of steam turbine technology in marine propulsion. This brought fame and fortune to Parsons and his colleagues.

Other noteworthy efforts include the work of *Barnaby* and *Thornycroft* (1898), who were studying problems that evolved during the trials of the HMS Daring. This early work led to the classic study of the potency for damage due to bubble collapse by *Lord Rayleigh* [15.6]. *Minnaert* [15.7] then set the stage for understanding the acoustics of cavitation and gas-filled bubbles. The foundations of our knowledge of bubble dynamics and its interrelationship to cavitation are found in several papers by *Plesset* and his coworkers [15.1, 8, 9].

Because of the myriad of applications, cavitation research has been an active discipline for more than a century. The complex physics involved have dictated the development of sophisticated experimental techniques, innovative analytical methods and, more recently, the development of powerful numerical schemes for addressing the investigation of a broad variety of phenomena. An enormous quantity of literature has been written on the subject of cavitation. There are several reviews of the topic, e.g., *Eisenberg* [15.10], *Wu* [15.11], *Acosta* and *Parkin* [15.12], *Plesset* and *Posperetti* [15.13], *Arndt* [15.14–16], *Blake* and *Gibson* [15.17], and *Rood* [15.18].

The state of the art is well summarized in the textbook by *Knapp* et al. [15.19] and more recently by *Young* [15.20], *Brennen* [15.21] and *Franc* and *Michel* [15.22]. There are other texts in languages other than English. Notable examples are the exhaustive treatise by *Anton* [15.23] in Romanian and the text by *Isay* [15.24] in German. Numerous symposia have been devoted to the topic in the United States and elsewhere.

In spite of significant advances in computational techniques, there is still a need for experimental research. In fact, recent numerical studies have underscored the need for new experimental data, utilizing advanced experimental techniques. Recently cavitation research has been carried out utilizing an interactive numerical/experimental approach. One of the aims of this chapter is to illustrate how numerical and experimental approaches are integrated in modern cavitation research. Emphasis will be placed on the special instrumentation requirements for cavitation research that are in addition to the instrumentation required in single-phase fluid mechanics.

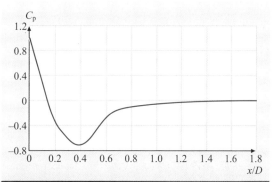

Fig. 15.3 Cavitation inception on a hemispherical-nosed body. The curve in the *top panel* is the pressure distribution on the body (Courtesy of A. Keller)

15.1.1 Fundamentals

Occurrence of Cavitation

Examples of cavitation as viewed in the laboratory are shown in Figs. 15.3, 4, 5, 6, 7. Limited cavitation in the form of a ring of bubbles surrounding a hemispherical-nosed body is shown in Fig. 15.3. Note that when Fig. 15.3a,b are compared, cavitation appears to occur in a region that is downstream of the calculated minimum-pressure point. The pressure distribution in Fig. 15.3a was calculated using potential flow theory. Actually cavitation is occurring at the trailing edge of a laminar separation bubble that is not accounted for in the inviscid theory. Figure 15.4 is an outtake of high-speed video of an individual cavitation event on a National Advisory Committee for Aeronautics (NACA) 0015 hydrofoil. In this case inception begins very close to the minimum-pressure point on the foil. As the resulting vapor bubble moves downstream into a region of higher pressure it continues to grow. Finally the growth is arrested and the bubble collapses at about a length of 3/4 chord from the leading edge. After the initial collapse, the bubble rebounds and collapses again several times before the trailing edge is reached. Figure 15.5 illustrates cavitation occurring away from the surface of a body in a turbulent jet. Figure 15.6 contains some information on propeller cavitation. Figure 15.6a is an observation of cavitation on a model propeller in the University of Tokyo water tunnel. Note that cavitation occurs in the blade tip vortices and in the hub vortex. Figure 15.6b is a view, from a different perspective, of propeller cavitation made in one of the water tunnels at the Marine Institute of the Netherlands (MARIN). Note that there is a tip vortex and blade surface cavitation. The fact that several different types of cavitation can occur simultaneously in practice makes the scaling of these physical phenomena in the laboratory very challenging. Figure 15.7 is an example of cavitation on a hydrofoil and is presented to indicate several salient features of the cavitation problem. More than one type of cavitation can occur simultaneously (sheet cavitation on the surface of the blade, vortex cavitation at the tips). It is also important to note that in many practical applications cavitation is a nonsteady phenomenon. In a marine propeller for example, the degree of cavitation varies with the angular position of each blade due to the periodic passage of each propeller blade through the ship's wake. This induces periodic variations in thrust and torque that in some cases can cause severe vibration.

These examples illustrate the complexity of the cavitation process. It is clear that cavitation occurs in regions of low pressure. For example the calculation of minimum pressure in Fig. 15.3a was made using potential flow theory that ignores viscous effects. A shear layer associated with the instability of the separated flow in a separation bubble induces pressure fluctuations that are responsible for the incipient cavitation. Classical poten-

Fig. 15.4 Bubble growth visualized with high-speed video. Flow is from left to right. Note that several rebounds occur after the initial collapse at about the 3/4 chord position

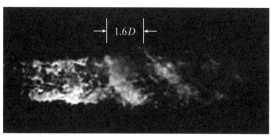

Fig. 15.5 Cavitating jet. Note the existence of a *smoke-ring*-like structure close to the nozzle ($m = 0$) and a helical structure ($m = 1$) further downstream [15.25]

tial flow theory argues that the minimum pressure in a flow must occur at a boundary. This is clearly not the case for the turbulent flow shown in Fig. 15.5, which contains an example of jet cavitation. Careful inspection of the cavitation process in turbulent flows also indicates that cavitation occurs in vortical structures. The influence of vorticity on cavitation is also evident in Fig. 15.5. However, careful inspection of sheet cavitation as seen in Fig. 15.6 will also indicate the existence of highly vortical structures at the trailing edge of sheet cavitation. These photos only hint at the complexity of the problem.

Cavitation Scaling
The fundamental parameter in describing the physics of the process is the cavitation index, defined by

$$\sigma = \frac{(p_0 - p_c)}{\frac{1}{2}\rho U_0^2},\qquad(15.1)$$

where p_0 and U_0 are a characteristic pressure and velocity, respectively, ρ is the density, and p_c is a critical pressure or cavity pressure depending on the application. Often σ is defined by setting p_c equal to the vapor pressure, p_v. Various hydrodynamics parameters such as lift and drag coefficient, torque coefficient, and efficiency, are assumed to be unique functions of σ when there is geometric similitude between the model and prototype. Generally speaking, these parameters are independent of σ above a certain critical value of σ. This critical value is often referred to as the incipient cavitation number. It should be emphasized that the point where there is a measurable difference in performance is not the same value of σ where cavitation can be first detected visually or acoustically.

The rationale for σ as a scaling parameter is as follows. Cavitation is normally assumed to occur when the minimum pressure in a flow is equal to the vapor pressure. For steady flow over a streamlined body the nondimensional minimum pressure coefficient is given by

$$C_{p_m} \equiv \frac{p_m - p_0}{\frac{1}{2}\rho U_0^2}.\qquad(15.2)$$

When $p_m = p_v$, the *incipient* value of σ is given by

$$\sigma_i = -C_{p_m}.\qquad(15.3)$$

We can think of σ_i as a performance boundary such that for $\sigma > \sigma_i$ there are no cavitation effects, while for $\sigma < \sigma_i$ cavitation effects such as performance degradation, noise, and vibration occur.

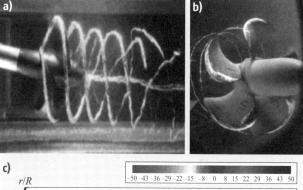

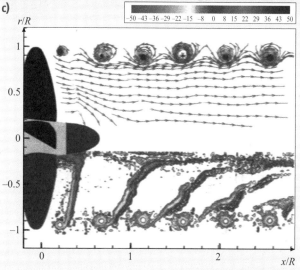

Fig. 15.6a–c Propeller cavitation. (**a**) Example of cavitation in the tip and hub vortices of a propeller (courtesy of H. Kato) (**b**) Example of tip vortex and surface cavitation on a propeller (courtesy of G. Kuiper) (**c**) Measurement of the trailing vortex structure behind a propeller using PIV (Courtesy of G. Felice)

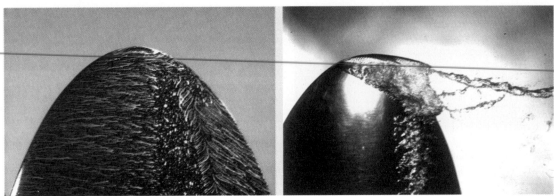

Fig. 15.7 Cavitation on a NACA $66_2 - 415$ hydrofoil. The photograph on *left* is an oil film visualization made in a wind tunnel. The photo on the *right* is an observation of cavitation on the same foil at the same angle of attack and Reynolds number. Note the existence of both surface cavitation (emanating from a separation bubble at about 60% chord) and tip vortex cavitation

Other definitions of the critical value of the cavitation index exist. Cavitation desinence refers to the conditions necessary to eliminate cavitation. This is denoted by the desinent cavitation number σ_d, which in general is higher than σ_i.

Inception Dynamics

The simple model of inception described by (15.3) is inadequate [15.14, 15]. This is generally an oversimplification since cavitation inception is governed by the single-phase flow characteristics (including turbulence) *and* the critical pressure, p_c. Hence a more general form of the inception cavitation index is given by

$$\sigma_i = -C_{p_m} + K \frac{\sqrt{\overline{p'^2}}}{\frac{1}{2}\rho U^2} - \frac{T}{\frac{1}{2}\rho U^2} \tag{15.4a}$$

where the second and third terms on the right-hand side of (15.4a) incorporate the effects of unsteadiness and bubble dynamics, respectively. The second term, which is proportional to the intensity of pressure fluctuations, is very important in free shear flows and boundary layers adjacent to smooth and roughened walls [15.14, 15, 26]. For example, the value of C_{p_m} in a turbulent jet is roughly equal to zero whereas the second term has values ranging from about 0.2 to 1.0, corresponding to the factor K being of order 10. The effects of turbulence are found to scale well with a shear stress coefficient defined as

$$C_f = \frac{\tau_0}{\frac{1}{2}\rho U^2} \quad \text{boundary layers} \tag{15.4b}$$

or

$$C_f = \frac{\overline{u_1 u_2}}{\frac{1}{2}\rho U^2} \quad \text{free-shear flows}. \tag{15.4c}$$

T is defined as the tensile strength of the liquid $(p_v - p_c)$, which can be an important factor in cavitation testing. It is generally accepted that cavitation inception occurs as a consequence of the rapid or explosive growth of small bubbles or nuclei that have become unstable due to a change in ambient pressure. These nuclei can be either imbedded in the flow or find their origins in small cracks or crevices at the bounding surfaces of the flow.

By considering the static equilibrium of a spherical bubble, the tension that a liquid can sustain before cavitating is found to depend on the size of nuclei in the flow

$$T = \frac{4S}{3R_n} \left[3 \left(1 + \frac{p_0 - p_v}{\frac{2S}{R_n}} \right) \right]^{-\frac{1}{2}} \tag{15.5a}$$

where S is the surface tension and p_0 is the free-stream pressure when the nucleus radius is R_n. Note that the tensile strength of the liquid T should not be confused with the surface tension S.

Equation (15.5a) implies that the critical pressure for cavitation inception approaches the vapor pressure when there is a sufficient supply of nuclei greater than approximately 100 μm. When the number of sufficiently large nuclei is small, the pressure required for cavitation inception can be *negative*, i.e., the flow is locally in *tension*. Measured nuclei size distributions vary greatly in various facilities. This leads to significant discrepancies in the measured value of σ_i.

Techniques for the measurement of cavitation nuclei have been developed over the past 30 years. A coherent review of nuclei measurement is given by *Billet* [15.28]. Many investigators have considered the details of how these nuclei exist. *Rood* [15.18] provides a review of these efforts. It is only recently that reliable measurements are possible. Most methods of cavitation nuclei measurement are tedious and require sophisticated equipment, as will be outlined subsequently.

Most of the early progress in understanding the details of the inception process has been made through consideration of the dynamic equilibrium of a spherical bubble containing vapor and non-condensable gas. The Rayleigh–Plesset equation describes this equilibrium [15.13]:

$$R\ddot{R} + \frac{3}{2}\dot{R}^2 = \frac{1}{\rho}\left[p_i - p_\infty(t) - \frac{2S}{R} - 4\mu\frac{\dot{R}}{R}\right],$$
(15.5b)

where R is the time-dependent bubble radius, p_i is the pressure inside the bubble and p_∞ is the externally applied pressure. However, as pointed out by Rood, *spherical bubbles passively convected by the mean flow* cannot adequately model real fluid effects on the inception process. Rood goes on to make a very strong argument that inception physics is crucially related to the interaction of bubble dynamics and vortical structures. The problem is complicated by non-spherical bubble deformation and complex modification of the interacting vortical structures. In addition, *Joseph* [15.29] has pointed out that the cavitation threshold is related to the *maximum* tensile stress in the liquid rather than the *minimum* thermodynamic pressure in the flow. These issues are underscored by the work of *Gindroz* and *Billet* [15.30] who were able to demonstrate the effect of the size/number distribution of nuclei on propeller cavitation. Particularly noteworthy for this discussion is their findings that tip vortex cavitation is particularly sensitive to nuclei content (water quality).

Influence of Dissolved Gas

Non-condensable gas in solution can also play a role in vaporous cavitation, since the size and number of nuclei in the flow are related to the concentration of dissolved gas. Under certain circumstances, cavitation can also occur when the lowest pressure in the flow is substantially *higher* than the vapor pressure. In this case bubble growth is due to *diffusion* of dissolved gas across the bubble wall. This can occur when nuclei are subjected to pressures below the saturation pressure for a relatively long period of time. *Holl* [15.31] suggested that *gaseous* cavitation could occur when the flow is locally supersaturated. He suggested an equilibrium theory such that

$$(p_m - p_v)_c \leq p_s = \beta C_g,$$
(15.6)

where p_s is the saturation pressure, β is Henry's constant and C_g is the concentration of dissolved gas. Henry's constant is a function of the type of gas in solution and the water temperature. As a rule of thumb $\beta \approx 6700$ Pa/ppm for air, when concentration is expressed in a mole/mole basis. In other words, water is saturated at one atmosphere when the concentration is approximately 15 ppm. Thus, for gaseous cavitation an upper limit on σ_i is given by

$$\sigma_i = -C_{p_m} + \frac{\beta C_g}{\frac{1}{2}\rho U_0^2}.$$
(15.7)

The experiments of *Holl* [15.31] indicate that both gaseous and vaporous cavitation can occur in the same experiment. This is an important consideration when examining the trends of experimental data such as shown in Fig. 15.28. The two types of cavitation are different physical processes and it is sometimes difficult to distinguish between them. Dissolved gas can also influence the measured values of hydrodynamic loads in cavitating flows [15.14, 15].

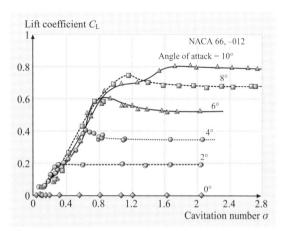

Fig. 15.8 Variation in the lift coefficient of a hydrofoil (after [15.27] with permission by Annual Reviews Inc.)

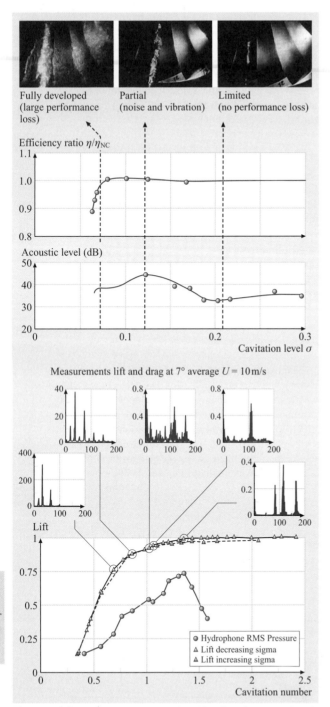

Fig. 15.9 (a) Schematic of the relationship of performance breakdown and noise with various stages of cavitation **(b)** Variation of lift, noise and spectral content of lift oscillations on a NACA 0015 hydrofoil (after [15.32])

Effects of Cavitation

Once cavitation occurs, a given flow field is significantly modified because the lowest pressure in the flow is typically limited to the vapor pressure. Thus

$$C_{\text{pm}} = -\sigma \quad \sigma \leq \sigma_i . \tag{15.8}$$

Since the lift coefficient of a hydrofoil scales with $-C_{\text{pm}}$, this parameter will decrease with decreasing σ, as shown in Fig. 15.8. The torque coefficient for a turbomachine has a similar trend.

Under some circumstances, inconsistencies are found when analyzing data similar to that shown in Fig. 15.8. This can sometimes be circumvented by measuring the cavity pressure p_c directly and defining the cavitation number in terms of the cavity pressure p_c:

$$\sigma \equiv \frac{p_0 - p_c}{\frac{1}{2}\rho U_0^2} . \tag{15.9}$$

The effect of cavitation on lift is directly related to the observed degradation of performance of turbomachinery due to cavitation. This is illustrated in Fig. 15.9. This figure is a composite of turbine model tests (courtesy of Voith Company) and acoustic data from *Deeprose* et al. [15.33]. The turbine head defines the cavitation number in this figure:

$$\sigma_T \equiv \frac{H_{\text{sv}}}{H} , \tag{15.10}$$

where H_{sv} is the net positive suction head [15.34] and H is the total head under which a given machine is operating. σ_T and σ are qualitatively equivalent. The accompanying photos in the figure clearly illustrate that cavitation occurs at much higher values of σ_T than at the point of performance degradation (as indicated by a reduction in efficiency). This illustrates that σ_T and σ are qualitatively equivalent.

Cavitation can also influence vortex dynamics in subtle ways. For example, *Young* and *Holl* [15.35] and *Belahadji* and *Michel* [15.36] found that cavitation strongly influenced the frequency of vortex shedding behind wedges. Because cavitation modifies the forcing frequency due to flow over a body, there is a possibility of unexpected hydroelastic vibration if a closer match between forcing frequency and a structural mode of vibration occurs. An even subtler feature is the *production* of vorticity at the trailing edge of attached cavitation [15.37, 38].

Cavitation is also basically a nonsteady phenomenon. For example, steady flow over a hydrofoil at a value of σ below σ_i results in a highly dynamic form of sheet cavitation [15.32, 39]. Figure 15.9a illustrates how the dynamics of lift oscillations can change dramatically with σ. Using a simple reentrant jet model, *Arndt* et al. [15.40] were able to show that for cavity lengths l less than about 75% of a chord length, an estimate of the frequency of oscillation is given by

$$\frac{fl}{U} = \frac{1}{4}\sqrt{1+\sigma}, \qquad (15.11)$$

where σ is defined by the pressure in the cavity (which may not be equal to the vapor pressure [15.14]). *Kawanami* et al. [15.41] have reviewed cavity oscillation data from several sources, including their own measurements. Equation (15.11) fits their published data very well. *Stinebring* [15.42] found a similar result for sheet cavitation on a zero caliber ogive. For more developed cavitation, the problem is more complex.

15.1.2 Types of Cavitating Flows

Shear Flows

Turbulent shear flows typically consist of vortical structures that are randomly distributed in space and time. Coherent structures in the flow can play an important role in the cavitation process. For example, *Daily* and *Johnson* [15.43] showed that cavitation occurred in a turbulent boundary layer when the wall pressure was greater than the vapor pressure. Cavitation nuclei are entrained into the middle of the boundary layer by large-scale vortical structures where cavitation was observed to occur. *Arndt* and *Ippen* [15.44] were able to measure the inception pressure in the cores of these vortical structures by observing the rate of bubble growth during inception with the aid of high-speed cinema photography. The observed growth rate was related to inception pressure utilizing the Rayeigh–Plesset equation. Instantaneous pressure drops of the order of ten times the root-mean-square (RMS) wall pressure were found. Further details can be found in *Arndt* and *George* [15.45] and *Arndt* [15.15].

Figure 15.5 contains an example of cavitation occurring in a turbulent jet. Two cylindrical modes of coherent structure (described by $\cos m\theta$) in the flow are evidenced by cavitation. A smoke-ring-like instability ($m = 0$) is evident close to nozzle, whereas further downstream a helical mode ($m = 1$) is evident. The calculations of *Michalke* [15.46] indicate that the growth rate of both modes is about equal. Apparently the helical mode dominates downstream. The axial wavelength of this mode is calculated to be about $1.6D$ where D is jet diameter, which fits very well with the observations. The tendency for coherent structures to form in turbulent jets was explored by *Chahine* and *Johnson* [15.47]. In developing a patented underwater cleaning device, they found that an underwater jet could be induced to *self-resonate* at a Strouhal number of about 0.3 [15.48]. They showed that the energy content at this natural frequency could be dramatically amplified by creating a feedback mechanism with a resonant chamber. In this manner, the cavitation inception number can be increased by factors greater than two. The jet shear layer organizes into ring vortices that cavitate to form a toroidal cloud of bubbles that is highly erosive. The enhanced cavitation in these structured jets have important applications in deep hole drilling, cleaning, cutting and underwater sound generation. An interesting biological application is used by the *snapping shrimp* that relies on cavitation induced by a high-speed jet to stun his prey [15.49].

Although coherent structures in turbulent jets are important in the cavitation process [15.50], the apparent mechanisms are complex. The inception process appears to occur in micro-vortices that are the *debris* from the pairing process [15.51–53]. Apparently very high negative peaks in pressure are associated with the pairing process. This issue is more complex than previously thought. *Gopalan* et al. [15.54] found significant differences in the inception mechanism for naturally occurring jets and jets whose nozzle boundary had been tripped. Their experiments revealed that inception in a naturally occurring jet occurred in secondary, axially oriented vortices in the form of inclined cylindrical bubbles at axial distances (x/D) of 0.55 with cavitation indices of 2.5. The measured alignment of the cylindrical bubbles was equal to the direction of the measured principal strain. Upon tripping the nozzle boundary layer, inception was observed to occur at $x/D \approx 2$ as distorted *spherical* bubbles within the primary ring-like structures with an inception index of 1.7. These results indicate that cavitation inception can occur both in the form of distorted spherical bubbles in the primary vortical structures and within secondary axial vortices in the form of cylindrical bubbles. Tripping the boundary layer of the jet apparently suppresses the secondary vortex cavitation.

By observing the growth of injected bubbles [15.55], *Ran* and *Katz* [15.50] found that measured negative peaks in pressure are more than a factor of 10 higher than the RMS pressure, similar to the results of *Arndt* and *Ippen* [15.44] for a turbulent boundary layer. They also noted that the nuclei distribution in a jet was significantly different from uniform. Nuclei are entrained into low-pressure regions in the flow, enhancing the probability for cavitation. Similar ideas have been promulgated in the past (e.g., [15.56]), but this study is the first *quantitative* analysis of this factor.

Attached Cavitation

Various types of cavitation on surfaces can be found in practice, including bubble cavitation, sheet cavitation, cloud cavitation, and various forms of vortex cavitation, depending upon how the low-pressure regions are generated. In spite of considerable research, there are still many features of the problem that have not been properly explored. For example, inception studies are based on fully wetted flow properties, i.e., pressure distribution, turbulence level etc., in the absence of cavitation. Also, classical models of developed cavitation consider only cavitation number as the primary variable. What has not been given adequate attention is a class of partially cavitating flows in which there is an interaction between fluid turbulence and cavitation. For example, vortex generation at the trailing edge of sheet cavitation is a manifestation of the cavitation itself [15.37, 38]. Cavitating microstructures are created that are highly energetic and are responsible for significant levels of noise and erosion. This is an important finding since turbulence is normally attributed to being a factor in the inception process, but cavitation as a mechanism for turbulence generation has been given scant attention.

The details concerning the transition of sheet cavitation to cloud cavitation are still not understood and are beyond the scope of this review [15.38, 57, 58]. It is well known that the modeling of even partial, time-averaged cavities is not simple, due to the inverse character of the flow representation in the vicinity of the cavity and its wake. In addition, partial cavity models cannot explain the breakup of sheet cavitation at the trailing edge into detached cavitation clouds. The process is inherently unsteady even for steady free-stream conditions. Within a certain envelope of cavitation number, σ, and angle of attack, α, of a lifting surface, the process is also periodic [15.32]. This creates a modulation of the trailing cloud cavitation that is highly erosive and very noisy [15.59]. A typical view of sheet/cloud cavitation on a NACA 0015 hydrofoil is shown in Fig. 15.10a. Figure 15.10b contains a pictorial display of the various types of cavitating flow that were observed on this hydrofoil at various combinations of angle of attack α and cavitation number σ. Several different cavitating regimes occur depending on the combination of σ and α. The demarcation between *inception* and $-C_{\text{Pm}}$ (computed) varies such that σ_i is always less than $-C_{\text{Pm}}$, as

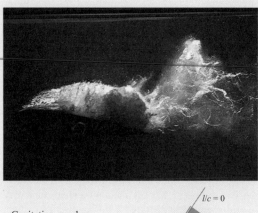

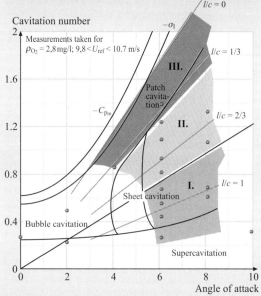

Fig. 15.10 Sheet/cloud cavitation. The figure on the *right* depicts the various types of dynamics that can occur over a range of σ and α. Note the lateral extent of the wake

expected. This illustrates that there are some effects of water tensile strength in these data. The solid black lines denote the results of visual inspection. It was noted that sheet cavitation could be subdivided into two regions such that at higher angles of attack ($\alpha \geq 5°$) the flow had a wider dynamic range. At low angle of attack, roughly less than 4°, only bubble cavitation occurred. A sample of bubble cavitation is shown in Fig. 15.4. At intermediate angles of attack and relatively high values of σ, cavitation inception is in the form of patchy cavitation.

Further lowering of the cavitation number results in sheet cavitation that is dominated by relatively low-frequency oscillations, $fc/U \leq 0.3$. At higher angles of attack, $\alpha \geq 5°$, a more-complex sequence of events occurs. The flow is still dominated by sheet cavitation. However, the characteristic frequencies of oscillation increase with increasing σ, as will be shown in subsequent plots. Sheet cavitation with large scale break-off of cavitation clouds is also observed.

Cavitation-induced lift oscillations have spectral characteristics that vary considerably over a range of $1.0 \leq \sigma/2\alpha \leq 8.5$, where α is the angle of attack, as shown in Fig. 15.11. The process was found to be highly dynamic. The amplitude of the fluctuations can exceed 100% of the steady-state lift and are associated with the periodic shedding of vortical clouds of bubbles into the flow. Three types of oscillatory behavior are noted [15.39]:

1. $1.0 \leq \sigma/2\alpha \leq 4.0$: a strong spectral peak exists at a Strouhal number, $fc/U = K$, that is independent of cavitation number.
2. $4.0 \leq \sigma/2\alpha \leq 6.0$: a higher-frequency, albeit weaker, spectral peak dominates. The frequency of this peak is almost a linear function of the cavitation number and corresponds to a constant Strouhal number, based on cavity length, of about 0.3.
3. $6.0 \leq \sigma/2\alpha \leq 8.5$: bubble/patch cavitation can occur. This induces a distinct, very low-frequency spectral peak.

Frequency data collected from high-speed video of the flow are also shown in Fig. 15.11. These data agree very well with the lift data. Note that at approximately $\sigma/2\alpha \approx 4$ there is a sharp transition from one type of frequency trend to the other. The transition that occurs at $\sigma/2\alpha = 4$ corresponds to a relative cavity length l/c of about 0.75, as predicted by the linearized analysis of *Watanabe* et al. [15.60]. Utilizing joint frequency–time analysis (JFTA), *Arndt* et al. [15.39] found that two mechanisms are at play and that they do not occur simultaneously.

The constant K is found to vary in the range 0.06–0.30. This raises an important issue concerning cavitation testing. The variation in the volume of cavitation gives rise to compliance in the test section of a water tunnel or other type of test loop (discussed subsequently). Thus the dynamic response of the test facility may play a role [15.61]. This issue is further described in *Franc* [15.62]. A series of photos showing how cavitation varies over a range of $\sigma/2\alpha$ is shown in Fig. 15.12.

Of interest to this discussion is the formation of highly vortical structures at the trailing edge of the sheet. Horseshoe vortices are clearly evident in the photos in Fig. 15.12. This phenomenon is well known. *Avellan* et al. [15.63] and *Yamaguchi* et al. [15.64] postulated the existence of these structures on the basis of flow instabilities that develop and become unstable on the surface of the sheet [15.65]. More-recent information indicates that baroclinic vortex generation is the primary factor [15.38].

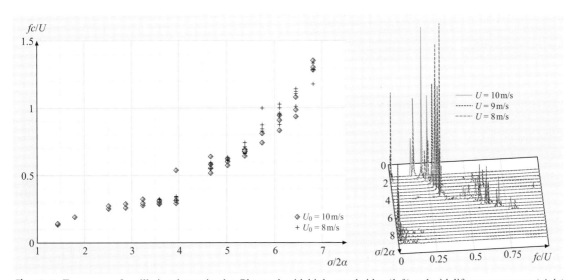

Fig. 15.11 Frequency of oscillation determined at Obernach with high-speed video (*left*) and with lift measurements (*right*) (after [15.39])

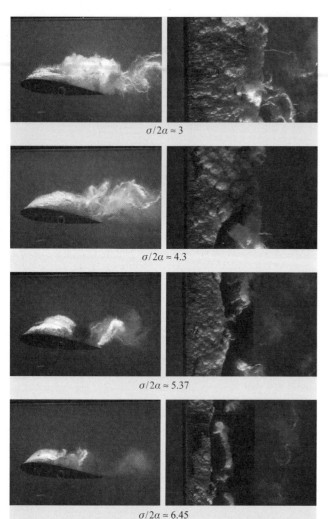

Fig. 15.12 Variation of sheet cloud cavitation with $\sigma/2\alpha$, Re $= 1.2 \times 10^6$

Developed Cavitation

When a vapor or gas-filled cavity is very long in comparison to the body dimensions, it is classified as a supercavity. Generally speaking, the shape and dimensions of vapor-filled and ventilated cavities (sustained by air injection) are the same when correlated with the cavitation number based on cavity pressure. The engineering importance of supercavitation relates especially to the design of very high-speed hydrofoil vessels as well as to the design of supercavitating propellers for very high-speed watercraft and supercavitating inducers for rocket pumps and other applications that require the pumping of highly volatile liquids. An example of supercavitation behind a sharp-edged disk is shown in Fig. 15.13. A detailed discussion of supercavitation can be found in *Knapp* et al. [15.19].

Ventilated cavities require a certain quantity of ventilation gas in order to be maintained. The issues involved are complex. This has been discussed recently by *Schauer* [15.67] and *Wosnik* et al. [15.68]. Figure 15.14 illustrates the relationship between air demand, cavitation number and cavity dimensions.

15.1.3 Cavitation Damage

The physics of cavitation damage is a complex problem. At the heart of the problem is the impulsive pressures created by collapsing bubbles [15.6]. Recent numerical techniques permit detailed examination of the collapse of individual bubbles [15.17, 69]. This work has been complemented by a wide variety of experimental stud-

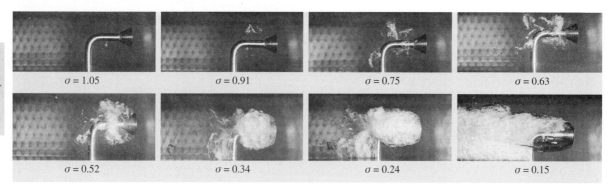

Fig. 15.13 Sequence of photos illustrating the transition from shear flow cavitation in the wake of a sharp-edged disk to fully developed supercavitation. Flow is from *right* to *left* (after [15.66])

ies [15.70–72]. All of these studies indicate that the final stages of collapse result in the formation of a microjet that can be highly erosive (Fig. 15.15). The collapse pressure is estimated to be greater than 1500 atmospheres.

Very little is known about the correlation between cavitation damage and the properties of a given flow field. However, it is important to bear in mind that cavitation erosion will scale with a high power of velocity at a given cavitation number and that cavitation erosion does not necessarily increase with a decrease in the cavitation index [15.73]. It has also been observed that the cavitation pitting rate is measurably reduced with an increased concentration of gas [15.73]. An important factor is that the pitting rate scales with a very high power of velocity (typically in the neighborhood of six). Since the velocity in turbomachinery passages is proportional to the square root of the head, this also implies that the magnitude of the erosion problem is more severe in high-head machinery.

Thiruvengadam [15.74] has analyzed a great deal of erosion data and has concluded that, for engineering purposes, the erosive intensity of a given flow field can be quantified in terms of depth of penetration per unit time $\dot{y}$ and the strength S_e of the material being eroded,

$$I = \dot{y} S_e .\tag{15.12}$$

The intensity I is a function of a given flow field. Many different forms of S_e have been tried. The most used value appears to be ultimate strength, which is basically a weighted value of the area under a stress-strain curve [15.75]. Although various materials have different rates of weight loss when subjected to the same cavitating flow, a normalized erosion rate versus time characteristic is often similar for a wide range of materials. Hence, a simplified theory allows for a rapid determination of I for a given flow by measuring $\dot{y}$ for a soft material in the laboratory. Service life for a harder material can then be predicted from the ratio of the strengths of the hard and soft materials.

Although the basic physics of the damage process in turbomachinery is complex, the essential features can be simulated by experiments with partially cavitating hydrofoils in a water tunnel [15.63, 76–80]. These studies indicate that maximum erosion occurs at the trailing edge of a cavity. The cavitation cloud at the trailing edge contains complex vortical structures that are highly erosive.

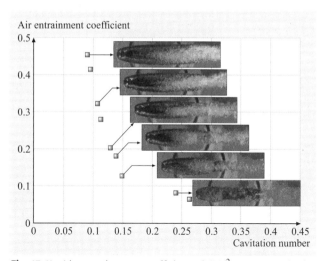

Fig. 15.14 Air entrainment coefficient Q/Ud^2 versus cavitation number for a ventilated cavity (after [15.67])

Partial Cavitation and its Relation to Erosion

In practical problems, the collective collapse of a cloud of bubbles is an important mechanism. *Hansson* and *Mørch* [15.81] suggested an energy-transfer model of concerted collapse of clusters of cavities. Because of mathematical difficulties this problem has not been studied in detail until recently [15.82, 83]. Earlier work [15.84] had already indicated the damage potential of a collapsing cloud of bubbles. Recent work supports this contention [15.85]. Very little is known about the correlation between cavitation damage and the properties of a given flow field. However, it is important to bear in mind that cavitation erosion will scale with

Fig. 15.15 Asymmetric bubble collapse near a surface (note the jet located in the bubble center). The bubble is approximately 0.2 cm in diameter

a high power of velocity at a given cavitation number and that cavitation erosion does not necessarily increase with a decrease in the cavitation index [15.73]. It has also been observed that the cavitation pitting rate is measurably reduced with a break-up of sheet cavitation at the trailing edge into detached cavitation clouds. The process is inherently unsteady, even for steady free-stream conditions. This creates a modulation of the trailing cloud cavitation that is highly erosive [15.40, 77, 78]. These details cannot be modeled with current numerical codes.

15.1.4 Facilities

Most cavitation observations and measurements are made in the laboratory. The exception to this is the recent development of cavitation monitoring techniques for hydroturbines [15.86]. Typical laboratory facilities include

1. Water tunnels;
2. Depressurized flumes;
3. Depressurized towing tanks;
4. Pump and turbine test loops;
5. Cavitation erosion test apparatus.

Water Tunnels

Water tunnels have been used for a wide variety of cavitation testing and research for about a century. The first known use of a water tunnel for cavitation research is due to *Parsons* in 1895 [15.20]. Since that time these facilities have grown in size and complexity. As of this writing (2004), the largest facility of its kind is the US Navy large cavitation channel [15.87,88]. A view of this facility is shown in Fig. 15.16. Much smaller facilities are more common in research laboratories, especially at universities. Examples of some smaller facilities are shown in Figs. 15.17, 18.

Important features necessary for cavitation tests include accurate, stable, independent control of pressure and velocity, measurement equipment for velocity, pressure, temperature, dissolved gas content and nuclei content and control, and photographic and video equipment. Because of the unsteady nature of cavitation and the extremely rapid physical processes that occur during bubble collapse and erosion, many laboratories are equipped with highly specialized high-speed video and photographic cameras that are capable of very high frame rates. Unsteady lift and drag measurements necessitate specially designed force balances [15.39]. A typical test section set up is shown in Fig. 15.19. Figure 15.20 illustrates the special monitoring equipment that is necessary for cavitation research.

Depressurized Flumes and Towing Tanks

A variety of facilities have been developed for studying cavitation phenomena in free surface flows. These include water tunnels with a variable pressure free surface test section, variable pressure towing tanks, and specialized variable pressure tanks for hydraulic model tests. It is important to note that cavitation testing in free surface flows can be especially demanding in terms of pressure

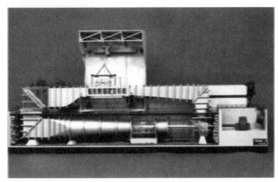

Fig. 15.16 Model of the US Navy large cavitation channel. This facility is approximately 10 storeys high and has a test section that is $3\,\text{m} \times 3\,\text{m} \times 16\,\text{m}$. The maximum flow velocity is approximately $16\,\text{m/s}$. The recirculating pump is powered by a $14\,000\,\text{hp}$ motor

1. Test section, 1270 × 190 SQ
2. Test section dome
3. Skimmer
4. 7° diffuser
5. Guide vane elbow
6. Axial flow pump, 150HP
7. Diffuser screens
8. Gas collector dome
9. Gas separator, 2134D
10. Honeycomb
11. Contraction and partial shape transaction
12. Nozzle
13. Booster bump

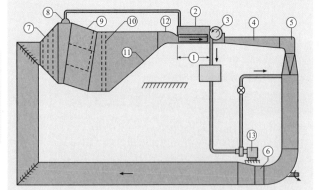

Fig. 15.17 The SAFL high-speed water tunnel at the University of Minnesota. The test section is $0.19\,\text{m} \times 0.19\,\text{m} \times 1\,\text{m}$. The maximum flow velocity is $30\,\text{m/s}$

control. Simultaneous modeling of cavitation number and Froude number requires scaling of the free surface pressure:

$$\frac{(p_0 - p_v)_m}{(p_0 - p_v)_p} = \frac{l_m}{l_p} \, . \tag{15.13}$$

In other words tests with a 1/25 model would have to be carried out at approximately (1/25)-th of an atmosphere.

The Vacutank at the Marin Institute, Netherlands (formerly NSMB) is a unique facility that was specially designed for ship propeller research using large ship models with simultaneous identity of Froude number and cavitation number. It differs from a standard towing tank because of its capability to maintain the pressure above the free surface in the range 40–1000 millibars. Broadly speaking the pressure must be reduced in proportion to the model scale; 40 millibars is the approximate pressure necessary for testing with a 1:30 model. The basin is 240 m in length, 18 m wide, and 8 m deep. A sketch of the facility is shown in Fig. 15.21. Ship models as large as 13 m in length can be studied in this facility, which has been recently upgraded (2003).

Pump and Turbine Test Loops

Pump and turbine test loops are similar in concept to water tunnels. Model testing is an important element in the design and development phases of turbine manufacture. Manufacturers own most of the laboratories that are equipped with model turbine test stands. However, there are independent laboratories available where relative performance evaluations between competing manufacturers can be carried out. An example is shown in Fig. 15.22, which is the Independent Turbine Test Facility at the Saint Anthony Falls Laboratory.

All test loops perform basically the same function. A model turbine is driven by high-pressure water from a head tank and discharges into a tail tank. The flow is recirculated by a pump, usually positioned well below the elevation of the model to ensure cavitation-free performance of the pump while performing cavitation testing with the turbine model. One important advantage of a recirculating turbine test loop is that cavitation testing can be done over a wide range of cavitation indices at constant head and flow, which is difficult, if not impossible, to accomplish in the field.

Cavitation Erosion Test Facilities

In many cases the service life of equipment and hydraulic structures subject to cavitation erosion can range from months to years. Because of the relatively lengthy

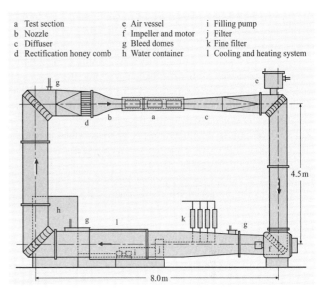

Fig. 15.18 The 30 cm-square water tunnel at the Versuchsanstalt für Wasserbau (VAO) in Obernach, Germany. The maximum flow velocity is about 16 m/s

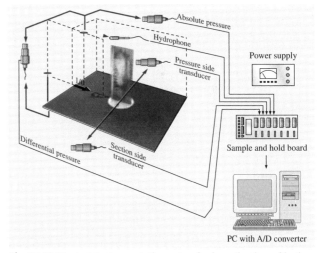

Fig. 15.19 Typical instrumentation setup for investigation of hydrofoil cavitation in the SAFL water tunnel

periods required to observe measurable erosion in the field, many different techniques have been developed in the laboratory to achieve significant time compression. The time compression factor achieved in accelerated erosion tests is as high as 10^5 [15.89]. Many of the devices used have little relationship to actual field conditions. For this reason they have typically been used for screening tests of different types of materials. Recent research is aimed at relating screening tests to predictions of service life in various applications [15.40]. The most commonly used device is the American Society for Testing and Materials (ASTM) vibratory apparatus. An oscillating horn produces a periodic pressure field that induces the periodic growth and collapse of a cloud of cavitation bubbles. A sample placed at the tip of the horn or immediately below it is easily eroded. The standard frequency of operation is 20 KHz, which produces a very high erosion rate due to the rapid recycling of the cavitation process.

As already mentioned new methods are being developed for measuring the erosion rate in the field. Usually the erosion rate is inferred from the measurement of noise or vibration. Measuring the impact pressure of the microscopic pressure jets caused by bubble collapse is a difficult task. Since the impact area is on the order of μm^2 and the duration of the impact is on the order of μs, advanced methods are needed to measure the impact. *Arndt* et al. [15.59] used a piezoelectric polymer, polyvinylidene fluoride (PVDF) to measure pressure impulses on a cavitating hydrofoil. *Soyama* and *Kumano* [15.90] used PVDF film to measure the impact energy of bubble collapse in a cavitating jet facility and estimated the threshold at which erosion will occur.

Quantifying cavitation erosion is a nontrivial endeavor as well. Perhaps the most commonly investigated parameters are pitting rate and mean depth of erosion penetration rate (MDPR), which is the number of pits caused by cavitation per unit time and the mean depth of erosion per unit time respectively. Determining the number of pits and average depth requires a microscope and well-positioned lighting since pit diameters are generally in the range 10–150 μm range (soft aluminum [15.19, 91]). In addition, other parameters include volume loss rate, maximum penetration depth, incubation period, and time until maximum damage rate. The last two give an indication of time before erosion actually occurs. In some cases, cavitation initially acts to harden the material, similar to shot peening. For this reason, an incubation period is often of interest, where the material is actually hardened. In addition to measuring cavitation erosion, the cavitating jet method is being investigated as a method of surface hardening materials.

As mentioned, it is necessary in an experimental investigation to compress the erosion time to a point where

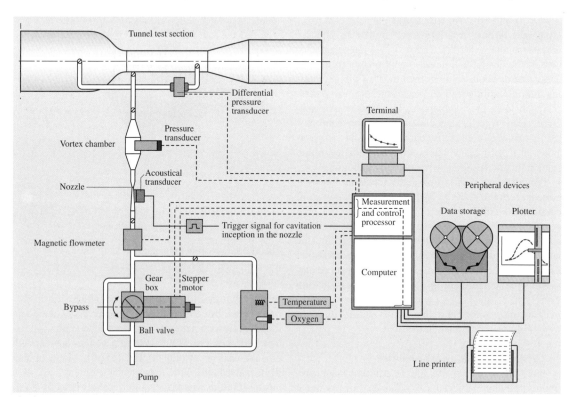

Fig. 15.20 Instrumentation for monitoring water quality at the water tunnel in Obernach, Germany

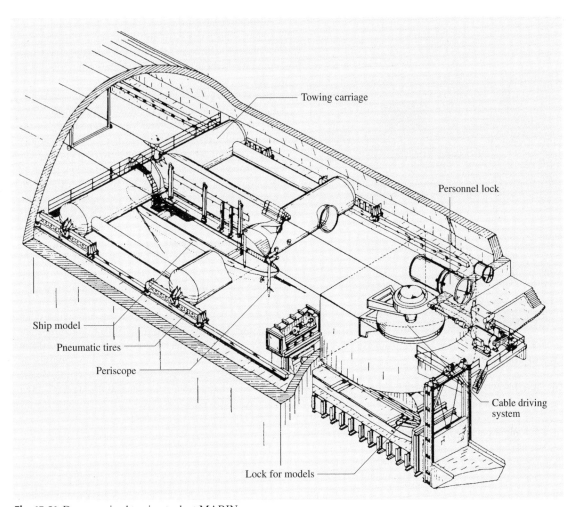

Fig. 15.21 Depressurized towing tank at MARIN

meaningful information can be acquired in a reasonable amount of time. While significantly increasing the erosion rate on a given material, these methods provide only a qualitative comparison of erosion rates, rather than a means for estimating erosion rate. The five most common methods are discussed: the vibratory method, the rotating disk method, the Venturi method, cavitating jets, and the liquid impingement method.

Venturi Method

The Venturi method has the least amount of time compression, but also most closely resembles flow conditions in practical applications. While a standard device does not exist, most Venturi erosion systems are a variation of the same principle. A test body is placed slightly downstream of the throat of a Venturi nozzle or in the test section of a high-speed water tunnel. The test body can either be a portion of the test section wall, or located in the center of the flow. The constriction of the throat can produce liquid velocities upwards of $100\,\mathrm{m/s}$, causing rapid erosion.

Numerous researchers [15.73, 92, 93] have found that pitting rate scales with velocity to the $\sim$6th power, despite significant variation in test section geometries. As a result, maintaining the same velocity from actual conditions to test specimen should result in identical erosion rates. Furthermore, a 25% increase in velocity will increase the pitting rate by nearly a factor of four. This trend will not increase indefinitely, however. *Belahadji* et al. [15.93] argue that the power should tend towards

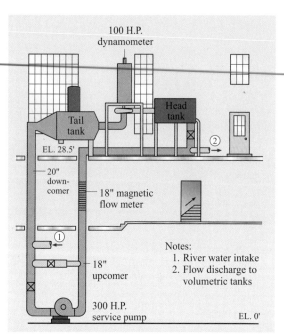

Fig. 15.22 Turbine test stand at the Saint Anthony Falls Laboratory (after [15.75])

unity as velocity increases and showed that the scaling does indeed break down at higher velocities.

In practice, a direct correlation between Venturi tests and actual applications is not always possible. These facilities are very expensive to build and operate. In addition, despite the power relation, the flow velocities necessary to perform tests that are analogous to the situation in the field in a reasonable amount of time are often still too high.

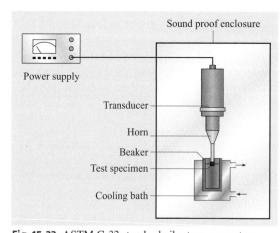

Fig. 15.23 ASTM G-32 standard vibratory apparatus

Vibratory Method

As mentioned previously, the vibratory method is perhaps the most common method used to carry out erosion studies. The vibratory method utilizes a rapidly oscillating disk to generate cavitation. The oscillation can be induced by amplified high-frequency sounds waves or a magnetostriction device. In either case, a specimen is made to oscillate at a frequency on the order of 10 kHz with an amplitude less than 100 μm in the desired liquid, creating cavitation and rapid erosion. The surface being examined can either be the oscillating disk (moving specimen), or a sample placed directly below the oscillating disk (stationary specimen). This method provides a low-cost alternative to the Venturi method. In addition, cavitation in various liquids for various materials can be studied with relative ease, since the specimens are naturally small and easy to manufacture, and the volume of liquid needed is much smaller. This method has been used to study cavitation in mercury and high-temperature liquid metals, a capability unique to this method.

The vibratory method is capable of compressing the erosion time anywhere from 300 to 700 times for steel [15.89]. Unfortunately, a correlation of erosion time between the vibratory method and field conditions does not exist. As a result, the method only allows for qualitative comparison. An extensive discussion on the vibratory method along with results for numerous metals is given by *Knapp* et al. [15.19].

Rotating Disk Method

The rotating disk employs a perforated, thin, rapidly rotating disk submerged in a fluid to create cavitation. Small holes are made in desired positions on the disk and inserts are placed either flush to the disk at the same radius or protruding perpendicular to the disk face. Under high rotationally speeds, cavitation bubbles are created by the holes and collapse on the inserts. The collapse of the cavitation bubbles results in erosion. The advantage of the rotating device is that it more closely mimics the flow patterns seen in rotating machinery, such as pump impellers, hydroturbines, and the like. The disadvantages, however, include a more complex flow field than the other methods.

Cavitating Jet Method

The cavitating jet method causes erosion by introducing a high-pressure jet into a low-pressure stagnation chamber where a specimen is placed at some distance from the nozzle. Vortex cavitation forms in the high-shear regions on either side of the jet core. As the jet propagates downstream and expands, periodic cloud cavities are observed [15.94]. The cavitating jet has some advantages over other methods. With relative ease, the upstream and downstream pressures of the jet can be varied to simulate a wide variety of cavitating conditions.

Impinging Jet Method

While the cause of erosion in the impinging jet method is not cavitation, the method is worthy of note. A liquid jet is made to impact the surface of a rotating wheel. Similar to the rotating disk, inserts are placed in the disk such that it crosses the path of the liquid jet(s). The disk is made to rotate very rapidly and the impact between the test specimen and a liquid jet of moderate velocity causes erosion similar to cavitation. This situation closely resembles erosion that occurs in Pelton-wheel-type hydroturbines. This method is capable of the most rapid erosion rates and can compress erosion time by as much as 9400 times [15.89].

Table 15.1 gives a comparison of the five methods as taken from results of the international cavitation erosion test (ICET). It is clear that the liquid jet method yields the highest erosion rates, followed by the rotating disk method. The vibratory and tunnel methods have

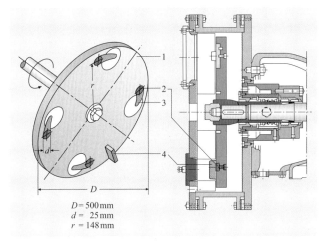

$D = 500\,\text{mm}$
$d = 25\,\text{mm}$
$r = 148\,\text{mm}$

Fig. 15.24 Rotating disk device used by KSB, 1 – rotating disk, 2 – cavitation generating holes, 3 – cavitating wake, 4 – stationary specimen (KSB Aktiengesellschaft, Frankental, Germany, http://www.imp.gda.pl/icet/KSB/KSB_rd.html)

similar erosion rates for most metals, and the cavitating jet has the lowest erosion rate. It should be noted that erosion rates are highly dependent on flow conditions

Table 15.1 Maximum erosion rates for various methods and various test metals (a: vibration amplitude, s: standoff distance) [Taken from the ICET results (http://www.imp.gda.pl/icet)]. ARMCO: American Rolling Mill Company

	Parameters	MDPR$_{\text{max}}$ (μm/min)				
		Cr–Ni steel	C steel	ARMCO Fe	Brass	Al alloy
Cavitation tunnel	$P_1 = 1030\,\text{kPa}$, $u = 30\,\text{m/s}$	0.05	0.06	0.15	0.25	2.35
	$P_1 = 103\,\text{kPa}$, $u = 14\,\text{m/s}$	0.15	0.32	0.33	0.74	2.01
Vibratory rig	(stationary) $a = 0.05\,\text{mm}$ $s = 0.35\,\text{mm}$	0.19	0.17	0.56	0.56	6.90
$F = 20\,\text{kHz}$	(moving) $a = 0.040\,\text{mm}$	0.45	0.39	1.83	1.42	6.10
Rotating disk	$P = 255\,\text{kPa}$, $u = 42.5\,\text{m/s}$	> 0.44	> 0.43	0.77	3.10	18.20
Cavitating jet	$P_1 = 19\,\text{MPa}$ $P_2 = 0.1\,\text{MPa}$, $s = 18\,\text{mm}$	0.03	0.14	0.09	0.17	1.38
Liquid jet	Jet vel.=6.75 m/s, sp. vel.=80 m/s	1.32	> 2.60	> 2.60	11.10	33.75

Table 15.2 Compression times of MDPR for aluminum and steel (From [15.89])

Experimental set up		MDPR (μm/h)		Compression time	
Method	Parameters	Al	St. G-X5	Al	St. G-X5
Venturi	60 m/s	4.35	0.0134	1.0	1.0
	80 m/s	24	0.072	5.5	5.4
	110 m/s	160	0.49	37	37
Magnetostriction oscillator $F = 20\,\text{Hz}$	Stationary specimen ($s = 1.2\,\text{mm}$)	–	2.51	–	187
	Moving specimen ($a = 0.010\,\text{mm}$)	537	4.24	123	316
	Moving specimen (0.015 mm)	847	6.77	195	505
	Moving specimen (0.025 mm)	1528	11.3	351	843
Jet cavitation	$P_1 = 250\,\text{bar}$, $P_2 = 1.3\,\text{bar}$, $s = 21\,\text{mm}$	1166	9.9	268	739
Drop impingement	jet vel. = 410 m/s $d = 1.2\,\text{mm}$	50 400	1260	11 586	94 030

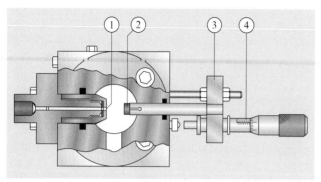

Fig. 15.25 ASTM G-134 standard cavitating jet apparatus, 1 – nozzle, 2 – specimen, 3 – specimen holder, 4 – micrometer head. (KSB, Frankental, Germany, http://www.imp.gda.pl/icet/KSB/KSB_rd.html)

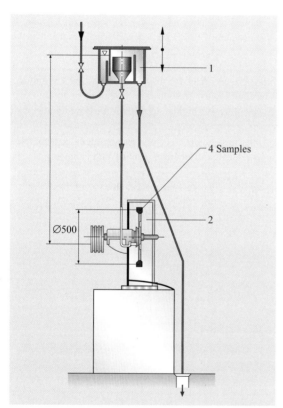

Fig. 15.26 Impinging jet device used by SIGMA, 1 – tank with controlled water level, 2 – test chamber. (SIGMA Research Institute Olomouc, Czechoslovakia, http://www.imp.gda.pl/icet/SIGMA/Sigma_li.html)

and it is difficult to generate analogous conditions between methods. Due in part to a lack of standards and difficulties in determining and correlating actual field conditions to test conditions, it has been argued that erosion tests should be performed on the actual field models [15.95].

Table 15.2 gives the erosion rates and compression time for various methods taken from *Durrer* [15.89]. Similar trends are noted in Table 15.1, except the compression time is significantly larger than seen in Table 15.2. It should also be noted that, while in both cases aluminum experiences the most severe erosion rates, the compression time between methods is the smallest. This further complicates attempts to scale erosion tests to the full-scale model where, in addition to the flow conditions, the difference in material plays a crucial role.

In general, however, the erosion rate decreases with increasing material strength. Various researches have investigated various scaling relations. *Syamala Rao* et al. [15.96] studied various methods and attempted to scale volume loss rate using different material properties. They found that results scaled best when scaling volume loss rate with the product of ultimate resilience and Brinell hardness. Figure 15.27 contains a plot of relative erosion rate using the concept developed by *Thiruvengadam* [15.74]. In this plot the erosive intensity expressed in terms of depth of penetration per unit time $\dot{y}$ is plotted versus time. Both the abscissa and the ordinate are normalized with respect to $\dot{y}_{max}$ and the time at which it occurs t_{max}. In this way a variety of different materials are shown to have the same relative time history of erosion. *Thiruvengadam* argues that erosion tests with a relatively soft material can be scaled up to prototype conditions with a harder material.

The relative erosion concept suggests another method for investigating erosion on a model or full scale is using soft metal inserts. If the proper scaling rates between metals can be determined, then softer metal inserted in cavitating equipment can be used to investigate erosion. This method requires drilling a hole in the model at the location of interest. Metal inserts are placed flush with the model surface, and cavitation tests are performed. Soft metal inserts placed in various locations on a turbine blade, for instance, give an indication of the relative erosion rate at a given location. With this knowledge, problematic areas can be treated to extend the life of the blade.

One of the simplest methods for erosion tests on full-scale equipment or scaled models can be done with a paint coating. Researchers found that a mixture of black stencil ink and thinner is capable of indicating areas where significant erosion will occur on a ship propeller in less than an hour. The international towing tank conference (ITTC) committee proposed technique using a soft surface made of stencil ink and ethyl alcohol for cavitation erosion tests. *Kadoi* and *Sasajima* [15.97] followed this procedure with some modifications to investigate erosion damage at different velocities on a model ship propeller. Recent tests at SAFL indicate that Dykem *Steel Blue* layout fluid also works well for qualitative visualization of cavitation erosion. Surface preparation, paint concentration, application method (brush versus spray) and drying time critically affect the paint erosion result, and need to be determined on a case-by-case basis.

While much knowledge has been gained concerning the cause and behavior of erosion due to cavitation, an accurate scaling procedure from test to field conditions is still not available. Results suggest that, in addition to flow and liquid characteristics, the material properties of the surface in question also play an important role. An acceptable scaling relation and theoretical explanation of cavitation erosion is a worthy endeavor and sorely needed.

15.1.5 Water Quality Measurements

Numerous studies highlight the importance of water quality monitoring when investigating cavitation. One well-known example is the ITTC test body that was tested at various international facilities. Cavitation inception was measured versus flow velocity. The results showed a large variation in measured inception value (Fig. 15.28). Differences in water quality are partly to blame for the huge discrepancy. Cavitation occurs when the pressure in a flow goes below a critical value, causing microscopic nuclei to expand explosively. A flow containing a large supply of nuclei will necessarily cavitate more readily than a flow with relatively few nuclei. An indication of the ability of a liquid to resist cavitation is the tensile strength. In theory, water without nuclei can withstand tensile stresses of approximately 10^4 atmospheres before cavitation occurs [15.99]. In practice, great care is needed even to produce water capable of withstanding more than two atmospheres.

Various factors influence the water quality of a given liquid. The presence of particulate matter in a liquid acts to decrease the surface tension of a bubble. *Church* [15.100] found that the surface tension of *clean bubbles* void of particle impurities on the liquid–gas interface is more than twice the value of *dirty bubbles* containing chemical and particle impurities. *Marschall* et al. [15.101] studied the effects of particle size on wa-

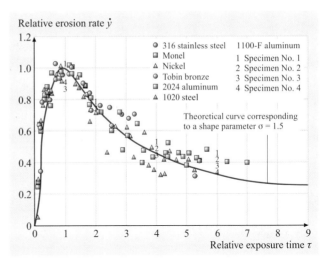

Fig. 15.27 Relative erosion rate versus dimensionless time for various metals

ter quality in degassed tap water and found that particle size, shape, and hydrophobic/hydrophilic characteristics influence the tensile strength. *Ma* [15.102] observed that tensile strength increased when water is stored under high pressure for a prolonged period of time. The high pressure, in addition to decreasing the mean nuclei size, causes gas bubbles to dissolve, decreases the number of available nuclei.

Numerous researchers have studied the effects of water quality on cavitation behavior. It is well known that

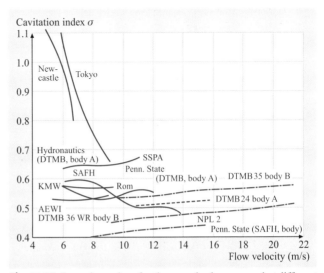

Fig. 15.28 Inception values for the same body measured at different water tunnels [15.98]

water quality effects influence the behavior of cavitation. *Van der Meulen* [15.103] studied water quality effects on inception and desinent cavitation on hemispherical-nosed bodies. In an extensive review of the subject, *Keller* [15.104] noted that not only is the value of the inception index affected, but water quality has an influence on the appearance of cavitation as well. *Kawakami* et al. [15.105] found that water quality can affect the lift dynamics on a cavitating hydrofoil. While it is well known that water quality affects cavitation behavior, a quantitative scale that accurately correlates cavitation behavior to water quality does not exist. In fact, it is unclear how the sum of various factors, such as gas content, particulate matter, and pressure history combine to characterize the tensile strength of a liquid. Indeed, even the tensile strength of a liquid can vary depending on how it is measured.

From a practical point of view, the tensile strength T, defined as $T = P_v - P_{crit}$, can be varied over a wide range in typical research facilities, and is a function of total gas content, system pressure and the history of the water. It should be emphasized that the relationship between tensile strength and preconditioning of the water is facility dependent. The tensile strength obtained as a result of these procedures is specific to the water tunnel being used. Different water tunnels require different procedures and a great deal of experience is necessary before water quality can be varied in a predictable manner. Care also has to be taken to ensure that there is not an excess of bubbles in the flow, which can result in pseudo-cavitation at local pressures considerably in excess of the vapor pressure of the liquid.

Gas Content Measurement

The van Slyke apparatus is used to measure total dissolved gas content of a liquid. The apparatus arose from an investigation of gas and electrolyte equilibria in blood in the early 1920s [15.106]. Developed by Donald van Slyke, the device was originally designed to measure gas content in blood. Because the device is very accurate, easy to use, and requires a small sample amount, it is ideal for water quality investigations as well. It utilizes cavitation to bring gas out of solution by creating a Torricelian vacuum above a sample of liquid. Figure 15.29 is a photograph of the modified van Slyke apparatus used at SAFL.

Numerous researchers have developed similar devices utilizing the same basic principle. *Numachi* [15.107] modified the van Slyke apparatus to allow for separate storage of released gas and water sample. Numachi's device also was easier to handle [15.108].

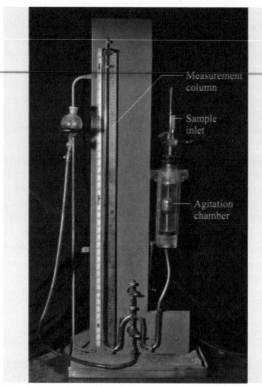

Fig. 15.29 Van Slyke apparatus

Schöneberger also made modifications to the van Slyke meter to more accurately measure gas content. Since each device is similar, only the van Slyke device will be discussed in detail. A $10\,\text{cm}^3$ sample is introduced and sealed in the apparatus. Lowering the mercury level creates a vacuum. The sample is mechanically shaken for approximately 10 min to ensure the dissolved gas is released from the sample. The partial pressure of released gas is measured and compared with the partial pressure of water vapor. The total dissolved gas is determined by the resulting pressure difference using the following relation.

$$Q = 2 \frac{273}{273+T} \frac{P_v - P_g}{7.6}, \quad (15.14)$$

where P_v denotes the partial pressure of the vapor and P_g is the partial pressure resulting from the released gas, and T is the fluid temperature in °C. Q is the total gas content in per milliliter. It should be noted that neither P_v nor P_g is measured, but rather it is the difference that is determined from the device. While capable of measuring the total dissolved gas in a liquid quickly, accurately, and over a wide range with a small sample, the large amount of mercury needed to operate the device has recently made it undesirable to use from an occupational health hazard point of view.

Recently, some mercury-free devices to measure gas content have been developed. Based on the same principle as the van Slyke meter, Brandt constructed and patented a device that uses an inflatable rubber membrane in place of mercury to create a vacuum; the amount of gas is determined volumetrically, therefore no pressure measurements are needed. *Heller* [15.109] developed a device that uses a bellows to create a vacuum and a pressure transducer to determine gas content. These devices, however, are not widely used or well known. Instead methods that infer total gas content from the concentration of other gases have become popular.

In an effort to replace the van Slyke meter, dissolved oxygen probes have been used to infer gas content. Most dissolved oxygen probes rely on an oxygen-permeable membrane and an electrolyte solution to determine dissolved oxygen in solution. Oxygen diffuses across a membrane into the solution where an anode and cathode are separated. The measured dissolved oxygen is a function of the current generated. While this method has been shown to be accurate in inferring total dissolved gas, caution should be taken when using this method. Unlike the van Slyke meter, this method requires regular calibration and maintenance. The solution and membrane must be periodically changed, or inaccurate reading will result. Also, it is incorrect to assume that the rate of oxygen diffusion into or out of a liquid is equal to that of other gases. The volume ratio of oxygen and nitrogen in air is not the same in water or other liquids under equilibrium conditions. In addition, the chemical oxygen demand of the tunnel water due to dissolved impurities is also a factor. While dissolved oxygen generally gives a qualitative indication of the amount of total dissolved gas, it should not be used as a method to infer total dissolved gas on a quantitative basis.

As already discussed, in addition to the difficulty of determining total dissolved gas, this parameter alone does not necessarily give a true indication of water quality. Since microscopic nuclei are needed for cavitation to occur, it is possible that a liquid with large total dissolved gas content will contain relatively few nuclei if exposed to high pressure for a sufficient amount of time. Under normal conditions, this is not the case. *Peterson* [15.110] found that the size and number of nuclei in a flow varies with gas content. *Arndt* and *Keller* [15.111] found that a doubling in gas content resulted in a tenfold increase in nuclei content.

Acoustic Techniques for Nuclei Measurement

Acoustic determination of nuclei content has many advantages in a variety of applications. The acoustic method is non-intrusive and does not require a transparent boundary between the instrument and fluid in question. The method is based on the acoustics of bubbles. In a simplified form, a bubble can be represented by a spring-bob system. The gas within the bubble acts as a spring to external forcing. The emission of an acoustic wave into a liquid medium containing bubbles will naturally cause the bubbles to oscillate. *Minnaert* [15.7] determined the resonance frequency of a bubble as a function of the radius and free-stream pressure.

$$\varpi_0 = \frac{1}{R_0}\sqrt{\frac{3\kappa p_0}{\rho}} \ ; \tag{15.15}$$

R_0 and p_0 denote the mean radius and pressure, respectively, ρ is the density of the liquid, and κ is an unknown constant to take into account the effects of heat conduction. When heat conduction effects are negligible, κ is unity. In theory, introducing an acoustic wave into a liquid will cause bubbles to oscillate. The resulting bubble oscillations will create acoustic waves in the liquid that are characteristic of the bubble size and number. The resulting energy scatter or absorption and attenuation of the sound wave by the bubbles are measured. This technique is termed resonance excitation or the attenuation method since it uses the resonant frequency of a bubble to determine the nuclei content in a flow.

Duraiswami et al. [15.112] proposed using a dispersive method to determine nuclei content. A dispersive method utilizes attenuation and the difference in the speed of sound between a liquid and gas. A pure liquid, without bubbles, will propagate sound at a different rate than a liquid with bubbles. By comparing the propagated sound speed with the theoretical value for a pure liquid, the nuclei content can be determined. *Chahine* and *Kalumuck* [15.113] used two hydrophones placed in a flow and recorded bursts of sound emitted from one and received by the other. By recording the interference caused by the bubbles, the bubble size distribution and number is determined.

Acoustic methods, however, have limitations. Acoustic methods tend to overestimate nuclei content for liquids with small nuclei, and underestimate nuclei content in liquids with large nuclei [15.112]. In addition, the accuracy is sensitive to the surrounding materials. Highly acoustically reflective surfaces can have an adverse effect on the signal acquired [15.114]. *Leighton* [15.115] notes that, when using the attenuation method, large bubbles in a liquid can emit a stronger signal than much smaller bubbles resonating at their natural frequency. *Oldenziel* [15.116] also found that the best results are attainable only when the concentration of bubbles is low. In addition, the attenuation method assumes spherical, linear bubble oscillation.

Optical Techniques for Nuclei Measurement

The simplest optical method uses photography. By illuminating a plane in the liquid and taking a photo, bubble size and number can be determined. This method is a simple, brute-force approach. It requires a camera with sufficient resolution, and individually measurement of the radius of each bubble in the image. In cases where nuclei content varies over a wide range, and where numerous conditions are studied, this method quickly became impractical. Recent advances in digital photography and image-processing techniques, however, have spurred interest in this method.

Holographic techniques have recently been developed along these lines. In holographic interferometry a laser beam is split into two coherent beams. One beam, the subject wave, is passed through the volume of interest. The other, the reference wave, bypasses the measurement volume and is directed onto a recording medium, where it is recombined with the subject wave. By recording the difference in amplitude and phase between the reference and subject wave, a hologram is produced.

Holographic images provide a high-resolution, three-dimensional image of a given flow sample. Once an image has been acquired and constructed, it is a matter of time to process and analyze the images. Holographic techniques provide many advantages. Direct visualization of a sample removes the need for calibration procedures. In addition to nuclei distribution, objects of all types are easily observed and distinguished, an advantage which is lacking in all other methods. In addition, holography can be used to measure objects as small as $5\,\mu m$. *O'Hern* [15.53] was able to examine nuclei and particle distribution in oceanic waters with great detail using a submergible holographic system. In addition to the cost of a holographic system, a considerable amount of time is needed to obtain meaningful information about water quality. As is the case with the photographic method, this technique is impractical for routine water quality measurements.

Keller [15.117] used light scattering to determine nuclei size in a flow. Using a HeNe gas laser to illuminate a control volume, a receiver lens is positioned perpendicular to the laser and light scattered by nuclei passing though the control volume is collected by the lens. The scattered light then passes through a photomultiplier and is collected by a pulse-height processor. The intensity of the scattered light is assumed to vary with nuclei size. By knowing the light intensity, the size spectrum can be determined. The procedure requires calibration using particles of known size and number. As a result, the accuracy of the method is largely dependent on the care and precision with which the calibration is performed. This method is accurate but has many limitations. The volume of the laser light/receiver path intersection limits the maximum size that can be measured. Keller's original apparatus was used to measure nuclei in the range of approximately $5-500\,\mu m$. In addition, only a small portion of the flow can be analyzed.

Along similar lines, phase Doppler anemometry (PDA) devices offer accurate nuclei size and velocity measurement without the need for calibration. PDA measures the phase shift created by particles passing through the probe volume and is capable of measuring particles in the range of approximately $1\,\mu m$ to $1\,mm$ [15.118]. The PDA has the advantage of measuring both nuclei content and liquid flow rate with high accuracy [15.119]. The sampling rate of the system is limited by the number of particles that pass through the probe volume in a given time, allowing for very high sampling rates. A sample, however, is taken only when a particle passes through the control volume, resulting in variable time separations between sampling. Like Keller's device the probe is not able to distinguish between solid particles and bubble nuclei, which can give potentially misleading results.

Cavitation Susceptibility Meters

Cavitation susceptibility meters (CSM) have evolved from a technique for determining nuclei content that measures the inception of cavitation in a known pressure field, and compares the pressure at inception with vapor pressure. Cavitation is caused in a pre-designed area of minimum pressure (C_{p_m}). The inception events for given flow conditions are observed and recorded. Knowing the pressure at inception, a formula can be derived from the equilibrium theory for the mean number and size of nuclei in the flow.

Schiebe [15.120] first proposed using mathematically derived axisymmetric bodies, known as *Schiebe bodies*, to measure cavitation inception. Using potential flow theory, a set of bodies with known value and location of C_{P_m} were developed and experimentally tested. An analytical model was developed where the cavitation occurrence rate is predicted for a flow with a given nuclei content. The body is tested in a flow, and the occurrence rate is acoustically measured. By adjusting the parameters of the analytical model until the occurrence rate matches that of the experiment, the nuclei distribution of the flow was determined.

While this method proved unsuccessful, an alternative use for CSM devices was proposed. Instead of measuring nuclei content, *Silberman* et al. [15.122] proposed generating a set of inception event curves for a given test body under different water qualities. These curves could then serve as a reference for future cavitation tests. By measuring cavitation events on the designed test body and comparing with previously generated curves, the tensile strength of the liquid can be determined. As a result, the focus of water quality monitoring becomes one of tensile strength of the test liquid, rather than nuclei content.

CSM devices using Venturi nozzles have been developed by numerous researchers [15.109, 123, 124] to measure the tensile strength of a liquid directly. Cavitation is caused in the throat of the nozzle by increasing the flow rate until inception is observed. The pressure and the flow rate through the nozzle are measured and the pressure in the throat is inferred. If cavitation is observed at pressures well below vapor pressure, the liquid is considered strong.

Oldenziel [15.123] first proposed using a Venturi nozzle to measure cavitation susceptibility. Inception in the nozzle throat was detected visually using a shadowgraph method. A light source and photodiode are placed on opposite sides of the nozzle. The explosion of a bubble causes light scattering and a change in voltage of the photodiode. *Lecoffre* and *Bonnin* [15.124] used a steel Venturi nozzle and a pressure sensor located on the downstream side of the nozzle to measure cavitation susceptibility. Many variations of the basic concept have been developed to measure water quality. One such device is discussed below.

Vortex Nozzle

One example of a modern susceptibility meter is the vortex nozzle. Detailed measurements by *d'Agostino* and *Acosta* [15.125] indicated that conventional Venturi devices suffer from inaccuracies due to viscous effects. To circumvent this problem, *Keller* [15.121] developed a so-called vortex nozzle for the measurement of tensile

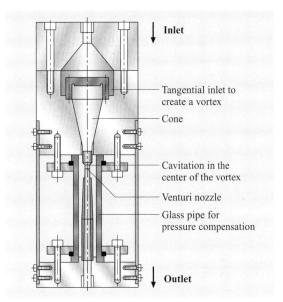

Fig. 15.30 Vortex nozzle susceptibility meter [15.121]

strength. The nozzle device consists of a vortex chamber followed by a Venturi nozzle. The vortex chamber superimposes a circulation on the flow to ensure that cavitation occurs at the center of the throat, away from the flow boundaries. The pressure drop across the nozzle is measured using pressure transducers. The throat pressure is calculated using an empirically derived equation. The throat pressure when inception occurs equals the critical pressure. With the critical pressure known, the tensile strength of the water is determined.

An empirical relation for the pressure in the throat is given by

$$P_\text{T} = P_1 - C\rho \left(\frac{Q}{A_\text{T}}\right)^2, \quad (15.16)$$

where the density and dimensions of the nozzle are assumed to be fixed, P_1 is measured at the inlet to the device and C is an experimentally derived calibration constant, and Q is the flow rate. Substituting $P_\text{T} = P_\text{v} - T$, where T is the tensile strength, and P_v is the liquid vapor pressure, the tensile strength is:

$$T = C\rho \left(\frac{Q}{A_\text{T}}\right)^2 - P_1 + P_\text{v}. \quad (15.17)$$

Q is a function of the pressure drop across the nozzle, a measured quantity, thus the tensile strength of a liquid can be determined and the average nuclei size can be inferred. In the above equation, P_1 and Q are determined for the instance when inception is first observed in the vortex nozzle. Inception can be measured visually or acoustically depending on the available instruments and material with which the nozzle is constructed.

In principle, an equation for the minimum pressure in the throat area could be obtained by calculating an idealized flow in the nozzle throat consisting of a uniform axial flow and a superimposed rotational vortex the nozzle throat is of interest. *Ma* [15.102] carried out such a calculation numerically and was able to correlate the tension measured in the vortex apparatus with the tension inferred from measurements of inception in tip vortex flow.

Regardless of the techniques used to measure tension, it should be noted that tensile strength is also a function of the static pressure of the liquid. This can be inferred from a simple equilibrium model. Under equilibrium conditions,

$$P_i - P_\infty - \frac{2S}{R} = 0, \qquad (15.18)$$

where R is the bubble radius, P_i is the internal bubble pressure, and P_∞ the surrounding liquid pressure. The internal bubble pressure is the sum of both vapor pressure and pressure due to incondensable gas. The pressure due to incondensable gas can in turn be determined by the ideal gas law.

$$P_i = P_v + P_g = P_v + \frac{M_g G \theta_g}{\frac{4}{3}\pi R^3}, \qquad (15.19)$$

where M_g, θ_g, and G denote the moles of gas, temperature of gas, and universal gas constant, respectively. Combining (15.18) and (15.19) and letting the R_c denote the critical radius at which the bubble becomes unstable, the tensile strength is written:

$$T = P_v - P_\infty = \frac{2S}{R_c} - \frac{M_g G \theta_g}{\frac{4}{3}\pi R_c^3}. \qquad (15.20)$$

It is impractical to measure the necessary quantities to determine the tensile strength of a liquid from the above equation. *Blake* [15.126] showed that, for isothermal expansion, the critical bubble radius can be written:

$$R_c = \left(\frac{9 M_g \theta_g G}{8 \pi S}\right)^{1/2}. \qquad (15.21)$$

Substituting (15.21) in (15.20) yields,

$$T = \frac{4S}{3R_c}. \qquad (15.22)$$

The ratio of the tensile strength of a liquid at two different liquid pressures is

$$\frac{T_1}{T_2} = \frac{R_{c2}}{R_{c1}}. \qquad (15.23)$$

Using the ideal gas law for the pressure inside the bubble, (15.23) can be written as

$$\frac{T_1}{T_2} = \left(\frac{P_1}{P_2}\right)^{1/3}. \qquad (15.24)$$

Thus, the tensile strength of the liquid scales with the cube root of the pressure. When determining the tensile strength of a liquid using any CSM device, it is important to note the pressure history and conditions under which the measurement is taken. In addition to reducing the bubble size, high pressure increases the gas saturation level of the liquid and causes gaseous nuclei to dissolve. While the theoretical estimate of the tensile strength of water is on the order of 10^4 atmospheres, this is never the case due to the large number of nuclei present in the liquid. When inception studies are made in a closed-circuit tunnel, it is necessary to insure that a continual supply of nuclei exists for cavitation to occur. The source of persistent nuclei in a flow has been the subject of much debate. *Fox* and *Hertzfeld* [15.127] proposed that an organic skin membrane forms around bubbles, preventing gas diffusion. Experimental evidence for this theory has been observed, but the model fails to explain why nuclei still persist in inorganic liquids and other liquids containing chemicals that should alter the behavior of the organic material. *Harvey* et al. [15.128] postulated that gas pockets become trapped in the crevices of surfaces and free-stream solid particles. These gas bubbles form concave surfaces with the surrounding liquid. The pressure in the bubble is lower than the surrounding liquid so the gas does not dissolve into the liquid. The surface tension caused by the hydrophobic surface acts to repel the liquid, establishing a stable condition for the trapped gas pocket. The pocket remains in a stable equilibrium until the vapor pressure is reached in the near vicinity. While no direct verification of the model has been made, it is generally accepted.

15.1.6 Cavitation Inception

Detection

The most common measurement in cavitation research is the determination of the conditions for inception. This may appear to be a trivial exercise, but measurements are fraught with difficulties. Although the definition of cavitation appears to be relatively simple and straightforward, there is some ambiguity in the experimental determination of σ_i. The inception process tends to be intermittent and difficult to quantify (see the Appendix in [15.129] for a simple model of intermittency). One factor is the relatively large range of nuclei sizes that tend to cavitate at different values of pressure. As the cavitation number is decreased, the cavitation events occur at an increasing rate. This behavior makes it difficult to define a concise point of inception. Numerous researchers have used a microphone or hydrophone to detect inception acoustically. The rapid expansion of a bubble produces easily detected acoustic waves in the audible range. Small hydrophones placed near the inception location provide an easy and accurate method for detecting cavitation inception. Often cavitation can be detected acoustically before it can be visually observed. This method is especially advantageous in cases where visual access to the test section is difficult or impossible. *Lecoffre* and *Bonnin* [15.124], *Ma* [15.102], and *Heller* [15.109] used a microphone to detect inception in a Venturi-type device in order to determine tensile strength.

As pointed out later, unexpected variations in σ_i occur because of factors not accounted for in the basic scaling law. In addition, there is a large source of error inherent to the various methods we use to characterize the onset of cavitation. Inception is usually defined by the visual appearance of bubbles. In the laboratory, testing is normally carried out in a water tunnel at constant velocity with the static pressure slowly lowered until cavitation is observed. An alternative scheme is to lower the pressure even further and then gradually raise it until the cavitation is extinguished (cavitation desinence). There is a discrepancy between σ_i defined by inception and that defined by desinence. This is referred to as a hysteresis effect and has been discussed in detail by *Holl* and *Treaster* [15.130].

Because of the noise associated with cavitation, it has been common for many years to use acoustic techniques to define the inception point. Sample acoustic data are shown in Fig. 15.31. Figure 15.31a is a presentation of acoustic data obtained in a plug valve, plotted in the form of noise intensity in various frequency bands as a function of the cavitation index. Inception is defined by the point where there is a rapid increase in noise level with further lowering of pressure. σ_i so measured depends on the frequency band of the data, the low-frequency noise being relatively insensitive to cavitation index. A different acoustic technique is illustrated in Fig. 15.31b where the number of cavitation

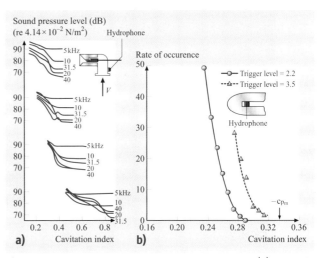

Fig. 15.31a,b Acoustic inception techniques [15.15]. (**a**) Acoustic intensity method. (**b**) Rate of occurrence

events per unit time is determined with a hydrophone and an electronic counter. To isolate the cavitation noise from background noise, a Schmidt trigger arrangement is used. In this situation, σ_i is defined by a rapid rise in the rate of occurrence. Unfortunately, the measured value of σ_i is sensitive to the threshold level of the Schmidt trigger, as shown in the diagram. While this method is useful to determine the first inception event, it has proven unable to measure the number of inception events accurately. *Pham* et al. [15.114] compared the number of inception events in a combination Venturi-center body-type device as measured acoustically, and

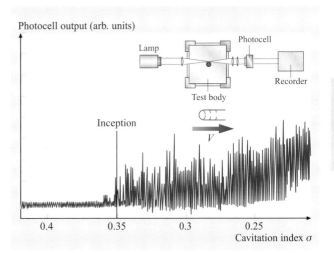

Fig. 15.32 Example of an optical detection technique [15.15]

Fig. 15.33 Tip vortex inception in weak water (*left*) and strong water (*right*). The cavitation number for inception in strong water is so low that sheet cavitation on the surface of the foil occurs before vortex cavitation inception [15.131]

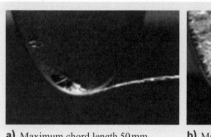
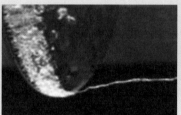
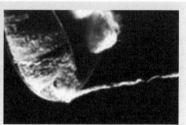

a) Maximum chord length 50 mm, $\sigma = 2.18$, $U = 11.0$ m/s, beginning tip vortex cavitation

b) Maximum chord length 100 mm, $\sigma = 2.18$, $U = 11.0$ m/s, developed cavitation

c) Maximum chord length 200 mm, $\sigma = 2.18$, $U = 11.0$ m/s, fully developed cavitation

Fig. 15.34a–c Size scale effect on a NACA 16-020 hydrofoil [15.104]. **(a)** Maximum chord length 50 mm, $\sigma = 2.18$, $U = 11.0$ m/s, beginning tip vortex cavitation **(b)** Maximum chord length 100 mm, $\sigma = 2.18$, $U = 11.0$ m/s, developed cavitation **(c)** Maximum chord length 200 mm, $\sigma = 2.18$, $U = 11.0$ m/s, fully developed cavitation

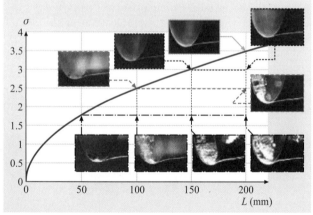

Fig. 15.35 Size effect on inception for tip vortex cavitation on a NACA 16-020 hydrofoil [15.104]

a photodiode to measure inception in a Venturi device. An example of an optical detection scheme is shown in Fig. 15.32. This method is based on the determination of scattered light that has been focused in the minimum-pressure region of a test body. With the use of a laser setup, several investigators have successfully used this technique. *Arndt* and *Keller* [15.131] used high-speed video tripped by a laser to compare vortex inception for a three-dimensional (3-D) hydrofoil in strong and weak water. By positioning a laser in the center of a trailing vortex, a high-speed digital video camera was set to record the instance a vapor bubble crossed the laser path. Using this technique, *Arndt* and *Keller* [15.131] found that inception occurs near the foil tip under weak water conditions, while inception occurs far downstream under strong water conditions (Fig. 15.33).

with high-speed video. They found that the acoustic counting technique overestimated the number of inception events by as much as seven times compared with the number seen using high-speed video. It may be for similar reasons that *Schiebe*'s [15.120] method was unsuccessful.

Visual detection of inception is an accurate method when the necessary tools are available. As previously mentioned, *Oldenziel* [15.123] used a light source and

The basic detection schemes are all subjective to some extent and can lead to discrepancies between data collected in different facilities. Unfortunately, there is no clear distinction between observed variations in σ_i due to the detection scheme used and variations in σ_i due to a fundamental difference in environmental factors. It should be emphasized at this point that cavitation inception as defined by a measurable reduction in the performance of a hydraulic machine has no relevance to this discussion since a state of fully developed cavitation is associated with measurable changes in performance.

Scale Effects

Keller [15.104], in an extensive review of the subject, pointed out that the first step in analyzing scale effects, defined as variations in σ_i with size and velocity, is to remove the ambiguity concerning the critical pressure, p_c in (15.1). As a historical note, Parsons used hot water in his 1895 water tunnel as a means of lowering the cavitation index, thereby allowing inception studies at relatively low velocity [15.132]. This is not recommended because of thermal effects [15.1]. He adjusted the tensile strength of the water to zero and methodically determined that there are other scaling factors for a variety of body shapes that vary as

$$\sigma_i = \sigma_0 \left(\frac{L}{L_0}\right)^{1/2} \left(\frac{v_0}{v}\right)^{1/4} \left[1 + \left(\frac{U_\infty}{U_0}\right)^2\right], \tag{15.25}$$

where L, v, U_∞ are characteristic length of the body, the viscosity and velocity, respectively. L_0, v_0 and U_0 are reference values and σ_0 is a constant. According to Keller, knowing σ_0 for a given body contour and type of cavitation the cavitation index can be predicted for every size, flow velocity and viscosity of the fluid. Keller also had a correlation for turbulence level, but it is felt by the authors that there is insufficient data to support this correlation fully. Although this correlation is supported by a very extensive set of experimental data, there is no theoretical underpinning to these results. A size scale effect is illustrated in Fig. 15.34.

This illustrates the appearance of cavitation with increasing size when cavitation number and velocity are held constant. A further indication of this effect is given in Fig. 15.35. Note the increase in σ_i with increasing size, in accordance with (15.25). Note also the increasing development of cavitation with size if σ is held constant.

Gas Content Effects

Care must be exercised in adjusting water quality to a low value of tensile strength. A relatively high gas content can obscure the basic physics that are being investigated. This is illustrated with some data originally presented and discussed by *Holl* [15.31], plotted in Fig. 15.37. These data illustrate the occurrence of two different types of cavitation on a NACA 16012 hydrofoil section. The lower set of data corresponds to band-type cavitation, which disappears uniformly across

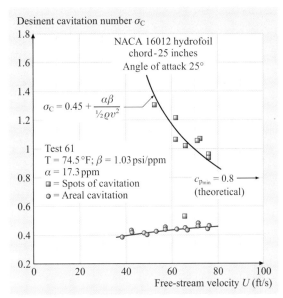

Fig. 15.36 Example of two types of cavitation on a NACA 16012 hydrofoil [15.31]

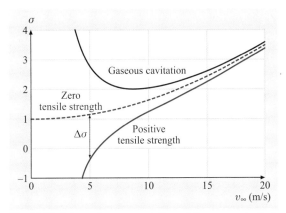

Fig. 15.37 Schematic representation of the gas content effect [15.104]

the span at the desinent value of the cavitation index. A second type, consisting of random spots, was also noted, having a much higher desinent cavitation index. The critical cavitation index for the former increases with velocity while it decreases for the latter type. Holl theorized that the former type was vaporous cavitation whereas the latter type was not vaporous cavitation at all, but instead a different type of bubble growth due to gaseous diffusion, very similar to the bubble formation occurring when the ambient pressure in a bottle of soda water is suddenly reduced upon opening. This so-called

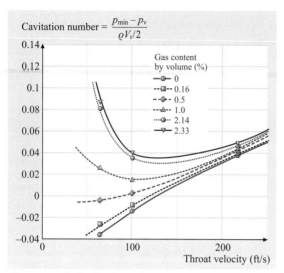

Fig. 15.38 Cavitation inception number for a Venturi device for several values of gas content [15.104]

gaseous cavitation follows a different scaling law, as given by (15.7). Sometimes the gas content effect is more difficult to distinguish, especially when vaporous cavitation is strongly dependent on velocity. This is shown schematically in Figs. 15.36, 37.

In summary, a review of previous laboratory experiments indicates that the details of the cavitation inception process are not adequately described by a single physical parameter such as the cavitation index. Previously overlooked details of the flow field can play a major role in the inception process and cavitation can occur when the local pressure is higher or lower than the vapor pressure. Ordinary liquids can sustain tension and more than one type of inception mechanism is possible on a given class of body. Thus, care must be taken to distinguish between the different forms of cavitation when attempting to arrive at scaling relationships.

15.1.7 Measurements in Cavitating Flows

In addition to classical measurements of flow necessary for understanding the inception process, it is necessary to make measurements in cavitating flows.

High-Speed Photography and Video

Due to the complexity of cavitating flows (essentially three-dimensional, nonstationary, turbulent two-phase flows with phase change) photography and cinematography are very important tools when studying cavitation phenomena. One look at any text on cavitation, e.g., *Knapp* et al. [15.19], makes it clear how much insight can be gained this way. As with all experimental techniques, *spatial and temporal resolution* are key. Time-resolved photography or cinematography can be achieved by very short illumination times while working in a totally dark environment. Good results have been obtained by letting the duration of a stroboscopic flash (typical strobe illumination time: $3\,\mu s$) determine the exposure time, examples are shown in Figs. 15.7, 10, 12, 13, 40, 46, 47. Another option is to use cameras capable of extremely short shutter times, either mechanically or electronically, such as high-speed complementary metal–oxide–semiconductor (CMOS)-based cameras and continuous illumination. Commercially available video equipment is capable of frame rates as high as 250 000 frames/s. Some laboratories are equipped with special cameras that are capable of frame rates as high as 2 000 000 frames/s. On the other hand, when combining long exposures (i. e., non-time-resolving shutter times) with continuous illumination, *time-averaged* pictures of cavitation can be taken. Spatial resolution is a function of the recording medium and the optical setup. It should be adjusted so that the smallest events of interest are sufficiently resolved.

Simultaneous Imaging of Different Views

Using mirrors in test sections with optical access from at least two sides, two (or more) views of cavitating flows can be imaged simultaneously. It is important that mirrors are placed so that the optical paths for each view are of equal length, allowing focusing on both at the same time. Figure 15.39 shows a configuration used in the SAFL high-speed water tunnel, Figures 15.40 and 15.41 show sample frames taken with a high-speed

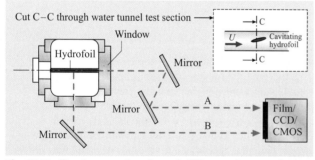

Fig. 15.39 Simultaneous imaging of different views, optical arrangement for photography/video of cavitating hydrofoils in SAFL high-speed water tunnel. Optical paths for side view (A) and bottom view (B) are of equal length

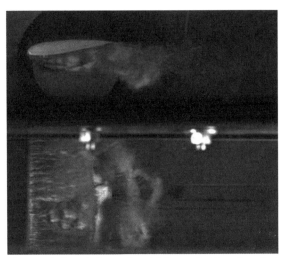

Fig. 15.40 Sample frame from high-speed video of cavitating NACA 0015 hydrofoil, simultaneous imaging of side and bottom view (2000 frames/s, shutter time $\approx 500\,\mu s$, $U = 8\,\text{m/s}$, $\sigma = 1.00$, $\alpha = 8°$)

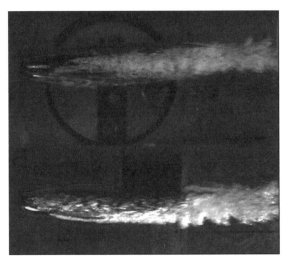

Fig. 15.41 Sample frame from high-speed video of a ventilated supercavitating body, simultaneous imaging of side and bottom view (2000 frames/s, shutter time $\approx 500\,\mu s$, $U = 7.8\,\text{m/s}$)

camera [2000 frames/s, field of view (FoV) 1024×1024 pixels].

Pressure Measurements in Erosion Studies

In viewing the current state of knowledge in this field, it becomes clear that further advances in cavitation research require a method for relating erosion rate to the amplitude of pressure transients striking the solid boundary during the collapse of cavitation bubbles. Methods have been developed for measuring in situ the impulsive pressures due to cavitation in both laboratory experiments and in the field. Various techniques have been explored in the past to measure the pressure transients both spatially and temporally. Most studies are made with conventional pressure transducers. However, the environment for pressure measurement equipment can be severe and it has been found that conventional pressure transducers often have unacceptably high failure rates. Recently, piezoelectric polyvinylidene fluoride (PVDF) polymer film has been used extensively in dynamic measurement and control. This film has been adopted for the measurement of impulse pressures in the study of cavitation erosion [15.40, 59].

Pressure transducers can be constructed in-house from piezoelectric film. The integral component of the film is a polarized PVDF polymer. This material is known for its high degree of piezoelectric activity. As outlined in *Arndt* et al. [15.59], the use of this technology for hydrofoil tests requires special development. One example of the basic measurement technique is shown in Fig. 15.42. In this example, a matrix of sensors was used to instrument an interchangeable plug on a hydrofoil for water tunnel tests. A grid of 14 piezoelectric pressure transducers was attached to a removable section of the suction side of the foil as shown in Fig. 15.43. Considerable effort was expended in developing the method

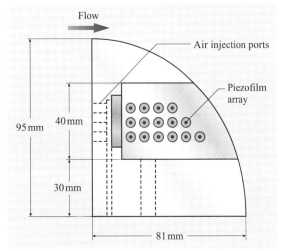

Fig. 15.42 View of a pressure film instrumented NACA 0015 hydrofoil. This setup was used to study the effects of air injection on minimizing erosion

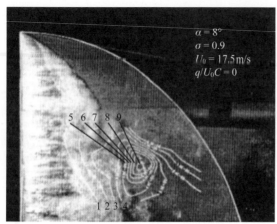

Fig. 15.43 View of sheet/cloud cavitation on the NACA 0015 hydrofoil shown to the left. ($U = 17/5 m/s$, $\sigma = 0.9$). Superimposed on the picture are isobars of mean square pressure measured with the piezometric film array

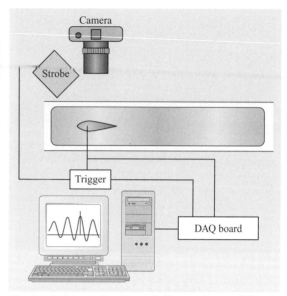

Fig. 15.44 Schematic of a typical setup for conditionally sampled photos

of layout, mounting, and architecture of the transducer matrix finally used.

Normally this technique is only suitable for transient response. Calibration can be achieved using a dropping ball technique or by sudden unloading of the film. The latter can be achieved with a special device whereby a pencil lead is gradually loaded until it breaks, suddenly releasing its pressure on the film. Inspection of the transient output form the transducer can be used to determine the frequency response, which can be quite high. Details concerning time and spatial resolution can be found in *Arndt* et al. [15.59].

A further example of using an array of piezoelectric film transducers is shown in Fig. 15.45. The test setup is shown in Fig. 15.19. Numerical simulations of cavitation cloud collapse indicate that cavity collapse occurs at approximately 75% of chord for the conditions tested. A plot of RMS pressure as a function of the composite parameter $\sigma/2\alpha$ and relative position on the foil indicates that the most intense erosion will take place when $\sigma/2\alpha = 3.0$ at a relative position of about $x/c = 0.75$, in agreement with the numerical simulations [15.39].

Conditionally Triggered Photography of Sheet Cavitation

As another application of pressure measurements, the transducer array cited above can be used to obtain phase-locked photos of the cavitation process. As already noted, sheet/cloud cavitation can be very periodic for a given set of operational parameters. Often it is desired to obtain a series of photographs to accompany data from an array of piezoelectric film transducers. This technique can provide information about the relationship between cavity location and pressure distribution. The best method to accomplish this would be to use a high-speed video camera that is synchronized to the data-acquisition system. Budgetary constraints often preclude the use of such a system. However, the periodicity of the flow can be exploited. By obtaining a series of photos with different delay times relative to the pressure signal a montage of photos can be obtained that can be used to construct a virtual film of the process.

To sample a pressure signal whose period and phase are correlated with the flow, the beginning of a period, as defined by the maximum of a pressure signal, can be sensed. Using a delay circuit to trigger a stroboscope a series of photos at relative times within a period of oscillation can be obtained. In principle, several photographs can be taken at each phase and ensemble-averaged. However, photographs taken at equal time delays are found to be remarkably consistent. It is recommended that the photographs be taken in total darkness. Typically a photograph is taken by manually opening the shutter, using a computer program to trigger two strobe lights simultaneously, and then the shutter is manually closed. By performing the procedure in this manner, the exposure, as well as the time that each photograph is taken, is determined by the firing of the strobes.

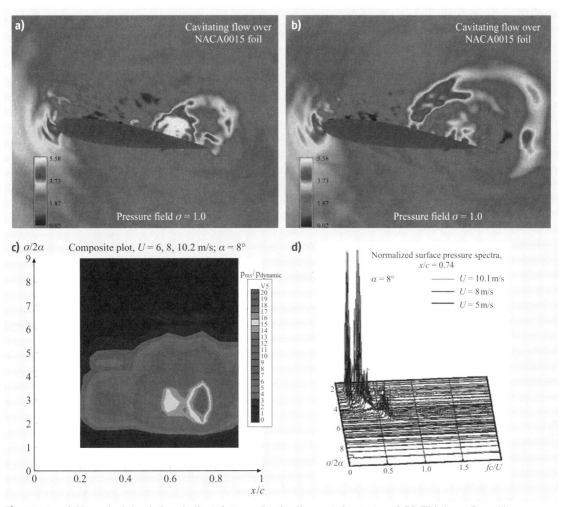

Fig. 15.45a–d Numerical simulations indicate intense cloud collapse at about $x/c = 0.75$. This is confirmed by measurements of pressure using an array of piezoelectric film transducers. Note the similarity of the pressure spectra to the lift spectra shown in Fig. 15.11 (**a**) Instant of cloud collapse (**b**) Radiation of pressure wave after collapse (**c**) RMS pressure versus $\sigma/2\alpha$ and x/c (**d**) Pressure spectra

A programmable triggering system affords much greater flexibility then a hard-wired triggering circuit. A technique in use at the St. Anthony Falls Laboratory utilizes a computer code originally developed at the Arizona State University for use with Metrabyte DAS-16 compatible data-acquisition boards that were modified to suit the current needs. A typical setup is shown schematically in Fig. 15.44. To remove the effect of errors resulting from the time delay introduced by the amplifier and the trigger circuit internal to the strobe lights, the light intensity was measured with a photocell, and recorded simultaneously with the other data. Postprocessing shows directly when the strobes lights fired relative to the pressure signal, and hence the time that the photograph was taken.

Typical results are shown in Fig. 15.46. Photos taken at two instances of time one corresponding to a positive peak in pressure and the other to a negative peak in pressure are shown. In general this method requires a considerable amount of fine tuning depending on the physics of the flow. Finally a third example is a study of the physics of a bubbly wake produced by a cavitating hydrofoil is shown in Fig. 15.47. Numerical simulations reveal that coherent vortex structures exist in the wake.

As a final note, other signals can be used to trigger stroboscopic lighting. *Arndt* and *Maines* [15.133] made measurements of nucleation in a tip vortex using a new system consisting of a laser beam as a *trip wire*, a threshold/delay circuit, strobe lighting and a standard 35 mm camera with an extension bellows for magnification. As a nucleus enters the vortex core and cavitates, it passes through a laser beam, scattering light that is picked up by photoamplifier, which in turn triggers the delay circuit. At the end of the delay strobe lighting is activated. By assembling a sequence of photos at several delay times, the equivalent of frame rates as high as 40 000 frames/s could be achieved. The same laser trip technique was used to trigger a high-speed video camera for observations at another laboratory with startling agreement between the data sets obtained in two different laboratories with two different visualization techniques [15.131] (Fig. 15.33).

Lift and Drag

Lift and drag measurements in a water tunnel are difficult even in non-cavitating flow. Most lift balances in use in small water tunnels are designed for steady conditions. A typical balance design incorporates the use of cantilever beam elements and strain gages to sense deflection. The challenge in force balance design is the need to eliminate *crosstalk* between lift and drag. Many balances have been designed in-house and are constructed with custom-made components. An example of such a balance is shown in Fig. 15.48a. This balance was successfully used at the Saint Anthony Falls Laboratory for many years [15.32]. Recently a variety of load cells have become available at reasonable cost. These provide the opportunity for constructing reliable balances at reasonable cost (Fig. 15.48b).

Measurements in cavitating flows place additional requirements on balance design. Cavitation often induces unsteady oscillations in the lift and drag that are often in a frequency range that is beyond the natural frequency of a typical force balance. *Kjeldsen* et al. [15.32] attempted to circumvent this problem using two flush-mounted pressure transducers at the base of hydrofoil, one placed on the pressure side and the other on the suction side. This provided a method to measure the instantaneous values of the pressure difference between the suction and pressure side of the foil. The main motivation for using this transducer configuration is to obtain an independent metric for fluctuating lift since the frequency response of a typical force balance is not adequate for resolving lift and drag dynamics in cavitating flow. Individual needs will of course vary. Kjeldsen et al. were able to confirm that Δp is proportional to C_l as expected, although this technique is not in general use. *Arndt* et al. [15.39] utilized a new force balance configuration that utilizes piezoquartz force sensors (Fig. 15.49). Although the frequency response of this type of force balance is high, it is generally not suitable for steady force measurements. *Arndt* et al. [15.39] found that, for the case of lift oscillations on a cavitating hydrofoil, the pressure difference technique and force balance technique gave similar results (Fig. 15.50).

Acoustic Measurements

Acoustic measurements in water tunnels are difficult because of reverberation. This is an acute problem in water tunnels because of the relatively long wavelengths of radiated sound. For example, a sound signal with a frequency of c/U would have a wavelength of ac/U, where a is the speed of sound and c is a typical length scale, e.g., the chord length. The long acoustic wavelengths relative to typical tunnel dimensions implies that the test environment is highly reverberant, necessitating very sophisticated acoustic calibration procedures. However, qualitative measurements can still be made effectively. An example is the use of an acoustic signal for inception detection (Fig. 15.31).

A limited number of examples of measurements of cavitation noise in water tunnels can be found in the literature [15.131, 134, 135]. For example, *Higuchi* et al. [15.135] were able to distinguish between tip vortex cavitation noise and surface cavitation noise by using two different hydrofoils designed such that surface cavitation occurred on one hydrofoil in the absence of tip vortex cavitation and tip vortex cavitation occurred in the absence of surface cavitation on the other. A special feature of this study was the use of an array of hydrophones to locate the inception location in a tip vortex. This is shown in Fig. 15.51. The experiment was made in a high-speed water tunnel. The acoustic signature was monitored by four Brüel and Kjær (B&K) model 8103 hydrophones (Fig. 15.52) positioned in a water-filled chamber separated from the test section by an almost acoustically transparent wall. A similar setup was used by *Barker* [15.136]. Both the pressure fluctuations due to the turbulent boundary layer on the test section wall and the water tunnel background noise were measured to be insignificantly low, compared to the cavitation noise. On the other hand, the test section was found to be highly reverberant.

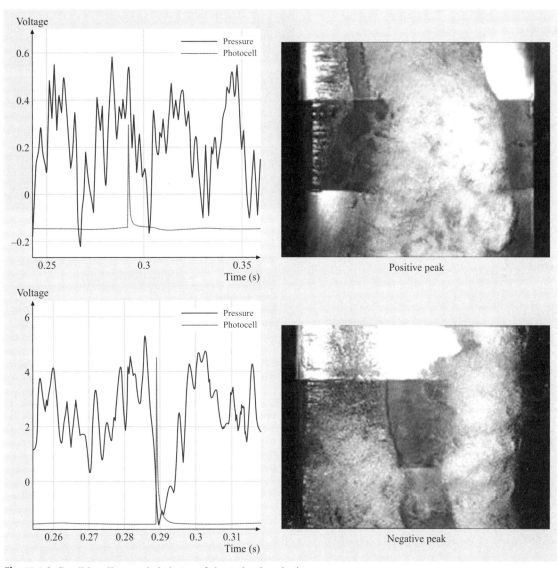

Fig. 15.46 Conditionally sampled photos of sheet cloud cavitation

The hydrophones were first calibrated in a large water tank where an acoustic free field could be simulated. Using a spherical projector, high frequency (typically a 60 KHz sine wave) tone bursts were generated and measured with the hydrophone array. Each hydrophone in the array was placed in the same relative position as in the water tunnel. In order to calibrate both the hydrophone array and the acoustic characteristics of the water tunnel itself, the projector was next placed in the test section in place of the hydrofoil. By triggering the digital oscilloscope with the projector pulse, the output from the individual hydrophones was simultaneously digitized. The initial pulses in these traces were the projector output itself via direct path, which had the same waveform as the near-field measurement of the projector signal. The subsequent pulses reflected by the tunnel walls could be clearly identified by this technique and were in qualitative agreement with a computer simulation of the reverberation in the tunnel.

Cavitation inception could be detected both visually and acoustically. Inception was marked by a distinct burst of noise in the flow field. By pre-triggering an

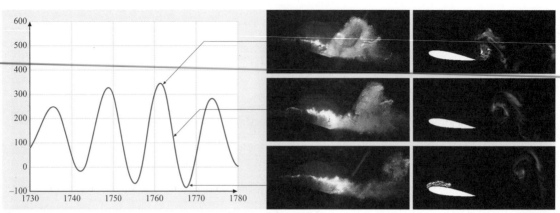

Fig. 15.47 Comparison of phase locked photos of cavitation (*left column*) with numerical simulations of incondensable gas that has come out of solution as a result of the cavitation [15.137]

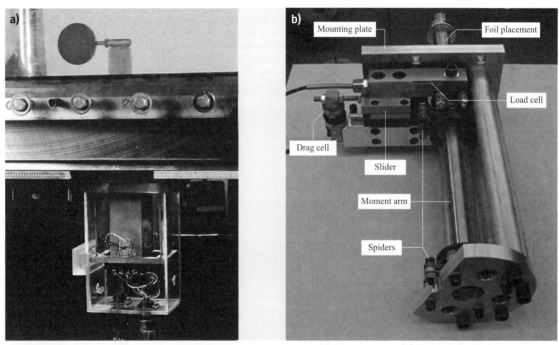

Fig. 15.48a,b Typical force balances (courtesy of Saint Anthony Falls Laboratory) (**a**) Force balance installed in the SAFL high-speed water tunnel. A sphere test for turbulence level is being performed. (**b**) New replacement balance for the one shown in (**a**). This balance utilizes off-the-shelf load cells

analog-to-digital (A/D) converter with the hydrophone signal itself, individual instantaneous pressure time traces at four different locations could be recorded. By measuring the time delay of arrival among the four hydrophones, the sound source location could be estimated by a triangulation method. The three time delays among the four time traces define three two-dimensional manifolds. The intersection of these manifolds is the location of the noise source. Given three time delays among four hydrophones as inputs, implicit simultaneous equations for the three coordinates of the sound source can be solved iteratively.

A technique for dealing with the effect of tunnel reverberation was studied for the case of surface cavitation

Fig. 15.49 Piezoquartz force balance used to measure fluctuating lift on a cavitating hydrofoil [15.39]

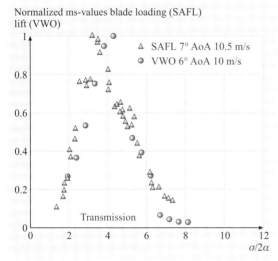

Fig. 15.50 Comparison of lift oscillation data obtained with Δp measurements made at SAFL with force balance measurements made at Obernach, Germany

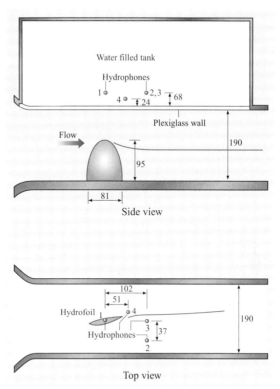

Fig. 15.51 Experimental setup for detecting the location of inception in a tip vortex [15.135] (All dimensions in mm)

Fig. 15.52 Typical commercially available hydrophones (courtesy of Brüel and Kjær)

inception. Since the individual pulses were recognizable in the signal, a Fourier analysis can be performed on the individual pulses. The waveform of a single bubble collapse in the free field was obtained by ensemble-averaging the acoustic cavitation inception data. Using this waveform, single-bubble noise in the presence of reflections from the tunnel walls could be simulated numerically. The computed effect of reverberation on the power spectrum was calculated by comparing the power spectrum of a simulated single-bubble collapse with the power spectrum of a simulated pulse with reflections. The spectrum of the reverberant signal was found to match very closely that of the original free-field pulse

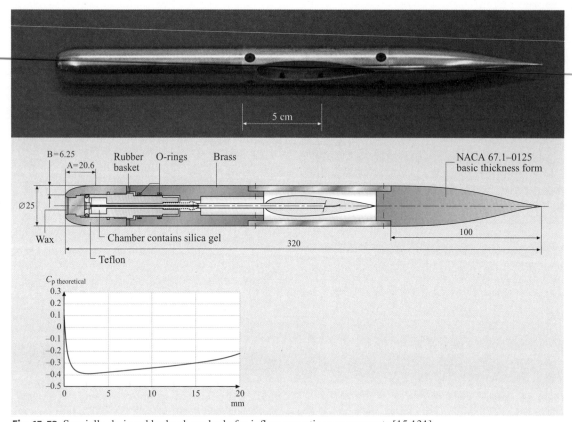

Fig. 15.53 Specially designed hydrophone body for inflow acoustic measurements [15.131]

when the spectra were normalized to maximum amplitudes. In the study by *Higuchi* et al. [15.135] the amplitude of the reverberant spectrum was approximately 7 dB higher than that of the free field in this example, and appears to scale with the number of reflections from each individual pulse. The result will differ in each facility but the lack of distortion of the inception noise spectra due to reverberation is an encouraging result.

Another technique for measuring inception noise was reported by *Arndt* and *Keller* [15.131]. The effects of nuclei content were studied in a series of comparative tests in a water tunnel and the Vacutank at the Marine Institute, Netherlands (Fig. 15.21). This investigation required careful planning because of the required remote operation in the Vacutank due to the very low operating pressures. Only values of σ greater than 1 could be achieved in both facilities. With the desire to use a fixed body rather than a model propeller, a series of sharp-edged disks of 2, 4, 8 and 16 cm diameter were manufactured. The disk geometry was selected because of its relatively high cavitation number. The four different sizes were selected not only to provide an overlap in Reynolds number in the two facilities, but also an overlap in time scales of the shear layer turbulence ($\tau \approx d/U$). At the time this was thought to be an important factor in the inception process. The relative disk thickness, shaft diameter etc. were fixed at the same values used in the classic study of *Kermeen* and *Parkin* [15.138]. Pressure ports were located in the supporting shaft of each disk so that the time-averaged pressure could be measured at various positions in the wake. The three largest disks were also fitted with accelerometers to allow an independent check on the acoustically measured cavitation event rate.

Inception observations were carried out both visually and acoustically in the water tunnel. These tests served to verify the hydrophone design and mounting system that had to be developed to perform remotely monitored inception tests in the Vacutank. The sensing element is a special hydrophone, designed and manufactured by the Physical Laboratories, TNO, model ZP 84 (early research utilized an Atlantic Research

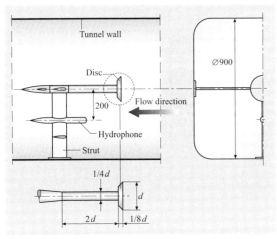

Fig. 15.54 Water tunnel installation for measuring inception noise [15.131]

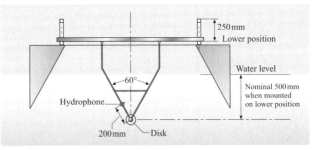

Fig. 15.55 Vacutank installation for measuring inception noise [15.131]

LC-60 hydrophone) that was housed in a specially designed streamlined body. Consideration had to be given to designing a housing that was free of cavitation and minimized non-cavitating flow noise. Several factors were taken into account in selecting the position of the hydrophone, including its location in the wake and minimization of the effects of reverberation in the water tunnel. The hydrophone and its position relative to the disks in both facilities are shown in Figs. 15.51, 53 and 15.54. Some of the results are given in *Arndt* [15.15].

As a final example, a similar setup to that used in *Higuchi* et al. [15.135] was used to measure acoustic radiation from sheet cloud cavitation by *Arndt* and *Levy* [15.139]. The results are shown in Fig. 15.56. Note the difference in the frequency content compared to the spectra of surface pressure fluctuations shown in Fig. 15.45.

Quantitative Laser Techniques for Cavitating and Bubbly Flows

Soon after the creation of the first laser at the Hughes Research Laboratory in May 1960 [15.140], the special properties of coherent light were being explored for flow measurement purposes. *Laser Doppler velocimetry* (LDV), which matured in the 1970s and 1980s, and *particle image velocimetry* (PIV), which came of age in the 1990s, are quantitative optical flow measurement techniques utilizing lasers.

The LDV technique [15.141], also known as *laser Doppler anemometry* (LDA) uses a small ellipsoidal measuring volume formed by crossing laser beams (the measuring volume is typically on the order of 0.1 × 0.1 × 0.5 mm, or smaller), in which the velocity of randomly arriving particles is measured at a point by the time (determined from Doppler burst frequency) it takes to travel a known distance (Doppler fringe spacing). A direct extension of LDV is *phase Doppler anemometry* (PDA), which can be used to determine the bubble or particle size by measuring the phase shift of the illuminating beam with multiple receivers (initially two, now typically three, receivers), as well as bubble or particle velocity [15.142].

For the PIV technique [15.143, 144], a seeded flow field is illuminated (exposed) twice (or more) with a laser light sheet with a known separation time between laser pulses, and particle images are recorded [on film, a charge-coupled (CCD) or CMOS device. If the recording medium is digital, the technique is referred to

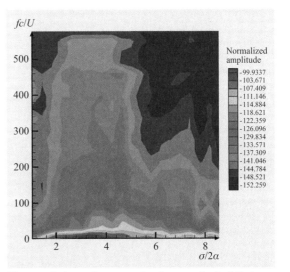

Fig. 15.56 Acoustic spectra for the same hydrofoil shown in Fig. 15.19. Note the relatively high frequency content in the range $2 \leq \sigma/2\alpha \leq 6$

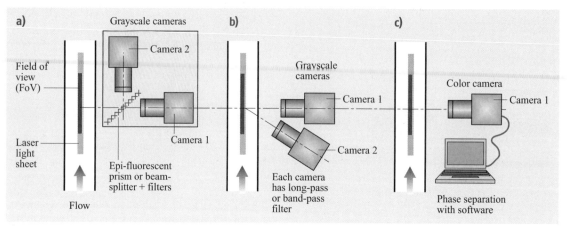

Fig. 15.57 Typical PIV camera arrangements for phase separation by color

as digital PIV (DPIV)]. For intermediate particle concentrations statistical correlation-based methods can be used to determine the displacement in small interrogation areas, typically resulting in thousands of velocity vectors in a plane per captured image pair.

The fundamentals of these techniques, implementation details and recent advances (e.g., high-frame-rate PIV, holographic PIV, and micro-PIV, micro-LDV) are discussed in Chap. 5. Nowadays, complete LDV and PIV systems including software are commercially available from at least half a dozen vendors, and several free software packages for analysis exist. When used in cavitating or bubbly flows, these techniques face a special set of problems and challenges, which are addressed here.

Particle Image Velocimetry

In a two-phase flow containing liquid and bubbles, the light from the laser sheet illuminating the object plane will be Mie-scattered by both the seeding particles and the bubbles. Typically the size of the seeding particles is one to tens of micrometers (μm), whereas the bubbles caused either by vaporous/gaseous cavitation or artificial ventilation are much larger, typically tens to a few hundreds of micrometers. In order to measure velocity fields, the liquid and gas phase need to be effectively separated in the PIV images. Several methods to separate phases have been developed; they are summarized briefly below, and their applicability to cavitating flows is discussed.

Phase Separation by Color through Laser-Induced Fluorescence. In this technique the liquid phase is seeded with microscopic particles coated or imbedded with a fluorescent dye. The dye is chosen so that the particles fluoresce at a wavelength (color) different from the incident light, e.g., in the yellow–orange range when illuminated with a green laser. Typical lasers currently used in PIV are frequency-doubled Nd:YAG (neodymium-doped yttrium aluminum garnet) lasers emitting light at 532 nm (green). Both phases will scatter the incident laser light, but only the liquid-phase seeding particles will fluoresce [laser-induced fluorescence (LIF)]. The liquid and gas (bubble) phases can now be separated in two ways: before image recording by using two synchronized grayscale cameras and color filters, or after image recording by using one color camera. If two grayscale PIV cameras are used, optical filters, e.g., long-pass filters (*long-pass* refers to the wavelength of the light; these filters are also called *low-pass* filters, referring to the frequency), and possibly a beam splitter or an epi-fluorescent prism are required. Some camera arrangements are shown in Fig. 15.57. In arrangement a,

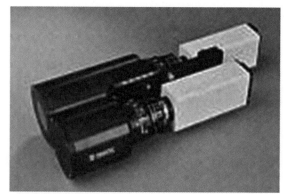

Fig. 15.58 Commercially available two-phase PIV camera mount (courtesy of Dantec Dynamics)

both grayscale cameras have the same field of view. An epi-fluorescent prism splits the beam, transmitting only the fluorescent wavelength to camera 2. Alternatively, a simple beam splitter in combination with a long-pass filter on camera 2 and an interference band-pass filter on camera 1 centered on the laser light wavelength can be used. In arrangement b, the two cameras have a different field of view, and a mapping function is required to *de-warp* the image of the off-axis camera 2. Long- and band-pass filters are also required to separate the phases. In arrangement c, a single color camera is used, and the phase separation is done after recording via software. Two-camera lens assemblies of this kind specifically designed for two-phase flow PIV/LIF have recently become commercially available (Fig. 15.58).

Phase separation by color was first successfully implemented by *Sridhar* et al. [15.145] in a cavitating jet facility while recording onto 35 mm color film. The experiment was conducted at very low, non-cavitating velocities (0.6 m/s) and 100 μm bubbles were injected at very low gas void fractions for demonstration purposes.

A more complex version of phase separation by color was proposed by *Towers* et al. [15.146] for autocorrelation PIV (single frame, double pulse). Here one phase was also seeded with fluorescent particles, but multiwavelength illumination and recording was used to obtain separate, directionally resolved velocity vector fields for both phases of a two-phase flow via autocorrelation processing. Since most newer PIV cameras are capable of cross-correlation through frame-straddling (double frame, double pulse) one does not have to go to such complexity to avoid directional ambiguity anymore.

Phase Separation by Particle Image Size. Bubbles caused by vaporous/gaseous cavitation or artificial ventilation are typically one to two orders of magnitude larger than the seeding particles. For the photometric parameters of a typical PIV setup, the image size of a seeding particle with a diameter of one to tens of micrometers will be dominated by diffraction-limited imaging. When imaging bubbles with a diameter of hundreds of micrometers, the contribution of geometric optics to the effective particle diameter is one to two orders of magnitude larger than for the seeding particles, and the contribution of diffraction-limited imaging is also larger due to the smaller aperture (higher f#) required by high-intensity reflections from the bubbles. This will lead to different size diameters on the imaging medium, allowing phase separation by image size. For example, in the bubbly wake behind the closure region of a ventilated supercavity, *Fontecha* [15.147] found the effective particle image diameters for 3 μm TiO_2 seeding particles and 300 μm bubbles to be one and six pixels, respectively.

The simplest way of separating the phases by particle image size is to threshold the grayscale PIV image, converting them to binary format: every pixel above a certain intensity value (the *threshold*) is set to 1, every pixel below is set to 0. Now regions of white (binary value 1) can be discriminated by size, identifying them as seeding particles or bubbles. This method was used by *Hassan* et al. [15.148,149]. *Hassan* et al. [15.148] solved the problem of having a grey corona around bright bubbles by additional local thresholding. They determined that the optimal local thresholding function to properly outline bright bubbles was $1/r^2$, where r is the distance to the bubble center. *Gui* and *Merzkirch* [15.150] and *Lindken* et al. [15.151, 152] developed a digital masking technique to separate the differently sized particle images of the two phases. Determination of these masks also required pixel intensity thresholding and size discrimination.

Phase Separation by Particle Image Intensity (Brightness). Laser illumination and camera aperture (the photometric parameters) can be adjusted so that seeding particle images will not exceed a certain grayscale value. Pixels with higher grayscale values would then be associated with a bubble image. *Sakakibara* et al. [15.153] and *Easson* and *Jakobsen* [15.154] employed this difference in image brightness in two-phase flows where solids were the disperse phase and air was the carrier phase. This technique should in principle work well for bubbly liquid flows created by cavitation events, since the larger bubbles scatter (reflect) incident light at higher intensity than the seeding particles. A problem may arise if the intensities of the light scattered by the particles and by the bubbles are orders of magnitude apart, e.g., in sheet/cloud cavitation or ventilated flows, and cannot be simultaneously imaged onto a single camera (cf. discussion of practical aspects below).

Other Phase-Separation Techniques. Phase separation based on object image shape was used by *Oakley* et al. [15.155] for individual large bubbles, and for dispersed large, solid particles by *Kiger* and *Pan* [15.156]. *Kiger* and *Pan* [15.156] eliminated the seeding particles by using a spatial median filter (seeding particles are discarded as high-frequency noise; this only works for low concentrations of the dispersed phase). The seeding particle images can then be processed separately by subtracting the filtered images from the originals. Correlation peak properties can be used for phase separation if there is a significant velocity difference between the two phases, as demonstrated in a bubble column by *Delnoij* et al. [15.157], albeit with prior knowledge of the liquid and gas velocities.

Combined Techniques. *Khalitov* and *Longmire* [15.158] developed a two-parameter phase discrimination by object image size *and* brightness. An object-weighted centroid vector (centroid position) is also determined and can later be used for particle tracking. A combination of phase separation by color *and* correlation peak widths applied to a bubbly jet in a box was reported by *Arnardottir* [15.159]. *Chaine* and *Nikitopulos* [15.160] used a single color PIV camera in a dispersed, bubbly jet. Phases were separated after recording by evaluating the green and red channel separately, and additional filtering by intensity was used to eliminate interference from seeding particles and *phantom* bubbles.

Of all the techniques introduced above, it is expected that phase separation by color through laser-induced fluorescence using two grayscale cameras with filters will become the dominant PIV technique in cavitating flows in the intermediate term. The main advantage of this technique is that it allows the capture of bubbles of all sizes and that the PIV image for the carrier liquid phase is uncontaminated by the bubble phase. The main drawback is currently cost: two grayscale cameras are needed and suitable fluorescent particles can be very expensive.

Practical Issues when Using PIV in Cavitating Flows. All of the phase-separation techniques introduced above were developed in dilute bubbly liquid (or sometimes air/solid) flows. While they should in principle work for cavitating flows, additional difficulties arise, especially since bubbly flows resulting from cavitation can have much higher local void fractions, i.e., vapor *pockets* or bubble clustering. Some of the difficulties are:

- Cavitation bubbles are typically larger than seeding particles and will scatter light more effectively. Their images will appear much brighter (and larger) than seeding particles on the recording medium. This can cause a problem when using a single camera at higher void fractions, since the light scattered by the seeding particles may be too weak in comparison to the bubble reflections to produce a usable signal. The proper selection of seeding particles and the use of higher sensitivity (12-bit or higher) CCD cameras can somewhat compensate for this. For a primer on the scattering characteristics of particles, i.e., scattering cross-section and intensity of light scattered in the vicinity of 90° from the incident light sheet [15.161].
- Bubbles must be dispersed enough to provide usable particle images for cross-correlation. This may not be the case in the initial stages of cloud shedding in vaporous cavitation or in the vicinity of the closure region behind a ventilated cavity. If the void fraction is high, bubbles may cluster or coalesce.
- The illuminating laser light sheet may be attenuated if a large number of bubbles is present.
- A bubble illuminated with a laser light sheet may appear as two adjacent bright spots on the image plane [15.155, 160]. The brighter one is due to reflection, the dimmer one is due to refraction, then internal reflection, and refraction again. If the photometric parameters cannot be adjusted to avoid this, additional post-recording treatment becomes necessary.
- Seeding particles provide additional nucleation sites for cavitation inception. If water quality is critical to the cavitating flow being investigated, this may affect the results.

Laberteaux and *Ceccio* [15.162, 163] investigated partial cavities on two-dimensional wedges without and with spanwise variation and a NACA 0009 hydrofoil with PIV. The flow was seeded with fluorescent latex particles with an average diameter of 30 μm and individual double-pulsed images (with image shift to resolve directional ambiguity) were recorded onto single frames of 35 mm film through a filter blocking the laser light wavelength. Thus velocity vector fields of only the liquid phase around partial cavities were obtained. A high-speed, cinemagraphic multi-frame PIV system was also developed, which was able to record up to 10 000 frames/s, also filtered before recording. This system also employed double pulses onto a single frame, but with no image shifting for the second pulse, therefore it was only useful in situations where there was no flow reversal. The video PIV data was used to obtain phase-averaged velocity vector fields of only the liquid phase for flows with periodic cavitation cycles.

Gopalan and *Katz* [15.38] investigated the flow structure in the closure region of attached cavitation, and used phase discrimination by color, further developing the technique developed by *Sridhar* et al. [15.145]. One Kodak ES-4 four-megapixel digital camera was used and alternatively fitted with a long-pass or band-pass filter to measure the velocity in the water and bubble phases, respectively.

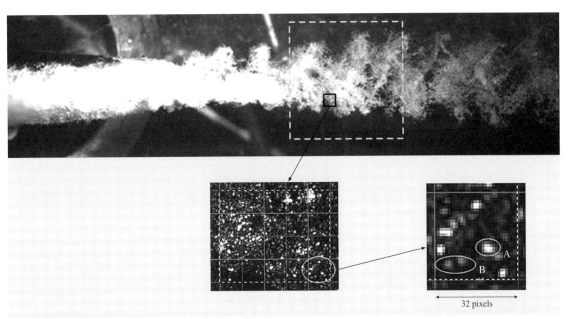

Fig. 15.59 Bubbly wake resulting from the collapse of an axisymmetric, ventilated supercavity (Cavitator diameter $d = 10$ mm, $Re_d = 90\,000$, $\sigma \approx 0.14$, exposure time 3 μs). The *dashed square* shows the approximate location of the PIV field of view. Bottom: Close-ups of a sample PIV image capture

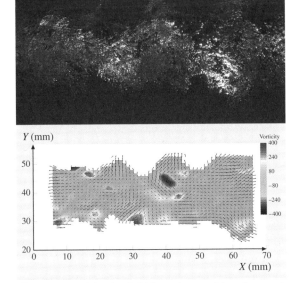

Fig. 15.60 PIV in the bubbly wake resulting from collapse of axisymmetric, ventilated supercavity using a single camera (1024×512 pixel). *Top*: PIV image capture (one image of a pair). *Bottom*: bubble velocity vector field and vorticity contours computed from this image pair

Iyer and *Ceccio* [15.164] studied the influence of developed cavitation on the dynamics of a turbulent shear layer with planar DPIV. The flow was seeded with 80 μm fluorescent latex particles, and a long-pass filter was used to record onto a single four-megapixel digital camera. Thus vector fields for only the liquid phase were obtained, while also measuring the bubble void fraction.

When considering PIV measurements in larger research water tunnels, the large total water volume requiring seeding may make the use of expensive fluorescent particles impractical because of the high cost. In this case, other, inexpensive non-fluorescing seeding particles and photometric parameters should be chosen carefully to achieve simultaneous imaging of bubbly and liquid phases. Recent tests at SAFL have shown that hollow glass spheres (mean diameter 11 μm, density 1100 kg/m^3), used as a non-conducting lightweight filler in the electronics industry or as paint additives, or silver-coated hollow glass spheres (mean diameter 14 μm, density 400–900 kg/m^3), used as a conducting lightweight filler, may be used in vaporous cavitation. Discrimination by particle image size, particle image intensity or a combined technique can then be used to separate the phases. If the light scattered by the seeding particles is too weak compared to the bubble reflections

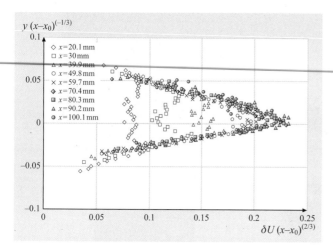

Fig. 15.61 Bubble PIV velocity data in the closure region of an axisymmetric, ventilated supercavity (cavitator diameter $d = 15$ mm, $\mathrm{Re}_d = 100\,000$, $\sigma \approx 0.2$). Average of 800 captures

velocity scales for the axisymmetric turbulent single-phase wake [15.168]. After cavity closure, the turbulent bubbly wake quickly rearranges itself into a similarity state, which in this case is achieved about four cavitator diameters downstream of the cavity closure (here the virtual origin was chosen at the point of cavity closure).

When computing velocity vector fields by cross-correlating bubble images, *Fontecha* [15.147] found the Hart algorithm to be more accurate and produce a larger number of valid vectors than the standard fast Fourier transform (FFT) algorithm. The Hart correlation method [15.169] identifies particles (bubbles) based on image intensity gradients, not just absolute intensity. It is thus well suited for varying intensity, relatively poor-quality PIV images of cavitation bubbles. PIV images that are slightly overexposed are often better suited for intensity gradient compression, since the saturated particle image centers have low gradients and will be discarded. Only the image regions with high intensity gradients are retained, and the remaining compressed, or *sparse*, image data results in a more efficient correlation algorithm. The Hart correlator also avoids the *intensity weighting* of the FFT correlation, where high-intensity pixels (particles) will contribute more to the cross-correlation than low-intensity pixels (particles). The compression rate must be chosen carefully to not lose correlation information.

to register on the recording medium, the bubble velocity field itself can still be obtained with a standard PIV system, using the bubbles as *particles*.

This technique was used to obtain PIV velocity data in the bubbly wake of an axisymmetric, ventilated supercavity by *Wosnik* et al. [15.68, 165–167] and *Fontecha* [15.147]. A sample strobe-illuminated picture of this flow is shown in Fig. 15.59, including the approximate location of the PIV field of view and close-ups of a sample PIV image capture down to a 32×32 pixel interrogation area. Standard PIV images were recorded from reflections of the air bubbles. The signal is usable for cross-correlation, as long as enough bubbles are present in the interrogation area. A sample PIV image capture and the corresponding velocity field from cross-correlation are shown in Fig. 15.60. Note that Figs. 15.59 and 15.60 show different realizations of this flow. Contrast and brightness of the PIV image were increased for reproduction purposes. The average percentage of validated velocity vectors pre-filtering will decrease significantly for the cavitating case (30–80% at the center of the bubbly wake) compared to the non-ventilated, non-cavitating single-phase case ($> 95\%$) at the same Reynolds number, due to bubble reflection and refraction, varying bubble density, vapor pockets and bubble *clusters*. The number of validated vectors also drops off moving radially outward from the center of the wake. However, the *bubble PIV* data shows good quantitative results, Fig. 15.61 shows velocity data in the region immediately following cavity collapse. The bubble velocity data are seen to scale well with length and

Phase-Locked PIV

For quantitative measurements in quasiperiodic cavitation processes it is often advantageous to obtain phase-locked or phase-averaged data. These data are especially useful for comparison with numerical results, where typically time-resolved information about cyclic events is available. When attempting to trigger a standard PIV system to obtain data phase-locked with cavitation events, the following problem can be encountered: pulsed Nd:YAG lasers tend to flash at a constant frequency, when triggered externally at unevenly spaced intervals the energy delivered per laser pulse may fluctuate significantly between flashes. A common work-around for this problem is to both record the signal documenting the quasiperiodic flow event (e.g., a pressure transducer on the suction side of a hydrofoil) and the TTL signal for PIV laser 1 and acquire a large number of image captures (typically thousands) at a constant laser pulsing frequency. The time stamp of each image capture can then be compared to the transducer signal, and the image captures can be binned according to their phase in the cyclic cavitation event. Obtaining phase-locked PIV data of cavitating flows is further complicated by the fact that cavitation processes that may look quasiperiodic in two dimensions, e.g., sheet/cloud cavitation under certain conditions on certain foil shapes, are inherently three-dimensional and affected by the experimental configuration, e.g., the presence of walls. Preliminary studies with hig- speed video and/or (phase-locked) photography to determine the feasibility of phase-locked or phase-averaged PIV are strongly suggested.

High-Frame-Rate PIV

Most of the experimental work to date on cavitating flows has either been whole-field observational and qualitative, i.e., it consists of high-speed photography and/or video, or single-point quantitative, e.g., velocity measurements with a laser Doppler velocimeter (LDV) and/or bubble size measurements with a phase Doppler anemometer (PDA). More recently, whole-field, quantitative measurements have been made using particle image velocimetry (PIV) [15.67, 147, 165]. Unfortunately, standard PIV systems, while providing high-resolution, are limited to low repetition rates (on the order of several Hz). This is due on one hand to the low repetition rates of their components (typically, Nd:YAG lasers: 10–20 Hz, digital cameras: 15–30 frames/s), and on the other hand to dataflow bottlenecks inherent to the system design [cameras have no *onboard* memory, particle images have to be acquired through, e.g., the peripheral component interface (PCI) bus]. Standard PIV systems are therefore limited to instantaneous *snapshots* and statistical information, but can shed little light on the evolution of flow structures (occurring at inverse time scales up to several hundred Hz).

The first application of time-resolved particle image velocimetry in cavitation research was by *Vogel* and *Lauterborn* [15.170], who combined PIV and high-speed photography to investigate the flow around cavitation bubbles during their collapse near a solid boundary. With a drum camera designed for use in high-speed photography and an argon laser pulsed by an acousto-optical deflector, they were able to achieve a limited number of frames at a temporal resolution of 10 kHz.

In contrast, recently commercialized, high-repetition rate (*time-resolved*) PIV systems are capable of recording particle images at up to 2000 frames/s at full 1024×1024 pixel resolution (or at higher framing rate and a reduced field of view: 4000 frames/s at 1024×512 resolution, 8000 frames/s at 1024×256, etc.). Since two frames are necessary to obtain a velocity vector field by cross-correlation, the actual PIV recording frequency would be 1 kHz (2 kHz, 4 kHz, etc.). This is sufficient

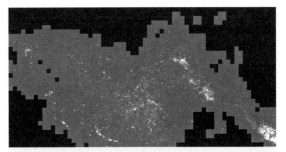

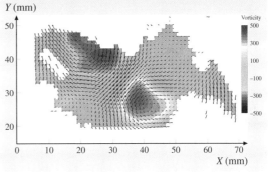

Fig. 15.62 PIV in the wake of a cavitating NACA 0015 hydrofoil, angle of attack= $-8°$, cavitation number= 1.1. *Top*: Adaptive processing mask obtained from sliding intensity thresholding overlaid on original image. *Bottom*: Velocity vectors and vorticity contours, from cross-correlation of bubbles

to temporally resolve cavitation events in most laboratory experiments. Typically, high-speed diode-pumped Nd:YLF lasers are used, which have dual heads and can achieve repetition rates of up to 10 kHz. The energy per pulse decreases with increasing frequency for these lasers, therefore a practical limit for most applications is set at about 5 kHz. An example of a currently commercially available high-speed Nd:YLF laser is the NewWave Research Pegasus, which delivers 10 mJ/pulse at its energy maximum near 2 kHz.

Figure 15.62 shows a sample frame from an experiment that quantitatively recorded the large-scale cavitating structures in the wake of a hydrofoil using high-speed, temporally resolved PIV [15.167]. With the given optical setup, flow speed and recording rate of 2000 Hz the mean flow traveled less than 5% of the field of view between captures, giving sufficient spatio-temporal resolution to track the large-scale cavitating vortices. The gas phase (cavitation) was separated with a simple, but effective adaptive masking algorithm based on a sliding intensity thresholding filter matched to the PIV interrogation spot size (left). Using bubbles as tracer particles, the velocity vector field for the gas phase in the cavitation cloud is obtained from cross-correlation (right).

Time-resolved PIV experiments will provide researchers with a wealth of information typically only obtained from model-free, direct numerical simulation (DNS), but at Reynolds numbers not attainable by DNS in the near to intermediate future. Comparison with synchronized high-speed video will bridge the gap between the vast body of qualitative observations and quantitative methods. High-repetition-rate PIV will enable researchers to fully correlate numerical models with observed cavitation. This has tremendous potential for integrated experimental/numerical research on cavitating flows, since it can provide the crucial time-resolved quantitative information needed for numerical modeling efforts.

Laser Doppler Velocimetry

Even though the LDV technique has been used in fluid mechanics for more than 30 years, its applicability to bubbly gas–liquid flow is the subject of ongoing discussion. Several studies have shown what can be measured under certain, well-defined conditions, but the applicability of LDV to real, practical two-phase flow situations is still being questioned. Generally speaking, the effect bubbles have on LDV measurements depends on their size and the volume fraction of the gas phase, as well as LDV parameters such as working distance and laser beam diameter.

In a high-void-fraction flow with *larger bubbles* (diameters larger than a few tenths of a millimeter) the LDV data rate will decrease compared to a single-phase liquid case. This is due to the lower probability of simultaneous beam penetration through the two-phase medium and the gaps in the signal when a bubble passes through the measurement volume itself [15.171, 172]. Distinguishing between the signal from the seeding particles in the liquid phase and the bubbles can be accomplished in certain cases for bubbly flow with larger bubbles. Methods include, phase discrimination on burst amplitude [15.173], phase discrimination on pedestal amplitude [15.174], controlling trigger level of signal intensity, liquid-phase seeding concentration [15.175], or a power-spectrum curve-fit procedure [15.176, 177]. In general, LDV in bubbly flow is feasible as long as the distances between the transmitting and receiving optics and the measuring volume are small enough to obtain useable Doppler signals from the liquid.

Bubbles resulting from cavitation events or remnants thereof, e.g., in the wake of sheet/cloud cavitation

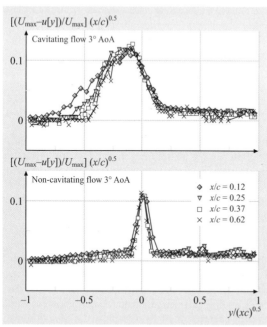

Fig. 15.63 Mean velocity in wake of a 2-D NACA 0015 hydrofoil, LDV measurements with and without cavitation. Data are scaled in similarity variables for turbulent plane wakes. Coordinate x is the distance from the trailing edge, c is chord length [15.32]

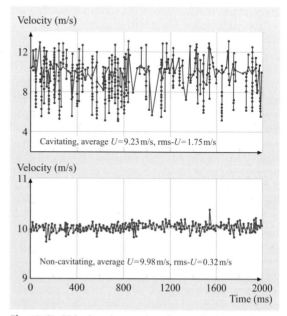

Fig. 15.64 Velocity time series for cavitating and non-cavitating wakes (LDV)

(Fig. 15.12) typically have diameters on the order of a hundred micrometers or smaller and will act as additional seeding. As a result the LDV data rate of valid signals can increase significantly, especially compared to just using water impurities for *natural* seeding.

Kjeldsen et al. [15.32] found the LDV data rate for cavitating conditions in the wake of hydrofoils at small angle of attack to be more than 10 times higher than for non-cavitating conditions. The mean velocity data from LDV measurements with and without cavitation in the wake of a two-dimensional (2-D) NACA 0015 hydrofoil are shown in Fig. 15.63 [15.32]. The data are scaled in similarity variables for turbulent plane wakes. It is evident that the rate of spreading of the wake is much larger under cavitating conditions, which is consistent with visual observations and the results of other investigations [15.178]. Cloud cavitation sheds large vortical structures containing bubbles into the wake. These clouds of bubbles extend much further in the cross-stream direction than the viscous wake associated with non-cavitating flow. While the non-cavitating LDV signal resembles a typical turbulent wake signature, the cavitation signal is skewed towards lower velocities (Fig. 15.64). The bubbles are counted more efficiently than naturally occurring LDV seeds in the flow, therefore the average velocity will contain more weight from the shed bubbles. This effect on the LDV measurements still needs to be quantified. However, the observation that the wake spreading rate is much greater under cavitating conditions is still correct.

The strong negative fluctuations in velocity are due to the imprint left by the periodic passage of bubble clouds, and this cyclical increase in the data rate can be exploited further. The bubble shedding is rooted in a cyclic cavitation process on the foil surface, and is characteristic of the dynamics of the cavitating hydrofoil. Plotting LDV data rate versus time, where data rate is defined as the time elapsed between two acquired valid samples by the LDV system, fitting a curve that is spaced uniformly in time with the help of interpolation functions, and processing the result with an FFT algorithm, a power spectrum of the LDV data rate can be calculated. This *LDV frequency-domain fingerprint* is shown in Fig. 15.65. Note how closely it matches the power spectrum of close-to-simultaneous lift measurements. It should be emphasized that the LDV measurement is a quasi-point measurement, and will give important information of the cavitation shedding dynamics unmatched by a lift/drag balance and pressure transducers at the base of the foil, all of which will measure a global value. All these features also appear to be cap-

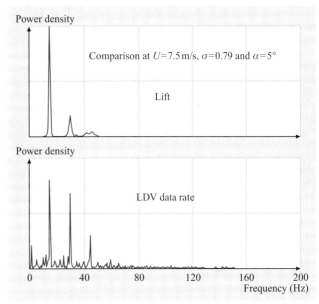

Fig. 15.65 Comparison of lift dynamics with the FFT of the LDV data rate in the wake of a cavitating hydrofoil. The fundamental frequency corresponds to a Strouhal number of $fc/U = 0.15$

tured by numerical simulations [15.137]. By integrating numerical and experimental research, an approach toward calculating bubbly wakes of cavitating hydrofoils is emerging.

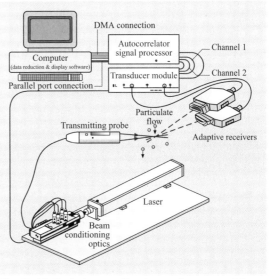

Fig. 15.66 Typical setup of hardware of a commercial two-receiver PDA system

As previously mentioned, the phase Doppler anemometry (PDA) is an extension of the LDV measurement principle, and can be used to measure bubble velocity and size. As a bubble passes through the fringes of the probe volume, it scatters light. A receiving lens located at an off-axis collection angle projects a portion of the fringe pattern onto two or three detectors. Each detector produces a Doppler burst signal with a frequency proportional to the particle velocity. The phase shift between the Doppler burst signals from two different detectors is proportional to the size of the bubbles. For a sketch of the PDA principle and details of the PDA technique, see Chap. 23 or [15.142]. A typical two-receiver PDA setup is shown in Fig. 15.66 [15.179]. The size range of bubbles that can be measured is limited on one hand by the amount of light they scatter (a function of bubble size, available laser power, light collecting optics) and on the other hand by the Gaussian nature of the laser beam and the far-field condition. For a two-receiver system the measurable bubble diameter range is approximately 1–500 μm, for a three-receiver system it is larger. Note that a two-receiver system is limited to measuring phase shifts of up to 360° without ambiguity. The bubble size range can be extended by reducing the distance between the two receivers, however, this will also reduce the measurement resolution. A three-receiver system provides both a large measurable size range and high resolution. State-of-the-art PDA receivers have three prealigned receiving apertures integrated into a single fiber-optic probe. For a two-receiver PDA system the largest bubbles one is trying to measure should be smaller than the measuring volume, otherwise the two-receiver system cannot distinguish between reflection and refraction. The polarization vector of the illumination laser beam should be vertical (0°), since the main scattering mechanism for bubbles is reflection. *Wang* [15.118] reports PDA measurements in a high-speed water tunnel in the wake of a cavitating hydrofoil with a two-receiver system. Here the PDA receiver angle was set to 90° in order to avoid light path compensation due to the change of index of refraction in water, as shown in Fig. 15.67. *Burdin* et al. [15.180] used a three-receiver PDA to characterize the acoustic cavitation cloud caused by an ultrasonic horn at 20 kHz (vibratory apparatus, Sect. 15.1.4).

Several experimental studies aimed at understanding how bubbles modify the turbulence in the carrier

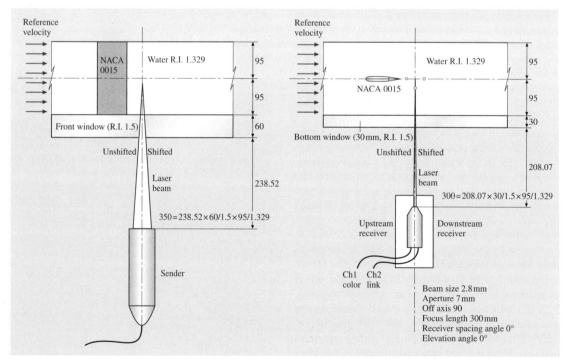

Fig. 15.67 Cavitation bubble measurement in a hydrofoil wake with a two-receiver PDA in a water tunnel; the receiver angle is 90° [15.118]

phase in an axisymmetric liquid jet were carried out and provide good examples of LDV/PDA application to bubble–liquid two-phase flows. *Sun* [15.181–183] used LDV to measure mean and fluctuating velocity components in both the liquid and the gas phase in a bubbly jet ($\mathrm{Re}_{D,\mathrm{liquid}} = 8500$–$9400$) with gas void fractions of up to 9%. For the continuous-phase measurements amplitude discrimination was used to discard signals from bubbles. Bubble sizes were approximately 1 mm, measured with a photographic technique. A surfactant was used to increase the surface tension and reduce bubble size and bubble coalescence. *Kumar* et al. [15.184] used LDV in a bubbly jet ($\mathrm{Re}_{D,\mathrm{liquid}} = 4700$–$9100$) with gas void fractions as high as 20% at the lower Reynolds number. Bubble sizes were measured with a photographic technique and ranged from 0.6 to 2 mm. Only the liquid-phase velocities were measured. *Stanley* [15.185] and *Stanley* and *Nikitopoulos* [15.186] used a phase Doppler approach to measure liquid and bubble velocities and the bubble size distribution in the developing region of a bubbly jet.

Void Fraction Measurements

Cavitating flows are two-phase flows, either single-fluid two-phase for vaporous cavitation or two-fluid two-phase for artificial, gas-ventilated cavitation or dissolved gas coming out of solution. An important quantity describing two-phase flows is the void fraction, usually defined as the gas volume to total volume. Similar to cavitation nuclei measurement, the instantaneous global void fraction can be determined in a very simple way with photography, by taking a picture of an illuminated plane in the flow to determine bubble size and number. Multiple pictures can be used to obtain an estimate of average global void fraction. Advances in CCD/CMOS technology and image-processing techniques have recently created renewed interest in these non-intrusive photographic methods.

For cavitating flows, a continuous measurement of the local average void fraction is easily accessible by quasi-point-averaging or small-volume-averaging void fraction probes. Note that all sufficiently miniaturized local void fraction probes may be arranged in groups of two or more to give additional information on interface (bubble) velocity, and in certain cases bubble size in the previous section.

Conductivity Probes. All conductivity probes consist of two electrodes with a potential difference immersed in the two-phase mixture. Thus the current flow is a direct measurement of the conductivity of the fluid between the electrodes, which is determined by the relative amounts of liquid and vapor/gas. Current can then be calibrated as a function of void fraction. Since the conductivity probe is really a phase indicator and reacts to the passing of phase interfaces, void fraction is often more accurately referred to as *interfacial area concentration*. Conductivity probes were first proposed by *Neal* and *Bankoff* [15.188]. They have been used successfully for many years and a great variety of them exists, for a review of the many common ones see *Jones* [15.189]. A miniaturized four-sensor conductivity probe designed by *Kim* et al. [15.187] is shown in Fig. 15.68, note that the measurement area is less than $0.2\,\mathrm{mm}^2$.

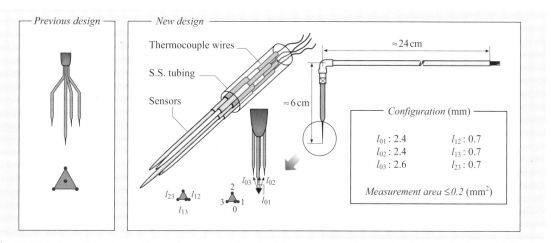

Fig. 15.68 Schematics of an older and a new miniaturized four-sensor tubing conductivity probe for void fraction (interfacial area concentration) measurement [15.187]. (Note: S.S. = stainless steel)

Optical Fiber Probes. An optical probe is sensitive to the change in refractive index and thus also responds to the passage of a liquid–gas interface. If water is present at the tip of the probe most of the light will enter the water due to the relatively small difference in refractive index between the probe material (glass fiber) and water. If a gas bubble covers the probe tip most of the light is reflected back into the probe and can be detected on the other end using a photomultiplier. With a suitable sampling frequency and thresholding, the signal can be converted to a binary time series with binary states of liquid $[b(t_i) = 0]$ or bubble/gas $[b(t_i) = 1]$. The local void fraction η is then simply the residence time of the gas state divided by total sampling time, or for equal sampling intervals Δt_s, the number of samples indicating gas state divided by total number of samples

$$\eta = \frac{\sum_{i=1}^{N} b(t_i)}{N} \,. \tag{15.26}$$

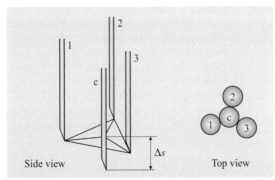

Fig. 15.69 Schematic of an optical four-sensor probe [15.190]

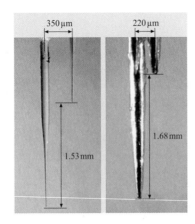

Fig. 15.70 Dimensions of conductivity and optical void fraction probes [15.191]

Mudde and *Saito* [15.190] used an in-house fabricated optical glass fiber probe [15.192] to measure void fraction, bubble size and velocity in a bubble column and bubbly pipe flow. The design of the probe is shown in Fig. 15.69. The polymethyl meta acrylate fibers have a diameter of 0.2 mm and are the tips are cut at an angle of 30° to the fiber axis.

Le Corre et al. [15.191] compared a miniaturized four-sensor conductivity probe [15.187] and a minituarized four-sensor optical probe (developed by the French Atomic Energy Commission – CEA, Grenoble) and showed that probe design greatly affects the accuracy of the measurements. To get an appreciation for the size of theses probes, photographs are shown in Fig. 15.70. While probes miniaturized to this degree will work well in *conventional* two-phase flows, e.g., bubble columns, it is questionable how long a micrometer-sized fiber-optic probe would last under cavitating conditions with cavitation bubbles collapsing onto its tip.

Grayscale Interpretation of PIV Images. A new technique to determine void fraction from PIV measurements of a cavitating flow has recently been developed at SAFL [15.67, 165]. Regular PIV images were obtained in the wake of an axisymmetric, ventilated supercavity, using the ventilation gas bubbles as seeding *particles*. The average grayscale value for a series of raw PIV images was calculated, and the background noise was subtracted. Assuming that the intensity of the reflected light is proportional to the local void fraction, the general shape of the void fraction distribution versus radial position $\eta^*(r)$ for any downstream location within the PIV field of view is known. However, the void fraction profiles are only qualitatively correct due to the fact that the light sheet was created by a Gaussian beam. The actual magnitude of the void fraction can be determined from the mass conservation integral for the gas phase

$$\dot{m} = \int_A k_1 \eta^*(r) \rho_{\text{gas}} U_{\text{gas}}(r) \, dA \,, \tag{15.27}$$

where $\dot{m}$ is the ventilation gas injection rate measured with a flow meter (rotameter), and k_1 is a correction factor for light sheet intensity variation due to the Gaussian beam. The ventilation gas density ρ_{gas} can be estimated from pressure measurements in the water tunnel and the ideal gas law. The velocity profiles of the gas phase $U_{\text{gas}}(r)$ were determined from PIV cross-correlation algorithms using air bubbles as particles. The correction or *calibration* factor k_1 can then be computed by numerically integrating equation (15.27), thus obtaining the quantitatively correct average void fraction profiles from $\eta = k_l \eta^*$.

An example of the void fraction obtained in this way is shown in Fig. 15.71. Note that the correction factor for the intensity variation of the light sheet, k_l, was assumed to be only a function of downstream position and independent of radial position r, i.e., the rays in the laser light sheet were approximated to be parallel.

Other Void Fraction Techniques and Probes. Several other techniques to determine void fraction exist. They include hot-film anemometry, which in principle has the advantage of also being able to measure phase-interface velocity. However, the fragility of hot-film probes limits their usefulness in cavitating flows. Gamma densitometers systems can be used to obtain line-averaged and cross-section-averaged void fraction. Both of these techniques were used successfully by *Kirouac* et al. [15.193] in a column filled with R-134a, simulating steam–water two-phase flow. *Shamoun* et al. [15.194] used a light extinction technique in bubbly flow. Analysis of how collimated, monochromatic light is scattered by a dispersion of gas bubbles provides direct measurement of the line average of the interfacial area density, which can be related to line-averaged void fraction by an image-processing algorithm. Note that the local average void fraction can also be determined from phase Doppler (PDA, laser diffraction) measurements described pre-

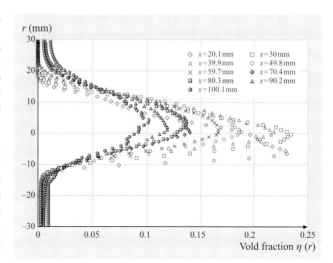

Fig. 15.71 Void fraction calculated from PIV images, in a turbulent bubbly wake immediately behind the collapse of a ventilated supercavity. $Re_d = 10^6$, $\sigma \approx 0.2$. Average of 800 vector fields

viously [15.195]. Three additional techniques for void fraction measurement (and bubble sizing) used primarily in oceanic environments are high-frequency backscatter measured with multiple- and single-frequency sonar, acoustical resonators and acoustic pulse propagation sensors. These have been compared to conductivity probes by *Vagle* and *Farmer* [15.196].

15.2 Wave Height and Slope

In this section surface elevation and slope measurements in time and space are discussed along with their complementary frequency-domain counterparts amplitude/energy and phase spectra. The discussion is divided into four subsections, each of which is subdivided into subtopics. The main subsections are: temporal point measurements at fixed geometric position; one-dimensional spatial measurements (measurements along a line); two-dimensional spatial measurements (measurements throughout an area); and special surface elevation measurements for large laboratory or field use. As several techniques described in each of the first three categories use similar physical principles, there will be some repetition in the discussion as the number of dimensions is increased.

This section is not a review paper in that it does not present all relevant references or give a complete histor-

ical perspective for each measurement method. Rather, it is written in the tradition of a handbook, as a guide for experimentalists. Therefore, general discussions of the appropriate measuring techniques are presented with sufficient references to clarify the basic principles of the methods. For more details on a particular topic, the reader should consult the entire open literature; the references cited herein can be used as a starting point for this process.

One last note, radar measurements have not been included in these discussions, although they are certainly valuable in the field. This decision was the result of the difficulty encountered in conducting radar measurements in the laboratory due to reflections, re-reflections, and such. For field measurements and large-scale experiments, these techniques afford endless possibilities and should not be discarded.

15.2.1 Temporal Point Measurements at Fixed Geometric Positions

If one could travel back in time, one would see that the first wave height measurements were made in the field eons ago by mariners and beach-goers who estimated this quantity for their logs as well as for fun. This first method for estimation of wave height is by direct physical observation and suffers from individual bias and the inability to estimate length (as well as time to determine wave period). To demonstrate the variability with which individuals estimate wave climate in the nearshore, one of the authors (MP) conducted an experiment at Duck, North Carolina while at the Coastal Engineering Research Center [15.197]. This report documents a confidence interval of ±0.23 m on wave heights that varied roughly over 0.30–1.0 m during the experiment. One can easily see that estimating wave height is nontrivial, and this technique can be dismissed summarily.

Perhaps the simplest measurement that does not incorporate an electronic instrument is the wave staff, possibly coupled with an imaging system. Early systems used single-frame film cameras or film movie cameras that offered time series rather than single snapshots. More recently, of course, film cameras have been supplanted by video/digital imaging of various types. In this most basic technique, the wave staff was usually the equivalent of a survey rod – a straight, long rod with tick marks inscribed along its length. In the laboratory, where positioning of the rod is more straightforward (i.e., from the ceiling, walls, floors, etc.), this simple method provides reasonably good results. Problems include the disturbance due to the in situ nature of the staff, the resolution of the imaging device, which may be marginal especially if a large dynamic range is required, and the labor-intensive data-reduction process needed to determine the time series of surface elevations. (Several other instruments that provide direct output of the surface elevation time series and that are still reasonably inexpensive will be discussed below.) One important advantage of the primitive wave staff used in conjunction with video is that, in the case of breaking waves, where other more-sophisticated instruments such as those discussed below give questionable results, one can observe the phenomenon directly and glean useful information not otherwise available [15.198]. Lastly, it is noted that in the laboratory one can utilize a grid along a channel wall, for example, to obtain spatial information in addition to temporal data.

In addition to the prosaic wave staff described above, many authors ascribe this name to more-advanced instruments such as the capacitance wave probe and the resistance wave probe, likely the most frequently used instruments for measuring liquid surface elevation at a prescribed location in a wave basin, flume, or tank. These instruments were first introduced in the late 1940s and 1950s [15.199–201]. In fact, their popularity is such that they are available commercially. Note that they are often used in arrays, linear or 2-D, to provide spatial information, as discussed in the following subsections. In the capacitance probe, an insulated single wire is oriented vertically through the liquid and extended into the air, thus forming one side of a capacitor, while the water comprises the other. For a fixed probe the capacitance varies with changes in water level. The insulation can be a glass tube in which a non-insulated wire is placed or it can be Teflon, for example. Some conductivity of the liquid (usually water) is required, and is generally present naturally due to impurities. Conversely, in the resistance probe, two non-insulated wires are mounted vertically, parallel and adjacent to one another, and the resistance between them varies as the liquid/water level changes. In principle, both of these instruments exhibit essentially linear electrical response.

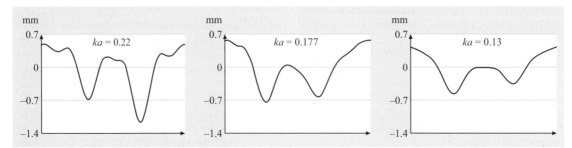

Fig. 15.72 A 9.8 Hz wave time series (Wilton's first ripple) measured at three downstream locations from the wave generator with steepness (wavenumber times stroke) 0.146. The ordinate spans 2.1 mm while the abscissa represents one wave period, 0.102 s (courtesy of Phys. Fluids)

For a discussion of their use in physical models, see for example *Hughes* [15.202].

In practice, resistance and capacitance gages have some differences and many similarities. The resistance gage is more robust and hence is frequently used in oceanic applications while the capacitance probe is more fragile and hence is more often used in laboratory investigations. Capacitance gages require simpler circuitry and, for multiple spatial measurements, their crosstalk interference is lower than that of resistance gages [15.203]. In addition, capacitance gages typically offer less blockage of the flow, as a stiff insulator can be used, thus removing the need for a strut of some sort. Electronics that cover a sufficiently large frequency range are available commercially; however, it is apparent that liquid receding from the probe may limit the accuracy of the measurements especially, for low frequencies and larger amplitudes or for high-frequency and relatively small-amplitude conditions. For gravity–capillary and capillary waves where wave heights are small (i. e., on the of order of millimeters) the meniscus effects on the intrusive wires can be significant [15.204], and these instruments should be avoided, or at least the smallest-diameter wire possible should be used, and again capacitance gages are the better choice. For an indication of the accuracy that can be achieved see Fig. 15.72 [15.205] where gravity–capillary waves in the internal resonance regime are shown; the capacitance wave probe diameter was 0.5 mm. Using either of these probe types is problematic in breaking waves where multivalued surface profiles and bubbly flows are found. In fact, this latter statement is true for any in situ probe, and more-advanced imaging techniques are required.

Pressure measurements via some type of transducer are another common method of measuring wave height indirectly as a function of time at a fixed spatial position. In this method, a pressure sensor mounted at a fixed depth below the mean free surface is used to measure the fluctuations in pressure induced by the passage of the wave, and theory is used to relate the pressure fluctuations to the surface elevations. Though highly accurate pressure measurement devices now exist, several related problems arise. First, the transducer sensing area in common with the fluid is of finite size and therefore yields spatially averaged readings that may be problematic for short wavelengths. Second, since the transducer's diaphragm/interface with the fluid necessarily displaces to provide the measurement, sensor–flow interactions occur that may affect the measurement. These two errors may not be a major problem for field measurements, but can hamper measurements on small scales in the laboratory. Last, since the general surface/interfacial wave theory problem remains unsolved in closed form, there is a level of approximation that accompanies this method and significant errors may arise depending on the wave steepness and surface complexity. A primary advantage of this measurement is its less-intrusive nature – it does not pierce the water surface and so does not affect the flow there. The seminal work in this area appeared in the literature in the mid 1940s by researchers at the Scripps Institution of Oceanography [15.206] and the University of California [15.207]. Obviously, present-day pressure transducers (e.g., those using quartz or fiber optics) are much more accurate, but still suffer from most of the pitfalls mentioned.

Yet another class of point measurement techniques uses sound to probe the surface elevation, usually from below. Perhaps the simplest example is the echo sounder in which a short-duration sound pulse is emitted vertically from below and the measured time of flight for the wave to travel to the surface and return to the sensor, traveling at an estimated sound speed in water, yields the surface height once the factor of two is taken into account. One such commercially available sensor is the acoustic surface tracker (Nortek, USA). These types of gages have several positive features as well as some disadvantages. As with the pressure transducer, these techniques are remote, do not pierce the surface, and can yield reasonable measurements; however, when used from below, they may suffer from many shortcomings, including the relatively high power required to generate the highly attenuated high-frequency acoustic waves necessary to maintain a small-cone-angle interrogation beam and thus a small surface footprint; data-processing problems associated with the extraneous reflection and re-reflection from surfaces other than the one that is to be measured; and difficulties when measuring steep surface slopes. Individual echo sounders are not commonly used in practice to measure surface elevations. However, two present-day reincarnations of this technique are the upward-looking acoustic Doppler current profiler (ADCP), which is capable of measuring depth-varying currents and thus inferring surface elevation via water wave theory, possibly including propagation direction [15.208], and side-scan and upward-looking sonar, which can measure orbital velocities (from which one can again construct the surface via a wave theory) and the sea surface spectrum [15.209]. Both of these instruments emit sound bursts and measure the Doppler shift in the sound waves reflected from particles found naturally in the water. In *Strong* et al. [15.208], the authors discuss the use of the ADCP in obtaining wave

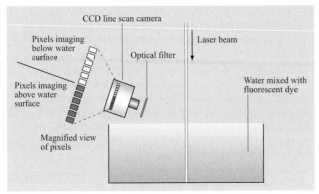

Fig. 15.73 Laser wave height gage. The optical filter blocks specular reflections of the laser light but passes the light from the fluorescing dye. (After *Liu* and *Lin* [15.210])

ments provide much more information than the others discussed thus far, but they are expensive, a wave theory must be invoked and an inversion performed, scatterers must be present over the water depth, and bubbles present due to strong winds or breaking in extreme seas may dominate the return and affect the results. For small-scale laboratory efforts, these instruments are probably not a good choice. The aforementioned side-scan sonar technique uses a somewhat comparable approach.

Major improvements in measuring surface elevations in the laboratory can be attributed directly to the advent of lasers. Using lasers to view and probe the surface, to obtain two-dimensional velocity fields, obtain three-component velocities over a small volume, and facilitate other research has increased steadily in popularity, and for good reason. These essentially non-intrusive devices in conjunction with viewing and forming optics are extremely effective and have proven to be the system of choice and the state of the art. Throughout this section on wave height measurement, the use of the laser is highlighted. The first laser-based device discussed herein is essentially a laser-based wave staff [15.210, 211]. The laser beam is oriented vertically and originates from above the water surface. A camera, in this case one with as little as a single line of pixels, in an elevation view records the intersection of the laser beam and the water surface with a small look-down angle from the horizontal and with the line of pixels aligned with the laser beam (Fig. 15.73). The water is mixed with a fluorescent dye whose spectral excitation peak overlaps the wavelength of light produced by the laser. (Often fluorescein dye is chosen for green–blue laser lines over rhodamine dyes as the former affects surface tension less and is safer to use.) A long-wavelength-pass optical cutoff filter is placed in front of the camera lens to remove specular reflections/refractions of the laser beam from the water surface while passing the light from the fluorescing dye. Each line of pixel data captured by the camera is processed to determine the pixel closest to the transition from the low light levels above the water surface to the high light levels at and below the water surface. By recording an image of an object with known distance markings (i.e., a precise target), a calibration is determined and the pixel location of the transition from air to water is converted to elevation. Limits on accuracy concern the diameter of the laser beam and the angle of inclination of the camera relative to the horizontal, with smaller-diameter beams and shallower angles yielding more-accurate results, see the discussion of laser light sheet methods in Sect. 15.2.2. Though the measurement is more accurate at shallower viewing angles, the inter-

height spectra, including directional information. The authors show reasonable agreement with both a gage that measures pressure and two components of horizontal current (PUV, discussed in Sect. 15.2.4 along with the ADCP) and a heave–pitch–roll (HPR) buoy (discussed below and in Sect. 15.2.4). These acoustic devices can be used in large laboratories, although they were developed primarily for use in the field. In the data analysis, cross-spectra are obtained that include amplitude and phase information as a function of direction; hence wave height is available for a complicated sea surface. These instru-

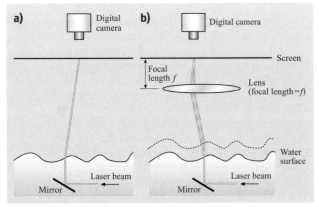

Fig. 15.74a,b Laser slope gage. (a) The height between the water surface and the screen must be large compared to the wave height so that the position of the beam on the screen can be converted directly to the water surface slope. (b) Because the screen is placed one focal length above the lens, the position of the refracted beam on the screen depends only on the angle of inclination of the laser beam relative to vertical. The trajectories of the beam with two water surface elevations are shown

section of the laser beam and the water surface may not be visible to the camera when the surface shape is highly three dimensional, as in breaking waves.

In another relatively simple method using a laser beam, the temporal variation of the free-surface slope is measured. Slope methods are particularly useful for measuring capillary waves where, though the amplitudes are quite small, the slopes may be large. In this method, the laser beam is oriented vertically from above (or below) and the direction of the beam after refraction through the interface is measured and related to the local surface slope in two directions. Perhaps the simplest method of measuring the direction of the refracted beam is to fix a translucent screen horizontally in the air with the laser beam directed vertically from below the water surface, and to view the beam position on the screen with a video imager from above the screen. A simple setup is shown in Fig. 15.74a. As can be seen in the figure, the position of the beam in the imager (once calibrated by imaging a target of known size or a precise grid) can be converted to the angle of the beam relative to the vertical if the height of the screen and the instantaneous height of the water surface are known. The difficulty of knowing the water elevation is overcome approximately if the vertical distance between the water surface and the screen is much larger than the wave height. Also, as shown in Fig. 15.74b, a lens can be placed between the screen and the water surface with the screen at the focal plane of the lens. In this case, all beams with the same slope relative to the vertical, regardless of the water surface height, will be focused at the same point on the screen. Likewise, the spot on an opaque screen can be viewed from below, though the correction may be more difficult. If one is interested in obtaining the surface height record from the measured surface slope, one can integrate the slope record if the surface height at some reference time and the wave phase speed are known. The phase speed is a theoretically well-known quantity in a linear monochromatic sea; however, in a physical experiment, problems arise computing this value accurately (even for monochromatic waves) because the phase speed is a function of the wave height for finite-amplitude waves and the wave height is, of course, not known. One of the first investigators to use this technique to measure surface slope was *Cox* [15.212], though at that time he used a pinhole in a telescope to focus the light on a photocell, and an incandescent light source rather than a laser. Other problems Cox experienced included elevation effects on the light intensity. *Sturm* and *Sorrell* [15.204] presented findings using a similar approach. In their work, they identified a limiting slope that can be measured due to total internal reflection of the laser light, and they avoided the problem of not knowing the surface elevation by requiring the distance of the screen (in their case a photo cell) to the water surface to be much larger than the wave height. In addition, they invoked linear wave theory and used monochromatic waves of low steepness to demonstrate the effectiveness of the method, and under those restrictions, the method works well. It is noted that this technique also has a footprint (the beam waist at the water surface) that for shorter wavelengths may affect the data significantly. Lastly, *Saylor* [15.213] is mentioned as an investigator who used total internal reflection to advantage for measuring the surface slope beneath capillary waves of large steepness.

The literature on surface slope measurement using lasers is voluminous, and as mentioned previously, it is not explored adequately here as this is not the purpose of this handbook. A discussion of several additional techniques is included in the two subsequent subsections, however. Extensions of these methods to obtain additional information are also possible. Using light reflection from the water surface, *Wu* et al. [15.214] developed a technique that enabled the measurement of surface curvature distribution concurrent with surface slope in one surface direction, and with it they measured laboratory-generated wind waves. Subsequently, *Wu* [15.215] further enhanced and generalized the technique to facilitate measurements in both surface directions. *Barter* et al. [15.216] discuss a method that measures surface slope and surface elevation simultaneously, the latter by measurements of the attenuation of the beam. This seems difficult to implement accurately due to water clarity issues. Lastly, we mention the contribution by *Jahne* et al. [15.217] that discusses the use of a telecentric imaging system to facilitate simultaneous measurements of the optical slope and surface elevation at the same geometric point on the surface. This novel technique greatly improves surface measurement capabilities at the expense of a more-complex system.

One additional point worth mentioning with regard to laser point measurements is whether the laser should be oriented propagating from liquid to gas, or gas to liquid. *Jahne* et al. [15.218] present an argument on the grounds of sensitivity and range that it is better to have the laser beam originating in the air and entering the water from above. We feel that the direction decision should be based on more-practical considerations; in particular the steepness of the waves to be measured plays an important role.

Buoy- or float-mounted accelerometers (or inclinometers) with onboard compasses represent yet another type of surface elevation measuring device. Typically, these are station-moored so that their degrees of freedom are limited. One popular version of this is the heave–pitch–roll buoy that measures the three degrees of freedom that comprise its name. (Two similar instruments that have been developed to obtain directional spectra and are discussed subsequently are the PUV gage and the differential pressure gage.) This method is a point measurement of surface elevation when heave acceleration only is recorded or available, and is known generally by the generic name *waverider* buoy. This instrument, of course, has several deficiencies that include motions other than heave due to imperfect mooring, and the fact that it is a floating body and thus exhibits its own dynamics. Although gyroscopic accelerometers can and are used, they have response times that affect the accuracy of these instruments, and waveriders are less accurate in very steep wave conditions where accelerations are large. These packages are very popular for nearshore measurement and are readily available commercially. Due to their deficiencies, they are not recommended for most laboratory uses.

One of the first to investigate a so-called proximity probe, in this case through the air to the liquid surface was *Killen* [15.220]. This instrument used sound; however, since then proximity probes that use other sources have been developed. Advantages include their essentially non-intrusive nature and hence their transportability; they can be moved easily along the surface (for example in towing tank experiments) or held stationary without affecting the liquid surface. Typically, their disadvantages are that the error increases as the slope increases, that standoff distance is an issue, and that they may have a large *footprint* on the surface so that the elevation obtained no longer approaches a point measurement. Though useful in measuring liquid surfaces, these instruments are extremely popular for use in measuring solid surfaces. In addition, several variations of these instruments are available commercially, and they can be used for longer wavelengths in the laboratory.

To conclude this subsection, a few simple ideas of other types of probes are mentioned. For example, if one uses fluorescing dyes dissolved in the liquid/water to make it nearly opaque (and one is not concerned with surface tension, perhaps due to longer wavelengths being investigated), a simple column of illuminated light-emitting diodes (LEDs) oriented vertically can be used as a wave staff by video recording the column over time and later counting the number of diodes that are

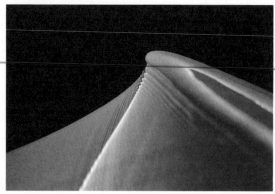

Fig. 15.75 Laser-induced-fluorescence photograph of a gravity wave that is about to form a weak spilling breaker from *Liu* and *Duncan* [15.219]. The wavy line forming the boundary between the dark upper region and the lower, orange region is the intersection of the laser light sheet and the water surface. Light variations in the bright region of the image are due to refraction of the light sheet as it enters the water and due to refraction along the line of sight of the camera as it views the radiant nonuniform light sheet through the curved water surface between the camera and the light sheet

visible in the image. Also with this nearly opaque fluorescent dye-laden liquid, another possibility is to use laser light from above, and view the single spot with two imagers with known geometric position in space, and use stereo pairs to obtain the spot's location. This technique is discussed for an instrument with multiple laser beams in Sect. 15.2.3. Finally, another technique discussed below that can be used also for point measurements, is the laser line-scan imaging method.

15.2.2 One-Dimensional Spatial Measurements

In many instances, one is interested in knowing the wave/surface shape along a horizontal line. The temporal history of this shape can be used for a variety of purposes, including the determination of the wavelength, the dispersion relationship for water waves, and the spatiotemporal power spectrum. Perhaps the simplest method to obtain this height distribution is to arrange an array of wire gages in a straight line, see for example *Mitsuyasu* [15.221] and *Wang* and *Hwang* [15.222]. The method is, of course, limited by the length of the array, which determines the longest wavelength that can be measured, and the spacing of the wire gages within it, which determines the shortest wavelength that can be measured. It is a useful and robust method that can be implemented especially in the field, as described in *Wang* and *Huang* [15.222].

Laser-based measurement methods are particularly well suited to one-dimensional spatial measurements. One obvious method is, of course, to use an array of laser wave staffs in the same manner as an array of wire gages. However, rather than employ the laser method as a discrete set of wave staffs, the laser beam is usually expanded optically into a light sheet and an area-based video camera views (from the side with a small angle from the horizontal) the fluorescence induced at the air–water interface by the light sheet. In a typical application, the laser beam is spread into a sheet either by using cylindrical lenses or by a rotating polygonal block with mirrored surfaces. The light source is usually mounted above the water surface to maximize the intensity of the laser light and thus the light emitted by the dye at the water surface. In the camera images, the region corresponding to the air is dark while the region corresponding to the water is bright (Fig. 15.75). The boundary between the dark upper region and the light lower region is the water surface profile formed by its intersection with the light sheet. At the time of this writing, most measurements of this type utilize digital still or movie imagers rather than film cameras. Temporal resolution is limited by the maximum frame rate of the camera and the brightness of the image incident on the camera's sensor, which in turn is determined by the laser power, the width and thickness of the light sheet, the lens number f, and the concentration and light conversion efficiency of the dye. If the light sheet is created with a rotating mirror, the scan rate of the rotating mirror must be set to include several laser beam scans in each image exposure to insure uniform lighting. Spatial resolution is limited by the number of pixels imaging a given distance in the plane of the light sheet and several optical effects. One optical effect is the combination of the thickness of the light sheet and the viewing angle of the camera. As the light sheet becomes thicker or the camera look-down angle increases (with zero defined as horizontal) the thickness of the light sheet on the flat water surface as seen in the camera image increases and the change in image light intensity across the light sheet–water surface intersection is reduced. This weaker light-intensity gradient decreases the clarity of the apparent water surface in the image and thus reduces the accuracy of its measured location. Another limitation is blockage effects produced by wave features interrupting the line of sight from the camera to the light sheet–water surface intersection, or due to the shading of parts of the water surface by other fractions of the water surface. As mentioned in the above discussion of laser wave staffs, these blockage effects only become unavoidable in highly three-dimensional surfaces such as during energetic wave breaking. Since the ratio of surface height to length rarely exceeds about 1/7, typical imagers, which have about the same number of pixels vertically and horizontally, are poorly utilized with say a wavelength distributed over the full width of the image but the height distributed over only a seventh of the image height. *Perlin* et al. [15.223] used a system of cylindrical lenses to stretch the wave height in the image so that it extended over additional pixels, thus increasing the vertical resolution without modifying the horizontal resolution.

In principle, the single-point laser slope gage can be adapted to measure one-dimensional spatial distributions of slope by scanning the beam at a high rate in a plane perpendicular to the water surface and measuring the deflection of the beam by the methods described in the previous section. However, these scanning laser slope gage methods have been reserved primarily for two-dimensional spatial measurements of slope and are discussed in the following section.

15.2.3 Two-Dimensional Spatial Measurements

Techniques for measuring two-dimensional spatial distributions of water surface height suffer from limited accuracy in the vertical direction and/or limited resolution in the plane of the mean water surface. The most rudimentary of these methods is simply a two-dimensional array of point measurement devices and the most straightforward of these is an array of wire gages. Each gage has the accuracy of an individual wire gage, but only a limited number of gages can be placed in a given area of the water surface. A method using essentially an array of laser wave staff devices has been developed by *Carneal* et al. [15.224]. The device used a 10×10 array of low-power laser diodes that point toward the surface from above with a small angle relative to the vertical. The intersection of these laser beams and the water surface (the water is mixed with a high concentration of fluorescent dye) was observed with a single digital camera that had a 1000×1000 pixel array. The camera viewed the water surface from a relatively small angle from vertical, on the opposite side of the vertical from the direction of the laser beams. The image positions of the illuminated points at the intersection of the laser beams and the water surface are used (through calibration) to determine the three-dimensional positions of these points on the water surface, thus providing a discrete representation of the water surface at each instant in time. This relatively new method has similarities to stereo methods discussed below. Its accuracy, particularly in height, is determined essentially by the number of pixels that the beam–water surface intersection points move due to the maximum change in surface height. Thus, its accuracy in the vertical is much lower than the single laser wave staff where the water surface varies over the entire column of pixels. Similar to the wire gage array, this method provides data at only a discrete set of points within the measurement area. In spite of these difficulties, this method is non-intrusive and thus well suited to a number of applications where the use of wire gages would be difficult or impossible.

Another interesting method for determining the spatial distribution of surface height is stereo photography. In this method, two cameras separated by a distance d (or a single camera with a split-screen stereo lens) view the water surface from above. The images from the two cameras are recorded simultaneously and the overlapping parts of the images are analyzed to obtain the surface height distribution. To maximize the region of image overlap, the cameras are typically pointed toward the same portion of the water surface. In stereo analysis, the position of a physical feature in three-dimensional space is determined from the relative change in position of that physical feature between the two images. The primary shortcoming of stereo measurements of water surfaces is that the image features used for analysis are typically reflections, say illuminated regions of high water surface slope or curvature, rather than physical objects. Thus, the three-dimensional position of the reflection point on the free surface may be different for each camera because of their different viewing angles. This introduces errors in the stereo analysis and the calculated distribution of surface height. This method has been used by *Banner* et al. [15.225] and is discussed by *Jahne* et al. [15.218]. Recently, *Wanek* and *Wu* [15.226] have developed a three-camera system oriented at an oblique angle to the water surface and performed a thorough evaluation of the method by comparison with simultaneous capacitance wire gage measurements. *Grant* et al. [15.227] also developed a stereo method; however, rather than use reflections from the water surface as points for stereo analysis, they marked points on the surface with a 10×10 array of laser beams created with a holographic lenslet array acting on the beam of a single pulsed Nd:YAG frequency-doubled laser. The intersection of the beams with the water surface was viewed with a split-view stereo camera. No fluorescent dye was used; rather the intersections of the beams and the water surface were made visible by scattering from naturally occurring solid particles in the water. The images were processed to obtain the three-dimensional (3-D) positions of the beam intersection points with the free surface. Of course, as in the multi-laser-beam single-camera method discussed in the previous paragraph, the spatial resolution is limited by the number of laser beams and the vertical resolution by the number of pixels of image displacement between the two images of the same physical feature.

Another optical wave-height method uses moiré fringes. A moiré projection method is described by *Grant* et al. [15.228]. In this method, light is projected on the water surface through a rectilinear grating. The water is mixed with fluorescent dye and the projected light pattern, which is distorted by the shape of the free surface, is viewed through a camera with an identical grating at the focal point of its lens. Fringes created by the interference of the two grids are then counted to find the surface height along each fringe contour. This method is not trivial to implement and its accuracy in height and resolution in the horizontal is limited by the number of fringes in the image. However, it too can be useful when non-intrusive methods are required.

Techniques for measurement of surface slope throughout an area of the free surface are used extensively in the laboratory and the field. The slope imaging methods described in the following paragraph are the current state of the art and provide highly accurate slope measurements with high spatial resolution in the measurement area. Before considering these slope imaging methods, the scanning laser slope gage is discussed. This instrument is capable of obtaining the spatiotemporal spectrum of the water surface slope over a small surface area and has been used by a number of researchers in both the laboratory and the field. The principles of these devices are essentially the same as the laser point slope gage described above; however, in the scanning devices the beam is traversed at a rapid rate over a rectangle or circle on the mean water surface. Typically the laser beam projects upward from below the water surface and a system of rotating or oscillating mirrors and lenses in a submersible housing is used to traverse the beam over the area of interest while keeping the beam trajectory vertical. The receiver optics are placed above the surface. As in the point slope gage, a lens with a diffuser screen placed above and at its focal plane results in a direct calibration of beam position to surface slope without knowing the instantaneous surface height at the point where the beam passes through the water surface. The position of the beam on the diffuser screen is tracked with a photosensitive position sensor or even a high-speed digital camera. Such a device was developed by *Bock* and *Hara* [15.229]; it scanned a circle of diameter 154 mm every 14.4 ms with 237 slope readings per revolution (a spacing between readings of 2.04 mm). A similar device described by *Li* et al. [15.230] scanned an 81 mm × 81 mm square pattern with 36 × 36 measurement points every 146 ms. These scanning rates and distances between samples determine the frequency and wavenumber limits of the spectra produced. Another critical parameter, as in any laser slope gage, is the diameter of the beam at the water surface. As the surface wavelength approaches this diameter the instantaneous slope measurement becomes inaccurate.

Slope imaging methods based on reflection and refraction have become a common technique for laboratory measurements in recent years. These methods are capable of measuring time sequences of the surface slope distribution throughout an area of the surface. Unlike the scanning slope methods, slopes are measured simultaneously at each pixel in the image covering the measurement area allowing one to obtain the temporal evolution of the slope field with high spatial and temporal resolution. The appropriate theory has been reviewed by *Jahne* et al. [15.218]. In the present article, only the refraction methods are described since they have been used most often and since the principles of the two types of devices are similar. In the refraction methods, a light source with spatially varying intensity or color is placed underwater at a depth such that its presence does not significantly affect the waves. A still or movie camera is placed above the water surface. The principles for this refraction method were first introduced by *Cox* [15.212], though he used an optical sensor to measure the slope at only one point. Lighting systems with spatially varying white-light intensity can be used to measure one component of surface slope over a two-dimensional patch of the surface. One such system was developed by *Jahne* and *Riemer* [15.231] in which the light source (a modification of the device developed by *Keller* and *Gotwols* [15.232]) consisted of a glass box filled with a suspension of latex particles in water (Fig. 15.76a). The light was provided by a row of lamps directed horizontally at one end of the box. The light scattering caused by the particles creates an exponentially decaying light intensity in the direction of the

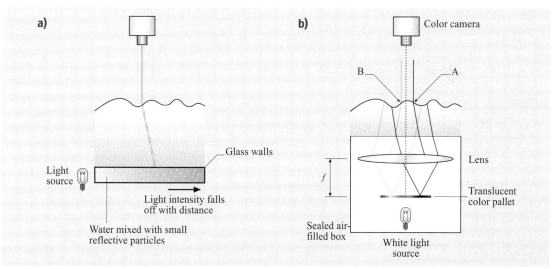

Fig. 15.76a,b Schematics of slope imaging devices. **(a)** System for measuring one component of slope (after the devices reported in *Jahne* and *Riemer* [15.231] and *Keller* and *Gotwols* [15.232]). The light box at the bottom of the schematic produces white light whose intensity varies with position in the plane of the page along the upper surface of the box. The light intensity is constant in the direction normal to the page. The camera is placed far above the water surface so that it receives only the vertical light rays. **(b)** System for measuring two components of slope (after the device reported in *Zhang* and *Cox* [15.233]). The lighting system produces beams of light that approach the water surface with continuously varying angle and a different color for each angle. The blue and yellow beams are shown in the figure. The point A (B) on the water surface has a slope such that it turns a light ray from the blue (yellow) beam to the vertical. The camera is placed far above the water surface so that it receives only the vertical light rays

light projection (say the downwind direction in a wind wave tank) and a uniform light intensity in the crosswind direction. Each elemental area of the water surface receives light rays from a wide range of points from the light box, and as each of these light rays passes through the surface it is bent away from the local surface normal by refraction. Typically, the camera is placed a large distance above the tank (i.e., a distance many times the maximum wave height) so that it receives only the light rays that are nearly vertical. Thus, tracing the light ray backward from the camera, each pixel images a small area of the water surface. At the water surface, this light ray is turned toward the local surface normal and is directed to a specific point on the surface of the light box. The light intensity received at the pixel is determined by the intensity at the corresponding point on the light box, thus encoding surface slope by light intensity. With this white-light box, only the slope component in the direction of the light gradient can be determined. Of course, both components of slope can be determined in sequential experiments by rotating the direction of the light gradient 90° about a vertical axis. This can also be facilitated by using separate sets of lamps on two perpendicular sides of the box in separate experiments.

This slope imaging method can be extended to simultaneous measurement of both components of surface slope by using a light source with spatially varying color and a color-sensitive camera, *Zhang* and *Cox* [15.233] and *Zhang*, *Dabiri* and *Gharib* [15.234]. The system used by *Zhang* and *Cox* [15.233] is depicted in Fig. 15.76b. The light source apparatus consists of lamps creating an upward-projecting cone of white light, a horizontal translucent screen with a color pattern, and a large-diameter lens with its axis oriented vertically and the lens placed one focal length above the screen. It is essential that each point on the screen have a different color. Since the screen is at the focal point of the lens, all light rays from any point on the screen leave the lens in a parallel beam whose direction from the vertical is determined by the position of the point on the screen relative to the vertical optical axis of the lens. Each point on the water surface is illuminated by all the different beams with direction encoded by color. According to the local slope, only one of these beams is deflected vertically. The camera is then placed high above the water surface and the pixel corresponding to a given point on the water surface receives the vertical ray whose color is then an encoded measure of the local slope. Both the gray-level and color versions of these slope imaging devices produce a wealth of spatiotemporal information about the surface slope field and are being used primarily to explore the physics of wind wave systems in both the laboratory and the field.

15.2.4 Special Surface Elevation Measurements for Large Laboratory or Field Use

In many cases, wave height records, particularly those measured in irregular or random seas, are expressed on a spatiotemporal frequency-by-frequency basis (i.e., in the temporal frequency and spatial frequency/wavenumber domain) via Fourier transform techniques. By Fourier transforming a time series and/or an instantaneous surface profile, one obtains a complex representation of the amplitude/energy and phase, or equivalently the real and imaginary parts of the wave frequency or wavenumber. These spectra are useful for the study of the physics of water waves for many practical scenarios (for example, the determination of a transfer function to convert sea-surface elevation spectra to roll spectra in a linear systems approach for an offshore platform design), and, most importantly for the discussion at hand. That is, to properly utilize and analyze the data recorded by PUV gages (which measure the pressure and two horizontal components of velocity, hence the acronym), ADCPs, and such, the amplitude and phase spectra (a frequency-by-frequency representation of the *linear* superposition of waves in the record) are required. As the topic of this handbook is experimental fluid mechanics, it is not our purpose to present these data analysis techniques here. Instead, the interested reader is referred to the many excellent texts that discuss the Fourier transformation process for surface waves as well as for any other temporal series and/or spatial information. As a starting point, the texts by *Bendat* and *Piersol* [15.235] and *Papoulis* [15.236] are recommended, though many other outstanding texts are available. Specific to the area of water waves, texts by *Ochi* [15.237, 238] are available, again among myriad others.

The focus of this subsection is to identify some of the instrumentation available to obtain data in large laboratory settings that more closely resemble the field. Instruments discussed include: those that provide a time series of pressure and the two horizontal components of velocity concurrently (i.e., PUV gages and variations thereof) which may be transformed to yield surface height frequency spectra with directional information; slope array buoys that can be used to determine surface height frequency spectra in time and space (i.e., wavenumber magnitude, wave frequency, and direction); ADCP devices; global positioning system (GPS) signals that may be used to obtain surface height spectra; and multi-point/grid surface height measurements by wave staff that can be used to approximate the amplitude and phase as a function of wavenumber vector, wave frequency, and direction. The techniques and algorithms that determine the various types of spectra from temporal and spatial series are continuing to evolve, as are the methods used in the laboratory and field to obtain the data series.

The first class of instrumentation to be discussed measures several of the dependent variables of linear (inviscid) wave theory at one location; from these data reconstruction of the *directional* temporal frequency spectrum is possible. Of course all these methods have drawbacks regarding the approximations required and these drawback result in differences between the computed and actual spectra. Perhaps the most fundamental approximation, as mentioned above, is the use of linear wave theory to obtain the spectra. As an example of such a ubiquitous recorder, the PUV wave gage can be used to calculate the directional frequency spectrum, say $S_{\eta\eta}(\omega, \theta)$ at a horizontal location (where η is the surface height, ω is the angular frequency, and θ is the wave propagation direction); here we neglect the phase spectrum, as is customary. PUV hardware units and their attendant analysis software are available commercially and may be used as standalone or shore-wired devices. Pressure is typically measured with a quartz pressure transducer while the two components of velocity are measured via an electromagnetic current sensor (e.g., those manufactured by Marsh–McBirney) or an acoustic current sensor (e.g., those manufactured by Falmouth Scientific). Both current sensors yield measurements averaged over a spatial volume on the order of $1\,\mathrm{cm}^3$. Knowing the mass density, the mean of the measured pressure yields the sensor depth required by the wave theory. These instruments provide the information required to obtain the first three components of a Fourier expansion in *direction* (i.e., with amplitude and phase). Variations of these instruments are the UVW (three components of velocity) gage, the heave–pitch–roll (HPR) buoy actually developed ealier, and the differential pressure gage [15.239]. A typical PUV-type gage utilizes an expansion to obtain the wave propagation direction based on the seminal work by *Longuet-Higgins*, *Cartwright*, and *Smith* [15.240] in which an HPR buoy was investigated. In their formulation, auto- and cross-spectra of surface elevation, and surface slope in each of the two directions were used to obtain expressions for the first five coefficients of the real-valued Fourier-expansion approximation to the directional spread (for a complex representation, amplitude and phase, the first three coefficients are thus available). More recently differential global positioning system (GPS) technology has been used to measure buoy motions and shown to be equivalent [15.241]. Essentially the same expansion is used for the various gages mentioned, though different Fourier directional coefficients result. In fact if one measures pressures at other positions locally in space, additional terms in the expansion are available providing presumably improved estimates of the directional temporal frequency spectrum.

At the cost of a more-expensive instrument than those discussed hitherto, the acoustic Doppler current profiler (ADCP) is well suited for obtaining the directional surface elevation spectrum as well as the water-particle velocity spectra in discrete bins over the water depth. Typically these units include a pressure transducer to obtain mean depth and tidal information, and were originally developed to obtain the current profile as a function of depth. For a complete description of one such commercially available product see the information from RD Instruments [15.242]. As a current profiler the instrument works as follows: four acoustic beams are paired in orthogonal planes at fixed orientation with respect to the vertical (typically 20° or 30°), known as a Janus configuration. The beam-axis component of velocity is measured and cell-averaged. The cell mean current is obtained by subtracting the measured velocities of the opposite beams. Using a cross-correlation under the assumption of stationary wave fields, the directional spectrum may be determined, though a larger computational effort is required than if simply using them for current profiling. One important advantage of the ADCP compared to the PUV-type devices discussed previously is its ability to determine velocity profiles to the surface. As the orbital velocities beneath waves decay exponentially in intermediate and deep water over vertical distances scaled with the wavelength and since the instrument is necessarily positioned below the main orbital motion, the ability to measure currents to the surface facilitates quantifying the higher-frequency part of the spectrum. Obviously these instruments along with the PUV-type are useful in field operations where hazardous ice flows exist, or perhaps seaway traffic is high. In the laboratory, for example in a large, deep basin where a model is to be used, these instruments may provide the additional advantage that they can be bottom-mounted and no longer represent a surface obstacle.

In concluding this subsection, one possible novel technique on the horizon is that of using GPS signal reflections, see *Zuffada* et al. [15.243]. They discuss the possibility of measuring sea height and surface roughness with global positioning system reflections from the sea surface. A review of the literature at this juncture indicates that this technique is still under development and that the associated difficulties remain largely unresolved.

15.3 Sediment Transport Measurements

15.3.1 A Brief Introduction to Sediment Transport

Sediment transport is a broad subject with many intricacies. Many books have been written about sediment transport [15.244, 245], and the reader is encouraged to explore these books for more detail on this complex subject. Nevertheless, a brief discussion about some of the basics of sediment transport is warranted herein to help clarify some of the issues surrounding measurements associated with sediment transport.

Sediment transport measurements are difficult for a number of reasons:

- Natural sediment is opaque, making optical penetration difficult.
- Sediment transport is strongly dependent on flow properties, and instrument intrusion can be a problem, especially near the bed.
- Most sediment transport occurs in complex physical environments, such as near bed forms and boundaries.

Sediment transport is extremely important in fluvial, oceanic, and Aeolian (wind-blown) geomorphology. For the sake of brevity, waterborne sediment transport will be the focus of discussion herein. Even so, many of the methods described can also be applied (perhaps in slightly different forms) to Aeolian sediments.

Traditionally, sediment transport has been subdivided into three components: bed load, suspended load, and wash load. Bed load is the mass flux of sediment particles that travel along the bed by rolling, bouncing, and sliding. Flow drag pushes the particles along the bed, and momentum exchange between the particles and the bed is often significant. Bed load is extremely difficult to measure accurately, and while new, innovative techniques have been proposed, none have proven effective. The primary problem is that bed load travels in a thin layer in the part of the flow with the strongest velocity gradient. Thus, it is difficult to accurately measure the velocity and number of all mobile particles that make up the bed load.

Suspended load consists of sediment that is entrained into the flow by turbulence. Sediment suspensions are not true suspensions, but consist of particles that settle out of the flow in the absence of turbulence. Suspended particles have a finite fall velocity. What distinguishes bed load from suspended load is that suspended particles may travel many grain diameters from the bed, and do not come into contact with the bed as frequently as bed load particles. Like bed load, the suspended sediment concentration at the interface between the bed and the water is generally very important and also very difficult to measure accurately [15.246], but for the most part, suspended load is less difficult to measure than bed load.

Wash load consists of very fine sediment particles that do not easily settle once they are entrained into the flow, even in the absence of turbulence. The wash load is nearly a true suspension. Because of this, wash load particles are finer than particles found in the bed. Since wash load particles are fine, they are usually also cohesive. We will limit our discussion of the theory of suspended sediment primarily to noncohesive sediment because, while cohesive sediment is very important, it is also very complicated. Many of the suspended load measurement techniques described in this chapter can be applied to wash load sediment measurements. However, there are issues associated with particle breakdown and flocculation that must be considered, and in situ measurements of cohesive sediment are often preferable [15.247].

Suspended Load Transport

The ultimate goal of sediment transport work is often to determine rates of erosion and/or deposition of sediment. In order to find either of these quantities mass flow rate of sediment must be measured. To demonstrate the importance of sediment mass flux for suspensions of fine sand, we begin with the mass conservation equation

$$\frac{\partial c}{\partial t} + \frac{\partial F_{si}}{\partial x_i} = 0, \quad (15.28)$$

in which c is sediment concentration, t is time, x_i represents the three principle coordinate directions, and F_{si} is the flux of suspended sediment in the i-th direction. For fine sediment with a terminal fall velocity v_s the flux can be written as:

$$F_{si} = (u_i - v_s \delta_{i3}) c, \quad (15.29)$$

where u_i is the flow velocity and the third principle direction is in the vertical. Essentially, (15.28) and (15.29) state that suspended sediment follows the flow exactly except for in the vertical direction. In the vertical direction the velocity of the sediment is the difference between the flow velocity and the terminal fall velocity of the sediment.

If the velocities and concentrations are decomposed into time-averaged and fluctuating components and back-substituted into (15.28) and (15.29), subsequent Reynolds averaging produces the equation:

$$\frac{\partial \overline{c}}{\partial t} + \frac{\partial \overline{F}_{si}}{\partial x_i} = 0 \qquad (15.30)$$

in which

$$\overline{F}_{si} = (\overline{u}_i - v_s \delta_{i3})\overline{c} + \overline{u'_i c'} \,. \qquad (15.31)$$

The overbars in (15.30) and (15.31) indicate time-averaged quantities and the primes indicate fluctuations about the time average. The flux given by (15.31) has three components: advection by the mean velocity $(\overline{u}_i \overline{c})$, fall velocity of sediment in the vertical direction $(-v_s \delta_{i3} \overline{c})$, and the turbulent Reynolds flux of sediment $(\overline{u'_i c'})$. Consider the vertical component of the flux given by (15.31). The fall velocity of the sediment always transports the sediment towards the bed. If the channel bed is flat and nearly horizontal, $\overline{u}_3$ is zero near the bed. In this case, suspended sediment transport only occurs if the Reynolds flux term is nonzero (i.e., suspended sediment transport cannot be sustained in the absence of turbulence).

For closure, the Reynolds flux is often modeled following Prandtl's mixing-length hypothesis:

$$\overline{u'_i c'} = -D_d \frac{\partial \overline{c}}{\partial x_i} \,. \qquad (15.32)$$

Here D_d is the kinematic eddy diffusivity of the flow. For fine sediment, it is often assumed that the eddy diffusivity and the eddy viscosity are equal. Then, the eddy diffusivity can be obtained directly from knowledge of the flow field.

Solving (15.30) through (15.32) for any given flow requires a flow model and boundary conditions. For open channel flows, boundary conditions are applied at the free surface and near the bed. The free surface boundary condition is simply that the flux of sediment at the gas–fluid interface is zero. The boundary condition at the bed has historically taken on two forms: specifying the near-bed concentration, and specifying the entrainment (vertical Reynolds flux) of sediment from the bed. Entrainment is perhaps more difficult to measure than concentration, but for uniform flows, entrainment can be directly calculated from concentration measurements using (15.31). Near-bed concentration and entrainment are usually correlated to the average bed shear stress (or shear velocity) by equations. Examples include *Smith* and *McLean* [15.248], *van Rijn* [15.249], and *García* and *Parker* [15.250]. For nonequilibrium flows, it is more appropriate to use entrainment as the boundary condition so that particle settling and particle entrainment can be simultaneously modeled in (15.31).

Bed Load Transport

Bed load is sediment that is transported by saltation (rolling, bouncing, and sliding) adjacent to the bed. For bed load composed of fine sediment, the saltation layer is very thin. Unlike suspended load, bed load can be sustained in the absence of turbulence. In practice, bed load is usually calculated using equations that correlate bed load with the average dimensionless bed shear stress (Shields stress)

$$\tau^* = \frac{\tau}{\rho g R D} \,, \qquad (15.33)$$

where τ is the bed shear stress, ρ is the density of the fluid, g is the gravitational acceleration, R is the submerged specific gravity of the sediment ($\rho_s/\rho - 1$), ρ_s is the density of the sediment, and D is the median diameter of the sediment. The critical shear stress – the minimum dimensionless shear stress at which sediment resting on the bed will begin to move – is also important. Most bed load functions give the dimensionless bed load q^* as a function of shear stress raised to the power of 1.5. The dimensionless bed load, or Einstein number, is defined as

$$q^* = \frac{q_b}{D\sqrt{gRD}} \,, \qquad (15.34)$$

where q_b is the volume flux of bed load per unit width of bed. As an example, a frequently used bed load formula is the *Meyer-Peter* and *Muller* [15.251] formula

$$q^* = 8\left(\tau^* - \tau_c^*\right)^{3/2} \,, \qquad (15.35)$$

where τ_c^* is obtained by substituting the critical shear stress into (15.33). The reader should be aware that *Wong* [15.252] has found a discrepancy in this formula and provides a viable correction. Many others [15.253–256] have developed relations between bed load and bed shear stress; most equations rely on the critical shear stress as well.

Shear Stress

As discussed, existing bed load and suspended load theories rely heavily on accurate measurement of bed shear stress. Average bed shear stress has long been the correlate of choice because it is easy to estimate for uniform channels and it occurs where sediment transport begins and ends – at the interface between the fluid and the bed. In open channel flow, the momentum principle is often applied to determine average bed shear stress. For example, in steady, uniform flow the momentum equation can be used to show

$$\tau_b = \sqrt{\rho g R_h S}, \qquad (15.36)$$

where τ_b is the bed shear stress, ρ is the density of the water, g is the gravitational acceleration, R_h is the hydraulic radius of the channel, and S is the energy slope of the channel. For transport in an infinite fluid, velocity measurements can be used to get the time-averaged velocity profile, which can be used in conjunction with known velocity distributions to assess the average bed shear stress.

Determining average bed shear stress is relatively easy for uniform flows with flat beds, but determining instantaneous, local shear stress is not easy, especially if the bed is not flat. Sediment transport is strongly influenced by local variations of shear stress and turbulence bursts, and measuring spatial and temporal variations of shear stress is important. Current methods of measuring local shear stress (hot film sensors, chemical rate sensors, etc.) are impractical above movable beds. Several methods for estimating shear stress from related measurements have been suggested. *Dewey* and *Crawford* [15.257] suggest that bed shear stress can be indirectly estimated from near-bed measurements of the time-averaged velocity profile, the Reynolds stress, or the turbulent dissipation rate. However, in complex flows these estimates are often inaccurate [15.258]. Other measurements that have been introduced as a replacement for average bed shear stress include turbulent kinetic energy [15.259] and velocity [15.260] near the bed. Measurement of bed shear stress, and bed shear stress substitutes, above movable beds is the subject of ongoing research.

Summary

In summary, important measurements in sediment transport include concentration, mean and instantaneous sediment flux, and bed shear stress. It is clear that the fluxes and concentrations of greatest importance are often located near the bed. In fact, this is the most difficult part of the flow in which to measure concentrations and fluxes because concentration gradients and velocity gradients are very high, requiring good spatial resolution. In addition, the presence of high concentrations of suspended sediment makes most techniques difficult to apply.

The fundamental measurement associated with sediment transport is sediment mass flux, or mass flow rate of sediment. Measuring the instantaneous mass flux requires either a direct time-based measurement of mass or simultaneous measurement of sediment concentration and velocity. In addition, the sediment size distribution is often important. When measuring fluxes of suspended sediment, it is common practice to measure concentration only and to assume that the mean velocity of the sediment and the flow are equivalent. For bed load measurements, however, the only proven methods require timed collection of bed load samples. This is because the bed load layer is extremely thin, and simultaneous measurements of bed load sediment concentration and velocity are at this time impossible.

15.3.2 Methods of Measuring Suspended Sediment Transport

For useful concentration measurements, measuring devices must have good spatial resolution and a small

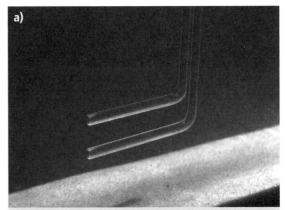

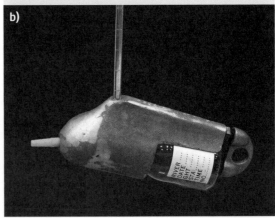

Fig. 15.77a,b Two common manual sampling devices: **(a)** a laboratory system with two manual sampling tubes, **(b)** a US DH-48 depth-integrating bottle sampler

Table 15.3 Properties of suspended sediment samplers. Properties given in the table are estimates based on experience, published literature, and manufacturers specifications. Actual properties may vary with design and implementation

Characteristic	Manual samplers	Optical backscatter	Single-frequency acoustic backscatter (ABS)	LISST	FBRM
Typical maximum sampling rate	≈ 0.1 Hz or less	≈ 10 Hz	≈ 10 Hz	Up to 4 Hz – generally less. 1 Hz cited by *Wren* et al. [15.261]	≈ 0.1 Hz Concentration dependent
Typical estimated measuring volume dimensions	Circular inlet Diameter: 0.3 to 0.7 cm	Cone Half angle: 15° Cone length: ≈ 2.5 cm Cone volume: ≈ 1 to 3 cm^3	Frustum of a cone Frustum height: ≈ 0.5 cm Frustum radius: ≈ 1 to 5 cm Frustum volume: ≈ 2 to 40 cm^3	Small cylindrical volume Length 2.5 cm Volume: ≈ 1 cm^3	Circular annulus Diameter: ≈ 1 cm Volume: ≈ 1 cm^3
Intrusiveness	Intrusive	Moderately intrusive	Non-intrusive	Intrusive	Intrusive
Optimal/typical size range	Silts and sands	Clays and silts	Sands	2 to 400 µm	1 to 1000 µm
Maximum concentration (by volume)	Very large	$\approx 2\%$ (depends on sediment size/type)	$\approx 2\%$	$\approx 0.2\%$	$\approx 2\%$
Sediment size distribution	Simultaneously measures size distribution and concentration	Sensor can only measure concentration of the sediment distribution used in its calibration	Sensor can only measure concentration of the sediment distribution used in its calibration	Measures size distribution and concentration	Measures size distribution – may take substantial time to gather enough data
Overall accuracy of concentration measurements	20 to 100% pump-type isokinetic samplers have highest accuracy	10 to 30%	10 to 30%	20%	Unknown
Ease of use	Easy to implement but time-consuming. Calibration unnecessary but isokinetic requirement adds a level of complexity	Easy to use with proper data logging. Calibration is required and may be time consuming	Moderately difficult to use with proper data logging. Calibration is required and may be time-consuming	Relatively complex	Relatively complex
Relative initial cost	Inexpensive	Moderately inexpensive	Moderately expensive	Expensive	Expensive

measuring volume. For two-dimensional flows the spanwise and lengthwise dimensions of the measuring volume are not critical, but height of the volume should be small relative to the thickness of the boundary layer. A small measuring volume height is necessary since most vertical distributions of suspended sediment are nonlinear. In unsteady flows and in flows in which the response of sediment to turbulence is to be measured, a good time response may also be required.

The most widely used methods of measuring suspended sediment transport are relatively old because they are inexpensive and do not require extensive calibration. In fact, traditional methods of sediment measurement are generally the standard by which new techniques are calibrated. Recent reviews of technology available to measure suspended sediment concentrations include those of *White* [15.246] and *Wren* et al. [15.261]. Another useful review, albeit somewhat older, is provided by *van Rijn* and *Schaafsma* [15.262].

Typical characteristics of three of the more widely used methods of measuring sediment concentration and size distribution – manual sampling, optical backscatter (OBS), and acoustic backscatter (ABS) – are given in Table 15.3. Two relatively new technologies that allow rapid measurement of nonuniform sediment size distributions, laser in situ scattering and transmissometry (LISST) and focused beam reflectance measurement (FBRM), are also shown in the table. These newer technologies are costly, flow intrusive, and difficult to implement. Nevertheless, their ability to measure both concentration and size distribution makes them worthy of discussion. Note that specifications given in Table 15.3 are estimates and vary somewhat with instrument design and implementation.

Manual Sampling

Of the three most widely used sampling techniques, only manual sampling provides direct, simultaneous measurements of concentration and sediment size range. Unfortunately, manual sampling is flow intrusive, requires sample processing (samples must be dried and weighed), and has poor temporal resolution; optical and acoustic techniques offer some advantage over manual sampling in these areas.

Manual sampling has been utilized extensively for measuring sediment concentration in sediment transport research [15.264–269]. There are many different types of manual samplers [15.262, 270], two of which are shown in Fig. 15.77. Although there is variation in ease of use, efficiency, and accuracy, most samplers operate similarly. Manual samplers generally consist of a small tube through which sediment/water mixtures are pumped, siphoned, or forced by the flow. The suspensions are carried into a container where they are deposited until sediment concentrations and size ranges can be determined by sieving and weighing the samples. Usually, the sampler nozzle is aligned with the flow, and samples are pumped at the same velocity as the unobstructed flow velocity – a process referred to as isokinetic sampling. Sampling tube diameters are typically 3–7 mm. Since volumetric sediment concentrations are measured directly, results gathered with the sampler are relatively accurate, but maintaining isokinetic sampling is not always easy. Errors result if the sampling velocity is not the same as the flow velocity because the sediment has a higher density than the fluid, and any redirection of streamlines leads to a change in sampled concentration. *Winterstein* and *Stefan* [15.269] and *Bosman* et al. [15.263] demonstrated that the accuracy of manual samplers varies with sampling rate, sediment size, and the angle of the sampler relative to the flow velocity. On the other hand, correction factors, such as those shown in Fig. 15.78 [15.263], can be used to adjust concentration measurements as long as the unobstructed flow velocity is known. In Fig. 15.78, α is the trapping efficiency (the ratio of the sampled to the true concentration), u_s is the sampling velocity, and u_c is the unobstructed flow velocity. If the unobstructed flow velocity is unknown, it is generally best to sample at a high rate. Figure 15.78 demonstrates that sampling at a velocity that is higher than the flow velocity may result in acceptable error. For example, the accuracy is reduced by less than 25% for 0.45 mm sand when intake velocities are two to three times the flow velocity. Reductions in accuracy are even smaller for fine sand. *Hay* [15.271] and others have suggested that sampling at velocities of as much as four times the flow velocity results in acceptable concentration measurements.

For most flows of interest, manual samplers have good spatial resolution. Furthermore, manual samplers can be used to measure suspensions of nonuniformly sized sediment distributions. Although the samplers do obstruct the flow, they are often small in comparison to

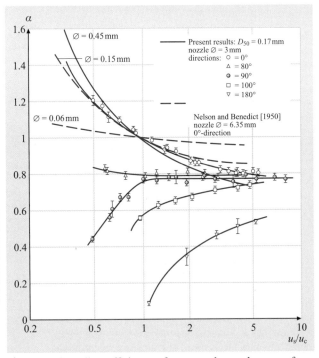

Fig. 15.78 Sampling efficiency of a manual sampler as a function of sampler orientation, sediment size, and sampling velocity (after [15.263])

the scales of interest. However, of the most widely used methods, manual sampling is one of the most intrusive. Manual samplers have three main drawbacks. First, it has already been mentioned that the unobstructed flow velocity must be known in order to sample the flow accurately. In most cases, measurements of suspended sediment concentration are most important near the bed, but this is also the very region where it is most difficult to gather an isokinetic sample or to determine the unobstructed flow velocity. Second, manual samplers have poor temporal resolution. It typically takes a minute or more to gather a manual sample [15.265]. Thus, manual sampling is not suitable for finding correlations between instantaneous concentration and velocity (e.g., measurement of the Reynolds flux) or shear stress. Finally, measurements are inconvenient and require postprocessing to assess size ranges and sediment concentrations.

Aligning manual samplers with the flow is important but not critical. *Nelson* and *Benedict* [15.272] suggested that a 20° misalignment of the intake has negligible influence on the accuracy of concentration measurements. Figure 15.78 further demonstrates this. For isokinetic samples, a measurement error of only 20% is expected for a 90° misalignment when 0.17 mm sand suspensions are sampled.

A depth-integrating bottle sampler, a field version of the manual sampler, is shown in Fig. 15.77. The depth-integrating sampler is intended to be drawn through the water column at a constant rate in order to determine the total suspended load in a river. Bottle samplers are widely used and have the same characteristics as laboratory samplers, except that they have a larger profile. The fill rate of most bottle samplers is dependent on depth (because of hydrostatic pressure variation) and flow velocity. To accommodate isokinetic sampling, some bottle samplers have a variety of nozzle sizes. Nozzle sizes are selected to achieve the appropriate sampling velocity [15.270].

Optical Techniques

A number of optical techniques have been utilized to measure suspended sediment concentrations. The majority of optical samplers can be divided into two categories: optical transmission [15.273–275] and optical backscatter [15.276]. Optical transmission devices consist of one or more transmitter/detector pairs. The transmitter is typically an infrared light-emitting diode (LED) and the detector is a photoresistor or photodiode. High concentrations of suspended material reduce the amount of light that travels between the transmitter and the detector.

Optical backscatter (OBS) is a more widely used method for measuring suspended sediment concentrations. In the case of OBS sensors, an emitter transmits light into a small measuring volume, and light reflected from sediment suspended within the sampling volume is measured by a photodetector. The device is a point measurement device, and is relatively easy to implement. The output of the sensors increases linearly with concentration. A typical calibrated sensor can be used to detect sand concentrations of up to $50\,g/l$ and is accurate to $\pm 0.1\,g/l$ according to the D&A Instrument Company [15.277]. Errors associated with calibration and implementation reduce the accuracy significantly.

Compared to other electrical devices, transmissometers and OBS sensors are relatively cheap and have high temporal response of about 10 Hz [15.278, 279]. Optical sensors have two shortcomings. First, they are somewhat intrusive for small-scale flows. A typical OBS sensor is approximately 3 cm in diameter and 20 cm in length, and has a measuring volume on the order of $3\,cm^3$, which is located a small distance from the sensor. Orientation of the sensor can be easily altered, reducing sensor intrusion, but the sensors are point measurement devices, and if an entire vertical profile is desired, multiple sensors are required. Second, although *Green* and *Boon* [15.280] propose a method of deciphering size distribution from OBS measurements, standard OBS sensors can only measure concentrations of homogeneous suspensions [15.246]. The sensors are calibrated by assuming that increasing reflection intensity is associated with increasing concentration. Variations in grain size cause a breakdown in this assumption. Furthermore, even if concentration measurements were unaffected by particle size, it would be desirable to be able to determine the size gradation of any given sample. *Sutherland* et al. [15.281] also reports some sensitivity of OBS sensors to sediment darkness or hue, such that even if sediment size is constant, a change in the sediment color may lead to a change in sensor calibration.

White [15.246] suggests that OBS sensors do not work as well near the bed because they interfere with the flow, and near the water surface, natural light can interfere with OBS measurements. Using OBS sensors is relatively simple as long as they are properly calibrated. In situ calibration is best (using isokinetic samplers), and the sensors must be calibrated with the same sediment that they will be measuring. OBS sensors are particularly susceptible to biofouling (becoming covered with algal slime) if used for more than a few days [15.282]. The effects of biofouling can be avoided by periodically cleaning the OBS lens. It is possible to automate cleaning [15.283] but this adds cost to the system.

Beach et al. [15.278] tested a fiber-optic backscatter (FOBS) sensor. The FOBS sensor is less intrusive than a standard OBS sensor, with a tip diameter of only 4 mm. Its small size reduces instrument flow interference near the bed. Apart from their size, FOBS sensors have characteristics that are similar to those of standard OBS sensors. Beach et al. [15.278] reported a concentration measurement range of 0–200 g/l for the FOBS sensor that they used. Figure 15.79 shows the approximate dimensions and measuring volume of a FOBS sensor. The extant of the measuring volume varies somewhat with concentration, but it is estimated to be about $3\,\text{cm}^3$ in size. Puleo et al. [15.284] used an array of FOBS sensors to measure vertical concentration profiles of 0.44 mm sand in the swash zone of a beach. Near the bed, the sensors were spaced at 1 cm intervals; sensors that were below bed level had saturated output and were used to identify bed elevation. Recently, fiber-optic backscatter (FOBS) sensors have also been used to measure sediment deposition rates Ridd et al. [15.283]. Ridd et al. were able to measure sediment deposition rates with a resolution of $0.01\,\text{mg}\,\text{cm}^{-2}$ at an estimated accuracy of 5% in still water.

Laser-based devices can be used to measure size, concentration and particle velocity [15.246]. Two relatively new laser based techniques, focused beam reflectance measurement (FBRM) and laser in situ scattering and transmissometry (LISST) can be used to determine the size distribution of suspensions of nonuniform particles in situ. The FBRM uses a focused, rotating laser to measure particle size range [15.285–287]. When the focused laser intercepts a particle, light is reflected to a photosensor. Since laser rotation speed is known and the time of light reflection is measured, the chord length of the intercepted particle can be determined. Assuming all of the particles are spherical, sediment size distribution can be statistically determined. After integrating over a period of time, the entire size range in the vicinity of the sensor can be determined.

LISST devices rely on transmission and refraction of light by suspended particles to determine both sediment concentration and size distribution [15.288, 289]. A collimated laser is focused through the sampling volume towards a group of concentric photo-optic sensor rings. The laser and the central sensor can be used as a transmissometer, detecting concentration. The outer sensor rings sense light diffraction – a particle-size-dependent property. The maximum measurable concentration is

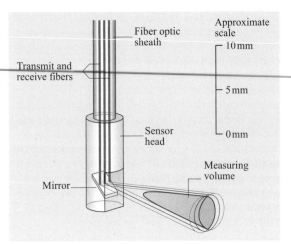

Fig. 15.79 Illustration of a fiber-optic backscatter (FOBS) sensor and measuring volume (shown as the *shaded region*). Volumes illuminated by the single transmit fiber and sensed by the dual receive fibers are also depicted. (After [15.278])

limited to about 0.2% by volume, although this limit can be adjusted with design changes. Typical accuracy of concentration measurements is on the order of 20% over the range of sediment sizes that the LISST is designed for [15.290]. Drawbacks of the FBRM and LISST are flow intrusiveness (both are significantly larger than OBS sensors) and cost. Furthermore, the response time of these devices, though in general better than manual sampling, is still slow compared to OBS and acoustic devices. Nevertheless, many field researchers choose to use a single LISST or FBRM device in conjunction with OBS sensor arrays or acoustic sensors, simply to determine the size ranges of suspended sediments [15.291]. Both FBRM and LISST require some degree of calibration, and like all optical sensors, both devices are susceptible to biofouling.

In laboratory studies, velocity and concentration fields have been simultaneously determined for suspended sediment transport of uniform sediment using particle tracking velocimetry techniques. This requires discrimination of flow tracers and sediment particles. There are two methods of discriminating between flow tracers and suspended sediment. The first method is to seed the flow with fluorescent tracers, and to use a filter to eliminate all but the light emitted from the tracers [15.292]. The second method is to discriminate based on particle size [15.293], something that can only be done if the sediment and the flow tracers are significantly different in size. Size distributions of suspended sediment can also be determined using photographic techniques [15.294]

Acoustic Techniques

Over the last two decades, there have been significant advances in the use of acoustics to measure suspended sediment. Acoustic sensors have been successfully applied both in the field [15.295, 296] and in the laboratory [15.297, 298]. Like optical transducers, acoustical transducers have been developed that are based on transmission [15.262, 299], but backscatter devices have been favored because of their ability to measure entire concentration profiles. *Thorne* and *Hanes* [15.300] provide a review of acoustic backscatter devices (ABS), which have the potential to simultaneously measure concentration profiles, velocity profiles, and bed topography. ABS devices have some distinct advantages over manual sampling and optical methods: first, they measure entire concentration profiles nearly non-intrusively; second, data collection is very rapid; and third, once calibrated, an ABS device can be operated without significant sample processing.

Figure 15.80 depicts how ABS devices measure concentration. ABS operation is initiated when the device transmits a pressure pulse into the water column. The pressure pulse travels through the water column at the speed of sound (≈ 1500 m/s in water). As it travels through the water column, a fraction of each pulse is reflected back to the transducer by particles suspended in the water. The strength of the return signal is directly related to the sediment concentration. It takes more time for reflections to return from scatterers that are more distant from the transducer. Thus, an entire profile of the sediment concentration can be resolved by measuring the strength of the return signal as a function of time. The amount of time required to gather an entire profile equals the time that it takes for the acoustic pulse to travel from the transducer to the most remote measuring volume and back. Since the speed of sound in water is so high, gathering one concentration profile only takes a fraction of a second. However, pressure waves backscattered from sediment particles are Rayleigh distributed, and multiple profiles must be averaged to measure concentration with any degree of accuracy [15.279, 300]. Typical ABS devices gather profiles at 100 Hz. Averaging enough profiles to reduce concentration uncertainty to about 25% reduces temporal response to 5–10 Hz, depending on the ABS device configuration (short profiling ranges allow higher sampling rates).

ABS sampling volumes are shaped like the frustum of a cone. The radius of the sampling volume increases

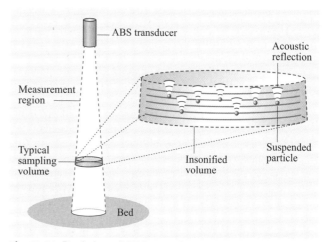

Fig. 15.80 Depiction of ABS operation. One of the sampling volumes is enlarged to illustrate acoustic reflections from individual particles. The measurement region is composed of many such sampling volumes which become active as the acoustic pulse travels through them

with distance from the transducer because the acoustic pulse spreads as it propagates through the water. The radius of the sampling volume of a typical device may vary between 1 and 5 cm over a 1 m profile. If small sampling volumes are desired, the device should be installed close to the region of interest [15.298]. The height of the sampling volume is dependent on the duration of the transmitted pulse. An estimate of the shortest reliable pulse is about five transducer cycles. For a 2 MHz ABS transducer this translates to a minimum sampling volume height of about 0.2 cm; so fairly good spatial resolution can be achieved with ABS devices. More often, sampling volume heights are on the order of 0.5 cm. The height of the sampling volume also affects the ability of the ABS device to measure near-bed concentrations. The bed produces a very strong acoustic reflection, making it impossible to measure bed load concentrations.

For the same concentration, different sizes of suspended sediment will return different signal strengths to the ABS transducer. Consequently, if nonuniform distributions of sediment are present, a strong signal could indicate low concentrations of course sediment, high concentrations of fine sediment, or some combination of the two. Multifrequency ABS systems have been used to simultaneously determine concentration and sediment size [15.296, 301]. However, it has only been possible to simultaneously discern the concentration and mean sediment size of unimodal sediment distributions. Nonuniform size distributions still cannot be resolved with multifrequency ABS Systems [15.246]. Furthermore, *Lynch* et al. [15.302] state that acoustic devices are limited to measuring particle sizes of greater than 25 μm. Incidentally, acoustical and optical backscatter devices can also be used in combination to simultaneously estimate suspended sediment particle size and concentration with reasonable accuracy [15.302].

ABS systems consist of a transducer/amplifier combination and a high-speed data-acquisition system. The data acquisition system must be able to capture the amplitude of acoustic return signals with frequencies of up to 5 MHz. For instance, *Admiraal* and *García* [15.298] used a 20 MHz data acquisition board with a 2.25 MHz transducer. One alternative is to use an envelope detector to detect only the peaks of the acoustic return pulse. Using this approach, the transducer output can be sampled at a lower rate. However, in some cases, filtering the transducer output can reduce spatial resolution.

Equations describing acoustic backscattering by suspended sediment have been given by *Hay* [15.271], *Thorne* and *Campbell* [15.303], and *Thorne* et al. [15.304]. According to *Admiraal* and *García* [15.298], the voltage output v of a typical ABS for measurements in the far field is

$$\langle v^2 \rangle = C \frac{M}{r^2} e^{-4\alpha_t r}, \tag{15.37}$$

where M is the mass of scatterers per unit volume of fluid, r is the distance from the transducer to the measuring volume, and C is a constant determined by calibration. According to *Thorne* et al. [15.304], measurements are in the far field as long as

$$r > \frac{\varepsilon \pi a_t^2}{\lambda}, \tag{15.38}$$

where ε is approximately 2, a_t is the radius of the transducer, and λ is the acoustic wavelength. The acoustic signal is attenuated as it travels through the water, both by particles in the water and by the water itself. The total attenuation coefficient α_t is given by

$$\alpha_t = \alpha_w + \frac{1}{r} \int_0^r \alpha_s \, dr, \tag{15.39}$$

where α_w is the attenuation of the acoustic pulse by the water and α_s is the attenuation of the acoustic pulse by sediment suspended in the water. Attenuation of sound by water is given by *Fisher* and *Simmons* [15.305] as:

$$\begin{aligned}\alpha_w &= (55.9 - 2.37T + 4.77 \times 10^{-2} T^2 - 3.48 \times 10^{-4} T^3) \\ &\quad \times 10^{-15} f^2 (1 - 3.84 \times 10^{-4} p + 7.57 \times 10^{-8} p^2),\end{aligned} \tag{15.40}$$

where f is the frequency of the transmitted signal in Hz, T is the water temperature in °C, and p is the absolute pressure in atm. According to *Hay* [15.271], the attenuation due to particles in the water is negligible when concentrations are less than 1% by volume. For high concentrations of suspended sediment, attenuation due to the sediment is dependent on the concentration, making concentration profiles more difficult to determine. The attenuation can be corrected for by first computing the concentration closest to the sensor, then calculating the corresponding attenuation, then calculating the concentration slightly further from the sensor (using the previously calculated attenuation correction), and so on until the entire concentration profile is resolved. The attenuation due to the suspended sediment has been shown to follow the relation [15.271, 306]:

$$\alpha_s = \zeta M, \tag{15.41}$$

where ζ is a constant dependent on the sediment characteristics.

To find the constants C and ζ in (15.37) and (15.41) the ABS device must be calibrated. Once C and ζ are known, an acoustic pulse can be transmitted, the return signal can be measured as a function of time, and (15.37) through (15.41) can be solved to get the mass concentration as a function of distance from the transducer.

While acoustic sensors have been primarily used to measure suspended sediment concentration, it is also possible to use the Doppler shift of the return signal to measure the sediment velocity or flux [15.297, 307, 308]. Likewise, acoustic devices designed for measuring flow velocities can also be used to measure sediment concentration. Examples include acoustic Doppler current profilers [15.291] and cross-correlation velocity profilers [15.309]. *Gartner* [15.291] compared suspended sediment concentration measurements made with acoustic Doppler current profilers (ADCP) and OBS devices. Measurements from the two instruments were well correlated in some locations, but near the bed, measurements did not agree. Nevertheless, the ADCPs did provide useful information about the suspended sediment concentration profile. *Holdaway* et al. [15.310] had some success using an ADCP to measure suspended sediment concentrations. In their study, they compared concentrations measured using the signal strength of an ADCP with transmissometer measurements. *Holdaway* et al. [15.310] demonstrated that the ADCP provides useful concentration information, but requires in situ calibration and independent information about the local particle size distribution.

Acoustic devices are very sensitive to coarse sediment, while optical sensors are more sensitive to fine sediment [15.262, 279]. Even so, the presence of fine air bubbles or particles other than the sediment particles can result in erroneous acoustic concentration measurements, particularly for low concentrations of suspended sediment [15.246, 279, 298]. The possibility that particulates or air bubbles are contaminating the output can be eliminated by verifying that the output of the device is zero when no suspended sediment is present. *Osborne* et al. [15.279] did a side-by-side comparison of OBS and ABS devices and found that agreement between the two devices was quite good. Concentration measurements using the two devices were within 10% of each other for time scales of several minutes – in their experiments it was not practical to compare instantaneous measurements.

Calibration of Acoustical and Optical Sensors

In order to use acoustical and optical measuring devices, the devices must first be calibrated. All parts of the sensor must be underwater, and temperature is important. It is best if these devices are calibrated in situ, most likely using manual samplers. If it is not possible to calibrate the device in situ, it can also be calibrated in a laboratory facility. The facility in which the device is calibrated should have a known or measurable suspended sediment distribution. *Hay* [15.271] lists a number of calibration methods, including the sediment-laden jet that he used to calibrate his acoustic sensor. *Thorne* and *Hanes* [15.300] give an example of another calibration facility – a calibration tower. Thorne and Hanes emphasize that the tower must be run for a long period of time prior to calibration to allow microbubbles to escape.

It is difficult to keep coarse sediment uniformly suspended so *Admiraal* and *García* [15.298] built a recirculating duct that maintained high velocities in all places where the sediment was present. High velocities kept the sediment in suspension at all times. A holding tank allowed microbubbles trapped in the water to be released, preventing the bubbles from causing erroneous measurements. The sediment in the duct bypassed the holding tank so that the concentration in the duct remained nearly constant throughout each test. Measurements with the ABS device confirmed that the suspended sediment concentration was uniform for the entire cross section of the duct. A sediment sampler situated just downstream of the location where ABS profiles were gathered was used to calibrate the ABS device.

Appropriate calibration requires a similar range of concentrations in the calibration facility and the test facility. Furthermore, the sediment attenuation correction for concentration profiles is nonlinear, and *Hay* [15.271] points out that a time-averaged attenuation correction will erroneously overcorrect the acoustic output voltage if there are large concentration fluctuations. This can be particularly problematic if concentration fluctuations are substantially different in the calibration facility and the test facility. One way of overcoming this problem is to parametrically estimate instantaneous attenuation and to use it to correct measured voltages before averaging.

Optical sensors must also be calibrated. However, since most optical devices are for point measurements, calibrating them is less difficult. A typical field method for calibrating OBS sensors is to gather bed material at the field site, mix it with water in a bucket, and keep it in suspension while multiple readings are taken with the sensor. Samples of the suspension in the bucket are collected for later analysis (drying and weighing), and the process is repeated with samples that are increasingly dilute [15.291]. Since OBS sensors are likely to be used with fine sediment, keeping the sediment in suspension throughout the calibration process is not so difficult.

15.3.3 Bed Load Sediment Measurements

As noted previously, bed load is much more difficult to accurately measure than suspended load. The main reason for this is that bed load is transported as a thin layer near the bed in a region with strong velocity gradients. Most optical, acoustic, and electronic sensors cannot resolve at scales necessary to distinguish the thin layer of bed load from the stationary layer of sediment just below it. Furthermore, bed load concentrations are often high, and optical and acoustic devices have difficulty penetrating through the high concentrations of sediment. Consequently, accepted methods of measuring bed load are nontechnical and have been around for many years. Needless to say, there is a need to improve spatial and temporal resolution of bed load measurements before many of the most important sediment transport questions can be answered. So much of the sediment travels very near to the bed, that it is difficult to resolve some very elementary questions. For now, only time- and spatially integrated bed load can be measured with a reasonable degree of accuracy. For the most part, this is done using sediment traps or pressure-difference samplers.

Pressure-Difference Samplers

The most commonly used bed load sampling devices are pressure-difference samplers. The Arnhem sampler, which dates back to 1937, was one of the earliest and widely used devices of this type [15.244, 270]. A more contemporary pressure-difference sampler is the Helley–Smith bed load sampler [15.312]. The Helley–Smith sampler was designed for use with coarse sands and fine gravels (2–10 mm in diameter) but has been regularly used to measure bed load composed of finer sediments [15.313]. Boat-deployed and handheld versions of the Helley–Smith sampler are available and are shown in Fig. 15.81. The sampler nozzle has two square openings at the front and rear. The downstream opening is larger than the upstream opening, and when the sampler is placed on the bed of a stream or river, a pressure drop is created inside the nozzle between the inlet and outlet.

The pressure drop causes the flow to accelerate through the nozzle, and sediment that enters through the front of the nozzle is carried into a bag attached to the rear of the nozzle. Sediment that is not fine enough to escape through the mesh bag becomes trapped and can later be dried, sieved, and weighed. According to *Edwards* and *Glysson* [15.270], a newer version of the sampler has a nozzle outlet to inlet ratio of 1.40 instead of the traditional ratio of 3.22. Both area ratios are accepted by federal agencies, pending further investigation. The optimal area ratio likely depends on sediment size, as coarse sediment requires higher velocities to push it through the nozzle, and fine sediment will be over-sampled if nozzle velocities are too high. Each bed load sample is a time-integrated (non-instantaneous) measurement.

For bed load samplers, sampling efficiency is defined as the ratio of the measured bed load to the actual bed load [15.312]. Sampling efficiencies of between 100 and 160% have been reported for the Helley–Smith sampler [15.270]. Researchers have found that fine and medium sands are typically oversampled by the original Helley–Smith sampler because the sampler was designed for coarse sand and fine gravels. For example, *Emmett* [15.311] found sampling efficiencies of about 150% and 93% for sediment size ranges of 0.25–0.5 mm and 0.5–16 mm, respectively. Emmett hypothesized that the high efficiency of the finer sediment might have been caused by inadvertent trapping of suspended load by the device. The amount of suspended sediment trapped by the device can be substantial since suspended load concentrations are highest near the bed.

The measuring efficiency of pressure-difference samplers should be used to correct measurements – especially for fine sediment which may result in an efficiency of 150%. Ignoring the sampling efficiency will cause a bias error. In addition to trapping suspended sediment, there are other sources of error associated with pressure-difference samplers. A first source of error is due to placement of the sampler. The lower lip of the nozzle must be flush with the bed. The presence of bedforms may prevent the sampler from sitting flat on the bed, leading to local scour and an erroneous bed load measurement. Moreover, inadvertently pushing the nozzle into the bed will cause an overestimate of the local bed load. Second, misalignment of the sampler with the mean flow direction can cause greatly reduced trapping efficiencies. *Gaudet* et al. [15.313] found trapping efficiencies could be reduced by nearly 50% for misalignments of only 10°. These misalignments may be caused by carelessness, but also by channel geometry and bedform effects.

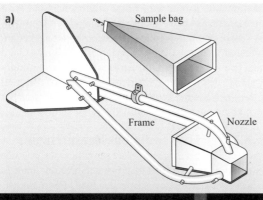

Fig. 15.81a,b The Helley–Smith pressure-difference sampler: **(a)** boat-deployed sampler (after [15.311]) and **(b)** handheld sampler

Bed load can vary significantly in the streamwise and transverse directions, so multiple measurements are necessary to measure the total bed load in a stream. For bed load composed of 1.15 mm sediment, *Helley* and *Smith* [15.312] found that the standard deviations of individual bed load measurements were extremely large – typically the same size as the measured bed load itself. It is difficult to assess the accuracy of the bed load sampler for individual measurements because even in uniform flow, bed load varies a great deal due to the presence of bedforms and other factors, both temporally and spatially. Thus, assessing how much of the standard deviation is due to natural variability of the bed load and how much is due to inaccuracy of the device is not entirely possible. *Hubbell* et al. [15.314] found that although relative uncertainty of bed load samplers could be high for low bed load rates, measurement uncertainty was generally small. However, they also demonstrated that actual bed load rates can vary significantly in the streamwise direction (for instance, in the presence of dunes), resulting in poor estimates of the time integrated total bed load if only one sample was gathered at each position along a transect. It has often been assumed that when the sampler is properly used the accuracy of individual samples is relatively good for coarse sand and fine gravel, but recent studies have even questioned this assumption [15.315, 316].

For field measurements, *Gaweesh* and *van Rijn* [15.317] demonstrated that individual pressure-difference sampler measurements may have an uncertainty of 50% in the presence of bedforms because the variation of bed load on the bedform is as high as 50% and the sampling location is often random. The uncertainty of the total bed load in a river can be reduced to less than 20% by gathering multiple bed load measurements across the river transect [15.299, 317]. The required number of samples varies with river characteristics and flow conditions; but *Gaweesh* and *van Rijn* [15.317] recommend dividing the river into seven subsections and gathering 25 samples per subsection. The samples gathered in each subsection should be gathered over the streamwise length of the bedforms in that subsection so that a representative measure of the average bed load is obtained [15.317]. This is clearly a very time-consuming process.

Measuring total bed load with a Helley–Smith sampler requires the following steps:

1. The sampler is lowered until it rests on the bed – the lower lip of the nozzle should be flush with the bed, but the sampler should not be pushed into the bed.
2. A sample is collected for a fixed amount of time. The collection time can be determined by collecting test samples at locations where the bed load is thought to be the highest. Choose the collection time such that the bag is never filled to more than 40% of its capacity [15.270] otherwise the pressure drop across the nozzle will be reduced. The collection time that is finally selected should be used for all samples. According to *Edwards* and *Glysson* [15.270], a collection time of less than 60 seconds is usually preferred.
3. *Emmett* [15.311] suggests that samples should be gathered at 20 equally spaced lateral locations to satisfactorily estimate the total bed load. For practicality, *Edwards* and *Glysson* [15.270] recommend collecting 40 bed load samples for each transverse (two at each lateral location). For accuracy, other researchers recommend more samples [15.299, 317]. Specific strategies for collecting the samples are given by *Edwards* and *Glysson* [15.270].
4. The bed load per unit width at a sampling location is determined by dividing the mass of the sample by the width of the sampler nozzle and the collection time. If multiple samples are gathered for a sampling location, the resulting bed load per unit width measurements must be averaged.
5. The total bed load is obtained by multiplying each bed load per unit width by the width of the river subsection that it represents and then summing all of the subsection bed loads.

Properly using a pressure-difference sampler is costly and time-consuming. The samples collected with the sampler must be postprocessed, making on-site data checks impractical.

Sediment Trapping

In laboratory settings, although instantaneous and local bed load is difficult to measure, time-averaged bed load (or total load) can be accurately measured simply by capturing the sediment in a trap at the end of the test flume for weighing [15.314, 318]; in this case, the only concern is whether or not the trap affects the measurements being gathered.

In the field, pit- or slot-type samplers are simply pits or depressions made in the stream bed (Fig. 15.82). Bed load falls into the depressions where it is trapped. The trapped sediment must be weighed in situ [15.319, 320] or removed for weighing in some fashion, possibly by means of a conveyor or pump [15.311]. Well-designed pit- and slot-type bed load traps capture nearly 100% of coarse bed load [15.244, 315] but may also catch some suspended load. Fine bed load material may be sampled at a lower efficiency because turbulence within the trap can resuspend some of the finer material, carrying it out of the trap [15.315]. Pit-type samplers are often permanent, as they are built into the stream bed. They can be used to accurately measure total bed load, but since they affect the bathymetry, they are not suitable for measuring spatial or temporal variations associated with bedforms or channel geometry. Permanent sediment traps are costly to install and operate.

Acoustic Measurements

In the future, it may become possible to use acoustics to measure bed load non-intrusively with good spatial and temporal resolution. Acoustics could provide easier measurements in rivers that are deep and where traditional bed load measurements are difficult. Application of acoustics to measure bed load is only in its infancy, and acoustics have only been used to measure bed load velocities [15.321]. In fact, the bed load layer is quite thin, but covers a wide velocity range. It is difficult to say what velocity is actually being measured by acoustic pulses reflected from the bed, and bed load velocity measurements may be biased [15.322]. At this time, only the relative magnitude of bed load velocities at various locations on a river transect can be gathered with sufficient confidence. However, even this information is useful, as it can be used to appropriately plan bed load collection with pressure-difference samplers.

15.3.4 Total Load Measurements

Subdividing the sediment load into suspended load and bed load is not always necessary; sometimes it is sufficient to determine only the total sediment load in a river. For instance, the longevity of a large reservoir is based on the total sediment load into the reservoir because all of the sediment settles in the reservoir, regardless of the mode of transport in the river upstream of the reservoir. When a location can be identified that puts the entire sediment load into suspension, that location is a good place to measure the total load since it is easier to measure suspended load accurately than it is to measure bed load. According to *Graf* [15.244], all of the sediment load may be forced into suspension at rapids or constrictions (bridges), or a turbulence flume may be constructed to force all of the sediment into suspension [15.323]. Once in suspension, the sediment may be sampled using isokinetic samplers or other suspension measurement techniques.

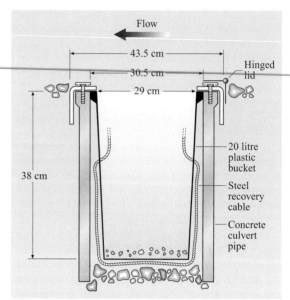

Fig. 15.82 A pit-style trap. The trap is cylindrical to eliminate directional bias (after [15.315])

15.3.5 Other Measurement Techniques

As described by *White* [15.246] there are perhaps two primary areas of sediment transport measurement that need improvement: sampling sediment fluxes near the bed, and sampling ranges of particle sizes. Traditional techniques partially overcome these issues, but are flow intrusive, have low temporal resolution, and require difficult sample processing. Although suspended sediment concentration can in some cases be measured with relatively high temporal and spatial resolution, the same is not true of bed load.

Only the most widely used sediment transport measurement techniques have been discussed in this chapter. There are a number of less commonly used techniques available that the reader should be aware of. These techniques include: nuclear radiation backscatter or transmission; hydrostatic pressure-difference measurements; sediment impact measurements; spectral reflectance; and tracking natural or seeded tracer particles that have magnetic, radiation, or fluorescence properties. Most of these techniques are not widely used either because they are too costly, too difficult to implement (particularly the tracking and nuclear techniques), or too inaccurate. The reader is encouraged to explore the review articles written by *White* [15.246] and *Wren* et al. [15.261] to find out more about these techniques.

References

15.1 M.S. Plesset: Physical effects on cavitation and boiling, Proc. 1st Symp. Naval Hydrodynamics, ed. by F.S. Sherman (Academic, Washington 1957) 297–323

15.2 R.E.A. Arndt, R. Voigt, J.P. Sinclair, P.R. Rodrigue: Cavitation erosion in hydroturbines, J. Hydraul. Eng. **115**, 1297–1315 (1989)

15.3 K.M. Kalumuck, G.L. Chahine: The use of cavitating jets to oxidize organic compounds in water, J. Fluids Eng. **122**, 465–470 (2000)

15.4 C. Gong, D.P. Hart: Ultrasound induced cavitation and sonochemical yields, J. Acoust. Soc. Am. **5**, 1–8 (1998)

15.5 L.C. Burrill: Sir Charles Parsons and cavitation – 1950 Parsons Memorial Lecture, Trans. Inst. Marine Eng. **63**, 149–167 (1951)

15.6 L. Rayleigh: On the pressure developed in a liquid during the collapse of a spherical cavity, Phil. Mag. **34**, 94–98 (1917)

15.7 M. Minnaert: On musical air-bubbles and sounds of running water, Phil. Mag. **16**, 235–248 (1933)

15.8 M.S. Plesset, S.A. Zwick: The growth of vapor bubbles in superheated liquids, J. Appl. Phys **25**(4), 493–500 (1954)

15.9 P.S. Epstein, M.S. Plesset: On the stability of gas bubbles in liquid-gas solutions, J. Chem. Phys. **18**(11), 1505–1509 (1950)

15.10 P. Eisenberg: Mechanics of cavitation. In: *Handbook of Fluid Dynamics*, ed. by V.L. Streeter (McGraw Hill, New York 1961) pp. 12.2–12.24, , Section 12

15.11 T.W. Wu: Cavity and wake flows, Annu. Rev. Fluid Mech. **4** (1972)

15.12 A.J. Acosta, B.R. Parkin: Cavitation inception – A selective review, J. Ship Res. **19**, 193–205 (1975)

15.13 M.S. Plesset, A. Prosperetti: Bubble dynamics and cavitation, Annu. Rev. Fluid Mech. **9**, 145–185 (1977)

15.14 R.E.A. Arndt: Cavitation in fluid machinery and hydraulic structures, Annu. Rev. Fluid Mech. **13**, 273–328 (1981)

15.15 R.E.A. Arndt: *Recent Advances in Cavitation Research*, Adv. Hydrosci., Vol. 12 (Academic, San Diego 1981) pp. 1–77

15.16 R.E.A. Arndt: Cavitation in vortical flows, Annu. Rev. Fluid Mech. **34**, 143–175 (2002)

15.17 J.R. Blake, D.C. Gibson: Cavitation bubbles near boundaries, Annu. Rev. Fluid Mech. **19**, 99–123 (1987)

15.18 E.P. Rood: Review – mechanisms of cavitation inception, J. Fluids Eng. **113**, 163–175 (1991)

15.19 R.T. Knapp, J.W. Daily, F.G. Hammit: *Cavitation* (McGraw-Hill, New York 1970)

15.20 F.R. Young: *Cavitation* (McGraw-Hill, London 1989)

15.21 C.E. Brennen: *Cavitation and Bubble Dynamics* (Oxford University Press, New York 1995)

15.22 J.P. Franc, J.M. Michel: *Fundamentals of Cavitation* (Kluwer Academic, Dordrecht 2004)

15.23 I. Anton: Cavitatia. Editura Academei (Polytechnical Institute of Timisoara, Bucharest 1984), R-79717

15.24 W.H. Isay: *Kavitation* (Hansa Schroedter, 1981)

15.25 R. Taghavi: *Cavitation Inception in Axisymmetric Jets*. Ph.D. Thesis (University of Minnesota, Minneapolis 1985)

15.26 R.E.A. Arndt: *Vortex Cavitation, Fluid Vortices* (Kluwer Academic, Dordrecht 1995) pp. 731–782

15.27 R.W. Kermeen: Water tunnel tests of NACA 66_1-012 hydrofoil in noncavitating and cavitating flows, Calif. Inst. Tech. Hydrodynamics Lab. Rep. **47-7** (1956)

15.28 M.L. Billet: Cavitation nuclei measurements - a review, Cavitation and Multiphase Flow Forum (ASME, New York 1985)

15.29 D.D. Joseph: Cavitation and the state of stress in a flowing liquid, J. Fluid Mech. **366**, 367–378 (1998)

15.30 B. Gindroz, M.L. Billet: Influence of the nuclei on the cavitation inception for different types of cavitation on ship propellers, J. Fluids Eng. **120**, 171–178 (1998)

15.31 J.W. Holl: An effect of air content on the occurrence of cavitation, J. Basic Eng. **82**, 941–946 (1960)

15.32 M. Kjeldsen, R.E.A. Arndt, M. Effertz: Spectral characteristics of sheet/cloud cavitation, J. Fluids Eng. **122**, 481–487 (2000)

15.33 W.M. Deeprose, N.W. King, P.J. McNulty, Pearsall: Cavitation noise, flow noise and erosion, Proc. Conf. Cavitation Inst. Mech. Eng., London (1974)

15.34 G.F. Wislicenus: *Fluid Mechanics of Turbomachinery*, Vol. 1 (Dover, New York 1965)

15.35 J.O. Young, J.W. Holl: Effects of cavitation on periodic wakes behind symmetric wedges, J. Basic Eng. **88**, 163–176 (1966)

15.36 B. Belahadji, J. Michel: Numerical and experimental study of cavitating vortices in the turbulent wake, Proc. Third Int. Symp. Cavitation, Grenoble, ed. by J.M. Michel, H. Kato (1998)

15.37 K.R. Laberteaux, S.L. Ceccio: Flow in the Closure Region of Closed Partial Attached Cavitation, Proc. Third Int. Symp. Cavitation, Grenoble, ed. by J.M. Michel, H. Kato (1998) 197–202

15.38 S. Gopalan, J. Katz: Flow structure and modeling issues in the closure region of attached cavitation, Phys. Fluids **12**, 895–911 (2000)

15.39 R.E.A. Arndt, C.C.S. Song, M. Kjeldsen, J. He, A. Keller: Instability of partial cavitation: A numerical/experimental approach. In: *Proc. 23rd Symp. Naval Hydrodynamics*, ed. by E. Rood (Academic, Washington 2000)

15.40 R.E.A. Arndt, C. Ellis, S. Paul: Preliminary investigation of the use of air injection to mitigate cavitation erosion, J. Fluids Eng. **117**, 498–592 (1995)

15.41 Y. Kawanami, H. Kato, H. Yamagushi: Three-dimensional characteristics of the cavities formed on a two-dimensional hydrofoil, Trans Int. Symp. on Cavitation, ed. by J.M. Michel, H. Kato (1998) 191–196

15.42 D.R. Stinebring: *Scaling of Cavitation Damage. MS Thesis* (Pennsylvania State University, Pennsylvania 1976)

15.43 J.W. Daily, V.E. Johnson: Turbulence and boundary layer effects on cavitation inception from gas nuclei, Trans ASME **78**, 1695–1706 (1956)

15.44 R.E.A. Arndt, A.T. Ippen: Rough surface effects on cavitation inception, J. Basic Eng. **90**, 249–261 (1968)

15.45 R.E.A. Arndt, W.K. George: Pressure fields and cavitation in turbulent shear flows. In: *12th Symp. Naval Hydrodynamics*, ed. by R. Cooper (Academic, Washington 1978) pp. 327–339

15.46 A. Michalke: On spatially growing disturbances in an inviscid shear layer, J. Fluid Mech. **23**, 521–544 (1965)

15.47 Chahine, V.E. Johnson: Mechanics of self-resonating cavitating jets, Jets and Cavities-Int. Symp. **31**, 21–33 (1985)

15.48 C.S. Crowe, F.H. Champagne: Orderly structure in jet turbulence, J. Fluid Mech. **48**, 547 (1971)

15.49 M. Versluis, B. Schmitz, A. von der Heydt, D. Lohse: On the sound of snapping shrimp, 53rd Annu. Meeting Am. Phys. Soc. Div. Fluid Dynamics, Washington (2000), Video Presentation

15.50 B. Ran, J. Katz: Pressure fluctuations and their effect on cavitation inception within water jets, J. Fluid Mech. **262**, 223–263 (1994)

15.51 K.K. Ooi: Scale effects on cavitation inception in submerged water jets: A new look, J. Fluid Mech. **151**, 367–390 (1985)

15.52 J. Katz, T.J. O'Hern: Cavitation in large scale shear flows, J. Fluids Eng. **108**, 373–376 (1986)

15.53 T.J. O'Hern: *Cavitation Scale Effects: I. Nuclei Distributions in Natural Waters, II. Cavitation Inception in a Turbulent Shear Flow. Ph.D. Thesis* (California Institute of Technology, Pasadena 1987)

15.54 S. Gopalan, J. Katz, O. Knio: The flow structure in the near field of jets and its effect on cavitation inception, J. Fluid Mech. **398**, 1–43 (1999)

15.55 K.K. Ooi, A.J. Acosta: The utilization of specially tailored air bubbles as static pressure sensors in a jet, J. Fluids Eng. **106**, 459–465 (1983)

15.56 R. Taghavi, R.E.A. Arndt: *Cavitation in various types of shear flow, Cavitation in Hydraulic Structures and Turbomachinery, ASME FED-25* (ASME, New York 1985) pp. 129–134

15.57 T. Leger, S.L. Ceccio: Examination of the flow near the leading edge of attached cavitation – Part 1. Detachment of two-dimensional and axisymmetric cavities, J. Fluid Mech. **376**, 61–90 (1998)

15.58 T. Leger, L.P. Bernal, S.L. Ceccio: Examination of the flow near the leading edge of attached cavitation – Part 2. Incipient breakdown of two-dimensional and axisymmetric cavities, J. Fluid Mech. **376**, 91–113 (1998)

15.59 R.E.A. Arndt, C. Ellis, S. Paul: Application of piezoelectric film in cavitation research, J. Hydraul. Eng. **123**, 539–548 (1997)

15.60 S. Watanabe, Y. Tsujimoto, J.-P. Franc, J.-M. Michel: Linear analyses of cavitating instabilities, Proc. Third Int. Symp. Cavitation, Grenoble, ed. by J.M. Michel, H. Kato (1998)

15.61 B. Svingen, M. Kjeldsen, R.E.A. Arndt: Dynamics of closed circuit hydraulic model loops, Proc. ASME Fluids Eng. Summer Meeting, Montreal (2002)

15.62 J.P. Franc: Partial cavity instabilities and re-entrant jet, CAV 2001: Fourth Int. Symp. Cavitation, ed. by C.E. Brennen, R.E.A. Arndt, L.S. Ceccio (California Institute of Technology, Pasadena 2001), http://cav2001.library.caltech.edu/

15.63 F. Avellan, P. Dupont, M. Farhat: Cavitation erosion power, Proc. Cavitation **91.116**, 135–140 (1991)

15.64 H. Yamaguchi, M. Tanaka, H. Kato: A numerical study on the mechanism of vortex generation downstream of a sheet cavity on a two dimensional hydrofoil, Cavitation and Multiphase Flow Forum, Vol. 109 (ASME, New York 1991)

15.65 M. Kjeldsen: Theoretical and experimental investigations of the instability of an attached cavity, Proc. ASME Fluids Eng. Div. Summer Mtg., Vancouver (1997), Paper FEDSMS97-3268

15.66 R.E.A. Arndt: *Cavitation Near Surfaces of Distributed Roughness. Ph.D. Dissertation* (Massachusetts Institute of Technology, Boston 1967)

15.67 T.J. Schauer: *An Experimental Study of a Ventilated Supercavitating Vehicle. MS Thesis* (University of Minnesota, Minnesota 2003)

15.68 M. Wosnik, T.J. Schauer, R.E.A. Arndt: Experimental investigation of the turbulent bubbly wake in a ventilated flow. In: *Advances in Turbulence X.*, ed. by H.I. Andersson, P.-A. Krogstad (CIMNE, Barcelona 2004) pp. 657–660

15.69 E.A. Brujan, G.S. Keen, A. Vogel, J.R. Blake: The final stage of the collapse of a cavitation bubble close to a rigid boundary, Phys. Fluids **14**, 85–92 (2002)

15.70 W. Lauterborn, H. Bolle: Experimental investigations of cavitation-bubble collapse in the neighborhood of a solid boundary, J. Fluid Mech. **72**, 391–399 (1975)

15.71 Y. Tomita, A. Shima: Mechanisms of impulsive pressure generation and damage pit formation by bubble collapse, J. Fluid Mech. **169**, 535–564 (1986)

15.72 A. Vogel, W. Lauterborn, R. Timm: Optical and acoustic investigations of the dynamics of laser-produced cavitation bubbles near a solid boundary, J. Fluid Mech. **206**, 299–338 (1989)

15.73 D.R. Stinebring, R.E.A. Arndt, J.W. Holl: Scaling of cavitation damage, J. Hydronautics **111**(3), 67–73 (1977)

15.74 A. Thiruvengadam: *Scaling laws for cavitation erosion* (Hydronautics Inc., Laurel 1971)
15.75 R.E.A. Arndt: Hydraulic turbines. In: *Hydropower Engineering Handbook*, ed. by J. Gulliver, R.E.A. Arndt (McGraw-Hill, New York 1991) pp. 4.1–4.67
15.76 P. Bourdon, R. Simoneau, F. Avellan, M. Farhat: Vibratory Characteristics of Erosive Cavitation Vortices Downstream of a Fixed Leading Edge Cavity, Proc. 15th IAHR Symp., Belgrade (1990)
15.77 F. Avellan, P. Dupont, I. Ryhming: Generation mechanism and dynamics of cavitation vortices downstream of a fixed leading edge cavity, Proc. 17th Symp. on Naval Hydrodynamics, ed. by E. Rood (Academic, Washington 1988)
15.78 P.A. Abbot, R.E.A. Arndt, T.B. Shanahan: Modulation noise analysis of cavitating hydrofoils, ASME FED. 176. In: *Proc. Symp. Bubble Noise and Cavitation Erosion in Fluid Systems*, ed. by R.E.A. Arndt (ASME, New York 1993)
15.79 Q. Le, J.P. Franc, J.M. Michel: Partial cavities: Global behavior and mean pressure distribution, J. Fluids Eng. **115**, 253–248 (1993)
15.80 Q. Le, J.P. Franc, J.M. Michel: Partial cavities: Pressure pulse distribution around cavity closure, J. Fluids Eng. **115**, 249–254 (1993)
15.81 I. Hansson, K.A. Mørch: The dynamics of cavity clusters in ultrasonic (vibratory) cavitation erosion, J. Appl. Phys. **51**, 4651–4658 (1980)
15.82 A. Prosperetti, N.Q. Lu, H.S. Kim: Active and passive acoustic behavior of bubble clouds at the ocean's surface, J. Acoust. Soc. Am. **93**(6), 3117–3127 (1993)
15.83 T. Colonius, C.E. Brennen, A.T. Preston: A numerical investigation of unsteady bubbly cavitating nozzle flows, Phys. Fluids **14**, 300–311 (2002)
15.84 L. van Wijngaarden: On the collective collapse of a large number of cavitation bubbles in water, Proc 11th Int. Cong. of Appl. Mech., ed. by H. Gortler (Springer, Berlin, Heidelberg 1964)
15.85 Y.C. Wang, C.E. Brennen: Shock development in the collapse of a cloud of bubbles, ASME Cavitation and Multiphase Flow Forum 153 (ASME, New York 1994)
15.86 P.A. Abbot, D.W. Morton, T.B. Shanahan: Hydroturbine cavitation detection using advanced acoustic emission techniques, Hydroacoustic Facilities and Experimentation Techniques, ASME Winter Annual Meeting, Atlanta (ASME, New York 1991)
15.87 J. Wetzel, R.E.A. Arndt: Hydrodynamic design considerations for hydroacoustic facilities: I flow quality, J. Fluids Eng. **116**(2), 324–331 (1994)
15.88 J. Wetzel, R.E.A. Arndt: Hydrodynamic design considerations for hydroacoustic facilities: II pump design factors, J. Fluids Eng. **116**(2), 332–337 (1994)
15.89 H. Durrer: Cavitation erosion and fluid mechanics, Sulzer Tech. Rev. **3**, 55–61 (1986)
15.90 H. Soyama, H. Kumano: The fundamental threshold level – a new parameter for predicting cavitation erosion resistance, J. Testing Evaluation **30**(5), 421–431 (2002)
15.91 B. Vyas, C.M. Preece: Stress produced in a solid by cavitation, J. Appl. Phys. **47**, 5133–5138 (1976)
15.92 R. Simoneau, A. Archer: *Transposition of Cavitation Marks on Different Hardness Metals*. ASME Fluids Engineering Division Summer Meeting, FEDSM97-3300 (ASME, New York 1997)
15.93 B. Belahadji, J.P. Franc, J.M. Michel: A statistical analysis of cavitation erosion pits, J. Fluid Eng. **113**, 700–706 (1991)
15.94 H. Soyama, T. Ikohagi, R. Oba: Observation of the cavitating jet in a narrow watercourse, ASME FED Cavitation Multiphasec Flow **194**, 79–82 (1994)
15.95 J. Steller: International cavitation erosion test and quaititative assessment of material resistance to cavitation, Wear **233–235**, 51–64 (1999)
15.96 B.C. Syamala Rao, N.S. Lakshmana Rao, K. Seetharamiah: Cavitation erosion studies with venturi and rotating disk in water, J. Basic Eng. **92**, 563–579 (1970)
15.97 H. Kadoi, T. Sasajima: Cavitation erosion prediction using a "soft surface", Int. Shipbuilding Prog. **25**, 141–150 (1978)
15.98 H. Lindgren, C.A. Johnsson: Cavitation inception on headforms, ITTC comparitive experiments, Proc. 11th Towing Tank Conf., Tokyo (1966) 219–232
15.99 M.S. Plesset: *Cavitating Flows, California Institute of Technology, Rep. 85-46* (California Institute of Technology, Pasadena 1969)
15.100 C.C. Church: A method to account for acoustic microstreaming when predicting bubble growth rates produced by rectified diffusion, J. Acoust. Soc. Am. **84**, 1758–1764 (1988)
15.101 H.B. Marschall, K.A. Morch, A.P. Keller, M. Kjeldsen: Cavitation inception by almost spherical solid particles in water, 4th Int. Symp. on Cavitation, Pasadena, ed. by C.E. Brennen, R.E.A. Arndt, L.S. Ceccio (2001)
15.102 D. Ma: *Experimental Studies of Water Quality Effects on Tip Vortex Cavitation*. MS Thesis (University of Minnesota, Minnesota 1994)
15.103 J.H.J. van der Meulen: Incipient and desinent cavtitation on hemispherical nosed bodies, Int. Shipbuilding Prog. **12**, 21–32 (1972)
15.104 A.P. Keller: Cavitation Scale Effects: Empirically Found Relations and the Correlation of Cavitation Number and Hydrodynamic Coefficients, 4th Int. Symp. on Cavitation, Pasadena, ed. by C.E. Brennen, R.E.A. Arndt, L.S. Ceccio (2001)
15.105 D.T. Kawakami, Q. Qin, R.E.A. Arndt: Can water quality affect the lift dynamics of cavitating hydrofoils?, 5th Int. Symp. on Cavitation, Osaka, ed. by Y. Tsujimoto (2003)
15.106 A.B. Hastings: *Biographical Memoirs: Donald Dexter van Slyke* (National Academy of Sciences, Washington 1976) pp. 308–361

15.107 F. Numachi: Über die Kavitationsentstehung mit besonderem Bezug auf der Luftgehalt des Wassers, Ing-Arch. **7**, 396–409 (1936)

15.108 W. Heller: *Hydrodynamische Effekte unter besonderer Berücksichtigung der Wasserqualität* (Technische Universität Dresden, Dresden 2004)

15.109 W. Heller: A new approach for determination of cavitation sensitivity of water, 4th ASME JSME Joint fluid Engineering Conference, Honolulu (2003)

15.110 F.B. Peterson: Hydrodynamic cavitation and some considerations of the influence of free gas content. In: *9th Symp. on Naval Hydrodynamics*, ed. by R. Cooper (Academic, Washington 1972)

15.111 R.E.A. Arndt, A.P. Keller: Free gas content effects on cavitation inception and noise in a free shear flow, Proc. IAHR Symp. Two Phase Flow and Cavitation in Power Generation Systems, Grenoble (1976) 3–16

15.112 R. Duraiswami, S. Prabhukumar, G.L. Chahine: Bubble size measurement using an inverse acoustic scattering method, J. Acoust. Soc. Am. **104**, 2699–2717 (1998)

15.113 G.L. Chahine, K.M. Kalumuck: Development of a near real-time instrument for nuclei measurement: The ABS acoustic bubble spectrometer, 4th ASME JSME Joint Fluid Eng. Conf., Honolulu (ASME, New York 2003)

15.114 T.M. Pham, J.M. Michel, Y. Lecoffre: Dynamical nuclei measurement: On the development and the performance evaluation of an optimized centerbody meter, J. Fluids Eng. **119**, 744–750 (1997)

15.115 T.G. Leighton: *The Acoustic Bubble* (Academic, San Diego 1994)

15.116 D.M. Oldenziel: *Bubble Cavitation in Relation to Liquid Quality*. Ph.D. Thesis (Technical University Twente, Twente 1979)

15.117 A.P. Keller: Influence of the cavitation nucleus spectrum on cavitation inception, investigated with a scattered light counting method, J. Basic Eng. **94**, 917–925 (1972)

15.118 H. Wang: *Experimental Study of Water Bubbly Hydrofoil Wakes*. MS Thesis (University of Minnesota, Minnesota 2004)

15.119 H. Tanger, E.A. Weitendorf: Applicability tests for the phase Doppler anemometer for cavitation nuclei measurements, J. Fluids Eng. **114**, 443–449 (1992)

15.120 Schiebe F.R.: Measurement of the Cavitation Susceptibility of Water Using Standard Bodies. St. Anthony Falls Hydraulic Laboratory, Rep. 118 (1972)

15.121 A.P. Keller: A vortex-nozzle cavitation susceptibility meter in routine application in cavitation inception measurements, Proc. Euromech. Colloquium 222 – Unsteady Cavitation and Its Effects (1987)

15.122 Silberman E., Schiebe F.R., Mrosla E.: The Use of Standard Bodies to Measure the Cavitation Strength of Water. St Anthony Falls Hydraulic Laboratory, Rep. 141 (1973)

15.123 D.M. Oldenziel: Measurements on the cavtiation susceptibility of water, 5th Conf. on Fluid Machinery Budapest **2**, 737–748 (1975)

15.124 Y. Lecoffre, J. Bonnin: Cavitation Test and Nucleation Control. In: *International Symposium on Cavitation Inception, ASME Winter Annual Meeting, New York*, ed. by W.B. Morgan, B.R. Parkin (ASME, New York 1979) pp. 141–147

15.125 L. d'Agostino, A.J. Acosta: On the design of cavitation susceptibility meters, Proc. 20th Am. Towing Tank Conf. **1**, 307–350 (1983)

15.126 F.G. Blake: *The Onset of Cavitation in Liquids I*, Acoustics Res. Lab., Tech Memo No 12 (Harvard Univ., Cambridge 1949)

15.127 F.E. Fox, K.F. Herzfeld: Gas bubbles with organic skin as cavitation nuclei, Ac. Soc. Am. **26**, 984–989 (1954)

15.128 E.N. Harvey, W.D. McElroy, A.H. Whiteley: On cavitation formation in water, J. Appl. Phys. **18**, 162–172 (1947)

15.129 R.E.A. Arndt, A.P. Keller: Water quality effects on cavitation inception in a trailing vortex, J. Fluids Eng. **114**, 430–438 (1992)

15.130 J.W. Holl, A.L. Treaster: Cavitation hysteresis, J. Basic Eng. **88**, 199–212 (1966)

15.131 R.E.A. Arndt, A.P. Keller: A case study of international cooperation: 30 years of collaboration in cavitation research, Proc. 4th ASME-JSME Joint Fluids Eng. Conf., Honolulu (2003), Available on CD

15.132 A. Richardson: *The Evolution of the Parsons Steam Turbine* (Engineering, London 1911)

15.133 R.E.A. Arndt, B.H. Maines: Nucleation and bubble dynamics in vortical flows, J. Fluids Eng. **122**, 488–493 (2000)

15.134 W.K. Blake, M.J. Wolpert, F.E. Geib: Cavitation noise and inception as influenced by boundary layer development on a hydrofoil, J. Fluid Mech. **80**(4), 617–640 (1977)

15.135 H. Higuchi, R.E.A. Arndt, M.F. Rogers: Characteristics of tip vortex noise, J. Fluids Eng. **111**, 495–501 (1989)

15.136 S.J. Barker: Measurements of radiated noise in the Caltech high-speed water tunnel, Part II: Radiated noise from cavitating hydrofoils, GALCIT (1975), Final Report on ONR contract Hydrodynamic Radiated Noise

15.137 R.E.A. Arndt, C.C.S. Song, Q. Qin: Experimental and numerical investigations of cavitation, 22nd IAHR Symposium on Hydraulic Machinery and Systems, Stockholm (2004)

15.138 R.W. Kermeen, B.R. Parkin: Incipient cavitation and wake flow behind sharp edged disks, CIT Hydrodynamics Lab Rep. **84-5** (1957)

15.139 R.E.A. Arndt, M. Levy: Acoustic radiation from cavitating hydrofoils, Proc. 7th Int. Congress on Noise and Vibration, Garmisch (1999)

15.140 T.H. Maiman: Stimulated optical radiation in ruby, Nature **187**, 493–494 (1960)

15.141 F. Durst, A. Melling, J.H. Whitelaw: *Principles and Practice of Laser Doppler Anemometry* (Academic, London 1981)

15.142 H.E. Albrecht, M. Borys, N. Damaschke, C. Tropea: *Laser Doppler and Phase Doppler Measurement Techniques* (Springer, Berlin, Heidelberg 2003)

15.143 R.J. Adrian: Particle-imaging techniques for experimental fluid mechanics, Annu. Rev. Fluid Mech. **23**, 261–304 (1991)

15.144 J. Westerweel: Fundamentals of digital particle image velocimetry, Meas. Sci. Technol. **8**, 1379–1392 (1997)

15.145 G. Sridhar, B. Ran, J. Katz: Implementation of particle image velocimetry to multi-phase flow, Cavitation Multiphase Flow Forum **109**, 205–210 (1991)

15.146 D.P. Towers, C.E. Towers, C.H. Buckberry, M. Reeves: A colour PIV system employing fluorescent particles for two-phase flow measurements, Meas. Sci. Technol. **10**, 824–830 (1999)

15.147 L.G. Fontecha: *PIV measurements in the wake of a supercavitating body. MS Thesis* (Chalmers University of Technology, Gothenburg 2004)

15.148 Y.A. Hassan, T.K. Blanchat, C.H. Seeley, R.E. Canaan: Simultaneous velocity measurements of both components of a two-phase flow using particle image velocimetry, Int. J. Multiphase Flow **18**, 371–395 (1992)

15.149 Y.A. Hassan, W. Schmidl, J. Ortiz-Villafuerte: Investigation of three-dimensional two-phase flow structure in a bubbly pipe flow, Meas. Sci. Technol. **9**, 309–326 (1998)

15.150 L. Gui, W. Merzkirch: Phase separation of PIV measurements in two-phase flow by applying a digital mask technique, ERCOFTAC Bull. **30**, 45–48 (1996)

15.151 R. Lindken, L. Gui, W. Merzkirch: Velocity measurements in multiphase flow by means of particle image velocimetry, Chem. Eng. Tech. **22**, 202–206 (1999)

15.152 R. Lindken, W. Merzkirch: Velocity measurements of liquid and gaseous phase for a system of bubbles rising in water, Exp. Fluids **29**, S194–S201 (2000)

15.153 J. Sakakibara, R.B. Wicker, J.K. Eaton: Measurements of the particle-fluid velocity correlation and the extra dissipation in a round jet, Int. J. Multiphase Flow **22**, 863–881 (1996)

15.154 W.J. Easson, M.L. Jakobsen: Slippage measurements in two-phase flows using particle image velocimetry, Institute of Physics Optical Group Half Day Conference Optics and Optical Diagnostics in Combustion (1996)

15.155 T.R. Oakley, E. Loth, R.J. Adrian: A two-phase cinematic PIV method for bubbly flows, J. Fluids Eng. **119**, 707–712 (1997)

15.156 K.T. Kiger, C. Pan: PIV technique for the simultaneous measurement of dilute two-phase flow, J. Fluids Eng. **122**, 811–818 (2000)

15.157 E. Delnoij, J. Westerweel, N.G. Deen, J.A.M. Kuipers, W.P.M. van Swaaij: Ensemble correlation PIV applied to bubble plumes rising in a bubble column, Chem. Eng. Sci. **54**, 5159–5171 (1999)

15.158 D.A. Khalitov, E.K. Longmire: Simultaneous two-phase PIV by two-parameter phase discrimination, Simultaneous two-phase PIV by two-parameter phase discrimination, Exp. Fluids **32**(2), 252–268 (2002)

15.159 M. Arnardottir: *Fundamentals of PIV Applied to Two-Phase Water-Bubble Flow. MS Thesis* (Aalborg University, Aalborg 2001)

15.160 G. Chaine, D.E. Nikitopoulos: Multiphase digital particle image velocimetry in a dispersed, bubbly, axisymmetric jet, Proc. ASME 2002 FEDSM, FEDSM2002-31435, Montreal (ASME, New York 2002)

15.161 A. Melling: Tracer particles and seeding for particle image velocimetry, Meas. Sci. Technol. **8**, 1406–1416 (1997)

15.162 K.R. Laberteaux, S.L. Ceccio: Partial cavity flows – Part 1. Cavities forming on models without spanwise variation, J. Fluid Mech. **431**, 1–41 (2001)

15.163 K.R. Laberteaux, S.L. Ceccio: Partial cavity flows – Part 2. Cavities forming test objects with spanwise variation, J. Fluid Mech. **431**, 43–63 (2001)

15.164 C.O. Iyer, S.L. Ceccio: The influence of developed cavitation on the flow of a turbulent shear layer, Phys. Fluids **14**, 3414–3431 (2002)

15.165 M. Wosnik, T.J. Schauer, R.E.A. Arndt: Experimental Study of a Ventilated Supercavitating Vehicle, Conference paper for CAV 2003-Fifth International Symposium on Cavitation, Osaka, ed. by Y. Tsujimoto (2003)

15.166 M. Wosnik, L. Fontecha, R.E.A. Arndt: Measurements in high void-fraction bubbly wakes created by ventilated supercavitation, Proc. ASME-FEDSM2005, ASME Fluids Eng. Div. Summer Meeting, Houston (2005), paper FEDSM2005-77200

15.167 M. Wosnik, R.E.A. Arndt: Measurements in high void-fraction turbulent bubbly wakes created by axisymmetric ventilated supercavitation, Proc. Sixth Int. Symp. Cavitation CAV2006, Wageningen (2006)

15.168 P.B.V. Johansson, W.K. George, M.J. Gourlay: Equilibrium similarity, effects of initial conditions and local Reynolds number on the axisymmetric wake, Phys. Fluids **15**, 603–617 (2003)

15.169 D.P. Hart: High-speed PIV analysis using compressed image correlatoion, J. Fluids Eng. **120**, 463–470 (1998)

15.170 A. Vogel, W. Lauterborn: Time-resolved particle image velocimetry used in the investigation of cavitation bubble dynamics, Appl. Opt. **27**(9), 1869–1876 (1988)

15.171 R.F. Mudde, J.S. Groen, H.E.A. van den Akker: Application of LDA to bubbly flows, Nucl. Eng. Des. **184**, 329–338 (1998)

15.172 J.S. Groen, R.F. Mudde, H.E.A. van den Akker: On the application of LDA to bubbly flow in the wobbling regime, Exp. Fluids **27**, 435–449 (1999)

15.173 T. Börner, W.W. Martin, H.J. Leutheuser: Comparative measurements in bubbly two-phase flow using laser Doppler and hot-film anemometry, Chem. Eng. Commun. **28**, 29–43 (1984)

15.174 S.L. Lee, F. Durst: On the motion of particles in turbulent duct flow, Int. J. Multiphase Flow **8**, 125–146 (1982)

15.175 S. So, H. Morikita, S. Tagaki, Y. Matsumoto: Laser Doppler velocimetry measurement of turbulent bubbly channel flow, Exp. Fluids **33**, 135–142 (2002)

15.176 H.R.E. van Maanen: *Retrieval of Turbulence and Turbulence Properties Form Randomly Sampled Laser Doppler Anemometry Data with Noise. Ph.D. Thesis* (Delft University of Technology, Delft 1999)

15.177 S. Guet, H.R.E. van Maanen, R.F. Mudde: Feasibility of LDA measurements in high void fraction bubbly flow, 11th Int. Symp. on Applications of Laser Techniques to Fluid Mechanics, Lisbon (2002)

15.178 A. Kubota, H. Kato, H. Yamaguchi, M. Maeda: Unsteady structure measurement of cloud cavitation on a foil section using conditional sampling technique, J. Fluids Eng. **111**, 204–210 (1989)

15.179 A.A. Naqwi, R. Menon, L.M. Fingerson: An adaptive phase/Doppler system and its applications including particle sizing in submicron and nanometer ranges, Exp. Fluids **20**, 328–334 (1996)

15.180 F. Burdin, N.A. Tsochatzidis, P. Guiraud, A.M. Wilhelm, H. Delmas: Characterisation of the acoustic cavitation cloud by two laser techniques, Ultrason. Sonochem. **6**, 43–51 (1999)

15.181 T.J. Sun: *A Theoretical and Experimental Study of Non-Condensible Turbulent Bubbly Jets. Ph.D. Thesis* (Pennsylvania State University, Pennsylvania 1985)

15.182 T.Y. Sun, G.M. Faeth: Structure of turbulent bubbly jets – I. Methods and centerline properties, Int. J. Multiphase Flow **12**, 99–114 (1986)

15.183 T.Y. Sun, G.M. Faeth: Structure of turbulent bubbly jets – II. Phase property profiles, Int. J. Multiphase Flow **12**, 115–126 (1986)

15.184 S. Kumar, D.E. Nikitopoulos, E.E. Michaelides: Effect of bubbles on the turbulence near the exit of a liquid jet, Exp. Fluids **7**, 487–494 (1989)

15.185 K.N. Stanley: *Non-Intrusive Characterization of a Dispersed Bubbly Axisymmetric Jet. Ph.D. Thesis* (Louisiana State University, Baton Rouge 2001)

15.186 K.N. Stanley, D.E. Nikitopoulos: Dispersed bubbly axisymmetric jets, Proc. of the 4th Int. Conference on Multiphase Flows, New Orleans (2001)

15.187 S. Kim, X.Y. Fu, X. Wang, M. Ishii: Development of the miniaturized four-sensor conductivity probe and the signal processing scheme, Int. J. Heat Mass Transfer **43**, 4101–4118 (2000)

15.188 L.G. Neal, S.G. Bankoff: A high resolution resistivity probe for determination of local void properties in gas liquid flow, AIChE J. **9**, 490–494 (1963)

15.189 O.C. Jones: Two-phase flow measurement techniques in gas-liquid systems. In: *Fluid Mechanics Measurements*, ed. by R.J. Goldstein (Hemisphere, Washington 1983) pp. 479–558

15.190 R.F. Mudde, T. Saito: Hydrodynamical similarities between bubble column and bubbly pipe flow, J. Fluid Mech. **437**, 203–228 (2001)

15.191 J.-M. Le Corre, E. Hervieu, M. Ishii, J.-M. Delhaye: Benchmarking and improvements of measurement techniques for local-time-averaged two-phase flow parameters, Exp. Fluids **35**, 448–458 (2003)

15.192 A. Cartellier: Simultaneous void fraction measurement, bubble velocity, and size estimate using a single optical probe in gas-liquid two-phase flows, Rev. Sci. Instrum. **63**, 5442–5453 (1992)

15.193 G.J. Kirouac, T.A. Trabold, P.F. Vassallo, W.E. Moore, R. Kumar: Instrumentation development in two-phase flow, Exp. Thermal Fluid Sci. **20**, 79–93 (1999)

15.194 B. Shamoun, M. El Beshbeeshy, R. Bonazza: Light extinction technique for void fraction measurements in bubbly flow, Exp. Fluids **26**, 16–26 (1999)

15.195 F. Burdin, P. Guiraud, A.M. Wilhelm, H. Delmas: Implementation of the laser diffraction technique for acoustic cavitation bubble investigations, Part. Part. Syst. Charact. **19**, 73–83 (2002)

15.196 S. Vagle, D.M. Farmer: A comparison of four methods for bubble size and void fraction measurements, IEEE J. Oceanic Eng. **23**, 211–222 (1998)

15.197 M. Perlin: *Statistical Analysis of Visual Wave Observations and Gage/Radar Measurements, Coastal Engr Res. Center, USACE, Misc Paper 84-6* (U.S. Army Corps of Engineers, Washington 1984)

15.198 A.V. Babanin, I.R. Young, M.L. Banner: Breaking probabilities for dominant surface waves on water of finite constant depth, J. Geophys. Res. **106**(C6), 11659–11676 (2001)

15.199 TM6: An Ocean Wave Measuring Instrument, Tech. Memo 6, Beach Erosion Board 1-52 (1948)

15.200 W.S. Campbell: *An Electronic Wave-Height Measuring Apparatus*, US Navy DW Taylor Model Basin, Rep. 859 (U.S. Dept. of the Navy, Washington 1953)

15.201 L.B. Wilner: Variable capacitance liquid level sensors, Rev. Sci. Inst. **31**(5), 501–507 (1960)

15.202 S.A. Hughes: *Physical Models and Laboratory Techniques in Coastal Engineering* (World Scientific, Singapore 1993)

15.203 F.T. Korsmeyer: *The Capacitance Wave Probe, NA & ME Dept Rep* (Univ. Michigan, Ann Arbor 1980)

15.204 G.V. Sturm, F.Y. Sorrell: Optical wave measurement technique and experimental comparison with conventional wave height probes, Appl. Opt. **12**(8), 1928–1933 (1973)

15.205 M. Perlin, C.-L. Ting: Steep gravity-capillary waves within the internal resonance regime, Phys. Fluids A **4**(11), 2466–2478 (1992)

15.206 W.H. Munk: *Measurement of Waves from Pressure Fluctuations at Ocean Bottom, Scripps Inst. Ocean, Wave Proj Rep No. 5* (The Scripps Institution of Oceanograhphy, La Jolla 1944)

15.207 R.G. Folsom, A.B. White, M.F. Killory: *Deep Water Wave Measurements, I, II, III, IV* (Univ. of Calif. Fluid Mech. Lab, Berkeley 1945, 1946)

15.208 B. Strong, B. Brumley, E.A. Terray, G.W. Stone: The Performance of ADCP-Derived Directional Wave Spectra and Comparison with other Independent Measurements, Proc. MTS/IEEE Oceans Conf., Providence (Institute of Electronic and Electrical Engineers, Piscataway 2000)

15.209 M.V. Trevorrow: Measurement of ocean wave directional spectra using Doppler side-scan sonar arrays, J. Atmos. Oceanic Technol. **12**(3), 603–616 (1995)

15.210 H.-T. Liu, J.-T. Lin: On the spectra of high-frequency wind waves, J. Fluid Mech. **123**, 165–185 (1982)

15.211 J.H. Duncan, A.A. Dimas: Surface ripples due to steady breaking waves, J. Fluid Mech. **329**, 309–339 (1996)

15.212 C.S. Cox: Measurements of slopes of high-frequency wind waves, J. Marine Res. **16**(3), 199–225 (1958)

15.213 J.R. Saylor: Internal reflection beneath capillary water waves: A method for measuring wave slope, Appl. Opt. **36**(6), 1121–1129 (1997)

15.214 J. Wu, J.M. Lawrence, E.S. Tebay, M.P. Tulin: A multiple purpose optical instrument for studies of short steep water waves, Rev. Sci. Instrum. **40**(9), 1209–1213 (1969)

15.215 J. Wu: Directional slope and curvature distributions of wind waves, J. Fluid Mech. **79**(3), 463–480 (1977)

15.216 J.D. Barter, K.L. Beach, P.H.Y. Lee: Collocated and simultaneous measurement of surface slope and amplitude of water waves, Rev. Sci. Instrum. **64**(9), 2661–2665 (1993)

15.217 B. Jahne, M. Schmidt, R. Rocholz: Combined optical slope/height measurements of short wind waves: Principle and calibration, Meas. Sci. Technol. **16**, 1937–1944 (2005)

15.218 B. Jahne, J. Klinke, S. Waas: Imaging of short ocean wind waves: A critical theoretical review, J. Opt. Soc. Am. A **11**(8), 2197–2209 (1994)

15.219 X. Liu, J.H. Duncan: The effects of surfactants on spilling breaking waves, Nature **421**, 520–523 (2003)

15.220 J.M. Killen: The sonic surface-wave transducer, Am. Towing Tank Conf., Univ. Calif., Berkeley (Defense Technical Information Center, Ft. Belvoir 1959)

15.221 H. Mitsuyasu, Y.-Y. Kuo, A. Masuda: On the dispersion relation of random gravity waves – Part 2: An experiment, J. Fluid Mech. **92**, 731–749 (1979)

15.222 D.W. Wang, P.A. Hwang: The dispersion relation of short wind waves from space-time wave measurements, J. Atmos. Oceanic Technol. **21**, 1936–1945 (2004)

15.223 M. Perlin, H. Lin, C.-L. Ting: On parasitic capillary waves generated by steep gravity waves: An experimental investigation with spatial and temporal measurements, J. Fluid Mech. **255**, 597–620 (1993)

15.224 J.B. Carneal, C.W. Baumann, P. Atsavapranee, J.H. Hamilton, J. Shan: A global laser rangefinder profilometry system for the measurement of three-dimensional wave surfaces, Proc. of REDSM05, 2005, ASME Fluids Eng. Div. Summer Meeting and Exhib., Houston (Am. Soc. Mech. Eng., Washington 2005)

15.225 M.L. Banner, I.S.F. Jones, J.C. Trinder: Wavenumber spectra of short gravity waves, J. Fluid Mech. **198**, 321–344 (1989)

15.226 J.M. Wanek, C.H. Wu: Automated trinocular stereo imaging system for three-dimensional surface wave measurement, Ocean Eng. **33**, 723–747 (2006)

15.227 I. Grant, Y. Zhao, G.H. Smith, J.N. Stewart: Split-screen, single-camera, laser-matrix, stereogrammetry instrument for topographical water wave measurements, Appl. Opt. **34**(19), 3806–3809 (1995)

15.228 I. Grant, N. Stewart, I.A. Padilla-Perez: Topographical measurements of water waves using the projection moire method, Appl. Opt. **29**(28), 3981–3983 (1990)

15.229 E.J. Bock, T. Hara: Optical measurements of capillary-gravity wave spectra using a scanning laser slope gauge, J. Atmos. Oceanic Technol. **12**, 395–403 (1995)

15.230 Q. Li, M. Zhao, S. Tang, S. Sun, J. Wu: Two-dimensional scanning laser slope gauge: Measurements of ocean-ripple structures, Appl. Opt. **32**(24), 4590–4597 (1993)

15.231 B. Jahne, K. Riemer: Two-dimensional wave number spectra of small-scale water surface waves, J. Geophys. Res. **95**(C7), 11531–11546 (1990)

15.232 W.C. Keller, B.L. Gotwols: Two-dimensional optical measurement of wave slope, Appl. Opt. **22**, 3476–3478 (1983)

15.233 X. Zhang, C.S. Cox: Measuring the two-dimensional structure of a wavy water surface optically: A surface gradient detector, Exp. Fluids **17**, 225–237 (1994)

15.234 X. Zhang, D. Dabiri, M. Gharib: Optical mapping of fluid density interfaces: Concepts and implementations, Rev. Sci. Instrum. **67**(5), 1858–1868 (1996)

15.235 J.S. Bendat, A.G. Piersol: *Random Data: Analysis and Measurement Procedures* (Wiley, New York 1986)

15.236 A. Papoulis: *Probability, Random Variables and Stochastic Processes*, 2nd edn. (McGraw Hill, New York 1984)

15.237 M.K. Ochi: *Applied Probability and Stochastic Processes* (Wiley-Interscience, New York 1990)

15.238 M.K. Ochi: *Ocean Waves: The Stochastic Approach* (Cambridge Univ Press, Cambridge 1998)

15.239 K.R. Bodge: *The Design, Development, and Evaluation of a Differential Pressure Gauge Directional Wave Monitor, Coastal Engr Res Center, USACE, Misc Rep 82-11* (U.S. Army Corps of Engineers, Washington 1982)

15.240 M.S. Longuet-Higgins, D.E. Cartwright, N.D. Smith: *Observations of the Directional Spectrum of Sea Waves Using the Motions of a Floating Buoy*, Ocean Waves Spectra (Prentice-Hall, Englewood Cliffs 1963)

15.241 H.E. Krogstad, S.F. Barstow, S.E. Aasen, I. Rodriguez: Some recent developments in wave buoy measurement technology, Coastal Eng. **37**(3-4), 309–329 (1999)

15.242 RD Instruments: *Waves User's Guide, Pub No. 957-6148-00* (RD Instruments, San Diego 2001)

15.243 C. Zuffada, G. Hajj, J.B. Thomas: *Sea Height and Roughness with GPS Reflections*, http://www.nasatech.com/Briefs/Jun02/NP020943.html (NASA, Washington 2002)

15.244 W.H. Graf: *Hydraulics of Sediment Transport* (McGraw-Hill, New York 1971)

15.245 S.M. Yalin: *Mechanics of Sediment Transport* (Pergamon, Oxford 1972)

15.246 T.E. White: Status of measurement techniques for coastal sediment transport, Coastal Eng. **35**(1998), 17–45 (1998)

15.247 J.M. Phillips, D.E. Walling: An assessment of the effects of sample collection, storage, and resuspension on the representativeness of measurements of the effective particle size distribution of fluvial suspended sediment, Water Resour. **29**, 298–308 (1995)

15.248 J.D. Smith, S.R. McLean: Spatially averaged flow over a wavy surface, J. Geophys. Res. **83**, 1735–1746 (1977)

15.249 L.C. van Rijn: Sediment transport – Part II. Suspended load transport, J. Hydraul. Eng. **110**(11), 1613–1641 (1984)

15.250 M. García, G. Parker: Entrainment of bed sediment into suspension, J. Hydraul. Eng. **117**(4), 414–435 (1991)

15.251 E. Meyer-Peter, R. Muller: Formulas for bedload transport, Proc. 2nd Cong. International Association of Hydraulic Res., Stockholm (IAHR, Delft 1948) 39–64

15.252 M. Wong: Does the bedload equation of Meyer-Peter and Muller fit its own data?, Int. Association of Hydraulic Research Congress, Thessaloniki (IAHR, Delft 2003)

15.253 H. Einstein: *The Bed-Load Function for Sediment Transportation in Open Channel Flows*, Technical Bulletin 1026, Soil Conservation Service (U.S. Dept. of Agriculture, Washington 1950)

15.254 M.S. Yalin: An expression for bedload transportation, J. Hydraul. Div. ASCE **89**(HY3), 221–250 (1963)

15.255 F. Engelund, J. Fredsoe: A sediment transport model for straight alluvial channels, Nordic Hydrol. **7**, 293–306 (1976)

15.256 R. Fernandez Luque, R. van Beek: Erosion and transport of bed sediment, J. Hydraul. Res. **14**(2), 127–144 (1976)

15.257 R.K. Dewey, W.R. Crawford: Bottom stress estimates from vertical dissipation rate profiles on the continental shelf, J. Phys. Oceanography **18**(8), 1167–1177 (1988)

15.258 R.L. Soulsby, K.R. Dyer: The form of the near-bed velocity profile in tidally accelerating flow, J. Geophys. Res. **86**(C9), 8067–8074 (1981)

15.259 K.W. Bedford, O. Wai, C.M. Libicki, R.I.II. van Evra: Sediment entrainment and deposition measurements in long island sound, J. Hydraul. Div. ASCE **113**(10), 1325–1342 (1987)

15.260 M.W. Schmeeckle, J.M. Nelson: Direct numerical simulation of bedload transport using a local, dynamic boundary condition, Sedimentology **50**(2), 279–301 (2003)

15.261 D.G. Wren, B.D. Barkdoll, R.A. Kuhnle, R.W. Derrow: Field techniques for suspended-sediment measurement, J. Hydraul. Eng. **126**(2), 97–104 (2000)

15.262 L.C. van Rijn, A.S. Schaafsma: Evaluation of measuring instruments for suspended sediment, Int. Conf. Measuring Techniques of Hydraulics Phenomena in Offshore, Coastal, and Inland Waters, London (British Hydraulic Research Association, Cranfield 1986) 401–423

15.263 J.J. Bosman, E.T.J.M. van der Velden, C.H. Hulsbergen: Sediment concentration measurement by transverse suction, Coastal Eng. **11**, 353–370 (1987)

15.264 V.A. Vanoni: Transportation of suspended sediment by water, ASCE Trans. **111**(3), 67–102 (1946)

15.265 R.P. Apmann, R.R. Rumer: Diffusion of sediment in developing flow, J. Hydraul. Div. ASCE **96**(1), 109–123 (1970)

15.266 H.E. Jobson, W.W. Sayre: Vertical transfer in open channel flow, J. Hydraul. Div. ASCE **96**(3), 703–724 (1970)

15.267 N.L. Coleman: Velocity profiles with suspended sediment, J. Hydraul. Res. **19**(3), 211–229 (1981)

15.268 L.C. van Rijn: *Entrainment of Fine Sediment Particles, Development of Concentration Profiles in a Steady, Uniform Flow Without Initial Sediment Load*, Report on Model Investigation, Delft Hy-

15.269 T.A. Winterstein, H.G. Stefan: *Suspended Sediment Sampling in Flowing Water: Laboratory Study of the Effects of Nozzle Orientation, Withdrawal Rate and Particle Size*, External Memorandum M-168, St. Anthony Falls Hydraulic Laboratory (University of Minnesota, Minneapolis 1983)

15.270 T.K. Edwards, G.D. Glysson: *Field Methods for Measurement of Fluvial Sediment: U.S.G.S. Techniques of Water-Resources Investigations* (USGS, Reston 1999)

15.271 A.E. Hay: Sound scattering from a particle-laden turbulent jet, J. Acoust. Soc. Am. **90**(4), 2055–2074 (1991)

15.272 M.E. Nelson, P.C. Benedict: Measurement and analysis of suspended sediment loads in streams, Proc. ASCE **76**(31), 1–28 (1950)

15.273 E.T. Baker, J.W. Lavelle: The effect of particle size on the light attenuation coefficient of natural suspensions, J. Geophys. Res. **89**(C5), 8197–8203 (1984)

15.274 J.P.-Y. Maa: Laboratory measurements of instantaneous sediment concentration under waves, IEEE J. Oceanic Eng. **13**(4), 299–302 (1988)

15.275 N.J. Clifford, K.S. Richards, R.A. Brown, S.N. Lane: Laboratory and field assessment of an infrared turbidity probe and its response to particle size and variation in suspended sediment concentration, Hydrol. Sci. J. **40**(6), 771–791 (1995)

15.276 J.P. Downing, R.W. Sternberg, C.R.B. Lister: New instrumentation for investigation of sediment suspension in the shallow marine environment, Marine Geol. **42**, 19–34 (1981)

15.277 D&A Instrument Company: *OBS-3 Product Information* (D&A Instrument Company, Port Townsend 2005)

15.278 R.A. Beach, R.W. Sternberg, R. Johnson: A fiber optic sensor for monitoring suspended sediment, Marine Geol. **103**, 513–520 (1992)

15.279 P.D. Osborne, C.E. Vincent, B. Greenwood: Measurement of suspended sand concentrations in the nearshore: Field intercomparison of optical and acoustic backscatter sensors, Continental Shelf Res. **14**(2/3), 159–174 (1994)

15.280 M.O. Green, J.D. Boon III: The measurement of constituent concentrations in nonhomogeneous sediment suspensions using optical backscatter sensors, Marine Geol. **110**, 73–81 (1993)

15.281 T.F. Sutherland, P.M. Lane, C.L. Amos, J. Downing: The calibration of optical backscatter sensors for suspended sediment of varying darkness levels, Marine Geol. **162**, 587–597 (2000)

15.282 D.H. Schoellhamer: Biological interference of optical backscatterance sensors in Tampa Bay, Florida, Marine Geol. **110**, 303–313 (1993)

15.283 P. Ridd, G. Day, S. Thomas, J. Harradence, D. Fox, J. Bunt, O. Renagi, C. Jago: Measurement of sediment deposition rates using an optical backscatter sensor, Estuarine, Coastal Shelf Sci. **52**, 155–163 (2001)

15.284 J.A. Puleo, R.A. Beach, R.A. Holman, J.S. Allen: Swash zone sediment suspension and transport and the importance of bore-generated turbulence, J. Geophys. Res. **105**(C7), 17021–17044 (2000)

15.285 D.J. Law, A.J. Bale, S.E. Jones: Adaptation of focused beam reflectance measurement to in-situ particle sizing in estuaries and coastal waters, Marine Geol. **140**, 47–59 (1997)

15.286 J.M. Phillips, D.E. Walling: Calibration of a Par-Tec 200 laser back-scatter probe for in situ sizing of fluvial suspended sediment, Hydrol. Processes **12**, 221–231 (1998)

15.287 A.R. Heath, P.D. Fawell, P.A. Bahri, J.D. Swift: Estimating average particle size by focused beam reflectance measurement (FBRM), Part. Part. Syst. Char. **19**, 84–95 (2002)

15.288 Y.C. Agrawal, H.C. Pottsmith: Laser diffraction particle sizing in STRESS, Cont. Shelf Res. **14**, 1101–1121 (1994)

15.289 Y.C. Agrawal, H.C. Pottsmith: Instruments for particle size and settling velocity observations in sediment transport, Marine Geol. **168**, 89–114 (2000)

15.290 Sequoia Scientific Inc.: *LISST-100 Product Information* (Sequoia Scientific, Redmond 2005)

15.291 J.W. Gartner: Estimating suspended solids concentrations from backscatter intensity measured by acoustic Doppler current profiler in San Francisco Bay, California, Marine Geol. **211**, 169–187 (2004)

15.292 D.M. Admiraal, R. Musalem, M.H. García: A Study of self-formed ripples using Particle Image Velocimetry (PIV), 4th IAHR Symposium on River, Coastal, and Estuarine Morphodynamics, Barcelona (IAHR, Delft 2003)

15.293 I. Nezu, R. Azuma: Turbulence characteristics and interaction between particles and fluid in particle-laden open channel flows, J. Hydraul. Eng. **130**(10), 988–1001 (2004)

15.294 S.C. Knowles, J.T. Wells: In situ aggregate analysis camera (ISAAC): A quantitative tool for analyzing fine-grained suspended material, Limnol. Oceanogr. **43**(8), 1954–1962 (1998)

15.295 C. Libicki, K.W. Bedford, J.F. Lynch: The interpretation and evaluation of a 3 MHz acoustic backscatter device for measuring benthic boundary layer sediment dynamics, J. Acoust. Soc. Am. **85**(4), 1501–1511 (1989)

15.296 A.E. Hay, J. Sheng: Vertical profiles of suspended sand concentration and size from multifrequency acoustic backscatter, J. Geophys. Res. **97**(C10), 15661–15677 (1992)

15.297 C. Shen, U. Lemmin: Application of an acoustic particle flux profiler, J. Hydraul. Res. **37**(3), 407–419 (1999)

15.298 D.M. Admiraal, M.H. García: Laboratory measurement of suspended sediment concentration using

15.299 M.G. Kleinhans, W.B.M. Ten Brinke: Accuracy of cross-channel sampled sediment transport in large sand-gravel-bed rivers, J. Hydraul. Eng. **127**(4), 258–269 (2001)

an Acoustic Concentration Profiler (ACP), Exp. Fluids **28**, 116–127 (2000)

15.300 P.D. Thorne, D.M. Hanes: A review of acoustic measurement of small-scale sediment processes, Continental Shelf Res. **22**, 603–632 (2002)

15.301 A.M. Crawford, A.E. Hay: Determining suspended sand size and concentration from multifrequency acoustic backscatter, J. Acoust. Soc. Am. **94**(6), 3312–3324 (1993)

15.302 J.F. Lynch, J.D. Irish, C.R. Sherwood, Y.C. Agrawal: Determining suspended sediment particle size information from acoustical and optical backscatter measurements, Continental Shelf Res. **14**(10/11), 1139–1165 (1994)

15.303 P.D. Thorne, S.C. Campbell: Backscattering by a suspension of spheres, J. Acoust. Soc. Am. **92**(2), 978–986 (1992)

15.304 P.D. Thorne, P.J. Hardcastle, R.L. Soulsby: Analysis of acoustic measurements of suspended sediments, J. Geophys. Res. **98**(C1), 899–910 (1993)

15.305 F.H. Fisher, V.P. Simmons: Sound absorption in sea water, J. Acoust. Soc. Am. **62**(3), 558–564 (1977)

15.306 P.D. Thorne, G.P. Holdaway, P.J. Hardcastle: Constraining acoustic backscatter estimates of suspended sediment concentration profiles using the bed echo, J. Acoust. Soc. Am. **98**(4), 2280–2288 (1995)

15.307 L. Zedel, A. Hay: A coherent Doppler profiler for high-resolution particle velocimetry in the ocean: Laboratory measurements of turbulence and particle flux, J. Atmos. Oceanic Technol. **16**, 1102–1117 (1999)

15.308 C. Shen, U. Lemmin: A two-dimensional acoustic sediment flux profiler, Meas. Sci. Technol. **8**, 880–884 (1997)

15.309 K.F. Betterridge, P.D. Thorne, P.S. Bell: Assessment of acoustic coherent Doppler and cross-correlation techniques for measuring near-bed velocity and suspended sediment profiles in the marine environment, J. Atmos. Oceanic Technol. **19**, 367–380 (2002)

15.310 G.P. Holdaway, P.D. Thorne, D. Flatt, S.E. Jones, D. Prandle: Comparison between ADCP and transmissometer measurements of suspended sediment concentration, Continental Shelf Res. **19**(3), 421–441 (1999)

15.311 W.W. Emmett: *A Field Calibration of the Sediment-Trapping Characteristics of the Helley–Smith Bed Load Sampler*, US Geological Survey Professional Paper 1139 (USGS, Washington 1980)

15.312 E.J. Helley, W. Smith: Development and calibration of a pressure-difference bedload sampler, US Geol. Survey Open-File Rep. **73-108**, 18 (1971)

15.313 J.M. Gaudet, A.G. Roy, J.L. Best: Effect of orientation and size of Helley–Smith sampler on its efficiency, J. Hydraul. Eng. **120**(6), 758–767 (1994)

15.314 D.W. Hubbell, H.H. Stevens Jr., J.V. Skinner, J.P. Beverage: New approach to calibrating bed load samplers, J. Hydraul. Eng. **111**(4), 677–694 (1985)

15.315 S.M. Sterling, M. Church: Sediment trapping characteristics of a pit trap and the Helley–Smith sampler in a cobble gravel bed river, Water Resour. Res. **38**(8), 10.1029/2000WR000052 (2002)

15.316 K. Bunte, S.R. Abt, J.P. Potyondy, S.E. Ryan: Measurement of coarse gravel and cobble transport using portable bedload traps, J. Hydraul. Eng. **130**(9), 879–893 (2004)

15.317 M.T.K. Gaweesh, L.C. van Rijn: Bed-load sampling in sand-bed rivers, J. Hydraul. Eng. **120**(12), 1364–1384 (1994)

15.318 K.T. Lee, Y.L. Liu, K.H. Cheng: Experimental investigation of bedload transport processes under unsteady flow conditions, Hydrol. Processes **18**, 2439–2454 (2004)

15.319 D.W.T. Jackson: A new, instantaneous aeolian sand trap design for field use, Sedimentology **43**, 791–796 (1996)

15.320 B.O. Bauer, S.L. Namikas: Design and field test of a continuously weighing, tipping-bucket assembly for Aeolian sand traps, Earth Surf. Processes Landforms **23**, 1171–1183 (1998)

15.321 C.D. Rennie, R.G. Millar, M.A. Church: Measurement of bed load velocity using an acoustic Doppler current profiler, J. Hydraul. Eng. **128**(5), 473–483 (2002)

15.322 C.D. Rennie, R.G. Millar: Measurement of the spatial distribution of fluvial bedload transport velocity in both sand and gravel, Earth Surf. Processes Landforms **29**, 1173–1193 (2004)

15.323 P.C. Benedict, M.L. Albertson, M.Q. Matejka: Total sediment load measured in turbulence flume, Trans. ASCE, Vol. 120 (1955) 457–488, Paper 2750

16. Aerodynamics

Vehicle aerodynamics has its roots in aviation, and the same is true for the related experimental techniques. The wind tunnel is the key carry-over element. Wind tunnels have been adapted stepwise to meet the specific needs of vehicle aerodynamic studies. Three areas of development must be mentioned: With vehicles a much greater blockage can be tolerated; to simulate the relative motion between vehicle and road or track different belt systems and mechanisms to reproduce wheel rotation have turned out to be useful; for tests in various climates a variety of specialized smaller wind tunnels have been developed. Measurement techniques are very much the same as in aeronautics at low Mach number and increasingly particle image velocimetry (PIV) is being applied to study the complex flow fields around vehicles. Nevertheless, flow visualization and photogrammetry still remains an indispensable tool for a quick evaluation of different geometries, also for dynamic flow conditions.

16.1 **Ground Vehicle Aerodynamics**................1043
    16.1.1 Vehicles in Their Natural Environment 1043
    16.1.2 Simulation of the Drive on Road and Track 1044
    16.1.3 Wind Tunnels for Ground Vehicles ..1044
    16.1.4 Wind-Tunnel Design.....................1047
    16.1.5 Equipment for Automotive Wind Tunnels.........1054
    16.1.6 Limits of Simulation 1058
    16.1.7 Typical Vehicle Wind Tunnels 1063
    16.1.8 Tests with Moving Models and Full-Scale Vehicles 1068
    16.1.9 Measurement Technique 1070
    16.1.10 Support by Computational Fluid Dynamics ..1080

16.2 **Short-Duration Testing of High Enthalpy, High Pressure, Hypersonic Flows** 1081
    16.2.1 Working Principle of Shock Tubes/Tunnels and Shock Expansion Tubes/Tunnels 1082
    16.2.2 Measurement Techniques..............1095
    16.2.3 Typical Applications of Shock Tunnel and Shock Expansion Tunnel Facilities 1111

16.3 **Bluff Body Aerodynamics**........................1125
    16.3.1 Flow Physics, Facilities, and Approach..............................1125
    16.3.2 How Bluff is a Bluff Body? 1128
    16.3.3 Base Pressure, Drag, Lift, and Strouhal Number 1133
    16.3.4 Overview of Vortex Shedding Regimes 1135
    16.3.5 Concluding Remarks 1145

References 1146

16.1 Ground Vehicle Aerodynamics

16.1.1 Vehicles in Their Natural Environment

The environment to which a ground vehicle is exposed during cruise is sketched in Fig. 16.1. The flow around it and its thermal loading depend on various conditions:

- vehicle speed
- grade of road or track
- strength and gustiness of wind
- rain or snow
- road dirt
- sun load

As can be seen in Fig. 16.1 the flow field past a vehicle on a road or track is made up of two components: the airflow equivalent to a vehicle's forward motion, and the other originating from the natural wind, which forms a turbulent boundary layer above the rough soil. The size of the turbulent eddies is on the same order of magnitude as the vehicle's length. As a consequence, the resulting oncoming flow is far from uniform. On the

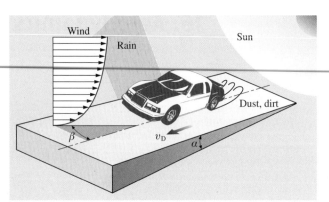

Fig. 16.1 A vehicle in its real environment

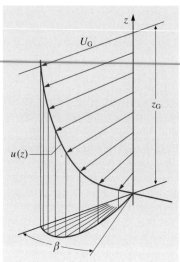

Fig. 16.2 The skewed wind profile of the boundary layer over the ground

contrary, as sketched in Fig. 16.2, its velocity profile is skewed. Due to the gustiness of the wind, the magnitude and direction of the velocity vary randomly with locus and time. Additionally, this turbulence can be increased by a superimposed unsteady flow field resulting from the vehicle's motion through the wakes of other vehicles, and from obstacles in the surroundings such as trees, bridges, and houses.

In addition, the temperature field above a road is not necessarily homogeneous. Intense sunlight will heat the surface of the road more than the surrounding air, generating a temperature boundary layer above the road.

Altogether, both fields around the vehicle, air flow and temperature, are very inhomogeneous and nonstationary.

16.1.2 Simulation of the Drive on Road and Track

Most of the development work necessary for high-quality aerodynamics and thermal properties of a vehicle is carried out in wind tunnels. According to the requirements of the different category of tests – aerodynamic, aero-acoustic, thermal – various types of test facilities have been developed in which air is blown past a vehicle. They all together are called wind tunnels, and are all in the focus of what follows.

Specific experiments must be carried out with moving models, and of course with real vehicles. These are described in Sect. 16.1.8. Typical examples are: investigating the pressure pulse when trains meet, when they pass through a station, or when they enter a tunnel. Another example is the response of a car to a sudden crosswind gust, and to the wakes of passing vehicles. Such tests are performed with scale models catapulted crosswise through a wind tunnel, or with self-propelled small-scale trains on track, with scale models and full-scale vehicles towed in water tanks, and with real vehicles in crosswind sections – and, of course, under real environmental conditions on the road and track. In the latter two cases the vehicle has to deal with other vehicles.

In any case, on a test bed the drive on a road or track is only *simulated*, never *reproduced* exactly. As is typical, this simulation is associated with deviations from reality, and it is not easy to quantify all the errors resulting therefrom.

Measurement techniques applied in vehicle aerodynamics will be treated in Sect. 16.1.9; they are very similar to those used in aircrafts aerodynamics.

16.1.3 Wind Tunnels for Ground Vehicles

Configuration Considerations

The manner in which the real flow on a road or track is represented in a wind tunnel is a more or less severe simplification. As depicted in Fig. 16.3, the relative motion is reversed: the vehicle is at rest and is subjected to moving air. However, with ground vehicles this would be an incomplete simulation, because the relative motion between the vehicle and ground, and the rotation of the wheels should not be overlooked. How this requirement is fulfilled will be outlined in Sect. 16.1.5.

Different from reality, in a wind tunnel the on-coming flow is made as uniform as possible, and the turbulence level is very low. This *ideal* flow is a carryover from aircraft aerodynamics, where it is valid for flight at high altitude. As a means of making results from different wind tunnels comparable, this may have its justification. However, extra expense for realizing extremely low turbulence is not reasonable for vehicle aerodynamics. On the contrary, the turbulence level should be high. And in fact, quite recently, one automotive wind tunnel was equipped with a device for artificially increasing turbulence. A description will follow in the section on turbulence (Sect. 16.1.5).

The climate inside a passenger compartment depends not only on the temperature of the air inside but also on direct radiation from the sun and diffuse radiation from the environment. The radiation from sunlight – intensity, spectrum, direction – can be simulated with sufficient accuracy; diffuse radiation is normally neglected. Furthermore, the comfort of passengers depends on the humidity of the surrounding air. Accordingly, this property of air also has to be reproduced. Humidity, in turn, interacts with components of the air conditioning system. Therefore, dewing and icing must be considered when the evaporator and its housing are tested.

In bad weather a vehicle is exposed to water from various directions: natural rain, and dirty water whirled up by the vehicle's own or other vehicles' wheels. Simulating rain in a wind tunnel is comparatively easy. A definite volume flow of water is sprayed into the undisturbed airstream ahead of the vehicle. Experiments on soiling – real or simulated – should be carried out in wind tunnels only with extremely small quantities of *dirt*, since it is difficult to protect installed probes and pressure taps from dirt, and cleaning costs are very high for a large wind tunnel.

During thermal tests the vehicle's engine must be loaded according to the road load. A chassis dynamometer must be installed, suited for front-, rear- and all-wheel drive. During the (full-scale) model phase of vehicle development the heat fluxes in the engine – water, oil, etc. – must be simulated by external boilers.

Not only flow and temperature fields but also the acoustic environment must be simulated. For cars, buses and wagons interior wind noise is an important criterion for comfort. For trains external wind noise is also important. Generally, wind tunnels are very noisy. The tunnel sound field, which is largely determined by its fan, masks the sound field produced by the flow around a vehicle. Specifically designed aero-acoustic wind tunnels have overcome this problem. They make it possible to discriminate wind noise from tunnel noise *objectively*, and allow for *subjective* assessment of the wind noise spectrum.

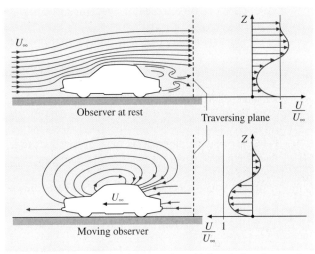

Fig. 16.3 Relative motion between the vehicle, wind, and ground on the road and in a wind tunnel

Evolutionary Development of Vehicular Wind Tunnels

In principle, vehicle wind tunnels resemble so-called *normal* wind tunnels that are common in aeronautics, and were in fact derived from them. However, with regard to their fluid-dynamic properties the test objects in both cases differ significantly from each other. Generally aircrafts are well-streamlined bodies, and the flow around them is mainly attached. They produce *lift*, which causes the airstream of a wind tunnel to bend. On the other hand, ground vehicles are *bluff* bodies, and the flow past them is characterized by separation. They experience high *drag*, and tend to *block* the jet of a wind tunnel. Vehicles with wings that produce high downwards force – like Formula 1 cars – also bend the airstream. This the more as vehicles are preferably tested at full scale in wind tunnels whose test sections, according to the *classical* layout rules, are by far too small. In addition, vehicle wind tunnels must also be suited for aero-acoustic investigations, and equipped for thermal tests.

Early ground-vehicle aerodynamics made use of wind tunnels that had been designed for aircraft. The scale of the vehicle models was as small as 1 : 10; later larger scales were preferred: in Europe 1 : 4 and 1 : 5, in USA 3 : 8 (1 : 3.75). Originally, the ground was simulated by the mirror image of the model. Later, when it was recognized that this was neither physically correct (vortices forming at the underside and at the rear oscillated from the model to the image and vice versa) nor practical (two models and a wind tunnel of twice the original size were needed), a rigid ground board was applied.

When automobile designers turned to full-scale models, as shapes in small scale are difficult to assess aesthetically, vehicle aerodynamics had to follow. W. Kamm in 1938 built the first full-scale automotive wind tunnel in the world, which is still operational. This wind tunnel (and several others built thereafter) was designed according to the rules for wind tunnels for aircrafts: accordingly, the blockage ratio, the key parameter that determines the overall dimensions of the tunnel, had to be about 5%.

Later, wind-tunnel development progressed down two routes:

1. while vehicle aerodynamicists were originally interested in forces and moments – mainly drag, and those components of the resulting air force controlling directional stability – other items later attracted attention, namely thermal problems such as engine cooling and air conditioning inside the passenger compartment, wind noise, and soiling. The corresponding tests led to particular test beds, so-called thermal wind tunnels, which were further specialized to hot and cold tunnels.
2. for pure aerodynamic testing at full scale, smaller wind tunnels were built with blockage ratios of up to 20%. Low-noise, so-called aero-acoustic, wind tunnels were developed, ground simulation was improved, and rotation of the wheels was implemented.

Despite the lack of enthusiasm of the designers to perform their modeling in a small scale, wind-tunnel tests with small-scale models are still carried out today, preferably in the early development phase of a new vehicle, and in research. The scale 1 : 2.5, the classical scale applied by mechanical engineers, was rediscovered for wind-tunnel models. Provided that the wind speed of a tunnel is high enough this scale allows testing at high enough Reynolds numbers – and with heavy-duty vehicles even for full-scale Reynolds numbers – without coming into conflict with Mach-number effects.

Selecting the Model Scale and Appropriate Wind Tunnel

Most of the work performed during the modeling phase is done in a wind tunnel. Water tunnels only serve as a supplement for special purposes, and are very well suited for flow visualization.

The question of which model scale to be selected – either in a wind or water tunnel – has to be considered with care, and in cooperation with design staff. The advantages of a small scale – low cost for both model and wind tunnel, and easy handling – must be weighed against the risk of errors due to insufficient fidelity of the model and too low a Reynolds number.

Detailed optimization and fine-tuning of cars must be made at full scale, and as soon as drivable prototypes are on hand, it has to be decided whether individual tests should be carried out on the road or in a wind tunnel. Tests on the road and track offer the advantage of realistic conditions. However, their drawbacks should not be ignored: test conditions are rarely repeatable, confidentiality is difficult to secure, and making measurements is more difficult. The advantages of the wind tunnel are obvious: faster and more-precise measurement techniques, and the possibility of unraveling different effects, such as noise from the drive train, tires, and wind around the body. However, as cannot be emphasized often enough,

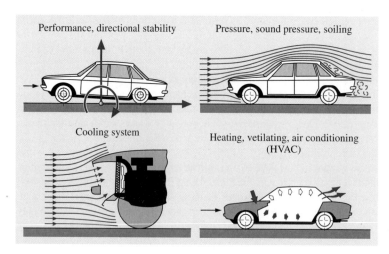

Fig. 16.4 The four main objectives of vehicle aerodynamics

the limitations of the wind tunnel should not be overlooked: deficiencies in simulation, high cost, and limited availability.

With respect to its objectives, vehicle aerodynamics can be classified into four categories, as illustrated in Fig. 16.4. The two sketches in the upper row represent the management of the outer flow, while those in the lower row symbolize thermal management and the related inner flow. Both the outer and inner flows are closely linked to each other. Their individual simulation requirements can be fulfilled in two kinds of facilities.

1. All physical parameters that have an influence according to Fig. 16.4 are simulated in one and the same facility. This leads to a large climatic aero-acoustic wind tunnel.
2. Only those parameters are reproduced that are significant during the specific test, all others being reproduced with less expense; this requires several specialized test beds:
   - full-scale aero-acoustic wind tunnel
   - hot and cold tunnels
   - chassis dynamometer with fan
   - small-scale wind tunnel.

Both routes – *all in one* and *several specialized tunnels* – have been followed. However, experience has shown the latter to be superior: investment and operational cost are lower, and instead of one test bed three to four are available – a number which is needed anyhow.

In any case, the more that physical parameters are neglected during the simulation, the greater the effort necessary to assure that the results achieved on a test bed will be confirmed on the road. Comparative measurements among the various test facilities and with the road must be carried out to uncover and quantify the deficiencies of any specific kind of simulation.

16.1.4 Wind-Tunnel Design

Configuration Considerations

As sketched in Fig. 16.5, there are two ways to lay out the air path in a wind tunnel:

1. closed return; known under the brand *Göttingen*, the first of which was designed by Ludwig Prandtl;
2. open, no return; designed by Gustave Eiffel and in his honor called an *Eiffel* tunnel.

In order to remain independent of environmental conditions, tunnels of the latter type are generally erected in a closed building, either in an existing one or one specifically designed to host the wind tunnel. In the latter case a flow pattern called *open return* is produced.

The pros and cons of both types of wind tunnels are discussed in detail in [16.1]. In vehicle aerodynamics the *Göttingen* type is clearly preferred. All hot and cold tunnels, and with only a few exceptions, also the full-scale aero and aero-acoustic vehicle wind tunnels, work with a closed air circuit. Examples of both will be presented in Sect. 16.1.7.

Properties of the Essential Components

When planning a test, the *user* of a wind tunnel is looking for answers to the following five questions:

1. What kind of wind tunnel do I need?
2. How do the expected results correlate with those from road tests? What are the specific deficits of the simulation, and can they be tolerated or adjusted?
3. How many wind tunnel hours are needed?
4. At what time is the specific wind tunnel needed available?
5. What are the costs of the test campaign?

Answering the first question appears to be trivial: it must be derived from the purpose of the test. What kind of information is needed: aerodynamic, aero-acoustic, thermal properties? The answer to the second question depends on the characteristics of the following components of the tunnel:

1. Test section (size, open or closed flow, kind of ground simulation)
2. Nozzle (including the settling chamber)

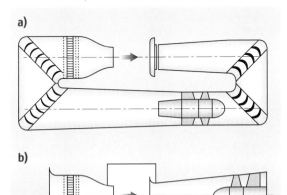

Fig. 16.5a,b The two types of air circuit of wind tunnels: (a) closed return, *Göttingen type*, (b) open, no return, *Eiffel type*

3. Collector (for tunnels with an open test section)
4. Chamber surrounding an open test section (known as the *plenum*)
5. Heat exchanger, sun load, rain and snow (in climatic tunnels).

In the following, these key components will first be discussed with no regard to the purpose for which the wind tunnel is specialized. Subsequently, in a separate section, examples of existing vehicle wind tunnels will be given. All other technical details, such as the design of the diffusers, turning vanes in the bends, the fan, fan drive and control, will be touched on only insofar as they are different from wind tunnels for aeronautical purposes [16.2].

The nature of the remaining three questions above is predominantly economic. The items to look at are:

1. How many data points can be acquired per unit time?
2. What are the costs per data point acquired?

A high hourly price can be more than compensated by fast test techniques, data acquisition, and evaluation.

Test Section. The test section is the point were the test engineer meets the wind tunnel. Together with its adjacent components, namely the nozzle, collector, and plenum, it is depicted in Fig. 16.6. The properties of the test section are decisive for the conception of his experiment and the assessment of the results. Subsequently, they will be discussed with special attention to bluff bodies such as automobiles and trains.

Two properties of the test section are characteristic: the magnitude of the airstream's cross section, A_N, which is measured at the nozzle exit, and the nature of its circumferential boundaries.

The nozzle cross section has to be seen in relation to the dimensions of the test object which, in vehicle aerodynamics, are characterized by its frontal area A, as defined in Fig. 16.7. The quotient of frontal area A and nozzle cross section A_N is called the blockage ratio $\varphi = A/A_N$, which, of course, strictly speaking is not a property of the wind tunnel. However, this relation emphasizes that it is not the size of the tunnel per se that is essential but its relation to the size of the model.

In order to provide kinematic similarity of the flow in a wind tunnel to that on road, the blockage ratio φ – which is zero on the road or track – should be as small as possible. However, cost considerations – construction and operation – demand a blockage ratio as large as feasible. However, what *feasible* means in this context is still a matter of discussion.

When the first full-scale automotive wind tunnels were built, the value of the blockage ratio φ was carried over from aeronautic wind tunnels, where $\varphi = 0.05$ was a target value. With a frontal area of $A = 2\,\mathrm{m}^2$ typical for cars, this would have led to a nozzle cross section of $A_N = 40\,\mathrm{m}^2$. Indeed, full-scale wind tunnels were built with nozzle cross sections of $30-40\,\mathrm{m}^2$ and above, the largest by General Motors in USA, with $A_N = 52.6\,\mathrm{m}^2$.

Later, comparative measurements and investigations on test section dimensions resulted in the finding that a larger blockage ratio could be well tolerated; $\varphi = 0.1$ seems to be a reasonable limit for cars. Several wind tunnels have been designed according to this number,

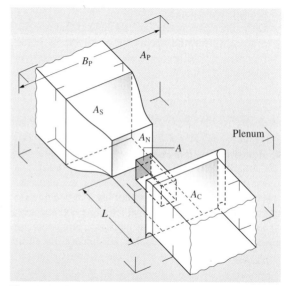

Fig. 16.6 The main geometric parameters of the test section of a wind tunnel

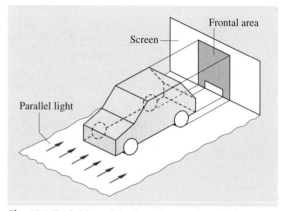

Fig. 16.7 Definition of the frontal area of a vehicle

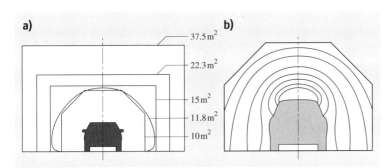

Fig. 16.8a,b Size and shape of the cross section of the airstream (after Janssen); **(a)** various sizes, **(b)** shape matched to the isobars around a car

having a cross section of $A_N \approx 20\,\text{m}^2$. Yet, even $\varphi \approx 0.2$ is used; the smallest full-scale wind tunnel has only $A_N = 10\,\text{m}^2$.

However, this consideration has to be seen in light of the fact that most car companies have vans and *sports utility vehicles* (SUVs) in their line-up, which have frontal areas of up to $3\,\text{m}^2$.

Not only the *magnitude* of the nozzle cross section A_N but also its circumferential *contour* has an effect on the flow field past the test object. Some typical cross sections of automotive wind tunnels are compiled in Fig. 16.8. Most frequently the shape of an open-jet cross section is rectangular. The ratio of the height H to the width B is about $H/B \approx 0.6$. Sometimes the corners are rounded off or chamfered. Very small full-scale wind tunnels that are exclusively used for cars have non-rectangular cross sections. A. Morelli [16.3], when designing the $11\,\text{m}^2$ wind tunnel at Pininfarina, selected a semicircle with its lower edges rounded off. His intuition was to create a cross section that was similar to the shape of the frontal area of a car. By cutting off the corners of a rectangle, L. J. Janssen, designer of the first aero-acoustic wind tunnel on behalf of BMW ($A_N = 10\,\text{m}^2$), intended to approximate the shape of the isobars around a car.

The other characteristic property of a test section is the nature of the circumferential boundary of the airstream. Three types of test section can be distinguished: open, closed, and slotted walls. The closed test section has been further developed to specific configurations such as streamlined and adaptive walls. All five types of jet boundaries are sketched in Fig. 16.9. The ground floor, which is typical for a vehicle wind tunnel, is *not* a component of the test section, although physically bound to it. Be it stationary or moveable, it is part of the test set-up representing either the road or rails.

Kinematically, the flow around a test object in a jet of finite dimensions is not exactly the same as in free air. The nature of the difference depends on the boundaries of the jet, as can be seen from the streamlines sketched in Fig. 16.10. In an open-jet test section the air flowing around a vehicle has room to make way. In comparison to free air the streamlines are further apart from each other and so the speed of the air at any given point in the vicinity of a vehicle is somewhat lower than in an airstream of infinite dimensions.

In contrast, in a closed test section the streamlines are constrained by the walls, and so the local speed in the vicinity of a vehicle is raised. In both cases, open and closed test sections, the effect of the jet's boundaries become more pronounced the nearer they are to the model, i.e., the larger the blockage ratio φ. The motivation for the various types of test-section boundaries shown in Fig. 16.9 is to minimize the effect of the limited tunnel size, thereby permitting a blockage ratio as large as possible.

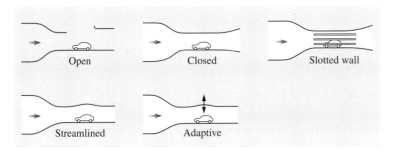

Fig. 16.9 Various test section boundaries

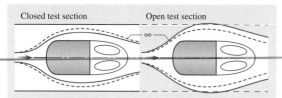

Fig. 16.10 Effect of the airstream boundary on the flow past a (bluff) body in comparison to free flow; *left*: closed test section, streamlines are quenched, *right*: open test section, streamlines are expanding

Both kinds of jet boundaries, open and closed, have their merits and drawbacks with respect to the physical properties of the wind tunnel, and the peculiarities of daily work; they will be discussed below. The corrections that compensate for the physical differences compared to the flow in free air are outlined in Sect. 16.1.6.

Frequently, the flow in an empty *open* test section is said to be comparable to that in an ideal *free jet*. The constant ambient pressure is impressed on its flow field, and, as a logical result, the static pressure along its axis is constant. Such a property would be highly advantageous for experiments with bluff bodies because, when measuring the drag of long bodies with large vertical surfaces at the front and the rear, even a small axial pressure gradient in the test section may lead to a significant error.

However, in most wind tunnels with an open test section the static pressure is anything but constant, as shown by the comparison of several wind tunnels in Fig. 16.11 from [16.4]. Generally, the pressure gradient along the tunnel axis is negative downstream of a nozzle exit. Near the center of the test section the pressure gradient goes through zero, and further downstream it becomes positive, being steepest immediately in front of the collector.

The second considerable advantage previously attributed to an open test section is a lower absolute value of correction. However, this argument no longer holds. For a long time, the only correction considered was *blockage correction*, and this in fact is small for an open jet. However, blockage (precisely, solid blockage) is only one of a series of interference effects that have to be considered, and so should not be looked at in isolation.

Finally, the third – and real – advantage of an open test section is its good accessibility, which facilitates model installation and the placement of probes, permits out-of-flow acoustic measurements, and allows easy photography of the flow.

The greatest drawback of an open test section is its limited usable length. The jet coming out of the nozzle mixes with the quiescent air in the surrounding plenum. The width of the jet's core (in which the set wind speed U_∞ can be found) decreases rapidly. Furthermore, turbulent mixing processes at the jet's boundary are more

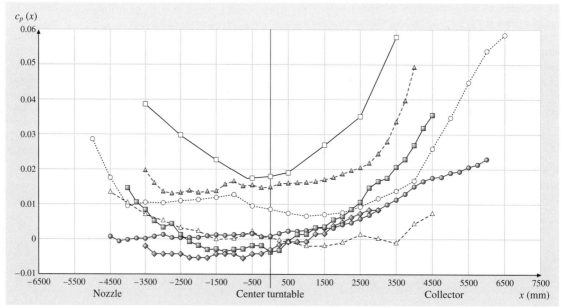

Fig. 16.11 Static pressure coefficient $c_p(x)$ along the axis of several automotive wind tunnels with an open test section (after [16.4])

intense than in a turbulent boundary layer adjacent to a wall, and this mixing causes higher losses than the friction on a plane wall. Therefore the power requirement of an open test section is higher than for a closed one. Other disadvantages of the open jet are its susceptibility to low-frequency oscillations (surge) and its unconstrained sound radiation, which may be regarded as a nuisance by wind-tunnel personnel. Finally, for climatic wind tunnels the plenum must be included within the air-conditioned area; hence large wall panels have to be insulated.

A specific problem arises when racing cars are tested in an open test section. Their rear wings produce a high down-force (negative lift). The related upwash can be so strong that it bends the airflow behind a car upwards and over the collector mouth. Air behind the car is then drawn in from the sides and this may significantly alter the flow pattern at the rear end of the car.

The negative properties of an open test section are corresponding positive properties of the closed. An advantage of a *closed* test section is its large usable length. The negative pressure gradient due to pressure loss at the walls can be compensated for by slightly increasing the tunnel cross section in the flow direction. However, this compensation is only correct for one specific configuration, namely for an empty test section; with a model installed this is only an approximation. A further advantage of a closed test section is the stability of its flow, i.e., the problem of pumping does not exist.

A disadvantage of a closed test section is its sensitivity to blockage; its correction is approximately twice that of a free jet (and is of opposite sign). Furthermore, at a high yaw angle β, the lateral deflection of the airstream may be so strong that the related adverse pressure gradients on the adjacent sidewalls of the test section may lead to boundary-layer separation there.

The *slotted-wall* test section is an attempt to combine the advantages of the open and closed test sections, and at the same time eliminate the drawbacks of both. The slots in the walls give ambient-pressure access to the flow inside the test section; thus the pressure along the axis is (almost) constant. The solid part of the walls prevents the jet from mixing with the ambient air; the core of the jet therefore remains usable for a greater length.

One means of avoiding the large corrections necessary for blockage in a closed test section is to *streamline* its walls. This idea is based on the fact that the flow pattern at a certain distance, even from a bluff body like a vehicle (and its wake), in the so-called far field, is only weakly dependent on the individual details of the model's shape. The far field is mainly determined by the overall parameters of the test object: its length, height, and width (and thus some kind of fineness ratio that can be derived from them). If the tunnel walls are shaped according to the free-air flow pattern of an average mid-sized vehicle, the flow around smaller and larger vehicles will only be slightly distorted, in any case far less than by parallel walls.

The disadvantage of a test section with streamlined but fixed walls is that its contour does not fit vehicles whose main dimensions differ significantly from the one used to define the contours of its walls. This, for example, will be the case if a box-type light truck is put into a test section that has been configured for passenger cars. A possibility for overcoming this problem is offered by *adaptive* walls, albeit at the price of great expense.

Nozzle. The function of the nozzle is fourfold, it:

1. accelerates the flow
2. makes the velocity distribution over the cross section of the flow more uniform
3. reduces the intensity of the turbulence in the airstream
4. serves to measure the wind speed in the test section

The requirements (specifications) for the flow properties for vehicle wind tunnels have not yet been formulated on a rational basis. However, the following data are typical for existing tunnels:

1. local deviations from average wind speed $\Delta u/U_\infty = (u - U_\infty)/U_\infty \leq \pm 0.5\%$;
2. angularity in pitch and yaw $\alpha, \beta \leq \pm 0.2°$;
3. turbulence level $\mathrm{Tu} = \sqrt{\overline{u'^2}}/U_\infty \leq 0.5\%$.

These quantities are determined by two geometric properties of a nozzle: the contraction ratio and contour of the walls. The contraction ratio κ is defined as the relation of the nozzle's entrance cross section to its outlet: $\kappa = A_S/A_N$, where A_S is the cross section of the nozzle's entrance, and A_N that at its exit.

According to experience the effect of the contraction ratio κ on the flow quality is:

$$\frac{\Delta u}{U_\infty} \sim \frac{1}{\kappa}, \quad \mathrm{Tu}_x \sim \frac{1}{\kappa}.$$

However, the contraction ratio κ determines not only the quality of the flow but also the size and thus the cost of a wind tunnel, and so it should not be larger than necessary.

From experience it is known that a turbulence level of 0.5% is achieved with a contraction ratio of $\kappa = 4$, a value which had already been proposed by L. Prandtl.

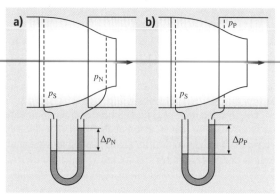

Fig. 16.12a,b Measuring the wind speed U in an open jet wind tunnel; (**a**) nozzle method, (**b**) plenum method

The *wind speed* in a test section is determined in the *classical* manner by measuring the difference Δp of static pressure between the entrance of the nozzle (i. e., in the settling chamber) and its exit. The dynamic head q_∞ within the jet leaving the nozzle is

$$q_\infty = \frac{\rho}{2} U_\infty^2 = k \Delta p\,, \tag{16.1}$$

where k is a nozzle factor to be determined by measuring (and averaging over the jet's cross section, excluding the mixing zone at its boundaries) the dynamic head in an empty test section. Generally, the numerical value of the nozzle factor k is greater than one, because the wind speed in the settling chamber is not exactly zero, and hence the local static pressure there is (slightly) lower than the total pressure.

In a wind tunnel with an *open* test section there exist two possibilities for the position of the *downstream* pressure tap, and therefore there are two possibilities to measure the wind speed. The first, the so-called *nozzle method*, is the same as for a closed test section. The location of the downstream pressure tap is close to the exit of the nozzle. The related pressure signal is: $\Delta p_N = p_{SC} - p_N$; see the left-hand side of Fig. 16.12. Alternatively, in the *plenum method*, the downstream pressure tap is located in the plenum surrounding the open jet; see the right-hand side of same figure. If the plenum is well vented, as is normally the case, this pressure p_P is equal to the atmospheric pressure. The pressure difference is designated $\Delta p_P = p_{SC} - p_P$.

For a long time there has been an intense dispute among wind-tunnel operators about which method should be preferred. However, comparative measurements [16.6] have clearly proven the superiority of the plenum method. The major reason for the inferiority of the nozzle method is the influence of the model on the pressure tap at the nozzle exit. For bluff bodies this effect is strong, and as shown in Fig. 16.13 is dependent on the model position (the distance between the nozzle exit and the front of the model).

This value can be taken as reasonable for automotive wind tunnels with a Göttingen circuit. For Eiffel tunnels, where the flow upstream of the nozzle is not disturbed by a fan, diffuser, turning vanes etc., an even smaller contraction ratio is sufficient: $\kappa = 2\text{--}3$. How turbulence can be increased for special investigations will be discussed in Sect. 16.1.5.

The shape of a nozzle – length and contour – should be selected so as to provide a rectangular velocity profile at its exit. Even more importantly, the angle between the jet and the geometrical axis of the test section should be small, because the flow around bluff bodies close to the ground is very sensitive to differences in the angle of attack (or pitch). Therefore a tolerance of $\alpha \leq \pm 0.5°$, as recommended in [16.5], is rather generous; $\alpha \leq \pm 0.2°$ seems to be more reasonable. The sensitivity of vehicles to small deviations in yaw near $\beta = 0$ is less pronounced.

In small wind tunnels, such as thermo tunnels, where the blockage ratio is extremely high ($\varphi \approx 0.3\text{--}0.4$), the wind speed is measured in a different way; details will be described in Sect. 16.1.6 together with wind-tunnel corrections.

Wind speed must be controllable at all levels, even at low speed. For instance, driving slowly uphill on a long steep grade represents one of the serious criteria for dimensioning the radiator and cooling fan (the other being maximum speed). The wind from the vehicle's forward motion as well as the radiator's fan contribute

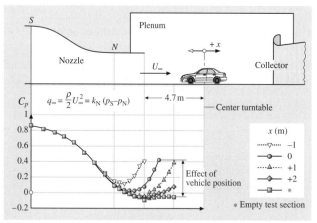

Fig. 16.13 The effect of model position on the pressure distribution inside the wind-tunnel nozzle [16.1]

to the cooling effect; therefore, accurate control of the (low) wind speed is important.

Technically, two solutions are available for adjusting wind speed: variable pitch of the fan's blades at constant fan speed, and continuously variable fan speed with fixed fan blades. Today, the latter is used more widely. Variable pitch permits a very quick change of wind speed. However, this advantage is offset by a high noise level at *all* wind speeds, including idle. For test engineers and craftsmen who have to work on a model in a tunnel between two runs while the fan is idling, this is rather unpleasant.

Collector. The collector's function is to gather the air transported by the free jet of an open test section and to guide it into the first diffuser. However, this task is accompanied by the following two phenomena:

- the static pressure along the axis of the test section is increasing in front of the collector, as shown in Fig. 16.11;
- the flow in the test section may pulsate at low frequency.

This has four major consequences:

1. drag is measured too low,
2. pressures and forces may be altered (the Katzmeier effect),
3. the pulsation will modulate the flow noise around the test vehicle,
4. the pulsation may grow to an extent at which the structure of the wind-tunnel building is endangered.

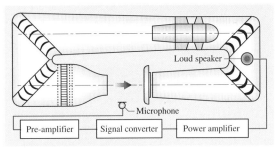

Fig. 16.15 Closed-loop control to attenuate low-frequency oscillations (booming) (after [16.7])

Despite significant effort, satisfactory measures against the pressure increase in front of the collector has not yet been found. Perhaps *active* devices, for example suction at the inlet, will lead to a solution.

The physics of the pumping in an open-jet wind tunnel – called *wind-tunnel buffeting* – has not yet been completely deciphered. There are two potential resonance effects:

- Vortices are shed from the nozzle mouth, as sketched in Fig. 16.14, and move downstream where they impinge on the collector mouth. The pressure pulse caused there travels back to the nozzle and triggers the formation of the next vortex.
- A standing wave is generated within the wind-tunnel duct.

In any case Seifert wings that break up these vortices, widely used in aeronautical wind tunnels, are not suitable for aero-acoustic wind tunnels because of the

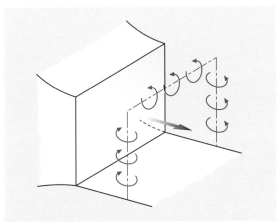

Fig. 16.14 Formation of ring vortices in the shear layer which builds up at the border of the free jet

Fig. 16.16 Array of loud speakers in an open chamber in the wall of the wind-tunnel tube behind the first corner (photo Audi AG)

high-frequency noise they generate. Therefore, vehicle wind tunnels are equipped with Seifert wings that can be removed during acoustic measurements.

An alternative is described in [16.8]: an active resonance control. Its use in a wind tunnel can be seen in the schematic in Fig. 16.15. It consists of an *out-of-flow* microphone that picks up the pressure fluctuations in the plenum, an array of loudspeakers mounted in a chamber behind the first turning vane open to the tunnel wall, see Fig. 16.16, and a time delay for phase adjustment between the microphone input and loudspeaker output. Low-frequency pulsations are attenuated by 20 dB without introducing any additional noise into the tunnel.

A drawback of the first implementation of this system was that it had to be adjusted to each new test configuration, because the frequency and amplitude of the excited resonance not only depend on the flow speed but also on the model. For this reason an improved, self-adjusting system has been developed [16.7].

Plenum. The plenum surrounding the free jet of an open test section, as sketched in Fig. 16.6, must be wide enough to avoid interference of the jet with the walls of the plenum. According to [16.9] drag is underpredicted if the volume of the plenum is less than a specific minimum value. Its main dimensions should not fall below the values compiled in Table 16.1.

Here A is the frontal area of the test object, b_F is its width, and B_P is the width of the plenum. The height and width of a plenum have to be large enough to keep the crane and traversing gear out of the flow. The room on both sides of the jet must be wide enough to keep test equipment such as an acoustic mirror or microphone array out of flow.

Anechoic Measures. Generally, if no specific precautions have been taken, normal wind tunnels are so loud that they are not suited for aero-acoustic investigations. In particular subjective assessment of noise level and character inside a car, such as speech recognition, is not possible. The noise level of anechoic wind tunnels is 20–40 dB(A) lower than in normal wind tunnels.

The guiding principles for an aero-acoustic wind tunnel are to avoid the generation of all noise, and to attenuate any remaining generated noise. Three essential measures have to be taken to build a low-noise wind tunnel:

- installation of a low-noise fan (low noise has higher priority than high efficiency); this requires three measures:

Table 16.1 Dimensions of the plenum related to the frontal area and width of the vehicle investigated

Cross section of plenum, related to vehicle frontal area A_P/A	> 45
Width of plenum, related to the width of the vehicle B_P/b_F	> 8

- low rpm to keep the tip speed and corresponding tonal noise low;
- a large number of blades to shift the first blade-passing frequency and thus the corresponding harmonics to high frequencies, where noise is easier to attenuate;
- limit broadband noise due to the boundary layer on the blades to the same level as tonal noise.
- fan noise has to be trapped in front of and behind the fan; the best suited elements for this task are turning vanes, designed as a *scene silencer*;
- the walls of the plenum have to be covered with broadband noise absorbers.

Technical details are given in Sect. 16.1.7.

Control of Air Temperature. In wind tunnels not dedicated to thermal tests the air temperature must be controlled only insofar as plasticine models do not loose their consistency. The air temperature should be kept below 30 °C. Control of the air temperature should not be performed by an in-stream cooler because of its additional power requirement – and thus additional noise. Air exchange or cooling of the air in a bypass are proper approaches.

This is not the case in thermal wind tunnels where air temperature is the key parameter. Uniformity across a test section should be ±1 K. Because of the cross-flow nature of the tunnel's heat exchanger this specification is not easy to meet. The heat exchanger must be operated under *saturated* conditions, i.e., the volume flux of coolant has to be very high. Furthermore, not only steady-state operation has to be considered when the heat exchanger is laid out. Tunnel time is rare and expensive; therefore the performance of the cooler should permit the desired temperature to be reached quickly. To keep the pressure loss of the air flow through the cooler low the cooler must be placed in a location where the cross section of the tunnel duct is large. The end of the large diffuser following the fan appears to be more suitable than inside the settling chamber. Disturbances in the air flow caused by the cooler (wakes from its tubes and struts) then have a longer flow path to level out.

16.1.5 Equipment for Automotive Wind Tunnels

Depending on the test spectrum for which the facility is designed it must be possible to reproduce or at least simulate a variety of physical processes or effects. These are compiled in Table 16.2 and discussed in the following text, along with their production methods.

Ground Floor

When the motion of a vehicle on the road or track is reversed in a wind tunnel, as indicated in Fig. 16.3, the flow between the vehicle's underside and the ground requires special attention. On the road, as drawn in Fig. 16.17a, a channel flow develops under the vehicle, with a no-slip condition at both walls. The upper wall, the underside of the vehicle, is moving, and the lower wall, the road or track, is at rest. When the flow reversal is correctly reproduced, the floor of the test section has to move, as sketched in Fig. 16.17c and, at the same time, the wheels of the vehicle must rotate.

Because of the high effort necessary to fulfill this condition – details will be discussed below – in many vehicle wind tunnels the relative motion between the vehicle and ground is not reproduced, and consequently the resulting velocity profile underneath the vehicle (Fig. 16.17b) differs from reality. However, measurements carried out with standard passenger cars have shown that the differences between wind tunnel and road are confined to a very thin layer immediately above the ground [16.10]. For this reason, for many tests one can refrain from reproducing the relative motion between the vehicle and the ground.

However, technically it is feasible to reproduce the relative motion, and the same holds for the rotation of

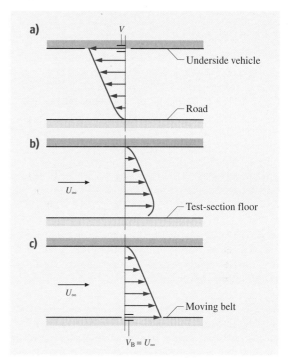

Fig. 16.17a–c Flow underneath a car; (**a**) on the road, (**b**) in a wind tunnel with a rigid (non-moving) floor, (**c**) in a wind tunnel with a moving floor

the wheels. The proper means for doing this is a moving belt that spans the entire width of the test section on which the wheels are rolling [16.11]. However, for the time being, belts able to carry the full weight of a vehicle are not available. Therefore this technique is used in special applications only, for instance for monoposto race cars. There the wheels are mechanically separated from the vehicle and held by slender struts (with an internal balance) from outside. Suspension of the (heavy) vehicle's body is performed by a sting from above, or sometimes from the rear; because of interference effects between the sting and vehicle both are debatable solutions.

To avoid this effort simpler simulation techniques are used; commonly applied methods are compiled in Fig. 16.18:

- a simple ground board, sometimes called a *rigid ground floor*;
- cut-off of the oncoming boundary layer by a *scoop* (left part Fig. 16.18a);
- boundary-layer removal by *basic* or *distributed* suction (Fig. 16.18c,d);

Table 16.2 Physical processes or effects and measures to reproduce or simulate them in a wind tunnel

Physical effect	Measure to perform it
Relative motion to the road	Blowing, suction, moving ground
Turbulence and boundary layer	Mesh, swiveling flaps
Tractive force	Chassis dynamometer
Sun radiation	Bulbs with spectrum of sun
Heat flux	External boilers for air, oil, and water
Humidity	External steam generator
Rain	Water spray system
Dirt	Water (plain or colored)
Geometry	Traversing gear

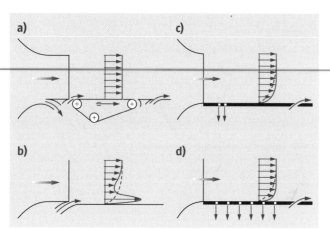

Fig. 16.18a–d The major possibilities to simulate the relative motion underneath a car; (**a**) moving belt with scoop, (**b**) *basic suction* either through a sheet of porous material or through slot; (**c**) tangential blowing, (**d**) distributed suction

- filling of the velocity deficit by tangential blowing (Fig. 16.18b);
- narrow belt as above, supplemented by short belts in front of the frontwheels (*T-belt* [16.12]).

Single measures may be combined, as can be seen in Fig. 16.18a, where a scoop and tangential blowing support the moving floor to further reduce the thickness of the oncoming boundary layer. All the aforementioned techniques can be combined with wheel rotation. The pros and cons of these techniques are discussed in [16.13].

Despite many investigations, to date no fully satisfactory solution for the simulation of the road or track has been found. Furthermore, it has turned out to be impossible to establish a general rule for correcting or compensating for the various negative effects.

Chassis Dynamometer

During thermal tests the power of the engine must be braked by a chassis dynamometer. To avoid overheating of the tires during long tests, the diameter of the rollers should be not smaller than 2 m. In order to cope with front- and rear-wheel-drive vehicles the rollers must be movable along the axis of the wind tunnel. Two dynamometers are needed for four-wheel-drive vehicles. For thermal brake tests the tractive power at high speed must be close to the braking power. However, because abraded break material soils the wind tunnel and may be toxic, braking tests are rarely carried out in a wind tunnel anymore.

Turbulence

Traditionally, as mentioned already in Sect. 16.1.4, vehicle wind tunnels are designed to have an airstream with low *turbulence*. However, vehicles generally operate in an environment with high turbulence, which has two origins: the natural wind, and from the wakes of other vehicles and of trees, buildings, and other obstacles along the roadside. This turbulence has at least two effects that bother for the passengers: it modulates the spectrum of wind noise, and it shakes the car.

To be able to generate reproducible turbulence with respect to spectrum, scale, and its distribution above

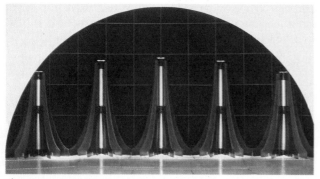

Fig. 16.19 Swiveling flaps to generate a highly turbulent boundary layer; design A. Cogotti; multiple exposures (photo Pininfarina)

Fig. 16.20 Sunlight simulation with a matrix of lamps assembled in a panel that can be tilted; climatic chamber Modine GmbH (photo Media Fotograf)

the ground, the wind tunnel of Pininfarina was equipped with flaps as shown in Fig. 16.19. They are used in a fixed position when the wake of a preceding vehicle is simulated, or in a swiveling motion to generate specific turbulence effects; for further details see [16.14, 15].

Sun Load

To simulate sun load the intensity of radiation must be $1000\,\text{W/m}^2$, the solar value. A homogeneity of $\pm 10\%$ is said to be sufficient. Light sources with a spectrum close to the sun's are commercially available. Two technical solutions are in use. In the older one (Fig. 16.20) the lamps are arranged in a matrix and assembled in a large overhead panel that can be turned to the side to expose a test object to radiation from an oblique angle. The newer version, see Fig. 16.21, make use of single lamps assembled in two frames that can be moved and swiveled. The lamps are equipped with flaps with which the radiation can be quickly switched on and off to simulate, for example, entering or leaving a tunnel.

Rain, Snow, and Dirt

Rain is simulated by water that is sprayed into the oncoming air through nozzles that are mounted on a tube crossing the test section horizontally. To make the water streaks and drop deposition on the vehicle visible a fluorescent liquid is added to the water. With ultraviolet illumination flow patterns such as those shown in Fig. 16.22 become visible.

Fig. 16.22 Water streaks on a vehicle's body made visible with sprayed water doped with an ultraviolet (UV)-sensitive pigment, and UV light (photo BMW AG)

In practice *snow* could be produced using commercially available snow guns. However, they need a long distance to freeze the spray. Therefore in wind tunnels an array of snow nozzles is used, which is positioned in front of the vehicle, similar to the rain nozzles. A mixture of cold water and high-pressure air (5–10 bar) is ejected through a nozzle (Fig. 16.23) into the ambient air. Depending on the water content, pressure, and temperature the quality of the snow can be varied between *wet* (adhesive, particle size $45\,\mu\text{m}$, density $450\,\text{kg/m}^3$) and *dry* (particle size $20\,\mu\text{m}$, density $200\,\text{kg/m}^3$).

Considering the sensibility of the test equipment and the high cost of cleaning the simulation of *dirt* has to be performed with care. Generally dirt is simulated by water, either blended with a fluorescent liquid or with whiting. Realistic dirt tests have to be performed on a specially prepared test track.

Fig. 16.21 Sunlight simulation with distributed movable spot lights; climatic wind tunnel Behr GmbH (photo Behr GmbH)

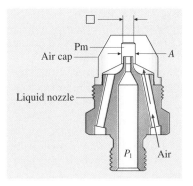

Fig. 16.23 Cross section through a spray nozzle of a snow-*gun* (drawing after RTA Rail Tec Arsenal Fahrzeugversuchsanlage GmbH)

Traversing Gear

Every aero-acoustic wind tunnel should be equipped with a traversing gear covering the entire test section. This device should be committed together with the construction of the wind tunnel because it is urgently needed during the calibration tests. In thermo tunnels, where such a device is generally not needed, these calibration tests are performed with a temporary set up.

16.1.6 Limits of Simulation

General Remarks

The effectiveness of all aerodynamic measurements on a vehicle have to be proved on the road or track. In a wind tunnel a real cruise can only be simulated, never reproduced exactly. The limitations of simulation fall into two categories:

- violation of the law of similarity
- deficits of the wind-tunnel technique.

In the next two sections both will be only briefly discussed; for more details see [16.1].

Experiments with Small-Scale Models

When considering whether a wind-tunnel test should be carried out at full or small scale the following questions have to be answered:

- How can we assure that the results from small-scale tests will agree with those at full scale?
- What wind tunnels are available?
- What is the time advantage compared to a full-scale test?
- What is the cost advantage of small-scale testing?

The first is the key question. If no satisfactory agreement with full-scale testing can be expected, the remaining questions are meaningless. Another question is: what, in this context, is *satisfactory*? Is it sufficient to assess shape *modifications* with regard to drag (or other forces and moments)? Or will the data be used to *compare* two or more design variants?

Coincidence of results from small- and full-scale testing is achieved if exact similarity is present with respect to the:

- geometry of the model. However, regardless of the skill of the model maker, not every detail can be reproduced in small scale. Therefore an *exact* similarity is not possible.

- kinematics of the flow. This will be achieved, when the Reynolds number is the same in both cases, i.e., at full (1) and small (2) scale:

$$Re_1 = Re_2,$$
$$\frac{U_1 \cdot l_1}{\nu_1} = \frac{U_2 \cdot l_2}{\nu_2},$$

where U is the velocity of the undisturbed oncoming flow, and l is a typical geometrical dimension, generally the length of the vehicle or, as discussed below, a specific radius r.

Since generally small- and full-scale tests are carried out in air the above formula results in the following rule: to attain kinematic similarity the wind speed in small-scale tests has to be increased by the same factor as the dimensions of the model are scaled down.

For two reasons this rule is not easy to observe exactly:

- the maximum wind speed in the model wind tunnel may not be high enough;
- if the oncoming speed exceeds a critical value, Mach-number effects become noticeable.

The latter is critical in vehicle aerodynamics because bluff bodies such as vehicles are sensitive to compressibility effects. While generally $Ma = 0.3$ is said to be the limit up to which flow can be treated as incompressible, for vehicles this value is $Ma = 0.2$. For further details see [16.16].

Generally, when testing road vehicles, the maximum Reynolds number in small-scale tests is two to five times smaller than at the full scale. In order to assess the effect of the Reynolds number, so-called Reynolds series are run, as demonstrated in Fig. 16.24. There, the drag coefficient c_D is plotted versus Reynolds number Re. Two different vehicles were investigated: in the upper part of the diagram a constant c_D was achieved at $Re = 5 \times 10^6$, which compares pretty well with full-scale data. The vehicle in the lower part of the diagram was completely different; c_D = constant was not achieved at all, neither at small nor full scale. The latter example may serve as a warning: the effect of the Reynolds number on the drag coefficient can be significant and is hard to predict.

There is one exception to this: the radius of the leading edges of a rectangular block, like a bus, or the cabin of a truck. For such a configuration a so-called *optimum radius* exists; the definition of *optimum* is explained with the aid of Fig. 16.25: if the leading edges of a block are rounded off in increments Δr, starting with sharp edges ($r = 0$), c_D will decrease, at first almost linearly with a steep gradient, and then more moderately with bigger radii, finally reaching a constant value. The *optimum radius* is the smallest radius r_{opt}, where $c_D = \text{const}$. As seen in the figure, its value depends on the Reynolds number: the higher the Reynolds number, the smaller the optimum radius [16.1]. From many experiments such as that displayed in Fig. 16.25a a general relation has been derived:

$$\text{Re}_{(r/b)\text{opt}} = \frac{U_\infty \cdot r_{\text{opt}}}{\nu} = 1.3 \times 10^5 \, .$$

This number is close to the critical Reynolds-number of a circular cylinder, where the Reynolds-number is formed with the diameter.

The sensitivity to Reynolds number of other details typically found in the geometry of vehicle bodies, such as tapering of the rear (*boat tailing*), and rounding of the C-pillars, have not yet been investigated in depth.

The attitude of vehicle aerodynamicists to working with small-scale models has undergone several changes. In early vehicle aerodynamics they had no choice because the only wind tunnels available were those from aeronautics. Initially they were pretty small; a scale of 1 : 10 was used for cars, and even smaller for trains. Later, when large wind tunnels became available, full-scale investigations were preferred – if tests were made at all, because vehicle development was interrupted for some years.

After 1945, this sequence repeated itself. Aerodynamic development started with models at a scale of 1 : 5 and 1 : 4. Only the large wind tunnel of Forschungsinsti-

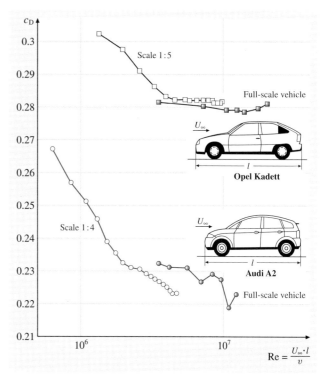

Fig. 16.24 Drag coefficient c_D versus Reynolds number Re for two different passenger cars, comparison between small- full-scale vehicles. *Top*: Opel Kadett, *bottom*: Audi A2 (data from Adam Opel GmbH and Audi AG)

tut für Kraftfahrwesen und Fahrzeugmotoren Stuttgart (FKFS, now belonging to Daimler Chrysler) allowed for full-scale models and cars, and was used mainly for final tests. When the vehicle manufacturers began to erect their own wind tunnels they built them for full-scale vehicles, and small-scale testing was pushed into the background again, more so as designers prefer full-scale models to assess their appearance. If they are fully equipped (e.g., with a *real* underbody), design models can be (and often are) used for wind-tunnel tests.

Some vehicle manufacturers made use of small-scale models in the early phase of development. However, these were displaced again, this time by numerical models. In some cases a scale of 1 : 2.5 was preferred. Models this large can be tested at the full-scale Reynolds number (in a full-scale wind tunnel) without conflicting with the Mach-number.

A full-scale Reynolds number can also be achieved with small models in a high-pressure (100 bar) or cryogenic wind tunnel (100 K) [16.16].

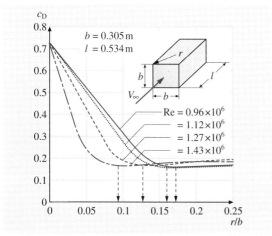

Fig. 16.25 Effect of Reynolds number on the value of the *optimum* front radius of a prismatic block with rectangular cross section (data from [16.16])

Wind-Tunnel Corrections

When driven on the road or track a vehicle moves through a space that is unlimited in all directions (free air), with the ground below (the road or track) being the only limiting surface. The dimensions of a wind tunnel's airstream, however, are finite. Its cross section and its usable length are limited, and the static pressure along its axis is not constant. The proximity of the airstream's boundaries to the test object modifies the flow around it compared to that in an unlimited space. The angularity of the flow, the distribution of velocity and pressure around a vehicle, and at least in principle, the shear stresses on its surface, are all modified and so, consequently, are the forces and moments acting on the vehicle. The smaller a wind tunnel relative to a vehicle, the larger these discrepancies are.

There are two possibilities to cope with the systematic errors resulting from these effects:

- corrections (some prefer the term *adjustments*), which are performed a posteriori, i.e., *after* the measurement. This is standard practice in large wind tunnels;
- correcting the setting of the wind speed *during* the test. This is practised in small (mainly thermal) wind tunnels.

The basic premise of wind-tunnel corrections is that the shape of the flow is assumed to remain unchanged by the finite size of the airstream's dimensions. In other words, the shape of the streamlines, the shape of the pressure distribution, and the locations and kinds of separation are all the same as in free air. The only effect of the finite dimensions is that the velocity of the oncoming flow is altered. By modeling the various perturbations resulting from the finite dimensions of the airstream this speed change is computed.

The development of these so-called *classical* corrections was performed in three steps. In the first, the perturbations of the flow field caused by the boundaries were identified and analyzed. In the second step, each perturbation was modeled and computed. Finally, the effects were superimposed, which means that it was assumed that the perturbations do not interact with each other (a linear approach).

Originally, this kind of corrections was carried over from aeronautics, where they had been developed for configurations with attached flow, i.e., slender bodies such as wings, nacelles and fuselages. Furthermore, they were limited to small blockage ratios, i.e., small models in comparatively large wind tunnels [16.2].

However, for vehicles both conditions are not fulfilled because:

- generally vehicles are bluff bodies, and their flow is characterized by separation and reattachment;
- commonly, blockage is high; i.e., large models are investigated in comparatively small wind tunnels.

Therefore it is questionable whether it is justified to apply these classical corrections for vehicles. However, because they have been developed so far that single perturbation effects can be discriminated [16.4], they are still applied – if applied at all.

The axial pressure gradient, dp/dx, within the airstream is taken into account differently; it is assumed that the axial pressure gradient with a model in the test section is the same as that in the empty tunnel. Its effect is horizontal buoyancy and will be computed as such, and the drag force will be corrected by adding the force resulting therefrom.

Already in Fig. 16.10 it was indicated that the effect of a test section's boundary on the flow pattern around a vehicle is opposite for open and closed test sections. In the former the streamlines around a car are wider apart from each other than in free air, while in the latter they are squeezed together. If this deformation of streamlines due to the boundaries of the jet were the only effect, the corrections for both kinds of test section could be treated simultaneously. However, a number of other perturbations exist, and these depend on the type of test section. Therefore it is appropriate to treat the corrections for each type separately.

The overall configuration of a car in a wind tunnel is asymmetric with respect to the tunnel floor. This floor (the ground plane) simulates the road and therefore should be interpreted as part of the test set up, not of the tunnel. When the model and ground plane are reflected about the ground plane, as shown in Fig. 16.26, the resultant configuration can be viewed as a test chamber

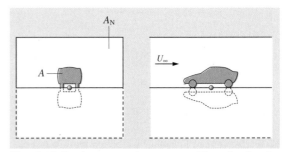

Fig. 16.26 *Duplex* test section, generated by reflecting the real test section in the floor

with double the cross section with a centrally mounted test object made up of the model itself and its mirror image, including the road or track. The advantage of this so-called *duplex test section with duplex test object* is easier formulation of the numerical models by which the perturbation effects are quantified, even more so if the model is replaced by a body of revolution and the cross section of the wind tunnel is assumed to be circular.

Closed test section: in a closed test section three kinds of perturbation are present:

- solid blockage
- wake blockage
- horizontal buoyancy.

The first two have the same effect as if the speed of the measured undisturbed oncoming flow, U_m, is superimposed by an additional speed u. When the nondimensional coefficients, such as the drag coefficient c_D, are computed, the corrected speed, namely U_corr, has to be used:

$$c_\mathrm{D} = \frac{D}{\frac{\rho}{2} U_\mathrm{corr}^2 A},$$
$$U_\mathrm{corr} = U_\mathrm{m} + u = U_\mathrm{m} + \varepsilon U_\mathrm{m} = U_\mathrm{m}(1+\varepsilon),$$
$$q_\mathrm{corr} = q_\mathrm{m}(1+\varepsilon)^2,$$

where ε is the so-called blockage coefficient, which is made up from the solid, ε_SB, and wake blockage, ε_WB:

$$\varepsilon = \varepsilon_\mathrm{SB} + \varepsilon_\mathrm{WB}.$$

Horizontal buoyancy adds a value $\Delta c_\mathrm{D\,HB}$ to the drag coefficient c_D. How it is affected by the (near) wake of the body is still a matter of discussion.

Finally, the corrected drag coefficient $c_\mathrm{D\,corr}$ is computed as follows:

$$c_\mathrm{D\,corr} = \frac{c_\mathrm{D\,m} + \Delta c_\mathrm{D}}{q_\mathrm{corr}/q_\mathrm{m}} = \frac{c_\mathrm{D\,m} + \Delta c_\mathrm{D\,HB}}{(1+\varepsilon)^2},$$

where the index m is the label for the measured values.

So far this calculation is strictly formal. To evaluate this equation, the blockage coefficient ε must be determined from the essential geometric parameters of the test section and the model, such as the blockage ratio φ and the length of the model. To date this has preferably been done using analytical models of the test section and model. Because only the far field is affected, the model is replaced by a body of revolution with the same thickness ratio.

Open test section: for a long time, vehicle aerodynamicists held the opinion that a wind tunnel with a free jet does not need corrections at all. And indeed, for larger wind tunnels the *classical* corrections are so small that they can be neglected. Furthermore, it was common understanding that the axial pressure gradient in the airstream of the empty test section is zero. This is because, according to textbooks, a free jet has the same pressure as the plenum into which it is ejected. However, comparative measurements disclosed that this is not true, and Fig. 16.11 is a striking result.

Following [16.4], compared to the closed test section several additional perturbations have to be taken into consideration. These are (Fig. 16.27)

1. jet expansion ε_JE
2. jet deflection ε_JD
3. nozzle blockage ε_N
4. collector blockage ε_C.

As before, the correction for the dynamic pressure of the undisturbed oncoming flow is the result of an addition:

$$\frac{q_\mathrm{corr}}{q_\mathrm{m}} = (1 + \varepsilon_\mathrm{JE} + \varepsilon_\mathrm{JD} + \varepsilon_\mathrm{N} + \varepsilon_\mathrm{C})^2.$$

The physical effects 1 to 4 are explained and modeled in [16.4].

The pressure gradient $\mathrm{d}p/\mathrm{d}x$ (horizontal buoyancy) is treated as for the closed test section. However, its value, and consequently its perturbation effect, is much

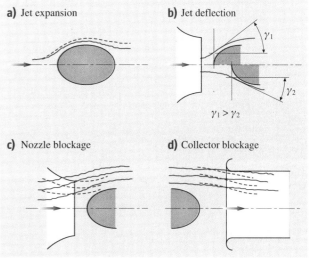

Fig. 16.27 Four major flow disturbances of the flow past a model in an open test section (after [16.4])

higher, and in some wind tunnels it may override all others.

The state of the art in computational fluid dynamics (CFD) makes it possible to surmount the restrictions of the *classical* corrections: low blockage, slender bodies, and questionable assumptions regarding the effect of the near wake. Making use of the panel method, the test object and test section, including the nozzle and the diffuser, can be modeled as shown in Fig. 16.28.

This method of correction is based on the assumption that the differences between the flow fields in free air, $\varphi = 0$, and with finite blockage, $\varphi \neq 0$, is the same in a real flow as in an inviscid flow. The wake can be modeled according to one of the methods described in [16.16]. As long as only *global* corrections are required, as described above, the model of the test object including its wake can be described using a rather coarse panel structure. Computation time will be extremely short; online and real-time computation of the corrections is possible. Beyond this, correction of the flow in specific locations can be carried out, for instance for the flow angle at the rear wing of a racimg car, which is affected by the final dimensions of the airstream.

Such models have been developed for the closed test section [16.17], but not yet for the open. There the nozzle and the collector have to be included in the model, and the edge of the free jet must be modeled by imaging the model plus its near wake.

Correction of the yawing angle is generally not performed in vehicle aerodynamics. If necessary, the method can be taken from aeronautics, especially from slender wings. The yaw angle β of a vehicle corresponds to the angle of attack α of a wing.

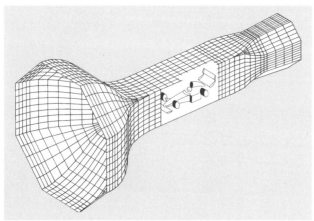

Fig. 16.28 Panel model for computing wind tunnel corrections: nozzle, test section plus model, first diffuser

Thermal tests: when tests are performed in a thermal wind tunnel, a posteriori correction is not possible. The effective wind speed must already *during the test* be the same as the circumferential speed of the rollers of the dynamometer. This is ensured by an *implicit correction*. Several suggestions have been made for how this can be carried out [16.1, 18, 19]:

- The effective wind speed is that which follows from the pressure at the stagnation point of the test object, which is the Pitot pressure p_0 of the oncoming flow

$$U_{\infty\,\text{eff}} = \sqrt{\frac{2(p_0 - p_\infty)}{\rho}}.$$

Of course, the Pitot pressure is the same across the oncoming flow. However, the advice to measure it at the stagnation point of the test object is a clear and reproducible one, more so as it must be ensured that the Pitot tube is approached at a zero yaw angle.

- The vertical velocity profile in front of the test car is measured on the road (or in a large wind tunnel). Comparison with the same profile measured in the thermal wind tunnel and a linear regression results in a nozzle factor k:

$$k = \frac{\overline{U}_{\infty\,\text{set}}}{U_\infty}.$$

This method is preferred for experiments with cooling systems. From the measured velocity profile this segment is taken for the computation of k in front of the radiator inlet.

- As an alternative the cooling flow duct itself can be used as a flow meter and be calibrated on the road.
- When radiators of buses are tested, their air intake is on one side of the vehicle at the rear. The wind speed is set equal to the road speed (and thus the circumferential speed of the dynamometer) at the sides of the vehicle.
- Comparative pressure distributions are measured in the longitudinal center section of the hood and in the scuttle on the road and in the thermal tunnel, and a blockage factor k is determined via linear regression. Generally the result is different from that determined with the velocity profile in front of the vehicle. This kind of adjustment is applied when climatic tests are performed.

- When climatic tests are performed with (long) train coaches the wind speed on the sides of the coaches has to be set as for the radiator test with buses.

16.1.7 Typical Vehicle Wind Tunnels

Classification of Wind Tunnels for Vehicle Aerodynamics

In the motor industry and research organizations a large variety of wind tunnels are in operation. Although their objects, namely cars and light-duty trucks, are very similar, the variety of the design of wind tunnels is surprising. In order to provide a survey they will be classified into four categories:

- large full scale
- small full scale
- thermal
- small scale.

Each will be characterized by one example. Additionally, some smart design details will be emphasized. For further details see [16.1], where the major full-scale wind tunnels worldwide are compiled, and some are described. In Europe, vehicle aerodynamicists prefer an open test section, and this type is gradually gaining preference in the US. All but one of the wind tunnels discussed have an open test section.

Large Full-Scale Aerodynamic and Aero-Acoustic Wind Tunnels

A typical large full-scale vehicle wind tunnels is that of the Institut für Verbrennungsmotoren und Kraftfahrwesen (IVK), at the University of Stuttgart. It was erected in 1988, when the wind tunnel built by Kamm in 1939 had to be given to Daimler-Benz AG. The air circuit (*Göttingen* type) together with is main dimensions are compiled in Fig. 16.29. The cross section of the jet is 22.5 m².

Originally, this wind tunnel was built as a purely aerodynamic facility, and as such it came into service in 1988; the maximum wind speed was 270 km/h ($\approx$ 168 mph). However, provision was made for later improvements to expand the testing possibilities of the tunnel to include wind-noise investigations [16.20]. This upgrade was performed in 1993 with a novel silencing concept. The air path (Fig. 16.29) shows how the fan noise is *locked in*. Both cross legs, (1) and (2), together with their turning vanes act as U-bend silencers. The plenum (3) is covered with noise-absorbing elements laid out for two frequency ranges. In the low range, 80–200 Hz, the fan noise is damped by the novel membrane absorber sketched in Fig. 16.30a. The middle and high frequencies are damped by porous polyester foam (Fig. 16.30b). The A-weighted self-noise level (SPL) measured *out of flow* and plotted versus tunnel speed (Fig. 16.31) is a proof of the effectiveness of this concept. The IVK aero-acoustic wind tunnel is now one of the quietest full-scale wind tunnels ever built (Fig. 16.31). The maximum wind speed of the tunnel was only reduced by 7 km/h by the damping measures.

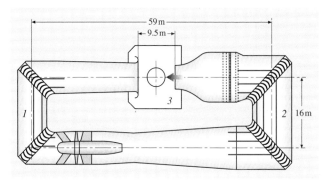

Fig. 16.29 Aero-acoustic wind tunnel at IVK (after [16.20]). Cross section $A_N = 22.5$ m², maximum wind speed $V_{max} = 257$ km/h, SPL at 140 km/h = 69 dB(A), *five-belt system*, rotating wheels

Another outstanding property of the IVK wind tunnel is its ground simulation, called the *five-belt system*. It is made up of a *narrow* belt (*narrow* compared to a belt that spans the entire width of the test section) moving between the wheel tracks and four *mini belts*, one under each wheel. Together with a system of tangential blowing and basic suction the boundary layer underneath the vehicle can be almost completely removed, and the wheels can be rotated, both up to the full wind-tunnel speed. J. Potthoff, the designer of this system and a similar one for the small-scale wind tunnel of IVK, has presented a description [16.21] and thoughts for future development.

Small Full-Scale Aero-Acoustic Wind Tunnels

The first small full scale wind tunnel ($A_N = 11.75$ m²) built for vehicle aerodynamics was designed on behalf of the Italian designer Pininfarina by *Morelli* (1971) [16.3], followed by the aeroacoustic wind tunnel ($A_N = 10$ m²) designed on behalf of BMW by L. J. Janssen (1988), both with an air path according to Eiffel.

In 1999 the small full-scale aero acoustic wind tunnel at Audi AG became operational [16.22]. Its air path is reproduced in Fig. 16.32 ($A_N = 11$ m²). Based on the correction method developed by *Mercker* and *Wiedemann* [16.4] the dimensions of the test section were fixed as to minimize corrections. The turning vanes in

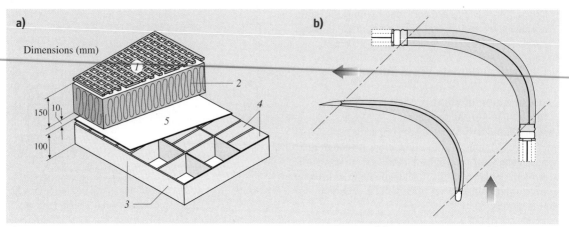

Fig. 16.30a,b Wind noise attenuation in the aero-acoustic wind tunnel at FKFS; (**a**) broadband absorber, (**b**) corner vanes covered with closed damping foam [16.20]

the 2nd and 3rd corner are designed as scene silencers to lock in the noise of the fan. The fan itself was designed to low noise generation [16.23]. As can be recognized from Fig. 16.31, this wind tunnel is the most silent at all.

Over the years the above mentioned wind tunnel of Pininfarina was converted into an aeroacoustic one, and upgraded. Due to design limitations it had not been possible to lock in the fan noise. Therefore a special fan was designed according to rules mentioned in Sect. 16.1.4. Noise level is now not as low as in newer aero acoustic tunnels; however, it is low enough to perform aero acoustic investigations of all kind. To increase top wind speed from 150 km/h to 260 km/h the semi-circle arch in the open return path was *filled* with 13 axial blowers (Fig. 16.33).

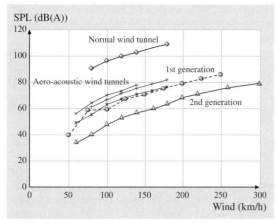

Fig. 16.31 Measured out-of-flow SPL of several open jet wind tunnels (source Audi AG)

Thermal Wind Tunnels

For each new vehicle two thermo systems have to be developed and tested:

- engine cooling, radiator and several heat exchangers
- heating, ventilation, and air conditioning (HVAC).

For the former a hot climate is needed, for the latter hot and cold, including further climate parameters such as humidity, sunlight, rain, and snow. Accordingly, two types of wind tunnels have been built:

- purely hot tunnels
- hot and cold tunnels.

With regard to their size (cross section A_N), three kinds of wind tunnels can be distinguished:

- $10-12 \, m^2$ for full-size trucks and buses;
- $6 \, m^2$ for cars;
- $\leq 4 \, m^2$ for radiator tests with cars.

Generally the air path for all these tunnels is a closed return (*Göttingen*) type, as shown in Fig. 16.34, and the test section is open. Comparative tests confirmed the aforementioned finding that ground simulation can be simple: a rigid floor is sufficient.

Representative of the first category are the two Fiat thermo tunnels [16.24]. They are part of a wind-tunnel center including a $31 \, m^2$ normal wind tunnel. One of the two $12 \, m^2$ tunnels is laid out for high temperatures and sunlight simulation (hot tunnel), the other for cold environment (cold tunnel). Equipped with dynamometers, this pair of wind tunnels is well suited for all kinds of thermal tests.

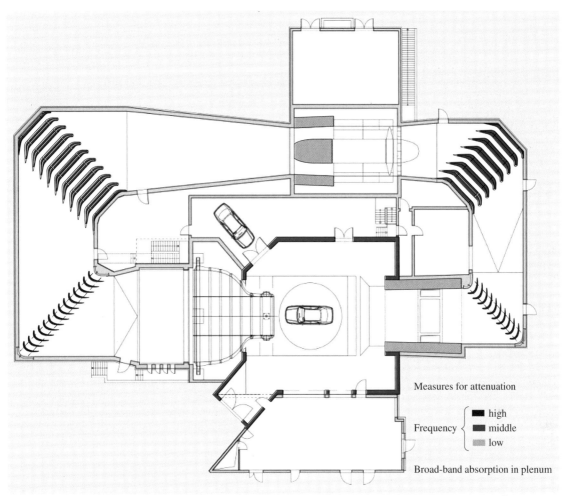

Fig. 16.32 Aero-acoustic wind tunnel of Audi AG; design *Wiedemann* [16.22]; cross section $A_N = 11\,\mathrm{m}^2$, maximum wind speed $V_{\max} = 300\,\mathrm{km/h}$, OSPL at $140\,\mathrm{km/h} = 57\,\mathrm{dB}(A)$, moving floor, rotating wheels.

For thermal tests with cars a cross section of $12\,\mathrm{m}^2$ is more than sufficient. From the investigations mentioned in [16.25, 26] and from experience $A_N = 6\,\mathrm{m}^2$ has proven to be sufficient. However, in order also to be able to test full-size trucks and buses, several of these thermal tunnels are equipped with nozzles that allow for different cross sections: $6\,\mathrm{m}^2$ for cars and $10–12\,\mathrm{m}^2$ for large vehicles. The loss of top speed with the greater nozzle area can be accepted because of the speed limit of these larger vehicles. Two types of nozzles have been developed for this purpose:

- nozzles with a flexible upper wall (Fig. 16.35)
- tandem nozzles (Fig. 16.36).

A particular vehicle type comes with a variety of cooling systems (engine, oil, intercooler, condenser, exhaust gas, etc.). To develop these different systems requires a large effort. However, many of the related tests can be performed at ambient temperature. This work is performed in wind tunnels or with blowers with a nozzle cross section of ca. $2\,\mathrm{m}^2$. Only fine-tuning is done in a hot tunnel.

For tests with *rolling stock* for railways and street cars a large wind tunnel is needed. Coaches are about 30 m long. The only way to simulate the flow on all four sides of the coach is in a *closed* test section. In its design a compromise has to be made between reproducing the axially constant flow velocity and a zero axial pressure

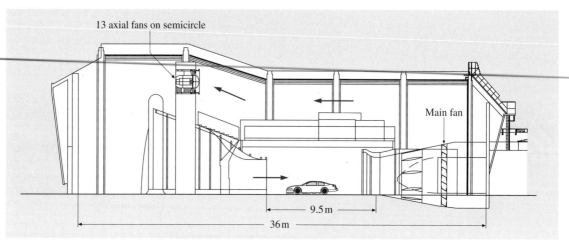

Fig. 16.33 Wind tunnel of Pininfarina, design *Morelli* [16.3]; upgrade *Cogotti* [16.27]; cross section $A_N = 11.75 \text{ m}^2$, maximum wind speed $V_{max} = 260 \text{ km/h}$, OSPL at 140 km/h = 77 dB(A), moving floor, rotating wheels

gradient. The air-conditioning units' air inlet and the outlet of the used air must be at the same pressure as on the track, where it is atmospheric.

Two such wind tunnels exist at Rail Tec Arsenal (RTA), Vienna. A view into a test section is reproduced with Fig. 16.37. The cross section of both test sections is $A_N = 16 \text{ m}^2$. The length of the large tunnel is 100 m, enough to host a train of three coaches and a complete street-car train. The length of the smaller tunnel (30 m) is sufficient for articulated buses (maximum length 18.75 m) and full-size trucks. One side wall of the test section is equipped with lamps at an angle of 30° (standard for railway testing). Intensity can be varied from 250 to 1000 W/m². Both tunnels are equipped with dynamometers (for data see the figure caption). The maximum wind speed is 300 km/h for the large and 120 km/h for the smaller tunnel.

Small-Scale Wind Tunnels for Automobiles

There are many wind tunnels that are well suited for testing small-scale vehicle models. Several are in academia, where they are used for this purpose only occasionally. Others, which were specifically designed for vehicle aerodynamics, are in research institutes. Some existing small-scale wind tunnels have served as pilot

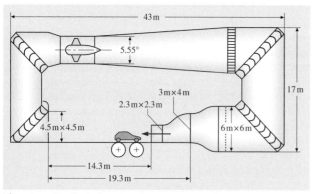

Fig. 16.34 Typical climatic wind tunnel (data from Modine GmbH) cross section $A_N = 5.3/12 \text{ m}^2$, maximum wind speed $V_{max} = 250$ or 130 km/h, four-roller dynamometer

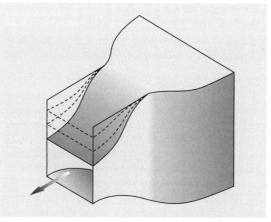

Fig. 16.35 Adjustable wind tunnel nozzle of the climatic wind tunnel of Behr GmbH; cross section $A_N = 2.0/8.0/10.0 \text{ m}^2$, maximum wind speed $V_{max} = 139/100/80 \text{ km/h}$ four roller dynamometer, temperature $-30 < T < +50\,°\text{C}$; sunlight 1000 W/m²

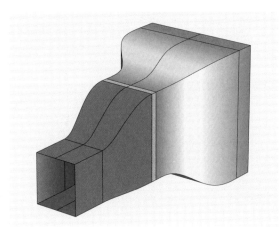

Fig. 16.36 *Tandem nozzle* of the climatic wind tunnel (graphic after Modine GmbH)

Fig. 16.37 Test section of one of the two large climatic wind tunnels of RTA Rail Tec Arsenal, Wien (photo RTA); cross section $A_N = 16\,\text{m}^2$, length 100/30 m, maximum wind speed V_{max} 300/120 km/h, dynamometer 850/250 km/h temperature $-50 < T < +60\,°\text{C}$; sunlight 1000 W/m^2

tunnels during the design of a full-scale tunnel. Because of geometrical similarity, simulation deficits such as blockage are the same as in the full-scale tunnel. Differences occur only in the boundary layer of the ground floor. With such model tunnels it is reliable taking over results developed at small scale to the full scale.

Reduced-scale wind tunnels *can* be built at very low cost; the quarter-scale tunnel at the Motor Industry Research Association (MIRA), shown in Fig. 16.38, is a striking example [16.28, 29]. Also the operating costs of such facilities are low, and flexibility is high. Both arguments are reasons why small-scale testing is coming into fashion again.

Small-Scale Wind Tunnels for Trains

Generally, aerodynamic investigations with trains are carried out in normal wind tunnels, which are otherwise used in aviation. A scale of 1 : 10 is common. Because the test sections of these tunnels are relatively short, only short trains can be investigated, the head (locomotive) and two to three coaches. Tests at higher Reynolds numbers are carried out in a cryogenic tunnel.

A wind tunnel specially laid out for testing trains is shown in Fig. 16.39. It is the *soufflerie à veine longue* at Saint-Cyr-L'École, France [16.30]. The model scale is 1 : 20, and for special tests (pressure measurements) 1 : 10. The height of the ceiling of the closed test section

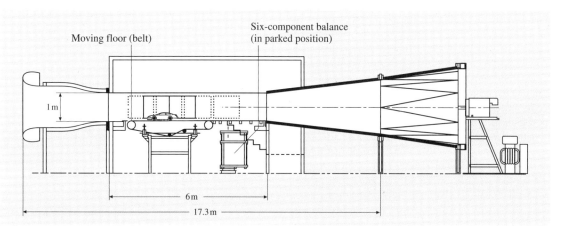

Fig. 16.38 Small scale wind tunnel of MIRA, Nuneaton, UK (graphic MIRA); cross section $A_N = 2.12\,\text{m}^2$, maximum wind speed V_{max} 90 mph, moving floor

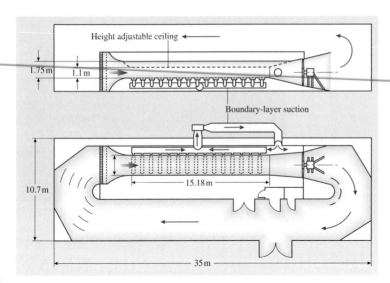

Fig. 16.39 La souflerie à veine longue in Saint-Cyr-L'École, France [16.30]

can be reduced from 1.75 m to 1.1 m to increase the wind speed from 40 m/s to 60 m/s. The axial pressure loss is compensated by the angle of the roof. Ground simulation was improved by boundary-layer suction. Later the first 6 m were equipped with a moving belt with rails and sleepers.

Tests with pantographs are performed with their originals in full scale, mainly in aero-acoustic wind tunnels from the automobile industry.

16.1.8 Tests with Moving Models and Full-Scale Vehicles

Small-Scale Models

Model track: as discussed, the boundary layer developing on a wind-tunnel floor impairs the simulation of the flow underneath a vehicle, more so with decreasing ground clearance. Other tests, for example the meeting of two trains or a train entering a tunnel, cannot be simulated under dynamic conditions in a wind tunnel at all. Model tracks for small-scale train models are suitable for such tests [16.31].

Catapult: unsteady effects can also be studied by *shooting* the model – train or car – crosswise through the jet of a wind tunnel [16.32] (Fig. 16.40). In order to keep the sections of the track needed to accelerate and break the model short, the balance inside the model must be able to bear a high acceleration. In any case, the test signal will be superposed by large amounts of noise from the spring–mass system formed by the model and balance, and must be filtered.

Towing tank: another possibility to avoid an impairing ground-floor boundary layer or to study transient effects is offered by a towing tank, in which a fully immersed model is towed through a water basin [16.33]. The speed must only be 1/15 of the speed in a wind tunnel. Meeting of trains, and entering a crosswind gust have been studied [16.34].

Full-Scale Cars and Trains

Aerodynamic drag: the aerodynamic drag coefficient of a vehicle can also be measured in a coast-down test [16.35] (this recent reference may well serve as an introduction to the subject of fuel economy, and

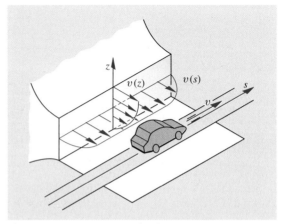

Fig. 16.40 Cross-wind test with small-scale model in a wind tunnel, schematic

in specific, to coast-down testing, the related literature, and the official regulations). It must be conducted on a long, straight, level test track. Several methods have been developed for the consideration of the influence of natural wind, although it is better yet to perform the test with no wind at all. The test vehicle is first accelerated to its top speed and then allowed to coast freely by disengaging the engine. The change of vehicle speed is continuously recorded versus time.

The deceleration is caused by the aerodynamic drag D_A and mechanical resistance D_M according to

$$m(1+f)\frac{dV(t)}{dt} = D_A + D_M ,$$

where m is the vehicle mass (kg), f takes into account the rotating masses, $V(t)$ is the vehicle speed (m/s) as a function of time t, and D_A is the aerodynamic drag (N)

$$D_A = c_D \frac{\rho}{2} V^2 A .$$

D_M is the mechanical resistance (N); it is composed of the tire rolling resistance, and the resistances of the driveline and bearings:

$$D_M = \mu_R m g ,$$

where μ_R is the coefficient of rolling resistance, which is a function of speed $\mu_R = \mu_R(V)$; m is the vehicle mass (during the test, which means including the driver and gasoline) and g is acceleration due to gravity. f is obtained from the equation of motion of the rotating components

$$f = \frac{I_d/r_d^2 + I_0/r_0^2}{m} ,$$

where I_d is the moment of inertia of the rotating components of the driveline including the wheels of the driven axle (Nms^2), I_0 is the same value of the non-driven axle (Nms^2) and r_d and r_0 are the dynamic rolling radii of the tires on the driven and non-driven axles (m).

The main difficulty with coast-down tests consists of accurately measuring the rolling resistance of the wheels. Its measurement on a drum leads to erroneous results. Considerable differences are found (either on the outside or inside surface of the drum) compared to on a level road. Furthermore, an additional effect is caused by the vehicle's suspension geometry (i.e., wheel camber, toe-in).

There are three possibilities for taking rolling resistance into consideration:

- rolling resistance is measured on a moving belt;
- it is measured with a trailer designed for investigating single wheels where all the wheel parameters (load, tire pressure, camber, etc.) of influence can be set [16.36];
- it is measured by towing the test car, which is covered by a shrouding trailer [16.37].

Alternatively, the determination of the mechanical drag D_M can be circumvented:

- when only increments of aerodynamic drag – due to additional parts such as spoilers or panels on the underside – have to be determined;
- when an artificial drag increment (with lift and pitching moment unchanged) can be produced and quantified by wind-tunnel measurements.

Crosswind sensitivity: the crosswind sensitivity of a vehicle is not easy to assess because the driver *must* be included. The traditional method is to subject the vehicle to an artificially generated crosswind gust during a straight-ahead drive on a test track [16.38]. Figure 16.41 shows a schematic illustration of such a test. Usually, the vehicle's lateral deviation $y(t)$ from the initial straight-ahead direction is considered as a characteristic measure.

This kind of crosswind tests can be performed in two ways:

1. The driver does not apply any steering correction. The steering wheel is held either in its original position (fixed control) or it remains untouched (free control).
2. The driver attempts to minimize the lateral deviation y of the vehicle by counteraction with the steering wheel.

The first method considers the reaction of the vehicle alone. On a first glance it seems to be well suited for

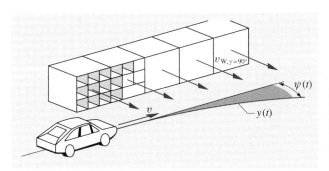

Fig. 16.41 Cross-wind test with real car, schematic

comparative tests. However, it does not have much to do with reality.

The second method includes the driver's reaction; thus the test result also depends on the driver. This procedure has the great advantage that the closed control loop made up by the driver, vehicle, and road is reproduced and is very close to a real-life situation. At the same time, the driver is acting as a sensor – by seeing the deviation and feeling the accelerations (yaw and roll) – and as an actuator – by operating the steering wheel.

Different crosswind facilities are in operation by the automobile manufacturers. Regarding their length, size, position, and number of blowers, and wind speed, they differ significantly from each other. As a result, the test data obtained in different facilities are hard to compare.

An error frequently incurred with crosswind tests, especially when carried out by the press, is that the yawing angle is set too high: $\beta \approx 45°$. Such high yawing angles are rather unlikely to occur at the *high* road speeds where such a crosswind can affect the vehicle and be dangerous. Yawing angles on the order of $\beta = 20°$ (to 30°) are considered more appropriate for realistic crosswind sensitivity comparisons among vehicles.

A realistic assessment of the effect of wind on the system made up of the vehicle and driver can only be achieved by driving under real conditions (stochastically changing wind) on public roads. Over time, several procedures have been developed that include the driver in the assessment (see [16.39] for the most recent).

Also *cooling tests* and *break tests* are carried out on road. Load is *produced* either in the mountains or with a break trailer on a level road.

16.1.9 Measurement Technique

The measurement technique applied in vehicle aerodynamics makes use of the same physical principles as in other areas of fluid dynamics. Differences exist with regard to technical details and in application, and only those will be considered here. The quantities to be measured are compiled in Table 16.2.

Forces and Moments

In vehicle aerodynamics forces and moments are stated in a *vehicle fixed* coordinate system, as sketched in Fig. 16.42. The reference point is the position on ground where the longitudinal centerline crosses the perpendicular line halfway along the wheel base ($a/2$). Instead of quoting the moments of yaw, pitch and roll as such, they are declared as force increments on the front and rear wheels, which makes communication with vehicle dynamicists easier.

To measure the resulting forces and moments produced by the flow and to reduce them to components, specific balances have been developed. In comparison to those used in aviation the following differing specifications exist:

- forces in horizontal and vertical direction are of the same order of magnitude;
- the resulting force and moment are reduced in a vehicle-fixed system;
- the balance swiveling $\beta = \pm 180°$.

Time in a vehicle wind tunnel is rare and very expensive. The balance must be designed for quick model change, and the measuring time must be short. Frequently, manipulations on the underside of the vehicle have to be executed. Therefore the balance should be completed by a device to lift the vehicle (1.80 m) and to relocate it exactly. Some balances are designed to be equipped with a moving belt and mini-belts to rotate the wheels.

According to the manner in which the vehicle is connected to the balance three types can be distinguished (Fig. 16.43):

1. *Wire suspension*: this *classical* technique borrowed from aviation is no longer applied in vehicle aerodynamics as assembling the model is too slow. However, it still offers several advantages. The thin wires disturb the flow far less than struts or stings, and their drag can easily be calibrated out. Furthermore, the suspension makes it possible to allow the vehicle's wheel roll on a broad, moving belt with reduced and controlled wheel load.
2. *Sting*: the vehicle is held by a rigid sting, either from above or from the rear. Flow is disturbed to varying degrees, especially separation at the rear. This technique is widely used for racing cars. Wheels

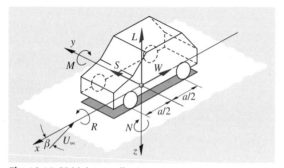

Fig. 16.42 Vehicle coordinate system

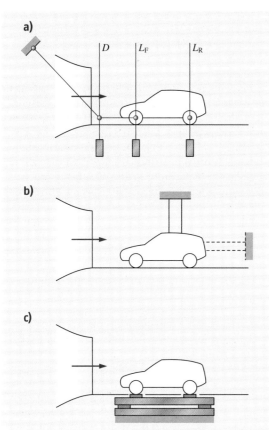

Fig. 16.43a–c Three possibilities to assemble a model to a wind-tunnel balance, schematic; (**a**) *traditional* wire suspension; (**b**) with sting from top or rear; (**c**) on platform

are mechanically disconnected from the vehicle's body, held from outside by covered two-component balances or stings, and can roll on the moving belt.

3. *Floor balance*: the vehicle rests with its four wheels on four small platforms that can be adjusted for the wheel base and track. This type is preferred in vehicle aerodynamics, and allows for the integration of devices to rotate the wheels.

Traditionally, floor balances were designed to separate the forces and moments in the six degrees of freedom mechanically [16.40]. One quality measure of a balance was how well the cross-effects were eliminated. Today this is no longer practiced; the decoupling is performed strictly numerically.

Three different designs of balances are in use (Fig. 16.44):

1. *Floating-frame balance, seven components*: vertical and horizontal components are mechanically decoupled. The frame is floating on hydrostatic bearings; drag and side forces on the front and rear axle are measured via connecting struts. The lift balances are mounted inside the frame, one for each wheel. Horizontal displacement of the test object (elastokinematic elements, tires) has no effect on the measurement.
2. *Platform balance, six components*: the platform is held by six struts that are connected with force-measuring elements. Horizontal forces move the test object and have an effect on the force distribution. The movement is measured optically and corrected for. The six forces are not mechanically decoupled.

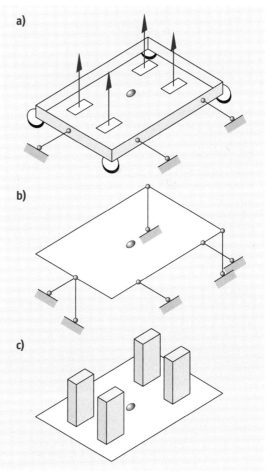

Fig. 16.44a–c Different designs of platform balance, schematic; (**a**) floating frame; (**b**) rigid platform; (**c**) four/two columns

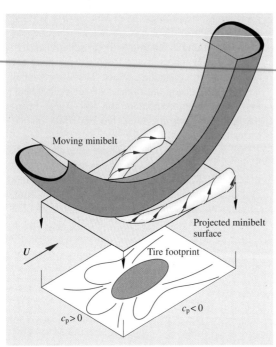

Fig. 16.45 Pressure distribution around a wheel, needed for wheel-pad correction, schematic

the footprint (Fig. 16.45), and has an effect on lift. By measuring the pressure field on the pads an empirical correction can be applied.

Pressure

When it is not permissible to drill a hole into the sheet metal of the body of the test object, pressure is measured by thin disc-like probes (so-called *bed bugs*), as shown in Fig. 16.46, which are fixed to the surface of the body by an adhesive. If more than one probe is used attention must be paid not to impair the local flow due to the plastic hoses connecting the probe to the pressure transducer.

Pressure-sensitive paints (PSP) have not yet gained acceptance in routine vehicle aerodynamics for three main reasons:

- because of the relatively low flow speeds, pressure differences to ambient are low
- the expenditure for preparing the model is high
- calibration needs a lot of time.

To scan fields of pressure and velocity in the space around the vehicle so-called multi-hole probes are used. The most prominent example is the 14-hole probe developed by A. *Cogotti* [16.41] (Fig. 16.47). Local values for the velocity $v(z, y)$ and the loss of total pressure in a plane $x = \text{const}$ behind the test object give an indication of where on the vehicle drag is generated. In the case of Fig. 16.48 upwind and downwash were produced with the flaps seen in Fig. 16.19 in specific stationary positions, leading to the vorticity shown on the diagrams on the right-hand side of Fig. 16.48.

Velocity

Vehicle aerodynamicists like to measure the air velocity with a vane anemometer; a collection is depicted

The interaction is registered during calibration and stored in a matrix.

3. *Column balance, 4×6 components*: The test object rests on four columns that are assembled on a rigid, common ground plate and can be adjusted to the wheel base and track. For testing motor cycles only two are needed. Each column is equipped with a complete six-component balance.

In any case, each wheel rests on a small rectangular surface – be it a rigid pad or a minibelt – which is inevitably wider and longer than the footprint of the tire. Thus it is subjected to the pressure field around

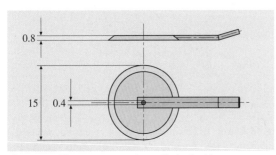

Fig. 16.46 Flat pressure probe, called a *bed bug*

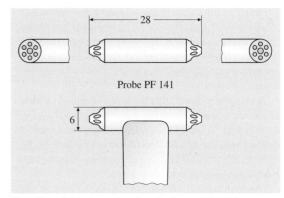

Fig. 16.47 14-hole probe after A. *Cogotti* [16.41]

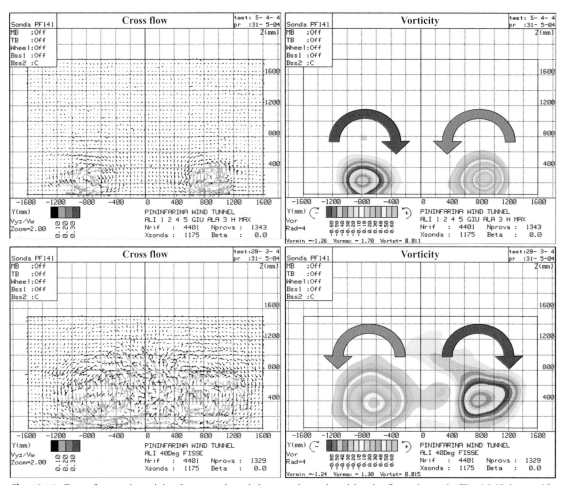

Fig. 16.48 Cross flow and vorticity for upwash and downwash produced by the flaps shown in Fig. 16.19 in specific steady-state positions (after A. Cogotti)

in Fig. 16.49. Often they are handheld to detect local flow streaks, for instance in front of breaks. Probes with a diameter of 80–100 mm but very thin are used to measure the flow distribution in front of (or behind) a radiator or condenser.

Measuring the flow velocity in discrete loci but continuous in time is the domain of laser Doppler anemometry (LDA). Using three pairs of laser beams all three components of the velocity vector can be measured. Its application is described in [16.42]. If continuous information of speed in a plane is needed particle image velocimetry (PIV) is well suited [16.43]. It allows recording the complete flow field in a plane within a few microseconds. Thus it makes it possible to get insight into an unsteady flow field, such as the vortex filed in a plane $x = $ const behind a car.

Air Flow Rate

The air flowing through a vehicle's passenger compartment serves as a transport medium for mass and heat. It is its task to control thermal comfort and hygiene. Apertures designed into a body are:

- inlet: for cars generally in front of the windshield
- outlet: for cars at the rear; formerly at the C-pillars, now in a region of moderate (low) pressure, for instance at both sides, hidden behind the rear bumper.

Besides these designed openings there are other unwanted ones, so-called leaks, where air may come in or go out.

Generally, the air flow rates through all these apertures depend on the flow field around the vehicle, i.e., on the driving speed. Measurement of this volume flow is

Fig. 16.49 Vane anemometers, diameter (from *left* to *right*): 11, 22, and 88 mm

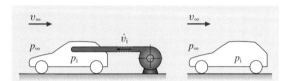

Fig. 16.50 Test set up to measure the air flow rate through a passenger compartment, schematic; *left*: outside auxiliary blower to measure the extraction line; *right*: air flow test

carried out in a wind tunnel in two steps, as summarized in the schematic in Fig. 16.50:

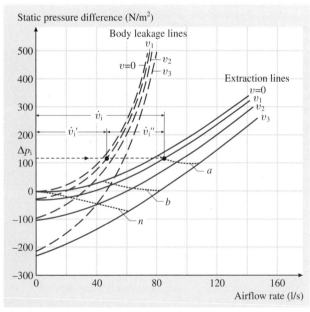

Fig. 16.51 Measurement of air-flow rate, extraction lines

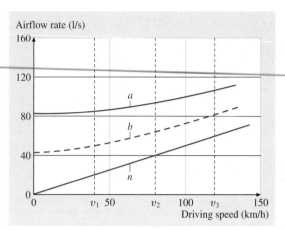

Fig. 16.52 Air flow rate versus driving speed for three different fan stages

1. measuring the *extraction curves* with an external blower
2. determination of the air flow rate using these curves

This technique is attributed to Wallis [16.44], however it was in fact developed independently in several places. The reference in both cases is the interior pressure $\Delta p_i = p_i - p_\infty$.

In order to measure the extraction curve all *inlet* openings are sealed; the *exit* openings remain open. While the wind is on, air is fed into the compartment through a hole in one of the windows from an external blower and hose. The static pressure inside the compartment, $\Delta p_i = p_i - p_\infty$, is measured as a function of the volume flow $\Delta p_i = f(\dot{V})$. The volume flow of the blower $\dot{V}$ is measured using a flow meter inside the tube. This measurement is repeated for several wind speeds U_∞ and plotted in a graph (Fig. 16.51).

In a second step the external blower and the supply hose are removed, and the inlet apertures are opened. Now the interior pressure Δp_i is measured again for the same wind speeds U_∞ as above, $\Delta p_i = f(U_\infty)$. All important settings of the control flaps of the HVAC system and speeds of its fan are investigated. For each set of $(n, \Delta p_i, U_\infty)$ the volume flow can be read as a function of wind speed, and is plotted as shown in Fig. 16.52.

If these two steps are repeated with the outlet apertures sealed the leakage air flow rate can also be measured.

Flow Visualization

Flow visualization serves as an important aid in experimental vehicle aerodynamics, in wind tunnels and on

Fig. 16.53 Traditional flow visualization with wool tufts (photo Volkswagen AG)

the road or track. Traditionally wool tufts were used, and they are sometimes still used even today, in various versions:

- tufts on the surface of the body, as shown in Fig. 16.53; this technique, although preparation of the model is lavish, was applied to localize separation; it is also suitable for tests on the road and track;
- tufts in an orthogonal screen (positioned behind the vehicle) to visualize longitudinal vortices;
- a tuft on a single probe, to detect the local flow direction, for instance in front of breaks.

A useful technique to visualize the flow on the surface is the application of a liquid film of petroleum and gasoline containing colored or luminescent pigments. Figure 16.54 displays a typical result. The liquid is painted (or sprayed) onto the surface, and the wind is quickly accelerated and kept on until the film dries. However, such painted images must be interpreted carefully,

Fig. 16.54 Flow visualization with emulsion of Schlemmkreide (photo Daimler Chrysler AG)

Fig. 16.55 Flow visualization with *smoke*; top: smoke introduced into the near wake; bottom: smoke introduced into the undisturbed oncoming flow (photo Volkswagen AG)

especially in regions of separated flow, for example, at the vertical areas of the vehicle body, where gravity affects the paths of the particles.

A device frequently used is a smoke generator. By feeding thin smoke streaks into an airstream, the path of the air flow becomes visible. Generally, an alcohol/water mixture is evaporated in the generator. An optical dense nontoxic vapor – frequently called *smoke* – develops, and is injected into the airstream with a long thin (heated) stem at two different locations:

- into the *undisturbed* flow ahead of the vehicle; streamlines become visible, as depicted in the lower part of Fig. 16.55;
- into the *separated* flow, where it fills the entire near wake up to the separation point, as seen in the upper part of Fig. 16.55.

Good illumination is a prerequisite for flow visualization. To generate a plane of light, light from a high-performance laser (5–24 W) must be first led through a cylindrical lens and then deflected with a mirror in such a way that the vehicle can be illuminated in each desired section across its full width, length or height. The thickness of the light sheet should be between a few millimeters and several centimeters.

Alternatively, the flow in a region of separation can be investigated with a bubble generator. Helium-filled soap bubbles are introduced into the air flow and photographed. If the exposure time is well chosen, individual flow paths are visible (Fig. 16.56). This method is well suited to small-scale models, but is occasionally also applied to full-size cars.

Fig. 16.56 Flow inside near wake made visible with He-filled soap bubbles (photo Adam Opel AG)

Fig. 16.57 Flow visualized by H-bubbles generated with a wire in a water tunnel (photo Daimler Chrysler AG)

A simple and very sensitive method to detect lines of separation is offered by the application of talcum. The location where separation is expected to occur is sprayed with a very thin oil film before the test. During the test talcum is blown into the zone of separated flow. It deposits where the flow separates, and a line of separation becomes clearly visible. Because talcum is dirt for a wind tunnel, this technique should only be used for special problems. However, on the road and track it can be used without restriction.

Water tunnels are extremely well suited for flow visualization. Formerly, methods in which flow patterns on the surface became visible were preferred. Later, methods in which the three-dimensional flow was made visible were introduced. One possibility is to add a thin colored streak of liquid (milk) to the flow; the most prominent example is Reynolds' experiment from 1883, with which he made the transition from laminar to turbulent flow in a pipe visible. However, adding a colored liquid to the flow of a tunnel with a closed circuit has the disadvantage that the colors dilute the water and make it quickly unusable.

Instead, the effect of electrolysis can be used. A thin wire is spanned across the flow and a pulsed direct-current (DC) voltage applied. Hydrogen bubbles are generated and carried away with the flow. If the wire is isolated, equidistant bubble streaks become visible, similar to the smoke streaks in a wind tunnel. As an example Fig. 16.57 depicts the shear layer over an open cabriolet. Pictures like this can be used for comparison with CFD results. If the wire is spanned in an area of separation similar pictures as with the He bubbles can be produced.

However, although very attractive pictures can be produced this way, water tunnels are barely used in the course of vehicle development anymore.

Flow Noise

The measurement of flow noise has two objectives [16.45, 46]:

- *Exterior noise*: the typical example is a high-speed train. To get authorization to run a train the operator has to prove that the train conforms with noise regulations. Specifically, the sound pressure level must be below 90 dB(A) at a distance of 25 m. To ensure this, during the development of the train the noise sources have to be localized and methods to attenuate them have to be developed.
- *Interior noise*: such as the noise inside the passenger compartment of a car, a bus or a wagon. Interior noise is an important matter of comfort. In this case not only objective data have to be recorded. The noise must be assessed according to subjective criteria and speech recognition.

What makes this task delicate is that interior noise can only be measured when a prototype exists. However, once a project is in the prototype phase only minor modifications can be performed. Therefore it is necessary to know how exterior noise sources and interior noise level correlate. Only then can measures be developed to keep interior noise at a level that is assessed as comfortable.

Four measuring systems are on hand to determine the data:

- a single microphone
- two microphones built into the auditory canals of an artificial head
- an acoustic mirror with one or more microphones
- a microphone array.

The latter two systems have been developed to localize sound sources on *moving* objects such as trains or cars. Of course they can also be used in a wind tunnel, however, only in those with an open test section.

Single microphone: a microphone is an extremely sensitive pressure sensor, transforming an alternating pressure into an alternating-current (AC) voltage. According to *George* [16.47, 48] in the frequency range of interest the measurement signal should vary by not more than ± 2 dB with frequency. Condenser microphones meet this prerequisite, provided that the diameter of the membrane is matched to the frequency range under consideration. This linearity is lost as the wavelength of the sound signal approaches the diameter of the membrane. Microphones with a small diameter are suitable for high frequencies. Inside a passenger compartment a frequency range of 20 Hz to 12 kHz must be covered. The appropriate microphone diameter is 1/2" (12.7 mm).

Because of pseudo-noise generated by the flow around a microphone, in-flow measurements with a microphone are hardly ever performed. Even a well-streamlined nose cone, which was formerly used for this purpose, cannot overcome this effect. Instead, noise close to a surface is measured with so-called surface microphones with the dimensions of a small coin: outer diameter 20 mm, thickness 2 mm. A typical test set up is shown in Fig. 16.58, where the sound pressure on the surface of a side window close to the wake of the side mirror is measured. The result is shown in Fig. 16.59. The difference in sound pressure level with and without mirror is compared for two different mirrors: the *noisy* mirror, lower diagram, increases the wind noise by approximately 20 dB compared to the case without a mirror, while the *silent* adds only 2–10 dB to the (exterior) noise.

When measurements are carried out in rooms with only very low wind speed, such as inside a passenger compartment, a definite and repeatable position of the microphone must be observed. Admittedly, because of the multiple reflections at the walls the noise field is diffuse, and the level of sound pressure is independent of the microphone's position. However, very close to the walls, which themselves are noise sources, this is no longer true. As shown in Fig. 16.60 the sound pressure level increases when the microphone approaches the side window. The left ear of the driver is within the near field of the side window [16.49]. Therefore, this position was formerly selected as the standard for comparison of interior noise inside a passenger car. A typical result is shown in Fig. 16.61. In this case, the interior noise was measured twice: with the exterior mirror in place and without. The mirror is perceptible in the high-frequency range.

Artificial head: to subjectively assess the interior noise more like a driver or passenger the single microphone can be replaced by an artificial head (Fig. 16.62). The noise is recorded by two microphones built into the auditory canals of this head model. Subsequent to the test in the vehicle the noise is analyzed, reproduced binaurally and listened to with a head set. Different noise signals can be compared by quickly switching from one to another, and the effect of specific frequency bands can be assessed. The target is to tailor an agreeable spectrum, and measures are elaborated with which it can be reproduced inside the car, a task known as *sound engineering*.

Acoustic mirror: noise sources, either stationary or moving, can be localized with an acoustic mirror. Its working principle evolves from Fig. 16.63 [16.50]. The mirror is made up from an ellipsoid of revolution. One focus is directed at the object, while the microphone is located in the other, where the sound waves emitted from the object arrive. The sound pressure level is between 20–40 dB higher than that measured with a single microphone without a mirror, although the amplification depends on frequency and is lowest at low frequencies.

The surface under investigation must be scanned point by point. More than one point can be measured at a time when several microphones are placed at the focus of the mirror (Fig. 16.64). However, their amplification is not the same, and must be calibrated before the test.

Fig. 16.58 Test set up with surface microphones according to Brüel and Kjaer (Photo courtesy Forschungsinstitut für Kraftfahrwesen und Fahrzeugmotoren Stuttgart (FKFS))

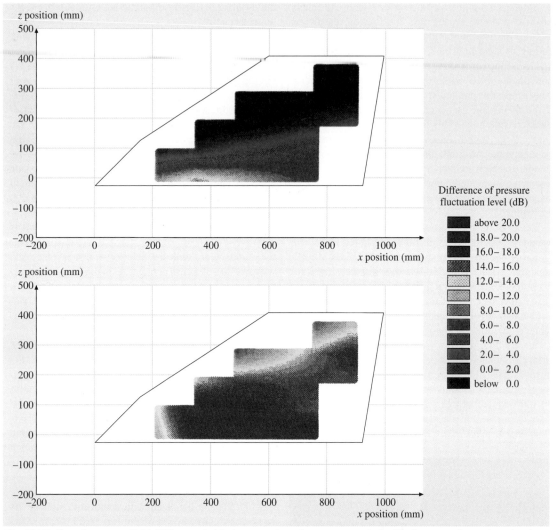

Fig. 16.59 Sound-pressure field difference behind an exterior mirror as compared with the case of no mirror in place; *bottom*: *noisy mirror*, *top*: *silent mirror* [courtesy Forschungsinstitut für Kraftfahrwesen und Fahrzeugmotoren Stuttgart (FKFS)]

When applied in a wind tunnel the mirror must be placed out of flow, and two effects have to be observed:

- the sound wave emitted by a source located on the object is transported by the airstream of the wind tunnel. The source of sound appears to be further downstream;
- refraction of the waves at the airstream's boundary make the noise source appear bigger than it is.

Microphone arrays: noise sources on moving objects can be localized with a phased array of microphones, as described in [16.51]. Only with these can the movement of a sound source be considered. Even narrow-band tones are correctly reproduced, because the Doppler effect can be regained. Figure 16.65 shows a schematic of the installation: 15 microphones produce one-dimensional (1-D) sensitivity.

Geometrical Quantities

In vehicle aerodynamics several geometrical quantities must be measured:

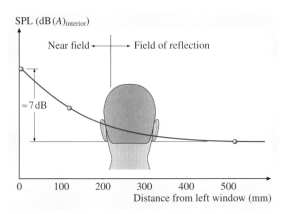

Fig. 16.60 Increase of SPL at the left ear of the driver when coming close to the side window (after [16.49])

- the frontal area of the vehicle (Fig. 16.7), which is the reference area for nondimensional coefficients
- the exact coordinates of the test object's surface; each shape modification must be documented consistently with the computer-aided design (CAD) dataset for communication with the design team
- stationary and unsteady deformations of body parts (e.g., hoods, doors or ballooning of the roof of a cabriolet).

Optical methods have been developed for these purposes. Of course, these are not fluid-dynamic methods, but they are indispensable tools for the vehicle aerodynamicists.

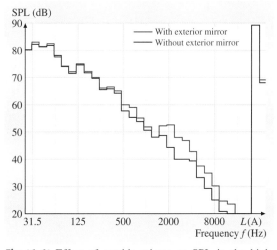

Fig. 16.61 Effect of outside mirror on SPL in the high-frequency range (after [16.45])

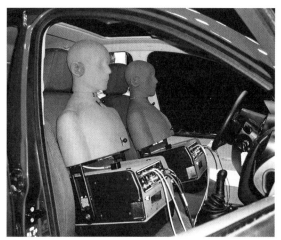

Fig. 16.62 Artificial head (photo Ford Werke GmbH)

Traditionally the frontal area is determined from the vehicle's shadow image, which is generated with a flash light positioned at a large distance. The contour projected onto a screen is drawn by hand and the area is determined with a planimeter. The divergence of the light beam is corrected for by comparing the shadow image of a rectangular plane of a precisely given magnitude (i.e., $2\,\text{m}^2$).

Newer methods make use of *edge tracking*. The disadvantage of these methods is that each measured point has to be (automatically but nevertheless time-consumingly) focused, because not all these points are in the same plane. The accuracy of $\pm 0.3\%$ is said to be sufficient. The time required to measure one frontal area is about 45 min.

A much faster method has also been developed [16.52]. The outline of the vehicle's shadow image is projected onto a screen with a 50 mm-diameter laser beam that is moved by the wind tunnel's traversing gear.

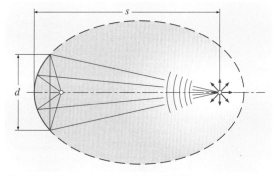

Fig. 16.63 Acoustic mirror (after [16.50])

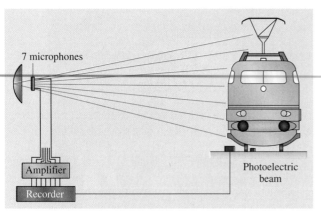

Fig. 16.64 Array or seven microphones close to the focal point of an acoustic mirror

The contour is tracked with a frequency of 10 Hz, and a photograph of each point is taken by a digital camera. This event is repeated as long as the contour is closed. The photo, which is taken by a digital camera oblique to the laser beam, is inverted and the pixels inside the contour are counted and multiplied by their area. The time needed to measure one frontal area can be reduced to 8 min, with accuracy remained the same.

With trains one goes without measuring the frontal area; it is *set* to $10 \, m^2$.

In the course of developing the body's shape in a wind tunnel numerous shape modifications are carried out. Their geometry has to be documented. While this was formerly performed by taking off hand-made patterns or metering on a measuring plate, today use is made of photogrammetry and triangulation, as described in detail in [16.53, 54].

Three kinds of tasks have to be performed:

1. static measurement of (many) single points
2. dynamic measurement of (many) single points
3. static measurement of free-formed surfaces.

For the first of these a single (digital) camera is needed. The object, which must be prepared with (at least) three marks, is photographed by hand from several directions. For the second, for instance to measure the lift off of a cabriolet's roof (*ballooning*), two cameras, mounted on a common console, are needed. For the third case two cameras are also needed. In addition, a projector, which is located between the cameras on the same console, projects a grid onto the object.

The registration of the surface data of a complete model at a scale of 1 : 4 can be done in 20–30 min, while for a full-scale model 3 h are needed. The accuracy is ± 0.1 mm.

16.1.10 Support by Computational Fluid Dynamics

To a large extent aerodynamic development work is based on empiricism. This is all the more relevant for optimizing the shape of a vehicle, a process which is performed by *trial and error*. All the information you get after a shape modification is a set of six coefficients. If, for instance, c_D was reduced, you will continue to modify the shape in the same direction. If not, you take the previous modification back and try your luck at another location. In neither case do you obtain information about *why* the shape change had a positive or negative effect on the flow.

Deeper insight into the flow physics can be gained by performing detailed measurements in the flow field, for instance a wake traverse, as shown with Fig. 16.48. However, this kind of measurement is a rare exception; time does not allow its performance very often.

In this situation computational fluid dynamics (CFD) is gaining importance. Once the basic equations are solved you can get almost any information you want – velocity vectors, pressure, shear stress, separation of flow and reattachment – throughout the entire space simply by appropriate postprocessing. All this information can be used as a rationale for the next stage of shape modification.

A survey of the related state of the art of CFD in vehicle aerodynamics was recently given in [16.55].

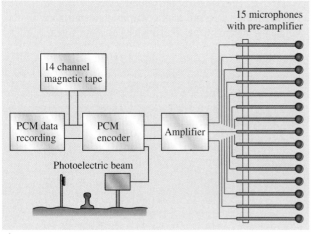

Fig. 16.65 Microphone array (after [16.51])

16.2 Short-Duration Testing of High Enthalpy, High Pressure, Hypersonic Flows

In hypervelocity flows the speed of the considered fluid is much larger than the speed of sound. Commonly the hypersonic flow regime is considered to start above a Mach number of $M = 5$. Ground-based testing of such flows is performed in many different types of facilities. The reason for this is the large range of flow conditions and phenomena encountered in hypersonic flight and the fact that no single facility can simulate all relevant flow parameters simultaneously. Therefore, in hypersonic testing, partial simulation of the complete flow situation concentrating on selected flow phenomena are performed in different types of facilities. Examples are Mach–Reynolds number simulation in cold hypersonic ground-based test facilities, verification and qualification of hot structures of space vehicles in arc-heated test facilities, and the investigation of the influence of the chemically reacting flow past an entry or re-entry vehicle on its aerodynamic behavior in shock tunnels or shock expansion tunnels. Comprehensive overviews of ground-based testing of hypersonic flows are given by, e.g., *Lukasiewicz* [16.56], and *Lu* and *Marren* [16.57].

One possibility to increase the Mach number in ground-based facilities is by reducing the free-stream temperature, i. e., the free-stream speed of sound. In this case, high Mach numbers can be achieved while the free stream velocity is significantly lower than the actual flight velocity. However, characteristic of high-Mach-number hypersonic flight with $M \approx 10$ and higher is that the kinetic energy of the flow is large enough that high-temperature effects such as vibrational excitation or dissociation of the fluid molecules occur in the flow past hypersonic vehicles. For such hypersonic, high-velocity flows the term hypervelocity flow is used. The high flow velocities and subsequently the high-temperature effects are not duplicated in cold hypersonic ground-based test facilities. During the re-entry flight of a space vehicle in the Earth's atmosphere or the interplanetary atmospheric entry of space vehicles or meteorites, speeds in excess of 6 km/s are achieved. Considering a flow with this speed in a test section with an area of 1 m^2 and a density of 0.003 kg/m^3, a power requirement of 300 MW already results. Therefore, continuous flow facilities are not a practical way to generate such high-enthalpy, hypersonic flows. Additionally, the correct simulation in ground-based testing of the chemical relaxation length of the dissociation reactions of the fluid molecules occurring, for example, behind the strong bow shock in front of the nose of a re-entry vehicle, requires the duplication of the flight binary scaling parameter, the product of the free-stream density ρ and a characteristic flow length L [16.58]. Consequently, the smaller the scale of the wind tunnel model is chosen, the higher the free-stream density or pressure needs to be. Considering that the flight trajectory range of a re-entry vehicle from low Earth orbit in about 70 km altitude, where the highest heat flux typically occurs, the atmospheric density is approximately 10^{-4} kg/m^3. Using a geometrical scaling factor of 30, a free-stream density in the ground-based facility of 0.003 kg/m^3 is required. If a flow with this free-stream density and a velocity of 6 km/s is generated by expansion in a convergent–divergent hypersonic nozzle from a reservoir at rest without adding energy, a total specific enthalpy of about 23 MJ/kg and a nozzle reservoir pressure on the order of 90 MPa is required. This results in a nozzle reservoir temperature of about 10 000 K. It is clear that such conditions can only be achieved in impulse facilities with short flow duration. The most successful types of facility that are able to generate high-enthalpy and high-pressure hypersonic flows are shock tunnels and shock expansion tunnels with typical test times of approximately 5 ms and less. The principle of these facilities is to store the energy over a long period of time, therefore reducing the necessary power requirement and subsequently releasing the stored energy rapidly. Due to the high flow speeds, test times on the order of a few ms are still sufficient for the development of a steady flow over a model. According to *Hornung* [16.59], a reasonable, conservative correlation of the necessary test time to establish a steady flow is

$$\tau = 20 \frac{L}{U_\infty} \,,$$

where L is the model length and U_∞ is the free-stream velocity. For a test using the aforementioned flow condition and a 0.3 m-long wind tunnel model, the required test time would be 1 ms. The high-pressure, high-velocity flows that can be generated in shock tunnels and shock expansion tunnels make these facilities not only suitable for the investigation of space vehicle aerothermodynamics but also for studying complete air-breathing propulsion systems, particularly supersonic combustion ramjets (scramjets) at flight Mach numbers of $M_\infty = 8$ and above.

In the present chapter, the basic working principle of different types of shock tunnels and shock expansion tunnels, the special aspects of measurement techniques in such short-duration facilities with test times of approximately 5 ms and less and some selected typical applications in existing facilities will be discussed. In the different types of facilities discussed here, the wind tunnel model is stationary, i.e., it is not moving. So-called hypervelocity range facilities in which the model is free flying at high speeds or combinations of hypervelocity range facilities and, e.g., shock tunnel facilities, so-called counterflow range facilities are not discussed here. Details of these facilities can be found, e.g., in [16.56, 57].

16.2.1 Working Principle of Shock Tubes/Tunnels and Shock Expansion Tubes/Tunnels

Basic Principle of Ideal Shock Tubes

The basic layout of a diaphragm-driven shock tube is shown in Fig. 16.66. It consists of two chambers of equal and constant cross section separated by a diaphragm. Both chambers are filled with gases at different conditions p_4, T_4, m_4, γ_4 and p_1, T_1, m_1, γ_1, where p is the pressure, T the temperature, m the molecular weight and γ the ratio of specific heats. Initially the gases are at rest in both chambers. As a convention for the present discussion of shock and shock expansion tubes and tunnels, the left section is called the driver or high-pressure section, containing the driver gas, and the right part the driven or low-pressure section, containing the test gas. After instantaneous removal of the diaphragm, a shock wave is moving into the driven section and the head of a centered expansion wave is moving into the high-pressure region.

In 1860, this basic principle of shock tubes was investigated theoretically by the mathematician Bernhard *Riemann* in Göttingen, Germany [16.60]. The first practical application of a diaphragm-driven shock tube was performed in 1899 by Paul Vieille in Paris, France, studying the establishment and propagation of discontinuous waves in columns of air traveling faster than the speed of sound [16.61]. In Vieille's shock tube with a length of 6 m and a diameter of 22 mm, shock wave speeds of up to 600 m/s were measured by expanding pressurized air from 2.74 MPa to atmospheric pressure using diaphragms of paper or collodion sheets.

The measured propagation speeds agreed well with ideal shock tube theory.

After this first application of a shock tube this efficient apparatus was more or less forgotten until about 40 years later. Since the 1940s, shock tubes have been extensively used for investigations related to gas dynamical phenomena such as shock wave propagation, shock wave reflection at obstacles or shock wave refraction at contact surfaces, research of detonation waves, studies related to aerodynamics, aerothermodynamics, determination of the physical and chemical kinetics properties of high-temperature gases and applications in laser technique, medicine or calibration of fast acting measurement gauges. Descriptions of their working principle and discussions of shock tube applications are given by, e.g., *Lukasiewicz* [16.62], *Glass* and *Hall* [16.63], *Oertel* [16.64], *Lukasiewicz* [16.56], *Anfimov* [16.65] or *Takayama* [16.66].

The flow field developing after diaphragm removal in the vicinity of the diaphragm location, i.e., when the waves have not yet reached the left and right boundary, is shown in an x–t wave diagram (Fig. 16.67), where x is the streamwise coordinate with its origin at the position of the diaphragm location, and t is the time with $t = 0$ at diaphragm rupture. The solution of this Riemann or shock tube problem can be obtained as an exact solution of the one-dimensional, time-dependent Euler equations for thermally and calorically perfect gases. A discussion of this solution is given, e.g., by *Sod* [16.67] in connection with the investigation of different finite-difference methods for the numerical solution of systems of nonlinear, hyperbolic conservation laws. It is interesting to note in this context that the approximate solution of Riemann problems with generalized initial conditions on either side of the grid cell interfaces (or diaphragm in the context of shock tubes) are a core element of many modern

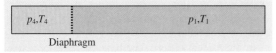

Fig. 16.66 Schematic layout of a diaphragm-driven shock tube

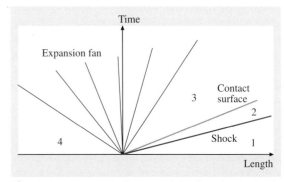

Fig. 16.67 Wave (x–t) diagram of the flow developing in a constant-area shock tube

upwind computational fluid dynamics schemes [16.68]. These schemes are based on a numerical method applying the exact solution of the Riemann problem at the cell interface developed by *Godunov* [16.69]. The numbers used in Figs. 16.66 and 16.67 denote distinct regions of the flow. Region 2 contains the shock compressed test gas, while in region 3 the driver gas processed by the unsteady expansion wave is contained. The test and driver gas are separated by a contact surface.

The initial filling conditions of the driver and driven section uniquely define the properties of the developing unsteady expansion and shock wave. The pressure and velocity obtained in region 3 by expansion of the driver gas is given by e.g. [16.70]

$$\frac{u_3}{a_1} = \frac{a_4}{a_1}\left(\frac{2}{\gamma_4 - 1}\right)\left[1 - \left(\frac{p_3}{p_1}\frac{p_1}{p_4}\right)^{(\gamma_4-1)/2\gamma_4}\right].$$

In order subsequently to generate a graphical solution of the shock tube flow, the pressure and speed of sound in regions 3 and 4 are nondimensionalized with their respective quantities in region 1. The velocity in region 3 is normalized by the speed of sound in region 1, a_1. A graphical interpretation of this relation is plotted in the left part of Fig. 16.68 for $p_4/p_1 = 1000$ and different speed of sound ratios a_4/a_1 and $\gamma_4 = 1.4$. The pressure and velocity obtained in region 2 by shock compression of the test gas can be obtained by the relation

$$\frac{u_2}{a_1} = \frac{1}{\gamma_1}\left(\frac{p_2}{p_1} - 1\right)\left(\frac{\frac{2\gamma_1}{\gamma_1+1}}{\frac{p_2}{p_1} + \frac{\gamma_1-1}{\gamma_1+1}}\right)^{1/2}.$$

This relation is plotted in the right part of Fig. 16.68. At the contact surface in Fig. 16.67 the conditions $p_2 = p_3$ and $u_2 = u_3$ must be fulfilled. Combining these boundary conditions with the above two equations results in the relation

$$\frac{p_4}{p_1} = \frac{p}{p_1}\left(1 - \frac{(\gamma_4 - 1)\left(\frac{a_1}{a_4}\right)\left(\frac{p}{p_1-1}\right)}{\sqrt{2\gamma_1\left[2\gamma_1 + (\gamma_1+1)\left(\frac{p}{p_1-1}\right)\right]}}\right)^{-2\frac{\gamma_4}{(\gamma_4-1)}}.$$

Graphically, the solution of the shock tube problem can be obtained by the point of intersection of the curve in the right part of Fig. 16.68 with one of the curves in the left part of the same figure. This is shown in Fig. 16.69 using $a_4/a_1 = 3$.

The velocity, temperature, density, and pressure distribution of the flow in a constant-area shock tube at a point in time after diaphragm rupture is shown in Fig. 16.70. The position of the incident shock, contact surface, and head of the unsteady expansion wave are denoted by *a*, *b* and *c*, respectively. It can be seen that the pressure and velocity distributions are continuous at the contact surface while the temperature and density show a discontinuity.

After reflection of the incident shock wave at the right end wall of the shock tube, the test gas is brought to rest in region 0 (Fig. 16.71). Subsequently, the reflected shock wave penetrates the contact surface. Depending on the local conditions, three types

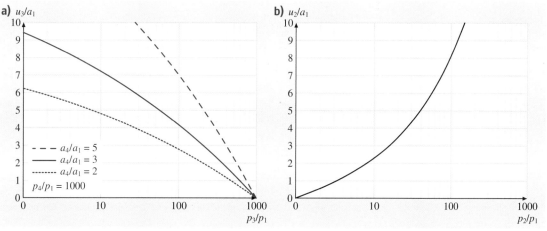

Fig. 16.68a,b Determination of the pressure and velocity in region 3 by separate consideration of processing the driver gas in region 4 by the unsteady expansion for different ratios of $a_4/a_1 = \sqrt{T_4/T_1}$ (**a**) and compression of the test gas in region 1 by the shock wave (**b**); $p_4/p_1 = 1000$ and $\gamma_4 = \gamma_1 = 1.4$

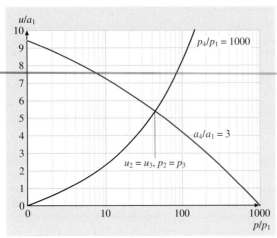

Fig. 16.69 Determination of the pressure and velocity in region 3 with $u_2 = u_3$, $p_2 = p_3$ at the contact surface; $a_4/a_1 = 3$

of shock wave/contact surface interaction can be differentiated. If $p_0/p_2 = p_5/p_3$ the reflected shock wave penetrates the contact surface without reflection, and the contact surface is brought to rest. This case is called a tailored interface condition, and the pressure remains constant in region 0 (Fig. 16.72). If $p_0/p_2 > p_5/p_3$ the undertailored interface condition is obtained, and an expansion wave is generated which lowers the pressure in region 0. For the case of an overtailored

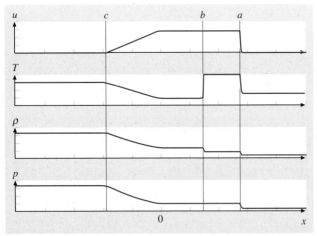

Fig. 16.70 Velocity, temperature, density, and pressure distribution of the flow in a constant-area shock tube at a constant point in time after diaphragm rupture; position of the incident shock wave (*a*), contact surface (*b*), and head of the unsteady expansion wave (*c*)

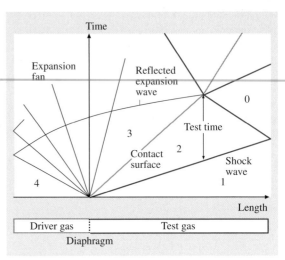

Fig. 16.71 Wave (x–t) diagram of the flow in a constant area shock tube after the expansion wave and the incident shock wave were reflected at the end walls

interface condition ($p_0/p_2 < p_5/p_3$), a shock wave propagates into region 0, increasing the pressure in region 0. Due to the fact that the shock compressed and heated slug of gas in region 0 is used in shock tube research for, e.g., the determination of physical or chemical kinetics properties of high-temperature gases or in reflected shock tunnel operation as the reservoir driving the flow in the nozzle and test section, shock tube operation in tailored interface mode is most desirable.

For a given driver section condition, a tailored interface condition can only be obtained for one particular driven section filling condition. For this reason different types of driver gas conditions and driver gas mixtures are used to achieve tailored interface conditions for different stagnation enthalpies [16.56].

The head of the unsteady expansion wave is reflected at the left wall of the driver section and is subsequently accelerated by interaction with the centered expansion wave (Fig. 16.71). Due to the fact that the head of the reflected expansion wave has a higher velocity than the incident shock wave, the lengths of the driver and driven sections must be chosen such that the arrival of the reflected expansion at the end of the shock tube is delayed as much as possible. If for example the shock processed gas contained between the shock and the contact surface is used as test flow, the reflected expansion wave should not intersect with the reflected shock wave before this intersects with the contact surface. In this way, the nominally available

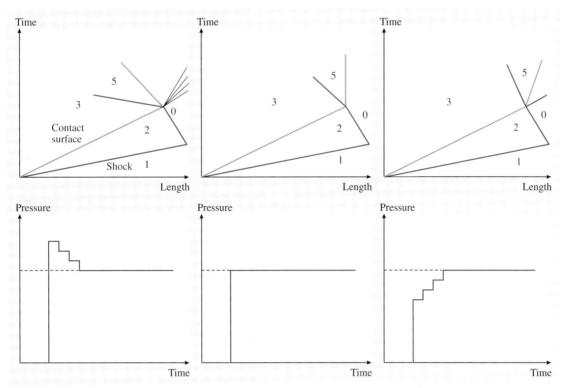

Fig. 16.72 Shock wave/contact surface interactions for undertailored (*left*), tailored (*middle*) and overtailored (*right*) interface condition; wave (*x–t*) diagram (*upper row*) and time history of pressure in region 0 (*lower row*)

time of constant flow conditions in region 2 can be maximized.

Shock Tunnels – Ideal Operation

If air is used as the test gas, the flow generated in shock tubes in region 2 can be used to duplicate stagnation enthalpies and pressures that occur during the re-entry flight of a space vehicle [16.56]. However, the Mach number is limited to $M \approx 3$. To overcome this limitation and to be able to generate hypersonic flows, the shock tunnel was proposed by *Hertzberg* [16.71]. Since then, many types of shock tunnels have been developed and are operated in laboratories around the world. Overviews of the facilities and their application can be found in, e.g. [16.64, 72, 73]. The basic principles of the two different ways of shock tunnel operation, the straight-through shock tunnel and the reflected shock tunnel will be discussed in the subsequent sections.

Straight-Through Shock Tunnel Operation. The basic layout and wave diagram of a shock tunnel operating in straight-through mode are shown in Fig. 16.73. At the end of the shock tube, a divergent nozzle is attached, in which the supersonic flow behind the incident shock is accelerated to hypersonic Mach numbers. The time increment to start the nozzle is reduced by initial evacuation of the nozzle and test section. For this reason, a secondary, thin diaphragm is placed at the end of the shock tube. The steady test flow is established in the test section after the incident shock wave, a contact surface and a system of waves has reached the nozzle exit. The wave traveling with the local velocity $u - a$ is an upstream facing wave that is initiated by a mismatch of the post incident shock wave condition and the condition generated by the nozzle expansion. Due to the fact that this wave is moving into a supersonic flow, it is swept downstream. The test time is finished after the $u + a$ wave generated by the arrival of the shock tube contact surface or the reflected expansion wave reaches the nozzle exit. In Fig. 16.73, the length of the facility has been chosen such that both waves reach the nozzle entrance at the same time. In their study and utilization of a straight-through shock tunnel *Stalker* and *Mud-*

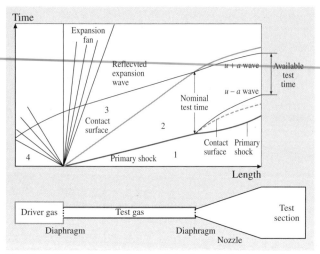

Fig. 16.73 Wave diagram for a straight-through shock tunnel

sen sufficiently small such that the incident shock wave is almost completely reflected, compressing and heating the test gas again. The stagnant slug of test gas, generated by the shock reflection, is subsequently expanded through the hypersonic nozzle. The principal layout and wave diagram of a reflected shock tunnel are shown in Fig. 16.74. In contrast to the shock tunnel and shock tube configurations discussed above, the facility sketched in Fig. 16.74 has different driver section and driven section diameters. This area change at the driver–driven tube intersection generates a steady expansion until the condition $M = 1$ is reached at the throat section followed by an unsteady expansion wave. For a given pressure difference between the driver and driven section, a steady expansion provides a higher velocity increase in subsonic flow than an unsteady expansion [16.56, 75]. In supersonic flow the opposite applies. Therefore, the area change is a means of improving the shock tube and shock tunnel performance.

For an ideal shock tube performance, i. e., not considering effects such as viscosity or multidimensional flows, the test flow quality and resulting test time depends on several aspects. The first is the wave pattern that is obtained when the reflected shock intersects with the contact surface. Based on the discussion on basic principles of ideal shock tubes, it is clear that the shock tunnel operation should be tuned in such a way that a tailored interface condition is obtained, resulting in a nozzle reservoir with constant conditions. Further, the shock tunnel geometry should be chosen such that the reflected expansion wave does not reach the nozzle entrance before the contact surface. An additional flow feature that influences the available test time is the nozzle starting process. Similar to the flow establishment in the divergent nozzle of the straight-through shock tunnel, a wave system has to pass through the nozzle before a steady flow is established. This process was studied by a quasi-one-dimensional analysis by *Smith* [16.76]. The waves resulting from this analysis are shown in Fig. 16.74. The incident shock wave (*a*) is followed by a contact surface (*b*), an upstream-facing secondary shock wave (*c*), and the upstream head of an unsteady expansion (*d*). Compared to the straight-through mode, significantly larger test times of the order of several ms can be achieved with a reflected shock tunnel.

ford [16.74] obtain test times in the order of 10 μs. More details concerning the straight-through shock tunnel can be found in, e.g. [16.56, 64].

Reflected Shock Tunnel Operation. The reflected shock tunnel is characterized by a convergent–divergent nozzle that is attached to the end of the shock tube. Similar to the straight-through shock tunnel, a thin secondary diaphragm is placed at the nozzle entrance in order to allow evacuation of the nozzle, test section and vacuum tank before the run. The nozzle entrance diameter is cho-

Techniques for Total Specific Enthalpy Augmentation in Reflected Shock Tunnels. For reflected shock tunnels, a good approximation to the total specific enthalpy of the nozzle reservoir condition is given by

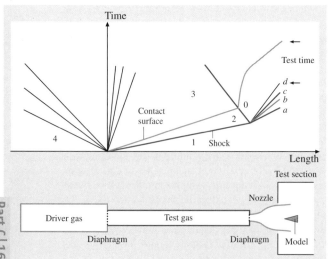

Fig. 16.74 Schematic and wave (*x*–*t*) diagram for a reflected shock tunnel

$h_0 = u_S^2$, where u_S denotes the speed of the incident shock wave [16.59]. Introducing the shock Mach number M_S and still assuming that the gases are thermally and calorically perfect, the relation between the initial pressure and speed of sound ratio between the driver and driven section reads:

$$\frac{p_4}{p_1} = \frac{2 \cdot \gamma_1 \cdot M_S^2 - (\gamma_1 - 1)}{\gamma_1 + 1}$$
$$\times \left[1 - \frac{\gamma_4 - 1}{\gamma_1 + 1} \cdot \frac{a_1}{a_4} \cdot \left(M_S - \frac{1}{M_S} \right) \right]^{-\frac{2\gamma_4}{\gamma_4 - 1}}.$$

This correlation is plotted in Fig. 16.75. Considering high-enthalpy flows, as a first approximation, the total specific enthalpy is approximately equal to the kinetic energy of the flow. Assuming the case that an air flow in the test section with a velocity of $u_\infty = 6\,\text{km/s}$ should be generated, a total specific enthalpy of $18\,\text{MJ/kg}$ ($h_0 \approx 0.5 u_\infty^2$) would be required. Due to practical reasons, air at room temperature is used in the driven section of reflected shock tunnels and consequently an approximate shock wave speed of $4.2\,\text{km/s}$ and a shock Mach number of $M_s = 12.5$ results. According to Fig. 16.75, using a pressure ratio of $p_4/p_1 = 2000$, this shock Mach number requires a ratio of speeds of sound of about $a_4/a_1 = 8$.

When increasing the shock Mach number in air such that the post-shock temperature exceeds an approximate value of $800\,\text{K}$, inner degrees of freedom of the test gas molecules are excited and, at temperatures above approximately $2500\,\text{K}$, oxygen molecules start to dissociate. Therefore, the assumption of a constant specific heat ratio is no longer valid. The influences of these high-temperature effects must be considered for the design and performance prediction of shock tunnels and they will be addressed later. Here the discussion is restricted to reveal basic dependencies.

A high ratio of a_4/a_1, i.e., a high value of a_4 (a_1 is fixed by the choice of using air at room temperature as test gas) can be obtained by reducing the molecular weight m_4 or by increasing the temperature of the driver gas. The first point is fulfilled by using hydrogen or helium as the driver gas. With this choice of driver gas, in conventional reflected shock tunnels, a ratio of $a_4/a_1 \approx 3$ can be achieved. The second way to increase the speed of sound ratio requires heating of the driver gas. Several heating techniques to increase the incident shock Mach number and the stagnation enthalpy have been developed and will be presented in the subsequent sections.

Electrically Heated Shock Tunnels. The driver gas can be heated by external or internal electrical resistance heaters. External heaters can be used to heat a light driver gas up to a temperature of $800\,\text{K}$. Using helium, this results in $a_4/a_1 \approx 5$, for hydrogen $a_4/a_1 \approx 6.2$. External heating must be applied over a long period of time and is limited by the material strength behavior of the high-pressure driver tube with increasing temperature. For this reason internal heaters with heating duration of about one minute were applied to minimize the tube heating [16.56]. Examples of electrically heated shock tunnels are the LENS I facility at the Calspan-UB Research Center Buffalo, New York, USA [16.77] and the Aachen shock tunnel TH2 at the Rheinisch–Westfälische Technische Hochschule (RWTH) Aachen University, Germany [16.78]. For the LENS I tunnel, test section velocities of up to $5\,\text{km/s}$ ($h_0 \approx 14\,\text{MJ/kg}$) using a hydrogen driver and in TH2 velocities of up to $3.6\,\text{km/s}$ ($h_0 \approx 7\,\text{MJ/kg}$) using helium as driver gas are reported.

Deflagrative Combustion Driven Shock Tunnels. In deflagrative combustion-driven shock tunnels, a stoichiometric mixture of hydrogen and oxygen diluted by helium or a helium–nitrogen mixture is utilized to create a hot driver gas [16.79]. The amount of helium dilution of the hydrogen–oxygen mixture is on the order of 70% by volume. An example of this type of facility is the 16-Inch Shock Tunnel of the National Aeronautics and Space Administration (NASA) Ames Research Center [16.80]. In this facility the combustion is initiated by wires extending over the length of

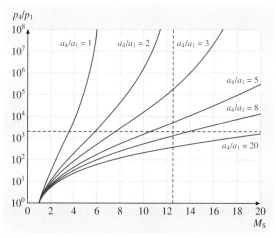

Fig. 16.75 Shock Mach number M_S as a function of p_4/p_1 and a_4/a_1 for a perfect monatomic driver ($\gamma_4 = 5/3$) and diatomic test gases ($\gamma_1 = 1.4$)

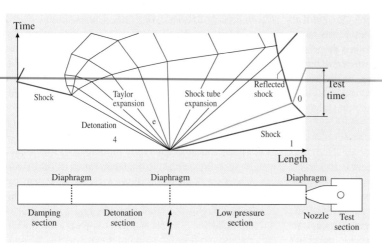

Fig. 16.76 Schematic and wave (x–t) diagram of a detonative combustion-driven shock tunnel operated in backward or upstream propagation mode including damping section

the driver section and heated by the discharge of a capacitor bank. The subsequent deflagrative combustion raises the pressure and temperature in the driver gas. For this facility, which has been applied to hypersonic airbreathing propulsion research, shock Mach numbers of $M_s \approx 10$ and total specific enthalpies of $h_0 \approx 12$ MJ/kg at nozzle reservoir pressures of $p_0 \approx 350$ bar have been reported [16.81]. In principle, the performance capability of deflagrative combustion driven shock tunnel exceeds that of electrically heated shock tunnels and detonative combustion-driven shock tunnels [16.82]. However, in the high-enthalpy regime the true performance of this type of facility is unknown and a limitation concerning the achievable driver pressure exists because for driver operating conditions with post-combustion pressures higher than about 400 bar, deflagrative combustion is difficult to maintain and local detonations may occur that could result in large driver pressure fluctuations [16.79].

Detonative Combustion-Driven Shock Tunnels. In detonative combustion-driven shock tunnels, the conventional driver is replaced by a detonation section that is filled with a gaseous reactive mixture. Typically, stoichiometric mixtures of oxygen and hydrogen are used as the driver gas, with dilution by helium or argon. Two main principles are applied in detonative combustion-driven shock tunnels. These are the back-

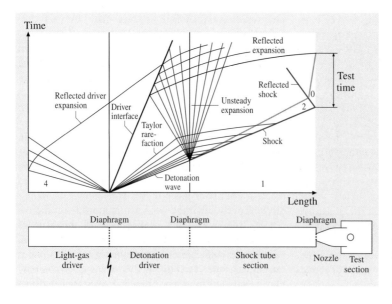

Fig. 16.77 Schematic and wave (x–t) diagram of a detonative combustion-driven shock tunnel operated in forward or downstream propagation mode including a shock-induced detonation technique

ward (or upstream) and the forward (or downstream) propagation mode [16.83]. The schematic and wave diagram of a detonative combustion-driven shock tunnel operated in backward or upstream mode is shown in Fig. 16.76. The detonation is initiated at the location of the main diaphragm and it moves backward, i.e., upstream toward the left end wall of the driver section. The flow velocity behind the detonation is directed in the same direction as the detonation wave. Due to the Taylor expansion that follows the detonation, the burnt driver gas is decelerated and brought to rest. This quasi-steady driver condition, obtained behind the characteristic labeled (e) in Fig. 16.76, acts similar to a high-temperature driver of a conventional shock tunnel and the wave processes in the driven section and the nozzle develop as discussed before. At the left end of the driver section, a damping section is added. This section is filled with low-pressure nitrogen. Initially, the detonation section is separated from the damping section by a thin diaphragm. After reaching the damping section, the detonation is converted into a shock wave, which attenuates in the damping section and is subsequently reflected at the end wall. The addition of this section is necessary in order to avoid the high mechanical loads on the tube when the detonation wave would be reflected at the end of the detonation section [16.83]. At the driver–damping section interface a rarefaction wave is generated that moves downstream and finally interacts with the reflected shock wave in the driven section (Fig. 16.76).

In the forward (or downstream) propagation mode, the detonation is initiated at the upstream end of the detonation section as indicated in Fig. 16.77 [16.84]. For the discussion of the basic working principle of the downstream propagation mode, the light gas driver shown in Fig. 16.77 should be ignored. After initiation, the burnt gas and the detonation wave move downstream towards the main diaphragm. As the detonation wave hits the main diaphragm, the hot accelerated combustion products, which yield an effective unsteady driver condition, drive the incident shock wave in the shock tube [16.84]. In the downstream propagation mode, the incident shock wave is attenuated by interaction with the Taylor expansion following the detonation wave, leading to unsteady flow conditions in the nozzle reservoir if a facility operation as sketched in Fig. 16.77 is used.

Due to the fact that with the forward propagation mode the kinetic energy behind the detonation wave can be used to drive the incident shock wave, higher shock Mach numbers can be achieved with this mode compared to the backward mode, where the driver gas is first accelerated toward the left end of the driver section thereby wasting some of the chemical energy of the detonation. This potential performance gain in the forward mode resulted in the development of different techniques to reduce and control the extent of the rarefaction caused by the Taylor expansion. The shock-induced detonation technique for example utilizes a light driver gas attached at the upstream end of the detonation section as shown in Fig. 16.77 [16.79]. This additional driver is used to generate a shock wave in the combustible driver gas mixture that subsequently initiates the detonation wave. The interface between the light driver gas and the detonation wave processed driver gas acts as a gaseous piston that leads to a higher pressure level behind the detonation wave as would be obtained if the detonation is initiated at a closed end wall. Further means of increasing the performance of tunnels operated in the forward mode are to increase the length of the detonation driver to decrease the flow gradients behind the detonation wave or to generate additional waves in the detonation driver to decrease the effect of the Taylor expansion, e.g., by an area change at the driver–driven section interface or the utilization of a cavity ring or a throat section in the detonation driver [16.83].

Examples of detonative combustion-driven shock tunnels are the TH2-D tunnel of the RWTH Aachen University, Germany, the JF-10 and the BFJ-60 tunnels of the Institute of Mechanics of the Chinese Academy of Sciences, Beijing [16.83] and the HYPULSE facility of the Alliant Techsystems Inc. – General Applied Science Laboratories (ATK–GASL) in Ronkonkoma, USA [16.79]. Using the backward (TH2-D) and forward propagation mode (JF-10), total specific-enthalpy and nozzle reservoir pressure conditions of 14.6 MJ/kg, 220 bar (resulting in a free-stream velocity of 4.9 km/s) and 19.6 MJ/kg, 800 bar respectively, are reported [16.83]. The available test time range is approximately 2–4 ms. According to the relations given in the framework of the discussion of the total specific enthalpy augmentation, these conditions would result in shock Mach numbers of $M_s \approx 11$ and $M_s \approx 13$. Concerning the ratio of the speed of sound in the driver and driven section, an effective value of $a_4/a_1 \approx 8$ can be achieved with detonative combustion-driven shock tunnels.

Electric-Arc-Driven Shock Tunnel. In electric-arc-driven shock tunnels, an electric discharge heats the driver gas hydrogen or helium to temperatures of about 8000 K for hydrogen, and about 20 000 K for helium. As in conventional shock tunnels, the driver gas is separated from the test gas by a diaphragm that bursts due to the rise in pressure within the driver during the capacitor-bank discharge, initiating the shock tube flow as discussed in previous sections. An example of this type of facility is the Electric Arc Shock Tunnel (EAST) at the NASA Ames Research Center [16.85–87]. The facility can be used in shock tube as well as shock tunnel mode. Due to its ability to generate very strong shock waves in the shock tube – in hydrogen shock wave speeds up to 50 km/s are reported – EAST is used to investigate the effects of radiation and ionization that occur during very high-velocity atmospheric entries such as that performed by the Galileo probe, which entered the atmosphere of the planet Jupiter in 1995.

The energy for the electric arc which heats the constant volume driver is supplied by a 1530 μF capacitor bank capable of storing 1.25 MJ when charged to a maximum of 40 kV. The arc, which is generated by the discharge of the capacitor bank through electrodes in the driver section, is contained within an insulated chamber and is initiated by a tungsten wire that extends along the driver. In the first phase of the discharge, the electric current passes mainly through the tungsten wire heating it up. This heat is transferred to the driver gas causing ionization and thereby allowing a portion of the electrical current to pass through the generated plasma. Subsequently, after reaching a temperature of about 4000 K, the tungsten wire explodes and vaporizes and an arc forms in the driver. Approximately 50 μs after the discharge was started, the diaphragm is fully open and the shock tube flow is initiated [16.86]. During the discharge process, a maximum current of approximately 600 kA is reached.

In the shock tunnel mode the EAST shock tube is connected to a 760 mm-diameter test section by a convergent–divergent nozzle. Using nitrogen as the test gas, tailored interface conditions were obtained at $7 \leq M_s \leq 19$, resulting in total specific enthalpies of $7 \, \text{MJ/kg} \leq h_0 \leq 40 \, \text{MJ/kg}$. Using helium heated up to 8000 K as the driver gas, this results in $a_4/a_1 \approx 15$. The nozzle reservoir pressure varies from 44 bar (40 MJ/kg) to 20 bar (7 MJ/kg). The test time ranges from 0.5 ms at the highest total specific enthalpy to 1.5 ms at the lowest total specific enthalpy [16.85].

Computational investigations of the potential operating conditions of EAST in reflected and straight-through shock tunnel mode using air as test gas show that shock Mach numbers of up to $M_s \approx 16$ would allow testing at equivalent flight velocities of up to 8700 m/s. For the reflected shock tunnel mode, the nozzle reservoir pressure drops when increasing the total specific enthalpy. In the equivalent flight velocity regime above 5500 m/s it ranges from about 400 bar at 14 MJ/kg to approximately 170 bar at 36 MJ/kg [16.87].

Free-Piston-Driven Shock Tunnels. In a free-piston-driven shock tunnel, the conventional driver of a shock

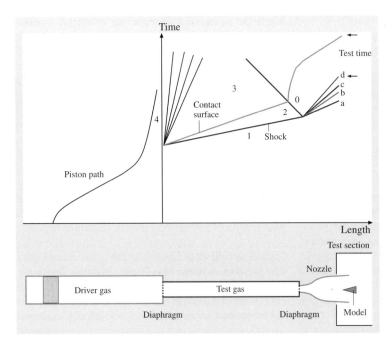

Fig. 16.78 Schematic and wave (x–t) diagram of a free-piston-driven shock tunnel; the waves a, b, c, and d are explained in the section related to reflected shock tunnel operation

tunnel is replaced by a free piston driver. This concept was proposed by *Stalker* [16.88]. A schematic and wave diagram of this type of facility is shown in Fig. 16.78. Free-piston-driven shock tunnels consist of an air buffer, a compression tube, separated from an adjoining shock tube via the primary diaphragm, and a subsequent nozzle, test section and vacuum tank. The high-pressure air stored in the air buffer is utilized to accelerate a heavy piston down the compression tube. During this quasi-adiabatic compression and heating of the light driver gas (typically helium or a helium argon mixture), the piston reaches a maximum velocity in the order of 300 m/s. The driver gas temperature increases with the driver gas volumetric compression ratio. In principle, there is no limit to the compression ratio used and therefore, there is also no limit to the stagnation enthalpy that can be generated [16.73].

When the main diaphragm burst pressure is reached it ruptures and the wave process, as in a conventional reflected shock tunnel, is initiated. The trajectory of the piston is chosen in such a way that, after the main diaphragm rupture, the pressure and temperature of the driver gas is maintained approximately constant. This is achieved by selecting the velocity of the piston at diaphragm rupture, and therefore the subsequent movement of the piston, such that it compensates for the loss of the driver gas flowing into the shock tube. For this reason, in contrast to the constant-volume driver of conventional shock tunnels, the free piston driver is a constant-pressure driver. Due to the large forces occurring during the operation of the free piston driver, the compression tube, shock tube, nozzle assembly is allowed to move freely in the axial direction. An inertial mass placed at the compression tube/shock tube junction can significantly reduce the recoil motion of the facility during operation. The test section and the vacuum tank remain stationary. A sliding seal is used at the nozzle/test section interface. In principle no limit on the achievable stagnation enthalpy exists for free-piston-driven shock tunnels. However, in real facility operation, effects such as contamination of the test gas impose upper limits. Tailored interface operation at shock Mach numbers of $M_s = 18.5$ in air has been reported [16.73, 88]. This results in a total specific enthalpy of about 40 MJ/kg.

Considering the different types of high-performance shock tunnel drivers, the free-piston technique appears to be the most developed and most distributed. After a series of free-piston-driven shock tunnels, T1–T3 (Australian National University, Canberra), T4 (The University of Queensland, Brisbane) were built in Australia between the early 1960s and the late 1980s [16.73, 89], this type of facility was implemented in a number of different institutions worldwide. The largest ones of these are the T5 at the Graduate Aeronautical Laboratories of the California Institute of Technology, USA [16.90], the High Enthalpy Shock Tunnel Göttingen (HEG) of the German Aerospace Center [16.91] and the High Enthalpy Shock Tunnel (HIEST) of the Japan Aerospace Exploration Agency (JAXA) at Kakuda [16.92]. The latter uses the largest nozzle with an exit diameter of 1200 mm. The maximum stagnation enthalpy used is 25 MJ/kg at a 1500 bar nozzle reservoir pressure, resulting in a test time of 2 ms.

Performance Considerations of Real Shock Tunnels.
The deviation from the ideal shock tunnel performance as discussed for the straight-through shock tunnel and for the reflected shock tunnel is caused by various phenomena, e.g. [16.56].

The opening of the main diaphragm has so far been considered as an instantaneous process. However, due to the high pressure ratios p_4/p_1 the diaphragm usually deforms before rupture followed by a gradual opening process of the diaphragm, e.g. [16.64]. Both effects result in an initially two-dimensional flow with a curved shock wave and contact surface. The shock wave propagates across the shock tube, resulting in repeated reflections at the wall and the center line. The interaction of the shock and the reflected waves cause the shock front to become planar within several tube diameters downstream of the diaphragm location. However, the bulging of the diaphragm and the gradual opening causes that the contact surface becomes and stays curved [16.93, 94]. For high initial pressure ratios across the diaphragm, it was found that the measured shock wave speed can exceed the shock speed predicted by one-dimensional model-

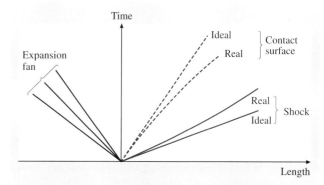

Fig. 16.79 Wave x–t diagram of shock tube performance for ideal (inviscid) and real (viscous) flow

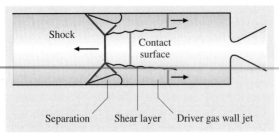

Fig. 16.80 Schematic of the reflected shock wave/boundary layer interaction which leads to shock bifurcation and the development of a driver gas wall jet

ing of the ideal shock tube. These higher than ideal shock speeds can be partially explained by the wave processes that occur during the gradual opening of the diaphragm [16.93].

A large impact on the real shock tunnel performance have viscous effects in the shock tube and the hypersonic nozzle. The effects of the growing boundary layer within the shock tube can be explained, for example, by the model developed by *Mirels* [16.95]. The wall boundary layer developing behind the moving shock acts as a sink that removes mass from the region between the shock wave and the contact surface. This causes deceleration of the shock wave and acceleration of the contact surface (Fig. 16.79). A limiting condition is reached at which the shock wave and the contact surface move at the same speed when the mass flow through the shock wave is equal to the mass flow in the boundary layer past the contact surface. Based on the ideal shock tube performance, the available test time as indicated in Fig. 16.71 would only depend on the length of the shock tube if the driver would be chosen sufficiently long such that the arrival of the reflected expansion waves is sufficiently delayed. However, the means of gaining test time by extending the shock tube length is limited due to the boundary-layer development.

For reflected shock tunnels, early driver gas contamination is an important issue that is also linked to the viscous flow developing in the shock tube. After reflection of the incident shock wave at the shock tube end wall, it intersects with the shock tube sidewall boundary layer. Due to this shock wave–boundary layer interaction, the reflected shock wave is bifurcated, and when it interacts with the contact surface a wall jet of driver gas is generated (Fig. 16.80), which arrives prematurely at the nozzle entrance where it contaminates the test gas [16.58]. Various devices have been developed to detect the arrival of driver gas in light-gas-driven facilities [16.73]. A discussion of this issue for detonative combustion-driven facilities is given [16.79].

The expansion of the test gas in hypersonic nozzles to high free-stream Mach numbers generates relatively thick boundary layers at the nozzle wall. This needs to be taken into account in the nozzle design process because the effective nozzle area ratio is reduced by viscous effects. Furthermore, the wave pattern dominating the nozzle starting process is influenced by the interaction with the nozzle wall boundary layer and multidimensional flow effects [16.96].

At high shock Mach numbers, the behavior of the test gas air deviates from that of an ideal gas due to the excitation of vibrational degrees of freedom of the molecules and chemical reactions that are generated in the shock heated gas. In the shock tube flow the test gas can be considered as being in thermal and chemical equilibrium. The largest differences between modeling the test gas air in the shock tube as an ideal gas or as a high-temperature equilibrium gas occur at high shock Mach numbers and low shock tube filling pressures. The temperatures in region 2 (behind the incident shock wave) and region 0 (behind the reflected shock wave) are significantly reduced due to the high-temperature effects [16.97].

When operating reflected shock tunnels at high total specific enthalpies, the temperature in the nozzle reservoir can be as high as $10\,000$ K. If air is used as the test gas the molecular species oxygen and nitrogen are highly dissociated. This high-temperature, high reservoir slug of test gas is subsequently expanded in a convergent–divergent nozzle to hypersonic velocities. In the first

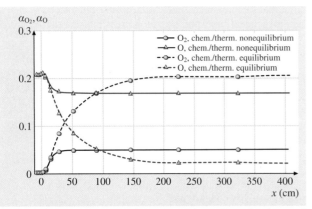

Fig. 16.81 Computed concentration of molecular and atomic oxygen along the center line of the HEG nozzle; the flow is modeled in chemical and thermal equilibrium as well as nonequilibrium conditions

part of the expansion, the resident time of a particle in a certain flow environment is long enough that the flow remains in thermal and chemical equilibrium. However, with increasing flow velocity a condition is reached that the characteristic time to reach equilibrium is larger than the residence time at the corresponding flow condition. Therefore, the chemical relaxation process, i. e., the recombination of dissociated species freezes. An example of this process is shown in Fig. 16.81 for the High Enthalpy Shock Tunnel Göttingen [16.98]. In this figure, the computed mass fraction of molecular and atomic oxygen is plotted as function of the stream wise direction along the nozzle center line, starting from the nozzle reservoir until the nozzle exit for a high-enthalpy condition. In the nozzle reservoir, the molecular oxygen is fully dissociated. Due to the decreasing temperature in the nozzle expansion, the oxygen atoms recombine. The different flow models assuming the flow in chemical/thermal equilibrium or in chemical/thermal nonequilibrium reveal the freezing point in the nozzle expansion about 25 cm downstream of the nozzle throat. The corresponding translational/rotational temperature and vibrational temperature distributions along the nozzle center line are shown in Fig. 16.82. The applied model to describe the thermal relaxation process shows that similar to the chemical relaxation, the thermal relaxation also freezes approximately 25 cm downstream the nozzle throat. Due to the partly dissociated free-stream flow which is generated in reflected high-enthalpy shock tunnels, the calibration process of these facilities is very complex and the knowledge of the free stream is important for the evaluation of the data obtained in high-enthalpy shock tunnels. An approach that strongly couples detailed diagnostics and computational fluid dynamics tools is required [16.99].

Shock Expansion Tubes/Tunnels

In reflected shock tunnels, the heating of the test gas is performed by the incident and reflected shock wave resulting in very high nozzle reservoir temperatures at high total specific enthalpy operating conditions. In practice this imposes limitations on shock tunnel operation at very high enthalpies. Shock expansion tubes/tunnels [16.100–103] overcome these limitations by avoiding the stagnant flow region in the nozzle reservoir.

The wave x–t diagram of an ideal shock expansion tube is shown in Fig. 16.83. The set up of this type of facility is similar to a shock tube with the addition of a light secondary diaphragm that separates the driven section into two parts: an intermediate section following the driver section and an acceleration section further downstream. Similar to the shock tube operation, the driver section is initially filled to high pressure with a low-molecular-weight gas such as helium. The shock tube is filled with the desired test gas to a pressure that is generally sub-atmospheric. The acceleration section (region 10 in Fig. 16.83) is filled with the acceleration gas at a lower pressure than the initial filling pressure of the shock tube. Typically, the same gas is used in both the shock tube and acceleration section. The flow is initiated by bursting the primary diaphragm leading to the wave processes as discussed in Sect. 16.2.1. Upon strik-

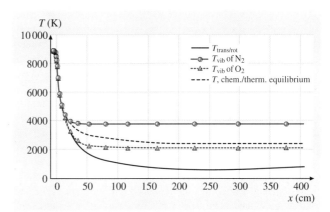

Fig. 16.82 Numerical prediction of the translational/ rotational temperature and vibrational temperatures of molecular nitrogen and oxygen along the nozzle center line of the HEG nozzle resulting from different flow models (chemical and thermal equilibrium and nonequilibrium)

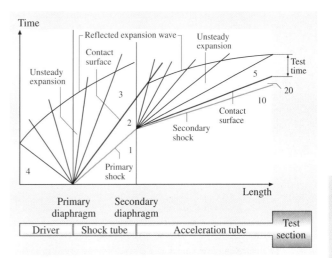

Fig. 16.83 Wave x–t diagram of an ideal shock expansion tube

ing and rupturing the secondary diaphragm, the incident (primary) shock wave acquires a higher Mach number as it enters the acceleration section (secondary shock wave) generating the flow in region 20. In order to equilibrate the pressure and velocity from region 2 to 20 an unsteady expansion is generated that accelerates the test gas of region 2 to the high velocity in region 5. The test section is located at the exit of the acceleration section. The available test time is the period between the arrival of the acceleration gas/test gas interface and the first wave that disrupts the uniform test flow. This can be caused by the arrival of the tail of the unsteady expansion wave generated after bursting of the secondary diaphragm. A second wave that can be responsible for terminating the uniform test flow is the reflected expansion wave which is generated when the leading wave of the unsteady expansion intersects with the driver gas–test gas interface. As indicated in Fig. 16.83, in an optimum shock expansion tube configuration, both limiting waves arrive simultaneously at the test section location. Adjustments to achieve a particular operating condition of the shock expansion tube are made by varying the driver gas speed of sound and pressure, as well as the shock tube and acceleration section filling pressures, and by changing the relative lengths of the shock tube and acceleration tube sections by changing the location of the secondary diaphragm.

The difference between a shock expansion tube and a reflected shock tunnel is the unsteady expansion from region 2 to 5. In a shock tunnel all the energy is added to the flow by heating of the incident and reflected shock wave. Subsequently, the test gas is expanded to high Mach numbers by means of a steady expansion in a convergent–divergent nozzle. The steady expansion is a constant total specific enthalpy process and is described by

$$H = h + \frac{1}{2}u^2 = \text{const},$$

or $dH = 0$. In a shock expansion tube, the incident (primary) shock wave adds only a fraction of the final enthalpy of the flow and the test gas is not stagnated. It is processed by the unsteady expansion which is generated after the secondary diaphragm burst. For this process,

$$2\frac{a}{(\gamma - 1)} + u = \text{const}$$

applies. From this follows that

$$dH = -(M-1)dh.$$

Since the incident shock wave is sufficiently strong to establish supersonic flow ($M > 1$) in the test gas upstream of the unsteady expansion and since for an expanding flow $dh < 0$, dH is always positive. Therefore, it follows that the total specific enthalpy and total pressure are amplified by the unsteady expansion. The total pressures achievable in the test section are far in excess of the initial pressure of the driver gas. The shock expansion tube process adds more energy per unit mass of test gas than the reflected shock tunnel process and is therefore better suited for very high-enthalpy ground testing [16.59]. However, the available test time is generally approximately one order of magnitude shorter than that for reflected shock tunnels when comparing facilities of similar size. In a shock expansion tube, the wind tunnel model is placed either inside or at the end of the acceleration tube. Therefore, the core flow and model size is limited by the tube diameter. Similar to straight-through shock tunnels, a divergent nozzle can be added to the end of the shock expansion tube to expand the hypersonic flow to a larger area. This configuration is termed the shock expansion tunnel, e.g. [16.79, 102].

For the former NASA Langley Research Center shock expansion tube/tunnel with a 15.24 cm inner diameter tube, using an unheated helium driver to accelerate test gases such as dry air or carbon dioxide to test section velocities of approximately 5.6 km/s, test times of 100–300 µs are reported [16.102]. The large 0.6 m inner diameter of the LENS X shock expansion tunnel at the Calspan-UB Research Center Buffalo, New York, USA allows test times of approximately 2–3 ms to be achieved at a flow velocity of about 3.5 km/s [16.104].

Techniques for Total-Specific-Enthalpy Augmentation in Shock Expansion Tubes/Tunnels. Since the first part of a shock expansion tube or tunnel is identical to a shock tube or tunnel, the same type of different driver techniques as used for shock tunnels can be applied for shock expansion tubes/tunnels to increase the shock Mach number in the shock tube. An overview of possible driver techniques is given in [16.82]. In the subsequent sections, examples of these applications are given.

Detonative Combustion–Driven Shock Expansion Tube/Tunnel. The forward running detonation driver of the HYPULSE facility of ATK-GASL in Ronkonkoma, USA [16.79] is used to operate the facility as a reflected shock tunnel as well as a shock expansion tube/tunnel. The set up of the different tube sections of the facility is flexible such that optimum component sizing for its different operating modes can be achieved. For shock expansion tube/tunnel operation, the ratio of shock to acceleration tube length is chosen in such a way that the reflected expansion waves and the secondary expansion wave all arrive simultaneously at the end of the acceleration tube. The facility with an inner tube diameter of 15.24 cm is used to generate hypersonic flows up to a total specific enthalpy of 12–22 MJ/kg. For these conditions, test times of 0.25–0.55 ms are achieved.

Free-Piston-Driven Shock Expansion Tubes/Tunnels. In order to study the flow past spacecrafts at superorbital velocities (above 8 km/s) in various atmospheres of the solar system, a series of free-piston-driven shock expansion tubes/tunnels was developed at the University of Queensland, Australia [16.73, 105, 106]. These facilities basically utilize the free piston driver as described in the section related to free piston driven shock tunnels. Modifications of this driver concept by using a two-stage piston or the addition of a compound driver to the basic shock expansion tube set up shown in Fig. 16.83 were investigated in order to increase the performance of superorbital shock expansion tubes [16.106]. The X3 facility is the largest of the superorbital shock expansion tubes/tunnels at The University of Queensland [16.107]. The inner diameter of the acceleration tube is 18.3 cm. The operation of X3 relies on a two-stage free-piston compression process, to generate a high-temperature compressed driver gas. Typical test times of the order of 300 μs are achievable at test section flow velocities of about 8.4 km/s. For the smaller X2 facility with an inner accelerometer tube diameter of 8.5 cm, total specific enthalpy conditions of 60 MJ/kg (10.3 km/s) using nitrogen as the test gas are reported. The test time for this condition is of the order of 50 μs [16.108]. The X1 facility has an accelerometer tube diameter of 3.7 cm. Electron density investigations were performed in an air test flow at a total specific enthalpy of 110 MJ/kg (12 km/s) with a test time of 20 μs [16.109].

Performance Considerations of Real Shock Expansion Tubes/Tunnels. Similar to shock tubes, there are a number of non-ideal processes during operation of the shock expansion tube that cause the actual operating cycle to deviate from the ideal. In addition to the phenomenon related to the opening of the primary diaphragm and the viscous attenuation of the incident (primary) shock wave as it progresses down the shock tube, the non-ideal rupture of the secondary diaphragm is an important feature of shock expansion tube operation.

In the distance–time (x–t) wave diagram of an ideal shock expansion tube as shown in Fig. 16.83 it is assumed that the secondary diaphragm ruptures instantaneously on impact by the incident (primary) shock wave, and that its mass contributes no inertia to the test gas for subsequent acceleration. However, in practice the diaphragm rupture requires a finite period of time, resulting in a reduction of the available test time [16.59]. Furthermore, the incident shock wave is reflected off the diaphragm traveling upstream into the oncoming test gas. For air as the test gas, the entropy generated may be sufficient to cause significant oxygen dissociation that can only be eliminated by recombination in the subsequent unsteady expansion. Investigations of this issue and additional references can be found, e.g., in [16.93, 110].

Further, viscous attenuation of the secondary shock wave and growth of the boundary layer along the acceleration tube walls influence the test gas conditions and test section flow. At total specific enthalpy conditions on the order of 22 MJ/kg, the level of test flow dissociation is reduced in shock expansion tubes/tunnels when compared with reflected shock tunnels. However, for extremely high total specific enthalpies ranging from approximately 60–100 MJ/kg, nonequilibrium chemistry effects during the unsteady expansion have to be considered for molecular test gases. Using, e.g., air as test gas at these high total specific enthalpy conditions, results in a highly dissociated free stream.

16.2.2 Measurement Techniques

Pressure

Surface pressure measurements in short-duration facilities require different approaches than for continuously running facilities. Because the measurement time is short, pressure transducers with fast response times have to be used. Additionally the susceptible area of the transducer has to be installed close to the surface to minimize the filling time of the tubing system in front of the susceptible area. The most commonly used pressure gauges in short-duration hypersonic ground-based test facilities are based on the piezoelectric and on the piezoresistive effect. Piezoelectric pressure transducers are explained in detail in Sect. 8.2. The piezoresistive effect leads to an increase of the resistance of a semiconductor when a pressure load is applied. Optical fibre pressure transducers, which are based on different principles including intensity modulation, interferometry, polarization effects, refractive index changes, reflectometry and fibre Bragg grating are currently under

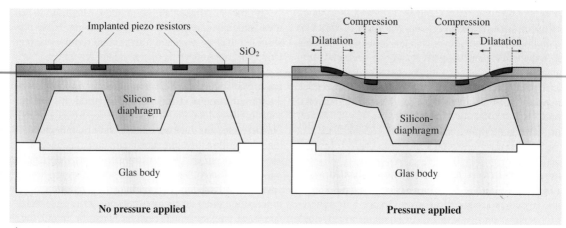

Fig. 16.84 Schematic of the Kulite piezoresistive pressure transducer

development. A review of micromachined pressure gauges is given by *Eaton* and *Smith* [16.111]. A recent development of an optical pressure gauge based on a Fabry–Pérot interferometer [16.112] with a natural frequency above 1 MHz enables data capture over a bandwidth exceeding 100 kHz. This results in response times that are suitable for short-duration hypersonic testing.

Miniaturized fast piezoresistive pressure transducers are manufactured by Kulite. Depending on the pressure range, the Kulite XCEL-100 pressure gauge with a diameter of 2.4 mm has a natural frequency between 240 kHz and 1 MHz and a bandwidth of 20 kHz. Figure 16.84 shows the schematic of a Kulite pressure transducer. It uses a silicon element for the mechanical diaphragm structure, and the sensing element itself is an integral part of the silicon element. The piezoresistors are formed within the silicon diaphragm by either diffusion or implantation of atoms. Two of these resistors are positioned on the silicon diaphragm such that they experience a compressive strain, and two are positioned where they experience a tensile strain. They are connected such, forming a fully active Wheatstone bridge. These transducers may be manufactured at very small sizes: diameters of the housing can be in the range 1–2 mm.

To achieve fast response times, it is necessary to position the susceptible area of the transducer as close as possible to the model surface. As shown in Fig. 16.85, small holes are drilled in the model which act as pressure tappings. The gauge is placed directly behind these tappings. The influence of the pressure tapping shape and manufacturing quality on the measured signal is discussed in Sect. 4.1. In Fig. 16.85 the schematic of two different types of pressure gauge installations as used in short-duration facilities is shown.

The pressure transducers are colored dark brown. Typical tapping diameters range from 0.5 to 1.5 mm. The schematic on the left-hand side of Fig. 16.85 shows an installation suited for tappings perpendicular to the flow. This installation is not suited for stagnation regions, where the flow may directly stagnate on the pressure transducer leading to high heat flux on the susceptible area. A stagnation point installation is shown on the right-hand side of Fig. 16.85. Here the susceptible area of the transducer is protected by a stagnation plate. Examples of pressure measurements using both configurations are given in Fig. 16.86.

This test has been conducted in a small shock tube. Two transducers have been installed in the end wall, where they measure the pressure rise caused by the incident shock wave. The presence of the stagnation plate extends the rise time for the used tapping diameter and cavity size by about 0.15 ms.

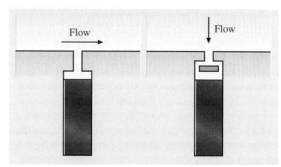

Fig. 16.85 Installation of pressure transducers into a model wall (transducers shown in dark brown)

Surface Heat Flux

In short-duration high-enthalpy, hypersonic ground-based test facilities the most common way to measure wall heat flux is by utilizing thin-film gauges or thermocouples. Due to the harsh environment in these facilities with high wall shear stresses during the starting process of the flow, the most robust and reliable technique to measure heat flux is the application of surface-mounted coaxial thermocouples. The principle of operation of thermocouples is the Peltier–Seebeck or thermoelectric effect [16.113].

In order to allow a detailed instrumentation of wind tunnel models, miniaturized thermocouples with diameters as small as 0.4 mm are used. Figure 16.87 shows a schematic view of a miniaturized coaxial ultrafast thermocouple manufactured by Medtherm Corporation [16.114]. Coaxial thermocouples are built with a core consisting of one conductive material separated by insulation from the outer ring made of a different material. Possible material combinations and their official Instrument Society of America (ISA) code are given in Table 16.3 together with their application range. The electrical connection on the top surface is realized either by coating the surface with one of the materials or simply by sanding of the transducer surface. In either case the mass of the electrical connection in comparison to the mass of the transducer body is small leading to response times of thermocouples in the order of a few μs. Additionally, the heat flux into the transducers is dominated by the thermocouple body and the junction itself is negligible due to its low mass. The sensitivity of thermocouples range from $1\,\mu V/K$ to $70\,\mu V/K$. The thermocouple inherits the advantage that the shape of the transducer can be adopted to any wall curvature by

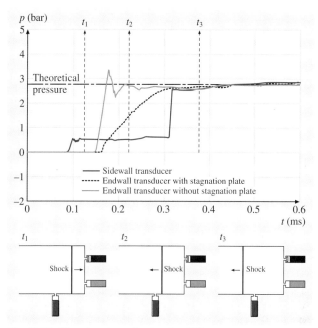

Fig. 16.86 Rise-time investigation for different pressure gauge installations in shock tube walls

sanding. In case of transducer failure, it can be repaired in situ by sanding the surface again.

Alternatively to thermocouples, in areas with less-harsh flow environments, such as wake flows or surfaces at small angles of attack, thin-film gauges can be used to measure the temperature history in order to evaluate the heat flux (see e.g. [16.115, 116]). Thin-film gauges consist of a nonconductive substrate on which a metallic film with a typical thickness of less than $1\,\mu m$ is met-

Table 16.3 Possible material combinations and application ranges of thermocouples

ISA code	Material A (+)	Material B (−)	Application range
B	Platinum 30% rhodium	Platinum 6% rhodium	1640 K to 1970 K
C	95% W5Re tungsten 5% rhenium	W26Re tungsten 26% rhenium	1920 K to 2585 K
E	Chromel	Constantan	365 K to 1170 K
J	Iron	Constantan	365 K to 1030 K
K	Chromel	Alumel	365 K to 1530 K
N	Nicrosil	Nisil	920 K to 1530 K
R	Platinum 13% rhodium	Platinum	1140 K to 1720 K
S	Platinum 10% rhodium	Platinum	1250 K to 1720 K
T	Copper	Constantan	70 K to 620 K

Chromel* – 90% nickel, 10% chromium; alumel* – 95% nickel, 2% manganese, 2% aluminium, 1% silicon; constantan – 55% copper, 45% nickel; nisil+ – 95.6% nickel, 4.4% silicon; nicrosil+ – 84.1% nickel, 1.3% silicon, 14.6% chromium
* Trademarks of Hoskins Manufacturing Company
+ Trademarks of Harrison Alloys Inc.

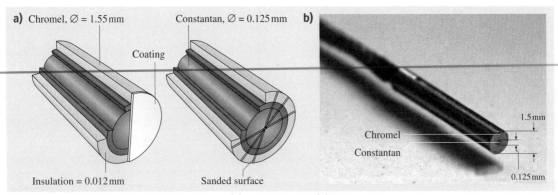

Fig. 16.87a,b MedTherm type E thermocouple: (**a**) schematic, and (**b**) photograph

allized. The working principle of these gauges is based on the dependence of the electrical resistance of the metallic film on temperature. For the metallic film noble metals are used because they exhibit the largest sensitivity. Again, the mass of the susceptible area is very small compared to the body in which the heat is conducted. Therefore, similar response times as for thermocouples are obtained.

As an example (Fig. 16.88) a thin-film gauge as manufactured by the French–German Research Institute of Saint Louis (ISL) is shown. The measuring point is a thin platinum film, which is metallized on a glass body. The electrical connection to the measurement wires is realized by a silver coating of the upper and lower third of the glass body. The typical diameter of these gauges is in the order of millimeters. The temperature sensitivity and signal-to-noise ratio of thin-film gauges are superior to thermocouples. However, disadvantages are the missing robustness in harsh flow environments usually found in short-duration facilities, the large effort of adapting the susceptible area to curved surfaces and the missing possibility to repair the gauge in situ.

Two assumptions are used to evaluate the heat flux from the measured temperature traces using coaxial thermocouples or thin-film gauges. The first assumption is that the measurement is dominated by one-dimensional heat conduction in the sensor body, and the second is that the gauge itself acts as a semi-infinite body [16.117]. To ensure that the first assumption is valid, only the transducer face should be exposed to the flow, and the combination of wind-tunnel model wall material and transducer material has to be chosen in an appropriate way. The validity of the semi-infinite body assumption requires $x \geq \sqrt{\alpha t}$ [16.118], where x is the length of the transducer, α the diffusivity and t the time. The diffusivity α is defined as the ratio of the heat conductivity κ and the product of density ρ and specific heat capacity c, $\alpha = \kappa/(\rho c)$. For thermocouples of ISA code E with a length of 10 mm the time before the heat reaches the end of the transducer is approximately 3 s. This period of time is orders of magnitude higher than the typical test times in short-duration facilities of a few ms. If the two assumptions mentioned can be regarded as valid, the heat conduction problem in the transducer body is described by the following differential equation:

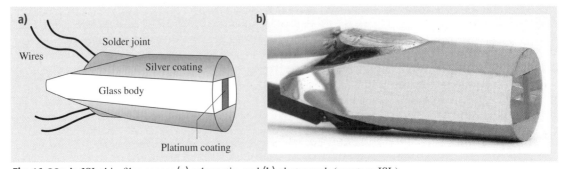

Fig. 16.88a,b ISL thin-film gauge: (**a**) schematic, and (**b**) photograph (courtesy ISL)

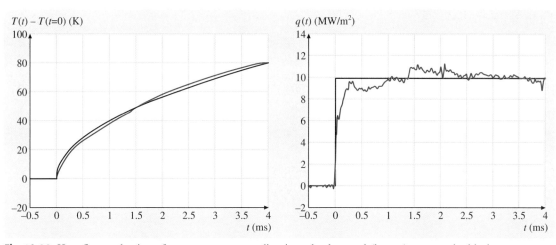

Fig. 16.89 Heat flux evaluation of a temperature recording in a shock tunnel (brown) compared with the temperature evolution based on a stepwise heat transfer load (black)

$$\frac{\partial^2 T}{\partial x^2} = \frac{1}{\alpha}\frac{\partial T}{\partial t}. \tag{16.2}$$

Here, T is the temperature minus a suitable reference temperature, x is the coordinate normal to the wind tunnel model wall and t the time. With the boundary conditions that the heat flux into the model wall, $\dot{q}_W(x=0) = -\kappa(\partial T/\partial x)$ and $T(x=\infty) = 0$, the following equation can be derived for the evaluation of the heat flux

$$\dot{q}_W(t) = \sqrt{\frac{\rho c \kappa}{\pi}}\left(\frac{T(t)}{\sqrt{t}} + \frac{1}{2}\int_0^t \frac{T(t)-T(\tau)}{(t-\tau)^{3/2}}\,d\tau\right), \tag{16.3}$$

as shown by *Schultz* and *Jones* [16.117].

For the evaluation of the wall heat flux by numerical integration of digitally stored temperature data, this relationship is replaced by a discretized form such as

$$\dot{q}_W(t_n) = 2\sqrt{\frac{\rho c \kappa}{\pi}}\sum_{j=1}^n \frac{T_j - T_{j-1}}{\sqrt{t_n - t_j} + \sqrt{t_n - t_{j-1}}},$$

proposed by *Cook* et al. [16.119] or an alternative scheme described by, e.g., *Kendall* et al. [16.120].

In Fig. 16.89 an example obtained with the data evaluation method of *Cook* et al. [16.119] is presented. In the left plot of Fig. 16.89, the temperature development measured by a type E thermocouple installed in the stagnation point of a sphere positioned in a hypersonic flow with total specific enthalpy of $h_0 = 15\,\mathrm{MJ/kg}$ is indicated by the brown line. In the right plot of Fig. 16.89 the resulting heat flux versus time (brown line) is shown. If the heat flux into the wall is given by an ideal step function, the exact solution for the temperature rise at $x=0$ resulting from (16.2) is

$$T_W(x=0, t) = \frac{2\dot{q}_W}{\sqrt{\pi}}\sqrt{\frac{t}{\rho c \kappa}}. \tag{16.4}$$

Using a constant heat flux into the transducer body of $\dot{q}_W = 10\,\mathrm{MW/m^2}$ as approximation of the measured heat flux evolution (the black line in the right plot of Fig. 16.89), the parabolic temperature history pro-

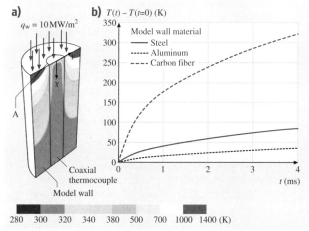

Fig. 16.90a,b Unsteady finite-element analysis of the heat conduction process in a type E thermocouple and the adjacent model wall for a wall heat flux of $10\,\mathrm{MW/m^2}$; axisymmetric temperature distribution at $t = 4\,\mathrm{ms}$ for a carbon-fibre model wall (**a**) and time evolution of the temperature at the interface (A) between the thermocouple and the model wall for different wall materials (**b**)

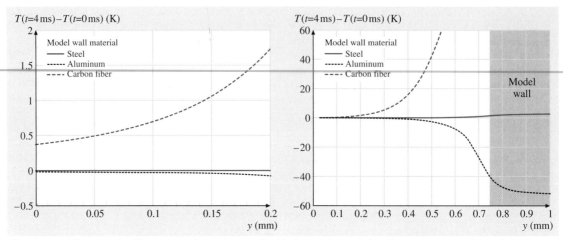

Fig. 16.91 Difference of radial computed temperature distribution at the model wall and thermocouple surface obtained for different thermocouple/wall material combinations compared to the combination chromel/chromel for which no temperature difference between the thermocouple and wall exists

file shown as black line in the left plot of Fig. 16.89 is obtained.

To ensure that the measurement is not affected by conduction in the wall tangential direction, the thermocouple type should be chosen in such a way that the material properties of the model wall match those of the thermocouple at the measurement position. The influence of different thermocouple/model wall material combinations on the measured temperature distribution is shown in Fig. 16.90 and Fig. 16.91. These results were obtained by numerical investigation of the unsteady heat conduction process in a type E thermocouple with 1.5 mm diameter and a length of 15 mm, installed in model walls of different materials, using Ansys (CAD-FEM GmbH). The thermocouple is modeled as a solid cylinder consisting of chromel. At the top surface a constant heat flux of $q_w = 10\,\mathrm{MW/m^2}$ is applied. The temperature evolution at the contact point between the thermocouple and the model wall is given in Fig. 16.90b for three different model wall materials. Depending on the material, large differences of the temperature evolution are obtained.

The difference of the radial temperature distribution at the model wall and thermocouple surface obtained for the thermocouple/wall material combination chromel/chromel, i.e. a combination for which no temperature difference between the thermocouple and the model wall exists, and the material combinations chromel/steel, chromel/aluminium and chromel/carbon fibre are plotted in Fig. 16.91. On the left-hand side of Fig. 16.91, a zoom of the core region of the thermocouple is shown. The right-hand side of Fig. 16.91 shows the distribution up to a radius of 1 mm. In case of a coated thermocouple with dimensions as shown in Fig. 16.87, the temperature measurement takes place at $y = 0{-}0.0625\,\mathrm{mm}$, the location of the chromel/constantan junction. For a thermocouple which is adapted to a three-dimensional surface or repaired in situ by sanding, the chromel/constantan junction is not precisely defined and the complete thermocouple surface may act as the measurement location. As can be seen from Fig. 16.91, depending on the thermocouple/wall material combination the assumption of one-dimensional heat conduction can be violated.

From (16.3) it becomes evident that the error in the estimation of the heat flux is directly proportional to the error of the value of $\sqrt{\rho c \kappa}$. To determine this value accurately two procedures may be applied. The contact procedure is used for thermocouples and thin-film gauges and is based on the fact that under the assumption of one dimensional heat conduction, the contact temperature of two media is only dependent on their temperature before contact and on the thermal properties ($\sqrt{\rho c \kappa}$) of both media. Knowing the thermal properties of one media allows to determine the gauge properties. A possible set up is a mechanically driven device that generates an instantaneous contact between the gauge and a liquid surface. Such a method is discussed in [16.121]. The electric discharge calibration can only be applied for thin-film gauges. It utilizes the effect that with an electric discharge, the thin film itself can be heated with a known and constant heat flux. Following (16.4), the measurement of the temperature of the thin-film gauge results in the determination of $\rho c \kappa$. This procedure is described in e.g. [16.122].

Forces and Moments

The ability to perform integrated force and moment measurements in ground-based testing facilities is an important part of the design and development of hypersonic vehicles. For a force balance to operate in short-duration ground-based test facilities with test times of order ms or less, it would become necessary for the balance to have an extremely short response time. However, given the short test times and the response times of force measurement techniques for conventional wind tunnels, e.g. [16.123], static equilibrium between the model and support structure is rarely established, i.e., it may only be obtained using unrealistically small models. Therefore, it becomes necessary to use specialized measurement methods that account for the dynamic response of the system. The basic force measurement techniques that have been applied in short-duration flows are outlined in the subsequent paragraphs. An overview of these techniques is given in [16.124].

Free-Flight Force Measurement Techniques. By using a free-flying (or a very weakly constrained) model, it is possible to directly measure the motion of a model in a wind tunnel. This can be achieved by measuring model accelerations (e.g. [16.125, 126]) or by visualization techniques to record the time history of model displacements (e.g. [16.127]). From these measurements, the applied forces and moments can be recovered.

The experiments performed by *Naumann* et al. [16.125] involved a model instrumented with accelerometers that was released by a fast-opening chuck just before the test flow arrives. The model used was 140 mm in length, and small quartz accelerometers with built-in amplifiers from PCB (model 309A) were used. Motion during the first few ms was small. Therefore, thin wires from the accelerometers were required such that the motion of the model was not disturbed. This technique overcomes the assumptions regarding balance stiffness, as there are no connections between the model and the support structure. By using an inertial matrix of the model and the measured accelerations, the applied forces can be determined via Newton's law (e.g. [16.128]). It is necessary to have the model equipped with six or more accelerometers to realize a six-component force balance and the inertial matrix must be determined in each individual case. The advantage of this technique is that it is relatively simple, avoiding expensive devices such as telemetry and complicated isolating supports. Further, once the inertial matrix is known, the data evaluation procedure is straightforward.

Other investigations by *Kussoy* and *Horstmann* [16.127] relied on recording photographically the trajectory of a free flying model. Since the model's displacement of its center of mass was much larger than any displacement due to model flexibility, reasonable results were obtained. *Bernstein* and *Scott* [16.129] attempted to account for model flexibility in free flying models. This was necessary as the facility used had a shorter test time, resulting in smaller displacements and therefore model flexibility was an issue. Laser interferometry was used for the detection of small displacements. The model was 120 mm long and the resulting displacement data was fitted to parabolic curves. Hence, it was assumed that the model had constant acceleration and thus failed to overcome the underlying problem of model flexibility.

The next two methods to measure forces and moments which will be discussed, the strain gauge force balance and the stress wave force balance, are based on utilizing a model/support structure assembly.

The basic features of the response of a structure as function of time after a load has been applied can be discussed by looking at a simple one degree of freedom system. If damping is neglected, such a system of mass M on a spring of stiffness k obeys the equation of motion

$$M\ddot{u} + ku = F(t) ,$$

where u is the displacement, $\ddot{u}$ the acceleration (second time derivative of the displacement), and $F(t)$ the loading. Assuming a sudden application of a constant load F_0, the solution to the equation of motion is

$$u = \frac{F_0}{k}[1 - \cos(\omega t)] ,$$

with $\omega = \sqrt{k/M}$ and the fundamental natural frequency $f = \omega/(2\pi)$. These relations show, that the natural frequency of the system can be increased by increasing the stiffness and reducing the mass.

Strain Gauge Force Balances. Strain gauge force balances are used in short-duration test facilities for measurements of forces and moments. In combination with a model of sufficiently high stiffness and low moments of inertia, natural frequencies of approximately 1 kHz can be achieved. Although acceleration compensation can be used to account for support flexibility, it generally remains an issue. Thus, in order to avoid flexible models, they are generally small, un-instrumented and lightweight.

Jessen and *Grönig* [16.130] designed a six-component strain gauge force balance for use in the RWTH Aachen University Shock Tunnel in Germany. This facility has a test time of 2–6 ms. The design resulted in a compromise between high stiffness and sufficient strain sensitivity (i. e., models of high stiffness and low moments of inertia were utilized). The balance was designed to work without acceleration compensation. Because the balance is part of the sting it is virtually independent of the model and may be calibrated statically with a high accuracy. However, strong coupling between the six force components was seen, resulting in extensive calibration being required to obtain accurate results. Since interference between components is significant, a third-order description of the system was used to calibrate the force balance. The relation between the signal and load is given by

$$S_i = R_{oi} + \sum_{j=1}^{6} a_{ij} Z_j + \sum_{j=1}^{6} \sum_{k=j}^{6} b_{ijk} Z_j Z_k + \sum_{j=1}^{6} C_{ij} Z_j^3 ,$$
$$i = 1, 2, 3, \ldots, 6 ,$$

where S_i is the signal of the strain gauge bridge for component i, R_{oi} is the signal of bridge i without load, Z_j the total load acting on component j, a_{ij} the calibration factors for linear terms, b_{ijk} the calibration factors for square terms, and c_{ij} the calibration factors for cubic terms.

This balance was also investigated further by *Störkmann* et al. [16.131]. The balance outputs were improved with compensation of low frequency oscillations but no attempt was made to account for model flexibility. Compensation was performed using two methods – sting-based acceleration compensation, which compensated only for frequencies less than the first resonant frequency of the system, and model-based compensation in which six accelerometers were used to measure the acceleration of the model. Experiments were conducted on three different models – a conical model, a model of the Apollo command module and a model of the elliptical aerodynamic configuration (ELAC) I (a delta wing configuration). Good results were obtained with the conical and Apollo models as they were sufficiently stiff, however, the ELAC I model was quite flexible. Consequently the balance outputs for this model contained large oscillations.

As discussed in Sect. 16.2.3, a similar type of strain gauge force balance with acceleration compensation was used in the HIEST shock tunnel of JAXA, Japan by *Itoh* et al. [16.132] for the investigation of the aerothermodynamic characteristics of the H-II Orbiting Plane (HOPE) space vehicle.

Stress Wave Force Measurement Technique. The stress wave force balance relies on the ability to measure the dynamic response of the model and supporting structure, and consequently any effects such as model flexibility and mass distribution are accounted for (e.g. [16.124]). Instead of using a static calibration as done for the strain gauge force balances, stress wave force balances require a dynamic calibration of the complete model, balance, and support structure assembly.

Upon flow arrival, stress waves propagate through the model at the speed of sound of the material and subsequently enter a stress bar that is instrumented with a strain gauge to record the time history of strain. If the system is linear, the resulting strain signal $y(t)$ can be related to the applied aerodynamic load $u(t)$ via an impulse response function $g(t)$ as described by the convolution integral,

$$y(t) = \int_0^t g(t-\tau) u(\tau) d\tau .$$

The aerodynamic force in an experiment can be determined by the deconvolution of the strain signal with the impulse response function. The impulse response function is determined either through experimental calibration or through finite element analysis. However, in order to reduce errors due to modeling approximations it is usually preferable to determine the impulse response function experimentally.

It has been found that deconvolution is best performed in the time domain using the algorithm of *Prost* and *Goutte* [16.133]. Additionally, beginning the deconvolution technique using the time history of Pitot pressure scaled in magnitude to the expected force component level improves the accuracy and stability of the process [16.134].

Sanderson and *Simmons* [16.135] first demonstrated the technique by measuring the drag force on a 15° conical model, 200 mm in length in a reflected shock tunnel. The model was made from aluminum and attached to a 2 m-long hollow brass bar. This stress bar or sting, was instrumented with strain gauges to record the axial strain time history. The test time was approximately 1 ms and the aerodynamic load was subsequently found through a numerical deconvolution process. The measured drag was found to be in good agreement with theoretical calculations.

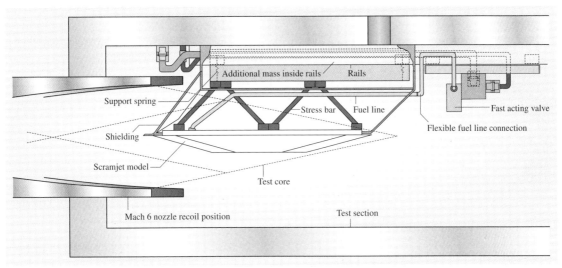

Fig. 16.92 Schematic of scramjet model and external stress wave force balance in the test section of the T4 shock tunnel (after *Robinson* et al. [16.136])

The technique was then extended to measure multiple components of force by *Mee* et al. [16.134]. This was done on a 15° cone with an internal balance arrangement. The lift, drag, and pitching moments were measured by short, stiff stress bars instrumented with strain gauges. The technique has subsequently been applied to a wide variety of models, including models with simultaneous measurements of pressure and heat flux. The major disadvantage with this technique is that aerodynamic shielding and vibration isolation of the support structure is required to separate the stress waves generated in the model from that of the test section environment. Calibration of the model is usually performed using an instrumented impact hammer or via a cut-weight technique. In order to maximize the performance of stress wave force balances, an individual design for each configuration to be tested is recommended.

Stress wave force balances have been developed with external as well as internal stress bar arrangements. A three-component stress wave force balance used by *Robinson* et al. [16.136] to measure the thrust, lift, and pitching moment of a supersonic combustion ramjet (scramjet) model in the T4 shock tunnel of The University of Queensland, Australia is shown in Fig. 16.92. Four stress bars are located above the model. They connect the scramjet model to a base plate. The bars are instrumented to measure axial strain using piezo-resistive strain gauges. The heavy base plate is suspended from the test section by light support springs to provide vibration isolation of the balance from the test section when the tunnel is operated. An aerodynamic shield, shown in the upper section of Fig. 16.92, isolates the elements of the force balance from aerodynamic forces so that only forces on the scramjet model are detected.

The internal three-component force balance applied by *Robinson* and *Hannemann* [16.137] in the High Enthalpy Shock Tunnel Göttingen (HEG) is shown in Fig. 16.93. Four short, stiff stress bars are mounted on a sting and each bar is instrumented with semiconductor strain gauges to measure the time history of strain.

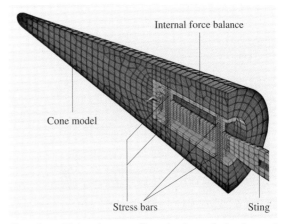

Fig. 16.93 Finite-element discretization of the internal stress wave force balance – cone model set up used in the HEG shock tunnel

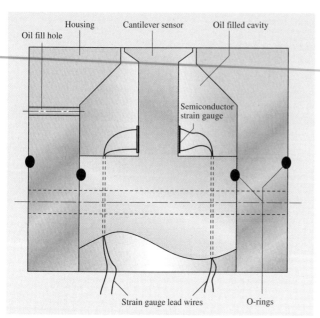

Fig. 16.94 VPI skin friction gauge design (after *Novean* et al. [16.138])

In summary, force measurements in test times of the order of ms in duration are a challenging and difficult area. Of the methods outlined, each has its own advantages and disadvantages, and the selection of the technique to use is somewhat dependent on the test time capability of the facility, the model to be tested, and the aims of the test campaign. For a very stiff or very large model for which the application of the strain gauge or stress wave technique is not feasible, the free-flight measurement technique is preferable.

A high natural frequency is particularly important for strain gauge force balances because a static calibration is performed for this type of balance, and a model with high natural frequency allows that within the available test time the mean values of the measured forces and moments approach the static value as close as possible. To obtain mean level information, approximately five cycles of the lowest natural frequency are required.

If the goal of a test campaign is to use many different models with slight geometry changes and for cost-efficiency reasons the models are manufactured from plastic, which is lightweight, then a strain gauge force balance with acceleration compensation might be chosen. In facilities with test times of 1 ms and below, and if, for example, a heavy model is required with simultaneous pressure and heat flux measurements, then a stress wave force balance may be most suitable.

Skin Friction

Indirect measurement of skin friction or the use of the Reynolds analogy have several disadvantages, including lack of direction sensitivity, a dependency on assumptions regarding the nature of the flow field, and lower precision with respect to direct measurements. Due to the importance of accurate skin friction measurement for supersonic flows and the prevalence of short-duration facilities for supersonic testing, skin friction gauges with high-frequency response have been developed. An overview of the application of skin friction gauges in short-duration facilities is given, e.g., in [16.140]. Typically, the gauge is mounted with the sensing face flush with the instrumented body's surface such that the skin friction force acts on this sensing face. A piezoelectric skin friction gauge was used in the shock tunnels of the Cornell Aeronautical Laboratory (now Calspan) by *Holden* [16.141]. The gauge consisted of two piezoceramic beams in bending mode supporting a measurement diaphragm flush with the model surface. The natural frequency of the gauge was approximately 5 kHz [16.140].

In the case of the miniature cantilever gauge design of the Virginia Polytechnic Institute (VPI) [16.138], the sensor face lies at the end of a cantilever beam, which is instrumented at the base with two strain gauges mounted at 180° to one another. A schematic of the VPI gauge is shown in Fig. 16.94. The wall shear force from the passing flow minutely deflects the sensor head leading to a change in the resistances of the semiconductor strain

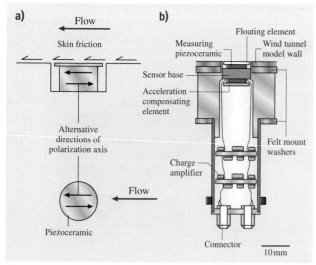

Fig. 16.95a,b UQ skin friction gauge design: (**a**) principle of operation; (**b**) design layout (after *Goyne* et al. [16.139])

gauges, which are incorporated in a Wheatstone bridge configuration. The cavity between the sensing element and the housing is filled with either a silicon oil or a silicon rubber compound. The rubber-filled variant was designed to reduce maintenance, although it leads to increased manufacturing and modeling complexity. These gauges had a natural frequency of 20 kHz and the uncertainty in skin friction measurement was estimated at 10% for the complex supersonic flows considered.

The piezoceramic skin friction gauge design developed at The University of Queensland (UQ), Australia is shown in Fig. 16.95 [16.139]. In this case, the sensing face is a floating element that is directly attached to a piezoceramic ring. The polarization direction of the piezoceramic is mounted parallel to the skin friction force direction. Acceleration compensation is achieved through an acceleration compensating element that is mounted inside the gauge. The element and associated piezoceramic for both the floating element and the acceleration compensating element are of the same dimensions and mounted similarly. The gauge is calibrated for transient shear and acceleration in separate bench tests and pressure calibration is achieved during experiments using adjacent pressure transducers. The UQ gauges had a lowest natural frequency of approximately 40 kHz and the uncertainty in skin friction measurement was estimated at $\pm 47\%$, $\pm 16\%$, and $\pm 7\%$ for laminar, transitional and turbulent boundary layers, respectively [16.142].

Optical Techniques

General Application of Optical Techniques. As in conventional ground-based facilities, in hypersonic short-duration test facilities optical techniques can provide a large range of data with the advantage of being able to obtain information without physically disturbing the flow. Without going into a detailed description, an overview of the major optical measurement techniques used in impulse facilities are summarized in this section (see also Chap. 21).

Laser-induced fluorescence (LIF) was applied to measure temperatures in the free stream of a high-enthalpy shock tunnel [16.143, 144].

Planar laser-induced fluorescence (PLIF) is a frequently used optical imaging method in short-duration ground-based facilities. A laser is tuned to a certain frequency such that it excites a chosen energy level in a probed molecule in the test flow. The tuned laser light is formed into a sheet and passed through the flow. An image of the fluorescence can be recorded which is directly related to the concentration of the probed molecule within the laser sheet. Commonly probed molecules in hypersonic flows include nitric oxide (NO), which is often naturally occurring in shock tunnels, and hydroxide (OH), which is present in oxygen–hydrogen combustion flows. This method can also be extended to measure temperature and velocity in hypersonic flows. The measurement of species concentrations, vibrational temperatures, and velocities has been performed in short-duration test facilities by, e.g., *Wollenhaupt* [16.145], *O'Byrne* et al. [16.146], *Palma* et al. [16.147], *Ben-Yakar* et al. [16.148], and *Kovachevich* et al. [16.149].

Tunable diode laser absorption (TDLAS) is used to measure translational/rotational temperatures, gas velocities, and concentrations of various molecules and atoms (e.g. [16.150–154]).

The application of temperature and pressure-sensitive paint in hypersonic short-duration test facilities is still under development. A discussion of the application of these techniques is given by *Nakakita* and *Asai* [16.155], and *Ohmi* et al. [16.156].

A method to obtain planar velocity visualization, called Doppler picture velocimetry (DPV), has been developed and used for short-duration measurements at the French–German Research Institute of Saint Louis (ISL) [16.157]. A specially designed Michelson interferometer transforms the Doppler shift of the scattered light of the flow particles into a shift in light intensity which is subsequently converted into a flow velocity.

The coherent anti-Stokes Raman spectroscopy (CARS) uses two pulsed laser beams with known pulse and Stokes frequencies. These beams are focused to a point generating an output with an anti-Stokes frequency. The analysis of this output allows information such as temperature and relative species densities to be obtained. This point measurement method has been applied in short-duration test facilities by, e.g., *O'Byrne* et al. [16.158], and *Danehy* et al. [16.159].

Flow tagging is a non-intrusive laser-based method for direct velocity measurements. The method is based upon laser-enhanced ionization velocimetry in which a tagged region is created by two-step excitation of sodium and subsequent collisional ionization. The achieved depletion of neutral atoms is then interrogated by planar laser-induced fluorescence, providing a measurement of the velocity. Velocities in the range of 9 km/s have been measured [16.160].

Electron-beam fluorescence is based on the fact that the inelastic collisions between electrons and gas molecules produce a radiative emission that is characteristic for the gas state properties. By creating a high-energy electron beam and passing it through a gas, the emission can be initiated and analyzed. While the intensity of this emission is related to the number density and to features of the electron beam, the spectral characteristics are related to the composition, density and temperature of the gas. General introductions into the technique are given by, e.g., *Muntz* [16.161], and *Bütefisch* and *Vennemann* [16.162]. Applications in shock tunnel environment are reported by *Shelton* and *Cassady* [16.163], and *Muntz* et al. [16.164].

Phase-Step Holographic Interferometry. Interferometry may be used as a technique to measure the variation of the refractive index of a gaseous flow in the test section of a short-duration facility. This information can be used to evaluate the density distribution of the investigated flow field. Holographic interferometry does not require special machining or manufacturing of test section windows, mirrors or lenses with high precision, because any imperfections in these components are automatically balanced out by the holographic two step procedure. Therefore, related to the application in short-duration ground-based test facilities, this technique has replaced the classical and labor-intensive Mach–Zehnder interferometry.

Determination of the Refractive Index by Interference Phenomena. The absolute speed c_0 at which light travels in absolute vacuum is constant. In any kind of gaseous media the speed of light c will be lower. The ratio of the two speeds defines the refractive index

$$n(\rho) = \frac{c_0}{c} = 1 + K^\lambda \rho \,.$$

The Gladstone–Dale relation describes that in a gaseous media consisting of one species the refractive index depends on the density ρ and the Gladstone–Dale constant K^λ [16.165]. K^λ is weakly dependent on the wavelength and is specific for each gas. For gas mixtures the refractive index is given by a linearly composed Gladstone–Dale constant

$$n(\rho) = 1 + \rho \sum_{i=1}^{S} K_i^\lambda \xi_i \,,$$

where K_i^λ are the Gladstone–Dale coefficients for the individual gas species, ξ_i are the species mass fractions, S is the number of species, and λ is the wavelength of the laser light used. The definition of a linearly composed Gladstone–Dale constant applies to general gas mixtures including chemically reacting, excited and ionized species.

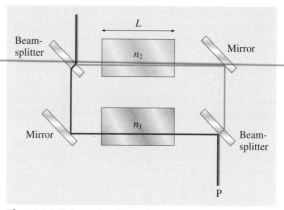

Fig. 16.96 Schematic of interference experiment

The basic principle of interferometry is shown in Fig. 16.96. Two rays of coherent light interfere in point P. Each ray passes through zones with different refractive index, which leads to a time shift Δt due to the different speeds of light in the two zones:

$$\Delta t = \frac{L}{c_2} - \frac{L}{c_1} = \frac{L}{c_0}(n_2 - n_1) \,.$$

With the time shift the difference in the optical path length evaluates to

$$\Delta l = c_0 \Delta t = L(n_2 - n_1) \,.$$

If the difference in the optical path length equals the wavelength, then the phase shift between both rays equals 2π. Therefore, it follows that:

$$\frac{\Delta l}{\lambda} = \frac{\phi}{2\pi} = \frac{L}{\lambda}(n_2 - n_1) = \frac{K^\lambda L}{\lambda}(\rho_2 - \rho_1) \,.$$

As will be derived later, the intensity, which is measured at the point P, is dependent on the phase shift between the two rays:

$$I \propto \cos \phi \,.$$

In summary it is clear that the measurement of the intensity or phase shift ϕ at point P is connected directly to the density difference in the two regions.

Using a Mach–Zehnder interferometer, which is shown schematically in Fig. 16.97a, the density in the test section can be evaluated if the density distribution is known at a reference point.

The schematic emphasizes that any imperfection that influences the beams, disturbs the interference in the measurement plane F. To avoid this, the measurement can be performed in two steps (Fig. 16.97b,c). In the first step, the light beam (grey) passing through the evacuated test section is recorded and the subsequently light

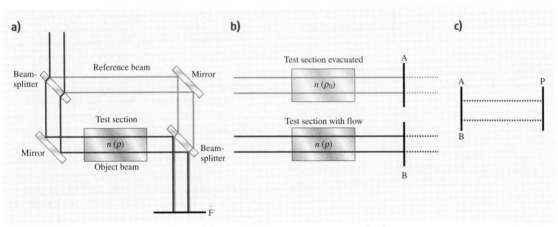

Fig. 16.97a–c Principle of Mach–Zehnder (**a**) and holographic interferometry (**b,c**)

beam (brown) passing through the test section with the flow is recorded in the second step. Both beams are then reconstructed starting from the recording planes A and B are brought to interference in the measurement plane F. Any imperfection within the optical set up is present in both beams, because they pass through the same optical components. As will be shown later, these imperfections cancel out.

To store and reconstruct both beams, a holographic storage technique has to be used. Therefore, this two-step measurement technique is called holographic interferometry.

Holographic Recording. Photoactive media is sensitive to the amount of light energy it receives during the exposure time. An electromagnetic wave such as light can be described by specifying the temporal and spatial dependence of its electric intensity vector E. A more complete description requires specification of the magnetic intensity H, the electric displacement D, and the magnetic induction B. The present consideration is restricted to E because photoactive media respond primarily to the electric field. Let $E_O(x, y, t)$ be the electrical field strength of an object light wave

$$E_O(x, y, t) = A_O(x, y) e^{i[\varphi_O(x,y)+\omega t]},$$

with amplitude A_O, phase φ_O, and circular frequency ω within an area $F(x, y)$ at the time t on the photoactive medium. Then the energy density is

$$W_{\text{phot}} = \frac{E_O E_O^*}{Z} t_B = \frac{A_O^2(x, y)}{Z} t_B,$$

where t_B is the exposure time, and Z is the characteristic impedance of the volume in which the light wave propagates. $E_O^*(x, y, t)$ is the complex conjugate value of $E_O(x, y, t)$. The energy density in this equation no longer contains any information about the phase distribution $\varphi_O(x, y)$ of the object wave.

If an interference of the object wave $E_O(x, y, t)$ with a coherent background in the form of a second light wave,

$$E_R(x, y, t) = A_R(x, y) e^{i[\varphi_R(x,y)+\omega t]}$$

with the same wavelength is generated (Fig. 16.98a), the energy in any point of the photoactive medium is given by

$$W_{\text{holo}} = \frac{t_B}{Z}(E_O + E_R)(E_O^* + E_R^*)$$
$$= \frac{t_B}{Z}[A_O^2 + A_R^2 + 2 A_O A_R \cos(\varphi_O - \varphi_R)].$$

Using this holographic recording technique, it is possible to store the phase-shift between the object and the reference wave. Depending on the phase-shift $\varphi_O - \varphi_R$, a distribution of high and low energy density on the area F (photoactive material) is obtained, leading to a strong

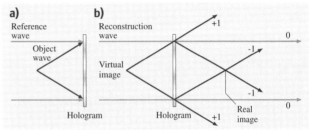

Fig. 16.98a,b Holographic recording (**a**) and holographic reconstruction (**b**) for a point light source

or weak exposure of the photoactive material. The resulting recorded interference pattern is called a hologram.

Holographic Reconstruction. The recorded object wave can be reconstructed by illuminating the developed hologram with a reconstruction wave

$$E_{RC}(x, y, t) = A_{RC}(x, y) e^{i[\varphi_{RC}(x,y) + \omega t]},$$

as shown in Fig. 16.98b.

The reconstruction wave is identical to the reference wave. During the passage of the reconstruction wave through the holographic plate, it is damped depending on the exposure of the holographic plate. Behind the holographic plate a wave is formed that is an identical copy of the original object wave with respect to the distribution of amplitude and phase. The following energy distribution is formed behind the hologram (the exposure time and impedance are neglected)

$$E_{RC} W_{\text{holo}} \sim A_{RC} e^{i(\varphi_R + \omega t)} (A_O^2 + A_R^2)$$
$$+ A_O A_R A_{RC} e^{i(\varphi_O + \omega t)}$$
$$+ A_O A_R A_{RC} e^{i(2\varphi_R - \varphi_O + \omega t)}.$$

The result is a system of three waves. The first term is the reconstruction wave with a modification in the amplitude. The second term is the reconstructed object wave, and the third wave also contains information on the object wave. Discussing the function of the hologram in terms of a diffraction grating, the three waves are identical with the zeroth and first diffraction order of the reconstruction wave. The unchanged reconstruction wave is of zeroth diffraction order. The reconstructed object wave (*direct wave*) is of the order $+1$, while the third wave (*conjugate wave*) is of the order -1.

Application of Holographic Interferometry. An example of the set up of a holographic interferometer, as used at the High Enthalpy Shock Tunnel Göttingen (HEG) of the German Aerospace Center (DLR), is discussed here. In Fig. 16.99, a schematic of this set up is shown. The grey line shows the light path of the reference beam and the brown line shows the light path of the object beam. The object beam passes through the test section and is brought to interference with the reference beam on the holographic plate; here the interference pattern is recorded as discussed in the previous sections. To achieve interference between the object and the reference beams, a light source with sufficient coherence length is needed. In this set up, a seeded Nd:YAG laser from Innolas (model Spitlight 300), emitting light at 532 nm is used. It has a coherence length larger than 1 m. The optical path length of one of the beams is about 15 m. The alignment of the optical path length difference between the two beams in the range of this coherence length allows to set up a functioning optical system.

As discussed above, the first exposure of the holographic plate with one reference beam is done prior to the run in the HEG and a second exposure with the other reference beam is done during the test time. After the chemical treatment of the holographic plate, two reconstruction waves are created in a separate reconstruction unit. The reference waves used in the reconstruction unit are identical to the reference waves used for both exposures. Each of the reconstruction wave forms a system of waves as shown in the previous section. Both reconstructed object waves (described by second term in previous equation) are aligned in such a way that they interfere on a charge-coupled device (CCD) camera behind the holographic plate. Let $R_1(x, y)$ and $R_2(x, y)$ be the reconstructed object waves:

$$R_1(x, y) = A_{O1} A_{R1} A_{W1} e^{i(\varphi_{O1} + \omega t)},$$
$$R_2(x, y) = A_{O2} A_{R2} A_{W2} e^{i(\varphi_{O2} + \omega t)}.$$

Then the resulting energy distribution of the interference pattern on the CCD camera is:

$$W_{\text{HI}}$$
$$= (R_1 + R_2)(R_1^* + R_2^*)$$
$$= (A_{O1} A_{R1} A_{RC1})^2 + (A_{O2} A_{R2} A_{RC2})^2$$
$$+ 2 A_{O1} A_{R1} A_{RC1} A_{O2} A_{R2} A_{RC2} \cos(\varphi_{O1} - \varphi_{O2}).$$

It is obvious from the last term that the camera sees the phase distribution of the light during the experiment φ_{O2}

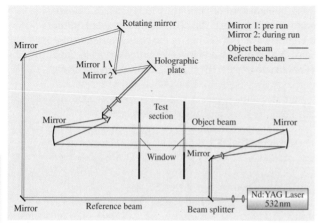

Fig. 16.99 Schematic set up of the HEG holographic interferometry system

relative to the phase distribution of the original object wave φ_{O1} before the experiment in the form of an intensity distribution. This intensity $I(x, y)$ on each pixel of the CCD camera is defined by

$$I(x, y) = I_0\{1 + m(x, y)\cos[\phi(x, y)]\},$$

where m is a contrast function and I_0 is the background intensity. These terms contain all the amplitude modulations within the interference pattern. $\phi(x, y) = (\varphi_{O1} - \varphi_{O2})$ is the phase-shift distribution. For the evaluation of the fringe distribution, the phase-shift technique is used. Here the optical path of one of the reconstruction waves is lengthened or shortened by a piezoelectric mirror, thus adding up a known phase shift with respect to the reconstruction wave. Three individual phase distributions are generated:

$$I_1(x, y) = I_0\{1 + m(x, y)\cos[\phi(x, y) + \Theta_1]\},$$
$$I_2(x, y) = I_0\{1 + m(x, y)\cos[\phi(x, y) + \Theta_2]\},$$
$$I_3(x, y) = I_0\{1 + m(x, y)\cos[\phi(x, y) + \Theta_3]\}.$$

This system of equations defined $\phi(x, y)$. The background intensity and the contrast function can be determined and thus eliminated from the interference pattern.

As already shown, the phase shift $\phi(x, y)$ is coupled to the density difference $\Delta\rho$ by

$$\frac{\phi(x, y)}{2\pi} = \int \left(\frac{\Delta\rho(x, y, z) - \rho_\infty}{\lambda} \sum_{i=1}^{S} K_i^\lambda \xi_i(x, y, z) \right) dz,$$

where K_i^λ are the Gladstone–Dale coefficients of the gas species, ξ_i are the species mass fractions, S is the number of species, and λ is the wavelength of the laser light. The two-dimensional projection of $\phi(x, y)$ contains line of sight integrated information of the three dimensional flow field. The density difference distribution may be reconstructed from the measurements only if the measured flow field is two dimensional or axisymmetric. In the case of a general three-dimensional flow field, the measurement of the two-dimensional phase shift distribution may be compared to computed line-of-sight integrated phase-shifts, resulting from three-dimensional CFD.

Holographic Interferometry Using Two Wavelengths. When a gas is heated to a sufficiently high temperature, some electrons are excited to the point where they separate from the nucleus and create a plasma. The refractive index of the plasma is the sum of the refractive indices of the atoms, ions, and electrons weighted by their number densities. The refractive index of the heavy particles is described by the Gladstone–Dale relation and it is of similar order of magnitude for the different species of a gas. The refractive index of electrons is given by

$$n_e = \sqrt{\left(1 - \frac{N_e e^2 \lambda^2}{2\pi m_e c^2}\right)}.$$

Here N_e is the number density, n_e the refractive index, e the charge, and m_e the mass of the electron. Evaluation of the constants leads to

$$n_e = \sqrt{(1 - 8.92 \times 10^{-16} \lambda^2 N_e)}.$$

With $\sqrt{1-x} \approx 1 - (1/2)x$ for $x \ll 1$, the equation above can be approximated by

$$n_e - 1 = -4.46 \times 10^{-16} \lambda^2 N_e.$$

It is important to note that the refractive index of electrons is strongly dependent on λ. Further, the contribution of electrons to the refractive index of a considered gas is an order of magnitude higher than the contributions of atomic and molecular species, and is of opposite sign. The electrons therefore dominate the refractive index of moderately and highly ionized plasmas. Because of this strong dispersion, a measurement of the refractive index of the electrons at two different wavelengths at the same time allows to determine the electron density directly from

$$n(\lambda_1) - n(\lambda_2) \cong n_e(\lambda_1) - n_e(\lambda_2)$$
$$= -4.46 \times 10^{-16} (\lambda_1^2 - \lambda_2^2) N_e.$$

Reviews of the initial efforts using this technique are reported by *Zaidel'* et al. [16.166], and *Jahoda* et al. [16.167]. Application of two-wavelength holographic interferometry in a flow created by a superorbital shock expansion tube has been performed by *McIntyre* et al. [16.109].

High-Speed Flow Visualization. The principal of flow visualization methods may be subdivided into methods that observe changes in the refractive index, methods that observe particles or tracing elements which have been added to the flow, and methods that introduce energy into the flow converting selected portions of the flow into luminous or visible areas. Most of the high-speed flow visualization (HSFV) systems rely on the first method and are based on the shadowgraph and Schlieren technique, for which the fundamentals are given, e.g., by *Schardin* [16.168], and *Merzkirch* [16.165]. A comprehensive review of shadowgraph or Schlieren techniques is given by *Settles* [16.169]. Further details may also be found in Chapts. 6,11. A comprehensive review

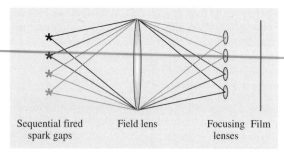

Fig. 16.100 Schematic of the Cranz–Schardin camera principle

about the current state of high-speed photography and photonics is given by *Ray* [16.170].

Light Sources. To visualize intermittent flow phenomena, it is important to use adequate light sources. If the required exposure times can be achieved by the use of either mechanical shutters for film cameras or by controlling the exposure time of the light-sensitive element in digital cameras itself, continuous light sources may be used.

Continuous light sources that may be used are tungsten–halogen lamps, compact arc lamps, xenon flash tubes and light-emitting diodes (LEDs). Details and more information about continuous light sources are given by *Ray* [16.170], and *Settles* [16.169].

The oldest intermittent light sources are spark gaps, which have already been used by *Toepler* [16.171] and *Mach* and *Salcher* [16.172]. In these early days of flow visualization, flash light durations from 0.1–1 μs were achieved. Modern Nanolite spark flash lamps, which are based on the design by *Fischer* [16.173], achieve flash durations of 5 ns with 20 kHz repetition rates [16.174]. Lasers and laser diodes offer pulse duration times in the ns range with high repetition rates up to 250 kHz and usually an order of magnitude higher energy than spark gaps. Semiconductor light sources such as LEDs can be pulsed in the sub-μs range.

The choice of the light source depends strongly on the application. Lasers and laser diodes offer high energies. However, optical artefacts are introduced, due to the monochromatic and coherent nature of their light. The laser illumination exhibits a binary Schlieren cut-off, and fringes appear due to diffraction which diminish the optical quality of the pictures. Generally, the use of coherent light sources transforms the classical Schlieren set up into a Schlieren interferometer. Broadband or incoherent white-light sources provide better Schlieren and shadowgraph images, but may not be used in certain applications. Especially in hypervelocity flow environments incorporating large energies, the resulting high level of self-illumination (usually broadband) overlays the recorded flow pictures. Therefore, using a narrow-band light source like a laser in combination with a filtering system may be the only solution to achieve satisfactory flow visualization.

High-Speed Cameras. To record the light representing the flow picture, numerous systems and techniques using photographic film or modern electronic cameras have been used and the development especially in the electronic domain is a quickly developing process. Digital cameras are based on either charge-coupled devices (CCDs) or recently on the complementary metal oxide semiconductor (CMOS) technology. A review of analogue and digital camera systems is given, e.g., by *Holzfuss* [16.175].

The main question is to achieve sufficiently short exposure times. This can be achieved by either using an adequate intermittent light source with sufficient energy or by using a continuous light source that is interrupted by either a shutter in the optical path, or by the camera system itself. Modern cameras allow short exposure times using electronic shutters.

One of the key developments for high-speed photography was the Cranz–Schardin camera. In Fig. 16.100, a schematic of the principle is shown. The framing rate is controlled by electronically firing the spark gaps one after another. The optical set up allows to record the individual frames at different locations on the film and no moving parts are required.

Other camera types used for high-speed flow visualization are the rotating drum (Fig. 16.101) and rotating-mirror/prism (Fig. 16.102) cameras. In the first case, the film is transported at high speed to record the

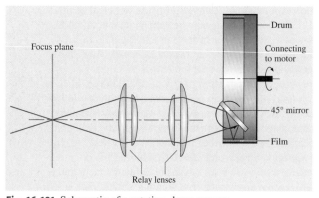

Fig. 16.101 Schematic of a rotating-drum camera

sequential images on different locations of the film. In the latter case a fast rotating prism or mirror is deflecting the light to different locations on the stationary film. Rotating drum cameras reach film speeds up to 300 m/s allowing to record up to 200 images with 50 kHz. To achieve these high speeds the drum housing is evacuated to reduce friction effects. One of the fastest rotating-mirror cameras is manufactured by Cordin. The framing rate is 25 million frames per second. A small turbine rotates the center mirror at 20 000 rotations per second.

The availability of fast and sensitive digital elements led to the development of rotating-mirror systems, where the film is replaced by a number of CCD or CMOS elements.

Example of a High-Speed Flow Visualization Schlieren Set Up. The set up shown in Fig. 16.103 is used at the HEG of the German Aerospace Center. It uses a Z-path layout for the object light path with spherical mirrors (Halle SDH4300) with a diameter of 300 mm and a focal length of 1500 mm.

The light source for this set up is a diode pulsed Nd:YAG laser from Lightwave Electronics (Model 612) emitting at 532 nm. It has the capability to operate at pulse rates up to 50 kHz. The laser light is widened by a telescope system (L1 and L2) and then expanded by a lens (L3) onto the spherical main mirror (H1), which aligns the beam to a parallel light-bundle of diameter 300 mm through the test section. On the other side of the test section the beam is collimated by the second spherical main mirror (H2). With the lens L4, the test section is focused onto the film plane (A) and a razor blade (R) is used at the focal point formed by the lens L4. The images are recorded with a Cordin rotating-drum camera (Model 318) on black-and-white film (Kodak TMAX 100). The camera consists of a rotating drum, which is equipped with 35 mm film stripes of 1 m length. The shutter of the camera remains open during the experiment. The narrow pulse width of the laser has a pseudo-framing function during the process of image recording on the moving film. More information about the HEG HSFV system is given by *Martinez* et al. [16.176].

One application of high-speed flow visualization is to investigate if the flow phenomena that are studied in short-duration ground-based facilities reach a steady state during the available test time. In Fig. 16.104 one such example – the establishment of the hypersonic, high-enthalpy flow past a cylinder – is shown.

The experiment was conducted in HEG at a total specific enthalpy of 15 MJ/kg using air as test gas.

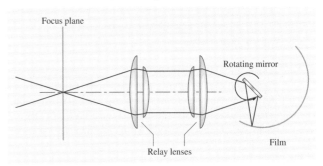

Fig. 16.102 Schematic of a rotating-mirror camera

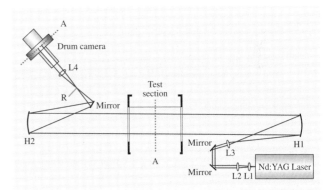

Fig. 16.103 Schematic of a high-speed flow visualization Schlieren set up using a drum camera

Fig. 16.104 High-speed flow visualization of the establishment of the hypersonic, high-enthalpy flow past a cylinder in the HEG

16.2.3 Typical Applications of Shock Tunnel and Shock Expansion Tunnel Facilities

In the subsequent sections, typical applications of shock tunnel and shock expansion tunnel facilities are summarized. The main objective of this overview is to highlight selected work and to refer the interested reader to references which cover the corresponding research in more detail.

Hypersonic, High-Enthalpy Flows Past Generic Flow Configurations

When considering the re-entry trajectory of a space vehicle returning from low Earth orbit into the atmosphere, the most critical point concerning the heating loads on the vehicle is found in the continuum flow region in approximately 70 km altitude. In this region of the re-entry path, the velocity of the vehicle is approximately 6 km/s and the flow past the vehicle is accompanied by strong shock waves, leading to high temperatures that cause dissociation reactions. The fundamental influence of the thermal and chemical relaxation processes caused by these high-temperature effects on the external aerodynamics, i.e., the pressure distribution, flap efficiency, shock–shock and shock–boundary layer interactions and on the heating loads can be investigated by looking at the flow past basic generic flow configurations which are especially designed in order to focus on one of these effects. Additionally, these studies are well suited to validate the ground-based facility performance, measurement techniques and computational fluid dynamics (CFD) codes. A summary of the effects occurring in hypervelocity aerodynamics with chemical nonequilibrium is given by *Stalker* [16.58].

Pioneering work related to the investigation of relaxation processes in the shock layer of generic bodies was performed in the T3 free-piston-driven shock tunnel of the Australian National University in the 1970s. Many of the tests were performed using nitrogen as test gas at total specific enthalpies of 22 MJ/kg and above. One aspect of chemical nonequilibrium generated behind the shock wave of a wedge is that the attached shock wave is no longer straight, as obtained in a perfect gas, but curved towards the wedge surface. The measurement of the shock curvature was used by *Kewley* and *Hornung* [16.177] in order to determine the dissociation rate of nitrogen at high temperatures. The investigation of the relaxation processes on shock detachment on a wedge showed that the detachment distance of the bow shock wave increases gradually with the wedge angle in the case of a relaxing flow and more rapidly in a perfect gas flow [16.178].

In the HEG of DLR, holographic interferometry to measure the density distribution in the shock layer as well as surface pressure and heat flux measurements were applied in conjunction with detailed CFD investigations to study the flow in the shock layer of a cylinder placed with its axis transverse to the flow [16.179]. The experimental data which was obtained at total specific enthalpies of approximately 12–22 MJ/kg for air as the test gas served as a basis for the validation of physico-chemical models used in CFD codes. The cylinder model was mounted on the nozzle centerline (Fig. 16.105a). The large shock stand-off distance generated by this

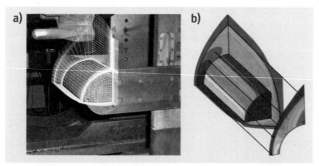

Fig. 16.105a,b Cylinder model in the HEG test section including the grid used for the three-dimensional flow-field computations (**a**) and line of sight reconstruction of the phase-shift distribution from flow-field computations (**b**)

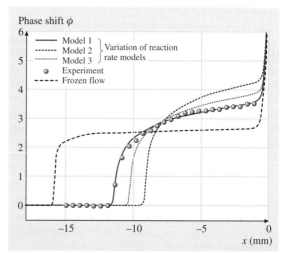

Fig. 16.106 Experimentally and numerically determined phase shift in the shock layer of a cylinder – the influence of using different chemical models is shown

model is advantageous for the investigation of the gas properties in the shock layer using optical measurement techniques. The phase shift distribution in the shock layer of the cylinder which can be correlated to the density difference between the free stream density and the density in the shock layer, was measured and simultaneously reconstructed utilizing three-dimensional flow field computations (Fig. 16.105b). The computations were performed using different chemical reaction rate models. Based on the comparison of the different computed phase-shift distributions and the measurements, the most suitable chemical reaction rate model for the present flow conditions can be selected (Fig. 16.106). The plot in Fig. 16.106 also includes the phase shift distribution resulting from the assumption of frozen chemistry in the shock layer, i.e., infinitely slow reaction rates. It is obvious that the various flow models cause large differences in the shock stand-off distance.

The interaction between a shock wave and a separated region in a hypersonic flow is a very important flow feature which appears for example at aerodynamic control surfaces of re-entry vehicles. In addition to the correct prediction of the pressure distribution on the control surface, the heat transfer and the length of the separation needs to be predicted correctly in the design process of a re-entry vehicle. Generic configurations such as different types of double-cone configurations or hollow cylinder flare configurations were used in the large-energy national shock (LENS) facility at the Calspan–University at Buffalo Research Center (CUBRC) to provide a benchmark dataset at well-calibrated test conditions with fully laminar hypersonic flow [16.180, 181]. These tests were performed using nitrogen as test gas at total specific enthalpies of approximately 4 MJ/kg. Numerical computations of these flow fields using continuum mechanical flow solvers as well as particle methods such as the direct simulation Monte Carlo method were performed by, e.g., *Candler* et al. [16.182]. In these numerical investigations, issues such as the influence of vibrational freezing in the free stream of the ground-based facility or slip flow boundary conditions on the model were examined.

When a weak oblique shock wave impinges on the strong bow shock wave ahead of a blunt body, it is known that the shock–shock interaction pattern can cause extremely high local surface pressure and heat flux on the blunt body. Therefore, the understanding of the phenomena associated with this type of flow is of particular interest for the design of hypervelocity vehicles. The influence of high-temperature effects on the shock–shock interaction generated when an oblique shock wave interacts with the bow shock of a circular cylinder was investigated by *Sanderson* [16.183] in the T5 free-piston-driven shock tunnel of Graduate Aeronautical Laboratories, California Institute of Technology (GALCIT), USA and by *Schnieder* [16.184] in the HEG of the German Aerospace Center.

The shock–boundary layer interaction in hypersonic, high-enthalpy flow including high-temperature effects on wedge-type compression ramps were investigated in the T3 free-piston-driven shock tunnel of the Australian National University in Canberra by *Mallinson* et al. [16.185], in the T5 free-piston-driven shock tunnel by *Davis* [16.186] and *Davis* and *Sturtevant* [16.187], and in the HEG by *Martinez Schramm* and *Eitelberg* [16.188]. These tests covered a total specific enthalpy range of 3–22 MJ/kg.

High-Enthalpy Flows Past Re-entry Configurations

The X-38 program was a joint effort of the US National Aeronautics and Space Administration (NASA) and the European Space Agency (ESA). The objective of this program was to demonstrate the technologies required for the development of the future crew return vehicle (CRV) for the International Space Station (ISS), providing the crew members with the capability to return to Earth in the event of an emergency. Within the X-38 program, the intention was that key design and operational aspects should be validated by flying a full-scale technology demonstrator. The determination of the aerodynamics and aerothermodynamics of the X-38/CRV was based on a close cooperation between ground-based testing and CFD [16.189]. Investigations in the HEG were performed which represented the first experiments using the X-38 configuration in the high total specific enthalpy flow regime up to 22 MJ/kg ($u_\infty = 6000$ m/s) for pressure and heat flux measurements [16.190]. The stainless-steel 1 : 24-scale wind tunnel model of the X-38, with a length of $l = 360$ mm, mounted in the HEG test section, is shown in Fig. 16.107a. The positions of the thermocouples used to present the normalized heat flux distribution along a streamwise cut including the body flap in Fig. 16.108 are indicated in Fig. 16.107b. The set of heat flux gauges consists of six transducers on the fore body and seven transducers on the flap. The heat flux measurements and the computed heat fluxes have been normalized using the stagnation point heat flux on X-38. Additionally to the normalized heat flux, the standard deviation of the measured data during the test time window is shown in the Fig. 16.108. The normalized measured and computed heat flux on the 20°

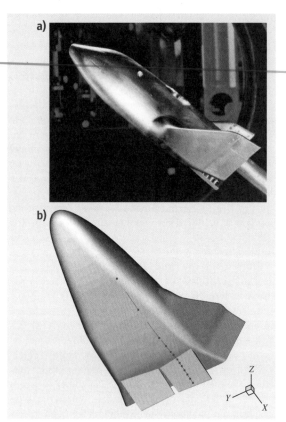

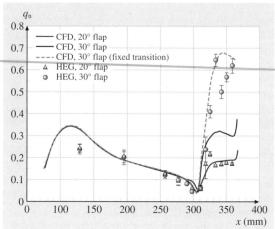

Fig. 16.108 Normalized heat flux distributions on the X-38 model; test gas: air at a total specific enthalpy of 22 MJ/kg

Fig. 16.107a,b X-38 model in the HEG test section (**a**) and locations of the heat flux gauges used in Fig. 16.108 (**b**)

and 30° deflected body flap using HEG operating condition I (22 MJ/kg) is shown in Fig. 16.108. Except at the first gauge position, where the measured heat flux is significantly lower than the computed one, the comparison between CFD and experiment along the fore body is good. In the reattachment region the measured heat flux distribution shows a clear peak that is not reproduced by the laminar flow computation. At the peak, the heat flux is about 25% higher than that measured further downstream on the flap. It is obvious that in the separated flow region the standard deviation is higher than at neighboring gauge positions located further upstream and downstream. This behavior was consistently reproducible during the test campaign and is regarded as being related to oscillations of the flow in this region. In the reattachment region and further downstream the measured heat flux on the 30° flap is approximately two times higher than the one predicted by the laminar flow computation. Therefore, a computation was performed assuming transitional flow past X-38 with fixed transi-

Fig. 16.109 HOPE 2.5% model equipped with force balance in the HIEST test section (from *Itoh* et al. [16.132])

tion in front of the body flap hinge line. This result approximately matches the heat flux level at reattachment. However it does not correctly reproduce the measured drop and subsequent increase of heat flux on the flap.

From an engineering point of view the significant increase of the heat flux for the 30° deflected body flap is a very important result indicating that in spite of the fact that the unit free stream Reynolds number of the HEG operating condition would suggest that the flow past the model should be laminar, the flow in the separated flow may be governed by flow instabilities or transitional/turbulent flow. This behavior was also identified on other vehicle configurations and is still in the focus of current research programs.

The aerothermodynamics of the Japanese space vehicle HOPE was investigated in the High Enthalpy Shock Tunnel (HIEST) of the Japan Aerospace Exploration Agency (JAXA) at Kakuda [16.132, 191]. The aim of these tests was to prove chemical nonequilibrium effects on the aerodynamics of the HOPE vehicle. In particular, the research was focused on the investigation of the nonequilibrium effect of oxygen dissociation by force measurements in HIEST. The tests were conducted at stagnation enthalpies of 8–24 MJ/kg in which oxygen dissociation probably dominates the nonequilibrium aerodynamic characteristics, and the binary scaling parameter was chosen such that the nonequilibrium dissociating flow around HOPE could be duplicated. The wind tunnel model had a length of $l = 400$ mm. The forces and pitching moment were measured by a strain gauge force balance with aid of acceleration compensation for the low frequency vibrations of the bending motion of the sting and the pitching motion of the model. The influence of nonequilibrium effects is most pronounced for the pitching moment. The pitching moment coefficient C_m of HOPE is plotted in Fig. 16.110 as function of the stagnation enthalpy. The measured data is compared with the respective values of the HOPE reference data base which was determined by cold hypersonic testing and nonequilibrium CFD results. The dissociation rate of nitrogen is much smaller than that of oxygen and it was found that except at the nose region almost no dissociation of nitrogen occurs. Except for the subsonic region at the nose where the pressure increases due to dissociation, on the main part of the body and wing, the pressure decreases with oxygen dissociation. Regarding the pitching moment, both the increase in nose pressure and the decrease in fuselage pressure can increase C_m. This effect is revealed in Fig. 16.110. Compared to a perfect gas flow which results in a pitching moment coefficient which is independent of the stagnation enthalpy, the measurements in HIEST and the numerical predictions, modeling the nonequilibrium flow past HOPE, result in an increasing C_m with increasing stagnation enthalpy. In addition, at a total specific enthalpy of 14 MJ/kg, the measurements in HIEST show a peak in the C_m distribution. Analysis of the ratio of the flow characteristic time and the relaxation time of oxygen dissociation led to the suggestions that this peak may be caused by the chemical nonequilibrium effect of oxygen dissociation, which for the considered flow becomes most critical at this moderate stagnation enthalpy condition.

The successful flight of the atmospheric re-entry demonstrator (ARD) in late 1998 marked the first civil atmospheric re-entry and recovery achieved by European nations. Launched on an Ariane 5 from the Guiana space port in Kourou, the capsule reached a maxi-

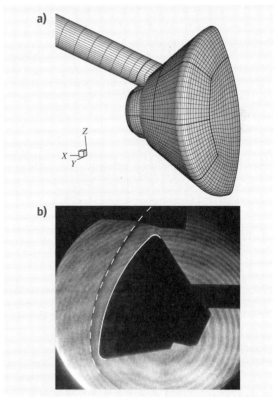

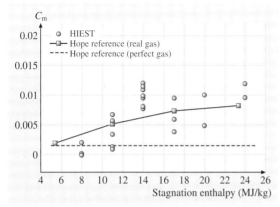

Fig. 16.110 Pitching moment coefficient as function of stagnation enthalpy for the HOPE space vehicle (after *Itoh* et al. [16.132])

Fig. 16.111a,b Surface grid for the ARD wind-tunnel model (**a**), and Schlieren image showing the measured and computed (*dashed line*) bow shock shape in HEG (**b**)

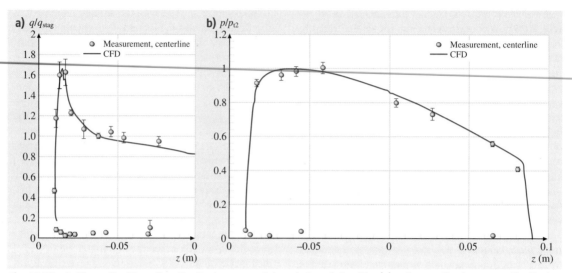

Fig. 16.112a,b Normalized heat flux distributions around the windward shoulder (**a**) and normalized pressure distribution along the centerline (**b**) of the ARD model; $h_0 = 12\,\text{MJ/kg}$

mum altitude of 830 km before descent to re-entry and splashdown within 5 km of its target point. The project was conducted primarily to validate existing and future ground-based experimental and numerical aerothermodynamic models, to qualify the design of thermal protection materials, and to help assess navigation, guidance and control system performance [16.192]. A post-flight simulation campaign for the ARD capsule has been conducted and coordinated by the prime contractor EADS Launch Vehicles. As part of the campaign, a range of numerical and experimental research work was performed at DLR to reproduce aero heating and aerodynamic data for the capsule in hypersonic flight [16.193]. The scale-model experiments were performed in the HEG using air as test gas at total specific enthalpy conditions of 12–22 MJ/kg. One objective of the work was to help refine and cross-check the available numerical and experimental ground-based simulation facilities through use of a single geometry. In HEG, Schlieren visualization and both surface pressure and heat flux results were recorded. Parallel to these experiments, three dimensional flow field computations were conducted. This combined effort provided results and insight in the advantages and deficiencies of the numerical and experimental techniques that would not be achievable otherwise. The CFD simulations, for example, allowed the effects of flow nonuniformity and the presence of experimental artefacts in the test facility to be assessed. Likewise, comparison with experimental results allowed identification of possible defects in the numerical solutions. Resembling the geometry of the Apollo capsules, the ARD capsule has a spherical nose and an axisymmetric shape. In Fig. 16.111a the grid around the ARD scale model is shown. It is designed to capture the full flow field around the model mounted on a sting in the shock tunnel. The grid allows assessment of the impact of sting and leeward flow on the forebody flow field. The substantial bluntness of the forebody supports a large subsonic region encompassing most of the windward shock layer, even at high Mach numbers. The size of the subsonic region presented challenges for obtaining good measurements in the short-duration test flows,

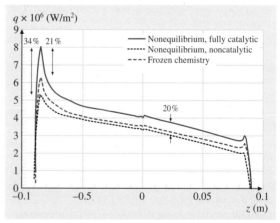

Fig. 16.113 Effect of chemistry models on the centerline heat flux profile on the ARD configuration; $h_0 = 22\,\text{MJ/kg}$

and for maintaining accuracy and stability of CFD simulations. All experiments were performed with the ARD scale model mounted at a 20° angle of attack and no yaw. In Fig. 16.112a, the normalized heat flux measurements made along the model centreline around the windward shoulder are plotted together with the computed data. On the windward shoulder the maximum heat flux occurs, and it can be seen from Fig. 16.112a that this peak is well reproduced by CFD. In Fig. 16.112b, the measured and computed normalized surface pressure is shown along the centreline of the ARD configuration. Several CFD computations were performed to estimate the influence and importance of thermochemical processes in the flow field past the ARD capsule in HEG. For an inflow corresponding to a 22 MJ/kg total specific enthalpy HEG operating condition, Fig. 16.113 shows the computed heat flux results along the centreline produced under different chemistry assumptions. The simulations presented were all performed assuming that the gas species remained in thermal equilibrium, since numerical tests showed that thermal nonequilibrium effects were negligible. The figure shows that the default chemistry model – nonequilibrium (finite-rate) chemical reactions with a fully catalytic model surface – results in significantly more heat flux than the frozen-flow assumption. A maximum difference of 21% was observed at the windward shoulder, while a difference of about 14% occurs over most of the remainder of the model forebody. The assumption of a noncatalytic ARD surface also results in a lower heat flux compared to the default chemistry model, differing by a maximum of 34%. The difference in heat flux results can be explained by considering the chemical processes occurring throughout the flow. Under finite-rate reaction chemistry, the molecular dissociation occurring in the shock layer means that some of the gas thermal energy is absorbed as species formation energy. This process is manifested in a much cooler shock layer and smaller shock stand-off, compared to non-reacting (frozen) flow. When the incident flow subsequently impinges on a fully catalytic model surface, species recombination causes the heat of formation to be released and results in an increased heat transfer to the body. In contrast, if the surface is noncatalytic or partially catalytic, then only some of the heat of formation will be recovered immediately with the remainder being swept downstream in the atoms or molecules which have not recombined fully. This process results in the lower heat flux observed for the nonequilibrium–noncatalytic simulation. In the case of frozen flow, where no dissociation or recombination occurs, flow energy acquired from shock heating is stored purely in thermal form and then transferred to the surface by conduction. The simulated heat flux for frozen flow is lower than for the nonequilibrium–fully catalytic case, because the sudden release of energy due to species recombination does not occur.

As discussed e.g. by *McIntyre* et al. [16.194] aerocapture is the process in which an interplanetary vehicle acquires sufficient drag for orbit insertion in a single pass through a planet's atmosphere. This process requires that the vehicle enters the lower altitudes, where the atmospheric density, and therefore the drag generated, is sufficiently large. Significant heat flux can result, and this necessitates the development of suitable heat shields, the design being dependent on the entry speed and type of atmosphere. Current aerocapture studies focus on two approaches, the use of the aeroshell and the ballute. In the former, the drag is generated by a forward-facing ablative heat shield on the vehicle. The

Fig. 16.114 Interferometric flow visualization of the flow past a toroidal ballute in the X2 shock expansion tube; flow from left, H_2-Ne test gas representing the atmosphere of Neptune, equivalent velocity 12 km/s (from *McIntyre* et al. [16.194])

latter approach involves the deployment of a large towed balloon/parachute (ballute) assembly behind the space vehicle that generates equivalent drag at lower densities, and thus indirectly reduces the heat flux on the vehicle.

A configuration consisting of a spherical spacecraft followed by a toroidal ballute similar to the one shown in Fig. 16.114 was investigated in the GALCIT T5 free-piston-driven shock tunnel by *Rasheed* et al. [16.195]. The test gases studied were carbon dioxide, nitrogen and hydrogen to simulate actual future missions to Mars, Titan and Neptune, respectively. The carbon dioxide runs were performed with reservoir pressures ranging from 5 MPa to 30 MPa and total specific enthalpies ranging from 12 MJ/kg to 23 MJ/kg. The nitrogen runs were performed with nominal reservoir pressures from 4 MPa to 10 MPa and nominal total specific enthalpies from 23 MJ/kg to 26 MJ/kg. The hydrogen runs were performed with reservoir pressures from 3 MPa to 25 MPa and total specific enthalpies ranging from 27 MJ/kg to 80 MJ/kg. The data collected for each run consisted of a resonantly enhanced shadowgraph and heat flux data on the ballute ring and the spacecraft. The results in T5 showed that the flow past the space vehicle/ballute configuration is steady. The experimental simulation of the flows in the Mars and Titan atmospheres could be used to permit extrapolation to the flight Reynolds numbers of planned missions with reasonable confidence.

An experimental investigation of hypersonic, high-enthalpy flow over the configuration shown in Fig. 16.114 was conducted in the X2 shock expansion tube facility of The University of Queensland, Australia [16.194]. Imaging using near-resonant holographic interferometry (Fig. 16.114) showed that the flows were steady and not subject of choking except when the opening of the toroid was blocked. Measurements at moderate total specific enthalpy conditions (15–20 MJ/kg) in nitrogen and carbon dioxide showed peak heat fluxes of around $20\,\mathrm{MW/m^2}$ on the toroid. At higher enthalpies ($> 50\,\mathrm{MJ/kg}$) in nitrogen, carbon dioxide, and a hydrogen–neon mixture, heat fluxes above $100\,\mathrm{MW/m^2}$ were observed.

Laminar Turbulent Transition in High-Enthalpy Flows

The knowledge of the transition from laminar to turbulent boundary layer flow on a entry or re-entry vehicle is an important point in the aerothermodynamic design of such vehicles, because, e.g., the prediction of heat flux depends critically on this process. Concerning laminar to turbulent boundary layer transition, one of the important differences between low-speed and hypersonic flows is that the dominant instability mode in the latter is the second or *Mack* [16.196] mode, in which the boundary layer acts as a waveguide for acoustic noise, where selected frequencies are trapped and amplified, eventually leading to transition. The second important difference is that in high-enthalpy hypersonic flows the relaxation processes associated with vibrational excitation and dissociation provide mechanisms for damping acoustic waves and may therefore be expected to affect the second mode [16.197].

It was pointed out by *Hornung* [16.197], that a large part of the experimental work on the problem of stability and transition at high Mach number has been done in cold, hypersonic ground-based test facilities. These investigations, together with the linear stability analysis by *Mack* [16.196] provide a substantial basis for understanding the path to transition in cold hypersonic flow, however, they are not able to capture the high-temperature effects that occur in hypervelocity flows.

An extensive experimental program on transition in hypervelocity flows was conducted at GALCIT in the T5 free-piston-driven shock tunnel in which the focus was specifically on the regime where relaxation processes associated with vibrational excitation and dissociation are important. A summary of this work is given by *Hornung* [16.197].

As pointed out earlier, the typical test time of free-piston-driven shock tunnels at high-enthalpy conditions is approximately 1–2 ms. Due to the fact that the frequencies of the most strongly amplified modes are typically 1–3 MHz, the short test time is not a serious limitation. The configuration used for the transition experiments in T5 was a slender cone (Fig. 16.115). The flow over this configuration has the advantages that the pressure gradient is zero, and that it is free of side-effects. A 5° half-angle cone was selected in order to be able to compare the new high-enthalpy results with those from cold hypersonic testing. The velocity in the T5 test section ranged up to 6 km/s, and the transition location on the cone as a function of the total specific enthalpy was determined from the distinct rise in heat flux that coincides with the transition process. The Reynolds number at transition, evaluated at the reference temperature [16.197] and based on the distance from the cone tip to the transition location, versus the total specific enthalpy of the flow is shown for different test gases in Fig. 16.116. It is obvious that a significant increase in transition Reynolds number with increasing total specific enthalpy is obtained. This increase is slightly larger in air than in nitrogen. Tests with helium, which behaves like a perfect gas in the T5 total specific enthalpy conditions, revealed that even at 15 MJ/kg, the same transition Reynolds number as at low total specific enthalpy was obtained. Carbon dioxide, which exhibits strong vibrational and dissociational effects in the used enthalpy range lead to a significantly larger transition Reynolds number. These results show that a dramatic transition delay, which is completely absent at low speeds, is evident at high enthalpy, and that the magnitude of the phenomenon and the enthalpy at which it sets in are different for different gases. Linear stability computations performed by *Johnson* et al. [16.198] agree with the trends observed in T5 and illustrate how strongly thermochemical nonequilibrium effects can influence the growth rate of disturbances (Fig. 16.117). The mechanism responsible for this transition delay is shown to be the damping of the acoustic second-mode instability by relaxation processes. In additional tests in T5 it was demonstrated that transition could be further delayed by up to a factor of two by suitable blind porosity of the surface [16.197].

Experiments to investigate the process of boundary layer transition in hypervelocity boundary layers were performed in the T4 free-piston-driven shock tunnel at The University of Queensland, Australia by *Mee* and *Goyne* [16.200], *Mee* [16.201]. An array of flush-mounted thin-film heat flux gauges was used to measure surface heat flux and to detect the location and extent of the transitional region on a 1500 mm-long 120 mm-wide flat plate that formed one of the inner walls of a duct that captured the core flow of the test section. The thin-film gauges, manufactured at The University of Queensland, consisted of a nickel film, 1 mm long × 0.3 mm wide × 1 μm thick, deposited onto a fused-quartz cylinder of 2.1 mm diameter. The gauges and associated instrumentation were estimated to have response times of 2 μs. The experiments were performed in a test gas of air at nozzle supply enthalpies of 6 and 12 MJ/kg. At the lower total specific enthalpy condition the unit Reynolds number (based on free-stream conditions) was varied from 1.7×10^6 m^{-1} to 4.9×10^6 m^{-1}. At the high-enthalpy condition, the unit Reynolds number was 1.6×10^6 m^{-1}. A typical heat flux distribution along the test surface with a smooth surface is shown in Fig. 16.118. The measurements indicate the transition of the boundary layer from a laminar region to a turbulent region along the test surface.

The detailed analysis of the measurements obtained in T4 by the individual thin-film gauges showed that transition takes place through initiation, growth and merger of turbulent spots in the considered flat plate hypervelocity boundary layer. The rates of initiation and growth of turbulent spots appear to be the factors which

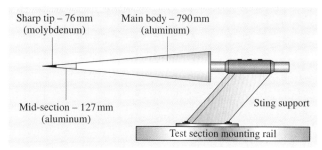

Fig. 16.115 Slender cone model used for the transition experiments in T5;. the main body is hollow and instrumented with thermocouple surface heat flux gauges (after *Hornung* et al. [16.199])

will determine the length of the transition region It was found that the spanwise extent of spots can be identified and spreading rates can be determined. Further experiments in T4 showed that earlier transition can be promoted by using both isolated and distributed roughness elements. Using distributed roughness elements located 100 mm downstream of the plate leading edge with a ratio of roughness element height to boundary layer thickness of 1.1 leads to the result in Fig. 16.119. The distributed roughness elements caused earlier transition at the low- and high-enthalpy condition used for the testing. An isolated rounded roughness element with a height 50% of the boundary-layer thickness was found

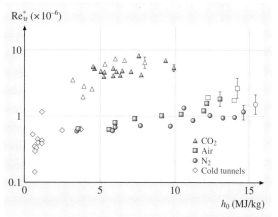

Fig. 16.116 Transition Reynolds number evaluated at reference conditions as function of total specific enthalpy for different test gases. *Open symbols* correspond to cases where the flow was laminar to the end of the cone. The references for the cold tunnel data are given by *Hornung* [16.197]; the carbon dioxide results show a large transition delay relative to the nitrogen and air results (after *Hornung* (2006) [16.197])

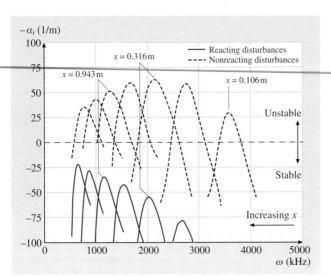

Fig. 16.117 Results of linear stability calculations with thermochemical nonequilibrium at the conditions of T5 (4.0 MJ/kg in carbon dioxide); the diagram shows growth rate of disturbances as function of disturbance frequency at several distances along the cone; the *dotted curves* show the results at the same conditions as for the continuous curves but with the rate processes turned off [16.198] (after *Hornung* [16.197])

to be capable of producing a turbulent patch immediately behind the element.

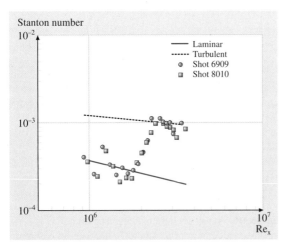

Fig. 16.118 Dimensionless mean heat flux along the test surface resulting from two different T4 runs [$h_s = 6.2$ MJ/kg, Re $= 2.6 \times 10^6$ m^{-1} (based on freestream conditions)]; the *continuous* and *dashed lines* indicate laminar and turbulent heat flux levels obtained from empirical correlations (after *Mee* [16.201])

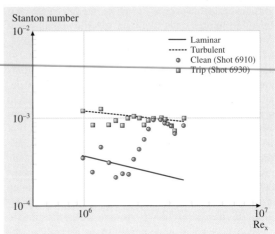

Fig. 16.119 Dimensionless mean heat flux along the test surface [$h_s = 6.2$ MJ/kg, Re $= 2.6 \times 10^6$ m^{-1} (based on freestream conditions)] for the smooth surface (*circles*) and distributed roughness elements located 100 mm downstream of the plate leading edge (*squares*); the *continuous* and *dashed lines* indicate laminar and turbulent heat flux levels obtained from empirical correlations (after *Mee* [16.201])

The same test geometry as used by *Mee* [16.201], was used by *Goyne* [16.140], and *Goyne* et al. [16.142] to conduct skin friction measurements in laminar, transitional and turbulent boundary layers in addition to pressure and heat flux measurements. The tests were performed in a total specific enthalpy range of 4–13 MJ/Kg and a Reynolds number (based on the distance from the leading edge) range of 0.16–21×10^6. For the transitional boundary layer, the characteristics of turbulent spots were investigated and similar results were obtained using heat flux and skin friction measurements.

Supersonic Combustion Ramjet Research

The goals to reduce the specific transport costs and to increase the reliability and flexibility of new transportation systems require that future space programs develop partly or fully reusable vehicles with improved propulsion systems. The supersonic combustion ramjet (scramjet), an air-breathing propulsion concept, is one option for a new and advanced type of engine to be operated at flight Mach numbers above $M = 8$. The efficiency gains of such engines potentially allow improvements in size of payload, cost/kg and reliability of future launch vehicles.

Recently significant advances have been made in developing scramjet engines, including The University of Queensland's supersonic combustion flight experiment HyShot [16.202] with its first successful flight performed in July 2002 and the flights of NASA's X-43A in March and November 2004 as part of the Hyper-X programme [16.203, 204]. Supersonic combustion had not occurred in flight until the performance of HyShot and it had not been used in an engine which propelled a flight vehicle until the X-43A flight of March 2004 [16.205]. Ground-based testing in reflected shock tunnels and a shock expansion tunnel represent significant contributions to these flight experiments.

Test techniques and strategies for large scale scramjet testing in large reflected shock tunnels are discussed by *Deiwert* et al. [16.207]. For the Ames 16-Inch deflagrative combustion-driven shock tunnel, large- to full-scale scramjet combustor test capabilities over the flight Mach number range 12 to 16 are reported.

The extensive scramjet research conducted in Australia in the T3 and T4 free-piston-driven shock tunnels was summarized by *Paull* and *Stalker* [16.208] and *Stalker* et al. [16.205]. The research programmes conducted in these facilities include the investigation of the thrust/drag performance of an integrated axisymmetric scramjet configuration consisting of intake, combustor and thrust nozzle, shown in Fig. 16.120 with a stress wave force balance by *Paull* et al. [16.206]. The model consists of an axisymmetric center body, with six combustion chambers and associated intakes arranged about its periphery. These intakes consist of compression ramps formed by the splitters which separate the combustion chambers. The thrust coefficient determined for this configuration as function of the total specific

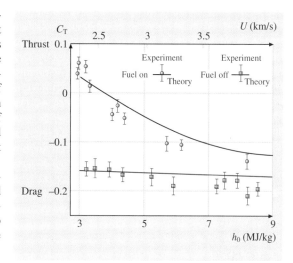

Fig. 16.121 Performance of the axisymmetric integrated scramjet model of Fig. 16.120; measured thrust coefficient for fuel off (*squares*) and fuel on (*circles*) versus total specific enthalpy and test section free-stream velocity; the lines indicate the calculated thrust coefficients [16.205, 206]

enthalpy of the flow and the test section free-stream velocity is shown in Fig. 16.121. The measured data points for fuel-on and fuel-off experiments are plotted together with two continuous lines indicating calculated axial forces. The important result of this investigation is that the difference between the fuel off and the fuel on data clearly shows the generation of thrust and for stagnation enthalpies below approximately 3.5 MJ/kg, positive net thrust could be demonstrated in a ground-based test facility.

The HyShot supersonic combustion flight experiment was led by The University of Queensland in Australia with contributions from a number of international partners [16.202]. The scramjet payload was launched by an unguided sounding rocket from the Woomera launch range in Australia on a highly parabolic trajectory to an altitude in excess of 328 km. The scramjet experiment was conducted during re-entry, and consisted of a double-wedge intake with two back-to-back constant-area combustion chambers, one fueled with hydrogen at an equivalence ratio of about 0.33, and the second unfueled (Fig. 16.122). In the fueled combustor, gaseous hydrogen was injected perpendicularly to the flow through a series of four portholes (Fig. 16.123a). Supersonic combustion data was successfully collected at flight Mach numbers of approximately $M = 7.8$–7.95 in an altitude range from 35 km down to 29 km [16.209].

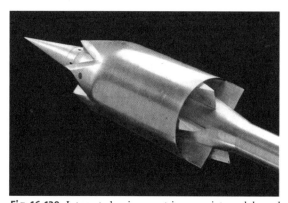

Fig. 16.120 Integrated axisymmetric scramjet model used in the T4 free-piston-driven shock tunnel of The University of Queensland, Australia for net thrust measurements [16.206] (picture courtesy A. Paull, The University of Queensland)

As part of the HyShot flight programme, extensive ground-based testing of the supersonic combustion experiment was performed in the T4 free-piston-driven shock tunnel. These experiments included pre- as well as post-flight test campaigns. An experimental and numerical post-flight analysis was also conducted at the German Aerospace Center, and the European Space Research and Technology Center (ESTEC) of the European Space Agency (ESA) [16.210, 211]. The experiments were performed in the HEG. The flow path geometry of the wind tunnel model used in the HEG HyShot post-flight analysis was identical to the flight configuration. A comparison between the measured and calculated surface pressure distribution in the combustion chamber for a HEG operating condition which duplicates the HyShot flight condition at an altitude of 32.5 km is shown together with the surface pressure of the fueled combustor measured in flight in Fig. 16.123b. The scatter bars of the HEG measurements indicate the root-mean-square (RMS) values of the pressure during the test time. The fuel-off static surface pressure obtained from CFD analysis compares well with the wind-tunnel experiment. Further, the pressure gain due to fuel injection in the HEG measurements is obvious. While some issues concerning the quantitative comparison of the ground-based, flight and CFD data are not yet fully resolved, the post flight analysis in HEG confirmed the interpretation of the flight data that supersonic combustion was successfully established.

A modification of the HyShot configuration used in the first and second flight of the programme was developed at the Japan Aerospace Exploration Agency (JAXA) using the High Enthalpy Shock Tunnel (HIEST) [16.212]. In this configuration, the porthole injection is replaced be a *HyperMixer* injection which utilizes generators of longitudinal vortices to enhance the supersonic air flow fuel mixing and the combustion process.

Further scramjet research performed in HIEST is presented by Itoh et al. [16.132]. An integrated scramjet configuration of 2 m length was investigated in the total specific enthalpy range of 3.3–7.5 MJ/kg which corresponds to the flight Mach number range of 7–13. The model was suspended by very thin wires and could move freely in the test section. Therefore, using the 13 accelerometers integrated on the model, the thrust/drag characteristic of the engine as a function of the fuel equivalence ratio and the stagnation enthalpy was investigated.

As part of the NASA Hyper-X programme, the X-43A (Fig. 16.124), a small-scale research vehicle to provide flight data for a hydrogen-fueled, airframe-integrated scramjet engine at flight Mach numbers

Fig. 16.122 HyShot supersonic combustion flight experiment configuration, used during the first two flights of the programme, mounted on top of a two-stage sounding rocket at the Woomera launch range in Australia (Photo courtesy The University of Queensland)

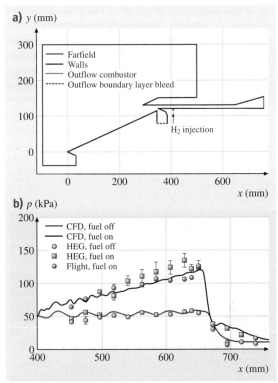

Fig. 16.123a,b HyShot supersonic combustion flight experiment configuration, used during the first two flights of the programme, geometry of the fueled combustor including the boundary conditions for the CFD analysis (**a**) and computed and measured pressure distributions in the fueled and unfueled combustor; flight altitude: 32.5 km (**b**)

of 7 and 10 was developed [16.204]. In addition, aerodynamic, thermal, structural, guidance, flush-air-data-system, and other data were to be obtained. The 3.66 m-long research vehicles were dropped from the NASA Dryden B-52, rocket-boosted to the test point by a modified Pegasus first stage, separated from the booster, and then operated in autonomous flight. The flight tests were conducted at an altitude of approximately 30 km.

Tests of the Hyper-X scramjet engine flow path have been conducted in the HYPULSE shock tunnel facility of ATK–GASL in Ronkonkoma, USA at conditions duplicating the stagnation enthalpy at flight Mach numbers of 7, 10, and 15 [16.213, 214]. For the tests at Mach number 7 and 10, HYPULSE was operated as a reflected shock tunnel; at Mach number 15, HYPULSE was operated as a shock expansion tunnel. A data base for the operation of the scramjet was obtained by variation of a number of flow parameters including the fuel equivalence ratio from lean to rich and the fuel composition from pure hydrogen to mixtures of 2% and 5% silane in hydrogen by volume to aid ignition.

Comparisons of the ground test data with selected data from the successful $M = 10$ flight showed good agreement for HYPULSE test runs in which the hydrogen fuel was augmented with silane to prompt ignition and combustion. Comparisons of tare data acquired without fueling the flow in both ground test and flight operation also exhibited very good agreement, confirming that scramjet testing in the ms time scale available in shock tunnels does provide meaningful data [16.214], thereby expanding the ground test regime for hypersonic air-breathing propulsion concepts.

Hot Model Technique

The requirement of ground-based testing at high total enthalpies and high pressures to properly duplicate high temperature effects leads to facilities with typical test times in the order of ms. Therefore, in these facilities the ratio of total temperature to wall temperature of a wind tunnel model can be approximately simulated, however, the absolute temperature of the model surfaces stays at room temperature. In order to overcome this partially, heated models were used in the shock tunnel facilities.

The influence of wall temperature on the shock wave/boundary layer interaction on a ramp configuration in hypersonic flow was investigated in the shock tunnel TH2 at the RWTH Aachen University, Germany [16.215]. For these investigations, TH2 was operated with the conventional and the heated helium

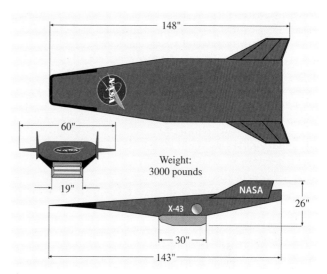

Fig. 16.124 X-43A vehicle geometry; measures in inches (after *Voland* et al. [16.204])

driver. The model surface was heated by electric heater elements. Actively cooled gauges to measure pressure and temperature (coaxial thermocouples) were integrated in a model and color Schlieren and infrared thermography were additionally applied. The ramp model consisted of a sharp leading edge flat plate at a 0° angle of attack and a second plate at 15° angle of attack. The total specific enthalpy was varied between 1.68 MJ/kg (total temperature 1520 K) and 4.12 MJ/kg (total temperature 3630 K) and the wall temperature could be varied between 293 K and 840 K. Using pressure measurements, the influence of the wall temperature and the stagnation conditions on the size of the separation bubble generated in the hinge line region was demonstrated for a flow that was laminar before the separation and possibly transitional at reattachment. At constant total temperature and increasing wall temperature, the size of the separation bubble increased. Increasing the total temperature at constant wall temperature led to a reduction of the size of the separation bubble. Furthermore, the pressure level in particular at reattachment increased with increasing wall temperature.

In atmospheric re-entry flight with a high level of aerodynamic heating, the utilization of low-catalytic surface materials, such as SiO_2-based materials, is desired in order to minimize the increase in vehicle surface temperature by reducing the recombination of atomic species at the wall. From experiences of ground and flight tests, it has been suspected that surface catalysis may strongly depend on changes in surface temperature and density. However, the detailed characterization of surface catalycity is still an actual research topic, mainly due to the lack of experimental data at sufficiently high density equivalent to flight conditions, since ground tests have been conducted mainly in low-density facilities such as arc-heated wind tunnels and plasma wind tunnels. High-density flows can be generated in high-enthalpy shock tunnels, however, the short available test duration is a severe drawback. Starting with room temperature, the wall temperature of test models in shock tunnels increases by only a few degrees. Therefore, an experimental research program to study the influence of the wall temperature on wall catalysis was conducted in the HIEST of JAXA utilizing a preheated wind tunnel model [16.132].

The test model was a 373 mm-long 260 mm-wide flat plate with blunt leading edge (Fig. 16.125). Silicon dioxide, as a material with low catalycity, was used to coat one half of the flat plate and silver (Ag) was used to coat the second half, providing the reference for a highly catalytic material. Heat flux measurements were performed using 10 thermocouples mounted in line on both sides. The surface catalycity effect was investigated by comparing the heat fluxes measured on the SiO_2 surface with those obtained with the Ag coating or with CFD results. Heaters were installed in the model which allowed wall temperature adjustment from 300 K to 1000 K. The tests in HIEST were carried out at 14 MJ/kg total specific enthalpy and 20 MPa nozzle reservoir pressure. From CFD analysis, it was found that most of the oxygen molecules should be dissociated in the high recovery temperature region in the hypersonic boundary layer near the wall, and the increment in heat flux due to recombination by wall catalysis was estimated to be 40% of the total heat flux. The experiments were conducted at wall temperatures of up to 800 K and the results are summarized in Fig. 16.126. The measured heat flux is normalized by the fully catalytic CFD results and subsequently averaged over the flat plate and plotted as function of the inverse of the wall temperature. A strong temperature dependence of the surface catalysis of the SiO_2-coated surface was observed from room temperature up to 570 K, as the normalized heat flux increased with wall temperature. It is pointed out by *Itoh* et al. [16.132], that the temperature dependence of surface catalysis appears at a relatively low temperature in the tests performed in HIEST compared to data obtained in lower-density wind tunnel tests where this dependence is found in the 400–800 K range. It is concluded that this early appearance of the temperature dependence of surface catalysis may be associated with the higher-density operation in the shock tunnel.

The same model as shown in Fig. 16.125 was also tested in HEG of DLR by *Ueda* et al. [16.216]. As part of this comparative study performed in two large free-piston-driven shock tunnels, the same trend concerning the temperature dependence of the surface catalycity of SiO_2 as first observed in HIEST was also found in HEG.

One concept for improving the efficiency of scramjet engines is that of injecting the fuel on the intake.

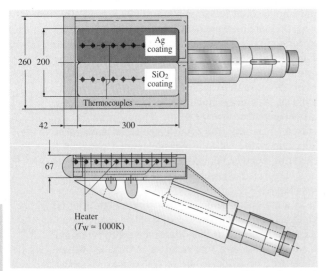

Fig. 16.125 Heated flat plate model used in HIEST (after *Itoh* et al. [16.132])

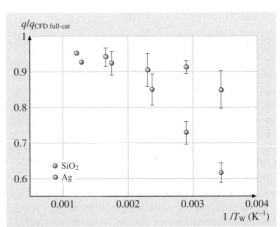

Fig. 16.126 Temperature dependence of the catalytic behavior of the SiO_2 surface obtained in HIEST; the *data points* show averaged values over the plate (after *Itoh* et al. [16.132])

The fuel injection before the combustion chamber allows the fuel to mix with the air prior to entering the combustion chamber and the length of the combustion chamber required for complete combustion is therefore reduced. Extensive testing of this concept has been conducted in the T4 shock tunnel facility at The University of Queensland, Australia. In contrast to real flight conditions, the wall temperature will only marginally increase during the available test time in T4. Therefore, realistic wall temperatures similar to those achieved in flight and the influence of the wall temperature on premature ignition on the intake can only be investigated by using a preheated model. For this reason, cold wall intake experiments were compared with experiments using a heated intake model with wall temperatures up to 700 K for total specific enthalpy conditions of 3.6 MJ/kg [16.149]. In the considered boundary layers on the intake, temperatures exist close to the wall that are, for both the unheated and heated model, significantly higher than the wall temperatures. These peak temperatures are above the ignition temperature of hydrogen. Therefore, the intake fuel injection must be performed in such a way that the fuel does not penetrate this region close to the wall downstream of the injection location. Planar laser-induced fluorescence using hydroxide (OH) radicals as the target species to identify regions where combustion occurs and pressure measurements on the intake were performed. No hydroxide was observed on the intake when the wall temperature was varied from 300 K to 700 K. Furthermore, no static pressure increase was observed along the intake model. These factors indicate that no premature combustion of hydrogen was occurring for the investigated flow fields.

16.3 Bluff Body Aerodynamics

16.3.1 Flow Physics, Facilities, and Approach

Bluff body wakes are complex; they involve the interactions of three shear layers in the same problem, namely a boundary layer, a separating free shear layer, and a wake. As has been remarked by *Roshko* [16.217], "the problem of bluff body flow remains almost entirely in the empirical, descriptive realm of knowledge", although our knowledge of this flow is extensive. In fact the recent surge of activity on wakes, over the last two decades, from experiment, direct numerical simulation and analysis has yielded a wealth of new understandings.

In this section, we shall briefly present the *experimental* approaches to studying bluff-body flows. However, rather than simply list the approaches and techniques used to study such flows, we shall present the different quantitative and qualitative approaches that have been employed, while at the same time presenting some of the fascinating, rich, and sometimes beautiful, physical phenomena that have been found in bluff body flows, as one progressively increases the Reynolds number (Reynolds number = Re = UD/ν, where U is the free-stream velocity, D is the width of the body and ν is the kinematic viscosity). We shall also present some of the key nondimensional parameters as a function of Re, and functional relationships to represent these variations in practice.

The nominally *two-dimensional* vortex-shedding process from bluff bodies has been described in a number of review papers over the last 50 years (*Rosenhead* [16.218]; *Morkovin* [16.219]; *Berger* and *Wille* [16.220]; *Oertel* [16.221]; *Coutanceau* and *Defaye* [16.222]; *Roshko* [16.217]; and *Williamson* [16.223]). In experiments, significant steps forward in our understanding of wake vortex dynamics have come from the many recent studies of three-dimensional (3-D) phenomena, which have led to some new explanations of longstanding controversies which were hitherto assumed to have two-dimensional origins (see the recent review of *Williamson* [16.223]).

Bluff-body wake flows have direct engineering significance. The alternate shedding of vortices in the near wake, in the classical *vortex street* configuration, leads to large fluctuating pressure forces in a direction transverse to the flow, and may cause structural vibrations, acoustic noise, or resonance, which in some cases can trigger failure. A recent review of vortex-induced vibrations arising from such effects can be found in *Williamson* and *Govardhan* [16.224]. The classical view of a *vortex street* in cross section is shown in Fig. 16.127, where regions of concentrated vorticity are *shed* into the downstream flow from alternate sides of the body (and with alternate senses of rotation), giving the appearance of an upper row of negative vortices and lower row of positive vortices. Interestingly, it will be shown later that these particular cross-sectional photographs actually contain useful information regarding the distribution of the *three-dimensional* vortex structure (see the figure caption).

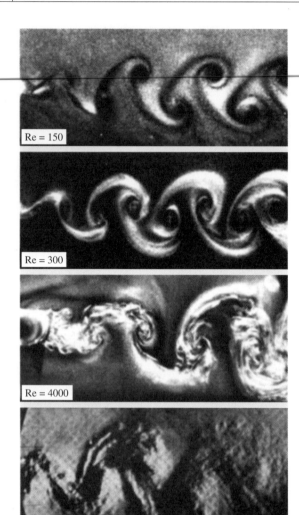

Fig. 16.127 Visualization of laminar and turbulent vortex streets. These photographs show the development of Karman vortex streets over a wide range of Re. Streamwise vorticity, in the *braid* between Karman vortices, is indicated by the white regions, and is visible for Re = 300 up to the highest Re = 270 000. The aluminium flake visualizations are from *Williamson* [16.223, 227]. The Schlieren photograph at Re = 270 000 is from *Thomann* [16.228]

rescence (LIF) technique, examples of which are shown later in Figs. 16.137, 138. Surface particle techniques, using aluminium flakes on the water surface were originally used to excellent effect by Prandtl in the 1920s, and have also been employed more recently by *Williamson* and *Roshko* [16.225], for example. Subsurface particle visualization will be seen here from a study by *Mills* et al. [16.226] in Fig. 16.129. Hydrogen bubble wires in water, employed by Rockwell and his group at Lehigh University have also been extremely effective, and an example may be observed in Fig. 16.139. Rockwell at Lehigh University, and Gharib at Caltech, have also been pioneers in the development of the particle image velocimetry technique (PIV) in the field of research concerning bluff bodies. An excellent example of this approach can be found from Rockwell's group, showing near-wake vortex dynamics in Fig. 16.140.

In air, visualization has been conducted using smoke from arrays of tubes, but also from the very effective smoke wire technique introduced in the 1950s, and optimized by Hassan Nagib and Tom Corke at IIT (*Cimbala* et al. [16.229]; *Corke* et al. [16.230]; and *Goldstein* [16.231], which is also an excellent reference for several of these techniques). In Fig. 16.127, we also see an example of a Schlieren photograph of a vortex street at Re = 270 000, where the variation of density in the flow is exploited in this case. In essence, there are a number of flow visualization techniques, which can lead to new discoveries in these flows, as well as to beautiful images of significant fluid flow phenomena.

Facilities in use for studies of bluff-body aerodynamics include not only wind tunnels, but also water facilities. For example, in our vortex dynamics and wake studies at Cornell, we utilize very low-turbulence wind tunnels, along with a recirculating water channel, which is indispensable for providing continuous flow. However, the use of a computer-controlled XY towing tank is extremely powerful, in that quite unsteady motions to bodies (periodic or transient) can be imparted in two degrees of freedom to bodies in the large (8 m × 1.25 m × 1.25 m) facility. One of the other key aspects of such a facility is the fact that one can study flow structures far downstream of a body, by waiting long times after the body has passed the region of interest. A further facility often used over the last few decades is the closed-circuit water tunnel, whose test section is enclosed, unlike the water channel. Other facilities for high-speed flows include shock tubes, which are described elsewhere in this book. A good resource for discussion of all the above facilities for use in bluff body aerodynamics, and the

These visualizations in Fig. 16.127 come from a thin light sheet shining on aluminium flakes suspended in water in the wake of a towed vertical circular cylinder. Visualizations of the cross sections of such vortices are often made in water, using ordinary dye, as well as using fluorescent dye employing the laser-induced fluo-

Fig. 16.128a–c Pressure measurements for a cylinder with laminar or turbulent boundary layer separation. A laminar boundary layer separates at an angle of around 82° (see (**a**)) leading to the pressure distribution from *Fage* and *Falkner* [16.232] in (**b**). At a higher Re, as plotted by *Roshko* [16.233] in (**b**), the separation point shifts to higher angles (typically 110–120°), leading to a thinner wake, with weaker vortices, and a much lower drag C_D. In (**c**) we observe the mean and fluctuating surface pressures at Re = 140 000 (after *Cantwell* and *Coles* [16.234]); the vertical lines show the maximum and minimum instantaneous pressures found over several thousand cycles, and it is therefore quite surprising that the mean pressure remains so uniform over the rear (or *base*) of the cylinder

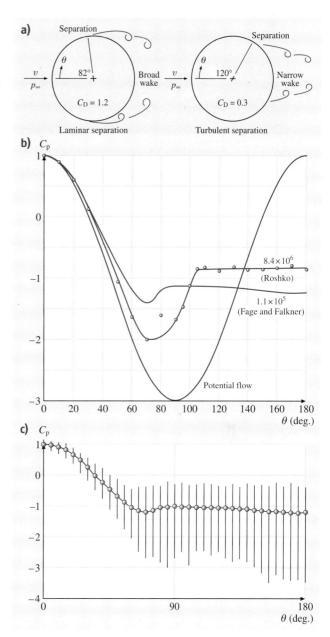

various relevant techniques, is the book by *Goldstein* [16.231].

To return to the flow structure from bluff bodies, the alternating formation of vortices mentioned earlier, were the origin of *Strouhal*'s [16.235] classical measurements of the sound frequency produced by translating cylindrical rods through air, and for the Aeolian tones, which are produced by the wind blowing over a wire or a string in an *Aeolian harp*. In a historical review by *Rott* [16.236], he discusses the later contributions of *Lord Rayleigh* [16.237] in normalizing Strouhal's frequency data using the *Strouhal* number

$$S = \frac{fD}{U}, \tag{16.5}$$

where f is the frequency of vortex shedding, D is diameter, and U is flow velocity. A great deal of impetus in this flow was triggered by the classical work of *von Karman* in 1911 [16.238], who not only analyzed the stability of vortex street configurations, but established a theoretical link between the vortex street structure and the drag on the body. This work came about from experiments conducted by Hiemenz (within Prandtl's laboratory in Göttingen), who had initially interpreted wake oscillations from a cylinder as an artifact of the experimental arrangement. However, von Karman viewed the wake oscillations and alternate generation of vortices as an *intrinsic* phenomenon, and went on to investigate their linear stability. Some descriptive understanding of the near-wake vortex formation process can be found from *Gerrard* [16.239] and from *Perry* et al. [16.240]. Gerrard suggested that a forming vortex draws the shear layer (of opposite sign) from the other side of the wake across the wake centreline, eventually cutting off the supply of vorticity to the growing vortex.

A great many experimental measurements have been carried out for the bluff-body wake including Strouhal numbers, coefficients of lift and drag, base pressure (i.e., the pressure at a point right at the downstream end of a body), separation points, surface shear stress, wake velocity measurements such as mean and fluctuation velocity profiles and Reynolds stresses, and estimates of the length and width of the *vortex formation* region.

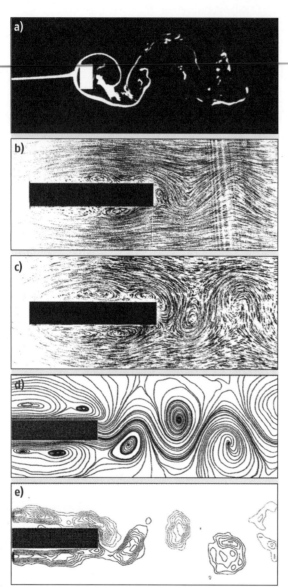

Fig. 16.129a–e Flow around rectangular cross-sections, indicated by various visualization techniques. In (**a**) the ratio of streamwise length to width of the body (l/h) is near a critical value which maximizes the drag [16.241], and we see a classic dye technique. In (**b**) and (**c**), the flow around a long rectangular cylinder is made clear using particle streaklines, and shows the usefulness of employing different reference frames (fixed with respect to the body and the wake vortices, respectively). In (**d**) and (**e**), we observe streamlines and vorticity from measurements using PIV. The images (**b**) to (**e**) are from *Mills* et al. [16.226]

like a streamlined body, such as an airfoil, where the width of the body decreases gradually towards the trailing edge, a bluff body has a more dramatic reduction of body width in the after-portions, and a larger adverse pressure gradient, leading to boundary layer separation. The subsequent rolling up of the separated vorticity in the free shear layers, causes the formation of concentrated local regions of vorticity, known as wake vortices. It is these vortices, and their associated low-pressure centers, in proximity to the rear of the body, which yield very large fluctuating pressures behind a bluff body, but also a surprisingly uniform region of low pressure behind the body (Fig. 16.128b). By integrating such a pressure distribution around the body, we can understand the large drag of a bluff body to be due to the difference between the high-pressure region in the vicinity of the front stagnation point, and the low uniform pressure in the rear (or *base*) region of the body. The drag coefficient is given by

$$C_\mathrm{D} = \frac{\mathrm{Drag}}{1/2 \rho U^2 DL} \,, \tag{16.6}$$

where ρ is the fluid density, D is the width of the body and L is the span length. Obviously, the *real* pressure distribution differs markedly from the potential flow distribution, where the front and rear velocity field (or pressures) are symmetric fore and aft, yielding zero drag.

In the case of the circular cylinder, we see in the pressure distribution of Fig. 16.128, that if the boundary layer remains laminar around the frontal portions of the body (for roughly Re < 200 000), then the separation point for the layer occurs at an angle $\theta \approx 80°$ from the front stagnation point (see the point of inflection in the early classic measurements of *Fage* and *Falkner* [16.232], at Re = 110 000). If the Reynolds number is increased to 8.4×10^6, in Fig. 16.128, then the boundary layer on the body becomes turbulent prior to separation, and remains attached for longer against the adverse pressure gradient. The layer only separates at around $\theta \approx 110°$, and the relatively weaker and smaller wake vortices exert

Such measurements may be found in a number of well-known reviews in the literature that were mentioned earlier. We shall mention several of these measurements in Sect. 16.3.2, as a function of increasing Reynolds number.

16.3.2 How Bluff is a Bluff Body?

A bluff body is one for which the boundary layer separates from the sides of the body at some point, typically from a point near the maximum width of the body. Un-

Table 16.4 Drag coefficients and Strouhal numbers for common two- and three-dimensional shapes. One interesting point may be made using the example of a semicircular cylinder. When the flat face is upstream, the drag is high, and the Strouhal number is low, and it is roughly equivalent to a flat plate. When the curved face is upstream, the drag is lower, and the Strouhal number is higher, and now the body is roughly equivalent to the circular cylinder. We note that, as the drag C_D increases, the frequency S decreases, which is a trend first clearly pointed out by *Roshko* [16.242]

2-D BODY	C_D	S	2-D BODY	C_D	S
Circular cylinder	1.2	0.21	Square cylinder	2.1	0.13
Semi-circular cylinder	1.2	0.21	Square cylinder at 45°	1.6	0.15
	1.7	0.16	Rectangular cylinder (2:1)	1.8	0.08
Flat plate	2.0	0.16	Rectangular cylinder (1:2)	2.4	0.14
Half tube	1.2	0.21	90° wedge	1.6	0.18
	2.3	0.13		2.0	0.13

3-D BODY	C_D	S	3-D BODY	C_D	S
Sphere	0.47	0.19	60° cone	0.80	0.19
Hemisphere	0.4	–	90° cone	1.2	0.16
Hemisphere	1.2	–	Disk	1.1	0.13

far less *base suction* on the rear of the body, pulling it downstream. The ensuing dramatic reduction of drag is known as the *drag crisis*, and will be discussed in the next section.

The degree of *bluffness* was addressed by Roshko in a classic paper on bluff-body wakes (*Roshko* [16.242]), and he made it clear that, even if different bodies had the same width, they could generate different base pressures (the roughly uniform low pressure behind the body), as a result of divergence of the separating shear layers to quite different degrees, depending on their shape. We have compiled a set of drag coefficients and Strouhal numbers for a number of two-dimensional shapes, and three-dimensional objects in Table 16.4, for Reynolds numbers around 10 000. The large variation of C_D values is evident. An illustrative example (Table 16.4) is given by the two half-cylinder shapes. When the flat face is downstream, one can imagine that the shape is more streamlined, and the separating shear layers will emerge from the body roughly as they do for the full circular cylinder, giving $C_D = 1.2$, and a vortex frequency or Strouhal number of $S = 0.21$. If, on the other hand, the flat face is upstream, the shear layers emerge from the body sides at a large angle normal to the free stream,

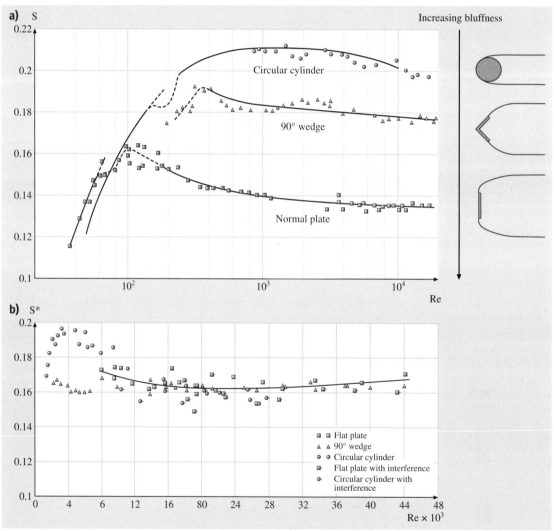

Fig. 16.130a,b Strouhal numbers (S) for different shapes, versus Re, and the introduction of a *wake* Strouhal number (S^*) by *Roshko* [16.242]. His idea was to normalize the Strouhal number by the *wake width* and the *wake velocity* close to separation, as shown in **(b)**, rather than scaling the Strouhal number simply on the body width and free-stream velocity as in **(a)**

the wake vortices will be larger and stronger, forming in a wider formation region behind the body. The drag is now significantly higher, $C_D = 1.7$, and the larger width of vortex formation corresponds with a lower normalized vortex shedding frequency, $S = 0.16$. As pointed out by Roshko, the shedding frequency scales with the width of the wake, rather than the physical width of the body itself.

We present a further example, which illustrates how the streamwise length of a body has an effect on the drag, even if the front face is kept a constant shape. If one studies the drag of a family of rectangular cross sections, as in Fig. 16.129a, then one finds the existence of a critical value of L/h (streamwise length/width). *Bearman* and *Truesdell* [16.241] showed that C_D increases from 2.0 to 3.0, as the length increases from a normal flat plate ($L/h \approx 0$) to the critical ratio $L/h = 0.6$. As the length of the body is stretched further, the drag diminishes, ultimately reaching a lower value than for the flat plate. Again, the proximity of the separating shear layers, and their roll up into strong wake vortices, close to the rear surface of the body is key to dictating the pressure distribution and drag of the body.

In Fig. 16.129b,c, we also present visualizations of long rectangular (nominally two-dimensional) shapes [16.226], which use particle streak photography. Clear differences can be found if one uses the reference frame fixed with respect with the body or fixed with the vortices, as illustrated also in *Williamson* and *Roshko* [16.225] and *Perry* et al. [16.240]. Examples from particle image velocimetry (PIV) are also shown in Fig. 16.129, showing the streamlines and vorticity contours using this powerful quantitative technique, described comprehensively elsewhere in this handbook. The technique shows clearly the shedding of vortices from the front face (leading-edge separation), which then travel along the sides of the cylinder, and largely influence the phase and formation of trailing-edge vortex formation.

As mentioned earlier, these ideas concerning the degree of *bluffness* of a body were first formulated by *Roshko* in [16.242]. He measured the Strouhal numbers for a number of bodies, as a function of Re, as seen in Fig. 16.130. He was essentially the first to note that *bluffer* bodies diverged the flow more, yielding a lower base pressure (more base *suction*), and lower vortex shedding frequencies. For example, in order of increasing bluffness, we have: circular cylinder → wedge → normal flat plate. From dimensional analysis, he proposed a Strouhal number based on the physical scales of the near-wake formation to collapse frequency data

from different body shapes. He suggested that the shedding frequency would scale with the wake width (L^*) rather than simply the body dimensions (D), and on a relevant velocity scale (U^*) for the vortex formation in the near wake, rather than simply on the free-stream velocity (U). He put forward a *wake* Strouhal number:

$$S^* = \frac{fL^*}{U^*} \approx \text{const.} \qquad (16.7)$$

The wake width L^* for different bodies was found using free-streamline theory, while the velocity scale was taken to be that velocity just outside the sepa-

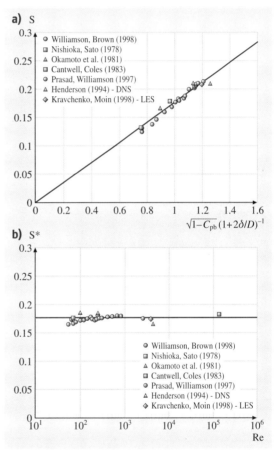

Fig. 16.131a,b Collapse of frequency data for a circular cylinder. This approach uses a wake length scale involving the body diameter plus the shear layer thickness (evaluated near separation). This scaling collapses the *wake* Strouhal number (S^*) well for a particular body, in this case a circular cylinder, over a wide regime of Re (as shown in *Williamson* and *Brown* [16.243]). (References are found in therein)

ration point (U_s) which, to a good approximation, is calculated from the base pressure coefficient. (Following a boundary layer approximation, $1/2 U_s^2$ is the flux of vorticity shed into the wake.) This *wake* Strouhal number, shown in Fig. 16.130b, and other such *universal* Strouhal numbers that have been put forward, for example by *Bearman* [16.244] and by *Griffin* [16.245], resulted in a very reasonable collapse of the shedding-frequency data for many different bluff bodies. These results confirm the merit of considering the characteristic scales of wake formation, rather than simply body dimensions (and free-stream velocity), to generate an effective Strouhal number representing vortex shedding frequency.

Such ideas also provide a basis for a functional relationship for S–Re measurements for a given bluff body. It is possible to consider different length and velocity scales upon which the shedding frequency might depend. *Bauer* [16.247] measured the vortex shedding frequencies behind flat plates parallel to the flow (whose cross sections were bullet-shaped). He normalized the Strouhal numbers by a characteristic lengthscale equal to twice the boundary layer displacement thickness (δ^*) plus the plate (or *bullet*) thickness (i.e., $D + 2\delta^*$). A subsequent study by *Eisenlohr* and *Eckelmann* [16.248] for many different plates showed a good collapse of their frequency data when it was plotted using such a Strouhal number.

Concerning the circular-cylinder wake, consistent with the above discussion, *Williamson* and *Brown* [16.243] use the concept that the frequency will scale with $(D+2\delta)^{-1}$, where δ is a characteristic shear layer *vorticity* thickness, so that in (16.7), they have $L^* = (D+2\delta)$, and $U^* = (U_s) =$ velocity measured at separation, giving:

$$S^* = \left(\frac{fD}{U}\right)\left(\frac{U}{U_s}\right)\left(1+\frac{2\delta}{D}\right). \tag{16.8}$$

It can be shown [16.242] that U_s is related to the base pressure coefficient (to a good approximation) by

$$\frac{U_s}{U} = \sqrt{1-C_{pb}}, \tag{16.9}$$

giving an expression for the Strouhal number as

$$S = S^*\sqrt{1-C_{pb}}\left(1+\frac{2\delta}{D}\right)^{-1}. \tag{16.10}$$

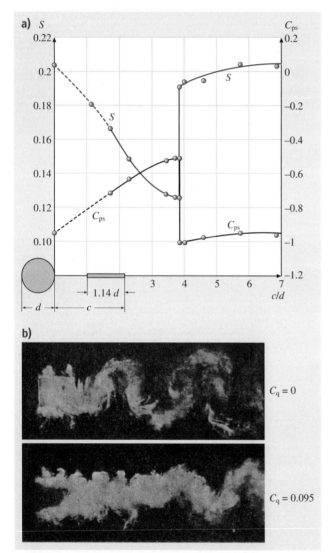

Fig. 16.132a,b Effect of a wake *splitter plate* on the base pressure coefficient (C_{ps}) and Strouhal number (S). Effect of *base bleed* on the wake structure. Both of these classic techniques, introduced respectively by *Roshko* [16.242] and by *Bearman* [16.246], can dramatically influence the wake flow structure as well as the base pressure, vortex shedding frequency, and also the drag of a bluff body. The principal effect is to stretch the vortex formation region, and to shift the low-pressure vortices further downstream from the base region of the body

Experimental support for the use of the Strouhal number S^* in (16.10) comes from Fig. 16.131a, showing S plotted against $\sqrt{1-C_{pb}}(1+2\delta/D)^{-1}$. We note that the data lies closely along a straight line, whose gradient yields the best-fit value for $S^* = 0.176$. One should note that the available data from the literature for this plot includes the wide range of Re from 55 up to 140 000. A plot of calculated S^* values versus Re in Fig. 16.131b gives another indication of the constancy of S^* over this large range of Re, and thus suggests that it is a reasonable means to collapse frequency data for this body.

Regarding the primary wake instability, a revealing experiment was conducted by *Roshko* [16.242], who studied the effect of a splitter plate (parallel to the free stream) located downstream of a bluff body at high Re. He found that, by bringing such a plate closer to the cylinder, he could interfere with the vortex shedding instability within a critical distance from the body, which caused a jump decrease in both the shedding frequency and base *suction* ($-C_{pb}$), as shown in Fig. 16.132a. A downstream shift of the low-pressure vortices reduced the suction near the base of the cylinder. The ensuing wider and longer *vortex formation* region exhibits a lower shedding frequency.

One may define a vortex *formation length* [16.246, 251] as that point downstream of the body where the velocity fluctuation level has grown to a maximum (Fig. 16.142b) and thereafter decays downstream. *Bearman* [16.246], also using splitter-plate wake interference, made the discovery that the base suction was very closely inversely proportional to the formation length, which has often been assumed since that time. It has generally been found in these studies that an *increase* in formation length (L_F/D) is associated with a *decrease* in the level of velocity fluctuation maximum u'_{rms}/U_∞ (or Reynolds stress maximum) and a decrease in the base suction. *Bearman* [16.244], on the other hand, used *base bleed* to bleed a rate of fluid flow out of the base region of a body, to alter the wake formation process. This is readily evident from his smoke visualizations for zero or finite bleed rates in Fig. 16.132b, where even small bleed rates show remarkable control of the downstream vortex formation.

It is also relevant in the discussion of the vortex-shedding regimes to consider that, if one averages over large time (compared to the shedding period), one can define a *mean recirculation region* in the wake, which is symmetric and closed. This was discussed in *Roshko* and *Fiszdon* [16.252], and in *Roshko* [16.217], where he linked the recirculation *bubble* length with the base pressure C_{pb}, and the Reynolds stresses ($\rho u'v'$), in an effective model of bluff-body flow.

16.3.3 Base Pressure, Drag, Lift, and Strouhal Number

In this section, we present the important variations of lift and drag, Strouhal number, and base pressure as a function of Reynolds numbers. At this point, we restrict

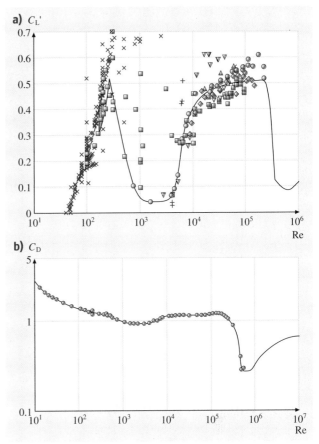

Fig. 16.133a,b Fluctuating lift coefficient (C'_L) and mean drag coefficient (C_D) as a function of Re. One can observe that the *drag crisis*, seen as the sharp reduction of drag at Re $\approx$ 200 000, is coincident with a sharp reduction in fluctuating lift. The lift variation and its measurement is discussed comprehensively in *Norberg* [16.249], where the symbols are defined in his Fig. 2. The drag variation includes classic data from *Wieselsberger* [16.250], which has been remarkably *robust* in the face of modern techniques. Data at the highest Re $\approx 10^6 - 10^7$ come from *Roshko* [16.233]

ourselves to discussion of the circular cylinder alone, since many of the variations exhibit similar features for the different bluff-body shapes.

Drag and lift forces are measured using many different techniques, which are discussed at length in the review of lift force measurements by *Norberg* [16.249], and in the book of *Goldstein* [16.231]. These techniques include, for example, pressure distribution measurements (described well in the early 1928 paper by *Fage*[16.253]), use of strain gauges, linear variable displacement transducers (LVDTs), and piezo-

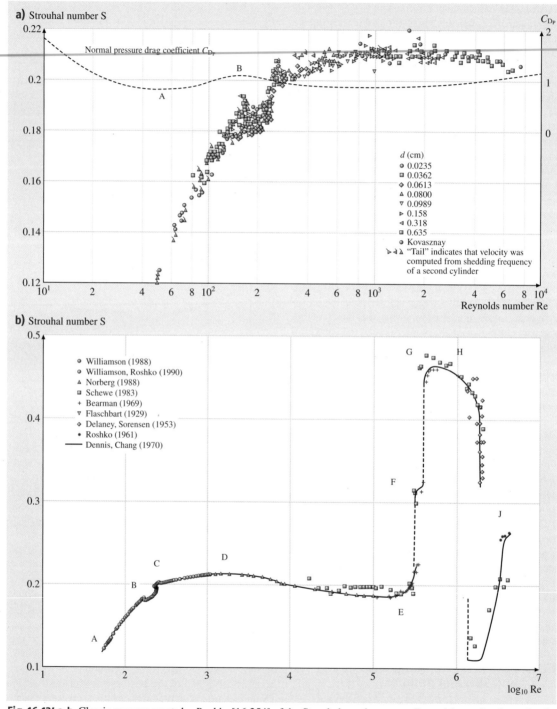

Fig. 16.134a,b Classic measurements by *Roshko* [16.254] of the Strouhal number versus Reynolds number for a circular cylinder in (**a**), and an updated plot showing the key regimes in (**b**) compiled for this handbook (References are found in [16.223])

electric transducers. Vortex shedding frequencies are generally measured using LDV (laser Doppler velocimetry), and hot-wire anemometers, or possibly by noting the lift force spectrum. The base pressure coefficient: $C_{pb} = (2pb - ps)/\rho U^2$ (where pb is the base pressure, ps is the static pressure of the free stream), is generally measured using a small pressure hole (tap) at the base of the bluff body.

It is immediately clear from Fig. 16.133 that the lift and drag variations are significantly different as a function of Re. The unsteady lift coefficient, which shows large variations, is influenced by the types of instability in the near wake, as Reynolds numbers are varied. These instabilities will be discussed in the following section. In brief, the rise of C_L to the first major peak at around $Re \approx 190$ is associated with increasingly energetic laminar vortex shedding in the near wake. The second broader peak at much higher $Re \approx 10\,000 - 200\,000$, corresponds with the energetic separated shear layer instability, which results in high shear stresses and stronger unsteady wake vortices in the near wake.

The drag force, on the other hand, shows remarkably little of the large fluctuations that characterize the lift. The presented data points in Fig. 16.133b are from some classic measurements by *Wieselsberger* [16.250], and indicate the reduction of C_D until around $Re \approx 1000$. In fact, the drag varies roughly as $C_D \approx 1/Re$ at the lowest $Re \ll 1$. Because it is extremely difficult to measure drag force on thin wires at very low Re, any fine variations in C_D around $Re \approx 100-200$, due to the onset of the energetic laminar vortex shedding, are not evident. An effective approach to determine the subtle variations in C_D, is to use direct numerical simulations, as completed to very high resolution by *Henderson* [16.255].

It is remarkable that the drag, and the Strouhal number in Fig. 16.134, remains almost constant at around $C_D \approx 1.2$, and $S \approx 0.20$, over a very wide regime of Re, from 1000 to 200 000, and it is usefully assumed by practising engineers that the flow field is reasonably independent of Reynolds numbers over this wide regime. On the other hand, the base pressure and lift force exhibit changes indicative of the shear layer instability becoming more energetic as Re increases. We go on to discuss the vortex instabilities in these flows in the next section.

An important effect, mentioned earlier, is the sharp drop in drag at $Re \approx 200\,000$, which is caused by the transition of the attached boundary layer into a turbulent state. The turbulent layer is more able to withstand the adverse pressure gradient around the cylinder, and separates later, causing weaker smaller vortices and less base *suction* contributing to the reduced drag. This significant effect is called the *drag crisis*, and in some cases the purposeful roughening of a bluff body will actually induce early transition and an advantageous reduction in drag.

The variation of Strouhal number was first defined by Roshko over a wide range of Reynolds numbers [16.242], and indicated a change in regime at around $Re \approx 300$, as seen in Fig. 16.134a, leading to the well-known S–Re relationships:

$$S = 0.212 - \frac{4.494}{Re}, \; Re \approx 50-150; \quad (16.11)$$

$$S = 0.212 - \frac{2.692}{Re}, \; Re \approx 300-2000. \quad (16.12)$$

A more-complete S–Re variation is shown in Fig. 16.134b, showing the existence of a number of regimes, and along with the base pressure measurements, will form the basis of the flow regime discussions in the next section.

A highly accurate representation of the variation of Strouhal numbers with Re was put forward by *Williamson* and *Brown* [16.243], based on the form of (16.10), noting that the shear layer thickness will scale on $1/\sqrt{Re}$, as for the attached boundary layer. In fact they have put forward the expression:

$$S = \left(A + \frac{B}{\sqrt{Re}} + \frac{C}{Re} + \ldots \right). \quad (16.13)$$

An excellent fit for the laminar (parallel) shedding regime, based on a truncation of the series (9), using data from *Williamson* [16.256], gives

$$S = \left(0.285 - \frac{1.390}{\sqrt{Re}} + \frac{1.806}{Re} \right), \; Re < 190. \quad (16.14)$$

If we apply the $\sqrt{Re}$ formulation to the frequency data from *Williamson* [16.227] for higher Re, we find:

$$S = 0.2234 - \frac{0.3490}{\sqrt{Re}}, \; Re \approx 260-1300. \quad (16.15)$$

For higher regimes of Re, the curve fits of *Fey* et al. [16.257] may be effectively used, as follows:

$$S = 0.204 - \frac{0.3364}{\sqrt{Re}}, \; Re \approx 1300-5000; \quad (16.16)$$

$$S = 0.1776 - \frac{2.2023}{\sqrt{Re}}, \; Re \approx 5000-200\,000. \quad (16.17)$$

For the complexity of the S variation for $Re > 200\,000$, it is probably worth using the plot in

Fig. 16.134b directly. Accurate data for vortex shedding frequencies (S) as well as base pressures (C_{pb}), for a variation of aspect ratios (cylinder length/diameter), are given by *Norberg* [16.258].

16.3.4 Overview of Vortex Shedding Regimes

It is particularly revealing as a key for this section to consider the plot of base pressure coefficient (C_{pb}) as a function of Reynolds number in Fig. 16.135. In contrast to some of the other parameters of the flow, the base pressure responds sensitively to the changes in flow instabilities and phenomena throughout the Reynolds number range. An incisive overview of the flow regimes was first given in *Roshko* and *Fiszdon* [16.252]. This was updated by *Roshko* [16.217] and *Williamson* [16.223], stimulated by the completion of the previously elusive data at low laminar-shedding Reynolds numbers by *Williamson* and *Roshko* [16.259]. In some of these studies, it was found convenient to refer to a *base suction* coefficient ($-C_{pb}$), rather than the base pressure itself.

Regarding the plot of base pressure in Fig. 16.135, it should be mentioned that this comes from experiments using a smooth cylinder in good flow quality (turbulence levels typically around 0.1%), and also from the simulations of *Henderson* [16.255]. It is known that roughness, turbulence levels (as well as the character of turbulence spectra), cylinder aspect ratio, end conditions, and blockage affect the transitions, although the trends remain the same. The first definition of flow regimes based on measurements of velocity fluctuation, spectra and frequency was given by *Roshko* [16.254]. He found a *stable* (periodic) laminar vortex shedding regime for Re = 40–150, a transition regime in the range Re = 150–300, with an *irregular* regime for Re = 300–10 000, where velocity fluctuations showed distinct irregularities. Similar regimes were confirmed by *Bloor* [16.263].

Referring now to the plot of base suction coefficient versus Re in Fig. 16.135, and following the lead of *Roshko* [16.217] and *Williamson* [16.223], we shall define the various shedding regimes with respect to the letters marked on this plot, and exhibit some of the principal results and experimental approaches that have been taken.

Regime up to A: Laminar Steady Regime (Re < 49)

At Re below around 49, the wake consists of a steady recirculation region of two symmetrically placed *vortices* on each side of the wake, whose length grows as the Reynolds number increases. This trend has been shown experimentally by *Taneda* [16.264], *Gerrard* [16.265], and *Coutanceau* and *Bouard* [16.266], and is supported by the computations of *Dennis* and *Chang* [16.267]. However, from analysis and computation in (constrained to be) steady two-dimensional (2-D) flow, it has proven surprisingly difficult to define the variation

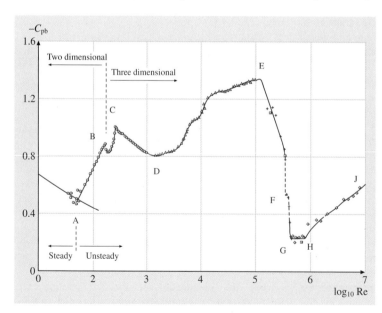

Fig. 16.135 Plot of base suction coefficients ($-C_{pb}$) over a large range of Reynolds numbers. A plot of base suction coefficient is particularly useful as a basis for discussion of the various flow regimes. The base suction coefficient (negative of base pressure coefficient, or $-C_{pb}$) is surprisingly sensitive to the process of vortex formation in the near wake, which itself is affected strongly by the evolution of various two- and three-dimensional wake instabilities, as Reynolds numbers are varied. The sources of the data, in order of increasing Re, data are from computations of *Henderson* [16.255]; and experiments of *Williamson* and *Roshko* [16.259], *Norberg* [16.258], *Bearman* [16.260], *Flaschbart* [16.261], *Shih* et al. [16.262]

of bubble shape with Reynolds number as Re becomes large [16.268], as discussed in *Roshko* [16.217]. It appears that the asymptotic formula for $C_D(Re)$ for this steady wake is not yet available. It should be noted that, as the length of the steady wake bubble increases, due to the viscous stresses, so the recent DNS (direct numerical simulation) computations of *Henderson* [16.255] and the measurements of *Thom* [16.269] clearly show a decrease in the base suction.

Regime A–B: Laminar Vortex Shedding Regime (Re = 49 to 190)

In this regime, the variation of base suction with Re shows a sharp deviation in trend from the steady wake regime discussed above (Fig. 16.135). The recirculation region develops instabilities, initially from the downstream end of the bubble, whose strength and amplification grows with Re. This effect may be measured by a monotonic increase in the amplitude of maximum wake velocity fluctuations with Re, and a gradual movement of the instability maximum (or formation length) upstream towards the cylinder.

The onset of the wake instability near to Re = 47–49 has been found to be a manifestation of a Hopf bifurcation, and the flow represents a dynamical system described by a Stuart–Landau equation [16.270]. As the wake instability becomes amplified, the Reynolds stresses in the *near*-wake region increase, the formation length decreases, and there is a consistent increase in the base suction. There is also an increase in the unsteady forces, as shown from computations [16.255], but not yet detected in experiment at these low Re.

Some early observations and measurements were made which showed or suggested that vortices can shed at some oblique angle to the axis of the cylinder in what we now term *oblique* shedding [16.220, 271], although towing-tank experiments by *Hama* [16.272] demonstrated only parallel shedding. A further phenomenon, which we shall see is directly related to this, is the phenomenon of discontinuities in the relationship between the Strouhal and Reynolds numbers (see Fig. 16.136a in the laminar (oblique) shedding regime), as first detected very clearly by *Tritton* [16.273] near to Re = 75 [16.274], and subsequently the source of a great deal of debate over a period of 30 years.

Experimental measurements of the Strouhal frequency over the period 1878–1978 showed a scatter of the order of 20% even among the modern experiments [16.223]. This scatter was present despite the fact, as pointed out by *Roshko* [16.275], that "the quantities

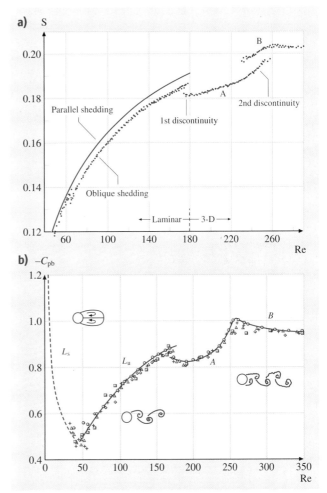

Fig. 16.136a,b Strouhal number (S) and base suction coefficient ($-C_{pb}$) over the laminar and wake transition regimes of Reynolds number. We can see quite sharp changes of mode, caused by: the inception of vortex shedding; the first jump to a mode A three-dimensional instability; and a second jump to mode B instability (after *Williamson* [16.227], *Williamson* and *Roshko* [16.259])

involved (U, D, ν and f) could be rather easily measured to better than 1% accuracy". ν is the kinematic viscosity. Many explanations were put forward over the years, although it was finally shown that, in the absence of certain effects, such as free-stream shear and cylinder vibration, a discontinuity in the S–Re relation can be caused by the unexplained phenomenon described earlier, namely *oblique shedding*. The S–Re discontinuity, originally observed by *Tritton* [16.273], is caused by a changeover from one mode of oblique shedding to another oblique mode, as Re is increased [16.256, 276–280]. It is shown that the particular boundary conditions at the spanwise ends of the cylinder dictate the angle of shedding over the whole span, even for a cylinder that is hundreds of diameters in length. *Gerich* and *Eckelmann* [16.281] had earlier shown that a region close to the ends of a cylinder (about 10 diameters in length) can be influenced in a *direct* manner, causing a cell of lower-frequency shedding to appear near to the ends.

In the towing tank and wind tunnel experiments of *Williamson* [16.256, 276], it was found that the oblique vortices formed a periodic *chevron* pattern, as shown in Fig. 16.137a. The technique used to observe the vortices clearly involves washing off fluorescent dye from the horizontal cylinders in a towing tank, and employing laser-induced fluorescence (LIF). Over each half span, the oblique angle is dictated by the end conditions in that half. It was then found that there are several means to achieve parallel shedding by manipulating the end conditions, as in Fig. 16.137b. With parallel shedding, the Strouhal–Reynolds number curve is completely continuous, as shown in Fig. 16.136. (An ingenious approach by *Leweke* and *Provansal* [16.282] has involved the study of vortex shedding past a torus, where there are no ends to the curved cylinder.) It was also shown [16.276] that one may define a *universal* Strouhal curve, in the sense that the experimental oblique-shedding data (S_θ) can be closely collapsed onto the parallel-shedding curve (S_0) by the transformation

$$S_0 = \frac{S_\theta}{\cos\theta}. \qquad (16.18)$$

Some control of vortex shedding angles has been made at higher Re by *Prasad* and *Williamson* [16.283], although the techniques are not as effective as in the laminar shedding regime.

Regime B–C: 3-D Wake-Transition Regime (Re ≈ 190 to 260)

The significant visualizations of *Hama* [16.272] showed that the instability in the wake-transition regime takes the form of a three-dimensional waviness on the primary Karman vortices, and the formation of what *Gerrard* [16.265] later calls "fingers of dye". It is now known that these dye *fingers* are associated with vortex loops and streamwise vorticity, in similarity with other free shear flows.

The transition to three-dimensionality in the wake can conveniently be described with reference to the measurements of the Strouhal–Reynolds number in Fig. 16.136a, and to the changes in base pressure in Fig. 16.136b, where it may be observed that the transition, originally described by *Roshko* in 1954 [16.254], actually involves two discontinuous changes (see *Williamson* [16.284]). At the first discontinuity the Strouhal frequency drops from the laminar curve to one corresponding to a *mode A* 3-D shedding, at close to Re = 180–190. This discontinuity is hysteretic, and the exact critical Re depends on whether the flow speed is increased or decreased. We see the inception of vortex loops (in a *mode A* instability), and the formation of streamwise vortex pairs due to the deformation of primary vortices as they are shed, at a wavelength of around 3–4 diameters, as shown in Fig. 16.138a. (The modes of 3-D instability in this figure are visualized using the LIF technique, in the same manner as found in Fig. 16.137.)

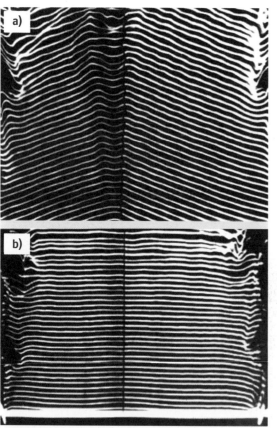

Fig. 16.137a,b Oblique and parallel shedding are controlled by the end boundary conditions for cylinders of even several hundreds of diameters in length. In (**a**), we have the *chevron* pattern of shedding, showing the existence of a phase *shock* in the center span, while in (**b**), we can induce parallel shedding [16.276]

Although there is a large range of reported critical Re for wake transition, Re = 140–194, it should be noted that it is influenced by the end boundary conditions. Floquet stability analysis of *Barkley* and *Henderson* [16.285] yields a critical Reynolds number of 188.5. *Miller* and *Williamson* [16.286] found that non-mechanical end conditions (using suction tubes downstream of the body) can yield rather *clean* end conditions, and they find that the laminar regime for parallel shedding could be extended experimentally up to $Re_{crit} = 194$.

At the second discontinuous change in the S–Re relation in Fig. 16.136a, there is a gradual transfer of energy from *mode A* shedding to a *mode B* shedding over a range of Re from 230–250. The latter mode comprises finer-scale streamwise vortices (Fig. 16.138b), with a spanwise length scale of around one diameter. There is evidence to show that the origin of the mode A instability is based on a core elliptic instability of the primary vortex cores, whereas the mode B structures evolve from an instability of the *braid* region of vorticity between the primary vortices. The origin of these instabilities are discussed from experiment in *Williamson* [16.227], and in *Leweke* and *Williamson* [16.288], and from numerical simulation in *Thompson* et al. [16.289], *Barkley* and *Henderson* [16.285], and *Thompson* et al. [16.290] and *Zhang* et al. [16.291].

The large intermittent low-frequency wake velocity fluctuations, originally monitored by *Roshko* [16.254] and then by *Bloor* [16.263], have been shown to be due to the presence of large-scale spot-like *vortex dislocations* in this transition regime [16.287]. These are caused by local shedding-phase dislocations along the span. The base suction and Strouhal frequency continue to increase in this regime, but follow curves at a lower level than may be extrapolated from the laminar shedding regime.

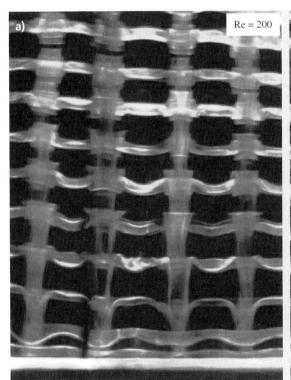 

Fig. 16.138a,b Modes A and B three-dimensional instabilities. **(a)** Mode A instability. This is associated with the inception of streamwise vortex loops. This example for Re = 200, corresponds with a spanwise wavelength: $\lambda/D = 4.01$, which is remarkably close to the maximum growth rate from Floquet analysis (after *Barkley* and *Henderson* [16.285]) **(b)** Mode B instability. This is associated with the formation of finer-scale streamwise vortex pairs. λ/D is roughly 1.0. Re = 270. Note that both photographs are to the same scale (after *Williamson* [16.227, 287])

Regime C–D: Increasing Disorder in the Fine-Scale Three-Dimensionalities

The peak in the base suction close to C, at Re = 260, is associated with a peak in the Reynolds stresses in the near wake, and a particularly ordered three-dimensional streamwise vortex structure in the near wake [16.227]. At this point, the primary wake instability behaves remarkably like the laminar shedding mode, with the exception of the presence of the fine-scale streamwise vortex structure. As Re is then increased towards point D in the plot of base pressure in Fig. 16.137, the fine-scale three-dimensionality becomes increasingly disordered, and this appears to cause a reduction in the two-dimensional Reynolds stresses, a consistent reduction in base suction, and an increasing length of the formation region [16.227, 293]. The increased length of the formation region (the downstream distance over which the vortices form), as one increases Re from 270 to 1360, can be seen clearly in Fig. 16.139, using the hydrogen bubble visualization technique of *Unal* and *Rockwell* [16.293].

Regime D–E: Shear-Layer Transition Regime (Re = 1000 to 200 000)

In this regime, the base suction increases again, the 2-D Reynolds stress level increases, the Strouhal number gradually decreases (Fig. 16.134), and the formation length of the mean recirculation region decreases [16.294], all of which are again consistent variations. These trends are caused by the develop-

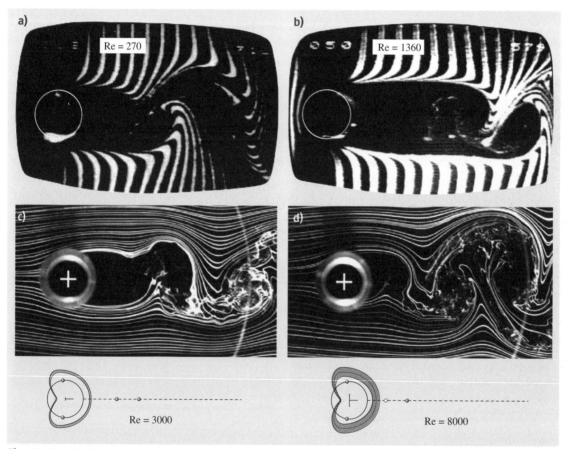

Fig. 16.139a–d Visualizations of the wake structure as the Reynolds number is varied. Hydrogen bubble images show the lengthening of the vortex formation region as Re increases from 270 to 1360, as shown by *Lin* et al. [16.292]. Smoke wire images from *Norberg* [16.249] show the diminishing vortex formation region as Re increases further from 3000 to 8000, as the stresses from the shear layer instability cause the region to shrink. Diagrams show mean pressure distributions by *Norberg* [16.249]

ing instability of the separating shear layers from the sides of the body. As noted by *Roshko* [16.217], this might be called the Schiller–Linke regime, after those who discovered it and who associated this regime with an increase in base suction and drag, while the turbulent transition point in the separating shear layers moves upstream, as Re increases. The decrease in formation length in the present regime (for Re increasing from 3000 to 8000), is well demonstrated by the smoke wire visualizations of *Norberg* [16.249], in Fig. 16.139c,d.

Physically, the shear layer vortices take on the appearance of the vortices so often observed in mixing layers, as shown from smoke wire visualization in Fig. 16.140, but in this case the length of the layer is obviously limited by the streamwise extent of the formation region. Evidence of shear layer vortices are seen in the earlier Fig. 16.127 at Re = 4000, and are found to amalgamate in the near wake into the Karman vortices, as seen in the PIV visualization of *Lin* et al. [16.292] in Fig. 16.140 for Re = 10 000. *Unal* and *Rockwell* [16.295] have also shown that the vortex formation can arise from the exponential amplification of the disturbance kinetic energy in accord with linear stability theory applied to the separating shear layer. Shear layer vortices generally appear for Re > 1200, but have been observed below this Re, possibly induced by vibration of the test body in some cases.

Although the shear layer transition was considered by *Roshko* [16.254], it was not until *Bloor* [16.263] that the frequency of the shear layer instability waves were studied. She demonstrated that the shear layer instability frequency scaled approximately with $\sqrt{\text{Re}}$, by considering the thickness and velocity of the separating laminar boundary layer. There is no question that this is of the right order, although even a cursory glance at the actual data points from these investigations does not precisely support the $\text{Re}^{0.5}$ variation (look carefully at the frequency data in Fig. 16.141). For clarity, the frequencies are defined by f_{sl}, which is the shear layer frequency, and f_k, which is the Karman vortex frequency. The frequency of shear layer vortices has sometimes been referred to as the Bloor–Gerrard frequency, in honor of the work of Bloor and Gerrard, who first investigated the separating shear layer instabilities, in the particular context of the bluff-body wake.

In the case of the shear layer frequency, a careful reevaluation of all data that could be found from a number of investigators, shows that the exponent p in the expression $(f_{sl}/f_k) = A\,\text{Re}^p$ is significantly greater than 0.5 in every case. All the compiled data are plotted in

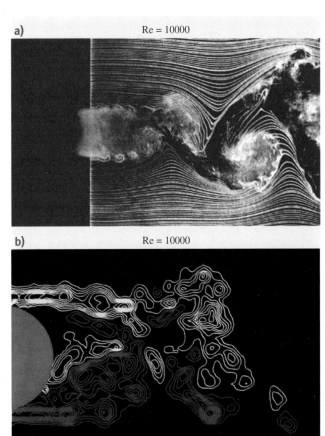

Fig. 16.140a,b Clear images of the shear layer instability in the near wake at Re = 10 000. In (**a**) we see the smoke wire technique, and a clear observation of the shear layer vortices, contrasting with the larger wake vortices [16.283]. In a similar flow in (**b**) the shear layer vorticity, amalgamating into Karman wake vortices, is shown using the PIV technique [16.292]

Fig. 16.141. The least-squares best fit to the data is

$$\left(\frac{f_{sl}}{f_k}\right) = 0.0235\,\text{Re}^{0.67}$$

and the line for $\text{Re}^{0.5}$, although clearly of the right order, would not closely fit the measured data. One might note that the data extends over a large range from Re = 1200 up to Re = 100 000. On a dimensional basis, one expects that the shear layer frequency will scale with a characteristic velocity and length scale in the form:

$$\left(\frac{f_{sl}}{f_k}\right) \sim \frac{U_{\text{sep}}}{\theta_{sl}},$$

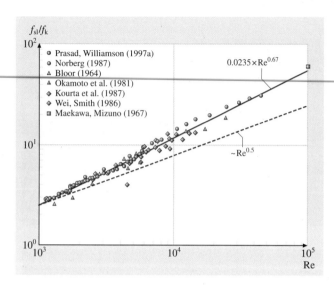

Fig. 16.141 Variation of normalized shear layer frequency with Reynolds number. The plot includes data from all the investigators up to 1997 who measured the shear layer frequency. By looking along the lines from the *left*, with one's nose right down on the page, the fact that the trend is higher than $Re^{0.5}$ is rather more obvious, and yields the best fit as: $(f_{sl}/f_k) = 0.0235\,Re^{0.67}$ (References are found in *Prasad* and *Williamson* [16.283])

where U_{sep} is the velocity outside the boundary layer at the separation point, and θ_{sl} is the momentum thickness of the separated shear layer. Bloor suggested, very reasonably, that the momentum thickness scales with the boundary layer thickness just before separation, which itself scales as $Re^{-0.5}$. If one notes that Strouhal number $S = f_k D/U$, one finds

$$\left(\frac{f_{sl}}{f_k}\right) \sim Re^{0.5}\frac{(1-C_{pb})^{0.5}}{S} ,$$

where we have used the relationship $U_{sep}/U = (1-C_{pb})^{0.5}$. If one assumes, over a large range of Re, that C_{pb} is approximately constant then one finds the expression used by Bloor

$$\left(\frac{f_{sl}}{f_k}\right) \sim Re^{0.5} .$$

Clearly this is of the same order as the measurements. However, the fact that the experimental power law for a comprehensive plot of all the previous studies follows $Re^{0.67}$ was shown by *Prasad* and *Williamson* [16.297] to be due to the fact that the characteristic lengths and velocities are somewhat different to the values assumed above. In particular it was found that there is indeed a variation of base pressure C_{pb} over a range of Re (see the range between D–E in Fig. 16.135), and also that the movement of the transition point upstream as Re increases affects the shear layer frequency. Both of these influences yield an increase in the exponent, above 0.5. One finds from the analysis that approximately $(f_{sl}/f_k) \sim Re^{0.70}$, which would explain the experimentally observed shear layer frequency.

It is important to introduce briefly the types of turbulence intensities that have been measured in the near

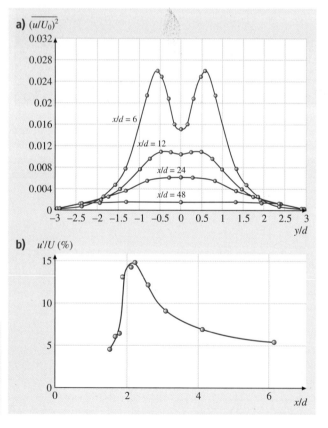

Fig. 16.142a,b Typical velocity fluctuation measurements made in the near wake using hot wire anemometry. In (**a**), *Roshko* [16.254] measured the streamwise turbulence intensity profiles at specific downstream stations, showing the characteristic twin peaks associated with the two rows of vortices moving downstream. In (**b**), *Bloor* and *Gerrard* [16.296] show the typical streamwise intensity profile along the wake centreline, indicating one of the measures of formation length as given by the location of the peak intensity ($x/d \approx 2$ in this case)

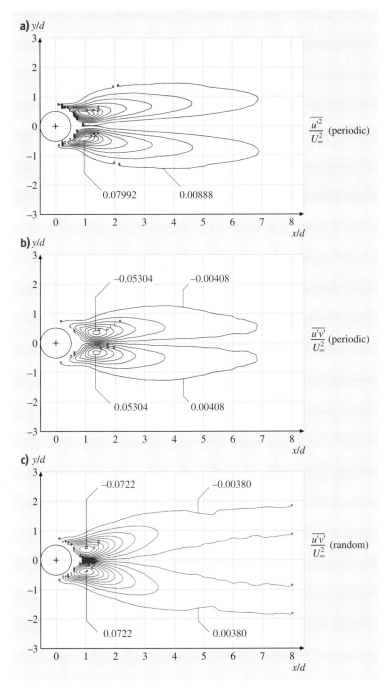

Fig. 16.143a–c Contours of Reynolds stress extracted by an ingenious method involving a flying hot wire. The previous classical profiles in Fig. 16.142 may be related to the streamwise Reynolds stress contours of (**a**) here. Measurements of Reynolds shearing stresses by *Cantwell* and *Coles* [16.234] in (**b**), and (**c**) show the periodic and random stress components (these combine to give the total shearing stress), so one has some idea of how much stress is due to the periodic vortex shedding. Re = 140 000

wake of the cylinder. *Roshko* [16.254] and *Kovasznay* [16.298] were the first to measure the streamwise turbulence intensity using the hot wire anemometer. In Fig. 16.142, Roshko plotted various transverse profiles across the wake, exhibiting the characteristic twin peaks. The intensity reaches a peak and decays downstream, as shown from the centreline measurements of *Bloor* and *Gerrard* [16.296] in Fig. 16.142b. More-detailed measurements of turbulence quantities have been made extensively since that time, and we shall focus here briefly on some of the data of *Cantwell* and *Coles* [16.234] in Fig. 16.143. Their contours of streamwise normal Reynolds stress and Reynolds shear stress show the typical peaks on either side of the near wake. These measurements, made possible with an ingenious flying hot wire, have been subdivided usefully into a periodic and random component.

It is particularly interesting to note from Fig. 16.144 that the periodic component of the Reynolds shear stress remains remarkably similar between Re = 3900 (from PIV of *Govardhan* and *Williamson* [16.300]), and the much higher Re = 140 000 of *Cantwell* and *Coles*. On the other hand, the *random* component of the shear stresses are more than double at the high Reynolds numbers. Govardhan and Williamson observed that a significant portion of the remnant of the stresses, over and above the periodic component, is caused by fluctuations due to the shear layer instability, rather than only representing random fluctuations. Finally, we include here a plot of Reynolds stresses coming from LDV measurements of *Djeridi* et al. [16.299], completing a figure showing all three techniques, namely

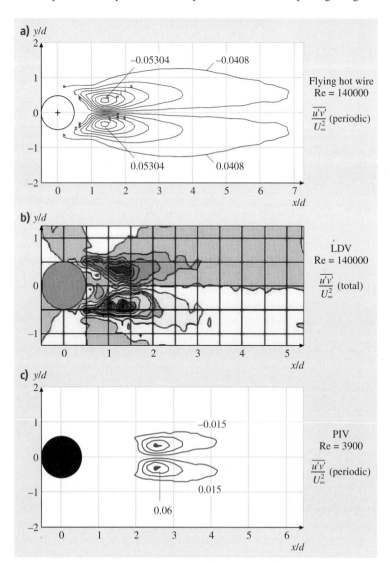

Fig. 16.144a–c Comparison of stress measurements using flying hot wire, LDV, and PIV methods. The data in (**a**) and (**b**) are at the same Re = 140 000, although they measure the periodic and total stresses [16.234, 299]. It is interesting that the PIV data for much lower Re = 3900 in (**c**) show quite comparable peak values of the *periodic* shearing stresses at Re = 140 000, although *Govardhan* and *Williamson* [16.300] find that the *random* stresses are far smaller at the lower Re. It appears that much of the energy of the *random* stresses at high Re are those caused by the shear layer instability, and so might not be considered all random

hot wire, LDV and PIV, producing a similar set of data.

The level of roughness and the level of free-stream turbulence, amongst other influences, can modify the Reynolds number for the inception of the drag crisis, as indicated by Fig. 16.145. A more turbulent free stream, or a rougher cylinder, will induce an earlier transition to boundary-layer turbulence on the body, and to a dramatic reduction in drag.

Regime E–F–G: Asymmetric Reattachment Regime (or Critical Transition)

In this regime, the base suction and the drag decrease drastically, associated with a separation–reattachment bubble, causing the revitalized boundary layer to separate much further downstream on the body sides (at the 140° line) and with a much reduced width of downstream wake than for the laminar case. There is a most interesting phenomenon which occurs at point F in Fig. 16.135, and this corresponds with a separation–reattachment bubble on *only one side* of the body, as discovered by *Bearman* [16.260], and shown by *Schewe* [16.303] to be bistable, causing rather large mean lift forces ($C_L \approx 1$). The effect on the base pressure and Strouhal number of this single *bubble* on one side, or the presence of both bubbles, is shown in Figs. 16.134b and 16.135, where the single-bubble configuration is associated with point F in these plots.

Regime G–H: Symmetric–Reattachment Regime (or Supercritical Regime)

In this regime, the flow is symmetric with two separation–reattachment bubbles, one on each side of the body. Some fluctuations are detected in the wake at large Strouhal numbers of around 0.4 [16.260], which is consistent with the relatively thin wake in this regime (one expects that the frequency will scale roughly inversely with the wake width). According to *Roshko* [16.217], the considerably higher Reynolds stresses of the boundary layer following the separation bubble allow the boundary layer to survive a greater adverse pressure gradient than in the post-critical regime (see below), where transition finally occurs before separation.

Regime H–J: Boundary-Layer Transition Regime (or Post-Critical Regime)

The increase in Reynolds numbers through the various regimes to this point is associated with a sequence of fundamental shear flow instabilities, following the order

- wake transition
- shear layer transition
- boundary layer transition.

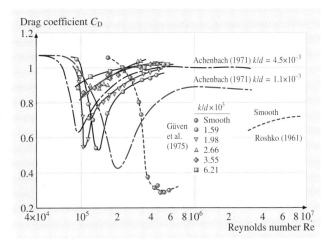

Fig. 16.145 The effect of surface roughness in stimulating an earlier *drag crisis*. Increasing roughness induces a lower critical Re at which one finds the drop in drag associated with laminar-turbulent transition in the attached boundary layer. However, it is apparent that high roughness then leads to a higher drag, after the crisis, than for smooth cylinders (see also *Shih* et al. [16.262]). (References are found in [16.301, 302])

The effect of an increase in Re up to this particular regime (H–J) is to move the turbulent transition point further upstream, until at high enough Re, the boundary layer on the surface of the cylinder itself becomes turbulent. It was generally assumed that, after this point, the downstream wake would be fully turbulent, and it was not expected that coherent vortices would be observed. However, in 1961, *Roshko* [16.233] was able to demonstrate the surprising result that periodic vortex shedding is strongly in evidence even in this flow regime, in what one might call a *Roshko street*, in honor of this important discovery. Separation occurs further upstream, yielding higher drag and base suction, and a wider downstream wake than in the previous regime. The drag in Fig. 16.133b and the base pressure of Fig. 16.135 both start to increase again when the vortex shedding resumes after the drag crisis in this regime discovered by Roshko. More-recent data up to very high Re $\approx 10^7$ may be found in *Shih* et al. [16.262].

16.3.5 Concluding Remarks

Despite the fact that the wake of a bluff body does not easily admit analytical approaches, it is exceedingly rich in flow phenomena. Over the last few years, there has been a surge of experimental discoveries concerning several aspects of bluff-body wakes, but particularly three-dimensional aspects. These activities have been matched by new understandings of wake flows coming from analysis and experiments and computations. Cellular shedding, vortex dislocations, oblique shedding, phase shocks and expansions, vortex loops, are all three-dimensional vortex dynamics phenomena that are becoming understood and that influence the variation of important parameters such as base pressure, drag, lift and vortex frequency.

Facilities to study bluff-body aerodynamics utilize both air and water as the fluid medium, and employ a large set of techniques to measure the velocity, vorticity, pressure, and the forces on bluff bodies, and to visualize such flows with a battery of methods.

References

16.1 W.-H. Hucho (Ed.): *Aerodynamik des Automobils*, 5th edn. (Vieweg, Wiesbaden 2005), in German

16.2 J.B. Barlow, W.H. Rae, A. Pope: *Low Speed Wind Tunnel Testing* (Wiley, New York 1999)

16.3 A. Morelli: *General layout characteristics and performance of a new wind tunnel for aerodynamic and functional tests on full-scale vehicles*, SAE-paper 710 214 (SAE, Warrendale 1971)

16.4 E. Mercker, J. Wiedemann: *On the correction of interference effects in open jet wind tunnels*, SAE SP-1145 (SAE, Warrendale 1996) pp. 1–15

16.5 SAE: *Aerodynamic testing of road vehicles, testing methods and procedures*, SAE J2084 (SAE, Warrendale 1993)

16.6 B.C. Nijhof, G. Wickern: *Reference static and dynamic pressures in automotive wind tunnels*, SAE-SP-1786 (SAE, Warrendale 2003) pp. 15–31

16.7 F. Evert, H. Miehling: *Active suppression of buffeting at the Audi AAWT: Operational experiences and enhancements of the control scheme*, SAE-SP-1874 (SAE, Warrendale 2004) pp. 227–238

16.8 G. Wickern, W. von Heesen, S. Wallmann: *Wind tunnel pulsations and their active suppression*, SAE SP-1524 (SAE, Warrendale 2000) pp. 223–236

16.9 K.-R. Deutenbach: *Influence of plenum dimensions on drag measurement in 3/3-open-jet automotive wind tunnels*, SAE SP-1078 (SAE, Warrendale 1985) pp. 198–198

16.10 W.-H. Hucho, L.J. Janssen, G. Schwarz: *The wind tunnel's ground plane boundary layer – its interference with the flow underneath cars*, SAE-Paper 750 066 (SAE, Warrendale 1975)

16.11 E. Mercker, H.W. Knape: *Ground simulation with moving belt and tangentail blowing for full-scale automotive testing in a wind tunnel*, SAE-Paper 890 367 (SAE, Warrendale 1989)

16.12 A. Cogotti: *The T-Belt – The New Moving Ground System of the Pininfarina Wind Tunnel*, SAE-Paper 2007-01-1044 (SAE, Warrendale 2007)

16.13 J. Wiedemann: Road simulation requirements and techniques. In: *Progress in Vehicle Aerodynamics – Advanced Experimental Techniques*, ed. by J. Wiedemann, W.-H. Hucho (Expert, Renningen 2000) pp. 48–63

16.14 A. Cogotti: Generation of a controlled turbulent flow in an automotive wind tunnel and its effect on car aerodynamics and aeroacoustics. In: *Progrogress in Vehicle Aerodynamics III – Unsteady Flow Effects*, ed. by J. Wiedemann, W.-H.. Hucho (Expert, Renningen 2004) pp. 150–176

16.15 G. Carlino, A. Cogotti: *Simulation of transient phenomena with the turbulence generation system in the pininfarina wind tunnel*, SAE-Paper 2006-01-1031 (SAE, Warrendale 2006)

16.16 W.-H. Hucho: *Aerodynamik der stumpfen Körper* (Vieweg, Wiesbaden 2002)

16.17 J. Katz: *Integration of computational methods into automotive wind tunnel testing*, SAE-Paper 890 601 (SAE, Warrendale 1989)

16.18 E.Y. Ng, P.W. Johnson, S. Watkins, L. Grant: *Wind-tunnel tests of vehicle cooling performance at high blockage*, SAE SP-1524 (SAE, Warrendale 2000) pp. 25–34

16.19 J.W. Yen, W.R. Martindale, E.G. Duell, S.A. Arnette: *Determining blockage correction in climatic wind tunnels using CFD*, SAE SP-1786 (SAE, Warrendale 2003) pp. 143–153

16.20 R. Künstner, J. Potthoff, U. Essers: *The aeroacoustic wind tunnel of the Stuttgart university*, SAE SP-1078 (SAE, Warrendale 1995) pp. 31–47

16.21 J. Potthoff: Future road simulation techniques. In: *Progress in Vehicle Aerodynamics – Advanced Experimental Techniques*, ed. by J. Wiedemann, W.-H. Hucho (Expert, Renningen 2000) pp. 64–82

16.22 J. Wiedemann, G. Wickern, B. Ewald, C. Mattern: *Audi aero-acoustic wind tunnel*, SAE-Paper 930300 (SAE, Warrendale 1993)

16.23 J. Wiedemann: The design of wind-tunnel fans for aeroacoustic testing. In: *Topics in Wind Noise – Automobile Wind Noise and ist Measurement, Part II*, SAE SP-1457, ed. by J.R. Callister, A.R. George (SAE, Warrendale 1999) pp. 1–12

16.24 G. Antonucci, G. Ceronetti, A. Costelli: *Aerodynamic and climatic wind tunnels in FIAT research center*, SAE-Paper 770 392 (SAE, Warrendale 1977)

16.25 R. Buchheim, D. Schwabe, H. Röhe: Der neue 6 m^2-Klimawindkanal von Volkswagen, Teil 1, ATZ **88**, 211–218 (1986), in German

16.26 R. Buchheim, D. Schwabe, H. Röhe: Der neue 6 m²-Klimawindkanal von Volkswagen, Teil 2, ATZ **88**, 389–392 (1986), in German

16.27 A. Cogotti: *Upgrade of the pininfarina wind tunnel – the new "13-fan" drive system*, Draft 06B-406 (SAE, Warrendale 2006)

16.28 G.W. Carr: The MIRA quarter-scale wind tunnel, MIRA Rep. **11** (1961)

16.29 G.W. Carr: An improved moving-ground system in the MIRA model wind tunnel, MIRA Rep. **2** (1987)

16.30 M. Bernard: : La souflerie á veine longue de l'Institute Aérotechnique de Saint-Cyr-L'Ecole, Rev. Gén. Chemins de Fer **January**, 704–711 (1973)

16.31 R.G. Gawthorpe: Wind effects on ground transportation, J. Wind Eng. Indust. Aerodyn. **52**, 73–92 (1994)

16.32 N. Kobayashi, M. Yamada: *Stability of a one box type vehicle in a cross-wind – an analysis of transient aerodynamic forces and moments*, SAE-Paper 881 878 (SAE, Warrendale 1988)

16.33 L. Larsson, L.U. Nilsson, A. Berndtsson, L. Hammar, K. Knutson, H. Danielson: *Study of ground simulation-correlation between wind-tunnel and water-basin tests of a full-scale car*, SAE-Paper 890 368 (SAE, Warrendale 1989)

16.34 H. Neppert, R. Sanderson: *Untersuchungen zur Zugbegegnung. Bauwerk- und Personenpassage im Wasserkanal. Seminar Schnellbahnaerodynamik* (DFVLR-AVA, Göttingen 1978), in German

16.35 G. Sovran, D. Blaser: *Quantifying the potential impacts of regenerative braking on a vehicle's tractive-fuel consumption for the U.S.*, European, and Japanese driving schedule, SAE-paper 2006-01-0664 (SAE, Warrendale 2006)

16.36 K.-L. Haken: Messung des Rollwiderstands unter realen Bedingungen. In: *Kraftfahrwesen und Verbrennungsmotoren*, ed. by M. Bargende, J. Wiedemann (Expert, Renningen 1999) pp. 488–504, in German

16.37 J.C. Kessler, S.B. Wallis: *Aerodynamic test techniques*, SAE-Paper 660 464 (SAE, Warrendale 1966)

16.38 E. Széchényi: Crosswind and its Simulation. In: *Progress in Vehicle Aerodynamics – Advanced Experimental Techniques*, ed. by J. Wiedemann, W.-H. Hucho (Expert, Renningen 2000) pp. 83–96

16.39 A. Wagner: *Ein Verfahren zur Vorhersage und Bewertung der Fahrerreaktion bei Seitenwind*, Dissertation (Universität Stuttgart, Stuttgart 2003), in German

16.40 L. Polanski, W. Matich, J.T. Kutney Sr.: *A new multicomponent wind-tunnel balance for automotive wind tunnels*, SAE SP-1078 (SAE, Warrendale 1995) pp. 137–145

16.41 A. Cogotti: Flow field measurements and their interpretation. In: *Progress in Vehicle Aerodynamics – Advanced Experimental Techniques*, ed. by J. Wiedemann, W.-H. Hucho (Expert, Renningen 2000) pp. 97–120

16.42 J. Schmitt, K. Wilharm: Measurement of flow fields with LDA. In: *Progress in Vehicle Aerodynamics – Advanced Experimental Techniques Wiedemann J.*, ed. by W.-H. Hucho (Expert, Renningen 2000) pp. 121–130

16.43 J. Kompenhans, M. Raffel, L. Dieterle, H. Richard, T. Dewhirst, H. Vollmers, K. Ehrenfeld, C. Willert, K. Pengel, C. Kähler, O. Ronneberger: Measurement of flow fields with particle immage velocimetry (PIV). In: *Progress in Vehicle Aerodynamics – Advanced Experimental Techniques*, ed. by J. Wiedemann, W.-H. Hucho (Expert, Renningen 2000) pp. 131–157

16.44 S.B. Wallis: *Ventilation systems aerodynamics – a new design method*, SAE-Paper 710 036 (SAE, Warrendale 1971)

16.45 M. Helfer: Localization of sond sources. In: *Progress in Vehicle Aerodynamics – Advanced Experimantal Techniques*, ed. by J. Wiedeman, W.-H. Hucho (Expert, Renningen 2000) pp. 170–182

16.46 U. Widmann, C. Zörner, N. Lindener: Aeroacoustic measurements and their correct physiological assessment. In: *Progress in Vehicle Aerodynamics – Advanced Experimantal Techniques*, ed. by J. Wiedeman, W.-H. Hucho (Expert, Renningen 2000) pp. 183–200

16.47 A.R. George (Ed.): *Automobile wind noise and its measurement, Part I*, SAE SP-1184 (SAE, Warrendale 1996)

16.48 A.R. George (Ed.): *Automobile wind noise and its measurement, Part II*, SAE SP 1457 (SAE, Warrendale 1999)

16.49 W. Dobrzynski: Windgeräuschquellen am Kraftfahrzeug. In: *Akustik und Aerodynamik des Kraftfahrzeuges*, ed. by S.R. Ahmed (Expert, Renningen 1995) pp. 48–73, in German

16.50 M. Helfer: Beurteilung von Hohlspiegelmikrophonen zur Schallortung. In: *Akustik und Aerodynamik des Kraftfahrzeuges*, ed. by S.R. Ahmed (Expert, Renningen 1995) pp. 142–151, in German

16.51 B. Barsikow, W.F. King., E. Pfizenmaier: Wheel/rail noise generated by a high-speed train with a line array of microphones, Sound Vibration **118**, 337–342 (1987)

16.52 W. Weiher, B. Schwab: Modernisierung einer Windkanal-Traversieranlage – Schnelle Vermessung einer Fahrzeugstirnfläche. In: *Virtuelle Instrumente in der Praxis*, ed. by R. Jamal, H. Jaschinski (Hüthig, Heidelberg 2005) pp. 94–97, in German

16.53 K. Wilharm: Schnelle optische Formerfassung bei der aerodynamischen Formoptimierung, VDI Berichte **1470**, 283–295 (1999), in German

16.54 P. Dannhäuser: Einsatz optischer 3D-Messtechnik in der Aerodynamikentwicklung. In: *Kraftfahrwesen und Verbrennungsmotoren, Internationales Stuttgarter Symposium*, ed. by M. Bargende, H.-C. Russ, J. Wiedemann (Expert, Renningen 2005) pp. 490–504, in German

16.55 J. Wiedemann, W.-H. Hucho (Eds.): *Progress in Vehicle Aerodynamics – Numerical Methods* (Expert, Renningen 2006)

16.56 J. Lukasiewicz: *Experimental Methods of Hypersonics* (Marcel Dekker, New York 1973)

16.57 F.K. Lu, D.E. Marren (Eds.): *Advanced Hypersonic Test Facilities*, Progress in Astronautics and Aeronautics, Vol. 198 (AIAA, Reston 2002)

16.58 R.J. Stalker: Hypervelocity aerodynamics with chemical nonequilibrium, Ann. Rev. Fluid Mech. **21**, 37–50 (1989)

16.59 H.G. Hornung: Experimental hypervelocity flow simulation, needs, achievements and limitations. In: *First Pacific International Conference on Aerospace Science and Technology* (Cheng-Kung Univ., Taiwan 1993)

16.60 B. Riemann: *Über die Fortpflanzung ebener Luftwellen von endlicher Schwingungsweite*, Vol. 8 (Königliche Gesellschaft der Wissenschaften, Göttingen 1860), in German

16.61 P. Vieille: Sur les discontinuités produites par la détente brusque des gas comprimés, C. R. Acad. Sci. **129**, 1228–1230 (1899)

16.62 J. Lukasiewicz: *Shock Tube Theory and Applications* (National Research Council, Ottawa 1950), Report MT-10

16.63 I.I. Glass, J.G. Hall: Shock Tubes. In: *Handbook of Supersonic Aerodynamics NAVORD Rep. 1488*, Vol. 6 (Springer, Berlin, Heidelberg 1959), Sect. 18

16.64 H. Oertel: *Stoßrohre* (Springer, Berlin, Heidelberg 1966), in German

16.65 N. Anfimov: TSNIIMASH Capabilities for aerogasdynamical and thermal testing of hypersonic vehicles. In: *AIAA 17th Aerospace Ground Testing Conference*, AIAA paper 92-3962 (AIAA, Reston 1992)

16.66 K. Takayama: Summary of forty years continuous shock wave research at interdisciplinary shock wave research center, Tohoku University. In: *Proc. 24th International Symposium on Shock Waves*, ed. by Z. Jiang (Springer, Berlin, Heidelberg 2004)

16.67 G.A. Sod: A survey of several finite difference methods for systems of nonlinear hyperbolic conservation laws, J. Comput. Phys. **27**, 1–31 (1978)

16.68 E.F. Toro: *Riemann Solvers and Numerical Methods for Fluid Dynamics. A Practical Introduction*, 3rd edn. (Springer, Berlin, Heidelberg 2006)

16.69 S.K. Godunov: A difference scheme for numerical computation of discontinuous solutions of the equations of fluid dynamics, Matem. Sbornik **47**, 271–306 (1959)

16.70 J.D. Anderson Jr.: *Modern Compressible Flow with Historical Perspective*, 2nd edn. (McGraw-Hill, New York 1990)

16.71 A. Hertzberg: A shock tube method of generating hypersonic flows, J. Aerosp. Sci. **18**(12), 803–805 (1951)

16.72 I.I. Glass: Over forty years of continuous research at UTIAS on nonstationary flows and shock waves, Shock Waves **1**, 75–86 (1991)

16.73 R.J. Stalker: Modern development in hypersonic wind tunnels, Aeronaut. J. **January**, 21–39 (2006)

16.74 R.J. Stalker, N.R. Mudford: Unsteady shock propagation in a steady flow nozzle expansion, J. Fluid Mech. **241**, 525–548 (1992)

16.75 E.L. Resler, S.-C. Lin, A. Kantrowitz: The production of high temperature gases in shock tubes, J. Appl. Phys. **23**(12), 1390–1399 (1952)

16.76 C.E. Smith: The starting process in a hypersonic nozzle, J. Fluid Mech. **24**(4), 625–640 (1966)

16.77 M.S. Holden, R.A. Parker: LENS hypervelocity tunnels and application to vehicle testing at duplicated flight conditions. In: *Advanced Hypersonic Test Facilities*, Progress in Astronautics and Aeronautics, Vol. 198, ed. by F.K. Lu, D.E. Marren (AIAA, Reston 2002)

16.78 H. Olivier, M. Habermann, M. Bleilebens: Use of shock tunnels for hypersonic propulsion testing. In: *35th AIAA/ASME/ASEE Joint Propulsion Conference and Exhibit*, AIAA paper 99-2447 (AIAA, Reston 1999)

16.79 R.SM. Chue, C.-Y. Tsai, R.J. Bakos, J.I. Erdos, R.C. Rogers: NASA's HYPULSE facility at GASL – a dual mode, dual driver reflected-shock/expansion tunnel. In: *Advanced Hypersonic Test Facilities, Progress in Astronautics and Aeronautics*, Vol. 198, ed. by F.K. Lu, D.E. Marren (NASA, Wshington 2002)

16.80 D.W. Bogdanoff, H.A. Zambrana, J.A. Cavolowsky, M.E. Newfield, C.J. Cornelison, R.J. Miller: Reactivation and upgrade of the NASA Ames 16 inch shock tunnel; Status Report. In: *AIAA 30th Aerospace Sciences Meeting and Exhibit*, AIAA paper 92-0327 (AIAA, Reston 1992)

16.81 J.A. Cavolowsky, M.P. Loomis, M.E. Newfield, T.C. Tam: Flow characterization in the NASA Ames 16-inch shock tunnel. In: *28th AIAA/SAE/ASME/ASEE Joint Propulsion Conference*, AIAA paper 92-3810 (AIAA, Reston 1992)

16.82 R.J. Bakos, J.I. Erdos: Options for enhancement of the performance of shock-expansion tubes and tunnels. In: *AIAA 33rd Aerospace Sciences Meeting and Exhibit*, AIAA paper 95-0799 (AIAA, Reston 1995)

16.83 H. Olivier, J. Zonglin, H.R. Yu, F.K. Lu: Detonation-driven shock tubes and tunnels. In: *Advanced Hypersonic Test Facilities*, Prog. Astronaut. Aeronaut., Vol. 198, ed. by F.K. Lu, D.E. Marren (AIAA, Reston 2002)

16.84 F.K. Lu, D.R. Wilson, R.J. Bakos, J.I. Erdos: Recent advances in detonation techniques for high-enthalpy facilities, AIAA J. **38**(9), 1676–1684 (2000)

16.85 J.O. Reller, N.M. Reddy: Analysis of the flow in a 1-MJ electric-arc shock tunnel, NASA TN **D-6865**, 2160–2165 (1972)

16.86 S.P. Sharma, C. Park: Operating of a 60- and 10-cm electric arc-driven shock tube – Part I: The driver, J. Thermophys. Heat Transfer **4**(3), 259–265 (1990)

16.87 S.P. Sharma, C. Park: Operating of a 60- and 10-cm electric arc-driven shock tube – Part II: The driven section, J. Thermophys. Heat Transfer **4**(3), 266–272 (1990)

16.88 R.J. Stalker: A study of the free-piston shock tunnel, AIAA J. **5**(12), (4-1)–(4-10) (1967)

16.89 R.J. Stalker: Shock tunnel for real-gas hypersonics, AGARD CP **428** (1987)

16.90 H.G. Hornung: Performance data of the new free-piston shock tunnel at GALCIT. In: *AIAA 17th Aerospace Ground Testing Conference*, AIAA paper 92-3943 (AIAA, Reston 1992)

16.91 G. Eitelberg, T.J. McIntyre, W.H. Beck, J. Lacey: High enthalpy shock tunnel in Göttingen, AIAA paper **92-3955** (1992)

16.92 K. Itoh, S. Ueda, T. Komuro, K. Saito, M. Takahashi, H. Miyajima, K. Koga: Design and Construction of HIEST (High enthalpy shock tunnel). In: *Proc. Int. Conference on Fluid Engineering*, Vol. 1 (JSME, Tokyo 1997) pp. 353–358

16.93 P. Petrie-Repar: *Numerical Simulation of Diaphragm Rupture*, Dissertation (University of Queensland, Queensland 1997)

16.94 H.G. Hornung, J.J. Quirk: Two effects of diaphragm bulge in shock tubes. In: *Advances in Fluid Mechanics and Turbomachinery*, ed. by H.J. Rath, C. Egbers (Springer, Berlin, Heidelberg 1998)

16.95 H. Mirels: Test time in low pressure shock tubes, Phys. Fluids **6**(9), 1201–1214 (1963)

16.96 H.O. Amann: Experimental study of the starting process in a reflection nozzle, Phys. Fluids **12**, 150–153 (1969), Suppl. I

16.97 B. Esser, H. Grönig, H. Olivier: High enthalpy testing in hypersonic shock tunnels. In: *Advances in Hypersonic, Defining the Hypersonic Environment*, Vol. 1, ed. by J.J. Berlin, J. Périaux, J. Ballmann (Birkhäuser, Boston 1989) pp. 182–258

16.98 K. Hannemann, V. Hannemann, S. Brück, R. Radespiel, G.SR. Sarma: Computational modeling for high enthalpy flows, Syst. Anal. Model. Simul. **34**, 253–277 (1999)

16.99 K. Hannemann, M. Schnieder, B. Reimann, J. Martinez Schramm: The influence and the delay of driver gas contamination in HEG. In: *21th AIAA Aerodynamic Measurement Technology and Ground Testing Conference*, AIAA paper 2000-2593 (AIAA, Reston 2000)

16.100 E.L. Resler, D.E. Bloxsom: *Very High Mach Number Flows by Unsteady Flow Principles* (Cornell Univ. Graduate School of Aeronautical Engineering Monograph, Ithaca 1952)

16.101 R.L. Trimpi: A preliminary theoretical study of the expansion tube, a new device for producing high-enthalpy short-duration hypersonic gas flows, NASA Tech. Rep. TR **R-133** (1962)

16.102 C.G. Miller: A critical examination of expansion tunnel performance, AIAA paper **78-768** (1978)

16.103 A. Paull, R.J. Stalker: Test flow disturbances in an expansion tube, J. Fluid Mech. **245**, 493–521 (1992)

16.104 I. Nompelis, G.V. Candler, T.P. Wadhams, M.S. Holden: Numerical simulation of high-enthalpy experiments in the LENS-X expansion tube facility. In: *42nd AIAA Aerospace Sciences Meeting and Exhibit*, AIAA paper 2004-1000 (AIAA, Reston 2004)

16.105 A. Paull, R.J. Stalker, I. Stringer: Experiments on an expansion tube with a free piston driver, 15th Aerodynamic Testing Conf. (American Institute of Aeronautics and Astronautics, Washington 1988) 173–178, Tech. Paper A88-37907 15-09

16.106 R.G. Morgan: A review of the use of expansion tubes for creating superorbital flows. In: *35th AIAA Aerospace Sciences Meeting and Exhibit*, AIAA paper 1997-279 (AIAA, Reston 1997)

16.107 R.G. Morgan: Development of X3, a superorbital expansion tube. In: *38th AIAA Aerospace Sciences Meeting and Exhibit*, AIAA paper 2000-0558 (AIAA, Reston 2000)

16.108 T.I. McIntyre, A.I. Bishop, T.N. Eichmann, H. Rubinsztein-Dunlop: Enhanced flow visualization with near-resonant holographic interferometry, Appl. Opt. **42**(22), 4445–4451 (2003)

16.109 T.I. McIntyre, M.J. Wegener, A.I. Bishop, H. Rubinsztein-Dunlop: Simultaneous two-wavelength holographic interferometry in a superorbital expansion tube facility, Appl. Opt. **36**, 8128–8134 (1997)

16.110 G. Wilson: Time-dependent quasi-one-dimensional simulations of high enthalpy pulse facilities. In: *AIAA 4th International Aerospace Planes Conference*, AIAA paper 92-5096 (AIAA, Reston 1992)

16.111 W.P. Eaton, J.H. Smith: Micromachined pressure sensors: Review and recent development, Smart Mater. Struct. **6**, 530–539 (1997)

16.112 S. Watson, W.N. MacPherson, J.S. Barton, J.D.C. Jones, A. Tyas, A.V. Pichugin, A. Hindle, W. Parkes, C. Dunare, T. Stevenson: Investigation of shock waves in explosive blasts using fibre optic pressure sensors, Meas. Sci. Technol. **17**, 1337–1342 (2006)

16.113 F. Lieneweg: *Handbuch Technische Temperaturmessung* (Vieweg, Braunschweig 1976), in German

16.114 Medtherm Corporation: *Coaxial Thermocouple Probes*, Bulletin 500 (Medtherm, Huntsville 2000)

16.115 H. Oertel: *Kurzzeitphysik*, Wärmeübergangsmessungen (Springer, Berlin, Heidelberg 1967), in German

16.116 C.G. Miller: Comparison of thin-film resistance heat-transfer gages with thin-skin calorimeter gages in conventional hypersonic wind tunnels, NASA TM **83197** (1981)

16.117 D.L. Schultz, T.V. Jones: Heat transfer measurements in short duration facilities, AGARD Rep. **165** (1973)

16.118 R.H. Eaves, C.T. Kidd: Miniature co-axial surface thermocouples for heat transfer rate measurements in hypersonic wind tunnels. In: *41st Supersonic Tunnel Association Meeting* (1974)

16.119 W.J. Cook, E.J. Felderman: Reduction of data from thin-film heat transfer gauges: A concise numerical technique, AIAA J. **4**(3), 561–562 (1966)

16.120 D.N. Kendall, W.P. Dixon, E.H. Schulte: Semiconductor surface thermocouples for determining heat transfer rates, IEEE Trans. Aerosp. Electron. Syst. **AES-3**(4), 596–603 (1967)

16.121 C. Jessen: *Messung von Druck, Temperatur und Kraft an Modellen im Stosswellenkanal. Dissertation* (RWTH Aachen, Aachen 1993)

16.122 G.T. Skinner: *A New Method of Calibrating Thin Film Gauge Backing Materials*, Rep. CAL-105 (Cornell University, Buffalo 1962)

16.123 B.FR. Ewald: Review article multi-component force balances for conventional and cryogenic wind tunnels, Meas. Sci. Technol. **11**, R81–R94 (2000)

16.124 M. Robinson: *Simultaneous Lift, Moment and Thrust Measurements on a Scramjet in Hypervelocity Flow*, Dissertation (University of Queensland, Queensland 2003)

16.125 K.W. Naumann, H. Ende, G. Mathieu, A. George: Millisecond aerodynamic force measurement with side jet model in the ISL shock tunnel, AIAA J. **31**(6), 1068–1074 (1993)

16.126 M. Takahashi, S. Ueda, T. Komuro, K. Sato, H. Tanno, K. Itoh: Development of a new force measurement method for scramjet testing in a high enthalpy shock tunnel. In: *9th AIAA International Space Planes and Hypersonic Systems and Technologies Conference*, AIAA paper 1999-4961 (AIAA, Reston 1999)

16.127 M.I. Kussoy, C.C. Horstmann: Cone drag in rarefied hypersonic flow, AIAA J. **8**(2), 315–320 (1970)

16.128 F.P. Beer Jr, E.R. Johnston, E.R. Eisenberg, W.E. Clausen, D.F. Mazurek, P.J. Cornwell: *Vector Mechanics for Engineers: Statics and Dynamics*, 8th edn. (McGraw-Hill, New York 2007)

16.129 L. Bernstein, G.T. Scott: A laser-interferometric trajectory-following system for determining forces on free flying models in a shock tunnel. In: *Proc. 13th Int. Symp. Shock Waves*, ed. by C.E. Treanor, J.G. Hall (State Univ. of New York Press, Albany 1981) pp. 150–158

16.130 C. Jessen, H. Grönig: A six-component balance for short duration hypersonic facilities. In: *New Trends in Instrumentation for Hypersonic Research*, ed. by A. Boutier (Kluwer Academic, Dordrecht 1993) pp. 295–305

16.131 V. Störkmann, H. Olivier, H. Grönig: Force measurements in hypersonic impulse facilities, AIAA J. **36**(3), 342–348 (1998)

16.132 K. Itoh, S. Ueda, H. Tanno, T. Komuro, K. Sato: Hypersonic aerothermodynamic and scramjet research using high enthalpy shock tunnel, Shock Waves **12**, 93–98 (2002)

16.133 R. Prost, R. Goutte: Discrete constrained iterative deconvolution algorithms with optimized rate convergence, Signal Process. **7**(3), 209–230 (1984)

16.134 D.J. Mee, W.J.T. Daniel, J.M. Simmons: Three-component force balance for flows of millisecond duration, AIAA J. **34**(3), 590–595 (1996)

16.135 S.R. Sanderson, J.M. Simmons: Drag balance for hypervelocity impulse facilities, AIAA J. **29**(12), 2185–2191 (1991)

16.136 M. Robinson, D.J. Mee, A. Paull: Lift, pitching moment and thrust measurement on a fuelled scramjet. In: *Proc. 23rd International Symposium on Shock Waves*, ed. by F. Lu (2001), on CD-ROM

16.137 M. Robinson, K. Hannemann: Short duration force measurements in impulse facilities. In: *25th AIAA Aerodynamic Measurement Technology and Ground Testing Conference*, AIAA paper 2006-3439 (AIAA, Reston 2006)

16.138 M.G. Novean, J.A. Schetz, R.D.W. Bowersox: Direct measurements of skin friction in complex supersonic flows, AIAA paper **97-0394** (1997)

16.139 C.P. Goyne, R.J. Stalker, A. Paull: Shock-tunnel skin-friction measurement in a supersonic combustor, J. Propuls. Power **15**(3), 699–705 (1999)

16.140 C.P. Goyne: *Skin Friction Measurements in High Enthalpy Flows at High Mach Number. Dissertation* (University of Queensland, Queensland 1998)

16.141 M.S. Holden: An experimental investigation of turbulent boundary layers at high Mach number and Reynolds numbers, Cornell Aero. Lab. Rep. **AB-5072-A-1**, NASA CR-112147 (1972)

16.142 C.P. Goyne, R.J. Stalker, A. Paull: Skin-friction measurements in high-enthalpy hypersonic boundary layers, J. Fluid Mech. **485**, 1–32 (2003)

16.143 M. Wollenhaupt, M. Rosenhauer, T. Müller, J. Jourdan, J. Scholz, S. Hartung, W.H. Beck: NO laser-induced fluorescence studies for the application of single-shot two-line thermometry to HEG. In: *Proc. 21st International Symposium on Shock Waves*, ed. by A.F.P. Houwing (Panther, Fyshwick 1997)

16.144 J. Scholz, S. Hartung, J. Jourdan, M. Wollenhaupt, W.H. Beck: Temporally resolved NO laser-induced fluorescence investigations for the application of

16.144 two-line thermometry to high enthalpy flows. In: *Proc. 22nd Int. Symp. Shock Waves*, ed. by G.J. Ball, R. Hillier, G.T. Roberts (Southampton Univ. Media, Southampton 1999)

16.145 M. Wollenhaupt: Einzelpuls Zwei-Linien-Thermometrie mit planarer laserinduzierter Fluoreszenz an NO-Molekülen in Hochenthalpieströmungen, DLR-FB **97-23** (1997), in German

16.146 S.O. O'Byrne, P.M. Danehy, A.F.P. Houwing: Non-intrusive temperature and velocity measurements in a hypersonic nozzle flow, 22nd AIAA Aerodynamic Measurement Technology and Ground Testing Conference (AIAA, Reston 2002)

16.147 P.C. Palma, P.M. Danehy, A.F.P. Houwing: Flourescence imaging of rotational and vibrational temperature in a shock tunnel nozzle flow, AIAA J. **41**(9), 1722–1732 (2003)

16.148 A. Ben-Yakar, M.G. Mungal, R.K. Hanson: Time evolution and mixing characteristics of hydrogen and ethylene transverse jets in supersonic cross-flows, Phys. Fluids **18**(2), 026101–026117 (2004)

16.149 A.L. Kovachevich, K.M. Hajek, T.J. McIntyre, A. Paull, M. Abdel-jawad: Imaging of hydrogen fuel injection on the intake of a heated wall scramjet. In: *42nd AIAA/ASME/SAE/ASEE Joint Propulsion Conference and Exhibit*, AIAA paper 2006-5039 (AIAA, Reston 2006)

16.150 A.K. Mohamed, D. Henry, D. Bize, M. Ory, W.H. Beck, M. Carl, U. Frenzel: MSTP phase 2 synthesis report: Infrared diode laser absorption spectroscopy in the HEG shock tube facility, ONERA RTS **14/7301 PY** (1997)

16.151 O. Trinks, W.H. Beck: Application of a diode-laser absorption technique with the D2 transition of atomic Rb for hypersonic flow-field measurements, Appl. Opt. **37**(30), 7070–7075 (1998)

16.152 W.H. Beck, O. Trinks, A.K. Mohamed: Diode laser absorption measurements in high enthalpy flows: HEG free stream conditions and driver gas arrival. In: *Proc. 22nd Int. Symp. Shock Waves*, ed. by G.J. Ball, R. Hillier, G.T. Roberts (Southampton Univ. Media, Southampton 1999)

16.153 M.A. Oehlschlaeger, D.F. Davidson, R.K. Hanson: Temperature measurement using ultraviolet absorption of carbon dioxide behind shock waves, Appl. Opt. **44**, 6599–6605 (2005)

16.154 T.C. Hanson, D.F. Davidson, R.K. Hanson: Shock tube measurements of water and n-dodecane droplet evaporation behind shock waves. In: *43rd Aerospace Sciences Meeting*, AIAA paper 2005-0350 (AIAA, Reston 2005)

16.155 K. Nakakita, K. Asai: Pressure-sensitive paint application to a wing-body model in a hypersonic shock tunnel, 22nd AIAA Aerodynamic Measurement Technology and Ground Testing Conference (2002)

16.156 S. Ohmi, H. Nagai, K. Asai, K. Nakakita: Effect of TSP layer thickness on global heat transfer measurement in hypersonic flow, 44th AIAA Aerospace Sciences Meeting and Exhibit (2006)

16.157 F. Seiler, M. Havermann, A. George, F. Leopold, J. Srulijes: Planar velocity visualization in high-speed wedge flow using Doppler picture velocimetry (DPV) compared with particle image velocimetry (PIV), J. Visual. **6**(3), 253–262 (2003)

16.158 S. O'Byrne, P.M. Danehy, A.D. Cutler: $N_2/O_2/H_2$ dual-pump CARS: Validation experiments, Proc. 20th Int. Congress on Instrumentation in Aerospace Simulation Facilities (2003)

16.159 P.M. Danehy, S. O'Byrne, A.D. Cutler, C. Rodriguez: CARS) as a probe for hydrogen-fuel/air mixing, JAN-NAF APS/CS/PSHS/MSS Joint Meeting (2003)

16.160 B.N. Littleton, A.I. Bishop, T.J. McIntyre, P.F. Barker, H. Rubinsztein-Dunlop: Flow tagging velocimetry in a superorbital expansion tube, Shock Waves **10**, 225–228 (2000)

16.161 E.P. Muntz: The electron beam fluorescence technique, AGARDograph **132**, 111 (1968)

16.162 K.A. Bütefisch, D. Vennemann: The electron beam technique in hypersonic rarefied gas dynamics, Prog. Aerosp. Sci. **15**, 217–260 (1974)

16.163 D.P. Shelton, P.E. Cassady: Electron beam density measurement in a hypersonic shock tunnel flow-field, 17th Aerospace Ground Testing Conference (AIAA, Reston 1992)

16.164 E.P. Muntz, F.M. Lufty, I.D. Boyd: The study of reacting, high energy flows using pulsed electron-beam fluorescence, 27th Fluid Dynamics Conference (AIAA, Reston 1996)

16.165 W. Merzkirch: *Flow Visualization* (Academic, New York 1974)

16.166 A.N. Zaidel', G.V. Ostrovskaya, Y.I. Ostrovskii: Plasma diagnostics by holography (a review), Sov. Phys. Tech. Phys. **13**, 1153–1164 (1969)

16.167 F.C. Jahoda, R.A. Jeffries, G.A. Sawyer: Fractional-fringe holographic plasma interferometry, Appl. Opt. **6**, 1407–1410 (1967)

16.168 H. Schardin: Das Toeplersche Schlierenverfahren: Grundlagen für seine Anwendung und quantitative Auswertung, VDI (Verein Deutscher Ingenieure) Forschungsheft **367**, 1–32 (1934), in German

16.169 G.S. Settles: Schlieren and shadowgraph techniques. In: *Visualizing Phenomena in Transparent Media*, ed. by R.J. Adrian, M. Gharib, W. Merzkirch, D. Rockwell, J.H. Whitelaw (Springer, Berlin, Heidelberg 2001)

16.170 S.F. Ray (Ed.): High speed photography and photonics, SPIE Monograph **PM120** (2002)

16.171 A. Toepler: Optische Studien nach der Methode der Schlierenbeobachtung, Poggendorfs Annal Phys. Chem. **210**(6), 194–217 (1867), in German

16.172 E. Mach, P. Salcher: *Optische Untersuchung der Luftstrahlen* (Akad. Wiss., Wien 1889), in German

16.173 H. Fischer: Simple submicrosecond light source with extreme brightness, J. Opt. Soc. Am. **47**(11), 981–984 (1957)

16.174 HSPS: *Website* (High-Speed Photo.-Systeme, Wedel 2006), www.hsps.com

16.175 J. Holzfuss: Analoge und digitale Hochgeschwindigkeitskinematographie – eine Übersicht (Analogue and digital high speed cinematography – a review), TM – Technisches Messen **68**(11), 499–506 (2001), in German

16.176 J. Martinez Schramm, S. Karl, K. Hannemann: High speed flow visualization at HEG. In: *New Results in Numerical and Experimental Fluid Mechanics IV*, Notes on Numerical Fluid Mechanics and Multidisciplinary Design, Vol. 87, ed. by C. Breitsamter, B. Laschka, H. Heinemann (Springer, Berlin Heidelberg 2004) pp. 229–235

16.177 D.J. Kewley, H.G. Hornung: Non-equilibrium dissociating nitrogen flow over a wedge, J. Fluid Mech. **64**(4), 725–736 (1974)

16.178 H.G. Hornung, G.H. Smith: The influence of relaxation on shock detachment, J. Fluid Mech. **93**(2), 225–239 (1979)

16.179 S. Karl, J. Martinez Schramm, K. Hannemann: High enthalpy cylinder flow in HEG: A basis for CFD validation, 33rd AIAA Fluid Dynamics Conference (2003)

16.180 M.S. Holden: Experimental studies of laminar separated flows induced by shock wave/boundary layer and shock/shock interaction in hypersonic flows for CFD validation, AIAA paper **2000-0930** (2000)

16.181 M.S. Holden, J.K. Harvey: Comparisons between DSMC and Navier–Stokes solutions on measurements in regions of laminar shock wave boundary layer interaction in hypersonic flows, AIAA paper **2002-0435** (2002)

16.182 G.V. Candler, I. Nompelis, M.-C. Druguety, M.S. Holden, T.P. Wadhams, I.D. Boyd, W.-L. Wang: CFD validation for hypersonic flight: Hypersonic double-cone flow simulations, AIAA J. **2002-0581** (2002)

16.183 S.R. Sanderson: *Shock Wave Interaction in Hypervelocity Flow*, Dissertation (California Institute of Technology, Pasadena 1995)

16.184 M. Schnieder: *Wechselwirkung einer starken und einer schwachen Stoßwelle in reagierender Hochenthalpieströmung*, Dissertation, DLR-FB 98-31 (RWTH Aachen, Aachen 1999), in German

16.185 S.G. Mallinson, S.L. Gai, N.R. Mudford: The interaction of a shock wave with a laminar boundary layer at a compression corner in high-enthalpy flows including real gas effects, J. Fluid Mech. **342**, 1–35 (1997)

16.186 J.P. Davis: *High-Enthalpy Shock/Boundary-Layer Interaction on a Double Wedge*, Dissertation (California Institute of Technology, Pasadena 1999)

16.187 J.P. Davis, B. Sturtevant: Separation length in high-enthalpy shock/boundary-layer interaction, Phys. Fluids **12**, 2661–2687 (2000)

16.188 J. Martinez Schramm, G. Eitelberg: Shock boundary layer interaction in hypersonic high enthalpy flow on a double wedge. In: *Proc. 22nd Int. Symp. Shock Waves*, ed. by G.J. Ball, R. Hillier, G.T. Roberts (Southampton Univ. Media, Southampton 1999)

16.189 T.J. Horvath, S.A. Berry, N.R. Merski, S.M. Fitzgerald: X-38 experimental aerothermodynamics, 34th AIAA Thermophysics Conference (AIAA, Reston 2000)

16.190 K. Hannemann, J. Martinez Schramm, S. Brück, J.MA. Longo: High enthalpy testing and CFD rebuilding of X-38 in HEG, Proc. 23rd International Symposium on Shock Waves, ed. by F. Lu (2001), on CD-ROM

16.191 K. Itoh, T. Komuro, K. Sato, H. Tanno, S. Ueda: Hypersonic aerodynamic research of HOPE using high enthalpy shock tunnel, AIAA/NAL-NASDA-ISAS 10th International Space Planes and Hypersonic Systems and Technologies Conference (AIAA, Reston 2001)

16.192 D. Isakeit, P. Watillon, A. Wilson, C. Cazaux, G. Bréard, T. Leveugl: *The Atmospheric Reentry Demonstrator*, European Space Agency Rep. BR-138 (1998)

16.193 I.A. Johnston, M. Weiland, J. Martinez Schramm, K. Hannemann, J. Longo: Aerothermodynamics of the ARD: Postflight numerics and shock-tunnel experiments, 40th AIAA Aerospace Sciences Meeting and Exhibit (AIAA, Reston 2002)

16.194 T.J. McIntyre, I. Lourel, T.N. Eichmann, R.G. Morgan, P.A. Jacobs, A.I. Bishop: Experimental expansion tube study of the flow over a toroidal ballute, J. Spacecr. Rockets **41**(5), 716–725 (2004)

16.195 A. Rasheed, K. Fujii, H.G. Hornung, J.L. Hall: Experimental investigation of the flow over a toroidal aerocapture ballute, 19th Applied Aerodynamics Conference (AIAA, Reston 2001)

16.196 L.M. Mack: *Boundary-Layer Stability Theory*, Special Course on Stability and Transition of Laminar Flow, AGARD Rep. 709 (1984)

16.197 H.G. Hornung: Hypersonic real-gas effects on transition. In: *IUTAM Symposium on One Hundred Years of Boundary Layer Research*, Proc. IUTAM Symposium, ed. by G.E.A. Meier, K.R. Sreenivasan (Springer, Berlin, Heidelberg 2006)

16.198 H.B. Johnson, T. Seipp, G.V. Candler: Numerical study of hypersonic reacting boundary layer transition on cones, Phys. Fluids **10**, 2676–2685 (1998)

16.199 H.G. Hornung, P.H. Adam, P. Germain, K. Fujii, A. Rasheed: On transition and transition control in hypervelocity flow, Proc. Ninth Asian Congress of Fluid Mechanics (2002)

16.200 D.J. Mee, C.P. Goyne: Turbulent spots in boundary layers in a free-piston shock tunnel flow, Shock Waves **6**, 337–343 (1996)

16.201 D.J. Mee: Boundary layer transition measurements in hypervelocity flows in a shock tunnel, 39th AIAA Aerospace Sciences Meeting (AIAA, Reston 2001)

16.202 A. Paull, H. Alesi, S. Anderson: The HyShot flight program and how it was developed, AIAA/AAAF 11th International Space Planes and Hypersonic Systems and Technologies Conference (AIAA, Reston 2002)

16.203 V. Rausch, C. McClinton, J. Sitz: Hyper-X program overview, Proc. Int. Symp. Airbreathing Engines (1999), ISABE 99-7213

16.204 R.T. Voland, L.D. Huebner, C.R. McClinton: X-43A hypersonic vehicle technology development, Acta Astronaut. **59**, 181–191 (2006)

16.205 R.J. Stalker, A. Paull, D.J. Mee, R.G. Morgan, P.A. Jacobs: Scramjets and shock tunnels – the Queensland experience, Prog. Aerosp. Sci. **41**, 471–513 (2005)

16.206 A. Paull, R.J. Stalker, D.J. Mee: Experiments on supersonic ramjet propulsion in a shock tunnel, J. Fluid Mech. **296**, 150–183 (1995)

16.207 G.S. Deiwert, J.A. Cavolowsky, M.P. Loomis: Large scale scramjet testing in the Ames 16-inch shock tunnel, 18th AIAA Aerospace Ground Testing Conference (AIAA, Reston 1994)

16.208 A. Paull, R.J. Stalker: Scramjet testing in the T3 and T4 hypersonic impulse facilities, Scramjet Propulsion, Vol. 189, ed. by E.T. Curran, S.N.B. Murthy (2000) 1–46

16.209 N.E. Hass, M.K. Smart, A. Paull: Flight data analysis of HyShot 2, 13th AIAA/CIRA International Space Planes and Hypersonic Systems and Technologies Conference (AIAA, Reston 2006)

16.210 A.D. Gardner, K. Hannemann, J. Steelant, A. Paull: Ground testing of the HyShot supersonic combustion flight experiment in HEG and comparison with flight data. In: *40th AIAA Joint Propulsion Conference*, AIAA paper 2004-3345 (AIAA, Reston 2004)

16.211 S. Karl, K. Hannemann, J. Steelant, A. Mack: CFD analysis of the HyShot supersonic combustion flight experiment configuration, 14th AIAA/AHI Space Planes and Hypersonic Systems and Technologies Conference (AIAA, Reston 2006)

16.212 T. Sunami, K. Itoh, H. Tanno, T. Komuro, K. Sato, M. Kodera, K. Fujimura, T. Maehara, T. Narita: Mach 8 firing tests of a hypermixer scramjet at HIEST, Proc. of the 35th Fluid Dynamics Conference Japan (2003) 123–126

16.213 R.C. Rogers, A.T. Shih, C.-Y. Tsai, R.O. Foelsche: Scramjet tests in a shock tunnel at flight Mach 7, 10, and 15 conditions, 37th AIAA/ASME/SAE/ASEE Joint Propulsion Conference and Exhibit AIAA-2001-3241 (2001)

16.214 R.C. Rogers, A.T. Shih, N.E. Hass: Scramjet development tests supporting the Mach 10 flight of the X-43, AIAA/CIRA 13th International Space Planes and Hypersonics Systems and Technologies Conference (AIAA, Reston 2005)

16.215 M. Bleilebens: *Einfluß der Wandtemperatur auf die Stoß/Grenzschicht-Wechselwirkung an einer Rampe im Hyperschall*, Dissertation (RWTH Aachen, Aachen 2004), in German

16.216 S. Ueda, M. Weiland, K. Itoh, K. Hannemann: Comparative experiment on the surface catalycity in two high enthalpy shock tunnels, 24th AIAA Aerodynamic Measurement Technology and Ground Testing Conference (AIAA, Reston 2004)

16.217 A. Roshko: Perspectives on bluff body aerodynamics, J. Wind Eng. Indust. Aerodyn. **49**, 79 (1993)

16.218 L. Rosenhead: Vortex systems in wakes, Adv. Appl. Mech. **3**, 185 (1953)

16.219 M.V. Morkovin: Flow around a circular cylinder – kaleidoscope of challenging fluid phenomena, Proc. ASME Symp. on Fully Separated Flows (Philadelphia 1964) 102–118

16.220 E. Berger, R. Wille: Periodic flow phenomena, Annu. Rev. Fluid Mech. **4**, 313 (1972)

16.221 H. Oertel: Wakes behind blunt bodies, Annu. Rev. Fluid Mech. **22**, 539 (1990)

16.222 M. Coutanceau, J.-R. Defaye: Circular cylinder wake configurations: A flow visualisation survey, Appl. Mech. Rev. **44**, 255 (1991)

16.223 C.H.K. Williamson: Vortex dynamics in the cylinder wake, Annu. Rev. Fluid Mech. **28**, 477 (1996)

16.224 C.H.K. Williamson, R. Govardhan: Vortex-induced vibration, Annu. Rev. Fluid Mech. **36**, 413 (2004)

16.225 C.H.K. Williamson, A. Roshko: Vortex formation in the wake of an oscillating cylinder, J. Fluid Struct. **2**, 355 (1988)

16.226 R. Mills, J. Sheridan, K. Hourigan: Particle image velocimetry and visualization of natural and forced flow around rectangular cylinders, J. Fluid Mech. **478**, 299 (2003)

16.227 C.H.K. Williamson: Three-dimensional wake transition behind a cylinder, J. Fluid Mech. **328**, 345 (1996)

16.228 H. Thomann: Measurements of the recovery temperature in the wake of a cylinder and of a wedge at Mach numbers between 0.5 and 3, Aero Res. Inst. Sweden (FFA) Rep. No. **84**, 1–30 (1959)

16.229 J.M. Cimbala, H.M. Nagib, A. Roshko: Large structure in the far wakes of two-dimensional bluff bodies, J. Fluid Mech. **190**, 265 (1988)

16.230 T. Corke, D. Koga, R. Drubka, H. Nagib: A new technique for introducing controlled sheets of smoke streaklines in wind tunnels, IEEE Pub. **77**, 74 (1974), CH 1251-8 AES

16.231 R.J. Goldstein: *Fluid Mechanics Measurements* (Taylor Francis, Washington 1996)

16.232 A. Fage, V.M. Falkner: An experimental determination of the intensity of friction on the surface of an airfoil, Proc. R. Soc. **129**, 378 (1930)

16.233 A. Roshko: Experiments on the flow past a circular cylinder at very high Reynolds number, J. Fluid Mech. **10**, 345 (1961)

16.234 B. Cantwell, D. Coles: An experimental study of entrainment and transport in the turbulent near-wake of a circular cylinder, J. Fluid Mech. **136**, 321 (1983)

16.235 V. Strouhal: Über eine besondere Art der Tonerregung, Annal. Phys. Chem. (Leipzig), Neue Folge **5**, 216 (1878), in German

16.236 N. Rott: Lord Rayleigh and hydrodynamic similarity, Phys. Fluids **4**, 2595 (1993)

16.237 Lord Rayleigh: Aeolian tones, Phil. Mag. **XXIX**, 433 (1915)

16.238 T. von Karman: Über den Mechanismus des Widerstands, den ein bewegter Körper in einer Flüssigkeit erfährt, Göttinger Nachr. Math. Phys. Kl. **509**, 324–330 (1911), in German

16.239 J.H. Gerrard: The mechanics of the vortex formation region of vortices behind bluff bodies, J. Fluid Mech. **25**, 401 (1966)

16.240 A.E. Perry, M.S. Chong, T.T. Lim: The vortex shedding process behind two-dimensional bluff bodies, J. Fluid Mech. **116**, 77 (1982)

16.241 P.W. Bearman, D.M. Trueman: An investigation of the flow around rectangular cylinders, Aero. Q. **23**, 229 (1972)

16.242 A. Roshko: On the wake and drag of bluff bodies, J. Aeronaut. Sci. **22**, 124 (1955)

16.243 C.H.K. Williamson, G.L. Brown: A series in $1/\sqrt{Re}$ to represent the Strouhal-Reynolds number relationship for the cylinder wake, J. Fluid Struct. **12**, 1073 (1998)

16.244 P.W. Bearman: On vortex street wakes, J. Fluid Mech. **28**, 625 (1967)

16.245 O.M. Griffin: Universal similarity in the wakes of stationary and vibrating bluff structures, J. Fluids Eng. **103**, 52 (1981)

16.246 P.W. Bearman: Investigation of the flow behind a two-dimensional model with a blunt trailing edge and fitted with splitter plates, J. Fluid Mech. **21**, 241 (1965)

16.247 A.B. Bauer: Vortex shedding from thin flat plates parallel to the free stream, J. Aeronaut. Sci. **28**, 340 (1961)

16.248 H. Eisenlohr, H. Eckelmann: . In: *Exp. Heat Transfer, Fluid Mech. and Thermod.*, ed. by R.K. Shah, E.N. Ganic, K.T. Yang (Elsevier, Amsterdam 1988) pp. 264–268

16.249 C. Norberg: Fluctuating lift on a circular cylinder: Review and new measurements, J. Fluid Struct. **17**, 57 (2003)

16.250 C. Wieselsberger: Recent statements on the laws of liquid and air resistance, Phys. Z. **22**, 321 (1921)

16.251 O.M. Griffin, S.E. Ramberg: The vortex street wakes of vibrating cylinders, J. Fluid Mech. **66**, 553 (1974)

16.252 A. Roshko, W. Fiszdon: On the persistence of transition in the near wake, Problems of Hydrodynamics and Continuum Mech. , SIAM, 606–616 (1967)

16.253 A. Fage: The air flow around a circular cylinder in the region where the boundary layer separates from the surface, Aero. Res. Council R&M **1179** (1928)

16.254 A. Roshko: On the development of turbulent wakes from vortex sheets, NACA Rep. **1191** (1954)

16.255 R.D. Henderson: Details of the drag curve near the onset of vortex shedding, Phys. Fluids **7**, 2102 (1995)

16.256 C.H.K. Williamson: Oblique and parallel modes of vortex shedding in the wake of a circular cylinder at low Reynolds numbers, J. Fluid Mech. **206**, 579 (1989)

16.257 U. Fey, M. Koenig, H. Eckelmann: A new Strouhal-Reynolds number relationship for the circular cylinder in the range $47<Re<2\times10^5$, Phys. Fluids **10**, 1547 (1998)

16.258 C. Norberg: An experimental investigation of the flow around a circular cylinder: Influence of aspect ratio, J. Fluid Mech. **258**, 287 (1994)

16.259 C.H.K. Williamson, A. Roshko: Measurements of base pressure in the wake of a cylinder at low Reynolds numbers, Z. Flugwiss. Weltraum. **14**, 38 (1990)

16.260 P.W. Bearman: On vortex shedding from a circular cylinder in the critical Reynolds number regime, J. Fluid Mech. **37**, 577 (1969)

16.261 O. Flaschbart: Messungen an ebenen und gewölbten Platten, Erg. Aerodyn. Versuch. Göttingen **4**, 96–100 (1932)

16.262 W.C.L. Shih, C. Wang, D. Coles, A. Roshko: Experiments on flow past rough circular cylinders at large Reynolds numbers, J. Wind Eng. Indust. Aerodyn. **49**, 351 (1993)

16.263 M.S. Bloor: The transition to turbulence in the wake of a circular cylinder, Fluid Mech. **19**, 290 (1964)

16.264 S. Taneda: Downstream development of wakes behind cylinders, J. Phys. Soc. Jpn. **14**, 843 (1959)

16.265 J.H. Gerrard: The wakes of cylindrical bluff bodies at low Reynolds number, Philos. Trans. R. Soc. A **288**, 351 (1978)

16.266 M. Coutanceau, R. Bouard: Experimental determination of the main features of the viscous flow in the wake of a circular cylinder in uniform translation: steady flow, J. Fluid Mech. **79**, 231 (1977)

16.267 S.C.R. Dennis, G.Z. Chang: Numerical solutions for steady flow past a circular cylinder at Reynolds numbers up to 100, J. Fluid Mech. **42**, 471 (1970)

16.268 B. Fornberg: Steady viscous flow past a circular cylinder up to Reynolds number 600, J. Comp. Phys. **61**, 297 (1985)

16.269 A. Thom: The flow past circular cylinders at low speeds, Proc. R. Soc. A **141**, 651 (1933)

16.270 M. Provansal, C. Mathis, L. Boyer: Benard-von Karman instability: Transient and forced regimes, J. Fluid Mech. **182**, 1 (1987)

16.271 E. Berger: Transition of the laminar vortex flow to the turbulent state of the Karman vortex street behind an oscillating cylinder at low Reynolds number, Jahrbuch Wiss. Ges. L. R., 164–172 (1964), in German

16.272 F.R. Hama: Three-dimensional vortex pattern behind a circular cylinder, J. Aeronaut. Sci. **24**, 156 (1957)

16.273 D.J. Tritton: Experiments on the flow past a circular cylinder at low Reynolds numbers, J. Fluid Mech. **6**, 547 (1959)

16.274 M. Gaster: Vortex shedding from circular cylinders at low Reynolds numbers, J. Fluid Mech. **46**, 749 (1971)

16.275 A. Roshko: *Modelling, Past and Future*, Comment paper for Whither Turbulence Workshop, Cornell University, ed. by J.L. Lumley (Cornell Univ., Ithaca 1989) pp. 486–489

16.276 C.H.K. Williamson: Defining a universal and continuous Strouhal-Reynolds number relationship for the laminar vortex shedding of a circular cylinder, Phys. Fluids **31**, 2742 (1988)

16.277 H. Eisenlohr, H. Eckelmann: Vortex splitting and its consequences in the vortex street wake of cylinders at low Reynolds number, Phys. Fluids A **1**, 189 (1989)

16.278 M. Koenig, H. Eisenlohr, H. Eckelmann: The fine structure in the S-Re relationship of the laminar wake of a circular cylinder, Phys. Fluids A **2**, 1607 (1990)

16.279 M. Hammache, M. Gharib: A novel method to promote parallel shedding in the wake of circular cylinders, Phys. Fluids A **1**, 1611 (1989)

16.280 M. Hammache, M. Gharib: An experimental study of the parallel and oblique vortex shedding from circular cylinders, J. Fluid Mech. **232**, 567 (1991)

16.281 D. Gerich, H. Eckelmann: Influence of end plates and free ends on the shedding frequency of circular cylinders, J. Fluid Mech. **122**, 109 (1982)

16.282 T. Leweke, M. Provansal: The flow behind rings: bluff body wakes without end effects, J. Fluid Mech. **288**, 265 (1995)

16.283 A. Prasad, C.H.K. Williamson: The instability of the shear layer separating from a bluff body, J. Fluid Mech. **333**, 375 (1997)

16.284 C.H.K. Williamson: The existence of two stages in the transition to three-dimensionality of a cylinder wake, Phys. Fluids **31**, 3165 (1988)

16.285 D. Barkley, R.D. Henderson: Three-dimensional Floquet stability analysis of the wake of a circular cylinder, J. Fluid Mech. **322**, 215 (1996)

16.286 G.D. Miller, C.H.K. Williamson: Control of three-dimensional phase dynamics in a cylinder wake, Exp. Fluids **18**, 26 (1994)

16.287 C.H.K. Williamson: The natural and forced formation of spot-like "vortex dislocations" in the transition of a wake, J. Fluid Mech. **243**, 393 (1992)

16.288 T. Leweke, C.H.K. Williamson: Three-dimensional instabilities in wake transition, Eur. J. Mech. B Fluids **17**, 571 (1998)

16.289 M. Thompson, K. Hourigan, J. Sheridan: Three-dimensional instabilities in the cylinder wake. In: *Int. Colloq. Jets, Wakes and Shear Layers*, ed. by K. Hourigan (CSIRO, Melbourne 1994)

16.290 M.C. Thompson, T. Leweke, C.H.K. Williamson: The physical mechanism of transition in bluff body wakes, J. Fluid Struct. **15**, 607–616 (2001)

16.291 H. Zhang, U. Fey, B.R. Noack, M. Koenig, H. Eckelmann: On the transition of the cylinder wake, Phys. Fluids **7**, 1 (1995)

16.292 J.-C. Lin, J. Towfighi, D. Rockwell: Instantaneous structure of near-wake of a cylinder: On the effect of Reynolds number, J. Fluid Struct. **9**, 409 (1995)

16.293 M.F. Unal, D. Rockwell: On vortex shedding from a cylinder, Part 1. The initial instability, J. Fluid Mech. **190**, 491 (1988)

16.294 L. Schiller, W. Linke: Druck und Reibungswiderstand des Zylinders bei Reynolds'schen Zahlen 5000 bis 40000, Z. Flugtech. Motorluft. **24**, 193 (1933), in German

16.295 M.F. Unal, D. Rockwell: On the role of shear layer stability in vortex shedding from cylinders, Phys. Fluids **27**, 2598 (1984)

16.296 M.S. Bloor, J.H. Gerrard: Measurements on turbulent vortices in a cylinder wake, Proc. R. Soc. **294**, 319 (1966)

16.297 A. Prasad, C.H.K. Williamson: Three-dimensional effects in turbulent bluff body wakes, J. Fluid Mech. **343**, 235 (1997)

16.298 L.S.G. Kovasznay: Hot wire investigation of the wake behind cylinders at low Reynolds numbers, Proc. Roy. Soc. **4**(198), 174 (1949)

16.299 H. Djeridi, M. Braza, R. Perrin, G. Harran, E. Cid, S. Cazin: Near-wake turbulence properties around a circular cylinder at high Reynolds number, Flow Turb. Combust. **71**, 19 (2003)

16.300 R. Govardhan, C.H.K. Williamson: Mean and fluctuating velocity fields in the wake of a freely-vibrating cylinder, J. Fluids Struct. **15**, 489 (2001)

16.301 E. Achenbach: Influence of surface roughness on cross-flow around a circular cylinder, J. Fluid Mech. **46**, 321 (1971)

16.302 O. Güven, V.C. Patel, C. Farrell: Surface roughness effects on the mean flow pst circular cylinders, Iowa Inst. Hydraulic Res. Rep. **175** (1975)

16.303 G. Schewe: On the force fluctuations acting on a circular cylinder in crossflow from subcritical up to transcritical Reynolds numbers, J. Fluid Mech. **133**, 265 (1983)

17. Atmospheric Measurements

A selected group of common instruments used to measure the atmosphere is described in this chapter. Typical atmospheric measurements include winds, temperature, pressure, humidity, dew point, moisture, radiation, visibility, cloud heights, lightning, gaseous composition, aerosols, and precipitation. Atmospheric sensors may measure at a point (in space) or remotely (with a given distance from the volume of air being measured). As an example of a common wind sensor, sonic anemometers measure wind speed in two or three directions using differences in the propagation speed of sound in different directions across moving air. The surface heat budget, e.g., solar radiation, albedo, the Earth's long-wave radiation, heat fluxes into the soil, and latent/sensible heat fluxes are major forcing effects for atmospheric motions. Standard measurement methodologies are presented in Sect. 17.1.3. Some of the most frequently used measuring instruments for atmospheric dispersion are reviewed in Sect. 17.2. Major field experiments have recently used arrays of both *bag* samplers and fast response sensors to track the dispersion of plumes of trace gases in urban areas. Remote sensing instruments, both active and passive, are becoming increasingly available and robust. Remote sensing instruments may be ground-based or satellite based. While the flexibility and power of modern remote sensing instruments is impressive, there is frequently an inversion challenge associated with data interpretation.

17.1	**Point Measurements** 1159
	17.1.1 Wind Velocity 1159
	17.1.2 Temperature, Relative Humidity, and Pressure 1162
	17.1.3 Fluxes 1164
	17.1.4 Vertical Profiling 1166
17.2	**Dispersion Measurements** 1167
	17.2.1 Tracers 1167
	17.2.2 Bag Samplers 1168
	17.2.3 Fast-Response Sensors 1168
	17.2.4 Recent Major Field Experiments 1168
17.3	**Remote Sensing** 1169
	17.3.1 Lidar 1169
	17.3.2 Ceilometers 1172
	17.3.3 Sodar 1172
	17.3.4 Radar Profilers 1174
	17.3.5 RASS 1174
17.4	**Satellite Measurements** 1175
References 1178	

Measurement of atmospheric flow and airborne material at different scales and with varying space–time resolutions is of great utility for meteorologists, atmospheric physicists, communication engineers, military planners, and teams subjected to major technological risks. Low space–time-resolution atmospheric data are used for forecasting and emergency planning, whereas high-resolution data are employed in atmospheric flow and dispersion research. Typical atmospheric measurements include winds, temperature, pressure, humidity, dew point, moisture, radiation, visibility, cloud heights, lightning, gaseous composition (e.g., CO_2, pollutants), aerosols, and precipitation. Over the past few decades, major technological advances have been made with regard to atmospheric sensors (detectors) as well as sensor platforms (which support a battery of sensors such as satellites, aircraft and towers). The advent of *fast-response* (typically up to 100 Hz) and *mean-value* sensors for direct and remote sensing measurements and progress in data acquisition, calibration, management, analysis, and quality-assurance techniques have enabled measurements with ever-increasing reliability and accuracy. Direct sensors make in situ measurements at a given location whereas remote sensors measure wave signals that are generated or modified by atmospheric phenomena at some distance away. The remote sensors that emit a particular type of waves (e.g., sound, light, microwaves) from a transmitter and receive the modified signal at a receiver are called *active* remote sensors whereas those that capitalize on receiving stray

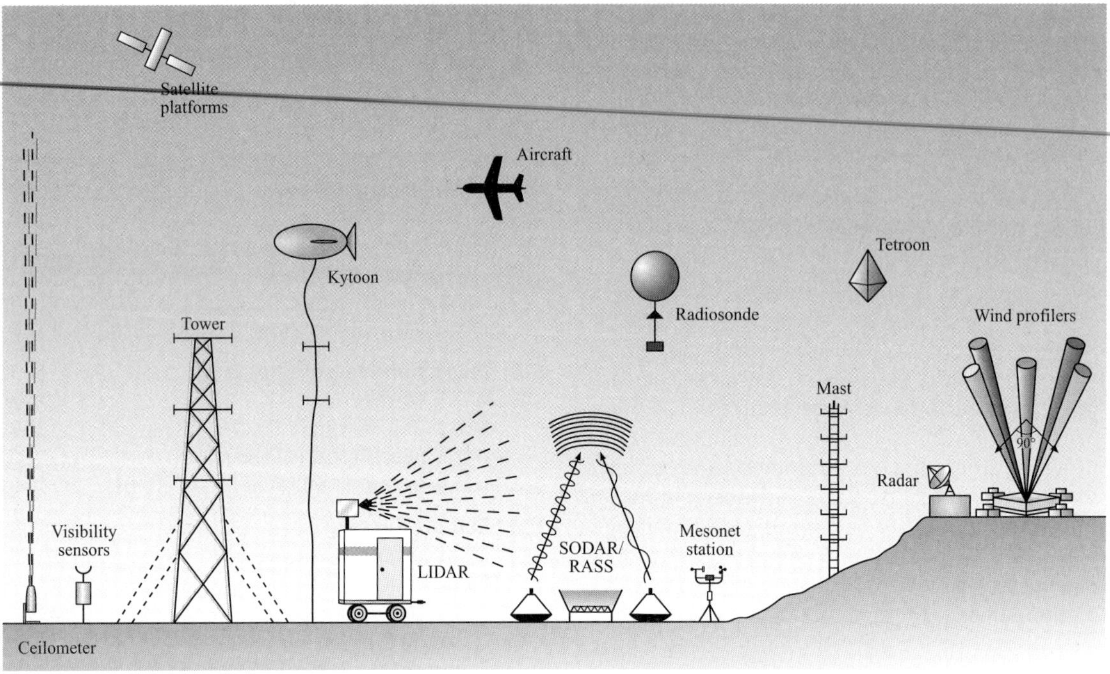

Fig. 17.1 A schematic of an atmospheric flow sensor deployment in a field campaign

waves emitted or reflected by atmospheric constituents or the Earth's surface are called passive remote sensors. Direct sensors have high space–time resolution, but they may also modify local wind fields to yield potentially unrepresentative point values. Remote sensing offers wide-area coverage (sometimes over periods of many years with the same instrumentation, as in the case of satellites) and physically nonintrusive measurements that are often more cost-effective than surface instruments, but usually their spatial resolution is limited, their view may be obscured by the clouds and gasses, and they may require convoluted calibration techniques.

Figure 17.1 shows a schematic of an idealized field campaign with an array of atmospheric instruments and instrument platforms. The meteorological mast, with a typical height of 10–50 m, is used to place point instruments at various heights. Sometimes much higher permanent masts (e.g., 300 m in Boulder, and 213 m in Cabauw) are used. Instruments associated with masts include: sensors for wind velocity (e.g., sonic, cup, hot-wire and vane anemometers), refractive index (scintillometers), visibility (transmissometers, backscatter meters, cameras, nephelometers) and falling precipitation (disdrometer); thermometers (thermistors, thermocouples); barometers (aneroid and capacitive elements); hygrometers (hygristors, psychrometers);

rain/snow gauges (bucket and electrically heated gauges, snowboards and pillows); and radiometers (net radiometers, pyranometers, pyrheliometers, diffusometers and pyrgeometers). Most of these instruments are multipurpose and measure more than one variable. Smaller portable mesonet stations are also employed, which consist of temperature, humidity, rain, pressure, radiation, and turbulent flux measurements at 2 m and 10 m. The flow around larger towers can be significantly disturbed by the structure, and thus instruments are deployed on booms projecting several meters outward of the tower, directed at different directions with the hope that one of the sensors will be facing the upwind direction.

Kytoons are helium balloons tethered to a winch on the ground. They are portable *towers* with instruments suspended along the tether at different heights. Tethers are available up to ≈ 2 km height, although air-traffic restrictions limit their use to less than 800 m, depending on the airspace. They can be used as instrument *towers* at a fixed balloon elevation or as profilers when the balloon is ascending or descending. The use of kytoons is limited because of their limited carrying capacity and difficulties at wind speeds above 5 m/s. Other profiling equipment includes radiosondes or rawinsondes (helium-filled expendable balloons carrying instrument packages) and tetroons (constant-pressure, balloons with a tetrahedron shape and suspended instruments, which theoretically stay at a constant altitude). Such platforms typically measure temperature, humidity, and pressure information (known as raobs) and are commonly used in operational weather forecasting.

Active remote sensors shown in Fig. 17.1 include radars (radio detection and ranging), sodars, sodars (which employ sound instead of radio waves), lidars (which employ light), and ceilometers (cloud-height measuring devices based on lidar technology). Typically they emit wave pulses, and the time and strength of signal returns allow the calculation of the distance to, and magnitude of, the quantity being measured. They can be used either on the ground or on an aircraft, but their size and large power requirements require the use of larger aircraft. Ground remote sensors can scan vertical swaths at a fixed azimuthal angle called a range–height indicator (RHI) or at a fixed polar angle known as a plan–position indicator (PPI). Volume scans can be obtained using a combination of the two techniques.

A wide range of aircraft platforms are used, including ultralights, gliders, single- and multiengine propeller or turboprop aircraft, helicopters, jets, and cabin-class civilian/military aircraft. The instruments are typically mounted on special booms projecting outside the boundary layer or on special pods to minimize aircraft-induced disturbances. The high speeds (typically 50–100 m/s) of aircraft and limitations of sensor response make it necessary to obtain large space–time segments in order to yield statistically stable turbulence statistics, which is difficult in practice. More recently, satellites have become useful tools for atmospheric observations, with satellite sensors providing routine data with resolution down to a few meters.

A number of point and remote sensors, and sensor platforms are briefly discussed in this chapter. Given the sophistication, complexity, and variety of the available atmospheric sensors and sensor platforms, a comprehensive review of them all is not feasible in such a limited space, and thus only selected commonplace atmospheric flow-measurement tools are described.

17.1 Point Measurements

Measurements taken at a fixed point or volume in space are known as point measurements. A typical setup for point measurements consists of a tower equipped with an array of sensors. Some of the quantities of interest are the wind speed and direction, temperature, solar radiation, relative humidity, pressure, and turbulent fluxes. In this section an overview of most commonly used point measurement sensors is given.

17.1.1 Wind Velocity

Standard meteorological convention determines the U wind component along the east–west direction, the V component along the south–north direction, and the W component vertically upward (against gravity). Wind direction is given in angular degrees, clockwise from the north (e.g., an easterly wind has a direction of $90°$). Note that some sensor manufacturers use different conventions, based on the instrument axis. If this is the case, one of the first steps in postprocessing and quality control should be correcting for wind direction (and its components) to the standard convention. To measure the mean wind, so-called cup and propeller anemometers are used (Figs. 17.2, 3). Their sampling rate is on the order of one second and data is usually stored as 1–30 min averages. The output is a voltage or frequency (magnet-activated reed switch), depending on the model, and they are very simple to use in combination with any standard data logger with a voltage or pulse-count input.

These types of anemometers have threshold speed of ≈ 0.5 m/s and an accuracy of ≈ 0.1 m/s, which is sufficient for routine (operational) meteorological or cli-

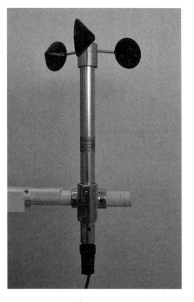

Fig. 17.2 The Met-one three-cup anemometer

matological measurements. Turbulence measurements, however, require higher sampling rates, lower thresholds with higher accuracy, and greater resolution. For this reason cup and propeller anemometers are of limited utility for research-grade measurements.

Sonic anemometers are widely used for wind and temperature measurements. They have superior performance (≈ 0.01 m/s and ≈ 0.01 K resolution with ≈ 0.05 m/s and ≈ 0.05 K accuracy), fast response (up to ≈ 60 Hz) and robustness with no moving parts. The electronics are self-contained within the probe, allowing it to be operated as a tower-mounted instrument capable of withstanding hostile environmental conditions. These devices measure the time that an emitted sound wave travels through the air between two ultrasonic transducers (hence the name *sonic*). If the air is in motion, the velocity will influence the speed of sound wave by accelerating or decelerating it. The time of flight of the first emitted signal is equal to

$$t_1 = \frac{s}{c + u_{\text{air}}}, \qquad (17.1)$$

and the time of flight of the second signal (emitted from the opposite transducer) is

$$t_2 = \frac{s}{c - u_{\text{air}}}, \qquad (17.2)$$

where s, c, and u_{air} are the distance between the transducers, the speed of sound, and the wind speed component along the transducers axis, respectively [17.1]. From this information it is straightforward to obtain the along axis wind component u_{air} as

$$u_{\text{air}} = \frac{s}{2}\left(\frac{1}{t_1} - \frac{1}{t_2}\right). \qquad (17.3)$$

Thus, only the airspeed component along the axis of the ultrasonic transducers is measured. Sonics are equipped with either two (*two-dimensional*) or three (*three-dimensional*) pairs of ultrasonic transducers, which may have orthogonal or nonorthogonal axes. In the latter case, the measured wind speed components are transformed to and expressed as those based on the orthogonal axes. Further, the speed of sound can be found as

$$c = \frac{s}{2}\left(\frac{1}{t_1} + \frac{1}{t_2}\right). \qquad (17.4)$$

The speed of sound in moist air is a function of temperature and humidity, which can be written as

$$c^2 = \gamma R T_{\text{v}}, \qquad (17.5)$$

where γ, R, and T_{v} are the ratio of the specific heat of dry air at constant pressure to that at constant volume, the gas constant of dry air, and the virtual air temperature (the temperature that dry air would have if its pressure and specific volume were the same as a sample of moist air), respectively ($\gamma = 1.4$, $R = 287.04$ J/(K kg)). In anemometer operation, the virtual temperature is calculated using

$$T_{\text{v}} = \frac{c^2}{\gamma R}. \qquad (17.6)$$

If barometric pressure measurements are available independently, the virtual potential temperature can be calculated through

$$\theta_{\text{v}} = \frac{T_{\text{v}}}{\left(\frac{P}{P_0}\right)^{R/c_{\text{p}}}}, \qquad (17.7)$$

which is a conserved quantity and is critical for sensible heat flux calculation. Here P, P_0, and c_{p} are the pressure, the pressure at the sea level ($P_0 = 1000$ mbar), and the specific heat at constant pressure, respectively. While most sonic anemometers have comparable accuracy in wind speed measurements, temperature accuracy varies significantly from one make to another, and is reflected in the instrument price. The cheapest sonic anemometers do not even have the capability to measure temperature.

Typically 10 or 20 Hz sampling rates are used for sonics, which is sufficient for turbulence calculations. In postprocessing (which can be performed in real time by a datalogger), the first step is the separation of the mean wind ($\overline{U}, \overline{V}, \overline{W}$, and $\overline{\theta_{\text{v}}}$) from the fluctuating components (u', v', w', θ'), selecting a proper averaging period that usually varies from several seconds to several hours, depending on the type of analysis. Additional filtering or trend removal (detrending) techniques may be applied. Fluctuating components are used for the calculations of turbulent intensities in the forms of variances ($\overline{u'^2}, \overline{v'^2}$, and $\overline{w'^2}$) and Reynolds stresses and fluxes in the forms of covariances ($\overline{u'v'}, \overline{u'w'}, \overline{v'w'}$, and $\overline{w'\theta'}$).

Fig. 17.3 The RM Young propeller anemometer

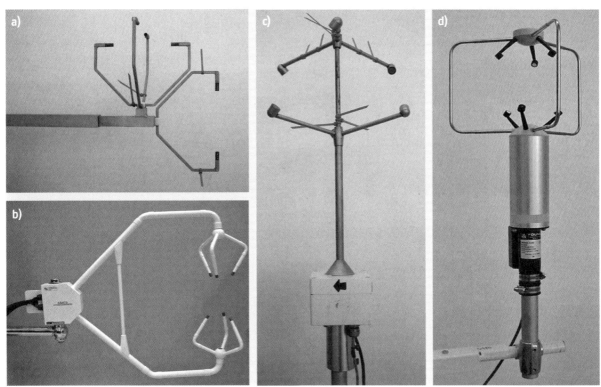

Fig. 17.4a–d Sonic anemometers: (**a**) Applied Technologies Sx, (**b**) Campbell Scientific CSTAT3, (**c**) Metek USA 1, (**d**) RM Young 81000. The wire straps are mounted on the Applied and Metek devices as bird protection – they are recommended for any type of sensor

Three-dimensional sensors have significant advantages over two-dimensional ones, as access to the vertical velocity enables calculations of quantities such as the friction velocity

$$u_* = \sqrt{\overline{u'w'}^2 + \overline{v'w'}^2}, \quad (17.8)$$

and the sensible heat flux

$$H_S = \rho c_p \overline{w'\theta'_v}, \quad (17.9)$$

which are crucial for developing atmospheric turbulence parameterizations.

Several sonic anemometer designs exist. For measurements at sites with a prevailing dominant wind direction, probes like those from Applied Technologies Inc. (A, K, S_x, V and V_x styles) or the Campbell Sci. CSAT3 probe are usually employed. Careful supporting base designs help minimize flow distortions due to tower interference while permitting a wide unobstructed coverage of the vertical component. These probes are also known for their ability to yield an unobstructed vertical component. Probes such as the Metek USA1 or RM Young 81 000 allow almost unobstructed measurements of horizontal wind components from any direction, but the supporting base of these types of probes obstructs the vertical wind component. Figure 17.4 shows photographs of these different probe types.

Superior environmental resistance for these probes is achieved by using ultraviolet (UV)-stabilized thermoplastic, stainless-steel, and anodized aluminum components, thereby isolating the sonic anemometer signal from the background noise. Conversely, sonic anemometers do not produce noise that could hamper the operation of nearby instruments, unless ultrasound measurements are made in the proximity of the sonic probe. Sonic probes, however, are not resistant to birds who find sonics arms convenient resting places; to repel them plastic wire straps can be carefully installed without obstructing the instrument operation. Sonics also cannot be operated during rain. Ice on the transducers, which usually occurs in the early morning or when the temperature drops below the freezing point, completely

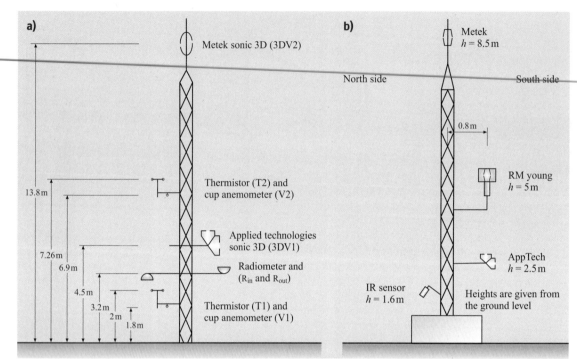

Fig. 17.5 (a) Schematic of the tower equipped with two sonic anemometers, two cup anemometers, two thermistors, and radiometers used during the VTMX2000 campaign. (b) Tower setup used for the Urban Canyon measurements during the JU2003 campaign. This tower was equipped with three sonic anemometers and an IR sensor for surface temperature measurements

disables the anemometer operation unless heaters are installed.

Examples of possible instrumented tower configurations are shown in Fig. 17.5. The schematic in Fig. 17.5a presents a configuration that was used for the evaluation of eddy diffusivities of momentum and heat

$$K_M = -\frac{\overline{u'w'}}{\left(\frac{\partial U}{\partial z}\right)} \quad \text{(momentum)}, \quad (17.10)$$

$$K_H = -\frac{\overline{w'\theta'}}{\left(\frac{d\theta}{dz}\right)} \quad \text{(heat)}, \quad (17.11)$$

when the wind direction was constant with height. Here the fluxes of momentum ($\overline{u'w'}$) and heat ($\overline{w'\theta'}$) measured by the lower sonic (height 4.5 m) are used together with gradients of the mean wind ($\partial U/\partial z$) and temperature ($d\theta/dz$) measurements of two cup anemometers and two thermistors placed below and above the sonic (heights ≈ 2 m and ≈ 7 m). The fluxes are also measured by a second sonic placed at the top of the tower. In the configuration presented on Fig. 17.5b, fluxes are measured at three levels and the gradients of mean values can be calculated using the two sonics.

Hot wires are sometimes used for atmospheric velocity measurements at very high frequencies. Given their high sensitivity and fragility, constant care and protection need to be applied. For example, sand grains carried by the wind can easily break a hot wire. This type of sensors is also very susceptible to noise and has a limited measurement range in terms of flow magnitude and direction. Therefore, hot wires are mainly used under controlled laboratory conditions and only for some specific turbulence-related atmospheric boundary-layer measurements.

17.1.2 Temperature, Relative Humidity, and Pressure

Ordinary thermistors are usually employed for atmospheric temperature measurements. They are inexpensive, conveniently available in different varieties, adaptable, and versatile. Thermistor circuits can have reasonable output voltages – compared to the milli-

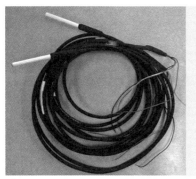 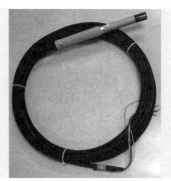 

Fig. 17.6 From the left: soil thermistors (M107), a relative humidity senor (HMP45C), and an infrared sensor (4000.4ZL)

volt outputs of thermocouples. In field applications, proper radiation shielding is needed to limit direct solar radiation loading. To reduce the influence of solar heating further, the radiation shield can be aspirated by drawing air across the measurement junction using a high-powered fan. This, however, creates an increased power demand, which may present a problem for remote weather stations that operates solely on solar power. When fast measurements (> 1 Hz) are needed, thermocouples are used. By their very nature, thermocouples are less stable and more sensitive to noise. Usually a signal gain is required for their operation, whence any noise picked up by the thermocouple will be amplified. When a fast response is needed, fine wire thermocouples are frequently used but the drawbacks stated in Sect. 17.1.1 for hot wires also apply to such fine wire thermocouples. Owing to their extremely small wire diameter, the solar loading on them is negligible, thus eliminating the need for radiation shielding.

Contact sensors used for surface measurements tend to influence the surface temperature being measured (e.g., by shading the surface from solar radiation), but maintaining a constant thermal contact with the surface is difficult. Therefore, infrared (IR) sensors are widely used for surface (skin) temperature measurements. The latter are noncontact instruments that sense emitted infrared radiation and calculate surface temperature via Stefan's law:

$$J = \varepsilon \sigma T^4, \qquad (17.12)$$

where J is the measured IR radiation, ε is the surface emissivity ($0 \leq \varepsilon \leq 1$), σ is the Stefan–Boltzmann constant ($\sigma = 5.67 \times 10^8 \, \text{W} \, \text{m}^{-2} \, \text{K}^{-4}$), and T is the surface temperature. The emissivities for different surface types can be found in handbooks [17.2]. For greater accuracy, the sensor body temperature is used to correct the target temperature. The IR sensor output voltage is proportional to the surface temperature.

The relative humidity (RH) is defined as the ratio of the actual amount of moisture in the atmosphere to the amount that the atmosphere can hold. RH is an important quantity in weather prediction for the estimation of precipitable water and evaluating the virtual temperature. It is also used to estimate evapotranspiration and to predict the health of plants. Water vapor is a strong greenhouse gas that acts to trap infrared radiation reflected from the Earth. Modern RH sensors are based on a principle that relates the changes in the capacitance of a thin polymer film when it absorbs water molecules. Such sensors yield a voltage output proportional to the relative humidity. Temperature compensation is required in computing final RH values, and thus a temperature sensor is needed in RH instruments. The accuracy of these measurements is 2–3%, depending on the absolute value of RH (the higher the RH, the lower the accuracy). As for temperature sensors, RH sensors also need a radiation shield. RH can also be obtained from wet- and dry-bulb temperature measurements. Other quantities that measure atmospheric water vapor are the absolute humidity (the mass of water vapor in a given volume of air), the mixing ratio (the ratio of the mass of water vapor contained in a sample of moist air to the mass of dry air in the sample), the specific humidity (the ratio of the mass of water vapor in a sample of moist air to the total mass of the sample), and the dew point (the temperature at which a parcel of air saturates). Figure 17.6 shows examples of temperature, RH, and IR sensors.

The barometric pressure (the pressure of the atmosphere) is one of the most important parameters for weather prediction, given that the movement of pressure fronts indicates the trajectory of weather fronts. A conventional barometer consists of a sealed glass tube, evacuated, and placed in an open pool of mercury; the pressure is measured directly as the height of mercury in the tube. Modern barometers use electronic pressure transducers fitted with a pressure-sensitive diaphragm, and a sensor whose electrical (or resonant) properties change with the shape of the diaphragm. These sensors have an accuracy of ≈ 0.3 hPa. The barometric pressure is expressed as either the *absolute barometric pressure* (exerted by the atmosphere relative to a vacuum), or as the *corrected barometric pressure* (the absolute barometric pressure at any level adjusted to be the measurement that would result if it was done at the sea level).

17.1.3 Fluxes

Calculation of sensible heat fluxes using sonic anemometers measurements was described in Sect. 17.1.1. Sensible heat flux is transmitted from the surface into the air by conduction, and the breakdown of the conduction layer causes convection to transport the heat flux to higher air layers. Supplementary heat transport comes through water-vapor condensation or evaporation in the form of latent heat flux. To calculate the latent heat flux, it is necessary to measure water-vapor fluctuations in air. A commonplace instrument for this purpose is the krypton hygrometer (Fig. 17.7), which measures the absorption of ultraviolet light (two lines at 123.58 nm and 116.49 nm) and produces a voltage output that is a function of the instantaneous deviation of water-vapor density (ρ'_v) from the mean. The response rate of a krypton hygrometer can be as high as 100 Hz. When used in conjunction with a sonic anemometer, or any other instrument capable of measuring fluctuations of the vertical velocity at the same rate as the hygrometer, the latent heat flux (LE) can be calculated via

$$\text{LE} = L_V \overline{w' \rho_{v'}}, \qquad (17.13)$$

where L_V is the latent heat of vaporization, and $\overline{w' \rho'_v}$ is the covariance between the vertical velocity and the vapor density.

The next important component of surface energy balance is the flux through the surface, usually referred to as the soil (surface) heat flux. Although our interest is flux at the air–soil surface, it is impossible to measure this directly as the flux sensor cannot be placed at the surface due to strong solar heating. Therefore, a soil heat flux plate is placed at some depth (usually 5–50 cm, depending on the soil type and amount of solar heating) to measure the thermal energy flux directly at that depth. For the surface flux calculation, it is necessary to determine the energy change in the soil layer above the soil heat flux sensor. This is achieved by placing an array of temperature sensors in the upper soil layer while the surface temperature is measured by an IR thermometer; see Fig. 17.10b for a schematic. Now the surface heat flux (SHF) can be calculated using a control-volume analysis as

$$\text{SHF} = \rho_s c_s \frac{\text{d}}{\text{d}t} \int_0^h T(z) \text{d}z + \text{HF}, \qquad (17.14)$$

where ρ_s is the soil density, c_s is the soil heat capacity, $T(z)$ are the temperatures measured in the upper soil layer, h is the depth of the soil heat flux sensor, and HF is the flux measured by the soil heat flux sensor. Modern soil heat flux sensors are equipped with a heater that performs self-calibration, during which it also determines the exact soil heat conductivity. Self-calibration

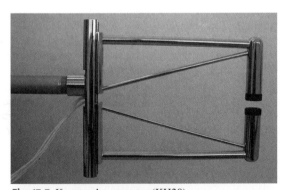

Fig. 17.7 Krypton hygrometer (KH20)

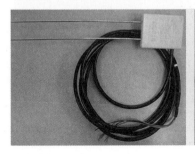
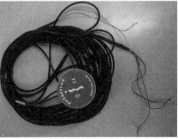

Fig. 17.8 Water content reflectometer (CS625), and soil heat flux plate (HFP01SC)

is recommended every 3–5 h. Since the soil density ρ_s, and heat capacity c_s are strongly dependent on the soil moisture content, some type of soil water content sensor, such as a *water-content reflectometer*, ought to be used simultaneously. The latter determines the volumetric water content of soils averaged over a certain depth (usually $\approx 30\,\text{cm}$) by measuring the dielectric constant of the soil. The reflectometer consists of two rods connected to a circuit board. Rods inside the soil act as antennae that transmit and receive electromagnetic signals produced by a multivibrator on the circuit board. The travel time of the signal in the probe rods depends on the dielectric permittivity of the soil, whilst the dielectric permittivity depends on the water content. Pictures of a water-content reflectometer and a soil heat flux plate are given in Fig. 17.8.

Solar radiation, its albedo, as well as the Earth's own long-wave radiation are major forcing effects for atmospheric motions and detailed measurements of these are necessary for complete energy analysis. A pyranometer measures the solar radiation R_S (short wave, $\approx 0.3\text{--}3\,\mu\text{m}$), while a pyrgeometer measures far-infrared radiation R_IR ($\approx > 3\,\mu\text{m}$). A typical setup for measuring the radiation balance consists of four sensors: a pair consisting of a pyranometer and a pyrgeometer, which face upward toward incoming solar and far-infrared radiation from the sky, and another pair, which face downward to measure the reflected solar and long-wave radiation emitted from the surface. The net radiation is given as the difference between the incoming

($\downarrow$) and outgoing radiation ($\uparrow$):

$$R_\text{net} = R_S^\downarrow + R_\text{IR}^\downarrow - R_S^\uparrow - R_\text{IR}^\uparrow . \qquad (17.15)$$

As an example, five days of measurements of all four components and net radiation are given in Fig. 17.9a. These measurements were taken above a grassy area in Oklahoma City during JU2003 campaign [17.3]. Figure 17.9b shows Kipp and Zonen CNR1, Net Radiometer which has all four sensors in one assembly.

Albedo is defined as the ratio of reflected and incoming solar radiation

$$A = \frac{R_S^\uparrow}{R_S^\downarrow} . \qquad (17.16)$$

Furthermore we can calculate effective sky and ground (or *skin*) temperatures as

$$T_\text{Sky} = \left(\frac{R_\text{IR}^\downarrow}{\sigma}\right)^{\frac{1}{4}} , \quad T_\text{Ground} = \left(\frac{R_\text{IR}^\uparrow}{\sigma}\right)^{\frac{1}{4}} , \qquad (17.17)$$

where σ is Stefan–Boltzmann constant (5.67×10^{-8} $\text{Wm}^{-2}\text{K}^{-4}$).

There are net radiometers that are capable of measuring directly the difference between incoming and outgoing radiation in the spectral range $0.25\text{--}60\,\mu\text{m}$ or even $0\text{--}100\,\mu\text{m}$. These are cheaper but do not give information about long and short wave components (e.g.

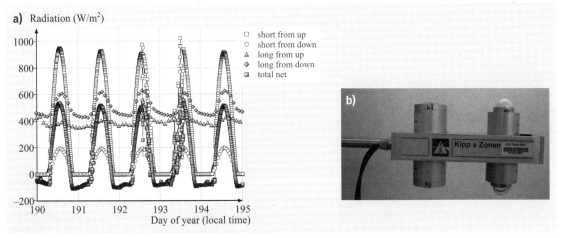

Fig. 17.9 (a) A five-day cycle of incoming and outgoing short- ($0.3\text{--}3\,\mu\text{m}$) and long-wave ($5\text{--}50\,\mu\text{m}$) radiation components and total net radiation measured during JU2003. (b) Kipp and Zonen CNR1, net radiometer. Note that the pyranometers (right side of the instrument) are covered with transparent glass hemispheres to allow undisturbed passage of solar radiation. This instrument was used for the measurements presented in (a)

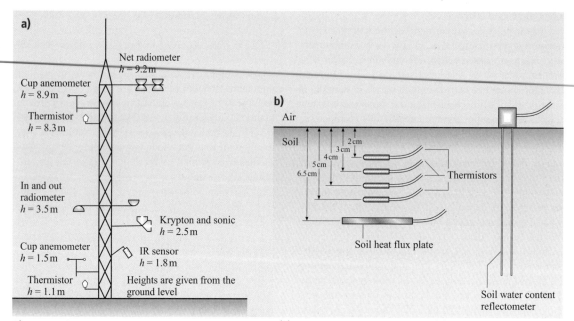

Fig. 17.10a,b Energy balance station used during JU2003: (**a**) instruments above the ground, (**b**) underground instruments for measuring the soil heat flux

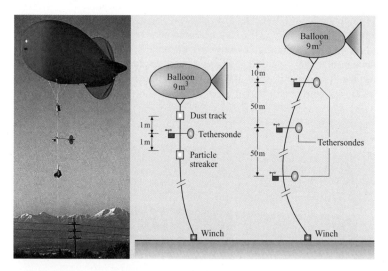

Fig. 17.11 Tethered balloon and schematic of the setup used during the VTMX2000 campaign

calculations of albedo, surface and sky temperatures are not possible).

Upon measuring these energy components, attempts can be made to close the energy budget using:

$$R_{net} = H_S + LE + SHF. \tag{17.18}$$

Note that advection is not taken into account in (17.18). It should also be borne in mind that the measured radiation represents an average over a relatively large area, whilst sensible, latent, and soil heat flux are essentially point values (covering a relatively small area). If horizontal homogeneity cannot be assumed (e.g., due to uneven land use or roughness changes), the use of several sets of instruments spanning a larger surface area is recommended. A schematic of the energy balance setup used during the JU2003 campaign is shown in Fig. 17.10.

17.1.4 Vertical Profiling

Along with point measurements it is useful to acquire vertical profiles of meteorological quantities such as potential temperature, wind speed and direction, relative humidity, aerosol concentration, chemical composition, and pressure. To a certain extent, these profiles can be obtained using remote sensing techniques (Sects. 17.3 and 17.4), but at relatively coarse resolution. In high-resolution (≈ 1 m) applications, tethersondes, radiosondes, and dropsondes are used for vertical profiling. A tethersonde system consists of sondes with sensors and radio transmitters, a helium balloon, an electrical winch, and a ground radio receiver. Several sondes can be used at different heights, subject to weight constraints. An electrical winch drives the tethered balloon up and down (*profile mode*) or simply keeps it at a desired height (*tower mode*). While in the profile mode, for accurate profiles, the winch speed should be commensurate with the sensor response time. A simple method of verifying that the winch speed is acceptable is to compare the consistency of two consecutive profiles (taken during balloon ascent and descent). On the other hand, by using too low of a winch speed, some small-scale phenomena may be missed, especially if the profiling height reaches several hundreds meters.

In balloon operations it should be noted that, although the winch's potentiometer setting is the same during the ascent and descent, the former velocity will be higher than the latter, given the aiding role of the balloon lift during the ascend. This problem can be remedied by incorporating a winch controller that measures actual balloon (tether line) velocity. Figure 17.11 shows a tethered balloon that was deployed in the vertical transport and mixing (VTMX) 2000 field campaign [17.4, 5]. Three instruments can be seen attached to the tether line: a DusTrack for sampling particulate matter (the bright box closest to the balloon); a tethersonde with sensors for measuring wind speed, wind direction, temperature, relative humidity, and pressure (in the middle, with wings in the tail); and a streaker for particle collection for the purpose of chemical composition analysis (dark box below the tethersonde). Radiosondes (also known also as rawinsondes) and dropsondes essentially have the same function as the tethersondes. Radiosondes are smaller helium-filled balloons released from the ground with a single (disposable) module that collects data during ascent. Conversely, dropsondes are released from aircraft and take measurements during their descent (see Sect. 17.1).

Owing to the deflection of the balloon and instruments by horizontal mean winds, profiles taken by these profiling systems may not be strictly vertical. Modern dropsondes and radiosondes are equipped with global positioning system (GPS) devices so that their exact trajectory can be mapped. Experience suggests that the upper limit of winds for safe tethered-balloon operation is about 5 m/s.

17.2 Dispersion Measurements

Understanding the movement and dispersion of either gaseous or aerosolized releases in the atmosphere is an important problem in atmospheric science. The need to protect people and inform rescue and mitigation efforts in the event of catastrophic atmospheric releases is a major motivation behind this area of research. Examples of the application of dispersion-related science include responses to the Three-Mile Island nuclear accident, the Chernobyl nuclear accident, chemical attacks in the middle-eastern conflicts, and the September 11, 2001 attack on the Twin Towers in New York. Numerous field experiments and modeling efforts have focused on this problem. Some of the most frequently used measuring instruments for field experiments on atmospheric dispersion are described below.

17.2.1 Tracers

Two of the most commonly used tracers in atmospheric field experiments are sulfur hexafluoride (SF_6) and perfluorocarbon (PFT) tracer gases. Both SF_6 and PFT are innocuous gases and are also frequently used in the medical and industrial fields. Both are stable, colorless, and odorless. In the quantities typically used, SF_6 has no known environmental or health effects. Studies in both Europe and the United States have used tracer gases since the 1960s. Many common manufacturing processes utilize SF_6. For example, the making of tennis balls, shoes, foam insulation, double-pane windows, and electrical equipment all frequently utilize SF_6. Consequently, however, it may be the case that background levels of SF_6 in urban areas may be large enough to swamp the relatively small amount of SF_6 released in a dispersion experiment. Therefore, it is important to perform background testing of SF_6 around the area of a proposed dispersion experiment to determine if ambient levels are too high. In these cases, which are usually in urban or industrial environments, it may be recommended to use PFTs instead of SF_6. Such deliberations are in progress, for example, at the time of writing for dispersion experiments being planned in the New York City area.

17.2.2 Bag Samplers

Bag samplers are one of a variety of names given to samplers containing bags which are filled by a small pump according to a schedule programmed into a microprocessor. An example is shown in Fig. 17.12. The boxes are usually battery operated and are therefore portable. Typically 12–18 sample bags are filled before the bags must be removed by hand and taken to a laboratory where a chemical analysis of the tracers is performed. The device used to perform the analysis is a gas-chromatography electron capture detector (GC-ECD).

17.2.3 Fast-Response Sensors

Several other types of instruments can provide analysis of tracer gas concentrations. The National Oceanic and Atmospheric Administration (NOAA)'s Field Research Division has developed a fast-response GC-ECD that produces data at about 1–2 Hz. Currently, the NOAA uses modified Scientech Inc. ECD devices (the TPA-4000) as the basis of their fast-response analyzers. As in the previous section, ECD stands for electron capture device and designates a type of gas chromatograph. These instruments allow atmospheric plumes of SF_6 to be sampled actively in real-time with van-mounted gas analyzers. There are also several instruments that rely on some form of infrared spectroscopy. The basic idea of infrared spectroscopy is that different compounds have different vibrational states to their molecules. Vibrational states that correspond to particular frequencies of infrared light absorb infrared energy and leave an absorption deficit in the transmitted light energy. Consequently, many compounds can be identified by matching their infrared absorption patterns with those in a database. Mobile Fourier-transform infrared (FTIR) instruments operate using infrared spectroscopy principles. Recently, the *Pentagon Shield* experiments used

Fig. 17.12 Sample boxes for the collection of SF_6 gas, designed by the Lawrence Livermore National Laboratory

FTIR, bag samplers, and the TPA-4000 gas analyzers to measure SF_6. Likewise, the Thermo Electron Corporation's MIRAN SapphIRe portable infrared ambient analyzer operates on the same spectroscopy principles and has been used in several major atmospheric dispersion field experiments including the VTMX/urban dispersion experiment and the joint urban dispersion experiment in 2003 [17.4, 6]. Analysis time for the MIRAN SapphIRe is 20–165 s depending on the number of wavelengths used. Typical response times are 18 s to achieve 90% of the reading. The device weighs about 10 kg and can operate on batteries for about four hours.

17.2.4 Recent Major Field Experiments

There has been a high level of interest and participation in field experiments in the last four years. In particular, the Department of Energy (DOE) VTMX/urban dispersion experiment and the joint urban dispersion experiment were two of the largest field experiments conducted with a focus on urban dispersion.

The VTMX/urban dispersion experiment was conducted in Salt Lake City in the fall of 2000. It was actually a combination of two experimental campaigns. The purpose behind the combination was to extend the reach and density of the relevant data for each experiment. The VTMX experiment was designed to improve the understanding of vertical transport and mixing processes in urban meteorological basins – particularly at night. Some of the science issues motivating the study are mentioned briefly here [17.4]. One science question concerned cold pools and the nature of the interaction between cold pools, drainage flows, and larger-scale synoptic forcing. Of particular interest were the patterns of pollutant dispersal throughout the valley. How do nocturnal flow phenomena at various scales in an urban valley affect the patterns and dispersion of tracers? How can improvement in the fundamental understanding of mixing processes lead to better mesoscale, large-eddy, and direct numerical modeling? A wide variety of instruments, using both in situ and remote sensing, were deployed as part of the campaign. The distribution of the instruments was an important characteristic of the VTMX campaign – there was a basin-wide distribution and corresponding basin-wide science focus. Remote sensing instruments deployed were radar wind profilers [some with radio acoustic sounding systems (RASS)], sodars, backscatter lidars, several coherent Doppler lidars, and a Raman water-vapor lidar. Tethered balloons and rawinsondes provided vertical profiles of wind velocity, temperature, and humidity. Three-dimensional sonics were also deployed, and a number of surface data stations already existed as a part of the MesoWest network. For the tracer releases, Brookhaven National Laboratory provided perfluorocarbon tracers (PFTs). Some releases were downtown, while other locations were near the slopes or mountain valley openings. An array of instruments to measure both particles in the air and CO were also deployed to study pollution accumulation within the valley. The NOAA/ATDD (atmospheric turbulence and diffusion division) LongEZ instrumented aircraft measured the horizontal structure of wind, temperature, and turbulence on several nights. A microbarograph array was operated to capture perturbations in the surface pressure potentially associated with gravity waves.

The urban dispersion campaign associated with VTMX was specifically designed to provide data to improve and test atmospheric dispersion models. Central to the focus and design of the experiment was a dense array of instruments (both to measure tracers and meteorological variables) deployed close to a downtown release location. The type of tracer used was SF_6 (some PFTs were released downtown by VTMX researchers), and all the various gas measurement devices examined above were deployed: FTIR, GC-ERD, bag samplers, and Miran gas analyzers. A total of 201 samplers were used: 56 PFT samplers, 105 SF_6 samplers, and 40 SF_6/PFT samplers [17.6]. In addition to a dense array of instruments in several blocks around the release site, samplers were located around arcs at two, four, and six kilometers from the release site. The intention of the deployments was to provide a dataset of concentrations that could track a short-duration release around individual buildings, through city blocks, and continuing through the urban area out to tens of kilometers from the release site. A hierarchy of dispersion models could be tested with this data, from LES (large eddy simulation) models close to the release source point, to simpler models using urban roughness parameterizations. There were 10 intensive observation periods (IOPs) with nocturnal releases.

In July 2003, the Department of Defense Defense Threat Reduction Agency (DTRA) and the US Department of Homeland Security (DHS) conducted a joint study in Oklahoma City, known as the Joint Urban 2003 (JU2003). The list of instruments deployed was similar to that of the VTMX/urban dispersion experiment with some additional notes. To the knowledge of the authors, it was the first time that two, matched, coherent Doppler lidars were deployed in a coordinated fashion in a major experiment. The lidar teams experimented with several strategies to explore and optimize the combined gathering of data for regions between the two lidars. For example, data were specifically taken for the purpose of 4DVAR (four-dimensional variational data assimilation) analysis. Intersecting RHI scans between the two lidars provided a unique opportunity to explore the concept of *virtual towers*. A helicopter-based remote SF_6 sensor was also deployed, providing images of SF_6 dispersal. Indoor tracer measurements were also conducted to measure the transport of SF_6 through the building envelope. The first six IOPs occurred during daylight hours, in contrast to the VTMX/urban dispersion experiment outlined above. The final four IOPs took place at night.

17.3 Remote Sensing

17.3.1 Lidar

A variety of types of lidar are currently used to probe the atmosphere. In the following, an introduction to the principles of coherent Doppler, backscatter, and differential absorption lidar (DIAL) is provided.

The general principle of coherent Doppler lidar is that pulses of laser light of a given frequency are sent

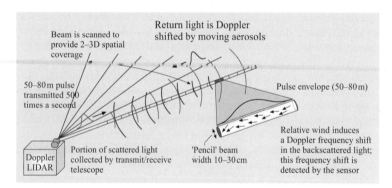

Fig. 17.13 Principle of operation of coherent Doppler lidar (courtesy of *Lockheed Martin Coherent Technologies*)

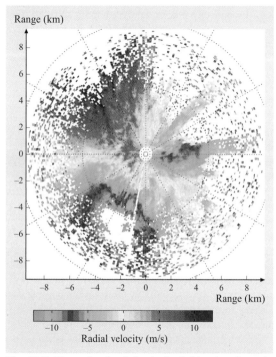

Fig. 17.14 Example of radial velocity fields taken from the VTMX field experiment in October 2000 in Salt Lake City. A full 360° PPI is shown

out into the atmosphere and returns are collected. Ambient aerosols scatter the laser light and a fraction of the light returns to the receiver/emitter. After a single pulse is emitted, the lidar waits for returns for a time corresponding to the outer length of its range. Modern coherent Doppler lidar may have high pulse-repetition frequencies (PRF). For example, Arizona State University's lidar emits 500 pulses per second. Because the returns are collected in discrete periods or *bins*, a distance can be associated with each time *bin* according to the distance that light can travel in the time measured. Therefore, the data is broken into range gates. For example, a modern coherent Doppler lidar may have 100 range gates, each 65 m long, for a total range of 6.5 km.

Doppler lidars have recently become available in commercial form, for example, Lockheed Martin Coherent Technologies' *WindTracer* (http://www.lockheedmartin.com/coherent). The principles of operation of this type of lidar are illustrated in Fig. 17.13. (The physical details described are for Arizona State University's Doppler lidar built by Lockheed Martin Coherent Technologies Inc.) Five hundred pulses of infrared laser light are sent into the atmosphere each second. The laser pulses are shaped like rods typically 50–80 m long and 10–30 cm in diameter, depending on the distance from the lidar. In addition to the range resolution based on discretely measuring the time of return for backscattered light, aerosols along the path of the beam impart a Doppler frequency shift proportional to velocity of the aerosols in the direction parallel to the laser beam. Therefore, both the range and radial velocity are measured. By scanning the laser beam, the two- and three-dimensional structure of the radial velocities and aerosol levels can be obtained using the RHI and PPI modes described in Sect. 17.1.

A typical example of a PPI radial velocity field is given in Fig. 17.14. The location of the lidar is at the center of the graphic. Blue colors represent velocities moving towards the lidar and orange–red colors indicate velocities away from the lidar. The lidar deployment was part of the DOE VTMX and urban dispersion experiment in October 2000 in Salt Lake City. Note that on the right side of the plot a drainage flow (darker blue) coming down from the surrounding Wasatch mountain range can be seen. This was a frequent occurrence in the evening hours during the VTMX experiment. The blue patterns to the left and down from the center and the brown/orange to the left and up from the center are evidence of a larger scale up the valley flow (from south to north).

Examples of RHI plots are shown in Figs. 17.15, 16. Both plots are from the JU2003 deployment of the Arizona State University lidar in Oklahoma City. These scans used the same look direction at different times and were chosen to show examples of the quiescent boundary layer (Fig. 17.15) and more-active conditions (Fig. 17.16).

While the Doppler frequencies measured with a coherent Doppler lidar are known with reasonably well-defined accuracy, estimating concentrations of aerosols from the strength of the return signal is fraught with difficulties. An important distinction should be made between the power signal received back at the lidar after a pulse is transmitted and the concentrations of aerosols. One can view the lidar equation (17.19) in order to understand this problem more carefully. This version of the equation is valid for single elastic backscatter of the laser light off aerosols and is appropriate for the coherent Doppler lidar.

$$P(r) = \frac{C}{r^2} \beta(r) e^{-2 \int_0^r k(r') dr'}, \quad (17.19)$$

where r is the distance of a range gate from the lidar, $P(r)$ is the power received from the range r, $\beta(r)$ is the volumetric backscatter coefficient of the atmosphere at range r, $k(r)$ is the corresponding extinction coefficient at range r, and C is the lidar constant (a function of the physical characteristics of the device).

Note that the power $P(r)$ is a function of both the extinction coefficient (i.e., the integrated degradation in the signal caused by absorption of light along the beam path) and the backscatter at the location r. The backscatter is dependent on the concentration of the aerosols; however, it is also dependent on the size distribution, shape, and composition of the aerosols. Therefore, there are significant difficulties associated with inversion of the lidar equation in order to obtain concentration. Furthermore, simplifications such as the use of Mie theory, which assumes spherically shaped particles, to relate backscatter to concentration may be poor approximations. Notice that in the lidar equation (17.19), both the backscatter and extinction are unknown. Therefore, to relate the power $P(r)$ to the backscatter requires some additional information or assumptions to be supplied. Some researchers have proposed various relationships between backscatter and extinction. Even if backscatter could be obtained, there are significant challenges in relating backscatter to concentration. Formally, one must know the composition, shape, and size distribution of

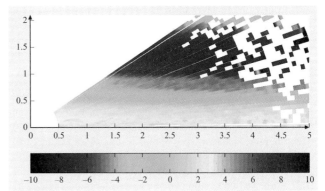

Fig. 17.15 RHI plot of winds in Oklahoma City during the JU2003 Experiment, 7/15/03 at 10:40 UTC. Note the rather quiescent boundary layer and compare with Fig. 17.16 below. Range away from the lidar is in kilometers directly below the plot, and the color bar shows how color relates to radial velocities

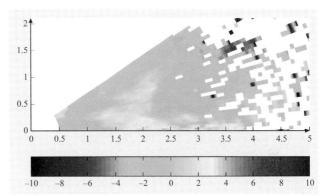

Fig. 17.16 RHI plot of winds in Oklahoma City during the JU2003 experiment, 7/14/03 at 20:20 UTC. Note the more-active boundary layer compared with Fig. 17.15. Range away from the lidar is in kilometers directly below the plot, and the color bar shows how color relates to radial velocities

aerosols at all the sampling locations. It may be possible to measure the characteristics of the aerosols at one or more points with in situ instrumentation, but the question of the validity of assuming the homogeneity of these characteristics over the lidar sampling area remains.

Backscatter lidars receive range-gated light returns similar to coherent Doppler lidar except a frequency shift cannot be measured. Backscatter lidars are relatively simple, typically monochromatic with only one transmitter and receiver channels. The same issues relating to difficulties inverting the equation describing the power returned to the receiver apply. It is usually implied that a backscatter lidar sends out a given frequency pulse and receives back that same frequency, i.e., light is elastically backscattered without a frequency shift (as opposed to Raman lidar).

Differential absorption lidar (DIAL) measurements are made with at least two different wavelengths of laser light. The most typical use of DIALs is to monitor trace gases. One wavelength is chosen so that it is absorbed maximally by the given trace gas, and the other is chosen so that it is not absorbed. The ratio of the power returned between the two types of pulses (that absorbed and not) is related to the amount of the target trace gas along the path. DIAL is also sometimes used to obtain atmospheric temperature. In this case, a target gas that does not vary significantly across the atmosphere is chosen, and the relational dependence between absorption and density, pressure, and temperature is utilized. The DIAL technique can be very sensitive and can be quite useful for monitoring pollution gases in urban areas. DIALs that detect ozone gas have been used to map the severity and extent of ozone events in major cities. It is particularly useful for this application to mount the DIAL on an aircraft.

17.3.2 Ceilometers

Laser ceilometers are pulsed, backscatter laser systems without scanning capability and without the ability to detect the Doppler shift from the returns. The returns are range-gated so that the distance (or ceiling) of the strong backscatter signal from clouds can be measured. Owing to the lack of a need for scanning motors

Fig. 17.17 The ceilometer deployed during the DOE VTMX urban dispersion experiment in Salt Lake City 2000. This small, lightweight device was provided by *Vaisala*, and is the CT25K model

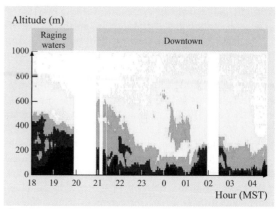

Fig. 17.18 The output of the ceilometer (backscatter signal) showing aerosol layers during DOE VTMX/Urban Dispersion Experiment in Salt Lake City 2000

and simpler signal-processing requirements, ceilometers are quite compact compared to coherent Doppler lidars (Fig. 17.17).

The data produced by ceilometers is $P(r)$, similar to the coherent Doppler lidar described above. Subject to the same caveats noted in the previous section, the output is frequently given in terms of aerosol backscatter. In this case, assumptions with regard to extinction are implied. The strength of ceilometers is their robust ability to accurately determine (within tens of meters typically) ceiling heights of cloud layers. Off-the-shelf ceilometers that are easy to operate and need not be attended for long periods can be purchased. In recent years, ceilometers have been used in larger experiments to locate layers of aerosols in the boundary layer and to help determine boundary layer heights. For example, the authors have experience with deployment of a *Vaisala* CT25k during the DOE VTMX/urban dispersion experiment in Salt Lake City [17.7]. Usual vertical resolutions are 10–30 m. During the VTMX experiment, the ceilometer showed evidence of nocturnal layers and occasional wavy behavior within these layers (Fig. 17.18).

17.3.3 Sodar

The general principle of sound detection and ranging (sodar) is that acoustic energy scatters back off temperature variations in the atmosphere. Acoustic pulses are sent out into the atmosphere, usually at frequencies of 1000–4500 Hz. For reference, the human ear can respond to a frequency range of 20–20 000 Hz. Consequently, sodar signals are frequently audible. The corresponding wavelengths of sodar frequency ranges are in the tens of centimeters. Since sound waves interact with inhomogeneities in the atmosphere similar to their own size, temperature variations in the range of 10–30 cm typically cause backscatter to occur. The backscatter is range-gated in the same manner as for coherent Doppler lidar, i.e., the time of return for the backscatter signal is translated into distance using a speed constant (the speed of sound for sodar vis-à-vis the speed of light for lidar). The Doppler shift of the returning signal is analyzed to provide the radial velocity of the sampling volume. Three directions or beams are required to recover the full velocity vector using simple geometrical arguments. This can be accomplished with three, physically separate, sound drivers, or with a phased array of many speakers (some arrays with more than 100). By precisely controlling the phase of the speakers, electronic steering can be accomplished.

Two main types of sodars exist: monostatic and bistatic. Monostatic sodars have the receiver and transmitter collocated, while in bistatic sodars there is a separation between the receiver and transmitter. Most commercial systems use the monostatic approach because of efficiencies of cost and construction associated with collocated receivers and transmitters. Sound will backscatter off of both mechanical and thermal inhomogeneities caused by atmospheric turbulence. However, only the thermal inhomogeneities backscatter at $180°$. Therefore monostatic systems rely on backscatter off thermal variations in the atmosphere. Bistatic sodars can use backscatter off both mechanical and thermal variations because the separation between the receiver and transmitter implies that smaller backscatter angles will still reach the receiver. In principle, this should result in more-robust returns, but the aforementioned efficiency of monostatic systems appears to dominate sodar design decisions.

Another important difference between monostatic and bistatic sodars is in the sampling volume. Monostatic devices have all receivers and transmitters in one location and point towards different volumes. Obviously, some degree of horizontal homogeneity of winds is implied when using monostatic sodars. This is due to the fact that different observation directions for sampling a (horizontally homogeneous) region of winds must be used in order to obtain different components of the velocity. A full velocity vector requires three observation directions. Bistatic sodars may have one, upward-looking, collocated receiver/transmitter, and two additional transmitters angled to overlap the central beam. Consequently, the assumption of homogeneity need only apply to the overlapping volume (three intersecting cones). Monostatic Doppler sodars usually perform better than phased-array systems.

Fig. 17.19 Sodar deployed by the U.S. Army Dugway Proving Grounds during the DOE VTMX/Urban Dispersion Experiment (Courtesy C. Biltoft)

Sodars typically have ranges from the tens of meters up to 1000 m with vertical resolutions of 10–50 m depending on the configuration and manufacturer. Minisodars often have a maximum vertical range of a few hundred meters. General characteristics of sodars that are important to keep in mind when comparing with other measurement devises are:

1. the relative low cost to operate and purchase
2. known accuracies and much previous experience
3. good near the ground, but somewhat limited performance in upper air measurements
4. sound contamination must be considered in choosing a location
5. signal in audible range, so people around the deployment may be annoyed by operating the sodar, and
6. ground clutter can cause difficulties by biasing the recovered wind toward zero, i.e., nearby trees or buildings may cause problems.

In the DOE VTMX/urban dispersion experiment (Fig. 17.19), for example, it was found that initially a sodar on a high-rise building in downtown Salt Lake City (surrounded by other buildings) was experiencing echoes from nearby buildings. The solution was to reorient the noisier side of the sodar away from neighboring high-rise buildings.

17.3.4 Radar Profilers

Radio detection and ranging (radar) profilers are similar to sodars in principle, except that they use radio rather than sound waves. Three to five near-vertical observation directions are used to obtain radial velocities using the Doppler shift of backscattered radio waves. The frequencies of radar profilers have been adjusted compared to other radars to have enhanced interaction with refractive-index regularities caused by turbulence. Refractive-index fluctuations are advected by the wind, and because they can be detected by the radar, an estimate of the wind moving along the radio beam can be obtained. Similar to Doppler lidar and sodar, different observation directions and the assumption of wind homogeneity over a measurement area can be used to produce velocity vectors with simple geometrical arguments. Figure 17.20 shows an NOAA 915 MHz radar profiler deployed in Salt Lake City for the DOE VTMX/urban dispersion experiment in 2000.

Several characteristics of radar profilers should be kept in mind relative to other types of profilers. Radar profilers typically have much higher vertical ranges than sodars. The NOAA 915 MHz radar in Fig. 17.20, for example, has a range from 150–4000 m and vertical resolutions in the range of 10–50 m. Like sodar, radar profilers are commercially available – with boundary or lower-atmospheric ranges up to 3–4 km – to tropospheric profilers that can measure up to 16 km vertically. Extra vertical range is usually achieved at the expense of vertical resolution. Secondly, the radio energy from radar is not as focused as the energy from a lidar. This allows increased backscatter from the sides of a much broader observation direction. Consequently, ground clutter can prevent useful data from being obtained near the ground. Radar profilers are typically deployed in tandem with sodars (Fig. 17.21). The latter performs well near the ground, whereas radar profilers can provide high vertical range. Third, it should be recognized that with higher vertical range comes much more extensive assumptions of homogeneity of winds associated with the reconstruction of radial velocities from three different observation directions into velocity vectors. At 10 km above the Earth's surface, obtaining wind vectors may imply an assumption of wind homogeneity on the 3 km scale. Lastly, radar technology is mature and these devices typically operate well in an unattended mode.

Fig. 17.20 The 915 MHz phased-array radar and Doppler sodar (*right*) deployed by NOAA's Air Resources Laboratory in Idaho Falls. These instruments were deployed as part of the DOE VTMX/urban dispersion experiment in Salt Lake City 2000

17.3.5 RASS

Early versions of radio acoustic sounding systems (RASS) were tested in the 1960s. The result of a RASS is the virtual temperature. Because there is a known relationship between the speed of sound and the virtual

Fig. 17.21 Example of co-located remote sensing instruments during the joint urban dispersion experiment 2003 in Oklahoma City (*from left*: sodar, ceilometer, radar). (Photo courtesy of Lawrence Livermore National Laboratory)

temperature in the atmosphere Sect. 17.3), measurements of the speed of sound at various levels in the atmosphere can be translated into virtual temperatures at each level. Collocated acoustic and radar profilers, as seen in Figs. 17.20, 21, can be used to obtain the speed of sound by backscattering radio energy off vertically propagating sound waves. Sound waves moving through the atmosphere change the refractive dielectric constant and may backscatter radio energy of an appropriate wavelength. By varying the acoustic frequency of a sodar, one can find the optimal frequency so that the radio waves from a given radar will backscatter strongly off the sodar's acoustic waves.

Time-of-arrival and Doppler shift information provide vertical range and speed of sound measurements. Early in the development of RASS technology, it was not easy to use the radar in a mode to acquire both profiles of radial velocities (i.e., as a radar profiler) and to track the vertical propagation of an acoustic wave. With more development, this has become possible and consequently RASS devices have access to both radial velocities and the acoustic propagation speeds at the same time. As a result, radial velocities can be used to improve the accuracy of the speed of sound measurements. The measured Doppler shift in a RASS is a projection onto the observation direction of a vector sum of the acoustic wave velocity vector and the wind vector. If the radial wind velocity is known, it can be subtracted off and a more-accurate virtual temperature can be obtained.

Some of the limitations of RASS are related to the advection of acoustic waves by the background wind and to distortion by turbulence. Therefore, the acoustic wave, which is intended to be located directly above the radar, can be off center due the background wind, and consequently, the vertical range may be comprised. However, beam-steering techniques that allow maximum ranges of up to 20 km to be obtained have been developed [17.8]. Depending on the frequency of the radar used, the appropriate corresponding acoustic signal may be attenuated in different ways. For example, ultrahigh frequency (UHF) RASS is more strongly limited vertically than very-high frequency (VHF) RASS. Another characteristic of RASS is that they are frequently very loud and need to be placed away from people. Since the output is virtual temperature, one would require additional information about moisture to obtain standard temperature. For this purpose, one might envision a collocated water-vapor DIAL lidar and a RASS.

17.4 Satellite Measurements

Technological breakthroughs in the 1990s have led to extraordinary advances of remote sensing capabilities for the 21st century and hence to the emergence of a new topical area known as satellite meteorology, which deals with the study of the Earth's atmosphere and oceans using data obtained from remote sensing devices onboard satellites. The enhanced remote probing capabilities are utilized by dozens of satellites continuously collecting data from multiple vantage points, allowing scientists from different continents to collect and share data. There are two main types of meteorological (weather) satellites, defined by their orbital configurations that essentially determine the surveillance characteristics: geostationary and polar-orbiting satellites. The former move in prograde orbits [GEO (Geostationary Earth Orbit); $\approx 35\,000$ km high] above the equator whereas the latter circle the Earth from pole to pole twice a day. Some examples of the geostationary satellites are the US geostationary operational environmental satellites (GOES) (two in service, GOES-12 and GOES-10, looking at the east and west coasts), the European METEOSAT, and the Japanese Geostationary Meteorological Satellite (GMS). Weather forecasting and presentations are based on these satellites. Operationally two polar-orbiting satellites are employed, one satellite passing south to north (ascending; e.g., the US Aqua satellite) and the other vice versa (descending; e.g. the US Terra satellite). These can be placed in sun-synchronous orbits, which place the ascending node at a constant solar time, permitting images to be taken at a given location received at the same time each day. These satellites are used for special purposes, such as monitoring of snow and ice sheets, but because of the viewing geometry their images need to be assembled as mosaics. Also of utility are the polar operational environmental satellites (POES) on low-Earth orbits (LEO ≈ 880 km) that travel from pole to pole on a shorter time scale (orbital time ≈ 1 h 42 min). These satellites collect data in a swathe as the Earth rotates beneath them, and yield high-resolution images and atmospheric profiles.

Flown on board these satellites are radiometers that measure either reflected or emitted electromagnetic radiation by the Earth's surface or atmospheric constituents in a series of discrete spectral bands. The measurements are transmitted back to the Earth, usually in the form of an image. Since each type of surface material on the Earth and each type of particle in the atmosphere has its own unique spectral signatures, the data can provide valuable information about the target. Two types of radiometers are used: imagers (which provide information on visible light from the sun reflected back to space by the Earth's surface or clouds), and sounders (which measure the amount of radiation emitted). Most commonly used sensors belong to the categories of visible, infrared (IR), water-vapor, and microwave sensors.

Visible sensors operate in a spectral range of $0.55-0.99\,\mu m$, and the amount of shortwave radiation detected therein is a function of the reflectivity of the target; for example, cloud tops and the Earth's surface. The images so obtained are similar to black-and-white photographs of the Earth. The brightest and whitest elements of these images indicate the most-reflective surfaces, such as thick clouds or fresh snow cover, whereas the darkest parts indicate the least reflective surfaces, such as the ocean surface (nearly black). Differences in the shading of clouds usually relate to cloud thickness. Land surfaces tend to appear gray. Some of the interesting features that may be visible in visual images include the sun glint or glitter (a large bright region due to the reflection of the sun off a large water surface), shadows (due to clouds when the sun angle is low or overshooting tops in tall thunderstorm clouds), terminators (shadow from the dark side of the earth), terrain features (lakes, rivers, mountains, snow cover or differences in reflectivity due to large forests), and snow cover (depending on the age and thickness of snow and ice).

IR sensors measure radiation at wavelengths of $10.3-12.6\,\mu m$, wherein cold objects appear white and hot surfaces black. For example, interplanetary space beyond the limb of the planetary disk appears white on IR images because of the extremely cold temperatures of space. This feature can be used to distinguish IR imagery from visible images. The IR sensor band is within a reasonably transparent *window* region of the atmosphere, and in cloud-free regions the long-wave radiation from the radiating Earth's surface can be detected by IR sensors both during the day and night. The Stefan–Boltzmann law determines the amount of IR radiation and its spectral quality, and thus the absolute temperature of the radiating objects (e.g., land, water, and clouds) can be measured remotely using IR sensors. Such imagery can also be used to distinguish low clouds (which are relatively warm and appear gray, as in the case of fog) from high clouds. Thick cold clouds, like the tops of thunderstorms, appear bright white. Methods have been developed to estimate the physical thickness of the clouds and the temperature and altitude of the cloud base [17.9] using satellite data and other information sources. A milky appearance over an otherwise cloud-free region may indicate an exceptionally cold air mass. Surface features can be detected in IR imagery by subtle shading contrasts resulting from differences in the surface temperatures. For example, cloud-free ocean regions appear more uniform because of more-uniform sea surface temperatures, whereas large differences in surface temperature over continents produce images with dark regions over hot deserts and lighter regions over mountainous terrain.

Water-vapor imagery is a valuable tool for weather analysis and forecasting, because it represents flow patterns of the upper troposphere. Water vapor is transparent to visible and IR wavelengths and thus are invisible to IR satellite imagery, but it is a very efficient absorber and emitter in the $6.5-6.9\,\mu m$ band (which is technically an IR band, but operating at a different band from the usual IR sensors). Thus radiometers sensitive to this band can be used to detect water vapor in the atmosphere (typically displaying water-vapor concentrations of $5-9\,km$). Black indicates low amounts of water vapor and milky white indicates high concentrations. Bright white regions correspond to cirrus clouds. In the middle latitude regions, zones with strong contrast in the amount of water vapor often indicate the presence of a jet stream.

Some satellites have a scanning microwave radiometer that senses the microwave (far infrared) radiation emitted by the Earth with wavelengths in the vicinity of $1.5\,cm$. In essence, a microwave detector can penetrate clouds and distinguish between ground and ice or snow surfaces. However, these sensors offers poor spatial resolution.

Spectral data about a target across a range of energy levels can be obtained using multispectral remote sensors, which is comprised of an array of sensors, each tuned to a particular channel or band of wavelengths. For example, weather satellite sensors are designed to probe the visible, near-infrared, and thermal infrared radiation from various surface materials and atmospheric components. The radiometers on land-use satellites such as Landsat and Spot are engineered to provide multispectral data that aids in measuring the spectral differences between various surface materials. Different land surface types such as concrete, asphalt, crops, meadow, forest, water, and desert all exhibit unique spectral signatures. An example of images taken from the advanced space-borne thermal emission and reflection radiome-

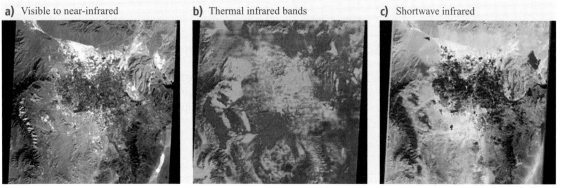

Fig. 17.22a–c Images obtained by ASTER at different wavelength bands (Courtesy of Will Stefanov, ASU). (**a**) Visible to near-infrared, (**b**) thermal infrared bands, (**c**) shortwave infrared

ter (ASTER) instrument onboard the Terra satellite in the visible to near-infrared (15 m/pixel resolution), short-wave infrared (30 m/pixel), and thermal infrared (90 m/pixel) is shown in Fig. 17.22. Using the visible to near-infrared band, it is possible to obtain information on major land-cover classes, vegetation health, soil properties, and soil contamination. Thermal infrared bands are used for evaluating surface energy balances, regional climate models, anthropogenic heat sources, and heat island development. Shortwave infrared bands are used for studies on urban surface materials, rooftop materials, energy use, fugitive dust production, and metal contamination.

The geosynchronous imaging Fourier-transform spectrometer (GIFTS) with ≈ 1600 channels in the thermal infrared spectral region, which is slated to be launched before 2010, is expected to provide new capabilities for measuring turbulent fluxes at the land surface.

Satellite-borne instruments are an invaluable tool for monitoring numerous environmental phenomena, such as fires, stack plumes, and air pollution. Wind vectors from satellite images are calculated using two techniques: tracking moisture gradients using (6.7 μm) water-vapor images, and detecting a cloud and tracking its motion through a sequence of images; the latter method is commonly applied to IR-window and visible images. Observations at a wavelength of approximately 4 μm are used to detect fog at night, discriminate between water clouds, ice clouds, and snow, detect fires and volcanic eruptions, and to meas-

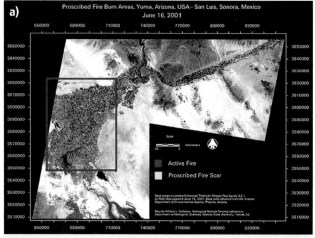

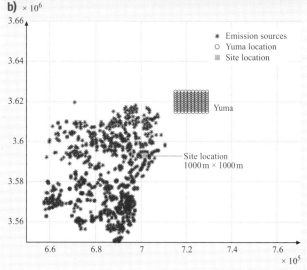

Fig. 17.23 (**a**) Fire scars in the Yuma/San Luis area, obtained by ETM+, and (**b**) the processed image of the boxed area using a pattern recognition technique

ure the temperature of the ocean during the night. The fires and smoldering fire scars appear as *hot spots*. This is evident from Fig. 17.23a, which was constructed using the reflectance in the visible through shortwave infrared portions of the spectrum of images acquired by the enhanced thematic mapper plus (ETM+) instrument orbiting onboard the Landsat 7 satellite (30 m/pixel ground resolution). In this case, the locations of agricultural burns in the Yuma-San Luis airshed surrounding the US–Mexico border were monitored to evaluate the plume pathways using dispersion models. Figure 17.23b shows extracted fire scars using a special algorithm. Once active fires and smoldering regions can be identified, respective emission factors can be used to model the dispersion of contaminants from agricultural fires. Such contaminants are believed to have health impacts on US–Mexico border communities.

References

17.1 Campbell Scientific Inc.: *CSAT3 Three Dimensional Sonic Anemometer, Instruction Manual* (Campbell, NorthLogan 1998)

17.2 T.R. Oke: *Boundary Layer Climates*, 2nd edn. (Methuen, New York 1987)

17.3 M. Princevac, R. Calhoun, D. Zajic, J.E. Holeman, R. Heap, H.J.S. Fernando: Arizona State University's contribution to the Joint Urban 2003 experiment: an overview. In: *Proc. Tenth Asian Congress of Fluid Mechanics* (University of Paradeniya, Paradeniya 2004)

17.4 J.C. Doran, J.D. Fast, J. Horel: The VTMX 2000 campaign, Bull. Am. Meteorol. Soc. **83**(4), 537–554 (2002)

17.5 P. Monti, H.J.S. Fernando, M. Princevac, W.C. Chan, T.A. Kowalewski, E.R. Pardyjak: Observations of flow and turbulence in the nocturnal boundary layer over a slope, J. Atmos. Sci. **59**(17), 2513–2534 (2002)

17.6 K.J. Allwine, J.H. Shinn, G.E. Streit, K.L. Clawson, M. Brown: Overview of URBAN 2000; a multiscale field study of dispersion through an urban environment, Bull. Am. Meteorol. Soc. **83**(4), 521–536 (2002)

17.7 R. Bowen, R. Calhoun, J. Rasanen: Ceilometer Boundary Layer Measurements During the DOE/VTMX/URBAN Field Experiment in Salt Lake City. In: *Fourth Symposium on Urban Environment* (Norfolk 2002)

17.8 N. Matuura, H. Masuda, S. Inuki, S. Kato, S. Fukao, T. Sato, T. Tsuda: Radio acoustic measurement of temperature profile in the troposphere and stratosphere, Nature **323**, 426–428 (1986)

17.9 G.R. Diak, W.L. Bland, J.R. Mecikalski: A note on first estimates of surface insolation from GOES-8 visible satellite data, Agricult. For. Meteorol. **82**, 219–226 (1993)

18. Oceanographic Measurements

This chapter describes various methods to measure the oceanographic variables that are dynamically significant. After a brief overview of the various motions and their temporal and spatial scales, the challenges of making measurements in a vast, inhospitable, and unforgiving environment are described. Then point measurements (pressure, temperature, salinity, sound speed, density, and velocity), Lagrangian measurements (floats and dye dispersion), and remote sensing methods (acoustic and electromagnetic) are described. Because many of the practical problems of oceanographic measurements are associated with the complete measurement system, examples of these and illustrative case studies of several experiments are given.

18.1 **Oceanography** 1179
    18.1.1 Oceanographic Processes 1179
    18.1.2 Challenges of the Environment 1181

18.2 **Point Measurements** 1182
    18.2.1 Pressure 1182
    18.2.2 Temperature 1182
    18.2.3 Salinity 1183
    18.2.4 Sound Speed 1185
    18.2.5 Density 1185
    18.2.6 Velocity 1186

18.3 **Lagrangian Techniques** 1188
    18.3.1 Surface Drifters 1188
    18.3.2 Deep Ocean Floats 1189
    18.3.3 Dispersion 1190

18.4 **Remote Sensing** 1192
    18.4.1 Acoustic Doppler Current Profilers (ADCP) 1192
    18.4.2 Acoustic Tomography 1195
    18.4.3 Acoustic Correlation Techniques 1199
    18.4.4 Electromagnetic Sensing Using the Geomagnetic Field 1199
    18.4.5 Surface Current Mapping Using Radar Backscatter 1201
    18.4.6 Satellites 1202

18.5 **Measurement Systems** 1203
    18.5.1 Expendable Sensors 1203
    18.5.2 HPIES 1203
    18.5.3 Moored Profiler 1204
    18.5.4 TriSoarus Towed Vehicle 1204
    18.5.5 Turbulence and Small-Scale Structure 1205
    18.5.6 Floats 1205
    18.5.7 Gliders 1207
    18.5.8 Autonomous Undersea Vehicles (AUVs) 1207
    18.5.9 LEO-15 1207

18.6 **Experiment Case Studies** 1208
    18.6.1 Hawaii Ocean Mixing Experiment (HOME) 1209
    18.6.2 Fluxes, Air–Sea Interaction, and Remote Sensing (FAIRS) 1212

References 1214

18.1 Oceanography

The oceans cover 70% of the Earth and profoundly affect the planet and our lives. The ocean is the flywheel of the climate, and if we hope to predict climate change, we must intimately understand the processes within and bounding the ocean, from the smallest scales of turbulent mixing to the largest scales of global ocean circulation (Fig. 18.1). The majority of the world's population lives nearby the sea and is directly affected by the marine weather, coastal ocean ecosystems, and associated hazards such as storm surges and tsunamis. Ocean ecosystems and the *environmental services* they provide, from food to carbon sequestration, are directly affected by ocean dynamics over the full range of scales. As governments and citizens increasingly appreciate the importance of knowing the present and future states of the ocean, more effort is being placed on measuring, observing, and understanding the ocean.

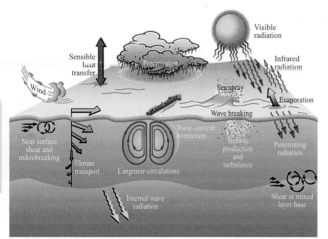

Fig. 18.1 Some of the many physical processes occurring in the ocean (courtesy of J. Doucette, Woods Hole Oceanographic Institution)

18.1.1 Oceanographic Processes

Ocean processes are active over a multitude of time and space scales (Fig. 18.2). Most are affected by the two major distinguishing characteristics of ocean flow: stratification and the Earth's rotation. Stratification can inhibit turbulence and, because it is determined both by temperature and salinity, double diffusion can occur. Rotation causes the Ekman spiral, the turning of the current vector with depth given a surface wind stress. The combination of the stratification and rotation is responsible for internal waves, ranging from nearly horizontal motions just above the inertial frequency to nearly vertical motions close to the buoyancy frequency. Planetary waves arise from the change in Coriolis force with latitude; examples are eastward-propagating equatorial Kelvin waves and westward-propagating Rossby waves.

Ocean flow is almost geostrophic, i. e., the Coriolis force almost balances the horizontal pressure gradient force. The kinematic geostrophic equations relating the horizontal velocity components u and v to the pressure gradient are:

$$u = -\left(\frac{1}{f\rho}\right)\left(\frac{\partial p}{\partial y}\right) \quad \text{and}$$
$$v = \left(\frac{1}{f\rho}\right)\left(\frac{\partial p}{\partial x}\right), \qquad (18.1)$$

where ρ is density and f is the Coriolis parameter ($2\Omega \sin\varphi$, where Ω is the Earth's angular velocity and φ is the latitude). These equations do not include dynamics. Much of the interesting fluid mechanics is in the *residual* ageostrophic, time-dependent flow. While often (but not necessarily) small, the latter have profound effects on the flow.

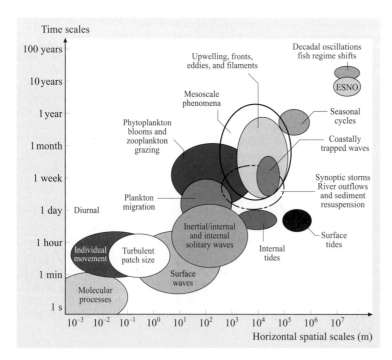

Fig. 18.2 Time and space scales of oceanic processes

Two major challenges confront the oceanographer making measurements: the limitations of the instrumentation, and the difficulty in obtaining adequate spatial and temporal sampling. This chapter primarily addresses the first, but it is necessary to note that sampling errors are the largest source of errors in oceanography. As one wise oceanographer has noted, the first century of modern oceanography was one of undersampling [18.1]. Another states: "The absence of evidence was taken as evidence of absence". The great difficulty of observing the ocean meant that when a phenomenon was not observed, it was assumed it was not present. The more one is able to observe the ocean, the more complexity and subtlety appears [18.2].

When planning, executing, and analyzing the results of an experiment, it is crucial to find means to overcome the problems of undersampling.

18.1.2 Challenges of the Environment

The dynamic air–sea interface, high pressures at depth, corrosion, and biofouling are just some of the challenges of making ocean measurements that face the experimental fluid mechanist. Much ocean work is still done from ships at the mercy of the weather; no matter how good the ship, operations are sometimes impossible. In fair weather, the ship is always moving, complicating the deployment and recovery of equipment. Making measurements on either side of the air–sea interface is faced with problems of spray, salt deposition, dynamic loading, mechanical fatigue, marine growth, and biofouling of sensors.

The combination of high pressures at depth and saltwater corrosion is challenging to overcome. The high pressures encountered deploying instrumentation at any significant depth dictate the geometry of instrument housings, which are typically either spheres or cylinders. Spheres are the *ideal* shape, if they can be made easily. Cylinders are often more convenient to fit circuit boards and to maintain. Squat-aspect-ratio cylinders with internal bracing are used occasionally for large internal components (such as commercial ethernet routers).

To minimize or reduce corrosion requires proper selection of materials. If possible, instrument housings and platforms must be made of materials that are either inherently noncorrosive (e.g., glass, plastics, titanium) or are protected against corrosion. Glass has been used for decades and, if used appropriately, is robust and inexpensive, even to full ocean depth; the 375 mm-diameter, 25 mm-wall-thickness glass sphere is a well-proven instrument housing. Various plastics are adequate for shallow depths less than 600 m. If metal housings are used, attention must be paid to the galvanic properties and galvanic connections with other metals; if current flows, the case will corrode and eventually fail. However, if one accepts a ≈ 0.2 mm/y loss of material, mild steel can be used. Carbon fiber housings are starting to be used for large pressure cases (e.g., for cameras) or for deep use requiring specific geometrical shapes (e.g., gliders). An extreme environment for instrument materials is deep hot vents with effluent temperature $> 350\,^\circ\text{C}$, pH ≈ 2, and precipitating chemicals.

O-rings are typically used to seal the endcaps of pressure cases or through-case penetrators/connectors; however, if a lifetime longer than 10 years is required, materials such as polyethylene should be considered. A note of caution: O-rings and connectors are the number one reason for instrument failure.

Yet another major problem is biofouling. In the shallow-water photic zone, marine life can grow very rapidly and completely cover an instrument; at depth, bacterial films (and eventually mats) are pervasive. There are no panaceas for this problem; some remedies include special paints and chemicals (already severely regulated in 2005), copper shutters, use of heat, ultraviolet light (e.g., a useful byproduct of some camera lights), conductive coatings on optical surfaces, and special coatings developed for biomedical use that resist biologically active molecules.

The absence of easily available power and communications severely hamper in situ measurements. Often instruments must use less than one watt of power (often milliwatts) and a preset sampling schedule. Fortunately, the electronics and data storage part of an instrument are becoming more efficient so the challenge is more in making the actual sensor/transducer more efficient. In an effort to remove the power and communication restriction, various national and international programs are being initiated with the goal of providing power and communications infrastructure. This will be accomplished with sea-floor cable systems with nodes (up to 10 kW and 1 Gb/s), or with deep-sea buoys with power and communication capability being supplied to the seafloor for surrounding sensor networks. Satellites are now providing accurate navigation [e.g., the global positioning system (GPS)] and modest communications (Iridium). Newer and better batteries, solar cells, fuel cells, and in situ power generation may provide solutions.

The greatest challenge is perhaps associated with cost and logistics. Day rates for US academic ships range from $10 000 for a 40 m coastal ship to $40 000 for an 85 m global-class ship with a remotely operated vehicle. These ships are relatively slow (12 kt) so transit times to remote locations and associated cost can be high. Ocean observing satellite costs are much higher, though usually the user does not pay the full cost. New paradigms centered on sustained long-term presence (e.g., cabled systems) will have their own logistical and cost challenges. These problems often outweigh those associated with actual measurements.

18.2 Point Measurements

Fundamental oceanographic point measurements are pressure, temperature, salinity, and sound speed; density is usually calculated as an empirical function of these (usually the first three). Pressure, density, and velocity are state variables in the dynamical equations. Various sensors used to make point measurements are described here. It may be useful for the reader to compare this section with Sect. 17.1, which discusses point measurements in the atmosphere.

18.2.1 Pressure

Pressure is a fundamental, and fortunately, routine measurement in oceanography. The SI unit of pressure is the pascal (Pa, $1\,N/m^2$); often the unit decibar is used because 1 dbar (10^4 Pa) is very nearly equivalent to the hydrostatic pressure of 1 m of water.

The measurement of pressure is used primarily for two purposes: for inferring depth and as a vertical coordinate for other measurements, and for use in the dynamical context. A strain gage is the simplest and least-expensive instrument for measuring pressure; accuracy is about $\pm 1\%$ of full scale.

A typical precision pressure transducer used for full-ocean-depth work is shown in Fig. 18.3. This transducer uses a very thin quartz crystal beam, electrically induced to vibrate at its lowest resonant mode. The oscillator is attached to a Bourdon tube that is open at one end to the ocean environment. As the pressure increases the Bourdon tube uncurls, stretching the quartz crystal and increasing the vibrational frequency. Conversely, a reduction in pressure causes the Bourdon tube to curl more tightly, thereby compressing the quartz crystal and lowering the vibrational frequency. These quartz frequency changes can be measured very precisely and the frequency changes are then converted into the corresponding changes in pressure. A small temperature dependence is accounted for using an independent thermistor. The sensor response is inherently rapid, and is limited in practice to how precise a measurement is desired and the frequency measurement method; they are often used in profiling systems (described below) sampling at 24 Hz. To measure pressure on the seafloor in the open ocean for the purpose of tsunami detection, averaging times greater than a minute or so are required. In this case, the transducer is sensitive to changes in pressure of less than one millimeter of water [18.3]. Long-term drift can be minimized and characterized by aging the device at high pressure for several months in the laboratory.

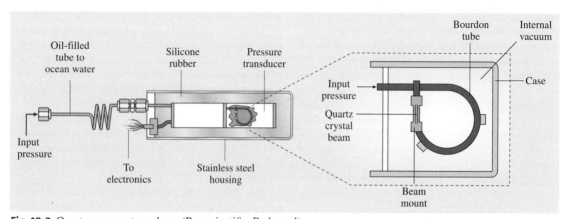

Fig. 18.3 Quartz pressure transducer (Paroscientific, Redmond)

18.2.2 Temperature

Temperature is now routinely measured with a high degree of accuracy using thermistors. One such temperature sensor consists of a glass-coated thermistor bead that is pressure-protected inside a 0.8 mm-diameter thin-walled stainless-steel tube (Sea-Bird Electronics, Bellevue). Exponentially related to temperature, the thermistor resistance is the controlling element in an optimized Wien bridge oscillator circuit. The resulting sensor frequency is inversely proportional to the square root of the thermistor resistance, and is in the range 2–6 kHz, corresponding to temperatures of $-5\,^\circ\text{C}$ to $+35\,^\circ\text{C}$. These sensors have a drift of less than $0.001\,^\circ\text{C}$ per six-month period and the time response is ± 0.010–0.065 s. Careful calibration procedures yield an overall transfer accuracy against a platinum reference thermometer within $0.0002\,^\circ\text{C}$.

18.2.3 Salinity

Salinity is a measure of the quantity of dissolved salts in ocean water. Seawater is a complex mixture of many ions with 11 major constituents making up more than 99.99% of all dissolved materials. These constituents are very well mixed through the world's oceans and are found in nearly constant proportions. This means that, although the salinity may vary from place to place, the major constituents are present in the same relative proportions. Salinity is now measured as practical salinity units (psu). The United Nations Educational, Scientific, and Cultural Organization (UNESCO) practical salinity scale of 1978 (PSS78) defines salinity in terms of an electrical conductivity ratio (relative to *standard* seawater), and so is dimensionless. Nominally, a salinity of 35 psu means that 35 kg of salt is dissolved per 1000 kg of seawater. Open-ocean salinities are generally 32–37 psu.

Conductivity (reciprocal resistivity) is an intrinsic property of seawater from which salinity and density may be derived. As conductivity is a strong function of temperature, effects due to the latter must be accounted for.

Oceanographic sensors cannot measure conductivity directly; instead they measure conductance, i. e., the voltage produced in response to the flow of a known electrical current. Conductivity is calculated from the conductance measured by the sensor using a scale factor or cell constant that depends on the ratio of length and cross-sectional area of the sampled water volume in which the electrical current actually flows. In certain types of cells, the length-to-area ratio corresponds very closely to the physical dimensions of the cell. However, in most cases, the sample volume cannot be directly measured to the necessary accuracy.

The determination of conductivity derives from:

$$R = \rho \frac{L}{A}, \tag{18.2}$$

where R is the resistance (1/conductance), ρ is the resistivity (1/conductivity), L is the length of sampled water volume, and A is the cross-sectional area of the sampled water volume. Note the equal importance of sample geometry and measured resistance in the determination of conductivity. (Conductance has units of siemens (S), equivalent to reciprocal ohms or volts/ampere. Conductivity has units of S/m.)

There are three main engineering problems: how to make an electrical connection to the water (difficult but not limiting), how to measure the resistance of the water to which the connection has been made (not difficult), and how to maintain a stable cell geometry (the limiting problem). The first problem is addressed by using electrodes or transformer coupling (inductive). In the latter case, one transform core is used to couple to the water, and the second measures the induced current. For the former, the electrodes must be stable and low resistance (e.g., platinum). The second problem is easily solved with reasonable resistance measuring circuits. The last

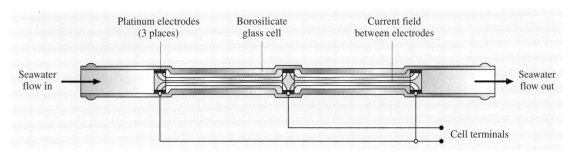

Fig. 18.4 A three-electrode conductivity cell (Sea-Bird Electronics, Bellevue)

problem, though, requires stable geometry of the sensing electric field.

In some sensors, the electric field is contained entirely within the geometry of the cell; in others, the field extends out into the free field and can be affected by nearby objects (e.g., the instrument case).

For field work the three-electrode conductivity cell has its sensing electric field entirely within the cell (Fig. 18.4). Because of its geometry (4 mm internal diameter) seawater must to be pumped through it; combined with a temperature sensor in the same fluid circuit and a fixed-flow-rate pump, the sensor and flow time constants are well known and the density can be accurately calculated without any *spiking*. Slowly leaching anti-foulant chemicals can easily be inserted into the fluid circuit to prevent any biological growth and attendant geometry changes. The measurement range is 0.0–7.0 S/m, the initial accuracy is 0.0003 S/m, the stability is 0.0003 S/m per month, and the resolution is 0.00004 S/m at 24 Hz.

A typical inductive cell is a simple open cylinder with two transformer windings around it. The electric field extends outside the cylinder significantly. An alternating voltage in the audio frequency range is applied to one transformer. This induces a voltage in the seawater circuit. The resulting current is directly proportional to the conductivity of the seawater and is measured by the second transformer. To overcome the often ill-defined geometry of such a cell, a new sensor has been developed (Fig. 18.5). It consists of two ceramic tubes mounted parallel to each other. The ends of each tube are fixed to ceramic boxes. Each tube has a pair of toroidal transformers fitted coaxially over them. The two cores act as charge pumps pulling the current lines from one end reservoir down the opposite tube. This situation is very similar in operation to guard electrodes used on the open ends of a tube electrode sensor. Because the cells are large and freely flushing, pumping is not required. Because the electric field is internal to the sensor (thus the name, nonexternal inductive conductivity sensor (NXIC)), anti-foulant chemical can be placed around the ends of the sensor.

To calibrate conductivity cells in situ to obtain the best possible salinity accuracy (currently $\approx \pm 0.002$ psu), water samples are collected and then compared with International Association for the Physical Sciences of the Oceans (IAPSO) standard seawater in the laboratory. An accepted commercial standard for in-laboratory measurement of water samples is the Guideline Instruments, Ltd. *Autosal*. The main feature of this instrument is a four-electrode conductivity cell immersed in a high-precision temperature-controlled water bath that allows a sequential measurement of standards and samples. Standard seawater is required to calibrate the instrument at regular intervals. The electric field is contained entirely within the cell and is therefore not affected by external factors.

A new, smaller and more-portable unit has been developed recently [18.4]. A dual inductive cell system removes the need for a highly stable bath temperature and gives direct reading of the conductivity ratio of the sample against standard seawater. A single master oscillator is used to drive the coil which transects both of the sample-containing rings of fluid, inducing the same proportion of flux in both measuring cells. Each ring has an independent pick-up coil, and the signals from these give the raw outputs. The dual cell removes the need for highly stable bath temperatures, and the cells are surrounded by a well-stirred oil bath to ensure thermal uniformity. Every sample is standard-

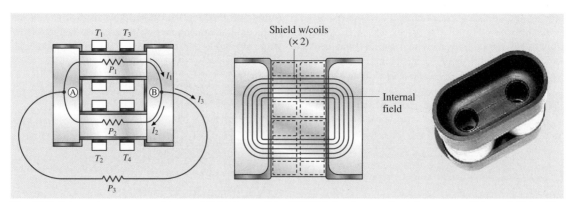

Fig. 18.5 Schematic, sensor, and internal field lines of the NXIC inductive conductivity sensor (Falmouth Scientific, Cataumet)

ized, and the standard itself is enclosed in glass to preserve the integrity of the reference measurement for weeks (and reduces cost). The instrument complements laboratory salinometers; it was developed as a collaboration between RBR Ltd., Guideline, and the Marine Hydrophysical Institute, Ukraine.

18.2.4 Sound Speed

The speed of sound in seawater is a strong function of temperature and pressure and a weak function of salinity. In some cases it is easier or more economical to measure temperature, pressure, and the speed of sound to then determine salinity. In other cases the speed of sound is desired directly or to infer temperature. In the acoustic measurement of water velocity, the speed of sound often enters into the equations.

A simple equation for the speed of sound in seawater is that of Mackenzie:

$$C(T, S, D) = 1448.96 + 4.591T - 5.304 \times 10^{-2} T^2 \\ + 2.374 \times 10^{-4} T^3 + 1.340(S - 35) \\ + 1.630 \times 10^{-2} D + 1.675 \times 10^{-7} D^2 \\ - 1.025 \times 10^{-2} T(S - 35) \\ - 7.139 \times 10^{-13} TD^3 \,, \quad (18.3)$$

where T is temperature in degrees Celsius, S is salinity in psu, D is depth in meters; the range of validity is temperature 2–30 °C, salinity 25–40 psu, and depth 0–8000 m [18.5].

If this is linearized one obtains

$$\delta C = 4.59 \delta T + 1.34 \delta S + 0.0163 \delta D \,, \quad (18.4)$$

where δ represents a small perturbation. The nominal sound speed in the ocean is 1500 m/s. For a typical ocean temperature change of 1 °C, the sound speed changes roughly 4 m/s; for a 1 psu change, the change in sound speed is roughly 1 m/s; for every 1000 m increase in water depth, the sound speed increases by 16 m/s. Typical salinity changes in the ocean are much smaller than 1 psu and in many underwater acoustics applications the effects of salinity can be ignored. For precise work, one should reference the UNESCO standard equation [18.6].

For *point* measurements of sound speed, time-of-flight sound velocimeters have a specified accuracy of ± 0.05 m/s, with a response time of ≈ 0.1 ms and a path length of 10–30 cm (Applied Microsystems, Ltd.). It is shown below that elements of acoustic tomography are in essence large sound velocimeters.

Acoustics plays a significant role in oceanographic measurements as will become evident below. A recent text presents the fundamentals of acoustical oceanography including studies of the near-surface ocean, bioacoustics, ocean dynamics, and the ocean bottom [18.7].

18.2.5 Density

Density is an empirical function of temperature, salinity, and pressure. Because much of the ocean circulation is governed by small density differences, density is a crucial measurement and attention must always be paid to precision and accuracy. Seawater density changes from roughly 1028 kg/m^3 at the surface to 1054 kg/m^3 at a depth of 6000 m. At the sea surface temperature ranges from -1.8 °C to ≈ 32 °C with corresponding densities of 1028 kg/m^3 and 1021 kg/m^3. While extremes of salinity are ≈ 34 psu to 37 psu, most ocean water is in the range 34.6–34.8 psu with a corresponding change in density of ≈ 0.17 kg/m^3.

Because salinity is difficult to measure directly in situ, electrical conductivity, sensitive to temperature and salinity, is measured instead; an independent measurement of temperature is relatively easy, permitting the salinity signal to be extracted. Density is commonly measured using a conductivity–temperature–

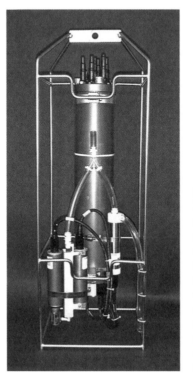

Fig. 18.6 Profiling conductivity–temperature–density (CTD) unit with pressure and pumped temperature, conductivity, and dissolved oxygen sensors (Sea-Bird Electronics, Bellvue)

depth/pressure (CTD) sensor package; an example is shown in (Fig. 18.6). This one is deployed on an electrical–mechanical wire from a ship in a profiling mode; other smaller ones are placed on moorings and mobile platforms (e.g., floats; see Sect. 18.3.2). This type of sensor package has been the major workhorse of the oceanographic community during the second half of the 20th century (reversing thermometers and water-sampling bottles were used before).

It is necessary to note that all three variables T, S, and P must be measured as near to simultaneously as possible, or with known time offsets, otherwise aphysical readings (density spiking) are obtained. In practice this means having the sensors as close together as possible and often pumping is used to create a predefined flowrate/time delay between sensors.

There is ongoing work to try to use optical methods to determine the index of refraction and thus the density directly. To date, though, the sensors either have excellent high-speed precision (but also respond to turbulent velocity signals, making it an optical *shear probe* as well) with little longer-term absolute accuracy, or are slow and have marginal absolute accuracy.

18.2.6 Velocity

There are three primary means of making point measurements of velocity: mechanical, acoustic, and electromagnetic. Mechanical current meters use rotors, vanes, and propellers, with compass cards to determine geomagnetic reference. They suffer from less than adequate response (e.g., stalling at low speed), low reliability, and high operating and maintenance costs. Because these are being phased out of use, the focus here is on the latter two methods.

The proceedings of the most recent quadrennial Institute of Electrical and Electronics Engineers (IEEE) Oceanic Engineering Society conference on current measurement technology [18.8] is a recommended source of the latest information on this and related topics. An earlier review of the subject is given by *Collar* [18.9].

Acoustic Current Meters

Acoustic current meters are of two basic types: time of flight and Doppler (Fig. 18.7). (There are many conceptual similarities between these instruments and their atmospheric equivalents, e.g., sonic anemometers; see Chap. 17.) Acoustic time-of-flight/travel-time (ATT) systems are based on the simple notion that sound travels faster with a flow than against; the differential travel time is a direct measurement of the flow velocity. At the same time, the average of the two travel times in each direction gives the sound speed. The general expression for travel time is the integral along a ray path Γ with arc length s:

$$t = \int_\Gamma \left(\frac{ds}{C + u\tau} \right) , \quad (18.5)$$

where C is the sound speed, and $u\tau$ is the velocity component projected onto the ray path.

Assuming that the current speed u is much less than a nominal background sound speed C_0 and a straight path of length R along the flow direction, one obtains for the differential and average travel times, respectively:

$$\delta t = -2u \left(\frac{R}{C_0^2} \right) \quad \text{and} \quad t = T_0 - \left(\frac{R}{C_0^2} \right) \delta C , \quad (18.6)$$

where T_0 is the nominal travel time along the path and δC is a sound speed (temperature) perturbation. For path lengths on the order 10 cm, a 1 cm/s resolution implies a 1 ns differential travel-time resolution. This can be achieved in three possible ways: direct measurement of broadband pulse travel time, phase difference, or frequency difference techniques. A representative ATT is shown in (Fig. 18.7). The supporting cage is meant for inline mounting in an oceanographic mooring. The four sensor heads are at the bottom. The specified accuracy is 2% of the reading or ±1 cm/s up to 3 m/s, with a resolution of 0.1 mm/s. Absolute heading and tilt accuracy

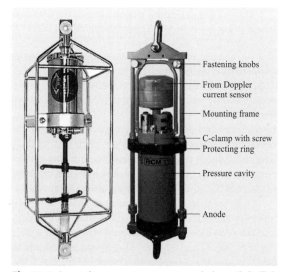

Fig. 18.7 Acoustic current meters: travel time (*left*, Falmouth Scientific 3-D ACM, Cataumet) and Doppler (*right*, Aanderaa RCM-11, Bergen)

are 2°, measured with a three-axis flux gate system and two accelerometers. The maximum sample rate is 2 Hz. A thermistor measures water temperature to obtain the sound speed.

The need for a mechanical supporting structure presents two problems. To prevent vibration from contaminating the velocity measurement, the structure must be rigid. To prevent structure-generated flow perturbations from contaminating the result, the redundant paths can be selectively sampled to include only those on the *upstream* side.

In custom applications these techniques have been extended to 3 m path length, operating for five years under a surface buoy without a problem. The latter highlights an advantage of acoustic techniques in ocean observations – the robust nature of the hardware and insensitivity to calibration problems.

The acoustic Doppler current meter transmits an acoustic pulse and measures the frequency shift of energy reflected or scattered off particles in the water. It does not matter if the particle impedance is higher (as for sediment particles) or lower (as for air or gases); both will result in a reflection.

The water velocity is determined from the measured Doppler shift f_D, the frequency difference between the transmitted and received signals using the equation

$$u = \frac{1}{2} c \frac{f_D}{f} . \qquad (18.7)$$

To obtain a velocity vector, data from several beams are combined. The beam geometry varies, some diverge and assume locally homogeneous flow while others cross to measure in a small volume, much like a laser Doppler velocimeter.

Because the frequencies used are typically quite high in these short-range applications (e.g., 2 MHz) and the Doppler shifts are small (30 Hz for 1 cm/s) there is no problem with resolving low velocities as they are small perturbations about the carrier. When particle scattering strength is low, the signal-to-noise ratio will be low, but it will be unbiased. Ambient acoustic noise at these frequencies is due to thermal fluctuations only and is white. The conversion of the measured Doppler shift to velocity depends only on the speed of sound, which is adequately known with an independent measurement sensor (transmit and receive circuits share a common oscillator, so errors in drift cancel). Again, because there are no moving parts, the instrument is inherently reliable. If in a moored configuration, the motion of the mooring could influence the interpretation of the measurements.

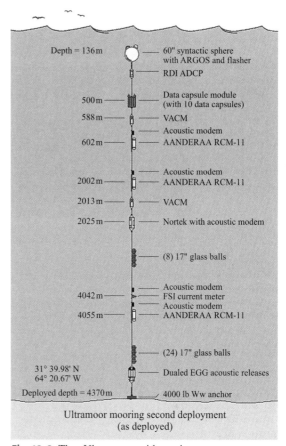

Fig. 18.8 The Ultramoor with various current meters, acoustic modems, and data capsules for data return. RDI (RD Instruments, Inc.), VACM (vector averaging current meter)

In an example of an oceanographic mooring with many current meters (Fig. 18.8), the mooring reaches from the seafloor 4370 m deep to close to the surface (a subsurface mooring avoids wave stresses and associated motion). This particular mooring is part of the Ultramoor program at the Woods Hole Oceanographic Institution, specifically meant to test data capsules and acoustic modems [18.10].

Electromagnetic Current Meters

A conducting fluid moving through a magnetic field generates an electric field that is dependent on the fluid velocity – a statement of Faraday's law of magnetic induction. In current meters, the magnetic field is generated by an alternating-current (AC) excitation in the instrumentation. In larger-scale applications, ocean flow

Fig. 18.9 An electromagnetic current meter in a frame for bottom use (InterOcean S-4, San Drego)

through the Earth's magnetic field generates a voltage often measured using submarine cables (see below). For the ocean case, the voltage E in the water is the product of the water flow velocity u, the magnitude of the magnetic field B, and the length of the conducting path L:

$$E = uBL \ . \tag{18.8}$$

In a common electromagnetic current meter (Fig. 18.9) the magnetic field intensity is generated by a circular coil, internal to the instrument, driven by a precisely regulated alternating current. Two orthogonal pairs of titanium electrodes located symmetrically on the equatorial ring of the instrument sense the respective voltages and the separation defines the path; data from a flux gate compass is used to calculate the current vector with respect to north.

The flow field around the sensor heads is of critical importance. The grooved surface of the housing produces stable hydrodynamic characteristics by inducing a fully turbulent boundary layer with uniform and stable vortex shedding through the range of use. This ensures a stable and linear calibration. (Recall the sensitivity of electric field lines to geometry as discussed in Sect. 18.2.3). The specifications are: range 0–350 cm/s, accuracy 2% of reading or ±1 cm/s, sampling rate 5 Hz, resolution 0.04 cm/s, noise 0.05 cm/s root-mean-square (rms) for 10 s averages, and direction ±2°. Because of the AC excitation there is no zero-velocity threshold. Biofouling can affect the calibration, though. Such sensors do not lend themselves to turbulence measurements because of the weighted volumetric averaging.

18.3 Lagrangian Techniques

The most basic Lagrangian data is position as a function of time, from which velocity can be calculated. This is true whether a parcel of dye is being followed or a float is following a flow. In the latter case, the sensors the float platform carries (typically) contribute moving point measurements. The interpretation of the data can be complicated by this space–time mix of information.

18.3.1 Surface Drifters

The design of surface drifters is driven by the need to faithfully follow the flow in a known and repeatable way, and at the same time is influenced by considerations of cost, ease of storage and deployment (around the world in large numbers, from ships and planes), robustness, and longevity. Systematic evaluations of the different surface drifter designs of the 1980s concluded that the key to accurate water following was to increase the ratio of drogue to float size. Much iterative work was done to determine optimal designs, and there is some convergence.

For the open ocean the surface velocity programme (SVP) drifter has a spherical plastic float, a wire tether, a holey sock drogue (a 4 m-long canvas cylinder centered at 15 m depth), and battery power sufficient for two years of operation (Fig. 18.10). SVP drifters can carry sensors that measure temperature and salinity, wind direction, ambient sound (to calculate wind speed and rain rate), upwelling light, and atmospheric pressure; most though only carry sea surface temperature and atmospheric pressure sensors. Advanced research and global observation satellite (ARGOS) receptions (for data transfer and positioning) vary from 6 to 14 per day, depending on latitude and storm conditions. About 7500 SVP drifters have been deployed since 1988.

A drifter originally developed for the coastal ocean dynamics experiment in the 1980s has had greatest use in coastal and marginal sea deployments. It can be deployed from a small plane, helicopter, or boat. Location is by GPS or ARGOS. Data transmission is to ARGOS or a local paging system on land. Two types are manufactured, one with a spherical float and a radar corner reflector-style drogue, the other with a cruciform drogue

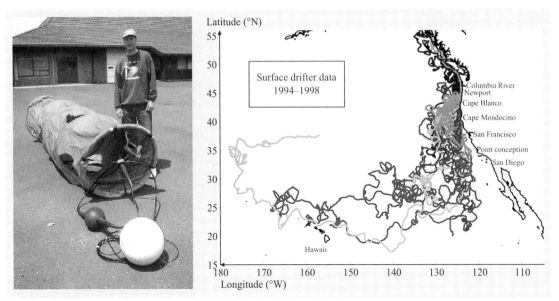

Fig. 18.10 (*left*) An SVP drifter. (right) Paths of surface drifters released off the Oregon coast in spring and early summer (red, yellow, and brown), late summer and fall (green and light blue), and winter (dark blue). (courtesy of J. Barth)

with four small floats. Temperature and salinity sensors have been placed on these drifters. A power recharging option is available, making them redeployable after recovery. About 2500 drifters of this type have been deployed in the past 10 years.

18.3.2 Deep Ocean Floats

The use of neutrally buoyant and profiling floats has been a cornerstone of oceanography for the last half century, and this will continue. In the late 1950s swallow

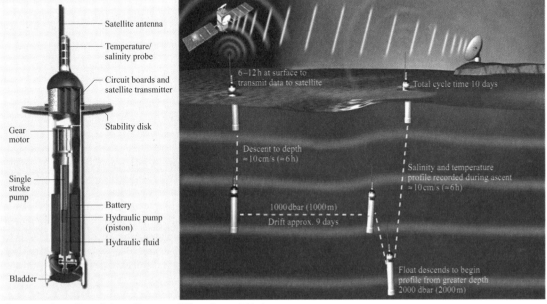

Fig. 18.11 Argo float (16 cm diameter, 130 cm long, 26 kg)

floats misbehaved by moving at *high* speed (10 cm/s) in seemingly random directions rather than slowly in a straight line (baffling the crew on the sailboat trying to track them acoustically), thus awakening oceanography to the presence of mesoscale ocean *weather*.

The key to constructing such floats is to ensure that the float's density, and especially the floats compressibility, accurately matches that of the seawater in which they float. Doing this not only lets them float at the desired depths and faithfully follow fluid accelerations, but also minimizes power needed to profile. Early floats were aluminum scaffolding tubing; in some cases, special hulls of glass (a large *test tube*) or numerically machined, ribbed cylinders were used. In any case, careful ballasting of each individual float (grams out of many kilograms) is required.

There are two basic modes of operation for freely drifting autonomous floats: acoustically tracked and profiling (and combinations). Profiling floats such as Argo (Fig. 18.11) typically come to the surface once every 10 days to telemeter temperature and salinity data and to obtain a satellite navigation fix. During the time submerged the float is typically untracked. These floats change and adjust their buoyancy using a piston to move oil in and out of a bladder exposed to the seawater. Their lifetime is about four years or 150 profiles. The goal of the international Argo program is to seed the ocean with 3000 floats to obtain on average one float every 300 km (750 floats/y); at the end of 2004 there were 1500 floats in operation around the world. This is a major effort being undertaken by the international community. While the initial emphasis was on temperature and salinity, additional sensors are being added for dissolved oxygen, horizontal electric field for velocity, and hydrophones for wind and rain (and seismic and marine mammal sound), to name a few.

A widely used and accurate method to study the ocean in motion employs the deep sound, or sound fixing and ranging (SOFAR), channel to track neutrally buoyant drifters over large horizontal distances, on the order of 10×10^3 km. Three or more anchored sound sources provide an acoustic navigation system whereby precisely timed acoustic signals [in this case, an 80 s-long 1.5 Hz-wide frequency-modulated (FM) sweep at 260 Hz] spread out radially from each of the sources. Given the arrival times at a drifting receiver equipped with a synchronous clock, the receiver's position can be determined to an accuracy of a few kilometers or better. Since the mid 1980s floats have been deployed in many different projects worldwide to study ocean currents.

These floats, also known as RAFOS floats (SOFAR spelled backwards), can be deployed to drift at any depth, although greater acoustic ranges are possible if both the sound sources and the floats are not too far from the sound channel axis. The floats record the arrival times in their microprocessor memory. At the end of each float's mission underwater, it surfaces and telemeters all its data back to shore. Typical missions last from months to several years. Special float designs have been developed whereby a float surfaces briefly to telemeter its data before returning to depth to continue its underwater mission. To make the floats mimic fluid motion accurately, they can be configured to drift with the water both horizontally and vertically; these are known as isopycnal floats. In regions of high shear, profiling floats can give erroneous estimates of velocity if they are based solely on surface fixes. Similarly, when trying to measure abyssal currents, estimates can be severely compromised if the float comes to the surface. In these cases it is advantageous to remain submerged. Some systems are *turned around* using fixed receivers and drifting sources (SOFAR floats).

One illustrative application of navigated floats was in the Atlantic climate change experiment (ACCE, Fig. 18.12) where the float provided in situ observations of dissolved oxygen. Here the sensors were only briefly activated, once daily, and thus operated over the entire lifetime of the floats. From a Lagrangian viewpoint, interpreting the variability of dissolved oxygen presents the difficulty of separating the biological from the physical processes. However, there are times when the nature of the variability clearly indicates that one process dominates the other.

18.3.3 Dispersion

Tracer release experiments use methods to estimate directly the dispersion and mixing across density interfaces (diapycnal diffusivity) in the ocean. Although logistically challenging, the method is straightforward: a passive tracer or dye is injected at depth (on a density surface), typically from a vessel, and an initial survey is done to map the tracer distribution. The region is resampled at discrete time intervals, and the goal for each survey is to map a large enough region to account for all of the tracer. The measurements are used to estimate the time- and space-averaged diapycnal diffusivity for the region. (Dispersion in the atmosphere is discussed in Sect. 17.2.

For deep-ocean tracer release experiments lasting several years, sulfur hexafluoride (SF_6) has been

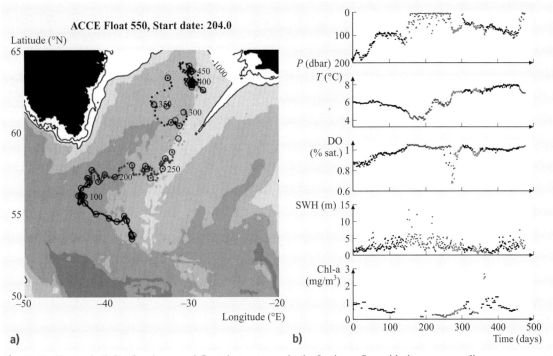

Fig. 18.12 The track (*left*) of an isopycnal float that outcrops in the Irminger Sea with the corresponding measurements of pressure, temperature, and dissolved oxygen (*right*) (courtesy of T. Rossby)

the tracer of choice [18.11–13]. Its background concentration in the marine environment is low, it is chemically conserved, and it is harmless to marine organisms [18.14]. However, SF_6 is becoming a useful anthropogenic tracer in studies of the pathways from the atmosphere into the ocean interior, and therefore replacement tracers are being sought. Trifluoromethyl sulfur pentafluoride (CF_3SF_5) has similar properties to SF_6 [18.15], but its background in the atmosphere is too low for it to be useful as a transient tracer; it may be a viable alternative to SF_6. The method of injection is the same for both gases; the gas is forced through atomizing orifices so that droplets dissolve before sinking about one meter below the release density surface. The gas is pumped continuously to an injector sled towed behind a slowly moving ship. The sled tracks a density surface using automatic feedback control from a CTD sensor mounted on the sled. Subsequent sampling uses a towed vertical array of water samplers and temperature–pressure sensors; samples are analyzed using a gas chromatograph equipped with an electron capture detector. The minimum detectable amount for both gases is about 0.01 femtomoles. The achievable sampling rate using shipboard gas chromatograph is of order 200 samples/d.

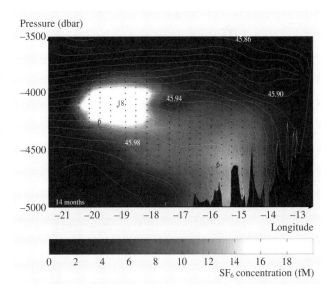

Fig. 18.13 SF_6 concentration 14 months after release in the Brazil Basin [18.16]

The Brazil basin tracer release experiment provides an example of the method [18.16]. (Fig. 18.13) shows the concentration of SF_6 14 months after release in early 1996. This data, when coupled with independent measurements of the dissipation rate of turbulent kinetic energy, support the hypothesis that the tides running over the rough bathymetry generate internal waves that propagate upwards into the water column where the associated shear intensifies, inducing turbulent mixing events.

Fluorescent dyes are used for experiments where higher sampling rates, on the order of 5 Hz, are required to resolve processes that have time scales of hours to days [18.17]. Although the minimum detectable level in this mode is about a million times higher than for SF_6, these dyes are superior for small volumes (about one cubic kilometer) and short durations (five days). Fluorescein, Rhodamine-WT, and Rhodamine-B are examples of commonly used dyes. Rhodamine-B is a suspected carcinogen; the other two are both nontoxic. Fluorescein is preferable to Rhodamine-WT in that it is less expensive and the background levels are much lower, but it decays rapidly in sunlight whereas WT is stable.

18.4 Remote Sensing

Remote sensors measure wave signals that are generated or modified by ocean processes at some distance. For the ocean case there are two classes: in situ and external. In the first, acoustics and electromagnetics are usually implied, often simply extending the range of what were called point measurements. The latter usually refers to the electromagnetic sensing of the sea surface, whether using radars from land or other sensors probing the ocean surface from space satellites. In atmospheric measurements, sonic detection and ranging (SODAR), light distance and ranging (LIDAR), and atmospheric GPS tomography are conceptually similar remote sensing methods (Sect. 17.3).

18.4.1 Acoustic Doppler Current Profilers (ADCP)

Acoustic Doppler current profilers (ADCPs) are multibeam Doppler SONARs used to measure ocean currents. They have several advantages over point current meters: they measure a profile of currents, have no moving parts (and so are not susceptible to stalling or mechanical fouling), and are nonintrusive.

ADCPs have come into widespread use since the technology was first introduced [18.18, 19] and different types of profilers have evolved for different applications. The range, resolution, and accuracy of the measurements depend upon the frequency of the ADCP and on sampling and environmental factors; some of the main considerations are discussed below, although it is impossible to be comprehensive here. ADCPs are used from a variety of platforms: current meter moorings, ships (through the hull, over the side, or towed), autonomous undersea vehicles (AUVs), in bottom-mounted configurations, or lowered through the water on a wire, to name a few. Because the ADCP measures only the velocity of the water relative to the instrument, the velocity of the instrument relative to the Earth must also be measured accurately. Hence it is important to include the measurement platform as part of the overall ADCP system.

An ADCP transmits a pulse of high-frequency sound (a *ping*) along multiple acoustic beams and estimates the Doppler shift in frequency of the sound reflected from scatterers in the water over successive time intervals (range gates) corresponding to the distance (range) from

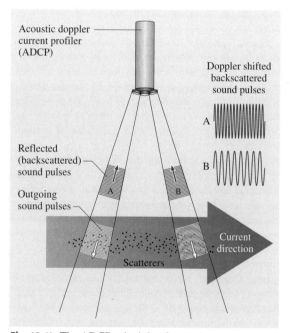

Fig. 18.14 The ADCP principle of operation

the transducer to the scatterers (Fig. 18.14). At the frequencies commonly used (from $\approx 10\,\text{kHz}$ to $\approx 10\,\text{MHz}$), the scatterers are primarily plankton and small particles that are assumed to be passively advected by the currents. The Doppler shift in frequency is proportional to the along-beam velocity of the scatterers relative to the transducer (18.7),

$$f_\text{D} = \frac{2 f_\text{t} u_\text{b}}{C}, \qquad (18.9)$$

where f_D is the Doppler shift, f_t is the transmit frequency, u_b is the velocity component along the beam, and C is the speed of sound in seawater. The beam geometry and the transducer orientation are used to convert beam velocities into Earth components: east, north, and vertical. A minimum of three beams is required to determine a three-dimensional velocity vector. In practice, a four-beam configuration is common, spaced at 90° in azimuth and at a fixed angle with respect to vertical. Figure 18.14 illustrates the principle for one beam pair. A current flowing from left to right will have a component towards the transducer along beam A and away from the transducer along beam B. If the current is uniform over the distance spanned by the beams, the Doppler shifts of the transmit frequency will be of equal size but opposite sign. Two beams determine the horizontal and vertical velocity in the plane defined by the beams. A second pair is used to determine the orthogonal component of horizontal velocity and a redundant estimate of vertical velocity.

A basic assumption of the technique is that the scatterers are present and that they are passively advected by the currents. In the ocean the distribution of scatterers varies both geographically and with depth. Scatterers are often not uniformly distributed with depth; they are layered, alternating from high concentrations to low, and there is typically a dramatic decrease at depths exceeding about 1000 m. Scattering layers can introduce bias in ADCP current measurements in a number of ways. First, they may swim, and a prime example is diel vertical migration at sunrise and sunset [18.20]. Second, very strong scattering layers serve as a hard reflector, like the bottom, and can bias the current measurement. Detecting bias and editing it out is one of the primary data quality screenings that is applied in post processing.

ADCPs can be categorized by type of acoustic pulse. Pulses can be coded (broadband) or uncoded (narrow band). A narrow-band ADCP transmits pulses that are gated sinusoids and measures the Doppler frequency shift of the echoes. Narrow-band ADCP bandwidth and accuracy are inversely proportional to the pulse duration. A broadband ADCP transmits pulses modulated with pseudorandom codes and estimates the Doppler phase shift of the echoes in the range ±180° [18.21]. Broadband bandwidth and accuracy are directly proportional to code length. The ambiguity velocity is the largest velocity that can be measured by a broadband ADCP, as phase wraps of 360° cannot be distinguished. Ambiguity velocity is an important issue for systems mounted on platforms that move at great speeds relative to ocean currents, e.g., vessel mounted, as the ADCP needs to resolve the velocity of the water relative to the ship (5–6 m/s) even though the ocean currents of interest rarely exceed 1 m/s. Increasing the maximum ambiguity velocity requires decreasing the code length and therefore bandwidth and accuracy. Broader bandwidth also reduces the maximum range and increases the chance of acoustic interference from other devices. Hence, although a broadband ADCP has the potential for better short-term accuracy than a narrowband ADCP, there are other tradeoffs that need to be considered for any particular application.

The basic transducer styles are monolithic and phased array, and either type can be driven with broadband or narrow-band pulses. A monolithic piston transducer consists of a single piezoelectric ceramic disk that generates one beam. A phased-array transducer consists of an array of small ceramics wired such that alternate columns or rows can be driven with different time delays, generating two or four beams simultaneously. Transducer style is often an issue when seeking to maximize the velocity profiling range. Lower-frequency SONARs get better range because the attenuation and absorption of sound in seawater is lower at lower frequency. Because the angular beam width varies inversely with the product of the frequency and the transducer width, a lower frequency corresponds to a larger transducer size for a fixed beam width. Large transducers are more expensive to manufacture and more cumbersome to deploy. At low frequency, a phased-array transducer can be more practical than a multibeam monolithic style transducer because it is smaller, generating multiple beams from a single transducer. It is more complicated and expensive to manufacture, however, and variations among the ceramic elements will degrade the beam pattern.

The limiting factor affecting the short-term accuracy of ADCPs is the theoretical bound on the random error [18.22], but the overall accuracy of the current measurement is determined by the sum of all random and bias errors of the ADCP system, i.e., ADCP plus platform motion. To use a specific example, a commonly used commercial ADCP is a 150 kHz narrow-band ADCP manufactured by RD Instruments. At 8 m vertical resolution, the single-ping standard deviation of horizontal current from this instrument is approximately 11 cm/s. Vector-averaging over 600 s intervals reduces the standard deviation to less than 1 cm/s [18.23]. Hence, in the absence of any system biases, the predicted error for averaged currents using this instrument mounted on a rigid platform is quite small compared to the variance of the currents being observed. When used to profile currents from a ship underway, however, the errors are dominated by the accuracy of determining the ship's motion. The ship-relative currents measured by the ADCP must be rotated into Earth coordinates using a measurement of ship's heading, and the ship's velocity over the Earth must be subtracted. Because the ocean current is a small residual between the rotated ADCP measurement and the ship speed, the calculation is particularly sensitive to errors in heading and/or transducer alignment. A heading bias of one degree for a ship steaming at 5 m/s can introduce an error of 9 cm/s in the cross-track component of horizontal velocity. Heading bias errors, formerly the most limiting error in shipboard profiling, have been largely eliminated by the use of GPS attitude measurements [18.24, 25]. With excellent GPS position and heading measurements, the accuracy of ocean currents from shipboard ADCP measurements can achieve an accuracy of 2 cm/s for 600 s averages [18.26].

The ADCP has numerous advantages over point current meters, and it is used routinely on moorings and in bottom-mounted configurations. Direction and orientation are determined from an internal magnetic compass and pendulum tilt sensors. In addition to profiling currents, the ADCP provides a remote sensing measure of the scattering layer, useful in estimating biomass [18.28]. High-frequency ADCPs can be used to estimate turbulence [18.29, 30].

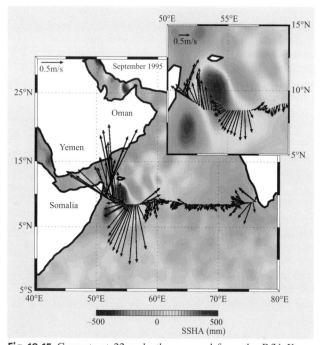

Fig. 18.15 Currents at 22 m depth measured from the R/V *Knorr* during the 1995 Arabian Sea southwest monsoon, overlaid on sea surface height anomalies from altimetry [18.27]. The Great Whirl is the large anticyclonic eddy centered at 55°E (positive sea surface height anomaly)

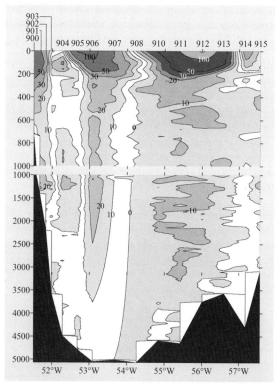

Fig. 18.16 Cross track component of current measured from the R/V *Knorr* during the 1995 Arabian Sea southwest monsoon. The section merges current observations in the upper 400 m made from continuous underway shipboard ADCP and lowered ADCP observations made at stations located at positions indicated by numbers along top axis. The Great Whirl is seen to have significant current strength to about 2000 m depth

Shipboard profiling was one of the earliest uses of the ADCP [18.31, 32]. The advent of GPS with its greatly improved accuracy of position and heading measurements helped to establish the shipboard ADCP as a standard survey tool. Horizontal resolution is 3 km (for typical ship speeds); range is limited by the frequency. A 150 kHz ADCP (e.g., (Fig. 18.15)) profiles to 400 m depth in good scattering conditions.

The lowered ADCP is an adaptation to extend the profiling range to full ocean depth [18.33, 34]. A self-contained ADCP such as used on a mooring is mounted on a package usually containing a CTD and lowered from the surface to the ocean floor. A set of overlapping profiles of horizontal currents are obtained, each with a range of about 100–200 m that are first differenced, averaged, and gridded to form a full depth profile of horizontal velocity shear. The shear or relative velocity profile is referenced in a similar manner to the shipboard ADCP, using accurate GPS. Figure 18.16 shows a vertical section of cross-track currents measured from R/V *Knorr* along the track shown in (Fig. 18.15). The upper 400 m was measured by shipboard ADCP; at greater depths the currents were measured by lowered ADCP observations made at the fixed station locations indicated along the top.

18.4.2 Acoustic Tomography

Sound travels faster in warm than in cold water. By measuring the travel time of sound over a known path, the sound speed and thus the temperature can be determined. Sound also travels faster with a current than against. By measuring the reciprocal travel times in each direction along a path, the absolute water velocity can be determined. Each acoustic travel time represents the path integral of the sound speed (temperature) and water velocity (18.5). As the sound travels along a particular ray path, it inherently averages these properties of the ocean, heavily filtering along-path horizontal scales shorter than the path length. Over a 1000 km range, a depth-averaged temperature change of 10 m°C is easily measured as a 20 ms travel-time change [18.35].

In ocean acoustic tomography travel time data from a multitude of paths crossing at many different angles are used to reconstruct the sound speed (temperature) and velocity fields (Fig. 18.17). The projection slice theorem requires views from many angles to fill in Fourier wavenumber space, and thus be able to map physical space via an *inverse*. The many crossing paths determine the spatial resolution, and differences in travel times give the strength of the perturbations.

Figure 18.17 illustrates an important characteristic of tomographic systems. For N instruments the number of paths and corresponding data is $N_{\text{paths}} = 1/2N(N-1)$. For example, there are eight instruments in the figure, resulting in 28 paths (not all shown); if one more instrument is added 36 paths results with an increase of eight data for only one additional instrument. This quadratic growth of data with the number of instruments is one reason why instruments only on the periphery can obtain high resolution in the interior.

An oceanic vertical sound speed profile and a set of acoustic rays for 620 km range are shown in (Fig. 18.18). Because of the competing effects of decreasing temperature and increasing pressure with depth, a sound speed minimum occurs at about 1 km depth. This deep-ocean sound channel (mentioned above in the navigation context), or waveguide [with an index of refraction of C/C_0,

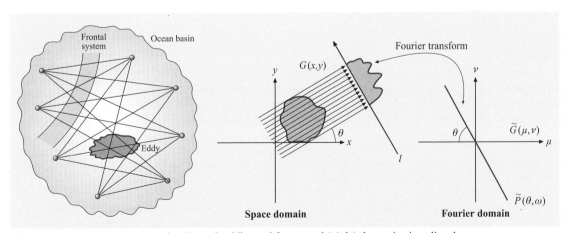

Fig. 18.17 (*left*) Acoustic paths going through eddies and fronts, and (*right*) the projection slice theorem

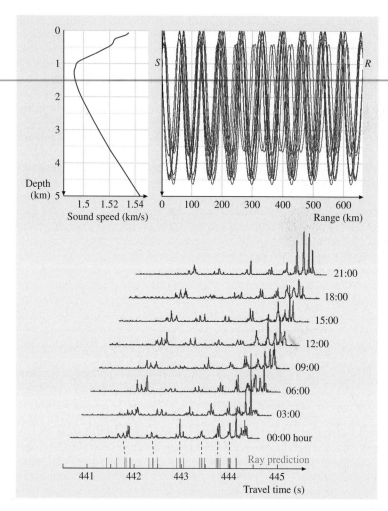

Fig. 18.18 Sound speed profile, acoustic eigenrays, and predicted and measured travel times for a range of 620 km

where C_0 is the minimum sound speed ≈ 1490 m/s at a depth of 1000 m in (Fig. 18.18)], causes acoustic energy to be trapped in the water column so that it can propagate great distances without interacting with the ocean bottom. The sound channel also results in multiple acoustic paths, called *eigenrays*, between an acoustic source and a receiver. The eigenrays have different travel times because they traverse water with different sound speeds and the path lengths are different. The acoustic receiver detects the arrival of multiple pulses, and each pulse can be identified with a predicted eigenray. The figure shows the measured predicted arrival patterns for a 620 km path south of Bermuda. The World Ocean Atlas [18.36] was used to make the predictions. Recorded arrivals are slightly earlier than predicted, suggesting that temperatures were above their climatological mean.

Recasting (18.5) and (18.6) results in the perturbation expression:

$$\delta t = \frac{-R}{C}\left(\frac{\delta C}{C} - \frac{u}{C} - \frac{\delta R}{R}\right), \quad (18.10)$$

where δt is the travel-time perturbation due to small perturbations in sound speed δC, range δR, and water velocity u along the path. As mentioned above, for a 10 m°C temperature perturbation ≈ 0.04 m/s sound speed perturbation over a 1000 km path, with $C = 1500$ m/s, the perturbation travel time is ≈ 20 ms, which is measured easily when averaged over several transmissions in one day.

Clearly precise timekeeping and instrument positioning are required, on the order of a millisecond and one meter for experiments with ranges > 100 km. The former is accomplished using either a GPS timing sig-

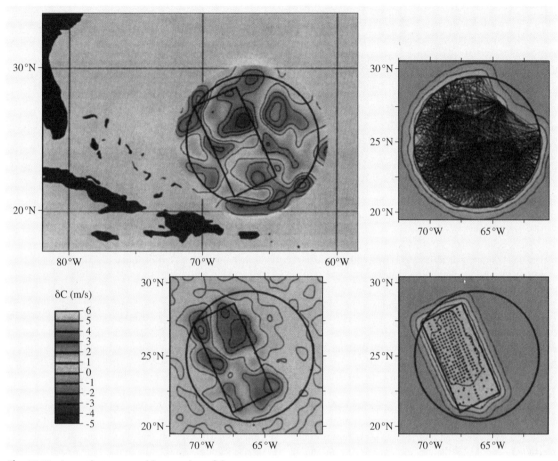

Fig. 18.19 Acoustic tomographic mapping of the ocean

nal if available, or a low-power crystal clock combined with a low-duty-cycle high-power rubidium frequency standard. If the source is moored on the bottom, possibly swaying in the *ocean breeze* 4 km above the bottom, precise navigation is usually accomplished with a local long-baseline high-frequency (10 kHz) acoustic navigation system (a pinger on the mooring ranging off transponders on the seafloor).

Most acoustic sources used for oceanography are limited by transducers and power considerations to source levels < 195 dB re 1 µPa (260 W). To obtain a high enough signal-to-noise ratio, pseudorandom codes are transmitted over long duration at relatively low power; the received signals are replica-correlated with a version of the transmitted signal with the result a single peak. Code processing gains are usually on the order of 30–40 dB, with the limit determined by ocean fluctuations that limit the coherent averaging times. Various frequencies are used depending on range: 20 Hz and 75 Hz for a multi-megameter, 250 Hz for <1 Mm, 400 Hz for ≈ 300 km, and 5 kHz for 20 km.

Since it was first proposed in the late 1970s [18.37, 38], ocean acoustic tomography has evolved into a remote sensing technique employed in a wide variety of physical settings [18.39]. In the context of long-term oceanic climate change, acoustic tomography provides integrals through the mesoscale and other high-wavenumber noise over long distances. In addition, tomographic measurements can be made without risk of calibration drift; these measurements have the accuracy and precision required for large-scale ocean climate observation. Trans-basin acoustic measurements offer a signal-to-noise capability for observing ocean climate variability that is not readily attainable by an ensemble of point measurements. On a regional scale, tomography has been employed to observe active convection,

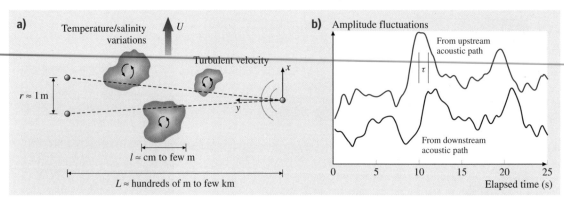

Fig. 18.20 (a) Acoustic propagation from a single source to two receivers. (b) Amplitude fluctuations at the two receivers show a time lag τ that corresponds to the time it takes for turbulence structures to cross the acoustic paths (courtesy of D. Di Iorio)

to measure changes in integrated heat content, to observe the mesoscale with high resolution, to measure depth-independent barotropic currents in a unique way, and to observe directly oceanic relative vorticity. The remote sensing capability has proven effective for measurements under ice in the Arctic and in regions such as the Strait of Gibraltar, where conventional in situ methods may fail. Coastal, shallow-water tomography is possible, though acoustic bottom interaction poses challenges. The acoustic paths are sensitive antennae for specific wavenumbers of the ocean wave field; the measurement technique is the only accurate in situ method to measure open-ocean barotropic and baroclinic (depth-dependent) tides [18.40]. This work led to a renaissance in the study of tides and ocean mixing (Sect. 18.6.1).

One of the original goals of tomography was to observe synoptically the mesoscale variability of the ocean. The acoustic mid-ocean dynamics experiment: moving ship tomography (AMODE-MST) obtained high-resolution, nearly synoptic three-dimensional maps of the ocean over a large area [18.41]. During 50 days in June and July 1991, a ship circumnavigated a 1000 km-diameter array of six moored acoustic sources, deploying a vertical receiving array every 3 h every 25 km to receive the signals from the sources (Fig. 18.19). The travel-time data along the many acoustic paths crossing at many different angles were then used to reconstruct the ocean sound speed (temperature) field in a way very analogous to a medical computed tomography (CT) scan. Sound speed perturbations at 700 m depth on the order of 1 m/s $\approx 0.25\,^\circ\text{C}$ were mapped [(Fig. 18.19), top left]. The acoustic travel time data were inverted to produce this field. For independent verification and comparison purposes, air-expendable bathythermograph (AXBT) data (July 18–22) were collected and mapped [(Fig. 18.19), lower left]. The panels on the right show the line-integral or point sampling that was used to obtain the maps on the left, with estimated errors. The two estimates of the sound speed are within the uncertainties of each method. This experiment demonstrated that high-resolution maps could be obtained with this technique in the ocean, and that the navigation challenges could be overcome.

A limiting case of tomography is an inverted echosounder (IES, single-layer reflection tomography), which sits on the bottom and measures the seafloor-to-surface round-trip travel time. The measured travel time τ varies primarily due to changes in the heat content of the water column [18.42]. The recognition that heat content is well correlated with thermocline depth or dynamic height across the thermocline led to the use of τ to predict these variables for dynamics studies [18.43]. Arrays of IESs can be used to determine horizontal variations of geopotential height anomaly derived from the specific volume anomaly profiles; from these full-water-column (first mode) baroclinic shear profiles relative to zero velocity at the bottom can be calculated. The accuracy of this approach, compared to direct current observations, is excellent [18.44]. Further, a statistical regression technique has recently been developed based on the *gravest empirical mode* (GEM) [18.45]. In ocean regions with stable T–S relationships (most of the world's oceans) and where there is a dominant gravest vertical mode of low-frequency variability (most of the nonequatorial oceans), τ can be combined with information on the vertical structure of the water column (the GEM, derived from CTD profiles) to yield time series of full-water-column profiles of temperature, salinity, and specific-volume anomaly. The limit of how much independent information can be obtained from one observable, though, must be recognized.

The major source of *noise* for large-scale tomography experiments are sound speed (temperature) fluctuations caused by internal waves. There are ongoing efforts to invert this *signal* to obtain the parameters of internal wave models [18.46].

Inverting travel times in general for three-dimensional ocean temperature and velocity is an involved process that benefits greatly if it is done in the context of a data-assimilating numerical ocean circulation model. Data assimilation, starting with weather nowcasting and forecasting decades ago, is just coming into its stride in oceanography, using Kalman filtering and adjoint/variational techniques [18.47]. The assimilation of acoustic travel times directly into numerical ocean circulation models requires the specification of the measurement functional relating model parameters to the measured travel times. Only recently have data-assimilating ocean models had the vertical resolution for the forward acoustic problem, the calculation of the travel-time path integral. The wider use and acceptance of tomography will depend on the data-assimilation community using the acoustic data.

18.4.3 Acoustic Correlation Techniques

When sound propagates through the ocean, the wavefront is affected by turbulent fluctuations in the sound speed field (caused by temperature, salinity, and water velocity fluctuations), producing amplitude and phase variations; this is acoustic scintillation. If there is a larger-scale current advecting these small-scale fluctuations through the sound field, the received sound field carries information about the mean current. By transmitting sound through the water volume of interest along two paths, and cross correlating the two received signals, one can obtain the mean current perpendicular to the two paths (Fig. 18.20). This inverse process assumes certain forms for the spectrum of the turbulent fluctuations, e.g., isotropic turbulence.

Figure 18.21 shows other possible geometries integrating the simple acoustic travel-time concepts mentioned above, using reciprocal transmissions and multiple transceivers to obtain mean flow shear, currents, and sound speed both perpendicularly and along the paths, and vorticity. With appropriate amplitude weighting of acoustic beams, one can determine some range dependence. Measurements have been made with these scenarios in tidal channels [18.48].

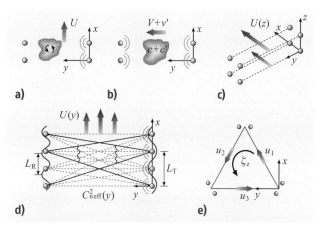

Fig. 18.21 (a) Parallel acoustic paths between two sources and two receivers. (b) Reciprocal transmission, where all transducers transmit and receive. (c) Two-dimensional arrays for shear and turbulent anisotropy measurements. (d) Linear arrays for spatial filtering of flow and turbulence parameters along the path. (e) Triangular array with paired transducers at each corner for horizontal velocity, turbulence, and vorticity measurements (courtesy of D. Di Iorio)

The acoustic scintillation flow meter (ASFM) has been commercialized (ASL Environmental Science). One version measures flow in hydroelectric dam pipes, and another measures flow in channels, e.g., the Second Narrows, Port of Vancouver. An ASFM system has been operating successfully in the Fraser River, on the British Columbia coast, since 1991.

18.4.4 Electromagnetic Sensing Using the Geomagnetic Field

The low-frequency (< 0.5 cycles per day) horizontal electric fields (HEFs) in the ocean are dominated by motional induction caused by the motion of conductive seawater through the Earth's permanent magnetic field. For the case of a flat-bottomed ocean with the length scale of the flow much larger than the water depth, a theoretical relationship between the HEF and the horizontal water velocity can be derived [18.49,50]; the motionally induced voltage E measured over a horizontal distance L caused by the ocean flow though the Earth's magnetic field B_z is (18.8)

$$E = V^* B_z L , \qquad (18.11)$$

where V^* is the vertically averaged flow, weighted by the conductivity σ

$$V^* = \frac{\int (\sigma V) \, dz}{\int \sigma \, dz} . \qquad (18.12)$$

Fig. 18.22 Florida Strait transports (*solid line* in Sv) compared with the NAO (*dashed line*, rescaled and unitless [18.51])

cal relationship between HEFs and barotropic currents have been reported for point seafloor HEF measurements [18.52], for moving platforms [18.53], and for seafloor cables [18.54].

Fixed Sensors (Bottom Arrays, Cables)

Bottom instrument packages with two crossed electrode pairs with 6 m separation have been used to measure V^* and infer the barotropic velocity. A mid-latitude B_z of $36\,\mu\mathrm{T}$ and V^* of 1 cm/s (typical of mid-gyre open ocean) gives a voltage of $2\,\mu\mathrm{V}$. Even very stable silver–silver chloride electrodes drift much more than this; to overcome this drift, a water chopper switch is used to periodically reverse the polarity of the measurement. New designs are in progress to simplify the chopper. Given the simplicity and robustness of this method of measuring the barotropic flow, it will likely be used more in new long-term ocean observatories.

Many submarine cables have and are being used to measure the motionally induced voltages generated by water flow. An example of a short cable is the one crossing the Ria de Aveiro lagoon in Portugal, where a large tidal flow occurs [18.55]. The cable crossing the lagoon is 280 m long, so the measured (tidal peak) value of $E = 10\,\mathrm{mV}$ gives a $V^* = 1.0\,\mathrm{m/s}$ (using the value of B_z above). Combining this with the average depth of 15 m gives a flow rate or transport of $4 \times 10^{10}\,\mathrm{m}^3/\mathrm{s}$. When tidal signals are removed, the weaker, residual flows are relevant in studying seasonal variations caused by river run-off, wind-forced currents, etc.

One of the first demonstrated links between ocean circulation and a climatic index, aside from the El Niño/Southern Oscillation, is provided by what may be the longest time series of ocean transport in existence. Measurements of the horizontal electric potential across the Florida Straits are directly related to the volume transport of the Florida Current; they were initiated in 1969 and have been measured since 1981 on inactive and powered submarine telecommunications cables, with calibration factors of 41 and 67 mV S/v and estimated uncertainties of 0.7 and 1.9 Sv, respectively (1 Sverdrup = $10^6\,\mathrm{m}^3/\mathrm{s}$). This long effort has demonstrated a significant relationship (correlation $R = 0.75$ at 95% CI) between the transport of the Florida Current and an index of short-term climate variability over the North Atlantic called the North Atlantic Oscillation (Fig. 18.22; [18.51]). Only such long records of spatially averaged quantities are suitable for statistical analysis.

Cable voltages are monitored around the world on an opportunistic basis using retired submarine telecom-

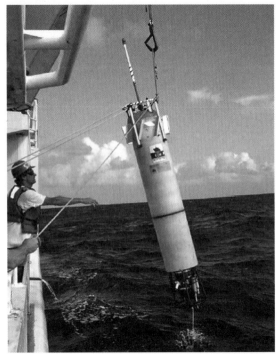

Fig. 18.23 Absolute velocity profiler (AVP); electrodes are in the black ring (courtesy of T. Sanford)

To a first approximation V^* is the vertically averaged ocean velocity over the horizontal path L. The perturbation influence of mild and strong topography has also been considered.

If the conductivity profile is known from independent data, and if there is independent information about the vertical shear of the horizontal currents, then the small conductivity bias is easily removed from the HEF-derived water velocities to yield the depth-averaged current – the transport, often called the barotropic velocity. Successful demonstrations of the theoreti-

munications cables. The reader is referred to the most recent proceedings of the quadrennial conferences on scientific uses of submarine cables [18.56].

Moving Sensors

When the measurement electrodes are moving through the water at velocity V, then the measured voltage is

$$E = (V - V^*)B_z L \,. \tag{18.13}$$

For the case of electrodes towed behind a ship, often called the geomagnetic electro-kinetograph (GEK), near-surface currents can be measured. Two electrodes, one behind the other in the tow direction, measure velocity perpendicular to path. With accurate ship navigation (and velocity), the absolute V^* can be obtained [18.57]. This technique, though known for a long time, is little used.

Free-falling probes are often used to obtain vertical profiles of velocity, especially for turbulence and fine structure measurements where no vibration is permissible, (Fig. 18.23). An ADCP on the nose measures water velocity, with an absolute velocity reference provided by the fixed bottom, sensed by the ADCP.

18.4.5 Surface Current Mapping Using Radar Backscatter

Radar backscatter measurements over the ocean in the 1950s showed obvious Doppler shifts [18.58, 59] (see references in a recent review [18.60]). Under Bragg scattering conditions, surface waves with a length half that of the radio waves constructively reflect energy, in a manner analogous to X-ray diffraction from atoms in a crystal lattice. The frequency shift relative to the carrier in the absence of a current is

$$f = \pm\sqrt{(g/\pi\lambda)} \,, \tag{18.14}$$

where λ is the radio wavelength and is related to the advancing and receding phase velocity of the surface waves. A mean current displaces both of the corresponding peaks in the frequency spectrum by an amount equal to the associated Doppler shift. In this case the waves act as strongly reflecting passive tracers that produce a Doppler shift. The radar antenna beam pattern is usually such that a number of azimuthal beams are formed. The radar signal returns are range-gated, so that radial velocity is obtained as a function of the azimuth angle. With two or more radar stations horizontally offset providing geometrically independent data sets, a vector map of velocity can be formed. High-frequency (HF) radars of 3–30 MHz, and more commonly 12–25 MHz (25–12 m wavelength), are used. The lower frequencies can reach 200 km ranges with range bins of 5–10 km, while the higher frequencies reach 50 km with range bins of 1–3 km.

The current this method senses is not just the surface current but some average of the current in the upper water column because of the interaction between the exponentially decaying surface gravity wave orbital velocities and the vertical profile of the current. The weighting factor is $\exp(-2kz)$, the attenuation of wave orbital velocity with depth. For various reasonable, assumed current profiles (influenced by geostrophic currents, mixed layer processes, and Stokes drift), the sampling depths are all

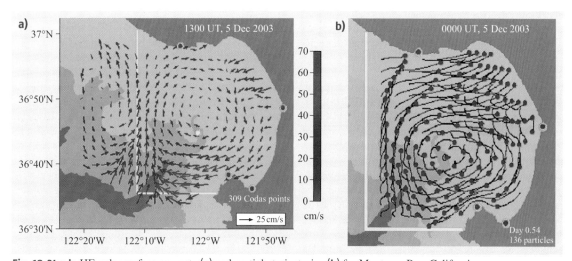

Fig. 18.24a,b HF radar surface currents (**a**) and particle trajectories (**b**) for Monterey Bay, California

less than about one meter, but greater than an electromagnetic skin depth of a few centimeters. It may be possible to use different frequency radars (sensitive to different surface waves and their depth sampling) to extract depth-dependent velocity information. In the future it is not unreasonable to expect accuracies to drop from the present ≈ 10 cm/s to 2–5 cm/s, the typical precision due to the frequency resolution of the Doppler spectrum, given further instrument development and verification.

Figure 18.24 shows recent measurements made in Monterey Bay, California, based on data from a four-site HF radar network. Data from these systems are processed in real time. A normal-mode analysis mapping of the raw data makes it possible to produce continuously updated surface particle trajectories. Currents, particularly during the summer months, are now known to be driven in nearly equal proportions by eddies and upwelling filaments, sea breeze winds, and tidal fluctuations. The two-dimensional current maps available from the HF radar network make it possible to observe and separate the effects of these multiple driving forces.

Recent measurements such as these have led to much broader acceptance of radar data within the oceanographic community (see the collection of articles [18.61]). As of 2004, there are approximately 100 active antennae sites in the US, most funded for research. A contiguous HF radar system spanning the east coast of the US has recently been demonstrated. HF radar will likely be one of the first components of the nascent operational coastal observing systems.

18.4.6 Satellites

Besides the intrinsic value of their data products, satellite observations are of use to oceanographers to provide an additional context for their in situ measurements (Sect. 17.4). Several satellites relevant to oceanography research are described here.

The first oceanographic satellite was Seasat, launched in 1978. While it operated for only 105 days, it demonstrated many of the sensing modalities of subsequent satellites: radar altimetry to measure sea surface height, microwave scatterometry to measure wind speed and direction, scanning multichannel microwave radiometry to measure the sea surface temperature, visible and infrared radiometry to identify cloud, land, and water features, and synthetic aperture radar (SAR) to observe the global surface wave field and polar sea-ice conditions. Seasat launched an entirely new mode of oceanography, one that for the first time had a global perspective.

The topography experiment (TOPEX)/Poseidon and Jason-2 altimeters measure sea surface height to an accuracy of ± 4 cm with a 10 day repeat cycle, nominal equatorial track spacing of 315 km, and an orbital period of ≈ 2 h. This allows the mapping changes in surface geostrophic currents as well as (time invariant) seafloor topography (seamounts affect local gravity and thus the sea surface height). The Gulf Stream, for example, has a 1 m sea surface height difference across it; a typical mesoscale eddy yields a 10 cm signal. The accuracy is a true tour de force, given that the total range is 1300 km (3 parts in 10^8).

Scatterometers today provide most of the operational observations of ocean wind speed and direction. Because the area and amplitude of centimeter-wavelength surface (capillary) waves is a function of wind speed, radar backscattering at the same wavelengths can be used to infer wind speed; by observing at different angles, the wind direction can be inferred. QuikScat, launched in 1999, maps out a 1800 km swath during each orbit and provides $\approx 90\%$ coverage of the ocean each day. Estimated accuracy is ± 1 m/s with a wind-vector resolution of 25 km.

Sea surface temperature has been determined since 1978 using the US National Oceanic and Atmospheric Administration (NOAA) polar-orbiting advanced very high-resolution radiometer (AVHRR). The instrument uses a scanning mirror/telescope to obtain a 2700 km-wide swath and views the Earth roughly twice a day. Absolute accuracy of $\pm 0.5\,°C$ is obtained when compared with in situ data (such as from ships and buoys). Due to the eruption of Mt. Pinatubo in 1991 and an increase of aerosols in the atmosphere, modifications had to be made to the original algorithm; this was determined only years after the eruption. The inconsistency in the data was a significant problem in understanding global climate change. The recently deployed moderate-resolution imaging spectroradiometer (MODIS) on the Terra and Aqua satellites is an optical scanner that views the Earth in 36 channels with spatial resolution ranging from 250 m to 1 km; several of these channels overlap directly with the AVHRR channels.

The tropical rainfall measuring mission, launched in 1997, uses a precipitation radar with a horizontal resolution at the ground of about 4 km and a swath width of 220 km. It provides vertical profiles (250 m resolution for vertical viewing) of rain and snow from the surface up to a height of about 20 km, detecting rain rates down to about 0.7 mm/h. At intense rain rates, where the attenuation effects can be strong, new data-processing methods have been developed that help correct for this effect.

Aquarius is a planned National Aeronautics and Space Administration (NASA) mission (≈ 2009) to measure sea surface salinity. The basis of the measurement is the dependence of the dielectric constant of seawater on salinity at microwave frequencies. The dielectric constant determines the surface emissivity, and this determines the brightness temperature, the measurable parameter. The baseline mission with an 8 day repeat cycle with a 350 km swath would provide 100 km resolution with an accuracy of 0.2 psu when averaged over a month. This will permit a reliable seasonal cycle to be measured.

18.5 Measurement Systems

Much of the challenge in oceanography is fielding the instrumentation; to reduce *delivery costs* it is usually the case that multiple sensors are combined into a suite of sensors supporting a common measurement objective. Here several examples are given: expendable sensors, a fixed bottom instrument package, a towed one, and several mobile platforms. In oceanography, there is always a tension between expendable, low-cost, low-accuracy probes and more-permanent, high-cost, and accurate sensor packages. Lastly, a description of the cabled, general-purpose long-term ecosystem observatory (LEO-15) is given; this example illustrates a trend to long-term sustained observations.

18.5.1 Expendable Sensors

In the early days of modern oceanography (World War II until the 1960s), bathythermographs (BTs) were deployed to 300 m from underway ships to measure temperature profiles (to obtain sound speed for antisubmarine warfare). In a small torpedo-shaped tow fish, a U-shaped xylene-filled copper tube swung an arm with a stylus across a glass slide coated with a thin layer of gold. Being somewhat awkward and limited in capability, these were replaced by expendable bathythermographs (XBTs). These similarly shaped devices contain a thermistor in the heavy lead nose and a spool of very fine, two-conductor wire in the tail. A corresponding spool of wire in the deck portion of the system assures no tension is on the wire. These free-falling probes can reach 1800 m, and cost about \$50 each. Accuracy is $\approx 0.1\,°C$ and depth is determined from a fall rate equation. There are now expendable sound velocimeters, CTDs, and (electromagnetic) current meters (more expensive); all are deployable from aircraft also.

18.5.2 HPIES

With the advent of sustained long-term ocean observatories, it will be important to choose sensors for their proven oceanographic utility, their resistance to degradation, and their ability to measure oceanographic variables remotely (an advantage that enhances reliability). An HPIES sensor suite combines a horizontal electrometer (HEM), a bottom pressure sensor, and an inverted echosounder measuring seafloor-to-sea surface acoustic travel time (Fig. 18.25). These measurements are integrating variables [18.62] because they are dominated by the currents and thermal variability with the largest vertical scales (and sometimes the largest horizontal scales). As such, these variables are ideal for observing the meso- to gyre-scale variations of the ocean circulation that are most likely related to climate change.

Arrays of IESs yield four-dimensional views (x, y, z, and t) of water properties and low-frequency (< 0.5 cycle per day) baroclinic (depth-dependent) current shear. Arrays of HEMs yield three-dimensional views (x, y, and t) of conductivity-weighted barotropic velocity. The HEM-IES combination is greater than the sum of the parts, because it yields four-dimensional views of absolute water velocity. The accuracy of the derived absolute velocities has been demonstrated in an experiment in the Sub-Antarctic Front. Like the horizon-

Fig. 18.25 A pressure inverted echosounder unit (PIES, courtesy of D. R. Watts)

tal electric field variations, low-frequency pressure variations are related to an integral (in this case, horizontal) of water velocity fluctuations, and it can be argued from geostrophy that bottom pressure will be more sensitive to, if not dominated by, the motions with largest horizontal scales. As such, bottom pressure is a good partner with IES and HEM arrays, and thus HPIES. Among other applications, bottom pressure has been employed in studies of the variability of strong frontal currents (e.g., the Gulf Stream [18.63] and the Antarctic Circumpolar Current [18.64]), and studies of the kinematics and dynamics of atmospherically forced barotropic variability [18.65]. The combination of bottom pressure and two-component barotropic velocity gives the energy flux vector, useful in any dynamical context, as well as being a substantial improvement over the single pressure sensors presently used for tsunami detection.

In the future, IES acoustic transceiver arrays can also be used in a tomographic mode, precisely measuring travel times along acoustic ray paths between nodes. Given typical sound speed profiles one can expect at least surface reflected rays connecting bottom instruments up to a separation of 50 km in the deep ocean. With an array the number of paths grows quadratically. The travel time data can be inverted to produce the horizontal fields of temperature, velocity, and vorticity, complementing the more-local IES, HEF, and bottom pressure measurements. Higher-order statistics of the acoustic signals can be inverted to produce internal wave parameters (e.g., travel-time variance and internal wave strength).

18.5.3 Moored Profiler

Fixed sensors alone cannot provide all the required sampling; nor can any single sensing system. The moored profiler (Fig. 18.26) is one new tool that offers a vast improvement over fixed-depth sensors. By crawling up and down a standard mooring wire, it measures time series of temperature, salinity, and velocity profiles. Measuring entire profiles instead of a handful of discrete depths enables calculation of shear, strain, and turbulent dissipation (from overturns and shear parameterizations), while covering the entire water column. The system has been deployed in over 20 experiments with successes far outnumbering failures; the number of present and planned deployments indicates its reliability.

Moored profilers have already resulted in substantial scientific advances in our understanding of variability in the eastern Atlantic, internal tides and mixing near Hawaii [18.66] and on the Virginia continental slope, the variability of the Kuroshio extension [18.67],

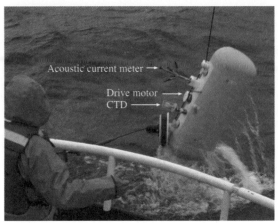

Fig. 18.26 A McLane moored profiler being recovered in the Puget Sound (courtesy of M. Alford)

and the evolution of arctic thermohaline layers and double-diffusive layers in the Barbados staircase. Future developments to the platform will likely include increasing profiling speed from 0.25 m/s to 0.4 m/s to reduce tidal aliasing, adding payload capability (e.g., more sensors), and providing docking to external power and communications (either via a seafloor cabled observatory system or a large battery pack and acoustic modem, for instance).

18.5.4 TriSoarus Towed Vehicle

TriSoarus is a towed, undulating vehicle for high-resolution, three-dimensional surveys of the upper ocean. TriSoarus began life as a SeaSoar vehicle, but has been heavily modified to increase profiling range

Fig. 18.27 Trisoarus towed undulating vehicle (courtesy of C. Lee)

on unfaired cable. Large symmetric foil section wings and a gravity-driven roll-stabilizing aileron significantly enhance range and stability compared to a stock Sea-Soar. The actuator and telemetry system have been upgraded by adapting Triaxus components (another *boxkite* towed vehicle). Thus, the two vehicles have identical subsea and topside electronics suites, down to the software control algorithm. Instrument payloads are similar. TriSoarus provides flexible support for a wide range of payloads, including physical, optical, biological and chemical sensors, and the system has been designed to ease the integration of new instruments. The basic sensor suite includes a CTD, chlorophyll fluorometer, transmissometer and dissolved oxygen sensor, with a fiber-optic tow cable providing high-bandwidth telemetry. At a typical tow speed of 8 knots, the vehicle profiles from the surface to 200 m, providing along-track resolutions of 2 km or better.

The TriSoarus has been used in the Adriatic Sea to measure the response to three primary driving forces:

1. wintertime Bora winds
2. weaker, along-basin Sirocco winds
3. seasonal buoyancy input from the Po River and other sources;

all govern the evolution of coastal filaments, eddies, and fronts in the northern Adriatic.

18.5.5 Turbulence and Small-Scale Structure

Seawater density is determined primarily by temperature and salinity, and the different molecular properties of these two scalars give rise to many interesting phenomena. Differential diffusion of heat and salt has the potential to radically alter large-scale ocean circulation due to freshwater mixing in polar regions. These phenomena are also important in estuaries where salt stratification is crucial to mixing. Anywhere there is warm salty water below cold fresh water, double-diffusive salt fingering can occur.

To study such phenomena, the MARLIN towed vehicle carries a fast salinity probe, which resolves conductivity and temperature on sub-centimeter scales (Fig. 18.28). Hundreds of kilometers of salinity data were obtained from MARLIN in 2002 during the Hawaii ocean mixing experiment (Sect. 18.6.1 below).

Turbulence measurements must be made on platforms that are free of vibration; this has driven the development of free-falling microstructure measurement platforms. Turbulent velocity fluctuations are

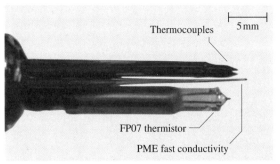

Fig. 18.28 The MARLIN fast salinity probe (courtesy of J. Nash)

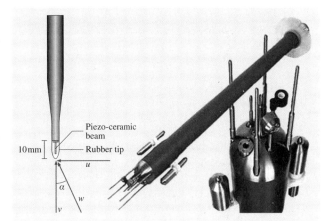

Fig. 18.29 Shear probe and microstructure profiler (Alec Electronics, Kobe)

measured with a shear probe (Fig. 18.29), which is the standard sensor for turbulence measurements. The probe's sensing element consists of a parabolically shaped, axially symmetric flexible rubber tip. For small angles of attack, the oncoming flow with velocity W produces a hydrodynamic lift force proportional to the cross-stream velocity component. A piezoceramic beam, located at the center of the rubber tip, translates the lift force into an electrical signal. Differentiating this produces a signal proportional to the rate of change of cross-stream velocity du/dt. Under Taylor's frozen-field assumption, the velocity shear is then $du/dz = W\,du/dt$.

18.5.6 Floats

Float platforms are increasingly used to carry a payload of diverse sensors dependent on the application. A recent float developed at the Applied Physics Labo-

Fig. 18.30 Four MLF2s ready for Hurricane Isidore after Puget Sound testing in July 2002 (courtesy of E. D'Asaro)

ratory is the mixed layer float-2 (MLF2, (Fig. 18.30)). Earlier experience showed that, although much could be learned from temperature and pressure measurements, these alone were insufficient to diagnose the dynamics of oceanic turbulent boundary layers. Measurements of shear, stratification, and surface waves are necessary. Furthermore, accurate ballasting of the floats under conditions of strong mixing in general requires knowledge of the density of the seawater, so salinity measurements were needed. Also, there is a great potential for studying biological phenomena using optical sensors. Finally, advances in satellite communications offered two-way data transmission so that floats can be controlled and their mission parameters modified remotely. These factors dictated the construction of a larger float with sufficient power (a larger battery pack; solar cells for duration) and the capability of easily accommodating a variety of sensors.

MLF2 displaces 50 L and has the ability to surface repeatedly by controlling 750 cm^3 of active volume control with an extruding piston. It has a cloth drogue, but this can be folded and unfolded as many times as desired. The MLF2 mission consists of periods of Lagrangian drift typically 12 h long. Between these the float surfaces, uses GPS to determine its position, and transmits its position and some of its data using Iridium. It then waits for instructions before beginning another drift period.

The float can carry a large instrument suite, including a Doppler SONAR, altimeter, CTDs, accelerometers, a photosynthetic active radiation (PAR) sensor, fluorometer, and horizontal electric field sensors. MLF2 has been used in a study of the wintertime North Pacific mixed layer in 2000 and in studies of upwelling off Oregon in 2000 and 2001. An air-deployment package was used to deploy MLF2s into hurricanes during 2002–2004.

The sensors allow much better understanding of the behavior and accuracy of floats. For example, in a stratified fluid, the dominant cause of drag on the float is the radiation of internal waves. This internal wave drag is much larger than the normal form drag at low speeds

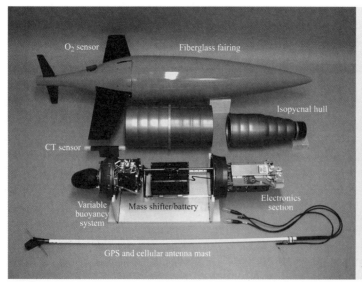

Fig. 18.31 Seaglider: (*left*) components and (*right*) being deployed in 2004 in the north Pacific

and causes the floats to be Lagrangian for frequencies higher than about $N_B/30$, where N_B is the buoyancy frequency. Because most mixing events in a stratified fluid happen on time scales of about N_B, this means that floats can be good instruments to measure mixing rates in the ocean interior. These results are described in a recent paper [18.68].

18.5.7 Gliders

Gliders are autonomous mobile platforms that use changes in buoyancy combined with wings to provide a horizontal motive force. As with Argo and other floats, oil is pumped in and out of an external bladder to effect the buoyancy change. There are three presently produced: the Spray (Scripps Institution of Oceanography), the Slocum (Webb Research Corp.), and the Seaglider (University of Washington) (Fig. 18.31); many variants are in progress. Gliders fill a niche between floats and powered vehicles, providing a directed platform with long duration.

The Seaglider [18.69] changes its attitude by mechanically shifting its battery pack fore and aft and side to side. When it surfaces, a combined GPS/Iridium antenna extends about one meter above the water. GPS provides position and time synchronization. Iridium provides real-time 2400 baud two-way communications; missions are routinely controlled from shore and new software can be uploaded. Typical sensors are temperature, conductivity, pressure, and dissolved oxygen. Because buoyancy/volume and drag are significant issues, interfacing new sensors is nontrivial.

Figure 18.31b shows a Seaglider in the process of being launched from an oceanographic vessel for a cross-Pacific mission that lasted 6 months, made 600 dives and covered 3 Mm at 1/2 kt using an average of 1/2 W. This particular glider carries a RAFOS acoustic receiver; ranges to 1900 km were obtained.

18.5.8 Autonomous Undersea Vehicles (AUVs)

AUVs are now used for short-duration (typically 1 day at 3 kt) missions, carrying a variety of sensors ranging from simple CTDs to very sophisticated bottom mapping SONARs. An example of a small one (19 cm diameter, 160 cm long, 37 kg) is the REMUS (remote environmental monitorung units) vehicle (Fig. 18.32). A standard configuration carries a small up- and down-looking ADCP, CTD, sidescan SONAR, and light-scattering sensors; other sensors that have been used (not all at

Fig. 18.32 REMUS AUV (Hydroid, Pocasset)

once) include an acoustic Doppler velocimeter, a shear probe, a plankton pump, up and down looking radiometers, a DIDSON (dual frequency identification sonar) *acoustic lens*, GPS, and deployable acoustic navigation transponders. These are still rather expensive (250 000 US$) and require skilled technical support. Two other vehicles are the autonomous benthic explorer (ABE, Woods Hole Oceanographic Institution) intended for deep use around topography (e.g., measuring vertical heat flux above hot vent fields), and the Dorado (Monterey Bay Aquarium Research Institute), a large (21 in-diameter) vehicle that has been used under the ice in the Arctic. The latter application will likely be a major driver for AUVs and gliders, as a means of directed sampling. The military is a major driver behind glider and AUV development.

18.5.9 LEO-15

The long-term ecosystem observatory (LEO) off the US New Jersey coast (Fig. 18.33) is a forerunner of ocean observing systems of the future. The core is a cabled node in 15 m of water, installed in 1996. Many in situ sensors and platforms are connected to this central core: winched CTD, ADCP, benthic acoustic stress sensor (BASS) tripod, electromagnetic current meter, optical backscatter, wave height and period, temperature, and fluorescence, video cameras, optical profiling node with chlorophyll, dissolved organic matter, fluorescence, spectral light backscatter, and particle size, and a REMUS AUV docking station. Around it are many autonomous instruments (some on land): moorings with temperature, salinity, and current sensors, IMET (improved meteorological) meteorological buoy, HF radar buoy extending the range of the shore-based system, hyperspectral radiometers to provide ground truth for new satellite sensors, gliders and AUVs, land IMET sensors, and SODAR for wind profiles (Chap. 17). These sen-

Fig. 18.33 Long-term ecosystem observatory (LEO-15) instrumentation used for the 2000–2001 experiment (courtesy of O. Schofield and S. Glenn)

sor systems are complemented by ships, aircraft, and satellites.

A major goal of this integrated ocean observatory is to develop a real-time capability for rapid environmental assessment and physical and biological forecasting in coastal waters. To this end, observational data from all sources are collected in real time, permitting adaptive sampling of episodic events and assimilation into ocean forecast models. This measurement-rich situation provides a unique opportunity to refine and verify models. To date, the complex sampling shown in the figure can only be sustained for one or two months at a time.

18.6 Experiment Case Studies

Actual oceanographic field experiments are usually expensive and often last a number of years, with analysis taking a decade or more. There is always a variety of measurements made using diverse technologies that facilitate novel approaches to the science questions. Two cases illustrate the complexity.

18.6.1 Hawaii Ocean Mixing Experiment (HOME)

The ocean's steady-state thermohaline structure is determined by downward diffusion balanced against upward currents. An outstanding oceanographic problem is that independent measurements of the ocean's vertical temperature structure and the amount of turbulent mixing are not consistent. Measurements find approximately one tenth of the mixing necessary to agree with the warm temperatures detected in the ocean's depth. A long-standing suggestion is that mixing does not occur uniformly over the entire ocean, but is concentrated in the vicinity of rough topography (Sect. 18.3.3). The Hawaii ocean-mixing experiment (HOME) has been an effort to test this theory with observations and computational models of mixing along the Hawaiian Ridge [18.70].

The Hawaiian Ridge is an attractive place for a mixing experiment because the topography is steep and

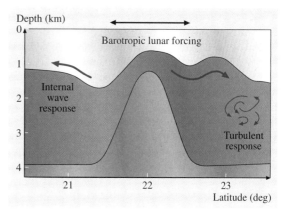

Fig. 18.34 Schematic of energy cascade near abrupt topography like the Hawaiian Ridge

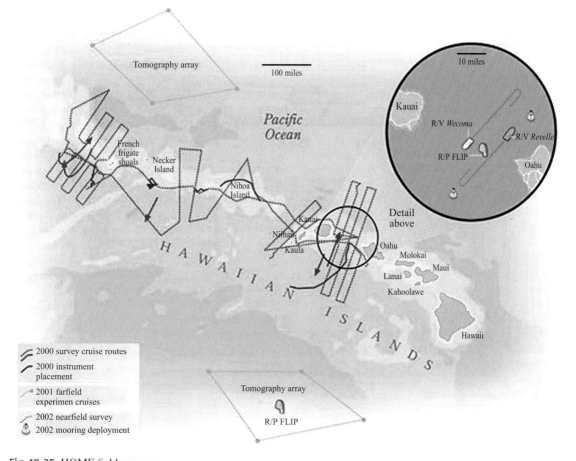

Fig. 18.35 HOME field program

energy is available for mixing in the form of tides, which strike the ridge perpendicularly (Fig. 18.34). Recent inferences from satellite altimeter data suggest that half of the 2 TW required for ocean mixing can be derived from the tides. Mixing is envisioned as a cascade from barotropic surface tides, through baroclinic internal tides and breaking internal waves, to turbulence.

Observing the tides-to-turbulence cascade in the ocean is a considerable challenge. Spatial scales range from megameters (the size of ocean basins and major topographic features) down to centimeter scales, where molecular diffusion occurs. The HOME project used an integrated program of modeling and observation; the latter includes components to survey the entire ridge, as well as to study far- and near-field regimes (Fig. 18.35). The numerical modeling of surface and internal tides and the analysis of historical data completed the experimental approach.

Two major challenges were to isolate baroclinic variability associated with the tides from a background of energetic processes and to extrapolate the results obtained from HOME to other regions of the ocean where the tides are an important energy source for open ocean mixing. In both regards, numerical models provide the means for interpretation of a comprehensive but undersampled set of observations, and data assimilation provides an optimal way to combine all available data into dynamically consistent estimates of barotropic and baroclinic energy fluxes, and hence to constrain the amount of tidal energy available for localized mixing.

Models can also be used to determine the topographic and oceanic conditions favorable for energetic tide-induced mixing, and thus experiment planning. Finally, models provide a dynamical context in which to synthesize and generalize the results, providing the opportunity to evaluate various turbulent closure schemes against direct and integrated measures of dissipation and diapycnal mixing.

Historical data from moored current meters, CTDs, and ADCPs (collected for other purposes) were analyzed for the presence of signatures of fine-scale mixing activity, with the results being grouped according to depth, isopycnal, proximity to topography, or other environmental variables. This work helped to determine the locations and ubiquity of regions where strong mixing activity appears to be common irrespective of energy source and this has guided the subsequent deployment of field resources.

The survey component-quantified the geographic variability of turbulent mixing (shear, stratification, and dissipation) in the vicinity of the Hawaiian Ridge. Five cruises in 2000 were conducted measuring temperature, salinity, horizontal and vertical velocity, turbulent energy dissipation, optical attenuation, acoustic backscatter, and seafloor topography. Horizontal resolution was as fine as O(1 m), with ranges of roughly 1000 km along the ridge and 100 km across the ridge. The finest vertical resolution was O(1 cm), ranging over the entire ocean depth. In an extensive survey along the 3000 m isobath, the absolute velocity profiler (AVP, (Fig. 18.23)) was used to estimate the semidiurnal internal tide energy-flux radiating away from the ridge, confirming energy was originating at the ridge rather than passing through it. Other measurements were made using SeaSoar (measuring stratification and shear to 300 m), the advanced microstructure profiler (AMP), and towed body (MARLIN, (Fig. 18.28)) capable of making measurements of turbulence and fine structure to depths of 3400 m, and moorings with ADCPs and other instrumentation. Model energy-flux hot and cold spots were consistent with these and other observations though with some differences in magnitude and direction.

The survey results guided the selection of sites for the near- and far-field components. The former included many of the same measurements as for the survey, concentrated on the Kaena Ridge, a representative mixing *hot spot*, on the northwest side of Oahu bordering the Kauai Channel. The far-field experiment consisted of tomographic arrays and pressure sensors and horizontal electrometers on the north and south sides of the ridge. These were used to estimate the first two terms of the

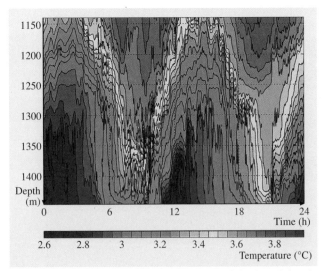

Fig. 18.36 Moored temperature as a function of depth and time in the eastern Kauai Channel near the Kaena Ridge [18.70]

equation:

> Energy lost from the barotropic tide
> = (baroclinic tide radiative energy)
> + (local dissipation).

R/P FLIP (floating instrument platform) (Fig. 18.39) was included in both the near- and far-field components to obtain high-resolution vertical profiles and Doppler SONAR data.

In combination, these measurements provided striking evidence that energy is being transferred from the incident tide to internal waves that then break, giving rise to turbulence. This occurs mainly at local *hot spots*. A modest amount of energy radiates as baroclinic tides into the ocean interior; most appears to be dissipated locally. This conversion from the tides to turbulence represents a cascade of energy from the very large tidal scales of 1000 s of km, down to the relatively tiny scale of turbulent dissipation, approximately 1 cm. A mooring near the Kaena Ridge (Fig. 18,36) equipped with temperature (and other) sensors showed peak-to-peak internal wave displacements of 300 m (compared with the 1 m surface tide) during spring tide at the semidiurnal period [18.70].

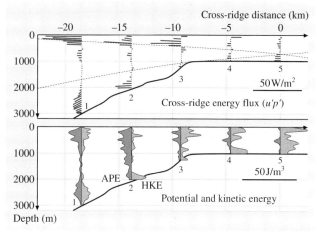

Fig. 18.37 Energy flux and potential and kinetic energy as a function of depth and range on the Kaena Ridge (courtesy of J. Nash et al.)

Baroclinic generation occurs over the steep upper flanks of the ridge and forms beams that propagate upward and downward. Superposition of waves from the northern and southern flanks of the ridge (and traveling in opposite directions) produce a standing-wave pattern over the

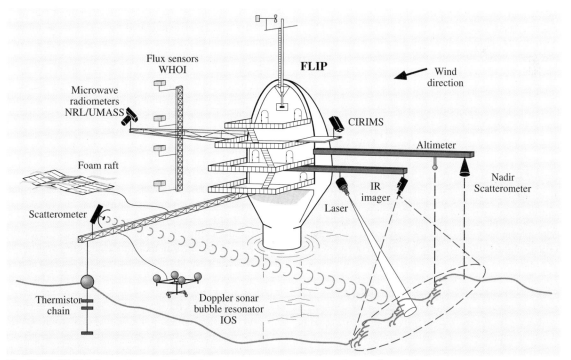

Fig. 18.38 FAIRS focused on coordinated measurements to improve understanding of air–sea interaction and its effects on remote sensing

Fig. 18.39 R/P FLIP drifted freely so the hull rotated to remain in the same orientation to the prevailing wind

Fig. 18.40 FAIRS infrared sensors (courtesy of A. Jessup)

ridge top. The standing wave has no net energy flux, but the interference of waves produces regions of high and low available potential energy-to-horizontal kinetic energy (APE:HKE) ratio that correspond to the ridge flanks (high APE) and ridge top (low APE) (Fig. 18.37).

18.6.2 Fluxes, Air–Sea Interaction, and Remote Sensing (FAIRS)

To better characterize and quantify the transfer of momentum, heat, and gas across the air–sea boundary, measurements of their underlying physical mechanisms were made using remote sensing instruments as part of the fluxes, air–sea interaction, and remote sensing (FAIRS) experiment (Fig. 18.38). This was a 32 day excursion aboard the research platform R/P FLIP off the coast of Monterey, California. FLIP is a unique 100 m vessel that is towed to sea and is literally flipped up on end so that only the top 20 m is above the water surface (Fig. 18.39). The result is a stable platform for housing scientists and their instruments, ideal for making air–sea interaction measurements because of its low flow distortion and high stability during rough conditions. The experiment has provided an extensive suite of complementary measurements that combine remote sensing, marine meteorology, and physical oceanography [18.71].

Microbreaking Waves and Infrared Imagery

The most obvious phenomena at the air–sea interface are ocean surface waves, especially large-scale breaking waves, or whitecaps. Less obvious is microscale wave breaking, or microbreaking, which occurs when small, wind-generated waves about 10 cm to 1 m long gently break. These short, wind-driven waves increase to a height of a few centimeters until they become too steep to support themselves and consequently *break* and overturn the water right at the surface. This process transfers momentum into the upper ocean as turbulence and mixing that enhances the transfer of heat and gas between the ocean and atmosphere.

The remote sensing instruments mounted on the decks and booms of FLIP included an infrared imager, microwave radiometers, microwave radars, and video cameras. A passive technique used the infrared imager to take a picture of the ocean's surface temperature, concentrating on the small but measurable temperature variations that occur when a breaking wave causes water from below to be mixed to the surface. An active technique used a CO_2 laser to heat a patch of water; the patch was then monitored by the imager to time its decay. During rough seas when the surface was broken frequently and repeatedly by microbreaking, the heated patch cooled down in less that a second. During very calm conditions, the patch remained visible in the infrared image for several seconds. This length of time for the heated patch to revert to the surrounding surface temperature is used to quantify the flux of heat at the surface as well as the amount of turbulent mixing.

Parasitic Capillary Waves and Microwave Backscatter

One characteristic of a microbreaking wave is that small capillary waves are generated on its forward face and travel along with the wave as it breaks. These capillary waves ride along on the steep front face of microbreakers, and thus they appear tilted with respect to the look angle of (microwave) radar; this tilting effect may explain why more reflected energy is observed than is predicted by current models. By combining infrared and microwave techniques, the role of microbreaking

Fig. 18.41 Foam generator raft produces a steady foam by forcing air through a matrix of gas-permeable tubing

in surface fluxes and radar backscatter can be better understood.

Whitecap Emissivity and Microwave Radiometry

Microwave radiometers are currently used to measure wind speed from satellites by exploiting the dependence of ocean surface emissivity on surface roughness. Because the wind direction signal is so small, the emissivity of foam generated by large-scale breaking waves becomes significant. To characterize the effect of foam on radiometer measurements, a 3×7 m raft made of gas-permeable tubing generated a uniform area of sea foam that filled the footprint of the microwave radiometer (Fig. 18.41); in this way a degree of *laboratory control* was brought to the field experiment. By measuring the microwave signal from the sea surface with and without foam, the foam's emissivity could be determined [18.72].

Surface Roughness and Air–Sea Fluxes

Over the ocean, direct measurements of turbulent air–sea fluxes are difficult due to platform motion, flow distortion, and the effects of sea spray. Instead, marine meteorologists have used flux–profile relationships that relate the turbulent fluxes of momentum, heat, and moisture (or mass) to the more easily measured profiles of velocity, temperature, and water vapor (or other gases), respectively (Sect. 17.2.1). These relationships are required for indirect methods such as the bulk aerodynamic, profile, and inertial dissipation methods that estimate the fluxes from mean, profile, and high-frequency spectral measurements, respectively. The flux–profile or flux–gradient relationships are also used extensively in numerical models to provide lower boundary conditions to *close* the model by approximating higher-order terms from low-order variables. Direct covariance techniques, gradient techniques combined with Monin–Obukhov similarity theory, and inertial dissipation methods are all used to infer fluxes. However, the use of land-based measurements to infer ocean surface fluxes has been questioned, particularly close to the ocean surface where wave-induced forcing can affect the flow. Therefore, the applicability of Monin–Obukhov similarity theory to ocean–atmosphere surface layers is a current topic of research.

During FAIRS, micrometeorological flux measurements were used to improve the parameterizations of heat and momentum fluxes [18.73]. Direct measurement of the atmospheric fluxes along with profiles of water vapor and temperature were performed near the end of the 18 m port boom (Fig. 18.40). Turbulent fluxes of momentum, heat, and water vapor were measured by sonic anemometers, sonic thermometers, and infrared hygrometers, respectively. A mast supported a profiling system that moved a suite of sensors between 5 m and 15 m above the mean ocean surface. The moving sensors were referenced against a fixed suite of sensors to remove naturally occurring variability during the profiling periods. The measurements of heat and momentum flux, combined with infrared and microwave measurements, quantify the correlation of air–sea fluxes with surface roughness, ranging in scales from dominant gravity waves to capillary waves and microbreaking.

Sea-to-Air Dimethylsulfide Fluxes

One of the consequences of wave breaking and sea surface renewal is that soluble gases are transferred between the ocean and the atmosphere. Dimethylsulfide (DMS) plays an important role in atmospheric chemistry and climate through its oxidation to sulfuric acid and subsequent formation of cloud condensation nuclei. Although the disequilibrium can be measured fairly easily for most gases, direct measurements of the gas transfer velocity are quite difficult. However, if the transfer velocity for one gas is known, the transfer velocity for other gases can be fairly easily calculated.

DMS fluxes were measured using two different techniques, the gradient flux (GF) approach and relaxed eddy accumulation (REA). In the GF technique, air samples are obtained at different heights from the surface, and the gradient is related to the gas flux through simultaneous micrometeorological measurements and

Monin–Obukhov similarity theory. In the REA, two samples are taken at the same height, one when the vertical wind has a significant upward component and one when the vertical wind has a downward component; the flux is proportional to the concentration difference between the two samples and the variance in the vertical wind. The gas transfer results using both GF and REA were consistent with previous measurements, with the REA values being somewhat higher than those from GF [18.74]. This measurement provided an opportunity for one of very few in situ comparisons of different methods of gas transfer measurement, and the first one for DMS. Measurements of the concentration of DMS in surface seawater were also performed, allowing estimation of the gas transfer velocity and comparison to the heat transfer velocities measured using the active infrared technique described above.

Turbulent Velocity and Enhanced Dissipation

A unique surface-tracking float was used to link measurements of the turbulent velocity field beneath the surface to the signature of turbulence on the surface (Fig. 18.42). Subsurface acoustic sensors were used to examine the link between bubble size distributions and near-surface turbulence. The float supported three Doppler SONARs, two resonators and a thermistor and pressure sensor. The velocity profiles, resolved on the centimeter scale, yielded wavenumber spectra directly. Energy dissipation was inferred from the spectra, consistent with an inertial subrange. The bubble size distribution was measured with a resonator placed close to the acoustic Doppler profiler. Video recordings of the float allowed verification of wave breaking events and provided estimates of the size and propagation speed of whitecaps. In addition, the directional surface wave field was monitored with a suite of Doppler SONARs mounted on the hull of FLIP. Measurements of the thermal structure of the surface layer of the ocean were acquired with two chains of self-recording thermistors to a depth of 70 m.

Simultaneous video recordings of wave breaking events during which turbulent velocity and bubble fields

Fig. 18.42 A surface-following float provided subsurface measurements of turbulent velocity and bubbles generated by large-scale breaking waves

were measured show that turbulence and dissipation in the ocean surface layer is greatly enhanced due to wave breaking [18.75]. Dissipation rates associated with a breaking event are up to five orders of magnitude greater than background rates. Bubbles of all sizes are rapidly injected beneath breaking waves and decay slowly after approximately one wave period. The large velocities that occur during breaking are sufficient to counteract the rise in velocity of larger bubbles only briefly. However, the more-persistent turbulent velocity that remains after breaking may keep small bubbles in suspension for a significant time. These and other observations provide insight into the turbulent transport in the near-surface layer and the dissipation of wave energy; with these measurements it will be possible to determine if infrared techniques can be used (more simply) to estimate near-surface dissipation rates.

References

18.1 W. Munk: *Oceanography Before, and after, the Advent of Satellites, Oceanography and Society*, ed. by D. Halpern (Elsevier, Amsterdam 2000) p. 1

18.2 C. Wunsch: What is the thermohalic circulation?, Science **298**, 1179–1180 (2002)

18.3 M. Eble, F. Gonzalez: Deep-ocean bottom pressure measurements in the northeast Pacific, J. Atmos. Ocean. Tech. **8**, 221–233 (1991)

18.4 I. Shkvorets, F. Johnson: *Proc. IEEE Oceans 2004 OTO'04 Conference* (Kobe, Japan 2004)

18.5 K. Mackenzie: Nine-term equation for sound speed in the oceans, J. Acoust. Soc. Am. **70**, 807–812 (1981)

18.6 G. Wong, S. Zhu: Speed of sound in seawater as a function of salinity, temperature, and pressure, J. Acoust. Soc. Am. **97**, 1732–1736 (1995)

18.7 H. Medwin: *Sounds in the Sea* (Cambridge University Press, Cambridge 2005)

18.8 J. Rizoli (Ed.): *Proc. IEEE/OES Seventh Working Conference on Current Measurement Technology, San Diego, California* (The Printing House, Stoughton 2003)

18.9 P. Collar: *A Review of Observational Techniques and Instruments for Current Measurement in the Open Sea, Report No. 304* (Institute of Oceanographic Sciences, Surrey 1993)

18.10 D. Frye, N. Hogg, C. Wunsch: A long duration mooring for ocean observation, Sea Technol. **45**, 29–39 (2004)

18.11 J. Ledwell, E. Montgomery, K. Polzin, L. St. Laurent, R. Schmitt, J. Toole: Evidence for enhanced mixing over rough topography in the abyssal ocean, Nature **403**, 179–182 (2000)

18.12 J. Ledwell, A. Watson, C. Law: Mixing of a tracer in the pycnolcline, J. Geophys. Res. **103**, 21499–21529 (1998)

18.13 K. Polzin, J. Toole, J. Ledwell, R. Schmitt: Spatial variability of turbulent mixing in the abyssal ocean, Science **276**, 93–96 (1997)

18.14 A. Watson, J. Ledwell, S. Sutherland: The Santa Monica Basin tracer experiment: Comparison of release methods and performance of perfluorodecalin and sulfur hexafluoride, J. Geophys. Res. **96**, 8719–8725 (1991)

18.15 W. Sturges, T. Wallington, M. Hurley, K. Shine, K. Sihra, A. Engel, D. Oram, S. Penkett, R. Mulvaney, C. Brenninkmeijer: A potent greenhouse gas identified in the atmosphere: SF5CF3, Science **289**, 611–613 (2000)

18.16 J. Ledwell, K. Polzin, L.St. Laurent, R. Schmitt, J. Toole: Evidence for enhanced mixing over rough topography in the abyssal ocean, Nature **403**, 179–182 (2000)

18.17 M. Sundermeyer, J. Ledwell: Lateral dispersion over the continental shelf: Analysis of dye-release experiments, J. Geophys. Res. **106**, 9603–9621 (2001)

18.18 F. Rowe, J. Young: An ocean current profiler using Doppler sonar, IEEE Proc. Oceans 79, 292–297 (IEEE Proc. Oceans, New York 1979)

18.19 R. Pinkel: Observation of strongly nonlinear motion in the open sea using a range-gated Doppler sonar, J. Phys. Oceanogr. **9**, 675–686 (1979)

18.20 C. Wilson, E. Firing: Sunrise swimmers bias acoustic Doppler current profiles, Deep Sea Res. Part **39**, 885–892 (1992)

18.21 R. Pinkel, J. Smith: Repeat sequence coding for improved precision of Doppler sonar and sodar, J. Atmos. Ocean. Technol. **9**, 149–163 (1992)

18.22 K. Theriault: Incoherent multibeam Doppler current profiler performance. Part I: Estimate variance, IEEE J. Ocean. Eng. **11**, 7–15 (1986)

18.23 T. Chereskin, A. Harding: Modeling the performance of an acoustic Doppler current profiler, J. Atmos. Ocean. Technol. **10**, 41–63 (1993)

18.24 B. King, E. Cooper: Comparison of ship's heading determined from an array of GPS antennas with heading from conventional gyrocompass measurements, Deep Sea Res. **40**, 2207–2216 (1993)

18.25 G. Griffiths: Using 3D GPS heading for improving underway ADCP data, J. Atmos. Ocean. Technol. **11**, 1135–1143 (1994)

18.26 T. Chereskin, C. Harris: *Shipboard Acoustic Doppler Current Profiling During the WOCE Indian Ocean Expedition: I10 , SIO-97-14* (Scripps Institution of Oceanography, La Jolla 1997)

18.27 T. Chereskin, D. Wilson, L. Beal: The Ekman temperature and salt fluxes at 8 30' N in the Arabian Sea during the 1995 southwest monsoon, Deep-Sea Res. **49**, 1211–1230 (2002)

18.28 C. Flagg, S. Smith: On the use of the acoustic Doppler current profiler to measure zooplankton abundance, Deep-Sea Res. **36**, 455–474 (1989)

18.29 A. Gargett: Velcro measurement of turbulence kinetic energy dissipation rate ε, J. Atmos. Ocean. Technol. **16**, 1973–1993 (1999)

18.30 M. Stacey, S. Monismith, J. Burau: Measurements of Reynolds stress profiles in unstratified tidal flow, J. Geophys. Res. **104:10**,)933–949 (1999)

18.31 L. Regier: Mesoscale current fields observed with a shipboard profiling acoustic current meter, J. Phys. Oceanogr. **12**, 880–886 (1982)

18.32 T. Joyce, D. Bitterman, K. Prada: Shipboard acoustic profiling of upper ocean currents, Deep-Sea Res. **29**, 903–913 (1982)

18.33 E. Firing, R. Gordon: *Deep ocean acoustic Doppler current profiling, Proc. IEEE Fourth International Working Conference on Current Measurements*, Clinton, MD (Ocean Engineering Society, Arlington 1990)

18.34 J. Fischer, M. Visbeck: Deep velocity profiling with self-contained ADCPs, J. Atmos. Ocean. Technol. **10**, 764–773 (1993)

18.35 W. Munk, P. Worcester, C. Wunsch: *Ocean Acoustic Tomography* (Cambridge Univ Press, Cambridge 1995)

18.36 http://www.nodc.noaa.gov/OC5/WOA05/pr_woa05.html

18.37 W. Munk, C. Wunsch: Abyssal recipes II: Energetics of tidal and wind mixing, Deep-Sea Res. **26**, 123–161 (1979)

18.38 W. Munk, C. Wunsch: Observing the ocean in the 1990s, Phil. Trans. R. Soc. London **307**, 439–464 (1982)

18.39 B. Dushaw, G. Bold, C.-S. Chiu, J. Colosi, B. Cornuelle, Y. Desaubies, M. Dzieciuch, A. Forbes, F. Gaillard, J. Gould, B. Howe, M. Lawrence,

18.39 J. Lynch, D. Menemenlis, J. Mercer, P. Mikhalevsky, W. Munk, I. Nakano, F. Schott, U. Send, R. Spindel, T. Terre, P. Worcester, C. Wunsch: *Observing the ocean in the 2000's: A strategy for the role of acoustic tomography in ocean climate observation*, Observing the Oceans in the 21st Century , C. Koblinsky and N. Smith (GODAE Project Office and Bureau of Meteorology, Melbourne 2001)

18.40 B. Dushaw, G. Egbert, P. Worcester, B. Cornuelle, B. Howe, K. Metzger: A TOPEX/POSEIDEN global tidal model (TPX0.2) and barotropic tidal currents determined from long-range acoustic transmissions, Prog. Oceanogr. **40**, 337–367 (1997)

18.41 The AMODET Group(T. Birdsall, J. Boyd and B. Cornuelle B. and Howe and R. Knox and J. Mercer and K. Metzger and R. Spindel and P. Worcester): Moving ship tomography in the northwest Atlantic Ocean, EOS Trans. AGU. **75**, 17–23 (1994)

18.42 D. Watts, H. Rossby: Measuring dynamic heights with inverted echo sounders: Results from MODE, J. Phys. Oceanogr. **7**, 345–358 (1977)

18.43 S. Garzoli, A. Bianchi: Time-space variability of the local dynamics of the Malvinas-Brazil confluence as revealed by inverted echo sounders, J. Geophys. Res. **92**, 1914–1922 (1987)

18.44 C. Meinen, D. Watts: Vertical structure and transport on a transect across the North Atlantic Current near 42N: Time series and mean, J. Geophys. Res. **105:21**, 891–869 (2000)

18.45 D. Watts, C. Sun, S. Rintoul: Gravest empirical modes determined from hydrographic observations in the Subantarctic Front, J. Phys. Oceanogr. **31**, 2186–2209 (2001)

18.46 J. Colosi, T. Birdsall, C. Clark, J. Colosi, B. Cornuelle, D. Costa, B. Dushaw, M. Dzieciuch, A. Forbes, B. Howe, D. Menemenlis, J. Mercer, K. Metzger, W. Munk, R. Spindel, P. Worcester, C. Wunsch: A review of recent results on ocean acoustic wave propagation in random media: Basin scales, IEEE J. Ocean. Eng. **24**, 138–155 (1999)

18.47 I. Fukumori: A partitioned Kalman filter and smoother, Mon. Weather Rev. **130**, 1370–1383 (2002)

18.48 D. Di Iorio, A. Gargett: *Sounds in the Sea: From Ocean Acoustics to Acoustical Oceanography*, ed. by H. Medwin (Cambridge Univ Press, Cambridge 2005)

18.49 T. Sanford: Motionally induced electric and magnetic fields in the sea, J. Geophys. Res. **76**, 3476–3492 (1971)

18.50 A. Chave, D. Luther: Low-frequency, motionally-induced electro-magnetic fields in the ocean, Part 1: Theory, Geophys J. Res. **95**, 7185–7200 (1990)

18.51 M. Baringer, J. Larsen: Sixteen years of Florida Current transport at 27 N, Geophys. Res. Lett. **28**, 3179–3182 (2001)

18.52 A. Chave, D. Luther, J. Filloux: Observations of the boundary current system at 26.5 N in the subtropical North Atlantic Ocean, Phys J. Oceanogr. **27**, 1827–1848 (1997)

18.53 T. Sanford: *Proc. of the IEEE Third Working Conference on Current Measurement, Airlie, VA* (IEEE, New York 1986)

18.54 J. Larsen: Transport and heat flux of the Florida Current at 27N derived from cross-stream voltages and profiling data: Theory and observations, Phil. Trans. Roy. Soc. London **338**, 169–236 (1992)

18.55 R. Nolasco, F. Monteiro Santos, A. Soares, N. Palshin, P. Represas, J. Dias: Measurements of motional induction voltage in the Ria de Aveiro lagood (Portugal). In: *Third Int. Workshop of Scientific Use of Submarine Cables and Related Technologies* (IEEE, Piscataway 2003) pp. 127–132

18.56 J. Kasahara: *Proc. Scientific Submarine Cable 2003 Workshop*, ed. by A. Chave (University of Tokyo, Tokyo 2003), IEEE, Piscataway, NJ, 2003

18.57 T. Sanford, R. Drever, J. Dunlap, W. Johns: *Proc. of the IEEE Fifth Working Conference on Current Measurement*, ed. by S. Anderson, G. Appell, A. Williams (William S. Sullwold, Taunton 1995)

18.58 D. Crombie: Proposal for a new Æther drift experiment, Nature **175**, 681–682 (1955)

18.59 D.E. Barrick: *HF Radio Oceanography-Arevier, Boundary-Layer Meteorology* (Springer, Berlin 1978), Historical Archive 4,1-4

18.60 C. Teague, J. Vesecky, D. Fernandez: Hf radar instruments, past to present, Oceanography **10**, 40–44 (1997)

18.61 High frequency radars for coastal oceanography, Oceanography **10**(2) (1997)

18.62 D. Luther, A. Chave: *Observing integrating variables in the ocean, Proc. 7th 'Aha Huliko'a Hawaiian Winter Workshop on Statistical Methods in Physical Oceanography* (University of Hawaii, Honolulu 2003)

18.63 D. Watts, X. Qian, K. Tracey: Mapping abyssal current and pressure fields under the meandering Gulf Stream, J. Atmos. Ocean. Technol. **18**, 1052–1067 (2001)

18.64 C. Hughes, M. Smithson: Bottom pressure correlations in the south Atlantic, Geophys. Res. Lett. **23**, 2243–2246 (1996)

18.65 D. Luther, A. Chave, J. Filloux, P. Spain: Evidence for local and nonlocal barotropic responses to atmospheric forcing during BEMPEX, Geophys. Res. Lett. **17**, 949–952 (1990)

18.66 M. Alford, M. Gregg, M. Merrifield: Structure, propagation and mixing of energetic baroclinic tides in Mamala Bay, Oahu, Hawaii, J. Phys. Oceanogr. **36**, 997–1018 (2006)

18.67 H. Yamamoto, T. Hatayama, Y. Yoshikawa, M. Fukasawa: The first science deployment of a McLane Moored Profiler, JAMSTECR **47**, 1–11 (2003)

18.68 E.D. Asaro: Performance of autonomous Lagrangian floats, Atmos J. Ocean. Technol. **20**, 896–911 (2003)

18.69 C. Eriksen, T. Osse, R. Light, T. Wen, T. Lehman, P. Sabin, J. Ballard, A. Chiodi: Seaglider: A long-range autonomous underwater vehicle for oceanographic research, IEEE J. Ocean. Eng. **26**, 424–436 (2001)

18.70 D. Rudnick, T. Boyd, R. Brainard, G. Carter, G. Egbert, M. Gregg, P. Holloway, J. Klymak, E. Kunze, C. Lee, M. Levine, D. Luther, J. Martin, M. Merrifield, J. Moum, J. Nash, R. Pinkel, L. Rainville, T. Sanford: From tides to mixing along the Hawaiian Ridge, Science **301**, 355–357 (2003)

18.71 A. Jessup: The FAIRS Experiment, IEEE Geosci. Rem. Sens. Soc. Newsletter **123**, 12–16 (2002)

18.72 M. Aziz, S. Reising, W. Asher, L. Rose, P. Gaiser, K. Horgan: Effects of air-sea interaction parameters on ocean surface microwave emission at 10 and 37 Ghz, IEEE Trans. Geosci. Remote Sens. **43**, 1763–1774 (2005)

18.73 W. Asher, A. Jessup, M. Atmane: Oceanic application of the active controlled flux technique for measuring air-sea transfer velocities of heat and gases, J. Geophys. Res. **109**, C08S12 (2004), DOI: 10.1029/2003JC001862

18.74 E. Hintsa, W. Dacey, W. McGillis, J. Edson, C. Zappa, H. Zeelink: Sea-to-air fluxes from measurements of the atmospheric gradient of dimethylsulfide and comparison with simultaneous relaxed eddy accumulation measurements, J. Geophys. Res. **109**, C01026 (2004), DOI: 10.1029/2002JC001617

18.75 J. Gemmrich, D. Farmer: Near-surface turbulence in the presence of breaking waves, , J. Phys. Oceanogr. **34**, 1067–1086 (2004)

19. Microfluidics: The No-Slip Boundary Condition

The no-slip boundary condition at a solid–liquid interface is at the center of our understanding of fluid mechanics. However, this condition is an assumption that cannot be derived from first principles and could, in theory, be violated. In this chapter, we present a review of recent experimental, numerical and theoretical investigations on the subject. The physical picture that emerges is that of a complex behavior at a liquid/solid interface, involving an interplay of many physicochemical parameters, including wetting, shear rate, pressure, surface charge, surface roughness, impurities and dissolved gas.

In Sect. 19.1 we present a brief history of the no-slip boundary condition for Newtonian fluids, introduce some terminology, and discuss cases where the phenomenon of slip (more appropriately, this may often be *apparent slip*) has been observed. In Sect. 19.2 we present the different experimental methods that have been used to probe slip in Newtonian liquids and summarize their results in the form of tables. A short presentation of the principle and results of molecular dynamics simulations is provided in Sect. 19.3, as well as remarks about the relation between simulations and experiments. We then present in Sect. 19.4 an interpretation of experimental and simulation results in light of both molecular and continuum models, organized according to the parameters

19.1	**History of the No-Slip Condition** 1220
	19.1.1 The Previous Centuries 1220
	19.1.2 Terminology 1220
	19.1.3 Traditional Situations Where Slip Occurs 1221
	19.1.4 Newtonian Liquids: Slip or No-Slip . 1222
19.2	**Experimental Methods** 1222
	19.2.1 Indirect Methods......................... 1222
	19.2.2 Local Methods............................. 1224
19.3	**Molecular Dynamics Simulations** 1226
	19.3.1 Principle 1227
	19.3.2 Results.. 1227
	19.3.3 Interpretation in the Continuum Limit................. 1228
19.4	**Discussion: Dependence on Physical Parameters** 1228
	19.4.1 Surface Roughness 1228
	19.4.2 Dissolved Gas and Bubbles............ 1229
	19.4.3 Wetting...................................... 1230
	19.4.4 Shear Rate.................................. 1232
	19.4.5 Electrical Properties..................... 1233
	19.4.6 Pressure 1233
19.5	**Perspective** 1234
	References 1235

upon which slip has been found to depend. We conclude in Sect. 19.5 by offering a brief perspective on the subject.

The vast majority of problems in the dynamics of Newtonian fluids are concerned with solving, in particular settings, the Navier–Stokes equations for incompressible flow

$$\rho(\partial_t + \boldsymbol{u} \cdot \nabla)\boldsymbol{u} = -\nabla p + \mu \nabla^2 \boldsymbol{u} \,, \quad \nabla \cdot \boldsymbol{u} = 0 \,. \tag{19.1}$$

The list of problems for which this task has proven to be difficult is long. However, most of these studies assume the validity of the no-slip boundary condition, i.e., that all three components of the fluid velocity on a solid surface are equal to the respective velocity components of the surface. It is only recently that controlled experiments, generally with typical dimensions microns or smaller, have demonstrated an apparent violation of the no-slip boundary condition for the flow of Newtonian liquids near a solid surface.

We present in this chapter a tentative summary of what is known about the breakdown of the no-slip condition for Newtonian liquids and discuss methods and results of experiments, simulations and theoretical models. This topic is of fundamental physical interest and has potential practical consequences in many areas of engineering and applied sciences where liquids interact with small-scale systems [19.1,2], including flow in porous media, microfluidics, friction studies and biological fluids. Furthermore, since viscous flows are relevant to the study of other physical phenomenon, such as the hydrophobic attraction in water, a change in the boundary condition would have significant quantitative impact on the interpretation of experimental results [19.3–5]. The present chapter complements previous work [19.6–8].

19.1 History of the No-Slip Condition

19.1.1 The Previous Centuries

The nature of boundary conditions in hydrodynamics was widely debated in the 19th century and the reader is referred to [19.9, 10] for historical reviews. Many of the great names in fluid dynamics have expressed an opinion on the subject at some point during their careers, including Bernoulli, Euler, Coulomb, Darcy, Navier, Helmholtz, Poisson, Poiseuille, Stokes, Hagen, Couette, Maxwell, Prandtl and Taylor. In his 1823 treatise on the movement of fluids [19.11], Navier introduced the linear boundary condition (also proposed later by *Maxwell* [19.12]), which remains the standard characterization of slip used today: the component of the fluid velocity tangent to the surface, $u_\parallel$, is proportional to the rate of strain, (or shear rate) at the surface,

$$u_\parallel = \lambda n \cdot \left[\nabla u + (\nabla u)^T\right] \cdot (1 - nn), \qquad (19.2)$$

where n denotes the normal to the surface, directed into the liquid. Alternatively, the right-hand side of this boundary condition, when multiplied by the shear viscosity of the liquid, states that the tangential component of the surface velocity is proportional to the surface shear stress. The velocity component normal to the surface is naturally zero as mass cannot penetrate an impermeable solid surface, $u \cdot n = 0$. In (19.2), λ has the unit of a length, and is referred to as the slip length. For a pure shear flow, λ can be interpreted as the fictitious distance below the surface where the no-slip boundary condition would be satisfied (Fig. 19.1). Note that on a curved surface the rate of strain tensor is different from the normal derivative of the tangential component of the flow so all the terms in (19.2) need to be considered [19.13].

A century of agreement between experimental results in liquids and theories derived assuming the no-slip boundary condition (i.e., $\lambda = 0$) had the consequence that today many textbooks of fluid dynamics fail to mention that the no-slip boundary condition remains an assumption. A few monographs however discuss the topic. In his classic book [19.14, p. 576], *Lamb* realizes that no-slip is the most probable answer but leaves the possibility open for extraordinary cases:

> *It appears probable that in all ordinary cases there is no motion, relative to the solid, of the fluid immediately in contact with it. The contrary supposition would imply an infinitely greater resistance to the sliding of one portion of the fluid past another than to the sliding of the fluid over a solid.*

Similarly, *Batchelor* [19.15, p. 149] offers two paragraphs where the question is discussed in detail, including the role of molecular effects in smoothing out discontinuities. He also mentions the importance of an experimental validation of the no-slip condition:

> *The validity of the no-slip boundary condition at a fluid–solid interface was debated for some years during the last century, there being some doubt about whether molecular interactions at such an interface lead to momentum transfer of the same nature as that at a surface in the interior of a fluid; but the absence of slip at a rigid wall is now amply confirmed by direct observations and by the correctness of its many consequences under normal conditions.*

19.1.2 Terminology

We introduce here some useful terminology that is used throughout this chapter.

Phenomenon of Slip
Refers to any situation in the dynamics of fluids where the value of the tangential component of the velocity

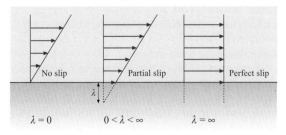

Fig. 19.1 Interpretation of the (Maxwell–Navier) slip length λ

appears to be different from that of the solid surface immediately in contact with it.

Molecular Slip (Also Intrinsic Slip)
Refers to the possibility of using hydrodynamics to force liquid molecules to slip against solid molecules. Such a concept necessarily involves large forces [19.7]. Let us denote by σ a typical molecular length scale and by A the Hamaker constant for the intermolecular forces. Molecular slip will occur when intermolecular interactions $\mathcal{O}(A/\sigma)$ are balanced by viscous forces $\mathcal{O}(\mu\sigma^2\dot{\gamma})$, where μ is the shear viscosity of the liquid and $\dot{\gamma}$ the shear rate; this can only happen for a very large shear rates $\dot{\gamma} \approx A/\mu\sigma^3 \approx 10^{12}\,\text{s}^{-1}$, where we have taken the viscosity of water $\mu = 10^{-3}\,\text{Pa s}$, and typical values $A \approx 10^{-19}\,\text{J}$ and $\sigma \approx 0.3\,\text{nm}$.

Apparent Slip
Refers to the case where there is a separation between a small length scale a where the no-slip condition is valid and a large length scale $L \gg a$ where the no-slip condition appears to not be valid. Well-known examples of such apparent slip include electrokinetics [19.16] (in this case a is the the thickness of the double layer) and acoustic streaming [19.15] (in this case a is the thickness of the oscillatory boundary layer). Similarly, a liquid flowing over a gas layer displays apparent slip (Sect. 19.4).

Effective Slip
Refers to the case where molecular or apparent slip is estimated by averaging an appropriate measurement over the length scale of an experimental apparatus.

19.1.3 Traditional Situations Where Slip Occurs

The phenomenon of slip has already been encountered in three different contexts.

Gas Flow
Gas flow in devices with dimensions that are on the order of the mean free path of the gas molecules shows significant slip [19.17]. An estimate of the mean free path is given by the ideal gas formula, $\ell_m \approx 1/(\sqrt{2}\pi\sigma^2\rho)$, where ρ is the gas density (here taken as the number of molecules per unit volume); for air under standard conditions of temperature and pressure, $\ell_m \approx 100\,\text{nm}$ and, in general, ℓ_m depends strongly on pressure and temperature. The possibility of gas slip was first introduced by *Maxwell* [19.12]. He considered the flow of an ideal gas and assumed that a percentage $(1-p)$ of the wall collisions were specular whereas a percentage (p) were diffuse. Such an assumption allows an exchange of momentum between the gas and the wall. The corresponding slip length is given by

$$\frac{\lambda}{\ell_m} = \frac{2(2-p)}{3p} \,. \tag{19.3}$$

The case of a rough surface with only specular reflections was considered in [19.18]. In general, a Knudsen number defined as the ratio of the mean free path to the system size $\text{Kn} = \ell_m/L$ is used to characterize the boundary condition for gas flow, with slip being important when $\text{Kn} \gtrsim 0.1$ [19.19].

Non-Newtonian Fluids
The flows of non-Newtonian fluids such as polymer solutions show significant apparent slip in a variety of situations, some of which can lead to slip-induced instabilities. This is a topic with a long history and is of tremendous practical importance, and we refer to [19.20–28] and references therein for an appropriate treatment.

Contact Line Motion
In the context of Newtonian liquids, molecular slip has been used as a way to remove singularities arising in the motion of contact lines, as reviewed in [19.29, 30]. Solving the equations of motion with a no-slip boundary condition in the neighborhood of a moving contact line leads to the conclusion that the viscous stresses and the rate of energy dissipation have non-integrable singularities. It was first suggested [19.31] that a local slip boundary condition for the flow would remove the singularity, and indeed it does [19.32–34]). Furthermore, since the slip length appears via a logarithmic factor in a condition involving the contact line speed, it has virtually no influence on macroscopic quantities such as force and pressure drops on scales larger than the capillary length $\ell_c = (\gamma/\rho g)^{1/2}$ [19.33], where γ is the liquid surface tension, ρ the liquid density and g gravity. Consequently, macroscopic measurements on moving droplets cannot be used in general to deduce the exact slip law [19.35]. The slip length could also become velocity dependent due to a combination of microscopic roughness and contact angle hysteresis [19.36]. Such local slip near contact lines was confirmed by early molecular dynamics simulations [19.37]. In a different context but with a similar purpose, slip was used to remove the singularity in the mobility of particles sliding near solid surfaces [19.38, 39].

19.1.4 Newtonian Liquids: Slip or No-Slip

The development of the surface force apparatus in the 1970s [19.40–42] (Sect. 19.2.1) has allowed for more than 30 years of precise probing down to the nanometer scale of both structure and dynamics of many Newtonian liquids against mica [19.43–53]. Experimental methods have included squeeze and/or shear flow for a variety of polar and nonpolar liquids displaying a wide range of wetting conditions and shear rates. With the exception of the flow of toluene over C_{60} (Fullerene)-coated mica [19.54], these studies have confirmed the validity of the no-slip boundary condition and the bulk rheological behavior down to a few nanometers. At smaller length scales, an increase in viscous resistance has been reported, with qualitative differences between the behavior of water and other nonpolar liquids [19.44,49,53,55]. The conclusions have been confirmed by molecular dynamics simulations [19.56] and are consistent with studies of flow in capillaries with diameters of tens of nanometers [19.57,58].

In this context, the large number of recent published experiments reporting some form of (apparent) slip with $\lambda \approx 1$ nm–1 μm in the flow of Newtonian liquids is surprising [19.59–85], and has allowed the rediscovery of a few early studies reporting some degree of slip [19.86–90]. In part, this chapter is an attempt to describe and interpret these more recent experimental results.

19.2 Experimental Methods

As will be discussed below, a large variability exists in the results of slip experiments so it is important first to consider the different experimental methods used to measure slip, directly or indirectly. In these setups, surface conditions may usually be modified by polymer or surfactant adsorption or by chemical modification. Two broad classes of experimental approaches have been used so far: indirect and local methods.

19.2.1 Indirect Methods

Indirect methods assume (19.2) to hold everywhere in a particular configuration and infer λ by measuring a macroscopic quantity. Such methods therefore report effective slip lengths, and they have been the most popular so far. If the effective slip length is λ, then a system size L at least comparable $L \sim \lambda$ is necessary in order for slip to have a measurable impact.

Pressure Drop Versus Flow Rate
This standard technique is used in many studies [19.64, 65,73,87,89], where the main results are summarized in Table 19.1 (other indirect methods for estimating the slip length are summarized in Table 19.2). The dependence of these results on the size of the system was studied in [19.91], where there was some evidence that λ increased with the size of the system. A known pressure drop Δp is applied between the two ends of a capillary or a microchannel and the flow rate $Q = \int u \, dS$ is measured. A slip boundary condition leads to a flow rate $Q(\lambda)$ larger than the no-slip one Q_{NS} by a factor that varies with the ratio of the slip length to the system size; e.g., for a circular pipe of radius a, we get

$$\frac{Q(\lambda)}{Q_{NS}} = 1 + \frac{4\lambda}{a} \,. \tag{19.4}$$

Using this method, we also note that two groups have reported a larger resistance than expected with the no-slip condition in microchannels [19.92] and for flow through small orifices [19.93]. Their results are not well understood but might be due to electrokinetic effects or flow instabilities.

Drainage Versus Viscous Force
This technique consists in imposing the motion (steady or oscillatory) of a curved body perpendicular to a solid surface, and measuring the instantaneous resistive force, which may be compared with that from a model of the fluid motion in the gap, assuming no-slip or slip boundary conditions [19.94–96]. This method is similar in principle to the pressure drop versus flow rate method, with the difference that, here, the pressure and velocity fields are unsteady. The two most common narrow-gap geometries are either a sphere of radius a close to a planar surface or two crossed cylinders of radius a. For both cases, the viscous force F opposing the motion has the form [19.97]

$$F = -f^* \frac{6\pi \mu a^2 V}{D} \,, \tag{19.5}$$

where V is the instantaneous velocity of the moving body, D the minimum distance between the two surfaces, and f^* the slip factor. If the no-slip boundary condition is valid, $f^* = 1$, otherwise when there is slip, $f^* < 1$

Table 19.1 Summary of slip results for pressure drop versus flow rate experiments. The following symbols are used in this table: –: unknown parameter; DDS: dimethyldichlorosilane; TMS: trimethylchlorosilane; CTAB/CTA(+): cetyltrimethyl ammonium bromide; PVP: polyvinylpyridine; OTS: octadecyltrichlorosilane; CCl$_4$: tetrachloromethane; SDS: sodium dodecyl sulfate; pp: peak to peak; rms: root mean square; L: slip independent of shear rate; NL: shear rate dependent

	Surfaces	Liquids	Wetting	Roughness	Shear rates	Slip length	L/NL
Schnell [19.89]	Glass+DDS	Water	–	–	$10^2-10^3\,\text{s}^{-1}$	1–10 μm	L
Churaev [19.87]	Quartz+TMS	Water	70–90°	–	$1\,\text{s}^{-1}$	30 nm	NL
		Mercury	115–130°	–	$10^3-10^4\,\text{s}^{-1}$	70 nm	NL
		CCl$_4$	Complete	–	–	No-slip	–
		Benzene	Complete	–	–	No-slip	–
Kiseleva [19.73]	Quartz+CTA(+)	CTAB solutions	70°	–	$10^2-10^3\,\text{s}^{-1}$	10 nm	L
Cheng [19.64]	Glass+photoresist	Water	–	5 Å (pp)	$10^2-10^4\,\text{s}^{-1}$	No-slip	–
		Hexane	–			10 nm	L
		Hexadecane	–			25 nm	L
		Decane	–			15 nm	L
		Silicon Oil	–			20 nm	L
Cheikh [19.63]	Poly(carbonate)+PVP	SDS solutions	< 90°	–	$0-10^5\,\text{s}^{-1}$	20 nm	L
Choi [19.65]	Silicon	Water	≈ 0°	11 Å (rms)	$10^3-10^5\,\text{s}^{-1}$	0–10 nm	NL
	Silicon+OTS	Water	≳ 90°	3 Å (rms)		5–35 nm	NL

and depends on the slip lengths on both surfaces. The calculation for $f^* = f_{\text{slip}}$ in the case of equal slip lengths is given by [19.39]

$$f_{\text{slip}} = \frac{D}{3\lambda}\left[\left(1+\frac{D}{6\lambda}\right)\ln\left(1+\frac{6\lambda}{D}\right)-1\right], \quad (19.6)$$

and has been extended to account for two different slip lengths [19.97] and for the case of any curved bodies [19.98]. Note that when $D \ll \lambda$, f_{slip} goes to zero as $f_{\text{slip}} \sim D \ln(6\lambda/D)/3\lambda$, so that the viscous force (19.5) only depends logarithmically on D; this is a well-known result in the lubrication limit [19.99].

Two different experimental apparatus have been used to measure drainage forces, the surface force apparatus (SFA) and the atomic force microscope (AFM). The SFA was invented to measure non-retarded van der Waals forces through a gas, with either a static or dynamic method [19.41, 42], and was extended in [19.40] to measure forces between solid surfaces submerged in liquids. More recently it has been used by many groups to measure slip in liquids, with results summarized in Table 19.3. This technique usually uses interferometry to report the separation distance between the smooth surfaces. The moving surface is attached to a spring system of known properties so the difference between imposed and observed motions allows a calculation of the instantaneous force acting on the surfaces.

The AFM was invented by *Binnig* et al. [19.100] and has also been used for many investigations of slip, with experimental results summarized in Table 19.4. A flexible cantilever beam (typically, a few microns wide and hundreds of microns long) with a small (tens of microns) attached colloidal sphere is driven close to a surface, either at its resonance frequency or at fixed velocity, and the deflections of the beam are measured. Since the mechanical properties of the beam are known, deflection measurements can be used to infer the instantaneous drainage force on the colloidal particle.

Sedimentation

This experimental method was used in [19.60], with their results summarized in Table 19.2. The sedimentation speed under gravity of spherical particles of radius a is measured. If the particles are small enough, their motion will occur at small Reynolds number; in that case, the sedimentation velocity with a slip length λ, $v(\lambda)$, is larger than its no-slip counterpart, v_{NS}, according to

$$\frac{v(\lambda)}{v_{\text{NS}}} = \frac{1+3\lambda/a}{1+2\lambda/a}. \quad (19.7)$$

Streaming Potential

This is the experimental technique employed in [19.66], with the results summarized in Table 19.2. A pressure drop is applied to an electrolyte solution between the two ends of a capillary and creates a net flow. Since the surfaces of the capillary acquire in general a net charge in contact with the electrolyte, the net pressure-driven flow

Table 19.2 Summary of alternative experimental methods to infer slip. The symbols used in this table are given in Table 19.1, with additional symbols as: S: sedimentation; FR: fluorescence recovery; PIV: particle image velocimetry; SP: streaming potential; FC: fluorescence cross-correlations; DETMDS: diethyltetramethyldisilazan; FDS: perfluorodecanetrichlorosilane; STA: stearic acid (octadecanoic acid); CDOS: chlorodimethyloctylsilane; Va: vacuum; PDMS: polydimethylsiloxane; KCl: potassium chloride; NaCl: sodium chloride

	Surfaces	Liquids	Wetting	Roughness	Shear rates	Slip length	L/NL
S: *Boehnke* [19.60]	Silica	Propanediol	$\approx 0°$	–	$1\,\mathrm{s}^{-1}$	No-slip	–
		Propanediol+Va		–		$1\,\mu\mathrm{m}$	–
		PDMS	–	–		No-slip	–
	Silica+DETMDS	Propanediol	70–80°	–		No-slip	–
		Propanediol+Va		–		$1\,\mu\mathrm{m}$	–
		PDMS	–	–		No-slip	–
FR: *Pit* [19.71, 76, 77]	Sapphire	Hexadecane	Complete	4 Å (rms)	$2\text{–}10^4\,\mathrm{s}^{-1}$	175 nm	L
	Sapphire+FDS		65°			No-slip	–
	Sapphire+OTS		40°			400 nm	L
	Sapphire+STA		25°			350 nm	L
PIV: *Tretheway* [19.79, 80]	Glass	Water	$\approx 0°$	–	$10^2\,\mathrm{s}^{-1}$	No-slip	–
	Glass+OTS		120°	2 Å		$0.9\,\mu\mathrm{m}$	–
PIV: *Joseph* [19.72]	Glass	Water	$\approx 0°$	5 Å (rms)	$10^2\,\mathrm{s}^{-1}$	50 nm	–
	Glass+OTS		95°			No-slip	–
	Glass+CDOS		95°			50 nm	–
SP: *Churaev* [19.66]	Quartz	KCl solutions	–	2 nm (pp)	$10^5\,\mathrm{s}^{-1}$	No-slip	–
	Quartz+TMS	KCl solutions	80–90°	25 nm (pp)		5–8 nm	–
FC: *Lumma* [19.74]	Mica	Water	–	15 nm (pp)	$10\,\mathrm{s}^{-1}$	0.5–0.86 μm	–
	Glass	Water	5–10°			0.6–1 μm	
		NaCl solutions				0.2–0.6 μm	–

creates an advection-of-charges current that results in a surplus of ions on one end of the capillary, and a deficit in the other end. If the two ends of the capillary are not short-circuited, a net steady-state potential difference, termed the streaming potential, exists between the two ends of the capillary and is such that the current due to advection of net charge near the solid surfaces is balanced by the conduction countercurrent in the bulk of the electrolyte [19.16, 101–103]. If the fluid experiences slip at the wall (and if the ζ-potential is unchanged by the treatment of the surface), a larger current will occur, hence a large potential difference $\Delta V(\lambda)$ given by

$$\frac{\Delta V(\lambda)}{\Delta V_{NS}} = 1 + \lambda\kappa\,, \tag{19.8}$$

where κ is the Debye screening parameter, which gives the typical distance close to the surface where there is a net charge density in the liquid, $\kappa^{-1} = (\epsilon_r\epsilon_0 k_B T/2e^2 n_0)^{1/2}$ [19.16]. Here ϵ_r is the dielectric constant of the liquid, ϵ_0 the permittivity of vacuum, k_B Boltzmann's constant, T the temperature, e the electron charge and n_0 the number density of ions in the bulk of the solution.

19.2.2 Local Methods

All of these methods have the disadvantage that the slip boundary condition (19.2) was not verified directly, but instead was estimated via the assumed effect of slip on some other measured macroscopic parameters. A few techniques have been introduced that try to alleviate this indirect estimation of slip.

Particle Image Velocimetry (PIV)

This method was proposed to investigate slip in [19.72, 79, 80]; we have summarized their results in Table 19.2 [19.104–106]. Let us consider a pressure-driven flow between two parallel plates with separation distance $2h$. In this case, a non-zero slip length λ leads to a velocity field

$$U_{\text{slip}}(z) = -\frac{h^2}{2\mu}\frac{\mathrm{d}p}{\mathrm{d}x}\left(1 - \frac{z^2}{h^2} + \frac{2\lambda}{h}\right)\,, \tag{19.9}$$

Table 19.3 Summary of slip results for experiments using the surface force apparatus (SFA). The symbols used in this table are given in Tables 19.1 and 19.2, with additional symbols as: HDA: 1-hexadecylamine; OTE: octadecyltriethoxysilane; PPO: polystyrene (PS) and polyvinylpyridine (PVP), followed by coating of OTE; PVP/PB: polyvinylpyridine and polybutadiene; PVA: polyvinylalcohol; OMCTS: octamethylcyclotetrasiloxane; av: average; th: polymer thickness. Note that many entries in this table, including the largest slip lengths, are from the same group (S. Granick, U. Illinois)

	Surfaces	Liquids	Wetting	Roughness	Shear rates	Slip length	L/NL
Chan [19.43]	Mica	OMCTS	–	–	$10-10^3 \text{ s}^{-1}$	No-slip	–
		Tetradecane	–	–		No-slip	–
		Hexadecane	–	–		No-slip	–
Israelachvili [19.47]	Mica	Water	–	–	$10-10^4 \text{ s}^{-1}$	No-slip	–
		Tetradecane	–	–		No-slip	–
Horn [19.46]	Silica	NaCl solutions	45°	5 Å (av)	$10-10^3 \text{ s}^{-1}$	No-slip	–
Georges [19.45]	6 surfaces [19.45]	9 liquids [19.45]	–	0.2–50 nm (pp)	$1-10 \text{ s}^{-1}$	No-slip	–
Baudry [19.59]	Cobalt	Glycerol	20–60°	1 nm (pp)	$1-10^4 \text{ s}^{-1}$	No-slip	–
	Gold+thiol		90°			40 nm	L
Cottin-Bizonne [19.68]	Glass	Glycerol	< 5°	1 nm (pp)	$1-10^4 \text{ s}^{-1}$	No-slip	–
	Glass+OTS	Glycerol	95°			50–200 nm	L
		Water	100°			50–200 nm	L
Zhu [19.82]	Mica+HDA	Tetradecane	12°	≈ 1 Å (rms)	$10-10^5 \text{ s}^{-1}$	0–1 µm	NL
	Mica +OTE	Tetradecane	44°			0–1.5 µm	NL
		Water	110°			0–2.5 µm	NL
Zhu [19.84]	Mica+OTS	Water	75–105°	6 nm (rms)	$10-10^5 \text{ s}^{-1}$	No-slip	–
		Tetradecane	12–35°	6 nm (rms)		No-slip	–
	Mica+.8 PPO	Water	85–110°	3.5 nm (rms)		0–5 nm	NL
		Tetradecane	21–38°	3.5 nm (rms)		0–5 nm	NL
	Mica+.2 PPO	Water	90–110°	2 nm (rms)		0–20 nm	NL
		Tetradecane	–	2 nm (rms)		0–20 nm	NL
	Mica+OTE	Water	110°	0.2 nm (rms)		0–40 nm	NL
		Tetradecane	38°	0.2 nm (rms)		0–40 nm	NL
Zhu [19.83]	Mica+PVP/PB	Tetradecane	–	≈ 1 nm (th)	$10-10^5 \text{ s}^{-1}$	No-slip	–
	Mica+PVA	Water	–		$10-10^5 \text{ s}^{-1}$	0–80 nm	NL
Zhu [19.85]	Mica	n-Alkanes	Complete	–	$10-10^5 \text{ s}^{-1}$	No-slip	–
	Mica+HDA	Octane	–	–	$10-10^5 \text{ s}^{-1}$	0–2 nm	NL
		Dodadecane	–	–		0–10 nm	NL
		Tetradecane	12°	–		0–15 nm	NL
Cottin-Bizonne [19.67]	Glass	Dodecane	≈ 0°	1 nm (pp)	10^2-10^4 s^{-1}	No-slip	–
		Water	≈ 0°			No-slip	–
	Glass+OTS	Dodecane	–			No-slip	–
		Water	105°			20 nm	L

which shows that a change in the condition at the boundary has a bulk effect: the (no-slip) Poiseuille flow is augmented by a plug flow. The idea of PIV is then to use small particles as passive tracers in the flow to measure the velocities of the particles with an optical method and check whether the velocities extrapolate to zero at the solid surface. Since small particles have large diffusivities, results need to be averaged to extract the advective part of the tracer motion. Also, the particles in general move relative to the fluid owing to hydrodynamic interactions [19.94], so care is needed when interpreting the measured velocities.

Near-Field Laser Velocimetry
Using Fluorescence Recovery

This is the experimental technique proposed in [19.71, 76, 77], and we have summarized their results in Table 19.2. In this method, the velocity field of small

Table 19.4 Summary of slip results for experiments using the atomic force microscope (AFM colloidal probe). The symbols used in this table are given in Tables 19.1, 2, 3, with additional symbols as: KOH: potassium hydroxide; HTS: hexadecyltrichlorosilane

	Surfaces	Liquids	Wetting	Roughness	Shear rates	Slip length	L/NL
Craig [19.69]	Silica+gold+thiols	Sucrose sol.	40–70°	6 Å (rms)	$10–10^6$ s^{-1}	0–15 nm	NL
Bonaccurso [19.62]	Mica/glass	NaCl solutions	Complete	1 nm (rms)	$10^2–10^6$ s^{-1}	8–9 nm	L
Sun [19.78]	Mica/glass	1-propanol	< 90°	1 nm (rms)	$10^2–10^6$ s^{-1}	10–14 nm	–
Bonaccurso [19.61]	Silicon/glass	Sucrose sol.	Complete	7 Å (rms)	$10^2–10^6$ s^{-1}	0–40 nm	NL
	Silicon/glass+KOH			4 nm (rms)		80 nm	NL
				12.1 nm (rms)		100–175 nm	NL
Neto [19.75]	Silica+gold+thiols	Sucrose sol.	40–70°	6 Å (rms)	$10–10^6$ s^{-1}	0–18 nm	NL
Vinogradova [19.81]	Silica/glass	NaCl solutions	Complete	3 Å (rms)	$10–10^5$ s^{-1}	No-slip	–
	Polystyrene	NaCl solutions	90°	2.5 nm (rms)		4–10 nm	L
Henry [19.70]	Silica/mica	Water	Complete	–	$10^2–10^5$ s^{-1}	80–140 nm	NL
	Silica/mica+CTAB	CTAB solutions	> 90°	–		50–80 nm	NL
Cho [19.107]	Borosilicate+HTS	Octane	13°	3 Å (rms)	$10^2–10^5$ s^{-1}	No-slip	–
		Dodecane	32°			No-slip	–
		Tridecane	35°			10 nm	–
		Tetradecane	37°			15 nm	–
		Pentadecane	39°			10 nm	–
		Hexadecane	39°			20 nm	–
		Cyclohexane	25°			10 nm	–
		Benzene	32°			50 nm	–
		Aniline	64°			50 nm	–
		Water	97°			30 nm	–
		Benzaldehyde	62°			20 nm	–
		Nitrobenzene	63°			10 nm	–
		2-nitroanisole	70°			No-slip	–

fluorescent probes is measured close to a nearby surface. An intense laser illuminates the probes and renders them non-fluorescent (photobleaching). Monitoring the fluorescence intensity in time using evanescent optical waves allows to obtain an estimate of the slip length. Note that the fluorescence intensity evolves in time due to both convection (part which depends on slip) and molecular diffusion, so a careful analysis is needed. Also, because of the fast diffusion of molecular probes, the method is effectively averaging over a diffusion length (typically $\approx 1\,\mu$m), which is much larger than the evanescent wavelength.

Fluorescence Cross-Correlations

This is the most recent experimental method, proposed by [19.74], and their results are also summarized in Table 19.2. Fluorescent probes excited by two similar laser foci are monitored in two small sample volumes separated by a short distance. Cross-correlation of the fluorescence intensity fluctuations due to probes entering and leaving the observation windows allows to determine both the flow direction and intensity. The measured velocities are averaged over the focal size of microscope and the characteristics of the excitation laser.

19.3 Molecular Dynamics Simulations

Molecular dynamics (MD) simulations are useful theoretical tools in the study of liquids [19.56, 108, 109] that have been extensively used to probe boundary conditions. We summarize below the principle of the technique and discuss the interpretation of their results in the continuum limit.

19.3.1 Principle

MD simulations integrate numerically Newton's law of motion for single atoms (or molecules)

$$m_i \frac{d^2 r_i}{dt^2} = \sum_j F_{ij} \,, \quad (19.10)$$

where m_i is the atomic mass, r_i the position of atom i, and F_{ij} the interatomic (or intermolecular) force between atoms i and j, that is $F_{ij} = -\nabla_i V_{ij}$, where V_{ij} is the interaction potential. Potentials used in simulations range from the Lennard–Jones two-body potential

$$V_{ij} = \epsilon \left[\left(\frac{\sigma}{r_{ij}}\right)^{12} - c_{ij} \left(\frac{\sigma}{r_{ij}}\right)^6 \right] \,, \quad (19.11)$$

where ϵ is an energy scale, σ the atomic size, and r_{ij} the distance between atoms i and j, to more-realistic potentials including many-body or orientation-dependent interactions [19.56, 108, 109]. The set of (19.10) are integrated in time, with appropriate numerical cut-offs, and with specified boundary conditions and initial conditions. Usually initial positions are random and initial velocities are taken from a Boltzmann distribution. It is also possible to modify (19.10) slightly to model evolution at constant temperature either by coupling the system of atoms to a heat bath or by a proper rescaling of the velocities at each time step. Interactions with a solid can occur by adding different wall atoms, either fixed on a lattice or coupled to a lattice with a large spring constant, to allow momentum transfer from the liquid but prevent melting. The constants (c_{ij}) in (19.11) allow variation of the relative intermolecular attraction between liquids and solids, which therefore mimics wetting behavior. Using a simple additive model [19.125], the case of a partially wetting fluid with contact angle θ_c can be modeled with

$$\cos \theta_c = -1 + 2 \frac{\rho_S c_{LS}}{\rho_L c_{LL}} \,, \quad (19.12)$$

where ρ_S (ρ_L) is the solid (liquid) density and c_{LS} and c_{LL} are, respectively, the liquid–solid and liquid–liquid intermolecular constants. Finally, two types of flow can be driven. In the first kind, atoms that constitute the wall(s) are driven at a constant velocity and the bulk liquid has a Couette flow profile. In the second kind, each liquid atom is acted upon by a body force and the liquid has a Poiseuille flow profile.

19.3.2 Results

The method described above has been used to study slip in different types of liquids [19.37, 110–124], with results summarized in Table 19.5. Early simulations showed no-slip except near contact lines [19.37, 111]. More recent investigations have reported that molecular

Table 19.5 Summary of MD simulation results for Lennard-Jones liquids with N liquid atoms. List of symbols: HML: heavy-mass lattice; NN: fixed atoms of carbon nanotube; FL: fixed lattice; BF: flow driven by a body force; CF: Couette flow; CL: contact line

	Solid	Flow	N	Wetting	$\frac{k_B T}{\epsilon}$	Results
Koplik [19.37]	HML	BF	1536	0–79°	1.2	No-slip except at CL
Heinbuch [19.110]	FL	BF	915	Complete	0.8–2	$-2\sigma \lesssim \lambda \lesssim 0$
Thompson [19.111]	FL	CF	672–5376	0–90°	1.4	No-slip except at CL
Koplik [19.112]	HML	BF/CF	1536–8000	0–80°	1.2	$\lambda \approx 0$–10σ
Thompson [19.113]	HML	CF	672	$\lesssim 90°$	1.1	$\lambda \approx 0$–2σ
Sun [19.114]	HML	BF	7100	–	1	No-slip except for frozen wall
Thompson [19.115]	FL	CF	1152–1728	0–140°	1.1	$\lambda \approx 0$–60σ
Barrat [19.116]	FL	BF/CF	10 000	90–140°	1	$\lambda \approx 0$–50σ
Jabbarzadeh [19.117]	HML	CF	–	Complete	9	$\lambda \approx 0$–10 nm
Cieplak [19.118]	HML	BF/CF	–	–	1.1	$\lambda \approx 0$–15σ
Fan [19.119]	HML	BF	3800–21 090	Complete	–	$\lambda \approx 0$–5σ
Sokhan [19.120]	NN	BF	2000	–	–	$\lambda \approx 0$–5 nm
Cottin-Bizonne [19.121]	FL	CF	–	110–137°	1	$\lambda \approx 2$–57σ
Galea [19.122]	HML	CF	6000	Complete	1	$-3\sigma \lesssim \lambda \lesssim 4\sigma$
Nagayama [19.123]	FL	BF	2400	0–180°	–	$\lambda \approx 0$–100 nm
Cottin-Bizonne [19.124]	FL	CF	–	110–137°	1	$\lambda \approx 0$–150σ

slip increases with decreasing liquid–solid interactions [19.116, 118, 123], liquid density [19.112, 113], density of the wall [19.115], and decreases with pressure [19.116]. The model for the solid wall, the wall–fluid commensurability and the molecular roughness were also found to strongly influence slip [19.114–117, 122, 126]. Related investigations include contact line motion [19.127] (and references therein), motion of a sliding plate [19.128], and the validity of the Stokes drag formulae at small scales [19.129].

19.3.3 Interpretation in the Continuum Limit

Results from MD simulations can sometimes be difficult to interpret in the continuum limit. First, for computational reasons, simulations to date are limited to tens of thousands of atoms, which restricts the size and time scale of the simulated physical system. The three control parameters in the simulations are the molecular/atomic mass m, the interaction energy ϵ, and the molecular/atomic size σ. Consequently, lengths are measured in units of σ (≈ 3 Å) and times in units of the molecular time scale $\tau \sim \sqrt{m\sigma^2/\epsilon}$ ($\approx 10^{-12}$ s). Simulated systems are therefore limited to tens of nanometers, and time scales to nanoseconds. The consequence of this observation is that MD simulations always probe systems with much higher shear rates than any experimental setup. For example, in MD simulations of Couette flow, the typical wall velocity is $U \sim \sigma/\tau$, corresponding to typical shear rates $\dot{\gamma} \sim \sigma/\tau h$ where h is the typical length scale of the simulation box, usually a few tens of σ. Consequently, $\dot{\gamma} \approx 10^{11}$ s^{-1}, which is orders of magnitude larger than experimental shear rates. Note that this does not apply to investigations inferring slip length from equilibrium simulations [19.126, 130].

A second significant issue in interpreting results of MD simulation was pointed out by Brenner and Ganesan in the case of particle diffusion near a solid surface [19.131], and concerns the scale separation between molecular and continuum phenomena. The idea is that the correct boundary condition in the continuum realm should arise asymptotically as a matching procedure between the outer limit of the inner (molecular) system and the inner limit of the outer (continuum) system. By doing so, the change in the physical behavior within a few intermolecular length scales of the surface is explicitly taken into account, which allows one to make a distinction between *conditions at a boundary* and *boundary conditions*. As a consequence, slip lengths should not be measured literally at the molecular scale but arise as the extrapolation, at the boundaries, of the far field hydrodynamic results, a procedure which is not always performed appropriately.

19.4 Discussion: Dependence on Physical Parameters

Having described the different methods by which slip is investigated, we present in this section a discussion of both experimental and simulation results and compare them with theoretical models. The discussion is organized according to the physical parameters upon which slip has been found to depend.

19.4.1 Surface Roughness

Roughness Influences Resistance
Be it at the molecular size [19.122] or on larger scales [19.45, 61, 76, 84, 117], roughness and geometrical features have been observed to influence the behavior at liquid–solid interfaces. Not only does roughness leads to an ambiguity as to the exact location of the surface, but it impacts the dynamics of the nearby fluid, leading experimentally either to an increase [19.45, 76, 84, 117] or a decrease [19.61] of the liquid friction.

Roughness Decreases Slip
The physical idea for a roughness-induced resistance is straightforward: on the roughness length scale, a flow is induced that dissipates mechanical energy and therefore resists motion. For the same reason, a bubble with a local no-shear surface rises at a finite velocity in a liquid. More generally, geometrical features of size a on a surface can be solely responsible for a large resistance on large scales $L \gg a$, independently of the details of the local boundary condition on the scale a. This feature was first recognized by *Richardson* [19.132] who assumed a periodic perfectly slipping surface shape and performed asymptotic calculations for the limit $a/L \to 0$; in this limit the no-slip boundary condition was recovered [19.133, 134]. The calculation was revisited by *Jansons* [19.135] who considered a small fraction c of roughness elements of size a with a local no-shear condition on an otherwise perfectly slipping surface. Equating the viscous force associated with the disturbance flow created by the defects

$\mathcal{O}(\mu\dot{\gamma}d^2)$ to the local Stokes drag on a defect, $\mathcal{O}(\mu a u_s)$, where $d \sim a/c^{1/2}$ is the typical distance between defects and u_s is the fluid velocity near the defects, leads to an effective slip length for the surface $\lambda = u_s/\dot{\gamma}$ given by

$$\lambda \sim \frac{a}{c}. \qquad (19.13)$$

When c is of order one, all these length scales (a, d, λ) are of the same order and results of [19.132] are recovered. Recently more-rigorous results were derived in [19.136].

When the boundary condition is locally that of no-slip, roughness shifts the position of the effective surface into the liquid. Calculations have been made for periodic and random surfaces [19.137–140] and are related to earlier work on the boundary conditions for porous materials [19.141, 142] and the Laplace equation [19.143].

Roughness-Induced Dewetting

The interaction of roughness with surface energies can lead to the spontaneous dewetting of a surface and the appearance of a super-hydrophobic state, as proposed in [19.121, 124]. In that case, roughness could increase slip by producing regions of gas–liquid interface at the solid boundary. Let us consider for illustration a surface S covered with a fraction c of roughness elements of height a in a liquid at pressure p and let us denote by $r > 1$ the ratio of real to apparent surface area. In that case, the change in free energy ΔG to dewet the apparent area $(1-c)S$ between the roughness elements arises from surface energies and work done against the liquid

$$\Delta G = r(1-c)S(\gamma_S - \gamma_{LS}) + (1-c)S(pa + \gamma), \qquad (19.14)$$

where γ is the surface tension of the liquid, γ_S that of the solid and γ_{LS} the liquid–solid interfacial tension. Consequently, using Young's law, $\gamma \cos\theta_c = \gamma_S - \gamma_{LS}$, we see that dewetting is energetically favorable when $p < -\gamma(1 + r\cos\theta_c)/a$, which, for a given value of the pressure, will occur if the surface is hydrophobic ($\cos\theta_c < 0$) and a is small enough. The super-hydrophobic state is therefore due to a combination of geometry and wetting characteristics.

This idea is related to the so-called fakir droplets [19.144–147] and to more-general drag reduction mechanisms found in nature using gas bubbles [19.148]. Trapped bubbles in rough surfaces were studied by [19.32] in the context of contact line motion and are probably responsible for the apparent slip lengths reported in [19.149–151] for flow over fractal surfaces, and possibly other studies as well (see also [19.91]).

A similar mechanism was quantified experimentally using trapped bubbles in rough silicon wafers [19.152] (see also the calculations in [19.153]) and show promise of decreasing turbulent skin-friction drag [19.154].

19.4.2 Dissolved Gas and Bubbles

Slip Depends on Dissolved Gas
! The amount of slip has been observed experimentally to depend on the type and quantity of dissolved gas. It is reported in sedimentation studies [19.60] that slip was not observed in vacuum conditions but only when the liquid sample was in contact with air. Furthermore, the study in [19.6] showed that tetradecane saturated with CO_2 leads to results consistent with no-slip but significant slip when saturated with argon, whereas the opposite behavior was observed for water. Similar results were reported in [19.155]. More generally, slip results in non-wetting systems are found to depend strongly on the environment in which the experiment is performed [19.67].

Flow Over Gas: Apparent Slip
The results above, together with experiments showing dependence of slip on the absolute value of the pressure [19.155], and spatially varying velocity fields [19.80], hint at the possibility of flow over surface-attached gas pockets or bubbles (see also the discussion in [19.91]). Recent results in [19.67] also point at the possibility of flow over gas pockets associated with the contamination of hydrophobic surfaces by nanoparticles. We also note that the group of Steve Granick reported a contamination of their previous ostensibly smooth mica surfaces by platinum nanoparticles [19.156], possibly affecting some of their experimental results in [19.82–85].

The idea of a flow over a gas layer was first mentioned in [19.157] and revisited in [19.158] as a possible explanation for the attraction between hydrophobic surfaces in water: the attraction could be due to the hydrodynamic correlated fluctuations of the gas interfaces, analogous to the Bjerknes force between two pulsating bubbles. Detailed theoretical considerations have shown that it would be favorable for water between two hydrophobic surfaces to vaporize [19.159]. Flow of binary mixtures have also been shown to phase separate by the sole action of intermolecular forces [19.160].

It is clear that flow over a layer of gas will lead to apparent slip. Since stress must be continuous at a liquid–gas interface, a difference of shear viscosities will lead to a difference of strain rates. If a liquid of viscosity μ_1 flows over a layer of height h with viscosity

μ_2, the apparent slip length for the flow above is [19.97]

$$\frac{\lambda}{h} = \frac{\mu_1}{\mu_2} - 1 ; \qquad (19.15)$$

where $\mu_1/\mu_2 \approx 50$ for a gas–water interface. Three differences exist however between a flow over a gas layer and flow over a set of bubbles:

1. The gas in bubbles recirculates, which decreases the previous estimate (19.15) by about a factor of four;
2. No-slip regions located between the bubbles will also significantly decrease the apparent slip lengths [19.91, 132, 135, 161–163] on the effect of nonuniform slip lengths);
3. Bubbles are in general not flat, which decreases the previous estimates even further (similar to the effect of roughness on a shear flow).

When the gas layer is in the Knudsen regime ($\sigma \ll h \ll \ell_m$), the shear stress in the liquid, $\mathcal{O}(\mu\dot{\gamma})$, is balanced by a purely thermal stress in the gas, $\mathcal{O}(\rho u_s u_{th})$, where ρ is the gas density, u_{th} the thermal velocity $u_{th} = \mathcal{O}(\sqrt{k_B T/m})$ (m is the mass of a gas molecule) and u_s is the liquid velocity at the interface. Balancing these two contributions leads to an apparent slip length, $\lambda = u_s/\dot{\gamma}$, given by [19.164]

$$\lambda \sim \frac{\mu}{\rho u_{th}} , \qquad (19.16)$$

which is independent of h and can be as large as microns. Note that the slip length given by (19.16) increases with the viscosity of the liquid.

Nanobubbles in Polar Liquids?

Over the last four years, many groups have reported experimental observation of nanobubbles against hydrophobic surfaces in water [19.165–175], with typical sizes ≈ 10–100 nm and large surface coverage (see also the reflectivity measurements in [19.176, 177]). The nanobubbles disappear when the liquid is degassed. Similar bubbles could be responsible for slip measurement in some of the experiments to date (see also [19.178, 179]). How could the formation of such bubbles be explained? Thermal fluctuations lead to bubble sizes $a \sim \sqrt{k_B T/\gamma}$, which are on the order of the molecular length. It has been proposed that shear might induce bubbles, but the mechanism is not clear [19.164]. An alternative scenario could be a local decrease in pressure near hydrophobic surfaces due to intermolecular forces (see (4.23) in [19.29]).

The second important issue for nanobubbles is their stability against dissolution. A spherical gas bubble of radius a, diffuses into the liquid on a timescale [19.180, 181]

$$\tau \sim \frac{M p_0 a^2}{D c_0 R T} \left(1 + \frac{p_0 a}{\gamma}\right) , \qquad (19.17)$$

where M is the molar mass, p_0 the far-field pressure, D the diffusion coefficient of the gas in the liquid, T the temperature, c_0 the saturated gas concentration in the liquid (mass per unit volume) at pressure p_0, γ the liquid surface tension and R the absolute gas constant; note that this estimate is independent of surface tension for sufficiently small bubbles. For a 10 nm bubble, $\tau \approx 10\,\mu s$ and τ becomes a few hours when $a \approx 100\,\mu m$. It has therefore been argued that the existence of such small bubbles can only be explained if the liquid is supersaturated with gas [19.165]. In many pressure-driven flow experiments at small scale, a high-pressure gas in contact with the liquid is used to induce the flow; for example pressured gas at 10 atm is used to drive motion in [19.65], equivalent to the internal pressure of a 100 nm bubble. However, the resulting equilibrium between a gas bubble and a supersaturated solution is well-known to be unstable as any perturbation either grows without limit or dissolves away. A possible resolution to the stability problem of such bubbles might come from intermolecular forces in the gas which become important when bubbles reach small radii and large pressures [19.182].

19.4.3 Wetting

Slip Depends on Wetting Properties

It was recognized early that friction at the liquid–solid boundary should be a function of the physicochemical nature of both the solid and the liquid [19.9]. In particular, the wetting properties have been found to play a crucial role in many experiments. Wetting of solids by liquids is reviewed in [19.29, 183] and is quantified by the spreading coefficient, $S = \gamma_S - \gamma - \gamma_{LS}$, which is the difference in surface energy between a dry solid surface and the same surface wet by a liquid layer (γ_S, γ and γ_{LS} are the solid, liquid and liquid–solid interfacial energies, respectively). When $S > 0$, the solid is completely wet by the liquid and when $S < 0$ the wetting is partial. In the latter case, a small liquid droplet on the solid surface would take the shape of a spherical cap with a contact line at an angle θ_c to the solid, where θ_c is the equilibrium contact angle, and is give by Young's law, $\gamma \cos \theta_c = \gamma_S - \gamma_{LS}$. The surface is said to be hydrophobic if $\theta_c > 90°$, and in that case the nucleation of small bubbles in the liquid should occur preferentially on the surface.

Slip has been measured for systems in complete wetting [19.61, 62, 70, 71, 76, 77] and partial wetting [19.59, 60, 63, 65–75, 78–82, 84, 85, 87, 107]. The amount of slip has usually been found to increase with contact angle, either systematically [19.82] or only for nonpolar liquids [19.107]. All these results are summarized in the plot displayed in Fig. 19.2 (left), which shows however, overall, a poor correlation between slip and contact angle.

The Tolstoi Theory

It appears that Tolstoi was the first to try to quantify the importance of surface energies on slip at the molecular level [19.184–186]. Using concepts from macroscopic thermodynamics at the molecular scale, Tolstoi considered the relation between surface energies and molecular mobility (hence diffusivity) near a solid surface by calculating the work it takes for molecules to make room for themselves in the liquid, and how that changes near a boundary. The molecular diffusivity D is given by the product of the molecular scale σ and a velocity $D \sim \sigma V$. The molecular velocity is $V \sim \sigma/\tau$, where τ is the typical time scale for hopping from one molecular position to the other. This is typically an inverse molecular frequency corrected for the energy it takes to create a void of size σ, which is similar to the surface energy $\gamma \sigma^2$. Near a solid, this energy involves the solid and liquid/solid interfacial energies, hence the possibility of having a different molecular mobility close to a surface. In the case of complete wetting, the Tolstoi model leads to the no-slip boundary condition within ± one molecular layer in the liquid, but in the case of partial wetting, molecules near the surfaces are found to have larger mobilities, leading to a slip length [19.184, 186]

$$\frac{\lambda}{\sigma} \sim \exp\left[\frac{\alpha \sigma^2 \gamma (1 - \cos\theta_c)}{k_B T}\right] - 1, \quad (19.18)$$

where α is a dimensionless geometrical parameter of order one, γ is the liquid surface tension and θ_c the equilibrium contact angle. The estimate (19.18) increases with the contact angle and can be orders of magnitude above the molecular length.

Intrinsic (Molecular) Slip

Another theory at the molecular scale uses the fluctuation-dissipation theorem and Green–Kubo relations to derive slip lengths from equilibrium thermodynamics [19.126, 130, 187]. Using Onsager's hypothesis

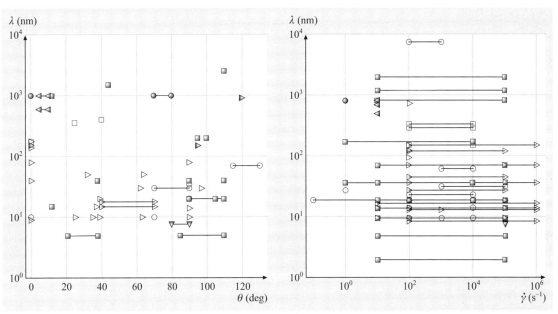

Fig. 19.2 Experimental variation of the slip length λ with the liquid–solid contact angle θ (*left*) and the typical experimental shear rate $\dot{\gamma}$ (*right*) for the experimental results summarized in Tables 19.1, 2, 3, 4: pressure-driven flow (○), sedimentation (●), fluorescence recovery (□), PIV (▶), streaming potential (▼), fluorescence cross-correlations (◀), SFA (■), and AFM (▷). When a solid line is drawn, the experimental results are given for a range of contact angles and/or shear rates. Furthermore, when the value of the contact angle is unknown, the results are not reported

of linear regression of fluctuations, i.e., that small fluctuations around equilibrium can be described by the same equations that describe the relaxation from non-equilibrium, leads to a formula for the time-dependent momentum correlation function in the liquid, function of both the slip length and the wall position. The boundary condition is found to be applied about one molecular layer inside the liquid and the slip length is given by

$$\frac{\lambda}{\sigma} \sim \frac{D^*}{S_t c_{LS}^2 \rho_c \sigma^3},\quad (19.19)$$

where $D^* = D_\parallel/D_0$, $D_\parallel$ is the collective molecular diffusion coefficient, D_0 the bulk diffusivity, S_t the structure factor for first molecular layer (both D^* and S_t are dimensionless numbers of order unity), ρ_c the fluid density at the first molecular layer (number per unit volume), and c_{LS} the dimensionless liquid–solid coefficient of the Lennard–Jones potential (19.11). In the case of complete wetting, the slip length is essentially zero as soon as the roughness is a few percent of the molecular size, but in a non-wetting situation the slip length can be up to two orders of magnitude above molecular size and increases with contact angle [19.187]. As the macroscopic contact angle goes to 180°, the slip length diverges as $\lambda/\sigma \sim 1/(\pi - \theta_c)^4$. The theoretical predictions are found to agree very well with MD simulations [19.116, 187] and the approach was extended to polymer solutions in [19.188]. The details of the molecular slip mechanism were subsequently studied in [19.189], which showed that slip at low shear rates occurs by localized defect propagation and switches to slipping of whole molecular layers for larger rates.

19.4.4 Shear Rate

Slip Depends on Shear Rates
In many investigations, slip was observed to depend on the shear rate at which the experiment or simulation was performed (Fig. 19.2, right). When that is the case, the slip boundary condition, (19.2), becomes nonlinear $\lambda = \lambda(\dot{\gamma})$, and we refer to this situation as NL in Tables 19.1, 2, 3, 4. Shear-dependent slip was reported experimentally in [19.61, 65, 69, 70, 75, 82–85, 87], with the strongest dependence to date in [19.82]. When such a dependence is not observed, (19.2) is linear (L) and the slip length is a property of the liquid–solid pair. Two drainage experiments have also reported linear boundary conditions, with force proportional to velocity in (19.5), but with a profile, for small separations between the surfaces, which differs from that given by a uniform slip length model [19.59, 68].

Most MD simulations report boundary conditions in the linear regime, except in [19.110, 123], where the magnitude of slip was found to depend on the magnitude of the driving force, and in [19.115] where simulations give slip lengths that diverge at high shear rates $\lambda/\sigma \sim (1 - \dot{\gamma}/\dot{\gamma}_c)^{-1/2}$.

The Leaking Mattress Model
A mechanical model for shear-dependent slip in drainage experiments was proposed in [19.190], based on the assumption that a layer of gas bubbles is present on the solid surface. Since drainage experiments are unsteady, the bubble sizes will be a function of time in response to pressure variations in the liquid (by the combination of compression and diffusion), hereby modifying the amount of liquid which is necessary to drain out at each cycle of the oscillation, and therefore modifying the viscous force on the oscillating surface. This idea of flow over a *leaking mattress* leads to a frequency-dependent decrease in the viscous force, $f^*(\omega)$ in (19.5), given by

$$\frac{f^*(\omega)}{f_{\text{slip}}} = \frac{1}{1 + \left[\delta k_1 + \frac{(\omega \delta k_2)^2}{1 + \delta k_1}\right]},$$

$$k_1 = \frac{n a_0^2 I(\theta_c) c_0 \sqrt{\omega \kappa}}{\pi \rho_0 p_0 D a},$$

$$k_2 = \frac{c_0 h_0}{\pi c_\infty p_0 D a}\left[1 + \frac{c_\infty (D - 2h_0)}{\rho_0 h_0}\right],$$

(19.20)

where f_{slip} is the slip factor due to flow over bubbles, given by (19.6) (i.e., f_{slip} is the zero-frequency force decrease, associated with a slip length, λ, which describes the effective resistance of the covered solid surface), and the other terms quantify the dynamic response of bubbles and contribute to an additional force decrease. In (19.20), a is the curvature of the surface, c_∞ is the far-field dissolved gas concentration (mass per unit volume), p_0 is the far-field liquid pressure, c_0 the dissolved gas concentration in equilibrium with gas at pressure p_0, ρ_0 the gas density at pressure p_0, ω the frequency of oscillation of the drainage experiment, D the minimum distance between the two surfaces, μ the liquid shear viscosity, n the number of bubbles per unit area, a_0 the equilibrium radius of curvature of the bubbles, κ the diffusivity of the gas in the liquid, I a geometrical shape factor of order unity, h_0 the mean bubble height on the surface and $\delta = 12\pi\mu a^2 f_{\text{slip}}/D$. Note that $f^*(\omega)$ increases with the liquid viscosity and the curvature of the surfaces. The results of this model compare well with the experiments of [19.82].

The Critical Shear-Rate Model

An empirical model for shear-dependence inspired by the data in [19.82] was also proposed in [19.191]. Slip is assumed to occur locally with a constant slip length λ as soon as the local shear rate reaches a critical value $\dot{\gamma}_c$; below this critical value, the no-slip boundary condition is assumed to remain valid. This model has therefore slip confined to an annular region around the narrow gap where shear rates are the highest. With the two fitting parameters ($\lambda, \dot{\gamma}_c$), the model can reconcile various experimental data, except for small separations of the surfaces.

Viscous Heating

In a steady flow, the rate of dissipation of mechanical energy (which, for a Newtonian fluid, depends on the square of the shear rate) is equal to the rate of change of internal energy due to changes of temperature. Since the viscosity depends on temperature, high shear rates could lead to the possibility of viscous heating [19.192], a flow-induced reduction in viscosity, which could be interpreted as an apparent slip. Assuming a traditional exponential law for the viscosity $\mu = \mu_0 \exp[-\beta(T - T_0)/T_0]$ and flow in a circular capillary of radius a, the apparent slip length due to viscous heating would be [19.91]

$$\frac{\lambda}{a} \sim \frac{\beta}{T_0} \left(\frac{\nu}{\kappa_T}\right) \frac{(\dot{\gamma}a)^2}{c_p}, \tag{19.21}$$

where T_0 is the reference temperature, β a dimensionless coefficient of order one, ν the fluid kinematic viscosity, κ_T the fluid thermal diffusivity and c_p the specific heat. Although viscous heating can be neglected in most experiments to date, it has the potential of becoming important at higher shear rates ([19.193] on the issue of temperature variations).

19.4.5 Electrical Properties

Apparent Slip Depends on Ionic Strength and Polarity

When probing slip in electrolyte solutions and polar liquids, the amount of slip was found to vary with electrical properties. Sedimentation experiments reported that slip was only observed for polar liquids [19.60]. Fluorescence-correlation measurements reported slip lengths of the same order as the screening length κ^{-1}, which decrease with the ionic strength of the solution [19.74]. Drainage experiments also reported that when liquids are polar, slip fails to increase with hydrophobicity but increases with the dipolar moment of the liquid [19.107]; this result was interpreted as a disruption by the drainage flow of the local liquid cohesive energy arising from dipole-image dipole interactions close to the surface [19.107]. Finally, we note that the morphology of nanobubbles has also been observed to depend on pH [19.173] (see also the discussion in [19.194]).

Electrostatic-Induced Averaging

When using small tracers to probe the fluid velocity, electrical effects need to be carefully taken into account. The first issue concerns measurement close to the surface. If the surface and the particles are similarly charged, particles will be repelled electrostatically and will not come within a distance $\sim \kappa^{-1}$ of the surface (if charges are opposite, tracers will stick to the the surface). This effect will therefore increase the average velocity of the particles when compared to what would be expected if electrical effects were not considered. If the averaging window has a height $h > \kappa^{-1}$ above the surface, then the mean flow velocity will appear to be increased by a factor $(1 + 1/\kappa h)$, which, if interpreted as a slip length would give an apparent value given by

$$\frac{\lambda}{h} \sim \frac{1}{\kappa h - 1}. \tag{19.22}$$

Apparent Slip Due to Charged Tracer

The other potential problem with experimental methods using small tracers is the influence of the streaming potential on their motion [19.195]. If the particles are charged, their velocity will also include an electrophoretic component in response to the flow-induced potential difference; moreover, if they are sufficiently charged, this velocity will be able to overcome the electro-osmotic back-flow in the bulk and the particles will move faster than the local liquid, leading to an apparent slip length. If we consider the case of a pressure driven flow between two parallel plates separated by a distance $2h$, the resulting apparent slip length is given by [19.195]

$$\frac{\lambda}{h} \sim \frac{\zeta_w(q\zeta_p - \zeta_w)(\epsilon_r \epsilon_0)^2}{2\sigma_e \mu h^2 + (\epsilon_r \epsilon_0 \zeta_w)^2 \kappa h}, \tag{19.23}$$

where ζ_w is the wall zeta potential, ζ_p the particle zeta potential, q a dimensionless factor of order one depending on the ratio of the screening length to the particle size, μ the shear viscosity of the liquid, σ_e the electrical conductivity of the liquid and ϵ_r its dielectric constant. In the case of low-conductivity electrolytes, such apparent slip length can be as large as hundreds of nanometers.

19.4.6 Pressure

Apparent Slip Depends on Pressure
In the velocimetry experiments of [19.155], the measured slip length was found to decrease with the value of the absolute pressure; for water, the no-slip boundary condition was recovered when the absolute pressure reached 6 atm. Such results are another hint at the likely role of surface-attached bubbles, presumably decreasing in size with an increase in pressure.

Slip Due to Pressure Gradients
The possibility of surface slip due to gradients in liquid pressure was proposed in [19.157] using arguments from equilibrium thermodynamics. The idea is that the chemical potential of a liquid molecule depends on pressure, so a pressure gradient leads to a gradient in chemical potential, hence a net force F on the liquid. Assuming $F \ll k_B T/\sigma$, a molecular model of diffusion under force allows us to get the net surface velocity and estimate the slip length. For a circular pipe of radius a, the slip length is predicted to be

$$\lambda \sim \frac{\mu D_S}{\rho a k_B T}, \qquad (19.24)$$

where D_S is the diffusion coefficient for molecules close to the surface and ρ the molecular density of the liquid (number per unit volume). For regular liquids such as water, the result (19.24) leads to molecular size slip length and suggests that the only way to get larger slip lengths would be for liquid molecules close to the surface to slip over a gas gap [19.157].

19.5 Perspective

Because of the great advances in micro- and nanofabrication technologies, the ability to engineer slip could have dramatic influences on flow since the viscous-dominated motion can lead to large pressure drops and large axial dispersion. As was shown in this chapter, the small-scale interactions between a liquid and a solid leads to extremely rich possibilities for slip behavior, with dependence on factors such as wetting conditions, shear rate, pressure, surface charge, surface roughness and dissolved gas.

We conclude this chapter by presenting a perspective summarizing the interpretation and use of the slip boundary condition (19.2) to describe the motion at a liquid–solid boundary.

- Physically, there is a difference between three different types of slip:
  a) Microscopic slip at the scale of individual molecules,
  b) actual continuum slip at a liquid–solid boundary (i. e., beyond a few molecular layers) and
  c) apparent (and effective) slip due to the motion over complex and heterogeneous boundaries.
- From a practical standpoint however, the distinction is not important. Whether it is real slip or apparent slip due to the interplay of many physical parameters, we have seen in this chapter that a large number of (generally small) experimental systems display some form of reduced resistance to fluid motion.
- The (apparent) slip lengths reported experimentally span many orders of magnitude, from molecular lengths up to hundreds of nanometers. The impact of slip on systems with typical dimensions larger than tens of microns will therefore likely be limited (unless the surfaces have been specifically designed to display super-hydrophobic properties).
- Molecular theories are able to predict intrinsic slip lengths of up to tens of nanometers for hydrophobic systems, suggesting that any measurement of larger slip is affected by factors other than purely fluid dynamical.
- The parameters that contribute to apparent slip include roughness-induced dewetting, the amount and nature of dissolved gas, contamination by impurities and viscous heating. Other parameters that influence the magnitude of apparent slip include contact angle, shear rate, electrical properties and pressure.
- Finally, although it is usually assumed that slip only occurs on hydrophobic surfaces, a large variety of hydrophilic surfaces with different wetting properties have be shown to be prone to slip (Fig. 19.2).

Other more-complex behaviors remain to be understood, including dependence of the results on the molecular shape and size [19.64, 71, 85, 122], probe size [19.74], or viscosity [19.69, 75]. The development of alternative direct experimental methods would allow for a more precise quantification of slip phenomena.

Similarly, it might be valuable to reproduce some of the experiments discussed above in degassed and clean environments to quantify the influence of dissolved gas on apparent slip. Answers to these questions will probably allow for precise engineering of slip in small-scale systems.

References

19.1 T.M. Squires, S.R. Quake: Microfluidics: Fluid physics on the nanoliter scale, Rev. Mod. Phys. **77**, 977–1026 (2005)

19.2 H.A. Stone, A.D. Stroock, A. Ajdari: Engineering flows in small devices: Microfluidics toward a lab-on-a-chip, Annu. Rev. Fluid Mech. **36**, 381–411 (2004)

19.3 O.I. Vinogradova: Possible implications of hydrophobic slippage on the dynamic measurements of hydrophobic forces, J. Phys. Cond. Mat. **8**, 9491–9495 (1996)

19.4 O.I. Vinogradova: Implications of hydrophobic slippage for the dynamic measurements of hydrophobic forces, Langmuir **14**, 2827–2837 (1998)

19.5 O.I. Vinogradova, R.G. Horn: Attractive forces between surfaces: What can and cannot be learned from a jump-in study with the surface forces apparatus?, Langmuir **17**, 1604–1607 (2001)

19.6 S. Granick, Y.X. Zhu, H. Lee: Slippery questions about complex fluids flowing past solids, Nature Mat. **2**, 221–227 (2003)

19.7 P. Tabeling: Slip phenomena at liquid–solid interfaces., C. R. Physique **5**, 531–537 (2004)

19.8 O.I. Vinogradova: Slippage of water over hydrophobic surfaces, Int. J. Mineral Process. **56**, 31–60 (1999)

19.9 S. Goldstein: Note on the condition at the surface of contact of a fluid with a solid body. In: *Modern Development in Fluid Dynamics*, Vol. 2, ed. by S. Goldstein (Clarendon, Oxford 1938) pp. 676–680

19.10 S. Goldstein: Fluid mechanics in first half of this century, Ann. Rev. Fluid Mech. **1**, 1–28 (1969)

19.11 C.L.M.H. Navier: Mémoire sur les lois du mouvement des fluides, Mémoires de l'Académie Royale des Sciences de l'Institut de France **VI**, 389–440 (1823)

19.12 J.C. Maxwell: On stresses in rarefied gases arising from inequalities of temperature, Phil. Trans. R. Soc. Lond. **170**, 231–256 (1879)

19.13 D. Einzel, P. Panzer, M. Liu: Boundary condition for fluid flow – curved or rough surfaces, Phys. Rev. Lett. **64**, 2269–2272 (1990)

19.14 H. Lamb: *Hydrodynamics* (Dover, New York 1932)

19.15 G.K. Batchelor: *An Introduction to Fluid Dynamics* (Cambridge Univ. Press, Cambridge 1967)

19.16 W.B. Russel, D.A. Saville, W.R. Schowalter: *Colloidal Dispersions* (Cambridge Univ. Press, Cambridge 1989)

19.17 E.P. Muntz: Rarefied-gas dynamics, Ann. Rev. Fluid Mech. **21**, 387–417 (1989)

19.18 L. Bocquet: Slipping of a fluid on a surface of controlled roughness, C. R. Acad. Sci. Ser. II **316**, 7–12 (1993)

19.19 M. Gad-el-Hak: The fluid mechanics of microdevices – The Freeman Scholar Lecture, J. Fluids Eng. **121**, 5–33 (1999)

19.20 F. Brochard, P.G. de Gennes: Shear-dependent slippage at a polymer/solid interface, Langmuir **8**, 3033–3037 (1992)

19.21 P.G. de Gennes: Viscometric flows of tangled polymers, C. R. Acad. Sci. Paris B **288**, 219–220 (1979)

19.22 M.M. Denn: Extrusion instabilities and wall slip, Ann. Rev. Fluid Mech. **33**, 265–287 (2001)

19.23 L. Léger, E. Raphael, H. Hervet: Surface-anchored polymer chains: Their role in adhesion and friction, Adv. Polymer Sci. **138**, 185–225 (1999)

19.24 Y. Inn, S.Q. Wang: Hydrodynamic slip: Polymer adsorption and desorption at melt/solid interfaces, Phys. Rev. Lett. **76**, 467–470 (1996)

19.25 A.M. Kraynik, W.R. Schowalter: Slip at the wall and extrudate roughness with aqueous solutions of polyvinyl alcohol and sodium borate, J. Rheol. **25**, 95–114 (1981)

19.26 K.B. Migler, H. Hervet, L. Léger: Slip transition of a polymer melt under shear stress, Phys. Rev. Lett. **70**, 287–290 (1993)

19.27 W.R. Schowalter: The behavior of complex fluids at solid boundaries, J. Non-Newtonian Fluid Mech. **29**, 25–36 (1988)

19.28 S.Q. Wang: Molecular transitions and dynamics at polymer/wall interfaces: Origins of flow instabilities and wall slip, Adv. Polymer Sci. **138**, 227–275 (1999)

19.29 P.G. de Gennes: Wetting – statics and dynamics, Rev. Mod. Phys. **57**, 827–863 (1985)

19.30 E.B.V. Dussan: Spreading of liquids on solid surfaces – Static and dynamic contact lines, Ann. Rev. Fluid Mech. **11**, 371–400 (1979)

19.31 C. Huh, L.E. Scriven: Hydrodynamic model of steady movement of a solid/liquid/fluid contact line, J. Colloid Int. Sci. **35**, 85–101 (1971)

19.32 L.M. Hocking: Moving fluid interface on a rough surface, J. Fluid Mech. **76**, 801–817 (1976)

19.33 E.B.V. Dussan: Moving contact line – slip boundary condition, J. Fluid Mech. **77**, 665–684 (1976)

19.34 E.B.V. Dussan, S.H. Davis: Motion of a fluid-fluid interface along a solid surface, J. Fluid Mech. **65**, 71–95 (1974)

19.35 J. Eggers, H.A. Stone: Characteristic lengths at moving contact lines for a perfectly wetting fluid: The

19.35 influence of speed on the dynamic contact angle, J. Fluid Mech. **505**, 309–321 (2004)

19.36 K.M. Jansons: Moving contact lines at nonzero capillary number, J. Fluid Mech. **167**, 393–407 (1986)

19.37 J. Koplik, J.R. Banavar, J.F. Willemsen: Molecular dynamics of Poiseuille flow and moving contact lines, Phys. Rev. Lett. **60**, 1282–1285 (1988)

19.38 A.M.J. Davis, M.T. Kezirian, H. Brenner: On the Stokes–Einstein model of surface diffusion along solid surfaces: Slip boundary conditions, J. Colloid Int. Sci. **165**, 129–140 (1994)

19.39 L.M. Hocking: Effect of slip on motion of a sphere close to a wall and of two adjacent spheres, J. Eng. Math. **7**, 207–221 (1973)

19.40 J.N. Israelachvili, G.E. Adams: Measurement of forces between two mica surfaces in aqueous-electrolyte solutions in range 0 to 100 nm, J. Chem. Soc. Faraday Trans. I **74**, 975–1001 (1978)

19.41 J.N. Israelachvili, D. Tabor: The measurement of van der Waals dispersion forces in the range 1.5 to 130 nm, Proc. R. Soc. Lond. A **331**, 19–38 (1972)

19.42 D. Tabor, R.H.S. Winterton: The direct measurement of normal and retarded van der Waals forces, Proc. R. Soc. Lond. A **312**, 435–450 (1969)

19.43 D.Y.C. Chan, R.G. Horn: The drainage of thin liquid films between solid surfaces, J. Chem. Phys. **83**, 5311–5324 (1985)

19.44 M.L. Gee, P.M. McGuiggan, J.N. Israelachvili, A.M. Homola: Liquid to solid-like transitions of molecularly thin films under shear, J. Chem. Phys. **93**, 1885–1906 (1990)

19.45 J.M. Georges, S. Millot, J.L. Loubet, A. Tonck: Drainage of thin liquid films between relatively smooth surfaces, J. Chem. Phys. **98**, 7345–7360 (1993)

19.46 R.G. Horn, D.T. Smith, W. Haller: Surface forces and viscosity of water measured between silica sheets, Chem. Phys. Lett. **162**, 404–408 (1989)

19.47 J.N. Israelachvili: Measurement of the viscosity of liquids in very thin films, J. Colloid Int. Sci. **110**, 263–271 (1986)

19.48 J.N. Israelachvili, P.M. McGuiggan, A.M. Homola: Dynamic properties of molecularly thin liquid films, Science **240**, 189–191 (1988)

19.49 J. Klein, E. Kumacheva: Confinement-induced phase transitions in simple liquids, Science **269**, 816–819 (1995)

19.50 J. Klein, E. Kumacheva: Simple liquids confined to molecularly thin layers, I: Confinement-induced liquid-to-solid phase transitions, J. Chem. Phys. **108**, 6996–7009 (1998)

19.51 E. Kumacheva, J. Klein: Simple liquids confined to molecularly thin layers, II: Shear and frictional behaviour of solidified films, J. Chem. Phys. **108**, 7010–7022 (1998)

19.52 U. Raviv, S. Giasson, J. Frey, J. Klein: Viscosity of ultra-thin water films confined between hydrophobic or hydrophilic surfaces, J. Phys. Cond. Mat. **14**, 1–9 (2002)

19.53 U. Raviv, P. Laurat, J. Klein: Fluidity of water confined to subnanometre films, Nature **413**, 51–54 (2001)

19.54 S.E. Campbell, G. Luengo, V.I. Srdanov, F. Wudl, J.N. Israelachvili: Very low viscosity at the solid-liquid interface induced by adsorbed C_{60} monolayers, Nature **382**, 520–522 (1996)

19.55 S. Granick: Motions and relaxations of confined fluids, Science **253**, 1374–1379 (1991)

19.56 M.O. Robbins, M.H. Muser: Computer simulations of friction, lubrication and wear. In: *Handbook of Modern Tribology*, ed. by B. Bhushan (CRC, Boca Raton 2000) pp. 717–765

19.57 J.L. Anderson, J.A. Quinn: Ionic mobility in microcapillaries, J. Chem. Soc. Faraday Trans. I **68**, 608–612 (1972)

19.58 T.K. Knudstrup, I.A. Bitsanis, G.B. Westermann-Clark: Pressure-driven flow experiments in molecularly narrow, straight pores of molecular dimension in mica, Langmuir **11**, 893–897 (1995)

19.59 J. Baudry, E. Charlaix, A. Tonck, D. Mazuyer: Experimental evidence for a large slip effect at a nonwetting fluid-solid interface, Langmuir **17**, 5232–5236 (2001)

19.60 U.C. Boehnke, T. Remmler, H. Motschmann, S. Wurlitzer, J. Hauwede, M. Th. Fischer: Partial air wetting on solvophobic surfaces in polar liquids, J. Colloid Int. Sci. **211**, 243–251 (1999)

19.61 E. Bonaccurso, H.S. Butt, V.S.J. Craig: Surface roughness and hydrodynamic boundary slip of a Newtonian fluid in a completely wetting system, Phys. Rev. Lett. **90**, 144501 (2003)

19.62 E. Bonaccurso, M. Kappl, H.S. Butt: Hydrodynamic force measurements: Boundary slip of water on hydrophilic surfaces and electrokinetics effects, Phys. Rev. Lett. **88**, 076103 (2002)

19.63 C. Cheikh, G. Koper: Stick-slip transition at the nanometer scale, Phys. Rev. Lett. **91**, 156102 (2003)

19.64 J.T. Cheng, N. Giordano: Fluid flow through nanometer-scale channels, Phys. Rev. E **65**, 031206 (2002)

19.65 C.-H. Choi, K. Johan, A. Westin, K.S. Breuer: Apparent slip flows in hydrophilic and hydrophobic microchannels, Phys. Fluids **15**, 2897–2902 (2003)

19.66 N.V. Churaev, J. Ralston, I.P. Sergeeva, V.D. Sobolev: Electrokinetic properties of methylated quartz capillaries, Adv. Colloid Int. Sci. **96**, 265–278 (2002)

19.67 C. Cottin-Bizonne, B. Cross, A. Steinberger, E. Charlaix: Boundary slip on smooth hydrophobic surfaces: Intrinsic effects and possible artifacts, Phys. Rev. Lett. **94**, 056102 (2005)

19.68 C. Cottin-Bizonne, S. Jurine, J. Baudry, J. Crassous, F. Restagno, É. Charlaix: Nanorheology: An investigation of the boundary condition at hydrophobic and hydrophilic interfaces, Eur. Phys. J. E **9**, 47–53 (2002)

19.69 V.S.J. Craig, C. Neto, D.R.M. Williams: Shear-dependent boundary slip in an aqueous Newtonian liquid, Phys. Rev. Lett. **87**, 054504 (2001)

19.70 C.L. Henry, C. Neto, D.R. Evans, S. Biggs, V.S.J. Craig: The effect of surfactant adsorption on liquid boundary slippage, Physica A **339**, 60–65 (2004)

19.71 H. Hervet, L. Léger: Flow with slip at the wall: From simple to complex fluids, C. R. Physique **4**, 241–249 (2003)

19.72 P. Joseph, P. Tabeling: Direct measurement of the apparent slip length, Phys. Rev. E **71**, 035303 (2005)

19.73 O.A. Kiseleva, V.D. Sobolev, N.V. Chuarev: Slippage of the aqueous solutions of cetyltrimethylammonium bromide during flow in thin quartz capillaries, Colloid J. **61**, 263–264 (1999)

19.74 D. Lumma, A. Best, A. Gansen, F. Feuillebois, J.O. Rädler, O.I. Vinogradova: Flow profile near a wall measured by double-focus fluorescence cross-correlation, Phys. Rev. E **67**, 056313 (2003)

19.75 C. Neto, V.S.J. Craig, D.R.M. Williams: Evidence of shear-dependent boundary slip in Newtonian liquids, Eur. Phys. J. E **12**, S71–S74 (2003)

19.76 R. Pit, H. Hervert, L. Léger: Friction and slip of a simple liquid at a solid surface, Trib. Lett. **7**, 147–152 (1999)

19.77 R. Pit, H. Hervert, L. Léger: Direct experimental evidence of slip in hexadecane: Solid interfaces, Phys. Rev. Lett. **85**, 980–983 (2000)

19.78 G. Sun, E. Bonaccurso, V. Franz, H.S. Butt: Confined liquid: Simultaneous observation of a molecularly layered structure and hydrodynamic slip, J. Chem. Phys. **117**, 10311–10314 (2002)

19.79 D.C. Tretheway, C.D. Meinhart: Apparent fluid slip at hydrophobic microchannel walls, Phys. Fluids **14**, L9–L12 (2002)

19.80 D.C. Tretheway, C.D. Meinhart: A generating mechanism for apparent fluid slip in hydrophobic microchannels, Phys. Fluids **16**, 1509–1515 (2004)

19.81 O.I. Vinogradova, G.E. Yakubov: Dynamic effects on force measurements, 2. Lubrication and the atomic force microscope, Langmuir **19**, 1227–1234 (2003)

19.82 Y. Zhu, S. Granick: Rate-dependent slip of Newtonian liquid at smooth surfaces, Phys. Rev. Lett. **87**, 096105 (2001)

19.83 Y. Zhu, S. Granick: Apparent slip of Newtonian fluids past adsorbed polymer layers, Macromolecules **35**, 4658–4663 (2002)

19.84 Y. Zhu, S. Granick: Limits of the hydrodynamic no-slip boundary condition, Phys. Rev. Lett. **88**, 106102 (2002)

19.85 Y. Zhu, S. Granick: No-slip boundary condition switches to partial slip when fluid contains surfactant, Langmuir **18**, 10058–10063 (2002)

19.86 R. Bulkley: Viscous flow and surface films, Bur. Stand. J. Res. **6**, 89–112 (1931)

19.87 N.V. Churaev, V.D. Sobolev, A.N. Somov: Slippage of liquids over lyophobic solid surfaces, J. Colloid Int. Sci. **97**, 574–581 (1984)

19.88 P. Debye, R.L. Cleland: Flow of liquid hydrocarbons in porous Vycor, J. Appl. Phys. **30**, 843–849 (1959)

19.89 E. Schnell: Slippage of water over nonwettable surfaces, J. Appl. Phys. **27**, 1149–1152 (1956)

19.90 J. Traube, S.-H. Whang: Über Reibungskonstante und Wandschicht, Z. Physikal. Chem. A **138**, 102–122 (1928)

19.91 E. Lauga, H.A. Stone: Effective slip in pressure-driven Stokes flow, J. Fluid Mech. **489**, 55–77 (2003)

19.92 J. Pfahler, J. Harley, H. Bau, J. Zemel: Liquid transport in micron and submicron channels, Sensors Actuators **A21–A23**, 431–434 (1990)

19.93 T. Hasegawa, M. Suganuma, H. Watanabe: Anomaly of excess pressure drops of the flow through very small orifices, Phys. Fluids **9**, 1–3 (1997)

19.94 J. Happel, H. Brenner: *Low Reynolds Number Hydrodynamics* (Prentice Hall, Englewood Cliffs 1965)

19.95 B.N.J. Perrson, F. Mugele: Squeeze-out and wear: Fundamental principles and applications, J. Phys.: Condens. Mat. **16**, R295–R355 (2004)

19.96 O. Reynolds: On the theory of lubrication and its application to Mr Beauchamp Tower's experiments, including an experimental determination of the viscosity of olive oil, Phil. Trans. R. Soc. Lond. **177**, 157–234 (1886)

19.97 O.I. Vinogradova: Drainage of a thin liquid-film confined between hydrophobic surfaces, Langmuir **11**, 2213–2220 (1995)

19.98 O.I. Vinogradova: Hydrodynamic interaction of curved bodies allowing slip on their surfaces, Langmuir **12**, 5963–5968 (1996)

19.99 A.J. Goldman, R.G. Cox, H. Brenner: Slow viscous motion of a sphere parallel to a plane wall, I. Motion through a quiescent fluid, Chem. Eng. Sci. **22**, 637–651 (1967)

19.100 G. Binnig, C.F. Quate, C. Gerber: Atomic force microscope, Phys. Rev. Lett. **56**, 930–933 (1986)

19.101 D. Burgeen, F.R. Nakache: Electrokinetic flow in ultrafine capillary slits, J. Phys. Chem. **68**, 1084–1091 (1964)

19.102 R.J. Hunter: *Zeta Potential in Colloid Science: Principles and Applications* (Academic, New York 1982)

19.103 C.L. Rice, R. Whitehead: Electrokinetic flow in a narrow cylindrical capillary, J. Phys. Chem. **69**, 4017–4023 (1965)

19.104 S. Jin, P. Huang, J. Park, J.Y. Yoo, K.S. Breuer: Near-surface velocimetry using evanescent wave illumination, Exp. Fluids **37**, 825–833 (2004)

19.105 J. Yamada: Evanescent wave Doppler velocimetry for a wall's near field, Appl. Phys. Lett. **75**, 1805–1806 (1999)

19.106 C.M. Zettner, M. Yoda: Particle velocity field measurements in a near-wall flow using evanescent wave illumination, Exp. Fluids **34**, 115–121 (2003)

19.107 J.H. Cho, B.M. Law, F. Rieutord: Dipole-dependent slip on Newtonian liquids at smooth solid hydrophobic surfaces, Phys. Rev. Lett. **92**, 166102 (2004)

19.108 M.P. Allen, D.J. Tildesley: *Computer Simulation of Liquids* (Clarendon, Oxford 1987)
19.109 J. Koplik, J.R. Banavar: Continuum deductions from molecular hydrodynamics, Annu. Rev. Fluid Mech. **27**, 257–292 (1995)
19.110 U. Heinbuch, J. Fischer: Liquid flow in pores – Slip, no-slip, or multilayer sticking, Phys. Rev. A **40**, 1144–1146 (1989)
19.111 P.A. Thompson, M.O. Robbins: Simulations of contact line motion – Slip and the dynamic contact-angle, Phys. Rev. Lett. **63**, 766–769 (1989)
19.112 J. Koplik, J.R. Banavar, J.F. Willemsen: Molecular dynamics of fluid flow at solid-surfaces, Phys. Fluids **1**, 781–794 (1989)
19.113 P.A. Thompson, M.O. Robbins: Shear flow near solids – Epitaxial order and flow boundary conditions, Phys. Rev. A **41**, 6830–6837 (1990)
19.114 M. Sun, C. Ebner: Molecular dynamics study of flow at a fluid-wall interface, Phys. Rev. Lett. **69**, 3491–3494 (1992)
19.115 P.A. Thompson, S.M. Troian: A general boundary condition for liquid flow at solid surfaces, Nature **389**, 360–362 (1997)
19.116 J.L. Barrat, L. Bocquet: Large slip effect at a nonwetting fluid-solid interface, Phys. Rev. Lett. **82**, 4671–4674 (1999)
19.117 A. Jabbarzadeh, J.D. Atkinson, R.I. Tanner: Effect of the wall roughness on slip and rheological properties of hexadecane in molecular dynamics simulation of Couette shear flow between two sinusoidal walls, Phys. Rev. E **61**, 690–699 (2000)
19.118 M. Cieplak, J. Koplik, J.R. Banavar: Boundary conditions at a fluid-solid interface, Phys. Rev. Lett. **86**, 803–806 (2001)
19.119 X.J. Fan, N. Phan-Thien, N.T. Yong, X. Diao: Molecular dynamics simulation of a liquid in a complex nano channel flow, Phys. Fluids **14**, 1146–1153 (2002)
19.120 V.P. Sokhan, D. Nicholson, N. Quirke: Fluid flow in nanopores: Accurate boundary conditions for carbon nanotubes, J. Chem. Phys. **117**, 8531–8539 (2002)
19.121 C. Cottin-Bizonne, J.L. Barrat, L. Bocquet, E. Charlaix: Low-friction flows of liquid at nanopatterned interfaces, Nature Mat. **2**, 237–240 (2003)
19.122 T.M. Galea, P. Attard: Molecular dynamics study of the effect of atomic roughness on the slip length at the fluid-solid boundary during shear flow, Langmuir **20**, 3477–3482 (2004)
19.123 G. Nagayama, P. Cheng: Effects of interface wettability on microscale flow by molecular dynamics simulation, Int. J. Heat Mass Transfer **47**, 501–513 (2004)
19.124 C. Cottin-Bizonne, C. Barentin, E. Charlaix, L. Boequet, J.L. Barrat: Dynamics of simple liquids at heterogeneous surfaces: Molecular dynamics simulations and hydrodynamic description, Eur. Phys. J. E **15**, 427–438 (2004)
19.125 J.N. Israelachvili: *Intermolecular and Surface Forces* (Academic, London 1992)
19.126 L. Bocquet, J.L. Barrat: Hydrodynamic boundary conditions and correlation functions of confined fluids, Phys. Rev. Lett. **70**, 2726–2729 (1993)
19.127 J.B. Freund: The atomic detail of a wetting/dewetting flow, Phys. Fluids **15**, L33–L36 (2003)
19.128 J. Koplik, J.R. Banavar: Corner flow in the sliding plate problem, Phys. Fluids **7**, 3118–3125 (1995)
19.129 M. Vergeles, P. Keblinski, J. Koplik, J.R. Banavar: Stokes drag and lubrication flows: A molecular dynamics study, Phys. Rev. E **53**, 4852–4864 (1996)
19.130 L. Bocquet, J.L. Barrat: Hydrodynamic boundary conditions, correlation functions, and Kubo relations for confined fluids, Phys. Rev. E **49**, 3079–3092 (1994)
19.131 H. Brenner, V. Ganesan: Molecular wall effects: Are conditions at a boundary "boundary conditions"?, Phys. Rev. E **61**, 6879–6897 (2000)
19.132 S. Richardson: On the no-slip boundary condition, J. Fluid Mech. **59**, 707–719 (1973)
19.133 J.F. Nye: A calculation on sliding of ice over a wavy surface using a Newtonian viscous approximation, Proc. R. Soc. Lond. A **311**, 445–467 (1969)
19.134 J.F. Nye: Glacier sliding without cavitation in a -linear viscous approximation, Proc. R. Soc. Lond. A **315**, 381–403 (1970)
19.135 K.M. Jansons: Determination of the macroscopic (partial) slip boundary condition for a viscous flow over a randomly rough-surface with a perfect slip microscopic boundary condition, Phys. Fluids **31**, 15–17 (1988)
19.136 J. Casado-Diaz, E. Fernandez-Cara, J. Simon: Why viscous fluids adhere to rugose walls: A mathematical explanation, J. Diff. Eq. **189**, 526–537 (2003)
19.137 M.J. Miksis, S.H. Davis: Slip over rough and coated surfaces, J. Fluid Mech. **273**, 125–139 (1994)
19.138 I.V. Ponomarev, A.E. Meyerovich: Surface roughness and effective stick-slip motion, Phys. Rev. E **67**, 026302 (2003)
19.139 K. Sarkar, A. Prosperetti: Effective boundary conditions for Stokes flow over a rough surface, J. Fluid Mech. **316**, 223–240 (1996)
19.140 E.O. Tuck, A. Kouzoubov: A laminar roughness boundary condition, J. Fluid Mech. **300**, 59–70 (1995)
19.141 S. Richardson: Model for boundary condition of a porous material, Part 2., J. Fluid Mech. **49**, 327–336 (1971)
19.142 G.I. Taylor: Model for boundary condition of a porous material, Part 1., J. Fluid Mech. **49**, 319–326 (1971)
19.143 K. Sarkar, A. Prosperetti: Effective boundary conditions for the Laplace equation with a rough boundary, Proc. R. Soc. Lond. A **451**, 425–452 (1995)
19.144 J. Bico, C. Marzolin, D. Quere: Pearl drops, Europhys. Lett. **47**, 220–226 (1999)

19.145 A.B.D. Cassie, S. Baxter: Wettability of porous surfaces, Trans. Faraday Soc. **40**, 546–551 (1944)

19.146 T. Onda, S. Shibuichi, N. Satoh, K. Tsujii: Super-water-repellent fractal surfaces, Langmuir **12**, 2125–2127 (1996)

19.147 R.N. Wenzel: Resistance of solid surfaces to wetting by water, Ind. Eng. Chem. **28**, 988–994 (1936)

19.148 D.M. Buschnell: Drag reduction in nature, Ann. Rev. Fluid Mech. **23**, 65–79 (1991)

19.149 K. Watanabe, S. Ogata: Drag reduction for a rotating disk with highly water-repellent wall, JSME Int. J. Ser. B **44**, 556–560 (1998)

19.150 K. Watanabe, Y. Udagawa, H. Udagawa: Drag reduction of Newtonian fluid in a circular pipe with a highly water-repellent wall, J. Fluid Mech. **381**, 225–238 (1999)

19.151 K. Watanabe, Yanuar, H. Mizunuma: Slip of Newtonian fluids at solid boundary, JSME Int. J. Ser. B **44**, 525–529 (1998)

19.152 J. Ou, B. Perot, J.P. Rothstein: Laminar drag reduction in microchannels using ultrahydrophobic surfaces, Phys. Fluids **16**, 4635–4643 (2004)

19.153 C.Y. Wang: Flow over a surface with parallel grooves, Phys. Fluids **15**, 1114–1121 (2003)

19.154 T.G. Min, J. Kim: Effects of hydrophobic surface on skin-friction drag, Phys. Fluids **16**, L55–L58 (2004)

19.155 D. Tretheway, S. Stone, C. Meinhart: Effects of absolute pressure and dissolved gases on apparent fluid slip in hydrophobic microchannels, Bull. Am. Phys. Soc. **49**, 215 (2004)

19.156 Z.Q. Lin, S. Granick: Platinum nanoparticles at mica surfaces, Langmuir **19**, 7061–7070 (2003)

19.157 E. Ruckenstein, P. Rajora: On the no-slip boundary condition of hydrodynamics, J. Colloid Int. Sci. **96**, 488–491 (1983)

19.158 E. Ruckenstein, N. Churaev: A possible hydrodynamic origin of the forces of hydrophobic attraction, J. Colloid Int. Sci. **147**, 535–538 (1991)

19.159 K. Lum, D. Chandler, J.D. Weeks: Hydrophobicity at small and large length scales, J. Phys. Chem. B **103**, 4570–4577 (1999)

19.160 D. Andrienko, B. Dunweg, O.I. Vinogradova: Boundary slip as a result of a prewetting transition, J. Chem. Phys. **119**, 13106–13112 (2003)

19.161 J.R. Philip: Flows satisfying mixed no-slip and no-shear conditions, Z.A.M.P. **23**, 353–372 (1972)

19.162 J.R. Philip: Integral properties of flows satisfying mixed no-slip and no-shear conditions, Z.A.M.P. **23**, 960–968 (1972)

19.163 A.A. Alexeyev, O.I. Vinogradova: Flow of a liquid in a nonuniformly hydrophobized capillary, Colloids Surf. A **108**, 173–179 (1996)

19.164 P.G. de Gennes: On fluid/wall slippage, Langmuir **18**, 3413–3414 (2002)

19.165 P. Attard, M.P. Moody, J.W.G. Tyrrell: Nanobubbles: The big picture, Physica A **314**, 696–705 (2002)

19.166 M. Holmberg, A. Kuhle, J. Garnaes, K.A. Morch, A. Boisen: Nanobubble trouble on gold surfaces, Langmuir **19**, 10510–10513 (2003)

19.167 N. Ishida, T. Inoue, M. Miyahara, K. Higashitani: Nano bubbles on a hydrophobic surface in water observed by tapping-mode atomic force microscopy, Langmuir **16**, 6377–6380 (2000)

19.168 S.T. Lou, J.X. Gao, X.D. Xiao, X.J. Li, G.L. Li, Y. Zhang, M.Q. Li, J.L. Sun, X.H. Li, J. Hu: Studies of nanobubbles produced at liquid/solid interfaces, Mat. Charac. **48**, 211–214 (2002)

19.169 A.C. Simonsen, P.L. Hansen, B. Klosgen: Nanobubbles give evidence of incomplete wetting at a hydrophobic interface, J. Colloid Int. Sci. **273**, 291–299 (2004)

19.170 R. Steitz, T. Gutberlet, T. Hauss, B. Klösgen, R. Krastev, S. Schemmel, A.C. Simonsen, G.H. Findenegg: Nanobubbles and their precursor layer at the interface of water against a hydrophobic substrate, Langmuir **19**, 2409–2418 (2003)

19.171 M. Switkes, J.W. Ruberti: Rapid cryofixation/freeze fracture for the study of nanobubbles at solid–liquid interfaces, Appl. Phys. Lett. **84**, 4759–4761 (2004)

19.172 J.W.G. Tyrrell, P. Attard: Images of nanobubbles on hydrophobic surfaces and their interactions, Phys. Rev. Lett. **87**, 176104 (2001)

19.173 J.W.G. Tyrrell, P. Attard: Atomic force microscope images of nanobubbles on a hydrophobic surface and corresponding force-separation data, Langmuir **18**, 160–167 (2002)

19.174 X.H. Zhang, X.D. Zhang, S.T. Lou, Z.X. Zhang, J.L. Sun, J. Hu: Degassing and temperature effects on the formation of nanobubbles at the mica/water interface, Langmuir **20**, 3813–3815 (2004)

19.175 X.H. Zhang, H. Jun: Nanobubbles at the solid/water interface, Prog. Chem. **16**, 673–681 (2004)

19.176 T.R. Jensen, M.O. Jensen, N. Reitzel, K. Balashev, G.H. Peters, K. Kjaer, T. Bjornholm: Water in contact with extended hydrophobic surfaces: Direct evidence of weak dewetting, Phys. Rev. Lett. **90**, 086101 (2003)

19.177 D. Schwendel, T. Hayashi, R. Dahint, A. Pertsin, M. Grunze, R. Steitz, F. Schreiber: Interaction of water with self-assembled monolayers: Neutron reflectivity measurements of the water density in the interface region, Langmuir **19**, 2284–2293 (2003)

19.178 O.I. Vinogradova, G.E. Yakubov, H.J. Butt: Forces between polystyrene surfaces in water-electrolyte solutions: Long-range attraction of two types?, J. Chem. Phys. **114**, 8124–8131 (2001)

19.179 G.E. Yakubov, H.J. Butt, O.I. Vinogradova: Interaction forces between hydrophobic surfaces. Attractive jump as an indication of formation of "stable" submicrocavities, J. Phys. Chem. B **104**, 3407–3410 (2000)

19.180 P.S. Epstein, M.S. Plesset: On the stability of gas bubbles in liquid–gas solutions, J. Chem. Phys. **18**, 1505–1509 (1950)

19.181 S. Ljunggren, J.C. Eriksson: The lifetime of a colloid-sized gas bubble in water and the cause of the hydrophobic attraction, Colloids Surf. A **130**, 151–155 (1997)

19.182 R.A. Wentzell: Van der Waals stabilization of bubbles, Phys. Rev. Lett. **56**, 732–733 (1986)

19.183 P.-G.F. de Gennes Brochard-Wyart, D. Quéré: *Capillarity and Wetting Phenomena: Drops, Bubbles, Pearls, Waves* (Springer, Berlin, Heidelberg 2004)

19.184 T.D. Blake: Slip between a liquid and a solid – D.M. Tolstoi (1952) theory reconsidered, Colloids Surf. **47**, 135–145 (1990)

19.185 J. Frenkel: *Kinetic Theory of Liquids* (Dover, New York 1955)

19.186 D.M. Tolstoi: Molecular theory for slippage of liquids over solid surfaces, Doklady Akad. Nauk SSSR **85**, 1089 (1952), in Russian

19.187 J.L. Barrat, L. Bocquet: Influence of wetting properties on hydrodynamic boundary conditions at a fluid/solid interface, Faraday Disc. **112**, 119–127 (1999)

19.188 N.V. Priezjev, S.M. Troian: Molecular origin and dynamic behavior of slip in sheared polymer films, Phys. Rev. Lett. **92**, 018302 (2004)

19.189 S. Lichter, A. Roxin, S. Mandre: Mechanisms for liquid slip at solid surfaces, Phys. Rev. Lett. **93**, 086001 (2004)

19.190 E. Lauga, M.P. Brenner: Dynamic mechanisms for apparent slip on hydrophobic surfaces, Phys. Rev. E **70**, 026311 (2004)

19.191 H. Spikes, S. Granick: Equation for slip of simple liquids at smooth solid surfaces, Langmuir **19**, 5065–5071 (2003)

19.192 J. Gavis, R.L. Laurence: Viscous heating in plane and circular flow between moving surfaces, I&EC Fundamentals **7**, 232–239 (1968)

19.193 W. Urbanek, J.N. Zemel, H.H. Bau: An investigation of the temperature dependence of Poiseuille numbers in microchannel flow, J. Micromech. Microeng. **3**, 206–208 (1993)

19.194 N.F. Bunkin, O.A. Kiseleva, A.V. Lobeyev, T.G. Movchan, B.W. Ninham, O.I. Vinogradova: Effect of salts and dissolved gas on optical cavitation near hydrophobic and hydrophilic surfaces, Langmuir **13**, 3024–3028 (1997)

19.195 E. Lauga: Apparent slip due to the motion of suspended particles in flow of electrolyte solutions, Langmuir **20**, 8924–8930 (2004)

20. Combustion Diagnostics

Combustion processes consist of a complex multidimensional interaction between fluid mechanics and chemical kinetics. A comprehensive experimental analysis needs therefore measurements of flow and scalar fields. These measurements need to be performed in-situ with high temporal and spatial resolution as well as high accuracy and precision. In addition, any disturbances during the measurement should be avoided. These requirements are fulfilled best by laser-optical techniques. Whereas flow fields are commonly measured by methods like laser Doppler or particle imaging velocimetry discussed elsewhere, the focus of this chapter is on scalar measurements based on spectroscopy. Scalars of interest are temperatures, chemical species concentrations, or rate of mixing between fuel and oxidant. Following an introduction, Sect. 20.3 presents the interconnection between experimental analysis and numerical simulation of combustion processes. In Sect. 20.4, various spectroscopic techniques are described exemplary in their application to different fields of combustion research. The chapter concludes with aspects on future developments in combustion diagnostics.

20.1	Basics	1242
20.2	Laser-Based Combustion Diagnostics	1243
20.3	Experimental Data Devoted to Validation of Numerical Simulations and Modeling	1244
	20.3.1 General Remarks	1244
	20.3.2 Submodels and Their Specific Demands for Validation	1245
	20.3.3 Example: Generic Turbulent Flame	1246
20.4	Application of Laser-Based Techniques	1247
	20.4.1 Detection Sensitivity, Selectivity and Resolution	1247
	20.4.2 Laminar Flames	1249
	20.4.3 Turbulent Combustion	1251
	20.4.4 Engine Combustion	1265
	20.4.5 Diagnostics for Stationary, Large-Scale Combustion Processes	1284
20.5	Conclusions	1299
References		1300

Abbreviations and Acronyms

AM	amplitude modulation
AS	absorption spectroscopy
CA	crank angle
CARS	coherent anti-Stokes Raman scattering
CCD	charge-coupled device
CFD	computational fluid dynamics
CL	chemiluminescence
CLD	chemiluminescence detector
CMD	count median diameter
CRDLAS	cavity-ring-down laser-absorption spectroscopy
DFB-DL	distributed feedback diode lasers
DFG	difference-frequency generation
DFWM	degenerate four-wave mixing
DNS	direct numerical simulation
EGR	exhaust gas recirculation
ELIF	excimer-laser-induced fragmentation fluorescence
FARLIF	fuel/air ratio LIF
FWHM	full width at half maximum
FIR	far IR
FMS	frequency-modulation spectroscopy
FTIR	Fourier transform infrared
HCCI	homogeneous charge compression ignition
IC	internal combustion
ICCD	intensified CCD
ICLAS	intra-cavity laser-absorption spectroscopy
IR	infra red
LAS	laser-absorption spectroscopy
LES	large-eddy simulation
LDV	laser Doppler velocimetry
LIF	laser-induced fluorescence
LII	laser-induced incandescence

MS	mass spectrometry	REMPI	Resonance-enhanced multiphoton ionization
MIR	mid IR		
NDIR	nondispersive infrared photometer	RMS	root mean square
NIR	near IR	SFG	sum frequency generation
NIR-DL	near-IR diode laser	SHG	second harmonic generation
OD	optical density	TEM	transmission electron microscopy
OPO	optical parametric oscillator	TDC	top dead center
PAH	polycyclic aromatic hydrocarbons	TDL	tunable diode laser
PDA	phase Doppler anemometry	TDLAS	tunable diode laser absorption spectroscopy
PDF	probability density function		
PIV	particle image velocimetry	THz	terahertz
PS	polarization spectroscopy	TiRe-LII	time-resolved LII
QCL	quantum cascade laser	UV	ultraviolet
RANS	Reynolds-averaged Navier–Stokes equations	VCSEL	vertical cavity surface emitting laser
		WMS	wavelength-modulation spectroscopy

20.1 Basics

Combustion is the oldest and one of the most successful technologies of mankind. An early important spectroscopic observation in a combustion process was made in 1859 by Robert Wilhelm *Bunsen* and Gustav Robert *Kirchhoff* [20.1], who recognized the origin of some of the famous dark lines in the solar spectrum first seen by Wollaston and indexed in 1814 by Josef Fraunhofer. Bunsen and Kirchhoff realized that it could not happen by chance that 60 of the Fraunhofer lines coincided with 60 iron emission lines measured in the hot nonluminous flame gases of the Bunsen burner. Their experimental success, however, was based on Isaac Newton, who introduced the word spectrum into physics. After Bunsen and Kirchhoff, spectroscopy in flames became a rapidly growing field. However, the meaning of all the observed lines was not known until 1913 when Niels Bohr formulated his model of the atom. A few years later, Albert *Einstein* [20.2] published his famous paper *Zur Quantentheorie der Strahlung* in which he derived Planck's law and Bohr's rule by discussing the possible elementary process of energy and momentum exchange through radiation. Despite Einstein's clear description of the stimulated emission, the practical application of this principle was achieved only many years later in 1955, first in the microwave region with the maser (an acronym for microwave amplification by stimulated emission of radiation) [20.3] and in 1960, in the visible spectral region with the laser (an acronym for light amplification by stimulated emission of radiation laser) [20.4]. The principle of stimulated emission has now been realized in all states of the matter: solids, liquids, gases and free electrons. This allows the generation of coherent radiation from the far-infrared to the X-ray region. Compared to flame spectroscopy with conventional light sources [20.5] the introduction of tunable lasers and the development of nonlinear optical techniques [20.6] greatly expanded the possibilities of combustion spectroscopy. Now the spectroscopic states of atoms and molecules in combustion process can be observed non-intrusively with high temporal, spectral and spatial resolution. An excellent overview can be found in the book by *Eckbreth* [20.7] and *Kohse-Höinghaus* and *Jeffries* [20.8].

Combustion processes consist of a complex multidimensional and time-dependent interaction of a large number of elementary chemical reactions with different transport processes for mass, momentum and energy. To increase the rate of chemical conversion in practical applications, turbulent flow conditions are mostly applied. Similar to the laminar case, turbulent reactive flows can be described by solving the conservation equations for total species mass, momentum (Navier–Stokes equations) and enthalpy [20.9]. However, solving the conservation equations by direct numerical simulation (DNS) is, even in the days of modern parallel computing, a very demanding task. For a realistic system of liquid hydrocarbon oxidation in an internal combustion engine one would need more than 10^{21} computing steps. Therefore, at present and for the near future direct numerical simulation of three-dimensional turbulent reactive flows in technical combustion systems will not be possible. Instead, turbulence has to be modeled either by the use of Reynolds-averaged Navier–Stokes (RANS) or spatially filtered conservation equations (large-eddy simulation, LES). Reaction rates can be evaluated with the help of probability density functions (PDF). These models can then be validated by multidimensional non-intrusive laser measurements. Quantities of interest are multifarious and include temperature, chemical species concentration, flow velocities, or their turbulent fluctuations. In addition, such laser techniques can be applied directly in practical combustion systems to study details of the complex interaction of chemical kinetics and transport processes and act as sensors in active control loops to improve fuel efficiency and reduce the environmental effects of combustion.

20.2 Laser-Based Combustion Diagnostics

Laser-based combustion diagnostics as well as other optical diagnostic methods extensively exploit the direct interaction of light and matter, providing an important tool to observe the spectroscopic states of molecules and atoms with high spectral and spatial resolution in time regimes from hours to the femtosecond (10^{-15} s) scale, using optical power densities from picowatts (10^{-12} W) to multi-terawatts (10^{12} W) per square centimeter. A main feature of optical techniques is the non-intrusive in-situ investigation of the sample without significant interference between the sample and the measuring device itself. This is especially important for reactive and aggressive environments such as combustion processes where probe sampling techniques are of questionable value. In all cases light with a certain intensity and temporal and spectral distribution is sent onto or through a sample. The energy re-emitted (in most cases in the form of light) from the interaction region of the irradiated sample is observed and used to characterize the composition and/or structure of the sample. This is done by a detailed analysis of the intensity, spectral composition or angular and temporal distribution of the re-emitted light. Only absorption techniques determine the losses of the incoming light along the line of sight.

Excellent reviews about the advancement of laser-based combustion diagnostics are presented in [20.10] and [20.8]. In general, laser combustion diagnostics can be classified into flow field measurements, scalar field measurements and combined flow–scalar diagnostics. Flow field measurements rely either on Doppler shift (laser Doppler velocimetry, LDV), Doppler broadening (filtered Rayleigh scattering), tracking of particles (particle imaging velocimetry, PIV) or spatial structures of molecular tracers (molecular tagging velocimetry and gaseous image velocimetry). As these flow field diagnostics are discussed in detail in Chap. 5 here the focus is on measurement of scalars such as temperature and chemical species concentrations. However, it needs to be pointed out that high-temperature environments in flames demand for an adaptation of these afore mentioned flow diagnostics. For example, oil droplets commonly used as seed material tracking the turbulent flow in cold PIV or LDV applications are not feasible in flames due to their evaporation. Instead, solid metal-oxide particles with high melting points are used. These materials are more difficult to seed into the turbulent flow. It must be checked that the presence of the seed particles does not impart on the thermokinetic state of the flame. Otherwise, chemical kinetics might be affected by the seed material and the technique is not anymore non-intrusive. For this reason, seeding densities must be fairly low. In consequence, spatial resolution such as in PIV applications or data rates in LDV especially in hot parts of the flame are often lower than in cold flow applications.

Laser-induced fluorescence (LIF) with excitation in the ultraviolet (UV) is frequently used for the detection of minor combustion-related species such as NO, OH, CH, C_2, HCHO. Fluorescing organic molecules have found interest as fluorescent tracers that allow the quantitative observation of fluid mixing processes as well as fuel concentration in combustion devices [20.11, 12]. Further techniques using laser radiation include: optical absorption [20.13, 14] and cavity ring-down [20.15] of CH_3, and two-photon LIF of atoms [20.16]. Raman scattering is frequently used to simultaneously measure the concentration of major combustion species and temperature [20.17]. Nonlinear techniques such as coherent anti-Stokes Raman scattering (CARS) [20.18], degenerate four-wave mixing (DFWM) [20.19] and polarization spectroscopy (PS) [20.20] have been applied for combustion diagnostics. Elastic scattering off droplets is frequently used for observing the droplet distribution in spray systems and Rayleigh scattering from the gas phase is used to measure gas density, and hence, temperature [20.21]. Particle sizes and concentrations are observed by laser-heating the particles and detection of the subsequent incandescence in a process called laser-induced incandescence (LII) [20.22]. The variety of species and techniques that can be probed with laser-based techniques offers the option for simultaneous measurements of several species with similar experimental apparatus, but this in turn can cause reduced selectivity due to interfering signal contributions.

The advantage of laser-based techniques is that they do not intrude into the observed object and that in most cases, information is gained from a well-defined volume element. Whereas absorption techniques typically accumulate information over the line of sight, several techniques generate a signal from a well-defined point in space. Furthermore, the use of pulsed lasers and fast detectors allows time-specific data acquisition.

Several techniques can be expanded to measure combustion-relevant quantities not only in single points in space and time but also to investigate concentrations along the laser beam (one-dimensional spatial resolution), across planes that are defined by laser beams that are expanded to a so-called *light sheet* (two-dimensional spatial resolution). In few cases these techniques are also applied to three-dimensional measurements by combining several two-dimensional images into information that covers a restricted three-dimensional volume element.

Time-dependent measurements are often carried out for point and line-of-sight measurements. With the advent of high-repetition-rate solid-state lasers and fast multi-frame cameras techniques with one- and two-dimensional spatial resolution can also be carried out in a time-resolved way as high-repetition-rate imaging.

Another way to increase the dimensionality of the technique is the simultaneous application of multiple techniques (i.e., multiple species concentrations) or the determination of multiple characteristics of one quantity (i.e., determination of concentration, temperature and/or pressure from the analysis of line-shapes, integral and time-resolved measurements). This was done by wavelength multiplexing or fast wavelength tuning of absorption measurements. Raman scattering per se has the potential to observe multiple species and temperature at the same time. Other experiments require the combination of multiple lasers and/or multiple detectors to gain simultaneous information about multiple species and properties.

In the following sections examples of these various diagnostics approaches for combustion will be presented. The chapter will be structured according to the types of combustion systems rather than the different techniques. Clearly, many techniques can be applied to different combustion systems. This will be mentioned in the corresponding subsections and backed by giving appropriate references.

20.3 Experimental Data Devoted to Validation of Numerical Simulations and Modeling

20.3.1 General Remarks

Combustion technology reached a level of development where further improvements rely on complementary methods. The classical trial-and-error approach based on experimental experience was enhanced by numerical simulation of turbulent combustion processes. A comprehensive model that describes all relevant aspects of turbulent combustion taking place in applications such as internal combustion (IC) engines, furnaces, aero- or stationary gas turbines consists of various submodels that are based on conservation equations, elementary chemical reactions and closures such as physical models for transport processes and the equations of state. Due to the complexity of the underlying chemical and physical processes as well as their mutual interaction, simplifying assumptions are necessary. Because it is not granted a priori that these underlying assumptions are valid for a specific problem, experimental validation is imperative.

A thorough experimental validation should consist of each submodel as well all mutual interdependencies between the various physical and chemical processes [20.23]. This is an enormous task and can in practice only be fulfilled piecewise. Figure 20.1 sketches the basic architecture of a mathematical model suitable for numerical simulations. The different submodels call for individual

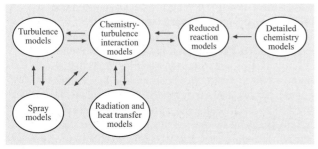

Fig. 20.1 Various submodels that build up a comprehensive model suitable for the numerical simulation of turbulent combustion processes

test rigs and specific diagnostic methods as outlined below.

Laser-based optical diagnostics have emerged in the past two decades as superior methods for studying combustion processes due to their low intrusiveness compared to probe techniques, their high temporal and relatively high spatial resolution. However, their application requires optical access that is not always easy to realize. Especially environments that mimic industrial processes such as internal combustion (IC) engines, gas turbines or furnaces, high pressure, dusty conditions, thermal radiation, and rapid contamination of optical accesses pose a challenge to these methods. Therefore, experiments devoted to the validation of numerical simulations are usually performed under somehow idealized conditions and are briefly discussed in the subsequent paragraphs. Furthermore, inlet and boundary conditions that have a significant impact on any process downstream of a mixing device (nozzle, intake port, etc.) in a burning chamber can be more easily controlled or measured with sufficient precision in test rigs that have been especially designed for validation purposes.

20.3.2 Submodels and Their Specific Demands for Validation

Figure 20.1 shows the modules (submodels) that comprise a comprehensive model describing turbulent combustion processes. Herein, the model for detailed chemistry describes combustion on the basis of elementary reactions. For the combustion of hydrocarbons, detailed chemical reaction mechanisms encompass hundreds of elementary reactions and species [20.9] and are therefore far too large to be used directly in computational fluid dynamics (CFD). Consequently, reduced reaction models are derived from detailed reaction mechanisms that contain a few reactive scalars only [20.24–26]. For the development and validation of chemistry models, a large and reliable experimental database is required. To circumvent problems linked to turbulence, chemical kinetics experiments are carried out for well-characterized boundary conditions such as laminar low-pressure flames [20.27, 28], flow reactors, shock-tube experiments [20.29] or rapid compression machines [20.30]. A survey of experimental methods to measure the chemical rate coefficients of single elementary reactions necessary for a detailed chemical reaction mechanism can be found in [20.31].

Turbulence models describe the properties of the flow field. The state of the art is to use the Reynolds-averaged Navier–Stokes equations (RANS) is used in most CFD models [20.32]. Within this approach, averaged conservation equations are solved for the statistical moments, assuming isotropic conditions. Unclosed terms are modeled on the basis of one- or two-equation models [20.33]. As a result, the CFD model predicts mean values and fluctuations in a pointwise manner. No spatially -correlated information is obtained. For this reason, experimental validation of RANS-based models can be restricted to quantitative, single-pulse, and pointwise measurements of species concentrations, temperature, and velocities. Of special importance are the experimental determination of well-characterized inlet conditions and the dissipation rate of turbulent kinetic energy based on two-point correlations.

Large-eddy simulation (LES) is an approach that has recently been extended to describe turbulent combustion. The conservation equations are spatially filtered before they are numerically solved on a spatial grid. Phenomena occurring on a scale smaller than the filter width are accounted for by subgrid-scale models [20.34]. The implicit LES combustion approach has various advantages. With increasing computational power, an increasing fraction of turbulent structures will be resolved, and the importance of an elaborated subgrid model will decrease. Unsteady effects can be described owing to the temporal resolution of the LES approach, and spatially correlated information is available. As a consequence, the requirements for the experimental validation increase. In addition to quantitative pointwise measurements, spatially correlated information is needed for validation purposes, such as turbulent length scales in different spatial directions [20.35] or gradients of scalar quantities such as temperature [20.36] or mixture fraction [20.37, 38]. It is worth noting, however, that direct numerical simulation (DNS) of the conservation equations is of growing importance for submodel development and validation [20.39]. Within this approach the grid resolution resolves all turbulent structures but – at least for three-dimensional (3-D) applications – it still relies on reduced chemical reaction models. DNS is a tool that ideally supplements experimental approaches.

In turbulent flames, a strong mutual influence of chemical reactions and the turbulent flow [20.40] is evident. Various chemistry–turbulence interaction models [20.41–43] exist. The state of the art is the commonly used flamelet and presumed probability density function (PDF) approach. However, a more comprehensive approach is to derive a PDF transport equation that may be solved by Monte Carlo methods or by unsteady flamelet modeling [20.44]. This approach shows great potential to describe, for example, flame extinction but is computationally expensive. To develop and validate submodels for the chemistry–turbulence interaction, quantitative experimental information for various simultaneously measured quantities is necessary, such as the main species, temperature, velocity, and reactive (radical) species. This requires sophisticated optical techniques, especially the combination of reliable velocity measurements simultaneous to main species concentration measurements. Although the combination of Raman, Rayleigh, and laser-induced fluorescence (LIF) [20.45] has already been developed to a high degree, the additional simultaneous measurement of gas velocities is at an early stage. Early approaches are discussed, for example, in [20.46–49] and show great potential.

In the case of liquid fuels, a submodel is required that describes breakup and the formation of a dense thin spray, as well as the evaporation and turbulence modulation [20.50]. The numerical and experimental investigation of this complex phenomenon is an ongoing task. Laser-based techniques such as spontaneous Raman scattering suffer on the one hand from optical density (multiple scattering in dense sprays, strong extinction, liquid core) as well as large local variations in number density at the interface between the liquid and gas phases, which necessitate a large dynamic range of the detector. However, for the development and validation of spray submodels, quantitative information regarding the liquid and gas phase is required. For the liquid phase, droplet diameters, droplet velocities and droplet size distribution need to be measured. For the gas phase, the local fuel–air ratio, temperature distribution, and gas velocities must be recorded. While for stationary conditions pointwise measurements such as phase Doppler anemometry (PDA) [20.51] are appropriate to characterize the liquid phase, the level of complexity rises if the spray characteristics for unsteady conditions (e.g., pressure oscillations that give rise to time-dependent spatial droplet distributions) are investigated. For this purpose, at least two-dimensional techniques, such as the interferometric particle imaging technique [20.52], are required. In addition, laser-induced fluorescence (LIF) schemes have been used in combination with Mie scattering to determine Sauter mean diameters [20.53, 54].

In many practical applications, it is necessary to account for radiation effects, particularly to predict thermal stress in walls correctly. The state of the art is to treat radiation either by integral methods [20.55–57] or by differential techniques. To account for the spectral characteristics of radiation, gray gas models [20.58] or spectral line models [20.59] are used. In open flames thermal radiation can be measured directly by pyrometers. Convective heat transfer to walls in the case of enclosed combustion can be measured spectroscopically as well, using a combination of thermographic phosphors and filtered Rayleigh scattering [20.60] or LIF [20.61].

20.3.3 Example: Generic Turbulent Flame

To ensure that a comprehensive model is valid for a certain range of operational conditions and capable of predicting trends correctly, a series of flames must be investigated where general parameters such as Reynolds number and fuel composition are varied. The burner configuration should exhibit some important aspects that

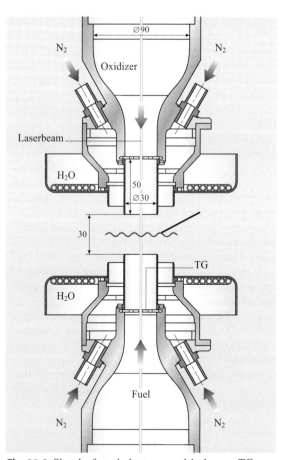

Fig. 20.2 Sketch of a turbulent opposed-jet burner. TG: perforated plates. The laser beam exciting Raman/Rayleigh scattering was directed along the symmetry axis of the burner. Laser radiation for LDV, PIV and LIF diagnostics entered the region of interest from the side

may arise in a similar manner within a practical realization. To demonstrate some of these special needs of experiments devoted to validation of numerical simulations, some recent measurements on a turbulent opposed jet burner are briefly outlined.

Opposing jets are a generic configuration in many industrial applications where fuel and oxidizing air impinge on each other, mix on a molecular level, and finally burn. The progress of chemical reactions in this turbulent mixing layer, however, depends on the mixing time, which can be quantified by the scalar dissipation rate [20.44]. If the residence time compared to typical finite-rate chemical time scales is too short, chemical reactions may even be extinguished. Flame extinction is indeed a problem in various practical combustion processes such as lean, low-NO_x gas turbines or direct-injection IC engines. The advantage of a generic setup such as the turbulent opposed jet burner is that the flow conditions leading to extinction can be well controlled. The location where extinction most probably starts is limited to a rather small volume. Therefore, data recording can be concentrated to the most important region, cutting down experimental effort enormously.

Figure 20.2 shows the cross section of a burner that allows to study these effects by detailed laser diagnostics. Details can be found in [20.37, 62]. Air emanating from the upper nozzle impinges on a rich methane/air mixture outside the flammability limits from the lower nozzle. The momentum of both streams is equal. Turbulence is enhanced by perforated plates located downstream of the contraction of each nozzle. A stagnation plane is formed which is located at the half nozzle distance (15 mm). The nozzles (30 mm diameter) are surrounded by a 60 mm-wide nitrogen co-flow to prevent mixing with ambient air. Optical access is possible along the burner center line through the central holes of the perforated plates and from the side perpendicular to the burner axis. The Reynolds number of the flows can be varied from stable towards extinguishing conditions and the equivalence ratio of the fuel can be varied.

To characterize the inflow conditions, the flow field between the two nozzles and the scalar field, the following diagnostics have been applied:

- Hot wire anemometry: nozzle exit velocity profiles and turbulent time scales [20.62].
- Two-dimensional laser Doppler velocimetry (LDV): radial and axial profiles of mean velocity, turbulent kinetic energy and Reynolds stress components [20.63].
- Planar LIF and particle imaging velocimetry (PIV), in part simultaneously: OH radical distribution during stable and extinguishing conditions, flame orientation, flow field conditioned on flame front [20.64, 65].
- Picosecond time-resolved LIF: absolute mean OH profiles, OH time series to characterize OH statistics [20.66].
- One-dimensional Raman/Rayleigh scattering: mean and fluctuations of major chemical species concentrations, temperature and mixture fraction, scalar dissipation rate [20.37, 63].

This database allows detailed comparisons with results from numerical simulations. Control and knowledge of the inlet boundary conditions as well as flow and scalar field are essential. Of special interest in this configuration are finite-rate chemistry effects due to the turbulence–chemistry interaction and turbulent mixing. The data have been used for comparison with Monte Carlo PDF models in combination with a RANS approach [20.63] and combustion LES [20.67]. Some of the results are discussed in Sect. 20.4.3 and in [20.37, 63].

20.4 Application of Laser-Based Techniques

20.4.1 Detection Sensitivity, Selectivity and Resolution

The detection sensitivity attainable in combustion environments is an essential parameter for the application of the various laser-based techniques. These limits cannot be stated generally since they depend on the line strength of the species to be detected, system parameters such as absorption length, response time or optical output power [20.68], while especially for in-situ measurements temperature and pressure in the probe region and disturbing effects of the in-situ beam path (dust, background emission) are most important factors for system-specific detection limits. A figure of merit for absorption spectrometers is the minimum detectable absorption change (MDA = $\Delta I / I_0$). In a precisely controlled absorption cell 10^{-3} is achievable with non-coherent light sources on a routine basis, while the

Method	Species	Transition	Energy of transition (cm^{-1})	p (total) (bar)	T (K)	Environment	Detection limit (abs.) (cm^{-3})	Detection limit (ppm)	Ref.
LAS	HO$_2$	$2\nu_1$-band	6623.8	6.7 (3)	295	photolysis cell	3 (13)[b]	16.8	[20.69]
	NO	(3, 0)-band	5524	1	1040	H$_2$/air flame	2.7 (15)	100	[20.70]
	CH$_4$	$2\nu_3$-band	6048	1	300	absorption cell	1.8 (14)	7	[20.71]
	NH	$A^3\Pi - X^3\Sigma(0, 0)$, $R_2(8)$	29762	1	2100	NH$_3$/N$_2$/O$_2$ flat flame, $\phi = 1.28$	3.0 (13)	7.9	[20.72]
CRDLAS	OH	$A^2\Sigma - X^2\Pi(0, 0)$	32500	0.04	1800	CH$_4$/air flame	2.0 (10)	0.12	[20.73]
	CH$_3$	$^rR(6, 6)$	3224.42	0.05	1400	CH$_4$/air flame	1.5 (13)	57.6	[20.74]
ICLAS	HCO	$A - X(09^00)-(00^10)$	16260	0.047	1800	CH$_4$/O$_2$/N$_2$-flame	1.4 (11)	0.57	[20.75]
LIF	OH	$A^2\Sigma - X^2\Pi(0, 0)$	32500	1	2000		25.6 (11)	0.07	[20.76]
		$A^2\Sigma - X^2\Pi(0, 0)$	32500	9.2	1700	C$_2$H$_6$/O$_2$/N$_2$-flame	1 (14)	2.5	[20.77]
	NO	$A^2\Sigma - X^2\Pi(0, 0)$	44247.8	1	2000		8 (11)	0.22	[20.7]
DFWM	OH	$A^2\Sigma - X^2\Pi(0, 0)$	32500	1	1700	premixed CH$_4$/air flame	2.0 (13)	4.66	[20.78]
	OH	$A^2\Sigma - X^2\Pi(0, 0)$	32500	1	2200	premixed CH$_4$/air flame	7.0 (13)	21.1	[20.79]
	NH	$A^3\Pi - X^3\Sigma(0, 0)$	29762	1	2100	NH$_3$/O$_2$/N$_2$ = 2.1/1.5/1.0			[20.80]
	CH$_4$	ν_3, Q(5)(1, 0)	3017.5	1 (−6)	300	CH$_4$/N$_2$	1.5 (11)	6174	[20.81]
	HF	ν_1, R(5)(1, 0)	4000	1 (−3)	300	HF/He	1.0 (10)	0.41	[20.82]
	CH$_3$	$3s^2A'_1 - 2p^2A''_2$	46185	1	1600	CH$_4$/N$_2$/O$_2$ flame, $\Phi = 1.25$	3.0 (14)	65	[20.83]
	C$_2$	$d^3\Pi_g - a^3\Pi_u$	19354	1	3000	C$_2$H$_2$/O$_2$ flame (welding torch)	5 (11)	0.19	[20.84]
CARS	C$_2$	Q(10)(1, 0)	1611.7	1	2500	C$_2$H$_2$/O$_2$ flame (welding torch)	1.0 (10)	0.003	[20.85]
	OH	$O_1(7.5)(1, 0)$	3065.3	1	≈ 1800	H$_2$/air flame (premixed)	1.0 (13)	2.5	[20.86]
	CO	Q(10)(1, 0)	2143	1	2000	flame	4.8 (16)	13000	[20.87]
	CO	Q(10)(1, 0)	2143	1	2000	flame	7.3 (16)	20000	[20.88]
	OH	$Q_1(7.5)(1, 0)$	3560	0.0106	300	HNO$_3$/He	1.4 (15)	5400	[20.89]
	NH$_2$	$\nu_1, /2\nu_4, (1, 0)$	3210	0.0005	300	NH$_3$	1.2 (12)	99	[20.90]
REMPI	1, 1 − DCE	$p^{-3}d(1+1)$	33863.9	–	5	molecular beam	6 ppbv[a]		[20.91]
	C$_6$H$_6$	$S_0 - S_1(1+1)$	38610		5	molecular beam	0.09 ppbv[a]		[20.91]
	C$_2$H$_4$	3 + 1	68181		5	molecular beam	10 ppbv[a]		[20.92]
PS	C$_2$	$d^3\Pi_g - a^3\Pi_u(0, 0)$	19357	1	1800	C$_2$H$_2$/O$_2$-flame	1.0 (12)	0.19	[20.93]
	OH	$A^2\Sigma - X^2\Pi(0, 0)$	32500	1	2100	propane/air flame	1.0 (13)	2.9	[20.94]

[a] Mass-spectrometric inlet system with molecular-beam sampling; the detection limits are referred to the concentration levels in the original sample mixture
[b] $1(11) = 1 \times 10^{11}$
[c] Mostly estimated temperatures

Table 20.1 Detection limits for minor, combustion-relevant species of various linear and nonlinear laser-spectroscopic techniques discussed in this review. The spectroscopic transitions probed for each species, their transition frequency (in cm^{-1}), total pressure p, an estimate of the temperature T, and the sample environment are given. The detection limits are approximate values, either determined directly or extrapolated from the respective experimental data ◄

best laser-based spectrometers provide MDAs in the 10^{-8} range, resulting in a dynamic range of 10^2–10^7. Table 20.1 is a (far from complete) collection of experimental work in the literature that stated the detection limits of combustion-relevant minor species for the various techniques introduced here at different pressures and environments to give a general impression of their practical applicability.

Species selectivity is another feature that has improved greatly through the introduction of lasers, since high radiance and small source linewidth are available simultaneously. For the case of continuous-wave (CW) lasers (diode and dye lasers) the laser line width can be of the order of 10 MHz, compared to molecular linewidths in the gas phase of several 100 MHz under low-pressure conditions and several GHz at atmospheric pressure. Ultimate selectivity can be ensured, if a spectral region or so-called *spectral window*, can be found where the species under investigation is the only (or at least the dominant) absorber.

Spatial and temporal resolution is another attribute that was improved significantly by the advent of lasers in diagnostics. Using the extremely short pulses of Q-switched or mode-locked lasers, the shortest turbulent time scales such as Kolmogorov time scales (in the μs regime) can be resolved even for very high turbulence levels. In contrast, the spatial resolution of laser diagnostics is in general insufficient to resolve the smallest fluid flow structures (Kolmogorov scales) or the smallest scalar structures (Batchelor scales) in turbulent flames. The latter is even worse for high-Schmidt-number flows. The spatial resolution is limited by the extension of the laser beam profile in the probe volume. Laser operation in the TEM$_{00}$ transversal mode using short focal lengths and large beam diameters through the last convex lens are typical measures taken to reduce the probe volume size. Constraints resulting from the high flame temperatures, flame enclosure, optical damage of windows used for optical access or gas breakdown, pulsed laser operation etc. typically limit the attainable spatial resolution to 100 μm at best in each direction of space. For high Reynolds numbers this is much larger than the smallest flow and scalar scales. Therefore under-resolving probe volumes act as spatial filters. Depending on the quantity, this under-resolution can cause a bias [20.95].

20.4.2 Laminar Flames

Low-Pressure Flames

Laminar premixed flames at low pressures on a flat-flame burner constitute an ideal experimental arrangement to study the interaction of elementary chemical combustion reactions. Experimental data on temperature as well as on concentration profiles for stable and unstable species are used to validate and further develop mathematical models that predict these profiles by numerical solution of the underlying conservation equations, including convection and molecular transport processes.

Figure 20.3a shows such an arrangement used for a low-pressure CH$_4$/O$_2$/NO flame. Absolute concentration profiles of methyl and hydroxyl radicals as well as nitric oxide are measured by laser absorption spectroscopy. CH radicals are detected by LIF. In this example (Fig. 20.3b) the shape and absolute values of the concentration profiles for OH and CH$_3$ radicals as well as the initial reduction of nitric oxide are predicted well by the models. Further improvements in sensitive absorption spectroscopy in flames can be obtained by using long-path absorption techniques such as cavity-ringdown laser-absorption spectroscopy CRDLAS [20.96] and intracavity absorption.

Resonance-enhanced multiphoton ionization (REMPI) – combined with molecular-beam sampling mass spectrometry – is a highly selective and sensitive technique for combustion diagnostics and environmental chemical analysis [20.91]. Detection sensitivities in the lower ppb to ppt (per volume) range for chloroethylenes, toluene and p-chlorphenol have been obtained in a helium carrier gas, whereas – due to cross sensitivities and nonresonant multiple-species ionization – this limit increases into the percent range in sample mixtures drawn directly from real-world industrial devices. To increase selectivity with an appropriate choice of excitation wavelengths various electronic transitions in large molecules can be accessed in one- or multiphoton transitions from which ionization is accomplished in a second step at the same or different laser frequency. Portable instruments have been designed that offer sufficient flexibility in switching between different laser wavelengths for the excitation of numerous unsaturated hydrocarbons, radicals and polycyclic aromatic hydrocarbons (PAH) for monitoring in real time. An example

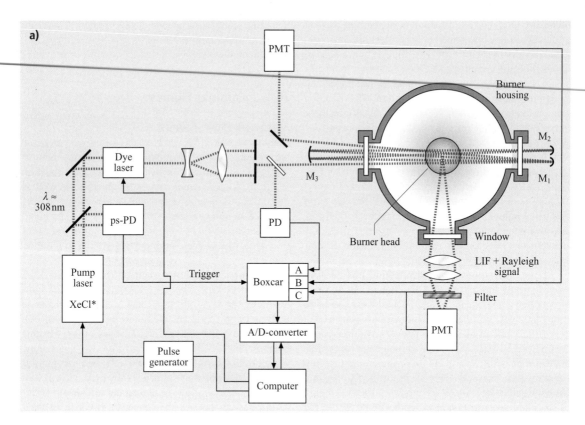
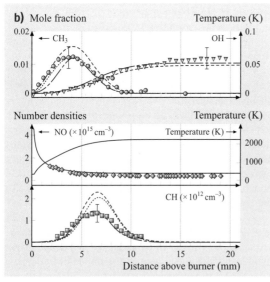

Fig. 20.3 (a) Experimental setup for laser absorption and LIF spectroscopy in low-pressure flames using tunable dye lasers and a long-path White-mirror arrangement. PD: photodiode, (ps-PD: picosecond resolution) **(b)** Comparison of measured and calculated concentration profiles in a laminar $CH_4/O_2/NO$ premixed flame at 1.33×10^3 Pa. Models: GRI 2.11 (*dashed*), Warnatz (*dash-dotted*) and Lindstedt (*dotted*) [20.97]

is the extraction and sensitive online REMPI detection of automotive exhaust gas pollutants at different engine loads and speeds [20.92].

Laser-ionization mass-spectrometric (MS) detection has also been successfully employed for spatially-resolved species analysis in low-pressure sooting flames to study the chemical kinetic mechanisms responsible for soot precursors and particle growth. Most soot precursors possess absorption maxima at 200–300 nm with ionization thresholds of 6–9.5 eV. Special scavenging techniques allow the sampling and subsequent REMPI/MS analysis of possible radical precursors that could act as important building blocks for the growth history of soot particles and carbon cages such as C_{60}, C_{70} in flame gases. *Homann* et al. [20.98] determined correlations between the number of hydrogen and carbon atoms contained in highly-condensed isomers of polycyclic aromatic hydrocarbons (PAHs) as precursors of C_{60} and their formation and destruction in low-pressure sooting flames. These authors claim the C_{60} cage is the final product in the chain of hydrogen abstraction, folding, condensation and isomerization of smaller fragments of five- and six-membered aromatic-ring condensates [20.99]. The detailed study of the fragmentation behavior, ionization efficiencies and absolute calibration of a variety of PAHs is essential for quantitative comparisons with numerical modeling of soot in flames [20.100].

Higher Pressure Flames

In laminar flames spatially-resolved temperature measurements using coherent anti-Stokes Raman scattering (CARS) are possible with high precision. Figure 20.4 shows a comparison of the temperature readings deduced from the pure rotational spontaneous Raman spectrum (squares) and the Q-branch vibrational CARS spectrum of molecular nitrogen (triangles) acquired at the same location and under the same operating conditions of a stable high-pressure premixed methane/air flame. Both data sets agree within the 3% limit of precision estimated from the spontaneous Raman data [20.101]. The pure rotational spontaneous Raman spectra exhibit well-resolved lines even at the highest pressures investigated. The spectrally dense CARS

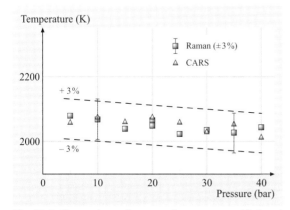

Fig. 20.4 Comparison of measured temperatures (deduced from fitting spectral shapes) in a methane/air flame at various pressures. Spectra were obtained from either spontaneous pure rotational Raman scattering (*squares*) or Q-branch vibrational CARS (*triangles*) measurements of molecular nitrogen at the same location in the post flame gases [20.101]. Shown as *dashed lines* are the ±3% temperature measurement accuracies inherent in fitting results from the spontaneous Raman data

Q-branch of N_2 suffers from partial collisional collapse of its rotational structure at the band origin, which usually degrades the temperature measurement accuracy at low temperatures (300–700 K) and even higher pressures [20.102]. Temperature measurements using pure rotational CARS have also been carried out in laminar flames and are often an alternative if higher precision is required in the low-temperature (300–870 K) higher-pressure (1–15 bar) regime [20.103].

For the detection of minor species fully resonant nonlinear techniques, such as DFWM, polarization spectroscopy (PS) and resonance-enhanced CARS are necessary to achieve sufficient detection sensitivity. DFWM spectra of electronic transitions have been recorded and gas temperatures determined in flames from OH, NH [20.80], NO [20.104], CN [20.105], CH [20.106], and in cell experiments from NO_2 [20.107], HCO [20.108]. Using polarization spectroscopy, important radicals in flames such as OH and NO were analyzed by *Nyholm* et al. [20.109] and *Löfstedt* et al. [20.110], respectively. The potential of polarization spectroscopy for two-dimensional imaging of these species and for single-shot temperature measurements was also shown [20.94, 109]. Because the total angular momentum of the photon–molecule system is conserved during the interaction process, a proper choice of beam polarizations can be exploited to excite P-, Q- or R-branch transitions selectively. Careful modeling of the PS spectral structure for OH and NH is essential for in-situ temperature measurements in atmospheric-pressure flames [20.20]. The intensity and collision frequency dependence of PS signal intensities has been modeled using the rigorous approach of direct numerical integration (DNI) of the Liouville equation for the density operator [20.111], which promises to address these effects more closely in practical measurement situations. To increase the Raman resonant susceptibility in a CARS interaction process, Stokes, pump and the generated anti-Stokes signal frequency should be tuned close to single-photon-allowed electronic transitions in the studied molecule. This makes the experimental approach rather complicated, since up to three narrow-band tunable laser systems must be employed. Nevertheless, spatially-resolved measurements of C_2 and OH distributions in pressurized flames have been performed by *Attal* et al. with high detection sensitivies [20.85, 86].

20.4.3 Turbulent Combustion

Fundamental studies of flames are based on various experimental setups tailored to select certain aspects of combustion. Practical combustion processes traditionally rely on non-premixed, turbulent combustion for reasons of efficiency and safety. However, demands for low NO_x emissions require the use of lean premixed combustion such as in stationary gas turbines. Detailed information about chemistry–turbulence interaction and complex recirculating flows is necessary to invent models suitable for simulating practically relevant flames. The understanding of the underlying processes is best verified through direct comparison of numerical simulation results with experimental measurements of properties such as temperature and species concentrations. Furthermore, detailed knowledge of the underlying flow field patterns is essential.

Laser-based techniques are ideal for studying turbulent processes since they do not disturb the flow and, using pulsed lasers, they are fast enough to resolve the smallest relevant timescales. Statistical investigations of correlations between temperature and the concentrations of different reactive species have been investigated using point measurements based on Rayleigh and Raman scattering and LIF diagnostics. Quantitative comparisons of the thermokinetic states in turbulent flames with idealized representations, such as steady strained laminar flames, perfectly stirred reactors, or adiabatic equilibrium are thus possible. Comparisons with direct numerical simulation (DNS) calculations on the other hand give insight into the validity of some of the basic assumptions upon which turbulent combustion models are built and allow the evaluation of the predictive accuracy, strengths and limitations of a wide variety of combustion models.

Revealing the structure of turbulent flames has attracted increasing interest and has been addressed using laser imaging for temperature, mixture fraction and reactive species such as OH and CH [20.113]. Simultaneous measurements of multi-scalars, yielding scalar dissipation rates, are critical for the validation of turbulent combustion models describing molecular mixing and extinction processes [20.37]. Simultaneous measurements of Rayleigh scattering and fuel Raman imaging provide information about temperature distribution and mixing. Both can be used to further quantify OH-LIF data by correcting for temperature and collisional quenching effects. These measurements (see Fig. 20.5) showed the correlation between the reaction zone width and jet Reynolds numbers and addressed the influence of local dissipation rate on flame extinction [20.112]. A review article that covers this topic in detail is given by *Masri* et al. [20.114].

With careful spectral filtering Raman measurements are possible even in dense media involving strong Mie scattering from small droplets. By forming a thin light sheet from the output beam of a XeF excimer laser at 351 nm *Decker* et al. [20.115] were able to image the Raman-scattered light from a dense cryogenic jet of liquid nitrogen onto a charge-coupled device (CCD) camera to study the evaporation and breakup of the fluid nitrogen exiting through a small (1.9 mm diameter) nozzle as employed in rocket engines, and performed temperature and density measurements from the Stokes/anti-Stokes intensity ratio and the Rayleigh intensity, respectively.

There are numerous applications of CARS as a temperature measurement technique in turbulent combustion processes. From the Q-branch spectrum of nitrogen, temperature measurement accuracies of 2% or better within a single laser pulse have been obtained [20.116], and applications have spread into diverse fields of technical combustion systems such as liquid-fuel combustors and swirl burners, IC engine combustion [20.117], gas turbines and jet propulsion devices [20.118–120] etc. High-resolution line-shape measurements of combustion-relevant species such as N_2 [20.121], O_2 [20.122], H_2 [20.123], CO [20.124], CO_2 [20.125], and NO [20.126] have improved modeling of CARS spectra – an important prerequisite for

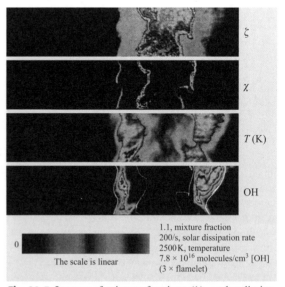

Fig. 20.5 Images of mixture fractions (ξ), scalar dissipation rates (χ), temperatures (T) and OH concentrations in turbulent CH_4/air diffusion flames obtained by 2-D Raman, Rayleigh and LIF spectroscopy [20.112]

precise temperature measurements through spectral simulation. Figure 20.6 shows temperature profiles from single-shot broadband N_2-CARS thermometry across the shear layer region of a supersonic air-fed combustor fueled with hydrogen [20.119]. The left-hand side of the graph depicts the mean temperature across the shear layer that is formed after injection of the fuel hydrogen into the supersonic air flow; it slowly decreases as the measurement point moves upstream. The necessity to do single-pulse measurements is clearly demonstrated. In the temperature histograms on the right-hand side of the picture the fluctuation in this parameter is illustrated as a function of position, which gives a more-detailed picture of the mixing properties and burning behavior of the combustor.

Multidimensional Diagnostics

While two-dimensional images yield a wealth of structural information [20.127, 128] and provide insight into the combustion process (Fig. 20.7), in many cases more information is desirable.

Multispecies measurements are of major interest to fully describe the thermochemical state of the observed volume element and to understand correlations between different species concentrations and potentially their local variations. In some cases linear combinations of different scalars correlate well with quantities such as heat-release rate that are otherwise difficult to measure [20.128]. For multispecies point measurements Raman scattering has been applied especially for the investigation of turbulent flames [20.17]. These techniques have been expanded for line measurements [20.37, 95, 129, 130] and allow – under special conditions – spatial multidimensional detection with limited resolution [20.131]. For two-dimensional imaging with high resolution, however, laser-induced fluorescence has been used to observe either different species at the same time in the same volume element [20.128, 132, 133], or to use the photophysical properties of fluorescent molecules to obtain information about more than one quantity (i.e., the concentration and the ambient conditions) [20.134, 135]. This will be shown in the following sections for organic tracers that allow not only concentration measurements but also the observation of temperature and local oxygen concentrations.

Turbulence is a three-dimensional process and therefore, three-dimensional measurements are desirable. Furthermore, the temporal behavior of concentrations and temperatures is of interest to understand turbulent phenomena, ignition, combustion instabilities, and transient phenomena. Imaging with high repetition rates

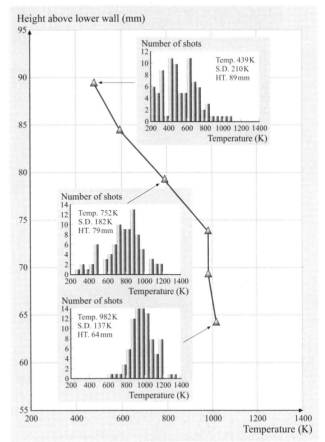

Fig. 20.6 Temperature distribution deduced from H_2 Q-branch CARS spectra across the shear layer of a supersonic hydrogen-fueled combustor [20.119]

that allows movie-like observation of two-dimensional scalar fields has been developed [20.136–138]. These techniques have become especially interesting since combustion modeling moved beyond the calculation of temporally-averaged quantities. For comparison with the results of large-eddy simulations (LES) and direct numerical simulations (DNS) both structural information and the temporal variation of structures is required.

Simultaneous Multispecies Detection in Turbulent Flames

Multi-Scalar 1-D Raman/Rayleigh Measurements In Turbulent Opposed Jet Flames. Simultaneous Raman/Rayleigh measurements allow the measurement of all main species (concentrations $> \approx 0.5-1\%$) and temperature. In hydrocarbon-fueled combustion, inter-

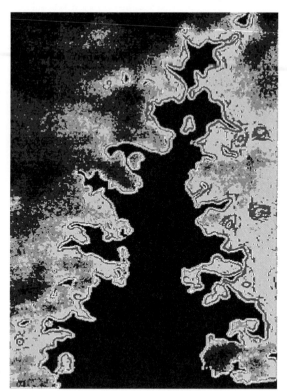

Fig. 20.7 OH-LIF distribution in a turbulent 150 kW flame. Size of the visualized field: 150 mm² × 110 mm² [20.140]

ference with LIF needs to be avoided. Therefore, the use of intense, pulsed 532 nm radiation from a frequency-doubled neodymium-doped yttrium aluminium garnet (Nd:YAG) laser is beneficial [20.139]. Figure 20.8 shows the experimental setup for multi-scalar measurements in turbulent partially-premixed opposed-jet flames [20.37, 67]. The burner is presented in Fig. 20.2 (Sect. 20.3.3).

A pulsed frequency-doubled Nd:YAG laser with a pulse energy of 900 mJ and a pulse duration of 10 ns was used to excite Raman/Rayleigh scattering. Two pulse stretchers were employed to lower the intensity below the threshold where optical breakdown occurred. The stretched pulse was focused ($f = 1100$ mm) from bottom to top along the vertically orientated symmetry axis of the turbulent opposed jet burner, passing through the central holes of the turbulence generating plates. In this arrangement the laser beam intersected the flame front on average perpendicularly and beam-steering effects were mostly avoided.

Perpendicular to the laser beam, inelastically scattered Raman photons within ±1.8 mm of the focal point were imaged on the entrance slit of a 310 mm imaging spectrograph. For this purpose, a specially designed achromatic corrected imaging lens system with a diameter of 150 mm and $f_{\#}$-numbers of 2.0 and 4.0 at the object and image side, respectively, was used. At the exit plane of the imaging spectrometer, ro-vibrational Raman bands of the main species (H_2, O_2, N_2, CO, CO_2, H_2O, and CH_4) were recorded spectrally and spatially resolved with a Gen IV intensified CCD (ICCD) camera. Rayleigh scattering was imaged onto a second ICCD camera with a perpendicular detector arrangement.

Binning the Raman signals in the spatial direction parallel to the laser beam resulted in a probe volume of $350 \times 360 \times 110\,\mu m^3$; the first dimension is along the laser beam and thereby across the reaction zone, the second is determined by the measured beam waist of 360 μm at the focal point and the last is due to the 220 μm width of the spectrograph entrance slit. This width is a compromise between the resolution necessary to discriminate the CO from the N_2 Raman signal and maximization of the signal. The overall resolution was limited by the diameter of the beam waist and exceeded the required resolution only slightly. Limitations in resolution are well-known phenomena in time-resolved Raman scattering due to the extremely weak signal of the scattering process.

Exemplarily, a partially-premixed flame is considered here. Air emanates from the upper nozzle and a rich methane/air premixed gas mixture (equivalence ratio 2.0) outside the flammability limit emanates from the lower nozzle. In the transient mixing layer both gas streams form an ignitable mixture and a partially premixed turbulent flame can be stabilized. As the most important quantities, the mixture fraction f and the temperature T along the center line are shown in Fig. 20.9. For comparison, this figure includes the non-reacting (isothermal) and combusting case.

In comparison to non-reacting conditions, the mean mixture fraction is almost unaltered by chemical reactions and heat release. The mixture fraction fluctuations, however, are very high for the non-reacting case and are significantly decreased, most likely by viscous effects, in the combusting case. Very close to the stagnation point a stoichiometric mean mixture fraction of $f_{st} = 0.51$ and a maximum mean temperature of approximately 1250 K were observed. Intense spatial fluctuations of the reaction zone cause the relatively small mean temperature. Depending on the instantaneous flow field with different local strain rates, the maximum temperatures vary between 600 and 2050 K.

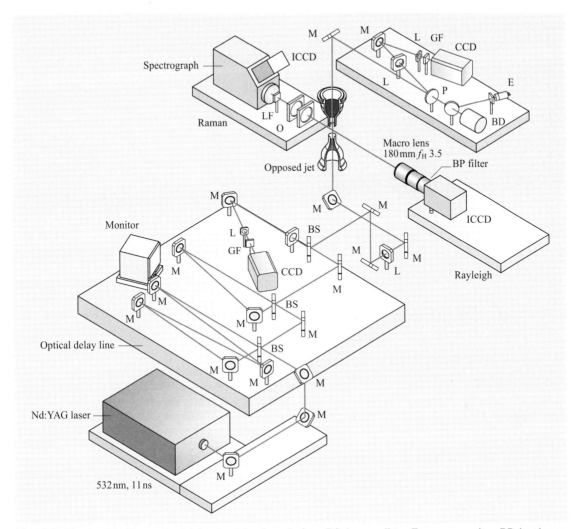

Fig. 20.8 Sketch of the Raman/Rayleigh setup. M: mirror, L: lens, BS: beam splitter, E: energy monitor, BP: band pass

Figure 20.10 shows mole concentrations and the corresponding fluctuations conditioned on mixture fraction for all main species except N_2. In general, the flame shows the typical features of non-premixed combustion as obvious from comparison to flamelet calculations employing a strain rate of $200\,\text{s}^{-1}$. It is clear from the figure that, on the lean side, laminar flamelet calculations are very similar to the results measured in the partially-premixed turbulent flame. At the rich side, however, deviations between the flamelet calculations and the turbulent flame data are apparent, especially for CO, H_2, and H_2O. Note that even small concentrations of CO and H_2 are detectable using the advanced technique applied.

Simultaneous Measurement of OH, HCHO and Temperature in Turbulent Flames. In the past decade much work has been aimed at the visualization of flame-front positions and structures in turbulent flames. Most of this effort has concentrated on measurements of the distribution of OH because of its high abundance in flames and the coincidence of OH transitions with high-power excimer laser wavelengths [20.127]. Measurements of flame front positions using OH, however, can give little information about important chemical reaction paths and fuel consumption rates. Recent investigations indicated that the distribution of HCO correlates well with peak heat-release rates for premixed flames of a certain range of equivalence ratios $\phi = 1.0\text{--}1.2$ [20.128, 141]. Due to

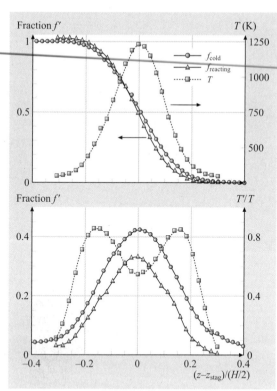

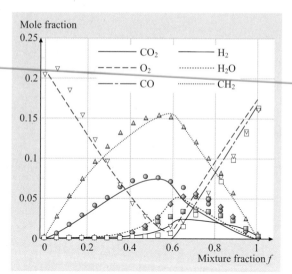

Fig. 20.10 Mole fraction conditioned on mixture fraction. *Symbols* represent experimental values obtained by Raman/Rayleigh spectroscopy in the turbulent opposed-jet burner, *lines* represent laminar flamelet calculations using a strain rate of $200\,\mathrm{s}^{-1}$

Fig. 20.9 Mean and fluctuation of the mixture fraction f and temperature T along the center line. Temperature fluctuations are normalized by the maximum temperature $T = 1250\,\mathrm{K}$

its low abundance, measurements of the HCO distribution via LIF have to apply phase sampling in acoustically forced unsteady flows to increase signal-to-noise ratios. Single-shot LIF measurements of HCO distributions in turbulent flows, however, do not seem to be feasible. Instead, *Paul* et al. [20.128] showed that the product of OH and HCHO concentrations is directly proportional to the reaction rate of $\mathrm{HCHO} + \mathrm{OH} \rightarrow \mathrm{H_2O} + \mathrm{HCO}$ and therefore yields an estimate for the production rate of HCO. The established link between HCO mole fraction and heat release therefore makes the product $I_\mathrm{LIF}(\mathrm{OH}) \times I_\mathrm{LIF}(\mathrm{HCHO})$ a good choice for flame front determination. Due to the high abundance of HCHO and OH in the flame-front region and their large fluorescence cross-sections, this approach enables single-shot measurements in highly turbulent reactive flows.

Whereas OH-LIF is frequently applied for measurements in different combustion media [20.142], efficient HCHO imaging have suffered from the lack of suitable high-power laser sources. As well as dye laser systems [20.143], broadband XeF and frequency-tripled Nd:YAG lasers have been used, which, however, access only weak transitions [20.144]. We took advantage of a coincidence of the side band of the XeF excimer laser at 353 nm with the strong 4^1_0 vibronic transition of formaldehyde [20.145, 146]. By combining two tunable excimer lasers (KrF and XeF), simultaneous measurements of OH and HCHO distributions by LIF are feasible on a single-shot basis. Rayleigh-scattered light provides information about the temperature distribution, which will be useful when addressing the quantification of LIF signal intensities.

Figure 20.11 shows an experiment in which a field measuring $20 \times 16\,\mathrm{mm}^2$ a distance of 2 mm above the burner exit in a turbulent natural gas/air Bunsen flame was investigated. A plane centered along the axis of the Bunsen flame is illuminated with the light of two tunable excimer lasers operated with KrF (248 nm) and XeF (353 nm). The LIF and scattering signal were detected with three ICCD cameras.

The results (Fig. 20.12) show strong OH LIF intensities throughout the burned gases within the observed area. Separate regions of burned gases are found within the fresh-gas zone, which might be separated from the flame front by the turbulent flow. The HCHO-LIF signal on the other hand is found within the fresh-gas

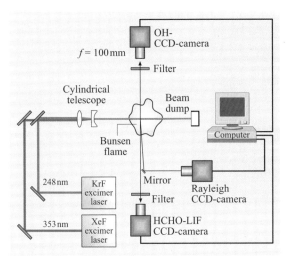

Fig. 20.11 Experimental setup for simultaneous measurements of HCHO and OH LIF and temperature in a turbulent flame

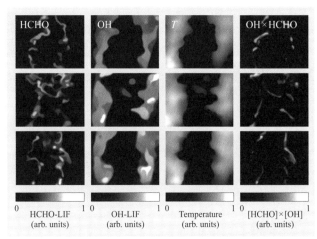

Fig. 20.12 Simultaneous measurement of formaldehyde (HCHO), OH LIF and temperature via Rayleigh scattering. The three *rows* show independent sets of images. The *right column* shows the linear combination of OH and HCHO LIF [20.132]

zone close to the flame front with little overlap with the OH distribution. Extended areas within the fresh gases showing increased HCHO signal levels were also found in areas with elevated temperatures. Here, several millimeters away from the flame front, cool-flame chemistry producing formaldehyde has already started. The H, CHO and OH images show reasonably good signal-to-noise ratio, allowing for the calculation of the product of the OH- and HCHO-LIF distributions as suggested by *Paul* et al. [20.128]. To reduce the influence of shot noise when calculating the LIF product images a nonlinear anisotropic diffusion algorithm was applied [20.147, 148]. The results are shown in the right column in Fig. 20.12 for three typical single-shot situations. The resulting images show sharp contours within the flame-front region. Extended areas with OH-LIF signal, as found on the hot side of the flame front as well as larger areas with HCHO-LIF signal within the fresh gases are suppressed. When comparing the resulting flame-front images to the temperature fields it is found that the product of the OH- and HCHO-LIF signals closely follows iso-lines within the temperature field.

Simultaneous Measurements of OH and NO Concentrations and Temperature in a Swirl Flame. The highly turbulent reactive flow field in practical-scale swirl flames is a challenge for both experiments and numerical simulations. Experiments aim to provide accurate data for model validation. The subject of the investigation described in this paragraph is the TECFLAM standard swirl burner [20.142]. Figure 20.13 shows the burner geometry and the position of the planes observed by laser imaging. The annular slot of the central fuel nozzle of the burner has an inner diameter of 20 mm, whereas its annulus is 3 mm wide. Swirl is provided to the air flow by a movable block swirl generator [20.149], which is set to an effective swirl number of $S = 0.9$. The burner is operated with a thermal load of $P = 150$ kW burning natural gas at an equivalence ratio of $\phi = 0.83$, which corresponds to a flow of 15 m^3/h natural gas and 180 m^3/h air. The Reynolds numbers of fuel and airstream, based on the exit bulk velocity,

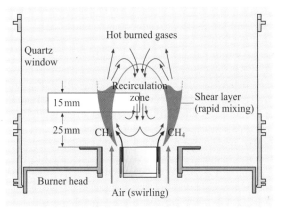

Fig. 20.13 Flow field and measurement position in the TECFLAM swirl flame [20.132, 133]

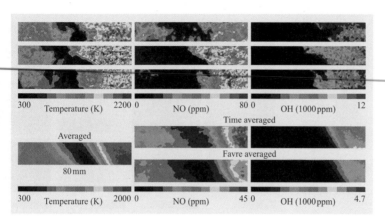

Fig. 20.14 Simultaneous measurements of temperature and NO and OH concentration in the reactive flow field of the TECFLAM swirl burner. *Upper frames*: instantaneous measurements, *lower frames*: averages [20.133]

are $Re_{gas} = 8000$ and $Re_{air} = 42\,900$, respectively. The combustion chamber is confined by a water-cooled housing (72–88 °C) to ensure constant boundary conditions. Optical access to the chamber is realized by four quartz windows ($100 \times 100\,\text{mm}^2$).

In the swirl flame a similar setup as in Fig. 20.11 was used to assess OH and NO concentration and the temperature distribution. One of the laser beams is frequency-converted in a hydrogen Raman shift cell [20.150] to provide 225.25 nm radiation for the excitation of the NO A-X(0,0) band at energy densities of $2 \times 10^6\,\text{W/cm}^2$. The NO-LIF signal is separated via a reflection band-pass filter (transmission: $248 \pm 10\,\text{nm}$) to detect light emitted from the A-X(0,1), (0,2) and (0,3) transitions. The second laser is triggered with a short delay (200 ns) to prevent crosstalk between the NO-LIF and Rayleigh signals. Rayleigh-scattered light is separated via a dielectric mirror (high reflectance at 248 nm at 45°). Achromatic UV lenses are used to focus the signal onto the chips of ICCD cameras. For simultaneous measurements of OH concentration and temperature the laser used for Rayleigh scattering was tuned to the OH transition ($P_2(8)$ within the OH A-X(3,0) band, laser energy density: $5 \times 10^7\,\text{W/cm}^2$) and the OH-LIF signal was detected using reflection band-pass filters centered at 295 nm. Since OH is a strongly diagonal molecule the strongest emission occurs from the (3,3) vibrational band (and after vibrational relaxation from the (2,2), (1,1), and (0,0) bands). The fluorescence contribution of OH LIF detected at 248 nm together with Rayleigh-scattered light is generally negligible compared to the Rayleigh signal [20.151]. However, in peak temperature zones where total number densities and hence Rayleigh signal are low OH LIF contributes to the overall signal, resulting in additional uncertainties of −5% in the temperature images.

The measurements presented here were carried out 25–40 mm (Fig. 20.13) above the burner exit. For calibration of the LIF signal intensities a *McKenna* burner was installed in the burner housing and investigated with an identical optical setup. The burner was operated on stoichiometric methane/air mixtures where resulting OH concentrations and temperatures are documented [20.152]. Additional calibration measurements have been carried out in lean flames ($\phi = 0.83$) doping known quantities of NO (300–800 ppm) into the fresh gases. The increase in signal intensities provided the necessary data for calibrating the NO-LIF measurements [20.153] in the swirl burner measurements.

Simultaneous measurements of temperature, OH and NO concentration allow one to quantify the LIF signal intensities by correcting for the temperature dependence of the ground-state population and collisional energy-transfer processes on a single-shot basis. Furthermore, by correcting for local total number densities, local concentrations can be calculated from number densities, allowing the computation of time-averaged concentration fields. In Fig. 20.14, three sets of simultaneously recorded temperature, NO and OH distribution fields are shown. The position of the observed area is depicted in Fig. 20.13. It can be seen that NO is present throughout the whole burned gas area. NO is formed within the hot burned gases on both sides of the fresh-gas intake and accumulated by recirculation. In the outer recirculation zone (the left part of the figure) peak NO concentrations are around 35 ppm, whereas in the inner recirculation zone peak NO concentrations in the range of 70–80 ppm are reached due to the higher temperatures. OH on the other hand is mainly found on the inner side of the fresh-gas intake. Due to the lean air–fuel mixture and the strong turbulence, OH is distributed across a relatively broad zone behind the flame front. Local peak OH concentrations are found to be 8000 ppm in the super-equilibrium range. Below the single-shot images, averaged images for the same measurement position are presented. For the NO and OH concentration fields, both, Favre- and time-averaged images are shown. Since both species are mainly present within the hot burned gases, the influence of density variations is restricted. However, in the burned gases, the time-averaged concentrations are higher than the Favre averages by up to 15%.

Three-Dimensional LIF Diagnostics

Two-dimensional images yield information about cross sections in turbulent combustion media. They can therefore not distinguish between isolated areas (i. e., *islands* of burning gases) and structures that are connected via the third dimension. Information on the propagation of a structure perpendicular to the measured plane is missing. Furthermore, gradients measured by these planar techniques are only a projection of the real gradients onto the observed line or plane.

Time-Averaged Three-Dimensional Measurement of Fuel Distribution. In isobaric, isothermal systems the use of tracer LIF is straightforward. When seeding acetone into the gas flow, the local LIF intensity is a direct measure of the local concentration. In later chapters we will describe how temperature and pressure effects must be corrected for more-general applications.

Figure 20.15 shows the setup for a measurement of fuel distributions in the combustion chamber of

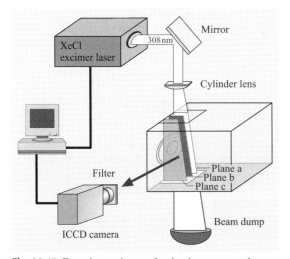

Fig. 20.15 Experimental setup for the time-averaged measurements of fuel–air mixing in a swirl burner of a stationary gas turbine [20.11]

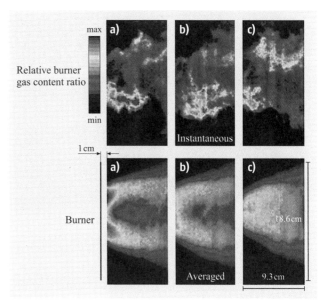

Fig. 20.16 Measurements of the fuel distribution (doped with acetone) in three parallel planes in the mixing chamber. The fuel nozzle is 1 cm to the left end of the images. *Upper row*: randomly chosen single-shot images, *lower row*: averaged fuel distribution [20.11]

a stationary gas turbine (thermal power of the investigated segment ≈ 1 MW). The gaseous fuel flow was simulated with air, seeded with acetone and the fluorescence was successively measured in three adjacent planes. Instantaneous and time-averaged images are shown in Fig. 20.16. The time-averaged concentration distributions from the adjacent planes were then used to reconstruct the average three-dimensional fuel distribution (Fig. 20.17) assuming cylindrical geometry. The strong variation in local concentrations (upper row in Fig. 20.16) indicates, however, that an instantaneous view into the three-dimensional structure would deviate significantly from the average distribution.

Instantaneous 3-D Measurement of Temperature and Flame-Front Position. For instantaneous observation of three-dimensional structures in turbulent flows several approaches have been developed and applied to turbulent mixing processes and flames. Parallel alignment of multiple detection planes either using subsequent detection of the imaged planes with a single camera and a long sweeping pulse or CW laser beams and scanning mirrors, or using high-repetition-rate imaging with the need for laser and camera *clusters* are the methods that have been used to observe extended volumes within turbulent flows [20.142, 154–156]. The spatial resolution in

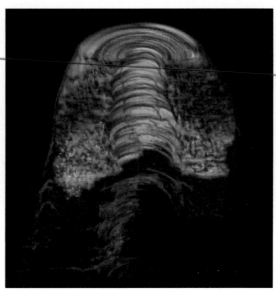

Fig. 20.17 Visualization of the averaged three-dimensional fuel distribution in the nonreactive swirl flow [20.157]

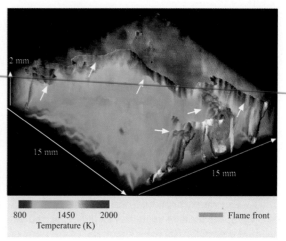

Fig. 20.19 Three-dimensional visualization of the temperature field and the flame front position (gray surface marked with *arrows*) in the reactive flow of a technical scale swirl flame at a swirl number of $S_0 = 0.9$ and an equivalence ratio of 0.83 [20.142]

the z-dimension (perpendicular to the observed planes), however, is restricted. Figure 20.18 shows the setup for instantaneous measurements of OH-LIF and temperature in three parallel planes within the reactive flow field of the TECFLAM swirl flame (Fig. 20.13).

The beams of two tunable excimer (tuned to an OH A-X(0,3) transition at 248.45 nm) are directed through the flame. Via a beam splitter a third beam is generated from one of the lasers, optically delayed and used to illuminate the third plane. Three cameras are aimed to detect the Rayleigh signal with dielectric mirrors as bandpass filters ((248±10) nm). Three additional cameras record the OH-LIF signal, which is spectrally separated from scattered light by WG280 long-pass filters.

Figure 20.19 presents a three-dimensional visualization of the flame front embedded in the turbulent temperature field of size 15×15 mm^2 at $x/D = 1.5$ and $r/D = 0.5$. The visualization is based on ray-tracing algorithms (in collaboration with IWR, Heidelberg). The flame-front position is evaluated for each individually measured OH concentration field by determining the position with the steepest gradient in OH concentration. Between the three adjacent planes the flame-front positions and temperatures are interpolated to obtain a 3-D representation.

Instantaneous 3-D Measurement of OH LIF with Crossed Light Sheets. An alternative attempt to image three-dimensional volume elements is the use of crossed detection planes (Fig. 20.20) [20.158–160]. A quasi-instantaneous detection is provided with two cameras [20.157]. In special mirror arrangements detection is possible with a single laser and camera [20.161]. One of the benefits of this method is the increasing resolution towards the crossing line and a small cylindrical volume of several hundred micrometers to approximately 1 mm around it. This geometry is ideal

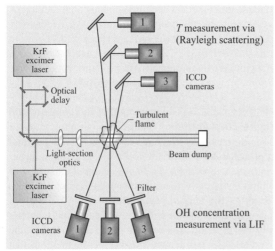

Fig. 20.18 Experimental setup of instantaneous 3-D measurements of OH LIF and temperature in a turbulent flame

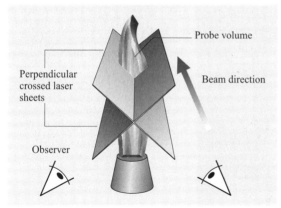

Fig. 20.20 Arrangement of the crossed light sheets for instantaneous three-dimensional observation of OH LIF around the crossing line

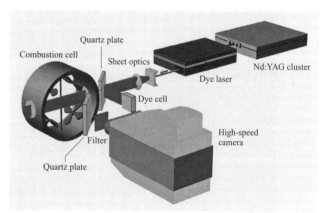

Fig. 20.22 Experimental setup for high-repetition-rate OH PLIF in the turbulent combustion cell

for the combination with one-dimensional (1-D) Raman line measurements which can provide temperature profiles and multispecies distributions in a limited part of the crossing line [20.162]. The resulting three-dimensional data matrix can also be visualized using a 3-D visualization tool that assigns not only color but also transparency values to the given intensity values [20.163]. The results for OH-LIF measurement in a lean Bunsen flame are presented in Fig. 20.21. To the left of the reactive zone is the unburned fresh gas in the center of the burner above the nozzle, while the right side shows the surrounding air around the flame cone. The depth of the visualized 3-D volume is 2 mm.

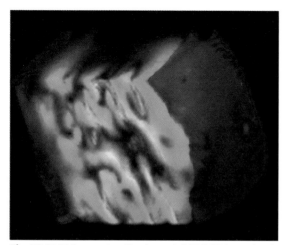

Fig. 20.21 Volume visualization of OH LIF intensity in a lean Bunsen flame [20.129, 161]

Multiple-Timestep Imaging
Tracking of Flame Kernel Development After Spark Ignition. Many practical combustion processes rely on electrical spark ignition in their initial phase. During the breakdown, arc and glow phases of an electrical discharge, a plasma is formed. At the interface of the plasma kernel and the ambient fuel–air mixture transfer of heat and radicals causes the formation of an initial flame kernel. Depending on the local fuel–air ratio and the turbulence levels the flame kernel may either grow in time, leading to a successful ignition, or be extinguished. To understand and predict the fate of an initial flame kernel requires the flame kernel development to be tracked temporally. This can be achieved by sequential OH LIF imaging.

A multiple laser and detector system [20.136] was used to record time sequences of OH-LIF images from single combustion events in a closed vessel using premixed methane/air mixtures [20.164]. The laser source of the high-repetition-rate imaging system is a cluster of four Nd:YAG lasers (BMI/CSF-Thomson) that are fired in series with a short time delay to generate a burst of four laser pulses. The separation between pulses was varied between 250 μs and 3 ms, depending on the fuel and turbulence intensity used in the experiment. The output from the four Nd:YAG lasers are combined to a single beam and frequency doubled to 532 nm. The pulse energy from each of the lasers was around 500 mJ. The Nd:YAG lasers pumped a dye laser (Continuum ND60) operating near 566 nm on rhodamine 590 dye. The output from the dye laser was frequency doubled to 283 nm and tuned to excite OH at the $Q_1(8)$ transition in the (1,0) band of the $A^2\Sigma^+ \leftarrow X^2\Pi$ system. The laser energy at 283 nm was around 2 mJ/pulse for a pump pulse

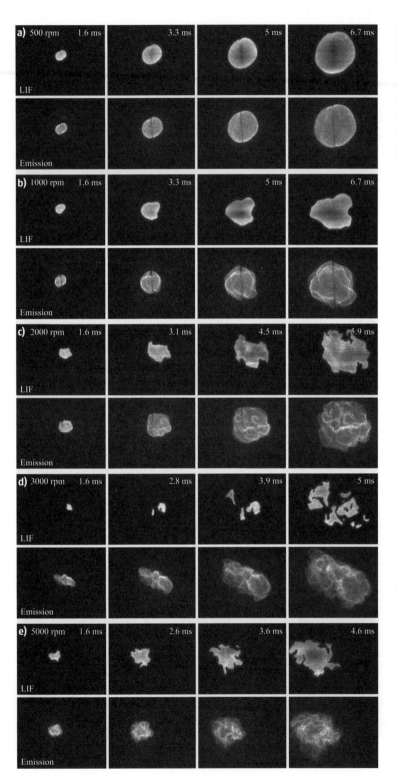

Fig. 20.23 Flame propagation for individual events measured by OH PLIF and flame emission for various rotor speeds as indicated in each row and stoichiometric methane/air mixtures. Times in each image denote the time after ignition. Turbulence intensities associated with the increasing rpm as stated in the figures are 0.31, 0.62, 1.3, 2.01, and 3.31 m/s

separation of 2 ms. The laser light was formed to a sheet, 50 mm in height and 0.2 mm in width, using a cylindrical telescope, and passed through the cell 1 mm in front of the electrodes (Fig. 20.22).

A high-speed camera consisting of eight individual intensified CCD units coupled to a single optical input (Imacon 468, DRS Hadland) was used to detect the fluorescence. An additional image intensifier was fitted to the optical input to achieve UV sensitivity. The first four CCD units were triggered by the individual lasers, and used to capture the four OH LIF images. The remaining four CCD units were used to capture flame chemiluminescence shortly after each LIF exposure. The delay between the LIF and chemiluminescence exposures was 2 μs, which can be considered simultaneous on the time scale of the studied process. An achromatic quartz lens ($f = 100$ mm, $f_\# = 2$) was fitted to the camera. A UG11 filter was used in combination with a long-pass filter, transmitting above 295 nm, was used to reject scattered laser light and transmitting fluorescence in the (1,1) and (1,0) bands of OH.

An external pulse generator synchronized the ignition system to the laser and detector system, while allowing the laser system to operate continuously at 10 Hz, to maintain stable operation. The ignition system consists of two units: the first unit produces an electrical breakdown between two electrodes (1 mm diameter Tungsten wire, conically shaped, 1 mm gap) and thereby an electrically conducting plasma channel. The energy of the breakdown was kept as low as possible and the released energy in this phase is below the minimum ignition energy. The second unit is a constant-current power supply giving rise to currents and voltages of approximately 1 A and 100 V, respectively, typical for an arc phase. In this study, an arc-phase duration of 200 μs is chosen. To achieve a constant current, the voltage is adjusted adaptively, leading to a reproducibility in energy deposition of better than 2%. Therefore, conditions for initially spherical flame kernels generated by the ignition system are highly reproducible as proven in [20.165] for initially quiescent gas mixtures.

The cylindrical constant-volume combustion vessel has a diameter of 39 cm and a height of 51 cm, resulting in a volume of approximately 55 l. Three optical access ports allow for LIF measurements. Four high-speed rotors of 144 mm in diameter are situated 200 mm from the electrode gap. Rotor axes are perpendicular to their respective neighbors. Rotor speeds are synchronized and can be varied between 0 and 5000 rpm. In the vicinity of the electrode gap a homogeneous isotropic turbulent flow field of variable intensity is generated.

Figure 20.23 shows data corresponding to single ignition events in stoichiometric CH_4/air mixtures subjected to different degrees of turbulence. The sequences were obtained using four consecutive laser shots fired in rapid succession with stated timings referring to time after ignition. The top rows of pictures are two-dimensional (2-D) LIF images of OH; the bottom rows are OH chemiluminescence images. Figure 20.23a corresponds to moderate levels of turbulence ($u' = 0.62$ m/s). It can be seen that first wrinkling (deviations from a spherical kernel) appears around 1.6 ms, which is a consequence of turbulence chemistry interaction although essentially the flame retains its near-laminar character. On the regime diagram [20.44] this corresponds to the laminar wrinkled flame front region. On the emission images a shadow appears indicating the position of the electrodes. Increasing turbulence levels to $u' = 2.01$ m/s (3000 rpm) results in a large change of flame shape. The flame now shows significant internal structure due to distortion by vortices. Here, the advantage of a planar light-scattering technique to assess flame structure is clearly seen: whereas emission images may still be interpreted as compact and connected flames the LIF images on the other hand show formation of cusps and handles or possibly islands.

From OH sequences, as shown in Fig. 20.23, a variety of parameters can be deduced to characterize the temporal evolution of the flame kernel. Here only the degree of wrinkling as a consequence of the turbulence–

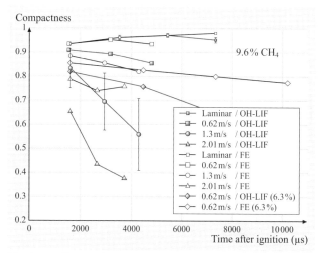

Fig. 20.24 Temporal evolution of compactness for stoichiometric methane–air mixtures subjected to various degrees of turbulence. FE: flame emission

chemistry interaction is briefly discussed. The degree of wrinkling and its effect on flame propagation can be quantified using the flame compactness factor K, which is defined as

$$K = \frac{4\pi A_t}{C_t^2} = \frac{C_s^2}{C_t^2}, \qquad (20.1)$$

where A_t is the area. OH LIF is detected within the observed cross section, and C_t is its circumference. C_s is the circumference of a spherical laminar flame of the same area A_t as the turbulent kernel. In this definition, spherical flame kernels have a compactness factor of $K = 1$ whereas wrinkled flames have $K < 1$. For example $K = 0.5$ would increase the circumference of the flame by a factor of 1.41. Figure 20.24 shows the temporal evolution of K for stoichiometric CH_4/air mixtures at various degrees of turbulence where K is calculated both from emission and from PLIF measurements. It is evident that line-of-sight-averaged chemiluminescence measurement is not a suitable tool to assess flame wrinkling: taking the $v' = 2.01$ m/s case as an example (3000 rpm rotor speed), large discrepancies between the two measurement techniques are visible. The LIF data indicate how strongly K is affected by turbulence and is especially strongly influenced if the mixture is lean. In contrast, in the laminar case (black line) both the emission data and the LIF data yield essentially the same results as expected. Note that the apparent increase in K for the laminar case is due to the decreasing effect of heat conduction to the electrodes as the flame kernel grows in size. The standard deviations of 30 individual measurements per point is indicated for the $v' = 0$ and $v' = 1.3$ m/s cases.

Flow Diagnostics Close to the Wall with Two-Time-Step Imaging. Flow-field imaging is most frequently carried out via particle imaging velocimetry (PIV) or laser Doppler velocimetry where particles are added to the flow. Close to surfaces, these techniques, however, fail due to the interaction of the particles with the surface. Different flow-tagging techniques have been proposed to overcome these problems. They are based either on Doppler shifts of molecular tracers [20.166] or on the laser-induced generation of a molecular tracer within the flow that can be observed for a certain time after its generation (tagging). This tagging is based on photodissociation of vibrationally hot H_2O [20.167], vibrationally excited O_2 [20.168, 169], NO formed from photodissociation of *tert*-butyl nitrate [20.170] or from photolytically generated oxygen atoms with subsequent reaction with nitrogen [20.171]. We discuss here photodissociation of NO_2 with subsequent LIF imaging of nitric oxide and expand this technique to allow the investigation of the temporal development of the turbulent flows by either photolytically tagging in two subsequent time steps with a single detection of both marked volumes or with one tagging and subsequent detection of each individual flow situation after two variable delays.

In an optically accessible flow channel (60×5 mm^2 cross-section, mean flow velocities of $v = 74$ m/s associated with a Reynolds number of Re $= 24\,500$ at room temperature, the flow seeded with 1200 ppm NO_2) a frequency-tripled Nd:YAG laser is focused to a line where the NO tag is generated. The resulting *line* of NO is then observed at two time steps with various delays (up to 300 µs) with LIF imaging. The frequency-tripled laser light of a double-cavity Nd:YAG laser is used to pump a midband optical parametric oscillator (OPO). It is tuned to generate a signal wavelength at 450 nm which is then extra-cavity frequency-doubled in a beta barium borate (BBO) crystal. The resulting UV beam has a pulse energy of approximately 1 mJ and a spectral width of ≈ 4 cm^{-1}. It therefore excites a large portion of the NO A − X(0, 0) band head. NO-LIF signals are discriminated against scattered light by a UG5 filter and focused onto the chip of an CCD-intensified interline camera. The image intensifier ensures short detection gate times for both frames. The maximum repetition rate of 200 ns of the double-frame camera is no restriction for our experiments. The optical setup is shown in Fig. 20.25.

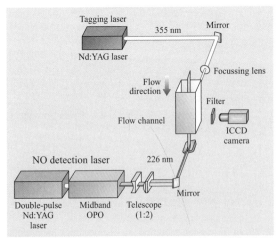

Fig. 20.25 Optical setup for the laser-based investigations in the flow channel

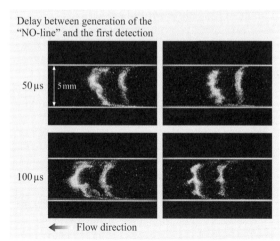

Fig. 20.26 Observation of a photolytically generated structure at two time steps after generation. Delay between generation and first detection: 50–100 µs, delay to the second detection: 75 µs. The flow is from *right to left*, the *horizontal lines* mark the position of the channel walls

Figure 20.26 shows the result of the double-frame imaging for NO lines generated at the right side of the frames (the flow is towards the left). The interaction of the flow with the boundary layer at the walls (horizontal lines) and with turbulence elements in the main flow is clearly seen.

Multiple-Timestep Imaging in Spray Diagnostics.
Techniques using simultaneous detection of Mie scattering and LIF signals have been used successfully for two-dimensional drop-size imaging in practical spray configurations. The fluorescence tracers commonly used, however, prevent the application in realistic fuels like gasoline and diesel because of strong interference with fluorescing compounds of the fuels. Therefore, the use of a fluorescent tracer is proposed that can be excited in the red spectral range, where the absorption by commercial fuels is minimized. We use a frequency-doubled diode-pumped Nd:YAlO$_3$ laser-oscillator power-amplifier system to generate pulse bursts at 671 nm with 30 ns pulses (≈ 1 mJ each) and a 30 kHz repetition rate. At this wavelength an organic tracer (rhodamine 800) with high fluorescence quantum yield is excited. Fluorescence and elastically scattered light is separated via dielectric filters and detected on two slow streak cameras. The fluorescence emitted at around 700 nm originates from the liquid phase only and the LIF/Mie signal ratio is a measure of the local Sauter mean diameter.

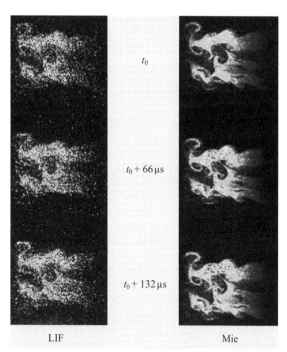

Fig. 20.27 LIF and Mie signal distributions in the spray of a nebulizer. Three images with a temporal separation of 66 µs are shown

Figure 20.27 shows a time series of Mie and LIF signal in the spray field of a medical nebulizer. The high-repetition-rate pulse bursts of the laser allow the observation of the spray development and evaporation. Further experiments will be carried out in diesel sprays. First measurements show that the tracer system with excitation in the red can be used in commercial diesel fuel without any significant fluorescent background from the fuel [20.172].

20.4.4 Engine Combustion

In-cylinder laser-based combustion diagnostics in internal combustion (IC) engines focus on three main fields:

1. the observation of the mixture preparation before ignition, which includes fuel injection and evaporation, fuel–air mixing, pre-combustion temperature distributions, and the distribution of residual inert gases from the previous engine cycle and intentionally recirculated exhaust gas
2. the observation of the ignition and combustion process including flame propagation and potential flame

quenching, also considering post-flame temperatures
3. the observation of pollutant formation, namely nitrogen oxides (NO_x) and soot

In the following paragraphs these different topics are treated separately.

Different strategies for optically accessible IC engines have been used over the last years. The design of optical engines requires compromises between the needs of the diagnostics and the level of comparability to production-line engines. Best optical accessibility is obtained with transparent cylinders and extended windows that are inserted into the piston (Fig. 20.28). The thickness of pressure-resistant quartz windows, however, usually prevents the operation of the neighboring cylinder and changes in-cylinder heat transfer considerably due to the extended quartz surfaces. More-advanced concepts sacrifice optical access by using smaller optical sapphire windows that can be thin enough to fit into a multi-cylinder engine block with minor modifications (Fig. 20.29). Recently, micro-optical laser-based probes using endoscope optics and fibers have been developed for in-cylinder investigation in IC engines [20.175–177].

Analysis of Mixture Formation

There are a number of quantities that would be desirable to measure with non-intrusive in-situ techniques to characterize the mixing process in the fresh gas completely. These quantities include:

- Fuel concentration
- Fuel–air ratio
- Temperature
- Fuel composition (i.e., the concentration of individual components)
- Residual gas concentration

To visualize the mixing process and to facilitate the interpretation of the results these quantities should be imaged in at least two dimensions with temporal resolution faster than the time scale of mixing and chemical reaction. A recent review summarizes the state of the art [20.11].

Mixing processes of interest can be categorized according to the level of difficulty in terms of quantitative imaging measurements. They are listed here in order of increasing complexity.

- Constant pressure versus temporally fluctuating pressure
- Constant temperature versus temporally fluctuating (spatially homogeneous) temperature
- Homogeneous temperature versus spatially varying temperature
- Temperature approaches the stability limit of the tracer

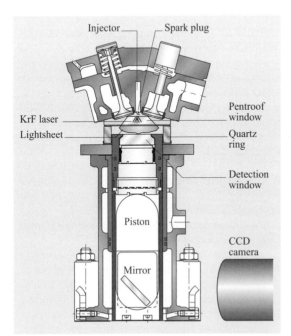

Fig. 20.28 Basic setup of an optical engine with access to the combustion chamber through a quartz cylinder and a piston window [20.173]

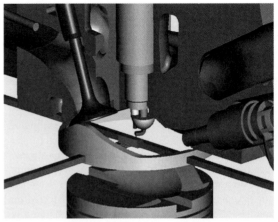

Fig. 20.29 Advanced concept for an optical engine with narrow sapphire ring that allows the operation of full quasi-production-line four-cylinder engines [20.174]

- Homogeneous versus inhomogeneous bath gas composition (mainly oxygen concentration is relevant)
- Single-component fuel versus multicomponent fuel
- Non-fluorescing versus fluorescing fuel
- Homogeneous (gas phase) versus heterogeneous (two-phase flows)

The ideal tracer should behave exactly like the fluid it is added to (i.e., the fuel or the desired component of a multicomponent fuel) in terms of droplet formation, evaporation, convection, diffusion, reactivity, and reaction rate. It is obvious that these requirements can not be met in full. However, practical tracers are often very similar to the fuel or are components that are present in commercial fuels. Therefore, in some situations, tracers must not be understood as being *added* to the fuel. Rather, the other fluorescing substances are replaced by non-fluorescing compounds. The modification of the system must be kept to a minimum and the influence of the tracers on a given experimental situation must be critically reviewed.

Ideally, the tracer should yield LIF signal intensities that are directly proportional to the desired quantity and should not be influenced by the ambient conditions. Unfortunately, signals from all fluorescent tracers show at least some dependence on local temperature, pressure and bath gas variation. Therefore, in experiments where ambient conditions change in time or space, the underlying interdependencies with the tracer signal must be understood in order to obtain quantitative results.

Tracers for Homogeneous Gas-Phase Systems. Tracer-based LIF techniques have been used for experimental studies in fluid mechanics and combustion for several years. Tracers are single components (molecules or atoms) with highly defined spectral behavior that represents the local concentration of the fluid of interest or that allow to remotely measure a quantity of interest (i.e., temperature or pressure). Typically, compounds are chosen that yield strong enough LIF signal intensities to allow two- (or even three-)dimensional visualization of the desired quantity with sufficient temporal resolution to freeze the motion.

There are two opposite cases that require the application of tracers for the measurement in fluids. First, the components of the fluid *do not (or only weakly) fluoresce*. This is the case for air and typical exhaust gases (H_2O [20.178], CO_2 [20.179]). At room temperature these species are only excited in the vacuum-UV region and only at high temperatures do their spectra extend into the spectral range that is of practical use for combustion diagnostics. The resulting signal in O_2 [20.180] and CO_2 [20.179] is then strongly temperature dependent and the practical use for concentration measurements is limited. In the second case the fuel contains *too many fluorescing compounds*. This is true for commercial fuels. Their fluorescence has been used to obtain qualitative and semiquantitative measurements of fuel vapor concentrations [20.181]. However, because all these compounds have different physical properties in terms of volatility and transport as well as in terms of their spectral response on variations in ambient conditions, the overall signal cannot be quantified. In both situations it is desirable to add a single tracer that can be selectively observed within the fluid. In the case of fuels (gasoline or Diesel-type fuels) this often means replacing the fluorescing compounds of the fuel by non-fluorescing compounds leaving only one, or adding an additional compound that was not part of the original mixture.

Different *classes* of molecules have been used as tracers. The choice of potential tracers is driven by the desire to add a minimum concentration of foreign material that yields a maximum LIF signal intensity while not perturbing the system under study. To provide high enough seeding concentrations especially at low temperatures (room temperature) the tracer must have a sufficient vapor pressure. While the main part of the following section focuses on the fuel-like hydrocarbon-based tracers, we include other concepts in the following overview.

Atoms have large absorption cross sections and are candidates that emit strong fluorescence upon excitation in the UV and the visible. The atomization of the material, however, requires the high temperatures that are present in flames. Some metal salts (such as thallium or indium chloride) can be dissolved in the fuel. In the flame front metal atoms are then generated that can be used to measure temperature in the burned gas [20.182, 183]. The strong transition moments in atoms allow the use of extremely low (and therefore non-perturbing) seeding levels. The strong transitions, in turn, are easily saturated. Excitation laser intensities are therefore limited and signals are weak, despite their favorable spectroscopic properties. For the observation of fuel distributions prior to combustion this class of fluorescing species is not suitable.

Small Inorganic Molecules. Di- and triatomic inorganic molecules are frequently used in combustion diagnostics. While unstable radicals such as OH, CH, C_2 that appear during the combustion process can be used for flame-front localization and combustion diagnostics [20.10], they are not suitable for observations in the mixing process prior to ignition. Strongly fluorescing stable species, however, are potentially interesting as tracers for the airflow. NO has been used despite its toxicity for studies in gaseous mixing processes [20.184, 185] and its spectroscopy is well understood for a wide range of possible applications [20.186]. OH and NO can also be photolytically generated in flow systems. While not suited for studying mixing on a large scale, these flow-tagging techniques give detailed insight into the fluid motion within the lifetime of the generated species. OH has been generated following photodissociation of vibrationally hot water [20.167], NO has been produced from NO_2 photolysis [20.187] and from O_2 photolysis in air [20.188].

Molecular oxygen has been used to *trace* the air flow [20.189] and to measure temperatures during mixture formation in a Diesel engine [20.180]. It was also used for flow tagging following the excitation of higher vibrational states by stimulated Raman scattering [known as Raman excitation plus laser-induced electronic fluorescence (RELIEF) [20.169]].

Iodine was applied as a fluorescing tracer that can be excited and detected in the visible spectral range [20.190]. Its toxicity, corrosiveness and the difficulty of seeding iodine at constant rates limits its practical applicability. SO_2 can be excited at various wavelengths below 390 nm and subsequently emits fluorescence from the UV to the violet [20.191–193]. The (corrosive and toxic) gas (boiling point: $-10\,°C$) can either be doped into the flow or generated in a flame from sulfur-containing precursors [20.194]. The latter application was suggested to mark residual burned gases in internal engine combustion and to visualize their mixing with fresh air and fuel. SO_2 fluorescence is strongly quenched by many molecules, including N_2 [20.192, 195–198]. Its applicability in high-pressure environments is therefore restricted. The LIF properties of high-temperature CO_2 upon excitation in the UV [20.179] offer the potential for new diagnostics for the observation of mixing of hot burned gases with air and fuel.

Organic Molecules. In contrast to the excitation of atoms and di- and several triatomic molecules, polyatomic organic molecules have a high density of states and therefore show broadband absorption spectra with excitation possible at various wavelengths that are often accessible with standard laser sources. The organic tracers are chemically close relatives of hydrocarbon fuels.

Some of the molecules that are attractive tracers are present in commercial fuels at the few percent level. Therefore, relatively high tracer concentrations can be applied without significantly disturbing the combustion process.

The chemical similarity between tracer and fuel has the additional advantage that the tracer disappears (burns) together with the fuel close to the flame front. In measurements with limited spatial resolution (≈ 1 mm) this is a good match to identify and visualize the position of reactive zones. However, since the reaction kinetics of the tracer is not perfectly identical to that of the fuel, tracers are not in general suitable for high-resolution measurements close to the flame front, because their concentration may not represent the fuel concentration accurately in this zone.

Fluorescing organic tracers come in different sizes and structures with different volatilities. According to their boiling points as a first criterion they can be used to represent different volatility classes of multicomponent fuels [20.199]. At the same time, care must be taken to avoid distillation processes that separate fuel and tracer due to non-ideal boiling behavior during the fuel evaporation [20.200–202].

Aromatic hydrocarbons are typical components of commercial fuels. These species are responsible for the strong absorption in the UV and the subsequently emitted fluorescence [20.203]. Single-ring aromatics such as benzene, toluene and xylene are part of gasoline fuels on the percent level while two-ring aromatics such as naphthalene and its derivatives are present in Diesel fuels. They typically have high fluorescence quantum yields (toluene: $\phi = 0.17$, benzene: $\phi = 0.22$, fluorene: $\phi = 0.66$, dimethyl anthracene: $\phi = 0.82$) and their absorption and emission spectra shift towards the red with increasing size of the aromatic system. The wide variety of molecular sizes (and therefore boiling points) makes this class of molecules attractive as tracers that can be adjusted to the boiling behavior of the fuel or that are representative for boiling classes in multicomponent fuels. For seeding room-temperature gas flows compounds larger than benzene and toluene have too low vapor pressures.

A major drawback of aromatic tracers is the strong quenching by oxygen. The signal intensities therefore do not only depend on the tracer but also (inversely) on the oxygen concentration. This effect was taken advantage of by interpreting the signal as proportional to fuel–air ratios, which are of major practical interest.

The aromatic compounds in commercial fuels have been used for qualitative and semiquantitative measurements, both in the vapor [20.204] and liquid phase [20.205, 206]. Benzene as a fuel tracer is typically avoided because of its carcinogenic effects. Toluene is less toxic and not considered carcinogenic. Therefore, it has been most frequently chosen as a fuel tracer [20.207] and recent publications have shed more light on the dependence of its fluorescence on p, T and n_{O_2} [20.208, 209]; α-methyl naphthalene was investigated by LIF, see e.g. [20.210], because it is part of a model two-component fuel that is used in experimental and numerical studies as a substitute for diesel or JP8 fuel. Like naphthalene, it is used in combination with N,N,N',N'-tetra-methyl-p-phenylenediamine (TMPD) in exciplex studies to visualize liquid and vapor phases simultaneously [20.211].

Aliphatic Compounds. Typical saturated aliphatic hydrocarbons such as alkanes and saturated alcohols do not fluoresce. They have their first absorption bands in the vacuum-UV region and excitation typically leads to photodissociation. Nonsaturated hydrocarbons with extended conjugated systems that would have useful spectroscopic properties are unstable and tend to polymerize.

Fluorescing aliphatic candidates contain chromophores that allow excitation into stable states that subsequently fluoresce. This class of molecules includes ketones (R_2CO), aldehydes ($R-CHO$) and amines (R_3N, where R is a saturated aliphatic hydrocarbon chain). The (conjugated) combination of chromophores (as in diketones $R-CO-CO-R$) typically shifts the absorption and fluorescence spectra to the red.

Ketones are the most frequently used class of fluorescent tracers. Their properties have been extensively studied [20.212–218] and they have been applied to various practical situations [20.219–222]. The high vapor pressure makes acetone (bp: 56 °C) an ideal tracer for gaseous flows [20.223, 224]. 3-pentanone (bp: 100 °C) [20.213, 225, 226] or mixtures of 3-pentanone and 3-hexanone [20.200] have been suggested as tracers that mimic the boiling and transport properties of gasoline. In most of those cases, *iso*-octane was substituting gasoline. This has the advantage that *iso*-octane is non-fluorescent and as a single component is more amenable to detailed modeling studies. For Diesel fuels, the use of even larger ketones was suggested, although they turned out to have limited stability at high temperatures. The larger aliphatic chains enhance the reactivity of the carbonyl group.

The smallest aldehyde, formaldehyde (HCHO), tends to polymerize and is therefore difficult to handle as a dopant. The next largest homologous molecule, acetaldehyde (CH_3CHO), has been used as a tracer in internal combustion engines [20.227]. Its low boiling point (21 °C) allows high seeding concentrations. Its spectral properties are comparable to those of acetone. Because acetone is considered less harmful, acetaldehyde is not frequently used as a tracer substance. Aldehydes with different molecular weight have been used to trace different boiling fractions in multicomponent fuels [20.228].

The fluorescence quantum yields ($\phi = 0.0008-0.0018$ for acetaldehyde depending on the excitation wavelength and $\phi = 0.002$ for acetone [20.229, 230]) are small compared to those of aromatic compounds in the absence of oxygen. In realistic situations where fuel is mixed with air, however, the resulting signals are comparable. Hexafluoroacetone has been suggested as an alternative that has fluorescence quantum yields about an order of magnitude stronger than in acetone [20.231, 232]. Because of its toxicity, however, this substance has not been further considered as a tracer for practical applications.

Biacetyl ($CH_3-CO-CO-CH_3$) is frequently used as a fluorescent tracer. With a boiling point of 88 °C the room-temperature vapor pressure is too low for many applications. On the other hand, the boiling point is significantly below that of typical fuels. This intermediate position, its stability and its strong odor might be the reasons why biacetyl, despite its attractive spectroscopic properties and its extensively studied photophysics [20.233–235], is less popular than ketones. For seeding oxygen-free flows, biacetyl is especially attractive because of its high phosphorescence quantum yield (15%, but strongly quenched by O_2).

Amines do fluoresce upon UV excitation. Some studies have covered ethylamine [20.236], and N,N-dimethylaniline [20.237]. Strong quenching by oxygen often hinders practical applications and an unpleasant (fish-like) smell and toxicity may deter people from using this class of tracers. Amines have found some attention as part of exciplex systems TMPD [20.211, 238], triethylamine (TEA) [20.239], and diethylmethylamine (DEMA) [20.240]. TEA and DEMA do not contribute significantly to the overall gas-phase signal but play an important role in forming the excited complex that is responsible for a shift of the liquid-phase fluorescence relative to that of the monomers.

Analysis of Air–Fuel Ratio. Depending on the system under study various approaches have been used for determining the local fuel–air equivalence ratio

ϕ. In situations where the local oxygen concentration can be predicted, the standard procedure relies on the application of a fluorescent tracer that is considered independent of oxygen (i.e., acetone or 3-pentanone). As long as the fuel concentration is reasonably low (not close to a directly-injected fuel spray) and as long as oxygen is homogeneously distributed in the fresh air (no exhaust gas recirculation) and as long as temperature is known, air–fuel ratios can be determined directly [20.220, 225, 241]. With a simultaneous measurement of temperature the results are less dependent on these assumptions [20.134]. In areas with high fuel concentration, the air partial pressure is reduced by the presence of fuel vapor. An iterative data evaluation can then be applied.

Another approach relies on the measurement of a single LIF signal that depends on both the fuel and the oxygen concentration [fuel–air ratio by laser-induced fluorescence (FARLIF)] [20.207]). FARLIF takes advantage of the strong quenching of toluene (or other tracer) LIF by O_2 in order to use toluene as a fluorescent tracer to measure ϕ directly [20.207]. Toluene fluorescence in air at room temperature is proportional to ϕ for $p > 3$ bar with 248 nm excitation. In IC engine applications the decrease of total fluorescence with increasing crank angle (i.e., increasing p and T) has been observed using excitation at 248 nm [20.242, 243]. Recent studies at elevated temperatures revealed a strong T dependence of the fluorescence quantum yield, which changes by more than two orders of magnitude within the temperature range of the compression stroke in engines, while the relative strength of O_2 quenching depends on the O_2 number density itself [20.208, 244, 245]. With these new findings FARLIF must be reconsidered and can only be applied within a limited range of conditions [20.246].

Further measurement strategies rely on the measurement of a number of signals (typically two) that depend on the fuel and the oxygen concentration. Mixtures of two fluorescent tracers, one of which is strongly quenched by O_2, have been applied in isothermal flows [20.135] and in an IC engine [20.242]. The combinative effects of oxygen and temperature variation are not yet fully understood. Therefore, measurements in systems with an inhomogeneous temperature distribution may be problematic. Additional information is currently being obtained [20.208, 209].

Finally, Raman scattering has been used for fuel–air measurements in pointwise measurements [20.241], line [20.247–249] and imaging measurements [20.250, 251]. These measurements require a detailed knowledge of the temperature-dependent shape of the Raman spectra and therefore require detailed calibration experiments or a detailed understanding of the underlying spectroscopy [20.252]. The influence of liquid droplets has been evaluated [20.248].

Multispecies Measurements during Mixture Formation in IC Engines. Knowledge of the spatial distribution of temperature and local fuel–air equivalence ratios in the combustion chamber in IC engines prior to ignition is of major interest when modeling engine combustion and modifying combustion chamber geometries. Especially in modern engines with stratified load and systems with exhaust gas recirculation that both strongly influence ignition and flame development, temperature and fuel concentration inhomogeneities are present.

2-D temperature distributions between 300 and 1000 K relevant for pre-combustion conditions can barely be assessed with other laser spectroscopic techniques developed for two-dimensional combustion thermometry so far. Rayleigh scattering has been employed for the determination of temperature fields in flames [20.140]. Engine measurements have been performed as well [20.253] but problems might occur in realistic engine geometries due to strong elastic scattering off surfaces. Since Rayleigh signal intensities are a function of a species-dependent scattering cross-section this method is not applicable in non-homogeneous systems where local effective Rayleigh cross-sections are unknown. Techniques based on atomic fluorescence [20.182, 254] and OH LIF [20.255] are only applicable to post-flame gases. NO-LIF thermometry [20.184, 185] would be an alternative because NO, when doped into the fresh gases, is present in both unburned and burned gases. This requires, however, an additional dopant and some uncertainty is induced in the fresh gases by the presence of NO_2 formed by reaction of NO with air.

For quantitative measurements of fuel vapor concentrations ketones such as 3-pentanone are frequently used because their evaporation properties are similar to those of common model fuels such as *iso*-octane [20.237]. Furthermore, the influence of collisional quenching, mainly by molecular oxygen, is greatly reduced compared to aromatic compounds since the lifetime of the excited states is limited by rapid intersystem crossing. 3-pentanone possesses an absorption feature at 220–340 nm with the peak near 280 nm at room temperature [20.213]. The absorption spectrum exhibits a temperature-induced shift towards longer wavelengths of about 10 nm per increase of 100 K (Fig. 20.30). Upon excitation in this region, fluorescence is emitted between 330 and 550 nm, with a spectral distribution that is almost independent of the absorbed wavelength. This spectral shift of the absorption, albeit undesired for concentration measurements with single-wavelength excitation, can be used for measuring temperature, e.g., when 3-pentanone is seeded to nonfluorescing model fuels, as the fluorescence intensity is a function of the absorption coefficient for a given excitation wavelength, and thus, of temperature. After excitation at two different wavelengths the ratio of the fluorescence signal intensities reflects the local temperature. This was first described by *Großmann* et al. [20.213] and later applied to temperature measurements using acetone as a tracer [20.256]. Since this temperature measurement is based on the ratio of signal intensities it is independent of local tracer concentrations and therefore allows measurements of 2-D temperature distributions in nonhomogeneous systems.

For the organic tracers the fluorescence signal S for weak laser excitation is given by the following equation:

$$S_{fl}(\lambda, p, c, T) = cI_{Laser}(\lambda)VN\sigma(\lambda, p, x, T) \times \phi(\lambda, p, x, T), \quad (20.2)$$

where I_{Laser} is the local laser pulse energy in the detection volume V, N is the number density and σ the molecular absorption cross section of the marker, and ϕ its fluorescence quantum yield. σ and ϕ depend on the excitation wavelength λ, the pressure p, the mixture composition x, and the temperature T. The factor c captures geometrical arrangements and detection optics properties. If the same volume V is excited by two laser pulses of different wavelengths (with a short temporal delay to ensure separate detection of the induced fluorescence), the ratio

$$\frac{S_1(\lambda_1, p, T)}{S_2(\lambda_2, p, T)} = \frac{I_{Laser_1}(\lambda_1)}{I_{Laser_2}(\lambda_2)} \frac{\sigma_1(\lambda_1, p, x, T)}{\sigma_2(\lambda_2, p, x, T)}$$

$$\times \frac{\phi_1(\lambda_1, p, x, T)}{\phi_2(\lambda_2, p, x, T)} \quad (20.3)$$

depends on the pressure- and temperature-induced variation in absorption cross-sections and fluorescence quantum yields whereas it is independent of the observed volume V, the number density N of the detected species and the detection efficiency c. Pressure and possible composition effects can be corrected for using data provided by *Großmann* [20.213] and *Ossler* [20.215] by including a pressure-dependent ratio π_1/π_2 that includes effects on σ and ϕ. Then, the remaining ratio of both signals normalized to the respective laser energy is a function $F(T)$ of temperature only.

$$\frac{S_1(\lambda_1, p, T)/I_{Laser_1}(\lambda_1)}{S_2(\lambda_2, p, T)/I_{Laser_2}(\lambda_2)} \frac{\pi_2}{\pi_1}$$

$$= \frac{\sigma_1(\lambda_1, T)}{\sigma_2(\lambda_2, T)} \frac{\phi_1(\lambda_1, T)}{\phi_2(\lambda_2, T)} = F(T). \quad (20.4)$$

A strong temperature dependence of the ratio, and hence high accuracy, can be expected when λ_1 and λ_2 are selected on opposite sides of the absorption maximum. We apply a combination of 248 and 308 nm excitation. The temperature dependence of the ratio $F(T)$ for these conditions is shown in Fig. 20.31 [20.257]. With the local

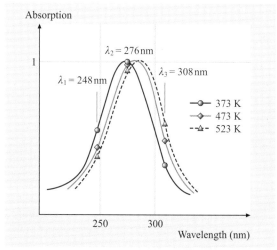

Fig. 20.30 Temperature dependence of the absorption spectra of 3-pentanone. The spectra are normalized, and the total integral increases with temperature [20.213]

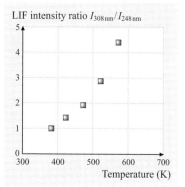

Fig. 20.31 Temperature dependence of the ratio of LIF intensities in 3-pentanone with 308 and 248 nm excitation [20.134]

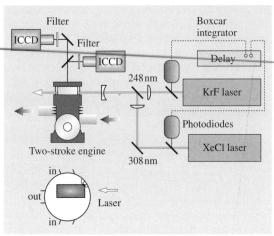

Fig. 20.32 Experimental setup for the simultaneous measurements of fuel concentration, temperature and air–fuel ratio in a two-stroke engine. The *lower figure* shows the position of the measurements displayed in Fig. 20.33

temperatures now determined from the signal ratio the fluorescence signals from either excitation wavelength can be corrected for temperature effects to calculate the number density N of tracer molecules. A calibration with a known number density of tracer molecules at a known temperature allows one to scale N to absolute number densities. With the ideal gas law the local *mole fraction* can then be calculated. The fuel–air equivalence ratio can then be determined based on the assumption that the cylinder contains fuel and air only and that the oxygen concentration is homogeneous within the air. There are limitations to these assumptions in systems with lots of recirculated exhaust gases. Techniques to overcome this problem will be discussed below.

The engine measurements [20.134] (Fig. 20.32) were conducted in a modified production-line single-cylinder two-stroke engine (bore: 80 mm, stroke: 74 mm, compression ratio: 8.6). The original cylinder head was replaced by a quartz ring with a height of 4 mm to allow for the entrance and exit of the laser sheets, and a full-size cylindrical quartz window on top through which fluorescence was monitored. The engine was carburetor-fueled with *iso*-octane doped with 10% (v/v) 3-pentanone. For the measurements described here the equivalence ratio was kept at $\phi = 0.62$ with an ignition timing at $-20°$ crank angle (CA) with respect to top dead center (TDC) at a speed of 1000 rpm. Two excimer lasers, operated with KrF (248 nm) and XeCl (308 nm), were fired with a fixed delay of 150 ns to prevent crosstalk between the LIF signals. The laser pulse energies were adjusted to no more than 50 mJ within a 20×0.5 mm^2 light sheet. The signals were directed via a metal-coated beam splitter to two ICCD cameras that were equipped with $f_\# = 2$, $f = 100$ mm achromatic UV lenses.

Figure 20.33 shows the LIF intensities upon 308 nm excitation for three single-shot images (18° CA after

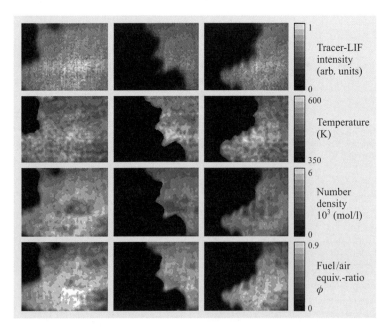

Fig. 20.33 Measurements of the tracer LIF intensity (first row), temperature (second row), fuel number density (third row) and fuel–air equivalence ratio (fourth row) in a two stroke engine. The columns show three randomly chosen data sets for instantaneous measurements [20.134]

TDC). The correction for temperature effects was performed on a per-pixel basis. The temperature-corrected LIF images were then calibrated to absolute number densities using the measurements under perfectly mixed conditions where the fuel concentration was calculated from the ratio of measured air flow and fuel consumption. The resulting images give absolute fuel number densities (third row). Since pressure gradients within the cylinder can be safely neglected temperature images allow the calculation of total number densities, which enable the calculation of fuel–air equivalence ratio distributions (fourth row). This series of images shows how the actual equivalence ratio corresponds to the measured LIF intensity.

Simultaneous Measurement of Fuel and Oxygen Concentration. In future highly energy-efficient combustion systems such as stratified-charge engines or homogeneous-charge compression-ignition (HCCI) engines, high concentrations of recycled exhaust gases strongly dilute the fuel–air mixture. In these situations, ignitability, flame speed, and auto-ignition susceptibility are governed by fuel and oxygen concentrations (number densities) rather than equivalence ratios. Whereas several techniques for measuring fuel concentrations and equivalence ratios are available (measurements for example based on the assumption that for tracers that are strongly quenched by oxygen – within a limited range [20.258] – the LIF signal is proportional to the equivalence ratio [20.207]. For the limitations of this assumptions under combustion-engine conditions see [20.246, 259]), we report here a technique that allows the simultaneous measurement of oxygen concentrations with a suitable tracer combination of aromatic and ketonic tracers. Aromatic molecules are efficiently quenched by molecular oxygen [20.260]. The influence of O_2 quenching on aliphatic ketones on the other hand is greatly reduced [20.213]. Aromatic and ketonic compounds have different LIF emission spectra upon excitation in the UV. Whereas single-ring aromatic compounds such as toluene display a narrow emission band peaking around 280 nm, the emission of ketones like 3-pentanone appears between 320 and 450 nm with a broad maximum at around 370 nm (Fig. 20.34). This difference has been used previously for the detection of the two different tracers when simultaneously observing the distribution of different volatility classes of multi-component fuels [20.199]. Here, we use this two-tracer approach for measuring oxygen concentrations.

For weak excitation the LIF signal intensity S_{fl} is given above. Molecular oxygen is regarded as the only relevant species quenching toluene fluorescence, while 3-pentanone fluorescence is considered independent of oxygen concentration. For a mixture of 3-pentanone (3P) and toluene (tol) the ratio of signal intensities $S_{\mathrm{fl}}^{\mathrm{R}}$ (20.5) is a function of the oxygen concentration only [$f(n_{O_2})$] since all the other factors are either constant (like absorption cross sections due to constant pressure and temperature) or cancel (like the local laser intensity and detection volume):

$$S_{\mathrm{fl}}^{\mathrm{R}} = \frac{S_{\mathrm{fl}}^{\mathrm{3P}}}{S_{\mathrm{fl}}^{\mathrm{tol}}} = \mathrm{const} \times \frac{\phi_{\mathrm{fl}}^{\mathrm{3P}}}{\phi_{\mathrm{fl}}^{\mathrm{tol}}} = f(n_{O_2}). \quad (20.5)$$

The fluorescence quantum yield ϕ_{fl}^{i} for species i is

$$\phi_{\mathrm{fl}}^{i} = \frac{k_{\mathrm{fl}}^{i}}{k_{\mathrm{tot}}^{i} + \tilde{k}_{\mathrm{q}}^{i} n_{\mathrm{q}}} \quad (20.6)$$

with the rate coefficient for spontaneous emission k_{fl}^{i} and the sum rate of all collision-independent de-excitation processes k_{tot}^{i}. $\tilde{k}_{\mathrm{q}}^{i}$ is the rate coefficient for collisional quenching with the quencher (i.e., O_2) number density n_{q}.

For atmospheric pressure in 3-pentanone, as in all aliphatic ketones, quenching by O_2 can be neglected (because $\tilde{k}_{\mathrm{q}}^{\mathrm{3P}} n_{\mathrm{q}} \ll k_{\mathrm{tot}}^{\mathrm{3P}}$). With the Stern–Volmer coefficient for toluene quenched by O_2 ($k_{\mathrm{SV}}^{\mathrm{tol},O_2} = \tilde{k}_{\mathrm{q}}^{\mathrm{tol},O_2}/k_{\mathrm{tot}}^{\mathrm{tol}}$) the above equation can be rewritten [20.135]:

$$S_{\mathrm{fl}}^{\mathrm{R}} = \mathrm{const} \frac{k_{\mathrm{fl}}^{\mathrm{3P}}(k_{\mathrm{tot}}^{\mathrm{tol}} + \tilde{k}_{\mathrm{q}}^{\mathrm{tol}} n_{\mathrm{q}})}{k_{\mathrm{tot}}^{\mathrm{3P}} k_{\mathrm{fl}}^{\mathrm{tol}}} = c_1 \left(1 + k_{\mathrm{SV}}^{\mathrm{tol},O_2} n_{\mathrm{q}}\right). \quad (20.7)$$

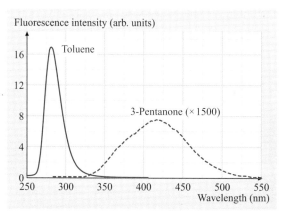

Fig. 20.34 Emission spectra of toluene and 3-pentanone with excitation at 248 nm. The signal of both species can be separated with appropriate filters. In pure nitrogen (shown here) the toluene signal dominates by far (3-pentanone signal enhanced by a factor of 1500) [20.135]

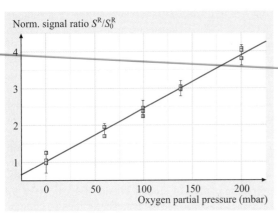

Fig. 20.35 Signal ratio of 3-pentanone and toluene LIF normalized to the ratio measured in pure nitrogen [20.135]

The signal ratio of 3-pentanone versus toluene LIF therefore increases linearly with the oxygen number density. The offset, given by c_1, can be obtained from a single-point calibration. An experimental validation is shown in Fig. 20.35. This simple approach assumes that the toluene and 3-pentanone signals are detected independently without any spectral overlap and that oxygen is the only quenching species. Further corrections may be added in order to reduce the systematic errors caused by these assumptions [20.135].

The Stern–Volmer coefficients and absorption and fluorescence spectra have been measured in a static cell where evaporated tracers were combined with various pressures of air and nitrogen. The imaging experiments were carried out in an optically accessible quartz cell with square cross section ($40\,\text{mm}^2 \times 40\,\text{mm}^2$). Two metal tubes (exit diameter 1 mm) were inserted 5 mm apart with the gas streams mixing at an angle of approximately 30°. The tubes were supplied with various mixtures of nitrogen and oxygen controlled by mass-flow controllers. The gas mixtures were bubbled through a blend of the liquid tracers (3-pentanone and toluene, 50:50 per volume), which were kept at a constant temperature of 293 K. The gaseous tracer concentration was therefore constant in both gas streams with a fixed ratio of both tracer compounds determined by the vapor pressure of the binary mixture. The volume of the liquid mixture was large enough to ensure negligible changes in its composition over time. For the experiments presented here one stream was supplied with pure nitrogen and the other contained variable oxygen concentrations. The gas exit velocities were set to 5–15 m/s. The mixing cell was purged with an additional nitrogen flow that was quiescent compared to the velocities of the other two gas streams. For calibration of the LIF signal ratio, a single gas stream with identical tracer concentrations but at various well-known oxygen concentrations was used. For imaging, the 248 nm laser beam was expanded to a light sheet 0.4 mm thick and 30 mm high and directed through the quartz cell 2 mm below the tube exits. The LIF signal was detected simultaneously with two ICCD cameras equipped with appropriate filters for observing each tracer separately.

Results of the two-tracer imaging technique in an arrangement of interacting N_2 and air gas jets within a quiescent N_2 atmosphere are shown in Fig. 20.36. The first box Fig. 20.36a shows the individual pictures taken in the toluene and 3-pentanone spectral region in the

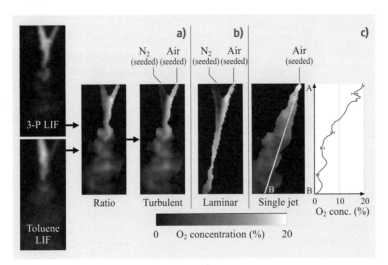

Fig. 20.36 Results of the O_2 imaging measurements in the mixing chamber [20.135]. For details see the text

two interacting turbulent gas jets. The 3P-LIF image shows increased signal intensities in the nitrogen jet due to some spectral overlap from the strong toluene signal. This error is corrected for by quantifying the relative contribution of toluene LIF to the 3-pentanone detection channel in additional experiments where the flow is seeded with toluene only. The second image shows that the ratio of signal intensities is not directly a function of oxygen number density [according to (20.7)] since the left jet (containing no O_2) yields higher signal intensities compared to the mixing region. After correcting for energy transfer processes [20.135], this artifact disappears (right image in Fig. 20.36a) and the resulting signal is calibrated using a measurement in pure air. Figure 20.36b shows the oxygen distribution in similar flows at a slower exit velocity at which the two laminar gas flows stay mostly separated. Figure 20.36c shows the O_2 distribution in an air jet injected into a quiescent N_2 atmosphere where the dilution of O_2 by entrainment of N_2 is clearly visible. The line plot shows the oxygen concentration along the line in the 2-D image including error bars to demonstrate the sensitivity of the technique. Since only the incoming jets were seeded by tracers, the outer region does not yield any information in these measurements and is shown in black.

Quantitative Nitric Oxide Diagnostics in High-Pressure Combustion

Detection of nitric oxide using LIF is of particular interest because NO is one of the most important combustion-generated pollutants. Oxides of nitrogen in the atmosphere contribute to photochemical smog and are precursors to acid rain. Exhaust from aircraft engines can add to the destruction of stratospheric ozone while automobile exhaust adds NO/NO_2 to our urban atmosphere. A significant amount of this nitrogen oxide is directly caused by combustion of biomass and fossil fuels [20.262]. As regulations for the restriction of NO become more stringent, much effort and research is underway to develop new and improved combustion systems that meet these standards. To optimize the combustion processes involved, a coupled understanding of physical and chemical processes and empirical diagnostic methods to map out the formation of the pollutants are required.

Choice of NO Detection Schemes for High-Pressure Flames. Two-dimensional LIF measurements of trace species require the coincidence of strong electronic transitions with the emitted wavelength of intense light

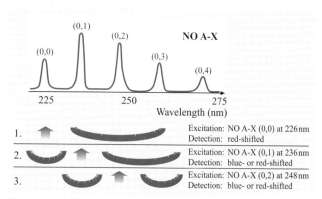

Fig. 20.37 Different options for NO LIF detection with excitation in the A − X band

sources; therefore, excimer lasers are a convenient source for NO excitation, but the limited wavelength tuning enables only a few excitation schemes. The NO D-X(0,1) system can be probed with 193 nm radiation, e.g., from an ArF excimer laser [20.263], the A-X(0,0)

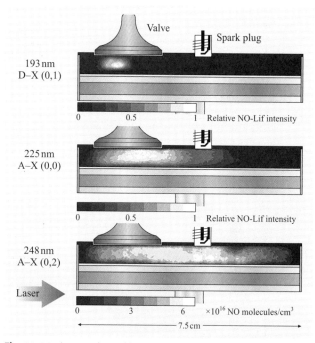

Fig. 20.38 Attenuation of laser light at different wavelengths during NO imaging with 193, 225 and 248 nm excitation. All three frames show the engine under identical operating conditions. Only with the longest wavelength (248 nm) was the expected symmetrical distribution of NO in the cylinder observed ($\phi = 1$ propane/air flame, actual pressure: 13 bar) [20.253, 261]

band at 225 nm with a H_2 Raman-shifted KrF excimer laser [20.150], or the A-X(0,2) band at 248 nm with a tunable KrF excimer laser [20.264]. Other measurements have been carried out with dye lasers and OPO systems in the (0,0) [20.265, 266] and (0,1) [20.267, 268] bands.

The NO D-X(0,1) system at 193 nm works for combustion processes where strongly absorbing species are present at low number densities only. In high-pressure combustion systems drastic attenuation of laser and signal light by hot CO_2, H_2O and O_2 limits their applicability [20.261, 270]. Therefore, we focus here on three NO detection schemes with A-X excitation, as depicted in Fig. 20.37.

In lean and stoichiometric high-pressure flames the selectivity of NO detection is perturbed by interfering hot-bands of O_2 that overlap the NO excitations. At pressures of a few bar, the quantum yield of O_2 LIF is dominated by predissociation, and the NO quantum yield decreases with increasing pressure due to collisional quenching while the O_2-LIF quantum yields remains constant. It is therefore crucial to choose NO excitation schemes that minimize the simultaneous excitation of O_2.

The NO A-X(0,0) system at 225 nm has been successfully used for measurements in high-pressure methane/air flames [20.271] and in engines fueled with propane under conditions up to 20 bar [20.272]. Three different strategies have been studied and applied to combustion diagnostics. Whereas $DiRosa$ et al. [20.265] uses the combined $P_1(23.5)$, $Q_1+R_{12}(14.5)$, $Q_2+R_{12}(20.5)$ feature at 226.03 nm, $Laurendeau$ et al. [20.266] prefer the $Q_2(26.5)$ transition at 225.58 nm. Within the tuning range of the Raman-shifted KrF excimer $Sick$ [20.272] chooses the $R_1+Q_1(21.5)$ line at 225.25 nm. We use measurements in a methane/air flames at 1–60 bar to compare the O_2 LIF interference and signal intensities of these three approaches.

Optical absorption in high-pressure combustion environments increases strongly at short wavelengths [20.261, 270] as a result of the transmission properties of hot CO_2 and H_2O [20.269, 273]. In high-pressure media, where the number density of absorbers is high, it is therefore desirable to use the longest possible wavelengths for excitation and detection. Excitation of the NO $A-X(0,2)$ near 248 nm [20.264, 274] exploits the strong KrF laser sources and has significantly less optical absorption in the engine; the feasibility of this approach is demonstrated with detailed spectroscopic studies [20.153, 274] and in-cylinder engine measurements (Fig. 20.38) [20.253, 275, 276]. In addition, this strategy allows the detection of NO fluorescence in the $A-X(0, 1)$ band, which is blue of the excitation light. This is a significant advantage as it suppresses LIF interference from intermediate hydrocarbon compounds. Unfortunately, the thermal population of the $v''=2$ level at post-flame temperatures is small, leading to a rather weak LIF signal intensity.

The availability of optical parametric oscillator systems (OPOs) which provide high laser pulse energies throughout the UV (up to 30 mJ/pulse) allows convenient excitation of the NO $A-X(0, 1)$ transition, and $Jamette$ et al. applied this scheme to measurements in an spark ignition (SI) engine [20.267].

A detailed analysis of the properties of the different excitation/emission schemes in the NO $A-X$ has been performed based on spectroscopic measurements in laminar high-pressure flames [20.153, 186, 277–280]. The results were used to develop the $LIFSim$ spectral simulation code that allows the simulation of absorption and emission spectra as well as the pressure and temperature influence in user-specified gas mixtures [20.174, 281].

Influence of Attenuation by Hot CO_2. Experiments in engines and high-pressure flames showed that hot exhaust gases strongly attenuate UV laser and signal light [20.261]. Figure 20.39 shows the strong wavelength dependence of the transmission of laser light in hot combustion gases. It was shown that this effect is mainly due to the shift of vacuum-UV absorption features of CO_2 towards longer wavelengths with increasing temperature [20.269]. For H_2O a similar

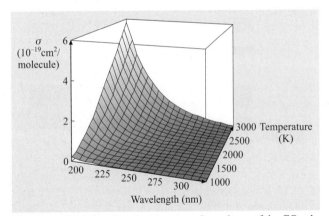

Fig. 20.39 Wavelength and temperature dependence of the CO_2 absorption cross sections [20.269]. A parameter fit to this data for the calculation of temperature- and wavelength-dependent absorption cross sections can be found in [20.261]

Table 20.2a,b Transmission at NO excitation wavelengths in two typical high-pressure combustion situations due to hot CO_2 (CO_2 and H_2O) absorption: (**a**) high-pressure flame: diameter 8 mm, 40 bar, 2300 K and (**b**) engine cylinder: diameter 7 cm, 13 bar, 2400 K [20.261]

	High-pressure flame, diameter: 8 mm	Engine cylinder, diameter: 7 cm
193 nm	13.7% (0.4%)	0.02% (10^{-6}%)
226 nm	62.4% (58.1%)	11.5% (9.1%)
237 nm	75.4% (73.8%)	26.6% (24.8%)
248 nm	84.4% (83.9%)	44.3% (43.4%)

behavior can be found at short wavelengths. It is, however, negligible at $\lambda > 230$ nm [20.269]. Transmission in typical high-pressure combustion situations is calculated (based on an empirical fit to the data in Fig. 20.40 [20.261]) for the different NO excitation wavelengths and summarized in Table 20.2. These data show that in many practical applications the longest possible wavelengths must be used for NO excitation despite the relatively weak fluorescence intensity that can be expected upon excitation from the excited vibrational levels.

Temperature Dependence of the Different NO Excitation Strategies. LIF diagnostics of practical systems are greatly simplified if the signal is relatively insensitive to the local gas temperature. The Boltzmann population of each of the vibrational manifolds produces quite different temperature sensitivities. In the temperature range relevant to post-flame gases (1500–2500 K) the signal intensity after excitation in the (0,0) band decreases with increasing temperature whereas in the (0,1) and (0,2) bands signals are strongly increasing. Minimal variation of the ro-vibrational population with temperature is found for the intermediate rotational levels in the (0,0) band. The (0,1) band has minimal pressure dependence since the chosen excitation line is within a very dense part of the spectrum where at elevated pressures neighboring lines contribute significantly to the overall signal. The simulated temperature dependence of the NO-LIF signal is shown in Fig. 20.40 for $p = 10$ bar for the candidates with minimal O_2 LIF interference for

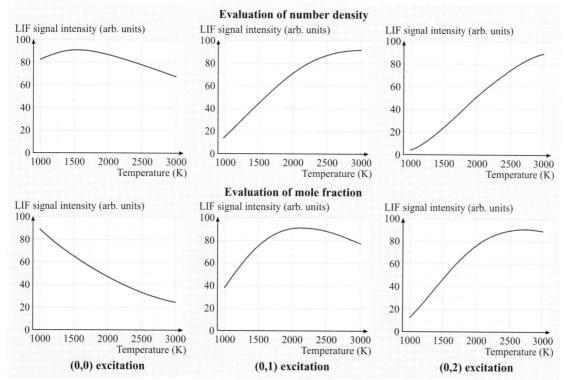

Fig. 20.40 Simulated variation of NO LIF signal with temperature for three excitation strategies. Rotational transitions with minimal O_2 LIF interference have been chosen [20.186]. $p = 10$ bar

each vibrational band. The data in Fig. 20.40 include the temperature dependence of Boltzmann population, spectral overlap of laser and NO transition, and collisional quenching. Note this temperature dependence can be mitigated with the judicious choice of measuring either number density or mole fraction. Excitation in the (0,0) band has weak temperature dependence of *number density*; however, excitation of the (0,1) or (0,2) bands can provide *mole fraction* measurements that are weakly temperature dependent at combustion temperatures.

There are three practical schemes for NO A − X LIF, which involve exciting transitions in the (0,0), (0,1) and (0,2) bands. In systems where excitation in the NO A − X(0, 0) band is not hindered by laser and signal absorption an NO absorption feature at 226.03 nm was found to be the best choice in terms of spectral purity and signal intensity. A − X(0, 1) excitation with (0,0) detection provides strong signal, small variation of signal with gas temperature when evaluating mole fraction, and good suppression of O_2-LIF interference. Any scheme exciting or detecting (0,0) in high-pressure hydrocarbon combustion, however, is plagued by attenuation of the excitation or fluorescence light. Excitation of A − X(0, 1) with (0,2) detection provides significantly stronger signals with less temperature variation than strategies using (0,2) excitation, and the interference from O_2 LIF is comparable for (0,1) and (0,2) excitation. However, practical engine diagnostics have hydrocarbon fluorescence interference red of the excitation light. The strategy of exciting NO A − X(0, 2) and detection of (0,1) provides LIF signal blue-shifted compared to the excitation light. This advantage may outweigh the lower signal and greater temperature sensitivity for practical diagnostics applications.

Quantitative Multi-Line Temperature Imaging. Temperature imaging is often performed as a two-line technique that probes the relative population of two ro-vibrational ground states in molecules such as NO [20.282, 283] or OH [20.255, 284]. This technique typically requires a calibration with measurements in an environment with known temperature and is affected by background signal. With simultaneous application of two excitation wavelengths, instantaneous temperature imaging is possible. Alternatively, scanning techniques have been applied that record the fluorescence excitation spectrum over an extended range of excitation wavelengths and evaluates absolute temperatures – independent of signal background – from the shape of the spectra [20.271]. We use the latter approach to measure temperature fields in stationary flames [20.185]. For the multi-line temperature imaging, 2-D LIF excitation scans were performed over a broad spectral range in the NO A − X(0, 0) band. For each individual pixel, LIF intensities were extracted versus excitation wavelengths, yielding experimental excitation spectra (Fig. 20.41).

The LIFSim spectral simulation code [20.281, 285] was then used to fit simulated spectra to the experimental data for each pixel. This yields absolute temperatures without the requirement for calibration. Simultaneous fit of background signal and LIF strength makes this technique robust against both broadband interference and laser and signal attenuation. This evaluation therefore yields temperature and semiquantitative NO concentrations that then can be calibrated by adding known amounts of NO to the fresh gases according to [20.153]. The broadband background that was recently attributed to CO_2 LIF [20.179] is measured at the same time yielding spatial distributions of CO_2 LIF intensities [20.179, 286]. The temperature information in turn can be used to calculate the local laser and signal attenuation due to hot CO_2 based on the data shown in Fig. 20.39 assuming a homogeneous concentration of CO_2 in the burned gases. These corrections have been applied for the measurement of quantitative NO concentration distributions in the (unseeded) methane/air high-pressure flame.

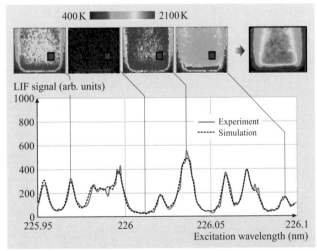

Fig. 20.41 Schematic of multi-line 2-D temperature measurement strategy based on a single point. NO A − X(0, 0) excitation at 5 bar. The images show an area of 11 mm^2 × 13 mm^2 above the burner exit [20.185]

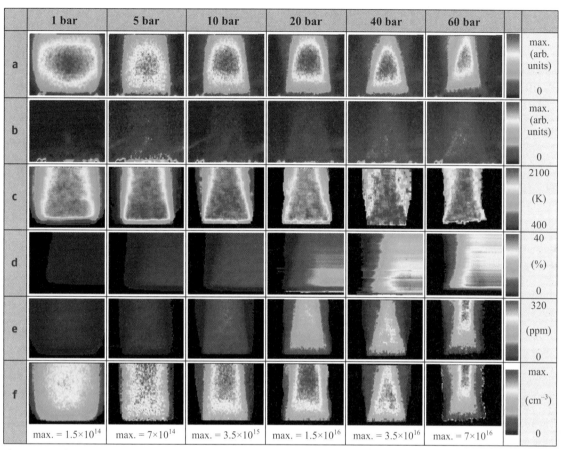

Fig. 20.42a–f Imaging results from a methane/air flame ($\phi = 0.95$, without seeding). **(a)** Raw NO-LIF images (not corrected for laser sheet inhomogeneities) **(b)** Raw off-resonant images **(c)** Absolute temperatures **(d)** Total attenuation due to laser and signal absorption by CO_2 and H_2O **(e)** Quantitative NO mole fractions (calibrated by NO-addition) **(f)** Quantitative NO number densities (calibrated by NO-addition method)

Figure 20.42 shows raw NO-LIF images, corrections performed for quantification, and evaluated NO concentrations. For maximum selectivity (i.e., minimum O_2 LIF interference), excitation was performed at 226.03 nm as discussed in the introduction section (Fig. 20.42a). Remaining interference was assessed by tuning the laser off-resonance to 224.78 nm (Fig. 20.42b). The temperature-sensitivity of the LIF signal was accounted for by using LIFSim. In many cases, temperature insensitivity is advantageous when conducting concentration imaging because of fluctuation of temperature within the view of the image. The temperature fields that were acquired using the multi-line technique are shown in Fig. 20.42c. A correction for collisional quenching was applied. Collisional quenching scales linearly with pressure when the gas composition is unchanged; however, due to the fact that gas composition and collisional cross-section are all temperature dependent, correction for this temperature effect is required. Figure 20.42d shows the corrections applied for signal and laser beam attenuation. Recent experiments have shown that in high-pressure and high-temperature combustion environments, the UV transmission is reduced by absorption from major species such as CO_2 and H_2O. The attenuation of both the laser beam and the LIF signal due to CO_2 and H_2O require calibration and have been accounted for in the imaging processing stage. The corrected and calibrated NO concentrations are shown in Fig. 20.42e (mole fractions) and f (number densities). Recent measurements report the application of multi-line NO-LIF thermometry also to sooting high-pressure flames and spray flames [20.276, 287–289].

Quantitative NO Diagnostics in Engines. Minimizing the generation of the pollutant nitric oxide is of particular interest when improving the environmental acceptance of internal combustion engines. While the release of many unwanted exhaust compounds like unburned hydrocarbons and carbon monoxide has been drastically reduced within the last years, additional effort is necessary to further reduce the NO release in order to meet future legislative regulations. Especially in modern lean-burning engines, inhomogeneous combustion conditions make it difficult to control the pollutant formation. Furthermore, standard catalytic converters can not be used since the overall air–fuel ratio deviates significantly from unity. Exhaust gas recirculation (EGR) has been used successfully in diesel and in gasoline engines to reduce the raw NO formation mainly due to decreased combustion temperatures. A detailed understanding of the elementary physical and chemical steps involved as well as their coupling is necessary to allow a mathematical description of the combustion process. Models with the ability of predicting optimized operating conditions, however, need input and feedback from preferably quantitative in-cylinder experiments.

For in-cylinder NO imaging diagnostics we use NO excitation in the $A - X(0, 2)$ band with a tunable KrF excimer laser at 247.94 nm and subsequent detection of the fluorescence light emitted from the $A - X(0, 1)$ transition around 237 nm. This technique allows the effective suppression of interfering LIF signals due to hot O_2 and CO_2 and partially burned hydrocarbons [20.274, 291, 292]. Therefore, selective detection of NO is possible even in inhomogeneous combustion environments like in direct injecting gasoline [20.275] and Diesel engines [20.293]. Furthermore, with this scheme the influence of laser beam attenuation is minimized [20.261, 270] (Fig. 20.38).

Figure 20.43 shows experimental and numerical results for the NO concentration distribution in a single-cylinder engine fueled with *iso*-octane air equipped with a production-type Volvo cylinder head [20.173, 294]. Care is taken to provide a homogeneous air/fuel mixture in these cases. Figure 20.44 shows results of spatially averaged NO concentration versus detection time in crank angle in a direct-injection engine with variable amounts of synthetic recirculated exhaust gases (variable mixtures and concentrations of N_2, CO_2, H_2O added to the intake air) [20.275, 290]. The experiments presented here were carried out in an optically accessible gasoline engine featuring a direct-injection cylinder head (BMW) and a Bosch injection system. A significant reduction of peak NO concentrations from 1200 ppm to 200 ppm with increasing EGR rate can be observed. A significant influence on the composition of the *synthetic exhaust gas* added to the fresh in-cylinder gases was found that correlates well with the heat capacity of the gas mixture [20.290]. For comparison with exhaust gas NO_x concentrations, results from the chemiluminescence (CLD) measurements

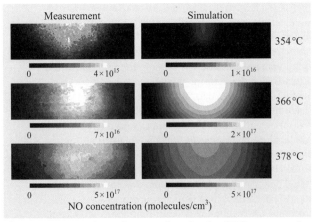

Fig. 20.43 Comparison of measured an simulated NO concentration distributions in an IC engine fueled with iso-octane/air [20.173]

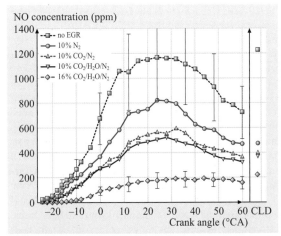

Fig. 20.44 Crank-angle-resolved spatially averaged NO concentrations for the different EGR conditions (see legend) in the BMW engine with gasoline direct injection. CLD NO_x concentration measurements are included in the graph [20.290]

have been included in Fig. 20.44. It should be kept in mind that the LIF images represent the NO concentration present in the plane defined by the position of the laser beam whereas the exhaust gas measurements represent averaged concentrations after homogeneously mixing the burned gases during the expansion and exhaust stroke. The comparison shows a good agreement of exhaust-gas concentrations and concentrations of the in-cylinder LIF measurements in the observed plane.

Soot Diagnostics

Soot formation has been a major issue in combustion research over the last few decades as soot is emitted by many practical combustion processes, e.g., power plants and IC engines. Soot is the result of incomplete combustion of hydrocarbons if not enough oxygen is locally available for full oxidation. Soot consists mainly of carbon but may have a wide range of properties and composition [20.295]. Transmission electron microscopy (TEM) shows that flame-generated soot consists of primary particles which have a typical size between 5 and 100 nm. Primary particles form aggregates that contain up to several hundred primary particles and have a fractal structure.

Laser-induced incandescence (LII) has proven to be a powerful tool for soot diagnostics. It was observed for the first time by *Week* and *Duley* in 1973 when they heated aerosols of carbon black and alumina with a CO_2 laser [20.296]. They set up the first energy balance describing the heating and cooling of particles and suggested that the method could be used for particle sizing. *Eckbreth* observed the incandescence as an interference in Raman scattering experiments in sooting propane diffusion flames [20.297]. In 1984 *Melton* set up the first detailed energy- and mass balance for LII and suggested to use LII as a tool for quantitative measurements of soot volume fractions [20.298]. Since the early 1990s LII has been used successfully in a variety of systems for measuring volume fractions of soot in laminar diffusion flames [20.299–303], laminar premixed flames [20.22, 304, 305], and turbulent flames [20.299, 306–308]. Particle sizes have been deduced from the temporal behavior of the LII signal taking the ratio of the LII signal at two different delay times after the signal peak [20.309, 310] as well as using the entire decay curve in point measurements [20.311–315]. In exhaust gases particle sizes have been measured with LII as well [20.316, 317]. Particle sizes were obtained assuming monodisperse particles [20.311, 313, 314] as well as considering a polydisperse lognormal particle-size distribution [20.312, 315].

The LII signal is due to thermal radiation from particles that are heated by an intense pulse laser. Subsequent cooling occurs due to several heat loss channels until, typically after a few hundreds of nanoseconds, the particle temperature reaches ambient gas temperature again. LII is basically used for two applications: Determination of the volume fraction of particles and particle sizing in the submicron region. Though LII has been used to investigate non-carbon particles [20.318–320] with increasing interest in recent years with the upcoming nanosciences, the main focus has been in the field of soot diagnostics. In the following overview LII will be discussed with respect to soot.

Soot volume fractions measured with LII are based on the detection of the temporally integrated LII signal over a certain range of wavelengths. This allows the expansion of the laser beam into a sheet and the detection of the signal light with a fast gated intensified CCD camera equipped with appropriate detection filters. In this way two-dimensional images of the soot distribution can be obtained, even in turbulent flames. As the emitted radiation of the laser-heated soot is close to that of a black body, corrected by the emissivity, a wide choice of detection wavelengths is available. Experimental and theoretical studies have shown [20.321, 322] that detection towards longer wavelengths minimizes the influence of a variation in particle size and ambient gas temperature. However, in most practical environments a detection in the blue around 400 nm is preferred to obtain a better discrimination of LII signal against flame luminosity.

A large impact on the accuracy is caused by the choice of the temporal detection gate. First, after the laser heating smaller particles cool faster than large ones because small particles have a larger surface-to-volume ratio than large particles. If long or even delayed (relative to the laser pulse) detection gates are used the integrated signal will underestimate small particles [20.299, 321]. Second, with increasing pressure the LII signal decays faster due to an increased heat conduction with increasing pressure [20.323]. If calibration is carried out at one single pressure, long or delayed detection gates should be avoided. Hence, short and prompt, i.e., with the start of the laser pulse, detection gates of 50 ns or less should be used.

In LII, several excitation wavelengths have been used, mainly the fundamental, 1064 nm, and second harmonic, 532 nm, of pulsed Nd:YAG lasers. Light at 532 nm is easier to handle as it is visible and lower laser fluences are necessary to heat up the soot particles as the absorption cross-section of soot at 532 nm is roughly twice the value at 1064 nm. However, 1064 nm causes less interference from electronically-excited C_2 [20.22, 324] and less broadband interference from PAH [20.325]. For low laser fluences the LII intensity rises monotonically with laser fluence as peak particle temperatures increase. Increasing laser fluence starts to vaporize the soot during the laser pulse and the maximum particle temperature remains constant. In this regime, where maximum particle temperature is reached but only minor vaporization occurs, the variation of LII with laser fluence is small, resulting in a so-called plateau region. This plateau region is preferred in soot volume fraction measurements as the LII signal is relatively independent of laser fluence. This is important in strongly absorbing environments. With higher fluences, the soot concentration is significantly reduced by vaporization resulting in reduced LII signals.

To obtain absolute soot volume fractions LII signals need to be calibrated. Therefore, the accuracy of LII is strongly influenced by the accuracy of the calibration method. A common method is the use of light extinction. However, this technique depends on the knowledge of the refractive index of soot. Different values have been reported for this value [20.326]. A method that is independent of an external calibration is the calibration with absolute light intensity [20.327]. The current status in LII is reviewed in [20.328].

Numerous models have been developed to predict the temporal behavior of the LII signal [20.298, 310, 329–332]. The basis for most models is an energy and mass balance between absorption of laser energy and heat loss due to vaporization of material from the surface, heat conduction to the surrounding gas and radiation:

$$q_{abs} = q_{int} + q_{evap} + q_{cond} + q_{rad}, \quad (20.8)$$

where q_{abs} is the flux of laser energy absorption, q_{int} the flux of increase of internal energy, q_{evap} the flux of energy loss by evaporation, q_{cond} the flux of heat conduction, and q_{rad} accounts for the cooling due to radiation. Various submodels for the different processes have been suggested in the past. Solving the resulting differential equation numerically yields the particle temperature as a function of time which is then turned into LII signal intensities using Planck's law. A more detailed model for soot particles including thermal annealing and oxidation has been developed by *Michelsen* [20.333]. However, several aspects in modeling LII are still related to large uncertainty. The evaporation term is usually based on equilibrium thermodynamics, e.g., describing the vapor pressure of carbon using the Clausius–Clapeyron equation. Large differences exist between experiment and model results at high laser fluences when significant evaporation of soot particles takes place. Hence, most experiments for obtaining particle sizes are carried out using low laser fluences. Still, if the heat transfer is dominated by conduction, which is the case for conditions at atmospheric and higher pressure, uncertainties remain in the value of the thermal accommodation coefficient. There are different approaches to determine the accommodation coefficient [20.319, 334]. However, in these studies the coefficient is treated more like a general calibration factor than assessing the true value of accommodation for molecular energy transfer. A comparison of the model approaches is published in [20.335]. A web-based simulation tool for LII signal, LIISim is available [20.336].

In-Cylinder Diesel Particle Sizing. The engine used for in-cylinder LII is a single-cylinder, two-stroke Diesel engine with a displacement volume of $250\,\text{cm}^3$. A custom-designed cylinder head provides the required optical access, which can be preheated or cooled, respectively, to a suitable operating temperature of approximately $80\,°\text{C}$. The laser beam axis passes through the center of the combustion chamber through two fused silica windows. The optical axis of the detection system is arranged perpendicular to the laser beam and has access to the combustion chamber by a third window at the top side of the chamber. Time-resolved signals

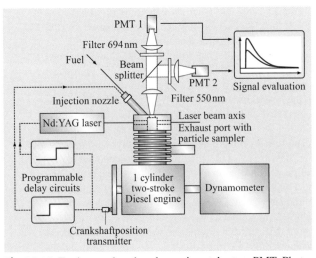

Fig. 20.45 Engine test bench and experimental setup, PMT: Photomultiplier

were detected at 550 and 694 nm. The irradiated test volume resulting from the present geometry is about 0.6 cm^3. The injection nozzle used is a Bosch common rail system, also located in the cylinder head. The engine is motored by an electrical asynchronous motor at a constant speed of 1500 min^{-1}. For time-resolved LII (TiRe-LII) measurements, the engine was fired for some individual cycles only. In this way, the windows could be kept as clean as possible and the thermal load on the cylinder head was low. The Diesel engine was equipped with a pressure transducer and a supplementary crankshaft position transmitter. All experiments were performed at an injection crank angle of 23° CA before TDC and an air/fuel equivalence ratio of $\phi = 0.26$. The particles were heated by an Nd:YAG laser at 1064 nm with a laser fluence of 0.10 J/cm^2. Finally, a thermophoretic particle sampler was located in the exhaust gas manifold to get particle probes for further analysis by transmission electron microscopy (TEM). The system is shown in Fig. 20.45.

In contrast to the laminar and well-defined conditions in the high-pressure burner the LII signals in the Diesel engine were taken in turbulent conditions with strongly varying pressures and temperatures. A series of normalized particle emission signals during cooling obtained from single shot experiments at six different crank angles is shown in Fig. 20.46. Particle cooling strongly depends on the engine crank angle and becomes longer with increasing crank angle. The signals of Fig. 20.46 and others were evaluated in terms of particle size by fitting lognormal distribution functions to the measured curves under the variation of the count mean diameter (CMD) and geometric standard deviation σ_g.

For the evaluation of the particle radiation signals in terms of particle size, it is important to know the respective mean combustion chamber pressure p_g and gas temperature T_g. For the present engine conditions, the pressure varied from nearly 80 bar at 0° CA to close to 1 bar at 100° CA. Gas temperatures changed in this crank angle range from 2000 K to 1500 K, respectively. The pressure was directly measured by a pressure transducer and the temperature was evaluated both from two-color pyrometry without laser heating and by calculating an individual combustion cycle. With these two parameters it was possible to evaluate the TiRe-LII signals shown in Fig. 20.46.

The results are shown in Fig. 20.47. CMD is in the range of 30 to 75 nm, increases up to a crank angle of about 10° CA and decreases again towards a value of about 30 nm at 100° CA after TDC. This behavior can be explained by particle formation and subsequent particle

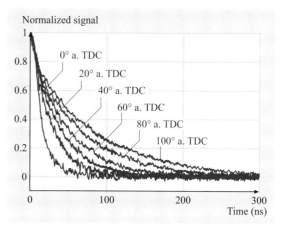

Fig. 20.46 Normalized TR LII signals at six crank angles

oxidation; σ_g is constant at a value of 1.1 up to a crank angle of 70° CA and then increases towards about 1.32. The two circled values of CMD and σ_g, which are shown at the right edge of the diagram, are results of the TEM analysis of the exhaust gas primary particles. The agreement between the TiRe-LII measured sizes at 100° CA after TDC and the TEM determined primary particle sizes in the exhaust gas is quite good.

Additional Species and Techniques for In-Cylinder Measurements

Rayleigh scattering can be used as an alternative method for gas-phase measurements of the distribution of large molecular species (i.e., hydrocarbon fuels) in the presence of small molecular species (i.e., air). Rayleigh scattering has been frequently used to measure gas density (and hence temperature) in homogeneously mixed

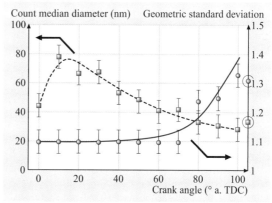

Fig. 20.47 CMD and σ_g evaluated from the signals in Fig. 20.46

systems where the effective cross-section for elastic scattering does not vary spatially [20.337]. It has also been used to measure relative temperature distributions in the post-flame gases in engines with simplified geometry [20.253]. Because of the linear dependence of the Rayleigh signal on number density and laser intensity, the interpretation of the signal is straightforward. As long as a single component dominates the effective Rayleigh cross-section in a mixing system, Rayleigh scattering also allows the measurement of species number densities. Hydrocarbons have a much larger cross-section for elastic scattering than air. Therefore, the overall scattering intensity can be used for measuring the local fuel density. With this technique, fuel vapor density has been measured in internal engine combustion [20.338–341] and in the mixing field of atmospheric-pressure natural gas flames [20.142]. In the case of single-component substitute fuels effects due to preferential evaporation (i.e., spatial separation of different fuel components that might have different scattering cross-sections) are avoided. The cross-sections for elastic scattering can be calculated based on data from [20.342–344] using dispersion relations from [20.345, 346]. Fuel dominates the scattering signal in the case of n-decane in air by a factor of $\sigma_{\text{Ray,fuel}}/\sigma_{\text{Ray,air}} = 120$. As long as no background scattering off surfaces and particles is present the total signal can be directly correlated with fuel number densities.

Rayleigh scattering, however, can be strongly influenced by scattering of laser light off walls and particles. It is therefore necessary to assess the background level that comes from these additional processes. In environments with lots of scattering off surfaces or particles (small enclosed housings, presence of fuel droplets, sooting environment) the standard technique fails. Filtered Rayleigh scattering [20.347] has been demonstrated under these conditions. Here, a narrowband laser is used for excitation that overlaps with a narrow line filter. Only slightly frequency-shifted signal (shifted by Doppler broadening due to the thermal velocity of the scattering molecules) is detected.

Raman Scattering. Valuable information for most major species in engines can again be obtained from spontaneous Raman scattering. *Grünefeld* et al. [20.249] performed spontaneous Raman imaging along a line with noise levels of a few per cent in a single shot using the beam of a KrF excimer laser at 248 nm with spectral as well as spatial dispersion of the scattered light accomplished by mounting an intensified CCD (ICCD) camera in the exit plane of a spectrograph. The larger scattering cross-sections in the ultraviolet allow imaging of the scattered light with smaller collection optics and therefore, measurements in technically relevant devices such as engines and boilers. By taking advantage of the highly polarized Raman and Rayleigh scattered light in comparison with unwanted laser excited fluorescence and stray light from enclosure walls, they were able to monitor major species in a *iso*-octane spray flame burner and a four-cylinder in-line engine by subtracting spectra taken with mutually perpendicular polarization of the exciting laser beam.

CARS. The most accurate spatially-resolved temperature measurements in engines can be performed with CARS as long as beam steering effects which severely degrade the CARS signal generation process can be tolerated. Therefore, precise cycle-resolved thermometry is possible in the compression stroke prior to ignition [20.117], in evaporating sprays [20.348] and in the expansion stroke when the flame front has traversed through the focal volume of the beams [20.349]. In any case the unstable resonator enhanced (USED) CARS phase-matching beam geometry [20.118] is preferred in such measurements to minimize beam-steering effects.

20.4.5 Diagnostics for Stationary, Large-Scale Combustion Processes

Efficient control and optimization of large-scale industrial combustion systems such as power plants, incinerators or heavy industries, is of considerable ecological and economical interest. In order to detect and correct deviations from the optimal process conditions detailed physical and chemical information – preferably from within the main process volume – is needed to permit feedback to the process by fast active-control loops. This places high demands on sensors for the gaseous major species and trace species concentrations, gas temperature, gas residence time and other process parameters. An efficient use of feedback loops for process control also requires sensors with sufficient temporal resolution and close coupling of the sensor response to the process of interest. Apart from sufficient sensitivity, selectivity and response times, a sensor for industrial process control applications must offer important additional features like compactness, long sensor life time, good long-term stability, reliable and simple calibration, ease of use, and finally low cost of ownership with respect to the possible savings in the process [20.10, 350].

In power plants and other stationary industrial combustion processes most parameters and especially gas-species concentrations are predominantly measured with extractive sensors. These are based on a multitude of different detection principles ranging from spectrally broadband optical sensors like nondispersive infrared (NDIR) photometers, fluorometric NO detectors, or Fourier transform infrared (FTIR)-based spectrometers to non-optical principles like paramagnetic or ion-conductive O_2 analyzers, metal-oxide-semiconductors and expensive multispecies gas chromatographs or mass spectrometers [20.351]. Common to these devices is that they always require extraction of a gas sample, which is typically done well behind the main reaction zone via a single access point in the flue gas duct. Prior to analysis the gas sample is filtered to remove particulates, usually cooled well below room temperature to remove water, avoid condensation and protect the sensor from the hot gas, and than transported to the main analyzer via a sampling tube that is sometimes up to 100 m long. This sampling process significantly affects the usefulness of the species signal for process control and creates numerous difficulties that can range from slow temporal response, systematic errors due to effects like adsorption, condensation, chemical reaction and fractionation during the sampling process and reduced representativeness due to the local sampling. Furthermore, many of this sensors often suffer – especially in the case of complicated gas mixtures – from a lack of specificity, caused for instance by the low spectral resolution of the optical filters used in some of these instruments. Extractive sensors in general and NDIR in particular also have a rather high cost of ownership due to the requirement for frequent cleaning of the sampling lines and routine calibrations with reference gases. Finally, due to the heterogeneity in the extractive sensor principles and the risk of fractionation during the sampling process it is in most cases impossible to perform a stringent comparison of, e.g., multiple species concentrations with high temporal resolution, which would be most interesting for monitoring and optimization of the combustion chemistry.

Avoidance of these problems requires a single, rather universally applicable, non-intrusive in-situ measurement technique for species concentrations, temperatures and flow parameters, which provides sufficient selectivity, resolution and speed, which completely avoids probe sampling and the need for costly calibration procedures and which is capable of simultaneously measuring nearly unrestricted combinations of process parameters within the same detection volume. Furthermore, in order to facilitate a widespread industrial use, the measurement technique should be rather independent of the type, size, and specific process conditions (temperature, pressure, particulate load of the gas) and, e.g., be applicable to a direct measurement even within large combustion chambers, with lateral dimensions in the order of tens of meters.

High-Resolution Laser-Absorption Spectroscopy

High-resolution laser absorption spectroscopy (LAS) [20.6, 352] is probably the only analytical technique that provides a simultaneous solution to most if not all of the above mentioned requirements. In absorption spectroscopy (AS), one of the oldest techniques of non-intrusive investigation of gaseous media, radiation directed straight through the measurement zone towards a radiation detector is being absorbed and this loss is monitored and analyzed for its dependence on wavelength.

AS is extremely versatile as every species of interest will show internal degrees of freedom which may be optically excited via the absorption process. Atomic species (and homonuclear molecules) maybe detected via electronic transitions in the ultraviolet (UV) and visible (VIS) spectral range, whereas heteronuclear molecules offer several important additional detection pathways via radiative excitation

1. of rotational levels in the far-infrared spectral range (FIR, $\lambda = 0\text{--}100\,\mu\text{m}$)
2. of vibrational and rotational sublevels in the mid IR (MIR, $\lambda = 10\text{--}3\,\mu\text{m}$) and
3. via vibrational overtone and combination bands in the near IR (NIR, $\lambda = 1\text{--}3\,\mu\text{m}$) [20.353–356].

This is a significant advantage to diagnostic techniques based on light emission like fluorescence, phosphorescence or chemiluminescence, which cannot be applied to every analyte as radiative transitions are prerequisite for those techniques [20.357]. Additionally, emission techniques often suffer from problems like radiative quenching, predissociation and other non-radiative de-excitation mechanisms. Thus, even though emission-based sensors are quite simple, inexpensive and robust, extensive pressure-, species- and device-dependent calibration information is needed in order to deliver absolute process data. Especially the low spectral resolution of chemiluminescence techniques provides only limited specificity, making it difficult to separate the signal from fluctuating background radiation.

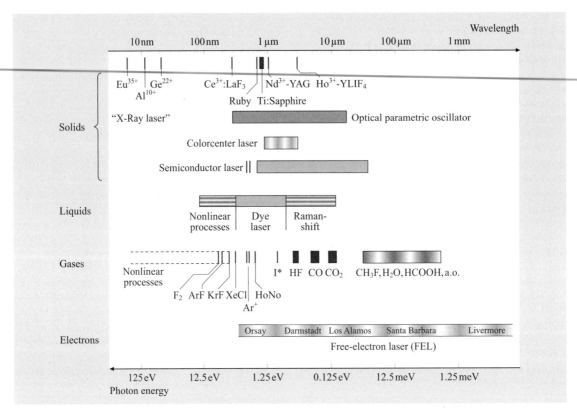

Fig. 20.48 Wavelength range of various available laser light sources

Further unique properties of AS that are important to combustion diagnostics are the possibility to determine absolute absorber densities and temperatures without the need for calibration and the linear sensitivity scaling with absorption path length. The line-of-sight character of AS may be, to a certain extend, a disadvantage compared to laser-induced emission-based techniques, because it prevents spatially resolved measurements in complicated geometries. But on one hand this restriction can be lifted for many cases using tomographic techniques [20.358, 359]. On the other hand this path-averaging effect, as it permits much more representative results than extractive or point-sampling techniques, proves to be very important for the surveillance of the very large measurement volumes frequently found in stationary combustion processes, reaching up to 20 m in diameter [20.10, 350, 357, 360].

Laser Sources for Absorption Spectroscopy

The analytical potential of AS could be significantly enhanced through its combination with laser excitation. The high spectral resolution of a laser light source maximizes the chemical selectivity especially in complex gas mixtures. Its high spectral power density minimizes or even completely removes the influence of detector noise, one of the major limits when working with broadband light sources, thereby significantly enhancing the spectrometer sensitivity. Furthermore, it is the high spectral power density that is an enabling laser feature for non-intrusive diagnostics in strongly luminescent processes, as it often permits one to neglect the influence of the strong thermal emission that occurs in many combustion processes.

Additionally, by taking advantage of the good laser beam quality it is possible to drastically reduce the size of the optical access ports to the process, and even more important, to considerably increase the maximum absorption path lengths, thereby permitting a significant increase in the sensitivity. Using laser sources in combination with simple single-path setups typical in-situ absorption path lengths in the order 1–5 m [20.361, 362] are used in combustion applications, but up to 20 m have been demonstrated in large industrial combustion processes [20.363, 364]. If even longer path lengths are needed, e.g., for the detection of minor species concentrations, and provided that optical disturbances such as light scattering by particles, are not too strong, it is possible to apply special, ruggedized multipass optics of the *Herriott* [20.365, 366] or *White* type [20.367, 368] to fold the beam path. Using that approach, path lengths in the order of 100 m and beyond have been achieved and used for gas-phase analysis in multiphase flows, e.g., to study the dynamics of ice cloud formation [20.369, 370]. The same technique is also advantageous if the measurement region has to be spatially confined. Such an approach was realized for the investigation of fire-suppression systems to study the gas-phase composition within water sprays [20.371].

When comparing laser sources for absorption spectroscopy rapid and, if possible, continuous tunability of the laser wavelength [20.10, 372] is one of the key features for combustion diagnostics. Rapid wavelength-tuning capabilities of many laser sources endow LAS with high temporal resolution and simultaneously permit a high degree of robustness against optical disturbances. On the other hand continuous wavelength tuning allows the recovery of entire absorption line shapes. This provides not only enhanced signal stability via suppression of spurious baseline effects, but more importantly the possibility of using line-shape analysis to access all the parameters that affect line area, shape and position (e.g., gas pressure, temperature, velocity, electric and magnetic fields etc. [20.6]) experimentally. Additionally, and most importantly for industrial applications, complete coverage of the absorption profile allows a calibration-free determination of the absolute absorber concentration by extraction of the integrated line strength according to the Lambert–Beer law [20.352, 361, 362, 364, 371].

Laser Selection and Spectral Coverage. Another key feature for laser selection is spectral coverage [20.10]. In general, the impressive versatility of LAS is based upon the very large spectral coverage provided by modern laser technology, spanning nearly seamlessly from the UV-VIS (e.g., via excimer, dye and solid-state lasers), over the mid infrared (e.g., CO_2 and CO gas lasers) [20.372, 373] to the far-infrared and THz region (via masers, quantum-cascade diode lasers) [20.374–376] (Fig. 20.48).

Nevertheless, it must be kept in mind that most individual laser types are limited to a relatively narrow spectral range and that many of them feature only discontinuous or even no tuning capabilities at all, thus having only rather specialized applications.

Typical examples are MIR gas lasers like the CO_2 and the infrared HeNe laser ($\lambda \approx 10.6\,\mu m$ resp. $3.39\,\mu m$). Both are based on quite mature and thus robust technologies, but their applicability for gas sensing relies on accidental coincidences with strong absorption lines. As the single emission line of the IR-HeNe coincides nicely with the very strong C—H stretch of the ro-vibrational transitions of most hydrocarbons it is quite frequently used for CH_4 monitoring [20.377, 378]. Combined with special IR-transmitting fibers, robust, fast, and sensitive absorption sensors with up to 2 kHz time resolution have been demonstrated and for instance used to study local mixing effects in CH_4 combustion [20.379].

The CO_2 laser is substantially more flexible owing to its high output power (> 1 W) and the multitude of discretely accessible lines. This enabled the realization of important sensing applications of hydrocarbons, ammonia and other species [20.380–382] and led to one of the first commercialized industrial in-situ species sensors for closed-loop control applications in large power plants: a laser in-situ ammonia sensor [20.383, 384] used for optimization of NO_x removal via the selective catalytic reduction (SCR) or selective non-catalytic reduction (SNCR) process [20.358, 385].

Despite these successful applications, only very few laser principles feature the possibility to cover a substantial part of the UV–VIS–IR spectral range with a single lasing mechanism, which is advantageous as the development efforts and cost are substantially lower when new species and hence new laser wavelengths must be incorporated. The two most important members of this laser class are semiconductor diode lasers and solid-state lasers, which provide this special property especially in combination with nonlinear crystal-based wavelength-mixing techniques like optical parametric generation (OPG), as well as second-harmonic, sum-frequency, and difference-frequency generation (SHG, SFG and DFG) [20.386].

However, practical combustion applications impose severe restrictions on the laser choice simultaneously requiring advantageous technical properties like room-temperature operation, long device life time, compactness, high electrical efficiency, and low cost, to name a few. As many of these laser systems are still in a *developmental* stage they have to mature significantly to be routinely applicable to the harsh boundary conditions often found in stationary combustion systems. As a consequence, the use of nonlinear laser-light generation techniques seems currently, despite turbulent developments, due to complexity, robustness, cost and space restrictions limited to the mid-term future, which leaves the diode laser as the premier choice with regard to general applicability.

The continuously *tunable diode laser* (TDL) is one of the most powerful laser sources for combustion diagnostics. Developed in the 1960s, soon after the first laser demonstration by *Maiman* [20.4], TDLs in the mid IR quickly became an important working horse in analytical basic research because they combined the three most important properties for analytical problems.

- Excellent selectivity due to their very high spectral resolution ($10^{-3}-10^{-4}$ cm^{-1})
- Very good sensitivity,
  - due to the access to the strong absorption lines in the molecular fingerprint region in the mid infrared, in combination with their
  - rapid, versatile and continuous wavelength-tuning properties. Latter permitting the development of advanced noise-reduction techniques [20.387], which provide significant sensitivity enhancement; in some cases even up to the shot-noise limit, the ultimate limit for absorption techniques [20.388].
- Excellent flexibility, due to their near-complete coverage of the MIR ($\lambda = 4-20\,\mu$m).

The versatility of MIR-TDL is widely documented in literature [20.389, 390] and first experiments deploying

their advantageous spectroscopic properties date back as far as in the 1970s, when a CO detection was first realized for environmental applications [20.391] and then quickly transferred to combustion problems [20.392–394]. However, despite significant long-term development efforts the major drawback of these so-called *lead salt* or IV–VI lasers [20.395] remained their need for cryogenic cooling to liquid-nitrogen or even liquid-helium temperatures, which severely hindered the widespread transfer into industrial combustion diagnostics. Nevertheless, MIR-TDLs are still frequently applied and of significant importance for basic [20.396] and combustion research [20.397].

The availability of room-temperature operated near infrared diode lasers (NIR-DL) in the AlGaAs and InGaAs material system ($\lambda = 780-1550$ nm) with relatively high optical output power (3–200 mW) significantly changed this situation, especially for diagnostics in large stationary combustors, because NIR-DL, which were initially developed and mass-produced for data storage and fiber-based data transmission, for the first time combined wide spectral coverage with superior spectroscopic and technical specifications. As a consequence, the NIR spectral range became quite interesting for absorption-based combustion diagnostics, first, to take advantage of the NIR-DLs high spectral quality, simple operation and low costs, and second, partially because the lower absorption coefficients of molecular overtone and combination bands in the NIR (Fig. 20.49) could be compensated by the longer absorption path commonly available in stationary combustion systems.

In recent years the wavelength coverage and thus the versatility of TDLAS has improved even further. As data storage capacities of optical discs improved, the available wavelength band for room temperature DL was first extended to the red ($\lambda > 630$ nm) and meanwhile even to the blue and near UV (450–350 nm) [20.398], with a small gap remaining in the 460–620 nm range. Simultaneously, the increased interest in industrial sensing applications in combination with a strong growth of the demand for security applications for military countermeasures and public safety provided the driving force for the development of promising new diode lasers in the short wave IR, the mid IR and even the far-infrared (FIR) or Terahertz (THz) spectral ranges. The short wave IR at 1.8–2.8 μm, which is important for combustion diagnostics, as this range is a good compromise between stronger absorption lines without too severe interference from hot water or CO_2, could be very recently accessed with relatively mature room-temperature devices, that are based on the InSb material system [20.362, 399, 400].

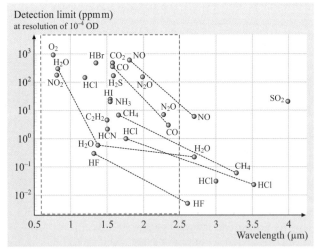

Fig. 20.49 Selection of detectable species and achievable sensitivities using NIR diode lasers. Vibrational overtone transitions of some species are connected with dashed lines. Sensitivities in ppm at one meter absorption path length are calculated assuming a detectable fractional absorption $\Delta I/I_0$ of 10^{-4} OD

Another major wavelength expansion could be derived from the invention of a completely new semiconductor laser type, the so called quantum-cascade lasers (QCL) [20.401]. QCLs permit not only surprisingly high output powers of up to watts, of at least the same if not higher importance is the possibility of operation at or near room temperature in the MIR, as well as the access to the THz region [20.376], though only at cryogenic temperatures.

This increasing flow of new semiconductor developments has already lead to a revival of absorption spectroscopy, which will proceed and expand as THz and especially room-temperature MIR sources will mature and as hybrid technologies like diode laser pumped solid state lasers improve nonlinear wave mixing techniques. This will provide spectroscopic sources with very high spectral resolution, very wide range tunability and a spectral coverage extending from the UV to the THz-region thus encompassing electronic, rovibrational as well as pure rotational transitions, making a TDLAS even more versatile.

In-Situ Tunable Diode-Laser Absorption Spectroscopy (TDLAS)

The development of in-situ laser diagnostics suitable for permanent use in stationary combustion processes impose very high demands on the spectrometer and its components. Currently the NIR spectral region offers the best boundary conditions for such applications, as all components, lasers, (fiber) optics as well as detectors, are highly developed due to their mass production for communication and data-storage technology. Commercial NIR-DL, provide near-complete spectral coverage from 635 to 1650 nm, which was recently expanded to 2600 nm by use of new InSb materials. With regard to industrial sensors NIR-DL offer a unique combination of positive properties like low cost, tiny size, room-temperature operation, very long laser lifetime of 10^5 hours and more, single-mode output power of up to several 100 mW, and excellent spectral resolution of 10^{-3} nm. Their sole disadvantage, to cover only relatively weak transitions in the visible and near infrared range (Fig. 20.49, e.g., overtone and combination vibrational bands, magnetic dipole transitions, line strength 10^{-23} to 10^{-21} cm/molecule), is of minor importance for stationary combustion sensors, as the species concentrations to be monitored are often quite high and the absorption lengths in full-scale combustors are rather large.

The versatile tuning properties of NIR-DL are most important for sensing applications. Monolithic diode lasers provide a wide range but slow tuning by laser temperature (typically 2–5 nm within a few 100 s), as well as a very fast and continuous tunability by laser current modulation ($10^3 - 10^6$ cm^{-1}/s over $\Delta \nu = 1-3$ cm^{-1}). Complete absorption line shapes may thus be captured even on µs time scales. The use of entire line shapes improves the stability of the data evaluation as time dependent baseline distortions may be separated. Furthermore by evaluation of the spectral line profile, it is possible to simultaneously extract absorber concentration as well as the physical boundary conditions in the measurement volume. The high tuning speed on the other hand is quite important for in-situ measurements in harsh environments as it permits to *freeze* all disturbances in the measurement volume, which is a prerequisite for their efficient removal [20.362].

Current tuning, however, simultaneously generates an unavoidable side effect: the amplitude modulation (AM) of the laser output power. AM tends to hide the weak molecular absorption signal in a strongly sloping background (Fig. 20.50) and also limits the usable dynamic range of the data acquisition system. The figure of merit, the specific AM, $dP/d\lambda$, i.e., the change in output power per wavelength interval, is usually best for vertical-cavity surface-emitting lasers (VCSELs), and worst for high data rate telecommunication distributed feedback diode lasers (DFB-DL), which on the other hand provide highest spectral purity [20.388]. However, hybrid diode lasers like external cavity diode lasers (XC-DL), which use antireflection-coated laser chips and

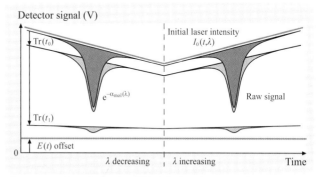

Fig. 20.50 Generic in situ TDLAS signal trace and measurement principle of direct tunable diode-laser spectroscopy. The disturbances [Tr(t), E(t)] have to be compensated before applying Beer's law. Tr(t) indicates the unspecific loss of the initial laser light (I_0) due to scattering and refraction, $E(t)$ is the level of background light and gray-body radiation reaching the detector. The *line area* is after disturbance correction proportional to the number density of absorbers

externally mechanically tuned optics to select the laser wavelength, may be used to avoid AM [20.402]. Moreover, XC-DL access the complete gain profile of the semiconductor thus providing the widest tuning ranges (20 nm in the red, up to 200 nm in the NIR) [20.403] However, increased cost, size, and susceptibility to mechanical instability, and reduced tuning speed, severely hinders their use in combustion applications.

The basic setup of an absorption spectrometer is quite simple: the beam of a diode laser is collimated, directed through the measurement volume, collected by a spherical mirror and projected onto a photo detector. As the laser wavelength is continuously scanned over the absorption line, the photodetector signal is amplified, filtered, digitized by an analog-to-digital (AD) converter, transferred to a computer, and evaluated for narrow band losses by a curve fitting algorithm, which, after removal of the AM and other disturbances, usually extracts the area of the absorption line by applying the Lambert–Beer law, here written for homogeneous conditions along the beam path:

$$I(\lambda) = I_0(\lambda) \exp\left[-S(T)g(\lambda - \lambda_0)NL\right].$$

The Lambert–Beer law relates the detected intensity $I(\lambda)$ to the initial intensity $I_0(\lambda)$ (describing the current- and thus wavelength-dependent AM) and the molecular absorption signal of a well-resolved individual absorption line. The latter is described by the temperature-dependent line strength $S(T)$ of this line, the number density N of absorbers, the absorption path length L, and an area normalized ($= 1$) function $g(\lambda - \lambda_0)$, describing only the shape of the absorption line, which is centered at wavelength λ_0. Using this normalization, this equation can be solved to yield the absolute absorber density N,

$$N = \frac{1}{S(T)L} \int \ln\left(\frac{I_0(\lambda)}{I(\lambda)}\right) d\lambda \rightarrow X = N\frac{k_B T}{p}$$

or the absorber concentration X by applying the ideal gas law and the measured temperature T and total pressure p.

It is important to note, that this equation does not contain any calibration parameters and that $S(T)$ is a molecular parameter, which is independent of the spectrometer configuration. Thus the absolute absorber density N, respectively the absolute absorber concentration X, can be directly inferred from the measured line area without any calibration. Only L, T and $S(T)$, and the total pressure p, must be known/measured. Furthermore, given a spectrally well-isolated absorption line, the line profile integration yields that N is independent of the line shape, i.e., all line-broadening effects and the total pressure. This property to avoid any calibration as well as pressure-broadening effects, is very advantageous for in-situ measurements in combustion devices or processes with variable pressure or temperature, and is hardly achievable with other measurement principles, thus giving absorption spectroscopy a rather unique position.

However, in order to take advantage these properties, it has to be kept in mind, that the normalization of the line area only holds in wavelength space. Hence, the absorption signal has to be transferred from time space to wavelength space by taking care of the dynamic tuning behavior, $d\lambda/dt$, of the diode laser. For very slow, quasistatic current tuning the conversion factor is just a constant. But, for higher tuning speeds, which are needed for fast measurements or, as we will see, under harsh boundary conditions in combustors, the tuning becomes highly nonlinear, which results in asymmetrically distorted line shapes in time space. The line fitting process thus requires that $d\lambda/dt$ is determined experimentally and taken into account. This is frequently realized using solid or air-spaced etalons as relative frequency markers [20.402]. As the systematic errors in N or X are dominated by errors in $S(T)$ and $d\lambda/dt$, it is most important to determine $d\lambda/dt$ at least with the accuracy of $S(T)$. The use of compact, solid etalons, within the instrument, may therefore, despite their robustness, generate too large uncertainties, due to refractive index errors, temperature effects, and reduced resolution. More precise are large air spaced etalons, whose frequency markers depend to a very large extend on the spatial separation of the etalon mirrors only.

In-situ laser absorption spectroscopy in combustion environments requires not only the consideration of the spectroscopic effects of temperature or pressure variations on the molecular signal. Furthermore, great care has to be taken to correct the detector signal for various strong disturbances that are often unavoidable under combustion conditions [20.362, 363, 404]; most important are strong fluctuations of the overall transmission $Tr(t)$ of the measurement path, which are caused by broadband absorption and scattering by particles, or beam steering due to refractive-index fluctuations. Additionally, there is frequently an intense thermal background radiation $E(t)$ from the combustion zone or radiating particles that increases the detector signals. Both effects may be changing rapidly in time and have to be included in an extended version of Beer's law:

$$I(\lambda) = I_0(\lambda) Tr(t) \exp[-S(T)g(\lambda - \lambda_0)NL] + E(t)$$

An exaggerated sketch of a typical in-situ TDLAS signal trace is shown in Fig. 20.50, indicating the additive emission and the multiplicative transmission term.

The relative signal-to-disturbance ratio depicted in Fig. 20.50 is however far from reality, as NIR absorption signals only rarely yield more than 5% absorption, while the transmission losses, e.g., in large coal combustors [20.363] or dense water sprays in fire research can [20.371, 405] exceed 99.99%, and the thermal background in coal plants may become 10 times stronger than the transmitted laser light. As the disturbances are often orders of magnitudes larger than the molecular absorption signals, they may bury the net signal completely. Effective disturbance suppression, compensation and removal strategies are thus essential in order to extract the narrow-band absorption and compute the desired combustion parameters.

The basic idea to effectively separate and correct the molecular absorption for the transmission and emission effects takes advantage of two important features of diode lasers: their fast wavelength tunability and – taking advantage of a mistake – their simultaneous current-induced output power modulation. By tuning the laser much faster than the transmission and emission fluctuations, both disturbances are effectively *frozen*, thus can be assumed to be constant during the wavelength scan and corrected by subtracting the offsets and dividing the offset-corrected scan through the baseline function [20.362]. The AM thus serves as a *transmission sensor*, while the emission term can be isolated from the DC coupled detector signal by quickly switching off the laser at the beginning of each scan. Hence, it is still possible to avoid calibrations and extract absolute absorber densities directly from the DC coupled detector signal by exploiting the extended Lambert–Beer law, provided that $Tr(t)$ and $E(t)$ can be extracted from the unprocessed detector signal. Additionally rapid phase-locked averaging of the wavelength scans, equivalent to using a comb filter at the modulation frequency and its overtones, may be used for further noise reduction and improvement of the detection limits [20.362, 406].

The described experimental concept, to directly capture and evaluate the DC coupled detector signal, including all offsets and disturbances, is termed direct absorption spectroscopy (DAS). Though the simplest approach to LAS, it was proved that DAS is capable of routinely achieving fractional optical resolutions, $\Delta I/I_0$, of 10^{-3}–5×10^{-5} in the in-situ signal, depending on the complexity of the disturbances and the measurement environment [20.362, 364, 407, 408]. Other more-complicated double-modulation techniques such as wavelength modulation spectroscopy (WMS) [20.387, 406, 409–414] or frequency modulation spectroscopy (FMS) [20.397, 415, 416] or dual-path approaches such as balanced detection [20.388, 417] are reported to provide one to two orders higher resolution, with $\Delta I/I_0$ better than 10^{-5}–10^{-6} being reported under well-controlled laboratory conditions or in extractive measurements. Application of these techniques to in-situ measurements in combustion scenarios, however, considerably increases their technical complexity and cost, which limits their widespread use. Especially the treatment of pressure and temperature dependent line shape effects, as well as the removal of transmission fluctuations is not straightforward with these techniques, so that sensor calibration if often required and most of the resolution advantage is lost due to incomplete disturbance correction.

In comparison, DAS has a under certain circumstances reduced sensitivity, but it also provides several highly important features, i.e., a relatively simple and robust setup, rapid response, and particularly the possibility to perform a calibration-free measurement of absolute gas concentrations while simultaneously avoiding line broadening effects. Hence, DAS offers a good compromise between sensitivity and complexity, and therefore is of great value for in-situ combustion diagnostics in technical processes.

Application of In Situ TDLAS

TDLAS has been extensively used over the last decades for combustion diagnostics [20.10, 350, 357, 360]. Its versatility can be nicely demonstrated by the long list of *accessible process parameters* using NIR-TDLs. Most important and most frequently employed is the detection of species concentrations, which has been realized for many light molecules (via their molecular overtone and combination bands (Fig. 20.49), and even many atomic species [20.418, 419], (e.g., most alkali [20.402] and earth alkali atoms) via their electronic transitions. Nearly as important to combustion diagnostics is the measurement of gas temperatures, which may be determined either via two-line thermometry [20.364] from the line strength ratio of a suitable line pair or from Doppler broadening [20.420]. Collisional broadening of the line shape has been used for pressure measurements [20.420]. Even the flow speed of the absorber may be extracted using tracer methods [20.402, 421] or from the Doppler shift of the line profile and combined with the species concentration to derive mass flux or thrust [20.420, 422]. The residence time distribution of the product gases in a reactor [20.402] is another important parameter which in high-temperature flows could only be measured by radioactive tracer methods. Using diode lasers that parameter could be inferred by seeding the flow with a tracer molecule [20.402, 421] and detecting the release of the tracer. Continuous seeding has been used in a pulsed flow [20.421, 423], as well as pulsed seeding in a continuous flow [20.402].

Multi-Parameter Spectrometers. One particular strength of laser absorption spectroscopy is the possibility for multi-parameter sensors with near arbitrary combinations of measurement parameters [20.10, 350, 357]. By optically superimposing the required number of different laser beams, usually one per measurement parameter, a multicolor laser beam is formed and used to enable simultaneous multi-parameter detection in the same detection volume via a single pair of optical access ports. Beam superposition can be realized by free space optics [20.364], or particularly elegant and robust by fiber-multiplexing several diode lasers into a single-mode glass fiber [20.424–427], taking advantage of the high-quality fiber-optical components developed for telecommunication. Depending on the spectral separation of the lasers needed for the individual parameters different wavelength multiplexing schemes maybe used:

- *Spectral wavelength multiplexing* may be used for well-separated wavelength channels. In this case the absorption signals are acquired by simultaneously scanning all lasers and dispersing the multi-wavelength beam after passage of the absorption path onto multiple detectors by use of a grating [20.424–427] or dichroic beam splitters [20.361, 364].
- Alternatively, for insufficient spectral separation, multi-parameter measurements may be realized by *temporal wavelength multiplexing* [20.408, 428], taking advantage of the rapid amplitude modulation possible with diode lasers. Time multiplexing means sequentially scanning each individual laser of a laser array and detecting the transmitted light with a single detector.
- Finally there is the possibility of *modulation frequency multiplexing* [20.410], which has only be demonstrated in combination with double modulation techniques like WMS. Here all lasers are operated simultaneously, but at different modulation frequencies. The light of all the lasers is directed onto a single detector and the respective components are separated by narrow-band electrical filters, e.g., lock-in amplifiers.

Due to the scan range limitations of monolithic lasers multiple-parameter detection has mostly been performed with a combination of several individual diode lasers. However, with the persistent, telecommunication-driven progress in external-cavity diode lasers [20.403, 429] and multi-section distributed-feedback (DFB) and distributed-Bragg-reflector (DBR) lasers [20.428, 430] laser systems continuously scanning up to 100 nm have become available, which will simplify the design of time-multiplexed multi-parameter spectrometers significantly.

Laboratory-Scale Combustion Control Applications. For large stationary combustion systems such as power plants, incinerators and heavy industries, a tight monitoring and control of combustion is of great importance to achieve a high fuel efficiency and reduced pollutant emission. The development of adequate sensors and fast active control loops suffered for a long time from the complex boundary conditions found in full-scale combustors. Thus the first successful TDLAS-based closed-loop combustion control experiments have been realized in well-controlled laboratory environments using small atmospheric burners with a few cm absorption path [20.424, 431]. This first setup consisted of two fiber-coupled wavelength-multiplexed DFB-DLs, which were rapidly scanned (1 kHz) across two water absorption lines in the 1.3 μm overtone band to measure water concentration (from peak height) and gas temperature (from peak ratio) 6 mm above the burner surface with 2 ms time resolution. This permitted a closed-loop temperature control from 900 to 2400 K by adjusting the fuel flow with a proportionally acting voltage-controlled solenoid valve thereby achieving a temperature deviation of only 1% (1σ) between 1800 and 2100 K. The next H_2O/temperature sensor generation [20.432] (with $\Delta l = 10$ cm) used an adaptive controller and acoustic forcing to optimize within 100 ms the performance of an atmospheric 5 kW non-premixed, C_2H_4–air, annular dump combustor (ADC) through improved fuel–air mixing achieved by adjusting amplitude and relative phase between air and fuel flow forcing. System scale-up ended with a 50 kW pulsed ADC waste incinerator, where CO emissions dropped from 2500 ppm to below 200 ppm by active optimization of the relative phase between primary and secondary air flow based on temperature measurements with 100 μs resolution in a 18 cm-wide flue gas duct [20.433]. Recently TDLAS control was also applied to modern propulsion systems, so-called pulse detonation engines (PDE). PDEs, essentially narrow, few cm-inner-diameter steel tubes, generate thrust by discontinuously igniting premixed gas. To optimize fuel consumption, maximize specific impulse and reduce temporal propulsion fluctuations it is important to avoid overfilling of the PDE and to adapt the ignition timing to the changing boundary conditions in the PDE (i.e. temperature, residual gas). Using fixed-wavelength DFB-DLs and standard lock-in detection C_2H_4 mixture fraction and temperature (for $T = 300-900$ K) was determined (after calibration) from the peak height/peak ratio of the C_2H_4 Q-branch multiplet at 1.62 μm [20.434]. Then, a simple closed-loop on–off controller adapted the fuel loading and ignition timing. This reduced the shot-to-shot thrust fluctuations significantly, clearly demonstrating the advantages of TDL-based control.

TDLAS Application to Stationary Combustion Systems

Sensor scale-up from rather short absorption paths in the order of 10 cm and laboratory-like conditions into industrial combustion environments (i.e. power plants with combustion chambers up to 100 times larger) has long been hampered by the large, quite variable and only rarely characterized disturbances present in large stationary combustors. Careful characterization of the optical and optomechanical properties of the in-situ absorption paths in these combustors has increased the knowledge about these problems in a way that robust and highly efficient correction and data evaluation algorithms for direct absorption spectroscopy could be developed, which can cope with the extreme disturbances levels, and extract weak molecular absorption signals despite severe light losses of more than 99.99% and background radiation signals, which may be more than a 10-fold stronger than the transmitted laser light [20.352, 361, 362, 364, 371]. The development of these procedures extended the working range of in-situ absorption spectroscopy drastically, so that despite the moderate laser output powers of 1–100 mW it became possible to detect and quantify specific absorptions on the order of a few percent with an optical resolution of $10^{-3}-10^{-4}$ and with a temporal resolution of 0.1–30 s.

Another key development which proved indispensable for the realization of industry grade in-situ spectrometers, was the design of a new computer-controlled optomechanical setup, which, in spite of strong disturbances caused by the in-situ path, is able to ensure permanent optical access to the measurement volume by automatically controlling, optimizing, and stabilizing the optical alignment of the spectrometer. Combined with various techniques to suppress the in-situ disturbances, this optomechanical setup could compensate for strong temperature-induced deformations of the combustion chambers, which before prevented continuous measurements during large combustor load changes [20.364]. Further, in cases of extremely poor visibility, this system could also be used to perform an automatic, computer-controlled initial alignment of the in-situ spectrometer [20.363].

All these critical developments are about to change the situation, and to permit the scale-up of TDLAS sensors to full-size combustion processes thus opening up new possibilities to optimize stationary combustion process. As a result of this progress NIR-DL-based in-situ TDLAS has been recently applied to an increasing range of industrial combustion processes (including grate-fired waste incinerators [20.361], rotary kilns [20.362], box-fired power plants [20.363] and even high-pressure coal combustion [20.408]) and an extensive variety of fuel types (encompassing coal, natural gas, biomass, household and special waste) thus impressively demonstrating the versatility and flexibility achievable by TDLAS. In order to give further insight in the potential of NIR-TDLAS for applications in stationary combustion processes, selected example applications will be described in further detail below, which indicate that NIRDL-based absorption sensors are now most promising for real-world applications, ranging from mobile and severely space-restricted sensors to fast, non-intrusive, in-situ species and temperature sensors working in high-temperature and high-pressure environments like full-scale power plants.

Waste Incinerators. The incineration of house hold as well as special waste is in many developed countries the most strictly regulated combustion process, which in combination with the high variability of the calorific value and water content of the waste creates a strong need for active combustion control to adapt the waste feed, the waste mixing or the amount of air/oxygen supplied to the process. To monitor the process conditions, enable process optimization via feedback control and ensure a highly efficient and environmentally safe waste incineration numerous optical techniques have been combined in an advanced combustion control system (CCS) and were applied to an industrial 20 MW_{th} waste incinerator with a 4 m-diameter combustion chamber, depicted in Fig. 20.51.

Efficient NO reduction according to the SNCR process required the injection of precise amounts of NH_3, which could be effectively monitored by the very first in-situ laser absorption sensor (LISA) based on a $^{13}CO_2$ laser and differential absorption spectroscopy [20.383]. In addition, the primary air injection, grate movement and fuel input are under closed-loop control through a scanning IR camera (TACCOS) [20.435] measuring the waste-bed temperature. Finally in order to control the secondary air injection actively a fast NIR-DL-based in-situ sensor for the simultaneous in-situ detection of O_2 and H_2O directly in the combustion chamber was successfully developed and tested [20.361].

Strong disturbances caused by the in-situ path such as transmission fluctuations and thermal background emission, which interfere with the molecular absorption signal had to be suppressed. The laser signals, which provided a linear response with a sensitivity of about 0.3 vol. % and a temporal resolution of 1 s were verified by extractive reference O_2 and H_2O sensors, which probed the cold flue gas duct. As the laser spectrometer completely avoided gas transport the laser signals were up to 60 s faster than the extractive reference signals, which is of major importance for the combustion control loop. A follow-up version of this sensor could be further improved by fiber-coupling of the laser and by the development of an inexpensive, compact and robust, purely analog data evaluation [20.436], with the potential for low-cost production. In the meantime even fully fiber-coupled TDLAS spectrometers [20.437] have been demonstrated, which enable the important expansion of TDLAS to highly space-restricted applications like internal combustion engines or high-pressure gas turbines.

Gas–Fired Power Plants. Owing to their short turn-on time large, gas-fired power plants are frequently used to compensate for demand fluctuations in the power distribution network. But on rare occasions ignition delays in the sequentially ignited multi-burner system, presumably due to excessive humidity levels inside the combustion chamber, are not detected because of a lack of suitable rapid response security sensors for CH_4. In these cases disastrous explosions causing damage of $10 to 100 million dollar have been experienced, even though UV flame sensors and extractive CH_4 detectors have been installed. In the future, such accidents can be avoided with the help of the first laser spectrometer [20.364] for the simultaneous in-situ measurement of the gas temperature and all majority species (CH_4, H_2O, O_2, CO_2) in large stationary combustors which has been developed recently (Fig. 20.52). Employing two DFB-DL at 760 nm and 1.65 μm to detect O_2, CH_4 and CO_2 and a Fabry–Pérot DL at 812 nm to extract absolute H_2O concentrations it is now possible to determine representative, spatially integrated species concentrations from within the main combustion chamber. The temperature signal that is needed to extract absolute species concentrations, which was not available before, is simultaneously derived from the line strength ratios within the group of rotational transitions of H_2O near 0.81 μm via two-line thermometry. These transitions have the advantage of coinciding with high-output-power diode lasers (≈ 35 mW), which increase the spectrometers tolerance to the substantial light losses often found in the large technical combustors.

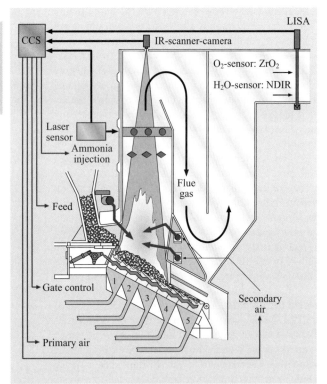

Fig. 20.51 Intersection through a 20 MW incinerator for household waste in which the in situ detection of O_2 and H_2O has been performed. A selection of important sensors (in situ diode-laser sensor, in situ laser ammonia sensor, LISA, and IR-scanner-camera, TACCOS) as well as the possible actors that are connected to the combustion control system (CCS) are indicated (further information in [20.361, 383, 435])

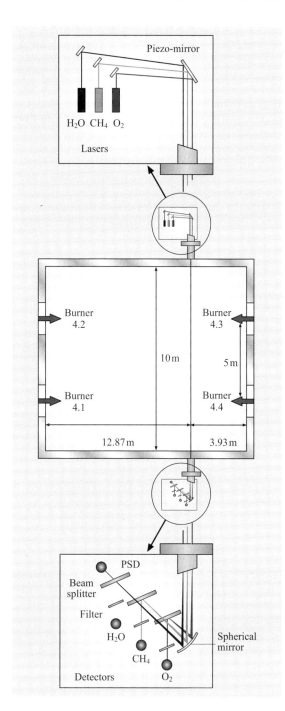

Fig. 20.52 Experimental setup of the spectrally multiplexed multi-parameter in situ diode-laser spectrometer used for the detection of CH_4, H_2O, O_2, CO_2 and gas temperature in the combustion chamber of a 1000 MW gas-fired power plant. Only the four of the 20 burners located in the plane of measurement are indicated. The optical setup containing lasers, optics, and detectors has been enlarged for clarity [20.364]

alignment control loop. Using this multi-parameter spectrometer it was possible to analyze the full start-up and power-down procedure of a 1000 MW gas-fired power plant with 20 50 MW burners and a 10 m furnace diameter and achieve to a time resolution of 1.6 s and a minimum detectable absorption better than 10^{-3} OD. This resulted in a CH_4 detectivity of about 100 ppmV and a dynamic range of more than two orders of magnitude, as well as a relative temperature resolution of ± 10 K and a total temperature range of 300–1300 K. The ability to monitor H_2O as the possible cause, as well as CH_4 as the consequence of an ignition delay offers new possibilities to ensure a safe ignition procedure of large-scale multi-burner gas-fired combustion systems. The device also demonstrates the excellent expandability of multiplexed in-situ diode laser spectrometers and is of high interest to investigate the fuel to air ratio as well as the local chemistry in combustion processes.

Coal–Fired Power Plants. CO, another important combustion species is one of the key control parameters for combustion processes and one of the major pollutants of fossil-fuel combustion. Furthermore it is closely linked to reducing conditions in the combustor, which are responsible for high-temperature corrosion effects as well as fouling and slagging of the combustor containment. CO in-situ sensors are therefore of high interest and intensively studied. In the near-infrared range CO may be detected via two overtone transitions, the 2ν and 3ν vibrational bands at 2.3 and 1.5 µm. The 2.3 µm band has long been inaccessible due to the lack of suitable lasers, while the 1.5 µm band directly coincides with the wavelength range of telecommunication lasers. Using the high optical power (over 20 mW) of the 1.5 µm telecom lasers, the long absorption paths available in full-size power plants and combining it with highly efficient and robust algorithms for the suppression of in-situ disturbances it was recently shown for the first time that CO can be detected even in the very large combustion chamber (20 m diameter) of a 700 MW_{th} lignite-fired power plant. [20.363]. This offers new diagnostic opportunities

Further, to allow permanent measurements during start-up conditions it was necessary to compensate for strong temperature-induced wall deformations, which was accomplished with a new active laser beam

to investigate the slagging and fouling process in power plants and their dependence on the composition of the combustion gases. In addition by spectrally multiplexing the 1.5 μm CO laser with a 35 mW 813 nm FP-DL it was possible to access a H_2O overtone transition, which permitted a simultaneous in-situ measurement of H_2O and gas temperature. A major problem which had to be solved was the compensation of beam steering and severe optical losses (Fig. 20.53) of 99.9–99.999% by dust. This zero-visibility condition also required a method to permit the initial spectrometer alignment, which was ensured by an automatic feedback-controlled beam alignment based on motorized bending mirrors and a special phase-locked extraction algorithm to separate the weak laser light from the strong background radiation.

Rotary Kiln Combustors. Batch-fired rotary kilns (Fig. 20.54), used for the combustion of special wastes, are known to generate rapid and strong intermittent CO peaks, which have to be kept below a safety limit, thereby restricting the total waste throughput and the plant efficiency. Advanced control concepts to reduce the CO peaks and improve the throughput demand a very fast CO detection, preferentially directly through the rotary kiln. This could be realized for the first time [20.404] using a modified version of the 1.5 μm CO spectrometer described above. [20.363]. However, first measurements using standard personal computer (PC)-based data-acquisition (DAQ) cards showed that the strong, rapid and burst-like transmission fluctuations lead to saturation and clipping in the detector signal and systematic errors in the measured CO concentration. A new high-speed data-acquisition system based on a fast digital signal processor (DSP) removed these problems by enabling – at modulation frequencies of up to 3 kHz – a real-time evaluation of the individual, unaveraged absorption scans on a ms time scale, so that distorted scans could be rejected. Furthermore the DSP ensured a much higher data throughput close to 100% and allowed the transmission information of each scan to be used to realize an automatic real-time

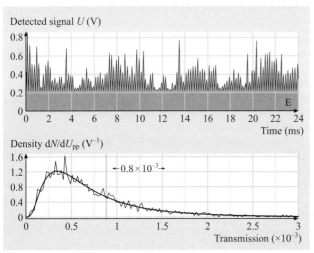

Fig. 20.53 Illustration of the transmission disturbances found in the combustion chamber of a 700 MW coal-fired power plant. *Top*: DC coupled in situ detector signal over a high number of wavelength scans (each sharp spike corresponds to an up and a down scan). The amplitude modulation of the signal is modified by strong transmission losses along the absorption path. In addition, the detector signal is increased by thermal background emission E. *Bottom*: distribution of transmission levels found over a longer time period, indicating an average in situ transmission of the measurement path (vertical line) of 8×10^{-4} [20.363]

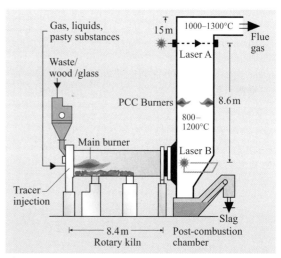

Fig. 20.54 Simplified cross section of the THERESA plant (Thermische Entsorgung spezieller Abfälle) [20.362], a rotary kiln fired incinerator for special waste. A secondary high-temperature combustion section [the post-combustion chamber (PCC)] is attached to the kiln exit. TDLAS was used in the process to realize a simultaneous CO/O_2 detection through the kiln (path A, Fig. 20.55), and at the kiln exit (path B, Fig. 20.56), as well as simultaneous O_2/H_2O/Temperature in the PCC. Multi-section gas residence-time measurements using binary alkali tracers were realized in the same process by injecting the tracer next to the main burner in the kiln and detecting the tracer at the exit of the kiln (path B, Fig. 20.57) and the exit of the PCC (path C, Fig. 20.57)

gain adaptation of the detector amplifiers, which significantly improved the usable dynamic range of the AD converter.

A comparison of both systems revealed that the relative systematic errors, that could be removed completely by the DSP system, can be larger then 1000% under special conditions in the case of a 8 m-long rotary kiln with 3 MW_{th} thermal power. Further enhancement of the DSP-based spectrometer was realized by spectrally multiplexing the 1.5 μm and 760 nm diode laser to access the R23-CO and the R23R23 O_2 transition for the first simultaneous CO/O_2 detection within a rotary kiln [20.404]. A 1.9 s response time yielded an optical resolution of $3-7 \times 10^{-4}$ and detection limits of 450 ppm CO resp. 2000 ppm O_2, which was sufficient to experimentally reveal the close anti-correlation of the intermittent CO and O_2 peaks (Fig. 20.55).

However, the approach of using the 3ν CO transitions is not feasible in smaller combustors or at low-ppm CO concentrations often found in modern combustion systems. To address this sensor deficiency new 2.3 μm distributed-feedback diode lasers (DFB-DL) that access the R30 line in the CO-2ν-band have been employed recently to develop a fast and highly sensitive in-situ CO absorption spectrometer suitable for hazardous waste incinerators [20.362, 404]. Additionally spectrally multiplexing the 2.3 μm DL with a 760 nm DFB-DL accessing the R17Q18/R19R19 O_2 lines again enabled a simultaneous in-situ detection of CO and O_2, which is most interesting for control strategies requiring the coverage of a wide range of fuel/air ratios. These new spectrometers were successfully tested over periods of up to two weeks in a 3.5 MW_{th} special waste incinerator. The absorption path ($l = 2.5$ m) was located at the rotary kiln exit in the lower part of the post combustion chamber with temperatures of 800–1000 °C (Fig. 20.54). Direct absorption spectroscopy using the DSP-based data evaluation enabled calibration-free species detection. With 1 s data-acquisition time a fractional absorption resolution (1σ) of 1.2×10^{-4} for CO (6×10^{-5} for O_2) could be achieved corresponding to significantly improved detection limits of 6.5 ppm CO (250 ppm O_2) [20.362]. This sensitivity and time resolution was high enough to detect even under lean-fuel conditions and with a relatively short absorption path small quasiperiodic stoichiometry changes of ±15 ppm CO caused by the periodic fuel feed of the rotary kiln (Fig. 20.56), thus nicely demonstrating the great potential of this device to establish an active combustion control system to optimize waste throughput in special waste incinerators.

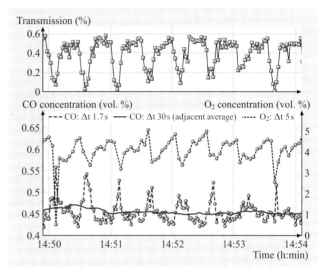

Fig. 20.55 Temporal evolution of the CO and O_2 concentration (*lower* and *middle traces* with *symbols*) and the broadband transmission (*top trace*) within the THERESA rotary kiln (Fig. 20.55) measured with 1.56 and 760 nm DFB-DL. While all signals vary synchronously with the waste feed cycle of 30 s, O_2 and CO are clearly anti-correlated, as expected from combustion chemistry. For comparison the flat *lower line* corresponds to a simulated signal for an extractive CO sensor [20.404]

High-Pressure Coal Combustion. The danger of severe corrosion damage in the high-temperature gas turbines needed for future high-efficiency pressurized coal combustion generates an immediate demand for optical in-situ diagnostic tools that are compatible with high-temperature high-pressure conditions. In particular alkali halide compounds have to be detected, but are spectroscopically difficult to access. ELIF [20.438–440], excimer-laser-induced fragmentation fluorescence is one of very few non-intrusive techniques capable of direct monitoring of alkali compounds in an industrial environment. Here an excimer laser photodissociates gaseous alkali compounds within the flue gas duct and generates electronically excited alkali atoms [Na(3^2P), K(4^2P)], which are then readily detected by their fluorescence in the visible region via a fiber-coupled two species detection system [20.438]. The measured signals are converted to absolute concentrations by a calibration of the complete detection apparatus under known thermodynamic conditions. A detection limit of 0.2 ppb could be established for both alkalis at 10 bar total pressure and 800 °C gas temperature.

As an alternative to the experimentally demanding, costly and calibration dependent, direct detection of

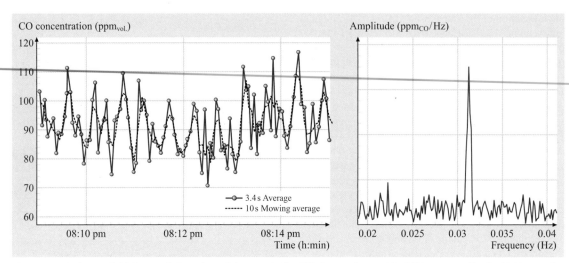

Fig. 20.56 *Left panel*: high-speed detection of fast, periodic CO fluctuations with ±17 ppm amplitude (±10 ppm for 10 s average) measured with 2.3 μm DFB-DL at the rotary kiln exit of the THERESA plant (Fig. 20.54). A fast Fourier transform (FFT) of the fluctuations over a 2 h period (*right panel*) shows a single prominent peak, with the same frequency as the feeding cycle of the incinerator [20.362]

the alkali compounds a diode-laser-based high-pressure compatible in-situ absorption spectrometer to monitor the alkali atoms (Li, K, Rb, Cs), which are produced by thermal dissociation of the compounds, was developed and tested in the flue gas duct of a 1 MW$_{th}$ coal combustor at temperatures of 1400 K and pressures up to 15 bar [20.408]. In spite of severely broadened absorption lines, transmission fluctuations, and strong background emission it was shown for the first time in a practical combustion system that a fast and highly sensitive in-situ detection of alkali atoms is possible via their D-lines in the 670–850 nm spectral range. Crucial to the realization of these spectrometers was the use and detailed characterization of widely tunable vertical-cavity surface-emitting lasers (VCSEL). Despite the low VCSEL output power of only a few 100 μW it was possible to realize the first VCSEL-based in-situ detection of alkali-metal atoms in a technical combustor and achieve detection limits of 1.4 pptm potassium. This demonstrates the applicability of TDLAS to high-pressure in-situ gas analysis and the possibility of low-cost absorption sensors for the surveillance of hot gas filters needed for the protection of flue gas turbines in future coal-fired combined-cycle power plants.

Gas Residence-Time Measurements. Up to now the residence time distribution of the product gases in high-temperature production processes could only be

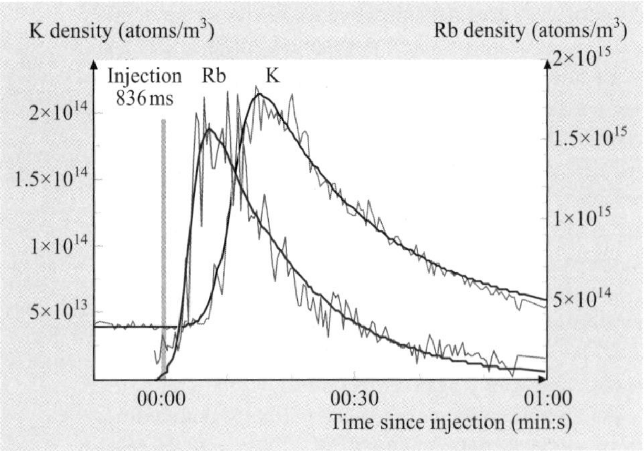

Fig. 20.57 Simultaneous residence-time distribution measurements in different plant segments of the THERESA rotary kiln (Fig. 20.54) using a binary tracer made from a mixture of K and Rb salts; the tracer mixture was injected at $t = 0$ over a period of 836 ms at the head of the rotary kiln (Fig. 20.54). The two species were measured at two locations within the plant: Rb atoms at the exit of the rotary kiln – 8.5 m from the injection site (*left peak* at $t = 9$ s); K atoms at the end of the PCC – 8.6 m downstream of the first measurement location (peak to the *right* at $t = 15$ s). Since K is part of the fuel its signal has – in contrast to Rb – an offset even without a tracer injection. Concentration distribution functions are fitted to the two traces [20.402]

measured by the use of radioactive tracers. In addition to the complicated regulations that have to be fulfilled for the manipulation of radioactive substances, their use inevitably causes a contamination of the plant and therefore hinders a frequent inspection of this important process parameter. Using the fast-response in-situ spectrometers for alkali atoms [20.408] it was possible to develop a new and promising alternative technique for the in-situ measurement of gas residence time distributions in high-temperature processes (patent pending), which uses cheap and harmless alkali compounds as temperature stable tracer species. These compounds can be effectively monitored via the alkali atoms produced by thermal dissociation near the detection volume after the compounds pass through the process volume. During the first successful demonstration of this technique [20.402] in a 3 MW$_{th}$ rotary kiln followed by a high-temperature post-combustion chamber (PCC; Fig. 20.54) it could be shown that an injection of only 100 mg of the tracer compound near the kiln entrance was sufficient in such a large process to effectively determine the residence time distribution. This high efficiency results from the extremely high detectivity of the laser in-situ absorption spectrometer of 80 ppq potassium (ppq = 1 in 10^{15}) or 4×10^{11} K atoms/m^3 at 1200 K [20.402]. Using binary K/Rb tracer mixtures and a multispecies spectrometer detecting the atoms at different locations in the plant it was possible to realize the simultaneous characterization of the transport properties of two consecutive plant segments (Fig. 20.57) located 8.4 m/17.0 m from the tracer discharge location and determine the mean residence times to (15 ± 1) s and (26 ± 1) s, which met the expected values. With this new technique it is feasible for the first time to realize a permanent online measurement and verification of the residence time distribution of a combustion process, which is particularly important for the optimization of chemical reactors and processes such as waste incinerators with statutory limits for the residence time of the combustion products.

20.5 Conclusions

Despite intensive research for alternative energy sources further economic growth will be closely linked to an extensive use of combustion processes for mobility, energy conversion, and generation of industrial process heat. The drastic increase of combustion fuel consumption has, however, now reached a level where the annual worldwide demand corresponds to amounts of fossil fuel that has been accumulated during a period of 1 million years of the Earth's history. The limited supply of fossil fuels and the detrimental environmental effects of the global use of combustion require pollutant emission to be minimized and total energy conversion efficiency and process performance to be optimized. Demands are similarly important for industrial combustion-based production processes, which are indispensable for high-volume products such as steel, glass, cement, and which require simultaneous optimization of combustion parameters, product quality, and product costs.

After four decades of research and development in the area of laser-based concepts for combustion diagnostics, many of the spectroscopic methods have matured now from qualitative to quantitative techniques. Its capabilities go far beyond the observation of two-dimensional distribution of signal intensities. The instantaneous simultaneous measurement of concentration distribution of several scalars (species concentration and temperature) gives insight in correlation of species, i.e., in turbulent combustion processes. The spatial combination of several observed planes on the other hand allows a visualization of three-dimensional concentration distributions and yields a unique view into turbulent nonreactive and reactive flow systems. The calculation of absolute (three-dimensional) species concentration gradients and scalar dissipation rates is of major importance for combustion modeling. Finally, multiple-time-step imaging yields movie-like insight into fast processes and therefore allows for the first time the investigation of highly dynamic processes in real time.

The interpretation of measured signal intensities in practical combustion situations is not trivial. High pressure, high temperature and a variety of sources of interfering signal and attenuation of laser and signal light hinders the selective and quantitative measurement. The spectroscopic properties of the species under study, therefore, must be well examined. We presented the required studies in high-pressure laminar flames for the example of nitric oxide. This information enables quantitative measurements even in harsh environments such as in-cylinder gasoline and Diesel engine measurements with direct injection. Advanced laser-based imaging diagnostics turns out to be an important tool for the development and optimization of modern combustion devices that can fulfill the future requirements in terms of energy efficiency and pollutant minimization.

Laser diagnostics techniques similar to those discussed here are also used to study diverse applications beyond combustion, including atmospheric chemistry, plasma chemistry, and chemical vapor deposition and material synthesis. Even biological applications of sophisticated laser techniques are now being pursued in some combustion diagnostics groups since the coupling of elementary chemical reactions with various transport processes is also the basis of living systems.

References

20.1 G.R. Kirchhoff, R.W. Bunsen: Chemische Analyse durch Spectralbeobachtungen, Poggend. Annal. **110**, 161–176 (1860), in German
20.2 A. Einstein: Zur Quantentheorie der Strahlung, Phys. Z. **18**, 21–128 (1917), in German
20.3 J.P. Gordon, H.H. Zeiger, C.H. Townes: The maser – new type of microwave amplifier, frequency standard and spectrometer, Phys. Rev. **99**, 1264–1274 (1955)
20.4 T.H. Maiman: Stimulated optical radiation in ruby, Nature **187**, 493–494 (1960)
20.5 A.G. Gaydon: *The Spectroscopy of Flames* (Chapman Hall, London 1967)
20.6 W. Demtröder: *Laserspektroskopie. Grundlagen und Techniken*, 3rd edn. (Springer, Berlin, Heidelberg 2000), in German
20.7 A.C. Eckbreth: *Laser Diagnostics for Combustion Temperature and Species*, 2nd edn. (Gordon and Breach, Amsterdam 1996)
20.8 K. Kohse-Höinghaus, J.B. Jeffries (Eds.): *Applied Combustion Diagnostics* (Taylor Francis, New York 2002)
20.9 J. Warnatz, U. Maas, R.W. Dibble: *Combustion*, 3rd edn. (Springer, Berlin, Heidelberg 2001)
20.10 J. Wolfrum: Lasers in combustion: From basic theory to practical devices, Proc. Combust. Inst. **27**, 1–42 (1998)
20.11 C. Schulz, V. Sick: Tracer-LIF diagnostics: Quantitative measurement of fuel concentration, temperature and air/fuel ratio in practical combustion situations, Prog. Energy Combust Sci. **31**, 75–121 (2005)
20.12 H. Zhao, N. Ladommatos: Optical diagnostics for in-cylinder mixture formation measurements in IC engines, Prog. Energy Combust. Sci. **24**, 297–336 (1998)
20.13 T. Etzkorn, J. Fitzer, S. Muris, J. Wolfrum: Determination of absolute methyl- and hydroxyl-radical concentrations in a low pressure methane-oxygen flame, Chem. Phys. Lett. **208**, 307–310 (1993)
20.14 D.F. Davidson, A.Y. Chang, M.D. DiRosa, R.K. Hanson: A CW laser absorption diagnostic for methyl radicals, J. Quant. Spectrosc. Radiat. Transfer **49**, 559–571 (1993)
20.15 P. Zalicki, R.N. Zare: Cavity ring-down spectroscopy for quantitative absorption measurements, J. Chem. Phys. **102**, 2708 (1995)
20.16 W.K. Bischel, B.E. Perry, D.R. Crosley: Detection of fluorescence from O and N atoms induced by two-photon absorption, Appl. Opt. **21**, 1419–1429 (1982)
20.17 R.S. Barlow, C.D. Carter, R.W. Pitz: Multi-scalar diagnostics in turbulent flames. In: *Applied Combustion Diagnostics*, ed. by K. Kohse-Höinghaus, J.B. Jeffries (Taylor Francis, London 2003)
20.18 K. Müller-Dethlefs, M. Pealat, J.-P.E. Taran: Temperature and hydrogen concentration measurements by CARS in an ethylene-air bunsen flame, Ber. Bunsenges. Phys. Chem. **85**, 803–807 (1981)
20.19 A. Dreizler, T. Dreier, J. Wolfrum: Thermal grating effects in infrared degenerate four-wave mixing for trace gas detection, Chem. Phys. Lett. **233**, 525–532 (1995)
20.20 A.A. Suvernev, A. Dreizler, T. Dreier, J. Wolfrum: Polarization-spectroscopic measurement and spectral simulation of OH ($A^2\Sigma$-$X^2\Pi$) and NH ($A^3\Pi$-$X^3\Sigma$) transitions, Appl. Phys. B **61**, 421–427 (1995)
20.21 R.W. Pitz, J.A. Wehrmeyer, J.M. Bowling, T.S. Cheng: Single pulse vibrational Raman scattering by a broadband KrFexcimer laser in a hydrogen/air flame, Appl. Opt. **29**, 2325–2332 (1990)
20.22 R.L. Vander Wal: Laser-induced incandescence: Development and characterization towards a measurement of soot-volume fraction, Appl. Phys. B **59**, 445–452 (1994)
20.23 A. Dreizler, J. Janicka: Diagnostic challenges for gas turbine combustor model validation. In: *Applied Combustion Diagnostics*, ed. by K. Kohse-Höinghaus, J.B. Jeffries (Taylor Francis, New York 2002)
20.24 M.D. Smooke: *Reduced Kinetic Mechanisms and Asymptotic Approximations for Methane-Air Flames*, Lecture Notes in Physics (Springer, Berlin, Heidelberg 1991)
20.25 N. Peters, F.A. Williams: The asymptotic structure of stoichiometric methane-air flames, Combust. Flame **68**, 185–207 (1987)
20.26 U. Maas, S.B. Pope: Simplifying chemical kinetics: intrinsic low-dimensional manifolds in composition space, Combust. Flame **88**, 239–264 (1992)
20.27 J.B. Jeffries, G.P. Smith, D.E. Heard, D.R. Crosley: Comparing LIF measurements and computer models of low pressure flame chemistry, Ber. Bunsenges. Phys. Chem. **96**, 1410–1418 (1992)

20.28 U. Meier, U. Kienle, I. Plath, K. Kohse-Höinghaus: Two-dimensional laser-induced fluorescence approaches for the accurate determination of radical concentrations and temperature in combustion, Ber. Bunsenges. Phys. Chem. **96**, 1401–1410 (1992)

20.29 K. Fieweger, R. Blumenthal, G. Adomeit: Shock-tube investigations on the self-ignition of hydrocarbon-air mixtures at high pressures, Proc. Combust. Inst. **25**, 1579–1585 (1994)

20.30 R. Minetti, M. Carlier, M. Ribaucour, E. Therssen, L.R. Sochet: A rapid compression machine investigation of oxidation and auto-ignition of n-heptane: measurements and modelling, Combust. Flame **102**, 298–309 (1995)

20.31 G.P. Smith: Diagnostics for detailed kinetic modeling. In: *Applied Combustion Diagnostics*, ed. by K. Kohse-Höinghaus, J.B. Jeffries (Taylor Francis, New York 2002) pp. 501–517

20.32 K.N.C. Bray: The challenge of turbulent combustion, Proc. Combust. Inst. **26**, 1–26 (1996)

20.33 W.P. Jones, B.E. Launder: The prediction of laminarization with a two-equation model of turbulence, Int. J. Heat Mass Transfer **15**, 301–314 (1972)

20.34 J. Janicka, A. Sadiki: Large eddy simulation of turbulent combustion systems, Proc. Combust. Inst. **30**, 537–548 (2005)

20.35 C. Schneider, A. Dreizler, J. Janicka: Fluid dynamical analysis of atmospheric reacting and isothermal swirling flows, Flow Turbul. Combust. **74**, 103–127 (2005)

20.36 G.H. Wang, N.T. Clemens, P.L. Varghese: High-repetition-rate measurements of temperature and thermal dissipation in nonpremixed turbulent jet flames, Proc. Combust. Inst. **30**, 691–700 (2005)

20.37 D. Geyer, A. Kempf, A. Dreizler, J. Janicka: Scalar dissipation rates in isothermal and reactive turbulent opposed-jets: 1D-Raman/Rayleigh experiments supported by LES, Proc. Combust. Inst. **30**, 681–689 (2005)

20.38 C.S. Barlow, A.N. Karpetis: Scalar length scales and spatial averaging effects in turbulent piloted methane/air jet flames, Proc. Combust. Inst. **30**, 673–680 (2005)

20.39 T.J. Poinsot, D. Veynante: *Theoretical and Numerical Combustion* (Edwards, Philadelphia 2005)

20.40 R.S. Barlow: Laser diagnostics and their interplay with computations to understand turbulent combustion, Proc. Combust. Inst. **31**, 49–76 (2007)

20.41 N. Peters: Laminar flamlet concepts in turbulent combustion, Proc. Combust. Inst. **21**, 1231–1250 (1986)

20.42 R.W. Bilger, S.B. Pope, K.N.C. Bray, J.F. Driscoll: Paradigms in turbulent combustion research, Proc. Combust. Inst. **30**, 21–42 (2004)

20.43 S.B. Pope: Computations of turbulent combustion: Progress and challenges, Proc. Combust. Inst. **23**, 591–612 (1990)

20.44 N. Peters: *Turbulent Combustion* (Cambridge Univ. Press, Cambridge 2000)

20.45 R.S. Barlow, C.D. Carter, R.W. Pitz: Multiscalar diagnostics in turbulent flames. In: *Applied Combustion Diagnostics*, ed. by K. Kohse-Höinghaus, J.B. Jeffries (Taylor Francis, New York 2002) pp. 384–407

20.46 G. Grünefeld, A. Gräber, A. Diekmann, S. Krüger, P. Andresen: Measurement system for simultaneous species densities, temperature and velocity double-pulse measurements in turbulent hydrogen flames, Combust. Sci. Technol. **135**, 135–152 (1998)

20.47 R.B. Miles, W.R. Lempert: Two-dimensional measurement of density, velocity, and temperature in turbulent high-speed air flows by UV Rayleigh scattering, Appl. Phys. B **51**, 1–7 (1990)

20.48 R.W. Dibble, W. Kollmann, R.W. Schefer: Conserved scalar fluxes measured in a turbulent nonpremixed flame by combined laser Doppler velocimetry and laser Raman scattering, Combust. Flame **55**, 307–321 (1984)

20.49 A. Nauert, A. Dreizler: Conditional velocity measurements by simulatenously applied laser Doppler velocimetry and planar laser-induced fluorescence in a swirling natural gas/air flame, Z. Phys. Chem. **219**, 635–648 (2005)

20.50 S. Geiss, A. Dreizler, Z. Stojanovic, A. Sadiki, J. Janicka: Experimental investigation of turbulence modification in a non-reactive two-phase flow, Exp. Fluids **36**, 344–354 (2004)

20.51 H.-E. Albrecht, M. Borys, N. Damaschke, C. Tropea: *Laser Doppler and Phase Doppler Measurement Techniques* (Springer, Berlin, Heidelberg 2003)

20.52 N. Damaschke, H. Nobach, C. Tropea: Optical limits of particle concentration for multidimensional particle sizing techniques in fluid mechanics, Exp. Fluids **32**, 143–152 (2002)

20.53 M.C. Jermy, D.A. Greenalgh: Planar dropsizing by elastic fluorescence scattering in sprays too dense for phase Doppler measurements, Appl. Phys. B **71**, 703–710 (2000)

20.54 I. Düwel, J. Schorr, J. Wolfrum, C. Schulz: Laser-induced fluorescence of tracers dissolved in evaporating droplets, Appl. Phys. B **78**, 127–131 (2004)

20.55 H.C. Hottel, A.F. Sarofim: *Radiative Heat Transfer* (McGraw Hill, New York 1997)

20.56 A.L. Crosbie, J.B. Farrell: Exact formulation of multiple scattering in a three-dimensional cylindrical geometry, J. Quant. Spectrosc. Radiat. Heat Transfer **31**, 397–416 (1984)

20.57 J.T. Farmer, J.R. Howell: Monte Carlo prediction of radiative heat transfer in homogeneous, anisotropic, nongray media, J. Thermophys. Heat Transfer **8**, 133–139 (1994)

20.58 A. Soufiani, E. Djavdan: A comparison between weigthed sum of gray gases and statistical

20.58 narrow-band radiation models for combustion applications, Combust. Flame **97**, 240–250 (1994)

20.59 R. Koch, S. Wittig, B. Noll: The harmonic transmission model: A new approach to multidimensional radiative transfer calculation in gases under consideration of pressure broadening, Int. J. Heat Mass Transfer **34**, 1871–1880 (1991)

20.60 J. Brübach, J. Zetterberg, A. Omrane, Z.S. Li, M. Aldén, A. Dreizler: Determination of surface normal temperature gradients using thermographic phosphors and filtered Rayleigh scattering, Appl. Phys. B **84**, 537–541 (2005)

20.61 T. Fuyuto, H. Kronemayer, B. Lewerich, W. Koban, K. Akihama, C. Schulz: Laser-based temperature imaging close to surfaces with toluene and NO-LIF, J. Phys.: Conference Series **45**, 69–76 (2006)

20.62 D. Geyer: 1D Raman/Rayleigh experiments in a turbulent opposed-jet. PhD Thesis (TU Darmstadt, Darmstadt 2004)

20.63 D. Geyer, A. Dreizler, J. Janicka, A.D. Permana, J.-Y. Chen: Finite rate chemistry effects in turbulent opposed flows: comparison of Raman/Rayleigh measurements and Monte Carlo PDF simulation, Proc. Combust. Inst. **30**, 711–718 (2005)

20.64 S.K. Omar, D. Geyer, A. Dreizler, J. Janicka: Investigation of flame structures in turbulent partially premixed counter-flow flames using laser-induced fluorescence, Prog. Comput. Fluid Dyn. **4**, 241–249 (2004)

20.65 B. Böhm, D. Geyer, K.K. Venkatesan, N.M. Laurendeau, M.W. Renfro: Simulatenous PIV/PTV/OH-PLIF imaging: Conditional flow field statistics in partially-premixed turbulent opposed jet flames, Proc. Combust. Inst. **31**, 709–717 (2007)

20.66 K.K. Venkatesan, N.M. Laurendeau, M.W. Renfro, D. Geyer, A. Dreizler: Time-resolved measurements of hydroxyl in stable and extinguishing partially-premixed turbulent opposed-jez flames, Flow Turbulence Combustion **76**, 257–278 (2006)

20.67 D. Geyer, A. Kempf, A. Dreizler, J. Janicka: Turbulent opposed jet flames: a critical benchmark for combustion LES, Combust. Flame **143**, 524–548 (2005)

20.68 K. Shimoda: Limits of sensitivity of laser spectrometers, Appl. Phys. **1**, 77–86 (1973)

20.69 C.A. Taatjes, D.B. Oh: Time-resolved wavelength modulation spectroscopy measurements of HO_2 kinetics, Appl. Opt. **36**, 5817–5821 (1997)

20.70 M.G. Allen: Diode laser absorption sensors for gas dynamic and combustion flows, Meas. Sci. Technol. **9**, 545–562 (1998)

20.71 S. Schäfer, M. Mashni, J. Sneider, A. Miklos, P. Hess, V. Ebert, K.-U. Pleban, H. Pitz: Sensitive detection of methane with 1.65 μm diode laser by photoacoustic and absorption spectroscopy, Appl. Opt. **66**, 511–516 (1998)

20.72 M.S. Chou, A.M. Dean, D. Stern: Laser absorption measurements on OH, NH and NH_2 in NH_3/O_2 flames: Determination of an oscillator strength for NH_2, J. Chem. Phys. **76**, 5334–5340 (1982)

20.73 S. Cheskis, I. Derzy, V.A. Lozovsky, A. Kachanov, D. Romanini: Cavity ring-down spectroscopy of OH radicals in low-pressure flames, Appl. Phys. B **66**, 377 (1998)

20.74 J.J. Scherer, D.J. Rakestraw: Cavity ring-down laser absorption spectroscopy detection of formyl (HCO) radical in a low pressure flame, J. Chem. Phys **265**, 169 (1997)

20.75 J.J. Scherer, D.J. Rakestraw: Cavity ringdown laser absorption spectroscopy detection of formyl (HCO) radical in a low-pressure flame, Chem. Phys. Lett. **265**, 168–176 (1997)

20.76 K. Kohse-Höinghaus, U. Meier, B. Attal-Trétout: Laser-induced fluorescence study of OH in flat flames of 1-10 bar compared with resonance CARS experiments, Appl. Opt. **29**, 1560–1569 (1990)

20.77 C.D. Carter, G.B. King, N.M. Laurendeau: Saturated fluorescence measurements of the hydroxyl radical in laminar high-pressure $C_2H_6/O_2/N_2$ flames, Appl. Opt. **31**, 1511–1522 (1992)

20.78 T. Dreier, D. Rakestraw: Measurement of OH rotational temperatures in a flame using degenerate four wave mixing, Opt. Lett. **15**, 72–74 (1990)

20.79 H. Bervas, B. Attal-Tretout, S. Le Boiteux, J.P. Taran: OH detection and spectroscopy by DFWM in flames: comparison with CARS, J. Phys. B **25**, 949–969 (1992)

20.80 T. Dreier, D.J. Rakestraw: Degenerate four-wave mixing diagnostics on OH and NH radicals in flames, Appl. Phys. B **50**, 479–485 (1990)

20.81 G.J. Germann, R.L. Farrow, D.J. Rakestraw: Infrared degenerate four-wave mixing spectroscopy of polyatomic molecules: CH_4 and C_2H_2, J. Opt. Soc. Am. B **12**, 25–32 (1995)

20.82 R.L. Vander Wal, B.E. Holmes, J.B. Jeffries, P.M. Danehy, R.L. Farrow, D.J. Rakestraw: Detection of HF using infrared degenerate four-wave mixing, Chem. Phys. Lett. **191**, 251–258 (1992)

20.83 R.L. Farrow, M.N. Bui-Pham: Degenerate four-wave mixing measurements of methyl radical distributions in hydrocarbon flames: Comparison with model predictions, Proc. Combust. Inst. **26**, 975–983 (1996)

20.84 C.F. Kaminski, I.G. Hughes, P. Ewart: Degenerate four-wave mixing spectroscopy and spectral simulation of C_2 in an atmospheric pressure oxy-acetylene flame, J. Chem. Phys. **106**, 5324–5332 (1997)

20.85 B. Attal, D. Débarre, K. Müller-Dethlefs, J.P.E. Taran: Resonance-enhanced coherent anti-Stokes Raman scattering in C_2, Rev. Phys. Appl. **18**, 39–50 (1983)

20.86 B. Attal-Trétout, P. Berlemont, J.P. Taran: Three-colour CARS spectroscopy of the OH radical at triple resonance, Mol. Phys. **70**, 1–51 (1990)

20.87 L.A. Rahn, P.L. Mattern, R.L. Farrow: A comparison of coherent and spontaneous Raman combustion diagnostics, Proc. Combust. Inst. **18**, 1533–1542 (1981)

20.88 A.C. Eckbreth, R.J. Hall: CARS concentration sensitivity with aqnd without nonresonant background suppression, Combust. Sci. Technol. **25**, 175–192 (1981)

20.89 T. Dreier, J. Wolfrum: Detection of the free OH ($X^2\Pi$) radical by CARS spectroscopy, J. Chem. Phys. **80**, 975–976 (1984)

20.90 T. Dreier, J. Wolfrum: Detection of free NH_2 (X^2B_1) radicals by CARS spectroscopy, Appl. Phys. B **33**, 213–218 (1984)

20.91 T.N. Tanada, J. Velazques, N. Hemmi, T.A. Cool: Detection of toxic emissions from incinerators, Ber. Bunsenges. Phys. Chem. **97**, 1516–1526 (1993)

20.92 C. Weickhard, U. Boesl: Time-resolved trace analysis of exhaust gas by means of laser mass spectrometry, Ber. Bunsenges. Phys. Chem. **97**, 1716–1719 (1993)

20.93 K. Nyholm, M. Kaivola, C.G. Aminoff: Polarization spectroscopy applied to C_2 detection in a flame, Appl. Phys. B **60**, 5–10 (1995)

20.94 K. Nyholm: Measurements of OH rotational temperatures in flames by using polarization spectroscopy, Opt. Commun. **111**, 66–70 (1994)

20.95 R.S. Barlow, A.N. Karpetis: Scalar length scales and spatial averaging effects in turbulent piloted methane/eir jet flames, Proc. Combust. Inst. **30**, 673–680 (2005)

20.96 J.J. Scherer, J.B. Paul, A. O'Keefe, R.J. Saykally: Cavity ringdown laser absorption spectroscopy: History, development, and application to pulsed molecular beams, Chem. Rev. **97**, 25 (1997)

20.97 W. Juchmann, H. Latzel, D.-I. Shin, G. Peiter, T. Dreier, H.-R. Volpp, J. Wolfrum, R.P. Lindstedt, K.M. Leung: Absolute radical concentration measurements and modeling of low pressure $CH_4/O_2/NO$ flames, Proc. Combust. Inst **27**, 469–476 (1998)

20.98 M. Bachmann, J. Griesheimer, K.-H. Homann: Thermal and chemical influences on the soot mass growth, Proc. Combust. Inst. **25**, 635–643 (1994)

20.99 M. Hausmann, K.H. Homann: Scavenging of hydrocarbon radicals from flames with dimethyldisulfide II. Hydrocarbon radicals in fuel-rich low-pressure flames of acetylene, ethylene, 1,3-butadiene and methane with oxygen, Ber. Bunsenges. Phys. Chem. **101**, 651–667 (1997)

20.100 J. Ahrens, A. Keller, R. Kovacs, K.-H. Homann: Large molecules, radicals, ions and soot particles in fuel-rich hydrocarbon flames. Part III: REMPI mass spectrometry of large flame PAHs and fullerenes and their quantitative calibration through sublimation, Ber. Bunsenges. Phys. Chem. **102**, 1823–1839 (1998)

20.101 M. Woyde, W. Stricker: The application of CARS for temperature measurements in high pressure combustion systems, Appl. Phys. B **50**, 519–525 (1990)

20.102 T. Dreier, M. Ridder, G. Schiff, A. Saur, A.A. Suvernev: Determination of temperature from N_2 and O_2 CARS spectra at very high pressure, Proc. Combust. Inst. **25**, 1727–1734 (1994)

20.103 L. Martinsson, P.-E. Bengtsson, M. Aldén, S. Kröll: Applications of rotational CARS for temperature measurements at high pressure and in particle-laden flames. In: *Temperature: Its Measurements and Control in Science and Industry*, ed. by J.F. Schooley (American Institute of Physics, New York 1992) pp. 679–684

20.104 R.L. Vander Wal, R.L. Farrow, Rakestraw: High-resolution investigation of degenerate four-wave mixing in the $\gamma(0,0)$ band of nitric oxide, Proc. Combust. Inst. **24**, 1653–1659 (1992)

20.105 S.J. Tsay, K.G. Owens, K.W. Aniolek, D.L. Miller, N.P. Cernansky: Detection of CN by degenerate four-wave mixing, Opt. Lett. **20**, 1725–1727 (1995)

20.106 S. Williams, R.N. Zare, L.A. Rahn: Reduction of degenerate four-wave mixing spectra to relative populations II. Strong-field limit, J. Chem. Phys. **101**, 1093–1107 (1994)

20.107 A.P. Smith, G. Hall, B.J. Whitaker, A.G. Astill, D.W. Neyer, P.A. Delve: Effect of inert gases on the degenerate four-wave mixing spectrum of NO_2, Appl. Phys. B **60**, 11–18 (1995)

20.108 G. Hall, A.G. Suits, B.J. Whitaker: Resonant degenerate four wave mixing detection of HCO, Chem. Phys. Lett. **203**, 277–282 (1993)

20.109 K. Nyholm, R. Fritzon, M. Aldén: Two-dimensional imaging of OH in flames by use of polarization spectroscopy, Opt. Lett. **18**, 1672–1674 (1993)

20.110 B. Löfstedt, R. Fritzon, M. Aldén: Investigation of NO detection in flames using polarization spectroscopy, Appl. Opt. **35**, 2140–2146 (1996)

20.111 T. Reichardt, R.P. Lucht: Theoretical calculation of line shapes and saturation effects in polarization spectroscopy, J. Chem. Phys. **109**, 5830–5843 (1998)

20.112 J.B. Kelman, A.R. Masri: Reaction zone structure and scalar dissipation rates in turbulent diffusion flames, Appl. Opt. **36**, 3506–3514 (1997)

20.113 Q.V. Nguyen, P.H. Paul: The time evolution of a vortex-flame interaction observed via planar laser-induced fluorescence imaging of CH and OH, Proc. Combust. Inst. **26**, 357–364 (1996)

20.114 A.R. Masri, R.W. Dibble, R.S. Barlow: The structure of turbulent nonpremixed flames revealed by Raman-Rayleigh-LIF measurements, Prog. Energy Combust. Sci. **22**, 307–362 (1996)

20.115 M. Decker, A. Schik, U.E. Meier, W. Stricker: Quantitative Raman imaging investigations of mixing phenomena in high-pressure cryogenic jets, Appl. Opt. **37**, 5620–5627 (1998)

20.116 M. Péalat, P. Bouchardy, M. Lefebvre, J.P. Taran: Precision of multiplex CARS temperature measurements, Appl. Opt. **24**, 1012–1022 (1985)

20.117 R.P. Lucht, D. Dunn-Rankin, T. Walter, T. Dreier, S.C. Bopp: Heat transfer in engines: Comparison of

20.117 CARS thermal boundary layer measurements and heat flux measurements, SAE Techn. Paper Ser. 910722 (1991)

20.118 A.C. Eckbreth, G.M. Dobbs, J.H. Stufflebeam, P.A. Tellex: CARS temperature measurements in augmented jet engine exhausts, Appl. Opt. **23**, 1328–1339 (1984)

20.119 A.C. Eckbreth, T.J. Anderson, J.A. Shirley: Laser Raman diagnostics for propulsion systems development, Ber. Bunsenges. Phys. Chem. **97**, 1597–1608 (1993)

20.120 R. Bédué, R. Gastebois, R. Bailly, M. Pealat, J.P. Taran: CARS measurements in a simulated turbomachine combustor, Combust. Flame **57**, 141–153 (1984)

20.121 B. Lavorel, G. Millot, J. Bonamy, D. Robert: Study of rotational relaxation fitting laws from calculation of SRS N_2 Q-branch, Chem. Phys. **115**, 69–78 (1987)

20.122 G. Millot, R. Saint-Loup, S. Santos, R. Chaux, H. Berger, J. Bonamy: Collisional effects in the stimulated Raman Q-branch of O_2 and O_2–N_2, J. Chem. Phys. **96**, 961–971 (1992)

20.123 L.A. Rahn, R.L. Farrow, G.J. Rosasco: Measurement of the self-broadening of the H_2 Q(0-5) Raman transitions from 295 to 1000 K, Phys. Rev. A **43**, 6075–6088 (1991)

20.124 J.P. Looney, G.J. Rosasco, L.A. Rahn, W.S. Hurst, J.W. Hahn: Comparison of rotational relaxation rate laws to characterize the Raman Q-branch spectrum of CO at 295 K, Chem. Phys. Lett. **161**, 232–238 (1989)

20.125 B. Lavorel, G. Millot, G. Fanjoux, R. Saint-Loup: Study of collisional effects on bandshapes of the $v_1/2v_2$ Fermi dyad in CO_2 gas with stimulated Raman spectroscopy. III. Modeling of collisional narrowing and study of vibrational shifting and broadening at high temperature, J. Chem. Phys. **101**, 174–177 (1994)

20.126 W. Kiefer: Non-linear Raman spectroscopy and its chemical applications. In: *NATO Advanced Study Insitutes Series C: Mathematical and Physical Sciences*, ed. by W. Kiefer, D.A. Long (Reidel, Dordrecht 1982) pp. 241–260

20.127 R. Suntz, H. Becker, P. Monkhouse, J. Wolfrum: Two-dimensional visualization of the flame front in an internal combustion engine by laser-induced fluorescence of OH radicals, Appl. Phys. B **47**, 287–293 (1988)

20.128 P.H. Paul, H.N. Najm: Planar laser-induced fluorescence imaging of flame heat release, Proc. Combust. Inst. **27**, 43–50 (1998)

20.129 A. Hoffmann, C. Schulz, J. Ruckwied, D. Malcherek, T. Dreier, R. Schießl, U. Maas: Three-dimensional measurement of OH-concentration gradients in a turbulent flame by simultaneous laser induced fluorescence and Raman scattering. In: *European Combustion Meeting* (The Combustion Institute, Orléans 2003)

20.130 P.C. Miles: Raman line imaging for spatially and spectrally resolved mole fraction measurements in internal combustion engines, Appl. Opt. **38**, 1714–1732 (1999)

20.131 G. Grünefeld, M. Schütte, P. Andresen: Simultaneous multiple-line Raman/Rayleigh/LIF measurements in combustion, Appl. Phys. B **70**, 309–313 (2000)

20.132 S. Böckle, J. Kazenwadel, T. Kunzelmann, D.-I. Shin, C. Schulz, J. Wolfrum: Simultaneous single-shot laser-based imaging of formaldehyde, OH and temperature in turbulent flames, Proc. Combust. Inst. **28**, 279–286 (2000)

20.133 S. Böckle, J. Kazenwadel, C. Schulz: Laser-diagnostic multi-species imaging in strongly swirling natural gas flames, Appl. Phys. B **71**, 741–746 (2000)

20.134 S. Einecke, C. Schulz, V. Sick: Measurement of temperature, fuel concentration and equivalence ratio fields using tracer LIF in IC engine combustion, Appl. Phys. B **71**, 717–723 (2000)

20.135 W. Koban, J. Schorr, C. Schulz: Oxygen distribution imaging with a novel two-tracer laser-induced fluorescence technique, Appl. Phys. B **74**, 111–114 (2002)

20.136 C.F. Kaminski, J. Hult, M. Aldén: High repetition rate planar laser induced fluorescence of OH in a-turbulent non-premixed flame, Appl. Phys. B **68**, 757–760 (1999)

20.137 C.F. Kaminski, M.B. Long: Multidimensional diagnostics in space and time. In: *Applied Combustion Diagnostics*, ed. by K. Kohse-Höinghaus, J.B. Jeffries (Taylor Francis, London 2003) pp. 224–251

20.138 I. Düwel, J. Schorr, P. Peuser, P. Zeller, J. Wolfrum, C. Schulz: Spray diagnostics using an all solid-state Nd:YAlO$_3$ laser and fluorescence tracers in commercial gasoline and Diesel fuels, Appl. Phys. B **79**, 249–254 (2004)

20.139 W. Meier, O. Keck: Laser Raman scattering in fuel-rich flames: background levels at different excitation wavelengths, Meas. Sci. Technol. **13**, 741–749 (2002)

20.140 A. Buschmann, F. Dinkelacker, M. Schäfer, T. Schäfer, J. Wolfrum: Measurement of the instantaneous detailed flame structure in turbulent premixed combustion, Proc. Combust. Inst. **26**, 619 (1996)

20.141 H.B. Najm, P.H. Paul, C.J. Mueller, P. Wyckoff: On the adequacy of certain experimental observables as measurements of flame burning rate, Combust. Flame **113**, 312–322 (1998)

20.142 T. Landenfeld, A. Kremer, E.P. Hassel, J. Janicka, T. Schäfer, J. Kazenwadel, C. Schulz, J. Wolfrum: Laserdiagnostic and numerical studies of strongly swirling natural-gas flames, Proc. Combust. Inst. **27**, 1023–1030 (1998)

20.143 D.I. Shin, G. Peiter, T. Dreier, H.-R. Volpp, J. Wolfrum: Spatially resolved measurements of CN, CH,

NH and H$_2$CO concentration profiles in a domestic gas boiler, Proc. Combust. Inst. **28**, 319–325 (2000)

20.144 N. Garland: Assignment of formaldehyde laser-induced Fluorescence spectrum from the Sandia engine experiment, SRI International MP **84-033** (1984)

20.145 D. Grebner, J. Hein, W. Triebel, A. Burkert, J. König, C. Eigenbrod: 2D-fast-scan cool-flame diagnostic using formaldehyde-LIF excited by XeF excimer laser radiation. In: *Proc. Joint Meeting of the British German and French sections of the Combustion Institute Nancy* (The Combustion Institute, Nancy 1999) pp. 485–487

20.146 S. Böckle, J. Kazenwadel, T. Kunzelmann, D.-I. Shin, C. Schulz: Single-shot laser-induced fluorescence imaging of formaldehyde with XeF excimer excitation, Appl. Phys. B **70**, 733–735 (2000)

20.147 J. Weickert: *Anisotropic Diffusion in Image Processing* (Teubner, Stuttgart 1998)

20.148 H. Scharr, B. Jähne, S. Böckle, J. Kazenwadel, T. Kunzelmann: Flame front analysis in turbulent combustion. In: *22. DAGM Symposium* (Deutsche Arbeitsgemeinschaft Mustererkennung, Kiel 2000) pp. 325–332

20.149 N. Fricker, W. Leuckel: The characteristics of swirl-stabilized natural gas flames. Part 3: the effect of swirl and burner mouth geometry on flame stability, J. Inst. Fuel **49**, 152–158 (1976)

20.150 T. Dreier, A. Dreizler, J. Wolfrum: The application of a Raman-shifted tunable KrF excimer laser for laser-induced fluorescence combustion diagnostics, Appl. Phys. B **55**, 381–387 (1992)

20.151 W. Ketterle, M. Schäfer, A. Arnold, J. Wolfrum: 2D single shot imaging of OH radicals using tunable excimer lasers, Appl. Phys. B **54**, 109–112 (1992)

20.152 E.W. Kaiser, K. Marko, D. Klick, L. Rimai, C.C. Wang, B. Shirinzadeh, D. Zhou: Measurement of OH density profiles in atmospheric-pressure propane-air flames, Combust. Sci. Tech. **50**, 163–183 (1986)

20.153 C. Schulz, V. Sick, U. Meier, J. Heinze, W. Stricker: Quantification of NO A-X(0,2) LIF: Investigation of calibration and collisional influences in high-pressure flames, Appl. Opt. **38**, 434–1443 (1999)

20.154 B. Yip, D.C. Fourguette, M.B. Long: Three-dimensional gas concentration and gradient measurements in a photoacoustically perturbed jet, Appl. Opt. **25**, 3919–3923 (1986)

20.155 J.H. Frank, K.M. Lyons, M.B. Long: Technique for three-dimensional measurements of the time development of turbulent flames, Opt. Lett. **16**, 958–960 (1991)

20.156 J. Nygren, J. Hult, M. Richter, M. Aldén, M. Christensen, A. Hultqvist, B. Johansson: Three-dimensional laser induced fluorescence of fuel distributions in an HCCI engine, Proc. Combust. Inst. **29**, 679–685 (2002)

20.157 A. Hoffmann, F. Zimmermann, H. Scharr, S. Krömker, C. Schulz: Instantaneous three-dimensional visualization of concentration distributions in turbulent flows with crossed-plane laser-induced fluorescence imaging, Appl. Phys. B **80**, 125–131 (2005)

20.158 S.S. Sattler, F.C. Gouldin, N.T. Boertlein: Combined crossed-plane imaging and stereo-particle image velocimetry. In: *Third Joint Meeting of the U.S. Sections of the Combustion Institute* (The Combustion Institute, Chicago 2003)

20.159 D.A. Knaus, F.C. Gouldin: Measurement of flamelet orientations in premixed flames with positive and negative Markstein numbers, Proc. Combust. Inst. **28**, 367–373 (2000)

20.160 D.A. Knaus, F.C. Gouldin, D.C. Bingham: Assessment of crossed-plane tomography for flamelet surface normal measurements, Combust. Sci. Technol. **174**, 101–134 (2002)

20.161 A. Hoffmann, F. Zimmermann, and C. Schulz: Instantaneous three-dimensional visualization of concentration distributions in turbulent flows with a single laser. In: *First International Conference on Optical and Laser Diagnostics* (London, 2002), in press

20.162 A. Karpetis, R. Barlow: Measurements of flame orintation and scalar dissipation in turbulent hydrocarbon flames. In: *Third Joint Meeting of the U.S. Sections of The Combustion Institute* (The Combustion Institute, Chicago 2003)

20.163 C. Dartu: *Visualization of Volumetric Datasets* (Universität Heidelberg, Heidelberg 2001)

20.164 C. Kaminski, J. Hult, M. Aldén, S. Lindenmaier, A. Dreizler, U. Maas, M. Baum: Spark ignition of turbulent methane/air mixtures revealed by time resolved laser induced fluorescence and direct numerical simulations, Proc. Combust. Inst. **28**, 399–405 (2000)

20.165 A. Dreizler, S. Lindenmaier, U. Maas, J. Hult, M. Aldén, C.F. Kaminski: Characterisation of a spark ignition system by planar laser-induced fluorescence of OH at high repetition rates and comparisons with chemical kinetics calculations, Appl. Phys. B **70**, 287–294 (2000)

20.166 R.B. Williams, P. Ewart, A. Dreizler: Velocimetry of gas flows using degeneratre four-wave mixing, Opt. Lett. **19**, 1486–1488 (1994)

20.167 R.W. Pitz, J.A. Wehrmeyer, L.A. Ribarov, D.A. Oguss, F. Batliwala, P.A. DeBarber, S. Deutsch, P.E. Dimotakis: Unseeded molecular flow tagging in cold and hot flows using ozone and hydroxyl tagging velocimetry, Meas. Sci. Technol. **11**, 1259–1271 (2000)

20.168 R. Miles, C. Cohen, J. Conners, P. Howard, S. Huang, E. Markovitz, G. Russell: Velocity measurements by vibrational tagging and fluorescent probing of oxygen, Opt. Lett. **12**, 861 (1987)

20.169 R.B. Miles, W. Lempert, B. Zhang: Turbulence structure measurement by RELIEF flow tagging, Fluid. Dyn. Res. **8**, 9–17 (1991)

20.170 S. Krüger, G. Grünefeld: Gas-phase velocity field measurements in dense sprays by laser-based flow tagging, Appl. Phys. B **70**, 463–466 (2000)

20.171 T. Elenbaas, N.M. Sijtsema, R.A.L. Tolboom, N.J. Dam, W. van de Water, J.J. ter Meulen: *Characterization of turbulence by air photolysis and recombination tracking (APART)*, AIAA 2002-0694 (American Institute of Aeronautics and Astronautics, Inc., 2002)

20.172 I. Düwel, P. Peuser, P. Zeller, J. Wolfrum, C. Schulz: Laser-based spray diagnostics in commercial fuels using a new all-solid-state laser system. In: *ILASS Europe* (Nottingham 2004)

20.173 G. Josefsson, I. Magnusson, F. Hildenbrand, C. Schulz, V. Sick: Multidimensional laser diagnostic and numerical analysis of NO formation in a gasoline engine, Proc. Combust. Inst. **27**, 2085–2092 (1998)

20.174 W.G. Bessler, M. Hofmann, F. Zimmermann, G. Suck, J. Jakobs, S. Nicklitzsch, T. Lee, J. Wolfrum, C. Schulz: Quantitative in-cylinder NO-LIF imaging in a realistic gasoline engine with spray-guided direct injection, Proc. Combust. Inst. **30**, 2667–2674 (2005)

20.175 M.J. Hall, P. Zuzek, R.W. Anderson: Fiber optic sensor for crank angle resolved measurements of burned gas residual fraction in the cylinder of an SI engine, SAE Techn. Paper Ser. 2001-01-1921 (2001)

20.176 R. Reichle, C. Pruss, W. Osten, H.J. Tiziani, F. Zimmermann, C. Schulz: Acquisition of combustion parameters close to the ignition spark with a fiber optic sensor. In: *Optical measurement systems for industrial inspection IV*, Vol. 5856, ed. by W. Osten, C. Gorecki, E.L. Novak (SPIE-The International Society for Optical Engineering, Bellingham, WA 2005) pp. 158–168

20.177 M. Richter, B. Axelsson, M. Aldén: Engine diagnostics using laser-induced fluorescence signals collected through an endoscopic detection system, SAE Techn. Paper Ser. 982465 (1998)

20.178 H. Neij, M. Aldén: Application of two-photon laser-induced fluorescence for visualization of water vapor on combustion environments, Appl. Opt. **33**, 6514–6523 (1994)

20.179 W.G. Bessler, C. Schulz, T. Lee, J.B. Jeffries, R.K. Hanson: Carbon dioxide UV laser-induced fluorescence in high-pressure flames, Chem. Phys. Lett. **375**, 344–349 (2003)

20.180 A.A. Roller, M. Decker, V. Sick, J. Wolfrum, W. Hentschel, K.-P. Schindler: Non-Intrusive temperature measurements during the compression phase of a DI Diesel engine, SAE Techn. Paper Ser. 952461 (1995)

20.181 M.C. Drake, T.D. Fransler, D.T. French: Crevice flow and combustion visualization in a direct-injection spark-ignition engine using laser imaging techniques, SAE Techn. Paper Ser. 952454 (1995)

20.182 J.E. Dec, J.O. Keller: High speed thermometry using two-line atomic fluorescence, Proc. Combust. Inst. **21**, 1737–1745 (1986)

20.183 C.F. Kaminski, J. Engström, M. Aldén: Quasi-instantaneous two-dimensional temperature measurements in a spark ignition engine using two-line atomic fluorescence, Proc. Combust. Inst. **27**, 85–93 (1998)

20.184 B.K. McMillin, J.L. Palmer, R.K. Hanson: Temporally resolved, two-line fluorescence imaging of NO temperature in a transverse jet in a supersonic cross flow, Appl. Opt. **32**, 7532–7545 (1993)

20.185 W.G. Bessler, C. Schulz: Quantitative multi-line NO-LIF temperature imaging, Appl. Phys. B **78**, 519–533 (2004)

20.186 W.G. Bessler, C. Schulz, T. Lee, J.B. Jeffries, R.K. Hanson: Strategies for laser-induced fluorescence detection of nitric oxide in high-pressure flames: III. Comparison of A-X strategies, Appl. Opt. **42**, 4922–4936 (2003)

20.187 C. Orlemann, C. Schulz, J. Wolfrum: NO-flow tagging by photodissociation of NO_2. A new approach for measuring small-scale flow structures, Chem. Phys. Lett. **307**, 15–20 (1999)

20.188 N. Dam, R.J.H. Klein-Douwel, N. Sijstema, J. ter Meulen: Nitirc oxide flow tagging in unseeded air, Opt. Lett. **26**, 36–38 (2001)

20.189 M.P. Lee, R.K. Hanson: Calculations of O_2 absorption and fluorescence at elevated temperatures for a broadband argon-fluoride laser source at 193 nm, J. Quant. Spectros. Radiat. Transfer **36**, 425–440 (1986)

20.190 B. Hiller, R. Hanson: Properties of the iodine molecule relevant to laser-induced fluorescence experiments in gaseous flows, Exp. Fluids **10**, 1–11 (1990)

20.191 K. Greenough, A. Duncan: The fluorescence of sulfur dioxide, J. Am. Chem. Soc. **83**, 555–560 (1961)

20.192 H. Mettee: Fluorescence and phosphorescence of SO_2 vapor, J. Chem. Phys. **49**, 1784–1793 (1968)

20.193 S.J. Strickler, D. Howell: Luminescence and radiations transitions in sulfur dioxide gas, J. Chem. Phys. **49**, 1947–1951 (1968)

20.194 V. Sick: Exhaust-gas imaging via planar laser-induced fluorescence of sulfur dioxide, Appl. Phys. B **74**, 461–463 (2002)

20.195 T. Rao, S. Collier, J. Calvert: Primary photophysical processes in the photochemistry of sulfur dioxide at 2875 Å, J. Am. Chem. Soc. **91**, 1609–1615 (1969)

20.196 T. Rao, S. Collier, J.G. Calvert: The quenching reactions of the first excited singlet and triplet states of sulfur dioxide with oxygen and carbon dioxide, J. Am. Chem. Soc. **91**, 1616–1621 (1969)

20.197 R. Caton, A. Duncan: Lifetime of the lowest triplet state of sulfur dioxide, J. Am. Chem. Soc. **90**, 1945–1949 (1968)

20.198 A. Lozano: *Laser-excited luminescent tracers for planar concentration measurements in gaseous*

20.199 H. Krämer, S. Einecke, C. Schulz, V. Sick, S.R. Nattrass, J.S. Kitching: Simultaneous mapping of the distribution of different fuel volatility classes using tracer-LIF and NIR tomography in an IC engine, SAE Transactions, J. Fuels Lubricants **107**, 1048–1059 (1998)

20.200 D. Han, R.R. Steeper: An LIF equivalence ratio imaging technique for multicomponent fuels in an IC engine, Proc. Combust. Inst **29**, 727–734 (2002)

20.201 M.-T. Lin, V. Sick: Mixture evaporation characteristics prediction for LIF measurements using PSRK (Predictive Soave-Redlich-Kwong) equation of state, SAE Transactions, J. Fuels Lubricants **111**, 1490–1499 (2002)

20.202 M.-T. Lin, V. Sick: Is toluene a suitable LIF tracer for fuel film measurements?, SAE Techn. Paper Ser. 2004-01-1355 (2004)

20.203 J. F. Le Coz, C. Catalano, T. Baritaud: Application of laser-induced fluorescence for measuring the thickness of liquid films on transparent walls. In: *7th International Symposium on Applications of Laser Techniques to Fluid Mechanics II* (Lisbon 1994)

20.204 T. Fansler, D. French, M.C. Drake: Fuel Distributions in a Firing Direct-injection Spark-ignition Engine Using Laser-Induced Fluorescence Imaging, SAE Techn. Paper Ser. 950110 (1995)

20.205 D. A. Greenhalgh, D. F. Bryce, R. D. Lockett, and S. C. Harding: Development of planar laser induced fluorescence for fuel: Application to gas turbine combustion. In: *AGARD PEP Symposium on Advanced non-intrusive instrumentation for propulsion engines* (Brussels 1997)

20.206 W. Hentschel, B. Block, T. Hovestadt, H. Meyer, G. Ohmstede, V. Richter, B. Stiebels, A. Winkler: Optical diagnostics and CFD-simulations to support the combustion process development of the Volkswagen FSI direct injection gasoline engine, SAE Techn. Paper Ser. 2001-01-3648 (2001)

20.207 J. Reboux, D. Puecheberty: A new approach of PLIF applied to fuel/air ratio measurement in the compressive stroke of an optical SI engine, SAE Techn. Paper Ser. 941988 (1994)

20.208 W. Koban, J.D. Koch, R.K. Hanson, C. Schulz: Oxygen quenching of toluene fluorescence at elevated temperatures, Appl. Phys. B **80**, 777–784 (2005)

20.209 W. Koban, J.D. Koch, V. Sick, N. Wermuth, R.K. Hanson, C. Schulz: Predicting LIF signal strength for toluene and 3-pentanone under engine-related temperature and pressure conditions, Proc. Combust. Inst. **30**, 1545–1553 (2005)

20.210 S.A. Kaiser, M.B. Long: The effect of temperature and quenching on laser-induced fluorescence of naphtalenes as JP-8 tracers. In: *Chemical and Physical Processes in Combustion – Technical Meeting of the Eastern States Section of the Combustion Institute* (Pennsylvania State University, University Park 2003)

20.211 L.A. Melton: Spectrally separated fluorescence emissions for Diesel fuel droplets and vapor, Appl. Opt. **22**, 2224–2226 (1983)

20.212 A. Lozano, B. Yip, R.K. Hanson: Acetone: A tracer for concentration measurements in gaseous flows by planar laser-induced fluorescence, Exp. Fluids **13**, 369–376 (1992)

20.213 F. Großmann, P.B. Monkhouse, M. Ridder, V. Sick, J. Wolfrum: Temperature and pressure dependences of the laser-induced fluorescence of gas-phase acetone and 3-pentanone, Appl. Phys. B **62**, 249–253 (1996)

20.214 L. Yuen, J. Peters, R. Lucht: Pressure dependence of laser-induced fluorescence from acetone, Appl. Opt. **36**, 3271–3277 (1997)

20.215 F. Ossler, M. Alden: Measurements of picosecond laser-induced fluorescence from gas-phase 3-pentanone and acetone: Implications to combustion diagnostics, Appl. Phys. B **64**, 493–502 (1997)

20.216 J.D. Koch, R.K. Hanson: Temperature and excitation wavelength dependencies of 3-pentanone absorption and fluorescence for PLIF applications, Appl. Phys. B **76**, 319–324 (2003)

20.217 J.B. Gandhi, P.G. Felton: On the fluorescent behavior of ketones at high temperatures, Exp. Fluids **21**, 143–144 (1996)

20.218 J.D. Koch, R.K. Hanson: Ketone photophysics for quantitative PLIF imaging, AIAA paper 2001-0413 (2001)

20.219 N.P. Tait, D.A. Greenhalgh: 2D laser induced fluorescence imaging of parent fuel fraction in nonpremixed combustion, Proc. Combust. Inst. **24**, 1621–1628 (1992)

20.220 D. Wolff, H. Schlüter, V. Beushausen, P. Andresen: Quantitative determination of fule air mixture distributions in an internal combustion engine using PLIF of acetone, Ber. Bunsenges. Phys. Chem. **97**, 1738–1741 (1993)

20.221 M. Berckmuller, N. Tait, R. Lockett, D. Greenhalgh: In-cylinder crank-angle-resolved imaging of fuel concentration in a firing spark-ignition engine using planar laser-induced fluorescence, Proc. Combust. Inst. **25**, 151–156 (1994)

20.222 M. Dawson, S. Hochgreb: Liquid Fuel Visualization Using Laser-Induced Fluorescence During Cold Start, SAE Techn. Paper Ser. 982466 (1998)

20.223 R.M. Green, L.D. Cloutman: Planar LIF observations of unburned fuel escaping the upper ring-land crevice in an SI engine, SAE Techn. Paper Ser. 970823 (1997)

20.224 R. Bryant, J.F. Driscoll: Acetone laser induced fluorescence for low pressure, low temperature flow visualization, Exp. Fluids **28**, 417–476 (2000)

20.225 H. Neij, B. Johansson, M. Aldén: Development and demonstration of 2D-LIF for studies of mixture preparation in SI engines, Combust. Flame **99**, 449–457 (1994)

20.226 A. Arnold, A. Buschmann, B. Cousyn, M. Decker, V. Sick, F. Vannobel, J. Wolfrum: Simultaneous imaging of fuel and hydroxyl radicals in an in-line four cylinder SI engine, SAE Trans. **102**, 1–9 (1993)

20.227 A. Arnold, H. Becker, R. Suntz, P. Monkhouse, J. Wolfrum, R. Maly, W. Pfister: Flame front imaging in an internal-combustion engine simulator by laser-induced fluorescence of acetaldehyde, Opt. Lett. **15**, 831–833 (1990)

20.228 J.C. Swindal, D.P. Dragonetti, R.T. Hahn, P.A. Furman, W.P. Acker: In-cylinder charge homogeneity during cold-start studied with fluorescent tracers simulating different fuel distillation temperatures, SAE Techn. Paper Ser. 950106 (1995)

20.229 D.A. Hansen, E.K.C. Lee: Radiative and nonradiative transitions in the first excited singlet state of simple linear aldehydes, J. Chem. Phys. **62**, 3272–3277 (1975)

20.230 D.A. Hansen, E.K. Lee: Radiative and nonradiative transitions in the first excited singlet state of symmetrical methyl-substituted acetones, J. Chem. Phys. **62**, 183–189 (1975)

20.231 A. Gandini, K. Kutschke: Primary process in photolysis of hexafluoroacetone vapour, 2. Fluorescence and phosphorescence, Proc. R. Soc. London Ser. A **306**, 511 (1968)

20.232 A. Gandini, K.O. Kutschke: The effect of mercury on the triplet state of hexafluoroacetone, Ber. Bunsenges. Phys. Chem **72**, 296–301 (1968)

20.233 H. Okabe, W.A. Noyes: The relative intensities of fluorescence and phosphorescence in biacetyl vapor, J. Am. Chem. Soc. **79**, 810–806 (1957)

20.234 H.W. Sidebottom, C.C. Badcock, J.G. Calvert, B.R. Rabe, E.K. Damon: Lifetime studies of the biacetyl excited singlet and triplet states in the gas phase at 25°C, J. Am. Chem. Soc. **94**, 13–19 (1972)

20.235 J.B. Liu, Q. Pan, C.S. Liu, J.R. Shi: Principles of flow field diagnostics by laser-induced biacetyl phosphorescence, Exp. Fluids **6**, 505–513 (1988)

20.236 P. Pringsheim: *Fluorescence and Phosphorescence* (Interscience, New York 1949)

20.237 T. Itoh, A. Kakuho, H. Hishinuma, T. Urushihara, Y. Takagi, K. Horie, M. Asano, E. Ogata, T. Yamashita: *Development of a New Compound Fuel and Fluorescent Tracer Combination for Use with Laser Induced Fluorescence* (Nissan Motor Co./University of Tokyo, Tokyo 1995)

20.238 P.G. Felton, F.V. Bracco, M.E.A. Bardsley: On the quantitative application of exciplex fluorescence to engine sprays, SAE Techn. Paper Ser. 930870 (1993)

20.239 A.P. Fröba, F. Rabenstein, K.U. Münch, A. Leipertz: Mixture of triethylamine and benzene as a new seeding material for the quantitative two-dimensional laser-induced exciplex fluorescence imaging of vapor and liquid fuel inside SI engines, Combust. Flame **112**, 199–209 (1998)

20.240 J.B. Ghandhi, P.G. Felton, D.F. Gajdezko, F.V. Bracco: Investigation of the fuel distribution in a two-stroke engine with an air-assisted injector, SAE Techn. Paper Ser. 940394 (1994)

20.241 M. Richter, B. Axelsson, K. Nyholm, M. Aldén: Real-time calibration of planar laser-induced fluorescence air-fuel ratio measurements in combustion environments using in situ Raman scattering, Proc. Combust. Inst **27**, 51–57 (1998)

20.242 D. Frieden, V. Sick, J. Gronki, C. Schulz: Quantitative oxygen imaging in an engine, Appl. Phys. B **75**, 137–141 (2002)

20.243 J.C. Sacadura, L. Robin, F. Dionnet, D. Gervais, P. Gastaldi, A. Ahmed: Experimental investigation of an optical direct injection SI engine using FARLIF, SAE Techn. Paper Ser. 2000-01-1794 (2000)

20.244 W. Koban, J.D. Koch, R.K. Hanson, C. Schulz: Absorption and fluorescence of toluene vapor at elevated temperatures, Phys. Chem. Chem. Phys. **6**, 2940–2945 (2004)

20.245 W. Koban, J.D. Koch, V. Sick, N. Wermuth, R.K. Hanson, C. Schulz: Predicting LIF signal strength for toluene and 3-pentanone under engine-related temperature and pressure conditions, Proc. Combust. Inst. **30**, 1545–1553 (2005)

20.246 W. Koban, C. Schulz: Toluene as a tracer for fuel, temperature and oxygen concentrations, SAE Techn. Paper Ser. 2005-01-2091 (2005)

20.247 M. Schütte, H. Finke, G. Grünefeld, S. Krüger, P. Andresen, B. Stiebels, B. Block, H. Meyer, W. Hentschel: Spatially resolved air/fuel ratio and residual gas measurements by spontaneous Raman scattering in a firing direct injection fasoline engine, SAE Techn. Paper Ser. 2000-01-1795 (2000)

20.248 B. Mewes, G. Bauer, D. Brüggemann: Fuel vapor measurements by linear Raman spectroscopy using spectral discriminattion from droplet interferences, Appl. Opt. **38**, 1040–1045 (1999)

20.249 G. Grünefeld, V. Beushausen, P. Andresen, W. Hentschel: Spatially resolved Raman scattering for multi-species and temperature analysis in technically applied combustion systems: Spray flame and four-cylinder in-line engine, Appl. Phys. B **58**, 333–342 (1994)

20.250 R. A. L. Tolboom, N. M. Sijtsema, N. J. Dam, J. J. t. Meulen, J. M. Mooij, J. D. M. Maassen: 2D stoichiometries from snapshot Raman measurements, in 40th AIAA Aerospace Sciences Meeting. AIAA paper 2002-0400

20.251 D.C. Kyritsis, P.G. Felton, F.V. Bracco: On the feasibility of quantitative, single-shot, spontaneous Raman imaging in an optically accessible engine cylinder, SAE Techn. Paper Ser. 1999-01-3537 (1999)

20.252 D. Geyer: 1D Raman/Rayleigh experiments in a turbulent opposed jet, PhD Thesis (TU Darmstadt, Darmstadt 2004)

20.253 C. Schulz, V. Sick, J. Wolfrum, V. Drewes, M. Zahn, R. Maly: Quantitative 2D single-shot imaging of

NO concentrations and temperatures in a transparent SI engine, Proc. Combust. Inst. **26**, 2597–2604 (1996)

20.254 J. Engström, J. Nygren, M. Aldén, C.F. Kaminski: Two-line atomic fluorescence as a temperature probe for highly sooting flames, Opt. Lett. **25**, 1469–1471 (2000)

20.255 A. Arnold, B. Lange, T. Bouché, Z. Heitzmann, G. Schiff, W. Ketterle, P. Monkhouse, J. Wolfrum: Absolute temperature fields in flames by 2D-LIF of OH using excimer lasers and CARS spectroscopy, Ber. Bunsenges. Phys. Chem. **96**, 1388–1392 (1992)

20.256 M.C. Thurber, F. Griseh, R.K. Hanson: Temperature imaging with single- and dual wavelength acetone planar laser-induced fluorescence, Opt. Lett. **22**, 251–253 (1997)

20.257 N.P. Tait, D.A. Greenhalgh: PLIF imaging of fuel fraction in practical devices and LII imaging of soot, Ber. Bunsenges. Phys. Chem. **97**, 1619–1625 (1993)

20.258 J. Kazenwadel, W. Koban, T. Kunzelmann, C. Schulz: Fluorescence imaging of natural gas/air mixing without tracers added, Chem. Phys. Lett. **345**, 259–264 (2001)

20.259 W. Koban, C. Schulz: FAR-LIF: Myth and Reality. In: *European Combustion Meeting* (The Combustion Institute, Louvain-la-Neuve 2005)

20.260 K. Kikuchi, C. Sato, M. Watabe, H. Ikeda, Y. Takahashi, T. Miyashi: New aspects on fluorescence quenching by molecular oxygen, J. Am. Chem. Soc. **115**, 5180–5184 (1993)

20.261 C. Schulz, J.B. Jeffries, D.F. Davidson, J.D. Koch, J. Wolfrum, R.K. Hanson: Impact of UV absorption by CO_2 and H_2O on NO LIF in high-pressure combustion applications, Proc. Combust. Inst. **29**, 2725–2742 (2002)

20.262 C.T. Bowman: Control of combustion-generated nitrogen oxide emissions: Technology driven by regulation, Proc. Combust. Inst. **24**, 859–878 (1994)

20.263 A.M. Wodtke, M. Huwel, H. Schlüter, G. Meijer, P. Andresen, H. Voges: High-sensitivity detection of NO in a flame using a tunable ArF laser, Opt. Lett. **13**, 910–912 (1988)

20.264 C. Schulz, B. Yip, V. Sick, J. Wolfrum: A laser-induced fluorescence scheme for imaging nitric oxide in engines, Chem. Phys. Lett. **242**, 259–264 (1995)

20.265 M.D. DiRosa, K.G. Klavuhn, R.K. Hanson: LIF spectroscopy of NO and O_2 in high-pressure flames, Comb. Sci. Tech. **118**, 257–283 (1996)

20.266 W.P. Partridge Jr., M.S. Klassen, D.D. Thomsen, N.M. Laurendeau: Experimental assessment of O_2 interferences on laser-induced fluorescence measurements of NO in high-pressure, lean premixed flames by use of narrow-band and broadband detection, Appl. Opt. **34**, 4890–4904 (1996)

20.267 P. Jamette, P. Desgroux, V. Ricordeau, B. Deschamps: Laser-induced fluorescence detection of NO in the combustion chamber of an optical GDI engine with A-X(0,1) excitation, SAE Techn. Paper Ser. 2001-01-1926 (2001)

20.268 W.G. Bessler, C. Schulz, T. Lee, J.B. Jeffries, R.K. Hanson: Laser-induced-fluorescence detection of nitric oxide in high-pressure flames with A-X(0,1) excitation. In: *Proc. Western States Section of the Combustion Institute Spring Meeting* (The Combustion Institute, Oakland 2001)

20.269 C. Schulz, J.D. Koch, D.F. Davidson, J.B. Jeffries, R.K. Hanson: Ultraviolet absorption spectra of shock-heated carbon dioxide and water between 900 and 3050 K, Chem. Phys. Lett. **355**, 82–88 (2002)

20.270 F. Hildenbrand, C. Schulz: Measurements and simulation of in-cylinder UV-absorption in spark ignition and Diesel engines, Appl. Phys. B **73**, 165–172 (2001)

20.271 A.O. Vyrodov, J. Heinze, M. Dillmann, U.E. Meier, W. Stricker: Laser-induced fluorescence thermometry and concentration measurements on NO A-X (0,0) transitions in the exhaust gas of high pressure CH_4/air flames, Appl. Phys. B. **61**, 409–414 (1995)

20.272 A. Bräumer, V. Sick, J. Wolfrum, V. Drewes, M. Zahn, R. Maly: Quantitative two-dimensional measurements of nitric oxide and temperature distributions in a transparent square piston SI engine, SAE Techn. Paper Ser. 952462 (1995)

20.273 J.B. Jeffries, C. Schulz, D.W. Mattison, M. Oehlschlaeger, W.G. Bessler, T. Lee, D.F. Davidson, R.K. Hanson: UV absorption of CO_2 for temperature diagnostics of hydrocarbon combustion applications, Proc. Combust. Inst. **30**, 1591–1599 (2005)

20.274 C. Schulz, V. Sick, J. Heinze, W. Stricker: Laser-induced fluorescence detection of nitric oxide in high-pressure flames using A-X(0,2) excitation, Appl. Opt. **36**, 3227–3232 (1997)

20.275 F. Hildenbrand, C. Schulz, M. Hartmann, F. Puchner, G. Wawrschin: In-cylinder NO-LIF imaging in a realistic GDI engine using KrF excimer laser excitation, SAE Techn. Paper Ser. 1999-01-3545 (1999)

20.276 M. Hofmann, H. Kronemayer, B.F. Kock, H. Jander, C. Schulz: Laser-induced incandescence and multi-line NO-LIF thermometry for soot diagnostics at high pressures. In: *European Combustion Meeting* (The Combustion Institute, Louvain-la-Neuve 2005)

20.277 W.G. Bessler, C. Schulz, T. Lee, D.I. Shin, M. Hofmann, J.B. Jeffries, J. Wolfrum, R.K. Hanson: Quantitative NO-LIF imaging in high-pressure flames, Appl. Phys. B **75**, 97–102 (2002)

20.278 T. Lee, W.G. Bessler, H. Kronemayer, C. Schulz, J.B. Jeffries, R.K. Hanson: Quantitative temperature measurements in high-pressure flames with multi-line nitric oxide (NO)-LIF thermometry. In: *4th Joint Meeting of the U.S. Sections of the Combustion Institute* (Drexel University, Philadelphia 2005)

20.279 W.G. Bessler, C. Schulz, T. Lee, J.B. Jeffries, R.K. Hanson: Strategies for laser-induced fluorescence detection of nitric oxide in high-pressure flames. I. A-X(0,0) excitation, Appl. Opt. **41**, 3547–3557 (2002)

20.280 W.G. Bessler, C. Schulz, T. Lee, J.B. Jeffries, R.K. Hanson: Strategies for laser-induced fluorescence detection of nitric oxide in high-pressure flames. II. A-X(0,1) excitation, Appl. Opt. **42**, 2031–2042 (2003)

20.281 W.G. Bessler, C. Schulz, V. Sick, J.W. Daily: A versatile modeling tool for nitric oxide LIF spectra. In: *3rd Joint Meeting of the US Sections of the Combustion Institute* (The Combustion Institute, Chicago 2003)

20.282 M. Tamura, J. Luque, J.E. Harrington, P.A. Berg, G.P. Smith, J.B. Jeffries, D.R. Crosley: Laser-induced fluorescence of seeded nitric oxide as a flame thermometer, Appl. Phys. B **66**, 503–510 (1998)

20.283 W. Bessler, F. Hildenbrand, C. Schulz: Vibrational temperature imaging using two-line laser-induced fluorescence of seeded NO. In: *Laser Applications to Chemical and environmental analysis*, OSA Technical Digest Series, Vol. 3 (Opt. Soc. Am., Washington 2000) pp. 149–151

20.284 R. Cattolica: OH rotational temperature from two-line laser-excited fluorescence, Appl. Opt. **20**, 1156–1166 (1981)

20.285 J.W. Daily, W.G. Bessler, C. Schulz, V. Sick: Role of non-stationary collisional dynamics in determining nitric oxide LIF spectra, AIAA 2004-0389 (2004)

20.286 T. Lee, W.G. Bessler, C. Schulz, M. Patel, J.B. Jeffries, R.K. Hanson: UV planar laser-induced fluorescence imaging of hot carbon dioxide in a high-pressure flame, Appl. Phys. B **79**, 427–430 (2004)

20.287 H. Kronemayer, W. Bessler, C. Schulz: Gas-phase temperature measurements in evaporating sprays and spray flames based on NO multiline LIF, Appl. Phys. B **81**, 1071–1074 (2005)

20.288 T. Lee, W.G. Bessler, H. Kronemayer, C. Schulz, J.B. Jeffries: Quantitative temperature measurements in high-pressure flames with multi-line nitric oxide (NO)-LIF thermometry, Appl. Opt. **31**, 6718–6728 (2005)

20.289 I. Düwel, H.-W. Ge, H. Kronemayer, R.W. Dibble: Experimental and numerical characterization of a turbulent spray flame, Proc. Combust. Inst. **31**, 2247–2255 (2007)

20.290 W.G. Bessler, C. Schulz, M. Hartmann, M. Schenk: Quantitative in-cylinder NO-LIF imaging in a direct-injected gasoline engine with exhaust gas recirculation, SAE Techn. Paper Ser. 2001-01-1978 (2001)

20.291 C. Schulz, J. Wolfrum, V. Sick: Comparative study of experimental and numerical NO profiles in SI combustion, Proc. Combust. Inst. **27**, 2077–2084 (1998)

20.292 F. Hildenbrand, C. Schulz, V. Sick, H. Jander, H.G. Wagner: Applicability of KrF excimer laser induced fluorescence in sooting high-pressure flames, in VDI Flammentag Dresden, VDI Berichte **1492**, 269–274 (1999)

20.293 F. Hildenbrand, C. Schulz, J. Wolfrum, F. Keller, E. Wagner: Laser diagnostic analysis of NO formation in a direct injection Diesel engine with pump-line nozzle and common-rail injection systems, Proc. Combust. Inst. **28**, 1137–1144 (2000)

20.294 F. Hildenbrand, C. Schulz, V. Sick, G. Josefsson, I. Magnusson, Ö. Andersson, M. Aldén: Laser spectroscopic invesitigation of flow fields and NO-formation in a realistic SI engine, SAE Techn. Paper Ser. 980148 (1998)

20.295 A. D'Alessio: Laser light scattering and fluorescence diagnostics of rich flames produced by gascons and liquid fuels. In: *Particulate carbon: Formation during combustion*, ed. by D.C. Siegla, G.W. Smith (Plenum, New York 1981) p. 207

20.296 R.W. Weeks, W.W. Duley: Aerosol-particle sizes from light emission during excitation by TEA CO_2 laser pulses, J. Appl. Phys **45**, 4661–4662 (1973)

20.297 A.C. Eckbreth: Effects of laser-modulated particle incandescence on Raman scattering diagnostics, J. Appl. Phys. **48**, 4473–4479 (1977)

20.298 L.A. Melton: Soot diagnostics based on laser heating, Appl. Opt. **23**, 2201–2208 (1984)

20.299 T. Ni, J.A. Pinson, S. Gupta, R.J. Santoro: Two-dimensional imaging of soot volume fraction by the use of laser-induced incandescence, Appl. Opt. **34**, 7083–7091 (1995)

20.300 D.J. Bryce, N. Ladommatos, H. Zhao: Quantitative investigation of soot distribution by laser-induced incandescence, Appl. Opt. **39**, 5012–5022 (2000)

20.301 C.R. Shaddix, K.C. Smyth: Laser-induced incandescence measurements of soot production in steady and flickering methane, propane, and ethylene diffusion flames, Combust. Flame **107**, 418–452 (1996)

20.302 P.O. Witze, S. Hochgreb, D. Kayes, H.A. Michelsen, C.R. Shaddix: Time-resolved laser-induced incandescence and laser elastic-scattering measurements in a propane diffusion flame, Appl. Opt. **40**, 2443–2452 (2001)

20.303 B. Quay, T.-W. Lee, T. Ni, R.J. Santoro: Spatially resolved measurements of soot volume fraction using laser-induced incandescence, Combust. Flame **97**, 384–392 (1994)

20.304 J. Appel, B. Jungfleisch, M. Marquardt, R. Suntz, H. Bockhorn: Assessment of soot volume fractions from laser-induced incandescence by comparison with extinction measurements in laminar, premixed, flat flames, Proc. Combust. Inst. **26**, 2387–2395 (1996)

20.305 B. Axelsson, R. Collin, P.-E. Bengtsson: Laser-induced incandescence for soot particle size and

20.306 H. Geitlinger, T. Streibel, R. Suntz, H. Bockhorn: Two-dimensional imaging of soot volume fractions; particle number densities; and particle radii in laminar and turbulent diffusion flames, Proc. Combust. Inst. **27**, 1613–1621 (1998)

(continuation) volume fraction measurements using on-line extinction calibration, Appl. Phys. B **72**, 367–372 (2001)

20.307 T.R. Meyer, S. Roy, V.M. Belovich, E. Corporan, J.R. Gord: Simultaneous planar laser-induced incandescence, OH planar laser-induced fluorescence, and droplet Mie scattering in swirl-stabilized spray flames, Appl. Opt. **44**, 445–454 (2005)

20.308 H. Bockhorn, H. Geitlinger, B. Jungfleisch, T. Lehre, A. Schön, T. Streibel, R. Suntz: Progress in characterization of soot formation by optical methods, Phys. Chem. Chem. Phys. **4**, 3780–3793 (2002)

20.309 S. Will, S. Schraml, A. Leipertz: Comprehensive two-dimensional soot diagnostics based on laser-induced incandescence, Proc. Combust. Inst. **26**, 2277–2284 (1996)

20.310 S. Will, S. Schraml, K. Bader, A. Leipertz: Performance characteristics of soot primary particle size measurements by time-resolved laser-induced incandescence, Appl. Opt. **37**, 5647–5658 (1998)

20.311 B. Axelsson, R. Collin, P.-E. Bengtsson: Laser-induced incandescence for soot particle size measurements in premixed flat flames, Appl. Opt. **39**, 3683–3690 (2000)

20.312 T. Lehre, B. Jungfleisch, R. Suntz, H. Bockhorn: Size distributions of nanoscaled particles and gas temperatures from time-resolved laser-induced-incandescence measurements, Appl. Opt. **42**, 2021–2030 (2003)

20.313 D. Woiki, A. Giesen, P. Roth: Time-resolved laser-induced incandescence for soot particle sizing during acetylene pyrolysis behind shock waves, Proc. Combust. Inst. **28**, 2531–2537 (2000)

20.314 R.L. Vander Wal, T.M. Ticich, A.B. Stephens: Can soot primary particle size be determined using laser-induced incandescence?, Combust. Flame **116**, 291–296 (1999)

20.315 B.F. Kock, T. Eckhardt, P. Roth: In-cylinder sizing of Diesel particles by time-resolved laser-induced incandescence (TR-LII), Proc. Combust. Inst. **29**, 2775–2781 (2002)

20.316 D.R. Snelling, G.J. Smallwood, R.A. Sawchuk, W.S. Neill, D. Gareau, D.J. Clavel, W.L. Chippior, F. Liu, Ö.L. Gülder: In-situ real-time characterization of particulate emissions from a Diesel engine exhaust by laser-induced incandescence, SAE Techn. Paper Ser. 2000-01-1994 (2000)

20.317 S. Schraml, S. Will, A. Leipertz: Performance characteristics of TIRE-LII soot diagnostics in exhaust gases of Diesel engines, SAE Techn. Paper Ser. 2000-01-2002 (2000)

20.318 R.L. Vander Wal, T.M. Ticich, J.R. West Jr.: Laser-induced incandescence applied to metal nanostructures, Appl. Opt. **38**, 5867–5879 (1999)

20.319 B.F. Kock, C. Kayan, J. Knipping, H.R. Orthner, P. Roth: Comparison of LII and TEM sizing during synthesis of iron particle chains, Proc. Combust. Inst. **30**, 1689–1697 (2005)

20.320 T. Lehre, R. Suntz, H. Bockhorn: Time-resolved two-color LII: size distributions of nano-particles from gas-to-particle synthesis, Proc. Combust. Inst. **30**, 2585–2593 (2005)

20.321 R.L. Vander Wal: Laser-induced incandescence: Detection issues, Appl. Opt. **35**, 6548–6559 (1996)

20.322 B. Mewes, J.M. Seitzman: Soot volume fraction and particle size measurements with laser-induced incandescence, Appl. Opt. **36**, 709–717 (1997)

20.323 M. Hofmann, W.G. Bessler, C. Schulz, H. Jander: Laser-induced incandescence (LII) for soot diagnostics at high pressure, Appl. Opt. **42**, 2052–2062 (2003)

20.324 P.-E. Bengtsson, M. Aldén: Soot-visualization strategies using laser techniques, Appl. Phys. B **60**, 51–59 (1995)

20.325 C. Schoemaecker Moreau, E. Therssen, X. Mercier, J.F. Pauwels, P. Desgroux: Two-color laser-induced incandescence and cavity ring-down spectroscopy for sensitive and quantitative imaging of soot and PAHs in flames, Appl. Phys. B **78**, 485–492 (2004)

20.326 K.C. Smyth, C.R. Shaddix: The elusive history of m = 1.57 − 0.56i for the refractive index of soot, Combust. Flame **107**, 314–320 (1996)

20.327 D.R. Snelling, G.J. Smallwood, F. Liu, Ö.L. Gülder, W.D. Bachalo: A calibration-independent LII technique for soot measurement by detecting abolute light intensity, Appl. Opt. **44**, 6773–6785 (2005)

20.328 C. Schulz, B.F. Kock, M. Hofmann, H.A. Michelsen, S. Will, B. Bongie, R. Suntz, G. Smallwood: Laser-induced incandescence: Recent trends and current questions, Appl. Phys. B **83**, 333–354 (2006)

20.329 D. R. Snelling, F. Liu, G. J. Smallwood, L. Ö. Gülder: Evaluation of the nanoscale heat and mass transfer model of LII: Prediction of the excitation intensity. In: 34th National Heat Transfer Conference (Pittsburgh 2000)

20.330 B.F. Kock, B. Tribalet, C. Schulz, P. Roth: Two-color time-resolved LII applied to soot particle sizing in the cylinder of a Diesel engine, Combust. Flame **147**, 79–92 (2006)

20.331 D.L. Hofeldt: Real-time soot concentration measurement technique for engine exhaust streams, SAE Techn. Paper Ser. 930079 (1993)

20.332 H. Bladh, P.-E. Bengtsson: Characteristics of laser-induced incandescence from soot in studies of a time-dependent heat and mass-transfer model, Appl. Phys. B **78**, 241–248 (2004)

20.333 H.A. Michelsen: Understanding and predicting the temporal response of laser-induced incandescence

20.334 D.R. Snelling, F. Liu, G.J. Smallwood, L. Ö. Gülder: Determination of the soot absorption function and thermal accomodation coefficient using low-fluence LII in a laminar coflow ethylene diffusion flame, Combust. Flame **136**, 180–190 (2004)

20.335 H.A. Michelsen, F. Lin, B.F. Kock, H. Bladh, A. Boiarcinc, M. Charwoth, T. Dreier, R. Hadef, M. Hofmann, J. Reimann, S. Will, P.-E. Bengtsson, H. Bockhorn, F. Foncher, K.-P. Geigle, C. Mounaïm-Rousselle, C. Schulz, R. Stirn, B. Tribalet, S. Suntz: Modeling laser-induced incandescence of soot: A summary and comparison of LII models, Appl. Phys. B **87**, 503–521 (2007)

20.336 M. Hofmann, B. Kock, C. Schulz: A web-based interface for modeling laser-induced incandescence (LIISim). In: Proceedings of the European Combustion Meeting, Chania, Greece, 2007 The Combustion Institute

20.337 R.W. Dibble, R.E. Hollenbach: Laser Rayleigh thermometry in turbulent flames, Proc. Combust. Inst. **18**, 1489–1499 (1981)

20.338 C. Espey, J.E. Dec, T.A. Litzinger, D.A. Santavicca: Planar laser Rayleigh scattering for quantitative vapor-fuel imaging in a Diesel jet, Combust. Flame **109**, 65–86 (1997)

20.339 Ö. Andersson, R. Collin, M. Aldén, R. Egnell: Quantitative imaging of equivalence ratios in DME sprays using a chemically preheated combustion vessel, SAE Techn. Paper Ser. 2000-01-2785 (2000)

20.340 C. Schulz, J. Gronki, S. Andersson: Multi-species laser-based imaging measurements in a Diesel spray, SAE Techn. Paper Ser. 2004-01-1917 (2004)

20.341 S. Vogel, C. Hasse, J. Gronki, S. Andersson, N. Peters, J. Wolfrum, C. Schulz: Numerical simulation and laser-based imaging of mixture formation, ignition and soot formation in a Diesel spray, Proc. Combust. Inst. **30**, 2029–2036 (2005)

20.342 A.I. Vogel: Physical properties and chemical constitution of aliphatic hydrocarbons, J. Chem. Soc. **9**, 133–142 (1946)

20.343 R.C. Weast: *Handbook of Chemistry and Physics*, 54th edn. (CRC, Cleveland 1973)

20.344 Landolt-Börnstein: *Eigenschaften der Materie in ihren Aggregatzuständen. Tabellen* (Springer, Berlin, Heidelberg 1962)

20.345 J.R. Partington: *An Advanced Treatise on Physical Chemistry*, Vol. 9 (Longmans, Green, London 1953)

20.346 W.C. Gardiner, Y. Hidaka, T. Tanzawa: Refractivity of combustion gases, Combust. Flame **40**, 213–219 (1981)

20.347 D. Hoffman, K.-U. Münch, A. Leipertz: Two-dimensional temperature determination in sooting flames by filtered Rayleigh scattering, Opt. Lett. **21**, 525–527 (1996)

20.348 F. Beyrau, A. Bruer, T. Seeger, A. Leipertz: Gas-phase temperature measurement in the vaporizing spray of a gasoline direct-injection injector by use of pure rotational coherent anti-Stokes Raman scattering, Opt. Lett. **29**, 247–249 (2004)

20.349 D. Klick, K.A. Marko, L. Rimai: Broadband single-pulse CARS spectra in a fired internal combustion engine, Appl. Opt. **20**, 1178–1185 (1981)

20.350 K. Kohse-Höinghaus, R.S. Barlow, M. Aldén, J. Wolfrum: Combustion at the focus: Laser diagnostics and control, in Proc. Combust. Inst. **30**, 89–123 (2005)

20.351 P.T. Moseley, J.O.W. Norris, D.E. Williams: *Techniques and Mechanisms in Gas Sensing* (Adam Hilger, New York 1991)

20.352 V. Ebert, J. Wolfrum: Absorption Spectroscopy. In: *Optical Measurements – Techniques and Applications*, 2nd edn., ed. by F. Mayinger, O. Feldmann (Springer, Berlin, Heidelberg 2001) pp. 227–266

20.353 G. Herzberg: *The Spectra and Structures of Simple Free Radicals: An Introduction to Molecular Spectroscopy* (Dover, New York 1988)

20.354 G. Herzberg: *Spectra of Diatomic Molecules*, Vol. I, 2nd edn. (Krieger, Malobar, FL 1989)

20.355 G. Herzberg: *Infrared and Raman Spectra of Polyatomic Molecules*, Vol. II (Krieger, Malobar, FL 1991)

20.356 G. Herzberg: *Electronic Spectra of Polyatomic Molecules*, Vol. III (Krieger, Malobar, FL 1991)

20.357 J. Wolfrum, T. Dreier, V. Ebert, C. Schulz: Laser-based combustion diagnostics. In: *Encyclopedia of Analytical Chemistry*, ed. by R.A. Meyers (Wiley, Chichester 2000) pp. 2118–2148

20.358 V. Ebert, R. Hemberger, W. Meienburg, J. Wolfrum: In-situ gas analysis with infrared lasers, Ber. Bunsenges. **97**, 1527–1534 (1993)

20.359 R. Villarreal, P. Varghese: Frequency-resolved absorption tomography with tunable diode lasers, Appl. Opt. **44**, 6786–6795 (2005)

20.360 V. Ebert, C. Schulz, H.R. Volpp, J. Wolfrum, P. Monkhouse: Laser diagnostics of combustion processes: from chemical dynamics to technical devices, Israel J. Chem. **39**, 1–24 (1999)

20.361 V. Ebert, J. Fitzer, I. Gerstenberg, K.U. Pleban, H. Pitz, J. Wolfrum, M. Jochem, J. Martin: Simultaneous laser-based in situ detection of oxygen and water in a waste incinerator for active combustion control purposes, Proc. Combust. Inst. **27**, 1301–1308 (1998)

20.362 V. Ebert, H. Teichert, P. Strauch, T. Kolb, H. Seifert, J. Wolfrum: High sensitivity in-situ CO-detection in a 3 MW$_{th}$ rotary kiln for special waste incineration using new 2.3 μm distributed feedback diode lasers, Proc. Combust. Inst. **30**, 1611–1618 (2005)

20.363 H. Teichert, T. Fernholz, V. Ebert: Simultaneous in situ measurement of CO, H_2O, and gas temperatures in a full-sized coal-fired power plant by near-infrared diode lasers, Appl. Opt. **42**, 2043–2051 (2003)

20.364 V. Ebert, T. Fernholz, C. Giesemann, H. Pitz, H. Teichert, J. Wolfrum, H. Jaritz: Simultaneous

20.364 diode-laser-based in situ detection of multiple species and temperature in a gas-fired power plant, Proc. Combust. Inst. **28**, 423–430 (2000)

20.365 D.R. Herriott: Folded optical delay lines, Appl. Opt. **4**, 883–889 (1965)

20.366 D.R. Herriott, H. Kogelnik, R. Kompfner: Off-axis paths in spherical mirror interferometers, Appl. Opt. **3**, 523–526 (1964)

20.367 J.U. White: Long optical path of large aperture, J. Opt. Soc. Am. **32**, 285–288 (1942)

20.368 J. White: Very long optical path in air, J. Opt. Soc. Am. **66**, 411–416 (1976)

20.369 V. Ebert, H. Teichert, C. Giesemann, H. Saathoff, U. Schurath: Fibre-coupled in-situ laser absorption spectrometer for the selective detection of water vapor traces down to the ppb-level, Tech. Mess. **72**, 23–30 (2005)

20.370 W. Gurlit, R. Zimmermann, C. Giesemann, T. Fernholz, V. Ebert, J. Wolfrum, U. Platt, J.P. Burrows: Lightweight diode laser spectrometer CHILD (compact high-altitude in-situ laser diode) for balloon-borne measurements of water vapor and methane, Appl. Opt. **44**, 91–102 (2005)

20.371 H.E. Schlosser, J. Wolfrum, V. Ebert, B.A. Williams, R.S. Sheinson, J.W. Fleming: In situ determination of molecular oxygen concentrations in full-scale fire-suppression tests using tunable diode laser absorption spectroscopy, Proc. Combust. Inst. **29**, 353–360 (2002)

20.372 L.F. Mollenauer, J.C. White: *Tunable Lasers* (Springer, Berlin, Heidelberg 1987)

20.373 F.K. Kneubühl, M.W. Sigrist: *Laser* (Teubner, Stuttgart 1991)

20.374 L. Ajili, G. Scalari, D. Hofstetter, M. Beck, J. Faist, H. Beere, G. Davies, E. Linfield, D. Ritchie: Continuous-wave operation of far-infrared quantum cascade lasers, Electron. Lett. **38**, 1675–1676 (2002)

20.375 L. Ajili, J. Faist, H. Beere, D. Ritchie, G. Davies, E. Linfield: Loss-coupled distributed feedback far-infrared quantum cascade lasers, Electron. Lett. **41**, 419–420 (2005)

20.376 R. Köhler, A. Tredicucci, F. Beltram, H.E. Beere, E.H. Linfield, A.G. Davies, D.A. Ritchie, R.C. Iotti, F. Rossi: Terahertz semiconductor-heterostructure laser, Nature **417**, 156–159 (2002)

20.377 T. Tsuboi, N. Arimitsu, D. Ping, J.M. Hartmann: Light absorption by hydrocarbon molecules at 3.392 μm of He-Ne laser, Jpn. J. Appl. Phys. **24**, 8–13 (1985)

20.378 W.G. Mallard, W.C. Gardiner: Absorption of the 3.39 μm He-Ne laser line by methane from 300 to 2400 K, Quant. Spectrosc. Radiat. Transfer **20**, 135–149 (1978)

20.379 R.K. Mongia, E. Tomita, F.K. Hsu, L. Talbot, R.W. Dibble: Use of an optical probe for time-resolved in situ measurement of local air-to-fuel ratio and extent of fuel mixing with applicators to low NO_x emissions in premixed gas turbines, Proc. Combust. Inst. **26**, 2749–2755 (1996)

20.380 S.A. Trushin: Photoacoustic air pollution monitoring with an isotopic CO_2 laser, Ber. Bunsenges. Phys. Chem. Chem. Phys. **96**, 319–322 (1992)

20.381 P.K. Cheo, C. Zhiping, C. Li, Z. Yi: Applications of a tunable CO_2 sideband laser for high-resolution spectroscopic measurements of atmospheric gases, Appl. Opt. **32**, 836–841 (1993)

20.382 D.E. Cooper, T.F. Gallagher: Frequency modulation spectroscopy with a CO_2 laser: results and implications for ultrasensitive point monitoring of the atmosphere, Appl. Opt. **24**, 710–716 (1985)

20.383 W. Meienburg, H. Neckel, J. Wolfrum: In-situ measurement of ammonia concentration in industrial combustion systems, Proc. Combust. Inst. **23**, 231–236 (1990)

20.384 A. Arnold, H. Becker, R. Hemberger, W. Hentschel, W. Ketterle, M. Köllner, W. Meienburg, P. Monkhouse, H. Neckel, M. Schäfer, K.P. Schindler, V. Sick, R. Suntz, J. Wolfrum: Laser in-situ monitoring of combustion processes, Appl. Opt. **29**, 4860–4872 (1990)

20.385 R. Hemberger, S. Muris, K.U. Pleban, J. Wolfrum, I.S. Zaslonko, P. Kilpinen, A.E.J. Akanetuk: An experimental and modeling study of the selective noncatalytic reduction of NO by ammonia in the presence of hydrocarbons, Combust. Flame **99**, 660–668 (1994)

20.386 P. Günter: *Nonlinear Optical Effects and Materials*, Springer Ser. Opt. Sci, Vol. 72 (Springer, Berlin, Heidelberg 2000)

20.387 J.A. Silver: Frequency modulation spectroscopy for trace species detection: Theory and comparison among experimental methods, Appl. Opt. **31**, 707–717 (1992)

20.388 P. Vogel, V. Ebert: Near shot noise detection of oxygen in the A-band with vertical-cavity surface-emitting lasers, Appl. Phys. B **72**, 127–135 (2001)

20.389 M. Tacke: Lead salt lasers. In: *Long Wavelength Infrared Emitters Based on Quantum Wells and Superlattices*, ed. by M. Helm (Gordon+Breach, Amsterdam 2000) pp. 347–396

20.390 R. Grisar, G. Schmidke, M. Tacke, G. Restelli: *Monitoring of Gaseous Pollutants by Tunable Diode Lasers* (Kluwer, Dordrecht 1989)

20.391 R.T. Ku, J.O. Sample, E.D. Hinkley: Long-path monitoring of atmospheric carbon monoxide with a tunable diode laser system, Appl. Opt. **14**, 854–861 (1975)

20.392 S.M. Schoenung, R.K. Hanson: CO and temperature measurements in a flat flame by laser absorption spectroscopy and probe techniques, Combust. Sci. Technol. **24**, 227–237 (1981)

20.393 P.L. Varghese, R.K. Hanson: Collision width measurements of CO in combustion gases using a tunable diode laser, J. Quant. Spectrosc. Radiat. Transfer **26**, 339–347 (1981)

20.394 R.K. Hanson: Absorption spectroscopy in sooting flames using a tunable diode laser, Appl. Opt. **19**, 482–484 (1980)

20.395 M. Tacke, J.R. Meyer, A.R. Adams: Lead-salt lasers, Philos. T. Roy. Soc. A **359**, 547–566 (2001)

20.396 P. Werle: Spectroscopic trace gas analysis using semiconductor diode lasers, Spectrochim. Acta A **52**, 805–822 (1996)

20.397 J.H. Miller, S. Elreedy, B. Ahvazi, F. Woldu, P. Hassanzadeh: Tunable diode-laser measurement of carbon monoxide concentration and temperature in a laminar methane-air diffusion flame, Appl. Opt. **32**, 6082–6089 (1993)

20.398 U. Gustafsson, G. Somesfalean, J. Alnis, S. Svanberg: Frequency-modulation spectroscopy with blue diode lasers, Appl. Opt. **39**, 3774–3780 (2000)

20.399 C. Lin, M. Grau, O.D. and, M.-C. Amann: Low threshold room-temperature continuous-wave operation of 2.24–3.04 μm GaInAsSb/AlGaAsSb quantum-well laser, Appl. Phys. Lett. **84**, 5088–5090 (2004)

20.400 V. Ebert, C. Giesemann, J. Koeth, H. Teichert: New room-temperature 2.3 μm DFB-diode lasers: First spectroscopic characterization and CO-detection. In: *Laser Applications to Chemical and Environmental Analysis OSA TEchnical Digest Series* (Opt. Soc. Am., Washington 2004), Paper TuF-98

20.401 J. Faist, F. Capasso, D.L. Sivco, C. Sirtori, A.L. Hutchinson, A.Y. Cho: Quantum cascade laser, Science **264**, 553–555 (1994)

20.402 E. Schlosser, J. Wolfrum, L. Hildebrandt, H. Seifert, B. Oser, V. Ebert: Diode laser based in situ detection of alkali atoms: development of a new method for determination of residence-time distribution in combustion plants, Appl. Phys. B **75**, 237–247 (2002)

20.403 D. Wandt, M. Laschek, K. Przyklenk, A. Tünnermann, H. Welling: External cavity laser diode with 40 nm continuous tuning range around 825 nm, Opt. Commun. **130**, 81–84 (1996)

20.404 V. Ebert, H. Teichert, C. Giesemann, T. Fernholz, E. Schlosser, P. Strauch, H. Seifert, T. Kolb, J. Wolfrum: Rapid multi-parameter in-situ spectrometer for sensitive detection of CO, O_2, H_2O and temperature in industrial rotary kilns. In: *17. External TECFLAM Seminar Joint Research Project Combustion Control and Simulation*, ed. by W. Meier (AG TECFLAM, Stuttgart 2003) pp. 122–149

20.405 A.R. Awtry, J.W. Fleming, V. Ebert: Simultaneous diode laser-based in-situ measurement of liquid water content and oxygen mole fraction in dense water mist environments, Opt. Lett. **31**, 900–902 (2006)

20.406 T. Fernholz, H. Teichert, V. Ebert: Digital, phase-sensitive detection for in situ diode-laser spectroscopy under rapidly changing transmission conditions, Appl. Phys. B **75**, 229–236 (2002)

20.407 V. Ebert, J. Fitzer, I. Gerstenberg, K.-U. Pleban, H. Pitz, J. Wolfrum, M. Jochem, J. Martin: Simultaneous laser-based in-situ-detection of oxygen and water in a waste incinerator for active combustion control purposes, Proc. Combust. Inst. **27**, 1301–1308 (1998)

20.408 E. Schlosser, T. Fernholz, H. Teichert, V. Ebert: In situ detection of potassium atoms in high-temperature coal-combustion systems using near-infrared-diode lasers, Spectrochim. Acta A **58**, 2347–2359 (2002)

20.409 M. Grabrysch, C. Corsi, F.S. Pavone, M. Inguscio: Simultaneous detection of CO and CO_2 using a semiconductor DFB diode laser at 1.578 μm, Appl. Phys. B **65**, 75–79 (1997)

20.410 D.B. Oh, M.E. Paige, D.S. Bomse: Frequency modulation multiplexing for simultaneous detection of multiple gases by use of wavelength modulation spectroscopy with diode lasers, Appl. Opt. **37**, 2499–2501 (1998)

20.411 L.G. Blevins, B.W. Peterson: Obtaining and interpreting near-infrared wavelength modulation absorption signals from hot fire gases: Practical issues. In: *Combustion Institute Technical Meeting Eastern States Section* (Raleigh, NC 1999) pp. 85–88

20.412 L. G. Blevins, W. M. Pitts: Carbon monoxide measurement using a near-infrared tunable diode laser. In: *Annual Conference on Fire Research: Book of Abstracts* (1998) pp. 21–22

20.413 K. Muta, M. Tanoura, H. Honda: In-situ measurement of CO by tunable diode laser absorption spectroscopy in a large scale waste test furnace. In: *Proc. International Laser Sensing Symposium* (Fukui 1999)

20.414 V. Ebert, K. U. Pleban, J. Wolfrum: In-situ oxygen-monitoring using near-infrared diode lasers and wavelength modulation spectroscopy. In: *Laser Applications to Chemical and Environmental Analysis, OSA Technical Digest Series* (Opt. Soc. Am., Washington, 1998) p. 206–209

20.415 J.J. Nikkari, J.M. Di Iorio, M.J. Thomson: In situ combustion measurements of CO, H_2O and temperature with a 1.58 μm diode laser an two-tone frequency modulation, Appl. Opt. **41**, 446–452 (2002)

20.416 M.G. Allen, B.L. Upschulte, D.M. Sonnenfroh, W.J. Kessler, P.A. Mulhall: Overview of diode laser measurement in large-scale test facilities, Proc. Aerodynamic Measurement Technology and Ground Testing Conference **21**, 2452–2454 (2000)

20.417 D.M. Sonnenfroh, M.G. Allen: Observation of CO and CO_2 Absorption near 1.57 μm with an external-cavity diode laser, Appl. Opt. **36**, 3298–3300 (1997)

20.418 C.E. Wieman, L. Hollberg: Using diode lasers for atomic physics, Rev. Sci. Instrum. **62**, 1–20 (1991)

20.419 J.C. Camparo: The diode laser in atomic physics, Contemp. Phys. **26**, 443–477 (1985)

20.420 L.C. Philippe, R.K. Hanson: Laser diode wavelength-modulation spectroscopy for simultaneous measurement of temperature, pressure, and velocity in shock-heated oxygen flows, Appl. Opt. **32**, 6090–6103 (1993)

20.421 S.T. Sanders, D.W. Mattison, L. Ma, J.B. Jeffries, R.K. Hanson: Wavelength-agile diode laser sensing strategies for monitoring gas properties in optically harsh flows: Application in cesium-seeded pulse detonation engine, Opt. Express **10**, 505–514 (2002)

20.422 M.F. Miller, W.J. Kessler, M.G. Allen: Diode laser-based air mass flux sensor for subsonic aeropropulsion inlets, Appl. Opt. **35**, 4905–4912 (1996)

20.423 Z.C. Owens, D.W. Mattison, E.A. Barbour, C.I. Morris, R.K. Hanson: Flowfield characterization and simulation validation of multiple-geometry PDEs using cesium-based velocimetry, Proc. Combust. Inst. **30**, 2791–2798 (2005)

20.424 E.R. Furlong, D.S. Baer, R.K. Hanson: Combustion control using a multiplexed diode-laser sensor system, Proc. Combust. Inst. **26**, 2851–2858 (1996)

20.425 R.M. Mihalcea, D.S. Baer, R.K. Hanson: Diode laser sensor for measurements of CO, CO_2, and CH_4 in combustion flows, Appl. Opt. **36**, 8745–8752 (1997)

20.426 R.M. Mihalcea, D.S. Baer, R.K. Hanson: Diode-laser absorption sensor system for combustion monitoring and control applications. In: *AIAA Paper* 97-3356 (1997)

20.427 D.S. Baer, R.K. Hanson, M.E. Newfield, N.K.J.M. Gopaul: Multiplexed diode-laser sensor system for simultaneous H_2O, O_2, and temperature measurements, Opt. Lett. **19**, 1900–1902 (1994)

20.428 M.G. Allen: Diode laser absorption sensors for gasdynamic and combustion flows, Meas. Sci. Technol. **9**, 545–562 (1998)

20.429 D. Wandt, M. Laschek, A. Tünnermann, H. Welling: Continuously tunable external-cavity diode laser with a double-grating arrangement, Opt. Lett. **22**, 390–392 (1997)

20.430 A.P. Larson, L. Sandström, S. Höjer, H. Ahlberg, B. Broberg: Evaluation of distributed Bragg reflector lasers for high-sensitivity near-infrared gas analysis, Opt. Eng. **36**, 117–123 (1997)

20.431 X. Zhou, X. Liu, J.B. Jeffries, R.K. Hanson: Diode Laser Sensors for Combustion Control. In: *AIAA Paper* 2003-1010 (2003)

20.432 E.R. Furlong, D.S. Baer, R.K. Hanson: Real-time adaptive combustion control using diode-laser absorption sensors, Proc. Combust. Inst. **27**, 103–111 (1998)

20.433 E.R. Furlong, R.M. Mihalcea, M.E. Webber, D.S. Baer, T.P. Parr, R.K. Hanson: Diode-laser sensor system for closed-loop control of a 50-kW incinerator, AIAA Paper 97-2833 (1997)

20.434 L. Ma, S.T. Sanders, J.B. Jeffries, R.K. Hanson: Monitoring and control of a pulse detonation engine using a diode-laser fuel concentration and temperature sensor, Proc. Combust. Inst. **29**, 161–168 (2002)

20.435 F. Schuler, F. Rampp, J. Martin, J. Wolfrum: TACCOS - a thermography-assisted combustion control system for waste incinerators, Combust. Flame **99**, 431–439 (1994)

20.436 V. Ebert, J. Fitzer, I. Gerstenberg, H. Pitz, K.U. Pleban, J. Wolfrum, M. Jochem, J. Martin: Online monitoring of water vapour with a fiber coupled NIR-diode laser spectrometer, VDI Berichte **1366**, 145–154 (1998)

20.437 J.T.C. Liu, J.B. Jeffries, R.K. Hanson: Diode laser absorption diagnostics for measurements in practical combustion flow fields. In: *AIAA Paper* 2003-4581 (2003)

20.438 F. Greger, K.T. Hartinger, P.B. Monkhouse, J. Wolfrum, H. Baumann, B. Bonn: In situ alkali concentration measurements in a pressurized fluidized bed coal combustor by excimer laser induced fragmentaiton fluorescence, Proc. Combust. Inst. **26**, 3301–3307 (1996)

20.439 K.T. Hartinger, S. Nord, P.B. Monkhouse: Quenching of fluorescence from Na(3^2P) and K(4^2P) Atoms following photodissociation of NaCl and KCl at 193 nm, Appl. Phys. B **64**, 363–367 (1997)

20.440 R.C. Oldenburg, S.L. Baughcum: Photofragment fluorescence as an analytical technique: application to gas-phase alkali chlorides, Anal. Chem. **58**, 1430–1436 (1986)

21. Electrohydrodynamic Systems

In this chapter we review briefly the fundamentals of electrohydrodynamics (EHD), the characteristic EHD dimensionless numbers and the techniques to measure conductivity and electric field, as well as the peculiarities imposed by charging of particles in the classical fluid-mechanical methods for measuring velocity and visualizing fluid flows.

We begin with a brief review of the basic equations, followed by an examination of the physical mechanisms that govern fluid flow through the relevant dimensionless numbers related to electric forces. However, the main emphasis is put on the description of the experimental methods, used to measure the fundamental EHD magnitudes. First, we discuss the basic mechanisms of conductivity, how to measure it, and how to obtain reproducible I–V characteristics. This section also includes a discussion of the techniques to control ion injection. This is followed by a section dedicated to the measurement of mobility. Then we describe the Kerr effect, and how it can be used to measure the electric field in liquids. The last section is dedicated to a description of the difficulties we encounter in the classical techniques of laser Doppler anemometry, and visualization techniques in EHD flows, and how they may be overcome, at least partially. We hope that this chapter will be useful, not only to EHD researchers, but also to practising fluid dynamicists, and to chemical and electrical engineers who need to understand and apply the principles and experimental techniques of EHD in their work.

21.1 **Equations** 1318
    21.1.1 Electrical Equations 1318
    21.1.2 Mechanical Equations 1318
    21.1.3 Temperature Equation 1319

21.2 **Fluid Statics and Dynamics in EHD** 1320
    21.2.1 Charge Decay 1320
    21.2.2 EHD Statics and Dynamics.............. 1321

21.3 **Experimental Methods in EHD** 1322

21.4 **Conductivity** 1323
    21.4.1 Conduction Mechanisms in Liquids . 1323
    21.4.2 Ohmic Versus Non-Ohmic Regime: Practical Estimation of the Transition Voltage 1324
    21.4.3 Unipolar Injection 1325
    21.4.4 Determination of the Conductivity.. 1325

21.5 **Mobility** 1327

21.6 **Electric Field Measurement: Kerr Effect** ... 1328

21.7 **Velocity** 1329
    21.7.1 Laser-Doppler Anemometry........... 1329

21.8 **Visualization** 1330
    21.8.1 Shadowgraph and Schlieren 1330
    21.8.2 Tracer Particles............................ 1331

References 1331

Electrohydrodynamic (EHD) systems are hydrodynamical systems subjected to electric fields. Electrical currents are weak, thus, in contrast to magnetohydrodynamics, the liquids are either good insulators (oils, alcohols, purified water), or weak electrolytes (aqueous solutions of conductivities comprised between 10^{-5} and $1\,\mathrm{S/m}$). Maxwell equations are reduced to the quasi-electrostatic equations. The electrical force term plays a crucial role in Navier–Stokes equation. Also temperature, through the variations induced in mobility, conductivity and permittivity and eventually through Joule heating, plays an important role in electrothermal flows.

21.1 Equations

We present the electrical, mechanical and thermal equations that govern the dynamics of EHD of liquids (for details see [21.1]). We consider a system of characteristic size l and characteristic fluid velocity u_0.

21.1.1 Electrical Equations

Most systems of interest have linear constitutive relations for the displacement vector $\boldsymbol{D} = \epsilon \boldsymbol{E}$, and the field, $\boldsymbol{H} = \boldsymbol{B}/\mu$, where $\boldsymbol{E}$, $\boldsymbol{B}$ are the electric and magnetic field, and ϵ and μ are the permittivity and permeability, respectively. A condition that must be met by an electromagnetic system in order to be quasi-electrostatic is that the magnetic field due to the total current density in the system, regardless of its origin, i.e., injection, dissociation, current due to particles, or displacement current, must satisfy the condition $cB_0 < E_0$, where B_0 and E_0 are the characteristic values of the magnetic and electric field, respectively, and c is the velocity of light in the system. This is equivalent to stating that the electric energy density dominates over the magnetic energy density, $(1/2)\epsilon E^2 \gg (1/2)B^2/\mu$.

The scale for B_0 comes from the fourth Maxwell's equation,

$$\nabla \times \boldsymbol{H} = \boldsymbol{J} + \frac{\partial \boldsymbol{D}}{\partial t} \tag{21.1}$$

and it can originate from the conduction, the displacement current or both. In the first case it is $B_0 \sim \mu l J_0$, and we must have

$$\frac{cB_0}{E_0} \sim \frac{cl\mu J_0}{E_0}, \tag{21.2}$$

and consequently,

$$J_0 \ll E_0 \sqrt{\frac{\epsilon}{\mu}} \frac{1}{l}. \tag{21.3}$$

For typical systems in the laboratory $l \approx 10^{-2}$ m, $\mu \approx \mu_0 \approx 10^{-6}$ H/m and $\epsilon \approx 10^{-11}$ F/m, and we must have $J_0 \ll 3E_0$ in SI units. In insulating liquids this condition is amply satisfied in practical situations as J_0 is in the range of microamperes per square centimeter, and E_0 is several kV per centimeter.

If the displacement current dominates, $B_0 \sim (l/c^2)\omega E_0$, with ω the frequency of the electric field. Then we have

$$\frac{cB_0}{E_0} \sim \frac{l\omega}{c}, \tag{21.4}$$

and we must satisfy $l\omega/c \ll 1$, which is obviously true (for typical systems in the laboratory) for frequencies below microwave frequencies.

Notice that the ratio of the displacement to the conduction current is given by $\epsilon \omega E_0 / J_0$. Therefore for ω less (larger) than $\epsilon J_0 / E_0$ the conduction (displacement) current dominates. For the particular case of an ohmic liquid, $\boldsymbol{J} = \sigma \boldsymbol{E}$, this transition value for ω is σ/ϵ.

The condition $cB_0 \ll E_0$ automatically implies that the electric field is irrotational. In effect,

$$\frac{\partial \boldsymbol{B}/\partial t}{\nabla \times \boldsymbol{E}} \ll \frac{\omega l}{c} \ll 1. \tag{21.5}$$

Therefore, in electrohydrodynamics, Maxwell equations reduce to

$$\nabla \cdot \boldsymbol{D} = q, \quad \nabla \times \boldsymbol{E} = 0, \quad \frac{\partial q}{\partial t} + \nabla \cdot \boldsymbol{J} = 0, \tag{21.6}$$

where q denotes the volume charge density.

We must complement these equations with the jump conditions at interfaces, i.e., the continuity of the tangential component of the electric field, and the conservation equation for surface charge density, which is equal to the jump of the normal component of the displacement, $q_s = [\epsilon E_n]$,

$$\frac{\partial q_s}{\partial t} + u_n \boldsymbol{n} \cdot \nabla q_s + (\nabla \cdot \boldsymbol{n}) u_n q_s$$
$$+ \nabla_s \cdot \boldsymbol{K}_s + \boldsymbol{n} \cdot [\boldsymbol{J}] - u_n [q]$$
$$= 0, \tag{21.7}$$

where $\boldsymbol{n}$ points from phase 1 to phase 2, $\nabla_s = \nabla - \boldsymbol{n}(\boldsymbol{n} \cdot \nabla)$, $\boldsymbol{K}_s$ is the surface current density, and $[f] = f_2 - f_1$ is the jump of f across the surface (for an alternative derivation to the one given in [21.1] see Appendix A in [21.2]).

21.1.2 Mechanical Equations

For fluid velocities much smaller than the velocity of sound, the fluid may be considered incompressible. Then, the mechanical equations reduce to

$$\nabla \cdot \boldsymbol{u} = 0, \tag{21.8}$$

for mass conservation, and to the Navier–Stokes equation for momentum conservation

$$\rho \frac{d\boldsymbol{u}}{dt} = -\nabla p_0 + \eta \nabla^2 \boldsymbol{u} + \boldsymbol{f}_e + \rho \boldsymbol{g}, \tag{21.9}$$

where ρ is the mass density, $\boldsymbol{u}$ the fluid velocity, p_0 the thermodynamic pressure in the absence of the electric field, η the fluid viscosity, f_e the electrical force density, and the last term is the gravitational force per unit volume with g the gravity acceleration.

The electrical force density is $f_e = \nabla \cdot T^e$, with T^e given by

$$T^e = \left\{ -E \cdot D + \int_0^D \left[E - \rho \left(\frac{\partial E}{\partial \rho} \right)_{T,D} \right] dD \right\} + DE. \tag{21.10}$$

For an electrically linear medium the components of T^e are

$$T^e_{ij} = \epsilon E_i E_j - \frac{1}{2}(1-a) E_k E_k \delta_{ij} \tag{21.11}$$

where $a = (\rho/\epsilon)(\partial \epsilon/\partial \rho)_T$ is the electrostriction parameter. In this case we recover the classical expression for the electric force density

$$f_e = \nabla \cdot T^e = qE - \frac{1}{2}E^2 \nabla \epsilon + \nabla \left(\frac{1}{2} \epsilon a E^2 \right). \tag{21.12}$$

The term qE, called the Coulomb force, is the force per unit volume on a medium containing free electric charge. This is the strongest EHD force term and usually dominates when direct-current (DC) electric fields are present. The next term, called the dielectric force, is due to the force exerted on a nonhomogeneous dielectric liquid by an electric field. This is usually weaker than the Coulomb force and only dominates when an alternating-current (AC) electric field with a period much shorter than the charge relaxation time and/or the ionic transit time, is applied. The last term, called the electrostrictive force is the gradient of a scalar and is treated as a modification to the fluid pressure.

21.1.3 Temperature Equation

The derivation of the temperature T equations is lengthy [21.1, 3]. The result is

$$\rho \left\{ c_V - T \frac{D^2}{2\rho} \left[\frac{\partial^2}{\partial T^2} \left(\frac{1}{\epsilon} \right) \right]_\rho \right\} \frac{dT}{dt}$$
$$= \frac{T}{\rho} \left(\frac{\partial p}{\partial T} \right)_{\rho,D} \frac{d\rho}{dt} - \nabla \cdot J'_0 + \Phi$$
$$+ E \cdot J' + T \left[\frac{\partial}{\partial T} \left(\frac{1}{\epsilon} \right) \right]_\rho D \cdot \frac{dD}{dt}, \tag{21.13}$$

where c_V is the specific heat at constant volume, J'_0, J' are the heat flux and current density measured in the rest frame of the fluid element, respectively, and $\Phi = 2\eta e_{ij}^2$ with $e_{ij} = (1/2)[(\partial u_i/\partial x_j) + (\partial u_j/\partial x_i)]$ is the heat generation due to viscous dissipation. If we neglect thermoelectric effects, the heat flux is given by $J'_0 = -\tilde{k} \nabla T$, with $\tilde{k}$ the thermal conductivity in the presence of the electric field. The dependence of the thermal conductivity on the electric field has not been reported in the literature, and is probably negligible, therefore we take $\tilde{k} = k$, where k is the usual thermal conductivity. The equation for temperature may be also written in terms of the pressure p instead of the mass density ρ as the independent variable for the entropy, since most processes in liquids take place at constant pressure [21.3].

For moderate heating the equation of state for the mass density may be approximated by $\rho = \rho_0[1 - \alpha(T - T_0)]$, with ρ_0 being the reference density at temperature T_0 and α the volume expansion coefficient evaluated at the reference temperature. The permittivity ϵ is also assumed to vary linearly with temperature, $\epsilon = \epsilon_0[1 - e_1(T - T_0)]$, with e_1 being the relative derivative of the permittivity with respect to temperature. At ambient temperature $e_1 \approx 10^{-3}$. For frequencies below the microwave region, which are typical in EHD, the molecular polarization is in equilibrium with the field, and dielectric heating is not present. Since under the Boussinesq approximation the fluid properties η, k, c_V and α are assumed to be constants, and the contribution from viscous dissipation is neglected, the temperature equation for processes at constant pressure reduces to

$$\rho_0 c_p \frac{dT}{dt} = k \nabla^2 T + E \cdot J', \tag{21.14}$$

where c_p is the specific heat at constant pressure.

For highly insulating liquids, the currents are of the order of μA or less, and the contribution of Joule heating to the energy equation may be neglected. The energy equation is decoupled from the electrical and mechanical equations. Therefore, the solution depends only upon the boundary and initial conditions imposed on the temperature T. We have

$$\rho c_p \left(\frac{\partial T}{\partial t} + u \cdot \nabla T \right) = k \nabla^2 T. \tag{21.15}$$

However, the temperature cannot be ignored as its distribution affects the dynamic behavior of the system through the buoyancy and electric forces. In addition the temperature also affects the charge density distribution through the variations induced in the permittivity, the conductivity and the mobility. Reference [21.4] has a discussion of these effects in some detail.

21.2 Fluid Statics and Dynamics in EHD

The importance of the principal electrical and mechanical effects depends on the ratio of charge decay and electrical and mechanical time constants [21.5], or equivalently on the pressure or forces ratios of mechanical and electrical origin [21.6], or alternatively through the dimensionless EHD equations [21.1]. In this section we discuss the effect that electrical terms have on the statics and dynamics of fluids, but first we need to discuss the evolution of the charge density.

21.2.1 Charge Decay

The evolution of the charge density is given by

$$\frac{\partial q}{\partial t} + \nabla \cdot \boldsymbol{J} = 0 . \tag{21.16}$$

Ohmic Regime

For liquids in the ohmic regime $\boldsymbol{J} = \sigma \boldsymbol{E} + q\boldsymbol{u}$, where $\boldsymbol{u}$ is the liquid velocity. Then,

$$\frac{\mathrm{d}q}{\mathrm{d}t} + \frac{\sigma}{\epsilon} q = 0 , \tag{21.17}$$

where $\mathrm{d}/\mathrm{d}t = \partial/\partial t + \boldsymbol{u} \cdot \nabla$. The solution is $q(t) = q(0)\mathrm{e}^{-t/\tau_\mathrm{e}}$. Therefore, in the local rest frame of the fluid, the charge relaxes exponentially with a characteristic time $\tau_\mathrm{e} = \epsilon/\sigma$. However, even if the interfaces are initially neutral, charges will accumulate at them because for steady conditions the normal current density has to be continuous, $\sigma_1 E_{\mathrm{n}1} = \sigma_2 E_{\mathrm{n}2}$, which implies that $q_\mathrm{s} = \epsilon_2 E_{\mathrm{n}2} - \epsilon_1 E_{\mathrm{n}1}$ will in general be different form zero. The charging time τ is a combination of the relaxation time of each phase and geometrical factors. An example of the importance of this relaxation time in an EHD problem is the electrified bouncing ball (see, for example [21.7]).

In AC electric fields of period $T = 2\pi/\omega$ much smaller than the relaxation time the charge density at interfaces cannot change, and consequently the average of the Coulomb electric surface forces is equal to zero. Therefore, when we want to avoid the effect of charges the common practice is to use fields of frequency much larger than the charge relaxation time of any of the phases present in the system. For a practical example see [21.8].

Unipolar Injection Regime

Metallic point electrodes, sharp edges, and fine wires submerged in insulating fluids have intense local electric fields at their surfaces because of the enhancing effects of curvature. This originates electron injection, either from the metal to molecules in the fluid or viceversa. As a result, ions are created and move under the influence of the electric field. For the sake of simplicity, consider that only one ionic species, taken to be positive, is injected into a perfectly insulating liquid $\sigma = 0$. The constitutive law for the current density is

$$\boldsymbol{J} = qK\boldsymbol{E} - D\nabla q + q\boldsymbol{u} , \tag{21.18}$$

where $K\boldsymbol{E}$ is the velocity of the ions relative to the fluid, called the drift velocity, and K is the ion mobility.

Let us compare the diffusion term to the drift term. It is

$$\frac{D\nabla q}{qK\boldsymbol{E}} \sim \frac{D}{KV} = \frac{0.025}{V} , \tag{21.19}$$

where we have used Einstein's relation $D/K = k_\mathrm{B} T/e$ (for an ambient temperature of $T \approx 300\,\mathrm{K}$). Consequently, diffusion will be negligible compared to drift unless the potential difference is of the order of 0.025 V. Usually this will only happen in thin layers, either boundary layers or internal layers, where diffusion is required to match the outer or mainstream solutions of the diffusion-free problem. For example, it will be important in the dynamics of the charged double electrical layers in electrolytes. However, for highly insulating liquids diffusion may be safely neglected. Then, the charge conservation equation can be written as

$$\frac{\partial q}{\partial t} + (\boldsymbol{u} + K\boldsymbol{E}) \cdot \nabla q + \frac{K}{\epsilon} q^2 = 0 , \tag{21.20}$$

or equivalently $\mathrm{d}/\mathrm{d}t(1/q) = K/\epsilon$, when we move with the ion velocity $K\boldsymbol{E} + \boldsymbol{u}$. This equation has the exact solution $q = q_0/(1 + t/\tau_\mathrm{C})$ with $\tau_\mathrm{C} = \epsilon/Kq_0$ being the algebraic bulk relaxation time. This solution is valid on $\mathrm{d}\boldsymbol{r}/\mathrm{d}t = K\boldsymbol{E} + \boldsymbol{u}$, so that unless a given element of liquid can be traced via a particle line to a source of charge, it will support no bulk charge density. The algebraic charge relaxation time τ_C depends on the initial level of charge, and after several decay times, the charge density reaches a limiting value ϵ/Kt independent of the initial conditions. It is important to note that charge relaxation due to Coulomb repulsion is much slower than the classical exponential relaxation time. This explains why insulating fluids may accumulate charges easily. A classical example is fuel, to which we must add electrostatic additives that increase their conductivity and facilitate the reduction of their charges in order to avoid electrostatic hazards.

The ratio between Coulomb repulsion and drift, $(K/\epsilon)q^2/K\boldsymbol{E}\cdot\nabla q$, is given by the dimensionless number $C = \tau_\mathrm{i}/\tau_\mathrm{C} = q_0 l^2/\epsilon V$ where $\tau_\mathrm{i} = l^2/KV$ is the transit time, i.e., the time for an ion to cross the typical length l. For low (large) values of C the characteristic time that enters into the problem is the transit time τ_i (Coulomb repulsion time τ_C). Small values of C indicate that the electric field is determined by the external power supply connected to the electrodes in the fluid. On the contrary, for high values of C the electric field is strongly dependent on the charges that are either inside or on the surface of the fluids.

The quotient u/KE is the ratio of the transit time to the mechanical convection time. This ratio is called the electric Reynolds number Re, in analogy to the Reynolds number, which is the ratio of viscous time to inertial time. For high values of this number the charge may be considered to be attached to the liquid, i.e., the liquid behaves like a good insulator. On the contrary, for small values of Re, the liquid may be considered to be a conductor.

Mixed Model

It is quite common to consider the mobility model with a background liquid conductivity

$$\boldsymbol{J} = q K \boldsymbol{E} + \sigma \boldsymbol{E} + q \boldsymbol{u} . \tag{21.21}$$

The ratio of conduction current to mobility current is $\sigma E/Kq E \sim \sigma/Kq_0 = \tau_\mathrm{C}/\tau_\mathrm{e} = C_0$. Large values of C_0 indicate that the liquid is basically a conductor, the opposite being true for $C_0 \ll 1$. Another way to write this important number is to take into account that $q_0 \sim \epsilon V/l$ (from Gauss's law). Then $C_0 = \sigma l^2/K\epsilon V$, which implies that for large systems or low values of the applied voltage the liquid will behave as an ohmic conductor.

21.2.2 EHD Statics and Dynamics

There may exist solutions to the EHD equations that are time independent. The simplest one is the hydrostatic solution $\boldsymbol{u} = 0$. For this solution to be possible, the electric stresses must be normal to any fluid interface present in the system, and in the bulk it must satisfy

$$\nabla \times \left(q\boldsymbol{E} - \frac{1}{2} E^2 \nabla \epsilon + \rho \boldsymbol{g} \right) = 0 . \tag{21.22}$$

The interfacial conditions are always met for perfectly insulating liquids, since for them it is $q_\mathrm{s} = 0$, and for perfect liquid conductors because the electric field is always normal to the interface. As previously indicated, for semi-insulating liquids there is a necessity to use AC fields of high enough frequency if we want to impose zero Coulomb electric force density at the interfaces. In general, these conditions, plus the condition in the bulk, are met only in highly symmetrical geometries (planar, cylindrical or spherical layers). Let us then consider the general case of dynamics.

Interfacial Dynamics

In the classical boundary conditions we have to add, to the jump across the interface of the mechanical stress tensor, the jump of electrical stresses $[\boldsymbol{T}_\mathrm{e} \cdot \boldsymbol{n}]$. The order of magnitude of these forces acting at the interface, is γ/R for capillary forces, where γ is the surface tension and R the minimum radius of curvature; $[\epsilon E_\mathrm{n} E_\mathrm{t}] \sim [\epsilon E_0^2]$ for the electrical terms; $[\rho]gl$ for the gravitational force, where l is the length scale for the interface along the gravity acceleration. Therefore, the system behavior is determined by the following two dimensionless numbers: the Bond number $\mathrm{Bo} \equiv [\rho]gLR/\gamma$, defined by the ratio between the gravitational and the capillary forces; the electric bond number $B_\mathrm{e} \equiv [\epsilon E_0^2]R/\gamma$, which is the analog using the electrical force instead of the gravitational force; and the number $W_\mathrm{e} \equiv [\epsilon E_0^2]/[\rho]gL$, defined by the ratio of the electrical force to the gravitational one. This parameter governs the equilibria when capillary effects are small compared to the electrical and gravitational forces. When only the normal stresses are different from zero and $\mathrm{Re} \gg 1$, the fluid may be considered as ideal, and the electric Bond number and W_e or a combination of both will govern the dynamics of the system. Well-known examples are linear and nonlinear waves and instabilities in horizontal interfaces and jets [21.2, 9]. Viscosity must always be considered when the liquids are neither perfectly conducting nor perfect insulators, since it is the only mechanism that balances the electrical stresses at the interface. The classical Ohnesorge number $\mathrm{Oh} \equiv [\rho\nu]u_0 R/\gamma h_0$, which compares the viscous terms with the capillary terms, enters into the problem.

An example of a steady-state solution is *Quincke's rotor*. This is a solid, highly insulating, cylinder immersed in a viscous, conducting, liquid. An electric field, uniform and of value E_0 at large distances of the cylinder, is imposed perpendicular to its axis. Due to the symmetry of the system, there is an equilibrium state, but this state is unstable. A slight rotation produces an asymmetrical distribution of charge. In this case, there is a net torque on the cylinder and it begins to rotate. Assuming that the system is viscous dominated, the angular velocity may be estimated as

$$\rho v \frac{u_0}{R} \sim \rho v \Omega \sim q_s E_s . \tag{21.23}$$

The tangential electric field is, in this system, of the same order as the external field, whereas the surface charge density can be obtained from the surface charge conservation equation. For steady conditions the latter can be written as $\Omega(\partial q_s/\partial\theta) + [\sigma E_r] = 0$, and from here, the order of magnitude for q_s is $q_s \sim (\sigma E_0)/\Omega$. Finally, we obtain

$$\Omega^2 \sim \frac{\sigma E_0^2}{\rho v} = \frac{1}{t_{\text{ev}} \tau_{\text{e}}} \tag{21.24}$$

where $t_{\text{ev}} = \rho v/\epsilon E_0^2$ is the electroviscous relaxation time (see below), and $\tau_{\text{e}} = \sigma/\epsilon$ is the charge decay time. For corn oil ($\epsilon \approx 3.1 \times 10^{-11}$, $\sigma \approx 5 \times 10^{-11}$, $\rho v \approx 5 \times 10^{-2}$ in SI units) and an applied electric field around $1\,\text{kV/cm}$, the resulting rotation period is of the order of one second. This phenomenon has been observed in practice, and is responsible for the rotation of small particles immersed in insulating liquids and subjected to DC fields [21.10–12].

Bulk Dynamics

The dynamics is governed by Navier–Stokes equation. In addition to the classical nondimensional numbers in fluid mechanics, we have to consider new electromechanical parameters defined as the ratios of the Coulomb force, or dielectric force, to the pressure, viscous or inertial forces. These numbers may be written in a variety of ways depending on the physical mechanisms that determine the velocity field, the charge density and the gradient of permittivity. Since $q_0 \sim \epsilon E/l$, the ratio of the Coulomb term to the dielectric term of the electric force is given by $\epsilon/(l\nabla\epsilon)$. In general, the Coulomb force is typically much larger than the dielectric force; for example, when the variation of the permittivity is induced by a temperature gradient it is $\nabla\epsilon = 10^{-3}\Delta T \epsilon$, and therefore this ratio is equal to $10^3/l\Delta T$. For temperature differences, which are typically of a few degrees Celsius, and for laboratory systems this ratio is much greater than one. In general, only for those situations for which q_0 is negligible, and $\nabla\epsilon$ different from zero, is the dielectric force the dominant electric force.

Consider the transient motion of a liquid driven solely by the electrical forces. In the viscous regime $\eta \nabla^2 \boldsymbol{u} \sim f_{\text{e}}$. Equating both terms, we find a natural time scale $\tau_{\text{ev}} = \eta/\epsilon E^2$, called the electroviscous scale, which plays a significant role in these transient viscous regimes. This same ratio tells us that for steady conditions the typical scale for the velocity is $u_0 = \epsilon V^2/(\eta l)$.

In inertial flows, the most frequently encountered in nature or in industrial processes, the acceleration term is comparable to the electric force. In other terms, the electrical energy is converted to kinetic energy of the liquid. This implies that $(1/2)\epsilon E^2 \simeq (1/2)\rho u_0^2$, and consequently that the typical liquid velocity is given by $u_0 \simeq (\epsilon/\rho)^{1/2} E_0 = K_{\text{h}} E_0$. The liquid velocity is proportional to the electric field, and the constant $K_{\text{h}} = (\epsilon/\rho)^{1/2}$ is called the hydrodynamic mobility. It is normal to define a dimensionless number $M = (\epsilon/\rho)^{1/2}/K$ as the ratio of two mobilities, the hydrodynamic mobility, and K, the true ionic mobility. Taking into account *Walden's rule*, i.e., $K\eta \sim$ const., we see that M is proportional to viscosity. Since $u \simeq MKE$ and for liquids M may vary from 4 to 400, ions are entrained by the motion of the liquid, contrary to the case of gases where M is less than one. For charged liquids, the time of flight of ions may be much larger than the algebraic charge relaxation time or the ionic transit time, and the final charge distribution may be strongly affected by the fluid motion. As a consequence, the electric field is modified, which in turn changes the fluid velocity, thus making these types of problems particularly difficult. For further discussion of the effect of electric forces on the dynamics of the system see the references cited at the beginning of this section.

Electrothermal flows are those flows due to the variations in the mobility K, conductivity σ, and permittivity ϵ due to a thermal gradient, either externally imposed or originating through Joule heating. These variations give rise to variations in the charge density, and therefore in the electrical force, thus affecting fluid flow. For the underlying physics of these subtle flows we refer the reader to [21.4] for the case of insulating liquids, and to [21.13] for the case of weak electrolytes in microfluidics.

21.3 Experimental Methods in EHD

The experimental study of EHD systems is besieged by several fundamental difficulties. One is the extreme difficulty of determining the nature of ions and to predict their density distributions due to chemical complexity. The result is that the constitutive equations for the current density are rarely known. Therefore, we must work with simplified models. Well-known examples of such models are the conduction and the mobility models previously discussed. Another important problem is the difficulty of measuring charge density, which is highly desirable as electrical forces depend on it. A third difficulty is avoiding or minimizing charging effects on particles used as tracers, as well as electrothermal flows when we use Schlieren techniques.

The determination of the nature of ions has been partially solved in gases using spectrography. In liquids the unavoidable presence of minute traces of impurities that have a huge effect on conductivity makes the problem untractable, unless

1. we dissolve a known salt or a compound, such that the dominant ions are known,
2. we inject massively one type of ion, or
3. we keep the liquid recirculating through a electrodialysis cell.

Let us discuss then how we model conduction, injection and the current density and the problems encountered in its measurement. Then, we will discuss the determination of the electric field, and finally the measurement of the velocity field and its visualization.

21.4 Conductivity

The experimental determination of the conductivity of low-conductivity liquids is one of the most challenging issues that researchers in electrohydrodynamics have to confront. The purpose of this section is to describe the difficulties often encountered in measuring the conductivity and how they can be overcome, or, at least, their effects reduced.

21.4.1 Conduction Mechanisms in Liquids

Low-conductivity liquids, such as hydrocarbons, do not self-ionize, and the charge carriers are ions originating from the spontaneous dissociation of tiny amounts of ionizable substances present in the liquid. In the case of practical grade liquids the ionizable substances spring from the unavoidable impurities. It is, however, a common practice to add some ionophores in order to increase and stabilize the conductivity of these liquids.

For a weak electrolyte AB dissolved in a liquid the following chemical equilibrium determines the number of dissociated ions:

$$AB \rightleftharpoons A^+ + B^- . \tag{21.25}$$

There is a constant associated with the forward direction, the dissociation constant K_D, whereas for the backward reaction the constant is the recombination constant K_R. The dissociation constant can be expressed as (see [21.1] and references therein)

$$K_D = 3e \frac{(K_+ + K_-)}{4\pi\epsilon a^3} e^{-U/k_B T} \tag{21.26}$$

and the recombination constant as

$$K_R = e \frac{(K_+ + K_-)}{\epsilon} \frac{1}{1 - e^{-U/k_B T}} \tag{21.27}$$

where $U = e^2/(4\pi\epsilon a)$, a is the minimum distance of approach between the centers of the ions, and K_+ and K_- are the mobilities of the cations and anions, respectively.

The number of ions in equilibrium is obtained from the balance

$$K_R n_+ n_- = K_D c \tag{21.28}$$

where c is the concentration of salt, n_+ is the concentration of positive ions and n_- is the concentration of negative ions. Hence

$$n_+ = n_- = n_0 = \sqrt{\frac{K_D c}{K_R}} = A e^{-e^2/(4\pi\epsilon a k_B T)} . \tag{21.29}$$

This number is very sensitive to the value of the dielectric constant ϵ, which appears in the argument of the exponential. This fact explains the dependence of the conductivity on the dielectric constant.

If the applied electric field is not very high, this equilibrium is not perturbed and determines the number of charge carriers. Also, the liquid volume remains electrically neutral and the electric field acting on the charge carrier is the applied field. Therefore the regime of conduction is ohmic and the current density becomes

$$j = e(n_+ K_+ + n_- K_-)E = \sigma E \tag{21.30}$$

where σ is the conductivity

$$\sigma = e(K_+ + K_-)n_0 . \qquad (21.31)$$

Dependence of the Conductivity on the Dielectric Constant

The effect of the dielectric constant on the equilibrium (21.25) is determined by its role on the exponential of (21.29). For low dielectric constants the exponential is small and so is the density of charge carriers. In this way the density of charge carriers is an increasing, and very sensitive, function of the dielectric constant. For example, a change in ϵ from 2 to 20 results in an increase of 10 orders of magnitude in n_0. The physical origin of this effect is the screening of the electric field of an ion by the molecular dipoles of the liquid. This screening is almost absent in low-polar liquids, and the ion pairs are tightly bound in that case.

Equation (21.29) allows one to obtain an empirical law for the dependence of the conductivity on the dielectric constant [21.14]:

$$\sigma \propto n_0 \propto e^{-e^2/(4\pi\epsilon a k_B T)} \rightarrow \log \sigma = A - \frac{B}{\epsilon} , \qquad (21.32)$$

where A and B are constants. This correlation fits a large number of experimental data reasonably well (Fig. 21.1).

Dependence of the Conductivity on the Concentration of Solute

For a weak electrolyte in a low-polar liquid, (21.29) and (21.31) yield a dependence of σ on c of the type

$$\sigma \propto \sqrt{c} . \qquad (21.33)$$

And this is the case for a great number of salts when added to a low-polar liquid such as tri-isoamyl-ammonium picrate (TIAP) or tetra-butyl-ammonium chloride (TBA) in hydrocarbons.

However, the addition of surfactants (for example, AOT, di-2-ethylhexyl sulfosuccinate) may give place to another kind of dependence [21.15]. In effect, surfactants are compounds whose molecules have a long non-polar tail (usually an alkyl radical) and a polar head. When surfactants are dissolved in a non-polar medium they tend to form inverse micelles. These are groups of molecules, formed from 10–20 molecules, with their polar heads oriented towards the center of the group and the non-polar tails pointing towards the solvent volume. The environment in the core of the micelle is a polar one, and it is a convenient place for an ion that otherwise could not be stable in the volume of the solvent.

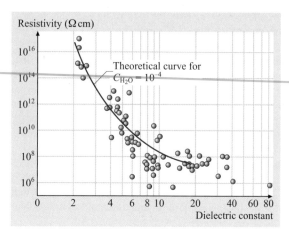

Fig. 21.1 Resistivity as a function of dielectric constant for different liquids (after [21.14])

The formation of micelles, which takes place above a certain concentration referred to as the critical micellar concentration, helps to increase the conductivity. However, in this case the kinetics of charge carrier formation is different to the case of bare ions (21.25). The equilibrium is replaced by

$$M + M \rightarrow M^+ + M^- \qquad (21.34)$$

because the anion and the cation reside in the interior of their respective micelles. If c is the concentration of the surfactant, the conductivity is in this case proportional to c:

$$\sigma \propto c , \qquad (21.35)$$

which is the same behavior as that expected for a strong electrolyte in an aqueous media.

21.4.2 Ohmic Versus Non-Ohmic Regime: Practical Estimation of the Transition Voltage

The ohmic regime, in which the electric current density is proportional to the applied electric field, is established only if the electric field scarcely perturbs the thermodynamic equilibrium between the charge carriers and the neutral molecules that originate them. Associated to this thermodynamic equilibrium there is a characteristic time τ_t, which is the time needed to recover the equilibrium after the perturbation. This thermodynamic time is one half the electric relaxation time of the liquid $\tau_e = \epsilon/\sigma$ [21.1]. On the other hand the transit time of a charge carrier between two electrodes a distance d apart is $\tau_i = l^2/(KV)$. This is also the time a local perturbation of the equilibrium needs to propagate across the gap between the electrodes.

The voltage above which the ohmic regime is no longer expected is estimated putting $C_0 = 1$, i.e., equating the time of transit and the electric relaxation time:

$$\frac{\epsilon}{\sigma} = \frac{l^2}{KV}. \tag{21.36}$$

This gives

$$V_s = \frac{\sigma l^2}{K\epsilon} \tag{21.37}$$

for the transition voltage, known as the saturation voltage.

Well above the saturation voltage, the ions are swept from the volume at a rate that prevents their recombination. The current is then limited by the rate at which the ions are created in the liquid volume and becomes independent of the applied voltage. This current is the saturation current and its value is

$$j_s = eK_D cd. \tag{21.38}$$

However, this is not the end of the story and at even higher voltages the field-enhanced dissociation [21.16] gives place to an increase in the current that is a nonlinear function of the applied voltage.

The notion of saturation voltages deserves two commentaries. First, the transition between ohmic and non-ohmic regime is a continuous one, from a clear linear relation between the electric current and the applied voltage to a nonlinear relation that depends on the kind of process that dominates the electric conduction above saturation. There is, in general, no discontinuity in the experimental current–voltage characteristics and no unambiguous way of assigning a definite voltage to the transition.

Secondly, if the voltage is applied during very short periods of time, compared to the dielectric relaxation time, the instantaneously measured current should correspond to the stationary ohmic value. This is due to the fact that the equilibrium has not yet been appreciable disturbed in that case.

21.4.3 Unipolar Injection

One of the most important mechanisms of charge transport above the saturation voltage is charge injection. This is also one of the most common process for originating space charge in the volume of insulating liquids. Under non-controlled circumstances charge injection is erratic, non-steady and unipolar or bipolar. There are several techniques to control injection.

The electron injection technique [21.17] uses an electron beam in a vacuum chamber. The electron beam sweeps the liquid surface in both directions analogously to a television set. The voltage of a grid above the liquid surface enables one to measure the surface voltage. This technique has been successfully applied to the study of EHD instabilities. However it only works on a free surface and in a vacuum.

Corona discharge from a needle or a blade can be used to induce injection over a free liquid surface. The injection in the liquid is strong, whether the corona current is space charge limited or not, because the mobility of ions in liquids is much smaller than the mobility of ions in air. The major drawbacks of this technique are that there is no direct way to measure the voltage on the liquid surface and that the injection is not uniform. The major advantage is that it is very easy to implement [21.18].

Ion exchange membranes are widely used in electrodialysis. They are also of interest for injection experiments. Covering the electrodes with ion exchange membranes creates a reservoir of ions suitable for injection when the electric field is applied. In this way a strong injection is usually obtained [21.19, 20].

The use of additives increases the conductivity of the liquid, but it may help to induce and stabilize injection. For example, the salt tri-isoamyl-ammonium picrate (TIAP) has often been used as such an additive. The level of injection achieved with this technique is low to moderate [21.21]. The addition of iodine with electrodes made from titanium has also turned out to be a good technique to obtain stable and reproducible injection [21.22].

Finally sharp electrodes immersed in the liquid may cause stable injection at high voltages without the use of any additive. Electrohydrodynamic plumes emanating from the sharp electrode have been observed in this way [21.23].

21.4.4 Determination of the Conductivity

Current–Voltage Characteristics

The most widely used method to measure the conductivity of low-conducting liquids is to obtain a current–voltage characteristic and estimate the conductivity from the linear part of the curve. For very low-conducting liquids the gap between the electrodes has to be small enough in order to have a detectable current for moderate applied voltages. For example a liquid of conductivity of the order of 10^{-12} S/m will yield a current of the order of picoamperes when subjected to a 100 V difference across a gap of 1 mm and 25 cm^2 of electrode surface. For this reason a very sensitive electrometer or picoammeter is required.

Concerning the cell design several precautions must be taken into account [21.25]:

- Since the liquid to be tested may have a very low conductivity the electric field is not totally confined to the liquid volume. This is the reason why a guard ring is desirable. This ensures that all the measured current originated in a region of uniform field.
- It is convenient to have a variable gap between the electrodes. One of the problems one usually confronts when measuring the conductivity is to assure that the conduction regime is ohmic; only then it is meaningful to extract a value of conductivity. In the ohmic regime, the current must decrease as the inverse of the electrode gap. This provides an additional criterium, besides the linearity with the applied voltage, to decide whether the ohmic regime is established or not. On the other hand a variable electrode gap facilitates mobility measurements, as we will see below.
- Materials are also very important. Chemical reaction at the electrodes may be the source of additional charge carriers. Although stainless steel is widely used as a material for the electrodes it may be interesting to have a set of electrodes made from different materials. Also the insulating parts are of concern. Although Teflon has often been used, because of its outstanding insulating properties, its porosity makes it unsuitable for parts in contact with the liquid, as it can be a reservoir for impurities.

After several current–voltage curves have been taken, and the correct dependence with voltage and distance have been found, the estimated conductivity should be used to determined from (21.37) the saturation voltage, confirming that one has been working well below this voltage. Figure 21.2 shows a typical current–voltage characteristics where a linear part is distinguishable at low voltages. After the saturation voltage, I increases faster than ohmically, a fingerprint of an injection process.

Commercially Available Apparatuses

There are commercially available apparatuses that work properly up to conductivities of the order of 10^{-8} S/m. However no one goes beyond this value. To the knowl-

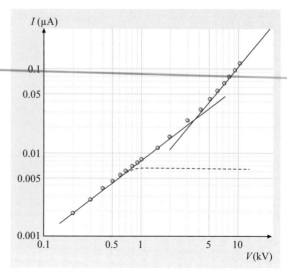

Fig. 21.2 I–V characteristics. From the linear part the conductivity can be estimated. The horizontal line corresponds to the saturation voltage. The steeper slope above the saturation voltage is characteristic of an injection process (after [21.24])

edge of the authors, the only commercially available conductivimeter is the one manufactured by IRLAB based on a method developed in Grenoble by *Tobazéon* et al. [21.26, 27].

This apparatus exploits the fact that the current measured just after the voltage has been established corresponds to the ohmic value, provided the time at which the measure is achieved is small compared to the transit time of ions. The liquid is placed between two concentric cylindrical electrodes and subjected to a square-wave voltage of amplitude 10 V and frequency 0.5 Hz. The conduction current is measured during a short time interval during each period and averaged over a suitable number of periods. In this way their builders claim that the apparatus is able to measure conductivities as low as 10^{-15} S/m.

Some Additional Drawbacks

On top of the already commented difficulties some additional problems are encountered when measuring the conductivity of low-conducting liquids. The two major problems are stability and reproducibility.

By stability we mean the constancy of the measured value in time. Usually this is not the case and the conductivity of the liquid may vary in time by as much as an order of magnitude. The reasons for this behavior are to be found in the sensitivity of the conductivity to the presence of impurities. Even a small amount of substance can change the conductivity notably.

In addition, the obtained values are not reproducible. This is also related to the chemical composition of the sample. Only a very careful purification of the liquid, followed by complete control of the additives and their purity can give a reproducible value.

We believe that, in general, for practical insulating-grade liquids it is impossible to give a value of the conductivity with less than 30% error.

21.5 Mobility

The mobility of charge carriers is another of the important physical properties in electrohydrodynamics. Two methods, closely related, are of practical importance in the measurement of mobility.

The first method is the so-called time-of-flight method [21.29, 30] (Chap. 3.7.3). This method is based on the measurement of the time the ions need to cover the distance between two electrodes. In [21.30] the ions are injected from a point and directed towards the collecting electrode through two consecutive grids. Initially the voltage of the different electrodes, the two grids, the collector and the injector, are arranged in such a way that the ions go from the injector to the collector and a constant current is established in the collector circuit. Then the voltage is switched in such a way that no more ions enter the drift space, i.e., the space between the second grid and the collector, whereas the voltage difference between this second grid and the collector is maintained. The time elapsed from switching off the voltage until the extinction of the collector current is the time of flight of the ions from the grid to the collector. From this time the velocity of the ions, and from it the mobility, can be calculated.

The second method makes use of the current transient observed after a step voltage is applied between two parallel electrodes. In principle two types of transient

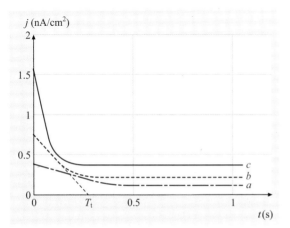

Fig. 21.3 Transient current after a step voltage has been applied. The voltage is greater than the saturation voltage and there is no other source of ions except dissociation in volume (after [21.15])

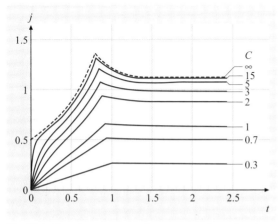

Fig. 21.4 Transient current after a step voltage has been applied when injection is taking place from one of the electrodes (after [21.28])

Table 21.1 The mobility of several ions in different solvents

	K (m^2/Vs)
TIAP in cyclohexane	1.3×10^{-8}
Bu$_4$NPi in cyclohexane	1.4×10^{-8}
TIClO$_4$ in cyclohexane	1.2×10^{-8}
DAP in cyclohexane	1.05×10^{-8}
TIAP in hexane	4×10^{-8}
TIAP in benzene	3.3×10^{-8}
AOT in cyclohexane	6×10^{-9}
CrAc in cyclohexane	3×10^{-9}

TIAP: tri-isoamyl-ammonium picrate, Bu$_4$NPi: tetra-butyl-ammonium picrate, TIClO$_4$: tri-isoamyl-ammonium perchlorate, DAP: dodecylamine propionate, AOT: sodium di-2-ethylhexyl sulfosuccinate, CrAc: alkyl-salicylate chromate

are expected if the applied voltage is above saturation. Figure 21.3 is typical of a volume conduction above saturation: the current is a decreasing function of time that reaches a stationary value corresponding to the saturation current. In this case the time of transit τ_i is the time of flight of the ions, $\tau_i = l^2/(KV)$. Figure 21.4 is typical of an injection process [21.28]. In this case it is the time at which the maximum is reached that corresponds to the time of flight of ions. This peak is well defined only for strong injection and cannot be well defined for weak injections, resulting in an ambiguous measurement.

The major drawback of this second method is that often the transit currents are not easy to interpret and do not fall clearly into one of the types mentioned. This can be due to various reasons. First it is possible that more than one type of carrier dominates the conduction process, producing a blurring of the curves. Secondly other process may be taking part: field-enhanced conduction and bipolar injection, for instances. In order to guarantee that what it is measured is a mobility it is very convenient to make use of a measurement cell with a variable gap. In this way the correct dependence of the time of flight on the voltage and the distance can be tested.

Table 21.1 shows the value of the mobility for various ions in several solvents. This table has been elaborated from different sources, mainly [21.21, 31, 32]. Another table can be found in Chap. 3.7.3 (Table 3.33) of this handbook.

21.6 Electric Field Measurement: Kerr Effect

To our knowledge, charge density in liquids has only been indirectly determined through measurements of the electric field. To measure the latter, we use the fact that many dielectric liquids become birefringent when stressed by a high electric field so that an incident linearly or circularly polarized light propagating through the liquid becomes elliptically polarized. This electro-optic effect, known as the Kerr effect, depends quadratically on the transversal components of the electric field. If we denote by E the applied electric field, then

$$\Delta n = (n_\parallel - n_\perp)\lambda B E_T^2, \quad (21.39)$$

where $n_\parallel$ and $n_\perp$ are the refraction indexes along the parallel and perpendicular directions of the electric field, respectively, λ is the free-space wavelength of light, E_T is the component of the electric field in the plane perpendicular to the direction of light, and B is the Kerr constant. Table 21.2 gives the value of the Kerr constant for several liquids.

Suppose that the electric field is constant and we know its direction. Then, with a source of white light, and the help of a polarizer, we may select a linear wave propagating in the direction perpendicular to the electric field, with electric vector e, making an angle α with E. After crossing the cell of length l, the phase shift between the extraordinary (parallel) and the ordinary (transversal) rays is given by $\phi = 2\pi B l E^2$, and the wave is elliptically polarized. Placing an analyzer that only lets linearly polarized waves with electric field vector forming an angle β with E pass, the ratio of intensities is given by

$$\frac{I}{I_0} = \cos^2(\alpha - \beta) - \sin 2\alpha \sin 2\beta \sin^2 \frac{\phi}{2}. \quad (21.40)$$

For a polarizer–analyzer set such that $(\alpha - \beta) = \pi/2$ the first term vanishes. Choosing the polarizer such that $\alpha = \pi/4$ ($\beta = 3\pi/4$) or $\alpha = 3\pi/4$ ($\beta = \pi/4$) the phase shift ratio of the transmitted to the incident light is

$$\frac{I}{I_0} = \sin^2 \frac{\phi}{2} \quad (21.41)$$

Table 21.2 List of Kerr constants for various liquids ($\lambda = 5890$ Å and $20\,°C$) (from [21.33, 34])

Liquid	Kerr constant B (10^{-14} m/V^2)	Dielectric constant
Carbon disulfide	3.59	2.6
Carbon tetrachloride	0.0823	2.2
Benzene	0.434	2.3
Acetone	18.13	20.7
Phenyl iso-thiocyanate	101.3	11.5
Nitrobenzene	362.8	36.1
n-Ethyl acetamide	445.1	135
Ethylene carbonate[1]	166.9	89.6
Propylene carbonate[2]	111.3	69
o-Dichlorobenzene	47.4	7.5
p-Chlorotoluene	25.6	6.4
Quinoline	16.7	9
Ethyl salicylate	21.8	8.6

[1] $45\,°C$, $\lambda = 5460$ Å
[2] $23\,°C$, $\lambda = 5460$ Å

with the result that whenever $\phi = 2n\pi$ with $n = 1, 2 \ldots$, or equivalently, $E^2 = n/Bl$ we have extinction of the transmitted light. If we monotonously increase the voltage we observe successive extinctions of the light, which allows us to determine the electric field. This extinction will be global if the electric field is uniform, or local otherwise. This technique has been developed and successfully applied to EHD systems [21.32, 35–37].

21.7 Velocity

The measurement of velocity in electrohydrodynamic flows has special peculiarities that make it more difficult than in hydrodynamic flows. To start with some of the standard techniques, like hot-wire or hot-plate anemometry, are not of use because the probes will not sustain the high electric fields that are always present in EHD flows and would modify the field distribution.

21.7.1 Laser-Doppler Anemometry

In laser-Doppler anemometry a laser beam is split into two beams that are made to interfere inside the liquid. The region of measurement is the volume where the fringes of interference are clearly distinguished. When a particle crosses the fringes it scatters light with a modulated amplitude. The frequency of this modulation is the velocity of the particle divided by the separation between the fringes. A photomultiplier detects the scattered light and an electronic circuit extracts the frequency of the modulation. The signal-to-noise ratio is high if the particle diameter is of the order of the fringe separation, typically of the order of 1 μm. For a more rigorous and detailed description see [21.38].

Some successes have been obtained using laser-Doppler anemometry [21.39–41]. However the common practice of seeding the liquid with particles, unavoidable to produce a sufficient amount of scattered light from the volume of measurement, has to be analyzed in detail. Two requirements, closely related, must be fulfilled: the particles should follow the liquid flow, that is, their drift velocity with respect to the liquid must be negligible. On the other hand the seeding particles should not alter the electric field distribution, otherwise their presence will modify the flow structure.

Let us analyze the factors that could induce a drift velocity

1. Common to all hydrodynamic flows is the ability of the particles to respond to a step change in the liquid flow. This is determined by the time scale [21.42]:

$$\tau_1 = \frac{2a^2 \rho_p}{9\eta}, \qquad (21.42)$$

where a is the radius of the particle, ρ_p its density, and η the liquid viscosity. This time scale is obtained by balancing the viscous and inertial force on an accelerated particle. For particles smaller than 1 μm in size this time is less than 1 μs, and therefore does not represent any limitation.

2. If the density of the particles differs from that of the liquid sedimentation takes place. The typical velocity of sedimentation is

$$u = \frac{2(\rho_p - \rho)a^2 g}{9\eta} \qquad (21.43)$$

which is of order of 1 μm/s for particles in the micrometer range [21.42]. As we have mentioned the velocity of electrohydrodynamic flows is of order of $\sqrt{\epsilon/\rho}E$, where the factor $\sqrt{\epsilon/\rho}$ is called the electrohydrodynamic mobility and is of order 10^{-7} m^2 V/s. For an electric field of order 10^6 V/m the velocity due to the EHD flow is well above the sedimentation of particles.

3. The major concern for stationary flows is the possible charge of the particles. The motion of charged particles in electric fields is referred to as electrophoresis. The electrophoretic mobility of charged particles is of order of $\epsilon \zeta/\eta$, where ζ is the zeta potential [21.43]. This mobility is to be compared with the electrohydrodynamic mobility. For zeta potentials of the order of 100 mV and liquids of viscosity 1 cP, the electrophoretic mobility is of order 10^{-9} m^2 V/s, whereas the electrohydrodynamic mobility is of order 10^{-7} m^2 V/s. Although this may not be a problem in regions where the flow is vigorous and well established, it may limit the use of laser-Doppler anemometry in EHD flows near the onset of the flow or in certain regions.

4. If the particles are uncharged, the dielectric force may come into play. In the case of a spherical particle this force is [21.44]

$$F = 2\pi^3 \epsilon \, \mathrm{Re}[K(\omega)] \nabla^2 E, \qquad (21.44)$$

where $K(\omega)$ is a function of the frequency and is not very different from one. From the balance of this force and the viscous drag the mobility of the particle is estimated as

$$\frac{a^2 \epsilon E}{\eta l} \,, \qquad (21.45)$$

where l is a typical length scale for the variation of the electric field (typically the distance between the electrodes). Taking 10^6 V/m for the electric field, 1 mm for l and $\eta \approx 10^{-3}$ the dielectric force leads to a mobility of order 10^{-11} m^2 V/s for a 1 μm particle. Therefore this effect is negligible in the presence of other effects, although it may be noticeable for uncharged particles.

Particles made from PMMA (poly-methyl methacrylate), TiO$_2$ or glass have been used successfully by various authors. However it is important to remember, especially in the case of PMMA or other latex particles, that the presence of functional groups in the particle surface and of additives in the liquid may drastically change the particle charge.

Besides the use of laser-Doppler anemometry to measure the liquid velocity, this technique has been used with success for the determination of the electrophoretic mobility of colloidal particles [21.45]. Indeed some commercially available apparatuses for evaluating the mobility and size of colloidal particle make use of this technique.

21.8 Visualization

Along with velocimetry, flow visualization is sometimes very useful in determining the flow structure. Let us briefly recall some of the most common techniques of flow visualization used in electrohydrodynamics.

21.8.1 Shadowgraph and Schlieren

In shadowgraph and Schlieren visualization the variation of the index of refraction determines the light pattern. In the shadowgraph technique a parallel beam of light crosses the sample and is observed or photographed on a screen. The differences in index of refraction induce different curvatures of the light rays, thus producing regions of shadow or brightness on the screen. In the Schlieren technique the beam that emanates from the sample is focused by a lens and a knife edge placed at the focal point cuts off half of the beam. This procedure increases the contrast on the screen. Actually in the Schlieren technique the intensity of light on the screen depends on the gradient of the index of refraction of the corresponding zone, whereas in shadowgraph the intensity depends on the second derivative of the index of refraction. For this reason, the Schlieren method is usually preferred (details can be found in [21.46]).

In EHD flows the gradient of the index of refraction has to be induced externally. Usually this is achieved by heating one of the electrodes. Precautions have to be taken in order not to induce noticeable natural convection or electrothermal flows. For natural convection the velocity induced by a temperature difference ΔT may be estimated from the balance between the inertial term and the buoyancy term in the Navier–Stokes equation:

$$\rho \frac{u^2}{l} \sim \rho \beta \Delta T g \qquad (21.46)$$

which yields

$$u = (g \beta l \Delta T)^{1/2} \,, \qquad (21.47)$$

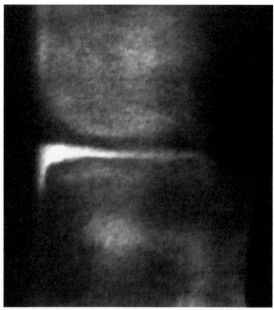

Fig. 21.5 Schlieren photograph of an EHD plume in silicone oil. The distance between the blade and the plane electrodes is 16.4 mm. The applied tension is 24 kV

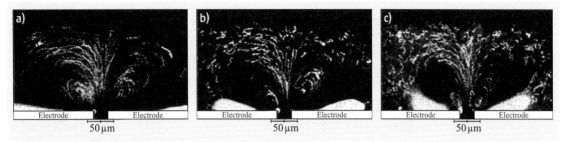

Fig. 21.6 Visualization of an EHD flow between microelectrodes using fluorescent particles (after 21.47)

where β is the coefficient of volumetric thermal expansion. Taking $\beta = 10^{-4}\,\text{K}^{-1}$ and $l = 1$ mm the velocity estimated is less than 0.01 m/s for differences of temperature of the order of 1 K. This values is below the expected velocities for EHD flows. Problems arise in large containers, where the velocity induced by natural convection may be non-negligible.

Schlieren techniques have been successful in the study of EHD plumes [21.23, 48]. Figure 21.5 shows a EHD plume originating from a blade at 24 kV [21.23]. The liquid is silicone oil of viscosity $5 \times 10^{-5}\,\text{m}^2/\text{s}$ and the distance between the blade and the opposite electrode is 16.8 mm.

21.8.2 Tracer Particles

The use of particles for visualization in EHD flows suffers from the same problems as seeding in laser-Doppler anemometry, which have already been discussed. However the use of fluorescent particles has opened a new field in flow visualization. The light from fluorescent particles is visible through optical microscopes even at magnifications at which the particles themselves are not visible. This fact allows the use of smaller particles, thus assuring that the particles follow the flow.

Fluorescent particles are typically made from a polymer, such as polystyrene, with a fluorescent dye added. There is a huge variety of fluorescent dyes that excite and emit at different wavelengths. Fluorescent visualization usually requires a microscope with suitable illumination corresponding to the wavelength that excites the dye, and filters that select the wavelength emitted by the particles. Figure 21.6 shows the visualization of flow in a microdevice using latex spheres, 557 nm in diameter, that excite in the blue region of the spectrum and emit green light [21.47].

References

21.1 A. Castellanos (Ed.): *Electrohydrodyamics* (Springer, Berlin, Heidelberg 1998), Chaps. 1–4
21.2 A. Castellanos, A. González: Nonlinear hydrodynamics of free surfaces, IEEE T. Dielect. El. In. **5**, 334–343 (1998)
21.3 A. Castellanos: Entropy production and the temperature equation in electrohydrodynamics, IEEE T. Dielect. El. In. **10**, 22–26 (2003)
21.4 F. Pontiga, A. Castellanos: Physical mechanisms of instability in a liquid layer subjected to an electric field and a thermal gradient, Phys. Fluids A **6**, 1684–1701 (1994)
21.5 J.R. Melcher: *Continuum Electromechanics* (MIT Press, Cambridge 1981)
21.6 J.M. Crowley: Dimensionless ratios in electrohydrodynamics. In: *Handbook of Electrostatics*, ed. by J.S. Chang, J.M. Crowley, A.J. Kelly (Marcel-Dekker, New York 1995) pp. 99–119
21.7 A. Khayari, A.T. Pérez: Nonlinear dynamics of a bouncing ball driven by electric forces, Int. J. Bifurcat. Chaos **13**(10), 2959–2975 (2003)
21.8 A. Ramos, H. González, A. Castellanos: Experiments on dielectric liquid bridges subjected to electric fields, Phys. Fluids A **6**, 3206–3208 (1994)
21.9 J.R. Melcher: *Field-Coupled Surface Waves* (MIT Press, Cambridge, MA 1963)
21.10 L. Lobry, E. Lemaire: Viscosity decrease induced by a DC electric field in a suspension, J. Electrostat. **47**(1–2), 61–69 (1999)
21.11 E. Lemaire, L. Lobry: Reverse electrorheological effect: A suspension of colloidal motors, Int. J. Mod. Phys. B **15**(6–7), 780–787 (2001)
21.12 E. Lemaire, L. Lobry: Chaotic behavior in electrorotation, Physica A **314**(1–4), 663–671 (2002)
21.13 A. Castellanos, A. Ramos, A. González, N.G. Green, H. Morgan: Electrohydrodynamics and dielectrophoresis in microsystems: scaling laws, J. Phys. D **36**, 2584–2597 (2003)

21.14 P.K. Watson: Space charged limited currents in liquid dielectrics. In: *Electrohydrodyamics*, ed. by A. Castellanos (Springer, Berlin, Heidelberg 1998), Chap. 10

21.15 A. Denat, B. Gosse, J.P. Gosse: Electrical conduction of solutions of an ionic surfactant in hydrocarbons, J. Electrostat. **12**, 197–205 (1982)

21.16 L. Onsager: Deviations from Ohm's law in weak electrolites, J. Chem. Phys. 2, 599–615 (1934), in The collected works of Lars Onsager (World Scientific, Singapore 1996)

21.17 P.K. Watson, T.M. Clancy: Electron injection technique for investigating processes in insulating liquids and solids, Rev. Sci. Instrum. **36**(2), 217–222 (1965)

21.18 F. Vega, A.T. Pérez: Corona-induced electrohydrodynamic instabilities in low conducting liquids, Exp. Fluids **34**, 726–735 (2003)

21.19 R. Tobazéon: Sur l'electrodialyse de liquides non polaires ou faiblement polaires, C. R. Acad. Sci. **282**, C153–C156 (1976)

21.20 N.J. Félici, R. Tobazéon: Charge carrier elimination and production by electrodyalitic polymers in contact with dielectric liquids, J. Electrostat. **11**, 135–161 (1981)

21.21 A. Denat, B. Gosse, J.P. Gosse: Ion injections in hydrocarbons, J. Electrostat. **7**, 220–225 (1979)

21.22 A.I. Zhakin: Classic theories of ion recombination and dissociation in liquids. In: *Electrohydrodyamics*, ed. by A. Castellanos (Springer, Berlin, Heidelberg 1998), Chap. 6

21.23 A.T. Pérez, P.A. Vázquez, A. Castellanos: Dynamics and linear stability of charged jets in dielectric liquids, IEEE T. Ind. Appl. **31**(4), 761–767 (1995)

21.24 F.MJ. McCluskey, P. Atten, A.T. Pérez: Heat Transfer enhancement by electroconvection resulting from an injected space charge between parallel plates, Int. J. Heat Mass Tran. **34**(9), 2237–2250 (1991)

21.25 M. Medrano, A.T. Pérez, C. Soria-Hoyo: Design of a conductivity meter for highly insulating liquids, J. Phys. D: Appl. Phys. **40**, 1477–1482 (2007)

21.26 M. Hilaire, C. Marteau, R. Tobazéon: Apparatus developed for measurement of the resistivity of highly insulating liquids, IEEE T. Electr. Insul. **23**(4), 779–787 (1988)

21.27 R. Tobazéon, J.C. Filippini, C. Marteau: On the measurement of the conductivity of highly insulating liquids, IEEE T. Dielect. El. In. **1**(6), 1000–1004 (1994)

21.28 P. Atten, J.P. Gosse: Transient of one-carrier injections in polar liquids, J. Chem. Phys. **51**(7), 2804–2811 (1969)

21.29 H.T. Davis, S.A. Rice, L. Meyer: On the kinetic theory of simple dense fluids – XI. Experimental and theoretical studies of positive ion mobility in liquid Ar, Kr, and Xe, J. Chem. Phys. **37**(5), 947–956 (1962)

21.30 B.L. Henson: Mobility of positive ions in liquefied argon and nitrogen, Phys. Rev. **135**(4A), A1002–A1008 (1964)

21.31 A. Alj, J.P. Gosse, B. Gosse, A. Denat, M. Nemancha: Influence de la nature du surfactant ionique sur la conduction électrique de ses solutions dans le cyclohexane, Rev. Phys. Appl. **22**, 1043–1053 (1987)

21.32 A. Denat: *Etude de la conduction eléctrique dans les solvants non polaires*, Ph.D. Thesis (Université Joseph-Fourier, Grenoble 1982)

21.33 D.E. Gray (Ed.): *American Institute of Physics Handbook* (McGraw-Hill, New York 1957)

21.34 J.C. Lacroix: *Etude et realisation d'une cellule de Kerr a champ electrique tournant*, Ph.D. Thesis (Université Joseph-Fourier, Grenoble 1970)

21.35 R. Tobazéon: *Etude du transfert convectif de charges electriques par un jet de liquide isolant et application a la generation de tensions elevés*, Ph.D. Thesis (Université Joseph-Fourier, Grenoble 1973)

21.36 M. Zahn: Optical, electrical and electromechanical meaurement methodologies of field, charge and polarization in dielectrics, IEEE T. Dielect. El. In. **5**, 627–650 (1998)

21.37 D. Filipovic, P. Osmokrovic, Z. Lazarevic: Electro-optical Kerr effect in liquid dielectrics, Mater. Sci. Forum **413**, 197–200 (2002)

21.38 L.E. Drain: *The Laser Doppler Technique* (Wiley, New York 1980)

21.39 R. Disselnkotter, K. Barner: Seeding of electrohydrodynamic convection flows for light scattering experiments, J. Electrostat. **19**, 323–336 (1987)

21.40 F.MJ. McCluskey, A.T. Pérez: The electrohydrodynamic plume between a line source of ions and a flat plate, IEEE T. Electr. Insul. **27**(2), 334–341 (1992)

21.41 P. Atten, M. Haidara: Electrical conduction and EHD motion of dielectric liquids in a knife-plane electrode assembly, IEEE T Electr. Insul. **20**(2), 187–198 (1985)

21.42 R.J. Adrian: Laser velocimetry. In: *Fluid Mechanics Measurements*, ed. by R.J. Goldstein (Springer, Berlin, Heidelberg 1983)

21.43 R.J. Hunter: *Foundations of Colloid Science*, 2nd edn. (Oxford Univ. Press, Oxford 2001)

21.44 T.B. Jones: *Electromechanics of Particles* (Cambridge Univ. Press, New York 1991)

21.45 J.P. Dalbiez, K. Tabti, P.J. Derian, M. Drifford: Vélocimétrie Doppler sous champ électrique: technique et application à l'étude de la mobilité électrophorétique des colloïdes et des polyélectorlytes, Rev. Phys. Appl. **22**, 1013–1024 (1987)

21.46 R.J. Goldstein: Optical systems for flow measurement: Shadowgraph, Schlieren and interferometric techniques. In: *Fluid Mechanics Measurements*, ed. by R.J. Goldstein (Springer, Berlin, Heidelberg 1983)

21.47 N.G. Green, A. Ramos, A. González, H. Morgan, A. Castellanos: Fluid flow induced by nonuniform

AC electric fields in electrolytes on microelectrodes – III. Observation of streamlines and numerical simulation, Phys. Rev. E **66**, 026305–026315 (2002)

21.48 B. Malraison, P. Atten, A.T. Pérez: Panaches chargés résultant de l'injection d'ions dans un liquide isolant par une lame ou une pointe placée en face d'un plan, J. Phys. III **4**, 75–85 (1994)

Part D Analysis and Post-Processing of Data

22 Review of Some Fundamentals of Data Processing
Holger Nobach, Göttingen, Germany
Cameron Tropea, Darmstadt, Germany
Laurent Cordier, Poitiers, France
Jean-Paul Bonnet, Poitiers, France
Joël Delville, Poitiers, France
Jacques Lewalle, Syracuse, USA
Marie Farge, Paris, France
Kai Schneider, Marseille, France
Ronald J. Adrian, Tempe, USA

23 Fundamentals of Data Processing
Holger Nobach, Göttingen, Germany
Cameron Tropea, Darmstadt, Germany

24 Data Acquisition by Imaging Detectors
Bernd Jähne, Heidelberg, Germany

25 Data Analysis
Bernd Jähne, Heidelberg, Germany
Michael Klar, Heidelberg, Germany
Markus Jehle, Heidelberg, Germany

This final part of the Springer Handbook is a reference source about signal and data processing techniques commonly encountered in fluid mechanics. These topics have been complemented by a section discussing data acquisition by imaging detectors, a topic becoming increasingly important for optical measurement techniques. These are all subjects that in their development are not naturally associated with fluid mechanics; hence **Part D** attempts to collect information from many diverse sources and present it conveniently to the fluid mechanics researcher. Topics covered in this part include fundamental topics of signal- and data-processing transforms (Fourier, Hilbert, wavelet, etc.), proper orthogonal decomposition and stochastic estimation. This is followed by a discussion of estimator expectation and variance and the influence of noise on these quantities. The Cramér–Rao lower bound (CRLB) is introduced and developed for several common signal processing examples from fluid mechanics. Imaging detectors and measures of their performance are then discussed in detail before closing with a chapter on image processing and motion analysis, two topics especially relevant to the particle image velocity (PIV) measurement technique.

22. Review of Some Fundamentals of Data Processing

This chapter is devoted to reviewing some fundamental transforms and analysis procedures commonly used for both signal and data processing in fluid mechanics measurements. The chapter begins with a brief review of the Fourier transform and its digital counterpart the discrete Fourier transform. In particular its use for estimating power spectral density is discussed in detail. This is followed by an introduction of the correlation function and its relation to the Fourier transform. The Hilbert transform completes the introductory topics. The chapter then turns to a rigorous presentation of the proper orthogonal decomposition (POD) in the context of the approximation theory and as an application of singular value decomposition (SVD). The relationship between POD and SVD is discussed and POD is described in a statistical setting using an averaging operation for use with turbulent flows. The different POD approaches are briefly introduced, whereby the main differences between the classical POD and the snapshot POD are highlighted. This section closes with a presentation of the POD as a generalization of the classical Fourier analysis to inhomogeneous directions. The chapter continues with a discussion of conditional averages and stochastic estimation as a means of studying coherent structures in turbulent flows before moving in a final section to a comprehensive discussion of wavelets as a combination of data processing in time and frequency domain. After first introducing the continuous wavelet transform and orthogonal wavelet transform their application in experimental fluid mechanics is illustrated through numerous examples.

22.1	**Fourier Transform**	1337
22.2	**Correlation Function**	1342
22.3	**Hilbert Transform**	1344
22.4	**Proper Orthogonal Decomposition: POD**	1346
	22.4.1 Basics	1346
	22.4.2 POD: An Approximation Method	1347
	22.4.3 The Proper Orthogonal Decomposition (POD)	1352
	22.4.4 The Different POD Approaches	1357
	22.4.5 Classical POD or Direct Method	1359
	22.4.6 Snapshot POD	1360
	22.4.7 Common Properties of the Two POD Approaches	1361
	22.4.8 POD and Harmonic Analysis	1362
	22.4.9 Typical Applications to Fluid Mechanics	1364
	22.4.10 POD Galerkin	1365
	22.4.11 Evaluative Summary of the POD Approach	1369
22.5	**Conditional Averages and Stochastic Estimation**	1370
	22.5.1 Conditional Averages	1370
	22.5.2 Stochastic Estimation	1373
22.6	**Wavelet Transforms**	1378
	22.6.1 Introduction to Wavelets	1378
	22.6.2 Continuous Wavelet Transform	1378
	22.6.3 Orthogonal Wavelet Transform	1383
	22.6.4 Applications in Experimental Fluid Mechanics	1387
References		1395

22.1 Fourier Transform

The Fourier transform (FT) is an integral transform with orthogonal sinusoidal basis functions of different frequencies. The result represents the frequency spectrum of the signal. Depending on the characteristics of the original (time) signal, different variants of the transform are defined.

A continuous periodic signal $x(t) = x(t+T)$ with the period T can be decomposed into an infinite series of sinusoidal functions (*Fourier series*), whose linear combination reproduces the original function

$$x(t) = x(t+T)$$

$$= \sum_{k=-\infty}^{\infty} a_k \cos\left(\frac{2\pi kt}{T}\right) + \sum_{k=-\infty}^{\infty} b_k \sin\left(\frac{2\pi kt}{T}\right). \quad (22.1)$$

The Fourier coefficients a_k and b_k are given by

$$a_k = \frac{1}{T} \int_0^T x(t) \cos\left(\frac{2\pi kt}{T}\right) dt,$$

$$b_k = \frac{1}{T} \int_0^T x(t) \sin\left(\frac{2\pi kt}{T}\right) dt. \quad (22.2)$$

Using

$$e^{ix} = \cos(x) + i\sin(x), \quad (22.3)$$

the Fourier coefficients a_k and b_k can be combined to a complex value $\underline{c}_k$ and the Fourier series can be extended easily for complex function $\underline{x}(t)$ yielding

$$\underline{x}(t) = \underline{x}(t+T) = \sum_{k=-\infty}^{\infty} \underline{c}_k \exp\left(\frac{2\pi ikt}{T}\right) \quad (22.4)$$

with

$$\underline{c}_k = \frac{1}{T} \int_0^T \underline{x}(t) \exp\left(-\frac{2\pi ikt}{T}\right) dt. \quad (22.5)$$

Note that, even for a real function $x(t)$, the Fourier coefficients are complex values. The real part is associated with the cosine function and the imaginary part with the sine function. Thus, the real part represents contributions to the signal, which are symmetric about zero and the imaginary part describes the asymmetric contributions.

By using $T \to \infty$, for a continuous complex signal $\underline{x}(t)$ with finite energy content the superposition of the Fourier series becomes

$$\underline{x}(t) = \int_{-\infty}^{\infty} \underline{X}(f) \exp(2\pi i f t) df \quad (22.6)$$

with

$$\underline{X}(f) = \int_{-\infty}^{\infty} \underline{x}(t) \exp(-2\pi i f t) dt. \quad (22.7)$$

Signal with finite energy fulfill $\int_{-\infty}^{\infty} |\underline{x}(t)| dt < \infty$. This implies that the signal is nonperiodic.

The result of the decomposition $\underline{X}(f)$ is called the *continuous Fourier transform* (CFT). It is a continuous, infinite, non-periodic, complex frequency spectrum, which fulfills the Plancherel theorem

$$\int_{-\infty}^{\infty} |\underline{X}(f)|^2 df = \int_{-\infty}^{\infty} |\underline{x}(t)|^2 dt, \quad (22.8)$$

indicating the conservation of energy by the Fourier transform. Note that upper- and lower-case notation will be used for the frequency and time domains, respectively.

A finite series of complex values $\underline{x}_n = \underline{x}(t = n\Delta t_s)$ with $n = 0, 1, \ldots, N-1$, sampled at equal time intervals and over the time duration $0 \leq t < T = N\Delta t_s$ can be decomposed into a finite sum of complex Fourier coefficients $\underline{X}_k$, yielding the discrete Fourier transform (DFT). The DFT is defined as

$$\underline{X}_k = \underline{X}(f = k\Delta f)$$

$$= \mathrm{FT}(\underline{x}_n) = \sum_{n=0}^{N-1} \underline{x}_n \exp\left(-i\frac{2\pi nk}{N}\right)$$

$$k = 0, 1, \ldots, (N-1) \quad (22.9)$$

and its inverse transform as

$$\underline{x}_n = \mathrm{FT}^{-1}\{\underline{X}_k\} = \frac{1}{N} \sum_{k=0}^{N-1} \underline{X}_k \exp\left(+i\frac{2\pi nk}{N}\right),$$

$$n = 0, 1, \ldots, (N-1), \quad (22.10)$$

where n is the data sample index at time intervals of Δt_s and with the corresponding sample frequency of f_s. The spectral coefficients are computed for the equally spaced frequencies given by

$$f_k = \frac{k}{N\Delta t_s} = \frac{kf_s}{N}, \quad k = 0, 1, \ldots, (N-1). \quad (22.11)$$

The frequency spacing of the resulting Fourier coefficients is therefore

$$\Delta f_s = \frac{1}{N\Delta t_s} = \frac{1}{T} = \frac{f_s}{N}. \quad (22.12)$$

This is also the lowest frequency that can be resolved.

The power spectral density (PSD) is given by the squared magnitude of the spectral coefficients

$$S_k = S(f = f_k) = \frac{1}{Nf_s} \underline{X}_k^* \underline{X}_k = \frac{1}{Nf_s} |\underline{X}_k|^2,$$

$$k = 0, 1, \ldots, (N-1). \quad (22.13)$$

This function represents the distribution of the total signal power between the frequencies 0 and f_s. The term density is used because power per frequency bandwidth Δf_s is being considered. The PSD is symmetric about $k = N/2$ and has a periodicity every N samples. Therefore, an alternative representation is the use of negative and positive frequencies. For this case, all values of $k \geq N/2$ are interpreted as negative frequency values and the spectrum is symmetric about $k = 0$. In this case, the function is known as the two-sided spectrum. The one-sided PSD simply considers the symmetry of the two-sided spectrum around 0 and yields the spectral distribution between $k = 0$ and $k = N/2$. It is given by

$$G_k = G(f = f_k) = \frac{2}{Nf_s} X_k^* X_k = \frac{2}{N^2 \Delta f_s} X_k^* X_k,$$

$$k = 0, 1, \ldots, \frac{N}{2}. \qquad (22.14)$$

Note that the total power of this spectrum is not the same because $f = 0$ and $f = f_s/2$ are also doubled.

The maximum resolvable frequency is half the sampling frequency $f_{max} = f_s/2 = f_{N/2}$ (the Nyquist

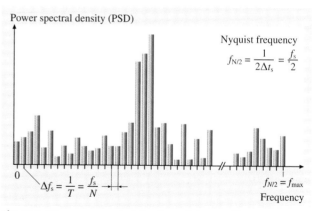

Fig. 22.1 The power spectral density and the sampling parameters

frequency) and the resolution is determined by the data set duration $\Delta f_s = 1/T$. Graphically the PSD and the parameters involved in computing it are shown in Fig. 22.1.

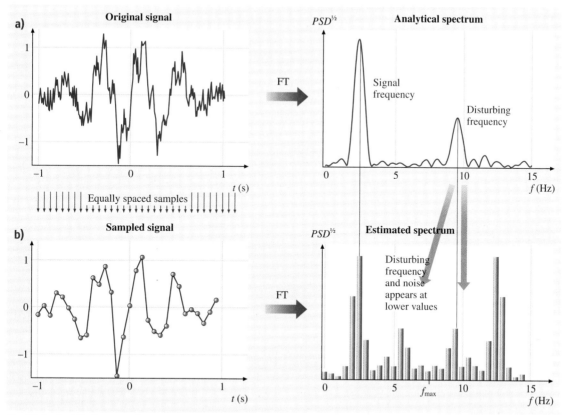

Fig. 22.2a,b Aliasing error in a spectrum due to signal frequencies occurring above the Nyquist frequency. (**a**) Original signal and spectrum, (**b**) sampled signal and falsified spectrum

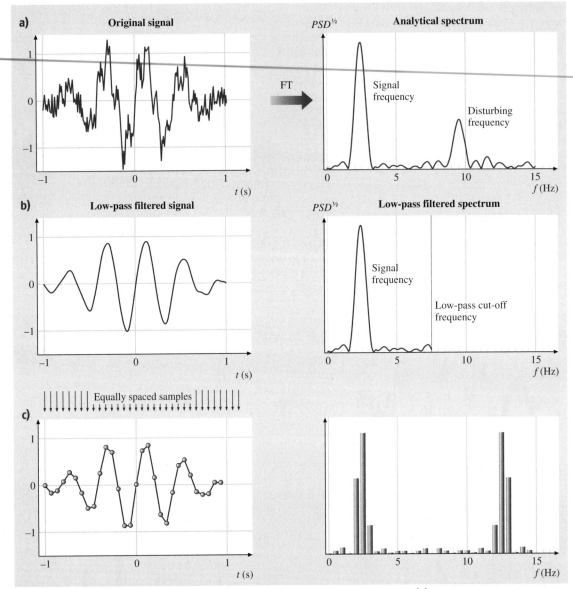

Fig. 22.3a–c Elimination of the aliasing error by use of a low-pass, anti-aliasing filter. (**a**) Original signal and spectrum, (**b**) low-pass filtered signal and spectrum, (**c**) sampled signal and non-aliased spectrum

Two properties of the DFT deserve particular attention. Since the time between the sample points is not infinitely small, the power in the signal at frequencies above f_{max} will appear in the PSD at lower frequencies, an effect known as aliasing. This falsifies the spectrum at the lower frequencies. An example of aliasing is given in Fig. 22.2. The signal in Fig. 22.2a contains two frequency peaks at 2.4 Hz and 9.5 Hz, as can be seen in the correct spectrum illustrated in Fig. 22.2a. By sampling the signal at 15 Hz, the maximum resolvable frequency is 7.5 Hz and thus, the Nyquist criterion is not fulfilled for the signal power at 9.5 Hz. The spectral portion above 7.5 Hz is mirrored about the Nyquist frequency and results in an additional peak at 5.5 Hz (Fig. 22.2b). Furthermore the signal noise at frequencies above f_{max} also increases the noise level at frequencies below f_{max}.

Aliasing errors in estimates of PSD are avoided by applying an analog anti-aliasing, low-pass filter with a sharp cut-off at half the sampling frequency. This procedure is illustrated in Fig. 22.3, using the same signal as used in Fig. 22.2. Before sampling the signal, a low-pass filter removes the frequencies higher than the Nyquist frequency f_{max} (Fig. 22.3b). The spectrum of the filtered and sampled signal in Fig. 22.3c contains no additional frequency peak. Furthermore, the noise level is reduced to the same as in the original signal in Fig. 22.3a.

Besides the periodicity after every N samples of the DFT in (22.9) and the PSD in (22.13), the inverse DFT in (22.10) also has a period of N samples. This effectively means that the DFT perceives and acts on an infinite juxtaposition of the input data record and the inverse DFT effectively transforms an infinite juxtaposition of the spectrum. This is illustrated in Fig. 22.4 for a time series.

If the beginning and end of the record do not merge smoothly into one another, sudden amplitude jumps are perceived, which give rise to additional frequency components in the spectrum. These *end effects* are unimportant for records of long duration; however, they deserve attention with short records. These effects are diminished by applying window functions in the time domain. Window functions scale the input data amplitude and force a tapering to zero at the beginning and end of the signal [22.1].

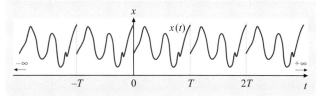

Fig. 22.4 Implicit periodicity of acquired signal when processing using the finite-length DFT

A further consequence of a finite input record duration is spectral broadening. A spectrum of an infinitely long sine wave is a delta function at the signal frequency. A finite-length sine wave yields however a broadened peak, in which the peak width is inversely proportional to the input signal duration. This process is graphically illustrated in Fig. 22.5. The spectrum of an infinite sine wave is a delta function at the signal frequency (Fig. 22.5a). A finite duration sine wave can be viewed as the product of an infinite sine wave with a rectangular window of duration T (Fig. 22.5b). The spectrum of

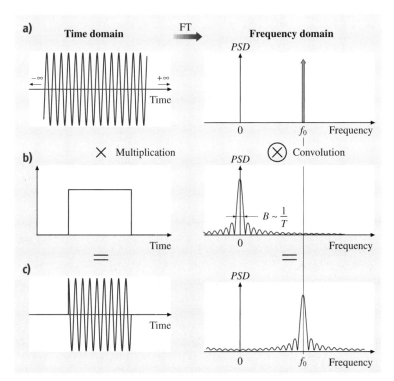

Fig. 22.5a–c A multiplication of two signals in the time domain is equivalent to a convolution in the frequency domain. This can be used to explain spectral broadening due to finite record lengths. (**a**) Infinite sine function and related spectrum, (**b**) rectangular function and related spectrum, (**c**) finite sine function and related spectrum

the finite sine wave will therefore be the convolution of the delta function with the magnitude of a sinc function, the transform of a rectangular window (Fig. 22.5c).

This can be easily illustrated using the following relations. If a signal $y(t)$ is given in the time domain as the product of two other signals, $x(t)$ and $h(t)$

$$y(t) = x(t)h(t) \tag{22.15}$$

then the Fourier transform of $y(t)$ is given by the convolution of the Fourier transforms of $x(t)$ and $h(t)$ [22.2].

$$\underline{Y}(f) = \underline{X}(f) \otimes \underline{H}(f) = \int_{-\infty}^{\infty} \underline{X}(\alpha) \underline{H}(f - \alpha) \, d\alpha \, . \tag{22.16}$$

The power spectral density of $y(t)$ is then

$$S_k = \frac{1}{Nf_s} \underline{Y}_k^* \underline{Y}_k \, , \quad k = 0, 1, \ldots, (N-1) \, . \tag{22.17}$$

An obvious consequence of spectral broadening is that the resolution of distinct signal frequencies in the PSD can be improved by sampling a longer portion of the signal.

In practical implementations of the DFT, (22.9) is not used directly but rather a recursive form known as the fast Fourier transform (FFT) is used. There are many realizations of the FFT, but they share one feature in common, namely, that they normally operate on 2^n points: sample records are restricted to values such as 16, 32, 64, 128, ... The calculation time of the DFT implemented with (22.9) increases with N^2. The FFT algorithm reduces the computation time to the order of $N \log N$. Algorithms exist for FFTs using other record lengths, especially prime number decompositions; however, these are not in widespread use.

A commonly used technique with the FFT is that of zero padding. Without changing the spectral content of the signal, zero padding forces the FFT algorithm to estimate the spectrum at additional frequencies between zero and $f_{\max}$, thus improving the resolution. This is easily seen by examining a signal doubled in length by adding zeros. Instead of (22.9) the transform becomes

$$\underline{X}_k = \sum_{n=0}^{2N-1} \underline{x}_n \exp\left(-\mathrm{i}\frac{2\pi nk}{2N}\right),$$

$$k = 0, 1, \ldots, (2N-1) \, . \tag{22.18}$$

However, since $\underline{x}_n = 0$ for $n = N, N+1, \ldots, (2N-1)$, this can be written as

$$\underline{X}_k = \sum_{n=0}^{N-1} \underline{x}_n \exp\left(-\mathrm{i}\frac{2\pi n \, (k/2)}{N}\right),$$

$$k = 0, 1, \ldots, (2N-1) \tag{22.19}$$

which is identical to the N-point transform for every other k value. However now $\underline{X}_k$ is also computed at intermediate k values. The spectral content of the signal has in no way been altered, but with the intermediate estimates, interpolation of peak locations can be improved. Zero padding can also be used to extend input data records up to a length of 2^n values, in preparation for an FFT.

22.2 Correlation Function

The (temporal) autocorrelation function of a signal $x(t)$ is defined as

$$R(\tau) = E[x(t)x(t+\tau)] \, . \tag{22.20}$$

It is a symmetric function. Principally, the information available in spectral domain is also available in the correlation domain, since the autocorrelation function $R(\tau)$ forms a Fourier transform pair with the power spectral density (Wiener–Khinchine relation).

$$S(f) = \mathrm{FT}\{R(\tau)\} = \int_{-\infty}^{\infty} R(\tau) \, d\tau \, , \tag{22.21}$$

$$R(\tau) = \mathrm{FT}^{-1}\{S(f)\} = \int_{-\infty}^{\infty} S(f) \, df \, . \tag{22.22}$$

In digital form this can be expressed as

$$R_n = R(\tau = n\Delta\tau) = \frac{f_s}{N} \sum_{k=0}^{N-1} S_k \exp\left(+\mathrm{i}\frac{2\pi nk}{N}\right),$$

$$n = 0, 1, \ldots, N-1 \, , \tag{22.23}$$

$$S_k = S(f = f_k) = \frac{1}{f_s} \sum_{n=0}^{N-1} R_n \exp\left(-\mathrm{i}\frac{2\pi kn}{N}\right),$$

$$k = 0, 1, \ldots, N-1 \, , \tag{22.24}$$

where $\Delta\tau = \Delta t_s$ is the time lag interval. The autocorrelation function is by definition symmetric about $\tau = 0$, as the auto spectral density is about $f = 0$, yielding the alternative expressions

$$R_n = R(\tau = n\Delta\tau)$$
$$= \frac{f_s}{2N} \sum_{k=-N/2}^{N/2-1} G_{|k|} \exp\left(+\mathrm{i}\frac{2\pi nk}{N}\right)$$
$$= \frac{f_s}{2N}\left[G_0 + (-1)^n G_{N/2} \right.$$
$$\left. + 2\sum_{k=1}^{N/2-1} G_k \cos\left(\frac{2\pi nk}{N}\right)\right],$$
$$n = 0, 1, \ldots, \frac{N}{2}, \tag{22.25}$$
$$G_k = G(f = f_k)$$
$$= \frac{2}{f_s} \sum_{n=-N/2}^{N/2-1} R_{|n|} \exp\left(-\mathrm{i}\frac{2\pi kn}{N}\right)$$
$$= \frac{2}{f_s}\left[R_0 + (-1)^k R_{N/2} \right.$$
$$\left. + 2\sum_{n=1}^{N/2-1} R_n \cos\left(-\frac{2\pi kn}{N}\right)\right],$$
$$k = 0, 1, \ldots, \frac{N}{2}. \tag{22.26}$$

With the mean removed, the autocorrelation function is known as the autocovariance function; however, these two terms will be used interchangeably, always assuming a mean-free input signal.

A computation of R_n using the FFT, first to compute the PSD and then to transform to the correlation domain, exhibits a speed advantage that increases with increasing data record length $N \log N$ compared to N^2 for a direct calculation of the correlation function shown below. However, there are some subtle drawbacks of using the estimate of (22.25). The most important of these is the so-called wrap-around error [22.2], which has its origins in the finite-length DFT, (22.9). The inherent periodicity in time, which is implied by (22.9) and illustrated in Fig. 22.4, essentially means that the correlation function computed according to (22.25) assumes an infinite juxtaposition of the input signal in time. The derived autocorrelation function will also be based on this assumption and is, therefore, known as the circular autocorrelation. This error is avoided by first padding the input signal with zeros at the beginning and end of the original signal to double its length.

The autocorrelation function will exhibit a periodicity at the same period as the original signal. For instance, the autocorrelation of an infinite sine wave will be an

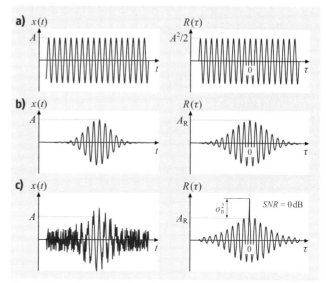

Fig. 22.6a–c Input signal and autocorrelation function. **(a)** Sine wave, **(b)** noise-free Doppler signal, **(c)** noisy Doppler signal

infinite cosine wave, as illustrated in Fig. 22.6a. Thus, the signal frequency can be estimated by measuring the elapsed time over one or more zero crossings of the autocorrelation function (period timing).

The autocorrelation of a Gaussian-windowed sine wave centered around $t = 0$, as shown in Fig. 22.6b, will yield as a correlation function a cosine wave with an amplitude decay directly related to the window width. Of particular interest is the effect of signal noise on the correlation function. As illustrated in Fig. 22.6c, the contribution of signal noise can be found entirely in the first coefficient of the autocorrelation function, i.e., at $\tau = 0$. This is because the signal noise has no inherent time scale, meaning that it is completely random and not correlated with itself over any length of time. This last property of the autocorrelation function is particularly interesting for signal processing, because it provides a means of separating the noise effects from the signal, thus, improving the estimation of signal frequency and other signal parameters.

The correlation function can also be computed directly. The estimator

$$R_n = R(n\Delta\tau) = \frac{1}{N} \sum_{i=0}^{N-1} x_i x_{i+n},$$
$$n = 0, 1, \ldots, (N-1) \tag{22.27}$$

with periodic boundary conditions ($x_{n\pm N} = x_n$) yields the same estimate of the correlation function as obtained with the FFT. Due to the assumption of the periodicity of the signal, it has systematic errors, as described above. The estimator

$$R_n = R(n\Delta\tau) = \frac{1}{N} \sum_{i=0}^{N-n-1} x_i x_{i+n} ,$$

$$n = 0, 1, \ldots, (N-1) \qquad (22.28)$$

is equivalent to the FFT-based estimation with zero padding (doubling the signal length). This estimator has systematic errors due to the decreasing number of products in the sum for increasing time lags $n\Delta\tau$, while the sum is always divided by N. Dividing the sum by the number of products in the sum yields the estimator

$$R_n = R(n\Delta\tau) = \frac{1}{N-n} \sum_{i=0}^{N-n-1} x_i x_{i+n} ,$$

$$n = 0, 1, \ldots, (N-1) , \qquad (22.29)$$

which is unbiased. On the other hand, this estimator has a time-lag-dependent estimation variance, which increases with time lag. The alternative estimator

$$R_n = R(n\Delta\tau) = \frac{\sum_{i=0}^{N-1} x_i^2 \sum_{i=0}^{N-n-1} x_i x_{i+n}}{\sqrt{\sum_{i=0}^{N-n-1} x_i^2} \sqrt{\sum_{i=0}^{N-n-1} x_{i+n}^2}} ,$$

$$n = 0, 1, \ldots, (N-1) \qquad (22.30)$$

first calculates the correlation coefficient using a normalization with estimates of the variance based on the same x_i as used for the summation with products $x_i x_{i+n}$. Then the correlation coefficient is denormalized to a correlation function using a variance estimate based on all available signal samples. This estimator has a small estimation variance. However, this estimator is only asymptotically bias-free for a sufficiently large N.

22.3 Hilbert Transform

The Hilbert transform of a function $x(t)$ is defined by

$$y(t) = \mathrm{HT}\{x(t)\} = \frac{1}{\pi} \int_{-\infty}^{\infty} \frac{x(\tau)}{t-\tau} \mathrm{d}\tau \qquad (22.31)$$

and is an integral transform, where the Cauchy principal value is taken in the integral. The function $y(t)$ is produced by passing $x(t)$ through a filter with the transfer function

$$H(f) = -\mathrm{i}\,\mathrm{sgn}(f) . \qquad (22.32)$$

A singularity exists at the value $f = 0$, which, however, does not cause any computational problems. On the other hand, the infinite integral causes problems for signals that are not mean-free. Thus, when processing signals with the Hilbert transform, it is necessary first to remove the mean.

The magnitude and phase of $H(f)$ are

$$|H(f)| = 1 , \qquad (22.33)$$

$$\arg\{H(f)\} = -\frac{\pi}{2}\mathrm{sgn}(f) . \qquad (22.34)$$

The inverse of the Hilbert transform is given by

$$x(t) = \mathrm{HT}^{-1}\{y(t)\} = -\mathrm{HT}\{y(t)\} = \frac{1}{\pi} \int_{-\infty}^{\infty} \frac{y(\tau)}{\tau-t} \mathrm{d}\tau . \qquad (22.35)$$

Table 22.1 Some sample Hilbert transform pairs

$x(t)$	$y(t) = \mathrm{HT}\{x(t)\}$
Const.	Defined as 0
$ax_1(t) + bx_2(t)$	$ay_1(t) + by_2(t)$
$x(at)$	$y(at)$
$x(t-t_0)$	$y(t-t_0)$
$\int_{-\infty}^{\infty} x(t)x(t-\tau)\mathrm{d}t$	$\int_{-\infty}^{\infty} y(t)y(t-\tau)\mathrm{d}t$
$a \sin bt$	$-a \cos bt$
$a \cos bt$	$a \sin bt$
$\delta(t-a)$	$\frac{1}{\pi}\frac{1}{t-a}$

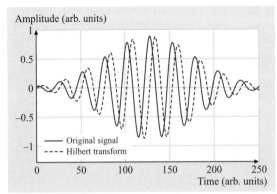

Fig. 22.7 A sample signal and its Hilbert transform illustrating the $-90°$ phase shift

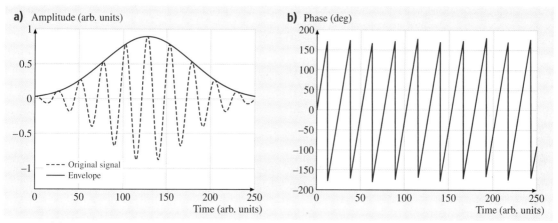

Fig. 22.8 (a) Input signal and computed envelope amplitude, (b) instantaneous phase of the input signal

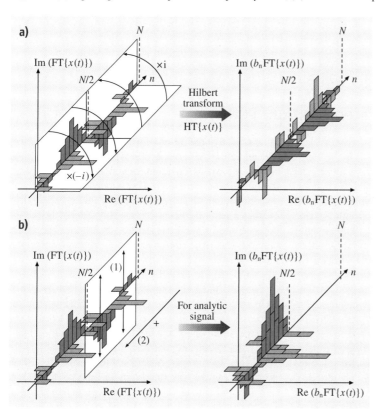

Fig. 22.9 (a) Fourier coefficients modified for the Hilbert transform, (b) modified coefficients for analytical signal

Some typical examples of Hilbert transform pairs are given in Table 22.1.

A sample signal and its Hilbert transform are shown in Fig. 22.7. For a given input signal $x(t)$ the Hilbert transform is the signal $y(t)$ shifted by $-90°$ in phase for all frequencies.

An analytical (complex) function for a given input signal $x(t)$ can be defined as

$$\underline{z}(t) = x(t) + \mathrm{i} \mathrm{HT}\{x(t)\}, \tag{22.36}$$

which has spectral values only for frequencies larger than or equal to zero ($f \geq 0$). Its Fourier transform is zero for all negative frequencies, or in the discrete case for all frequencies $f \geq N/2$. This analytical signal can be used to derive the signal envelope $A(t)$ and the instantaneous signal phase $\varphi(t)$,

$$A(t) = |\underline{z}(t)|, \qquad (22.37)$$

$$\varphi(t) = \arg\{\underline{z}(t)\}. \qquad (22.38)$$

The envelope and phase of the Doppler-like signal from Fig. 22.7 are shown in Fig. 22.8.

The calculation of the Hilbert transform for a discrete signal of finite length $x_n = x(t = t_n)$, $t_n = n/f_s$, $n = 0, 1, \ldots, (N-1)$ can be performed in the frequency domain using the discrete Fourier transform or its fast implementation, the FFT

$$\aleph\{x_n\} = \mathrm{FT}^{-1}\{b_n \mathrm{FT}\{x_n\}\} \qquad (22.39)$$

with

$$b_n = \begin{cases} -\mathrm{i} & \text{for } 0 \leq n < N/2 \\ \mathrm{i} & \text{for } N/2 \leq n < N \end{cases}. \qquad (22.40)$$

The analytical signal defined in (22.36) can be obtained as

$$\underline{z}_n = \underline{z}(t = t_n) = \mathrm{FT}^{-1}\{b_n \mathrm{FT}\{x_n\}\} \qquad (22.41)$$

with

$$b_n = \begin{cases} 2 & \text{for } 0 \leq n < N/2 \\ 0 & \text{for } N/2 \leq n < N \end{cases}. \qquad (22.42)$$

These expressions are illustrated graphically in Fig. 22.9, in which the real and imaginary Fourier coefficients of a real input signal are shown. The modified coefficients used in the inverse transform to obtain the Hilbert transform (22.39) are shown in Fig. 22.9a, and the modified coefficients used in the inverse transform to obtain the analytical function (22.41) are shown in Fig. 22.9b. From this figure it becomes apparent that the Hilbert transform can be implemented using very simple operations in combination with the Fourier transform.

22.4 Proper Orthogonal Decomposition: POD

22.4.1 Basics

Collecting very large amounts of data by numerical simulations or experimental approaches is a common situation in almost any scientific field. There is therefore a great need to have specific postprocessing techniques able to extract from these large quantities of high-dimensional data, synthetic information essential to understand and eventually to model the processes under study. The proper orthogonal decomposition (POD) is one of the most powerful method of data analysis for multivariate and non linear phenomena. Essentially, POD is a linear procedure that takes a given collection of input data and creates an orthogonal basis constituted by functions estimated as the solutions of an integral eigenvalue problem known as a Fredholm equation (22.60). These eigenfunctions are by definition (22.58) characteristic of the most probable realizations of the input data. Moreover, it can be shown that they are optimal in terms of representation of the energy present within the data (Sect. 22.4.3).

Historical Background of POD
Historically, the proper orthogonal decomposition was introduced in the context of turbulence by *Lumley* [22.3] as an objective definition of what was previously called *big eddies* by *Townsend* [22.4] and which is now widely known as coherent structures (CS, see [22.5] for a detailed discussion of CS and an overview of their detection methods). According to Yaglom [22.6], the POD is a natural idea to replace the usual Fourier decomposition in nonhomogeneous directions. The POD method was then introduced for different purposes independently by several scientists, in particular, by *Kosambi* [22.7], *Loève* [22.8, 9], *Karhunen* [22.10], *Pougachev* [22.11], and *Obukhov* [22.12, 13]. This technique then became known under a variety of names: Karhunen–Loève decomposition or expansion, principal component analysis [22.14] or hotelling analysis [22.15], and singular value decomposition [22.16]. Naturally, the proper orthogonal decomposition has been used widely in studies of turbulence but other popular applications involve random variables [22.17], image processing such as characterization of human faces [22.18], signal analysis [22.19], data compression [22.20], and more recently optimal control [22.21, 22].

From a mathematical point of view, the proper orthogonal decomposition is just a transformation that diagonalizes a given matrix A and brings it to a canonical form $A = U \Sigma V^\dagger$, where Σ is a diagonal matrix (see the paragraph on *singular value decomposition* for a complete description). The mathematical content of POD is therefore classical and is based on the spectral theory of compact, self-adjoint operators [22.23]. Two geometric interpretations of this mathematical procedure are discussed later.

POD and Turbulent Flows

A complete literature review on applications of POD to turbulence is far beyond the scope of this handbook: good reviews can be found in *Holmes* et al. [22.24], *Delville* et al. [22.25] and in the appendix of *Gordeyev* [22.26]. In the following, we provide a brief reminder of the insight that can be gained from the use of POD for eduction and modeling of the coherent structures observed in most turbulent flows.

For our purposes, it is sufficient to bear in mind [22.27] that CS identification has to be done for at least two reasons: firstly, from an energetic point of view because the relative energy content of the CS compared with the total turbulent energy can be from 10% (for boundary layers, far jets) up to 20% (far wakes, plane mixing layers) or 25% (near wakes or jets) [22.28]; secondly, because the dynamical properties of CS play an essential role in mixing processes, drag, noise emission, etc. For these reasons, the idea of controlling turbulent flows by means of influencing their coherent structures seems promising [22.29, 30].

Several characteristics of the proper orthogonal decomposition technique, as introduced by *Lumley* [22.3], are quite attractive in terms of CS identification. Firstly, compared to many other classical methods used for large-scale identification (flow visualization, conditional methods, VITA (Variable Integration Time Average), pattern recognition analysis), no a priori is needed for the eduction scheme. CS are defined in an objective and unique manner as the flow realization that possesses the largest projection onto the flow field (22.58). Secondly, the POD yields an optimal set of basis functions in the sense that no other decomposition of the same order captures an equivalent amount of kinetic energy. Up to now, POD is only presented as a data analysis method that takes as input an ensemble of data, obtained from physical experiments or from detailed numerical simulations, and extracts basis functions optimal in terms of the representativeness of the data. For illustrative purposes of the ability of the proper orthogonal decomposition to educe CS, POD is applied in *Cordier* and *Bergmann* [22.31] to a database obtained by large-eddy simulation of a three-dimensional plane turbulent mixing layer. However, proper orthogonal decomposition can also be used as an efficient procedure to compute low-dimensional dynamical models of the CS. Reduced-order modeling by POD is based on projecting the governing equation of motion onto subspaces spanned by the POD basis functions (Galerkin projection) yielding a simple set of ordinary differential equations (ODEs). Finally, due to the optimality of convergence in terms of kinetic energy of the POD functions, only a small number of POD modes are necessary to represent the dynamical evolution of the flow correctly. Reduced-order modeling based on POD has recently received an increasing amount of attention for applications to optimal control problems for partial differential equations [22.32–35]. In *Cordier* and *Bergmann* [22.31], a low-order model based on POD is developed for the incompressible unsteady wake flow behind a circular cylinder at a Reynold's number of 200. In particular, it was demonstrated how the control action could be incorporated into the low-dimensional model.

22.4.2 POD: An Approximation Method

In this chapter, we decide to follow the view of *Chatterjee* [22.36] and to introduce the singular value decomposition and its generalization, the proper orthogonal decomposition (Sect. 22.4.3), in the general context of approximation theory [22.37].

Suppose we want to approximate a possibly vector-valued function $u(\bm{x}, t)$ over some domain of interest $\mathcal{D} = \Omega \times [0; T]$ as a finite sum in the separated-variables form:

$$u(\bm{x}, t) \simeq \sum_{k=1}^{K} a^{(k)}(t) \phi^{(k)}(\bm{x}) \,. \quad (22.43)$$

For simplicity and because it will be the case in fluid mechanics applications, $\bm{x}$ can be viewed as a spatial coordinate and t as a temporal coordinate.

Our expectation is that this approximation becomes exact as $K \longrightarrow +\infty$. The representation (22.43) is clearly not unique. A classic way to solve this approximation problem is to use for the basis functions $\phi^{(k)}(\bm{x})$, functions given a priori, for example Fourier series, Legendre polynomials or Chebyshev polynomials. An alternative approach could be to determine the functions $\phi^{(k)}(\bm{x})$ that are naturally intrinsic for the approximation of the function $u(\bm{x}, t)$. As will be explained in the following, this particular approach corresponds to the proper orthogonal decomposition (POD).

An additional difficulty is that a different sequence of time functions $a^{(k)}(t)$ corresponds to each choice of basis functions $\phi^{(k)}(\bm{x})$. So, given $\phi^{(k)}(\bm{x})$, how can we determine the coefficients $a^{(k)}(t)$? Suppose we have chosen orthonormal basis functions, i. e.,

$$\int_\Omega \phi^{(k_1)}(x)\phi^{(k_2)}(x)\,\mathrm{d}x = \delta_{k_1 k_2}\,, \qquad (22.44)$$

where

$$\delta_{k_1 k_2} = \begin{cases} 0 & \text{for } k_1 \neq k_2, \\ 1 & \text{for } k_1 = k_2 \end{cases}$$

is the Kronecker delta symbol, then:

$$a^{(k)}(t) = \int_\Omega u(x,t)\phi^{(k)}(x)\,\mathrm{d}x\,.$$

Therefore for orthonormal basis functions, $a^{(k)}(t)$ depends only on $\phi^{(k)}(x)$ and not on the other ϕ. So for selecting the function $\phi^{(k)}(x)$, it would be useful to use the orthonormality condition.

Moreover, while an approximation to any desired accuracy can always be obtained if K can be chosen large enough, we may like to find, once and for all, a sequence of orthonormal functions $\phi^{(k)}(x)$ in such a way that the approximation for each K is as good as possible in a least-squares sense. Now consider that we can measure (experimentally or numerically) at N_t different instants of time, M realizations of $u(x,t)$ at M different locations $x_1, x_2, \ldots, x_M$. The approximation problem (22.43) is then equivalent to finding the orthonormal functions $\{\phi^{(k)}(x)\}_{k=1}^K$ with $K \leq N_t$ that solve:

$$\min \sum_{i=1}^{N_t} \| u(x,t_i) - \sum_{k=1}^K [u(x,t_i), \phi^{(k)}(x)]\phi^{(k)}(x) \|_2^2\,, \qquad (22.45)$$

where $\|\cdot\|_2$ define the norm associated with the usual L^2 inner product $(.,.)$. Remind that, for any vector $y \in \mathbb{R}^M$, we have

$$y = \begin{pmatrix} y_1 \\ \vdots \\ y_M \end{pmatrix} \Longrightarrow \|y\|_2 = (y,y)^{1/2} = \sqrt{y^\mathrm{T} y}$$

$$= \sqrt{y_1^2 + \ldots + y_M^2}\,. \qquad (22.46)$$

The practical method of solving the minimization problem (22.45) is to arrange the data set $\mathcal{U} = \{u(x,t_1), \ldots, u(x,t_{N_t})\}$ in an $M \times N_t$ matrix A called the snapshot data matrix

$$A = \begin{pmatrix} u(x_1,t_1) & u(x_1,t_2) & \cdots & u(x_1,t_{N_t}) \\ u(x_2,t_1) & u(x_2,t_2) & \cdots & u(x_2,t_{N_t}) \\ \vdots & \vdots & \vdots & \vdots \\ u(x_M,t_1) & u(x_M,t_2) & \cdots & u(x_M,t_{N_t}) \end{pmatrix}$$

$$A \in \mathbb{R}^{M \times N_t}\,. \qquad (22.47)$$

Each column $A_{:,i} \in \mathbb{R}^M$ of the snapshot data matrix represents a single snapshot $u(x,t_i)$ of the input ensemble $\mathcal{U}$. We note that, if the snapshot data are assumed to be linearly independent (this will be the case in particular for the snapshot POD method for reasons explained in Sect. 22.4.6), the snapshot data matrix has full column rank.

The solutions of the minimization problem (22.45) are given by the truncated singular value decomposition of length K of the matrix A. For this reason, the singular value decomposition of a matrix is reviewed below. The relationship between the proper orthogonal decomposition and the singular value decomposition is addressed later in Sect. 22.4.2.

Singular Value Decomposition (SVD)

Definition of SVD. Let A be a general complex $M \times N_t$ matrix. The singular value decomposition (SVD) of A is the factorization [22.16]:

$$A = U\Sigma V^\dagger\,, \qquad (22.48)$$

where U and V are (non-unique) unitary $M \times M$ and $N_t \times N_t$ matrices, respectively, i.e. $UU^\dagger = I_M$ and $VV^\dagger = I_{N_t}$, and $\Sigma = \mathrm{diag}(\sigma_1, \ldots, \sigma_r,)$ with $\sigma_1 \geq \sigma_2 \geq \ldots \geq \sigma_r \geq 0$ where $r = \min(M,N_t)$. The rank of A equals the number of nonzero singular values it has. Here, $V^\dagger$ denotes the adjoint matrix of V defined as the conjugate transpose of V. Remind that for a unitary matrix $A^{-1} = A^\dagger$. If $A \in \mathbb{R}^{M \times N_t}$ then $V^\dagger = V^\mathrm{T}$, and V is said to be orthogonal.

The σ_i are called the singular values of A (and of $A^\dagger$), the first r columns of $V = (v_1, v_2, \ldots, v_{N_t})$ are the right singular vectors, and the first r columns of $U = (u_1, u_2, \ldots, u_M)$ are the left singular vectors. Since the singular values are arranged in a specific order, the index i of the i-th singular value will be called the singular value number.

Geometric Interpretations of SVD.
Geometric Structure of a Matrix. By definition of a matrix, an $M \times N_t$ matrix A is a linear operator that maps

vectors from an N_t-dimensional space, say $\mathcal{E}_{N_t}$, to an M-dimensional space, say $\mathcal{E}_M$. Imagine the unit sphere in $\mathcal{E}_{N_t}$ (the set of vectors of unit magnitude). Multiplication of these vectors by the matrix A results in a set of vectors that defines an r-dimensional ellipsoid in $\mathcal{E}_M$, where r is the number of nonzero singular values. The singular values $\sigma_1, \sigma_2, \ldots, \sigma_r$ are the lengths of the principal radii of that ellipsoid (Fig. 22.10). Intuitively, the singular values of a matrix describe the extent to which multiplication by the matrix distorts the original vector. Moreover, since the matrix V is unitary, (22.48) becomes $AV = U\Sigma$. The consequences are that the directions of these principal radii are given by the columns of U and the pre-images of these principal radii are the columns of V. A second geometric interpretation is given in the next section.

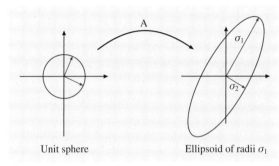

Fig. 22.10 Geometric interpretation of the SVD of the matrix A: image by A of a unit sphere

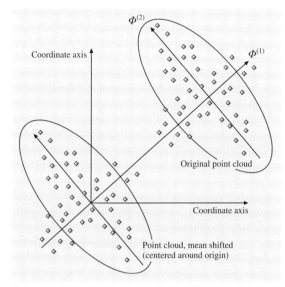

Fig. 22.11 Geometric interpretation of the SVD of the matrix A: phase-space rotation

Due to the interpretation of the matrix A in terms of linear algebra, it is now obvious that the 2-induced norm of A is σ_1

$$\|A\|_2 = \max_{\|x\|=1} \|Ax\|_2 = \sigma_1 \,. \tag{22.49}$$

SVD as a Phase-Space Rotation. A second geometric interpretation may be attributed to SVD applications. We now view the $M \times N_t$ matrix A as a list of coordinates of M points denoted $P_1, P_2, \ldots, P_M$ in an N_t-dimensional space. Each point P_i is represented in Fig. 22.11 by a diamond. For any $k \leq N_t$, we seek a k-dimensional subspace for which the mean square distance of the points, from the subspace, is minimized, i.e., we search a vector $\boldsymbol{\Phi}^{(1)}$ (Fig. 22.11) such that $\sum_{i=1}^{M} |\boldsymbol{H_i P_i}|^2$ is minimized, where H_i are the orthogonal projection of P_i onto the line of direction vector $\boldsymbol{\Phi}^{(1)}$. This mathematical procedure can be geometrically interpreted (Fig. 22.11) as a rotation of the phase space from the original basis into a new coordinate system whose orthogonal axes coincide with the axes of inertia of the data. This formulation of the SVD problem corresponds exactly to the way principal component analysis is commonly introduced in the literature [22.14].

When the singular value decomposition is used for data analysis, the SVD algorithm is generally applied to a matrix deduced from the snapshot matrix A by subtracting from each column of A the mean of that column. This mean shift ensures that the M-point cloud is now centered around the origin of the coordinate (Fig. 22.11).

Relationships Between SVD and Eigenvalue Problems

In this section, we present how the singular values and the right and left singular vectors of a rectangular matrix A can also be computed by solving symmetric eigenproblems with, e.g., the matrices $A^\dagger A$ or $AA^\dagger$, instead of computing the SVD of A. In this case, $A^\dagger A$ and $AA^\dagger$ represent a finite-dimensional version of the two-point space–time correlation R introduced in Sect. 22.4.3. The results of this section will be used later.

Let $A = U\Sigma V^\dagger$ be a singular value decomposition of $A \in \mathbb{R}^{M \times N_t}$. Then $A^\dagger A = V\Sigma U^\dagger U\Sigma V^\dagger = V\Sigma^2 V^\dagger$, where Σ^2 is a diagonal matrix. Since $A^\dagger A$ is an Hermitian matrix, its eigenvalue decomposition can be written: $A^\dagger A = W\Lambda W^{-1} = W\Lambda W^\dagger$, where W is an $N_t \times N_t$ unitary matrix. By comparing the two expression of A, we conclude that $\Sigma^2 = \Lambda$, and $W = V$. In other words $\sigma_i = \sqrt{\lambda_i}$, and (V, Λ) is the eigenvector–eigenvalue decomposition of $A^\dagger A \in \mathbb{R}^{N_t \times N_t}$.

The same development applied to the matrix $AA^\dagger$ leads to $AA^\dagger = U\Sigma V^\dagger V \Sigma U^\dagger = U\Sigma^2 U^\dagger = W\Lambda W^\dagger$, so (U, Λ) is the eigenvector–eigenvalue decomposition of $AA^\dagger \in \mathbb{R}^{M \times M}$.

At this point, we remark that the eigenvalue problem associated with $A^\dagger A$ is more practical to solve than the eigenvalue problem associated with $AA^\dagger$ in cases where the input collection N_t is significantly smaller than the number of coefficients needed to represent each item of the collection M. This remark explains why two different POD approaches exist: the classical POD (Sect. 22.4.5) and the snapshot POD (Sect. 22.4.6).

Lower-Rank Approximation to A

Given $A \in \mathbb{R}^{M \times N_t}$, the computation of a matrix $X \in \mathbb{R}^{M \times N_t}$ with $\text{rank}(X) = k < \text{rank}(A)$ such that an appropriate norm of the error $E = A - X$ is minimized is a classical problem. This problem can be solved explicitly if we take as the norm the Frobenius norm, defined as the square root of the sums of squares of all the elements and denoted by $\|\cdot\|_F$ or any unitarily invariant norm. The solution is given by the Eckart–Young theorem [22.38] which states that

$$\min_{\text{rank}(X) \leq k} \|A - X\|_F = \|A - A_k\|_F = \sqrt{\sum_{i=k+1}^{r} \sigma_i^2(A)}, \tag{22.50}$$

where

$$A_k = U \begin{pmatrix} \Sigma_k & 0 \\ 0 & 0 \end{pmatrix} V^\dagger = \sigma_1 u_1 v_1^\dagger + \ldots + \sigma_k u_k v_k^\dagger$$

where Σ_k is the matrix obtained by setting $\sigma_{k+1} = \sigma_{k+2} = \ldots = \sigma_r = 0$ in Σ.

For example, the 2-norm defined by (22.49) can be used. In this case, the Eckart–Young theorem (22.50) yields [22.39]:

$$\min_{\text{rank}(X) \leq k} \|A - X\|_2 = \|A - A_k\|_2 = \sigma_{k+1}(A).$$

Remark: This theorem establishes a relationship between the rank k of the approximant, and the $(k+1)$-th largest singular value of A. Therefore, if the singular values decrease is fast, we can hope to find an approximant with small rank (see the section on *Examples of Image Processing by SVD* below).

Relationship Between POD and SVD

Here, we discuss the close relationship between POD and SVD. Our presentation follows the view of Fahl [22.34] but similar treatments can be found in Atwell and King [22.40]. The reader is referred to Volkwein [22.41] for the mathematical demonstrations.

Suppose that each member of the input collection $\mathcal{U}$ defined in Sect. 22.4.2 can be written in terms of n-th-order finite-element basis functions $\{\varphi^{(j)}(x)\}_{j=1}^n$, i.e.,

$$u(x, t_i) = u^n(x, t_i) = \sum_{j=1}^{n} u^{(j)}(t_i) \varphi^{(j)}(x),$$

where the superscript n denotes a high-order finite-element discretization.

The inner product can then be defined by:

$$(u, v)_{\mathcal{M}} = u^T \mathcal{M} v, \tag{22.51}$$

where $\mathcal{M} \in \mathbb{R}^{n \times n}$ is the finite-element mass matrix and $u, v \in \mathbb{R}^n$ are the finite-element coefficient vectors for a given t_i. Employing a Cholesky factorization $\mathcal{M} = \mathcal{M}^{1/2}(\mathcal{M}^{1/2})^T$, the $\mathcal{M}$ inner product (22.51) can be transformed to the standard Euclidean inner product (22.46) such that the condition

$$\|u\|_{\mathcal{M}} = (u, u)_{\mathcal{M}}^{1/2} = \|(\mathcal{M}^{1/2})^T u\|_2$$

holds. The minimization problem (22.45) can then be reformulated for the $\mathcal{M}$ inner product as

$$\min \sum_{i=1}^{N_t} \|u^n(x, t_i) - \sum_{k=1}^{K} [u^n(x, t_i), \phi^{(k)}(x)]_{\mathcal{M}} \phi^{(k)}(x)\|_{\mathcal{M}}^2, \tag{22.52}$$

where the POD basis functions $\{\phi^{(k)}\}_{k=1}^K$ are assumed to be in the linear space spanned by the finite-element basis functions $\{\varphi^{(j)}(x)\}_{j=1}^n$, i.e.,

$$\phi^{(k)}(x) = \sum_{j=1}^{n} \phi_j^{(k)} \varphi^{(j)}(x).$$

To reformulate the minimization problem (22.52) in a matrix approximation context, let $\Phi \in \mathbb{R}^{n \times K}$ denote a matrix collecting the finite-element coefficients of the unknown POD functions. Since for any matrix $\hat{A} \in \mathbb{R}^{n \times N_t}$, $\sum_{i=1}^{N_t} \|\hat{A}_{:,i}\|_2^2 = \|\hat{A}\|_F^2$, where $\|\cdot\|_F$ denotes the Frobenius norm defined earlier (22.50), the problem (22.52) is equivalent to solving:

$$\min_{Z \in \mathbb{R}^{n \times K}} \|\hat{A} - ZZ^T \hat{A}\|_F^2 \quad \text{s.t.} \quad Z^T Z = I_K \tag{22.53}$$

with $\hat{A} = (\mathcal{M}^{1/2})^T A$ and $Z = (\mathcal{M}^{1/2})^T \Phi$, where $Z \in \mathbb{R}^{n \times K}$.

Equation (22.53) indicates that we are looking for a K dimensional subspace with orthogonal matrix Z such that $X = ZZ^T \hat{A}$ is the best approximation to $\hat{A}$ compared to all subspaces of dimension K. According to the Eckart–Young theorem (22.50), the solution to problem (22.53) is given by a truncated singular value decomposition of $\hat{A}$ of length K

$$\hat{A}_K = U_K \Sigma_K V_K^T, \tag{22.54}$$

where U_K and V_K correspond to the first K columns of U and V, respectively. Finally, comparing $\hat{A}_K$ and the form of X, we find that the matrix Φ solves

$$(\mathcal{M}^{1/2})^T \Phi = U_K \in \mathbb{R}^{n \times K}. \tag{22.55}$$

The finite-element coefficients of the POD basis functions can then be computed by solving the linear system (22.55) where the left singular vectors U of $\hat{A} = U \Sigma V^T$ can be obtained directly as the eigenvalues of the previous $\hat{A} \hat{A}^T$ matrix. However, as was previously remarked at the end of Sect. 22.4.2, when N_t is significantly smaller than n it is more practical to solve the eigenvalue problem $\hat{A}^T \hat{A}$. It follows that, in this case, the right singular vectors V of $\hat{A}$ are obtained and U must be deduced from V by the equation $U = \Sigma^{-1} \hat{A} V$.

Remark: The eigenvalue problems can be solved with the library LAPACK [22.42] and efficient algorithms for POD computations based on Lanczos iterations can be found in *Fahl* [22.34].

Examples of Image Processing by SVD

As an illustration of the SVD process for computing low-rank approximations to data matrices, consider a time-independent problem where the input collection consists of greyscale images. In Fig. 22.14a and Fig. 22.15a the *clown* picture and the *trees* picture from MatLab are considered. These images can be represented by means of a 200×330, and a 128×128 matrix, respectively, each entry (pixel) having a value between 0 (white) and 1 (black) in 64 levels of gray. Both matrices have full rank, i.e., 200 and 128, respectively. Their numerical ranks however are much lower. The singular values of these two images are shown in Fig. 22.12 on a semi-log scale; both sets of singular values fall off rapidly, and hence low-rank approximations with small errors are possible.

By comparing the spectrum of the two singular value plots, we can determine that the relative error for approximants of the same rank is greater for the *clown* image than for the *trees* image. Thus the *trees* image is easier to approximate.

The Eckart–Young theorem states that, for any matrix A of rank N, an approximation of rank $k \leq N$ of the matrix A can be obtained by

$$A = \sigma_1 u_1 v_1^\dagger + \sigma_2 u_2 v_2^\dagger + \ldots + \sigma_k u_k v_k^\dagger.$$

Thus using the singular value decomposition, one can obtain a high-fidelity model, perhaps with large k. In order to obtain a lower-rank representation of these images, singular modes corresponding to small singular values are neglected. So if the spectrum of the singu-

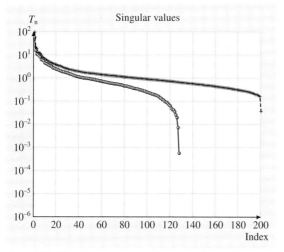

Fig. 22.12 Singular values for the *clown* image (*crosses*) and the *trees* image (*open circles*)

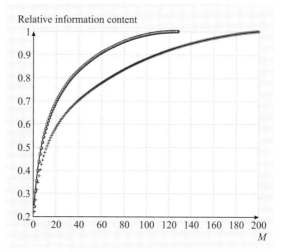

Fig. 22.13 Relative information content for the *clown* image (*crosses*) and the *trees* image (*open circles*)

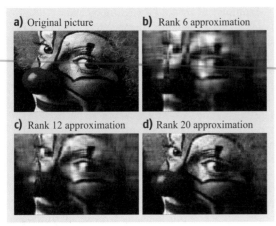

Fig. 22.14a–d Approximations of the *clown* image from MatLab by images of lower rank. (**a**) Original picture. (**b**) Rank 6 approximation. (**c**) Rank 12 approximation. (**d**) Rank 20 approximation

lar values decays fast, one can choose a cutoff value $M \ll N$ and carry out an approximation of A with a reduced number of singular modes. To make this idea more precise, one can define the *relative information content* (RIC) of the singular value decomposition of A by

$$\text{RIC}(M) = \frac{\sum_{i=1}^{M} \sigma_i}{\sum_{i=1}^{N} \sigma_i} \,. \tag{22.56}$$

If the low-rank approximation is required to contain $\delta\%$ of the total information contained in the original image, the dimension M of the subspace D_M^{SVD} spanned by the M first singular modes is determined by

$$M = \arg\min\{\text{RIC}(M); \text{RIC}(M) \geq \delta\} \,. \tag{22.57}$$

Another way to measure the degree of order of the signal u contained in the snapshot data matrix A is to define a global entropy [22.43] as

$$H(u) = -\frac{1}{\log N} \sum_{k=1}^{N} p_k \log(p_k)$$

with

$$p_k = \frac{\sigma_k}{\sum_{i=1}^{N} \sigma_i} \,.$$

If the information is uniformly distributed among all the modes, the entropy is maximal and equal to one.

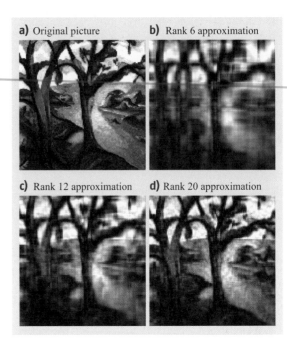

Fig. 22.15a–d Approximations of the *trees* image from MatLab by images of lower rank. (**a**) Original picture. (**b**) Rank 6 approximation. (**c**) Rank 12 approximation. (**d**) Rank 20 approximation

Similarly, if there is only one nonzero singular value, then the entropy is zero. At intermediate states, $H(u)$ keeps increasing as more modes become necessary to represent the data.

In Fig. 22.13, the relative information content for the *clown* image and the *trees* image are shown. The same result as previously mentioned for the two images when the singular values spectrum was discussed is evidenced. For a given number of singular modes, say $M = 20$, respectively, 60% and 70% of the information content of the original *clown* image and *trees* image are contained in the approximation. This clearly demonstrates that the *trees* image is easier to approximate by a lower-rank image than the *clown* image.

Lastly, we present in Figs. 22.14 and 22.15, clockwise from the top, the original picture, and approximants of rank 6, rank 20, and rank 12, for the *clown* image and *trees* image, respectively.

22.4.3 The Proper Orthogonal Decomposition (POD)

This section introduces the proper orthogonal decomposition in the spirit of *Holmes* et al. [22.24], as a technique that can contribute to a better understanding of turbulent flows. Here, POD is not reduced to an advanced processing method that allows the extraction of coherent structures from experimental or numerical data. Rather, POD is used to provide a set of basis functions with which to identify a low-dimensional subspace on which to construct a dynamical model of the coherent structures by projection on the governing equations. This idea was first applied in *Aubry* et al. [22.29] to model the near-wall region of a turbulent boundary layer and more recently by *Ukeiley* et al. [22.30] to study the dynamics of the coherent structures in a plane turbulent mixing layer.

The Fredholm Equation

Let $\{u(X), X = (x, t_n) \in \mathcal{D} = \mathbb{R}^3 \times \mathbb{R}^+\}$ denote the set of observations (also called *snapshots*) obtained at N_t different time steps t_n over a spatial domain of interest Ω ($x = (x, y, z) \in \Omega$). These snapshots could be experimental measurements or numerical solutions of velocity fields, vorticity fields, temperatures, etc. taken at different time steps and/or different physical parameters, for example Reynolds number [22.44]. The underlying problem is to extract from this ensemble of random vector fields a coherent structure. Following *Lumley* [22.3], a coherent structure is defined as *the deterministic function which is best correlated on average with the realizations $u(X)$*. In other words, we look for a function Φ that has the largest mean square projection onto the observations $|(u, \Phi)|^2$. Since it is only the parallelism between Φ and the observations that is of interest, the dependence on the amplitude of Φ must be removed. One way is to normalize the amplitude of Φ. It is then natural to look at a space of functions Φ for which the inner product exists, i. e., to impose Φ to be an element of $L^2(\mathcal{D})$, the collection of square-integrable functions defined on the flow region $\mathcal{D}$. Finally, in order to include the statistics, we must maximize the expression:

$$\frac{\langle|(u, \Phi)|^2\rangle}{\|\Phi\|^2}$$

in some average sense (temporal, spatial, ensemble, or phase average), denoted here by $\langle \cdot \rangle$, to be specified for each application. The choice of the average operator is at the heart of the different POD approaches and a detailed discussion of this point is postponed to Sect. 22.4.4.

Hence, mathematically, the function Φ corresponds to the solution of the constrained optimization problem:

$$\max_{\Psi \in L^2(\mathcal{D})} \frac{\langle|(u, \Psi)|^2\rangle}{\|\Psi\|^2} = \frac{\langle|(u, \Phi)|^2\rangle}{\|\Phi\|^2} \quad (22.58)$$

with respect to

$$(\Phi, \Phi) = \|\Phi\|^2 = 1 .$$

Here $(.,.)$ and $\|\cdot\|$ denote the usual L^2 inner product and L^2 norm over $\mathcal{D}$:

$$(u, \Phi) = \int_{\mathcal{D}} u(X) \cdot \Phi^*(X) \, dX$$

$$= \sum_{i=1}^{n_c} \int_{\mathcal{D}} u_i(X) \Phi_i^*(X) \, dX ; \quad \|u\|^2 = (u, u) ,$$

where the superscript asterisk indicates the complex conjugate and n_c is the number of vectorial components of $u(X)$. Note that L^2 seems to be a natural space in which to do fluid mechanics since it corresponds to flow having finite kinetic energy, but the choice of other norms for the POD basis computation is possible, see Sect. 22.4.4 for a discussion.

The maximization problem (22.58) can be cast in an equivalent eigenvalue problem. To see this, let us define the operator $\mathcal{R} : L^2(\mathcal{D}) \longrightarrow L^2(\mathcal{D})$ by

$$\mathcal{R}\Phi(X) = \int_{\mathcal{D}} R(X, X')\Phi(X') \, dX' ,$$

where $R(X, X') = \langle u(X) \otimes u^*(X') \rangle$ is the two-point space–time correlation tensor ($\otimes$ is the dyadic product). Then, straightforward calculations reveal that

$$(\mathcal{R}\Phi, \Phi)$$
$$= \left(\int_{\mathcal{D}} \langle u(X) \otimes u^*(X') \rangle \Phi(X') \, dX', \Phi(X) \right)$$
$$= \int_{\mathcal{D}} \int_{\mathcal{D}} \langle u(X) \otimes u^*(X') \rangle \Phi(X') \, dX' \Phi^*(X) \, dX$$
$$= \left\langle \int_{\mathcal{D}} u(X) \Phi^*(X) \, dX \int_{\mathcal{D}} u^*(X') \Phi(X') \, dX' \right\rangle$$
$$= \langle|(u, \Phi)|^2\rangle \geq 0$$

Furthermore, it follows that:

$$(\mathcal{R}\Phi, \Psi) = (\Phi, \mathcal{R}\Psi) \text{ for any } (\Phi, \Psi) \in [L^2(\mathcal{D})]^2 .$$

We suppose that the probabilistic structure of the ensemble of observations is such that the average and integrating operations can be interchanged [22.45]. Then $\mathcal{R}$ is linear and self-adjoint, i. e., $\mathcal{R}^\dagger = \mathcal{R}$, where the adjoint of $\mathcal{R}$, $\mathcal{R}^\dagger$, is defined by

$$(\mathcal{R}u, v) = (u, \mathcal{R}^\dagger v)$$

for all $u \in L^2(\mathcal{D})$ and $v \in L^2(\mathcal{D})$.

nonnegative operator on $L^2(\mathcal{D})$, i.e. $(\mathcal{R}\boldsymbol{u}, \boldsymbol{u}) \geq 0$ for all $\boldsymbol{u} \in L^2(\mathcal{D})$. Consequently, spectral theory applies [22.23, 46] and guarantees that the maximization problem (22.58) admits a solution, equal to the largest eigenvalue of the problem

$$\mathcal{R}\boldsymbol{\Phi} = \lambda \boldsymbol{\Phi}, \tag{22.59}$$

which can be written as a Fredholm integral eigenvalue problem

$$\sum_{j=1}^{n_c} \int_{\mathcal{D}} R_{ij}(\boldsymbol{X}, \boldsymbol{X}') \Phi_j(\boldsymbol{X}') \, d\boldsymbol{X}' = \lambda \Phi_i(\boldsymbol{X}). \tag{22.60}$$

The properties of the empirical eigenfunctions $\Phi_i(\boldsymbol{X})$ obtained by solving the Fredholm equation (22.60) are fully discussed in Sect. 22.4.3. Here, it is sufficient to make some comments shedding light on the constraints linked to the POD method.

In (22.60), the integral $\int_{\mathcal{D}} \cdot \, d\boldsymbol{X}'$ is over the entire domain of interest $\mathcal{D}$. The consequence is that the two-point correlation tensor R_{ij} has to be known over all $\mathcal{D}$. Therefore, the data volume to handle can be very important (several gigabytes are not rare) and sometimes data compression is necessary to reduce the data storage requirements (see [22.31] for an example). Due to the important size of the data sets necessary to apply POD, renewed interest in POD only appeared in the 1990s, explained by the great advances in numerical simulation capability and measurement techniques.

Remark: An alternative approach for finding the solution to maximization of (22.58) is by directly solving a classical problem in the calculus of variations. Since $(\mathcal{R}\boldsymbol{\Phi}, \boldsymbol{\Phi}) = \langle |(\boldsymbol{u}, \boldsymbol{\Phi})|^2 \rangle$, the problem (22.58) is equivalent to determining the $\boldsymbol{\Phi}$ that maximizes λ, where

$$\lambda = \frac{\langle |(\boldsymbol{u}, \boldsymbol{\Phi})|^2 \rangle}{(\boldsymbol{\Phi}, \boldsymbol{\Phi})} = \frac{(\mathcal{R}\boldsymbol{\Phi}, \boldsymbol{\Phi})}{(\boldsymbol{\Phi}, \boldsymbol{\Phi})}. \tag{22.61}$$

Using the calculus of variations, $\boldsymbol{\Phi}$ is determining by imposing the condition $dF(\epsilon)/d\epsilon|_{\epsilon=0} = 0$ with

$$F(\epsilon) = \frac{[\mathcal{R}(\boldsymbol{\Phi} + \epsilon \boldsymbol{\Upsilon}), (\boldsymbol{\Phi} + \epsilon \boldsymbol{\Upsilon})]}{[(\boldsymbol{\Phi} + \epsilon \boldsymbol{\Upsilon}), (\boldsymbol{\Phi} + \epsilon \boldsymbol{\Upsilon})]}$$
$$= \frac{(\mathcal{R}\boldsymbol{\Phi}, \boldsymbol{\Phi}) + \epsilon(\mathcal{R}\boldsymbol{\Phi}, \boldsymbol{\Upsilon}) + \epsilon(\mathcal{R}\boldsymbol{\Upsilon}, \boldsymbol{\Phi}) + \epsilon^2 (\mathcal{R}\boldsymbol{\Upsilon}, \boldsymbol{\Upsilon})}{(\boldsymbol{\Phi}, \boldsymbol{\Phi}) + \epsilon(\boldsymbol{\Phi}, \boldsymbol{\Upsilon}) + \epsilon(\boldsymbol{\Upsilon}, \boldsymbol{\Phi}) + \epsilon^2 (\boldsymbol{\Upsilon}, \boldsymbol{\Upsilon})}.$$

This leads one to verify for any $\boldsymbol{\Upsilon}$ the condition

$$(\mathcal{R}\boldsymbol{\Phi}, \boldsymbol{\Upsilon}) = \lambda(\boldsymbol{\Phi}, \boldsymbol{\Upsilon}),$$

which is equivalent to finding the eigenvalue of the eigenvalue problem (22.59).

Properties of the POD Basis Functions
Eight main properties can be derived.

1. For a bounded integration domain $\mathcal{D}$, Hilbert–Schmidt theory applies [22.46] and assures us that there is not one, but a denumerable infinity of solutions of (22.60). Then, the Fredholm equation (22.60) has a discrete set of solutions satisfying:

$$\sum_{j=1}^{n_c} \int_{\mathcal{D}} R_{ij}(\boldsymbol{X}, \boldsymbol{X}') \Phi_j^{(n)}(\boldsymbol{X}') \, d\boldsymbol{X}' = \lambda^{(n)} \Phi_i^{(n)}(\boldsymbol{X}), \tag{22.62}$$

where $\lambda^{(n)}$ and $\Phi_i^{(n)}$ denote, respectively, the POD eigenvalues and POD eigenvectors or eigenfunctions of order $n = 1, 2, 3, \ldots, +\infty$. Each new eigenfunction is sought as the solution problem of the maximization problem (22.58) subject to the constraint of being orthogonal to all previously found eigenfunctions. Hence, by construction, the eigenfunctions are mutually orthogonal but they can be chosen orthonormal (see item 4). Any d-fold degenerate eigenvalue is associated with d linearly independent eigenfunctions.

2. If $\mathcal{R}$ is a self-adjoint and nonnegative operator then all eigenvalues are real and positive:

$$\lambda^{(1)} \geq \lambda^{(2)} \geq \lambda^{(3)} \geq, \ldots \lambda^{(+\infty)} \geq 0 \tag{22.63}$$

and the corresponding series converges:

$$\sum_{n=1}^{+\infty} \lambda^{(n)} < +\infty.$$

3. The eigenfunctions $\boldsymbol{\Phi}^{(n)}$ form a complete orthogonal set, which means that almost every member (except possibly on a set of measure zero, see [22.47]) of the snapshots can be reconstructed in the following way:

$$u_i(\boldsymbol{X}) = \sum_{n=1}^{+\infty} a^{(n)} \Phi_i^{(n)}(\boldsymbol{X}). \tag{22.64}$$

4. The eigenfunctions $\boldsymbol{\Phi}^{(n)}$ can be chosen to be mutually orthonormal:

$$\sum_{i=1}^{n_c} \int_{\mathcal{D}} \Phi_i^{(m)}(\boldsymbol{X}) \Phi_i^{*(n)}(\boldsymbol{X}) \, d\boldsymbol{X} = \delta_{mn}$$

where $\delta_{mn} = \begin{cases} 0 & \text{for } m \neq n; \\ 1 & \text{for } m = n. \end{cases} \tag{22.65}$

Since $\mathcal{R}$ is a self-adjoint operator, orthogonality is verified necessarily. On the other hand, the choice of orthonormality for the eigenfunctions is rather arbitrary because they are determined relative to a real multiplicative constant. Hence, it is numerically equivalent to impose:

$$\sum_{i=1}^{n_c} \int_{\mathcal{D}} \Phi_i^{(m)}(X) \Phi_i^{*(n)}(X) \, \mathrm{d}X = \lambda^{(m)} \delta_{mn} \quad (22.66)$$

for the eigenfunctions $\Phi_i^{(m)}(X)$ and the condition $\langle a^{(n)} a^{*(m)} \rangle = \delta_{mn}$ for the projection coefficient $a^{(n)}$ or to impose for the eigenfunctions the orthonormality condition (22.66) and the orthogonality condition (22.69) for the coefficients. For numerical reasons, it is easier to use (22.66) for *classical POD*, and (22.65) for *snapshot POD* [22.48].

5. The random coefficients $a^{(n)}$, projections of u onto Φ, are then calculated by using the orthonormality of the eigenfunctions Φ:

$$a^{(n)} = (u, \Phi) = \sum_{i=1}^{n_c} \int_{\mathcal{D}} u_i(X) \Phi_i^{*(n)}(X) \, \mathrm{d}X \,. $$

$$(22.67)$$

6. The two-point space–time correlation tensor R_{ij} can be decomposed as a uniformly convergent series [22.23]:

$$R_{ij}(X, X') = \sum_{n=1}^{+\infty} \lambda^{(n)} \Phi_i^{(n)}(X) \Phi_j^{*(n)}(X') \,.$$

$$(22.68)$$

This result is known as Mercer's theorem.

7. The diagonal representation of the tensor R_{ij} combined with the decomposition of u onto the eigenfunctions Φ and their orthogonality assure that the coefficients $a^{(n)}$ are mutually uncorrelated and that their mean square values are the eigenvalues themselves.

$$\langle a^{(n)} a^{*(m)} \rangle = \delta_{mn} \lambda^{(n)} \,. \quad (22.69)$$

Proof: This assertion derives directly from the representation of $R_{ij}(X, X')$, given in equation (22.68):

$$R_{ij}(X, X')$$
$$= \langle u_i(X) u_j^*(X') \rangle$$
$$= \left\langle \sum_{n=1}^{+\infty} a^{(n)} \Phi_i^{(n)}(X) \sum_{m=1}^{+\infty} a^{*(m)} \Phi_j^{*(m)}(X') \right\rangle$$
$$= \sum_{n=1}^{+\infty} \sum_{m=1}^{+\infty} \langle a^{(n)} a^{*(m)} \rangle \Phi_i^{(n)}(X) \Phi_j^{*(m)}(X') \,.$$

But we know from the Mercer's theorem that

$$R_{ij}(X, X') = \sum_{n=1}^{+\infty} \lambda^{(n)} \Phi_i^{(n)}(X) . \Phi_j^{*(n)}(X') \,,$$

and so, since the $\Phi^{(n)}(X)$ are an orthonormal family in $L^2(\mathcal{D})$, we see that $\langle a^{(n)} a^{*(m)} \rangle = \delta_{mn} \lambda^{(n)}$.

8. Finally, Mercer's theorem and the orthonormality of $\Phi^{(n)}$ lead to

$$\sum_{i=1}^{n_c} \int_{\mathcal{D}} R_{ii}(X, X) \, \mathrm{d}X = \sum_{n=1}^{+\infty} \lambda^{(n)} = E \,. \quad (22.70)$$

If $u(X)$ is a velocity field, then E corresponds to the turbulent kinetic energy (TKE) integrated over the domain $\mathcal{D}$. In the same way, if $u(X)$ is a vorticity field, as in *Sanghi* [22.49], this relation leads to the system enstrophy. So, whatever variable is considered for the POD, the eigenvalues $\lambda^{(n)}$ obtained by solving the Fredholm equation (22.60) are always homogeneous to energy but are not strictly speaking energy. Thinking of the POD eigenvalues as energy in a general mechanical context is incorrect in principle and may lead to misleading results. The interpretation of this equation is that every structure of order (n) makes an independent contribution to the TKE. Then, the amplitude of the eigenvalues $\lambda^{(n)}$ measure the relative importance of the different structures present within the flow.

Optimality of the POD Basis

Suppose that we have a signal $u(X)$ with $u \in L^2(\mathcal{D})$ and an approximation u^a of u with respect to an arbitrary orthonormal basis $\Psi^{(n)}(X)$, $n = 1, 2, \ldots, +\infty$. One can write:

$$u_i^a(X) = \sum_{n=1}^{+\infty} b^{(n)} \Psi_i^{(n)}(X) \,.$$

Equations (22.70) and (22.69) clearly state that, if the $\Psi_i^{(n)}(X)$ have been nondimensionalized, $\langle b^{(n)} b^{*(n)} \rangle$ represents the average energy in the n-th mode. The following lemma establishes the notion of the optimality of the POD approach.

Lemma. Let $\{\Phi^{(1)}(X), \Phi^{(2)}(X), \ldots, \Phi^{(\infty)}(X)\}$ denote an orthonormal set of POD basis elements, and $\{\lambda_1, \lambda_2, \ldots, \lambda_\infty\}$ denote the corresponding set of eigenvalues. If

$$u_i^P(X) = \sum_{n=1}^{+\infty} a^{(n)} \Phi_i^{(n)}(X)$$

denotes the approximation to u with respect to this basis, then for any value of N [22.24]:

$$\sum_{n=1}^{N} \langle a^{(n)} a^{*(n)} \rangle = \sum_{n=1}^{N} \lambda^{(n)} \geq \sum_{n=1}^{N} \langle b^{(n)} b^{*(n)} \rangle \ .$$

Proof: It is straightforward [see the proof of (22.69)] to show that the kernel R_{ij} can be expressed in terms of $\Psi^{(n)}$, $n = 1, \ldots, +\infty$ as

$$R_{ij}(X, X') = \sum_{n=1}^{+\infty} \sum_{m=1}^{+\infty} \langle b^{(n)} b^{*(m)} \rangle \Psi_i^{(n)}(X) \Psi_j^{*(m)}(X') \ .$$

Therefore, the projection of the kernel R_{ij} in an N-dimensional space spanned by $\{\Psi^{(n)}\}_{n=1}^{N}$ can be written in matrix form as

$$R = \begin{pmatrix} \langle b^{(1)} b^{*(1)} \rangle & \cdots & \langle b^{(1)} b^{*(N)} \rangle & 0 & \cdots & 0 \\ \langle b^{(2)} b^{*(1)} \rangle & \cdots & \langle b^{(2)} b^{*(N)} \rangle & 0 & \cdots & 0 \\ \vdots & \vdots & \vdots & \vdots & & \vdots \\ \langle b^{(N)} b^{*(1)} \rangle & \cdots & \langle b^{(N)} b^{*(N)} \rangle & 0 & \cdots & 0 \\ 0 & \cdots & 0 & 0 & 0 & 0 \\ \vdots & \vdots & \vdots & \vdots & & \vdots \end{pmatrix} \ .$$

The proof finally relies on a result for linear operators [22.50, p. 260] that states that the sum of the first N eigenvalues of a self-adjoint operator is greater than or equal to the sum of the diagonal terms in any N-dimensional projection of it:

$$\sum_{n=1}^{N} \lambda^{(n)} \geq \text{Tr}(R) = \sum_{n=1}^{N} \langle b^{(n)} b^{*(n)} \rangle \ .$$

This lemma establishes that, among all *linear* decompositions, the POD is the most efficient, in the sense that, for a given number of modes, N, the projection on the subspace spanned by the N leading eigenfunctions will contain the greatest possible kinetic energy on average. The reader must remember that optimality of the POD functions is obtained only with respect to other linear representations.

Model Reduction Aspects

The energetic optimality of the POD basis functions suggests that only a very small number of POD modes, say M, may be necessary to describe any signal $u(X)$ of the input data efficiently. The choice of M is then an important and critical task and adequate criteria for choosing M must be introduced.

Let N_{POD} denote the number of POD modes obtained by solving the Fredholm equation (22.60). The truncation error $\epsilon(M)$ due to the use of M instead of N_{POD} POD basis functions in representing the input data is given by

$$\epsilon(M) = \left\| u(X) - \sum_{n=1}^{M} [u(X), \Phi^{(n)}(X)] \Phi^{(n)}(X) \right\|^2$$

$$= \left\| \sum_{n=M+1}^{N_{\text{POD}}} [u(X), \Phi^{(n)}(X)] \Phi^{(n)}(X) \right\|^2 \ . \quad (22.71)$$

It is immediate to deduce from (22.71) the equivalent forms for the two particular approaches of POD described in Sect. 22.4.4. The reader is referred to *Fahl* [22.34], where ϵ is defined for the snapshot POD.

The quantity $\epsilon(M)$ measures the accumulated squared error in representing the input snapshots, due to neglecting the POD basis elements that correspond to small POD eigenvalues.

However, in practice, this criterion is never used and the choice of M is rather based on heuristic considerations. As indicated in point 8 of Sect. 22.4.3 on the properties of the POD basis functions, $\sum_{i=1}^{M} \lambda^{(i)}$ represents in some sense the average energy contained in the first M POD modes. For turbulent flows, it corresponds exactly to the average turbulent kinetic energy. Therefore, to capture most of the energy contained in the N_{POD} POD modes, it suffices to choose M so that $\sum_{i=1}^{M} \lambda^{(i)} \simeq \sum_{i=1}^{N_{\text{POD}}} \lambda^{(i)}$. By definition, the ratio $\sum_{i=1}^{M} \lambda^{(i)} / \sum_{i=1}^{N_{\text{POD}}} \lambda^{(i)}$ yields the percentage of the total kinetic energy in the N_{POD} POD modes that is contained in the first M POD basis functions. For a predefined percentage of energy, δ, the dimension M of the subspace spanned by the M first POD functions is chosen such that the condition

$$\frac{\sum_{i=1}^{M} \lambda^{(i)}}{\sum_{i=1}^{N_{\text{POD}}} \lambda^{(i)}} \geq \delta \quad (22.72)$$

holds [22.22, 31, 34]. The criterion (22.72) is equivalent to the one based on the relative information content used in Sect. 22.4.2 for the singular value decomposition (22.57). The POD reduced basis subspace is defined as $D_M^{POD} = \text{span}\{\Phi^{(1)}, \Phi^{(2)}, \ldots, \Phi^{(M)}\}$.

To this point we have only discussed the model reduction associated with using POD basis functions in the approximation of the input collection. Dynamical models based on POD have not been discussed.

Nevertheless, the optimal energetic convergence of the POD basis functions suggests that only a very small number of modes may be necessary to describe the dynamics of the system. Therefore, starting from data issued from high-dimensional models (experimental data or detailed simulations), it seems conceivable that POD modes can be efficiently used in Galerkin projections that yield low-dimensional dynamical models. Even though there are no theoretical guarantees of optimality in dynamical modeling, this method was already used in many cases, for turbulent flows or optimal control of fluids and reasonable to excellent models were obtained. The presentation of this approach can be found in *Cordier* and *Bergmann* [22.31]. For turbulent flows, POD is used to build low-dimensional models that address the role of coherent structures in turbulence generation [22.29, 30]. For the optimal control of fluids, POD is used to obtain reduced-order models of dynamics that reduce the computational complexity associated with high-complexity models such as the Navier–Stokes equations [22.21, 22, 32, 34]. In the control literature [22.40], several philosophies exist for using a reduced basis obtained by applying POD in low-order control design. A *reduce-then-design* approach involves reduction of the system model before control design, and the *design-then-reduce* approach, in which full-order model design is followed by full-order control design, and then control order reduction.

As a partial conclusion, note that the reduced-order models based on POD belong to a wider class of approximation methods, called singular-value-based methods by *Antoulas* and *Sorensen* [22.51]. These authors recently reviewed the state of the art in the area of model reduction of dynamical systems and distinguish three broad categories of approximation methods: singular-value-based methods, Krylov-based methods, and iterative methods combining aspects of both the SVD and Krylov methods. Since, the strengths and weaknesses of these methods are different, new insights can certainly be gained by applying these approximation methods to fluid flow control. For example, the reader is referred to *Allan* [22.52] for an application of the Krylov subspace method to derive an optimal feedback control design for driven cavity flow.

22.4.4 The Different POD Approaches

Except for the inner product, defined as the standard L^2 inner product for simplicity of presentation, the POD was derived in Sect. 22.4.3 in a general setting. The fundamental questions of the choice of

- the input collection
- the inner product
- the averaging operation $\langle \cdot \rangle$ (spatial or temporal)
- the variable X (spatial $\boldsymbol{x} = (x, y, z)$ or temporal t)

were not discussed. This section demonstrates that different orthogonal decompositions can be obtained depending, for example, on the way the averaging operator $\langle \cdot \rangle$ is defined to calculate the kernel of the Fredholm equation (22.60). In what follows, only two methods, classical POD (Sect. 22.4.5) and snapshot POD (Sect. 22.4.6), will be fully described. The reader is referred to *Aubry* et al. [22.53] for a presentation of the generalization of these two methods called *the biorthogonal decomposition*.

Choice of Input Collection

Choosing an input collection is a vital part of the proper orthogonal decomposition process since the POD basis only reflects information provided by the input collection. Indeed, the POD algorithm tries to remove *redundant* information (if any) from the database. As a consequence, the ability of POD modes to approximate any state of a complex system is totally dependent on the information originally contained in the snapshot set used to generate the POD basis. The POD eigenfunctions are intrinsically linked to the input data used to extract them. This is the source of the method's strengths as well as its limitations: extrapolation of the POD functions to different geometries or control parameters (Reynolds number, ...), can be difficult to undertake (see [22.54, Chap. 4.6, p. 254] for a discussion).

When the POD basis is used for model reduction, an input collection of time snapshots is frequently chosen (Sect. 22.4.6). Typically, this data set comes from experimental measurement or numerical computations. Hence the data have some error associated with them. Therefore it is important to study the effect of these errors, assimilated to infinitesimal perturbations, on the outcome of the POD model-reduction procedure. This fundamental question has only recently been investigated theoretically, by *Rathinam* and *Petzold* [22.55]. These authors introduced the POD sensitivity factor as a nondimensional measure of the sensitivity of the resulting projection to perturbations in data. They found that the POD sensitivity is relevant in some applications while it is not in others. These theoretical results still need to be illustrated by realistic examples issued from fluid mechanics, for example. Now, consider the ideal case with no perturbations in the input collection. Choosing a time snapshot input collection relevant for dynamical system description remains a difficult task because there is no definitive way to decide how many snapshots are necessary to capture the information content of the system, how long numerical simulations or experiments should be run to generate sufficiently resolved snapshots, and which initial conditions should be used. The reader should remember that the input collection corresponds to solutions belonging to the attractor of a dynamical system such as the Navier–Stokes equation in fluid mechanics. If this attractor is ergodic, the initial conditions are forgotten as time proceeds [22.24].

For control problems based on reduced-order models, an open question is how to incorporate control information in the model implicitly. A simple solution, and one that is generally used, is to generate snapshots using a variety of control inputs to excite system dynamics that arise when a control is applied [22.56]. Recently, *Burkardt* et al. [22.57] demonstrate that centroidal Voronoi tessellations could be one method of intelligent sampling in the control parameter space that can be employed for generating good snapshots set.

Choice of Inner Product and Norm

So far, POD was described in the context of the standard L^2 inner product for reasons of simplicity and, more importantly, because it corresponds to the general case for fluid flow applications for reasons explained later in Sect. 22.4.4. However, in a few cases it may be useful to use a different inner product to obtain different notions of optimality.

L^2 Inner Product. Let $L^2(\Omega)$ be the Hilbert space of square-integrable complex-valued functions defined on Ω. Square-integrable means that the functions $f(x)$ belonging to $L^2(\Omega)$ satisfy

$$\|f\| = (f, f)^{1/2} = \left[\int_\Omega |f|^2 \, \mathrm{d}x\right]^{1/2} < +\infty \, .$$

For vector-valued functions $\boldsymbol{u}$, such as the velocity field in a fluid flow, the inner product on $L^2(\Omega)$ is defined by

$$(\boldsymbol{u}, \boldsymbol{v}) = \int_\Omega \left(u_1 v_1^* + u_2 v_2^* + u_3 v_3^*\right) \mathrm{d}x \, ;$$

$$\|\boldsymbol{u}\|^2 = (\boldsymbol{u}, \boldsymbol{u}) \, , \qquad (22.73)$$

where Ω denotes the spatial domain occupied by the fluid. Moreover, its kinetic energy is proportional to $\|\boldsymbol{u}\|^2$. Therefore, L^2 is a natural space in which to do fluid mechanics since it corresponds to flow having finite kinetic energy. This is the reason why the L^2 inner product is the most commonly used to define the proper orthogonal decomposition.

H^1 Inner Product. Let $H^1(\Omega)$ be the Sobolev space of functions that, along with their first derivatives, belong to $L^2(\Omega)$.

In *Iollo* [22.58], it is found that the low-order model developed for the Euler equations by a straightforward Galerkin projection (see Sect. 22.4.10 for a description of the method) was unstable. Therefore, *Iollo* et al. [22.59] proposed a way to improve the numerical stability of the low-order model developed by Galerkin POD by redefining the norms involved in the POD definition as

$$(u, v)_\epsilon = \int_\Omega uv \, \mathrm{d}x + \epsilon \int_\Omega (\nabla u \cdot \nabla v) \, \mathrm{d}x \, , \qquad (22.74)$$

where ϵ is a parameter to take into account different metrics. Numerical experiments demonstrate the definite benefit of employing the H^1 formulation in the POD. Even though the use of the H^1 inner product seems beneficial for the robustness of the reduced-order model, we believe that it has not been given sufficient attention in the literature.

Inner Product for Compressible Flow. For compressible flow configurations, the velocity $\boldsymbol{u} = (u, v, w)$ and thermodynamic variables (e.g., density ρ, pressure p, enthalpy h) are dynamically coupled. This introduces questions of whether to treat the thermodynamic variables separately from the velocity, or together as a single vector-valued variable (e.g., $\boldsymbol{q}(x) = (\rho, u, v, w, p)(x)$). For a scalar-valued POD, where separate POD modes are computed for each flow variable, the standard L^2 inner product defined by (22.73) can be used [22.60]. For vector-valued POD, where all the flow variables are written as a single vector $\boldsymbol{q}$, the standard inner product:

$$(\boldsymbol{q}_1, \boldsymbol{q}_2)$$
$$= \int_\Omega (\rho_1 \rho_2 + u_1 u_2 + v_1 v_2 + w_1 w_2 + p_1 p_2) \, \mathrm{d}x$$

may not be a sensible choice for dimensional reasons. Of course, one could simply nondimensionalize the variables, but then the sense in which projections are optimal is rather arbitrary and depends on the nondimensionalization. *Rowley* [22.60] sought an inner product for compressible flow, which makes intuitive sense in that the *energy* defined by the induced norm is a meaningful physical quantity. For a two-dimensional configuration, Rowley introduced a vector-valued variable $q = (u, v, a)$, where u and v are the velocities and a is the local sound speed, and defined a family of inner products as

$$(q_1, q_2)_\epsilon = \int_\Omega \left(u_1 u_2 + v_1 v_2 + \frac{2\epsilon}{\gamma(\gamma-1)} a_1 a_2 \right) dx , \quad (22.75)$$

where γ is the ratio of specific heats and ϵ is a parameter. If $\epsilon = \gamma$ then the induced norm gives $\|q\|^2 = 2h_0$, i.e., twice the total enthalpy of the flow, and if $\epsilon = 1$ then the induced norm gives twice the total energy of the flow.

In spite of the proposals made by Rowley, the choice of the inner product best adapted to aeroacoustics still remains an open question.

To summarize: a POD approach is linked first to the definition of an inner product, which depends on the quantity to be investigated. When very different parameters have to be considered at the same time in this decomposition, weighting or proper normalization has to be considered in such a way that the behavior of these parameters is properly taken into account.

22.4.5 Classical POD or Direct Method

This approach was originated by *Lumley* [22.3]. In this case, the average $\langle \cdot \rangle$ is temporal:

$$\langle \cdot \rangle = \frac{1}{T} \int_T \cdot \, dt$$

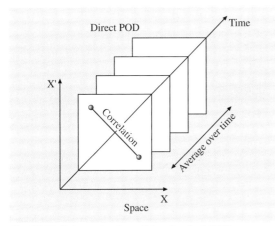

Fig. 22.16 Schematic view of the classical POD

and is evaluated as an ensemble average, based on the assumptions of stationarity and ergodicity. On the other hand, the variable X is assimilated to the space variable $x = (x, y, z)$ defined over the domain Ω.

Figure 22.16 describes schematically the principle of the classical POD.

The corresponding eigenvalue problem is easily deduced from (22.62) by replacing the domain of integration $\mathcal{D}$ by Ω and the variable X by x. The integral Fredholm equation to be solved is then given by

$$\sum_{j=1}^{n_c} \int_\Omega R_{ij}(x, x') \Phi_j^{(n)}(x') dx' = \lambda^{(n)} \Phi_i^{(n)}(x) , \quad (22.76)$$

where $R_{ij}(x, x')$ is the *two-point spatial correlation tensor* defined as

$$R_{ij}(x, x') = \frac{1}{T} \int_T u_i(x, t) u_j(x', t) dt$$

$$= \sum_{n=1}^{N_{\text{POD}}} \lambda^{(n)} \Phi_i^{(n)}(x) \Phi_j^{(n)*}(x')$$

with T being a sufficiently long period of time for which the space–time signal $u(x, t)$ is known and with N_{POD} being the number of POD modes, i.e., the size of the eigenvalue problem (22.76). Note that the eigenfunctions arising from this decomposition are purely spatial.

Discussion of the Size of the Eigenvalue Problem

Given M, the number of spatial points of the snapshots data, and n_c, the number of components of the variable u used for the decomposition, $N_{\text{POD}} = M \times n_c$, $M = N_x \times N_y \times N_z$, where N_x, N_y and N_z are the number of nodes of the experimental or numerical grid, respectively, in the X, Y and Z directions. Now, suppose one performs a detailed numerical simulation or one employs a modern measurement technique such as particle image velocimetry in fluid mechanics. In each case, a large number of grid points M can be obtained and the size of the POD problem can then quickly become too large to be solved with a good numerical precision even with numerical library dedicated to this kind of problem, like the ARPACK library [22.61].

Nevertheless, as will be demonstrated in Sect. 22.4.8, the POD method can be viewed as the generalization of the harmonic decomposition to the inhomogeneous directions. So one way to take into account this size constraint with the classical POD approach is to decompose the flow directions in homogeneous and inhomogeneous directions as was done in general in experimental approaches [22.25, 30].

Now, suppose that the number of ensemble members deemed adequate for a description of the process is N_t with $N_t \ll M$ (the question of determining N_t is not addressed), then even if the eigenvalue problem can be accurately solved, time can be saved in solving an eigenvalue problem of size N_t. This remark is at the heart of the method of snapshots.

22.4.6 Snapshot POD

The snapshot POD method, suggested by *Sirovich* [22.62–64], is the exact symmetry of the classical POD. The average operator $\langle . \rangle$ is evaluated as a space average over the domain Ω of interest:

$$\langle \cdot \rangle = \int_\Omega \cdot \, d\mathbf{x}$$

and the variable X is assimilated to time t.

The principle of the snapshot POD method is schematically described in Fig. 22.17.

The Discrete Eigenvalue Problem

To derive the discrete eigenvalue problem corresponding to the snapshot POD, we assume that $\boldsymbol{\Phi}$ has a special form in terms of the original data:

$$\boldsymbol{\Phi}(\mathbf{x}) = \sum_{k=1}^{N_t} a(t_k) \mathbf{u}(\mathbf{x}, t_k), \quad (22.77)$$

where the coefficients $a(t_k)$, $k = 1, \ldots, N_t$ are to be determined so that $\boldsymbol{\Phi}$ given by (22.77) provides a maximum for (22.58), i.e., is the solution of equation (22.76), written here for convenience as

$$\int_\Omega R(\mathbf{x}, \mathbf{x}') \boldsymbol{\Phi}(\mathbf{x}') \, d\mathbf{x}' = \lambda \boldsymbol{\Phi}(\mathbf{x}). \quad (22.78)$$

More exactly, the properties of the span of the POD eigenfunctions guarantee that such a development exists [22.24].

The two-point spatial correlation tensor $R(\mathbf{x}, \mathbf{x}')$ is estimated under stationarity and ergodicity assumptions as:

$$R(\mathbf{x}, \mathbf{x}') = \frac{1}{T} \int_T \mathbf{u}(\mathbf{x}, t) \otimes \mathbf{u}^*(\mathbf{x}', t) \, dt$$

$$= \frac{1}{N_t} \sum_{i=1}^{N_t} \mathbf{u}(\mathbf{x}, t_i) \otimes \mathbf{u}^*(\mathbf{x}', t_i).$$

Substituting this expression of R and the decomposition (22.77) of $\boldsymbol{\Phi}$ into (22.78), we obtain

$$\sum_{i=1}^{N_t} \left[\sum_{k=1}^{N_t} \frac{1}{N_t} \left(\int_\Omega \mathbf{u}(\mathbf{x}', t_k) \cdot \mathbf{u}^*(\mathbf{x}', t_i) \, d\mathbf{x}' \right) a(t_k) \right]$$
$$\times \mathbf{u}(\mathbf{x}, t_i)$$
$$= \lambda \sum_{k=1}^{N_t} a(t_k) \mathbf{u}(\mathbf{x}, t_k)$$

and we conclude that a sufficient condition for the coefficients $a(t_k)$ to be a solution of (22.78) is to verify that

$$\sum_{k=1}^{N_t} \frac{1}{N_t} \left[\mathbf{u}(\mathbf{x}', t_k), \mathbf{u}^*(\mathbf{x}', t_i) \right] a(t_k) = \lambda a(t_i),$$
$$i = 1, \ldots, N_t. \quad (22.79)$$

This can be rewritten as the eigenvalue problem

$$CV = \lambda V, \quad (22.80)$$

where

$$C_{ki} = \frac{1}{N_t} \int_\Omega \mathbf{u}(\mathbf{x}, t_k) \cdot \mathbf{u}^*(\mathbf{x}, t_i) \, d\mathbf{x} \quad \text{and}$$

$$V = [a(t_1), a(t_2), \ldots, a(t_{N_t})]^T.$$

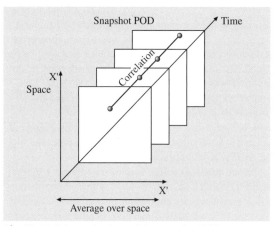

Fig. 22.17 Schematic view of the snapshot POD

Note that, in order for (22.79) to be a necessary condition, one needs to assume that the observations $u(x, t_i), i = 1, \ldots, N_t$ are linearly independent.

Since C is a nonnegative Hermitian matrix, it has a complete set of orthogonal eigenvectors

$$V^{(1)} = \left[a^{(1)}(t_1), a^{(1)}(t_2), \ldots, a^{(1)}(t_{N_t}) \right]^T ,$$
$$V^{(2)} = \left[a^{(2)}(t_1), a^{(2)}(t_2), \ldots, a^{(2)}(t_{N_t}) \right]^T , \ldots ,$$
$$V^{(N_t)} = \left[a^{(N_t)}(t_1), a^{(N_t)}(t_2), \ldots, a^{(N_t)}(t_{N_t}) \right]^T$$

along with a set of eigenvalues $\lambda^{(1)} \geq \lambda^{(2)} \geq \ldots \geq \lambda^{(N_t)} \geq 0$. Now, for reasons of simplicity, we can impose that the projection coefficients $a(t_k), k = 1, \ldots, N_t$ verify the same orthogonality conditions as for the classical POD. Then we can normalize the temporal eigenfunctions $V^{(i)}$ by requiring that

$$\frac{1}{N_t}(V^{(n)}, V^{(m)}) = \frac{1}{N_t} \sum_{k=1}^{N_t} a^{(n)}(t_k) a^{(m)*}(t_k)$$
$$= \lambda^{(n)} \delta_{nm} .$$

It is now easy to check that, if the POD eigenfunctions $\boldsymbol{\Phi}^{(n)}(x)$ are not estimated via (22.77) but as

$$\boldsymbol{\Phi}^{(n)}(x) = \frac{1}{N_t \lambda^{(n)}} \sum_{k=1}^{N_t} a^{(n)}(t_k) u(x, t_k) , \quad (22.81)$$

then the spatial modes are orthonormal

$$\int_\Omega \boldsymbol{\Phi}^{(n)}(x) \cdot \boldsymbol{\Phi}^{(m)}(x) \, dx = \delta_{nm} .$$

The Continuous Eigenvalue Problem

So far, the snapshot POD method was presented as by Sirovich in his original work [22.62–64]. Therefore the eigenvalue problem (22.80) is discrete and not continuous, as was defined the eigenvalue problem (22.76) derived for the classical POD. However, deducing an integral Fredholm equation from (22.80) is immediate; we obtain

$$\int_T C(t, t') a^{(n)}(t') \, dt' = \lambda^{(n)} a^{(n)}(t) , \quad (22.82)$$

where $C(t, t')$ is the *two-point temporal correlation tensor* defined as

$$C(t, t') = \frac{1}{T} \int_\Omega u_i(x, t) u_i(x, t') \, dx$$
$$= \frac{1}{T} \sum_{n=1}^{N_{\text{POD}}} a^{(n)}(t) a^{(n)*}(t') .$$

where in this definition, the summation over i is implicit.

The main properties of the snapshot POD are the following:

1. The eigenfunctions are purely time dependent.
2. No cross-correlations appear in the kernel.
3. The homogeneity hypothesis is not required to lower the size of the eigenvalue problem.
4. Linear independence of the snapshots is assumed.
5. The size of the eigenvalue problem (22.82) is $N_{\text{POD}} = N_t$. Then, as was mentioned in Sect. 22.4.5, the snapshot POD drastically reduces the computational effort when M, the number of spatial points in the snapshots data, is much greater than N_t. For this reason, whenever this condition is fulfilled, the snapshot POD is preferred.

22.4.7 Common Properties of the Two POD Approaches

General Properties

Whatever the particular method used to determine the spatial and temporal POD eigenfunctions, they fulfil the following properties:

1. Each space–time realization $u_i(x, t)$ can be expanded into orthogonal basis functions $\Phi_i^{(n)}(x)$ with uncorrelated coefficients $a^{(n)}(t)$:

$$u_i(x, t) = \sum_{n=1}^{N_{\text{POD}}} a^{(n)}(t) \Phi_i^{(n)}(x) .$$

2. The spatial modes $\boldsymbol{\Phi}^{(n)}(x)$ are specified to be orthonormal:

$$\int_\Omega \boldsymbol{\Phi}^{(n)}(x) \cdot \boldsymbol{\Phi}^{(m)}(x) \, dx = \delta_{nm} .$$

3. The temporal modes $a^{(n)}(t)$ are orthogonal:

$$\frac{1}{T} \int_T a^{(n)}(t) a^{(m)*}(t) \, dt = \lambda^{(n)} \delta_{nm} .$$

Incompressibility and Boundary Conditions

The spatial basis functions $\Phi_i^{(n)}(x)$ can be calculated from the velocities $u_i(x, t)$ and the coefficients $a^{(n)}(t)$ by integrating over a sufficiently long period of time T and normalizing by the eigenvalues $\lambda^{(n)}$:

$$\Phi_i^{(n)}(x) = \frac{1}{T\lambda^{(n)}} \int_T u_i(x, t) a^{(n)*}(t) \, dt \, . \tag{22.83}$$

The POD basis functions are then represented as linear combinations of instantaneous velocity fields. All the properties of the snapshots that can be written as linear and homogeneous equations pass directly to the POD basis functions. For example, if the snapshots are divergence free, then we obtain divergence-free POD basis functions

$$\nabla \cdot u = 0 \Longrightarrow \nabla \cdot \Phi^{(n)} = 0 \quad \forall n = 1, \ldots, N_{\text{POD}} \, .$$

If the snapshots satisfy homogeneous Dirichlet boundary conditions then we also obtain POD basis functions satisfying homogeneous boundary conditions.

Snapshot or Classical POD?

As presented in Sect. 22.4.5, 4.6, two different POD approaches exist: the classical POD and the snapshot POD; how then can we choose the pertinent method for each practical configuration? The answer is mainly determined by the particular data set available for the evaluation of the kernels.

On the one hand, data obtained by numerical simulations like direct numerical simulation or large-eddy simulation can be highly resolved in space and time but due to cost considerations only a very short time sample is simulated. In the same vein, a good spatial resolution can be obtained by particle image velocimetry, but associated with a poor temporal resolution. These two configurations, characterized by a moderate time history and high spatial resolution, correspond to situations for which the two-point temporal correlation tensor $C(t, t')$ is statistically well converged.

On the other hand, experimental approaches such as hot-wire anemometry or laser Doppler anemometry provide a well-defined time description but with limited spatial resolution. These measurements techniques enabled long time histories and moderate spatial resolution. Therefore, the two-point spatial correlation tensor $R_{ij}(x, x')$ is statistically well converged.

In conclusion, the data issued from an experimental approach will generally be treated using the classical method and data issued from numerical simulations by the snapshots method. An exception is the case of data sets obtained from particle image velocimetry.

22.4.8 POD and Harmonic Analysis

As long as the domain $\mathcal{D}$ defined in (22.60) is bounded, the Hilbert–Schmidt theory applies [22.46], and all the properties stated in Sect. 22.4.3 hold. It is thus necessary to pay special attention to flow directions assumed to be homogeneous, stationary or periodic.

A First Approach: Homogeneity in One Direction

As a first approach, we can assume, for example, that the spatial direction OX_3 is homogeneous (a generalization including other directions is straightforward). If OX_3 is homogeneous then the two-point correlation $R(x, x')$ depends only on the difference $r_3 = x'_3 - x_3$ of the two coordinates in the OX_3 direction:

$$R_{ij}(x_1, x'_1, x_2, x'_2, x_3, x'_3, t, t') \\ = R_{ij}(x_1, x'_1, x_2, x'_2, t, t'; r_3) \, .$$

Splitting the space–time variable $X = (x_1, x_2, x_3, t)$ into a homogeneous variable x_3 and an inhomogeneous variable $\chi = (x_1, x_2, t)$, the integral Fredholm equation (22.60) is written

$$\sum_{j=1}^{n_c} \int_{\mathcal{D}'} \int_{-\infty}^{+\infty} R_{ij}(\chi, \chi'; x_3 - x'_3) \Phi_j(\chi', x'_3) \, d\chi' \, dx'_3 \\ = \lambda \Phi_i(\chi; x_3) \, . \tag{22.84}$$

Under the homogeneity hypothesis, we may develop the spatial eigenfunction Φ_l in a Fourier series decomposition as:

$$\Phi_l(\chi; r_3) = \sum_{k_3=-\infty}^{+\infty} \hat{\Phi}_l(\chi; k_3) \exp(2\pi i k_3 r_3) \tag{22.85}$$

and introduce Π_{ij}, the Fourier transform of R_{ij} in the direction OX_3:

$$\Pi_{ij}(\chi, \chi'; k_3) \\ = \int_{-\infty}^{+\infty} R_{ij}(\chi, \chi'; r_3) \exp(-2\pi i k_3 r_3) \, dr_3 \\ = \int_{-\infty}^{+\infty} R_{ij}(\chi, \chi'; -r_3) \exp(2\pi i k_3 r_3) \, dr_3 \, , \tag{22.86}$$

where k_3 is the spatial wavenumber associated to r_3.

Substituting expression (22.85) in (22.84), we first obtain

$$\sum_{j=1}^{n_c}\sum_{k_3=-\infty}^{+\infty}\int_{\mathcal{D}'}\int_{-\infty}^{+\infty}R_{ij}(\chi,\chi';-r_3)\hat{\Phi}_j(\chi',k_3)$$
$$\times\exp[2\pi i k_3(x_3+r_3)]\,\mathrm{d}\chi'\,\mathrm{d}r_3$$
$$=\sum_{k_3=-\infty}^{+\infty}\lambda(k_3)\hat{\Phi}_i(\chi;k_3)\cdot\exp(2\pi i k_3 x_3)\,. \quad (22.87)$$

Then, replacing the two-point correlation R_{ij} by his Fourier transform Π_{ij} defined by (22.86), the Fredholm equation becomes

$$\sum_{k_3=-\infty}^{+\infty}\left[\sum_{j=1}^{n_c}\int_{\mathcal{D}'}\Pi_{ij}(\chi,\chi';k_3)\hat{\Phi}_j(\chi',k_3)\,\mathrm{d}\chi'\right]$$
$$\times\exp(2\pi i k_3 x_3)$$
$$=\sum_{k_3=-\infty}^{+\infty}\lambda(k_3)\hat{\Phi}_i(\chi;k_3)\cdot\exp(2\pi i k_3 x_3)\,. \quad (22.88)$$

Finally, the uniqueness of the Fourier series coefficients implies that the Fredholm equation (22.60) is equivalent to:

$$\sum_{j=1}^{n_c}\int_{\mathcal{D}'}\Pi_{ij}(\chi,\chi';k_3)\hat{\Phi}_j(\chi',k_3)\,\mathrm{d}\chi'$$
$$=\lambda(k_3)\hat{\Phi}_i(\chi;k_3)\,. \quad (22.89)$$

The conclusion is that the homogeneity hypothesis in the OX_3 direction decouples the initial POD problem into a set of lower-dimensional problems. For each Fourier wavenumber k_3 the eigenvalue problem to solve is:

$$\sum_{j=1}^{n_c}\int_{\mathcal{D}'}\Pi_{ij}(\chi,\chi')\hat{\Phi}_j(\chi')\,\mathrm{d}\chi'=\lambda\hat{\Phi}_i(\chi)\quad\forall k_3\,. \quad (22.90)$$

Another key result is that, in each homogeneous (or stationary) direction, harmonic functions are solutions of the integral Fredholm equations. Then, as a first approximation, the proper orthogonal decomposition can be viewed as the generalization of the harmonic decomposition to the inhomogeneous directions.

This result is especially useful in systems where the domain $\mathcal{D}$ is of higher dimension. For example, in the study of the three-dimensional turbulent plane mixing layer realized via the classical POD in *Delville* et al. [22.25] and *Ukeiley* et al. [22.30], we appeal to homogeneity in the spanwise (x_3) and streamwise (x_1) directions. Selecting the finite domain $[0,L_1]\times[0,L_3]$ in these variables, we use a mixed Fourier–empirical decomposition of the form:

$$u(x,t)$$
$$=\sqrt{L_1 L_3}\int_{-\infty}^{+\infty}\int\sum_{n=1}^{N_{\text{POD}}}a_{k_1,k_3}^{(n)}(t)\Phi^{(n)}(x_2;k_1,k_3)$$
$$\times\exp[2\pi i(k_1 x_1+k_3 x_3)]\,\mathrm{d}k_1\,\mathrm{d}k_3\,.$$

As detailed in *Delville* et al. [22.25], time is mapped to the streamwise direction through Taylor's hypothesis.

The vector-valued eigenfunctions $\Phi^{(n)}(x_2;k_1,k_3)$ are obtained by solving a Fredholm equation analogous to (22.60) in which the kernel R_{ij} is replaced by the cross-spectral tensor $\Psi_{ij}(x_2,x_2';k_1,k_3)$ defined as the streamwise and spanwise transform of the cross-correlation tensor. More details are given in *Cordier* [22.65]; see also *Delville* et al. [22.25] and *Ukeiley* et al. [22.30].

Phase Indetermination

The *phase indetermination* is one of the most important limitations of the POD. This indetermination is due to the use of two-point correlations and, as will be demonstrated in the following, appears only for directions where an harmonic decomposition is used.

Suppose that the eigenfunction $\hat{\Phi}_j(\chi';k_3)$ is a solution of (22.89). Then it can easily be proven that every function $\hat{\Phi}_j(\chi';k_3)\theta(k_3)$, where $\theta(k_3)$ is a random phase function, will also be a solution. The phase information between the different modes is lost, and the eigenfunctions $\hat{\Phi}_i(\chi;k_3)$ are known up to an arbitrary function $\theta(k_3)$, which needs to be determined. In particular, for classical POD, it is impossible to obtain directly a description of the preferred modes in the physical space. However, a description of the dominant modes can be obtained by using a complementary method called *the shot-noise theory* fully described in *Herzog* [22.66], and *Moin* and *Moser* [22.67]. The reader is referred to *Delville* et al. [22.25] for an application of the shot-noise theory to recover the dominant modes of a three-dimensional turbulent mixing layer from the POD eigenfunctions determined by classical POD.

An alternative way is to build, from the dominant POD eigenfunctions, a low-order model by use of a Galerkin projection of the governing equations onto the POD modes, leading to a low-order dynamical system described by a set of ordinary differential equations. In this case, these equations themselves drive the missing spectral phase information. This approach has been successfully addressed for the wall region of a turbulent boundary layer in *Aubry* et al. [22.29] and for a plane turbulent mixing layer in *Ukeiley* et al. [22.30]. Naturally, this kind of low-order model is particularly suited for active flow control studies.

22.4.9 Typical Applications to Fluid Mechanics

POD in the Mixing Layer

In this section, we briefly present a few results obtained from the application of classical POD to experimental measurements obtained in a two-dimensional plane turbulent mixing layer [22.68]. The streamwise and transverse instantaneous velocity components are determined by using a rake of 12 equally spaced X-wires uniformly distributed according to the transverse y-direction of the flow. In the present study, only the spatial direction y (mean gradient, inhomogeneous) and the time delay τ are considered. The flow being stationary, a Fourier transform has to be used in the time direction before applying POD (Sect. 22.4.8). Of course, this approach is limited by the fact that the three-dimensional aspect of the flow cannot be assessed; only one slice of the flow is viewed here. Hence the full three-dimensional behavior of the flow is not analyzed. However, useful information on the global organization of the flow can be outlined.

Following the classical POD method described in Sect. 22.4.5, the dominant structure of the flow can be determined from the following Fredholm equation:

$$\sum_{j=1}^{n_c} \int_{\Omega} \Psi_{ij}(y, y'; f) \Phi_j^{(n)}(y'; f) \mathrm{d}y'$$
$$= \lambda^{(n)}(f) \Phi_i^{(n)}(y; f), \qquad (22.91)$$

where n_c is the number of velocity components on which POD is performed, and where the cross-spectrum $\Psi_{ij}(f)$ is the temporal Fourier transform of the two-point space–time correlations

$$R_{ij}(y, y'; \tau) = \langle u_i'(y, t) u_j'(y', t + \tau) \rangle .$$

The Fourier transform of the velocity can be retrieved from the eigenfunctions Φ_i

$$\hat{u}_i'(y; f) = \sum_{n=1}^{N_{\mathrm{POD}}} a^{(n)}(f) \Phi_i^{(n)}(y; f) ,$$

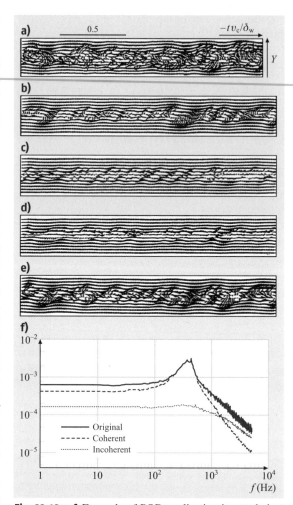

Fig. 22.18a–f Example of POD application in a turbulent plane mixing layer. A rake of 12 X-wire probes is used to measure the instantaneous distribution of two velocity components across the mixing layer. (**a**) Original measurements. (**b**) First POD mode. (**c**) Second POD mode. (**d**) Third POD mode. (**e**) Contribution of the first three POD modes. (**f**) Energy spectrum of the transverse velocity component. *Original* corresponds to the hot-wire measurements, *Coherent* to the contribution of the first three POD modes and *Incoherent* to the remaining POD modes

where $a^{(n)}(f)$ is found from

$$a^{(n)}(f) = \int_{\Omega} \hat{u}_i'(y; f) \Phi_i^{(n)}(y; f) \mathrm{d}y . \qquad (22.92)$$

In terms of structure identification, the participation of any POD mode in the instantaneous flow field can be considered. Figure 22.18 shows, for a selected sample, an example of the instantaneous velocity field plotted in a frame moving with a constant convection velocity U_c using a Taylor hypothesis based on U_c for the longitudinal direction. Here, this convection velocity is chosen arbitrarily as the average of the two velocity streams of the mixing layer. The organization, which can be visually detected from the original velocities (Fig. 22.18a), is relatively well reconstructed by the first POD mode (Fig. 22.18b). However, the spatial extent of the events in the y direction is generally underestimated, and higher modes (Fig. 22.18c,d) are needed to improve this (Fig. 22.18e).

The power spectra deduced from the original, coherent, and incoherent contributions to the transverse velocity field are superimposed on Fig. 22.18f. The coherent flow field has a similar spectral distribution to the original (total) flow until the frequency 500 Hz, while the incoherent flow is uncorrelated and exhibits a flat energy spectrum in this frequency domain. We then see that the POD coherent component captures the strongest frequency peak, lying at around 400 Hz and associated with the two-dimensional Kelvin–Helmholtz instability. At high frequencies, the spectral slope obtained for the incoherent spectrum is similar to that deduced from the original spectrum, while the coherent spectrum exhibits a different slope. The spectral content of the small scales has then been well reconstituted with the POD incoherent field. Finally, the background fluctuations exhibit a spectrum corresponding to a quasihomogeneous equilibrium turbulence. This figure also shows that POD coherent structures are not localized in the spectral domain. They correspond to multiple scales and differ from the usual approaches of Fourier filtering in time or space.

The Complementary Technique in the Mixing Layer

In order to perform the projection required to obtain the random coefficients $a^{(n)}(f)$ from the POD, it is necessary to have knowledge of the flow field at all points in space (22.92). Depending on the measurement apparatus, this is not always possible. For example, it is sometimes difficult to operate several hot-wire probes simultaneously. On the other hand, the determination of the correlation tensor used for applying POD only requires a two-point measurement procedure. In this case, the linear stochastic estimation (see Sect. 22.4.6 for further details), also based on two-point correlations, allows the estimation of the raw data on which POD can be used as a structure identification process (by retaining only a small number of modes). This approach, called

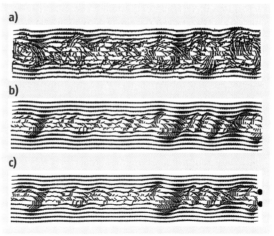

Fig. 22.19a–c Application of the complementary technique to the plane mixing layer. (**a**) Original measurements. (**b**) Direct application of POD. (**c**) Application of the complementary technique (two points of reference were used for the linear stochastic estimation)

a complementary method, was introduced by *Bonnet* et al. [22.69].

Mathematically, the stochastic estimates of the random coefficients are now calculated from:

$$a^{(n)}_{\text{est}}(f) = \int_\Omega u'_{i,\text{est}}(y;f) \Phi^{(n)}_i(y;f) \, dy \,,$$

where $u'_{i,\text{est}}(y;f)$ is either a single or multipoint linear stochastic estimate of the velocity field and $\Phi^{(n)}_i(y;f)$ is obtained from the original POD eigenvalue problem. The estimated velocity can be reconstructed in Fourier space by

$$\hat{u}'_i(y;f) = \sum_{n=1}^{N_{\text{POD}}} a^{(n)}_{\text{est}}(f) \Phi^{(n)}_i(y;f) \,,$$

and then inversely transformed to obtain $\hat{u}'_i(y,t)$. The example plotted in Fig. 22.19 illustrates this procedure applied to the same data as used in Fig. 22.18.

22.4.10 POD Galerkin

General Methodology

Before going into the details of the application of the Galerkin projection for the POD eigenfunctions, we first recall the basic ideas of Galerkin projection.

Galerkin projection is a special case of *weighted residual methods* [22.70]. These methods are dedicated to solve functional equations, such as ordinary (ODE) or partial (PDE) differential equations, or integral equations. For example, consider the equation

$$\mathcal{L}(u) = 0 \tag{22.93}$$

defined in a domain Ω, where $\mathcal{L}$ is some differential operator. The Galerkin method is an approximation to the true solution of (22.93) sought by weighting the residual of the differential equation. Assume that U is an approximate solution to (22.93). Substitution of U for u in (22.93) results in a nonzero residual $r = \mathcal{L}(U) \neq 0$.

The best approximation for U is that which reduces the residual r to the least value at all points in the domain Ω. The weighted residual method enforces the condition

$$R_j = \int_\Omega w_j r \, d\Omega = 0, \tag{22.94}$$

where the R_j are the weighted residual integrals and w_j are the weighting functions. In the Galerkin method, the weighting functions are chosen to be the same as the basis functions used in the expansion of the approximate solution. Hence, if U is approximated by the expansion

$$U(x) = \sum_{i=1}^{\infty} \chi_i \varphi_i(x), \tag{22.95}$$

where $\varphi_i(x)$ are the basis functions and χ_i are the coefficients to be determined, then the weighting functions are selected as $w_j = \varphi_j(x)$, $j = 1, \ldots, +\infty$. The fact that the unknown u solution of (22.93) is a member of an infinite-dimensional space is a practical difficulty. The discretization of the Galerkin procedure then consists of truncating the sum in (22.95) at a finite index $i_{\max}$, thus rendering the problem a finite dimensional one. Therefore, (22.94) becomes

$$\int_\Omega \varphi_j(x) \mathcal{L}(U) \, d\Omega = 0, \quad j = 1, \ldots, i_{\max}. \tag{22.96}$$

If the definition (22.73) of the inner product is introduced, then equation (22.96) further simplifies to

$$\left\{ \mathcal{L} \left[\sum_{i=1}^{i_{\max}} \chi_i \varphi_i(x) \right], \varphi_j \right\} = 0, \quad j = 1, \ldots, i_{\max}. \tag{22.97}$$

Finally, the Galerkin Projection is equivalent to impose the $i_{\max}$ scalar products defined by equation (22.97) to vanish.

For such an approach to work [22.71], the two following requirements should be satisfied:

1. the function space $\{\varphi_i\}$ of the basis functions $\varphi_i(x)$ must be complete and for practical reasons to become clear in the next section, an orthonormal basis is especially desirable.
2. the basis functions $\varphi_i(x)$ must meet the boundary conditions of the problem.

Reduced Order Models Based on POD

From the properties of the POD eigenfunctions $\boldsymbol{\Phi}^m(x)$ as described in Sect. 22.4.3 one can immediately see that these eigenfunctions are particularly suited for Galerkin projection.

The Navier–Stokes equations for incompressible flows can be written symbolically as

$$\frac{\partial u}{\partial t} = F(u) \text{ with } u = u(x, t) x \in \Omega \text{ and } t \geq 0, \tag{22.98}$$

where F is a differential operator that contains only spatial derivatives and where Ω is the spatial domain under study. The differential equation (22.98) is mathematically well posed if the system is completed by initial conditions:

$$u(x, t = 0) = u_0(x) \tag{22.99}$$

and boundary conditions. Here, we decide to follow the viewpoint of *Fahl* [22.34] and to formulate the differential equation (22.98) in the general context of boundary control problem for fluid flows. Hence, we assume that the boundary of the domain Γ, can be split into two parts such that Γ_c denotes that part of the boundary where the control is applied and $\Gamma \setminus \Gamma_c$ is the part of the boundary that is not controlled. Then, we can complete the (22.98) with the Dirichlet boundary conditions:

$$u(x, t) = h[x, t; d(t)] \text{ with } x \in \Gamma \text{ and } t \geq 0, \tag{22.100}$$

where d is the control input. More precisely, the boundary conditions can be written as

$$h[x, t; d(t)] = \begin{cases} \gamma(t) c(x) & x \in \Gamma_c, t \geq 0, \\ g(x) & x \in \Gamma \setminus \Gamma_c, t \geq 0, \end{cases} \tag{22.101}$$

where $\gamma(t)$ can be interpreted as the temporal variation of a prescribed control action defined by $c(x)$, $x \in \Gamma_c$. For example, this formulation corresponds to flow control by blowing and suction along a portion of the boundary as considered in *Joslin* et al. [22.72].

The reduced-order model is then derived by Galerkin projection of the PDE (22.98) onto the POD subspace of dimension N_{POD}. The first step is to insert in (22.98) the development of u on the POD basis $\boldsymbol{\Phi}^{(m)}$:

$$u(x,t) = \sum_{m=1}^{N_{POD}} a^{(m)}(t) \boldsymbol{\Phi}^{(m)}(x)$$

to obtain (using the notation $\dot{a} = \mathrm{d}a/\mathrm{d}t$)

$$\sum_{m=1}^{N_{POD}} \dot{a}^{(m)}(t) \boldsymbol{\Phi}^{(m)}(x) = F\left(\sum_{m=1}^{N_{POD}} a^{(m)}(t) \boldsymbol{\Phi}^{(m)}(x)\right) . \tag{22.102}$$

The set of spatial eigenfunctions $\boldsymbol{\Phi}^{(m)}(x)$ being a basis, the right-hand side of (22.102) can be written as a linear combination of $\boldsymbol{\Phi}$:

$$F\left(\sum_{m=1}^{N_{POD}} a^{(m)}(t) \boldsymbol{\Phi}^{(m)}(x)\right)$$
$$= \sum_{n} \mathcal{F}^{(n)}\left(a^{(1)}, a^{(2)}, \ldots\right) \boldsymbol{\Phi}^{(n)}(x) .$$

Finally, the Galerkin projection of (22.98) onto the POD eigenfunctions is evaluated as

$$\left(\boldsymbol{\Phi}^{(n)}, \frac{\partial u}{\partial t}\right) = \left(\boldsymbol{\Phi}^{(n)}, F(u)\right)$$

for $n = 1, \ldots, N_{Gal}$,

where N_{Gal} is the number of Galerkin modes kept in the projection. From the orthonormality of the eigenfunctions $\boldsymbol{\Phi}^{(m)}(x)$, the PDE (22.98) is replaced by a set of ODEs defined by

$$\dot{a}^{(n)}(t) = \mathcal{F}^{(n)}\left(a^{(1)}(t), \ldots, a^{(n)}(t)\right) ,$$

with $n = 1, \ldots, N_{Gal}$, \hfill (22.103)

where

$$\mathcal{F}^{(n)} = \left[\boldsymbol{\Phi}^{(n)}, F\left(\sum_{m=1}^{N_{Gal}} a^{(m)}(t) \boldsymbol{\Phi}^{(m)}(x)\right)\right] .$$

(The left-hand side of (22.103) is estimated as

$$\left(\boldsymbol{\Phi}^{(n)}, \sum_{m} \dot{a}^{(m)} \boldsymbol{\Phi}^{(m)}(x)\right)$$
$$= \sum_{m} \dot{a}^{(m)}(t) \left(\boldsymbol{\Phi}^{(n)}, \boldsymbol{\Phi}^{(m)}\right) .$$

Hence, as previously noted at the beginning of this section, an orthonormal basis is especially desirable because we avoid inverting an $N_{Gal} \times N_{Gal}$ matrix to solve for $\dot{a}^{(n)}(t)$)

The functions $\mathcal{F}^{(n)}$ are linear if F is a linear operator and, in our case, due to the convective terms in the Navier–Stokes equations, $\mathcal{F}^{(n)}$ are usually quadratic functions of $a^{(n)}$. In *Aubry* et al. [22.29] and *Ukeiley* et al. [22.30], the low-order dynamical system (22.103) have cubic terms because the velocity field is decomposed into mean and fluctuating components ($u = \langle u \rangle + u'$), where the mean is slowly varying in time. The mean may then be described in terms of the fluctuations u', which give rise to Reynolds stresses and then to cubic equations for (22.103). Another enhancement to the basic theory described so far is the modeling of energy transfer to the higher modes (corresponding to the dissipative scales of the flow) neglected in the Galerkin projection procedure. In *Aubry* et al. [22.29], the influence of the missing scales is simply parameterized by a generalization of the Heisenberg spectral model in homogeneous turbulence [22.73] and in *Ukeiley* et al. [22.30] the mean velocity is split into a steady and time-dependent part by choosing cutoff wavenumbers. Other methods for improving the accuracy of the POD reduced-order model are discussed at the end of this section.

To obtain a well-posed mathematical problem, one needs to add a set of initial conditions to the reduced-order model (22.103) and to make sure that the problem (22.103) matches the original boundary conditions (22.100). The initial conditions can be directly inferred from the conditions (22.99) imposed to the original problem

$$a^{(n)}(t=0) = a_0^{(n)} , \text{ where } a_0^{(n)} = \left(u_0(x), \boldsymbol{\Phi}^{(n)}\right) .$$

For the boundary conditions, the answer is not so direct because it depends mainly of the type of boundary conditions applied, homogeneous or nonhomogeneous boundary conditions. For this reason, this question is postponed to the next subsection.

To sum up, combining the Galerkin projection methodology and the optimality of convergence of POD eigenfunctions (Sect. 22.4.3), we demonstrate that high-dimensional models represented by partial differential equations (PDE) can be replaced by low-dimensional dynamical models of nonlinear ordinary differential equations (ODE).

Problem of Boundary Conditions

According to (22.83), when the Navier–Stokes equations (NSE) are subjected to homogeneous Dirichlet boundary conditions (for example, $h(x, t; c(t)) = 0$), the POD basis functions inherit these boundary conditions. Therefore, the reduced-order models developed by Galerkin projection of the NSE on the POD subspace are equivalent to the original NSE. More precisely, we can only argue that the system of ODEs (22.103) are mathematically equivalent to the original problem (22.98). We are sure that there exists a solution of the reduced-order model (22.103) that lies on the manifold that is defined by the boundary conditions of the original problem. However, the fundamental question if whether a flow can be represented exactly by a finite-dimensional basis of POD eigenfunctions and the question of stability of that manifold are still not fully answered. Issues concerning the stability and the accuracy of a Galerkin projection are discussed in *Iollo* [22.58] and *Iollo* et al. [22.59]. The possible connections between the stability properties of the manifold and the stability properties of the physical phenomenon are addressed in *Rempfer* [22.71]. The expansion coefficients $a_{DS}^{(n)}(t), n = 1, \ldots, N_{Gal}$ that are solutions of the dynamical system (22.103) can then be used to compute the reduced-order solution

$$u_{DS}(x, t) = \sum_{n=1}^{N_{Gal}} a_{DS}^{(n)}(t) \Phi^{(n)}(x),$$

where the subscript DS denotes the dynamical system approximation to the original solution $u(x, t)$ of the NSE.

Now we consider the NSE with nonhomogeneous Dirichlet boundary conditions defined by (22.101). Assume that the snapshots data $u(x, t_i), i = 1, \ldots, N_t$ of the input ensemble $\mathcal{U}$ satisfy the required (nonhomogeneous) boundary conditions. Due to (22.83), the POD basis functions are no longer suitable to use in a Galerkin projection. The solution of this problem is to transform the actual problem to a problem with homogeneous boundary conditions.

When $h(x, t; d(t))$ does not depend on time t, for example $h(x, t; d(t)) = g(x)$ for all $x \in \Gamma$ and $t \geq 0$, *Sirovich* [22.62] suggests overcoming this difficulty by computing the POD basis functions for the fluctuations around the mean flow field. Given N_t time snapshots, the mean velocity $\langle u(x, t) \rangle = \frac{1}{N_t} \sum_{i=1}^{N_t} u(x, t_i)$ is first computed as an ensemble average. The POD eigenfunctions are then estimated using the modified input data $\mathcal{U}' = \{u(x, t_1) - \langle u(x, t) \rangle, \ldots, u(x, t_{N_t}) - \langle u(x, t) \rangle\}$ (see [22.35] for an example). Due to its construction, the mean flow $\langle u(x, t) \rangle$ is a solenoidal field and satisfies the prescribed nonhomogeneous boundary conditions. Furthermore, each modified snapshot $u(x, t_i) - \langle u(x, t) \rangle$ is also divergence free, but satisfies homogeneous Dirichlet boundary conditions. In the case of time-independent nonhomogeneous boundary conditions, the reduce- order solution can be computed as

$$u_{DS}(x, t) = \langle u(x, t) \rangle + \sum_{m=1}^{N_{Gal}} a_{DS}^{(m)}(t) \Phi^{(m)}(x),$$

where the coefficients $a_{DS}^{(n)}(t), n = 1, \ldots, N_{Gal}$ are the solutions of the dynamical system (22.103).

For boundary control problems, the more interesting case is the one where the Dirichlet boundary conditions $h(x, t; d(t))$ defined by (22.101) depend on time t. In order to match these boundary conditions, *Graham* et al. [22.56], *Ravindran* [22.21, 22] and *Fahl* [22.34] propose searching for the reduced-order solution of the low-order dynamical system (22.103) as

$$u_{DS}(x, t) = \langle u(x, t) \rangle + \gamma(t) u_c(x)$$

$$+ \sum_{m=1}^{N_{Gal}} a_{DS}^{(m)}(t) \Phi^{(m)}(x),$$

where $u_c(x), x \in \Omega$ is a reference flow field, describing how the control action $\gamma(t) c(x), x \in \Gamma_c, t \geq 0$ influences the flow and satisfying the following boundary conditions

$$\gamma(t) u_c(x) = \begin{cases} \gamma(t) c(x), & x \in \Gamma_c, t \geq 0, \\ 0 & x \in \Gamma \setminus \Gamma_c, t \geq 0. \end{cases}$$

Similar to the procedure presented for the time-independent case, a mean velocity $\langle u(x, t) \rangle$ is first computed as the ensemble average of the modified input data defined as $\mathcal{U}' = \{u(x, t_1) - \gamma(t_1) u_c(x), \ldots, u(x, t_{N_t}) - \gamma(t_{N_t}) u_c(x)\}$. Afterward, the POD basis functions are estimated with the input collection $\mathcal{U}'' = \{u(x, t_1) - \gamma(t_1) u_c(x) - \langle u(x, t) \rangle, \ldots, u(x, t_{N_t}) - \gamma(t_{N_t}) u_c(x) - \langle u(x, t) \rangle\}$.

Since $(u(x, t_i) - \gamma(t_i) u_c(x))|_{\Gamma_c} = 0$ and $\langle u(x, t) \rangle$ matches all other nonhomogeneous boundary conditions, the POD basis functions satisfy homogeneous boundary conditions on the whole domain. This approach is used in *Bergmann* et al. [22.35] to incorporate the boundary control for the cylinder wake into the POD-based reduced-order model.

Accurate Model Reduction

When the POD-based reduced-order model (22.103) is integrated in time, it yields a set of predicted time histories for the mode amplitudes $a^{(i)}(t)$, which can be compared with the POD temporal eigenfunctions. However, it is now well known that, when the equations (22.103) are integrated in time with initial conditions obtained from corresponding direct numerical simulations, a gradual drift from the full-state solution to another erroneous state may arise, prohibiting a correct description of the long-term dynamics [22.74]. Even worse, in some cases, the short-term dynamics of the POD model may not be sufficiently accurate to be used as a surrogate model of the original high-fidelity models. Essentially, two sources of numerical errors can be identified. First, the pressure term is often neglected in the Galerkin projection. In many closed flows, it can be demonstrated rigorously that the contribution of the pressure term is exactly zero. For convectively unstable shear layers, as the mixing layer or the wake flow, *Noack* et al. [22.75] proved that neglecting the pressure term may lead to large-amplitude errors in the Galerkin model, requiring the introduction of a pressure-term representation [22.75, 76]. The second source of numerical errors is the truncation involved in the POD Galerkin approach. Indeed, since only the most energetic POD modes are kept, the POD model is not sufficiently dissipative to prevent erroneous time amplifications of its solution. This problem is similar to that of large-eddy simulation, where the energy transfers between the resolved scales and the subgrid scales have to be modeled [22.77]. Recently, Karniadakis employed the same dissipative model, called the spectral vanishing viscosity model (SVVM), to formulate alternative Large Eddy Simulation (LES) approaches [22.78] and to improve the accuracy of POD flow models [22.74]. In *Bergmann* et al. [22.35], the POD model is stabilized by the introduction of a time-dependent eddy-viscosity model estimated for each POD mode as the solution of an auxiliary optimization problem (see [22.79] for a description). This approach can be viewed as a calibration procedure of the POD-Galerkin system similar to the methods recently introduced in *Galletti* et al. [22.76] for the pressure model or in *Couplet* et al. [22.80] for the polynomial coefficients of the system.

22.4.11 Evaluative Summary of the POD Approach

The proper orthogonal decomposition is a powerful and elegant method of data analysis aimed at obtaining low-dimensional approximations of high-dimensional processes. For turbulent flows, the POD approach by itself is neither a theory nor a closure method. However, a better understanding of the role of coherent structures in turbulence generation can be gained with low-order dynamical systems developed by Galerkin projection of the governing equations onto the POD basis functions (see [22.29], for example). On the other hand, the recent invention of microelectromechanical systems has generated substantial interest for control methods for fluid dynamics. The design of reduced-order controllers for fluid system is essential for real-time implementation and the POD method is particularly suited for deriving reduced-order models [22.21, 22, 34, 40].

Among the *advantages* related to the proper orthogonal decomposition, the following points can be underlined.

- The method is objective, methodic, and rigorous: a mathematical framework is provided by the Hilbert–Schmidt theory.
- The POD is a linear method but no linear hypothesis is imposed on the process. The fact that this approach always looks for linear or affine subspaces instead of curved submanifolds makes it computationally tractable. However the POD does not neglect the nonlinearities of the original vector field. If the original dynamical system is nonlinear, then the resulting reduced-order POD model will also typically be nonlinear.
- The POD basis functions are optimal in terms of energy.
- The efficiency of POD increases with the level of inhomogeneity of the process. This method is therefore particularly suited to the analysis of turbulent shear flows. Moreover, as the generalization of Fourier methods to inhomogeneous directions, POD is complementary to harmonic methods.
- Combined with the Galerkin projection procedure, POD provides a powerful method for generating lower-dimensional models of dynamical systems that have a very large or even infinite-dimensional space.

Among the *disadvantages* related to POD are the following.

- The technique requires knowledge of a two-point correlation tensor over a large number of points. Its use is therefore limited by the size of the data sets, which can quickly becomes huge [22.31].
- Due to the use of two-point correlations, phase indetermination appears for directions where an harmonic decomposition has to be used. For classical POD, in particular, complementary techniques are necessary to obtain a description of the preferred modes in the physical space.

- By definition, POD selects structures that are best correlated with the entire fluctuating velocity field. It is therefore not helpful if we want to study a region S that contains only a small fraction of the total kinetic energy. Fortunately, in this type of situation, it is then possible to use the extended POD (EPOD) approach recently introduced by *Borée* [22.81]. Using POD modes computed in the subdomain S only, this method provides a decomposition of the velocity field in the whole domain. This decomposition is such that the extended mode of order p gives the only local contribution to the velocity field that is correlated with the projection of the velocity field onto mode POD of order p in S.
- The nature of the POD basis functions is intrinsically linked to the flow configurations from which they have been derived. The same argument can be used to explain the energetic optimality of the POD basis functions. Therefore, a POD basis determined with a set of realizations of the flow model for a specified control input can perfectly reproduce the dynamics of the flow for a fixed system and may not be sufficient when the system is under the action of a control. In these cases, the POD basis needs to be improved by an adaptive procedure [22.21, 34].

22.5 Conditional Averages and Stochastic Estimation

22.5.1 Conditional Averages

Conditional averaging is a natural way of refining averages to pertain to a more precisely defined set of conditions. It is used extensively in the study of *coherent structures* in turbulent flow. In this section the concepts underlying conditional averaging will be presented briefly, coherent structure concepts will be reviewed, and the method of *conditional eddies* will be explained.

Conditional Averaging Concepts

Let $\tilde{q}$ be any random quantity, and denote the mean value of $\tilde{q}$ by $\langle \tilde{q} \rangle$, where the brackets denote an averaging operation. As usual, we separate $\tilde{q}$ into a sum of its mean and its fluctuation about the mean:

$$\tilde{q} = \langle \tilde{q} \rangle + q \,. \tag{22.104}$$

The conditional average of $\tilde{q}$ given that a set of events $E = \{E_1, \ldots, E_M\}$ occurs is denoted by $\langle \tilde{q} | E \rangle$. The conditional average of the total is equal to the unconditional average plus the conditional average of the fluctuation,

$$\langle \tilde{q} | E \rangle = \langle \tilde{q} \rangle + \langle q | E \rangle \,. \tag{22.105}$$

Hereafter, we shall deal mostly with fluctuations, bearing in mind that the mean component can always be incorporated.

Suppose that $\tilde{q}$ and E are random functions over time. The unconditional time average of $\tilde{q}$ would be found by integrating over its values at all times and dividing by the total time. (Ideally, this would be done in the limit of the total time approaching infinity, but in practice the total time is merely large.) The conditional average for a given E can be found by summing over all points in time at which the event E occurs, and dividing by the fraction of the total time for which E occurs. That is, we would restrict attention to just the set of times for which E occurs. To illustrate this further, suppose that $\tilde{q}$ is the heat flux at a point on a surface and E_1 is the event $\{w(x, t) \geq 0\}$, where w is the vertical velocity at a location x above the surface. The conditional average would be found by sampling the heat flux whenever the vertical velocity is positive. To normalize the average of these samples properly, the averaging time would have to be the fraction of the total time during which the vertical velocity is positive.

As another example, suppose that $\tilde{q}$ is the weight of an individual, $E_1 = h$ is the individual's height, and the averaging brackets represent an ensemble average over a large population of individuals. Then, the conditionally averaged weight given the height would be found first by selecting all of the individuals having a certain height, then finding the average of that group's weight. If the total population in the ensemble were N, and if a group equal to 10% of N had a height of (1.8 ± 0.01) m, then the conditional average would be found by summing the heights of all individuals in the group and dividing by $0.1N$. The conditional averaging processes for area averages, line averages, or combinations of any of the various averages are analogous.

By specifying that random events have certain values, conditional averages diminish uncertainty, and by specifying more event information, the uncertainty diminishes further. For example, by specifying the height of an individual, the conditional weight becomes more representative than the average over all individuals. By specifying the height and age, the group becomes even more specific, and the conditional average becomes even more representative. One expects the variance about the conditional average to continue to diminish as more data are added, but this is only true if the data are correlated with the quantity being averaged. Thus, the conditional weight of an individual is more likely to be representative if the events are genetic or dietary in nature, but not if they are astronomical or otherwise unrelated to a person's weight.

In the context of fluid mechanics, one way we use conditional averages is to improve measurements in the presence of uncontrolled random variations. For example, suppose that laminar flow moves over a surface whose position with respect to a fixed sensor is slowly changing. Let the distance from the surface to the sensor be $y_s(t)$ and the velocity measured by the sensor be $\tilde{u}$. One could represent the velocity at distance y above the surface by the unconditional average, $\langle \tilde{u} \rangle$, but the measurement would be contaminated by the motion of the surface. By sampling the velocity at the sensor only when $y_s(t) = y$, the conditional average $\langle \tilde{u} | y_s(t) = y \rangle$ could be formed, and it would represent the velocity at y more accurately.

Coherent Structures

A second way we use conditional averages is to evaluate the average behavior of flow fields during certain important events, such as the occurrence of strong Reynolds stresses or strong pressure fluctuations. These averages then tell us something about the state of the flow that creates these strong events. In the study of turbulent shear flows, the best-known conditional averaging method is quadrant analysis. One uses this method to explore the mechanisms by which Reynolds shear stresses $\langle u_1 u_2 \rangle$ are created in shear flows such as wall turbulence. (u_1 is streamwise and u_2 is wall-normal.) The flow is classified according to the quadrant in which the instantaneous values (u_1, u_2) fall

$$E_1 = \{u_1 > 0, u_2 > 0\}; \quad E_2 = \{u_1 < 0, u_2 > 0\};$$
$$E_3 = \{u_1 < 0, u_2 < 0\}; \quad E_4 = \{u_1 > 0, u_2 < 0\}.$$
(22.106)

By taking conditional averages given these events one can find the mean fraction of the total Reynolds stress associated with each quadrant, the mean velocity in each quadrant, the mean rate of turbulent energy transport associated with each quadrant, etc. Events in the second quadrant are *ejections* that move low-streamwise-momentum fluid away from the wall, and events in the fourth quadrant are called *sweeps*, which transport high-momentum fluid towards the wall. Both types of events contribute to a negative value of $\langle u_1 u_2 \rangle$, so knowing the conditional average values of other flow properties during these events is useful.

Associating the occurrence of a conditional event with a physical structure in the flow is a more-challenging problem. This problem arises when one wishes to determine the physical form of *coherent structures*. These structures are characteristic, three-dimensional, vortical motions that occur repeatedly and contribute substantially to the mean behavior of the flow. They are random, but not so random as to make them unrecognizable – they are the order within the chaos. Coherent structures are often large scale, but they need not be. Examples of large-scale structures are the spanwise roll vortices that occur in turbulent shear layers. Within the rolls smaller coherent structures occur in the form of *braids* that have smaller vortex cores and a significant component of vorticity in the direction of the mean flow. In turbulent boundary layers the *bulges* are large-scale coherent structures, and the near-wall *quasi-streamwise vortices* are small scale. Even isotropic turbulence, which is the least structured of all turbulent flows, contains small coherent structures in the form of *worms* – slender vortices whose diameters are of the order of 10 Kolmogorov lengths.

Coherent structures are important because they provide a conceptual means of reducing the complexity of turbulence to manageable units. To implement the concept one would like to be able to decompose a flow field into its various coherent structural elements so as to study their various forms, functions, and interactions. To do so, one must have a mathematically unambiguous definition by which the coherent structures can be recognized and isolated. Unfortunately, there is no such definition. The properties of coherent structures listed above may be common to coherent structures, but they do not constitute a complete definition. For example, the quadrant events in (22.106) cannot be associated with a single coherent structure because many different coherent structures could contain regions whose velocity vectors fall in a given quadrant. Sometimes, it is mistakenly asserted that coherent structures are defined by their coherence in space. However, the continuity of fluid motion virtually guarantees that *any continuous fluid motion is spatially coherent*, so spatial coherence can only be one aspect of coherent structures, not a defining property.

It is more useful to define coherent structures as *motions that have a spatial pattern that persists in time*, as measured by temporal coherence functions having long time scales. Long temporal coherence implies that the patterns would be characteristic because they would contribute significantly to the mean statistics of the flow by virtue of dominating long segments of the time averages. Unfortunately, a definition based on temporal coherence would require analysis of the flow patterns in space and time, and this is too difficult for current experimental methods, and too data-intensive for current computational simulations.

For the time being, one is forced to return to the idea of defining coherent structures by conditional averages using intelligently selected events that recognize the coherent structure. This was the basic idea of quadrant analysis. Quadrant 2 events were associated with ejections of low-momentum fluid upwards, and quadrant 1 events were associated with sweeps of high-momentum fluid towards the wall. If these statements are taken to be the definitions of ejection and sweep, then there is no ambiguity. But, to go further and attribute detailed structural behavior to the occurrence of such events, one needs much more evidence. For example, quadrant 2 ejections need not coincide with the turbulent burst phenomenon, which is defined in terms of a rapid eruption of dye in flow visualizations of near-wall low-speed streaks. Eddies of many different forms could possess regions in which the velocity vector falls into one quadrant or another, without qualifying as bursts or sweeps. The root cause of this ambiguity is that there are many possible conditional events, and no guarantees that any of them unambiguously identifies coherent structure of a particular type.

Conditional Eddies

One approach to resolving the conundrum outlined above is to employ conditional averages, but to refrain from attempting identification using ad hoc events. Instead, one concentrates on events that occur naturally

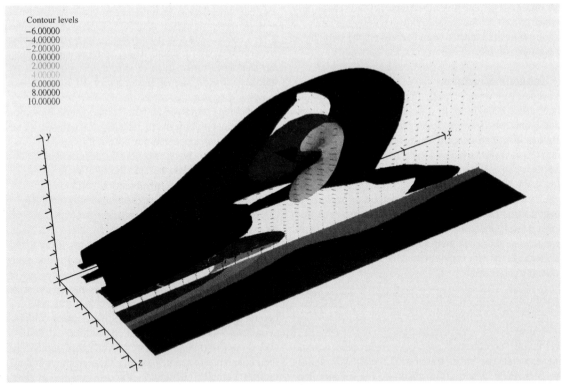

Fig. 22.20 Conditional eddy $\langle u(x,t)|u(x_1,t)\rangle$ in turbulent channel flow given a second quadrant event vector. The colored surfaces are contours of constant magnitude of the vorticity of the conditional eddy $\nabla \times \langle u(x,t)|u(x_1,t)\rangle = \langle \nabla \times u(x,t)|u(x_1,t)\rangle$. The event vector is located $y_1 = 49$ viscous wall units above the lower wall of the $\mathrm{Re}_\tau = 180$ channel flow. The carpet map on the wall depicts the wall shear stress associated with the conditional eddy. The conditional average is found by an approximation that will be discussed later

in the fundamental equations governing the statistics of turbulent flow [22.82]. These equations depend on the particular form of statistical theory one is interested in, but for a broad class of theories it can be shown that the equations governing the probability density function for an event E naturally involve conditional averages of the form $\langle u(x,t)|E \rangle$. From the probability density function equation one can derive equations for any statistical moment of E, so understanding velocity fields defined by $\langle u(x,t)|E \rangle$ is fundamental to understanding the closure problem for the E-theory.

The quantity $\langle u(x,t)|E \rangle$ defines a velocity vector field that we refer to as the *conditional eddy*. An important class of conditional eddies is the set defined by events of the form

$$E_N = \{c_1 \leq u(x_1, t) < c_1 + dc_1$$
$$\text{and } c_2 \leq u(x_2, t) < c_2 + dc_2 \text{ and } \ldots$$
$$\text{and } c_N \leq u(x_N, t) < c_N + dc_N\}. \quad (22.107)$$

This event occurs in the Lundgren–Monin equation for the probability density function (PDF) of the velocities at N points. The simplest equation in the Lundgren–Monin hierarchy is the one-point PDF from which equations for all one-point moments (mean, root mean square, and higher-order moments) are derivable. This equation involves the simplest conditional eddy

$$\langle u(x,t)|c_1 \leq u(x_1,t) < c_1 + dc_1 \rangle.$$

Note that quadrant analysis is a special case of this conditional eddy, found by averaging the event velocity vector over each of the quadrants in the (u_1, u_2) plane and all values of u_3. This conditional eddy is quite powerful as a means of recognizing structure. In isotropic turbulence, it takes the form of a vortex ring centered on the event vector [22.83]. In wall turbulence, if the event vector is given values corresponding to the most probable second-quadrant ejection event, the conditional eddy takes the form of the hairpin vortex, shown in Fig. 22.20. This velocity pattern corresponds quite well with patterns observed in experimental flows.

Another type of conditional eddy is the *conditional eddy given local kinematics*. The event defining this eddy consists of the velocity at x_1 and the velocity gradient tensor ∇u at x_1, which completely specifies the kinematics at that point. In particular, from the velocity gradient tensor, one can specify the vorticity. In isotropic turbulence, the conditional eddy found by specifying the vorticity looks like a slender vortex whose diameter is of the order of the Taylor microscale and whose length is of the order of the integral length scale, corresponding in shape and dimensions to a turbulent worm. In homogeneous turbulent shear flow the conditional eddy given the vorticity resembles a hairpin vortex [22.84].

22.5.2 Stochastic Estimation

In general, the analysis of data by the process of *stochastic estimation* refers to the approximation of one or more random variables in terms of available data, which may also be random. We shall see that it is intimately related to conditional averaging, and that it will ultimately provide a powerful practical tool for determining conditional eddies from data produced by physical and numerical simulation experiments. To clarify the differences between conditional averaging and stochastic estimation the development of stochastic estimation concepts in this section will begin with a fresh problem statement.

Estimation of Random Processes

In the context of structure in flows that are turbulent or otherwise random, the estimated variable is typically the velocity vector field, which we decompose into the sum of its mean $U(x,t)$ and its fluctuating part $u(x,t)$. As in our discussion of conditional averaging, suppose that the data are associated with the occurrence of certain events at one or more points in the field. Their totality is referred to as the *event data vector* E. We shall denote the estimate of $u(x,t)$ in terms of the event data by $\hat{u} = F(E, x, t)$.

The data can take many forms, but in the simplest case E is one or more components of the velocity vector at a single point in the flow, $u_1 = u(x,t)$. Then the estimate would represent a best representation of the velocity field that occurs around the point x_1 when the velocity at x_1 is equal to the value specified for u_1. Note that this value can be varied to explore different states of the flow. Note also that the estimate of the *total* velocity field is just the mean velocity plus the estimate of the *fluctuating* velocity field.

It can be proven that, in general, the least mean-square error estimate of $u(x,t)$ given the data E is the conditional average of $u(x,t)$ given E, denoted by $\langle u(x,t)|E \rangle$. That is, of all possible estimates in terms of the event data, $F(E, x, t)$, the mean-square error $e = \langle |u - F|^2 \rangle$ is a minimum when $F = \langle u|E \rangle$.

This result is so important that it is derived here in detail. The condition for the error to be a minimum is that the first variation $\delta e = 0$. Taking the variation δF_i

with respect to all possible functions of the event data gives

$$\delta e = \langle 2(u_i - F_i)(-\delta F_i) \rangle \quad (22.108a)$$
$$= -2\langle\langle(u_i - F_i)|E\rangle\delta F_i\rangle . \quad (22.108b)$$

Equation (22.108b) follows from (22.108a) by writing the average as the average over the random event data E of the conditional average given E. Since E is fixed inside the conditional average, $\delta F_i(E)$ is also fixed, and it can be taken outside of the conditional average. [The conditional average given E averages over fluctuations in u, while the unconditional (outer) average averages over fluctuations in E.] In order for (22.108b) to vanish for arbitrary variations δF_i, the conditional average must vanish identically, leading to

$$F_i = \langle u_i | E \rangle \quad (22.109)$$

or

$$\hat{u} = \text{best estimate of } u \text{ given the data } E = \langle u | E \rangle . \quad (22.110)$$

This simple but elegant result is quite general. We could replace the velocity field u with *any* quantity q and arrive at the same result.

An estimate of u in terms of event data E must become increasingly erroneous as the location x is moved away from the locations of the event data, because u will become uncorrelated with the event data for large spatial (or temporal) separations. The conditional average $\langle u(x,t)|E\rangle$ manifests this property by reducing to the unconditional average $\langle u \rangle$ (which equals zero if u is a fluctuation) as x is removed from the vicinity of the event data.

Linear and Nonlinear Mean–Square Estimation

In general, $\langle u|E\rangle$ is a nonlinear function of the components of E. However,, under the condition that the components of u and E are joint normally distributed, it is well known that $\langle u|E\rangle$ is a linear function of E [22.85]. Often, this property is applied approximately to non-normal random variables by postulating an estimate of u in the form of a linear combination of the event data. Then one speaks of *linear mean-square estimation*. This subject is discussed thoroughly in the literature of stochastic theory [22.85, 86]. Let

$$\hat{\hat{u}}_i = \text{linear estimate of } u_i = \sum_{j=1}^{M} L_{ij} E_j . \quad (22.111)$$

The mean-square error of this estimate is

$$\left\langle \left(u_i - \sum_{j=1}^{M} L_{ij} E_j\right)^2 \right\rangle = \min ,$$
$$i = 1, 2, 3 \text{ and } j = 1, \ldots, M . \quad (22.112)$$

The mean-square error of the linear estimate must be greater than or equal to the error of the conditional average. The necessary condition for minimization of the mean-square error is the *orthogonality principle* which states that the errors $u_i - \sum_{j=1}^{M} L_{ij} E_j$ are statistically orthogonal to the data

$$\left\langle \left(u_i - \sum_{j=1}^{M} L_{ij} E_j\right) E_k \right\rangle = 0 ,$$
$$i = 1, 2, 3 \text{ and } j, k = 1, \ldots, M . \quad (22.113)$$

Equation (22.113) follows from (22.112) by setting the derivative with respect to L_{ik} of the mean-square error equal to zero. Simple manipulation of (22.113) leads to an $M \times M$ system of linear algebraic equations for the coefficients for each value of i

$$\sum_{j=1}^{M} \langle E_j E_k \rangle L_{ij} = \langle E_k u_i \rangle ,$$
$$j, k = 1, \ldots, M , \quad i = 1, 2, 3 . \quad (22.114)$$

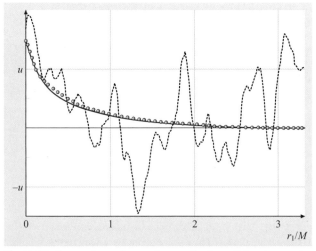

Fig. 22.21 Comparison of a random realization of the streamwise velocity $u(x')$ measured in grid turbulence (*dashed line*), the conditional average of the velocity given the value at $x' = x$ (*solid line*), and the linear stochastic estimate (*open circles*)

The coefficients in these equations are the correlations between each pair of event data and between each event datum and the quantity being estimated.

Figure 22.21 illustrates the behavior of the conditional average and the linear estimate. In this example, the quantity being estimated is the streamwise velocity component in an experimental grid turbulence flow field. The event is that the velocity component is equal to $1.5\sigma_1$ at $x' = x$, where σ_1 is the root mean square of u. A single realization of $u(x')$ that satisfies this event is plotted as a dashed line, and the conditional average and the linear estimate are shown as a solid line and round symbols, respectively. At $x' = x$ the realization and the estimates each equal the given event. As $x' - x$ increases, the conditional average and the linear estimate decay to zero because they lose correlation with the event. The conditional average must approach the unconditional average, which vanishes, and the linear estimate follows suit. As they vanish, the root-mean-square error increases with increasing x' until it asymptotes to the unconditional root-mean-square value σ_1.

This illustrates the general rule that estimation of a random variable does not significantly reduce the error unless the variable being estimated is highly correlated with the data, e.g., in the vicinity of the data and correlated with it. Providing information about data that have little physical relationship to the process will not reduce the error, nor will providing physically relevant data that are too far removed from the point of estimation in space or time.

Estimation of Conditional Averages

In addition to estimating $\boldsymbol{u}$ as a function of $\boldsymbol{E}$, one can also *estimate the conditional average* $\langle \boldsymbol{u}|\boldsymbol{E}\rangle$. The conditional average is a deterministic function of random data $\boldsymbol{E}$, so it toe is a random variable. It is natural to expand $\langle \boldsymbol{u}|\boldsymbol{E}\rangle$ in a Taylor series about $\boldsymbol{E} = 0$, and truncate at some order [22.82, 83]. When the series contains only first-order terms, we refer to this as a *linear stochastic estimation* (LSE) to distinguish it from linear mean-square estimation. The equations for *linear stochastic estimation* of the i-th component of $\langle \boldsymbol{u}|\boldsymbol{E}\rangle$ are

$\check{u}_i = $ linear stochastic estimate of $\langle u_i|\boldsymbol{E}\rangle$

$$= \sum_{j=1}^{M} \check{L}_{ij} E_j \,, \quad (22.115)$$

where M is the number of event data, and $\check{L}_{ij}$ is a function of $\boldsymbol{x}$ and the positions of the event data. The estimation coefficients $\check{L}_{ij}$ are chosen so that the mean-square error

$$\left\langle \left(\langle u_i|\boldsymbol{E}\rangle - \sum_{j=1}^{M} \check{L}_{ij} E_j \right)^2 \right\rangle = \min \,,$$

$i = 1, 2, 3$ and $j = 1, \ldots, M$. (22.116)

As above, the necessary condition for minimization is the *orthogonality principle*, which states that the errors $\langle u_i|\boldsymbol{E}\rangle - \sum_{j=1}^{M} \check{L}_{ij} E_j$ are statistically orthogonal to the data

$$\left\langle \left(\langle u_i|\boldsymbol{E}\rangle - \sum_{j=1}^{M} \check{L}_{ij} E_j \right) E_k \right\rangle = 0 \,,$$

$i = 1, 2, 3$ and $j, k = 1, \ldots, M$. (22.117)

The resulting equations for $\check{L}_{ij}$,

$$\sum_{j=1}^{M} \langle E_j E_k\rangle \check{L}_{ij} = \langle E_k u_i\rangle \,,$$

$j, k = 1, \ldots, M \,, \quad i = 1, 2, 3 \,,$ (22.118)

are identical to those for L_{ij} in (22.114). Thus, *the linear stochastic estimate of* $\langle \boldsymbol{u}|\boldsymbol{E}\rangle$ *and the linear mean-square estimate of* $\boldsymbol{u}$ *are numerically equal*. The principal difference is one of interpretation. In particular, while the mean-square error of the linear mean-square estimate of $\boldsymbol{u}$ must be large when $\boldsymbol{u}$ is uncorrelated with $\boldsymbol{E}$, (due, for

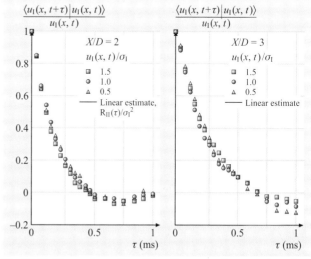

Fig. 22.22 Comparison of conditionally averaged streamwise velocity on the centerline of the shear layer in a round turbulent jet with the linear stochastic estimate of the conditional average. (After *Adrian* et al. [22.87])

example, to a large separation between x and the location of the event data), the error of the linear stochastic estimate of $\langle u|E\rangle$ may be small because $\langle u|E\rangle$ also vanishes as the separation becomes large. The results in Fig. 22.21 illustrate this behavior clearly if we consider the linear estimate to be an estimate of the conditional average.

The linear estimate of the conditional average is surprisingly good for a variety of different types of turbulent flows, including isotropic turbulence, pipe flow, and jet flow [22.88], and for a variety of different types of variables, including velocity, pressure and Reynolds stresses [22.87]. As an example, conditional averages of the streamwise velocity at $t+\tau$ in a turbulent jet given the value of the velocity at t are compared to the linear stochastic estimate in Fig. 22.22. For this simple case the reader can verify that the linear estimate is

$$\check{u}_1(\tau) = \text{lin. est.}\langle u_1(t+\tau)|u_1(t)\rangle$$
$$= \frac{\langle u_1(t)u_1(t+\tau)\rangle}{\langle u_1^2(t)\rangle} u_1(t) \ . \tag{22.119}$$

The averages are conditioned on three different values of the velocity at zero time and the linear estimate is calculated from the time-delayed autocorrelation.

Usually, the accuracy of linear estimation is good, even for large values of the event data. In situations where the linear estimate falls short of the accuracy required, the estimation can be improved by postulating $F(E)$ to be a nonlinear function such as a quadratic or a higher-order polynomial. Often, higher-order terms have little effect on probable values of the events, but sometimes they improve accuracy [22.89–91]. Interestingly, *Brereton* [22.92] has also suggested a Laurent-type expansion that contains negative powers of the event data and showed improved estimation for low-amplitude events. Ultimately, the accuracy of an estimate must rest in direct comparison of the estimate with the conditional average.

It was noted earlier that the most basic conditional eddy in the hierarchy of conditional averages in the PDF equations is $\langle u(x,t)|u(x_1,t)\rangle$. The field shown in Fig. 22.20 was presented as an approximation to a conditional eddy of this type. It can now be explained that the approximation was found by the stochastic estimate

$$\text{lin. est.}\langle u_i|u_1\rangle = \check{L}_{ij}u_{1j}, \quad i,j = 1,2,3 \tag{22.120}$$

in which $\check{L}_{ij}$ is found by solving

$$\langle u_{1j}u_{1k}\rangle \check{L}_{ij}(x_1,x) = R_{ik}(x_1,x), \quad i,j,k = 1,2,3 \ , \tag{22.121}$$

where

$$R_{ik}(x_1,x) = \langle u_i(x_1,t)u_k(x,t)\rangle \tag{22.122}$$

is the two-point spatial correlation with a fixed reference point x_1. In matrix form

$$\begin{pmatrix} \langle u_{11}u_{11}\rangle & \langle u_{11}u_{12}\rangle & \langle u_{11}u_{13}\rangle \\ \langle u_{12}u_{11}\rangle & \langle u_{12}u_{12}\rangle & \langle u_{12}u_{13}\rangle \\ \langle u_{13}u_{11}\rangle & \langle u_{13}u_{12}\rangle & \langle u_{13}u_{13}\rangle \end{pmatrix} \begin{pmatrix} \check{L}_{i1} \\ \check{L}_{i2} \\ \check{L}_{i3} \end{pmatrix}$$
$$= \begin{pmatrix} R_{i1}(x_1,x) & R_{i2}(x_1,x) & R_{i3}(x_1,x) \end{pmatrix} \tag{22.123}$$

A common and useful procedure is to use event data consisting of velocity vectors $u_\alpha = u_\alpha(x_\alpha,t)$ at a set of N space points $\{x_\alpha\}$, $\alpha = 1, \ldots, N$. Then, E contains $M = 3N$ data,

$$E = \{u_{11}, u_{12}, u_{13}, u_{21}, \ldots, u_{N1}, u_{N2}, u_{N3}\} \ , \tag{22.124}$$

and $\langle E_j E_k\rangle$ involves only second-order two-point spatial correlations.

$$R_{jk}(x_\alpha, x_\beta) \equiv \langle u_j(x_\alpha,t)u_k(x_\beta,t)\rangle \ ,$$
$$\alpha, \beta = 1, \ldots, N \ . \tag{22.125}$$

The correlation on the right-hand side of (22.118) involves correlations of the form

$$\langle E_k u_i\rangle = \langle u_l(x_\alpha,t)u_i(x,t)\rangle = R_{li}(x_\alpha,x) \ ,$$
$$k = 1, \ldots, M; \quad i, l = 1,2,3 \ . \tag{22.126}$$

Thus, when the event data consist of velocity vectors, knowledge of the two-point spatial correlations of the velocity suffices to determine the linear estimate.

Given the estimate of a conditional eddy velocity field, the estimate of a derivative of the conditional eddy can be found by differentiating the estimate with respect to x. We have seen this already for the vorticity field in Fig. 22.20. A more-complex application that also illustrates the use of multipoint velocity data is shown in Fig. 22.23. The data consist of a grid of velocity vectors lying in a plane located above a wall in channel flow. The linearly estimated velocity field, given these data, is differentiated at the wall to get the viscous wall shear stress, shown in color contours. Comparison with the actual viscous wall shear stress shows close agreement, although the estimate is smoothed somewhat with respect to the actual stress field.

If the event vector contains derivatives of the velocity field, $(\partial u_i/\partial x_j)$ at the points x_α the resulting two-point correlations that appear in (22.122) contain correlations between velocities and derivatives, and derivatives and

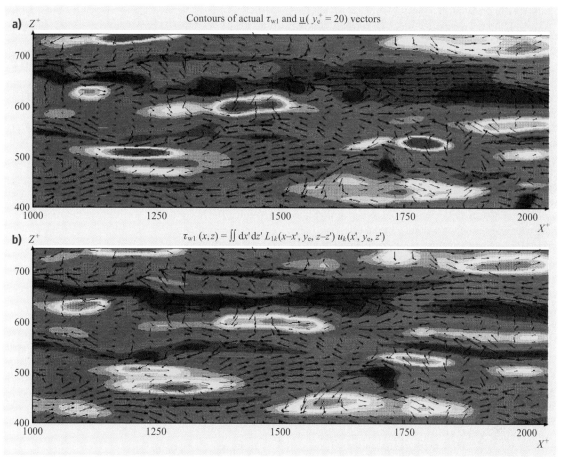

Fig. 22.23 (a) Instantaneous streamwise wall shear stress and velocity vectors on a plane $y+ = 20$ viscous wall units above the wall in direct numerical simulation of turbulent channel flow. **(b)** Linear estimate of the conditional average of the wall shear stress given the same set of velocity vectors on the $y+ = 20$ plane. (After *Bagwell* et al. [22.93])

derivatives. Since each type of correlation can be expressed in terms of R_{ij} and its derivatives, it follows that knowledge of the two-point spatial correlation function again suffices to determine fully the linear estimate of u in terms of velocity data and derivatives of any order. This is a powerful result, as it states that very complicated conditional averages can be approximated once R_{ij} is known.

Conditional averages play three important roles. First, they are used in the study of coherent flow structures in turbulence. Second, they appear naturally in PDF theories. These theories define conditional eddies in the form $\langle u(x,t)|E\rangle$, where E consists of a set of velocities at N points. Third, conditional averages play a fundamental role in estimation theory by virtue of being the best least mean-square estimate given the available data. Linear and nonlinear stochastic estimation can be used to estimate either the random variable u or its conditional average. Estimation of the random variable produces a mean-square error that increases as the correlation of the variable with the data decreases. However, estimation of the conditional average does not suffer from this problem. Thus, one must be careful to distinguish clearly between estimation of the variable or the conditional average. The linear estimate of a conditional average is exact for joint normal random variables, but this should not be taken to imply that joint normal distribution is a necessary condition for accuracy, or that accuracy implies that the variable must be joint normal. Experimental evidence shows that stochastic estimation works well for non-joint-normal variables.

Stochastic estimation offers two huge advantages over conditional averaging. First, calculating the conditional average requires finding realizations of the variable and the data in which the data assume the prescribed values of interest. When the event vector has high dimension, the occurrence of the event is highly improbable, and one must take immense amounts of data to build up a stable conditional average. For example, if an N-point conditional eddy is to be evaluated for a single set of velocity vector values, one would have to sift through an N-dimensional space of events to find the realizations in which the prescribed values of the events occur. In contrast, the stochastic estimate requires evaluation of the correlations between the data and the variable being estimated, and it is straightforward to calculate such two-point correlations, especially from data produced by direct numerical simulations or particle image velocimetry. Secondly, once the correlations have been found, the stochastic estimate, linear or nonlinear, gives the conditional average for *any* set of event values. Evaluation for new values only requires plugging them into the linear form. In contrast, evaluating the conditional average for a new event vector would require sorting through all of the data to find realizations that satisfied the new event.

The linear estimate of a conditional eddy offers a third attractive feature. Namely, it involves only the two-point second-order spatial correlation tensors. These are fundamental quantities in the statistical description of turbulent structure, but the correlation tensor in not easily interpreted in terms of the underlying eddy structure. The linear estimate repackages the correlation tensor information into a vector field that is much more readily interpreted.

22.6 Wavelet Transforms

22.6.1 Introduction to Wavelets

Conventional data processing is performed in either the time domain (moments, correlations, etc.) or the frequency domain (power spectra, etc.). Wavelet processing combines the two, allowing the definition of local spectral properties and the ability to zoom in on local features of the signal. To simplify notations, we use *time-frequency*, with an implied reference to time series for the data; the alternative use of data is straightforward. Similarly, the *signal* is simple terminology for scalar or vector field.

A wavelet is a basis function (an elementary building block to analyze or synthesize the signal) characterized by

- its shape and its amplitude, to be selected by the user
- its scale (frequency or size) and location (time or location) relative to the signal, spanning a range of interest to study a given phenomenon.

Wavelet coefficients are the scalar products of the signal with all dilated and translated wavelets. The set of wavelet coefficients thus obtained is indexed by position and scale (always positive) in the wavelet half-plane.

We will first present the continuous wavelet transform, then the discrete and orthogonal wavelet transforms. Additional details can be found in standard references [22.94–96]. We will discuss their implementation and how to use them to analyze signals measured in laboratory experiments or generated numerically. We will also explain how the wavelet transform can be used to extract coherent structures form a velocity or vorticity field. An overview of techniques can be found in [22.97] and in the archival literature.

22.6.2 Continuous Wavelet Transform

Definition of a Wavelet
Let us assume that a wavelet $\psi(t)$ is given.

Notations: Fourier Transform. For reference and to establish notation, we define the Fourier transform of $\psi(t)$ as

$$\hat{\psi}(\omega) = \int_{-\infty}^{\infty} \psi(t) e^{-2i\pi\omega t} \, dt \, ,$$

with its inverse

$$\psi(t) = \int_{-\infty}^{\infty} \hat{\psi}(\omega) e^{2i\pi\omega t} \, d\omega \, .$$

The energy spectrum is defined on the basis of the Parseval relation

$$\frac{1}{2} \int_{-\infty}^{\infty} |\psi(t)|^2 \, dt = \int_{-\infty}^{\infty} E_F(\omega) \, d\omega \, .$$

The subscript 'F' refers to Fourier, to distinguish it from wavelet spectra; in the applications, we will also use the compensated spectrum $\omega \cdot E$ to display dominant scales.

Regardless of its scale and magnitude, a function ψ is admissible as a wavelet if and only if

$$c_\psi = \int_0^\infty |\hat{\psi}(\omega)|^2 \frac{d\omega}{|\omega|} < \infty, \quad (22.127)$$

for which it is sufficient that its mean vanishes

$$\int_{-\infty}^\infty \psi(t)\,dt = 0. \quad (22.128)$$

Optionally, higher moments $\int t^n \psi(t)\,dt$ may also vanish. In addition, it is required that ψ asymptote to zero fast enough for large t

$$|\psi(t)| < \frac{C}{1+|t|^p} \quad (22.129)$$

with $C > 0$ and $p \in \mathbb{N}$. Other desirable properties include good localization (steep decay) in both the time and frequency domains. These conditions are not very restrictive, and many different wavelets have been used in the literature; only a few examples (and reasons for their selection) can be treated here, but the rationale for wavelet selection can be inferred from the applications discussed in Sect. 22.6.4.

Concerning the Choice of Wavelet. Some of the trade-offs implied by the choice of wavelet are apparent from a few examples. A difference of two Dirac functions separated by a time τ, used to build structure functions, has an extended Fourier spectrum. The Haar function also has compact support (Fig. 22.24) (compact support means that the subset of the time domain over which the function is nonzero is closed and bounded), and an extended oscillatory spectrum ($1/\omega^2$ decay) (Fig. 22.25) as a consequence. The Mexican-hat and Morlet wavelets (formulae in next section) have an exponential drop-off in both time and frequency, and are preferred in many cases. The trade-off between them is that the Morlet wavelet has narrow spectral bandwidth and an extended time domain, while the Mexican-hat wavelet is more localized in the time domain at the expense of a wider spectral content. Perfect localization in both time and frequency is impossible (due to the Heisenberg uncertainty principle): each wavelet represents a compromise, the pros and cons of which should be reflected in the interpretation of the results.

General Formulae

The continuous wavelet transform of a function $u(t)$ (assumed to have zero mean and finite energy) is defined as a convolution

$$\tilde{u}(a, t) = \int_{-\infty}^\infty \left[u(t') \frac{1}{\sqrt{a}} \psi^* \left(\frac{t-t'}{a} \right) \right] dt', \quad (22.130)$$

where the asterisk superscript denotes complex conjugation; $\tilde{u}(a, t)$ is the wavelet coefficient at time t and scale a. The integral measures the comparison of the local shape of the signal and the shape of the wavelet. The dilation factor a acts as a zoom, so large features of the original signal appear at large values of a, while short-duration events appear at small a. By centering the scaled

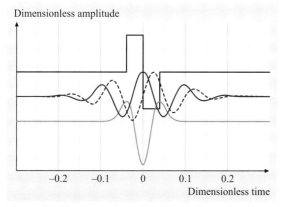

Fig. 22.24 Three wavelets of similar scale (with arbitrary vertical offsets). From top to bottom: Haar, Morlet (real (*solid*) and imaginary (*dashed*) parts), and Mexican hat

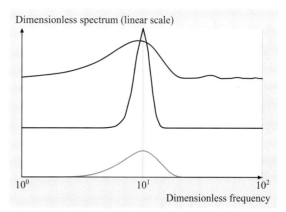

Fig. 22.25 The Fourier spectra $E_F(\omega)$ of the three wavelets of Fig. 22.24 (arbitrary vertical offsets). From top to bottom: Haar, Morlet, and Mexican hat

wavelet at time t, some time localization around t and some spectral localization around scale a is achieved.

Two interpretations of the variables t and a are consistent with t/a being dimensionless. We can assume that both t and a have dimensions of time; alternatively, t can be nondimensionalized by the wavelet's centroid frequency ω_ψ for $a = 1$,

$$\omega_\psi = \frac{\int_0^\infty \omega |\hat{\psi}(\omega)| \, d\omega}{\int_0^\infty |\hat{\psi}(\omega)| \, d\omega} .$$

in which case the dilation factor a is also dimensionless. We use them indifferently.

The value of the wavelet transform as an analytical tool rests on the next two formulae. First, the inverse transform

$$u(t) = \frac{1}{c_\psi} \int_{-\infty}^{\infty} \int_0^\infty \left[\tilde{u}(a, t') \frac{1}{\sqrt{a}} \psi\left(\frac{t-t'}{a}\right) \right] \frac{da}{a^2} \, dt' \quad (22.131)$$

requires a convolution in time as well as integration over a. It expresses the fact that no information is lost in the transform, and that the signal can be interpreted as made of a linear combination of wavelets. Second, there is a Parseval relation

$$\int_{-\infty}^{\infty} [u(t) v^*(t)] \, dt = \frac{1}{c_\psi} \int_{-\infty}^{\infty} \int_0^\infty \tilde{u}(a,t) \tilde{v}^*(a,t) \frac{da}{a^2} \, dt , \quad (22.132)$$

and in particular

$$\int_{-\infty}^{\infty} |u(t)|^2 \, dt = \frac{1}{c_\psi} \int_{-\infty}^{\infty} \int_0^\infty |\tilde{u}(a,t)|^2 \frac{da}{a^2} \, dt , \quad (22.133)$$

which shows that the energy of the signal is conserved in the time–frequency domain. Thus, the squared wavelet coefficients $\tilde{u}$ provide an instantaneous power spectrum, useful for studying transient or intermittent signals.

The Energy Distribution. For statistically stationary signals, one classically considers the modulus of its Fourier transform [i.e., the spectrogram or energy spectrum $E_F(\omega)$]. Since the phase of the Fourier coefficients is thus lost, no information on the local structure of the signal can be retrieved, as the time or space information is encoded by the phases of all Fourier coefficients. Therefore the classical energy spectrum, based on the Fourier transform, is not ideal to analyze statistically nonstationary or inhomogeneous signals. The wavelet transform extends the concept of energy spectrum so that one can define a local energy spectrum. Several related algorithms are illustrated in Sect. 22.6.4.

Variants. Wavelets can be normalized in such a way as to have unit energy at all scales. This option is favored by many authors, and is associated with the $\sqrt{a}$ and c_ψ factors in the formulae above. Alternatively, changes in variables and different normalizations (as used later) give simpler expressions of inverse transforms and spectra. It is recommended that the user verify the consistency of formulae and software of different provenance by performing wavelet transforms of simple test signals.

Examples: Continuous Wavelet Transforms and Interpretation

In the examples illustrated below, the wavelet transform will be applied to an artificial signal (Fig. 22.26), so that specific experimental conditions need not be described. The signal combines features encountered in various applications in fluid mechanics. We superpose a modulated wave, isolated events in the same frequency range, and continuous and intermittent wave and noise packets (some of the latter correlated to the phase of the primary wave, e.g., at times 2, 7, and 13 of the primary wave).

Implementation: Convolution. Since the wavelet transform at each scale a is a convolution of the signal with the scaled wavelet, the operation is efficiently implemented as a multiplication in Fourier space, i.e.,

$$\tilde{u}(a,t) = \sqrt{a} \int_{-\infty}^{\infty} \left[e^{2i\pi\omega t} \hat{\psi}^*(a\omega) \hat{u}(\omega) \right] d\omega ,$$

which involves the Fourier transform of the signal, multiplication by the (precalculated) transform of the scaled wavelet, and inverse Fourier transform. This relation also shows that the wavelet transform is a band-pass filtered signal, with $\hat{\psi}(a\omega)$ describing the shape of the filter. Each wavelet coefficient combines the information in a temporal vicinity (weighted by the wavelet shape) and in a spectral vicinity (weighted by its Fourier spectrum); it is this time–frequency compromise that makes wavelets so versatile with intermittent data.

Discretization. In practice, the signal $u(t)$ of duration T is sampled at constant intervals on a grid $t_i = iT/N$, $i \in [0, N-1]$ with $N = 2^J$. This corresponds to a Nyquist frequency $\omega_{\max} = N/2T$, and a spectrum of discrete decreasing frequencies, for which a logarithmic progression $\omega_j = \omega_{\max} a_0^{-j}$ is typically used. The frequency ratio a_0 is 2 (one per octave) for orthogonal wavelets; for continuous wavelets, a ratio $a_0 = 2^{1/3}$ was used for most figures, with a finer resolution $a_0 = 2^{1/9}$ where required. The lowest Fourier frequency $\omega_0 = 1/T$ does not necessarily apply to the wavelet transform, although it has been adopted in the figures. The Fourier frequencies vary linearly, with $\omega_k = k\omega_0$. For a signal u, the transform is sampled in the wavelet half-plane according to the expression

$$\tilde{u}(a_j, t_i) = \sqrt{a_j} \sum_{k=-N/2}^{N/2} \left[e^{2i\pi\omega_k t_i} \hat{\psi}^*(a_j\omega_k)\hat{u}(\omega_k) \right] ,$$

where $a_j = a_0^j$ and $\hat{u}(\omega_k) = \Delta t \sum_{n=0}^{N-1} u(t_n) e^{-2\pi i t_n \omega_k}$ and can be efficiently computed using the FFT. Note that frequencies ω_k have been centered around $\omega = 0$ and that $\hat{u}(-\omega) = \hat{u}^*(\omega)$ since u is real-valued.

Discretization Error. Sampling of the continuous wavelet transform at discrete points entails some loss of information. This loss can be treated in practice as an approximation for which errors can be kept as small as desired on a suitably fine grid. Bounds on the errors can be found in the mathematical theory of frames. Generally, using the time-series sampling interval and a frequency ratio of 2–5 per octave yields sufficient accuracy for many applications.

Mexican-Hat Wavelets. Starting from the normalized Gaussian filter

$$F_\sigma(t) = \frac{1}{2\sqrt{\pi\sigma}} e^{-\frac{t^2}{4\sigma}} , \qquad (22.134)$$

in which $\sigma = a^2/2$, its derivative with respect to σ (second time derivative) gives the Mexican-hat wavelet:

$$\psi_2(\sigma, t) = \frac{\mathrm{d}F_\sigma}{\mathrm{d}\sigma} = \left(\frac{t^2}{4\sigma^2} - \frac{2}{\sigma} \right) F_\sigma(t) = \frac{\mathrm{d}^2 F_\sigma}{\mathrm{d}t^2} . \qquad (22.135)$$

It is a difference of Gaussian filters of different scales (band-pass filtering), divided by the scale difference. Rather than energy normalization, this variant favors the relation to Gaussian filtering and the simple formulae below, including the compatibility equation. The Fourier transform of ψ_2 is

$$\hat{\psi}_2(\sigma, \omega) = -4\pi^2 \omega^2 e^{-4\pi^2\omega^2\sigma} . \qquad (22.136)$$

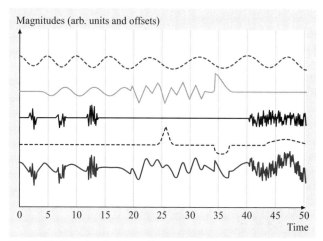

Fig. 22.26 The signal $u(t)$ used in examples (*bottom*), and its various contributions (*top*); arbitrary offsets and magnitude

For a signal u, the Mexican-hat wavelet transform is written

$$\tilde{u}_2(\sigma, t) = \int u(t')\psi_2(t - t', \sigma)\mathrm{d}t' = \frac{\mathrm{d}^2}{\mathrm{d}t^2}(F_\sigma * u) . \qquad (22.137)$$

Then the energy of the signal is given by

$$\int \frac{|u|^2}{2}\mathrm{d}t = \iint 2\, |\sigma \tilde{u}_2|^2 \frac{\mathrm{d}\sigma}{\sigma} \mathrm{d}t \qquad (22.138)$$

and the inverse transform by either one of the relations

$$u(t) = -\int_0^\infty \sigma \tilde{u}_2(\sigma, t) \left[\frac{\mathrm{d}\sigma}{\sigma} \right] \qquad (22.139)$$

$$= 4 \iint \sigma \tilde{u}_2(\sigma, t')\psi_2(t - t', \sigma)\mathrm{d}\sigma\, \mathrm{d}t' . \qquad (22.140)$$

The relations show that $\sigma \tilde{u}_2$ is physically relevant.

For plotting purposes, the conversion from σ to an equivalent frequency is needed. Taking the wavelet transform of $\cos(2\pi\omega t)$, two simple conversions can be adopted: the peak of the compensated energy spectrum corresponds to $\omega\sqrt{\sigma} = 1/\pi\sqrt{8}$, or the centroid of the spectrum coincides with $\omega\sqrt{\sigma} = 1/2\pi$. The second alternative is adopted here, with the largest ω equal to the Nyquist frequency of our discrete signal.

The example signal (Fig. 22.26) and its Mexican-hat wavelet map are shown in Fig. 22.27.

The largest wavelet coefficients (red for minima, blue for maxima) identify individual extrema of the

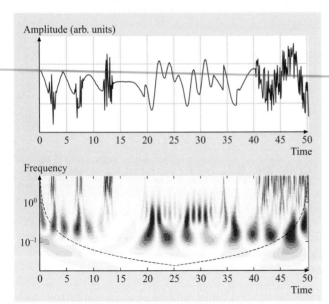

Fig. 22.27 Mexican hat wavelet map $\sigma\widetilde{s}_2$ (*bottom*) and signal (*top*). The cone of influence of end-points is also shown

signal. The near-periodicity of one contribution to the signal is visible in the energetic contributions along a horizontal strip of the map. The noise is intermittently distributed in the top third of the map, corresponding to larger frequencies. Additional features are discussed in connection with the Morlet transform.

Cone of Influence and End Effects. For a wavelet located at a fixed time, increasing its scale from some very small value gradually brings a larger part of the signal into the wavelet's view, generating the cone of influence of the wavelet. Signal values within the cone of influence affect the wavelet coefficient at that scale. If the wavelet is located near either end of a signal, it will see whatever information is located beyond the end-points, or the lack of information; the corresponding wavelet coefficients are of questionable value. Various techniques such as wrap-around (implied periodicity of the signal), mirror symmetry or zero padding do not entirely eliminate this problem, and results should be interpreted accordingly. Periodicity is used throughout the examples below. The cone of influence is the envelope, for some suitably low threshold (2% of maximum in this case), of the wavelet coefficients $\tilde{u}$ of a Dirac function located at position t_0 or t_{N-1}. It is recommended that the cone of influence of end-points be shown as a reminder that the wavelet coefficients within the cone should be used with caution.

Morlet Wavelet. The Morlet wavelet is complex-valued, and consists of a Fourier wave train inside a Gaussian envelope of width z_0/π:

$$\psi_{M,z_0}(t) = \left(e^{2i\pi t} - e^{-\frac{z_0^2}{2}}\right)e^{-2\pi^2 t^2/z_0^2}. \quad (22.141)$$

The envelope factor z_0 controls the number of oscillations in the wave packet; a value of $z_0 = 5$ is generally adopted, with the result shown on Fig. 22.24. The correction factor $e^{-z_0^2/2}$, making the wavelet admissible, is very small for $z_0 \geq 5$ and often neglected. The Fourier transform is

$$\hat{\psi}_{M,z_0} = \frac{z_0}{2\sqrt{\pi}}e^{-\frac{z_0^2}{2}(1+\omega^2)}\left(e^{-z_0^2\omega} - 1\right). \quad (22.142)$$

There is apparently no closed-form expression of c_ψ, but numerical integration presents no difficulty and yields the values shown in Table 22.1.

Instead of the usual dilation factor a, we adopt its inverse $\omega = 1/a$, which is its own equivalent frequency. For normalization, we adopt

$$\Psi_M(t\omega) = \frac{\omega}{\sqrt{c_\psi}}\psi_{M,z_0}(t\omega). \quad (22.143)$$

Then, the three basic formulae are, for the transform

$$\tilde{u}_M = \int u(t')\Psi_M^*[(t-t')\omega]\,dt', \quad (22.144)$$

for the energy of the signal

$$\int dt\,\frac{u^2}{2} = \frac{1}{2}\iint |\tilde{u}_M|^2\,\frac{d\omega}{\omega}\,dt, \quad (22.145)$$

and for the inverse transform

$$u(t) = \iint \Psi_M[(t-t')\omega]\tilde{u}_M(t',\omega)\frac{d\omega}{\omega}\,dt. \quad (22.146)$$

Figure 22.28 shows the norm and phase of the Morlet transform of our signal.

Comparison and Interpretation. Similarities and differences between Figs. 22.27 and 22.28 are equally informative.

The cone of influence of the end-points is wider for the Morlet wavelet: this is not surprising, since the wavelet captures several oscillations at a given frequency, whereas the Mexican-hat wavelet only covers one-and-a-half periods. The better frequency resolution of Morlet is counterbalanced by its broader temporal resolution. The wavelet coefficients within the cone of influence of the end-points are necessary for energy conservation and signal reconstruction, but should not blindly be interpreted as physical events (although the presence of oscillations at time $t > 40$ and of a maximum at $t = 47$ are not in doubt in the example).

Local spectral intensity (as measured by the wavelet coefficients) is concentrated in the same frequency band around $\omega \approx 0.3$. In the time–frequency resolution trade-off, the Morlet wavelet gives a much narrower frequency spread for the dominant wave, whereas the Mexican-hat wavelet identifies individual extrema of the signal in time. In the Morlet case, one weak energetic band at lower frequency is associated with the signal modulation, and at yet a lower frequency with end-effects (wrap-around creates periodicity).

At times 42 to 50, and frequencies larger than 0.6, we see a scattering of events that is due to the interplay of random number generation and the wavelet's reproducing kernel (see the next paragraph).

Interpretation: Reproducing Kernel. The mapping of $u(t)$ from the time axis to the wavelet half-plane (time and scale) shows that the continuous wavelet representation is redundant, which implies that the wavelet coefficients are not independent of one another. The various patterns observed in the wavelet coefficients reflect a combination of the features of the signal and of the wavelet used for the analysis. Mathematically, this is shown by substituting the expression for the inverse transform in the direct transform relation; simple rearrangement yields

$$\tilde{u}(a,t) = \iint K(a,t,a',t'')\tilde{u}(a',t'')\,\mathrm{d}a'\,\mathrm{d}t'' \,,$$

where the reproducing kernel K is

$$K = \frac{1}{c_\psi \sqrt{a a'^{5/2}}} \\ \times \int \left[\psi\left(\frac{t'-t''}{a'}\right) \psi^*\left(\frac{t-t'}{a}\right)\right] \mathrm{d}t' \,.$$

The reproducing kernel provides structure to the field of wavelet coefficients through the integral equation above, by which the wavelet coefficient at any point captures information from its vicinity as weighted by K, itself dependent on the wavelet shape.

Alternatively, it can be shown that, in the case of the Mexican-hat wavelet, the structure of the field of coefficients is such that

$$\partial_\sigma \tilde{u}_2 = \partial_{tt}^2 \tilde{u}_2 \,.$$

This is a compatibility condition for $\tilde{u}(\sigma, t)$ to be the Mexican-hat transform of a signal $u(t)$. (Note that we

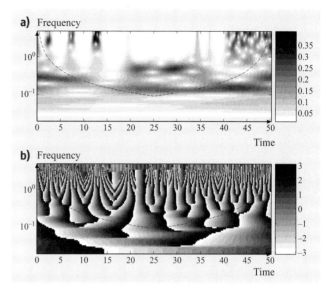

Fig. 22.28a,b Norm (**a**) and phase (**b**) of the Morlet transform of our signal. Also shown is the cone of influence of the end-points

plot $\sigma \tilde{u}_2$, which obeys a related partial differential equation.) The compatibility equation for the Morlet wavelet is much more complicated.

Thinking of the analyzing wavelet as a template for features of the signal, the main difference between the Morlet and Mexican-hat shapes is the local periodicity of the former as opposed to the central minimum (with two side maxima for admissibility) of the latter. Large wavelet coefficients are associated with a good match of shape at the right scale and location, but the pattern recognition is far from perfect. The Mexican-hat wavelet responds to any local maximum or minimum of the signal, and will do so over a range of scales; the Morlet wavelet is better suited to respond to a sequence of maxima and minima, of which the location is harder to pinpoint but the frequency is more narrowly determined.

22.6.3 Orthogonal Wavelet Transform

Discrete Wavelets
It is possible to obtain a discrete set of quasi-orthogonal wavelets by sampling the dilation and time axes a and b. For the scale a we use a logarithmic discretization, i.e., a is replaced by $a_j = a_0^{-j}$ where a_0 is the sampling rate of the $\log a$ axis [$a_0 = \Delta(\log a)$] and where $j \in \mathbb{Z}$ is the scale index. The position b is discretized linearly, i.e., b is replaced by $t_{ji} = ib_0 a_0^{-j}$, where b_0 is the sampling rate of the position axis at the largest scale and where $i \in \mathbb{Z}$ is the position index. Note that the sampling rate

of the position varies with scale, i.e., for finer scales (increasing j and hence decreasing a_j) the sampling rate increases. Accordingly we obtain discrete wavelets

$$\psi_{ji}(t) = \frac{1}{\sqrt{a_j}} \psi\left(\frac{t - t_{ji}}{a_j}\right) \tag{22.147}$$

and the corresponding discrete decomposition formula is

$$\tilde{u}_{ji} = \langle \psi_{ji}, u \rangle = \int_{-\infty}^{\infty} u(t)\psi_{ji}^*(t)\,\mathrm{d}t \tag{22.148}$$

with $\langle u, v \rangle = \int_{-\infty}^{+\infty} u(t)v^*(t)\,\mathrm{d}t$ being the scalar product in energy norm, i.e., for signals of finite energy such as those encountered in fluid mechanics.

The discrete reconstruction formula is

$$u(t) = C \sum_{j=-\infty}^{+\infty} \sum_{i=-\infty}^{+\infty} \tilde{u}_{ji} \psi_{ji}(t) + R(t), \tag{22.149}$$

where C is a constant and $R(t)$ is a residual, both of which depend on the choice of the wavelet and the sampling of the scale and position axes. For the particular choice $a_0 = 2$ (which corresponds to a scale sampling by octaves) and $b_0 = 1$ we have the dyadic sampling, for which there exist special wavelets ψ_{ji} that form an orthonormal basis, i.e.,

$$\langle \psi_{ji}, \psi_{j'i'} \rangle = \delta_{jj'}\delta_{ii'}, \tag{22.150}$$

where δ denotes the Kronecker symbol. This means that the wavelets ψ_{ji} are orthogonal with respect to their translates by discrete steps $2^{-j}i$ and their dilates by discrete steps 2^{-j} corresponding to octaves. In this case the reconstruction formula is exact with $C = 1$ and $R = 0$, but the discrete wavelet transform has lost the invariance by translation and dilation of the continuous one.

Orthogonal Wavelets

The construction of orthogonal wavelet bases and the associated fast numerical algorithms are based on the mathematical concept of multiresolution analysis (MRA). The underlying idea is to consider approximations u_j of the signal $u(t)$ at different scales j. The amount of information needed to go from a coarse approximation u_j to a higher-resolution approximation u_{j+1} is then described using orthogonal wavelets. The orthogonal wavelet analysis can thus be interpreted as decomposing the signal into approximations of the signal at coarser and coarser scales (i.e., for decreasing j) where the differences between the approximations are encoded using wavelets.

The coarse-graining at a given scale is done by filtering the signal with the scaling function ϕ. As a filter, the scaling function ϕ does not have vanishing mean but is normalized $\int_{-\infty}^{\infty} \phi(t)\,\mathrm{d}t = 1$. We construct translated and dilated versions of ϕ

$$\phi_{ji}(t) = 2^{j/2}\phi(2^j t - i). \tag{22.151}$$

This basis ϕ_{ji} is orthonormal at a given scale j with respect to its translates by steps $i/2^j$

$$\langle \phi_{ji}, \phi_{ji'} \rangle = \delta_{ii'} \tag{22.152}$$

but not to its dilates, in contrast to wavelets. In general there is no explicit expression for the scaling function. However, the scaling function satisfies a so-called refinement equation:

$$\phi_{j-1,i}(t) = \sum_{n=-\infty}^{+\infty} h_{n-2i} \phi_{jn}(t) \tag{22.153}$$

with the coefficients $h_i = \langle \phi_{ji}, \phi_{j-1,0} \rangle$. These coefficients determine the scaling function completely. Equation (22.153) implies that the approximation of a coarser scale can be described by linear combinations of the signal at finer scales, which corresponds to a nestedness of the approximation space.

The associated wavelet ψ is a linear combination of the scaling function ϕ,

$$\psi_{ji}(t) = \sum_{n=-\infty}^{+\infty} g_{n-2i} \phi_{jn}(t) \tag{22.154}$$

with coefficients $g_n = \langle \phi_{jn}, \psi_{j-1,0} \rangle$. These filter coefficients g_n are computed from the filter coefficients h_n using the relation

$$g_n = (-1)^{1-n} h_{1-n}. \tag{22.155}$$

As in the continuous case, the wavelets have at least vanishing mean, and also possibly vanishing higher-order moments, i.e.,

$$\int_{-\infty}^{+\infty} t^m \psi(t)\,\mathrm{d}t = 0 \quad \text{for } m = 0, \ldots, M-1. \tag{22.156}$$

Now we consider approximations of the signal $u(t)$ at two different scales j

$$u_j(t) = \sum_{i=-\infty}^{+\infty} \tilde{u}_{ji} \phi_{ji}(t) \tag{22.157}$$

and $j-1$

$$u_{j-1}(t) = \sum_{i=-\infty}^{+\infty} \bar{u}_{j-1,i}\phi_{j-1,i}q(t) \quad (22.158)$$

with the scaling coefficients

$$\bar{u}_{ji} = \langle u, \phi_{ji}\rangle \quad (22.159)$$

which correspond to local averages of the signal u at position $i2^{-j}$ and at scale 2^{-j}. The difference between the two approximations is described by the wavelets

$$u_j(t) - u_{j-1}(t) = \sum_{i=-\infty}^{+\infty} \tilde{u}_{j-1,i}\psi_{j-1,i}(t) \quad (22.160)$$

with the wavelet coefficients

$$\tilde{u}_{ji} = \langle u, \psi_{ji}\rangle, \quad (22.161)$$

which correspond to local differences of the signal at position $(2i+1)2^{-(j+1)}$ between approximations at scales 2^{-j} and $2^{-(j+1)}$.

Iterating the two-scale decomposition (22.160), any signal $u(t)$ of finite energy can be expressed as a sum of a coarse-scale approximation at a reference scale j_0, and their successive differences (the details needed to go from one scale j to the next smaller scale $j+1$ for $j = j_0, \ldots, J-1$),

$$u(t) = \sum_{i=-\infty}^{+\infty} \bar{u}_{j_0 i}\phi_{j_0 i}(t) + \sum_{j=j_0}^{J}\sum_{i=-\infty}^{+\infty} \tilde{u}_{ji}\psi_{ji}(t). \quad (22.162)$$

In numerical applications the sums in theis equation have to be truncated in both scale j and position i. The truncation in scale corresponds to a limitation of u to a given finest scale J, which is in practice imposed by the available sampling rate. Due to the finite length of the available data the sum over i becomes also finite. The decomposition (22.162) is orthogonal, as, by construction,

$$\langle \psi_{ji}, \psi_{lk}\rangle = \delta_{jl}\delta_{ik}, \quad (22.163)$$
$$\langle \psi_{ji}, \phi_{lk}\rangle = 0 \quad \text{for } j \geq l \quad (22.164)$$

in addition to (22.152).

Fast Wavelet Transform

Starting with a signal u given at the finest resolution 2^{-J}, i.e., we know u_J and hence the coefficients $\bar{u}_{Ji}$ for $i \in \mathbb{Z}$, the fast wavelet transform computes its wavelet coefficients $\tilde{u}_{ji}$ by successively decomposing each approximation u_J into a coarser-scale approximation u_{J-1}, plus their differences, encoded by the wavelet coefficients. The algorithm uses a cascade of discrete convolutions with the filters h_n (low pass filter) and g_n (band pass), followed by down-sampling, i.e., we retain one coefficient out of two.

- Initialization: given $u \in L^2(\mathbb{R})$ and $\bar{u}_{Ji} = u\left(\frac{i}{2^J}\right)$ for $i \in \mathbb{Z}$.
- Decomposition: for $j = J$ to 1, step -1, do

$$\bar{u}_{j-1,i} = \sum_{n \in \mathbb{Z}} h_{n-2i}\bar{u}_{jn} \quad (22.165)$$

$$\tilde{u}_{j-1,i} = \sum_{n \in \mathbb{Z}} g_{n-2i}\bar{u}_{jn}. \quad (22.166)$$

The inverse wavelet transform is based on successive reconstructions of a fine-scale approximation u_j from the coarser-scale approximation u_{j-1}, plus the differences between approximations at scale $j-1$ and the smaller scale j that are encoded by $\tilde{u}_{j-1,i}$. The algorithm uses a cascade of discrete convolutions with the filters h_n and g_n, plus up-sampling. The up-sampling is obtained by adding zeros between two successive coefficients.

- Reconstruction: for $j = 1$ to J, step 1, do

$$\bar{u}_{ji} = \sum_{n=-\infty}^{+\infty} h_{i-2n}\bar{u}_{j-1,n} + \sum_{n=-\infty}^{+\infty} g_{i-2n}\tilde{u}_{j,n}. \quad (22.167)$$

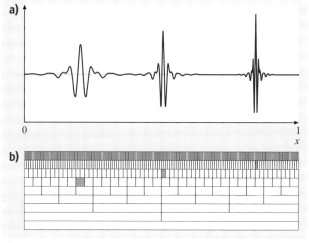

Fig. 22.29a,b Orthogonal quintic spline wavelets $\psi_{j,i}(t) = 2^{j/2}\psi(2^j t - i)$ at different scales and positions: $\psi_{5,6}(t)$, $\psi_{6,32}(t)$ $\psi_{7,108}(t)$ (**a**) and corresponding wavelet coefficients (**b**)

Examples of Orthogonal Wavelets

Orthogonal wavelets (constituting an MRA) are typically defined by their filter coefficients h_n and for most no explicit expression for ψ is available. In the following we give filter coefficients of h_n for typical orthogonal wavelets. The filter coefficients of g_n can be obtained using the quadrature relation between the two filters (22.155).

- Haar D1 (one vanishing moment):
$$h_0 = \frac{1}{\sqrt{2}},$$
$$h_1 = \frac{1}{\sqrt{2}}.$$

- Daubechies D2 (two vanishing moments):
$$h_0 = \frac{(1+\sqrt{3})}{4\sqrt{2}} = 0.482962913145,$$
$$h_1 = \frac{(3+\sqrt{3})}{4\sqrt{2}} = 0.836516303736,$$
$$h_2 = \frac{(3-\sqrt{3})}{4\sqrt{2}} = 0.224143868042,$$
$$h_3 = \frac{(1-\sqrt{3})}{4\sqrt{2}} = -0.129409522551.$$

- Coiflets: coefficients, too numerous to appear here, are listed in other sources [22.98]; wavelets are shown on Fig. 22.30.

As an example (Fig. 22.31), we develop the signal $u(t)$, sampled on $N = 2^J$ points, as an orthonormal wavelet series from the largest scale $a_{\max} = 2^0$ to the smallest scale $a_{\min} = 2^{-J}$:

$$u(t) = \sum_{j=0}^{J-1} \sum_{i=0}^{2^j-1} \tilde{u}_{ji} \psi_{ji}(t). \quad (22.168)$$

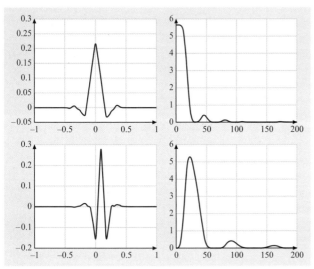

Fig. 22.30 Orthogonal wavelets Coiflet 12. *Top*: scaling function $\phi(t)$ (*left*) and $|\hat{\phi}(\omega)|$. *Bottom*: wavelet $\psi(t)$ (*left*) and $|\hat{\psi}(\omega)|$

Comparing the orthogonal wavelet transform (Fig. 22.31) with the continuous versions (Figs. 22.27 and 22.28) shows qualitative agreement in the distribution of coefficients. The cone of influence depends on the filter length L of g and h (22.153, 155). On the finest scale $L/2 - 1$ coefficients on the left and on the right maybe be influenced by boundary effects. At larger scales, the number of influenced coefficients increases accordingly (22.165, 166). Note that for the Haar wavelet ($L = 2$), no boundary effects are present.

We note two other differences. First, the orthogonal wavelet uses very few coefficients to describe the large-scale features. For a signal of N points, the highest resolution (Nyquist) requires $N/2$ points, the next highest $N/4$, and so forth, so that a total of N wavelet coefficients for all scales are needed for an exact reconstruction. This contrasts with the $N\dot{M}$ continuous wavelet coefficients, where M is the number of discretized frequencies for an approximate reconstruction. Although the $N \cdot M$ number can be improved upon, this justifies the use of orthogonal wavelets whenever the

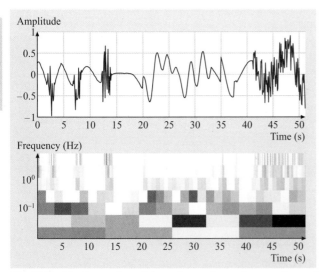

Fig. 22.31 Absolute value of the orthogonal wavelet transform of our signal using Coiflet 12 wavelets

economy of representation is critical (numerical simulation, data storage and transmission). Second, the subtle variations in frequency in the signal are not readily diagnosed from the orthogonal transform, because of the scale discretization into octaves.

22.6.4 Applications in Experimental Fluid Mechanics

In this section we describe some diagnostics based on the continuous and orthogonal wavelet transform. The diagnostics can be local (properties of single coefficients) or collective (properties of groups of coefficients or statistics based on them). While our purpose is to illustrate the strengths of wavelet-based methods, we also caution about some possible misuses. We will illustrate the ideas on our artificial signal which contains, by construction, a number of the observed features that might be difficult to isolate with other techniques; and mention a few published results from a rapidly evolving archival list. This is about data interrogation: from the selection of the wavelet to the processing algorithms, each subsection should be read as an example of a question made possible by wavelet analysis, and the corresponding quantitative answer.

Energy Distribution
Energy Maps. The total energy of the signal, normalized per unit time, is

$$E = \frac{1}{T}\int_0^T \frac{u^2}{2}\,\mathrm{d}t \ . \tag{22.169}$$

The Parseval relations for Fourier and wavelets (Mexican hat, Morlet, and orthogonal take the subscripts 'F', '2', 'M', and 's', respectively) give the expressions

$$E = \frac{1}{T}\int E_\mathrm{F}\,\mathrm{d}\omega = \frac{1}{T}\int E_2\,\mathrm{d}\omega$$
$$= \frac{1}{T}\int E_\mathrm{M}\,\mathrm{d}\omega = \frac{1}{T}\sum_{i,j} E_\mathrm{s}(\omega_j, t_i) \ . \tag{22.170}$$

On this basis, we can define the local scalogram (or energy distribution in time and frequency), and, after time integration/summation, the global scalogram (or mean energy spectrum) as an alternative to the Fourier spectrum.

We remark that, for mean spectra or for local spectra, many distinct distributions of energy are possible, adding to the same total energy. The interpretation of the various options, and their relation to the Fourier spectrum, are discussed in the following subsections.

Local Spectra. Since each scale 2^{-j} of the wavelet ψ_j is inversely proportional to the mean wavenumber $\omega_j = \omega_\psi 2^j$. The local wavelet spectrum is then defined as

$$E_\mathrm{s}(\omega_j, t_i) = |\tilde{u}_{ji}|^2 \frac{2^j}{\omega_j} \ . \tag{22.171}$$

By measuring $E_\mathrm{s}(\omega_j, t_i)$ at different positions t_i, one can study how the energy spectrum depends on local flow conditions and estimate the contribution to the overall Fourier energy spectrum of different components of the flow. For example, one can determine the scaling of the energy spectrum contributed by coherent structures, such as isolated vortices, and the scaling of the energy spectrum contributed by the unorganized part of the flow. The spatial variability of the local wavelet spectrum $E_\mathrm{s}(\omega_j, t_i)$ measures the flow's intermittency and lends itself to conditional sampling.

A comparison of the Mexican-hat and Morlet compensated spectra is shown on Fig. 22.32.

It was verified that the integral adds up to the signal's energy for each case. While the spatial and spectral energies are similar in some respects, the details reflect the properties of the corresponding wavelets. The most obvious common feature is that most of the energy is concentrated in a small region of the wavelet half-plane: most wavelet coefficients contribute negligible energy to the signal. This is the basis for wavelet compression algorithms.

Mean Spectra. When one integrates the local wavelet spectrum over time, one gets the global wavelet spec-

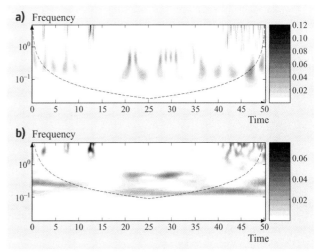

Fig. 22.32a,b Compensated local spectra of our test signal with the Mexican-hat (ωE_2, (**a**)) and Morlet (ωE_M, (**b**)) wavelets

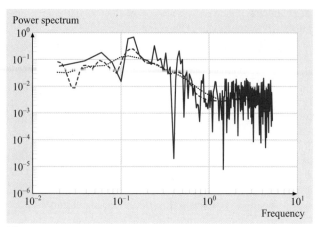

Fig. 22.33 Fourier (E_F, *solid line*), and mean Mexican-hat (E_2, *dotted line*) and Morlet (E_M, *dashed line*) mean power spectra

trum. The Fourier spectrum (Fig. 22.33) appears very noisy, which is due to the brevity of our signal (512 points) combined with its various intermittent features. The power spectrum $E(\omega)$ shows the dominant mid frequencies; at high frequencies, we have a combination of noise and corrections for nonsinusoidal low-frequency content and for intermittent events.

The Morlet spectrum is a smoother alternative to the Fourier spectrum, obtained by squaring the wavelet coefficients, and summing them at each frequency. The dominant lower frequencies stand out, but the familiar Fourier scatter is absent: this is a consequence of the wavelet coefficients being band-pass-filtered Fourier coefficients, with the spectrum of the wavelet as the filter shape. Depending on the application, such drastic smoothing can be an advantage or a weakness of the mean wavelet spectra compared to the Fourier spectra.

The wider spectrum of the Mexican-hat wavelet leads to further smoothing, so much that the low-frequency peaks are no longer distinct, but merge into a broader clump. This illustrates the time–frequency resolution compromise associated with wavelet selection.

Relation to Fourier Spectrum. Although the wavelet transform analyzes the flow into wavelets rather than complex exponentials, it has been shown that the global wavelet spectrum converges to the Fourier spectrum provided that the analyzing wavelets have enough cancellations. More precisely the global wavelet spectrum (with $\tilde{E}$ standing for either E_F, E_2 or E_M) is

$$\tilde{E}(\omega) = \int_0^{+\infty} \tilde{E}(\omega, t)\, dt \qquad (22.172)$$

gives the correct Fourier exponent for a power-law Fourier energy spectrum $E_F(\omega) \propto \omega^{-\beta}$ if the analyzing wavelet has at least $m > (\beta-1)/2$ vanishing moments. Thus, the steeper the energy spectrum, the more vanishing moments the analyzing wavelet should have. The inertial range in turbulence has a power-law behavior. The ability to correctly characterize power-law energy spectra is therefore an important property of the wavelet transform (which is related to its ability to detect and characterize singularities).

The global wavelet spectrum $\tilde{E}(\omega)$ is a smoothed version of the Fourier spectrum $E_F(\omega)$. This can be seen from the following relation between the two spectra

$$\tilde{E}(\omega) = \frac{1}{C_\psi \omega_\psi} \int_0^{+\infty} E(\omega') |\hat{\psi}\left(\frac{\omega_\psi \omega'}{\omega}\right)|^2 d\omega' \qquad (22.173)$$

which shows that the global wavelet spectrum is an average of the Fourier spectrum weighted by the square of the Fourier transform of the wavelet ψ shifted at wavenumber ω. Note that the larger ω is, the larger the averaging interval, because wavelets are bandpass filters with $\frac{\Delta\omega}{\omega}$ constant. This property of the global wavelet spectrum is particularly useful for turbulent flows. Indeed, the Fourier spectrum of a single realization of a turbulent flow is too oscillatory to clearly detect a slope, but this is not an issue for the smoother global wavelet spectrum.

For instance, the real-valued Mexican-hat wavelet has only two vanishing moments and thus can correctly measure energy spectrum exponents up to $\beta < 5$. In the case of the complex-valued Morlet wavelet (22.141), only the zeroth-order moment is zero, but the higher m-th-order moments are very small ($\propto \omega_0^{2m+1} e^{(-\omega_0^2/2)}$) provided that ω_0 is larger than 5. Therefore the Morlet wavelet transform gives accurate estimates of the power-law exponent of the energy spectrum at least for approximately $\beta < 7$ (if $\omega_0 = 6$). There is also a family of wavelets with an infinite number of cancellations

$$\hat{\psi}_k(\omega) = \alpha_k \exp\left[-\frac{1}{2}\left(\omega^2 + \frac{1}{\omega^{2k}}\right)\right], \quad k \geq 1, \qquad (22.174)$$

where α_k is chosen for normalization. These wavelets can therefore correctly measure any power-law energy spectrum, and detect the difference between a power-law energy spectrum and a Gaussian energy spectrum $E(\omega) \propto e^{-(\omega/\omega_\psi)^2}$. For instance in turbulence this wavelet can identify the changing spectral slopes around the Kolmogorov microscale and into the dissipative range of the spectrum.

Ridges. In some applications, the signal may be primarily oscillatory, but with modulation of the frequency and amplitude. Extracting these oscillations and quantifying the modulation can be done from the continuous wavelet map. One algorithm consists of identifying the ridges of the energy distributions, i.e., the coordinates of the points in time at which the compensated energy spectrum has a local spectral maximum. In Fig. 22.34, we plot the ridge lines of the Mexican-hat and Morlet energy distributions. The Morlet wavelet is a better template for localized oscillations, and emphasizes the locally periodic structure of the signal. In contrast, although local periodicity is visible on the Mexican-hat energy map, the ridge lines do not capture the continuity of wave trains. While remedies exist for this shortcoming of the Mexican-hat wavelet, the Morlet wavelet is a more natural tool for this application.

Even when the frequency modulation is very small, the relevant physics can be detected through related modulations in the amplitude and or phase discontinuities along the ridge line. This technique has been successfully used to unravel the interactions between modes in shear layers.

Lines of Modulus Maxima. A skeleton of the continuous wavelet transform can also be obtained by mapping its modulus maxima at each frequency. In the case of real-valued wavelets, these alternate with the lines of zero crossing (sign change). The skeleton combines information about the signal and the wavelet (reproducing kernel), as shown in Fig. 22.35. For a given signal, the particulars of the skeleton are clearly wavelet dependent, and should be interpreted with caution. However, statistically the skeleton provides useful partitions of the wavelet half-plane.

While modulus maxima and zero-crossing lines are topologically equivalent, the values of the extrema on the former contain more information. At any given frequency, the wavelet transform can be approximated between extrema by interpolating a smooth monotonic function. Thus, the knowledge of the wavelet coefficients along the lines of modulus maxima is sufficient to reconstruct the wavelet map, and thereby the signal. In conjunction with thresholding of the wavelet coefficients, this can be used for data compression and for computing multifractal statistics.

Intermittent Fluctuations. One of the advantages of the time–frequency representation is the ability to define and compute error bars to supplement the mean spectra. In intermittent cases, the (squared) wavelet coefficients

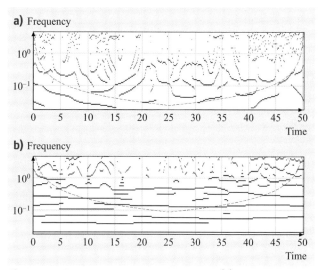

Fig. 22.34a,b Ridge lines of the Mexican-hat (**a**) and Morlet transforms (**b**), showing the frequency modulation of the energetic contributions to the signal

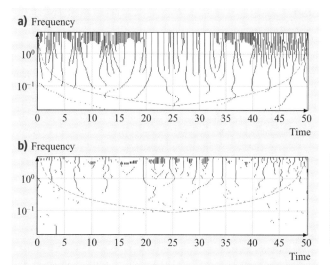

Fig. 22.35a,b Lines of modulus maxima for the Mexican-hat (**a**) and Morlet (**b**) wavelets

show considerable variability in time, and a corresponding standard deviation can be calculated at each scale. The result, using additive rather than logarithmic departures from the mean, is shown in Fig. 22.36.

The error bars are taken as covering the mean, plus or minus three standard deviations. The lower end of the error bars corresponds to negative energies, and is assimilated to zero; the profile of the top of the error

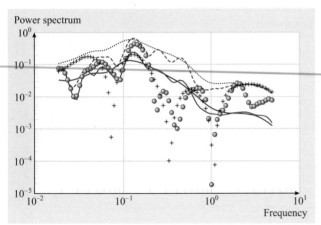

Fig. 22.36 Mean Morlet and Mexican-hat mean power spectra (*solid lines*), as in Fig. 22.33, and their error bar profile as *dashed line* (Mexican hat) and *dotted line* (Morlet); also the local spectra at $t = 45$: Mexican hat (*circles*) and Morlet (*crosses*)

bars is shown. It is seen that the error bars can be larger than the mean by a factor of 3–4 at low frequencies, and considerably more at high frequencies, for our signal. A larger bracket around the mean spectrum has been interpreted as an indicator of intermittency in homogeneous turbulence.

The superposition of the local spectra (scaled per unit duration of signal) is also informative. First, there is considerable difference between the Morlet and Mexican-hat local energies Fig. 22.32) – as was obvious from the wavelet maps (Figs. 22.28 and 22.27). At $\omega \approx 0.15$, the local energies contribute to one of the dominant peaks: the smallest dominant frequency is represented at this time. The larger Morlet energy is indicative of an oscillation at this frequency. In contrast, at $\omega \approx 0.045$, the larger Mexican-hat energy content captures the large, single bump better than Morlet. At $\omega \approx 3$–4, the double peak of Morlet coefficients is matched by a much larger single peak of the Mexican-hat coefficients: this combines the random noise and the reproducing kernel for this particular instant, and is very different at nearby points.

Also noteworthy is that the Mexican-hat energies around $\omega = 4$ lie outside the three-standard-deviation bracket. In general, this can be attributed to several possible causes. First, the difference in magnitude between Morlet and Mexican-hat energies indicates that an individual event, rather than an oscillation, is present in this vicinity; such an event could be a statistical outlier. Second, in this instance, the mean spectrum and the standard deviation are ensemble statistics, underestimat-

ing the (intermittent) noise energy content per unit time. This points to the need for conditional statistics.

Intermittency Measures. We now turn to intermittency, which is defined as localized bursts of high-frequency activity. This means that intermittency is a phenomenon localized in both physical space and spectral space, and thus a suitable basis for representing intermittency should reflect this dual localization.

To measure intermittency we use the space-scale information contained in the wavelet coefficients to define scale-dependent moments and moment ratios. Useful diagnostics to quantify the intermittency of a signal $u(t)$ are the moments of its wavelet coefficients at different scales j,

$$M_{p,j}(u) = 2^{-j} \sum_{i=0}^{2^j - 1} |\tilde{u}_{ji}|^p . \tag{22.175}$$

Note that $E_j = 2^j M_{2,j}$.

The sparsity of the wavelet coefficients at each scale is a measure of intermittency, and it can be quantified using ratios of moments at different scales,

$$Q_{p,q,j}(u) = \frac{M_{p,j}(u)}{[M_{q,j}(u)]^{p/q}} . \tag{22.176}$$

Classically, one chooses $q = 2$ to define typical statistical quantities as a function of scale. Recall that for $p = 4$ we obtain the scale-dependent flatness $F_j = Q_{4,2,j}$. It is equal to 3 for Gaussian white noise at all scales j, which proves that this signal is not intermittent. The scale-dependent skewness, hyperflatness, and hyperskewness are obtained for $p = 3, 5$ and 6, respectively. For intermittent signals $Q_{p,q,j}$ increases with j.

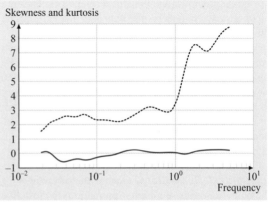

Fig. 22.37 Scale dependence of the skewness (*solid line*) and flatness (*dashed line*)

On Fig. 22.37, we show the frequency-dependent skewness and kurtosis of our test data, calculated based on the Mexican-hat coefficients. The skewness hovers around zero, and the kurtosis around 3 at low frequencies. The larger kurtosis values at higher frequencies confirms the intermittency of the noise interspersed in the data.

Various techniques have been used to separate outlier events from a background signal. One can mention the local intermittency measure (LIM) and large dissipative events.

Higher Moments and Scaling

The idea in this group of applications is that any line in the wavelet transform matrix can be treated as a time trace (band-pass-filtered data) with statistics in its own right. Auto- and cross-correlations can be found in the literature. The spectral error bars described already are another example.

Scaling describes the relation between moments of a signal at various scales. In particular, anomalous scaling relates to departures from Kolmogorov scaling for high-order moments of turbulent fields. Because of their intermittent and spectral content, wavelet coefficients can be used effectively to study anomalous scaling. By comparing results before and after filtering certain types of events, it has been possible to associate anomalous scaling with certain classes of events, such as coherent structures or highly dissipative eddies.

Relation to Structure Functions. Since its introduction to turbulence by Kolmogorov, the second-order structure function has been a frequently used diagnostic tool for the study of a turbulent signal $u(t)$. It is easy to measure and, moreover, Kolmogorov's theory of homogeneous and isotropic turbulent predicts its scaling exponent, which explains its wide use. It is defined as

$$S_2(\tau) = \int_{-\infty}^{\infty} [u(t+\tau) - u(t)]^2 \, dt \, . \tag{22.177}$$

In this paragraph we link the scale-dependent moments of the wavelet coefficients to the structure functions and show that the global wavelet spectrum corresponds to the second-order structure function. Furthermore, we prove that the structure functions are *blind* to some scaling exponents and propose a way to overcome this limitation.

We first remark that the increments of a signal, also called the modulus of continuity, can be seen as its wavelet coefficients using the difference of Diracs (DOD) wavelet

$$\psi^\delta(t) = \delta(t+1) - \delta(t) \, . \tag{22.178}$$

We thus obtain

$$u(t+\tau) - u(t) = \tilde{u}_{t\tau} = \langle u, \psi^\delta_{t\tau} \rangle \tag{22.179}$$

with $\psi^\delta_{t\tau}(t) = 1/\tau\{\delta[(t-t')/\tau + 1] - \delta[(t-t')/\tau]\}$. Note that in this case the wavelet is normalized with $1/\tau$ normalization rather than $1/\sqrt(\tau)$. The p-th-order structure function $S_p(l)$ therefore corresponds to the p-th-order moment of the wavelet coefficients at scale τ,

$$S_p(\tau) = \int_{-\infty}^{\infty} [u(t+\tau) - u(t)]^p \, dt = \int_{-\infty}^{\infty} (\tilde{u}_{t\tau})^p \, dt \, . \tag{22.180}$$

As the DOD wavelet has only one vanishing moment (its mean), the exponent of the p-th-order structure function in the case of a self-similar behavior is limited by p, i.e., if $S_p(\tau) \propto \tau^{\zeta(p)}$ then $\zeta(p) < p$. To be able to detect larger exponents one has to use increments with a larger stencil, or wavelets with more vanishing moments, i.e., $\int t^m \psi(t) \, dt = 0$ for $m = 0, 1, \ldots, M-1$.

We now concentrate on the case $p = 2$, i.e., the energy norm. Equation (22.172) gives the relation between the global wavelet spectrum $\tilde{E}(\omega)$ and the Fourier spectrum $E(\omega)$ for an arbitrary wavelet ψ. For the DOD wavelet we find, since $\hat{\psi}^\delta(\omega) = e^{i\omega} - 1 = e^{i\omega/2}(e^{i\omega/2} - e^{-i\omega/2})$ and hence $|\hat{\psi}^\delta(\omega)|^2 = 2(1 - \cos \omega)$, that

$$\tilde{E}(\omega) = \frac{1}{C_\psi \omega} \int_0^\infty E(\omega') \left[2 - 2\cos\left(\frac{\omega_\psi \omega'}{\omega}\right) \right] d\omega' \, . \tag{22.181}$$

Setting $\tau = \omega_\psi/\omega$ and comparing with (22.177) we see that the wavelet spectrum corresponds to the second-order structure function, i.e.,

$$E_s(\omega) = \frac{1}{C_\psi \omega} S_2(\tau) \, . \tag{22.182}$$

These results show that, if the Fourier spectrum behaves like $\omega^{-\alpha}$ for $\omega \to \infty$, $\tilde{E}(\omega) \propto \omega^{-\alpha}$ if $\alpha < 2M + 1$, where M is the number of vanishing moments of the wavelets. Consequently, we find for $S_2(\tau)$ that $S_2(\tau) \propto \tau^{\zeta(p)} = \left(\frac{\omega_\psi}{\omega}\right)^{\zeta(p)}$ for $\tau \to 0$ if $\zeta(2) \leq 2M$. In the present case we have $M = 1$, i.e., the second-order structure function can only detect slopes smaller than 2, corresponding to an energy spectrum with slopes shallower than -3. Thus we find that the usual structure function gives spurious results for sufficiently smooth signals.

Wavelet-based calculation of structure functions and their statistics (including anomalous scaling, above) has been documented.

Nonlinear Filtering

Nonlinear filtering is the common approach to denoising, data compression, coherent structure eduction, and associated conditional statistics. It consists in separating from a wavelet map any event meeting a physically relevant criterion, and reconstructing a filtered signal suitable for further processing. The quality of the results depends on the choices of wavelet and of event definition. In Fourier filtering, the presence of a spectral gap is highly desirable. In the wavelet representation, objective criteria for separating flow events need the same careful formulation. Different events, with different spectral and temporal signatures, might be captured with different wavelets; but a wavelet coefficient is not a flow structure, in the usual sense, unless the experimental definition of the event is reflected in the filtering algorithm.

Whether based on local intermittency measures, on being surrounded by a temporal–spectral gap, or related energy-based criteria, some thresholding is usually involved in the filtering scheme, and threshold sensitivity needs to be documented. One notable exception, which has been used for coherent structure eduction, is the mathematically proven optimality of some wavelets to separate a coherent signal from Gaussian noise, without adjustable thresholding parameters. This has been used effectively in two- (2-D) and three-dimensional (3-D) turbulence.

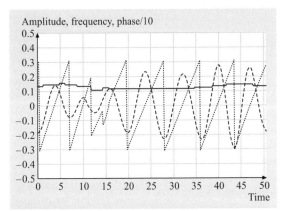

Fig. 22.38 The dominant oscillations (*dashed line*), its local frequency (*solid line*) and its phase (*dotted line*). The phase in $(-\pi, \pi)$ is divided by 10 for plotting convenience

Modulated Oscillations. For our signal, the mean spectra do not indicate a unique dominant periodicity, but several weak mid-frequency peaks can be identified. The Morlet wavelet, a template for local periodicity (over a time frame of only a few periods), enables us not only to extract the modulation, but also to quantify the frequency shifts already visible on Fig. 22.34. The algorithm consists of

1. searching, at each time, for the peak energy within range of scales (excluding the noisy high frequency band, for example)
2. recording the frequency and phase of the wavelet coefficients
3. canceling the wavelet coefficients outside this dominant mode (a spectral bracket of $\pm 10\%$ of peak local frequency has been retained here)
4. performing the inverse wavelet transform of the remaining coefficients to obtain a *filtered* signal. The result is shown on Fig. 22.38.

Comparing with Fig. 22.26, we observe that the clean part of the signal, between times 15 and 40, is extracted together with amplitude corrections caused by the isolated events of matching scale. The dominant frequency is seen to vary slightly, matching one of the ridge lines on Fig. 22.34, and increasing slightly towards both ends of the plot. Thus, depending on

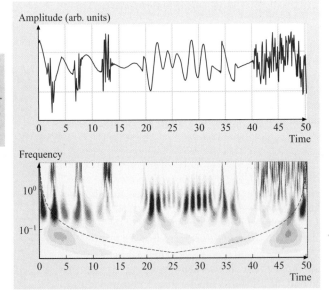

Fig. 22.39 The signal obtained by subtracting the dominant oscillations (Fig. 22.38) from the original signal (Fig. 22.26) and its Mexican-hat wavelet transform

the context, this oscillation could indicate a physical event. In contrast, in the first third of the signal, the ridge lines are broken, and we also see phase dislocations and erratic amplitude in Fig. 22.38. This happens because of interference from the oscillations at nearby frequencies, and from end-effects, so no carrier wave can be isolated unambiguously in this region. Additional physical insight, different wavelets and/or more-elaborate algorithms might modify this cautious conclusion.

Noteworthy is the relatively complex nature of the filter, made possible by the time–frequency representation. Furthermore, the small frequency shifts are easily observed, a trademark of continuous (Morlet in particular) wavelet transforms; for actual anemometric and acoustic traces, they have been associated, e.g., with subtle shear layer resonances. Finally, the phase of this nonperiodic signal is clearly defined away from dislocations, and can be correlated with other features. Similar ideas have been used in relation to vortex shedding and atmospheric oscillations.

Selective Filtering. Complementing the above viewpoint, the wavelet coefficients excluding the modulated oscillation can be retained, and the inverse Morlet transform removes the oscillation from the signal. A similar procedure has been carried out to characterize background turbulence as distinct from coherent structures. The result is shown on Fig. 22.39. The higher-frequency wave packet, the isolated bumps, and the noise are left intact, and only the faintest trace of the main oscillation is observed in the filtered signal. The difference on the time trace is not huge (weaker low frequencies between times $t = 15$ and 35, and the weaker bump at $t = 6$, are the more visible differences), but is quite clear on the Mexican-hat wavelet transform compared with Fig. 22.27.

This filtered signal can be treated as any signal, and its Mexican-hat wavelet transform is also shown on Fig. 22.39. The isolated events, notably the discontinuity at $t = 34$, stand out. Several points are illustrated in this context.

1. The selective removal of groups of wavelet coefficients meeting specific criteria designed by the analyst amounts to local surgery on the signal.
2. Because of their different shapes, the Morlet and Mexican-hat wavelets emphasize different features (local periodicity and isolated extrema, respectively). Other wavelets can be used, e.g., for edge detection.
3. Physical events (e.g., the discontinuity at time $t = 34$) can be multiscale events, extending spectrally over several orders of magnitudes; conversely, events of similar shape, location and/or scale (as were seeded in our signal on Fig. 22.26) may not have distinct wavelet signatures. Spectrally or temporally overlapping groups of coefficients should not be interpreted as indicative of distinct physical objects without corroborating additional insight specific to the signal at hand.

Several filtering variants can extract coherent structures. In addition to the denoising approach mentioned above, several wavelet-based eduction schemes have been used successfully. One example is shown in Fig. 22.40 [22.99], for the decomposition of the vector field in a mixing layer into coherent eddies and unstructured turbulence. In this case, both coherent and unstructured fields carried a significant fraction of the turbulent energy and of the turbulent stresses.

Conditional Spectra. Statistics such as energy or kurtosis provide a global summary of the data. Conditional statistics do the same for any subset of the data that meets user-defined criteria. This approach has been used in the case of unsteady flows encountered in turbomachinery, where periodic wake passing affects the boundary-layer development and transition and the relevant turbulent time scales. Similar steps are taken here with our test

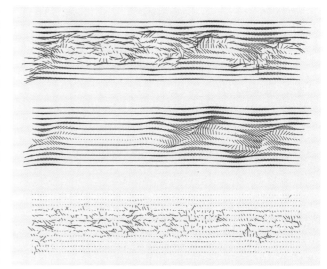

Fig. 22.40 A velocity field in a turbulent mixing layer (*top*), and its decomposition into intermittent coherent (*middle*) and unstructured fluctuations (*bottom*) (after [22.99])

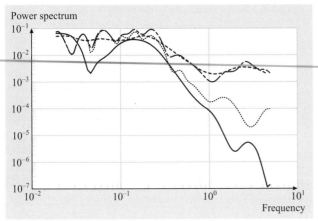

Fig. 22.41 Conditional power spectra over the first 15 units of test data, separating the contributions from data with the phase of the modulated oscillation between $-\pi/5$ and $\pi/5$ ('part 1'), and the other phases ('part 2'). Mexican hat spectrum of part 1 (*dashed line*) and part 2 (*solid line*); Morlet spectrum of part 1 (*dash-dot line*) and part 2 (*dotted line*)

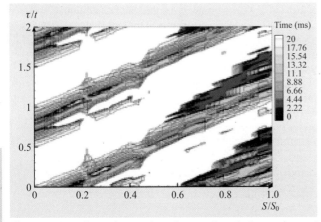

Fig. 22.42 Dominant time scales of unsteady turbulence triggered in the boundary layer over a curved plate by periodic wakes; s/s_0 is the dimensionless chord coordinate; time is nondimensionalized with the wake-passing half-period (after [22.100])

Note that the (phase-related) condition and the (energy-related) statistic can be obtained with different wavelet transforms.

The result is shown in Fig. 22.41. As far as determining the dominant frequencies of part 1 of the signal, the Mexican-hat and Morlet spectra agree both for the low-frequency and noisy contributions, separated by a spectral gap. For part 2 of the signal, they also concur that there is no noise. This might serve as an indication that noise is triggered in some way by the modulated oscillation, or that they have a common driving mechanism. However, one should be careful that conditional statistics imply the use of an indicator function (value 1 when the conditioning criterion is met, 0 otherwise), which introduces a cone of influence for each transition. Therefore, quantitative spectral statistics would be accurate only at frequencies higher than the dominant frequencies of the indicator function.

These ideas have been implemented in the study of unsteady boundary-layer transition (Fig. 22.42) [22.100]. The wakes of guide vanes are experienced as nearly periodic disturbances traveling along the rotor blade surfaces, triggering transition. The characterization of three types of turbulence (between wakes, in wakes, and in the boundary layer) evolving along the blade surface, could lead to better models of these complex flows.

Further Applications

Some general applications of wavelets overlap partially with fluid dynamics, and can be mentioned briefly.

Multidimensional Wavelets. Beyond time series, experimental and numerical data fields can be multidimensional, in space (e.g., PIV data) or space and time. Multidimensional wavelet analysis has been used in this context, and examples were presented above. Distinct transforms are normally executed in space and time; for 2-D or 3-D fields, multidimensional wavelets have been used successfully. They can be constructed as tensor products of one-dimensional wavelets, or as derivatives of multidimensional (sometimes anisotropic) filters. They have been successfully applied to the study of scaling in isotropic turbulence, of anisotropy in turbulent shear flows, to the eduction of coherent structures in 2-D and 3-D numerical data, among others. Generally speaking, many tools developed in the image processing literature are potentially relevant.

Denoising. Denoising has been used to separate unstructured turbulence (the noise) from the coherent structures. In this context, noise is not necessarily a loss in signal

data. We restrict the computation of mean spectra to the filtered signal (Fig. 22.39) subject to the condition that the phase of the modulated oscillations over the first 15 s of data is between $-\pi/5$ and $\pi/5$ (Fig. 22.38), called part 1; other phases in the same time span contributed to part 2. For comparison, all spectra are normalized to a unit time of their respective samples. A statistical quantity of interest might be the dominant time scale (peak of compensated spectrum) associated with the condition.

quality due to extraneous random interference. Depending on the context, the noise may be undesirable or physically relevant: what matters here is that the two contributions to the signal have different wavelet signatures, either locally (e.g., the magnitude of the wavelet coefficients) or statistically (e.g., Gaussian energy distribution). Denoising is one particular instance of nonlinear filtering.

Denoising and two-dimensional fields are combined in Fig. 22.43. PIV data in the near wake of a circular disk normal to the uniform flow were obtained in an axial plane, and the tangential vorticity was calculated by finite difference. The combination of experimental and numerical noise can be alleviated by denoising. The orthogonal wavelet denoising scheme [22.98] has been adapted to the Mexican-hat wavelet, resulting in the vorticity distribution shown on Fig. 22.43.

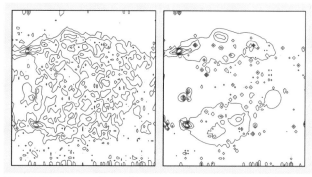

Fig. 22.43 Contour lines of instantaneous tangential vorticity in an axial section of the wake behind a circular disk, calculated from PIV data. *Right*: raw data; *left*: after continuous wavelet denoising; the same contour levels are used for both plots (data courtesy of H. Higuchi and R. P. Bigger)

Data Compression. Most of the energy of the signal is concentrated in a relatively small number of wavelet coefficients, as seen on Fig. 22.32. Independently of the context-dependent interpretation of energetic events as coherent structures, as in POD (Sect. 22.5), the remaining weak fluctuations may be ignored or modeled to yield a lower-dimensional system. Further economy of representation is achieved by keeping only modulus-maxima lines (the skeleton, without the surrounding *flesh*). Because of the commercial impact for data storage and transmission, this subfield has evolved very rapidly. Dynamical system modeling based on modulus maxima may be achievable.

Detection of Transitions. Transitions come in various guises, and several options should be considered. The asymmetric first derivative of a Gaussian (related to the Mexican hat, which is the second derivative) has been used for edge detection; in nearly periodic signals, phase dislocations (Morlet) have also been used successfully. The image-processing literature is too extensive to cite.

Resolution of Singularities, Fractal Signals. The ability of wavelets to zoom into ever-smaller domains of the signal can also be used to characterize singularities. It has been shown that the scaling exponent of the wavelet coefficients is related to the strength of the singularity.

Additional Remarks. Nonuniform sampling intervals in time series complicate the use of fast algorithms for wavelet transforms. As for Fourier processing, methods such as interpolation and resampling can be used.

As a pattern-recognition technique, the use of a single wavelet shape and simple algorithms is usually not satisfactory, because the value of the wavelet coefficients is determined not only by the shape, but also the amplitude of fluctuations. Some successful applications are emerging in the flow control area. While wavelets are usually too simple to serve as pattern-recognition tools, a collection of wavelet-based diagnostics can decrease the number of false-positive matches. Neural nets have been used to process multiple wavelet-based criteria, and applications to fluid mechanics appear likely in the future.

References

22.1   S.L. Marple Jr.: *Digital Spectral Analysis* (Prentice-Hall, Englewood Cliffs 1987)

22.2   J.S. Bendat, A.G. Piersol: *Random Data: Analysis and Measurement Procedures* (Wiley, New York 1986)

22.3   J.L. Lumley: The structure of inhomogeneous turbulence. In: *Atmospheric Turbulence and Wave Propagation*, ed. by A.M. Yaglom, V.I. Tatarski (Nauka, Moscow 1967) pp. 166–178

22.4   A.A. Townsend: *The Structure of Turbulent Shear Flow*, 2nd edn. (Cambridge Univ. Press, Cambridge 1976)

22.5   R. Adrian, J.P. Bonnet, J. Delville, F. Hussain, J. Lumley, O. Metais, C. Vassilicos: *CISM/ERCOFTAC*

Advanced Course: Eddy Structure Identification Techniques in Free Turbulent Shear Flows (Springer, Berlin, Heidelberg 1996)

22.6 J.L. Lumley: *Stochastic Tools in Turbulence* (Academic, New York 1970)

22.7 D.D. Kosambi: Statistics in function space, J. Indian Math. Soc. **7**, 76–88 (1943)

22.8 M. Loève: Fonctions aléatoires du second ordre, Comptes Rend. Acad. Sci. **220**, 295–300 (1945)

22.9 M. Loève: *Probability Theory* (Van Nostrand, New York 1955)

22.10 K. Karhunen: Zur Spektraltheorie stochastischer Prozesse, Ann. Acad. Sci. Fenn. A1 **34**, 1–7 (1946), in German

22.11 V.S. Pougachev: General theory of the correlations od random functions, Izv. Akad. Nauk. SSSR Mat. **17**, 401–402 (1953)

22.12 A.M. Obukhov: Energy distribution in the spectrum of a turbulent flow, Izv. A. N. SSSR Geogr. Geophys. **4-5**, 453–466 (1941)

22.13 A.M. Obukhov: Statistical description of continuous fields, Trudy Geophys. Int. Akad. Nauk. SSSR **24**, 3–42 (1954)

22.14 I.T. Joliffe: *Principal Component Analysis* (Springer, New York 1986)

22.15 H. Hotelling: Analysis of a complex statistical variables into principal components, J. Educ. Psychol. **24**, 417–441 (1933)

22.16 G.H. Golub, C.F. Van Loan: *Matrix Computations*, 2nd edn. (Johns Hopkins Univ. Press, Baltimore 1990)

22.17 A. Papoulis: *Probability, Random variables, and Stochastic Processes* (McGraw-Hill, New York 1965)

22.18 M. Kirby, L. Sirovich: Application of the Karhunen–Loève procedure for the characterization of human faces, IEEE T. Pattern Anal. **12**(1), 103–108 (1990)

22.19 V.R. Algazi, D.J. Sakrison: On the optimality of the Karhunen–Loève expansion, IEEE Trans. Inform. Theory **15**, 319–321 (1969)

22.20 C.A. Andrews, J.M. Davies, G.R. Schwartz: Adaptative data compression, Proc. IEEE **55**, 267–277 (1967)

22.21 S.S. Ravindran: Reduced-order adaptive controllers for fluid flows using POD, J. Sci. Comput. **15**(4), 457–478 (2000)

22.22 S.S. Ravindran: A reduced-order approach for optimal control of fluids using proper orthogonal decomposition, Int. J. Numer. Meth. Fluids **34**, 425–448 (2000)

22.23 R. Courant, D. Hilbert: *Methods of Mathematical Physics*, Vol. 1 (Wiley, New-York 1953)

22.24 P. Holmes, J.L. Lumley, G. Berkooz: *Turbulence, Coherent Structures, Dynamical Systems and Symmetry*, Cambridge Monographs on Mechanics (Cambridge Univ. Press, Cambridge 1996)

22.25 J. Delville, L. Ukeiley, L. Cordier, J.-P. Bonnet, M. Glauser: Examination of large-scale structures in a turbulent mixing layer, Part 1. Proper orthogonal decomposition, J. Fluid Mech. **391**, 91–122 (1999)

22.26 S. Gordeyev: *Investigation of coherent structure in the similarity region of the planar turbulent jet using POD and wavelet analysis*, Ph.D. Dissertation (University of Notre Dame, Notre Dame 1999)

22.27 J.-P. Bonnet, J. Delville: Coherent structures in turbulent flows and numerical simulations approaches. In: *Lecture Series 2002-04 on Post-Processing of Experimental and Numerical Data*, ed. by P. Millan, M.L. Riethmuller (Von Karman Institute for Fluid Dynamics, Bruscelles 2002)

22.28 H.E. Fiedler: Control of free turbulent shear flows. In: *Flow Control: Fundamental and Practices*, Lecture Notes Phys., Vol. 53, ed. by M. Gad-el-Hak, A. Pollard, J.-P. Bonnet (Springer, Berlin, Heidelberg 1998) pp. 336–429

22.29 N. Aubry, P. Holmes, J.L. Lumley, E. Stone: The dynamics of coherent structures in the wall region of a turbulent boundary layer, J. Fluid Mech. **192**, 115–173 (1988)

22.30 L. Ukeiley, L. Cordier, R. Manceau, J. Delville, M. Glauser, J.-P. Bonnet: Examination of large-scale structures in a turbulent mixing layer, Part 2. Dynamical systems model, J. Fluid Mech. **441**, 67–108 (2001)

22.31 L. Cordier, M. Bergmann: Two typical applications of POD: Coherent structures eduction and reduced order modelling. In: *Lecture Series 2002-04 on Post-Processing of Experimental and Numerical Data*, ed. by P. Millan, M.L. Riethmuller (Von Karman Institute for Fluid Dynamics, Bruscelles 2002)

22.32 M. Hinze: *Optimal and instantaneous control of the instationary Navier–Stokes equations, Accreditation to supervise research dissertation* (Berlin University, Berlin 2000)

22.33 S. Volkwein: *Optimal and suboptimal control of partial differential equations: Augmented Lagrange-SQP methods and reduced order modeling with proper orthogonal decomposition, Accreditation to supervise research dissertation* (Graz University, Graz 2001)

22.34 M. Fahl: *Trust-Region methods for flow control based on reduced order modeling*, Ph.D. Dissertation (Trier University, Trier 2000)

22.35 M. Bergmann, L. Cordier, J.-P. Brancher: Optimal rotary control of the cylinder wake using POD reduced order model, Phys. Fluids **17**(9), 1–21 (2005)

22.36 A. Chatterjee: An introduction to the proper orthogonal decomposition, Curr. Sci. **78**(7), 808–817 (2000)

22.37 T.J. Rivlin: *An Introduction to the Approximation of Functions* (Dover, New York 1981)

22.38 N.J. Higham: Matrix nearness problems and applications. In: *Applications of Matrix Theory*, ed. by M.J.C. Glover, S. Barnett (Clarendon, Oxford 1989) pp. 1–27

22.39 L. Hubert, J. Meuleman, W. Heiser: Two purposes for matrix factorization: A historical appraisal, SIAM Rev. **42**, 68–82 (2000)

22.40 J.A. Atwell, B.B. King: Reduced order controllers for spatially distributed systems via proper orthogonal decomposition, Virginia Tech. ICAM **99-07-01** (1999)

22.41 S. Volkwein: *Proper Orthogonal Decomposition and Singular Value Decomposition*, Tech. Rep. Institut für Mathematik 153 (Graz University, Graz 1999)

22.42 E. Anderson, Z. Bai, C. Bischof, S. Blackford, J. Demmel, J. Dongarra, J. Du Croz, A. Greenbaum, S. Hammerling, A. McKenney, D. Sorensen: LAPACK User's Guide (SIAM third edn. 1999)

22.43 N. Aubry: On the hidden beauty of the proper orthogonal decomposition, Theor. Comp. Fluid Dyn. **2**, 339–352 (1991)

22.44 E.A. Christensen, M. Brøns, J.N. Sørensen: *Evaluation of POD-based decomposition techniques applied to parameter-dependent non turbulent flows*, DCAMM Rep. 573 (Technical University of Denmark, Arhus 1998)

22.45 G. Berkooz: *Turbulence, coherent structures, and low dimensional models*, Ph.D. Dissertation (Cornell University, Ithaca 1991)

22.46 F. Riesz, B.S. Nagy: *Functional Analysis* (Ungar, New York 1955)

22.47 G. Berkooz, P. Holmes, J.L. Lumley: The proper orthogonal decomposition in the analysis of turbulent flows, Annu. Rev. Fluid Mech. **25**, 539–575 (1993)

22.48 D. Rempfer, H.F. Fasel: Evolution of three-dimensional coherent structures in a flat-plate boundary layer, J. Fluid Mech. **260**, 351–375 (1994)

22.49 S. Sanghi: *Mode interaction models in near-wall turbulence*, Ph.D. Dissertation (Cornell University, Ithaca 1991)

22.50 R. Temam: *Infinite-Dimensional Dynamical Systems in Mechanics and Physics* (Springer, New York 1988)

22.51 A.C. Antoulas, D.C. Sorensen: *Approximation of Large-Scale Dynamical Systems: An Overview*, Tech. Rep. (Rice University, Houston 2001)

22.52 B.G. Allan: *A reduced order model of the linearized incompressible Navier–Stokes equations for the sensor/actuator placement problem*, ICASE Rep. 2000-19 (NASA, Washington 2000)

22.53 N. Aubry, R. Guyonnet, R. Lima: Spatio-temporal analysis of complex signals: Theory and applications, J. Stat. Phys. **64**(3/4), 683–739 (1991)

22.54 J. Delville, L. Cordier, J.-P. Bonnet: Large-scale structure identification and control in turbulent shear flows. In: *Flow Control: Fundamental and Practices*, Lecture Notes Phys., Vol. 53, ed. by M. Gad-el-Hak, A. Pollard, J.-P. Bonnet (Springer, Berlin, Heidelberg 1998) pp. 199–273

22.55 M. Rathinam, L.R. Petzold: A new look at proper orthogonal decomposition, SIAM J. Numer. Anal. **41**(5), 1893–1925 (2001)

22.56 W.R. Graham, J. Peraire, K.Y. Tang: Optimal control of vortex shedding using low order models, Part 1: Open-loop model development, Int. J. Numer. Meth. Eng. **44**(7), 945–972 (1999)

22.57 J. Burkardt, M.D. Gunzburger, H.-C. Lee: *Centroidal Voronoi Tessellation-Based Reduced-Order Modeling of Complex Systems*, Tech. Rep. (Florida State University, Tallahassee 2004)

22.58 A. Iollo: Remarks on the approximation of the Euler equations by a low order model, INRIA Res. Rep. **3329**, 1–28 (1997)

22.59 A. Iollo, S. Lanteri, J.-A. Désidéri: Stability properties of POD-Galerkin approximations for the compressible Navier–Stokes equations, INRIA Res. Rep. **3589**, 1–30 (1998)

22.60 C.W. Rowley: *Modeling, simulation and control of cavity flow oscillations*, Ph.D. Dissertation (California Institue of Technology, Pasadena 2002)

22.61 R.B. Lehoucq, D.C. Sorensen, C. Yang: ARPACK Users' Guide: Solution of Large-Scale Eigenvalve Problems with Implicity Restarted Arnoldi Methods (SIAM 1998)

22.62 L. Sirovich: Turbulence and the dynamics of coherent structures, Part 1: Coherent structures, Q. Appl. Math. **45**(3), 561–571 (1987)

22.63 L. Sirovich: Turbulence and the dynamics of coherent structures, Part 2: Symmetries and transformations, Q. Appl. Math. **45**(3), 573–582 (1987)

22.64 L. Sirovich: Turbulence and the dynamics of coherent structures, Part 3: Dynamics and scaling, Q. Appl. Math. **45**(3), 583–590 (1987)

22.65 L. Cordier: *Etude de systèmes dynamiques basés sur la décomposition orthogonale aux valeurs propres (POD), Application à la couche de mélange turbulente et à l'écoulement entre deux disques contra-rotatifs*, Ph.D. Dissertation (Poitiers University, Poitiers 1996)

22.66 S. Herzog: *The large scale structure in the near-wall of turbulent pipe flow*, Ph.D. Dissertation (Cornell University, Ithaca 1986)

22.67 P. Moin, R.D. Moser: Characteristic-eddy decomposition of turbulence in a channel, J. Fluid Mech. **200**, 471–509 (1989)

22.68 J. Delville: Characterization of the organization in shear layers via the proper orthogonal decomposition, Appl. Sci. Res. **53**, 263–281 (1994)

22.69 J.P. Bonnet, D. Cole, J. Delville, M. Glauser, L. Ukeiley: Stochastic estimation and proper orthogonal decomposition: Complementary techniques for identifying structures, Exp. Fluids **17**, 307–314 (1994)

22.70 C.A.J. Fletcher: *Computational Techniques for Fluid Dynamics* (Springer, New York 1991)

22.71 D. Rempfer: Investigations of boundary layer transition via Galerkin projections on empirical eigenfunctions, Phys. Fluids **8**(1), 175–188 (1996)

22.72 R.D. Joslin, M.D. Gunzburger, R.A. Nicolaides, G. Erlebacher, M.Y. Hussaini: A self-contained, automated methodology for optimal flow control

22.73 J.O. Hinze: *Turbulence*, 2nd edn. (McGraw-Hill, New York 1975)

22.74 S. Sirisup, G.E. Karniadakis: A spectral viscosity method for correcting the long-term behavior of POD model, J. Comp. Phys. **194**, 92–116 (2004)

22.75 B.R. Noack, P. Papas, P.A. Monkewitz: The need for a pressure-term representation in empirical Galerkin models of incompressible shear flows, J. Fluid Mech. **523**, 339–365 (2005)

22.76 B. Galletti, A. Bottaro, C.-H. Bruneau, A. Iollo: Accurate model reduction of transient and forced wakes, Eur. J. Mech. B **26**(3), 354–366 (2007)

22.77 P. Sagaut: *Large-Eddy Simulation for Incompressible Flows – An Introduction* (Springer, Berlin, Heidelberg 2005)

22.78 G.S. Karamanos, G.E. Karniadakis: A spectral vanishing viscosity method for large eddy simulations, J. Comp. Phys. **162**, 22–50 (2000)

22.79 M. Bergmann: *Optimisation aérodynamique par réduction de modèle POD et contrôle optimal, Application au sillage laminaire d'un cylindre circulaire*, Ph.D. Dissertation (Institut National Polytechnique de Lorraine, Nancy 2004)

22.80 M. Couplet, C. Basdevant, P. Sagaut: Calibrated reduced-order POD-Galerkin system for fluid flow modelling, J. Comp. Phys. **207**, 192–220 (2005)

22.81 J. Borée: Extended proper orthogonal decomposition: A tool to analyse correlated events in turbulent flows, Exp. Fluids **35**, 188–192 (2003)

22.82 R.J. Adrian: On the role of conditional averages in turbulence theory. In: *Turbulence in Liquids*, ed. by G. Patterson, J. Zakin (Science, Princeton 1977) pp. 322–332

22.83 R.J. Adrian: Conditional eddies in isotropic turbulence, Phys. Fluids **22**, 2065–2070 (1979)

22.84 R.J. Adrian, P. Moin: Stochastic estimation of organized turbulent structure: Homogeneous shear flow, J. Fluid Mech. **190**, 531–559 (1988)

22.85 A. Papoulis: *Probability, Random Variables and Stochastic Theory*, 2nd edn. (McGraw-Hill, New York 1984)

22.86 R. Deutsch: *Estimation Theory* (Prentice-Hall, New York 1965) p. 269

22.87 R.J. Adrian, B.G. Jones, M.K. Chung, Y. Hassan, C.K. Nithianandan, A.T.C. Tung: Approximation of turbulent conditional averages by stochastic estimation, Phys. Fluids A **1**, 992–998 (1989)

22.88 D.R. Cole, M.N. Glauser, Y.G. Guezennec: An application of the stochastic estimation to the jet mixing layers, Phys. Fluids A **4**, 192 (1992)

22.89 T.C. Tung, R.J. Adrian: Higher-order estimates of conditional eddies in isotropic turbulence, Phys. Fluids **23**, 1469–1470 (1980)

22.90 H.L. Pécseli, J. Trulsen: A statistical analysis of numerically simulated plasma turbulence, Phys. Fluids B **1**, 1616–1636 (1989)

22.91 Y.G. Guezennec: Stochastic estimation of coherent structure in turbulent boundary layers, Phys. Fluids A **1**, 1054 (1989)

22.92 G.J. Brereton: Stochastic estimation as a statistical tool for approximating turbulent conditional averages, Phys. Fluids A **4**, 1046–2054 (1992)

22.93 T.G. Bagwell, R.J. Adrian, R.D. Moser, J. Kim: Improved approximation of wall shear stress boundary conditions for large eddy simulations. In: *Near Wall Turbulent Flows*, ed. by R. So, C.B. Speziale, B.E. Launder (Elsevier, Amsterdam 1993) pp. 265–275

22.94 I. Daubechies: *Ten Lectures on Wavelets* (SIAM, 1992)

22.95 M. Farge: Wavelet transforms and their applications to turbulence, Annu. Rev. Fluid Mech. **24**, 395–457 (1992)

22.96 S. Mallat: *A Wavelet Tour of Signal Processing* (Academic, San Diego 1998)

22.97 J.C. van den Berg (Ed.): *Wavelets and Physics* (Cambridge Univ. Press, Cambridge 1999)

22.98 D. Donoho, M.R. Duncan, X. Huo, O. Levi: Wavelab, http://www-stat.stanford.edu/ wavelab (1999)

22.99 J. Lewalle, J. Delville, J.P. Bonnet: Decomposition of Mixing Layer Turbulence in coherent structures and background fluctuations, Flow Turb. Comb. **64**, 301–328 (2000)

22.100 M.T. Schobeiri, K. Reid, J. Lewalle: Effect of unsteady wake passing frequency on boundary layer transition, experimental investigation and wavelet analysis, J. Fluids Eng. **125**, 251–266 (2003)

23. Fundamentals of Data Processing

In this chapter the fundamentals of statistical parameter estimation are reviewed for applications typical in experimental fluid mechanics. The chapter begins with a review of the probability density function and its moments and continues with common estimators for the mean and variance of stationary random processes. A brief introduction to signal noise is given as a prelude to a rigorous discussion of the Cramér–Rao Lower Bound (CRLB). The CRLB represents the lower bound of variance of unbiased estimators of a parameter. This concept is deepened using illustrations from the laser Doppler, phase Doppler and PIV measurement techniques. The chapter closes with a short discussion about the propagation of errors in a measurement chain.

23.1 Statistical Principles 1399

23.2 Stationary Random Processes 1401

23.3 Estimator Expectation and Variance 1402
    23.3.1 Estimators for the Mean 1402
    23.3.2 Estimators for Higher-Order Statistics 1404

23.4 Signal Noise 1406

23.5 Cramér–Rao Lower Bound (CRLB) 1408
    23.5.1 Laser Doppler and Phase Doppler Signals 1409
    23.5.2 Particle Imaging 1414

23.6 Propagation of Errors 1416

References 1417

23.1 Statistical Principles

The first concept to be introduced is the probability distribution function $P(x)$, which is the probability assigned to a set of points k, such that the random variable $x(k)$ satisfies $x(k) \leq x$, where x is some fixed value.

$$P(x) = \text{Prob}[x(k) \leq x]. \qquad (23.1)$$

This set of points $x(k) \leq x$ is a subset of all points satisfying $x(k) \leq \infty$. Thus

$$P(-\infty) = 0, \quad P(\infty) = 1. \qquad (23.2)$$

The *probability density* function (PDF) $p(x)$ is defined by the relation

$$p(x) = \lim_{\Delta x \to 0} \left(\frac{\text{Prob}[x < x(k) \leq x + \Delta x]}{\Delta x} \right). \qquad (23.3)$$

Thus,

$$p(x) \geq 0, \qquad (23.4)$$

$$\int_{-\infty}^{\infty} p(x)\,dx = 1, \qquad (23.5)$$

$$P(x) = \int_{-\infty}^{x} p(\xi)\,d\xi\,; \quad \frac{dP(x)}{dx} = p(x). \qquad (23.6)$$

The next concept to be discussed is that of expected values. The expected value for any real, single-valued, continuous function $g(x)$ of the random variable $x(k)$ is given by

$$E\{g[x(k)]\} = \int_{-\infty}^{\infty} g(x) p(x)\,dx \qquad (23.7)$$

in particular, for $g(x) = x$, the mean value of $x(k)$ is obtained by

$$E[x(k)] = \mu_x = \int_{-\infty}^{\infty} x p(x)\,dx \qquad (23.8)$$

and for $g(x) = x^2$, the mean square value of $x(k)$ is obtained by

$$E[x^2(k)] = \psi_x^2 = \int_{-\infty}^{\infty} x^2 p(x)\,dx\,. \qquad (23.9)$$

The quantities defined in (23.8) and (23.9) are also known as the first and second *moments* of the random variable $x(k)$. Note that often $\bar{x}$ is used instead of μ_x for the mean value of $x(k)$. Furthermore, often the variance of $x(k)$, σ_x^2, is used rather than the mean square value,

$$\sigma_x^2 = \psi_x^2 - \mu_x^2 = \int_{-\infty}^{\infty} (x - \mu_x)^2 \, p(x) \, dx \, . \quad (23.10)$$

The standard deviation σ_x of $x(k)$ is the square root of the variance. Equation (23.10) is one example of the more general r-th-order *central moment*

$$\mu_r = \int_{-\infty}^{\infty} (x - \mu_x)^r \, p(x) \, dx \quad (23.11)$$

which quantifies deviations of $x(k)$ about its mean value.

Similar expressions can be written for the bivariate case, in which two random variables $x(k)$ and $y(k)$ are considered. The joint probability function is defined by

$$P(x, y) = \text{Prob}\,[x(k) \leq x \text{ and } y(k) \leq y] \quad (23.12)$$

and the associated joint probability density function by

$$p(x, y)$$
$$= \lim_{\substack{\Delta x \to 0 \\ \Delta y \to 0}} \left(\frac{\text{Prob}\begin{bmatrix} x < x(k) \leq x + \Delta x \\ \text{and } y < y(k) \leq y + \Delta y \end{bmatrix}}{\Delta x \Delta y} \right) \quad (23.13)$$

yielding also

$$p(x, y) \geq 0 \, , \quad (23.14)$$

$$\int_{-\infty}^{\infty} \int_{-\infty}^{\infty} p(x, y) \, dx \, dy = 1 \, , \quad (23.15)$$

$$P(x, y) = \int_{-\infty}^{y} \int_{-\infty}^{x} p(\xi, \eta) \, d\xi \, d\eta \, ,$$

$$\frac{\partial^2 P(x, y)}{\partial x \partial y} = p(x, y) \, . \quad (23.16)$$

The two random variables are said to be statistically independent if

$$p(x, y) = p(x) p(y) \, . \quad (23.17)$$

The expected value of any real, single-valued, continuous function $g(x, y)$ of two random variables $x(k)$ and $y(k)$ is given by

$$E\,[g(x, y)] = \int_{-\infty}^{\infty} \int_{-\infty}^{\infty} g(x, y) p(x, y) \, dx \, dy \quad (23.18)$$

One special example is when $g(x, y) = [x(k) - \mu_x] \times [y(k) - \mu_y]$, where μ_x and μ_y are the respective mean values. The expected value is known as the covariance

$$C_{xy} = E\left\{[x(k) - \mu_x][y(k) - \mu_y]\right\}$$
$$= E\,[x(k) y(k)] - E\,[x(k)]\, E\,[y(k)]$$
$$= \int_{-\infty}^{\infty} \int_{-\infty}^{\infty} [x(k) - \mu_x][y(k) - \mu_y] p(x, y) \, dx \, dy \, .$$
$$(23.19)$$

The correlation coefficient is then defined by

$$\rho_{xy} = \frac{C_{xy}}{\sigma_x \sigma_y} \quad (23.20)$$

which lies between -1 and $+1$.

Data processing deals with the *estimation* of relevant process statistics from the primary measurement quantities. The term *estimation*, rather than determination or computation, is used, since in almost all cases, the physical process has a stochastic part, meaning that the result of an estimation is a random variable (even an exact replication of the experiment would yield a slightly different answer). The procedure or computational algorithm used to obtain the estimation is known as the *estimator*.

Estimators are evaluated on the basis of three properties. First, the expected value of the estimation should be equal to the parameter being estimated

$$E\left(\hat{\phi}\right) = \phi \, . \quad (23.21)$$

If this is true, the estimator is *unbiased*. Note that an estimator is often signified by the hat symbol. Second, the mean square error of the estimator should be smaller than for any other possible estimator.

$$E[(\hat{\phi}_1 - \phi)^2] \leq E[(\hat{\phi}_i - \phi)^2] \, . \quad (23.22)$$

In this case the estimator $\hat{\phi}_1$ is said to be *efficient*. Note that the smallest possible estimation variance for any unbiased estimator is given by the Cramér–Rao lower bound (CRLB). Finally, the estimate should converge to the parameter being estimated for a large sample number or for a long observation time

$$\lim_{N \to \infty} \text{Prob}(|\hat{\phi} - \phi| \geq \varepsilon) = 0 \, . \quad (23.23)$$

For an arbitrarily small $\varepsilon > 0$, the estimator is said to be *consistent*. A sufficient condition to meet this requirement is

$$\lim_{N \to \infty} [(\hat{\phi} - \phi)^2] = 0 \, . \quad (23.24)$$

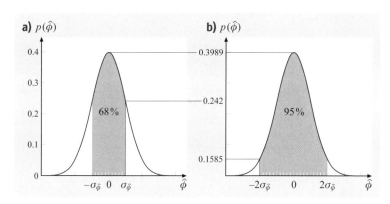

Fig. 23.1a,b Gaussian (normal) distribution illustrating confidence limits. (**a**) For $\pm\sigma$ (68%), (**b**) For $\pm 2\sigma$ (95%)

The mean square error used above can be expanded to yield

$$E[(\hat{\phi}-\phi)^2] = E\{[\hat{\phi}-E(\hat{\phi})+E(\hat{\phi})-\phi]^2\}$$
$$= E\{[\hat{\phi}-E(\hat{\phi})]^2\}$$
$$+ E\{[E(\hat{\phi})-\phi]^2\}. \quad (23.25)$$

Hence, the mean square error is the sum of two parts: the first part is a variance term that describes the random part of the error

$$\mathrm{var}(\hat{\phi}) = E\{[\hat{\phi}-E(\hat{\phi})]^2\} = E(\hat{\phi}^2) - E^2(\hat{\phi}), \quad (23.26)$$

which can be made arbitrarily small by increasing the sample size. The second part is the square of a bias term describing the systematic portion of the error

$$b^2(\hat{\phi}) = E\{[E(\hat{\phi})-\phi]^2\}. \quad (23.27)$$

This part is not influenced directly by the sample size and can arise from many sources, often found outside of the data processing. Often special calibration procedures are required to quantify such errors; however, these will not be considered further here. In fact, the bias error will be assumed to be negligible in the following discussion.

Under these conditions and for a *small* normalized random error

$$\varepsilon = \frac{\sigma(\hat{\phi})}{\phi} = \frac{\sqrt{\mathrm{var}(\hat{\phi})}}{\phi}, \quad (23.28)$$

the probability density function for the estimates, $p(\hat{\phi})$, can often be approximated by a Gaussian distribution with the mean value $E(\hat{\phi}) = \phi$ and a standard deviation $\sigma_{\hat{\phi}} = \varepsilon\phi$

$$p(\hat{\phi}) = \frac{1}{\varepsilon\phi\sqrt{2\pi}} \exp\left(\frac{-(\hat{\phi}-\phi)^2}{2(\varepsilon\phi)^2}\right). \quad (23.29)$$

Probability statements about the bounds in which future estimates $\hat{\phi}$ will lie can thus be made as follows

$$\mathrm{Prob}[\phi(1-\varepsilon) \le \hat{\phi} < \phi(1+\varepsilon)] \approx 0.68,$$
$$\mathrm{Prob}[\phi(1-2\varepsilon) \le \hat{\phi} < \phi(1+2\varepsilon)] \approx 0.95, \quad (23.30)$$

since for a Gaussian distribution $\pm\sigma$ or $\pm 2\sigma$ about the mean contains, respectively, 68% or 95% of the probability mass, as sketched in Fig. 23.1.

This leads directly to the concept of *confidence intervals*, i.e., the interval in which the true value will lie with a given probability (valid for small ε).

$$\hat{\phi}(1-\varepsilon) \le \phi \le \hat{\phi}(1+\varepsilon) \quad \text{with 68\% confidence},$$
$$\hat{\phi}(1-2\varepsilon) \le \phi \le \hat{\phi}(1+2\varepsilon) \quad \text{with 95\% confidence}. \quad (23.31)$$

The value of ε can be estimated directly from the sampled data, as discussed in the next section.

23.2 Stationary Random Processes

Given some random phenomena, any single time history of this function is called a sample function. The collection of all possible sample functions, possibly an infinite number, is known as a *random process* or *stochastic process*.

The mean value (first moment) of the ensemble of sample functions at time t_1 is then the arithmetic mean over the instantaneous values of the sample functions at time t_1, as illustrated in Fig. 23.2. A correlation or joint moment of the process at two different times can be computed by taking the ensemble average of the product of instantaneous values at two times t_1 and $t_1+\tau$. These values can be written as

$$\mu_x(t_1) = \lim_{N\to\infty} \frac{1}{N} \sum_{k=1}^{N} x_k(t_1), \qquad (23.32)$$

$$R_{xx}(t_1, \tau) = \lim_{N\to\infty} \frac{1}{N} \sum_{k=1}^{N} x_k(t_1) x_k(t_1+\tau), \qquad (23.33)$$

where R_{xx} is known as the autocorrelation function.

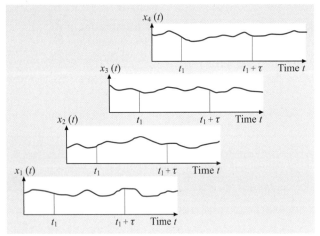

Fig. 23.2 Ensemble of sample functions defining a random process

A random process is known as *weakly stationary* when the value defined by (23.32) is independent of t_1 and the autocorrelation is only a function τ, the variance is limited and the mean value is constant. The process is known as *strongly stationary* when also the entire probability density function is independent of t_1. Otherwise the process is *instationary*.

Generally, however, statistics of a stationary random process are not computed over an ensemble of sample functions but over a time average. For example,

$$\mu_x(k) = \lim_{T\to\infty} \frac{1}{T} \int_0^T x_k(t)\,\mathrm{d}t = \mu_x, \qquad (23.34)$$

$$R_{xx}(\tau, k) = \lim_{T\to\infty} \frac{1}{T} \int_0^T x_k(t) x_k(t+\tau)\,\mathrm{d}t = R_{xx}(\tau). \qquad (23.35)$$

If these values do not differ from those in (23.32) and (23.33), then the process is said to be *ergodic*, in which case the index k is dropped. All stationary processes encountered in fluid mechanics can be considered ergodic.

Note that the covariance function is simply the autocorrelation function with the mean removed and the cross-covariance function is the cross-correlation function with the product of the means removed

$$\begin{aligned} C_{xx}(\tau) &= R_{xx}(\tau) - \mu_x^2, \\ C_{xy}(\tau) &= R_{xy}(\tau) - \mu_x \mu_y. \end{aligned} \qquad (23.36)$$

23.3 Estimator Expectation and Variance

In many cases the expectation and variance of an estimator can be derived analytically and several examples are given below. For more complicated quantities, this is not always possible and other strategies can be followed. The jackknife algorithm will be introduced as one such approach.

23.3.1 Estimators for the Mean

The first estimator to be examined is the mean value. The most common sample mean estimator is given by

$$\hat{\mu}_x = \frac{1}{N} \sum_{i=1}^{N} x_i, \qquad (23.37)$$

where x_i are individual samples of the process x. Instead of $\hat{\mu}_x$ the alternative expression $\bar{x}$ is also commonly used.

However, the difference between the estimated value and the true mean value can be shown better using $\hat{\mu}_x$ and μ_x respectively. The estimator of the mean value given in (23.37) is non-biased, since $E(\hat{\mu}_x) = \mu_x$ [23.1]. The mean square error, or variance, of this estimator is then given by

$$\mathrm{var}(\hat{\mu}_x) = \sigma_{\hat{\mu}_x}^2 = E\left[(\hat{\mu}_x - \mu_x)^2 \right]. \qquad (23.38)$$

Substituting (23.37) into (23.38) leads to

$$\begin{aligned} \sigma_{\hat{\mu}_x}^2 &= E\left[\left(\frac{1}{N} \sum_{i=1}^{N} x_i - \mu_x \right)^2 \right] \\ &= \frac{1}{N^2} E\left[\left(\sum_{i=1}^{N} x_i - \mu_x \right)^2 \right]. \end{aligned} \qquad (23.39)$$

If the condition $E(x_i x_j) = 0$ is satisfied, i.e., consecutive samples are uncorrelated or statistically independent, (23.39) can be further reduced to

$$\sigma_{\hat{\mu}_x}^2 = \frac{1}{N^2} E\left[\sum_{i=1}^{N}(x_i - \mu_x)^2\right]$$
$$= \frac{\sigma_x^2}{N}, \quad (23.40)$$

which states that the variance of the mean estimator decreases with increasing number of samples.

This analysis has been performed for an estimator based on discrete samples x_i; however, a similar analysis could be made for a mean estimator based on the continuous signal $x(t)$

$$\hat{\mu}_x = \frac{1}{T}\int_0^T x(t)\,dt, \quad (23.41)$$

which differs from the true mean μ_x, since the integral is performed only over a finite time T. The variance of this estimator becomes

$$\mathrm{var}\,(\hat{\mu}_x) = \sigma_{\hat{\mu}_x}^2 = E\left[(\hat{\mu}_x - \mu_x)^2\right] = E\left(\hat{\mu}_x^2\right) - \mu_x^2.$$
$$(23.42)$$

In terms of the autocovariance function, this can be written as [23.1]

$$\sigma_{\hat{\mu}_x}^2 = \frac{1}{T}\int_{-T}^{T}\left(1 - \frac{|\tau|}{T}\right) C_{xx}(\tau)\,d\tau \quad (23.43)$$

for a stationary random process. For small τ only C_{xx} remains in the integral and for large τ, C_{xx} goes to zero, thus the integral can be expressed as

$$\sigma_{\hat{\mu}_x}^2 = \frac{2\sigma_x^2 T_x}{T} \quad (23.44)$$

with the integral time scale

$$T_x = \frac{1}{\sigma_x^2}\int_0^\infty C_{xx}(\tau)\,d\tau. \quad (23.45)$$

As pointed out by *George* [23.2], if the results given by (23.40) and (23.44) are equated, the condition for statistically independent samples can be obtained, namely

$$N = \frac{T}{2T_x}. \quad (23.46)$$

This is graphically represented in Fig. 23.3 and leads to two very insightful interpretations.

- Samples are statistically independent if they are separated by a period of the least $2T_x$ in time.
- Segments of the continuous signal $2T_x$ in length contribute to the mean estimate as one, statistically independent sample.

The manifestation of this relation is that sampling a signal with time intervals less than $2T_x$ will not accelerate the convergence of the mean estimator. At this point, the difference between *data* and *information* should become very clear. New information (with respect to the mean estimate), comes only every $2T_x$ time periods.

Equation (23.44) makes a statement about the necessary observation or measurement time to achieve a given statistical uncertainty (variance of the mean estimator). However, to use this equation the integral time scale, as defined using the autocovariance function, must be known beforehand. Moreover, the integral time scale may change by orders of magnitude, e.g., in flow fields between different points of a single velocity profile. Often, however, a simple estimate of T_x suffices. This will be illustrated with the following example of how (23.44) can be used in practice.

The example chosen is a velocity measurement in the recirculation zone of a backward-facing step water flow. In a preliminary measurement the local variance of the velocity fluctuations is estimated to be $0.2\,\mathrm{m}^2/\mathrm{s}^2$ at point A (Fig. 23.4). The requirement is that the mean velocity at point A be determined to within $\pm 0.04\,\mathrm{m/s}$ with 95% confidence.

The integral time scale of the velocity fluctuations can be estimated from appropriate velocity and length scales, in this case $U_0 = 2\,\mathrm{m/s}$ and x_R, which is approximately $8H$ or $0.4\,\mathrm{m}$. Thus, $T_u = x_R/U_0 = 0.2\,\mathrm{s}$. Note that the subscript u for the integral time scale is used, since the process being measured is the velocity u. Assuming a normal distribution for the scatter of the estimates, the probability of being within $\pm\sigma_{\hat{u}}$ of the true mean value would be about 68%. This would increase to the required 95% for $\pm 2\sigma_{\hat{\mu}_u}$.

$$2\sigma_{\hat{\mu}_u} = 0.04\,\mathrm{m/s}, \quad \sigma_{\hat{\mu}_u}^2 = 0.0004\,\mathrm{m}^2/\mathrm{s}^2. \quad (23.47)$$

Equation (23.44) can now be solved for the required measurement time to fulfill this condition

$$T = \frac{2\sigma_u^2 T_u}{\sigma_{\hat{\mu}_u}^2} = 200\,\mathrm{s}. \quad (23.48)$$

Note that this calculation has been performed independent of the choice of measurement technique. In fact, no measurement technique can shorten the necessary observation time given in (23.48), since this

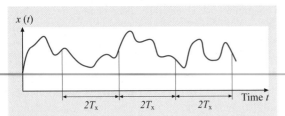

Fig. 23.3 Graphical interpretation of statistical independence of consecutive samples of a continuous process

describes the fundamental statistical behavior of a random process.

In practice, it is unusual to make such calculations prior to every measurement. It is more convenient to display the current measured mean velocity online, accumulated over all samples up to that time, and then to allow the user to terminate the measurement when the fluctuations of the mean are below an acceptable level. Indeed, from the necessary measurement duration, and from the fluctuation level of the mean, a rough estimate of the integral time scale can often be made. This technique of user intervention does not lend itself to automation, so that still a third approach is often used, in which a fixed number of samples is used for each point, whereby the number is chosen to be very large to ensure sufficient convergence for all measurement points. In many flows there are regions where data rates decrease dramatically, e.g., near walls. In such cases there is often no choice but to accept a higher degree of statistical uncertainty, since otherwise the data collection time becomes excessive.

Alternatively, (23.40) could have been used if the velocity data were available in discrete form at regular time intervals. Assuming the sample rate was not faster than every $2T_u$, the number of samples required to insure the requested accuracy would be

$$N = \frac{\sigma_u^2}{\sigma_{\hat{\mu}_u}^2} = \frac{0.2\,\text{m}^2/\text{s}^2}{0.0004\,\text{m}^2/\text{s}^2} = 500\,. \tag{23.49}$$

This discussion puts into perspective expressions such as 'high' or 'low' data rates or 'many' or 'few' samples. The data rate, or the number of samples, must always be considered with respect to the integral time scale of the process at the particular measurement point. This explains the preferred use of data *density* rather than data *rate*. It should also be apparent that, for the same Reynolds number, measurements performed in air flows will typically be much shorter in duration than in water flows, given the same target accuracy. The reason for this lies in the fact that, for the same Reynolds number, the integral time scale of an air flow is generally shorter.

Further guidelines for reporting measurement uncertainties can be found in *Kline* and *McClintock* [23.3], *Kline* [23.4] or *Moffat* [23.5, 6].

23.3.2 Estimators for Higher-Order Statistics

In the study of turbulence, statistics of not only the mean velocity but also of higher-order moments are required. General formula for the estimator variance for higher-order statistics have been given by *Stuart* and *Ord* [23.7] and *Kendall* and *Stuart* [23.8]. *Benedict* and *Gould* [23.9] have summarized their results in the following manner.

An unbiased estimator of the r-th central moment μ_r (23.11) is given by

$$\hat{\mu}_r = \frac{1}{N} \sum_{i=1}^{N} (x_i - \hat{\mu}_x)^r \tag{23.50}$$

in which the true mean has been replaced by the sample mean (23.37). Strictly this estimator is unbiased only for $r = 1$, however this also applies for higher moments when N is large. The sampling variance of $\hat{\mu}_r$ is given by

$$\text{var}(\hat{\mu}_r) = \sigma_{\hat{\mu}_r}^2 = \frac{1}{N}(\mu_{2r} - \mu_r^2 + r^2 \mu_{r-1} \mu_2 - 2r\mu_{r+1}\mu_{r-1})\,, \tag{23.51}$$

where terms of order N^{-2} and higher have been neglected. 95% confidence intervals are then $\hat{\mu}_r \pm 2\sigma_{\hat{\mu}_r}$. Note that (23.51) uses $\mu_1 = \mu_0 = \mu_{-1}$ and $\mu_2 = \sigma_x^2$. Furthermore, it uses the exact central moments μ_r,

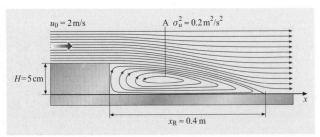

Fig. 23.4 Sketch of example backward facing step flow. x_R is the mean reattachment length

which are actually unknown. However if N is suitably large, typically $N = 100$, these can be replaced by the central moment sampling statistics, $\hat{\mu}_r$, for practical computations.

Similarly, the mixed central moment

$$\mu_{r,s} = \int_{-\infty}^{\infty} \int_{-\infty}^{\infty} (x - \mu_x)^r (y - \mu_y)^s p(x)p(y) \,dx\,dy \quad (23.52)$$

can be estimated using

$$\hat{\mu}_{r,s} = \frac{1}{N} \sum_{i=1}^{N} (x_i - \hat{\mu}_x)^r (x_i - \hat{\mu}_y)^s \quad (23.53)$$

which exhibits the variance

$$\begin{aligned}
\operatorname{var}(\hat{\mu}_{r,s}) &= \sigma_{\hat{\mu}_r}^2 \\
&= \frac{1}{N} \big(\mu_{2r,2s} - \mu_{r,s}^2 + r^2 \mu_{2,0} \mu_{r-1,s}^2 \\
&\quad + s^2 \mu_{0,2} \mu_{r,s-1}^2 + 2rs \mu_{1,1} \mu_{r-1,s} \mu_{r,s-1} \\
&\quad - 2r\mu_{r+1,s} \mu_{r-1,s} - 2s\mu_{r,s+1} \mu_{r,s-1} \big) \,.
\end{aligned} \quad (23.54)$$

Note that $\mu_{10} = \mu_{01} = 0$, $\mu_{r,-1} = \mu_{-1,s} = 0$, $\mu_{20} = \sigma_x^2$ and $\mu_{02} = \sigma_y^2$. Equation (23.54) can be simplified for normally distributed processes, since then all odd moments are zero and the second, fourth, sixth and eighth moments are 1, 3, 15 and 105 times σ_x^2, respectively.

The variances of the most common statistics in turbulence research are summarized in Table 23.1, for both an arbitrary and a normal distribution of the process. Note that the formulas given in Table 23.1 are multiplied by N. The u and v velocity components have been used for illustration. As an example, the variance of the mean estimator is given as $\sigma_u^2 N^{-1}$, which agrees with (23.40).

The expressions in Table 23.1 all assume statistical independence between samples, as specified by (23.46). If the sample rate is too high to insure statistical independence, the total number of samples N must be adjusted so that the total observation time yields the desired confidence bounds, according to (23.44). Furthermore, it should be noted that turbulence quantities are seldom normally distributed, so that the simplifications given in Table 23.1 can lead to significant errors if normality is not previously established.

For more-complex estimators, there exist several resampling algorithms with which the uncertainty of the measured quantity can be estimated. In particular the jackknife algorithm will be discussed, as first introduced by *Tukey* [23.10]. Notes on its practical implementation are given by *Efron* and *Tibshirani* [23.11] and an evaluation of its potential with laser Doppler data is given by *Benedict* and *Gould* [23.9]. This algorithm also assumes statistical independence in the data set $x = (x_1, x_2, \ldots, x_N)$ when computing some statistical estimator. The jackknife samples

$$x_{\text{jack},i} = (x_1, x_2, \ldots, x_{i-1}, x_{i+1}, \ldots, x_N) \quad (23.55)$$

are obtained by leaving out in turn one of the data samples. The jackknife samples are then used to compute N estimates $\hat{\phi}_{\text{jack},i}$ with $i = 0, 1, \ldots, (N-1)$. The jackknife variance for $\hat{\phi}$ is then given by

$$\operatorname{var}(\hat{\phi})_{\text{jack},i} = \frac{N-1}{N} \sum_{i=1}^{N} (\hat{\phi}_{\text{jack},i} - \mu_{\hat{\phi}_{\text{jack}}})^2 \,, \quad (23.56)$$

Table 23.1 Estimator variances multiplied by N [23.9]

Statistic	Variance for any distribution	Normal assumption
μ_u	σ_u^2	σ_u^2
σ_u	$\frac{\mu_4 - \sigma_u^4}{4\sigma_u^2}$	$\frac{\sigma_u^2}{2}$
$R_{uv} = \mu_{1,1}$	$\mu_{2,2} - \mu_{1,1}^2$	$(1 + \rho_{uv}^2)\sigma_u^2 \sigma_v^2$
$\rho_{uv} = \frac{\mu_{1,1}}{\sigma_u \sigma_v}$	$\rho_{uv}^2 \left[\frac{\mu_{2,2}}{\mu_{1,1}^2} + \frac{1}{4}\left(\frac{\mu_{4,0}}{\sigma_u^4} + \frac{\mu_{0,4}}{\sigma_v^4} + \frac{2\mu_{2,2}}{\sigma_u^2 \sigma_v^2} \right) - \left(\frac{\mu_{3,1}}{\mu_{1,1}\sigma_u^2} + \frac{\mu_{1,3}}{\mu_{1,1}\sigma_v^2} \right) \right]$	$(1 - \rho_{uv}^2)^2$
σ_u^2	$\mu_4 - \sigma_u^4$	$2\sigma_u^4$
μ_3	$\mu_6 - \mu_3^2 - 6\mu_4 \sigma_u^2 + 9\sigma_u^6$	$6\sigma_u^6$
$\mu_{2,1}$	$\mu_{4,2} - \mu_{2,1}^2 + \sigma_u^4 \sigma_v^2 + 8\sigma_u^2 \mu_{1,1}^2 - 2\sigma_u^2 \mu_{2,2} - 4\mu_{1,1}\mu_{3,1}$	$2(1 + 2\rho_{uv}^2)\sigma_u^4 \sigma_v^2$
μ_4	$\mu_8 - \mu_4^2 - 8\mu_5 \mu_3 + 16\mu_3^2 \sigma_u^2$	$96\sigma_u^8$

where

$$\mu_{\hat{\phi}_{\text{jack}}} = \frac{1}{N} \sum_{i=1}^{N} \hat{\phi}_{\text{jack},i} . \quad (23.57)$$

The 95% confidence interval for the estimator is given by $\hat{\phi} \pm 2\text{var}((\hat{\phi})_{\text{jack}})^{1/2}$.

The jackknife algorithm requires N^2 calculations per variance estimate. This computational load can be greatly reduced if the programming is modified specifically for each statistic to be studied. For example, if the mean square of the velocity fluctuations σ_u^2, is being studied, the jackknife sample can be written

$$\hat{\phi}_{\text{jack},i} = \hat{\sigma}_{u,\text{jack},i}^2 = \frac{1}{N-1} \sum_{\substack{j=1 \\ j \neq i}}^{N} (u_j - \hat{\mu}_{u,\text{jack},i})^2 . \quad (23.58)$$

This equation can be rewritten as

$$\hat{\sigma}_{u,\text{jack},i}^2 = \frac{1}{N-1} \left[\sum_{\substack{j=1 \\ j \neq i}}^{N} u_j^2 - 2\hat{\mu}_{u,\text{jack},i} \sum_{\substack{j=1 \\ j \neq i}}^{N} u_j \right. \\ \left. + (N-1)(\hat{\mu}_{u,\text{jack},i})^2 \right] . \quad (23.59)$$

Each term in the brackets is summed only once over all $j = 1, \ldots N$ and then decremented by u_j^2 and u_j respectively for each jackknife replication.

It can be shown theoretically that the jackknife is biased high on its estimation of uncertainty and thus, it will never underestimate the uncertainty of a statistic.

23.4 Signal Noise

Noise is essentially any amplitude deviation of an individual realization of a signal from its ideal model. Sources of noise are manifold, e.g., shot noise or thermal noise in any electronics, additional generation-recombination and modulation noise in semiconductors, the photon noise for optical components or quantization noise.

The power of signal fluctuations σ_s^2 put into relation with the power of noise fluctuations σ_n^2 is known as the signal-to-noise ratio (SNR) and is generally expressed in decibels:

$$\text{SNR/dB} = +10 \log \left(\frac{\sigma_s^2}{\sigma_n^2} \right) . \quad (23.60)$$

The estimation of the SNR from a given signal segment is often required in the detection/validation step of signal processing to indicate whether a result can be expected to be reliable or not. Unfortunately, the estimation of the SNR directly from the signal is complicated since the signal fluctuations and the noise are superimposed. However, noise contributions in the system are usually considered to be spectrally white. This refers to the fact that the total noise power is distributed evenly over all frequencies up to the upper bandwidth of the system. Still, the spectral distributions of the useful signal and the noise are superimposed. However, if the bandwidth of the expected signal model is limited, the noise power can be derived from the frequency spectrum.

As an example, in Fig. 23.5, a laser Doppler signal, a noise signal and the summation of the two in time, spectral and correlation domain is illustrated. It becomes obvious from Fig. 23.5 that the power spectral density (PSD) or the autocorrelation function (ACF) offer excellent means to monitor SNR and to determine whether a particle signal is present or not.

An idealized graphical interpretation of SNR is given in Fig. 23.6, which shows schematically the PSD of a Doppler signal logarithmically scaled. The SNR is given by the ratio of the areas A to B. A more detailed estimation procedure is given by *Tropea* [23.12]. The noise appears as a base line floor of width Δf, the bandwidth of the system, and of amplitude σ_n^2/f_s. Any filtering of the signal, for instance using a low-pass filter, will directly decrease area B and thus increase the SNR, since more of the noise is removed. The use of a band-pass filter to increase SNR increases the reliability of the signal detection, since the SNR acceptance threshold can be chosen higher. In contrast, the variance of the frequency estimation remains constant because the peak in the spectrum still has the same width. Indeed, such adjustable input filters are usually an integral part of any Doppler signal processor. On the other hand there is a danger in filtering with too narrow a bandwidth, since in general the signal frequency is not known a priori. This can lead to truncation of the velocity distribution and to a bias of the estimated moments.

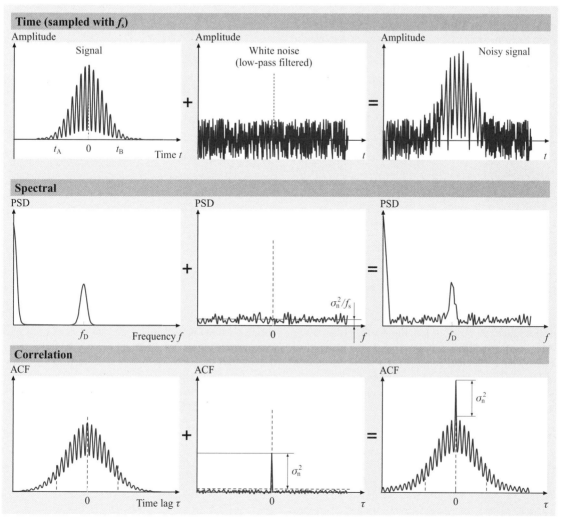

Fig. 23.5 Representation of a laser Doppler signal, a noise signal and a combination of the two in time, spectral and correlation domain

The SNR can also be estimated from the autocorrelation function. Noise, being fully stochastic and having zero correlation duration, appears only in the first autocorrelation coefficient, i.e., $R(\tau = 0)$. The statistical scatter (error) of the autocorrelation coefficients increases with SNR for every $\Delta\tau$, given a finite number of samples. Thus, the SNR can be estimated by comparing the amplitude of the autocorrelation function at $\tau = 0$ to the maximum peak amplitude of the remaining periodicity, exemplary shown in Fig. 24.6c, Chap. 24 for a high-pass filtered burst signal with added noise. If the frequency of the periodicity f has already been determined, the amplitude of the signal A_R

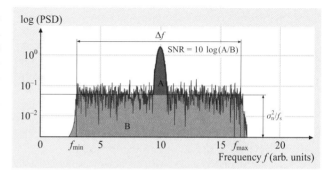

Fig. 23.6 Graphical representation of SNR using the power spectral density (PSD)

(Fig. 24.6, Chap. 24) can be estimated by fitting a cosine wave to the measured correlation function at points removed from $\tau = 0$. This can be computed using the expression

$$A_R \approx \frac{R(n\Delta\tau)}{\cos(2\pi f n \Delta\tau)} \tag{23.61}$$

from which the variance of the noise portion of the signal can be computed

$$\sigma_n^2 = R(0) - A_{R^2} \,. \tag{23.62}$$

The index n should ideally be chosen at the first maximum or minimum removed from $\tau = 0$.

The SNR is then given as

$$\text{SNR/dB} = 10 \log\left(\frac{A_{R^2}}{\sigma_n^2}\right). \tag{23.63}$$

The presence of noise in the signal can have different effects on the estimation of signal parameters or statistics. First of all, noise can directly affect the signal processing and cause systematic errors of the parameter extraction from individual signals. Furthermore, noise can disturb the signal detection, which often leads to a biased selection of signal realizations. In both cases, the derived signal statistics are biased. But even without systematic errors, the possible accuracy of the parameter extraction is limited, known as the Cramér–Rao lower bound (CRLB), which is discussed in more detail in the following section.

23.5 Cramér–Rao Lower Bound (CRLB)

The goal of analyzing an acquired signal is to derive several signal parameters according to a given model representing the physical basis of the signal generating process, e.g., a signal frequency, phase or amplitude. In practical cases, the recorded signals are equidistantly sampled and time limited, so that the amount of available information is also finite. Furthermore, the signal is influenced by noise and this introduces uncertainty into any parameter determined from the signal. The calculation of signal parameters is therefore called *estimation*, since it contains a random component.

The true values of the parameters to be estimated are seldom known and different estimation algorithms (*estimators*) will also yield different results. Therefore, it is of interest to quantify the accuracy of each estimator statistically. To begin with, the *expectation* of the estimator should be equal to the true value, i.e., non-biased. Second, the estimator should be *efficient*, meaning that it uses all available information to estimate the required parameter as accurately as possible. In Sect. 23.1, features of estimators are discussed in more detail. The efficiency of an estimator is quantified by its *variance*. While the bias should be zero, the finite amount of information yields a lower bound of achievable accuracy and thus, a finite variance. For unbiased estimators this lower bound of variance is given by the CRLB [23.13, 14].

No unbiased estimator can obtain estimates with a variance smaller than the CRLB, thus this quantity can be used to evaluate the performance of a specific algorithm. On the other hand, the CRLB gives no information about how an algorithm should process a measured signal to reach this lower bound. However, based on estimation theory, and closely related to the CRLB, the maximum-likelihood (ML) estimator can be derived. If any unbiased estimator reaches the CRLB, then the ML estimator will also reach it, at least asymptotically [23.13, 15].

For a signal

$$x(t = t_i) = x_i = m_i + n_i, \quad i = 0, 1, \ldots, (N-1) \tag{23.64}$$

with

$$\boldsymbol{x} = \begin{pmatrix} 0 \\ x_1 \\ \vdots \\ x_{N-1} \end{pmatrix}, \quad \boldsymbol{m} = \begin{pmatrix} m_0 \\ m_1 \\ \vdots \\ m_{N-1} \end{pmatrix},$$

$$\boldsymbol{n} = \begin{pmatrix} n_0 \\ n_1 \\ \vdots \\ n_{N-1} \end{pmatrix}, \quad \boldsymbol{x} = \boldsymbol{m} + \boldsymbol{n} \tag{23.65}$$

consisting of the model signal $\boldsymbol{m}$ of known type, the noise $\boldsymbol{n}$ and the unknown (scalar) parameter a, the CRLB is

given by

$$\sigma_a^2 \geq \frac{1}{E\left[\left(\frac{\partial \ln p(x,a)}{\partial a}\right)^2\right]} = -\frac{1}{E\left[\frac{\partial^2 \ln p(x,a)}{\partial a^2}\right]}, \quad (23.66)$$

where $p(x, a)$ is the joint probability density function, (Sect. 23.1), of the measured signal x for a given parameter a. Since a is normally a vector, (23.66) is the inverse of a matrix, the Fisher information matrix $\mathbf{J}$, whose typical element is given by

$$J_{ij} = E(H_{a_i} H_{a_j}) = -E(H_{a_i a_j}) \quad (23.67)$$

with

$$H_{a_i} = \frac{\partial \ln p(x,a)}{\partial a_i}, \quad a = \begin{pmatrix} a_0 \\ a_1 \\ \vdots \\ a_{A-1} \end{pmatrix} \quad (23.68)$$

The bound of the i-th unknown element of the parameter vector a is given by the i-th diagonal element with index ii of the inverse Fisher information matrix

$$\sigma_{a_i}^2 \geq \left(\mathbf{J}^{-1}\right)_{ii}, \quad (23.69)$$

where no summation is implied. For uncorrelated and signal independent noise with power σ_n^2, and with a Gaussian distribution, the joint probability density function $p(x, a)$ becomes

$$p(x, a) = \left(\frac{1}{2\pi\sigma_n^2}\right)^{\frac{N}{2}} \exp\left[-\frac{1}{2\sigma_n^2} \sum_{i=0}^{N-1} (x_i - m_i)^2\right] \quad (23.70)$$

and the elements of the Fisher information matrix become [23.16]

$$J_{ij} = \frac{1}{\sigma_n^2} \sum_{k=0}^{N-1} \left(\frac{\partial m_k}{\partial a_i} \frac{\partial m_k}{\partial a_j}\right). \quad (23.71)$$

To derive the lower bounds for a given Doppler burst this can be calculated and inverted, at least numerically.

23.5.1 Laser Doppler and Phase Doppler Signals

As an example, the lower bounds for the estimation of the frequency ω and the phase φ will be derived from a laser Doppler-like signal. Since the parameter vector a must contain *all* unknown parameters, including those that are not estimated (hidden parameters), a constant amplitude of unity during the observation time is assumed for simplification. The time-dependent signal

$$x(t) = m(t) + n(t) \quad (23.72)$$

is composed of the model signal

$$m(t) = \cos(\omega t + \varphi) \quad (23.73)$$

and the time-dependent noise $n(t)$. The measured signal after sampling is therefore

$$x_i = x(t = t_i) = \cos(\omega t_i + \varphi) + n_i, \\ i = 0, 1, \ldots, (N-1) \quad (23.74)$$

and each sample is a function of the two model parameters $x_i(\omega, \varphi)$ and the noise. The sampling times are given by $t_i = i/f_s$. The noise n is uncorrelated and Gaussian distributed. The parameter vector is

$$a = \begin{pmatrix} \omega \\ \varphi \end{pmatrix}. \quad (23.75)$$

The derivatives of the model-parameter-dependent samples $m_i = \cos(\omega t_i + \varphi)$ are

$$\frac{\partial m_i}{\partial \omega} = -t_i \sin(\omega t_i + \varphi),$$
$$\frac{\partial m_i}{\partial \varphi} = -\sin(\omega t_i + \varphi). \quad (23.76)$$

The Fisher information matrix becomes

$$\mathbf{J} = \frac{1}{\sigma_n^2} \\ \times \begin{pmatrix} \sum_{i=0}^{N-1} t_i^2 \sin^2(\omega t_i + \varphi) & \sum_{i=0}^{N-1} t_i \sin^2(\omega t_i + \varphi) \\ \sum_{i=0}^{N-1} t_i \sin^2(\omega t_i + \varphi) & \sum_{i=0}^{N-1} \sin^2(\omega t_i + \varphi) \end{pmatrix}. \quad (23.77)$$

The inverse of the Fisher information matrix is

$$\mathbf{J}^{-1} = \frac{\sigma_n^2}{\det(\mathbf{J})} \\ \times \begin{pmatrix} \sum_{i=0}^{N-1} \sin^2(\omega t_i + \varphi) & -\sum_{i=0}^{N-1} t_i \sin^2(\omega t_i + \varphi) \\ -\sum_{i=0}^{N-1} t_i \sin^2(\omega t_i + \varphi) & \sum_{i=0}^{N-1} t_i^2 \sin^2(\omega t_i + \varphi) \end{pmatrix}. \quad (23.78)$$

As an example, a numerical simulation was performed for the (true) parameters $\omega = 2$, $\varphi = 1.1$, $f_s = 10$ and

$N = 256$. For this and subsequent examples in this section, frequencies have been non-dimensionalized using 2π. Thus, for $\omega = 2$, $f_s = 10$, $\omega = 2$, $f_s = 10$ corresponds to 10π samples per cycle. The noise power was varied logarithmically in 25 steps from e^{-10} to e^{+10}, which corresponds to 25 equal steps of SNR, expressed in dB. The noise power has been normalized with the signal variance, thus $\sigma_n^2 = 1$ corresponds to SNR $= 0$ dB. For each noise level, 1000 independent realizations were generated. The individual signals were processed by a least mean square estimation routine, which for Gaussian distributed noise is equal to the maximum-likelihood estimation. The Fisher information matrix and its inverse were calculated to be

$$\mathbf{J} = \frac{1}{\sigma_n^2} \begin{pmatrix} 130 & 1653 \\ 1653 & 28356 \end{pmatrix}, \quad (23.79)$$

$$\mathbf{J}^{-1} = \sigma_n^2 \begin{pmatrix} 0.029787 & -0.001737 \\ -0.001737 & 0.000137 \end{pmatrix}. \quad (23.80)$$

In Fig. 23.7a a sample signal with SNR $= 15$ dB is illustrated. The results presented in Fig. 23.7b show that the maximum-likelihood estimator meets the calculated CRLB. Furthermore, a threshold noise power $\sigma_{n,\max}^2$ can be seen for the frequency estimate. Above this limit, the noise dominates the spectrum and the algorithm estimates the frequency randomly from the entire frequency range. The frequency of the threshold depends not only on the signal characteristics, but also on the capability of the estimation procedure to find the correct peak in the spectrum. The phase range is limited by $\pm\pi$. Therefore, the variance of the phase estimation is also limited.

In the case of phase Doppler signals, the Doppler frequency and the phase difference between two signals

$$x_i = \cos(\omega t_i + \varphi_x) + n_{x,i}, \quad i = 0, 1, 2\ldots, (N-1), \quad (23.81)$$

$$y_i = \cos(\omega t_i + \varphi_y) + n_{y,i}, \quad i = 0, 1, 2\ldots, (N-1), \quad (23.82)$$

with independent noise components n_x and n_y are of interest. To derive the CRLB for the phase difference, it is convenient to rewrite these signals as

$$x_i = m_{x,i} + n_{x,i} = \cos(\omega t_i + \varphi) + n_{x,i},$$
$$i = 0, 1, 2\ldots, (N-1), \quad (23.83)$$

$$y_i = m_{y,i} + n_{y,i} = \cos(\omega t_i + \varphi + \Delta\varphi) + n_{y,i}$$
$$i = 0, 1, 2\ldots, (N-1). \quad (23.84)$$

Since the signals are of the same length with independent noise components, the joint probability density function $p(\mathbf{x}, \mathbf{y}, \mathbf{a})$ now becomes

$$p(\mathbf{x}, \mathbf{y}, \mathbf{a}) = \left(\frac{1}{2\pi\sigma_n^2}\right)^N \exp\left(-\frac{1}{2\sigma_n^2} \sum_{i=0}^{N-1} \left[(x_i - m_{x,i})^2 \right.\right.$$
$$\left.\left. + (y_i - m_{y,i})^2\right]\right) \quad (23.85)$$

with

$$m_{x,i} = \cos(\omega t_i + \varphi), \quad (23.86)$$
$$m_{y,i} = \cos(\omega t_i + \varphi + \Delta\varphi), \quad (23.87)$$

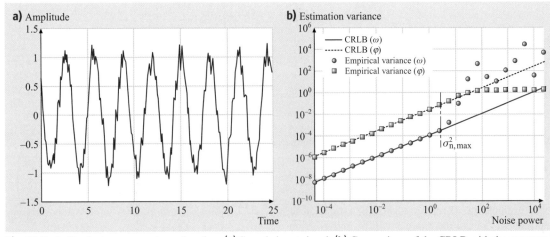

Fig. 23.7a,b Single-tone parameter estimation. (**a**) Sample input signal. (**b**) Comparison of the CRLB with the computed variance for frequency and phase estimates

and the elements of the Fisher information matrix become

$$J_{ij} = \frac{1}{\sigma_n^2} \sum_{k=0}^{N-1} \left(\frac{\partial m_{x,k}}{\partial a_i} \frac{\partial m_{x,k}}{\partial a_j} + \frac{\partial m_{y,k}}{\partial a_i} \frac{\partial m_{y,k}}{\partial a_j} \right). \tag{23.88}$$

The vector of unknown parameters is

$$\boldsymbol{a} = \begin{pmatrix} \omega \\ \varphi \\ \Delta\varphi \end{pmatrix}. \tag{23.89}$$

Note that φ is included in the parameter vector since it is unknown, even though it is not used. The derivatives of m_x and m_y are

$$\frac{\partial m_{x,i}}{\partial \omega} = -t_i \sin(\omega t_i + \varphi), \tag{23.90}$$

$$\frac{\partial m_{x,i}}{\partial \varphi} = -\sin(\omega t_i + \varphi), \tag{23.91}$$

$$\frac{\partial m_{x,i}}{\partial \Delta\varphi} = 0, \tag{23.92}$$

$$\frac{\partial m_{y,i}}{\partial \omega} = -t_i \sin(\omega t_i + \varphi + \Delta\varphi), \tag{23.93}$$

$$\frac{\partial m_{y,i}}{\partial \varphi} = -\sin(\omega t_i + \varphi + \Delta\varphi), \tag{23.94}$$

$$\frac{\partial m_{y,i}}{\partial \Delta\varphi} = -\sin(\omega t_i + \varphi + \Delta\varphi). \tag{23.95}$$

The Fisher information matrix becomes

$$\mathbf{J} = \frac{1}{\sigma_n^2} \begin{pmatrix} P_2 + Q_2 & P_1 + Q_1 & Q_1 \\ P_1 + Q_1 & P_0 + Q_0 & Q_0 \\ Q_1 & Q_0 & Q_0 \end{pmatrix} \tag{23.96}$$

with

$$P_k = \sum_{i=0}^{N-1} t_i^k \sin^2(\omega t_i + \varphi), \tag{23.97}$$

$$Q_k = \sum_{i=0}^{N-1} t_i^k \sin^2(\omega t_i + \varphi + \Delta\varphi). \tag{23.98}$$

The inverse of the Fisher information matrix is

$$\mathbf{J}^{-1} = \sigma_n^2 \begin{pmatrix} P_2 + Q_2 & P_1 + Q_1 & Q_1 \\ P_1 + Q_1 & P_0 + Q_0 & Q_0 \\ Q_1 & Q_0 & Q_0 \end{pmatrix}^{-1}. \tag{23.99}$$

As an example, a numerical simulation was performed for the (true) parameters $\omega = 2$, $\varphi = 1.1$, $\Delta\varphi = -0.8$, $f_s = 10$ and $N = 256$. The noise power varied logarithmically in 25 steps from e^{-10} to e^{+10}. For each noise level 1000 independent realizations were generated. The individual signals (Fig. 23.8a) were processed by a maximum-likelihood estimation routine. The Fisher information matrix and its inverse were

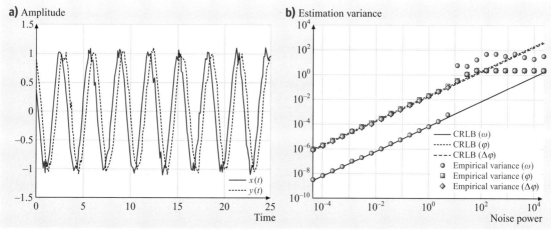

Fig. 23.8a,b Single-tone parameter estimation from two orthogonal signals. (**a**) Sample input signals. (**b**) Comparison of the CRLB with computed variance for frequency, phase and phase difference estimates

calculated to be

$$J = \frac{1}{\sigma_n^2} \begin{pmatrix} 55536 & 3261 & 1608 \\ 3261 & 258 & 128 \\ 1608 & 128 & 128 \end{pmatrix}, \quad (23.100)$$

$$J^{-1} = \sigma_n^2 \begin{pmatrix} 0.000070 & -0.000894 & 0.000008 \\ -0.000894 & 0.019063 & -0.007794 \\ 0.000008 & -0.007794 & 0.015536 \end{pmatrix}. \quad (23.101)$$

The results presented in Fig. 23.8b show that the maximum-likelihood estimator meets the calculated CRLB. Again, a threshold noise power can be seen for the frequency estimate and the phase range is limited by $\pm\pi$. Note that the CRLB and the empirically derived estimation variance of the phase φ and the phase difference $\Delta\varphi$ are different.

The expressions of the CRLB derived above are not convenient for practical use in setting up a signal processor. Explicit expressions of the CRLB are required. To derive these, a set of two orthogonal signals x and $\tilde{x}$ with independent noise components n and $\tilde{n}$ are considered.

$$x_i = m_i + n_i, \quad i = 0, 1, \ldots, (N-1), \quad (23.102)$$

$$\tilde{x}_i = \tilde{m}_i + \tilde{n}_i, \quad i = 0, 1, \ldots, (N-1), \quad (23.103)$$

with

$$m_i = A \cos(\omega t_i + \varphi), \quad i = 0, 1, \ldots, (N-1), \quad (23.104)$$

$$\tilde{m}_i = A \sin(\omega t_i + \varphi), \quad i = 0, 1, \ldots, (N-1), \quad (23.105)$$

where additionally the amplitude A is unknown. The joint probability density function becomes

$$p(x, \tilde{x}, a) = \left(\frac{1}{2\pi\sigma_n^2}\right)^N \exp\left\{-\frac{1}{2\sigma_n^2} \sum_{i=0}^{N-1} \left[(x_i - m_i)^2 + (\tilde{x}_i - \tilde{m}_i)^2\right]\right\} \quad (23.106)$$

and the elements of the Fisher information matrix are

$$J_{ij} = \frac{1}{\sigma_n^2} \sum_{k=0}^{N-1} \left(\frac{\partial m_k}{\partial a_i} \frac{\partial m_k}{\partial a_j} + \frac{\partial \tilde{m}_k}{\partial a_i} \frac{\partial \tilde{m}_k}{\partial a_j}\right). \quad (23.107)$$

The vector of unknown parameters is

$$a = \begin{pmatrix} \omega \\ \varphi \\ A \end{pmatrix}. \quad (23.108)$$

Using the fact that $m_i^2 + \tilde{m}_i^2 = A^2$, the Fisher information matrix can be expressed explicitly as

$$J = \frac{1}{\sigma_n^2} \begin{pmatrix} A^2 \sum_{i=0}^{N-1} t_i^2 & A^2 \sum_{i=0}^{N-1} t_i & 0 \\ A^2 \sum_{i=0}^{N-1} t_i & A^2 \sum_{i=0}^{N-1} 1 & 0 \\ 0 & 0 & N \end{pmatrix} = \frac{A^2}{6\sigma_n^2 f_s^2}$$

$$\times \begin{pmatrix} N(N-1)(2N-1) & 2N(N-1)f_s & 0 \\ 2N(N-1)f_s & 6Nf_s^2 & 0 \\ 0 & 0 & \frac{6Nf_s^2}{A^2} \end{pmatrix}. \quad (23.109)$$

The zero elements in this matrix indicate that the amplitude can be estimated completely independent of the frequency and the phase. Thus, the amplitude can be presumed to be known without changing the lower bounds of the frequency and phase estimator variance. The inverse of the Fisher information matrix becomes

$$J^{-1} = \frac{2\sigma_n^2}{A^2 N^2 (N^2 - 1)}$$

$$\times \begin{pmatrix} 6Nf_s^2 & -3N(N-1)f_s & 0 \\ -3N(N-1)f_s & N(N-1)(2N-1) & 0 \\ 0 & 0 & \frac{A^2 N(N^2-1)}{2} \end{pmatrix} \quad (23.110)$$

leading to the CRLB for the frequency

$$\sigma_\omega^2 \geq \frac{12\sigma_n^2 f_s^2}{A^2 N(N^2 - 1)}. \quad (23.111)$$

This is the CRLB for two signals with independent noise components. For only one signal, the information content is approximately one half, leading to

$$\sigma_\omega^2 \geq \frac{24\sigma_n^2 f_s^2}{A^2 N(N^2 - 1)}. \quad (23.112)$$

Using

$$\text{SNR} = \frac{A^2}{2\sigma_n^2}, \quad (23.113)$$

the CRLB for ω can be expressed as [23.17–19]

$$\sigma_\omega^2 \geq \frac{12 f_s^2}{N(N^2 - 1)\text{SNR}}, \quad (23.114)$$

or, if $\omega = 2\pi f$ is used, this variance reduces to

$$\sigma_f^2 \geq \frac{3 f_s^2}{\pi^2 N(N^2 - 1)\text{SNR}}. \quad (23.115)$$

This expression was derived assuming that the noise is spectrally white. Any filtering used to reduce the signal noise violates this assumption and (23.115) no longer strictly holds. Thus, while (band-pass) filtering may improve the SNR, the estimator variance may not be reduced.

For the case of phase Doppler signals, a second signal pair is required.

$$x_i = m_{x,i} + n_{x,i} \quad m_{x,i} = A\cos(\omega t_i + \varphi),$$
$$i = 0, 1, \ldots, (N-1), \quad (23.116)$$
$$\tilde{x}_i = \tilde{m}_{x,i} + \tilde{n}_{x,i} \quad \tilde{m}_{x,i} = A\sin(\omega t_i + \varphi),$$
$$i = 0, 1, \ldots, (N-1), \quad (23.117)$$
$$y_i = m_{y,i} + n_{y,i} \quad m_{y,i} = A\cos(\omega t_i + \varphi + \Delta\varphi),$$
$$i = 0, 1, \ldots, (N-1), \quad (23.118)$$
$$\tilde{y}_i = \tilde{m}_{y,i} + \tilde{n}_{y,i} \quad \tilde{m}_{y,i} = A\sin(\omega t_i + \varphi + \Delta\varphi),$$
$$i = 0, 1, \ldots, (N-1). \quad (23.119)$$

The joint probability density function becomes

$$p(x, \tilde{x}, y, \tilde{y}, a) = \left(\frac{1}{2\pi\sigma_n^2}\right)^{2N}$$
$$\exp\left(-\frac{1}{2\sigma_n^2}\sum_{i=0}^{N-1}\left[(x_i - m_{x,i})^2 + (\tilde{x}_i - \tilde{m}_{x,i})^2\right.\right.$$
$$\left.\left. + (y_i - m_{y,i})^2 + (\tilde{y}_i - \tilde{m}_{y,i})^2\right]\right) \quad (23.120)$$

and the elements of the Fisher information matrix are

$$J_{ij} = \frac{1}{\sigma_n^2}\sum_{k=0}^{N-1}\left(\frac{\partial m_{x,k}}{\partial a_i}\frac{\partial m_{x,k}}{\partial a_j} + \frac{\partial \tilde{m}_{x,k}}{\partial a_i}\frac{\partial \tilde{m}_{x,k}}{\partial a_j}\right.$$
$$\left. + \frac{\partial m_{y,k}}{\partial a_i}\frac{\partial m_{y,k}}{\partial a_j} + \frac{\partial \tilde{m}_{y,k}}{\partial a_i}\frac{\partial \tilde{m}_{y,k}}{\partial a_j}\right). \quad (23.121)$$

The vector of unknown parameters is

$$a = \begin{pmatrix} \omega \\ \varphi \\ \Delta\varphi \\ A \end{pmatrix}. \quad (23.122)$$

The Fisher information matrix becomes

$$\mathbf{J} = \frac{1}{\sigma_n^2}\begin{pmatrix} 2A^2\sum_{i=0}^{N-1}t_i^2 & 2A^2\sum_{i=0}^{N-1}t_i & A^2\sum_{i=0}^{N-1}t_i & 0 \\ 2A^2\sum_{i=0}^{N-1}t_i & 2A^2\sum_{i=0}^{N-1}1 & A^2\sum_{i=0}^{N-1}1 & 0 \\ A^2\sum_{i=0}^{N-1}t_i & A^2\sum_{i=0}^{N-1}1 & A^2\sum_{i=0}^{N-1}1 & 0 \\ 0 & 0 & 0 & 2N \end{pmatrix},$$
$$(23.123)$$

$$\mathbf{J} = \frac{A^2}{6\sigma_n^2 f_s^2}$$
$$\times \begin{pmatrix} 2N(N-1) & 6N & 3N & 0 \\ \times(2N-1) & \times(N-1)f_s & \times(N-1)f_s & \\ 6N(N-1)f_s & 12Nf_s^2 & 6Nf_s^2 & 0 \\ 3N(N-1)f_s & 6Nf_s^2 & 6Nf_s^2 & 0 \\ 0 & 0 & 0 & \frac{12Nf_s^2}{A^2} \end{pmatrix}. \quad (23.124)$$

The inverse of the Fisher information matrix is

$$\mathbf{J}^{-1} = \frac{\sigma_n^2}{2A^2 N}\begin{pmatrix} \frac{12f_s^2}{N^2-1} & -\frac{6f_s}{N+1} & 0 & 0 \\ -\frac{6f}{N+1} & \frac{5N-1}{N+1} & -2 & 0 \\ 0 & -2 & 4 & 0 \\ 0 & 0 & 0 & A^2 \end{pmatrix} \quad (23.125)$$

leading to the following CRLBs [23.20]

$$\sigma_\omega^2 \geq \frac{6\sigma_n^2 f_s^2}{A^2 N(N^2-1)}, \quad (23.126)$$
$$\sigma_\varphi^2 \geq \frac{\sigma_n^2(5N-1)}{2A^2 N(N+1)}, \quad (23.127)$$
$$\sigma_{\Delta\varphi}^2 \geq \frac{2\sigma_n^2}{A^2 N}. \quad (23.128)$$

Note that these are the lower bounds for the four signals with independent noise components. If only the two phase Doppler signals are given, then the information content is one half and the bounds become

$$\sigma_\omega^2 \geq \frac{12\sigma_n^2 f_s^2}{A^2 N(N^2-1)} = \frac{6f_s^2}{N(N^2-1)\mathrm{SNR}}, \quad (23.129)$$
$$\sigma_\varphi^2 \geq \frac{\sigma_n^2(5N-1)}{A^2 N(N+1)} = \frac{5N-1}{2N(N+1)\mathrm{SNR}}, \quad (23.130)$$
$$\sigma_{\Delta\varphi}^2 \geq \frac{4\sigma_n^2}{A^2 N} = \frac{2}{N\mathrm{SNR}}. \quad (23.131)$$

The CRLB is exactly half of the value for a Doppler signal, since the frequency information content in the

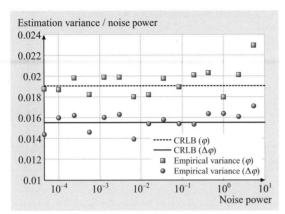

Fig. 23.9 Variance of the maximum-likelihood estimator, normalized by the noise power for phase and phase difference. Comparison to the respective CRLBs (Simulation parameters as used in Fig. 23.8)

two phase Doppler signals is twice as large. The CRLB for the phase difference between the two phase Doppler signals is lower than that for the absolute phase by a factor of 0.8, for large N. This can be seen in Fig. 23.9, which presents the estimator variance results from Fig. 23.8b, normalized by the noise power.

All of the above derivations were based on a signal model of constant amplitude. In fact, laser Doppler and phase Doppler signals exhibit an amplitude described by a Gaussian envelope and for this case, the derivation of CRLB is somewhat more tedious. Results for the frequency, amplitude, arrival time and residence time are presented by *Høst-Madsen* and *Gjelstrup* [23.21].

23.5.2 Particle Imaging

The task of deriving the two-dimensional position from images of small particles is addressed. For simplification, a sampled two-dimensional Gaussian function

$$z(x,y) = A \exp\left\{-\eta\left[(x-\xi)^2 + (y-\psi)^2\right]\right\} \quad (23.132)$$

with the particle position $[\xi, \psi]$, the maximum intensity amplitude A and the parameter η, defining the width of the particle image, is used for the intensity profile of the particle image. The originally continuous intensity profile is given at discrete positions (pixels) z_{ij}. For simplicity, an ideas sampling is assumed with $z_{ij} = z(x_i, y_j)$.

To derive the particle position from its sampled image, several methods can be applied such as the centroid method or a Gaussian fit. For the given condition with a sampled Gaussian intensity profile, the centroid method is biased while the Gaussian fit is bias-free. However, all methods are affected by noise.

Assuming a Poisson-distributed fluctuation of the intensity values due to the photon noise as in *Wernet* and *Pline* [23.22], the elements of the Fisher information matrix become

$$J_{ij} = \sum_k \sum_l \frac{1}{z_{kl}} \frac{\partial z_{kl}}{\partial a_i} \frac{\partial z_{kl}}{\partial a_j}. \quad (23.133)$$

Note that the limits of the sums are not given here. A summation over all pixels that contribute to the particle image is assumed. Since the particle images are usually much smaller then the dimensions of the observed imaging area, the limits of the sums can be dropped off as long as the particle images do not overlap. The parameter vector is given by

$$\boldsymbol{a} = \begin{pmatrix} A \\ \eta \\ \xi \\ \psi \end{pmatrix}. \quad (23.134)$$

The element $J_{\xi\xi}$ of the Fisher information matrix then becomes

$$\begin{aligned} J_{\xi\xi} &= \sum_k \sum_l \frac{1}{z_{kl}} \left(\frac{\partial z_{kl}}{\partial \xi}\right)^2 \\ &= \sum_k \sum_l z_{kl}[-2\eta(x_k-\xi)]^2 \\ &= \sum_k \sum_l 4A\eta^2(x_k-\xi)^2 \exp\{-\eta[(x_k-\xi)^2 \\ &\quad + (y_l-\psi)^2]\}. \end{aligned} \quad (23.135)$$

A separation in two sums yields

$$J_{\xi\xi} = 4A\eta^2 \left\{\sum_k (x_k-\xi)^2 \exp\left[-\eta(x_k-\xi)^2\right] \right. \\ \left. \left|\sum_l \exp\left[-\eta(y_l-\psi)^2\right]\right\}\right.. \quad (23.136)$$

Approximating the sums by integrals yields

$$J_{\xi\xi} = \frac{4A\eta^2}{\Delta x \Delta y} \left\{\int (x-\xi)^2 \exp\left[-\eta(x-\xi)^2\right] dx \right. \\ \left. \left|\int \exp\left[-\eta(y-\psi)^2\right] dy\right\}\right. \quad (23.137)$$

with the sampling intervals Δx and Δy, which are one pixel ($\Delta x = \Delta y \equiv 1$). Assuming furthermore an unbounded particle image and therefore integrals within the range $[-\infty, \infty]$, the first integral becomes $\sqrt{\pi/(4\eta^3)}$ and the second $\sqrt{\pi/\eta}$, yielding

$$J_{\xi\xi} = 2\pi A \, . \tag{23.138}$$

Assuming that the parameters of the parameter vector a can be estimated independently, the appropriate element of the inverse Fisher information matrix can be approximated by

$$\left(J^{-1}\right)_{\xi\xi} \approx \left(J_{\xi\xi}\right)^{-1} = \frac{1}{2\pi A} \tag{23.139}$$

and finally the CRLB for estimating the component ξ of the particle position in the x-direction becomes

$$\sigma_\xi^2 \geq \frac{1}{2\pi A} \, . \tag{23.140}$$

Similarly, for estimating the component ψ of the particle position in the y-direction is

$$\sigma_\psi^2 \geq \frac{1}{2\pi A} \, . \tag{23.141}$$

In Fig. 23.10 the estimation variances are shown for the centroid method, the Gaussian fit and an alternative maximum-likelihood estimator (MLE), obtained empirically based on a computer simulation. The deviations of the two estimated components $\hat{\xi}$ and $\hat{\psi}$ from the correct values ξ and ψ are combined to a common deviation

$$\sqrt{(\hat{\xi} - \xi)^2 + (\hat{\psi} - \psi)^2} \tag{23.142}$$

as well as the CRLB

$$\sigma_\xi^2 + \sigma_\psi^2 \geq \frac{1}{\pi A} \, . \tag{23.143}$$

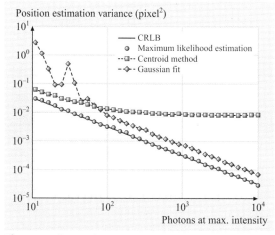

Fig. 23.10 CRLB for the position estimation of particle images for Poisson-distributed noise

As expected the MLE obviously meets the CRLB, while the Gaussian fit has a significantly larger estimation variance. This is a result of the different image information used for the two estimators. While the MLE uses the entire particle image for its position estimation, the position estimate of the Gaussian fit is based only on the nine central pixels of the particle image. However, extending the number of pixels to be used for the fit is not useful. On the one hand this would reduce the estimation variance. But on the other hand, systematic errors due to overlaps of particles images would increase.

Compared to those both estimators, the centroid method is dominated by systematic errors for large numbers of photons (low noise). However, for small number of photons (high noise) the centroid method is more robust then the Gaussian fit.

Due to the common use of powerful pulsed lasers the accuracy of the position estimation is nowadays not limited by the number of photons but by thermal noise and the discrete gray values of the images. Both can be approximated by Gaussian-distributed noise [23.23]. For a Gaussian-distributed noise with the variance σ_n^2 the elements of the Fisher information matrix become

$$J_{ij} = \frac{1}{\sigma_n^2} \sum_k \sum_l \frac{\partial z_{kl}}{\partial a_i} \frac{\partial z_{kl}}{\partial a_j} \, . \tag{23.144}$$

The element $J_{\xi\xi}$ of the Fisher information matrix then becomes

$$\begin{aligned} J_{\xi\xi} &= \frac{1}{\sigma_n^2} \sum_k \sum_l \left(\frac{\partial z_{kl}}{\partial \xi}\right)^2 \\ &= \frac{1}{\sigma_n^2} \sum_k \sum_l z_{kl}^2 [-2\eta(x_k - \xi)]^2 \\ &= \frac{1}{\sigma_n^2} \sum_k \sum_l 4A^2\eta^2 (x_k - \xi)^2 \exp\{-2\eta[(x_k - \xi)^2 \\ &\quad + (y_l - \psi)^2]\} \, . \end{aligned} \tag{23.145}$$

A separation in two sums yields

$$J_{\xi\xi} = \frac{4A^2\eta^2}{\sigma_n^2} \left\{ \sum_k (x_k - \xi)^2 \exp\left[-2\eta(x_k - \xi)^2\right] \right. \\ \left. \times \left| \sum_l \exp\left[-2\eta(y_l - \psi)^2\right] \right| \right\} \, . \tag{23.146}$$

Approximating the sums by integrals yields

$$J_{\xi\xi} = \frac{4A^2\eta^2}{\sigma_n^2 \Delta x \Delta y} \left\{ \int (x - \xi)^2 \exp\left[-2\eta(x - \xi)^2\right] dx \right. \\ \left. \times \left| \int \exp\left[-2\eta(y - \psi)^2\right] dy \right| \right\} \tag{23.147}$$

with the sampling intervals Δx and Δy, which again are 1 pixel ($\Delta x = \Delta y \equiv 1$). Assuming an unbounded particle image and therefore integrals within the range $[-\infty, \infty]$, the first integral becomes $\sqrt{\pi/(32\eta^3)}$ and the second $\sqrt{\pi/(2\eta)}$, yielding

$$J_{\xi\xi} = \frac{\pi A^2}{2\sigma_n^2} \qquad (23.148)$$

Assuming that the parameters of the parameter vector a can be estimated independently, the appropriate element of the inverse Fisher information matrix can be approximated by

$$\left(J^{-1}\right)_{\xi\xi} \approx \left(J_{\xi\xi}\right)^{-1} = \frac{2\sigma_n^2}{\pi A^2} \qquad (23.149)$$

and finally, the CRLB for estimating the component ξ of the particle position in the x-direction becomes

$$\sigma_\xi^2 \geq \frac{2\sigma_n^2}{\pi A^2} . \qquad (23.150)$$

Similarly, for estimating the component ψ of the particle position in the y-direction is

$$\sigma_\psi^2 \geq \frac{2\sigma_n^2}{\pi A^2} . \qquad (23.151)$$

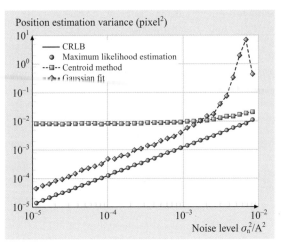

Fig. 23.11 CRLB for the position estimation of particle images for Gaussian-distributed noise

In Fig. 23.11 the results of the computer simulation are shown for Gaussian-distributed noise. The MLE again meets the CRLB, while the Gaussian fit has a larger estimation variance due to the reduced information used for the estimation procedure. The centroid method again is limited by systematic errors for low noise levels, but is more robust than the Gaussian fit for high noise levels.

23.6 Propagation of Errors

The concepts of stochastic and systematic errors for a given measurement quantity have already been introduced in (23.26) and (23.27), respectively. If a derived quantity y depends on several individual measurement quantities x_i, the question arises as to the measurement error in y

$$y = f(x_1, x_2, \ldots, x_n) . \qquad (23.152)$$

The propagation of errors from the quantities x_i to y is treated separately for systematic and stochastic errors.

The resulting systematic error in y is found by using a first-order Taylor expansion

$$\delta y = \frac{\partial f}{\partial x_1}\delta x_1 + \frac{\partial f}{\partial x_2}\delta x_2 + \ldots + \frac{\partial f}{\partial x_n}\delta x_n , \qquad (23.153)$$

where δx_i are the systematic errors for each measurement quantity x_i and δy is the overall systematic error.

Note that all δx_i quantities are signed and, as such, systematic errors may be compensating in nature.

Stochastic errors are treated in the mean square, leading to the relation

$$\sigma_y = \sqrt{\left(\frac{\partial f}{\partial x_1}\sigma_{x_1}\right)^2 + \left(\frac{\partial f}{\partial x_2}\sigma_{x_2}\right)^2 + \ldots + \left(\frac{\partial f}{\partial x_n}\sigma_{x_n}\right)^2} \qquad (23.154)$$

where the individual estimator variances $\sigma_{x_i}^2$ have been evaluated using techniques described in the previous section. This formula assumes that all of the individual stochastic errors are normally distributed and that the standard deviations are all evaluated with the same confidence intervals. An extensive discussion of error propagation can be found in *Kline* and *McClintock* [23.3], *Kline* [23.4] and *Moffat* [23.5].

References

23.1 J.S. Bendat, A.G. Piersol: *Random Data: Analysis and Measurement Procedures* (Wiley, New York 1986)

23.2 W.K. George Jr.: Processing of random signals, Proc. Dyn. Flow Conf. (1978) 20–63

23.3 S.J. Kline, F.A. McClintock: Describing uncertainties in single-sample experiments, Mech Eng **75**, 3–8 (1953)

23.4 S.J. Kline: The purposes of uncertainty analysis, ASME J. Fluid. Eng. **107**, 153–160 (1985)

23.5 R.J. Moffat: Using uncertainty analysis in the planning of an experiment, ASME J. Fluid. Eng. **107**, 173–182 (1985)

23.6 R.J. Moffat: Describing uncertainties in experimental results, Exp. Thermal Fluid Sci. **1**, 3–17 (1988)

23.7 A. Stuart, J.K. Ord: *Kendall's Advanced Theory of Statistics* (Edward Arnold, London 1994)

23.8 M. Kendall, A. Stuart: *The Advanced Theory of Statistics*, Vol. Charles Griffen (Charles Griffen, London 1963)

23.9 L.H. Benedict, R.D. Gould: Towards better uncertainty estimates for turbulence statistics, Exp. Fluids **22**, 129–136 (1996)

23.10 J.W. Tukey: Bias and confidence in not-quite large samples, Ann. Math. Statist. **29**, 614 (1958)

23.11 B. Efron, R.J. Tibshirani: *An Introduction to the Bootstrap* (Chapman Hall, New York 1993)

23.12 C. Tropea: Performance testing of LDA/PDA signal processing systems, Int. Conf. on Laser Anemom - Adv. and Appl., Swansea, UK: paper II 4 (1989)

23.13 M. Kendall, A. Stuart: *The Advanced Theory of Statistics*, Vol. 2 (Charles Griffen, London 1963)

23.14 A. Papoulis: *Probability, Random Variables and Stochastic Processes* (McGraw-Hill, New York 1988)

23.15 S.M. Kay: *Fundamentals of Statistical Signal Processing: Estimation theory* (Prentice Hall, Englewood Cliffs 1993)

23.16 A.D. Whalen: *Detection of Signals in Noise* (Academic, New York 1971)

23.17 K. M. Ibrahim, D. Werthimer, W. D. Bachalo: Signal Processing Considerations for Laser Doppler and Phase Doppler Applications. Appl. of Laser Techn. Fluid. Mech. 1990 291-316

23.18 D.C. Rife, R.R. Boortstyn: Single tone parameter estimation from discrete time observations, IEEE Trans. on Inform. Theory **20**, 591–596 (1974)

23.19 T. Wriedt, K.A. Bauckhage, A. Schöne: Application of Fourier analysis to phase Doppler-signals generated by rough metal particles, IEEE Trans. Instrum. Meas. **38**, 984–990 (1989)

23.20 A. Høst-Madsen, K. Andersen: Lower bounds for estimation of frequency and phase of Doppler signals, Meas. Sci. **6**, 637–652 (1995)

23.21 A. Høst-Madsen, P. Gjelstrup: New processing methods for LDA signals, Proc. 8th Int. Symp. on Appl. of Laser Techn. to Fluid Mech., Lisbon, Portugal: paper 31.3 (1996)

23.22 M.P. Wernet, A. Pline: Particle displacement tracking technique and Cramer-Rao lower bound error in centroid estimates from CCD imagery, Exp. Fluids **15**, 295–307 (1993)

23.23 J. Westerweel: Theoretical analysis of the measurement precision in particle image velocimetry, Exp. Fluids **29**, S3–S12 (2000)

24. Data Acquisition by Imaging Detectors

Imaging sensors convert radiative energy into an electrical signal and such sensors are available that cover the wide spectrum from gamma rays to the infrared. They accumulate an electrical signal during the exposure time and convert all the signals of an array of detectors into a time-serial analog or digital data stream. The dominate and most successful devices to perform this task are charge coupled devices (CCD). However directly addressable imaging sensors on the basis of CMOS fabrication technology are becoming more and more promising because the image acquisition, digitalization and preprocessing can be integrated on a single chip; hence yielding very fast frame rates. This chapter provides a comprehensive survey of the available imaging sensors, details the parameters that control their performance and gives practical tips to select the best camera for different imaging tasks.

24.1 Definitions 1419

24.2 Types of Detectors 1420
    24.2.1 Quantum Detectors 1420
    24.2.2 Thermal Detectors 1420

24.3 Imaging Detectors 1421
    24.3.1 The Charge-Coupled Device 1421
    24.3.2 CMOS Imaging Sensors 1421
    24.3.3 CCD Sensor Architectures 1422
    24.3.4 Standard Interfaces for Digital Cameras 1425

24.4 Performance of Imaging Sensors 1426
    24.4.1 Responsivity 1426
    24.4.2 Quantum Efficiency 1426
    24.4.3 Signal Irradiance Relation 1427
    24.4.4 Dark Current 1427
    24.4.5 Noise-Equivalent Exposure 1427
    24.4.6 Saturation Equivalent Exposure ... 1427
    24.4.7 Dynamic Range 1428
    24.4.8 Photon-Noise-Limited Performance 1428
    24.4.9 Linear Noise Model for Image Sensors 1429
    24.4.10 Signal-to-Noise Ratio 1429
    24.4.11 Spectral Sensitivity 1431
    24.4.12 Nonuniform Responsivity 1433
    24.4.13 Artifacts and Operation Errors 1433

24.5 Camera Selection 1435
    24.5.1 Measurements at Low Light Levels 1435
    24.5.2 Measurements with High Irradiance Variations ... 1435
    24.5.3 Precise Radiometric and Geometric Measurements 1435

References 1436

24.1 Definitions

An imaging sensor converts an *image*, i.e., the spatially varying irradiance projected on the image plane by an optical system into a digital data stream that can be stored and processed by a digital computer. This involves a number of steps Fig. 24.1:

- *Conversion*. The incident radiative energy is converted into an electrical charge, voltage, or current. In order to form an image, an array of detectors is required. Each of the detector elements integrates the incident irradiance over a certain area.
- *Storage*. The generated charges must be integrated and thus stored over a certain time interval, the *exposure time*. In order to control illumination, the sensor should accumulate charges only during a time interval that can be controlled by an external signal. This feature is called an *electronic shutter*.
- *Read-out*. The exposure, the accumulated electrical charges must be read out. This step essentially converts the spatial charge distribution (parallel data) into one (or multiple) sequential data stream that is then processed by one (or multiple) output circuit.
- *Signal conditioning*. In a suitable electrical amplifier, the read-out charge is converted into a voltage

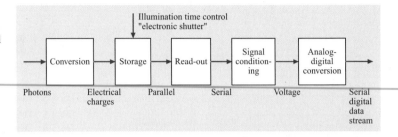

Fig. 24.1 Chain of processes that takes place in an imaging sensor that converts incident photons into a serial digital data stream

and by appropriate signal conditioning reduced from distortions. In this stage, also the responsivity of the sensor can be controlled.

- *Analog–digital conversion* In a final step, the analog voltage is converted it into a digital number for input into a computer.

24.2 Types of Detectors

For imaging detectors, basically two types of detectors are utilized: quantum detectors and thermal detectors.

24.2.1 Quantum Detectors

The term *quantum detector* refers to a class of detectors where the absorption of the smallest energy unit for electromagnetic radiation, the *photon*, triggers the detection of radiation. This process causes the transition of an electron into a higher excited state. Three variants are possible, which lead to three subclasses of quantum detectors.

Photoemissive Detectors
By the absorption of the photon, the electron receives enough energy to be able to leave the detector and become a *free electron*. This effect is known as the (extrinsic) *photo effect*. The photo effect can be triggered only by photons below a critical wavelength (i.e., above a certain energy of the photon that is sufficient to provide enough energy to free the electron from its binding).

Photoemission of electrons is utilized in *vacuum photo tubes* and *photomultiplier tubes (PMT)*. PMTs are sensitive enough to count single photons, since the initial electrons generated by photon absorption are accelerated to hit another dynode with sufficient energy to cause the ejection of multiple secondary electrons. This process can be cascaded to achieve high gain factors. Because PMTs have short response times that may be less than 10^{-10} s, individual photons can be counted as short current impulses. Such a device is known as a *photon-counting device*.

Radiation detectors based on the photo effect can be sensitive only in a quite narrow spectral range. The lowest wavelength is given by the minimum energy required to free an electron. For higher photon energies, the material may become more transparent, leading to a lower probability that the photon is absorbed.

Photovoltaic and Photoconductive Detectors
Semiconductor devices that utilize the inner photo effect have largely replaced imaging detectors based on photoemission. A thorough understanding of these devices requires knowledge of condensed matter physics. Thus, discussion of these devices here is superficial and concentrates on the basic properties directly related to imaging detectors. The semiconductor devices have in common with the photoemissive detectors that they have a threshold energy and thus a minimum frequency of radiation is required. This is due to the fact that electrons must be excited from the valence band across a band gap to the conduction band. In the valence band, the electrons cannot move and thus cause no further effects.

In a suitable photodetector material, the conduction band is empty. The absorption of a photon excites an electron into the conduction band, where it can move rather unrestricted, thus increasing the conductance of the detector material. Detectors operating in this mode are called *photoconductive detectors*.

In an appropriately designed detector, the generation of an electron builds up an electric tension. Under the influence of this tension, an electric current can be measured that is proportional to the rate with which the absorbed photons generate electrons. Detectors of this type are known as *photovoltaic detectors*.

24.2.2 Thermal Detectors

Thermal detectors respond to the temperature increase resulting from the absorption of incident radiation. The delivered signal is proportional to the temperature increase. The common feature of thermal detectors is the wide spectral range to which they can be made sensitive. They do not, however, reach the sensitivities of quantum detectors, since the radiation detection is based on secondary effects. The three most widely used secondary effects for imaging detectors are thermoelectricity, pyroelectricity, and thermoconductivity.

Thermoelectric Effect
Two separate junctions of two dissimilar metals at different temperatures generate a voltage proportional to the temperature difference between them. One junction must be kept at a reference temperature, while the other be designed to absorb the electromagnetic radiation with minimum thermal mass and good thermal insulation. Such a device is known as a *thermopile*.

Pyroelectric Effect
Pyroelectric materials have a permanent electrical polarization. Changes in the temperature cause a change in the surface charges. Pyroelectric detectors can only detect changes in the incident radiant flux and thus require a chopper. Furthermore, they lose their pyroelectric behavior above a critical temperature, the Curie temperature.

Thermoconductive Effect
Some materials feature large changes in electric conductivity with temperature. A device measuring radiation by change in conductance is called a *bolometer*. Recently, microbolometer arrays have been manufactured for uncooled infrared imagers in the 8–14 μm wavelength range that show a noise equivalent temperature difference (NEΔT) of about 100 mK [24.1].

24.3 Imaging Detectors

The photovoltaic effects are most suitable for imaging detectors. Instead of moving the photoinduced electrons through the conduction band, they can be captured by additional potential walls in cells. In these cells, the electrons can be accumulated for a certain time.

Given this basic structure of a quantum detector, it is obvious that the most difficult problem of an imaging device is the conversion of the spatial arrangement of the accumulated charges into an electric signal. This process is referred to as *read-out*.

24.3.1 The Charge-Coupled Device

The invention of the *charge-coupled device (CCD)* in the mid 1970s was a breakthrough for semiconductor imaging devices. A CCD is effectively an analog shift register.

By an appropriate sequential change of potential, charge can be moved across the imaging sensor. A so-called four-phase CCD is shown in Fig. 24.2. Potential walls separate the charge packets. First, a potential wall to the left of the three-cell-wide charge-storage area is raised, confining the charges to two cells. Subsequently, a potential barrier to the right is lowered so that the charge pockets move to the right by one cell. Repeating this procedure, the charge packets can be moved all the way across the array. The charge transfer process can be made efficient enough that only a negligible fraction of electrons are *lost* during the whole transfer. The charge transfer mechanism is the base for one- and two-dimensional arrays of photosensors.

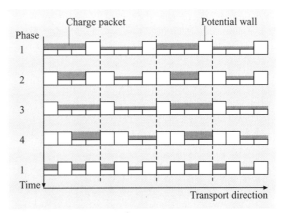

Fig. 24.2 Principle of the charge-transfer mechanisms of the charge-coupled device (CCD)

Table 24.1 Spectral sensitivity of different CCD sensors or focal plane areas. The last column lists either the quantum efficiency or the number of electrons generated per kV for X-ray detectors

Sensor type	Spectral range	Typical sensitivity
Scintillator, glass-fiber coupled to Si-CCD	20–100 keV	0.5 e/keV
Fiber-coupled X-ray-sensitive fluorescence coating on Si-CCD	3–40 keV	5 e/keV
Direct detection with beryllium window on Si-CCD	3–15 keV	0.3 e/keV
Specially treated, thinned, back-illuminated Si-CCD	<30 nm 30 eV–8 keV	
Si-CCD with UV fluorescence coating and MgF_2 window	0.12–1.0 μm	< 0.4
Thinned, back-illuminated Si-CCD	0.25–1.0 μm	< 0.80
Standard silicon CCD or CMOS sensor	0.35–1.0 μm	0.1–0.65
GaAs focal plane array	0.9–1.7 μm	
Pt:Si focal plane array	1.4–5.0 μm	0.001–0.01
InSb focal plane array	1.0–5.5 μm	< 0.85
HgCdTe (MCT) focal plane array	2.5–5.0 or 8.0–12.0 μm (tunable)	
GaAs/AlGaAs quantum well photodetectors (QWIP)	3.0–5.0 or 8.0–10.0 μm (tunable)	Low
Antimonide superlattice (SL)	3.0–10.0 μm (tunable)	

24.3.2 CMOS Imaging Sensors

Active Pixel Sensors

Despite the success of CCD imaging sensors, there are valuable alternatives [24.2]. With the rapidly decreasing sizes of transistors on semiconductors, it is also possible to give each pixel its own preamplifier and possibly additional circuits. Such a pixel is known as an *active pixel sensor (APS)*. In this case transfer of the accumulated photocharge is no longer required. The voltage generated by the circuits of the APS is just sensed by appropriate selection circuits. In contrast to a CCD sensor, it is possible to read out an arbitrary part of the sensor without any speed loss and with the same pixel clock rate.

APS technology is very attractive because the imaging sensors can be produced in standard complementary metal oxide semiconductor (CMOS) technology and additional analog and digital circuits can be integrated onto the chip. This opens the way to integrate the complete functionality of a camera on a single chip including analog-to-digital converters.

High-Speed CMOS Sensors

CMOS sensors offer the significant advantage that massively parallel read-out circuits can easily be added to the sensor. Therefore high-speed cameras are dominated today by CMOS sensors Table 24.1. The frame rates for high-speed sensors are not limited by the CMOS technology itself but rather by the transfer of the digital image data (Sect. 24.3.4).

24.3.3 CCD Sensor Architectures

Frame Transfer

Frame-transfer CCD sensors (Fig. 24.3) use the photo accumulation sites also as charge transfer registers. At the end of the exposure time, the whole frame is transferred across the whole illuminated area into an optically isolated frame storage area, where it is read out row by row with another horizontal shift register, while in the meantime the next exposure takes place in the illuminated frame array. During the transfer phase, the sensor must not be illuminated. Therefore illumination with either a flashlight or a mechanical shutter is required. The required dark period is rather short compared to the readout time for the whole area because it is approximately equal to the time required to read out just one row.

For large-area CCD sensors, the extra storage area for the frame transfer can no longer be afforded. Then,

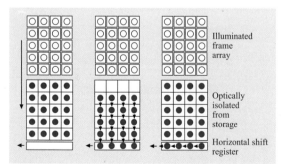

Fig. 24.3 Frame integration with a frame-transfer CCD sensor

Table 24.2 Selection of monochrome CCD imaging sensors, C: charge saturation capacity in electrons, eNIR: enhanced NIR sensitivity, FR: frame rate in s^{-1}, ID: image diagonal in mm, QE: peak quantum efficiency, Sony (ICX...) and Kodak (KAI...) sensors

Chip	Format $H \times V$	FR	ID	Pixel size $H \times V (\mu m)$	Comments
Interlaced American video (EIA standard)					
ICX278AL 1/4"	768 × 494	30	4.56	4.75 × 5.55	eNIR
ICX258AL 1/3"	768 × 494	30	6.09	6.35 × 7.4	eNIR
ICX248AL 1/2"	768 × 494	30	8.07	8.4 × 9.8	eNIR
ICX422AL 2/3"	768 × 494	30	11.1	11.6 × 13.5	
Interlaced European video (CCIR standard)					
ICX279AL 1/4"	752 × 582	25	4.54	4.85 × 4.65	eNIR
ICX259AL 1/3"	752 × 582	25	6.09	6.5 × 6.25	eNIR
ICX249AL 1/2"	752 × 582	25	8.07	8.6 × 8.3	eNIR
ICX423AL 2/3"	752 × 582	25	10.9	11.6 × 11.2	
Progressive scanning interline					
ICX098AL 1/4"	659 × 494	30	4.61	5.6 × 5.6	
ICX424AL 1/3"	659 × 494	30	6.09	7.4 × 7.4	
ICX074AL 1/2"[a]	659 × 494	40	8.15	9.9 × 9.9	C 32k, QE 0.43 at 340 nm
ICX414AL 1/2"	659 × 494	50	8.15	9.9 × 9.9	C 30k, QE 0.40 at 500 nm
ICX075AL 1/2"	782 × 582	30	8.09	8.3 × 8.3	
ICX204AL 1/3"	1024 × 768	15	5.95	4.65 × 4.65	
ICX205AL 1/2"	1360 × 1024	9.5	7.72	4.65 × 4.65	C 13k
ICX285AL 2/3"	1360 × 1024	10	11.0	6.45 × 6.45	C 18k, QE 0.65 at 500 nm
ICX085AL 2/3"[a]	1300 × 1030	12.5	11.1	6.7 × 6.7	C 20k, QE 0.54 at 380 nm
ICX274AL 1/1.8"	1628 × 1236	12	8.99	4.4 × 4.4	
ICX625ALA 2/3"	2456 × 2058	15	11.0	3.45 × 3.45	
KAI-0340DM 1/3"	640 × 480	200	5.92	7.4 × 7.4	C 20k, QE 0.55 at 500 nm
KAI-1010M	1008 × 1018	30	12.9	9.0 × 9.0	QE 0.37 at 500 nm
KAI-1020M	1000 × 1000	49	10.5	7.4 × 7.4	C 42k, QE 0.45 at 490 nm
KAI-2001M	1600 × 1200	30	14.8	7.4 × 7.4	C 40k, QE 0.55 at 480 nm
KAI-4020M	2048 × 2048	15	21.4	7.4 × 7.4	C 40k, QE 0.55 at 480 nm
KAI-11002M	4008 × 2672	3	43.3	9.0 × 9.0	C 60k, QE 0.50 at 500 nm
KAI-16000M	4872 × 3248	3	43.3	7.4 × 7.4	C 30k
Electron-multiplying frame-transfer CCD					
E2V CCD97-00fi	512 × 512	40	11.6	16.0 × 16.0	C 130k, QE 0.46 at 720 nm
E2V CCD97-00bi	512 × 512	40	11.6	16.0 × 16.0	C 130k, QE 0.92 at 570 nm, thinned and back illuminated
TI TC285SPD-30	1004 × 1002	13	11.3	8.0 × 8.0	C 70k, QE 0.60 at 680 nm
In situ storage high-speed image sensors (ISIS)					
Shimadzu IS-CCD	312 × 260	10^6	26.9	66.3 × 66.3	C 25k, stores up to 100 frames on chip, 13% fill factor for photogate

[a] no longer available. Sources: http://www.framos.de, http://www.kodak.com/global/en/digital/ccd/, http://www.pco.de, http://www.e2v.com, http://www.schimadzu.com, and [24.3]

the frame is directly read out. This variant of frame-transfer sensors is called a *full-frame transfer* sensor. It has the disadvantage that the chip must not be illuminated during the whole read-out phase. This means that the exposure time and the read out time must not overlap. Most scientific-grade CCD sensors and those used in consumer and professional digital cameras are full-frame transfer sensors.

Interline Transfer

Because frame-transfer sensors are not suitable for continuous illumination and high frame rates, the most common scan mode is the more complex *interline transfer*. Such a sensor has charge storage sites between the lines of photosensors. At the end of an exposure period, the charges collected at the photosensors are transferred to these storage sites. Then a two-stage transfer chain follows. First, the charge packets are shifted down the vertical interline storage registers. The charge package in the lowest cell of the vertical shift register is then transferred to a horizontal shift register at the lower end of the CCD chip, where a second transfer of an image row is shifted out to a charge-sensitive amplifier to form a time-serial analog video signal.

Electronic Shutter

A very useful feature is the *electronic shutter*. With electronic shuttering, the exposure time can be limited to shorter times than a field or frame duration. This is achieved by draining the accumulating charges at the beginning of the exposure time. Accumulation (and thus the exposure) starts when the draining is stopped and lasts until the end of the normal exposure time. Electronic shutter times can be as short as a few µs.

High Speed with In Situ Storage

At the current state of the art, the serial read-out speed of CCD sensors is limited to several 10 MHz. Even with up to four output taps, maximal pixel rates are only about 100 MPixel/s. A 1024×1024 sensor with 1000 frames/s, however, would require 1000 MPixel/s.

This principle limitation can be overcome if multiple charge storages are integrated for each sensor element on the chip. The price to be paid is a rather short image sequence and a small fill factor for the photosensitive area. Recently a 312×260 IS-CCD sensor has been introduced that can capture 100 consecutive images at 1 000 000 frames/s ([24.3], Table 24.2). This chip has large pixels ($66-66\,\mu m^2$) with a fill factor of 13% and a maximum effective pixel transfer rate into the on-chip charge storage of 80 000 MPixel/s.

Microlens Arrays

Interline transfer sensors have the disadvantage that only a rather small fraction, typically 20–30%, of the sensor element is light sensitive because the main area of the sensor is required for the interline storage areas and other circuits. Arrays of microlenses can overcome this disadvantage. With such an area, each photosensor is covered by a microlens, effectively enlarging the light-collecting area and thus sensitivity by a factor of 2–3.

This enhanced sensitivity comes at the price of a smaller acceptance cone for incoming light. Therefore, an imaging sensor with microlenses may show a lower than expected sensitivity with high-aperture lenses (f-numbers smaller than 2) and show a stronger fall-off towards the edge of the array for wide-angle lenses with short focal lengths. In addition, microlenses limit UV sensitivity. For many applications, the advantages of microlens arrays overcome their disadvantages. They have boosted the effective *quantum efficiency* of interline transfer CCD sensors to 65% (Fig. 24.11a).

Color and Spectral Sensors

For color imaging, either three images sensors are required – one each for red, green, and blue – in a three-CCD camera, or sensor elements with different color sensitivities are used by integration of a color filter above the photosensing array and below the microlens. The most common color filter set up is the Bayer pattern. In a 2×2 pixel area, the upper left and lower right pixel have a green filter, the upper right a red, and the lower left a blue filter. Microlenses and filters make a sensor element in a solid-state imager a complex electrooptical system (Fig. 24.4).

Imaging sensors with custom color sensitivities are possible in principle, but very expensive. Thus the standard approaches either use several monochrome sensors with dichroitic beam splitters for simultaneous acquisition or a single camera with a filter wheel or a con-

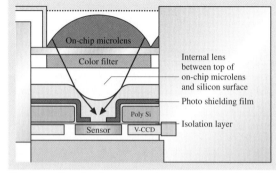

Fig. 24.4 Sketch of a sensor element of a modern CCD imager with a microlens and a color filter

Table 24.3 Selection of CMOS imaging sensors. C: charge saturation capacity in electrons, FR: frame rate in s^{-1}, PC: pixel clock in MHz, QE: peak quantum efficiency, HDRC: high dynamic range camera

Chip	Format $H \times V$	FR	PC	Pixel size $H \times V$ (μm)	Comments
Linear response					
Micron[a] MT9V403	656 × 491	200	66	9.9 × 9.9	QE 0.32 at 520 nm
Fillfactory[b] IBIS54-1300	1280 × 1024	30	40	6.7 × 6.7	QE 0.30–0.35 at 600 nm, C 60k
Fillfactory[b] IBIS4-4000	2496 × 1692	4.5		11.4 × 11.4	C 150k
Fast-frame-rate linear response					
Fillfactory[b] LUPA1300	1280 × 1024	450	40	12.0 × 12.0	16 parallel ports
Micron[a] MT9M440	2352 × 1728	240	80	7.0 × 7.0	16 parallel 10-bit ports
Micron[a,c] MT9M413	1280 × 1024	600	80	12.0 × 12.0	QE 0.27 at 520 nm, C 63k, 10 parallel 10-bit ports
Photron APX-RS[d]	1024 × 1024	3000	?	17.0 × 17.0	?
Logarithmic response					
IMS, Stuttgart HDRC[e]	640 × 480	30	10	12.0 × 12.0	
PhotonFocus EMPHIS-300[f,g]	748 × 480	60	20	10.6 × 10.6	C 200k
PhotonFocus[f,g]	1024 × 1024	150	80	10.6 × 10.6	QE 0.29 at 600 nm, C 200k

[a] http://www.framos.de, [b] http://www.fillfactory.com, [c] http://www.pco.de, [d] http://www.photron.com, [e] http://www.ims-chips.de, [f] http://www.photonfocus.com, [g] linear response at low light levels with adjustable transition to logarithmic response

trollable bandpass filter for consecutive acquisition of different color channels. Spectroscopic imaging or hyperspectral imaging uses a dispersive element, such as a grid or a prism, to map the wavelength onto one spatial coordinate of the sensor. In this way, however, spectral data can only be taken along one spatial coordinate. The acquisition of spectral images requires scanning in the missing spatial direction.

24.3.4 Standard Interfaces for Digital Cameras

Unfortunately, there is no unique standard available to transfer the serial stream of digital image data from a camera to a computer. Besides proprietary digital image data communication, four standard digital links are used to connect digital cameras to computers.

Camera Link
The oldest standard is the Camera Link standard. It is based on the serial communication protocol known as Channel Link and uses several such serial channels to transfer image data and control signals. Depending on the number of serial channels used in parallel (base, medium or full configuration) 24, 48 or 64 bits can be transported with clock rates of 20–60 MHz. At a clock rate of 40 MHz, this results in data rates of 120, 240, and 320 MB/s.

Camera Link is the fastest available standard. Standard personal computer (PC) bus systems such as the peripheral component interconnect (PCI) bus can only cope with the base configuration; the medium and full configuration require faster bus systems such as the recently introduced PCI express bus. This fact underlines a serious handicap of high-speed imaging. While affordable high-speed image sensors are available with serial pixel rates of more than 1000 Mpixel/s (Table 24.3), standard peripheral bus systems cannot handle these fast data streams. This problem makes high-speed imaging expensive. An intermediate storage system is required and limits the continuously recordable image sequences currently to several GB of image data.

Another big disadvantage of the Camera Link standard is that only the physical layer including the connectors are standardized. The way in which the camera is controlled, e.g., the exposure time or trigger mode, is still proprietary and requires the adaption of software drivers for each manufacturer or even camera.

Firewire and USB2
These two standards connect digital cameras to personal computers via standard bus systems, which are used for many other peripheral devices. Their advantage is their low cost. Even the power supply for the camera is provided by the bus systems. The disadvantages are the low maximum throughput of about 30–40 MBytes/s

and the short connection lengths of 5 m without bus repeaters.

While with the Firewire or IEEE1394 buses at least a standard is defined for handling the basic functionality of cameras (DCAM, http://www.1394ta.org), standardization does not exist at all for USB2. With the IEEE1394b extension of the Firewire standard, some disadvantages have been removed; the maximum throughput is doubled and with optical fiber connections large distances between the camera and PC can be bridged.

GigE Vision and GenICam
The latest development is the GigE Vision standard that takes advantage of the high data rates possible with gigabit Ethernet networking connections. A significant advantage of this new standard is the fact that the vision industry has finally agreed to develop a generic programming interface to handle digital cameras independent of the interface technology. Camera Link, IEEE1394 DCAM, and GigE Vision are supported by this standard (http://www.emva.org, http://www.GenICam.org).

24.4 Performance of Imaging Sensors

This section introduces the basic terms that describe the response of detectors to incident radiation. Most of theses quantities characterize an individual sensor, but some also describe the nonuniformity of an array of sensors in an imaging sensor [24.4–6].

24.4.1 Responsivity

The basic quantity that describes the sensitivity of a sensor to radiation is called its *responsivity* R. The responsivity is given as the ratio between the flux Φ incident on the detector area and the resulting signal s:

$$s = R\Phi . \tag{24.1}$$

A good detector shows a constant responsivity over a wide range of radiative fluxes. A constant responsivity means that the signal is directly proportional to the incident flux. The units of responsivity are either A/W or V/W, depending on whether the signal is measured as an electrical current or voltage, respectively.

24.4.2 Quantum Efficiency

The term to describe the responsivity of a sensor is the *quantum efficiency* (QE). It relates to the particulate nature of electromagnetic radiation and the electric signal. The quantum of electromagnetic radiation, the photon, carries an energy $h\nu$, where h is Planck's constant and ν is the frequency of the radiation. The quantum of electric charge is the elementary charge e. So the quantum efficiency η is the ratio of induced elementary charges e, N_e, and the number of incident photons N_p:

$$\eta = N_e/N_p . \tag{24.2}$$

For low-energy radiation, the quantum efficiency is always lower than one for photosensors where one photon can generate at most one elementary charge. Not all incident photons generate a charge unit, because some are reflected at the sensor surface, some are absorbed in a region of the sensor where no electric charges are collected, and some are transmitted through the sensor without being absorbed. For high-energy radiation, e.g., ultraviolet (UV) and X-ray radiation absorbed in silicon, many charge units can be generated, depending on the energy of the radiation and the properties of the sensor material.

The elementary relation between the quanta for electromagnetic radiation and electric charge can be used to compute the responsivity of a detector. The radiative flux measured in photons is given by

$$\Phi_e = h\nu \, dN_p \, dt . \tag{24.3}$$

The resulting current I is given as the number of elementary charges per unit time multiplied by the elementary charge

$$I = e\frac{dN_e}{dt} , \tag{24.4}$$

and the responsivity as

$$R(\lambda) = \frac{I}{\Phi_e} = \eta(\lambda)\frac{e}{hc}\lambda \approx 0.8066\eta(\lambda)\lambda \left[\frac{A}{W\mu m}\right] . \tag{24.5}$$

The responsivity increases linearly with wavelength, as radiation is quantized into smaller units for larger wavelengths. Thus, a detector for infrared radiation in the 3–5 μm range is intrinsically about 10 times more sensitive than a detector for light. For practical purposes, it is important to realize that the responsivity of a detector is weakly dependent on many parameters. Among others, these include the angle of the incident light, the temperature of the detector, fatigue, and aging.

24.4.3 Signal Irradiance Relation

As illustrated in Fig. 24.1, the irradiation collected by a photosensor is finally converted into a digital number g at each sensor element of a sensor array. Thus the overall gain of the sensor element can be expressed by a single digital gain constant K that relates the number of collected charge units to the digital number g:

$$g = KN_e \, . \tag{24.6}$$

The digital gain constant K is dimensionless and indicates the number of digits per unit charge. An ideal charge unit counting sensor has a digital gain constant K of one.

The incident radiative flux Φ_e is received by the sensor for an exposure time t on an area A. Therefore the received radiant energy (energy-based Q_e or photon-based N_p) can be related to the irradiance incident on the sensor by

$$Q_e = \Phi_e t = AEt \quad \text{or} \quad N_p = At\frac{\lambda}{hc}E \, . \tag{24.7}$$

Using (24.2) and (24.6), the relation between the incident irradiance and the digital signal g is

$$g = K\frac{A}{hc}\eta(\lambda)\lambda E(\lambda)t \, . \tag{24.8}$$

Normally a sensor does not receive monochromatic radiation so that (24.8) must be integrated over the spectral irradiance E_λ:

$$g = K\frac{At}{hc}\int_{\lambda_1}^{\lambda_2}\eta(\lambda')\lambda' E_\lambda(\lambda')\,\mathrm{d}\lambda' \, . \tag{24.9}$$

24.4.4 Dark Current

Generally, a detector generates a signal s_0 even if the incident radiative flux is zero. In a current-measuring device, this signal is called the *dark current*. A sensor with a digital output will show a digital signal g_0 without illumination.

Thermal energy can also excite electrons into the conduction band and thus gives rise to a *dark current* even if the sensor is otherwise perfect. The probability for thermal excitation is proportional to $\exp[-\Delta E/(k_B T)]$, where ΔE is the energy difference across the band gap.

The appropriate method to limit the dark current is cooling. Because of the exponential dependency of the dark current on the absolute temperature, the dark current decreases strongly with temperature, by about a factor of two per 7–10 K. Imaging sensors cooled to liquid-nitrogen temperatures can be illuminated for hours without noticeable dark current.

24.4.5 Noise-Equivalent Exposure

The dark signal has not only a direct-current (DC) component but also a randomly fluctuating component with a standard deviation σ_0 caused by the both the statistical fluctuations of the dark current and the noise of the electronic circuits. Using the responsivity, σ_0 can be converted into an equivalent radiative flux known as the *noise-equivalent exposure* or NEE,

$$H_0 = \sigma_0/R \, . \tag{24.10}$$

The NEE is related to the certain frequency band to which the detector is responding and essentially gives the minimum exposure that can be measured with a detector in that band.

24.4.6 Saturation Equivalent Exposure

A photosensor is also limited to a maximum signal. For a digital output this is simply the largest digital number that is delivered by the analog-to-digital converter (ADC), which is $g_s = 2^D - 1$ for an ADC with D bits. As with the NEE, a *saturation-equivalent exposure* or SEE can be defined by

$$H_s = \frac{g_s}{R} \, . \tag{24.11}$$

There is also a physical limit for the maximum signal of a photosensor because it can only store a limited number of photoelectrons (the *full-well capacity*). This limit determines the maximum possible exposure. For an ideal sensor with a quantum efficiency of one, the maximum number of electrons is equal to the number of

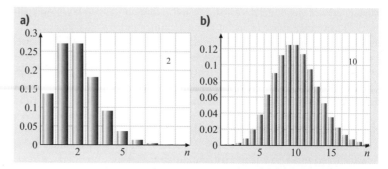

Fig. 24.5 Poisson distribution (24.14) describing the counting statistics for photons for values N as indicated. With increasing $\bar{n}$, the distribution quickly becomes symmetrical and approaches the Gaussian distribution (24.16)

photons that can be detected in one exposure by a sensor element.

If we know the quantum efficiency and the electron capacity, it is easy to compute the saturation exposure H_s of a sensor element:

$$H_s = \eta^{-1} N_s h\nu, \qquad (24.12)$$

where N_s is the full-well capacity of the sensor, and $h\nu$ is the energy of the photon. For light at the frequency of maximum sensitivity of the eye ($\nu = 5.4 \times 10^{14}$ Hz, $\lambda = 555$ nm), with an electron capacity of $N_e = 100\,000$, and a quantum efficiency of 80%, the saturation exposure is about 4×10^{-14} J. For pixels with a size of $10\,\mu\text{m} \times 10\,\mu\text{m}$, this translates into a saturation irradiance of 10^{-6} W/cm^2 for an exposure time of 40 ms.

24.4.7 Dynamic Range

The *dynamic range* (DR) of an image sensor is the ratio of the maximum output signal g_s to the standard deviation of the dark signal σ_0 when the sensor receives no irradiance. The dynamic range is often expressed in units of decibels (dB) as

$$\text{DR} = 20 \log \frac{g_s}{\sigma_0}. \qquad (24.13)$$

A dynamic range of 60 dB thus means that the saturation signal is 1000 times larger than the standard deviation of the dark signal.

24.4.8 Photon-Noise-Limited Performance

An ideal detector would not introduce any additional noise. Even then, the detected signal is not noise-free, because the generation of photons itself is a random process. Whenever the detector related noise is below

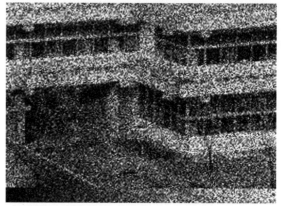

Fig. 24.6 A low-light image taken with a maximum of 10 photons

the photon noise, a detector is said to have *photon-noise-limited* performance. Photon generation is (like radioactive decay) a random process with a *Poisson distribution*. The probability density function (PDF) is given by

$$p_P(n) = \frac{N^n \exp(-N)}{n!}. \qquad (24.14)$$

This equation gives the probability that n photons are detected in a certain time interval when on average N are detected. The variance of the Poisson distribution σ_n^2,

$$\sigma_n^2 = \sum_n p_P(n)(n-N)^2 = N, \qquad (24.15)$$

is equal to the average value N.

For low mean values, the Poisson PDF is skewed with a longer tail towards higher values Fig. 24.5. However, even for moderate N, the Poisson distribution quickly converges to the continuous *Gaussian distribution* (also known as the *normal distribution*) given by

$$p_n(x) = \exp{-\frac{(x-N)^2}{2N}} \qquad (24.16)$$

with an equal mean value and variance. Figure 24.6 shows an example of low-light images taken with a low number of photons.

Note that, if a sensor element can store even 1 000 000 electrons, the standard deviation of the irradiance measurement is not better than 0.1%, even if the only noise source of the sensor is the photon noise. CCD sensors have an electron capacity of 10 000–500 000 electrons; only devices sensitive in the infrared range with larger sensor elements have a significantly higher capacity of up to 10 million electrons.

24.4.9 Linear Noise Model for Image Sensors

The photosignal for a single pixel is Poisson distributed as discussed in the previous section. The electronic circuits add a number of other noise sources. For practical purposes, it is however only important to know that these noise sources are normal distributed and independent of the photon noise and have a variance of σ_d. According to the laws of error propagation, the noise variances add up linearly. Therefore the noise variance of the total number of generated charge units $N = N_0 + N_e$ is given by

$$\sigma_N = \sigma_d^2 + \sigma_{N_e}^2 = \sigma_{N_0}^2 + N_e \,. \qquad (24.17)$$

According to (24.18), the digital signal is given as

$$g = g_0 + KN \,. \qquad (24.18)$$

An arbitrary signal g_0 is added here to account for an offset in the electronic circuits. Then the variance σ_g^2 can be described by only two terms as

$$\sigma_g^2 = K^2(\sigma_d^2 + N_e) = \sigma_0^2 + K(g - g_0) \,. \qquad (24.19)$$

The term $\sigma_0^2 = K^2 \sigma_d^2$ includes the variance of all non-signal-dependent noise sources. The units of σ_d and σ_d are number of electrons (e$^-$) and digits (DN). Equation (24.19) predicts a linear increase of the noise variance with the measured gray value g. In addition it can be used to measure the absolute gain factor K (photon transfer method [24.4]).

24.4.10 Signal-to-Noise Ratio

The responsivity is not a good measure for a sensor, because a sensor with a high responsivity can be very noisy.

The precision of the measured irradiance rather depends on the ratio of the received signal and its standard deviation σ_g. This term is known as the *signal-to-noise ratio (SNR)*:

$$\text{SNR} = \frac{g - g_0}{\sigma_g} = \frac{g'}{\sigma_g} \,. \qquad (24.20)$$

The inverse SNR σ_g/g' gives the relative resolution of the irradiance measurements. A value of $\sigma_g/g' = 0.01$ means, e.g., that the standard deviation of the noise level corresponds to a relative change in the irradiance of 1%.

The discussion in the previous section has shown that the most important quantity of an image sensor is the relation of the signal quality expressed by the signal-to-noise ratio to the irradiation H. Using (24.18) and (24.19), this relation can be expressed in the simplest way by

$$\text{SNR} = \frac{N_e}{\sqrt{\sigma_d^2 + N_e}} = \frac{N_p}{\sqrt{\eta^2 \sigma_d^2 + \eta N_p}} \,. \qquad (24.21)$$

At high irradiation ($Kg' \gg \sigma_0^2$) this equation reduces to

$$\text{SNR} = \sqrt{\frac{N_p}{\eta(\lambda)}} \qquad (24.22)$$

and at low irradiation to

$$\text{SNR} = \frac{N_p}{\eta(\lambda)\sigma_d} \,. \qquad (24.23)$$

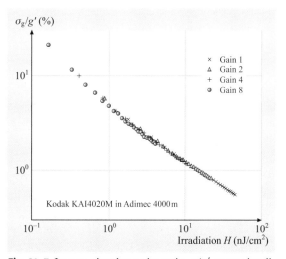

Fig. 24.7 Inverse signal-to-noise ratio σ/g' versus irradiation for an Adimec A4000m with a Kodak KAI-4020M sensor at four different gains as indicated

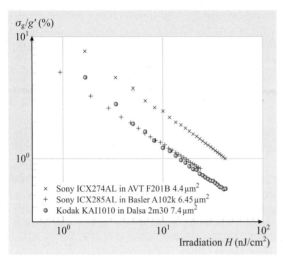

Fig. 24.8 Influence of pixel size on the relation between the inverse signal-to-noise ratio σ/g' and the irradiation

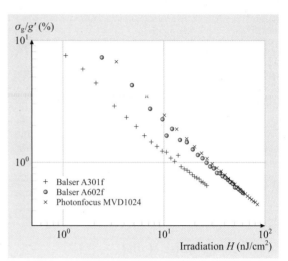

Fig. 24.10 Comparison of interline CCD (Basler A301f) and CMOS (Basler A602f, Photofocus MVD 1024) sensors with pixels of the same size

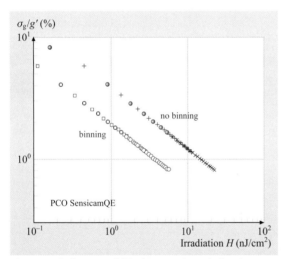

Fig. 24.9 Influence of binning on the relation between the inverse signal-to-noise ratio σ/g' and the irradiation

The number of photons perceived by a sensor element can be computed from the irradiance, area of the sensor element, and the exposure time according to (24.7).

These equations give surprisingly simple and general answers for the most important quality parameters of imaging sensors. They clearly show that the SNR increases at high irradiance levels with the square root of the quantum efficiency and the area of the sensor element. The SNR does not depend on the amplification factor K at all. Increasing the amplification only increases the signal but not its quality.

Only recently, the European Machine Vision Association (EMVA, http://www.emva.org/) established the EMVA1288 standard for a comparable characterization of cameras and imaging sensors. The results of some measurements of the SNR as a function of the irradiation are discussed in the following. First, Fig. 24.7 proves that the gain does not influence the relation between the SNR and irradiation at all. At higher gain lower irradiations are reached, but at the price of a worse SNR. The values at the highest possible irradiation give the best SNR that is possible with a given camera.

Figure 24.8 demonstrates that sensors with large pixel areas generally result in a higher SNR because larger pixel have a larger full-well capacity Sect. 24.4.6.

The sensitivity of an imaging sensor can be significantly increasing by a technique called *binning* Fig. 24.9. Here the charges of pixels of a small neighborhood, e.g., a 2×2 neighborhood, are collected together. This reduces the spatial resolution in horizontal and vertical direction by a factor of two, but increases the sensitivity by a factor of four. Note that binning does not generally increase the SNR.

Finally, Fig. 24.10 compares CCD and CMOS sensors with the same pixel size. The two CMOS sensors show about two times lower sensitivity, because these two sensors do not use microlens arrays to increase the

Table 24.4 Imaging sensors for the infrared (IR), C: full-well capacity in millions of electrons [Me], IT: integration time, NETD: noise-equivalent temperature difference, QE: peak quantum efficiency

Chip	Format $H \times V$	FR	PC	Pixel size $H \times V (\mu m)$	Comments
Near-infrared (NIR)					
Indigo[1] InGaAs	320 × 256	345		30 × 30	0.9–1.68 μm, C 3.5 Me
Mid-wave infrared (MWIR)					
AIM[2] PtSi	640 × 486	50	12	24 × 24	3.0–5.0 μm, NETD < 75 mK at 33 ms IT
Indigo[1] InSb	320 × 256	345		30 × 30	2.0–5.0 μm, C 18 Me
Indigo[1] InSb	640 × 512	100		25 × 25	2.0–5.0 μm, C 11 Me
AIM[2] HgCdTe	256 × 256	800	20	40 × 40	3.0–5.0 μm, NETD < 20 mk at 1.2 ms IT, C 14 Me
AIM[2] HgCdTe	384 × 288	120	20	24 × 24	3.0–5.0 μm, NETD < 20 mk at 2 ms IT
AIM[2]/IaF FhG[3] QWIP	640 × 512	30	18	24 × 24	3.0–5.0 μm, NETD < 15 mk at 20 ms IT
Long-wave infrared (LWIR)					
AIM[2] HgCdTe	256 × 256	200	16	40 × 40	8–10 μm, NETD < 20 mk at 0.35 ms IT
Indigo[1] QWIP	320 × 256	345		30 × 30	8.0–9.2 μm, C 18 Me, NETD < 30 mk
AIM[2]/IaF FhG[3] QWIP	256 × 256	200	16	40 × 40	8.0–9.2 μm, NETD < 8 mk at 20 ms IT
AIM[2]/IaF FhG[3] QWIP	640 × 512	30	18	24 × 24	8.0–9.2 μm, NETD < 10 mk at 30 ms IT
Dual band					
AIM[2] QWIP	388 × 284	100	80	40 × 40	4.8 and 8.0 μm peak, NETD < 25 mk
Uncooled sensors					
Indigo[1] Microbolometer	320 × 240	60		30 × 30	7.0–14.0 μm, NETD < 120 mk
FLIR[1] ThermaCAM 640	640 × 480	30		?	7.5–13.0 μm, NETD < 80 mk

Sources: [1] http://www.flir.com, [2] http://www.aim-ir.de, [3] http://www.iaf.fhg.de

light-collecting area. However, the SNR is significantly better. This indicates that these two CMOS sensors have a higher full-well capacity than the corresponding interline CCD sensor with the same pixel area.

24.4.11 Spectral Sensitivity

The band gap of the semiconductor material used for the detector determines the minimum energy of the photon to excite an electron into the conduction band and thus the maximum wavelength that can be detected. For silicon, the threshold given by the band gap is at about $1.1\,\mu\mathrm{m}$. Thus, imaging detectors based on silicon are also sensitive in the near-infrared. Other materials are required to sense radiation at larger wavelengths (Table 24.1). The response for shorter wavelengths largely depends on the design. Standard imagers are sensitive down to about 350 nm. Then, the photons are absorbed in such a short distance that the generated electrons do not reach the accumulation regions in the imagers. With appropriate design techniques, however, imaging silicon sensors sensitive for ultraviolet radiation and even X-rays can be made (Table 24.1).

An ideal photodetector has a *quantum efficiency* of unity. This means that each photon irradiating the sensor generates an electron. Real devices come quite close to this ideal response. Even standard commercial devices have peak quantum efficiencies of about 0.65 (Fig. 24.11a) while scientific-grade devices may reach quantum efficiencies of up to 0.95 (Fig. 24.11b). This can be achieved by illuminating a thinned sensor from the back and by using appropriate antireflection coating. In this way it is also possible to extend the sensitivity far into the ultraviolet.

Modern CCD sensors do not show much variation in the spectral sensitivity in the visible range (Fig. 24.11). There are, however, significant differences in the ultraviolet and infrared spectral range.

The sensitivity in the ultraviolet (UV) range is generally low but there are special sensors available with enhanced UV sensitivity (Fig. 24.11). Note that the UV sensitivity is also influenced by the material used for the glass window protecting the CCD sensor. Standard CCD sensors do not use a quartz glass window and thus typically cut off the sensitivity for wavelengths below 360 nm.

All sensors are also sensitive in the near-infrared (near-IR) range, but the sensitivity does not extend beyond 1000 nm, because silicon becomes transparent at larger wavelengths. There is a general tendency for the IR sensitivity to decrease with the size of the pixels. There are special sensors available with enhanced IR sensitivity (Fig. 24.11). Many camera vendors, however, integrate an infrared cut-off filter to limit the sensitivity of the CCD sensor to the visible range.

The spectral sensitivity of CMOS sensors (Table 24.3) typically differs somewhat from that of CCD sensors. The peak sensitivity is shifted from about 550 nm to about 650 nm and the sensitivity is higher in the infrared. Thus, one cannot generally say that current CMOS sensors are less sensitive than CCD sensors. A direct comparison of the Sony ICX074AL CCD sensor with the Micron MT9V403 CMOS sensors, e.g., shows

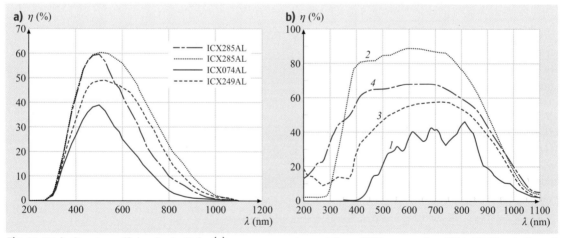

Fig. 24.11a,b Quantum efficiencies in % for **(a)** some commonly used Sony CCD sensors (see Table 24.2) as indicated, data courtesy of PCO AG, Kelheim, Germany. **(b)** Scientific-grade CCD sensors as indicated (data from Scientific Imaging Technologies, Inc. (SITe), Beaverton, Oregon)

that the CMOS sensor is about two times less sensitive at 550 nm but more sensitive at wavelengths larger than 750 nm.

Sensors that are sensitive in the mid (3.0–5.0 μm) and long (8.0–14.0 μm) infrared range for thermal sensing require special and expensive semiconductor materials with lower band gaps (Tables 24.1, 24.4, [24.7]). In order to limit the dark current (Sect. 24.4.4), it is also required to cool the sensors down to cryogenic temperatures (60–100 K). Only thermal detector array-based thermoconductivity (bolometer arrays) and pyroelectric sensors can be operated at room temperatures. However, these sensors are typically one order of magnitude less sensitive. The noise-equivalent temperature difference (NEDT) is only 70–100 mK, while it is 7–20 mK for quantum detectors.

24.4.12 Nonuniform Responsivity

For an imaging sensor that consists of an array of sensors, it is important to characterize the nonuniformity in the response of the individual sensors since it significantly influences the image quality. Both the dark signal and the responsivity can vary. The nonuniformity of the dark signal becomes evident when the sensor is not illuminated and is often referred to as the *fixed pattern noise* (FPN). This term is a misnomer. Although the nonuniformity of the dark current appears as noise, it is that part of the dark signal that does not fluctuate randomly but is a static pattern. The correct expression is *dark signal nonuniformity* (DSNU). When the sensor is illuminated, the variation in the responsivity of the individual sensors leads to an additional component of the nonuniformity. This variation of the image sensor under constant irradiance is called the *photoresponse nonuniformity* (PRNU).

If the response of the image sensors is linear, the effects of the nonuniform dark signal and responsivity can be modeled with (24.1) as

$$\mathbf{G} = \mathbf{R} \cdot \mathbf{H} + \mathbf{G}_0 , \tag{24.24}$$

where all terms in the equation are matrices.

The photoresponse nonuniformity σ_R is often simply given by the standard deviation:

$$\sigma_R^2 = \frac{1}{MN-1} \sum_{m=0}^{M-1} \sum_{n=0}^{N-1} (r_{m,n} - \overline{R})^2$$

$$\text{with} \quad \overline{R} = \frac{1}{MN} \sum_{m=0}^{M-1} \sum_{n=0}^{N-1} r_{m,n} . \tag{24.25}$$

24.4.13 Artifacts and Operation Errors

Given the complexity of modern imagers, the quality of the acquired images can seriously suffer by misadjustments or by simply operating them incorrectly. It can also happen that a vendor delivers a sensor that is not well adjusted. In this section, we therefore show some common misadjustments and discuss a number of artifacts limiting the performance of imaging sensors.

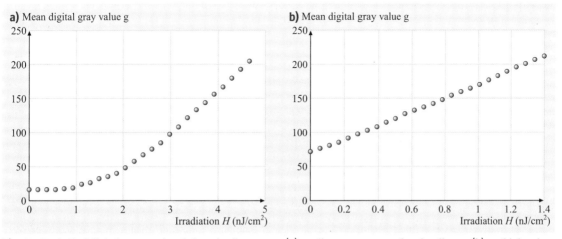

Fig. 24.12a,b Bad digital camera signals by misadjustments: **(a)** nonlinear response at low irradiance, **(b)** too high values for the dark image

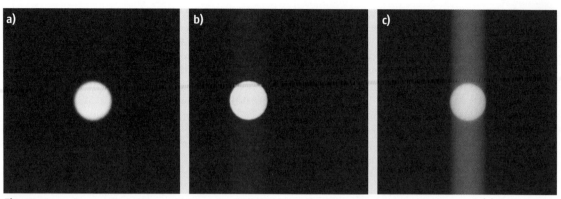

Fig. 24.13a–c Smear effects observed by constant illumination and a short exposure time of 20 μs with (**a**) PCO pixelfly scientific (Sony ICX285AL sensor, diameter of spot: about 80 pixel); (**b**) PCO pixelfly qe (Sony ICX285AL sensor, diameter of spot: about 80 pixel); (**c**) Basler A602f CMOS camera (Micron MT9V403 sensor, diameter of spot: about 54 pixel)

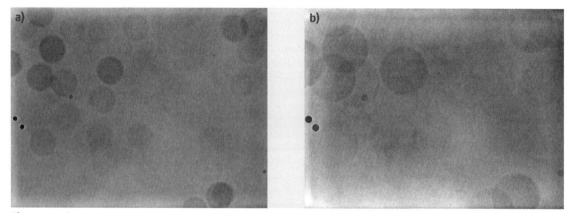

Fig. 24.14a,b Small dirt particles on the cover glass of a CCD sensor and the IR cut-off filter made visible by illuminating the sensor with uniform diverging light corresponding to an aperture of (**a**) 10.6, and (**b**) 6.1. The images are contrast enhanced with ranges that cover ±2% and ±1% change in intensity, respectively

Misadjustments

A serious misadjustment is a strong nonlinearity shown in (Fig. 24.12a). The trouble with such nonlinearity is that it can barely be detected by just observing images of the camera. In order to discover such a misadjustment, a sophisticated linearity measurement is required. It is much easier to detect and avoid too high an offset value, as shown in (Fig. 24.12b). Especially disturbing is an underflow. Then the digital gray values are zero below a critical irradiation and it appears that the dark image shows no noise.

Blooming and Smear

When short exposure times in the μs range are used with constant illumination, an artifact known as *smear* can be observed. This effect is caused by the tiny residual light sensitivity of the interline storage areas. At exposure times equal to the read-out time, this causes no visible effects. When the exposure time is only a small fraction of the read-out time (e.g., 20 μs versus 100 ms), additional charges are accumulated when a charge packet is transferred through an illuminated area. Therefore vertical streaks are observed when a small area is illuminated (Fig. 24.13a,b). In contrast to CCD sensors, CMOS sensors do not show smear effects (Fig. 24.13c).

Another artifact, known as *blooming*, occurs in overexposed areas, where a part of the generated charges leaks to neighboring pixels. Modern CCD imaging sensors include antiblooming circuits so that this effect

Dirt on the CCD Cover Glass

A trivial but ubiquitous flaw is dirt on the cover glass of the CCD sensor. Dirt can significantly contribute to the inhomogeneity of the sensor responsivity (PRNU, see Sect. 24.4.12. The images in Fig. 24.14 were taken by removing the lens from the camera and illuminating the chip with an integrated sphere. In this way, the dirt is directly projected onto the chip surface. The visibility of dirt in images strongly depends on the f-number. At lower f-numbers it is less visible because the projection onto the chip is very blurred due to the more-convergent light beams.

Cleaning of the window of a CCD imager is very difficult. Therefore the first rule is to keep dirt from settling on the window by removing the cover or lens of the camera only when required and then only in a clean environment.

24.5 Camera Selection

There is no generally valid answer to the question of what is the best imaging sensor. In Sect. 24.4 we have seen that there are many different quantities that describe the quality of an imaging sensor. The demands of the application determine which of these features are the most important. In the following, a number of typical demands in applications and their implication for camera selection are discussed.

24.5.1 Measurements at Low Light Levels

Intuitively, one would argue that in this case, a camera with a high responsivity is required. What really counts, however, is the signal-to-noise ratio (SNR) at *low irradiation* levels, see (24.23). This signal-independent dark noise depends most significantly on the read-out frequency of the pixels. A higher read-out frequency requires a higher bandwidth of the electronic circuits, which implies a higher noise variance. Modern CCD sensors, such as the Sony ICX285AL (Table 24.2), show σ_d values of only a few electrons (about 4–12) even for high read-out frequencies.

Dark noise levels below one electron can only be achieved with imaging sensors that amplify the generated charge unit in some way, so-called *intensified CCD sensors* (ICCD).

The newest development in this direction are electron-multiplying circuits (EM-CCD) that are integrated as a last stage of the charge-transfer chain before the charge is converted into a voltage (Table 24.2). In this way, the charge is multiplied but not the dark noise, leading to a suppression of σ_d proportional to the charge gain factor A. The price to be paid is that the randomness of the gain multiplication introduces an excess noise factor e, which is about 1.4. Once the dark noise is low compared to the photon noise, it appears as if the quantum efficiency of an EM-CCD would be $1.4^2 \approx 2$ times smaller than for a non-charge-multiplying sensor, see (24.22).

If long exposure times can be afforded, even low irradiance results in a high irradiation of the sensor. In this case the limiting factor for good image quality is a low *dark current* (Sect. 24.4.4). This requires cooling of the sensor. The Sensicam QE and Pixelfly QE cameras from PCO, for example, use the same Sony ICX285AL CCD chip. The lower dark current of the Sensicam QE, where the CCD chip is cooled down to $-12\,°C$, allows exposure times of up to 1000 s, whereas the much higher dark current of the uncooled chip degrades the image quality of the Pixelfly QE at exposure times as low as 10 s.

24.5.2 Measurements with High Irradiance Variations

Many applications are plagued by high irradiation variations. If the illumination conditions cannot be set up to avoid such large dynamic ranges, a good sensor requires not only a low σ_d but also a high *full-well capacity* so that a high *dynamic range* is the most important quality parameter Sect. 24.4.7.

24.5.3 Precise Radiometric and Geometric Measurements

If small irradiance differences or high-precision irradiance values must be measured, a high SNR is the most important parameter (Sect. 24.4.10). The SNR of the CCD and CMOS sensors included in (Table 24.2, 3) does not exceed values of 200 close to the saturation illumination. This means that the relative standard deviation of a single irradiance measurement at a single pixel is no better than 0.5%.

A critical parameter to compare irradiances measured at different pixels is the PRNU Sect. 24.4.12. For CCD sensors the standard deviation of the PRNU is typically 0.25–0.35%, for CMOS sensors it is no better than 0.5%. Therefore the PRNU does not significantly degrade the accuracy of irradiance measurements for single measurements. However, when either spatial or temporal averaging is performed this is no longer the case, and one must correct the PRNU by suitable calibration measurements.

References

24.1 P.W. Kruse: *Uncooled Thermal Imaging Arrays, Systems, and Applications* (SPIE, Bellingham, WA 2001)

24.2 P. Seitz: *Solid-State Image Sensing, Handbook of Computer Vision and Applications*, ed. by B. Jähne, H. Haußecker, P. Geißler (Academic, Sam Diego 1999)

24.3 T. Etoh, D. Poggemann, G. Kreider, H. Mutoh, A. Theuwissen, A. Ruckelshausen, Y. Kondo, H. Maruno, K. Takubo, K. Soya, K. Takehara, T. Okinaka, Y. Takano: An image sensor which captures 100 consecutive frames at 1 000 000 frames/s , IEEE T. Electron. Dev. **50**, 144–151 (2003)

24.4 J.R. Janesick: *Scientific Charge-Coupled Devices* (SPIE, Bellingham, WA 2001)

24.5 G. C. Holst: *CCD Arrays, Cameras, and Displays, 2nd ed.* (SPIE, Bellingham, WA 1998)

24.6 A. J. P. Theuwissen: *Solid-State Imaging with Charge-Coupled Devices* (Kluwer, Dordrecht 1995)

24.7 G. Gaussorgues: *Infrared Thermography* (Chapman Hall, London 1994)

25. Data Analysis

From the beginning of science, visual observation has played a major role. At that time, the only way to document the results of an experiment was by verbal description and manual drawings. The next major step was the invention of *photography* more than one and a half centuries ago, which enabled experimental results to be documented objectively. In experimental fluid mechanics, flow visualization techniques gave direct insight into complex flows, but it was very difficult and time consuming to extract quantitative measurements from photographs and films.

Nowadays, we are in the middle of a second revolution sparked by the rapid progress in both photonics and computer technology. Sensitive solid-state cameras are available that acquire digital image data, and standard personal computers and workstations have become powerful enough to process these data. These technologies are now available to any scientist or engineer. As a consequence, image processing has expanded and continues to expand rapidly from a few specialized applications into a standard scientific tool.

This chapter gives a brief presentation of some of the most important general image processing

- 25.1 **Image Processing** 1437
  - 25.1.1 Sampling and Quantization 1437
  - 25.1.2 Radiometric Corrections 1440
  - 25.1.3 Geometric Corrections 1442
  - 25.1.4 Averaging and Noise Suppression ... 1445
  - 25.1.5 Edge and Line Extraction 1451
  - 25.1.6 Direction and Orientation 1453
  - 25.1.7 Local Wavenumber and Local Phase 1458
  - 25.1.8 Multiscale Processing 1461
- 25.2 **Motion Analysis** 1464
  - 25.2.1 General Considerations on Motion Analysis 1464
  - 25.2.2 Correlation-Based Velocity Analysis 1469
  - 25.2.3 Least-Squares Matching 1473
  - 25.2.4 Tracking Techniques 1474
  - 25.2.5 Optical-Flow-Based Velocity Analysis 1481
- **References** 1488

techniques that are required to process image data in experimental fluid mechanics. The second section (Sect. 25.2) deals with motion analysis. The most important methods are introduced and classified according to the fundamental principles, assumptions and approximations upon which they are based.

25.1 Image Processing

25.1.1 Sampling and Quantization

Computers process digital numbers. Therefore, the final steps of digital image formation are *digitization* and *quantization*.

Sampling Theorem
Digitization means sampling the gray values at a discrete set of points, which can be represented by a matrix. Sampling may already occur in the sensor that converts the collected photons into an electrical signal. A CCD camera already has a matrix of discrete sensors. Each sensor is a sampling point on a two-dimensional (2-D) grid.

Sampling not only leads to a reduction in resolution, but also to a loss of information as structures of about the scale of the sampling distance and finer will be lost. It also introduces considerable distortions if fine structures are sampled. Figure 25.1 shows a simple example. Digitization is simulated by taking only every fourth pixel in

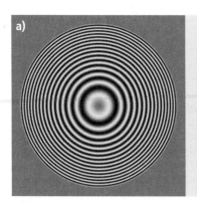 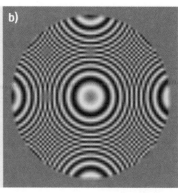

Fig. 25.1a,b Explanation of the moiré effect with a ring test pattern: (**a**) original pattern with structures that are sampled four times per wavelength at the edge of the ring; (**b**) Pattern first downsampled using only every fourth column and row and then upsampled using linear interpolation to the original size

every fourth row. This kind of image distortion is called the *moiré* effect. The same phenomenon, called *aliasing*, is known for one-dimensional signals, especially time series.

The sampling theorem states under which conditions these distortions can be avoided and, even more, whether a complete reconstruction of the sampled continuous image is possible; if the two-dimensional (2-D) spectrum $\hat{g}(k_1, k_2)$ of a continuous image function $g(x_1, x_2)$ is band-limited, i. e.,

$$\hat{g}(\boldsymbol{k}) = 0 \quad \forall |k_{1,2}| \geq k_{\max 1, \max 2}, \quad (25.1)$$

then it can be reconstructed exactly from samples with a distance

$$\Delta x_w = \frac{1}{2k_{\max 1, \max 2}}. \quad (25.2)$$

In other words, at least two samples per wavelength are required. The maximum wavenumber that can be sampled without errors is called the *Nyquist* or *limiting wavenumber* $k_{\max}$. Often dimensionless wavenumbers $\tilde{k} = k/k_{\max}$ that are scaled by the *Nyquist wavenumber* and are confined to an interval $[-1, 1]$ are used.

Standard Sampling

An array of photosensitive sensor elements does not perform a point sampling. Rather the average over the light-sensitive area of each sensor element is taken. This corresponds to a convolution of the image signal by the spatial distribution of the light sensitivity. In the best case, the whole area of the sensor element is equally sensitive. This is known as *standard sampling*. It is a kind of *regular sampling*, because each point in the continuous space is equally weighted.

The averaging over the light-sensitive area causes spatial blurring of the signal and thus some band limitation. However, this is not sufficient to avoid moiré effects. Convolution by a box function $\Pi(x/\Delta x)$ of the width Δx in the spatial domain is equivalent to a multiplication by

$$\mathrm{sinc}(\tilde{k}/2) = \frac{\sin \pi \tilde{k}/2}{\pi \tilde{k}/2}$$

in the Fourier domain. At the Nyquist wavenumber $\tilde{k} = 1$, the Fourier transform of the box function is still $2/\pi$. The first zero crossing occurs only at twice the Nyquist wavenumber. The band limitation is worse with a real imaging sensor, when only a fraction of the sensor element area is light sensitive.

Because of the insufficient band limitation by the imaging sensor, other means have to be taken in order to avoid moiré effects. The best situation is when the imaged object is bandlimited itself. Additional band limitation can also be introduced by the optical system.

Reconstruction from Samples

Reconstruction is performed by a suitable *interpolation of the sampled points*. Generally, the interpolated points at continuous positions $g_r(\boldsymbol{x})$ are calculated from the sampled values $g(\boldsymbol{r}_{m,n})$ on a regular grid

$$\boldsymbol{r}_{m,n} = [m\Delta x_1, n\Delta x_2]^{\mathrm{T}} \quad \text{with} \quad m, n \in \mathbb{Z}. \quad (25.3)$$

weighted with suitable factors depending on the distance from the interpolated point:

$$g_r(\boldsymbol{x}) = \sum_{m,n} h(\boldsymbol{x} - \boldsymbol{r}_{m,n}) g_s(\boldsymbol{r}_{m,n}). \quad (25.4)$$

From the sampling theorem it can be inferred that an *exact* reconstruction of the continuous image is possible when the original continuous image meets the sampling theorem and the transfer function of the *interpolation kernel* $h(\boldsymbol{x})$ is a box function:

$$\hat{g}_r(\boldsymbol{k}) = \Pi(\tilde{k}_1/2, \tilde{k}_2/2) \hat{g}(\boldsymbol{k}). \quad (25.5)$$

Then the ideal interpolation function is the inverse Fourier transform of the box function, a sinc function:

$$h(\boldsymbol{x}) = \text{sinc}(x_1/\Delta x_1)\text{sinc}(x_2/\Delta x_2) \,. \quad (25.6)$$

Oversampling

Unfortunately, the ideal interpolation function only decreases like $1/x$ towards zero. Therefore, correct interpolation requires many sampling points and is therefore inefficient.

More-efficient solutions to the interpolation problem can be obtained if the sampling theorem is *overfilled*, i.e., $\hat{g}(\boldsymbol{k})$ is already zero before the Nyquist wavenumber is reached. Then $\hat{h}(\boldsymbol{k})$ can have any value in the region where $\hat{g}$ vanishes without introducing errors. This freedom can be used to construct an interpolation function that decreases more quickly in the spatial domain, i.e., has a minimum-length interpolation mask. We can also start from a given interpolation formula. Then the deviation of its Fourier transform from a box function tells us to what extent structures will be distorted as a function of the wavenumber.

The principle of *oversampling* is not only of importance for the construction of effective interpolation functions. It is also essential for the design of any type of precise filter with small filter masks (Sects. 25.1.4, 25.1.5). Generally, the rate of oversampling, which increases the number of data points, and the requirements of the filter design must be balanced. Practical experience shows that a sample rate between three and six samples per wavelength, i.e., 1.5- to 3-fold oversampling, is a good compromise.

Quantization and Resolution

Computer can only handle digital numbers. Therefore continuous number are mapped onto a limited number Q of discrete gray values (*quantization*):

$$[0, \infty[\xrightarrow{Q} \{g_0, g_1, \ldots, g_{Q-1}\} = G \,.$$

Quantization always introduces errors, as the true value g is replaced by one of the quantization levels g_q. If the quantization levels are equally spaced with a distance Δg and if all gray values are equally probable, the variance introduced by the quantization is given by

$$\sigma_q^2 = \frac{1}{\Delta g}\int_{g_q-\Delta g/2}^{g_q+\Delta g/2}(g-g_q)^2 dg = \frac{1}{12}(\Delta g)^2 \,. \quad (25.7)$$

The standard deviation σ_q is about 0.3 times the distance between the quantization levels Δg.

With respect to the quantization, the question arises of the accuracy to which we can measure a gray value. At first glance, the answer to this question seems to be trivial and given by (25.7): the maximum error is half a quantization level and the mean error is about 0.3 quantization levels.

But what if we measure the value repeatedly? This could happen if we take many images of the same object or if we have an object of a constant gray value and want to measure the mean gray value of the object by averaging over many pixels.

For repeated measurements, the error of the mean value decreases with the number N of measurements according to

$$\sigma_{\text{mean}} \approx \frac{1}{\sqrt{N}}\sigma \,, \quad (25.8)$$

where σ is the standard deviation of the individual measurements and N is the number of measurements taken. If 100 measurements are taken, the error of the mean should be just about a tenth of the error of the individual measurement.

Taking quantization into account, however, averaging requires a more-detailed analysis. If no noise is present, the same quantized value is always measured. Then the result cannot be more accurate than the individual measurements.

However, if the measurements are noisy, we would obtain different values for each measurement. The probability for the different values reflects the mean and variance of the noisy signal, and because we can the distribution, we can estimate both the mean and the variance.

As an example, a standard deviation of the noise equal to the quantization level is discussed. Then, the standard deviation of an individual measurement is about three times larger than the standard deviation due to the quantization. However, already with 100 measurements, the standard deviation of the mean value is only 0.1, or 3 times lower than that of the quantization.

As in images we can easily obtain many measurements by spatial averaging, there is the potential to measure mean values with standard deviations that are much smaller than the die standard deviation of quantization in (25.7).

The accuracy is also limited, however, by other, systematic errors. The most significant source is the unevenness of the quantization levels. In a real quantizer, such as an analog-to-digital converter, the quantization levels are not equally distant but show systematic deviations that may be up to half a quantization interval. Thus, careful investigation of the analog-to-digital converter is required to estimate what really limits the accuracy of the gray value measurements in images.

25.1.2 Radiometric Corrections

The first image processing steps include two classes of operations: point and geometric operations. Essentially, these two types of operations modify the *what* and *where* of a pixel.

Point operations modify the gray values at individual pixels depending only on the gray value and possibly on the position of the pixels. Generally, such a kind of operation is expressed by

$$G'_{mn} = P_{mn}(G_{mn}) . \tag{25.9}$$

The indices of the function P denote the possible dependence of the point operation on the position of the pixel.

Homogeneous point operators are the same for all pixel and can be implemented via look-up tables. They are a very useful tool for inspecting images. As the look-up table operations work in real time, images can be manipulated interactively. If only the output look-up table is changed, the original image content remains unchanged. Here, some typical tasks are demonstrated.

Evaluating and Optimizing Illumination

With the naked eye, is is very hard to estimate the homogeneity of an illuminated area (Fig. 25.2a). We need to mark gray scales such that absolute gray levels become perceivable for the human eye. If the radiance distribution is continuous, it is sufficient to use equidensities. This technique uses a staircase-type homogeneous point operation by mapping a certain range of gray scales onto one. This point operation is achieved by zeroing the p least-significant bits with a *logical AND operation*:

$$q' = P(q) = q \wedge \overline{(2^p - 1)} , \tag{25.10}$$

where $\wedge$ denotes the logical (bitwise) *and* and the overline denotes *negation*. This point operation limits the resolution to $Q - p$ bits and, thus, 2^{Q-p} quantization levels. Now, the jump between the remaining quantization levels is large enough to be perceived by the eye and to see contour lines of equal absolute gray scale in the image (Fig. 25.2). Another way to mark absolute gray values is the so-called *pseudocolor image*. With this technique, a gray level q is mapped onto a *red–green–blue (RGB) triple* for display. As color is much better

Fig. 25.2 (a) and **(b)** (contrast-enhanced, gray scale 184–200): edges artificially produced by a staircase look-up table (LUT) with a step height of 1.0 and 2.0 make contours of constant irradiance easily visible

recognized by the eye, this helps reveal absolute gray levels.

Detection of Underflow and Overflow

Under- and overflows of the gray values of a digitized image often go unnoticed and cause serious bias in further processing, for instance, for mean gray values of objects or the center of gravity of an object. In most cases, such areas cannot be detected directly. They may only become apparent in textured areas when the texture is bleached out. Over- and underflow are detected easily in histograms by strong peaks at the minimum and/or maximum gray values (Fig. 25.3). With pseudocolor mapping, the few lowest and highest gray values could be displayed, for example, in blue and red, respectively. Then, gray values dangerously close to the limits

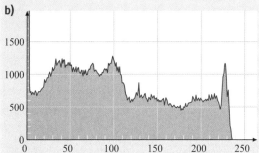

Fig. 25.3a,b Detection of underflow and overflow in digitized images by histograms: (**a**) image with underflow and (**b**) its histogram

immediately pop out of the image and can be avoided by correcting the illumination, lens aperture or gain and offset of the camera.

Noise Variance Equalization

The variance of the noise of a linear image sensor (Sect. 26.4.9) is not constant but depends on the image intensity g according to

$$\sigma_g^2(g) = \sigma_0^2 + Kg \,. \tag{25.11}$$

Many statistical image analysis procedures, however, are based on gray-value-independent, normally distributed, additive noise. Because brighter regions in the image have a larger variance, their influence is generally overestimated, while darker regions still contains valid information with less statistical uncertainty, which is not adequately used.

The nonlinear gray value transform

$$h(g) = \gamma g_{\max} \frac{\sqrt{\sigma_0^2 + Kg} - \sigma_0}{\sqrt{\sigma_0^2 + Kg_{\max}} - \sigma_0} \,, \tag{25.12}$$

maps the gray values into the interval $[0, \gamma g_{\max}]$ and the standard deviation

$$\sigma_h = \frac{\gamma K g_{\max}/2}{\sqrt{\sigma_0^2 + K g_{\max}} - \sigma_0} \tag{25.13}$$

is independent of the gray value.

The nonlinear transform becomes particularly simple for an ideal imaging sensor with no dark noise ($\sigma_0 = 0$). Then a square-root transform must be applied to obtain an intensity-independent noise variance:

$$h(g) = \gamma \sqrt{g g_{\max}} \quad \text{and} \quad \sigma_h = \frac{\gamma}{2} \sqrt{K g_{\max}} \,. \tag{25.14}$$

Correction for Inhomogeneous Illumination

Every real-world application has to contend with *uneven illumination* of the observed scene. Even if a lot of effort is spent optimizing the illumination setup, it is still very hard to obtain perfectly even object irradiance. A nasty problem is caused by small dust particles in the optical path, especially on the glass window close to the charge-coupled device (CCD) sensor. Because of the distance of the window from the imager, these particles – if they are not too large – are blurred to such an extent that they are not directly visible. However, they still absorb some light and thus cause a drop in the illumination level in a small area. These effects are not easily visible in a scene with high contrast and many details, but become very apparent in the case of a uniform background (Fig. 25.2a,b). Some imaging sensors, especially complementary metal oxide semiconductor (CMOS) sensors, also show considerable uneven sensitivity of the individual photoreceptors, which adds to the nonuniformity of the image. These distortions can severely limit the quality of the images. These effects make it more difficult to separate an object from the background, and introduce systematic errors for subsequent image processing steps.

It is possible to correct for these effects if we know the nature of the distortion and can take suitable reference images. A simple *two-point radiometric calibration* of an imaging sensor can be applied for every sensor with a linear response. The following reference images are taken: firstly, a dark image ***B*** without any illumination and secondly a reference image ***R*** with an object of constant radiance, e.g., by looking with the camera into an *integrating sphere*. Then, a normalized image corrected for both an inhomogeneous

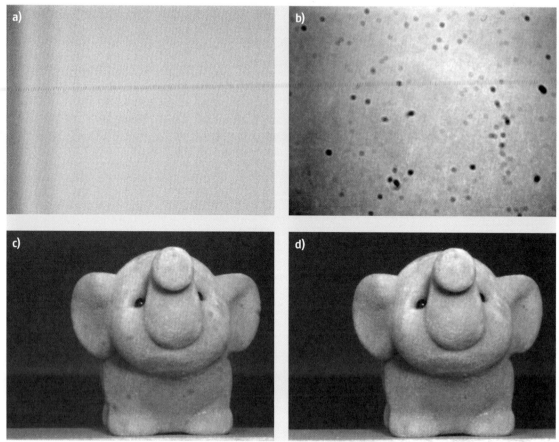

Fig. 25.4a–d Contrast-enhanced (**a**) dark image and (**b**) reference image for a two-point radiometric calibration of a CCD camera with analog video output. Two-point radiometric calibration with the dark and reference image: (**c**) original image and (**d**) calibrated image; in the calibrated image the dark spots caused by dust are no longer visible

dark image and inhomogeneous sensitivity is given by

$$G' = c \frac{G - B}{R - B} \,. \tag{25.15}$$

Figures 25.4a,b show a contrast-enhanced dark image and reference image of a CCD camera. Typical signal distortions can be observed. The signal oscillation at the left edge of the dark image results from an electronic interference, while the dark blobs in the reference image are caused by dust on the glass window in front of the sensor. The improvement due to the radiometric calibration according to (25.15) can clearly be seen in Figs. 25.4c,d.

Often, the quantity to be measured by an imaging sensor is related in a nonlinear way to the measured gray value. In such cases a more-complex *nonlinear radiometric calibration* is required.

25.1.3 Geometric Corrections

Geometric transforms of images are required to correct for geometric distortions introduced during the image formation process or if scaling, rotating or any other kind of geometric transform of images is required. These *geometric operations* modify only the position of a pixel. A pixel located at the position x is relocated to a new position x'. These operations include two major steps. In most applications, the mapping function is not given explicitly but must be derived from corresponding points. When an image is warped by a geometric transform, the pixels in the original and warped images almost never fall onto each other. Thus, it is required to interpolate gray values at these pixel from neighboring pixels.

Forward and Inverse Mapping

Geometric transforms define the relationship between the points in two images. This relation can be expressed in two ways. Either the coordinates of the output image x' can be specified as a function of the input coordinates x or vice versa:

$$x' = M(x) \quad \text{or} \quad x = M^{-1}(x'), \qquad (25.16)$$

where M specifies the mapping function and M^{-1} is its inverse. The two expressions in (25.16) give rise to two principal kinds of spatial transformation: *forward* and *inverse mapping*.

With *forward mapping*, a pixel of the input image is mapped onto the output image (Fig. 25.5a). Generally, a pixel of the input image lies between the pixels of the output image. With forward mapping, it is not appropriate just to assign the value of the input pixel to the nearest pixel in the output image (point-to-point or nearest-neighbor mapping), as it may happen that the transformed image contains holes as a value is never assigned to a pixel in the output image or that a value is assigned more than once to a point in the output image. An appropriate technique distributes the value of the input pixel to several output pixels. The easiest procedure is to regard pixels as squares and to take the fraction of the area of the input pixel that covers the output pixel as the weighting factor. Each output pixel accumulates the corresponding fractions of the input pixels which – if the mapping is continuous – add up to cover the whole output pixel.

With *inverse mapping*, the coordinates of a point in the output image are mapped back onto the input image (Fig. 25.5b). It is obvious that this scheme avoids holes and overlaps in the output image as all pixels are scanned sequentially. Now, the interpolation problem occurs in the input image. The coordinates of the output image in general do not hit a pixel in the input image but lie in between the pixels. Thus, its correct value must be interpolated from the surrounding pixels. Generally, inverse mapping is a more-flexible technique, as it is easier to implement various types of interpolation techniques.

Ideal Interpolation

The basis of interpolation is the sampling theorem. The problem is related to the fact that the reconstruction of the continuous image from the sampled image in practice is quite involved and can be performed only approximately because the ideal continuous interpolation mask h, the sinc function, requires too many operations.

Any approximate solution should still reproduce the grid points and not depend on any other grid point (the *interpolation condition*):

$$h(x_{m,n}) = \begin{cases} 1 & m = 0, n = 0 \\ 0 & \text{otherwise} \end{cases}. \qquad (25.17)$$

Any interpolation mask must, therefore, as for the ideal interpolation mask, have zero crossings at all grid points except the zero point, where it is 1.

The ideal interpolation function in (25.6) is separable. Therefore, interpolation can be as easily formulated for higher-dimensional images. We can expect that all solutions to the interpolation problem will also be separable. Consequently, we need only discuss the one-dimensional (1-D) interpolation problem. Once it is solved, we also have a solution for the multidimensional interpolation problem.

An important special case is the interpolation to intermediate grid points halfway between the existing grid points. This scheme doubles the resolution and image size in all directions in which it is applied. Then, the continuous interpolation kernel reduces to a discrete convolution mask. As the interpolation kernel (25.5) is separable, we can first interpolate the intermediate points in a row in the horizontal direction before we apply vertical interpolation to the intermediate rows. In linear interpolation three dimensions, a third 1-D interpolation

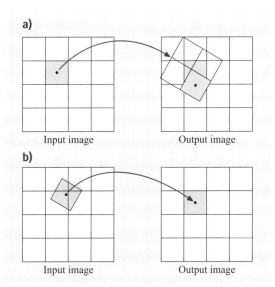

Fig. 25.5a,b Illustration of (**a**) forward mapping and (**b**) inverse mapping for spatial transformation of images

is added in the z or t direction. The interpolation kernels are the same in all directions. We need the continuous kernel $h(x)$ only at half-integer values for $x/\Delta x$.

Linear Interpolation

Linear interpolation is the classic approach to interpolation. The interpolated points lie on pieces of straight lines connecting neighboring grid points. In order to simplify the expression, we use in the following normalized spatial coordinates $\tilde{x} = x/\Delta x$. We locate the two grid points at $-1/2$ and $1/2$. This yields the *interpolation equation*

$$g(\tilde{x}) = \frac{g_{1/2} + g_{-1/2}}{2} + (g_{1/2} - g_{-1/2})\tilde{x} \quad (25.18)$$

for $|\tilde{x}| \leq 1/2$. The continuous interpolation mask for linear interpolation is

$$h_1(\tilde{x}) = \begin{cases} 1 - |\tilde{x}| & |\tilde{x}| \leq 1 \\ 0 & \text{otherwise} \end{cases} \quad (25.19)$$

Linear interpolation introduces serious distortions: while low wavenumbers (and especially the mean value $\tilde{k} = 0$) are interpolated correctly, high wavenumbers are slightly reduced in amplitude, resulting in some degree of smoothing. Furthermore spurious high wavenumbers are introduced, because the first derivative is discontinuous at the grid points.

Spline-Based Interpolation

Besides of its limited accuracy, linear and higher-order polynomial interpolation has another significant disadvantage: the interpolated curve is already discontinuous in its first derivative at the grid points. This is due to the fact that, for each interval between grid points, another polynomial is taken. Thus, only the interpolated function is continuous at the grid points but not the derivatives.

Splines avoid this disadvantage by additional constraints for the continuity of derivatives at the grid points. From the wide classes of splines, *B-splines* prove to be most useful for interpolation. As B-splines are separable, it is sufficient to discuss the properties of 1-D B-splines. From the background of image processing, the easiest access to B-splines is their convolution property. The kernel of a P-order B-spline curve is generated by convolving the box function $P+1$ times with itself:

$$\beta_P(\tilde{x}) = \underbrace{\Pi(\tilde{x}) * \ldots * \Pi(\tilde{x})}_{(P+1) \text{ times}} \quad (25.20)$$

with the Fourier transform (the *transfer function*)

$$\hat{\beta}_P(\hat{k}) = \left(\frac{\sin \pi \tilde{k}/2}{(\pi \tilde{k}/2)}\right)^{P+1} . \quad (25.21)$$

The B-spline function itself is not a suitable interpolation function. The transfer function decreases too early, indicating that B-spline interpolation performs too much averaging, and the B-spline kernel does not meet the interpolation condition (25.17) for $P > 1$. B-splines can only be used for interpolation if the discrete grid points are first transformed in such a way that a following convolution with the B-spline kernel restores the original values at the grid points. This transformation is known as the *B-spline transformation* and is constructed from the condition:

$$g_P(x) = \sum_n c_n \beta_P(x - x_n) \quad \text{with} \quad g_P(x_n) = g(x_n) . \quad (25.22)$$

If centered around a grid point, the B-spline interpolation kernel is unequal to zero for only three grid points. The coefficients $\beta_3(-1) = \beta_{-1}$, $\beta_3(0) = \beta_0$, and $\beta_3(1) = \beta_1$ are $1/6$, $2/3$, and $1/6$, respectively. The convolution of this kernel with the unknown B-spline transform values c_n should result in the original values g_n at the grid points.

The B-spline coefficients can be computed very efficiently by a recursive filter that is applied first in the forward and then in the backward direction with the following recursion [25.1]:

$$g'_n = g_n - (2 - \sqrt{3})(g'_{n-1} - g_n) ,$$
$$c'_n = g'_n - (2 - \sqrt{3})(c_{n+1} - g'_n) . \quad (25.23)$$

The whole operation takes only two multiplications and four additions.

The B-spline interpolation is applied after the B-spline transformation. In the continuous case this yields the *effective transfer function*

$$\hat{\beta}_I(\tilde{k}) = \frac{\sin^4(\pi \tilde{k}/2)/(\pi \tilde{k}/2)^4}{(2/3 + 1/3 \cos \pi \tilde{k})} . \quad (25.24)$$

Essentially, the B-spline transformation performs an amplification of high wavenumbers (at the Nyquist wavenumber $\tilde{k} = 1$ by a factor 3). This compensates for the smoothing of the B-spline interpolation to a large extent.

For an image enlargement by a factor of two, the intermediate points are given by a convolution with the mask

$$[1 \ 23 \ 23 \ 1]/48 . \quad (25.25)$$

25.1.4 Averaging and Noise Suppression

In a region with totally independent pixel gray values, nothing can be recognized. *Spatial coherency* either in one or two dimensions is required in order to recognize lines and regions, respectively (Fig. 25.6). Averaging within regions of constant gray values (an object of interest) appears to be a central tool in image processing. In addition, the deviation of the gray values of the pixel in the neighborhood gives a quantitative measure as to how well a region of constant gray values is encountered in the neighborhood. This approach is very much the same as that used for any type of measurements. A single measurement is meaningless. Only repeated measurements give us a reliable estimate both of the measured quantity and its uncertainty. In image processing, averaging needs not necessarily be performed by taking several images, although this is a very useful procedure. Because spatial information is obtained with images, spatial averaging offers an alternative.

Many objects do not show a distinct constant gray value, but we can still recognize them if the pattern they show differs from the background. After suitable preprocessing such an image can be converted into a *feature image*. Then, the feature image can be handled in the same way as a gray scale image for simple objects.

An important general question for smoothing is how it can be computed efficiently. This question is of special importance if we want to analyze coarse features in the images that require averaging over larger distances and thus large smoothing masks (or if we apply smoothing to higher-dimensional images such as volumetric images or image sequences).

While averaging works well within regions, it is questionable at the edges of an object. When the filter mask contains pixels from both the object and the background, averaged values have no useful meaning. These values cannot be interpreted as an object-related feature since this depends on the fraction of the object pixels contained in the mask of the smoothing operator. Therefore, smoothing techniques that stop or at least diminish averaging at discontinuities are discussed. Such an approach is not trivial as it requires the detection of the *edges* before the operation can be applied. Obviously, such advanced smoothing techniques need to analyze the local neighborhoods in more detail and adapt the smoothing process in one or the other way to the structure of the local neighborhood.

Box Filters

The simplest type of averaging filters is the *box filter* or *running mean*. It averages $R \times R$ pixel around a central pixel and writes this average to the central pixel. This procedure is repeated for all pixels of an image and is mathematically a convolution of the image with an $R \times R$ mask of equal coefficients with the value $1/R^2$.

The *spatial variance* of a 1-D box filter with R coefficients is given by

$$\sigma_x^2 = \frac{1}{12}(R^2 - 1) . \tag{25.26}$$

The *standard deviation* σ of smoothing increases approximately linearly with the size of the mask. The box filter is separable. Higher-dimensional box filters result from a cascaded application of the 1-D box filter in all directions.

In the Fourier domain, convolution reduces to a multiplication of the Fourier transformed image $\hat{g}(k)$ by the Fourier transform of the smoothing mask, which is known as the transfer function. The transfer function of the box filter

$$^R\hat{r}(\tilde{k}) = \frac{\sin(R\pi\tilde{k}/2)}{R \sin(\pi\tilde{k}/2)} \tag{25.27}$$

is one at the wavenumber zero (preservation of the mean value) and decreases for small wavenumbers proportionally to the wavenumber squared:

$$^R\hat{r} \approx 1 - \frac{R^2 - 1}{24}(\pi\tilde{k})^2 . \tag{25.28}$$

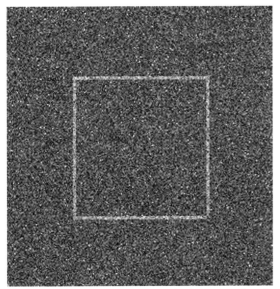

Fig. 25.6 Spatial coherency is required to recognize objects. No object can be recognized in a region that contains only randomly distributed gray values

The attenuation of large wavenumbers is not steeper than $\propto \tilde{k}^{-1}$. This is only a very weak attenuation of high wavenumbers, which renders box filters useless for many applications. With the exception of the ^{2}R box filter (the box filter [1 1]/2), the transfer function does not decrease monotonically; rather it shows significant oscillations. This leads to the following disadvantages. First, certain wavenumbers are eliminated:

$$^{2R}\hat{r}(\tilde{k}) = 0 \quad \forall \tilde{k} = \frac{n}{R/2}, \quad 1 \leq n \leq R. \qquad (25.29)$$

Note that the transfer function for the highest wavenumber $\tilde{k} = 1$ vanishes only for even-sized box filters. Second, the transfer function becomes negative in certain wavenumber ranges. This means a 180° phase shift and a contrast inversion.

Because the box filter simply computes the average of $R \times R$ pixels (*running mean*), the variance of the averaged pixel reduces to

$$\frac{\sigma'}{\sigma} = \frac{1}{R}, \qquad (25.30)$$

provided that the input pixels are statistically independent (*white noise*).

A box filter is isotropic only for small wavenumbers. Generally, it has the other significant disadvantage that it is strongly nonisotropic. Structures along the axes are attenuated much less than in the direction of the diagonals.

The only big advantage of the box filter is that it can be computed very efficiently as a recursive filter according to the following equation:

$$g'_n = g'_{n-1} + \frac{1}{2R+1}(g_{n+R} - g_{n-R-1}). \qquad (25.31)$$

This recursion can be understood by comparing the computations for the convolution at neighboring pixels. When the box mask is moved one position to the right, it contains the same weighting factors for all pixels except for the last and the first pixel. Thus, we can simply take the result of the previous convolution, (g'_{n-1}), subtract the first pixel that just moved out of the mask (g_{n-R-1}) and add the gray value at the pixel that just came into the mask (g_{n+R}). In this way, the computation of a box filter does not depend on its size; the number of computations is of $\mathcal{O}(R^0)$. Only one addition, one subtraction, and one multiplication are required per pixel.

Thus the box filter is a fast filter with bad properties. Cascading box filters avoid or at least diminish many of the disadvantages of box filters. Since individual box filters can be computed independently of their size, cascading them still remains independent of the size. The computational effort can be balanced against the remaining anisotropy and other distortions of the filter.

Binomial Filter

Binomial filters are built by cascading the simplest and most elementary smoothing mask

$$\boldsymbol{B} = \frac{1}{2}[1 \ 1], \qquad (25.32)$$

taking the mean value of the two neighboring pixel. Cascading this mask R times results in the filter mask

$$\frac{1}{2^R} \underbrace{[1 \ 1] * [1 \ 1] * \ldots * [1 \ 1]}_{R \text{ times}}. \qquad (25.33)$$

Only odd-sized masks should be applied if the resulting smoothed image should lie on the same grid.

Some examples of the resulting filter masks are:

$$\boldsymbol{B}^2 = [1 \ 2 \ 1]/4$$
$$\boldsymbol{B}^4 = [1 \ 4 \ 6 \ 4 \ 1]/16$$
$$\boldsymbol{B}^8 = [1 \ 8 \ 28 \ 56 \ 70 \ 56 \ 28 \ 8 \ 1]/256. \qquad (25.34)$$

The coefficients of the mask are the values of the *binomial distribution*. The iterative composition of the mask by consecutive convolution with the 1/2[1 1] mask is equivalent to the computation scheme of *Pascal's triangle* (Table 25.1).

The standard deviation σ of the binomial mask B^R is generally given by

$$\sigma^2 = \frac{R}{4}. \qquad (25.35)$$

Table 25.1 Computation of binomial coefficients using Pascal's triangle. R denotes the order of the binomial, f the scaling factor 2^{-R}, and σ^2 the variance, i.e., the effective width of the mask.

R	f		σ^2
0	1	1	0
1	1/2	1 1	1/4
2	1/4	1 2 1	1/2
3	1/8	1 3 3 1	3/4
4	1/16	1 4 6 4 1	1
5	1/32	1 5 10 10 5 1	5/4
6	1/64	1 6 15 20 15 6 1	3/2
7	1/128	1 7 21 35 35 21 7 1	7/4
8	1/256	1 8 28 56 70 56 28 8 1	2

The standard deviation increases only with the square root of the mask size. For high R the mask size ($R+1$ coefficients) is therefore much larger than the standard deviation $\sqrt{R}/2$.

Two- and higher-dimensional binomial filters can be composed by cascading filters along the corresponding axes: The smallest odd-sized mask ($R=2$) of this kind is a 3×3 binomial filter in 2-D:

$$\frac{1}{4}\begin{bmatrix}1 & 2 & 1\end{bmatrix} * \frac{1}{4}\begin{bmatrix}1\\2\\1\end{bmatrix} = \frac{1}{16}\begin{bmatrix}1 & 2 & 1\\2 & 4 & 2\\1 & 2 & 1\end{bmatrix}$$

The transfer function of $\mathcal{B}^R$ is then given as the R-th power:

$$\hat{b}^R(\tilde{k}) = \cos^R(\pi\tilde{k}/2), \quad (25.36)$$

which can be approximated for small wavenumbers by

$$\hat{b}^R(\tilde{k}) \approx 1 - \frac{R}{8}(\pi\tilde{k})^2. \quad (25.37)$$

The transfer function decreases monotonically and approaches zero at the largest wavenumber. The smallest mask $\mathcal{B}^2$ has a halfwidth of $\tilde{k}/2$. For larger masks, both the transfer function and the filter masks quickly approach the Gaussian distribution with an equivalent variance. Larger masks result in smaller half-width wavenumbers according to the *uncertainty relation*.

The noise suppression factors of a 2-D binomial mask for uncorrelated pixels is given by

$$\frac{\sigma'}{\sigma} = \frac{(2R)!}{4^R(R!)^2} \approx \left(\frac{1}{R\pi}\right)^{1/2}\left(1 - \frac{1}{8R}\right). \quad (25.38)$$

A direct computation of a $(R+1)\times(R+1)$ filter mask requires $(R+1)^2$ multiplications and $(R+1)^2 - 1$ additions. If we decompose the binomial mask into elementary smoothing masks $1/2\,[1\ 1]$ and apply this mask in horizontal and vertical directions R times each, we only need $2R$ additions. All multiplications can be handled much more efficiently as shift operations. Despite the efficient implementation of binomial smoothing filters $\mathcal{B}^R$ by cascaded convolution with $\mathcal{B}$, the number of computations increases drastically because the smoothing distance σ is only proportional to the square root of R according to (25.35). Doubling σ quadruples the number of computations.

Multistep Averaging

The problem of slow large-scale averaging originates from the small distance between the pixels averaged in the elementary $\mathcal{B} = 1/2\,[1\ 1]$ mask. This problem can be overcome if the same elementary averaging process is used with more-distant pixels:

$$\mathbf{B}_{2x} = \frac{1}{4}[1\ 0\ 2\ 0\ 1], \quad \mathbf{B}_{2y} = \frac{1}{4}\begin{bmatrix}1\\0\\2\\0\\1\end{bmatrix}. \quad (25.39)$$

The subscripts in these masks denote the stepping width and coordinate direction. The standard deviation of these filters is proportional to the distance between the pixels. The most efficient implementations are multistep masks along the axes. Because of separability, this approach can be applied to image data of arbitrary dimensions.

The problem with these filters is that they perform subsampling. Consequently, they are no longer smoothing filters for larger wavenumbers. Used individually, these filters are not useful.

Cascaded multistep binomial filtering with recursive step doubling

$$\underbrace{\mathcal{B}^R_{2^{S-1}x}\cdots\mathcal{B}^R_{8x}\mathcal{B}^R_{4x}\mathcal{B}^R_{2x}\mathcal{B}^R_{x}}_{S\text{ times}} \quad (25.40)$$

leads to a significant performance increase for large-scale smoothing. For normal separable binomial filtering, the number of computations is proportional to σ^2 [$\mathcal{O}(\sigma^2)$]. For multistep binomial filtering it depends only logarithmically on σ [$\mathcal{O}(\mathrm{ld}\,\sigma^2)$].

The standard deviation of smoothing is

$$\sigma^2 = \underbrace{R/4 + R + 4R + \ldots + 4^{S-1}R}_{S\text{ times}} = \frac{R}{12}(4^S - 1)$$

$$(25.41)$$

and the transfer function is

$$\prod_{s=0}^{S-1}\cos^R(2^{s-1}\pi\tilde{k}). \quad (25.42)$$

Thus, for S steps only RS additions are required, while the standard deviation grows exponentially with $\approx \sqrt{R/12}\cdot 2^S$.

With the parameter R, the degree of isotropy and the degree of residual inhomogeneities in the transfer function can be adjusted. For the most efficient implementation with $R = 2$ ($\mathbf{B}^2 = [1\ 2\ 1]/4$ in each direction), the residual side peaks at high wavenumbers with maximal amplitudes up to 0.08 still cause significant disturbances. With the next larger odd-sized masks

($R = 4$, $\boldsymbol{B}^4 = [1\ 4\ 6\ 4\ 1]/16$ in each direction) these residual side peaks at high wavenumbers are suppressed well below 0.005.

Nonlinear Averaging

Linear smoothing filters cannot distinguish between a useful feature and noise. This property can be best demonstrated in the Fourier space (Fig. 25.7). White noise is added to the image. Because of the linearity of the Fourier transform, the Fourier transform of the white noise adds directly to the Fourier transform of the image.

Any linear filter operator works in such a way that the Fourier transform of the image is multiplied by the Fourier transform of the filter. The result is that at each wavenumber the noise level and the image features are attenuated by the same factor. Thus, nothing has improved at all. The signal-to-noise ratio is just the same. The noise level is reduced but so is the signal.

It is obvious that more-complex approaches than linear filtering are required. Common to all these approaches is that in one or the other way the filters are dependent on the context. Therefore a kind of control strategy is an important part of adaptive filtering that tells us which filter or in which way a filter has to be applied at a certain point in the image. In the following sections, some classes of nonlinear filter techniques are discussed.

Problem-Specific Nonlinear Filters. This approach is the oldest one and works if a certain specific type of distortion is present in an image. A well-known example is the *median filter*, which can excellently remove single-distorted pixels with minimum changes to the image features. This approach, of course, only works for the type of distortion it is designed for. A median filter is excellent for removing a single pixel that has a completely incorrect gray value because of a transmission or data error; it is less well suited, however, to the reduction of white noise.

Weighted Averaging. So far, each pixel was treated equally assuming that the information it carries is of equal significance. While this seems to be a reasonable first approximation, it is certain that it cannot be generally true. Already during image acquisition, the sensor area may contain bad sensor elements that lead to erroneous gray values at certain positions in the image. Likewise, transmission errors may occur so that individual pixels may carry wrong information. In one way or another we may attach a certainty measurement to each data point.

Once a certainty measurement has been attached to a pixel, it is obvious that standard convolution operators are no longer a good choice. Instead, the weight we attach to the pixel has to be considered when performing any kind of operation with it. Each pixel enters the convolution sum with a weighting factor associated with it. This kind of approach is called *normalized convolution*. Thus, normalized convolution requires two images. One is the image to be processed, the other is an image with the weighting factors:

$$G' = \frac{\boldsymbol{H} * (\boldsymbol{W} \cdot \boldsymbol{G})}{\boldsymbol{H} * \boldsymbol{W}}, \qquad (25.43)$$

where $\boldsymbol{H}$ is any convolution mask, $\boldsymbol{G}$ is the image to be processed, and $\boldsymbol{W}$ is the image with the weighting factors. A normalized convolution with the mask $\boldsymbol{H}$ essentially transforms the set of the image $\boldsymbol{G}$ and the weighting image $\boldsymbol{W}$ into a new image $\boldsymbol{G}'$ and a new weighting image $\boldsymbol{W}' = \boldsymbol{H} * \boldsymbol{W}$, which can undergo further processing. *Standard* convolution can be regarded as a special case of normalized convolution where all pixels are assigned the same weighting factor and a weighting image is not required, since the factor remains constant.

Actually, this type of approach seems very natural to a scientist or engineer, as they are used to qualifying any data by a measurement error, which is then used in any further evaluation of the data. Normalized convolution applies this common principle to image processing.

The power of this approach is related to the fact that there are varied possibilities for the definition of the certainty of the measurement; it does not only have to be related to a direct measurement error of a single pixel. If we are, for example, interested in computing an estimate

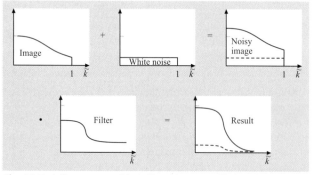

Fig. 25.7 A linear smoothing filter does not distinguish between useful features and noise in an image. It reduces the feature and the noise amplitudes equally at each wave number so that the signal-to-noise ratio remains the same

of the mean gray value in an object, we could devise a kind of certainty measurement that analyzes neighborhoods and attaches low weighting factors where we suspect an edge so that these pixel do not contribute much to the mean gray value or feature of the object. In a similar way, we could, for instance, also check how likely the gray value of a certain pixel is if we suspect some distortion by transmission errors or defective pixels. If the certainty measurement of a certain pixel is below a critical threshold, it is effectively replaced by a weighted value from the surrounding pixels.

Adaptive Filtering. *Adaptive filters* in the narrower sense use a different strategy than normalized convolution. Now, the filter operation itself is made dependent on the neighborhood. Adaptive filtering can best be explained by a classical application, the suppression of noise without significant blurring of image features. The basic idea of adaptive filtering is that in certain neighborhoods a smoothing operation can be applied without blurring structures. If, for instance, the neighborhood is flat, it can be assumed that this is an area within an object of constant features and thus an isotropic smoothing operation can be to this pixel to reduce the noise level. If an edge is present in the neighborhood, some smoothing is still possible, namely along the edge. In this way, some noise is removed but the edge is not blurred. With this approach, we need a kind of large filter set of directional smoothing operations. Because of the many filters involved, it appears that adaptive filtering might be a very computational-intensive approach; this is indeed the case if either the coefficients of the filter to be applied have to be computed for every pixel or if a large set of filters has to be used. With the discovery of *steerable filters* [25.2], however, adaptive filtering techniques have become attractive and computationally much more efficient.

With this approach a small set of base filters is used to compute a set of filtered images. Then, these images are interpolated using parameters that depend on the adjustable parameters. In operator notation this reads

$$\mathcal{H}(\alpha) = \sum_{p=1}^{P} f_p(\alpha) \mathcal{H}_p \qquad (25.44)$$

where $\mathcal{H}_p$ is the p-th filter and $f_p(\alpha)$ is a scalar function of the steering parameter α. Two problems must be solved to use steerable filters. First, and most basically, it is not clear that such a filter base H_p exists at all. Second, the relation between the steering parameter(s) α and the interpolation coefficients f_p must be found. If the first problem is solved, we mostly get the solution to the second for free.

A simple example of a steerable filter is directional smoothing. A directional smoothing filter is to be constructed with the following transfer function

$$\hat{h}_\theta(k, \phi) = f(k) \cos^2(\phi - \theta). \qquad (25.45)$$

In this equation cylinder coordinates (k, ϕ) are used in the Fourier domain. The filter in (25.45) is a *polar separable filter* with an arbitrary radial function $f(k)$. This radial component provides an isotropic bandpass filtering.

The steerable angular term is given by $\cos^2(\phi - \theta)$. Structures oriented into the direction θ remain in the image, while those perpendicular to θ are completely filtered out. The angular width of the directional filter is $\pm 45°$.

Using elementary trigonometry it can be shown that this filter can be computed from only two base filters in the following way:

$$\hat{h}_\theta(k, \phi) = \frac{1}{2} + \frac{1}{2}[\cos(2\theta)\hat{H}_0(k, \phi) \\ + \sin(2\theta)\hat{H}_{\pi/4}(k, \phi)] \qquad (25.46)$$

with the filter base

$$\hat{h}_0(k, \phi) = f(k) \cos^2 \phi, \\ \hat{h}_{\pi/4}(k, \phi) = f(k) \sin^2(\phi). \qquad (25.47)$$

The two base filters are directed towards $0°$ and $45°$. The directional filter $\hat{h}_\theta$ can be steered into any direction between $-90°$ and $90°$.

Using separable filters, a polar separable directional filter can be approximated only in a limited wavenumber range. Thus $f(k)$ must be a bandpass filter. The following filter set turns out to be a good approximation. It uses only binomial filters along the axes ($\mathcal{B}_x$ and $\mathcal{B}_y$) and diagonals ($\mathcal{B}_{x-y}$ and $\mathcal{B}_{x+y}$) with equal variance:

$$\mathcal{H}_0 = \frac{1 - \mathcal{B}_x^{2R} \mathcal{B}_y^{2R} - \left(\mathcal{B}_x^{2R} - \mathcal{B}_y^{2R}\right)}{2},$$

$$\mathcal{H}_{\pi/4} = \frac{1 - \mathcal{B}_x^{2R} \mathcal{B}_y^{2R} - \left(\mathcal{B}_{x+y}^R - \mathcal{B}_{x-y}^R\right)}{2}. \qquad (25.48)$$

For small wavenumbers the transfer function of the filter steered in the direction θ agrees with the required form (25.45). Thus it is not surprising that this filter

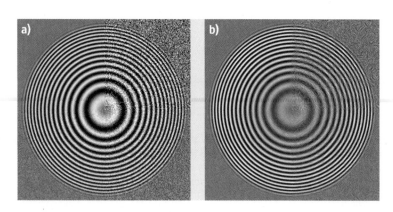

Fig. 25.8 (**a**) Ring test image with an amplitude of 100 superposed by zero mean normal distributed noise with a standard deviation of 10, 20, and 40 in three quadrants. (**b**) Image (**a**) after two iterations of steerable directional smoothing in the direction of constant gray values

can be used well to remove noise in images with directed gray value structures when this filter is steered to smooth in the direction of constant gray values (Fig. 25.8).

Nonlinear Diffusion Filters. In recent years, a whole new class of image processing operators has been investigated, known as *diffusion filters* [25.3]. Diffusion is a transport process that tends to level out concentration differences and thus work like a smoothing filter. Diffusion processes govern the transport of heat, matter, and momentum. To apply a diffusion process to an image, the gray value g is regarded as the concentration of a chemical species. The elementary law of diffusion states that the flux induced by a concentration difference is against the direction of the concentration gradient and proportional to it:

$$\boldsymbol{j} = -D\nabla g,\qquad (25.49)$$

where the constant D is known as the *diffusion coefficient*. Using the continuity equation

$$\frac{\partial g}{\partial t} + \nabla \boldsymbol{j} = 0,\qquad (25.50)$$

the nonstationary diffusion equation is

$$\frac{\partial g}{\partial t} = \nabla(D\nabla g).\qquad (25.51)$$

For the case of a homogeneous diffusion process (D does not depend on the position), the equation reduces to

$$\frac{\partial g}{\partial t} = D\Delta g.\qquad (25.52)$$

The general solution to this equation is equivalent to a convolution with a smoothing mask. A spatial Fourier transform, which results in

$$\frac{\partial \hat{g}(\boldsymbol{k})}{\partial t} = -D|\boldsymbol{k}|^2 \hat{g}(\boldsymbol{k}),\qquad (25.53)$$

reduces the equation to a linear first-order differential equation with the general solution

$$\hat{g}(\boldsymbol{k}, t) = \exp(-D|\boldsymbol{k}|^2 t)\hat{g}(\boldsymbol{k}, 0),\qquad (25.54)$$

where $\hat{g}(\boldsymbol{k}, 0)$ is the Fourier-transformed image at time zero. Multiplication of $\hat{g}(\boldsymbol{k}, 0)$ in the Fourier space with the Gaussian function $\exp(-|\boldsymbol{k}|^2/(2\sigma_k^2))$ with $\sigma_k^2 = 1/(2Dt)$ as given by (25.54) is equivalent to a convolution with the same function but of reciprocal width. Thus,

$$g(\boldsymbol{x}, t) = \frac{1}{2\pi\sigma^2(t)} \exp\left(-\frac{|\boldsymbol{x}|^2}{4Dt}\right) g(\boldsymbol{x}, 0).\qquad (25.55)$$

Equation (25.55) establishes the equivalence between a diffusion process and convolution with a Gaussian kernel. In the discrete case, the Gaussian kernel can be replaced by binomial filters.

Given the equivalence between convolution and a diffusion process, it is possible to adapt smoothing to the local image structure by making the diffusion constant dependent on the position (*inhomogeneous diffusion*) and/or the direction (*anisotropic diffusion*).

To avoid smoothing of edges, it appears logical to attenuate the diffusion coefficient there. Thus, the diffusion coefficient is made dependent on the strength of the edges as given by the magnitude of the gradient

$$D(g) = D(|\nabla g|).\qquad (25.56)$$

Perona and *Malik* [25.4] used the following dependency of the diffusion coefficient on the magnitude of the gradient:

$$D = D_0 \frac{\lambda^2}{|\nabla g|^2 + \lambda^2},\qquad (25.57)$$

where λ is an adjustable parameter. For small gradients $|\nabla g| \ll \lambda$, D approaches D_0; for high gradients $|\nabla g| \gg \lambda$, D tends to zero.

As simple and straightforward as this idea appears, it is not without problems. Depending on the functional form of D on ∇g, the diffusion process may become unstable, resulting even in steeping of the edges. A safe way to avoid this problem is to use a regularized gradient obtained from a smoothed version of the image [25.3].

Inhomogeneous diffusion has one significant disadvantage: it stops diffusion completely and in all directions at edges, leaving them noisy. Edges are, however, only blurred by diffusion perpendicular to them while diffusion parallel to edges is even advantageous since it stabilizes the edge.

An approach that makes diffusion independent of the direction of edges is known as anisotropic diffusion. With this approach, the flux is no longer parallel to the gradient. Therefore, the diffusion can no longer be described by a scalar diffusion coefficient as in (25.49). Now, a *diffusion tensor* is required:

$$\boldsymbol{j} = -\boldsymbol{D}\nabla g = - \begin{pmatrix} D_{11} & D_{12} \\ D_{12} & D_{22} \end{pmatrix} \begin{pmatrix} \partial g/\partial x \\ \partial g/\partial y \end{pmatrix} . \quad (25.58)$$

The properties of the diffusion tensor can best be seen if the symmetric tensor is brought into its principal axis system by a rotation of the coordinate system. Then, (25.58) reduces to

$$\boldsymbol{j} = -\begin{pmatrix} D_{x'} & 0 \\ 0 & D_{y'} \end{pmatrix} \begin{pmatrix} \partial g/\partial x' \\ \partial g/\partial y' \end{pmatrix} = -\begin{pmatrix} D_{x'} \partial g/\partial x' \\ D_{y'} \partial g/\partial y' \end{pmatrix} . \quad (25.59)$$

Now, the diffusion in the two directions of the axes is decoupled. The two coefficients on the diagonal $D_{x'}$ and $D_{y'}$ are the *eigenvalues* of the diffusion tensor. In analogy to isotropic diffusion, the general solution of the anisotropic diffusion can be written

$$\hat{g}(\boldsymbol{x}, t) = \frac{1}{2\pi\sigma_{x'}(t)\sigma_{y'}(t)} \exp\left(-\frac{x'^2}{4D_{x'}t}\right)$$
$$\times \exp\left(-\frac{y'^2}{4D_{y'}t}\right) g(\boldsymbol{x}, 0) \quad (25.60)$$

in the spatial domain, provided that the diffusion tensor does not depend on the position.

This means that anisotropic diffusion is equivalent to cascaded convolution with two 1-D Gaussian convolution kernels that are steered into the directions of the principal axes of the diffusion tensor.

If one of the two eigenvalues of the diffusion tensor is significantly larger than the other, diffusion occurs only in the direction of the corresponding eigenvector. Thus the gray values are smoothed only in this direction. Implementation details for nonlinear diffusion filters are given in [25.3, 5].

25.1.5 Edge and Line Extraction

Averaging filters suppress structures with high wavenumbers. Edge detection requires a filter operation that emphasizes the spatial changes in signal values and suppresses areas with constant values. Derivative operators are suitable for such an operation in the one-dimensional case. The first derivative shows an extreme at the edge (maximal positive or negative steepness), while the second derivative crosses zero (vanishing curvature) where the edge has its steepest ascent or descent. Both criteria can be used to detect edges.

In higher dimensions the description of signal change is more complex. In 2-D images edges, corners, lines, and local extremes can be distinguished as relevant features for image processing. At an *edge*, we have a large change of the signal value perpendicular to the direction of the edge, but in the direction of the edge the change is low. However, if the curvature perpendicular to the gradient is high, the edge becomes a *corner*. A *line* is characterized by low first- and second-order derivatives along the line and a maximal curvature perpendicular to the direction of the line. *Local extremes* are characterized by zero first-order derivatives, but large curvatures in all directions.

In three dimensions, i. e., *volumetric images*, there can be *surfaces* with a strong first-order change in the direction perpendicular to the surface and low slopes and curvatures in the two directions within the surface. At an edge, there are low signal changes only in the direction of the edge, while at a corner the signal changes in all directions. All the local features described in multidimensional signals can be well represented with first- and second-order derivatives.

First-Order Derivation, Gradient

A p-th-order *partial derivative* operator corresponds to multiplication by $(2\pi \mathrm{i}k)^p$ in wavenumber space. The first-order partial derivatives into all directions of a W-dimensional signal form the W-dimensional *gradient vector*:

$$\nabla = \left(\frac{\partial}{\partial x_1}, \frac{\partial}{\partial x_2}, \ldots, \frac{\partial}{\partial x_W}\right)^\mathrm{T} \circ\!\!-\!\!\bullet \; 2\pi \mathrm{i}\boldsymbol{k} . \quad (25.61)$$

The magnitude of the gradient vector,

$$|\nabla| = ||\nabla||_2 = (\nabla^T \nabla)^{1/2} = \left[\sum_{w=1}^{W} \left(\frac{\partial}{\partial x_w} \right)^2 \right]^{1/2}, \quad (25.62)$$

is invariant to rotation of the coordinate system and thus a good measure for edge strength.

First-order discrete differences are the simplest approximation to compute the gradient vector. For the first partial derivative in the x direction, the symmetric difference is the most useful, with the convolution mask

$$\boldsymbol{D}_{2x} = 1/2[1 \ 0 \ -1] \quad (25.63)$$

and the transfer function

$$\hat{d}_{2x} = i \sin(\pi \tilde{k}_x). \quad (25.64)$$

Second-Order Derivation, Curvature

Second-order derivatives can be used to detect edges as zero crossings and lines and other second-order features by measurement of the *curvature*. All possible combinations of second-order partial differential operators of a W-dimensional signal form a symmetric $W \times W$ matrix, known as the *Hessian matrix*:

$$\boldsymbol{H} = \begin{pmatrix} \frac{\partial^2}{\partial x_1^2} & \frac{\partial^2}{\partial x_1 x_2} & \cdots & \frac{\partial^2}{\partial x_1 x_W} \\ \frac{\partial^2}{\partial x_1 x_2} & \frac{\partial^2}{\partial x_2^2} & \cdots & \frac{\partial^2}{\partial x_2 x_W} \\ \vdots & \vdots & \ddots & \vdots \\ \frac{\partial^2}{\partial x_1 x_W} & \frac{\partial^2}{\partial x_2 x_W} & \cdots & \frac{\partial^2}{\partial x_W^2} \end{pmatrix}. \quad (25.65)$$

It is always possible to find a coordinate transform $\boldsymbol{R}$ into the *principal coordinate system* so that the Hessian matrix becomes diagonal. In two dimensions this is

$$\boldsymbol{H}' = \begin{pmatrix} \frac{\partial^2}{\partial x'^2} & 0 \\ 0 & \frac{\partial^2}{\partial y'^2} \end{pmatrix}. \quad (25.66)$$

The gradient has only one nonzero component in the principal coordinate system. This is not the case for curvatures. Generally, *all* curvatures are nonzero in the principal coordinate system.

There are two curvature parameters that are invariant to a rotation of the coordinate system. The first is the trace of this matrix, i.e., the sum of the diagonal, called the *Laplacian operator* or the *mean curvature* and denoted by Δ:

$$\Delta = \mathrm{tr}\,\boldsymbol{H} = \sum_{w=1}^{W} \frac{\partial^2}{\partial x_w^2} \circ\!\!-\!\!\bullet\ -4\pi k^2. \quad (25.67)$$

The second invariant is the Gaussian curvature, which is equal to the determinant of the Hessian matrix:

$$\det \boldsymbol{H} = \frac{\partial^2}{\partial x'^2} \frac{\partial^2}{\partial y'^2} = \frac{\partial^2}{\partial x^2} \frac{\partial^2}{\partial y^2} - \left(\frac{\partial^2}{\partial x \partial y} \right)^2. \quad (25.68)$$

The simplest discrete approximations of second-order derivative filters are the following second-order differences:

$$\boldsymbol{D}_x^2 = \begin{bmatrix} 1 & -2 & 1 \end{bmatrix}$$

$$\boldsymbol{L} = \begin{bmatrix} 0 & 1 & 0 \\ 1 & -4 & 1 \\ 0 & 1 & 0 \end{bmatrix}. \quad (25.69)$$

Regularized Edge Detection

The edge detectors discussed so far are still poor performers, especially in noisy images. Because of their small mask sizes, they are most sensitive to high wavenumbers. At high wavenumbers there is often more noise than signal in images. Thus an optimum edge detector is tuned to the scale (wavenumber range) with the maximum signal-to-noise ratio. Consequently, we must design filters that perform a derivation in one direction but also smooth the signal in all directions.

Smoothing is particularly effective in higher-dimensional signals because it does not blur the edge in all directions perpendicular to the direction of the gradient. Derivative filters that incorporate smoothing are also known as *regularized edge detectors* because they result in robust solutions for the ill-posed problem of estimating derivatives from discrete signals.

2×2 Cross-Smoothing Operator. The smallest cross-smoothing derivative operator has the following 2×2 masks

$$\frac{1}{2} \begin{bmatrix} 1 & -1 \\ 1 & -1 \end{bmatrix} \quad \text{and} \quad \frac{1}{2} \begin{bmatrix} 1 & 1 \\ -1 & -1 \end{bmatrix}. \quad (25.70)$$

There is nothing that can be optimized with this small filter mask. Because of the imperfect approximation, the direction of the gradient computed with this operator has errors of up to $5°$ at large wavenumbers ($\tilde{k} = 0.5$) and the computed magnitude of the gradient depends on the direction of the edge [25.5].

Sobel Edge Detector. The *Sobel operator* is the smallest difference filter with an odd number of coefficients that averages the image in the direction perpendicular to the differentiation:

$$\frac{1}{8}\begin{bmatrix} 1 & 0 & -1 \\ 2 & 0 & -2 \\ 1 & 0 & -1 \end{bmatrix}, \quad \frac{1}{8}\begin{bmatrix} 1 & 2 & 1 \\ 0 & 0 & 0 \\ -1 & -2 & -1 \end{bmatrix}. \qquad (25.71)$$

The errors in the magnitude and direction of the gradient are similar to the 2 × 2 cross-smoothing difference operator.

Optimized Regularized Edge Detectors. An optimized regularized derivative operator with about a 10 times lower error in the estimate of the direction of edges is [25.6]:

$$\frac{1}{32}\begin{bmatrix} 3 & 0 & -3 \\ 10 & 0 & -10 \\ 3 & 0 & -3 \end{bmatrix}, \quad \frac{1}{32}\begin{bmatrix} 3 & 10 & 3 \\ 0 & 0 & 0 \\ -3 & -10 & -3 \end{bmatrix}. \qquad (25.72)$$

Similar optimizations are possible for larger-sized regularized derivative filters [25.6].

25.1.6 Direction and Orientation

A local neighborhood could also contain more-complex patterns than edges and constant regions. If it contains oriented patterns, it is denoted as a simple neighborhood or *linear symmetry* [25.7]. In 2-D images this could be edges between objects of constant intensity, oriented patterns or, in a space–time image, objects moving with a constant velocity (Fig. 25.9). Although the three examples refer to entirely different image data, they have in common that the local structure is characterized by an orientation,

Simple Neighborhood in the Spatial Domain

A local neighborhood with ideal local orientation is characterized by the fact that the gray value only changes in one direction. In all other directions it is constant. If the coordinate system is oriented along the principal directions, the gray values become a 1-D function of only one coordinate. Generally, we will denote the direction of local orientation with a unit vector $\bar{n}$ perpendicular to the lines of constant gray values. Then, a simple neighborhood is mathematically represented by

$$g(x) = g(x^T \bar{n}), \qquad (25.73)$$

where we denote the scalar product simply by $x^T \bar{n}$. Equation (25.73) is also valid for image data with more

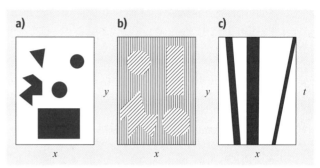

Fig. 25.9a–c Three different interpretations of local structures in 2-D images: (**a**) edge between uniform object and background; (**b**) orientation of pattern; (**c**) orientation in a 2-D space–time image indicating the velocity of 1-D objects

than two dimensions. The projection of the vector x onto the unit vector $\bar{n}$ makes the gray values depend only on a scalar quantity, the coordinate in the direction of $\bar{n}$ (Fig. 25.10). The gradient lies in the direction of $\bar{n}$.

Representation in the Fourier Domain

A simple neighborhood also has a special form in Fourier space. If the whole image is described by (25.73), i.e., $\bar{n}$ does not depend on the position, then the Fourier transform must be confined to a line. The direction of the line is given by $\bar{n}$:

$$g(x^T \bar{n}) \quad \circ\!\!-\!\!\bullet \quad \hat{g}(k)\delta[k - \bar{n}(k^T \bar{n})], \qquad (25.74)$$

where k denotes the coordinate in the Fourier domain in the direction of $\bar{n}$. The argument in the δ function is only zero when k is parallel to $\bar{n}$. If (25.74) is restricted to a local neighborhood around x_0, this corresponds to a multiplication of $g(x^T \bar{n})$ by a window function $w(x - x_0)$ in the spatial domain. The size and shape of the neighborhood is determined by the window function. A window function that gradually decreases to zero

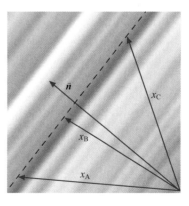

Fig. 25.10 Illustration of a linear symmetric or simple neighborhood. The *grey values* depend only on a coordinate given by a unit vector $\bar{n}$

diminishes the influence of pixels as a function of their distance from the outer pixel. Thus,

$$w(\boldsymbol{x}-\boldsymbol{x}_0) \cdot g(\boldsymbol{x}^T \bar{\boldsymbol{n}}) \quad \circ\!\!-\!\!\bullet$$
$$\hat{w}(\boldsymbol{k}) * \hat{g}(\boldsymbol{k})\delta[\boldsymbol{k} - \bar{\boldsymbol{n}}(\boldsymbol{k}^T \bar{\boldsymbol{n}})] \,, \qquad (25.75)$$

where $\hat{w}(\boldsymbol{k})$ is the Fourier transform of the window function.

The limitation to a local neighborhood thus blurs the line in Fourier space to a *sausage-like* shape. Because of the reciprocity of scales between the two domains, its thickness is inversely proportional to the size of the window. Thus the accuracy of the orientation estimate is directly related to the ratio of the window size to the wavelength of the smallest structures in the window.

The Structure Tensor

A suitable representation should be able to determine a unique orientation (given by a unit vector $\bar{\boldsymbol{n}}$) and to distinguish constant neighborhoods from neighborhoods without local orientation.

Such a representation can be introduced by the following optimization strategy to determine the orientation of a simple neighborhood. The optimum orientation is defined as the orientation that shows the least deviations from the directions of the gradient. A suitable measure for the deviation must treat gradients pointing in opposite directions equally. The squared scalar product between the gradient vector and the unit vector representing the local orientation $\bar{\boldsymbol{n}}$ meets this criterion:

$$(\nabla g^T \bar{\boldsymbol{n}})^2 = |\nabla g|^2 \cos^2\left[\angle(\nabla g, \bar{\boldsymbol{n}})\right] \,. \qquad (25.76)$$

This quantity is proportional to the cosine squared of the angle between the gradient vector and the orientation vector and is thus maximal when ∇g and $\bar{\boldsymbol{n}}$ are parallel or antiparallel, and zero if they are perpendicular to each other. Therefore, the following integral is maximized in a two-dimensional local neighborhood:

$$\int w(\boldsymbol{x}-\boldsymbol{x}')\left[\nabla g(\boldsymbol{x}')^T \bar{\boldsymbol{n}}\right]^2 d^2 x' \,, \qquad (25.77)$$

where the window function w determines the size and shape of the neighborhood around a point $\boldsymbol{x}$ at which the orientation is averaged. The maximization problem must be solved for each point $\boldsymbol{x}$. Equation (25.77) can be rewritten in the following way:

$$\bar{\boldsymbol{n}}^T J \bar{\boldsymbol{n}} \to \max \qquad (25.78)$$

with

$$J = \int w(\boldsymbol{x}-\boldsymbol{x}')[\nabla g(\boldsymbol{x}') \nabla g(\boldsymbol{x}')^T] d^2 x' \,,$$

where $\nabla g \nabla g^T$ denotes an outer (Cartesian) product. The components of this symmetric 2×2 tensor, named the

Table 25.2 Eigenvalue classification of the structure tensor in 2-D images

Condition	Rank(J)	Description
$\lambda_1 = \lambda_2 = 0$	0	Both eigenvalues are zero. The mean squared magnitude of the gradient ($\lambda_1 + \lambda_2$) is zero. The local neighborhood has constant values.
$\lambda_1 > 0, \lambda_2 = 0$	1	One eigenvalue is zero. The values do not change in the direction of the corresponding eigenvector. The local neighborhood is a simple neighborhood with ideal orientation (straight edge or 1-D texture).
$\lambda_1 > 0, \lambda_2 > 0$	2	Both eigenvalues are unequal to zero. The gray values change in all directions. In the special case of $\lambda_1 = \lambda_2$, we speak of an isotropic gray value structure as it changes equally in all directions.

Table 25.3 Eigenvalue classification of the structure tensor in 3-D (volumetric) images

Condition	Rank(J)	Description
$\lambda_1 = \lambda_2 = \lambda_3 = 0$	0	The gray values do not change in any direction; constant neighborhood.
$\lambda_1 > 0, \lambda_2 = \lambda_3 = 0$	1	The gray values change only in one direction. This direction is given by the eigenvector to the nonzero eigenvalue. The neighborhood includes a boundary between two objects (surface) or a layered texture. In a space–time image: constant motion of a spatially oriented pattern (*planar wave*).
$\lambda_1 > 0, \lambda_2 > 0, \lambda_3 = 0$	2	The gray values change in two directions and are constant in a third (edge or extruded texture). In a space–time image: constant motion of a spatially distributed pattern. The eigenvector to the zero eigenvalue gives the direction of the constant gray values.
$\lambda_1 > 0, \lambda_2 > 0, \lambda_3 > 0$	3	The gray values change in all three directions.

structure tensor, are

$$J_{pq}(\mathbf{x}) = \int_{-\infty}^{\infty} w(\mathbf{x} - \mathbf{x}') \left(\frac{\partial g(\mathbf{x}')}{\partial x'_p} \frac{\partial g(\mathbf{x}')}{\partial x'_q} \right) \mathrm{d}^2 x' \,. \tag{25.79}$$

These equations indicate that a tensor is an adequate first-order representation of a local neighborhood. More-complex structures such as structures with multiple orientations cannot be distinguished.

By a rotation of the coordinate system, this can be brought into a diagonal form. Then, (25.78) reduces to

$$J' = (\bar{n}'_1, \bar{n}'_2) \begin{pmatrix} J'_{11} & 0 \\ 0 & J'_{22} \end{pmatrix} \begin{pmatrix} \bar{n}'_1 \\ \bar{n}'_2 \end{pmatrix} \,. \tag{25.80}$$

A unit vector $\bar{n}' = (\cos\theta \sin\theta)$ in the direction θ gives the values

$$J' = J'_{11} \cos^2 \theta + J'_{22} \sin^2 \theta \,.$$

Without loss of generality, we assume that $J'_{11} \geq J'_{22}$. Then, it is obvious that the unit vector $\bar{n}' = (1\ 0)^\mathrm{T}$ maximizes (25.80). The maximum value is J'_{11}. In conclusion, this approach not only yields a tensor representation for the local neighborhood but also shows the way to determine the orientation. Essentially, (25.78) constitutes an *eigenvalue problem*. The eigenvalues λ_w and eigenvectors $\boldsymbol{e}_w$ of a 2×2 matrix are defined by

$$\mathbf{J} \boldsymbol{e}_w = \lambda_w \boldsymbol{e}_w \,. \tag{25.81}$$

An eigenvector $\boldsymbol{e}_w$ of $\mathbf{J}$ is thus a vector that is not turned in direction by multiplication by the matrix $\mathbf{J}$ but is only multiplied by a scalar factor, the eigenvalue λ_w. This implies that the structure tensor becomes diagonal in a coordinate system that is spanned by the eigenvectors (25.80). For a symmetric matrix the eigenvalues are all real and nonnegative, and the eigenvectors form an orthogonal basis. According to the *maximization problem* formulated here, the eigenvector to the maximum eigenvalue gives the orientation of the local neighborhood.

Classification of Local Neighborhoods

The power of the tensor representation becomes apparent if we classify the eigenvalues of the structure tensor. The classifying criterion is the number of eigenvalues that are zero. If an eigenvalue is zero, this means that the gray values in the direction of the corresponding eigenvector do not change. The number of zero eigenvalues is also closely related to the rank of a matrix. The *rank* of a matrix is defined as the dimension of the subspace for which $\mathbf{J}\boldsymbol{k} \neq \boldsymbol{0}$. The space for which is $\mathbf{J}\boldsymbol{k} = \boldsymbol{0}$ is denoted as the *null space*. The dimension of the null space is the dimension of the matrix minus the rank of the matrix and is equal to the number of zero eigenvalues. We will perform an analysis of the eigenvalues for two and three dimensions. In two and three dimensions, we can distinguish the cases summarized in Tables 25.2 and 25.3, respectively.

In practice, it will not be checked whether the eigenvalues are zero but below a critical threshold that is determined by the noise level in the image.

Orientation Vector

With the simple convolution and point operations discussed in the previous section, we computed the components of the structure tensor. In two dimensions, we can readily solve the eigenvalue problem. The orientation angle can be determined by rotating the inertia tensor into the principal axes coordinate system. The orientation angle is given by

$$\tan 2\theta = \frac{2J_{12}}{J_{22} - J_{11}} \,. \tag{25.82}$$

Because $\tan 2\theta$ is gained from a quotient, we can regard the dividend as the y and the divisor as the x component of a vector and can form the *orientation vector* $\boldsymbol{o}$, as introduced by Granlund [25.8]:

$$\boldsymbol{o} = \begin{pmatrix} J_{22} - J_{11} \\ 2J_{12} \end{pmatrix} \,. \tag{25.83}$$

The argument of this vector gives the orientation angle and the magnitude a certainty measure for the local *orientation*. The term orientation is used in all cases where an angle range of only $180°$ is required. It is not possible to distinguish between patterns that are rotated by $180°$. Orientation is still, of course, a *cyclic* quantity.

The orientation vector can be well represented as a color image. It appears natural to map the certainty measure onto the luminance and the orientation angle as the hue of the color. Our attention is then drawn to the bright parts in the images where we can distinguish the colors well. The darker a color is, the more difficult it becomes to distinguish the different colors visually. In this way, our visual impression coincides with the orientation information in the image.

Structure Operator and Coherency

The orientation vector reduces local structure to local orientation. From three independent components of the symmetric tensor only two are used. When we fail to observe an orientated structure in a neighborhood, we do not know whether no gray value variations or distributed orientations are encountered. This information is included in the not yet used component of the tensor, $J_{11} + J_{22}$, which gives the mean square magnitude of the gradient. Consequently, a well-equipped structure operator also needs to include the third component. A suitable linear combination is

$$s = \begin{pmatrix} J_{11} + J_{22} \\ J_{22} - J_{11} \\ 2J_{12} \end{pmatrix}. \tag{25.84}$$

This structure operator contains the two components of the orientation vector and, as an additional component, the mean square magnitude of the gradient, which is a rotation-invariant parameter. Comparing the latter with the magnitude of the orientation vector, a constant gray value area and an isotropic gray value structure without a preferred orientation can be distinguished. In the first case, both squared quantities are zero; in the second only the magnitude of the orientation vector. In the case of a perfectly oriented pattern, both quantities are equal. Thus their ratio seems to be a good *coherency measure* c_c for local orientation:

$$c_c = \frac{\sqrt{(J_{22} - J_{11})^2 + 4J_{12}^2}}{J_{11} + J_{22}} = \frac{\lambda_1 - \lambda_2}{\lambda_1 + \lambda_2}. \tag{25.85}$$

The coherency ranges from 0 to 1. For ideal local orientation ($\lambda_2 = 0$, $\lambda_1 > 0$) it is 1, for an isotropic gray value structure ($\lambda_1 = \lambda_2 > 0$) it is 0.

A color representation of the structure tensor requires only two slight modifications compared to the color representation for the orientation vector. First, instead of the length of the orientation vector, the squared magnitude of the gradient is mapped onto the intensity. Second, the coherency measure (25.85) is used as the saturation. In the color representation for the orientation vector, the saturation is always one. The angle of the orientation vector is still represented as the hue.

In practice, a slight modification of this color representation is useful. The squared magnitude of the gradient shows variations too large to be displayed in the narrow dynamic range of a display screen with only 256 luminance levels. Therefore, a suitable normalization is required. The basic idea of this normalization is to compare the squared magnitude of the gradient

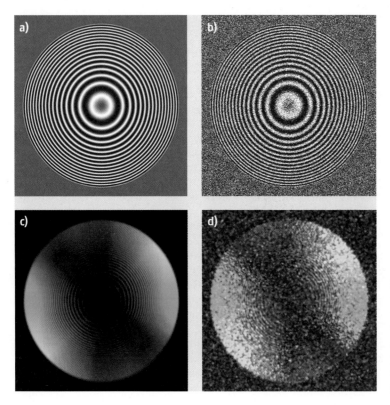

Fig. 25.11a–d Demonstration of the computation of the structure tensor with the ring test pattern: (**a**) ring without noise, (**b**) ring with a signal-to-noise ratio of 2, (**c**) color presentation of the structure tensor computed from (**a**), (**d**) the same from (**b**)

with the noise level. Once the gradient is well above the noise level it is regarded as a significant piece of information. This train of thoughts suggests the following normalization for the intensity I:

$$I = \frac{J_{11} + J_{22}}{(J_{11} + J_{22}) + \gamma \sigma_n^2}, \quad (25.86)$$

where σ_n is an estimate of the standard deviation of the noise level. This normalization provides a rapid transition of the luminance from 1, when the magnitude of the gradient is larger than σ_n, to 0 when the gradient is smaller than σ_n. The factor γ is used to optimize the display.

Implementation

The structure tensor can be computed straightforwardly as a combination of *linear convolution* and *nonlinear point operations*. The partial derivatives in (25.79) are approximated by discrete derivative operators. The integration weighted with the window function is replaced by a convolution with a smoothing filter that has the shape of the window function. If we denote the discrete partial derivative operator with respect to the coordinate p by the operator $\mathcal{D}_p$ and the (isotropic) smoothing operator by $\mathcal{B}$, the local structure of a gray value image can be computed with the structure tensor operator

$$\mathcal{J}_{pq} = \mathcal{B}(\mathcal{D}_p \cdot \mathcal{D}_q). \quad (25.87)$$

The equation is written in an operator notation. Pixel-wise multiplication is denoted by the dot to distinguish it from successive application of convolution operators.

These operators are valid in images of any dimension $W \geq 2$. In a W-dimensional image, the structure tensor has $W(W+1)/2$ independent components, hence three in 2-D, six in 3-D, and ten in 4-D images. These components are best stored in a multichannel image with $W(W+1)/2$ components.

The smoothing operations consume the largest number of operations. Therefore, a fast implementation must, in the first place, apply a fast smoothing algorithm. A fast algorithm can be established based on the general observation that higher-order features always show a lower resolution than the features from which they are computed. This means that the structure tensor can be stored on a coarser grid and thus in a smaller image. A convenient and appropriate subsampling rate is to reduce the scale by a factor of two by storing only every second pixel in every second row.

The accuracy of the orientation angle depends strongly on the implementation of the derivative filters. The straightforward implementation of the algorithm using the standard derivative filter mask $(1/2)[1\ 0\ -1]$ or the *Sobel operator* results in surprisingly high errors, with a maximum error in the orientation angle of more than $7°$ at a wavenumber of $\tilde{k} = 0.7$. The error depends on both the wavenumber and the orientation of the local structure. For orientation angles in the direction of axes and diagonals, the error vanishes.

The error in the orientation angle can be suppressed significantly if better derivative filters are used. The little extra effort invested in optimizing the derivative filters thus pays off in an accurate orientation estimate (Fig. 25.11b). A residual angle error of less than $0.5°$ is sufficient for almost all applications. The various derivative filters discussed for edge and line extraction give the freedom to balance computational effort with accuracy.

Even with a low signal-to-noise ratio, the orientation estimate is still correct if a suitable derivative operator is used. With increasing noise level, the coherency decreases and the statistical error of the orientation angle estimate increases (Fig. 25.11d).

Energy Tensor

Recently, a phase-invariant extension of the structure tensor was proposed, named the *energy tensor*, $\mathbf{E}$, de-

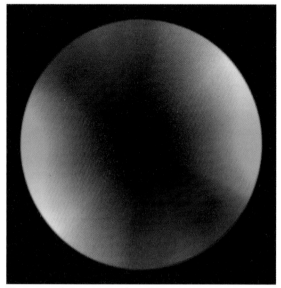

Fig. 25.12 Color presentation of the energy tensor (compare with Fig. 25.11c) [25.9]

fined as

$$\mathbf{E} = \nabla g \cdot \nabla g^T - g \cdot \mathbf{H}g$$

$$= \begin{pmatrix} g_x^2 - gg_{xx} & g_xg_y - gg_{xy} \\ g_xg_y - gg_{xy} & g_y^2 - gg_{yy} \end{pmatrix}, \quad (25.88)$$

where $\mathbf{H}$ is the Hessian matrix. The energy tensor can be computed accurately using the filter optimization techniques described in Sect. 25.1.5 if the second-order derivative filters are computed by consecutive application of first-order filters.

The energy tensor requires no averaging. Firstly, this constitutes a significant saving in terms of number of computing operations. Secondly, the energy tensor gives better results. This can be demonstrated by computing the structure tensor and energy tensor of the ring test pattern in Fig. 25.11. While the averaging of the structure tensor is not sufficient at large wavelengths (small wavenumbers) close to the center of the ring pattern, this effect does not show up in the energy tensor (Fig. 25.12).

25.1.7 Local Wavenumber and Local Phase

Local structure is not only characterized by orientation but also by a local scale that can be represent by a local wavenumber, i.e., the number of periods of a structure per unit length. The determination of the amplitude, phase, wavenumber, and orientation is, of course, a central image processing task for any type of technique delivering fringe patterns.

Phase

The key to determining the local wavenumber is the *phase* of the signal. Consider the one-dimensional periodic signal

$$g(x) = g_0 \cos(kx). \quad (25.89)$$

The argument of the cosine function is known as the phase of the periodic signal

$$\phi(x) = kx. \quad (25.90)$$

Thus the phase is a *linear function* of the position and the wavenumber. The wavenumber of the periodic signal is given by the first-order spatial derivative of the phase signal

$$\frac{\partial \phi(x)}{\partial x} = k. \quad (25.91)$$

Hilbert Filter and Analytic Signal

The key to determining the phase is an operator that delays the signal by a phase of 90°. This operator would convert the $g(x) = g_0 \cos(kx)$ signal into a $g'(x) = -g_0 \sin(kx)$ signal. Using both signals, the phase of $g(x)$ can be computed by

$$\phi(g(x)) = \arctan\left(\frac{-g'(x)}{g(x)}\right). \quad (25.92)$$

As only the ratio of $g'(x)$ and $g(x)$ goes into (25.92), the phase is indeed independent of amplitude. Together with the signs of the two functions $g'(x)$ and $g(x)$, the phase can be computed over the full range of 360°.

Thus the phase of a signal is determined by a linear operator that shifts the phase of a signal by 90°. Such an operator is known as the *Hilbert filter* $\mathbf{H}$ or *Hilbert operator* $\mathcal{H}$ and has the transfer function

$$\hat{h}(k) = \begin{cases} i & k > 0 \\ 0 & k = 0 \\ -i & k < 0 \end{cases}. \quad (25.93)$$

The magnitude of the transfer function is 1, as the amplitude remains unchanged. As the Hilbert filter has a purely imaginary transfer function, it must have odd symmetry to generate a real-valued signal. Therefore positive wavenumbers are shifted by 90° ($\pi/2$) and negative wavenumbers by $-90°$ ($-\pi/2$). A special situation is given for zero wavenumber, for which the transfer function is 0. A signal with zero wavenumber is a constant and can be regarded as a cosine function with infinite wavenumber sampled at the phase zero. Consequently, the Hilbert-filtered signal is the corresponding sine function at phase zero, i.e., zero.

Because of the discontinuity of the transfer function of the Hilbert filter at the origin, its point spread function is of infinite extent

$$h(x) = -\frac{1}{\pi x}. \quad (25.94)$$

The convolution with (25.94) can be written

$$g_h(x) = \frac{1}{\pi} \int_{-\infty}^{\infty} \frac{g(x')}{x' - x} dx'. \quad (25.95)$$

This integral transform is known as the *Hilbert transform* [25.10].

Because the convolution mask of the Hilbert filter is infinite, it is impossible to design an exact discrete Hilbert filter for arbitrary signals. This is only possible if we restrict the class of signals to which it is applied. Thus the following approach is taken to design an effective implementation of a Hilbert filter.

First, the filter should shift the phase by precisely $\pi/2$. This requirement comes from the fact that we cannot afford an error in the phase because it includes the position information. A wavenumber-dependent phase shift would cause wavenumber-dependent errors. This requirement is met by any convolution kernel of odd symmetry.

Second, the requirements for a magnitude of 1 can be relaxed if the Hilbert filter is applied to a band-passed signal. Then, the Hilbert filter must only show a magnitude of one in the pass-band range of the bandpass filter used. This approach avoids discontinuities in the transfer function at the wavenumber 0 and thus results in finite-sized convolution kernels. Optimized Hilbert filters are discussed by Jähne [25.11].

A real-valued signal and its Hilbert transform can be combined into a complex-valued signal by

$$g_a = g - ig_h .\tag{25.96}$$

This complex-valued signal is denoted as the *analytic function* or analytic signal. According to (25.96) the analytic filter has the point spread function

$$a(x) = 1 + \frac{i}{\pi x} \tag{25.97}$$

and the transfer function

$$\hat{a}(k) = \begin{cases} 2 & k > 0 \\ 1 & k = 0 \\ 0 & k < 0 \end{cases} .\tag{25.98}$$

Thus all negative wavenumbers are suppressed. Although the transfer function of the analytic filter is real, it results in a complex signal because it is asymmetric. For a real signal no information is lost by suppressing the negative wavenumbers. They can be reconstructed as the Fourier transform of a real signal is Hermitian. The analytic signal can be regarded as just another representation of a real signal with two important properties. The magnitude of the analytic signal gives the *local amplitude*

$$|a|^2 = g^2 + g_h^2 .\tag{25.99}$$

and the argument the *local phase*

$$\phi = \arg(a) = \arctan\left(\frac{-g_h}{g}\right) .\tag{25.100}$$

The original signal and its Hilbert transform can be obtained from the analytic signal using (25.96) by

$$g(x) = [g_a(x) + g_a^*(x)]/2$$
$$g_h(x) = i[g_a(x) - g_a^*(x)]/2 .\tag{25.101}$$

To determine the local wavenumber, the first spatial derivative of the phase signal is computed (25.91). This derivative has to be applied in the same direction as the Hilbert or quadrature filter has been applied. However, direct computation of the partial derivatives is not advisable, because of the inherent discontinuities in the phase signal. A phase computed with the inverse tangent restricts the phase to the main interval $[-\pi, \pi[$ and thus inevitably leads to a *phase wrapping* from π to $-\pi$ with the corresponding discontinuities.

If only the local wavenumber is of interest, this problem can be avoided by computing the phase derivative directly from the derivatives of g and g_h [25.12]. The result is

$$k = \frac{\partial}{\partial x} \arctan(-g_h/g) = \frac{g_h \partial g/\partial x - g \partial g_h/\partial x}{g^2 + g_h^2} .\tag{25.102}$$

This procedure to compute the phase gradient also eliminates the need to use trigonometric functions and is, therefore, significantly faster.

Phase in Higher-Dimensional Signals and the Monogenic Signal

The concept of the analytic signal makes it possible to extend the ideas of local phase into multiple dimensions. The transfer function of the analytic operator uses only the positive wavenumbers, i.e., only half of the Fourier space. If we extend this partitioning to multiple dimensions, we have more than one choice to partition the Fourier space into two half-spaces. Instead of the wavenumber, we can take the scalar product between the wavenumber vector k and any unit vector $\bar{n}$ and suppress the half-space for which the scalar product $k\bar{n}$ is negative:

$$\hat{a}(k) = \begin{cases} 2 & k\bar{n} > 0 \\ 1 & k\bar{n} = 0 \\ 0 & k\bar{n} < 0 \end{cases} .\tag{25.103}$$

The unit vector $\bar{n}$ gives the direction in which the Hilbert filter is to be applied. The definition (25.103) of the transfer function of the analytic signal implies that the Hilbert operator can only be applied to directionally filtered signals. This results from the following considerations. For one-dimensional signals we have seen that a discrete Hilbert filter does not work well for small wavenumbers. In multiple dimensions this means that a Hilbert filter does not work well if $\hat{k}\bar{n} \ll 1$. Thus no wavenumbers near an orthogonal to the direction of the Hilbert filter may exist, in order to avoid errors.

This fact makes the application of Hilbert filters and thus the determination of the local phase in higher-dimensional signals significantly more complex. It is not sufficient to use isotropic band-pass-filtered images. In addition, the band-pass-filtered images must be further decomposed into directional components. At least as many directional components as the dimensionality of the space are required.

The extension of the Hilbert transform from a 1-D signal to higher-dimensional signals is not satisfactory because it can only be applied to directionally filtered signals. For wavenumbers close to the separation plane, the Hilbert transform does not work. What is really required is an isotropic extension of the Hilbert transform.

A vector-valued extension of the analytic signal meets both requirements. It is known as the *monogenic signal* and was introduced to image processing by Felsberg and Sommer [25.13]. The monogenic signal is constructed from the original signal and its *Riesz transform*. The transfer function of the Riesz transform is given by

$$\hat{h}(k) = i\frac{k}{|k|}. \quad (25.104)$$

The magnitude of the vector h is 1 for all values of k. The Riesz transform is thus isotropic. It also has odd symmetry because

$$\hat{h}(-k) = -\hat{h}(k). \quad (25.105)$$

The Riesz transform can be applied to a signal of any dimension. For a 1-D signal it reduces to the Hilbert transform.

For a 2-D signal the transfer function of the Riesz transform can be written using polar coordinates as

$$\hat{h}(k) = i(\cos\theta, \sin\theta)^T. \quad (25.106)$$

The transfer function is similar to the transfer function for the gradient operator (25.61). The convolution mask or point spread function (PSF) of the Riesz transform is given by

$$h(x) = -\frac{x}{2\pi|x|^3}. \quad (25.107)$$

The original signal and the signal convolved by the Riesz transform can be combined for a 2-D signal to the 3-D monogenic signal as

$$g_m(x) = (p, q_1, q_2)^T \quad (25.108)$$

with $p = g$, $q_1 = h_1 * g$, and $q_2 = h_2 * g$. The local amplitude of the monogenic signal is given as the norm of the vector of the monogenic signal as in the case of the analytic signal (25.99):

$$|g_m|^2 = p^2 + q_1^2 + q_2^2. \quad (25.109)$$

The monogenic signal does not only give an estimate for the *local phase* ϕ as the *analytic signal* does. The monogenic signal also gives an estimate of the *local orientation* θ by the following relations:

$$p = a\cos\phi,$$
$$q_1 = a\sin\phi\cos\theta,$$
$$q_2 = a\sin\phi\sin\theta. \quad (25.110)$$

Therefore the monogenic signal combines the estimation of local orientation and local phase. This is of high significance for image processing because the two most important features of a local neighborhood, the local orientation and the local wavenumber can be estimated in a unified way.

It is significantly more complex to compute the local wavenumber from the *monogenic signal*, because there are three signals in two dimensions. From (25.110) we obtain two different equations for the phase:

$$\phi_1 = \text{arccot}\left(\frac{p\cos\theta}{q_1}\right), \quad \phi_2 = \text{arccot}\left(\frac{p\sin\theta}{q_2}\right). \quad (25.111)$$

It is necessary to combine these equations because each of them gives no result for certain directions. The solution is use the *directional derivative*. When differentiating the phase in the direction of the wavenumber vector, the magnitude of the wavenumber vector is obtained:

$$k = \frac{\partial\phi}{\partial \bar{k}} = \cos\theta\frac{\partial\phi_1}{\partial x} + \sin\theta\frac{\partial\phi_2}{\partial y}. \quad (25.112)$$

The terms $\cos\theta$ and $\sin\theta$ can be obtained from (25.110):

$$\cos^2\theta = \frac{q_1^2}{q_1^2 + q_1^2} \quad \text{and} \quad \sin^2\theta = \frac{q_2^2}{q_1^2 + q_1^2}. \quad (25.113)$$

Then the magnitude of the wavenumber vector results in

$$k = \frac{p(q_{1x} + q_{2y}) - q_1 p_x - q_2 p_y}{p^2 + q_1^2 + q_1^2}. \quad (25.114)$$

The components of the wavenumber vector $k = (k\cos\theta, h\sin\theta)$ can be computed by combining (25.114) and (25.112).

Quadrature Filters

Quadrature filters is an alternative approach to getting a pair of signals that differ only by a phase shift of

90° ($\pi/2$). It is easiest to introduce the complex form of the quadrature filters. Essentially, the transfer function of a *quadrature filter* is also zero for $k\bar{n} < 0$, like the transfer function of the analytic filter. However, the magnitude of the transfer function is not one but can be any arbitrary real-valued function $h(k)$:

$$\hat{q}(k) = \begin{cases} 2h(k) & k\bar{n} > 0 \\ 0 & \text{otherwise} \end{cases}. \quad (25.115)$$

The quadrature filter thus also transforms a real-valued signal into an analytical signal. In contrast to the analytical operator, a wavenumber weighting is applied. From the complex form of the quadrature filter, we can derive the real quadrature filter pair by observing that they are the part of (25.115) with even and odd symmetry. Thus

$$\hat{g}_+(k) = [\hat{q}(k) + \hat{q}(-k)]/2,$$
$$\hat{g}_-(k) = [\hat{q}(k) - \hat{q}(-k)]/2. \quad (25.116)$$

The even and odd part of the quadrature filter pair show a phase shift of 90° and can thus also be used to compute the local phase.

Quadrature filters can also be designed on the basis of the monogenic signal. These quadrature filters have one component more than the dimension of the signal. The transfer function is

$$[\hat{h}(k), \mathrm{i}k\hat{h}(k)/|k|]^\mathsf{T}. \quad (25.117)$$

The best-known quadrature filter pair is the *Gabor filter*. A Gabor filter is a bandpass filter that selects a certain wavelength range around the center wavelength k_0 using the Gauss function. The complex transfer function of the Gabor filter is

$$\hat{g}(k) = \begin{cases} \exp\left(|k - k_0|^2 \sigma_x^2/2\right) & kk_0 > 0 \\ 0 & \text{otherwise} \end{cases}. \quad (25.118)$$

If $|k_0|\sigma_x > 3$, (25.118) can be approximated by

$$\hat{g}(k) \approx \exp\left(-|k - k_0|^2 \frac{\sigma_x^2}{2}\right). \quad (25.119)$$

Using the relations in (25.116), the transfer function for the even and odd component are given by

$$\hat{g}_\pm(k) = \left[\exp\left(-|k - k_0|^2 \frac{\sigma_x^2}{2}\right) \pm \exp\left(-|k + k_0|^2 \frac{\sigma_x^2}{2}\right)\right]. \quad (25.120)$$

The point spread function of these filters is

$$g_+(x) = \cos(k_0 x) \exp\left(-\frac{|x|^2}{2\sigma_x^2}\right),$$
$$g_-(x) = \mathrm{i}\sin(k_0 x) \exp\left(-\frac{|x|^2}{2\sigma_x^2}\right), \quad (25.121)$$

or combined into a complex filter mask:

$$g(x) = \exp(\mathrm{i}k_0 x) \exp\left(-\frac{|x|^2}{2\sigma_x^2}\right). \quad (25.122)$$

25.1.8 Multiscale Processing

The powerful concept of neighborhood operations is only the starting point for image analysis. This class of operators can only extract local features at scales of at most a few pixels distance. It is obvious that images contain information also at larger scales. To extract object features at these larger scales, we need correspondingly larger filter masks. The use of large masks, however, results in a significant increase in computational costs. If we use a mask of size R^W in a W-dimensional image the number of operations is proportional to R^W. Thus a doubling of the scale leads to a four- and eightfold increase in the number of operations in 2- and 3-D images, respectively.

The explosion in computational cost is only the superficial expression of a problem with deeper roots. The more important question is at which scale can a certain feature in an image be detected in an optimal way? This scale depends, of course, on the characteristic scales contained in the object to be detected. Optimal processing of an image thus requires the representation of an image at different scales.

If an $N \times N$ image is represented on a grid in the spatial domain, we do not have any information at all about the wavenumbers contained at that point in the image. We know the position with an accuracy of the grid constant Δx, but the local wavenumber at this position may be anywhere in the range of the possible wavenumbers from 0 to $N\Delta k = 1/\Delta x$.

In the wavenumber representation, each pixel represents one wavenumber with a wavenumber resolution of $\Delta k = 1/(N\Delta x)$. However, any positional information is lost, as one point in the wavenumber space represents a periodic structure that is spread over the whole image. Thus the positional uncertainty is $N\Delta x$.

This discussion shows that the representation of an image in either the spatial or wavenumber domain constitutes one of two opposite extremes. Either the spatial or the wavenumber resolution is maximal, but the resolution in the other domain is completely lost. A multiscale image representation requires a type of joint resolution that allows for a separation into different wavenumber ranges (scales) but still preserves as much spatial resolution as possible.

This can be done in the most efficient way in a *multigrid representation*. The basic idea is simple. While the representation of fine scales requires the full resolution, coarse scales can be represented at lower resolution. This leads to a scale space with smaller and smaller images as the scale parameter increases.

Gaussian Pyramid

If we want to reduce the size of an image, we cannot just *subsample* the image by taking, for example, every second pixel in every second line. This would disregard the *sampling theorem*. For example, a structure that is sampled three times per wavelength in the original image would only be sampled one and a half times in the subsampled image and thus appear as an aliased pattern. Consequently, we must ensure that all structures that are sampled fewer than four times per wavelength are suppressed by an appropriate smoothing filter to ensure a properly subsampled image. This means that size reduction must go hand in hand with appropriate smoothing.

Generally, the requirement for the smoothing filter can be formulated as

$$\hat{B}(\tilde{\mathbf{k}}) = 0 \quad \forall \tilde{k}_p \geq \frac{1}{r_p}, \qquad (25.123)$$

where r_p is the subsampling rate in the direction of the p-th coordinate.

The combined smoothing and size reduction can be expressed in a single operator by using the following notation to compute the $(q+1)$-th level of the Gaussian pyramid from the q-th level:

$$\mathbf{G}^{(0)} = \mathbf{G}, \quad \mathbf{G}^{(q+1)} = \mathcal{B}_{\downarrow 2} \mathbf{G}^{(q)}. \qquad (25.124)$$

The number next to the ↓ in the index denotes the subsampling rate. The 0-th level of the pyramid is the original image.

Repeated smoothing and subsampling operations result in a series of images called the *Gaussian pyramid*. From level to level, the resolution decreases by a factor of two; the size of the images decreases correspondingly. Consequently, we can think of the series of images as being arranged in the form of a pyramid, as illustrated in Fig. 25.13.

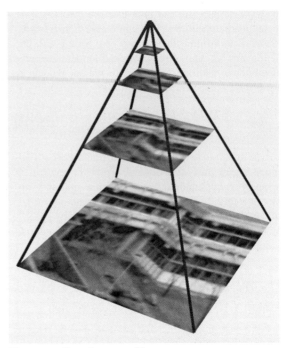

Fig. 25.13 Gaussian pyramid

The pyramid does not require much storage space. Generally, if we consider the formation of a pyramid from a W-dimensional image with a subsampling factor of 2 and M pixels in each coordinate direction, the total number of pixel is given by

$$M^W \left(1 + \frac{1}{2^W} + \frac{1}{2^{2W}} + \ldots\right) < M^W \frac{2^W}{2^W - 1}. \qquad (25.125)$$

For a two-dimensional image, the whole pyramid needs only one third more space than the original image, and for a three-dimensional image only one seventh more. The computation of the pyramid is equally effective. The *same* smoothing filter is applied to each level of the pyramid. Thus the computation of the *whole* pyramid only needs 4/3 and 8/7 times more operations than for the first level of a two-dimensional and three-dimensional image, respectively.

The pyramid brings large scales into the range of local neighborhood operations with small kernels. Moreover, these operations are performed efficiently. Once the pyramid has been computed, neighborhood operations on large scales in the upper levels of the pyramid are much more efficient than for finer scales because of the smaller image sizes.

The Gaussian pyramid constitutes a series of low-pass-filtered images in which the cut-off wavenumbers decrease by a factor of two (an octave) from level to level. Thus only the coarser details remain in the smaller images (Fig. 25.13). Only a few levels of the pyramid are necessary to span all possible wavenumbers. For an $N \times N$ image we can compute at most a pyramid with $\mathrm{ld}\,N + 1$ levels. The smallest image consists of a single pixel.

Laplacian Pyramid

From the Gaussian pyramid, another pyramid type can be derived, the *Laplacian pyramid*, which leads to a sequence of band-pass-filtered images. In contrast to the Fourier transform, the Laplacian pyramid only leads to a coarse wavenumber decomposition without a directional decomposition. All wavenumbers, independent of their direction, within the range of about an octave (a factor of two) are contained in one level of the pyramid.

Because of the coarse wavenumber resolution, we can preserve good spatial resolution. Each level of the pyramid only contains matching scales, which are sampled a few times (two to six) per wavelength. In this way, the Laplacian pyramid is an efficient data structure that is well adapted to the limits of the product of wavenumber and spatial resolution set by the *uncertainty relation*.

In order to achieve this, we subtract two levels of the Gaussian pyramid. This requires an upsampling of the image at the coarser level. This operation is performed by an *expansion operator* $\uparrow_2$. The degree of

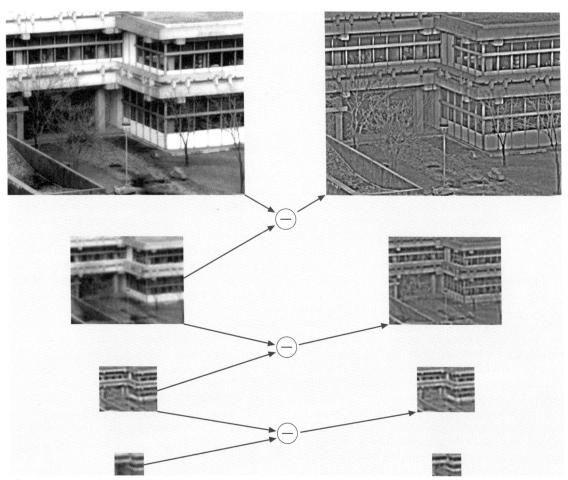

Fig. 25.14 Construction of the Laplacian pyramid (*right column*) from the Gaussian pyramid (*left column*) by subtracting two consecutive planes of the Gaussian pyramid

expansion or upsampling is denoted by the figure after the $\uparrow$ in the index, in a similar notation as for the *reduction operator* (25.124).

The expansion is significantly more difficult than the size reduction as the missing information must be interpolated. For a size increase of two in all directions, first every second pixel in each row must be interpolated and then every second row. Interpolation is discussed in detail in Sect. 25.1.3. Using this notation, the generation of the p-th level of the Laplacian pyramid can be written

$$L^{(p)} = G^{(p)} - \uparrow_2 G^{(p+1)}, \quad L^{(P)} = G^{(P)}. \quad (25.126)$$

The Laplacian pyramid is an effective scheme for a *bandpass decomposition* of an image. The center wavenumber is halved from level to level. The last image of the Laplacian pyramid $L^{(P)}$ is a low-pass-filtered image $G^{(P)}$ containing only the coarsest structures.

The Laplacian pyramid has the significant advantage that the original image can be reconstructed quickly from the sequence of images in the Laplacian pyramid by recursively expanding the images and summing them. The recursion is the inverse of the recursion in (25.126). In a Laplacian pyramid with $p+1$ levels, the level p (where the counting starts from zero) is the coarsest level of the Gaussian pyramid. Then the level $p-1$ of the Gaussian pyramid can be reconstructed by

$$G^{(P)} = L^{(P)}, \quad G^{(p-1)} = L^{(p-1)} + \uparrow_2 G^p. \quad (25.127)$$

This is just an inversion of the construction scheme for the Laplacian pyramid. This means that, even if the interpolation algorithms required to expand the image contain errors, they affect only the Laplacian pyramid and not the reconstruction of the Gaussian pyramid from the Laplacian pyramid, as the same algorithm is used. The recursion in (25.127) is repeated with lower levels until level 0, i.e., the original image, is reached again. As illustrated in Fig. 25.14, finer and finer details become visible during the reconstruction process. Because of the progressive reconstruction of details, the Laplacian pyramid has also been used as a compact scheme for image compression. Nowadays, more-efficient schemes are available on the basis of wavelet transforms, but they operate on principles that are very similar to those of the Laplacian pyramid [25.14, 15].

25.2 Motion Analysis

25.2.1 General Considerations on Motion Analysis

Image-based whole-field velocimetry methods are used to measure the flow field of a fluid, based on the analysis of an image sequence visualizing the flow under consideration. A large number of different methods have been developed and successfully applied during the past two decades, with both camera/computer-hardware and image processing algorithms being constantly improved upon in terms of accuracy, spatial and temporal resolution, and dynamic range. In general, all these methods estimate the flow velocity by determining the displacements of some kind of image features in a number of successive frames (at least two). In a computer vision context, this displacement field is called the *optical flow* $f(x, t)$.

Computing optical flow based on motion analysis in general is one of the major issues of computer vision, with applications not restricted to fluid flow but including any kind of dynamic processes and scenes. In addition, photogrammetrists use matching methods that are closely related to motion analysis, e.g., to establish correspondences between two stereo images to compute a disparity map or to locate target patterns within an image. Accordingly, there is a huge amount of methods, algorithms, and publications spread out through the computer vision, photogrammetry and fluid mechanics literature. The vast terminology for the different methods might confuse the unfamiliar reader, particularly because notions like optical flow, image matching, image correlation, and tracking are not always used consistently by the different communities (or even within the same community).

The methods differ in the following aspects:

- What kind of image features are used?

    A1 Single particles, i.e., discrete features,
    A2 particle patterns, i.e., patterns of discrete features,
    A3 continuous features.

- What kind of input data is used for the velocity estimation?

    B1 Spatial information, i.e., positions of features in the image plane,

Table 25.4 Examples of different approaches to image-based velocity analysis

Method	Reference	Features	Data	Calculation
Standard PIV	*Willert* and *Gharib* [25.16], *Westerweel* [25.17]	A2	B1,B3	C1
Correlation-based tracking, correlation imaging velocimetry	*Fincham* and *Spedding* [25.18]	A2	B1,B3	C1
Image correlation velocimetry, adaptive least-squares matching	*Tokumaru* and *Dimotakis* [25.19], *Gruen* [25.20]	A2,A3	B1,B3	C2
Multi-grid PIV with deformable windows	*Scarano* and *Riethmuller* [25.21]	A2	B1,B3	C1
Hybrid PIV/PTV	*Cowen* and *Monismith* [25.22], *Bastiaans* et al. [25.23]	A1,A2	B1,B3	C1,C3
Two-frame tracking	*Baek* and *Lee* [25.24], *Ohmi* and *Li* [25.25]	A1	B1,B3	C3
Four-frame tracking	*Hassan* and *Canaan* [25.26], *Malik* et al. [25.27]	A1	B1,B2,B3	C3
Kalman filtering	*Takehara* et al. [25.28]	A1	B1,B2,B3	C3
Optical flow techniques	*Jähne* et al. [25.29]	A2,A3	B1,B2,B3,B4	C2

B2 temporal information, i.e., more than two frames are used,

B3 intensity, i.e., gray values of features,

B4 intensity gradients, i.e., local gray value differences.

- What is the computational approach to solve the motion correspondence problem?

C1 Cross-correlation,

C2 least-squares optimization (linear or nonlinear),

C3 discrete tracking techniques (kinematic models, combinatorial optimization, Kalman filters).

Algorithms using almost any combination of image features, input data, and calculation method can be found in the literature. Some examples are compiled in Table 25.4. A common classification is to divide the methods into two major groups: *region-based methods* and *feature-based methods*.

Region-based methods estimate the motion of gray value patterns within small image patches, so-called *interrogation areas* or *interrogation windows*. The most common region-based method used in fluid-mechanic applications is *particle image velocimetry* (PIV) (Sect. 25.2.2), whereas in computer vision and photogrammetry, *optical flow techniques* (Sect. 25.2.5) and *least-squares matching* (Sect. 25.2.3) are frequently used. Since in all these methods, the image is divided into a regularly spaced array of interrogation windows, the result is a displacement field on a regular grid (which may be the pixel grid itself, as in optical flow methods of computer vision).

In contrast, feature-based methods try to identify single objects in the image, segment them from the background, and follow their motion throughout an image sequence. Thus, feature-based methods yield randomly spaced velocity information, depending on the distribution of objects in the image. The most important feature-based method for flow visualization is *particle-tracking velocimetry* (PTV) (Sect. 25.2.4), where individual tracer particle images are the objects to be tracked.

In general, both classes of methods have advantages and disadvantages, which are outlined in the following sections. Note that there are also hybrid methods that try to combine the advantages of region- and feature-based approaches, and consequently have better performance in many cases.

To summarize, four important groups of methods can be distinguished: correlation-based analysis (Sect. 25.2.2), least-squares matching (Sect. 25.2.3), tracking techniques (Sect. 25.2.4), and optical flow methods (Sect. 25.2.5). In the remainder of this section, some general aspects of motion analysis, which apply equally to all the different approaches, are outlined.

Dynamic Range, Sampling Theorem, and Subpixel Accuracy

Dynamic Range. An important quantity characterizing the potential of a motion estimator is its *dynamic range* DR, which is defined as the ratio of the maximum to the minimum displacement that can be measured:

$$DR = \frac{\xi_{\max}}{\xi_{\min}}. \tag{25.128}$$

Obviously, a dynamic range as high as possible is desirable, in particular with regard to the measurement of turbulent flows, which may contain strong velocity fluctuations.

The fundamental limits on the dynamic range of a digital imaging method are related to the discrete nature of the image data. The measurement of large displacements is limited by the *temporal sampling theorem* [25.29], while the measurement of small displacements is limited by the maximum *subpixel accuracy* that can be achieved, which in turn depends on the sampling and quantization of the image intensity [25.29]. Approaches to overcome the limitations of large displacements and increase the dynamic range are outlined below. If such an approach is used together with a subpixel-accurate determination of small displacements (see below), a dynamic range of 100–1000 can be achieved using standard equipment.

Sampling Theorem and Motion Correspondence. To estimate an object's velocity given two successive image frames, the *motion correspondence problem* has to be solved, i.e., a unique correspondence between two images of the same object in the two successive frames has to be established. This can only be achieved, if the temporal *sampling theorem* is valid. In simple words, the (temporal) sampling theorem (or *Nyquist criterion*) states, that the motion between two images, i.e., the optical flow f, should be less than half the smallest local spatial scale $\lambda_{g,\min}$ of the image intensity $g(\mathbf{x}, t)$:

$$|f|\Delta t = |f| \cdot (1 \text{ frame}) < \frac{1}{2}\lambda_{g,\min} , \qquad (25.129)$$

where Δt is the time interval between two successive images, in units of frames (thus, $\Delta t = 1$), and f is the optical flow in units of pixel/frame. The sampling theorem puts a fundamental limit on the relation between the size and intensity structure of an object and its motion, i.e., on the relation between the spatial and temporal intensity gradients. Given just two images, the motion of an object can only be recovered unambiguously if (25.129) is valid. In this case, the motion correspondence problem can be solved. Otherwise, temporal aliasing occurs [25.29], and the problem of motion estimation becomes ill-posed, i.e., there is no unique solution.

As a consequence of the sampling theorem, there is a maximum allowed displacement that can be recovered by any region-based method. For example, consider a differential optical flow technique (Sect. 25.2.5). In this case, there has to be a unique relation between the spatial and temporal gray value gradients. To estimate the motion of a single particle with a symmetrical Gaussian intensity distribution (as typical for PIV and PTV particle images), the maximum allowed displacement corresponds roughly to the standard deviation of the Gaussian radius of the particle. As a second example, for a quadratic PIV interrogation window of length L, the maximum displacement corresponds roughly to $L/4$, assuming that the intensity distribution or particle density within the window is homogeneous and sufficiently large. The latter result is known as the one-quarter rule in the PIV literature [25.30]. Obviously, the smallest spatial scale within a PIV interrogation window depends on the particle distribution and density within that window. Generally, an optimal density of about 10 particles per interrogation window is recommended in the PIV literature. Assuming a homogeneous distribution, the particles form a periodic intensity pattern of wavelength $\lambda = L/2$. Hence, the one-quarter rule follows from the sampling theorem. Note that these limits are not strict but should be considered as more or less accurate estimates, since the actual spatial frequency content of the image depends on the particle distribution. The latter is a stochastic quantity, with varying values for different interrogation windows within one image.

The situation is a bit different for feature-based tracking methods. As a simple example, if only one object is visible in the image, its motion can be tracked with the only restriction that it stays within the field of view, since in any case two successive images of the object can be related to each other unambiguously. In this case, the wavelength of the spatial image structure corresponds to the size of the image (or twice this size). However, such a case is of limited practical importance, since there will always be more than one object to be tracked. With increasing object or particle density, the spatial image scales become smaller and the motion correspondence becomes more difficult, which again is a manifestation of the sampling theorem. Still, tracking algorithms are able to track motions violating the sampling theorem. However, the latter is only possible, if further information is used (apart from two successive images). For example, a common assumption is that object trajectories are smooth, i.e., the direction and speed of an object does not change abruptly between two frames. In this case, it is possible to use information from previous frames as input to a motion model and predict the position of the object in the next frame by extrapolation. If the model provides a good description of the actual motion, much larger displacements can be handled compared to approaches using only two frames without any modeling.

Subpixel Accuracy. The subpixel accuracy of a velocity estimator determines the minimum displacement that can be measured. Since a digital image provides a sampled version of the original intensity distribution of the physical image, with gray values defined on an integer grid (pixel positions), the position of an object, e.g., a particle image or the correlation peak resulting from the cross-correlation of two PIV images, can only be determined with an accuracy of ± 0.5 pixel. To achieve higher accuracy, some kind of subpixel interpolation has to be carried out. One way to do this is to use a model of the intensity distribution of the object and determine the best fit of this model to the image data in a least squares sense. The most common model in PIV and PTV is a Gaussian distribution, since it provides a very good approximation to both the image of a single tracer particle and the displacement peak in a PIV correlation. The subpixel-accurate coordinates are introduced as parameters of the model and determined using a least-squares algorithm.

Another common approach to achieve subpixel accuracy in PIV and optical flow techniques is to warp the original images according to an estimated flow model. The warping is carried out iteratively, and a refined techniques of the velocity field is computed in each iteration. Since the warped image will generally be defined on noninteger pixel positions, warping requires a precise method to interpolate gray values; see, e.g., Sect. 25.1.3.

In applying subpixel interpolation, one should keep the following considerations in mind. To compute a subpixel-accurate position within an image, the information contained in the image intensity, i. e., in the gray values, is translated into geometric information, i. e., position in the image. This translation is based on certain assumptions concerning radiometric aspects of the imaging process. One such assumption is the Gaussian intensity distribution mentioned above. Further important assumptions, which are often taken for granted, are the linearity and homogeneity of the sensor and a homogeneous illumination. If any of these assumptions is violated, subpixel accuracy will deteriorate or even become meaningless. Thus, it is very important to take into account the radiometric properties of the cameras and illumination, if very high accuracy is required. For example, if the cameras suffer from strong fixed pattern noise, a radiometric correction should be applied to the images. Even if the image data is perfect and all the assumptions are valid, the result of the subpixel interpolation may still be biased. For example, one source of bias in PIV evaluation is the so-called peak-locking effect.

As a general limit, for typical 8 bit images with 256 gray levels, one can expect a (theoretical) maximum subpixel accuracy of the order of magnitude of 0.01 pixel, given optimal image data, a good object model, and an unbiased estimator. Note that it may be very difficult to actually achieve such ideal circumstances in real PIV or PTV applications, where measurement errors are typically in the range of 0.05–0.2 pixel.

Hierarchical Multigrid Approaches

As explained in the previous section, the maximum displacement that can be determined by region-based approaches such as PIV and differential optical flow methods is limited by the smallest spatial scales of the underlying image structure. However, images also contain information at larger scales than the neighborhood size of the interrogation windows. The basic idea of iterative, hierarchical methods is to start the estimation of the optical flow at the largest image scales, which enable the determination of large displacements in a first iteration. This first estimation may be applied to warp the second image back along the estimated displacement field and refine the estimation at smaller spatial scales. An efficient implementation of such a coarse-to-fine strategy is a *Gaussian image pyramid* (Sect. 25.1.8), which is basically a multigrid representation of an image at different spatial scales. The efficiency of Gaussian pyramids is due to the reduction of the linear image size by a factor of two at each level of the pyramid. This reduction makes the large-scale information in the image available to small filter masks. However, at the same time the image becomes increasingly blurred. Hence, we have to take care in applying Gaussian pyramids to PIV images, since the small particle images may soon be completely smoothed out. Large-scale information can only be obtained if there is a certain fraction of larger particles in the image or the particle density varies locally. Hierarchical PIV approaches are often realized by starting with large interrogation windows and iteratively decreasing the size of the interrogation windows instead of decreasing the image size as in a Gaussian pyramid.

Modeling of Displacement Fields

Given two successive images $g_0 = g(\mathbf{x}, t_0)$ and $g_1 = g(\mathbf{x}, t_1)$ of a flow field, the displacement field $\boldsymbol{\xi}(\mathbf{x}, t)$ can be thought of as the transformation, or mapping, of the spatial image intensity field from the first image to the second. The optical flow is the time derivative of this mapping: $\boldsymbol{f}(\mathbf{x}, t) = \partial_t \boldsymbol{\xi}(\mathbf{x}, t)$. Within a local neighborhood N centered at $\mathbf{x}_0$, the displacement field may be approximated by a Taylor expansion

$$\xi(x,t) = \xi(x_0, t) + (x - x_0)\nabla \xi(x_0, t)$$
$$+ \frac{1}{2!}[(x-x_0)\nabla]^2 \xi(x_0, t) + \ldots$$

Taking into account only the first-order terms, the equivalent formulation for the optical flow reads

$$f(x,t) = \begin{pmatrix} a_1 & a_2 \\ a_3 & a_4 \end{pmatrix} \begin{pmatrix} x - x_0 \\ y - y_0 \end{pmatrix} + \begin{pmatrix} a_5 \\ a_6 \end{pmatrix} \quad (25.130)$$
$$= A(x - x_0) + t . \quad (25.131)$$

In this first-order approximation, the displacement field consists of a constant shift t and a linear (affine) deformation of the local neighborhood, described by the matrix A. Note that in such a formulation, the spatial derivatives of the flow field are introduced as parameters:

$$a_1 = \frac{\partial f_x}{\partial x}, \quad a_2 = \frac{\partial f_x}{\partial y},$$
$$a_3 = \frac{\partial f_y}{\partial x}, \quad a_4 = \frac{\partial f_y}{\partial y}. \quad (25.132)$$

This offers the possibility to estimate spatial velocity gradients without performing explicit differentiation of the velocity field. Thus, important hydromechanic quantities like the in-plane vorticity $\omega_z = \partial_x f_y - \partial_y f_x$ and the rate-of-strain tensor S can be directly estimated, since

$$A = \begin{pmatrix} 0 & -\frac{\omega_z}{2} \\ \frac{\omega_z}{2} & 0 \end{pmatrix} + S, \quad (25.133)$$

with

$$S = \begin{pmatrix} \partial_x f_x & \frac{1}{2}(\partial_y f_x + \partial_x f_y) \\ \frac{1}{2}(\partial_y f_x + \partial_x f_y) & \partial_y f_y \end{pmatrix}. \quad (25.134)$$

Similar to this spatial modeling of the displacement fields, the temporal evolution of the motion of a single particle along its Lagrangian trajectory around the point x_0 may be approximated using a Taylor expansion in time:

$$\xi(x_0, t) = \xi(x_0, t_0) + v(t - t_0) + \frac{1}{2}a(t - t_0)^2 + \ldots \quad (25.135)$$

This kind of modeling is frequently applied in particle-tracking algorithms, see Sect. 25.2.4.

A more-detailed discussion of the modeling of flow fields is given by *Jähne* et al. [25.29].

Confidence Measures, Validation and Postprocessing

As any measurement technique, the result of a velocity estimation should not only supply the velocity field, but also a *measure of confidence*. To enable a reliable interpretation of the velocity field, gross errors have to be detected and removed. The optical flow methods discussed in Sect. 25.2.5 yield confidence measures as an integral part of the result. In PIV, typically the ratio of the tallest to the second tallest correlation peak is used to detect unreliable measurements. Based on such confidence measures, questionable measurements are detected and removed from the velocity field, which is typically done in a postprocessing step after the velocity field has been computed. However, in iterative methods, where the results strongly depend on the quality of the velocity estimates in previous iterations, the validation should be done after each iteration.

After the erroneous vectors have been removed, the resulting gaps in the velocity field may be filled by applying an interpolation technique, e.g., *adaptive Gaussian windowing* (AGW) [25.31]. Such interpolation techniques can also be used to interpolate the randomly distributed velocity vectors resulting from a PTV technique to a regular grid. Basically, the interpolation corresponds to a convolution of the velocity field using a special convolution kernel, e.g., a Gaussian in the AGW. To account for the varying uncertainty of the computed velocity vectors, a *normalized convolution* may by computed, where pixels with suspicious information (as indicated by their confidence measure) are given a low weighting factor in the convolution sum. For further information on interpolation and convolution techniques, refer to [25.29].

3-D Motion Estimation

Most of the methods discussed in this chapter refer to the case of 2-D motion estimation within a plane, i.e., the image plane. However, all of these methods can easily be extended to the case of 3-D motion estimation within a volume in space. From an algorithmic point of view, there is no principal difference between, e.g., computing a cross-correlation in 2-D and in 3-D. Optical flow algorithms and tracking methods can also be applied to the 3-D case simply by adding a further dimension. The challenge in 3-D motion estimation is rather a technological one: the acquisition of 3-D image data. Most approaches to 3-D velocity measurement are based on *stereoscopic* or *multi-view imaging* using two or more views of the same flow scene to recover the 3-D velocity field. The most prominent method applied to flow measurement is stereoscopic PIV [25.32]. Some 3-D PTV approaches are discussed in Sect. 25.2.4.

The basic new ingredient of 3-D methods as compared to 2-D methods is a *geometric camera calibration*. This calibration is necessary because perspective effects have to be taken into account in the evaluation of stereo images. The task of the stereo evaluation is to establish stereoscopic correspondences between two different views of the same scene. Thus, in addition to the motion correspondence problem (temporal correspondence), the *stereo correspondence* problem (spatial correspondence) has to be solved: given two views of the same scene, e.g., a flow field visualized by tracer particles, a unique correspondence between the particle images in the two views has to be found. The camera calibration provides the geometric relationship between the two views, the so-called *epipolar geometry*. If this relationship is known, the stereo correspondence problem can be solved much easier and faster. Further, the calibration also provides the necessary geometric information to compute the 3-D position of an object by triangulation of two or more views.

Stereo algorithms can be implemented very efficiently and transparently in terms of *projective geometry*. The (projective) geometry of multiple views and its implications for motion analysis are extensively discussed in the computer vision and photogrammetry literature [25.33]. Camera calibration is a classic topic of photogrammetry [25.34].

Another powerful but experimentally very elaborate approach is holographic imaging [25.35]. More information on this and other 3-D flow visualization methods can be found in [25.36].

25.2.2 Correlation-Based Velocity Analysis

In this section, approaches to velocity analysis based on the computation of cross-correlation coefficients are discussed. These approaches belong to the region-based methods. The focus is on particle image velocimetry (PIV), which is the method that is most often applied in fluid mechanics applications.

Standard Digital PIV Analysis

Basic Principle. Particle image velocimetry (PIV) is a technique to determine the two-component displacement vectors of tracer particle patterns in a 2-D plane (light sheet) within a flow. The result is a snapshot of the Eulerian flow field. The displacements are found by dividing two subsequent frames of a PIV sequence, $g_1 = g(x, t_1)$ and $g_2 = g(x, t_2)$ into interrogation windows, typically of a size of 16×16 or 32×32 pixels, and computing the cross correlation coefficient $r(x, s)$ of two corresponding windows

$$r(x, s) = \frac{\langle g_1(x') g_2(x' - s) \rangle}{\sqrt{\langle g_1^2(x') \rangle \cdot \langle g_2^2(x') \rangle}}, \quad (25.136)$$

using the abbreviation

$$\langle a(x') \rangle = \int_{-\infty}^{\infty} w(x - x') a(x') \mathrm{d}^2 x',$$

where the weight function $w(x - x')$ represents the size of the interrogation window and it is assumed that the local mean values over the interrogation windows have been subtracted from g_1 and g_2. The correlation coefficient is computed for a given 2-D range of displacements s of the interrogation window, resulting in a so-called *correlation plane*.

Computation of Velocity. Because the direct evaluation of the cross correlation coefficient (25.136) is computationally very expensive, it is usually computed using fast Fourier transform (FFT) methods, because in Fourier space the double summation is replaced by a simple pointwise multiplication. Once the correlation plane has been determined, the correct displacement is given by the maximum correlation peak. Thus, the optical flow is approximated as

$$f(x, t) \approx \frac{1}{\Delta t} \arg\max r(x, s), \quad (25.137)$$

where Δt is the time difference between the two successive images. Subpixel accuracy is achieved by centroiding or fitting a Gaussian to the correlation peak. Usually, in both methods only three neighboring correlation values in each direction are used (three-point estimators). Depending on the image quality and the evaluation method, the accuracy of the displacement estimation is of the order of 0.01–0.1 pixels and the dynamic range of the method is of the order of 100–1000.

Velocity Postprocessing. To compute an instantaneous velocity field, interrogation windows are distributed on a regular grid and evaluated by cross-correlation. The result is the instantaneous Eulerian velocity field. Since PIV is a statistical evaluation method, this vector field will contain a certain amount of spurious vectors (outliers), which result from interrogation windows containing an insufficient number of particle images.

These outliers have to be removed prior to any further evaluation of the velocity field. Some methods for outlier removal and interpolation of the resulting gaps are mentioned in Sect. 25.2.1. After this postprocessing, higher-order quantities such as vorticity, divergence, or rate of strain may be computed.

Limitations. The basic approach to PIV as discussed in this section suffers from a number of shortcomings that limit the accuracy, dynamic range, and spatial resolution of PIV. The main origin of these shortcomings is the fixed, finite size or shape of the interrogation window used in the correlation analysis, which effectively acts as a spatial low-pass filter on the estimated velocity field. Another source of error stems from the spatial discretization of digital particle images. In detail, the following limitations exist:

- *In-plane loss of pairs.* Particles may enter or leave the finite interrogation window between subsequent frames, in particular those that are moving faster than the mean velocity within the window. Thus, fast particles will not contribute to the correlation peak, since they do not have a matching partner within the interrogation window. This results in a bias of the estimated velocity towards lower values.
- *Velocity gradients.* Spatial velocity gradients within the interrogation window also contribute to the in-plane loss of pairs and thereby reduce the signal-to-noise ratio in the correlation plane since not all particles within an interrogation window correlate equally well due to their nonuniform motion. As a rule of thumb, the degradation of the PIV result becomes significant if the displacement of tracer particles due to local flow gradients gets larger than the image diameter of the particles.
- *Out-of-plane loss of pairs.* Particles may enter or leave the light sheet along the optical axis during the time of two successive exposures. Such particles are only visible in one of the images and do not have a matching partner. Again, this results in a reduction of the signal-to-noise ratio. The out-of-plane loss of pairs is a principal physical limitation of PIV that can only be overcome by adjusting the experimental parameters, e.g., the thickness of the light sheet or the frame rate of the cameras.
- *Computational aspects.* To reduce the computational load, the correlation is often computed in the Fourier domain using FFT methods. However, the necessary assumption of the periodicity of the image data within the interrogation window introduces inaccuracies compared to a direct spatial cross-correlation, which is, in principal, more accurate [25.37].
- *Peak-locking or pixel-locking.* The discrete nature of the PIV images introduces a bias towards integer displacements in the subpixel evaluation of the displacement peak. Peak-locking is the result of a biased subpixel estimation, if the input data (i.e., correlation values) is distributed asymmetrically around the maximum peak. The degree of peak-locking depends on the size of the particle images.

Apart from peak-locking, all these limitations are a consequence of the finite extent of the interrogation windows. The size of the window is given by a trade-off between dynamic range and accuracy on one hand and spatial resolution on the other hand. Large interrogation windows can resolve large motions and provide good accuracy due to a high signal-to-noise ratio, given that the window contains only weak velocity gradients. Large windows are also more robust to outliers. On the other hand, smaller windows provide a better spatial resolution and are less affected by velocity gradients, e.g., shear flows or strong vortices. However, to enable a reliable evaluation of the cross-correlation, the windows must contain a sufficient number of particle images and thus must have a certain minimum size, which depends on the particle density.

The limitations imposed by fixed interrogation windows can also be explained by looking at the spatial Taylor expansion of the velocity field (25.130). The standard PIV approach can only compute a straight shift of the interrogation windows between two frames. The velocity field within the interrogation window is assumed to be constant. This corresponds to a zeroth-order expansion of the velocity field. Linear effects like rotation, shear and dilation, or higher-order deformations are not accounted for. Due to the spatial averaging over the interrogation window, flow scales smaller than the window size cannot be recovered.

To summarize, the accuracy, spatial resolution, and dynamic range of the standard PIV method are coupled by the size of the interrogation window. The performance of PIV depends on three main factors: particle size and density, the size of the interrogation window, and the presence of velocity gradients. Particle size and density can be controlled during the setup of the experiment and are not discussed further. Recommendations for optimal settings are given in the PIV literature [25.30]. In the following sections, some advanced PIV methods are discussed. The goal of these methods is to overcome the limitations of the standard PIV approach to increase the accuracy, resolution, and dynamic range. Towards this end, the latter three performance measures have to be decoupled. Most of the advanced methods rely on the following three major ideas: iterative methods instead of a single-pass evaluation to refine the solution, hierarchical multigrid approaches to resolve both large and small motions, and higher-order approximations of the velocity field to account for velocity gradients and higher-order deformations.

Advanced Digital PIV Analysis
Multiple-Pass Interrogation with Window Shifting. To reduce the in-plane loss of pairs, a discrete integer window offset determined in a first interrogation pass is introduced before doing a second interrogation using the shifted window. The increased number of matched particle pairs results in an increased signal-to-noise ratio of the correlation peak. Iterations of the window shifting may be carried out until the displacement determined in the final iteration is below one pixel. Due to the discrete window shifting, the result still suffers from peak-locking, which can be reduced by applying continuous window shifting.

Correlation-Based Tracking. In the standard PIV approach, the interrogation windows in the first and second frame are of the same size and at the same location within the image. This is the major reason for the low-velocity bias error due to the in-plane loss of pairs. A simple modification to eliminate this error is to use a larger interrogation window in the second frame, centered around the smaller window in the first frame [25.18]. In this case, the correlation coefficient has to be computed directly in the spatial domain for all displacements of the small window within the large window. *Fincham* et al.[25.18] have termed this approach *correlation image velocimetry* to distinguish it from the standard correlation-based interrogation using equally sized windows. In a number of more-recent references, the approach using differently sized windows is referred to as *correlation-based tracking*, since the particle pattern defined by the small interrogation window is tracked within a search area defined by the large interrogation window.

Multiple Passes with Decreasing Window Size. The optimal interrogation window size for PIV depends on the local flow conditions and the local seeding particle density, which means that it is rarely constant from one region of the flow to another. Thus, instead of using fixed window sizes, the size of the window should be dynamically adapted to the local flow conditions. A simple way to implement this idea is to refine the correlation interrogation in an iterative way by starting with large windows and decreasing the window size during the course of the iterations [25.38]. In such a multigrid approach, the maximum in-plane displacement is decoupled from the interrogation window size, which increases the dynamic range without decreasing the spatial resolution. The displacements computed with larger interrogation windows can be used as predictions for further interrogations with smaller windows to shift the windows according to the prediction before the next interrogation is calculated. Thus, a high signal-to-noise ratio can be maintained also with small interrogation windows. The size of the interrogation windows may be decreased down to a correlation of single particles.

Since in such iterative methods, the quality of the final result depends on the results of previous iterations, in particular the first iteration, validation methods (Sect. 25.2.1) should be applied after each iteration. Since the first iteration will generally be a standard PIV correlation and as such suffer from all the basic limitations mentioned above, more-sophisticated methods have been developed for the first iteration [25.39].

Deformable Windows and Higher-Order Approximations of the Displacement Field. In the standard PIV evaluation, interrogation windows with fixed size and shape are used, and the velocity field is assumed to be constant within the windows, which is a zeroth-order approximation of the velocity field (Sect. 25.2.1). To account for variations of the velocity within the windows due to velocity gradients and higher-order effects, deformations of the interrogation windows of the particle images have to be considered, corresponding to a higher-order approximation of the velocity field. Towards this end, [25.40, 41] introduced the *particle image distortion* technique: they use fixed interrogation windows, but apply an iterative deformation of the images to compensate for in-plane loss of pairs. In each iteration, the image area within the interrogation window is deformed according to the displacement field calculated in the previous iteration. To compute the deformed images, some kind of image interpolation has to be applied, e.g., bilinear interpolation or spline interpolation (Sect. 25.1.3). Care has to be taken in the interpolation step not to spoil the accuracy gain due to the window deformation with an inaccurate interpolation scheme. A further advantage of the window deformation using image interpolation is the possibility to introduce continuous window offsets, which reduces the peak-locking effect.

Second-Order Correlation. As an effective method to suppress false correlation peaks and amplify the correct one, [25.42] introduced the second-order correlation method, which is simply a multiplication of the correlation plane of an interrogation area by the correlation plane of one or more neighboring interrogation areas (overlapping by, e.g., 50%). Thus, it is a correlation of the correlation. Since any peak that does not appear in both planes is eliminated, correlation anomalies are suppressed, resulting in more-reliable and accurate velocity estimates. Unlike statistical PIV postprocessing methods to remove spurious vectors, which rely on the accuracy and similarity of neighboring vectors, errors are directly eliminated in the correlation data. The second-order correlation may be applied together with any of the PIV methods discussed in this section to validate the results already during the computation step.

Super-Resolution PIV, Hybrid PIV/PTV Methods. Clearly, the maximum amount of information contained in a PIV image is the motion of the individual particles. The number of particles within an image, i.e., the particle density, defines the maximum spatial resolution of the velocity field that can be achieved. The approach of particle-tracking velocimetry (Sect. 25.2.4) is to actually exploit the maximum resolution by identifying the individual particles and measuring their motion. However, such an approach is not feasible in the evaluation of PIV images, since the particle density is much higher than in PTV images, which gives rise to ambiguities in the temporal correspondence analysis of the particle motion that cannot be resolved without further information. The idea of super-resolution PIV respectively hybrid PIV/PTV is to combine PIV and PTV. The goal is to increase the spatial resolution and accuracy of PIV and overcome the averaging and gradient-biasing effects of the standard PIV interrogation by tracking the individual particles within the interrogation windows [25.22]. The initial result of a coarse PIV interrogation is used in a predictor step to direct the particle-matching algorithm in the right direction and thereby reduce the size of the search area. With a smaller search area located near the correct match partner, the probability of ambiguities is reduced. If a unique match is established, the particle velocity is calculated from the two positions in the successive images by a finite-difference scheme.

3-D/3-C PIV

Several methods have been proposed to extend the PIV technique towards measurements of full three-component (3-C) vectors respective measurements within a three-dimensional (3-D) volume in space. Stereoscopic PIV is the method that is applied most frequently.

Stereoscopic PIV. Stereoscopic PIV enables the measurement of 3-C vectors within a plane in space. Hence, it is a 2-D/3-C method. For a review, see [25.32]. The basic idea is to use two cameras observing the light sheet, and to compute the third velocity component (i.e., the out-of-plane motion) from the disparity map between the two particle images. Further, stereoscopic PIV also offers the possibility to eliminate perspective errors, which may contaminate the in-plane measurements if perspective effects are strong, i.e., when the lateral dimensions of the object plane are comparable to its depth.

Stereoscopic PIV systems can be arranged in two configurations. *Translational systems* have parallel optical axes, whereas in *rotational systems* the two optical axes are arranged enclosing a convergence angle α. Both arrangements have advantages and disadvantages [25.32].

3-C vectors are obtained by mapping the displacements from each image plane to the object plane and combining them to obtain the third component. There are three different approaches [25.32]:

1. *Geometric reconstruction.* A priori knowledge of the complete recording geometry is necessary. This information is used to perform an explicit ray tracing of the projection rays. This method is tedious and not very accurate, since the necessary geometric parameters (e.g., stereo baseline, depth of the measurement plane) often cannot be measured with sufficient accuracy.
2. *2-D calibration.* A calibration is performed using one image of a calibration target, which has to coincide exactly with the plane of the light sheet during flow measurements. A general polynomial transformation (typically up to second or third order to account for lens distortions) between the object plane and the image planes of the two cameras is estimated, based on the known correspondences between object and image points of the calibration target. The final step of determining the 3-C velocity uses reconstruction equations that still require some knowledge of the geometry such as the separation between the

lenses, the object distance, or the angular orientation of the cameras to the object plane.
3. *3-D calibration*. A full 3-D geometric camera calibration is performed, using several images of translated calibration planes. To compute 3-C vectors, explicit knowledge of the system geometry is not required. General higher-order polynomial transformations are also frequently applied in the 3-D calibration. Instabilities related to overparameterization might be introduced if the measurements are noisy, since typically ≈ 40 free parameters are calibrated for each camera. The application of photogrammetric pinhole-camera models and self-calibration methods in stereoscopic PIV is a rather recent development [25.43].

Defocusing PIV. [25.44] introduced defocusing PIV as a method to obtain 3-D/3-C velocity fields. A volume illumination is applied, and the defocus principle is used to identify three-dimensional particle locations. *Pereirra* et al.[25.45] use a similar technique to obtain full 3-D information. A volumetric cross-correlation is computed to estimate the velocity field.

Multiplane Stereoscopic PIV. The idea of multiplane stereoscopic PIV is to use several light sheets in different depths to obtain flow information from a number of different planes within a 3-D volume. The planes may be illuminated either simultaneously or sequentially. In the former case, several stereo camera setups are used to acquire the images. For details, see [25.46]. A recent variant of multiplane stereoscopic PIV is the XPIV method [25.47]. It combines stereoscopic PIV, multisheet illumination and defocusing PIV. The latter is applied to separate the different depth planes which are all projected simultaneously into the same camera.

Photogrammetric PIV. *Pereirra* et al. [25.48] describe a photogrammetric PIV system. The principle is similar to that of a 3-D particle-tracking velocimetry setup (Sect. 25.2.4). Three cameras are used to acquire images of the flow. The 3-D particle positions are reconstructed by triangulation, based on a geometric camera calibration that is performed prior to the flow measurements. The 3-D/3-C flow field is obtained by computing the volumetric cross-correlation of the particle positions in subsequent frames. The only difference between 3-D particle-tracking and photogrammetric PIV is that the former tracks single particles in 3-D, whereas the latter computes the cross-correlation of 3-D interrogation areas, i. e., volumetric, spatial particle patterns.

Holographic PIV. In contrast to all the methods discussed so far, holographic PIV [25.35] requires a volumetric illumination with *coherent* light to record holograms of the flow. The 3-D/3-C flow field is recovered by interrogating the holograms with coherent light beams. In principle, holographic PIV is superior to all the other methods, but the experimental setup and the data evaluation is very complex. For these reasons, holographic PIV does currently not provide the ability to collect large data bases for statistical analyses. Hence, the application of holographic PIV is limited to relatively simple flow configurations.

Further Reading
Different PIV methods are reviewed by *Adrian* [25.49]. Information on autocorrelation PIV including film-based acquisition and optical evaluation methods can be found in *Kean* and *Adrian* [25.50, 51]. The theory of cross-correlation PIV is developed in [25.52]. The fundamentals of digital PIV are discussed by *Westerweel* [25.17], *Willert* and *Gharib* [25.16] and in the books by *Westerweel* [25.53] and *Raffel* et al. [25.30]. The latter gives a large number of references to further information on PIV.

25.2.3 Least-Squares Matching

Basic Principle
Least-squares matching is an alternative approach to maximizing the cross-correlation between two image patches to estimate the interframe motion. Like correlation techniques, it also belongs to the region-based methods of motion estimation. Given two successive images $g_1 = g(\mathbf{x}, t_1)$ and $g_2 = g(\mathbf{x}, t_2)$, an interrogation window is selected in the first frame, and a larger search area centered around this interrogation window is selected in the second frame. The displacement of the interrogation window is calculated by *minimizing* a distance measure that quantifies the dissimilarity between two image regions. This distance measure is given by the *sum-of-squared differences* (SSD, the *least squares*) of the gray values within the interrogation window between the first and second frame:

$$d(\mathbf{x}, \mathbf{s}) = \int_{-\infty}^{\infty} w(\mathbf{x} - \mathbf{x}')[g_1(\mathbf{x}') - g_2(\mathbf{x}' - \mathbf{s})]^2 \mathrm{d}^2 x' , \quad (25.138)$$

where the weight function $w(\mathbf{x} - \mathbf{x}')$ represents the size of the interrogation window. The optical flow is approximated as

$$f(\mathbf{x}, t) \approx \frac{1}{\Delta t} \arg \min d(\mathbf{x}, \mathbf{s}), \quad (25.139)$$

where Δt is the time difference between the two successive images. Subpixel accuracy may be achieved using the same methods as in correlation-based approaches, e.g., by fitting a Gaussian function to the (inverse) displacement peak.

Least-squares matching is also referred to as *image correlation velocimetry* [25.19], *adaptive least-squares correlation* [25.20], and the *minimum quadratic differences* (MQD) method [25.54].

Relation to Other Region-Based Approaches

Least-squares matching is closely related to the other region-based approaches of motion estimation, namely optical flow techniques, and correlation-based analysis.

The similarity between least-squares matching and differential optical flow techniques (Sect. 25.2.5) is revealed by approximating $g_2(\mathbf{x}' - \mathbf{s})$ in (25.138) by a Taylor expansion about $\mathbf{s} = 0$ and skipping all terms above first order. The resulting expression is the gradient-based formulation of the optical flow. For the case of subpixel motions, the equivalence of first-order differential optical flow estimation and least-squares matching using bilinear interpolation is shown by [25.55].

Gui and *Merzkirch* [25.54, 56] discuss the relation between least-squares matching and correlation techniques. Expanding the squared term in (25.139) shows that this expression contains the (negative) cross-correlation coefficient as used in a PIV evaluation. But in addition, there is a term accounting for nonuniformities in the particle image distribution and nonuniform illumination. *Gui* and *Merzkirch* [25.56] show that this term is responsible for the superiority of the least-squares matching as compared to conventional correlation-based methods.

Advanced Least-Squares Matching

The real strength of least-squares matching is revealed if it is combined with similar advanced evaluation methods as outlined in Sect. 25.2.2. In particular, iterative approaches using a coarse-to-fine strategy together with a higher-order approximation of the displacement field (Sect. 25.2.1) are widely used and show good performance [25.19]. Radiometric effects, i.e., intensity changes, may also be included in the model. The implementation of such methods within a general least-squares parameter estimation framework provides the ability to estimate higher-order quantities such as vorticity and rate of strain directly. Towards this end, these quantities are introduced as parameters to be estimated within the least-squares optimization. Thus, no explicit finite differences of the velocity field, which are very sensitive to numerical errors and noise, have to be calculated. In addition, no explicit differentiation of the image data is required. In most cases, only a few parameters are extracted from the optimization (e.g., six parameters for a general 2-D affine transformation), but many more data points are used for the computation (e.g., 256 pixels in a 16×16 interrogation window). Due to this strong overdetermination, the least-squares matching is quite immune to image noise. Furthermore, the precision and the reliability of the estimated parameters can be easily assessed by a covariance analysis of the least-squares result. Overparameterization can be avoided by selecting a model of the displacement fields based on the significance of the computed parameters. Various models can be used for different interrogation windows, making the method adaptive to the local image structure.

25.2.4 Tracking Techniques

In this section, various tracking techniques are reviewed. Tracking techniques are feature-based motion estimation algorithms: individual tracer particle images are the features that are tracked throughout an image sequence. The most important tracking method used to measure fluid flow is *particle-tracking velocimetry* (PTV). This method is described in detail in this section.

Besides PTV, some region-based methods are also referred to as tracking techniques: *correlation-based tracking*, *least-squares tracking* or *Kanade–Lucas tracker* (KLT) tracking. The latter methods consider the particle or gray value patterns within the interrogation windows as features that are tracked. Correlation-based tracking ([25.18] and Sect. 25.2.2) is a special PIV evaluation mode, where the interrogation window in the second frame (the search window) is larger than that in the first frame, as opposed to the standard correlation-based interrogation mode of PIV where both windows are of equal size. Thus, in this method, the correlation coefficient is used as a tracking criterion. Least-squares tracking was discussed in Sect. 25.2.3. The KLT tracker [25.57] is a differential optical flow method based on the early work of *Lucas* and *Kanade* [25.58] (Sect. 25.2.5). This method performs tracking in the sense that individual image regions are automatically selected and tracked if the image structure within the regions is sufficient to compute the optical flow.

Particle-Tracking Velocimetry

Basic Principle. The basic idea of particle-tracking velocimetry (PTV) is to identify single particle images within an image, segment them from the background, and track them along their trajectories throughout an image sequence. Thus, PTV is a feature-based approaches to motion estimation. A PTV algorithm has to solve the following three tasks: *particle segmentation*, determination of *particle position*, and *particle matching*, i.e., solving the motion correspondence problem. Since in most applications, the particles cannot be distinguished from each other reliably (e.g., by their shape or intensity), the latter task is the most difficult, due to ambiguities occurring especially for high particle densities. Thus, the particle density in PTV applications is generally lower than in PIV applications.

As a simple example, consider the following optimal conditions for PTV: high image contrast (bright particles on a dark, noise-free background), low particle density, and small interframe motions. The latter two conditions imply that the mean distance between particles is much larger than their interframe motion. In this case, a very simple PTV approach may be used: segment the particles by a global intensity threshold, determine their position by centroiding, and match each particle in the first frame to that particle in the second frame that is closest to its position in the first frame (i.e., its nearest neighbor). However, since the optimal conditions assumed above will rarely be given in real applications, more-sophisticated algorithms are needed, in particular for particle segmentation (see below) and the tracking (Sect. 25.2.4).

Differences to Region-Based Approaches. The major differences between tracking techniques and region-based motion estimation are the following:

- *Temporal scope:* To increase the reliability of the tracking, most PTV techniques use more than two successive frames to establish the temporal correspondences (*multi-frame tracking*), e.g., three-frame tracking [25.59], four-frame tracking [25.26], five-frame tracking [25.60]. Note that there are also two-frame tracking techniques [25.24] as well as techniques that try to find the optimal set of trajectories by taking into account their complete (visible) length within a global optimization [25.61]. Most correlation-based approaches and least squares techniques try to establish a matching between only two frames, while in optical flow techniques, more than two frames may also be used, e.g., for a more-accurate computation of temporal gradients or to stabilize the results by temporal smoothing [25.62].
- *Particle density and spatial resolution:* Since PTV aims at identifying individual particles and finding corresponding match partners in the next frame, the particle density is generally lower than in PIV, which results in a lower spatial resolution of the underlying flow field. On the other hand, the resolution given by the particle seeding is fully exploited without any averaging effects as in PIV (Sect. 25.2.2), and therefore with a higher accuracy. For example, the motion of two particles located within one PIV interrogation window can be resolved individually. Thus, the local spatial resolution is higher than in PIV.
- *Spatio-temporal distribution of velocity vectors:* Feature-based approaches compute the *Lagrangian representation* of a flow field, i.e., the result of such algorithms is a set of trajectories of the tracked objects. Thus, the velocity information is given at random locations depending on the tracer distribution and density. In contrast to region-based methods, no dense, instantaneous flow field defined on a regular grid is computed (which is the *Eulerian representation* of a flow field). On the other hand, tracking methods allow the motion of individual particles to be followed in time, enabling e.g., a Lagrangian study of diffusion, which is not possible using PIV results.
- *Large motions:* Since feature-based methods do not exploit the relation of spatial and temporal intensity gradients for velocity estimation, they are less sensitive to violations of the sampling theorem. Thus, feature-based methods are more suitable for the handling of larger motions, given that the motion correspondence problem can be solved using one of the advanced methods outlined below.

Particle Segmentation. In the particle segmentation step, a decision has to be made for each pixel of the image, whether it belongs to a particle or to the image background. Thus, for an image $g(x, t)$, the segmented image $g_s(x, t)$ is given by the following operation:

$$g_s(x, t) = \begin{cases} 1 & : \quad g(x, t) \in \text{object(particle)} \\ 0 & : \quad g(x, t) \in \text{image background} \end{cases}.$$
(25.140)

The result of the segmentation is a binary image in which the particle images are marked with the value 1 and the background is marked with the value 0.

Particle segmentation is rather difficult because most particle images do not have a bimodal gray value histogram. Therefore, a simple segmentation by a global intensity threshold is not feasible. Particle images do not have a uniform (mean) gray value for the following reasons:

- *Image noise*: Image noise introduces false positives and false negatives, especially in the segmentation of low-contrast particles.
- *Motion blur*: While small and slow particles are imaged as bright circular spots, the shape of the image of a faster particle is elongated in the direction of its motion due to the integration time of the camera. Faster particles cover a larger area in the image, i.e., the irradiance is distributed over a larger number of pixels. Thus, faster particles have a lower intensity and may appear as faint objects with gray values close to the image background.
- *Inhomogeneous illumination:* Several factors may contribute to an inhomogeneous illumination of the images, e.g., the general intensity distribution of the light source, glow of dirty water, or particles reflecting light to their neighbors. In particular, for 3-D applications using a volume illumination, inhomogeneities have to be taken into account.
- *Out-of-focus imaging*: In applications with volume illumination, particles may be out-of-focus, which reduces the image contrast.

As a result of all these factors, the (mean) gray values of particle images vary locally and may cover the complete range from the background noise level to the saturation of the sensor. Many authors have pointed out that the particle segmentation is one of the most crucial steps in a PTV algorithm, since it is the dominating factor controlling both the reliability and the accuracy of the tracking [25.63].

Another difficulty is introduced by overlapping particles. Especially in 3-D applications using volume illumination, particle images may partly overlap, or particles may be completely occluded by others. Particle occlusion is a principal physical limitation of 3-D techniques using a volume illumination. A trade-off has to be made between the seeding density (i.e., spatial resolution of the flow field) and the depth of view. Different techniques to resolve overlapping particle images are discussed by [25.64, 65].

Particle Position. Once the particle images have been segmented from the background, various methods are available for the subpixel-accurate determination of the position of individual particles: centroiding, least squares fits, e.g., in the form of a Gaussian three-point estimator as applied in PIV, fit of extended models taking into account the motion of the particles or template matching by cross-correlation.

The performance of the different methods depends on the properties of the images, e.g., particle size and density, particle contrast, and image noise. In general, errors in the determination of the particle positions are introduced by the discretization, noise in the signal (photon shot noise), and noise related to the image acquisition. The discretization errors for typical particle images are of the order of magnitude of $0.01-0.1$ pixel [25.29]. *Wernet* and *Pline* [25.60] show that the Cramer–Rao lower bound for the error in determining the position of a Gaussian-shaped particle is 0.015 pixel. In practical applications with additional noise sources related to the image acquisition, the mean centroid estimation error is generally larger ($0.1-0.2$ pixel). If highest accuracy is desired, the bias towards integer positions becomes important (peak-locking as discussed in Sect. 25.2.2). In this case, iterative weighted least-squares methods have to be applied to compute an unbiased estimate.

Computation of Velocity. If a unique particle match is found, a first-order approximation of the particle's velocity may be computed by subtracting the positions of the particle in the two successive frames and dividing by the frame period of the camera. Assuming that the latter is given free of error, the absolute error in estimating the particle displacement is $\sqrt{2}$ times the particle position error. Therefore, errors of individual velocity vectors may be higher than in PIV. However, *Wernet* and *Pline* [25.60] have shown that better accuracy than with PIV can be achieved by averaging the PTV results over an area of the size of a typical PIV interrogation window. Higher-order finite differences or spline fits to particle trajectories may also be used to achieve a better quality of the velocity results. Further, the errors in the particle positions of two nearby particles in successive frames are typically highly correlated. The systematic error cancels out in the difference of the two positions. Hence, the error in the velocity estimation will be lower then the conservative estimate given above.

Velocity Postprocessing. Techniques to remove outliers in the velocity field and interpolate the resulting gaps as well as techniques to interpolate randomly spaced data to a regular grid have been mentioned in Sect. 25.2.1. Similar techniques may be applied to PTV data, where outliers have to be defined relative to a particle trajectory. *Malic* and *Dracos* [25.66] present dedicated interpolation schemes for three-dimensional velocity fields from scattered data using Taylor expansions. Besides the interpolation to regular grids, the final PTV results (i.e., the flow trajectories) enable a Lagrangian analysis of the flow field.

Limitations. An important parameter describing the difficulty of particle tracking is the particle spacing displacement ratio r_p [25.27]:

$$r_p = \frac{\Lambda_0}{\Lambda_t}, \qquad (25.141)$$

where Λ_0 is the average distance between the particles and Λ_t is the average particle displacement between two successive frames. Tracking is easy, if $r_p \gg 1$. In this case, even simple nearest-neighbor approaches such as the one described above may yield good results. The tracking difficulty increases for $r_p \approx 1$, and tracking becomes virtually impossible for $r_p \ll 1$. In the latter case, the probability of ambiguities in the particle matching is very high. The motion correspondence problem cannot be solved if no additional information about the particles or their motion (e.g., size, shape, color, intensity, direction of motion) is available. The fundamental reason is the violation of the sampling theorem (Sect. 25.2.1) in the case of high particle density and large motion.

Thus, for a given particle density, the tracking difficulty is related to the maximum possible displacement between two frames, which depends on the frame rate, image magnification, and the flow field under investigation. For a reliable tracking, Λ_t should be small, i.e., the frame rate of the camera should be sufficiently high. On the other hand, a larger Λ_t yields a lower relative error of the velocity vector, since the absolute error in determining particle positions is independent of Λ_t. The basic idea to enable reliable particle tracking for values of $r_p \approx 1$ and smaller is to take into account additional information about the flow and use this information to guide the particle matching. Towards this end, most PTV approaches introduce a motion model (Sect. 25.2.1). Some advanced PTV techniques based on motion models are discussed in Sect. 25.2.4.

Another difficulty in PTV stems from the fact that the number of particles in two successive frames may not be equal. Even if the number of particles is equal, the assumption that the same set of physical particles is visible in both images is not valid. Instead, a PTV algorithm has to handle the following events:

- Entry respective trajectory initialization: Particles may enter the field of view. These particles do not have a correspondence partner in the previous frame.
- Exit respective trajectory termination: Particles may leave the field of view. These particles do not have a correspondence partner in the following frame.
- False positives and false negatives: Image noise may result in spurious particle images at locations where actually no particle is visible. Segmentation failures may result in a loss of particles, e.g., particles are not segmented due to their low intensity.
- Overlapping particles and occlusion: Especially for high particle densities, an overlap and occlusion handling has to be introduced.

Since all these events and their consequences have to be considered for each particle in each image of a sequence, PTV algorithms tend to be quite complex. In the following sections, some advanced PTV techniques are discussed.

Advanced Techniques to Solve the Motion Correspondence Problem

As discussed in the previous section, the most difficult step of a PTV algorithm is to find the unique, correct solution to the *motion correspondence* problem: all particles in an image have to be associated to their correct matching partners in the next frame. This association is difficult for higher particle densities, since the probability of *ambiguities* becomes large, i.e., there will be several possible matching partners within a search region in the next frame. The basic idea of advanced PTV algorithms is to use additional *a priori* knowledge and assumptions about the flow field. Based on such information, a model of the flow can be introduced. Further, kinematic constraints can be used to reduce the number of corresponding particles and thus the number of ambiguities in the matching. The reduction of the number of ambiguities becomes possible for two reasons. First, using a model of the flow field enables the *prediction* of the future position of particles by extrapolation. Thus, the search region for the correspondence search may be located at this predicted position. Second, if the flow model is a good approximation of the actual flow field, there will be a high probability that the predicted position is already at the correct location of the match partner. Thus, the size of the search region may be reduced. If there is only one particle within the predicted search region, a unique match is established. If still several matching partners remain, one of them may be chosen as the correct match according to certain criteria, which are again based on additional information or assumptions on the flow.

The main assumption that is made in many tracking algorithms is the *smoothness of the flow field*, which is based on the physical principle of inertia. Due to inertia, the motion of an object will not change abruptly between two frames, given that the frame rate is sufficiently high. Inertia is related to the *temporal* smoothness of particle trajectories. In addition, *spatial smoothness* (or spatial *coherence*) of the velocity field may also be assumed. In particular, for incompressible, viscous flows, the velocity vectors within a spatial neighborhood will vary smoothly. Thus, velocity vectors next to each other will be similar in speed and direction.

In addition to the basic smoothness assumptions, any a priori knowledge about the flow field may also be incorporated into the tracking. For example, if the flow is known to have a mean bulk velocity, this velocity can be used as an offset in the tracking algorithm. Search radii defining the area where matching particles are supposed to be found can be defined based on *hydromechanic knowledge*, e.g., maximum expected velocities or turbulence scales such as the Kolmogorov scales or Taylor microscales [25.27]. Such considerations are particularly important for the initialization of trajectories, since for particles entering the field of view, no velocity vector is available that can be used to predict their next position.

Another approach to resolve ambiguities in the correspondence analysis is to take into account several possible matches and defer the decision of the correct match on later frames. Such approaches are realized using techniques of *statistical data association* or *combinatorial optimization*. These techniques can also deal with particle occlusion and therefore resolve crossing trajectories. This is possible because a larger temporal scope is taken into account when solving the correspondence problem, e.g., a temporal neighborhood of three previous and three future frames. The correspondence is solved by finding an optimal set of trajectories within this temporal neighborhood, where optimality is expressed, e.g., in terms of trajectory smoothness.

Finally, some remarks concerning the optimal choice of thresholds and other tracking parameters, e.g., size of search regions, shall be made. In many cases, the optimal parameters depend on the flow conditions, i.e., particle density and flow velocity. However, the latter are not constant throughout the whole image, but there may be significant variations of these quantities within a single image. Obviously, optimal performance of a tracking algorithm cannot be achieved using a fixed set of parameters. Instead, the parameters should adapt to the local flow conditions. For example, it does not make sense to use a search region based on a global maximum velocity constraint within a region of the image where the velocity is very small and the particle density is very high. In this case, a search region based on a maximum velocity constraint will be much too large. It may also be advantageous to adapt the shape of the search region to the flow conditions. For example, if there is a prominent main flow direction, the search region should be elongated along this direction.

In the following subsections, some implementations of the ideas developed in this section are presented.

Two-Frame Tracking. The simplest two-frame tracking technique performs a nearest neighbor search based on a minimum-velocity constraint. *Hering* et al. [25.67] use the spatial overlap of particle images to identify the nearest neighbor. The overlap is caused by overlapping integration times of the even and odd fields in an interlaced camera frame. If non-interlaced cameras are used, the overlap may be created artificially by a morphological dilation. The approach is limited to low particle densities.

The performance of two-frame tracking can be increased towards higher particle densities by including a spatial coherence constraint and requiring velocity vectors within a spatial neighborhood to be similar [25.25].

Multiframe Tracking. Multiframe techniques are based on the assumption of trajectory smoothness. They use a model of the particle motion to predict the particle positions in the next frame. This model is given by the Taylor expansion of the particle trajectory equation (25.135). The difference in these techniques is the degree of approximation in the Taylor expansion. *Three-frame techniques* use the actual frame and one previous frame to compute a first-order approximation of the particle velocity. The resulting velocity vector is used to predict the particle position in the next frame, assuming that the velocity stays constant. If several match candidates are found within a neighborhood around the predicted particle position, the particle with the smallest distance to the predicted position is chosen. This choice corresponds to a *minimum acceleration constraint* on the particle motion. In a similar manner, higher-order terms in the Taylor expansion may be taken into account to improve the accuracy of the predicted particle position [25.27].

From the point of view of implementation, imposing the temporal smoothness constraint is actually a matter of defining search areas around predicted particle positions. In most applications, circular areas are used. The radii of these areas are chosen according to the kinematic constraints imposed by the flow model or according to prior hydromechanic knowledge about the flow, for example, to initialize a trajectory, the search radius in the second frame may be defined by the maximum expected velocity. The centers of the search areas in the third and fourth frame are predicted by extrapolating the model, while their radii may be chosen according to the expected fluctuations in the velocity. The latter can be estimated from the Kolmogorov scales of the flow and the imaging parameters [25.27].

Combinatorial Optimization. The motion correspondence problem between two sets of features in two successive frames may be formulated as a *combinatorial optimization problem*. Based on such a formulation, results and algorithms developed in graph theory and operations research may be applied to PTV.

Given are two sets of particle coordinates P_1 and P_2 (particle images in the first frame at $t = t_1$ and particle images in the second frame at $t = t_2 = t_1 + \Delta t$):

$$P_1 = \{p_{1,i}, i = 1..N_1\}, \quad (25.142)$$

$$P_2 = \{p_{2,j}, j = 1..N_2\}, \quad (25.143)$$

where $p_{i,j} = (x_{i,j}, y_{i,j})$ are the coordinates of particle j in frame i. The particle matching between two successive frames may be described by an association matrix $\alpha = (\alpha_{ij})$, with $\alpha_{ij} = 1$ if particle i in the first frame is matched with particle j in the second frame and $\alpha_{ij} = 0$ elsewhere. The task is to find an *optimal assignment* between the elements of the first and those of the second set.

Optimality is expressed in terms of an objective function d that is linear in the associations between the two sets:

$$d = \sum_{i=1}^{N_1} \sum_{j=1}^{N_2} \alpha_{ij} c_{ij}, \quad (25.144)$$

where c_{ij} is the cost for the association α_{ij}. The costs c_{ij} are chosen according to the kinematic constraints discussed in the previous section, e.g., favoring smooth trajectories. The optimal assignment is determined by minimizing the objective function (25.144). This formulation of the two-frame tracking is equivalent to a so-called *bipartite graph matching* or *assignment problem*, which is a basic problem of combinatorial optimization occurring in many applications. Efficient algorithms to solve it have been developed [25.68]. For an example of a PTV application based on bipartite graph matching (see [25.69]). *Stellmacher* and *Obermayer* [25.70] present an interesting approach to simultaneously estimate particle correspondences and a local affine transformation by applying a combined discrete and continuous optimization method. This method has been originally proposed by *Gold* et al. [25.71] for solving point-matching problems in statistical pattern recognition.

If the temporal scope of the tracking is extended to three or more frames, the problem becomes a *multidimensional assignment problem*. Such problems are known to be NP-complete, i.e. there is no efficient algorithm to compute their solution [25.68]. However, approximate solutions can be found using greedy search techniques and other heuristics [25.61]. The complexity of the correspondence analysis is significantly increased due to the extended temporal scope. Therefore, combinatorial techniques are computationally expensive. On the other hand, these techniques are also able to resolve crossing trajectories and find an optimal set of trajectories in the presence of particle occlusions and dropouts due to segmentation failures.

Statistical Techniques. Statistical approaches to solve the motion correspondence problem (*statistical data association*) have been originally developed in the context of radar target tracking and surveillance, where particle tracking is referred to as multiple target tracking. A large number of different approaches and algorithms has been published. For an introduction see [25.72, 73]. A review of statistical data association techniques in the context of computer vision is given by *Cox* [25.74].

Statistical data association techniques are specifically developed to resolve ambiguities in motion correspondences, including events like track initiation and termination, particle occlusion, and false positives/negatives. Like the other advanced techniques discussed so far, most statistical techniques are also based on the two paradigms of prediction of an optimal search region and postponing assignment decisions by examining subsequent frames. The difference of statistical approaches is that they model the motion of a particle over time as a *stochastic process*. The optimal particle state according to the measurements and the underlying model is then computed within the framework of Bayesian inference [25.75]. Within this framework, the *probability distribution of the particle state* is propagated over time and updated by the measurements in each frame. A (stochastic) state vector is introduced, which describes the actual state of a particle, e.g., its position, velocity and acceleration. This state vector follows a probability distribution, the so-called *prior distribution*. When measurements of the particle state become available (e.g., its position resulting from the particle segmentation), this measurement information is converted into a *likelihood function* defined on the particle state space. Likelihood functions are presumed to contain all the relevant information in the observed data. They provide a probabilistically correct method of combining all types of sensor information and incorporating it into a tracker's estimate of the particle state. Note that this may include the detailed physics of the camera response and its noise characteristics. Bayes rule is applied to combine all this information and compute the *posterior distribution* on the target state by combining the prior distribution with the likelihood function. Finally, optimal particle associations are computed from these probability distributions. All statistical data association techniques may be formulated within this *Bayesian framework*.

The simplest statistical tracking technique is the *Kalman filter* [25.73, 76]. It is based on the assumptions of a linear model of particle motion (like (25.135)) and Gaussian distributions of particle state and measurement error. The Kalman filter computes the optimal position of a particle in the next frame, by predicting its position according to a model and combining this prediction with a measurement. Note that the prediction also includes the precision of the particle location. Optimality is achieved by taking the errors of the prediction and the measurement into account. For example, the shape of the search region around the predicted position may be chosen according to the covariance of the predicted position, which typically results in elliptical search regions instead of circular ones. Examples of Kalman filters applied in PTV are described in [25.28].

Simple Kalman filters have several drawbacks. They provide a statistically optimal estimate of particle position and thus can guide the tracking and implicitly reduce ambiguities. However, they cannot handle ambiguities explicitly. An extension of a Kalman filter with increased capabilities in the handling of ambiguities is the *multiple hypothesis tracker* (MHT) [25.73, 77]. In the case of ambiguities, the latter takes into account the k-best hypotheses (computed by a Kalman filter). Probabilities for all the hypotheses are computed and the most probable hypothesis is chosen. Thus, the MHT is a combined approach based on statistical reasoning and combinatorial optimization.

In complex tracking problems (such as turbulent motion), the assumptions of the Kalman filter, i.e., a linear model and Gaussian probability distributions, may be too simplistic. It is possible to relax these assumptions and work with general probability distributions of arbitrary shape as well as arbitrary nonlinear models. This general framework of *Bayesian multiple-target tracking* is discussed in the book by *Stone* et al. [25.75].

Hybrid Methods Using Velocity Estimation Techniques. Hybrid methods combining a PIV velocity estimation with the tracking of single particles [25.22, 63, 78] have already been discussed in Sect. 25.2.2. *Bastiaans* et al. [25.23] present a hybrid method combining a PIV prediction step with a tracking step based on combinatorial optimization. An important condition for hybrid PIV/PTV methods is the spatial coherence of the flow field, since only then the PIV prediction step will yield a reasonable estimate of the local velocity.

3-D/3-C Tracking

PTV can be extended to 3-D measurements using the same basic techniques as in 3-D PIV, namely stereoscopic or multicamera image acquisition. In addition so solving the motion correspondence problem (temporal correspondence), a 3-D PTV algorithm has to solve the *stereo correspondence problem* (spatial correspondence), i.e., to find corresponding particles in two or more images taken from different viewpoints. It is necessary to establish the stereoscopic correspondences of particles uniquely in order to compute their 3-D position by triangulation of the optical rays.

Basically, there are two different approaches to 3-D PTV. They differ in the order in which the spatial and temporal correspondences are solved. One approach first performs particle tracking in the image planes, resulting in a set of 2-D particle trajectories. Thus, the motion correspondence is solved first. Then, the stereo correspondences of the particle trajectories are analyzed to find the 3-D coordinates [25.29]. The other approach first solves the stereoscopic correspondences of single particles to compute their 3-D coordinates. Afterwards, particle tracking is performed in 3-D space [25.27, 64]. For this approach, three or more views of the flow field are needed in order to resolve stereoscopic ambiguities in the determination of the particle positions.

A recent approach is to combine temporal and spatial information and solve the correspondence analysis simultaneously [25.79].

The goal of 3-D PTV is to measure the three components (3-C) of the velocity vectors within a 3-D volume in space. Towards this end, a volume illumination has to be used instead of a light sheet that is typically used in 2-D applications. The volume illumination introduces a number of difficulties concerning the imaging of particles and the processing of such images:

- *Particle occlusion:* Particles located along the same optical ray occlude each other, which limits the particle density or the depth of view. A trade-off between these two parameters has to be chosen according to the goals of the measurement.
- *Projection of the 3-D scene:* A particle image obtained using volume illumination is the 2-D projection of a 3-D scene with a certain depth range. Depending on the particle density, this will increase the probability of spatial overlap of the particle images or trajectories crossing each other in the image plane. In addition, the assumption of spatial coherence of the projected flow field is no longer valid, since particles that are actually far away from each other in the 3-D volume may be located next to each other in the image.
- *Out-of-focus imaging:* If the illuminated volume is larger than the depth of field of the lens, some particles will be out of focus, resulting in a larger size and lower contrast of these particles. Such effects have to be considered in the particle segmentation.

Further Readings
Further information about tracking methods can be found in [25.81, 82] and [25.83] in the context of computer vision and in [25.36] in the context of flow measurement. In-depth textbooks on Kalman filtering in the context of radar applications are available [25.72, 73, 75]. Many ideas and techniques described there can also be successfully applied in visual tracking applications like PTV. *Cook* et al. [25.84] and *Nemhauser* and *Wolsey* [25.68] discuss algorithms to solve combinatorial optimization problems, similar to those arising in tracking applications with increased temporal scope.

25.2.5 Optical-Flow-Based Velocity Analysis

Optical flow can be considered as the distribution of apparent velocities of movement of brightness patterns in an image [25.85]. More precisely the optical flow is an approximation to the *two-dimensional* motion field of an image sequence. We obtain this motion field by projecting the three-dimensional velocities of object points in three-dimensional space onto the two-dimensional image plane [25.86].

In estimating the optical flow and in determining motion fields from the estimated optical flow, there arise a number of peculiarities and difficulties, which have to be taken into account:

- There may be regions in the images where no motion can be determined (the *black wall problem*) or where only the normal velocity component can be determined (the *aperture problem*). Figure 25.15 illustrates these situations. In order to obtain a dense flow field an interpolation or a regularization technique has to be applied. This is an intrinsic problem in optical flow computation, and it is addressed throughout the chapter.
- There may be regions in the image, where the motion is nontranslational. Especially for applications in fluid dynamics the nonrigid behavior of fluids has to be taken into account. For these situations appropriate models have been developed.
- For large displacements the temporal sampling theorem may be violated. Coarse-to-fine techniques

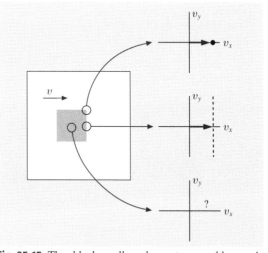

Fig. 25.15 The black wall and aperture problems. At the corners of the moving square, a unique velocity can be estimated. On the edges of the squares only the normal component of the velocity is determinable. In the center the motion is completely unconstrained (from [25.80])

applied on bandpass-filtered image sequences help to cope with this problem.
- In the presence of motion discontinuities, reflections or corrupted pixels, parameterized flow field models fail. Using a robust approach, described in the paragraph on *robust estimation* below, we can determine a correct motion field even in these situations.
- Brightness changes have to be taken into account. Illumination changes lead to a misinterpretation of the optical flow field. This can be nicely demonstrated by considering a rigid sphere with homogeneous surface reflectance, spinning around an axis through the center of the sphere (Fig. 25.16). If the surface is not textured and the illumination stays constant, the optical flow field would be zero over the entire sphere. If a directional light source moves around the same sphere the illumination changes would be falsely attributed to motion of the sphere surface. In many situations we can apply physical models of brightness variations to deal with brightness changes.

Optical flow methods can be classified as belonging to one of these groups:

- Differential techniques, which compute image velocity from spatiotemporal intensity derivatives.
- Frequency-based techniques, which use energy/phase information in the output of velocity tuned filters.
- Tensor-based techniques, which deduce the motion field from the local image brightness distribution represented as a structure tensor.

Though the attempts to estimate optical-flow look very different at first glance, all of these approaches are closely related. *Simoncelli* [25.80] showed that differential techniques are equivalent to frequency-based technique, provided that the derivatives and filters are chosen appropriately. The structure tensor can be constructed based on operations in the spatiotemporal domain [25.62], but it can also be obtained by linear combinations of outputs of filters in frequency-domain [25.87]. *Jähne* et al.[25.29] and *Barron* et al.[25.86] provide a well-founded overview on the field of optical-flow estimation and give a quantitative comparison of the results.

Though the method of optical-flow is quite common in computer vision, the extent of application in experimental fluid dynamics is relatively small, so far. Therefore in the following we will provide a detailed review of the optical flow methods classified above, and we will show, how to cope with each of the difficulties listed above. We conclude this chapter with a literature review.

Optical-Flow Methods

Differential Techniques. The general task is to determine the optical flow field $\boldsymbol{f} = (f_1, f_2)^{\mathrm{T}} = (\mathrm{d}x/\mathrm{d}t, \mathrm{d}y/\mathrm{d}t)^{\mathrm{T}}$ from the gray values of an image sequence. In the simplest case it is assumed, that the gray value $g(\boldsymbol{x}, t)$ along a path $\boldsymbol{x}(t)$ remains constant for all time:

$$g(\boldsymbol{x}(t), t) = \mathrm{const} .$$

By taking the temporal derivative on both sides and applying the chain rule one obtains:

$$\frac{\mathrm{d}g}{\mathrm{d}t} = \frac{\partial g}{\partial x}\frac{\mathrm{d}x}{\mathrm{d}t} + \frac{\partial g}{\partial y}\frac{\mathrm{d}y}{\mathrm{d}t} + \frac{\partial g}{\partial t} = 0 .$$

Writing down in vector notation using $\nabla g = (\partial g/\partial x, \partial g/\partial y)^{\mathrm{T}}$ this yields the brightness *constancy constraint equation* (BCCE):

$$(\nabla g)^{\mathrm{T}} \boldsymbol{f} + g_t = 0 . \tag{25.145}$$

Because the partial derivatives $g_x = \partial g/\partial x$, $g_y = \partial g/\partial y$ and $g_t = \partial g/\partial t$ are accessible by the application of a derivative filter, one obtains one constraint for the two-component flow field. Dealing with two unknowns in one equation we have an ill-posed problem. Graphically spoken the solution of (25.145) determines a *line*, containing all vectors that are possible candidates for the true optical flow vector (Fig. 25.15). Without further assumptions only the flow perpendicular to the constraint line can be estimated. This problem is commonly referred as the *aperture problem* of motion estimation.

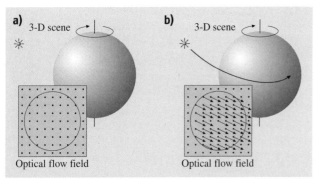

Fig. 25.16a,b Illumination changes and optical-flow. (**a**) Spinning sphere with fixed illumination leads to zero optical flow. (**b**) Moving illumination source causes apparent optical-flow field without motion of the sphere (after [25.29])

Solving (25.145) for points x_U in a sufficiently large neighborhood U around x we may get other constraint lines, so that we can determine the true optical flow vector by the intersecting point of the constraint lines. But the neighborhood must not be chosen too large, because it is not assured, that the motion is constant in a larger area. How large to choose the neighborhood is referred as the *generalized aperture problem*. One way to weaken the assumption of constant motion is finding a local parameterization of the flow field, so that one demands local coherency instead of local constancy. This leads us to the second constraint of differential optical flow estimation, the *spatial coherence constraint*.

The concept of optical flow originates from hydrodynamics. Grey values *flow* over the image plane, like volume elements flow in fluids. In hydrodynamics the principle of conservation of mass is formulated by the continuity equation, which reads in its differential form

$$\frac{\partial \rho}{\partial t} + \nabla(\boldsymbol{u}\rho) = \frac{\partial \rho}{\partial t} + \boldsymbol{u}\nabla\rho + \rho\nabla\boldsymbol{u} = 0 . \quad (25.146)$$

The three-dimensional velocity $\boldsymbol{u}$ of a fluid element with density ρ in three-dimensional space is apparently analogous to the two-dimensional optical flow $\boldsymbol{f}$ of a gray value g in two-dimensional space. The BCCE (25.145) corresponds to the continuity equation (25.146), if one drops $\rho\nabla\boldsymbol{u}$. Why do we have to drop the last term? Consider an object moving away from the camera. In this case the total brightness change dg/dx of a gray value on a path $x(t)$ is zero, because the irradiance in the image plane remains the same for an object moving perpendicular to the image plane. Because both g_t and ∇g are zero, we can apply (25.145). But we cannot apply (25.146), because the additional term $\rho\nabla\boldsymbol{u}$ or $g\nabla\boldsymbol{f}$ would *not* be zero, because the motion is not divergence free.

However, under certain conditions the use of a two-dimensional continuity equation instead of the BCCE can be motivated. This is the case, if one deals with 2-D transmittance images of a 3-D fluid flow, so that the imaged 2-D flow is the density weighted average of the physical 3-D flow [25.88,89]. The constraint on the data now becomes

$$g_x f_1 + g_y f_2 + g f_1 + g f_2 + g_t = 0. \quad (25.147)$$

Local Weighted Least Squares. Assuming the optical-flow to be constant within a small neighborhood, [25.58] proposed a local method to estimate the optical-flow. Goal is to minimize the squared left-hand side of the BCCE (25.145) in a local neighborhood U around x, which is given by the weighting (or window) function $w(x - x')$:

$$\hat{\boldsymbol{f}} = \arg\min_{\boldsymbol{f}} \int_{-\infty}^{\infty} w(\boldsymbol{x}-\boldsymbol{x}')[(\nabla g)^{\mathrm{T}}\boldsymbol{f} + g_t]^2 d\boldsymbol{x}' .$$

The weighting function in the simplest case is given by a box-filter (all points in the neighborhood are weighted equally), but better results can be achieved using a binomial filter. Standard least-squares minimization (setting the partial derivatives of the functional with respect to f_1 and f_2 to zero) yields the equation system

$$\underbrace{\begin{pmatrix} \langle g_x g_x \rangle & \langle g_x g_y \rangle \\ \langle g_x g_y \rangle & \langle g_y g_y \rangle \end{pmatrix}}_{\mathbf{A}} \underbrace{\begin{pmatrix} f_1 \\ f_2 \end{pmatrix}}_{\boldsymbol{f}} = -\underbrace{\begin{pmatrix} \langle g_x g_t \rangle \\ \langle g_y g_t \rangle \end{pmatrix}}_{\boldsymbol{b}} \quad (25.148)$$

with the abbreviation

$$\langle a \rangle = \int_{-\infty}^{\infty} w(\boldsymbol{x}-\boldsymbol{x}') a \, d\boldsymbol{x}' .$$

The solution of (25.148) is given by

$$\boldsymbol{f} = \mathbf{A}^{-1}\boldsymbol{b} ,$$

provided that the inverse of $\mathbf{A}$ exists, i. e.. the determinant of $\mathbf{A}$ is unequal to zero:

$$\det \mathbf{A} = \langle g_x g_x \rangle \langle g_y g_y \rangle - \langle g_x g_y \rangle^2 \neq 0 .$$

This is not the case, if all spatial derivatives in the local neighborhood are zero (the *black wall problem*), or if all gradients in the local neighborhood point into the same direction (the *aperture problem*).

Global Constraints. Instead of assuming local spatial constancy (and therefore coherence) by introducing a window function we can demand global spatial coherence. One can determine the optical-flow by minimizing the BCCE (25.145) over the entire image Ω. To make the problem well-posed an additional term, the regularizing spatial coherence constraint $\|e_S^2\|$, is introduced:

$$\hat{\boldsymbol{f}} = \arg\min_{\boldsymbol{f}} \int_{\Omega} [(\nabla g)^{\mathrm{T}}\boldsymbol{f} + g_t]^2 d\boldsymbol{x}' + \lambda^2 \|e_S^2\| .$$

$$(25.149)$$

The parameter λ controls the influence of the spatial coherence term. *Horn* et al.[25.85] propose global smoothness for the spatial coherence constraint:

$$\|e_S^2\| = \int_{\Omega} \|\nabla f_1(\boldsymbol{x}')\|^2 + \|\nabla f_2(\boldsymbol{x}')\|^2 d\boldsymbol{x}' .$$

There are other suggestions for $\|e_S^2\|$, which may be better suited to special kinds of problems (e.g., fluid flow analysis). To solve the minimization problem a variational approach can be adapted. Thus, the integral equation (25.149) can be solved by a system of Euler–Lagrange equations:

$$L_{f_1} - \frac{\partial}{\partial x} L_{f_{1x}} - \frac{\partial}{\partial y} L_{f_{1y}} = 0,$$

$$L_{f_2} - \frac{\partial}{\partial x} L_{f_{2x}} - \frac{\partial}{\partial y} L_{f_{2y}} = 0.$$

The integrand of (25.149) can be identified with the Lagrange function:

$$L = (g_x f_1 + g_y f_2 + g_t)^2 + \lambda^2 (f_{1x}^2 + f_{1y}^2 + f_{2x}^2 + f_{2y}^2),$$

L plugged into the Euler–Lagrange equations yields the diffusion–reaction system

$$((\nabla g)^T f + g_t) \nabla g - \lambda^2 \nabla^2 f = 0.$$

For the case, that there is a high spatial gray value variation (that means ∇g is large), the first summand dominates in the equation, and the optical flow is calculated using the BCCE. But if there is a *black wall problem*, the optical flow is calculated from the last summand, which states the Laplacian equation $\nabla^2 f = 0$.

The discretization can be performed using finite differences or finite elements. Once guaranteed, that the problem is well-posed, there exist a number of minimization schemes like Gauss–Jordan elimination or Gauss–Seidel iteration, which can be applied.

Frequency-Based Techniques. The concept of identifying sequences of 2-D images as 3-D spatiotemporal structures allows one to analyze motion in the corresponding spatiotemporal frequency domain (the Fourier domain). Let $g(x, t)$ be an image sequence of any pattern moving with constant velocity, causing the optical flow f at any point in the image plane, the resulting spatiotemporal structure can be described by

$$g(x, t) = g(x - ft). \tag{25.150}$$

The spatiotemporal Fourier transform $\hat{g}(k, \omega)$ of equation (25.150) is given by

$$\hat{g}(k, \omega) = \hat{g}(k) \delta(k^T - \omega), \tag{25.151}$$

where $\hat{g}(k)$ is the spatial Fourier transform of the pattern, and $\delta(\cdot)$ denotes Dirac's delta distribution. This equation states, that the three-dimensional Fourier spectrum of a pattern moving with constant velocity condenses to a plane in Fourier space. The plane equation in Fourier domain is given by the argument of the delta distribution in (25.151) and can be considered as an alternative formulation of the BCCE (25.145):

$$\omega(k, f) = k^T f. \tag{25.152}$$

Taking the derivatives of $\omega(k, f)$ with respect to k_x and k_y yields both components of the optical flow:

$$f = \nabla_k \omega(k, f).$$

Quadrature-filter techniques try to estimate the orientation of this plane by using velocity tuned filters in the Fourier-domain. A quadrature-filter pair is a real frequency selective filter together with its imaginary Hilbert transform (Sect. 25.1.7). Its transfer function can be written in complex notation $\hat{q}(k)$.

The most common quadrature-filter pair is the *Gabor filter*, which selects a certain spatiotemporal frequency region with a Gaussian window centered at (k_0, ω_0). Its complex transfer function is

$$\hat{G}(k, \omega) = \exp\left(-\frac{1}{2}\sqrt{(k-k_0)^2 + (\omega-\omega_0)^2} \sigma^2\right).$$

From this the spatiotemporal filter mask can be computed using the shift theorem:

$$G(x, t) = \frac{1}{(2\pi)^{3/2} \sigma^3} \exp[i(k_0 x + \omega_0 t)]$$

$$\exp\left[-\left(\frac{x^2 + y^2 + t^2}{2\sigma^2}\right)\right].$$

By applying this filter for different parameter sets (k, ω) on the original spatiotemporal image we get estimates of the spectral density (or *energy*) of the corresponding periodic image structure belonging to these parameter sets. Ideally, for a single translational motion, the responses of these filters are concentrated about a plane in k–ω-space, so that we are able to get the optical flow by a least-squares fit to the data. Another way to calculate the optical flow is by constructing a structure tensor composed of the filter outputs [25.90].

Tensor-Based Techniques. Optical-flow estimation can be formulated as orientation analysis in a three-dimensional spatiotemporal image. The concept of orientation analysis of a pattern in 2-D (Fig. 25.17a) can be generalized to 3-D (Fig. 25.17b). Any constantly moving gray value structure causes inclined patterns. Goal of tensor-based optical-flow estimation is to find the orientation of these patterns, provided that there exist any oriented patterns.

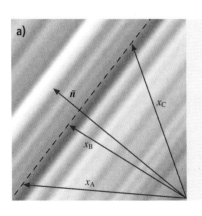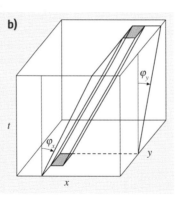

Fig. 25.17 (a) A simple neighborhood in 2-D. The grey values depend on one coordinate in the direction of the unit vector $\hat{n}$ only. **(b)** Space–time diagram with two spatial components and a temporal component (after [25.62])

Let $r = (r_1, r_2, r_3)^T$ be the vector pointing into the direction of constant brightness within the three-dimensional x–t domain. Once estimated r one obtains for the optical flow:

$$f = (f_1, f_2)^T = \frac{1}{r_3}(r_1, r_2)^T \qquad (25.153)$$

Here r points orthogonally to the spatiotemporal gradient vector $\nabla_{xt} g = (g_x, g_y, g_t)^T$. Therefore the scalar product between r and $\nabla_{xt} g$ has to vanish:

$$(g_x, g_y, g_t) \cdot (r_1, r_2, r_3)^T = r_3\left[(\nabla g)^T f + g_t\right] = 0\,.$$

We arrive at the well-known BCCE equation (25.145). Instead of the approach of [25.58], where the BCCE is minimized in a spatial neighborhood here one minimizes $\nabla_{xt} g \cdot r$ in a *spatiotemporal* neighborhood U, which is characterized by the window function $w(x - x', t - t')$.

$$\hat{r} = \arg\min_{r} \left\langle [\nabla_{xt} g \cdot r]^2 \right\rangle$$

using the abbreviation

$$\langle a \rangle = \int_{-\infty}^{\infty} w(x - x', t - t') a \, dx' dt'\,.$$

Under the assumption of constant r (that is, constant f) within U, we the minimization problem can be reformulated as

$$\hat{r} = \arg\min_{r} \left[r^T \langle \nabla_{xt} g \cdot \nabla_{xt} g^T \rangle r \right] \qquad (25.154)$$

$$= \arg\min_{r} r^T \mathbf{J} r\,, \qquad (25.155)$$

where $\mathbf{J}$ with its components $J_{pq} = \langle g_p g_q \rangle$ is the three-dimensional symmetric *structure tensor*, and g_p, $p \in \{x, y, t\}$, denotes the partial derivative along the coordinate p.

The structure tensor can be transformed into diagonal shape by means of rotation. Thus the principal axes of the structure tensor can be found by solving the eigenvalue problem

$$\mathbf{J} r = \lambda r\,.$$

The eigenvector to the corresponding *minimal* eigenvalue denotes the direction of constant brightness in the x–t domain, from which the optical flow can be calculated according to (25.153). From the rank of the structure tensor the type of motion can be deduced: Constant brightness (the *black wall problem*, rank($\mathbf{J}$) = 0), spatial orientation and constant motion (the *aperture problem*, rank($\mathbf{J}$) = 1), distributed spatial structures and constant motion (rank($\mathbf{J}$) = 2), and distributed spatial structures but no coherent motion (rank($\mathbf{J}$) = 3). The structure tensor technique is not only able to give an estimate for the optical flow, but is also able to present a confidence measure which assesses the quality of the estimate. A detailed analysis of the structure tensor technique and its practical application to optical flow computation can be found in [25.29]. The structure tensor technique can be formulated as a solution of the total least-squares (TLS) problem in a more-general way [25.91].

Implementation of the Structure Tensor. The expression for the structure tensor in (25.154) can be written explicitly as:

$$\mathbf{J} = \begin{pmatrix} \langle g_x g_x \rangle & \langle g_x g_y \rangle & \langle g_x g_t \rangle \\ \langle g_x g_y \rangle & \langle g_y g_y \rangle & \langle g_y g_t \rangle \\ \langle g_x g_t \rangle & \langle g_y g_t \rangle & \langle g_t g_t \rangle \end{pmatrix},$$

using the abbreviation

$$\langle a \rangle = \int_{-\infty}^{\infty} w(x - x', t - t') a \, dx' dt'\,.$$

The implementation of the structure tensor can be carried out very efficiently by standard image processing operations. Identifying the convolution with a smoothing operation (for example with the isotropic binomial operator $\mathcal{B}$), and the derivatives in the p-th respective q-th direction with edge detectors (for example the optimized *Sobel filters* $\mathcal{D}_p$ and $\mathcal{D}_q$ [25.92]), we can construct a structure tensor operator

$$\mathcal{J}_{pq} = \mathcal{B}(\mathcal{D}_p \cdot \mathcal{D}_q),$$

where the dot signals a pixel-wise multiplication.

Because smoothing in general comes along with a loss of information, the result after applying the structure tensor operator can be stored in a more compact sequence than the original data. In practise this can be handled by downsampling the resulting sequence by a factor of two, for instance. In order to find a procedure to perform the principal axis transformations efficiently, we bear in mind that we are dealing with very small and symmetric matrices. These types of data can be covered by using the numerical method of Jacobi transformations to find the eigenvalues and eigenvectors [25.93].

Improvements of Optical Flow Determination

In this section improvements of determining optical flow will be presented. These improvements will help to solve the difficulties mentioned in the introduction. Each improvement can be applied to more than one (but not necessarily to all) methods of determining optical flow.

Parameterization of 2-D Optical-Flow Fields. As mentioned in the introduction to this chapter the standard local optical-flow methods assume that the optical flow $f(x, t)$ is constant within the local neighborhood U around x. However, the optical flow may be expanded to a first order Taylor series in the vicinity of (x_0, t_0), in the way as in Sect. 25.2.1 (25.130). Then one is capable to estimate implicitly spatial gradients of the flow fields, for instance.

The BCCE supplemented by this parameterization yields the *extended brightness change constraint equation* (EBCCE):

$$(\nabla g)^\mathrm{T}(t + Ax + at) + g_\mathrm{t} = 0. \quad (25.156)$$

Coarse-to-Fine Techniques. The temporal sampling theorem, already noted in Sect. 25.2.1, states a theoretical upper limit for the magnitude of displacements that are able to be analyzed. Apparently the maximum determinable displacements are limited by the magnitude of the highest spatial wavenumbers, which are contained in the image.

Coarse-to-fine techniques or hierarchical multigrid approaches (Sect. 25.2.1) help to estimate large motions. By smoothing the image the high frequency content can be eliminated from the image sequence, so that one is able to estimate a coarse motion field. Then the motion can be *undone* by transforming the image back by means of the estimated coarse motion field. Now the higher-frequency content can be used to estimate a finer motion field. Added to the previously coarse motion field the fine motion field provides a more accurate approximation to the real motion field.

This procedure can be improved in several ways:

- If an estimation on a coarse level is incorrect, the fine-level estimate has no chance of correcting the errors. To fix this, we must have knowledge of the error in the coarse-level estimates. This suggests working in a probabilistic framework. Indeed, a *state evolution equation* and a *measurement equation* can be proposed similar to Kalman filtering [25.80].
- For more-accurate computation additional filters can be introduced, which slice the bandwidth in smaller pieces than these given by a dyadic pyramid structure. This results in a combined multiresolution and multiscale approach [25.94].
- The motion can be distributed very irregularly over the image plane. In these situations a selective multiresolution approach is suggested [25.95].

Robust Estimation. There are situations in which even parameterized flow field models fail to determine the optical flow correctly; these include the presence of multiple motions, such as motion discontinuities at boundaries (occluded multiple motions), or different motions being overlaid (transparent multiple motions) but also the presence of reflexes or corrupted pixels.

Least squares estimation tries to minimize a quadratic objective function $\rho(x)$:

$$\hat{f} = \arg\min_f \int_{-\infty}^{\infty} w(x-x')\rho[(\nabla g)^\mathrm{T} f + g_\mathrm{t}]\mathrm{d}x'$$

with $\rho(x) = x^2$. The influence function $\psi(x)$ of the objective function is defined as the derivative of ρ with respect to x:

$$\psi(x) = \frac{\partial \rho(x)}{\partial x}.$$

In the least-squares case the influence of data points increases linearly and without bound, so that outliers that do not fit to the model such as corrupted pixels have a great influence and distort the estimation of the correct optical flow dramatically. This is due to the fact that, by using a quadratic objective function, we inherently assumed that the residual errors are Gaussian and independently distributed within the neighborhood U.

To achieve a more-robust parameter estimation we have to replace the quadratic objective function by a suitable other function, which is referred to as an *M-estimator* in statistics. The influence function of an M-estimator has to be redescending, i.e., it has to approach zero for large residuals after an initial increase for small values. [25.96] proposed a commonly used M-estimator (Fig. 25.18), which reads together with its influence function

$$\rho(x,\sigma)=\frac{x^2}{\sigma+x^2}, \quad \psi(x,\sigma)=\frac{2x\sigma}{(\sigma+x^2)^2}, \quad (25.157)$$

where σ is a scale parameter.

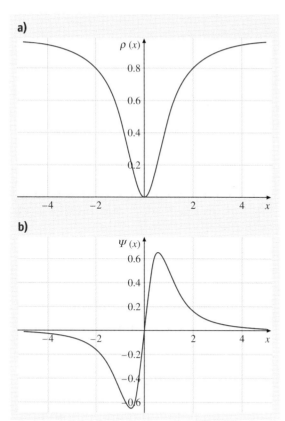

Fig. 25.18a,b An example for an M-estimator: (a) Geman and McClure norm (25.157). (b) Its derivative (after [25.29])

Given a robust formulation, there are numerous optimization techniques that can be employed to recover the motion estimates, and the most appropriate technique will depend on the particular formulation and choice of the ρ function. For detailed information about robust estimation the reader is referred to [25.97].

Dealing with Brightness Changes. In many situations the constraint of brightness constancy (25.145) is violated. In some cases we are able to find a physical model for the time-dependent brightness variation. So one can estimate both the correct optical flow field f and the parameters a of the underlying physical process. The approach of [25.98] constitutes an extension of the brightness change constraint equation to parameterized models of brightness variation, provided that these models are linear in a or can be linearized by a Taylor series expansion.

In Sect. 25.2.5 we have stated that for the case that the brightness $g(\boldsymbol{x}, t)$ is constant along a path $\boldsymbol{x}(t)$ for all times, we are able to derive a constraint on the optical-flow. Now we allow, that the brightness along the path may change according to a time-dependent parameterized function $h(g_0, t, \boldsymbol{a})$:

$$g(\boldsymbol{x}(t), t) = h(g_0, t, \boldsymbol{a}), \quad (25.158)$$

where $g_0 = g(\boldsymbol{x}(t_0), t_0)$ denotes the image at time t_0, and $\boldsymbol{a}$ is the parameter vector for the brightness change model. The total derivative on both sides of (25.158) yields the *generalized brightness change constraint equation* (GBCCE),

$$(\nabla g)^\mathrm{T} \boldsymbol{f} + g_t = \frac{\mathrm{d}}{\mathrm{d}t} h(g_0, t, \boldsymbol{a}),$$

which reduces to the well-known BCCE if h is constant.

In the following some models of brightness variation are presented:

Linear source terms. When sources are present, the brightness depends linearly on time: $h = qt$, where q denotes the source strength. The GBCCE becomes

$$(\nabla g)^\mathrm{T} \boldsymbol{f} + g_t = q .$$

Exponential decay. In relaxation processes the time-dependent brightness can be modeled by an exponential decay: $h = g_0 \exp(-\kappa t)$, where κ denotes the relaxation constant. By differentiating h with respect to t the exponential function reproduces itself, so that we can write for the GBCCE:

$$(\nabla g)^\mathrm{T} \boldsymbol{f} + g_t = -\kappa g .$$

Diffusion process. Fick's second law, which states that for isotropic diffusion the rate of change of the gray value is proportional to its Laplacian, tells us what the GBCCE looks like in this case, where the diffusion constant D is the proportional constant:

$$(\nabla g)^T f + g_t = D\nabla^2 g \,,$$

One can combine these physical models of brightness variation with various differential or tensor-based techniques: One has to replace the BCCE by the GBCCE, and the minimization has to be carried out over the optical flow f and over the parameters a simultaneously.

Literature Review

Each of the optical-flow methods (differential, frequency-based, or tensor-based approaches) presented in Sect. 25.2.5 was adapted to evaluation in the field of fluid mechanics. The term *optical flow* suggests its application to image sequences dealing with continuous tracer, such as heat or concentration. Indeed, most of the literature addresses continuous tracer. On the other hand, in practice fluid flow analysis is to a great extent particle based, which does not prevent optical-flow-based methods from being applied as well. Examples can be found in the literature.

Differential Techniques. *Ruhnau* et al.[25.94] applied the method of *Horn* et al.[25.85] (25.149) to images recorded with the conventional PIV technique. They used a coarse-to-fine strategy. The same authors replaced the global smoothness constraint by a regularizing coherence constraint, relying on the Stokes equation [25.99] and on the vorticity transport equation [25.100].

Instead of applying the BCCE, other authors used the 2-D continuity equation (25.147) as model. This was justified by the special kind of recording technique (such as transmittance imagery [25.88]) or data (such as satellite imagery [25.89]). The latter used a second-order div–curl regularization scheme instead of just assuming global smoothness. Moreover, they applied their scheme to PIV sequences [25.101].

Cohen et al. [25.95] applied a global method using a nonquadratic regularization technique to atmospheric and oceanographic image sequences. They have shown, that using an appropriate tessellation of the image according to an estimate of the motion field can improve optical flow accuracy and yields more reliable flows. This method defines a nonuniform multiresolution approach for coarse-to-fine grid generation.

Frequency-Based Techniques. *Larsen*[25.90] applied the local energy distribution to satellite images using the frequency-based techniques presented in Sect. 25.2.5. They used the optical-flow estimates together with confidence measures as an input for a regularization method based on the Markovian random-field approach.

Tensor-Based Techniques. *Garbe* et al. [25.102] estimated both velocity vector field and heat flux simultaneously in image sequences, recorded using infrared thermography. They expanded the structure tensor technique by a model including brightness changes.

Jehle and *Jähne* [25.103] estimated the wall shear stress in a medical engineering application directly, without previous computation of the velocity vector fields. The wall shear rate emerges as some components of the velocity gradient tensor, which is a 3-D generalization of the matrix **A**.

References

25.1 M. Unser, A. Aldroubi, M. Eden: Fast B-spline transforms for continuous image representation and interpolation, IEEE Trans. PAMI **13**, 277–285 (1991)

25.2 W.T. Freeman, E.H. Adelson: The design and use of steerable filters, IEEE Trans. PAMI **13**, 891–906 (1991)

25.3 J. Weickert: *Anisotropic Diffusion in Image Processing* (Teubner, Stuttgart 1998)

25.4 P. Perona, J. Malik: Scale-space and edge detection using anisotropic diffusion, IEEE Trans. PAMI **12**, 629–639 (1990)

25.5 B. Jähne: *Digital Image Processing*, 6th edn. (Springer, Heidelberg 2005)

25.6 B. Jähne, H. Scharr, S. Körgel: Principles of filter design. In: *Computer Vision and Applications*, Signal Processing and Pattern Recognition, Vol. 2, ed. by B. Jähne, H. Haußecker, P. Geißler (Academic, San Diego 1999) pp. 125–151

25.7 J. Bigün, G.H. Granlund: Optimal orientation detection of linear symmetry, ICCV'87 (IEEE, Washington 1987) 433–438

25.8 G.H. Granlund: In search of a general picture processing operator, Comput. Graph. Imag. Process. **8**, 155–173 (1978)

25.9 M. Felsberg, G.H. Granlund: POI detection using channel clustering and the 2D energy tensor, Pattern Recognition: 26th DAGM Symposium, Tübingen, Germany, LNCS, Vol. 3175 (Springer, Berlin 2004) 103–110

25.10 V.K. Madisetti, D.B. Williams: *The Digital Signal Processing Handbook* (CRC, Boca Raton 1998)

25.11 B. Jähne: *Handbook of Digital Image Processing for Scientific and Technical Applications*, 2nd edn. (CRC, Boca Raton 2004)

25.12 D.J. Fleet: *Measurement of Image Velocity* (Dissertation University of Toronto, Canada 1990)

25.13 M. Felsberg, G. Sommer: A new extension of linear signal processing for estimating local properties and detecting features. In: *Mustererkennung 2000, 22. DAGM Symposium, Kiel*, Informatik aktuell, ed. by G. Sommer, N. Krüger, C. Perwass (Springer, Berlin 2000) pp. 195–202

25.14 C.K. Chui (Ed.): *Wavelets: A Tutorial in Theory and Applications* (Academic, Boston 1992)

25.15 T. Acharya, P.-S. Tsai: *JPEG2000 Standard for Image Compression* (Wiley, New York 2005)

25.16 C.E. Willert, M. Gharib: Digital particle image velocimetry, Exp. Fluids **10**, 181–193 (1991)

25.17 J. Westerweel: Fundamentals of digital particle image velocimetry, Meas. Sci. Technol. **8**, 1379–1392 (1997)

25.18 A.M. Fincham, G.R. Spedding: Low cost, high resolution DPIV for measurement of turbulent fluid flow, Exp. Fluids **23**, 449–462 (1997)

25.19 P.T. Tokumaru, P.E. Dimotakis: Image correlation velocimetry, Exp. Fluids **19**, 1–15 (1995)

25.20 A.W. Gruen: Adaptive least squares correlation: a powerful image matching technique, S. Afr. J. Photogramm. Remote Sensing Cartogr. **14**(3), 175–187 (1985)

25.21 F. Scarano, M.L. Riethmuller: Advances in iterative multigrid PIV image processing, Exp. Fluids **29**, S51–S60 (2000)

25.22 E. Cowen, S. Monismith: A hybrid digital particle tracking velocimetry technique, Exp. Fluids **22**, 199–211 (1997)

25.23 R.J.M. Bastiaans, G.A.J. van der Plas, R.N. Kieft: The performance of a new PTV algorithm applied in super-resolution PIV, Exp. Fluids **32**, 346–356 (2002)

25.24 S.J. Baek, S.J. Lee: A new two-frame particle tracking algorithm using match probability, Exp. Fluids **22**, 23–32 (1996)

25.25 K. Ohmi, H.-Y. Li: Particle tracking velocimetry with new algorithms, Meas. Sci. Technol. **11**(6), 603–616 (2000)

25.26 Y.A. Hassan, R.E. Canaan: Full-field bubbly flow velocity measurements using a multiframe particle tracking technique, Exp. Fluids **12**, 49–60 (1991)

25.27 N.A. Malik, T. Dracos, D. Papantoniou: Particle tracking velocimetry in three-dimensional flows, Exp. Fluids **15**, 279–294 (1993), Part II: Particle tracking

25.28 K. Takehara, R.J. Adrian, G.T. Etoh, K.T. Christensen: A Kalman tracker for super-resolution PIV, Exp. Fluids **29**, S34–S41 (2000)

25.29 B. Jähne, H. Haussecker, P. Geissler: *Handbook of Computer Vision and Applications* (Academic, San Diego 1999)

25.30 M. Raffel, C. Willert, J. Kompenhans: *Particle Image Velocimetry: A Practicle Guide* (Springer, Heidelberg 1998)

25.31 G.R. Spedding, E.J.M. Rignot: Performance analysis and application of grid interpolation techniques for fluid flows, Exp. Fluids **15**, 417–430 (1993)

25.32 A.K. Prasad: Stereoscopic particle images velocimetry, Exp. Fluids **29**, 103–116 (2000)

25.33 R. Hartley, A. Zisserman: *Multiple View Geometry in Computer Vision* (Cambridge University Press, Cambridge 2000)

25.34 C.S. Slama: *Manual of Photogrammetry*, 4th edn. (American Society of Photogrammetry, Falls Church 1980)

25.35 K.D. Hinsch: Holographic particle image velocimetry, Meas. Sci. Technol. **13**, R61–R72 (2002)

25.36 T. Dracos: *Three-Dimensional Velocity and Vorticity Measuring and Image Analysis Techniques* (Kluwer Academic, Dordrecht 1996)

25.37 S.P. McKenna, W.R. McGillis: Performance of digital image velocimetry processing techniques, Exp. Fluids **32**, 2 (2002)

25.38 D.P. Hart: Super-Resolution PIV by Recursive Local-Correlation, J. Visual. **3**(2), 187–194 (2000)

25.39 H.J. Lin, M. Perlin: Improved methods for thin, surface boundary layer investigations, Exp. Fluids **25**, 431–444 (1998)

25.40 H.T. Huang, H.F. Fielder, J.J. Wang: Limitation and improvement of PIV, Exp. Fluids **15**, 168–174 (1993), Part I: Limitation of conventional techniques due to deformation of particle image patterns

25.41 H.T. Huang, H.F. Fielder, J.J. Wang: Limitation and improvement of PIV, Exp. Fluids **15**, 263–273 (1993), Part II. Particle image distortion, a novel technique

25.42 D.P. Hart: PIV error correction, Exp. Fluids **29**, 13–22 (2000)

25.43 B. Wienecke: Stereo-PIV using self-calibration on particle images, 5th International Symposium on Particle Image Velocimetry (Busan, Korea 2003)

25.44 C.E. Willert, M. Gharib: Three-dimensional particle imaging with a single camera, Exp. Fluids **12**, 353–358 (1992)

25.45 F. Pereirra, M. Gharib, D. Dabiri, M. Modarress: Defocusing PIV: a three component 3-D DPIV measurement technique, Exp. Fluids **29**, S78–S84 (2000), Application to bubbly flows

25.46 C. Kähler, J. Kompenhans: Fundamentals of multiple plane stereo particle image velocimetry, Exp. Fluids **29**, 70–77 (2000)

25.47 A. Liberzon, R. Gurka, G. Hetsroni: XPIV-Multiplane stereoscopic particle image velocimetry, Exp. Fluids **36**, 355–362 (2004)

25.48 A. Schimpf, S. Kallweit, J.B. Richon: Photogrammatic particle image velocimetry, 5th Int. Symp. on Particle Image Velocimetry (2003)

25.49 R.J. Adrian: Particle-imaging techniques for experimental fluid mechanics, Annu. Rev. Fluid Mech. **23**, 261–304 (1991)

25.50 R.D. Keane, R.J. Adrian: Optimization of particle image velocimeters, Meas. Sci. Technol. **1**, 1202–1215 (1990), Part I: Double pulsed systems

25.51 R.D. Keane, R.J. Adrian: Optimization of particle image velocimeters, Meas. Sci. Technol. **2**, 963–974 (1991), Part II: Multiple pulsed systems

25.52 R.D. Keane, R.J. Adrian: Theory of crosscorrelation analysis of PIV images, Appl. Sci. Res. **49**, 191–215 (1992)

25.53 J. Westerweel: *Digital Particle Image Velocimetry – Theory and Application* (Delft Univ. Press, Delft 1993)

25.54 L.C. Gui, W. Merzkirch: A method for tracking ensembles of particle images, Exp. Fluids **21**, 465–468 (1996)

25.55 C.Q. Davis, Z.Z. Karu, D.M. Freeman: Equivalence of subpixel motion estimators based on optical flow and block matching, Int. Symposium on Computer Vision (Coral Gables, Florida 1995)

25.56 L.C. Gui, W. Merzkirch: A comparative study of the MQD method and several correlation-based PIV evaulation algorithms, Exp. Fluids **28**, 36–44 (2000)

25.57 J. Shi, C. Tomasi: Good features to track, Computer Vision Pattern Recognition (1994)

25.58 B.D. Lucas, T. Kanade: An iterative image registration technique with an application to stereo vision, Imaging Understanding Workshop (1981) 121–130

25.59 D. Papantoniou, T. Dracos: Analyzing 3-D turbulent motions in open channel flow by use of stereoscopy and particle tracking, Adv. Turb. **2**, 278–285 (1989)

25.60 M.P. Wernet, A. Pline: Particle displacement tracking technique and Cramer-Rao lower bound error in centroid estimates from CCD imagery, Exp. Fluids **15**, 295–307 (1993)

25.61 C.J. Veenman, M.J.T. Reinders, E. Backer: Establishing motion correspondence using extended temporal scope, Artif. Intell. **145**(1–2), 227–243 (2003)

25.62 B. Jähne: *Digital Image Processing*, 5th edn. (Springer, Heidelberg 2002)

25.63 Y.G. Geuzennec, N. Kiritsis: Statistical investigation of errors in particle image velocimetry, Exp. Fluids **10**, 138–146 (1990)

25.64 H.G. Maas, A. Gruen, D. Papantoniou: Particle tracking velocimetry in three-dimensional flows, Exp. Fluids **15**, 133–146 (1993), Part I: Photogrammetric determination of particle coordinates

25.65 Y.G. Guezennec, R.S. Brodkey, N. Trigui, J.C. Kent: Algorithms for fully automated three-dimensional particle tracking velocimetry, Exp. Fluids **17**, 209–219 (1994)

25.66 N.A. Malik, T. Dracos: Interpolation schemes for three-dimensional velocity fields from scattered data using Taylor expansions, J. Comput. Phys. **119**, 231–243 (1995)

25.67 F. Hering, D. Wierzimok, C. Leue, B. Jähne: Particle tracking velocimetry beneath water waves, Exp. Fluids **23**(6), 472–482 (1997), Part I: Visualization and tracking algorithms

25.68 G. Nemhauser, L. Wolsey: *Integer and Combinatorial Optimization* (Wiley, New York 1999)

25.69 S.B. Dalziel: Decay of rotating turbulence: some particle tracking experiments,. In: *Flow Visualization and Image Analysis*, ed. by F.T.M. Nieuwstadt (Kluwer Academic, Dordrecht 1993)

25.70 M. Stellmacher, K. Obermayer: A new particle tracking algorithm based on deterministic annealing and alternative distance measures, Exp. Fluids **28**, 506–518 (2000)

25.71 S. Gold, A. Rangarajan, C.-P. Lu, S. Pappu: New algorithms for 2D and 3D point matching: pose estimation and correspondence, Pattern Recog. **31**, 1019–1031 (1998)

25.72 E. Brookner: *Tracking and Kalman Filtering made easy* (Wiley, New York 1998)

25.73 S. Blackman, R. Popoli: *Design and Analysis of Modern Tracking Systems* (Artech House, Boston 1999)

25.74 I.J. Cox: A review of statistical data association techniques for motion correspondence, Int. J. Comput. Vis. **10**(1), 53–66 (1993)

25.75 L.D. Stone, C.A. Barlow, T.L. Corwin: *Bayesian Multiple Target Tracking* (Artech House, Boston 1999)

25.76 G. Welch, G. Bishop: *An Introduction to the Kalman Filter*, Tech. Rep. TR 95-041 (Univ. North Carolina, Chapel Hill 2001)

25.77 I.J. Cox, S.L. Hingorani: An efficient implementation of Reid's multiple hypothesis tracking algorithm and its evaluation for the purpose of visual tracking, IEEE Trans. Pattern Anal. **18**(2), 138–150 (1996)

25.78 R.D. Keane, R.J. Adrian, Y. Zhang: Superresolution particle imaging velocimetry, Meas. Sci. Technol. **6**, 754–768 (1995)

25.79 J. Willneff, B. Lüthi: Particle tracking velocimetry measurements for lagrangian analysis of turbulent flows, 6th Conference on Optical 3-D Measurement Techniques, Vol. 2 (Zurich 2003) 191–198

25.80 E.P. Simoncelli: *Distributed representation and analysis of visual motion*, Ph.D. Thesis (MIT, Cambridge 1993)

25.81 O. Faugeras: *Three Dimensional Computer Vision: A Geometric Viewpoint* (MIT Press, Cambridge 1993)

25.82 B.F. Murray, D.W. Buxton: *Experiments in the Machine Interpretation of Visual Motion* (MIT Press, Cambridge 1990)

25.83 Z. Zhang, O. Faugeras: *3D Dynamic Scene Analysis*, 27 Springer Inform. Sci. (Springer, Heidelberg 1992)

25.84 W.J. Cook, W.H. Cunningham, W.R. Pulleyblank, A. Schrijver: *Combinatorial Optimization* (Wiley, New York 1998)

25.85 B.K.P. Horn, B.G. Schunk: Determining optical flow, Artif. Intell. **17**, 185–204 (1981)

25.86 J.L. Barron, D.J. Fleet, S.S. Beauchemin: Performance of optical flow techniques, Int. J. Comput. Vis. **12**(1), 43–77 (1994)

25.87 G.H. Granlund, H. Knutsson: *Signal Processing for Computer Vision* (Kluwer, Dordrecht 1995)

25.88 R. Wildes, M. Amabile, A.-M. Lanziletto, T.-S. Leu: Recovering estimates of fluid flows from image sequence data, Comput. Vis. Image Underst. **80**, 246–266 (2000)

25.89 T. Corpetti, E. Memin, P. Perez: Dense estimation of fluid flows, IEEE Trans. Pattern Anal. Machine Intell. **24**(3), 365–380 (2002)

25.90 R. Larsen: Estimation of dense image flow fields in fluids, IEEE T. Geosci. Remote Sens. **36**(1), 256–264 (1998)

25.91 S. van Huffel, J. Vandewalle: *The Total Least Squares Problem: Computational Aspects and Analysis* (SIAM, Philadelphia 1991)

25.92 H. Scharr: *Optimal Operators in Digital Image Processing*, Ph.D. Thesis (University of Heidelberg, Heidelberg 2000)

25.93 W.H. Press, S.A. Teukolsky, W. Vetterling, B. Flannery: *Numerical Recipes in C: The Art of Scientific Computing* (Cambridge Univ. Press, New York 1992)

25.94 P. Ruhnau, T. Kohlberger, C. Schnörr, H. Nobach: Variational optical flow estimation for particle image velocimetry, Exp. Fluids **38**, 21–32 (2005)

25.95 I. Cohen, I. Herlin: Non uniform multiresolution method for optical flow and phase portrait models: environmental applications, Int. Comput. Vis. **33**(1), 24–49 (1999)

25.96 S. Geman, D.E. McClure: Bayesian image analysis: An application to single photon emission tomography, Am. Statist. Assoc. Statist. Comput. Sect. (1984) 12–18

25.97 M.J. Black, P. Anandan: The robust estimation of multiple motions: parametric and piecewise-smooth flow fields, Comput. Vis. Image Understand. **63**, 75–104 (1996)

25.98 H.W. Haussecker, D.J. Fleet: Computing optical flow with physical models of brightness variation, IEEE Trans. Pattern Anal. **23**(6), 661–673 (2001)

25.99 P. Ruhnau, C. Schnörr: Optical Stokes flow: an image based control approach, Exp. Fluids **42**, 61–78 (2007)

25.100 P. Ruhnau, A. Stahl, C. Schnörr: On-line variational estimation of dynamical fluid flows with physics-based spatio-temporal regularization, 26th DAGM (2006), Pattern Recognition

25.101 T. Corpetti, D. Heitz, G. Arroyo, E. Memin, A. Santa-Cruz: Fluid experimental flow estimation based on an optical-flow scheme, Exp. Fluids **40**(1), 80–97 (2005)

25.102 C. Garbe, H. Spies, B. Jähne: Estimation of surface flow and net heat flux from infrared image sequences, J. Math. Imag. Vis. **19**, 159–174 (2003)

25.103 M. Jehle, B. Jähne: Direct estimation of the wall shear rate using parametric motion models in 3D, Lect. Notes Comput. Sci. **174**, 434–443 (2006)

Acknowledgements

B.3 Material Properties: Measurement and Data
*by William A. Wakeham, Marc J. Assael,
Abraham Marmur, Joël De Coninck, Terry D. Blake,
Stephanus A. Theron, Eyal Zussman*

The research leading to this article (JdeC, TDB) was partially supported by the Ministère de la Région Wallonne, Fonds National de la Recherche Scientifique, the Université de Mons-Hainaut, and Kodak Ltd.

B.4 Pressure Measurement Systems
by Beverley J. McKeon, Rolf H. Engler

This chapter was partly written while the author (BJM) was a Royal Society Dorothy Hodgkin Fellow in the Department of Aeronautics at Imperial College, London. The support of the Royal Society is gratefully acknowledged.

The editorial comments of John Foss were gratefully received and significantly improved this text.

B.5 Velocity, Vorticity, and Mach Number
*by Beverley J. McKeon, Geneviève Comte-Bellot,
John F. Foss, Jerry Westerweel, Fulvio Scarano,
Cameron Tropea, James F. Meyers, Joseph W. Lee,
Angelo A. Cavone, Richard Schodl,
Manoochehr M. Koochesfahani,
Yiannis Andreopoulos, Werner J.A. Dahm,
John A. Mullin, James M. Wallace,
Petar V. Vukoslavčević, Scott C. Morris,
Eric R. Pardyjak, Alvaro Cuerva*

Sect. 5.1. This section was partly written while Beverly McKeon was a Royal Society Dorothy Hodgkin Fellow in the Department of Aeronautics at Imperial College London. The support of the Royal Society is gratefully acknowledged.

The editorial comments of John Foss were gratefully received and significantly improved this text.

Sect. 5.2.1–5.2.7. Geneviève Comté-Bellot addresses her sincere thanks to John Foss, Cam Tropea, Olivier Marsden, Julien Weiss, Garimella R. Sarma, Jean-Christophe Béra, Arun Mangalam, and Thomas Castelain for their scientific exchanges and editing improvements. She has also a special remembrance of Stanley Corrsin who wrote a thorough review on HWA for Springer in 1963.

Sect. 5.3.3. This chapter is based on lectures presented by the authors at courses on PIV and related techniques held at the German Aerospace Establishment (DLR) in Göttingen and at the Von-Karman Institute (VKI) in Brussels. During these courses a large number of participants asked questions or initiated discussions, during the lectures, the breaks and the social events. It is these questions and discussions that helped the authors to revise and improve their lecture notes. The authors are therefore grateful to all participants of these courses.

Sect. 5.3.5. I would like to express my gratitude tho the members or former members of the engine measurement system group: Manfred Beversdorff, Wolfgang Förster, Toni Klemmer, and Elza Rymenants for their engaged cooperation and valuable contributions to the development of the LTV technique.

Sect. 5.5.1. The present contribution would not be possible without the assistance and data obtained by Drs. Anant Honkan and Juan Agui.

Sect. 5.7. The authors wish to thank Dr. Sanz-Andres for the extremely useful remarks and recommendations provided during the preparation of this document.

B.7 Temperature and Heat Flux
*by Tomasz A. Kowalewski, Phillip Ligrani,
Andreas Dreizler, Christof Schulz, Uwe Fey*

This is a summary of the work that the author (TAK) has been fortunate to share with his colleagues and students, first at the Max-Planck-Institut and presently at his home institution. In particular I would like to acknowledge the contribution of W. Hiller, St. Koch, C. Abegg, A. Cybulski, M. Rebow, K. Dekajlo and T. Michalek, to what has been a team effort over a number of years.

B.8 Force and Moment Measurements
by Klaus Hufnagel, Günter Schewe

The author of Sect. 8.1 would like to gratefully acknowledge the assistance of Mr. David Rival in proofreading the manuscript.

C.10 Measurements of Turbulent Flows
by Giovanni Paolo Romano, Nicholas T. Ouellette, Haitao Xu, Eberhard Bodenschatz, Victor Steinberg, Charles Meneveau, Joseph Katz

The author (GPR) acknowledges Profs. R.A. Antonia and A. Cenedese for introducing and accompanying him into the fascinating world of turbulence. The author would like to thank Professors F. Lalli, P. Monti, and G. Querzoli for reading the manuscript and for helpful suggestions and discussions. Many thanks also to Dr. M. Falchi.

This work (VS) was partially supported by grants from the Israel Science Foundation, the Binational US–Israel Foundation, the Minerva Foundation, and the Minerva Center for Nonlinear Physics of Complex Systems.

C.12 Wall-Bounded Flows
by Joseph C. Klewicki, William S. Saric, Ivan Marusic, John K. Eaton

Throughout the course of conducting boundary-layer stability experiments and constructing wind tunnels, I (WSS) have learned a great deal from my colleagues who freely discussed their own ideas and techniques with me and offered suggestions to improve my own work. Many of the ideas on experimental techniques in this paper came from these discussions. The short list of principal contributors contains Thomas Corke, James Kendall, Philip Klebanoff, Victor Levchenko, Hassan Nagib, Mark Morkovin, Werner Pfenninger, Ronald Radeztsky, Mark Reibert, Gregory Reynolds, and Shohei Takagi to whom I should like to express my sincere thanks.

C.16 Aerodynamics
by Wolf-H. Hucho, Klaus Hannemann, Jan Martinez Schramm, Charles H.K. Williamson

The author (CHKW) would like to acknowledge continued support over several years from the Ocean Engineering Division of the O.N.R., monitored by Tom Swean, presently under contract N00014-04-1-0031. The author is extremely grateful to the fluids team of Tim Morse, Matt Horowitz and Paolo Luzzatto Fegiz for help in preparing this chapter.

C.19 Microfluidics: The No-Slip Boundary Condition
by Eric Lauga, Michael P. Brenner, Howard A. Stone

We thank L. Bocquet, K. Breuer, E. Charlaix, H. Chen, C. Cottin-Bizonne, B. Cross, R. Horn, J. Israelachvili, J. Klein, J. Koplik, J. Rothstein, T. Squires, A. Steinberger, O. Vinogradova, and A. Yarin for useful feedback on an early draft of this chapter. Funding by the NSF Division of Mathematical Sciences, the Office of Naval Research and the Harvard MRSEC is acknowledged.

About the Authors

David M. Admiraal
University of Nebraska-Lincoln
Department of Civil Engineering
Lincoln, NE, USA
dadmiraal2@unl.edu

Chapter C.15, Sect. 15.3

Dr. Admiraal is an Associate Professor of Civil Engineering at the University of Nebraska. He earned his Ph.D. in Civil Engineering at the University of Illinois at Urbana-Champaign in 1999. His research interests include sediment transport and fluvial hydraulics, and his current research includes fish passage design and enhancement of energy dissipation in culverts.

Ronald J. Adrian
Arizona State University
Ira A. Fulton School of Mechanical
and Aerospace Engineering
Tempe, AZ, USA
rjadrian@asu.edu

Chapter D.22, Sect. 22.5

Ronald J. Adrian is the Ira A. Fulton Professor of Mechanical and Aerospace Engineering at Arizona State University. The methods to which he has made fundamental contributions are the laser Doppler velocimeter technique, particle image velocimetery and the stochastic estimation method. Dr. Adrian is a Fellow of the American Institute of Aeronautics and Astronautics, American Physical Society, American Society of Mechanical Engineers and the American Academy of Mechanics.

Nuri Aksel
Universität Bayreuth
Department of Applied Mechanics
and Fluid Dynamics
Bayreuth, Germany
tms@uni-bayreuth.de

Chapter C.9, Sect. 9.3

Dr. Aksel is Professor of Applied Mechanics and Fluid Dynamics at the University of Bayreuth. He obtained his Dr. degree from the University of Darmstadt in 1984, and his Habilitation degree from the University of Karlsruhe in 1990. In 1993 he became Professor of Fluid Dynamics at the Technical University of Chemnitz. Since 1998 he holds the Chair of Applied Mechanics and Fluid Dynamics at the University of Bayreuth. Dr. Aksel is on the board of the International Association of Applied Mathematics and Mechanics (GAMM) and is a member of The Society of Rheology. His research interests are dynamics of continua including dispersion rheology and hydrodynamics of liquid films.

Yiannis Andreopoulos
The City College of the City University
of New York
Department of Mechanical Engineering
New York, NY, USA
andre@ccny.cuny.edu

Chapter B.5, Sect. 5.5.1

Dr. Andreopoulos is The Michael Pope Professor of Energy of the School of Engineering at The City College of New York. He holds a Ph.D. and M.Sc. in Aeronautics from Imperial College and MS/BS degrees in Mechanical and Electrical Engineering from the National Technical University of Athens. His research interests are in experimental fluid mechanics and aerodynamics with emphasis in low and high speed turbulence, shock waves, flows in porous media, ski mechanics and environmental flows.

Roger E.A. Arndt
University of Minnesota
Saint Anthony Falls Laboratory
Minneapolis, MN, USA
arndt001@umn.edu

Chapter C.15, Sect. 15.1

Professor Roger E.A. Arndt received his Ph.D. from MIT. He has served on the faculties of Aerospace Engineering at Penn State University and Civil Engineering at the University of Minnesota. Several awards including the ASME Fluids Engineering Award have recognized his research in the areas of cavitation, aeroacoustics, alternate energy and aeration technology. His industrial experience includes rocket propulsion and high-speed marine vehicles. His experience in the design of research facilities for cavitation and hydroacoustics has lead to the design responsibility for two major hydroacoustic research facilities in the US and Germany.

About the Authors

Marc J. Assael

Aristotle University
Chemical Engineering Department
Thessaloniki, Greece
assael@auth.gr

Chapter B.3, Sects. 3.1, 3.4–3.6

Dr. Marc J. Assael received his Ph.D. from the Chemical Engineering Department at Imperial College, London. He is currently Professor of Thermophysical Properties in the Aristotle University of Thessaloniki. His research interests are in thermophysical properties. He has authored 2 books, 12 handbook chapters, and more than 180 technical papers in refereed journals and conference proceedings.

Howard A. Barnes

University of Wales Aberystwyth
Institute of Mathematical
and Physical Sciences
Ceredigion, UK
howard.barnes@ntlworld.com

Chapter C.9, Sect. 9.2

After 33 years in industrial research, Howard Barnes is now a research professor in the University of Wales. He has published four books and over 80 papers in rheology. He has been awarded a DSc by the University of Wales; is a Fellow of the Royal Academy of Engineering and has been honoured by the Queen by being made an Officer of the Order of the British Empire.

Terry D. Blake

Tring, UK
terrydblake@btinternet.com

Chapter B.3, Sect. 3.3.2

Dr. Terry Blake is a physical chemist. He obtained his Ph.D. at the University of Bristol where he developed the molecular-kinetic theory of dynamic wetting. After postdoctoral experience at the University of Texas and Imperial College London, he joined Kodak Limited Research and Development in 1971. There he carried out extensive research in the fundamentals of coating processes and lead research in surface and colloid science. He retired from Kodak in 2002 and is currently visiting professor at universities in the UK, Belgium and Australia where he maintains his interest in the dynamics of wetting.

Eberhard Bodenschatz

Max Planck Institute for Dynamics
and Self-Organization
Laboratory of Fluid Dynamics, Pattern
Formation and Nanobiocomplexity (LFPN)
Göttingen, Germany
eberhard.bodenschatz@ds.mpg.de

Chapter C.10, Sect. 10.2

Eberhard Bodenschatz received his doctorate from the University of Bayreuth in 1989. From 1992 until 2005, during his tenure at Cornell he was a visiting professor at the MPI for Polymer Research in Mainz and at UCSD. In 2003 he became a scientific member of the Max Planck Society and an adjunct Director at the MPI for Dynamics and Self-Organization (Göttingen). Since 2005 he is a Director at this Institute. He is an Adjunct Professor of Physics and of Mechanical and Aerospace Engineering at Cornell University and a Professor of Physics at the University of Göttingen. He is an Alfred P. Sloan Fellow, a Cottrell Scholar and a Fellow of the American Physical Society. His research expertise covers a broad range of investigations in fluid dynamics, and pattern formation in physics and biology.

Jean-Paul Bonnet

Université de Poitiers, ENSMA, CNRS
Laboratoire d'Etudes Aérodynamiques
Poitiers, France
Jean-paul.bonnet@univ-poitiers.fr

Chapter D.22, Sect. 22.4

Dr. Jean-Paul Bonnet is research Director at CNRS. He obtains a Ph.D. and a Doctorat d'Etat in Fluid Mechanics at the University of Poitiers in 1982. He is the Director of a federation of 6 CNRS laboratories in ME. He is senior member of French AAAF. His current research, inside the Laboratoire d'Etudes Aérodynamiques-LEA, is focussed on turbulent incompressible and supersonic flows mostly from experimental approaches; his studies includes fundamental turbulence for free turbulent shear flows and flow control by micro-jets for jet noise reduction and separation control on wings with strong partnerships with industry.

Michael P. Brenner

Harvard University
School of Engineering
and Applied Science
Cambridge, MA, USA
brenner@deas.harvard.edu

Chapter C.19

Michael Brenner is Gordon Mckay Professor of Applied Mathematics and Applied Physics at Harvard. He obtained his Ph.D. degree in Physics from the University of Chicago in 1994. Prior to joining Harvard he was on the faculty at MIT. His research covers a range of problems in applied mathematics, from fluid mechanics to engineering design to theoretical biology.

Ronald J. Calhoun

Arizona State University
Mechanical and Aerospace Engineering
Tempe, AZ, USA
Ron.Calhoun@asu.edu

Chapter C.17, Sects. 17.2–17.4

Dr. Calhoun is an Assistant Professor of Mechanical and Aerospace Engineering at Arizona State University. He obtained his Ph.D. from Stanford University in 1998 and worked as a physicist at Lawrence Livermore National Laboratory for several years. Dr. Calhoun's graduate work was in computational fluid dynamics and he now specializes in atmospheric flows, especially transport and dispersion problems. He leads a lidar group at ASU currently and has deployed instruments in four atmospheric science field campaigns.

Antonio Castellanos

Universidad de Sevilla
Sevilla, Spain
castella@us.es

Chapter C.21

Antonio Castellanos is Professor of Electromagnetism at the University of Seville, Spain. His current research interests are in the coupling of electric fields to fluids (electrohydrodynamics, EHD), in the control of bio-particles and liquids in microelectrode structures (AC electrokinetics and EHD in MEMS), in gas discharges (ozonizers, pollution control), and in the physics of cohesive granular media.

Angelo A. Cavone

NASA Langley Research Center
ATK Space Systems, Inc.
Hampton, VA, USA
angelo.a.cavone@nasa.gov

Chapter B.5, Sect. 5.3.4

Angelo Cavone is a Research Engineer with ATK Space Systems, Inc. He has been developing advanced electronics systems along with system control software for laser velocimetry applications for over 20 years. His expert support of Doppler global velocimetry research at NASA Langley Research Center has led to its many successful applications in NASA and U.S. Air Force wind tunnels. He is currently developing point Doppler velocimetry for turbulence power spectra measurements.

Teresa K. Chereskin

University of California–San Diego
Scripps Institution of Oceanography
La Jolla, CA, USA
tchereskin@ucsd.edu

Chapter C.18

Dr. Chereskin is an observational physical oceanographer who makes measurements at sea and analyzes data. Her research interests include the wind-driven ocean circulation, eastern boundary currents, Southern Ocean dynamics, and acoustic Doppler measurement techniques.

Geneviève Comte-Bellot

Ecole Centrale de Lyon
Centre Acoustique
Ecully, France
genevieve.comte-bellot@ec-lyon.fr

Chapter B.5, Sects. 5.2.1–5.2.7

Geneviève Comte-Bellot is Emeritus Professor in the Department of Fluid Mechanics, Acoustics and Energetics at the Ecole Centrale de Lyon, France. She is Corresponding Member of the French Academy of Sciences, Emeritus Member of the French Academy of Engineering and Associate Member of the U.S. National Academy of Engineering. She received her Ph.D. from the University of Grenoble in 1963. She continued her research work at the Johns Hopkins University studying isotropic turbulence. She has authored or coauthored several books, including a recent monograph on turbulence, and around 150 archival and conference papers.

Joël De Coninck

Université de Mons-Hainaut
Centre for Research
in Molecular Modelling
Mons, Belgium
joel.de.coninck@galileo.umh.ac.be

Chapter B.3, Sect. 3.3.2

Physicist and Mathematician, Joël De Coninck is now Professor in the University of Mons-Hainaut and invited professor in several foreign universities. His main domain of expertise is related to wetting, spreading and coating, combining experimental techniques and molecular modelling tools. He is now the head of the 'Centre for Research in Molecular Modelling', a multidisciplinary group of 25 researchers working in nanotechnology, biotechnology, surface and interface treatment and characterization and, also, image analysis. He has published extensively within these areas. He is member of several scientific associations and is entitled of several patents in these domains.

Laurent Cordier

CNRS, Université de Poitiers
Laboratoire d'Etudes Aérodynamiques
Poitiers, France
Laurent.Cordier@univ-poitiers.fr

Chapter D.22, Sect. 22.4

Laurent Cordier received his Ph.D. from Poitiers University (France). He is currently CNRS senior researcher at Laboratoire d'Etudes Aérodynamiques where he is mainly involved in flow control on wakes and channel flow configurations. His research interests are reduced-order modelling based on proper orthogonal decomposition, numerical optimization methods and closed-loop feedback control.

Alvaro Cuerva

Universidad Politécnica de Madrid
Instituto Universitario de
Microgravedad, Aeronáuticos
Madrid, Spain
acuerva@idr.upm.es

Chapter B.5, Sect. 5.7

The research activity of A. Cuerva is related to wind energy, sonic anemometry and rotor craft aerodynamics. He has several refereed publications and more than 20 international communications. He has been vice-Chairman of the International Network for Wind Energy Testing MEASNET and has participated as independent evaluator in the wind energy program of EU and the aerospace program of NASA.

Werner J.A. Dahm

The University of Michigan
Department of Aerospace Engineering
Ann Arbor, MI, USA
wdahm@umich.edu

Chapter B.5, Sect. 5.5.2

Professor Werner J.A. Dahm has been on the faculty of the Department of Aerospace Engineering at The University of Michigan since 1985. His research has been primarily in fluid dynamics, turbulent flows, turbulent mixing, combustion and microsystems. His Ph.D. is in Aeronautics from Caltech and he previously worked in industry as a Research Engineer in the Transonic Wind Tunnel Section of the Propulsion Wind Tunnel Facility at the USAF Arnold Engineering Development Center (AEDC). He is a Fellow of the Division of Fluid Dynamics (DFD) of the American Physical Society (APS), a Fellow of the American Institute of Aeronautics and Astronautics (AIAA). Professor Dahm is an author of over 170 journal articles, conference papers, and technical publications, a holder of two U.S. patents. He has served extensively on technical advisory and organizational committees for numerous technical conferences, and as a consultant for industry.

Joël Delville

CNRS, Université de Poitiers – CEAT
Laboratoire d'Etudes Aérodynamiques
Poitiers, France
joel.delville@lea.univ-poitiers.fr

Chapter D.22, Sect. 22.4

Joël Delville is a senior research engineer at CNRS in the field of turbulence, flow control and low order modelling. He has a Ph.D. from the French University of Poitiers in Fluid Mechanics. He has an extensive experience in experimental approaches, and multidimensional data analysis, proper orthogonal decomposition, stochastic estimation, low order modelling of complex flows.

Andreas Dreizler

Technische Universität Darmstadt
Fachgebiet für Energie- und
Kraftwerkstechnik
Darmstadt, Germany
dreizler@ekt.tu-darmstadt.de

Chapters B.7, C.20, Sect. 7.3

Dr. Dreizler is head of the experimental working group Technische Verbrennung within the Fachgebiet für Energie- und Kraftwerkstechnik at the Technische Universität Darmstadt. Focus of his research is on physical and chemical processes in turbulent reactive flows with an emphasis on turbulent combustion. He is an expert in development and application of laser diagnostic methods. His data sets of various target flames serve internationally as benchmark for validation of numerical simulations. He authored over 80 contributions in reviewed journals and reviewed conference proceedings. In 2002 he received the Adolf-Messer Award for outstanding contributions in the field of combustion science.

John K. Eaton

Stanford University
Department of Mechanical Engineering
Stanford, CA, USA
eaton@vk.stanford.edu

Chapter C.12, Sect. 12.4

Professor Eaton conducts research in turbulent flow fluid mechanics and heat transfer including particle-laden flows, turbulent boundary layers, gas turbine heat transfer, and complex internal flows. He has pioneered the use of Magnetic Resonance Velocimetry in complex turbulent flows. He has recently begun work in microscale flows including gas–liquid flows in fuel cells and nanofluid heat transfer. He is a Fellow of ASME and APS.

Volker Ebert

Universität Heidelberg
Physikalisch Chemisches Institut
Heidelberg, Germany
volker.ebert@pci.uni-heidelberg.de

Chapter C.20

Dr. Ebert is Associate Professor at the University of Heidelberg and head of a research group on laser-based in situ process diagnostics. His research concentrates on development of diode-laser-based multi-parameter spectrometers and their application as physico-chemical sensors for species concentration, temperature and gas residence times in industrial combustion and chemical production processes as well as fire suppression research. Additionally he cooperates with environmental research groups focusing on aerosol–cloud interaction and atmospheric transport processes. He has authored 47 articles in reviewed journals.

Yasuhiro Egami

Institute of Aerodynamics
and Flow Technology
German Aerospace Center
Göttingen, Germany
yasuhiro.egami@dlr.de

Chapter B.6, Sect. 6.4

Yasuhiro Egami received his Ph.D. degrees in Mechanical Engineering from Sendai's Tohoku University in 1997. Then he worked as a research associate at the Institute of Fluid Science, Tohoku University from 1997 to 1999 and was an invited researcher at Karlsruhe University in Germany during 1998 and 1999. Then, he joined the National Aerospace Laboratory of Japan (now JAXA) in the field of pressure-sensitive paint (PSP) and temperature-sensitive paint (TSP) from 1999 to 2003. Currently he is working at the German Aerospace Center (DLR) as a researcher since 2003. His current research interest is in the development of PSP/TSP technique for a cryogenic wind tunnel.

Rolf H. Engler

Institute of Aerodynamics
and Flow Technology
German Aerospace Center,
Experimental Methods
Göttingen, Germany
rolf.engler@dlr.de

Chapter B.4, Sect. 4.4

Dr. Rolf H. Engler received his Dipl. Ing. degree in 1969 at the Ing.-Academy of Wolfenbüttel in the field of air conditioning. After three years in industry, he started a study at the University of Göttingen and received his Dipl.-Physiker degree in 1979 and in 1986 the Dr. rer. nat. degree. Since 1991 he works on pressure- and temperature-sensitive paint techniques PSP and TSP in different wind tunnels of DLR and around Europe. Since 1995 he is the project leader of the PSP team at DLR Göttingen.

Marie Farge

Ecole Normale Supérieure
Laboratoire de Météorologie Dynamique
(IPSL–CNRS)
Paris, France
farge@lmd.ens.fr

Chapter D.22, Sect. 22.6

Marie Farge is Research Director at the Centre National de la Recherche Scientifique and is a member of the Laboratoire de Météorologie Dynamique which is connected to Ecole Normale Supérieure, Ecole Polytechnique and Université Paris VI. She obtained a M.S. degree in engineering from Stanford University, a Ph.D. in physics from Université Paris VII, and a Ph.D. in mathematics from Université Paris VI. In 1981 she was post-doctoral fellow at Harvard University. She is studying the nonlinear dynamics of turbulent flows using computer simulation. In 1988 she pioneered the use of wavelets in fluid mechanics. She received several prizes from CNRS, Cray Research, American Physics Society and the French Academy of Sciences.

Harindra J.S. Fernando

Arizona State University
Mechanical and Aerospace Engineering
Tempe, AZ, USA
j.fernando@asu.edu

Chapter C.17, Sect. 17.4

Dr. Fernando is a Professor of Engineering at Arizona State University and is also the Director of the Environmental Fluid Dynamics Program. His research spans experimental, theoretical, numerical and field experimental aspects of atmospheric and oceanic flows and their operational forecasting. He was a recipient of many awards, including the NSF Presidential Young Investigator Award and the Rieger Distinguished Award for environmental sciences. He is a fellow of the American Physical Society and the American Society of Mechanical Engineers.

Uwe Fey

Institute of Aerodynamics
and Flow Technology
German Aerospace Center
Göttingen, Germany
Uwe.Fey@dlr.de

Chapter B.7, Sect. 7.4

Uwe Fey studied physics at the University of Göttingen and received his Diploma in 1994. He continued his work at the Max-Planck Institut für Fluid Dynamik in Göttingen, where he received his Ph.D. in 1997. Thereafter he joined the Research Center Rossendorf (FZR) near Dresden where he worked on boundary-layer control by means of electromagnetic forces. In 2001, he returned to Göttingen and joined a company as an application engineer for optical measurement systems. Since 2002 he is a member of the PSP team at the Institute of Aerodynamics and Flow Technology of the German Aerospace Center (DLR). His main activities are the development, application and improvement of the TSP and PSP method for large, industry-scale cryogenic wind tunnels.

John F. Foss

Michigan State University
Mechanical Engineering
East Lansing, MI, USA
foss@egr.msu.edu

Chapters A.2, B.5, C.13, Sects. 2.1, 5.2.8, 5.6

Professor Foss received his BSME (1961), MSME (1962) and Ph.D. (1965) from Purdue University. He has been on the faculty at Michigan State University since 9/1964. He served as the NSF Program Director for Fluid Dynamics and Hydraulics (1998–2000). His research specialty is vorticity measurements. His research group addresses fundamental and applied problems in turbulent flows. The latter are primarily associated with automotive applications. He is a Fellow of ASME and the A.von Humboldt Stiftung and a Chartered Physicist of the IOP. He holds 7 patents involving fluid mechanics.

Marcelo H. García

University of Illinois
at Urbana-Champaign
Ven Te Chow Hydrosystems Laboratory,
Department of Civil and Environmental
Engineering
Urbana, IL, USA
mhgarcia@uiuc.edu

Chapter C.15, Sect. 15.3

Professor Garcia's research interests are in river and coastal morphodynamics, sediment transport mechanics and environmental hydraulics. He is the Chester and Helen Siess Professor of Civil Engineering and the Director of the Ven Te Chow Hydrosystems Laboratory at UIUC. He has received the Walter Huber Research Prize (1998) and the Karl Emil Hilgard Award (1996, 1999) from the American Society of Civil Engineers (ASCE), and the Arthur Thomas Ippen Award (2001) from the International Association of Hydraulic Engineering and Research (IAHR). Professor Garcia is a Member of the National Academy of Engineering of Argentina.

Klaus Hannemann

German Aerospace Center (DLR)
Institute of Aerodynamics and Flow
Technology, Spacecraft Section
Göttingen, Germany
Klaus.Hannemann@dlr.de

Chapter C.16, Sect. 16.2

Dr. Klaus Hannemann is head of the Spacecraft Section in the Institute of Aerodynamics and Flow Technology of the German Aerospace Center (DLR) in Göttingen. He obtained his doctoral degree in Mechanical Engineering from the Universität Karlsruhe in 1987. His main research activities are in the field of numerical and experimental aerothermodynamics of space vehicles including re-entry flows, rocket propulsion and high speed airbreathing propulsion.

Lutz Heymann

Universität Bayreuth
Lehrstuhl für Technische Mechanik
und Strömungsmechanik, Fakultät
für Angewandte Naturwissenschaften
Bayreuth, Germany
lutz.heymann@uni-bayreuth.de

Chapter C.9, Sect. 9.3

Dr. Lutz Heymann is senior scientist in the Department of Applied Mechanics and Fluid Dynamics. He studied Chemical Process Engineering at the Technical University Merseburg where he received his Dr. degree for work on thixotropic and viscoelastic behavior of suspensions in 1984. Prior to joining the University of Bayreuth in 1998 he worked in the Rheology Research Group of the Institute of Mechanics in Chemnitz and at the Technical University Chemnitz, Germany. His current research is focused on the rheology and rheometry of suspensions. Dr. Heymann is member of The Society of Rheology.

Bruce Howe

University of Washington
Applied Physics Laboratory
Seattle, WA, USA
howe@apl.washington.edu

Chapter C.18

Dr. Howe's work focuses on measurement systems and has included: laser Doppler velocimetry for air–sea interaction and atmospheric boundary layer experiments, ocean acoustic tomography and basin scale thermometry, and ionospheric and atmospheric tomography. Current interests include long-term cabled and buoy ocean observatories and mobile platforms, and the integration of acoustic systems with these to simultaneously support navigation, communications, and science.

Wolf-Heinrich Hucho

Schondorf, Germany
HuchoWHH@t-online.de

Chapter C.16, Sect. 16.1

Wolf-Heinrich Hucho studied Mechanical Engineering at Braunschweig Technical University. He received his Dr.-Ing. degree from Professor Schlichting. He started his industrial career as test engineer at Volkswagen, first as head of the climatic wind tunnel, later as head of power train research. Later on he was VP R& D in the supply industry. Since 1985 he is a freelancing technical author. He is editor of *Aerodynamics of Road Vehicles* and author of *Aerodynamics of Bluff Bodies* (in German).

Klaus Hufnagel

Technische Universität Darmstadt
Fluid Mechanics and Aerodynamics
Darmstadt, Germany
k.hufnagel@aero.tu-darmstadt.de

Chapter B.8, Sect. 8.1

Dr. Hufnagel is director of the wind tunnel facilities at Technische Universität Darmstadt. He started his career as research engineer in the cryobalance project in 1984 and obtained his doctoral degree from Technische Universität Darmstadt in 1995. Dr Hufnagel has 25 years experience in designing and building wind tunnel balances for tunnels all over the world.

Bernd Jähne

University of Heidelberg
Research Group Image Processing,
Interdisciplinary Center for Scientific Computing
Heidelberg, Germany
Bernd.Jaehne@iwr.uni-heidelberg.de

Chapters D.24, D.25, Sect. 25.1

Bernd Jähne studied physics in Saarbrücken and Heidelberg, Germany. He occupied a research professorship at the Scripps Institution of Oceanography, University of California in San Diego from 1988–1994 and heads now the research group Image Processing of the Interdisciplinary Center for Scientific Computing and the Institute for Environmental Physics at Heidelberg University. He is founder and head of the Heidelberg Image Processing Forum.

Markus Jehle

University of Heidelberg
Research Group Image Processing,
Interdisciplinary Center
for Scientific Computing
Heidelberg, Germany
Markus.Jehle@iwr.uni-heidelberg.de

Chapter D.25, Sect. 25.2

Markus Jehle has studied physics in Karlsruhe, where he has received his diploma in 2001. He received his Dr. degree in 2006 from the University of Heidelberg. His research interests include computer vision, image processing and flow visualisation, in particular spatio-temporal analysis of flows close to free water surfaces using optical-flow-based techniques.

Joseph Katz

The Johns Hopkins University
Department of Mechanical Engineering
Baltimore, MD, USA
katz@jhu.edu

Chapters C.10, C.14, Sect. 10.4

Joseph Katz's research focuses on experimental fluid mechanics and development of optical diagnostics techniques involving PIV and holography for laboratory and field applications, especially in the ocean. He has worked in bubble dynamics and cavitation, turbulent single and multiphase flows, flow structure and turbulence within turbomachines, turbulence in the bottom boundary layer of the coastal ocean and canopy flows. Using holography he has also studied swimming behavior of plankton both in the laboratory and in the ocean. He has received several awards including the 2004 ASME Fluids Engineering Award.

Damien Kawakami

University of Minnesota
Mechanical Engineering
Bloomington, MN, USA
kawa0036@umn.edu

Chapter C.15, Sect. 15.1

Mr. Kawakami investigated water quality effects on cloud cavitation as part of his Masters of Science research. Other research areas include cavitation erosion, super-cavitation, and cavitation instabilities. Presently, he works for the United States Navy, Department of Naval Reactors.

Saeid Kheirandish

Universität Karlsruhe (TH)
Institut für Mechanische
Verfahrenstechnik und Mechanik
Karlsruhe, Germany
Saeid.Kheirandish@mvm.uni-karlsruhe.de

Chapter A.1, Sect. 1.8

Saeid Kheirandish has received his Ph.D. from the Institute of Polymer Engineering, Technical University of Berlin, in 2005. His research areas of interest include theoretical and experimental rheology, polymer processing, and structure-rheology relationship in complex fluids. After finishing a period of research at the School of Chemical Engineering at the University of South Carolina, Columbia, he has joined the Institute of Mechanics at the Department of Chemical Engineering at the University of Karlsruhe as a postdoctoral fellow. His current research is focused on the elongational behaviour of microgel-containing structures, colloids and fluids with yield-stress.

Michael Klar

University of Heidelberg
Research Group Image Processing,
Interdisciplinary Center
for Scientific Computing
Heidelberg, Germany
Michael.Klar@iwr.uni-heidelberg.de

Chapter D.25, Sect. 25.2

Michael Klar studied physics in Heidelberg, where he received his diploma and Ph.D. in 2001 and 2005, respectively. His research interests include computer vision, photogrammetry and flow visualization, in particular endoscopic 3-D particle-tracking velocimetry for flow measurements in porous media. He is now with the Robert Bosch GmbH in Plochingen, Germany.

Joseph C. Klewicki

University of New Hampshire
Department of Mechanical Engineering
Durham, NH, USA
Joe.klewicki@unh.edu

Chapter C.12, Sects. 12.1, 12.2

Professor Klewicki studies complex and turbulent fluid flows, with an emphasis on wall-bounded flows. His on-going efforts are in the experimental study of boundary layer vortical motions, and flow field modelling (physical and analytical). He is a fellow of the ASME, and is currently Dean of the College of Engineering and Physical Sciences at the University of New Hampshire.

Manoochehr M. Koochesfahani

Michigan State University
Department of Mechanical Engineering
East Lansing, MI, USA
koochesf@egr.msu.edu

Chapter B.5, Sect. 5.4

Dr. Manoochehr Koochesfahani is Professor of Mechanical Engineering at Michigan State University. He received his Ph.D. in Aeronautics from the California Institute of Technology. His research interests include: fundamental studies of turbulent mixing and mixing control/enhancement, unsteady fluid mechanics and aerodynamics, micro- and nano-flows, and development of advanced optical diagnostics for fluid flow and mixing studies using molecule-based methods and quantum dots.

Tomasz A. Kowalewski

Polish Academy of Sciences
Institute of Fundamental Technological
Research (IPPT PAN), Department
of Mechanics and Physics of Fluids
Warszawa, Poland
tkowale@ippt.gov.pl

Chapter B.7, Sect. 7.1

Tomasz A. Kowalewski is a physicist with an extensive research background in the field of fluid mechanics. He earned his Ph.D. in 1982 and habilitation in 1996 from the Polish Academy of Sciences. Presently he is Associate Professor and head of department at the Institute of Fundamental Technological Research working mainly in experimental and computational fluid mechanics, thermally driven and free surface flows, flow associated with phase change phenomena, nanofibers and microfluidics.

Eric Lauga

Massachusetts Institute of Technology
Department of Mathematics
Cambridge, MA, USA
lauga@mit.edu

Chapter C.19

Eric Lauga is Assistant Professor of Applied Mathematics in the Department of Mathematics at Massachusetts Institute of Technology. He obtained his Ph.D. in Applied Mathematics from Harvard University in 2005 where he studied the mechanics of fluids on the micron scale. His research interests are in biological fluid mechanics, biophysics, nonlinear dynamics, and soft condensed matter.

Joseph W. Lee

NASA Langley Research Center
Advanced Sensing and Optical
Measurements Branch
Hampton, VA, USA
Joseph.W.Lee@nasa.gov

Chapter B.5, Sect. 5.3.4

Joseph Lee is a Senior Electronics Technician at the NASA Langley Research Center. Since joining NASA he has developed electronic, optical and data acquisition systems crucial to the implementation of laser velocimetry techniques in NASA and U.S. Air Force wind tunnels. His work on the design, installation and testing of Doppler global velocimetry systems and subsystems was the prime contributor to the successful implementation of the technique to wind tunnel applications.

Jacques Lewalle

Syracuse University
Mechanical Engineering
Syracuse, NY, USA
jlewalle@syr.edu

Chapter D.22, Sect. 22.6

Jacques Lewalle has applied the continuous wavelet transforms to data analysis, e.g. paper formation and nerve response, as well as fluid dynamics (2004 Lewis F. Moody Award of ASME/FED), and to the mathematical physics of the diffusion, Poisson and Navier–Stokes equations. His primary interest is in the multiscale interactions in incompressible turbulence and 3-D vorticity dynamics.

Phillip Ligrani

University of Oxford
Department of Engineering Science
Oxford, UK
phil.ligrani@eng.ox.ac.uk

Chapter B.7, Sect. 7.2

Phil Ligrani is Professor of Mechanical Engineering and Director of the Convective Heat Transfer Laboratory in the University of Utah. His current research areas include measurement techniques, convective heat transfer, heat transfer augmentation, internal cooling, turbulent boundary layers, transitional phenomena, film cooling, macro-scale and micro-scale pumping systems, and peptide suspension flows. He was awarded "Professor of the Year", ASME Fellow, an SCIES-AGTSR Faculty Fellowship, a Universität Karlsruhe Guest Professorship, a NASA Space Act Tech Brief Award, and the Carl E. and Jessie W. Menneken Faculty Award for Excellence in Scientific Research.

Abraham Marmur

Technion-Israel Institute of Technology
Department of Chemical Engineering
Haifa, Israel
marmur@technion.ac.il

Chapter B.3, Sect. 3.2

Professor Abraham Marmur received his Ph.D. in 1974 from the Technion – Israel Institute of Technology. Then he spent two years as a postdoc at the State University of New York at Buffalo. Later he was a visiting associate professor at the University of Wisconsin, Madison, and a visiting scientist at the IBM Almaden research center. Professor Marmur has been working in the field of interfacial phenomena for over twenty five years, has published over a hundred scientific papers, and has been consulting for major companies. At the Technion, Professor Marmur received awards for excellence in research and in teaching.

Jan Martinez Schramm

Institute of Aerodynamics
and Flow Technology
Department Spacecraft
Göttingen, Germany
Jan.Martinez@dlr.de

Chapter C.16, Sect. 16.2

Jan Martinez Schramm studied physics at the University of Göttingen where he graduated in 1999. He is a research scientist at the German Aerospace Center (DLR) in Göttingen. The main focus of his work is on short duration measurement techniques and experimental investigations of high-temperature, hypersonic flows in the free piston driven High Enthalpy Shock Tunnel Göttingen (HEG).

Ivan Marusic

University of Melbourne
Victoria, Australia
imarusic@unimelb.edu.au

Chapter C.12, Sect. 12.4

Professor Marusic conducts research primarily on turbulent boundary layers and high Reynolds number incompressible flows. He is a Packard Fellow of Engineering and Science, and a recipient of a National Science Foundation Career Award.

Beverley J. McKeon

California Institute of Technology
Graduate Aeronautical Laboratories,
Division of Engineering
and Applied Science
Pasadena, CA, USA
mckeon@caltech.edu

Chapters B.4, B.5, Sects. 4.1–4.3, 5.1

Beverley McKeon is an Assistant Professor of Aeronautics in the Graduate Aeronautical Laboratories, California Institute of Technology. She received her Ph.D. from Princeton University in 2003, then moving to Imperial College London, where in 2004 she was awarded a Royal Society Dorothy Hodgkin Fellowship. Her current research focuses on the control and fundamental physics of wall-bounded flows, such as reduction of drag, noise and structural loading or expansion of vehicle performance envelopes.

Gareth H. McKinley

Massachusetts Institute of Technology
Department of Mechanical Engineering
Cambridge, MA, USA
gareth@mit.edu

Chapter C.9, Sect. 9.1.3

Gareth McKinley is the School of Engineering Professor of Teaching Innovation within the Department of Mechanical Engineering at MIT. He is also the director of the Program in Polymer Science and Technology and Head of the Hatsopoulos Microfluids Laboratory. He is a co-founder and member of the Board of Directors of Cambridge Polymer Group. His research interests include extensional rheology, interfacial fluid dynamics and super-hydrophobicity, microrheology and microfluidics and the processing of nanocomposite materials.

Charles Meneveau

The Johns Hopkins University
Department of Mechanical Engineering
Baltimore, MD, USA
meneveau@jhu.edu

Chapter C.10, Sect. 10.4

Charles Meneveau is the Louis M. Sardella Professor of Mechanical Engineering at the Department of Mechanical Engineering and Director of the Center for Environmental and Applied Fluid Mechanics at The Johns Hopkins University. His area of research is focused on understanding and modeling hydrodynamic turbulence, and complexity in fluid mechanics in general. He combines experimental, computational and theoretical tools for his research, and places special empasis on the multiscale aspects of turbulence.

Wolfgang Merzkirch

Universität Essen
Lehrstuhl für Strömungslehre
Essen, Germany
wolfgang.merzkirch@uni-essen.de

Chapters B.6, C.11

Wolfgang Merzkirch is Professor Emeritus at Universität Essen, Germany. He was professor of fluid mechanics at Universität Essen and Universität Bochum, Germany, research associate at Ernst-Mach-Institut, Freiburg, Germany, and NASA Ames Research Center, USA. He is adjunct professor at Texas A&M University, USA, founding editor of Experiments in Fluids, author of the book *Flow Visualization*. His research areas are experimental fluid mechanics with emphasis on optical methods and flow visualisation. He received the Leonardo da Vinci Award at the 8th International Symposium on Flow Visualization and the Asanuma Award from the Visualization Society of Japan.

James F. Meyers

NASA Langley Research Center
Advanced Sensing and Optical
Measurements Branch
Hampton, VA, USA
james.f.meyers@nasa.gov

Chapter B.5, Sect. 5.3.4

James Meyers is a Distinguished Research Associate at the NASA Langley Research Center. During his 39 year career at NASA he has been instrumental in the advancement of the state-of-the-art in laser velocimetry techniques and their implementation to wind tunnel applications. His research in Doppler global velocimetry led to the first successful application of the technology to wind tunnel testing and formed the basis for subsequent developments.

Scott C. Morris

University of Notre Dame
Department of Aerospace and Mechanical
Engineering
Notre Dame, IN, USA
s.morris@nd.edu

Chapter B.5, Sect. 5.5.4

Professor Morris completed his Ph.D. in mechanical engineering from Michigan State University. He has been at the University of Notre Dame since 2002. Current research activities include overlapping areas of study within fundamental turbulence, aeroacoustics, aeromechanics, and turbomachinery.

John A. Mullin

Northrop Grumman Space Technology
Redondo Beach, CA, USA
John.mullin@ngc.com

Chapter B.5, Sect. 5.5.2

John Mullin earned his Ph.D. from the University of Michigan in 2004. Dr. Mullin's research is currently in the areas of fluid dynamics, reacting flows, and mixing including experimental and numerical applications in combustion-driven high-energy lasers, aerooptics, and launch vehicle/spacecraft contamination. Other research includes nonintrusive laser diagnostics to measure velocity and velocity gradient fields. He is a Senior Member of AIAA and a member of APS.

Klaas te Nijenhuis

Delft University of Technology
Department of Polymer Materials
and Polymer Engineering
Delft, The Netherlands
k.tenijenhuis@tnw.tudelft.nl

Chapter C.9, Sects. 9.1, 9.2, 9.4

Dr. Klaas te Nijenhuis is a retired physical chemist with an extensive research background in the field of thermo-reversible polymer networks. Furthermore he is interested in film formation of polymer latices as studied with the aid of light (UV-VIS) scattering. Finally he has a considerable experience in teaching rheology of polymeric fluids on academic and post academic level.

Holger Nobach

Max Planck Institute for Dynamics
and Self-Organization
Göttingen, Germany
holger.nobach@nambis.de

Chapters D.22, D.23, Sect. 22.1–22.3

Dr. Nobach received his doctorate in electrical engineering from the University of Rostock in 1997. During his postdoctoral research, between 1998 and 2000 on an industrial research program with Dantec Dynamics in Copenhagen and between 2000 and 2005 at the Technical University of Darmstadt, he developed measurement techniques for flow investigations. Since 2005 he is a scientist at the Max Planck Institute for Dynamics and Self-Organization in Göttingen, Germany and studies characteristics of turbulent flows with improved and extended optical measurement systems.

Jeffrey A. Odell

University of Bristol
Department of Physics
Bristol, UK
Jeff.odell@bristol.ac.uk

Chapter C.9, Sect. 9.5

Jeff Odell is an experimental physicist at the University of Bristol. He has worked for 30 years in the fields of polymer morphology, macromolecular dynamics and flow, especially elongational flow. Current research interests are microfluidics and microrheology, especially of biological fluids. He has over 120 publications in these fields. He is founder and managing director of MicroRheology Ltd.

Nicholas T. Ouellette

Max Planck Institute for Dynamics
and Self-Organization
Laboratory of Fluid Dynamics, Pattern
Formation and Nanobiocomplexity
(LFPN)
Göttingen, Germany
nicholas.ouellette@ds.mpg.de

Chapter C.10, Sect. 10.2

Nicholas T. Ouellette received his Ph.D. from Cornell University in 2006, where he used particle-tracking techniques to study the Lagrangian dynamics of intensely turbulent flows. He is currently a postdoctoral researcher at Haverford College, where he is working on instabilities and spatiotemporal chaos in quasi-two-dimensional flows and on the fluidic transport of asymmetric particles.

Ronald L. Panton

The University of Texas
Mechanical Engineering Department
Austin, TX, USA
rpanton@mail.utexas.edu

Chapters A.1, A.2, Sect. 1.1–1.7, 2.2

Ronald L. Panton is the J. H. Herring Professor of Mechanical Engineering at the University of Texas at Austin. After receiving a B. A. degree in Mathematics and a B.S. degree in Mechanical Engineering from Wichita State University, he practiced engineering at North American Aviation followed by active duty with the Air Force. As an officer in the Air Force he was assigned to the X-15 Research Plane project at Wright-Patterson AFB. Subsequently, he studied Mechanical Engineering at the University of Wisconsin (Madison), M.S. and the University of California (Berkeley), Ph.D. Following graduation he became a professor at Oklahoma State University, 1966-71, and then the University of Texas at Austin, 1971 to present. His research has been in a variety of fluid flow topics ranging from flow-acoustic interactions to molten solder behavior. Much of his work is in turbulent flow. He authored the textbook *Incompressible Flow* edited the monograph *Self-Sustaining Mechanisms of Wall Turbulence*.

Eric R. Pardyjak

University of Utah
Department of Mechanical Engineering
Salt Lake City, UT, USA
pardyjak@eng.utah.edu

Chapter B.5, Sect. 5.7

Pardyjak's primary area of research is in the Environmental Fluid Mechanics with a focus on urban boundary layer transport and dispersion processes. Pardyjak received his Ph.D. from Arizona State University in 2001 in the Environmental Fluid Dynamics Program. His research activities involve both experimental (large scale atmospheric experiments and laboratory measurements) and modelling aspects of transport and dispersion.

Marc Perlin

University of Michigan
Naval Architecture
and Marine Engineering
Ann Arbor, MI, USA
perlin@umich.edu

Chapter C.15, Sect. 15.2

Dr. Perlin is a Professor of Naval Architecture and Marine Engineering, Professor of Mechanical Engineering, and Professor of Civil and Environmental Engineering. He earned his Ph.D. in Engineering Mechanics from the University of Florida in 1989. His research areas include nonlinear water waves, contact-line dynamics, and drag-reduction. Energy dissipation and vorticity in steep and breaking waves are studied experimentally including the effects of wind. Viscous drag reduction using polymers, microbubbles, and air layers are investigated at near prototype Reynolds numbers. Contact-line dynamics in oscillatory flows and thin films in rotating cylinders are studied experimentally and numerically.

Marko Princevac

University of California Riverside
Department of Mechanical Engineering
Riverside, CA, USA
marko@engr.ucr.edu

Chapter C.17, Sect. 17.1

Marko Princevac has a Ph.D. in Mechanical Engineering from Arizona State University (2003) and is now an Assistant Professor of Mechanical Engineering at the University of California, Riverside. His research includes laboratory, theoretical, and field experimental aspects of small scale atmospheric flows. He is a member of the American Society of Mechanical Engineers (ASME) and the Air and Waste Management Association (AWMA).

Alberto T. Pérez

Universidad de Sevilla
Departamento de Electrónica
y Electromagnetismo, Facultad de Física
Sevilla, Spain
alberto@us.es

Chapter C.21

Alberto Pérez is Profesor Titular (Associate Professor) in the electronics and electromagnetism department since 1994. His current research deals with the electrohydrodynamics of liquids and suspensions. Problems under study are the instabilities induced by the corona discharge in a layer of dielectric liquid, the chaotic behavior of a conducting ball bouncing on an electrode and the conductivity and electrophoretic mobility of concentrated suspensions.

Giovanni Paolo Romano

University of Roma "La Sapienza"
Department Mechanics and Aeronautics
Rome, Italy
romano@dma.ing.uniroma1.it

Chapter C.10, Sect. 10.1

Giovanni Paolo Romano graduated in Physics in 1984 at the University of Roma "La Sapienza". At present is Full Professor of Experimental Fluid Mechanics at the same University (Department of Mechanics and Aeronautics). He is author of more than 90 scientific publications concerning fundamental and applied fluid-mechanics involving large and small scale turbulent structures (near-wall turbulence, jets, wakes, biomedical flows). His main interest is in the experimental investigation of turbulent fields by means of advanced optical techniques (LDA, PIV, LIF).

William S. Saric

Texas A&M University
Aerospace Engineering
College Station, TX, USA
saric@tamu.edu

Chapter C.12, Sect. 12.3

William S. Saric is the Stewart and Stevenson Professor of Aerospace Engineering at Texas A&M University. He is a member of the National Academy of Engineering and received the AIAA Fluid Dynamics Award and the SES G.I. Taylor Medal. He is a Fellow of AIAA, APS, and ASME. Most recently, he has conducted experimental, and flight research on stability, transition, and control of 2-D and 3-D boundary layers.

Fulvio Scarano

Delft University of Technology
Aerospace Engineering Department
Delft, The Netherlands
f.scarano@tudelft.nl

Chapter B.5, Sect. 5.3.3

Fulvio Scarano is Associate Professor in the Aerodynamics Section at the Aerospace Engineering Department of TU Delft. He obtained his Ph.D. in Aerospace Engineering from the University of Naples in collaboration with the von Kármán Institute for Fluid Dynamics, where he was also awarded the von Kármán Prize for his studies on advanced PIV image processing algorithms. His research focuses on particle image velocimetry developments and its applications in high-speed aerodynamics and turbulent shear flows. The group has recently developed the tomographic-PIV technique for the instantaneous measurement of the three-dimensional velocity field. He is member of several international scientific committees and is recipient of the Impulse to Innovation award from the Dutch Technology Foundation.

Günter Schewe

German Aerospace Center (DLR)
Institut für Aeroelastik
Göttingen, Germany
guenter.schewe@dlr.de

Chapter B.8, Sect. 8.2

Günter Schewe studied physics in Aachen and Göttingen where he received his Dr. degree in physics of fluids at the MPI für Strömungsforschung. He is now senior scientist at the DLR-Institut für Aeroelastik. He has worked in the fields of turbulent boundary layers, nonlinear effects in transonic flutter, very high Reynolds number flows and special measuring techniques for the mentioned fields.

Kai Schneider

Université de Provence
Centre de Mathématiques et d'Informatique
Marseille, France
kschneid@cmi.univ-mrs.fr

Chapter D.22, Sect. 22.6

Dr. Kai Schneider is Professor of Mechanics and Applied Mathematics at the Université de Provence, Marseille, France. He obtained his Master degree in 1993 and his Ph.D. degree in 1996 both from the Universität Kaiserslautern, Germany. In 2001 he obtained his habilitation from the Université Louis Pasteur, Strasbourg, France. His current research activities are focused on the development of multiscale techniques for scientific computing and their application for modeling and computing turbulent and reactive flows.

Richard Schodl

German Aerospace Center (DLR)
Institute of Propulsion Technology
Engine Measurement Systems
Köln, Germany
richard.schodl@dlr.de

Chapter B.5, Sect. 5.3.5

Dr. Richard Schodl is heading a research group in the DLR-Institute of Propulsion Technology. The group's current research involves the development and application of laser instrumentation for turbomachinery flow analysis. He obtained his Dr. degree in Mechanical Engineering from the RWTH Aachen in 1978 and is author of numerous publications, received national and international awards, holds various patents and was many times invited as a lecturer to international conferences. During the years he has collected extensive experience in the development and application of optical instrumentation for non-intrusive flow analysis.

Christof Schulz

Universität Duisburg-Essen
Institut für Verbrennung
und Gasdynamik
Duisburg, Germany
christof.schulz@uni-duisburg.de

Chapters B.7, C.20, Sect. 7.3

Dr. Schulz is a Professor at the University of Duisburg-Essen. His research focuses on the development and application of laser-based diagnostics techniques to combustion processes, specifically in-cylinder diagnostics of mixture formation, in-cylinder temperatures and pollutant formation in internal combustion engines. Additionally he works on kinetics of high-temperature reactions and on nanoparticle formation in gasphase reactions. He heads a research group of 25 people. He has authored over 70 reviewed journal articles and has received the BMW Scientific Award and the Freudenberg-Award (Heidelberg Academy of Sciences).

Stephen Spiegelberg

Cambridge Polymer Group
Boston, MA, USA
steve@campoly.com

Chapter C.9, Sect. 9.1.3

Dr. Stephen Spiegelberg is the president and co-founder of Cambridge Polymer Group, Inc. He received his BS in Chemical Engineering from the University of Wisconsin-Madison, and his Ph.D. in Chemical Engineering from the Massachusetts Institute of Technology, where he studied toughening mechanisms in brittle plastics. He was a post-doctoral fellow at Harvard University, where he worked with NASA engineers to design and test an extensional rheometer for microgravity applications. He is a co-inventor on a capillary break-up rheometer with inventors at Cambridge Polymer Group, where he also works with polymeric materials for biomedical applications.

Victor Steinberg

Weizmann Institute of Science
Department of Physics
of Complex Systems
Rehovot, Israel
victor.steinberg@weizmann.ac.il

Chapter C.10, Sect. 10.3

Dr. Victor Steinberg is The Harry De Jur Professor of Physics at the Department of Physics of Complex Systems of the Weizmann Institute of Science. He is the recipient of the Ulam Distinguished Scholarship at Los Alamos National Lab, USA in 1987, Alexander von Humboldt Research Award, Germany, 1993, and Municipal Professorial Chair of the University of Joseph Fourier 2000–2003. Currently his research activity includes hydrodynamics of polymer solutions and complex fluids, microfluidics of complex fluids, convective and hydrodynamic turbulence, hydrodynamics of fluid near the gas-liquid critical point, single polymer dynamics in various flows, and dusty plasma.

Howard A. Stone

Harvard University
Division of Engineering
and Applied Sciences
Cambridge, MA, USA
has@deas.harvard.edu

Chapter C.19

Howard A. Stone is Vicky Joseph Professor of Engineering and Applied Mathematics in the Division of Engineering and Applied Sciences at Harvard University. He received a Bachelor of Science degree from the University of California at Davis in 1982 and his Ph.D. from Caltech in 1988, with both degrees being in Chemical Engineering. Following a post-doctoral year in the Department of Applied Mathematics and Theoretical Physics at the University of Cambridge he joined the Harvard faculty. His research interests are centered in fluid dynamics and its applications in various traditional fields from engineering to chemistry and physics.

Stephanus A. Theron

Technion – Israel Institute of Technology
Department of Mechanical Engineering
Haifa, Israel
therons@tx.technion.ac.il

Chapter B.3, Sect. 3.7

Dr. Theron completed his Ph.D. degree in Mechanical Engineering at the Technion in 2004. He is an engineer with a background in control and electronic and electric systems. His Ph.D. research focussed on the electrospinning process and methods for the characterization of polymer solutions.

Cameron Tropea

Technische Universität Darmstadt
Fachgebiet Strömungslehre und Aerodynamik
Darmstadt, Germany
ctropea@sla.tu-darmstadt.de

Chapters B.5, D.22, D.23, Sects. 5.3.2, 22.1–22.3

Cameron Tropea received his Bachelors and Masters of Applied Science at the University of Toronto (Eng. Sci.) before moving to the University of Karlsruhe, Germany, where he received his Dr.-Ing. in 1982. After holding a researcher assistantship at the University of Erlangen-Nuremberg, a guest professorship at the University of Waterloo, Canada, and an industrial position, he moved to his current position as head of the Chair of Fluid Mechanics and Aerodynamics at the Technical University of Darmstadt, Germany. His research interests lie in the development of optical measurements techniques for flow velocity and particle characterization and their application to investigate complex turbulent flows and atomization and spray processes.

Oğuz Uzol

Middle East Technical University
Department of Aerospace Engineering
Ankara, Turkey
uzol@ae.metu.edu.tr

Chapter C.14

Dr. Oğuz Uzol received his Master of Science in Aeronautical Engineering in 1995 from the Middle East Technical University, Ankara, Turkey, his Ph.D. in Aerospace Engineering, from The Pennsylvania State University, University Park, PA in 2000. He was a post-doctoral fellow in the Department of Mechanical Engineering, Johns Hopkins University, Baltimore, MD. His current research mainly focuses on aircraft propulsion systems, gas turbines, turbomachinery flows, turbulence physics and modeling, particle image velocimetry and other experimental techniques.

Petar V. Vukoslavčević

University of Montenegro
Department of Mechanical Engineering
Podgorica, Montenegro
petarv@cg.ac.yu

Chapter B.5, Sect. 5.5.3

Dr. Petar Vukoslavčević is a Professor of Mechanical Engineering at the University of Montenegro and a member of the Montenegrin Academy of Science and Art. He works in the fields of thermal anemometry measurement methods and turbulence. He developed a nine sensor hot-wire probe to simultaneously measure, for the first time, all three vorticity and velocity components. He is currently working on the implementation of thermal anemometry measurements in variable temperature and supercritical turbulent fluid flows.

Manfred H. Wagner

Technische Universität Berlin
Polymertechnik/Polymerphysik
Berlin, Germany
manfred.wagner@tu-berlin.de

Chapter A.1, Sect. 1.8

Manfred H. Wagner is Professor for Polymer Engineering and Polymer Physics at the Technical University (TU) of Berlin. His scientific interests include constitutive equations for polymeric systems, the application of rheology to polymer processing, and structure-property relations for polymers. He is President of the European Society of Rheology (ESR), and Secretary of the International Committee on Rheology. The Institute of Materials, London, awarded him the Swinburne Award 2002.

William A. Wakeham

University of Southampton
Southampton, UK
vice-chancellor@soton.ac.uk

Chapter B.3, Sects. 3.1, 3.4–3.6

Dr. William A. Wakeham obtained his Ph.D. from the Physics Department of the University of Exeter. He is currently the Vice-Chancellor of the University of Southampton and its Executive Head. His research interests include intermolecular forces, statistical physics and thermophysical properties. He is the author/editor of 7 books and has published over 400 technical articles in refereed journals. Among several awards he is a Fellow of the Royal Academy of Engineering and is a recipient of the Yeram S Touloukian Award of ASME.

James M. Wallace

University of Maryland
Department of Mechanical Engineering
College Park, MD, USA
wallace@eng.umd.edu

Chapter B.5, Sect. 5.5.3

James Wallace received his D. Phil. in Engineering Science at Oxford University in 1969 and was on the research staff of the Max-Planck-Institut für Strömungsforschung in Göttingen from 1969 until he joined the faculty of the University of Maryland in 1975. He is currently a Professor of Mechanical Engineering there. Wallace does experimental research on turbulent shear flows, in particular with the development of techniques for measuring and analyzing velocity gradient fields. He is currently investigating scalar dispersion in shear flows with environmental and mixing applications, as well as turbulence in high temperature and supercritical flows. He became a Fellow of the American Physical Society in 1989 and received the Distinguished Service Award in the Engineering Sciences of the Washington Academy of Sciences (1984).

Jerry Westerweel

Delft University of Technology
Department of Mechanical Engineering
Delft, The Netherlands
J.Westerweel@wbmt.tudelft.nl

Chapter B.5, Sect. 5.3.3

Dr. Westerweel is the Anthoni van Leeuwenhoek Professor at Delft University of Technology since 2001 and holds the Fluid Dynamics Chair in the Department of Mechanical Engineering of Delft University since 2005. He earned his Ph.D. from Delft University of Technology in 1993. His areas of interest cover optical measurement techniques for quantitative measurements in flows, i.e. particle image velocimetry, planar laser-induced fluorescence, holography, turbulence and coherent flow structures, turbulent mixing and chemical reactions, disperse multiphase flows, microfluidics, and biological flows.

Charles H.K. Williamson

Cornell University
Mechanical and Aerospace Engineering
Ithaca, NY, USA
cw26@cornell.edu

Chapter C.16, Sect. 16.3

Charles H.K. Williamson is a Full Professor at Cornell, and he spent 16 years there, having before spent 6 years at Caltech, 2 years in industry, and before that he received a Ph.D. at Cambridge University. He has received 10 Teaching Prizes (not grants), including one National one in 1994 from the Keck Foundation, including the 1999 Weiss Presidential Fellowship, and the Carnegie Foundation's Professor of the Year 2006.

Jürgen Wolfrum

Ruprecht-Karls-Universität Heidelberg
Physikalisch-Chemisches Institut
Heidelberg, Germany
wolfrum@urz.uni-heidelberg.de

Chapter C.20

Jürgen Wolfrum holds the chair of Physical Chemistry at the Ruprecht-Karls-University of Heidelberg. His research interests are in microscopic dynamics of elementary chemical reactions, combustion kinetics, kinetics of atmospheric and heterogeneous catalytic reactions, laser diagnostics of technical combustion processes (Otto- and Diesel-engines, gas turbines, power plants, fluidized bed combustion, municipal waste incinerators), laser applications in biology and genomics. He is recipient of many prestigious awards (Philip-Morris-Prize, BMW Scientific Award, Max-Planck-Research Award, Karl Heinz Beckurts Award, Polanyi Medal, Bunsen-Denkmünze, a.o.), the chairman of several advisory boards and member of various Academies of Sciences.

About the Authors

Martin Wosnik

Alden Research Laboratory
Holden, MA, USA
mwosnik@aldenlab.com

Chapter C.15, Sect. 15.1

Martin Wosnik studied at TU Darmstadt and University at Buffalo (SUNY), where he received his Ph.D. in Mechanical Engineering. He worked at Chalmers University of Technology and St. Anthony Falls Laboratory, University of Minnesota, and is now Senior Flow Engineer at Alden Research Laboratory. His varied research interests are in turbulence, cavitation and bubbly flows, hydraulics and flow measurement techniques.

Haitao Xu

Max Planck Institute for Dynamics
and Self-Organization
Department of Fluid Dynamics, Pattern
Formation, and Nanobiocomplexity
Göttingen, Germany
haitao.xu@ds.mpg.de

Chapter C.10, Sect. 10.2

Dr. Haitao Xu is a research associate at the Max Planck Institute for Dynamics and Self-Organization, Göttingen, Germany. He received his Ph.D. in mechanical engineering from Cornell University in 2003. Before joining MPIDS in 2006, he was a post-doctoral associate at the Physics Department, Cornell University. His current research focuses on experimental investigation of the Lagrangian properties of high Reynolds number turbulence.

Alexander L. Yarin

University of Illinois at Chicago
Department of Mechanical
and Industrial Engineering
Chicago, IL, USA
ayarin@uic.edu

Chapter A.2, Sect. 2.3

Dr. Alexander Yarin is a Professor of Mechanical Engineering at the University of Illinois at Chicago. He is an applied physicist and has a Ph. D. and habilitation in physics and mathematics from the USSR Academy of Sciences in Moscow. His research interests are in hydrodynamics of flows with free surfaces (jets, films and droplets), rheology and hydrodynamics of non-Newtonian (e.g. polymeric) liquids, combustion and nanotechnology. Currently his research activities include electrospinning of polymer nanofibers and their applications, suspensions of nanoparticles and coating of nanoparticles in low-pressure plasma.

Eyal Zussman

Technion – Israel Institute of Technology
Department of Mechanical Engineering
Haifa, Israel
meeyal@tx.technion.ac.il

Chapter B.3, Sect. 3.7

Dr. Zussman is a Professor at the Department of Mechanical Engineering at the Technion-Israel Institute of Technology. He received his D.Sc. degree in Mechanical Engineering from the Technion in 1992. He worked as a post-doctoral associate at the Technical University of Berlin from 1992–1994. Dr. Zussman's main areas of research are nanotechnology and polymer processing, electrospinning, nanoassembly and materials characterization.

James H. Duncan

University of Maryland
Department of Mechanical Engineering
College Park, MD, USA
duncan@eng.umd.edu

Chapter C.15, Sect. 15.2

James H. Duncan is a Professor of Mechanical Engineering at the University of Maryland. His research interests include breaking water waves, fluid–structure interactions, bubbly flows, cavitation, and underwater explosions. Dr. Duncan received his Ph.D. from the Johns Hopkins University in 1979. He is a Fellow of the American Physical Society.

Daniel G. Nocera

Massachusetts Institute of Technology
Department of Chemistry
Cambridge, MA, USA
nocera@mit.edu

Chapter B.5, Sect. 5.4

Daniel G. Nocera is the W. M. Keck Professor of Energy at the Massachusetts Institute of Technology. He received his B.S. degree from Rutgers University in 1979 and his Ph.D. degree from Caltech in 1984. Nocera's expertise in excited state design led him to co-invent the Molecular Tagging Velocimetry (MTV) technique. He began his academic career at Michigan State University and joined the faculty of MIT in 1997. Among his awards, he is a Presidential Young Investigator, Alfred P. Sloan Fellow, and was elected to the American Academy of Arts and Sciences. In 2005, he was rewarded the Italgaz Prize for his fundamental contributions to the development of renewable energy at the molecular level.

Detailed Contents

List of Abbreviations	XXI
Nomenclature	XXV

Part A Experiments in Fluid Mechanics

1 Experiment as a Boundary-Value Problem
Ronald L. Panton, Saeid Kheirandish, Manfred H. Wagner 3
- 1.1 Thermodynamic Equations
  Ronald L. Panton 3
  - 1.1.1 Thermodynamics 4
- 1.2 Kinematic Equations
  Ronald L. Panton 5
- 1.3 Balance Laws and Local Governing Equations
  Ronald L. Panton 6
  - 1.3.1 Continuity 6
  - 1.3.2 Linear Momentum and Related Equations 6
  - 1.3.3 Angular Momentum 7
  - 1.3.4 Energy 7
  - 1.3.5 Entropy 7
- 1.4 Balance Laws and Global Governing Equations
  Ronald L. Panton 8
  - 1.4.1 Regions 8
  - 1.4.2 Leibnitz and Gauss Theorems 8
  - 1.4.3 Volume 8
  - 1.4.4 Mass 8
  - 1.4.5 Linear Momentum 8
  - 1.4.6 Total Energy 9
  - 1.4.7 Thermal Energy 10
  - 1.4.8 Mechanical Energy 10
  - 1.4.9 Entropy 10
- 1.5 Constitutive Equations
  Ronald L. Panton 10
- 1.6 Navier–Stokes Equations
  Ronald L. Panton 11
  - 1.6.1 Incompressible Flows 11
- 1.7 Discontinuities in Density
  Ronald L. Panton 11
  - 1.7.1 Normal Surface Discontinuity 11
  - 1.7.2 Fluid–Solid Boundary 12
  - 1.7.3 Interfaces with Surface Tension 12

	1.8		Constitutive Equations and Nonlinear Rheology of Polymer Melts	
			Saeid Kheirandish, Manfred H. Wagner	13
		1.8.1	Classical Theories	13
		1.8.2	Convected Derivatives and Differential Equations	16
		1.8.3	Microstructural Theories	17
		1.8.4	Conclusions	29
	References			30

2 Nondimensional Representation of the Boundary-Value Problem

John F. Foss, Ronald L. Panton, Alexander L. Yarin 33

	2.1		Similitude, the Nondimensional Prototype and Model Flow Fields	
			John F. Foss	34
		2.1.1	Governing Equations – Newtonian and Incompressible	34
		2.1.2	Boundary Conditions	35
		2.1.3	Initial Conditions	36
		2.1.4	Parameters that Influence the Solution to the Boundary-Value and/or the Initial-Value (BV/IV) Problem	36
		2.1.5	Governing Equations – Newtonian and Compressible	38
		2.1.6	Flows for Which U and L May Not Be Apparent	39
	2.2		Dimensional Analysis and Data Organization	
			Ronald L. Panton	42
		2.2.1	Variables, Function List, and Extra Information	42
		2.2.2	Dimensions and Scale Ratios	43
		2.2.3	Natural Scales and Repeating Variables	43
		2.2.4	Π Theorem	45
		2.2.5	Example with Rank Less than the Number of Dimensions	45
		2.2.6	Example with Redundant Dimensions	46
		2.2.7	Anatomy of a Nondimensional Variable	47
		2.2.8	Nonuniqueness of Scales	48
		2.2.9	Reference	48
		2.2.10	Scales Chosen for Experimental Purposes	48
		2.2.11	Nondimensional Variables Interpreted as Physical Ratios	50
		2.2.12	Scales Found from Boundary Conditions and Equations	50
		2.2.13	Limiting Cases	51
		2.2.14	Singular Perturbations	52
		2.2.15	Overlap Behavior and Composite Expansions	52
		2.2.16	Common Scales and Nondimensional Parameters	55
	2.3		Self-Similarity	
			Alexander L. Yarin	57
		2.3.1	General Causes of Self-Similar Behavior in Certain Situations in Fluid Mechanics and Heat Transfer	57
		2.3.2	Implications of Self-Similarity in Experimental Studies	58
		2.3.3	Particular Examples of Self-Similar Navier–Stokes Flows	59
		2.3.4	Particular Examples of the Boundary Layer Flows	62
		2.3.5	Gas Dynamics: Strong Explosion	79
		2.3.6	Free-Surface Flows	80
	References			82

Part B Measurement of Primary Quantities

3 Material Properties: Measurement and Data
William A. Wakeham, Marc J. Assael, Abraham Marmur, Joël De Coninck, Terry D. Blake, Stephanus A. Theron, Eyal Zussman 85

- 3.1 Density
  William A. Wakeham, Marc J. Assael 85
  - 3.1.1 Piezometers 86
  - 3.1.2 Bellows-Type Densimeters 87
  - 3.1.3 Vibrating-Element Densimeters 88
  - 3.1.4 Buoyancy-Type Densimeters 91
  - 3.1.5 Density Reference Values 95
  - 3.1.6 Tables of Density Values 96
- 3.2 Surface Tension and Interfacial Tension of Liquids
  Abraham Marmur 96
  - 3.2.1 Surface Tension of Pure Liquids 96
  - 3.2.2 Surface Tension of Liquid Solutions 98
  - 3.2.3 Interfacial Tension 99
  - 3.2.4 Implications of Surface and Interfacial Tension on Liquid–Fluid Systems 100
  - 3.2.5 Measurement of Surface Tension and Interfacial Tension .. 101
  - 3.2.6 Surface Tension Values for Liquids 105
- 3.3 Contact Angle
  Joël De Coninck, Terry D. Blake 106
  - 3.3.1 The Equilibrium Contact Angle 106
  - 3.3.2 Dynamic Contact Angle 112
- 3.4 Viscosity
  Marc J. Assael, William A. Wakeham 119
  - 3.4.1 Oscillating-Body Viscometers 119
  - 3.4.2 Vibrating Viscometers 122
  - 3.4.3 Torsional-Crystal Viscometer 124
  - 3.4.4 Capillary Viscometers 125
  - 3.4.5 Falling-Body Viscometers 127
  - 3.4.6 Viscosity Reference Values 131
  - 3.4.7 Tables of Viscosity Values 132
- 3.5 Thermal Conductivity and Thermal Diffusivity
  Marc J. Assael, William A. Wakeham 133
  - 3.5.1 Transient Methods for Thermal Conductivity 134
  - 3.5.2 Steady-State Methods for Thermal Conductivity 138
  - 3.5.3 Light-Scattering Methods for Thermal Diffusivity 141
  - 3.5.4 Thermal Conductivity Reference Values 146
  - 3.5.5 Tables of Thermal Conductivity Values 147
- 3.6 Diffusion
  Marc J. Assael, William A. Wakeham 147
  - 3.6.1 Diffusion in Liquids 149
  - 3.6.2 Diffusion in Gases 154

	3.6.3	Diffusion Reference Values	156
	3.6.4	Tables of Diffusion Coefficient Values	157

3.7 Electric and Magnetic Parameters of Liquids and Gases
Stephanus A. Theron, Eyal Zussman 158
    3.7.1 Introduction 158
    3.7.2 Dielectric Constant 159
    3.7.3 Electric Conductivity 160
    3.7.4 Broadband Measurement of the Conductivity and Dielectric Constant 166

References 169

4 Pressure Measurement Systems
Beverley J. McKeon, Rolf H. Engler 179

4.1 Measurement of Pressure with Wall Tappings
Beverley J. McKeon 180
    4.1.1 Cavity Shape, Connection and Alignment 181
    4.1.2 Finite-Area Effects 181
    4.1.3 Effect of Compressibility 182
    4.1.4 Effect of Finite Depth 183
    4.1.5 Condition of the Orifice Edge 184
    4.1.6 Correction for Distance from Measuring Point 184

4.2 Measurement of Pressure with Static Tubes
Beverley J. McKeon 185
    4.2.1 Effect of Geometry 185
    4.2.2 Effect of Hole Location 186
    4.2.3 Directional Sensitivity 186
    4.2.4 Effect of Turbulence 186

4.3 Hardware and Other Considerations
Beverley J. McKeon 187

4.4 Pressure-Sensitive Paint (PSP)
Rolf H. Engler 188
    4.4.1 Basics of PSP 188
    4.4.2 Paints 190
    4.4.3 Imaging Systems 200
    4.4.4 Processing 204
    4.4.5 Applications 206
    4.4.6 Concluding Remarks 208

References 209

5 Velocity, Vorticity, and Mach Number
*Beverley J. McKeon, Geneviève Comte-Bellot, John F. Foss,
Jerry Westerweel, Fulvio Scarano, Cameron Tropea, James F. Meyers,
Joseph W. Lee, Angelo A. Cavone, Richard Schodl,
Manoochehr M. Koochesfahani, Daniel G. Nocera, Yiannis Andreopoulos,
Werner J.A. Dahm, John A. Mullin, James M. Wallace,
Petar V. Vukoslavčević, Scott C. Morris, Eric R. Pardyjak, Alvaro Cuerva* 215

5.1	Pressure-Based Velocity Measurements		
	Beverley J. McKeon		216
	5.1.1	Measurement of Total Pressure Head with Pitot Tubes	218
	5.1.2	Dynamic Head from Separate Measurements of Total and Static Pressures	226
	5.1.3	Direct Measurement of Dynamic Head (Combined Pitot-Static and Other Probes)	226
	5.1.4	Measurement of Dynamic Head and Flow Direction (Multihole Probes)	228
5.2	Thermal Anemometry		
	Geneviève Comte-Bellot, John F. Foss		229
	5.2.1	Introduction	229
	5.2.2	Sensors	231
	5.2.3	Anemometer Electronics	258
	5.2.4	Calibration Procedures in Subsonic Flows	266
	5.2.5	Measurement of Velocity and Temperature Fluctuations	273
	5.2.6	Calibration Procedures in Compressible Flows	278
	5.2.7	Special Techniques	279
	5.2.8	A Comprehensive Technique for X-Array Calibration and Data Processing	283
5.3	Particle-Based Techniques		
	Cameron Tropea, Fulvio Scarano, Jerry Westerweel, Angelo A. Cavone, James F. Meyers, Joseph W. Lee, Richard Schodl		287
	5.3.1	Tracer Particles	287
	5.3.2	Laser Doppler Technique	296
	5.3.3	Particle Image Velocimetry	309
	5.3.4	Doppler Global Velocimetry	342
	5.3.5	Laser Transit Velocimetry	353
5.4	Molecular Tagging Velocimetry (MTV)		
	Manoochehr M. Koochesfahani, Daniel G. Nocera		362
	5.4.1	The Photochemistry of MTV: Molecular Tracers and Chemical Mechanisms	363
	5.4.2	The Experimental Implementation of MTV: Tagging Methods, Detection and Processing	373
	5.4.3	Examples of MTV Measurements	377
	5.4.4	Summary and Conclusions	382

5.5 Vorticity
Yiannis Andreopoulos, Werner J.A. Dahm, John A. Mullin, James M. Wallace, Petar V. Vukoslavčević, Scott C. Morris 382
    5.5.1 Optical Techniques in Strophometry – Vorticity Measurements Methods 383
    5.5.2 High-Resolution Dual-Plane Stereo PIV (DSPIV) 400
    5.5.3 Measurements of the Vorticity Vector and Other Velocity Gradient Tensor-Based Turbulence Properties 408
    5.5.4 Transverse Vorticity Measurements with a Four-Sensor Hot-Wire Probe 429

5.6 Thermal Transient Anemometer (TTA)
John F. Foss 434
    5.6.1 Operational Description 434
    5.6.2 Representative Results 435

5.7 Sonic Anemometry/Thermometry
Eric R. Pardyjak, Alvaro Cuerva 436
    5.7.1 Definition 436
    5.7.2 Measurement Principles 437
    5.7.3 Device Characteristics, Accuracy, and Limitations 439
    5.7.4 Data-Acquisition Requirements 444
    5.7.5 Use and Calibration Procedures 444
    5.7.6 Manufacturers and Costs 445
    5.7.7 Device Comparison 445

References 446

6 Density-Based Techniques
Wolfgang Merzkirch 473

6.1 Density, Refractive Index, and Optical Flow Visualization
Wolfgang Merzkirch 473

6.2 Shadowgraphy
Wolfgang Merzkirch 474

6.3 Schlieren Method
Wolfgang Merzkirch 476

6.4 Moiré Deflectometry
Yasuhiro Egami 478

6.5 Interferometry
Wolfgang Merzkirch 480

6.6 Optical Tomography
Wolfgang Merzkirch 485

References 485

7 Temperature and Heat Flux
Tomasz A. Kowalewski, Phillip Ligrani, Andreas Dreizler, Christof Schulz, Uwe Fey, Yasuhiro Egami 487

7.1 Thermochromic Liquid Crystals
Tomasz A. Kowalewski 488
    7.1.1 Materials 489

7.1.2 Illumination 491
7.1.3 Acquisition and Calibration 493
7.1.4 Examples 496
7.1.5 Conclusions 499

7.2 Measurements of Surface Heat Transfer Characteristics Using Infrared Imaging
Phillip Ligrani 500
7.2.1 Introduction and Background 500
7.2.2 Chapter Organization 501
7.2.3 Infrared Cameras 501
7.2.4 Overall Physical Arrangement 502
7.2.5 In Situ Calibration 503
7.2.6 Measurement of Surface Nusselt Numbers 503
7.2.7 Measurement of Surface Adiabatic Film Cooling Effectiveness 506
7.2.8 Measurement of Surface Heat Flux Reduction Ratio 508
7.2.9 Transient Surface Temperature Measurements 510
7.2.10 Boundary-Condition Information for Numerical Predictions 512
7.2.11 Summary and Conclusions 515

7.3 Temperature Measurement via Absorption, Light Scattering and Laser-Induced Fluorescence
Andreas Dreizler, Christof Schulz 515
7.3.1 Overview 516
7.3.2 Non-Resonant Techniques 516
7.3.3 Resonant Techniques 523
7.3.4 Nonlinear Techniques 535
7.3.5 Conclusions 537

7.4 Transition Detection by Temperature-Sensitive Paint
Uwe Fey, Yasuhiro Egami 537
7.4.1 Introduction 537
7.4.2 Surface Heat Transfer Processes 543
7.4.3 Temperatures in Laminar and Turbulent Boundary Layer 545
7.4.4 Transition Detection by Means of TSP 548
7.4.5 Comparison with Other Methods 551

References 553

8 **Force and Moment Measurement**
Klaus Hufnagel, Günter Schewe 563

8.1 Steady and Quasi-Steady Measurement
Klaus Hufnagel 564
8.1.1 Basics 564
8.1.2 Basic Terms of Balance Metrology 568
8.1.3 Mounting Variations 570
8.1.4 Strain Gauges 572
8.1.5 Wiring of Wheatstone Bridges 573
8.1.6 Compensation of Thermal Effects 575

	8.1.7	Compensation of Sensitivity Shift	576
	8.1.8	Strain Gauge Selection	578
	8.1.9	Strain Gauge Application	578
	8.1.10	Materials	579
	8.1.11	Single-Force Load Cells	580
	8.1.12	Multicomponent Load Measurement	582
	8.1.13	Internal Balances	582
	8.1.14	External Balances	586
	8.1.15	Calibration	590
8.2	Force and Moment Measurements in Aerodynamics and Aeroelasticity Using Piezoelectric Transducers *Günter Schewe*		596
	8.2.1	Basic Aspects of the Piezoelectric Force-Measuring Technique	598
	8.2.2	Typical Properties	602
	8.2.3	Examples of Application	607
	8.2.4	Conclusions	614
References			615

Part C Specific Experimental Environments and Techniques

9 Non-Newtonian Flows
Klaas te Nijenhuis, Gareth H. McKinley, Stephen Spiegelberg,
Howard A. Barnes, Nuri Aksel, Lutz Heymann, Jeffrey A. Odell 619

9.1	Viscoelastic Polymeric Fluids *Klaas te Nijenhuis, Gareth H. McKinley, Stephen Spiegelberg*		619
	9.1.1	Measurements in Shear Flow	627
	9.1.2	Rheogoniometers and Rheometers	635
	9.1.3	Elongational Flows	646
9.2	Thixotropy, Rheopexy, Yield Stress *Howard A. Barnes, Klaas te Nijenhuis*		661
	9.2.1	A History of Thixotropy	661
	9.2.2	Description of Thixotropic Phenomenon	665
	9.2.3	Typical Thixotropic Experiments	669
	9.2.4	Semi-Empirical Phenomenological Theories Used to Fit Experimental Data	670
	9.2.5	The Breakdown and Build-Up of Isolated Flocs	674
	9.2.6	Examples of Systems and Studies from the Literature	674
	9.2.7	Overall Conclusions	679
9.3	Rheology of Suspensions and Emulsions *Nuri Aksel, Lutz Heymann*		680
	9.3.1	Preliminaries	680
	9.3.2	Suspensions	683
	9.3.3	Emulsions	711

	9.4	Entrance Correction and Extrudate Swell	
		Klaas te Nijenhuis	720
		9.4.1 Correction for Entrance Effect: Bagley Correction	720
		9.4.2 Extrudate Swell or Die Swell	721
		9.4.3 Conclusions	723
	9.5	Birefringence in Non-Newtonian Flows	
		Jeffrey A. Odell	724
		9.5.1 The Molecular Origin of Birefringence	725
		9.5.2 Techniques for Birefringence Measurement	725
		9.5.3 Relation Between Birefringence and Molecular Strain	729
		9.5.4 Optical Properties of Macromolecules in Solution: Effects of Macroform and Microform Birefringence	731
		9.5.5 Example Calculation of the Theoretical Birefringence for Stretched Molecules of Atactic Polystyrene	731
		9.5.6 Conclusions	732
References			732

10 Measurements of Turbulent Flows

Giovanni Paolo Romano, Nicholas T. Ouellette, Haitao Xu,
Eberhard Bodenschatz, Victor Steinberg, Charles Meneveau, Joseph Katz .. 745

	10.1	Statistical Eulerian Description of Turbulent Flows	
		Giovanni Paolo Romano	746
		10.1.1 Basics of Measurements of Turbulent Flows	746
		10.1.2 Reynolds Decomposition and Equations	765
		10.1.3 Scales in Turbulent Flows	776
	10.2	Measuring Lagrangian Statistics in Intense Turbulence	
		Eberhard Bodenschatz, Nicholas T. Ouellette, Haitao Xu	789
		10.2.1 Image Processing	789
		10.2.2 Experimental Implementation	792
		10.2.3 Turbulent Relative Dispersion	797
		10.2.4 Concluding Remarks	799
	10.3	Elastic Turbulence in Viscoelastic Flows	
		Victor Steinberg	799
		10.3.1 Basics	799
		10.3.2 Elastic Turbulence in Swirling Flow Between Two Plates ...	804
		10.3.3 Elastic Turbulence in a Curved Channel: Dean Flow	821
		10.3.4 Conclusions	829
	10.4	Measurements for Large-Eddy Simulations	
		Joseph Katz, Charles Meneveau	830
		10.4.1 Large-Eddy Simulation and Data Requirements	831
		10.4.2 Arrays of Single-Point Instruments for Studies of SGS Dynamics	833
		10.4.3 Planar Particle Image Velocimetry (PIV) for SGS Dynamics and LES	835
		10.4.4 Case Studies and Sample Results Using Planar PIV Measurements	836

		10.4.5	Holographic PIV Measurements of SGS Dynamics	844
		10.4.6	Scalar Concentration Measurements for SGS Mixing and Combustion Studies	848
	References			849

11 Flow Visualization
Wolfgang Merzkirch 857

11.1	Aims and Principles of Flow Visualization		857
11.2	Visualizations of Flow Structures and Flow Direction		859
	11.2.1	The Addition of Tracer Materials as Initial Condition	859
	11.2.2	Dye Lines and Contours in Liquid Flow	860
	11.2.3	Smoke Visualization in Air Flows	862
	11.2.4	Hardware for Flow Visualization Experiments: Illumination, Recording, Confinements of the Flow Regime	863
	11.2.5	Enhancement of Image Quality: Fluorescence, Image Processing	865
	11.2.6	Direct Visualizations of Velocity Profiles	866
11.3	Visualization of Free Surface Flows		868
References	....		869

12 Wall-Bounded Flows
Joseph C. Klewicki, William S. Saric, Ivan Marusic, John K. Eaton 871

12.1	Introductory Concepts		
	Joseph C. Klewicki		871
	12.1.1	Governing Equations	871
	12.1.2	Brief Overview of Wall Flow Structure	872
	12.1.3	Scaling Ideas and Parameters	874
	12.1.4	Overview of Measurement Considerations	874
12.2	Measurement of Wall Shear Stress		
	Joseph C. Klewicki		875
	12.2.1	Methods for Determining the Time-Averaged Wall Shear Stress	876
	12.2.2	Time-Resolved Methods	882
12.3	Boundary-Layer Stability and Transition		
	William S. Saric		886
	12.3.1	The Process of Transition for Boundary Layers in External Flows	886
	12.3.2	Nomenclature of Linear Theory	886
	12.3.3	Basic Rules for Boundary-Layer Stability Experiments	887
	12.3.4	Experimental Techniques	887
	12.3.5	Wind Tunnel Environment	890
	12.3.6	T–S Measurements...	892
	12.3.7	Visualization Methods	896

12.4	Measurements Considerations in Non-Canonical Flows	
	John K. Eaton, Ivan Marusic	896
	12.4.1 Pressure Probe Measurements	897
	12.4.2 Turbulence Measurements	898
	12.4.3 Wall Shear Stress	901
	12.4.4 Planar and Whole-Field Measurements	902
References		902

13 Topological Considerations in Fluid Mechanics Measurements

	John F. Foss	909
13.1	A Companion Document	909
13.2	Utilization of Topological Considerations for Flow Field Analyses	910
	13.2.1 A *Tabbed* Jet	910
	13.2.2 A Conical Flame Holder Geometry	911
	13.2.3 Flow Past a Small-Aspect-Ratio *Protruding* Cylinder	912
	13.2.4 A Vortex Ring–Wall Interaction	913
	13.2.5 An Annular Jet	915
	13.2.6 Summary	917
References		918

14 Flow Measurement Techniques in Turbomachinery

	Oğuz Uzol, Joseph Katz	919
14.1	Background On Turbomachinery Flows	919
14.2	Non-Optical Measurement Techniques	921
	14.2.1 Data-Acquisition Techniques	921
	14.2.2 Non-Optical Measurement Techniques	928
14.3	Optical Measurement Techniques	933
	14.3.1 Applications of Laser Doppler Velocimetry (LDV)	934
	14.3.2 Applications of Particle Image Velocimetry (PIV)	936
	14.3.3 Applications of Laser Two-Focus Velocimetry (L2F)	945
	14.3.4 Applications of Doppler Global Velocimetry (DGV)	948
	14.3.5 Applications of Pressure-Sensitive Paint (PSP)	950
14.4	Concluding Remarks	950
References		951

15 Hydraulics

	Roger E.A. Arndt, Damien Kawakami, Martin Wosnik, Marc Perlin,	
	James H. Duncan, David M. Admiraal, Marcelo H. García	959
15.1	Measurements in Cavitating Flows	
	Roger E.A. Arndt, Damien Kawakami, Martin Wosnik	959
	15.1.1 Fundamentals	961
	15.1.2 Types of Cavitating Flows	967
	15.1.3 Cavitation Damage	970
	15.1.4 Facilities	972
	15.1.5 Water Quality Measurements	979
	15.1.6 Cavitation Inception	984
	15.1.7 Measurements in Cavitating Flows	988

15.2 Wave Height and Slope
Marc Perlin, James H. Duncan 1009
    15.2.1 Temporal Point Measurements
             at Fixed Geometric Positions 1009
    15.2.2 One-Dimensional Spatial Measurements 1014
    15.2.3 Two-Dimensional Spatial Measurements 1015
    15.2.4 Special Surface Elevation Measurements
             for Large Laboratory or Field Use 1018
15.3 Sediment Transport Measurements
David M. Admiraal, Marcelo H. García 1020
    15.3.1 A Brief Introduction to Sediment Transport 1020
    15.3.2 Methods of Measuring Suspended Sediment Transport 1022
    15.3.3 Bed Load Sediment Measurements 1029
    15.3.4 Total Load Measurements 1032
    15.3.5 Other Measurement Techniques 1032
References 1033

16 Aerodynamics
*Wolf-H. Hucho, Klaus Hannemann, Jan Martinez Schramm,
Charles H.K. Williamson* 1043
16.1 Ground Vehicle Aerodynamics
Wolf-H. Hucho 1043
    16.1.1 Vehicles in Their Natural Environment 1043
    16.1.2 Simulation of the Drive on Road and Track 1044
    16.1.3 Wind Tunnels for Ground Vehicles 1044
    16.1.4 Wind-Tunnel Design 1047
    16.1.5 Equipment for Automotive Wind Tunnels 1054
    16.1.6 Limits of Simulation 1058
    16.1.7 Typical Vehicle Wind Tunnels 1063
    16.1.8 Tests with Moving Models and Full-Scale Vehicles 1068
    16.1.9 Measurement Technique 1070
    16.1.10 Support by Computational Fluid Dynamics 1080
16.2 Short-Duration Testing
of High Enthalpy, High Pressure, Hypersonic Flows
Klaus Hannemann, Jan Martinez Schramm 1081
    16.2.1 Working Principle of Shock Tubes/Tunnels
             and Shock Expansion Tubes/Tunnels 1082
    16.2.2 Measurement Techniques 1095
    16.2.3 Typical Applications of Shock Tunnel
             and Shock Expansion Tunnel Facilities 1111
16.3 Bluff Body Aerodynamics
Charles H.K. Williamson 1125
    16.3.1 Flow Physics, Facilities, and Approach 1125
    16.3.2 How Bluff is a Bluff Body? 1128

		16.3.3	Base Pressure, Drag, Lift, and Strouhal Number	1133
		16.3.4	Overview of Vortex Shedding Regimes	1135
		16.3.5	Concluding Remarks	1145
	References			1146

17 Atmospheric Measurements
Harindra J.S. Fernando, Marko Princevac, Ronald J. Calhoun 1157

- 17.1 Point Measurements
  Marko Princevac 1159
  - 17.1.1 Wind Velocity 1159
  - 17.1.2 Temperature, Relative Humidity, and Pressure 1162
  - 17.1.3 Fluxes 1164
  - 17.1.4 Vertical Profiling 1166
- 17.2 Dispersion Measurements
  Ronald J. Calhoun 1167
  - 17.2.1 Tracers 1167
  - 17.2.2 Bag Samplers 1168
  - 17.2.3 Fast-Response Sensors 1168
  - 17.2.4 Recent Major Field Experiments 1168
- 17.3 Remote Sensing
  Ronald J. Calhoun 1169
  - 17.3.1 Lidar 1169
  - 17.3.2 Ceilometers 1172
  - 17.3.3 Sodar 1172
  - 17.3.4 Radar Profilers 1174
  - 17.3.5 RASS 1174
- 17.4 Satellite Measurements
  Ronald J. Calhoun, Harindra J.S. Fernando 1175
- **References** 1178

18 Oceanographic Measurements
Bruce Howe, Teresa K. Chereskin 1179

- 18.1 Oceanography 1179
  - 18.1.1 Oceanographic Processes 1179
  - 18.1.2 Challenges of the Environment 1181
- 18.2 Point Measurements 1182
  - 18.2.1 Pressure 1182
  - 18.2.2 Temperature 1182
  - 18.2.3 Salinity 1183
  - 18.2.4 Sound Speed 1185
  - 18.2.5 Density 1185
  - 18.2.6 Velocity 1186
- 18.3 Lagrangian Techniques 1188
  - 18.3.1 Surface Drifters 1188
  - 18.3.2 Deep Ocean Floats 1189
  - 18.3.3 Dispersion 1190

	18.4	Remote Sensing	1192
		18.4.1 Acoustic Doppler Current Profilers (ADCP)	1192
		18.4.2 Acoustic Tomography	1195
		18.4.3 Acoustic Correlation Techniques	1199
		18.4.4 Electromagnetic Sensing Using the Geomagnetic Field	1199
		18.4.5 Surface Current Mapping Using Radar Backscatter	1201
		18.4.6 Satellites	1202
	18.5	Measurement Systems	1203
		18.5.1 Expendable Sensors	1203
		18.5.2 HPIES	1203
		18.5.3 Moored Profiler	1204
		18.5.4 TriSoarus Towed Vehicle	1204
		18.5.5 Turbulence and Small-Scale Structure	1205
		18.5.6 Floats	1205
		18.5.7 Gliders	1207
		18.5.8 Autonomous Undersea Vehicles (AUVs)	1207
		18.5.9 LEO-15	1207
	18.6	Experiment Case Studies	1208
		18.6.1 Hawaii Ocean Mixing Experiment (HOME)	1209
		18.6.2 Fluxes, Air–Sea Interaction, and Remote Sensing (FAIRS)	1212
	References		1214

19 Microfluidics: The No-Slip Boundary Condition

Eric Lauga, Michael P. Brenner, Howard A. Stone 1219

	19.1	History of the No-Slip Condition	1220
		19.1.1 The Previous Centuries	1220
		19.1.2 Terminology	1220
		19.1.3 Traditional Situations Where Slip Occurs	1221
		19.1.4 Newtonian Liquids: Slip or No-Slip	1222
	19.2	Experimental Methods	1222
		19.2.1 Indirect Methods	1222
		19.2.2 Local Methods	1224
	19.3	Molecular Dynamics Simulations	1226
		19.3.1 Principle	1227
		19.3.2 Results	1227
		19.3.3 Interpretation in the Continuum Limit	1228
	19.4	Discussion: Dependence on Physical Parameters	1228
		19.4.1 Surface Roughness	1228
		19.4.2 Dissolved Gas and Bubbles	1229
		19.4.3 Wetting	1230
		19.4.4 Shear Rate	1232
		19.4.5 Electrical Properties	1233
		19.4.6 Pressure	1233
	19.5	Perspective	1234
	References		1235

20 Combustion Diagnostics
Christof Schulz, Andreas Dreizler, Volker Ebert, Jürgen Wolfrum 1241

- 20.1 Basics 1242
- 20.2 Laser-Based Combustion Diagnostics 1243
- 20.3 Experimental Data Devoted to Validation of Numerical Simulations and Modeling 1244
  - 20.3.1 General Remarks 1244
  - 20.3.2 Submodels and Their Specific Demands for Validation 1245
  - 20.3.3 Example: Generic Turbulent Flame 1246
- 20.4 Application of Laser-Based Techniques 1247
  - 20.4.1 Detection Sensitivity, Selectivity and Resolution 1247
  - 20.4.2 Laminar Flames 1249
  - 20.4.3 Turbulent Combustion 1251
  - 20.4.4 Engine Combustion 1265
  - 20.4.5 Diagnostics for Stationary, Large-Scale Combustion Processes 1284
- 20.5 Conclusions 1299
- **References** 1300

21 Electrohydrodynamic Systems
Antonio Castellanos, Alberto T. Pérez 1317

- 21.1 Equations 1318
  - 21.1.1 Electrical Equations 1318
  - 21.1.2 Mechanical Equations 1318
  - 21.1.3 Temperature Equation 1319
- 21.2 Fluid Statics and Dynamics in EHD 1320
  - 21.2.1 Charge Decay 1320
  - 21.2.2 EHD Statics and Dynamics 1321
- 21.3 Experimental Methods in EHD 1322
- 21.4 Conductivity 1323
  - 21.4.1 Conduction Mechanisms in Liquids 1323
  - 21.4.2 Ohmic Versus Non-Ohmic Regime: Practical Estimation of the Transition Voltage 1324
  - 21.4.3 Unipolar Injection 1325
  - 21.4.4 Determination of the Conductivity 1325
- 21.5 Mobility 1327
- 21.6 Electric Field Measurement: Kerr Effect 1328
- 21.7 Velocity 1329
  - 21.7.1 Laser-Doppler Anemometry 1329
- 21.8 Visualization 1330
  - 21.8.1 Shadowgraph and Schlieren 1330
  - 21.8.2 Tracer Particles 1331
- **References** 1331

Part D Analysis and Post-Processing of Data

22 Review of Some Fundamentals of Data Processing
Holger Nobach, Cameron Tropea, Laurent Cordier, Jean-Paul Bonnet, Joël Delville, Jacques Lewalle, Marie Farge, Kai Schneider, Ronald J. Adrian 1337

- 22.1 Fourier Transform
  Holger Nobach, Cameron Tropea 1337
- 22.2 Correlation Function
  Holger Nobach, Cameron Tropea 1342
- 22.3 Hilbert Transform
  Holger Nobach, Cameron Tropea 1344
- 22.4 Proper Orthogonal Decomposition: POD
  Laurent Cordier, Jean-Paul Bonnet, Joël Delville 1346
  - 22.4.1 Basics 1346
  - 22.4.2 POD: An Approximation Method 1347
  - 22.4.3 The Proper Orthogonal Decomposition (POD) 1352
  - 22.4.4 The Different POD Approaches 1357
  - 22.4.5 Classical POD or Direct Method 1359
  - 22.4.6 Snapshot POD 1360
  - 22.4.7 Common Properties of the Two POD Approaches 1361
  - 22.4.8 POD and Harmonic Analysis 1362
  - 22.4.9 Typical Applications to Fluid Mechanics 1364
  - 22.4.10 POD Galerkin 1365
  - 22.4.11 Evaluative Summary of the POD Approach 1369
- 22.5 Conditional Averages and Stochastic Estimation
  Ronald J. Adrian 1370
  - 22.5.1 Conditional Averages 1370
  - 22.5.2 Stochastic Estimation 1373
- 22.6 Wavelet Transforms
  Marie Farge, Jacques Lewalle, Kai Schneider 1378
  - 22.6.1 Introduction to Wavelets 1378
  - 22.6.2 Continuous Wavelet Transform 1378
  - 22.6.3 Orthogonal Wavelet Transform 1383
  - 22.6.4 Applications in Experimental Fluid Mechanics 1387
- References 1395

23 Fundamentals of Data Processing
Holger Nobach, Cameron Tropea 1399

- 23.1 Statistical Principles 1399
- 23.2 Stationary Random Processes 1401
- 23.3 Estimator Expectation and Variance 1402
  - 23.3.1 Estimators for the Mean 1402
  - 23.3.2 Estimators for Higher-Order Statistics 1404
- 23.4 Signal Noise 1406

23.5	Cramér–Rao Lower Bound (CRLB)		1408
	23.5.1	Laser Doppler and Phase Doppler Signals	1409
	23.5.2	Particle Imaging	1414
23.6	Propagation of Errors		1416
References			1417

24 Data Acquisition by Imaging Detectors
Bernd Jähne 1419

24.1	Definitions		1419
24.2	Types of Detectors		1420
	24.2.1	Quantum Detectors	1420
	24.2.2	Thermal Detectors	1420
24.3	Imaging Detectors		1421
	24.3.1	The Charge-Coupled Device	1421
	24.3.2	CMOS Imaging Sensors	1421
	24.3.3	CCD Sensor Architectures	1422
	24.3.4	Standard Interfaces for Digital Cameras	1425
24.4	Performance of Imaging Sensors		1426
	24.4.1	Responsivity	1426
	24.4.2	Quantum Efficiency	1426
	24.4.3	Signal Irradiance Relation	1427
	24.4.4	Dark Current	1427
	24.4.5	Noise-Equivalent Exposure	1427
	24.4.6	Saturation Equivalent Exposure	1427
	24.4.7	Dynamic Range	1428
	24.4.8	Photon-Noise-Limited Performance	1428
	24.4.9	Linear Noise Model for Image Sensors	1429
	24.4.10	Signal-to-Noise Ratio	1429
	24.4.11	Spectral Sensitivity	1431
	24.4.12	Nonuniform Responsivity	1433
	24.4.13	Artifacts and Operation Errors	1433
24.5	Camera Selection		1435
	24.5.1	Measurements at Low Light Levels	1435
	24.5.2	Measurements with High Irradiance Variations	1435
	24.5.3	Precise Radiometric and Geometric Measurements	1435
References			1436

25 Data Analysis
Bernd Jähne, Michael Klar, Markus Jehle 1437

25.1	Image Processing		
	Bernd Jähne		1437
	25.1.1	Sampling and Quantization	1437
	25.1.2	Radiometric Corrections	1440
	25.1.3	Geometric Corrections	1442
	25.1.4	Averaging and Noise Suppression	1445
	25.1.5	Edge and Line Extraction	1451
	25.1.6	Direction and Orientation	1453

		25.1.7 Local Wavenumber and Local Phase	1458
		25.1.8 Multiscale Processing	1461
25.2	Motion Analysis		
	Markus Jehle, Michael Klar		1464
		25.2.1 General Considerations on Motion Analysis	1464
		25.2.2 Correlation-Based Velocity Analysis	1469
		25.2.3 Least-Squares Matching	1473
		25.2.4 Tracking Techniques	1474
		25.2.5 Optical-Flow-Based Velocity Analysis	1481
References			1488

Acknowledgements 1493
About the Authors 1495
Detailed Contents 1513
Subject Index 1531

Subject Index

3C-Doppler L2F 361
3-D Navier–Stokes 361
3-pentanone 532
4-D MRV (4-D magnetic resonance velocimetry) 902
4-D magnetic resonance velocimetry (4-D MRV) 902

A

Abel inversion 485
ABS (acoustic backscatter) 1024
absolute 124
– velocity profiler (AVP) 1200, 1210
absorption 523, 525, 1243, 1276
– coefficient 524
– cross-section 532
– line 342, 343, 348, 349, 351, 352
– long-path 525
– thermometry 525
absorption spectroscopy (AS) 516, 526, 1285
– tunable diode laser 526
acac (acetylacetonate) 372
accelerometer 1095, 1101, 1102, 1122
– Buoy 1013
– float-mounted 1013
acetone 371, 532, 1259
acetonitrile (ACN) 167
acetylacetonate (acac) 372
ACF (autocorrelation function) 302, 306, 1342, 1406
ACN (acetonitrile) 167
acoustic 1185, 1192
– backscatter (ABS) 1024
– cavitation cloud 1006
– Doppler current meter 1187
– eigenrays 1196
– inception technique 985
– measurement 992
– mirror 1077
– pulse propagation sensor 1009
– scintillation 1199
– sea surface height 1202
– sound channel 1195
– spectra 997
– tomography 1185, 1195
– travel time integral 1186
– travel time path integral 1195
– waveguide 1195
acoustic Doppler current profiler (ADCP) 1011, 1028, 1192
– broadband 1193
– lowered 1195
– narrow band 1193
acoustic scintillation flow meter (ASFM) 1199
acoustical resonator 1009
active
– control loop 1284
– pixel sensor (APS) 1422
– resonance control 1054
adaptive
– filter 1449
– Gaussian windowing (AGW) 1468
– least-squares correlation 1474
– multipass technique 403
– test section 609
– wall 1049, 1051
ADC (analog-digital converter) 203, 1427
ADCP (acoustic Doppler current profiler) 1011, 1028, 1192
adjustable nozzle 1066
ADP (acoustic Doppler current profiler)
– broadband 1193
– lowered 1195
– narrow band 1193
adsorption layer thickness 688
advanced very high-resolution radiometer (AVHRR) 1202
adverse pressure gradient 897
aero acoustic wind tunnel 1063
aerodynamic 1082, 1112, 1113, 1115
– blockage 875
– drag 1069
– perturbations 235
aeroelasticity 609
aerosol generator 862
aerothermodynamic 1081, 1082, 1113, 1115
AFM (atomic force microscopy) 234, 1223
AFTRF (axial flow turbine research facility) 923
aggregate 1281

AGW (adaptive Gaussian windowing) 1468
air 1119
– conditioning 1046
– entrainment coefficient 971
– flow rate 1073
– injection 970
– path 1047
– velocity 1072
air-breathing propulsion 1081, 1088, 1120, 1123
air-expendable bathythermograph (AXBT) 1198
airfoil
– at high angle of attack 608
– model 601
– NACA 0012 610
– supercritical 611
air–fuel ratio 1269
aliasing 1340, 1438
alkali 1297
alkene 92
altimetry sea surface height 1202
AM (amplitude modulation) 1289
American Society for Testing and Materials (ASTM) 126
amplitude and phase spectra 1018
amplitude modulation (AM) 1289
– specific 1289
amplitude of fluctuation 969
analog-to-digital converter (ADC) 203, 1427
analytic signal 1459
analytical (complex) function 1345
anemometer
– calibration 266, 277
– constant-current (CCA) 231, 246, 248, 258
– constant-temperature (CTA) 231, 247, 248, 260
– constant-voltage (CVA) 231, 247, 248, 264
– hot wire 1135, 1143
– propeller 1160
– pulsed-wire 901
– sonic 1159
– stability 261, 262, 265
– vane 1072
anemometry
– thermal 901
angle of attack α 968

angular
- momentum 7
angularity 1051
anharmonicity 519
- correction 519
anisotropic diffusion 1450
anisotropy
- optical 724
annular jet 915
anti-Stokes
- line 517, 520
- Raman scattering 517
APE:HKE (available potential energy-to-horizontal kinetic energy) 1212
aperture problem 1481, 1483
apparent extensional viscosity 647, 650, 658
APS (active pixel sensor) 1422
Aquarius 1202
aqueous solutions of potassium chloride 156
arbitrary region 8
ARD (atmospheric re-entry demonstrator) 1115
Argo profiling float 1190
argon laser 356
array of laser device 1015
artificial
- head 1077
- ventilation 998
AS (absorption spectroscopy) 516, 526, 1285
- tunable diode laser 526
ASFM (acoustic scintillation flow meter) 1199
aspect ratio 684
assignment problem 791
ASTM
- G-32 standard vibratory apparatus 976
ASTM (American Society for Testing and Materials) 126
atmospheric re-entry demonstrator (ARD) 1115
atomic force microscopy (AFM) 234, 1223
atomization 293
atomizer 293
attached
- cavities 959
- flow 596
attachedcavitation 968
attenuation 1028, 1053
autocorrelation function (ACF) 302, 306, 1342, 1406

autocovariance function 1343
autonomous undersea vehicle (AUV) 1192
AUV (autonomous undersea vehicle) 1192
available potential energy-to-horizontal kinetic energy (APE:HKE) 1212
AVHRR (advanced very high-resolution radiometer) 1202
AVP (absolute velocity profiler) 1200, 1210
AXBT (air-expendable bathythermograph) 1198
axial
- flow turbine research facility (AFTRF) 923
- velocity gradient 880
- wavelength 967
axisymmetric
- submerged jet 64, 68, 74, 76
- ventilated supercavity 1001, 1002

B

background
- oriented schlieren 479
- scattering 520
bag sampler 1168
Bagley 721
- correction 721
- plot 720, 721
balance 1070
- external 597
- half model 597
- internal 600
- platform 597
- rigidity of a 596
Baldwin–Dewey diagram 240
bandpass decomposition 1464
bandwidth 262, 265
baroclinic flow 1198
barotropic tide mixing 1210
base
- bleed 1133
- pressure coefficient 1132, 1135, 1136
- suction coefficient 1136
basic suction 1055, 1063
Basset–Boussinesq–Oseen (BBO) 288
Batchelor prediction 799
Batchelor's scaling 799
bathythermograph (BT)
- expendable bathythermograph (XBT) 1203

Bayesian
- framework 1480
- multiple target tracking 1480
BBO (Basset–Boussinesq–Oseen) 288
BC (boundary condition) 35
BCCE (brightness constancy constraint equation) 1482
beam
- expansion 301
- separation 359
- superposition 1292
- waist 300
bed load 1020
bed shear stress 1021
bedforms 1030
Beer–Lambert law 523, 1290
Bénard convection 483
benchmark 499
bending frequency 609
Bernoulli equation 216, 285
biacetyl 371
bifurcation
- subcritical 610
bimodal suspension 691
binary scaling parameter 1081, 1115
binomial filter 1446
Biot number 250
bipolar injection 1328
bistatic 1173
BL (boundary layer) 57, 58, 62, 63
black wall problem 1481, 1483, 1484
blade passage 359
Blasius 887
- flow 62
blockage 1051
- coefficient 1061
- correction 1050
- ratio 1046, 1048, 1049
blowing 1055
bluff body 607, 1045, 1125, 1128, 1133
- flow 1125
- shape 1133
- two-dimensional 597
body force 34
Boger fluids 803
boiling 959
bolometer 1421
Boltzmann
- fraction 518, 529
- plot 531

– population 1278
– superposition principle 619, 624, 626, 627
bond number 118, 647, 653, 657, 1321
bootstrap method 118
boundary condition (BC) 35, 746
boundary layer (BL) 57, 58, 62, 63, 545, 896, 1092, 1095, 1112, 1113, 1119, 1123, 1125, 1128, 1132, 1141, 1145
– laminar 1105, 1118, 1119
– stability 886
– temperature difference 547
– transition 544, 1118, 1119
– transitional 1105, 1120
– turbulent 1105, 1118, 1119
Boussinesq approximation 1319
box filter 1445
Bragg cell 298, 360
break test 1070
breakdown
– optical 1254
breaking 718
– wave 1010
break-up time 656
bridge section 608
brightness constancy constraint equation (BCCE) 1482
Brinell hardness 978
broadband absorber 1064
Brownian motion 684, 797
B-spline 1444
– interpolation 1444
– transformation 1444
BT (bathythermograph)
– expendable bathythermograph (XBT) 1203
bubble 998
– cavitation 968
– cloud 959
– cluster 1002
– collapse 961, 971
– diameter range 1006
– dynamic 961
– growth 962
– liquid 1007
– patch cavitation 969
– size range 1006
– wake 1001, 1002
buffer layer 873
buffet frequency 597, 611
buffeting 1053
building aerodynamic 597

bulk modulus 633
Bunsen 1242
buoyancy 1319
– driven convective flow 378
– effect 237, 238
Burgers model 619, 623, 624, 626, 627
Burnett 87
burst amplitude 1004

C

CaBER 660
cable effect 266
CAD (crank angle degree) 381
caged fluorescent dye 368
calibration 519, 522, 523, 536, 602, 897
– matrix 603
– practical aspect 522
– target 403
– time scale 522
Calspan–University at Buffalo Research Center (CUBRC) 1113
camera 1108, 1110, 1111
– analogue 1110
– CCD 1108–1110
– CMOS 1110
– Cranz–Schardin 1110
– digital 1110
– electronic 1110
– high-speed 1110, 1111
– mirror 1111
– rotating drum 1110, 1111
– rotating mirror 1111
– rotating prism 1111
– rotating-mirror/prism 1110
capacitance
– effect 261, 266
– gage 1011
– wave probe 1010
capacitor bank 1088, 1090
capillary 1212
– break-up rheometer 647, 656
– flow 119
– master viscometer 126
– number 647, 652, 657, 712
– rheometer 636, 637, 643, 646
– wave 57, 80
carbon
– dioxide 121, 140, 1119, 1120
– sequestration 1179
carbon-fiber-reinforced plastic (CFK) 546

CARS (coherent anti-Stokes Raman scattering) 516, 536, 1243, 1251, 1252, 1284
– rotational 537
CARS (coherent anti-Stokes Raman spectroscopy) 1105
catapult 1068
cavitating
– jet 963
– jet method 976
cavitation 959, 997
– application 960
– cloud 968, 1002
– damage 960, 970
– desinence 963, 985
– erosion 960, 971
– erosion test apparatus 972
– erosion test facility 973
– in the tip 963
– inception 960, 964, 986, 988, 993
– inception detection 984
– index 963
– induced lift oscillation 969
– limited 960
– noise 992
– nuclei 965
– nuclei measurement 965
– pitting rate 971
– shedding dynamics 1005
– susceptibility meters (CSM) 982
– test 972
cavity
– length 967, 969
– oscillation 967
– pressure 963
– ring-down 1243
– ring-down laser absorption spectroscopy (CRLAS) 525
– ring-down laser-absorption spectroscopy (CRDLAS) 1249
CCA (constant-current anemometer) 231, 246, 248, 258
CCD (charge-coupled device) 203, 938, 997, 1110, 1421
CD (cyclodextrin) 372
ceilometer 1172
centrifugal compressor 361
CFD (computational fluid dynamic) 188, 1080, 1083, 1093, 1112, 1245
CFK (carbon-fiber-reinforced plastic) 546
CFT (continuous Fourier transform) 1338
chain stretch 18
channel flow 872
charge

– amplifier   602
– coupled device (CCD)   203, 938, 997, 1421
– decay time   1322
– injection   1325
– relaxation time   1319, 1320
chassis dynamometer   1045, 1056
chemical
– kinetics   1245
– reaction mechanism   1245
chevron pattern   1138
chord length   967
chromel   1100
circular cylinder   607, 613, 1132
circulation   430
Clauser
– chart   901
– plot   879, 880
– plot method   880
climate   1045
climatic tunnel   1066
closed test section   1049, 1051, 1061
closed-tube technique   154
cloud cavitation   968
CLSM (confocal laser scanning microscopy)   865
CMC (critical micelle concentration)   99
CMD   1283
CMOS   997
– camera   793
– imaging sensor   1422
coalescence   718
coarse-to-fine technique   1481, 1486
coast-down   1068
coating processes   116
coaxial
– cone and plate   635, 637
coaxial-cylinder
– cell   139, 140
cobra probe   897
coherent anti-Stokes Raman scattering (CARS)   516, 536, 1105, 1243, 1251, 1284
coherent structure (CS)   309, 967, 1346, 1371
– wavelet   1393
Coiflet   1386
coincidence   304
cold wire   274
collapsing bubble   970
collector   1047, 1048, 1050, 1053
– blockage   1061
collisional broadening   529, 530
color filter   998

column balance   1072
combinatorial optimization   1478, 1479
combined
– technique   1000
combustion   1087, 1089, 1105, 1123, 1125
– chamber   1121, 1125
– coal, high-pressure   1297
– control   1292
– deflagrative   1088
– detonative   1088, 1089, 1092
– oxygen–hydrogen   1105
– process, large-scale   1284
– supersonic   1081, 1120–1122
– turbulent   1251
commercial
– fuel   1267
complex
– flow   710
compliance   620, 623, 624, 627, 628, 633, 635
compressibility
– coefficient   4
compressible
– flow   38, 238, 242, 278, 386
compression   1083, 1095, 1113, 1121
– quasi-adiabatic   1091
– ratio   1091
– tube   1091
compressor   919
computational fluid dynamic (CFD)   188, 1080, 1083, 1093, 1112, 1245
concentrated suspension   690
concentration of dissolved gas coarse   965
concentric cylinder   629, 632
condenser microphone   1077
conditional
– averaging   1370
– eddy   1370, 1373
conditionally
– triggered photography   990
conductivity   160, 1317, 1319–1326
– Ohmic   162
– probe   1007
conductivity probes   1008
cone
– angle   900
cone and plate   628, 629, 634, 637–642, 646, 727
– rheometer   637
confidence interval   1401
confocal

– laser scanning microscopy (CLSM)   865
– microscope   824
conservation of mass   8
considère   654
constant
– fraction discriminator   358
constitutive equation   119
constraint release LMSF modelconcentration   23
contact
– angle   106
– angle hysteresis   112
– surface   1082–1086, 1091, 1092
contact-line
– friction   114
– velocity   112
continuity equation   6
continuous
– fluid motion   1371
– Fourier transform (CFT)   1338
– phase   680
continuum
– approximation   216
– hypothesis   683
contraction ratio   1051
controlled-shear-rate (CSR) experiments   696
controlled-shear-stress (CSS) experiments   696
cooling test   1070
corner vane   1064
correction   135
correlation
– coefficient   1400
– depth   315
– function   806, 825
– plane   1469
correlation-based
– tracking   1471, 1474
– velocity analysis   1469
correlator   354
Couette   636, 637, 646, 727, 728
– flow   635, 637
– geometry   636, 638
– instrument   637–639
Couette–Taylor (CT)   801
– flow   801
Coulomb
– force   1319, 1322
– repulsion time   1321
counter   358
counterflow range facility   1082
covariance function   1400
Cox–Merz
– rule   635, 695

Cramér–Rao lower bound (CRLB) 303, 1400, 1408
crank angle degree (CAD) 381
Cranz–Schardin camera 1110
CRDLAS (cavity ring-down laser-absorption spectroscopy) 1249
creaming 718
creep 619–621, 623, 624, 626, 627
crew return vehicle (CRV) 1113
critical
– capillary number 713
– cone angle 900
– micelle concentration (CMC) 99, 1324
– point 122
– pressure 963, 964
– region 139
– shear rate 694
– shear rate, profile 825
– shear stress 703
– state 610
– transition 1145
CRLAS (cavity ring-down laser absorption spectroscopy) 525
CRLB (Cramér–Rao lower bound) 303, 1400, 1408
cross correlogram 354
cross-correlation function 821
cross-flow 891
crossover frequency 720
crosswind
– facility 1070
– gust 1044
– sensitivity 1069
CRV (crew return vehicle) 1113
cryogenic
– Ludwieg tube 605, 613
– tunnel 1067
– wind tunnel 538, 542, 545, 548
CS (coherent structure) 309, 967, 1346, 1371
CSM (cavitation susceptibility meters) 982
CT (Couette–Taylor) 801
CTA (constant-temperature anemometer) 231, 247, 248, 260
CUBRC (Calspan–University at Buffalo Research Center) 1113
Currie approximation 16
curvature 1452
– of the surface 12
curvilinear
– channel 824
– flow 802, 804
cut-off frequency 262

CVA (constant-voltage anemometer) 231, 247, 248, 264
cycle-to-cycle variation 380
cyclodextrin (CD) 372
cylinder - protruding 912
cylindrical
– capillary 635
– hot-film 234, 247, 251

D

damping function 15
– cauchy tensor 15
– Finger tensor 15
– Green tensor 15
– section 1089
dark current 1427
dark signal nonuniformity (DSNU) 1433
DAS (direct absorption spectroscopy) 1291
data density 1404
Daubechies 1386
DBP (di-n-butyl phthalate) 117
Dean flow 821, 825
Deborah number 647, 652, 653
Debye length 687, 704
decorrelation 385
deflection 596
defocusing PIV 1473
deformation 1079
degeneracy factor 519
degenerate four-wave mixing (DFWM) 1243
degree of freedom
– heaving 610
– torsional 611
DEHS (di-ethyl-hexyl-sebacat) 293
denoising 1395
densimeter
– bellows-type 87
– buoyancy-type 92
– single-sinker 90, 92
– tuning-fork 89
– two-sinker 94
– vibrating-cylinder 89
– vibrating-element 88
– vibrating-tube 88, 89
– vibrating-wire 90
densimetric Froude number 37
density 3, 473, 1081, 1083, 1084, 1098, 1106, 1107, 1109, 1112, 1117, 1124
– electron 1095, 1109
– gases 95
– number 1106, 1109

– reference values 95
– steam 90
– water 90
depletion layer 688
depolarization 518
de-pressurized flumes 972
de-pressurized towing tanks 972
depth of penetration 971
depth-integrating sampler 1025
desinent 980
detached shock 216
detail optimization 1046
detection 521, 525
– 2f 527
– efficiency 519, 529, 530
– limit 1247
– low-noise 521
– selectivity 1249
– sensitivity 525, 1251
– spatial resolution 1249, 1259
– temporal resolution 1249
– two-color 535
detectors 793
detonation 1082, 1088, 1089
– driver 1089, 1094
– propagation mode, backward 1088
– propagation mode, downstream 1089
– propagation mode, forward 1088, 1089
– propagation mode, upstream 1089
– section 1088, 1089
Deutsches Institut für Normung (DIN) 126
developed cavitation 970
DFB (distributed feedback)constraint 526
DFG (difference-frequency generation) 1287
DFT (discrete Fourier transform) 1338
DFWM (degenerate four-wave mixing) 1243
DGV (Doppler global velocimetry) 345, 347, 348, 350–352
diagnostics
– flow 1264
– multidimensional 1253
– spray 1265
– three-dimensional 1259
diaphragm 1082, 1085, 1086, 1089, 1091, 1094–1096, 1104
– main 1089, 1091
– primary 1091, 1093, 1095

- rupture 1082–1084, 1091, 1095
- secondary 1085, 1086, 1093–1095
dichroism 725
die swell 720, 723
dielectric constant 158, 159, 167, 168
dielectric force 1319, 1322
Diesel 1265, 1267–1269, 1280, 1282
di-ethyl-hexyl-sebacat (DEHS) 293
difference-frequency generation (DFG) 1287
differential cross-section 519
differential optical flow methods 1465
differential pom-pom equation 20
differential pressure gage 1014
diffractive optical element (DOE) 864
diffuse radiation 1045
diffusion 965
- diaphragm-cell technique 149
- gases 154
- inhomogeneous 1450
- liquids 149
- reference values 156
- taylor-dispersion technique 149
diffusion coefficient 684
- calibration data for gas-phase 156
- hydrocarbons in supercritical CO_2 157
- inter 149
- intra 149
- mutual 149
- organic solutes in n-hexane 157
- organic solutes in toluene 157
- self 149
- tracer 149
diffusion filters 1450
diffusion of the vorticity 384
diffusion tensor 1451
digital PIV 319, 326, 997
digital spatial correlation 376
digitization 1437
dilatation 386
diluted suspensions 690
dimensional analysis 57, 688
dimethylformamide (DMF) 167, 168
dimethylsulfide (DMS) 1213
dimethylsulfoxide (DMSO) 167
DIN (Deutsches Institut für Normung) 126
di-n-butyl phthalate (DBP) 117

dipole moment 517
- induced by the incident electromagnetic field 517
- permanent 524
direct absorption spectroscopy (DAS) 1291
direct numerical simulation (DNS) 251, 423, 765, 774, 791, 1242, 1245, 1252
direct physical observation 1010
direction of flow 637
directional
- derivative 1460
- frequency spectrum 1019
director 725, 727, 728
dirt 1057
discrete Fourier transform (DFT) 1338
discriminator 357
disperse phase 680
dispersion
- ocean tracer 1190
- prism 361
dispersions 680
displacement thickness 888
displacement-correlation peak 322
dissipation rate 400, 407, 411
dissipation rate of turbulent kinetic energy 386
dissociation 1318, 1323
dissociation constant 1323
dissolved gas 965
dissolved oxygen 981
distributed feedback (DFB) 526
distributed suction 1055
distribution 1401
DLR (German Aerospace Center) 1108
DLVO theory 687
DMF (dimethylformamide) 167, 168
DMS (dimethylsulfide) 1213
DMSO (dimethylsulfoxide) 167
DNS (direct numerical simulation) 251, 423, 765, 774, 791, 1242, 1245
DOE (diffractive optical element) 864
Doi–Edwards (DE) model 18
Doppler
- broadened 516, 521
- broadening 1284
- burst 302
- burst frequency 997
- effect 389, 1078
- frequency 297

- frequency shift 361
- fringe spacing 997
- global velocimetry (DGV) 287, 342, 361, 933
- picture velocimetry (DPV) 1105
- shift 342, 343, 350, 351
- shifted frequency 351
- velocimetry 1264
- width 520
Dorodnitsyn–Howarth transformation 63, 77
down force 1045
down-force 1051
drag 1045, 1046
- coefficient 607, 614, 1128
- crisis 1129, 1135, 1145
- flow rheogoniometer 635
- flow rheometer 636
- fluctuations 608
drift
- correction procedure 602
- zero point 597
driven section 1082–1084, 1086, 1087, 1089, 1093
driver gas 1082–1084, 1087–1089, 1091, 1092, 1094, 1095
- argon 1088, 1091
- contamination 1091, 1092
- helium 1087–1091, 1093, 1094, 1123
- hydrogen 1087–1089
- nitrogen 1087, 1090
- oxygen 1087, 1088
driver section 1084, 1086, 1088–1090, 1093
drop volume concentration 715
droplet
- break-up 714
- deformation 713, 714
- distribution 1243
- spreading scaling laws 114, 117
DSNU (dark signal nonuniformity) 1433
DSPIV (dual plane particle image velocimetry) 388
dual plane particle image velocimetry (DSPIV) 388
duplex test section 1060, 1061
dye 860
dynamic
- behavior 603
- contact angle 112, 114
- friction 113
- mechanical measurements 628, 635
- moduli 627, 630–633, 635

- pressure 216
- pressure measurement 179
- range (DR) 1428, 1465
- shear moduli 634
- stall 378
- stall control 611
- viscosity 634
dynamics of the cavitating hydrofoil 1005

E

Earth 1113
- atmosphere 1081
- low Earth orbit 1112
EBCCE (extended brightness change constraint equation) 1486
echo sounder 1011
ecosystems
- ocean 1179
eddy
- diffusivities 1162
- viscosity 77, 831
eddy viscosity 771
edge extraction 1451
edge tracking 1079
edge-detection algorithms 115
effect of orifice shape 181
effective
- Hamaker constant 686
- measuring time 606, 613
- Péclet number 689
- solid volume concentration 690
- wind speed 1062
efficiency 522, 963
EGR (exhaust gas recirculation) 1280
EHD (electrohydro dynamics) 1317–1322, 1325, 1329, 1331
- plume 1331
eigenfrequency 603
eigenvalue problem 1455
eight-constant Oldroyd equation 17
Einstein coefficient 690
Einstein number 1021
ejection 1371
elastic
- deformation 612
- instability 801, 806, 808, 817, 818, 822, 824–826
- stress 799, 813, 817, 820
- stress tensor 820
- stress, nonlinear 801
- stress, turbulent 810
- suspension 610

- turbulence 799, 804, 810, 814–816, 821
- wing model 612
elastically scaled 612
elasticity 720, 723
electric Bond number 1321
electric conductivity 160
electric field 1317–1321, 1323–1326, 1328, 1329
electric fields 1319
electric force 1319, 1321
electric Reynolds number 1321
electrical
- conductivity 158, 168
- double layer 686
- force 1317, 1319, 1321, 1323
electrically conducting fluids 136
electrically insulating fluids 136
electrohydrodynamic 1317, 1329
- mobility 1329
- plume 1325
electrohydrodynamics (EHD) 1317, 1318, 1323, 1327, 1330
electron injection 1320, 1325
electronic noise in anemometers 259, 263, 266, 275
electronic shutter 1419
electron-multiplying CCD (EM-CCD) 1435
electrophoresis 1329
electrophoretic mobility 1329
electrostatic stabilization 687
electrosteric stabilization 688
electrostrictive force 1319
electrothermal flows 1317, 1322, 1323
electroviscous relaxation time 1322
electroviscous scale 1322
elementary reaction 1245, 1249
ELIF (excimer-laser-induced fragmentation) 1297
elongational
- flows 708
- viscosity 651, 652, 657, 658
EM-CCD (electron-multiplying CCD) 1435
emissivity 1163
emulsions 681, 711
EMVA (European Machine Vision Association) 1430
end losses 248, 277
end-pinching 714
energy 7
- mechanical 7
energy tensor 1457

energy transfer 517, 521, 529, 531
- rotational (RET) 529
- vibrational (VET) 529
engine 520, 534, 1244, 1252, 1270
- cooling 1046
- internal combustion 1265
- model 608
- multispecies measurement 1270
- optically accessible 1266
- pulse detonation 1293
ensemble-average 924
enstrophy 11, 400, 401, 407, 411, 774
enstrophy production rate 401, 407
enstrophy-production-rate 411
enthalpy 4, 1081, 1087, 1088, 1093, 1094, 1097, 1111–1113, 1118, 1119
- stagnation 1084, 1085, 1087, 1091, 1115, 1121, 1122
- total 1123
- total specific 1081, 1086–1095, 1099, 1111–1116, 1118–1125
entrance correction 721
entrance effect 720, 721
entropy 4
entry 1090, 1117
- interplanetary atmospheric 1081
- vehicle 1081, 1118
environmental services 1179
epipolar geometry 1469
equation of energy 134
equation of state 4
equations of motion 635, 637, 638, 640, 641
equilibrium 1092
- chemical 1092, 1093
- static 1101
- thermal 1092, 1093, 1117
equilibrium shear compliance 633
equivalence ratio 1273
equivalent temperature difference 1421
erosion rate 974
erosion rates 978
erosive intensity 971
ESA (European Space Agency) 1113
ESTEC (European Space Research and Technology Center) 1122
estimating erosion rate 975
estimator 1400
- consistent 1400
- efficient 1400
- higher order statistics 1404
- mean 1402

– signal-to-noise 1406
– standard deviation 1401
– unbiased 1400
– variance 1403
ETW (European transonic wind tunnel) 199, 548
Euler characteristic 909
Euler equations 1082
Eulerian 651, 652
– correlation function 815
– velocity 798
– viewpoint 5
Euler–Lagrange equations 1484
European Machine Vision Association (EMVA) 1430
European Space Agency (ESA) 1113
European Space Research and Technology Center (ESTEC) 1122
European transonic wind tunnel (ETW) 199, 548
evaporation 697
event data vector 1373
excimer-laser-induced fragmentation (ELIF) 1297
exciter
– electrodynamic 610
– force 610
exhaust gas recirculation (EGR) 1280
expansion 1081, 1083, 1084, 1092, 1093
– operator 1463
– steady 1086, 1094
– Taylor 1089
– unsteady 1083, 1086, 1094, 1095
– wave, centered 1082, 1084
– wave, reflected 1084–1086, 1092, 1094, 1095
– wave, secondary 1095
– wave, unsteady 1083, 1084, 1086, 1094
exposure time 1419
extended brightness change constraint equation (EBCCE) 1486
extended pom-pom (XPP) model 21
extensional
– flows 727, 729
– thickening 651
exterior noise 1076
extraction curve 1074
extrudate swell 720
extrudate swell ratio 721–723

F

far infrared (FIR) 502, 1285
Faraday cells 727
FARLIF 1270
Farris effect 692
fast Fourier transform (FFT) 306, 327
fast shutter valve 547
fault current 606
FBRM (focused beam reflectance measurement) 1024
feature image 1445
feature-based method 1465
feedback 610
Fermat's principle 474
FFT (fast Fourier transform) 306, 327
fiber optic backscatter 1026
fibre 357
Fick's law 148
Fick´s diffusion law 10
field-enhanced dissociation 1325
filament
– line 860
– stretching 647, 650–656, 658, 659
– stretching rheometer 653, 656, 658
film 997
filter
– median 1448
filtered Rayleigh scattering 345
filtering
– wavelet 1393
FIR (far infrared) 502, 1285
first and second normal stress coefficients 640
first normal stress coefficient 642
first normal stress difference 639, 642
first-order derivatives 1451
Fisher information matrix 1409
five-belt system 1063
fixed pattern noise (FPN) 1433
fixed region 8
Fizeau interferometry 876
flame 519
– atmospheric pressure 525
– Bunsen 536, 1256, 1261
– chemiluminescence 1263
– extinction 1247
– finite-rate chemistry effect 1247
– front 1255, 1259, 1260
– higher-pressure 1251
– high-pressure 532
– kernel 1261, 1263
– laminar 525, 1245, 1249
– low-pressure 530, 532, 1249
– opposed jet 1247, 1253
– sooting 520, 532, 1250
– swirl 1257
– turbulent 519, 1246, 1253, 1255
flamelet 1245, 1255
floating frame 1071
floating-element sensor 883
float-mounted accelerometers 1013
flocculation 718
flocculation factor 716
floor balance 1071
Florida Strait transport 1200
flow
– complex 710, 898
– noise 1076
– optical 1464
– random 819
– resistance 800, 806, 807, 816, 824
– seeding 350, 352
– separation 897
– separation onset 609
– structure 825
– tagging 1264
– tagging velocimetry 363
– turbulent 382
– visualization 857, 1074
– work 9
fluctuating pressure 614
fluctuating velocity field 1373
fluid 3
fluorescein-conjugated dextran 824
fluorescence 364
– lifetime 529
– quantum yield 528, 532, 535, 1271, 1276
fluorescent dye 290, 865, 998
fluorescent-conjugated dextran (FITCD) 829
flutter
– boundary 610
– control system 610
– oscillations 609
– torsional 611
flying hot wire 279, 900
FMS (frequency modulation spectroscopy) 1291
focused beam reflectance measurement (FBRM) 1024
force
– aerodynamic 609
– balance 994
– cutting 597
– fluctuating 596
– inertia 609

- measurement, free-flight technique 1103
- spectrum 609
- steady 596
- time averaged 596
- transducer 599
- unsteady 609
force balance 1101, 1103, 1114
- dynamic response 1102
- stiffness 1101
- strain gauge 1101, 1102, 1104, 1115
- stress wave 1101–1104, 1121
forced
- dewetting 112
- motion 610
- Rayleigh scattering 142, 143
- torsional oscillations 611
- wetting 112, 116, 117
forced vibrations 630
- nonresonance 630
- resonance 630
Fourier heat conduction 10
Fourier transform 1378
- discrete 1337
- fast (FFT) 1342
- zero padding 1342
Fourier's law 133
Fourier-transform rheology (FTR) 707
four-sensor conductivity probe 1008
four-sensor optical probe 1008
FPN (fixed pattern noise) 1433
fpolymethylmethacrylate (PMMA) 700
fractal 1395
fractional population 523, 524
free
- damped vibrations 630
- diffusion 151
- mixing layers 67, 73
- shear layer 1125, 1128
- surface flow 868
- surface slope 1013
- vibrations 630, 631
free-flight force measurement 1101
free-stream disturbances 891
free-surface flows 80
French Atomic Energy Commission 1008
French–German Research Institute of Saint Louis (ISL) 1105
frequency
- bending 602
- buffet 597
- natural 1101, 1104, 1105
- nondimensional buffet 612
- Nyquist 1339
- shift 609
frequency flat field 348, 352
frequency modulation spectroscopy (FMS) 1291
frequency of vortex shedding 966
frequency response 262, 265, 884
frequency-based techniques 1484
frequency-stabilized argon-ion laser 361
friction velocity 873, 1161
fringe model 297
fringe spacing 359
fringe-type laser velocimetry 342, 346, 351, 353
frontal area 1048, 1079
Froude number 37, 972
FTR (Fourier-transform rheology) 707
fuel
- commercial 1267
- liquid 1246
fuel–air ratio 1246, 1266
fundamental differential equation of thermodynamics 4
furnace 1244
fuzzy slotting technique 307

G

gabor filter 1461, 1484
gamma densitometers 1009
gas
- content effects 987
- content measurement 980
- dynamics 57, 58, 79
- residence-time 1298
- turbine 536, 1244, 1252, 1259
gaseous cavitation 965
Gauss theorem 8
Gaussian
- beam 300, 1008
- curve 356
- differentiating kernel 795
- filter 831
- image pyramid 1467
- pyramid 1462
GBCCE (generalized brightness change constraint equation) 1487
generalized brightness change constraint equation (GBCCE) 1487
geometric shape factor 628, 629
geometric transforms 1442

geometrical quantity 1078
geosynchronous imaging Fourier-transform spectrometer (GIFTS) 1177
German Aerospace Center (DLR) 1108
Germano identity 831
GIFTS (geosynchronous imaging Fourier-transform spectrometer) 1177
Gladstone–Dale constant 1106
Gladstone–Dale equation 473
glass compliance 633
global constraints 1483
global Doppler velocimetry 362
global positioning system (GPS) 1194
Godunov 1083
Gouy interferometer 153
GPS (global positioning system) signals 1018
grade of dispersity 681
gradient 1451
Grashoff number 236
grayscale interpretation of PIV images 1008
greedy-matching 791
ground board 1045
ground floor 1049, 1055
gustiness 1044

H

Haar 1386
Hagen–Poiseuille equation 125
hairpin vortex 311
half-span model 599
Hamaker constant 685
hard spheres 685
harmonic motion 517
head loss 10
heat
- radiation 544
- transfer 57, 62, 543, 544
- transfer coefficient 545
- transfer laws in subsonic flows 236
- transfer laws in supersonic flows 239
- transfer measurement 538
heat balance 230, 246
heat conduction 544, 1098, 1099
- one-dimensional 1098, 1100
- unsteady 1100
heat conductivity 1098
heat convection 544

heat exchanger 1054
heat flux 7, 1081, 1096–1100, 1103, 1104, 1112–1114, 1116–1120, 1124
– plate 1164
heave motion 610
heave-pitch-roll buoy 1014
helical mode 967
Helley–Smith bed load sampler 1030
HEM (horizontal electrometer) 1203
hemispherical
– nosed body 961
Hencky strain 647, 651–655, 658
Henri de Pitot 218
Henry's constant 965
Hessian matrix 1452
heterodisperse 681
HIEST (high enthalpy shock tunnel) 1115
high enthalpy shock tunnel (HEG) 1108
high enthalpy shock tunnel (HIEST) 1115
high frequency backscatter 1009
high image density PIV 320
high Reynolds number 797
– facility 607
high speed
– flow visualization 1111
– water tunnel 972
high temperature effect 1081, 1087, 1113, 1118, 1123
high turbulence intensity flow 899
high-frame-rate PIV 1003
high-pressure wind tunnel 607
high-repetition rate 1265
high-Reynolds-number turbulence 792
high-speed
– imaging 116
– photography and video 988
– video 969
high-temperature effect 1092
– dissociation 1081, 1112, 1118
– vibrational excitation 1081, 1118
Hilbert
– filter 1458
– transform 1344, 1458
histogram equalization 319
HITRAN 527
hole effect 642

holographic
– interferometry 138
– particle image velocimetry (HPIV) 388, 844, 902, 1473
– techniques 982
holography 864
– holographic plate 1108
– reconstruction 1108, 1109
– recording 1101, 1107, 1111
hoop stress 803, 812
Hopf bifurcation 1137
horizontal buoyancy 1060, 1061
horizontal electrometer (HEM) 1203
Horseshoe vortices 969
hot wire flying 1144
hot wire on the wall 884
hot-film anemometry 1009
hot-wire 894, 921
hot-wire anemometry (HWA) 309, 748, 899
HPR (heave, pitch, roll) buoy 1019
HTV (hydroxyl tagging velocimetry) 365, 369
humidity 1045
HWA (hot-wire anemometry) 309, 748, 899
hybrid PIV/PTV 1472
hydration 716
hydraulic
– analogy 869
– linear actuator 610
hydraulically smooth 882
hydrocarbons 88, 122
hydrodynamic
– loads 965
– mobility 1322
– theory 113
hydroelastic vibration 966
hydrofoil 962, 965, 968, 986, 1000, 1002
hydrogen-bubble technique 866
hydrophone 981, 992, 996
hydroturbine 976
hydroxide (OH) 1105
hydroxyl tagging velocimetry (HTV) 365, 369
hypersonic 1085, 1088, 1111–1113, 1118, 1123
– boundary layer 1124
– flight 1081, 1116
– flow 1081, 1085, 1094, 1095, 1099, 1105, 1113, 1118, 1123
– ground-based test facility 1081, 1095, 1097, 1118
– short-duration test facility 1105

– testing 1081, 1096, 1115, 1118
– vehicle 1081, 1101
hypervelocity 1081, 1110, 1112, 1113, 1118, 1119
– range facility 1082
hysteresis 106, 108, 111, 610, 807

I

IAPWS (International Association for the Properties of Water and Steam) 95
IC (internal combustion) 1245
ICCD (intensified CCD sensors) 1435
ICET (international cavitation erosion test) 977
ideal gas 4
ideal model 135
IEP (isoelectric point) 695
IES (inverted echosounder) 1198, 1203
IGV (inlet guide vanes) 930
illumination
– inhomogeneous 1441
image correlation velocimetry 1474
image distortion 351
image processing 116
impact of a weight 604
impulse response function 1102
inception 967, 980, 996
– mechanism 967
– of cavitation 982
– pressure 967
incipient cavitation number 963
incipient separation 897
incompressible
– flow 11, 34
– fluid 4
in-cylinder flow 380
index of refraction 1006
inductance effects 261
inertial matrix 1101
infrared 1163
infrared thermography (IRT) 539, 551, 896
inhomogeneity of the turbulence intensities 385
initial amplitude 610
initial condition (IC) 746
injected bubbles 967
injection 1318, 1323, 1325, 1326, 1328
ink 650, 656
inkjet printing 117
inlet guide vanes (IGV) 930

inner (viscous) length 404
inner (viscous) scale 402
inner flow 1047
inner frequency 875
in-plane loss of pairs 1470
in-plane SGS stresses 837
instability 610, 886
– elastic 801, 806, 808, 817, 822, 825, 826
instabilityelastic 818, 824
instantaneous wall gradient methods 884
insulation resistance 602
integral time scale 799
integrated experimental/numerical research on cavitating flows 1004
integrating sphere 1441
intense turbulence 793
intensified CCD sensors (ICCD) 1435
intensity 517
– flat field 349, 351, 352
– fluorescence 529
– LIF 528
– of anti-Stokes Raman scattering 517
– of Rayleigh scattering 517
– of Stokes Raman scattering 517
interaction 609
interface condition
– overtailored 1084
– tailored 1084, 1086, 1090
– undertailored 1084
interfacial
– area concentration 1007
– rheology 714
– shear viscosity 715
– tension 96, 99–101, 103
interference 602, 1278
– fringe 480
interferogram 480
interferometer 480
– differential 481
– holographic 481
– Mach–Zehnder (MZI) 481
– phase-stepping 484
– reference beam 480
– schlieren 480
– shearing 480, 483
interferometric particle imaging technique 1246
interferometric technique 137

interferometry 474, 1095, 1101, 1106
– Fabry–Pérot 1096
– holographic 1106–1109, 1112, 1118
– holographic, two wavelength 1109
– Mach–Zehnder 1106
– Michelson 1105
interior noise 1076
intermittency measures 1390
internal
– energy 4
– flow 878
– gravity wave 484
– piezo balance 600
internal combustion (IC) 1245
– engine 379
International Association for the Properties of Water and Steam (IAPWS) 95
international cavitation erosion test (ICET) 977
International Organization for Standards (ISO) 126
International Space Station (ISS) 1113
international towing tank conference (ITTC) 978
interpolation 1443
interpolation kernel 1438
interpolation of sampled points 1438
interrogation areas 1465
intersystem crossing (ISC) 364
intracavity etalon 361
inverse micelles 1324
inverted echosounder (IES) 1198, 1203
iodine absorption line 347
iodine cell 361
iodine vapor cell (IVC) 343, 345, 347–349, 351
ion injection 1317
ion mobility 166
ionic transit time 1319
ionization 528, 529
IRT (infrared thermography) 539, 551, 896
ISC (intersystem crossing) 364
ISL (French–German Research Institute of Saint Louis) 1105
ISO (International Organization for Standards) 126
isoelectric point (IEP) 695
isokinetic sampling 1024

ISS (International Space Station) 1113
iterative analysis 335
ITTC (international towing tank conference) 978
IVC (iodine vapor cell) 343, 345, 347–349, 351

J

jackknife algorithm 1402, 1405
Japan Aerospace Exploration Agency (JAXA) 1115
Japanese Industrial Standards (JIS) 126
Jason 1202
JAXA (Japan Aerospace Exploration Agency) 1115
jet deformation 1061
jet expansion 1061
jet propagating along a cone 71
JFTA (joint frequency–time analysis) 969
JIS (Japanese Industrial Standards) 126
joint frequency–time analysis (JFTA) 969

K

Kalman filter 1480
Karman vortex street 614, 863
Karman–Corrsin–Pao spectra 276
Karman–Saffman–Pao spectra 253
K-BKZ equation 15
Kerr effect 1317, 1328
kinematic similarity 1048
kinetic energy 7, 1081, 1087, 1089
kinetic-energy dissipation rate 401, 407, 411
King's law 237
Kirchhoff 1242
KLT-tracking 1474
Knudsen number 12, 216, 239
Kolmogorov
– equation 770
– length scale 793
– scale 404, 797, 874, 1249
– time scale 793
Kovasznay modes 245
Kramers law 236

L

L2F (laser two-focus velocimetry) 353, 921, 945

laboratory-generated wind waves 1013
lag time 307
Lagrangian 5
– experiment 789, 792
– particle tracking (LPT) 789
– time scale 793
– trajectory 789, 817
– turbulence 795
Lambert–Beer 523, 524, 1287
– law 1290
laminar boundary layer 545
laminar shear flow 806
laminar–turbulent transition 543
Langevin function
– inverse 730
lanthanide ion 371
LAOS (large-amplitude oscillatory shear) 707
Laplacian operator 1452
Laplacian pyramid 1463
large cavitation channel 972
large eddy simulation (LES) 423, 773, 831, 1243, 1245
large energy national shock (LENS) 1113
large full-scale vehicle wind tunnel 1063
large fullsize climatic wind tunnel 1067
large-amplitude oscillatory shear (LAOS) 707
large-scale flow 793
LAS (laser absorption spectroscopy) 523, 1249, 1285
laser 516, 520, 1082, 1101, 1105, 1110, 1111, 1125, 1242, 1264
– cluster 1261
– CO_2 1287
– coherence length 1108
– diode 1110
– distributed-Bragg reflector (DBR) 1292
– distributed-feedback (DFB) 1292
– dye 536, 1256, 1261
– excimer 1255, 1256
– external cavity diode 1289
– fiber-coupled 526
– frequency doubled 1261
– frequency monitor (LFM) 343
– frequency-doubled 1254, 1265
– frequency-tripled 1256, 1264
– HeNe 1287
– lead salt 1288
– longitudinal mode 520

– Nd:YAG 522, 1108, 1111, 1254, 1261, 1264
– near infrared diode (NIR-DL) 1288
– OPO 1264
– pulse width 1111
– pulsed 522
– quantum-cascade 1289
– quantum-cascade diode 1287
– semiconductor diode 1287
– solid state 1287
– telecommunication 1295
– transit velocimetry (LTV) 353
– tunable diode 516, 1288
– vertical cavity surface emitting 1298
– vertical-cavity surface-emitting 1289
– wavelength 1106, 1109
laser absorption spectroscopy (LAS) 523, 1249, 1285
– intra cavity (ICLAS) 525
– ring-down (CRLAS) 525
laser attenuation 1276
laser diodes 357
laser Doppler (LD) method 353
laser Doppler anemometry (LDA) 287, 389, 698, 748, 898, 997, 1073
laser Doppler velocimeter (LDV) 805
laser Doppler velocimetry (LDV) 287, 882, 921, 933, 997, 1004, 1243
– seed material 1243
laser in situ scattering and transmissometry (LISST) 1024
laser light sheet 342, 345, 348, 351, 997
laser point measurements 1013
laser slope gage 1012
laser sources 1286
laser speckle strophometry (LSS) 388
laser transit
– anemometer 346
– anemometry 342, 345
– velocimetry (LTV) 287, 354
laser transit (LT) 353
laser triangulator 612
laser two-focus velocimetry (L2F) 353, 921, 945
laser vibrometer 610
laser wave height gage 1012

laser-based combustion diagnostics 1243
– in situ 1243
– non-intrusive 1243
laser-based wave staff 1012
laser-enhanced ionization (LEI) 370
laser-induced fluorescence (LIF) 312, 516, 521, 528, 751, 848, 865, 998, 1014, 1105, 1126, 1138, 1243
– multi-line thermometry 516
– single-line thermometry 516
– tracer 516
– two-line thermometry 516
laser-induced incandescence (LII) 1243, 1281
laser-induced photochemical anemometry (LIPA) 363, 365
Laskin nozzle 293
latent heat 1164
lateral deviation 1069
law of similarity 1058
law of the wall 879
LCO (limit-cycle oscillation) 611, 613
LCT (liquid-crystal thermography) 551
LDA 296
LDA (laser Doppler anemometry) 287, 389, 698, 898, 997, 1073
LDPE (low-density polyethylene) 653
LDV (laser Doppler velocimetry) 287, 296, 882, 921, 933, 997, 1004, 1135, 1243
– data rate 1004
– measurements with cavitation 1004
– techniques 885
leading-edge separation 378
leaks 1073
least-squares
– matching 1465, 1473
– tracking 1474
LED (light-emitting diode) 1110
LEI (laser-enhanced ionization) 370
Leibnitz theorem 8
LENS (large energy national shock) 1113
LES (large eddy simulation) 423, 773, 831, 1243, 1245
LFM 347
lidar 1170

LIF (laser-induced fluorescence) 312, 516, 521, 528, 529, 751, 848, 865, 998, 1014, 1105, 1126, 1138, 1243, 1259
– model nontransient 529
– model transient 529
– thermometry multi-line 529, 530
– thermometry single-line 529, 530
– thermometry two-line 529, 530
lift 1045
– and drag 992
– and drag coefficient 963
– coefficient 611, 965, 1135
– curve slope 612
– drag measurements 992
– fluctuations 608, 614
– oscillations 966
light extinction technique 1009
light scattering 858
light sheet 864, 1015
light source 1110, 1111
– coherent 1110
– continuous 1110
– incoherent 1110
– intermittent 1110
– narrowband 1110
– semiconductor 1110
light-emitting diode (LED) 1110
LII (laser-induced incandescence) 1243, 1281
likelihood function 1480
limit-cycle oscillation (LCO) 611, 613
limiting cases 51
limiting wavenumber 1438
Lin equation 771
line strength 524
linear convolution 1457
linear mean-square estimation 1374
linear momentum and related equations 6
linear stability 886
linear stochastic estimation (LSE) 1375
linear symmetry 1453
linear variable displacement transducer (LVDT) 1133
line-of-sight
– method 474, 485, 858
– multipath 516
line-shape 524, 529, 531
– absorption 527
– function 524
– Gaussian 524

– Lorentzian 524
– Voigt 524
LIPA (laser-induced photochemical anemometry) 363, 365
liquid crystalline 728
liquid drop spreading 115
liquid flows 238, 269
liquid-crystal thermography (LCT) 551
liquid-crystalline fluid 724
liquid–gas interface 1008
LISST (laser in situ scattering and transmissometry) 1024
load cells 992, 994
local
– intermittency measure (LIM) 1391
– normalization 307
– phase 1458
– wavenumber 1458
– weighted least squares 1483
– Weissenberg number 819
Lodge 14
logarithmic layer 873
logicalAND 1440
longitudinal bulk modulus 633
longitudinal wave 632
long-wave infrared (LWIR) 1431
long-wavelength infrared band (LWIR) 502
Loschmidt cell 154
loss 523
– angle 620, 631
– modulus 620, 626, 631, 634, 635
– tangent 713
low image density PIV 320
low noise fan 1054
low-density polyethylene (LDPE) 653
lower convective Maxwell model 17
low-image-density PIV 320
low-noise 523
LSE (linear stochastic estimation) 1375
LSS (laser speckle strophometry) 388
LT (laser transit) 353
Ludwieg tube 605, 613
luminescence 541
LVDT (linear variable displacement transducer) 1133
LWIR (long-wave infrared) 1431
LWIR (long-wavelength infrared band) 502

M

Mach number 39, 236, 239, 243, 473, 612, 613, 1046, 1058, 1081, 1085, 1087, 1092, 1094, 1116, 1118, 1120–1122
– binomial expansion 216
– shock 1087–1092, 1094
Mach–Zehnder 1107
macrochannel 822
macroform 731, 732
macrolens 314
magnetic resonance imaging (MRI) 866, 902
magnetic susceptibility 158, 159
manual sampling 1024
marine propulsion 961
MARLIN 1205
mass averaged velocity 4
material derivative 6
material region 8
maximization problem 1455
maximum likelihood (ML) 1408
maximum packing fraction 691
maximum penetration depth 974
maximum-likelihood estimator (MLE) 1415
Maxwell model 16, 619, 621–627
Maxwell–Wiechert (MW) model 622, 626, 627
MC (methylene chloride) 168
MDA (minimum detectable absorption) 1247
mean curvature 1452
mean flow velocity 354
mean momentum equation 873
mean velocity 355
measure of confidence 1468
measurement 516
– 3D 1260
– concentration 1253, 1257, 1258
– effective Rayleigh scattering cross-section 516
– fuel distribution 1259
– heat-release 1255
– high-repetition-rate imaging 1244, 1253, 1259, 1261
– imaging 516
– line-of-sight 1244
– multispecies 1253
– one-dimensional 1244
– point 1244, 1252
– pointwise 516
– simultaneous 1243, 1253, 1255, 1257, 1258, 1265

– temperature 1253, 1255, 1257–1259
– two-dimensional 1244
– viscosity 124
measuring deflections 601
mechanical decoupling 601
median test 339
MEH-PPV [poly(2-methoxy, 5-(2'-ethyl-hexoxy)-1, 4-phenylene-vinylene)] 162
melt flow index (MFI) 721
memory function 14
MEMS 235, 236, 251, 271
MEMS (microelectromachined sensors) 233
MEMS sensors 884
mercury 122
M-estimator 1487
methylene chloride (MC) 168
Mexican-hat 1381
MFI (melt flow index) 721
MHT (multiple hypothesis tracker) 1480
micellar solutions 713
microchannel 822, 827
microelectromachined sensors (MEMS) 233
microemulsions 711
microfluidic device 822
microform 731, 732
microjet 971
microphone array 1078
microscopic particle image velocimetry 823
micro-vortices 967
mid-wavelength infrared band (MWIR) 502
Mie parameter 290
Mie scattering 289
Mie-scattering 998
miniemulsions 711
minimum acceleration constraint 1478
minimum detectable absorption (MDA) 1247
minimum quadratic differences 1474
MIR 1285
mirror image 1045
misalignment errors 883
mixed layer float 1206
mixed metal hydroxide (MMH) 676
mixture formation 1266
mixture fraction 1252, 1254
ML (maximum likelihood) 1408

MLE (maximum-likelihood estimator) 1415
MMA (monomethylacetamide) 167
MMH (mixed metal hydroxide) 676
MMT (monomethyltryptamine) 167
mobility 1317, 1319–1323, 1325–1329
mode A instability 1139
mode B instability 1139
model
– engine 600
– rigid 612
– testing 973
– track 1068
– wall mounted 597
moderately concentrated suspensions 693
modern instruments 136
modulation 297
modulus
– complex 621
moiré 474, 478
– deflection mapping 478
– effect 1437
– fringes 1016
– Ronchi method 478
molar averaged velocity 4
molecular dynamics 119
– interpretation in the continuum limit 1228
– principle 1226
– results of slip investigations 1227
molecular filter 342–344, 347
– technology 343, 345
– velocimetry 350, 352, 353
molecular mean free path 216
molecular stress function (MSF) model 22
molecular tagging velocimetry (MTV) 362, 882, 885, 1243
molecular weight 1082, 1087, 1093
molecular-kinetic theory 113
molten salts 122
moment fluctuations 611
moment of inertia 805
moment of momentum 7
monodisperse 681
monogenic signal 1459, 1460
monomethylacetamide (MMA) 167
monomethyltryptamine (MMT) 167
monostatic 1173
Monte Carlo method 1245
montmorillonite (MT) 676
Morlet 1382
most stable apparent contact angle (MSACA) 111

motion blur 1476
motion correspondence 1477
motored IC engine 380
moving belt 1055
moving boundary work 9
moving floor 1065
MRA (multiresolution analysis) 1384
MRI (magnetic resonance imaging) 866, 902
MSACA (most stable apparent contact angle) 111
MT (montmorillonite) 676
MTV (molecular tagging velocimetry) 362, 882, 885, 1243
multichannel analyzer 358
multiframe tracking 1475, 1478
multigrid analysis 335
multigrid window-shift technique 333
multihole pressure probe 897
multihole probe 1072
multihole probes
– asymmetric shocks 229
– calibrated modes 228
– calibration 229
– conical 228
– five-hole probe 229
– flow direction 228
– flow separation 229
– fluctuation intensity 229
– hemispherical wedge 228
– instantaneous yaw 229
– limitations on these techniques 229
– null reading 228
– nulling method 229
– seven-hole probe 229
– three-hole Cobra probe 228
multiline temperature imaging 1278
multimode differential pom-pom model 20
multimode pom-pom model 20
multipass cell 1287
– Herriott 1287
– White 1287
multiplane stereoscopic PIV 1473
multiple hypothesis tracker (MHT) 1480
multiple overheat method 274, 277
multiple-pass interrogation 1471
multiple-timestep imaging 1265
multiresolution analysis (MRA) 1384
multiscalar 521
multiscale approach 1486

multispecies measurement   1270
multistep averaging   1447
multiview imaging   1468
multiwavelength illumination   999
MWIR (mid-wavelength infrared band)   502

N

NACA 0012 airfoil   610
n-alcohols   93
n-alkanes   91
narrow belt   1056
NASA (National Aeronautics and Space Administration)   1113
National Aeronautics and Space Administration (NASA)   1113
natural convection   57, 58, 77, 134, 1330, 1331
Navier–Stokes equation   11, 57, 59, 61, 1242
Nd:YAG   998
near infrared (NIR)   502, 1285
nearest-neighbor matching   320
nebulizer   293
NEE (noise-equivalent exposure)   1427
net positive suction head   966
neutral stability   887
Newton   1242
Newton's law   1101
Newtonian
– fluid   800
Newtonian fluid   10, 119, 813
Newtonian solvent   800
NIR (near infrared)   502, 1285
nitric oxide (NO)   531, 1105, 1243, 1251, 1252, 1257, 1258, 1264, 1275, 1277
nitrogen   132, 1093, 1119
NMR (nuclear magnetic resonance)   698
NO (nitric oxide)   531, 1105, 1243, 1251, 1252, 1257, 1258, 1264, 1275, 1277
noble gases   146
node   909
noise   960
– signal   1406
noise variance equalization   1441
noise-equivalent exposure (NEE)   1427
noncanonical flows   896
noncondensable gas   965

nonequilibrium   1092, 1093, 1095, 1115, 1117
– chemical   1093, 1112, 1115
– dissociation   1115, 1117
– thermal   1093, 1117
– thermochemical   1119, 1120
nonintrusive   523, 537
nonintrusive measurements   898
nonisentropic compression   216
nonlinear
– effects   608
– optical technique   1242
– point operations   1457
– regression   356
nonlinearity (dynamic)   248, 263
nonlinearity (static)   268
non-ohmic regime   1324, 1325
nonresonance   630
nonresonant   516, 632
normal
– stress   636, 639, 641, 642, 645
– stress difference   635, 636, 705, 800
– stress differences/coefficients   646
normalized convolution   1448, 1468
normalized erosion rate   971
no-slip   *see* slip
nozzle   1047, 1048, 1050, 1051, 1065, 1084–1086, 1089, 1091–1093, 1112, 1119, 1121
– area ratio   1092
– blockage   1061
– convergent–divergent   1081, 1086, 1090, 1092, 1094
– design   1092
– divergent   1085, 1086, 1094
– expansion   1085, 1093
– factor   1052, 1062
– hypersonic   1081, 1086, 1092
– method   1052
– reservoir   1086, 1089, 1092, 1093
– reservoir pressure   1081, 1088–1091, 1124
– reservoir temperature   1081, 1093
– starting process   1086, 1092
– throat   1093
– wall boundary layer   1092
nuclear magnetic resonance (NMR)   698
nuclei   964, 979
– content   981, 982
– sizes   985
null space   1455
number density   519
numerical   136
– modeling   1004

numerical simulation   1005, 1244, 1252
– boundary condition   1245, 1247
– chemistry-turbulence interaction   1252
– inlet condition   1245
– submodel   1245
Nusselt number   236, 239, 247
Nyquist criterion   1340, 1466
Nyquist wavenumber   1438

O

oblique shedding   1137
OBS (optical backscatter)   1024
observation of elastic turbulence   824
occurrence of cavitation   962
ocean
– biofouling   1181
– climate   1197
– conductivity–temperature–density (CTD)   1185
– density spiking   1186
– earth rotation   1180
– geostrophy   1180
– internal waves   1180
– mesoscale weather   1190
– mooring   1186
– Rossby waves   1180
– salinity spiking   1184
– sea-floor cable systems   1181
– stratification   1180
ocean acoustic tomography
– adjoint methods   1199
– data-assimilating model   1199
– inversion   1198
– Kalman filtering   1199
– moving ship tomography (AMODE-MST)   1198
ocean glider   1207
ocean sensing
– electric field   1199
ocean sensors
– microstructure sensors   1205
– shear probe   1205
ODE (ordinary differential equation)   57
OH   1243, 1247, 1251, 1252, 1255, 1257, 1258, 1260, 1270, 1278
OH (hydroxide)   1105
ohmic   1324, 1325
– conductivity   161
– regime   1320, 1324, 1326
Ohnesorge number   1321

oil-film
- fringes 877
- interferometry (OFI) 876, 901
- visualization 964
Oldroyd-B model 800
one-point moment 1373
open test section 1049, 1050, 1052, 1061
OPG (optical parametric generation) 1287
OPO (optical parametric oscillator) 525, 1276
optical access 1245, 1247, 1263
optical backscatter (OBS) 1024
optical fiber probes 1008
optical flow visualization 473
optical measurement technique 1105, 1113
- coherent anti-Stokes Raman spectroscopy 1105
- colour schlieren 1123
- Doppler picture velocimetry 1105
- electron-beam fluorescence 1105
- flow tagging 1105
- high speed flow visualization 1109, 1110
- infrared thermography 1123
- laser-induced fluorescence 1105
- planar laser-induced fluorescence 1105
- pressure sensitive paint 1105
- schlieren 1109–1111, 1116
- schlieren-interferometer 1110
- shadowgraph 1109, 1110, 1118
- tunable diode laser absorption 1105
optical methods 385
optical parametric generation (OPG) 1287
optical parametric oscillator (OPO) 525, 1276
optical particle tracking 789, 792
optical paths 988
optical polarizability 729
optical probes 1008
optical technique 385
optical technique for nuclei measurement 982
optical wave-height method 1016
optimizing 1080
optimum radius 1058, 1059
ordinary differential equation (ODE) 57
organic molecules 532
orientation 1453, 1455
- vector 1455

orthogonality principle 1374, 1375
oscillating disk 976
oscillating half model 612
oscillating horn 974
oscillating hot-wire 902
oscillatory
- boundary-layer thickness 120
oscillatory behavior 969
Ostwald ripening 718
OTV (oxygen tagging velocimetry) 365
outer flow 1047
out-of-plane loss of pairs 1470
overflow 1440
overheat parameter 230, 245, 274
overheat stepping 244
oversampling 1439
overtailored 1085
oxygen 1092, 1093
oxygen tagging velocimetry (OTV) 365

P

PAA (polyacrylic acid) 168
PACA (practical advancing contact angle) 109
PAH (polycyclic aromatic hydrocarbon) 1249
paint coating erosion tests 978
palenstrophy coefficient 773
palladium porphyrins (PdOEP) 191
panel method 1062
Pao spectra 253
Papanastasiou 16
parallel plate
- infinitely extended 636, 637
parallel plates 629, 636, 637, 639, 640, 646
parallel shedding 1137, 1139
parallel wires 254
parallel-plate instrument 140
parameter
- estimation 1408
- estimator 1408
- expectation 1408
parameterization of 2-D optical-flow fields 1486
Parseval theorem 1378
Parseval wavelet 1380
partial cavitation 971
partial differential equation (PDE) 57
partial pressure 980

particle 681, 1281
- identification 790
- image 789, 790, 997
- image intensity 999
- image size 999
- image velocimetry (PIV) 489
- migration 682, 697
- path 5, 860
- position vector 5
- response time 289
- Reynolds number 683, 689
- segmentation 1475
- separation 797
- size 1243, 1281
- size distribution (PSD) 681
- tracking 789
- tracking algorithm 799
- tracking experiment 792
particle image thermometry (PIT) 492
Particle image velocimetry (PIV) 915
particle image velocimetry (PIV) 287, 309, 342, 345, 346, 352, 353, 362, 388, 487, 492, 497, 805, 835, 902, 921, 933, 997, 998, 1073, 1126, 1131, 1243, 1465
particle image velocimetry technique (PIV) 749
particle track 791
particle tracking system 794
particle tracking velocimetry (PTV) 320, 751, 805, 837, 1465, 1474
particle trajectories 789
particle trajectory 799
particle-finding algorithm 790
partition function 519, 529, 530
Pascal's triangle 1446
patchy cavitation 968
path-averaging 1286
paucidisperse 681
PCI (peripheral component interconnect) 494
PCL (polycaprolactone) 168
PDA receiver angle 1006
PDE (partial differential equation) 57
PDI (polydispersity index) 681
PDMS (polydimethylsiloxane) 700
PdOEP (palladium porphyrins) 191
PDR method 273
PDV (planar Doppler velocimetry) 948
peak detectability 339
peak-locking 1470
Peclet number 689, 783, 829

Subject Index

pedestal amplitude 1004
Peltier–Seebeck effect 1097
Pelton-wheel-type hydroturbines 977
perfect gas 4, 1112
– calorically 1082, 1087
– thermally 1082, 1087
performance degradation 963
performance of turbomachinery 966
peripheral component interconnect (PCI) 494
permeability 1318
permittivity 1317–1319, 1322
PET (poly(ethyleneterephthalate)) 117
PETW (pilot facility of ETW) 548
pH indicator 861
PHANTOMM (photoactivated non-intrusive tracing of molecular motion) 365
phase 1458
– angle 628, 631
– average 924
– discrimination 1004
– Doppler anemometry (PDA) 982, 997, 1005, 1246
– indicator 1007
– inversion 711
– inversion temperature (PIT) 717
– locked PIV 1002
– matching 536
– resolved velocity measurement 359
– separation 998, 999
– separation techniques 999
– shift 611, 1006, 1106, 1107, 1109, 1112, 1113
– step holographic interferometry 1106
– wrapping 1459
phosphorescence 364, 539, 541
phosphorescent supramolecule 373
photo effect 1420
photoactivated non-intrusive tracing of molecular motion (PHANTOMM) 365, 367
photochromic chemicals 365
photochromic dye 868
photochromism 366
photoconductive detectors 1420
photodetectors 728
photodissociation 528
photoemission 1420
photogrammetric PIV 1473
photogrammetry 790, 1080
photography 1437

photomultiplier tubes (PMT) 1420
photon 203
photon correlators 354
photon noise 1428
photon-correlation spectroscopy 141
photon-counting device 1420
photoresponse nonuniformity (PRNU) 1433
photovoltaic detector 1421
PIB (polyisobuthylene) 710
piezoelectric effect 598
– inverse 598, 611
piezoelectric film transducers 990
piezoelectric force transducer 598
– multicomponent 597
– one component 610
piezoelectric polyvinylidene fluoride (PVDF) 989
piezometer
– continuously-weighed 86
– expansion 87
– fixed-volume 86
– variable-volume 87
piezometric film 990
piezoquartz 992, 995
pilot facility of ETW (PETW) 548
pinhole 300
pipe flow 872
PIT (particle image thermometry) 492
pitching moment 1103, 1115
pitching-moment-slope curve 612
Pitot
– pressure 1062
– tubes 879
Pitot probes 218
– blockage 219
– blockage because of the probe body 226
– blockage effects of the strut on a wall pressure tapping 226
– calibration 220
– compressible flow 221
– constant displacement correction 224
– diameter ratio 221
– difference between measured and true velocity ΔU, at y_{cl} 223
– directional sensitivity 222
– displacement of the effective stagnation streamline 223
– displacement of the stagnation streamline 222
– effect of shear 222
– effect of turbulence 222

– effect of wall proximity 224
– flattened probes 225
– hemispherical nose design 221
– higher Reynolds numbers 221
– implied error in probe position 223
– inferred value of the von Kármán constant 226
– instantaneous yaw angle 222
– interference effect 226
– Kiel probe 219
– large-scale fluctuations and coherent structures 222
– MacMillan displacement correction 224
– MacMillan wall corrections 224
– mutual interference in supersonic flow 219
– offset of the probe centerline due to spatial integration 223
– Princeton displacement correction 224
– Princeton wall correction 225
– probe misalignment 219
– probe Reynolds number 220, 221
– probe tip geometry 219
– probe-induced bow shock 221
– ratio of inner to outer probe diameter 220
– ratio of inner to outer probe diameters 219
– Rayleigh supersonic Pitot formula 221
– round-nosed probe 221
– shear parameters α and β 223
– spatial averaging 219
– spatial integration 222
– square-ended flattened probes 221
– stagnation streamsurface 219, 220
– subsonic compressible flow 221
– supersonic flow 221
– temporal response 219
– the size of the characteristic scale relative to the probe diameter 222
– variations in the angle of a turbulent or unsteady flow 219
– Venturi effect 226
Pitot-static probes
– cantilever-type Pitot cylinder 227
– cones 227
– effect of yaw 227
– effects of pitch and yaw 227
– modified Prandtl 227
– Prandtl probe design 227
– ratio of inner to outer diameter 227

– shear effect 227
– tip and stem influence 227
– wall proximity effect 227
– wedges 227
– yaw sensitivity 227
PIV (particle image velocimetry) 287, 309, 342, 345, 346, 351–353, 362, 388, 487, 489, 492, 497, 805, 835, 902, 921, 933, 997, 998, 1073, 1126, 1131, 1243, 1465
pixel fill factor 326
pixel-locking 1470
planar Doppler velocimetry 342
planar Doppler velocimetry (PDV) 948
planar laser-induced fluorescence (PLIF) 848, 1105
planar submerged jet 66, 73, 77
planar wall jet 69
planposition indication (PPI) 1170
plan–position indicator (PPI) 1159
platform 1071
platinum porphyrins (PtOEP) 191
plenum 1048, 1050, 1054, 1063
– method 1052
PLIF (planar laser-induced fluorescence) 848, 1105
plunging tapes 116
PM (polarization modulation) 727
PMMA (fpolymethylmethacrylate) 700
PMT (photomultiplier tubes) 1420
Pockels 727
Pockels cell 360
POCS (projection onto convex sets) 309
POD (proper orthogonal decomposition) 235, 1346, 1352
– accurate model reduction 1368
– approximation method 1347
– basics 1346
– choice of inner product and norm 1358
– choice of input collection 1357
– classical or direct method 1359
– common properties approaches 1361
– complementary technique 1365
– different approaches 1357
– evaluative summary 1369
– Fredholm equation 1353
– Galerkin 1365
– harmonic analysis 1362
– model reduction aspects 1356
– optimality of basis 1355
– phase indetermination 1363

– problem of boundary conditions 1367
– properties of the basis functions 1354
– reduced order models based on 1366
– snapshot 1360
– typical applications to fluid mechanics 1364
point measurements 1009
point operations 1440
Poiseuille 727
– flow 644
– profile 62
polar separable filter 1449
polarizability 517
polarization 725–728
polarization modulation (PM) 727
polarization spectroscopy (PS) 1243
polarizers 727, 728
polarizing microscope 725
pollutant
– combustion-generated 1275
poly(ethyleneterephthalate) (PET) 117
polyacrylamide (PAA) 803
polyacrylic acid (PAA) 168
polycaprolactone (PCL) 168
polycyclic aromatic hydrocarbon (PAH) 1249
polydimethylsiloxane (PDMS) 700
polydisperse 681
polydispersity index (PDI) 681
polyethylene oxide 657
polyisobuthylene (PIB) 710
polymer molecule 818
polymer relaxation time 804, 807, 817
polymer solution 800, 803, 806, 809, 812, 817, 823, 827
polymer stress 816
polyurethane (PU) 168
polyvinyl alcohol (PVA) 168
polyvinylidene floride (PVDF) 974
pom-pom model 19
post-critical 1145
posterior distribution 1480
potential flow theory 962
power plant 527, 1292, 1295
– coal-fired 1295
– gas-fired 1294
power spectral density (PSD) 302, 1338, 1406
power spectrum of velocity 814
PPI (planposition indication) 1170

PPI (plan–position indicator) 1159
practical advancing contact angle (PACA) 109
practical issues PIV in cavitating flows 1000
practical receding contact angle (PRCA) 109
practical salinity units (psu) 1183
Prandtl mixing length 772
Prandtl number 39, 63
PRCA (practical receding contact angle) 109
predictive algorithm 791
predissociation 529, 1276
pre-shear, preconditioning 700
pressure 1072, 1081–1091, 1093, 1094, 1096, 1103–1105, 1112, 1113, 1115, 1116, 1118, 1120, 1122, 1123
– at the wall 180
– coefficient 889, 963
– diaphragm burst 1091
– distribution 1128, 1133
– gauge 642, 1095–1097
– gauge, micromachined 1096
– gauge, optical 1096
– gradient 1050, 1060, 1061
– measurement 1096
– mechanical 6
– Pitot 1102
– probe 897
– probe calibrations 179
– quartz crystal transducer 1182
– ratio 1087
– static 1122, 1125
– surface 1112, 1113, 1116, 1122
– tapping 1096
– total 1094
– transducer 989, 992, 1095, 1096, 1105
– transducer, optical fibre 1095
– transducer, piezoelectric 1095
– transducer, piezoresistive 1096
pressure measurement 989, 1011, 1095, 1123, 1125
– bank manometer 187
– configuration of the probes and accompanying hardware 187
– connections between probes and transducers 187
– diaphragm-type transducers 187
– digital output 187
– ellipsoidal-tipped probes 187
– fast response times 187
– high precision 187
– Liquid manometers 187

– multi-channel devices 187
– response time of the system 187
– sensitivity of the transducer 187
– sequential and simultaneous measurements 187
– zero drift 187
pressure–density relation 216
pressure-sensitive paint (PSP) 188, 203, 896, 921, 1072
Preston tube 879
– method 879
prestress 598
pretriggering 993
principal coordinate system 1452
principal strain rates 410
prior distribution 1480
PRNU (photoresponse nonuniformity) 1433
probability
– central moment 1400, 1404
– density distribution (PDF) 304
– density function (PDF) 356, 393, 753, 1243, 1245, 1373, 1399
– distribution function (PDF) 433, 810, 814, 819
– histogram 356
– joint density function 1400
probe volume geometry 358
process
– random 1401
projection onto convex sets (POCS) 309
propagation of errors 1416
propeller 961, 963
proper orthogonal decomposition (POD) 235, 1346, 1352
properties of a continuum 3
proximity probe 1014
PS (polarization spectroscopy) 1243
PSD (power spectral density) 302, 1338, 1406
pseudocolor image 1440
pseudo-noise 1077
PSP (pressure-sensitive paint) 188, 203, 896, 921, 1072
PtOEP (platinum porphyrins) 191
PTV (particle tracking velocimetry) 320, 751, 805, 1465, 1474
PU (polyurethane) 168
pulsed hot-wires 280, 281
pulsed-sound 894
pump 919
pump and turbine test loops 972, 973
pump impellers 976

pumping 1053
pure extensional flow 651
PUV gage 1014
PVA (polyvinyl alcohol) 168
PVDF (piezoelectric polyvinylidene fluoride) 989
PVDF (polyvinylidene floride) 974
PVDF film 974
pycnometer
– high-temperature 87
pyramid 1462
pyrometry 1283

Q

QMSF model 26
quadrature filter 1461, 1484
quantization 1437, 1439
– error 1439
quantum detector 1420
quantum efficiency 1426
quantum yield 1265
quarter-wave plate 356
quartz 596
quasistatic measurement 602, 606
quencher 364
quenching 528, 529, 1268, 1273, 1278, 1279
Quincke's rotor 1321

R

Rabinowitch equation 645
radar (radio detection and ranging) 1159, 1174
radar altimetry 1202
radial velocity 814
radiation 1057, 1246
radiation of sunlight 1045
radio acoustic sounding system (RASS) 1169, 1174
radio detection and ranging (radar) 1159, 1174
radiometer 1165
– correction 1440
radiosondes 1167
radius of gyration 803
RAFOS (sound fixing and ranging, reversed) 1190
rain 1045, 1048, 1057
rainfall 1202
Raman 521, 1252
– cross-section 522
– frequency-shift 520
– spontaneous 1251

Raman excitation plus laser-induced electronic fluorescence (RELIEF) 365, 366
Raman measurement
– multiscalar 1253
Raman scattering 516, 1243, 1253, 1284
– multispecies 516
– one-dimensional 516
random correlation term 322
random variable 1399
– expectation 1399
– variance 1400
range of the balance 608
range–height indicator (RHI) 1159
RANS (Reynolds-averaged Navier–Stokes) 771, 872
rate equation 529
rate of rotation 389
rate of shear 636, 637
rate of strain 6
rate of wetting 112
rate-of-strain tensor 400
Rayeigh–Plesset equation 967
Rayleigh 294, 1256
– cross-section 519, 522
– interferometric technique 151
– interferometry 152
– measurement, multiscalar 1253
– supersonic Pitot formula 216
Rayleigh scattering 516, 519, 1270, 1283
– along-line 522
– filtered 516, 520, 1243, 1284
– single-point 522
– two-dimensional plane 522
Rayleigh–Plesset equation 965
rear sting support 600
receptivity 886
recirculation region 1133, 1140
recombination 1093, 1095, 1117, 1124, 1325
– atomic species 1123
– constant 1323
recovery temperature 239
reduced frequency 36
reduced reaction model 1245
reduction operator 1464
reentrant jet 967
re-entry 1112, 1115, 1121
– flight 1081, 1085, 1123
– trajectory 1112
– vehicle 1081, 1113, 1118
reference conditions 86
reference point 1070
reflection 1006

reflectometer 1165
refraction 1006
refractive index 473, 724–726, 730, 732, 1008, 1095, 1106, 1109
– matching 861
refrigerants 86, 88, 94
regime diagram 1263
region-based methods 1465
registration error 317
regular sampling 1438
regularization 1453
relative humidity (RH) 1163
relative uniaxial elongational viscosity 709
relaxation 626, 1112, 1118, 1119
– chemical 1081, 1093, 1112
– modulus 14, 619, 622
– thermal 1093, 1112
– time 622, 623, 626, 627, 647, 651, 652, 656–658, 723, 1115
RELIEF (Raman excitation plus laser-induced electronic fluorescence) 365
repeller 610
reproducibility 605
reptation 18
residence time 304, 723
resistance gage 1011
resistance wave probe 1010
resolution 523, 597, 604, 1247
resonance 630, 631
– curve 631
– effect 604
– enhanced multiphoton ionization (REMPI) 1249
– frequency 631
– frequency of a bubble 981
resonant 523
responsivity 1426
retardation 725–728
– time 623, 624
Reynolds
– analogy 244, 1104
– shear stress 354, 355, 1371
– stress 816, 873, 1133, 1137, 1140, 1144, 1160
– stress symmetric tensor 758
Reynolds number 39, 236, 239, 607, 613, 793, 825, 1046, 1058, 1059, 1114, 1118, 1137
– effect 1059
– high 613
– transition 1118
– unit 1119
Reynolds-averaged Navier–Stokes (RANS) 830, 872, 1242, 1245

RH (relative humidity) 1163
rheogoniometer 635–637, 646
rheometer 635–637
– filament stretching 647, 650–656, 658, 659
– pressure driven 635, 636, 643
– slit 636, 646
Rheotens 660
RHI (range–height indicator) 1159
Richardson–Obukhov law 799
Riemann problem 1082
Riesz transform 1460
Rivlin–Saywers equation 15
RMS value 597, 609
road dirt 1043
robust estimation 1482
Rochon prism 356, 358
rolling resistance 1069
root load 612
rotary kiln combustors 1296
rotating
– disk method 976
– drum camera 1110
– frame 921
– machinery 976
– mirror camera 1111
– shaft 9
rotational diffusion coefficient 685
rotational friction coefficient 685
rotational systems 1472
rotor 919
rough wall flows 881
rouse time 18
rubber-like liquid constitutive equation 14
running mean 1445
Rutgers–Delaware rule 695

S

S+H (sample-and-hold) 306, 308
saddle 909
sample-and-hold (S+H) 306, 308
sampling
– ocean 1181
– point 1286
– process 1285
– theorem 1443, 1462
– vector averaging 1192
sandwich 628, 629
SAOS (small-amplitude oscillatory shear) 714
SAR (synthetic aperture radar) 1202
satellite 1175
– salinity 1202, 1203

saturation 529, 531
– collisional 536
– current 1328
– equivalent exposure (SEE) 1427
– voltage 1325, 1326
Sauter mean diameter 1246, 1265
scalar dissipation rate 1252
scale 1046
– effects 987
– integral time 1403
– similarity model 836
scaling 57, 80
– laws 114
– wavelet 1391
scatterometry 1202
scene silencer 1063
scene-silencer 1054
Scheimpflug condition 403
Scheimpflug configuration 315
Schiebe bodies 983
schlieren 474, 1323, 1330, 1331
– background oriented 477
– frequency filter 477
– interferometer 477
– method 476
– sharp-focusing system 478
– synthetic 477
Schrödinger equation 517
scoop 1055
scramjet 1121
sea surface temperature 1202
search volume 791
seawater
– electrical property 1183
second law of thermodynamics 7, 10
second normal stress coefficient 642
second normal stress difference 635, 642
second-harmonic generation (SHG) 1287
second-order correlation 1472
second-order derivatives 1452
sediment
– suspensions 1020
– transport 1020
– trapping 1031
sedimentation 682, 696
seeding particles 998, 1001
Seifert wing 1053
seismic mass 597
selectivity 1247
selfexcited
– flutter oscillations 610
– shock oscillations (buffet) 612
self-similar variable 58, 59, 72, 79

self-similarity   57–59, 82
sensible heat flux   1161
sensitivity   530, 598, 605, 1247
– coefficients in subsonic flows   240, 242, 259, 263, 265, 273
– coefficients in supersonic flows   243, 244, 246
sensors
– biofouling   1181
– conductivity cell   1183
– conductivity–temperature–density (CTD)   1185
– corrosion   1181
– integrating measurement   1203
sentmanat extension rheometer (SER)   653
separable K-BKZ equation   15
separated shear layer   608
separation bubble   1145
separation point   1128, 1131, 1142
SER (sentmanat extension rheometer)   653
setup
– for self-excited oscillations   610
– pitch   611
SFA (surface force apparatus)   1223
SFG (sum-frequency generation)   1287
SG (specific gravity)   85
SGS stresses   837
shadowgraph   474, 1330
– technique   1330
Shaft work   9
shape factor   889
sharp edged disks   996
shear
– compliance   708
– creep experiment   708
– flow   624, 626
– force   602
– layer   40, 432, 1141
– layer frequency   1142
– layer instability   1135, 1141, 1144
– modulus   619, 620, 622, 623
– rate   624, 625, 635, 637–646, 721–723, 804, 806, 807, 812
– stress   620, 625, 721, 723
– thickening   694
– wave   632, 633
sheet cavitation   962, 966, 968
sheet cloud cavitation   968, 970
SHG (second-harmonic generation)   1287
shields stress   1021
shift frequency   298
ship models   973

ship propeller   973
shock expansion tube   1082, 1093–1095, 1117, 1118
– free-piston-driven   1095
– ideal   1093, 1095
– superorbital   1095, 1109
shock expansion tunnel   1081, 1082, 1093–1095, 1112
– free-piston-driven   1095
– superorbital   1095
shock tunnel   1081, 1082, 1085–1087, 1089, 1090, 1093, 1094, 1099, 1102–1106, 1112, 1116, 1123, 1124
– deflagrative combustion-driven   1087, 1121
– detonative combustion driven   1088, 1089
– detonative combustion-driven   1088
– electrically heated   1087, 1088
– electric-arc-driven   1089
– free piston driven   1090, 1113, 1121
– free piston driven shock   1118
– free-piston-driven   1091, 1112, 1118, 1119, 1121, 1122, 1124
– high enthalpy   1091, 1093, 1103, 1105, 1108, 1115, 1122, 1124
– ideal, performance   1091
– performance   1086, 1091, 1092
– reflected   1084–1087, 1090–1095, 1102, 1121, 1123
– straight through   1085, 1086, 1090, 1091
– straight-through   1094
shock wave   79, 80, 475, 1082–1085, 1090–1092, 1095, 1112, 1113, 1123
– incident   1083–1087, 1089, 1092–1096
– reflected   1083, 1084, 1089, 1092, 1094
– secondary   1094, 1095
– secondary, upstream facing   1086
– speed   1082, 1087, 1090, 1091
shocktube   1082–1086, 1089–1097
– constant area   1082–1084
– diaphragm-driven   1082
– filling pressure   1092–1094
– ideal   1082, 1092
– ideal, performance   1086, 1092
– length   1082, 1092
shocktube performance   1091
side-scan and upward-looking sonar   1011

signal attenuation   1276
signal-to-noise ratio (SNR)   302, 1406, 1429
– estimation procedure   1406
similarity   409, 1058
similitude   34
simple shear flow   727, 728, 730
simulation   1058
single microphone   1076
single normal hot-wire   234, 241, 242, 251, 267, 268, 273
single realization   302
single-point equations   774
single-wire sensor resolution   875
singular points   909
singular value decomposition (SVD)   1348
singularities   1395
sink   61
skewed   1044
skewness   814
skin friction   1104
– coefficient   875
– force   1104
– gauge   1104
– gauge, piezoceramic   1105
– gauge, piezoelectric   1104
– measurement   1104, 1105, 1120
slip
– apparent   1221
– boundary condition   1219
– contact line motion   1221
– dependence on dissolved as and bubbles   1229
– dependence on pressure   1234
– dependence on shear rate   1231, 1232
– dependence on surface roughness   1228
– dependence on wetting properties   1230
– effective   1221
– electrical properties   1233
– experimental results   1223–1226, 1231
– gas flow   1221
– history   1220
– indirect experimental methods   1222
– length   12
– local experimental methods   1224
– mo,ecular dynamics results   1226
– molecular   1221
– Newtonian liquids   1222
– non-Newtonian fluids   1221
– phenomenon   1220

– ring   921, 923
– velocity   288
slit rheometer   636, 646
slope array buoys   1018
slope imaging devices   1017
slot correlation   307
slotted wall   1049
– test section   1051
Smagorinsky coefficient   831
small-amplitude oscillatory shear (SAOS)   714
small-scale
– model   1058
– wind tunnel   1066
– wind tunnel for trains   1068
smoke   862
– generator   1075
– tunnel   862
– wire   868
– wire visualization   1141
snapping shrimp   967
SNCR process   1294
snow   1048, 1057
snow nozzle   1057
SNR (signal-to-noise ratio)   302, 1406, 1429
soap bubble   1075
Sobel
– edge detection   1452
– filters   1486
– operator   1452, 1457
sodar (sound detection and ranging)   1159, 1172
SOFAR (sound fixing and ranging)   1190
soft sphere interactions   685
soil flux   1164
soil temperature   1163
soiling   1046
solar radiation   1165
sol-gel transition   713
solid blockage   1061
solid volume concentration   689
solid/liquid interfacial formation process   114
solid-liquid transition   703
Soloff method   316
soot volume fraction   1281
sound detection and ranging (sodar)   1159, 1172
sound engineering   1077
sound fixing and ranging (SOFAR)   1190
source   57, 61
space vehicle   1081, 1102, 1115, 1118

space–time diagram   827, 828
spark ignition   1261
spark tracer method   868
spatial
– and temporal resolution   988
– coherency   1445
– distribution of surface height   1016
– filtering device   357
– pattern   1371
– resolution   388, 403
– resolution (SR)   748
– resolution for derivatives   254, 276
– resolution for temperature spectra   275
– resolution for velocity spectra   252
– resolution of wires   251, 257
– smoothness   1478
– velocity fluctuation   807
spatial variance   1445
species concentration   1105, 1109
specific gravity (SG)   85
specific heat
– capacity   1098
– constant pressure   4
– constant volume   4
– ratio   1082, 1087
speckle photography   474, 479
– correlation techniques   479
– spatial filtering   479
spectra
– drag   608
– lift   608
– moment   611
spectral
– broadening   1341
– fingerprint   523
– overlap   529
– peak   969
– sensitivity   1432
spectrally dispersed   521
spectroscopy in flames   1242
spectrum   1242, 1387
– Fourier   1378
– wavelet   1387
speech recognition   1054
speed of light   1106
speed of sound   1081–1083, 1087, 1089, 1094, 1102, 1160
sphericity   682
spinability   649
spinning disc   41
splines   1444
splitter plate   1133
spontaneous dewetting   113
spontaneous wetting   113

spreading drops   117
spring
– bending   610
– cross   610
– torsion   610
square section cylinder   607
square-wave test   258, 262, 265, 275
squeeze flow   710
SSD (sum-of-squared differences)   1473
stability   610
stability of flow   695
stagnation enthalpy   1115
stagnation pressure   216
standard deviation   605, 1445
standard form of the general viscous fluid   13
standard sampling   1438
start-up experiment   708
static
– aeroelastic effects   612
– calibration equation for a shear stress sensor   884
– equilibrium   964
– pressure   216, 879
– pressure measurement   179
static tube   185
– a spherical form   185
– a static wedge   185
– angular sensitivity   186
– blockage   186
– bow shock   186
– cylindrical probes with rounded tips   185
– dependence of the behaviour on the mean velocity distribution   186
– directionally sensitive   186
– distribution of turbulent energy   187
– eddy size   187
– effects of turbulence   187
– ellipsoidal   185
– ellipsoidal nose   186
– fine-nosed probe   186
– hemispherical   185
– hemispherical tips   186
– high-subsonic freestream flows   186
– interaction of the shock with a laminar boundary layer   186
– local shocks   186
– location of the holes relative to the probe tip and stem   186
– Mach number approaches unity   186
– mutual interference   185

- ogival or conical 186
- Prandtl probes 186
- pressure coefficient 185
- rounded nose 186
- satandard design 186
- short-head design 185
- static disc 185
- subsonic flow 186
- supersonic flow 186
- tip shapes 185
- trip ring 186
- turbulence intensity 187
- Venturi effect 186
- viscosity 185
- yaw angles 186
stationary frame 921
statistical data analysis 354
statistical data association 1479
statistically independent 1403
steady flow rig 379
steady preload 597
steady shear flow 624–626
steerable filters 1449
Stefan's law 1163
stereo
- correspondence 1469
- correspondence problem 1480
- imaging 376
- measurements of water surfaces 1016
- photography 864
stereomatching 790
stereoscopic PIV 1472
steric stabilization 687
Stern–Volmer coefficient 1273, 1274
stimulated emission 1242
sting 1055, 1070, 1071
sting balance 601
stochastic process 1479
Stokes
- assumption 10
- derivative 6
- first problem 59
- law 127
- line 517, 520
- number 289, 793
- Raman scattering 517
- second problem 60
- wave 893
storage and loss moduli 628
storage and loss modulus 706
storage modulus 620, 626
straight through shock tunnel 1086

strain
- gauge balance 596
- hardening 651
- thickening 695
streak line 5
streakline 860
streamline 5, 860, 1051
streamlined wall 1049
stress
- elastic 799, 813, 815, 817, 820
- elastic, nonlinear 801
- elastic, turbulent 810
- relaxation 619–621, 626, 627, 723
- tensor 6, 635, 636, 800, 812, 816
- tensor, elastic 800
- wave force balance 1103
stretch the wave height 1015
stroboscopic flash 988
strong explosion 79
strong injection 1328
strophometry 388
Strouhal number 36, 597, 607, 614, 967, 969, 1127, 1131, 1133, 1137, 1142
structural force 609
structural parameter 693
structural relaxation time 720
structure function 382, 1391
structure functions 806, 815
structure tensor 1454
structure tensor operator 1486
structured jets 967
structuring in shear 702
sub-grid 945
subgrid-scale stress (SGS) 831, 838
subharmonics 608
submerged radial swirling jet 68, 75
subpixel accuracy 1466, 1467
subpixel interpolation 326
substantial derivative 6
substrate materials 271
suction 1055
sum-frequency generation (SFG) 1287
sum-of-squared differences (SSD) 1473
sun load 1043, 1048, 1057
sunlight 1044
sunlight simulation 1057
supercavitating inducers 970
supercavitating propellers 970
supercavitation 959, 970
supercritical 1145
supercritical state 156
superharmonics 608

super-hydrophobicity 111
super-resolution analysis 320
super-resolution PIV 1472
supersaturated 965
supersonic
- combustion 1122
- combustion ramjet (scramjet) 1081, 1103, 1120–1122, 1124
supramolecule 371
surface
- auto-spectra 1019
- catalysis 1123, 1124
- cavitation 963
- cross-spectra 1019
- curvature 1013
- discontinuity 11
- elevation 1009, 1019
- force apparatus (SFA) 1223
- hardening 974
- height frequency spectra 1018
- microphone 1077
- potential 686
- slope 1019
- slope measurement 1013
- stress 6
- temperature measurement 538
- tension 12, 38, 96, 97, 101, 103, 105–107, 647, 648, 650, 653, 654, 656, 657, 661, 964
- treatment 704
surfactant 98, 712
surge 1051
susceptibility 536, 537
suspended load 1020
suspended sediment transport 1021
suspensions 681
SVD (single value decomposition)
- and eigenvalue problems 1349
- examples of image processing 1351
- geometric interpretations 1348
- lower-rank approximation 1350
SVD (singular value decomposition) 1348
swell ratio 723
swept wing 612
swirling flow 792
swiveling flaps 1057
synchronizer 359
synchronizing electronics 359
synthetic aperture radar (SAR) 1202
synthetic schlieren 479

T

TACA (theoretical advancing contact angle) 109
tailored 1085
talcum 1076
tandem nozzle 1065, 1067
tangential blowing 1063
target molecules 531
Taylor
– Dean flow 801
– diffusion 798
– dispersion instrument 151
– dispersion technique 156
– microscale 793
– number 801
T-belt 1056
TCFB (two-color flow birefringence) 727
TDC (top dead center) 381
TDLAS (tunable diode-laser absorption spectroscopy) 1105, 1289
telecentric lenses 314
TEM (transmission electron microscopy) 1281
temperature 4, 1054, 1081–1084, 1087–1089, 1091, 1092, 1095, 1098–1100, 1105, 1106, 1109, 1112, 1118, 1124
– coefficient of resistance 230, 232, 243
– drifts 272
– field 1045
– fluctuation 273
– helium 599
– history 1097, 1099
– measurement 1100, 1101
– measurement, thermistor 1183
– profile 249
– rotational 1093
– sensitive coating 538
– sensitive molecule 541
– sensitive paint (TSP) 537, 896
– total 1123
– translational 1093
– translational/rotational 1093, 1105
– vehicle surface 1123
– vibrational 1093, 1105
– wall 1123–1125
temporal
– resolution 387, 519
– sampling theorem 1466
– spectra 807

tensile creep rate decay function 710
tensile strength 964, 968, 979, 980, 984
tensor-based techniques 1484
term value 519
ternary complex 373
test
– case 613
– gas 1083
– gas, air 1114
– section 1047, 1084–1087, 1090, 1091, 1094, 1095, 1103, 1106, 1108, 1111, 1113, 1118, 1121, 1122
– time 1081, 1085, 1086, 1089, 1090, 1092, 1094, 1095, 1098, 1101, 1102, 1104, 1108, 1111, 1113, 1118, 1122, 1123, 1125
tethered balloon 1166, 1167
tetrahydrofuran (THF) 168
theoretical advancing contact angle (TACA) 109
theoretical receding contact angle (TRCA) 109
thermal
– energy equation 7, 10
– inertia 231, 246
– quenching 541
– test 1054, 1062
– wind tunnel 1064
thermal conductivity 138
– alkenes 145
– critical point 137
– gases 148
– n-alcohols 146
– n-alkanes 144
– reference values 146
– refrigerants 147
– toluene 142
thermal diffusivity
– critical region 141
– light scattering methods 141
– molten salts 146
thermochromic liquid crystal (TLC) 487, 488, 490, 498, 539
thermocouple 1097–1100, 1113, 1124
– coated 1100
– coaxial 1097, 1098, 1123
– material, chromel 1100
– material, constantan 1100
– sensitivity 1097
– type E 1099
thermodynamic pressure 4
thermographic phosphor 538

thermographic phosphor thermography (TPT) 538
thermography 487, 488
thermometry 516
thermopile 1421
THF (tetrahydrofuran) 168
thin-film gauge 1097, 1098, 1100, 1119
– platinum 1098
thin-oil-film equation 876
three-component Doppler L2F system 362
three-component LT system 360
three-component-velocimeter 360
three-dimensional 897
– calibration 316
– flow 878
three-frame technique 1478
three-wire probe 234, 270
threshold for dynamic measurement 605
thrust of ejector engine 608
thymol blue method 867
THz 1287
time
– averaged wall shear stress 876
– compression factor 973
– constant 246, 247, 261, 275, 606
– history 604
– of-flight histogram 354
– of-flight measurement 354
– resolution (TR) 748
– resolved particle image velocimetry 1003
– resolved photography 988
– resolved PIV experiment 1004
– to-amplitude converter 358
tip
– streaming 714
– vortex 962, 963
– vortex cavitation 986
TiRe-LII 1283
TLC (thermochromic liquid crystal) 487, 488, 490, 498, 539
TLS (total least squares) 1485
Tollmien–Schlichting (TS) 547, 887
toluene 146, 534
tomography 864
– optical 474, 485
top dead center (TDC) 381
TOPEX (topography experiment) 1202
top-hat filter 831
topography experiment (TOPEX) 1202
torque 805, 816

torque coefficient 963, 966
torsion pendulum 630
torsional hydraulic actuator 611
total cross-section 520
total dissolved gas 980
total head 10, 966
total least squares (TLS) 1485
total load measurements 1032
total pressure 216
total temperature 242
total velocity field 1373
towed vehicle
– SeaSoar 1204
– TriSoarus 1204
– turbulence 1205
tower 1162
towing tank 1068
TPT (thermographic phosphor thermography) 538
tracer 532, 535, 1243, 1259, 1264, 1267, 1271
– aliphatic 1269
– aromatic hydrocarbon 1268
– atom 1267
– exciplex 1269
– gas 1168
– inorganic 1267
– ketones 1269
– material 859
– organic 1268
– particle 353, 789, 793
– photolytically generated 1268
tracking algorithm 793
tracking system 794
trajectory 792
transducer sensitivity 179
transfer function 603, 1444
transient technique 134
– hot-wire 135
transit time 304, 1324
transition 529, 886
– detection 538, 541, 545, 548, 551
– frequency 524
– hysteretic-encumbered 610
– range 610
– Reynolds number 1119
– to turbulence 886
transition moment 517, 518
– space-averaged 518
translating shaft 9
translational friction coefficient 684
translational systems 1472
transmission 361
transmission electron microscopy (TEM) 1281
transonic flow 611

transport aircraft wing 612
traversing gear 1058
TRCA (theoretical receding contact angle) 109
triangulation 1080
Trouton ratio 647, 649, 652–655, 709
TS (Tollmien–Schlichting) 547, 887
tube model 18
tunable diode-laser absorption spectroscopy (TDLAS) 1105, 1289
tunnel reverberation 994
turbine 919
– head 966
– model 966, 973
– test stand 976
turbomachinery 355, 919
turbulence 789, 804, 1044, 1056
– chemistry-turbulence interaction 1245
– dissipation rate 874
– elastic 799, 804, 810, 813–816, 821
– intensity 354, 355, 1143
– level 1045, 1051
– model 1245
– turbulence chemistry interaction 1263
– turbulence/chemistry interaction 1263
turbulent
– intensity 1160
– kinematic viscosity 772
– kinetic energy 837
– relative dispersion 797
– resistance 806
– shear flows 967
– structure 382
– thermal diffusivity 772
– wall-bounded flows 871
turbulent boundary layer 543, 545, 896, 967
– laminar 538
– pipe, channel flows 872
– transition 538
turning vane 1054
two infinitely extended parallel plates 635
two-bulb technique 155
two-color flow birefringence (TCFB) 727
two-dimensional
– array of point measurement devices 1015
– assignment problem 792

– boundary layer flow 872
– probability density function 355
two-frame tracking 1478
two-line thermometry 527
two-phase flows 1007
two-photon LIF 1243
type E thermocouple 1099

U

U-bend silencer 1063
uncertainty relation 1447
underflow 1440
undertailored 1085
uniaxial elongational viscosity 709
unidirectional solidification 378
unipolar injection 1320, 1325
unstable limit cycle 611
unsteadiness 596
unsteady boundary-layer separation 377
unsteady, reverse flows 901
upgrading a wind tunnel 1066
upper convective Maxwell model 16
upwash 1051
UVW (three components of velocity) gage 1019

V

Vacutank at the Marin Institute, Netherlands (formerly NSMB) 973
vacuum tank 1086, 1091
validation 1244
– modeling 1244
– numerical simulation 1244
van der Waals forces 685
van Slyke apparatus 980
vapor filled 970
vapor pressure 959
vaporous cavitation 965
vaporous/gaseous cavitation 998
variable pressure towing tanks 972
VCSEL (vertical-cavity surface-emitting lasers) 1298
vector field variables 909
vehicle
– coordinate system 1070
– mass 1069
– speed 1043
velocity 5, 644, 1081, 1083–1087, 1089–1091, 1093–1095, 1105, 1112, 1117, 1118, 1121
– fluctuation 807, 1133

– gradient 409, 637
– gradient tensor 400, 401
– gradient/vorticity measurement techniques 387, 391
– hypersonic 1092
– profile 644, 1044
– superorbital 1095
– time series for cavitating and non-cavitating wakes (LDV) 1004
velocity measurements
– accuracy of the transducer 217
– blockage and interference 217
– condensation 217
– convergence of the reading 217
– particulates 217
– Reynolds number 217
– sensitivity of the velocity to errors 217
– yaw acceptance 217
ventilated cavities 970
ventilation gas 970
Venturi
– center body-type device 985
– method 975
Versuchsanstalt Wasserbau (VAO) Obernach 973
vertical-cavity surface-emitting lasers (VCSEL) 1298
vibrating ribbon 892
vibration 517, 960, 962
– normal 517
vibratory apparatus 1006
vibratory method 976
video system 116
Vieille, Paul 1082
Virginia Polytechnic Institute (VPI) 1104
virtual temperature 1160
virtual tower 1169
viscoelastic
– liquid 800
viscoelasticity analyzer 804
viscometer
– capillary 125
– falling-body 127
– falling-cylinder 129
– falling-sphere 127
– high pressures 127
– liquids 126
– oscillating-body 119
– oscillating-cup 122
– oscillating-cylinder 122
– oscillating-disk 120, 121
– oscillating-sphere 122
– torsional-crystal 124
– vibrating 122

– vibrating-wire 122, 123
viscosity 619, 624, 625, 627, 632, 634–636, 639–641, 644, 646, 722, 723
– absolute measurement 121
– alkenes 130
– complex 620, 634, 635
– gases 132, 133
– hydrocarbon 131
– intrinsic 690
– n-alcohol 131
– n-alkane 129
– noble gase 128
– ratio 712
– reference value 131
– refrigerant 132
– relative 689
– specific 690
– standard 126
– water 126, 131
viscous
– diffusion scale 404
– dissipation 7
– flow 113
– no-slip condition 12
– relaxation time 685
– sublayer 396
viscous sublayer 882
viscoussublayer 873
visibility 301
visually 996
void
– fraction 1007
– fraction from PIV measurements 1008
– fraction measurements 1007
Voigt–Kelvin model 619, 621, 623, 624, 627
volume flow 1073
volume of the region 8
volume region 8
volume-average diameter 681
von Karman constant 772
von Karman integral method 879
vortex
– cavitation 962, 968
– dislocation 1139
– force 813
– formation 1133
– generation 968
– nozzle 983
– ring 913
– street 1125
vortex shedding 597
– frequency 608
– instability 1133

vortical flow 382
vortical structures 963, 967
vorticity 6, 7, 60, 382, 383, 400, 429, 774, 813, 818, 820, 825
– flux 384
– instantaneous 383
VPI (Virginia Polytechnic Institute) 1104

W

Wagner equation 15
waist diameter 358
wake
– blockage 1061
– factor 881
– formation 1132
– instability 1133, 1137, 1140
– layer 873
– velocity 1127, 1137
– vortex 1127
– width 1131, 1145
Walden's rule 1322
wall 1051
– flow structure 872
– hot-film 235, 271, 279
– normal positioning uncertainties 882
– shear stress 901
– slip 697
– slip, slip velocity 698
– vorticity flux 400
wall tapping
– angled 181
– burrs 184
– calibration 181
– cavity depth 183
– cavity geometry 181
– chamfered 181
– commercial standard 181
– compressibility 182
– curvature of the streamlines 181
– difference between local and wall pressure introduced in by turbulent fluctuations 185
– directional sensitivity 181
– dust 184
– error in the measured pressure 180
– finite size 181
– generation of Mach waves 183
– high Mach number 183
– high tapping Reynolds number 182
– iezometer 180
– large tappings 182

– lengthscale of pressure variation 182
– method to produce small tappings 184
– minimum deep tapping 183
– pressure coefficient 182
– radius of curvature 181
– radius on the edge of the hole 184
– radiused 181
– rapid spatial variation of pressure 182
– shallow tappings 183
– slot-type 181
– small 181
– system of eddies 183
– system of vortices 181
– zero-error condition 181
wash load 1020
waste incinerator 1293
water 146
water quality 974, 979
water quality measurements 979
water tunnel 1046
water tunnels 972
wave
– gage 1019
– height 1009
– propagation 630, 632
– staff 1010
wave diagram 1082, 1084–1086, 1088–1091, 1095
wavefunction 517
wavelength modulation spectroscopy (WMS) 527, 1291
wavelength multiplexing
– modulation frequency 1292
– spectral 1292
– temporal 1292
wavelength multiplexing 527, 1244
wavelet
– admissible 1379
– coherent structures 1393
– compatibility equation 1383
– compression 1395

– conditional spectra 1393, 1394
– cone of influence 1382
– convolution 1380
– decomposition 1385
– denoising 1394
– discrete 1383
– discretization 1380
– discretization error 1381
– dominant scale 1394
– energy distribution 1387
– fast algorithm 1385
– filtering 1392, 1393
– fractal 1395
– generalities 1378
– inverse transform 1380
– local spectrum 1387
– mean spectrum 1387, 1388
– Mexican-hat 1381
– modulated oscillation 1392
– modulus maxima 1389
– moment 1390, 1391
– Morlet 1382
– multidimensional 1394
– orthogonal 1384
– Parseval 1380
– reconstruction 1385
– reproducing kernel 1383
– ridge 1389
– scaling 1391
– scaling coefficient 1385
– scaling function 1384
– singularity 1395
– transform 1379
wavenumber vector 1019
waveplates 728
waverider buoy 1014
weak injection 1328
Weber number 38, 40
Weissenberg effect 800
Weissenberg number 800, 801, 828
wetting 106, 110
Wheatstone bridge 1096, 1104
white noise 1446
whiting 1057
Wiener–Khinchine relation 1342

wind 1043
– noise 1045, 1046
– speed 1052, 1058
wind tunnel 1044, 1046
– blow-down 613
– boundary layer 605
– cryogenic 598
– high pressure 596
– high-pressure 607
– low speed 596
– transonic 596, 609
window function 1341
window shifting 1471
windows 726–728
wind-tunnel corrections 1060
wing/engine combination 599, 608
wire
– arrays 235
– aspect ratio 232
– materials 232
– suspension 1070, 1071
wireless telemetry 923
WMS (wavelength modulation spectroscopy) 527, 1291
Wollaston prism 483
wool tuft 1074
work of adhesion 99, 100
wrap-around error 1343

X

xanthan gum 657
X-array 283, 430
X-wire 234, 255, 269, 900

Y

yawed wire 242
yawing angle 1062, 1070
yield stress 693
Young equation 106
Young–Laplace equation 101, 115

Z

zero-shear viscosity 634